Y0-CCU-535

Poisonous and Venomous Marine Animals of the World

(REVISED EDITION)

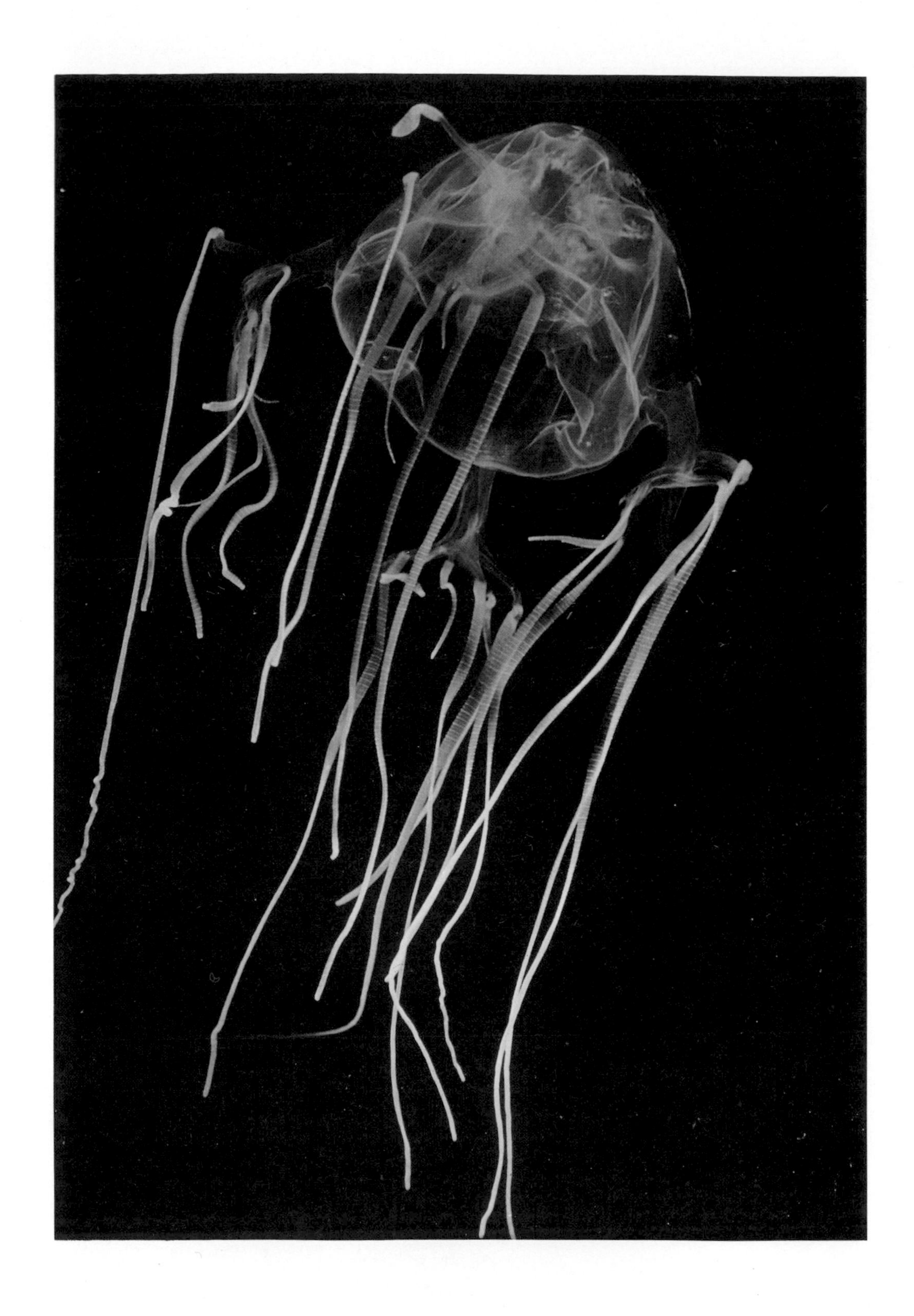

Chironex fleckeri Southcott. Living specimen taken near Cairns, Queensland, Australia. This is believed to be the finest photograph ever taken of the living organism. Specimen measured 7.5 cm across the bell. (K. Gillett)

by BRUCE W. HALSTEAD, M.D.
Director, International Biotoxicological Center,
World Life Research Institute
with the editorial assistance of LINDA G. HALSTEAD

Poisonous and Venomous Marine Animals of the World

(REVISED EDITION)

(This new edition incorporates the original three-volume work: "Part One–Invertebrates" and "Part Two–Vertebrates" in a single volume)

THE DARWIN PRESS, INC.
Princeton, New Jersey

Library of Congress Cataloging in Publication Data

Halstead, Bruce W.
Poisonous and venomous marine animals of the world.

Includes bibliographies and indexes.
1. Dangerous marine animals. 2. Poisonous
animals. I. Halstead, Linda G. II. Title.
QL100.H33 1978 591.6'9'09162 72-177977
ISBN 0-87850-012-X

The author acknowledges the generous support of the

Deputy Chief of Staff for Research and Development
Department of the Air Force

Office of the Surgeon General
Department of the Army

Bureau of Medicine and Surgery
Department of the Navy

that made the publication of this book possible.

Graphic Design: Albert McGrigor

A Darwin® book

Printed in the United States of America

by Science Press
Ephrata, Pennsylvania

Contents

[EDITOR'S NOTE: References to page numbers of plates in the back of this monograph appear in italics in the margins.]

Maps

Foreword to the First Edition

A complete survey of a broad field of science dating back to the dawn of history is rarely attempted in this day, but the present work is an example of such a study. It is unlikely that anyone but Dr. Halstead would have attempted or could have accomplished the task. It is particularly appropriate that this monograph on the poisonous and venomous marine animals of the world should have been undertaken by Bruce W. Halstead, whose name has been linked with the subject for several decades and whose publications and personal enthusiasm have provided the initial inspiration to many of the investigators who have begun to explore the field in recent years.

The scope of the present work is truly comprehensive, and an effort has been made to include references to all published work which has appeared since antiquity. Persons carrying out research on these animals and their toxins know well the difficulties of surveying the literature, difficulties due in part to the appearance of much important information in sources other than research publications. No single abstract journal even begins to cover this field, but Dr. Halstead has done the job for us and provided the comprehensive background essential to new studies.

The relative lack of attention given until recently to poisonous and venomous marine animals is somewhat mystifying in the light of the widespread occurrence of such creatures and of the often remarkable biological actions of their poisons and venoms. Even now there are a host of such animals known to be capable of inflicting severe injuries in man, and yet chemical and pharmacological studies are nonexistent. Possible roles of most of these toxins in the ecology of the animals have not been investigated in spite of the obvious importance of this aspect. The present monograph will surely lead to an intensification of research effort in this fascinating realm and to discoveries of great scientific and medical importance.

Dr. Halstead has written the definitive work in an important but neglected area of biology, and his treatise will be the major reference source for decades to come. We and others who will follow will be forever in his debt.

PAUL R. SAUNDERS, Ph.D., *Chairman* *
Department of Biological Sciences
University of Southern California
Los Angeles, Calif.

* (Now deceased)

Preface

The preparation for the writing of this monograph on *Poisonous and Venomous Marine Animals of the World* was begun in 1943. It was originally intended to limit the scope of the work to fishes, but later it became evident that there was a need for expanding the project to include other marine animals as well. At the time the program was instituted, the only publications generally available in this field were Marie Phisalix's *Animaux venimeux et venins* (1922) and E. N. Pawlowsky's *Gifttiere und Ihre Giftigkeit* (1927). Yoshio Hiyama's notable *Poisonous Fishes of the South Seas,* published in Japanese in 1943, originally intended for the operations of the Japanese Imperial Navy, and the biotoxic experiences encountered by American troops during World War II further pointed up the need for an exhaustive review of the entire subject of marine biotoxicology. Preliminary studies revealed a vast scattered literature, foreign and domestic, written over centuries of time. However, a glaring confusion in nomenclature, lack of information on the identity of etiological agents, and vast gaps in zoogeographical, morphological, pharmacological, and chemical data soon appeared. Before a marine biotoxicological monograph could be attempted, it required field investigations, recollections of specimen materials from localities in which the toxic organisms were originally described, surveys among numerous public health agencies, contacts with foreign medical organizations, examination of military and historical medical records, and the development of an enormous international correspondence with groups concerned with biotoxications. The preliminary phases of this project continued for a period of 14 years. During this time more than 30 major expeditions and field trips were made to such areas as Mexico, Central and South America; San Benitos, Cocos, Galapagos, Hawaii, Marshall, Phoenix, Line, Caroline, Mariana, Ryukyu, Bonin, Taiwan, and Philippine Islands; Europe, Japan, Thailand, India, Pakistan, Saudi Arabia, Egypt; Red Sea, Mediterranean Sea, Caribbean Sea, and the tropical Atlantic Ocean. Several thousand histological preparations were made of the venom organs of marine invertebrates, fishes, and reptiles. More than 25,000 fish samples were screened for toxicity, the specimens having been collected through several organizations from many different parts of the world.

The methods of obtaining data, specimen material, illustrations, translations, and other services were numerous, varied, and frequently unorthodox. One thing that always frustrated the government auditor was the list of food and clothing purchases which appeared on the invoices sent to countries where, at that time (1945-1950), foreign currency could not be accepted. So, we sent food and clothing in exchange for technical services or fish specimens.

The literature disclosed considerable confusion as to the nature of some marine biotoxins: whether they were contracted in the food chain of the fish or were merely the result of bacterial contamination; whether the oral fish poisons were large or small molecular substances; whether they were water or fat soluble. The phylogenetic distribution of toxic marine organisms was an aspect of the project which presented numerous questions. No one had established a standardized screening procedure for testing for marine biotoxins, and the necessity of trying to develop a reliable detection method proved to be a problem of no mean proportion. There were questions as to whether or not stingrays and many other kinds of fishes had a true venom apparatus. In perspective, one might say that the subject of poisonous marine organisms was one of confusion fraught with many questions but few answers. The studies that were instituted, needless to say, did not resolve all of the questions. However, the investigations did provide a degree of understanding of the scope of the problem and permitted a systematic approach to the subject matter.

Initial support for this work was obtained through a grant from the National Institutes of Health (Grant No. RG-2366) and the Office of Naval Research (Contract No. NONR-205-00) in 1949. Research during this early period was further implemented by assistance provided by the Pacific Oceanic Fishery Investigations of the U.S. Fish and Wildlife Service in our studies in the Line and Phoenix Islands. The funds from these early grants provided much of the specimen material and epidemiological data, and permitted some of the early screening work which formed the foundation for a considerable portion of the present monograph. One of the major essentials in order to proceed with the preparation of this monograph was the translation of many foreign publications. We were assisted in this phase of our work by the Translating Unit of the National Institutes of Health, the Army Language School in Monterey, and numerous foreign scientists, particularly in Japan, and more recently by Dr. R. E. Asnis of Philadelphia. Expeditions sponsored by the Army Chemical Corps aided in the procurement of specimen materials which would otherwise not have been available. Work at the Marine Biological Laboratory of the Atomic Energy Commission at Eniwetok provided also specimen materials and field data. Our studies in the eastern Pacific were facilitated by expeditions sponsored by Capt. George Allan Hancock of the Allan Hancock Foundation, and Woodrow G. Krieger of the Douglas Oil Co. Several expeditions to the central Pacific and Red Sea were made possible through the cooperation of the Office of Scientific Research, and the Science Division, Directorate of Science and Technology of the U.S. Air Force.

The first sponsorship concerned directly and solely with the writing of the monograph was obtained through the School of Aviation Medicine of the U.S. Air Force at Randolph Field, Texas, 1952, Contract No. AF 18(600)-451, followed by, among others, AF 49(638)-187 from the Office of Scientific Research, U.S. Air Force. The first draft of the monograph was completed in the fall of 1957; but with my entry into active duty with the Navy, efforts toward publishing the monograph came to a temporary standstill. Upon joining the staff of the U.S. Naval Medical School, National Naval Medical Center, Department of the Navy, under the administration of Rear Adm. C. B. Galloway, I was able to continue active work on the monograph. During this period, much additional material was included in the work. In 1958, the monograph was recommended by the Armed Forces Publications Committee for publication. However, the necessity for costly color illustrations and the budgetary restrictions prevented its publication until 1962, at which time work was again taken up to complete the project. On this occasion the updating and publication were sponsored jointly by the Office of the Surgeon General, Department of the Navy, and School of Aerospace Medicine, Department of the Air Force—the project being monitored by the latter organization under Contract No. AF 41(609)-1702. Because of the time lapse and continuing accumulation of data, the manuscript required comprehensive revision. Finally, in 1963, the actual process of printing the monograph was begun by the U.S. Government Printing Office.

The primary purpose of the original 3-volume monograph was to provide a systematic organized source of technical data on marine biotoxicology, covering the total world literature from antiquity to modern times. Toxic marine algal literature was limited to that which had a direct bearing on poisonous marine animals. [The coverage of Russian and Oriental literature leaves something to be desired; but nevertheless, over a period of more than two decades, an intensive effort has been made to bring this literature to light.] With few exceptions, all the literature referred to has been examined in its original form and was translated into English by competent technical translators. The exceptions have been so indicated by NSA (not seen by author) at the end of each reference. The literature upon which this book is based has been deposited in the International Biotoxicological Center of the World Life Research Institute, Colton, Calif.

A phylogenetic arrangement utilizing a historical approach has been adopted because it seemed to serve the objectives of the monograph to a greater degree than other methods that were examined. The accurate scientific identity of the organism is imperative if the medical, pharmacological, and chemical data are to be meaningful. The phylogenetic arrangement has been followed except for the listing of families and genera, in which case an alphabetic arrangement has been adopted in order to assist the nontaxonomist in finding scientific names. The "Other Names" given within the "Lists of Toxic Species" are not synonymies in the taxonomic sense, but are synonyms of scientific names which appear in the biotoxicological literature. [In this revised edition, representative lists of toxic marine organisms are often only included. Each species was selected on the basis of its commonality or uniqueness within a geographical area, its toxicity, or its medical importance.] In some instances, it was impossible to make an accurate taxonomic diagnosis of the species discussed. References are given to the general taxonomic literature of each group upon which the systematic literature, as used in this book, is based. An attempt was made to utilize systematic works which are generally considered to be standards in their field, even though they may not represent the latest taxonomic revision of that particular group. In dealing with the various phylogenetic groups, recognized systematists have been consulted. The literature references in the sections listing the organisms reported to be toxic to man are generally limited to those articles which refer to the toxicity of the organism rather than its biology or taxonomy.

This monograph is intended to serve several disciplines having varied interests. The identity of all of the organisms discussed has been fully documented by both illustrations and references to more complete taxonomic literature which will lead one to keys and descriptive data making possible a definitive identification. This information is of vital concern to all who are working with toxic marine organisms. The illustrations will also prove useful to anyone concerned with the preparation of health education materials, preventive medicine, or survival manuals. Biologists will find the sections dealing with nomenclature, geographical distribution, and biology of the organism to be of particular interest. Epidemiologists, public health workers, and allied groups will find the public health sections and the information on mechanisms of intoxication helpful. The clinical characteristics, pathology, treatment, and prevention have been dealt with at length for the benefit of physicians and first aid workers. Toxicologists, physiologists, and pharmacologists will be concerned with those portions of the chapters which deal with the physiological effects of the poisons on laboratory animals. Unfortunately, in most instances the chemistry of marine biotoxins is unknown, but the information that is available has been summarized. Swimmers and divers inhabiting the marine environment will do well to heed some of the admonitions relating to venomous marine creatures. This is one area where the old proverb "an ounce of prevention is worth a pound of cure" really pays rich dividends.

Work in marine biotoxicology continues at the World Life Research Institute. Since 1970, venomous research has been conducted under contract with the Oceanic Biology Program, Office of Naval Research, under Contract No. 14-67-C-0379, and the Arctic Program, Office of Naval Research, U.S. Navy. One field study was conducted in the Arctic in order to study the toxic properties of arctic marine mammals. Several field investigations were conducted in the West Indies and in Mauritius under sponsorship of the Food and Agriculture Organization of the United Nations, working in cooperation with the World Health Organization. The World Life Research Institute was recently (1970) designated by the WHO, FAO, and UNESCO as an international Reference Centre in marine toxicology.

The appointment of the author to membership in the Joint Group of Experts on the Scientific Aspects of Marine Pollution (GESAMP) of the United Nations by the World Health Organization directed attention to the possible relationship of marine pollutants to naturally-occurring biotoxins. This particular aspect of marine biotoxicology is an interesting one because it bears out many of the theories which have been propounded by numerous workers over the past several hundred years. Many persons believe that certain types of toxicity cycles, such as ciguatera poisoning, can be triggered by man-induced pollutants. The conclusion of the GESAMP study revealed the need for much greater research in this area. Subsequent work with the Scientific Committee on Problems of the Environment (SCOPE) of the International Council of Scientific Unions (ICSU) further pointed up the need for greater attention to all types of marine environmental poisons. One of the significant developments with SCOPE was the establishment of a Commission on Ecotoxicology in order that the international scientific community may deal with these problems.

The purpose of this new edition of *Poisonous and Venomous Marine Animals of the World* is to provide an abridged, updated work within a single volume—for scientist, specialist, and layman alike. When the author and publisher foresaw the need for a handy, practical edition of the monograph that could be used in the field if necessary, they wished to retain the original format and important information. Although some of the ancillary material was deleted, the essential work has been retained and the results of new investigations added. As of this writing, I have visited more than 150 countries in quest of biotoxicological data for use in this work. Thus, this revised edition incorporates a substantial amount of new material and the book has been reset in its entirety. We hope it will continue to be the standard compendium in marine biotoxicology and medicine for years to come.

While visiting the University of Okayama, Okayama, Japan, in 1953, I obtained from Dr. Akira Yokoo two small vials of tetrodotoxin crystals—the result of more than 50 years of combined chemical and pharmacological research on the part of the Japanese. These crystals were brought back to the United States as part of an intensive U.S. Army Chemical Corps study to determine the molecular structure and pharmacology of tetrodotoxin. This study incorporated the tremendous contributions of Japanese chemists and pharmacologists who had published extensively on the subject. The basic work of the Army Chemical Corps in the United States did more to focus on the importance of this biotoxin than any other single scientific effort. The history of tetrodotoxin is a spectacular example of how an extremely dangerous food poison, of certain military importance, has metamorphosed into one of the most valuable modern research tools providing unique insights into biophysical and neurophysiological phenomena. This gestation period has now taken over 80 years and the end is not yet in sight. After having worked in the field of marine biotoxicology for the past 30 years and having experienced the agony and the ecstasy of this floundering embryonic science, I feel obliged to make the following commentary: I am appalled at the reluctance of international, governmental, grant-in-aid organizations and foundations to sponsor research in this field to any significant degree, and the general myopia of the scientific community-at-large to move ahead in a positive manner in a field that offers the scientist an almost infinite and spectacular array of fascinating biodynamic compounds.

The problems and unanswered questions in biotoxicology are many. Nevertheless, biotoxicological research offers exciting possibilities and will profoundly affect the health and economy of mankind in the future.

BRUCE W. HALSTEAD

Colton, Calif.
January, 1978

Acknowledgments

One of the pleasures of authorship is the opportunity to acknowledge the generosity and cooperation of those who have made the writing possible. Thanks are due to the following organizations and persons who provided substantial logistics, financial support through gifts, grants, or contracts, or other forms of research support: School of Aerospace Medicine, Air Force Office of Scientific Research, and Science Division, Directorate of Science and Technology, Department of the Air Force; Office of Naval Research, Bureau of Medicine and Surgery, Department of the Navy; Chemical and Radiological Laboratories, Office of the Surgeon General, Department of the Army; National Institutes of Health, U.S. Public Health Service; National Science Foundation; Armed Forces Institute of Pathology; U.S. Coast Guard; Pacific Oceanic Fishery Investigations, U.S. Fish and Wildlife Service; Van Camp Laboratories; Food and Agriculture Organization, United Nations; American Philosophical Society; William Waterman Fund of the Research Corporation of America; Woodrow Krieger, Douglas Oil Co.; Anglo Egyptian Oil Co.; University of Cairo; Mr. and Mrs. Earl S. Webb, Mr. William Carter, Mrs. E. Larson, Miss Margaret Freer, Dr. Iner S. Ritchie, Dr. James Kuninobu, and Dr. James Slayback; and W. J. Voit Rubber Corp.

The following organizations provided technical data or literature dealing with poisonous and venomous marine organisms: National Library of Medicine; Pacific Science Board, National Research Council; U.S. National Museum, Smithsonian Institution; Edward Rhodes Stitt Library, National Naval Medical Center; Philippine Bureau of Fisheries; Japanese Ministry of Welfare and Health; Fisheries Agency, Ministry of Agriculture and Forestry; Tokyo Metropolitan Office; Headquarters of the Far East Command, and 406th General Medical Laboratories, U.S. Army; Zoological Institute of the U.S.S.R., Leningrad; Fisheries Research Board of Canada; Australian Museum; Division of Fish and Game, Hawaii; Department of Public Health, Hawaii; Marine Laboratory, University of Miami; Vernier Radcliffe Memorial Library, Loma Linda University; Hawaii Marine Laboratory, University of Hawaii; Bernice P. Bishop Museum; Shellfish Sanitation Branch, U.S. Public Health Service; Fishery Products Laboratory, Fisheries Experimental Commission of Alaska; Defense Documentation Center, U.S. Department of Defense; Henry E. Huntington Library and Art Gallery, San Marino, Calif.; New York Public Library; New York Academy of Medicine.

Logistics, finances, data, or other forms of support were provided by the following organizations: Arctic Health Research Center, U.S. Public Health Service; Arctic Institute of North America; Arctic Program, Geography Branch, Office of Naval Research (Contract No. N00014-67-C-0379), Department of the Navy; Department of Ichthyology, California Academy of Sciences; Public Health Laboratories, Alaska Department of Health and Welfare.

For assistance in providing technical translations of scientific articles in languages other than English, I am indebted to the following: Mr. John Fletcher, Mrs. Hope Norris, Miss Rosemary Roberts, and their associates at the Translating Unit, National Institutes of Health, U.S. Public Health Service; U.S. Army Language School; and Miss Tatiana Boldyreff, translator, National Naval Medical Center, U.S. Navy.

I appreciate the assistance provided by the following persons who have criticized portions of the manuscript, submitted data, or furnished scientific specimens:

Tokiharu Abe, Central Fisheries Station of Japan, Tokyo, Japan
Sidney Anderson, Noroton, Connecticut
James Atz, Malverne, N.Y.
Dorothy Ballantine, Laboratory of the Marine Biological Association, Plymouth, England
Albert H. Banner, Hawaii Marine Laboratory, University of Hawaii, Honolulu, Hawaii
John H. Barnes, Cairns, Queensland, Australia
Isobel Bennett, University of Sydney, Sydney, Australia
Dorothea Berry, Library, University of California, Riverside, Calif.
S. Stillman Berry, Redlands, Calif.
Rolf L. Bolin, Hopkins Marine Station, Stanford University, Pacific Grove, Calif.
Max C. Brewer, Arctic Research Laboratory, Office of Naval Research, Department of the Navy, University of Alaska, Barrow, Alaska
Vernon E. Brock,* U.S. Fish and Wildlife Service, Honolulu, Hawaii
Edwin H. Bryan, Jr., Pacific Science Information Center, Bernice P. Bishop Museum, Honolulu, Hawaii
Mariano N. Castex, S.J., Institute of Biology, College of the Immaculate Conception, Sante Fe, Argentina
Jose del Castillo, School of Medicine, University of Puerto Rico, San Juan, P.R.
Fenner A. Chace, Jr., U.S. National Museum, Washington, D.C.
Eliza P. Chugg, General Library, University of California, Berkeley, Calif.
Eugenie Clark, Cape Haze Marine Laboratory, Sarasota, Fla.
Preston E. Cloud, Jr., U.S. Department of Interior, Geological Survey, Washington, D.C.
Doris M. Cochran, U.S. National Museum, Washington, D.C.
Frank Cox, Jr.
Charles E. Cutress, U.S. National Museum, Washington, D.C.

* Deceased

E. Yale Dawson,* Natural History Museum, San Diego, Calif.
Elisabeth Deichmann, Museum of Comparative Zoology, Harvard University, Cambridge, Mass.
M. W. De Laubenfels,* Oregon State University, Corvallis, Oreg.
Lillian Dempster, California Academy of Sciences, San Francisco, Calif.
Sylvia A. Earle, Cape Hare Marine Laboratory, Sarasota, Fla.
Robert Endean, University of Queensland, St. Lucia, Brisbane, Queensland, Australia
Vittorio Erspamer, Instituto de Farmacologia, Universita di Parma, Parma, Italy
J. C. Fardon, Institutum Divi Thomae, Cincinnati, Ohio
H. B. Fell, Victoria University of Wellington, Wellington, New Zealand
John E. Fitch, California State Fisheries Laboratory, Terminal Island, Calif.
W. I. Follett, California Academy of Sciences, San Francisco, Calif.
A. Fraser-Brunner, British Museum (Natural History), London, England (Formerly)
Frederick A. Fuhrman, Max C. Fleischmann Laboratories of the Medical Sciences, Stanford University, Stanford, Calif.
John S. Garth, University of Southern California, Los Angeles, Calif.
Issac Ginsburg, U.S. Fish and Wildlife Service, Washington, D.C.
William Gosline, University of Hawaii, Honolulu, Hawaii
E. W. Gudger,* American Museum of Natural History, New York, N.Y.
Cadet Hand, University of California, Berkeley, Calif.
Olga Hartman, University of Southern California, Los Angeles, Calif.
W. D. Hartman, Peabody Museum of Natural History, Yale University, New Haven, Conn.
Albert W. Herre,* Stanford University, Stanford, Calif.
Ralph T. Hinegardner, University of Southern California, Los Angeles, Calif. (Formerly)
Yoshio Hiyama, University of Tokyo, Tokyo, Japan
Carl L. Hubbs, Scripps Institution of Oceanography, La Jolla, Calif.
S. H. Hutner, Haskins Laboratories, New York, N.Y.
Libbie Hyman, American Museum of Natural History, New York, N.Y.
Edmund C. Jaeger, Riverside, Calif.
Eugene T. Jensen, Shellfish Sanitation Branch, U.S. Public Health Service, Washington, D.C.
Rune Johansson, San Bernardino, Calif.
Witold L. Klawe, Inter-American Tropical Tuna Commission, La Jolla, Calif.
Alan J. Kohn, University of Washington, Seattle, Wash.
P. L. Kramp, University Zoological Museum, Copenhagen, Denmark
Charles E. Lane, Marine Laboratory, University of Miami, Coral Gables, Fla.
David J. Lee, School of Public Health and Tropical Medicine, the University, Sydney, Australia

* Deceased

Kwan-ming Li, University of Hong Kong, Hong Kong
George C. McLain, McLain's Camera Center, Anchorage, Alaska
Zvonimir Maretić, Medical Faculty, General Hospital, Zagreb, Croatia, Yugoslavia
Basil J. Marlow, Australian Museum, Sydney, Australia
Tom C. Marshall, Department of Harbours and Maine, Queensland, Australia
Edgar J. Martin, San Francisco State College, San Francisco, Calif.
Norman T. Mattox,* University of Southern California, Los Angeles, Calif.
John Maxwell, Library, University of California, Riverside, Calif.
Karl F. Meyer,* University of California Medical Center, San Francisco, Calif.
Robert R. Miller, University of Michigan, Ann Arbor, Mich.
Philip O'B. Montgomery, Jr., Southwestern Medical School, University of Texas, Dallas, Tex.
Reid Moran, Natural History Museum, San Diego, Calif.
Jean Morice, Institut Scientifique et Technique des Pêches Maritimes, Société d'Aide Technique et de Coopération, Saint-Barthélemy, Antilles Françaises
Harry S. Mosher, Stanford University, Stanford, Calif.
Allen C. Mueller, Miami Seaquarium, Miami, Fla.
M. A. Newman, Vancouver Public Aquarium, Vancouver, Canada
Jorge A. Novis, Faculdade de Medicina da Universidade da Bahia, Salvador, Bahaia, Brazil
Robert Owen, Entomology Laboratory, Trust Territory of the Pacific Islands, Koror, Western Caroline Islands
Ye. N. Pavlovskiy,* Zoological Institute, Leningrad, U.S.S.R.
Elizabeth C. Pope, Australian Museum, Sydney, Australia
A. Prakash, Fisheries Research Board of Canada, St. Andrews, New Brunswick, Canada
John E. Prince, School of Aerospace Medicine, U.S. Air Force, Brooks Air Force Base, Tex.
Luigi Provasoli, Haskins Laboratories, New York, N.Y.
John E. Randall, Bernice P. Bishop Museum, Honolulu, Hawaii
Harold A. Rehder, U.S. National Museum, Washington, D.C.
K. Reich, the Hebrew University, Jerusalem, Israel
H. A. Reid, Penang General Hospital, Malaysia
Robert Retherford, International Underwater Enterprises, Honolulu, Hawaii
Findlay E. Russell, School of Medicine, University of Southern California, Los Angeles County General Hospital, Los Angeles, Calif.
Edward J. Schantz, U.S. Army Biological Laboratories, Fort Detrick, Frederick, Md.
Waldo L. Schmitt, U.S. National Museum, Washington, D.C.
William C. Schroeder, Museum of Comparative Zoology, Harvard University, Cambridge, Mass.

*Deceased

Leonard P. Schultz, U.S. National Museum, Washington, D.C.
Mary Sears, Woods Hole Oceanographic Institution, Woods Hole, Mass.
Donald P. De Silva, Marine Lab., University of Miami, Coral Gables, Fla.
R. V. Southcott, South Australian Museum, Adelaide, South Australia
Donald Squires, U.S. National Museum, Washington, D.C.
Don H. Strode, Alaska Department of Fish and Game, Juneau, Alaska
Nobuo Tamiya, University of Tokyo, Hongo, Tokyo, Japan
Margaret Titcomb, Bernice P. Bishop Museum, Honolulu, Hawaii
Takashi Tomoyose, Ryukyu Fisheries Institute, Okinawa
Tsuchih Tu, Medical Center, University of Alabama, Birmingham, Ala.
M. W. F. Tweedie, Raffles Museum and Library, Singapore, Malaysia
James C. Tyler, Academy of Natural Sciences, Philadelphia, Pa.
Richard Wallen, Alaska Department of Fish and Game, Juneau, Alaska
Boyd Walker, University of California, Los Angeles, Calif.
Talbot H. Waterman, Gibbs Research Laboratory, Yale University, New Haven, Conn.
Holman R. Wherritt, Alaska Native Health Area Office, U.S. Public Health Service, Anchorage, Alaska
Bayard Wiebert, Phoenix, Ariz.
D. P. Wilson, Laboratory of the Marine Biological Association, Plymouth, England
R. B. Woodward, Department of Chemistry, Harvard University, Cambridge, Mass.
J. C. Yaldwyn, Australian Museum, Sydney, Australia
Hung Chai Yang, Taiwan Fisheries Research Institute, Kaohsiung Hsien, Taiwan

The research and preparation of this monograph was conducted at the School of Tropical and Preventive Medicine, Loma Linda University, Loma Linda, Calif.; the Walla Walla College Biological Station, Anacortes, Wash.; the U.S. Naval Medical School, National Naval Medical Center, Bethesda, Md.; and the World Life Research Institute, Colton, Calif.

I am particularly grateful for the continued interest, guidance, and support provided by Rear Adm. C. B. Galloway, MC, USN, Commanding Officer, National Naval Medical Center, and Brig. General Jack Bollerud, MC, USAF, of the Directorate of Science and Technology, Department of the Air Force, for without their assistance this publication might never have become a reality.

Gratitude is also expressed to my colleagues of the Loma Linda University who provided constructive criticism on various sections of this monograph; namely, Drs. F. Rene Modglin, Jack Zwemer, Vernon Bohr, Roscoe Bartlett, Millard Smith, Robert Macomber, Don R. Goe, and Donald W. Schall. For the preparation of histological sections I am indebted to Myrna Chitwood and David Howard. Bibliographical assistance was rendered by Leona Carscallen, Florence Hill, Jean Bunker, Norman Bunker, Erma Brown, and Linda G. Halstead.

The enormous task of tabulations and typing, which went through numerous editions, was the work of Gladys Titus, Lois Mitchell, Rosa-Lynne Wilson, and Roberta L. Barry. I am especially grateful to Mrs. Wilson for her excellent work in typing the final manuscript and checking thousands of names and bibliographic references.

The exacting task of editing the initial manuscript and much of the proof was the work of Ada Turner, Madelynn Haldeman, and Dr. and Mrs. Ernest Booth. I am indebted to Dr. Merlin Neff,* and Mr. C. A. Oliphant, La Sierra College, and Dr. Thomas Little, Loma Linda University, who did the final editing. However, I accept full responsibility for whatever errors may appear in the book.

The excellent photographs and artwork that appear in this monograph are the work of the following persons: Robert Ames, Shigeru Arita, Harry Baerg, Lou Barlow, Isobel Bennett, Julie Booth, Ralph Buchsbaum, Ron Church, George Coates, Padre S. Giacomelli, Keith Gillett, Neil Hastings, Anthony Healy, Yoshio Hiyama, Tokumitsu Iwago, Robert Kendall, Robert H. Knabenbauer, T. Komai, Robert Kreuzinger, Toshio Kumada, David J. Lee, George Lower, Frank McNeill, F. G. Myers, Don Ollis, Coles Phinizy, Allan Power, Ellis Rich, Paul R. Saunders, Mitsuo Shirao, Robert Straughan,* John Tashjian, K. Tomita, Henry Vester, and Douglas P. Wilson.

The editorial suggestions of Mr. Paul H. Oehser and Mr. John Lea, Editorial and Publications Division, Smithsonian Institution, were most helpful.

I am delighted to have my daughter Linda in the editorship of this new volume.

The talented efforts of Mr. Frank Mortimer and his associates of the U.S. Government Printing Office for their contribution to the typography and design of the monograph are greatly appreciated. The design of the original monograph was largely the work of Robert McKendry. The composition and printing of the present volume was carried out under the supervision of Edward H. Breisacher. Thanks are also due to Darrell R. Crawford of the Directorate of Administrative Services, Publishing Division, Department of the Air Force, and Dr. John E. Prince, Contract Monitor, Cellular Biology Unit, School of Aerospace Medicine, Brooks Air Force Base, for their assistance in the preparation of the original three-volume work. We have retained much of the work of Dr. Donovan Courville, to whom we are deeply indebted.

Above all, I am indebted to my teacher and friend Howard Walton Clark (1923 to 1941), formerly Curator of Fishes, California Academy of Sciences, San Francisco, Calif., who took time to introduce me as a youth to the wonders of the aquatic world, and to Wilbert McLeod Chapman,* who showed me the need for research on marine biotoxicology.

Finally, but by no means least, I am forever in the debt of my family, my wife Joy, and children Linda, Sandra, David, Larry, Claudia, and Shari, all of whom wondered at times what had happened to Bruce, who was always hurried in the laboratory, library, or underwater.

Sincerely and gratefully acknowledged,
BRUCE W. HALSTEAD

* Deceased

Introduction

Biotoxicology, the science of plant and animal poisons, is concerned with a vast number of toxic chemical substances, many of which remain to be characterized. Biotoxins are of two major types: phytotoxins, or plant poisons, and zootoxins, or animal poisons. Biotoxins can be classified in a variety of ways, according to one's approach to the subject. However, for the purpose of this monograph only those biotoxins pertinent to the marine environment have been selected, and of these, emphasis has been placed on marine zootoxins. Only those toxic marine plants have been included which have a bearing on biotoxicity cycles in marine animals.

Marine zootoxins can be further subdivided into those that are poisonous to eat, the oral poisons, and those that are administered by means of a venom apparatus, the parenteral poisons. The administration of venoms entails mechanical trauma, whereas other types of poisons do not. The term "poisonous" may be used in the generic sense, referring to both oral and parenteral poisons; but it is more commonly used in the specific sense to designate oral poisons. Thus, *all venoms are poisons, but not all poisons are venoms.* Oral marine biotoxins are generally thought to be small molecular substances; whereas most venoms are believed to be large molecular substances, a protein, or in close association with one. A third general type of marine biotoxins (referred to as crinotoxins), which is found in flatworms, nemerteans, fishes, etc., are endogenous poisons produced by glands, but they are not accompanied by a venom-purveying structure. The crinotoxin is generally released into the adjacent environment by means of pores, in a manner somewhat comparable to the action of sweat glands. For the most part, little is known about the chemical or pharmacological nature of these poisons.

Little is known of the biological significance of marine biotoxins. However, biotoxins appear to serve as defensive or offensive mechanisms, in food procurement, or they may be accidentally contracted in the food chain of the organism. This is an area of ecological research which only now is beginning to receive attention.

The scientific importance of biotoxins is frequently misunderstood. The most popular attitude is that poisons are lethal substances, causing intoxications and death, and are therefore substances to be avoided. It follows that since these poisons are dangerous to humans, research is justified to determine their pharmacological and chemical properties with the hope of ultimately developing a useful antitoxin or of learning more about their clinical characteristics. However, there is another aspect of this subject which is equally deserving of consideration, and this relates to the exploration of new and little-known biologically active chemical structures. It is estimated that of the thousands of marine organisms known or believed to contain biotoxic substances, less than 1 percent of them have been examined for their biological activity. Of the total number that have been studied even in a cursory manner, less than a half dozen of them have been evaluated to the point of determining their chemical and pharmacological characteristics. The marine biotoxic substances that have been studied reveal fungicidal, growth-inhibitory, antiviral, antibiotic, antitumor, hemolytic, analgesic, cardioinhibitory, psychopharmacological, and numerous other types of biological properties. There is a common misconception that whatever chemical substances are needed they can and will be developed by the synthetic chemist, but the facts fail to completely justify such optimism. An appreciation of the fantastically potent level which characterizes many of these biotoxic substances can be obtained by examining Table 1. It should be noted that one of our most biologically active synthetic chemical agents, a war gas, is among the least potent materials in this list.

There are indications that some of the marine biotoxins yet to be explored may rank among the most toxic substances known. One such substance is

TABLE 1.—*Relative Toxicities of a Selected Group of Toxic Substances (after Mosher* et al., *1964)*

Toxin	*Minimum Lethal dose* (μg/kg)	*Source*	Molecular weight
		Protein	
Botulinus toxin A	0.00003	Bacterium: *Clostridium botulinum*	900,000
Tetanus toxin	0.00010	Bacterium: *Clostridum tetani*	100,000
Ricin	0.02000	Plant: caster bean, *Ricinus communis*	(Not given)
Palytoxin	0.15000	*Palythoa* spp.	33,000
Diptheria toxin	0.30000	Bacterium: *Corynebacterium diphtheriae*	72,000
Cobra neurotoxin	0.30000	Snake: *Naja naja*	(Not given)
Crotalus toxin	0.20000	Snake: rattlesnake, *Crotalus atrox*	(Not given)
		Nonprotein	
Kokol venom	2.70000	Frog: *Phyllobates bicolor*	400
Tarichatoxin	8.00000	Newt: *Taricha torosa*	319
Tetrodotoxin	8-20.00000	Fish: *Sphaeroides rubripes*	319
Saxitoxin	9.00000	Shellfish. Produced by dinoflagellate *Gonyaulax catenella* ingested by shellfish	372
Bufotoxin	390.00000	Toad: *Bufo vulgaris*	757
Curare	500.00000	Plant: *Chendodendron tomentosum*	696
Strychnine	500.00000	Plant: *Strychnos nux-vomica*	334
Muscarin	1,100.00000	Mushroom: *Amanita muscaria*	210
Samandarin	1,500.00000	Salamander: *Salamandra maculosa*	397
Dilsopropylfluorophosphate	3,000.00000	Synthetic war gas	34
Sodium cyanide	10,000.00000	Inorganic poison	49

the venom produced by the deadly sea wasp *Chironex fleckeri*. Human death times have been recorded of less than 1 minute after contact with the tentacles of this jellyfish. It is now known that numerous species of dinoflagellates produce violently toxic substances, but of the more than 1,000 known species the biotoxins of only 4 kinds have been examined to any extent.

Toxicity is a useful indicator of biological activity and, as such, provides important leads to other pharmacological properties of value. For example, in some instances, there appears to be a direct correlation between biotoxicity and antibiosis. In natural products research, of which biotoxicology is a segment, ethnobotany serves a useful purpose in providing leads on types of biological activity. Ethnobotanical studies reveal the ways in which plants have been utilized in a native economy—an important facet of human ecology. Knowledge of an ethnic use of a particular plant or animal substance frequently provides a useful starting point for types of pharmacological properties which might otherwise be overlooked. Unfortunately our knowledge of ethnic marine biology is generally not as good as that of ethnobotany. Although much valuable data can be obtained from surveys of ethnic groups and early historical or ethnological writings, many of the marine organisms that we would like to know most about are frequently inaccessible to native groups. Consequently it is helpful to employ ancillary methods in obtaining leads to useful pharmacological properties of an unknown marine biochemical substance. One such approach is to use biotoxicity as an indicator of general pharmacological properties. Biotoxins offer a spectrum of new or little-known molecular structures which the synthetic chemist may ultimately find possible to synthesize and manipulate into useful products. The importance of biotoxins in providing new molecular blueprints merits greater scientific attention than they have received to date.

The use of highly active biotoxins as molecular probes in neurophysiological research offers intriguing possibilities. Thus, biotoxins may be useful in elucidating the nature of both nervous activity and muscular function.

There is an appalling dearth of fundamental information on the morphology of venom organs of marine organisms ranging from protozoans throughout the phylogenetic series to reptiles. Satisfactory progress in biotoxicological research is dependent upon a balance of disciplinary approaches. Chemical and pharmacological advances are dependent upon a basic knowledge of the primary source of the venom, which is derived from "old-fashioned" descriptive morphological research.

The need for increased effort in oceanological research is recognized internationally, particularly from the viewpoint of developing new food resources in a world faced with a population explosion. In terms of our national security, greater study of the marine environment is imperative. In the international area more attention is being directed toward the utilization of the so-called "trash fishes." When one realizes that of the some 25,000 species of fishes recognized by ichthyologists, less than 13 kinds make up the bulk of the world's fisheries catches, it is time that a second look be taken to develop a broader spectrum of food fishes. However, with the increased development of shore fisheries, particularly in tropical areas, and the greater utilization of the "trash fishes," the ever-present threat of various kinds of ichthyotoxins will be encountered. Failure to properly assess this situation can result in the introduction of international public health problems, and the destruction of an otherwise profitable, meritorious, and greatly needed aspect of the fisheries industry. A comparable situation is to be found in studies relating to the utilization of planktonic organisms for food.

The important role of marine biotoxins and biotoxicology in general in the study of human ecology with reference to military operations has long been recognized by Russian military physicians and scientists. It is of particular significance that despite the political changes within that country their basic concepts in military medical geography have not altered appreciably since the middle of the 18th century. Marine biotoxins and their relationship to military medical geography, a segment of human ecology, is an aspect which brings up a fundamental area of marine ecological research which to date has been largely ignored in the United States; namely, the biogenesis of marine biotoxins. At present essentially nothing is known as to how or why marine biotoxins come into being. How and why do some dinoflagellates manufacture a violently toxic substance such as saxitoxin whereas other closely related species, and possibly individuals within the same species, do not? What are the precursor substances involved, and precisely what is the process by which the poison is produced? How is it that such a violently toxic nonprotein substance as tetrodotoxin, produced by the liver and ovaries of tropical marine puffer fish, is chemically identical with tarichatoxin, a poison found in the embryo of a California freshwater salamander? Mass mortalities of fishes are frequently caused by toxic dinoflagellates, and the appreciation of the magnitude of this problem exists in fisheries and public health circles. A more subtle and less dramatic problem in marine biotoxicology is ichthyosarcotoxism, caused by ciguatoxic and clupeotoxic fishes. However, the results and public health implications may be even more serious. Ichthyosarcotoxic fishes pose a serious threat because of the unpredictable nature of the occurrence of the poison in a fish population. In ciguatoxic fishes there is an enormous phylogenetic spread in which any marine fish species appears to be a potential candidate. Ichthyosarcotoxins may appear any time and in a variety of habitats without warning. The external appearance of the fish reveals no evidence of the presence of ciguatoxin, and the species involved may be an otherwise valuable food fish. In the Line Islands during 1943-45, food fishes which previously had been eaten with impunity suddenly caused violent biotoxications. Unfortunately the serious nature of this problem has not been given the recognition that it deserves by either fisheries or public health organizations.

Intensive ecological research is needed on the food web of ichthyosarcotoxic fishes. Not only is this information needed because of its fundamental importance in the biogenesis of biochemical agents in marine organisms, but also because of its importance as a public health threat, and its development as a potential military weapon system. Available unpublished data indicates that it may be possible to initiate toxicity cycles, and there is always the hope that they can be controlled. Certainly there is a need to be able to forecast the occurrence of ciguatoxic and clupeotoxic fishes. A critical evaluation of the environmental factors contributing to the establishment of the ichthyosarcotoxic biotope is desperately needed.

The preliminary food chain studies that have been conducted on ciguatoxic fishes strongly indicate that blue-green algae *Lyngbya*, and perhaps other related genera, are responsible for producing either ciguatoxin or biotoxic precursor substances. It is a well-known fact that marine algae have a phenomenal ability to concentrate and store chemical substances from seawater. Exactly what the elemental requirements of algae for the production of biotoxins are, and whether these substances could be artificially supplied, is not known. Whether certain types of blocking agents might be introduced artificially into the environment and thus abort the development of the ciguatoxic cycle is also not known. Preliminary pharmacological evaluation of *Lyngbya majuscula* have shown antimicrobial, antiviral, fungicidal, and other types of growth-inhibitory properties. Again, these are areas which need to be explored further. The intimate relationship of a great array of microorganisms and macroorganisms living in conjunction with sponges poses some fascinating problems regarding the microecological factors contributing to the biogenesis of biotoxins having antibiotic properties.

During the last five years there has been a proliferation of reports on outbreaks of human biotoxications on poisonous arthropods, especially crabs. Most of these reports have involved the tropical Indo-Pacific region, ranging from southern Polynesia westward to the Indian Ocean (Mauritius). The global importance of poisonous fishes is poignantly revealed by even a cursory examination of some of the interrelationships of fishes to man. Fishes contribute materially to the health and welfare of man as sources of food, pleasure, chemical and pharmaceutical agents, and other raw products having industrial and agricultural uses. Fishes may adversely affect the health of man and reduce his economic potential as transvectors of toxins, traumogenous organisms, vectors of bacterial, viral, or mycotic organisms, producers of electrical discharges, etc. However, fishes undoubtedly play their most important role as sources of human and animal nutrients.

Famine, pestilence, and war have been ever-present scourges since the dawn of human history, and there has been an intimate relationship among the three of them. Although these scourges have been rather clearly defined in the past, the solution to some of these problems has not been readily apparent. One of the basic elements of this triumvirate of afflictions is the constant threat of hunger in many parts of the world. It is the feeling of many economists that if the hunger problem could be eliminated, one of the major causative factors of war might be reduced. With the present population explosion, currently reported to be increasing at the rate of 2.5 percent, the problem of food supply is becoming a subject of vital concern. The world population is presently estimated to be in excess of 4 billion people, which imposes a total annual protein requirement that exceeds 60 million tons. It is further estimated that, generally speaking, about 24 million tons of animal protein are needed. However, since problems in distribution are prevalent, much larger supplies of animal protein are required to fill the need. The oceans annually produce a harvest of about 400 million tons of animal protein in a variety of species and sizes suitable for harvest and use by man and could provide a sustained harvest of about 100 million tons annually.

Since 1850, the world catch of fishes and shellfishes has progressively increased from about 1.5 million tons to about 50 million tons in 1964. Less than 10 percent of the total catch came from fresh water. The remainder were marine organisms. Analysis of the total catch revealed that about 85.9 percent of the organisms were fishes, and the remainder were shellfish (7.6 percent), whales (5.1 percent), and other aquatic animals (1.4 percent). Thus fishes play a dominant role in human protein food reserves of the sea.

A study of the food habits of man reveals that fishes constitute about 5 percent of the total protein eaten. The remainder is derived from plants (67 percent), meat and milk (24 percent), and poultry and eggs (4 percent). The use of fishes in the United States has increased from about 5,641 million pounds in 1948 to 12,032 million pounds in 1964. Present consumption exceeds 63.5 pounds per capita per year.

The United Nations Food and Agriculture Organization estimated in 1954 that if our fisheries resources were to keep pace with the population demands they would have to increase their production by about 88 percent by 1980 (Rao and Amarai, 1954). Walford (1958), in his stimulating book *Living Resources of the Sea*, indicated that merely to maintain the present status of global food requirements would demand an increase in fishery products of 8.5 million tons during the next 50 years, but if we were to attempt to meet the total world protein requirements, i.e., provide food to the peoples now starving, it would demand an increase of about 50 million tons during this same period of time. The more recent reports have shown that present world fisheries are increasing at a rate of about three times faster than the human population, or about 10 million tons of animal protein, or a little more than one-third of the world's basic need of this product. Despite these optimistic production figures, fisheries are continuing to be viewed as an important protein food source which must be further expanded and exploited to the fullest if they are ultimately to keep up with the world's projected needs.

Further evaluation of fisheries catches reveals that the dominant part of the total catch is fishes (about 86 percent), and that 88.4 percent of the fish catch comes from the sea. Studies indicate that the rapidly growing fisheries are not in northern or southern latitudes but rather are in tropical seas. It is anticipated that this trend will continue, with even greater efforts being directed to tropical insular areas.

Despite the fact that ichthyologists recognize about 25,000 species of fishes, less than 13 kinds make up the bulk of the world's fisheries catches. A few kinds of fishes are heavily exploited, some are fished but unexploited, and others are abundant but go unharvested (Schaefer and Revelle, 1959). At present the greatest single fish component is herring and anchovy (more than 50 percent in 1964). Cod, haddock, and hake contribute 15.5 percent; horse mackerel, sea perch, etc. 12 percent; tunas and mackerels 6.7 percent; flatfishes 3.4 percent; salmon and smelt 1.5 percent; sharks and rays 1 percent; all other kinds of fishes comprise 18.5 percent. In the future, greater attention will be focused on the possibilities of developing shore fisheries as well as high seas fisheries in undeveloped tropical regions, and efforts will be directed toward the utilization of more of the so-called "trash-fishes"—all of which will bring us headlong into the problem of poisonous fishes.

Although poisonous fishes are rapidly becoming recognized as a highly significant public health problem in some parts of the tropical world, their full significance in the marine ecosystem has not been fully defined. Nigrelli (1958) has commented on the well-known fact that herring cannot be trapped in nets coated with "Dutchman's baccy juice," a slimy growth consisting of diatoms and a species of brown alga. This phenomenon has been studied quite intensively by Hardy (1956) who has proposed the theory of animal exclusion, i.e., the distributional relationship between plants and animals is due to a modification in the vertical migrational behavior of animals in relation to dense concentrations of plants. Nigrelli has further suggested that the metabolic products of the diatoms are suspected of being responsible for this exclusion effect. Moreover, the distribution of marine animals also appears to be influenced by the presence or absence of certain animal species. For example, water with large populations of the arrow worm *Sagitta setosa* is invariably poor in other organisms including fishes, whereas water in which *S. elegans* predominates is often rich in phytoplankton, zooplankton, and fishes. Experimental evidence has shown that sea urchins and marine annelids reared in *S. setosa* water have a higher degree of abnormal development and lower viability than those reared with *S. elegans*. Studies have shown that negative effects are also exerted upon fishes and possibly other organisms by sponges and sea cucumbers. Similar inhibitory effects have been reported to be produced by the sipunculoid worm *Bonellia* (Ruggieri and Nigrelli, 1962) and by the starfish *Pycnopodia helianthoides* (Rio *et al.*, 1963). There is evidence which suggests that ichthyocrinotoxic fishes, i.e., fishes that secrete into the water toxic glandular substances but lack a venom apparatus, may exert a greater controlling influence in the development and distribution of fish populations than has been previously suspected (Thomson, 1963; Liguori *et al.*, 1963). It is becoming increasingly evident that the presence of toxic metabolites released by phytoplankton, zooplankton, benthic invertebrates, algae, and fishes may play important roles as regulatory mechanisms in fisheries dynamics. This is an area of biochemical oceanography about which we are still ignorant and in urgent need of exploration (Nigrelli, 1958).

Despite recognition of the serious public health threat of poisonous fishes to the economic development of commercial shore fisheries in tropical and subtropical regions, governmental research efforts on this subject are generally on a less than minimal basis. The Japanese outbreaks of Minamata disease in which algae, shellfishes, and fishes had ingested industrial wastes containing highly toxic organic mercurial compounds and thereby caused fatalities in the local human population stands as testimony of the price of ignorance regarding the complex biotoxic factors in the marine ecosystem. It is particularly noteworthy that these organic mercurials apparently did not destroy the marine organisms, but they very obviously unleashed the full fury of their toxic effects against man. As the effluent of industrial wastes along our cities increases in volume and chemical complexity, the potential danger to our fisheries is far more subtle and vicious than the obvious gross destruction of these resources. The Minamata outbreaks point up the ever-present danger of providing to the general public either acutely lethal, or what is even worse, chronically lethal, commercial fisheries products.

Oral fish biotoxications must not be taken lightly since the compounds involved are among some of the most powerful lethal substances known to toxicologists. At the time of this writing, another outbreak of ciguatera has suddenly manifested itself at Washington Island, Line Islands. According to all available reports poisonous fishes were never known to occur at Washington Island even though they have been previously found at all the other islands in this archipelago. The fishes involved were such valuable food species as *Caranx* spp., *Lutjanus* spp., *Scarus* spp., etc. The cause of the outbreak was attributed to the sinking of the freighter MS *Southbank* on December 26, 1964, which was believed to have been heavily loaded with metals and other substances which may have contributed to triggering the outbreak. At this point the cause of the outbreak is pure conjecture since no one has been able to evaluate the situation in the field and to correlate it with laboratory analyses. It is interesting, however, that the toxicity of the fishes did not appear until that portion of the ship containing the main cargo had been completely sunk in August 1965. Salvage crews had caught and eaten fishes with impunity in the immediate vicinity of the wreck up until the time that the main cargo area had been completely sunk. Then in August 1965, the fishes previously edible in the vicinity of the ship suddenly and without warning became poisonous, and the crew members who ate them became violently ill. It is the precipitous nature of these ciguatoxic outbreaks that accentuates the seriousness of the problem.

The triggering mechanism in ciguatoxications is unknown, but it is of vital concern that the factors involved be fully determined. Although it is frightening to contemplate that these toxicity cycles may some day be volitionally triggered, it is even more serious if we find ourselves in a position where we can do nothing to control the situation. Medical oceanography in all of its varied complexions must take its rightful place in our future considerations of the magnitude of problems relating to human ecology, environmental health, nutrition, water pollution, and fisheries biology.

Venomous fishes constitute a biological hazard to anyone invading their environment. They, like other forms of toxic fishes, offer a rich source of new and little-known highly active biochemical agents. Probably less than 5 percent of the venomous fish species that are known to exist have been studied even in a cursory manner. The phylogenetic spectrum of venomous fishes is broad, and the morphology of their venom organs is remarkably diversified. The fact that the pharmacology and chemistry of fish venoms are for the most part unknown is a regrettable commentary on the state of knowledge on this subject. Moreover, the basic chemical structure has never been determined for a single fish venom. There is some evidence that the venoms of stingrays, weevers, and scorpionfishes may be similar or have similar chemical components, but our present knowledge is much too meager to permit one to indulge in generalizations at this time. One serious deterrent to the furtherance of our biochemical knowledge of fish venoms is the prevalent misconception in some scientific circles that descriptive morphology is largely passé. In terms of fish venomology, certainly nothing is further from the truth. Until such fundamentals as the anatomical distribution of fish venoms have been determined, the pharmacological and chemical characterization of these compounds will continue to be unstudied.

The geographical distribution of venomous marine fishes follows somewhat the distributional pattern of fishes in general. Venomous marine fishes are present in greatest numbers in warm temperate and tropical waters. The variety of species decreases as one proceeds in the direction of either the cold northern or cold southern latitudes. Why venomousness appears to follow this particular zoogeographic distributional pattern is not known. It is doubtful that venomousness is related solely to temperature gradients, because some species of venomous stingrays, chimaeras, scorpaenids, etc. are found in cold deep waters of tropical and temperate latitudes. As one might expect, venomous fishes are most abundant in variety and number of species within the Indo-Pacific faunal belt. The West Indian fauna is probably the next richest in venomous fish fauna.

The greatest development in venomous fishes, particularly in reference to variety of species, is found in the family Scorpaenidae. This family also contains one of the most dangerous stingers, namely, the stonefish *Synanceja*. The genera *Pterois, Scorpaena, Notesthes, Minous, Inimicus, Centropogon*, and many other members of this great family are capable of producing extremely painful stings. Anatomically, the most highly developed venom apparatus in a fish is found in the toadfish family Batrachoididae. On the other hand, the potency of the venom of batrachoids does not appear to equal that of either the scorpaenids or trachinids.

Fish venoms differ markedly in their pharmacological and chemical properties from the toxins of other venomous animals. The investigation of fish venoms is difficult because of their instability. Some of these venoms are labile even at temperatures of 0° C, and much of the activity is lost even on lyophilization of freshly prepared crude extracts. In many species, the venomous stings must be stored at temperatures below -20° C if stability is to be assured. The study of fish venoms is further complicated by the fact that qualitative as well as quantitative differences in the chemical composition of these toxins may exist from one species to the next and from one individual to the next. A venom may even vary within the individual animal at different seasons of the year. Marine piscine venoms appear to be chemically less complex than those found in reptiles. Furthermore, it is very difficult to compare the work of one investigator with that of another because of the variations in methods of extracting the venom, in means of storage and preservation of the venom, and in techniques of bioassay.

The classification and naming of animal venom has resulted in considerable duplication and confusion in nomenclature. At present there is no completely satisfactory system. Most investigators have more or less followed a taxonomic scheme of naming marine animal poisons, i.e., the toxin is usually named after the generic or family name of the organism from which the poison was derived. Examples of this system of naming marine biotoxins are seen in the following: saxitoxin, after the mollusk *Saxidomus*; venerupin, after the bivalve *Venerupis*; murexine, after the gastropod *Murex*; asterotoxin, after the starfish *Asterias*; holothurin, after the seacucumber *Holothuria*; tetrodotoxin, after the pufferfish *Tetrodon*; suberitin, after the sponge *Suberites*; plototoxin, after the catfish *Plotosus*; ostracitoxin, after the trunkfish *Ostracion*; etc. Although this system has certain obvious drawbacks from a chemical or pharmacological viewpoint, it appears to be the most acceptable system that has evolved to date, and biotoxicologists should be encouraged to continue to use it until such time as more exacting information is available. A few investigators have followed other nomenclatorial methods such as naming the biotoxin after a little-known vernacular name. The use of vernacular names for naming biotoxins should be discouraged because they are merely of provincial significance and are meaningless to most scientific workers. Until the chemical and pharmacological properties of biotoxins have been completely characterized, we must be cautious about systematizing biotoxicological data. The present system of biotoxicological organization is far from satisfactory, but at the moment the phylogenetic approach appears to be the most acceptable one developed.

One very interesting and highly significant aspect of marine biotoxicology that has emerged into focus from this study is the recently established major category of crinotoxic organisms. Although it has been recognized since the time of Prokhoroff (1884) that the skin of certain fishes produces toxic secretions, it was Thomson (1963) who pointed out the need for establishing a new major category of poisonous fishes. Fishes producing toxic secretions by means of specialized glandular structures, but which lack a traumogenic organ, have now been classified as ichthyocrinotoxic organisms. It is now recognized that there are also crinotoxic invertebrates as well. Examples are to be found within the sponges, coelenterates, mollusks, nemerteans, etc. The mechanism of production of the crinotoxins and means of dispersal differ considerably from those found in most other types of toxic marine organisms. The toxic secretions of these animals are released into the water, and, as Nigrelli (1958, 1962) has suggested, toxic metabolites in seawater may play an important role as regulatory mechanisms in the population dynamics of marine organisms.

Chelonitoxication, or turtle poisoning, constitutes one of the more violent forms of oral marine biotoxications. Although marine turtle poisoning appears to have been known since the time of the T'ang Dynasty (A.D. 618-907), our knowledge of the nature of the poison and how it is derived through the food chain of the animal has progressed but little since that time. Outbreaks of marine turtle poisoning resemble ciguatoxications in that they are sporadic and appear without warning. Turtle poisoning differs from ciguatera in that the intoxications are characteristically more violent and the mortality rate is more than twice that of ciguatera. Why chelonitoxications appear to be largely limited to certain specific parts of the Malay Archipelago, southern India, and Sri Lanka is not known. Most of the evidence points to toxic algae as the primary causative agents, but the species involved are unknown. Thus marine turtles present some fascinating problems in biotoxic ecology which await investigation.

The problem of sea snake envenomations has caused considerable controversy regarding the incidence and seriousness of their bites and aggressiveness of their behavior. Recent studies have shown that sea snakes are capable of inflicting fatal bites and that some species are aggressive. Moreover, sea snake envenomations are far more common than was formerly recognized. One surprising thing is the meager amount of critical morphological data available regarding the nature of their venom organs. It is also regrettable that so little attention has been directed to either the pharmacology or the chemistry of sea snake venoms.

One of our major gaps in biotoxicological knowledge concerns the lack of specific information on the toxicity of arctic marine mammals. Hypervitaminosis A is one of the disorders which occurs in man after having eaten the livers of arctic animals, particularly the polar bear. For some unknown reason, very little effort has been directed toward toxicologically and chemically analyzing the livers of these arctic animals. Although the preponderance of evidence indicates that most of these liver intoxications are due to hypervitaminosis A, there remains a question as to whether certain other organotropic or neurotoxic factors are also present. Much greater attention needs to be given to the toxicity of marine animals in general, both in the arctic and antarctic regions.

Certain general observations have been made in preparing the subject matter for this monograph during the past 25 years which may be of interest to organizations and persons working in marine biotoxicology. Probably the most overwhelming impression has been the fantastic array of potent biologically-active substances which exist throughout the entire marine animal phylogenetic series. The constant finding of new biotoxic substances from some of the most unsuspected sources clearly indicates that we have barely scratched the surface of this subject. The sea is indeed a vast physiological medium which is yet to be tapped by man as a source of new nutrients and highly useful biochemical substances. Why industry and government have so consistently ignored marine organisms as sources of valuable therapeutic agents will remain as one of the enigmas of science, medicine, and technology for all time. If the biotoxicological findings of the past are any indication of the future, it will ultimately be shown that the biochemical wealth of the sea probably far outweighs that of the utilizable marine inorganic resources. How long mankind can continue this luxury of scientific and economic ignorance only future historians will be able to ascertain.

With the rapid advances of this present era of molecular biology and the popular vogues of DNA and RNA research, it is well to pause and take a good backward look down the meandering path from which biotoxicology has come. In evaluating the more recent pharmacological and chemical advances of our time in biotoxicology, one cannot help but be impressed with the fact that most of these highly sophisticated disciplines have had their genesis in the basic observations of naturalists, historians, morphologists, clinicians, and ecologists. In most instances in marine biotoxicology, the discovery of the presence of a highly active biochemical substance was not made by either the pharmacologist or the chemist, but was the result of documented observations of the "old-fashioned biologist." Several thousand years of human activity testify to the importance of these fundamental observations and strongly suggest the continued need for the efforts of the "antiquated biologist" if modern molecular biology is to make the strides in biotoxicology that it proposes for the future.

Perhaps a word or two is in order as to the "state of the art." Marine biotoxicology has progressed over the years by spurts and sputters. The pharmacology and chemistry of such biotoxins as paralytic shellfish poison (saxitoxin), murexine, tetrodotoxin, and holothurin have now been largely determined, and progress is being made on ciguatoxin and several other toxins. However, the field at large has progressed but little since the time of Louis Pasteur (1822-95). Within recent years there has been an increased awareness on the part of fisheries experts, biologists, and public health workers of some of the implications of these toxic substances in the marine environment. However, with the advent of severe budgetary restrictions and inflation, research efforts in marine biotoxicology have been drastically reduced. A major international collaborative effort involving several agencies of the United Nations system, university laboratories, and research institutes was regrettably discarded because of insufficient funding in 1972. With the steady increase of chronic degenerative diseases and the compelling need for more nutrients for a burgeoning, hungry population, these are economies that mankind can ill afford. It was therefore of considerable interest and vast import that the USSR announced in 1974 their intention of exploring global marine resources as part of a fifteen-year ocean program. The Soviets have stated that the program will include: 1) the study of world ocean resources; 2) the elaboration of problems connected with the protection of those resources; and 3) the utilization of the resources. One of the major elements of this new program will be the study of marine biological and chemical resources. It is interesting to note that this new ocean program will come under the Far Eastern Scientific Methodology Council in Vladivostok. This Council also houses the Institute of Biologically-Active Substances, which for many years has been energetically pursuing the field of terrestrial pharmacognosy. Undoubtedly the new Soviet program will actively pursue investigations in marine biotoxicology and marine pharmacognosy. Similar programs are already in progress in Australia and Japan.

One encouraging note on the research horizon has been the steady progression of the Food and Drug Symposia of the Marine Technology Society during the years 1967, 1969, 1972, and 1974. Especially noteworthy are the excellent publications *Marine Pharmacology* by Baslow (1969), *Marine Pharmacognosy* by Martin and Padilla (1973), and *Chemistry of Marine Natural Products* by Scheuer (1973). These volumes are of great assistance to anyone studying bioactive marine compounds. A valuable compilation of abstracts and summaries, *Literature on Drugs from the Sea* (1967-70), has been prepared by Auffhammer and Deichmann (1971). A major step forward in the commercial evaluation of marine biodynamic substances has been made by the Swiss pharmaceutical company of Hoffman-LaRoche which has established a multimillion dollar facility known as the Roche Research Institute of Marine Pharmacology, Dee Why, Australia. They have already isolated a number of interesting compounds which they are presently investigating.

* * *

Despite the reluctance of the exchequers of research monies to get their feet wet in the briny deep, research efforts in marine biochemistry continue to be agonized by persistent investigators. Man, in his never-ending quest for more nutrients and more effective therapeutic agents, will continue to turn his attention to the sea and thereby be forced to reckon with those potent marine biotoxins which may someday be seen to play a dual role of being able to preserve life as well as to destroy it.

LITERATURE CITED

AUFFHAMMER, I. W., and W. B. DEICHMANN

1971 Abstracts and summaries of the literature on drugs from the sea, 1967-1970. Univ. Miami Sea Grant Techn. Bull. No. 16. 191 p.

BASLOW, M. H.

1969 Marine pharmacology. Williams & Wilkins Co., Baltimore, Maryland. 286 p.

HARDY, A. C.

1956 The new naturalist. The open sea, its natural history. The world of plankton. Collins, London. 335 p.

LIGUORI, V. R., RUGGIERI, G. D., BASLOW, M. H., STEMPIEN, M. F., JR., and R. F. NIGRELLI

1963 Antibiotic and toxic activity of the mucus of the Pacific golden striped bass *Grammistes sexlineatus*. Am. Zool. 3(4): 302.

MARTIN, D. F., and G. M. PADILLA, *eds.*

1973 Marine pharmacognosy. Academic Press, New York. 317 p.

MOSHER, H. S., FUHRMAN, F. A., BUCHWALD, H. D., and H. G. FISCHER

1964 Tarichatoxin-tetrodotoxin: a potent neurotoxin. Science 144(3622): 110-111.

NIGRELLI, R. F.

1958 Dutchman's baccy juice or growth-promoting and growth-inhibiting substances of marine origin. Trans. N.Y. Acad. Sci., II, 20(3): 248-262.

1962 Antimicrobial substances from marine organisms. Introduction: the role of antibiosis in the sea. Trans. N.Y. Acad. Sci., II, 24(5): 496-497.

PROKHOROFF, P.

1884 On the poisonous character of certain lampreys. [In Russian] Vrach. St. Petersb. 5(4): 54-55.

RAO, K. K., and C. J. AMARAI

1954 Population and food requirements. Food Agr. Org. U.N., Meeting No. 22, p. 483-491.

RIO, G. J., RUGGIERI, G. D., STEMPIEN, M. F., JR., and R. F. NIGRELLI

1963 Saponin-like toxin from the giant sunburst starfish, *Pycnopodia helianthoides*, from the Pacific Northwest. Am. Zool. 3(4): 331.

RUGGIERI, G. D., and R. F. NIGRELLI

1962 Effects of *Bonellin*, a water-soluble extract from the proboscis of *Bonellia viridis*, on sea urchin development. Am. Zool. 2(4): 365.

SCHAEFER, M. B., and R. REVELLE

1959 Marine resources, p. 73-109, 6 figs. *In* M. R. Huberty and W. L. Flock [eds.], Natural Resources. McGraw-Hill, New York.

SCHEUER, P. J.

1973 Chemistry of marine natural products. Academic Press, New York. 201 p.

THOMSON, D. A.

1963 A histological study and bioassay of the toxic stress secretion of the boxfish, *Ostracion lentiginosus*. Ph.D. Dissertation. Univ. Hawaii. 194 p.

WALFORD, L. A.

1958 Living resources of the sea. Opportunities for research and expansion. Ronald Press Co., New York. 321 p., 23 figs.

Chapter I—HISTORY

History of Marine Zootoxicology

ANTIQUITY

The story of toxic marine organisms and their relationship to mankind dates back to the dawn of man's written language. Astronomy and medicine are the oldest sciences, and it is in the records of the latter that we find the earliest references to poisonous organisms. In the mythologies of the past, where fancy seems to blend with facts, poisons and poisonous organisms played a dominant role in the affairs of early man. To the ancient magician-physicians and priests, a working knowledge of plant and animal poisons was an important part of their armamentarium. Not only was it of value to know the means of treating a poisonous sting, but it was equally useful to understand the virtues of biological poisons that were able to cause the death of an enemy. The primitive mind considered poisons to be an indifferent substance—a mystic reagent. They believed that poisons were lethal to a guilty person, but harmless to the innocent (Sigerist, 1951). Because of the profound and little-understood physiological effects resulting from contact with plant and animal poisons, the ancients often surrounded biotoxins with an aura of mystery, superstition, and religion. Thus the poison-lore and demoniacal medicine served as foundation stones for modern biomedical sciences.

Babylon and Assyria. Mesopotamian medicine was based on magic. Gula, the consort of Nibibi, commonly called the Great Physician, was the goddess of potions and poisons. These ancient people were among the first to concern themselves with animal anatomy, and to attempt to codify in written form the responsibilities of the physician. The use of scarification, bloodletting, and cupping by using animal horns in the treatment of venomous stings and bites originated during this period. There appears to be little information available regarding their knowledge of specific poisonous animals; but, since they were keen observers of natural history, undoubtedly they were aware of many noxious pests (Garrison, 1929; Castiglioni, 1941; Mettler and Mettler, 1947).

Egypt. The cultural developments in Mesopotamia were paralleled by the civilization of the Nile Valley. The documentation of poisonous biological organisms is provided by the Egyptian medical papyri. The Smith Papyrus

(1600 B.C.) cites the use of charms against snake poison, adjuring "the vessels of the body not to take up the excretion of sickness." The Hearst Medical Papyrus (2000-1350 B.C.) written between the 12th and 18th dynasties, is a practicing physician's formulary and contains about 260 prescriptions, some of which are for various kinds of animal bites. Similar prescriptions, which for the most part contained restorative natural products rather than poisons, are to be found in the Ebers Papyrus, a noteworthy step in therapeutic advancement.

The hieroglyphics of several Fifth Dynasty (ca. 2700 B.C.) tombs (Ti, Méra, Ptah-hoptep, and Deir-el-Gebrawi) depict the poisonous puffer *Tetraodon stellatus* (Fig. 1). The puffer is represented by the hieroglyphic inscription "chept":

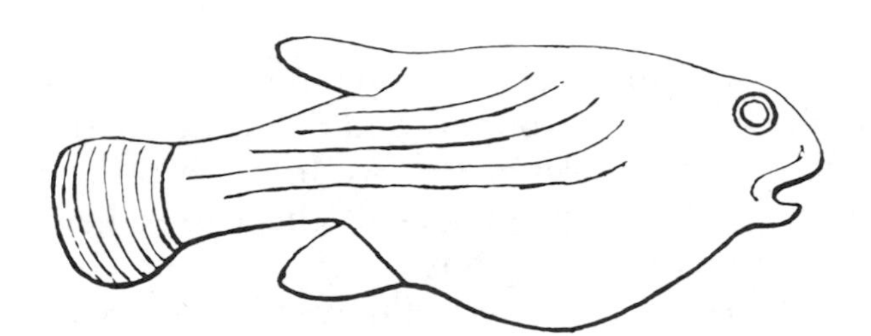

FIGURE 1.—Drawing of the inscription of the fahaka *Tetraodon stellatus*, from the tomb of the Egyptian pharaoh Ti, Fifth Dynasty. (From Gaillard)

Keimer (1955), who has made an intensive study of the natural history of ancient Egypt, states that *T. stellatus* was also called "špt" or "shepet," and was recognized as a poisonous species.

Israel. The biomedical history of ancient Israel, spanning a period of almost two millenniums, is dominated by the theocratic principles that guided the moral, social, and political thinking of the people. The Biblical record (Exodus 7: 19-21; 1491 B.C.) of the Egyptian water turning to "blood," causing the fish in the water to die, is said to be the earliest published record of the toxic effects of a dinoflagellate phenomenon (Kofoid, 1911; Hayes and Austin, 1951).

At least one of the Mosaic laws is believed to have been aimed directly at eliminating poisonous fishes from the Israelite diet (Macht, 1942); viz., "These ye shall eat of all that are in the waters; all that have fins and scales shall ye eat; And whatsoever hath not fins and scales ye may not eat; it is unclean unto you" (Deuteronomy 14: 9-10; ca. 1451 B.C.). This exhortation has a basis in scientific fact because many of the most deadly kinds of poisonous fishes are scaleless and some appear finless, to wit, the plectognaths and moray eels. It is noteworthy that more than two millenniums later American troops during World War II were to be given this same ancient admonition:

> All of the important fish with poisonous flesh belong to one large group, the Plectognathi, of which there are many kinds in the tropics. *All these fish lack ordinary scales* such as occur on bass, grouper, and sea trout. Instead, these poisonous fish are covered with bristles or spiny scales, strong sharp thorns, or spines, or are encased in a bony boxlike covering. Some of them have naked skin, that is, no spines or scales. Never eat a fish that blows itself up like a balloon.

The modern version as it appears in the booklet "Survival on Land and Sea" (Office of Naval Intelligence, U.S. Navy, 1944) is merely a more lengthy statement of the old Mosaic principle.

Persia and India. Few traces of Persian civilization or its system of medicine remain, and there is no information as to their knowledge of poisonous animals.

Generally, Indian medicine is less difficult to evaluate because its ancient views are adhered to in recent writings (Jolly, 1901), and thus continue in modern times (Nadkarni, 1954). The *Suśruta-saṁhitā*, written by Suśruta (ca. A.D. 300) was one of the original medical texts used by Brahmin priests, and is ranked as the basic medical text of Hindu medicine. Although the *Suśruta-saṁhitā* is mainly devoted to surgery and other aspects of medicine, particular attention is paid to aphrodisiacs and poisons, and antidotes for the bites of venomous snakes and other animals. Scarification, bloodletting, and cupping were apparently used by the Hindus in the treatment of venomous bites and stings.

China. The literature of ancient Chinese medicine is vast. In Chinese symbolism fishes hold an esteemed position since they are associated with the elementary character of man. This explains many of the legendary properties of fishes, some of which have a basis in scientific fact, while others do not. Fishes are generally symbols of health, reproduction, wealth, and prosperity. Poisonous properties of certain fishes have been known in China from the earliest times.

Read's (1939) work *Chinese Materia Medica of Fish Drugs*, is based on the ancient writings of Ch'en T'sang-chi of the T'ang Dynasty (A.D. 618-907), Jihhua of the Sung Dynasty (A.D. 960-1279), Li Shih-Chen of the Ming Dynasty (A.D. 1368-1644), and others. Read lists the fishes having toxic properties with their modern nomenclature.

Meng Hsien (A.D. 618) claimed that catfish and other scaleless fish were harmful to people, since they caused boils and scabies. The similarity between this statement and that found in the old Mosaic prohibitions is worthy of note.

Probably the earliest reference to ciguatoxic fish is that by Ch'en T'sangchi, written sometime during the T'ang Dynasty (A.D. 618-907), who claimed that the yellowtail is "a large poisonous fish fatally toxic to man." The yellowtail or amberjack *Seriola* is one of a variety of fishes that is commonly incriminated in ciguatera fish poisoning today. Puffer poisoning is recorded as far back as the time of K'ou Tsung-shih of the late Sung Dynasty (A.D. 960).

Japan. The earliest writings of Japanese medicine show the influence of old Chinese medicine. Modern Japanese authors (Suehiro, 1947) indicate that the toxic effects of puffer poison were known in both China and Japan from ancient times, but there does not appear to be any detailed record of this knowledge in Japan prior to A.D. 1500.

Greece. The dissociation of medicine from magic and religion was an achievement of the Greeks. The Homeric poems, the *Iliad* and *Odyssey*, written ca. 800 B.C., although never intended as medical writings, give us some insight as to the medical thinking of the period (Sigerist, 1951). Medicine had already become an art and the lay physician had come into his own. Specific medicaments were known and used. The Greeks at the time of the *Iliad* and *Odyssey* were aware that certain fishes were harmful, causing disturbances and weakness to the body, and they eliminated the noxious species from their dietary. Homer was acquainted with venomous snakes; he refers to Philoctetes, the legendary son of Poeas, king of the Malians of Mount Oeta, who was bitten by a snake on his journey to Troy and had to be left behind on the Island of Lemnos. Snakes were believed to be venomous not because of magic properties, but because they ate evil "pharmaka." Medicine in the ancient world reached its zenith in Greece during the millennium between 500 B.C. and A.D. 500. This creative period is symbolized by Hippocrates, the "Father of Medicine," whose name has come to represent the beauty, value, and dignity of medicine for all times. The relationship between disease and its etiology was propounded by Hippocrates. The toxicity of echinoderms is said to have been known to Hippocrates (Phisalix, 1922). His analytical approach to medical problems served as a solid basis for a scientific understanding of biotoxications.

Aristotle (384-322 B.C.), acclaimed the greatest biologist of antiquity, knew of the stinging abilities of jellyfishes (Storer and Usinger, 1957), and in his *Historia Animalium*, first published in Venice in 1476, he referred to the stingray and called it the trygon. Although he makes no specific references to envenomations by the stingray, Aristotle knew that it was armed with a sting (Gudger, 1943). He was also aware of the existence of the scorpionfish, which is called the sea hog (*Scorpaena porcus*) (Coutière, 1899).

Alexander the Great (356-323 B.C.) is said to have forbidden his soldiers to eat fish during conquests since he believed that some species caused disease.

The second century before Christ culminated in the development of "quasi-experimental pharmacology and toxicology." Mithridates VI, King of Pontus (120-63 B.C.), was one of the first to make the art of poisoning and preparation of antidotes an intensive study. He aspired to develop a universal antidote. His antidotes became known as "mithridates" and "theriacs." Thus, Mithridates is regarded as the originator of the concept of polyvalent drugs and sera. His concepts regarding the compounding of antidotes were actively followed by pharmacists through the beginning of the 18th century (Garrison, 1929).

Victor Nicander (175-135 B.C.), physician-poet, was one of the foremost writers of antiquity on the subject of venomous animals. He wrote several works in both prose and verse; his *Theriaca* is a hexameter poem on the nature of venomous animals and the wounds they inflict.

Little else on toxic marine animals appears in the ancient writings of this period until the time of Oppian of Corycus in Cilicia (ca. A.D. 172-210), who lived during the reign of Marcus Aurelius. Oppian apparently had an intimate knowledge of marine creatures and wrote enthusiastically about them. In his major work, *De Piscatus Libri V*, a fishing poem published in Latin in 1478, it is apparent that the writer was convinced that the spines of the stingray, weeverfish, scorpionfish, and dogfish were purveyors of venom (Grevin, 1571; Coutière, 1899; Gudger, 1943; Thompson, 1947).

Rome. Prior to the time of Pliny, little of a specific nature is found regarding the Romans' knowledge of poisonous marine organisms. According to Pliny, the people did without doctors before the advent of Greek medicine into Rome, depending mainly on medicinal herbs, superstitious rites, and religious observances (Garrison, 1929).

Pedanios Dioscorides (A.D. 50-100), a Greek army surgeon who served under Nero (A.D. 54-68), was the originator of the materia medica. He traveled in Italy, Greece, Asia Minor, Spain, and France where he collected a vast number of biological and mineralogical specimens. The data gathered provided the basis for his great materia medica, *De Universa Medicina*, which served medicine for about 1,600 years. Some of the earliest references to the toxicity and treatment for the intoxications caused by the sea hare, polychaete bristle worms, "sea vipers,"[1] and the stingray "Raia Pastinaca" are discussed by Dioscorides.

Gaius Plinius Secundus or Pliny the Elder (A.D. 29-79), born at Comum, Italy, is generally recognized as the greatest of the Latin naturalists. His *Naturalis Historia* is a vast and comprehensive encyclopedic work which covered the entire knowledge of nature. Books VIII to XI deal with zoology, and it is largely in these books that we find references to poisonous and venomous marine animals.

Pliny recommended that the sting of the stingray be ground up and used as a potion in the treatment of toothache and in obstetrics. In his section on *De Venenatis Marinus*, Pliny discusses the dangerous wounds inflicted by the venomous spines on the back and opercula of draco marinus and araneus, the weeverfishes. He also wrote on the toxicity of the nudibranch *Aplysia*, and he stated that the poison is so deadly that if a pregnant woman even looks at it she will immediately feel pain, become nauseated, and will abort. Any contact with the sea hare would cause death. Aside from references to the dangers from certain marine organisms, Pliny does not discuss the anatomy of their venom organs.

Although Pliny's biological writings were not as critical as some might desire, nevertheless they were monumental as a summary of the biological thinking that existed among the classical writers of antiquity. Pliny's work became not only the most popular natural history ever written, but it served as the foundation for biological thoughts through the Dark Ages, the Renaissance, and even to the close of the 17th century (Gudger, 1924).

[1] Whether the "sea viper" referred to here is a hydrophiid sea snake or one of the piscine eels is not known for certain.

Much has been written about the medical treatises of Claudius Galenus (Galen) (A.D. 131-201), but suffice to say he was the most voluminous of all the ancient writers. His works are of encyclopedic proportions; however, his remarks regarding poisonous animals are relatively brief. He discusses patients who had been bitten by venomous animals, and makes reference to stingrays, but in the main his ideas appear to have been borrowed from Dioscorides and Nicander.

Eastern Roman Empire—Byzantium. With the establishment of the Byzantine Empire (A.D. 476-732), the center of medical culture shifted to Constantinople, where Greek was the prevailing language. The natural sciences were not popular in Constantinople, and medicine likewise went into decline. Noteworthy medical writings of this period are few and poisonous animals are seldom mentioned.

Paulus of Aegina (A.D. 625-690) is the last great Byzantine physician. His best known work, an *Epitome* of medicine in seven books written in Greek, deals with various facets of medicine, hygiene, pathology, leprosy, skin diseases, burns, hemorrhages, surgery, and pharmacology. His fifth book is of particular interest to biotoxicologists since it is concerned with venomous stings, bites, and toxicology. Paulus provides the most complete account of military surgery of any writer of antiquity. However, he contributes nothing new to our knowledge of venomous marine organisms, but simply quotes the comments of Pliny, Nicander, and others.

Islam. After the death of Mohammad (A.D. 632), the medical concepts of Hippocrates, Aristotle, and Galenus found acceptance among the Arab peoples. During the course of their conquests in Damascus, Caesarea, and Alexandria, the Arabs had ample opportunity to become acquainted with the medical thinking of many classical writers. One of the oldest sources of Arabic medicine is a manuscript dealing with toxicology by the Persian Geber ibn Hajan (ca. A.D. 750-760) which is derived from Greek authors. The Arabs are said to have written at great length on poisons, including the effects of venoms, and their antidotes (Mettler and Mettler, 1947). A comprehensive review of Arabic toxicology has been written by Steinschneider (1871).

MEDIEVAL PERIOD

After the downfall of Rome came the Dark Ages when Western Europe passed into a period of intellectual decadence. Most historians agree that until the Renaissance there was neither induction nor experiment. The works of Aristotle, Pliny, Galenus, and others were accepted without cavil.

Among the few exceptions was Avicenna, or Ibn Sina (A.D. 980-1036), born near Bokhara, Persia. His great work, the *Q'anun*, or *Canon Medicinae*, is a compilation of his experiences in the practice of medicine together with a compendium of all that was known about medicine at that time. He deals with animal, plant, and mineral poisons; and for more than 500 years he was considered the outstanding authority on poisons and their antidotes. Avicenna recognized both oral and parenteral poisons, and he discussed venomous stings and their treatment. Some of his views were followed until the 18th century (Riesman, 1935).

The redactant thinking of the Medieval period is exemplified in the writings of Peter of Abanos (1250-ca. 1316). In his *De Venenis*, which was a 13th century composite of Arabic ideas, he divided all poisons into the three Arabic categories of mineral, vegetable, and animal—a simple classification of poisons which has met with popular acceptance down through the centuries. For the most part Peter's ideas were too confused with magic and demoniacal medicine to be of much value.

Undoubtedly, many others besides Avicenna wrote on poisons during this period. However, little else on the subject of plant and animal poisons has come to light.

MODERN PERIOD

The invention of printing in the latter part of the 15th century permitted the rapid diffusion of scientific knowledge. With the stimulating and refreshing thinking of such men as Bacon, Copernicus, Cusanus, Bruno, Leonicensus, Gesner, Aldrovandi, Rondelet, Delon, Da Vinci, and Vesalius, the modern period of scientific thought began.

1500-1699. The writers of the early Renaissance largely reflected the thinking of the classicals—Aristotle, Nicander, Pliny, and Dioscorides—in their discussions of toxic marine organisms. The numbers of poisonous organisms were essentially limited to those species which lived in the Mediterranean Sea. Knowledge of known toxic species had not increased for more than 2,000 years. The toxic species up until the Renaissance period consisted mainly of the usual "venomous" organisms: weeverfish, stingrays, scorpionfish, sea hares, jellyfish, and moray eels. Puffer poisoning had been known in ancient Egypt, China, and Japan, but apparently was of little or no concern to European writers. Although knowledge of ciguatoxic fishes is believed to date back to the T'ang Dynasty of China, awareness of the problem by European writers did not come about until the explorations of the New World by Old World discoverers. Most European physicians or biologists up to the end of the 17th century knew only one fish to be poisonous—the freshwater barbel, one of the causative agents of ichthyootoxism. Venomous sea snakes and poisonings by eating polar bear liver had been reported, but these views attracted little attention. Paralytic shellfish poisoning was known in Europe, but the nature of the disease was grossly confused. Knowledge of the nature of marine biotoxins continued to be a combination of mysticism, misconceptions, and generalizations. Intoxications were still accepted by the masses of the people as "acts of God." The treatment of venomous stings continued the same as it had been developed by the ancients. It consisted of scarification, cupping, and bloodletting, supplemented by the application of poultices, and the oral ingestion of alexipharmics, mithridates, and theriacs.

Progress in thinking and experimentation evolved slowly. During the 17th century only a small nucleus of courageous men were willing to fight against the superstition that continued to exist in most of the educated and upper levels of society. These noble investigators of the Renaissance provided important foundation stones upon which the systematists and experimentalists of later centuries were to build.

One of the earliest problems of this age to be reported in the European literature was fish roe poisoning, ichthyootoxism, due to ingesting toxic barbel roe. In 1491, Antonio Gazio (ca. 1450-1530) tested the edibility of barbel (members of the freshwater cyprinid genus *Barbus*) eggs by eating a small portion of them. According to the account published in *Corona Florida Medicinae sive De Conservatione Sanitatis*, in Venice in 1491, the experiment resulted in acute gastrointestinal disturbance. This is probably the first recorded experiment in piscine biotoxicology.

Born at Arona, Italy, Peter Martyr (1457-1526) was the first historian of America. In his *De Orbe Novo Decades Octo*, he recorded the presence of ciguatoxic fishes in the West Indies during the 16th century. Poisonous fishes in the West Indies, he believed, were the result of the fish eating the fallen fruit of poisonous trees growing along the water. Gudger (1930), who has thoroughly investigated this matter, is convinced that Martyr was referring to the toxic manchineel berry, *Hippomane mancinella* Linnaeus of the family Euphorbiaceae. The erroneous concept in the West Indies that the manchineel berry is the source of ciguatera poison has met with general acceptance among the laity, and has appeared with regularity in the scientific literature through the centuries since Martyr's time.

The noted biologist Guillaume Rondelet (1507-57) repeats many of the admonitions of the ancients concerning the venomous characteristics of stingrays in his work, *Libri de Piscibus Marinus*. Although treating the subject generally, Rondelet's description of weeverfish stings is probably the earliest record of an attempt to describe scientifically the venom apparatus of a fish. He was also the first to refer to dragonets (*Callionymus*) as venomous.

Konrad Gesner (1516-65), a physician in Zurich, is reputed to have been the greatest zoologist of the Renaissance (Nordenskiöld, 1935). His *Historia Animalium* is essentially a work in medical zoology since it discusses the medical significance of the organisms. Gesner lists the fishes which were recognized as venomous.

The great work of the Renaissance in biotoxicology was by Jacques Grevin (1538-70) (Fig. 2). Included in the first portion of his book, *Deux livres des venins*, were venomous reptiles, rats, salamanders, spiders, scorpions, flies, centipedes, lizards, stingrays, moray eels, weeverfish, sea hares (Fig. 3), electric rays, and the bites of rabid dogs, which at that time were thought to be poisonous. The second portion of his book is concerned with a variety of poisonous

DEVX
LIVRES DES
VENINS,

Aufquels il eſt amplement diſcouru des beſtes venimeuſes, theriaques, poiſons & contrepoiſons:

PAR

IAQVES GREVIN de Clermont en Beauuaiſis, Medecin à Paris.

ENSEMBLE,
Les œuures de Nicandre, Medecin & Poëte Grec, traduictes en vers François.

A ANVERS,
De l'Imprimerie de Chriſtofle Plantin.
M. D. LXVIII.
AVEC PRIVILEGE DV ROY.

FIGURE 2.—(Left) Jacques Grevin of Clermont at Beauvais (1538-70) father of modern biotoxicology. (Courtesy, National Library of Medicine). (Right) Title page of Grevin's *Deux livres des venins*, published by Christofle Plantin at Antwerp in 1568. (From the Library, World Life Research Institute)

DE LA VIVE, OV DRAGON MARIN. CHAPITRE XXXII.

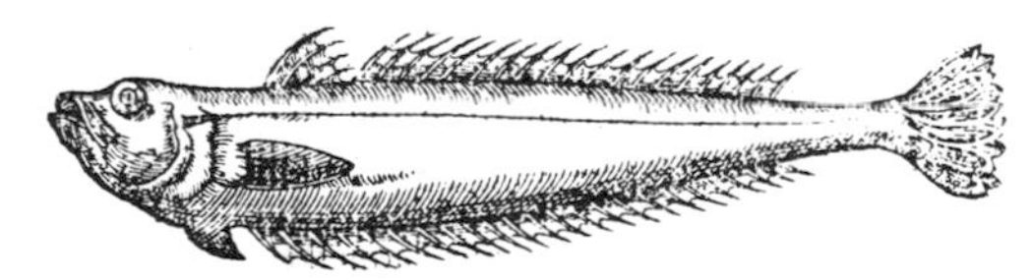

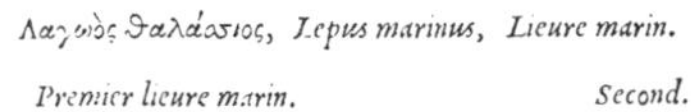

Δράκων θαλάσσιος, *Draco marinus*, *Viue.*

DE LA MVRENE. CHAP. XXX.

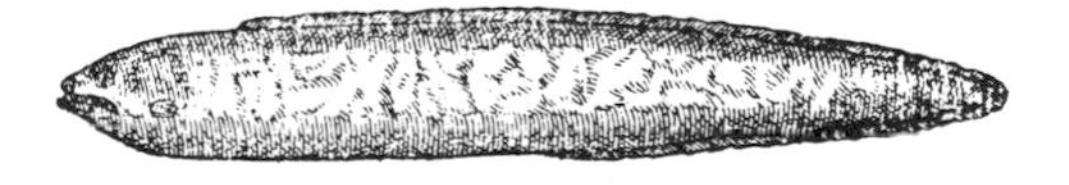

Μύραινα, *Murena*, *Murene.*

Λαγωὸς θαλάσσιος, *Lepus marinus*, *Lieure marin.*

Premier lieure marin.

Second.

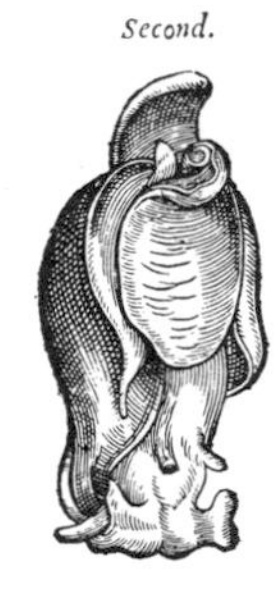

DE LA PASTENAQVE. CHAPIT. XXXI.

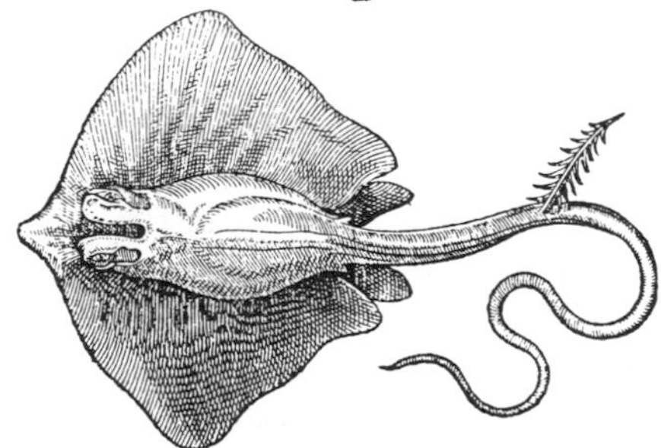

Τρυγών, *Paſtinaca*, *Paſtenaque.*

FIGURE 3.—Illustrations taken from Grevin's *Deux livres des venins*. (From Grevin)

plants. In this latter section the sea hare (*Aplysia*) and toads are included. Grevin undoubtedly provides the most comprehensive review and critique of his time on the subject of naturally-occurring poisons.

With Grevin a crude attempt was made to compile systematically and evaluate the total knowledge of toxic organisms which had accrued to that time. He provided the foundation and the historical departure point from which subsequent researchers began their investigations. Certainly Jacques Grevin can be aptly termed the Father of Modern Biotoxicology.

Some of our earliest references to fishes that are poisonous to eat concern the puffers and their allies, members of the piscine order Tetraodontiformes. According to Diego de Landa's *Relación de las Cosas de Yucatan* (1566), the Mayas were aware of the poisonous qualities of puffers in America. The oldest recorded experiment with puffer poison was a study conducted in Japan by Masamori Fukushima (1561-1624), who tested the toxicity of puffers by feeding samples of the flesh and internal organs to human prisoners. The record of these early experiments is said to have been reported in the *Busho Kanzaki* and the *Ipondo Yakusen* about 1716 (Suehiro, 1947).

Probably the most illustrious surgeon and physician of the Renaissance was Ambroïse Paré (1510-90). It was Paré who gave us the first truly medical commentary on the clinical characteristics of weeverfish stings. Paré wrote on the poisons produced by the moray eel and stingrays, and he also succumbed to the popular fallacy of the period that the discharge of the electric ray was a venom. A complete discussion of the writings of Paré on venomous fishes appears in *Oeuvres complètes d'Ambroise Paré*, first published at Paris in 1575.

Reports on the toxicity of polar bear liver date back to the explorations of the famous Dutch navigator Willem Barents (ca. 1540-97), who was in search of a northeast passage to eastern Asia. On his third voyage, he and his crew were forced to winter in the Arctic. In their desperation, some of the members of the expedition shot and ate polar bear liver and became violently ill.

The physicians Guilielmus Willem Piso (1611-78) and George Marcgrave (1610-44) are noteworthy because they provided reasonably accurate scientific descriptions of the organisms discussed, and because their travels marked the first major study of New World fishes. The medical work of Piso and the natural history of Marcgrave were combined in the *Historia Naturalis Brasiliae*. It is the first scientific commentary on the occurrence of venomous fishes in the Americas.

César de Rochefort (ca. 1600-90), lexicographer and author, published the *Histoire naturelle et morale des Iles Antilles, avec un dictionnaire caraïbe* (1658, 1666). Rochefort is one of the first to record the dangers resulting from the stings of the scorpionfish (*Scorpaena grandicornis*) in the West Indies (Phisalix, 1922). He also refers to fish having poisonous flesh, which produces vomiting, loss of consciousness, and death.

A remarkably comprehensive case history of an outbreak of barbel roe poisoning was published by Georgii Franci in 1683, entitled "Ova Barbi Comesta Noxia." This article is important because of the detailed account of the symptomatology and treatment of ichthyootoxism, and the valuable review of the Renaissance medical literature which it provides on the subject of barbel poisoning. The article exemplifies the awakening scientific attitude that was emerging during the latter part of the 17th century.

Thomas Bartolin (1616-80), a Danish physician, provides one of the earliest clinical descriptions of a weeverfish sting in his article "Acta Medica et Philosophica Hafniensia," Ann: 1674,1675,1676.

1700-1799. The discoveries of the 18th century were largely contributed by physicians, naturalists, explorers, historians, and missionaries on great voyages. Reliable scientific communication regarding biotoxic transvectors was made possible by means of Linnean systematics. Strange to say, the physician-ichthyologists of the 18th century made, for the most part, relatively minor contributions to the development of biotoxicology. The work of Trembley on the nematocyst apparatus, and Baster on the pedicellariae of sea urchins, were the glimmers of the coming dawn in morphological venomology that was to break forth so brilliantly during the 19th century. An increase in the knowledge of varieties of toxic marine organisms was observed, i.e., mollusks, arthropods, stingrays, sharks, catfishes, puffers, a variety of other tropical marine fishes, and marine turtles. There was also a decided increase in knowledge relating to the geography of biotoxications, e.g., fish poisoning in New Hebrides and New Caledonia; paralytic shellfish poisoning at Vancouver Island; puffer poisoning in Baja California, Japan, and China; turtle poisoning in Latin America; freshwater stingray envenomations in Brazil; and greater knowledge concerning the kinds of poisonous fishes in the West Indies. Knowledge of the geographical distribution of poisonous marine organisms was no longer confined to the environs of the Mediterranean Sea. Moreover, biomedical problems in the New World and tropic seas, and biotoxications were becoming of keen interest to European physicians and naturalists. Finally, it was a period in which fish poisoning became recognized as a valid clinical entity, and the symptomotology and nature of these biotoxications became more fully elucidated.

Probably no man of the age made a greater contribution to the development of an organized, systematic approach to biology, and to a significant extent to medicine, than did Carl von Linné, or Carolus Linnaeus (1707-78). His book, *Systema Naturae* (1758), marks the starting points of systematics or taxonomy, and nomenclature.[2] The establishment of a scientific methodology for naming and classifying organisms made possible accurate scientific communication regarding pathogenic and toxic agents.

Jesuit Padre Francisco Javier Clavijero (also written Francisco Saverio Clavigero) (1731-87), was an ardent student of Mexican history, and in his studies of Mexico, he brings to light one of the first known references to the

[2] Systematics, or taxonomy, is the science of classification of organisms, whereas nomenclature is concerned with a system of names. In the case of zoology, this name system is controlled by the International Rules of Zoological Nomenclature and in botany by the International Rules of Botanical Nomenclature.

poisonous puffer, or botete, in Baja California. The account of the two deaths involved was published by Clavijero in his *Historia de la Antigua o Baja California* in 1852.

The first report on poisonous Japanese puffers (Fig. 4) by a foreigner was written by Engelbrecht Kaempfer (1651-1716), a German physician-

FIGURE 4.—A poisonous Japanese puffer, fugu, or furube, as pictured in Kaempfer's *History of Japan* (1727). This is one of the earliest illustrations of a poisonous Japanese puffer to appear in occidental literature. (From Kaempfer)

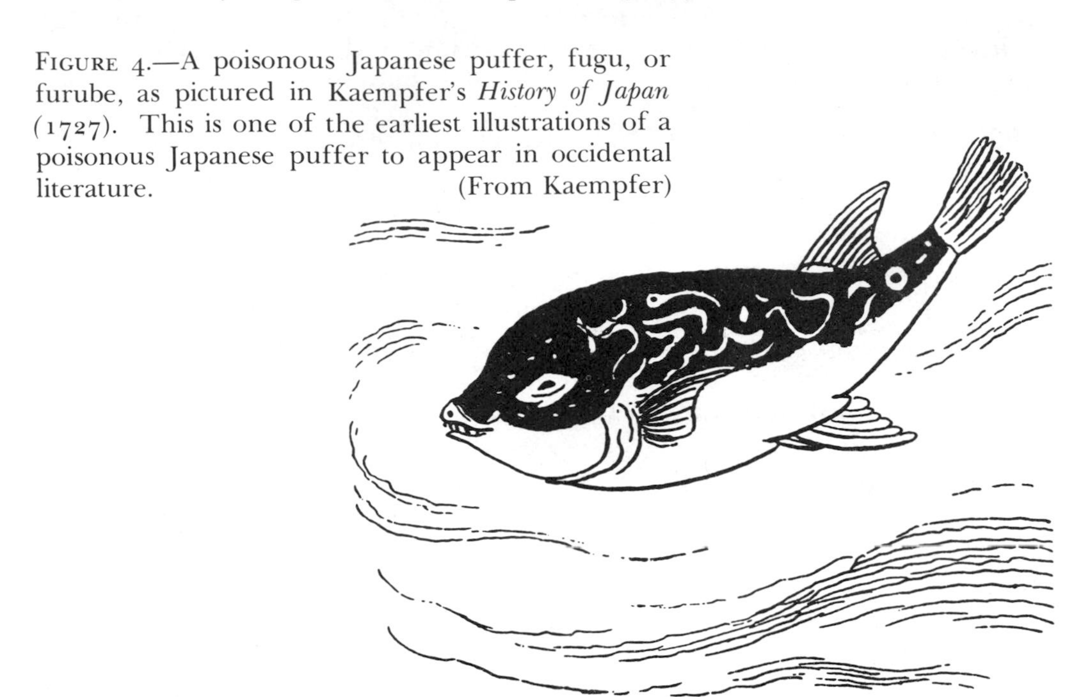

naturalist. Kaempfer's comments are of importance because they give some insight into the knowledge of puffer poisoning that existed among the Japanese during the 18th century.

The British Admiral George Anson (1697-1762), who made a round-the-world voyage between 1740-44, reported two intoxications from eating marine turtles in the Windward Islands, which were said to have occurred in 1697. His records appear in his *A Voyage Round the World, in the Years MDCCXL, I, II, III, IV*. This account is probably the earliest report in the literature regarding marine turtle poisoning.

Abraham Trembley (1700-84), a Swiss naturalist, published a monograph on freshwater coelenterates, *Mémoires pour servir à l'histoire d'un genre de polypes d'eau douce, à bras en forme de cornes,* in 1744, and was the first to describe and publish recognizable drawings of the nematocyst venom apparatus of these minute creatures. Students of marine venomology are indebted to Trembley, for he was the first to attempt to describe a very minute and complex venom apparatus that was later proven to be a lethal weapon system of certain marine jellyfishes.

The first report on the venomous spines of marine catfishes is that of the German ichthyologist Johann G. O. Richter (1754), who stated that the fin spines of the oceangoing "sea cat" were greatly feared by Spanish fishermen.

Although the toxic properties of certain types of echinoderms are believed to have been known as far back as the time of Hippocrates, nothing appears to have been said concerning the nature of their venom organs, the pedicellariae, until the time of Job Baster, who described the pedicellariae of sea urchins. His crude descriptions were published in his *Natuurkundige Uitspanningen* in 1762.

Clupeoid fish poisoning is a violent and lethal form of poisoning which has been known for several centuries, but has only recently been clinically characterized. One of the earliest reports on clupeoid fish poisoning was by the French physician Jean-Baptiste-René Pouppé Desportes (1704-48), who wrote on the subject in his *Histoire des maladies de S. Dominique*. In his report he refers to a violent form of fish poisoning occurring in the West Indies caused by the "cayeux," a small sardine-like fish which at that time was scientifically designated as *Clupea thryssa* (*see* Chisolm, 1808).

On their famous world voyage of 1774, the crew of His Majesty's ship, *Resolution*, under the command of Capt. James Cook (Fig. 5), almost terminated their trip prematurely at Port Sandwich, Malekua, New Hebrides, because of eating a fish described as most closely resembling the red pargo *Sparus pagrus* Linnaeus, a reddish-colored fish about 38 cm in length. This fish is now believed to have been *Lutjanus bohar*, the red snapper, one of the most common causes of ciguatera poisoning. The details of the incident are described by William Anderson (d. 1778), surgeon's mate on the voyage, in a letter entitled "An Account of Some Poisonous Fish in the South Seas," published in the *Philosophical Transactions of the Royal Society in London* in 1776. Anderson's article is generally recognized to be one of the most complete

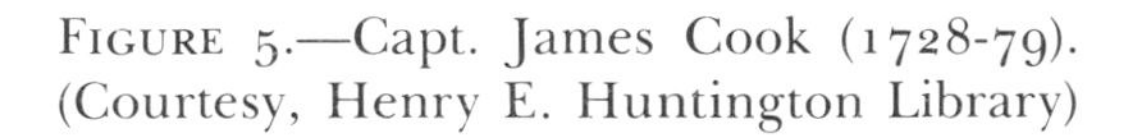

Figure 5.—Capt. James Cook (1728-79). (Courtesy, Henry E. Huntington Library)

clinical accounts of a typical case of ciguatera fish poisoning to appear during the 18th century and has since become a classic in the annals of marine biotoxicology.

In addition to the outbreak of fish poisoning caused by the "red pargo," which was written up so completely by their shipmate William Anderson, the father and son naturalists, Johann Reinhold Forster (1729-98) and Johann Georg Adam Forster (1754-94), also described the outbreak and mentioned a second poisoning due to a species of *Tetraodon* which occurred in New Caledonia on September 7, 1774, about 4 months after the first incident.

Capt. James Cook (1728-79) was sufficiently impressed with the outbreak of puffer poisoning which he incurred that he too comments on the incident in his book *A Voyage Towards the South Pole, and Round the World Performed in His Majesty's Ship the Resolution.*

Although poisonous fishes, now known to cause ciguatera fish poisoning, have been with man since antiquity, the clinical entity, or ciguatera syndrome, was first described by the Portuguese biologist Don Antonio Parra (ca. 1771) in his *Descripción de Diferentes Piezas de Historia Natural las mas del Ramo Marítimo, Representadas en Setenta y Cinco Láminas*, published in Havana, Cuba, in 1787. There was a tendency among Cuban authors of this period to substitute "S" for the usual "C" spelling in Spanish words because the "S" sound more adequately expressed the local Cuban pronunciation (Poey, 1866). Consequently, the early spelling of ciguatera was "Siguatera." The actual origin of the term "ciguatera" appears to be unknown, but is believed to have originated among the natives of Cuba, and dates back to at least the time of the Spanish conquests, and perhaps earlier. A Spanish dictionary published by the Real Academia of Spain (ca. 1866) defined the term ciguatera as "a disease contracted by persons who eat fish that is affected with disease or jaundice." As Poey (1866) later pointed out, there was a univalve mollusk *Turbo pica* in Cuba which the natives termed the "cigua." This mollusk was said to cause indigestion, and people who developed the ailment were called "ciguatos," or "enciguatados." The term was later expanded to include other forms of toxic marine life and hence the term "ciguatera"—the presently accepted clinical nomenclature for a specific form of fish poisoning. Parra's original description of "Siguatera" related the personal experience of his family after eating ciguatoxic fish.

The first outbreak of paralytic shellfish poisoning to be reported in the New World occurred on June 15, 1793. The incident is given in Capt. George Vancouver's (1757-98) *A Voyage of Discovery to the North Pacific Ocean, and Round the World*, published in 1798 and 1801. In his explorations along the west coast of North America, Vancouver visited many of the regions which in future years were to become the scene of several very serious outbreaks of paralytic shellfish poisoning.

By the latter part of the 18th century, fish poisoning was recognized as a valid clinical entity which should be investigated and for which effective means of therapy needed to be instituted. Recognition of this problem was presented to the Medical Society of London in a treatise by the English physician Edward Thomas entitled, "On the Poison of Fish," published in the *Memoirs of the Medical Society of London* in 1799. Thomas' article is of historical importance because it is the first general treatise, and remarkably complete for its time, on the subject of oral fish poisoning.

Thomas discussed the various causes of fish poisoning, with particular reference to the West Indies. He viewed with healthy skepticism the theory that copper banks served as the source of the poison. He noted also the distinct possibility that the poison originated in the food chain of the fish, but was not certain where. Thomas is one of the earliest references in medical literature on poisonous crabs. He stated that the land crab *Cancer terrestris* and the mountain crab of the West Indies became poisonous from eating the bark and leaves of the manchineel tree. Thomas listed the species of poisonous fishes in the West Indies, and he summarized the symptomatology of fish poisoning as known in the West Indies, but made no reference to the term ciguatera. Thomas urged the importance of being able to detect the difference between an edible and a poisonous fish and reaffirmed the old adage that "fish without scales are most liable to be affected by this poison; which seems to be the result of experience, and deserves particular attention from those who mean hereafter to prosecute the inquiry. . . ."

1800-1899. The 19th century is generally regarded as one of the most important epochs in the history of the natural sciences. Here is to be observed a climactic transition period—a shift from the era of natural philosophy to that of modern experimental science. Scientific progress, which had been slowed during the American (1775-83) and French (1789-99) Revolutions, gained momentum in the years that followed. Scientific thought of the 19th century reflected the vigorous intellectual, political, and social currents of the time. As political freedom was achieved, it helped produce rebellion against metaphysical dogma and other restrictive influences of human thought. Moreover, the access of the Third Estate to the universities, the rapid progress of industry, the development of urban centers, the remarkable development of the United States of America, the enormous increase in land and sea traffic, improvement in communication—all contributed to a greater exchange of ideas and scientific thought (Castiglioni, 1941). This is the period whose virility is so ably characterized by the thinking of such men of science as Cuvier, Lamarck, Humboldt, Purkinje, Goethe, Schleiden, Schwann, Helmholtz, Pasteur, Mayer, Liebig, Darwin, and many others of equal or greater importance. This brief reflection may assist in establishing perspective as we review the contributions of those who made up the warp and woof in the story of marine biotoxicology of the 19th century.

The article by the British physician Colin Chisholm entitled "On the Poison of Fish," published in *The Edinburgh Medical and Surgical Journal* of October 1808, is one of the more comprehensive treatises of this period on the poisonous fishes of the West Indies, and was regarded by his contemporaries as one of the most authoritative of its kind. The special value of Chisholm's article centers in his lengthy discussions of the various theories of the origin of fish poisons in nature and the methods of treatment in vogue during that era. Although clupeoid fish poisoning, or clupeotoxism, has been referred to by several of Chisholm's predecessors, insofar as is known, he was the first to present a detailed clinical case report of an actual human fatality.

The earliest reference to experimental research which attempted to determine the origin of oral fish poisons is that by Alexandre Moreau de Jonnes (1778-1870), who placed edible fish in water containing various minerals. His paper entitled "Sur les Poissons Toxicophores des Indes Occidentales," was published in 1821 in the Parisian *Nouveau journal de médicine, chirurgie, pharmacie*. The experiments of Moreau de Jonnes were unique for his time and constituted a milestone in experimental biotoxicology.

The writer of this period who gave the greatest impetus to the subject of marine biotoxicology at large and medical ichthyology in particular is Hermann Friedrich Autenrieth (1799-1874) (Fig. 6). Autenrieth's important biotoxicological work was entitled *Über das Gift der Fische mit Vergleichender Berücksichtigung des Giftes von Muscheln, Käse, Gehirn, Fleisch, Fett und Würsten, sowie der so Genannten Mechanischen Gifte*, published in 1833. It discusses all of

FIGURE 6.—Hermann Friedrich Autenrieth (1799-1874).
(Courtesy, Universitätsbibliothek Tübingen)

the then known poisonous and venomous fishes, including both freshwater and marine species, listing a total of some 125 species of toxic aquatic organisms. It included a thorough review of practically all the early literature on poisonous and venomous fishes, a list of all the species incriminated, symptomatology, treatment, and origin of the poisons in nature.

As microbiology progressed, with the development of the microscope by Anton van Leeuwenhoek (1632-1723), interest in matters relating to public health rapidly accelerated in the last quarter of the 18th century and continued at a fast pace during the 19th century. Thus, awareness of the problems of marine biotoxications came to the forefront, and these problems were given their due attention in the medical literature of the time. Recognition of the public health significance of biotoxications caused by fishes, shellfish, crabs, turtles, etc. is reflected in an important paper of this period entitled "Mémoire sur les empoisonnements par les huîtres, les moules, les crabes, et par certains poissons de mer et de rivière," which appeared in two parts, published in the *Annales d'hygiène publique et de médicine légale*, in 1851. The articles were written by two collaborators, Jean-Baptiste-Alphonse Chevallier (1793-1879) and Edouard Adolphe Duchesne (1804-69). Chevallier was a French pharmaceutical chemist who published extensively on drugs, toxicology, and various matters relating to public health. Duchesne, on the other hand, was a French botanist who, among varied activities, had published in 1836 his well-known *Répertoire des plantes utiles et des plantes vénéneuses du globe*. The two-part article by the collaborators covers much of the current medical literature on the subject and includes considerable information on the etiology, clinical characteristics, and treatment of marine biotoxications. For several decades the publication of Chevallier and Duchesne was generally referred to as an authoritative source for information on this subject.

A comparable article appeared in 1855, in the *Schweizerische Zeitschrift für Medizin und Chirurgie*, by the Swiss physician Dr. Konrad M. Meyer-Ahrens (1817-73) entitled "Von den Giftigen Fischen." This is particularly useful because of the clinical data it provides on freshwater ichthyootoxism, fish roe poisoning.

In 1866, the Cuban naturalist Felipe Poey y Alay (1799-1891) published a comprehensive treatise on fish poisoning entitled "Ciguatera, Memoria Sobre la Enfermedad Ocasionada por los Peces Venenosos," which appeared in the *Repertorio Físico-Natural de la Isla de Cuba*. Poey's paper is one of the most comprehensive on the subject of ciguatera for this entire period, and is especially rich in biological and clinical data pertinent to the poisonous fish fauna of the West Indies. Poey provides one of the earliest reports regarding the establishment of quarantine regulations governing the sale of potentially toxic fishes. Poey also discussed at great length the various theories concerning the origin of fish poisons in nature. He was one of the earliest writers to call attention to the spotty geographical distribution of ciguatoxic fishes, i.e., poisonous on one side of the island, but not on the other. The treatise of Poey is recommended to anyone interested in the history of ciguatera fish poisoning.

France was probably the first country to publish a naval preventive medicine manual which devoted a significant portion to the topic of poisonous plants and animals—terrestrial and marine. The book entitled *Traité d'hygiene navale*, written by Jean-Baptiste Fonssagrives (1823-84) is concerned with a variety of topics relating to naval hygiene, and includes 29 pages dealing with toxic jellyfishes, mollusks, crustaceans, fishes, reptiles, and plants. Most of the organisms discussed are illustrated. Fonssagrives' work marks the beginning of military medical research on marine biotoxicology in modern history.

The monumental work of the British physician Sir Joseph Bart Fayrer

(1824-1907), the *Thanatophidia of India*, is the earliest systematic work dealing strictly with venomous snakes and includes an excellent discussion of the hydrophiid sea snakes, together with two rare clinical reports on humans bitten by sea snakes. He lists 31 species of sea snakes for the Indian Ocean region. Fayrer's work has become a classic in venomology and has provided us with one of the most comprehensive early reports on venomous sea snakes.

On October 17, 1885, a mass biotoxication took place in Wilhelmshaven, Germany, which focused the attention of the medical world on the seriousness of poisonous shellfish. Although this was by no means the first such outbreak, this particular incident occurred at a time and in a place which attracted the attention of some of the finest scientific medical minds of the day. One such person was Rudolph Virchow (1821-1902), the founder of cellular pathology. Virchow, in his two articles "Ueber die Vergiftungen durch Miesmuscheln in Wilhelmshaven" (1885) and "Beiträge zur Kenntniss der Giftigen Miesmuscheln" (1886), contended that it was possible to observe certain external differences which characterized toxic mussels. The subject became quite controversial and of considerable interest to numerous researchers. Although Virchow was wrong in his assumptions, his recognition of the problem undoubtedly livened the interest of medical authorities in the subject of paralytic shellfish poisoning and gave impetus to research on this problem.

While experimental medicine and biology were rapidly developing in Europe and America, comparable investigations were beginning in Japan. Some of the earliest toxicological experiments on poisonous fishes with the use of test animals was conducted by a French physician Charles Remy, a member of the Faculty of Medicine at the University of Paris, while on a visit to Japan. His selection of poisonous puffer species was based on information supplied to him by a Doctor A. Goertz,[3] a physician in Yokohama, who had compiled a list of Japanese puffers incriminated in human intoxications. Remy recorded the first autopsies to be made on laboratory animals that had been killed by puffer poison. Remy's preliminary findings were reported in his article "Note sur les poissons toxiques du Japon," and his more extensive laboratory results in "Sur les poissons toxiques du Japon," both articles appearing in the *Comptes rendus des séances et mémoires de la Société de Biologie* of 1883. At this time Remy's work attracted considerable interest in Europe, America, and Japan, in the physiological effects of puffer poison, and apparently did much to stimulate the experimental work that quickly followed in his wake by Japanese investigators.

Medical geography in the true sense of the word had its beginning in Russia during the first part of the 18th century, and made steady progress in the years that followed. The high quality of Russian research in military preventive medicine and its appreciation of the importance of faunistic studies in medical geography is exemplified in the excellent work of Peter Nikolaevich

[3] There appears to be no further information available regarding the identity of this man.

FIGURE 7.—Peter Nikolaevich Savtschenko (1846-85).
(Courtesy, National Library of Medicine)

Savtschenko, also spelled Savchenko (1846-85) (Fig. 7), on poisonous and venomous marine fishes. Savtschenko clearly recognized the need and importance of basic medical-zoological data on toxic marine organisms, especially as they related to military preventive medicine. His *Atlas des poissons vénéneux*, published in 1886, contained a brief history of the subject, types of toxic fishes, clinical characteristics, and treatment. Savtschenko referred to "syguatera" poisoning, but failed to recognize the scope of the disease. His comments regarding the Polynesian antidotes, which consisted entirely of plant substances, reflect the general interest of Russian military medical geographers in ethnobotanical studies—an interest which has not diminished over the years. The remainder of the volume contained zoological descriptions and illustrations of 45 species of poisonous and venomous fishes from various parts of the world. The work of Savtschenko established a new standard of excellence in the presentation of military preventive medical data for use in field activities, and was without peer or counterpart for more than 60 years.

Experimental research on ichthyohemotoxic fishes began with the great Italian physiologist Angelo Mosso (1846-1910). In his general investigations on blood, Mosso studied the various properties of serum obtained from several kinds of eels, morays, and congers. Tasting eel serum, he noted that it produced a very unpleasant and irritating effect on the mucous membranes, a tightening of the throat, hypersalivation, and other untoward effects. He concluded that there was a toxic principle present, which was readily destroyed by heat or gastric juices. This principle Mosso termed "ichthyotoxicum." Mosso's work on the physiological properties of eel serum was carefully and critically done, and ultimately became a classic in its field. His major work on eel serum was first reported in the *Archives Italiennes de Biologie*, published in 1888. His articles continue to be referred to as an important advance in our knowledge of ichthyohemotoxins, fish serum poisons.

Serious scientific research on piscine venomology begins with the outstanding work of Louis Alphonse Bottard (1854- ?) entitled *Les poissons veni-*

meux, published in 1889. This was the first authoritative, systematic work on the venom organs of fishes. It also provided information on the clinical characteristics of the stings which these fishes produced, many of which Bottard had personally observed. An attempt was made to illustrate some of the fishes and their venom organs, but many of the drawings were relatively crude. Nevertheless the illustrations were among the first available on piscine venom organs.

Research on the venom organs of marine invertebrates progressed slowly. Rudolph Bergh (1824-1909) was the first to attempt a thorough anatomical description of the cone shell venom apparatus. Bergh's monumental work entitled "Beiträge zur Kenntniss der Coniden" was published in the *Nova Acta der Kaiserlich Leopoldinisch-Carolinisch Deutschen Akademie der Naturforscher* in 1895. Included in this work are descriptions of the venom organs of 34 species of cone shells. His descriptions are brief, but well illustrated, and deal largely with the gross morphology rather than histological details. The work of Bergh provided a firm groundwork for those who followed him in the study of cone shell anatomy.

Toward the end of the 19th century there developed in Japan an intense interest in the chemistry and pharmacology of puffer poison. This interest was generated to a large extent by some very formidable public health statistics which revealed that each year more than a hundred Japanese died as a result of ingesting one of their favorite epicurean delights—the deadly fugu. The most comprehensive report on the historical development of poisonous fish research in Japan is by Suehiro (1947). Serious scientific investigations of puffer poison seem to begin with the work of the famous Japanese zoologist Shinnosuke Matsubara, who in 1882 tested the toxicity of puffer eggs by feeding them to dogs and determining the results. His investigations were reported in the *Tokyo Gakugei Zasshi* of 1882 and 1885.

The physiological investigations on puffer poison by Kenji Osawa and his assistant S. Furukawa in 1885 determined the most toxic portions of the fish to be first, the ovary, then the liver, and to a lesser extent the testes. They said the cause of death was due to asphyxia caused by paralysis of the voluntary respiratory muscles which resulted in cessation of respiration.

The first research devoted solely to the isolation of puffer poison is said to have been done by a little-known Japanese chemist, Yashizawa, who in 1894 isolated two active principles from the ovaries of the puffers *Spheroides chrysops, S. rubripes*, and *S. prophyreus*, which he termed "tetrodonin," and a basic material termed "fugamin." The exact chemical nature of these substances is apparently not known (Suehiro, 1947), but they are said to have differed from the two substances later produced by Tahara.

This period draws to a close with the outstanding work of the renowned Japanese chemist Yoshizumi Tahara (1855-1935) (Fig. 8). The most complete article on Tahara's chemical studies was published in German in the *Biochemische Zeitschrift* of 1910, entitled "Über das Tetrodongift." This is the article to which most recent workers refer when they discuss Tahara's studies. In this article Tahara reviews his complete isolation procedures and characterization analyses. In addition, he alters some of his previous conclusions. Among other things he decided that tetrodonin was not a pure substance, but "probably that it consisted of nothing else but a crystalline nontoxic substance, mixed with tetrodotoxin to a greater or lesser extent." Two additional terms are presented, a crystalline base called "tetronine," and a "crystalline substance free from nitrogen," called "tetrodopentose." However, Tahara believed that these latter materials were decomposition products resulting from his isolation procedures and were not present in that form in the puffer. Tahara presented as the empirical formula of fugu or puffer poison, tetrodotoxin, $C_{16}H_{31}NO_{16}$.

Although Tahara's original tetrodotoxin is now believed to have contained only about 0.2 percent of the active principle and his empirical formula is no longer accepted, the importance of his work is not to be minimized. His researches were brilliant and he made a monumental contribution considering the crude equipment and the period in which he worked. Certainly no other person of his era contributed more toward the advancement of the chemistry of marine biotoxins than Yoshizumi Tahara, "a subject of the Emperor of Japan residing at 2 Yechizembori Nichome, Kyobashiku, Tokyo, Japan." One might further add that despite the frightening toxicity of the drug, there are undoubtedly thousands of Oriental physicians who over the decades could testify to the therapeutic usefulness of tetrodotoxin.

FIGURE 8.—Yoshizumi Tahara (1855-1935).
(Courtesy, National Library of Medicine)

Three valuable and highly useful biotoxicological monographs were published near the close of the 19th century. The first of these was by the German physician Otto Friedrich Bernard von Linstow (1842-1916), *Die Giftthiere und ihre Wirkung auf den Menschen*, published in 1894. In 1899, the French pharmacist Henri Coutière (1869-1952) published his 217-page doctoral thesis entitled *Poissons venimeux et poissons vénéneux*. During this same year Jacques Pellegrin (1873-1944) published his 113-page doctoral thesis in medicine entitled *Les poissons vénéneux*. All three of these monographs were destined to become basic works in the field of marine biotoxicology.

A noteworthy advance in echinoderm venomology is to be found in the contributions of Jacob Baron von Uexküll (1864-1944). His two important works were "Ueber Reflexe bei den Seeigeln," and "Die Physiologie der Pedicellarien," which appeared in the *Zeitschrift für Biologie* in 1896 and 1899 respectively. These two articles deal at length with the physiological mechanism of the globiferous pedicellariae and the toxic nature of their secretions. The works of Uexküll are of major importance to anyone studying the venom organs of sea urchins.

1900-1940. At the turn of the century there developed an intense interest by many of the French physiologists in zootoxins and their immunological properties. Aquatic organisms, both freshwater and marine, provided biologically active substances which proved to be well suited for investigations in this budding field of immunology. The ichthyohemotoxins of eels, fish venoms, and the nematocyst venoms of coelenterates attracted the attention of the early immunologists of this period. Among the more notable French workers of this epoch were Lucien Camus (1867-1934) and Marcel Eugene Emile Gley (1857-1930).

Although Camus and Gley are renowned for their research in other areas of physiology, they were particularly active in studying the toxicological properties of eel serum and the mechanism of protective immunization against the effects of eel ichthyohemotoxins in laboratory animals. The works of Camus and Gley on the immunological properties of eel serum were extensive and covered the period from 1898 to 1929.

Research on the toxin of the purple or hypobranchial gland of *Murex* shellfish was first conducted by the French physiologist Raphael Dubois. His studies were primarily concerned with animal pigments, but during the course of his investigations he encountered the toxic properties of extracts prepared from the purple gland of muricine shellfish. His most important work in this field was his "Recherches sur la pourpre et sur quelques autres pigments animaux," which appeared in the *Archives de Zoologie Expérimentale et Générale* in 1909. The work of Dubois attracted the attention of later physiologists and chemists and provided some of the groundwork for the final isolation and characterization of murexine by Erspamer and his associates several decades later.

A noteworthy but brief report on a little-known area of biotoxicological research is that of Josephine Azurem Furtado on the poisonous fishes of

Brazil. She published her findings in "Pesquizas ichthyologicas na Bahia do Rio de Janeiro" (1903), one of the few publications ever to appear on the poisonous fishes of Brazil.

The only investigative report dealing with the toxic gonadal extracts of sea urchins is by the French scientist Gustave Loisel (1903, 1904).

The first attempt to determine the immunological properties of weeverfish venom was by A. Briot of the Faculty of Sciences of Marseilles, France.

The work of the Italian investigator Antonio Porta of the University of Camerino, Italy, did much to stimulate the later anatomical researches of H. Muir Evans and others on venomous fishes. His single work, entitled "Ricerche Anatomiche Sull'apparecchio Velenifero di Alcuni Pesci," appeared in the *Anatomischer Anzeiger* of 1905.

The investigations of Friedrich Wolfgang Martin Henze (1873-1956) of the Medicinal Chemistry Laboratory of the University of Innsbruck, Austria, marked a milestone in the study of cephalopod venom in that he was the first to attempt to characterize chemically the toxic fraction of octopus saliva.

An important basic work to any student in echinoderm venomology is Eugénie Kayalof's *Étude des toxines des pédicellaires chez les oursins*, published in Paris in 1906. This was the first study to deal with some of the immunological properties of sea urchin venom (Henri and Kayalof, 1906).

The studies of Etienne de Rouville of the University of Montpellier, contains an extensive review of all the work that had been published to date and additional original data on the toxicological effects of cephalopod venom. He published his "Etudes physiologiques sur les glandes salivaires des céphalopodes" in the *Bulletin Mensuel de l'Académie des Sciences et Lettres de Montpellier* in 1910.

Two very useful general works on poisonous and venomous animals appeared during the first decade of the 20th century. In 1905, Albert Calmette (1863-1933), the renowned French physician-bacteriologist, published his *Venoms, Venomous animals and Antivenomous Serum-therapeutics* in both French (1907) and English (1908) editions. Although most of the 403-page book is concerned with venomous snakes and the physiology of envenomation and treatment, a section of the book is devoted to toxic marine invertebrates and fishes. Calmette's book was authoritative, well illustrated, and the most comprehensive medical treatise of its kind at that time.

A second valuable general work on toxic animals was *Die Giftigen Tiere*, written by the German zoologist Otto Taschenberg (1854-1922), professor of the Zoological Institute at Halle. The book, a well-illustrated volume of 325 pages, was published in Stuttgart in 1909. Although largely zoological in approach, it does contain valuable medical information as well.

Despite the numerous ancient reports on the toxicity of the nudibranch *Aplysia*, actual experimentation on the nature of their poisons was until recently the work of a single man, the German physician-pharmacologist Ferdinand Flury (1877-1947). Flury was an outstanding pharmacologist internationally recognized for his studies in industrial hygiene and toxicology and probably best known for his *Lehrbuch der Toxicologie*.

Catfish venoms are among the least known ichthyoacanthotoxins. One paper which has appeared on the properties of catfish acanthotoxins is the remarkable work of the Japanese physician Toyojiro Toyoshima (1885- ?) entitled a "Serological Study of the Toxin of the Fish *Plotosus anguillaris* Lacepede," which appeared in Japanese in the *Journal of the Japan Protozoological Society* in 1918. Toyoshima's work covered the biology of the fish, methods of extracting the venom, physiological effects of the poison on a variety of invertebrates and on fish, amphibians, reptiles, birds and mammals, the toxicology and histological changes at injection sites of the venom, the chemical and immunological properties of the poison, and attempts to develop an antitoxin. His work appears to have been carefully done and was an important contribution to piscine venomology. Unfortunately it was written in Japanese and published in a generally inaccessible journal. Consequently, his work has been largely overlooked by the scientific community at large.

Among the more notable works on the physiological effects of puffer poison are the papers of the Japanese physician Fusao Ishihara of the Tokyo Medical University. His three most comprehensive papers contain a large amount of useful information on the chemistry and pharmacology of puffer poison.

During the period of 1904 to 1918 the Polish physician Wladislaus Kopaczewski (1886-1959) published more than a dozen papers on the toxicity of eel serum, most of which appeared in the *Comptes rendus des seances de la Société de Biologie et de ses filiales* at Paris.

Another important series of contributions of this period is the work of the great Italian comparative physiologist-physician Filippo Bottazzi (1867-1941). His extensive research on the physiology of the posterior salivary glands of cephalopods has done much to increase our understanding of the function of the salivary venom glands of cephalopods.

Noteworthy among the researches of this epoch were the studies of Olympio Oliveira Ribeiro da Fonseca (1895-) of the Oswaldo Cruz Institute, Rio de Janeiro, Brazil. Fonseca's (1917, 1919) reports were largely confined to the toxicity of certain species of Brazilian puffers, and were especially appreciated since they dealt with a little-known poisonous fish fauna of the tropical Atlantic.

The most comprehensive publication that has appeared to date on zootoxicology is the two-volume *Animaux venimeux et venins* by Marie Phisalix (1861-1946), published by Masson, Paris, in 1922. Her monograph contains a vast compilation of biological, toxicological, and medical data, and is well illustrated throughout. Mme. Phisalix worked with many kinds of animal poisons, including those produced by arachnids, fishes, batrachians, lizards, and snakes. She was an indefatigable worker and published numerous articles on poisonous animals and the nature of their poisons, most of which are completely reviewed in her great monograph.

One of the outstanding leaders in venomological research was Yevgeniy Nikanorovich Pavlovskiy (also spelled Evgenii N. Pavloskii, or Eugenius N.

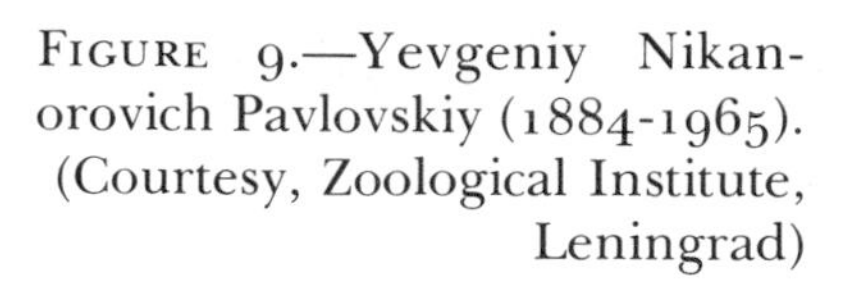

FIGURE 9.—Yevgeniy Nikanorovich Pavlovskiy (1884-1965). (Courtesy, Zoological Institute, Leningrad)

Pawlowsky) (1884-1965) (Fig. 9). The work of Academician Pavlovskiy and his associates at the Zoological Institute of the Leningrad Academy of Sciences demonstrated a remarkable degree of insight and appreciation of the overall public health and military importance of biotoxic agents. In the U.S.S.R. the study of biotoxins is part of a long-range program in medical geography, or human ecology, and constitutes a segment of technical data to be utilized as a part of the total national economy (Pavlovskiy, 1961).

Although Pavlovskiy is well known by microbiologists for his large number of contributions to the communicable disease field, he is equally recognized for his important contributions to venomology. Most of Pavlovskiy's publications in venomology appeared during the period from 1903 through 1929 and concern themselves with such organisms as venomous insects, scorpions, ticks, fishes, and reptiles. Moreover, he is one of the very few investigators to have worked on the poisonous freshwater fishes of central Asia. His greatest work on poisonous animals is his well-known *Gifttiere und Ihre Giftigkeit*, published in 1927. In 1963, Pavlovskiy published *Works on Experimental Zoology and Venomous Animals*, in Moscow (in Russian).

The work of the German physician Edwin Stanton Faust (1870-1929), although not as comprehensive as that of some of his predecessors, was nevertheless important to general biotoxicologists. Faust was a physician, worked actively as a pharmacologist, and among other things studied the poisons of toads, salamanders, and snakes. Among his better known general works are his *Die Tierischen Gifte* (1906), and *Vergiftungen Durch Tierische Gifte* (1927), in which he discusses various venomous invertebrates and fishes.

Interest in the ciguatera problem in the Caribbean region continued through the publications of Wilhelm Henrich Hoffman (1871-1950) of Cuba. His publications were concerned largely with the public health restrictions governing the sale of ciguatoxic fishes in Cuba and are among the very few in existence concerned with this aspect of the poisonous fish problem.

Much of our knowledge of the toxic properties of sea anemone venom is based on the work of two Rumanian microbiologists of the Faculty of Medicine of Bucharest, Jean Cantacuzène (1863-1943) and Nicholas L. Cosmovici (fl. 1925). They investigated the physiological effects of toxic extracts obtained from the nematocysts and tentacles of the sea anemone *Adamsia* which they tested on various crabs.

Outstanding among fish venomologists of this epoch is Harold Muir Evans (1866-1947). As a practicing physician and surgeon in the fishing port of Lowestoft, England, Evans came in frequent contact with fishermen who suffered from severe wounds inflicted by weeverfish (*Trachinus*), stingrays (*Trygon*), the spiny dogfish (*Squalus*), and others. Some of the most severe wounds that he encountered were weeverfish stings; consequently he directed much of his early efforts to the venom organs of *Trachinus* and the nature of their venom. Later he published on the venom organs of stingrays and the spiny dogfish (Fig. 10).

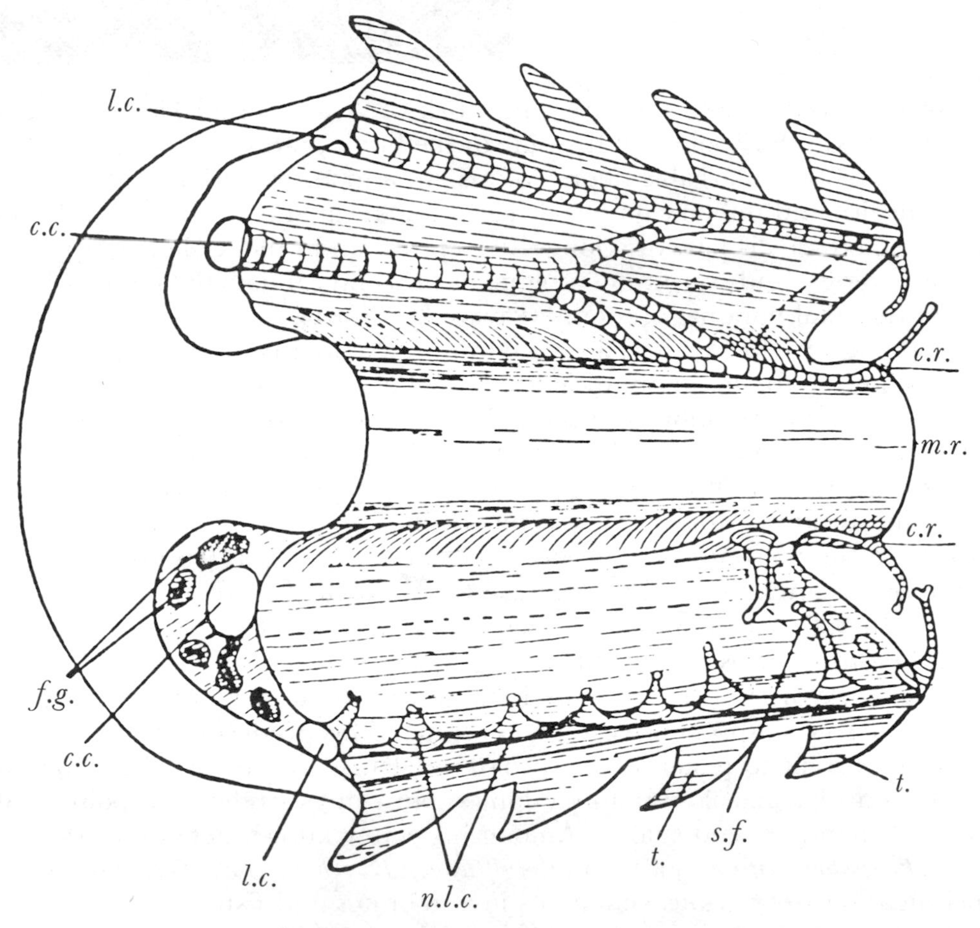

FIGURE 10.—One of Evans' (1916) earlier drawings of a cross section of the stingray sting in which he erroneously portrayed a series of venom canals and nipples. This is one of the first descriptions of a stingray venom apparatus. Evans later corrected his mistake. (From Evans, 1916)

Evans' most comprehensive technical work was "The Defensive Spines of Fishes, Living and Fossil, and the Glandular Structure in Connection Therewith, with Observations on the Nature of Fish Venoms," which appeared in the *Philosophical Transactions of the Royal Society of London* in 1923.

With the outbreaks of paralytic shellfish poisoning in the vicinity of San Francisco, Calif., during July 1927, in which more than 100 persons were involved and several deaths resulted, the problem came into sharp focus, and with it a demand for a solution. Under the direction of Karl Friederich Meyer (1884-1974) and Hermann Sommer (1899-1950) (Fig. 11) of the George Williams Hooper Foundation, University of California, working in cooperation with the California State Fish and Game Commission, a series of studies was conducted which provided the foundation upon which most recent research in this field has been established. The brilliant researches of Sommer and his

FIGURE 11.—Hermann Sommer (1899-1950). (Courtesy, George W. Hooper Foundation)

associates finally demonstrated that there was a direct correlation between the toxicity of the shellfish and the numbers of dinoflagellates *Gonyaulax catenella* present in the seawater. Their discovery of the primary cause of paralytic shellfish poisoning will remain in the annals of biotoxicology as one of the great breakthroughs in man's never ending inquiry into the mysteries of marine organisms and the biogenesis of their poisons.

In recounting the historical highlights of marine biotoxicology, one would be remiss in failing to mention the important researches of two Japanese physicians, Rikuichi Matsuo and Takashi Yasukawa. Both men investigated the poisonous fishes of Micronesia. Matsuo's work was entitled "A Study of the Poisonous Fishes of Jaluit Island." Yasukawa's investigations were limited to the poisonous fishes of the Mariana Islands, and specifically Tinian, Rota, and Saipan. He furnished a brief summary on the incidence of poisonous fishes based on fishermen's reports, ecology and distribution, stomach content analyses, bacteriology, and medical aspects.

The work of Theodore A. Maass is of great interest to biotoxicologists because it marks the first attempt to compile an exhaustive tabular handbook of all known toxic animals and their poisons. Descriptive material was reduced to a minimum and only the pertinent facts regarding the toxicological, pharmacological, and chemical properties were reported, together with abbreviated references to the literature. His work, entitled *Gift-tiere*, appeared in the *Tabulae Biologicae* series published by W. Junk at 's-Gravenhage in 1937, and consisted of 272 pages of tabular data.

Zénon Marcel Bacq (1903-) of Belgium has made numerous valuable contributions to our knowledge of the physiological effects of the posterior salivary gland secretions of the octopus on laboratory animals and has the distinction of being the only worker to have published on the pharmacological properties of extracts prepared from the tissues of nemertean worms.

The contributions of the American ichthyologist Eugene Willis Gudger (1905-51) to medical ichthyology are broad in scope and unusual in their coverage. The writings of Gudger provide a broader foundation to the overall subject of medical ichthyology than the publications of any other single author. Gudger's writings are scholarly and well documented. His works contain a great quantity of information on venomous stingrays, poisonous gempylid fishes, ciguatera poisoning, and other fishes dangerous to man.

The Danish zoologist Theodor Mortensen (1868-1952) is probably the greatest single contributor to our knowledge of the general morphology of the pedicellariae of sea urchins. His great work, *A Monograph of the Echinoidea*, has no peer in its field, and there are few counterparts to be found in other areas of marine zoology.

Although most of his researches were concerned with the physiological effects of Australian snake venoms, Charles Halliley Kellaway (1889-1952) also did some of the earliest studies on the central and peripheral actions of paralytic shellfish poisons. His major efforts in venomology were on the circulatory effects of snake venoms, and in particular on how these circulatory effects were secondary to the liberation of histamine from cells damaged by venom. This interest in the role of histamine in toxic and traumatic conditions was the central theme of his studies. In brief, it can be safely stated that the physiological investigations of Kellaway and his group on Australian snake venoms are some of the most comprehensive ever conducted.

The Australian physician-radiologist Hugo Flecker (1884-1957) is best known for his contributions on jellyfish and cone shell stings which occurred in the North Queensland area. During his activities with the Australian armed forces Flecker made numerous observations on jellyfish stings, and this, coupled with his keen interest in natural history, stimulated him to publish his valuable observations on "Irukandji stings" and other biotoxications. Flecker's writings are among the earliest reports on the deadly sea wasp, which Southcott subsequently named *Chironex fleckeri* in his honor.

Confirmation of the causative agent of paralytic shellfish poisoning as *Gonyaulax tamarensis* was largely the result of the excellent research of Alfreda B. Needler (1913-51) and her associates. Her report confirming the identification of *G. tamarensis* as the causative agent appeared in the *Journal of the Fisheries Research Board of Canada* for 1949.

World War II Period. The most exhaustive and outstanding biotoxicological research conducted during the World War II period was that done by the Japanese Government. With the expansion of Japanese naval operations in the South Pacific just prior to World War II, there was a rapid increase in biotoxic fatalities caused by the eating of poisonous tropical reef fishes, whereupon the Imperial Japanese Navy requested Toshio Kumada (1884-1953), president of the Nissan Fisheries Experiment Station, Odawara, Japan, to conduct an extensive investigation on poisonous fishes in the Japanese mandated islands of Micronesia. The fisheries station was operated by the Nippon Suisan Company, Ltd., one of the largest commercial fishing companies in Japan and one with extensive experience in fisheries operations in Micronesia. Because of the elderly condition of Kumada, Yoshio Hiyama, a fisheries biologist of the Tokyo Imperial University, was requested to take charge of the field investigations. Their studies were conducted largely in Saipan, Mariana Islands, and Jaluit, Marshall Islands, although other island areas were visited throughout Micronesia. The survey included interrogating natives, fishermen, local fish technologists, and physicians located in the islands. Despite numerous handicaps because of the lack of adequate field laboratory conditions, a great number of reef fishes of various sizes and kinds were captured and tested for their toxicity.

The results of Hiyama's study appeared in a series of special civilian and military handbooks, written in Japanese and profusely illustrated with magnificent color plates (Fig. 12). These manuals are undoubtedly the most superb survival handbooks ever published for military operations. The military editions, under the title of *Poisonous Fish*, were edited by Toshio Kumada and contained a map showing the geographical distribution, information on the kinds, habitat, and methods of detecting poisonous fishes, and information on the symptoms of fish poisoning and treatment. Despite the intense efforts to warn their troops of the dangers of eating poisonous fishes, Japanese technical sources estimate that they lost more than 400 military personnel from the eating of toxic fishes in Micronesia during World War II. The work of Hiyama will always remain as one of the great pioneering efforts in biotoxicological research, an attempt to elucidate the nature and dangers of poisonous fishes in the tropical Pacific.

During World War II, Allied troops encountered many types of noxious marine organisms in the region of the Great Barrier Reef and throughout Australasia. In an attempt to provide information on this subject, Gilbert Percy Whitley (1903-76) of Australia, prepared his well-illustrated bulletin on *Poisonous and Harmful Fishes*, published by the Commonwealth Council for Scientific and Industrial Research in 1943. The booklet was widely used by troops operating throughout the Australasian area.

Most of the scientific contributions of Albert William Christian Theodore Herre (1868-1962) to marine biotoxicology appeared as a series of unpublished U.S. Army Intelligence Reports containing considerable practical information regarding the identification of noxious species, methods of

FIGURE 12.—(Left) Japanese officer's military survival manual on poisonous fishes and other marine organisms. The original book was bound in silk. An edition for enlisted men as well as one for civilians was also published. (Below) Manual open to show folding color plates. (Library, World Life Research Institute)

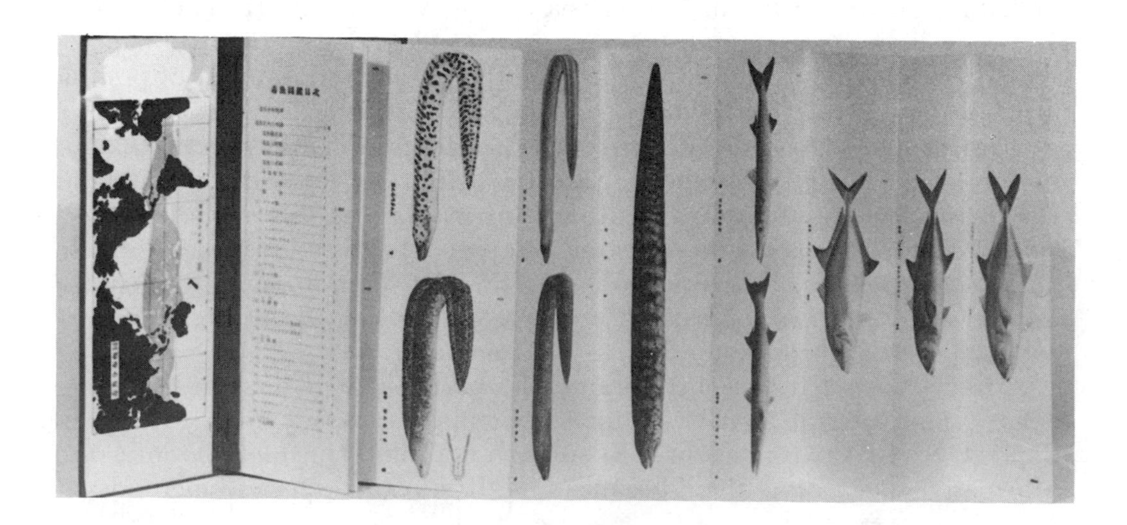

catching fishes, and dangers of contact with poisonous and venomous fishes in the Indo-Pacific regions. Unfortunately, the material was subsequently classified and never made available to the scientific community. However, Herre has published one of the very few reports to appear on the poisonous fishes of the Philippine Islands, entitled "Poisonous and Worthless Fishes—an Account of the Philippine Plectognaths." It appeared in the *Philippine Journal of Science* for 1924.

Herre published one of the most complete clinical accounts available on a sting from the Oriental catfish, *Plotosus anguillaris* (1949). His review of the scorpaenoid fishes of the Philippines and adjacent seas (1951), while primarily a taxonomic treatise, will prove helpful to any venomologist working with this group of fishes.

Although biotoxins were recognized to be of military importance prior to World War II, they were largely looked upon by the scientific community as being of academic curiosity rather than of practical significance. During World War II, military operations again emphasized the importance of having scientific facts available regarding the nature of these substances and the organisms that produce them. Steady improvement in perspective by the scientific community towards biotoxicology has been conspicuous during the postwar era. The result is a noticeable increase in recognition of the character of biotoxic marine organisms in the overall picture of oceanology, and their bearing on such areas as nutrition, military sciences, medicine, chemistry, and new drug development. The need for further research in marine biotoxicology has been clearly pointed up by the exemplary work of Walford entitled *Living Resources of the Sea*, published in 1958, and by the conference on *Biochemistry and Pharmacology of Compounds Derived from Marine Organisms* held and sponsored conjointly by the New York Academy of Sciences, the American Institute of Biological Sciences, and the Office of Naval Research, Department of the Navy, during April 1960 (Nigrelli, 1960).

Although the future appears hopeful, the field of biotoxicology has not made the progress that is needed. It is well to keep in mind that at present there is no organized effort on the part of any laboratory or any country to explore systematically the scantily known world of biotoxins. Thus, "There remains much land yet to be possessed."

Current Investigators in Marine Biotoxicology. The following list is based mostly on publication data or on a first-hand knowledge of those who are currently working in the field, even though their publications on the subject may be minimal. Every attempt has been made to make this list as comprehensive as possible. Any omissions are due to oversight rather than to intent. The author welcomes any information that may make this list more inclusive.

TABLE 1.—*List of Current Investigators in Marine Biotoxicology*

Name	Affiliation/Address*	Area of Interest
Abe, Tokiharu	Fisheries Agency, Tokai Regional Fisheries Research Laboratory, Kyobashi, Tokyo, Japan	Taxonomy of poisonous puffers
Ackermann, Dankwart	University of Würzburg, Germany	Chemistry of toxic coelenterates and mollusks
Akiba, Tomoichiro	Research laboratories, Chugai Pharmaceutical Co., Ltd., Toshima-ku, Tokyo, Japan	Venerupin poisoning
Alcala, A. C.	Biology Department, Silliman University, Dumaguete City, Philippine Islands	Crab poisons
Alender, Charles B.	California State College, Department of Biology, Long Beach, California	Echinoderm poisons
Anderson, James I. W.	Northeast Shellfish Sanitation Research Center, U.S. Public Health, Narragansett, Rhode Island	Marine biotoxins
Asan, Motokazu	Department of Fisheries, Faculty of Agriculture, Tohoku University, Sendai, Japan	Biochemistry of mollusks
Bagnis, Raymond	Medical Oceanographic Branch, Medical Research Institute Louis Malardé, Papeete, Tahiti	Fish poisons
Bannard, Robert A. B.	Chemical Laboratories, Defense Research Board, Ottawa, Canada	Biotoxicology of paralytic shellfish poison
Banner, Albert H.	Department of Zoology, University of Hawaii, Honolulu, Hawaii	Ciguatera poison in fishes
Barnes, John H.	General medical practice, Cairns, Queensland, Australia	Coelenterate stings, especially *Chironex*
Baxter, E. H.	Commonwealth Serum Laboratories, Parkville, Victoria, Australia	Immunochemistry of coelenterate and sea-snake venoms
Boylan, David B.	Hawaii Institute of Marine Biology, University of Hawaii, Honolulu, Hawaii	Chemistry of boxfish poisons
Burnett, J. W.	School of Medicine, University of Maryland, Baltimore, Maryland	Jellyfish venoms
Calton, G. J.	School of Medicine, University of Maryland,	Jellyfish venoms
Cameron, Ann M.	Department of Zoology, University of Queensland, Brisbane, Australia	Venomous marine animals
Castex, Mariano N.	Institute of Biology, College of the Immaculate Conception, Sante Fe, Argentina	Envenomations by freshwater stingrays
Ciereszko, Leon S.	Department of Chemistry, University of Oklahoma	Biochemistry of marine invertebrates
Cleland, John B.†	School of Medicine, University of Adelaide, Adelaide, South Australia	Stinging coelenterates and echinoderms
Courville, Donovan A.	School of Medicine, Loma Linda University, Loma Linda, California	Chemistry of marine biotoxins
Crone, H. D.	Australian Defence Scientific Service, Defence Standards laboratories, Melbourne, Victoria, Australia	Biochemistry of coelenterates
Doorenbos, Norman J.	School of Pharmacy, University of Mississippi, University, Mississippi	Fish poisons

* According to the most recent information available to the author.

† Deceased.

TABLE 1.—*List of Current Investigators in Marine Biotoxicology (continued)*

Name	Affiliation/Address*	Area of Interest
Endean, Robert	Department of Zoology, University of Queensland, Brisbane, Australia	Poisonous and venomous marine animals
Erhardt, J. P.	Division of General Biology and Ecology, Health Service Research Center, Armed Forces, Paris, France	Fish poisons
Erspamer, Vittorio	Institute of Pharmacology, University of Parma, Parma, Italy	Pharmacology of mollusk poisons
Fänge, Ragnar	Institute of Zoological Physiology, University of Göteborg, Göteborg, Sweden	Venomous mollusks
Fish, Charles J.	University of Rhode Island, Kingston, Rhode Island	Toxic marine animals
Freeman, Shirley E.	Australian Defence Scientific Service, Defence Standards Laboratories, Melbourne, Victoria Australia	Pharmacology of coelenterates and mollusks
Friess, Seymour L.	Physical Biochemistry Division, Naval Medical Research Institute, National Naval Medical Center, Bethesda, Maryland	Pharmacology of holothurin
Fuhrman, Frederick A.	Max C. Fleischmann Laboratories of the Medical Sciences, Stanford University, Stanford, California	Pharmacology of tarichatoxin, tetrodotoxin, and saxitoxin
Fukuda, Tokushi	Kagoshima University, Kagoshima, Japan	Puffer poisoning
Gardiner, J. E.	Department of Pharmacology, Faculty of Medicine, University of Singapore, Sepoy Lines, Singapore	Crab poisons
Ghiretti, Francesco	University of Sassari, Sassari, Italy	Physiology and biochemistry of cephalotoxin
Goldner, Ronald	School of Medicine, University of Maryland, Baltimore, Maryland	Jellyfish poisons
Goto, Toshio	Chemical Institute, Faculty of Science, Nagoya University, Nagoya, Japan	Chemistry of tetrodotoxin
Guinot, Danièle	Laboratoire de Zoologie (Arthropodes), Muséum National d'Histoire Naturelle, Paris, France	Toxicity of crustaceans, taxonomy of poisonous crabs
Halstead, Bruce W.	International Biotoxicological Center, World Life Research Institute, Colton, California	Biotoxicology
Hashimoto, Yoshiro	Laboratory of Fisheries Chemistry, Faculty of Agriculture, University of Tokyo, Hongo, Tokyo, Japan	Chemistry of poisonous marine animals
Hatano, Mutsuo	Laboratory of Food Chemistry, Faculty of Fisheries, Hokkaido University, Hokkaido, Japan	Chemistry of marine biotoxins
Helfrich, Philip	Hawaii Marine Laboratory, University of Hawaii, Honolulu, Hawaii	Poisonous fishes
Hessel, Donald W.	Bio-Laboratories, Colton, California	Chemistry of marine biotoxins
Hessinger, D. A.	Department of Biology, University of South Florida, Tampa, Florida	Coelenterate venoms

* According to the most recent information available to the author.

TABLE 1.—*List of Current Investigators in Marine Biotoxicology (continued)*

Name	Affiliation/Address*	Area of Interest
Huang, Chian L.	Department of Pharmacology, School of Pharmacy, Wayne State University, Detroit, Michigan	Pharmacology of marine biotoxins
Jullien, Antoine	Faculty of Sciences, and National School of Medicine and Pharmacy, University of Besançon, France	Toxicity of shellfish
Kaiser, Erich	Institute of Medicinal Chemistry, University of Vienna, Vienna, Austria	Chemistry of animal poisons
Kao, C. Y.	Department of Pharmacology, State University of New York, Brooklyn, New York	Pharmacology of tetrodotoxin and saxitoxin
Kawabata, Toshiharu	Division of Fish Poisoning and Botulism, Department of Food Control, National Institute of Health, Tokyo, Japan	Epidemiology of poisonous marine organisms
Keegan, Hugh L.	Department of Entomology, Medical General Laboratory (406), U.S. Army Medical Command, Tokyo, Japan	Toxic marine organisms
Keen, T. E. B.	Australian Defence Scientific Service, Defence Standards Laboratories, Melbourne, Victoria, Australia	Toxicology of coelenterates
Kohn, Alan J.	Department of Zoology, University of Washington, Seattle, Washington	Venomology of mollusks
Lakso, Jolean D. U.	Department of Pharmacology, University of California, Davis, California	Pharmacology of *Palythoa* poisons
Lane, Charles E.	Institute of Marine and Atmospheric Sciences, University of Miami, Miami, Florida	Coelenterate and mollusk venoms
Legeleux, Gilbert	Official pharmacist, Lombez (Gers), France	Poisonous fishes
Lenhoff, H. M.	Department of Developmental and Cell Biology, University of California, Irvine, California	Coelenterate venoms
Li, Kwan-ming	Hawaii Marine Laboratory, Department of Pharmacology, University of Hong Kong, Hong Kong	Pharmacology of fish poisons
Maretić, Zvonimir	Service for Infectious Disease and Epidemiology, General Hospital, Medical Center, Pula, Yugoslavia	Toxicology and medical aspects of venomous fishes
Marr, A. G. M.	Commonwealth Serum Laboratories, Parkville, Victoria, Australia	Immunochemistry of coelenterates
Martin, D. F.	Department of Chemistry, University of South Florida, Tampa, Florida	Dinoflagellate poisons
Martin, Edgar J.	Consultant in biochemical pharmacology, Berkeley, California	Toxins of sea anemones
McFarren, Earl F.	Food Chemistry, Milk and Food Research, Robert A. Taft Engineering Center, Cincinnati, Ohio	Poisonous fish and shellfish
McLachlan, Jack L.	Applied Biology, National Research Council, Ottawa, Canada	Toxic algae

* According to the most recent information available to the author.

TABLE 1.—*List of Current Investigators in Marine Biotoxicology (continued)*

Name	Affiliation/Address*	Area of Interest
McLaughlin, John J. A.	Haskins Laboratories, and Biology Department, Fordham University, New York	Dinoflagellate poisons
McNeill, Frank	Australian Museum, Sydney, Australia	Noxious marine invertebrates
Medcof, John C.	Fisheries Research Board of Canada, Biological Station, St. Andrews, New Brunswick	Paralytic shellfish poisoning
Michl, Heribert	Institute of Analytical Chemistry, University of Vienna, Vienna, Austria	Chemistry of zootoxins
Migita, Masao	Japan's Women's University, Tokyo, Japan	Marine biochemistry
Minton, Sherman A., Jr.	Department of Microbiology, School of Medicine, Indiana University, Bloomington, Indiana	Toxicology of venoms
Mir, G. N.	Material Science Toxicology Laboratories, University of Tennessee Medical Center, Memphis, Tennessee	Pharmacology of marine biotoxins
Modglin, Francis R.	Bio-Laboratories, Colton, California	Marine venomology
Mold, James D.	Organic Chemical Research, Liggett and Myers Tobacco Co., Durham, North Carolina	Chemistry of paralytic shellfish poison
Moore, Richard E.	Department of Chemistry, University of Hawaii, Honolulu, Hawaii	Chemistry of paralytic shellfish poison
Morice, Jean	Chief of the Laboratory, Institute of Marine Fisheries, Saint Barthélemy, French Antilles	Poisonous fishes
Mosher, Harry S.	Department of Chemistry, Stanford University, Stanford, California	Chemistry of taricha-toxin and tetrodotoxin
Nachmansohn, David	Department of Neurology, College of Physicians and Surgeons, Columbia University, New York	Neurophysiology of puffer poison
Niaussat, P.	Division of General Biology, Research Center of the Health Service of the Army, Ministry, of the Army, Clamart, France	Ciguatoxic fishes
Nigrelli, Ross F.	New York Aquarium; Laboratories of Marine Sciences, New York Zoological Society; Department of Biology, New York University, New York	Marine biochemistry and ecology
Ogura, Yasumi	Department of Pharmacology and Toxicology, Institute of Food Microbiology, Chiba University, Chiba, Japan	Pharmacology of puffer poison
Okonogi, Takashi	Pathology Department, Central Research Laboratories, Sankyo Company, Ltd., Tokyo, Japan	Immunology of sea-snake venoms
Pope, Elizabeth C.	Department of Marine Invertebrates, Australian Museum, Sydney, Australia	Noxious marine animals
Prakash, Anand	Fisheries Research Board of Canada, Biological Station, St. Andrews, New Brunswick	Toxic dinoflagellates
Pringle, Benjamin H.	Northeast Shellfish Sanitation Research Center, Public Health Service, Narragansett, Rhode Island	Paralytic shellfish poisoning

* According to the most recent information available to the author.

TABLE 1.—*List of Current Investigators in Marine Biotoxicology (continued)*

Name	Affiliation/Address*	Area of Interest
Provasoli, Luigi	Yale University, New Haven, Connecticut	Toxic dinoflagellates
Pugsley, Leonard I.	Food and Drug Directorate, Department of Health and Welfare, Ottawa, Ontario, Canada	Paralytic shellfish poisoning
Randall, John E.	Bernice P. Bishop Museum, Honolulu, Hawaii	Ciguatera and other noxious fishes
Rapoport, Henry	Department of Chemistry, University of California, Berkeley, California	Chemistry of paralytic shellfish poison
Rayner, Martin D.	Department of Physiology, School of Medicine, University of California, San Francisco, California	Pharmacology and physiology of fish and mollusk poisons
Reid, H. Alistair	Liverpool School of Tropical Medicine, Pembroke Place, Liverpool, England	Venomology of sea snakes
Riegel, Byron	Chemical Research, G. D. Searle & Co., Chicago, Illinois	Chemistry of paralytic shell-fish poison
Rodahl, Kaare	Institute of Work Physiology, Oslo, Norway	Arctic physiology, toxicity of polar bear and seal livers
Rosenberg, Philip	School of Pharmacy, University of Connecticut, Storrs, Connecticut	General toxicology, par-ticularly sea-snake venoms
Roux, Georges	Faculté de Médicine et de Pharmacie, Toulouse, France	Marine biotoxicology
Russell, Findlay E.	Laboratory of Neurological Research, School of Medicine, University of Southern California, Los Angeles, California	Zootoxicology
Saunders, Paul R.†	Department of Biological Sciences, Uni-versity of Southern California, Los Angeles, California	Pharmacology and chemistry of venomous marine animals
Schantz, Edward J.	U.S. Army Biological Laboratories, Fort Detrick, Frederick, Maryland	Paralytic shellfish poison
Scheuer, Paul J.	Department of Chemistry, University of Hawaii, Honolulu, Hawaii	Marine biotoxins
Shilo, Moshe	Laboratory of Microbiological Chemistry, Hadassah Medical School, Hebrew University, Jerusalem, Israel	Protozoan poisons
Sobotka, Harry H.	Department of Chemistry, Mt. Sinai Hospital, New York	Chemistry and physiological properties of holothurin
Southcott, Ronald V.	South Australian Museum, Adelaide, Australia	Human injuries by marine vertebrates
Stephenson, Norman R.	Physiology and Hormones Section, Food and Drug Directorate, Department of National Health and Welfare, Ottawa, Ontario, Canada	Bioassay of paralytic shellfish poison
Sutherland, Struan K.	Commonwealth Serum Laboratories, Park-ville, Melbourne, Australia	Immunochemistry of coelenterate venoms
Tamiya, Nobuo	Department of Biochemistry, Tohoku Uni-versity, Sendai, Japan	Chemistry of sea-snake venom
Tange, Yoshiyuki	Nippon Kokan Co., Yokohama, Japan	Fish venomology

* According to the most recent information available to the author.
† Deceased.

TABLE 1.—*List of Current Investigators in Marine Biotoxicology (continued)*

Name	Affiliation/Address*	Area of Interest
Teh, Y. F.	Department of Pharmacology, Faculty of Medicine, University of Singapore, Sepoy Lines, Singapore	Crab poisons
Thayer, Mary C. (Cobb)	Meteorological Department, Woods Hole Oceanographic Institution, Woods Hole, Massachusetts	Noxious marine animals
Tsuda, Tyosuki	Institute of Applied Microbiology, University of Tokyo, Bunkyo-ku, Tokyo, Japan	Chemistry of tetrodotoxin
Tu, Anthony T.	Department of Biochemistry, Colorado State University, Fort Collins, Colorado	Chemistry and toxicology of sea-snake venoms
Tu, Tsuchih	Department of Physiology and Biophysics, University of Alabama, Medical Center, Birmingham, Alabama	Pharmacology of sea-snake venoms
Turner, R. J.	Australian Defence Scientific Service, Defence Standards Laboratories, Melbourne, Victoria, Australia	Pharmacology of coelenterates and mollusks
Vellard, Jehan A.	Bolivian Institute for High Altitude Biology, Ministry of Public Health, La Paz, Bolivia	Venomous marine animals
Vick, James A.	Walter Reed Army Institute of Research, Washington, D.C.	Snake venoms
Watson, Michael	Idaho Department of Health, Laboratories Division, Statehouse, Boise, Idaho	Pharmacology of mollusks
Welsh, John H.	Department of Biology, Harvard University, Cambridge, Massachusetts	Coelenterate toxins
Wiener, Saul	Royal Melbourne Hospital, Melbourne, Victoria, Australia	Pharmacology of stonefish and coneshell venom
Winkler, Lindsay R.	Department of Sciences, College of the Desert, Palm Desert, California	Pharmacology of nudibranch poisons
Woodward, Robert B.	Department of Chemistry, Harvard University, Cambridge, Massachusetts	Chemistry of tetrodotoxin
Yamanouchi, Toshihiko	Department of Biology, Hanazono College, Kyoto, Japan	Pharmacology of holothurin
Yanagita, Tame Masa	Faculty of Science, Ochanomizu University, Otsuka, Bunkyo-ku, Tokyo, Japan	Morphology of coelenterate nematocysts
Yang, Hung Chai	Taiwan Marine Products Research Institute, Kaohsiung, Taiwan	Poisonous fishes
Yokoo, Akira	Department of Chemistry, Faculty of Science, University of Okayama, Japan	Chemistry of tetrodotoxin

* According to the most recent information available to the author.

LITERATURE CITED

BREASTED, J. H.
1930 The Edwin Smith surgical papyrus. Vol. I. Hieroglyphic transliteration, translation and commentary. Univ. Chicago Press, Chicago, Ill. 596 p., 18 figs., 8 pls.

CASTIGLIONI, A.
1941 A history of medicine. Engl. transl. by E. B. Krumbhaar. Alfred A. Knopf, New York. 1013 p., 433 figs.

CHISHOLM, C.
1808 On the poison of fish. Edinburgh Med. Surg. J. 4(16): 393-422.

COUTIÈRE, H.
1899 Poissons venimeux et poissons vénéneux. Thèse Agrég. Carré et Naud, Paris.

GAILLARD, C.
1923 Recherches sur les poissons représentés dans quelques tombeaux Égyptiens de l'ancien empire. Mém. Inst. Français Arch. Orient 51: 97-100; Figs. 56-57.

GARRISON, F. H.
1929 An introduction to the history of medicine. 4th ed. W. B. Saunders Co., Philadelphia. 996 p.

GREVIN, J.
1571 De venenis. Latin Transl. by J. Martin. 2 vols. Christofle Plantin. Antwerp.

GUDGER, E. W.
1924 Pliny's Historia Naturalis. The most popular natural history ever published. Isis (Bruxelles) 6(3): 269-281.
1930 Poisonous fishes and fish poisonings with special reference to ciguatera in the West Indies. Am. J. Trop. Med. 10(1): 43-55.
1943 Is the sting ray's sting poisonous? A historical résumé showing the development of our knowledge that it is poisonous. Bull. Hist. Med. 14: 467-504, 12 figs.

HAYES, H. L., and T. S. AUSTIN
1951 The distribution of discolored sea water. Texas J. Sci. 3(4): 530-541, 1 fig.

HENRI, V., and E. KAYALOF
1906 Etude des toxines contenues dans les pedicellaires chez les oursins. Compt. Rend. Soc. Biol. 60: 884-886.

HERRE, A. W.
1949 A case of poisoning by a stinging catfish in the Philippines. Copeia (3): 222.
1951 A review of the Scorpaenoid fishes of the Philippines and adjacent seas. Philippines J. Sci. 80(4): 381-482.

JOLLY, J.
1901 Medizin. *In* Grundr. d. Indo-Arischen Philol. und Altertumsk. Vol. III. No. 10. Trübner, Strassburg. (NSA)

KEIMER, L.
1955 Poisonous puffers in ancient Egypt. (Personal communication, June 2, 1955.)

KOFOID, C. A.
1911 Dinoflagellata of the San Diego region. 4. The genus *Gonyaula*, with notes on its skeletal morphology and a discussion of its generic and specific characters. Univ. Calif. (Berkeley) Publ. Zool. 8(4): 187-269, Figs. A-E, Pls. 9-17.

LOISEL, G.
1903 Les poisons des glandes génitales. Première note. Recherche et expérimentation chez l'oursin. Compt. Rend. Soc. Bull. 55: 1329-1331.
1904 Recherches sur les poisons genitaux de differents animaux. Compt. Rend. Acad. Sci. 139: 227-229.

MACHT, D. I.
1924 An experimental appreciation of Leviticus XI, 9-12 and Deuteronomy XIV, 9-10. Hebrew Med. J. 2: 165-170.

METTLER, C. G., and F. A. METTLER
1947 History of Medicine. Blakiston Co., Philadelphia. 1215 p., 16 illus.

NADKARNI, A. K.
1954 Dr. K. M. Nadkarni's Indian materia medica. 3d ed. Vol. I. Popular Book Depot, Bombay, India. 1319 p.

(NSA)—Not seen by author.

NIGRELLI, R. F., *ed.*
1960 Biochemistry and pharmacology of compounds derived from marine organisms. Ann. N.Y. Acad. Sci. 90(3): 615-950, figs.

NORDENSKIÖLD, E.
1935 The history of biology. New ed. L. B. Eyre (transl.). Tudor Publishing Co., New York. 629 p., 15.

PAVLOVSKIY, Y. N., *ed.*
1961 Geographic collection, vol. 14. Medical geography. [In Russian] Publishing House of the Academy of Sciences USSR. Leningrad. 200 p. (Engl. trans. by U.S. Dept. of Commerce JPRS: 15,633.)

PHISALIX, M.
1922 Animaux venimeux et venins. Vol. I. Masson et Cie., Paris. 656 p., 232 figs.

POEY, F.
1866 Ciguatera. Memoria sobre la enfermedad ocasionada por los peces venenosos. Repert. Fisico-Natural Isla Cuba (Havana) 2: 1-39.

READ, B. E.
1939 Chinese materia medica. Fish drugs. Peiping Nat. Hist. Bull. 136 p., 190 figs.

RIESMAN, D.
1935 The story of medicine in the middle ages. Paul B. Hoebner, Inc., New York. 402 p., 79 figs.

SIGERIST, H. E.
1951 A history of medicine. Vol. I. Primitive and archaic medicine. Oxford Univ. Press, New York. 564 p., 48 pls.
1961 A history of medicine. Vol. II. Early Greek, Hindu, and Persian medicine. Oxford Univ. Press, New York. 352 p., 85 figs.

STEINSCHNEIDER, M.
1871 Die toxicologischen schriften der araber bis ende. Arch. Pathol. Anat. Physiol. 52: 340-375, 467-503.

STORER, T. I., and R. L. USINGER
1957 General Zoology. 3d ed., McGraw-Hill Book Co., Inc., New York. 664 p., 36 figs.

SUEHIRO, Y.
1947 Poison of globefish. p. 140-159. Suehiro, Practice of fish physiology. [In Japanese] Takeuchi Bookstore, Tokyo.

THOMPSON, D. W.
1947 A glossary of Greek fishes. Oxford Univ. Press, London. 302 p.

Part One

INVERTEBRATES

Chapter II—INVERTEBRATES

Phylum Protozoa

ONE-CELLED ANIMALS

Protozoa are single-celled microscopic organisms, most of which are free-living and inhabit an aquatic environment. A few live in the body fluids of other animals. Most protozoa live as independent cells, but some are grouped as colonies. It is estimated that there are about 30,000 species.

The phylum Protozoa is generally divided into four classes:

MASTIGOPHORA: Flagellates, dinoflagellates, etc. *
SARCODINA: Amoebae, foraminifera, etc.
SPOROZOA: Sporozoans.
CILIATA: Ciliates.

Marine Protozoa poisonous to man are largely mastigophorans of the order Dinoflagellata. They abound in neritic waters and in the high seas, ranging from tropic to polar oceans. Dinoflagellates form an important part of the ocean plankton as synthetic producers of carbohydrates, proteins, and fats. In abundance, they are second only to diatoms, which, on occasion, they may surpass in the total mass of substances produced and in the rapidity of development (Kofoid and Swezy, 1921; Sverdrup, Johnson, and Fleming, 1942). During their periodic maxima, they may cause yellow, brown, green, black, red, or milky local discolorations of the sea. Since red is one of the most common discolorations, this phenomenon is frequently referred to as "red water," "red current," or "red tide." The "blooming" of these toxic plankton in excessive numbers frequently causes a mass mortality of the fishes and other animals living in the region involved. Considerable study has been made to develop means of predicting and controlling the red tide problem.

Useful reviews and bibliographies have been prepared by Galtsoff (1948), Gunter *et al.* (1948), Hayes and Austin (1951), Smith (1954), Ballantine and Abbott (1957), Brongersma-Sanders (1957), and Provasoli (1962). Since red tide and its associated problem of mass mortalities of marine animals go beyond the scope of this volume, the reader is referred to the foregoing authors for further information.

General reviews on the subject of paralytic shellfish poisoning have been prepared by Hill (1953), Meyer (1953), Dack (1956), and McFarren *et al.* (1956). Although broad and intensive research has been done on toxic dinoflagellates, study should be given to determining the number of toxic species; their role in the food chains of other marine organisms; improved methods of forecasting dinoflagellate blooms; factors affecting reproduction; more rapid methods of conducting field toxicity tests; characterization of the molecular structure of their poisons; and specific antidotes.

---

* Some authors include the dinoflagellates in the Kingdom Plantae, whereas others place them in the Kingdom Protista.

Research efforts in Canada have resulted in a review of paralytic shellfish poisonings in eastern Canada by Prakash, Medcof, Tennant (1971), and Medcoff (1971, 1972).

Paralytic shellfish poison is one of the most potent biological poisons known to man. If the origin of this biotoxin could be determined, elucidation of the biogenesis of many other small molecular poisons might be possible, thereby permitting control of these biotoxins. Produced by plant-animals and transvected by an array of mollusks, echinoderms, and arthropods, this poison affects the health and the economic welfare of man by entering his food chain.

The systematics and identifying characteristics of the dinoflagellates pertinent to this work have been discussed by Kofoid (1911), Kofoid and Swezy (1921), and Schiller (1933, 1937).

REPRESENTATIVE LIST OF MARINE PROTOZOANS REPORTED AS TOXIC TO MAN

Phylum PROTOZOA

Class MASTIGOPHORA: Flagellates, Dinoflagellates

Order DINOFLAGELLATA

Family GYMNODINIIDAE

2 *Gymnodinium breve*[1] Davis (Pl. 1, fig. a).

DISTRIBUTION: Gulf of Mexico and Florida.

SOURCES: Phillips and Brady (1953), Ray and Wilson (1957), McFarren *et al.* (1965), Ray and Aldrich (1965), Trieff *et al.* (1973), Martin and Padilla (1974).
OTHER NAME: *Gymnodinium brevis.*

Family PERIDINIIDAE

2 *Gonyaulax catenella* Whedon and Kofoid (Pl. 1, fig. b).

DISTRIBUTION: Pacific coast of North America.

SOURCES: Whedon and Kofoid (1936), Sommer *et al.* (1937), Sommer and Meyer (1941), Schantz (1957, 1960, 1961), Schantz *et al.* (1966), Fraser-Brunner (1973), Mebs (1973), Halstead (1974).

2, 3 *Gonyaulax tamarensis* Lebour (Pl. 1, fig. c; Pl. 2, figs. a, b).

DISTRIBUTION: Eastern coast of North America.

SOURCES: Sommer *et al.* (1937), Medcof *et al.* (1947), Needler (1949), Tennant, Naubert, and Corbeil (1955), Medcof (1960, 1972), Prakash (1967), Fraser-Brunner (1973), Halstead (1974).

---

[1] *G. breve* has not been incriminated as a dinoflagellate source of paralytic shellfish poison, but does produce a respiratory irritant (Lackey and Hynes, 1955).

Pyrodinium phoneus Woloszynska and Conrad (Pl. 1, fig. d). 2
DISTRIBUTION: Belgium.

SOURCES: Koch (1939), McFarren *et al.* (1956), Fraser-Brunner (1973), Halstead (1974).

The studies on red tide indicate that toxic dinoflagellates are far more extensive than is shown by the list given above, but these appear to be the more common species that have been incriminated in human intoxications.

BIOLOGY OF DINOFLAGELLATES AND RELATIONSHIP TO TOXIC SHELLFISH

Dinoflagellates and other phytoplankton are important as producers of the primary food supply of the sea. The production of these organisms is dependent upon radiant energy for the process of photosynthesis and for the presence of certain inorganic nutrient substances, phosphates, nitrates, silicates, and salts of heavy metals. Physical and biological factors that operate directly or indirectly upon the availability of these dissolved substances are also necessary for production. In addition, there are certain physical, chemical, and biological factors that directly affect the metabolism or the survival of the organisms themselves.

Growth control is expressed in *Leibig's law of the minimum,* which states that growth is limited by the factor that is present in minimal quantity. Conditions most favorable for the growth of dinoflagellates are found more often in coastal waters than far offshore, because the zone of decomposition lies close to the production zone that supplies the nutrients in abundance. If the water is deep and the shore steep-to, upwelling may occur with the same advantageous effect.

Plankton blooms are often associated with weather disturbances or weather alterations that may bring about changes in water masses or upwellings. No observations have been reported showing that seismic disturbances cause upwellings. Nutrient salts are less abundant in warm or tropical waters than in temperate or cool waters. Dinoflagellates can grow or flower at lesser minimum concentrations of these materials than can diatoms.

The dinoflagellates involved in human intoxications are largely members of the genus *Gonyaulax*. Members of this genus are enveloped by a cuticle or layer of cellulose composed of numerous plates. The shape of the organism varies according to the species, being spherical, polyhedra, fusiform, or elongated with stout apical and antapical prolongations, or in some instances, dorsoventrally flattened. The cuticle is marked by a descending, displaced equatorial girdle and by transverse and longitudinal furrows. The surface of the cuticle may be smooth or rugose, with major thickenings along the suture lines and plates forming a variable polygonal mesh, often with vermiculates, longitudinal, or spinulate elements. Dinoflagellates possess two flagella that arise about the middle of the body, one trailing in the body axis, and the other transverse to the axis. The nucleus is single, massive, and complex. Contained within the body are yellow to dark-brown chromatophores.

The nutrition of these organisms is said to be holophytic, saprozoic, holozoic, or mixtrophic. Ingestion of small organisms in an amoeboid manner has been observed. Since the fundamental function of nutrition in dinoflagellates overlaps that of both animals and plants, zoologists designate these organisms as Protozoa, and botanists classify them as unicellular algae. Perhaps a more appropriate term is that used by Hutner (1961), namely, "plant-animals."

According to Brongersma-Sanders (1948), temperature probably never limits the production of plankton, but usually affects its compostion. Dinoflagellates seem to prefer warmer waters; therefore, their maximum production occurs during the warm summer months. In her studies on *G. tamarensis*, Needler (1949) concluded that the principal physical factor is water temperature; the principal enemy is the ciliate *Favella ehrenbergi*; and the principal competitors are diatoms. Comparatively large numbers of *G. tamarensis* appeared anytime after the surface water reached 10° C, the greatest occurring in the latter half of August when *F. ehrenbergi* are rare. When many diatoms are present in late July or August, they may compete with *G. tamarensis* and so check production. Prakash and Medcof (1962) demonstrated that there is a clear-cut relationship between sunlight and toxicity production in *G. tamarensis*. They also pointed out that both summer and winter temperatures may be significant factors in dinoflagellate blooms and toxic production. Low temperatures favor encystment of the dinoflagellates, and they remain encysted until threshold encystment temperatures are reached. The overall factors to be considered are water turbulence, transparency, surface illumination, passive sinking of the phytoplankton themselves to depths beyond the photic zone, and grazing action of the zooplankton population.

Kudo (1954) states that asexual reproduction in dinoflagellates is by binary or multiple fission or budding in either the active or the resting stage, varying among different groups. Encystment is a common occurrence. *Gonyaulax* may form chains of two, four, and as many as eight individuals. Relatively little is known regarding the life histories of these organiams.

The possible relationship of paralytic shellfish poisoning to discolored areas of the sea was first pointed out in scientific literature by Lemouroux (cited by Chevallier and Duchesne, 1951*a, b*) and the consistent presence of "saprozoic protozoans" in some European mussels obtained in brackish water was first reported by Lindner (1888*a, b*).

Wolff's (1886) report that the poison was located primarily within the "liver" (the digestive gland) was later confirmed by the studies of Sommer *et al.* (1937), who observed that in sections stained with Ehrlich's hematoxylin, a yellow substance could be observed in the secondary hepatic tubules. This substance, always present whether the mussels were toxic or nontoxic, was absent in starved specimens. No other differences could be demonstrated in the digestive glands.

Certain unidentifiable plankton, found in considerable numbers in the stomachs of the toxic mussels, were either absent or present in very small amounts in the stomachs of nontoxic mussels. There was no histological evidence that gonadal development had any bearing on the matter, since ripe or ripening sex products were observed in the mantles of both toxic and nontoxic mussels.

The finding of plankton in the stomach contents of mussels stimulated Sommer and his group to undertake an investigation during the summer of 1931 to determine the specific planktonic forms used as food by the mussels. Diatoms and dinoflagellates were the dominant organisms present. Apparently the mussel *Mytilus californianus* exercises selectivity in regard to species of dinoflagellates ingested. Protozoans not exceeding certain dimensions were chosen. The results of 3 years of qualitative and quantitative experiments indicated that the yearly maxima of certain species of the genus *Gonyaulax* occurred preceding and during each poison period. Since there was no accompanying increase in the quantity of poison extracted from mussels at the times of these maxima, diatoms were excluded from further consideration as a poison source. *G. catenella* appeared to be the only species of dinoflagellate having a direct relationship to the toxicity of mussels. At times other species appeared in the collections in greater number than *G. catenella*, but their presence did not affect the degree of toxicity in the mussels. A poison isolated from *G. catenella* in July, 1933, was compared with the toxin from mussels and was found to be identical with it in solubility, stability, and mouse symptomatology.

The relationship between the toxicity of dinoflagellates and the toxicity of mussels within a given area is graphically shown in Fig. 1 as taken from Needler's paper (Sommer *et al.*, 1937; Needler, 1949).

Sommer's findings may be summarized as follows (the figures were given as tentative values):

3,000 *Gonyaulax* have a net weight of 100 micrograms.
3,000 *Gonyaulax* yield 15 micrograms of dry extract, which equals 15 percent of the net weight.
3,000 *Gonyaulax* yield 1 microgram of pure poison, which equals 1 mouse unit.
3,000 *Gonyaulax* yield 1 percent of their wet weight or 6.5 percent of the weight of the extractives, as pure poison.
The amount of poison in a single *Gonyaulax* varies, but may be taken as one three-thousandth of a mouse unit.

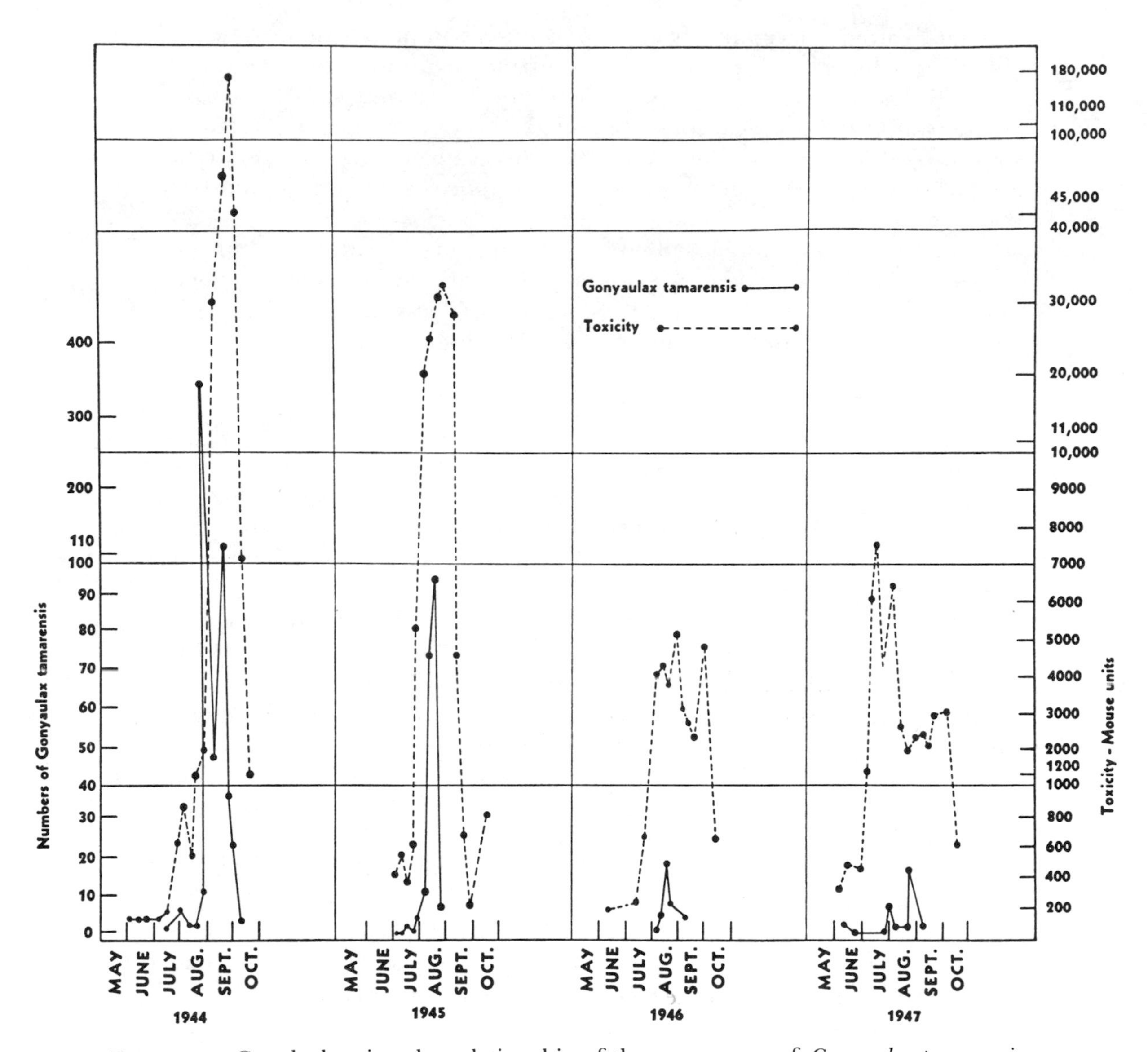

FIGURE 1.—Graph showing the relationship of the occurrence of *Gonyaulax tamarensis* to the toxicity of mussels, *Mytilus edulis*, at Head Harbour, Bay of Fundy, during 1944 through 1947. (Kreuzinger, from Needler, 1949)

Sommer's group experimented with nontoxic mussels that were submitted to a variety of temperature and oxygen tension conditions. These tests were continued until the death of the mussels appeared imminent; however, in no instance was a poison found to be present. Pronounced bacterial spoilage caused an increase in the rate of destruction of the poison. Mussels removed from their natural habitat and brought to the laboratory failed to show an increase in toxicity regardless of the environmental conditions to which they were subjected. Schmidtmann (1888) demonstrated that if atoxic mussels were transplanted to a contaminated area, they would become toxic within 3 to 4 days. Sommer *et al.* (1937) found that if mussels are starved in seawater they lose almost one-half their toxicity in about 10 days. Mussels that had turned practically atoxic in the laboratory became toxic again after ingestion of fresh, unfiltered seawaters.

With the use of curare, strychnine, and mussel poison, Thesen (1902) demonstrated that mussels were able to take up poison from the surrounding water. However Sommer *et al.* (1937) observed that mussels placed in water to which toxic mussel extract had been added would close their shells and refuse to open them until the water had been changed. Shellfish can be considered, therefore, resistant to comparatively large quantities of the poison. Metabolic studies failed to show any significant variation of the respiratory quotient in either toxic or nontoxic mussels. There was no increase in death rate.

Sommer *et al.* (1937) found that the Pacific coast mussels attain their maximal toxicity about the middle of July in localities south of San Francisco, but somewhat later northward. The poison was not found in lethal quantities between November and January. Of the two toxicity curves during each year, the first in early spring and the second in summer, the second toxicity peak is of primary public health importance. This toxicity, which generally occurs during the 2 weeks elapsing between two consecutive spring tidal periods, lasts for only a few days. However, when the peak of toxicity extends high above the danger line, the mussels remain toxic for as long as 1 month within a given locality. No clear relationship was observed in regard to tides. Mussels gathered from the lowest tidal areas, which are swept almost continually by the waves, were on the whole more toxic than those taken from upper tidal areas.

Further confirmation as to the origin of mussel poison was obtained from the experiments of Riegel *et al.* (1949*a*, *b*) in which 500,000 mouse units of poison were centrifuged from 5,000 liters of seawater taken from a red water area containing *G. catenella*.

Needler (1949) observed that the mussels became toxic when *Gonyaulax* appeared along the eastern Canadian coast. Once *Gonyaulax* was present, the toxicity level tended to lag behind the *Gonyaulax* counts, and the mussels retained their toxicity for some time after the dinoflagellates had disappeared. She noted that the dinoflagellates were in the process of spore formation when present in greatest number. Needler suggests that the spores may have sunk from the surface water, but remained in the mussel food and retained their power to make the mussels toxic.

Prakash (1963) found that there was a seasonal rise in shellfish toxicity in the Bay of Fundy coincident with an increased abundance of the marine dinoflagellate *G. tamarensis* and with the appearance of toxin in plankton extracts. Experiments with unialgal cultures of *G. tamarensis* isolated from the Bay of Fundy provided strong evidence that this dinoflagellate is the primary source of toxin in shellfish in this area and in the Gulf of Saint Lawrence. Evidence indicates that both toxic and nontoxic strains of *G. tamarensis* are widely distributed in the North Atlantic; therefore, the primary source or genesis of the poison remains a question.

Although the dinoflagellate *Prymnesium parvum* (Pl. 1, fig. e) has not been demonstrated as having any relationship to paralytic shellfish poisoning, Shilo and Aschner (1953) have pointed out a significant observation on the puzzling

2

lack of correlation between population density of *P. parvum* and the toxicity of the water. The pronounced lability of the prymnesium toxin to oxidizing agents, its ready adsorption by bottom soils, and its rapid destruction by micro-organisms may explain nontoxic pond waters. The presence of light was found to augment toxin production. The concentration of toxin observed, therefore, represents a dynamic equilibrium between toxin production and destruction. The dramatic increase in toxicity observed in fishponds with stable *Prymnesium* populations may represent either the withdrawal of a toxin-removing agency or the presence of a toxin-producing factor. This observation may have a bearing on some of the paralytic shellfish toxicity fluctuations.

Anderson (1960) and Berkeley (1962) suggested an examination of the crystalline style of clams for toxic plankton that may be ingested by *Saxidomus*, the Alaska butter clam. The style, a transluscent-yellow flexible rod containing a starch-reducing enzyme useful in digesting plankton, is found in the pyloric caecum of clams. A grinding and stirring action of the crystalline style against the gastric shield results in a softening and an eventual solution of its material with a concurrent release of polysaccharide-splitting enzymes. Peroxidase, one of the substances present in the style, and the food of the mollusk establish an oxidizing system. One of the products resulting from the oxidizing activity of the style system is glucosone, a product toxic to many animals. According to Anderson, examination of the style of *Saxidomus*, which revealed many adhering plankton, may be a more precise means of determining the exact species of toxic plankton upon which the clam has been feeding.

Berkeley (1962) found that seawater extracts prepared from plankton collected in the vicinity of clam beds were toxic to cultures of *Cristispira*. The biological constituents of the plankton did not seem to have any bearing on toxicity.

The determination of *G. catenella* as the source of the toxin of California mussels has been further confirmed by the axenic cultural and toxicological studies of Burke *et al.* (1960),[2] and the chemical research of Schantz (1960), and Schantz, Lynch, and Vayvada (1962). As Burke *et al.* (1960) have pointed out, previous studies provide good circumstantial evidence of dinoflagellates as the origin of the poison found in shellfish, but do not exclude the possibility of bacteria or other micro-organisms accompanying the dinoflagellate blooms as the actual source of the poison. Samples of the toxin obtained from axenic cultures of *G. catenella* and compared with California mussel poison revealed that both *G. catenella* and mussel toxin are dialyzable and have the same diffusion coefficients, pH stability, R_f values in paper chromatography, and color reactions with the Jaffe reagent. Their behavior is similar in ion-exchange adsorption, and their patterns are the same in electrochromatography. Finally, their physiological reactions are identical when the poisons are injected intraperitoneally into mice. Schantz also demonstrated that the poison found in the California mussel, the Alaska butter clam, and the toxin from cultured *G. catenella* are similar in their chemical and physical properties (Schantz *et al.*, 1962).

---

[2] Similar axenic studies have been conducted by Thomson, Laing, and Grant (1957) on freshwater forms *Anacystis, Microcystis,* and *Nostoc*.

REPRESENTATIVE LIST OF MOLLUSKS REPORTED TO TRANSVECT DINOFLAGELLATE POISONS

The systematics and identifying characteristics of the shellfish pertinent to this chapter have been discussed by Fitch (1953) and Abbott (1954).

Phylum MOLLUSCA

Class PELECYPODA: Bivalves

Family MACTRIDAE

Spisula solidissima (Dillwyn) (Pl. 3, fig. a). Solid surf clam, hen clam, Atlantic surf clam (USA). *4*

DISTRIBUTION: Labrador to North Carolina.

SOURCE: Gibbard and Naubert (1948).
OTHER NAME: *Mactra solidissima.*

Family MYACIDAE

Mya arenaria Linnaeus (Pl. 3, fig. b). Soft-shell clam, soft clam, long clam, mud clam, sand clam (USA), sand saper (England), strandmuschel (Germany). *4*

DISTRIBUTION: Atlantic coast of North America south to the Carolinas, Greenland, Britain, Scandinavia; Alaska, Japan, Vancouver Island, British Columbia. Introduced about 1865 into San Francisco Bay whence it has spread along the California and Oregon coasts.

SOURCES: Medcof *et al.* (1947), Medcof (1960).

Family MYTILIDAE

Mytilus californianus Conrad (Pl. 3, fig. c). California mussel (USA). *4, 5*

DISTRIBUTION: Unalaska, Aleutian Islands, eastward and southward to Socorro Island.

SOURCES: Meyer (1928, 1929), Sommer and Meyer (1937), Schantz *et al.,* (1961).

Mytilus edulis Linnaeus (Pl. 3, fig. d). Bay mussel, blue mussel, pile mussel (USA), moule (France), miesmuschel, seemuschel (Germany), tzu k' ts'ai (China), dagnje (Yugoslavia), mitili, pidocchi, marini (Italy), srounbag, bou zroug, bou chegg (Middle East), muslinge (Denmark), mosselen (Belgium), blaaskjael (Norway), mexilhao (Portugal and Brazil). *4*

DISTRIBUTION: Arctic Ocean to South Carolina, Alaska to Cape San Lucas, Baja California. Worldwide in temperate waters.

SOURCES: Sommer *et al.* (1937), Gibbard, Collier, and Whyte (1939), Evans (1970).

4 *Modiolus modiolus* (Linnaeus) (Pl. 3, fig. e). Northern horse mussel (USA).
DISTRIBUTION: Pacific coast of America from the Arctic Ocean to San Ignacio Lagoon, Baja California; circumboreal.

SOURCE: Medcof *et al.* (1947).

Family VENERIDAE

4 *Protothaca staminea* (Conrad) (Pl. 3, fig. f). Rock cockle, bay cockle, hard shell clam, common little neck, rock clam, ribbed carpet shell quahaug (USA).
DISTRIBUTION: Aleutian Islands to Cape San Lucas, Baja California.

SOURCE: Sommer and Meyer (1937).
OTHER NAME: *Paphia staminea.*

4 *Saxidomus giganteus* (Deshayes) (Pl. 3, fig. g). Smooth Washington clam, butter clam, money clam, giant saxidome (USA).
DISTRIBUTION: Sitka, Alaska, to San Francisco Bay, Calif.

SOURCES: Schantz (1960), Schantz *et al.* (1961), Price and Lee (1972).

4 *Saxidomus nuttalli* Conrad (Pl. 3, fig. h). Common Washington Butter clam, butter clam, money shell, giant saxidome, sand cockle (USA).
DISTRIBUTION: Humboldt Bay, Calif. to San Quentin Bay, Baja Calif.

SOURCES: Sommer and Meyer (1935, 1937).

MECHANISM OF INTOXICATION

Paralytic shellfish poisoning results from ingestion of shellfish, i.e., mussels, clams, scallops, etc., that have ingested toxic dinoflagellates. The poison accumulates in the digestive glands of mussels and clams. In some of the clams the gills may be quite toxic, whereas in *Saxidomus*, the butter clam, the poison is in the siphons. The distribution of the poison in the body of the animal appears to vary somewhat with the species of shellfish and with the season of the year. Although the whelk *(Buccinum undatum)* is not a plankton feeder, it may become toxic by feeding on bivalve mollusks, resulting in human fatalities (Medcof, 1972).

Respiratory irritation may result from the inhalation of toxic products contained in windblown spray from red tide areas of *Gymnodinium breve.*

MEDICAL ASPECTS

Clinical Characteristics

There are three clinical types of shellfish poisoning that may be concurrent in the same patient and should be considered in the differential diagnosis[3] (Linstow, 1894; Sommer and Meyer, 1935).

Gastrointestinal or Choleratic Shellfish Poisoning: This form is characterized

[3] Paralytic shellfish poisoning should also be distinguished from Japanese venerupin poisoning or Minamata disease, which are entirely different clinical entities.

by a relatively long incubation period of 10 to 12 hours and by such symptoms as nausea, vomiting, diarrhea, and abdominal pain. Usually the victim recovers within a short period of time. A variety of bacterial pathogens may cause this nonspecific type of food poisoning (Meyer, 1953).

Erythematous Shellfish Poisoning: Incubation period is relatively short, the onset of symptoms beginning in a few hours, varying with the individual. Symptoms are characteristic of an allergic reaction, viz., diffuse erythema, swelling, and urticaria that particularly affects the face and neck, but may subsequently involve the entire body. The rash is generally accompanied by severe pruritus. Headache, sensation of warmth, conjunctivitis, coryza, epigastric distress, dryness of the throat, swelling of the tongue, and dyspnea may be present. Inadequate preservation is thought to be a factor in this type of poisoning although fresh shellfish have also been involved. This form of the disease is thought to be on an allergic basis, but the exact nature of the disorder is not understood. Patients usually recover within a few days, but occasionally deaths have occurred.

Paralytic Shellfish Poisoning: This last type of intoxication, caused specifically by the dinoflagellate poison present in shellfish, has been designated variously as clam, mussel, or gonyaulax poisoning, paralytic shellfish poisoning, or mytilo-intoxication. The preferred clinical term is paralytic shellfish poisoning.

Paralytic shellfish poisoning may be diagnosed readily by the presence of pathognomonic symptoms which usually manifest themselves within 30 minutes. Initially, there is a tingling or burning sensation of the lips, gums, tongue, and face, with gradual progression to the neck, arms, fingertips, legs, and toes. The paresthesia later changes to numbness, so that voluntary movements are made with difficulty. In severe cases, ataxia and general motor incoordination are accompanied in most instances by a peculiar feeling of lightness, "as though one were floating in air." Constrictive sensations of the throat, incoherence of speech, and aphonia are prominent symptoms in severe cases. Weakness, dizziness, malaise, prostration, headache, salivation, rapid pulse, intense thirst, dysphagia, perspiration, anuria, and myalgia may be present. Gastrointestinal symptoms of nausea, vomiting, diarrhea, and abdominal pain are less common. As a rule, the reflexes are not affected. Pupillary changes are variable, and there may be an impairment of vision or even temporary blindness. Mental symptoms vary, but most victims are calm and conscious of their condition throughout their illness. Occasionally, patients complain that their teeth feel "loose or set on edge." Muscular twitchings and convulsions are rare.

In the terminal stages of the disease, the motor weakness and the muscular paralyses become progressively more severe. Seven (1958) reported a case in which there appeared to be myocardial involvement. Death occurs as a result of respiratory paralysis, usually within a period of 12 hours. The prognosis is good if the patient survives the first 12 hours. The clinical picture may be further complicated by the presence of one or more of the other forms of shellfish poisoning. In a series of 409 cases of paralytic shellfish poisoning collected by Meyer (1953), the case fatality rate was 8.5 percent. Repeated intoxications do not produce an immunity.

The clinical characteristics of shellfish poisoning have been discussed by Meyer, Sommer, and Schoenholz (1928) and Sommer and Meyer (1935, 1941).

The dinoflagellates are also capable of producing respiratory irritants (Galtsoff, 1948; Hayes and Austin, 1951; Phillips and Brady, 1953). Inhalation of toxic products contained in windblown spray from red tide areas of *Gymnodinium breve* outbreaks in Florida in 1947 irritated mucous membranes of the nose and throat, and caused spasmodic coughing, sneezing, and respiratory distress. Experiments demonstrated that samples of this water heated to a temperature of 80-90° C released a vapor that produced the same effects as noted above.

Pathology

The pathological findings resulting from paralytic shellfish poisoning are not of great significance. Autopsies usually reveal evidence of increased capillary permeability with edema, congestion, and petechial hemorrhages of the central nervous system and viscera. In pathological experiments on guinea pigs, rabbits, and kittens, Covell and Whedon (1937) found macroscopic evidence of softening and edema of the brain with moderate congestion of surface vessels. The lungs revealed multiple areas of hemorrhage and congestion of the abdominal viscera. Hemorrhages into the medulla of the adrenals and alveoli of the lungs occurred frequently. Microscopic examination revealed no alterations in the nerve cells of the central nervous system of acutely poisoned animals. The large nerve cells of the ventral horn of the spinal cord and ganglion cells of the medulla of the chronically poisoned animals showed alterations in certain cytologic constituents. The mitochondria appeared normal in the nerve cells, but revealed definite damage in the convoluted tubules of the kidneys. The Golgi apparatus of the small- and medium-sized ganglion cells of the spinal cord were condensed in one part and slightly dispersed in another part of the same cell. Most human autopsies reveal the routine pathological changes of an individual who has died in shock. The pathological findings of paralytic shellfish poisoning have been discussed by Linstow (1894), Meyer *et al.* (1928), and Takahashi *et al.* (1955).

TABLE 1.

Symptoms	Dosage[a]	
	Canadian Mouse Units (Medcof *et al.*, 1947)	Sommer's Mouse Units (Meyer, 1953)
Mild (Paresthesias, etc.)	2,000-10,000	15,000
Moderate (Paralysis)	10,000-20,000	.....
Severe (Respiratory distress, etc.)	30,000	22,000
Minimal Lethal Dose	.....	20,000-40,000

[a] Medcof's estimates are based on a series of 28 cases of shellfish poisoning, whereas Meyer's figures are based on a single outbreak involving 3 persons. The U.S. Food and Drug Administration has established the maximal human tolerance for paralytic shellfish poison as 1,200 Sommer mouse units per 100 g of shellfish meat, or about 2,500 mouse units per meal.

Relationship of Dosage to Symptomatology

The susceptibility to paralytic shellfish poison in humans appears to vary widely with the individual. The progression of severity of symptoms according to the increase in dosage has been variously estimated (*see* Table 1).

Results similar to those of Medcof *et al.* (1947) were obtained by Bond and Medcof (1958) in their evaluation of the epidemic of paralytic shellfish poisoning that occurred in New Brunswick in 1957. The effects of paralytic shellfish poison vary from one person to the next. Some persons seem to tolerate the poison more readily than other persons (Edwards, 1956).

Treatment

The treatment of paralytic shellfish poisoning is largely symptomatic. The poison has no specific antidote. Apomorphine is more effective than lavage in removing pieces of shellfish from the stomach. Since mussel poison is readily adsorbed on charcoal, Lloyd's reagent and similar adsorbents may be tried. Alkaline fluids are of value since the toxin is unstable in that medium. Diuresis may be instituted with 5 percent ammonium chloride.

Artificial respiration is an important adjunct and should be instituted promptly if there is any evidence of respiratory embarrassment. Experimentally this technique has been used with good results and is recommended (Sapeika, 1953; Murtha, 1960). Primary shock may be present and require attention.

Drug therapy has varying degress of success. The anticurare drugs, such as neostigmine, are useful in aiding artificial respiration. DL-amphetamine, epinephrine, ephedrine, and DMPP (1,1-dimethyl-4-phenylpiperazinium iodide) are also recommended (Murtha, 1960). To counteract the acetylcholine esterase-like inhibitory effect, Pepler and Loubser (1960) advised using oximes, such as pyridinealdoxime methiodide. Digitalis and alcohol are not recommended.

The treatment of paralytic shellfish poisoning has been discussed also by Meyer *et al.* (1928), Jordan (1931), and Sommer *et al.* (1937).

Prevention

The best precaution is strict adherence to local quarantine regulations (*see* Quarantine Regulations under Public Health Section). Despite the numerous folk tales as to methods of differentiating poisonous from edible mussels, there is still no reliable substitute for laboratory toxicity tests. Poisonous mussels cannot be detected by their appearance or smell, by discoloration of a silver object, or by placing garlic in the cooking water, etc. The usual methods of cooking, i.e., steaming, baking, boiling, and frying, do not remove the danger of intoxication, although Medcof *et al.* (1947) stated that cooking procedures may reduce the original poison content of the raw meat by more than 70 percent. Because of the necessary high temperatures, panfrying seems to be more effective in reducing the toxicity than any other cooking procedure. Müller (1932), Sommer and Meyer (1941), and others have recommended adding sodium bicarbonate (1 tablespoonful to a quart of water) and cooking the

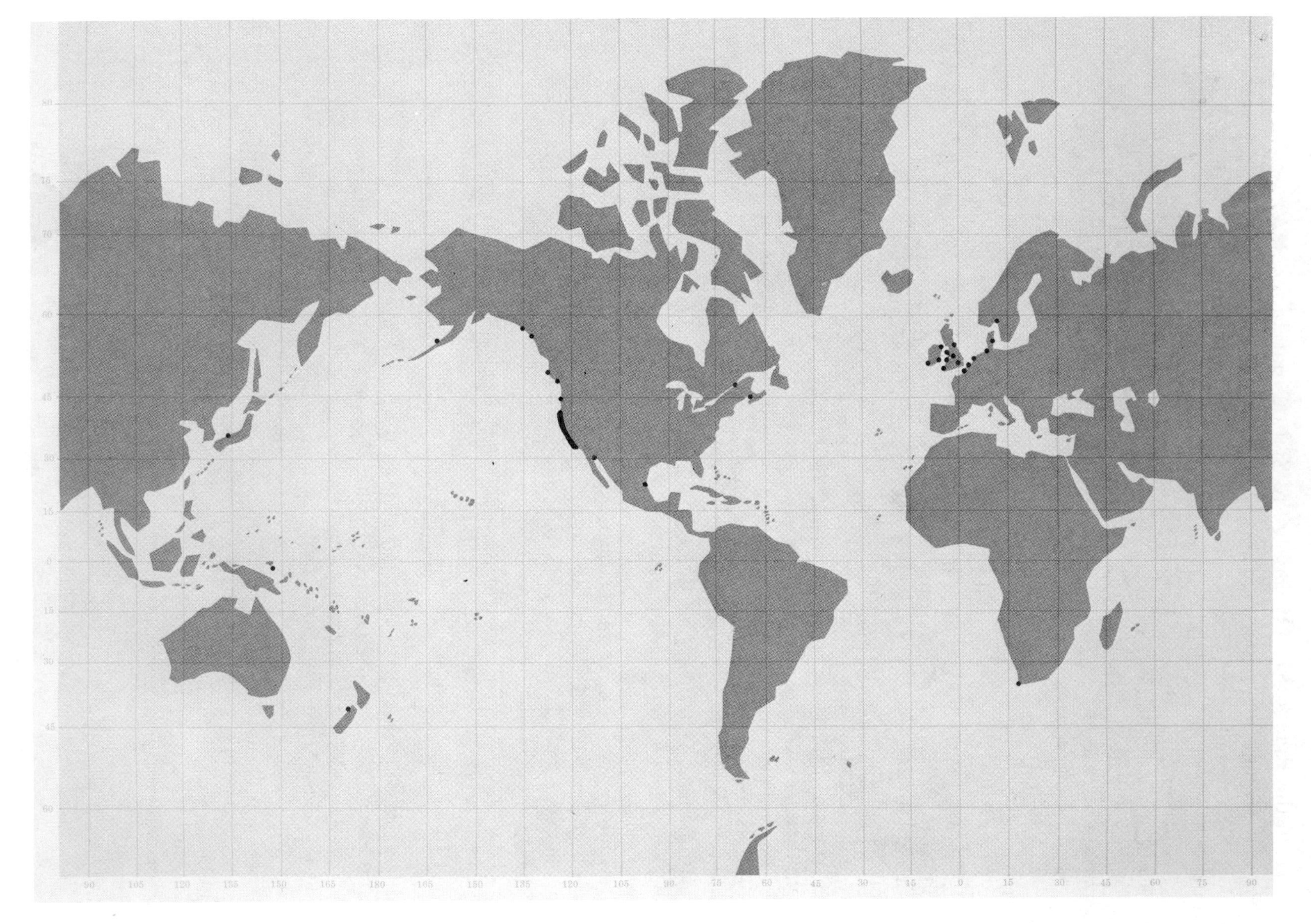

MAP 1

World distribution of paralytic shellfish outbreaks.

(L. Barlow)

mussels in this solution 20 to 30 minutes. This technique destroys about 85 percent of the poison, but detracts from the flavor. However, if the toxicity level is high, intoxication may still ensue.

The toxicity of shellfish varies somewhat according to the organ and species involved. In general, the digestive organs (dark meat) and gills of mussels *Mytilus*, the siphons of Alaska butter clams *Saxidomus*, and the bouillon or broth in which shellfish have been cooked should be discarded. Although generally harmless, the musculature (white meat) of shellfish should be washed thoroughly before cooking. The edibility of shellfish is not guaranteed merely because the shellfish are taken from a particular type of terrain or tidal level. The commercial canning process usually reduces the toxicity to a safe level, but if the shellfish have been collected from a dangerous area, toxicity checks should be made. A high mortality rate among sand crabs *Emerita analoga* found along the Pacific coast of North America during the warmer months indicates that the shellfish in the local area are toxic and should not be eaten (Sommer and Meyer, 1937; Gibbard *et al.*, 1939).

PUBLIC HEALTH ASPECTS

Until recent years, the consumption of shellfish was confined largely to maritime communities. With the advent of modern transportation and refrigeration systems, shellfish have become an important food item even in inland regions. The shellfish industry in the United States and Canada has a combined annual output of about 600 million pounds (Edwards, 1956). According to the Food and Agricultural Organization of the United Nations, the total world shellfish catch for the year 1953 amounted to more than 1,661 metric tons (Walford, 1958). The problems of shellfish sanitation and paralytic shellfish poisoning, therefore, take on significant proportions in the field of global public health.

Geography

Outbreaks of paralytic shellfish poisoning have occurred in widely scattered areas throughout the world. The areas incriminated thus far in human intoxications are: *North America*—Alaska, British Columbia, Washington, Oregon, California, New Brunswick, Nova Scotia, Maine; *Middle America*—Gulf of California, Vera Cruz, Mexico; *Europe*—Scotland, England, Ireland, France, Belgium, Germany, Denmark, Norway; *Africa*—South Africa; *Asia*—Japan; *Oceania*—Admiralty Islands, New Zealand. The worldwide distribution of paralytic shellfish poisoning is geographically illustrated in Map 1.

Incidence and Mortality Rate

There are no accurate figures regarding the overall incidence of paralytic shellfish poisoning. Meyer (1953) appears to have collected medical data on the largest number of outbreaks reported to date in his series of 409 cases that occurred over a period of years along the Pacific coast of North America. In this particular series there were 35 deaths making the case fatality rate 8.5 percent. McFarren *et al.* (1956) also list some of the cases of paralytic shellfish poisoning that occurred between 1798 and 1954. Numerous outbreaks have taken place in Europe over the years, but the statistics are so grossly incomplete that it is difficult to determine the mortality rate with any degree of accuracy.

Seasonal Changes

The period when toxicity levels are highest varies somewhat according to geographical location. Along the Pacific coast of the United States and Canada, all cases of paralytic shellfish poisoning have occurred between May 15 and October 26 (Meyer, 1953). Thus far, only one dangerous period, lasting a few days, has occurred each year. But mussels may be toxic to humans within a particular geographical area for a month. Although mussels can become toxic within a period of a few days, detoxification may take a month. Toxic mussels have not been detected on the Pacific coast (Washington, Oregon, California) from November through January (Sommer and Meyer, 1937). Thus, the adage that warns against the consumption of shellfish during the months without the letter "R" has some foundation. The seasonal toxicity of the Alaska butter clam *Saxidomus giganteus* differs from that of the California mussel *Mytilus californianus* and the Washington clam *S. nuttalli*. Chambers and Magnusson (1950) found that in some areas in Alaska, *S. giganteus* maintains dangerous toxicity levels throughout the year. The highest toxicity levels in shellfish in the New Brunswick and Nova Scotia areas occurred between mid-July and the latter part of September and reached an overall maximum in late August (Gibbard *et al.*, 1939; Gibbard and Naubert, 1948). The European and South African outbreaks took place during May through October (Meyer *et al.*, 1928; Sapeika, 1948).

Seasonal variation in the toxicity of shellfish and dinoflagellates has been given intensive study by several groups in both Canada and the United States. The public health aspects relating to seasonal fluctuation of shellfish toxicity have been discussed at length by the following: McFarren *et al.* (1956), U.S. Public Health Service (1959), Bond and Medcof (1958), and Medcof (1960).

Forecasting Toxicity Levels

The present system of predicting toxicity and establishing quarantine regulations carries certain inherent dangers that public health authorities are attempting to overcome. Shellfish become toxic as a result of their feeding upon toxic dinoflagellates. Therefore present testing and forecasting procedures are based on assays of secondary rather than of primary sources. Since present toxicity testing methods require several days, the dinoflagellate bloom of the shellfish beds may have increased severalfold between collection and analysis. Consequently, forecasting methods can be dangerously inaccurate.

Some workers have suggested that assays should be made directly on the dinoflagellate population itself. But as Bond and Medcof (1958) and others have shown, the dinoflagellate population itself is subject to violent and rapid changes that can take place within a matter of hours. The reliability of either method as an early warning device is therefore subject to question. Needler (1949) and Bond and Medcof (1958), noting that the year-to-year difference in water temperature is one of the factors determining abundance of dinoflagellates, and hence toxicity of the shellfish, recommended that future research on the improvemernt of forecasting techniques be directed toward those factors affecting dinoflagellate population fluctuations (Medcof, 1960; Prakash and Medcof, 1962).

Bates and Rapoport (1974) have developed a chemical assay for saxitoxin which is specific, rapid, routine, and reliable. It has been applied to a number of samples of *Saxidomus giganteus* and *Mytilus californianus*.

Quarantine Regulations

In an effort to suppress outbreaks of paralytic shellfish poisoning, the Shellfish Sanitation Section of the U.S. Public Health Service has established regulations governing the growing, harvesting, and processing of shellfish in the United States (*see* manual of recommended practice for *Sanitary Control of the Shellfish Industry*, U.S. Public Health Service, 1957, 1959, 1962). These regulations are revised from time to time and the latest revision should be consulted by anyone concerned with this aspect of the problem.

The Canadian Department of Health and Welfare and the U.S. Public Health Service have agreed that shellfish with toxicity levels up to 400 mouse units per 100 g are generally safe for human consumption. Areas are closed for commercial harvesting and marketing of raw shellfish as soon as this toxicity level is approached.

The quarantine level in British Columbia was set at 2.0 g of poison per whole clam or mussel (Pugsley, 1939). In August, 1936, the public health department for Nova Scotia proclaimed a mussel quarantine because of an outbreak resulting in the death of two persons. Japan also established shellfish quarantines in some areas (Hattori and Akiba, 1952).

Ecology

On the Pacific coast of North America, toxic mussels are generally taken along open unprotected coastline. Samples of shellfish taken from the mouths of bays or from open passages show a decrease in toxicity as compared with samples taken toward the head of bays or in protected areas (Sommer and Meyer, 1937; Furk, 1950; McFarren *et al.*, 1956). In Alaska, toxic clams are not generally found along the beaches of outside waters, but rather on or near open water of the wide straits characteristic of southeastern Alaska (Chambers and Magnusson, 1950). According to Sommer and Meyer (1941), it has never been necessary to quarantine bay mussels in California. Medcof *et al.* (1947) found that the most toxic shellfish in the Bay of Fundy came from areas where extreme tidal conditions brought an ocean environment close to shore. Very low toxicity was recorded from enclosed and sheltered inlets along the eastern Canadian coast. Although no outbreaks of human poisoning have been reported, Le Messurier (1935) obtained poisonous shellfish from Batemen's Bay and Lake Burrill, New South Wales. Most of the European outbreaks resulted from eating mussels that had been taken from a harbor, estuary, or brackish water areas (Sommer and Meyer, 1937). There is no evidence that the type of beach terrain is useful as an index in determining the toxicity of mussels (Waldichuk, 1958; Sparks *et al.*, 1962).

Effects of the Commercial Canning Process

The effects of the commercial canning process upon paralytic shellfish poison have been studied by Canadian public health workers. The process used by some Canadian canneries is as follows. The clams are placed in barrels with loose-fitting heads and are precooked with live steam at approximately atmospheric pressure for 15 to 20 minutes. They are then shucked (the mollusk meats cut and freed from the shells), and the siphons trimmed off. The meats are washed in warm fresh water, packed in cans, and covered with some of the remaining bouillon. The cans are sealed without being exhausted and are retorted at 121° C for 45 minutes.

The vinegar that seasons some of the remaining bouillon used to fill the cans alters the pH from 8.2 to 6.5. In the process of retorting, the poison is significantly attenuated both by the alkaline condition of steaming and by the slightly acid condition from the addition of vinegar. Since the poison is less stable in an alkaline than in an acid medium, canning without the adjustment of the pH would probably result in an even greater elimination of the poison.

Draining and discarding the bouillon is found to be a highly important factor in reducing the toxicity, and extending the period of retorting might reduce toxicity somewhat. But the procedures used in shucking have no important effect on toxicity (Medcof *et al.*, 1947; Gibbard and Naubert, 1948).

Pugsley (1939) found that in British Columbia, clam siphons were 20 to 40 times more toxic than the whole clam. Chambers and Magnusson (1950) obtained similar results in their studies on Alaskan clams. Pugsley stated that since canneries in British Columbia routinely remove the siphons from the clams and can only during the fall and winter months, toxicity does not present a serious threat to the industry.

The Department of National Health and Welfare of Canada adopted a routine sampling procedure for the testing of canned shellfish. No canned clams with toxicities exceeding 100 Canadian mouse units are allowed on the market. Such a product is usually obtained from raw shellfish having a toxicity of 1,000 mouse units or less (Medcof *et al.*, 1947). The Canadian Government subsequently adopted the 400 mouse unit quarantine level and now permits the marketing of clams (fresh, frozen, or canned) when the average score of representative samples from each shipment is less than 400 mouse units, and the scores of all samples are less than 2,000 mouse units per 100 g of shellfish.

TOXICOLOGY

Extraction Methods

Salkowski (1885), Virchow (1885), and Wolff (1886, 1887) were the first to develop toxicological methods for the testing of mussel poison. Alcoholic extracts and the pure juice from mussels were tested subcutaneously on rabbits and frogs. Crushed mussel meat was also injected subcutaneously into rabbits.

By selecting various parts of the mussel, i.e., foot, mantel, gills, digestive organs, and gonads, Wolff found that only the digestive organs ("liver") were toxic. Desiccation did not destroy the poison.

As a result of the outbreaks of paralytic shellfish poisoning in California in 1927, a series of toxicological experiments was conducted by Meyer *et al.* (1928), Sommer and Meyer (1937), and Meyer (1953). Initially, the mussels were fed to kittens and rabbits. Guinea pigs, unaffected by the ingestion of toxic mussels, developed symptoms from subcutaneous injection of tissue extracts. Because force feeding produced vomiting, dogs seldom yielded the desired symptoms. Transplantation of mussel livers into rabbits was also tried.

Subcutaneous or intraperitoneal injection of mice and rats gave the most satisfactory results. Sommer and his group developed a qualitative test consisting of intraperitoneal injection of alcoholic extracts into mice. Subsequently, they demonstrated that the addition of a small amount of hydrochloric acid removed additional poison from the shellfish. Whole mussels were placed in a meat grinder. Ethyl alcohol acidified with hydrochloric acid (congo red was used as an indicator) was then added, and the mixture was boiled for 1 to 3 hours under reflux. The liquid was filtered from the insoluble portion, evaporated *in vacuo* to a pasty brown mass, weighed out, and dissolved in a known sample of water. Additional poison was obtained when the alcohol-insoluble residue was exhausted with further quantities of acid ethanol. Later, because of the time factor, the test was greatly simplified: only the shellfish livers were used; methyl alcohol, because of its lower boiling point, was chosen as the solvent; the acid was omitted; and the evaporation was performed on the open water bath. The removal of fats and pigments with chloroform before injection produced more uniform aqueous solutions. This simplified method proved to be a useful qualitative procedure in testing large numbers of samples within a short period of time. To develop a quantitative procedure causing as little destruction of the toxic substance as possible, Sommer and his group substituted centrifugation for filtration, and the number of mussels per sample was reduced.

Testing Procedures

Several different methods have been used in the bioassay of dinoflagellate and paralytic shellfish poisons. Sommer and Meyer (1937) have described two toxicological screening procedures which, with some modification, have been widely adopted in this field. The details of conducting these tests are as follows.

Sommer-Meyer Standard Laboratory Test: The digestive glands are removed and macerated with acid methyl alcohol (4 ml of concentrated hydrochloric acid per liter). Several portions of the liquid are used until the residue becomes nearly colorless. The combined residue and liquids are poured into a centrifuge tube, placed in a water bath, brought to a boil, and then allowed to cool gradually to room temperature. Standing overnight permits most of the alcohol-insoluble substances to settle, and these are removed by centrifugation. The clear dark-colored supernatant or an aliquot part is poured into a weighed glass dish and evaporated on a boiling water bath. The residue, removed from the water bath when still of a pasty consistency, is triturated with several small portions of chloroform until the liquid comes off nearly colorless. The solvent is discarded each time. Extraction of the lipids and pigments is readily accomplished if the residue is of a pasty consistency. Floating particles are settled with the addition of ether. The residue is heated in the water bath for 1 or 2 minutes, being stirred constantly until it is quite viscous. Since it is hygroscopic, the residue should be weighed as soon as it is cold.

For injection purposes, a 10 percent distilled water solution is prepared from the residue. Usually it is not necessary to neutralize the poison for testing purposes; but if for any reason neutralization is attempted, it must be done with care, since even a slight alkaline reaction may destroy 50 percent of the poison within a short time. The chloroform extraction and subsequent heating remove the free hydrochloric acid introduced.

Sommer-Meyer Field Test: A simplification of the standard test was developed as a field method for the rapid examination of mussel beds and the establishment of quarantine measures. The digestive organs are removed from three mussels weighing approximately 50 g each. To the organs of each mussel is added 10 ml of 0.1 *N* hydrochloric acid. The supernatant is used for testing purposes.

Hashimoto and Migita (1950) recommended extraction with the use of methanol and acetic acid. This method readily extracts the toxin in large quantities without producing the toxic contaminants resulting from hydrochloric acid extraction.

Both the distilled water solution in the *Standard Laboratory Test* and the supernatant fluid in the *Field Test* are evaluated in the same manner. One ml of the extract is injected intraperitoneally into each of three or more laboratory mice. In their 1937 work, Sommer and Meyer defined the average lethal dose,

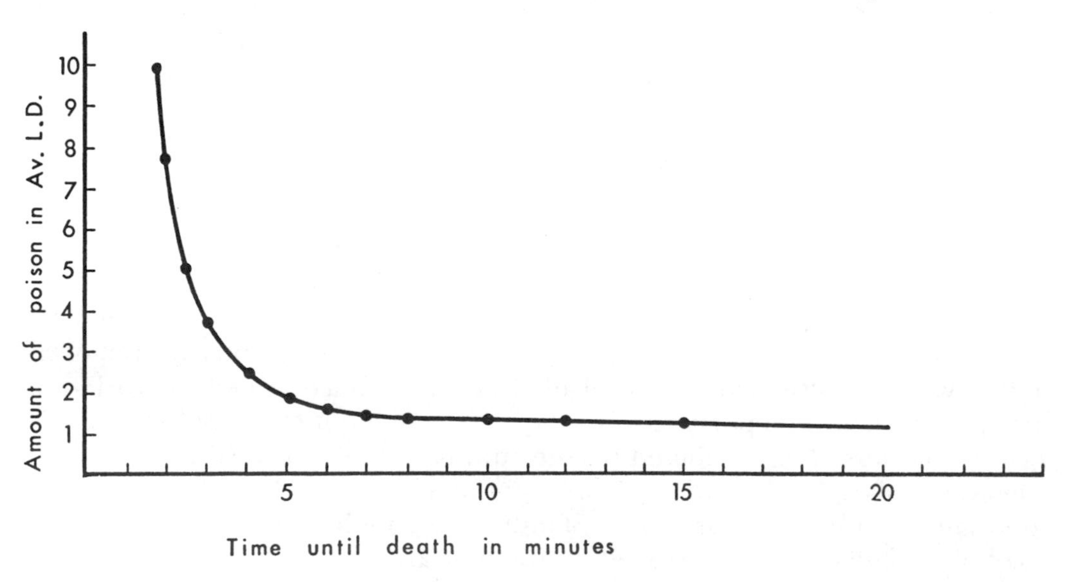

FIGURE 2.—Graph showing relationship of mouse death time to dose of paralytic shellfish poison. (After Sommer and Meyer, 1937)

or mouse unit, as the amount of shellfish poison that will kill a 20 g mouse in 10 to 20 minutes. The typical neurological symptoms are nervousness, ataxia, convulsions, respiratory distress, and paralysis (*see* Fig. 2). According to Meyer's (1953) most recent modification of the *Field Test*, if a 20 g mouse dies within 15 minutes of receiving 1 ml of the supernatant, the concentration of the poison is expressed as 10 mouse units per 50 g of mussel.

Sommer and Meyer (1937) found that in the mussel samples tested during a 9-year period, the average lethal dose for mice varied from 0.017 to approximately 60 mg. Meyer (1953) stated that the purest preparation obtained from toxic mussels has a toxicity of 4 Sommer mouse units per microgram.

Canadian Testing Methods: The toxicological methods used by Canadian workers (Medcof *et al.*, 1947) are a modification of the *Sommer-Meyer Field Test*. The meat is removed from the shellfish, washed in fresh water, drained 5 minutes, and then minced. A 100 g sample of the mince is suspended in 100 ml of 0.1 *N* hydrochloric acid and slowly brought to a boil. It is then cooled and made up to its original volume of 200 ml with distilled water.The pH is adjusted to between 4.0 and 4.5 by the addition of 5.0 *N* acid or 0.1 *N* sodium hydroxide. The supernatant liquid is clarified by settling or by centrifugation. One ml of the extract is injected intraperitoneally into each of the three mice weighing 18 to 22 g. If 1 ml of the supernatant proves fatal to the animal in 15 minutes, the extract is considered to contain 200 mouse units of poison. If the Canadian method is to be compared with the Sommer standard of mouse units per 500 g of mussel, the Canadian units must be divided by seven. Studies have shown that 30 percent of the total weight of the shellfish is meat. "Thus, 100 g of meat, as used by the Canadians, is equal to 350 g of total weight, or equal to (7) seven shellfish weighing 50 g each" (Meyer, 1953).

In an attempt to develop a stable reference standard for the bioassay of paralytic shellfish poison, Stephenson *et al.* (1955) investigated the various factors affecting bioassay results. They determined the toxicity of a shellfish extract, measured in mouse units and based on the mean death time of the test animals, by comparing it with that estimated by the LD_{50} obtained by means of a quantal response type of assay. The slope of the log dose-response line obtained with butter clam extracts was not significantly different from that found with scallop liver extracts. A refrigerated lyophilized extract of scallop liver was stable throughout a test period of nearly 3 years. They felt that such a preparation could be used as a reference standard for the bioassay of shellfish poison. Female mice were more susceptible to the paralytic poison than male mice. The LD_{50} per mouse was related directly to the average body weight of the test animals. The volume of the injection medium had no effect on either the magnitude of the LD_{50} or the slope of the probit regression line. A highly significant inverse relationship was demonstrated between the toxicity measured in mouse units, a procedure based on the mean death time of the test animals, and the LD_{50} per kg of mouse determined by an assay with an all-or-none response.

McFarren *et al.* (1956) found that the oral LD_{50} dose for rats varied with the strain and the weight, or both, of the rats used. A nonlethal oral dose of

TABLE 2.—LD_{50} *of Shellfish Poison for Various Animals (McFarren* et al., *1956)*

Animal	Oral LD_{50} per kilogram [a]	Estimated lethal dose for adult animal [a]
Mice	2,100	63
Rats	1,060	320
Monkeys	2,000 to 4,000	15,000
Cats	1,400	3,500
Rabbits	1,000	3,500
Dogs	1,000 approx.	11,000
Guinea Pigs	640	260
Pigeons	500	250
Man	500	36,500

[a] Mouse Units (Sommer or International).

paralytic shellfish poison rendered rats less sensitive to subsequent doses. They further observed that purified shellfish poison produced a higher LD_{50} than a crude extract of poison. This fact indicates that salt lowers the bioassay of clam meat extracts when mice are used (*see also* Schantz *et al.*, 1958; Wiberg and Stephenson, 1960). Wiberg and Stephenson (1960), who made a thorough study of the effect of metal ions on the toxicity of paralytic shellfish poison showed that the toxicity of the poison was reduced in the presence of Na^{+}, but is augmented when Ca^{++}, Ba^{++}, Sr^{++}, Mg^{++}, Mn^{++}, Co^{++}, Ni^{++}, Fe^{++}, and Fe^{+++} ions were present. A large number of monovalent, divalent, trivalent, and tetravalent metal ions had no influence on the toxicity at the level tested.

U.S. Public Health Service Bioassay Method: The first attempts to standardize bioassay methods were made in 1943 by the U.S. Public Health Service, the U.S. Fish and Wildlife Service, and the U.S. Food and Drug Administration. A tentative procedure was adopted for the bioassay of poisonous shellfish, and certain tolerances were recommended for guidance in closing shellfish operations.

On May 26, 1955, the U.S. Public Health Service sponsored a Canadian-United States Conference on Shellfish Toxicology in order to establish a uniform procedure for the bioassay of shellfish poison. It was at this Conference that Schantz and his associates announced the isolation of paralytic shellfish poison and recommended that the purified poison be adopted as a reference for bioassay.

Studies at that time, and since then, have shown that the poison from the California mussel *Mytilus californianus*, the Alaska butter clam *S. giganteus*, and the dinoflagellate *Gonyaulax catenella* appear to be identical in chemical, pharmacological, and biological properties (Schantz *et al.*, 1958; Schantz, 1960;

Murtha, 1960; Schantz *et al.*, 1961, 1962). The dihydrochloride salts are white solids soluble only in water, in methanol, and, to some extent, in ethanol. They have a toxicity of about 5.5×10^6 mouse units per gram, possess no absorption in the ultraviolet and visible range, and have a specific optical rotation of +130°. None of these properties changes when the poison is stored for 2 years: hence, the poison is sufficiently stable to serve as a reliable standard (Schantz *et al.*, 1958).

The 1955 Conference adopted purified poison as a tentative reference standard for a uniform bioassay and recommended that the results be reported in terms of the weight of the poison. The procedure for carrying out these recommendations was presented in the form of an "Interim Plan for Standardizing the Bioassay of Paralytic Shellfish Poison by Use of a Reference Standard" (Shellfish Sanitation Branch, U.S. Public Health Service, Washington, D.C.).

Schantz *et al.* (1958) and Schantz (1960), elaborating further on the "Interim Plan," discussed the basis for establishing certain standard conditions under which the micrograms of purified poison equivalent to one mouse unit can be determined and the usefulness of the reference standard in assaying poisonous clams. Their studies showed that in the determination of the conversion factor or CF value, as well as in the bioassay of an acid extract of clams, the relationship between mouse units and median death time can be used equally as well as an LD_{50} value; the pH of the solution injected into the mice should be between two and four; dilutions should be made so that the median death time falls between 5 and 7 minutes; the weight of the mice should be between 19 and 21 g and not over 23 g; a total of 10 mice should be used with each dilution in the determination of the CF value; and for the routine assay the injection of three mice is sufficient provided the median death time falls between 5 and 7 minutes. Recovery studies indicated that in the assay of low toxicity clams (about 400 mouse units per 100 g) the bioassay may underestimate the toxicity by as much as 60 percent. A 60 to 70 percent reduction of the bioassay was observed also when the ash from an equivalent amount of acid extract of clams or 0.5 percent sodium chloride was added to a solution of purified poison. Schantz and his coworkers felt that salt effects were responsible, at least in part, for the poor recovery in the bioassay of low toxicity clams.

Wiberg and Stephenson (1960) have shown that the difference in toxicity of paralytic shellfish poison for the three routes of administration is probably related to the rate of absorption. The acute LD_{50}'s for a highly purified preparation of paralytic shellfish poison in mice were determined for the oral (263 μg/kg), intraperitoneal (10.0 μg/kg), and intravenous (3.4 μg/kg) routes. Female mice were more susceptible than were males to lethal doses of the poison. Increases of pH of the injection medium or the addition of sodium ions reduced the intraperitoneal toxicity. These two effects were not additive and the sodium effect appeared to be the stronger. The sodium ion did not influence the oral or the intravenous toxicity. The use of the median death time as a criterion of toxicity is not reliable at pH levels above 4.0 or in the presence of sodium ions above 0.1 *M*. They found that in their evaluation of the bioassay procedure of the Association of Official Agricultural Chemists ("Interim Plan"), satisfactory results were obtained when the specified conditions were followed (McFarren, 1960).

The Association of Official Agricultural Chemists adopted this method of using purified shellfish poison as the reference standard for measuring the toxicity of poisonous shellfish. Essentially a modification of the Field Test of Sommer and Meyer (1937), this procedure undoubtedly will serve as an international standard in the future.

Details for a chemical determination of paralytic shellfish poison in clams have been published by McFarren *et al.* (1958).

Miscellaneous Testing Procedures: Although Sommer and Meyer's mouse intraperitoneal test has been accepted generally as a standard procedure, other methods have been used in testing for dinoflagellate poisons. Macht and Spencer (1941) introduced a phytopharmacological technique that employed the seedlings of *Lupinus albus*. The increment or growth of the single straight root of the seedling was measured after the plant had been grown for 24 hours in the dark at 15° C in the extract to be tested. Growth of the root was then compared with controls that were grown in plant physiological or Shive's solution. The index of growth or phytotoxic index was expressed by the ratio of X/N or average growth in the unknown solution over the growth of the controls.

Galtsoff (1948) prepared acidulated alcoholic extracts of the dinoflagellates *Gymnodinium* and *Evadne* and tested them on killifish *Fundulus heteroclytus*. One part of dried extract was placed in 1,000 ml of seawater. Test fish and controls, placed in 500 ml of water that was not aerated, were observed for the development of symptoms. Various other concentrations of dinoflagellate extracts were also used in this same manner. Galtsoff conducted toxicity tests by placing live *G. breve* in aquaria containing sheepshead, minnows, puffers, and mojarras. In each instance the fish died.

Cornman (1947) tested dinoflagellate extracts on the eggs of the sea urchin *Arbacia* and found that it retards cleavage.

Outbreaks of mass mortalities of brackish-water pond fishes in Israel have resulted in intensive investigations on the toxicity of *Prymnesium parvum*. Although not a known cause of paralytic shellfish poisoning or human intoxication, the poison produced by this plant-animal is of sufficient potency to be considered a potential offender. The toxicity of this organism has been studied by Otterstrøm and Steemann-Nielsen (1939), Prescott (1948), Reich and Rotberg (1958), Shilo and Aschner (1953), Shilo *et al.* (1953), Yariv (1958), Berman (1960), and Shilo and Rosenberger (1960). Relatively little is known about the chemistry and pharmacology of the poison. It is not only heat-labile, nondialyzable, and oxidation-sensitive, but also readily adsorbable on a variety of adsorbents, including pond-bottom soils. The toxin is destroyed at a pH level below 7.0. Previously the toxin was thought to be a protein (Shilo *et al.*, 1953), but recent studies indicate that it is a lipid (Shilo and Rosenberger, 1960).

Shilo and Aschner (1953) have developed a bioassay method using *Gambusia* minnows and tadpoles *Pana* or *Bufo* as test animals. The tadpole tests consisted of five animals immersed in 5 ml volumes in test tubes. Cell-free centrifugates of the toxic samples were used for the assay. The dilution medium was sodium chloride solution (0.16 percent), 9 volumes plus Sorensen's phosphate buffer pH 8, 1 volume, the mixture being adjusted to final pH value of 7.5-8.5. The test was conducted at a temperature of 15-25° C. Bacterial contamination was prevented with streptomycin (1 mg/ml). Assays were based on direct determinations of the MLD or on observations of the time of onset of toxic reactions, particularly the loss of equilibrium sense in minnows (equilibrium-loss time), or of tail curvature in tadpoles. Minnows, being more sensitive, were preferred for dilute quantities of the toxin; tadpoles, reacting more quickly, were preferred for rapid assay work. The prymnesium toxin unit (TU) was defined as the minimum amount of toxin per 20 ml that causes death of *Gambusia* minnows within 12 hours in standard test conditions, which are specified as two minnows about 1.5 cm long and immersed together in 20 ml test fluid in a 50 ml beaker having a 40 mm diameter. The relationship of toxin concentration to time of showing tail curvature (curvature time) in tadpoles is similar, in general, to that of equilibrium-loss time in *Gambusia* minnows. The reader is referred to Shilo and Aschner (1953), and Parnas (1963) for further details of their bioassay methods on *P. parvum* toxin.

Shilo and Rosenberger (1960) have successfully used erythrocytes, Ehrlich ascites cells, HeLa cells, human amnion cells (FL strain), and the Chang strain of liver cells in testing *P. parvum* toxin for biological activity. In all of the nucleated cells tested, the active compound was found to induce the formation of pseudopodia-like extrusions that was followed by a complete breakdown of the osmotic barrier, as shown by the uptake of trypan blue.

Ray and Wilson (1957) have worked with axenic cultures of *G. breve* and found that the dinoflagellate is directly responsible for the toxic effects on fish. Bacteria do not produce or directly contribute to the toxin production. Moreover, the toxicity is not dependent upon the presence of living *G. breve* organisms. The toxin of *G. breve* readily passes through a millipore membrane. Experiments showed that oxygen deficiency was not a factor. This finding confirms that of Galtsoff (1948). Cultures of *G. breve* and filtrates of the crude toxin were placed in aquaria containing the fishes *Membras vagrans, Mugil cephalus, Fundulus grandis, F. similis, Mollinesia latipinna,* and *Cyprinodon variegatus*. All the fishes died within 4 minutes to 19 hours. Similar studies have been performed by Starr (1958). The reader is referred to Ray and Wilson (1957) for details of dosage and experiment design.

Gates and Wilson (1960) have tested bacterized cultures of *Gonyaulax monilata* and found them to be toxic to mullet *Mugil cephalus*.

Using gobies *Gobius flavescens*, Abbott and Ballantine (1957) developed a bioassay for the toxicity of *Gymnodinium veneficum*. Tests were made by placing the gobies in aquaria containing cultures of the dinoflagellate or extracts of the poison. Results were determined by plotting time to complete loss of balance of the fish and time to death against dose in arbitrary units. Similar tests were conducted on coelenterates, annelids, arthropods, mollusks, echinoderms, tunicates, cephalocordates, fishes, amphibians, and mammals.

A chemical assay for saxitoxin, the paralytic shellfish poison, has been developed by Bates and Rapoport (1974). The technique involves alkaline hydrogen peroxide oxidation of saxitoxin to 8-amino-6-hydroxymethyl-2-imino-purine-3 (2*H*)-propionic acid, the fluorescence of which is measured at pH 5. This chemical assay is one hundred times more sensitive than the existing bioassay and eliminates various problems associated with the bioassay, particularly at low levels of toxin. The method was developed initially with pure saxitoxin and has been applied to a number of samples of the marine bivalves *Saxidomus giganteus* and *Mytilus californianus*.

Localization of the Poison in Shellfish

Wolff (1886), Sommer *et al.* (1937), Sommer and Meyer (1937), Medcof *et al.* (1947), and Meyer (1953) have shown that the greatest concentration of the poison in mussels *Mytilus* is in the digestive organs or so-called liver. Medcof *et al.* (1947) reported that in the bar clam *Mactra* and in the soft-shell clam *Mya* the gills are another important center of poison concentration. Toxicity studies for *Mya* reveal seasonal changes in the anatomical distribution of the poison. During the summer, the poison is concentrated primarily in the digestive glands; but during the fall and winter, the concentration is greatest in the gills. Apparently the gills are slow both in accumulating and in losing the poison. Scallops *Pecten* have the ability to retain the poison in high concentration in various other organs for long periods of time, but the adductor muscle, which is the part used for human consumption, has never been found to be toxic. Pugsley (1939) and Chambers and Magnusson (1950) found that the poison in the butter clams *Saxidomus* is concentrated primarily in the siphons. Since the toxicity varied greatly from month to month in the siphons, but did not fluctuate to any great extent in the bodies, the ratio of the toxicity of siphons to that of bodies is not constant (*see* Pl. 4).

Toxic Agents Other Than Paralytic Shellfish Poison

Whenever large amounts of crude shellfish extract are injected, the combined action of several substances may be responsible for the death of the animal. Sommer and Meyer (1937) determined that calcium and magnesium salts were most commonly encountered in mussel extracts. The average lethal dose for calcium is 20 mg and for magnesium salts, 10 mg. Symptoms produced by these salts may resemble those caused by sublethal doses of mussel poison. A quaternary ammonium salt has also been incriminated as a toxic agent in crude mussel extracts. In addition, there are believed to be other toxic substances whose chemical nature is poorly understood. Hashimoto and Migita (1950) demonstrated that methanol acidulated with hydrochloric acid "produces an artificial poison from some common components of shellfish."

PHARMACOLOGY

The pharmacological effects of paralytic shellfish poison have been described by Prinzmetal, Sommer, and Leake (1932), Kellaway (1935*a*, *b*), Sapeika

(1953), Fingerman *et al.* (1953), Murtha (1960), and Evans (1969*c*). Most of the early studies were based on a chemically nondefined aqueous solution of solvent fractionated acid alcohol extracts of whole mussel or mussel livers. A mouse lethal dose was the unit of potency, with a time factor often to 20 minutes. On a weight basis, about 0.1 mg of the dry crude extract per kilogram of mouse weight was found to be lethal. Susceptibility to the poison, in order of decreasing sensitivity, varied with the animal—mouse, rabbit, dog, frog. An acid stabilized solution apparently maintains its potency.

Murtha (1960) extracted the poisons from the mussel *Mytilus californianus* and the Alaska butter clam *Saxidomus giganteus* and purified them by the methods of Schantz *et al.* (1957). The extracts contained 5,000 to 6,000 mouse units per milligram of solids. The lethal dose of the material used was 3-4 mg/kg (15 to 20 mouse units/kg). Anesthetized and unanesthetized cats, dogs, and rabbits of mixed breed and of both sexes were used in his studies. He was unable to demonstrate any pharmacological differences between the two poisons.

The principal action of the toxin is on the central nervous system, i.e., respiratory and vasomotor centers, and the peripheral nervous system, i.e., neuromuscular junction, cutaneous tactile endings, and muscle spindles. Absorption occurs through the gastrointestinal tract, and rapid excretion of the active toxin occurs through the kidneys.

Effects on Respiration

A function of dose, the respiratory response varies from instantaneous failure to slowing. Dyspnea with short rapid exhalations or gaspings may occur. Electrophysiologic studies on the phrenic nerves indicate enhanced activity up to the time of failure. Kellaway (1935*b*) believed that the poison actually depressed the center, but that stimulation secondary to the resulting asphyxia accounted for the apparent enhanced activity. Junctional block of the respiratory muscles was also considered to be a factor. D'Aguanno (1959) reported that shellfish poison depressed respiration centrally with no curarization peripherally.

Effects on Circulation

Again the effects are variable, depending upon the animal and the dose. With a sublethal dose, the blood pressure falls initially, recovers, and then slowly, but steadily falls. Apparently the blood pressure change is due primarily, though not entirely, to vasodilation in the splanchnic area, mediated by way of the vasomotor center, because the changes are not found in decapitate or spinal animals (Kellaway, 1935*b*; Sapeika, 1953). The presence or absence of circulatory "buffering" mechanisms does not seem to influence the vascular responses. The heart does not stop beating even after the cessation of respiration, but apparently atrio-ventricular block occurs, along with occasional arrhythmias.

Murtha (1960) found that mussel poison had no major vascular action, and that peripheral vasodilatation was not an important factor that contributed to the marked cardiovascular collapse. Apparently the rapid fall in blood pressure that followed lethal doses of mussel poison was produced largely by the actions of the poison on the brain and partially by a direct depression of the myocardium.

The effect of gymnodin from *Gymnodinium breve* on the coagulation time of human blood plasma is described by Doig and Martin (1973). Its anticoagulant activity, for example, was comparable to 1/6 that of carrageenan and 1/70 that of heparin.

The effect of two toxins, prymnesin from *Prymnesium parvum* and gymnodin, on potassium transport across red cell membranes has been studied by Martin and Padilla (1974).

Effects on Nerve-Muscle

Paralytic shellfish poison is a potent neurotoxin capable of producing complete depression of both peripheral and reflex transmission (Murtha, 1960).

Kellaway (1935*a, b*) has described this neurotoxic effect as a typical curarization in frog nerve-muscle preparations that is followed by cessation of nerve conduction and finally by diminution of muscle conduction to direct stimulation. Concentration of 1:200,000 of the dry poison will produce curarization in 1 1/2 hours. Cutaneous tactile receptors and muscle spindles failed to respond after exposure to the toxin. Tactile receptors are paralyzed in 2 minutes with concentrations of 1:500,000 at 15° C, whereas the spindles failed in 58 minutes with similar concentrations at 8° C. The intoxicated muscle does not respond to the application of acetylcholine, and the paralysis is not reversible. The toxin has no apparent effect on isolated smooth muscle (Kellaway, 1935*b*; Sapeika, 1953).

Because of the action on the neuromuscular junction, dinoflagellate poison is sometimes compared to curare. However, these drugs differ in their action in that curare is believed to produce a competitive nonpolarizing block at the neuromuscular junction, whereas paralytic shellfish poison prevents the synaptic region from producing acetylcholine (Hutner and McLaughlin, 1958).

Bolton *et al.* (1959) found in their studies with frog nerve-muscle preparations that curare and botulinus toxin did not prevent muscle contraction on direct stimulation, but paralytic shellfish poison would modify such a reaction; however, the effects of paralytic shellfish poison are partially reversible. The absence of marked inhibition of cholinesterase activity at concentrations causing paralysis suggests that this effect is not dependent on excessive acetylcholine accumulation; therefore, they concluded that shellfish poison is not a cholinesterase inhibitor.

It is interesting to note that Murtha (1960) was unable to detect any significant pharmacological differences between mussel poison (*Mytilus californianus*) and clam poison (*Saxidomus giganteus*).

Comparative studies of the mechanism of saxitoxin and tetrodotoxin poisoning have been prepared by Evans (1969*a, b*). His comparison of paralytic shellfish poisons from *M. californianus, S. giganteus*, and *Gonyaulax catenella* suggests they are all identical (Evans, 1971; *see also* Kao and Fuhrman, 1967).

CHEMISTRY

Despite the lapse of time since Brieger (1885, 1888, 1889) and Salkowski (1885) undertook studies on the chemistry of mussel toxin, it was not until recently that the structure of the toxin was determined. Methods are now available, however, for the isolation of the purified toxin in quantities large enough to permit more detailed studies. Of equal importance to these techniques is a chemical assay method which permits the recovery of the toxin from low content sources.

The empirical formula ($C_{10}H_{19}N_7O_4Cl_2$) for the toxin hydrochloride seemed well established up until 1965. However, recent studies by Wong, Oesterlin, and Rapoport (1971) found that combustion analysis of saxitoxin, dried to a constant weight of 110° (10^{-5} mm) (with no loss of biological activity) were consistent with $C_{10}H_{15}N_7O_{32} \cdot 2$ HCL. The structural formula appears to be as follows:

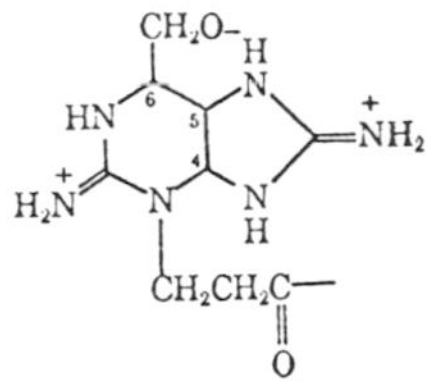

The toxin appears to exist in two tautomeric forms, one having a somewhat higher toxicity than the other. The color reactions, the specific rotation ($[\alpha]^{20}_{D} + 133°$), and the absorption spectra remain constant for the two forms. The structure must certainly contain a ring system with one or two guanidine groups as part of the molecule. Conjugate double bonds appear to be absent. The nature of the oxygen-carrying groups remains to be clarified.

Perhaps the greatest need at the present time is for a method of degradation that will yield large fragments of the molecule in quantities sufficient to permit identification. The work of Schuett and Rapoport (1962) has provided a start in this direction. The near-failure of the degradation methods tried suggests that the solution to the problem will be attained only as hypothetical molecules are assumed and as model molecules are synthesized for reaction studies.

Trieff *et al.* (1973) isolated the toxin from *Gymnodinium breve* by thin layer chromatographic procedure and studied its physico-chemical and toxicological characteristics. The LD_{50} of the crude toxin was about 2.2 mg/kg. A molecular weight of 468 was found by osmometry.

Shimizu *et al.* (1975) have isolated four different toxins from the clam *Mya arenaria* using Sephadex G-15 and high pressure ion exchange chromatography. One of the poisons was saxitoxin, but the other three appear to be new toxins. These same toxins were also isolated from the extract of cultured *Gonyaulax tamarensis*.

LITERATURE CITED

ABBOTT, B. C., and D. BALLANTINE
1957 The toxin from *Gymnodinium veneficum* Ballantine. J. Marine Biol. Assoc. U.K. 36: 169-189, 8 figs.

ABBOTT, R. T.
1954 American seashells. D. Van Nostrand Co., Inc., New York. 541 p., 40 pls., 100 figs.

ANDERSON, L. S.
1960 Toxic shellfish in British Columbia. Am. J. Public Health 50(1): 71-83. 5 figs.

BALLANTINE, D., and B. C. ABBOTT
1957 Toxic marine flagellates; their occurrence and physiological effects on animals. J. Gen. Microbiol. 16(1): 271-281, 3 figs.

BATES, H. A., and H. RAPOPORT
1974 A chemical assay for saxitoxin, the paralytic shellfish poison. Agric. Food Chem. 23(2): 237-239.

BERGMANN, F., PARNAS, I., and K. REICH
1963 Observations on the mechanism of action and on the quantitative assay of ichthyotoxin from *Prymnesium parvum* Carter. Toxicol. Appl. Pharmacol. 5: 637-649, 4 figs.

BERKELEY, C.
1962 Toxicity of plankton to *Cristispira* inhabiting the crystalline style of a mollusk. Science 135(3504): 664-665.

BERMAN, I.
1960 Formation of toxic principles in concentrated cell suspensions of *Prymnesium parvum*. M. Sci. Thesis. Hebrew Univ., Jersulem.

BOLTON, B. L., BERGNER, A. D., O'NEILL, J. J. and P. F. WAGLEY
1959 Effects of a shell-fish poison on end-plate potentials. Bull. Johns Hopkins Hosp. 105(4): 233-238, 4 figs.

BOND, R. M., and J. C. MEDCOF
1958 Epidemic shellfish poisoning in New Brunswick, 1957. Can. Med. Assoc. J. 79(19-24): 1-14.

BRIEGER, L.
1885 Ueber basische producte in der miesmuschel. Deut. Med. Wochschr. 11(53): 907-908.
1888 Zur kenntniss des tetanin und des mytilotoxin. Arch. Pathol. Anat. Physiol. 112: 549-551.
1889 Beitrag zur kenntniss der zusammensetzung des mytilotoxins nebst einer uebersicht der bisher in ihren haupteigenschaften bekannten ptomaine und toxin. Arch. Pathol. Anat. Physio. Ser. 9, 115(5): 483-492.

BRONGERSMA-SANDERS, M.
1948 The importance of upwelling water to vertebrate paleontology and oil geology. Verhandel. Koninkl. Ned. Akad. Wetenschap. (Tweede Sect) 45(4): 59, 68-72.
1957 Mass mortality in the sea, p. 941-1040, 7 figs. *In* J. W. Hedgpeth [ed.], Treatise on marine ecology and paleoecology. Vol. I. Geol. Soc. Am., Mem. 67. Waverly Press, Baltimore.

BURKE, J. M. MARCHISOTTO, J., MCLAUGHLIN, J. J., and L. PROVASOLI
1960 Analysis of the toxin produced by *Gonyaulax catenella* in axenic culture. Ann. N. Y. Acad. Sci. 90(3): 837-842. 1 fig.

CHAMBERS, J. S., and H. W. MAGNUSSON
1950 Season variations in toxicity of butter clams from selected Alaska beaches. U.S. Fish Wildlife Serv., Spec. Sci. Rept., Fish. No. 53, 19 p., 9 figs.

CHEVALLIER, A., and E. A. DUCHESNE
1815*a* Mémoire sur les empoisonnements par les huîtres, les moules, les crabes, et par certains poissons de mer et de rivière. Ann. Hyg. Publ. (Paris) 45(11): 387-437.
1851*b* Mémoire sur les empoisonnements par les huîtres, les moules, les crabes, et par certains poissons de mer et de rivière. Ann. Hyg. Publ. (Paris) 46(1): 108-147.

CLARK, R. B.
1968 Biological causes and effects of paralytic shellfish poisoning. Lancet 2(7571): 770-772, 1 fig.

COLBY, M.
1943 Poisonous marine animals in the Gulf of Mexico. Proc. Trans. Texas Acad. Sci. 26: 62-70.

CORNMAN, I.
1947 Retardation of Arbacia egg cleavage by dinoflagellate-contaminated seawater (red tide). Biol. Bull. 93: 205.
COVELL, W. P., and W. F. WHEDON
1937 Effects of the paralytic shell-fish poison on nerve cells. Arch. Pathol. 24(4): 411-418, 3 figs.
DACK, G. M.
1956 Poisonous plants and animals, p. 31-58. *In* G. M. Dack, Food poisoning. Rev. ed. Univ. Chicago Press, Chicago.
D'AGUANNO, W.
1959 Pharmacology of paralytic shellfish poison. Pharmacologist 1: 71.
DOIG, M. T., III, and D. F. MARTIN
1973 Anticoagulant properties of a red tide toxin. Toxicon 11: 351-355.
EDWARDS, H. I.
1956 The etiology and epidemiology of paralytic shellfish poisoning. J. Milk, Food, Technol. 19(12): 331-335.
EVANS, M. H.
1965 Cause of death in experimental paralytic shellfish poisoning. Brit. J. Exp. Pathol. 46(3): 245-253.
1969*a* Differences between the effects of saxitoxin (paralytic shellfish poison) and tetrodotoxin on the frog neuromuscular junction. Brit. J. Pharmacol. 36: 426-436.
1969*b* Mechanism of saxitoxin and tetrodotoxin poisoning. Brit. Med. Bull. 25(3): 263-267.
1969*c* Spinal reflexes in cat after intravenous saxitoxin and tetrodotoxin. Toxicon 7: 131-138.
1970 Paralytic shellfish poisoning in Britain. Mar. Poll. Bull. 1(12): 184-186.
1971 A comparison of the biological effects of paralytic shellfish poisons from clam, mussel, and dinoflagellate. Toxicon 9: 139-144.
FINGERMANN, M., FORESTER, R. H., and J. H. STOVER, JR.
1953 Action of shellfish poison on peripheral nerve and skeletal muscle. Proc. Soc. Exp. Biol. Med. 84: 643-646, 4 figs.
FITCH, J. E.
1953 Common marine bivalves of California. Calif. Fish Game, Fish Bull. No. 90, 102 p., 1 pl., 63 figs.
FRASER-BRUNNER, A.
1973 Danger in the sea. Hamlyn Publ. Group Ltd., New York. 128 p., illus.
FURK, D. M.
1950 Shellfish poisoning. Fish Res. Board Canada, Progr. Repts. Pacific Coast Sta. No. 82, p. 3-5, 1 fig.
GALTSOFF, P. S.
1948 Red tide. Progress report on the investigations of the cause of mortality of fish along the west coast of Florida. U.S. Fish Wildlife Serv., Spec. Rept. No. 46, 44 p., 9 figs.
GATES, J. A., and W. B. WILSON
1960 The toxicity of *Gonyaulax monilata* Howell to *Mugil cephalus*. Limnol. Oceanog. 5(2): 171-174.
GIBBARD, J. COLLIER, F. C., and E. F. WHYTE
1939 Mussel poisoning. Can. Publ. Health J. 30(4): 193-197, 1 fig.
GIBBARD, J., and J. NAUBERT
1948 Paralytic shellfish poisoning on the Canadian Atlantic coast. Am. J. Publ. Health 38(4): 550-553.
GUNTER, G., WILLIAMS, R. H., DAVIS, C. C., and F. G. SMITH
1948 Catastrophic mass mortality of marine animals and coincident phytoplankton bloom on the west coast of Florida, November 1946 to August 1947. Ecol. Monographs 18(3): 311-324.
HALEVY, S., SALITERNIK, R., and L. AVIVI
1971 Isolation of rhodamine-positive toxins from *Ochromonas* and other algae. Internatl. J. Biochem. 2(8): 185-192.
HALSTEAD, B. W.
1974 Marine biotoxicology, p. 212-239. *In* F. Coulston and F. Korte [eds.], Environmental quality and safety, vol. 3. Academic Press, Inc., New York.
HASHIMOTO, Y., and M. MIGITA
1950 On the shellfish poisons: I. Inadequacy of acidulated alcohols with hydrochloric acid as solvent. Bull. Japan Soc. Sci. Fish. 16(3): 77-85.
HATTORI, Y., and T. AKIBA
1952 Studies on the toxic substance in Asari (*Venerupis semidecussata*); II. Detection of toxic shellfish. [In Japanese, English summary] J. Pharm. Soc. Japan 72(4): 572-577.

HAYES, H. L., and T. S. AUSTIN
1951 The distribution of discolored sea-water. Texas J. Sci. 3(4): 530-541, 1 fig.

HILL, W. H.
1953 The occurrence and etiology of paralytic shellfish poisoning: I. A review (1790 to 1952). Minutes Pacific Coast Shellfish Com., January 1953. Appendix I. British Columbia Dept. Health, Victoria, B.C. (Unpublished).

HUTNER, S. H.
1961 Plant animals as experimental tools for growth studies. Bull. Borrey Botan. Club 88(5): 339-349.

HUTNER, S. H., and J. J. MCLAUGHLIN
1958 Poisonous tides. Sci. Am. 199(2): 92-98, 4 figs.

JORDAN, E. O.
1931 Poisonous plants and animals, p. 26-64. *In* E. O. Jordan, Food poisoning and food-borne infection. Univ. Chicago Press, Chicago.

KAO, C. Y., and F. A. FUHRMAN
1967 Differentiation of the actions of tetrodotoxin and saxitoxin. Toxicon 5: 25-34.

KELLAWAY, C. H.
1935*a* Mussel poisoning. Med. J. Australia 1: 399-401.
1935*b* The action of mussel poison on the nervous system. Australian J. Exp. Biol. Med. Sci. 13: 79-94. 8 figs.

KOCH, H. J.
1939 La cause des empoisonnements paralytiques provoqué par les moules. Assoc. Franc. Avan. Sci., Liege. 63d Session, p. 654-657.

KOFOID, C. A.
1911 Dinoflagellata of the San Diego region; IV. The genus *Gonyaulax*, with notes on its skeletal morphology and a discussion of its generic and specific characters. Univ. Calif. (Berkeley) Publ. Zool. 8(4): 187-269; Figs. A-E, Pls. 9-17.

KOFOID, C. A., and O. SWEZY
1921 The free-living unarmoured Dinoflagellata. Memoirs of the University of California. Vol. VI. Univ. California Press, Berkeley, Calif. 562 p., 388 figs., 12 pls.

KUDO, R. R.
1954 Protozoology. 4th ed. Charles C. Thomas, Springfield, Ill. 966 p., 376 illus.

LACKEY, J. B., and J. A. HYNES
1955 The Florida Gulf Coast "red tide." Eng. Progr. Univ. Florida Bull. Ser. 70, 9(2): 1-22, 6 figs.

LE MESSURIER, D. H.
1935 A survey of mussels on a portion of the Australian coast. Med. J. Australia 1(16): 490-492, 12 fig.

LINDNER, G.
1888*a* Über giftige miesmuscheln. Centralblatt Zent. Bak. Parasiten 3: 352-358.
1888*b* Beitrag zur kennzeichnung giftiger miesmuscheln und zur ermittelung der veranlassenden ursachen des muschelgiftes. Deut. Med. Wochschr. 9(49-50): 585-586, 597.

LINSTOW, O. V.
1894 Mollusca, p. 130-145; Figs. 10-19. *In* O. V. Linstow, Die gifttiere und ihre wirkung auf den menschen. August Hirschwald, Berlin.

MCFARREN, E. F.
1960 Collaborative studies of the chemical assay for paralytic shellfish poison. J. Assoc. Offic. Agri. Chemists 43: 544-547, 4 figs.

MCFARREN, E. F., SCHAFER, M. L., CAMPBELL, J. E., LEWIS, K. H., JENSEN, E. T., and E. J. SCHANTZ
1956 Public health significance of paralytic shellfish poison. A review of literature and unpublished research. Proc. Nat. Shellfisheries Assoc. 47: 114-141, 2 figs.

MCFARREN, E. F., SCHANTZ, E. J., CAMPBELL, J. E., and K. H. LEWIS
1958 Chemical determination of paralytic shellfish poison in clams. J. Assoc. Offic. Agri. Chemists 41(1): 168-177.

MCFARREN, E. F., TANABE, H., SILVA, F. J., WILSON, W. B., CAMPBELL, J. E., and K. H. LEWIS
1965 The occurrence of a ciguatera-like poison in oysters, clams, and *Gymnodinium breve* cultures. Toxicon 3: 111-123.

MACHT, D. I., and E. C. SPENCER
1941 Physiological and toxicological effects of some fish muscle extracts. Proc. Soc. Exp. Biol. Med. 46: 228-233.

MARTIN, D. F., and G. M. PADILLA
1974 Effect of *Gymnodinium breve* toxin on potassium influx of erythrocytes. Toxicon 12: 353-360.

MEBS, D.
1973 Chemistry of animal venoms, poisons and toxins. Experientia 29(11): 1328-1334.

MEDCOF, J. C.
1960 Shellfish poisoning—another North American ghost. Can. Med. Assoc. J. 82: 87-90.
1971 Winter variability in paralytic shellfish poison scores for Crow Harbour, New Brunswick. Fisheries Research Board of Canada (Manuscript Report Series No. 1163). 17 p.
1972 The St. Lawrence rough whelk fishery and its paralytic shellfish poison problems. Fisheries Research Board of Canada (Manuscript Report) Series No. 1201). 38 p.

MEDCOF, J. C., LEIM, A. H. NAUBERT, A. B., NEEDLER, A. W., GIBBARD, J., and J. NAUBERT
1947 Paralytic shellfish poisoning on the Canadian Atlantic coast. Bull. Fish. Res. Board Can. 75, 32 p.

MEYER, K. F.
1928 Mussel poisoning in California. Calif. Fish Game 14(3): 201-202, 1 fig.
1929 Ueber muschelvergiftungen in Kalifornien. Z. Fleisch Milchhygiene 39 (12): 210-212.
1953 Medical progress; food poisoning. New Engl. J. Med. 249(19): 765-773; (20): 804-812; (21): 843-852.

MEYER, K. F., SOMMER, H., and P. SCHOENHOLZ
1928 Mussel poisoning. J. Prev. Med. 2(5): 365-394.

MIGITA, M., and K. KANNA
1957 On the source of shell-fish poison at the Lake Hamana, 1. Attempts at experimentally rendering shellfish toxic. [In Japanese, English summary] Bull. Jap. Soc. Sci. Fish. 23(4): 215-221, 1 fig.

MOSHER, H. S.
1966 Non-protein neurotoxin. Science 51(3712): 860-861.

MÜLLER, H.
1932 Mussels and clams: a seasonal quarantine-bicarbonate of soda as a factor in the prevention of mussel poisoning. Calif. West. Med. 37(5): 327-328.
1935 Chemistry and toxicity of mussel poison. J. Pharmacol. Exp. Therap. 53(1): 67-89.

MURTHA, E. F.
1960 Pharmacological study of poisons from shellfish and puffer fish. Ann. N.Y. Acad. Sci. 90(3): 820-836, 9 figs.

NAKAZIMA, M.
1955 Studies on the toxicity of little-neck clam and oyster. [In Japanese] Bull. Biogeogr. Soc. Japan 16-19: 84-87.
1965 Studies on the source of shellfish poison in Lake Hamana, 2. Shellfish toxicity during the "Red Tide." Bull. Jap. Soc. Sci. Fish. 31(3): 204-207.
1965 Studies on the source of shellfish poison in Lake Hamana, 3. Poisonous effects of shellfish feeding on *Prorocentrum* sp. Bull. Jap. Soc. Sci. Fish. 31(4): 281-285, 1 pl.

NARAHASHI, T., HAAS, H. G., and E. F. THERRIEN
1967 Saxitoxin and tetrodotoxin: comparison of nerve blocking mechanism. Science 157(3795): 1441-1442.

NEEDLER, A. B.
1949 Paralytic shellfish poisoning and *Gonyaulax tamarensis*. J. Fish. Res. Board Can. 7(8): 490-504, 3 figs.

OTTERSTRØM, C. V., and E. STEEMANN-NIELSEN
1939 Two cases of extensive mortality in fishes caused by the flagellate *Prymnesium parvum* Carter. Rept. Danish Biol. Sta. 44: 1-24, 9 figs.

PARNAS, I.
1963 The toxicity of *Prymnesium parvum* (a review). Israel J. Zool. 12(1-4): 15-23.
1964 Pharmacological effects of *Prymnesium* toxin. (Personal communication to Dr. Russell, Nov. 20, 1964.)

PEPLER, W. J., and E. LOUBSER
1960 Histochemical demonstration of the mode of action of the alkaloid in mussel poisoning. Nature 188 (4753): 800.

PHILLIPS, C., and W. H. BRADY
1953 Sea pests—poisonous or harmful sea life of Florida and the West Indies. Univ. Miami Press, Miami, Fla. 78 p., 38 figs., 7 pls.

PHISALIX, M.
1922 Mollusques, p. 465-483; Figs. 188-193. *In* M. Phisalix, Animaux venimeux et venins. Vol. I. Masson et Cie., Paris.

PRAKASH, A.
1963 Source of paralytic shellfish toxin in the Bay of Fundy. J. Fish. Res. Board Can. 20(4): 983-996, 5 figs.
1967 Growth and toxicity of a marine dinoflagellate, *Gonyaulax tamarensis*. J. Fish. Res. Board Can. 24(7): 1589-1606.

PRAKASH, A., and J. C. MEDCOF
1962 Hydrographic and meteorological factors affecting shellfish toxicity at Head Harbour, New Brunswick. J. Fish. Res. Board Can. 19(1): 101-112, 4 figs.

PRAKASH, A., MEDCOF, J. C., and A. D. TENNANT
1971 Paralytic shellfish poisoning in eastern Canada. Fish Res. Board Can., Bull. 177. Ottawa, Canada. 87 p., 25 figs.

PRESCOTT, G. W.
1948 Objectionable algae with reference to the killing of fish and other animals. Hydrobiologia 1(1): 1-13.

PRICE, R. J., and J. S. LEE
1972 Paralytic shellfish poison and melanin distribution in fractions of toxic butter clam (*Saxidomus giganteus*) siphon. J. Fish. Res. Board Can. 29: 1657-1658.

PRINZMETAL, M., SOMMER, H., and C. D. LEAKE
1932 The pharmacological action of "mussel poisoning." J. Pharmacol. Exp. Therap. 46: 63-73.

PROVASOLI, L.
1962 Organic regulators of phytoplankton fertility. *In* L. Provasoli, The sea.

PUGSLEY, L. I.
1939 The possible occurrence of a toxic material in clams and mussels. Fish. Res. Board Can., Progr. Rept. Pacific Coast Sta. No. 40, p. 11-13.

RAY, S. M., and C. V. ALDRICH
1965 *Gymnodinium breve*: induction of shellfish poisoning in chicks. Science 148(3678): 1748-1749.

RAY, S. M., and W. B. WILSON
1957 Effects of unialgal and bacteria-free cultures of *Gymnodinium brevis* on fish and notes on related studies with bacteria. U.S. Fish Wildlife Serv., Fish. Bull. 53(123): 469-496.

REICH, K., and M. ROTBERG
1958 Some factors influencing the formation of toxin poisonous to fish in bacteria-free cultures of *Prymnesium*. Bull. Res. Council Israel 7B(3/4): 199-202.

RIEGEL, B., STANGER, D. W., WIKHOLM, D. M.
MOLD, J. D., and H. SOMMER
1949*a* Paralytic shellfish poisoning; IV. Bases accompanying the poison. J. Biol. Chem. 177(1): 1-6.
1949*b* Paralytic shellfish poisoning; V. The primary source of the poison, the marine plankton organism *Gonyaulax catenella*. J. Biol. Chem. 177(1): 7-11.

SALKOWSKI, E.
1885 Zur kenntniss des giftes der miesmuschel (*Mytilus edulis*). Arch. Pathol. Anat. Physiol. 102(3): 578-592.

SAPEIKA, N.
1948 Mussel poisoning. S. African Med. J. 22(10): 337-338.
1953 Actions of mussel poisons. Arch. Intern. Pharmacodyn. 93(2): 135-142, 4 figs.
1958 Mussel poisoning: a recent outbreak. S. African Med. J. 32(20): 527.

SCHANTZ, E. J.
1957 Purified shellfish poison for bioassay standardization, p. 17-36. *In* Conference on shellfish toxicology. U.S. Public Health Serv., Spec. Publ. (Unpublished.)
1960 Biochemical studies on paralytic shellfish poisons. Ann. N.Y. Acad. Sci. 90(3): 843-855.
1961 II. Some chemical and physical properties of paralytic shellfish poisons related to toxicity. J. Med. Pharm. Chem. 4(3): 459-468.

SCHANTZ, E. J., LYNCH, J. M., VAYVADA, G., MATSUMOTO, K., and H. RAPOPORT
1966 The purification and characterization of the poison produced by *Gonyaulax catenella* in axenic culture. Biochemistry 5(4): 1191-1194.

SCHANTZ, E. J., LYNCH, J. M., and G. VAYVADA
1962 Some chemical and physical properties of cultured *Gonyaulax catenella* poison.

SCHANTZ, E. J., MCFARREN, E. F., SCHAFER, M. L., and K. H. LEWIS
1958 Purified shellfish poison for bioassay standardization. J. Assoc. Offic. Agri. Chemists 41(1): 160-168.

SCHANTZ, E. J., MOLD, J. D., STANGER, D. W., SHAVEL, J., RIEL, F. J., BOWDEN, J. P., LYNCH, J. M., WYLER, R. S., RIEGEL, B., and H. SOMMER
1957 Paralytic shellfish poison; VI. A procedure for the isolation and purification of the poison from toxic clam and mussel tissues. J. Am. Chem. Soc. 79: 5230-5235.

SCHANTZ, E. J., MOLD, J. D., HOWARD, W. L., BOWDEN, J. P., STANGER, D. W., LYNCH, J. M., WINTERSTEINER, O. P., DUTCHER, J. D., WALTERS, D. R., and B. RIEGEL
1961 Paralytic shellfish poison; VIII. Some chemical and physical properties of purified clam and mussel poisons. Can. J. Chem. 39: 2117-2123, 3 figs.

SCHILLER, J.
1933 Dr. L. Rabenhorst's kryptogamenflora von Deutschland, Osterreich und der Schweiz; Dinoflagellatae. Vol. I. Acad. Verlags. Ges. (Leipzig). 617 p., 631 figs.
1937 Dr. L. Rabenhorst's kryptogamenflora von Deutschland, Osterreich und der Schweiz; Dinoflagellatae. Vol. II. Acad. Verlags. Ges. (Leipzig). 589 p., 612 figs.

SCHMIDTMANN, —
1888 Miesmuschelvergiftung zu Wilhelmshafen im Herbst 1887. Z. Med.-Beamte 1: 19-23, 49-53.

SCHUETT, W., and H. RAPOPORT
1962 Saxitoxin, the paralytic shellfish poison. Degradation to a pyrrolophrimidine. J. Am. Chem. Soc. 84: 2266.

SEVEN, M. J.
1958 Mussel poisoning. Ann. Internal. Med. 48: 891-897.

(NSA)—Not seen by author.

SHILO, M., and M. ASCHNER
1953 Factors governing the toxicity of cultures containing the phytoflagellate *Prymnesium parvum* Carter. J. Gen. Microbiol. 8(3): 333-343.

SHILO, M., ASCHNER, M., and M. SHILO
1953 The general properties of the exotoxin of the phytoflagellate *Prymnesium parvum*. Bull. Res. Council Israel 2(4): 446.

SHILO, M., and R. F. ROSENBERGER
1960 Studies on the toxic principles formed by the chrysomonad *Prymnesium parvum* Carter. Ann. N.Y. Acad. Sci. 90(3): 886-876, 7 figs.

SHIMIZU, Y., ALAM, M. *et al.*
1975 Presence of four toxins in red tide infested clams and cultured *Gonyaulax tamarensis* cells. Biochem. Biophys. Res. Commun. Marine Reprint 49. Sea Grant Advisory Service, University of Rhode Island. 66(2): 731-737.

SMITH, F. G.
1954 Red tide studies. Univ. Miami, Marine Lab., Prelim. Rept. January to June, 117 p., 35 figs.

SOMMER, H., and K. F. MEYER
1935 Mussel poisoning. Calif. West Med. 42(6): 423-426.
1937 Paralytic shell-fish poisoning. Arch. Pathol. 24(5): 560-598, 2 figs.
1941 Mussel poisoning — a summary. Calif. State Dept. Public Health, Weekly Bull. 21(14): 53-55.

SOMMER, H., WHEDON, W. F., KOFOID, C. A., and R. STOHLER
1937 Relation of paralytic shell-fish poison to certain plankton organisms of the genus *Gonyaulax*. Arch. Pathol. 24(5): 537-559, 2 figs.

SPARKS, A. K., SRIBHIBHADH, A., CHEW, K. K., and D. PEREYRA
1962 Ecology of paralytic shellfish toxicity in Washington. Res. Fish Contrib. 139: 39-42. (NSA)

STARR, T. J.
1958 Notes on a toxin from *Gymnodinium breve*. Texas Rept. Biol. Med. 16(4): 500-507.

STEPHENSON, N. R., EDWARDS, H. L., MACDONALD, B. F., and L. I. PUGSLEY
1955 Biological assay of the toxin from shellfish. Can. J. Biochem. Physiol. 33: 849-857.

SVERDRUP, H. U., JOHNSON, M. W., and R. H. FLEMING
1942 The oceans, their physics, chemistry and general biology. Prentice-Hall, New York. 1,087 p., 265 figs.

TAKAHASHI, T., SERISAWA, S., OKA, S., SAKITA, T., and H. HOSAKA
1955 Experimental cirrhosis of the liver induced by shellfish poison, with special reference to the pathogenesis of fibrosis. [In Japanese] Sogo Igaku 12: 81-86. (NSA)

TENNANT, A. D., NAUBERT, J., and H. E. CORBEIL
1953 An outbreak of paralytic shellfish poisoning. Can. Med. Assoc. J. 72: 436-439.

THESEN, J.
1902 Studien über die paralytische form von vergiftung durch muscheln (*Mytilus edulis* L.). Arch. Exp. Pathol. Pharmakol. 47: 311.

THOMSON, W. K., LAING, A. C., and G. A. GRANT
1957 Toxic algae; IV. Isolation of toxic bacterial contaminants. Dept. of National Defence, Canada. Rept. No. 51, 6 p.

TRIEFF, N. M., SPIKES, J. J., RAY, S. M., and J. B. NASH
1973 Isolation and purification of *Gymnodinium breve* toxin. p. 557-575. 6 figs. *In* A. de Vries and E. Kochva [eds.], Toxins of animal and plant origin, vol. 2. Gordon and Breach Science Publ., New York.

U.S. PUBLIC HEALTH SERVICE
1957 Sanitary control of the shellfish industry. Part II. Sanitation of the harvesting and processing of shellfish. Spec. Publ. No. 33, 26 p.
1959 Manual of recommended practice for sanitary control of the shellfish industry. Rev. ed. Part I. Sanitation of shellfish growing areas. Spec. Publ. No. 33, 36 p.
1962 Manual of recommended practice for sanitary control of the shellfish industry. Part III. Public health service appraisal of state shellfish sanitation programs. Spec. Publ. 39 p.

VIRCHOW, R.
1885 Ueber die vergiftungen durch miesmuscheln in Wilhelmshaven. Berlin. Klin. Wochschr. 22(48): 781-785.

WALDICHUK, M.
1958 Shellfish toxicity and the weather in the Strait of Georgia during 1957. Fish. Res. Board Can., Progr. Rept. Pacific Coast Sta. 112: 10-14.

Walford, L. A.
1958 Living resources of the sea. Ronald Press, New York. 321 p., 23 figs.

WHEDON, W. F., and C. A. KOFORD
1936 Dinoflagellata of the San Francisco region; I. On the skeletal morphology of two new species, *Gonyaulax catenella* and *G. acatenella*. Univ. Calif. (Berkeley) Publ. Zool. 41(4): 25-34, 16 figs.

WIBERG, G. S., and N. R. STEPHENSON
1960 Toxicologic studies on paralytic shellfish poison. Toxicol. Appl. Pharmacol. 2(6): 607-615.

WOLFF, M.
1886 Die ausdehnung des gebietes der giftigen miesmuscheln unter sonstigen giftigen seethiere in Wilhelmshaven. Verhandl. Berlin. Med. Ges. 17: 71-79, 1 figs.
1887 Ueber das erneute vorkommen von giftigen miesmuscheln in Wilhelmshaven. Arch. Pathol. Anat. Physiol. 110: 376-380, 2 pls.

WONG, J. L., OESTERLIN, R., and H. RAPOPORT
1971 The structure of saxitoxin. J. Am. Chem. Soc. 93(26): 7344-7345.

YARIV, J.
1958 Toxicity of *Prymnesium* cultures. Ph.D. Thesis. Hebrew Univ., Jerusalem.

YARIV, J., and S. HESTRIN
1959 Purification and properties of a cytotoxin formed by *Prymnesium parvum*. (Unpublished.)

(NSA)—Not seen by author.

Chapter III—INVERTEBRATES

Phylum Porifera

SPONGES

Sponges are multicellular animals of simple and loose organization. Generally, they have spicules of silica or calcium carbonate imbedded in their bodies for support and fibrous skeletons made of a horny substance called spongin; however, either or both of these may be lacking. Regarded as plants for many centuries, sponges received animal status from zoologists in 1835.

Because sponges lack a distinct enteron and the germ layers are not well established, the phylum Porifera is sometimes classed in a separate subkingdom, the Parazoa, or the Metazoa.

Prior to the advent of synthetics, the marine sponge industry was quite substantial. The annual world commercial sponge crop during its peak years is said to have been valued at more than $5 million (Potter, 1938). Production in the United States is at present less than 6 percent of what it was in 1936 (Walford, 1958). Sponge fishing is one of man's most ancient industries, the finest sponges having come from the Mediterranean area. Moore (1908) has written an excellent and well-illustrated article on the former sponge fishing industry.

There are approximately 4,000 species of sponges: about 1 percent (all members of a single family) inhabit fresh water; 10 percent are intertidal; and the remainder are marine or benthic.

The brevity of this chapter merely reflects the lack of our knowledge in this vast uncharted biochemical sea.

The systematics and identifying characteristics of the sponges pertinent to this chapter have been discussed by De Laubenfels (1932, 1936, 1950, 1953), Dickinson (1945), Pratt (1948), and Miner (1950).

REPRESENTATIVE LIST OF MARINE SPONGES REPORTED AS TOXIC

Phylum PORIFERA

Class DEMOSPONGIAE: Sponges

Family DESMACIDONIDAE

Fibulia nolitangere (Duchassaing and Michelotti) (Pl. 1, fig. a). *8*

DISTRIBUTION: West Indies.

SOURCES: De Laubenfels (1953), Fraser-Brunner (1973).
OTHER NAME: *Amphimedon nolitangere*.

9 *Microciona prolifera* (Ellis and Solander) (Pl. 2). Red moss, red or red beard sponge, oyster sponge (USA).

DISTRIBUTION: Cape Cod to South Carolina.

SOURCES: Corson and Pratt (1943), Nigrelli, Jakowska, and Calventi (1959), Jakowska and Nigrelli (1960), Fraser-Brunner (1973).

Family HALICLONIDAE

8 *Haliclona viridis* (Duchassaing and Michelotti) (Pl. 1, fig. b). Green sponge (USA).

DISTRIBUTION: West Indies.

SOURCE: Jakowska and Nigrelli (1960).

Family TEDANIIDAE

8 *Tedania ignis* (Duchassaing and Michelotti) (Pl. 1, fig. c). Fire sponge, scarlet sponge (USA).

SOURCES: Duchassaing and Michelotti (1864), Verrill (1907), De Laubenfels (1936, 1953), Fraser-Brunner (1973).

BIOLOGY

Sponges obtain food by propelling water through tiny pores in the body wall, thus capturing microorganisms and organic detritus that may be present. It is believed by some that they may utilize dissolved nutrients. Since there is no digestive tract, digestion is intracellular and takes place in the specialized collar cells, in the choanocytes, and in the amebocytes. Some of the products of digestion that are stored in amebocytes as glycogen, fats, glycoproteins, and lipoproteins are known as thesocytes. Excretory products are usually complex nitrogen bases such as agmatine, a guanidine derivative.

Because of their sessile habits and porous structure, sponges are a veritable hotel of living sea creatures, including other sponges, bryozoans, mollusks, coelenterates, annelids, crustaceans, echinoderms, fishes, blue-green algae, and cryptomonads. With the limited exception of certain nudibranchs, sponges are not used for food to any great extent by other marine animals. Despite their complex ecological relationships with other marine organisms, sponges appear to do an excellent job of resisting the action of the multitudinous organisms that enter the labyrinthine canals and cavities, and, in some instances, become part of their diet. Sponges constitute a rich source of biologically-active substances, including antibiotic and toxic materials (Nigrelli, Jakowska, and Calventi, 1959; Jakowska and Nigrelli, 1960).

MECHANISM OF INTOXICATION

Sponge poisoning is contracted from handling the sponge. There is no knowledge of human intoxications resulting from the ingestion of sponges.

MEDICAL ASPECTS

Clinical Characteristics

Very little information is available regarding the toxic effects of sponge extracts on man. Verrill (1907) and De Laubenfels (1936, 1953) pointed out that contact with the surface of the West Indian sponges, *Tedania ignis* and *Fibulia nolitangere*, results in immediate swelling and smarting of the skin. The reaction resembles that produced by poison oak *(Rhus)*. Similar lesions are said to be caused by *Hemectyon ferox* (Duchassaing and Michelotti, 1864). The dermatitis is believed to be the result of a chemical rather than a mechanical irritant, i.e., the spicules. Corson and Pratt (1943) reported a contact dermatitis known as "red moss" or sponge poisoning, which occurs frequently among oyster fishermen in Northeastern United States from handling *Microciona prolifera*, the red sponge. The initial symptoms consist of redness, a feeling of stiffness of the finger joints, and swelling of the affected part. Within a few hours, variable-sized blisters containing clear or purulent fluid may develop over the affected area. This dermatitis gradually spreads until it involves an extensive skin area and, if inadequately treated, continues for several months. The diagnosis of the disease can be confirmed by a patch test using a small piece of the sponge. Corson and Pratt's experiments suggest that red moss poisoning is probably due to a mechanical irritation produced by the spicules of the sponge, but this has not been definitely established. Nothing is known about the effects of the oral administration of toxic sponge extracts to man.

In the Tisza Valley, Szeged, Hungary, fishermen who came in contact with certain freshwater sponges were reported to have a similar type of dermatitis (Szentkirályi, 1937).

The ailment known as "sponge fishermen's disease," or "la maladie des pêcheurs d'eponges nus" as it is termed in some parts of the Mediterranean region is caused by contact with the stinging tentacles of the small coelenterate *Sagartia rosea* which is attached to the base of the sponge. The disease is not due to a chemical irritant produced by the sponge itself (Zervos, 1903, 1934, 1938; De Laubenfels, 1936; White, 1934). This disease is discussed elsewhere (*see* p. 110). Cleland and Southcott (1965) have discussed injuries caused by sponges in the Australian region at length, and Southcott (1976) has incriminated *Neofibularia mordens* and *Lissodendoryx sp.* as capable of inflicting a contact dermatitis.

Treatment

Soothing lotions and antiseptic dressings may be applied to the affected areas. De Laubenfels (1936) found dilute solutions of acetic acid to be effective. Antibiotic therapy may also be indicated.

TOXICOLOGY

Data on the toxic effects of sponge extracts are meager. Richet (1906*a*) was not only the first, but also one of the few men to test sponge extracts on laboratory animals. He prepared alcoholic extracts from the marine sponge *Suberites domunculus* and obtained a crude toxic product that he called suberitine. Richet found that dogs or rabbits developed vomiting, diarrhea, prostration, evidence of abdominal pain, intestinal hemorrhages, and respiratory distress if suberitine were injected intravenously. The lethal dose of the crude extract was said to be 10 mg/kg of animal weight in dogs. Autopsies of the animals revealed pericardial and intraperitoneal hemorrhages. If suberitine was administered orally, no untoward effects were observed. Moreover, if the sponge extract was heated for a period of time at temperatures higher than 80° C, the toxic principle was destroyed. In another series of experiments, using a wider spectrum of test animals—dogs, rabbits, guinea pigs, frogs, turtles, pigeons, etc. —Richet (1906*b*) found that the lethal dose of suberitine varied from 6 to 16 mg/kg of animal weight, depending upon the species and the individual.

As a result of the autopsy finding observed by Richet (1906*b*) in animals that had been killed by sponge extracts, Lassablière (1906) attempted to determine the effect of suberitine on globular resistance and hemolysis production. He concluded that dog red blood cells became susceptible to disruption with dilutions of suberitine; however, hemolysis did not occur.

De Laubenfels (1932) has made the field observation that if the sponge *Tedania toxicalis* is placed in a bucket of seawater with fish, mollusks, crabs, and worms, the other animals expire in less than an hour. When placed under similar conditions, other species of sponges were observed to be nontoxic. De Laubenfels believed that this species of sponge exuded a chemical substance that was toxic to most species of marine animals.

Zahl (1953) found aqueous extracts prepared from *T. ignis, Ircinia fasciculata, Spheciospongia vesparia, Dysidea etheria*, and *Callyspongia vaginalis*, to be toxic if injected intraperitoneally into white mice.

Halstead and Habekost (1954) prepared aqueous extracts from *Pseudosuberites pseudos*, a marine sponge that was obtained from Guaymas, Sonora, Mexico. Crude extracts were prepared by adding 1 ml of distilled water for each gram of sponge tissue, homogenizing the mixture in a Waring Blender, and centrifuging for 20 minutes at 2,000 rpm. One ml samples of the clear supernatant were injected into each of eight Swiss Webster mice weighing between 15 and 20 g. All the mice died within from 52 minutes to 21$^1/_2$ hours. The extracts showed some decrease in toxicity after storage for 3 to 6 months at a temperature of -5° C. Preliminary studies indicated that this particular poison is not soluble in methanol.

The antibiotic properties of aqueous extracts from the green sponge *Haliclona viridis* were investigated and found to be quite toxic to a variety of laboratory animals (Jakowska and Nigrelli, 1960). The extracts appeared microscopically as homogeneous cell populations. The active principle was heat-stable and could be extracted selectively from frozen and heat-dried sponges with a variety of organic solvents and chromatographic procedures. The authors did not indicate the exact nature of their methods. Aqueous extracts in dilutions from 10 to 100 ppm were found to be lethal to a variety of invertebrates, fish, amphibia, and mice. These same extracts were also observed to exhibit an antimicrobial effect on *Staphylococcus aureus* and *Escherichia coli*. There appeared to be a direct relationship between toxicity and antimicrobial activity.

No other reports have been found thus far on the toxic properties of marine sponges. However, Arndt (1928) has reported that aqueous extracts prepared from the freshwater sponges *Spongilla lacustris* (Linnaeus), *S. fragilis* Leidy, *Ephydatis fluviatilis* (Linnaeus), and *E. mulleri* (Lieberkuhn), are toxic when injected intraperitoneally, intracardially, or subcutaneously into white mice or guinea pigs.[1] Injections produced diarrhea, prostration, and respiratory distress. The extracts were observed to produce a weak hemolysis of sheep and guinea pig red blood cells. Desiccation for 2 years and temperatures up to 100° C did not destroy the toxic properties of the extract.

PHARMACOLOGY

Das, Lim, and Teh (1971) have reported the isolation of histamine and other histamine-like substances from the marine sponge *Suberites inconstans*, commonly found in the western tropical Pacific.

CHEMISTRY

(Unknown)

[1] According to De Laubenfels (1936), the sponges that Arndt called *Ephydatia* actually belong to the genus *Meyenia*.

LITERATURE CITED

ARNDT, W.

1928 Die spongien als kryptotoxische Tiere. Zool. Jahrb. 45: 343-360.

CLELAND, J. B., and R. V. SOUTHCOTT

1965 Injuries to man from marine invertebrates in the Australian region. National Health and Medical Research Council, Spec. Rept. Ser. (12): 1-282, 9 pls., 12 figs.

CORSON, E. F., and A. G. PRATT

1943 "Red moss" dermatitis. Arch. Dermatol. Syphil. (Chicago) 47(4): 574-579.

DAS, N. P., LIM, H. S., and Y. E. TEH

1971 Histamine and histamine-like substances in the marine sponge, *Suberites inconstans*. Comp. Gen. Pharmacol. 2(8): 473-475.

DE LAUBENFELS, M. W.

1932 The marine and fresh-water sponges of California. Proc. U.S. Nat. Mus., Publ. No. 2927, 81(4): 1-140, 79 figs.

1936 A discussion of the Dry Tortugas in particular and the West Indies in general, with material for a revision of the families and orders of the Porifera. (Papers from Tortugas Laboratory, vol. 30. Publ. No. 467.) 225 p., 22 pls.

1950 The Porifera of the Bermuda Archipelago. Trans. Zool. Soc. London 27(1): 1-154.

1953 A guide to the sponges of eastern North America. Marine Lab., Univ. Miami, Spec. Publ. 32 p., 8 figs.

DICKINSON, M. G.

1945 Sponges of the Gulf of California. Allen Hancock Pacific Expeditions. Reports, vol. 11, No. 1. Univ. So. Calif. Press, Los Angeles. 252 p., 97 pls.

DUCHASSAING, P., and G. MICHELOTTI

1864 Spongiaires de la mer caraïbe. Natuurk. Verhandel. Hollandsche. Maatsch. Wetenschap. Ser. 2, 21(2): 1-124, 25 pls.

FRASER-BRUNNER, A.

1973 Danger in the sea. Hamlyn Publ. Group Ltd., New York. 128 p., illus.

HALSTEAD, B. W., and R. C. HABEKOST

1954 Unpublished experiments on the toxicity of *Pseudosuberites pseudos* Dickinson.

JAKOWSKA, S., and R. F. NIGRELLI

1960 Antimicrobial substances from sponges. Ann. N.Y. Acad. Sci. 90(3): 913-916.

LASSABLIÈRE, P.

1906 Influence des injections intraveineuses de subéritine sur la résistance globulaire. Compt. Rend. Soc. Biol. 61: 600-601.

MINER, R. W.

1950 Field book of seashore life. G. P. Putnam's Sons, New York. 888 p. 275 pls.

MOORE, H. F.

1908 The commercial sponges and the sponge fisheries. U.S. Fish Wildlife Serv., Fish Bull. 28(1): 403-511, 56 pls.

NIGRELLI, R. F., JAKOWSKA, S., and I. CALVENTI

1959 Ectyonin, antimicrobial agent from the sponge, *Microciona prolifera* Verrill. Zoologica 44(4): 173-176. 1 pl.

POTTER, G. E.

1938 Textbook of Zoology. C. V. Mosby Co., St. Louis. 915 p., 440 figs., 15 pls.

PRATT, H. S.

1948 A manual of the common invertebrate animals (exclusive of insects). The Blakiston Co., Philadelphia. 854 p., 975 figs.

RICHET, C.

1906*a* De l'action toxique de la subéritine (Extrait aquex de *Suberites domuncula*.) Compt. Rend. Soc. Biol. 61: 598-600.

1906*b* De la variabilité de la dose toxique de subéritine. Compt. Rend. Soc. Biol. 61: 686-688.

SOUTHCOTT, R. V.
1976 Harmful sponges in South Australian seas, Hospitals Department of South Australia, Bull. (11): 15-18.

SZENTKIRÁLYI, S.
1937 Über eine durch süsswasserschwämme verursachte hauterkrankung der Tisze- (Theiss-) Fischer. Dermat. Wochschr. 104(20): 602-607, 9 figs.

VERRILL, A. E.
1907 The Bermuda Islands; V. Characteristic life of the Bermuda coral reefs. Trans. Conn. Acad. Arts. Sci. 12: 340.

WALFORD, L. A.
1958 Living resources of the sea. Ronald Press, New York. 321 p., 23 figs.

WHITE, R. P.
1934 The dermatergoses; or, occupational affections of the skin. 4th ed. H. K. Lewis Co., London. 716 p.

ZAHL, P. A.
1953 Toxic sponges of the Caribbean. (Personal communication, July 20, 1953.)

ZERVOS, S. G.
1903 La maladie des pêcheurs d'éponges. Semaine Méd. (Paris) 1903(25): 208-209.
1934 La maladie des pêcheurs d'éponges nus. Paris Méd. 93: 89-97, 16 figs.
1938 La maladie des pêcheurs d'éponges nus et l'anémone de la mer "actinion." Bull. Acad. Méd. (Paris) Ser. 3, 119: 379-395, 8 figs.

Chapter IV—INVERTEBRATES

Phylum Coelenterata[1]

HYDROIDS, JELLYFISHES, SEA ANEMONES, CORALS

The coelenterates or cnidarians are simple metazoans having primary radial, biradial, or radio-bilateral symmetry. They are composed essentially of two epithelial layers and one internal cavity—the gastrovascular cavity, or coelenteron, which opens only by the mouth. A dominant characteristic of the group is the presence of tentacles equipped with nematocysts. The group is further characterized by showing a remarkable degree of polymorphism, having an alternation of generations with a sexual and asexual phase, as well as a specialization of individual polyps as in the siphonophores. A single species may present a variety of forms of either the sessile polyp or the free-swimming medusoid type. For at least a part of their life span, most coelenterates are attached or sedentary.

The phylum Coelenterata is generally divided into three classes:

HYDROZOA: Hydroids and craspedote medusae. To this class belong the hydroids that are commonly found growing in plume-like tufts on rocks, seaweeds, and pilings. Small medusae are budded from these branching polyps. The order Siphonophora includes the venomous *Physalia*, which is a planktonic medusae. There are approximately 2,700 species of hydroids, most of which are marine.

SCYPHOZOA OR SCYPHOMEDUSAE: Jellyfish, acraspedote, or true medusae which do not possess a velum. This class includes the larger medusae having eight notches in the margin of the bell. A polyp stage is present but suppressed in this group. The deadly cubomedusae or sea wasps are examples of this class. There are about 200 species, entirely marine.

ANTHOZOA: Sea anemones, corals, and alcyonarians. The medusae stage is absent in this group, and many of the polyps are colonial. The corals are notable for their precipitation of calcareous skeletal structures and their role as reef builders. Important venomous examples are the anemones *Sagartia, Actinia*, and *Anemonia*. There are about 6,100 solitary or colonial species exclusively marine.

The majority of coelenterates are innocuous to man; however, the nematocyst apparatus, which is one of the morphological characteristics of all coelenterates, makes these animals of significance to venomologists.

A review of the extensive literature of coelenterate venomology impresses one with the broad phylogenetic spectrum of stinging members of this group noxious to man. Coelenterates become particularly impressive when one realizes that some of the cubomedusae can cause the death of an adult human within 4 to 6 minutes. Indeed, here are some extremely potent venoms that must

[1] This phylum is also designated as Cnidaria.

be reckoned with by both the clinician and the pharmacologist. Studies on the epidemiology of coelenterate stingings are meager and often inconclusive. Our knowledge of the exact nature and function of the nematocyst apparatus is limited largely to *Hydra, Physalia*, or one of several species of sea anemones. Many important groups of venomous coelenterate species have received little or no attention by modern investigators. Biochemical studies on coelenterate poisons are primarily the work of a very few scientists. Despite the enormous number of coelenterate stings which occur each year—most of them mild in nature—with few exceptions, we still do not have adequate methods of treating serious stings.

Attention is called to two great and indispensable works on the taxonomy of coelenterates, viz., *Medusae of the World* by Mayer (1910) and *Synopsis of the Medusae of the World* by Kramp (1961). The nomenclature as presented by Kramp has been adopted for the medusae listed in this chapter. Walsh and Bowers (1971) have prepared an excellent review of the Hawaiian zoanthids.

REPRESENTATIVE LIST OF MARINE COELENTERATES REPORTED AS VENOMOUS

Phylum COELENTERATA

Class HYDROZOA: Hydroids

Family MILLEPORIDAE

12 *Millepora alcicornis* Linnaeus (Pl. 1). Stinging coral, false coral, fire coral (USA, Australia), Karang gatal (Cocos-Keeling).

DISTRIBUTION: Caribbean Sea.

SOURCES: Southcott (1963), Fraser-Brunner (1973).

13 *Millepora dichotoma* Forskål (Pl. 2). Stinging coral, false coral, fire coral (USA, Australia), Karang gatal (Cocos-Keeling).

DISTRIBUTION: Red Sea, Gulf of Aden, and possibly eastward to the tropical Pacific.

SOURCE: Halstead (1957).

Family OLINDIADIDAE

14 *Gonionemus vertens* (Agassiz) (Pl. 3). Orange-striped jellyfish (USA).

DISTRIBUTION: *G. vertens* is found in north temperate Pacific and Atlantic Oceans, and the Mediterranean. However, reports of stings from *Gonionemus* are limited to the eastern coast of USSR.

SOURCES: Bari (1922), Aznaurian (1964), Brekhman (1969), Pigulevsky and Michaleff (1969), Fraser-Brunner (1973), Acad. Sci. USSR (1974).

14 *Olindioides formosa* (Goto (Pl. 4). Stinging medusa (USA).

DISTRIBUTION: Japan.

SOURCES: Aoki (1922), Faust (1924), Strong (1944).

Family PHYSALIIDAE

Physalia physalis (Linnaeus) (Pl. 6, fig. a; Pl. 9). Portuguese man-o-war, bluebottle (USA), caravelle, galère, physalie, vaisseau de guerre portugais, vaisseau portugais, vessie de mer (France), caravela (Portugal), schwimmpolypen (Germany). *16, 19*

DISTRIBUTION: Tropical Atlantic, occasionally as far north as the Bay of Fundy and the Hebrides, Mediterranean Sea.

SOURCES: Phillips and Brady (1953), Lane and Larsen (1965), Mebs (1973).
OTHER NAMES: *Physalia arethusa, Physalia caravella, Physalia pelagica* (part).

Physalia utriculus (La Martinière) (Pl. 5, fig. a; Pl. 7; Pl. 8). Portuguese man-o-war, bluebottle (USA, Australia), brûlant (French West Indies), benang-benang (Indonesia), katsueno-eboshi (Japan), morski mjehur (Yugoslavia). *15, 17, 18*

DISTRIBUTION: Indo-Pacific, as far north as southern Japan, Hawaii.

SOURCES: McNeill and Pope (1943*a, b*). Fish and Cobb (1954).
OTHER NAMES: *Physalia arethusa, Physalia megalista, Physalia pelagica.*

Family PLUMULARIIDAE

Aglaophenia cupressina Lamouroux (Pl. 10). Stinging hydroid (USA). *20*

DISTRIBUTION: Indian Ocean, Zanzibar to North Australia and the Philippines.

SOURCES: Saville-Kent (1900), Cleland (1912), Southcott (1963).
OTHER NAME: *Aglaophenia macgillivrayi.*

Lytocarpus philippinus (Kirchenpauer) (Pl. 11, figs. a, b). Feather hydroid (USA). *21*

DISTRIBUTION: Tropical and subtropical Pacific, Indian and Atlantic Oceans, Mediterranean Sea.

SOURCE: Southcott (1963).

Family RHIZOPHYSIDAE

Rhizophysa eysenhardti Gegenbaur (Pl. 12). *22*

DISTRIBUTION: Warmer waters of all oceans.

SOURCES: Southcott (1963), Cooke and Halstead (1970).

Class SCYPHOZOA: Jellyfishes

Family CARYBDEIDAE

Carybdea rastoni Haacke (Pl. 13). Sea-wasp (USA), mona, andonkurage lantern medusae (Japan), jimble (Australia). *22*

DISTRIBUTION: Tropical Pacific, Malayan Archipelago to South Australia, Marquesa and Hawaii, southern Japan.

SOURCES: Southcott (1956, 1958*a, b*, 1960, 1963).
OTHER NAME: *Charybdea rastoni.*

23 *Carukia barnesi* Southcott (Pl. 14).
DISTRIBUTION: Coast of Queensland, Australia.

SOURCES: Southcott (1967, 1970), Fraser-Brunner (1973).

Family CATOSTYLIDAE

24 *Catostylus mosaicus* (Quoy and Gaimard) (Pl. 15). Man-o-war, brown blubber, blubber, German blubber (Australia).
DISTRIBUTION: Brisbane to Melbourne, Australia, New Guinea, Philippines.

SOURCES: Pope (1947, 1953*a*, *b*).

Family CHIRODROPIDAE

25, 26, 27 *Chironex fleckeri* Southcott (Pls. 16,17, 18). Sea-wasp, box-jellies, indringa (Australia).
DISTRIBUTION: Northern Territory and North Queensland, Australia.

SOURCES: Southcott (1956, 1958*a*, *b*, 1959, 1960, 1963), Barnes (1965*a*, 1969), Baxter, Marr, and Lane (1968, 1973), Mebs (1973).

28 *Chiropsalmus quadrigatus* Haeckel (Pl. 19). Sea-wasp (USA).
DISTRIBUTION: Northern Australia, Philippines, Indian Ocean.

SOURCES: Southcott (1952, 1956, 1958*a*, *b*, 1959, 1963), Kingston and Southcott (1960), Fraser-Brunner (1973).

29 *Chiropsalmus quadrumanus* (Müller) (Pl. 20). Sea-wasp (USA).
DISTRIBUTION: Atlantic, North Carolina to Brazil, Indian Ocean, North Australia.

SOURCES: Southcott (1963), Phillips and Burke (1970).

Family CYANEIDAE

15, 29 *Cyanea capillata* (Linnaeus) (Pl. 5, fig. b; Pl. 21). Sea blubber, hairy stinger, sea nettle, hair jellyfish, lion's mane, molonga, snotty (USA, Australia), braendegoplen, den rød̸e, vandmand (Denmark).
DISTRIBUTION: North Atlantic and Pacific, southern coast of New England to the Arctic Ocean, France to Northern Russia, Baltic Sea; Alaska to Puget Sound, Japan, China, and Australia.

SOURCES: Phisalix (1922), Southcott (1960,1963), Mitchell (1962).
OTHER NAMES: *Cyanea annaskala, Cyanea artica*.

Family LYCHNORHIZIDAE

30 *Lychnorhiza lucerna* Haeckel (Pl. 22, fig. a). Agua viva (Brazil).
DISTRIBUTION: Coast of Brazil, French Guiana.

SOURCE: Machado (1943).
OTHER NAME: *Cramborhiza flagellata*.

Family NAUSITHOIDAE

Nausithoë punctata Kölliker (Pl. 22, fig. b, c; Pl. 23). Juvenile form is sometimes called "stinging alga." *30, 31*

DISTRIBUTION: All warm seas.

SOURCES: Fish and Cobb (1954), Southcott (1963).
OTHER NAMES: *Stephanoscyphus corniformis* Komai and *Stephanoscyphus racemosus* Komai (both juvenile forms of *N. punctata*).

Family PELAGIIDAE

Chrysaora quinquecirrha (Desor) (Pl. 5, fig. c). Sea nettle (USA), fosfore (Philippines). *15*

DISTRIBUTION: Azores and Woods Hole, New England, to the Tropics, West Africa, Indian Ocean, western Pacific from Malayan Archipelago to Japan, Philippines.

SOURCES: Light (1914*a, b*). Burnett *et al.* (1968), Phillips and Brady (1953), Rice and Powell (1970), Blanquet (1972).
OTHER NAME: *Dactylometra quinquecirrha.*

Pelagia noctiluca (Forskål (Pl. 24). Aguas vivas, aguas mas (Portugal), quallen (Germany), mauve stinger, mauve blubber (Australia). *32*

DISTRIBUTION: Warmer parts of Atlantic, Pacific, and Indian Oceans. West Indies, Florida to New England.

SOURCES: Southcott (1963), Fraser-Brunner (1973).
Other names: *Pelagia cyanella, Pelagia panopyra.*

Family RHIZOSTOMATIDAE

Rhizostoma pulmo (Macri) (Pl. 25). Cabbage blebs (USA). *32*

DISTRIBUTION: Europe, Norway to France, Mediterranean Sea, including Adriatic, Holland, Black Sea, Bosphorus.

SOURCES: Phisalix (1922), Southcott (1963).
OTHER NAMES: *Rhizostoma cuvieri, Phizostoma pulmo* var. *octopus.*

Class ANTHOZOA: Corals, Sea Anemones, Sea Pansies

Family ACROPORIDAE

Acropora palmata (Lamarck) (Pl. 26). Elk horn coral (USA). *33*

DISTRIBUTION: Florida Keys, Bahamas, West Indies.

SOURCE: Levin and Behrman (1941).
OTHER NAME: *Madapora palmata* (misspelled).

Family ACTINIIDAE

Actinia equina Linnaeus (Pl. 27, fig. a; Pl. 28, fig. a). *34*

DISTRIBUTION: Eastern Atlantic from Arctic Ocean to Gulf of Guinea, Mediterranean, Black Sea, Sea of Azov.

SOURCES: Pawlowsky and Stein (1929), Mebs (1973), Ferland and Lebez (1974), Sket *et al.* (1974).

34, 35, 36 *Anemonia sulcata* (Pennant) (Pl. 27, fig. b; Pl. 28, figs. c, d; Pl. 29). Sea anemone (USA).

DISTRIBUTION: Eastern Atlantic, from Norway and Scotland, to the Canaries, Mediterranean.

SOURCES: Giunio (1948), Béress and Béress (1971).
OTHER NAMES: *Actinia sulcata, Anthea cereus.*

41 *Physobrachia douglasi* Kent (Pl. 35, fig. c). Lumane (Samoa).

DISTRIBUTION: Polynesia westward through the Indian Ocean.

SOURCES: Zahl (1957), Cutress (1962).
OTHER NAME: *Gyrostoma kraemeri* (junior homonym).

Anthopleura elegantissima (Brandt).

DISTRIBUTION: California coast.

SOURCES: Lakso (1965), Lakso and Martin (1972).

Condylactis gigantea (Weinland).

DISTRIBUTION: Bermuda, West Indies.

SOURCE: Narahashi, Moore, and Shapiro (1969).

Family ACTINODENDRONIDAE

35, 37 *Actinodendron plumosum* Haddon (Pl. 28, fig. b; Pl. 30, figs. a, b). Hell's fire sea anemone, stinging anemone, actinarian (USA).

DISTRIBUTION: Tropical Pacific, Great Barrier Reef of Australia, Gilbert, Marshall, and Palau Islands.

SOURCES: Saville-Kent (1900), Cleland (1912, 1913), Hansen and Halstead (1971).
OTHER NAMES: *Actinodendron alcyonoideum, Actinodendron arboreum.*

Family ACTINODISCIDAE

38 *Rhodactis howesi* Kent (Pl. 31). Matalelei (Samoa).

DISTRIBUTION: Polynesia westward to the Indian Ocean.

SOURCE: Martin (1960, 1966*b*).

Family ALICIIDAE

38, 41 *Alicia costae* (Panceri) (Pl. 32; Pl. 35, fig. b).

DISTRIBUTION: Mediterranean.

SOURCE: Cutress (1962).

39 *Lebrunia danae* (Duchassaing and Michelotti) (Pl. 33, fig. a).

DISTRIBUTION: West Indies.

SOURCE: Cutress (1962).

39 *Triactis producta* Klunziger (Pl. 33, fig. b).

DISTRIBUTION: Red Sea.

Source: Levy, Masry, and Halstead (1970).

Family HORMATHIIDAE

Adamsia palliata (Bohadsch) (Pl. 27, fig. c; Pl. 34, fig. a). *34, 40*

Distribution: Norway to Spain, Mediterranean.

Source: Pawlowsky (1927).

Calliactis parasitica (Couch) (Pl. 34, fig. b). *40*

Distribution: English coast, Mediterranean.

Source: Nobre (1928). (Although this cannot be positively identified, Deichmann and Cutress believe that *Calliactis parasitica* is most likely to be the species to which Nobre refers.)
Other name: *Sagartia effecta.*

Family RENILLIDAE

Renilla muelleri Kölliker (Pl. 30, fig. c). Sea Pansy (USA). *37*

Distribution: Gulf of Mexico, southward to Brazil.

Source: Huang and Mir (1971).

Family SAGARTIIDAE

Corynactis australis Haddon and Duerden (Pl. 30, figs. c-d). *40*

Distribution: Australia.

Source: Pope (1963).

Sagartia elegans (Dalyell) (Pl. 27, fig. d). Rosy anemone (USA). *34*

Distribution: Iceland to Atlantic coast of France, Mediterranean Sea, and coast of Africa.

Sources: White (1934), Fraser-Brunner (1973).
Other name: *Sagartia rosea.*

Family STOICHACTIIDAE

Radianthus paumotensis (Dana) (Pl. 35, fig. a). Matamala samasama (Samoa). *41*

Distribution: Southern Polynesia westward to Micronesia.

Source: Cutress (1962).

Family ZOANTHIDAE

Palythoa toxica Walsh and Bowers. Limu-make-o-hana (Hawaii).

Distribution: Hawaiian Islands.

Source: Walsh and Bowers (1971).

Palythoa tuberculosa Esper (Pl. 32, fig. b). Toxic zoanthid, limu-make-o-hana (USA). *38*

Distribution: Tropical Pacific.

Sources: Hashimoto, Fusetani, and Kimura (1969), Moore and Scheuer (1971), Kimura, Hashimoto, and Yamazato (1972).
(*See* illustration of related species *P. australiensis*, Pl. 36.) *42*

BIOLOGY

Hydrozoa: Included within the Hydrozoa are the hydroids (e.g., *Aglaophenia*), stinging or hydroid corals (e.g., *Millepora*), and the free-floating siphonophores (e.g., *Physalia*). The hydroids are generally found attached to a substratum in shallow waters, from low tide down to depths of 1,000 m or more. The ecological conditions in which hydroids are found are extremely variable and they fluctuate according to the species. Some species have a wide distribution, whereas others are restricted to definite latitudes. Generally, hydroids are more abundant in temperate and cold zones. The form of the colony may be radically altered by such environmental factors as wave shock, currents, and temperatures. Colonies usually are small or moderate in size, but some species may attain a length of 2 m. Because of the sessile habits of hydroids, commensalism with other animals is of frequent occurrence. Since hydroids attach themselves to pilings, rafts, shells, rocks, or algae, and display their fine moss-like growth, they are sometimes mistaken for seaweed. Colonial hydroids are comprised of two kinds of polyps or zooids: the feeding hydranth that takes in food for the colony, and the reproductive polyp or gonangium. The nutritive polyps have a crown of tentacles and a central mouth that leads into the stomach cavity to which all the other polyps are connected by the coenosarc which encloses the common enteron. The nematocysts (nettle or stinging cells) are restricted to the tentacles. Food is procured by the use of the nematocysts.

Hydroids reproduce by budding; the free-swimming, solitary medusae that separate from the reproductive polyp produce ova that result in attached hydroids. Medusae are more difficult to obtain than are the plantlike hydroids, but they may be taken in fine-meshed plankton nets. The medusae may be likened to a tiny umbrella with a short handle, the manubrium, which contains the mouth. Tentacles provided with stinging cells hang from the vellum or margin of the umbrella. Medusae swim in a jerky fashion by spasmodic contractions of the umbrella. In the medusae, the sexes are separate.

The order Hydrocorallina or hydroid corals, of which *Millepora* is the best-known genus, is widely distributed throughout tropical seas in shallow water to a depth of 30 cm. Hydroid corals are important in the development of reefs, for they form upright, clavate, bladelike, or branching calcareous growths, or encrustations over other corals and objects. They vary in color from white to yellow-green; and, because of their variable appearance, are sometimes difficult to recognize. The order is characterized by a massive exoskeleton of lime carbonate, the surface of which is covered with numerous minute pores. There are two sizes of pores: the larger gastropores, which are 1 mm to 2 mm apart, and the smaller dactylopores, irregularly interspersed about the gastropores. The surface of the coral between the pores has a pitted appearance. The entire stony mass is traversed by a complex system of branched canals which communicate with the pores. From the gastropores protrude the feeding gastrozooids, equipped with a hypostome and capitate tentacles. Dactylozooids, extending from the dactylopores, are mouthless; however, they are provided with tentacles, which are believed to have a protective and tactile function. According to Hyman (1940), millepores have two or three types of powerful nematocysts located on the polyps, polyp bases, and in the general coenosarc. Apparently most of the *Millepora* have the ability to sting, but the venomousness of the sting varies from one species to the next (Boschma, 1948).

The order Siphonophora, of which *Physalia* is a pertinent example, are highly polymorphic free-swimming or floating colonies composed of several types of polypoid or medusoid individuals attached to a floating stem. Siphonophores are pelagic animals, inhabiting the surface of the sea. They depend largely upon currents, wind, and tides for their movement. They are widely distributed as a group, but most abundant in warm waters. The float or pneumatophore of *Physalia* is greatly enlarged, and it is represented by an inverted, modified, medusan bell, whereas the remainder of the coenosarc is correspondingly reduced. *Physalia* may attain a large size with a float 10 cm to 30 cm in length. From the underside of the float hang gastrozooids, dactylozooids, and the reproductive gonodendra with their gonophores or budding medusoids. The female gonophores are medusoid and may swim free, but the male reproductive zooid remains attached to the float. The gastrozooids or feeding polyps are without tentacles. Some of the tentacled dactylozooids are small; however, several of the large dactylozooids are equipped with very elongate "fishing" tentacles.

The number of fishing tentacles vary with the species of *Physalia*. In the Pacific form *P. utriculus* there is usually a single fishing tentacle; but in the Atlantic species *P. physalis* there are multiple fishing tentacles. Extending along the entire length of the large dactylozooid, a band of specialized tissue covers diverticulae of the gastrovascular cavity of the tentacle. These fishing tentacles or large dactylozooids may be found in the water to a depth of more than 30 m and, because of their transparent appearance, constitute a definite hazard to the unsuspecting swimmer. Upon contraction, the remainder of the tentacle shortens more completely than does the superficial band, and this causes the band to be thrown into loops and folds that are known as "stinging batteries." The nematocysts are contained in cnidoblasts located in the superficial epithelium of the battery. The toxin, a structureless fluid within the nematocyst capsule, bathes the surface of the nematocyst tubule (Lane, 1960). (*See* p. 100 for further details on the nematocyst apparatus.)

According to Parker (1932), scores of these fishing filaments may extend down into the water from a single *Physalia*. Parker has observed that the nematocyst heads occur at regular intervals along the side of the filament opposite the point from which the main muscle plate takes its origin. Each full-sized head contains about 500 large nematocysts and about 2,000 small ones. In one extended filament measuring 9 m in length, the nematocyst heads were distributed at intervals of approximately 3 cm apart. According to these figures, each fishing filament contained about 750,000 nematocysts. When one considers the large number of fishing filaments on each *Physalia*, he finds a formidable venom apparatus.

When the animal is moving through the water, the fishing tentacles undergo a continuous rhythmic movement, alternately contracting and relaxing. Thus there is a constant sampling of the water beneath the pneumatophore. If the tentacle brushes against a prey organism, the nematocysts are stimulated and they trigger the immediate release of the coiled nematocyst thread (Fig. 2).

The fully uncoiled thread may be several hundred times as long as the diameter of the parent capsule. The extreme length of the tubule, together with its chitinous barbs and spines, constitute a highly effective entanglement. If the tip of the cnidal thread penetrates the victim, the toxin is conveyed directly into the body of the prey through the hollow thread. Lane (1960) found that the thread can penetrate even a surgical glove.

Lane further observed that the magnitude of the response to contact with the victim is proportional to the area of contact between tentacle and prey. A small copepod may elicit the discharge of 20 to 50 adjacent nematocysts, whereas contact with a larger animal might evoke a discharge of several hundred thousand nematocysts. Gentle stimulation of the nematocyst results in a rapid release of the nematocyst thread, but does not dislodge the parent capsule from its position in the epithelium. Vigorous resistance by the prey results not only in greatly increasing the number of cnidae but also in dislodging many of them from the epithelium. Dislodged nematocysts are replaced by cnidoblasts that differentiate outside the stinging battery but subsequently come to occupy a definitive position in the battery epithelium.

It is interesting that the loggerhead turtle *Caretta caretta* has been reported to feed on *Physalia*. The potency of the toxin and the ability of the *Physalia* nematocysts to penetrate even a surgical glove make this a gastronomic feat of no small accomplishment (Lane, 1960).

Nudibranchs *Glaucus* and *Glaucilla* may feed on *Physalia*, consuming and storing the nematocysts, and later using them for their own defense (Thompson and Bennett, 1969).

Scyphozoa: All Scyphozoans or jellyfishes are marine and the majority are pelagic. A few species are known to inhabit depths of 2,000 fathoms or more. In the adult stage, most jellyfishes are free swimming. Because their swimming ability is relatively weak, jellyfishes are greatly influenced in their movements by currents, tides, and wind. Scyphozoans are widely distributed throughout all seas. Many medusae reveal that they are affected by light intensity in that they surface during the morning and late afternoons and descend during the midday and in the darkness, whereas others react in just the opposite manner. A descension is usually made during periods of stormy weather. Swimming is accomplished by rhythmic pulsations of the bell, and this action determines the vertical rather than the horizontal progress of the animal. Jellyfishes display a remarkable ability to withstand considerable temperature and salinity changes. They are carnivorous, some of the larger species being capable of capturing and devouring large crustaceans and fishes. Jellyfishes display a wide variety of sizes, shapes, and colors; many of them are semitransparent or glassy in appearance and often have brilliantly colored gonads, tentacles, or radial canals. In some species, they may vary in size from a few millimeters to more than 2 m across the bell, with tentacles more than 36 m in length, as in *Cyanea capillata*.

Cyanea has been aptly described as "a mop hiding under a dinner plate." It is a large, repulsive, slimy jellyfish. The upper surface of the disc is circular, almost flat, roughened, and varied in coloration, usually bluish or yellow-tinged. A multitude of threads several meters in length hang from the undersurface of the disc. *Cyanea* secretes a thick mucoid material which has a strong fishy odor.

Regardless of their size, jellyfishes are very fragile; many of them contain less than 5 percent of solid organic matter. Scyphomedusae have an eight-notch marginal bell, but lack a velum; the gonads are connected with the endoderm. Reproduction is by an alternation of generations, as in the hydroids, although the polyp stage is reduced. Jellyfishes have a complex system of branched radial canals, and numerous oral and marginal tentacles.

The cubomedusae are among the most venomous marine creatures known. The genera *Chironex, Chiropsalmus, Carybdea*, and *Chirodropus* contain some of the more dangerous species of the group. They range in size from a small grape to that of a large pear. Cubomedusae are widely distributed throughout all the warmer seas. They generally seem to prefer the quiet shallow waters of protected bays and estuaries, and sandy bottoms, although some species have been found in the open ocean. During the summer months, the immature forms which stay on the bottom reach maturity. The adults may then be found swimming at the surface. Light-sensitive cubomedusae, however, descend to deeper water during the bright sun of the middle of the day and come to the surface during early morning, late afternoon, and evening. Millions of cubomedusae have been observed hovering about 30 cm above the sandy bottom of some of the bays between Adelaide and Cape Jervis, Australia (Southcott, 1958*a*). Most of the specimens within the half-meter thick layer are found lying at an oblique angle, which is the customary position for the cubomedusae during such periods of inactivity. During the day the jellyfish ascend and gradually disperse in a zone ranging from the surface to a depth of 6 m of more.

Cubomedusae are strong and graceful swimmers, capable of moving along at a steady 2 knots, and are believed to feed mainly on fishes. They are very sensitive to water turbulence. During a flat calm period, they may appear in large numbers close to the beach in very shallow water (Barnes, 1960). They are most abundant during the warm summer months, but have been recorded at other times of the year.

Little is known regarding the developmental stages and life histories of the members of this group. Cubomedusae have four periradial stomach-pouches. They bear a superficial resemblance to the Hydromedusae in the shape of the boxlike bell and in the presence of an annular diaphragm that constricts the aperture of the bell-cavity (Mayer, 1910). One interesting feature of the cubomedusae is the presence of a highly developed eye, which has been well described in *Chironex* and in other species. On either side of the bell, a sense organ or rhopalium is set in a small niche containing both a position-sense organ and an optic apparatus of six eye-spots, the largest (the distal median eye-spot) having a biconvex lens. The other eye-spots have no lens (Southcott, 1956). Based on the research of Conant (1898) and Berger (1900), two useful publications have appeared that deal with the anatomy and physiology of some of the less dangerous members of the cubomedusae.

The tentacles of the cubomedusae are interradial and arise just above the bell-margin. Their four proximal parts or pedalia, which are tough and gelatinous, have a characteristic wing or spatulalike shape. A strong, outstanding pedalium containing a single or a cluster of four or more tentacles arises from each corner of the boxlike medusa. In the case of *Chironex* and *Chiropsalmus*, the pedalia are clawlike and may have as many as 12 tentacles attached to a single pedalium. The hollow tentacles, generally thick and strong, taper distally to a blunt point. The outer surface is covered with rings of nematocysts. The tentacles are highly contractile and during life may extend down into the water for a distance of more than a meter. Some of the differences in morphology between *Chironex* and *Chiropsalmus* have been well described by Southcott (*see* Fig. 3).

A noteworthy study on the symbiotic behavior between small fishes and jellyfishes has been published by Mansueti (1963).

Anthozoa: The class Anthozoa is comprised of two subclasses (Hyman, 1940): the Alcyonaria, which includes the soft corals, sea fans, sea pens, and sea pansies; and the Zoantharia, which includes the sea anemones and the true corals. There are no known published data on the biotoxicity of Alcyonarians. Some of the sea anemones and true corals are of concern to venomologists. The sea anemones are members of the order Actiniaria, whereas the true or stony corals belong to the Madreporaria (or Scleractinia; Vaughan, and Wells, 1943).

Actiniarians or sea anemones are one of the most abundant of seashore animals. There are approximately 1,000 species. Their bathymetric range extends from the tidal zone to depths of more than 2,900 fathoms. While they abound in warm tropical waters, yet some species inhabit arctic seas. Anemones vary in size from a few millimeters to a half meter or more in diameter. Most species are sessile and attached to objects of various kinds, but are nevertheless able to creep about to some extent. When they are covered by water and undisturbed, the body and tentacles are expanded; and because of their variety of colors they frequently have a flowerlike appearance. If the animal is irritated or the water recedes, the tentacles may be invaginated rapidly and the body contracted. The food of anemones consists of mollusks, crustaceans, other invertebrates, and fishes. Most anemones have a short cylindrical body and a flat oral disc margined with a variable number of tentacles around a slitlike mouth. The base or pedal disc of the anemone serves for attachment to objects. The mouth is connected to the enteron by a tubelike gullet. The body is internally divided into radial compartments by septa that are in multiples of six. The free inner margins of the septa within the enteron may bear convoluted septal filaments which are continued as threadlike acontia, bearing nematocysts. Occasionally, the acontia may be extended through pores in the body wall or through the mouth to aid in subduing prey. Nematocysts are also situated on the marginal tentacles. Reproduction may be either sexual or asexual by fission. There is no alteration of generations, and no medusal stage.

Madreporarians or true corals, requiring water temperatures of 20° C or higher, are confined largely to the torrid zone, although a few species do inhabit more temperate waters. The vertical distribution of corals ranges from the low tide zone to depths of about 20 fathoms. A few of the solitary corals are known to occur at 6,000 fathoms or more. Corals are a major constituent of living reefs. With calcareous algae, they dominate the reef in numbers and volume and provide the ecological niches essential to the existence of all other reef-dwelling animals and plants. Assuming that the oxygen supply and temperatures are favorable, the number of species of hermatypic reef corals is controlled largely by light intensity and radiant energy. This situation appears to be the result of the restrictive effect of the symbiotic zooxanthellae in the tissues of hermatypic corals. Temperature variation is a significant factor in determining the distribution of corals over reef flats, since corals do not flourish where the water circulation does not maintain equable temperatures and provide fresh supplies of nutrients. Since CaO_3 is generally present in supersaturated quantities in seawater of normal salinity, it is not a significant factor in the distribution of corals. The average coral colony contains three times as much plant as animal tissue, and most of the plant tissue is comprised of filamentous green algae in the skeleton (Odum and Odum, 1955).

The stony hexacorals, which include *Acropora, Astreopora, Goniopora*, and others, as a result of their wide range of ecological adaptability, their growth habits, and their near immunity to predators, tend to dominate reef communities. With the exception of certain fishes, living coral polyps are not a direct food supply for most marine organisms. This might suggest the presence of certain noxious chemical constituents within the tissues of the coral polyps. Corals are constantly hampered in their growth by such destructive forces as perforating, boring, and dissolving algae, sponges, mollusks, worms, and echinoids, and by fishes which nibble on them. Wave shock and the abrasive action of dislodged coral debris and sand exert further destructive effects.

Corals, small anemonelike polyps which have a reduced musculature, short tentacles, and no pedal disc, live in a stony cup having basal radial ridges (Fig. 1). The individual polyps connect with each other through the coenosarc.

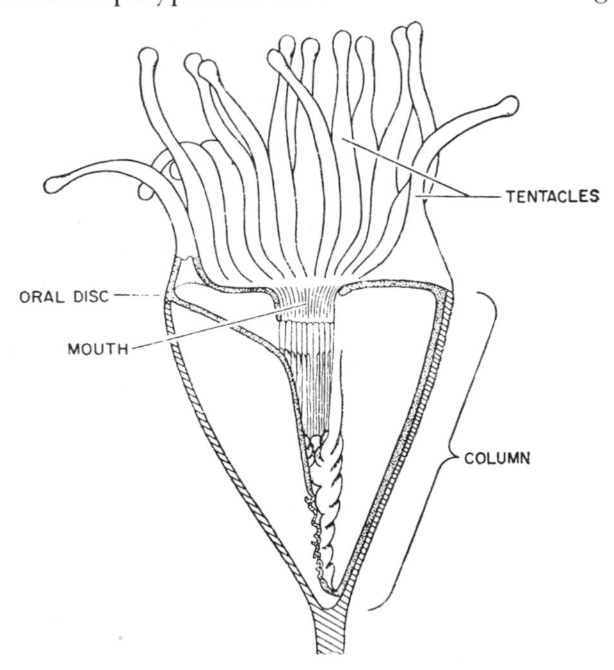

FIGURE 1.—Structure of a coral polyp. Diagrammatic. Sagittal section.
(R. Kreuzinger, modified from Vaughan and Wells)

Corals are carnivores with highly specialized feeding mechanisms. With the use of tentacles and cilia, they are able to capture and consume living zooplankton. The primary function of the cilia is to remove debris and wastes from the polyps' surface. Because of their sedentary habits and limited tentacular range, corals are largely dependent upon good water circulation to bring food within range of contact. The nematocysts, frequently of a complex type, are present in the tentacles, body wall, gullet, and septal filaments. Madrepores may be white, pink, yellow, green, brown, or purple in color. There are about 2,500 species, most of which are colonial. Wells (1957*a*) has published an excellent account on the ecology of corals and also has prepared a useful bibliography on the subject (Wells, 1957*b*).

MORPHOLOGY OF THE VENOM APPARATUS

The venom apparatus of coelenterates consists of the nematocysts located largely on their tentacles.

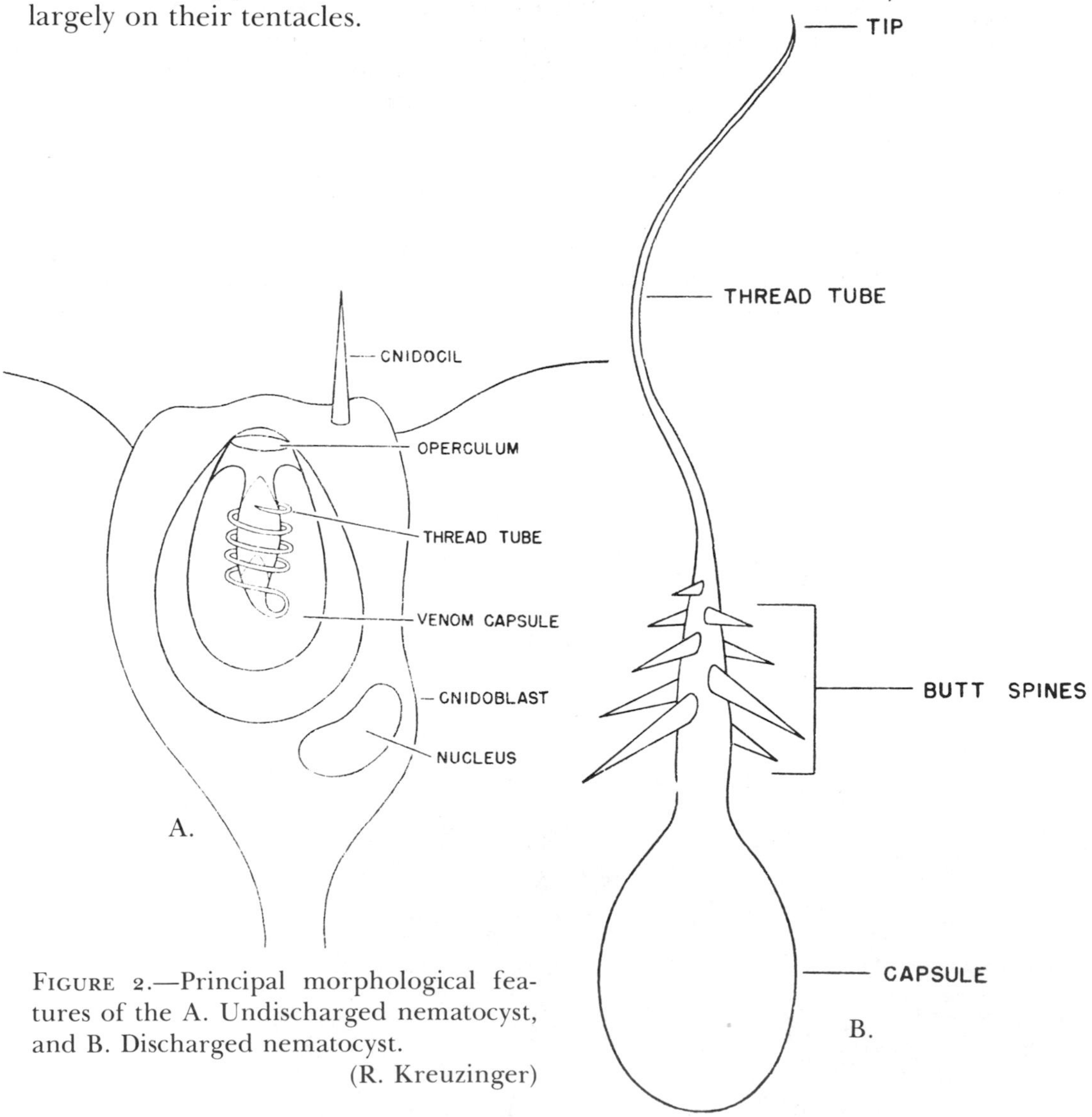

FIGURE 2.—Principal morphological features of the A. Undischarged nematocyst, and B. Discharged nematocyst.
(R. Kreuzinger)

Since a complete history of research on the nematocyst apparatus has been prepared by Weill (1934), attention need be directed to only two historically important papers on this subject. Trembley (1744) is said to have been the first to describe the nematocyst, and Moebius (1866) is said to have been the first to study its structure in detail. More recent investigations on the structure of coelenterate nematocysts have been conducted by Carlgren (1940, 1945), Pantin (1942), Yanagita (1959*a, b, c,* 1960*a, b, c*), Picken (1953), Yanagita and Wada (1959), and others. Some of the papers on this subject have not been included because they go far beyond the scope of this work. For the purposes of this discussion, the terminology as given by Weill (1934) and Hyman (1940) has been adopted.

The term nematocyst, or "stinging cell," is actually a misnomer since the structure is not a cell, but rather a cell organoid that is thought to be composed of keratin or a keratinlike substance (Brown, 1950; Picken, 1953).

Acording to Weill (1934) and Hyman (1940), 17 types of nematocysts are found in coelenterates. The following is Weill's classification of the various nematocyst types:

Weill's Classification of Nematocysts[2] (Pl. 37): *42, 43*

I. Tube closed at the end.
  A. *Desmonemes*, tube threadlike, forming a corkscrewlike coil, also called volvents.
  B. *Anacrophoric rhopalonemes*, tube club-shaped, without an apical appendix.
  C. *Acrophoric rhopalonemes*, tube club-shaped, with an apical appendix.

II. Tube open at the tip, without a butt (*haplonemes*).
  A. Tube of the same diameter throughout, *isorhizas*, also called glutinants.
    1. *Holotrichous isorhizas*, tube spiny throughout.
    2. *Atrichous isorhizas*, tube devoid of spines.
    3. *Basitrichous isorhizas*, tube spiny at base only.
  B. Tube slightly dilated toward the base, *anisorhizas*.
    1. *Homotrichous anisorhizas*, spiny throughout, spines equal.
    2. *Heterotrichous anisorhizas*, spiny throughout, spines larger at base.

III. Tube open at tip with a definite butt (*heteronemes*).
  A. Butt cylindrical, of the same diameter throughout (*rhabdoids*).
    1. *Microbasic mastigophores*, tube continued beyond the butt; butt not more than three times the capsule length.
    2. *Macrobasic mastigophores*, tube continued beyond the butt; butt four or more times the capsule length.
    3. *Microbasic amastigophores*, no tube beyond the butt; butt not more than three times the capsule length.

---

[2] This classification has been modified somewhat by Carlgren (1940) and Cutress (1955). Carlgren subdivided Weill's microbasic mastigophores into macrobasic *b* mastigophores and microbasic *p* mastigophores. Cutress accepts the modifications of Carlgren except for the elimination of two categories (microbasic and macrobasic amastigophores) and the addition of two new ones (microbasic *q* mastigophores and macrobasic *p* mastigophores).

4. *Macrobasic amastigophores*, no tube beyond the butt; butt four or more times the capsule length.

B. Butt dilated at the summit (*euryteles*).

1. *Homotrichous microbasic euryteles*, butt short, spines on butt of equal size.
2. *Heterotrichous microbasic euryteles*, butt short, spines on butt of unequal size.
3. *Telotrichous macrobasic euryteles*, butt long, with distal spines only.
4. *Merotrichous macrobasic euryteles*, butt long, spines elsewhere than at the ends.

C. Butt dilated at its base.

1. *Stenoteles*, also called penetrants.

The type of nematocyst(s) present varies from one coelenterate group to another. Weill has found that the location and nematocyst type follow a pattern of useful taxonomic characteristics in determining phylogenetic relationships. Since our knowledge of nematocyst types and their correlation with the venomous abilities of the organism is meager, it is impossible to draw any conclusion concerning them. The general distribution of nematocyst types among the coelenterate classes are as follows (Hyman, 1940):

Class	Nematocyst types
HYDROZOA	Rhopalonemes Desmonemes Isorhizas Anisorhizas Mastigophores Euryteles Stenoteles
SCYPHOZOA	Isorhizas Euryteles
ANTHOZOA	Isorhizas Mastigophores Amastigophores

Nematocysts are usually most abundant on tentacles, grouped on protuberances, circular or spiral ridges. They are also found in the epidermis of the oral region, and internal tentaclelike structures, i.e., the gastric filaments, the septal filaments, and the acontia. Nematocysts initially develop inside interstitial cells, termed cnidoblasts or nematocytes. Usually the developmental site of the nematocyst is some distance from the region in which it is finally utilized. The basal enlargements of the tentacles of medusae are regular developmental sites. Cnidoblasts containing developing nematocysts are transported by amoeboid movement through the body wall or by way of the gastrovascular cavity to their final destination in the ectodermal epithelium. The process of cnidogenesis has been discussed at length by Weill (1934) and Yanagita and Wada (1959).

The final placement of the adult nematocyst probably varies from one group of coelenterates to the next. It has been found (Yanagita, 1959*b*) that in some of the actinians the mature nematocysts are not embedded in cnidoblasts, but are situated freely each in a vacuolelike space within the supporting epithelial cells. The distal ends of the nematocysts are separated from the exterior by only a very thin layer of cortical cytoplasm of the surrounding epithelium.

The developing cnidoblast contains the capsule and within it the thread which first appears as a tube growing in from one pole of the capsule. The thread continues to lengthen and coil until it is developed fully. If the thread is the barbed type, the spines will begin to develop in the tube until it is filled. Last of all, the operculum, or plug or stopper, is formed at the pole where the tube joins the capsule. Now the nematocyst is developed fully and ready for discharge.

The structure of the nematocyst is remarkably complex considering its minute size, which ranges from 5μ to 1.12 mm in length. The largest nematocyst known is found in the siphonophore *Halistemma* which discharges a tube several millimeters in length. The essential features of the nematocyst are as follows (Pl. 38, fig. b): The capsulelike nematocyst is contained within the outer cnidoblast, or epithelium (as in the case of some actinians), which is fixed in the epidermis by a slender stalk connecting with the mesoglea. At one point on the outer surface of the cnidoblast projects the triggerlike cnidocil. The structure, placement, and occurrence of the cnidocil vary from one coelenterate group to the next, e.g., the cnidocil is apparently absent in the anthozoans (Weill, 1934; Pantin, 1942). Near the base of the cnidocil, in the periphery of the cnidoblast, is usually found a circlet of stiff rods (probably a circlet of a supporting nature), and frequently a basketwork of sinuous fibrile, which extends down into the stalk of the cnidoblast. Within the fluid-filled capsule is the hollow, coiled, thread tube containing the folded spines. Prior to discharge, the opening through which the thread tube is everted is closed by a lidlike device called the operculum, which in some of the actinians appears merely as a plug or stopper. The fluid in the capsule is the venom. *44*

Burnett and Sutton (1969), Sutton and Burnett (1969), and Burnett (1971*b*) have described the ultrastructure of the nematocysts of the sea nettle *(Chrysaora quinquecirrha)*.

The precise nature of the cnidocil, its mode of operation, and its excitation mechanism for discharging nematocysts have not been fully determined. Most observations indicate that the cnidocil, when present, serves as a triggering mechanism to discharge the contents of the nematocyst. Schulze (1922), Zick (1929, 1932), and others, have shown that the cnidocil is sensitive to physical contact. Parker and Van Alstyne (1932) have suggested that the cnidocil is primarily a chemoreceptor. Pantin's (1942) studies in *Anemonia sulcata* and his extensive review of the nematocyst apparatus in other coelenterates show that the cnidocil is sensitive to both mechanical and chemical stimuli, and that the discharge of the nematocyst is selective. In nature, excitation of the nematocyst appears to be caused largely by food. (*See also* Pls. 39-42.) *45-48*

The nematocyst is not wholly dependent upon the cnidocil as the mechanism of discharge, because in the anthozoans and possibly other coelenterates in which the cnidocil is absent, the nematocyst containing sensory, excitor, and effector elements seems to serve as an independent effector. The precise mechanisms involved are not understood. Two distinct sense organs, sensitive to both mechanical and chemical stimuli, perform the duties of a neuromuscular system. Thus the nematocyst serves as a double sense organ as well as an effector. No obvious analogies to this mechanism are found in the tissues of higher animals, and as Yanagita (1959*b*) has noted, this represents the simplest and perhaps the most primitive form of sensory-motor system. Studies in coelenterates fail to reveal any histological evidence of nerves running to the nematocyst apparatus (Parker and Van Alstyne, 1932; Weill, 1934; Pantin, 1942).

One of the most plausible explanations of the nematocyst discharge mechanism is presented in the extensive research of Yanagita (1959*a, b, c,* 1960*a, b, c*) and Yanagita and Wada (1959), in which they worked with the acontial nematocysts of the sea anemone *Diadumene luciae*. Nematocysts in the resting stage are embedded within a vacuole (probably the remnant of the cnidoblast) in the epithelium of the nematocyst band of the acontium with their distal ends separated from the exterior by a thin continuous hyaline protoplasmic sheet which is ciliated on the outer surface. During the excitation process, the surface of this protoplasmic sheet changes its mechanical or permeability properties under the action of ions, electric current, surfactants, etc. A weakening of consistency or an increase in permeability of that protoplasmic layer, particularly of its surface membrane, is all that is necessary for a nematocyst response to take place. However, the nematocyst response was found to vary according to the availability of the Cl^- ion. If Cl^- ions were absent, or present only in minimal quantities, the response was limited merely to extrusion of the intact nematocyst from the acontium, and the nematocyst failed to evert its stinging thread. On the other hand, if Cl^- ions were present in abundance, as is usually the case in seawater, the Cl^- ions came in contact with the distal tips of the nematocysts and they released the plugs, or stoppers. The explosive eversion of the stinging threads ensued as the result of built-in elastic energy, or pre-existent intracapsular pressure (Yanagita, 1959*b*).

Various theories have been propounded concerning the energy mechanism involved in the explosive eversion of the nematocyst thread. Most of these theories fall into three categories: (a) Elastic contraction of the capsular wall, (b) Swelling of the capsular contents, or (c) Active dilatation of the thread as the result either of an inherent elasticity or a swelling by rapid hydration. These theories have been reviewed in detail by Weill (1934), Kepner *et al.* (1951), Picken (1953,1957), and Yanagita (1959*a*).

After, or at the time of discharge, the exhausted nematocyst is extruded and a new nematocyst eventually takes its place.

Lane (1961*a, b*) believes that in the case of *Physalia*, the toxin is synthesized by gastrodermal cells, and passed through the mesoglea and into the nematocyst during the morphogenesis of the structure. The nature of the fluid contents within the nematocyst capsule is discussed in this chapter under the section on Toxicology (*see* p. 116).

The function of the nematocyst complex of *Gonionemus vertens* has been studied in elaborate detail by Westfall (1970).

MECHANISM OF INTOXICATION

Coelenterates inflict their injurious effects upon man by the use of their nematocyst apparatus. The venom is conveyed from the capsule of the nematocyst through the tubule into the tissues of the victim. Theoretically, any coelenterate equipped with a nematocyst apparatus is a potential stinger. The injurious effects may range from a mild dermatitis to almost instant death. The severity of the stinging is modified by the species of coelenterate, the type of nematocyst that it possesses, the penetrating power of the nematocyst, the area of exposed skin of the victim, and the sensitivity of the person to the venom.

Some of the sea anemones have been found to be poisonous to eat—particularly when raw. It is not known whether oral actinian intoxications are caused by their nematocyst poisons, or are the result of other noxious chemical substances contained in the tissues of their tentacles.

MEDICAL ASPECTS

Clinical Characteristics

The symptomatology of coelenterate stings varies according to the species, the site of the sting, and the person.

Hydrozoan Stings: This group includes stings resulting from contact with the members of coelenterate genera such as *Aglaophenia, Halecium, Liriope, Lytocarpus, Millepora, Olindias, Olindioides, Pennaria, Physalia,* and *Rhizophysa*. The stings produced by these organisms vary from a very mild stinging sensation, such as may be caused by some of the hydroids, to an extremely painful almost shocklike sensation, produced by *Olindias* and *Physalia*. The following effects have been reported as the result of contact with various types of hydrozoans, namely, a mild pricking sensation, nettlelike stinging sensation, redness of the skin, urticarial rash (may develop within minutes to hours after contact with the organism), haemorrhagic, zosteriform, or a generalized morbilliform rash, vesicle and pustule formation, and desquamation of the skin (Pl. 43, fig. a; Pls. 44-45). Abdominal pain, chills, fever, malaise, apprehension, and diarrhea have also been reported in hydroid stings.

49-51

Stings from *Olindioides* and *Physalia* can have severe effects and they may be accompanied by more generalized symptoms than are usually caused by some of the other hydrozoans. The effects most commonly reported are intense local pain extending along a pattern corresponding to the lymphatic drainage (i.e., axillary lymph nodes), and joint and muscle pain. The pain may be moderate to severe, sharp, shooting, or throbbing depending upon the individual. Some victims have compared it with an "electric shock-like sensation." Barnes (1960) who has experimentally inflicted himself and others with *P.*

utriculus stings states that the usual lesion is a discontinuous line of small papules, each surrounded by a small zone of erythema. There may be well-defined linear welts or scattered areas of punctate whealing and redness. Whealing is prompt, rarely massive, and it usually disappears within 2 hours. Erythema generally remains for about 24 hours. The intensity of the sting appears to vary depending upon the length of time that the tentacle remains in contact with the skin. This is true of most coelenterate stings. Lesions caused by stings from the multiple-tentacled Atlantic species *P. physalis* would probably be more extensive than those produced by the single-tentacled *P. utriculus*.

The above-mentioned signs and symptoms may be accompanied by headache, malaise, primary shock, collapse, faintness, pallor, weakness, cyanosis, nervousness, hysteria, chills and fever, muscular cramps, abdominal rigidity, nausea, vomiting, etc. Death has been reported from *Physalia* stings, but Cleland and Southcott (Southcott, 1963) have been unable to substantiate any deaths resulting from contacts with *Physalia*.

Hercus (1944) has reported a case of a corneal injury resulting from a *Physalia* sting. Ricord-Madiana (cited by Guérin, 1900) records the case of a Negro who poisoned an enemy by feeding him a soup prepared from the dry tentacles of *Physalia*. Phisalix (1922) reported fatalities in laboratory animals that had been fed extracts from *Physalia*. A dermatitis due possibly to *Hydra* has also been reported (Clemens, 1922).

Scyphozoan Stings: This group includes stings resulting from contacts with genera such as *Aurelia, Carybdea, Cassiopea, Catostylus, Chironex, Chirodropus, Chiropsalmus, Chrysaora, Cyanea, Linuche, Lobonema, Lychnorhiza, Nausithoë, Pelagia, Rhizostoma, Sanderia,* and *Tamoya*. As in the case of hydrozoan stings, the severity varies considerably according to the scyphozoan and the individual. Decisive factors in the severity of the case are the number of nematocysts encountered and the length of time that the victim remains in contact with the tentacles of the jellyfishes.

Stings from *Aurelia, Carybdea, Catostylus, Chirodropus, Linuche, Lobonema, Lychnorhiza, Nausithoë, Pelagia,* and *Tamoya* are usually relatively mild. Contact with the tentacles of these jellyfishes results in symptoms ranging from an immediate mild prickly or stinging sensation like that of a nettle sting, to an intense burning, throbbing, or shooting pain which may render the victim unconscious. In some cases, the pain is restricted to an area within the immediate vicinity of the contact, or again, it may radiate to the groin, abdomen, or armpit depending upon the initial site of the lesion. It may become somewhat generalized. Sometimes the localized pains may be followed by a feeling of numbness, or hyperesthesia. The area of skin coming in contact with the tentacles usually becomes erythematous, followed by a pronounced urticaria, blistering, swelling, and petechial hemorrhages. The stings may also cause redness and flushing of the face, increased perspiration, lacrymation, coughing, sneezing, and rhinitis. An asthmaticlike condition has been reported (Anon., 1924) as a result of inhaling dust containing dried jellyfish tentacles. Kitagawa (1943) reported a case of dermatitis resulting from the ingestion of a jellyfish.

Cassiopea, Catostylus, Chrysaora, Cyanea, Rhizostoma, and *Sanderia* are generally considered to be in the moderate-to-severe class. The symptoms are similar to those produced by the group mentioned above; however, they are accompanied frequently by generalized symptoms which can be more intense in character. Muscular cramps, severe whealing, mucoid expectoration, intense coughing, extreme mental depression, respiratory embarrassment, and a sensation of chest constriction are not uncommon. Loss of consciousness may occur, and if the victim is in the water, drowning may result (Pl. 46). *52*

Stings from *Chironex* and *Chiropsalmus* can be dangerous. Usually the effects consist of extremely painful localized areas of whealing, edema, and vesiculation, which later result in necrosis involving the full thickness of the skin. The initial lesions, caused by the structural pattern of the tentacles, are multiple linear wheals with transverse barring. The purple or brown tentacle marks form a whiplike skin lesion (Pls. 47-49). Painful muscular spasms, respiratory distress, a rapid weak pulse, prostration, pulmonary edema, vasomotor and respiratory failure, or death may result. The pain is said to be excruciating, with the victim frequently screaming and becoming irrational. Death may take place within from 30 seconds to 2 or 3 hours, but the usual time is less than 15 minutes. The cough and mucoid expectoration that are present in some of the other forms of jellyfish attacks are generally absent in *Chironex* and *Chiropsalmus* stings (Southcott, 1963). *53-55*

The etiological differential diagnosis of jellyfish stings is somewhat difficult to make as there is considerable overlapping in the spectrum of symptoms caused by the mild stingers in comparison with those caused by the extremely dangerous species. Barnes (1960) has prepared a table giving some of the more important clinical diagnostic characteristics of three important species (*see* Table 1).

Irukandji Sting Syndrome and Unidentified Jellyfish Stingings: Numerous jellyfish stingings in which there was lacking a positive identification of the causative agent have been reported from the vicinity of Cairns, Australia; Gulf of Carpentaria; Darwin; New Guinea; Fiji; and elsewhere (McNeill and Pope, 1943*a, b*, 1945; McNeill, 1945; Flecker, 1945*a, b*, 1957*a, b*; Southcott, 1952, 1959; Barnes, 1960). Lacking identification of the organisms, Southcott (1952), who personally observed more than a hundred of these cases, classified them according to their clinical characteristics. He found that they could be divided into two types, viz., *Type A Stings*—those without whealing, but with severe general effects, and *Type B Stings*—those with whealing, but without general effects. Type A stings are generally referred to in Australia as "Irukandji stings"—a term originally proposed by Flecker (1952*b*). The term Irukandji is derived from the tribal name of the aborigines who formerly inhabited the section of the Australian coast in which this type of sting is most commonly encountered. It is now used to designate a clearly defined clinical syndrome caused by jellyfish attacks. Irukandji does not refer to the causative agent.

Carukia barnesi was recently identified as the medusa responsible for Irukandji sting, *Type A* (Barnes, 1964; Southcott, 1967). The formal name

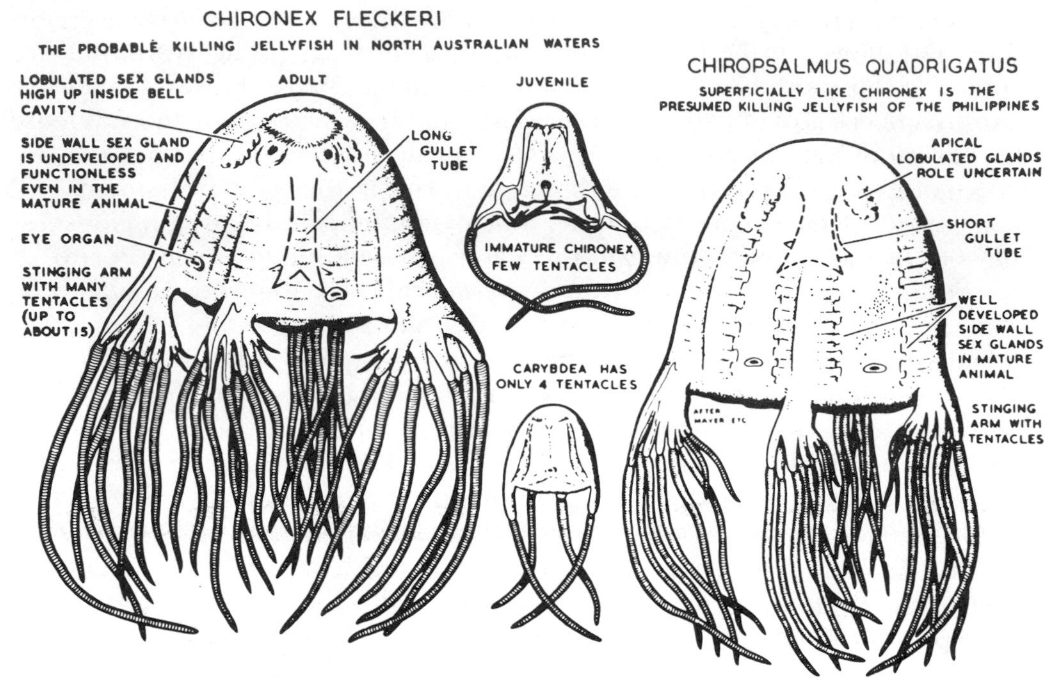

FIGURE 3.—Some of the significant morphological differences between the dangerous cubomedusae *Chironex fleckeri* and *Chiropsalmus quadrigatus*.

(Courtesy of R. V. Southcott)

TABLE 1.—*Clinical Characteristics of Some Important Coelenterate Stings (Barnes, 1960)*

Characteristics of Sting	*Physalia*	Cubomedusae	*Cyanea*
Wheal:			
Type	Single or replicated line	Multiple lines	Multiple lines
Width	Variable	3 to 7 millimeters	2 to 3 millimeters
Pattern	Not regular	Transverse bars	Zig-zag
Duration	1 to 4 hours	2 to 24 hours	½ to 1 hour
Pain:			
Type	Moderate	Severe	Burning
Duration	½ to 2 hours	Many hours	10 to 30 minutes
Necrosis or vesication	Rare	Usual	Rare

"carukiosis," based on the generic name *Carukia*, has been proposed in place of "Irukandji syndrome" or "Type A stinging" (Southcott, 1967).

Type B stings are probably caused by various jellyfishes, but Southcott (1959) believes that the cubomedusae are probably one of the most common causes—at least in the Australian area. Undoubtedly, as the identity of many of these organisms becomes better known, this classification will be modified or entirely discarded. In the meantime, it serves as a useful clinical designation for the bulk of the unidentified types of jellyfish stings, at least in Australia and adjacent areas.

Flecker (1952*b*) and Southcott (1963) have described the *Irukandji syndrome* as follows: The sting may occur on any part of the body, and may penetrate a bathing suit. The initial lesion is frequently invisible, insignificant, or comparable to a small "flea bite." In some instances, there may be a localized erythematous patch up to the size of the palm of the hand. At first, pain may be absent, or the initial pain may be comparable to that of a wasp sting; but after a delay of 5 to 60 minutes or longer, the pain may increase in severity, first at the site of the bite, and then as it spreads it may become excrutiating. This may be followed by intense backache, pain in the back of the thighs and knees, arthralgia, and headaches. Nausea is said to be a constant symptom. Abdominal pain is frequent and generally accompanied by vomiting. Sweating is profuse. Additional signs and symptoms may be restlessness, pyrexia, rapid irregular pulse, mucoid expectoration, coughing, hematemesis, hemoptysis, tightness of the chest, and prostration. Recovery generally occurs in 1 or 2 days.

The recovery period from jellyfish stings varies from a few hours to several weeks or longer depending upon the severity of the case. Gunn (1949) and Grupper and David (1958) reported cases in which there was subsequent disappearance of the subcutaneous fat in the vicinity of the lesion. Recurrent stings from coelenterates may produce a sensitivity to the toxin, and a subsequent sting may cause a fatal anaphylactic reaction in the victim (Berntrop, 1934; Giunio, 1948). Richet (1902*a*, 1903*c*, 1904*a, b*, 1905*a, b, c*) and Portier and Richet (1902*b, c*) experimentally demonstrated anaphylaxis in dogs with *Actinia* venom, while Binet, Burstein, and Lemaire (1951) produced anaphylaxis in dogs and guinea pigs with *Physalia* venom.

The so-called "status thymolymphaticus" appears to be present in individuals susceptible to stress. Such persons are probably more prone to fatalities resulting from coelenterate stings than normal individuals. This situation probably accounts for the high incidence of "status thymolymphaticus" among coelenterate fatalities (Wade, 1928; Berntrop, 1934). Another important cause of jellyfish sting fatalities is primary shock followed by drowning before the victim can be rescued.

Anthozoan Stings: Included within this grouping are those stings caused by sea anemones of the genera *Actinia, Actinodendron, Adamsia, Alicia, Anemonia, Calliactis, Lebrunia, Physobrachia, Rhodactis, Sagartia, Telmatactis*, and corals of the genera *Acropora, Astreopora*, and *Goniopora*.

Sea anemone stings tend to be more localized in their effects. There may be itching and a burning sensation at the sting site, accompanied by swelling and erythema, ultimately followed by local necrosis and ulceration. Severe sloughing of the tissues may occur, with a prolonged period of purulent discharge. Multiple abscesses have been reported. Localized symptoms may be accompanied by such generalized effects as fever, chills, malaise, abdominal pain, nausea, vomiting, headaches, a feeling of extreme thirst, and prostration. Sea anemone ulcers tend to be resistant to treatment and are slow to heal. As in the case of other types of coelenterate stings, they can be quite mild with little or no ill effects to the victim.

The "maladie des plongeurs" or the sponge fishermen's disease, as described by Zervos (1934), Phisalix (1922), Evans (1943), and Southcott (1963) is now known to be caused by a sea anemone, probably *Sagartia* or *Actinia*, which grows on the sponge. As Zervos (1934) has pointed out, nude fishermen diving for sponges locate them by feeling with their hands along the ocean bottom. After uprooting the sponges and cleaning them by removing stones and other encrusted debris, the fisherman places the sponges in a net sack which is suspended from his neck. During such activity, the sponge fishermen have ample opportunity to be stung by the sea anemones adhering to the sponge.

Coral cuts and stings are ill-defined problems. Although such cuts and ulcers are well known to most individuals working in tropical waters, the actual stinging ability of scleractinian or stony hexacorals is not well defined. The scleractinian corals are generally considered to be of minor significance among venomous coelenterates. However, there are a few genera which have members reputedly capable of stinging human beings, viz., *Acropora, Astreopora, Goniopora,* and *Plesiastrea*. The effects have been described as a distinct stinging sensation, followed by weeping of the lesion, wheal formation, and itching. If coral cuts or stings are left untreated, a superficial scratch may within a few days become an ulcer with a septic sloughing base surrounded by a painful zone of erythema. Cellulitis, lymphangitis, enlargement of the local lymph glands, fever, and malaise are commonly present. The ulcer may be quite disabling and usually the pain is out of proportion to the physical signs. If the ulcer occurs in the lower extremity, the patient may be unable to walk for weeks or months after the injury. Relapses, which occur without warning, are not uncommon.

The severity of coral lesions is probably due to a combination of factors—laceration of tissues by the razor-sharp exoskeleton of the coral, effects of the nematocyst venom, introduction of foreign materials into the wound (i.e., minute bits of calcium carbonate that come from the animal's exoskeleton and debris), secondary bacterial infection, and adverse climatic and living conditions. Coral ulcers are slow to heal. Coral stings and ulcers have been discussed by Levin and Behrman (1941), Halstead (1957, 1959), and Southcott (1963).

The clinical characteristics of coelenterate stings have been discussed by Zervos (1934), Cleland (1912, 1932), Pawlowsky and Stein (1929), Berntrop (1934), Tweedie (1941), Kitagawa (1943), McNeill and Pope (1943*a, b*), Evans (1945), Taft (1945), Phillips and Brady (1953), Fish and Cobb (1954), Halstead (1957), Keegan (1960), Kingston and Southcott (1960), and Payne (1961).

Sea Anemones Poisonous to Eat: Intoxications resulting from the ingestion of poisonous sea anemones in Samoa and other parts of the tropical Pacific have been reported by Farber and Lerke (1963), and Martin (1960). Sea anemones are commonly eaten by the natives of Samoa, but only certain species are eaten and they are generally cooked. Some of the others are considered poisonous to eat. *Rhodactis howesi* (matalelei) and *Physobrachia douglasi* (lumane) are generally considered to be poisonous when raw, but safe to eat when cooked. However, *Radianthus paumotensis* (matamala samasama) and one other species of *Radianthus* (matamalu uliuli) are considered to be poisonous either raw or cooked. Poisonings occurred most frequently in children who had eaten raw anemones when adults carelessly left their catch within the reach of the young folks. The children ate the raw anemones, mistaking them for other edible delicacies.

The initial symptoms were those of acute gastritis with nausea, vomiting, abdominal pain, cyanosis, and prostration. Shortly after ingesting the anemones, the victim sank into a stupor which lasted from 8 to 36 hours. During this period, the superficial reflexes were absent, but the blood pressure and pulse rate were normal. All the patients eventually went into prolonged shock, and finally died with pulmonary edema. Three children died as a result of ingesting raw sea anemones *Rhodactis howesi*. There is no known antidote.

Pathology

The literature on the autopsy findings in fatal cases of coelenterate stings has little to report. Wade (1928) records a single fatal case of jellyfish poisoning in the Philippine Islands. The identity of the jellyfish was unknown. The significant findings consisted of acute congestion of the viscera, serous edema of the lungs with diapedesis, acute toxic nephritis with diapedesis, persistent thymus, and moderate lymphoid hyperplasia of the pharynx, duodenum, ileum, and spleen. There was no evidence of an anaphylactic bronchiospasm. Scott (1921) stated that fatal cases of jellyfish stings clearly revealed intense congestion of meningeal and cerebral vessels, and engorgement of the right ventricle. Acute visceral congestion in man and experimental studies on dogs have also been reported by Nobre (1928) and Berntrop (1934). Pawlowsky and Stein (1929) obtained skin biopsies from persons exposed to the tentacles of *Actinia equina*. Pathological findings consisted of acute inflammatory changes of the skin.

Kingston and Southcott (1960) have reported on the autopsy findings of two fatal cases in Australia believed to be caused by a species of cubomedusae. One of the victims was a female, age 11, and the other was a male, age 38. Autopsy findings were similar in both instances. The lungs and air passages were filled with large quantities of frothy mucus. The abdominal organs, kidneys, and brain were congested, but otherwise they were normal. The heart was normal. In the case of the male, there was evidence of minute hemorrhages into the substance of the cerebral hemispheres. The skin of the legs of the child showed linear reddish-brown streaks about 6 mm in width where the jellyfish tentacles had adhered. Death of the child was estimated to have occurred in from one-half to 10 minutes, and less than 45 minutes in the adult. The child was wading in about 76 cm of water about 15 m from shore, in the vicinity of Tully, North Queensland; and the adult was standing in about 61 cm of water near Townsville, Queensland.

Small strips of affected skin, which included areas of whealing, were excised from the legs of the victim. Paraffin sections were prepared and stained with eosin and other staining techniques. Microscopic examination revealed numerous nematocysts on the surface of the skin, the threads of which had penetrated through to the dermis (Pl. 50). The stratum corneum was edematous and its layers separated. The Malpighian layer was thinned and the cells degenerate, having pyknotic nuclei. No pathological changes were observed in the deep layers. The surface nematocysts were demonstrable with suitable illumination and standard reticulum staining techniques. The nematocysts were identified as microbasic mastigophores, unusually large, having a distinctive elongate capsule; and they were identical in appearance with those obtained from the tentacles of the cubomedusae *Chironex fleckeri* and *Chiropsalmus quadrigatus*. Although the coroner's report indicated that the child had died of anaphylactic shock and pulmonary edema, Kingston and Southcott were unable to find any histological evidence of anaphylaxis, and they believed that death was a direct result of the jellyfish venom.

56

Diagnosis of Coelenterate Stings

Barnes (1960) and Southcott (1963), who have probably had more extensive clinical experience with jellyfish stings than any other investigators, have pointed out that the recognition of the jellyfish as the attacker is made by the companion or the victim. Frequently the jellyfish is not seen by anyone, and even if observed, it is seldom identified by the layman. Common names of jellyfishes are often misleading because the same name applies to several diferent unrelated species of organisms. Every effort should be made to capture the jellyfish and preserve it in 5 percent formalin solution. Lacking this, the rescuer should attempt to preserve a portion of the tentacle which may remain upon the victim. The material should be labeled, using a good quality paper and a soft pencil. The data should include the locality, date, name of the collector, and whatever other facts may appear to be pertinent, such as the tide, time of day, etc.

In order to make an accurate diagnosis, it is important to examine the skin wheals of the victim. The wheal should be scraped with the edge of a microscopic slide or a scalpel, and the material examined under a microscope for the presence of nematocysts. The nematocysts may also be obtained by microscopic examination of a strip of Scotch tape which has been pressed against the surface of the wheal. The nematocysts will adhere to the sticky side of the Scotch tape. Southcott has accomplished the same result by placing a thin layer of clear gum-mucilage on a slide, letting it dry, pressing the sticky surface of the slide against the wheals of the victim, and then examining the slide. Permanent mounts of the nematocysts can be made by placing them in glycerine jelly or embedding them in liquid plastic. Methylene blue can be used to stain the many nematocysts, but a stain is not necessary for identification.

While identification of the nematocysts from the skin of the victim will yield positive identification of a coelenterate sting, yet it will not give positive identification of the species. However, one can obtain much worthwhile information by utilizing this technique. Much additional research is needed on the cnidom of dangerous coelenterate species.

Treatment

Jellyfish tentacles that are adhering to the skin of the victim should be *immediately removed* with the use of sand, clothing, bathing towel, seaweed, gunny sacks, or other available materials. This is one of the most important steps, because as long as the tentacles are on the victim's skin, they continue to discharge their venom. Alcohol, sun lotion, oil, and other materials that are readily available should be applied promptly to the wheals or skin lesions to inhibit the further activity of adherent microscopic nematocysts. Numerous local remedies have been advocated in various parts of the world: sugar, soap, vinegar, lemon juice, papaya latex, ammonia solution, sodium bicarbonate, plant juices, boric acid solution, etc. These have been used with varying degrees of success (Hadley, 1941; Tweedie, 1941; Graham and O'Roke, 1943; Southcott, 1963). Topical or oral cortisone preparations are sometimes useful (Southcott, 1963). According to Waite (1951), oral antihistamines and topical antihistaminic creams alleviate urticarial lesions and symptoms. Morphine sulfate has been found to be most effective in relieving pain (Wade, 1928; Berntrop, 1934; Evans, 1945; Southcott, 1963). Severe stings may require epinephrine, 7 minims, subcutaneously, repeated as necessary. Intravenous hypertonic glucose solutions may also be useful. Muscular spasms can be relieved with the use of intravenous injections of 10 ml of 10 percent calcium gluconate (Stuart and Slagle, 1943; Southcott, 1963), or sodium amytal, intravenously (Frachtman and McCollum, 1945). Artificial respiration and oxygen may be required. Southcott (1963) has pointed out that a number of fatalities have occurred as a result of stings on the lower extremities, and he suggests that in such cases the immediate use of a tourniquet might save the life. Cardiac and respiratory stimulants, and other supportive measures, may be required.

Treatment must be directed toward the accomplishment of three primary objectives: relieving pain, alleviating the neurotoxic effects of the venom, and controlling primary shock. Excellent reviews on the treatment of coelenterate stings have been prepared by Southcott (1959) and Keegan (1960). Baxter, Marr, and Lane (1968, 1973) examined the feasibility of active or passive immunization against fatal envenomation by the sea wasp *Chironex fleckeri*. The potent antivenin, a globulin preparation, will protect laboratory animals against lethal effects and greatly reduce dermal necrosis. Their studies resulted in the development of an antivenin for humans. This antivenin is now produced by the Commonwealth Serum Laboratories, Melbourne, Australia.

The object of treatment with antivenin is to neutralize the venom as soon as possible. The antivenin should, therefore, be injected intravenously, preferably by infusion. Use of the intramuscular route may be justified if the person giving the antivenin is not trained in the technique of intravenous injection. The risk of serum reactions normally precludes the use of antivenin by lay persons, but if remote from medical aid and confronted with an emergency situation, intramuscular injection of antivenin is justifiable. It is important that any lay person provided with antivenin is fully acquainted with the technique of intramuscular injection and the necessary preparations. If injected directly into a vein, care must be taken to ensure that the antivenin is given slowly.

The contents of one container of Sea Wasp Antivenin constitutes the initial dose by the intravenous route. The contents of three containers should be given if the intramuscular route is to be used. Sea Wasp Antivenin is supplied in containers of 20,000 units. It should be stored in a refrigerator at 2°-10° C. It must not be frozen.

The possible uses of this antivenin against envenomation by other jellyfish is under investigation (Baxter and Marr, 1974).

Coral cuts should receive the following prompt treatment: cleansing the wound, removing of foreign particles, debriding if necessary, and applying antiseptic agents. Considerable difficulty can be prevented by promptly painting coral abrasions (even minor ones) with an antiseptic solution such as 2 percent tincture of iodine, etc. In severe cases, it may be necessary to give the patient bed rest with elevation of the limb, kaolin poultices, magnesium sulfate in glycerin solution dressings, and antibiotics (Halstead, 1959; Keegan, 1960). Levin and Behrman (1941) found roentgen therapy to be useful in resolving a chronic coral ulcer with keloidal formation.

Prevention

Bathers should exercise particular care when they swim in areas in which dangerous coelenterates such as *Physalia*, and especially the cubomedusae *Chiropsalmus* and *Carybdea*, are known to exist. It should be kept in mind that the tentacles of some species may trail a great distance from the body of the animal. Consequently, they should be given a wide berth. Tight-fitting long woolen underwear or rubber skindiving suits afford protection against attacks from these creatures. Jellyfishes washed up on the beach, even though appearing dead, may be capable of inflicting a serious sting. Since the tentacles of some jellyfishes may cling to the skin, a person should exercise care in removing the tentacles, otherwise additional stings will be received. Some species of coelenterates have powerful nematocysts capable of stinging through clothing and even surgical gloves; therefore a towel, rag or seaweed, stick, or a handful of sand should be used to remove the tentacles. If a person goes swimming soon after a storm in tropical waters infested with large numbers of jellyfish, he may receive severe multiple stings from remnants of damaged tentacles floating in the water. Upon being stung, the victim should make every effort to get out of the water as soon as possible in order to avoid the danger of drowning. Dilute ammonia and alcohol should be applied to the area affected, and other therapeutic measures should be instituted promptly.

Coral cuts and abrasions should be promptly painted with 2 percent tincture of iodine or some other suitable antiseptic solution.

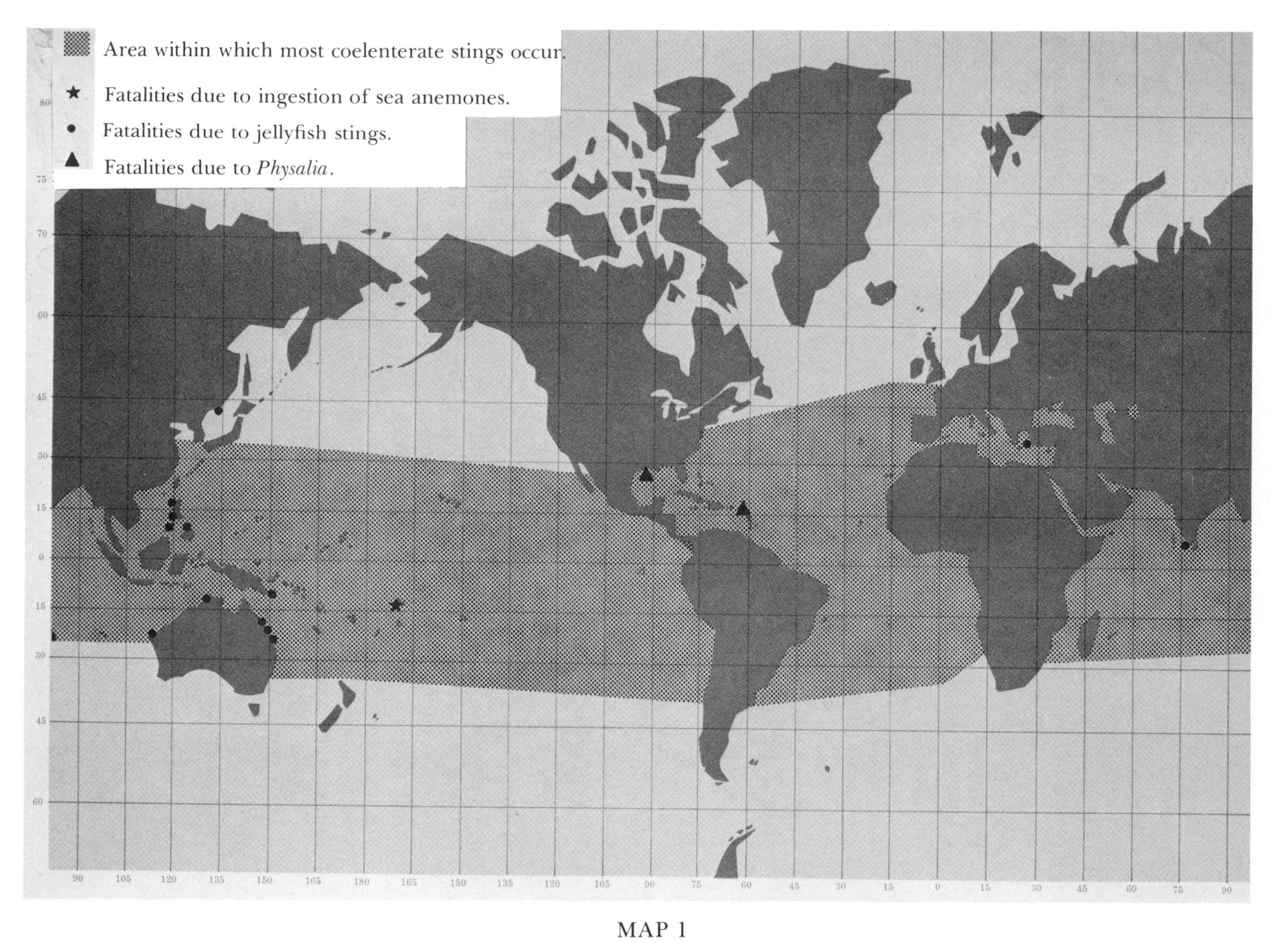

MAP 1

World distribution of reported outbreaks of coelenterate intoxications. (L. Barlow)

PUBLIC HEALTH ASPECTS

Our knowledge of the global public health aspects of coelenterate stings is exceedingly meager and is based largely on the scattered reports of clinicians and others who have published accounts of envenomations. Aside from the limited efforts of Australian workers, no attempt has been made to record the geographical distribution and incidence of coelenterate intoxications. The results of Southcott (1952, 1963) and Barnes (1960) clearly indicate that only a fraction of coelenterate stings that occur each year ever find their way into the medical literature. This is probably due to the fact that jellyfish stings are non-reportable public health illnesses, and fortunately most of the cases are relatively mild in nature.

Geographical Distribution

Theoretically, any coelenterate equipped with a nematocyst apparatus is capable of inflicting stings on man. Thus, in a general sense, the geography of coelenterate envenomations corresponds to that of the causative agents. Representatives of the phylum Coelenterata are to be found in every major oceanic area of the world.

The majority of species of greatest danger to man are found in tropical and warm temperate waters, which is a geographical area roughly corresponding to that of the distribution of corals (*see* Map 1). Major exceptions are the families Cyaneidae and Pelagiidae which are most abundant in temperate and colder waters. Fortunately, the colder water coelenterates do not constitute the public health hazard of some of the warmer representatives because the opportunity for exposure to man is less in the cooler latitudes.

Incidence

Barnes (1960) has provided the only statistics available on the incidence of jellyfish stings over a given period of time: Between November 1956 to May 1960, there were 116 known victims stung by jellyfishes in the vicinity of Cairns, North Queensland. The stingings took place between November and April with the greatest number occurring in December. In general, stingings are most common during the warmer months of the year, probably because weather conditions, water turbulence, and the number of bathers are at an optimum relationship for jellyfish attacks.

There is no way of determining with any degree of accuracy the actual annual incidence of coelenterate stings. On the basis of the published reports from 1836 through 1960, there have been >502 stings with >46 fatalities, giving a crude case fatality rate of about 9 percent.

TOXICOLOGY

Portier and Richet (1902*a, b, c*), Richet (1902*a, b, c*), Richet, Perret, and Portier (1902), and Richet and Portier (1936) were the first to attempt an evaluation on an experimental basis of the toxicity of coelenterate venoms. In their initial experiments, they prepared crude glycerinated toxic extracts which they termed "actinotoxin," made from the ground-up tentacles of *Anemonia sulcata*. These were injected intravenously into dogs and rabbits. Rabbits were found to be less sensitive to the venom than dogs. Similar extracts were prepared from the tentacles of *Physalia* (Portier and Richet, 1902*a*) and injected intramuscularly into pigeons.

Most of the early research of Portier and Richet was directed to a hypersensitivity phenomenon which they termed "anaphylaxis." When dogs were first administered a sensitizing dose, less than 0.15 ml/kg intravenously, of the crude actinotoxin, they became very sick but recovered. Then, if after a period of 2 or 3 weeks, these same dogs received a second dose of 0.08 to 0.25 ml intravenously, they immediately developed intense vomiting, pruritus, difficulty in respiration, foaming at the mouth, bloody diarrhea, and died within a period of several hours. During the course of the investigations, it was determined that three distinct fractions had individual chemical and physiological characteristics. These fractions they called "thalassin," "congestin," and "hypnotoxin."

Thalassin: This compound was first obtained from the tentacles of *A. sulcata* (Richet, 1902*b* to 1904*b*). Intravenous injection into dogs was found to produce sneezing, pruritus, urticaria, genital excitation, vasodilatation, edema of the face, nervousness, redness, coughing, rhinitis, rolling on the ground, conjunctivitis, "nervous depression," cardiac arrest, and death. Thalassin is especially characterized by its pruritogenic properties. It is active in dogs in dosages of 0.0001 g/kg of animal weight. The lethal dose in dogs is said to be 0.02 g/kg (Richet, 1905*c*). Richet (1904*c*, 1905*c*) also found that thalassin, or a thalassinlike substance, is widely distributed in shrimps, crabs, lobsters, amphipods, mussels, oysters, and the hydatid cysts of *Echinococcus granulosus*.

Congestin: This toxin was likewise first obtained from sea anemone tentacles by Richet (1903*a, c,* 1904*a, b,* 1905*a, b, c*). Intravenous administration of the substance into dogs resulted in vomiting, bloody diarrhea, evidence of abdominal pain, prostration, decrease in blood pressure, intense splanchnic vasodilatation, respiratory paralysis, and death. Congestin received its name because of the characteristic intense splanchnic congesting which it produces. This substance has also been referred to as "actinocongestin" (Richet, 1908). If the animal has previously received a sensitizing dose of the drug, a second injection given within a period of 2 to 3 months, or longer, results in anaphylaxis (Portier and Richet, 1902*b, c*; Richet, 1904*a, b,* 1905*a, b, c*). However, if small doses of thalassin are administered first, the animal is able to tolerate considerably larger doses of congestin. An initial injection combining thalassin and congestin when followed by an injection of congestin offers some protection, but mild anaphylaxis results. If dogs are administered congestin first and then thalassin, followed by an additional injection of congestin, lethal anaphylaxis results. A dose of 0.07 mg/kg of animal weight in dogs produces symptoms. The lethal dose is said to be 0.005 g/kg of animal weight in dogs, and 0.0009 g/kg in rabbits (Richet, 1904*a,* 1905*a*).

Hypnotoxin: This drug was first obtained by Portier and Richet (1902*a*, *b*) from the tentacles of *Physalia*. The authors believed that the toxin is present also in other coelenterates, but further research is needed before it can be established definitely. Hypnotoxin characteristically produces a central nervous system depression affecting both motor and sensory elements. Symptoms most commonly encountered are inactivity, coma, muscular paralysis, and cutaneous anesthesia when injected intramuscularly into pigeons, guinea pigs, ducks, and frogs. Death was reported from respiratory paralysis, but the lethal dosage is not known.

In unpublished experiments on *Physalia* tentacles, Bodansky (cited by Taft, 1945) has obtained results on hypnotoxin similar to those of Portier and Richet. Extracts of the tentacles injected into pigeons produced symptoms identical to those mentioned above. Preparations of the dried tentacles were found to retain their toxic properties for 2 years or longer. Hypnotoxin is normally produced in small quantities in the body tissues during muscular activity (Phillips and Brady, 1953).

Meduso-congestin: Dujarric de la Rivière (1915) isolated from the tentacles of the jellyfish *Rhizostoma cuvieri* a toxic substance which he termed "meduso-congestin." Crude aqueous extracts injected into rabbits and guinea pigs in dosages of 0.06 to 0.08 g/kg of animal weight produced localized edema, diarrhea, prostration, ruffling of the hair, respiratory distress, and death. Autopsies revealed an intense visceral congestion. Meduso-congestin is believed to be identical with the substance which Richet termed congestin.

Equinatoxin: Pawlowsky and Stein (1929) conducted 39 toxicological experiments on humans using various portions of *Actinia equina*. Where the tentacles were rubbed on the flexor surface of the forearm, painful urticarial reactions occurred. No attempt was made to isolate the toxin until very recently. A lethal protein from *A. equina*, called equinatoxin, was purified by fractional precipitation with acetone and by gel chromatography, with a molecular weight about 20,000 (Ferlan and Lebez, 1974). It is thermolabile, is hydrolyzed by pepsin and trypsin, and can act as an antigen. Following intravenous injection into rats the LD_{50} is 33.3 μg per kg. The pathophysiological actions of the toxin on the respiratory and cardiovascular systems were discussed by Sket *et al.* (1974).

Studies on other coelenterate venoms: Cosmovici (1925*a, b, c, d, e*) prepared aqueous extracts from the tentacles of the sea anemone *Adamsia palliata* and injected the material into the legs of crabs *Carcinus maenas*. He reports that the injections produced violent symptoms of pruritus, tetanic convulsions, paralysis, autotomy of the injected appendage, cardiac fibrillation, and, finally, what appeared to be sustained contraction of the heart muscle. He found that a potent extract (0.1 ml) would kill a crab in 15 to 30 minutes, according to the size of the crab. Additional aqueous extracts of *A. palliata* were tested on various other marine invertebrates, *Sagartia parasitica, Anemonia sulcata, Actinia equina, Strongylocentrotus lividus, Asterias glacialis, Sipunculus nudus*, and *Solen ensis*, and various cephalopods (Cantacuzène and Cosmovici, 1925). The researchers concluded that decapods were the most sensitive to the toxin, and that the cephalopods were the most resistant to the poison. In subsequent tests with *Adamsia palliata* poison on the crab *Eupagurus prideauxi* they found that this particular species was unaffected by the poison (Cantacuzène, 1925*a, b*). Therefore, it was concluded that the immunity of the crab *E. prideauxi* to the poison was due to the fact that it feeds on *A. palliata* and has established an immunity to it. Investigations of the gut contents of *E. prideauxi* revealed the nematocysts of *A. palliata* (Cantacuzène, 1926).

Further studies were conducted in an attempt to determine the nature of the immunity mechanism in *E. prideauxi* against the toxic effects of *A. palliata* extracts. It was found that if serial dilutions of extracts of *A. palliata* were placed in a solution of sheep red blood cells and lecithin, complete hemolysis occurred in all but the most dilute solutions of extracts. The author concluded from his experiments that the tissues and body fluids in *E. prideauxi* contain sufficient lecithin to produce lysis of *A. palliata* tissues, and he believed this to be the basis of the crab's immunity against *A. palliata* toxin.

Chemical evaluation of *A. palliata* extracts revealed the material to be proteinaceous in nature (Cantacuzène and Damboviceanu, 1934*b*). The extracts were then treated with trichloracetic acid to deproteinize them. When tested on crabs *C. maenas*, the deproteinized extracts were found to have the same toxic effects as the untreated toxic extracts (Cantacuzène and Damboviceanu, 1934*a*). It was concluded that the toxin from *Adamsia* loses none of its biological activity after deproteinization. Moreover, when mixed with lecithin, it has a hemolyzing effect on red blood cells which neither the poison alone nor the lecithin alone possess.

Fifty years after Richet and Portier's first experiments on anaphylaxis, Binet, Burstein, and Lemaire (1951) used the same procedures to prepare glycerinated extracts from the tentacles of *Physalia pelagica*. Intravenous injections of the extracts into dogs resulted in arterial hypotension which, it was found, could be prevented with the use of antihistamines. Successive injections of the extracts in guinea pigs resulted in anaphylaxis with signs of shock and finally death.

One of the difficulties encountered by previous investigators working on the toxicity of coelenterate venoms was the problem of separating the nematocysts from the tentacles and other tissues. In most instances the entire animal or the tentacles were macerated. This produced a toxic extract, without regard to its histological origin. Consequently, much of the earlier work on the true nature of nematocyst poisons has been brought into question. For example, Lane and Dodge (1958) and Lane (1961*a, b*) contend that thalassin and congestin as described by Richet and Portier are chemical constituents of the tentacles rather than the nematocysts.

Glaser and Sparrow (1909) were apparently the first to devise a method of separating the nematocysts from other tissues; their method was later modified by Phillips (1956), Phillips and Abbott (1957), Lane and Dodge (1958), and Lane (1960, 1961*a, b*). Nematocyst isolation techniques have been utilized during recent years, and have contributed to a better understanding of the chemistry and pharmacology of nematocyst poisons.

Phillips and Abbott (1957) developed a method of assaying the isolated nematocyst toxin of *Metridium senile* by determining the effect of the toxin on the righting reflex of the marine snail *Littorina planaxis*. When this snail is placed upside down in seawater, it rapidly rights itself and moves out of the water to a relatively dry place. However, it was found that the time required for this righting and withdrawal from the water could be prolonged by the addition of the toxin, and that the length of inhibition depends upon the concentration of the toxin. Excessively high doses of the toxin were able to extend the righting time and eventually kill the snail.

Lane and Dodge (1958) obtained isolated nematocysts by washing the animals in clean seawater, permitting the animal to autolyze at 4° C for 24 to 48 hours, diluting the mixture with several volumes of seawater, and passing it through graded screens of 24 and 115 meshes per inch. This removed most of the muscle and connective tissue and permitted passage of the undischarged nematocysts. The screened suspension was then allowed to settle overnight in the cold. The supernatant was decanted and discarded. The residue was then centrifuged at 300 to 400 g for 15 to 30 minutes. The supernatant solution was again discarded and the residue resuspended in seawater. These processes were continued until injection of 0.1 ml of the supernatant solution into the hemocoele of the fiddler crab *Uca* was made without effect. Most of the nematocysts obtained by this technique were generally undischarged. When the nematocysts were frozen and stored at -5° C, Lane and Dodge found that they could be made reactive after 20 weeks of frozen storage. They found that 3.4 liters of isolated fishing tentacles of *Physalia* yielded 60 g of packed wet "purified" nematocysts. They were able to liberate the contents of the isolated nematocysts by homogenization in a chilled Potter-Elvejhem homogenizer, using Amphibian Ringer's seawater, or distilled water as the dilutent. The sample was examined microscopically at intervals and homogenization was continued until about 90 percent of the capsules were fragmented. The homogenate was centrifuged at 600 g for 10 minutes to separate capsules and capsular fragments from the diluted capsular contents. The supernatant solution was described as cloudy, yellowish-white in color, and extremely toxic to crabs, fish, and small mammals. Since nematocysts may retain their reactivity for at least 2 weeks, precaution was taken to avoid exposure of the skin to contamination by any mixture, wet or dry, which contained undischarged nematocysts. Surfaces, clothing, and skin can be decontaminated with 95 percent ethanol.

Lane and Dodge (1958) and Lane (1960, 1961*a, b*) believe the nematocyst toxin of *Physalia* to be a highly labile protein complex. The poison was destroyed or denatured by heating to 60° C, by drying, or by treatment with ethyl ether, acetone, or ethanol. Activity could be preserved for at least 2 months when the material was stored at -5° C. The approximate lethal dose of toxin for mice, injected intraperitoneally, was found to be 0.037 ml/kg of a preparation which contained 0.201 percent total nitrogen. The toxin was shown to be devoid of hemolytic activity for fish erythrocytes. The poison appeared to affect the nervous system of fish, frogs, or mice, particularly the respiratory centers, before voluntary muscles. Injection of the crude toxin seemed to produce a general paralysis. Symptoms observed in mice included increased activity and tremors. After 10 minutes, there was ataxia, decreased muscle tone, flaccid paralysis, slowed and labored breathing, defecation, aphradisia, marked myosis, cyanosis, anoxic convulsions, and death. Survival time in mice was 1 to 48 hours, depending on the dose administered. Post mortem examination revealed that the lungs were ischemic; the heart was contracted, especially the left ventricle; hemorrhagic edema appeared in the peritoneal cavity; the skin of the nose and ears was very white; the cornea was cloudy; the colon had no formed stools; and the urinary bladder was empty. Localized changes in cardiovascular tone were observed in some test animals. *Physalia* toxin in the isolated heart of the clam elicited responses similar to those caused by acetylcholine.

Burkholder and Burkholder (1958) demonstrated antimicrobial activity in extracts prepared from sea whips, sea fans, and plexurid corals. The active principle is said to be present in the outer gray-purple of the cortex of the coral. No antimicrobial activity could be detected in the species of stony corals that were tested. The researchers were unable to determine whether the coral polyps or their associated zooxanthellae produce the antibiotic materials.

Lane, Coursen, and Hines (1961) have found that the crude toxin of *Physalia* withstands lyophilization without significant loss in toxicity for prolonged periods of time. They were able to separate *Physalia* toxin into component peptides by one-dimensional chromatography in 80 percent of aqueous *n*-propanol. It was observed that each of the resultant peptides retained considerable toxicity for the fiddler crab *Uca pugilator*. Further studies indicated that the crude toxin of *Physalia* nematocysts had no effects on the growth of 14 species of marine yeasts and 10 species of marine bacteria, or the ciliate protozoans *Paramecium caudatum* and *Tetrahymena gelli*.

The loggerhead turtle *Caretta caretta* is able to eat *Physalia* with apparent immunity. Wangersky and Lane (1960) thought that the immunity of the turtle to Physalia toxin might be the result of protective antibodies in the blood acquired by contact with *Physalia* early in life. However, subsequent turtle plasma-toxin precipitation tests failed to indicate that the loggerhead turtle possesses such blood immune bodies. The authors suggested that possibly the loggerhead turtle may possess localized tissue antibodies, or that the turtle is not susceptible to venoms of the type injected by *Physalia*.

The toxicity of the sea anemone *Rhodactis howesi* has been investigated by Martin (1960) and Farber and Lerke (1963). Martin prepared toxic aqueous extracts from homogenates of whole sea anemones and injected the material intraperitoneally into toads *Bufo marinus*. A maximum dose of 8 ml of homogenate was administered for each 100 g of toad. The toads showed no change in behavior or reaction for several hours after injection. Later they demonstrated a stuporous condition, respiratory distress, and most of them died within 48 hours. Their parotid glands turned ischemic immediately prior to death. Martin believed the poison to be of the paralytic type. Farber and Lerke (1963) fed toxic *R. howesi* to mice. The mice suddenly developed a mucous diarrhea and respiratory distress. The animals had a staggering gait and appeared to remain in a stuporous condition for prolonged periods of time. There was often intestinal hemorrhage, and death apparently resulted from respiratory failure. Post mortem findings consisted of subcutaneous and pulmonary hemorrhages, a moist sticky exudate on the intestines, an injected diaphragm, and pale soft kidneys. The toxin appeared to be slightly more active by mouth than by parenteral injection. However, most subsequent studies were based on intraperitoneal injections of four to six mice, weighing 18 to 20 g per mouse, using a dose of 0.1 ml of the homogenate. After experimenting with various solvents and buffered solutions, it was found that plain distilled water was a satisfactory extractant. The best aqueous homogenate had an MLD of 6.2 mg of whole coelenterate per mouse. Attempts were made, with a formalized fraction, to induce an immune response in mice and rabbits, but without positive results. The researchers believe that the toxic principle in *R. howesi* is a protein of fairly high molecular weight.

Aqueous extracts of the sea pansy *Renilla muelleri* from the Mississippi Gulf Coast were injected intraperitoneally in mice. Dosages of 100-250 mg/kg produce death within a period of 24 hours. The LD_{50} was 160 mg/kg. A dose of 10-20 mg/kg intravenously produced a fall in blood pressure and bradycardia in cats (Huang and Mir, 1971).

Martin (1963) has tested aqueous extracts prepared from a variety of California sea anemone species, viz., *Anthopleura elegantissima, A. xanthogrammica, Metridium senile, Corynactis californica, Tealia lofotensis, T. crassicornis,* and *T. coriacea*. Homogenates were prepared with a Waring blender, and the supernatant obtained by centrifugation. The extracts were dialyzed through commercial cellophane tubing, before being tested by intraperitoneal injection into white mice weighing 17 to 23 g. A dose level of 20 ml of extract per kilo of mice was used. All the extracts tested were found to be toxic; however, the potency of the extracts varied according to the species, with *A. elegantissima* and *A. xanthogrammica* being the most toxic. Toxicity was based on the survival time of the injected mice using an LD_{99}. All the sea anemone extract lost its toxicity when heated at 90° C for 90 minutes.

Emery and Grubb (cited by Martin, 1962) immunized rabbits with *R. howesi* extracts, and they obtained an immune serum which protected mice from injected extracts. Immune serums have also been prepared with *Physalia* extracts (Lane, 1960).

In considering the toxicity of coelenterates, one should keep in mind that the ability to produce a painful sting is not necessarily an indication of the amount of toxic substance present in the tentacles of a given species (Welsh, 1956). The process of stinging depends upon the ability of the nematocysts to penetrate skin, and this varies greatly with the species. To illustrate, Welsh pointed out that *A. sulcata* has penetrants that enter the skin of the finger sufficiently to cause a stickiness between the tentacle and the skin, but seldom produce pain. However, if a tentacle is placed against the tongue, a painful long-lasting burning sensation occurs, which indicates that the nematocysts are able to penetrate the delicate epithelium of the tongue.

There is an increasing amount of evidence that toxic substances are present in both the tentacles and nematocyst capsular fluid of coelenterates (Lane, 1960; Martin, 1962).

Some species of the genus *Palythoa* have been found to contain a potent toxin. The toxicity of a given species seems to vary with the individual. Some specimens are toxic and others are not. The poison is within the body of the animal and is not purveyed by means of its nematocyst apparatus. The poison derived from this animal has been termed by Scheuer (1964) as "palytoxin." Scheuer's work was reportedly done on the Hawaiian species *Palythoa vestius*. However, specimens taken from tidal pools near Hana, Maui, Hawaii were toxic whereas specimens taken elsewhere were not. The poison is highly hygroscopic and has not been crystallized. It is sensitive to both acid and base. Nitrogen is present in the molecule, but the ratio of nitrogen to carbon appears to be very low. There is no information available concerning the pharmacological properties of the poison.

Puffer, McIntosh, and Given (1970) obtained a toxic extract from the hydroid *Lytocarpus nuttingi* from Southern California, having an LD_{50} (crude extract) of 360 mg/kg when injected intravenously into mice. Hashimoto and Ashida (1973) isolated a toxic polypeptide from stony corals *Goniopora* spp. from the Ryukyu Islands. The toxin was reported lethal to mice at 0.5 mg/kg by the intraperitoneal injection. The toxin possessed hemolytic activity on rabbit blood cells.

Moore and Scheuer (1971) have isolated palytoxin from an unidentified species of *Palythoa*. The LD_{50} of the pure toxin is 0.15 g/kg by intravenous injection in mice. Yields have ranged from 5-270 ppm (Scheuer, 1972). It appears to be at least 100 times more active than the most active cardiac glycosides both in terms of lethal dose and minimum effective dose (Rayner, cited by Scheuer, 1972).

Baxter and Marr (1974) and Keen and Crone (1969*a, b*) investigated the lethal dermonecrotic and hemolytic properties of *Chironex* venom.

An examination was made of the penetration of the hemolytic fraction of tentacle extracts of *Chironex fleckeri* into monolayers of lipid, by Keen (1973). Hemolytic activity was shown to be inhibited by gangliosides *in vitro*. These findings suggested that lytic activity might be associated with an interaction between gangliosides and hemolytic protein.

Burnett and Calton (1973) separated the nematocysts of the sea nettle (*Chrysaora quinquecirrha*) from tentacular debris by zonal centrifugation. The nematocyst fluid was separated into 25 lethal components and one additional fraction which was capable of rupturing lysosomes. A sample of the supernatant free of nematocyst capsules and debris was found to produce an LD_{50} by IV in mice of 3.75 μl.

A toxin of low molecular weight was isolated from tentacles of the sea anemone *Anemonia sulcata* by Novak *et al.* (1973). Partially purified toxin was obtained by precipitation with ammonium sulphate, ion exchange chromatography on DEAE cellulose and chromatography on Sephadex G-50. The molecular weight of a partially purified toxin, probably a basic protein, was approximately 6000 as determined by gel filtration. Disc electrophoresis of the partially purified toxin revealed four proteins. Heating for 15 minutes at 90° C inactivated the toxin. The toxic activity remained unchanged over 12 months when stored in frozen solution at -26° C. The LD_{100} is 6 mg of the toxin per kg (rat) by IV injection. The toxin showed a cardiotropic action on the rat heart. When injected IV the toxin caused cardiac arrhythmia and fall of the arterial blood pressure.

That sea anemone (*Aiptasia pallida*) nematocyst venom exhibits phospholipase A activity on red cell membrane phospholipids was demonstrated by Hessinger and Lenhoff (1974). Their findings suggest that the underlying mechanism of nematocyst venom-induced hemolysis is due to the enzymatic action of venom phospholipase A on membrane phospholipids.

Much of the early work on the toxicology of coelenterate poisons has been reviewed by Phisalix (1922), Pawlowsky (1927), Thiel (1935), Sonderhoff (1936), Maass (1937), and Kaiser and Michl (1958).

PHARMACOLOGY

Work on the general pharmacological properties of coelenterate poisons is of recent origin and represents the findings of only a few investigators. Most of the research has been conducted on invertebrate organ systems and relatively little is known of the effects on mammals. With the exception of the work done by Phillips and Abbott (1957) and Lane *et al.* (1961), extracts have been prepared from homogenates of whole tentacular tissues. The Phillips and Lane groups worked with extracts prepared from isolated nematocysts.

Some of the principal actions of coelenterate poisons appear to relate to pain production, paralysis, and the urticariallike effects on the skin (welts, erythema, itching, edema, vesiculation, etc.).

Pain Production

It has been suggested by Welsh (1956, 1961) and Payne (1961) that pain production from coelenterate venom is probably due to the direct effect of 5-hydroxytryptamine, a known potent pain-producer and histamine releaser, on the pain receptor organs of the skin. The presence of 5-hydroxytryptamine in coelenterate tentacles has been well-established by the investigations of Welsh (1956, 1961), Mathias, Ross, and Schachter (1957, 1958), and Welsh and Moorhead (1959). However, Mathias *et al.* have questioned the quantity of 5-hydroxytryptamine actually present in the capsular fluid of the nematocyst.

Effects on Nerve Conduction

Welsh and Haskin (1939) found that autotomy, dropping of the legs, in crustaceans is a sensitive indicator of the activity of such substances as acetylcholine and epinephrine. Welsh (1956) assumed that nematocyst capsular fluid might act at some step in the conduction or transmission of nerve impulses; therefore, the autotomy reflex was chosen as an indicator of such action. He adopted the fiddler crab *Uca mordax* because of its tendency to drop a leg when grasped by the leg with a pair of forceps. Tentacular extracts were prepared from the siphonophore *Physalia*, and the sea anemones *Condylactis gigantea* and *Aiptasia* sp., by grinding whole tentacles with an equivalent volume of seawater, and the extract adjusted to a neutral pH value. Extracts were injected into the base of one of the walking legs in the amount of 0.02 to 0.05 ml, depending upon the size of the crab. The study showed that the tendency to drop or autotomize legs was reduced by the injections of tentacular extracts, and that the crabs exhibited clear signs of paralysis. The results led to the conclusion that a paralyzing agent of some type was present in the tentacular extract. Previous studies by Ackermann, Holtz, and Reinwein (1923) had shown that tetramethylammonium chloride, or tetramine, is a normal constituent of coelenterates and is also a paralysant having curarelike properties in vertebrates. It was also found that injected tetramethylammonium chloride and acetylcholine have similar actions to those of the tentacular extracts.

Further comparable experiments were made with the shore crab *Hemigrapsus nudus* as the test animal. Extracts were prepared from the tentacles of the anemone *Metridium dianthus*. In this instance, injections of tentacular extracts produced spontaneous autotomy of the walking legs and chelae, followed by degrees of paralysis dependent upon the dose. In addition, extracts were prepared from the jellyfish *Cyanea capillata* and injected into *H. nudus*, but spontaneous autotomies rarely occurred at any dilution. With dilutions up to 1:50, paralysis occurred, but this was less severe than with equivalent amounts of *Metridium* extract. It was also found that *H. nudus* is less subject to autotomizing than *U. mordax*. When tetramethylammonium chloride was injected into *H. nudus*, there was no spontaneous autotomy; however, there was a type of paralysis similar to that produced by tentacular extract.

Tetramethylammonium chloride, and a related compound called banthine, were found to be highly effective in blocking the autotomy-inducing and paralyzing actions of *Metridium* extract. Welsh has suggested that this is added evidence that a quaternary ammonium base similar to tetramethylammonium occurs in coelenterate tentacular extracts. However, as Welsh has pointed out, although tetramine is found in large amounts in coelenterates, this does not prove that it is the paralyzing agent in nematocysts.

Mathias *et al.* (1958) have worked with extracts prepared from the bodies and tentacles of *Actinia equina*, and they found them to produce marked contracture of the isolated frog rectus abdominis muscle. They believe that there are at least two paralyzing agents present in *Actinia* extracts. One is a dialyzable substance, soluble in alcohol, pharmacologically indistinguishable from tetramethylammonium. The other is a nondialyzable substance, insoluble in alcohol, and inactivated by chymotrypsin, which also produces a prolonged contracture of the frog rectus abdominis muscle. Twice as great a concentration of these two substances is found in the tentacles as in the body of the *Actinia*. The researchers also identified α-picolinic acid N-methyl betaine (homarine), which they found to be present in high concentrations. They were unable to detect acetylcholine and urocanylcholine (murexine) in any of their preparations. Extracts prepared from the tentacles of *Physalia* when injected into guinea pigs, rats, and rabbits, proved to be extremely toxic, but as yet the chemical constituents have not been identified.

Welsh and Prock (1958) have theorized that the toxic paralyzing effects of some of the coelenterate venoms may be the result of tetramine being linked with a protein which might produce a substance more toxic than tetramine alone. They based this idea partly on the evidence that two alkylated tetracovalent nitrogens, properly spaced in a molecule, can produce highly active junctional blocking agents such as curare and some of the synthetic curariform bis-quaternary substances. Protein denaturation by heat, or something else, might alter the spacing of the tetramines on the protein or release them, and thereby reduce the paralyzing action of the extract. In support of this idea, Welsh and Prock point out that nematocysts have a high affinity for methylene blue, a basic dye with two methylated nitrogens which through resonance may become tetracovalent. This would imply that there are molecules (probably protein) within the nematocyst contents that bind methylene blue and possibly other quaternary ammonium bases.

Narahashi, Moore, and Sahpiro (1969) have studied the effects of the toxin of the Bermuda sea anemone *Condylactis gigantea* on the interaction with nerve membrane ionic conductances.

Effects on Heart Muscle

Welsh (1956) tested *Metridium* and *Cyanea* tentacular extracts on the isolated cardiac ventricle of the horse clam *Schizothaerus nuttalli*. The horse clam's heart was used because it is sensitive to the quaternary ammonium compounds. Tetramine and its derivatives are known to decrease the amplitude and frequency of the heartbeat, whereas adrenaline and tyramine are known to be excitatory when applied to the isolated mollusk heart. The tested extracts of *Metridium* and *Cyanea* demonstrated a mixture of heart excitor and inhibitor substances, but there was a predominance of the excitor agent. Subsequent studies strongly indicated that the excitor substance was 5-hydroxytryptamine. *Metridium* extracts produced an increase in amplitude, frequency, and a tonic shortening of cardiac contractions, whereas *Cyanea* extracts caused an increase in frequency and more marked tonic shortening. It was suggested that such a difference could be the result of relatively larger amounts of the excitor substance in *Cyanea* tentacles.

Burnett and Goldner (1969, 1970*a*), and Burnett, Calton, and Cargo (1972) investigated the cardiovascular action of *Chrysaora quinquecirrha* to determine whether this agent was capable of injuring tissues other than skin. Cardiac conduction abnormalities and ischemic electrocardiographic changes occurred in rats shortly after intravenous injections of the toxin. Large doses of toxin produced a prompt increase in arterial blood pressure. Kleinhaus, Cranefield, and Burnett (1973) found that sea nettle *(C. quinquecirrha)* toxin produces the same effects as those produced by exposing cardiac fibers to a solution containing almost no Ca^{2+}. The authors suggest that the toxin in some way alters the effective concentration or the action of Ca^{2+} at the membrane of such fibers.

Turner and Freeman (1969) studied the effects of toxin extracts from *Chironex fleckeri* on the isolated, perfused guinea pig heart. Small doses of toxin resulted in reversible decreases in coronary flow, heart rate, and amplitude of contraction while larger doses produced irreversible changes. They further suggested that the toxin acts by altering membrane permeability (Freeman and Turner, 1969).

Freeman and Turner (1972) made a comparison of the pharmacologic properties of toxins isolated from *Chiropsalmus quadrigatus* and *Chironex fleckeri*. The effects of the toxins were studied in intact anesthetized animals and in a number of isolated organ preparations. Toxin extracted from *Chiropsalmus* tentacle was found to resemble the *Chironex* cardiotoxin (Keen, 1971). However, it was less stable and its activity differed in onset and duration of effects. Toxin from *Chiropsalmus* caused an initial hypertensive response in animals due to direct vasoconstriction. This was followed by hypotension and cardiac irregularities. Arterial pressure oscillations were frequently seen prior to death. It was not possible to elicit a carotid occlusion reflex during the action of the toxin. Ganglion blocking drugs prevented oscillations by blocking the efferent arm of the vasomotor reflex arc. Their studies concluded that *Chironex* appears to be the more dangerous species for man.

Two high weight protein toxins isolated from *Chironex fleckeri* tentacles exert their effects on the transmembrane potentials of the heart by different mechanisms. The cardiotoxic fraction appears to increase sodium permeability, whereas the hemolytic fraction appears to increase the leak component of membrane permeability (Freeman, 1974).

Gould and Burnett (1971) found *Chrysaora* toxin to increase the short-circuit across isolated frog skin.

Histamine Release

The common findings of wheals, redness, painful burning sensation, urticarial eruptions, etc. in coelenterate stings indicate the presence of histamine or a histamine-releaser substance. Uvnäs' (1960) investigations of the pharmacological properties of extracts prepared from the tentacles of *Cyanea capillata* revealed no histamine, but they did indicate a histamine-releasing agent. Welsh (1956) injected crabs with 0.05 ml of 1 percent histamine. This resulted in marked paralysis, but no spontaneous autotomy. If 0.05 ml of 1 percent pyribenzamine was administered, rapid autotomy resulted in 70 percent of the chelae, but no walking legs were dropped, and they could not be made to autotomize by crushing their legs. One-tenth of this dose of pyribenzamine clearly inhibited autotomy of the legs and antagonized the autotomizing effect of *Metridium* tentacular extract. Welsh (1961) concluded that a histamine releaser or 5-hydroxytryptamine, a histaminelike substance, may be a constituent of nematocyst toxins.

Miscellaneous

Payne (1961) has prepared tentacular extracts from *Chironex fleckeri* and assayed them on a rat uterus preparation. She found that an active principle in the extract caused the uterus to contract and remain in spasm for a period up to 2 hours or longer, despite repeated flushing of the organ-bath and forced relaxation of the muscle. The substance was heat labile and deteriorated slowly under moist cold, but it lasted a year or more if lyophilized. It is believed that this smooth muscle contractor is the main toxic principle released by *C. fleckeri* nematocysts and is the cause of the respiratory effects reported by clinicians (Flecker, 1952*a, b*; Southcott, 1963).

Crone and Keen (1969) and Keen and Crone (1969*a, b*) have shown that extracts from the tentacles of *C. fleckeri* contain at least two toxins. The first has only lethal activity and has a molecular weight of 150,000. The second has lethal, hemolytic, and dermonecrotic activity and has a molecular weight of 70,000 (Crone and Keen, 1971). The tentacle extracts and a toxin obtained by a milking technique were compared by chromatography and shown to be very similar if not identical in their biological activities.

Endean *et al.* (1969) isolated five types of nematocysts from *C. fleckeri*. Protein, carbohydrate, cystine-containing compounds and 3-indolyl derivatives were detected in all types of nematocysts. Saline extracts of the contents of nematocysts were highly toxic to prawns and fish and were capable of eliciting fatal systemic effects within seconds if administered to rats and mice intravenously. Saline extracts of toxic material from ruptured nematocysts elicited a powerful contracture of the striated musculature of barnacles and of skeletal, respiratory, and extravascular smooth musculature of rats. Their preliminary results indicated that the *C. fleckeri* nematocyst toxin is antigenic.

Saline extracts of intracapsular material from the nematocysts of *C. fleckeri* lost their activity against muscle if incubated with protease, heated at 42° C, exposed to strong salt solutions, acid solutions of pH 3 and below, alkaline solutions of pH 10.5 and above, iodine, sodium fluoride, sodium iodoacetate or sodium acid. Their activity against muscle was unimpaired in the presence of ouabain or by exposure to EDTA for 5 minutes. Hemolytic activity was exhibited by extracts; the extracts showed no phospholipase A or proteolytic activity (Endean and Henderson, 1969).

Lane *et al.* (1961) have found that *Physalia* nematocyst toxin can be separated into component peptides by one-dimensional chromatography. Isolated frog sciatic-gastrocnemius preparations were used by Larsen and Lane (1970) to investigate the actions of *P. physalis* toxin on the conduction of nerve impulses and on muscle contraction. Segments of sciatic nerves treated for 10 minutes with toxin solutions (1.0 mg/ml) failed to conduct impulses, although stimulation of the nerve distal to the treated segments elicited normal muscle contractions. Direct electrical stimulation of curarized gastrocnemius muscles was ineffective when toxin was present in the Ringer's solution bathing the muscle at a concentration of 1.0 mg/ml.

Heat inactivates the lethal effect of *Chrysaora quinquecirrha* and *Physalia physalis* but has no effect on their ability to interfere with the transfer of ionic calcium within various cellular compartments (Calton and Burnett, 1973*a, b*; Calton *et al.*, 1973). The muscolotoxic effect of these toxins is thought to result from their ability to prevent the uptake of calcium ions into skeletal muscle sarcoplasmic reticulum (Endean, cited by Calton and Burnett, 1973*a*).

Burnett and Goldner (1971) found *Chrysaora* venom to be antigenic to rabbits and guinea pigs, forming complement fixing antibodies and protective factors against lethal, hemolytic, and dermonecrotic factors.

CHEMISTRY

Welsh (1961) has provided a summary of the knowledge to date on the biological properties to be ascribed to the toxins of coelenterates. He regards the following points as established:

1. Of the quaternary bases found in coelenterates, only tetramine has a marked paralyzing action.
2. The production of pain and the release of histamine is due to 5-hydroxytryptamine, which in some coelenterates is present in the largest amounts in regions where nematocysts are most concentrated, though in others it may have a more general occurrence.
3. "The paralyzing and edema-producing actions of coelenterate toxins is due in large measure" to a toxin of protein composition. This does not eliminate the fact that tetramine is in some way related to the paralyzing action observed, but generally this substance is not present in amounts large enough to permit the conclusion that this is a major factor.

The properties of thalassin and congestin are the properties of tetramine and 5-hydroxytryptamine respectively, and need not be reviewed here. The most notable differences of coelenterate toxins from the properties of the toxin of protein nature are the solubilities in alcohol, glycerol, and the stability of heat.

Lane and coworkers have provided most of the more recent information relative to the properties of *Physalia* toxin (Lane and Dodge, 1958; Lane, 1960). The activity of the toxin is decreased markedly by heating to 60° C for 15 minutes, and by precipitation with acetone or by extracting with ether. The toxin is nondialyzable. Both before and after hydrolysis, it gives a positive ninhydrin test but negative Benedict's test. The toxin was found stable for at least 2 months when stored at -5° C. Precipitation of the toxin with alcohol, followed by redissolving in sea water, resulted in some loss of toxicity, but lethal doses on crabs did not produce death until after a 24-hour delay. However, the characteristic action on the legs was noted immediately. Adsorption on paper followed by elution released only the autotomy-producing activity. The lethal dose for mice by a solution of the toxin containing 0.201 percent total nitrogen was 2.1 ml/kilo when injected subcutaneously, and 0.037 ml/kilo when injected intraperitoneally. Death seemed to be due to respiratory failure. A dose of 0.5 ml of crude toxin containing 2.43 μg of nitrogen per ml was uniformly lethal when injected into the left ventral lymph sac of frogs. The crude toxin is strongly biuret-positive. The Molisch test is negative, suggesting the presence of little of no polysaccharide. The crude toxin withstands lyophilization without significant loss of toxicity. It is separable into component peptides by one-dimensional chromatography in 80 percent aqueous *n*-propanol or by electrophoretic methods. The crude toxin isolated by lyophilization assays between 15 and 16 nitrogen. The presence of phospholipase A (phosphatide acyl-hydrolase, EC 3.1.1.4) and phospholipase B (lysolecithin acyl-hydrolase, EC 3.1.1.5) has been demonstrated by Stillway and Lane (1971).

An extensive series of publications on the chemical characteristics of the venom of the sea nettle *(Chrysaora)* have been prepared by Burnett *et al.* (1968), Goldner *et al.* (1969), Stone, Burnett, and Goldner (1970), Burnett and Goldner (1970*b*), Burnett and Gould (1971*a, b*), Calton and Burnett (1973*a, b*), and Burnett and Calton (1974*a, b*).

Palytoxin has been estimated by Moore and Scheuer (1971) to have a formula of $C_{145}H_{264}N_4O_{78}$ and a molecular weight of 3300. Their studies showed that palytoxin contains no repetitive amino acids or sugar units.

Keen and Crone have contributed a series of studies which delineated the hemolytic, dermonecrotic, and chromatographic activity of *C. fleckeri* toxins (Keen and Crone, 1969*a, b*; Crone and Keen, 1969). It was concluded that the hemolytic activity of the extracts was due to a single component, a protein of molecular weight approximately 70,000. At least one other toxic component was present, which had a molecular weight of approximately 150,000.

Blanquet (1972) isolated two toxic proteins from toxin samples of *Chrysaora quinquecirrha*. The most toxic of these was analyzed for its amino acid content and found to contain high concentrations of glutamic and aspartic acid. No tyrosine could be detected. This protein has a molecular weight of at least 100,000.

Burnett and Calton (1974*b*) screened the venoms of the sea nettle *(Chryssaora)* and the Portuguese man-o-war *(Physalis)* for enzyme content.

Courville, Halstead, and Hessel (1958) published a brief résumé of some of the chemical literature dealing with coelenterate poisons; however, Kaiser and Michl (1958) have published the most comprehensive review on the biochemistry of coelenterate poisons. The reader is also referred to the recent work of Scheuer (1973) on coelenterate chemical constituents.

LITERATURE CITED

ACADEMY OF SCIENCES, USSR
1974 Studies on the venomous medusae *Gonionemus vertens vertens*. Trans. Acad. Sci. USSR, Far East. Sci. Cent., Inst. Mar. Biol. No. 2. 65 p., figs.

ACKERMANN, D., HOLTZ, F., and H. REINWEIN
1923 Reindarstellung and konstitutionsermittelung des tetramins, eines giftes aus Aktinia equina, Z., Biol. 79: 113-120.

ANONYMOUS
1924 Fish poisons. U.S. Nav. Med. Bull. 20(4): 466-467.

AOKI, T.
1922 Über medusenstichkrankheit. [In Japanese] Dermatol. Urol. (Fukuoka, Japan) 22(10): 835-891.

AZNAURIAN, M. S.
1964 The venomous Gonionemus jellyfish. [In Russian] Akas. Nauk USSR, Vladivostok. 34 p., 5 figs.

BARI, A. K.
1922 On a problem of the action of medusa's venom. News of the Physicians' Society of the Southern Ussuri, p. 3-9. (NSA)

BARNES, J. H.
1960 Observations on jellyfish stingings in North Queensland. Med. J. Australia 2: 993-999, 4 pls.
1963 Letter dated 7 November, 1963 (personal communication).
1964 Cause and effects of irukandji stingings. Med. J. Australia 1: 897-904, 3 figs.
1965*a* *Chironex fleckeri* and *Chiropsalmus quadrigatus*—morphological distinctions. N. Queensland Nt. 32 (137): 13-22.
1965*b* A diagnostic procedure for marine stings. Recovery of stinging capsules from skin of victims. Roy. Australian Nurs. Med. Newsletter 3(1): 15-16.
1967 Extraction of cnidarian venom from living tentacle, pp. 115-129, 7 figs. *In* F. E. Russell and P. R. Saunders [eds.], Animal toxins. Pergamon Press, New York.
1969 *Chironex fleckeri*. Image 31: 24-29, 11 figs.

BARI, A. K.
1922 On a problem of the action of medusa's venom. *News of the Physicians Society of the Southern Ussuri*, p. 3-9.

BAXTER, E. H., and A. G. MARR
1974 Sea wasp (*Chironex fleckeri*) antivenene: neutralizing potency against the venom of three other jellyfish species. Toxicon 12: 223-229.

BAXTER, E. H., MARR, A. G., and W. R. LANE
1968 Immunity to the venom of the sea wasp *Chironex fleckeri*. Toxicon 6: 45-50.
1973 Sea wasp (*Chironex fleckeri*) toxin —experimental immunity, p. 941-953, 8 figs. *In* A. de Vries and E. Kochva [eds.], Toxins of animal and plant origin, vol. 2. Gordon and Breach Science Publ., New York.

BÉRESS, L., and R. BÉRESS
1971 Reinigung zweier krabbenlähmender toxine aus der seeanemone *Anemonia sulcata*. Kieler Meeresforsch. 27(1): 117-127.

BERGER, E. W.
1900 Physiology and histology of the Cubomedusae. Mem. Biol. Lab. Johns Hopkins Univ. 4(4), 84 p., 3 pls.

BERNTROP, J. C.
1934 Over ziekteverschijnselen veroorzaakt door in de Noordzee voorkomende kwallen. Ned. Tijdschr. Geneesk. 78: 2084-2089.

BINET, L., BURSTEIN, M., and R. LEMAIRE
1951 Les effets hypotenseurs et anaphylactisants des extraits de *Physalie*. Compt. Rend. Acad. Sci. 233(10): 565-566.

BLANQUET, R.
1972 A toxic protein from the nematocysts of the scyphozoan medusa, *Chrysaora quinquecirrha*. Toxicon 10: 103-109.

BOSCHMA, H.
1948 The species problem in *Millepora*. No. 1. Zool. Verhandel. Rijksmus. Natuur. Hist. Leiden, 115 p., 13 figs., 15 pls.

BREKHMAN, I. I.
1969 A review of pharmacological and clinical research of biologically active substances of marine origin in the Soviet Far East (1949-1969), p. 359-367. *In* H. W. Youngken, Jr. [ed.], Food-drugs from the sea, proceedings. Marine Technology Society, Washington, D.C.

BROWN, C. H.
1950 Keratins in invertebrates. Nature 166 (4219): 439.

(NSA)—Not seen by author.

BURKHOLDER, P. R., and L. M. BURKHOLDER
1958 Antimicrobial activity of horny corals. Science 127(3307): 1174.

BURNETT, J. W.
1971*a* An electron microscopic study of two nematocystes in the tentacle of *Cyanea capillata*. Chesapeake Sci. 12(2): 67-71, 7 figs.
1971*b* An ultrastructural study of the nematocytes of the polyp of *Chrysaora quinquecirrha*. Chesapeake Sci. 12(4): 225-230, 3 figs.

BURNETT, J. W., and G. J. CALTON
1973 Purification of sea nettle nematocyst toxins by gel diffusion. Toxicon 11: 243-247.
1974*a* Sea nettle and man-o-war venoms: a chemical comparison of their venoms and studies on the pathogenesis of the sting. J. Invest. Derm. 62: 372-377.
1974*b* The enzymatic content of the venoms of the sea nettle and the Portuguese man-o-war. Comp. Biochem. Physiol. 47B: 815-820.

BURNETT, J. W., CALTON, G. J., and D. G. CARGO
1972 Recent investigations on the nature and action of sea nettle toxins. Derm. Proc. XIV Internat. Congr. (289): 768-769.

BURNETT, J. W., and R. GOLDNER
1969 Effects of *Chrysaora quinquecirrha* (sea nettle) toxin on the rat cardiovascular system. Proc. Soc. Exp. Biol. Med. 132(1): 353-356.
1970*a* Effect of *Chrysaora quinquecirrha* (sea nettle) toxin on rat nerve and muscle. Toxicon 8: 179-181.
1970*b* Partial purification of sea nettle (*Chrysaora quinquecirrha*) nematocyst toxin. Proc. Soc. Exp. Biol. Med. 133: 978-981.
1971 Some immunological aspects of sea nettle toxins. Toxicon 9: 271-277.

BURNETT, J. W., and W. M. GOULD
1971*a* Immunodiffusion—a technique for coelenterate polyp identification. Comp. Biochem. Physiol. 40A: 855-857.
1971*b* Further studies on the purification and physiological actions of sea nettle toxin. Proc. Soc. Exp. Biol. 138: 759-762.

BURNETT, J. W., PIERCE, L. H., NAWACHINDA, U., and J. H. STONE
1968 Studies on sea nettle stings. Arch. Dermatol. 98: 587-589, 1 fig.

BURNETT, J. W., STONE, J. H., PIERCE, L. H., CARGO, D. G., LAYNE, E. C., and J. S. SUTTON
1968 A physical and chemical study of sea nettle nematocysts and their toxin. J. Invest. Dermatol. 51(5): 330-336.

BURNETT, J. W., and J. S. SUTTON
1969 The fine structural organization of the sea nettle fishing tentacle. J. Exp. Zool. 172(3): 335-348, 10 figs.

CALMETTE, A.
1908 Venoms; venomous animals and antivenomous serum-therapeutics. John Bale, Sons & Danielsson, Ltd. London. 403 p., 125 figs.

CALTON, G. J., and J. W. BURNETT
1973*a* Sea nettle nematocysts: anatomy, toxicology, and chemistry. Proc. Food and Drugs from the sea symposium, Mar. Tech. Soc.: 147-270.
1973*b* The purification of Portuguese man-o-war nematocyst toxins by gel diffusion. Comp. Gen. Pharmacol. 4(15): 267-270.

CALTON, G. J., BURNETT, J. W., GARBUS, J., and S. R. MAX
1973 The effects of *Chrysaora* and *Physalia* venoms on mitochrondrial structure. Proc. Soc. Exp. Biol. Med. 143: 971-977, 4 figs.

CANTACUZÈNE, J.
1925*a* Immunité d'*Eupagurus prideauxii*, vis-à-vis des poisons de l'*Adamsia palliata*. Compt. Rend. Soc. Biol. 92: 1133-1136.
1925*b* Action toxique des poisons d'*Adamsia palliata* sur les crustacés décapodes. Compt. Rend. Soc. Biol. 92: 1131-1133.
1926 Activation des poisons de l'*Adamsia palliata* par la lécithine et pouvoir hémolytique. Compt. Rend. Sco. Biol. 95: 118-120.

CANTACUZÈNE, J., and N. COSMOVICI
1925 Action toxique des poisons d'*Adamsia palliata* sur divers invertébres marins. Compt. Rend. Soc. Biol. 92: 1464-1466.

Cantacuzène, J., and A. Damboviceanu
1934*a* Caractères biologiques de l'extrait des acconties d'*Adamsia palliata* après déprotéinisation. Compt. Rend. Soc. Biol. 117: 136-138.
1934*b* Caractères physio-chimiques du poison des acconties d'*Adamsia palliata*. Compt. Rend. Soc. Biol. 117: 138-140.

Carlgren, O.
1940 A contribution to the knowledge of the structure and distribution of the Cnidae in the Anthozoa. Lunds Univ. Arsskr. 36(3), 62 p.
1945 Further contributions to the knowledge of the cnidom in the Anthozoa especially in the Actiniaria. Lunds Univ. Arsskr. 41(9), 24 p.

Chu, G. W., and C. E. Cutress
1953 Human dermatitis caused by marine organisms in Hawaii. Proc. Hawaiian Acad. Sci. 1953: 9.

Cleland, J. B.
1912 Injuries and diseases of man in Australia attributable to animals (except insects). Australasian Med. Gaz. 32: 269-272.
1913 Injuries and diseases of man in Australia attributable to animals (except insects). J. Trop. Med. Hyg. 16: 25-31.
1932 Injuries and diseases in Australia attributable to animals (other than insects). Med. J. Australia 4: 157-166.

Clemens, W. A.
1922 Hydra in Lake Erie. Science 55 (1426): 445-446.

Conant, F. S.
1898 The Cubomedusae. Mem Biol. Lab. Johns Hopkins Univ. 4(1), 61 p., 8 pls.

Cooke, T. S., and B. W. Halstead
1970 Report of stingings by the coelenterate *Rhizophysa eysenhardti* Gegenbaur in California waters. Clin. Toxicol. 3(4): 589-594, 5 figs.

Cosmovici, N. L.
1925*a* L'action des poisons d'*Adamsia palliata* sur les muscles de *Carcinus moenas*. Compt. Rend. Soc. Biol. 92: 1230-1232, 1 fig.
1925*b* L'action des poisons d'*Adamsia palliata* sur le coeur de *Carcinus moenas*. Compt. Rend. Soc. Biol. 92: 1300-1302.
1925*c* Action convulsivante des poisons d'*Adamsia palliata* sur le *Carcinus moenas*. Compt. Rend. Soc. Biol. 92: 1466-1469, 4 figs.
1925*d* Autotomie chez *Carcinus moenas*, provoquée par les poisons d'*Adamsia palliata*. Compt. Rend. Soc. Biol. 92: 1469-1470.
1925*e* Les poisons de l'extrait aqueux des tentacules et des nématocystes d'*Adamsia palliata* sont-ils détruits par l'ébullition? Essais d'adsorption. Compt. Rend. Soc. Biol. 92: 1373-1374.

Courville, D. A., Halstead, B. W., and W. Hessel
1958 Marine biotoxins: isolations and properties. Chem. Rev. 58(2): 235-248.

Crone, H. D., and T. E. B. Keen
1969 Chromatographic properties of the hemolysin from the cnidarian *Chironex fleckeri*. Toxicon 7: 79-87.
1971 Further studies on the biochemistry of the toxins from the sea wasp *Chironex fleckeri*. Toxicon 9: 145-151.

Cutress, C. E.
1955 An interpretation of the structure and distribution of Cnidae in Anthozoa. Systematic Zool. 4(3): 120-137, 10 figs.
1962 Regarding venomous coelenterates. (Personal communication, Dec. 28, 1962).

Dujarric de la Rivière, R.
1915 Sur l'existence d'une médusocongestine. Compt. Rend. Soc. Biol. 78: 596-600.

Endean, R., and L. Henderson
1969 Further studies of toxic material from nematocysts of the cubomedusan *Chironex fleckeri* Southcott. Toxicon 7: 303-314.

Endean, R., Duchemin, C., McColm, D., and E. H. Fraser
1969 A study of the biological activity of toxic material derived from nematocysts of the cubomedusan *Chironex fleckeri*. Toxicon 6: 179-204, 19 figs.

Evans, H. M.
1943 Sting-fish and seafarer. Faber and Faber, Ltd., London. 180 p., 31 figs.
1945 Toxic properties of sting-ray's sting. Brit. Med. J. 4413: 165.

FARBER, L., and P. LERKE
1963 Observations on the toxicity of *Rhodactis howesi* (matamalu), p. 67-74, 3 figs. *In* H. L. Keegan and W. V. MacFarlane [eds.], Venomous and poisonous animals and noxious plants of the Pacific region. Pergamon Press, New York.

FAUST, E. S.
1924 Coelenterata (zoophyta), pflanzentiere, p. 1925-1928. *In* E. S. Faust, Handbuch Exp. Pharm. Berlin.

FERLAN, L., and D. LEBEZ
1974 Equinatoxin, a lethal protein from *Actinia equina*, 1. Toxicon 12: 57-61.

FISH, C. J., and M. C. COBB
1954 Noxious marine animals of the central and western Pacific Ocean. U.S. Fish Wildlife Serv., Res. Rept. No. 36, p. 17-20.

FLECKER, H.
1945*a* Injuries by unknown agents to bathers in North Queensland. Med. J. Australia 1: 417.
1945*b* Injuries by unknown agents to bathers in North Queensland. Med. J. Australia 1(4): 98.
1952*a* Fatal stings to North Queensland bathers. Med. J. Australia 1(2): 35-38.
1952*b* Irukandji sting to North Queensland bathers without production of weals but with severe general symptoms. Med. J. Australia 2: 89-91.
1952*c* Fatal stings to North Queensland bathers. Med. J. Australia 1: 458.
1957*a* Further notes on Irukandji stings. Med. J. Australia 1: 9.
1957*b* Injuries produced by marine organisms in tropical Australia. Med. J. Australia 2: 556.

FRACHTMAN, H. J., and W. T. MCCOLLUM
1945 Portuguese man-of-war stings: a case report. Am. J. Trop. Med. Hyg. 25(6): 499-500.

FRASER-BRUNNER, A.
1973 Danger in the sea. Hamlyn Publ. Group Ltd., New York. 128 p., illus.

FREEMAN, S. E.
1974 Actions of *Chironex fleckeri* toxins on cardiac transmembrane potentials. Toxicon 12: 395-404.

FREEMAN, S. E., and R. J. TURNER
1969 A pharmacological study of the toxin of a cnidarian, *Chironex fleckeri* Southcott. Brit. J. Pharmacol. 35: 510-520.
1972 Cardiovascular effects of cnidarian toxins: a comparison of toxins extracted from *Chiropsalmus quadrigatus* and *Chironex fleckeri*. Toxicon 10: 31-37.

GIUNIO, P.
1948 Otrovne ribe. Higijena (Belgrade) 1(7-12): 282-318, 20 figs.

GLASER, O. C., and C. M. SPARROW
1909 The physiology of nematocysts. J. Exp. Zool. 6(3): 361-382.

GOLDNER, R., BURNETT, J. W., STONE, J. S., and M. S. DILAIMY
1969 The chemical composition of sea nettle nematocysts. Proc. Soc. Exp. Biol. Med. 131(4): 1386-1388.

GOULD, W. M., and J. W. BURNETT
1971 Effects of *Chrysaora quinquecirrha* (sea nettle) toxin on sodium transport across frog skin. J. Invest. Dermatol. 57(4): 266-268.

GRAHAM, S. A., and E. C. O'ROKE
1943 Animals you should avoid eating, p. 56-57; Dangerous aquatic and marine animals, p. 70-73. *In* Graham and O'Roke, On your own. How to take care of yourself in wild country. A manual for field and service men. Univ. Minnesota Press, Minneapolis.

GRUPPER, C., and R. DAVID
1958 Dermatite provoquée par la méduse réaction d'abord hypertrophique, puis atrophique, persistante. Bull. Soc. Franc. Dermatol. Syphil. 65: 57-59.

GUÉRIN, P.
1900 De la toxicité des *Physalies*. Ann. Hyg. Med. Colon. 3: 265-271, 1 fig.

GUNN, M. A.
1949 Localized fat atrophy after jellyfish sting. Brit. Med. J. 2: 687, 1 fig.

HADLEY, H. G.
1941 The sea nettle. Med. Ann. District Columbia 10(5): 178-180.

HALSTEAD, B. W.
1957 Jellyfish stings and their medical management. U.S. Armed Forces Med. J. 8(11): 1587-1602, 11 figs.
1959 Dangerous marine animals. Cornell Maritime Press, Cambridge. 146 p. 88 figs.

HANSEN, P. A., and B. W. HALSTEAD
1971 The venomous sea anemone *Actinodendron plumosum* Haddon of South Vietnam. Micronesia 7(1-2): 123-136, 5 figs.

HASHIMOTO, Y., and K. ASHIDA
1973 Screening of toxic corals and isolation of a toxic polypeptide from *Goniopora* spp. Publ. Seto Mar. Biol. Lab. 20: 703-711.

HASHIMOTO, Y., FUSETANI, N., and S. KIMURA
1969 Aluterin: a toxin of filefish, *Aleutera scripta*, probably originating from a zoantharian, *Palythoa tuberculosa*. Bull. Jap. Soc. Sci. Fish. 35(11): 1086-1093, 5 figs.

HERCUS, J. P.
1944 An unusual eye condition. Med. J. Australia 1: 98-99.

HESSINGER, D. A., and H. M. LENHOFF
1974 Degradation of red cell membrane phospholipids by sea anemone nematocyst venom. Toxicon 12: 379-383.

HUANG, C. L., and G. N. MIR
1971 Toxicological and pharmacological properties of sea pansy, *Renilla mulleri*. J. Pharm. Sci. 60(11): 1620-1622.

HYMAN, L. H.
1940 The invertebrates: Protozoa through Ctenophora. Vol. I. McGraw-Hill Book Co., New York. 726 p., 221 figs.

JAQUES, R., and M. SCHACHTER
1954a The presence of histamine, 5-hydroxytryptamine and a potent, slow-contracting substance in wasp venom. Brit. J. Pharmacol. 9: 53-58.
1954b A sea anemone extract (thalassine) which liberates histamine and a slow-contracting substance. Brit. J. Pharmacol. 9: 49-52. 2 figs.

KAISER, E., and H. MICHL
1958 Die biochemie der tierischen gifte. Franz Deuticke, Wien. 258 p., 23 figs.

KEEGAN, H. L.
1960 Some venomous and noxious animals of the Far East. Med. Gen. Lab. (406), U.S. Army Med. Command, Japan. 46 p., 70 figs.

KEEN, T. E. B.
1971 Comparison of tentacle extracts from *Chiropsalmus quadrigatus* and *Chironex fleckeri*. Toxicon 9: 249-254.
1973 Interaction of the hemolysin of *Chironex fleckeri* tentacle extracts with lipid monolayers. Toxicon 11: 293-299.

KEEN, T. E. B., and H. D. CRONE
1969a Dermatonecrotic properties of extracts from the tentacles of the cnidarian *Chironex fleckeri*. Toxicon 7: 173-180.
1969b The hemolytic properties of extracts of tentacles from the cnidarian *Chironex fleckeri*. Toxicon 7: 55-63.

KEPNER, W. A., REYNOLDS, B. D., GOLDSTEIN, L., BRITT, G., ATCHESON, E. ZIELINSKI, Q., and M. B. RHODES
1951 The discharge of nematocysts of Hydra, with special reference to the penetrant. J. Morphol. 88: 23-47, 4 figs., 2 pls.

KIMURA, S., HASHIMOTO, Y., and K. YAMAZATO
1972 Toxicity of the zoanthid *Palythoa tuberculosa*. Toxicon 10: 611-617.

KINGSTON, C. W., and R. V. SOUTHCOTT
1960 Skin histopathology in fatal jellyfish stinging. Trans. Roy. Trop. Med. Hyg. 54(4): 373-384, 12 figs.

KITAGAWA, K.
1943 A case of dermatitis probably due to poison from eating jellyfish. [In Japanese] Acta Dermatol. (Kyoto) 41(4): 266.

KLEINHAUS, A. L., CRANEFIELD, P. F., and J. W. BURNETT
1973 The effects on canine cardia Purkinje fibers of *Chrysaora quinquecirrha* (sea nettle) toxin. Toxicon 11: 341-349.

KRAMP, P. L.
1961 Synopsis of the medusae of the world. J. Marine Biol. Assoc. U.K. Vol. 40, 469 p.

LAKSO, J. U.
1965 A study of the variability in toxicity of extracts made from the coelenterate *Anthopleura elegantissima* (Brandt). Thesis, Sacramento State College, 112 p.

LAKSO, J. U., and E. J. MARTIN
1972 Toxic stability of the anemone *Anthopleura elegantissima* (coelenterata) through annual seasons. Proc. West. Pharmacol. Soc. 15: 47-51.

LANE, C. E.
1960 The toxin of *Physalia* nematocysts. Ann. N.Y. Acad. Sci. 90(3): 742-750, 5 figs.
1961*a* Observations on the general biology of *Physalia*. Am. Zool. 1(3): 49.
1961*b* *Physalia* nematocysts and their toxin, p. 169-178, 3 figs. *In* H. M. Lenhoff and W. F. Loomis [eds.], The biology of Hydra and of some other coelenterates: 1961. University Miami Press, Coral Gables, Fla.

LANE, C. E., COURSEN, B. W., and K. HINES
1961 Biologically active peptides in *Physalia* toxin. Proc. Soc. Exp. Biol. Med. 107: 670-672, 1 fig.

LANE, C. E., and E. DODGE
1958 The toxicity of *Physalia* nematocysts. Biol. Bull. 115(2): 219-226, 2 figs.

LANE, C. E., and J. B. LARSEN
1965 Some effects of the toxin of *Physalia physalis* on the heart of the land crab *Cardisoma guanhumi* (Latreille). Toxicon 3: 69-71.

LARSEN, J. B., and C. E. LANE
1970 Direct action of *Physalia* toxin on frog nerve and muscle. Toxicon 8: 21-23.

LEVIN, O. L., and H. T. BEHRMAN
1941 Coral dermatitis. Arch. Dermatol. Syphilol. 44: 600-603.

LEVY, S., MASRY, D., and B. W. HALSTEAD
1970 Report of stingings by the sea anemone *Triactis producta* Klunzinger from Red Sea. Clin. Toxicol. 3(4): 637-643, 4 figs.

LIGHT, S. F.
1914*a* Another dangerous jellyfish in Philippine waters. Philippine J. Sci. 9: 291-295.
1914*b* Some Philippine Scyphomedusae, including two new genera, five new species, and one new variety. Philippine J. Sci. 9(3): 195-231, 16 figs.

MAASS, T. A.
1937 Gift-tiere. *In* W. Junk [ed.], Tabulae biologicae. Vol. XIII. N.V. Van de Garde & Co's Drukkerij, Zaltbommel, Holland. 272 p.

MACHADO, O.
1943 Catálogo sistemático dos animais urticantes e peconhentos do Brasil. Bol. Inst. Vital Brazil (25): 41-65.

MCNEILL, F. A.
1945 Injuries by unknown agents to bathers in North Queensland. Med. J. Australia 2(1): 29.

MCNEILL, F. A., and E. C. POPE
1943*a* A venomous medusa from Australian waters. Australian J. Sci. 5(6): 188-191, 2 figs.
1943*b* A deadly poisonous jellyfish. Australian Mus. Mag. 8(4): 127-131, 5 figs.
1945 Injuries by unknown agents to bathers in North Queensland. Med. J. Australia 1: 334-335.

MANSUETI, R.
1963 Symbiotic behavior between small fishes and jellyfishes, with new data on that between the stromateid, *Peprilus alepidotus*, and the scyphomedusa, *Chrysaora quinquecirrha*. Copeia (1): 40-80, 5 figs.

MARTIN, E. J.
1960 Observations on the toxic sea anemone *Rhodactis howesii* (Coelenterata). Pacific Sci. 14(4): 403-407.
1962 On the toxicity of sea anemone extracts. (Personal communication, Jan. 11, 1962.)
1963 Toxicity of dialyzed extracts of some California anemones (Coelenterata). Pacific Sci. 17: 302-304.
1966*a* The macromolecular toxin of sea anemones (Coelenterata). Proc. Galapagos Internat. Sci. Proj. 1966: 136-138.
1966*b* Anticoagulant from the sea anemone *Rhodactis howesii*. Proc. Soc. Exp. Biol. Med. 121: 1063-1065.
1968 Specific antigens released into sea water by contracting anemones (Coelenterata). Comp. Biochem. Physiol. 25: 169-176, 2 figs.

MATHIAS, A. P., ROSS, D. M., and M. SCHACHTER
1957 Identification and distribution of 5-hydroxytryptamine in a sea anemone. Nature 180: 658-659.
1958 Distribution of histamine, 5-hydroxytryptamine, tetramethylammonium, and other substances in coelenterates possessing nematocysts. J. Physiol. 142: 56-57.

MAYER, A. G.
1910 Medusae of the world. 3 vols. Carnegie Inst. Washington, Publ. No. 109. 735 p., 428 figs., 76 pls.

MEBS, D.
1973 Chemistry of animal venoms, poisons and toxins. Experientia 29(11): 1328-1334.

MITCHELL, J. H.
1962 Eye injuries due to jellyfish (*Cyanea annaskala*). Med. J. Australia 2(8): 303-305, 5 figs.

MOEBIUS, K.
1866 Über den bau, den mechanismus und die entwicklung der nesselkapseln einiger polypen und quallen. Abhandl. Naturw. Ver. Hamburg 5(1): 1-22, 2 pls.

MOORE, R. E., and P. J. SCHEUER
1971 Palytoxin: a new marine toxin from a coelenterate. Science 172: 495-498, 4 figs.

NARAHASHI, T., MOORE, J. W., and B. I. SHAPIRO
1969 Condylactis toxin: interaction with nerve membrane ionic conductances. Science 163: 680-681.

NOBRE, A. F.
1928 Animals venenosos de Portugal. [In Portuguese] Inst. Zool Univ. Pôrto. Vol. I p. 1-8.

NOVAK, V., SKET, D., CANKAR, G., and D. LEBEZ
1973 Partial purification of a toxin from tentacles of the sea anemone *Anemone sulcata*. Toxicon 11: 411-417.

ODUM, H. T., and E. P. ODUM
1955 Trophic structure and productivity of a windward coral reef community Eniwetok Atoll. Ecol. Monographs 25: 291-320, 12 figs.

PANTIN, C. F.
1942 The excitation of nematocysts. J. Exp. Biol. 19(3): 294-310, 3 figs.

PARKER, G. H.
1932 Neuromuscular activities of the fishing filaments of *Physalia*. J. Cellular Comp. Physiol. 1: 53-63, 2 figs.

PARKER, G. H., and M. A. VAN ALSTYNE
1932 The control and discharge of nematocysts, especially in *Metridium* and *Physalia*. J. Exp. Zool. 63(2): 329-344, 2 figs.

PAWLOWSKY, E. N.
1927 Gifttiere und ihre giftigkeit. Gustav Fischer, Jena. 515 p., 170 figs.

PAWLOWSKY, E. N., and A. K. STEIN
1929 Experimentelle untersuchung über die wirkung des actiniengiftes (*Actinia equina*) auf die menschenhaut. Arch. Dermatol. Syphil. 157: 647-656, 5 figs.

PAYNE, J. H.
1961 Cubomedusae in northern Australian waters. Proc. Pacific Sci. Congr. Pacific Sci. Assoc., 10th, Honolulu (Unpublished.)

PHILLIPS, C., and W. H. BRADY
1953 Sea pests—poisonous or harmful sea life of Florida and the West Indies. Univ. Miami Press, Miami, Fla. 78 p., 38 figs., 7 pls.

PHILLIPS, J. H.
1956 Isolation of active nematocysts of *Metridium senile* and their chemical composition. Nature 178: 932.

PHILLIPS, J. H., and D. P. ABBOTT
1957 Isolation and assay of the nematocyst toxin of *Metridium senile fimbriatum*. Biol. Bull. 113: 296-301.

PHILLIPS, P. J., and W. D. BURKE
1970 The occurrence of sea wasps (Cubomedusae) in Mississippi Sound and the northern gulf of Mexico. Bull. Mar. Soc. 20(4): 853-859.

PHISALIX, M.
1922 Animaux venimeux et venins. Masson & Cie., Paris. 656 p., 232 figs.

PICKEN, L. E.
1953 A note on the nematocysts of *Corynactis viridis*. Quart. J. Microscop. Sci. 94(3): 203-227, 9 figs., 2 pls.
1957 Stinging capsules and designing nature. New Biol. 22: 56-71, 3 figs., 7 pls.

PIGULEVSKY, S. V., and P. V. MICHALEFF
1969 Poisoning by the medusa *Gonionemus vertens* in the Sea of Japan. Toxicon 7: 145-149, 1 fig.

POPE, E. C.
1947 Some sea animals that sting and bite. Australian Mus. Mag. 9(5): 164-168, 5 figs.
1953*a* Sea lice or jellyfish? Australian Mus. Mag. 11(1): 16-21, 5 figs.
1953*b* Marine stingers. Australian Mus. Mag. 11(4): 111-115, 4 figs.
1963 Some noxious marine invertebrates from Australian seas, p. 91-102, 2 figs. *In* J. W. Evans [chmn.], Proceedings: first international convention on life saving techniques. Part III. Scientific Sect. Suppl. Bull. Post Grad. Comm. Med., Univ. Sydney.

PORTIER, P., and C. RICHET
1902*a* Sur les effets physiologiques du poison des filaments pêcheurs et des tentacules des coelentérés (hypnotoxine). Compt. Rend. Acad. Sci. 134: 247-248.

PORTIER, P., and C. RICHET—Continued
1902*b* De l'action anaphylactique de certains venins. Compt. Rend. Soc. Biol. 54: 170-172.
1902*c* Nouveaux faits d'anaphylaxie, où sensibilisation aux venins par doses réitérés. Compt. Rend. Soc. Biol. 54: 548-551.

PUFFER, H. W., MCINTOSH, M. E., and R. R. GIVEN
1970 Preliminary investigations of toxic material contained in the hydroid *Lytocarpus nuttingi*. Proc. West. Pharmacol. 13: 120-126.

RICE, N. E., and W. A. POWELL
1970 Observations on three species of jellyfishes from Chesapeake Bay with special reference to their toxin, 1. *Chrysaora (Dactylometra) quinquecirrha*. Biol. Bull. 139(1): 180-187.

RICHET, C.
1902*a* Des effets anaphylactiques de l'actinotoxine sur la pression artérielle. Compt. Rend. Soc. Biol. 54: 837-838.
1902*b* Du poison pruritogène et urticant contenu dans les tentacules des actinies. Compt. Rend. Soc. Biol. 54: 1438-1440.
1902*c* Phénomènes d'urtication produits par le poison des actinies. Semaine Méd. 22(51): 418.
1903*a* Des poisons contenus dans les tentacules des actinies (congestine et thalassine). Compt. Rend. Soc. Biol. 55: 246-248.
1903*b* De la thalassine, toxine cristallisée pruritogène. Compt. Rend. Soc. Biol. 55: 707-710.
1903*c* De la thalassine, considérée comme antitoxine cristallisée. Compt. Rend. Soc. Biol. 55: 1071-1073.
1904*a* Des effets prophylactiques de la thalassine et anaphylactiques de la congestine dans le virus des actinies. Compt. Rend. Soc. Biol. 56: 302-303.
1904*b* Nouvelles expériences sur les effets prophylactiques de la thalassine. Compt. Rend. Soc. Biol. 56: 775-777.
1904*c* De la thalassine pruritogène chez les crevettes *(Crangon)*. Compt. Rend. Soc. Biol. 56: 777-778.
1905*a* De l'action de la congestine (virus des actinies) sur les lapins et de ses effets anaphylactiques. Compt. Rend. Soc. Biol. 58: 109-112.
1905*b* De l'anaphylaxie après injections de congestine, chez le chien. Compt. Rend. Soc. Biol. 58: 112-115.
1905*c* Notizen über thalasin. Arch. Physiol. (Pflügers) 108:369-388.
1908 De la substance anaphylactisante où toxogénine. Compt. Rend. Soc. Biol. 64: 846-848.

RICHET, C., PERRET, A., and P. PORTIER
1902 Des propriétés chimiques et physiologiques du poison des actinies (actinotoxine). Compt. Rend. Soc. Biol. 54: 788-790.

RICHET, C., and P. PORTIER
1936 Recherches sur la toxine des coelentérés et les phénomènes d'anaphylaxie. Rés. Camp. Sci., Prince Albert I., Monaco (95), 23 p.

SAVILLE-KENT, W.
1900 The Great Barrier Reef of Australia. Its products and potentialities.

SCHEUER, P. J.
1964 The chemistry of toxins isolated from some marine organisms. Fortsch. Chem. Organ. Naturst. 22: 265-278.
1972 Recent developments in the chemistry of marine toxins. Naturwissenschaften 59: 545-556.
1973 Chemistry of marine natural products. Academic Press, New York. 201 p.

SCHULZE, P.
1922 Der bau und die entladung der penetrante bei *Hydra attenuata* Pallas. Arch. Zellforsch. 16: 383-438, 26 figs., 1 pl.

SCOTT, H. H.
1921 Vegetal and fish poisoning in the tropics, p. 790-798, 5 figs., 2 pls. *In* W. Byam and R. G. Archibald [eds.] The practice of medicine in the tropics. Vol. I. Henry Frowde and Hodder & Stoughton, London.

SKET, D., DRAŠLAR, K., and FERLAND, I., and D. LEBEZ
1974 Equinatoxin, a lethal protein from *Actinia equina*, 2. Pathophysiological action. Toxicon 12: 63-68.

SONDERHOFF, R.
1936 Über das gift der seeanemonen, I. Ein beitrag zur kenntnis der nesselgifte. Ann. Chem. 525(2-3): 138-150.

SOUTHCOTT, R. V.
1952 Fatal stings to North Queensland bathers. Med. J. Australia 1(18): 272-273, 1 fig.

SOUTHCOTT, R. V.—Continued
1956 Studies on Australian Cubomedusae, including a new genus and species apparently harmful to man. Australian J. Marine Freshwater Res. 7(2): 254-280, 3 pls., 23 figs.
1958*a* South Australian jellyfish. S. Australian Nat. 32(4): 53-59, 2 figs.
1958*b* The Cubomedusae—lethal jellyfish. Discovery 19: 282-285, 7 figs.
1959 Tropical jellyfish and other marine stingings. Military Med. 124(8): 569-579, 6 figs.
1960 Venomous jellyfish. Good Health (113): 18-23, 4 figs.
1963 Coelenterates of medical importance, p. 41-65, 6 figs. *In* H. L. Keegan and W. V. MacFarlane [eds.], Venomous and poisonous animals and noxious plants of the Pacific region. Pergamon Press, New York.
1967 Revision of some carybdeidae (Scyphozoa: cubomedusae) including a description of the jellyfish responsible for the "Irukandji syndrome." Australian J. Zool. 15: 651-671, 7 pls., 11 figs.
1970 Human injuries from invertebrate animals in the Australian seas. Clin. Toxicol. 3(4): 617-636, 13 figs.

STONE, J. H., BURNETT, J. W., and R. GOLDNER
1970 The amino acid content of sea nettle *(Chrysaora quinquecirrha)* nematocysts. Comp. Biochem. Physiol. 33: 707-710.

STILLWAY, L. W., and C. E. LANE
1971 Phospholipase in the nematocyst toxin of *Physalia physalis*. Toxicon 9: 193-195.

STRONG, R. P.
1944 Poisonous arthropods, fish and coelenterates, p. 1544-1551, 1 fig. *In* R. P. Strong, Stitt's diagnosis, prevention, and treatment of tropical diseases. Vol. II. Blakiston Co., Philadelphia.

STUART, M. A., and T. D. SLAGLE
1943 Jellyfish stings, suggested treatment, and report on two cases. U.S. Nav. Med. Bull. 41(2): 497-501.

SUTTON, J. S., and J. W. BURNETT
1969 A light and electron microscopic study of nematocytes of *Chrysaora quinquecirrha*. J. Ultrastructure Res. 28: 214-234.

TAFT, C. H.
1945 Poisonous marine animals. Texas Rept. Biol. Med. 3(3): 339-352.

THIEL, M. E.
1935 Über die wirkung des nesselgiftes der quallen auf den menschen. Ergeb. Fortschr. Zoo. 8: 1-35, 9 figs.

THOMPSON, T. E., and I. BENNETT
1969 *Physalia* nematocysts: utilized by mollusks for defense. Science 166: 1532-1533, 2 figs.

TREMBLEY, A.
1744 Mémoires pour servir à l'histoire d'un genre de polypes d'eau douce, à bras en forme de cornes. J. and H. Verbeck, Leiden. 324 p., 13 pls.

TURNER, R. J., and S. E. FREEMAN
1969 Effects of *Chironex fleckeri* toxin on the isolated perfused guinea pig heart. Toxicon 7: 277-286.

TWEEDIE, M. W.
1941 Poisonous animals of Malaya. Malaya Publishing House, Singapore. 90 p., 29 figs.

UVNÄS, B.
1960 Mechanism of action of a histamine-liberating principle in jellyfish *(Cyanea capillata)*. Ann. N. Y. Acad. Sci. 90(3): 751-759, 7 figs.

VAUGHAN, T. W., and J. W. WELLS
1943 Revision of the suborders, families, and genera of the Scleractinia. Geol. Soc. Am., Spec. Pap. No. 44, 363 p., 24 figs., 51 pls.

WADE, H. W.
1928 Post-mortem findings in acute jellyfish poisoning with sudden death in status lymphatiocus. Am. J. Trop. Med 8(3): 233-241.

WAITE, C. L.
1951 Medical problems of an underwater demolition team. U.S. Armed Forces Med. J. 2(7): 1325-1326.

WALSH, G. E., and R. L. BOWERS
1971 A review of Hawaiian zoanthids with descriptions of three new species. Zool. J. Linnean Soc. 50(2): 161-180, 8 figs., 8 pls.

WANGERSKY, E. D., and C. E. LANE
1960 Interaction between the plasma of the loggerhead turtle and toxin of the Portuguese Man-of-War. Nature 185 (4709): 330-331.

WEILL, R.
1934 Contribution à l'étude des cnidaires et de leurs nématocystes. Vols. 10-11. Trav. Stat. Zool. Wimbereux. 347 p., 427 figs.

WELLS, J. W.
1957a Coral reefs. Geol. Soc. Am., Mem. 67, (1): 609-631, 2 figs., 9 pls.
1957b Corals. Geol. Soc. Am., Mem. 67, (1): 1087-1104, 1 fig.

WELSH, J. H.
1956 On the nature and action of coelenterate toxins. Deep Sea Res., Suppl. 3: 287-297.
1961 Compounds of pharmacological interest in coelenterates, p. 179-186. *In* H. M. Lenhoff and W. F. Loomis [eds.], The biology of Hydra and of some other coelenterates: 1961. Univ. Miami Press, Coral Gables, Fla.

WELSH, J. H., and H. H. HASKIN
1939 Chemical mediation in crustaceans; III. Acetylcholine and autotomy in *Petrolisthes armatus* (Gibbes). Biol. Bull. 76: 405-415. (NSA)

WELSH, J. H., and M. MOORHEAD
1959 Identification and assay of 5-hydroxytryptamine in molluscan tissues by fluorescence method. (Abstr.) Science 129: 1491-1492.

WELSH, J. H., and P. B. PROCK
1958 Quaternary ammonium bases in the coelenterates. Biol. Bull. 115(3): 551-561, 2 figs.

WESTFALL, J. A.
1970 The nematocyte complex in a hydromedusan *Gonionemus vertens*. Z. Zellforsch. 110: 457-470, 10 figs.

WHITE, R. P.
1934 The dermatergoses or occupational affections of the skin. H. K. Lewis and Co., Ltd., London. p. 404-407.

YANAGITA, T. M.
1959a Physiological mechanism of nematocyst responses in sea-anemone, I. Effects of trypsin and thioglycolate upon the isolated nematocysts. Japan. J. Zool. 12(3): 360-375, 7 figs.
1959b Physiological mechanism of nematocyst responses in sea-anemone, II. Effects of electrolyte ions upon the isolated cnidae. J. Fac. Sci., Univ. Tokyo, Sect. IV, 8(3): 381-400, 10 figs.
1959c Physiological mechanism of nematocyst responses in sea-anemone, VII. Extrusion of resting cnidae—its nature and its possible bearing on the normal nettling response. J. Exp. Biol. 36(3): 478-479, 5 figs.
1960a Physiological mechanism of nematocyst responses in sea-anemone, III. Excitation and anaesthetization of the netting response system. Comp. Biochem. Physiol. 1: 123-139, 3 figs.
1960b Physiological mechanism of nematocyst responses in sea-anemone, IV. Effects of surface-active agents on the cnidae *in situ* and in isolation. Comp. Biochem. Physiol. 1: 140-154, 2 figs.
1960c The physiological mechanism of nematocyst responses in sea-anemone, V. The effects of lipoid solvents on the cnidae *in situ* and in isolation. Annotationes Zool. Japon. 33(4): 203-210.

YANAGITA, T. M., and T. WADA
1953 Discharge-inducing concentrations of acids and bases for the nematocysts of anemone. Nat. Sci. Rept. Ochanomizu Univ., Tokyo 4(1): 112-118, 4 figs.
1954 Effects of trypsin and thioglycollate upon the nematocysts of the sea anemone. Nature 173(4395): 171, 4 figs.
1959 Physiological mechanism of nematocyst responses in sea-anemone. VI. A note on the microscopical structure of acontium, with special reference to the situation of cnidae within its surface. Cytologia (Tokyo) 24: 81-97, 7 figs.

ZERVOS, S. G.
1903 La maladie des pêcheurs d'éponges. Semaine Méd. 25: 208-209.
1934 La maladie des pêcheurs d'éponges nus. Paris Méd. 93: 89-97, 16 figs.
1938 La maladie des pêcheurs d'éponges nus et l'anémone de la mer "actinion." Bull. Acad. Méd. 119: 379-395, 9 figs.

ZICK, K.
1929 Die wirkung der nesselkapseln auf protozoen. Zool. Anz. 83(11-12): 295-313, 9 figs.
1932 Die entladung der nesselkapseln protozoen. Zool. Anz. 98: 191-197.

(NSA)—Not seen by author.

Chapter V—INVERTEBRATES

Phylum Echinodermata

STARFISHES, BRITTLE STARS, SEA URCHINS, SEA CUCUMBERS

Echinoderms, characterized by their radial symmetry, have a body with usually five radii around an oral-aboral axis, comprised of calcareous plates which form a more or less rigid skeleton, or with plates and spicules embedded in the body wall. Spines and pedicellariae are present in the asteroids and echinoids, but are absent in some of the others. The coelom is complex and includes a water vascular system with tube feet. The digestive tract may or may not include an anus. The sexes are usually separate. Echinoderms, with the exception of a few planktonic holothurians, are all benthic, and all are marine. It is estimated that there are about 5,300 species in the phylum.

Usually the phylum is divided into two subphyla: the *Pelmatozoa*, which contains the sea lilies and their relatives, and the *Eleutherozoa*. Since no toxic organisms have been reported in the former group, no further mention will be made of them. However, the *Eleutherozoa* is known to contain a number of noxious members. The living *Eleutherozoa* is divided, according to Hyman (1955), and Storer and Usinger (1957), into the following classes:

ASTEROIDEA: Starfishes.

OPHIUROIDEA: Brittle stars. Toxicologically of minor importance.

ECHINOIDEA: Sea urchins, heart urchins, and sand dollars.

HOLOTHUROIDEA: Sea cucumbers.

The nomenclature as provided by Mortensen (1910, 1927, 1928, 1935, 1940*a*, *b*, 1943*a*, *b*, 1948, 1950, 1951*a*, *b*) has been adopted for the asteroids and echinoids, and Clark (1921, 1938, 1946) has been followed for the holothurians.

Excellent summaries on the general biology of echinoderms have been written by Ludwig and Hamann (1904), Cuénot (1948), Hyman (1955), and Nichols (1962).

The glimmer of information available on the toxicity of starfishes leads one to speculate that these creatures may offer an interesting and productive area of investigation for future research.

The Asteroidea or Starfishes

REPRESENTATIVE LIST OF STARFISHES REPORTED AS TOXIC

Phylum ECHINODERMATA

Class ASTEROIDEA: Starfishes

Family ACANTHASTERIDAE

58, 59 *Acanthaster planci* (Linnaeus) (Pls. 1-2). Venomous starfish (USA), crown of thorns starfish (Australia).

DISTRIBUTION: Indo-Pacific, Polynesia to Red Sea.

SOURCES: Fish and Cobb (1954), Pope (1963*a*, 1964), Brown (1972), Fraser-Brunner (1973), Halstead (1974).

Family ASTERIIDAE

60 *Aphelasteria japonica* (Bell) (Pl. 3, fig. a).

DISTRIBUTION: Japan.

SOURCE: Sawano and Mitsugi (1932).

60 *Asterias amurensis* Lütken (Pl. 3, fig. b).

DISTRIBUTION: Japan.

SOURCE: Sawano and Mitsugi (1932), Mebs (1973).
OTHER NAME: *Asterias rollestoni*.

Pycnopodia helianthoides (Brandt). Giant sunburst starfish (USA).

DISTRIBUTION: Central California to Alaska.

SOURCE: Rio, Stempien, and Nigrelli (1963).

Family ASTERINIDAE

60 *Asterina pectinifera* Müller and Troschel (Pl. 3, fig. c).

DISTRIBUTION: Japan.

SOURCE: Hashimoto and Yasumoto (1960).

Family ASTROPECTINIDAE

60 *Astropecten scoparius* Valenciennes (Pl. 3, fig. d).

DISTRIBUTION: Japan.

SOURCE: Sawano and Mitsugi (1932).
OTHER NAME: *Asteropecten scoparius*.

Family SOLASTERIDAE

61 *Solaster papposus* (Linnaeus) (Pl. 4). Sunstar, rose star (England), seesterne (Germany), étoile de mer (France).

DISTRIBUTION: Circumpolar, European seas south to the English Channel.

SOURCES: Parker (1881), Pawlowsky (1927), Halstead (1956).
OTHER NAME: *Crossaster papposus*.

BIOLOGY

Starfishes are free-living echinoderms that have a flat, star-shaped or pentagonal body, usually with a continuous disc that has five or more raylike extensions called arms. Tube feet are located in an open furrow along the underside of the arms. These arms contain the digestive glands and genital organs. Located on the upper, or aboral, surface are many blunt, calcareous spines. The stomach is large and sac-shaped, the intestine minute, and the mouth turns downwards. With the use of their tube feet, starfishes move about over the floor of the sea by slow, gliding movements. Usually the sucking discs are not used when walking over smooth bottom, but only when the starfish is climbing. Starfishes are voracious, for they eat other echinoderms, mollusks, and worms. It is believed that their habit of eating poisonous shellfish is the primary cause of their own toxicity (Sommer and Meyer, 1937). Pope (1963*b*) has suggested that the toxicity of *Acanthaster planci* may result in part to its feeding on the nematocysts of corals.

Because of their remarkably extensible mouth, starfishes are able to swallow relatively large animals. If the animal is too large to swallow, the starfish extrudes its stomach and digests the food outside its body. By exerting steady pressure with the use of its tube feet, it is able to pry open mollusks. Thus, starfishes are of economic significance to the oyster industry. Since starfishes are not generally used for food, their commercial value is negligible.

During early summer, eggs and sperm are released into the water where the fertilization of starfishes takes place. It is believed that starfishes usually take about 4 years to attain full size. Their ability to regenerate damaged parts of their body is phenomenal. Starfishes are among the most common of shore animals, and they range from the intertidal zone to great depths. They inhabit a wide variety of biotopes, rocks, sand, mud, coral, etc. Usually starfishes are negative to light and, therefore, seek shaded areas. An excellent discussion on the ecology of starfishes is credited to Hyman (1955).

MORPHOLOGY OF THE VENOM APPARATUS

Little is known as to the nature of the venom apparatus of *Acanthaster planci*, the venomous starfish. The following is a preliminary description based on a series of histological sections prepared from a large specimen of *A. planci*, taken at Heron Island, Great Barrier Reef, Australia, by K. Gillett. The spines were preserved in 10 percent formalin solution immediately after capture, sectioned without decalcification (6-8μ), and stained with hematoxylin and triosin, by D. Howard, Department of Pathology, School of Medicine, Loma Linda University.

A. planci attains a large size, up to 60 cm or more in diameter, and possesses 13 to 16 rays or arms. The outer surface of the entire body is covered by a series of large, pungent spines, which are completely enveloped in a thin layer of rugose integument. The spines may measure up to 6 cm or more in length, and tend to be somewhat less friable than their equivalent size in sea urchins.

Histological sections reveal that the spines of *A. planci* are covered by an integument consisting of an outer cuticle, an epidermis, and an underlying dermis, which surrounds the endoskeletal calcareous supportive structure of the spine (Pl. 2). The epidermal cells are tall, flagellated, and apparently of pseudostratified columnar epithelial type. The epidermal cells are interspersed

59

with neurosensory cells, pigment cells, and two types of glandular cells: an acidophilic cell having a densely granular cytoplasm, and a basophilic cell having a light-staining, vacuolated or reticulated cytoplasm—resembling a sebaceous type of cell. The acidophilic cells are peripheral in their distribution whereas the basophilic cells are for the most part situated adjacent to the basement membrane. There is observed at periodic intervals along the outer surface of the epidermal layer indentations or small accumulations of acidophilic glandular cells which lie wholly within the epithelium and contain their own small lumen. They have the appearance of a typical intraepithelial gland. It is believed that the venom is produced by these acidophilic cells. The basophilic cells are probably mucous-producing cells. However, the exact site of venom production remains for future research to determine. The basement membrane separates the epidermis from the underlying thick dermis of fibrillar connective tissue. Beneath the dermis is the endoskeletal structure of the spine. The areas of calcification appear to vary from one region to the next. The greatest density of calcification is in the peripheral portion. As one progresses to the center of the spine, the areas of calcification become less organized, trabeculated, and finally merge into a central unorganized core having almost the appearance of mucous-connective tissue.

MECHANISM OF INTOXICATION

Poisonous starfishes are believed to be toxic to eat. Contact with the slime of some species of asteroids may result in a contact dermatitis. In both cases the poison is thought to be produced by the glandular cells which are present in abundance in the epidermis of starfish.

Acanthaster planci is the only known venomous asteroid. The spines of this starfish are elongate, pungent, and covered by a venom-producing integument. Wounds produced by these spines can be extremely painful.

MEDICAL ASPECTS

Clinical Characteristics

Nothing is known about the symptomatology in humans caused by the ingestion of poisonous starfishes. Contact with the slime of *Marthasterias glacialis* may cause swelling if it touches the lips. Some species of starfishes (the exact identity of which is unknown) may cause a pruritic rash if their slime is brought into contact with the skin (Giunio, 1948; Clark, 1953). Swan (1889) cited a fatality resulting from a prick made by a starfish spindle, but the death appears to have resulted from a secondary bacterial infection rather than a toxin produced by the starfish.

Contact with the venomous spines of *Acanthaster planci* may cause an extremely painful wound, redness, swelling, protracted vomiting, numbness, and paralysis (Fish and Cobb, 1954; Pope, 1963*b*).

Treatment

Symptomatic.

Prevention

Avoid contact with dangerous species. The local inhabitants should be contacted for information concerning the identity of dangerous species. At present no specific data are available on this subject.

TOXICOLOGY

Relatively little is known concerning the toxicology of asterotoxin, or starfish poison. Parker (1881) cited the death of two cats that had ingested a sun star *Solaster papposus*. One cat became sick within 10 minutes after ingesting the starfish, and died about 5 minutes later. The second cat soon became very sick, screeched, developed ataxia, and died in violent convulsions within 2 hours after eating the material.

The toxic effects of starfish poison on oysters have been discussed by Sawano and Mitsugi (1932), and Imai, Hatanaka, and Oshima (1956). Extracts prepared from the alimentary tissues of the Japanese starfish species *Aphelasteria japonica, Asterias amurensis*, and *Astropecten scoparius*, act as a cardioinhibitor on oyster heart, *Ostrea circumpicta*. Akiya (1937), Ando (1954), and Ando and Hasegawa (1955) reported on the toxic effects of asterotoxin on fly larvae; and the growth-inhibitory effects on chicks have been presented by Whitson and Titus (1945), and Lee (1951).

The most recent report on the toxicology of asterotoxin is by Hashimoto and Yasumoto (1960). The toxin was prepared from an ethanol extract of dried specimens of the Japanese starfish *Asterina pectinifera*. It was found to be a saponin and highly toxic to killifish *Oryzias latipes*. The toxin had strong hemolytic properties, with a hemolytic index of about 7,000, as determined by the method of Fujita and Nishimoto (1952).

Chaet (1962) reported finding an autotomizing toxin in the coelomic fluid of the common starfish *Asterias forbesi* which had been scalded. When this fluid was injected into healthy starfish, it induced autotomy and caused death. Coelomic fluid from nonscalded ("normal") starfishes had no apparent effect when injected under otherwise identical conditions. Extracts obtained from heated hepatic and aboral tissue from the arms of normal starfishes produced toxic effects, but coelomic fluid and radial nerve tissue did not. It was concluded that the site of toxin production was in the epithelial lining. The toxin was also found to be poisonous to several other starfishes, a sea urchin, a marine annelid, and a horseshoe crab. The poison from scalded starfishes was found to be heat stable when autoclaved, and it retained its activity after being frozen for one year.

A lethal saponin fraction was isolated by Owellen *et al.* (1973) from *Asterias vulgaris* and found to have cytolytic activity against erythrocytes and lymphocytes.

CHEMISTRY

When Hashimoto and Yasumoto (1960) observed that starfishes kept in an aquarium produced the appearance of foam in the water when the animals died, they suspected that the toxic principle previously observed from ingestion of starfishes by cats and dogs was saponin. To test this theory, starfishes were subjected to the isolation procedure for saponins developed by Wall *et al.* (1952).

The crude toxin was brown in color, soluble in water and in aqueous alcohol, but not soluble in the fat solvents. Tests were positive for steroid and for carbohydrate. Aqueous solutions produced notable foaming, even in very dilute concentrations. Fehling's solution was reduced after hydrolysis, but not before. The hydrolysate, freed of hydrochloric acid by passing it through an ion-exchange column containing IR 4B resin, was reduced in volume and examined by paper chromatography. Four component sugars were found to be present, three of which had R_f values corresponding to glucose, xylose, and rhamnose.

The crude toxin was partially purified by dialysis. This step eliminated approximately 50 percent of the solids. The nondialyzable fraction contained 90 percent of the hemolytic activity. The fraction was dissolved in minimal methanol and passed through a column of activated alumina. This was followed by elution with absolute methanol until the eluate was found, on evaporation to dryness, to be free of any solid residues. A subsequent elution with *n*-butanol, however, gave further eluate fractions which revealed solid residues on evaporation. Residues from both eluates were found to contain toxic principles. From these observations, it was concluded that the crude saponin fraction contained two distinct toxins. The hydrolysate of the toxin was noted to resemble that of holothurin, obtained from the sea cucumber, from which glucose, xylose, glucomethylose, and 3-o-methylglucose had been previously identified by Chanley *et al.* (1959).

The Echinoidea or Sea Urchins

REPRESENTATIVE LIST OF SEA URCHINS REPORTED AS TOXIC

Phylum ECHINODERMATA

Class ECHINOIDEA: Sea Urchins

Family DIADEMATIDAE

62 *Diadema antillarum* Philippi (Pl. 5, fig. a). Black sea urchin, needle-spined urchin (USA), sea-needle, sea sting, sea-egg (Jamaica), cobbler (Barbados).

DISTRIBUTION: West Indies.

SOURCES: Phillips and Brady (1953), Halstead (1974).
OTHER NAME: *Centrechinus antillarum*.

Diadema setosum (Leske) (Pl. 5, figs. b, c). Long-spined or black sea urchin (USA), enor (Nauru). *62*

DISTRIBUTION: Indo-Pacific area, from Polynesia to east Africa, China, Japan, Bonin Islands.

SOURCES: Cleland (1912, 1932).
OTHER NAMES: *Centrechinus setosus, Diadema setosa.*

Echinothrix calamaris (Pallas) (Pl. 6, fig. a). Sea urchin (USA), jarum-jarum bĕlang (Malaysia). *63*

DISTRIBUTION: Indo-Pacific area, from Polynesia to east Africa, Red Sea, Australia, Japan.

SOURCES: Fish and Cobb (1954), Fraser-Brunner (1973).

Family ECHINIDAE

Paracentrotus lividus Lamarck (Pl. 6, fig. b). Sea urchin (USA, England), oursin (France), seeigel (Germany). *63*

DISTRIBUTION: Atlantic coast of Europe, from Ireland to west Africa, Mediterranean, Canaries, Madeira, Azores.

SOURCES: Uexküll (1899), Phisalix (1922), Pawlowsky (1927), Halstead (1974).
OTHER NAMES: *Toxopneustes lividus, Strongylocentrotus lividus.*

Family ECHINOTHURIDAE

Araeosoma thetidis (Clark) (Pl. 7; Pl. 8, fig. a). Tam o'shanter urchin (Australia). *64, 65*

DISTRIBUTION: East coast of Australia, northern New Zealand.

SOURCES: Mortensen (1935), Hyman (1955).

Phormosoma bursarium Agassiz (Pl. 8, fig. b; Pl. 9). Sea urchin (USA). *65, 66*

DISTRIBUTION: Indo-Pacific, from Hawaii to the Arabian Sea, Natal coast.

SOURCES: Mortensen (1935), Hyman (1955).

Family TOXOPNEUSTIDAE

Toxopneustes pileolus (Lamarck) (Pl. 10, figs. a, d; Pl. 11; Pl. 16, fig. h). Sea Urchin (USA), terong-terong lebar (Malaysia). *67, 68, 73*

DISTRIBUTION: Indo-Pacific area, Melanesia to east Africa, Japan.

SOURCES: Fujiwara (1935), Fish and Cobb (1954), Fraser-Brunner (1973), Halstead (1974). *See also: Toxopneustes roseus* Agassiz). Eastern tropical Pacific (Pl. 13a, b, c). *70*

Tripneustes gratilla (Linnaeus) (Pl. 12, fig. b). Sea urchin (USA), terong-terong bulat (Malaysia). *69*

DISTRIBUTION: Indo-Pacific area, Polynesia to east Africa, Japan, Australia.

SOURCES: Mortensen (1943*a, b*), Alender (1964).
(This species may be identical with *T. ventricosus*, and if so, the ova are suspect as being toxic during certain periods of the year.)

69 *Tripneustes ventricosus* (Lamarck) (Pl. 12, figs. c-d). White sea urchin (USA, West Indies), white sea-egg, edible sea-egg (Barbados).
DISTRIBUTION: West Indies, south to Brazil, West Coast of Africa.

SOURCES: Earle (1940), Halstead (1974).
OTHER NAME: *Tripneustes esculentus* (not *Echinus esculentus*).

BIOLOGY

Sea urchins are free-living echinoderms, having a globular, egg-shaped, or flattened body. The viscera are enclosed within a hard shell or test, formed by regularly arranged plates, carrying spines articulating with tubercles on the test. Between the spines are situated three-jawed pedicellariae, which are of interest to the venomologist and will be subsequently described in detail. In some species of sea urchins, the spines are also venomous. Tube feet are arranged in 10 meridian series rather than in furrows. A double pore in the test corresponds to each tube foot. The intestine is long and coiled, and an anus is always present. The gonads are attached by mesenteries to the inner aboral surface of the test. The mouth, situated on the lower surface, turns downward, and is surrounded by five strong teeth incorporated in a complex structure termed "Aristotle's lantern." Their power of regeneration is great; but autotomy, as observed in the asteroids, does not occur. By means of spines on the oral side of the test, sea urchins move slowly in the water. The tube feet are utilized to climb vertical surfaces. Some forms have the ability to burrow into crevices in rocks, while others cover themselves with shells, sand, and bits of debris.

Some urchins are nocturnal, hiding under rocks during the day and coming out to feed at night. Echinoids tend to be omnivorous in their feeding habits, ingesting algae, mollusks, foraminifera, and various other types of benthic organisms.

Sea urchins are dioecious, hermaphroditism occurring only as a rare anomally. Sexual dimorphism is generally absent. Spawning usually takes place during the spring and summer in the Northern Hemisphere, but somewhat earlier in the more southern latitudes. The reproductive periods of echinoids have been discussed at great length by Hyman (1955). Several species of European and tropical echinoids serve as important sources of food to man. Only the gonads are eaten, either raw or cooked. The bathymetric range of echinoids is great, extending from the intertidal zone to great depths.

MORPHOLOGY OF THE VENOM APPARATUS

The venom organs of sea urchins are of two types—spines and pedicellariae.

Spines: The spines of sea urchins are generally differentiated into two main sizes—the larger "primary spines," and the smaller "secondary spines." Sometimes the very small spines are termed "tertiary spines." They are also designated by their location on the surface of the test of the urchin, namely, "oral spines" (those situated on the ventral surface of the test), and "aboral spines" (those situated on the upper or dorsal surface of the test). Various other anatomical designations also are used for the naming of sea urchin spines, but since this aspect is beyond the scope of this present work, the reader is referred to Hyman (1955) for further information on the subject.

In general, sea urchin spines are mounted on tubercles of the test, having an indented base that fits over a tubercle which forms a ball and socket joint (Fig. 1). Above the base of the spine is a projecting circular ledge known as the milled ring. A circle of longitudinal muscle fibers extends from around the tubercle to the milled ring, which permits the spine to be moved in any direction. The shaft of sea urchin spines, although appearing to be smooth in some species, is always ornamented; and, in some instances it is quite rugose or thorny. The outside of the spine is covered by a thin layer of epithelium. The spines are composed of calcium carbonate, which occurs in the form of calcite intermingled with organic material. The calcareous substance has a fenestrated construction. The construction and details of arrangement vary throughout the Echinoidea, but are constant for each species. Thus the morphology of the spines and their arrangement on the test is a useful taxonomic tool in identifying the organism (Fig. 1).

In most echinoids the spines are solid, have blunt, rounded tips, and do not constitute a venom organ. However, members of the families Echinothuridae and Diadematidae possess long, slender, hollow, sharp spines, which are dangerous to handle. The acute tip and the retrose spinules, which are present in some instances, permit the spines to make easy entrance deep into the flesh. Because of their extreme brittleness, they readily break off in the wound. The spines of the echinothurids have a thin, fluted wall which is pierced by pores that connect the surface grooves with the axial cavity. The spines in *Diadema* are very long, frequently attaining a length of 300 mm or more. They are slender, usually verticillate, fragile, sharp, having a thin bark surrounding a large axial cavity (Fig. 2). Earle (1941), Taft (1945), Phillips and Brady (1953), and others, believed that the aboral spines of some of the diadematids secrete a venom; but this has not been experimentally demonstrated. However, the degree of pain connected with stings from diadematids would tend to support clinically their opinions.

The spines of some of the echinothurids, e.g., primary oral spines of *Phormosoma bursarium*, and secondary aboral spines of *Araesoma thetidis* and *Asthenosoma varium* (Fig. 3), are particularly suited as a venom organ because of their unique structure. In each instance the acute tip of the spines is encased within a venom sac, or "poison bag" as it has been previously termed (Mortensen, 1935; Hyman, 1955). The spinal venom organs of echinothurids appear to be developed best in the secondary aboral spines of *A. varium.* The spinal venom organs of *A. varium* are composed of connective and muscular fibers which enclose a venom sac which surrounds the tip of the

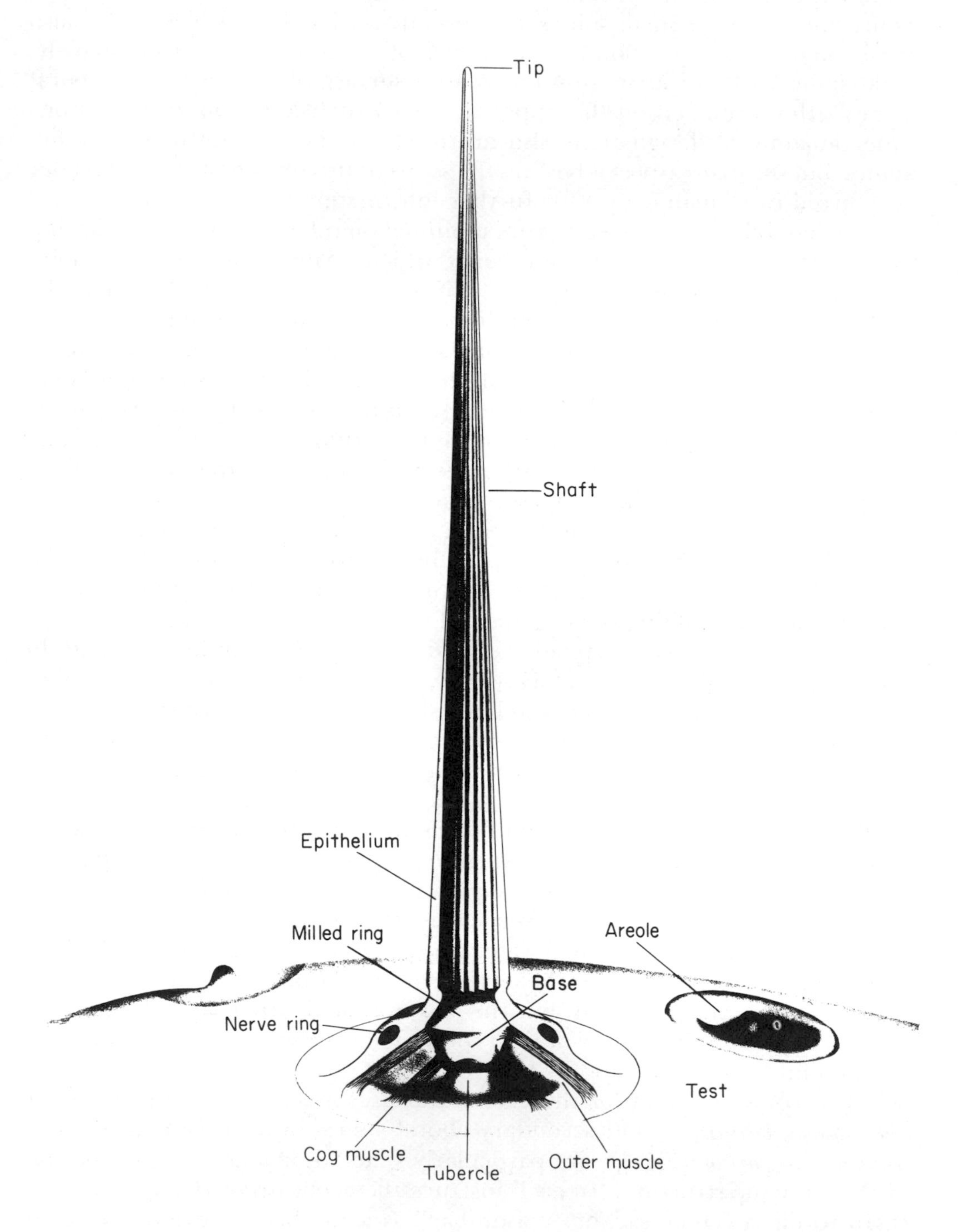

FIGURE 1.—Morphology of a typical sea urchin spine and its supporting structure. Semidiagrammatic. (R. Kreuzinger)

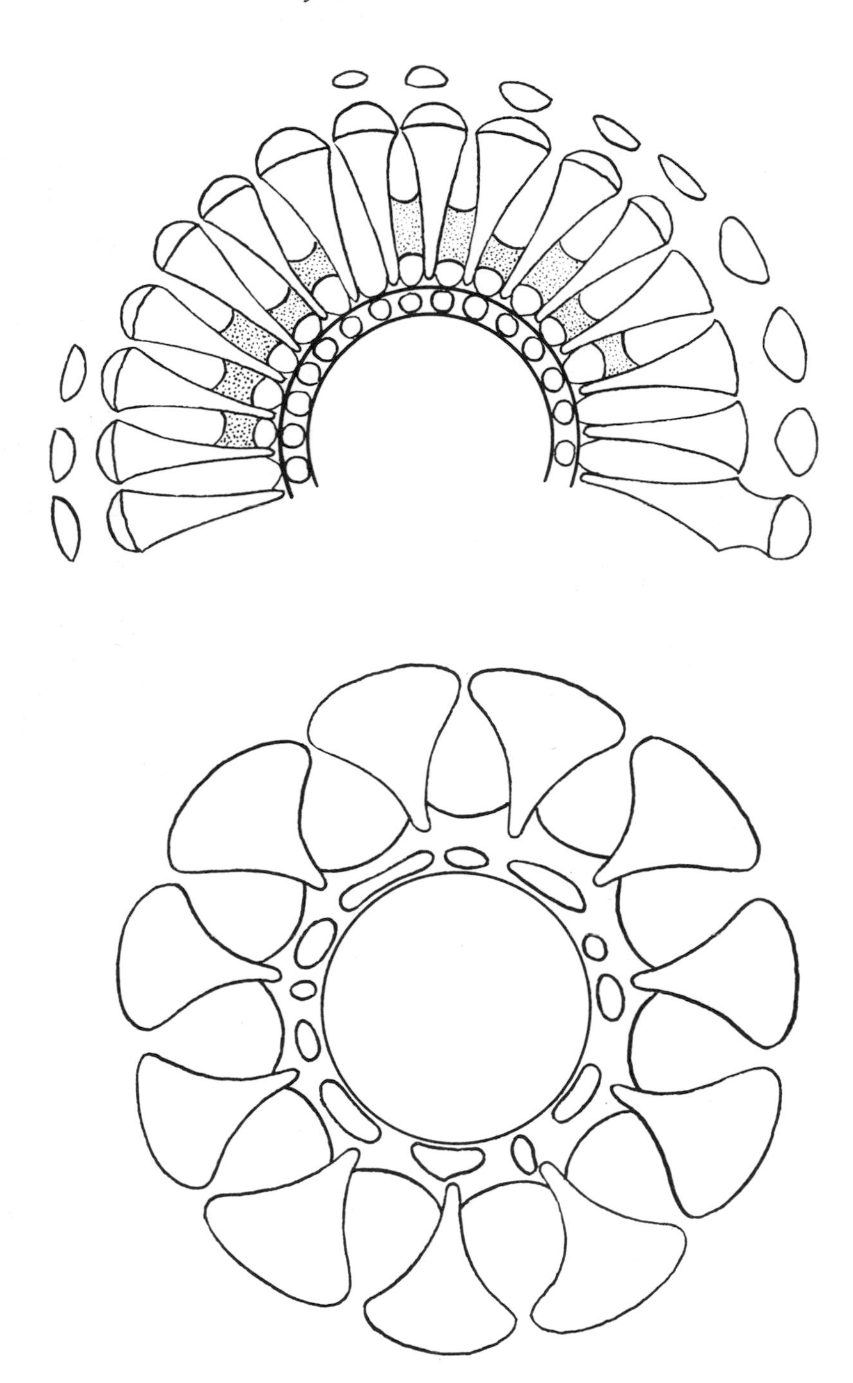

Figure 2.—Cross section of the hollow spines of two representative species of diadematid sea urchins *Diadema setosum* (Leske), × 125 (top), and *Echinothrix calamaris* (Pallas), × 225 (bottom). Note the large axial cavities of these spines.

(R. Kreuzinger, after Mortensen)

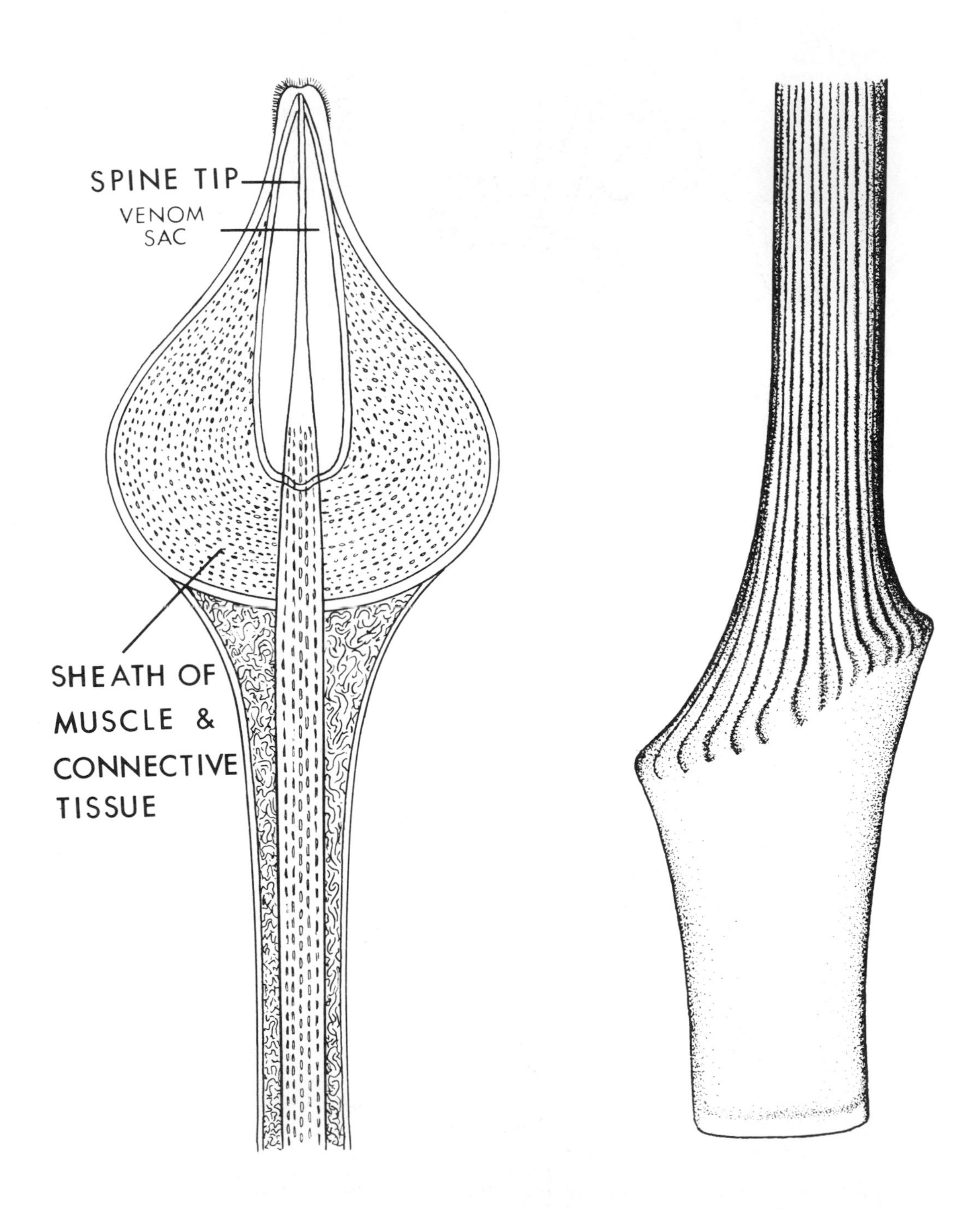

FIGURE 3.—Morphology of a venomous secondary aboral spine of the echinothurid sea urchin *Asthenosoma varium* Grube. The tip of the spine is completely encased within a sheath containing a venom sac (left). Base of primary oral spine (right). Semi-diagrammatic. (R. Kreuzinger, modified from Mortensen)

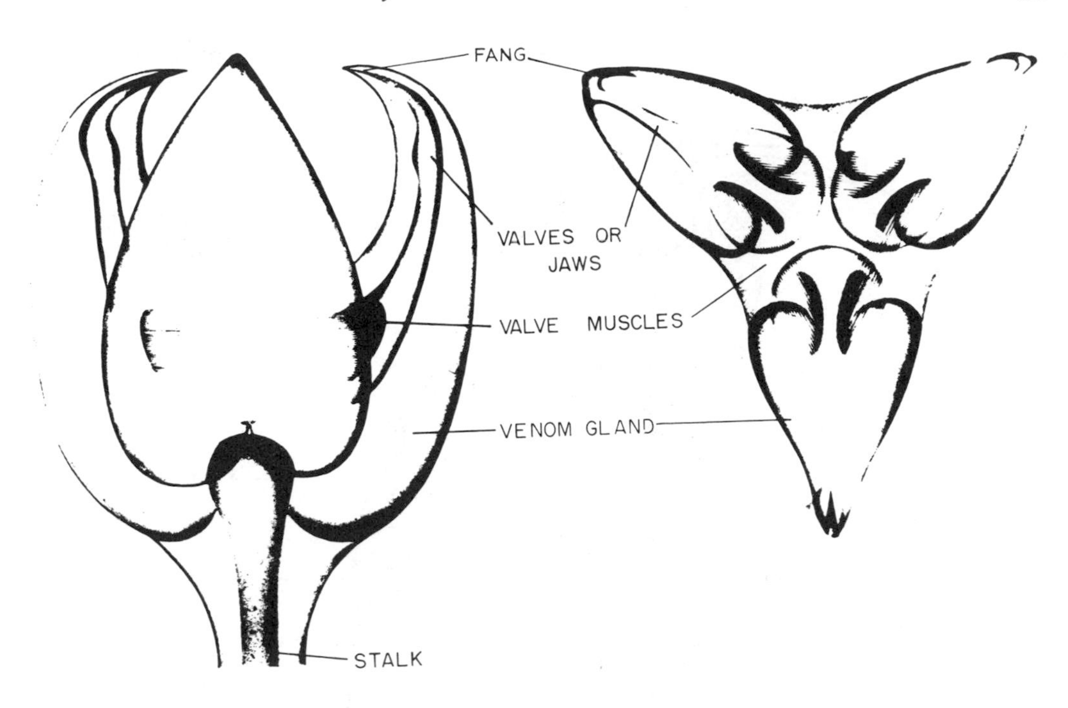

FIGURE 4.—Gross morphology of a venomous globiferous pedicellaria of a sea urchin. Details of the supporting stalk are not shown. (R. Kreuzinger, after Mortensen)

spine. The sac is filled with a toxin that supposedly is secreted by the epithelial lining of the sac.

Pedicellariae: The best general summarizations on the morphology of the pedicellariae of echinoderms have been given by Ludwig and Hamann (1901, 1902), Phisalix (1922), Mortensen (1927, 1928-51), Hyman (1955), and Nichols (1962). The following discussion is based largely upon their works.

Pedicellariae are small, delicate, seizing organs, which are found scattered among the spines of the test peculiar to the echinoderm classes, Echinoidea and Asteroidea (Fig. 4). They are comprised of two parts—a terminal, swollen, conical head, armed with a set of calcareous pincerlike valves or jaws, and a supporting stalk. The head is attached to the stalk either directly by the muscles, or by a long flexible neck. The head may consist of two or four, but more commonly three, calcareous valves or jaws (Fig. 5). The tip of these jaws is sometimes referred to as a fang. On the other side of each valve is a small elevation provided with fine sensory hairs. Contact with these sensory hairs causes the valves to close instantly. The pedicellariae of each species of sea urchin has its own characteristic skeleton. The skeletal elements of the stalk are formed by a calcareous rod composed of loose filaments united by five crosspieces. Generally this rod is dilated at its base into a larger cylindrical structure, which is connected by soft parts to a small tubercle on the shell. The length of the stalk varies according to the species and the type of pedicellariae. The soft parts of the pedicellariae consist of an outer epithelium, an

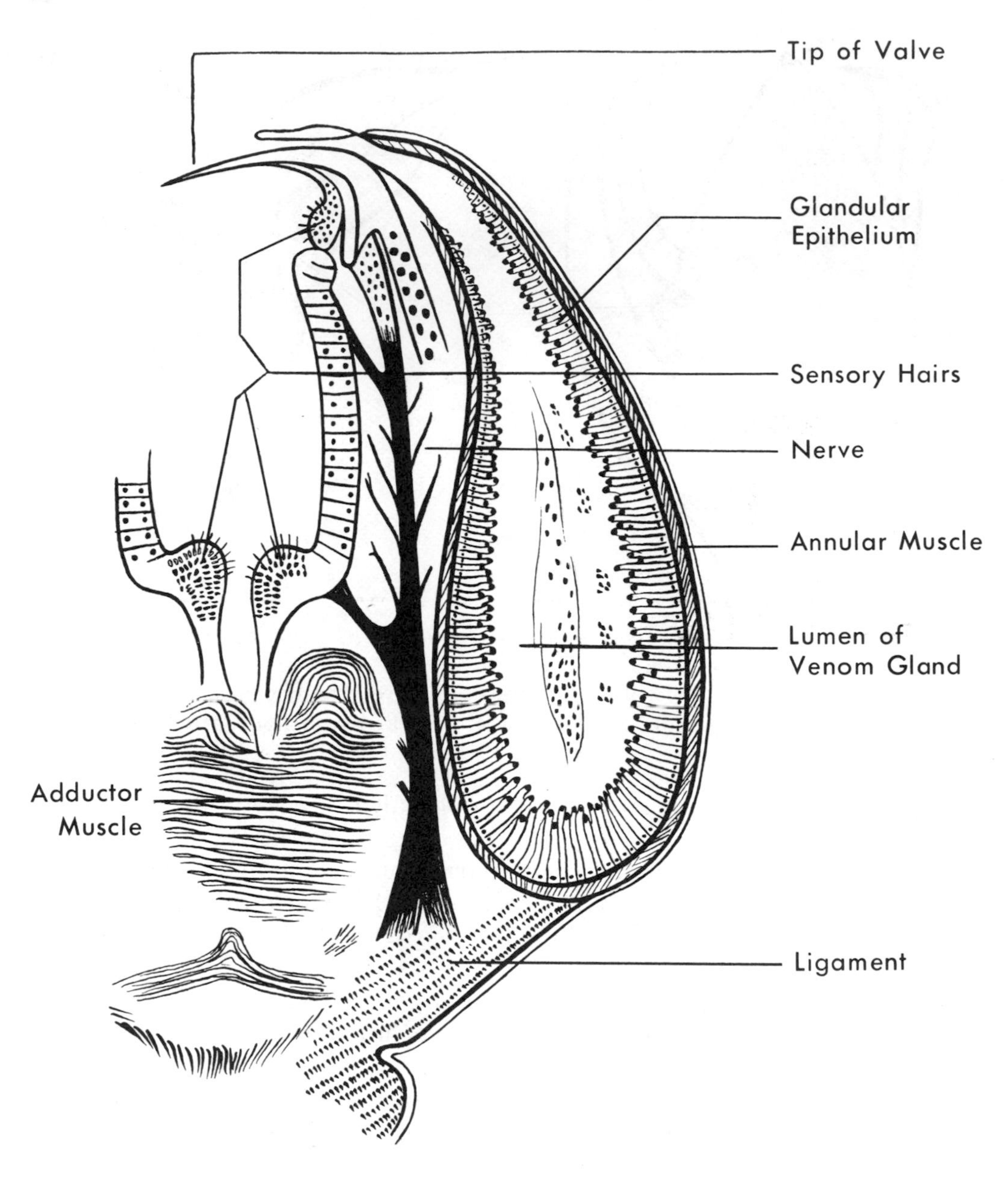

FIGURE 5.—Microscopic anatomy of a single valve of a venomous globiferous pedicellaria of a sea urchin. Semidiagrammatic. (R. Kreuzinger, after Hamann)

underlying layer of connective tissue, muscles, venom and mucous glands. These structures will be described in greater detail under the various types of pedicellariae.

Types of Pedicellariae: Pedicellariae have been described by a variety of terms which have caused some confusion. The following is a synonomy which will be helpful in deciphering the literature on this subject.

A Synonomy of Terms Used for the Various Types of Pedicellariae (Pl. 15)

71, 72

1. *Globiferous.*
   Synonyms: gemmiform, glandular, Pedicellaria gemmiforme, P. armé, P. globifera.
2. *Tridentate.*
   Synonyms: tridactyle, Pedicellaria tridactyle, P. inerme, P. tridens.
3. *Triphyllous.*
   Synonyms: trifoliate, Pedicellaria trifolié, P. triphylla.
4. *Ophiocephalous* (also *ophicephalous*).
   Synonyms: buccal, Pedicellaria ophicéphale. (There is some confusion in the literature between the ophiocephalous and triphyllous types; Phisalix, 1922, combines the two.)
5. *Dactylous.*
   Synonym: tetradactylous.

71-73

Globiferous Pedicellariae (Pl. 15, fig. a; Pl. 16; Pl. 17): Generally this type is found disseminated over the surface of the shell. It is more complex and varied in its structure than any other type. Hyman (1955) lists the dactylous pedicellariae as a variant of the globiferous type.

Globiferous pedicellariae attain their greatest development within the family Toxopneustidae. The head of large globiferous pedicellariae is globular in shape and comprised of three calcareous valves; and it may reach a size of 2 mm by 3 mm, with an overall height of 10 mm or more. Each of the valves has an enlarged trigonal-shaped base, which continues upward as a narrow stalk, and finally terminates in a sharp incurved spine or fang. The stalk in *Toxopneustes* is characteristically and unusually elongate. On the inner aspect of the lower third of the valve is a broad depression divided into two lateral fossae by a pronounced median ridge. The adductor muscles of the pedicellariae have their attachment within these lateral fossae. The adductor muscles of the pedicellariae of each head consist of three bundles of smooth muscle fibers. In addition, three abductor muscles are inserted on the outer aspect of the base of the valves, which causes them to open. The abductors are less developed than the adductors, but their tonicity is sufficient to keep the valves open when the adductors are relaxed. There also exists three flexor muscles which join the tip of the calcareous pedicellarial stalk to the base of the valves. The synergistic action produces retraction of the head and aids in the abduction of the valves. When the flexors are contracted singly, the head bends on the stalk in the direction of the contracted muscle. The flexor muscles are further strengthened by a connective tissue ligament which connects the base of each valve to the head of the calcareous stalk.

The outer surface of each valve is covered by a large gland which in *Toxopneustes* has two efferent venom ducts that empty in the vicinity of a small toothlike projection on the terminal fang of the valve. In some species, venom glands may also be found encircling the stalk (Figs. 4-5). Phisalix (1922) states that the gland is surrounded by a muscular membrane of flat smooth muscle fibers that are arranged in a circular fashion. This muscle has been termed the compressor muscle because of its function. An accurate and complete histological description of these glands seems to be lacking, but the venom, according to Foettinger (1881), is granular in nature. The cells appear to undergo dissolution, indicating a holocrine type of production; but this has not been determined with certainty. The nature of the venom is discussed in a subsequent section of this chapter. Overlaying the circular muscle is a thin layer of loose fibrous connective tissue. The entire structure is covered by a layer of pigmented epithelial tissue, whose exact nature has not been described. Sladen (1880) reported sensorial or "nervous papillae" on the inner aspect of each valve. These organs are composed of sensitive cells furnished with a rigid bristle, and are intercalated with supporting cells. Nerves are said to extend from these "papillae" to the muscles. Stimulation of these bristles causes the adductor muscles to contract. Our knowledge of the microscopic anatomy of the venom organs of echinoderms is incomplete and much additional research is needed.

Venomous pedicellariae, generally of the globiferous type, have been reported in the families Echinidae, Spatangidae, Strongylocentrotidae, and Temnopleuridae. However, as previously mentioned, they are most highly developed into a powerful venom apparatus in Toxopneustidae. Duncan (1889), cited by Mortensen (1935), says that certain members of the Arbaciidae have "large glands" on the pedicellarial stalk; however, Mortensen has not observed them. Members of the order Cidaroidea are also reported (Mortensen, 1928) to have globiferous pedicellariae, but they do not seem to be of any clinical significance to humans. Numerous anatomical variations are to be found in the globiferous pedicellariae of the various species of sea urchins. The reader is referred to the exhaustive works of Mortensen (1928-51) for further information on this subject. Examples of globiferous pedicellariae found in some of the sea urchins believed to be of medical significance to man are illustrated (Pls. 16-17).

73-74

Since venom glands have not been described for the remaining types of pedicellariae, they will be only briefly discussed.

72-73

Tridentate Pedicellariae (Pl. 15, fig. b): This is a nonvenomous type of pedicellariae usually having long slender valves, which in some forms may be very coarse. The blade of the valve is filled with a thick, reticulate calcareous meshwork. Often there are irregular, serrate crests in the blade. They may attain a length of 5 mm or more.

72-73

Triphyllous Pedicellariae (Pl. 15, fig. c): This is a nonvenomous type of pedicellariae, which has a head formed by three delicate, and sometimes very ornate serrate, fenestrated, calcareous lamellae. These are larger at their free extremity than at their base. The stalk is thin and formed almost entirely of two calcareous rods connected by small crosspieces. Triphyllous pedicellariae are generally small, and according to Phisalix (1922) seldom exceed a total length of 1.5 mm.

72-73

Ophiocephalous Pedicellariae (Pl. 15, fig. d): This nonvenomous type generally is found grouped in thick patches on the buccal membrane, but it may also occur elsewhere on the shell. They typically have very elaborate valves, with the edge of the triangular-shaped blade thickened and finely dentate with an arc under the articular surface of each valve. There is usually no distinct neck, for the head rests directly on the upper end of the stalk. Phisalix (1922) states that they seldom exceed a total length of 2 to 5 mm.

Dactylous Pedicellariae (Pl. 15, fig. e): This nonvenomous form is reputed to be the most delicate and elaborate of any of the pedicellariae, and is characteristic of the genus *Araeosoma* (Mortensen, 1935). It has three to five valves, with a long, slender, tube-shaped blade, terminating in a widened, fenestrated, gracefully curved part, the edge of which is distinctly serrate. The base of the valves is little developed; and the adductor muscles, if present, are very weak. Mortensen (1935) states that some dactylous pedicellariae have venom glands, but the pedicellariae are so weak that their function is questionable.

72-73

Function of Pedicellariae: One of the primary functions of pedicellariae is that of defense. When the sea urchin is at rest in calm water, the valves are generally extended, moving slowly about, awaiting their prey. When a foreign body comes in contact with valves, it is immediately seized. The pedicellariae do not release their hold as long as the object moves. If it is too strong to be held, the pedicellariae will continue to seize the object, even though it may be torn from the test. Detached pedicellariae may remain alive for several hours. The globiferous pedicellariae are essentially venom organs, used to immobilize their prey. Pedicellariae are also used by urchins to procure food, and to seize and deport detritus which may encumber the animal.

MECHANISM OF INTOXICATION

Intoxications from sea urchins may result from ingestion of their poisonous gonads, such as is the case in *Paracentrotus lividus*, *Tripneustes ventricosus*, and *Centrechinus antillarum*. However, in most instances sea urchin poisonings are due to stings from either their spines or pedicellariae.

MEDICAL ASPECTS

POISONING

Clinical Characteristics

During the reproductive season of the year, generally spring and summer (variable in some species as early as January or as late as November in others), the ovaries of certain species of sea urchins are reported to develop toxic products that are injurious to man (Phisalix, 1922; Pawlowsky, 1927; Earle, 1940, 1941; Manson-Bahr, 1950; Kaiser and Michl, 1958). The symptoms consist of general epigastric discomfort, nausea, vomiting, and diarrhea. Severe migrainelike attacks have been reported (Earle, 1940), which are believed to be on an allergenic basis. Little else is known regarding sea urchin oral intoxications in man.

Treatment

Symptomatic.

Prevention

Care should be taken to avoid eating the ova of known toxic species of sea urchins during the reproductive season of the year. Local inhabitants of the area—especially in tropical regions—should be contacted regarding the edibility of sea urchins.

ENVENOMATIONS

Clinical Characteristics

The hollow, elongate, fluid-filled spines of Echinothurid and Diadematid sea urchins are particularly dangerous to handle. Their sharp, needlelike points are able to penetrate the flesh with ease, and these produce an immediate and intense burning sensation. As the spines penetrate, they release a violet-colored fluid which causes discoloration of the wound. Intense pain is soon followed by redness, swelling, and aching sensations. Partial motor paralysis of the legs, slight anesthesia, edema of the face, and irregularities of the pulse have been reported (Pugh, 1913). According to Earle (1940, 1941), secondary infection is a frequent complication with some species. The pain usually subsides after several hours, but the discoloration may continue for 3 to 4 days.

There is a divergence of opinion as to whether the spines actually contain a venom. However, the clinical effects indicate that a poison of some type is present, since the pain far exceeds that produced by mere mechanical injury. According to Mortensen (1935), there is no doubt of the presence of venom in the spines of *Araeosoma* and *Asthenosoma*.

The clinical aspects of stings from the spines of sea urchins have been discussed by Cleland (1912), Pugh (1913), Mortensen (1935), Earle (1940, 1941), Taft (1945), Giunio (1948), Phillips and Brady (1953), and Fish and Cobb (1954).

Although numerous writers refer to the dangers of pedicellarial stings, there is little worthwhile clinical data. Oshima (1915) and Fujiwara (1935) state that a sting from the pedicellariae of *Toxopneustes pileolus* results in immediate and intense radiating pain, faintness, numbness, generalized paralysis, aphonia, respiratory distress, and death. The pain may diminish after about 15 minutes, and completely disappear within 1 hour, but paralysis may continue for 6 hours or longer (*see* section on Toxicology for further details). This subject has been discussed briefly by Mortensen (1943), Clark (1950), Fish and Cobb (1954), and Endean (1961).

Treatment

Insofar as the venom is concerned, sea urchin stings should be handled in a manner similar to any other venomous sting. However, attention is directed to the need for prompt removal of the pedicellariae from the wound When pedicellariae are detached from the parent animal, they frequently continue to be active for several hours. During this time they will introduce venom into the wound.

The extreme brittleness and retrose barbs of some sea urchin spines present an added mechanical problem. Nielly (1881) recommended that grease be applied, stating that this would allow the spines to be scraped off quite easily. Cleland (1912), Earle (1940), and others, are of the opinion that some sea urchin spines need not be removed, as they are readily absorbed. Absorption of the spines is said to be complete within 24 to 48 hours. However, Earle (1941) later pointed out that the spines of *Diadema setosum* are not readily absorbed, and months later roentgenological examination may reveal them in the wound. It is recommended that the spines of *Diadema* be removed surgically.

Prevention

No sea urchin having elongate, needlelike spines should be handled. Moreover, leather or canvas gloves and shoes do not afford protection, since the spines of *Diadema* and others are able to penetrate leather and canvas with ease. Care should be taken to avoid handling without gloves any tropical species of short-spined sea urchin because of the pedicellariae. Members of the family Toxopneustidae are especially dangerous, and should be avoided.

TOXICOLOGY

Gonadal Poison

Only the work of a single investigator is available on the toxicology of sea urchin gonads. Loisel (1903, 1904) prepared isotonic aqueous extracts from the testes and ovaries of *Paracentrotus lividus*, and injected them into the marginal ear vein of rabbits. A second series of acidulated aqueous extracts were prepared and injected into rabbits. The animals developed a loss of reflexes, exophthalmos, micturition, lacrymation, dilatation, tonic and clonic convulsions, respiratory distress, and muscular paralysis. The rabbits tested on the testicular extracts recovered, but the ovarian extracts proved to be lethal.

Pedicellarial Venom

Uexküll (1899) was the first, and one of the few, to attempt to study the toxicology of echinoid globiferous pedicellarial venom. Working on the venom of *Sphaerechinus granularis*, he found it to be a clear liquid, "having the consistency of a concentrated sugar solution." The poison coagulated into a white mass when it was spread over a slide. The venom changed into a granular mass, giving an acid reaction, when diluted with water before it dried. With these few observations on the physical properties of pedicellarial venom, Uexküll tested the poison on a series of marine animals by permitting the pedicellariae to sting them. He observed that when a small marine snail *Pleurobranchus meckeli* was placed on the sea urchin and stung by the pedicellariae, the snail immediately coiled up into a ball and fell off the urchin. It was found that a pedicellarial sting from *Sphaerechinus* on the octopus *Eledone moschata* resulted in a response similar to that produced by faradic stimulation. Small eels, when bitten by pedicellariae, became hyperactive and promptly died if stung in the medulla. If a frog's sciatic nerve was stung, it continued to conduct impulses for a short period of time, but then lost its conductivity at the site of the sting. A single pedicellarial bite was able to stop a frog's heart. A single sting on a frog's spinal cord resulted in generalized convulsions. Experiments with other sea urchin species yielded similar results; but in general it was found that the larger pedicellariae were more potent, and that a sea urchin having large pedicellariae, but few of them, was more dangerous than one having numerous small pedicellariae.

A different approach was used by Henri and Kayalof (1906), and Kayalof (1906), in their experiments with the globiferous, tridentate, and ophiocephalous pedicellariae of *Paracentrotus lividus, Arbacia lixula, Sphaerechinus granularis*, and *Spatangus purpureus*. The pedicellariae were removed from live sea urchins, macerated in sea water, and the solution injected into the visceral cavity of crabs, sea cucumbers, starfishes, fish, frogs, and lizards. Also, the solution was injected intravenously into octopuses and rabbits. From these studies it was concluded that all four types of pedicellariae produce toxic substances, which cause paralysis and death to the animal injected. Only sea cucumbers, starfishes, and frogs were not affected by the venom. An extract of macerations of 200 globiferous pedicellariae from *Sphaerechinus granularis* injected into a frog, and an additional 100 pedicellariae on the following day, resulted in no ill effect. They therefore assumed that frogs had a natural immunity to the poison. Crabs measuring 4 to 5 cm died within 15 to 20 minutes when injected with venom from 20 or more globiferous pedicellariae, but recovered if injected with less than that amount. Lizards and fish produced similar results. Rabbits weighing 1.5 kg died of respiratory paralysis from saline extracts prepared from 40 pedicellariae. Extracts from 50 pedicellariae of *Sphaerechinus* produced paralysis and death in an octopus. The venom was found to be somewhat heat stable, withstanding temperatures of 100° C for 15 minutes, with only slight attenuation of the toxin.

Henri and Kayalof (1906) also conducted a few simple immunological experiments. Rabbits were injected every 3 days with increasing doses of saline extracts of globiferous pedicellarial venom from *S. granularis*. After four injections the rabbits were able to tolerate the usual lethal dose of 40 pedicellariae. However, they found that holothurians, starfishes, and frogs apparently had a natural immunity to the pedicellarial venom. Frogs were able to tolerate venom extracts containing 200 pedicellariae without showing ill effects. The serum of the immunized rabbits did not protect other non-immunized animals such as rabbits, crabs, or fish against the effects of the poison; but if frog's serum was first injected into a crab, the crustacean was then protected from a subsequent injection of venom.

Stimulated by the investigations of Uexküll, and Henri and Kayalof, Lévy (1925) attempted to determine the hemolytic properties of pedicellarial venom. The pedicellariae were ground in physiological saline and filtered, and the clear extract used for testing. When the venom extract was used alone, no hemolysis resulted; but if lecithin, or "chicken vitellus," was added, the globiferous pedicellarial venom produced hemolysis. None of the other types of pedicellarial extracts resulted in hemolysis. Similar findings were obtained using horse, cow, and pig blood. Globiferous and tridentate pedicellarial venom from *Echinus esculentus* resulted in hemolysis, but no hemolysis was obtained from the ophiocephalous pedicellarial extract. Tests on *Psammechinus miliaris* were positive on tridentate extracts, but negative on the globiferous type. Similar tests were made on extracts prepared from the blood, spines, tube feet, testes, and ovaries of sea urchins, but they failed to produce hemolysis.

Fujiwara (1935) determined the toxicity of the venom of *Toxopneustes pileolus* by permitting the globiferous pedicellariae of living sea urchins to sting the shaven abdomens of mice. Stings by seven or eight pedicellariae resulted in respiratory distress (apparently the result of muscular paralysis) and a drop in body temperature in the mouse. Saline extracts prepared from 50 globiferous pedicellariae were injected subcutaneously into mice. Again the mice developed respiratory distress and a drop in body temperature. While handling a live specimen of *T. pileolus*, Fujiwara was stung by seven or eight globiferous pedicellariae. He suffered severe pain, respiratory distress, giddiness, paralysis of the lips, tongue, and eyelids, relaxation of the muscles in the limbs, difficulty in speech, and loss of control of his facial muscles. The pain disappeared in about an hour, but the facial paralysis continued for about 6 hours. Recovery was uneventful. Fujiwara's personal experience points up the highly toxic nature of the globiferous pedicellarial venom of this sea urchin.

Okada (1955), and Okada, Hashimoto, and Miyauchi (1955) performed a unique experiment in which they removed with forceps a globiferous ("giant") pedicellaria from *T. pileolus*, and permitted it to sting an oyster heart which was attached to a kymograph for recording the effects. They observed that when the pedicellaria came in contact with the heart there was a vigorous and positive closing action of the pedicellarial valves, and an ejaculation of a white milky fluid from the pedicellaria into the heart. This resulted in an immediate inhibition of cardiac pulsation followed by a sustained contraction of the heart from which it did not recover. Extraction of pedicellarial venom with 0.5 *M* KCl solution and injection of this extract into the heart also resulted in similar cardioinhibitory effects. Controls using injections of normal saline failed to elicit a similar response.

The résumés of Ludwig and Hamann (1904), Phisalix (1922), Pawlowsky (1927), and Brocq-Rousseu and Fabre (1942) on the toxic effects of pedicellarial venom are based largely on the works of Uexküll (1899) and Henri and Kayalof (1906).

PHARMACOLOGY

The first attempt to evaluate the general pharmacological properties of globiferous pedicellarial venom of sea urchins was made by Mendes, Abbud, and Umiji (1963). Saline extracts were prepared from homogenates of globiferous pedicellariae of *Lytechinus variegatus* and tested on accepted cholinergic effector systems, viz., guinea pig ileum, rat uterus, amphibian heart, longitudinal muscle of a holothurian, the protractor muscle of a sea urchin lantern, and the blood pressure of dogs. The response obtained was consistent with that of a dialyzable acetycholinelike substance which the researchers concluded to be in pedicellarial venom.

More recent studies (Feigen, Sanz, and Alender, 1966; Feigen *et al.*, 1968) have been done of the pedicellarial toxin of *Tripneustes gratilla*. The toxin was shown to release dialyzable pharmacologically active agents from ileal, pulmonary, and cardiac tissues of the guinea pig as well as from colonic and pulmonary preparations of the rat. Preliminary pharmacologic tests (Feigen, Sanz, and Alender, 1966) excluded acetylcholine but suggested that histamine, as well as several other substances, could be released by the reaction. Subsequent specific chemical analyses confirmed the presence of histamine. Later studies (Feigen *et al.*, 1968) showed that the interaction of crude sea urchin toxin, or its purified fractions, with certain substrates in the plasma produces dialyzable, heat-stable substances that have the capacity to initiate contractile reactions of the guinea pig ileum and the rat uterus. The same study discusses the immunological behavior of precipitating antibodies produced to the formalinized toxoids.

Alender, Feigen, and Tomita (1965) prepared a singular study on the isolation and characterization of toxin from the sea urchin *T. gratilla*. The results of their effort showed that the active principle of sea urchin toxin is a non-dialyzable, thermolabile protein. It is pH-stable, almost completely soluble in distilled water, and can be precipitated at relatively high potency in the presence of $^{2}/_{3}$ saturated ammonium sulfate in a yield amounting to 26 percent of the dry weight of the starting material.

CHEMISTRY

(Unknown)

The Holothuroidea or Sea Cucumbers

REPRESENTATIVE LIST OF HOLOTHURIANS REPORTED AS TOXIC

Phylum ECHINODERMATA

Class HOLOTHUROIDEA: Sea Cucumbers[1]

Family CUCUMARIIDAE

Afrocucumis africana (Semper) (Pl. 20, fig. a). Sea cucumber (USA). *77-78*

DISTRIBUTION: Japan.

SOURCES: Yamanouchi (1955), Nigrelli and Jakowska (1960).
OTHER NAME: *Pseudocucumis africana.*

Cucumaria echinata Von Marenzeller (Pl. 19). Sea cucumber (USA). *76*

DISTRIBUTION: Japan.

SOURCES: Yamanouchi (1955), Nigrelli and Jakowska (1960).

Pentacta australis (Ludwig) (Pl. 20, fig. h). Sea cucumber (USA). *77-78*

DISTRIBUTION: Tropical west Pacific.

SOURCES: Yamanouchi (1955), Nigrelli and Jakowska (1960).

Family HOLOTHURIIDAE

Actinopyga agassizi (Selenka) (Pl. 18, fig. a). Sea cucumber (USA). *75*

DISTRIBUTION: West Indies.

SOURCES: Hyman (1955), Yamanouchi (1955), Nigrelli and Jakowska (1960), Mebs (1973).

Actinopyga lecanora (Jaeger). Sea Cucumber (USA), stone fish (Australia).

DISTRIBUTION: Australia, Indian Ocean, Celebes, Micronesia, Ryukyu Islands.

SOURCES: Yamanouchi (1955), Nigrelli and Jakowska (1960).
OTHER NAME: *Holothuria lecanora.*

Holothuria argus (Jaeger) (Pl. 21, fig. a). Sea cucumber (USA), tiger fish (Australia). *78*

DISTRIBUTION: Indo-Pacific, Polynesia to the Indian Ocean, northward to the Ryukyus.

SOURCES: Yamanouchi (1955), Nigrelli and Jakowska (1960).
OTHER NAME: *Bohadschia argus.*

[1] The nomenclature as given by Clark (1946) has been largely followed in the presentation of this group.

75 *Holothuria tubulosa* Gmelin (Pl. 18, fig. c). Sea cucumber (USA).
DISTRIBUTION: Mediterranean, adjacent Atlantic coasts.

SOURCES: Yamanouchi (1955), Nigrelli and Jakowska (1960).

Family STICHOPODIDAE

75 *Stichopus chloronotus* Brandt (Pl. 18, fig. b). Sea cucumber (USA).
DISTRIBUTION: Indo-Pacific, Australia.

SOURCES: Yamanouchi (1955), Nigrelli and Jakowska (1960).

Family SYNAPTIDAE

78 *Euapta lappa* (Müller) (Pl. 21, fig. b). Sea cucumber (USA).
DISTRIBUTION: West Indies.

SOURCE: Nigrelli and Jakowska (1960).

Vernacular Names of Sea Cucumbers: Some species of holothurians do not have vernacular names, and are merely designated as sea cucumber, sea slug, nigger, cotton spinner (USA, England, Australia), holothurie, fieuse de coton, bêche-de-mer (France), biche-do-mar (Portugal), trepang, bêche-de-mer (Indo-Pacific area), and erico (Japan).

BIOLOGY

Sea cucumbers are free-living echinoderms, having an elongate wormlike or sausagelike body, without free arms, but with a series of tentacles circling the mouth, located at the anterior end of the body. The intestinal tract is long and looped, terminating in an anus at the posterior end. The skeleton consists of variable-sized, irregularly arranged plates embedded in the skin. Tube feet are present, but not situated in a furrow. In some species of sea cucumbers there are attached to the common stem of the respiratory trees a number of white, pink, or red tubules. These are the so-called Organs of Cuvier, or Cuvierian tubules (Fig. 6). If the sea cucumber is irritated, the Cuvierian tubules are emitted through the anus. Upon contact with the water, the tubules swell and elongate into sticky slender threads which serve to entangle any predator that attempts to annoy it. Only part of the tubules are emitted at any one time, and the autotomized tubules are soon replaced by new ones. In some species the Organs of Cuvier are quite toxic, containing large concentrations of holothurin. The role of these tubules is not completely known, but they are believed to serve as a protective mechanism for the sea cucumber.

Holothurians generally have separate sexes, but some are hermaphroditic. Sea cucumbers are sluggish creatures generally moving over the bottom of the sea by means of rhythmic contractions of the body. The tube feet are used principally as organs of attachment. The power of regeneration is great in holothurians. Under adverse conditions some species eviscerate, but are able to regenerate their organs within 10 to 25 days. Sea cucumbers sometimes camouflage themselves by covering their body with bits of sand, rocks, and debris. The food of sea cucumbers consists of bottom materials, plankton, and detritus, which are shoveled into the mouth or captured by the tentacles.

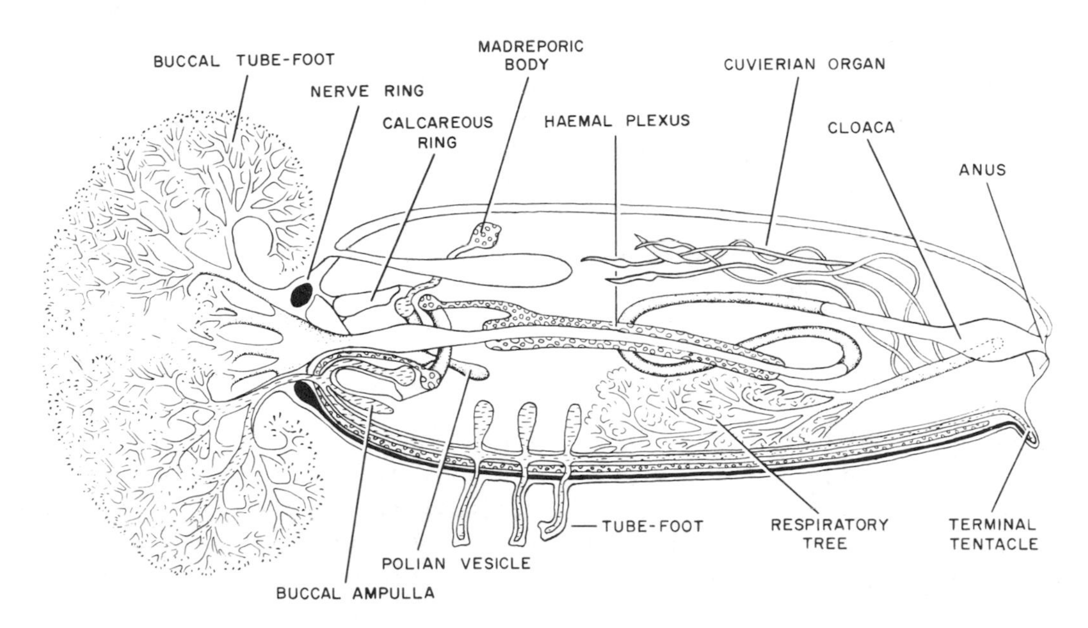

FIGURE 6.—Internal anatomy of a typical sea cucumber. Semidiagrammatic.
(R. Kreuzinger, after Nichols)

Holothurians live in a variety of habitats, viz., under rocks or coral, in crevices, mud, or seaweed, or on flat, open, sandy areas, etc. Some species live primarily in burrows, while others prefer open areas. They are cosmopolitan in their distribution, ranging in their vertical distribution from shallow water tidal areas to great depths.

Sea cucumbers are not used much as food by fish. However, they are an important item of food for man in the Indo-Pacific region, where they are sold under the name of "trepang" or "bêche-de-mer." The sea cucumbers are boiled to cause the animal to eviscerate and to shorten and thicken. After thorough drying, the trepang is ready for marketing. Trepang is used to flavor soups and stews.

MECHANISM OF INTOXICATION

Intoxications can reputedly result from ingestion of toxic sea cucumbers. Liquid ejected from the visceral cavity of some species may result in contact dermatitis or blindness. The poison of sea cucumbers is termed holothurin. Holothurin is said to be concentrated in the Organs of Cuvier.

MEDICAL ASPECTS

Clinical Characteristics

Little information is available regarding the clinical effects of holothurin in humans. Reported symptoms of dermal contact with sea cucumber poison are burning pain, redness, and a violent inflammatory reaction. If the fluid contacts the eyes of the victim, blindness may result. Ingestion of sea cucumber poison may be fatal (Cleland, 1913; Castellani and Chalmers, 1919; Fränkel and Jellinek, 1927).

Treatment

Symptomatic. Pharmacological studies suggest that anticholinesterase agents may be effective in the event of ingestion of holothurin (Friess, 1963). However, no reliable clinical data are presently available on this subject.

Prevention

Local inhabitants should be contacted before a person eats sea cucumbers, particularly in tropical areas. Simple, but reliable tests can be conducted by first feeding small portions of the sea cucumber to dogs or cats.

TOXICOLOGY

Although it has been known since the latter part of the 19th century that certain species of sea cucumbers are toxic, serious investigations in holothurian biotoxicology appear to have started with the work of Yamanouchi in 1929 when he observed that marine fish *Enedrias nebulosus, Girella punctata*, and others, when placed in aquaria contaminated with the tissue extracts of the sea cucumber *Holothuria vagabunda* died within a few minutes. Yamanouchi prepared scrapings from the body wall of *H. vagabunda*, removed small pieces from the body wall, ground them in a mortar with washed sand, boiled some of the material, and mixed other portions with solutions of calcium chloride. Then he tested the extracts on fishes. Unfortunately, most of Yamanouchi's (1942, 1943*a, b*) early works have been published in Japanese journals having very limited distribution.

Yamanouchi (1955) established a toxicity test by using as a test animal the freshwater fish *Oryzias latipes* which had a total length of about 2 cm. Small pieces of the body wall of the sea cucumber were removed, washed in seawater, boiled gently for 15 minutes, and then filtered. Ethanol (94 percent) was added to bring the solution up to 60 to 70 percent alcohol, filtered, and the salts discarded. The alcoholic solution was further distilled, condensed, and a large amount of absolute ethanol added to bring the condensed extract up to 94 percent alcohol. The sea cucumber poison, or holothurin, appeared as minute colorless crystals. The final separation from other contaminating salts was obtained by repeated heating, cooling, and filtration of the solution, since the holothurin is soluble in hot 94 percent ethanol, but precipitated when cold. The poison was found to kill *O. latipes* at 17° C in a solution of 1:200,000 to 1:250,000, or 0.0004-0.0005 percent in 24 hours. The poison was found to be most concentrated in the body wall of *H. vagabunda*, but present to a lesser extent in the viscera and body fluids. Yamanouchi estimated the entire animal to contain about 0.05 percent of the poison.

When Yamanouchi (1955) prepared aqueous extracts from *H. vagabunda*, he found they were toxic to sea anemones (*Actinia equina, Cavernularia obesa*), crustaceans (*Pachygrapsus crassipes, Metapenaeus sp., Eupagurus samuelis*), mollusks (*Aplysia kurodai*), earthworms (*Eisenia foetida*), marine fish (*Enedrias nebulosus, Girella punctata, Pomacentrus coelestis, Plotosus anguillaris*, and *Thallosoma cupido*), and freshwater fish (*Carasius auratus, O. latipes*). Freshwater fish were found to be more resistant to holothurin than marine forms. In most instances the animals died as a result of complete muscular paralysis. Intraperitoneal injections of aqueous extracts of holothurin in dosages of 0.4 to 3.0 mg/10 g mouse weight proved lethal to frogs (*Rana nigromaculata*). Extracts were also tested on mice by the oral, subcutaneous and intravenous routes, and the lethal dosage was found to be 4.0 mg/g, 0.7 mg/g, and 0.0075 mg/g, respectively. The terminal symptoms in mice consisted of a loss of reflexes and motor coordination, and finally a complete muscular paralysis. Autopsies of the mice showed a generalized hyperemia of the intestinal organs and hemorrhages from the lungs. Comparison of the hemolytic properties of holothurin with standardized solutions of commercial Merck's saponin on rabbit red blood cells revealed that holothurin was 10 times more potent than the commercial product.

Yamanouchi (1955) conducted toxicity tests on a larger number of species of sea cucumbers than any other investigator. In a series of 27 Indo-Pacific and Japanese holothurian species tested, 24 species proved to be toxic. (*See* p. 163, above, for partial list.) It is particularly noteworthy that several of the species found to be toxic to laboratory animals are generally considered to be edible and are commonly eaten throughout the tropical Pacific and in the Orient. However, it was found that the edible species were less toxic than those species considered to be dangerous for human consumption. Both *Stichopus japonicus* and *Cucumaria japonica*, which are commonly eaten raw in Japan, were also found to be toxic. Nevertheless, the literature reports no human intoxications from these two species. Yamanouchi explains the apparent discrepancy between laboratory toxicity results and human reactions to poisonous sea cucumbers on the following basis: Most "edible" species contain less poison than the toxic species. Moreover, sea cucumber poison is hydrolyzed by the gastric acids into nontoxic products. In the case of dried trepang, it is usually boiled for 1 hour and permitted to stand overnight in the water. This process is continued for at least 3 days. The poison is leached out and the water discarded, or it is attenuated by the boiling process.

76-77

Yamanouchi found the four most toxic Indo-Pacific species to be *H. atra, H. axiologa, S. variegatus*, and *Thelenota ananas*. The remaining species tested proved to be considerably less toxic. *H. vagabunda* is used in the Tokara Islands as a fish poison. The sea cucumbers are ground up, poured into a tidal pool, and later the paralyzed fishes are captured and eaten.

Most of the research in the United States on the toxicology of holothurin has been conducted by Ross F. Nigrelli and his associates. Crude aqueous extracts were prepared from the West Indian sea cucumber *Actinopyga agassizi* (Nigrelli, 1952). A dilution of 1:100,000 of the crude holothurin was found to kill the fish, *Cyprinodon baconi*, in 23 minutes. A dilution of 1:1,000 killed the fish *Carapus bermudensis* in 8 minutes. When injected subcutaneously, the crude holothurin solution was also demonstrated to be lethal to mice. Efforts to determine the distribution of the poison within the body of the sea cucumber revealed that the poison was most concentrated within the Organs of Cuvier

(Fig. 6). Nigrelli and Zahl (1952) tested crude extracts of holothurin on cultures of protozoans *Tetrahymena pyriformis, Euglena gracilis*, and *Ochromonas malhamensis*. Holothurin produced complete growth inhibition in dilutions of 2.2 mg of holothurin per 100 ml of nutrient liquid.

In an unpublished communication, Philips *et al.* (1956) indicated that solutions of holothurin are much less toxic to mammals when administered by mouth than when given parenterally. Toxicity in fish was believed to be through their gills. The studies of Philips *et al.* suggested that the hemolytic action of holothurin may be the major cause of death in poisoned vertebrates. They estimated the LD_{50} by the intraperitoneal route in mice, and by the intravenous route in dogs and cats to be between 2 to 5 mg/kg.

When Quaglio *et al.* (1957) tested crude aqueous extracts of holothurin on planarians (*Dugesia tigrina*), they found that the poison was lethal in concentrations between 0.00003 percent and 0.00001 percent. If the planarians were transvected, placed in 0.01 percent solution for 30 and 45 seconds and 1 minute, then removed and placed in fresh uncontaminated water, the anterior portions of the planarians regenerated within the usual period of time; but the posterior sections either disintegrated or failed to regenerate. The authors have suggested that holothurin may have antimetabolic properties.

Goldsmith, Osburg, and Nigrelli (1958) have shown that crude aqueous extracts of holothurin cause retardation of pupation when tested on the fruitfly *Drosophila melanogaster*.

Jakowska *et al.* (1958) showed that 0.2 ml of a 5 percent solution of crude holothurin injected in the dorsal lymph spaces of winterized frogs *Rana pipiens* caused 50 percent hemolysis, and stimulated hemopoiesis in the bone marrow.

Ruggieri, Ruggieri, and Nigrelli (1960) have shown that pre- and postfertilized sea urchin eggs (*Arbacia punctulata*) immersed for variable periods of time in aqueous extracts of holothurin in dilutions of 1,000 to 0.001 ppm resulted in various developmental abnormalities of the ova.

Nigrelli and Jakowska (1960) reviewed most of the published toxicological literature dealing with holothurin. They reported that holothurin extracts suppress root hair development in watercress, and cause necrosis of onion root tips. Their studies on the hemolytic properties of holothurin on rabbit red blood cells indicated that holothurin had greater hemolytic activity than commercial saponin.

Lasley and Nigrelli (1970) conducted experiments to show the effect of crude holothurin on leucocytes. Various concentrations of holothurin killed proportionate numbers of staphylococci as well as leucocytes.

Corbett (1971) anticipates the potential therapeutic uses of holothurins as a result of their demonstrated antimetabolic and anticholinergic properties.

PHARMACOLOGY

Data on the pharmacology of holothurin is quite limited, and what little is known is of recent vintage. Although Yamanouchi's (1942, 1943*a, b*, 1955) valuable work is quite extensive, it is limited largely to the toxicology of holothurin, and provides very little information concerning the general pharmacodynamics of sea cucumber poison. The early studies of Nigrelli (1952) and Nigrelli and Zahl (1952) were concerned with the general biological characteristics of the poison. Among other things, they noted that solutions of crude holothurin were capable of suppressing the growth of Sarcoma 180 in laboratory mice. Similar effects were observed in working with Krebs-2 ascites tumor in mice (Sullivan and Nigrelli, 1956; Nigrelli and Jakowska, 1960).

Friess *et al.* (1959, 1960) have studied the effects of purified holothurin on the propogated action potential in desheathed bullfrog sciatic nerve, single nerve fiber preparations from frog and toad, and on the rat and kitten nerve-diaphragm preparations. A purified fraction of holothurin was obtained using the method of Nigrelli *et al.* (1955). It was found that the steroidal glycoside holothurin had a powerful and irreversible action on both amphibian and mammalian nerve preparations, and it appeared to produce a direct contractural effect on muscle. Holothurin appeared to be comparable to the reference blocking agents cocaine, procaine, and physostigmine in potency on desheathed bullfrog sciatic nerve. However, it was in marked contrast to these reference agents in that the action of holothurin is quite irreversible on washing, and it did not alter impulse conduction velocity in the course of its attenuation of the impulse. Holothurin also displayed the same irreversibility of action with respect to the blockade of the action current in single fiber-single node nerve preparations from the toad and the frog. Holothurin irreversibly blocked the twitch response via either direct or indirect stimulation paths in the rat phrenic nerve-diaphragm. In addition, holothurin displayed a direct contractural effect on the muscle.

Friess (1963) has noted a very significant pharmacological finding as the result of investigations on holothurin. Purified holothurin (designated holothurin A) appeared to hit selected chemoreceptor targets hard and it simultaneously destroyed their capacity for evoking and controlling their linked physiological functions. However, the irreversibility in knockout of key chemoreceptors at the mammalian neuromuscular synapse by a $10^{-4} M$ of holothurin could be largely obviated by the presence of mere traces of certain protecting agents. It was found that a 10^{-10} to $10^{-9} M$ level of the classical anticholinesterase physostigmine would protect against much of the irreversible damage normally inflicted by holothurin at $10^{-4} M$, but this protection disappeared at higher concentrations of physostigmine. Also it was observed that the quaternary anticholinesterase neostigmine protected against the irreversible facets of holothurin-synapse interactions, but in a higher concentration range of the protecting agent.

Friess *et al.* (1970) observed that the saponin of purified holothurin A (H-) and its neutral desulfated derivative (DeH) acted upon the cat superior cervical ganglion preparations to produce irreversible inactivation on the ganglion. Friess *et al.* (1972) studied holothurin A from a Bahamian sea cucumber with regard to potency in the blockade of responses of phrenic nerve-diaphragm preparations from rat and guinea pig under the stresses of high environmental pressure or variable loading of the muscle.

The hemolytic and hemopoetic properties of holothurin were previously discussed in the section on toxicology.

CHEMISTRY

by DONOVAN A. COURVILLE

Certain species of sea cucumbers contain a poisonous principle to which the name holothurin has been applied. Nigrelli *et al.* (1955), in their investigation, used the Cuvier organs of *Actinopyga agassizi* as the source of the poison.

The toxin was submitted to elemental analysis by Chanley *et al.* (1955) to yield data suggesting an empirical formula $C_{50}H_{82}O_{26}S$.

The toxin is described as hygroscopic, soluble in water in all proportions, without water of hydration in the crystals, having an indefinite melting point which is complete at 206° C, nondialyzable through a collodion membrane, partly adsorbed on animal charcoal, nonreducing prior to hydrolysis but liberating reducing substances on acid hydrolysis, having a strong hemolytic action, and having foaming properties, even in dilute solutions. The substance isolated was termed "holothurigenin."

The ultraviolet absorption bands at 244, 238, and 255 mμ together with the color reaction with tetranitromethane, suggested a heteroannular diene with conjugate double bonds. The presence of conjugate double bonds was confirmed by the disappearance of the band at 244 mμ on catalytic hydrogenation to the dihydro derivative. The results of the investigation thus far suggested that "holothurigenin" was a trihydroxy-lactone-diene belonging to the terpenoid series. See work by Nigrelli *et al.* (1955), Chanley *et al.* (1955, 1959, 1960), Yamanouchi (1955), Matsumo and Yamanouchi (1961).

LITERATURE CITED

Akiya, R.
1937 Starfish lees is effective for extermination of flies. [In Japanese] Rept. Hokkaido Fish. Lab. (361): 10.

Alender, C. B.
1964 The venom from the heads of the globiferous pedicellariae of the sea urchin, *Tripneustes gratille* (Linnaeus). Ph.D. Thesis, Univ. Hawaii. 126 p.

Alender, C. B., Feigen, G. A., and J. T. Tomita
1965 Isolation and characterization of sea urchin toxin. Toxicon 3: 9-17.

Alender, C. B., and F. E. Russell
1966 Pharmacology, p. 529-543, 1 fig. *In* R. A. Boolootian [ed.], Physiology of Echinodermata. Interscience Publishers, New York.

Ando, Y.
1954 J. Hokkaido Fish. Sci. Inst. Ser. 7, 11(7): 29. (NSA)

Ando, Y., and O. Hasegawa
1955 Study of preventive effect of starfish lees to flies. [In Japanese] Rept. Hokkaido Hyg. Lab. 7: 67.

Brocq-Rousseu, D., and R. Fabre
1942 Toxalbumines animales, p. 9-10. *In* D. Brocq-Rousseu and R. Fabre, Les toxalbumines. Hermann & Cie, Paris.

Brown, T.
1972 Crown of thorns, the death of the Great Barrier Reef? Angus and Robertson Publishers, Sydney, Australia. 128 p., illus.

Castellani, A., and A. J. Chalmers
1919 Venomous animals: Protozoa to Arthropoda, p. 203-241; Figs. 17-26. *In* A. Castellani and A. J. Chalmers, Manual of tropical medicine. 3rd ed. Wm. Wood Co., New York.

Chaet, A. B.
1962 A toxin in the coelomic fluid of scalded starfish (*Asteria forbesi*). Proc. Soc. Exp. Biol. Med. 109(4): 791-794, 1 fig.

(NSA)—Not seen by author.

Chanley, J. D., Kohn, S. K., Nigrelli, R. F., and H. Sobotka
1955 Further chemical analysis of holothurin, the saponin-like steroid from the sea-cucumber. Part II. Zoologica 40: 99.

Chanley, J. D., Ledeen, R., Wax, J., Nigrelli, R. F., and Sobotka
1959 Holothurin. I. The isolation, properties, and sugar components of holothurin A. J. Am. Chem. Soc. 81: 5180-5183.

Chanley, J. D., Perlstein, J., Nigrelli, R. F., and H. Sobotka
1960 Further studies on the structure of holothurin. Ann. N.Y. Acad. Sci. 90(3): 902-905.

Clark, A. H.
1950 Echinoderms from the Cocos-Keeling Islands. Bull. Raffles Mus., Singapore (22): 53-67; Pl. 15.
1953 Poisonous echinoderms. (Personal communication, July 14, 1953.)

Clark, H. L.
1921 The echinoderm fauna of Torres Strait: its composition and its origin. Carnegie Institution of Washington, Publ. 214, p. 155-156.
1938 Echinoderms from Australia. Mem. Mus. Comp. Zool., Harvard, Vol. 55, 596 p., 64 figs., 28 pls.
1946 The echinoderm fauna of Australia. Its composition and its origin. Carnegie Institution of Washington, Publ. 566, 567 p.

Cleland, J. B.
1912 Injuries and diseases of man in Australia attributable to animals (except insects). Australasian Med. Gaz. 32: 269-272.
1913 Injuries and diseases of man in Australia attributable to animals (except insects). J. Trop. Med. Hyg. 16: 25-31.
1932 Injuries and diseases in Australia attributable to animals (other than insects). Med. J. Australia, 4, 1932: 157-166.

CORBETT, M. D.
1971 Holothurins—potential drugs from the sea. Bull. Bur. Pharm. Serv. 7 (7): 1-4, 2 figs. [University of Mississippi: School of Pharmacy].

CUÉNOT, L.
1948 Les échinids; Pédicellaires, p. 131-134. *In* P. P. Grasse [ed.], Traité de zoologie. Anatomie, systématique, biologie. Vol. 11, Echinodermes—stomocordes, procordés. Masson et Cie., Paris.

EARLE, K. V.
1940 Pathological effects of two West Indian echinoderms. Trans. Roy. Soc. Trop. Med. Hyg. 33(4): 447-452, 2 figs.
1941 Echinoderm injuries in Nauru. Med. J. Australia 2(10): 265-266.

ENDEAN, R.
1961 The venomous sea-urchin *Toxopneustes pileolus.* Med. J. Australia 1(9): 320.

FEIGEN, G. A., SANZ, E., and C. B. ALENDER
1966 Studies on the mode of action of sea urchin toxin, 1. Conditions affecting release of histamine and other agents from isolated tissues. Toxicon 4: 161-175, 6 figs.

FEIGEN, G. A., SANZ, E., TOMITA, J. T., and C. B. ALENDER
1968 Studies on the mode of action of sea urchin toxin, 2. Enzymatic and immunological behavior. Toxicon 6: 17-43, 15 figs.

FISH, C. J., and M. C. COBB
1954 Noxious marine animals of the central and western Pacific Ocean. U.S. Fish Wildlife Serv., Res. Rept. No. 36, pp. 21-23.

FOETTINGER, A.
1881 Sur la structure des pédicellaires gemmiformes de *Sphaerechinus granularis* et d'autres échinides. Arch. Biol. 2: 455-496.

FRÄNKEL, S., and C. JELLINEK
1927 Über essbare holothurin. Biochem. Z. 185: 389-391.

FRASER-BRUNNER, A.
1973 Danger in the sea. Hamlyn Publ. Group Ltd., New York. 128 p., illus.

FRIESS, S. L.
1963 Some pharmacological activities of the sea cucumber neurotoxin. A.I.B.S. Bull. 13(2): 41.

FRIESS, S. L., CHANLEY, J. D., HUDAK, W. V., and H. B. WEEMS
1970 Interactions of the echinoderm toxin holothurin A and its desulfated derivative with the cat superior cervical ganglion preparation. Toxicon 8: 211-219.

FRIESS, S. K., DURANT, R. C., FINK, W. L., and J. D. CHANLEY
1972 Effects of pressure and muscle loading on the toxic actions of echinoderm saponins in neuromuscular tissues. Toxicol. Appl. Pharmacol. 22: 115-127.

FRIESS, S. L., STANDAERT, F. G., WHITCOMB, E. R., NIGRELLI, R. F., CHANLEY, J. D., and H. SOBOTKA
1959 Some pharmacologic properties of holothurin, an active neurotoxin from the sea cucumber. Pharmacol. Exp. Therap. 126(4): 323-329, 3 figs.
1960 Some pharmacologic properties of holothurin A, a glycosidic mixture from the sea cucumber. Ann. N.Y. Acad. Sci. 90(3): 893-901, 2 figs.

FUJITA, M., and K. NISHIMOTO
1952 On the biological assay of Japanese Senega. [In Japanese, English summary], J. Pharm. Soc. Japan 72: 1645-1646.

FUJIWARA, T.
1935 On the poisonous pedicellaria of *Toxopneustes pileolus* (Lamarck). Annotationes Zool. Japon. 15(1): 62-67.

GIUNIO, P.
1948 Otrovne ribe. Higijena (Belgrade) 1(7-12): 282-318, 20 figs.

GOLDSMITH, E. D., OSBURG, H. E., and R. F. NIGRELLI
1958 The effect of holothurin, a steroid saponin of animal origin on the development of the fruit fly. (Abstr.) Anat. Record 130: 411-412.

HALSTEAD, B. W.
1956 Animal phyla known to contain poisonous marine animals, p. 9-27. *In* E. E. Buckley and N. Porges [eds.], Venoms. Am. Assoc. Adv. Sci., Washington, D.C.
1974 Marine biotoxicology, p. 212-239. *In* F. Coulston and F. Korte [eds.], Environmental quality and safety, Vol. 3. Academic Press, Inc., New York.

HASHIMOTO, Y., and T. YASUMOTO
1960 Confirmation of saponin as a toxic principle of starfish. Bull. Japan. Soc. Sci. Fish 26(11): 1132-1138.

HENRI, V., and E. KAYALOF
1906 Étude des toxines contenues dans les pédicellaires chez les oursins. Compt. Rend. Soc. Biol. 60: 884-886.

HYMAN, L. H.
1955 The invertebrates: Echinodermata. The coelomate Bilateria. Vol. IV. McGraw-Hill Book Co., New York. 763 p., 280 figs.

IMAI, T., HATANAKA, O., and S. OSHIMA
1956 Bull. Japan. Soc. Sci. Fish, Tokoku Branch 7, (1-2). (NSA)

JAKOWSKA, S., NIGRELLI, R. F., MURRAY, P. M., and A. M. VELTRI
1958 Hemopoietic effects of holothurin, a steroid saponin from the sea cucumber, *Actinopyga agassizi*, in *Rana pipiens*. (Abstr.) Anat. Record 132: 459.

KAISER, E., and H. MICHL
1958 Die biochemie der tierischen gifte. Franz Deuticke, Wien. 258 p., 23 figs.

KAYALOF, E.
1906 Étude des toxines des pédicellaires chez les oursins. These, Fac. Med. Univ. de Geneve, Paris. 57 p., 3 pls.

LASLEY, B. J., and R. F. NIGRELLI
1970 The effect of crude holothurin on leucocyte phagocytosis. Toxicon 8: 301-306.

LEE, C. F.
1951 Part III. Value of starfish meal—protein supplement for growth of rats and chicks and for egg production, p. 15-26, 2 figs. *In* C. F. Lee, Technological studies of the starfish, U.S. Wildlife Serv., Fish. Leaflet 391.

LÉVY, R.
1925 Sur les propriétés hémolytiques des pédicellaires de certains oursins réguliers. Compt. Rend. Acad. Sci. 181(2): 690-692.

LOISEL, G.
1903 Les poisons des glandes génitales. Premiére note. Recherches et expérimentation chez l'oursin. Compt. Rend. Soc. Biol. 55: 1329-1331.

(NSA)—Not seen by author.

1904 Recherches sur les poisons génitaux de différents animaux. Compt. Rend. Acad. Sci. 139: 227-229.

LUDWIG, H., and O. HAMANN
1901 Dr. H. G. Bronn's Klassen und ordnungen des thier-reichs, wissenschaftlich dargestellt in wort und bild. Echinoderm (Stachelhauter). Vol. 2. Sect. 3. Lieferung 41-43. C. F. Winter'sche Verlagshandlung, Leipzig, p. 1022-1030.
1902 Dr. H. G. Bronn's Klassen und ordnungen des thier-reichs, wissenschaftlich dargestellt in wort und bild. Echinoderm (Stachelhauter). Vol. 2. Sec. 3. Lieferung 44-48. C. F. Winter'sche Verlagshandlung, Leipzig, p. 1031-1042; Pls. II and IV.
1904 Dr. H. G. Bronn's Klassen und ordnungen des thier-reichs, wissenschaftlich dargestellt in wort und bild. Echinoderm (Stachelhauter). Vol. 2. Sect. 3. IV. Buch. Die seeigel. C. F. Winter'sche Verlagshandlung, Leipzig. 1,413 p., 18 pls., 15 figs.

MANSON-BAHR, P. H., *ed.*
1950 Animal poisons, p. 850-852. *In* P. H. Manson-Bahr [ed.], Manson's tropical diseases. A manual of the diseases of warm climates. 13th ed. Williams & Wilkins Co., Baltimore.

MATSUNO, T., and T. YAMANOUCHI
1961 A new triterpenoid sapogenin of animal origin (sea cucumber). Nature 191 (4783): 75-76.

MEBS, D.
1973 Chemistry of animal venoms, poisons and toxins. Experientia 29(11): 1328-1334.

MENDES, E. G., ABBUD, L., and S. UMIJI
1963 Cholinergic action of homogenates of sea urchin pedicellariae. Science 139 (3553): 408-409.

MORTENSEN, T.
1910 On some West Indian echinoids. U.S. Nat. Mus.,. Bull. 74, 31 p., 17 pls. (NSA)
1927 Handbook of the echinoderms of the British Isles. Oxford Univ. Press, Edinburgh. 471 p., 269 figs.
1928 A monograph of the Echinoidea. Cidaroidea. Vol. I. Carlsberg-Fund, C. A. Reitzel, Copenhagen. 551 p., 173 figs., 88 pls.

MORTENSEN, T.—Continued

1935 A monograph of the Echinoidea. Bothriocidaroida, Melonechinoida, Lepidocentroida, and Stirodonta. Vol. II. Carlsberg-Fund, C. A. Reitzel, Copenhagen. 647 p., 377 figs. 89 pls.

1940*a* A monograph of the Echinoidea. Aulodonta with additions to Vol. II (Lepidocentroida and Stirodonta). Vol. III. Part I. Carlsberg-Fund, C. A. Reitzel, Copenhagen. 370 p., 196 figs., 77 pls.

1940*b* Report on the Echinoidea collected by the U.S. Fisheries steamer *Albatross* during the Philippine expedition, 1907-10. U.S. Nat. Mus., Bull. 100, Vol. 14, pt. 1, 52 p., 3 figs., 1 pl.

1943*a* A monograph of the Echinoidea. Camarodonta. I. Orthopsidae, Glyphocyphidae, Temnopleuridae, and Toxopneustidae. Vol. III. Part 2. Carlsberg-Fund, C. A. Reitzel, Copenhagen. 553 p., 321 figs., 56 pls.

1943*b* A monograph of the Echinoidea. Camarodonta. II. Echinidae, Stronglyocentrotidae. Parasaleniidae, Echinometridae. Vol. III. Part 3. Carlsberg-Fund, C. A. Reitzel, Copenhagen. 446 p., 215 figs., 66 pls.

1943*c* A monograph of the Echinoidea. Holectypoida, Cassiduloida. Vol. IV. Part 1. Carlsberg-Fund, C. A. Reitzel, Copenhagen. 371 p., 326 figs., 14 pls.

1948 A monograph of the Echinoidea. Clypeastroida, Clypeastridae, Arachnoididae, Fibulariidae, Laganidae and Scutellidae. Vol. IV. Part 2. Carlsberg-Fund, C. A. Reitzel, Copenhagen. 471 p., 258 figs., 72 pls.

1950 A monograph of the Echinoidae, Spatangoida. I. Protosternata, Meridosternata, Amphisternata. I. Palaeopneustidae, Palaeostomatidae, Aeropsidae, Toxasteridae, Micrasteridae, Hemiasteridae. Vol. V. Part 1. Carlsberg-Fund, C. A. Reitzel, Copenhagen. 432 p., 315 figs., 25 pls.

1951*a* A monograph of the Echinoidea. Spatangoida. II. Amphisternata. II. Spatangidae, Loveniidae, Pericosmidae, Schizasteridae, Brissidae. Vol. V. Part 2. Carlsberg-Fund, C. A. Reitzel, Copenhagen. 593 p., 286 figs., 64 pls.

1951*b* A monograph of the Echinoidea. Index to Vols. I-V. Carlsberg-Fund, C. A. Reitzel, Copenhagen. 63 p.

NICHOLS, D.

1962 Echinoderms. Hutchinson University Library, London. 200 p., 26 figs.

NIELLY, M.

1881 Animaux et végétaux nuisibles, p. 709-710. *In* M. Nielly, Elements de pathologie exotique. Delahaye and Lecrosnier, Paris.

NIGRELLI, R. F.

1952 The effects of holothurin on fish, and mice with sarcoma 180. Zoologica 37: 89-90.

NIGRELLI, R. F., CHANLEY, J. D., KOHN, S. K., and H. SOBOTKA

1955 The chemical nature of holothurin, a toxic principle from the sea-cucumber (Echinodermata: Holothurioidea). Zoologica 40: 47-48.

NIGRELLI, R. F., and S. JAKOWSKA

1960 Effects of holothurin, a steroid saponin from the Bahamian sea cucumber (*Actinopyga agassizi*), on various biological systems. Ann. N.Y. Acad. Sci. 90(3): 884-892, 2 figs.

NIGRELLI, R. F., and P. ZAHL

1952 Some biological characteristics of holothurin. Proc. Soc. Exp. Biol. Med. 81: 379-380, 1 figs.

OKADA, K.

1955 Biological studies on the practical utilities of poisonous marine invertebrates. I. A preliminary note on the toxical substance detected in the trumpet sea-urchin, *Toxopneustes pileolus*. Records Oceanogr. Works Japan 2(3): 49-52, 3 figs.

OKADA, K., HASHIMOTO, T., and Y. MIYAUCHI

1955 A preliminary report on the poisonous effect of the *Toxopneustes* toxin upon the heart of oyster. Bull. Marine Biol. Sta. Asamushi, Tohoku Univ. 7(2-4): 133-140, 11 figs.

OSHIMA, H.
1915 Dokugase (poisonous sea-urchin). [In Japanese, English summary]; Dobut-sugaku Zasshi 27: 605.

OWELLEN, R. J., OWELLEN, R. G., GOROG, M. A., and D. KLEIN
1973 Cytolytic saponin fraction from *Asterias vulgaris.* Toxicon 11: 319-323.

PARKER, C. A.
1881 Poisonous qualities of the star-fish. Zoologist, 3, 5: 214-215.

PAWLOWSKY, E. N.
1927 Tiere mit giftigen drüsen und einem besonderen verwundungsapparat, p. 32-38, 15 figs. *In* E. N. Pawlowsky, Gifttiere und ihre giftigkeit. Gustav Fischer, Jena.

PHILIPS, F. S., GALLAGHER, T. F., BALIS, M. E. and S. S. STERNBERG
1956 Pharmacology of holothurin. (Personal communication, Jan. 18, 1956.)

PHILLIPS, C., and W. H. BRADY
1953 Sea pests—poisonous or harmful sea life of Florida and the West Indies. Univ. Miami Press, Miami, Fla. 78 p., 38 figs., 7 pls.

PHISALIX, M.
1922 Échinodermes, p. 77-99, 12 figs. *In* M. Phisalix, Animaux venimeux et venins. Vol. I. Masson et Cie., Paris.

POPE, E. C.
1963*a* Some noxious marine invertebrates from Australian seas, p. 91-102, 2 figs. *In* J. W. Evans [chmn.], Proceedings: first international convention on life saving techniques. Part III. Scientific Section. Bull. Post Grad. Comm. Med., Univ. Sydney.
1963*b* Australian venomous marine animals. (Personal communication, July 25, 1963.)
1964 A stinging by a crown-of-thorns starfish. Australian Nat. Hist. 14(11): 350.
1967 Venomous sea urchin in Sydney harbour. Australian Nat. Hist. 15(9): 289, 1 fig.

PUGH, W. S.
1913 Report of case of poisoning by sea urchin. U.S. Naval Med. Bull. 7(2): 245-255.

QUAGLIO, N. D., NOLAN, S. F., VELTRI, A. M., MURRAY, P. M., JAKOWSKA, S., and R. F. NIGRELLI
1957 Effects of holothurin on survival and regeneration of planarians. (Abstr.) Anat. Record 128: 604-605.

RANDALL, J. E.
1958 A review of ciguatera tropical fish poisoning, with a tentative explanation of its cause. Bull. Mar. Sci. Gulf Caribbean 8(3): 236-267, 2 figs.

RIO, G. J., STEMPIEN, M. F., JR., and R. F. NIGRELLI
1963 Saponin-like toxin from the giant sunburst starfish, *Pycnopodia helianthoides*, from the Pacific Northwest.

RIO, G. J., STEMPIEN, M. F., JR., NIGRELLI, R. F., and G. D. RUGGIERI
1965 Echinoderm toxins, 1. Some biochemical and physiological properties of toxins from several species of Asteroidea.

RUGGIERI, G. D., RUGGIERI, S. F., and R. F. NIGRELLI
1960 The effects of holothurin, a steroid saponin from the sea cucumber, on the development of the sea urchin. Zoologica 45: 1-16, 4 pls.

SAWANO, E., and K. MITSUGI
1932 Toxic action of the stomach extracts of the starfishes on the heart of the oyster. Sci. Rept. Tohoku Univ., 4, 7(1): 79-88, 16 figs.

SLADEN, W. P.
1880 On a remarkable form of pedicellaria, and the functions performed thereby. Together with general observations on the allied forms of this organ in the Echinidae. Ann. Mag. Nat. Hist., 5, 6(32): 101-114; Pls. 12-13.

SOMMER, H., and K. F. MEYER
1937 Paralytic shell-fish poisoning. Arch. Pathol. 24(5): 560-598, 2 figs.

STORER, T. I., and R. L. USINGER
1957 Phylum Echinodermata. Echinoderms, p. 339-350; Figs. 22-1 through 22-14. *In* E. J. Boell [ed.], General zoology. 3d ed. McGraw-Hill Book Co., Inc., New York.

SULLIVAN, T. D., and R. F. NIGRELLI
1956 Antitumorous action of biologics of marine origin. I. Survival of Swiss mice inoculated with Krebs-2 ascites tumor and treated with holothurin, a steroid saponin from the sea-cucumber *Actinopyga agassizi*. (Abstr.) Proc. Am. Assoc. Cancer Res. 2: 151.

SWAN, J. G.
1889 Fatal injury inflicted by a starfish. Bull. U.S. Fish Comm. 7(3): 34.

TAFT, C. H.
1945 Poisonous marine animals. Texas Rept. Biol. Med. 3(3): 339-352.

TASCHENBERG, O.
1909 Die giftigen tiere. Ferdinand Euke, Stuttgart. 325 p., 64 figs.

UEXKÜLL, J. VON
1899 Die physiologie der pedicellarien. Z. Biol. 37(19): 334-403, 2 pls.

WALL, M. E., KRIDER, M. M., ROTHMAN, E. S., and C. R. EDDY
1952 Steroidol sapogenins. I. Extraction, isolation, and identification. J. Biol. Chem. 198 (2): 533-543.

WHITSON, D., and H. W. TITUS
1945 The use of starfish meal in chick diets. Bull. Bingham Oceanogr. Coll. 9: 24-27.

YAMANOUCHI, T.
1942 Study of poisons contained in holothurians. [In Japanese] Teikoku Gakushiin Hokoku (17): 73.
1943a On the poison contained in *Holothuria vagabunda*. [In Japanese] Folia Pharmacol. Japon. 38(2): 115.
1943b Distribution of poison in the body of *Holothuria vagabunda*. [In Japanese] Zool. Mag. (Tokyo) 55(12): 87-88.
1955 On the poisonous substance contained in holothurians. Publ. Seto Marine Biol. Lab. 4(2-3): 183-203, 2 figs.

YASUMOTO, T., and Y. HASHIMOTO
1965 Properties and sugar components of asterosaponin A isolated from starfish. Agric. Biol. Chem. 29(9): 804-808.

YASUMOTO, T., TANAKA, M., and Y. HASHIMOTO
1966 Distribution of saponin in echinoderms. Bull. Jap. Soc. Sci. Fish. 32(8): 673-676, 1 fig.

YASUMOTO, T., WATANABE, T., and Y. HASHIMOTO
1964 Physiological activities of starfish saponin. Bull. Jap. Soc. Sci. Fish. 30(4): 357-364, 3 figs.

Chapter VI—INVERTEBRATES

Phylum Mollusca

SNAILS, BIVALVES, OCTOPUSES, ETC.

Mollusks are unsegmented invertebrates having a soft body and usually secreting a calcareous shell. A muscular foot is present which may be modified to serve various functions. Covering at least a portion of the body is a soft skin, the mantle, the outer surface of which secretes the shell. Respiration is by means of gills or a modified primitive pulmonary sac. Jaws are present in some species. In four of the five classes, food is obtained by the use of a rasp-like device called a radula. In the cone shells and a few others, the radula ribbon is lost and the teeth are modified into hollow, harpoonlike structures. In the bivalves, the radula is absent. It is estimated that there are 45,000 living species of mollusks.

The phylum Mollusca is generally divided into five classes:*

AMPHINEURA: Chitons. There are few data regarding the toxicity of chitons, which are included in this class. *Mopalia muscosa* is reported to transvect paralytic shellfish poisoning.

SCAPHOPODA: Tooth or tusk shells. No tooth or tusk shell is reputed to be toxic to man.

GASTROPODA: Univalve snails and slugs. The Gastropoda includes land, freshwater and marine snails, and slugs. Members of this class are characterized by a univalve shell or the lack of a shell. The body is usually asymmetrical in a spirally coiled shell (the slugs are an exception to this arrangement). Typically there is a distinct head, with one to two pairs of tentacles, two eyes, and a large, flattened, fleshy foot. Gastropods may be either monoecious or dioecious, and are mostly oviparous.

Most species of the three families Conidae, Turridae, and Terebridae, which constitute the suborder Toxoglossa, are believed to be provided with venom organs. However, of the toxoglossids, only the family Conidae has been incriminated in human intoxications (Sars, 1878; Clench, 1946; Hartley, 1954). Other members of the Gastropoda which are known to contain toxic species are *Aplysia, Creseis, Haliotis, Livona, Murex,* and *Neptunea*. It is estimated that there are 33,000 living species of gastropods.

---

* Malacologists have recently added the class Monoplacophora—primitively segmented, deepsea mollusks with a limpetlike shell. There are no reports available concerning their toxicity.

PELECYPODA: Bivalves (scallops, oysters, clams). The members of this class—clams, oysters, mussels, scallops, etc.—are characterized as having a shell of two lateral valves, usually symmetrical with a dorsal hinge, a ligament, and closed by one or two adductor muscles. They have a mantle of right and left lobes with the margins forming posterior siphons. These siphons control the flow of water into the mantle cavity. The foot is often hatchet shaped. They have no head, jaws, or radula. Most species are dioecious, producing veliger larvae or a glochidial stage. The genera *Modiolus, Mya, Mytilus, Protothaca, Saxidomus, Spisula*, etc. are transvectors of paralytic shellfish poisoning; however, these genera are discussed in Chapter II. Only the families Ostreidae and Veneridae are considered to be of direct concern to the subject matter of this chapter. Occasionally pelecypods have been involved in outbreaks of Minamata disease. It is estimated that there are 11,000 living species of Pelecypoda.

CEPHALOPODA: Squid, octopus, nautilus, cuttlefish. This class includes the nautilus, squid, cuttlefish, and octopus. The shell is external, internal, or absent. The head is large and contains conspicuous and well-developed eyes. The mouth, armed with horny jaws and a radula, is surrounded by 8 or 10 tentacles equipped with numerous suckers. Rapid movements can be produced by expelling water from the mantle cavity through the siphon. The sexes are separate. The toxicity of the salivary secretions of *Octopus* has been of principal interest to venomologists. There are about 650 species, all of which are marine.

Mollusks constitute the largest single group of biotoxic marine invertebrates of direct importance to man. Although dinoflagellates are the primary etiological agent of many mollusk biotoxications, mollusks serve as the major transvector through which these poisonings take place. During 1955-57 mollusks accounted for 1.9 million metric tons of invertebrates consumed in the world market, whereas the next highest consumer item was crustaceans at 0.8 million metric tons (Riedl, 1961). Thus, from the viewpoint of global public health and economics mollusks are an important item in human ecology. In terms of their biotoxicological order of importance within the phylum Mollusca, the pelecypods probably rank first in the number of intoxications which they cause each year. The next most important group is the gastropods, with the cephalopods ranking third. Historically mollusks have received greater attention by physicians and scientists than any other invertebrate group.

The systematics and identifying characteristics of the mollusks presented in this chapter are discussed by: Bergh (1895), Berry (1912, 1914), Rogers (1951), Abbott (1954), Kohn (1959*b*), Powell (1961), Warmke and Abbott (1961), Hanna (1963), and Riedl (1963).

Research on molluskan venomology is still in its infancy and much remains to be learned about the phylogenetic distribution of venom organs among mollusks. There is an increasing amount of evidence that venom organs are present in a far greater number of mollusks than was formerly suspected. This is indeed an area for much productive research in the future.

REPRESENTATIVE LIST OF WHELKS REPORTED AS POISONOUS[1]

Phylum MOLLUSCA

Class GASTROPODA: Snails, Slugs

Subclass PROSOBRANCHIA

Order ARCHAEOGASTROPODA

Family BUCCINIDAE

Babylonia japonica (Reeve). Japanese ivory shell (Japan).
Distribution: Japan.

SOURCES: Shibōta and Hashimoto (1970, 1971).

Neptunea antiqua (Linnaeus) (Pl. 1, fig. a). Red whelk, almond, buckie, mutlog, googawn, cuckoo shell (USA, England), barnagh (Ireland), trompetenschnecke, kinkhorn (Germany). *80*
DISTRIBUTION: Northern Europe.

SOURCES: Fänge (1957, 1958, 1960).

Neptunea arthritica (Bernardi) (Pl. 1, fig. b). *80*
DISTRIBUTION: Japan.

SOURCE: Asano and Ito (1960).

Family MURICIDAE

Thais haemastoma (Clench).
DISTRIBUTION: Southeast U.S., West Indies.

SOURCE: Huang and Mir (1972).

BIOLOGY

The whelks are a large and aggressive family of carnivorous mollusks that range from tropical to circumpolar seas. They have a vertical distribution which ranges from the littoral zone to great depths. Their shells come in a variety of shapes, sizes, and colors. Some are brilliantly colored whereas others are quite plain.

Whelks are commonly observed clambering about on rocks or plowing their way through mud, sand, or gravel, with the muscular foot largely buried in the substrate. When at rest, the mollusk retracts the foot and thereby closes the aperture with a horny operculum. Most whelks are more scavengers than active predators, feeding on dead fish and other scraps; but they tend to shun anything in an advanced state of decay. Because of their feeding habits they are a pest to lobster and crab fishermen, for they steal their bait at every opportunity. Some of the whelks feed on live mollusks by boring holes in their shells. At times they may cause considerable damage to oyster beds (Lovell, 1867). Egg laying takes place during the spring or early summer

[1] Species of mollusks involved in paralytic shellfish poisoning are not listed in this chapter (*see* Chapter II).

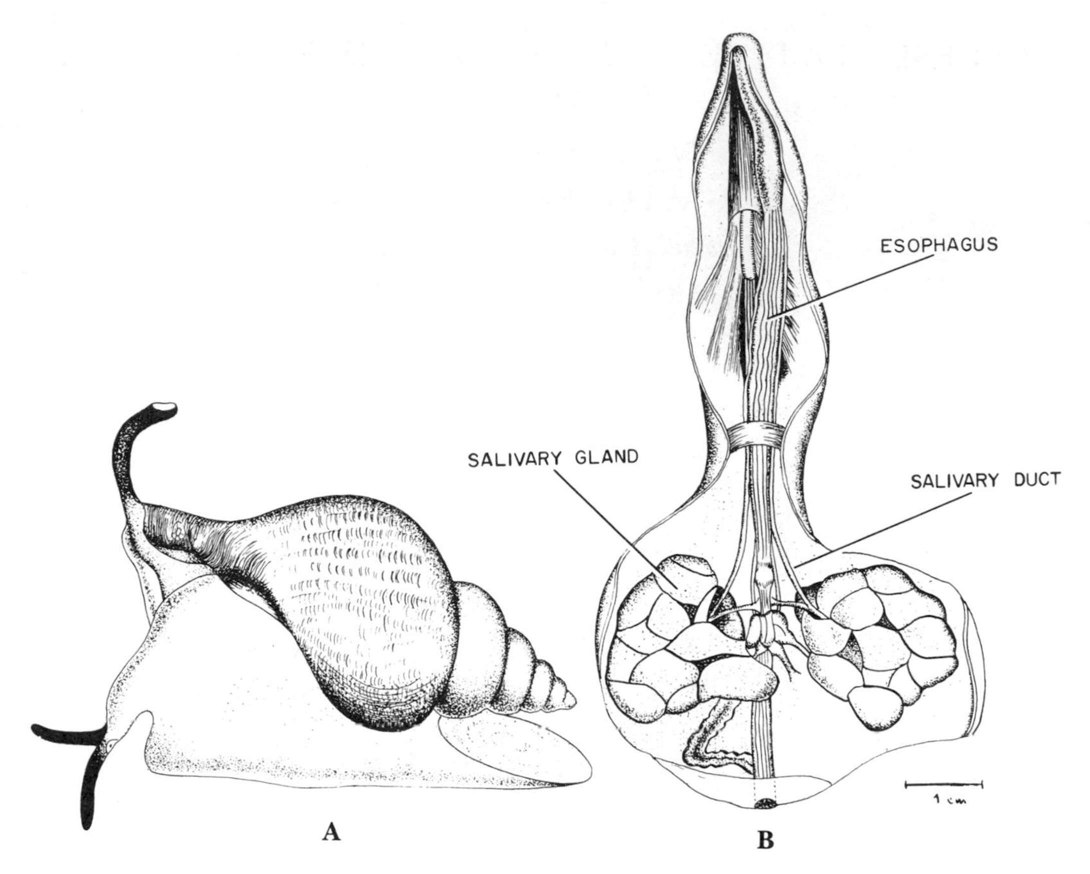

Figure 1.—A. *Neptunea antiqua* (Linnaeus), the common European red whelk. Drawing of living specimen. The salivary glands of this species produce a toxic secretion having potent muscle-contracting activity. Length about 10 cm. B. Salivary glands of *Neptunea antiqua*. The salivary glands of whelks consist of two lobulated yellowish-white masses situated one on either side of the esophagus at the level of the brain ganglia. Surrounding tissues have been removed to show anatomical relationships. The brain ganglia are seen, but not labeled, lying between the two salivary glands. Dorsal view.
(R. Kreuzinger, after Fänge)

months. Codfish, dogfish, and a variety of other fishes are predators to the whelk. Whelks are commonly used as fish bait.

MORPHOLOGY OF THE POISON GLAND

The whelk does not possess a true venom apparatus; however, the poison is concentrated in the salivary glands and apparently is not distributed elsewhere throughout the body of the mollusk. The following morphological description is based on the work of Dakin (1912) and Fänge (1960).

Gross Anatomy: The salivary glands of whelks consist of two lobulated yellowish-white masses situated one on each side of the esophagus at the level of the brain ganglia (Fig. 1B). The left gland lies more posterior and more dorsal than the right. Dorsally the surface of the glands are slightly convex and rather smooth. There are two salivary ducts, one from each gland, which originate among the many small branches that come from the glandular tissue. In the living animal, the glands are soft and friable, and fluid squeezed from the glands has a "fishy smell."

The salivary ducts run anteriorly, one on either side of the esophagus, and finally enter the pharynx in the vicinity of the radular apparatus. On the floor of the pharynx is the odontophore, which consists essentially of a tongue projecting into the pharyngeal space covered by a flexible rasp, a membrane set with teeth, known as the radula. The teeth are not modified to purvey venom such as is observed in *Conus* and other members of the Toxoglossa.

Microscopic Anatomy: Histological sections reveal that the salivary ducts break up in the gland, the branches dividing into finer tubules, and finally terminate through complete occlusion of their lumen. Sections through the ducts reveal ciliated columnar cells. The bulk of the glandular tissue is composed of large, intensely vacuolated cells which in *Neptunea antiqua* measure about 20μ in diameter. In most of these cells there is a trace of cytoplasmic contents and a distinct nucleus, and the remainder of the cell contains either one large vacuole or numerous secretory granules.

MECHANISM OF INTOXICATION

Whelk poison is believed to be restricted to the salivary glands of the mollusk. Intoxication results when these glands are ingested in whole shellfish either in the raw, cooked, or canned state. No true venom apparatus is present. Only two Japanese species, *Neptunea arthritica* (Bernardi) and *N. intersculpta* (Sowerby), have been involved in human intoxication; but other related species, *Argobuccinum oregonense* Redfield, *Buccinum leucostoma* (Lischke), and *N. antiqua* (Linnaeus), have been found to contain tetramine in their salivary glands and are to be considered with suspicion.

MEDICAL ASPECTS

Clinical Characteristics

Little is known regarding the clinical characteristics of whelk poisoning. Asano (1952) states that the principal symptoms are intense headaches, dizziness, nausea, and vomiting.

Tetramine is an autonomic ganglionic blocking agent and intoxication from this substance may result in a number of symptoms: nausea, vomiting, anorexia, weakness, fatigue, faintness, dizziness, photophobia, impaired visual accommodation, and dryness of the mouth. Some of these symptoms are a direct result of a drop in blood pressure and the decreased cardiac output. Frequently gastrointestinal motility is disturbed, and this results in diarrhea or, in some cases, constipation. Paralytic ileus is not uncommon.

Treatment

The treatment is largely symptomatic. Rest will help relieve some of the symptoms caused by the low blood pressure and decreased cardiac output.

Prevention

Asano (1952) recommends removing the salivary glands prior to the consumption of the mollusk.

PUBLIC HEALTH ASPECTS

Whelk poisoning is said to be a public health problem in Hokkaido, Japan, where numerous intoxications have resulted, but the details of these outbreaks are not known (Asano, 1952).

TOXICOLOGY

Asano (1952), and Asano and Ito (1959, 1960) prepared ethanol extracts from homogenized salivary glands of *Neptunea arthritica*. The alcohol was then distilled and the residue dissolved in water and defatted with ether. After the removal of the ether, the residue was diluted with water to make a solution of 1 g gland/ml water. The solution was then further diluted and 0.5 ml portions of each dilution were injected into mice to determine the degree of dilution that provided a minimal lethal dose. The scientists considered the smallest amount of poison that killed mice within 30 minutes after injection as the minimal lethal dose. They expressed toxicity in Mouse Units (MU), which they calculated with the following equation:

$$\mathrm{MU} = \frac{\text{VOLUME OF ORIGINAL SOLUTION (ml)} \times \text{body weight of mouse (g)}}{\text{weight of salivary gland (g)} \times \text{volume of injected solution (usually 0.5 ml)} \times \text{dilution degree of original solution at minimal lethal dose (1/N)}}$$

Results showed that there was some seasonal variation in the toxicity of the salivary glands tested, ranging from 240 MU in September to a maximum of <430 MU in July. All the tests were made on fresh shellfish which had been shucked, and each organ (salivary gland, hypobranchial gland, gonads, and mid-gut) was extirpated separately, weighed, and then extracted with ethanol. It was found that the poison was limited to the salivary glands.

Toxic symptoms in mice consisted of salivation, lacrimation, miosis, increased peristalsis, motor paralysis, respiratory failure, and death. The toxic effects from the poison are said to be 5 to 10 times weaker given by mouth than when administered by parenteral injection.

Toxicity tests were also made on carp *Cyprinus carpio* by muscular injection or by mouth through a polyethylene catheter. The resulting symptoms consisted of gradual loss of balance, inversion of the abdomen, delayed respiratory rate, clonic convulsions, and death.

PHARMACOLOGY

Tetramine produces typical curarelike effects in mammals and frogs. The vascular system shows a fall of blood pressure, with a slowing of the heart by peripheral vagus stimulation and peripheral vasomotor depression. The respiration is temporarily paralyzed by intravenous injections of tetramine. The paralysis is due to a curarelike effect of the phrenic nerves. Some believe that the respiratory action is entirely central, with the primary stimulation followed by depression of the medullary centers (Sollmann, 1949). Excretion of tetramine is very rapid in mammals, which accounts for the transient symptoms of food poisoning.

The toxin of *Babylonia japonica* was purified by Shibōta and Hashimoto (1970, 1971). The purified specimen evoked a minimum mydriasis in mice at a dose of about 0.02 > μ/g body weight.

The salivary gland extract of *Thais haemastoma* was found to be a powerful vasodilator and hypotensive agent (Huang and Mir, 1972). When the extract was administered intravenously to anesthetized cats at a dose of 5-20 mg/kg, it produced a sustained fall in blood pressure and a distinct bradycardia. The effect of the material on the blood pressure was partially blocked by atropine. The LD_{50} in mice was 43 mg/kg in the isolated rabbit heart; the extract produced a decrease in the rate and amplitude of cardiac contractions and a decrease in the rate of coronary outflow. The extract caused contractions of isolated guinea pig ileum and rabbit duodenum. These effects were not blocked by atropine.

CHEMISTRY

A toxic principle was isolated by Asano and Ito (1959) from the saliva of the marine gastropod *Neptunea arthritica*. The toxin was shown to possess a histamine-producing property which contracted the ileum of rabbits and guinea pigs, suggesting the characteristics of tetramine, previously identified as one of the toxins in coelenterates. Identity of the toxin with tetramine was convincingly demonstrated by comparison of the infrared spectra of the toxin picrate and toxin chloroaurate with those of the corresponding derivates of tetramine. The concentration of the toxin in *N. arthritica* was found to be five to eight times that in the coelenterate *Actinia equina* (Asano and Ito, 1960). As with coelenterates, other quaternary bases were found to be present, though the toxicity seemed to be lodged with the tetramine fraction. The same toxin was reported to be present in the salivary gland of *N. antiqua* by Fänge (1960).

An acetylcholinelike substance was isolated by Whittaker (1960) from the gastropod *Buccinum undatum*, which was found to have an identity with that of *Thais floridana*. The presence of an ester group in the active principle was demonstrated by the ferric hydroxamate reaction. Ultraviolet spectroscopy indicated a distinction from acetylcholine by the presence of a double bond. Reduction of the hydrolyzed acid of the ester yielded isovaleric acid. On the basis of these observations, it was concluded that the active principle was an α, β unsaturated derivative of isovalerycholine (senecioylcholine). This conclusion was confirmed by comparison of the infrared spectrum of the active principle with that of synthetic senecioylcholine.

The purified toxin from *Babylonia japonica* is thought to be a polyfunctional, highly oxygenated, complex bromo compound. The toxin was separated into two closely related components, but there was no clue as to the chemical structure of the toxin (Shibōta and Hashimoto, 1970,1971).

REPRESENTATIVE LIST OF CONE SHELLS REPORTED AS VENOMOUS

Phylum MOLLUSCA

Class GASTROPODA: Snails, Slugs

Subclass PROSOBRANCHIA

Order ARCHAEOGASTROPODA

Family CONIDAE

81 *Conus aulicus* Linnaeus (Pl. 2, figs a-b). Court cone (Australia, USA), tsuboima (Japan).

Distribution: Polynesia to Indian Ocean.

Sources: Cleland (1912, 1913), Fraser-Brunner (1973).
Other name: *Regiconus aulicus.*

81, 87, 89 *Conus geographus* Linnaeus (Pl. 2, figs. c, d; Pl. 8, fig. b; Pl. 10, fig. a). Geographer cone (Australia, USA), anboina (Japan).

Distribution: Indo-Pacific; Polynesia to East Africa.

Sources: Cleland (1912, 1942), Fish and Cobb (1954), Rice and Halstead (1968), Endean, Parish, and Gyr (1974).
Other name: *Rollus geographus.*

82 *Conus gloria-maris* Hwass (Pl. 3, fig. d). Glory of the sea (USA).

Distribution: Malay Archipelago, Philippines.

Sources: Miner (1923), Abbott (1950), Fraser-Brunner (1973).
This is an extremely rare and valuable shell. It is said that there are less than 100 specimens in existence. However, this species is believed to be capable of inflicting lethal wounds.

83 *Conus magus* Linnaeus (Pl. 4).

Distribution: Indo-Pacific.

Sources: Endean and Izatt (1965), Endean and Duchemin (1967), Freeman and Turner (1972), Endean, Gyr, and Parish (1974).

82, 85 *Conus marmoreus* Linnaeus (Pl. 3, figs. a-b; Pl. 6, fig. a). Marbled cone (Australia, USA).

Distribution: Polynesia to Indian Ocean.

Sources: Cleland (1912), Clench (1946), Kohn, Saunders, and Wiener (1960), Fraser-Brunner (1973).
Other name: *Coronaxis marmoreus.*

Conus striatus Linnaeus (Pl. 3, fig. c; Pl. 8, figs. a, c; Pl. 11, figs. a, b). Striated cone (USA). *82, 87, 90*

DISTRIBUTION: Indo-Pacific; Australia to East Africa.

SOURCES: Hiyama (1943), Endean, Izatt, and McColm (1967).

Conus textile Linnaeus (Pl. 5, figs. a-b; Pl. 7; Pl. 8, fig. d; Pl. 9, fig. b; Pl. 10, fig. b; Pl. 11, fig. c; Pl. 12, figs. c, d; Pl. 14, fig. d). Cloth-of-gold cone, textile cone, woven cone (Australia, USA), tagayasanminashi (Japan). *84, 86, 87, 88, 89, 90, 91, 93*

DISTRIBUTION: Polynesia to Red Sea.

SOURCES: Cleland (1912, 1942), Hiyama (1943), Clench (1946), Fraser-Brunner (1973).
OTHER NAME: *Darioconus textile.*

Conus tulipa Linnaeus (Pl. 5, figs. c-d; Pl. 6, fig. b). Tulip cone (Australia, USA), shiro anboino (Japan). *84, 85*

DISTRIBUTION: Polynesia to Red Sea.

SOURCES: Cleland (1912), Fish and Cobb (1954), Fraser-Brunner (1973).
OTHER NAME: *Tuliparia tulipa.*

BIOLOGY

The families Conidae, Turridae, and Terebridae constitute the suborder Toxoglossa. The members of the Toxoglossa are characterized by the possession of a venom apparatus, but only the family Conidae has been incriminated in human intoxications. There are about 400 species in the family Conidae, all of them members of the single genus *Conus*, and with few exceptions these are confined to tropical and subtropical seas. The species *C. aulicus, C. geographus, C. gloria-maris, C. marmoreus, C. omaria, C. striatus, C. textile,* and *C. tulipa* have a well-developed venom apparatus, and are capable of inflicting human fatalities (Hiyama, 1943; Abbott, 1950; Petrauskas, 1955; Kohn, 1958). Because of their size and the nature of their venom apparatus, numerous other species of cones are of potential danger to man. An annotated bibliography on poisoning by cone shells has been prepared by Brygoo (1972).

Cone shells are for the most part shallow-water inhabitants, since they range from tidal reef areas down to depths of several hundred meters. They are found in a variety of microhabitats. Some species are associated with the attached algae of coral reefs; others crawl about or under coral heads; still others prefer a sandy or coral rubble substrate. Those species of cones most dangerous to humans are found chiefly in the sand or rubble habitat. Althought not all sand dwellers have a bad reputation, they must be handled with caution. One of the largest of the cone shells is *C. leopardus* Röding. It attains a length of 18 cm to 23 cm; but its venom dart is one of the smallest in the genus, measuring only about 1 mm.

The cone shells are chiefly nocturnal in habit, burrowing in sand or under coral or rocks in the daytime and becoming active at night when feeding usually occurs. The feeding habits of cones have been studied intensively by Kohn (1955, 1956*a, b,* 1959*a, b*), and they are discussed at length in his excellent monograph on the ecology of the Hawaiian cone shells. Similar valuable studies have been conducted by Saunders and Wolfson (1961). The food of cones, varying somewhat with the species, consists of polychaete worms, other gastropods, pelecypods, octopuses, and small fishes.

Cones are predacious and they feed by injecting a venom into the prey by means of a detachable, dartlike radular tooth. The morphology of this venom apparatus is described at length elsewhere in this chapter (Fig. 2 A-B). Food detection is believed to be largely through the use of the osphradium, a chemoreceptory organ which is more highly specialized in the Conidae than any other group of mollusks (Kohn, 1956*a, b*). Vision is believed to play a minor role in food procurement. Water is taken in through the siphon and directed over the gill, the osphradium, and by the anus before leaving the mantle cavity. When food is detected, the cone becomes active and extends the proboscis about until contact is made with the prey. A single radular tooth is held within the lumen of the proboscis with its point slightly posterior to the aperture. Stimulation for release of the radular tooth has been observed to be tactile. Upon contact with the prey, the radular tooth is ejected and penetrates into the body of the victim. The effects of the venom are immediate and the victim is soon paralyzed. The mouth of the cone expands from a few millimeters in diameter to about 2 cm in some species. The impaled paralyzed organism, if small enough, is drawn into the mouth by rapid contraction of the proboscis. Finally, engulfment is completed by expansion of the buccal cavity around the victim (Kohn, 1956*a, b*). The buccal cavity remains distended for several hours after the ingestion of the prey, during which time the prey must be partially digested before it can pass on through the alimentary tract of the cone. For further details concerning the function of the venom apparatus of cones, see p. 194.

The activity of cones and their reactions to being handled vary greatly from one species to another. Some species tend to be timid and sluggish, and withdraw rapidly into their shell when disturbed. Several of the more tropical Indo-Pacific species, however, are quite active and when they are picked up they may extend their proboscis with their venom dart in place. *C. textile* may on occasion be very aggressive. Kohn (1955) cites a personal communication from R. Sheats, who on September 12, 1955, was diving in 25 m of water off Pearl Harbor, Oahu, Hawaii. He picked up a *C. textile* which, when brought to the surface, extended its proboscis about 2.5 cm. When the proboscis was stimulated with a piece of paper, a radular tooth was promptly ejected, penetrated the paper, and left a small drop of venom. This same process was repeated three times in the following 2 hours.

MORPHOLOGY OF THE VENOM APPARATUS

The following anatomical description of *Conus californicus* Hinds are based on unpublished data obtained from Ralph T. Hinegardner. These were collected in the vicinity of Point Fermin, Los Angeles County, Calif., and are supplemented by studies conducted by Bruce W. Halstead on the same species, and specimens of *C. textile* Linnaeus and *C. geographus* Linnaeus collected at Agana, Guam. All the specimens were collected fresh, fixed in Bouin's, Gilson's fluid, or in formalin, and processed in the routine manner for embedding in paraffin. Sections were stained with Mallory's triple, hematoxylin and eosin, and mounted in permount. Some of the radular teeth were treated with strong sodium hydroxide solution, soaked in concentrated silver nitrate solution for several days, washed in distilled water, placed in standard photographic developer, transferred through the alcohols and xylene, and mounted on a slide.

General: The venom apparatus of *Conus* consists of the venom bulb, venom duct, radular sheath, and radular teeth. The pharynx and proboscis, which are a part of the digestive system, also play an important role as accessory organs. The venom apparatus (Fig. 2 A-D) lies in a cavity posterior to the rostrum, on the dorsum of the animal. A thin layer of connective tissue covers a slightly thicker muscular layer, whose fibers run from the anterior body wall behind the rostrum to the posterior body wall behind the venom bulb.

Conus californicus

Gross Anatomy: The venom bulb (Pl. 9, fig. a), the most posterior portion of the venom apparatus, lies adjacent to the posterior body wall. It appears as a slightly curved, cucumber-shaped structure, about 7 mm × 3 mm in its greatest dimensions in the specimens examined. The bulb is held in place by a number of ventral and lateral connective tissue attachments which connect with the ventral body wall. The bulb has a gross appearance of a muscular rather than a glandular organ. As seen in cross section, the bulb is comprised of six layers of tissue, the details of which can be determined only by histological study. Gross dissection reveals a thick outer muscular layer and a relatively thin inner muscular layer which are separated by a connective tissue septum. By dissecting the outer muscular layer, it can be demonstrated that the muscle fibers of *Conus californicus* run the length of the bulb. The muscle fibers in the inner muscular layer are circular in their arrangement, and they run at right angles to the longitudinal axis of the bulb. *88*

The venom duct (Fig. 2C) originates at the right end of the venom bulb, passes sharply ventrad under the bulb, then over the esophagus, and to the left side of the cavity in which the venom apparatus is contained. The part of the duct under the bulb is flexible and generally flattened. After it reaches the left side of the body there is a constriction in the duct. The portion of the

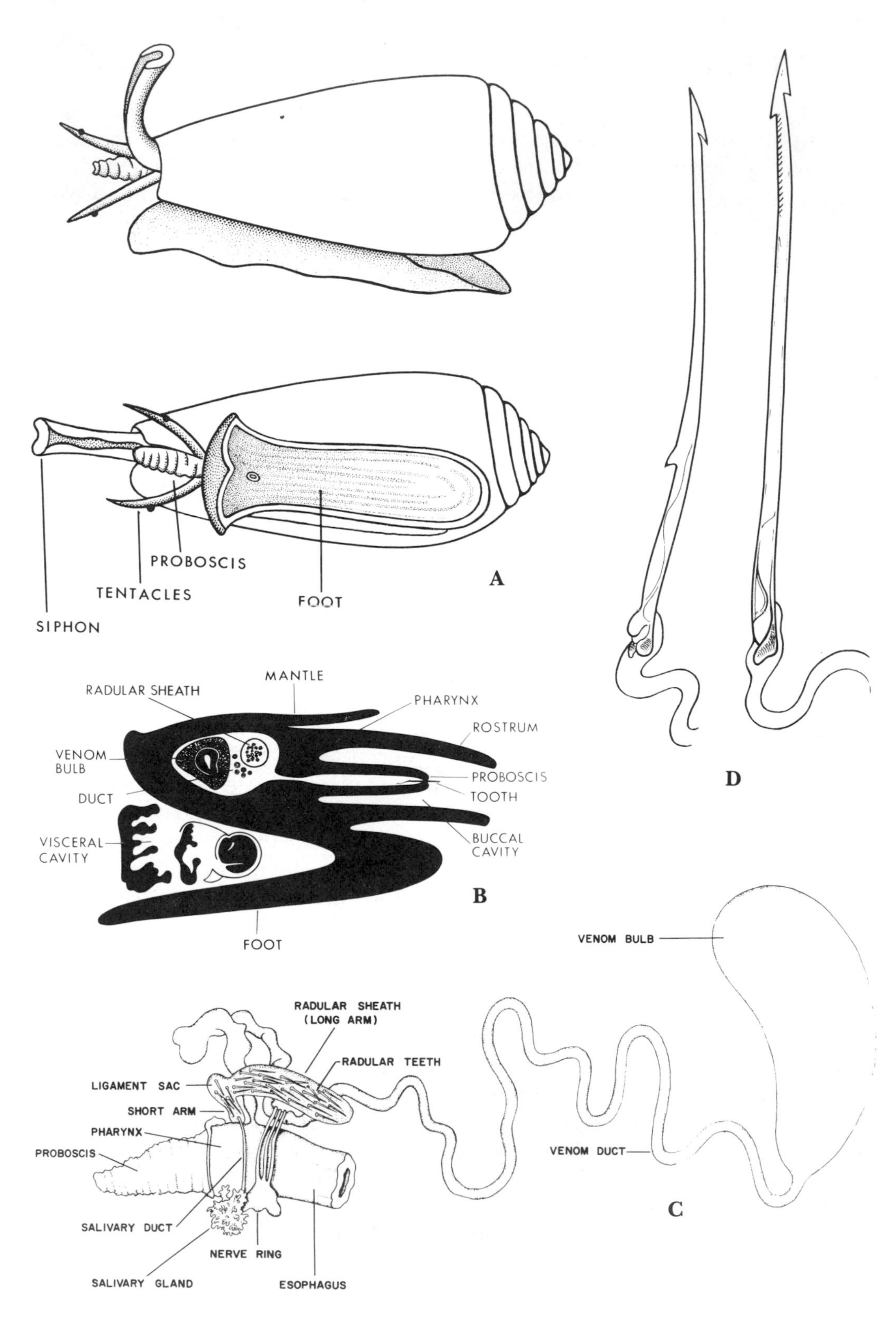
PROBOSCIS
TENTACLES
SIPHON
FOOT
A
MANTLE
RADULAR SHEATH
PHARYNX
ROSTRUM
VENOM BULB
PROBOSCIS
DUCT
TOOTH
VISCERAL CAVITY
BUCCAL CAVITY
B
FOOT
D
VENOM BULB
RADULAR SHEATH (LONG ARM)
RADULAR TEETH
LIGAMENT SAC
SHORT ARM
PHARYNX
PROBOSCIS
VENOM DUCT
C
SALIVARY DUCT
NERVE RING
SALIVARY GLAND
ESOPHAGUS

duct immediately following the constriction contains a light yellow granular material that is viscous and slimy. This viscous material has been found to be present in all the specimens examined to date, and is believed to be venom. The duct on the left side of the caivty forms three or four large loops, and then passes to the right side of the cavity, extending anteriorly on the same plane as the bulb, between the bulb and the radular sheath. On the right side of the cavity, at the level of the end of the bulb, the duct forms a large loop and then passes ventrally to the right side of the pharynx under the pharyngeal nerve ring. From this point it once more passes dorsad, makes another large loop, and then enters the right posterolateral aspect of the muscular pharynx just anterior to the pharyngeal nerve ring. Immediately before entering the pharynx the duct becomes constricted and transparent in its terminal 0.1 mm. In one of the largest specimens, the total length of the duct measured 85 mm.

The radular sheath lies anterior to the bulb, and is divided into an elongate dorsal arm and a short ventral one. The long arm is located just beneath the dorsal body wall, and extends about two-thirds of the distance between the right and left side of the cavity. The wall of the sheath is sufficiently transparent to make it possible to visualize the radular teeth contained within it. The long arm of the sheath is approximately of uniform diameter throughout its proximal two-thirds, but is slightly swollen in the distal third. Then it tapers sharply to a rounded point. The teeth in the long arm all point down and to the left at an angle of 45° with reference to the dorsal surface of the sheath. The sheath forms a 90° angle at the junction of the long and the short arm. At this junction there is a saclike enlargement on the outer side which contains a large number of the ligaments of the radular teeth. The short arm enters the pharynx ventral and at a point anterior to the entrance of the venom duct. The teeth in the short arm are directed toward the pharynx, or in the opposite direction to those in the long arm. In one specimen examined, the long arm contained about 60 teeth lying in two longitudinal rows, with the most distal ones most completely developed. The short arm contained about 10 to 12 teeth.

The radular teeth (Pl. 16) are in the form of a flat sheet of chitin rolled or supervoluted into a hollow tubelike structure. Troschel (1856-63), Bergh (1895), and Peile (1937) have demonstrated that a similar tooth formation *95*

FIGURE 2.—(Left) A. External features of a typical cone shell *Conus*. Note the relationship of the proboscis to the siphon. It is the muscular proboscis which is used to grasp the radular teeth and thereby inflict the sting. (R. Kreuzinger) B. Sagittal section through the venom apparatus of *Conus* showing some of the anatomical relationships and relative position of the venom apparatus within the body of the mollusk. Semidiagrammatic. (R. Kreuzinger, modified from Hermitte) C. Venom apparatus of *Conus californicus* showing relationship of venom organs to digestive tract. This arrangement is quite representative of most other species of this genus (O. Hon, after Hinegardner). D. Radular teeth of two representative species of venomous cone shells, *Conus marmoreus* × 220 (left), and *C. tulipa*, × 140 (right). (R. Kreuzinger, after Bergh)

is also found in other species of cones. In *C. californicus* the inner edge of the sheet has a series of protuberances, and the outer edge forms the barbs and hooklets of the tooth. Near the distal tip of the tooth the sheet of chitin becomes narrow and forms a groove rather than a complete tube. The base of the tooth is more intricate, in that the chitinous sheet instead of being straight-edged has a tonguelike projection extending from the base which folds over, but does not occlude the tubelike basal opening of the tooth. Attached to this lingual process is a short, solid, transparent ligamentous structure termed the "tooth ligament." The distal end of the ligament tapers to a fine point and attaches to the wall of the sheath.

The pharynx (Fig. 2C) is readily differentiated from the rest of the digestive tract by a large muscular ring which encircles it, and by an internal lining which differs from that of either the proboscis or esophagus. The muscular ring is flattened on the outer surface and convex on the inner. This ring serves as the wall for a major portion of the pharynx, and separates the entrance of the venom duct from that of the radular sheath. The venom duct enters the pharynx immediately posterior to the ring, whereas the radular sheath enters immediately anterior to the ring. It is believed that this is the site where the radular teeth are charged with venom.

The salivary gland (Fig. 2C) lies to the right of the pharynx and appears as a highly branched organ. This gland is attached to the short arm of the radular sheath by two small ducts. One duct passes over the pharynx to the left side of the short arm, the other under the pharynx to the right side. The nature of this organ as it relates to the venom apparatus is presently unknown. It may contribute to production of some substance having to do with the function of the radular teeth.

The proboscis (Fig 2C) is located a short distance anterior to the pharyngeal muscle ring, and when contracted, appears grossly as a cochleate muscular mass having a central orifice. In extended specimens, the proboscis may appear as an elongate cone-shaped mass extending into the rostral or buccal cavity. The proboscis is covered on the outside by a layer of circular muscle and connective tissue which is continuous with the lining of the rostrum. Adjacent to the layer of circular muscle is a longitudinal layer of muscle which extends through the length of the proboscis.

Microscopic Anatomy: The histological details of the venom bulb are best seen in cross section (Pl. 12, figs. a-b). The bulb appears to be round to ovoid in outline and is divided into four distinct zones. The outer zone is comprised of a thick longitudinal layer of smooth muscle covered by a thin cuticle. A septum of dense fibrous, connective tissue separates a thick circular layer of smooth muscle from the longitudinal layer. The triangular-shaped lumen of the bulb is lined by a single layer of simple epithelium which varies from a low cuboidal to columnar types. There is no histological evidence to indicate that the venom bulb is a secretory organ, and the term "venom gland" as used by previous authors is a misnomer. The bulb may possibly serve as a venom reservoir. The highly developed musculature provides the means for expel-

91

ling the venom through the duct and into the aperture of the hollow supervoluted radular teeth. Similar viewpoints regarding the function of the venom bulb have been expressed previously by Bouvier (1887), Hermitte (1946), and Hinegardner (1958), and more recently by Kohn *et al.* (1960).

The wall of the venom duct (Pl. 13), in cross section, consists of two distinct zones. The outer zone is comprised of two distinct layers of equal thickness of smooth muscle—an outer longitudinal and an inner circular one. Resting upon this inner circular layer is a broad zone of glandular epithelium showing distinct variations. The basal portion of this zone consists of variable-sized masses of closely packed red oval and spherical bodies. The masses taper toward the lumen. Basally, between these masses, one can see an occasional nucleus with a prominent nucleolus. The cytoplasmic outlines are not distinct, and they blend into long columns of closely packed blue and purple bodies which are approximately twice the size of the associated red ones. These columns are oriented parallel to one another and extend toward the lumen. The blue and purple bodies become more concentrated as they approach the lumen. Thin clefts appear to separate the adjacent columns. Appearing in the lumen are large masses of round purple bodies (which appear to be droplet bodies), and also patches of coagulated amorphous secretion. Histological evidence seems to indicate that the venom is produced by the glandular epithelium of the venom duct. The venom is probably produced by a holocrine type of secretion. Recent toxicological studies by Kohn *et al.* (1960) tend to confirm the histological evidence that the venom is produced in the venom duct.

92

A cross section of the distal third of the long arm of the radular sheath (Pl. 14, fig. c) appears as a tubular structure having a distinct wall and lumen. The wall consists of a band of moderately dense, fibrous, connective tissue upon which rests an epithelium of variable height. Along the midventral axis of the sheath, the epithelium appears as a bulbous longitudinal fold. In the thicker areas of the fold, large oval vesicularlike nuclei are seen. The epithelium of the wall and longitudinal fold is of a stratified or pseudostratified type, the exact nature of which has not been determined with certainty. The epithelial lining of the short arm of the radular sheath in the region entering the pharynx appears to be ciliated, as does the epithelium along parts of the longitudinal fold. The nuclei frequently stain the basal portions of the epithelium a golden brown color. In some regions the epithelium shows many distended clear areas. More distal sections reveal that the longitudinal epithelial fold occupies a major portion of the lumen, with cross sections of the scroll-like radular teeth concentrated dorsal to and on either side of the fold. Most of the teeth appear to originate in the distal half of the longitudinal fold. Thin immature teeth are found embedded in the cells or lying beside them. The epithelial cells around the embedded teeth are short and irregularly arranged. The tooth ligaments appear as blue-stained, finely fibrillar masses situated in the ventral expanse of the lumen on either side of the fold. In some places the ligamentous structures are seen concentrated in small bundles, some of

93

which are attached to teeth. The distal ends of the ligaments appear to fuse with the longitudinal fold. Apparently the primary function of the longitudinal epithelial fold is to give rise to the radular teeth.

The ciliated epithelial lining probably facilitates the ejection of the radular teeth from the sheath into the pharynx.

In sections taken from a more proximal region (middle third) of the long arm, the ventral longitudinal fold seems to diminish in size gradually and finally disappears (Pl. 14, fig. b). The wall of the sheath in the proximal region becomes thinner, and more mature radular teeth almost fill the entire lumen of the sheath.

93

Cross sections taken near the junction (proximal third) of the long and short arm of the radular sheath reveal a low-lying dorsal septum of connective tissue (Pl. 14, fig. a) which gradually increases in size in more distal sections of the short arm.

93

94

The pharynx is best seen in longitudinal sections (Pl. 15, figs. a-c) as two large outer smooth muscle masses which constrict the lumen of the digestive tract between the esophagus and the proboscis. The lumen of the pharynx is lined by ciliated pseudostratified epithelium arranged in longitudinal folds extending the entire length of the pharynx. In the lumen of the anterior part of the pharynx the epithelial lining alters its configuration and appears as folds convoluted at right angles to that of the pharynx. The pharyngeal epithelium is in marked contrast to the glandular epithelium of the esophagus. The venom duct can be seen entering the pharynx immediately posterior to the pharyngeal muscle. In longitudinal sections taken at lower levels, the short arm of the radular sheath can be seen entering the anterior pharynx.

94

In longitudinal sections, the proboscis (Pl. 15, fig. d) appears as an elongated, tapering, tubular organ having a wall and a lumen which opens to the outside of the anterior end. The point in the lumen where the convoluted epithelium of the anterior pharynx begins to uncoil marks the beginning of the proboscis. The wall of the proboscis is comprised of four distinct zones of tissue. The outer zone consists of a thick convoluted layer of pseudostratified epithelium having a striated border. Next, there is a layer of smooth muscle, which in some areas parallels the gyrations of the epithelial folds. If extended the muscle would be longitudinally oriented. Between the muscle layer and the epithelial lining of the lumen are myelinated nerve bundles. The lining of the lumen consists of ciliated pseudostratified epithelium similar to that of the pharynx. The lumen of the proboscis is filled with variable-sized spherical bodies which stain brown, red, and purple, and whose exact nature is not known.

Conus textile

The gross features of the venom apparatus of *Conus textile* are essentially the same as that of *C. californicus*. The major differences, aside from those which will be mentioned below, are quantitative. The specimen examined measured 19 mm by 4.5 mm. (Pl. 8, fig. d; Pl. 9, fig. b; Pl. 10, fig. b).

87, 88, 89

A noteworthy difference between the venom bulb of *C. textile* and that of *C. californicus* is the orientation of the outer layer of muscle fibers. In *C. textile* the muscle fibers are in a spiral configuration, whereas in the California cone they are longitudinal; also, the lumen of the bulb is round or ovoid rather than triangular as in *C. californicus* (Pl. 12, figs. c-d).

91

The arrangement of the venom duct is somewhat distinctive. The duct leaves the right end of the venom bulb, passes ventral to the left side of the body, then anterior to the bulb, where it forms a large coiled mass measuring 80 mm in length. This mass is anterior to the right end of the bulb and posterior to the elbow of the radular sheath, and it contains about 30 mm of duct. The duct next passes to the middle of the body cavity, and posterior to the distal end of the radular sheath, where it forms several small loops before passing ventrally, and under the right side of the pharyngeal nerve ring. A coil is then formed posterior and somewhat ventral to the proximal end of the long portion of the radular sheath. Finally, the duct enters the right lateral aspect of the pharynx posterior to the entrance of the short arm of the radular sheath. Throughout most of its length the duct is approximately 0.8 mm in diameter, except at the point it leaves the bulb where it is only 0.4 mm in diameter for a distance of about 2 mm, and at the point it enters the pharynx where it is less than 0.3 mm in diameter.

The general microscopic anatomy of the duct is similar to that described for *C. californicus* except that the colums of cells are smaller in width, appear to be more compact, and the granular material in the lumen is finer.

The gross and microscopic anatomical differences of the remaining venom organs in *C. textile* are primarily in size rather than in structure (Pl. 14, fig. d).

93

Conus geographus

The specimen examined measured 94 mm by 46 mm. The outer muscle fibers of the venom bulb spiral about the organ. The bulb tends to be less crescent-shaped than that of the other species (Pl. 10, fig. a).

89

The venom duct leaves the right end of the bulb, passes over the ventral floor of the cavity to the left anterior side where it forms several small coils containing 45 mm of duct. These loops coil in and around the salivary gland which is located in the left anterior corner of the cavity. The duct then returns to the right side of the cavity where a coiled mass is formed containing about 60 mm of duct. Finally, the duct loops under the pharyngeal nerve ring and enters the pharynx posterior to the radular sheath after forming one small kink containing 6 mm of tubing. The total length of the duct is somewhat over 160 mm and it has an average diameter of 1.0 mm.

The radular sheath is relatively larger than that seen in the two other species. The gross and microscopic anatomical differences of the remaining venom organs do not differ remarkably from that of *Conus californicus.*

For additional details regarding the anatomy of cones, the reader is referred to Bergh (1895), Shaw (1914), Alpers (1931), Taki (1937), Peile (1939), Jaeckel (1952), and Hinegardner (1958).

Function of the Venom Apparatus

A cone shell is able to inflict a sting only when the head of the animal is out of the shell. As the soft sluglike body emerges, the foot extends beyond both the anterior and posterior ends of the shell. The foot is truncated in front and bluntly pointed behind. Extending from the upper anterior part of the body is the elongate siphon which is believed to be both sensory and respiratory in function. However, as previously mentioned, it is believed that food detection is made largely through the osphradium aided by the proboscis. Vision probably plays a less significant role. The siphon may sometimes be observed extending some distance beyond the anterior margin of the shell. Ventral to the siphon, the tentacles and eyes protrude from the head near the base of the tubular rostrum, which contains the proboscis. Stinging is accomplished by extrusion of the long, tapered, fleshy, tubular proboscis from which emerges the radular teeth. The proboscis is a powerful muscular organ which can be retracted or extended at will. Garrett (1874-78), Clench (1946), and Kline (1956) have observed that during the act of stinging, a radular tooth, which had been previously released from the radular sac into the proboscis, extended from the anterior opening of the proboscis. Hermitte (1946) suggested that the proboscis is retracted, flattened against the collar of the rostrum, and then the tooth is pushed against the radular sac by muscular action, with the barbed point just past the surface, while the posterior end is gripped by a strong muscular collar near the tip of the prepharynx. Extension of the proboscis would carry it forward, snapping the anchoring ligament. Hermitte (1946) also suggested that the tooth from the radular sheath is released into the pharynx, and then grasped by inversion of the proboscis. The histological research of Hinegardner, however, indicated that a radular tooth is probably propelled by the action of the epithelial cilia (Fig. 2A-D), together with a peristaltic action from the radular sac, through the pharynx to the anterior opening of the proboscis where it is grasped ready for thrusting into the flesh of the prey (Pl. 7, fig. c).

86

Each radular tooth is used only once. The base of the tooth is held by the proboscis and serves as a harpoon (Fig. 2A-D). The impaled prey is withdrawn into the buccal cavity of the cone with the use of the tooth. If the prey is successfully captured, the tooth is swallowed with it; if not, the tooth is completely ejected, and a new one moves into position from the radular sheath (Kohn, 1956*b*).

It is believed that the venom produced in the venom duct is probably forced under pressure, by contraction of the venom bulb and duct, into the radular sheath, thus being forced into the lumen of the supervoluted radular teeth.

MECHANISM OF INTOXICATION

Cones inflict their sting by means of a venomous radular tooth. The radular teeth originate in the radular sheath where they reside until used. When needed, a single tooth passes from the sheath through the pharynx where it is charged with venom that is produced in the venom duct and purveyed to the hollow supervoluted radular tooth under pressure by the muscular venom bulb. The tooth then passes from the pharynx into the anterior opening of the proboscis where it is held ready to be plunged into the flesh of the victim. Most cone stings result from the careless handling of the mollusk by curious shell collectors.

MEDICAL ASPECTS

Clinical Characteristics

Stings produced by *Conus* are of the puncture wound variety. Localized ischemia, cyanosis, and numbness in the area about the wound, or a sharp stinging or burning sensation are usually the initial symptoms. The presence and intensity of the pain vary considerably from one individual to the next. Some persons state that the pain is similar to a wasp sting whereas others find it excruciating. Swelling of the affected part usually occurs. The numbness and paresthesias begin at the wound site and may spread rapidly involving the entire body, particularly about the lips and mouth. In severe cases paralysis of the voluntary muscles is initiated early, first by motor incoordination and followed by a complete generalized muscular paralysis. Knee jerks are generally absent. Aphonia and dysphagia may become very marked and distressing to the victim. Some patients complain of a generalized pruritis. Blurring of vision and diplopia are commonly present. Nausea may be present, but gastrointestinal and genitourinary symptoms are usually absent. The recovery period in less serious cases varies from a few hours to several weeks. Until fully recovered, victims complain of extreme weakness and tiring easily with the least amount of physical exertion. Coma may ensue, and death is said to be the result of cardiac failure.

The clinical characteristics of cone shell stings have been discussed by Garrett (1874-78), Cleland (1912), Flecker (1936), Yasiro (1939), Hermitte (1946), Abbott (1950), Hartley (1954), Petrauskas (1955), Kohn (1958), Keegan (1960), Hanna (1963), and Endean and Rudkin (1965).

Pathology

The only reference to an autopsy on a victim having succumbed to a cone shell sting is that by Flecker (1936), who states, "A *post mortem* examination showed that all the organs, heart, lungs, etc., were quite healthy."

Treatment

Symptomatic. Although cone shell stings do not produce a typical curarelike blocking effect, the use of neostigmine should be tried. Artificial respiration may be required.

Prevention

Since cone shells are often sought by collectors, a few words of precaution on the handling of them is pertinent. Gloves should be worn if possible. The animal should always be picked up by the posterior end of the shell. If the proboscis is extended from the pointed anterior end, the shell should be dropped immediately, for the proboscis is very extensible and may reach around within striking range of the hand of the holder. Usually, when the animal is disturbed it withdraws into the shell. However, it should not be held in the hand any longer than necessary, for the animal may soon relax and extend the proboscis to investigate its surroundings. It is noteworthy that most stings have occurred while the collector was attempting to scrape encrusted organic debris from the shell. It is possible that the scraping process may stimulate the cone to strike. At any rate, the handling of living cones is not without danger and all species should be treated with care.

PUBLIC HEALTH ASPECTS

Cone shell stings are of relatively minor public health significance in most parts of the world. Although the geography of cone shell envenomations corresponds to that of the geographical distribution of the causative agents, the greatest number of reported stings and fatalities have occurred in the Indo-Pacific area. The exact incidence is unknown since stingings of this type are generally not reported. The species most frequently involved in human envenomations are *Conus textile* and *C. geographus*. Species having caused fatalities are *C. textile, C. geographus, C. striatus,* and *C. aulicus*. However, a number of other closely related cone species are believed to be capable of producing fatalities. More than 28 cases, resulting in more than 8 deaths, have been reported in the literature. This gives a crude fatality rate of about 25 percent.

TOXICOLOGY

Hermitte (1946) appears to have been the first to have attempted animal toxicity experiments on the venom of *Conus*. Unfortunately, the alcoholic extracts which he used proved to be inactive and only negative results were obtained.

Kohn *et al.* (1960) have published an excellent preliminary report on some of the properties of cone venom. Various aqueous extracts were prepared from the venom ducts, venom bulbs, and radular sheaths of *C. textile, C. striatus, C. aulicus*, and *C. marmoreus*. The extracts from the venom ducts proved to be lethal when injected by several different routes into other mollusks, fishes, and mice. Extracts of venom bulbs and radular sheaths of *C. textile* and *C. aulicus* did not produce toxic effects when injected into snails and mice. Contents of *C. striatus* venom bulbs were lethal when injected into one mouse and several fishes.

In addition to the preceding cone species, aqueous extracts were prepared from the venom ducts and venom bulbs of one specimen of each of the following: *C. flavidus, C. leopardus*, and *C. maldivus*. However, none of the extracts demonstrated any toxic effects.

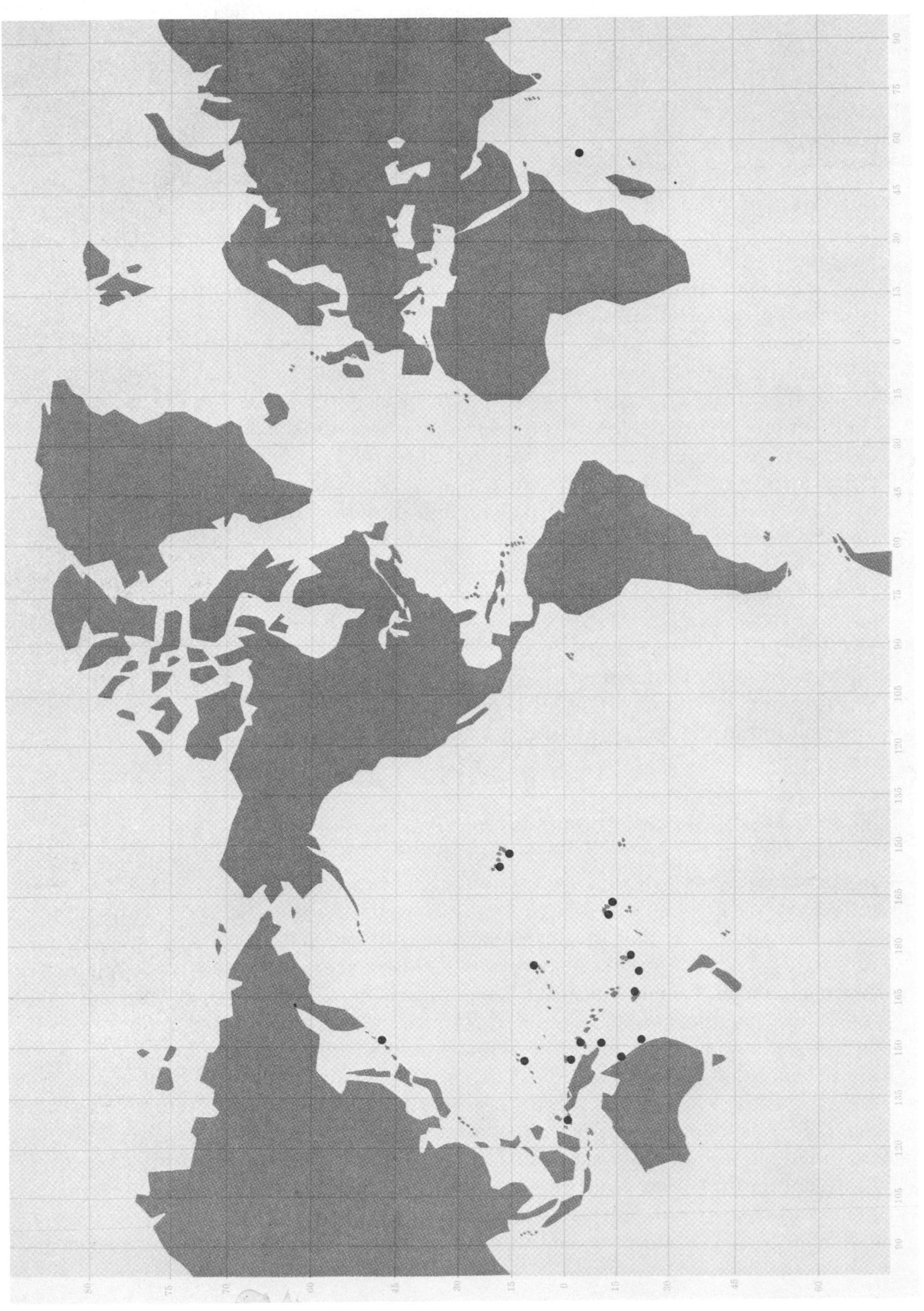

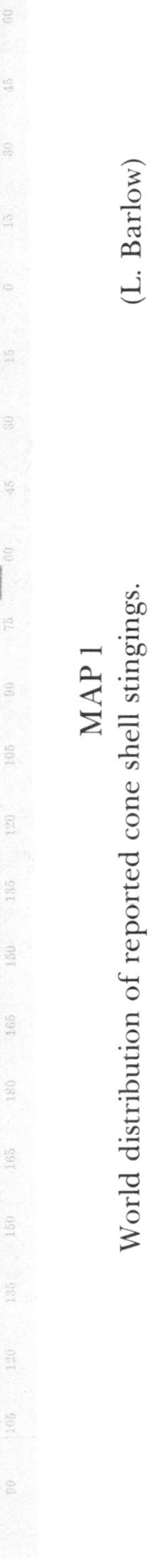

MAP 1
World distribution of reported cone shell stingings.
(L. Barlow)

The MLD of *C. textile* venom tested on *C. californicus* was <0.8 mm^3 of extract of whole venom duct. The MLD of *C. striatus* venom was <0.2 mm^3 of venom duct contents injected into fishes 50 mm to 65 mm in length. The LD_{50} of *C. striatus* venom was 0.5 to 1.0×10^{-4}mg of venom duct contents, tested intracerebrally, and 0.1 mg to 0.3 mg tested intravenously and intraperitoneally.

Toxic symptoms in fishes consisted of color change, ataxia, convulsions or quivering, and paresis. Toxic symptoms in mice consisted of ataxia, lacrimation, tonic spasms, dyspnea, hyperexcitability, violent scratching movements, flushing of tail and paws, sluggishness, paralysis, and coma prior to death. A single experiment with *C. striatus* venom failed to demonstrate any hemolytic properties on mouse erythrocytes.

The morphology of the venom apparatus of *C. magus* has been described in detail by Endean and Duchemin (1967). Endean and Izatt (1965) have shown that the venom of *C. magus* is more toxic to mice than any other cone shell venoms that have been studied to date. Amounts as low as 0.2 mg (wet wt.) of *C. magus* venom per kg are lethal to mice. Venom extracted from posterior regions of the venom duct of *C. magus* was lethal to mice and produced a contracture of isolated skeletal and smooth musculature of the rat in the absence of electrical stimulation. The effect was sustained and the musculature became paralyzed in the contracted state. Posterior duct venom increased the strength of contraction of the musculature of the ventricle of the toad but decreased its rate of contraction. Cardiac musculature was not paralyzed by the venom at concentrations tested. The effects produced by the venom in skeletal, smooth, or cardiac musculature could be reversed by washing out the venom. The MLD of LD_{50} were not determined. There is no information available concerning the chemical characteristics of the venom.

PHARMACOLOGY

The work of Kohn *et al.* (1960) on the pharmacology of cone venom is preliminary in nature and based on certain observations made on the toxicological properties observed in various test animals. They attributed the death of the mice to respiratory failure followed by cardiac arrest. Some of the toxic manifestations suggested that the principal action of the venom was due to interference with neuromuscular transmission, but the possibility of an action on the central nervous system could not be excluded.

Their chemical analyses showed the presence of quaternary ammonium compounds *n*-methlpyridinium, homarine, and γ-butyrobetaine, and unidentified indole derivative homogenates in the venom duct. One cone species gave a positive chromaffin reaction, suggesting the presence of amines. Not detected was 5-hydroxytryptamine; however, other indole amines were not excluded. These findings suggested the possibility that the quaternary ammonium compounds may compete for acetylcholine receptors at the motor-end plate, either preventing acetylcholine from exerting its depolarizing action as *d*-tubocurarine, or as decamethonium, producing continuous depolarization.

Thus acetylcholine, which is rapidly hydrolyzed by cholinesterase, is prevented from acting as a physiological depolarizing agent. In general, they found the responses to be similar to those caused by the curariform drugs.

Whyte and Endean (1962) have investigated the effects of the venom from *Conus textile* and *C. geographus* on the neuromuscular junction, using the phrenic nerve-diaphragm preparation. The venom was taken from various regions of the venom duct.

Venom obtained from *C. textile* was found to be inactive against rats and mice, and on isolated tissue preparations. However, *C. textile* venom was found to be toxic to other mollusks. The researchers have raised the question as to the toxicity to humans of this species. Whether their findings represent an individual, geographical, or seasonal fluctuation is a matter to be determined by further research. In the light of past fatality reports, it would be dangerous to assume at this juncture that *C. textile* is innocuous to humans.

Venom was also taken from various portions of the venom duct of *C. geographus*. It was found that the venom taken from the anterior portion of the venom duct was inactive, but venom from the posterior portion was highly potent. This finding was unexplained. Active *C. geographus* was found to cause muscular paralysis, but no effects could be observed on the neuromuscular junction from the venom alone. The venom was found to enhance the effect of liminal doses of tubocurarine on the neuromuscular junction, and the tubocurarine enhanced the muscle paralysis produced by small doses of venom. The muscle paralysis in entire animals resulted in respiratory paralysis. Organ bath studies demonstrated paralysis of diaphragm and other muscle. The lack of effect of eserine on the paralysis caused by the venom suggested that the phenomenon is not closely allied to the effect of tubocurarine on the neuromuscular junction.

The authors concluded that there is no rationale for the use of eserine or other neuromuscular-blocking agents in the treatment of cone stings on humans. Their findings indicate that artificial respiration may be highly beneficial.

A comparison of the activities of the venom of *C. geographus* with *C. magus* shows markedly different pharmacological properties (Endean, Parish, and Gyr, 1974). Whereas the activity of *C. geographus* venom appears confined to vertebrate skeletal muscle, the venom of *C. magus* elicits activity in most types of vertebrate muscle and nerves. Moreover, *C. geographus* venom elicits flaccid paralysis of vertebrate skeletal musculature whereas *C. magus* venom elicits a spastic paralysis of this musculature. These authors (Endean *et al.*, 1974) have conducted further pharmacological studies on the direct action of *C. magus* on vertebrate skeletal and cardiac musculature.

Songdahl and Lane (1970) tested the lyophilized venom of the alphabet cone *C. spurius atlanticus* on the crabs. The venom induced ataxia, muscle spasms, paralysis, and death (LD_{50} about 100 μg per crab).

Freeman and Turner (1972) summarized their pharmacological studies on *Conus* species as follows:

1. Toxins isolated from the venom apparatus of *C. magus* and *C. achatinus* have the same pharmacological properties, but differ from the toxins of several other piscivorous species of cone shells.

2. *C. magus* and *C. achatinus* toxins are heat labile at pH 8.5. A single lethal component with an approximate molecular weight of 10,000 was isolated from *C. achatinus* toxin by exclusion chromatography.

3. Animals died from a characteristic spastic paralysis after intravenous injection of the toxin.

4. Nerve transmission was unaffected by the toxin; skeletal muscle appeared to be the primary site of action. The toxin caused a persistent contracture of rat diaphragm muscle, and a dose-dependent decline in twitch tension. The contracture was potentiated by caffeine.

5. The decline in twitch tension was associated with depolarization of the cell membrane. The muscle recovered very slowly on washout of the toxin. The depolarization could be reversed by exposure of the preparation to 5 mM Na^+ solution or tetrodotoxin or saxitoxin.

6. Miniature end-plate potential frequency in the rat diaphragm decreased, as did the quantal content of the end-plate potential. The acetylcholine-induced contraction and depolarization of the chronically denervated rat diaphragm were increased by low doses of toxin and reduced by higher, depolarizing toxin doses.

7. Cardiac and smooth muscle were relatively resistant to the toxin. The isotonic contraction of the isolated perfused guinea-pig heart was increased by the toxins from both Conidae. The heart rate decreased.

8. The effects of the toxin are compared with those of batrachotoxin, which it somewhat resembles.

Turner and Freeman (1974) investigated the action of *Conus achatinus* toxin on mammalian skeletal muscle. They also discussed possible modes of action of the toxin and the antagonist compounds.

CHEMISTRY

Aside from the work of Kohn *et al.* (1960), nothing has been reported on the chemical properties of cone venom.

The results of toxicity experiments with venom that had been heated, dialyzed, and incubated with proteolytic enzymes, suggested a partial role of protein or polypeptides. The protein present in *Conus* venom may have been nontoxic, and the retention of considerable toxicity following heating and incubation with trypsin suggested the presence of other toxic agents. It is believed that the venom of *Conus* is comprised of a complex of active substances, including protein, quaternary ammonium compounds, and possibly, amines.

LIST OF ABALONE REPORTED AS POISONOUS

Phylum MOLLUSCA

Class GASTROPODA: Snails, Slugs

Subclass PROSOBRANCHIA

Order ARCHAEOGASTROPODA

Family HALIOTIDAE

Haliotis discus Reeve (Pl. 17). Abalone (USA), kurogai (Japan). *96*

DISTRIBUTION: Japan.

SOURCE: Hashimoto and Tsutsumi (1961).
OTHER NAME: *Haliotis discus hannai.*

Haliotis sieboldi Reeve. Abalone (USA), megai (Japan).

DISTRIBUTION: Japan.

SOURCE: Hashimoto and Tsutsumi (1961).

BIOLOGY

Abalones are widely distributed, with species occurring in California, Japan, Indian Ocean, Australia, Africa, and Europe. They dwell along outer surf-beaten coastal areas wherever they can get a foothold or beneath rocks in areas free from loose sand and mud. Abalones have a vertical distribution that is roughly within the extreme high and low tidal range.

The outer shell of the mollusk is covered by a rough, horny coating which quite frequently is hidden by a thick cover of algae and other growth. The spire of the shell is greatly flattened and the epipoda is bordered with a fringe and tentacles which project around the margin of the shell. Along the margin of the shell in older specimens there is a single row of holes through which long feelers may project, and from which water passing over the gills is discharged. The anus and the nephridial perforations lie beneath one of the more posterior openings. The living mollusk projects its head out from under the edge of the shell in the area where the row of holes terminates. The tip of the broad muscular foot is pointed backward from under the spiral. When undisturbed, the abalone moves along with a clumsy swinging gait. If disturbed, the mollusk is able to cling to a rock with remarkable power. However, it can be removed easily if taken unawares. Stories of abalone holding people until the tide comes in are believed to be fictitious. The lining of the shell is of a pearly iridescence. Abalone shell has been used extensively as a source of mother-of-pearl in the manufacture of buttons, ornaments, and trinkets. The muscular foot is used extensively in the preparation of soups, chowder, and as steaks. Some species of abalone attain large size—30 cm or more in length. A spawning female abalone may produce over 2 million eggs in a single season. Spawning takes place in California from the middle of February through the first of April. *Haliotis* is a strict vegetarian, feeding largely on sessile algae and some plankton (Ricketts and Calvin, 1952; Barnes, 1963).

The toxic principle found in the viscera of abalones is believed to originate through the food which they ingest—namely, certain species of seaweeds belonging to the genus *Desmarestia* (Hashimoto, Naito, and Tsutsumi, 1960).

MECHANISM OF INTOXICATION

Poisonings from abalone are the result of eating the viscera of the mollusk. The custom of eating the entire mollusk is practiced in Oriental countries. Elsewhere, only the muscular foot (*abalone steak*) is eaten. Although only two species of Japanese abalone, *Haliotis discus* Reeve and *H. sieboldi* Reeve, have been incriminated, the viscera of other species is suspect.

MEDICAL ASPECTS

Clinical Characteristics

The symptoms of abalone viscera poisoning have been described as a sudden onset of burning and stinging sensations over the entire body, followed by an urticarialike reaction consisting of a prickling sensation, itching, erythema, edema, and subsequent development of areas of skin ulceration (Hashimoto *et al.*, 1960). It has been observed that the skin lesions are limited to those parts of the body which are exposed to sunlight and that there is a distinct boundary between covered and uncovered parts of the body.

Abalone viscera poisoning is one of the rare instances in which dietary photosensitization in humans is known to be induced by a food of animal origin.

Treatment

The treatment is largely symptomatic. Abalone viscera should be promptly eliminated from the diet.

Prevention

Eliminate the viscera. Only the foot or "steak" of the mollusk should be eaten.

PUBLIC HEALTH ASPECTS

This problem is of minor public health significance since the disease is found only rarely and in very restricted parts of the Orient—chiefly in Okushiri Island, Hokkaido, and Iki Island, Nagasaki Prefecture, Japan. The actual incidence of the disease is unknown, but it is believed to have been very low during the past 60 years. Intoxications are most likely to occur in Japan during the spring months of the year.

TOXICOLOGY

Hashimoto *et al.* (1960), and Hashimoto and Tsutsumi (1961) fed cats, rats, mice, and rabbits, livers taken from *Haliotis discus* and *H. sieboldi* collected in May at Misaki, Kanagawa Prefecture, Japan. If the test animals were kept in a dark room, no appreciable symptoms developed; but when placed in sunlight, all the animals with the exception of the rabbit, demonstrated a dramatic response of erythema and pruritus in a few seconds. The animals became very irritable, scratching and rubbing areas of exposed skin, and frequently shaking their head. The portions of the skin not covered by hair became swollen and inflamed. There was increased lacrimation and salivation, intermittent convulsions, paralysis, and death in about 30 minutes in those cases in which the animals were highly sensitive to the toxin. Most of the rats recovered if they were promptly removed from the sunlight and kept in the dark. The rats subsequently showed a loss of appetite, loss of weight, accompanied by a lowering of body temperature. The swelling subsided in a few days leaving much of the skin in a necrotic state. Sclerosis and necrosis developed on the ears, back, and head, and there was a sloughing of the skin after about 3 weeks. The animals' ears remained malformed, but the naked skin areas of the back and head soon regenerated new hair growth.

The control animals showed no evidence of photosensitivity.

The abalones were dissected into various anatomical parts, the muscular foot, liver, and various other portions of the viscera, but the toxic principle was found only in the digestive gland or liver.

Photosensitization was found to be dependent upon the amount of abalone liver administered. The dosage was found to be the same for *H. discus* and *H. sieboldi*; nevertheless there was believed to be some variation in the concentration of photodynamic substance in the liver of the two species of *Haliotis* used because the experimental conditions varied somewhat. Under the experimental conditions which they used, the researchers found that any amount of *Haliotis* liver over 5 g/100 g of body weight of the test animal would induce photosensitivity.

PHARMACOLOGY

(Unknown)

CHEMISTRY

Hashimoto and Tsutsumi (1961) found that the photodynamic principle is fairly heat stable, i.e., not destroyed by boiling for 30 minutes; it can be kept frozen (-15° C to -20° C) without loss of activity for 10 months, and salting or placing it in a brine solution does not destroy it. It is soluble in methanol, ethanol, acetone, ether, chloroform, and acetic acid; but insoluble in water. Extracts appear dark brownish-green and show an intense flourescent red under ultraviolet light or sunlight.

Hashimoto and Tsutsumi believed that the photodynamic principle was a chlorophyll derivative obtained from the algae upon which the abalone fed.

REPRESENTATIVE LIST OF MUREX REPORTED AS POISONOUS

Phylum MOLLUSCA

Class GASTROPODA: Snails, Slugs

Subclass PROSOBRANCHIA

Order ARCHAEOGASTROPODA

Family MURICIDAE

97 *Murex brandaris* Linnaeus (Pl. 18, fig. a). Rock shell, murex (USA), rocher droite épine, biou cavellan (France), bulo maschio, bulo femina, garusola (Italy), rocher (Morocco).

Distribution: Mediterranean Sea and West Coast of Africa.

Sources: Dubois (1903, 1907), Roaf and Nierenstein (1907*a, b*).

97 *Murex trunculus* Linnaeus (Pl. 18, fig. b). Murex (USA), rocher fascie épineux, biou viouret (France), bulo maschio, bulo femina, garusola (Italy).

Distribution: Mediterranean Sea.

Sources: Dubois (1903), Emerson and Taft (1945).

Thais haemastoma (Clench)

Distribution: Southeast U.S., West Indies.

Source: Huang and Mir (1972).

98 *Thais lapillus* Linnaeus (Pl. 19, fig. a). Atlantic dogwinkle (USA).

Distribution: Southern Labrador to New York, United States; Norway to Portugal.

Sources: Dubois (1903, 1907), Roaf and Nierenstein (1907*a, b*).
Other name: *Purpura lapillus.*

BIOLOGY

The genus *Murex* contains about 300 species that inhabit tropical and temperate seas. Rock shells, or murex as they are commonly called, are essentially shallow water dwellers, ranging from the tidal zone to depths of 50 fathoms or more. Murex live in a variety of biotopes, varying from rocky tidepools to sandy bottom areas, and mud flats. Because of their carnivorous eating habits, some species of *Murex* are of considerable economic threat to oyster fisheries. Muricids are able to drill a hole through an oyster shell within 4 hours. The drilling process is accomplished with the help of an acid secreted by the borer's salivary gland and its sharp rasplike "tongue." The juices of the victim are sucked through the hole. *M. trunculus* is reportedly the mollusk used by the ancients in obtaining the Tyrian purple dye (Lovell, 1867). The pigment is produced by the so-called purple, or hypobranchial, gland which secretes a colorless or yellowish fluid. Upon a short exposure to sunlight the fluid changes to a brilliant violet or reddish color and gives off a penetrating fetid odor. The toxic substance which has been termed purpurine or murexine is also produced by this gland. Muricids are generally distinguished by their highly ornamental spine-processed whorls. *M. brandaris* and *M. trunculus* attain a length of about 8 cm, while *Thais lapillus* attains a length of about 5 cm.

MORPHOLOGY OF THE POISON GLAND

The poison of *Murex* is located in the hypobranchial, or purple, gland which is part of the skin and appears as a conspicuous folded glandular structure on the roof of the mantle cavity (Fig. 3A-B). If the mantle is incised and peeled back about the region of the rectogenital prominence, the inferior surface of the hypobranchial gland will be revealed. The hypobranchial gland in *M. trunculus* extends in the transverse plane from the branchia to the rectogenital protuberance, and in the longitudinal plane from the digestive gland to within about a half centimeter of the free margin of the mantle.

The hypobranchial gland appears as three narrow, elongate, parallel bands or zones, oriented in an anteroposterior direction. These zones are best seen in cross section (Fig. 3B). The two lateral, or marginal, bands have a milky appearance. Because of their adjacent anatomical relationships, they are termed the branchial zone and rectal zone. The medial band is termed the median zone, and is a pearly gray color which gradually merges into a yellow-green and finally into a violet-blue toward the anterior end of the gland. Erspamer (1947) has divided the median zone still further into an anterior median zone and a posterior median zone. The cells of the hypobranchial gland are comprised largely of glandular epithelium. Murexine, a neurotoxic secretion, is produced in the median zone of the hypobranchial gland. Unfortunately, little is known about the histochemistry of this gland; however, Grynfeltt (1911, 1913) is one of the few to have described its morphology.

The hypobranchial gland of *Murex* produces a yellowish viscid slime. Dubois (1903) has suggested that it serves as a venom gland. However, there is little evidence to substantiate this viewpoint as there is no accessory venom apparatus by which to introduce the venom into the prey. Fischer (1925) claims that it functions as a genital organ—a viewpoint also held by Jullien (1948*a, b*). Erspamer and Glässer (1957) have suggested that the neurotoxin murexine is used by the mollusk for food procurement and as a defense mechanism. Fretter and Graham (1962) believe that the hypobranchial gland is concerned mainly with the production of a viscid slime which cements particles together as they are being eliminated from the mantle cavity. Our knowledge of the function of the hypobranchial gland appears to be too meager at this time to draw any final conclusions.

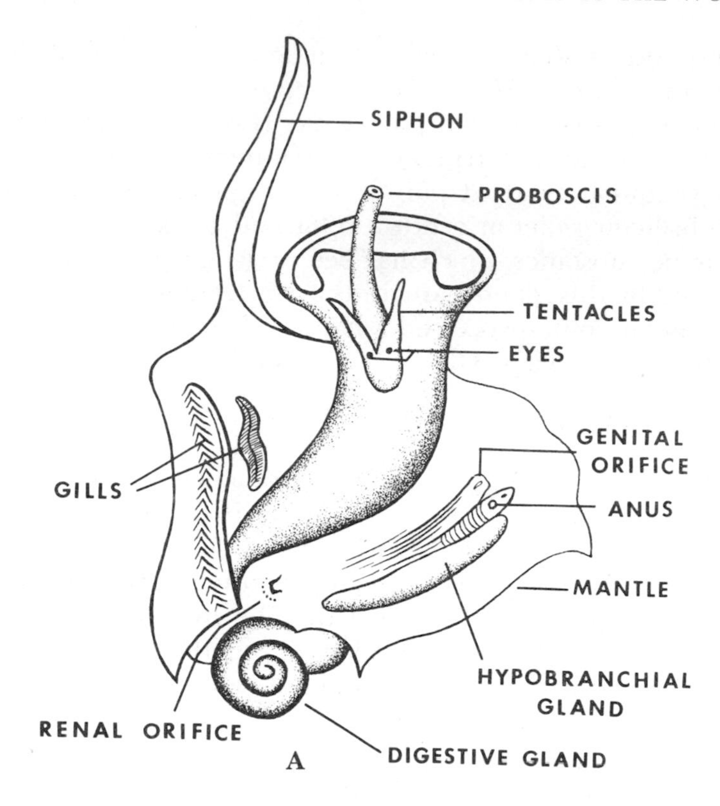

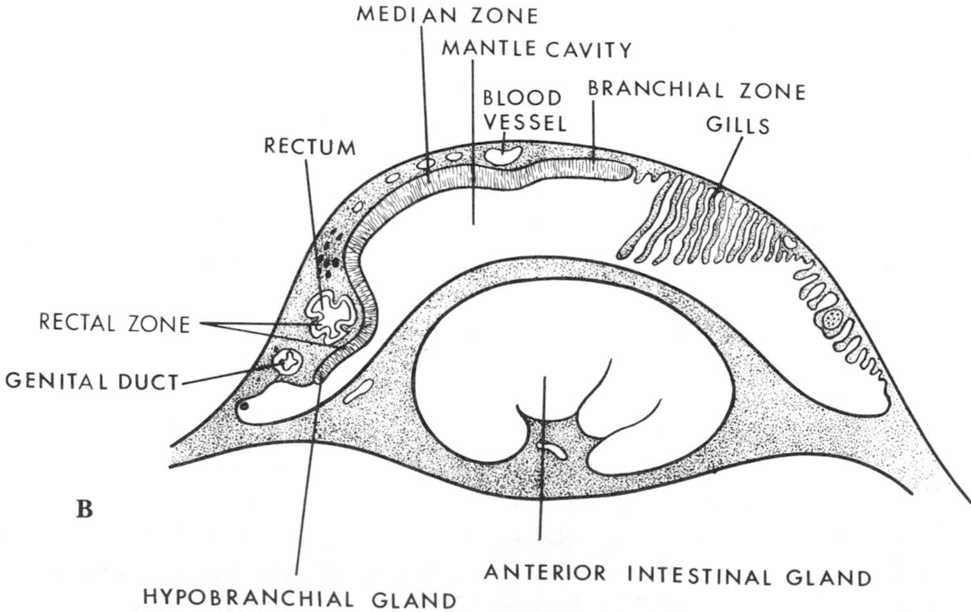

FIGURE 3.— A. Anatomy of *Murex*. The mantle has been peeled back to show the hypobranchial gland. Semidiagrammatic (R. Kreuzinger, modified from Fischer). B. Cross section through the body of *Murex* to show the various zones of the hypobranchial gland. Murexine is produced in the median zone. Semidiagrammatic.

(R. Kreuzinger, modified from Grynfeltt)

MECHANISM OF INTOXICATION

The subject of murex poisoning as it relates to man is still a moot one. Several species of *Murex* are commonly eaten in Europe (Lovell, 1867; Rogers, 1951). On the other hand, at least one serious outbreak of poisoning of undetermined character has occurred in man (Plumert, 1902). It would appear from the toxicological data which are available that murex poisoning by the oral route is generally unlikely. Since *Murex* does not possess a true venom apparatus, human contact with murexine via the parenteral route would not be possible under ordinary circumstances.

MEDICAL ASPECTS

Clinical Characteristics

Few references appear in the scientific literature on the toxicity of muricine shellfishes. Most of our knowledge on this subject has been derived from laboratory research. Plumert (1902) states that in 1900 at the Gulf of Trieste 43 persons were poisoned from eating *Murex brandaris*, and 5 of them died. Since the complete clinical characteristics of the outbreak are not given, the precise etiology of the outbreak cannot be determined. Dubois (1903), who has studied the toxic properties of *Murex*, states that *Murex* are eaten in great abundance in some parts of Europe without ill effects. Charnot (1945), on the other hand, states that *M. brandaris* may produce gastroenteritis, pruritus, convulsions, and death. No other data appear to be available regarding this subject.

Treatment

Symptomatic.

Prevention

Charnot (1945) believes that *Murex brandaris* should not be eaten. In general, the dangers of eating *Murex* are believed to be slight.

TOXICOLOGY

Extracts from the hypobranchial, or purple, gland of *Murex brandaris* and *M. trunculus* were first studied by Dubois (1903, 1907, 1909). Alcoholic extracts were prepared and tested by injecting them subcutaneously into frogs. This resulted in inactivity, muscular paralysis, convulsions, and death. Similar results were obtained with marine gobies and freshwater cyprinids; but the extracts proved to be inactive in warm-blooded vertebrates, dogs, rabbits, and guinea pigs. Dubois also demonstrated a similar substance to be present in the purple gland of *Purpura lapillus*, which he termed purpurine. He concluded that the purple gland is a venom gland which serves to capture prey or is used as a defense mechanism.

Roaf and Nierenstein (1907*a, b*) tested acidified aqueous extracts from the hypobranchial gland of *P. lapillus* by perfusing frog's blood vessels and by blood pressure experiments on rabbits. These tests revealed marked vasoconstriction and a drop of blood pressure in the rabbits. No deaths or other toxic effects were reported. Vincent and Jullien (1938*a, b, c*), Cardot and Jullien (1940), Jullien (1940, 1946, 1948*a, b*), Jullien and Bonnet (1941), Jullien, Garabadian, and Gibault (1941), Jullien, Jacquemain, and Ripplinger (1948), and Jullien and Ripplinger (1950) prepared extracts from the hypobranchial gland of *M. trunculus* and tested them on leeches, frogs, fishes, crustaceans, cephalopods, echinoderms, and rabbits. The toxic effects were similar to those produced by curare. Initially the symptoms in most of the organisms consisted of agitation, followed by muscular paralysis, and in some instances, death. Only the cephalopods and rabbits appeared to be unaffected by the extracts. The reactions varied considerably from extracts prepared from *M. brandaris.* In general, *M. brandaris* extracts appeared to be less toxic (Vincent and Jullien, 1938*c*; Bonnet and Jullien, 1941).

Erspamer and Dordoni (1947), and Erspamer (1948*a, b, c, d*) worked with purified salts of murexine, the active toxic principle obtained from the median salt zone of the hypobranchial gland of *M. trunculus, M. erinaceus*, and *M. brandaris*. They found that murexine was present in large concentrations in *M. trunculus* and *M. erinaceus*, and to a much lesser extent in *M. brandaris*. Murexine oxalate was found to have an LD_{100} of 300 mg/kg when given subcutaneously to white mice. The lethal dosage for intravenous injection was one-tenth to one-twentieth of those given subcutaneously. Death appeared to result from asphyxia and was frequently followed by fasciculatory and fibrillary contractions in nearly all the muscles. The heart continued to beat for some minutes after respiratory arrest. The extracts were found to be 3 to 5 times more toxic in the rat than in the mouse, and 10 times more toxic in the rabbit than in the mouse when given subcutaneously. Intravenous injections of strong extracts from the hypobranchial gland of *M. trunculus* resulted in a loss of consciousness, loss of superficial reflexes, and muscular paralysis; however, the test animals could be kept alive by artificial respiration. Tests of murexine picrate on leeches, batrachians, fishes, and rats resulted in a curariform effect.

Pasini, Vercellone, and Erspamer (1952) tested synthetic murexine on mice and found that it produced the same physiological responses. The LD was the same, 8.4 mg/kg, by intravenous administrations.

Erspamer and Ghiretti (1951) and Erspamer (1956) have also isolated 5-hydroxytrptamine from the hypobranchial gland of *Murex*. The 5-hydroxytryptamine was found to have a very low acute toxicity. The LD_{50} in mice by the intravenous route was 160 mg/kg, or $>$868 mg/kg subcutaneously. In the rat, the LD_{50} was 30 mg/kg intravenously, and 117 mg/kg subcutaneously. In chronic toxicity studies, it was found that the daily subcutaneous administration over a 30-to-120-day period in a dosage of 1 mg/kg did not affect the general health or systemic blood pressue, and did not produce any visceral pathology in the test animals. If the dosage was increased to 5 mg/kg to 15 mg/kg over a 10-to-40-day period, there was deterioration in the general health of the animal and renal damage with tubular necrosis was the outstanding microscopic finding.

PHARMACOLOGY

Dubois was the first to study the physiological effects of extracts from the hypobranchial glands of *Murex* and *Purpura*. He found that alcoholic extracts produced a generalized central paralysis in such cold-blooded vertebrates as the frog, and freshwater and saltwater fishes. Later studies revealed cardiovascular depression in addition to the central paralysis. Dubois believed the purple gland extracts had no adrenergic properties. Roaf and Nierenstein (1907*a, b*) claimed, on the other hand, that the hypobranchial gland of *Purpura* contained a substance chemically and physiologically similar to adrenalin. Vincent and Jullien (1938*a, b*) found in trichloracetic acid extracts of *Murex* hypobranchial gland a substance which caused acetylcholinelike contractions in the eserinized dorsal muscle of the leech. They considered the active principle to be an ester or a mixture of esters of choline, probably the acetic ester.

Later Cardot and Jullien (1940) tested noneserinized Ringer extracts of *M. trunculus* purple gland for their depression of excitability of nerve and muscle. They found that an injection of such extracts into a crab caused a profound muscular relaxation such as that produced by curare. Using the frog sciatic-gastrocnemius preparation, they also found that the total extracts of the gland provoked a strong curarization which consisted principally of an elevation of muscle chronaxie accompanied by a slight diminution of nervous chronaxie.

Erspamer and his associates have worked extensively on the pharmacology of murexine (Erspamer, 1946, 1947, 1948*a, b*, 1952, 1953; Erspamer and Dordoni, 1946, 1947; Erspamer and Glässer, 1957). They identified murexine chemically as urocanylcholine (β-[imidazole-4(5)]-acryl-choline), a naturally occurring choline ester of urocanic acid. They found that extracts of the hypobranchial organ of *M. trunculus*, when applied to an isolated frog rectus muscle, caused a strong contraction similar to that produced by acetylcholine. The activity on the noneserinized muscle was such that an extract of 1.0 g of fresh tissue was equivalent to 15,000 to 20,000 μg of acetylcholine. The frog rectus was found to be sensitive to dilutions of extract in the range of 1/35,000 to 1/70,000. Application of hypobranchial gland extracts from *M. trunculus* also caused strong acetylcholinelike contractions in the dorsal muscle of the leech. These contractions could be abolished by washing with Tyrode's solution. The above data were presumptive evidence that the active substance was quite similar to acetylcholine.

Extracts of the posterior portion of the median zone of the hypobranchial were not inactivated by contact with defibrinated calf or rabbit blood nor with rabbit brain extracts. Acetylcholine solutions (1:1,000 to 1:10,000) lose 99 percent of their activity after similar treatment. On the frog rectus and the dorsal muscle of the leech, extracts of *M. trunculus* and *M. brandaris* had less activity on eserinized preparations than on those not so treated. On the other hand, the activity of the extracts from *M. erinaceus* was increased two to three times in the presence of eserine.

It was found that extracts of hypobranchial gland from *M. trunculus* had very little effect on the isolated frog heart (the acetylcholine equivalent of 1.0 g of fresh posterior median zone from *M. trunculus* was less than 7 to 8 μg). Concentrations of 1:20 or greater caused a decrease in contractility and an irregular rhythm. This depression could be reversed by washing the organ with fresh Tyrode's solution, but it was unaffected by atropine.

Only relatively large doses (1:50) of extract had any stimulating effect on the large intestine of rats. The authors were uncertain as to whether or not this was a direct effect or a secondary stimulation resulting from the release of histamine. Extracts of a concentration of 1:30 to 1:10 had a slight depressive action on the frog lung preparation. Acetylcholine bromide (0.1 μg) had a definite stimulating effect on the frog lung.

When hypobranchial extracts were injected into the dorsal lymph sac of the frog, a curariform syndrome developed which included muscular weakness, loss of motor coordination, paralysis, disappearance of respiratory movements and reflexes, and pronounced mydriasis. Neither atropine nor prostigmine had any effect in preventing or attenuating the curariform syndrome.

In order to obtain a clue to the chemical nature of the active substance in the extracts, and thus an indication of the mechanism of its pharmacological action, Erspamer *et al.* subjected the extracts to various chemical procedures. They found that extracts after alkaline hydrolysis and subsequent acetylation caused acetylcholinelike contractions in the frog rectus muscle. The above treatment increased the acetylcholine equivalence of fresh extracts approximately 50 percent. The activity of acetylated extract applied to an eserinized frog rectus was markedly potentiated. The fact that acetylated extracts could be quickly and totally inactivated, not only by alkaline hydrolysis but also by contact with defibrinated ox blood or phosphate extracts of rabbit brain, strongly suggested susceptibility to cholinesterase, probably acetylcholinesterase. The activity of extracts acetylated without prior alkaline hydrolysis was essentially unchanged, indicating the absence of either choline or corresponding free cholines, at least in any significant amounts.

The activity of acetylized extracts on the dorsal muscle of the leech was essentially nil. However, after treatment with eserine the response of the leech preparation was increased about one thousandfold. It is also interesting to note that while unaltered extracts of *Murex* are about equally active on the frog rectus and the dorsal muscle of the leech, acetylcholine and the extracts submitted to alkaline hydrolysis and subsequent acetylation are at least 10 times more active on the leech preparation.

As previously mentioned, extracts of *M. trunculus* are essentially inactive on the frog heart except in very large doses (greater than 1:100). Acetylated extracts, on the other hand, markedly inhibit the heart at dilutions of 1:7,000 to 1:10,000. The action of the acetylated extracts is blocked by atropine.

Extracts acetylated without previous alkaline hydrolysis had the same activity on the heart as untreated extract.

Acetylation increases by 100 to 200 times the mild stimulating activity of unaltered extracts. This increased activity, resulting from acetylation, is blocked strongly by atropine. Whereas frog lung is slightly depressed by untreated extract, acetylation imparts a very intense contractive activity similar to that of acetylcholine extracts. After acetylation, the extracts manifest only about one-tenth the curariform activity in the intact frog. The paresis is of shorter duration.

When compared with acetylcholine, murexine is practically inactive on frog heart, frog lung, and rat large intestine (all of which are very sensitive to acetylcholine). Murexine is much less effective on the frog rectus, and much more effective on the leech muscle not treated with eserine.

Erspamer (1948*b*) in studying the effect of murexine on the central nervous system, found that small doses strongly excited the respiratory center, while large doses paralyzed it. The excitatory action was most evident in reversing the respiratory depression resulting from morphine or barbiturates. Erspamer found the bulbar vasomotor center to be influenced only slightly by murexine.

Erspamer also found that the frog heart was depressed only slightly by murexine, and that this effect was not altered by atropine. Although the direct action of murexine on the smooth muscle of blood vessels was not demonstrated, intravenous administration of murexine to the intact animal and the decapitated animal resulted in definite, but floating, hypertension. In the spinal cat, the hypertensive effect was thought to be due for the most part to adrenalin discharge both from the adrenals and from adrenergic nerve endings.

Murexine applied to the isolated eyes of the squid, octopus, frog, and toad produced a rapid and intense mydriasis, which in the squid was accompanied by an expansion of the iris chromatophores. In the intact squid, cat, and dog, murexine caused a transient myosis and a loss of iris pigmentation. Both these conditions in the squid were later reversed.

Murexine, even in large doses, had no effect on lacrimation in the dog or cat. However, a marked but transitory salivation was produced which was abolished by atropine. On the isolated intestine of the frog, pigeon, guinea pig, dog, cat, rabbit, rat, and on the isolated uterus of guinea pig, dog, cat, rabbit, and rat, murexine was only slightly stimulating. This excitation was blocked by atropine.

The intense contractions of frog uterus and the dorsal muscle of the leech induced by murexine were only slightly enhanced by eserine.

In summary, the basic pharmacological action of murexine on both vertebrates and invertebrates provoked a paralysis of the skeletal muscle. Murexine possessed marked neuromuscular blocking and nicotinic actions, but was almost devoid of muscarinic effects. The blocking action appeared to be of the depolarizing type. It was weaker, but qualitatively similar, to suxamethonium (Erspamer and Glässer, 1957).

Clinical trials of murexine were carried out by De Blasi and Leone (1955) on 47 patients, and by Ciocatto, Cattaneo, and Fava (1956) on 123 patients. The tests showed that murexine was a short-lasting muscle relaxant. The mean paralyzing dose in adult patients was approximately 1 mg/kg to 1.2 mg/kg, intravenously. Maximum effects were achieved in 45 to 60 seconds using a single intravenous dose, and the paralysis lasted 3 to 6 minutes. A satisfactory long-lasting muscular relaxation could be obtained by slow, intravenous infusion of a 1/1,000 solution of murexine in physiological saline. The side effects observed were attributed to the nicotinic actions of the drug. Muscular paralysis was always preceded and followed by muscular fasciculations. Hypersalivation was common, but it was blocked easily by atropine. A moderate rise in blood pressure was frequent. Persistence of pharyngeal and laryngeal reflexes, lacrimation, and sweating were noted.

Quilliam (1957), Holmstedt and Whittaker (1958), Keyl and Whittaker (1958), and Whittaker (1960) have also worked on the pharmacology of murexine. Their reports are essentially in agreement with the findings of Erspamer and his associates.

The pharmacological actions of a series of 19 synthetic murexinelike substances have been studied by Erspamer and Glässer (1958) to determine the relation between the chemical structure of the murexinelike substances and their pharmacological effects. Since these are synthetic agents, a discussion of these properties goes beyond the scope of this work.

CHEMISTRY

Murexine is the name given by Erspamer and Dordoni (1947) to the physiologically active compound found in large amounts in the median zone of the hypobranchial body of *Murex trunculus* and other related species of mollusks.

Considerable detail with reference to the chemical reactions of the toxin through a wide variety of reagents has been provided by Erspamer (1947). The observations from his series of tests led to the conclusion that two different components of the toxin fraction existed. These were referred to as purpurine (evidently identical to murexine) and an enteraminelike substance. The amounts of these two types were found to differ with species and even with the location in the hypobranchial organ. The "enteramine-like" factor was identified as enteramine in itself, otherwise known as 5-hydroxytryptamine (Erspamer and Ghiretti, 1951; Erspamer and Asero, 1951, 1952; Asero *et al.,* 1952; Erspamer, 1956).

The problem of determining the exact structure of murexine as an ester of choline was taken up by Erspamer and Benati (1953*a, b*). This toxic principle was found to form a dipicrate containing 67.1 to 67.7 percent picrate. By means of Feigl's hydroxylamine reaction, the toxin was shown to be an ester. This ester could be hydrolyzed in either an acid or alkaline medium. Acid hydrolysis-liberated choline was identified as the reineckate, and this identification was confirmed by the formation of acetylcholine on acetylation. Elementary analysis of the picrate indicated that murexine itself must have an empirical formula $C_{11}H_{18}O_3N_3$. Since choline accounted for only one nitrogen atom, the carboxylic acid part was believed to contain the two remaining nitrogen atoms. This carboxylic acid, derived by hydrolysis of murexine, was found to give an intense coupling reaction with diazonium salts. This excluded purines, pyrimidines, phenols, and indoles. The two nitrogen atoms suggested an imidazole derivative, and the carbon content suggested a 3-carbon acid derivative of imidazole. Adsorption of the color of bromine indicated a double bond. Identity to β-[imidazolyl-(4)]-acrylic acid was proven by coincidence of properties with the synthesized compound. Murexine was concluded then to have the structure of β-[imidazolyl-(4)]-acrylcholine, which is presumed to be formed by the conjunction of an intermediate of histidine metabolism with choline. Murexine is known otherwise as urocanylcholine.

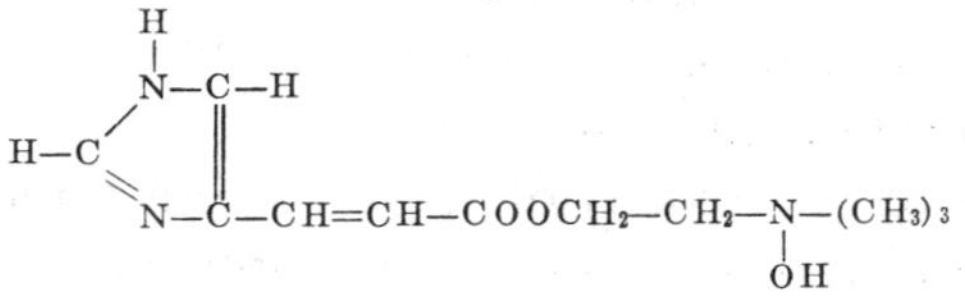

```
           H
           |
           N—C—H
          /  ||
      H—C    ||
          \\ ||
           N—C—CH=CH—COOCH2—CH2—N—(CH3)3
                                |
                                OH
```

The correctness of the above formula was confirmed by the synthesis of the toxin by Pasini *et al.* (1952). The synthesized toxin gave the same dipicrate as that obtained from the natural product and also showed the same pharmacological actions. From the hydrolysis products of the synthesized toxin, choline and urocanic acid were isolated.

The publications of Erspamer and his associates are particularly noteworthy in that they state that murexine constitutes one of the few instances in marine biotoxicology in which the chemistry of the poison has been completely characterized and synthesized. Certainly this represents a significant advance in the vast uncharted sea of marine biotoxins.

Brief general reviews on murex poison have been published by Calmette (1908), Maass (1937), and Kaiser and Michl (1958).

LIST OF TOP SHELLS REPORTED AS POISONOUS

Phylum MOLLUSCA

Class GASTROPODA: Snails, Slugs

Subclass PROSOBRANCHIA

Order ARCHAEOGASTROPODA

Family TROCHIDAE

Livona pica Linnaeus (Pl. 19, fig. b). West Indian top-shell (USA), cigua (Cuba). *98*

DISTRIBUTION: Southeast Florida and the West Indies.

SOURCES: Gudger (1930), Arcisz (1950).
OTHER NAME: *Turbo pica, Turbo nicobaricus.*

Family TURBINIDAE

Turbo argyrostomus Linné. Silver mouthed turban-shell (Japan).

DISTRIBUTION: Indo-Pacific.

SOURCE: Hashimoto *et al.* (1970).

The top shells are members of the gastropod family Trochidae. As their vernacular name implies, they are generally conical or top-shaped, spiral, and have a pearly luster on the inner surface of the shell. Some of the best-known members are the trochus shells of tropical seas which are widely used in the manufacture of bracelets, buttons, and pearl ornaments. The Trochidae is a large family and includes both littoral and deep-sea forms. The top shells are largely algae feeders.

Members of the Turbinidae have a spiral shell which is turban shaped, solid with a simple circular or oval aperture. They are generally vegetarian and inhabit the littoral of warm seas.

Our knowledge of the toxicity of *Livona pica* is limited to references such as Gudger (1930) and Arcisz (1950), who quote Poey as saying that the ingestion of "*Turbo pica,*" a marine snail, will give rise to indigestion and nervous disorders. The West Indian top shell is a popular seafood in the West Indies and Central America, where it abounds on rocks and coral reefs near the shore. It attains a length of about 10 cm. It is possible that in some localities these shellfish become toxic during certain seasons of the year, as do some of the West Indian marine fishes, since the toxin is believed to originate from the algae upon which they have been feeding. Unfortunately no other data are available regarding the toxicity of this organism.

Human intoxications following the ingestion of *Turbo argyrostomus* have been described by Hashimoto *et al.* (1970). The case histories resemble ciguatera poisoning, but the toxin is differentiated by a fat-soluble fraction. Toxicity tests of various specimens of *Turbo*, using mice with the water-soluble and fat-soluble fractions, revealed that the mid-gut gland and gut contents were more or less toxic, while the muscle was nontoxic. The water-soluble toxin was partially purified and found to be non-dialyzable, extractable with *n*-butanol, precipitable with acetone, and hemolytic.

LIST OF SEA HARES REPORTED AS POISONOUS

Phylum MOLLUSCA

Class GASTROPODA: Snails, Slugs

Subclass OPISTHOBRANCHIA

Order TECTIBRANCHIA

Family APLYSIIDAE

99 *Aplysia californica* Cooper (Pl. 20, fig. a). California sea hare (USA).

DISTRIBUTION: Monterey to San Diego, Calif.

Sources: Winkler (1961), Winkler and Tilton (1962), Winkler, Tilton, and Hardinge (1962), Watson and Rayner (1973).

Aplysia depilans Linnaeus (Pl. 20, fig. b). Sea hare (USA), seehase (Germany). *99*

Distribution: Coast of France, Atlantic Ocean, and throughout the Mediterranean Sea.

Source: Flury (1915).

Other name: *Aplysia limacina.*

Aplysia punctata Cuvier (Pl. 20, fig. c). Sea hare (USA), lievre de mer (Morocco), seehase (Germany). *99*

Distribution: Mediterranean Sea.

Source: Fayrer (1878).

Other name: *Aplysia punctata* (misspelled).

BIOLOGY

Sea hares inhabit coastal waters on bottoms covered with algae and eel grass, and they have a vertical range from shallow water tidal zones to depths of about 40 fathoms. They tend to be omnivorous, feeding chiefly on seaweed (Winkler and Dawson, 1963), but also devouring animal substances. Usually they are found singly, but during the spring they congregate in large numbers to oviposit. Despite the reputation given them by the ancients, they can be handled with impunity. If they are molested they discharge a violet fluid from their mantle. According to Tryon (1882) the fluid may have a nauseous odor, which probably contributed to their reputation. *Aplysia depilans* may attain a length of 25 cm, whereas some of the other species are considerably smaller. Sea anemones are among the very few to feed on sea hares (nudibranchs), apparently without ill effects (Winkler and Tilton, 1962).

MORPHOLOGY OF THE POISON GLAND

The distribution of the poison in the body of the sea hare has not been fully determined, but toxic substances have been reported from the following areas: (a) opaline gland (Flury, 1915); (b) purple gland (Flury, 1915); (c) digestive gland (Winkler, 1961; Winkler and Tilton, 1962; Winkler *et al.,* 1962). The morphology of these glands has been well described in an excellent monograph by Eales (1921) on the anatomy of *Aplysia.*

Opaline Gland (Fig. 4): This gland has been variously termed the poison gland, grape-shaped gland, and the gland of Bohadsch (Cunningham, 1883). It lies beneath the floor of the anterior part of the pallial cavity into which it discharges by numerous apertures. In some species the ductlets unite and discharge through a single external aperture. The gland has been described in one species as resembling a bunch of grapes. It consists of large unicellular vesicles whose contents in stained preparations generally resemble those of the mucous glands of the skin. The fluid secreted is milky and acrid in nature, and in some species possesses a nauseating odor. It was the secretion of this gland that gave *Aplysia* much of the evil reputation it had among ancient writers. The secretion of this gland is discharged less readily than that of the purple gland. It is believed that the secretions are both protective and excretory in function.

Purple Gland: The purple gland is situated on the underside of the free edge of the mantle shelf, and also opens into the pallial cavity. It is yellowish or brown in color, and has a somewhat pitted appearance due to the numerous ducts of the large unicellular glands that comprise the structure. If the living animal is irritated, the purple gland exudes a rich, reddish-purple dye mixed with mucus. The dye mixes slowly with the surrounding seawater and forms a protective screen which facilitates the escape of the sea hare.

Digestive Gland (Fig. 4): The digestive gland has been termed the liver or hepatopancreas. It is one of the glands of the alimentary canal, a large compact

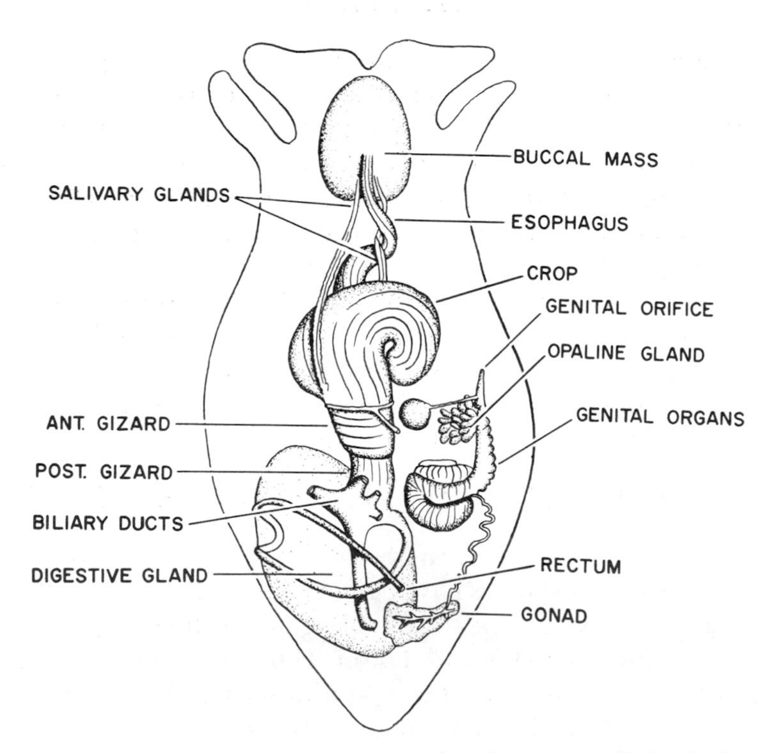

FIGURE 4.—Anatomy of *Aplysia* showing some of the anatomical relationships of the various glands which are believed to produce toxic secretions.
(R. Kreuzinger, modified from Eales)

structure, and makes up the greater part of the visceral mass. The posterior portion of the alimentary canal is embedded in the digestive gland. The outline of the gland is smooth and rounded, and divided into irregular lobes. Microscopically, the digestive gland consists of simple groups of cells lining central spaces with the spaces uniting to form ciliated channels—the bile ducts. The secretions of the digestive gland are discharged directly into the stomach by a varying number of ducts. The ducts are almost equal in diameter to the stomach itself. The ducts are ciliated in order to force the secretions into the alimentary canal. It is believed that the secretions are both digestive and excretory in nature.

MECHANISM OF INTOXICATION

The secretion of the sea hare *Aplysia* is reputed by ancient writers to be toxic if ingested. The toxic effects of *Aplysia* upon humans has not been confirmed.

MEDICAL ASPECTS

Clinical Characteristics

According to ancient writers (Grevin, 1568), extracts from *Aplysia* were frequently used by emperors to dispatch their political enemies. Even contact with the organism was reputed to produce death. Urticaria, severe inflammation, and depilation are said to result from contact with the secretions of *A. punctata* (Fayrer, 1878). Flury (1915) tested the secretions of *A. depilans* and found them to be bitter in taste, producing a burning sensation and slight irritation of the oral muscles, but no intoxication. Application of the secretion to the skin failed to produce irritation or depilation. The toxicity of nudibranchs to humans is still unknown. Flury (1915) claims that some species of *Aplysia* are eaten in tropical countries.

Treatment

Symptomatic.

TOXICOLOGY

Toxicological studies on the secretions of *Aplysia* were first conducted by Flury (1915), who worked on *A. depilans* and "*A. limacina*." According to Flury, the secretions of the sea hare are of three distinct types: (1) a colorless substance that is secreted from the surface of the body; (2) a whitish, strong-smelling, viscous substance secreted from the opaline gland through the genital pore; (3) a violet or purple substance from the so-called purple gland. Flury attempted to evaluate the toxicity of the surface secretion, and that from the purple gland.

The opaline secretion was obtained by carefully wiping the animal clean and then stroking it with the fingers. This process resulted in producing more secretion. The fluid was then injected into a series of coelenterates

(*Aiptasia, Anemonia, Rhopalonema, Hermiphora*), annelids (*Heteroneris, Asterope*), echinoderms (*Strongylocentrotus, Antedon*), mollusks (*Nasa, Pleurobranchaea*), arthropods (copepods, *Carcinus*), fishes (*Gobius, Motella, Crenilabrus, Serranus*), and frogs. All of the animals developed a muscular paralysis and some died. An injection of 0.1 ml into frogs resulted in hyperactive reflexes and pronounced muscular paralysis within a period of 25 minutes. Death usually occurred within 2 to 4 hours. A single injection of 1.0 ml into rabbits was the only test conducted on warm-blooded vertebrates and it was negative.

Flury found the opaline secretion produced a bitter taste and caused burning and irritation of the oral mucosa. When the secretion was placed on human skin it did not produce irritation or alopecia. When a drop of the secretion was placed in the eye of a rabbit, or a dog, it was found to produce an immediate irritation, causing the animal to close its eyes and manifest anxiety. No inflammatory symptoms were noted.

Injections of the purple secretion from *Aplysia* into frogs produced minor changes in the reflexes, and a brief, transient, weak paralysis.

Winkler (1961) prepared crude aqueous and acetone fractions from the digestive glands of the California sea hare *A. californica* and a related species, *A. vaccaria*. The extracts were bioassayed on frogs, chicks, mice, guinea pigs, rats, and kittens. The LD_{50} in mice were expressed in grams of digestive gland, and was found to be 0.0028 g/g in one batch and 0.036 g/g of mouse body weight in another batch when administered intraperitoneally. The LD_{50} for day-old chicks was about 25 percent more than for mice. The toxin also killed mice when given by a stomach tube at approximately 12 times the intraperitoneal dose. The signs and symptoms consisted of immediate hyperventilation, drooping of the ears, hypersalivation, hyperactivity, muscular twitchings, ataxia, loss of motor coordination, relaxation of the bladder sphincter muscles, respiratory paralysis, and death.

Watson (1973) obtained two separate lethal extracts, termed the ether-soluble toxin and water-soluble toxin, from the midgut glands of the Hawaiian sea hares, *A. pulmonica, Dolabella auricularia, Dolabrifera dolabrifera,* and *Stylocheilus longicauda*. Neither raw glands nor their extracts were orally toxic to mice, but both ether- and water-soluble fractions displayed potent effects following intraperitoneal injection. Signs following injection of the ether-soluble toxin included irritability, viciousness, and severe flaccid paralysis. The water-soluble toxin, in contrast, caused convulsions and respiratory distress.

PHARMACOLOGY

Flury (1915) stated that in the frog heart, the secretion from the body surface of *Aplysia* produced an arrhythmia followed by a bradycardia which terminated in complete diastolic stasis. Often the heart resumed beating after standing a few hours, but the new rate was relatively slow. Application of the milky secretion from the genital pore of *Aplysia* to the frog heart resulted within 1 to 2 hours in diastolic standstill following stages of arrhythmia and lowered rate. Atropine was without effect in relieving the cardiac depression produced by secretions of *Aplysia*.

Winkler *et al.* (1962) found aplysin, a purified acetone extract of the digestive glands of two California sea hares *A. californica* and *A. vaccaria*, to have the following pharmacological effects:

Cardiorespiratory System: There was an immediate lowering of blood pressure upon intravenous injection of aplysin in anesthetized dogs, but rapid recovery if it was given in small doses. The heart was initially irregular, but soon resumed a slower regular rate. Respiration was first stimulated and then depressed. In the frog, the heart was stopped in diastole.

Nervous System: The anterior cervical sympathetic ganglion of the cat was initially stimulated and then reversibly blocked.

Skeletal Muscle: Frog rectus abdominis muscle responded by contracture similar to the response to acetylcholine. The neuromuscular junction of the rat diaphragm was blocked with small doses. The block was monophasic and was antagonized by neostigmine. Chick semimembranosus muscle responded with contracture, but usually relaxed spontaneously before complete neuromuscular blocking occurred.

Smooth Muscle: Rabbit intestine smooth muscle became spastic with temporary cessation of peristalsis. The spasm could be immediately released or prevented by the addition of atropine.

Washing of any treated preparation resulted in almost immediate reversal.

Aplysin was not destroyed by serum or liver homogenates or by alkaline or acidic hydrolysis under the *in vitro* conditions of testing.

Measurable amounts of active aplysin were excreted, but it appeared that metabolic degradation occurred.

In tissue baths, aplysin produced a contractatory response of the external surface muscle of the sea anemone *Anthopleura xanthogrammica*; but no effects were observed on the internal wall of the enteron (Winkler and Tilton, 1962).

Watson and Rayner (1973) found that ether-soluble and water-soluble extracts from the midgut glands of *Aplysia californica* and *A. vaccaria* may have direct effects on the contractility of vascular smooth muscle. This activity was not mediated by alpha-adrenergic or cholinergic mechanisms. Sublethal doses of the crude ether-soluble extract produce hypertension when injected intravenously into anesthetized rats, whereas the crude water-soluble extract produces a transient hypotension, bradycardia, and apnea.

CHEMISTRY

Such data as are available on the chemistry of the toxic principle of *Aplysia* come to us from the studies by Flury (1915). He noted that the toxic principle is confined to the secretion which comes from the surface of the body of the animal when it is irritated, and that the secretions from the "genital pore" and the violet secretion from the purple gland are not toxic. He believed the toxic material to be related chemically to the terpenes because of its aromatic odor and absence of nitrogen.

In their work on aplysin, Winkler *et al.* (1962) have suggested that the active molecule of aplysin may possibly be a cholinester. One thing not in favor of it being a cholinester was the fact that pseudocholinesterases of the serum and liver did not hydrolyze aplysin; and it was not inactivated even by prolonged treatment with alkalies at 100° C. They noted, however, that effects similar to those of succinylcholine and aplysin are produced by decamethonium and other non-cholinesters. They also noted the similarity in some of the pharmacological properties of aplysin and murexine.

LIST OF PTEROPODS REPORTED AS TOXIC

Phylum MOLLUSCA

Class GASTROPODA: Snails, Slugs

Subclass PROSOBRANCHIA

Order PTEROPODA

Family CAVOLINIDAE

100 *Creseis acicula* Rang (Pl. 21). Straight needle-pteropod (USA).
DISTRIBUTION: Atlantic and Pacific Oceans.

SOURCE: Hutton (1960).

BIOLOGY

Pteropods, or sea butterflies, are small pelagic gastropods found in abundance in open seas throughout the world. They belong to the order Pteropoda, which is divided into two suborders, Thecosomata (those having shells) and the Gymnosomata (those without shells). Relatively little is known about the biology of pteropods. They are hermaphroditic. Their food consists largely of protozoans and microscopic algae. There is a total of about 60 species. Pteropods seldom exceed 2½ cm in total length.

The biology and systematics of pteropods have been discussed by Rang and Souleyet (1852), Cooke (1895), Pelseneer (1906), Tesch (1946, 1950), and Abbott (1954).

MEDICAL ASPECTS

Clinical Characteristics

Pteropod "stings" consist of a maculopapular rash which resembles that produced by some coelenterates; but no difficulties were reported after the initial stings. No other data are available.

Hutton (1960) reported a single outbreak in a shallow inshore area of the Gulf of Mexico, Redington Beach, St. Petersburg, Fla. The stings were said to be caused by the needlelike organism which penetrated the swimmers' bathing suits. Several hundred "needles" were said to have clung to the bathing suit of one swimmer. Whether the lesions were due to the mechanical effects or contact with a toxic substance is not known.

Treatment

The treatment is symptomatic.

Prevention

Avoid contact with the organism.

TOXICOLOGY

(Unknown)

PHARMACOLOGY

(Unknown)

CHEMISTRY

(Unknown)

LIST OF SHELLFISH REPORTED TO CAUSE CALLISTIN POISONING*

Phylum MOLLUSCA

Class PELECYPODA: Bivalves

Order FILIBRANCHIA

Family VENERIDAE

Callista brevisiphonata Carpenter (Pl. 22). Japanese callista (USA). *101*

DISTRIBUTION: Japan.

SOURCE: Asano (1954).

BIOLOGY

Callista generally prefer a protected sandy cove or bay. Spawning takes place during the late spring or summer months. Some species attain a length of 15 cm or longer.

---

* Exclusive of paralytic shellfish poisoning.

MECHANISM OF INTOXICATION

Poisoning results from eating the ovary of callista which contains a high concentration of choline. Shellfish are said to be toxic only during the spawning season of the year—May through September.

MEDICAL ASPECTS

Clinical Characteristics

The onset of symptoms varies from 30 minutes to 1 hour after ingestion of the shellfish. In some instances, the development of symptoms may be immediate, occurring while the victim is in the process of consuming the shellfish. The chief symptoms are itching, flushing of the face, urticaria, sensation of constriction of the chest, epigastric and abdominal pain, nausea, vomiting, dyspnea, cough, asthmatic manifestations, hoarseness, sensation of constriction, paralysis or numbness of the throat, mouth and tongue, thirst, hypersalivation, sweating, chills, and fever. Body temperature and pulse rate may be increased. There may be a slight drop in blood pressure, dilatation of the pupils, and leucocytosis.

The allergiclike manifestations that are present in this form of poisoning have been termed a "pseudoallergic reaction" (Asano, 1954).

Victims generally recover within 1 or 2 days. No fatalities have been reported.

Treatment

Symptomatic. The use of antihistamines has been recommended despite the fact that histamine does not appear to be the direct cause of the allergiclike reactions present in callistin shellfish poison.

Prevention

The shellfish are said to be safe to eat if the ovaries are eliminated. The ovaries can be identified by smearing a small amount of the whitish- or yellowish-colored gonads on a glass slide. If the shellfish is a female, one can observe the white granular-appearing ova. In the male, the gonadal substance appears as a milky paste. Cooking does not destroy the toxic properties of callistin shellfish poison. The degree of freshness of the shellfish is not a factor in the occurrence of the disease.

PUBLIC HEALTH ASPECTS

Callistin shellfish poisoning has been reported only from the vicinity of Mori, Hokkaido, Japan, where the ovaries were eaten with the shellfish during the spawning season—May through September.

Thus far, a total of 119 persons have been intoxicated (Asano, Takayanagi, and Kitamura, 1953). This is not a widespread public health problem.

TOXICOLOGY

Routine toxicological tests using mice, kittens, and dogs proved to be successful in demonstrating the toxicity of this poison (Asano *et al.*, 1953). Nine human volunteers were then fed 50 to 70 g samples taken from various parts of *Callista brevisiphonata* collected in the vicinity of Mori, Hokkaido. One of the nine patients developed a severe case of callistin shellfish poisoning with symptoms typical of those victims observed in the previous accidental outbreaks. One other patient developed very mild symptoms, and the remainder were either negative or doubtful.

The authors developed a method of skin testing using an allergen which they extracted from the shellfish with a Coca and Milford buffer solution. The liquid allergen, 0.05 ml, was injected intracutaneously with a tuberculin syringe into the skin of the inner forearm of human volunteers. Markedly positive reactions were obtained which resulted in wheals characteristic of those produced by allergens in the accidental cases of poisoning. Readings of the reactions were taken 20 minutes, 1 hour, and 10 hours after injection.

Asano (1954) also prepared allergen extracts and injected them subcutaneously into guinea pigs. Then after 3 weeks, intravenous injections were made on sensitized guinea pigs. Under usual circumstances, this should result in anaphylactic shock or some evidence of a hypersensitivity reaction if a true allergy were involved. However, in this case the only result was a slight quivering of the animal with a mild decrease of body temperature (less than 1° C). The animals were sacrificed and the plasma examined for the presence of free histamine according to Lubschez's method, but Asano was unable to detect any evidence that histamine was directly concerned with the production of the allergic manifestations.

No further toxicological data are available regarding callistin shellfish poison.

PHARMACOLOGY

Asano (1954) has shown that the causative agent in callistin shellfish poisoning is choline. Choline and its derivatives are said to have three distinct actions, varying somewhat from one derivative to the next (Sollmann, 1949):

Muscarinic Action: There is a parasympathetic stimulation which lowers blood pressure in nonatropinized cats.

Nicotinic Action: It stimulates the autonomic ganglia and affects many myoneural functions. It raises the blood pressure of atropinized cats and produces contracture of denervated gastrocnemius cats.

Epinephrine Mobilization: It results in dilatation of denervated and atropinized iris.

The toxicity of choline is said to be less severe than acetylcholine and histamine, which is suggested as the reason why some experimental animals are insensitive to the active substance in callistin shellfish poison (Asano, 1954).

Perfusion studies using isolated guinea pig ileum also failed to show evidence of a histaminic action.

CHEMISTRY

In view of the allergiclike clinical manifestations of the disease, Asano (1954) believed that callistin shellfish poisoning was due to either histamine (or a similar substance), or to choline (or one of its derivatives).

Histamine: Various ovarian extracts were prepared and fractionated by the Vickery-Leavenworth procedure, and then tested according to the Koessler-Hanke method. They were found to be negative for histamine and arginine fractions, but positive for the histidine fraction. Paper partition chromatography also failed to reveal any evidence of histamine.

Choline: The presence of choline ester-splitting enzymes in the ovaries of the *Callista brevisiphonata* were determined by the method of Stedman. Using the procedure of Kapfhammer-Bischoff, Asano obtained dark yellow-colored chloraurate crystals which he recrystallized three times from water and finally obtained the needlelike crystals of choline chloraurate. These he identified by using the melting point method. His determinations regarding the presence of choline were further substantiated by using the methods of Shaw and Glick. Choline was found to be relatively abundant in the ovary of *C. brevisiphonata*, with amounts ranging from 100 mg to 500 mg of choline per 100 g of dry ovarian material.

LIST OF SHELLFISH REPORTED TO CAUSE VENERUPIN POISONING

Phylum MOLLUSCA

Class PELECYPODA: Bivalves

Order FILIBRANCHIA

Family OSTREIDAE

101 *Crassostrea gigas* (Thunberg) (Pl. 23). Giant Pacific oyster, Japanese oyster, giant oyster, Pacific oyster (USA).

DISTRIBUTION: Japan, British Columbia to Morro Bay, California.

SOURCES: Akiba (1943, 1949), Akiba and Hattori (1949).

OTHER NAME: *Ostrea gigas.*

Order EULAMELLIBRANCHIA

Family VENERIDAE

98 *Dosinia japonica* (Reeve) (Pl. 19, fig. c). Japanese dosinia (USA).

DISTRIBUTION: Japan.

SOURCE: Akiba (1943).

Tapes semidecussata Reeve (Pl. 19, fig. d). Japanese littleneck, Japanese cockle (USA), asari (Japan). *98*

DISTRIBUTION: Japan, British Columbia to Elkhorn Slough, California.

SOURCES: Akiba (1943, 1949), Akiba and Hattori (1949).

BIOLOGY

Since venerupin poison appears to be limited to a single species of oyster, *Crassostrea gigas*, and certain shellfish members of the family Veneridae, *Tapes semidecussata, Dosinia japonica*, taken from specific areas in the Kanagawa and Schizuoka Prefectures, the present discussion on the biology of these mollusks will be limited to those aspects which appear to be concerned directly with their biology within these parameters.

Crassostrea: Toxic specimens of *Crassostrea* have been obtained largely from the deeper portions of the Yahei Shallows at the mouth of Hamana Lake, Schizuoka Prefecture, and in the vicinity of Nagai, Kanagawa Prefecture, Japan. The lake is a shallow saltwater slough, having a sand-mud bottom and poor tidal drainage. The shell tends to be long and thin, and in clusters if growing in the mud, or round and deep if growing singly on firm substratum. The Japanese oyster *C. gigas* spawns from June to August, whereas its toxic season occurs from January to April. A definite relationship has been found to exist between the presence of at least one species of dinoflagellate and the toxicity of the shellfish. The work of Masao Nakazima (1963) has determined a definite correlation between the presence of a poisonous dinoflagellate *Prorocentrum sp.* in the digestive gland of the toxic specimens of *Tapes semidecussata,* and the absence of the dinoflagellate in nontoxic specimens of *T. semidecussata*. Moreover, extracts of *Prorocentrum* poison, obtained from cultures of the dinoflagellate, produced hepatic lesions in mice similar to those caused by toxic shellfish. No outbreaks of venerupin poisoning have been reported from eating *C. gigas* (Akiba, 1949) taken from locations other than the Kanagawa or Schizuoka Prefectures. *C. gigas* attains a length of about 25 cm or more.

Tapes and Dosinia: The bivalves, Tapes and Dosinia, are generally found in coarse, sandy mud, or bays, sloughs, and estuaries, seldom more than 2 or 3 cm beneath the surface. Toxic specimens have been obtained only from certain restricted sloughs in Japan having a sand-mud bottom where the water is relatively deep and the tidal drainage is slow. *Tapes* spawns twice a year, during the spring and autumn. However, no relationship exists between the spawning season and the toxicity period, the latter reaching its peak between January and April. *D. japonica* attains a length of about 7 cm, and *T. semidecussata* a length of about 7.5 cm.

Two other species of bivalves, *Meretrix lusoria* and *Mactra veneriformis*, and a gastropod, *Batillaria multiformis*, were also collected in the Hamana Lake area but were found to be nontoxic (Akiba, 1949).

MECHANISM OF INTOXICATION

Venerupin shellfish poisoning results from the ingestion of poisonous bivalve shellfish taken from certain restricted areas in the Kanagawa and Schizuoka Prefectures of Japan during the period of January through April. Only two species have been incriminated in human intoxications: the Japanese oyster *Crassostrea gigas* and the asari *Tapes semidecussata*. The Japanese dosinia *Dosinia japonica* has also been found to be toxic, but apparently has not thus far been involved in any human intoxications. The poison is concentrated in the digestive gland or "liver" of the mollusk. The poison is believed to be derived from a dinoflagellate *Prorocentrum sp.* (Akiba, 1961).

MEDICAL ASPECTS

Clinical Characteristics

The symptoms of venerupin or asari shellfish poisoning usually develop in from 24 to 48 hours after ingestion of the toxic mollusks, but an incubation period is believed to extend up to 7 days (Akiba and Hattori, 1949). The initial symptoms are: anorexia, gastric pain, nausea, vomiting, constipation, headache, and malaise. Body temperature usually remains normal. Within 2 to 3 days nervousness, hematemesis, and bleeding from the mucous membranes of the nose, mouth, and gums develop. Halitosis is a dominant part of the clinical picture. Jaundice, petechial hemorrhages, and ecchymoses of the skin are generally present, particularly about the chest, neck, and upper portion of the arms and legs. Leucocytosis, anemia, retardation of blood clotting time, and evidence of disturbances in liver function have been noted. The liver is generally enlarged, but painless. In fatal cases the victim usually becomes extremely excitable, delirious, and comatose. There is no evidence of paralysis or other neurotoxic effects usually observed in paralytic shellfish poisoning. In the outbreaks that have been reported, there was found to be an average case fatality rate of 33.5 percent. The low mortality rates in 1949 and 1950 outbreaks were explained on the basis of an early diagnosis and prompt institution of medical care. In severe cases, death occurs within 1 week; in mild cases, recovery is slow, with the victim showing extreme weakness. The development of ascites is a frequent complication. According to Akiba and Hattori (1949), intoxication does not result in immunity.

The clinical characteristics of venerupin shellfish poisoning have also been discussed by Hashimoto, Kanna, and Shiokawa (1950), Hashimoto and Migita (1950), Hattori and Akiba (1952), and Dack (1956).

Pathology

According to Akiba (1943, 1949), and Akiba and Hattori (1949), autopsy findings consist of diffuse necrosis with hemorrhage and fatty degeneration of the liver. The heart, lungs, and gastrointestinal tract show congestion with hemorrhage.

Treatment

Treatment is symptomatic—bed rest, injections of intravenous glucose, and the administration of vitamins B, C, D, and insulin.

Prevention

Shellfish taken in the Schizuoka and Kanagawa Prefectures, Japan, should not be eaten during the period of January through April. Ordinary cooking procedures do not destroy the poison. Toxic shellfish cannot be detected by their taste or odor, since decomposition is not a factor in the production of the poison. The only certain method of determining poisonous shellfish at present is by preparing tissue extracts and testing them on laboratory animals.

PUBLIC HEALTH ASPECTS

Geographical Distribution

Venerupin shellfish poisoning is a public health problem restricted to the Kanagawa and Schizuoka Prefectures in Japan. There is no record of this type of shellfish poison occurring elsewhere despite the fact that the species of shellfish involved are found in other parts of Japan and have been introduced into the United States.

Incidence and Case Fatality Rate

Outbreaks of venerupin shellfish poisoning have been reported for 1889, 1941, 1942, 1949, and 1950 by Akiba (1943, 1949), Akiba and Hattori (1949), and Hattori and Akiba (1952). The overall case fatality rate was 33.5 percent.

Seasonal Changes

There is a definite seasonal fluctuation in the toxicity of the shellfish. All the human intoxications have taken place during the months of February and March. However, animal studies reveal that the toxicity of the shellfish begins to increase about October and rises rapidly to a dangerous level in early January, reaching a peak during February and March, after which the toxicity drops off precipitously during April. From June through September the toxin content in the shellfish is at its lowest level. Toxicity fluctuations are about the same for *Tapes semidecussata* and *Crassostrea gigas*—the only two species of shellfish involved in human intoxications (Akiba, 1949). Quarantine regulations have been published, and affected areas have been placed under quarantine for certain periods of time (Hattori and Akiba, 1952). The dangerous season is considered to be from the latter part of December through April.

Bacteriological Tests

Extensive bacteriological tests have been conducted for the usual types of human pathogens, but with negative results. Moreover, there has not been any evidence of putrefaction either by macroscopic observation or by chemical tests (Akiba and Hattori, 1949).

TOXICOLOGY

The toxicology of venerupin shellfish poison obtained from *Tapes semidecussata* and *Crassostrea gigas* has been reported by Akiba (1943, 1949) and Akiba and Hattori (1949).

Extraction Procedures: Whole bivalves were ground in a meat grinder, and the pulp mixed with two volumes of 80 percent ethyl or methyl alcohol, and extracted for an hour on the water bath under reflux. In some of the experiments hydrochloric acid was added to the alcohol until it was found that the acid was not only unnecessary for extraction, but destructive to the poison. The liquid was filtered, concentrated under vacuum to one-fifth the original volume, and then washed with an equal volume of ether. After removal of the ether layer, the liquid was evaporated by water bath until quite viscous. The residue was weighed, and distilled water added, so that 1.0 ml of extract contained the equivalent of 7.0 g of tissue. In subsequent experiments, it was found that the poison was located primarily in the digestive gland (liver). Digestive gland extracts were prepared in a manner similar to those from whole shellfish. An extract was routinely tested by injecting 1.0 ml intraperitoneally into laboratory mice. Extracts were also administered by mouth, intravenous, and subcutaneous injections into dogs, cats, and rabbits. In every instance toxic effects were observed—consisting of organotropic rather than neurotropic changes, viz., liver damage, and a hemorrhagic diathesis.

Feeding Tests: Whole cooked and raw bivalves were fed to dogs and cats. Animals fed 100 g/kg to 150 g/kg of body weight of nontoxic bivalves remained healthy; but animals fed more than 50 g/kg of body weight of toxic shellfish became ill within 1 or 2 days after feeding. Almost all sick animals died with typical symptoms such as vomiting, hematemesis, hemorrhagic diarrhea, leucocystosis, and marked retardation of coagulation time. Autopsies revealed liver damage and a hemorrhagic diathesis. In no cases were neurological symptoms present.

Thermoresistance: Shellfish which had been boiled for 1 hour were found to be lethal to both man and dogs. Heating at pH 5.4 from 30 minutes to 3 hours at 100° C to 119° C did not destroy the toxin. If the toxin was heated for 30 to 60 minutes at 115° C to 119° C, about 50 to 85 percent of the poison was destroyed. At pH 5.4 to 8.0, the toxin was not destroyed when boiled for 1 hour, but at pH 9.42 to 10.0, about 50 to 75 percent of the poison was destroyed when boiled for 30 minutes. The crude toxin has been found to retain its toxicity for more than 3 years either under refrigeration or at room temperature.

Antigenic Properties: The antigenic properties of the venerupin toxin were tested on mice. Ten mice were injected with one-half the MLD, whereas a second group of 10 mice were injected with one-fourth the MLD. Each mouse was injected three times at 4-day intervals. Both groups were injected with two MLD's of the poison 7 to 10 days after the last injection of the series. The inoculated mice were found to be just as sensitive to the toxin as the untreated controls, and all died without extension of survival time. Formal treatment failed to detoxify the poison.

Transplantation of Toxic Shellfish: It was found that if toxic shellfish were removed from Hamana Lake and placed in a tub of clean seawater, the toxin content dropped about 50 percent in from 1 to 2 weeks, and about 75 percent within 4 weeks. Poisonous oysters transplanted to nontoxic areas in the lake were observed to have reduced their toxicity about 90 percent. Moreover, nontransplanted oysters in the vicinity did not become contaminated, indicating that the origin of the toxin was not on an infectious basis. On the other hand, if nontoxic oysters were transplanted to a toxic zone in the lake, the bivalves became poisonous within 10 days.

Origin of Venerupin Poison: Early plankton studies in the Lake Hamana area failed to demonstrate any relationship to venerupin shellfish poisoning. Likewise, water pollution from chemical wastes did not appear to be a factor. The poison is presently believed to originate with the dinoflagellate *Prorocentrum sp.* (Akiba, 1961).

Anatomical Distribution of the Poison in Shellfish: Analysis of the various parts of the shellfish during the most toxic period of the 1949 outbreak revealed that the digestive gland of the shellfish contained 40 μ/g, whereas the remainder of the viscera contained only 0.57 μ/g. In this respect, the anatomical distribution of the results was similar to that observed in paralytic shellfish poisoning.

Relationship of Dosage to Symptomatology: The "toxic dose" was said to be 40 to 60 shellfish (150 g to 250 g), but 10 to 15 shellfish were harmless to adults. Those who ate more than 100 shellfish died. Children who ate 10 to 20 shellfish became severely intoxicated and most of them died.

Methods of Testing for Shellfish Toxicity: The early tests for venerupin shellfish poison utilized laboratory animals—in most instances mice. Since animal tests were slow and expensive, Hattori and Akiba (1952) developed a phthalein colorimetric test which they found to be inexpensive, rapid, and accurate. The test is conducted in the following manner: The digestive glands (liver) are excised, a given amount homogenized, and about 1 percent of the supernatant liquid removed for testing. To 10 ml samples of serial dilutions of the glandular solution are added 5 drops of phenophtalein reagent,[2] 2 drops of 30 percent acetic acid (pH should be 9.6 to 10.0) and 3 drops of 3 percent H_2O_2. The solution changes to a red color (perioxidase reaction) after 10 minutes, if the room temperature is 15° C to 25° C. If the room temperature is higher, the solution should be cooled or observed after 5 minutes for the red change. The amount of dilution at which the coloration occurs is multiplied by 1,000 and this is given as the phthalein value (P_1). Next, the original solution is heated for an hour in a boiling water bath, and the P_2 value is obtained in the same manner.

[2] Phenophtalein, 2 g, is dissolved in 100 ml of water, 20 g of KOH and 10 g of zinc dust are added, and the mixture is discolored by heating and then stored with a piece of zinc in a tightly stoppered colored bottle.

For example, when a 0.001 percent solution prepared from the 1 percent original solution showed a positive phthalein reaction, the P_1 value was 1. If the 1 percent original solution was heated for an hour in boiling water and showed a positive phthalein reaction, the P_2 value was 1,000. The end point of the positive reactions is determined by making comparisons with blank solutions such as is done in indicators. It is recommended that the coloration of the ring be checked by letting the solution stand for some time and then shake gently in order to obtain the correct color values.

Extensive testing over a 2-year period revealed that the P_1 value for nonpoisonous shellfish is usually between 1.0 to 10, but less than 1 if they are toxic. During February and March, the P_1 value is generally down to 0.1 to 0.5. The P_2 value for nonpoisonous shellfish is usually over 800 (rarely down to 600), but when toxic the P_2 value is usually around 20 to 40.

Shellfish are regarded as safe to eat if the P_1 value is larger than 1.0, and the P_2 value is between 800 to 1,000. Hattori and Akiba (1952) state that for the test to be reliable, both the P_1 and P_2 values must be used.

CHEMISTRY

The toxin responsible for certain symptoms of poisoning which result from eating Japanese *Tapes* and *Crassostrea* has been termed "venerupin." The study of its chemistry was initiated by Akiba (1943) following the mass poisoning in the Lake Hamana area of Japan in 1942, and most of the subsequent work on the chemistry of this toxin has been done by Akiba and his coworkers. Work has evidently not progressed far enough to permit any elucidation of the chemical nature of the toxin. Akiba and Hattori (1949) believed venerupin to be an amine. Akiba (1949) regarded it as containing a double bond since it readily absorbed bromine. Hashimoto and Migita (1950) saw a possible identity with the toxin called PIII, described by Sommer and Meyer (1937), in view of the apparent similarity of symptoms produced in animals.

LIST OF TRIDACNA CLAMS REPORTED AS TOXIC

Phylum MOLLUSCA

Class PELECYPODA: Bivalves

Order EULAMELLIBRANCHIA

Family TRIDACNIDAE

102 *Tridacna gigas* (Linnaeus) (Pl. 24). Tridacna clam (USA).
DISTRIBUTION: Indo-Pacific.

SOURCE: Halstead (1970). (Banner and Helfrich, 1964, list *Tridacna* sp.)

Tridacna maxima (Röding). Tridacna clam (USA).
DISTRIBUTION: Indo-Pacific.

SOURCES: Bagnis (1967), Banner (1967).

BIOLOGY

The family Tridacnidae includes the giant clams and their relatives, the larger shells reaching over a meter in length. They are recognized by the massive shells, and large irregular hinge teeth, set on a broad hinge-plate, and may be found burrowed into coral or lying on the bottom. The shape and color of the shell often blends into the background of coral rock but the giant clam is nonetheless quickly detected by the brilliant coloration of its siphonal edges and mantle.

MECHANISM OF INTOXICATION

Poisoning results from the eating of the shellfish. The poison is probably concentrated within the viscera of the animal.

MEDICAL ASPECTS

Clinical Characteristics

Tridacna clams, apparently of several different species, have been incriminated in human intoxications which clinically resemble ciguatera fish poisoning. The most complete account of tridacna clam poisoning is by Bagnis (1967) which involved 33 persons and a number of domestic animals that had eaten *Tridacna maxima* at Bora-Bora, Society Islands, French Polynesia. The clinical characteristics present included gastrointestinal vasomotor, and various neurological disturbances, including a loss of motor coordination. The symptoms present resembled those of ciguatera poisoning. The precise origin of the toxin was not determined, but was believed to have been involved in the food chain of the clam.

Treatment

The treatment is symptomatic.

Prevention

It is advisable to contact the natives regarding the toxicity of tridacna clams since the clams are usually edible in most areas.

TOXICOLOGY

(Unknown)

PHARMACOLOGY

(Unknown)

CHEMISTRY

(Unknown)

REPRESENTATIVE LIST OF CEPHALOPODS REPORTED AS TOXIC

Phylum MOLLUSCA

Class CEPHALOPODA: Cuttlefishes, Squids, and Octopuses

Order DECAPODA

Suborder SEPIOIDEA

Family SEPIIDAE

103 *Sepia officinalis* Linnaeus (Pl. 25). Common European cuttlefish (USA, England).

DISTRIBUTION: Europe.

SOURCE: Pawlowsky (1927).

Suborder TEUTHOIDEA

Family OMMASTREPHIDAE

103 *Ommastrephes sloani pacificus* (Steenstrup) (Pl. 26). Squid (USA).

DISTRIBUTION: Pacific Ocean.

SOURCE: Kawabata, Halstead, and Judefind (1957).

Order OCTOPODA

Suborder INCIRRATA

Family OCTOPODIDAE

103 *Eledone moschata* (Lamarck) (Pl. 27). Octopus (USA).

DISTRIBUTION: Mediterranean Sea, Red Sea.

SOURCES: Krause (1895), Livon and Briot (1905), Henze (1906), Calmette (1908), Fleig and Rouville (1910), Rouville (1910*a, b, c*), Pawlowsky (1927).

104 *Octopus apollyon* Berry (Pl. 28). Octopus (USA).

DISTRIBUTION: Alaska to Baja California.

SOURCES: Halstead (1949), Berry and Halstead (1954), Ballering *et al.* (1972).

105 *Octopus maculosus* Hoyle (Pl. 29, fig. a). Spotted octopus (USA), common ringed octopus, blue-banded octopus (Australia).

DISTRIBUTION: Indo-Pacific, Australia, Japan, Indian Ocean.

SOURCES: Mabbet (1954), Flecker and Cotton (1955), McMichael (1963), Trethewie (1965), Sutherland and Lane (1969).
OTHER NAMES: The species of octopus causing the death of the diver was provisionally considered to be *Octopus rugosus*, but was later determined, according to E. Pope (1963), as *O. maculosus*.[3] This is sometimes given as *Hapalochlaena maculosa*.

Octopus macropus Risso (Pl. 30, fig. a). Octopus (USA). *106*
DISTRIBUTION: Atlantic Ocean, Mediterranean Sea, Red Sea, Persian Gulf, Indian Ocean, Indonesia, China, Japan, and Australia.

SOURCES: Henze (1906), Mebs (1973).

Octopus vulgaris Lamarck (Pl. 30, fig. b). Octopus (USA), madako, mizudako, yanagidako (Japan). *106*
DISTRIBUTION: North Atlantic, Mediterranean Sea, Indian Ocean, Indo-Pacific region, Red Sea, West Indies, and South Africa.

SOURCES: Henze (1906), Mebs (1973).

BIOLOGY

Cephalopods are distributed widely in all temperate and tropical seas, but they are restricted to a few species in the Arctic and Antarctic Circles. True *Octopus* are primarily inhabitants of warmer waters. All cephalopods are marine and seem to have relatively little tolerance to extremes in salinity. *O. vulgaris* is one of the few members of this group to inhabit estuaries, but they do this only when the salinity approaches that of open seawater. They live best in water whose density exceeds that of normal seawater. Most cephalopods inhabit depths of less than 100 fathoms, but a few species live at great depths. Many octopuses are found in rock pools in the intertidal zone. On the whole, adult octopuses prefer rocky bottoms, but some are found in muddy and sandy areas. Cuttlefish and squid seem to have an affinity for the open sea, although some are found in shallow water.

---

[3] The exact identity of the Australian species of octopus, generally known as *Octopus maculosus*, which is the species believed to have been responsible for the only known octopus bite fatality, has been questioned. There are three closely allied species of spotted Australian octopuses, and they are: *O. maculosus* Hoyle, *O. fasciatus* Hoyle, and *O. lunulatus* Quoy and Gaimard. These three species might be confused with each other. According to Berry (1953) there remains some question regarding the identity and possibly the validity of these three species. More recent studies (March 1967) by Dr. Robert Endean's group at the University of Queensland have shown that *O. lunulatus* was probably responsible for the death of the Australian diver, since this species has a northern range. Moreover, *O. lunulatus* venom and *O. maculosus* venom have both been tested for their toxicity. Both venoms are very toxic, but *O. lunulatus* appears to be the more potent of the two. Several near-fatal envenomations have recently occurred from the bites of *O. maculosus*.

Cephalopods are largely predatory and carnivorous, feeding on Crustacea, bivalves, and occasionally fish. MacGinitie (1942), who has observed octopuses feeding in aquaria, states that an octopus may attack a crab larger than it is able to surround by its web. The octopus will spread its tentacles out on the glass, and hover over the victim in a tentlike fashion. When the crab is first seized, it will make a vigorous struggle with one cheliped grasping the edge of the web. However, within 20 seconds the cheliped will open wide in the manner assumed during a defensive attitude and then slowly close. Within a short time the abdomen of the crab will unbend in the manner characteristic of death, the appendages will quiver, and a slight brownish-colored fluid will issue from the branchial canals. Within another 45 seconds, the crab will appear to be dead. During this time the crab will not be attacked by the beak of the octopus. MacGinitie (1942) states:

> Crabs are no doubt killed by a secretion from the 'salivary' glands, of which the octopus has two pairs, one pair being fairly large. In order to eat the crab the octopus opened it at the dorsal juncture between carapace and abdomen, the place where the break comes when a crab molts. The octopus then pulled off the back of the crab, ate the viscera first, dropped the back, and then one by one pulled off the legs, cleaned out and ate the contents, and dropped the empty shell of each as it was finished.

Apart from the simpler reflexes involved in procuring food and mating, complex processes indicating memory and remarkable manipulative performances have been observed in cephalopods. When alarmed, cephalopods are able to move backward rapidly by ejecting powerful jets of water forward from the anterior opening of the funnel. The arms are stretched out horizontally in a straight line, with the visceral dome or head pointing forward. Octopuses also move about by gliding or creeping over the bottom. Usually, this latter method is used when the creature is in pursuit of food. Cephalopods will eject a dark cloud of "ink" when suddenly disturbed, which offers them excellent concealment and opportunity for escape. Octopuses tend to be solitary, irritable, pugnacious, whereas decapods are more gregarious. On occasion, octopuses are known to display the interesting habit of autophagy (Taki, 1936). Bartsch (1917) records the finding of an octopus 9 m in length. Giant squids, reputed to attain a length of 19 m or more, are very pugnacious and can inflict a serious wound with their horny beaks. Octopus and squid are considered to be a delicacy by many people. Octopuses that are to be used for laboratory purposes are best captured in fishing pots, which take the animal in an uninjured condition. Taki (1941) published a useful paper on the rearing of octopuses for experimental purposes.

See Lane's (1960) *Kingdom of the Octopus* for additional information on the biology of cephalopods.

MORPHOLOGY OF THE VENOM APPARATUS

The venom apparatus of cephalopods (Fig. 5) comprises the anterior and posterior salivary glands and the structures with which they are directly associated, namely, the salivary ducts, the buccal mass, and the mandibles or beak.

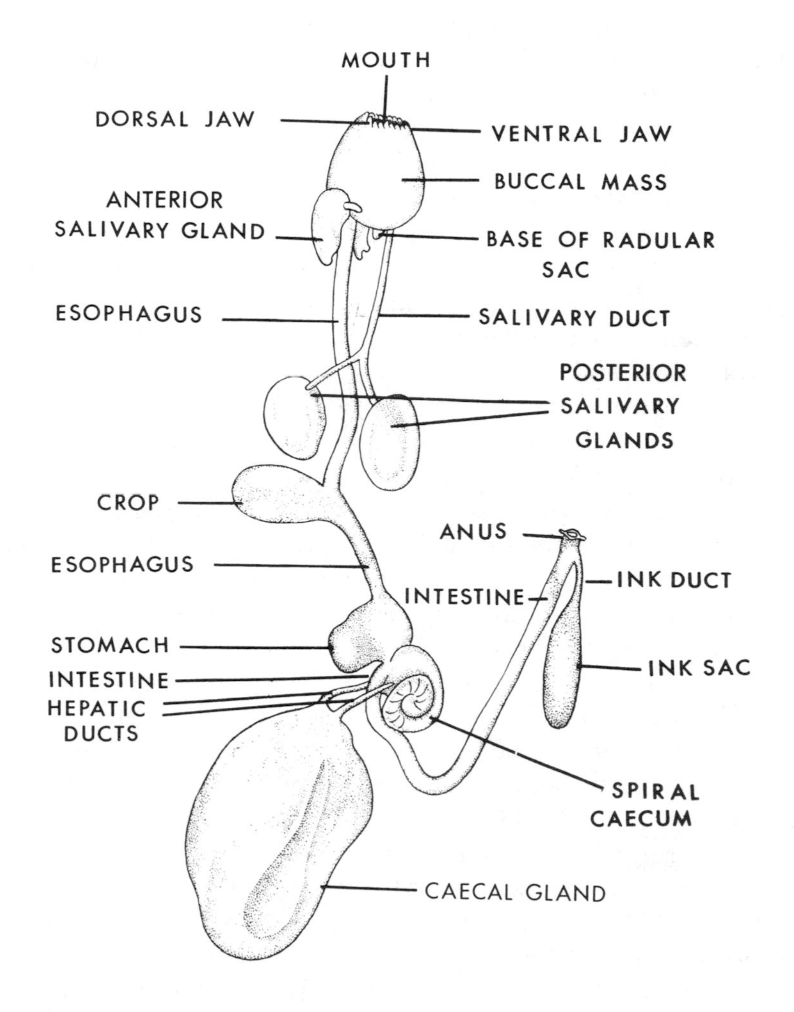

Figure 5.—Anatomy of the digestive tract of the octopus *Eledone*, showing the relationship of the salivary glands to the buccal mass and other internal organs. The toxic secretions are produced by the salivary glands. (R. Kreuzinger, after Isgrove)

Isgrove (1909) prepared an excellent, comprehensive anatomical work on the octopus *Eledone cirrhosa*. A similar treatise on the cuttlefish *Sepia officinalis* was written by Tompsett (1939). Wu (1940) prepared a well-illustrated, but diagrammatic and generalized, description of the anatomy of the octopus. Saito (1934) described the anatomy of *Octopus fang-siao*. The anatomy of the common octopus of northern Washington was described by Winkler and Ashley (1954). Lane (1960) included some useful anatomical drawings of cephalopds in his compendium.

The following description of the venom apparatus of cephalopods is based largely on the observations of *Eledone* of the order Octopoda, and *Sepia* of the order Decapoda, as made by Isgrove (1909) and Tompsett (1939).

Gross Anatomy: The venom apparatus is an intimate part of the digestive system. However, for the purposes of this work, discussion will be limited to those parts of the digestive system directly associated with venom production.

The mouth of the octopus is located in the center of the oral and anterior surface of the arms, surrounded by a circular lip fringed with fingerlike papillae (Fig. 6). The external surface of the lip is continuous with that of the web, but marked off from it by a deep groove. The edge of the web forms a kind of contractile outer lip. The mouth leads into a pharyngeal cavity with thick muscular walls, which is known as the buccal mass. It is surrounded and concealed by the muscular bases of the arms. The pharynx is furnished with two powerful dorsal and ventral chitinous jaws, which are shaped like a parrot's beak. The arrangement of the jaws differs from that of the parrot's in that the ventral one bites outside the dorsal, and is wider and larger. The jaws, which bite vertically with great force, tear the food as it is held by the suckers before it is passed on to the rasping action of the radula. The anterior edge of each jaw is thick, hard, and dark brown in color. The cutting edge is sharp, and a raised ridge some distance behind it provides attachment to the muscles working the jaws. Posteriorly, the jaws decrease in thickness, lighten in color, and finally become thin, colorless, and translucent. On the floor of the pharynx, slightly anterior to the midpoint, is the muscular tongue, which forms the anterior wall of the radular sac, at the base of which is the growing point of the radula. The radula appears as a broad, chitinous ribbon which extends over the upper and anterior surface of the tongue. This chitinous ribbon is responsible for the rasping action of the radula. Internally, the tongue is strengthened by two small cartilaginous strips which give it rigidity and also provide attachment for its muscles.

The duct from the posterior salivary glands passes above the sublingual gland on the ventral surface of the bulb and enters the buccal mass below the radular sac. The duct finally opens on the tip of the subradular organ which appears as an outgrowth in front of the tongue. The ducts from the anterior salivary glands are paired and open into the pharynx laterally and posteriorly.

The massive muscular wall of the buccal bulb is formed largely by the jaw and radular muscles. The buccal mass is attached anteriorly to the bases of the arms by a circular muscle band and posteriorly by two ligaments.

The pharynx is continued posteriorly into the esophagus. The esophagus appears as a narrow tube extending ventrally and posteriorly to the stomach, dorsally to the caecal gland or liver.[4] A narrow passage extends from the

[4] This portion of the digestive tract has been variously designated as the "liver," "pancreas," "hepatopancreas," "digestive gland," "mitteldarmdrüse," "glandular zone," etc. This glandular structure, as Arvy (1960) has pointed out, should not be termed the "liver" or "pancreas" since it is not analogous to either the liver or pancreas of vertebrates. Since the glandular mass is divided into two parts that are contained within the same capsule, Arvy has designated the anterior portion as the anterior caecal gland, and the other as the posterior caecal gland. The posterior caecal gland has no equivalent in other mollusks.

stomach, bifurcating, to give rise to the spiral caecum and the remainder of the intestinal tract. The relationships of these various structures are shown in Fig. 5.

Eledone has five salivary glands: an anterior pair, a posterior pair, and a single sublingual, or median, salivary gland. The anterior glands are closely

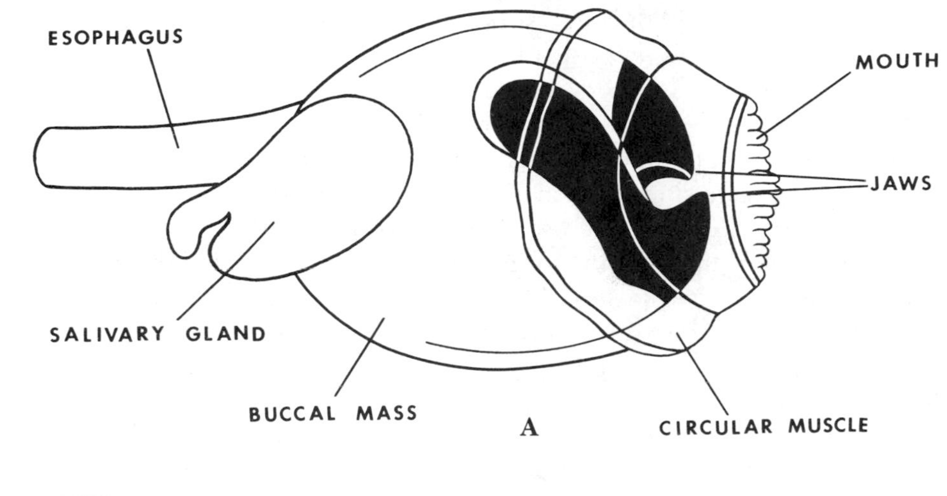

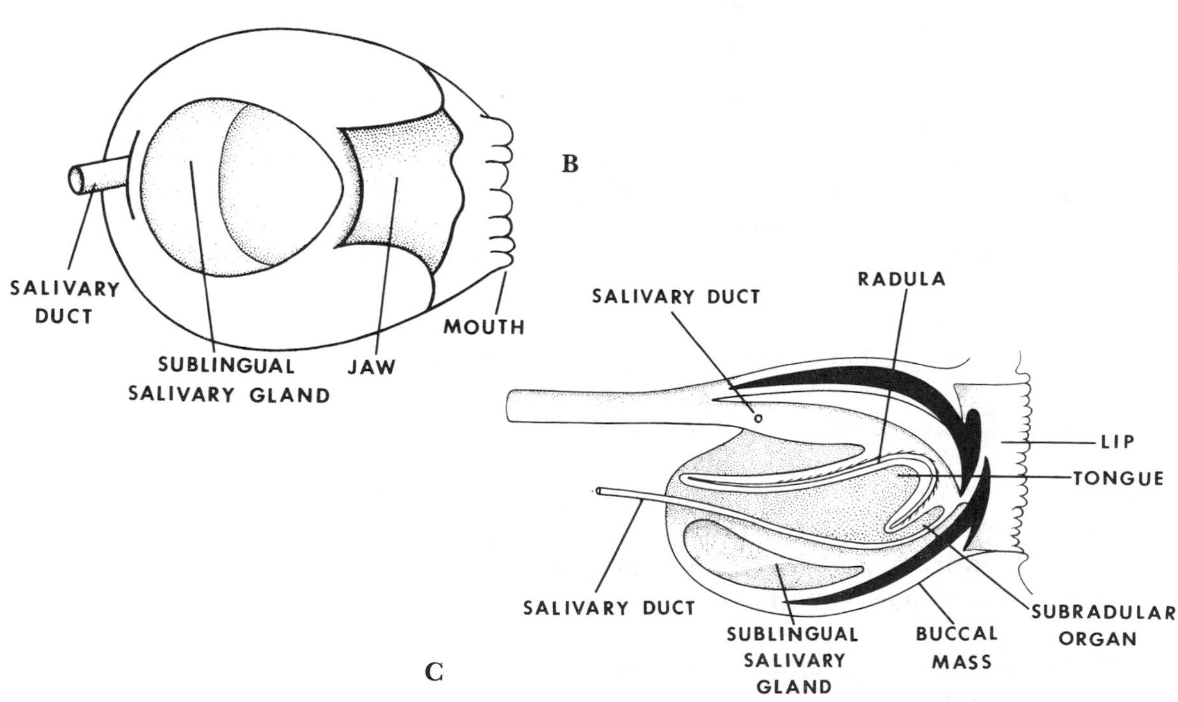

FIGURE 6.—A. Gross anatomy of the buccal mass of the octopus *Eledone*. (Side view.) B. Ventral view of buccal mass of *Eledone* showing the sublingual salivary gland. C. Sagittal section of the buccal mass of *Eledone* showing the openings of some of the salivary ducts. (R. Kreuzinger, after Isgrove)

applied to the external surface on the posterior aspect of the buccal mass. They are attached at the angle between the esophagus and the buccal mass. They are flattened, oval, bilobed posteriorly, and much smaller than the posterior pair. The anterior salivary duct, which is very short, leads from a slightly elevated ridge on the inner surface of the gland passing inwards to the pharynx. The duct is accompanied by an artery and a nerve.

The posterior glands are situated at the side of the crop which must be turned aside to fully expose the glands. These glands are attached to the visceral sac by a suspensory ligament. The glands are large and flattened. The posterior salivary ducts originate from the anterior inner aspect where there is a slight depression. After a short course the duct merges with the duct from the opposite side to form a single common posterior salivary duct which runs forward alongside the esophagus to the buccal mass.

The sublingual gland is located in the ventral wall of the buccal mass. It is oval and thickened posteriorly. All the glands are granular in appearance, soft and spongy in texture, and of a translucent whitish color.

The observations of Tompsett (1939) on *Sepia* indicate that the decapods follow a similar anatomical pattern.

Microscopic Anatomy: Isgrove (1909) presents a succinct description of the microscopic anatomy of the salivary glands of octopus which more or less summarizes the findings of the early octopodan anatomists:

> The three salivary glands all consist of glandular secretory tubules, embedded in a stroma of connective tissue. These tubules are closely adpressed in the anterior glands, but much further apart in the posterior glands, and branch dichotomously here. The secretory cells of the three glands are all similar, and are columnar with a basal nucleus. The secretion forms in globules in the anterior portion of the cell, and then falls into the lumen of the tubule. The secretion of these glands is a kind of mucus only, and contains no ferment whatever.

The following anatomical description of the salivary glands of *O. rubescens* Berry is based on histological sections taken from a male specimen having a spread of 25 cm. The octopus was captured by the author at Moss Beach, San Mateo County, Calif., in a tidepool, March 24, 1948. Tissue sections were prepared by Mrs. Myrtle Ernest of the Department of Pathology, and the anatomical description was written by Dr. F. Rene Modglin, Bio Laboratories, Colton, California. The slides were prepared in the routine manner, using paraffin as an embedding medium, cut at 8μ, and stained with hematoxylin and triosin.

Anterior Salivary Glands: These organs, in cross section, appear as a closely packed, circumscribed, glandular mass situated adjacent to the postero-lateral aspect of the buccal mass. The glandular elements are lined by a simple columnar epithelium containing either peripherally- or basally-placed dark granules in their cytoplasm, and having vesicular nuclei. No ductlike structures are seen. The intervening stroma is essentially of thin strands of fibrillar connective tissue, which forms partitions between neighboring acini. In some areas the lumina are dilated, and contain a weblike acellular protein coagulum. The entire gland is surrounded by a homogeneous, thin, refractile, acellular membrane.

Sublingual or Median Salivary Glands: In cross section, this organ appears as a semilunar shaped, circumscribed, glandular mass, embedded in the ventral side of the buccal mass posterior to the chitinous jaws. A thin sheet of smooth muscle overlays the gland and hides it from view when the ventral aspect of the bulb is examined. The sublingual gland resembles the anterior salivary gland, except that in the former some of the glandular cells have the typical clear swollen cytoplasm which characterize mucous cells. There is less stroma, and the gland is more densely packed. The glandular elements also form definite acini, and no ducts are observed.

Posterior Salivary Glands: The gland consists of fairly close-packed acini of glandular tissue, which appear clumped in two large triangular-shaped, circumscribed masses, one on either side of the muscular esophagus. The acini are mucous in nature, having large glandular cells with clear swollen cytoplasm, and an abundance of secretory granules. The acini are circumscribed by a hyaline band which is similar to a basement membrane. The intervening supporting tissue is a loose fibrillar connective tissue, which is sprinkled liberally with large mononuclear-type cells. Numerous small ductlike structures are seen lined by a tall, simple, columnar epithelium with peripherally placed, large, vesicular nuclei; while basally, there are striations and pink granules. The ducts also are surrounded by a hyaline band comparable to a basement membrane. The varying shapes of these acini and ducts suggest a convoluted arrangement. The entire gland is surrounded by a homogeneous, refractile, thin, acellular membrane. The salivary glands of cephalopods resemble the mucous glands of vertebrates, and appear to have a merocrine type of secretion.

Function of the Salivary Glands: The role of the salivary glands of cephalopods has attracted the attention of many anatomists and physiologists over the years. Krause (1897), Willem (1898), Falloise (1905-6), and others, have assigned them a digestive function; whereas Bourquelot (1885), Fredericq (1878), and Rouville (1910*a, b, c*) denied this, Cuénot (1905) and Henze (1906) believed them to be primarily excretory organs. The possibility of a hormonal function has also been considered, and they have been compared to the adrenal glands of vertebrates (Bacq and Ghiretti, 1951*a, b*). Most authors believe the salivary glands to be defensive organs or venom glands because of the toxic substances which they secrete (Krause, 1895, 1897; Lo Bianco, 1899; Livon and Briot, 1905, 1906; Henze, 1906; Baglioni, 1909; Rouville, 1910*a, b, c*; Erspamer, 1948*a, b, c*; Halstead, 1949; Ghiretti, 1953*a, b*; Berry and Halstead, 1954).

Recent research clearly indicates that the salivary glands of cephalopods have several functions. There is no question regarding their venomous properties; but, in addition, they appear to have a digestive role. Bacq, Fischer, and Ghiretti (1952) have shown that there is proteolytic activity in the salivary gland secretions; and Romanini (1952) has demonstrated the presence of hyaluronidase as a factor having intense mucinolytic properties. Arvy (1960) has shown in his histochemical studies that there is a serous secretion that has an acetylnaphtholesterase activity.

MECHANISM OF INTOXICATION

POISONING

Intoxications resulting from the ingestion of poisonous cephalopods have been caused by certain species of octopus and squid taken from specific areas in Japan. The nature of the poison is unknown, but there is no evidence that bacterial contaminants are involved.

ENVENOMATIONS

Cephalopods inflict their envenomations with the use of a well-developed apparatus (the beak and salivary glands). The sharp parrotlike beak produces the initial wound into which is introduced the toxic saliva or venom, cephalotoxin.

MEDICAL ASPECTS

POISONING

Clinical Characteristics

Intoxications resulting from the ingestion of cephalopods are extremely rare. Our knowledge of the symptomatology of this disease is derived from the single report by Kawabata *et al.* (1957) on a series of outbreaks of cephalopod intoxications which occurred in Japan from 1952 through 1955.

The incubation period varied from 10 to 20 hours. The predominant symptoms consisted of nausea, vomiting, abdominal pain, diarrhea, fever (38° to 39° C), headache, chills, weakness, and severe dehydration. Paralysis was present in 3 patients and 3 others developed convulsions, out of a total of 758 victims. Neurological symptoms were generally absent. Usually the clinical picture was that of a severe gastrointestinal upset, and recovery was within 48 hours.

Routine microbiological tests to determine the etiology of the outbreaks proved negative. There was no evidence of the presence of any histamine or histaminelike substances or any evidence of putrefaction. The only laboratory findings consisted of a leucocytosis with a decrease in lymphocytes and an increase in urobilinogen in the urine.

The mortality rate in the Japanese series of outbreaks was 0.84 percent.

Treatment

Symptomatic.

Prevention

The Japanese outbreaks occurred without warning and there was no evidence in the appearance or taste of the cephalopods as to their freshness. The species of cephalopods eaten were the squid *Ommastrephes sloani pacificus* Schauinsland, and the octopuses *Octopus vulgaris* Lamarck and *O. dofleini* (Wülker). All these species are commonly eaten and generally regarded as nontoxic in Japan and elsewhere. None of the usual laboratory tests have proved to be effective in demonstrating the toxicity of these cephalopods. The only valid test that has been found to date in determining the toxicity is by human volunteers. In the light of these findings it is recommended that before a person eats these species, a careful check with the local health or fisheries authorities be made regarding their edibility.

ENVENOMATIONS

Clinical Characteristics

Cephalopod lesions usually consist of two small puncture wounds produced by the sharp, parrotlike, chitinous jaws of the mollusk. Usually the pain is immediate and consists of a sharp burning or stinging sensation. It is sometimes described as similar to a bee sting, which at first is localized, but may later radiate to include the entire appendage. Within a few minutes a tingling or pulsating sensation develops in the area about the wound. There is some indication that coagulation time is retarded since bleeding is profuse and prolonged in most cephalopod bites. Swelling, redness, and heat usually develop about the wound. Some victims complain of an intense itching sensation about the affected area. Motor and severe sensory disturbances are generally absent.

However, McMichael (1963) reported several cases in which the symptoms consisted of numbness of the mouth and tongue, blurring of vision, difficulty in speech and swallowing, loss of tactile sensation, feeling of one's hands "floating up to my head and hitting my ears instead of my mouth." One victim complained of a sensation of being detached from his body. Complete muscular paralysis of the legs and arms was present in several cases, and loss of equilibrium was possible. In most instances healing of the wound was quite rapid and uneventful.

Mabbet (1954) and McMichael (1963) reported a fatality resulting from an octopus bite which occurred near East Point, Darwin, Australia. According to the account, the diver captured a small octopus which had a span of about 20 cm. The diver permitted the octopus to crawl over his arms and shoulders, and finally to the back of his neck where the animal remained for a few moments. It was later determined that during the period that the octopus was on his neck, a small bite was inflicted, producing a trickle of blood. A few minutes after being bitten, the victim complained of a sensation of dryness in his mouth and difficulty in swallowing. After walking a short distance up the beach from the scene of the accident, the victim began to vomit, became ataxic, weak, and developed pronounced respiratory distress and aphonia. The diver was rushed to a hospital and placed in a respirator where he died about 2 hours after being bitten. The identity of the species of octopus causing the fatality has been tentatively determined as *O. maculosus*. Other fatal cephalopod bites have also been discussed by Lane and Sutherland (1967) and Sutherland and Lane (1969).

Lane (1960) quotes Hornell (1917) as saying that some of the small octopuses near Madras, India, have a reputation of being venomous. Some of the Indian octopuses are said to be very active and pugnacious, inflicting a bite that is similar in its effects to the sting of a scorpion.

Human case reports of cephalopod bites are scarce. Montfort, in the earliest published record (cited by Bartsch, 1917), states that he was painfully bitten by an octopus. Halstead (1949), Berry and Halstead (1954), Hopkins (1964), Wittich (1968), and Snow (1970) have recorded a series of human case reports on octopus bites. Some of the larger squids are said to be capable of inflicting severe wounds with their powerful beaks (Duncan, 1941). The subject of octopus bites has been discussed at length by Lane (1960) in his excellent work *Kingdom of the Octopus*.

Treatment

Symptomatic. (*See* Hopkins, 1964.) Treat in the same manner as venomous fish stings. There are no effective antidotes. The Commonwealth Serum Laboratories of Melbourne, Australia have reported an attempt to produce neutralizing antibodies in rabbits (Sutherland and Lane, 1969).

Prevention

Avoid coming in contact with the beak of the octopus which is located in the center of the under surface of the web from which the tentacles project. The reckless handling or placing of even small specimens of octopuses against the bare skin or the chest or back should be avoided. Octopus bites can be dangerous, particularly if they occur in the upper part of the body.

PUBLIC HEALTH ASPECTS

POISONOUS CEPHALOPODS

Geographical Distribution

Outbreaks of intoxications caused by the ingestion of poisonous cephalopods is restricted to Japan. The octopuses responsible were captured in the vicinity of Hokkaido, and along the east coast of the northern tip of Honshu in a region called Sanriku, and in the vicinity of Sado Island. Octopuses in the vicinity of Sendai, Miyagi Prefecture, are said to be poisonous to eat during the month of September. The squids involved in the outbreaks were captured in the Yamagata Prefecture in the vicinity of Tobishima Island, Japan Sea, about 35 km northwest of Sakata City, and in the vicinity of Sado Island, off Niigata City, Niigata Prefecture.

Incidence

All the known outbreaks of cephalopod poisoning that have occurred in recent years in Japan have been reported by Kawabata *et al* (1957). In the largest series of outbreaks, the morbidity rate was determined as 27.1 percent, with a mortality rate of 0.84 percent.

There appeared to be a definite seasonal incidence since all the intoxications occurred during the months of June through September.

Etiology

The actual cause of the intoxications is unknown. All bacteriological tests for human pathogens, including *Clostridium botulinum* and *C. perfringens*, proved to be negative. See the original article for details of the tests (Kawabata *et al.*, 1957).

VENOMOUS CEPHALOPODS

Venomous cephalopods are not considered to be a public health problem.

TOXICOLOGY

POISONOUS CEPHALOPODS

Routine tests using extracts prepared from squid and octopuses taken in the Japanese outbreaks when used on mice and guinea pigs proved to be negative. When fresh squid were fed to 50 human volunteers, 15 of them developed pronounced gastrointestinal symptoms typical of those found in the accidental intoxications previously discussed. The quantity of squid fed to each person is unknown.

VENOMOUS CEPHALOPODS

The first reported experiments on the toxicity of cephalopod venom were by Lo Bianco (1888, 1909) who worked with *Octopus macropus* and *O. vulgaris*. He obtained octopus saliva, injected it into the gills of crabs, and observed the toxic effects. Krause (1897) extracted the saliva from *O. macropus* by dissecting out the posterior salivary glands and inserting a canula into the common posterior salivary duct. Secretory activity was produced by electrical stimulation of the nerve leading to the salivary glands. Stimulation resulted in producing 4 ml to 5 ml of salivary secretion within a period of 1 hour. Secretory activity usually ceased within 90 minutes after the death of the organism. A few drops of salivary secretion injected into the body or placed in contact with the gills of a crab produced instantaneous death. Within 5 to 10 minutes after injection, frogs developed complete muscular paralysis. The heart rate remained normal, but death usually resulted within a few hours. Krause found that injection of the toxin into rabbits produced varying effects; in some the results were negative, while in others death was immediate.

Briot (1905*a, b*) reported that inoculation of secretion from the anterior salivary glands had no effect on crabs, frogs, and rats. However, his studies on the posterior salivary glands produced results similar to those of Krause. Injections of the saliva into crabs resulted in convulsions of the extremities.

Paralysis soon followed, and death resulted in from 15 minutes to a few hours. The macerated salivary glands of decapods were observed to produce similar results. Both aqueous and glycerinated extracts were found to be active. Alcohol was said to precipitate the active constituent, and boiling for 10 minutes completely destroyed it. Even temperatures as low as 58° C for 1 hour were destructive to the toxin. Rats, frogs, and rabbits were found to be insensitive to the venom. The effect on fishes was questionable.

Baglioni (1909) used aqueous extracts that were prepared from the posterior salivary glands of *O. vulgaris*. When he injected 0.2 ml to 0.3 ml of the crude extracts into crabs, he found that it produced clonic convulsions, increased nervous excitability, motor coordination, paralysis, and finally death. Baglioni observed that small specimens of the fishes *Uranoscopus, Scorpaena*, and *Conger*, which had been bitten by *O. vulgaris*, developed respiratory distress, lost the position reflex, and developed violent jerking movements. With the exception of the *Conger*, all the fish died within 2 to 3 hours. When frogs *Rana esculenta* were injected intraperitoneally with 0.3 ml to 0.5 ml of crude salivary extract, the following signs of intoxication developed within from 10 to 20 minutes: increased nervous excitability and clonic convulsion. These were followed by profound torpor, respiratory distress, and eventual recovery. Baglioni concluded that the poison acted upon the central nervous system. Lo Bianco (1909) also tested the salivary secretion from *O. vulgaris* on crabs and obtained results similar to those of previous workers.

Early research was directed to the toxicity of the secretion from the posterior salivary glands of octopuses. Rouville (1910*a, b, c*) appears to have been the first to evaluate extracts from both the anterior and posterior salivary glands of *Eledone moschata*. Saline extracts were prepared from the salivary glands and injected into rabbits and anesthetized dogs. Injections of the posterior salivary extract in rabbits resulted in convulsive respiratory movements and death. Intravenous injections of the same extract into dogs resulted in a marked drop in blood pressure, slowing of the heart, vasodilatation, and an increased respiratory rate. Large doses of salivary extract in rabbits resulted in retarding the blood coagulation time. Extracts from the anterior salivary glands proved to be active in crabs and rabbits, but less toxic. Rouville's studies on the effect of the toxin on crabs are comparable to those of previous authors. Fleig and Rouville (1910) demonstrated in another series of experiments that salivary extracts from *O. vulgaris* are as toxic as those from *E. moschata*. They also determined that the poison was a specific glandular secretion rather than a toxic substance present in the blood of cephalopods. They therefore concluded that the salivary glands were true venom glands serving as organs of defense.

Research on octopus venom during the past two decades has been concerned largely with its pharmacological and chemical characteristics. The most recent work on the toxicology of cephalopod salivary venom is by Ghiretti (1959, 1960) who tested both salivary gland extracts and pure saliva from *Sepia officinalis, O. vulgaris, O. macropus*, and *E. moschata* on crabs, mollusks, and other invertebrates. Ghiretti has termed the active toxic principle in cephalopod saliva "cephalotoxin." He has taken as a measure of purification an arbitrary unit defined as the amount of toxin which paralyzes in 15 minutes a crab weighing 40 g to 50 g. The crab used was *Eriphia spinifrons.* He injected control animals with equal amounts of extract previously heated at 100° C for 5 minutes. "Specific activity was calculated as units per mgm. of protein." The protein content was determined colorimetrically with a Folin-phenol reagent as is shown by the method of Lowry *et al.* (1951). Cephalotoxin has also been found to be toxic for insects *Periplaneta americana*, and for crustacea *Pagurus, Squilla, Palinurus.* Ghiretti (1960) has used the esophagus and aorta of *Octopus* in determining the toxicity of cephalotoxin. His preparations have shown an LD_{50} of 0.33 mg/ml. Erspamer and Anastasi (1962) isolated an endecapeptide from the posterior salivary glands of the octopus *Eledone*, which when injected in large doses subcutaneously (300 μg/kg) resulted in hypotension and terminated in death due to heart failure.

Sutherland and Lane (1969) investigated the venom of *Octopus (Hapalochlaena) maculosus* on rabbits and mice. They found that 0.1 mg of the crude extract was the MLD for a 2 kg rabbit. A high concentration of hyaluronidase is present in the venom. There is believed to be at least two fractions of the venom, both non-antigenic and of low molecular weight. A single (25mg) octopus, with salivary glands weighing 300 mg, contains sufficient poison to produce flaccid paralysis in 750 kg of rabbits or ten humans weighing 75 mg.

Recently Ballering *et al.* (1972) devised a method of extracting saliva from *Octopus apollyon* without dissecting the octopus, by means of a plastic bag.

PHARMACOLOGY

POISONOUS CEPHALOPODS

Pharmacological investigations on poisonous squids and octopuses involved in the Japanese outbreaks were confined to the testing of methanol extracts on isolated strips of guinea pig intestines. No appreciable activity was detected that would indicate the presence of histamine or other toxic substances of a similar nature (Kawabata *et al.,* 1957).

VENOMOUS CEPHALOPODS

Prior to the work of Livon and Briot (1905, 1906), research on cephalopod saliva was primarily concerned with the general toxicology (most of the tests being conducted on crustaceans and, in some instances, frogs or other vertebrates). Livon and Briot, on the other hand, attempted to determine in a more precise manner the mode of action of cephalopod venom. Fixing a crab *Carcinus maenas* on its back, and attaching a marking stylus to a cheliped, they recorded the crab's normal activity under electrical stimulation with a kymograph. Next they injected aqueous extracts of posterior salivary glands of *Octopus vulgaris* and *Eledone moschata* into the musculature of the cheliped, and recorded the effects. Similar tests were made on the crab's central nervous system. From these studies they concluded that the cephalopod saliva does not affect the muscular system of the crab, but acts directly on the nerves. They were uncertain whether the site of action was central or peripheral.

Working with aqueous extracts from both the anterior and posterior salivary glands of *E. moschata*, Rouville (1910*a, b, c*) showed that both glands produced substances that were toxic to crabs *C. maenas* and rabbits. He concluded that in crabs the venom acted directly upon the neuromuscular system. Intravenous injections of octopus saliva in rabbits resulted in violent convulsions, respiratory distress, paralysis, and rapid death. Autopsy revealed diastolic cardiac arrest. Rouville believed, therefore, that octopus venom affected the respiratory and other medullary centers of the rabbit. Observing a distinct hypotensive effect, which persisted after severing the vagus nerve, he believed it to be due to both a cardiac effect and a peripheral vasodilatation. There was also evidence of hemolysis of the blood and an increased coagulation time.

The studies of Bottazzi (1916, 1918, 1919, 1921), and Bottazzi and Valentini (1924) dealt with the chemistry and the pharmacology of octopus venom. Bottazzi's pharmacological efforts were directed toward examining specific effects of octopus venom on blood pressure, the nervous system, and on smooth muscle in dogs. Aqueous and alcoholic extracts were prepared from the posterior salivary glands of *O. macropus* and injected into the femoral veins of dogs. This resulted in a prolonged marked drop in blood pressure, accompanied by a reduction in strength and an increased frequency of ventricular systoles. Also observed were respiratory distress, vomiting, general agitation and nervousness, convulsions, paralysis, and death. Autopsies revealed extensive subserous hemorrhages of the stomach, intestines, and throughout the omentum, as well as congestion of the spleen. Severe contractions of the intestinal musculature were noted. Octopus venom was also found to produce strong contractures of uterine smooth muscle. Bottazzi attributed the vasomotor effects largely to two constituents which he believed to be present in octopus venom, namely, histamine and tyramine. He also suggested that the smooth muscle contractions were probably due to the presence of *p*-oxyphenylethylamine, β-iminazolylethylamine, or choline.

With the use of perfusion techniques on isolated salivary glands from *O. macropus, O. vulgaris,* and *E. moschata*, Bacq (1951) and Bacq and Ghiretti (1951*a, b,* 1953) demonstrated that no spontaneous secretions are produced by the salivary glands. The secretory activity is dependent upon nervous stimulation such as occurs in the presence of prey or an enemy. Moreover, they were of the opinion that neither tyramine nor histamine, either alone or in combination, were able to account for all the neurotoxic effects produced by octopus saliva injected into crabs.

Erspamer and Ghiretti (1951), and Bacq, Fischer, and Ghiretti (1951) have shown that extracts and perfusates from the posterior salivary glands of *O. vulgaris, O. macropus*, and *E. moschata* have a powerful positive inotropic, chronotropic, and tonotropic action on molluscan heart, *Helix*, and *Octopus*. This action they ascribed to 5-hydroxytryptamine (also known as 5-HT, serotonin, enteramine).

Ghiretti (1960) demonstrated that neither tyramine nor any other amine present in the salivary glands, when injected in an amount equal to that present in a lethal quantity of saliva, is able to kill the animal. The early experiments of Livon and Briot (1906), and more recently confirming research by Ghiretti, indicate that the saliva of cephalopods contains several factors that are responsible for the various phases observed during the intoxication of a crab. Ghiretti (1959) has isolated a substance which he termed "cephalotoxin," and which is believed to be responsible for the paralyzing action in crustaceans. The chemical composition of cephalotoxin is as yet unknown, but is thought to be a glycoprotein (*see* Chemistry section for further discussion on this aspect). Cephalotoxin produces paralysis in crustaceans, the length of time required varying with the amount of poison and the species of crustacean. Neuromuscular preparations of the pincers and legs of crabs are unaffected by cephalotoxin. Isolated hearts of crabs and octopuses that were perfused with cephalotoxin showed a temporary increase in tonus followed by a decrease in amplitude and arrest in diastole. Cephalotoxin caused an increase in tonus and in frequency of the normal spontaneous rhythmic contractions of the esophagus and aorta of *Octopus*. Cephalotoxin has also been tested on frog neuromuscular preparations, frog esophagus and intestine, and rabbit gut. Cephalotoxin caused an increase in tonus and amplitude of spontaneous contractions in rabbit duodenum, but the effects were abolished by washing. Cephalotoxin was found to inhibit the respiration of rabbits.

Hartman *et al.* (1960) contributed a valuable paper on the isolation of pharmacologically active amines from the posterior salivary gland of *O. apollyon*. However, since they were primarily concerned with the chemical isolation of these substances, their results are reviewed in the section on Chemistry.

Erspamer (1949) isolated a pharmacologically active substance from the posterior salivary glands of the octopus *E. moschata*, which he termed "moschatin."[5] Moschatin, later called "eledoisin," was found to have a powerful vasodilator and hypotensive action, and it stimulated certain extravascular smooth muscle. Eledoisin also increased the permeability of the cutaneous vessels in the guinea pig. Moschatin, or eledoisin, is said to be absent in the salivary glands of *O. vulgaris* and *O. macropus* (Anastasi and Erspamer, 1962; Erspamer and Anastasi, 1962). They also reported an eledoisin-like substance from skin extracts of the South American amphibian *Physalaemus fuscumaculatus*, which they termed "physalaemin" (Erspamer, Bertaccini, and Cei, 1962).

Simon *et al.* (1964) found that aqueous extracts of the venom of *Octopus maculosus* injected into laboratory animals resulted in death due to respiratory paralysis. The authors believed this to be due to phrenic nerve blockage or to inhibition at the neuromuscular junction, or to both. The venom produced a marked hypotensive effect and bradycardia, but did not seem to affect the electrocardiogram. The venom produced deterioration in both nerve and muscle recordings in phrenic nerve-diaphragm preparations.

Trethewie (1965) injected aqueous extracts of *Octopus (H.) maculosus* intravenously into intact cats, the intracoronary route in isolated perfused heart of the cat, and applied it to isolated jejunum of the guinea pig, and the

---

[5] The name of this substance was later changed to "eledoisin" in order to avoid confusion with the plant alkaloid, moschatin, derived from *Achillea moschata* (Anastasi and Erspamer, 1962).

phrenic nerve diaphragm preparation of the rat. Respiratory failure followed by cardiac failure occurred in the cat. The coronary blood flow was little altered, and the heart rate was relatively unchanged in the isolated perfused heart. The jejunum responded with an immediate contraction followed by delayed relaxation. Neuromuscular transmission was inhibited in the diaphragm preparation, and there was a reversible direct depression of the muscle contractions of the diaphragm.

Freeman and Turner (1970) investigated maculotoxin, a potent venom secreted by *O. maculosus* of Australia. They found that it contains a neurotoxin which is pharmacologically similar to tetrodotoxin and saxitoxin. The poison causes hypotension, bradycardia, and respiratory paralysis in rabbits and rats. It appears to have a neuromuscular blocking action at low dosage levels, but at higher levels the muscle membrane becomes excitable. Animals can be resuscitated after a marginal lethal dose by artificial respiration, provided this is instituted before hypoxis becomes severe.

McDonald and Cottrell (1973) described some of the pharmacological properties of the toxin of *Eledone cirrosa* which showed some similarity to cephalotoxin.

CHEMISTRY

POISONOUS CEPHALOPODS

The only report dealing with this subject is by Kawabata *et al.* (1957). Samples of the toxic squid were examined for ammonia-nitrogen, trimethylene-nitrogen, and histamine. No significant differences were found between toxic and nontoxic squid.

VENOMOUS CEPHALOPODS

The secretion of the posterior salivary glands of the cephalopods contains a toxin which is used by the animal to kill its prey. Some of the more obvious physical and chemical properties of the secretion were described by Hyde as early as 1897. The significance of some of these observations has not become apparent until more recent times. What was earlier believed to be a single toxic principle in the secretion is now recognized as a composite of no less than three different chemical substances. The sequence of discoveries leading to the current views may be traced in the reports subsequent to that of Hyde.

Livon and Briot (1905) reported that the active substance in the secretion was precipitable by alcohol and was destroyed by heating for 10 minutes at 100° C or for 1 hour at 58° C. Henze (1906), on the other hand, reported that the toxic agent was extractable with alcohol and that it could be concentrated by heating for hours on a water bath without destruction. Treatment with phosphotungstic acid, silicotungstic acid, or calcium iodide, however, resulted in destruction. Henze regarded the toxin as an alkaloid because of its association with an unidentified impurity which was also believed to be alkaloidal in nature. It was also observed that taurine was present in the secretion in large amounts (Henze, 1910).

The isolation technique of Livon and Briot was modified by Rouville

(1910*a*) by extracting the dried and ground glands with normal saline solution instead of alcohol. These investigators noted that tyrosin precipitated from the secretion on standing.

Shortly thereafter (1913, 1929), Henze reported the isolation of a toxic substance from the secretion which gave a color reaction with iron chloride similar to that given by adrenalin. The toxin was identified as *p*-oxyphenylethylamine (tyramine), a substance chemically related to both adrenalin and tyrosin. The identification of the toxin was based on a comparison of the physiological action with that of the synthetic compound and by the coincidence of the melting point of the benzoyl derivative with that of the synthetic derivative. Henze concluded that *p*-oxyphenylethylamine was the only poisonous substance in the octopus salivary gland. The problem lay dormant until 1938 when Vialli and Erspamer reported evidence of the presence of a di-phenolic substance in the granules which exist in the salivary glands. Erspamer (1948*a, b, c, d*) described a substance in the saliva of *Octopus vulgaris* which, on irradiation with ultraviolet light in the presence of air, yielded a substance with intense adrenalinelike action. The parent substance was called "octopamine" while the irradiated relative with adrenalinelike properties was called "hydroxyoctopamine." In a later report by Erspamer (1952), evidence was presented to show that octopamine has the structure 1-*p*-hydroxyphenylethanolamine and the hydroxyoctopamine is the same as 1-noradrenaline. The proofs offered for the conclusions on chemical structure were based on the identities of the chromatograms of the toxin fraction with those obtained from synthetic compounds, and on the failure of new spots to appear on chromatograms of mixtures of the toxin fractions with the synthetic compounds. It was also shown that on irradiation, the spot for octopamine faded and a spot due to some catecholic substance appeared in its place. Identical pharmacological properties confirmed the conclusions of identity.

Evidence for the presence of a toxic substance in the saliva of octopuses, which was indolic in structure, in contrast to octopamine and hydroxyoctopamine which are phenolic, appeared in the report of Erspamer and Boretti (1950). It would seem that the reason the presence of such a toxin had been overlooked in earlier investigations was that the toxin has approximately the same R_f value as certain of the phenolic substances present, and thus its presence was masked in the development of the spots with reagents used for the detection of phenolic substances. When the development of the spots was accomplished by means of *p*-dimethylaminobenzaldehyde as well as reagents commonly used for detection of phenolic substances, the presence of indolic substances became apparent. Five spots representing substances of indolic structure were noted.

The first fraction was found to contain alteration products of enteramine (5-hydroxytryptamine); the second fraction was essentially enteramine picrate with only a slight contamination with tyramine picrate and some enteramine derivative; the third fraction was picrate of octopamine and tyramine with only traces of enteramine.

Ghiretti (1953*a, b*) demonstrated that octopus saliva differs in composition depending upon whether it is obtained by means of chemical, or by electrical stimulation. Stimulation mechanically or by electricity yields a saliva which is viscous and stringy in *O. vulgaris* and cloudy in *O. macropus*. When injected into crabs, it produced first a state of general excitement followed by paralysis. Chromatographic analysis of the saliva from *O. vulgaris* revealed the presence of tyramine, octopamine, hydroxytryptamine, and tyrosine; from *O. macropus*, only tyramine, octopamine, and tyrosine were present. Proteolytic activity was pronounced.

The excitement produced in the crab by the injection of mechanically stimulated saliva from octopuses may be explained by the physiological action of the active amines, i.e., tyramine, octopamine, and 5-hydroxytryptamine. These cannot, however, account for the subsequent paralysis. The logical conclusion is that this action results from a protein material in the saliva, probably of proteolytic enzyme nature. This fraction was called cephalotoxin by Ghiretti (1959).

Erspamer (1949) noted the presence of a toxic substance having powerful vasodilator and hypotensive action in the posterior salivary glands of *E. moschata* and *E. aldrovandi*, but it was absent in *O. vulgaris* and *O. macropus*. This toxin, first called moschatin, was later renamed eledoisin in order to avoid confusion with the vegetable alkaloid moschatin. This principle was identified as having a polypeptide configuration and the sequence of the amino acids in the purified product has been determined by Erspamer and Anastasi (1962), and Anastasi and Erspamer (1962). The sequence is pyroglutamyl-proline-serine-lysene-hydroxyaspartic acid-alanine-phenylalanine-isoleucine-glycine-leucene-methylamine. The structure has been confirmed by synthesis. Carboxypeptidase digestion failed to split any C-terminal amino acid and the fluorodinitrobenzene technique failed to reveal a free N-terminal amino acid. All amino acids appear to have the L-configuration. The isolation procedure involved absorption of the crude polypeptide in 95 percent alcohol on a column of alkaline alumina with subsequent elution with descending concentrations of ethanol. Eledoisin appears in the fraction eluted with 40 to 60 percent alcohol. Further purification was by ion-exchange chromatography on a column of Amberlite CG-50 in the H^+ form, followed by elution with *m*-ammonium acetate buffer at pH 8.4 and finally by means of countercurrent distribution between 0.5 *N* acetic acid and 1-butanol.

Following the beliefs of earlier investigators, researchers have referred to the physiologically active amines present in the salivary secretion of various species of octopuses as toxins. As pointed out by Ghiretti (1960), none of these amines is lethal to the crab in concentrations found in the salivary secretion. It seems essentially certain that one must look to a toxic principle of protein nature as the death-producing factor in the saliva of the octopus.

Hartman *et al.* (1960) summarized a considerable number of substances, including a number of enzymes, found or suspected to be present in the salivary glands of octopuses, some of which are also present in the salivary secretion. The part which each of these plays in the total defensive and offensive weapon system of octopuses remains to be elucidated.

Membrane studies of octopus neurotoxins (Sutherland, Broad, and Lane, 1970) indicated that the neurotoxins of *Octopus (H.) maculosus* have a molecular weight below 540 which would account for their total lack of antigenicity and indirectly their rapidity of action. As a result of their study, the investigators concluded that this is the lowest molecular weight toxin that any creature has injected into man with fatal results.

Croft and Howden (1972) isolated the toxin from *H. maculosus* in a chromatographically pure state. Maculotoxin was found to be a stable, highly polar compound having a relatively low molecular weight. It behaves as a cation under certain conditions and was shown to be chemically similar but distinct from tetrodotoxin.

Some of the chemical properties of the toxin of *Eledone cirrosa* have been described by McDonald and Cottrell (1973). They found it to be similar to cephalotoxin.

A preliminary study on the purification and composition of a toxin from the posterior salivary gland of *Octopus dofleini* has been conducted by Songdahl and Shapiro (1974). The toxin has a molecular weight of about 23,000. It consists of a single subunit with an isoelectric point of 5.2-5.3. The amino analysis shows a content of 31 aspartic and 27 glutamic residues in contrast to 10 lycine and 7 arginine. The toxic activities are resistant to proteolytic destruction but can be destroyed by boiling for 10 minutes.

Howden and Williams (1974) isolated histamine, tyramine, and serotonin from the venom of *H. maculosa*.

Jarvis *et al.* (1975) were unable to distinguish by chromatographic procedures any chemical differences between maculotoxin from *O. maculosus* and tetrodotoxin.

Subsequent studies by Sheumack, Howden, Spence, and Quinn (1978) have shown that maculotoxin from *Octopus (H.) maculosus* is chemically identical to tetrodotoxin. Pure maculotoxin was twice freeze-dried with D_2O then dissolved in 20 μl of 3 percent completely deuterated acetic acid (CD_3CO_2D) in D_2O. Its proton nuclear magnetic resonance spectrum (JEOL: 100 Mhz) showed a singlet at 2.72 (CD_2HCO_2D), a doublet centered on 2.98 (J=9.5 hertz), a multiplet with peaks at 4.62 and 4.88, a large proton peak at 5.39 (HOD), and a doublet centered on 6.14 parts per million (ppm) (J=9.5 hertz). This spectrum was identical with that of authentic tetrodotoxin examined under the same conditions. The pair of doublets at 2.98 and 6.14 ppm are the hallmarks of tetrodotoxin.

LITERATURE CITED

ABBOTT, R. T.

1950 The venomous cone shells. Sci. Counselor 13(4): 125-126, 153, 3 figs.

1954 American seashells. D. Van Nostrand Co., New York. 541 p., 40 pls., 100 figs.

AKIBA, T.

1943 Study of poisons of *Venerupis semidecussata* and *Ostrea gigas*. [In Japanese] Japan Iji Shimpo 1078: 1077-1082, 1 fig.

1949 Study of poisoning by *Venerupis semidecussata* and *Ostrea gigas* and their poisonous substances. [In Japanese, English summary], Nisshin Igaku 36(6): 1-24.

1961 Food poisoning due to oyster and baby clam in Japan, and toxicological effects of the toxic substance, p. 446-447. *In* Abstracts of symposium papers. Tenth Pacific Sci. Congr. Pacific Sci. Assoc. Honolulu, Hawaii.

AKIBA, T., and Y. HATTORI

1949 Food poisoning caused by eating asari (*Venerupis semidecussata*) and oyster (*Ostrea gigas*) and studies on the toxic substance, venerupin. Japan. J. Exp. Med. 20: 271-284, 1 fig.

ALPERS, F.

1931 Zur kenntnis der anatomie von *Conus lividus* Brug., besonders des darmkanals. (Fauna et anatomia ceylanica.) Jen. Z. Naturwiss. 65(2): 587-658, 40 figs.

ANASTASI, A., and V. ERSPAMER

1962 Occurrence and some properties of eledoisin in extracts of posterior salivary glands of *Eledone*. Brit. J. Pharmacol. 19(2): 326-336, 4 figs.

ARCISZ, W.

1950 Ciguatera: tropical fish poisoning. U.S. Fish Wildlife Serv., Spec. Sci. Rept., Fish. No. 27, 23 p.

(NSA)—Not seen by author.

ARVY, L.

1960 Histoenzymological data on the digestive tract of *Octopus vulgaris* Lamarck (Cephalopoda). Ann. N. Y. Acad. Sci. 90(3): 929-949, 10 figs.

ASANO, M.

1952 Studies on the toxic substances contained in marine animals. I. Locality of the poison of *Neptunea (Barbitonia) Arthritica* Bernardi. Bull. Japan. Soc. Sci. Fish. 17(8-9): 73-77, 2 figs.

1954 Occurrence of choline in the shellfish, *Callista brevisiphonata* Carpenter. Tohoku J. Agri. Res. 4(3-4): 239-250, 2 figs.

ASANO, M., and M. ITO

1959 Occurrence of tetramine and choline compounds in the salivary gland of a marine gastropod, *Neptunea arthritica* Bernardi. Tohoku J. Agri. Res. 10(2): 209-227, 8 figs.

1960 Salivary poison of a marine gastropod, *Neptunea arthritica* Bernardi, and the seasonal variations of its toxicity. Ann. N.Y. Acad. Sci. 90(3): 674-688, 9 figs.

ASANO, M., TAKAYANAGI, F., and T. KITAMURA

1953 Shellfish poisoning from *Callista brevisiphonata* Carpenter and its clinical symptoms. Tohoku J. Agri. Res. 3(2): 321-330, 1 fig.

ASERO, B., COLO, V., ERSPAMER, V., and A. VERCELLONE

1952 Synthèse des enteramins (5-Oxytryptamin). Ann. Chem. 576: 69-74, 3 figs.

BACQ, Z. M.

1951 Isolement et perfusion des glandes salivaires postérieures des céphalopodes octopodes. Arch. Intern. Physiòl. 59(3): 273-287, 6 figs.

BACQ, Z. M., FISCHER, P., and F. GHIRETTI

1951 Action de la 5-hydroxytryptamine chez les céphalopodes. Arch. Intern. Physiol. 59(2): 165-171, 3 figs.

1952 Action de la 5-hydroxytryptamine chez les céphalopodes. Arch. Intern. Physiol. 60: 565. (NSA)

Bacq, Z. M., and F. Ghiretti
1950 Un metodo di perfusione delle ghiandole salivari posteriori dei cefalopodi. Boll. Soc. Ital. Biol. Sper. 26(1): 775-776.
1951*a* Vasomotor phenomena in cephalopods. J. Physiol. 113: 525-527, 2 figs.
1951*b* La sécrétion externe et interne des glandes salivaires postérieures des céphalopodes octopodes. Arch. Intern. Physiol. 59(3): 288-314, 16 figs.
1953 Physiologie des glandes salivaires postérieurs des céphalopodes octopodes isolées et perfusées *in vitro*. Publ. Staz. Zool. Napoli 24(3): 267-277, 3 figs., 2 pls.

Baglioni, S.
1909 Sur l'action physiologique du poison des céphalopodes. Arch. Ital. Biol. 51: 349-352.

Bagnis, R.
1967 A propos de quelques cas d'intoxications par des mollusques du genre "Bénitier," dans une île de la Société. Bull. Soc. Pathol. Exotique 60(6): 580-592, 2 figs.

Ballering, R. B., Jalving, M. A., VenTresca, D. A., Hallacher, L. E., Tomlinson, J. T., and D. R. Wobber
1972 Octopus envenomation through a plastic bag via a salivary proboscis. Toxicon 10: 245-248.

Banner, A. H.
1967 Marine toxins from the Pacific, 1. Advances in the investigation of fish toxins, p. 157-165. *In* F. E. Russell and P. R. Saunders [eds.], Animal toxins, Pergamon Press, New York.

Banner, A. H., and P. Helfrich
1964 The distribution of ciguatera in the tropical Pacific. University of Hawaii, Honolulu. 48 p. (Hawaii Marine Laboratory Technical Report No. 3).

Barnes, R. D.
1963 Invertebrate zoology. W. B. Saunders Co., Philadelphia. 632 p.

Bartsch, P.
1917 Pirates of the deep—stories of the squid and octopus. Smithsonian Inst. Ann. Rept. 1916. Publ. No. 2464, p. 347-375, 19 pls., 9 figs.

Bartsch, A. F., and E. F. McFarren
1962 Fish poisoning: a problem in food intoxication. Pacific Sci. 16(1): 42-56.

Bergh, R.
1895 Beiträge zur kenntniss der coniden. Novo Acta Leopoldina 65(2): 69-214, 13 pls.

Berry, S. S.
1912 A review of the cephalopods of western North America. Bull. Bur. Fish 1910, 30(761): 269-336; Pls. 32-56, 18 figs.
1914 The cephalopoda of the Hawaiian Islands. Bull. Bur. Fish. 1912, 32 (789): 257-362; Pls. 45-55, 41 figs.
1953 Preliminary diagnoses of six west American species of octopus. Leaflets Malacol. 1(10): 51-58.

Berry, S. S., and B. W. Halstead
1954 Octopus bites—a second report. Leaflets Malacol. 1(11): 59-65, 1 fig.

Bonnet, A., and A. Jullien
1941 Toxicité comparée des extraits de la glande à pourpre chez *Murex trunculus* et *Murex brandaris*. Compt. Rend. Soc. Biol. 135: 958-960.

Bottazzi, F.
1916 Richerche sulla ghiandola salivare posteriore dei cefalopodi. I. Pubbl. Staz. Zool. Napoli 1: 59-146, 33 figs.
1918 Richerche sulla "ghiandola salivare posteriore dei cefalopodi. II. Pubbl. Accad. Nazl. Lincei 27(1): 190-196.
1919 Richerche sulla ghiandola salivare posterioire dei cefalopodi. II. Pubbl. Staz. Zool. Napoli 1: 69-148, 42 figs.
1921 Recherches sur la "glande salivaire postérieure" de l'*Octopus macropus*. Arch. Intern. Physiol. 18: 313-331, 4 figs.

Bottazzi, F., and V. Valentini
1924 Nuove ricerche sul veleno della "saliva" di *Octopus macropus*. Arch. Sci. Biol. (Bologna) 6: 153-168, 12 figs.

Bourquelot, E.
1885 Recherches sur les phénomènes de la digestion chez les mollusques céphalopodes. Arch. Zool. Exp. Gen. 3: 1-75.

Bouvier, E. L.
1887 Système nerveux, morphologie générale et classification des gastéropodes prosobranches. Ann. Sci. Nat. Zool. 7, 3: 1-510.

BRIOT, A.

1905*a* Sur le rôle des glandes salivaires des céphalopodes. Compt. Rend. Soc. Biol. 57:384-386.

1905*b* Sur le mode d'action du venin des céphalopodes. Compt. Rend. Soc. Biol. 57: 386.

BRYGOO, E. R.

1972 Bibliographie de l'envenimation par des mollusques de genre *Conus* Linné, 1758. Arch. Inst. Pasteur Madagascar 41(1): 123-133.

CALMETTE. A.

1908 Venoms. Venomous animals and antivenomous serum-therapeutics. John Bale, Sons & Danielsson, Ltd., London. 403 p., 125 figs.

CARDOT, H., and A. JULLIEN

1940 Action de la pourpre sur l'excitabilité du nerf et du muscle. Compt. Rend. Soc. Biol. 133: 521-523.

CHARNOT, A.

1945 La toxicologie au Maroc. Mem. Soc. Sci. Nat. Maroc. 1945(47): 86-97. (NSA)

CIOCATTO, E., CATTANEO, A., and E. FAVA

1956 Esperienze cliniche con un nuovo curarizzante: Murexina. Minerva Anestes. 22(6): 197-202.

CLELAND, J. B.

1912 Injuries and diseases of man in Australia attributable to animals (except insects). Australasian Med. Gaz. 32(11): 269-272.

1913 Injuries and diseases of man in Australia attributable to animals (except insects). J. Trop. Med. Hyg. 16: 25-31, 43-47.

1942 Injuries and diseases in Australia attributable to animals (insects excepted). Med. J. Australia 2(14): 313-320.

CLENCH, W. J.

1946 The poison cone shell. Occ. Pap. on Mollusks (Harvard) 1(7): 49-52, 1 pl.

COOKE, A. H.

1895 Molluscs, p. 64-65. *In* S. F. Harmer and A. E. Shipley [eds.], The Cambridge Natural History. Vol. III. Macmillan and Co., London.

COOPER, M. J.

1964 Ciguatera and other marine poisonings in the Gilbert Islands. Pacific Sci. 18(4): 411-440, 11 figs.

(NSA)—Not seen by author.

CROFT, J. A., and M. E. H. HOWDEN

1972 Chemistry of maculotoxin: a potent neurotoxin isolated from *Hapalochlaena maculosa*. Toxicon 10: 645-651.

CUÉNOT, L.

1905 Fonctions absorbantes et excrétrice du foie des céphalopodes. Arch. Zool. Exp. Gén. 7: 227-245. (NSA)

CUNNINGHAM, J. T.

1883 Note on the structure and relations of the kidney in *Aplysia*. Mitt. Zool. Sta. Neapel 4: 420-428, 1 pl.

DACK, G. M.

1956 Food poisoning. Rev. ed. Univ. Chicago Press, Chicago. 251 p.

DAKIN, W. J.

1912 Buccinum (the whelk). Liverpool Marine Biol. Comm., Mem. 20, 115 p., 6 figs., 8 pls.

DE BLASI, S., and U. LEONE

1955 Uso clinico di un nuovo curarizzante: la Murexina. Minerva Anestes. 21: 137-141.

DUBOIS, R.

1903 Sur le venin de la glande à pourpre des murex. Compt. Rend. Soc. Biol. 55: 81.

1907 Adrénaline et purpurine. Compt. Rend. Soc. Biol. 63: 636-637.

1909 Recherches sur la pourpre et sur quelques autres pigments animaux. Arch. Zool. Exp. Gén. 5, 2: 471-590, 2 figs.

DUNCAN, D. D.

1941 Fighting giants of the Humboldt. Nat. Geogr. Mag. 79: 373-400.

EALES, N. B.

1921 *Aplysia*. Liverpool Marine Biol. Comm., Mem. 24, 84 p., 7 pls.

EMERSON, G. A., and C. H. TAFT

1945 Pharmacologically active agents from the sea. Texas Rept. Biol. Med. 3(3): 302-338.

ENDEAN, R., and C. DUCHEMIN

1967 The venom apparatus of *Conus magus*. Toxicon 4: 275-284, 17 figs.

ENDEAN, R., GYR, P., and G. PARISH

1974 Pharmacology of the venom of the gastropod *Conus magus*. Toxicon 12: 117-129, 5 figs.

ENDEAN, R., and J. IZATT

1965 Pharmacological study of the venom of the gastropod *Conus magus*. Toxicon 3(2): 81-93.

Endean, R., Izatt, J., and D. McColm
1967 The venom of the piscivorous gastropod *Conus striatus*, p. 137-144. *In* F. E. Russell and P. R. Saunders [eds.], Animal toxins. Pergamon Press, New York.

Endean, R., Parish, G., and P. Gyr
1974 Pharmacology of the venom of *Conus geographus*. Toxicon 12: 131-138, 5 figs.

Endean, R., and C. Rudkin
1963 Studies of the venoms of some Conidae. Toxicon 1(2): 49-64.
1965 Further studies of the venoms of Conidae. Toxicon 2: 225-249, 3 figs.

Erspamer, V.
1946 Richerche chimiche e farmacologiche sugli estratti di ghiandola ipobranchiale di *Murex (Trybcularia) trunculus* (L.), e *Tritonalia erinacea* (L.), 1. Distribuzione e caratteristiche della purpurasi e delle purpurine. Pubbl. Staz. Zool. Napoli 20: 91-101, 2 figs.
1947 Richerche chimiche e farmacologiche sugli estratti di ghiandola ipobranchiale di *Murex trunculus, Murex brandaris*, e *Tritonalia erinacea*, 2. Reazioni chimiche colorate degli estratti. Arch. Intern. Pharmacodyn 74(2): 113-140, 2 figs.
1948*a* Richerche chimiche e farmacologiche sugli estratti di ghiandola ipobranchiale di *Murex brandaris* e *Tritonalia erinacea*, 4. Presenza negli estratti di enteramina o di una sostanze enteraminosimile. Arch. Intern. Pharmacodyn 76(3): 308-326, 5 figs.
1948*b* Osservazioni preliminari, chimiche e farmacologiche, sulla murexina. Experientia 4(6): 226-228, 2 figs.
1948*c* Active substances in the posterior salivary glands of Octopoda, 1. Enteramine-like substance. Acta Pharmacol. Toxicol. 4: 213-223, 5 figs.
1948*d* Active substances in the posterior salivary glands of Octopoda, 2. Tyramine and octopamine (oxyoctopamine). Acta Pharmacol. Toxicol. 4: 224-247, 9 figs.
1949 Ricerche preliminari sulla moschatina. Experientia 5: 79-81, 2 figs.
1952 Wirksame stoffe der hinteren speicheldrüsen der octopoden und der hypobranchialdrüse der purpurschnecken. Arzneimittel-Forsch. 2: 253-258, 8 figs.
1953 Murexine. Arch Exp. Pathol. Pharmakol. 218(½): 142-144.
1956 The enterochromaffin cell system and 5-hydroxytryptamine (enteramine, serotonin). Triangle 2(4): 129-138, 8 figs.

Erspamer, V., and A. Anastasi
1962 Structure and pharmacological actins of eledoisin, the active endecapeptide of the posterior salivary glands of *Eledone*. Experientia 18: 58.

Erspamer, V., and B. Asero
1951 L'enteramina, prodotto ormonale specifico del sistema enterochromaffine. Caratteristische chimiche e farmacologiche della sostanza pura, naturale e di sintesi. Ric. Sci. 21: 3-7. (NSA)
1952 Identification of enteramine, the specific hormone of the enterochromaffin cell system, as 5-hydroxyptamine. Nature 169: 800-801.

Erspamer, V., and O. Benati
1953*a* Isolierung des murexins aus hypobranchialdrüsenextrakten von *Murex trunculus* und seine identifizierung als β-[imidazolyl-4(5)]-acryl-cholin. Biochem. Z. 324: 66-73, 4 figs.
1953*b* Identification of murexine as β-[imidazolyl-(4)]- acryl-choline. Science 117(3033): 161-162.

Erspamer, V., Bertaccini, G., and J. M. Cei
1962 Occurrence of an eledoisin-like polypeptide (physalaemin) in skin extracts of *Physalaemus fuscumaculatus*. Experientia 18: 562-563.

Erspamer, V., and G. Boretti
1950 Identification of enteramine and enteramine-related substances in extracts of posterior salivary glands of *Octopus vulgaris* by paper chromatography. Experientia 6(9): 348, 2 figs.

Erspamer, V., and F. Dordoni
1946 Sulla murexina, nuova derivato colinico degli estratti di organo ipobranchiale di *Murex trunculus, Murex brandaris* e *Tritonali erinacea*. Ric. Sci. 16(8): 1114-1116.

(NSA)—Not seen by author.

ERSPAMER, V., and F. DORDONI—Continued
1947 Ricerche chimiche e farmacologiche sugli estratti di ghiandola ipobranchiale di *Murex trunculus, Murex brandaris* e *Tritonalia erinacea*. 3. senze negli estratti di una colina omologa: la murexina. Arch. Intern. Pharmacodyn. 74(3-4): 263-285, 9 figs.

ERSPAMER, V., and G. F. ERSPAMER
1962 Pharmacological actions of eledoisin on extravascular smooth muscle. Brit. J. Pharmacol. 19(2): 337-354, 12 figs.

ERSPAMER, V., and F. GHIRETTI
1951 The action of enteramine on the heart of molluscs. J. Physiol. 115(4): 470-481, 7 figs.

ERSPAMER, V., and A. GLÄSSER
1957 The pharmacological actions of murexine (urocanylcholine). Brit. J. Pharmacol. 12: 176-184, 7 figs.
1958 The pharmacological actions of some murexine-like substances. Brit. J. Pharmacol. 13: 378-384, 5 figs.

FALLOISE, A.
1905-6 Contribution à la physiologie comparée de la digestions. La digestion chez les céphalopodes. Arch. Intern. Physiol. 32: 282-305.

FÄNGE, R.
1957 An acetylcholine-like salivary poison in the marine gastropod *Neptunea antiqua*. Nature 180: 196-197.
1958 Paper chromatography and biological effects of extracts of the salivary gland of *Neptunea antiqua* (Gastropoda). Acta Zool. (Stockholm) 39: 39-46, 3 figs.
1960 The salivary gland of *Neptunea antiqua*. Ann. N.Y. Acad. Sci. 90(3): 689-694, 4 figs.

FAYRER, J.
1878 Venomous animals. Edinburgh Med. J. 23: 105-109.

FISCHER, P. H.
1925 Sur le rôle de la glande purpurigène des Murex et des pourpres. Compt. Rend. Acad. Sci. 180: 1369-1371.

FISH, C. J., and M. C. COBB
1954 Noxious marine animals of the central and western Pacific Ocean. U.S. Fish Wildlife Serv., Res. Rept. No. 36, p. 17-20.

FLECKER, H.
1936 Cone shell mollusc poisoning, with report of a fatal case. Med. J. Australia 1(14): 464-466.

FLECKER, H., and B. C. COTTON
1955 Fatal bite from octopus. Med. J. Austr. 2(9): 329-331, 2 figs.

FLEIG, C., and E. DE ROUVILLE
1910 Origine intra-glandulaire des produits toxiques des céphalopodes pour les crustacés. Toxicité comparée du sang, des extraits de glandes salivaires et d'extraits de foie des céphalopodes. Compt. Rend. Soc. Biol. 69: 502-504.

FLURY, F.
1915 Über das aplysiengift. Arch. Exp. Pathol. Pharmakol. 79: 250-263, 2 figs.

FRASER-BRUNNER, A.
1973 Danger in the sea. Hamlyn Publ. Group Ltd., New York. 128 p., illus.

FREDERICQ, L.
1878 Physiologie du poulpe commun. Arch. Zool. Exp. Gén. 7: 580-583.

FREEMAN, S. E., and R. J. TURNER
1970 Maculotoxin, a potent toxin secreted by *Octopus maculosus*. Hoyle Tox. Appl. Pharmacol. 16(3): 681-690, 4 figs.
1972 A myotoxin secreted by some piscivorous *Conus* species. Brit. J. Pharmacol. 1972: 329-343.

FRETTER, V., and A. GRAHAM
1962 British prosobranch molluscs. Roy. Society, London. 755

GARRETT, A.
1874-78 Annotated catalogue of the species of *Conus*, collected in the South Sea Islands. Quart. J. Conchology 1: 353-367.

GHIRETTI, F.
1953*a* Enteramina, octopamina e tiramina nelle secrezioni esterna ed interna delle ghiandole salivari posteriori dei cefalopodi octopodi. Arch. Sci. Biol. (Bologna) 37(5): 435-441, 4 figs.
1953*b* Les excitants chimiques de la secretion chez les céphalopodes octopodes. Arch. Intern. Physiol. 61(1): 10-21, 9 figs.
1959 Cephalotoxin: the crab-paralyzing agent of the posterior salivary glands of cephalopods. Nature 183: 1192-1193.
1960 Toxicity of octopus saliva against crustacea. Ann. N.Y. Acad. Sci. 90(3): 726-741, 11 figs.

GREVIN, J.
1568 Deux livres des venins. Christofle Plantin, Paris. 423 p.

GRYNFELTT, E.
1911 Sur la glande hypobranchiale de *Murex trunculus*. Biblio. Anat. 21: 181-209, 5 figs.
1913 Sur la genèse des boules picriphiles dans la glande hypobranchiale de *Murex trunculus*. Bull. Mens. Acad. Sci. Lett. Montpellier 5: 119-127, 2 figs.

GUDGER, E. W.
1930 Poisonous fishes and fish poisonings, with special reference to ciguatera in the West Indies. Am. J. Trop. Med. 10(1): 43-55.

HALSTEAD, B. W.
1949 Octopus bites in human beings. Leaflets Malacol. 1(5): 17-22.
1970 Poisonous and venomous marine animals of the world, vol. 3. Vertebrates. U.S. Government Printing Office, Washington, D.C. 1006 p., illus.

HANNA, G. D.
1963 West American mollusks of the genus *Conus*—II. Occ. Pap. Calif. Acad. Sci. (35), 103 p., 4 figs., 11 pls.

HARTLEY, T.
1954 Conidae. Malacological Club of Victoria, Publ. No. 6, 8 p., Figs. 1-7.

HARTMAN, W. J., CLARK, W. G., CYR, S. C., JORDAN, A. L., and R. A. LIEBHOLD
1960 Pharmacologically active amines and their biogenesis in the octopus. Proc. W. Pharmacol. Soc. 3: 106-122.

HASHIMOTO, Y., KANNA, K., and A. SHIOKAWA
1950 On shellfish poisons. II. Paralytic poison (preliminary report). [In Japanese, English summary] Bull. Japan. Soc. Sci. Fish. 15(12): 771-776, 2 figs.

HASHIMOTO, Y., and M. MIGITA
1950 On the shellfish poisons. I. Inadequacy of acidulated alcohols with hydrochloric acid as solvent. Bull. Japan. Soc. Sci. Fish. 16(3): 77-85.

HASHIMOTO, Y., NAITO, K., and J. TSUTSUMI
1960 Photosensitization of animals by the viscera of abalones, *Haliotis* spp. Bull. Japan. Soc. Sci. Fish. 26(12): 1216-1221, 1 pl.

HASHIMOTO, Y., KONOSU, S., SHIBŌTA, M., and K. WATANABE
1970 Toxicity of a turban-shell in the Pacific. Bull. Jap. Soc. Sci. Fish. 36(11): 1163-1171.

HASHIMOTO, Y., MIYAZAWA, K., KAMIYA, H., and M. SHIBŌTA
1967 Toxicity of the Japanese ivory shell. Bull. Japan. Soc. Sci. Fish. 33(7): 661-668, 2 figs.

HASHIMOTO, Y., and J. TSUTSUMI
1961 Isolation of a photodynamic agent from the liver of abalone, *Haliotis discus hannai*. Bull. Japan. Soc. Sci. Fish. 27(9): 859-866, 3 figs.

HATTORI, Y., and T. AKIBA
1952 Studies on the toxic substance in asari *(Venerupis semidecussata)*. 2. Detection of toxic shellfish. [In Japanese, English summary] J. Pharm. Soc. Japan 74(4): 572-577.

HENZE, M.
1906 Chemisch-physiologische studien an den speicheldrüsen der kephalopoden. Das gift und die stickstoffhaltigen substanzen des sekretes. Zentr. Physiol. 19(26): 986-990.
1910 Über das vorkommen des betains bei cephalopoden. Hoppe-Seylers Z. Physiol. Chem. 31: 253-255.
1913 *p*-oxyphenyläthylamin, das speicheldrüsengift der cephalopoden. Hoppe-Seylers Z. Physiol. Chem. 87: 51-58.
1929 Über den tyramin- und tyrosingehalt der speicheldrüse der cephalpoden. Zugleich methodisches zur mikrobestimmung der beiden substanzen. Hoppe-Seylers Z. Physiol. Chem. 182: 227-240, 1 fig.

HERMITTE, L. C.
1946 Venomous marine molluscs of the genus *Conus*. Trans. Roy. Soc. Trop. Med. Hyg. 39(6): 485-512, 5 pls.

HINEGARDNER, R. T.
1958 The venom apparatus of the cone. Hawaii Med. J. 17(6): 533-512, 6 figs.

HIYAMA, Y.
1943 Report of an investigation on poisonous fishes of the south Seas. [In Japanese] Nissan Fish. Exp. Sta. (Odawara, Japan), 137 p., 29 pls., 83 figs.

HOLMSTEDT, B., and V. P. WHITTAKER
1958 Pharmacological properties of β-diamethylacryloylcholine and some other β-substituted acryloylcholines. Brit. J. Pharmacol. 13: 308-314, 9 figs.

HOPKINS, D. G.
1964 Venomous effects and treatment of octopus bite. Med. J. Australia 1(3): 81-82.

HOWDEN, M. E. H., and P. A. WILLIAMS
1974 Occurrence of amines in the posterior salivary glands of the octopus *Hapalochlaena maculosa* (Cephalopoda). Toxicon 12: 317-320.

HUANG, C. L., and G. N. MIR
1972 Pharmacological investigations of salivary gland of *Thais haemastoma* (Clench). Toxicon 10(2): 111-117.

HUTTON, R. F.
1960 Marine dermatosis. Notes on "seabather's eruption" with *Creseis acicula* Rang (Mollusca: pteropoda) as the cause of a particular type of sea sting along the west coast of Florida. Arch. Dermatol. 82: 951-956, 4 figs.

HYDE, J. H.
1897 Beobachtungen über die secretion der sogenannten speicheldrüsen von *Octopus macropus*, 2. Biol. 35(17): 459-477, 2 figs.

ISGROVE, A.
1909 *Eledone*. Liverpool Marine Biol. Comm., Mem. 18, 105 p., 7 figs., 10 pls.

JAECKEL, S.
1952 Über vergiftungen durch *Conus*-arten (Gastr. pros.) mit einem beitrag zur morphologie und physiologie ihres giftapparates. Zool. Anz. 149(9-10): 206-216, 6 figs.

JARVIS, M. M., CRONE, H. D., FREEMAN, S. E., and R. J. TURNER
1975 Chromatographic properties of maculotoxin, a toxin secreted by *Octopus (Hapalochlaena) maculosus*. Toxicon 13: 177-181.

JOHANNES, R. E.
1963 A poison-secreting nudibranch (Mollusca: Opisthobranchia). Veliger 5(3): 104-105.

JULLIEN, A.
1940 Variations dans le temps de la teneur des extraits de glande à pourpre en substances actives sur le muscle de sangsue. Compt. Rend. Soc. Biol. 133: 524-527, 3 figs.
1946 Recherches sur les constituants et les propriétés de la pourpre. Ann.Sci. Franche-Comté, Litt., 37 p., 8 figs.
1948a La substance toxique de la glande à pourpre est-elle un dérivé de la choline. Compt Rend. Sco. Biol. 142: 101-102.
1948b Recherches sur les fonctions de la glande hypobranchiale chez *Murex trunculus*. Compt. Rend. Soc. Biol. 142: 102-103.

JULLIEN, A., and A. BONNET
1941 Toxicité de la pourpre en rapport avec la présence des substances à action stimulante sur le muscle de sangsue. Compt. Rend. Acad. Sci. 212: 932-934.

JULLIEN, A., GARABEDIAN, M. D., and R. GIBAULT
1941 Observations relatives aux propriétés pharmacologiques des constituants de la pourpre chez *Murex trunculus*. Compt. Rend. Soc. Biol. 135: 1636-1639, 1 fig.

JULLIEN, A., JACQUEMAIN, R., and J. RIPPLINGER
1948 Sur quelques propriétés de la glande à pourpre desséchée et vieillie chez *Murex trunculus*. Compt. Rend. Acad. Sci. 227: 1174-1176.

JULLIEN, A., and J. RIPPLINGER
1950 L'extrait desséché de glande à pourpre de *Murex trunculus* et son action biologique. Bull. Soc. Hist. Nat. Doubs (53): 29-30.

KAISER, E., and H. MICHL
1958 Die biochemie der tierischen gifte. Franz Deuticke, Wien. 258 p., 23 figs.

KARLING, T. G.
1966 On nematocysts and similar structures in turbellarians. Acta Zool. Fenn. 116: 3-28, 51 figs.

KAWABATA, T., HALSTEAD, B. W., and T. F. JUDEFIND
1957 A report of a series of recent outbreaks of unusual cephalopod and fish intoxications in Japan. Am. J. Trop. Med. Hyg. 6(5): 935-939.

KEEGAN, H. L.
1960 Some venomous and noxious animals of the Far East. Med. Gen. Lab. (406), U.S. Army Med. Command Japan. 46 p., 70 figs.

KEYL, M. J., and V. P. WHITTAKER
1958 Some pharmacological properties of murexine (urocanoylcholine). Br. J. Pharmacol. 13: 103-106,. 3 figs.

KLINE, G. F.
1956 Notes on the stinging operation of *Conus*. Nautilus 69(3): 76-78.

KOHN, A. J.
1955 Notes on *Conus* for monograph of poisonous and venomous marine animals. (Unpublished.)
1956*a* Feeding in *Conus striatus* and *C. catus*. Proc. Hawaiian Acad. Sci. 31st Ann. Meet., p. 13.
1956*b* Piscivorous gastropods of the genus *Conus*. Proc. Natl. Acad. Sci. 42(3): 168-171, 7 figs.
1958 Recent cases of human injury due to venomous marine snails of the genus *Conus*. Hawaii Med. J. 17: 528-532, 1 fig.
1959*a* The ecology of *Conus* in Hawaii. Ecol. Monographs 29: 47-90, 30 figs.
1959*b* The Hawaiian species of *Conus* (Mollusca: Gastropoda). Pacific Sci. 13: 368-401, 4 figs., 2 pls.

KOHN, A. J., SAUNDERS, P. R., and S. WIENER
1960 Preliminary studies on the venom of the marine snail *Conus*. Ann. N.Y. Acad. Sci. 90(3): 706-725, 7 figs.

KRAUSE, R.
1895 Die speicheldrüsen der cephalopoden. Centr. Physiol. 9(7): 273-277.
1897 Über bau und function der hinteren speicheldrüsen der octopoden. Sitzber. Deut. Akad. Wiss. Berlin 51: 1085-1098.

LANE, F. W.
1960 Kingdom of the octopus. The life history of the Cephalopods. Sheridan House, New York. 300 p., 13 figs., 46 pls.

LANE, W. R., and S. SUTHERLAND
1967 The ringed octopus bite: a unique medical emergency. Med. J. Aust. 2: 475-476.

LIVON, C., and A. BRIOT
1905 Le suc salivaire des céphalopodes est un poison nerveux pour les crustacés. Compt. Rend. Soc. Biol. 58: 878-880.
1906 Sur le suc salivaire des céphalopodes. J. Physiol. Pathol. Gén. 8: 1-9, 16 figs.

LO BIANCO, S.
1888 Notizie biologiche riguardanti specialmente il periodo di maturità sessuale degli animali del golfo di Napoli. Mitt. Zool. Sta. Neapel 8: 385-440; Figs. 1-32.
1899 Notizie biologiche riguardanti specialmente il periodo di maturità. Mitt. Sta. Zool. Neapel 13: 548-549.
1909 Notizie biologiche riguardanti specialmente il periodo di maturità sessuale degli animali del golfo di Napoli. Mitt. Zool. Sta. Neapel 19(4): 650-653.

LOVELL, M. S.
1867 The edible mollusks of Great Britain and Ireland with recipes for cooking them. Reeve and Co., London. 207 p., 12 pls.

MAASS, T. A.
1937 Gift-tiere. *In* W. Junk [ed.], Tabulae biologicae. Vol. XIII. N.V. Van de Garde & Co.'s Drukkerij, Zaltbommel, Holland. 272 p.

MABBET, H.
1954 Death of a skindiver. Australia Skin Diving and Spear-fishing Digest (December), p. 13, 17.

MCDONALD, N. M., and G. A. COTTRELL
1973 Purification and mode of action of toxin from *Eledone cirrosa*, p. 743-760, 5 figs. *In* A. de Vries and E. Kochva [eds.], Toxins of animal and plant origin, vol. 2. Gordon and Breach Science Publ., New York.

MACGINITIE, G. E.
1942 Notes on the natural history of some marine animals. Am. Midland Naturalist 19(1): 213-214.

MCMICHAEL, D. F.
1963 Dangerous marine molluscs, p. 74-80. *In* J. W. Evans [chmn.], Proc. first international convention on life saving techniques. Part III. Scientific Section. Suppl. Bull. Post Grad. Comm. Med., Univ. Sydney.

MEBS, D.
1973 Chemistry of animal venom, poisons and toxins. Experientia 29(11): 1328-1334.

MINER, R. W.
1923 The glory of the sea. Nat. Hist. 23(4): 325-328, 5 figs.

NAKAZIMA, M.
1963 Summary on the cause of shellfish poisoning in Hamana Lake. (Personal communication, May 23, 1963.)

PASINI, C., VERCELLONE, A., and V. ERSPAMER
1952 Synthèse des murexins (β-[imidazolyl-4(5)]-acryl-cholin). Justus Liebigs Ann. Chem. 578: 6-10, 2 figs.

PAWLOWKSY, E. N.
1927 Gifttiere und ihre giftigkeit. Gustav Fischer, Jena. 516 p., 170 figs.

PEILE, A. J.
1937 Some radula problems. J. Conchology 20: 292-304.
1939 Radula notes, VIII. Proc. Malacol. Soc. London 23: 348-355, 30 figs.

PELSENEER, P.
1906 The Gastropoda, p. 66-196. *In* E. R. Lankester [ed.], A treatise on zoology. Part V. Mollusca. Adam & Charles Black, London.

PETRAUSKAS, L. E.
1955 A case of cone shell poisoning by "bite" in Manus Island. Papua New Guinea Med. J. 1(2): 67-68.

PLUMERT, A.
1902 Über giftige seetiere im allgemeinen und einen fall von massenvergiftung durch seemuscheln im besonderen. Arch. Schiffs Tropenhyg. 6: 15-23.

POPE, E. C.
1963 Australian venomous marine animals. (Personal communication, July 1963.)
1968 Venomous ringed octopus. Australian Nat. Hist. 16(1): 16, 1 fig.

POWELL, A. W.
1961 Shells of New Zealand. 4th ed. Whitcombe & Tombs Ltd., New Zealand. 203 p., 36 pls.

QUILLIAM, J. P.
1957 The mechanism of action of murexine on neuromuscular transmission in the frog. Brit. J. Pharmacol. 12: 388-392, 7 figs.

RANDALL, J. E.
1958 A review of ciguatera tropical fish poisoning, with a tentative explanation of its cause. Bull. Marine Sci. Gulf Caribbean 8(3): 236-267, 2 figs.

RANG, P. C., and L. SOULEYET
1852 Histoire naturelle des mollusques ptéropodes. J. B. Ballière, Paris. 86 p., 15 pls.

RICE, R. D., and B. W. HALSTEAD
1968 Report of fatal cone shell sting by *Conus geographus* Linnaeus. Toxicon 5: 223-224.

RICKETTS, E. F., and J. CALVIN
1952 Between Pacific tides. 3d ed. Stanford Univ. Press, Stanford, Calif. 502 p., 134 figs., 46 pls.

RIEDEL, D.
1961 World fisheries, p. 41-75, 4 figs. *In* G. Borgstrom [ed.], Fish as food. Vol. I. Production, biochemistry, and microbiology. Academic Press, New York.

RIEDL, R.
1963 Fauna and flora der Adria. Ein systematischer meeresführer für biologen und naturfreunde. Paul Parey, Hamburg und Berlin. 640 p., 17 figs., 122 pls.

ROAF, H. E., and M. NIERENSTEIN
1907*a* The physiological action of the extract of the hypobranchial gland of *Purpura lapillus*. J. Physiol. 36: 5-8.
1907*b* Adrénaline et purpurine (reply to M. R. Dubois). Compt. Rend. Soc. Biol. 63: 773-774.

ROGERS, J. E.
1951 The shell book. A popular guide to a knowledge of the families of living mollusks, and an aid to the identification of shells native and foreign. Rev. ed. Charles T. Branford Co., Boston. 503 p., 87 pls.

ROMANINI, M. G.
1952 Osservazioni sulla ialuronidasi delle ghiandole salivari anteriori e posteriori degli octopodi. Pubbl. Staz. Zool. Napoli 23: 251-270, 3 figs.

ROUVILLE, E. DE
1910*a* Études physiologiques sur les glandes salivaires des céphalopodes et, en particulier, sur la toxicité de leurs extraits. Compt. Rend. Soc. Biol. 68: 834-836.
1910*b* Sur la toxicité des extraits des glandes salivaires des céphalopodes pour les mammifères. Compt. Rend. Soc. Biol. 68: 878-880.
1910*c* Études physiologiques sur les glandes salivaires des céphalopodes et, en particulier, sur la toxicité de leurs extraits. Bull. Mens. Acad. Sci. Lett. Montpellier 2: 125-147, 2 figs.

SAITO, I.
1934 Anatomy of *Octopus fang-siao* (d'Orbigny). [In Japanese] Zool. Mag. Tokyo 46: 6-14, 53-59. (NSA)

SARS, G. O.
1878 Bidrag til kundskaben om norges arktiske fauna, 1. Mollusca regionis arcticae norvegiae. Trykt Hos A. W. Brøgger, Christiania. 466 p., 18 pls.

SAUNDERS, P. R., and F. WOLFSON
1961 Food and feeding behavior in *Conus californicus* Hinds, 1844. Veliger 3(3): 73-76, 1 pl.

(NSA)—Not seen by author.

Shaw, H. O.
1914 On the anatomy of *Conus tulipa* Linn., and *Conus textile* Linn. Quart. J. Microscop. Sci. 60(1): 1-60, 6 pls., 12 figs.

Sheumack, D. D., Howden, M. E. H., Spence, I., and R. J. Quinn
1978 Maculotoxin: A neurotoxin from the venom glands of octopus *Hapalochlaena maculosa* identified as tetrodotoxin. Science 199: 188-189.

Shibōta, M., and Y. Hashimoto
1970 Purification of the ivory shell toxin. Bull. Japan. Soc. Sci. Fish. 36(1): 115-119, 3 figs.
1971 Further purification of the ivory shell toxin. Bull. Japan. Soc. Sci. Fish. 37(9): 936.

Simon, S. E., Cairncross, K. D., Satchell, D. G., Gay, W. S., and S. Edwards
1964 The toxicity of *Octopus maculosus* Hoyle venom. Arch. Int. Pharmocodyn. 149(3-4): 318-329, 4 figs.

Snow, C. D.
1970 Two accounts of the northern octopus, *Octopus doefleini*, biting scubadivers. Res. Rept. Fish Comm. Oregon 2(1): 103-104, 1 fig.

Sollmann, T.
1949 A manual of pharmacology and its applications to therapeutics and toxicology. 11th ed. W. B. Saunders Co., Philadelphia. 1,132 p.

Sommer, H., and K. F. Meyer
1937 Paralytic shell-fish poisoning. Arch. Pathol. 24(5): 560-598, 2 figs.

Songdahl, J. H., and C. E. Lane
1970 Some pharmacological characteristics of the venom of the alphabet cone, *Conus spurius atlanticus*. Toxicon 8: 289-292.

Songdahl, J. H., and B. I. Shapiro
1974 Purification and composition of a toxin from the posterior salivary gland of *Octopus dofleini*. Toxicon 12: 109-115, 4 figs.

Sutherland, S. K., Broad, A. J., and W. R. Lane
1970 Octopus neurotoxins: low molecular weight non-immunogenic toxins present in the saliva of the blue-ringed octopus. Toxicon 8(3): 249-250.

Sutherland, S. K., and W. R. Lane
1969 Toxins and mode of envenomation of the common ringed or blue-banded octopus. Med. J. Australia 1: 893-898, 7 figs.

Taki, I.
1936 Observations on autophagy in octopus. Annot. Zool. Japon. 15(3): 352-354, 2 figs.
1937 Zur morphologie und systematischen stellung von *Conus tuberculosus* Tomlin. [In Japanese, German summary], Zool. Mag. (Tokyo) 49(6): 218-231, 3 pls.
1941 On keeping octopods in an aquarium for physiological experiments, with remarks on some operative techniques. Venus 10(3-4): 140-156, 4 figs.

Tesch, J. J.
1946 The thecosomatous pteropods, 1. The Atlantic. Dana-Report No. 28. C. A. Reitzels Forlag, Copenhagen. 82 p., 35 figs., 8 pls.
1950 The Gymnosomata, 2. Dana-Report No. 36. Andr. Fr. Høst & Søn, Copenhagen. 55 p., 37 figs.

Toba, G.
1928 Venus 1: 24. (NSA)

Tompsett, D. H.
1939 Sepia. L.M.B.C. Mem. No. 32 (NSA)

Trethewie, E. R.
1965 Pharmacological effects of the venom of the common octopus *Hapalochlaena maculosa*. Toxicon 3: 55-59, 4 figs.

Troschel, F. H.
1856-63 Das gebiss der schnecken zur begründung einer natürlichen classification. Vol. I. Nicolaische Verlagsbuch Handlung. Berlin. 252 p., 20 pls.

Tryon, G. W., Jr.
1882 Manual of conchonology, 4. Nassidae, Turbinellidae, Volutidae, Mitridae. Acad. Nat. Sci., Philadelphia. 276 p., 58 pls.

Turner, R. J., and S. E. Freeman
1974 Factors affecting the muscle depolarization due to *Conus achatinus* toxin. Toxicon 12: 49-55, 2 figs.

Vialli, M., and V. Erspamer
1938 Ricerche istochimiche sulla ghiandola salivare posteriore di *Octopus vulgaris*. Mikrochemie 24: 253-261.

Vincent, D., and A. Jullien
1938*a* La teneur en acétylcholine du coeur des mollusques. Compt. Rend. Soc. Biol. 127: 334-336.

(NSA)—Not seen by author.

VINCENT, D., and A. JULLIEN—Continued

1938*b* Richesse de la glande à pourpre des *Murex* en esters de la choline. Compt. Rend. Soc. Biol. 127: 1506-1509, 3 figs.

1938*c* De la teneur des principaux organes de *Murex* en esters de la choline. Compt. Rend. Soc. Biol. 129: 602-603.

WARMKE, G. L., and R. T. ABBOTT

1961 Caribbean seashells. Livingston Publishing Co., Pennsylvania. 348 p., 19 maps, 44 pls.

WATSON, M.

1973 Midgut gland toxins of Hawaiian sea hares, 1. Isolation and preliminary toxicological observations. Toxicon 11: 259-267.

WATSON, M., and M. D. RAYNER

1973 Midgut gland toxins of Hawaiian sea hares, 2. A preliminary pharmacological study. Toxicon 11: 269-276.

WHITTAKER, V. P.

1960 Pharmacologically active choline esters in marine gastropods. Ann. N.Y. Acad. Sci. 90(3): 695-705, 8 figs.

WHYTE, J. M., and R. ENDEAN

1962 Pharmacological investigation of the venoms of the marine snails *Conus textile* and *Conus geographus*. Toxicon 1: 25-31, 3 figs.

WHYSNER, J. A., and P. R. SAUNDERS

1963 Studies on the venom of the marine snail *Conus californicus*. Toxicon 1: 113-122, 1 fig.

1966 Purification of the lethal fraction of the venom of the marine snail *Conus californicus*. Toxicon 4: 177-181, 3 figs.

(NSA)—Not seen by author.

WILLEM, V.

1898 Résumé de nos connaissances sur la physiologie des céphalopodes. Bull. Sci. France Belgique 30: 31-54, 8 figs.

WINKLER, L. R.

1961 Preliminary tests of the toxin extracted from California sea hares of the genus *Aplysia*. Pacific Sci. 15(2): 211-214, 2 figs.

WINKLER, L. R., and L. M. ASHLEY

1954 The anatomy of the common octopus of northern Washington. Walla Walla College Publ. Dept. Biol. Sci. Sta. (10): 1-30, 19 figs.

WINKLER, L. R., and E. Y. DAWSON

1963 Observations and experiments on the food habits of California sea hares of the genus *Aplysia*. Pacific Sci. 17(1): 102-105.

WINKLER, L. R., and B. E. TILTON

1962 Predation on the California sea hare, *Aplysia californica* Cooper, by the solitary great green sea anemone, *Anthopleura xanthogrammica* (Brandt), and the effect of sea hare toxin and acetylcholine on anemone muscle. Pacific Sci. 16(3): 286-290, 2 figs.

WINKLER, L. R., TILTON, B. E., and M. G. HARDINGE

1962 A cholinergic agent extracted from sea hares. Arch. Intern. Pharmacodyn. 137(1-2): 76-83, 5 figs.

WITTICH, A. C.

1968 Account of an octopus bite. Quart. J. Florida Acad. Sci. 29(4). (NSA)

WU, C. F.

1940 Anatomy of the octopus. Peiping Nat. Hist. Bull. 14: 147-152, 1 fig., 1 pl.

Chapter VII—INVERTEBRATES

Phyla Platyhelminthes, Rhynchocoela, Annelida, Arthropoda, Sipunculida, and Ectoprocta

Phylum PLATYHELMINTHES: Flatworms

The members of the phylum Platyhelminthes, or the flatworms, are characterized by having a soft body that is dorsoventrally flattened, and often elongate. Their integument consists of a ciliated epidermis, or they are covered by a cuticle. Some of the parasitic members have external suckers or hooks, or both. The digestive tract of flatworms is either incomplete or absent. They excrete body wastes by protonephridia with ducts. The body spaces of platyhelminthes are filled with parenchyma. The nervous system of flatworms is a simple type consisting of an anterior ganglia or nerve ring, and one to three longitudinal nerve cords. There are no skeletal, circulatory, or respiratory organs. There is no true segmentation. Usually they are monoecious, and fertilization is internal. Turbellarians are free living, commensalistic, or parasitic in their habits; they inhabit terrestrial, freshwater, and marine environments. It is estimated that there are about 10,000 species.

The phylum Platyhelminthes is comprised of three classes:

TURBELLARIA: The members of this class are mainly free living, although some species are commensalistic or parasitic in their existence. The turbellarians have a cellular or syncytial epidermis, usually provided with some cilia and rhabdoids. Their body is undivided, and with the exception of the Acoela, an intestine is present. Their life cycle is simple.

The class Turbellaria is further subdivided into five orders: *Acoela* is a group of primitive marine turbellarians having a mouth, but lacking an intestine. *Rhabdocoela* are small freshwater and marine turbellarians that have a complete unbranched saclike digestive tract. *Alloeocoela* are small turbellarians with a simple, bulbose, or plicate pharynx. The intestinal track may have short diverticula. *Tricladida* are large turbellarians of elongate form, having a plicate pharynx and an intestine with three highly diverticulated branches. The members of this group are marine, freshwater, and terrestrial. *Polycladida* are large marine turbellarians, mostly of broad flattened form; some, however, are elongate, having a plicate pharynx which opens into a main intestine from which numerous branches radiate to the periphery.

TREMATODA: These are the flukes. All the members of this group are either ecto- or endoparasitic, and therefore do not come within the scope of this work.

CESTODA: The tapeworms are all endoparasitic, and do not come within the scope of this work.

The Turbellaria are of interest to the marine biotoxicologist mainly because of the toxic secretions which are reputedly produced by the rhabdoids and some of the glands of these worms.

The systematics and biology of the Turbellaria pertinent to this chapter have been discussed by Lang (1884), Wilhelmi (1909), and Hyman (1951).

LIST OF FLATWORMS REPORTED AS TOXIC

Phylum PLATYHELMINTHES

Class TURBELLARIA: Free-Living Flatworms

Order POLYCLADIDA

Family LEPTOPLANIDAE

108 *Leptoplana tremellaris* Oersted (Pl. 1, fig. a). Flatworm (USA).

DISTRIBUTION: Atlantic coast of Europe, Irish Sea, North Sea, English Channel, Norway, Greenland, Mediterranean Sea, and Red Sea.

SOURCE: Arndt (1943).

Family PLANOCERIDAE

108 *Stylochus neapolitanus* (Delle Chiaje) (Pl. 1, fig. b). Flatworm (USA).

DISTRIBUTION: Mediterranean Sea.

SOURCE: Arndt (1943).

Family PSEUDOCERIDAE

108 *Thysanozoon brocchi* Grube (Pl. 1, fig. c). Flatworm (USA).

DISTRIBUTION: European seas, tropical Pacific, Indian Ocean.

SOURCE: Arndt (1943).

Order TRICLADIDA

Family BDELLOURIDAE

109 *Bdelloura candida* (Girard) (Pl. 2, fig. a). Flatworm (USA).

DISTRIBUTION: East coast of North America. Commensalistic on the horseshoe crab *Limulus*.

SOURCE: Arndt (1943).

Family PROCERODIDAE

109 *Procerodes lobata* (Schmidt) (Pl. 2, fig. b). Flatworm (USA).

DISTRIBUTION: Mediterranean and Black seas.

SOURCE: Arndt (1943).

BIOLOGY

Turbellarians are in general marine inhabitants, although representatives are found in fresh water and on land. They are benthic, preferring hard bottom, under rocks, shells, gravel, coral, or debris. Some of the smaller species are commonly found on sandy or muddy bottoms. A few of the acoels and polyclads are pelagic. However, most benthic turbellarians are littoral, and a few species have been found at great depths. Those species inhabiting shallow water areas, particularly tidal pools, are subjected to a variety of environmental conditions and temperatures.

Some of the acoels, rhabdocoels, and alloecoels harbor symbiotic chlorellae and xanthellae, which occur in the mesenchyme; they are believed to serve a useful function to the turbellarian.

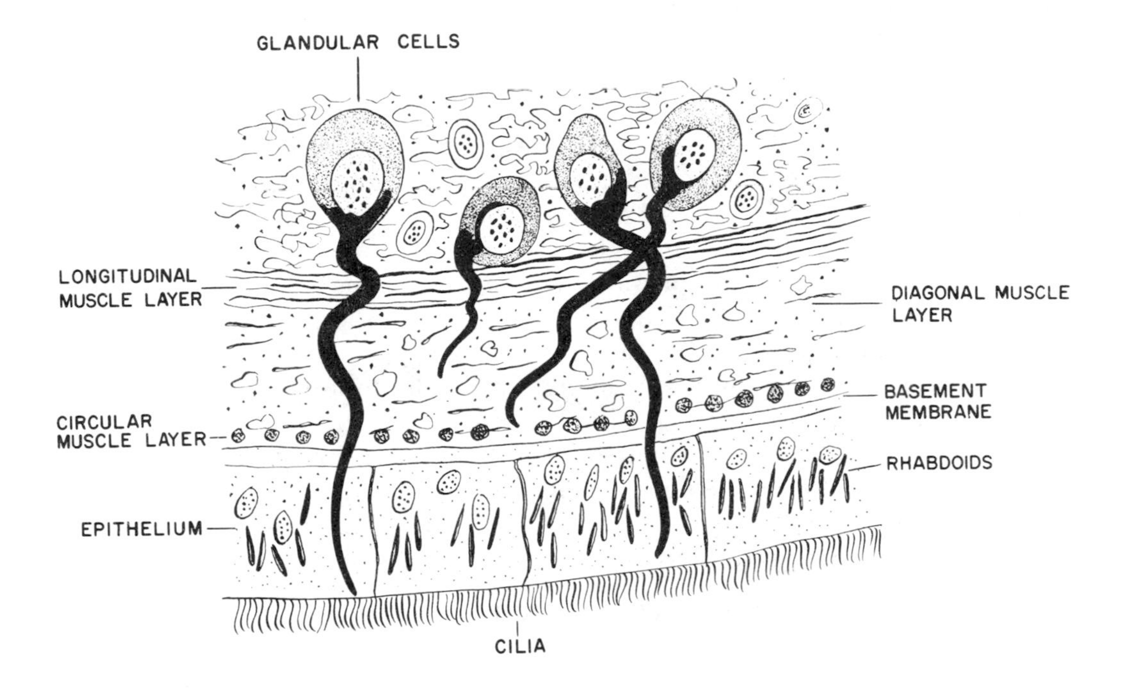

FIGURE 1.—Histological structure of the integument of the polyclad turbellarian, *Thysanozoon brocchi*, showing the glandular cells and rhabdoids which are believed to produce the poison present in some species of marine turbellarians. Semidiagrammatic. (R. Kreuzinger, after Lang)

The toxicity of turbellarians is thought to be part of their chemical protective mechanism which is used to ward off some of their enemies. With their soft, unarmed bodies and sometimes bright coloring, they certainly tend to attract predators. Turbellarian poisons are thought to be produced by either the rhabdoids, or by certain of the glands, or both. Although absent in some members of the Turbellaria, rhabdoids are peculiarly characteristic of turbellarian epidermis. Rhabdoids are rod-shaped bodies, arranged at right angles to the surface of the epidermis, which are secreted by epidermal or mesenchymal glands (Fig. 1). The different types of rhabdoids are designated as adenal rhabdites, epidermal rhabdites, rhammites, or chondrocysts, depending upon their anatomical location and morphology (Hyman, 1951). In general, rhabdoids consist of a shell which encloses a material of fluid consistency. The exact nature of rhabdoids is unknown. Prenant (1919) has suggested that rhabdoids are formed by nuclear breakdown, and that they consist of calcium nucleoproteinate, thought to be useless excreta in the process of being eliminated. Disagreeing with Prenant, most researchers are of the opinion that the rhabdoids secrete a slimy or adhesive material that serves the turbellarian as a protective coating against its enemies, or other inadverse conditions, such as during periods of cyst formation. It has been suggested that rhabdoidal secretions may be toxic (Hyman, 1951), but this lacks laboratory confirmation.

Turbellarians are richly provided with glands, most of which are unicellular in structure. Some of the glands are epidermal, but the majority are located in the mesenchyme. The frontal gland, consisting of a cluster of cyanophilous gland cells, is located near the brain ganglion with long necks that form a band and open by a pore or group of pores on the anterior tip of the body of the animal. The frontal gland is believed to be concerned with the capture of food and it may possibly secrete a toxic material. The so-called poison gland of turbellarians is thought to be a prostatoid, or part of the sexual apparatus of the organism (Hyman, 1951).

Nematocysts that originate from ingested coelenterate hydroids have also been found in the epidermis of turbellarians. After the nematocysts are released by the digestion of the hydroid tissue, they pass into the mesenchyme, and on to the epidermis. Some turbellarians utilize ingested nematocysts to ward off their prey (Hyman, 1951). See Chapter IV for information on the coelenterate nematocyst venom apparatus.

MECHANISM OF INTOXICATION

Although toxic substances have been reported in turbellarians, there is no history of human intoxications. It is assumed that if an intoxication were to result, it would be due to an accidental ingestion of the flatworm.

MEDICAL ASPECTS

Clinical Characteristics

Nothing is known about the clinical effects of turbellarian toxins in humans. Arndt and Manteufel (1925), Arndt (1943), and Hyman (1951) report that *Stylochus neapolitanus* and some of the large land planarians may produce an unpleasant astringent effect on the human tongue. The South American planarian *Polycladus gayi* is reputed to be fatal if eaten by horses and cattle.

Treatment

Symptomatic.

Prevention

Avoid eating turbellarians.

TOXICOLOGY

Initial investigations on crude tissue extracts prepared from freshwater turbellarians *Dendrocoelum lacteum, Polycelis nigra, P. cornuta, Planaria gonocephala, P. lugubris, Bdellocephala punctata*, by Arndt and Manteufel (1925), and the land planarian *Placocephalus kewensis* by Arndt (1925), revealed that they were toxic to a variety of laboratory animals. Arndt (1943) later prepared crude saline tissue extracts from the marine polyclads *Leptoplana tremellaris, Stylochus neapolitanus, Thysanozoon brocchi,* and *Yungia aurantiaca*, and the marine triclads *Procerodes lobata* and *Bdelloura candida*, all of which were taken from the Gulf of Naples, Italy. These extracts, when injected intracardially into guinea pigs, resulted in death of the animal in 1 to 40 minutes. The same extracts injected intraperitoneally into guinea pigs caused respiratory distress, but no immediate deaths. Although there were no immediate deaths in the test animals, several of them died within 13 to 24 days after the injections, but the cause of death could not be attributed to the toxin with any degree of certainty. Intraperitoneal injections of crude extracts from *B. candida* resulted in the death of the animals within 2 to 3 hours. Heating of the extract to 100° C for 1 minute destroyed its toxicity.

PHARMACOLOGY

The only data on the general pharmacological properties of turbellarian poison are by Arndt (1943), who perfused isolated frog heart with crude extracts from the polyclads *Thysanozoon brocchi, Stylochus neapolitanus, Leptoplana tremellaris* and *Yungia aurantiaca*, and the triclad *Procerodes lobata*, which were added to the isotonic frog Ringer's solution. The poison produced cardiac arrest, usually in systole, within 1 to 25 minutes. The effect could be reversed by perfusion with fresh Ringer's solution. Small amounts of the extracts caused an increase in the amplitude of the contraction, and this was followed by variable results: a return to normal, cardiac arrest in systole or diastole, which could be reversed by perfusion with fresh Ringer's solution. Atropine solutions failed to reverse the effects.

Arndt (1943) stated that extracts from all the marine turbellarians that he investigated showed toxic properties, but the mechanism of action of the poison seemed to differ from one species to the next.

CHEMISTRY

(Unknown)

Phylum RHYNCHOCOELA: Ribbon Worms

The phylum Rhynchocoela, sometimes referred to as Nemertinea, or the nemertines, nemertean or ribbon worms, is a small group of slender worms with soft, cylindrical-to-flattened, unsegmented bodies which are capable of great elongation and contraction. The body surface of these worms is covered with a glandular ciliated epithelium. A characteristic of this group is the possession of a highly eversible proboscis which is unarmed in the class Anopla, but may be armed in the members of the class Enopla. The mouth is located anteriorly on the ventral side; or the digestive tract may open through the proboscis pore, and thus a separate mouth is lacking. The tubular proboscis lies within a cavity called the rhynchocoel. Nemerteans differ chiefly from the turbellarians in that they possess a circulatory system, a greater differentiation of the digestive tract, and a definite anus. Some of the excretory products are eliminated by the protonephridea. There is no true segmentation. The nervous system consists of a well-developed brain and a pair of main lateral nerves and their accessory branches. The sexes are separate, but asexual reproduction may take place by fragmentation. Many of the nemerteans are beautifully colored red, brown, yellow, green, or white, or in combinations of these colors. Some are solid colored, whereas others are striped or cross-banded. Most nemerteans are free-living marine forms living in shallow water, coiled under stones, among algae, or in burrows. A few species are bathypelagic or benthic, while others are either freshwater or terrestrial. Several species are commensals, living in the mantle cavities of pelecypods, or among the gills or egg masses in crabs. It is estimated that there are about 500 nemertean species.

Extracts prepared from the tissues of certain species of nemerteans have been found to contain a potent toxic substance.

The biology and systematics of the nemerteans pertinent to this chapter have been discussed by McIntosh (1873), Coe (1943), Hyman (1951), Barnes (1963), and Riedl (1963).

REPRESENTATIVE LIST OF NEMERTEANS REPORTED AS TOXIC

Phylum RHYNCHOCOELA

Class ENOPLA

Order HOPLONEMERTINI

Family AMPHIPORIDAE

110 *Amphiporus lactifloreus* (Johnston) (Pl. 3, fig. a). Nemertean or ribbon worm (USA).

DISTRIBUTION: Arctic to the Mediterranean Sea.

SOURCES: Bacq (1937), Hyman (1951).

Family DREPANOPHORIDAE

Drepanophorus crassus (Quatrefages) (Pl. 3, fig. b). Nemertean or ribbon worm (USA). *110*

DISTRIBUTION: Cosmopolitan.

SOURCE: Bacq (1937).

Family EMPLECTONEMATIDAE

Paranemertes peregrina Coe.

DISTRIBUTION: San Pedro, California, to Alaska.

SOURCES: Kem, Abbott, and Coates (1971), Kem (1971).

Family LINEIDAE

Lineus lacteus Montagu. Nemertean or ribbon worm (USA).

DISTRIBUTION: South coast of England.

SOURCE: Bacq (1937).

BIOLOGY

Ribbon worms are found in a diversity of habitats, but the majority of them live under stones that lie on a muddy or sandy bottom within the tidal zone. Consequently, nemerteans are seldom seen by the casual onlooker. Some of the more common species may be found by the hundreds coiled in a tangled mass, hidden under a single stone; whereas others may be observed in great numbers hidden around the roots of coralline algae. A few of the larger species of nemerteans tend to be solitary in their habits and are sometimes located with difficulty. Algae serve as a favorite site of concealment for some of the smaller kinds of nemerteans. Another good collection area foı nemerteans is in the tangled mass of growth at the base of mussels along open coastal areas. Still other locations are to be found under rocks or in mud burrows.

Ribbon worms vary in length from a few millimeters to more than 30 m. However, some of the longer species have a diameter that is about the equivalent of a large thread.

Despite their apparent sedentary habits, nemerteans are quite active, and at times move with great rapidity. They are able to crawl over smooth surfaces with exquisite, and seemingly effortless, mobility. They are also able to float with ease on the surface of the water. When irritated, ribbon worms can propel themselves rapidly through the water with a serpentlike wiggle of the tail.

If given reasonable care, nemerteans are easily maintained in aquaria for many months and even years. They appear to have the capacity to live without food for extended periods of time. Nemerteans are said to be carnivorous and predacious. They have been observed feeding on mollusks and even fishes much larger than themselves. They are able to distend their mouths and digestive tracts to relatively enormous dimensions and thus capture their prey.

Nemerteans are also cannibalistic. Their normal diet consists of foraminifera and a variety of other microscopic organisms. One species, *Lineus gesserensis*, when sufficiently irritated, has the ability to turn itself inside out. Thus its viscera may be observed in this exposed state without requiring dissection.

MORPHOLOGY OF THE VENOM APPARATUS

It has now been established that a true venom apparatus is present in nemertean worms (Kem, 1971). A well-developed proboscis in the armed members of the Enopla (those species belonging to the order Hoplonemertini), demonstrates that a venom apparatus does exist. Ribbon worms have a long threadlike proboscis which can be extended explosively if they are roughly handled or otherwise stimulated. In most species of nemerteans, the proboscis is not connected with the digestive tract. In certain species of hoplonemerteans, the proboscis may be as long as the body, and it is armed at the tip with a tiny stylet which serves as a weapon of defense and offense (McIntosh, 1873).

At the extreme anterior end of the worm is a pore through which the proboscis can be extended or withdrawn (Fig. 2). The proboscis pore leads into a short canal known as the rhynchodaeum. The lumen of the rhynchodaeum is continuous with that of the proboscis proper. The structure may be compared to the result obtained from inverting a finger in a glove. The proboscis appears as a long tube, usually coiled, and lies within a fluid-filled cavity called the rhynchocoel. The armed proboscis is comprised of three portions: (1) an anterior thick-walled tube, (2) a short, middle bulbous portion which bears the stylet, and (3) a posterior blind tube. Attached to the posterior tip of the blind tube is a retractor muscle which has its origin in the wall of the posterior end of the rhynchocoel. The bulb is a muscular structure which serves as a diaphragm to separate the anterior tube from the posterior one. This diaphragm, bearing the stylet of the proboscis, is pierced by a slender canal—the only communication between the anterior and posterior tubes. At either side of the stylet are two or more lateral pouches or pockets which bear accessory stylets in various stages of development. (*See* Plate 4.)

111

The bulbous diaphragm contains many glandular cells whose secretion supplies the stylet base. It is suspected that the glandular cells produce a toxic substance (McIntosh, 1873).[1] Bacq (1937) states that amphiporine, obtained from nemertean worms, is not a venom and is not localized in the proboscis, but is scattered throughout the body of the worm. Nemerteans having an armed proboscis are reputed to be more toxic than the unarmed species.

[1] McIntosh (1873) cites several early workers who believed that the nemertean stylet apparatus served as a venom organ.

When the armed proboscis is extended, it everts as far as the stylet apparatus, which then occupies the tip of the everted proboscis and serves to pierce and hold the prey. Withdrawal of the proboscis is effected by the retractor muscle attached to the wall of the rhynchocoel or by some comparable muscular apparatus.

Excellent anatomical descriptions of nemertean worms have been provided by McIntosh (1873), Hyman (1951), and Barnes (1963).

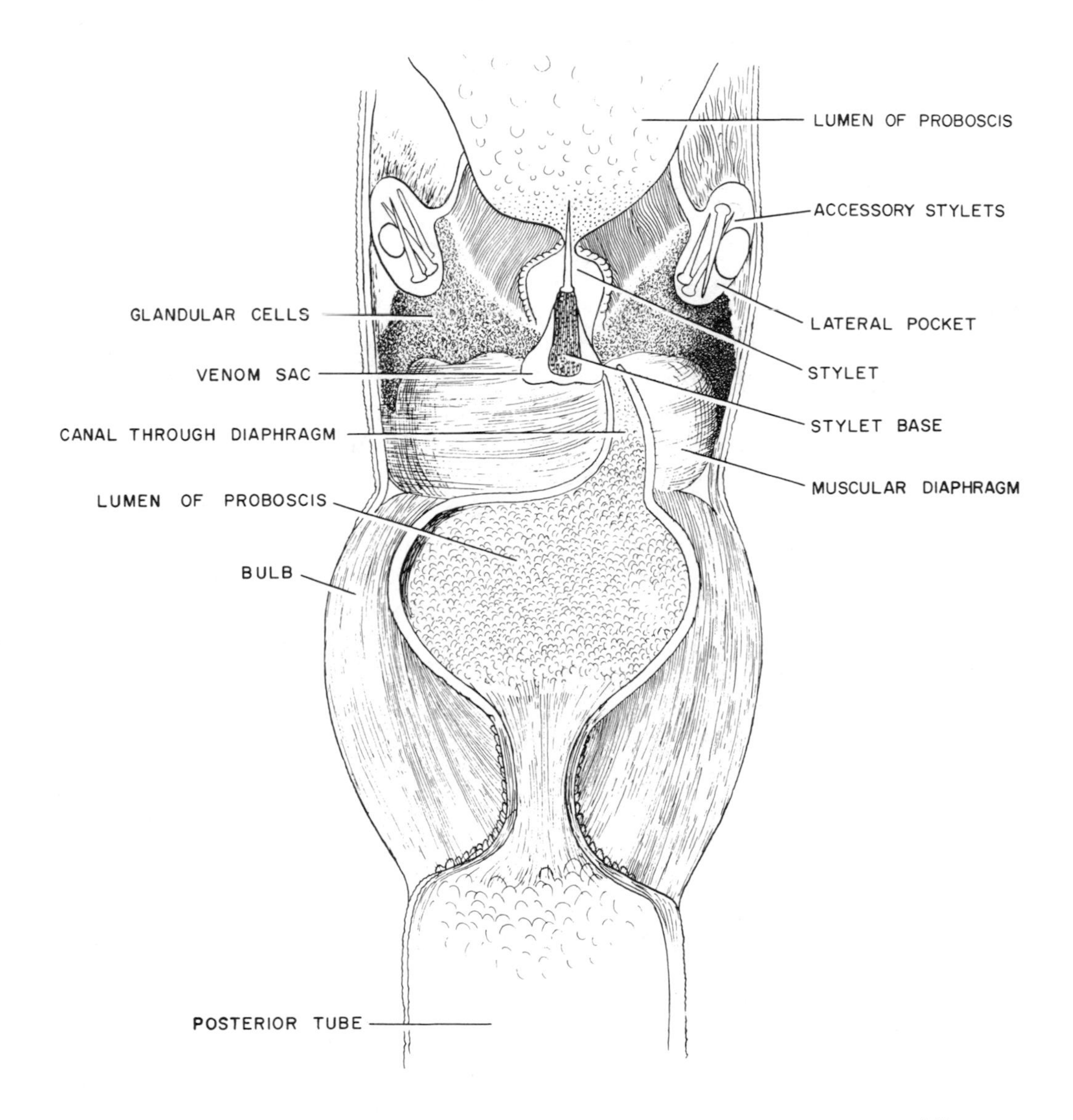

FIGURE 2.—Morphology of the stylet apparatus of nemertean worms. Semidiagrammatic. (R. Kreuzinger, after McIntosh)

MECHANISM OF INTOXICATION

There is no record of human intoxications from nemertean worms. The stylet of the nemertean proboscis is capable of penetrating the human skin, but the resulting effects are unknown. It can only be assumed that ingestion of nemertean worms containing amphiporine might produce toxic effects.

MEDICAL ASPECTS

Clinical Characteristics

(Unknown)

Treatment

Symptomatic.

TOXICOLOGY

Bacq (1937) prepared crude alcoholic extracts from the tissues of nemertean worms *Amphiporus lactifloreus, Drepanophorus crassus*, and other species, and injected them into frogs and crabs. When extracts were injected in the lymph sacs of frogs they caused increased respiration, dilated pupils, and extension of the limbs. The frogs recovered in a period of about 30 minutes. Injections into crabs resulted first in excitation, followed by prostration and death in some cases. Crabs that recovered did so in about 2 hours.

The poison amphiporine (so termed by Bacq in 1937) was not localized in any particular part of the worm. It appeared to be distributed throughout the body of the worm. Amphiporine did not appear to diffuse outside the body of the worm into the surrounding water. Bacq (1937) also isolated a second substance which he termed nemertine, from the nemerteans *Lineus lacteus* and *L. longissimus*. Although nemertine was found to serve as a nerve stimulant, it appeared to be less toxic than amphiporine.

Kem (1971) found that the nemertine toxin, anabaseine, which was isolated from *Paranemertes peregrina* Coe from San Juan Island, Washington, was widely distributed throughout the body of the worm, including the proboscis, viscera, and body integument. The anterior proboscis contains about seventy times the anabaseine necessary to paralyze an annelid its own size. Anabaseine is the only paralyzing constituent of the proboscis venom. This same poison was found in several other groups of nermertines that were studied. Two other paralytic substances were observed in other nemertine species, but the chemical nature of the poisons was not identified.

PHARMACOLOGY

Bacq (1937) tested crude extracts of amphiporine on the rectus abdominalis muscle of frogs and found they produced a sustained contraction on striated muscle. Perfusion of isolated frog heart *Rana esculenta* with amphiporine, using the technique of Straug, showed a release of the inhibitory inotropic action in cardiac muscle. Amphiporine was found to exert a paralyzing effect on sympathetic ganglia. Bacq concluded that amphiporine produced a nicotinic effect similar to that of acetylcholine. Aside from the statement that nemertine acted as a nerve stimulant on crustaceans, no other pharmacological properties have been reported for this substance.

King (1939) reported that a 1:500,000 dilution of crude amphiporine corresponded in its pharmacological activity to a 1:400,000 dilution of nicotine, but it differed in its chemical properties.

CHEMISTRY

Bacq (1937) observed a toxic principle in the marine nemertean worms *Amphiporus lactifloreus* and *Drepanophorus*, to which the name amphiporine was given. The toxin has an acetylcholine-like action on the rectus abdominalis muscle of the frog, but it is evidently not related structurally to acetylcholine or to the quaternary bases generally. Acetylcholine esterase has no action on the toxin. In comparing the results of various spot tests with the results of similar tests on nicotine, Bacq concluded that the toxin is an alkaloid related to nicotine. The toxic principle was described as stable to boiling in alkali or acid for 15 minutes, readily dialyzable, soluble in ethyl alcohol, methyl alcohol, chloroform, slightly soluble in ether, but insoluble in acetone.

A second substance, nemertine, less toxic and also distinct from amphiporine, was observed in extracts of *Lineus lacteus* and *L. longissimus*. This toxin was not related to nicotine and could be distinguished from amphiporine because it was less readily dialyzable, though the solubilities and stability were the same.

Kem, Abbott, and Coates (1971) isolated the toxic fraction from the proboscis venom of the nemertine *Paranemertes peregrina* Coe from San Juan Island, Washington. The toxin was identified as anabaseine, 2-(3-pyridyl)-3,4,5,6-tetrahydropyridine.

Phylum ANNELIDA: Segmented Worms

The phylum Annelida, or segmented worms, are organisms having an elongate, usually segmented body with paired setae. The outer covering of the body consists of a thin, nonchitinous cuticle. The digestive system is tubular, the coelom is large, and the vascular system is closed. Some species have well-developed chitinous jaws. Cosmopolitan in distribution, annelids are found in marine, freshwater, and terrestrial environments. Marine annelids also have a wide distribution bathometrically, and are for the most part benthic, either creeping or burrowing in mud, pilings, corals, or under rocks. Some are sedentary in their habits, building calcareous or fibrous tubes, whereas a few are pelagic. It is estimated that there are more than 6,200 species of annelids.

The phylum Annelida is comprised of three classes:

POLYCHAETA: Marine sandworms, tube worms, etc.

OLIGOCHAETA: Freshwater annelids and terrestrial earthworms.

HIRUDINEA: Leeches.

Very little is known regarding the toxicity of annelids. Several species of marine annelids have been reported as noxious to man because of their pungent, bristlelike setae; and at least one species *Glycera dibranchiata* is believed capable of inflicting venomous wounds with its jaws. A toxic substance has also been isolated from the tissues of certain polychaete worms.

Most of the annelids discussed in this section can be identified with the use of Pratt (1948), *A Manual of the Common Invertebrate Animals.* Useful reviews on the biology of annelids have been prepared by Fauvel *et al.* (1959), and Barnes (1963). An excellent comprehensive bibliography on the literature and a catalog of the polychaetous annelids of the world have been prepared by Hartman (1951, 1959).

REPRESENTATIVE LIST OF ANNELIDS REPORTED AS TOXIC

Phylum ANNELIDA

Class POLYCHAETA: Marine Annelids

Order ERRANTIA

Family AMPHINOMIDAE

112 *Chloeia flava* (Pallas) (Pl. 5, fig. a). Sea mouse, bristle worm (Australia, USA).
DISTRIBUTION: Indo-Pacific, Japan, China Sea, and Indian Ocean.

SOURCE: Tweedie (1941).

112 *Chloeia viridis* Schmarda (Pl. 5, fig. b). Sea mouse, bristle worm (USA).
DISTRIBUTION: West Indies, Gulf of California, Mexico south to Panama, Galapagos Islands, tropical America, both sides.

SOURCES: Steinbeck and Ricketts (1941), Phillips and Brady (1953), Halstead (1956).
OTHER NAME: *Chloeia euglochis.*

113 *Eurythoë complanata* (Pallas) (Pl. 6). Bristle worm (USA).
DISTRIBUTION: Circumtropical.

SOURCES: Pope (1947, 1963), Yaldwyn (1966).

114 *Hermodice carunculata* (Pallas) (Pl. 7). Bristle worm (USA).
DISTRIBUTION: Florida and the West Indies.

SOURCES: Phillips and Brady (1953), Penner (1970).

Family GLYCERIDAE

115 *Glycera dibranchiata* Ehlers (Pl. 8). Blood worm (USA, Canada).
DISTRIBUTION: North Carolina to northeast Canada.

SOURCE: Klawe and Dickie (1957).

Family LUMBRINERIDAE

Lumbriconereis heteropoda Marenzeller (Pl. 9). Polychaete worm (USA). *115*
DISTRIBUTION: Japan.

SOURCES: Nitta (1934, 1941), Hashimoto and Okaichi (1960).

BIOLOGY

The polychaetes are divided into two major groups: the Errantia, which includes most of the free-moving kinds, and the Sedentaria, or tube-dwelling and burrow-inhabiting species. The polychaetes that have been incriminated as toxic are largely errant forms.

Polychaetes have cylindrical bodies and are metameric, having numerous somites—each bearing a fleshy paddlelike appendage, or parapodia, that bears many setae. The head region has tentacles. There is no clitellum. Sexes are usually separate. There are no permanent gonads and fertilization is commonly external. Polychaetes have a trochophore larval stage, and there is asexual budding in some species.

Most polychaetes are free living; a few are extoparasitic. They have a bathymetric range from the tide line to depths of more than 5,000 meters. A few species are pelagic. Some of the burrowing worms feed on bottom detritus, whereas the tube dwellers subsist on plankton. Generally, polychaetes spend their existence crawling under rocks, burrowing in the sand or mud, in and around the base of algal growths; or they construct tubes, which they leave at periodic intervals in search of food. The majority of polychaetes range in size from 5 to 10 cm. However, some of the syllids are only 2 mm in length, whereas the giant Australian species *Onuphis teres* and *Eunice aphroditois* may attain a meter or more in length.

The most renowned members of the polychaete group are the palolo worms, members of the genus *Palola* of the tropical Pacific islands—Samoa, Fiji, etc. There is also a European species of *Palola* whose reproductive swarmings are similar to the Pacific variety. Palolo worms live in holes and crevices among rocks and coral growth on the ocean bottom. Each year on a certain day the worms come to the surface of the ocean in vast swarms in order to reproduce. In the Pacific area this takes place at dawn for 2 days in each of the months of October and November, the day prior to the day on which the moon is in its last quarter. The worms are present in greatest quantities on the second day. The posterior portion of the worm becomes modified to carry the gonads. On the morning of the swarming, the worm backs out of the burrow, and when the modified portion is fully protruded, it breaks off and wiggles to the surface. The anterior portion of the worm returns to the depths of its burrow. The worms are several centimeters in length. The males are light brownish in color while the females tend to be greenish. The natives relish palolo worms as food which they eat in a variety of ways—raw, boiled, fried in batter, scrambled with eggs, and in soup. In certain parts of the world, the natives insist that there is a definite relationship between the toxicity of some species of marine fishes and the swarming of palolo worms (Russell and Yonge, 1944; Burrows, 1945; McNeill, 1954; Dakin, Bennett, and Pope, 1960).

The giant Australian polychaetes *E. aphroditois* and *O. teres* have powerful chitinous jaws with which they are able to inflict a nasty bite. The blood worm *Glycera dibranchiata* has venom glands associated with its jaws and can inflict a painful bite.

Many polychaetes are very beautiful, colored red, pink, or green, or with a combination of colors and iridescence.

MORPHOLOGY OF THE VENOM APPARATUS

Setae: The members of the polychaete genera *Chloeia, Eurythoë, Hermodice*, and others, possess elongate pungent chitinous bristles or setae which project from the parapodia. The parapodia are a pair of lateral appendages extending from each of the body segments. The structure appears as a more or less laterally compressed fleshy projection of the body wall. Each parapodium is biramous; it consists of a dorsal portion, the notopodium, and a ventral part, the neuropodium. Each division of the parapodium is supported internally by one or more chitinous rods, termed acicula, to which are attached some of the parapodial muscles. Each of the distal ends of the two parapodial divisions are invaginated to form a setal sac or pocket in which the projecting setae are situated. Each seta is secreted by a single cell at the base of the setal sac. Generally the setae of polychaetes project some distance beyond the end of the parapodium. However, *Eurythoë* and *Hermodice* have the ability to retract or extend their setae to a remarkable degree. When the living worm is at rest, the setae appear to be quite short and barely in evidence; but when irritated, the setae are rapidly extended and the worm appears to be a mass of bristles.

The severity of symptoms reported in some of the clinical accounts lends credence to the belief that both *Eurythoë* and *Hermodice* possess venomous setae. The setae of both *E. complanata* and *H. carunculata* appear to be hollow, and at times seem to be filled with fluid. A seta of *E. complanata* has a series of retrorse spinules along the shaft, whereas the seta of *H. curunculata* is without spinules and has a needlelike appearance. The setae of *Chloeia* are said to be nonvenomous (Pope, 1963), but are listed as a "stinging" worm by others (Phillips and Brady, 1953; Steinbeck and Ricketts, 1941). Examination of histological sections of the parapodia of both species failed to reveal any glandular elements. However, the material examined was poorly preserved and it is quite possible that glandular structures might have been present and were not recognized. Final decision on this matter awaits further histological study.

Jaws: Glycera possesses a long tubular proboscis which can be extended to about one-fifth the length of the body. When the proboscis is retracted it occupies approximately the first 20 body segments. The pharynx is located behind the proboscis and bears four jaws or fangs, which are arranged equidistant around the pharyngeal wall. Attached to and immediately following the pharynx is an S-shaped esophagus. The proboscis and associated portions of the digestive tract lie free, unattached, in the coelom. An explosive extension of the pharynx results from the sudden contraction of the longitudinal muscles, which slide the pharynx forward and straighten out the esophagus.

The four fangs appear at the tip of the proboscis when the pharynx is in the extended position (Barnes, 1963).

Each jaw appears as a curved fang, to which is attached the venom gland. The duct and body of the gland lie within the pharyngeal wall and can be observed only by dissection. The gland empties its contents through the long slender duct which has its opening at the base of the fang.

MECHANISM OF INTOXICATION

Injurious effects from marine annelids are generally the result of careless handling. Bristle worms *Chloeia, Eurythoë*, and *Hermodice* inflict their injuries by means of their pungent, parapodial, bristlelike setae. *Eunice aphroditois, Onuphis teres, Glycera dibranchiata, G. ovigera*, and related species are capable of inflicting painful bites by their chitinous fangs or jaws, which in the case of the glycerids are associated with venom glands. The fangs of *Glycera* are situated at the extreme tip of the everted proboscis, and can be everted for striking or retracted with great rapidity. Nereistoxin, a neurotoxic substance present in the Japanese annelid *Lumbriconereis heteropoda* Marenz, is known to be toxic to warm-blooded animals and certain insects; but there are no records of this worm poisoning man.

MEDICAL ASPECTS

Clinical Characteristics

Bristle worm (*Chloeia, Eurythoë, Hermodice*) stings may result in an intense inflammatory reaction of the skin, consisting of redness, swelling, burning sensation, numbness, and itching. According to Mullin (1923), *Hermodice carunculata* is able to inflict a "paralyzing effect" with its setae. Contacts with the setae of bristle worms have been likened to handling the spines of prickly pear cactus or nettle stings. Severe complications may result which cause secondary infections, gangrene, and loss of the affected part. The clinical effects of bristle worm stings have also been discussed by Paradice (1924), Pope (1947, 1963), LeMare (1952), Phillips and Brady (1953), Halstead (1956, 1959), and Penner (1970).

Eunice aphroditois, Onuphis teres, Glycera dibranchiata, and *G. ovigera* are able to inflict painful bites. Whether *Eunice* and *Onuphis* have venom glands associated with their jaws is not known, but the glycerids are believed to secrete a venom at the time of biting. The bite has been compared to a bee sting, and it results in swelling, redness, and pain. Two to four minute oval lesions are present at the site where the jaws pierced the skin. There may be blanching of the skin immediately surrounding the lesions due to the effects of the venom. With the exception of severe itching, recovery is generally uneventful within a period of several days.

Treatment

The treatment of worm bites and bristle worm stings is largely symptomatic. There are no specific antidotes. Secondary infections may occur which require the use of antibiotic therapy. The setae of bristle worms can best be removed from the skin with adhesive tape, and then ammonia or alcohol should be applied to the area to alleviate the discomfort.

Prevention

Caution should be exercised in handling bristle worms, *Eunice, Onuphis*, and the glycerid worms. What appears to be a harmless polychaete may, if aggravated, suddenly become a noxious bristle worm, or one of the biting worms, and may suddenly extend its setae or biting jaws. Heavy rubber gloves are recommended in the handling of noxious polychaete worms.

TOXICOLOGY

Although Japanese fishermen have observed for many years that flies, ants, and other carnivorous insects die upon contact with the bodies of dead marine polychaetes, it remained for Nitta (1934) to demonstrate scientifically that at least one species of a Japanese marine annelid, *Lubriconereis heteropoda*, produces a toxic substance. This poison Nitta termed nereistoxin. Aqueous extracts of nereistoxin were tested on a variety of laboratory animals, including flies, fishes, mice, monkeys, etc. The poison affected primarily the nervous system and the heart. The lethal dose was found to be 0.38 mg/10 g for mice, and 1.8 mg/kg for rabbits when injected subcutaneously.

Hirayama, Matsue, and Komaki (1960) found that fishes were more susceptible than warm-blooded mammals to nereistoxin, and that the toxicity was enhanced when the medium was adjusted to the alkaline side. They also established a biological assay using killifish *Oryzias latipes* as the test animal. Similar studies were conducted by Okaichi and Hashimoto (1962*a, b*). Their procedure was as follows: 10 to 20 g of *L. heteropoda* were homogenized with a half weight of sand in a mortar. Three g of the homogenate were extracted with 25 ml of 0.1 *N* HCL in a boiling water bath for 15 minutes. The material was centrifuged, the residue washed with a small amount of water, and the washing added to the supernatant. The solution was then diluted to about 50 ml and adjusted to pH 7.6 with NaOH solution. Any amorphous precipitates that developed were eliminated by centrifugation. Six ml of M/15 Sörensen's phosphate buffer (pH 7.6) were added to the supernatant which was finally made up to 100 ml, and the material was then bioassayed.

The MLD for killifish *O. latipes* was about 0.3 ppm. The LD_{50} of nereistoxin hydrogen oxalate by intravenous injection in mice was 33.6 mg/kg. Neither a growth-promoting factor on *Streptococcus faecalis* colonies, nor an antagonizing effect to α-lipoic acid was observed. Tests were conducted on the housefly (*Musca domestica domestica*), American cockroach (*Periplaneta americana*), Azuki bean weevil (*Callosobruchus chinnensis*), larvae of *Hyphantria cunea*, and almond moth (*Ephestia catella*) to determine the insecticidal properties of nereistoxin. It was found that nereistoxin is less toxic than most of the commercial insecticides, but it possesses a rapid-acting anesthetic effect on insects.

PHARMACOLOGY

Little is known concerning the general pharmacological properties of nereistoxin. Nitta (1941) appears to be the only one who has studied to any extent the pharmacology of this annelid toxin. Nereistoxin is said to exert its effect principally on the nervous system. Rabbits and mice injected with nereistoxin developed myosis, increased motility of intestinal smooth muscle, lacrimation, and hypersalivation. Small doses of the toxin increased the cardiac rate, but large doses decreased it. However, the depressive effects were somewhat diminished by atropine. Nereistoxin appears to have a marked ganglionic blocking agent action on the central nervous system of insects (Hashimoto, Sakai, and Konishi, 1972).

CHEMISTRY

The only worm poison to have been analyzed chemically is nereistoxin. Taking his cue from the well-known observation that carnivorous insects die from feeding on the dead bodies of the marine annelid *Lumbriconereis heteropoda,* Nitta (1934) isolated from this worm the neurotoxic principle as the hydrogen oxalate (m.p. 178° C to 180° C). The free base was described as having an unpleasant odor, soluble in the fat solvents but insoluble in water. The empirical formula of nereistoxin was postulated as $C_5H_9NS_2$ by the analyses of the oxalate and picrate (m.p. 170° C). The toxin was found to react to various alkaloidal reagents, and to reduce Fehling's and ammonium silver nitrate solutions. When treated with potassium permanganate, the poison lost its toxicity. See also the summary by Cleland and Southcott (1965). Nereistoxin has now been synthesized and is used as an insecticide (Hashimoto, Sakai, and Konishi, 1972).

Phylum ARTHROPODA: Joint-Footed Animals

The phylum Arthropoda is the largest single group of the animal kingdom, for it contains more than 800,000 species. It carries the dubious honor of being the group about which we know the least in terms of its role in marine biotoxicology. Arthropods are characterized as having the body usually divided into a head, thorax, and abdomen of like or unlike somites, variously fused; and each segment bears a pair of jointed appendages. The exoskeleton is chitinous and is molted at intervals. The digestive tract is complete, divided into fore-, mid-, and hind-gut. The coelum is reduced. The body spaces serve as a hemocoel, and the heart is dorsally located. The nervous system is of a so-called modified ladder type with a ventrally situated cord. Arthropods inhabit a multitudinous variety of ecological niches.

The phylum Arthropoda is divided as follows:

- Subphylum TRILOBITA: Trilobites (extinct).
- Subphylum CHELICERATA: Arthropods having a body divided into a cephalothorax and an abdomen, with the first pre-oral appendages modified into chelicerae or claws. Antennae are absent.
  - Class MEROSTOMATA: Horseshoe crabs.
  - Class PYCNOGONIDA: Sea spiders.
  - Class ARACHNIDA: Scorpions, spiders, ticks, mites.
- Subphylum MANDIBULATA: Arthropods having a body divided into two parts (head and trunk) or three parts (head, thorax, and abdomen) and possessing antennae, mandibles, and maxillae as head appendages.

Class CRUSTACEA: Crustaceans.
Class INSECTA: Insects.
Class CHILOPODA: Centipedes.
Class DIPLOPODA: Millepedes.
Class SYMPHYLA: Symphylans or garden centipedes.
Class PAUROPODA: Pauropodans.
Etc.*

Although the literature has numerous reports on bacterial food intoxications and allergenic reactions from eating marine crustaceans, little is known about the biotoxic nature of marine arthropods. The only two groups of marine arthropods that have been incriminated as toxic to man are the class Merostomata (horseshoe crabs), and the class Crustacea (crustaceans).

The systematics and biology of the organisms pertinent to this chapter have been discussed by Smedley (1929, 1931), Light *et al.* (1954), Shipley (1958), Waterman (1958), and Barnes (1963).

REPRESENTATIVE LIST OF MARINE ARTHROPODS REPORTED AS TOXIC

Phylum ARTHROPODA

Subphylum CHELICERATA

Class MEROSTOMATA: Horseshoe Crabs

Family XIPHOSURIDAE

116 *Carcinoscorpius rotundicauda* (Latreille) (Pl. 10, fig. a). Horseshoe crab, kingcrab (USA), keroncho, belangkas padi (Malaysia), mangda fai, mada tuey, hera (Thailand).

DISTRIBUTION: Southern Philippines, Indonesia, Malaysia, Gulf of Siam, Bay of Bengal.

SOURCES: Smith (1933), Waterman (1953*b*)), Banner and Stephens (1966), Trishnananda *et al.* (1966).

116 *Tachypleus gigas* (Müller) (Pl. 10, fig. b). Horseshoe crab, Indo-Papuan kingcrab (USA), belangkas (Malaysia), mangda tale (Thailand).

DISTRIBUTION: East coast of Bay of Bengal, Indonesia to northern Vietnam, Torres Straits.

SOURCES: Tweedie (1941), Cleland and Southcott (1965).

Class CRUSTACEA: Crustaceans

Subclass CIRRIPEDIA: Barnacles

* The classification as presented has been purposely reduced for the purposes of this monograph.

Family SACCULINIDAE

Sacculina carcini Thompson (Pl. 12). Parasitic barnacle (USA, England). *118*
DISTRIBUTION: European seas.

SOURCE: Cantacuzène (1913).

Subclass MALACOSTRACA: Lobsters, Crayfish, Crabs

Family COENOBITIDAE

Birgus latro (Linnaeus) (Pl. 13). Coconut crab (USA), kaveu, aveu, vahi ha'ari (Tuamotu Islands, French Polynesia), matsukan (Japan). *118*
DISTRIBUTION: Indo-Pacific.

SOURCES: Hashimoto *et al.* (1967), Bagnis (1970).

Family XANTHIDAE

Atergatis floridus (Linnaeus) (Pl. 14, fig. c). *119*
DISTRIBUTION: Indo-Pacific.

SOURCES: Inoue *et al.* (1968), Konosu and Hashimoto (1968), Konosu *et al.* (1969), Hashimoto *et al.* (1969).

Carpilius convexus (Forskål). (Pl. 11, fig. a). *117*
DISTRIBUTION: Indo-Pacific.

SOURCES: Cooper (1964), Guinot (1967), Erhardt and Niaussat (1970).

Carpilius maculatus (Linnaeus) (Pl. 11, fig. b). *117*
DISTRIBUTION: Indo-Pacific.

SOURCE: Halstead and Cox (1973).

Demania alcalai Garth (Pl. 11, fig. c). *117*
DISTRIBUTION: Indo-Pacific.

SOURCE: Garth (1975).

Demania toxica Garth (Pl. 11, fig. d). *117*
DISTRIBUTION: Philippine Islands.

SOURCES: Alcalá and Halstead (1970), Garth (1971).

Eriphia sebana (Shaw and Nodder) (Pl. 14, fig. d). *119*
DISTRIBUTION: Indo-Pacific.

SOURCE: Mote, Halstead, and Hashimoto (1970).
OTHER NAME: *Eriphia laevimana.*

Lophozozymus pictor (Fabricius) (Pl. 11, fig. e). *117*
DISTRIBUTION: Indo-Pacific.

SOURCE: Teh and Gardiner (1970).

119 *Platypodia granulosa* (Rüppell) (Pl. 14, fig. b).
DISTRIBUTION: Indo-Pacific.

SOURCES: Hashimoto *et al.* (1967), Konosu *et al.* (1969), Hashimoto *et al.* (1969).

119 *Zozymus aeneus* (Linnaeus) (Pl. 14, fig. a).
DISTRIBUTION: Indo-Pacific.

SOURCES: Banner and Randall (1952), Konosu *et al.* (1968), Hashimoto *et al.* (1969), Konosu, Noguchi, and Hashimoto (1970), Mote, Halstead, and Hashimoto (1970).
OTHER NAME: *Zosimus aeneus*.

BIOLOGY

Merostomata: The horseshoe or king crabs are characterized by an arched cephalo-thorax, a horseshoe-shaped carapace, a wide, unsegmented abdomen possessing three-jointed chelicerae, pedipalpi, and six-jointed legs. They inhabit coastal waters on sandy or muddy bottoms at depths of from 2 to 6 fathoms where they lie buried. Annandale (1909) and Smith (1933) state that *Tachypleus* is essentially a marine species, whereas *Carcinoscorpius* prefers an estuarine environment and can live in almost fresh water. Horseshoe crabs are scavengers and feed on polychaete worms, mollusks, miscellaneous small marine animals, and algae which they obtain by plowing through sand and mud. Their usual method of locomotion is by crawling; however, they are capable of a clumsy, inverted swimming action by beating their gill books together. Reproduction occurs during late spring and early summer. At this time, they migrate toward the shore with a male clinging tightly to the back of the female. The eggs are laid in the shallow water in a series of shallow holes dug in the sand by the female. There are only three genera *Limulus, Carcinoscorpius*, and *Tachypleus*, and five species living today. One of these belongs to the North American western Atlantic coast and the Gulf of Mexico. The remainder are Asiatic, and range from Japan and Korea south through the Indo-Pacific region. Shuster (1950, 1953, 1957, 1960*a, b,* 1962) has written several interesting articles on the biology of horseshoe crabs.

Cirripedia: Most of the barnacles are sessile crustaceans, which, because of their external adult morphology, were formerly classed as mollusks. The free-living species are usually found attached to rocks, shells, coral, pilings, ship bottoms, and a variety of other objects. Some barnacles are parasitic and others are commensalistic on whales, turtles, fishes, and other animals. *Sacculina*, the only barnacle reported to be toxic, is parasitic to crustaceans. The sacculinids have become so highly specialized that they have lost all traces of their arthropod structure, at least during adulthood.

The developmental history of *Sacculina* is one of the most remarkable to be found in the annals of zoology. The adult is hermaphroditic. The young are hatched out in the mantle cavity of the adult as microscopic nauplius larvae, typical of the Cirripedia. Their early life is spent swimming near the surface of the sea, where later they become transformed into cypris larvae. After a period of free existence, the cypris seeks out a crab and with the use of its antennal hooks attaches itself to hair on any part of the crab's body. The antenna penetrates the base of the hair and gradually the cypris becomes fixed. The appendages of the cypris are gradually discarded, and a small mass of undifferentiated cells are passed down the hollow antenna into the body of the crab. The cell mass is then carried about the bloodstream of the crab until it reaches the vicinity of the intestine in the thorax. The cells then attach themselves to the intestine, just below the stomach, and proceed to develop rootlike extensions in all directions. The main bulk of the tumor mass, known as the central tumor, grows down toward the hind-gut. The tumor mass gradually undergoes a process of differentiation until all the adult organs have been developed in miniature. At this point the entire mass is surrounded by a sac, generally situated at the junction between the thorax and the abdomen of the crab. This stage is known as "*Sacculina interna.*" Gradually the *Sacculina* erodes a small depression in the crab's epithelium so that when the crab molts a little hole remains in the chitin the same size as the body of the *Sacculina*. Finally the parasite pushes its way through the hole and develops into the adult phase, "*Sacculina externa.*" The *Sacculina* produces a toxin which inhibits the growth of the crab, prevents further molting, and alters the sex of the crab. The life cycle of *Sacculina* is generally completed within a period of 3 or 4 years, and then it dies. If the crab survives parasitization, which it usually does not, the crab may then revert to its normal activity and form. Male crabs which are infected with this parasite assume the appearance of females. They develop a broad abdomen, smaller claws, and small hind legs on the hind part of the body. About 70 percent of infected males show signs of taking on female characteristics, but the female does not develop male characteristics. Both sexes of crabs tend to become sterile. However, if the male survives the transformation, it may continue to the point of laying eggs (Russell and Yonge, 1944).

Malacostraca: The Malacostraca, which includes about three-quarters of the known species of crustaceans, crabs, lobsters, shrimp, etc., are typically composed of 19 somites with the head fused to one or more thoracic somites. A carapace is commonly present, and the abdomen is equipped with appendages. Only five species of this large order have been incriminated as toxic, and it is to be regretted that little is known concerning the biology of four of them.

The sand crab *Emerita analoga* inhabits shifting, wave-washed sandy beaches of the Pacific coast of North America. When the waves sweep in, the crab may swim about for a short time; but it burrows rapidly into the sand when the water recedes. Sand crabs may sometimes be found buried under the sand some distance from the edge of the water, but usually not beyond the reach of the flood tide. Locomotion is always backward, whether swimming, burrowing, or crawling. Reproduction occurs during late spring or early summer. The female carries the eggs from 4 to 5 months until they are ready to hatch. Since *Emerita* lacks hard mandibles for chewing, it feeds largely on minute plants and animals. The fact that they feed on *Gonyaulax* accounts for their periodic toxicity. *Emerita* is not generally used for food in the United States, although an allied species is an article of diet on the coast of Peru (Johnson and Snook, 1935; MacGinitie and MacGinitie, 1949). Sand crabs attain a length of almost 50 mm.

Xanthid crabs have a carapace that is transversely oval or transversely hexagonal or subquadrate, rarely subcircular and almost always broader than long. The front of the carapace tends to be broad and is never produced in the form of a rostrum. The legs of the crab are of the ambulatory rather than swimming type. The family Xanthidae is widely represented throughout the tropical littoral.

MECHANISM OF INTOXICATION

Asiatic horseshoe crab intoxications reputedly result from eating the unlaid green eggs, flesh, or viscera during the reproductive season of the year. Like puffer fish, Asiatic crabs should be cooked by an expert, or better yet, eliminated from the diet. Despite their periodic toxicity, the large masses of green, unlaid eggs are highly esteemed by Asiatic peoples (Waterman, 1953*a, b*).

There are no recorded instances of intoxication in humans due to the eating of the sand crab *Emerita analoga* or the parasitic barnacle *Sacculina carcini*.

MEDICAL ASPECTS

Clinical Characteristics

The onset of symptoms in Asiatic horseshoe crab, or mimi poisoning, usually occurs within a period of 30 minutes. Initial symptoms consist of dizziness, headache, nausea, slow pulse rate, decreased body temperature, vomiting, abdominal cramps, diarrhea, cardiac palpitation, numbness of lips, paresthesias of lower extremities, and generalized weakness. More severe symptoms occur in rapid succession: aphonia; sensation of heat in mouth, throat, and stomach; inability to lift arms and legs; generalized muscular paralysis; trismus; hypersalivation; drowsiness; and loss of consciousness. The mortality rate is unknown, but is reportedly very high. Death, when it occurs, takes place within a period of 16 hours. The clinical characteristics of Asiatic horseshoe crab poisoning have been discussed by Smith (1933), Soegiri (1936), Waterman (1953*b*), and Banner and Stephens (1966).

Human intoxications from the parasitic barnacle *Sacculina* and the sand crab *Emerita* are unknown.

Reports of human biotoxications from marine crabs have appeared in the literature at infrequent intervals, but until recently there have been no serious investigations on the subject. Hashimoto *et al.* (1967) reported finding two species of xanthid crabs that have been involved in human intoxications in Ryukyu and Amami Islands. A total of 19 persons were poisoned. The victims developed paresthesias, muscular paralyses, aphasia, nausea, vomiting, and collapse, and two persons died within a period of several hours. The other patients recovered after several days. The treatment was apparently symptomatic. The crab responsible for the intoxications was identified as *Zozymus aeneus*. Screening tests revealed that both *Z. aeneus* and a second species *Platypodia granulosa* contained a paralytic poison showing a dose-death time curve similar to tetrodotoxin. The poison was found to be present in the muscles, viscera, and gills. The toxin was readily dialyzable, soluble in water and methanol, but insoluble in most fat solvents. Konosu and Hashimoto (1968) discovered a third species of a highly toxic xanthid crab *Atergatis floridus* which was taken in the Ryukyu Islands. Mote *et al.* (1970) incriminated another species of a toxic zanthid crab *Eriphia* in Palau, Western Caroline Islands. This species was reported to have been responsible for the death of two Japanese fishermen who died within a period of about two hours from eating this crab.

A human fatality due to a newly described species of crab, *Demania toxica*, has been reported by Alcalá and Halstead (1970). Symptoms prior to death included nausea, vomiting, hypersalivation, muscular exhaustion, convulsions, and respiratory distress. Halstead and Cox (1973) reported two deaths from the ingestion of the red-spotted crab *(Carpilius maculatus)*.

The coconut crab *Birgus latro* has been incriminated in a series of poisonings occurring at infrequent intervals at Miyako, Taketomi, and Ishigaki, Ryukyu Islands, over the past several decades (1935, 1947, 1948, 1958-59, 1962, etc.). The coconut crab is generally eaten with impunity in most parts of the tropical Pacific, but it is suspected that on occasion the coconut crab may ingest toxic plants and thereby develop poisonous flesh. One of the suspect plants has been identified as *Hernandia sonora* Linnaeus, a member of the family Hernandiacac. The clinical characteristics of coconut crab poisoning include violent gastrointestinal upset, headache, chills, joint aches, exhaustion, and muscular weakness. Deaths have been reported. There are presently no experimental toxicological data available (Hashimoto *et al.*, 1967; Bagnis, 1970).

The symptoms produced by poisoning from the crabs *Angatea* and *Xanthodes* are unknown. *Angatea* and *Xanthodes* are reputed to produce a high mortality rate because of their powerful toxicity (Serène, 1952). Stcherback (1907), LaFavre (1930), Biswas (1932), Jones (1939), and others have indicted crabs and lobsters as causative agents of human intoxications, but the identity of the crustaceans and the nature of the poisonings are undetermined. Most of these intoxications are believed to be the result of bacterial food poisoning.

Pathology

Soegiri (1936) appears to be the only one to report the pathological findings of a fatal case of Asiatic horseshoe crab poisoning. Autopsy findings consisted of mild visceral congestion, but nothing else of significance.

Treatment

Symptomatic, with no known antidote.

Prevention

Although Asiatic horseshoe crabs are eaten in many parts of southeast Asia, they should be avoided during the season of reproductivity. Special precaution should be taken against eating *Zozymus* and *Carpilius* since they are relatively large crabs and liable to be served on the table when caught. The crabs are common throughout the tropical Indo-Pacific and may be toxic throughout their range.

It is advisable to check with the local public health authorities or responsible fisheries people as to the seasonal edibility of tropical crabs. It is advisable to avoid eating tropical shore crabs unless there is reliable information available concerning their edibility. Most species of tropical shore crabs are reputedly safe to eat, but the few species that are toxic are sufficiently dangerous to warrant the precaution of checking with the local authorities.

TOXICOLOGY

Research on the toxicology of paralytic shellfish poison as it relates to the sand crab *Emerita* has been discussed by Sommer (1932) and Sommer and Meyer (1937).

Cantacuzène (1913) found that the blood of the crab *Carcinus maenas* when parasitized by *Sacculina* contained a substance that acted as an antibody (amboceptor) when brought in contact with an antigen produced by *Sacculina* and by complement (alexin) from guinea pig. His observations were based on a series of complement fixation tests which he conducted using extracts of *Sacculina* tissues, sacculinized crab's serum, guinea pig complement, sheep red blood cells, sheep antiserum, and saline solution. The tests showed only slight hemolysis; but if similar tests were conducted using normal crab's serum (a crab which had not been parasitized by *Sacculina*), there was total hemolysis.

It has been demonstrated that the toxicity of *Zozymus aeneus* varies greatly according to the individual and geographical location (Hashimoto *et al.*, 1969; Konosu *et al.*, 1970). For example *Z. aeneus* is edible in the New Hebrides, yet is responsible for deaths in Fiji. Some specimens have shown a toxicity level above 1,000 M.U./g. The marked regional variation of toxicity suggests that the toxin may be of an exogenous origin. Konosu *et al.* (1968) made a comparison of partially purified crab toxin to saxitoxin which showed them indistinguishable in paper and thin-layer chromatography. It was concluded that crab toxin may be identical or closely related to saxitoxin (Noguchi, Konosu, and Hashimoto, 1969). Inoue *et al.* (1968) found *Atergatis floridus* to be highly toxic. Further studies on the chemical properties of *Z. aeneus* have been conducted by Uwatoko-Setoguchi, Obo, and Hashimura (1969).

Teh and Gardiner (1970) showed that aqueous extract of *Lophozozymus pictor* was lethal when injected into mice. It was a potent depressant of respiration and blood pressure in the rat but had little effect on the response of the gastrocnemius muscle to sciatic nerve stimulation.

No other laboratory data are available regarding the toxicology, pharmacology, and chemistry of other arthropod poisons.

Phylum SIPUNCULIDA: Sipunculoids or Peanut Worms

The phylum Sipunculida is a group of unsegmented, wormlike invertebrates, having a body differentiated into an anterior slender introvert and a posterior plump cylindrical trunk. The terminal mouth is generally encircled by a number of tentacular outgrowths. There is a recurved digestive tract with a highly coiled intestine opening dorsally at the end of the introvert. The nervous system is of the annelid type. One or two nephridia are present. A definite circulatory system is absent. Sipunculoids are dioecious and produce a trochophore larva.

For many years sipunculoids were assigned an indefinite phylogenetic position. They were formerly classed as echinoderms, and later as annelids and included in the order *Gephyrea*, which also encompassed the unrelated priapuloids and echiuroids. Sipunculoids are now recognized to be a distinct group and are treated as a separate phylum (Barnes, 1963).

Sipunculoids, exclusively marine, are cosmopolitan in their distribution. They have a vertical range from the intertidal zone to depths of more than 5,000 m. There are about 250 species.

Only three species of sipunculoids have been reported as toxic members of the genera *Bonellia* and *Golfingia*.

At present, the phylum Sipunculida is arranged only in genera. No attempt has been made to assign the groups to families. The systematics and biology of the sipunculoids pertinent to this chapter have been discussed by Gerould (1913), Fisher (1952), Barnes (1963), and Riedl (1963).

REPRESENTATIVE LIST OF SIPUNCULOIDS REPORTED AS TOXIC

Phylum SIPUNCULIDA[2]

Bonellia viridis (Pl. 15). Sipunculoid, peanut worm (USA). *120*
DISTRIBUTION: Mediterranean Sea.

SOURCE: Baltzer (1924).

BIOLOGY

Sipunculoids are sedentary marine animals, living in burrows or crevices in sand, mud, gravel, coral, shells, at the base of algae, sponges, cavities in rocks, or wherever they can find a suitable protective environment. They crawl slowly about, attaching the tentacular crown to a solid object and then pulling the trunk forward. Much of their movement consists of running the introvert in and out. When in its natural habitat and left undisturbed, the sipunculoid extends the introvert out of the burrow and spreads the tentacles for feeding and respiration. If disturbed, the animal instantly retracts. Some species are said to be capable of exerting feeble swimming movements. Because of their tough outer covering, sipunculoids can withstand removal from the water without ill effects for limited periods of time. Little is known of the feeding habits of sipunculoids. The non-burrowing species are said to have a mucous-ciliary mode of feeding, living on minute particles and organisms. Their food consists largely of ciliates, foraminifera, turbellarians, tiny crustaceans, polychaete larvae, and a variety of other small animals. The burrowing species apparently feed on the substrate and detritus with which they come in contact. Sipunculoids are afflicted with the usual array of parasites and commensals characteristic of marine sedentary animals. Sipunculoids are commonly eaten by anemones, crabs, cephalopods, and fishes. In the Indo-Pacific region, they are also eaten by native peoples. Sipunculoids range in size from less than 2 mm to more than 60 cm in length. Peanut worms are for the most part drab in their coloration, which may be whitish, gray, yellowish, or a dull brown.

---

[2] At present there are no family designations in this phylum (Hyman, 1959).

MEDICAL ASPECTS

There are no known human intoxications.

TOXICOLOGY

Baltzer (1925*a, b*) has made an intensive study on the toxicity of female *Bonellia fulginosa* and *B. viridis*, and the effects of the poison on the larval development of the male. Sexual dimorphism in these two species of *Bonellia* is said to be quite pronounced. The free-swimming trochophore larvae attaches itself to the introvert of the female where it continues to remain in a symbiotic relationship with the female. With the use of methylene blue studies, Baltzer was able to demonstrate that the male received all its nourishment through the female. The attached male never attains a size comparable to that of the female, but remains a tiny degenerate form whose sole purpose is to fertilize the female. In examining the various tissues of the female *Bonellia*, Baltzer found that the skin of the introvert was quite toxic, but the other tissues were not. Toxicity tests were conducted on various animals: protozoa, annelid worms, crustaceans, tadpoles, and adult sipunculoids. Aqueous extracts obtained from ground-up *Bonellia* introvert tissue were added in aquaria in various concentrations, ranging mostly from 1:1,400 to 1:4,000. The test animals were then placed in the aquaria and their reactions observed. It was found that paramecia, flagellates, and other protozoans (freshwater and marine species) were rapidly paralyzed and soon died. Similar results were obtained in tests conducted on daphnia, freshwater nematodes, and tubifex annelids. Concentrations of the toxin in dilutions of 1:3,000 were also found to be lethal to adult male *Bonellia* that had been separated from the female. Tests on tadpoles resulted in paralysis and death. When tadpoles were placed in sublethal concentrations of the sipunculoid toxin over an extended period of time, their growth and development were inhibited.

Baltzer concluded from his studies that the poison of *B. fulginosa* and *B. viridis* is concentrated in the superficial layers of the introvert skin of the female, and that the poison possesses definite growth inhibitory properties. One question that remained unanswered is what is the role of sipunculoid poison in determining the sex of the *Bonellia* larvae? Baltzer felt that there may be a definite biochemical relationship between growth inhibition and sex determination. The effect of sipunculoid poison on tumor systems might prove to be a very productive area of research.

PHARMACOLOGY

(Unknown)

CHEMISTRY

(Unknown)

Phylum ECTOPROCTA: Moss Animals

The members of the phylum Ectoprocta were formerly included in the Bryozoa, or the moss animals; they form microscopic sessile branched, plant-like colonies, or low incrustaceans on rocks, shells, etc., or may resemble gelatinous mosses. The individual organisms are numerous, minute, each housed in a separate zooecium, having ciliated tentacles around the mouth. They have a complete U-shaped digestive tract. They lack nephridia and a circulatory system. The coelom is developed and lined. Sexually they are hermaphroditic. Bryozoans are free-living and found in both marine and freshwater. There are about 4,000 living species. The systematics and biology of the Ectoprocta are discussed by Hyman (1959) and Kluge (1962). A useful bibliography on this subject has been prepared by Osburn (1957).

The toxic effects of bryozoans have been reported in association with an allergic dermatitis in European trawler fishermen known as Dogger Bank Itch, or weed rash. This rash was first described by Bonnevie (1948) and the causative agent was identified as the bryozoan *Alcyonidium hirsutum*. Clinical reports indicate, however, that Danish fishermen from the Port of Esbjerg, Denmark, were suffering in the late 1930s from contact with a "seaweed" known as the "sea chervil," which was subsequently identified as *A. hirsutum*. In 1939, the Danish Workmen's Compensation Act included skin disorders caused by this organism. Guldager (1959) reported 95 cases at the fishermen's hospital in Esbjerg suffering from skin rashes caused by *A. hirsutum*.

Bryozoan infections are also reported to be common along the east coast ports of England, and especially among fishermen based at Lowestoft (Seville, 1957; Newhouse, 1966*a, b*). Bryozoan specimens taken from the Dogger Bank by English fishermen have been identified as *A. gelatinosum*.

LIST OF MOSS ANIMALS REPORTED AS TOXIC

Phylum ECTOPROCTA (BRYOZOA): Moss Animals

Class GYMNOLAEMATA

Family ALCYONIDIIDAE

Alcyonidium gelatinosum (Linnaeus) (Pl. 16). Sea-chervil, curly weed, amber weed (England). *120*

DISTRIBUTION: North Sea.

SOURCES: Bonnevie (1948), Newhouse (1966*b*), Halstead (1970).

Alcyonidium hirsutum (Fleming). Sea-chervil, curly weed, amber weed (England).

DISTRIBUTION: North Sea.

SOURCES: Bonnevie (1948), Newhouse (1966*a*), Halstead (1970).

BIOLOGY

Very little appears to be known about the biology of the two species of *Alcyonidium* involved. These organisms are said to flourish during the summer, but die off in the winter (Newhouse, 1966*a*). The organisms are usually taken in trawl nets which are towed over the sea bed by fishing trawlers. *Alcyonidium* are benthic organisms living attached to rocks or other objects on the sea floor. They are usually taken in relatively shallow waters.

MECHANISM OF INTOXICATION

Bryozoan infections are usually contracted as a result of handling or removing clusters of the organisms from trawl nets. Contacts are frequently made when mending nets in which *Alcyonidium* are entangled.

MEDICAL ASPECTS

Clinical Characteristics

The rash is most often seen in the antecubital fossa and flexor aspects of the forearms, wrists, extensor surfaces of the hands, fingers, and palms. Papules and blisters up to 3-4 mm are usually present in acute cases. The face may also be affected with a papulomacular rash of the forehead and cheeks, and a periorbital edema with scaling and desquamation of the eyelids. Patch tests using homogenates of the organism may produce very severe skin reactions. The dermatitis may be sufficiently intense as to be incapacitating.

The incidence among fishermen working in endemic areas was found to be about 7 percent. Continual exposure to *Alcyonidium* results in a chronic eczematoid dermatitis. With continued exposure the skin becomes highly sensitive to any further contact with the organism.

The clinical characteristics of Alcyonidium infections have been discussed by Bonnevie (1948), Guldager (1959), and Newhouse (1966*a, b*).

Treatment

The most important item is to discontinue exposure to the *Alcyonidium*. Treatment of the skin rash is symptomatic. Application of cremor fluocinolone and chlorpheniramine maleate to the affected areas has been recommended.

Prevention

Protective clothing and the use of gauntlet gloves have been found to be useful. Hypersensitive persons are sometimes required to discontinue fishing operations in endemic areas. Fishermen are urged to wash their hands, forearms, and faces regularly whether the rash is present or not.

TOXICOLOGY

(Unknown)

PHARMACOLOGY

(Unknown)

CHEMISTRY

(Unknown)

LITERATURE CITED

PLATYHELMINTHES

Arndt, W.

1925 Über die gifte der plattwürmer. Verhandl. Deut. Zool. Ges. 30: 135-145, 4 figs.

1943 Polycladen und maricole tricladen als giftträger. Mém. Estud. Museum Zool. Univ. Coimbra (148): 1-15.

Arndt, W., and P. Manteufel

1925 Die turbellarien als träger von giften. Z. Morphol. Oekol. Tiere 3: 344-357.

Hyman, L. H.

1951 The invertebrates: Platyhelminthes and Rhynchocoela; the acoelomate bilateria. Vol. 2. McGraw-Hill Book Co., New York. 550 p., 208 figs.

Lang, A.

1884 Die polycladen (seeplanarien) des Golfes von Neapel und der angrenzenden meeresabschnitte; eine monographie. Zoologische Station Neapel, Wilhelm Engelmann, Leipzig. 688 p., 54 figs., 39 pls.

Prenant, M.

1919 Recherches sur les rhabdites des turbellaries. Arch. Zool. Exp. Gen. 58: 219-250, 12 figs.

Wilhelmi, J.

1909 Fauna und flora des Golfes von Neapel und der angrenzenden meeresabschnitte. Herausgegeben von der zoologischen station zu Neapel. Monographie 32, Tricladen. Verlag von R. Friedländer & Sohn, Berlin. 405 p., 16 pls.

RHYNCHOCOELA

Bacq, Z. M.

1937 L' "amphiporine" et la "némertine" poisons des vers némertiens. Arch. Intern. Physiol. (Liége) 44(2): 190-204, 4 figs.

Barnes, R. D.

1963 Invertebrate zoology. W. B. Saunders Co., Philadelphia. 632 p.

Coe, W. R.

1943 Biology of the nemerteans of the Atlantic coast of North America. Trans. Connecticut Acad. Arts Sci. 35: 129-328, 4 pls.

Hyman, L. H.

1951 The invertebrates: Platyhelminthes and Rhynchocoela; the acoelomate bilateria. Vol. 2. McGraw-Hill Book Co., New York. 550 p., 208 figs.

Kem, W. R.

1971 A study of the occurrence of anabaseine in *Paranemertes* and other nemertines. Toxicon 9: 23-32, 5 figs.

Kem, W. R., Abbott, B. C., and R. M. Coates

1971 Isolation and structure of a hoplonemertine toxin. Toxicon 9: 15-22.

King, H.

1939 Amphiporine, and active base from the marine worm *Amphiporus lactifloreus*. J. Chem. Soc. (2): 1365-1366.

McIntosh, W. C.

1873 Monograph of the British annelids. Vol. I. Part I. The nemerteans. Ray Society, London. 96 p., 10 pls.

Riedl, R.

1963 Fauna und flora der Adria. Paul Parey, Hamburg und Berlin. 640 p., 221 figs., 8 pls.

ANNELIDA

Barnes, R. D.

1963 Invertebrate zoology. W. B. Saunders Co., Philadelphia. 632 p.

Burrows, W.

1945 Periodic spawning of 'palolo' worms in Pacific waters. Nature 155(3924): 47-48.

Cleland J. B., and R. V. Southcott

1965 Injuries to man from marine invertebrates in the Australian region. National Health and Medical Research Council, Spec. Rept. Ser. No. 12. 282 p., 10 pls.

Dakin, W. F., Bennett, I., and E. Pope

1960 Australian seashores. A guide for the beach-lover, the naturalist, the shore fisherman, and the student. Rev. ed. Angus and Robertson, Sydney. 372 p., 23 figs., 99 pls.

Fauvel, P., Avel, M., Harant, H., Grassé, P., and C. Dawydoff

1959 Embranchement des annelides. *In* P. Grassé [ed.], Traité de zoologie. Vol. 5. Part I. Masson et Cie., 686 p.

Halstead, B. W.

1956 Animal phyla known to contain poisonous marine animals, p. 9-27. *In* E. E. Buckley and N. Porges [eds.], Venoms. Am. Assoc. Adv. Sci., Washington, D.C.

1959 Dangerous marine animals. Cornell Maritime Press, Cambridge. 146 p., 88 figs.

HARTMAN, O.
1951 Literature of the polychaetous annelids. Vol. I. Bibliography. Edwards Brothers, Inc., Ann Arbor, Mich. 290 p.
1959 Catalogue of the polychaetous annelids of the world. Univ. Southern Calif. Press, Los Angeles 1: 1-353; 2: 354-628.
HASHIMOTO, Y., and T. OKAICHI
1960 Some chemical properties of nereistoxin. Ann. N.Y. Acad. Sci. 90(3): 667-673, 3 figs.
HASHIMOTO, Y., SAKAI, M., and K. KONISHI
1972 A new insecticide developed from nereistoxin. Food-Drugs from the Sea, Marine Tech. Soc., Proc. 1972, p. 129-138, 3 figs.
HIRAYAMA, K., MATSUE, Y., and Y. KOMAKI
1960 Agricult. (Japan) 8: 95. (NSA)
KLAWE, W. L., and L. M. DICKIE
1957 Biology of the bloodworm, *Glycera dibranchiata* Ehlers, and its relation to the bloodworm fishery of the Maritime Provinces. Fish. Res. Board Can., Bull. 115, 37 p., 18 figs.
LEMARE, D. W.
1952 Poisonous Malayan fish. Med. J. Malaya 7(1): 1-8, 1 pl.
MCNEILL, F.
1954 Palolo: Food worm of the Pacific. Australian Mus. Mag. 11(6): 173-174, 1 fig.
MULLIN, C. A.
1923 Report on some polychaetous annelids; collected by the Barbados-Antigua expedition from the University of Iowa in 1918. Univ. Iowa Studies Nat. Hist. 10(3): 39-45; Pls. 5-6.
NITTA, S.
1934 Über nereistoxin, einen giftigen bestandteil von *Lumbriconereis heteropoda* Marenz (Eunicidae). [In Japanese] J. Pharmacol. Soc. Japan 54: 648-652.
1941 Pharmakologische untersuchung des nereistoxins, das vom verf. im körper des *Lumbriconereis heteropoda* (Isome) isoliert wurde. [In Japanese] Tokyo J. Med. Sci. 55: 285-301, 18 figs., 1 pl.
OKAICHI, T., and Y. HASHIMOTO
1962*a* The structure of nereistoxin. Agr. Biol. Chem. (Tokyo) 26(4): 224-227.
1962*b* Physiological activities of nereistoxin. Bull. Japan. Soc. Sci. Fish. 28(9): 930-935, 1 fig., 1 pl.
PARADICE, W. E.
1924 Injuries and lesions caused by the bites of animals and insects. Med. J. Australia 1924: 650-652, 2 figs.
PENNER, L. R.
1970 Bristleworm stinging in a natural environment. Univ. Conn. Occ. Papers 1(4): 275-280, 3 figs.
PHILLIPS, C., and W. H. BRADY
1953 Sea pests—poisonous or harmful sea life of Florida and the West Indies. Univ. Miami Press, Miami, Fla. 78 p., 38 figs., 7 pls.
POPE, E. C.
1947 Some sea animals that sting and bite. Australian Mus. Mag. 9(5): 164-168, 5 figs.
1963 Some noxious marine invertebrates from Australian seas, p. 91-102, 2 figs. *In* J. W. Evans [chmn.], Proc. First International Convention on Life Saving Techniques. Part III. Scientific Section. Suppl. Bull. Post Grad. Comm. Med., Univ. Sydney.
PRATT, H. S.
1948 A manual of the common invertebrate animals (exclusive of insects). Rev. ed. The Blakiston Co., Philadelphia. 854 p., 974 figs.
RUSSELL, F. S., and C. M. YONGE
1944 The seas. Our knowledge of life in the sea and how it is gained. 2d ed. Frederick Warne & Co., Ltd., London. 379 p., 65 figs., 127 pls.
STEINBECK, J., and E. F. RICKETTS
1941 Sea of Cortez. A leisurely journal of travel and research. The Viking Press, New York. 598 p., 40 pls.
TWEEDIE, M. W.
1941 Poisonous animals of Malaya. Malaya Publishing House, Singapore. 90 p., 29 figs.
YALDWYN, J. C.
1966 Blistering from bristle-worm. Australian Nat. Hist. 15(3): 86, 1 fig.

ARTHROPODA

ALCALÁ, A. C., and B. W. HALSTEAD
1970 Human fatality due to ingestion of the crab *Demania* sp. in the Philippines. Clin. Toxicol. 3(4): 609-611.

(NSA)—Not seen by author.

ANNANDALE, N.
1909 The habits of Indian king crabs. Rec. Indian Mus. 3: 294-295.

BAGNIS, R.
1970 A case of coconut crab poisoning. Clin. Toxicol. 3(4): 585-588, 1 fig.

BANNER, A. H., and J. E. RANDALL
1952 Preliminary report on marine biology study of Onotoa Atoll, Gilbert Islands. Atoll. Res. Bull. (13): 43-62.

BANNER, A. H., and B. J. STEPHENS
1966 A note on the toxicity of the horseshoe crab in the Gulf of Thailand. Nat. Hist. Bull. Siam Soc. 21(3-4): 197-203.

BARNES, R. D.
1963 Invertebrate zoology. W. B. Saunders Co., Philadelphia. 632 p.

BISWAS, N. N.
1932 Poisoning after taking crabs. Indian Med. Record 52: 162.

CANTACUZÈNE, J.
1913 Observations relatives à certaines propriétes du sang de *Carcinus moenas* parasité par la sacculine. Compt. Rend. Soc. Biol. 74: 109-111.

CLELAND, J. B., and R. V. SOUTHCOTT
1965 Injuries to man from marine invertebrates in the Australian region. National Health and Medical Research Council. Spec. Rept. Ser. 12. 282 p., 10 pls.

COOPER, M. J.
1964 Ciguatera and other marine poisoning in the Gilbert Islands. Pac. Sci. 18(4): 411-440, 11 figs.

ERHARDT, J. P., and P. NIAUSSAT
1970 De léventuelle toxicité du décapode brachyoure *Carpilius convexus* Forskål, étude d'éxemplaires provenant de l'atoll de Clipperton. Cah. Pac. (14): 115-114, 2 pls.

GARTH, J. S.
1971 *Demania toxica*, a new species of poisonous crab from the Philippines. Micronesia 7(1-2): 179-183, 1 pl.
1975 *Demania alcalai*, a second new species of poisonous crab from the Philippines (Crustacea, Decapoda, Brachyura). Philippine J. Sci. 104(1-2): 1-6, 1 fig.

GUINOT, D.
1967 Les crabes comestibles de l'Indo-Pacifique. Editions de la Fondation Singer-Polignac, Paris. 145 p., 10 pls.

HALSTEAD, B. W., and K. W. COX
1973 A fatal case of poisoning by the red-spotted crab *Carpilius maculatus* (Linnaeus) in Mauritius. Proc. Roy. Soc. Arts Sci. Mauritius 4(2): 27-30, 1 pl.

HASHIMOTO, Y., KONOSU, S., INOUE, A. SAISHO, T., and S. MIYAKE
1969 Screening of toxic crabs in the Ryukyu and Amami Islands. Bull. Japan. Soc. Sci. Fish 35(1): 83-87, 1 fig.

HASHIMOTO, Y., KONOSU, S., YASUMOTO, T., INOUE, A., and T. NOGUCHI
1967 Occurrence of the toxic crabs in Ryukyu and Amami Islands. Oshima Branch, Kagoshima Prefectural office. 4 p., 3 pls.

HASHIMOTO, Y., KONOSU, S., YASUMOTO, T., and H. KAMIYA
1967 Investigation on toxic marine animals in the Ryukyu and Amami Islands, 8. A survey on coconut crab poisoning. [In Japanese] Tech. Rept. Lab. Marine Biochem., Fac. Agric., Univ. Tokyo. 7 p., 3 figs.

INOUE, A., NOGUCHI, T., KONOSU, S., and Y. HASHIMOTO
1968 A new toxic crab, *Atergatis floridus*. Toxicon 6: 119-123, 2 figs.

JOHNSON, M. E., and H. J. SNOOK
1935 Seashore animals of the Pacific coast. Macmillan Co., New York. 659 p., 700 figs.

JONES, H. M.
1939 Lobsters and gastro-enteritis. Some experiments on cooking and sterilisation. Lancet 2: 738-739.

KONOSU, S., and Y. HASHIMOTO
1968 Occurrence of toxic crabs in the Ryukyu and Amami Islands and similarity of the crab toxin to saxitoxin. South Pacific Comm., Seminar on Icthyosarcotoxism, Rangiro, French Polynesia. 6 p., 2 figs.

KONOSU, S., INOUE, A., NOGUCHI, T., and Y. HASHIMOTO
1968 Comparison of crab toxin with saxitoxin and tetrodotoxin. Toxicon 6: 113-117.
1969 A further examination on the toxicity of three species of xanthid crab. Bull. Japan. Soc. Sci. Fish. 35(1): 88-92, 2 figs.

KONOSU, S., NOGUCHI, T., and Y. HASHIMOTO
1970 Toxicity of a xanthid crab *Zosimus aeneus*, and several other species in the Pacific. Bull. Japan. Soc. Sci. Fish. 36(7): 715-719.

LAFAVRE, H. B.
1930 Report of a small outbreak of food poisoning on board the U.S.S. "Patoka" attributed to crawfish. U.S. Nav. Med. Bull. 28(2): 511-512.

LIGHT, S. F., SMITH, R. I., PITELKA, F. A., ABBOTT, D. P., and F. M. WEESNER
1954 Intertidal invertebrates of the central California coast. 2d ed. Univ. Calif. Press, Berkeley. 446 p., 138 figs.

MACGINITIE, G. E., and N. MACGINITIE
1949 Natural history of marine animals. McGraw-Hill Book Co., Inc., New York. 473 p., 282 figs.

MOTE, G. E., HALSTEAD, B. W., and Y. HASHIMOTO
1970 Occurrence of toxic crabs in the Palau Islands. Clin. Toxicol. 3(4): 597-607, 3 figs.

NOGUCHI, T., KONOSU, S., and Y. HASHIMOTO
1969 Identity of the crab toxin with saxitoxin. Toxicon 7: 325-326.

RUSSELL, F. S., and C. M. YONGE
1944 The seas. Our knowledge of life in the sea and how it is gained. 2d ed. Frederick Warne & Co. Ltd., London. 379 p., 65 figs., 127 pls.

SERÈNE, R.
1952 A report on the dangerous marine organisms of Indo-China. (Personal communication, Sept. 25,. 1952.)

SHIPLEY, A. E.
1958 Introduction to Arachnida, king-crabs, Tardigrada (water-bears) and Pentastohida, p. 253-279, 476-497. *In* S. F. Harmer and A. E. Shipley [eds.], The Cambridge Natural History. Vol. IV. Macmillan and Co., London.

SHUSTER, C. N., JR.
1950 III. Observations on the natural history of the American horseshoe crab, *Limulus polyphemus*. 3d Rept. Investigations of methods of improving the shellfish resources of Massachusetts. Woods Hole Oceanogra. Inst., Contr. No. 564, p. 18-23; Figs. 2-3.
1953 Odyssey of the horseshoe crab. Audubon Mag. 55(4): 162-163, 167.
1957 Xiphosura (with especial reference to *Limulus polyphemus*). Geol. Soc. Am., Mem. 67, 1: 1171-1174, 2 figs.
1960*a* Horseshoe "crabs." Estuarine Bull. 5(2): 3-9.
1960*b* Xiphosura. Encyclopedia Sci. Tech., McGraw-Hill Book Co., Inc. 14: 563-567, 7 figs.
1962 Serological correspondence among horseshoe "crabs" (Limulidae). Zoologica 47: 1-8; Pl. 1, 2 figs.

SMEDLEY, N.
1929 Malaysian king-crabs. Bull. Raffles Mus. 2: 73-78, 4 figs.
1931 Notes on king crabs (Xiphosura). Bull. Raffles Mus. 5: 71-74.

SMITH, H. M.
1933 A poisonous horseshoe crab. J. Siam Soc., Nat. Hist. Suppl. 9(1): 143-145.

SOEGIRI (Dr.)
1936 Een geval van mimi-vergifting. Geneesk. Tijdschr. Nederlandsch-Indië.

SOMMER, H.
1932 The occurrence of the paralytic shellfish poison in the common sand crab. Science 76(1981): 574-575.

SOMMER, H. W., and K. F. MEYER
1937 Paralytic shell-fish poisoning. Arch. Pathol. 24(5): 560-598, 2 figs.

STCHERBACK, A. E.
1907 Poisoning with lobsters. Acute polyneuritic ataxia, combined with acroneuritis, partial disturbance of the sensation of movements. Vrach. Gas. St. Petersb. 14: 285-317.

TEH, Y. F., and J. E. GARDINER
1970 Toxin from the coral reef crab *Lophozozymus pictor*. Pharmacol. Res. Comm. 2(3): 251-256.

TRISHNANANDA, M., TUCHINDA, C., YIPINSEI, T., and P. OONSOMBAT
1966 Poisoning following the ingestion of the horseshoe crab (*Carcinoscorpius rotundicauda*): report of four cases in Thailand. J. Trop. Med. Hyg. 69(9): 194-196, 2 figs.

TWEEDIE, M. W.
1941 Poisonous animals of Malaya. Malaya Publishing House, Singapore. 90 p., 29 figs.

UWATOKO-SETOGUCHI, Y., OBO, G., and S. HASHIMURA
1969 Purification and chemical properties of the toxin of crab. Acta Med. Univ. Kagoshima 11(1): 35-42.

WATERMAN, T. H.
1953*a* Xiphosura from Xuong-ha. Am. Scientist 41(1): 293-302, 3 figs.

WATERMAN, T. H.—Continued

1953*b* Poisonous Pacific horseshoe crabs. (Personal communication, Dec. 2, 1953.)

1958 On the doubtful validity of *Tachypleus hoeveni* Pocock, an Indonesian horseshoe crab (Xiphosura). Postilla Yale Peabody Mus. Nat. Hist. (36): 1-16, 3 pls.

SIPUNCULIDA

BALTZER, F.

1924 Über die giftigkeit der weiblichen Bonellia-Gewebe auf das Bonellia-Männchen und andere organismen und ihre beziehung zur bestimmung des geschlechts der Bonellia-Larve. Mitt. Naturf. Ges. Bern. 8: 1-20.

1925*a* Über die giftwirkung weiblicher Bonellia-Gewebe auf das Bonellia-Männchen und andere organismen und ihre beziehung zur bestimmung des geshlechts der Bonellienlarve. Mitt. Naturf. Ges. Bern. 8: 98-117.

1925*b* Über die giftwirkung der weiblichen Bonellia und ihre beziehung zur geschlechtsbestimmung der Larve. Rev. Suisse Zool. 32(7): 87-93.

BARNES, R. D.

1963 Invertebrate zoology. W. B. Saunders Co., Philadelphia. 632 p.

CHAET, A.

1955 Further studies on the toxic factor in Phascolosoma. Biol. Bull. 109. (NSA)

FISHER, W. K.

1952 The sipunculid worms of California and Baja California. U.S. Nat. Mus. 102: 371-450; Pls. 18-39, 1 fig.

GEROULD, J. H.

1913 The sipunculids of the eastern coast of North America. U.S. Nat. Mus. 44: 373-437; Pls. 58-62.

HYMAN, L. H.

1959 The protostomatous coelomates—phylum Sipunculida, p. 610-696, 29 figs. *In* E. J. Boell [cons. ed.],The Invertebrates: smaller coelomate groups Chaetognatha, Hemichordata, Pogonophora, Phoronida, Ectoprocta, Brachiopoda, Sipunculida, the coelomate Bilateria. Vol. V. McGraw-Hill Book Co., Inc., New York.

(NSA)—Not seen by author.

MAASS, T. A.

1937 Gift-tiere. *In* W. Junk [ed.], Tabulae biologicae. Vol. XIII. N. V. Van de Garde & Co's Drukkerij, Zaltbommel, Holland. 272 p.

PAWLOWSKY, E. N.

1927 Gifttiere und ihre giftigkeit. Gustav Fischer, Jena. 516 p., 176 figs.

RIEDL, R.

1963 Fauna und flora der Adria. Paul Parey, Hamburg und Berlin. 640 p., 221 figs., 8 pls.

ECTOPROCTA

BONNEVIE, P.

1948 Fishermen's "Dogger Bank itch." Acta Allergol. (Kbh.) 1: 40-46, 1 fig.

GULDAGER, A.

1959 Ugeskr. Laeg. 41: 1567. (NSA)

HALSTEAD, B. W.

1970 Poisonous and venomous marine animals of the world. Vol. 3, Addenda, p. 960. U.S. Government Printing Office, Washington, D.C.

HYMAN, L. H.

1959 The lophophorate coelomates—phylum Ectoprocta, p. 275-515, 84 figs. *In* E. J. Boell [cons. ed.], The invertebrates: smaller coelomate groups Chaetognatha, Hemichordata, Pogonophora, Phoronida, Ectoprocta, Brachiopoda, Sipunculida, the coelomate Bilateria. Vol. V. McGraw-Hill Book Co., Inc., New York.

KLUGE, G. A.

1962 Keys on the fauna of the USSR, no. 76. Bryozoan of the northern seas of the USSR. Akademiya Nauk Publishers, Moscow. [Engl. trans.] Amerind Publishing Co. Pvt. Ltd., New Delhi, 1975.

NEWHOUSE, M. L.

1966*a* Dogger Bank itch: a survey of trawlermen. Br. Med. J. 1: 1142-1145, 3 figs.

1966*b* Dogger Bank itch [abridged]. Proc. Roy. Soc. Med. 59(11): 1119-1120.

OSBURN, R. C.

1957 Marine Bryozoa, p. 1109-1112. *In* J. W. Hedgpeth [ed.], Treatise on marine ecology and paleoecology, Vol. 1, Ecology. Geol. Soc. Amer., Memoir 67, New York.

SEVILLE, R. H.

1957 Dogger Bank itch. Brit. J. Dermatol. 69: 92-93.

Part Two

VERTEBRATES

Introduction to Marine Vertebrates

CLASSIFICATION OF POISONOUS AND VENOMOUS MARINE VERTEBRATES

The relationships of toxic marine vertebrates and their position in the total framework of the Animal Kingdom can best be appreciated by a brief presentation of the major phylogenetic categories in which they occur. A further breakdown of these phylogenetic groups below the class level will be given in the chapters that follow.

The phylum Chordata comprises those organisms possessing a notocord, hence are termed chordates. The chordates are divided into two major divisions: (1) *Acrania*—those lacking both a cranium and vertebrae, and (2) *Craniata*—those possessing both a cranium and vertebrae, or the Vertebrata (fishes through mammals). A list of the major groups is as follows:

Phylum CHORDATA: Chordates.
- *ACRANIA:* Chordates lacking a cranium.
  - *Subphylum* HEMICHORDATA: Acorn Worms.
  - *Subphylum* UROCHORDATA: Sea Squirts.
  - *Subphylum* CEPHALOCHORDATA: Lancelets.
- *CRANIATA:* Chordates possessing both a cranium and vertebrae.
  - *Subphylum* VERTEBRATA: Vertebrates.
    - *Class* AGNATHA: Lampreys and Hagfishes.
    - *Class* CHONDRICHTHYES: Sharks, Rays, Skates, Chimaeras.
    - *Class* OSTEICHTHYES: Bony Fishes.
    - *Class* AMPHIBIA: Amphibians.
    - *Class* REPTILIA: Reptiles.
    - *Class* AVES: Birds.
    - *Class* MAMMALIA: Mammals.

Since there does not appear to be any biotoxicological information available regarding the members of the Acrania, they will of necessity be eliminated from further consideration in this work. The only two classes not coming within the area of interest to the marine biotoxicologist are the Amphibia and Aves.

CLASSIFICATION OF TOXIC FISHES

The following classification attempts to take into consideration the phylogenetic relationships of the etiological fishes, the clinical characteristics of the biotoxications, and the chemical nature of the poisons involved.

ICHTHYOTOXIC FISHES

Subphylum VERTEBRATA

I. POISONOUS FISHES[1]

Fishes which when ingested cause a biotoxication in humans due to a toxic substance present in the fish. It does not include fishes that may become accidentally contaminated by bacterial food pathogens.

A. *Ichthyosarcotoxic fishes.*—Those fishes that contain a poison within the flesh, i.e., in the broadest sense, musculature, viscera or skin, or slime, which when ingested by humans will produce a biotoxication. The toxins are oral poisons believed to be small molecular structures and are generally not destroyed by heat or gastric juices.

Class AGNATHA: Lampreys and hagfishes—causing cyclostome poisoning.

Class CHONDRICHTHYES: Sharks, rays, skates, chimaeras—causing elasmobranch and chimaera poisonings.

Class OSTEICHTHYES: Bony fishes.

1. *Ciguatoxic fishes*—causing ciguatera poisoning.[2]
2. *Clupeotoxic fishes*—causing clupeoid fish poisoning.
3. *Gempylotoxic fishes*—causing gempylid fish poisoning.
4. *Scombrotoxic fishes*—causing scombroid fish poisoning.
5. *Hallucinogenic fishes*—causing hallucinatory fish poisoning.
6. *Tetrodotoxic fishes*—causing puffer poisoning.

B. *Ichthyootoxic fishes.*—Those fishes that produce a poison which is generally restricted to the gonads of the fish. The musculature and other parts of the fish usually are edible. There is a definite relationship between gonadal activity and toxin production. Fishes in this group are mainly freshwater, but a few marine species have been incriminated.

Class OSTEICHTHYES: Bony fishes—producing fish roe poisoning.

C. *Ichthyohemotoxic fishes.*—Those fishes having poisonous blood. The poison is usually destroyed by heat and gastric juices.

Class OSTEICHTHYES: Bony fishes—producing fish blood poisoning.

---

[1] A general term used to designate any fish containing a poison exclusive of bacterial food poisons. The term *poisonous fish* as used in this particular classification is used in the restricted sense and as such pertains only to oral intoxications. In the broader sense any fish that contains a toxic substance, including venoms, is a poisonous fish.

[2] Moray eels are tentatively included within this group since recent chemical studies indicate that gymnothorax poisoning is caused by a neurotoxin that appears to be closely related, if not identical, with ciguatera poison. The clinical differences are believed to be the result of variations in quantity rather than quality of the poison.

II. ICHTHYOCRINOTOXIC FISHES—*Ostracion*, etc.

Those fishes that produce a poison by means of glandular structures, independent of a true venom apparatus; i.e, poison glands are present, but there is no traumagenic device.

III. VENOMOUS OR ACANTHOTOXIC FISHES.[3]

Those fishes that produce their poisons by means of glandular structures and are equipped with a traumagenic device to purvey their venoms. The poisons are parenteral toxins, usually large molecules, and are readily destroyed by heat or gastric juices.

Class CHONDRICHTHYES: Sharks, rays, and chimaeras—inflicting stings.

Class OSTEICHTHYES: Bony fishes—inflicting stings.

[3] The term "acanthotoxic" fishes is synonymous with "ichthyoacanthotoxic" fishes, and may be used interchangeably with it.

Chapter VIII—VERTEBRATES

Class Agnatha

Poisonous: LAMPREYS, HAGFISHES

The class Agnatha (cyclostomes, or lampreys and hagfishes) is a group of fishlike vertebrates having an eellike form, cartilaginous or fibrous skeleton, no definite jaws or bony teeth, and a cranium of a primitive type. There are no pelvic girdles, paired limbs, or true ribs. There are 6 to 14 pairs of gill pouches opening either directly into the pharynx or into a separate respiratory tube. Only a single nostril is present. The skin is scaleless. Development is oviparous, with or without a definite larval stage, and the sexes may or may not be separate. Because of their structural simplicity, cyclostomes are generally considered the most primitive of true vertebrates.

The systematics and biology of cyclostomes have been discussed by Day (1880-84), Jordan (1905), Joubin (1931), and Berg (1948-49). Excellent discussions on the Atlantic cyclostomes have been provided by Bigelow and Shroeder (1948), and by Brodal and Fänge (1963).

REPRESENTATIVE LIST OF CYCLOSTOMES REPORTED AS TOXIC

Phylum CHORDATA

Class AGNATHA

Order MYXINIFORMES: Hagfishes, Lampreys

Family MYXINIDAE

Myxine glutinosa Linnaeus (Pl. 1, fig. a). Atlantic hagfish (USA), borer, hagfish, poison ramper (England), piral eller pilal (Sweden). *122*

DISTRIBUTION: North Atlantic (Norwegian coast).

SOURCE: Engelsen (1922).

Family PETROMYZONTIDAE

Caspiomyzon wagneri (Kessler) (Pl. 2, fig. a). Caspian lamprey, nine-eye (USA), Caspian kaspiiskaya minoga (USSR). *123*

DISTRIBUTION: Caspian Sea, rivers of the Caspian basin.

SOURCE: Pawlowsky (1927).

123 *Lampetra fluviatilis* (Linnaeus) (Pl. 2, fig. b). River lamprey (USA), neva lamprey, rechnaya minoga, nevskaya, minoga (USSR).

DISTRIBUTION: Baltic and North Sea Basins, westward as far as Ireland and France. Enters the rivers from the sea.

SOURCES: Chevallier and Duchesne (1851), Kobert (1894, 1902), Phisalix (1922).

122 *Petromyzon marinus* Linnaeus (Pl. 1, fig. b). Sea lamprey, large nine-eyes (USA, England), meerneunauge, seelamprete (Germany), lamproie (France), pricke (Austria), lampreda (Italy), hafs-nejnoga (Sweden), morskaya minoga (USSR).

DISTRIBUTION: Coasts and rivers of both sides of the Atlantic, rivers of Mediterranean.

SOURCES: Chevallier and Duchesne (1851), Kobert (1902).

BIOLOGY

Family MYXINIDAE: The hags are exclusively marine fishes and are only occasionally encountered in brackish waters. They are very sensitive to salinity, generally requiring an optimum range of 3.2 to 3.4 percent and water temperature of less than 13° C. These tolerances tend to confine hagfishes to depths of 15 to 20 fathoms, except during the colder seasons of the year. Hags prefer a clay, mud, or sandy bottom, where they are believed to spend most of their time. Representatives of this group inhabit temperate and subtropical waters of the Atlantic and Pacific Oceans.

Since the hag is somewhat of a scavenger in its habits, its food consists of dead, disabled, or netted fishes. Polychaete worms form the main item of diet of some hagfishes. Hags are sometimes hauled on board ship still clinging to the side of a fish which they have just victimized. Their habit of attacking netted fishes has given them a bad reputation among haddock fishermen. Hagfishes are blind and locate most of their food by scent. They spawn throughout most of the year. The eggs are fewer than 30 in number, and generally they are deposited attached to a fixed object on the bottom of the sea. Hags do not pass through a larval stage.

The skin of the hagfish is richly supplied with large mucous cells. A large hagfish is said to be capable of filling a 2-gallon bucket with slime. The slime is reputed to be toxic.

Family PETROMYZONTIDAE: The members of the family Petromyzontidae are inhabitants of marine and fresh waters of the Northern Hemisphere. Some species are marine and anadromous, whereas others are confined to fresh waters. The habits of *Petromyzon marinus* are representative of the marine and anadromous species. Since lampreys are seldom seen in the open sea, little is known about this phase of their life. When encountered in the sea, they are usually close to the land, or in estuaries or other comparatively shallow water areas. Lampreys have been taken at 547 fathoms off Nantucket, Massachusetts. Life history studies indicate that lampreys tolerate a wide range of temperatures and salinities.

Adult lampreys are parasitic, feeding on the blood of other fishes. They attach themselves to the body of a host fish by sucking with the oral disc; by means of their horny teeth, they rasp through the skin and scales, and then suck the blood until the host is dry. Gage and Gage-Day (1927) have shown that the buccal secretion of lampreys contains an anticoagulant which facilitates the flow of blood. Mackerel, shad, cod, pollack, salmon, basking sharks, herring, swordfish, hake, sturgeon, and eels are favorite prey of lampreys. Larval lampreys feed largely on microscopic organisms.

P. marinus, an anadromous species, seeks a river having a gravelly bottom and rapid running water for spawning beds, and muddy or soft sandy bottom in quiet water for the larvae. In the New England states, migration up the rivers usually takes place during the spring and early summer months. The parents die soon after spawning. The larval lampreys remain in the parent stream, living in mud burrows for a period of 3 to 5 years. When they reach a length of 10 to 12 cm, they migrate to the sea, usually during the autumn months of the year.

MECHANISM OF INTOXICATION

Cyclostome poisoning is caused by eating the flesh, particularly skin and slime of raw or cooked hagfish and certain species of lampreys.

MEDICAL ASPECTS

Clinical Characteristics

Symptoms of cyclostome poisoning develop within a few hours after ingestion of the fish. The usual symptoms consist of nausea, vomiting, dysenteric diarrhea, tenesmus, abdominal pain, and weakness. Recovery generally takes place within a period of several days. Brief references to the clinical characteristics of cyclostome poisoning have been made by Chevallier and Duchesne (1851), Kobert (1894, 1902), Engelsen (1922), and Pawlowksy (1927).

Treatment

Symptomatic.

Prevention

Most cyclostome poisoning has been reported in the literature as being due to the failure of de-sliming the fish. Engelsen (1922), and others claim that if the fresh fish is covered with salt and left in a concentrated brine solution for several hours prior to cooking, the fish is then safe to eat. This procedure reputedly removes or detoxifies the poison. Cooking apparently does not always render the fish safe to eat.

TOXICOLOGY

Secretions from the buccal glands of various species of the genera *Petromyzon, Lampetra*, and *Entosphenus* delay or prevent coagulation when mixed with human blood. When lamprey buccal gland secretion was mixed with catfish *Amiurus* blood, it prevented coagulation and hemolyzed the blood (Gage and Gage-Day, 1927). There appear to be no reports on the toxicity of the slime or flesh. (*See also* Chapter XXIV on ichthyocrinotoxic fishes.)

PHARMACOLOGY

(Unknown)

CHEMISTRY

Except for such brief general references as Flössner and Kutscher (1925), Flössner and Miller (1933), and Kaiser and Michl (1958) to the chemistry of cyclostome poison, little else appears to be known on the subject. It has been suggested that the active principle of cyclostome poison is a biogenic amine. Gage and Gage-Day (1927) have worked on the anticoagulant and hemolytic properties of buccal-gland secretions of lampreys.

LITERATURE CITED

BERG, L. S.
1948-49 Freshwater fishes of the USSR and neighboring countries. [In Russian] 3 vols. Moscow. 1,381 p., 946 figs.

BIGELOW, H. B., and W. C. SCHROEDER
1948 Fishes of the western North Atlantic. Part I. Mem., Sears Found. Marine Res. 576 p., 106 figs.

BRODAL, A., and R. FÄNGE
1963 The biology of myxine. Universitelsforlaget, Oslo. 588 p.

CHEVALLIER, A., and E. A. DUCHESNE
1851 Mémoire sur les empoisonnements par les huitres, les moules, les crabes, et par certains poissons de mer et de rivière. Ann. Hyg. Publ. (Paris) 45 (11): 387-437.

DAY, F.
1880-84 The fishes of Great Britain and Ireland. 2 vols. Williams and Norgate, London. 179 pls.

ENGELSEN, H.
1922 Om giftfisk og giftige fisk. Nord. Hyg. Tidskr. 3: 316-325.

FLÖSSNER, O., and F. KUTSCHER
1925 Biochemische studien über *Petromyzon fluviatalis* L. Z. Biol. 82(69): 302-305.

(NSA)—Not seen by author.

FLÖSSNER, O., and P. V. MILLER
1933 Zur kenntnis des neosins. Z. Biol. 94(3): 307-311.

GAGE, S. H., and M. GAGE-DAY
1927 The anti-coagulating action of the secretion of the buccal glands of the lampreys (*Petromyzon, Lampretra,* and *Entosphenus*). Science 66(1708): 282-284, 2 figs.

JORDAN, D. S.
1905 A guide to the study of fishes. Vol. I. Henry Holt and Co., New York. 624 p., 393 figs.

JOUBIN, L. M.
1931 Faune ichthyologique de l'Atlantique Nord. Nos. 6-9. A. Fred Høst, Copenhague.

KAISER, E., and H. MICHL
1958 Die biochemie der tierischen gifte. Franz Deuticke, Wien. 258 p., 23 figs.

KOBERT, R.
1894 Compendium der praktischen toxikologie, p. 90-93. Ferdinand Enke, Stuttgart. (NSA)
1902 Ueber giftfische und fischgifte. Med. Woche (20): 209-212.

PAWLOWKSY, E. N.
1927 Gifttiere und ihre giftigkeit. Gustav Fischer, Jena. 515 p., 170 figs.

PHISALIX, M.
1922 Animaux venimeux et venins. Masson et Cie., Paris. 656 p., 232 figs., 4 pls.

Chapter IX—VERTEBRATES

Class Chondrichthyes

Poisonous: SHARKS, RAYS, SKATES, CHIMAERAS

The class Chondrichthyes comprises the sharks, rays, skates, chimaeras, and their relatives, all characterized by their highly developed jaws, their possession of fins and a general body structure developed along the familiar piscine plan. The skeleton is cartilaginous, and fertilization in most species is the internal type.

The class CHONDRICHTHYES, according to Lagler, Bardach, and Miller (1962), is divided into the following major groups:

Subclass ELASMOBRANCHII: Sharks, skates, and rays. This large subclass is further divided into the order Squaliformes (or Pleurotremata), which includes the various families of sharks; and the order Rajiformes (sometimes termed Hypotremata or Batoidei), which includes the sawfishes, skates, rays, mantas, and their relatives.

Subclass HOLOCEPHALI: Chimaeras. This subclass contains the order Chimaeriformes, and a single family.

Chondrichthyes includes both poisonous and venomous fish species, but only the ones poisonous for humans to eat will be discussed in this chapter.

Poisonous Sharks

Sharks are a group of fishes characterized by having a cartilaginous skeleton, primitive ribs, and a complete braincase or chondrocranium. The gill openings are elongate, lateral or partially so, and the gill pouches are supported by cartilaginous arches. Median and paired fins are present and supported by cartilaginous rods at their bases from which radiate unsegmented horny rays. The paired fins are attached to simple cartilaginous girdles. Portions of the pelvic fins in the males are modified to form a copulatory intromittent organ. The nostrils are paired and ventral in location. The mouth is ventral and equipped with well-developed jaws, the upper part not fused with the cranium.

The biology and systematics of the sharks listed in this section have been discussed by Günther (1870), Jordan and Evermann (1896), Garman (1913), Joubin (1929-31), Bigelow and Schroeder (1948), Schultz *et al.* (1953), and Smith (1961).

REPRESENTATIVE LIST OF SHARKS REPORTED AS POISONOUS

Phylum CHORDATA

Class CHONDRICHTHYES

Order SQUALIFORMES: Sharks

Family CARCHARHINIDAE

126 *Carcharhinus melanopterus* (Quoy and Gaimard) (Pl. 1, fig. a). Black-tip reef shark (USA, Australia), black shark (Natal), nilow (West Africa).

DISTRIBUTION: Tropical Indo-Pacific and Red Sea.

SOURCES: Smith (1947), Fish and Cobb (1954), Helfrich (1961).
OTHER NAME: *Carcharias melanopterus.*

126 *Carcharhinus amblyrhynchos* (Bleeker) (Pl. 1, fig. b). Gray reef shark (USA).

DISTRIBUTION: Indo-Pacific.

SOURCE: Banner and Helfrich (1964).
OTHER NAME: *Carcharhinus menisorrah.*

126 *Galeorhinus galeus* (Linnaeus) (Pl. 1, fig. c). Tope, oil shark, school shark, soupfin shark (USA, England, South Africa), milandre, chat marin (France), hundshai (Germany), canesca (Italy).

DISTRIBUTION: Mediterranean, eastern North Atlantic, South Atlantic, and Indian Ocean.

SOURCES: Autenrieth (1833), Chevallier and Duchesne (1851*a, b*) Hiyama (1943).
OTHER NAMES: *Galeus canis, Squalus galeus.*

126 *Prionace glauca* (Linnaeus) (Pl. 1, fig. d). Blue shark, great blue shark, blue whaler (USA, England), le bleu, squale bleu, chien de mer bleu (France), blauhai (Germany), yoshikiri (Japan), azulejo (Chile).

DISTRIBUTION: Cosmopolitan, in tropical, subtropical, and warm-temperate belts of all oceans.

SOURCES: Autenrieth (1833), Phisalix (1922), Hiyama (1943).
OTHER NAMES: *Carcharias glaucus, Charcarias glaucus, Squalus glaucus.*

Family DALATIIDAE

127 *Somniosus microcephalus* (Bloch and Schneider) (Pl. 2, fig. a). Greenland shark, sleeper shark, nurse shark, gurry shark (USA, England, Greenland), ondenzame (Japan), polyarnaya akula (USSR).

DISTRIBUTION: Arctic, North Sea, east to the White Sea, and west to the Gulf of St. Lawrence. There is also a closely related but distinct species in the Pacific (Bigelow and Schroeder, 1948).

SOURCES: Jensen (1914, 1948), Bigelow and Schroeder (1948), Halstead (1974).

Family HEXANCHIDAE

Hexanchus griseus (Bonnaterre) (Pl. 2, fig. b). Six-gilled shark, cow shark, gray shark, mud shark (USA), requin, griset, hexanche (France), cañabota (Spain), kagura (Japan), gato de mar (Latin America), squalo capopiatto (Italy). *127*

DISTRIBUTION: Atlantic, Pacific coast of North America, Chile, Japan, Australia, southern Indian Ocean, South Africa.

SOURCES: Coutière (1899), Phisalix (1922).
OTHER NAMES: *Heptanchus greslus, Notidanus griseus.*

Family ISURIDAE

Carcharodon carcharias (Linnaeus) (Pl. 2, fig. c). White shark, white pointer (USA, Australia), chien de mer, requin (France), pescecane (Italy), menschenhai (Germany), oshirozame, hitokuizame (Japan), tiburón blanco (Latin America). *127*

DISTRIBUTION: Cosmopolitan, in tropical, subtropical, and warm-temperate belts of all oceans.

SOURCES: Helfrich (1961), Bouder, Cavallo, and Bouder (1962).
OTHER NAME: *Squalus carcharias.*

Family SCYLIORHINIDAE

Scyliorhinus caniculus Linnaeus (Pl. 3, fig. a). Lesser-spotted cat shark, dogfish, rough dog (USA, England), kleingefleckter katzenhai (Germany), pesce gatta (Italy), petite rousette (France), röodhai (Norway). *128*

DISTRIBUTION: Atlantic Ocean northward from West Africa to the British Isles and Scandinavia; Mediterranean Sea.

SOURCES: Autenrieth (1833), Pawlowsky (1927), Hiyama (1943).
OTHER NAMES: *Scyllium canicula, Squalus catulus.*

Family SPHYRNIDAE

Sphyrna zygaena (Linnaeus) (Pl. 3, fig. b). Hammerhead shark (USA), requin marteau (France), pesce martello (Italy), crüz, medialuna, cornuda, pejecapelo (Peru), shumokuzame, kasebuka (Japan). *128*

DISTRIBUTION: Tropical to warm temperate belts of the Atlantic and Pacific Oceans.

SOURCES: Hiyama (1943), Bouder *et al.* (1962).

BIOLOGY

Family CARCHARHINIDAE: The requiem sharks constitute the largest family of sharks, having at least 13 genera and numerous species, which include the majority of modern sharks. Most carcharhinids are inhabitants of tropical or warm-temperate waters, entering the cooler zones only during the summer months, if at all. Some are circumtropical in distribution, whereas others are confined to narrow distributional areas. Most of the species are harmless, but a few have been incriminated as maneaters. Requiem sharks tend to be omnivorous in the broadest sense of the word, feeding on a great variety of fishes, sharks, stingrays, turtles, birds, sea lions, porpoises, gastropods, horseshoe crabs, crustaceans, carrion, garbage, etc. Many of them exhibit cannibalistic tendencies. Carcharhinids inhabit a diversity of biotopes, ranging from open seas and coastal waters to estuaries, inlets, river mouths, and lagoons. One species *Carcharinus nicaraguensis* is known only from Lake Nicaragua, a freshwater lake. On numerous occasions, carcharhinids have been observed pursuing schools of small fishes in shoals where the water was so shallow that the backs of the sharks were almost completely exposed to the air. Their development is either of the viviparous or ovoviviparous type. Some species may grow to be very large. *Galeocerdo* the tiger shark is said to attain a length of 9 m or more.

Family DALATIIDAE: The sleeper sharks constitute one of the small families of sharks. They have a range which extends from the Arctic waters of Greenland to South Africa and the Indo-Pacific. The single incriminated toxic member of this family *Somniosus microcephalus* is an inhabitant of Arctic and north temperate waters. Bigelow and Schroeder (1948) report it as feeding on a variety of fishes (skate, herring, salmon, capelin, sculpin, lumpfish, pollack, cod, haddock, wolffish, flatfish), seals, birds, mollusks, and coelenterates. It also feeds on carrion whenever available. It has a wide bathymetric range, from the surface to 660 fathoms or more. *Somniosus* is considered of commercial value in Norway, Iceland, and Greenland, where it is sought for its liver oil. The flesh is used for dogfood and, when dried, is occasionally eaten by humans. The flesh is said to be toxic when fresh (Jensen, 1914, 1948). Unlike most sharks, *S. microcephalus* is oviparous.

Family HEXANCHIDAE: Cow sharks are widely distributed, representatives of the family occurring throughout most temperate and tropical seas. Hexanchids are largely bottom dwellers and somewhat sluggish in their movements. Their bathymetric range is considerable. *Hexanchus griseus* has been taken at more than 700 fathoms, whereas other species seem to prefer shallow water. They are said to be voracious feeders, eating fishes and various crustaceans. *H. griseus* may reach a length of more than 4 m, but *Heptranchus perlo* is reputed to have a maximum length of about 2 m. Cow sharks are ovoviviparous in their development.

Family ISURIDAE: The only mackerel shark incriminated as toxic is the great white shark *Carcharodon carcharias*, which is cosmopolitan in all warm seas. This species has the reputation of being one of "the most voracious of fishlike vertebrates" (Bigelow and Schroeder, 1948), devouring practically intact sharks, large fishes, sea lions, turtles, and garbage. This species has been known to charge small boats without provocation and has a bad reputation as a maneater. Although typically an inhabitant of the open seas, it is not averse to entering shallow water to feed. *C. carcharias* attains a length of 5 m or more. Its development is ovoviviparous.

Family SCYLIORHINIDAE: The cat shark family is comprised of numerous species of small sharks inhabiting tropical and temperate seas in both shoal and deep water. Most of them are bottom dwellers, but little is known of their habits. Cat sharks are oviparous in their development. They seldom exceed a length of 1 m.

Family SPHYRNIDAE: The hammerhead sharks, as their name implies, are characterized by having the anterior portion of the head greatly flattened dorsoventrally and widely expanded laterally, in a hammerlike shape. The eyes are situated at the outer edges of the head. Members of the family are found in all warm seas. Hammerhead sharks are known to be maneaters. Members of this group may be found far offshore, swimming near the surface of the water, but not infrequently are encountered inshore, in bays and lagoons. Their diet consists largely of fish, crustaceans, squid, and barnacles. They attain a length of 4 m or more. Some species are viviparous whereas others are believed to be ovoviviparous.

MECHANISM OF INTOXICATION

Elasmobranch poisoning is most commonly caused by eating the liver of sharks. However, the musculature of some of the larger tropical sharks and the arctic *Somniosus microcephalus* may also be toxic. Tropical sharks are generally considered to be especially dangerous to eat.

MEDICAL ASPECTS

Clinical Characteristics

Intoxications caused by the ingestion of poisonous shark flesh are generally mild. The symptoms seldom exceed that of a mild gastroenteritis, predominated by a diarrhea. However, ingestion of toxic shark liver may result in a severe biotoxication. The symptoms usually develop within a period of 30 minutes and consist of nausea, vomiting, diarrhea, abdominal pain, anorexia, headache, prostration, rapid weak pulse, malaise, insomnia, cold sweats, oral paresthesias, and burning sensation of the tongue, throat, and esophagus. As time progresses, the neurological and other symptoms become more pronounced, resulting in extreme weakness, trismus, muscular cramps, sensation of heaviness of the limbs, blepharospasm, dilatation of pupils, spasmodic contractions of upper lids, hiccoughs, visual disturbances, tingling sensation of fingertips, joint aches, delirium, ataxia, incontinence, dysuria, desquamation, pruritus, respiratory distress, coma, and death. The recovery period varies from a day or two to several weeks. The mortality rate is unknown, but is believed to be relatively high.

Coutaud (1879) has published a detailed clinical account on a single outbreak of elasmobranch poisoning involving seven persons which occurred on the Isle of Pines, New Caledonia, in 1873. The victims had eaten the liver from an unidentified species of shark which had been captured in local waters. The symptoms were rapid in onset and very severe in three of the victims, one of whom finally died. Some of the shark liver was fed to a pig, and it also died. The symptoms in the remainder of the group were relatively mild and were not described in detail. In the severe cases the gastrointestinal and neurological disturbances were predominant.

The poison found in the Greenland shark *Somniosus microcephalus* appears to be present largely in the flesh of the shark rather than the liver. Jensen (1914) states that the fresh liver can be eaten without harm. The poison present in fresh shark flesh can be either deactivated or possibly destroyed by drying, which is not the case in ciguatera poison. The symptoms in humans caused by eating the flesh of the Greenland shark are similar to those found in dogs and are described in the section on toxicology. The predominant symptoms are of a gastrointestinal and neurological disorder, but appear to differ somewhat from the clinical picture observed in the usual elasmobranch biotoxications caused by eating the livers of some of the larger tropical sharks. Cooper (1964) has reported the death of several persons from eating the liver of the tiger shark, believed to be *Galeocerdo cuvieri*, in the Gilbert Islands. However, the flesh was said to be nontoxic.

From all the clinical evidence that has been examined thus far, it appears that sharks may contain at least three different types of poisons. There is no reason to believe that ciguatoxin does not occur in tropical sharks since this poison is found in a wide phylogenetic range of marine fishes. There is also clinical evidence that the poison found in the liver of large specimens of tropical sharks differs from that found in the flesh of the Greenland shark *Somniosus*. The belief that tropical shark liver poison differs from ciguatera poison is partially supported by the absence of the pathognomonic paradoxical sensory disturbance (temperature reversal sensation) which constitutes a striking part of the clinical picture of ciguatera poisoning. However, until such a time as adequate chemical and pharmacological analyses have been conducted on the poisons found in various shark species, one can merely speculate as to their nature.

Bouder *et al.* (1962) (*see also* Helfrich, 1961) have suggested that tropical shark liver poisoning is caused by hypervitaminosis A. However, the clinical characteristics of severe cases of elasmobranch poisoning are not consistent with hypervitaminosis A. The wide variety and severe neurological disturbances of elasmobranch poisoning can hardly be explained on the basis of excessive vitamin A intake alone, even if hypervitaminosis A were a factor. Moreover, Cooper (1964) has pointed out that in studies conducted on the vitamin A content of Gilbert Island sharks (Lonie, 1950), the quality of vitamin A was too low for commercial extraction.

Our knowledge of the symptoms produced by elasmobranch poisoning is based almost entirely upon the writings and observations of Autenrieth (1833), Coutaud (1879), Jensen (1914), and Bøje (1939). (*See also* Halstead, 1974).

Pathology

Coutaud (1879) has written the only published autopsy report on lethal biotoxications resulting from elasmobranch poisoning, and this was based on a single case. The victim had eaten the liver from an unidentified shark captured in the vicinity of the Isle of Pines, New Caldeonia. Aside from evidence of congestion of the viscera and meninges of the brain, no unusual pathological findings were noted. Similar findings have been reported by Bøje (1939) on a dog that had died from eating the Greenland shark *Somniosus.*

Treatment

There is no known antidote. Treatment is symptomatic.

Prevention

The flesh and liver of large tropical sharks should be avoided unless positively known to be safe to eat, and the only way that that can be determined is by feeding a piece of the meat to a dog, cat, or some other satisfactory test animal.

The toxic flesh of the Greenland shark *Somniosus* can be rendered safe to eat by cutting it into strips and hanging them up to dry in the open air. The strips should be hung in such a manner as to permit the juices to drain free of the meat. Detoxification is most complete when the meat is hung up during the winter and frozen. Later when it thaws the fluid drains off, and it should be permitted to dry thoroughly in the sun and wind. Meat that is incompletely dried during the summer months may form a superficial crust which prevents the inner portions from becoming dry. Poorly dried meat is said to be more toxic than fresh meat. Moreover, if dogs are fed both properly dried and fresh meat, there appears to be a synergistic effect, resulting in a more serious intoxication than if fresh shark meat were used alone. It is suggested by Bøje (1939) that the drying process "inactivates" or possibly destroys the poison. The toxicity of *Somniosus* seems to vary according to the locality, time of year, and individual specimen, so one must be cautious in assuming that because this shark is safe to eat in one place that it is safe to eat elsewhere and at other times. Ordinary cooking temperatures do not destroy the poison, but it is said to be water soluble and can be removed either by several washings and changes of water or by boiling and discarding the water.

TOXICOLOGY

The toxic effects resulting from ingestion of the flesh of the Greenland shark *Somniosus microcephalus* have been studied by Bøje (1939). He observed that dogs intoxicated by *Somniosus* walk with slow, stiff steps and develop hypersalivation, vomiting, explosive diarrhea, conjunctivitis, muscular twitching, turning of the eyes upward and outward, respiratory distress, tonic and clonic convulsions, and death. The diarrhea which is characteristic of Greenland shark-meat intoxications is usually absent in the more serious cases. Gradual habituation to toxic *Somniosus* flesh "sometimes gives a certain degree of immunity from the toxin." Physical work is said to precipitate the appearance of the symptoms. Dogs which have eaten poisonous shark meat without ill effect become sick when they pull a sledge. Exertion results in the dog's gasping for breath, perspiration, salivation, ataxia, and collapse. After resting, the dogs may partially recover, stagger to their feet, and continue on for a period of time.

Macht and Spencer (1941) and Macht (1942) tested aqueous extracts from the musculature of *Mustelus canis.* Extracts were injected intraperitoneally into laboratory mice which were observed for the development of symptoms. They also employed a phytopharmacological technique utilizing the seedlings of *Lupinus albus.* The increment or growth of the single straight root of the seedling was measured after the plant had been grown for 24 hours in the dark at 15° C in the extract to be tested. Growth of the root was then compared with controls which were grown in plant physiological (Shive's) solution. The index of growth or phytotoxic index was expressed by the ratio of X/N or average growth in the unknown solution over the growth of the controls. Toxic fishes usually gave a phytotoxic index of less than 70 percent. *M. canis* gave a reading of 62 percent.

Halstead and Schall (1958) found the liver and gonads of the gray reef shark *Carcharhinus menisorrah* of Palmyra Is., Line Islands, to be toxic. Aqueous extracts were prepared by homogenizing the tissues in a Waring blender, centrifuging, and injecting the supernatant intraperitoneally into white laboratory mice. The mice showed evidence of hypoactivity, ataxia, ruffling of the hair, lacrimation, diarrhea, paralysis, and finally death within a period of 36 hours.

PHARMACOLOGY

Research on the pharmacology of elasmobranch poison has not been reported. The suggestion has been made that the poison in the flesh of *Somniosus* may develop from the decomposition of urea present in the tissues. Bøje (1939) attempted to test this theory by feeding dogs codfish that had been mixed with large quantities of urea, but this experiment failed to produce any intoxication. The toxin appears to be primarily a parasympathomimetic substance. The physiological evidence lends some credence to Bøje's conjecture that the poison may be related to the choline present in muscle as a component of lecithin. The similarity of choline to muscarine has also been suggested as an explanation of the mode of action of the poison. The pharmacology of elasmobranch poison is still unknown. Li (1965) has isolated a biotoxin from the liver of the gray reef shark *Carcharinus menisorrah* and found that it exhibited pharmacological properties similar to ciguatoxin derived from the red snapper *Lutjanus bohar.* Cooper (1964) reports that symptoms caused by shark liver poisoning in the Gilbert Islands are the same as those for ciguatera but develop with greater rapidity and severity.

CHEMISTRY

(Unknown)

Poisonous Skates and Rays

The order Rajiformes, which includes the skates and rays, consists of a group of fishes which have the gills confined to the ventral surface of their body. The body proper is elongate, depressed, and almost elliptical in cross section. The pectoral fins have wide bases which are greatly extended and attached along the side of the body from the tip of the snout to the anterior margin of the pelvic fins. The anal fin is absent.

Skates and rays contain both poisonous and venomous species, but only the ones poisonous to eat are discussed in this section.

The systematics and identifying characteristics of the skates and rays listed as poisonous have been discussed by Garman (1913), Joubin (1929-31), Whitley (1940), Bigelow and Schroeder (1953), Smith (1961), and Lagler *et al.* (1962).

REPRESENTATIVE LIST OF SKATES AND RAYS REPORTED AS POISONOUS

Phylum CHORDATA

Class CHONDRICHTHYES

Order RAJIFORMES: Skates, Rays, Etc.

Family DASYATIDAE

129 *Dasyatis sayi* (LeSueur) (Pl. 4, fig. a). Blunt-nose stingray (USA), ráia (Brazil).

DISTRIBUTION: Western Atlantic from south Brazil to New Jersey.

SOURCES: Macht and Spencer (1941), Macht (1942).

Family MYLIOBATIDAE

129, 220 *Aetobatus narinari* (Euphrasen) (Pl. 4, fig. b). Spotted duck-billed ray, spotted eagle ray (USA, Australia, England), hihimanu, ihimanu (Hawaii), pagi manok, paol, banogan, taligmanok (Philippines), pari lang (Malaysia), eel-tenkee, currooway tiriki (India), bonnet skate, spotted ray, keppierog, pylstert (South Africa), impogo, kombedo (West Africa), raya (Latin America), ráia (Brazil).

DISTRIBUTION: Tropical seas.

SOURCE: Halstead and Bunker (1954).

Family RAJIDAE

129 *Raja batis* Linnaeus (Pl. 4, fig. c). Flapper skate (England), raya noriega (Spain), pocheteau blanc (France), razza bavosa (Italy), rochen (Germany).

DISTRIBUTION: Coasts of Europe, Baltic Sea, North Sea, Mediterranean Sea, South Africa.

SOURCE: Phisalix (1922).[1]
OTHER NAME: *Raia batis.*

Family TORPEDINIDAE

129 *Torpedo marmorata* Risso (Pl. 4, fig. d). Cramp fish, numb fish, marbled electric ray (USA, England), torpille marbrée (France), tremelga (Portugal), tremielga (Spain), marmorierter zitterroche (Germany), torpedine marezatta (Italy).

DISTRIBUTION: Atlantic Ocean, Indian Ocean, Mediterranean Sea.

SOURCES: Gley (1904),[1] Phisalix (1922),[1] Maass (1937).[1]
OTHER NAME: *Raja torpedo.*

---

[1] Authors state that only the blood or serum of the fish was found to be toxic. However, the toxicological data presented were insufficient to determine the exact nature of the poison.

BIOLOGY

Family DASYATIDAE: The members of the stingray family Dasyatidae are characterized by their absence of a well-developed caudal fin having cartilaginous medial supports, the absence of a distinct dorsal fin, and the absence of a slender median process directed forward from the pelvis. The anterior portions of the pectorals are continuous along the sides of the head rather than being separated off as subrostral lobes or fins. The tail in dasyatids is slender, tapering, and longer than the disc in most species. Typical of stingrays in general, the tail in most species is armed with one or more elongate, acute, serrated, venomous spines. These are some of the more pertinent characteristics which separate this family from the other members of the Rajiformes.

The dasyatids are inhabitants of the bottoms of bay and estuary flats, shoal lagoons, river mouths, or on patches of sand between coral heads, etc. They are sometimes difficult to detect because of their habit of lying partially buried in the mud or sand with only a portion of their head or body exposed. They feed by excavating the bottom with their pectoral fins, thereby exposing worms, mollusks, crustaceans, small fish, etc., which they ingest. Although stingrays remain inactive for long periods of time, they will at intervals swim actively about, using undulating motions of their pectorals to propel themselves through the water. Dasyatids are primarily warm-water fish, preferring subtropical and tropical latitudes. They are largely confined to shallow waters, but a few species occur in depths of 60 fathoms or more. They are ovoviviparous, with the embryos lying loose in the uterus without any physical connection with the mother. Dasyatids vary in size from about 30 cm to more than 2 m across the disc.

Family MYLIOBATIDAE: The members of the family Myliobatidae, the eagle rays, are characterized by their pectorals, which are either narrow opposite the eyes or entirely interrupted at that point, the head thus being conspicuously marked off from the rest of the body. The anterior portions of the pectorals unite across the front of the head below the tip of the snout, forming a subrostal lobe. The crown of the head is conspicuously elevated. The tail is much longer than the disc, slender, and armed with a venomous spine in some species, but absent in others. There is a small dorsal fin on the anterior part of the tail, but no caudal fin. These are some of the more pertinent characteristics which distinguish this family from other members of the Rajiformes.

Eagle rays are widely distributed in tropical and warm continental waters where they are encountered about off-lying islands, coastal areas, and at times in the open sea, but are commonly seen in shallow bays and estuaries over sand and mud flats. Eagle rays are fast graceful swimmers and appear to literally fly through the water beating their large pectoral fins. At times they will break the surface of the sea and skim for a short distance through the air. Eagle rays are bottom feeders, living chiefly on crustaceans and mollusks. Development is ovoviviparous. The larger species may attain a length of 5 m or more, including the tail.

Family RAJIDAE: Most of the skates are members of a single family—the Rajidae. Skates are characterized by their moderately slender tails, which are not more than twice as long as the body. They are rounded above, flattened below, and have a narrow dermal fold along each side. Two dorsal fins are present. The orbits are prominent and rise above the general level of the head. The front of the cranium is with or without a rostral projection. The skin of the upper surface of the disc and tail is more or less rough with small prickles, larger thornlike denticles, or both, but lack serrate venomous caudal spines. These are some of the more dominant characteristics which separate the skates from other members of the Rajiformes.

Skates are bottom fishes and are often observed in sand where they are able to bury themselves by using their pectoral fins. They prefer smooth, sandy, gravel bottoms, or mixtures of sand and shells. They frequently rise off the bottom in pursuit of prey and occasionally are seen at the surface. Skates swim by undulating their pectoral fins, and at times can move with great rapidity. They are primarily carnivores, feeding on crustaceans, worms, mollusks, and fishes, mainly at night. Skates are most abundant in depths of less than 100 fathoms, but have been recorded at depths greater than 1,500 fathoms. They are most common in temperate and boreal latitudes. The female lays her eggs over a period of several months in spring, summer, or fall. The egg capsules are horny or leathery in texture, comprised of a keratin-like material, quadrate in shape, range in color from a light auburn to dark brown or black, and may have a beautiful sheen to them. Skates attain a length of about 2 m. Some species of skates are known to have electric organs, but apparently are unable to produce a shock of any significance to man.

Family TORPEDINIDAE: The electric rays are characterized by their disc, subcircular to elongate, which is fleshier toward its margins and thicker than in most other disc-shaped batoids, and by their softer body. The tail is with or without lateral folds, sharply marked off from the body portion and broader in the base than in most other batoids. The attachment of the anterior portion of the pectorals extends forward to or beyond the level of the eyes. There are one or two well-developed fins. The caudal fin is well developed. The skin is soft and entirely naked. One of the major characteristics of the members of this group is the possession of two electric organs, one on either side of the anterior part of the disc between the anterior extension of the pectoral and the head, extending forward to about the level of the eye and posteriorly past the gill region to the vicinity of the pectoral girdle. These are some of the essential characteristics which separate this group from other members of the Rajiformes.

Electric rays are feeble swimmers and sluggish in their habits, spending most of their time on the bottom partially buried in sand or mud. Although most species seem to prefer shallow waters, some members are found at moderately great depths. Their food consists of crustaceans, mollusks, worms, other vertebrates, and fishes. Electric rays are inhabitants of temperate, subtropical, and tropical latitudes. They attain a length of about 2 m and a weight of over 91 kg. Development is ovoviviparous. Large specimens of electric rays have been known to produce up to 220 volts. A rested ray is said to be able to produce an electrical discharge sufficient to knock down and temporarily disable an adult man (Bigelow and Schroeder, 1953).

MECHANISM OF INTOXICATION

Biotoxications may result from ingestion of either the flesh or viscera of certain species of skates or rays. The viscera of tropical species are believed to be particularly dangerous.

MEDICAL ASPECTS

Clinical Characteristics

Little is known concerning the symptomatology of biotoxications from eating the flesh or viscera of batoids. It is believed that the symptoms are similar to those found in shark poisoning. The relationship of this form of intoxication to ciguatera is not known, and it is quite possible that ciguatoxin may occur in skates and rays inhabiting subtropical and tropical waters (Halstead, 1959).

Treatment

Symptomatic.

Pathology

(Unknown)

Prevention

Until more precise information is available the flesh and particularly the viscera of large tropical skates and rays should not be used for human consumption.

TOXICOLOGY

(Unknown)

PHARMACOLOGY

(Unknown)

CHEMISTRY

(Unknown)

Poisonous Chimaeras

Chimaeras, elephantfish, or ratfish, as they are sometimes called, are a group of cartilaginous fishes having on either side a single external gill opening covered over by an opercular skin fold with cartilaginous supports leading to a common branchial chamber into which the true gill clefts open. Externally, chimaeroids are more or less compressed laterally, tapering posteriorly to a slender tail. The snout is rounded or conical, extended as a long pointed beak or bearing a curious hoe-shaped proboscis. There are two dorsal fins. The first dorsal is triangular, usually higher than the second, and edged anteriorly by a strong sharp-pointed bony spine which serves as a venom apparatus. There are various other anatomical characteristics of this group which have been thoroughly discussed by Bigelow and Schroeder (1953).

Some of the more important works on the systematics of the chimaeras are by Joubin (1929-31), Bigelow and Schroeder (1953), Roedel (1953), Smith (1961), and Lagler *et al.* (1962).

LIST OF CHIMAERAS REPORTED AS POISONOUS

Family CHIMAERIDAE

130 *Chimaera monstrosa* Linnaeus (Pl. 5, fig. a). Chimaera, ratfish, rabbitfish (USA), seeroche (Germany), chimère (France), peixe rato (Portugal), quimera (Spain), khimery (USSR).

DISTRIBUTION: North Atlantic from Norway and Iceland to Cuba, the Azores, Morocco, Mediterranean Sea, South Africa.

SOURCES: Coutière (1899), Phisalix (1922), Halstead (1959).

130 *Hydrolagus colliei* (Lay and Bennett) (Pl. 5, fig. b). Ratfish (USA).

DISTRIBUTION: Pacific coast of North America.

SOURCE: Halstead and Bunker (1952).

BIOLOGY

Chimaeras have an extensive geographical range, extending from north temperate to south temperate waters. However, the specific range of the individual members of this order indicates a definite preference for cooler temperatures. The depth range extends from the surface down to 1,400 fathoms or more. The vertical distribution of *Hydrolagus colliei* differs from most other chimaeras. Goode and Bean (1895) stated that the Pacific ratfish may be seen swimming at the surface of the water in southeastern Alaska and British Columbia, but it generally inhabits deeper water southward. The usual depth range for this species is given as 5 to 60 fathoms, while most other chimaeras inhabit depths of 100 to 1,900 fathoms or more. Chimaeras are weak swimmers, propelling themselves largely by undulations of the posterior part of their body, second dorsal fin, and caudal. The large pectorals serve both as propelling and balancing organs. Chimaeras are somewhat nocturnal in their habits. They are not vigorous, struggle but little, and die soon after being removed from the water. They have well-developed dental plates, and can inflict a nasty wound. Dean (1906) has expressed the opinion that fishermen fear the jaws of the ratfish more than they do the formidable dorsal spine. He mentioned one case in which a ratfish bit a piece of flesh from the hand of a fisherman. Chimaeras tend to be omnivorous in their eating habits, the bulk of their diet consisting of small invertebrates, mollusks, a variety of crustaceans, and small fishes. They are said to bite on almost any bait and line. Fertilization is internal and is effected by the pelvic claspers. All chimaeras are oviparous, laying their eggs in large horny capsules. Chimaeras attain a length of about 20 cm.

MECHANISM OF INTOXICATION

Poisonings may result from ingesting the flesh, and the viscera of chimaeras is considered to be particularly dangerous to eat.

MEDICAL ASPECTS

Clinical Characteristics

There are no clinical reports available on oral chimaera biotoxications. Ingestion of the flesh is reputed to cause a stupefying effect or mental depression (Buniva, 1803). The viscera of the Pacific ratfish is also reported to be toxic (Halstead and Bunker, 1952). It is questionable that chimaeras become involved in the ciguatera form of ichthyosarcotoxism since their geographical distribution is largely beyond the usual range of ciguatera (Halstead, 1959).

Pathology

(Unknown)

Treatment

Symptomatic.

Prevention

Chimaeras should not be used for human consumption. Their viscera is believed to be particularly dangerous to eat.

TOXICOLOGY

The only published laboratory studies on the toxicity of the flesh of chimaeras are by Halstead and Bunker (1952). Aqueous extracts were prepared from the musculature and viscera of a large specimen of *Hydrolagus colliei* taken at the Walla Walla College Biological Station, Ship Harbor, Fidalgo Is., Wash., during June, 1950. The extracts were administered intraperitoneally into white Swiss-Webster mice. Only the extracts prepared from the oviducts were found to be moderately toxic. The remaining musculature and visceral extracts showed no evidence of toxicity.

PHARMACOLOGY

(Unknown)

CHEMISTRY

(Unknown)

LITERATURE CITED

Autenrieth, H. F.
1833 Ueber das gift der fische. C. F. Osiander, Tübingen. 287 p.

Banner, A. H., and P. Helfrich
1964 The distribution of ciguatera in the tropical Pacific. Hawaii Marine Lab. Tech. Rept. No. 3. Final Rept. NIII Contr. SA-43-ph-3741. 48 p.

Bigelow, H. B., and W. C. Schroeder
1948 Fishes of the western North Atlantic. Part I. Mem., Sears Found. Marine Res. 576 p., 106 figs.
1953 Fishes of the western North Atlantic. Part II. Mem., Sears Found. Marine Res. 588 p., 127 figs.

Bøje, O.
1939 Toxin in the flesh of the Greenland shark. Meddr. Grønland 125(5): 1-16.

Bouder, H., Cavallo, A., and M. J. Bouder
1962 Poissons vénéneux et ichtyosarcotoxisme. Bull. Inst. Océanogr. 59(1240), 66 p., 2 figs.

Buniva, C.
1803 Concernant la physiologie et la pathologie des poissons, suivi d'un tableau indiquant l'ictyographie subalpine. Mém. Acad. Sci. Litt. et Beaux-Arts Turin 12: 78-122.

Chevallier, A., and E. A. Duchesne
1851*a* Mémoire sur les empoisonnements par les huitres, les moules, les crabes, et par certains poissons de mer et de rivière. Ann. Hyg. Publ. (Paris) 45 (11): 387-437.
1851*b* Mémoire sur les empoisonnements par les huitres, les moules, les crabes, et par certains poissons de mer et de rivière. Ann. Hyg. Publ. (Paris) 46(1): 108-147.

Cooper, M. J.
1964 Ciguatera and other marine poisoning in the Gilbert Islands. Pacific Sci. 18(4): 411-440, 11 figs.

Coutaud, H.
1879 Observations sur sept cas d'empoisonnement par le foie de requin. M.D. Thèse 8. Fac. Méd. Paris. 47 p.

Coutière, H.
1899 Poissons venimeux et poissons vénéneux. Thèse Agrég. Carré et Naud, Paris. 221 p.

Dean, B.
1906 Chimaeroid fishes and their development. Carnegie Inst. Wash. Publ. 32. 194 p., 11 pls., 144 figs.

Fish, C. J., and M. C. Cobb
1954 Noxious marine animals of the central and western Pacific Ocean. U.S. Fish Wildlife Serv., Res. Rept. No. 36, p. 14-23.

Garman, S.
1913 The plagiostomia (sharks, skates, and rays). Mem. Mus. Comp. Zool. Harvard College. Vol. 36. 514 p., 75 pls.

Gley, E.
1904 Recherches sur le sang des Sélaciens. Action toxique du sérum de Torpille *(Torpedo marmorata)*. Compt. Rend. Acad. Sci. 138: 1547-1549.

Goode, G. B., and T. H. Bean
1895 Oceanic ichthyology. U.S. Nat. Mus., Spec. Bull. 553 p., 415 figs.

Günther, A.
1870 Catalogue of the fishes of the British Museum. Vol. 8. British Museum, London. 549 p.

Halstead, B. W.
1959 Dangerous marine animals. Cornell Maritime Press, Cambridge. 146 p., 88 figs.
1974 Marine biotoxicology, p. 212-239. *In* F. Coulston and F. Korte [eds.], Environmental quality and safety, vol. 3. Academic Press, Inc., New York.

Halstead, B. W., and N. C. Bunker
1952 The venom apparatus of the ratfish, *Hydrolagus colliei.* Copeia (3): 128-138, 1 pl., 4 figs.
1954 A survey of the poisonous fishes of Johnston Island. Zoologica 39(2): 61-77, 1 fig.

Halstead, B. W., and D. W. Schall
1958 A report on the poisonous fishes of the Line Islands. Acta Trop. 15(3): 193-233, 4 figs.

Helfrich, P.
1961 Fish poisoning in the tropical Pacific. Hawaii Marine Lab., Univ. Hawaii. 16 p., 7 figs.

HIYAMA, Y.
1943 Report of an investigation on poisonous fishes of the South Seas. [In Japanese] Nissan Fish. Exp. Sta. (Odawara, Japan), 137 p., 29 pls., 83 figs.

JENSEN, A. S.
1914 The selachians of Greenland. Mindeskrift Japetus Steenstrup. 30: 12-16.
1948 Contributions to the ichthyofauna of Greenland. Skrift. Univ. Zool. Mus. Copenhagen 9: 20-25.

JORDAN, D. S., and B. W. EVERMANN
1896 The fishes of North and Middle America. U.S. Nat. Mus., Bull. 47(1): 145, 149.

JOUBIN, L. M.
1929-31 Faune ichthyologique de l'Atlantique nord. Nos. 11-18. Andr. Fred. Høst & Fils, Copenhague.

LAGLER, K. F., BARDACH, J. E., and R. R. MILLER
1962 Ichthyology. Wiley and Sons, New York. 545 p., 14 figs.

LI, K.
1965 Ciguatera fish poison: a cholinesterase inhibitor. Science 147(3665): 1580-1581.

LONIE, T. C.
1950 Excess vitamin A as a cause of food poisoning. New Zealand Med. J. 49(274): 680-685.

MAASS, T. A.
1937 Gift-tiere. *In* W. Junk [ed.], Tabulae biologicae. Vol. XIII. N. V. Van de Garde & Co's Drukkerij, Zaltbommel, Holland. 272 p.

MACHT, D. I.
1942 An experimental appreciation of Leviticus XI. 9-12 and Deuteronomy XIV. 9-10. Hebrew Med. J. 2: 165-170.

MACHT, D. I., and E. C. SPENCER
1941 Physiological and toxicological effects of some fish muscle extracts. Proc. Soc. Exp. Biol. Med. 46(2): 228-233.

PAWLOWSKY, E. N.
1927 Gifttiere und ihre giftigkeit. Gustav Fischer, Jena. 515 p., 170 figs.

PHISALIX, M.
1922 Animaux venimeux et venins. 2 vols. Masson et Cie., Paris. 1: 1-659, 521 figs., 17 pls.; 2: 1-864, 521 figs., 17 pls.

ROEDEL, P. M.
1953 Common ocean fishes of the California coast. Calif. Dept. Fish Game, Fish. Bull. 91. 184 p., 175 figs.

SCHULTZ, L. P., HERALD, E. S., LACHNER, E. A., WELANDER, A. D., and L. P. WOODS
1953 Fishes of the Marshall and Marianas Islands. Vol. I Families from Asymmetrontidae through Siganidae. U.S. Nat. Mus., Bull. 202. 685 p., 74 pls., 89 figs.

SLOANE, H.
1707 A voyage to the islands Madera, Barbados, Nieves, S. Christophers and Jamaica. Vol. I. London. p. 22-29.
1725 A voyage to the islands Madera, Barbados, Nieves, S. Christophers and Jamaica. Vol. II. London. p. 275-291.

SMITH, J. L.
1961 The sea fishes of southern Africa. 4th ed. Central News Agency, South Africa. 580 p., 1,232 figs., 102 pls.

SMITH, R. O.
1947 Survey of the fisheries of the former Japanese Mandated Islands. U.S. Fish Wildlife Serv., Fish Leaflet 273. 105 p., 50 figs.

WHITLEY, G. P.
1940 The fishes of Australia. Part I. The sharks, rays, devil-fish, and other primitive fishes of Australia and New Zealand. Royal Zoological Society of New South Wales, Sydney. 280 p., 303 figs.

Chapter X—VERTEBRATES

Class Osteichthyes

Poisonous: CIGUATOXIC FISHES

The marine biotoxication generally referred to as "ciguatera" involves a broader phylogenetic spectrum of fishes and possibly other marine organisms than any other known type of human poisoning caused by the eating of marine animals. Because of the complexity and heterogeneity of etiological agents, ecological factors, and the somewhat elusive nature of the disease and the poison, any attempt to clearly delineate the total problem of ciguatera is presently beset with conflicting viewpoints, observations, and other difficulties. Needless to say, the problem is extremely complex and of vast importance to our future understanding of the total economy of the sea. The study of ciguatera poses some profound problems in the biogenesis of marine biotoxins. Ciguatera will become of increasing international importance to public health workers, fishery biologists, biochemists, nutritionists, and militarists in the future. The previous extensive literature on ciguatera has been concerned with only facets of the problem. This is the first attempt to bring together the total sum of published data on the subject of ciguatera and to present it in an organized manner. Despite this present effort, the problem of ciguatoxic fishes continues with questions of major importance unanswered.

The transvectors of ciguatera are believed to be restricted to marine fishes and possibly some marine invertebrates and algae. The disease is not known to occur in freshwater fishes. It appears that under proper circumstances any fish living in the sea may be a potential transvector of ciguatoxin. However, since the disease appears to be largely restricted to warm temperate, subtropical, and tropical regions, ciguatoxin is found most frequently in certain fish species, generally insular shore forms, although a few pelagic species have been incriminated.

The systematics of the fishes concerned with this chapter has been discussed by the following authors: Lacépède (1799-1804, 1819, 1833-35), Valenciennes (1835-50), Günther (1859-70), Day (1865, 1878-88, 1880-84), Jordan and Evermann (1896-1900, 1905, 1908), Fowler (1904, 1928, 1931, 1933, 1936, 1938, 1941, 1956, 1959), Jordan, Tanaka, and Snyder (1913), Meek and Hildebrand (1923, 1925, 1928), Joubin (1929-38), Kumada (1937), Longley and Hildebrand (1940, 1941), Schultz (1943), Herre (1953), Schultz *et al.* (1953), All Union Scientific Research Institute (1957), Gosline and Brock (1960), Schultz *et al.* (1960), and De Sylva (1963).

REPRESENTATIVE LIST OF FISHES REPORTED AS CIGUATOXIC[1]

Phylum CHORDATA

Class OSTEICHTHYES

Order CLUPEIFORMES (ISOSPONDYLI): Herrings, Anchovies, Etc.

Family ALBULIDAE

132 *Albula vulpes* (Linnaeus) (Pl. 1, fig. a). Bonefish, ladyfish, banana fish (USA), oio (Hawaii), te ikari (Gilbert Islands), bandang tjurorot (Indonesia).

DISTRIBUTION: Red Sea to Hawaii.

SOURCES: Ross (1947), Banner and Helfrich (1964), Bagnis *et al.* (1970).

Family CLUPEIDAE[2]

132 *Clupanodon thrissa* (Linnaeus) (Pl. 1, fig b). Sprat, thread herring, tassart, shad (USA), dorokui (Japan).

DISTRIBUTION: Indo-Pacific, China, Japan, Korea.

SOURCES: Chevallier and Duchesnc (1851*a, b*), Fonssagrives (1877), Savtschenko (1886), Hiyama (1943), Bagnis *et al.* (1970).
OTHER NAMES: *Clupea thrissa, Meletta thrissa*.

132 *Harengula humeralis* (Cuvier) (Pl. 2, fig. a). Red-ear sardine, sardine, pilchard, sprat, whitebill, pincer (USA), sardina, sardina de ley (West Indies).

DISTRIBUTION: Florida, Bermuda, West Indies to Brazil.

SOURCES: Corre (1865), Coutière (1899), Phisalix (1922), De Sylva (1963).
OTHER NAME: *Clupea humeralis*.

132 *Harengula ovalis* (Bennett) (Pl. 2, fig. b). Sardine (USA), tunsoy, sardina (Philippines).

DISTRIBUTION: Indo-Pacific, Red Sea.

SOURCES: Meyer-Ahrens (1855), Phisalix (1922), Ross (1947), Bagnis *et al.* (1970).
OTHER NAME: *Clupanodon venenosa, Clupea venenosa, Harengula punctata, Meletta venenosa*.

---

[1] The classification as herein presented follows that of Berg (1947), and Lagler, Bardach, and Miller (1962). Elasmobranch fishes that may possibly be ciguatoxic have not been included in this list. See Chapter IX on sharks, rays, etc.

[2] Some members of this family are also involved in clupeoid poisoning, but since clupeoid poisoning is a recent distinction, early references to toxic clupeoid fishes have been listed under both ciguatoxic and clupeotoxic categories. Some of the early references are too vague to determine the exact nature of the type of ichthyosarcotoxism.

Macrura ilisha (Buchanan-Hamilton) (Pl. 3, fig. a). Hilsa, sablefish (USA), palasoh, pulla, oolum (India), nga-tha-louk (Burma), matabelo (Indonesia), zoboor, soboor, shad (Iran). *133*

DISTRIBUTION: Persian Gulf, India, Sri Lanka, Burma, Vietnam.

SOURCES: Coutière (1899), Pawlowsky (1927), Hiyama (1943), Bagnis *et al.* (1970).
OTHER NAME: *Clupea ilisha*.

Nematolosa nasus (Bloch) (Pl. 3, fig. b). Gizzard shad, hairback herring, thread-finned gizzard shad, long-finned gizzard shad (USA, Australia), suagan, kabasi (Philippines), noonah, muddu candai (India), pla kōk, pla kup (Thailand), zobi, goaff, pahlowa, gooag (Iran). *133*

DISTRIBUTION: Indo-Pacific, India, Burma.

SOURCES: Autenrieth (1833), Coutière (1899), Bagnis *et al.* (1970).
OTHER NAME: *Clupea nasus*.

Opisthonema oglinum (LeSueur) (Pl. 4, fig. a). Atlantic thread herring, hairyback, bristle herring (USA), sargo, sargo de gato, sardinha large, tassart, sardinha sargo, caillen, machuelo (West Indies). *133*

DISTRIBUTION: West Indies, occasionally to Cape Cod.

SOURCES: Chisholm (1808), Orfila (1817), Fonssagrives (1877), Randall (1958).
OTHER NAMES: *Clupea thrissa, Clupea thryssa, Megalops thrissa, Meletta theissa, Meletta thrissa.*

Sardinella perforata (Cantor) (Pl. 4, fig. b). Sardine (USA), tembang, tamban (Malaysia), tamban (Philippines), lemuru (Indonesia). *133*

DISTRIBUTION: Indo-Pacific, Arabia, Persian Gulf.

SOURCES: Coutière (1899), Phisalix (1922), Bagnis *et al.* (1970).
OTHER NAMES: *Clupea perforata, Harangula perferata, Clupeonia perforata.*

Family ENGRAULIDAE

Thrissina baelama (Forskål) (Pl. 5). Anchovy (USA), dumpilas, tigi (Philippines), trich (Indonesia). *134*

DISTRIBUTION: Indo-Pacific, Red Sea, enters rivermouths.

SOURCES: Duméril (1866), Day (1871), Bagnis *et al.* (1970).
OTHER NAMES: *Anchovia baelama, Engraulis baelama, Thrissocles baelama.*

Order ANGUILLIFORMES (APODES): True Eels

Family MURAENIDAE

Gymnothorax flavimarginatus (Rüppell) (Pl. 6, fig. a). Moray eel (USA), puhikapa (Hawaii), dreb, jaunagi (Marshall Islands), kwatuma (east *134*

Africa), palang (South Africa), payangitan, malabanos, pinangitan, barason, buriwaran, indong, labung, ogdok, pananglitan, ubod (Philippines), dabea (Fiji), to'e (Samoa), hagman (Marianas), reef eel (Australia), tekaibiki, terebona, tebukimeri, teimone (Gilbert Islands), kiari, kuiru, makiki, takataka, hamurenga, tiohu, tavere (Tuamotus), utsubo (Japan), divi gal gulla, kabara gal gulla, tamil anjalai (Sri Lanka).

DISTRIBUTION: Indo-Pacific, east Africa.

SOURCES: Khlentzos (1950), Ralls and Halstead (1955), Bartsch and McFarren (1962), Banner and Helfrich (1964).

134 *Gymnothorax javanicus* (Bleeker) (Pl. 6, fig. b). Moray eel (USA), dabea (Fiji), to'e (Samoa), hagman (Marianas), puhi (Hawaii), dreb, jaunagi (Marshall Islands), tekaibiki, terebona, tebukimeri, teimone (Gilbert Islands), kiari, kuiru, makiki, takataka, hamurenga, tiohu, tavere (Tuamotus), malabanos, payangitan, pinangitan, barason, buriwaran, indong, labung, ogdok, pananglitan, ubod (Philippines), utsubo (Japan), divi gal gulla, kabara gal gulla, tamil anjalai (Sri Lanka), kwatuma, palang (South Africa).

DISTRIBUTION: Indo-Pacific, east Africa.

SOURCES: Hiyama (1943),[3] Fish and Cobb (1954), Halstead and Lively (1954), Yasumoto and Scheuer (1969), Rayner (1972).

134 *Gymnothorax meleagris* (Shaw and Nodder) (Pl. 6, fig. c). Moray eel (USA), dabea (Fiji), to'e (Samoa), hagman (Marianas), puhi (Hawaii), dreb, jaunagi (Marshall Islands), tekaibiki, terebona, tebukimeri, teimone (Gilbert Islands), kiari, kuiru, makiki, takataka, hamurenga, tiohu, tavere (Tuamotus), malabanos, payangitan, pinangitan, barason, buriwaran, indong, labung, ogdok, pananglitan, ubod (Philippines), utsubo (Japan), divi gal gulla, kabara gal gulla, tamil anjalai (Sri Lanka), kwatuma, palang (South Africa).

DISTRIBUTION: Indo-Pacific, Japan.

SOURCES: Hiyama (1943), Fish and Cobb (1954), Ralls and Halstead (1955).
OTHER NAME: *Lycodontis meleagris.*

135 *Gymnothorax undulatus* (Lacépède) (Pl. 7, fig. a). Moray eel (USA), dabea (Fiji), hagman (Marianas), to'e (Samoa), puhi (Hawaii), dreb, jaunagi (Marshall Islands), tekaibiki, terebona, tebukimeri, teimone (Gilbert Islands), kiari, kuiru, makiki, takataka, hamurenga, tiohu, tavere (Tuamotus), malabanos, payangitan, pinangitan, barason, buriwaran, indong,

---

[3] The illustration shown in Hiyama's (1943) work is mistakenly identified as *G. flavimarginatus*, but should be *G. javanicus*.

labung, ogdok, panaglitan, ubod (Philippines), utsubo (Japan), divi gal galla, kabara gal gulla, tamil anjalai (Sri Lanka), kwatuma, palang (South Africa).

DISTRIBUTION: Indo-Pacific, east Africa, Red Sea.

SOURCES: Hiyama (1943), Lin, Lin, and Yang (1953), Randall (1961).
OTHER NAME: *Lycodontis undulata.*

Muraena helena Linnaeus (Pl. 7, fig. b). Moray eel (USA), murry (England), muräne (Germany), murène (France), murena (Italy, Spain). *135, 258*

DISTRIBUTION: Eastern Atlantic Ocean and Mediterranean Sea.

SOURCES: Autenrieth (1833), Chevallier and Duchesne (1851*a, b*). Phisalix (1922), Pawlowsky (1927), Bagnis *et al.* (1970).
OTHER NAMES: *Gymnothorax helena, Muraena punctatus.*

Order BELONIFORMES (SYNENTOGNATHI): Needlefish, Halfbeaks, Flyingfishes, Etc.

Family BELONIDAE

Strongylura caribbaea (LeSueur) (Pl. 8, fig. a). Needlefish, houndfish (USA), agujón (West Indies). *135*

DISTRIBUTION: West Indies.

SOURCES: Fonssagrives (1877), Coutière (1899), Pellegrin (1899), Bagnis *et al.* (1970).
OTHER NAMES: *Belone caribaea, Belone carribaea.*

Family HEMIRAMPHIDAE

Hemiramphus brasiliensis (Linnaeus) (Pl. 8, fig. b). Ballyhoo, halfbeak (USA), balao, escribano (West Indies). *135*

DISTRIBUTION: Tropical Atlantic.

SOURCES: Cloquet (1821), Chevallier and Duchesne (1851*a, b*), Coutière (1899).
OTHER NAMES: *Belone brasiliensis, Esox brasiliensis.*

Order BERYCIFORMES (BERYCOMORPHI): Squirrelfishes, Soldierfishes, Etc.

Family HOLOCENTRIDAE

Myripristis murdjan (Forskål) (Pl. 9). Soldierfish (USA), uu, menpachi (Hawaii), mon (Marshall Islands), sagamilung (Marianas), iihi, maunauma (Tahiti), baga baga, suga-suga, tinik tinik (Philippines), malau, malautea, mamo, segasega (Samoa), te kungkung (Gilbert Islands), mas laut, sowangi (Indonesia), ittodai (Japan). *136*

DISTRIBUTION: Indo-Pacific.

SOURCES: Seale (1912*a, b*), Halstead and Schall (1958), Bagnis *et al.* (1970).

Order PERCIFORMES (PERCOMORPHI): Perchlike Fishes

Family ACANTHURIDAE

136 *Acanthurus chirurgus* (Bloch) (Pl. 10, fig. a). Surgeonfish (USA).
DISTRIBUTION: Tropical Atlantic.

SOURCES: Savtschenko (1886), Banner and Helfrich (1964).
OTHER NAME: *Acanthurus nigricans.*

136 *Acanthurus lineatus* (Linnaeus) (Pl. 10, fig. b). Lined surgeonfish (USA), maroa (Tahiti), katawa (Gilbert Islands), hidsung (Marianas), labahita, indangan, mangadlit, yaput (Philippines), venaki, pidja, abila (Indonesia), imim, ael, bir, diepro, tiebro (Marshall Islands), pone, palagi, nanife, ili ilia (Samoa), balagi (Fiji), surgeonfish (Australia).
DISTRIBUTION: Indo-Pacific, excluding Hawaii.

SOURCES: Jordan (1929), Cooper (1964), Banner and Helfrich (1964), Yasumoto *et al.* (1971).
OTHER NAME: *Hepatus lineatus.*

137 *Acanthurus nigrofuscus* (Forskål) (Pl. 11, fig. a). Surgeonfish, tang, doctorfish (USA), pakuikui (Hawaii), labahita, indangan, mangadlit, yaput (Philippines), vanaki, pidja, abila (Indonesia), te nimunai, te nrakerake (Gilbert Islands), imim, ael, bir, diepro, tiebro (Marshall Islands), apii, malto horopape, maroa, umeurne (Tahiti), pone, palagi, nanife, ili ilia (Samoa), balagi (Fiji), surgeonfish (Australia).
DISTRIBUTION: Indo-Pacific.

SOURCES: Hiyama (1943), Lin *et al.* (1953), Halstead and Schall (1958), Bagnis *et al.* (1970).
OTHER NAMES: *Acanthurus elongatus, Hepatus nigrofuscus.*

137 *Acanthurus olivaceus* Bloch and Schneider (Pl. 11, fig. b). Orange spot surgeonfish, tang (USA), ael (Marshall Islands), pakuikui (Hawaii), labahita, indangan, mangadlit, yaput (Philippines), vanaki, pidja, abila (Indonesia), te nimunai, te nrakerake (Gilbert Islands), apii, malto horopape, maroa, umeurne (Tahiti), pone, palagi, nanife, ili ilia (Samoa), balagi (Fiji), surgeonfish (Australia).
DISTRIBUTION: Indo-Pacific.

SOURCES: Hiyama (1943), Halstead and Bunker (1954*b*), Banner and Helfrich (1964).
OTHER NAME: *Hepatus olivaceus.*

Acanthurus triostegus (Linnaeus) (Pl. 11, fig. c). Convict surgeonfish, tang (USA), manini (Hawaii), kuban (Marshall Islands), koinawa (Gilbert Islands), shimahagi (Japan), five-banded surgeonfish (Australia). *137*

Distribution: Indo-Pacific.

Sources: Hiyama (1943), Halstead (1958), Banner and Helfrich (1964), Helfrich and Banner (1968), Helfrich, Piyakarnchana, and Miles (1968).

Acanthurus xanthopterus (Cuvier and Valenciennes) (Pl. 11, fig. d). Surgeonfish (USA), pualu (Hawaii), mako (Gilbert Islands), labahita, indangan, mangadlit, yaput (Philippines), vanaki, pidja, abila (Indonesia), imim, ael, bir, diepro, tiebro (Marshall Islands), apii, malto horopape, maroa, umeurne (Tahiti), pone, palagi, nanife, ili ilia (Samoa), balagi (Fiji). *137*

Distribution: Indo-Pacific.

Sources: Halstead and Bunker (1954*a*), Banner *et al.* (1963*b*), Cooper (1964), Helfrich and Banner (1968), Helfrich *et al.* (1968).
Other names: *Acanthurus crestonis, Acanthurus fuliginosus*.

Ctenochaetus striatus (Quoy and Gaimard) (Pl. 12, fig. a). Surgeonfish (USA), maito (Tahiti), te ribabui (Gilbert Islands), sazanamihagi (Japan), labahita, indangan, mangadlit, yaput (Philippines), vanaki, pidja, abila (Indonesia), imim, ael, bir, diepro, tiebro (Marshall Islands), pone, palagi, nanife, ili ilia (Samoa), balagi (Fiji). *138*

Distribution: Indo-Pacific excluding Hawaii, Red Sea.

Sources: Hiyama (1943), Halstead and Bunker (1954*a*), Helfrich and Banner (1963), Helfrich *et al.* (1968), Yasumoto *et al.* (1971).
Other name: *Ctenochaetus strigosus*.

Ctenochaetus strigosus (Bennett) (Pl. 10, fig. c). Surgeonfish (USA), kala (Hawaii), maito (Tahiti), te ribabui (Gilbert Islands), sazanami-hagi (Japan), labahita, indangan, mangadlit, yaput (Philippines), vanaki, pidja, abila (Indonesia), imim, ael, bir, diepro, tiebro (Marshall Islands), pone, palagi, nanife, ili ilia (Samoa), balagi (Fiji). *136*

Distribution: Indo-Pacific.

Sources: Hiyama (1943), Halstead and Bunker (1954*b*), Cooper (1964).
Other name: *Ctenochaetus striatus*.

Naso lituratus Bloch and Schneider (Pl. 12, fig. b). Unicornfish (USA), kala, kalaholo (Hawaii), balak (Marshall Islands), tabaku-tabaku (Philippines), ili ilia (Samoa), te nrakerake (Gilbert Islands), maito (Tahiti), galak karang (Indonesia). *138*

Distribution: Indo-Pacific, Red Sea.

Sources: Halstead and Bunker (1954*b*), Bartsch *et al.* (1959).

138 *Zebrasoma veliferum* (Bloch) (Pl. 12, fig. c). Sailfin tang (USA), kihikihi, api (Hawaii).

DISTRIBUTION: Indo-Pacific, Red Sea.

SOURCES: Hiyama (1943), Fish and Cobb (1954), Bouder, Cavallo, and Bouder (1962).

Family CARANGIDAE

139 *Caranx crysos* (Mitchill) (Pl. 13, fig. a). Blue runner, jack (USA), cojinuda (West Indies).

DISTRIBUTION: Tropical Atlantic.

SOURCES: Mowbray (1916), Arcisz (1950), Bagnis *et al.* (1970).
OTHER NAME: *Caranx chrysos.*

139 *Caranx hippos* (Linnaeus) (Pl. 13, fig. b). Crevalle, jack (USA), crevalle (West Indies).

DISTRIBUTION: Tropical Atlantic.

SOURCES: Cloquet (1821), Fonssagrives and de Méricourt (1861), Taschenberg (1909), Fish and Cobb (1954), Bouder *et al.* (1962).
OTHER NAMES: *Caranx carangus, Scomber carangus.*

139 *Caranx ignobilis* (Forskål) (Pl. 13, fig. c). Kingfish, jack, crevalle, cavalla (USA), ulua (Hawaii), rewa, lane (Marshall Islands), lupo (Samoa), paruku (Tuamotus), teaongo, tekuana (Gilbert Islands), saga (Fiji), talakitok, momsa, atoloy (Philippines).

DISTRIBUTION: Indo-Pacific.

SOURCES: Halstead and Schall (1958), Banner *et al.* (1963*b*), Helfrich and Banner (1968), Helfrich *et al.* (1968).

139 *Caranx latus* Agassiz (Pl. 13, fig. d). Horse-eye jack (USA), jurel (West Indies).

DISTRIBUTION: Tropical western Atlantic.

SOURCES: Moquin-Tandon (1860), Nielly (1881), Pellegrin (1899), Gilman (1942), De Sylva (1963), Larson and Rothman (1967).
OTHER NAME: *Caranx fallax.*

139 *Caranx lugubris* Poey (Pl. 13, fig. e). Jack, crevalle (USA), white ulua (Hawaii), lupo (Samoa), paruku (Tuamotus), teaongo, tekuana (Gilbert Islands), saga (Fiji), lane (Marshall Islands), talakitok, momsa, atoloy (Philippines), cavalla, cavalho, cocinero (Latin America), kingfish, cavally (South Africa), sante, tanet, caballa, bouebouesina, cotro, ogombo, gaoua, (west Africa).

Distribution: Circumtropical.

Sources: Coutière (1899), Halstead and Bunker (1954*a*, *b*), Banner and Helfrich (1964).
Other names: *Caranx ascensionis, Caranx frontalis.*

Caranx melampygus Cuvier (Pl. 14, fig. a). Blue crevally, jack, crevalle (USA), omilu, hishi, ulua (Hawaii), lupo (Samoa), paruku (Tuamotus), teaongo, tekuana (Gilbert Islands), saga (Fiji), lane, deltokrok (Marshall Islands), talakitok, momsa, atoloy (Philippines). *140*
Distribution: Indo-Pacific.

Sources: Hiyama (1943), Halstead and Schall (1956, 1958), Banner and Helfrich (1964), Helfrich *et al.* (1968), Niaussat, Gak, and Ehrhardt (1969).
Other name: *Caranx ascensionis.*

Caranx sexfasciatus Quoy and Gaimard (Pl. 14, fig. b). Jack, crevalle (USA), ulua, pake ulua, mempachi ulua, papio—young (Hawaii), lupo (Samoa), paruku (Tuamotus), teaongo, tekuana (Gilbert Islands), saga (Fiji), lane (Marshall Islands), talakitok, momsa, atoloy (Philippines). *140*
Distribution: Indo-Pacific.

Sources: Savtschenko (1886), Fish and Cobb (1954), Halstead and Schall (1958), Bartsch *et al.* (1959), Bouder *et al.* (1962).
Other names: *Caranx fallax, Caranx hippos, Caranx lessoni.*

Elagatis bipinnulatus (Quoy and Gaimard) (Pl. 14, fig. c). Rainbow runner (USA), kamanu, Hawaiian salmon (Hawaii), cojinua (Latin America), prodigal son, rainbow runner (South Africa), runner (Australia). *140*
Distribution: Circumtropical.

Sources: Mann (1938), O'Neill (1938), Helfrich (1963).
Other name: *Elongatus bipinnulatus* (misspelled).

Selar crumenophthalmus (Bloch) (Pl. 15). Bigeye scad, horse-eye jack, goggle-eye jack (USA), chicharro (West Indies), akule (Hawaii), matang, baka, bunutan, atoloy (Philippines), mijinga, korkofanha (west Africa). *140*
Distribution: Circumtropical.

Sources: Corre (1865), Phisalix (1922), Bouder *et al.* (1962).
Other names: *Caranx crumenophthalmus, Caranx plumieri, Trachurops crumenopthalmus.*

Seriola dumerili (Risso) (Pl. 16, fig. a). Amberjack, yellowtail (USA), serviola (Spain), kahala (Hawaii), sériola (France), ricciola (Italy), seriola fisch (Germany), peixe limão (Portugal), cojinua (Latin America). *141*
Distribution: Indo-Pacific, Mediterranean to West Indies.

SOURCES: Coutière (1899), Hoffmann (1927, 1929*a*, *b*), Banner and Helfrich (1964).
OTHER NAMES: *Seriola gigas, Seriola proxima, Zonichtys gigas.*

141 *Seriola falcata* Valenciennes (Pl. 16, fig. b). Almaco jack (USA), madregal (West Indies).
DISTRIBUTION: West Indies, north to the Carolinas.

SOURCES: Hoffmann (1927, 1929*a*, *b*), Gilman (1942), Randall (1958), Bagnis *et al.* (1970).

Family CORYPHAENIDAE

141 *Coryphaena hippurus* Linnaeus (Pl. 17, figs. a, b). Dolphin (USA, Australia, South Africa), mahi-mahi (Hawaii), dorado (Latin America), kui tou tao (Taiwan), akpei, apem, derefi (west Africa), koko (Marshall Islands), shiira (Japan), dourado macho (Portugal), gemeine bramen (Germany), hampuga (Spain), coryphene (France), lumpuga (Italy).
DISTRIBUTION: Pelagic, in all tropical and temperate seas.

SOURCES: Orfila (1817), Chisholm (1821), Pawlowsky (1927), Wheeler (1953), Helfrich (1963).
OTHER NAME: *Coryphaena dorado.*

Family GOBIIDAE

142 *Ctenogobius criniger*[4] (Cuvier and Valenciennes) (Pl. 18). Goby (USA), oopu (Hawaii), bia (Philippines), jibalan (Marshall Islands).
DISTRIBUTION: Indo-Pacific, Australia.

SOURCES: Corre (1865), Savtschenko (1886), Pawlowsky (1927), Herre (1953), Noguchi, Kao, and Hashimoto (1971).
OTHER NAMES: *Gobius criniger, Rhinogobius nebulosus.*

Family LABRIDAE

142 *Coris gaimardi* (Quoy and Gaimard) (Pl. 19). Wrasse, lazy fish (USA), hialea akilolo (Hawaii), jollol (Marshall Islands), bagondon, talad (Philippines).
DISTRIBUTION: Indo-Pacific.

SOURCES: Hiyama (1943), Randall (1958), Bouder *et al.* (1962).

142 *Epibulus insidiator* (Pallas) (Pl. 20). Wrasse (USA), teuianau (Gilbert Islands), mo (Marshall Islands), sling-jaw (Australia).
DISTRIBUTION: Indo-Pacific.

SOURCES: Hiyama (1943), Randall (1958), Bouder *et al.* (1962).

---

[4] Hashimoto and Noguchi (1971) have obtained a tetrodotoxin-like substance in *C. criniger* and have suggested the poison may be more closely related to tetradotoxin than ciguatoxin.

Family LUTJANIDAE

Aprion virescens Cuvier and Valenciennes (Pl. 21, fig. a). Streaker, blue snapper (USA), uku (Hawaii), laum, suzuki (Marshall Islands), aona, aomachi, omachi (Saipan), guntal (Philippines), kaakap (Indonesia), jobfish (Australia). *143*

Distribution: Indo-Pacific.

Sources: Hiyama (1943), Bouder *et al.* (1962), Banner and Helfrich (1964).

Gnathodentex aureolineatus (Lacépède) (Pl. 21, fig. b). Snapper (USA), maene (Tahiti), tinar (Marshall Islands). *143*

Distribution: Indo-Pacific.

Sources: Hiyama (1943), Banner and Helfrich (1964).
Other name: *Pentapus aurolineatus.*

Gymnocranius griseus (Temminck and Schlegel) (Pl. 21, fig. c). Snapper (USA), kanulo, labongan (Philippines), mijmij (Marshall Islands), seabream (Australia). *143*

Distribution: Indo-Pacific.

Source: Bartsch *et al.* (1959).
Other name: *Gymnocranius microdon.*

Lethrinus harak (Forskål) (Pl. 22, fig. a). Scavenger, snapper (USA), batarde (Mauritius), drijing, mamennie (Marshall Islands), emperor (Australia). *144*

Distribution: Indo-Pacific, Red Sea.

Sources: Wheeler (1953), Randall (1958).

Lethrinus mambo Montrouzier (*species inquirendae*). Scavenger, snapper (USA).

Distribution: New Caledonia.

Sources: Fonssagrives and de Méricourt (1861), Blanchard (1890), Hiyama (1943), Bouder *et al.* (1962).

Lethrinus miniatus (Forster) (Pl. 22, fig. b). Scavenger, grey snapper, sweetlips (USA), oeo (Tahiti), jalia (Marshall Islands), fuefuki (Japan), emperor (Australia), scavenger (South Africa). *144*

Distribution: Indo-Pacific, Japan, east Africa.

Sources: Jordan (1929), Hiyama (1943), Banner and Helfrich (1964), Helfrich *et al.* (1968).
Other names: *Lethrinella miniata, Lethrinus rostratus.*

Lethrinus variegatus Cuvier and Valenciennes (Pl. 22, fig. c). Snapper, scavenger, porgy (USA), meko (Tuamotus), teokaoka (Gilbert Islands), *144*

mameni, net, woeo (Marshall Islands), kutambak, sapingan, bukutat (Philippines), mati-hari (Indonesia), variegated emperor (Australia), scavenger (South Africa).
DISTRIBUTION: Indo-Pacific.

SOURCES: Hiyama (1943), Halstead and Schall (1958), Helfrich *et al.* (1968).
OTHER NAME: *Lethrinus semicinctus.*

145 *Lutjanus argentimaculatus* (Forskål) (Pl. 23, fig. a). Snapper (USA), nanue (Samoa), kanulo, labongan (Philippines), mangrove jack (Australia).
DISTRIBUTION: Indo-Pacific, Australia.

SOURCES: Coutière (1899), Phisalix (1922), Banner and Helfrich (1964).
OTHER NAME: *Mesoprion argentimaculatus.*

145 *Lutjanus aya* (Bloch) (Pl. 23, fig. b).[5] Red snapper (USA), pargo colorado, pargo guachinango (West Indies).
DISTRIBUTION: Western tropical Atlantic.

SOURCES: Gilman (1942), Arcisz (1950), Randall (1958), Bagnis *et al.* (1970).
OTHER NAME: *Lutjanus blackfordi.*

145 *Lutjanus bohar* (Forskål) (Pl. 23, fig. c). Red snapper (USA), haamea (Tahiti), jab, baan, pan (Marshall Islands), mumea (Samoa), anglais (New Caledonia), ingo (Gilbert Islands), mylah, tambak (Indonesia), aknulo, tabon (Philippines), bati, damu (Fiji), red bass (Australia), akamasu (Japan).
DISTRIBUTION: Indo-Pacific, Red Sea, East Indies.

SOURCES: Jordan (1929), Hiyama (1943), Halstead and Bunker (1954*a*), Randall (1958, 1961), Banner *et al.* (1963*a, b*), McFarren *et al.* (1965), Helfrich *et al.* (1968).

145 *Lutjanus gibbus* (Forskål) (Pl. 23, fig. d). Red snapper (USA), ikanibong (Gilbert Islands), jab (Marshall Islands), fuena, mimija (Okinawa), malai (Samoa), tuhara (Tahiti), damu (Fiji), maya-maya, tapako (Philippines), fuedokutarumi (Japan), chemise (Seychelles, Mauritius), mylah, tambak (Indonesia), paddletail (Australia).
DISTRIBUTION: Tropical Indo-Pacific.

SOURCES: Hiyama (1943), Halstead and Lively (1954), Helfrich *et al.* (1968).
OTHER NAME: *Genyoroge melanura.*

145 *Lutjanus janthinuropterus* (Bleecker) (Pl. 23, fig. e). Snapper (USA), hashi-yantobi (Japan), dark-tailed sea perch (Australia).

[5] There is considerable question regarding the relationship of *Lutjanus aya* and *L. blackfordi.* Some authors consider them to be distinct species whereas others do not.

Distribution: Indo-Pacific.

Sources: Hiyama (1943), Fish and Cobb (1954), Bouder *et al.* (1962).
Other name: *Lutjanus flavipes.*

Lutjanus jocu (Bloch and Schneider) (Pl. 24, fig. a). Dog snapper (USA), jocu (West Indies). *146*

Distribution: Florida and West Indies, south to Brazil.

Sources: Fonssagrives and de Méricourt (1861), Coutière (1899), Hoffmann (1927, 1929*a, b*), Randall (1958), Bagnis *et al.* (1970).
Other names: *Anthias jocu, Mesoprion jocu, Mesoprion yocu.*

Lutjanus kasmira (Forskål) (Pl. 24, fig. b). Snapper (USA, South Africa), tanda tanda (Indonesia), taape (Tahiti), blue-banded sea perch (Australia). *146*

Distribution: Indo-Pacific, South Africa.

Sources: Randall (1958), Banner and Helfrich (1964), Cooper (1964).

Lutjanus monostigmus (Cuvier and Valenciennes) (Pl. 24, fig. c). Snapper (USA), tairaira (Tahiti), jawajo, jeblo, kelikrok (Marshall Islands), akajamatobi (Ryukyus), baweina (Gilbert Islands), black-spotted sea perch (Australia). *146*

Distribution: Indo-Pacific, Red Sea.

Sources: Jordan (1929), Halstead and Schall (1958), Banner *et al.* (1960), Helfrich *et al.* (1968).
Other names: *Lutjanus fulviflamma, Lutjanus leioglossus, Lutjanus tuluttharmna* (no such species—erroneously translated from the Japanese—Banner and Helfrich, 1964; Banner, 1965).

Lutjanus nematophorus (Bleeker)[6] (Pl. 25, fig. a). Chinaman fish, snapper (USA, Australia).[7] *147*

Distribution: Australia.

Sources: Whitley (1943*a, b*), Cleland (1942), Wheeler (1953), Banner and Helfrich (1964).
Other name: *Paradicichthys venenatus.*

---

[6] According to Whitley (1943*a, b*), *Paradicichthys venenatus* Whitley is the adult form of *Lutjanus nematophorus* (Bleeker) as evidenced by the illustrations and legends presented in Whitley's report. Hence, *P. venenatus* can no longer be considered as valid since *L. nematophorus* was first described by Bleeker (1862-77).

[7] Some of the members of this family have also been incriminated as causing hallucinogenic fish poisoning. Consequently, some of these species will be found under both sections. The relationship of hallucinogenic fish poisoning to ciguatera—if any—is unknown at this time. See section on hallucinogenic fish poisoning, page 587.

147 *Lutjanus semicinctus* Quoy and Gaimard (Pl. 25, fig. b). Snapper (USA), tinaemea (Gilbert Islands).
DISTRIBUTION: Indo-Pacific.

SOURCES: Hiyama (1943), Fish and Cobb (1954), Bouder *et al.* (1962).

147 *Lutjanus vaigiensis* (Quoy and Gaimard) (Pl. 26, fig. a). Red snapper (USA), toau (Tahiti), jato, jeblo, jej, kelikrok (Marshall Islands), bawe, (Gilbert Islands), akana (Ryuykus), mos (western Carolines), akadokut-arumi (Japan), yellow-margined sea perch (Australia).
DISTRIBUTION: Indo-Pacific, East Coast of Africa, Australia.

SOURCES: Hiyama (1943), Dawson, Aleem, and Halstead (1955), Halstead (1958).
OTHER NAME: *Lutjanus flavipes.*

147 *Monotaxis grandoculis* (Forskål) (Pl. 26, fig. b). Bigeye burgy, snapper (USA), mu (Tahiti), kielotan, kie (Marshall Islands).
DISTRIBUTION: Indo-Pacific.

SOURCES: Hiyama (1943), Randall (1958, 1961), Helfrich *et al.* (1968).

Family MUGILIDAE

148 *Chelon vaigiensis* (Quoy and Gaimard) (Pl. 27, fig. a). Mullet (USA), aua (Samoa), akor, moreat, yol (Marshall Islands), balanak, bunsit, taguman (Philippines), diamond-scaled mullet (Australia).
DISTRIBUTION: Indo-Pacific.

SOURCES: Dawson *et al.* (1955), Bartsch and McFarren (1962).
OTHER NAME: *Mugil vaigiensis.*

148 *Mugil cephalus* Linnaeus (Pl. 27, fig. b). Common mullet, striped mullet (USA), cefalo (Italy), macho, machuto, lisa cabezuda (Latin America), gemeine meeräsche (Germany), cabot (France), tainha (Portugal), cabezudo (Spain).
DISTRIBUTION: Cosmopolitan.

SOURCES: Mills (1956), Helfrich (1961, 1963).

Family MULLIDAE

148 *Mulloidichthys auriflamma* (Forskål) (Pl. 28, fig. a). Goatfish, salmonet, surmullet (USA, Australia, South Africa), weke-ula (Hawaii), baybao, tubac, tuyo (Philippines), jome (Marshall Islands), tebaweina (Gilbert Islands), afolu i'a sina (Samoa).
DISTRIBUTION: Indo-Pacific, Red Sea.

SOURCES: Halstead and Bunker (1954*b*), Banner and Helfrich (1964).

Mulloidichthys samoensis (Günther) (Pl. 28, fig. b). Goatfish, salmonet, surmullet (USA, South Africa), wek'a'a (Hawaii), baybao, tubac, tuyo (Philippines), jome (Marshall islands), tebaweina (Gilbert Islands), afolu i'a sina (Samoa), gold-striped goatfish (Australia). *148*

DISTRIBUTION: Indo-Pacific.

SOURCES: Halstead and Bunker (1954*a, b*), Banner and Helfrich (1964).

Parupeneus chryserydros (Lacépède) (Pl. 29, fig. a). Goatfish, salmonet, surmullet (USA, Australia, South Africa), weke (Hawaii), baybao, tubac, tuyo (Philippines), jome (Marshall Islands), tebaweina (Gilbert Islands), afolu i'a sina (Samoa). *149*

DISTRIBUTION: Indo-Pacific, east Africa.

SOURCES: Nichols and Bartsch (1945), Halstead and Bunker (1954*b*), Bagnis *et al.* (1970).

OTHER NAMES: *Pseudupeneus chryserydros, Upeneus chryserydros.*

Upeneus arge Jordan and Evermann (Pl. 29, fig. b). Goatfish, salmonet, surmullet (USA, Australia, South Africa), weke pueo, weke pahula, nightmare weke, crazy surmullet (Hawaii), baybao, tubac, tuyo (Philippines), jome (Marshall Islands), tebaweina (Gilbert Islands), afolu i'a sina (Samoa). *149*

DISTRIBUTION: Indo-Pacific.

SOURCES: Titcomb and Pukui (1952), Helfrich (1961, 1963).

Family POMACENTRIDAE

Abudefduf septemfasciatus (Cuvier) (Pl. 30). Sergeant major, damselfish (USA), maomao (Hawaii), ulavapua, alala saga, mutu (Samoa), bakej (Marshall Islands), tebukibuki (Gilbert Islands), palata (Philippines), damselfish (Australia). *149*

DISTRIBUTION: Indo-Pacific, Australia, east Africa, China.

SOURCES: Halstead and Bunker (1954*a*), Bartsch *et al.* (1959), Bagnis *et al.* (1970).

Family SCARIDAE[8]

Chlorurus pulchellus (Rüppell) (Pl. 31, fig. a). Parrotfish (USA), lala (Marshall Islands), fuga (Samoa), pahoro (Tahiti), ulavi (Fiji), bontog, loro, mulmol (Philippines), kakatoi (Seychelles). *150*

---

[8] The parrotfish family frequently appears in the literature under the family Callyodontidae with the principal genus listed as *Callyodon*. However, most modern fish systematists have adopted the family name as Scaridae and the type genus as *Scarus*. The parrotfishes listed in this chapter are named according to Schultz' (1958) " Review of the parrotfishes Family Scaridae."

DISTRIBUTION: Indo-Pacific.

SOURCES: Bartsch *et al.* (1959), Bartsch and McFarren (1962).

150 *Scarus coeruleus* (Bloch) (Pl. 31, fig. b). Blue parrotfish (USA), loro (West Indies).

DISTRIBUTION: Florida, West Indies, Panama.

SOURCES: Chisholm (1808), Coutière (1899), Maass (1937), Bagnis *et al.* (1970).

150 *Scarus ghobban* Forksål (Pl. 31, fig. c). Parrotfish (USA, Australia), bera (Japan), fuga (Samoa), pahoro (Tahiti), alwon, noelok (Marshall Islands), ulavi (Fiji), bontog, loro, mulmol (Philippines), kakatoi (Seychelles).

DISTRIBUTION: Indo-Pacific, Japan, Red Sea, Australia.

SOURCES: Meyer-Ahrens (1855), Phisalix (1922), Bagnis *et al.* (1970).
OTHER NAMES: *Pseudoscarus pyrrhostethus, Scarus psittacus.*

151 *Scarus microrhinos* Bleeker (Pl. 32, fig. a). Parrotfish (USA, Australia), uha nanao (Tahiti), mao (Marshall Islands), fuga (Samoa), ulavi (Fiji), bontog, loro, mulmol (Philippines), kakatoi (Seychelles).

DISTRIBUTION: Indo-Pacific.

SOURCES: Smith (1947), Halstead and Lively (1954), Bouder *et al.* (1962).
OTHER NAME: *Callyodon microrhinos.*

151 *Scarus vetula* Bloch and Schneider (Pl. 32, fig. b). Queen parrotfish (USA), vieja (West Indies).

DISTRIBUTION: West Indies to Florida.

SOURCES: Corre (1865), Fonssagrives (1877), Gomez (1926), Bagnis *et al.* (1970).

Family SCOMBRIDAE[9]

152 *Acanthocybium solandri* (Cuvier) (Pl. 33, fig. a). Wahoo (USA, South Africa), ono (Hawaii), kamasusawara, okamasu (Japan), wahoo, peto (Australia), wahoo, poisson becune (Seychelles), peto (Latin America, Australia).

DISTRIBUTION: Circumtropical.

SOURCES: Bouder *et al.* (1962), Bartsch and McFarren (1962), Robinson (1962).

[9] See also the species and references listed in Chapter XIII on scombrotoxic fishes since scombroid fishes have been incriminated in both scombroid and ciguatera forms of ichthyosarcotoxism. Confusion has resulted in the past because the authors failed to distinguish between these two types of biotoxications.

Scomberomorus cavalla (Cuvier) (Pl. 33, fig. b). King mackerel, sierra, kingfish, cavalla (USA). *152*
DISTRIBUTION: Tropical Atlantic.

SOURCE: Bouder *et al.* (1962).
OTHER NAME: *Cybium caballa.*

Family SERRANIDAE

Anyperodon leucogrammicus (Cuvier and Valenciennes) (Pl. 34, fig. a). Rockcod, grouper, seabass (USA), kabro (Marshall Islands). *153*
DISTRIBUTION: Indo-Pacific.

SOURCES: Bartsch *et al.* (1959), Banner and Helfrich (1964), Helfrich *et al.* (1968).

Cephalopholis argus Bloch and Schneider (Pl. 34, fig. b). Spotted seabass, rockcod, black rock cod, spotted grouper, garrupa (USA, South Africa, Australia), roi (Tahiti), maloslos (western Carolines), kalemej (Marshall Islands), baraka, kugtung (Philippines), kuroganmo (Saipan). *153*
DISTRIBUTION: Indo-Pacific.

SOURCES: Autenrieth (1833), Halstead and Bunker (1954*a*), Helfrich *et al.* (1968).
OTHER NAMES: *Bodianus guttatus, Serranus guttatus.*

Cephalopholis fulvus (Linnaeus) (Pl. 33, fig. c). Coney (USA), guativere, fino, ouatilibi (West Indies). *152*
DISTRIBUTION: Florida, West Indies, southward to Brazil.

SOURCES: Corre (1865), Pawlowsky (1927), Bouder *et al.* (1962).
OTHER NAMES: *Epinephelus punctatus, Serranus punctatus, Serranus ouatalibi.*

Cephalopholis leopardus (Lacépède) (Pl. 34, fig. c). Grouper (USA), gatala (Samoa), rero (Tahiti), baraka, kugtung (Philippines), rockcod, grouper (Australia), vieille (Seychelles), grouper (South Africa). *153*
DISTRIBUTION: Indo-Pacific.

SOURCES: Hiyama (1943), Bouder *et al.* (1962).
OTHER NAME: *Epinephelus leopardus.*

Cephalopholis miniatus (Forskål) (Pl. 34, fig. d). Grouper (USA), gatala (Samoa), rero (Tahiti), baraka, kugtung (Philippines), rockcod, grouper (Australia, South Africa), vieille (Seychelles). *153*
DISTRIBUTION: Indo-Pacific.

SOURCE: Cooper (1964).

Epinephelus adscensionis (Osbeck) (Pl. 35, fig. a). Rock hind (USA), cabra mora (West Indies). *154*

DISTRIBUTION: Florida and West Indies southward to Brazil.

SOURCES: Fonssagrives and de Méricourt (1861), Bouder *et al.* (1962).
OTHER NAME: *Serranus nigriculus.*

154 *Epinephelus fuscoguttatus* (Forskål) (Pl. 35, fig. b). Grouper (USA), galiy (Palau), kuro (Marshall Islands), vieille (Seychelles), rockcod, grouper (Australia, South Africa).
DISTRIBUTION: Indo-Pacific.

SOURCES: Hiyama (1943), Wheeler (1953), Helfrich *et al.* (1968).
OTHER NAMES: *Epinephelus microdon, Serranus fuscoguttatus, Serranus microdon, Serranus lutra.*

154 *Epinephelus guttatus* (Linnaeus) (Pl. 35, fig. c). Red hind (USA), mero guajiro, cabrilla (West Indies).
DISTRIBUTION: South Carolina to Brazil.

SOURCES: Fonssagrives and de Méricourt (1861), Coutière (1899), Bouder *et al.* (1962).
OTHER NAMES: *Epinephelus maculosus, Serranus arara, Epinephelus catus, Serranus lunulatus, Serranus maculosus.*

156 *Epinephelus tauvina* (Forskål) (Pl. 37, fig. a). Black sea bass (USA), gatala (Samoa), kuro (Marshall Islands), tekuau (Gilbert Islands), baraka, kugtung, lapulapu (Philippines), mohata (Japan), rockcod, grouper (Australia, South Africa), vieille (Seychelles).
DISTRIBUTION: Indo-Pacific.

SOURCES: Helfrich (1963), Banner and Helfrich (1964), Helfrich and Banner (1968), Helfrich *et al.* (1968).

155 *Mycteroperca venenosa* (Linnaeus) (Pl. 36, fig. a). Yellowfin grouper (USA), cabrilla (West Indies).
DISTRIBUTION: Western tropical Atlantic.

SOURCES: Burrows (1815), Coutière (1899), Randall (1958), Bouder *et al.* (1962).
OTHER NAMES: *Epinephelus venenosus, Mycteroperca apua, Mycteroperca venenosa apua, Mycteroperca venenosa, Serranus cardinalis, Serranus rupestris, Serranus venenosus.*

156 *Plectropomus leopardus* (Lacépède) (Pl. 37, fig. b). Grouper (USA), tonu (Tahiti), ikuit (Seychelles), gatala (Samoa), kuro (Marshall Islands), tekuau (Gilbert islands), baraka, kugtung, lapulapu (Philippines), mohata (Japan), rockcod, grouper (Australia, South Africa).
DISTRIBUTION: Indo-Pacific, Japan, Red Sea.

SOURCES: Harry (1953), Banner and Helfrich (1964).
OTHER NAMES: *Epinephelus leopardus, Serranus leopardus.*

Plectropomus maculatus (Bloch) (Pl. 37, fig. c). Grouper (USA), loche saumonnée (New Caledonia), lapulapu (Philippines), julae (Marshall Islands), rockcod, grouper (Australia, South Africa), vieille (Seychelles). *156*
DISTRIBUTION: Indo-Pacific, Japan, east Africa, Red Sea.

SOURCES: Wheeler (1953), Bouder *et al.* (1962), Banner and Helfrich (1964).

Plectropomus oligacanthus Bleecker (Pl. 38, fig. a). Grouper (USA), loche saumonnée (New Caledonia), lapulapu (Philippines), julae (Marshall Islands), rockcod, grouper (Australia, South Africa), vieille (Seychelles). *157*
DISTRIBUTION: Indo-Pacific.

SOURCES: Hiyama (1943), Randall (1958), Bouder *et al.* (1962).
OTHER NAME: *Paracanthistius oligacanthus.*

Plectropomus truncatus Fowler and Bean (Pl. 38, fig. b). Grouper (USA), akajin (Ryukyus, Japan), taiyaw (Palau), tonu (Tahiti), loche saumonnée (New Caledonia), lapulapu (Philippines), julae (Marshall Islands), rockcod, grouper (Australia, South Africa), vieille (Seychelles). *157*
DISTRIBUTION: Indo-Pacific.

SOURCES: Hiyama (1943), Bouder *et al.* (1962), Cooper (1964).

Variola louti (Forskål) (Pl. 38, fig. c). Grouper (USA), hokahoka, maere (Society Islands), kaikbet (Marshall Islands), hoa (Tahiti), nbwele (Palau), nagaju akajin (Ryukyus), barahata (Japan). *157*
DISTRIBUTION: Indo-Pacific.

Sources: Coutière (1899), Fish and Cobb (1954), Helfrich *et al.* (1968).
OTHER NAMES: *Epinephelus louti, Serranus louti.*

Family SIGANIDAE

Siganus argenteus (Quoy and Gaimard) (Pl. 39, fig. d). Rabbitfish (USA), ellok, mole (Marshall Islands), lopauulu (Samoa), baliwis, palit (Philippines), spinefoot (Australia), rabbitfish, slimy, spiny (South Africa). *158*
DISTRIBUTION: Indo-Pacific, Red Sea.

SOURCES: Wheeler (1953), Randall (1958).
OTHER NAMES: *Eiganus rostratus, Teuthis rostreta.*

Siganus fuscescens (Houttuyn) (Pl. 39, fig. a). Rabbitfish (USA), ellok, mole (Marshall Islands), lopauulu (Samoa), baliwis, palit (Philippines), nai moun (China), aigo (Japan). *158*
DISTRIBUTION: Indo-Pacific.

SOURCES: Clark (1950), Lin *et al.* (1953), Kudaka (1960).
OTHER NAME: *Teuthis fuscescens.*

158 *Siganus lineatus* (Valenciennes) (Pl. 39, fig. b). Rabbitfish (USA), ellok, mole (Marshall Islands), lopauulu (Samoa), baliwis, palit (Philippines), golden-lined spinefoot (Australia), rabbitfish, slimy, spiny (South Africa).
DISTRIBUTION: Indo-Pacific.

SOURCE: Clark (1950), Bagnis *et al.* (1970).
OTHER NAME: *Teuthis lineata.*

155 *Siganus oramin* (Schneider) (Pl. 36, fig. b). Rabbitfish (USA), ellok, mole (Marshall Islands), lopauulu (Samoa), baliwis, palit (Philippines), golden-lined spinefoot (Australia), rabbitfish, slimy, spiny (South Africa).
DISTRIBUTION: Indo-Pacific, east Africa, Saudi Arabia.

SOURCES: Wheeler (1953), Randall (1958).
OTHER NAME: *Teuthis oramin.*

158 *Siganus puellus* (Schlegel) (Pl. 39, fig. c). Rabbitfish (USA), ellok, mole (Marshall Islands), lopauulu (Samoa), baliwis, palit (Philippines), spinefoot (Australia), rabbitfish, slimy, spiny (South Africa).
DISTRIBUTION: Indo-Pacific.

SOURCES: Bartsch *et al.* (1959), Bartsch and McFarren (1962).
OTHER NAME: *Teuthis puella.*

Family SPARIDAE

159 *Pagellus erythrinus* (Linnaeus) (Pl. 40, fig. b). Porgy (USA), red seabream (England), goyne youfief, ticotico (west Africa), pagel, breca, garapello (Spain), kleine rothbrassen, rothe goldbrassen (Germany), bica (Portugal), pagel rouge (France).
DISTRIBUTION: Black Sea, Mediterranean, and east Atlantic, from British Isles and Scandinavia to Azores, Canaries, and Fernando Po.

SOURCES: Autenrieth (1833), Knox (1888).
OTHER NAME: *Sparus erythrinus.*

159 *Pagrus pagrus* (Linnaeus) (Pl. 40, fig. c). Porgy (USA), braize, seabream (England), gemeine sackbrasse, grosse rothbrassen (Germany), pagro (Italy), pargo, pargoli (Spain), parguete, capatao (Portugal), pagre commun (France).
DISTRIBUTION: Eastern Atlantic, Mediterranean.

SOURCES: Corre (1865), Savtschenko (1886).
OTHER NAMES: *Pagrus vulgaris, Sparus pagrus.*

159 *Sparus sarba* (Forskål (Pl. 40, fig. a). Porgy (USA), bakoko, gaud-gaud (Philippines), bream, river perch, sandela, mbande (South Africa).

DISTRIBUTION: Indo-Pacific, Red Sea.

SOURCES: Savtschenko (1886), Knox (1888), Phisalix (1922), Bagnis *et al.* (1970).
OTHER NAMES: *Aurata psittacus, Chrysophrys sarba, Sparus psittacus.*

Family SPHYRAENIDAE[10]

Sphyraena barracuda (Walbaum) (Pl. 41). Great barracuda (USA), lupak (Philippines), luccio marino (Italy), copeton (Spain), brochet de mer, spet (France), pfeilhecht (Germany), dokukamasu (Japan), chikirukamasa (Ryukyus), ono (Tahiti), bicuda (Portugal), kaku (Hawaii), jure, jujukobe (Marshall Islands), barracuda, seapike (Australia), becime (Seychelles), scapihe, snoek, barracuda (South Africa). *160*

DISTRIBUTION: All tropical seas, with the exception of the eastern Pacific Ocean.

SOURCES: Catesby (1743), Phisalix (1922), Halstead (1951), De Sylva (1963), Prosvirov (1963), Helfrich *et al.* (1968).
OTHER NAMES: *Esox barracuda, Esox becuna, Perca major, Sphyraena becuna, Sphyraena picuda, Sphyraenidae barracuda, Umbra minor.*

Order TETRAODONTIFORMES (PLECTOGNATHI)[11]: Filefishes, Triggerfishes, Puffers, Etc.

Family ALUTERIDAE

Alutera monoceros (Linnaeus) (Pl. 42, fig. a). Unicorn filefish (USA), oili lepa, ohua (Hawaii), pareva (Tahiti), sensuri (Ryukyus), hoshinamihagi (Japan). *160*

DISTRIBUTION: All warm seas.

SOURCES: Chisholm (1821), Savtschenko (1866), Clark and Gohar (1953), Bouder *et al.* (1962).
OTHER NAMES: *Aluteros monoceros, Balistes monoceros, Monacanthus cinereus, Monacanthus monoceras.*

Alutera scripta (Osbeck) (Pl. 42, fig. b). Scrawled filefish (USA), barbuda (West Indies), leatherjacket (South Africa), abu'arabeiya (Suez), pareva (Tahiti), oili lepa, ohua (Hawaii), sensuri (Ryukyus), hoshinamihagi (Japan). *160*

---

[10] There is considerable question regarding the validity and synonomy of some of the species of the barracuda family Sphyraenidae. There is no single work that covers all the species listed, but Weber and De Beaufort (1922) and De Sylva (1963) are helpful.

[11] There is reason to believe that tetraodontiform (plectognath) fishes may at times serve as transvectors of both tetrodotoxin and ciguatoxin. However, final confirmation of this matter awaits further laboratory investigation. Most of the references listed under the tetraodontiform fishes included in this chapter apply to fishes suspected of having caused outbreaks of ciguatera poisoning in humans, but there is no way of knowing for certain at this time since many of the clinical reports are too vague to support a definitive diagnosis.

DISTRIBUTION: All warm seas.

SOURCES: Autenrieth (1833), Herre (1927), Hiyama (1943), Helfrich (1961).
OTHER NAMES: *Aleuteres scriptus, Balistes scriptus, Monacanthus scriptus, Osbeckia scripta.*

Family BALISTIDAE

161 *Balistapus undulatus* (Mungo Park) (Pl. 43, fig. a). Triggerfish (USA), red-striped papaco (Philippines), tebubutakataka (Gilbert Islands), kokiri karava (Tuamotus), red-lined triggerfish (Australia).
DISTRIBUTION: Indo-Pacific, China, Japan, Red Sea, Zanzibar, Mozambique.

SOURCES: Seale (1912*b*), Halstead and Schall (1958), Bouder *et al.* (1962).

161 *Balistes vetula* Linnaeus (Pl. 43, fig. b). Triggerfish (USA), ballesta (Spain), baliste (France), balestra (Italy), hornfisch, seebock (Germany).
DISTRIBUTION: Tropical Atlantic, Mediterranean, and Indian Ocean.

SOURCES: Cloquet (1921), Prokhoroff (1884), Bouder *et al.* (1962).

161 *Balistoides conspicillum* Bloch and Schneider (Pl. 43, fig. c). Triggerfish (USA), humuhumu (Hawaii), sumu (Samoa), tebubutakataka (Gilbert Islands), bub (Marshall Islands), mongarakawahagi (Japan), papakol, puggot (Philippines), big-spotted triggerfish (Australia).
DISTRIBUTION: Indo-Pacific, China, Japan.

SOURCES: Meyer-Ahrens (1855), Hiyama (1943), Fish and Cobb (1954), Bouder *et al.* (1962).
OTHER NAMES: *Balistes conspicillum, Balistes niger, Balistoides niger, Odonus niger.*

162 *Odonus niger* (Rüppell) (Pl. 44, fig. a). Triggerfish (USA), humuhumu (Hawaii), sumu (Samoa), tebubutakataka (Gilbert Islands), bub (Marshall Islands), mongarakawahagi (Japan), papakol, puggot (Philippines).
DISTRIBUTION: Indo-Pacific, Red Sea.

SOURCES: Herre (1924), Smith (1947), Randall (1958), Bartsch *et al.* (1959).
OTHER NAME: *Balistes erythrodon.*

162 *Pseudobalistes flavimarginatus* (Rüppell) (Pl. 44, fig. b). Triggerfish (USA), humuhumu (Hawaii), sumu (Samoa), tebubutakataka (Gilbert Islands), bub (Marshall Islands), mongarakawahagi (Japan), papakol, puggot (Philippines).
DISTRIBUTION: Indo-Pacific, Red Sea.

SOURCES: Seale (1912*b*), Hiyama (1943), Bouder *et al.* (1962).
OTHER NAME: *Balistes flavimarginatus.*

Family MONACANTHIDAE[12]

Amanses sandwichiensis (Quoy and Gaimard) (Pl. 44, fig. c). Filefish (USA), aimeo (Samoa), oili lepa, ohua (Hawaii), pareva (Tahiti). *162*

DISTRIBUTION: Indo-Pacific.

SOURCES: Seale (1912*b*), Halstead and Bunker (1954*b*), Bagnis *et al.* (1970).
OTHER NAMES: *Cantherines sandwichiensis, Monacanthus sandwichiensis.*

Stephanolepis setifer (Bennett) (Pl. 45, fig. b). Pygmy filefish (USA), lija (West Indies), komuki (Japan), gallo, cajo canario (West Africa), peleg, bitig (Philippines), leatherjacket (Australia). *163*

DISTRIBUTION: Tropical Atlantic, and Indo-Pacific Oceans, Japan.

SOURCES: Jouan (1867), Maass (1937), Bagnis *et al.* (1970).
OTHER NAMES: *Monacanthus komuki, Monacanthus setifer.*

Family OSTRACIONTIDAE

Acanthostracion quadricornis (Linnaeus) (Pl. 45, fig. c). Cowfish (USA), chapin, toro (West Indies), kounaye (west Africa), baiacú, taoca (Brazil). *163*

DISTRIBUTION: Western Atlantic, tropical coasts of the Americas and Caribbean.

SOURCES: Faust (1906, 1924, 1927*a, b*), Prosvirov (1963).
OTHER NAMES: *Lactophrys quadricornis, Lactophrys tricornis, Ostracion quadricornis.*

Lactoria cornuta (Linnaeus) (Pl. 45, fig. a). Trunkfish (USA), bill (Marshall Islands), pahu (Hawaii), umi-suzume, suzume-fugu (Japan), bacabaca, obuluk-sungyan, pegapega, tikung, patagan (Philippines), cowfish, boxfish (Australia), cowfish, trunkfish, oskop, seekoei, seevarkie (South Africa). *163*

DISTRIBUTION: Tropical Indo-pacific, Japan, South Africa.

SOURCES: Chevallier and Duchesne (1851*a, b*), Herre (1924), Pawlowsky (1927), Bouder *et al.* (1962).
OTHER NAME: *Ostracion cornutum.*

Ostracion meleagris Shaw (Pl. 45, fig. d). Trunkfish (USA), pahu (Hawaii), bill (Marshall Islands), obuluk, tabayong (Philippines), umi-suzume, suzume-fugu (Japan), cowfish, boxfish (Australia), cowfish, trunkfish, boxfish, oskop, seekoei, seevarkie (South Africa). *163, 269*

DISTRIBUTION: Indo-Pacific, Japan, South Africa.

SOURCES: Heẹre (1924), Helfrich (1961).
OTHER NAMES: *Ostracion lentiginosum, Ostracion lentiginosus, Ostracion sebae.*

[12] Adopting the classification used in this text, Randall (1964) has written a revision of the genera *Amanses* and *Cantherines* which should be consulted by anyone attempting to identify filefishes.

Order LOPHIIFORMES (PEDICULATI): Goosefishes, Frogfishes, Etc.

Family ANTENNARIIDAE

164 *Histrio histrio* (Linnaeus) (Pl. 46, fig. a). Sargassumfish, frogfish (USA), marbled anglerfish (Australia), toadfish, sargassumfish (South Africa).

DISTRIBUTION: All tropical seas.

SOURCES: Phisalix (1922), Maass (1937), Bagnis *et al.* (1970).
OTHER NAME: *Antennarius marmoratus.*

Family LOPHIIDAE

164 *Lophiomus setigerus* (Vahl) (Pl. 46, fig. b). Goosefish, angler, frogfish (USA), anko (Japan).

DISTRIBUTION: Indo-Pacific, Japan.

SOURCES: Nielly (1881), Savtschenko (1886), Fish and Cobb (1954).
OTHER NAME: *Lophius setigerus.*

BIOLOGY

Fishes incriminated in ciguatera poisoning are scattered over a wide phylogenetic range, having varied body shapes and biological characteristics. Their habitat, habits, feeding, and reproductive processes are as remarkably diversified as their morphology. Consequently it is difficult to make any stereotyped generalizations as to precisely what constitutes the "standard biology" of a ciguatoxic fish. Perhaps when the ciguatera problem has been more clearly defined, certain common biological denominators will emerge, but thus far our knowledge of this subject is much too meager to do more than provide a series of brief summaries on the biology of the fish families that have participated as transvectors of ciguatoxin. Unfortunately so little is known regarding the ecology of tropical reef fishes that it is impossible at this time to provide the reader with anything more than a very cursory review of this extremely important, but greatly neglected, aspect of marine biology.

Outstanding examples of the type of studies that are needed for many different groups of fishes, and most other marine organisms in general, are the following: Harry (1953) on ichthyological field data of Raroia Atoll, Tuamotu Archipelago; Hiatt and Strasburg (1960) on ecological relationships of the fish fauna on coral reefs of the Marshall Islands; Randall (1961) on a contribution to the biology of the convict surgeonfish of the Hawaiian Islands, *Acanthurus triostegus sandvicensis*; Randall and Brock (1960) on the ecology of epinepheline and lutjanid fishes of the Society Islands, with emphasis on food habits; De Sylva (1963) on the systematics and life history of the great barracuda *Sphyraena barracuda* (Walbaum). Other examples might be given, but for the most part, studies of this type come few and far between. For those desiring to pursue this aspect of marine biology their attention is directed to the excellent work by Fosberg and Sachet (1953) "Handbook for Atoll Research" and the publications that appear from time to time in the *Atoll Research Bulletin*, Pacific Science Board, National Academy of Sciences, Washington, D.C.

The following summarizations on the biology of fish families that have been incriminated as transvectors of ciguatoxin are presented in the same sequence as given in the "Representative List of Fishes Reported as Ciguatoxic," p. 326.

Order CLUPEIFORMES (ISOSPONDYLI)

Family ALBULIDAE: The ladyfishes or bonefishes are found in most warm seas and are common along sandy coasts. Occasionally they are seen in large schools. They prefer shallow water over low tidal flats. Their food consists of shellfish, crustaceans, and worms, which they take from the mud and grind up with their pavementlike teeth. Bonefish are considered a good game fish in some areas, and are eaten to a limited extent.

Family CLUPEIDAE: Herring are among the most cosmopolitan groups of fishes. Some species live in fresh water, whereas others enter rivers to spawn, but the majority are oceanic species. Herring are frequently found swimming in immense schools. They have been incriminated in both the ciguatera and clupeoid types of ichthyosarcotoxism. Two species, *Clupanodon thrissa* and *Clupea tropica*, have been reported as violently toxic species. Although these two species are considered to be valuable food fishes in most areas, they may be extremely poisonous at times elsewhere. The dentition of herring is small and feeble, and their food consists largely of planktonic organisms, shrimp, crustaceans, worms, etc. They are greatly valued as food fishes. The economic significance of herring has been noted by Jordan (1905), viz.: "As salted, dried, or smoked fish the herring is found throughout the civilized world, and its spawning and feeding grounds have determined the location of cities."

Family ENGRAULIDAE: Anchovies are small herringlike fishes, abundant in temperate and warm seas. Although generally found in the open seas, they enter bays and ascend rivers. They are characterized by their snout projecting beyond their very wide mouth. The flesh of some species is of excellent flavor, and they are commercially canned in large numbers. Anchovies are valuable bait fishes and serve as an important food source for larger fishes. They are largely plankton feeders.

Order ANGUILLIFORMES (APODES)

Family MURAENIDAE: Moray eels are a group of savage, moderate-sized marine fishes. Their bathymetric range extends from intertidal reef flats to depths of a hundred meters or more. They are found throughout temperate and tropical seas. Moray eels inhabit a large variety of ecological biotopes, in surge channels, coralline ridges, inter-islet channels, reef flats, and lagoon patch reefs. They are nocturnal in their habits, hiding in crevices, holes, and under rocks or coral during the day, and coming out at night. With the aid of a light one can observe morays wriggling over a reef flat at night in large numbers. They may often be seen thrusting their heads out of coral holes, their mouths slowly opening and closing, waiting for some unwary victim to pass by their lair. They are able to strike with great rapidity and ferocity. Their long, fanglike, depressible teeth are exceedingly sharp, and can inflict serious lacerations. Their powerful muscular development, tough leathery skin, and dangerous jaws make them formidable animals. Some of the larger morays attain a length of 3 m or more and may constitute a real hazard to divers. The flesh of some species is used by some peoples as food. The flesh is said to be agreeable, but oily and not readily digestible. Some species are violently poisonous to eat.

Order BELONIFORMES (SYNENTOGNATHI)

Family BELONIDAE: The needlefishes resemble the freshwater garpike in form, but they are quite distinct from it. They are generally found along coastal areas and frequently enter estuaries. Needlefishes are largely surface swimmers. The body of the needlefish is long, slender, and covered with small scales; sharp unequal teeth fill the long jaws. Some of the larger species, which may attain a length of 2 m, have been known to inflict fatal wounds with the thrust of their sharp beaks. Needlefishes are voracious feeders and will readily take a hook baited with a live fish.

Family HEMIRAMPHIDAE: The halfbeaks have been appropriately named, having a spearlike, much prolonged, lower jaw, and a greatly reduced upper one. Halfbeaks are widely distributed throughout warm and temperate seas. Their main food consists of planktonic organisms, vegetable matter and crustaceans, which they skim off the surface of the water. Most of the species are small and seldom exceed 30 cm in length. A few are oceanic, but the majority of them are shore fishes.

Order BERYCIFORMES (BERYCOMORPHI)

Family HOLOCENTRIDAE: The squirrelfishes or soldierfishes are characteristic species of rocky or coral flats in tropical seas. Their ctenoid scales are very hard and remarkably spiny. They are further characterized by the presence of a spine at the angle of the preopercle. The coloration of these fishes is usually a brilliant crimson, with or without stripes. Squirrelfishes are predacious, active and largely nocturnal in their habits, feeding on crustaceans, worms, and algae. During the day they may be observed hovering almost motionless in coral or rock crevices. At night they come out to feed on the reef flat or in the surf.

Order PERCIFORMES (PERCOMORPHI)

Family ACANTHURIDAE: Surgeonfishes or tangs are a group of shore fishes inhabiting warm seas. The principal genus *Acanthurus* is characterized by the presence of a sharp, lancelike, movable spine, which is located on each side of the caudal peduncle. Surgeonfishes are particularly common in surge channels and shoal areas in the vicinity of coral patch reefs. They are predominantly herbivores, feeding on fine filamentous algae of numerous species. Small fishes and invertebrates may be ingested. Most surgeonfishes are small to moderate in size. The eggs and larvae of at least some species are pelagic.

Family CARANGIDAE: Jacks, scads, and pompanos are a large group of swift-swimming, oceanic fishes, which are cosmopolitan in distribution. They are particularly common in the vicinity of coral reefs. Some of the Pacific species are especially noted for their long migrations in quest of food. They are mostly carnivorous in their eating habits. Many are desirable game fishes, and most of them are valued as food.

Family CORYPHAENIDAE: Dolphins are large, fast-swimming, beautifully colored, predatory, oceanic fishes, and found in all warm seas. They are usually observed swimming near the surface of the water and may enter bays or inlets. The flesh is fine-flavored, and they are an excellent game fish. They are carnivorous in their feeding habits.

Family GOBIIDAE: Gobies are excessively numerous in tropical and temperate zones. Usually they are found in lakes, brooks, swamps, bays, or in shallow coastal waters. Seldom, if ever, are they found in deep water. Some of them burrow in the mud, whereas others are found clinging on or under rocks. Most species are small in size, and some of them are equipped with a ventral fin modified into a sucker. The Asiatic mud skipper *Periophthalmus* and related species are able to skip over the rocks, among weeds, and out of water with the agility of a lizard. Few gobies are used for food. They are largely carnivorous, feeding on crustaceans, small fish, mollusks, worms, but at times include small fragments of algae.

Family LABRIDAE: Wrasses are characterized by their large, separate, conical teeth in the front of the jaws. They have an extensive geographical range, but most are concentrated in warmer water. Most species are shore forms, inhabiting rocky areas, coral reefs, and in amongst growths of marine algae. Some species are herbivorous, but most of them are carnivores. They are among the most beautiful and gaily colored of the reef creatures. Their quality as food fishes varies considerably from one species to the next.

Family LUTJANIDAE: Snappers are carnivorous, voracious, gamy, shore fishes, and abundant in all warm seas. Their food consists largely of smaller fishes. They are common in rocky, coral reef areas. Snappers are usually good food fishes, and of commercial value in some places. They take the hook readily. Certain members of the genus *Lutjanus* are among the most serious offenders as ciguatoxic fishes.

Family MUGILIDAE: Mullets are small to moderate-sized fishes with broad heads and a mouth fringed with very feeble teeth. They are tropical and temperate zone inhabitants of shallow bays, estuaries, and lagoons, although a few are confined to fresh water. Mullets are largely herbivores. They are frequently observed in schools swimming near the surface of the water. In some areas they are a valuable food fish and are often raised in large numbers in ponds. They have a phenomenal ability to jump over the cork line of a seine and give the fisherman a real test of his ability.

Family MULLIDAE: Surmullets or goatfishes are small to moderate-sized shore fishes of warm seas. A variety of forms may be seen swimming about coral reefs, many gaily colored. They are usually observed as a few scattered individuals or in small schools. With the use of their feelers they feel their way along a sandy bottom in quest of food. They are carnivorous, feeding largely on a variety of small animals. Surmullets serve as food for a variety of predatory fishes. Their flesh is usually excellent.

Family POMACENTRIDAE: Damselfishes are chiefly confined to coral reefs, although a few species are found in more temperate zones. They are most abundant around coral crevices, in holes and rocky areas, which they utilize for protection. Although gregarious, they do not form large schools. Damselfishes are omnivorous in their feeding, and although apparently preferring an herbivorous diet, they readily ingest any organism that is picked up in the process of their grazing. Most of these fishes are too small to be considered of food value.

Family SCARIDAE: Parrotfishes are similar in appearance to the wrasses, but differ in that they have their teeth fused into plates. They are shallow-water shore fishes and very common in coral reefs, lagoons, and rocky areas of tropical waters. Their food consists of algae and small animals that are ingested along with corals, which make up the major part of their diet. Deep gouges and scratches may be observed on coral rocks where small schools of parrotfishes have browsed. They thus contribute significantly to the formation of fine sand by means of returning the pulverized rock and skeletal material to the bottom as fecal components.

Family SCOMBRIDAE: The great family of true mackerels, which includes the tunas, are distinguished by their streamlined bodies, smooth scales, magnificent metallic coloration, and by the presence of a number of detached finlets behind the dorsal and anal fins. Mackerels are for the most part swift-swimming pelagic fishes and are one of our most valuable fisheries resources. Most species run in large schools. Some of these schools have been reported to be "a half mile wide and 20 miles long." Although there are a large number of species in the Scombridae, only a few members of the genera *Acanthocybium, Euthynnus, Sarda*, and *Scomberomorus* have been incriminated in ciguatera poisoning. They, like most of their relatives, are oceanic fishes, but during the reproductive season come in close to shore, and some species may at times be found in bays and lagoons. Apparently it is during this inshore migration period that they encounter ciguatoxin by means of their diet. They feed on a variety of foods: sardines, anchovies, plankton, etc.

Family SERRANIDAE: Seabass or groupers are robust, carnivorous, predacious shore fishes of tropical and temperate waters. A variety of biotopes are inhabited by this large family: coral reefs, rocks, sandy areas, and kelp. A few are found in fairly deep water, but most live in shallow water. Some attain great size. They are usually considered as good food fishes.

Family SIGANIDAE: Rabbitfishes are moderate-sized fishes, having compressed, ovate bodies with slippery skins and minute scales. They are found in tropical waters, traveling in dense schools, around coral reefs where they browse on marine algae. However, they are not obligatory herbivores and at times may shift to a fleshy diet.

Family SPARIDAE: Porgies are a group of shore fishes having a perch-like appearance. They are closely related to the Pomadasyidae. They are found on tropical and temperate coasts, usually confined to shallow waters in a variety of habitats. Small fishes, crustaceans, and other invertebrates comprise their diet. They are among the more important food fishes.

Family SPHYRAENIDAE: Barracuda are long, slender, swift-swimming, carnivorous, savage, and exceedingly voracious shore fishes of tropical and temperate waters. Their jaws contain long, knifelike, canine teeth. Some attain large size. They are common in lagoons, passageways, and coral reefs. Their flesh is of excellent quality, but the larger specimens of some species may be very toxic especially during the reproductive season.

Order TETRAODONTIFORMES (PLECTOGNATHI)

Family ALUTERIDAE: Filefishes are a group of small-sized, shallow-water shore fishes of temperate and tropical seas characterized by compressed lean bodies covered by shagreenlike prickles. The musculature is generally too meager to make these fishes useful as food. However, some natives do skin and eat them. The flesh is considered to be dry, bitter, and offensive to the taste. Filefishes are omnivorous, frequently making a meal of a single item if available in sufficient quantity. Corals appear to be a popular dietary item for filefishes.

Family BALISTIDAE: Triggerfishes are a group of small to moderate-sized shore fishes, characterized by their deep compressed body, covered with a thick layer of enlarged bony scales. The first two dorsal spines are modified into a triggerlike device. Triggerfishes are widely distributed throughout all warm seas. They seem to prefer shallow reef areas, although some are found in fairly deep water. They are omnivorous in their eating habits, ingesting corals, sponges, urchins, algae, and various other small organisms. Despite their slow swimming movement, they travel long distances by floating with the currents. Some species bite eagerly on almost any bait, whereas others are difficult to take.

Family MONACANTHIDAE: The family Aluteridae is sometimes included in the Monacanthidae. The habits of the fishes of these two families are essentially the same.

Family OSTRACIONTIDAE: Trunkfishes are one of the more peculiarly constructed groups of plectognaths. The body is enveloped within a bony box, comprised of six-sided scutes, leaving openings only for the jaws, fins, and tail. They live in tropical seas, and are frequently seen swimming slowly along among the corals in shallow water. Many of them are brilliantly colored. The "exoskeleton," which serves them well for protection, is also used by primitive peoples as a container for cooking them over a fire. Little is known of their habits.

Order LOPHIIFORMES (PEDICULATI)

Family ANTENNARIIDAE: Frogfishes are a group of small, misshapen, lumplike fishes, which are widely distributed throughout tropical seas. Their bodies are covered with a skin having prickles or dermal flaps, and on the snout is located a lurelike filament which is used to attract small fishes. They live at shallow depths, frequently near the surface of the water, drifting with floating seaweeds. By holding air in their mouths they are able to float without effort and are carried long distances by currents. They are also found in the vicinity of sand banks and coral reefs. They are not generally used for food.

Family LOPHIIDAE: Goosefishes have a large depressed grotesque head, which constitutes the bulk of the fish. The mouth is remarkably wide, having powerful jaw muscles and strong canine teeth. Dermal flaps are scattered about over the head. The first dorsal spine is modified to form a lurelike device which is used to attract smaller fishes near the mouth. Goosefishes lie motionless on the bottom for long periods of time waiting for their hapless victims. They prefer sandy or rocky areas in relatively shallow water. They attain a length of 1 m or more. They are of minimal value for food.

BIOGENESIS OF CIGUATOXIN

The biogenesis or process by which ciguatoxin originates in the body of the fish is a problem about which man has speculated for many centuries. The primitive mind from the days of ancient demoniacal medicine has attempted to explain the cause of poisonings—and no exception has been made in the case of ciguatera. Numerous theories have been propounded as to the origin of ciguatera poison, but most investigators have concluded that they are without scientific basis. These theories have been listed together with references to representative authors who discussed them and are presented in Table 1 without further comment.

Ciguatoxications have frequently been categorized as mere outbreaks of bacterial food poisoning. However, since bacterial food pathogens have no bearing on the ciguatera problem no attempt has been made to either list or discuss articles suggesting the bacterial origin of ciguatera poisoning. May it suffice to say that the symptomatology of ciguatera poisoning is not characteristic of a bacterial food poisoning. The reader is referred to the works of Meyer (1953) and Dack (1956) for comprehensive reviews of the subject of bacterial food poisonings and their clinical characteristics. Moreover, bacteriological studies such as those by Hiyama (1943), Arcisz (1950), Bartsch *et al.* (1959), De Sylva (1963) fail to provide any scientific support for the concept of the bacterial origin of ciguatera poisoning.

Algal Food Chain Theory: Those who have had the opportunity of studying the ciguatera problem in the field are unanimously agreed that the poison (or complex of poisons) known as ciguatoxin has its origin in the environment of

TABLE 1.—*Theories as to the Origin of Ciguatoxin*

THEORIES	REFERENCES*
Ingestion of manchineel fruit	Chisholm (1808), Burrows (1815), Duméril (1866), Gudger (1930), Bréta (1939), Gilman (1942), Arcisz (1950), Halstead (1951), Wheeler (1953), De Sylva (1963)
Ingestion of *Cocculus* berries, etc.	Duméril (1866), Arcisz (1950), Wheeler (1953)
Poisonous algae	Chisholm (1808), Burrows (1815), Savtschenko (1886), Dunlop (1917), Gudger (1930), Matsuo (1934), Yasukawa (1935), Bréta (1939), Gilman, (1942), Kawakubo & Kikuchi (1942), Hiyama (1943), Ross (1947), Arcisz (1950), Halstead (1951), Whitley (1954), Dawson *et al.* (1955), Habekost *et al.* (1955), Randall (1958, 1961), Halstead (1959), Kudaka (1960), Helfrich (1961), Bouder *et al.* (1962), Helfrich and Banner (1963, 1968), Cooper (1964), Helfrich *et al.* (1968)
Poisonous protozoans or plankton	Fonssagrives & de Méricourt (1861), Arcisz (1950), Halstead (1951), Mills (1956)
Sponges	Mills (1956)
Jellyfishes	Chisholm (1808), Burrows (1815), Duméril (1866), Gudger (1930), Bréta (1939), Arcisz (1950), Halstead (1951), Whitley (1954), Randall (1958)
Corals	Duméril (1866), Savtschenko (1886), Pellegrin (1899), Walker (1922), Pawlowsky (1927), Gudger (1930), Matsuo (1934), Yasukawa 1935), Bréta (1939), Kawakubo & Kikuchi (1942), Hiyama (1943), Arcisz (1951), Halstead (1951), Russell (1952), Wheeler (1953), Whitley (1954), Mills (1956), Randall (1958), Banner & Helfrich (1964)
Sea anemones *(Actinia)*	Bréta (1939)
Mollusks (including mussels, sea hares, and pteropods)	Savtschenko (1886), Jordan (1905), Gudger (1930), Kawakubo & Kikuchi (1942), Halstead (1951), Randall (1958)
Palolo worms or other sea worms	Gudger (1930), Halstead (1951), Randall (1958), Bouder *et al.* (1962)
Poisonous crabs or Crustacea	Yasukawa (1935), Hiyama (1943), Kudaka (1960)

* NOTE: A listing of the author's name does not necessarily imply that he endorses the theory. It merely indicates that the author has referred to it.

TABLE 1.—*Theories as to the Origin of Ciguatoxin (Continued)*

THEORIES	REFERENCES*
Starfishes (or echinoderms in general)	Hiyama (1943), Whitley (1954)
Sea cucumbers (holothurians)	Chisholm (1808), Burrows (1815), Jordan (1905)
Other poisonous fishes	Pawlowsky (1927), Bréta (1939), Kawakubo & Kikuchi (1942), Hiyama (1943)
Copper or copper compounds	Chisholm (1808), Burrows (1815), Duméril (1866), Savtschenko (1886), Gudger (1930), Bréta (1939), Gilman (1942), Arcisz (1950), Halstead (1951), Jones (1956), Bouder *et al.* (1962)
Other metallic products, dumping of war matériel, etc.	Burrows (1815), Savtschenko (1886), Van Zant (1914), Schnackenbeck (1943), Ross (1947), Dempster (1949), Russell (1952), Mills (1956), Jones (1956), Cavallo & Bouder (1961), Bouder *et al.* (1962)
Pumice	Whitley (1954), Mills (1956)
Contamination of water with wastes, factory refuse, or chemicals	Duméril (1866), Steinbach (1895), Pellegrin (1899), Dunlop (1917), Halstead (1951), Randall (1958), Cooper (1964)
Effect of moonlight or stars	Bryant (1912), Hutchins (1912)
Climate changes	Cloquet (1821)
Parasitic infestation	Behre (1949)
Pathological conditions of the fish	Chisholm (1808), Meyer-Ahrens (1855), Pellegrin (1899), Dunlop (1917)
Maturation process	Fonssagrives & de Méricourt (1861), Duméril (1866), Savtschenko (1886), Jordan (1905), Mills (1956)
Metabolic changes at the spawning season	Chisholm (1808), Meyer-Ahrens (1855), Fonssagrives & de Méricourt (1861), Savtschenko (1886), Pellegrin (1899), Jordan (1905), Dunlop (1917), Herre (1927), Pawlowsky (1927), Matsuo (1934), Gilman (1942), Ross (1947), Arcisz (1950), Halstead (1951), Randall (1958), Bouder *et al.* (1962), De Sylva (1963)

* NOTE: A listing of the author's name does not necessarily imply that he endorses the theory. It merely indicates that the author has referred to it.

the fish. It is further believed, based on a substantially increasing amount of field and laboratory evidence, that the poison is transmitted from the environment to the fish by means of food ingested by the fish (Hiyama, 1943; Habekost *et al.*, 1955; Dawson *et al.*, 1955; Randall, 1958; Bartsch *et al.*, 1959; Bartsch and McFarren, 1962; Halstead, Hessel, and Suchy, 1963; Cooper, 1964; McFarren *et al.*, 1965) (Fig. 1).

There are certain valid observations that lend further support to the foodchain theory, namely:

1. *The extreme variability of toxicity* in any given so-called "ciguatoxic fish species." Every species of fish incriminated in ciguatera poisoning has been found to be nontoxic and edible in some other locality. In other words, thus far no one has reported a single species of ciguatoxic fish that is universally poisonous all the time.

2. *The spotty geographical distribution* of ciguatoxic fishes. There appears to be no consistency in the incidence of poisonous fish species within a toxic fish zone. Some species may be extremely toxic within a given locality, whereas other species are completely innocuous in the same locality. Most toxic fishes are found in island areas, but are generally limited to specific localities around the island. Ciguatoxic fishes will be found in one part of the island, but absent in others. There is a definite spotty insular distribution of ciguatoxic fishes (Lagraulet *et al.*, 1973).

3. *Ciguatoxic fishes are generally bottom dwellers* or feed on bottom dwelling fishes. Although several pelagic fish species (i.e., scombroids, clupeoid fishes, etc.) have been incriminated in ciguatoxications, the fishes were usually taken close to shore where they had apparently been feeding on benthic fishes or some other benthic organism. Most ciguatoxic fishes are reef dwellers, but exceptions to this rule are known to occur.

The aforementioned observations are not new with this writing, but have been repeatedly reported over several centuries by numerous authors. These observations further raised the question as to the nature of the environmental source of the poison, all of which have led to the development of various theories that have attempted to explain the ciguatera phenomenon. Since one or more food items of the fish appear to be involved in the ecology of ciguatoxin production, a variety of poison sources have been suggested which include such things as manchineel berries, other poisonous land plants, marine algae, protozoans, plankton, sponges, jellyfishes, corals, sea anemones, mollusks, palolo worms, crustaceans, echinoderms, fishes, toxic metals, dumping of war materials, strychnine, pumice, factory wastes, and numerous other items of a similar varied nature.

A cursory review of the eating habits of ciguatoxic fishes shows that their eating habits are quite varied, but can be grouped into three categories: (*a*) herbivores, (*b*) carnivores, and (*c*) omnivores. Since some species of ciguatoxic fishes appear to be benthic obligate herbivores and ciguatoxin arises from an environmental food source, then one is led to the conclusion that one or more species of benthic algae participate at some point in ciguatoxin production (Fig. 1).

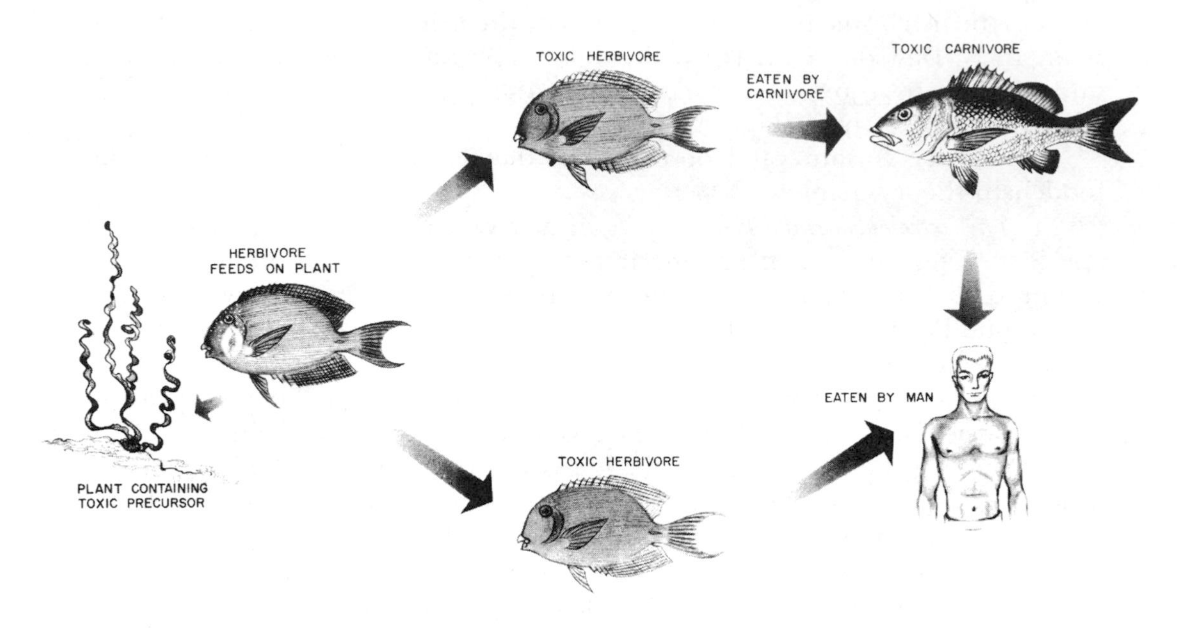

FIGURE 1.—Diagram depicting the biogenesis or transvectoring mechanism by which ciguatoxin is obtained through the food chain of fishes and thereby contracted by man. (R. Kreuzinger)

Several recent investigators (Banner *et al.,* 1963*a*; Helfrich and Banner, 1963; Cooper, 1964) have implied in their publications that the concept of the algal foodchain theory as a source of ciguatoxin is of recent origin (Randall, 1958), but such is not the case. Exactly when the algal food chain theory originated no one really knows, but apparently it was in existence as far back as the late 18th century and appears to have been propounded by native fishermen in the West Indies. At any rate, the British physician Colin Chisholm (d. 1825) in 1808 seems to have been among the first to have recorded the idea that West Indian fishes became poisonous as a result of eating toxic benthic marine algae. Chisholm gave quite a thorough review of the early theories regarding the origin of the poison, and his article will be found of historical interest to anyone studying this subject.

Apparently a similar belief has been present among the fishermen for many years in various islands of the tropical Pacific. Matsuo (1934), who investigated the poisonous fishes of the Marshall Islands, stated that: "It is widely believed by the natives that fish are poisonous because they eat poisonous seaweeds or corals in the sea." Similar beliefs among native fishermen have been reported by Banner (1959) in the Hawaiian Islands, and by Cooper (1964) in her studies on the ciguatoxic fishes of the Gilbert Islands.

During the summer of 1951, Halstead (1951) and his associates at the School of Tropical and Preventive Medicine, Loma Linda University, initiated a series of fish feeding experiments that were aimed at evaluating the algal foodchain hypothesis. The first effort attempted to determine if toxicity in a fish could be induced by feeding poisonous fish flesh to a nontoxic fish. For the purposes of this experiment a violently ichthyosarcotoxic puffer (*Arothron hispidus*), a nonciguateric species, was used as a source of the poison. Samples of the toxic puffer musculature and liver were fed to a nontoxic California cottid (*Leptocottus armatus*). Four *L. armatus* were used in the experiment. Two of them were fed toxic puffer flesh every other day for a period of 10 days. Two of the *L. armatus* were kept in a separate tank and used as controls where they were fed the flesh of the nontoxic California seabass *Paralabrax nebulifer*. Ten days later the controls and the test fish were killed. Aqueous homogenized extracts were made from the musculature of the *L. armatus* and injected intraperitoneally into four white mice. Three of the mice that had been injected with extracts prepared from the *L. armatus* that had been fed the toxic puffer flesh died within a period of 12 hours and one mouse was sick but recovered. All the mice that had been injected with the tissue extracts prepared from the control *L. armatus* remained asymptomatic. Further studies were planned utilizing the flesh of several ciguatoxic fish species as a source of poison and *P. nebulifer* as a test fish, but due to a lack of financial support it was not possible to complete this aspect of the food chain investigation. This preliminary study suggested that it was possible to induce ichthyosarcotoxicity in a nontoxic fish species.

Halstead and Bunker (1954*a, b*) believed, along with other previous investigators, that the source of ciguatoxin was to be found in marine algae. They further suggested that the original substance, however, might be either a toxin or in the form of a nontoxic precursor substance that is later metabolized into a toxic product. They postulated that the variation in the distribution of ciguatoxin within the body of the fish was probably governed by three principal factors, viz., venous drainage of the intestine, detoxication, and metabolic processes of the fish. When the active principle was ingested by the fish from the alga, it would be absorbed by the intestine, transported via the hepatic portal vein through the liver, and then distributed throughout the body of the fish. Concentrations of the toxin within a given organ might vary depending upon the rate of metabolism, detoxication, and the time interval after ingestion of the substance, but in general the intestine and liver would probably contain the greatest concentration of the poison.

Dawson *et al.* (1955) studied the alimentary tract contents of 21 species of reef fishes from Palmyra Island, Line Islands. The study included species from the following genera of fishes: *Acanthurus* (4 species), *Ctenochaetus* (1 species), *Zebrasoma* (1 species), *Chaetodon* (3 species), *Chelon* (1 species), *Abudefduf* (1 species), *Lutjanus* (3 species), *Arothron* (1 species), *Upeneus* (1 species), *Holocentrus* (1 species), *Lethrinus* (1 species), *Rhinecanthus* (1 species), and *Sufflamen* (1 species). Eleven species were found to be herbivorous, at least in part, consisting of the fish genera *Acanthurus, Ctenochaetus, Zebrasoma, Chaetodon, Abudefduf, Arothron,* and *Chelon*. All the herbivorous fish species were found to contain fish specimens that were toxic when tested by mouse intraperitoneal injections. Ten species were found to be carnivorous and consisted of the fish genera *Chaetodon, Holocentrus, Lethrinus, Lutjanus, Rhinecanthus, Sufflamen,* and *Upeneus*. Only two of the genera *Lethrinus* and *Upeneus* included nontoxic fish specimens. The remainder were found to be toxic. In a random series of 106 samples of gut contents from the entire group of fishes, it was found that algae were present in a larger number of poisonous fishes than any other type of food and that the benthic blue-green Cyanophyta alga *Lyngbya majuscula* Gomont was detected in a large majority of the samples of poisonous fishes. Because of the limited amount of material that was available for study it was not possible to arrive at any definite conclusions.

The frequent presence of the blue-green alga *L. majuscula* suggested the possibility that this alga might serve as a primary source of ciguatoxin or its phytochemical precursor. *L. majuscula* is an olive drab filamentous alga that grows in the form of long unbranched hairlike filaments one cell in diameter. Surveys conducted during April and May 1953 by Halstead and his associates at Palmyra Island revealed that large masses of *Lyngbya* were conspicuously growing on the warm inner parts of the broad reef flats. The filaments of *Lyngbya*, not having a holdfast, become entangled in other algae, coral, rubble, and sand. As the filaments proliferate and increase in length the strands are frequently broken up, become matted, and are carried to shore by waves. It was observed that *Lyngbya* was particularly abundant at the east end of the island where large numbers of surgeonfishes *Acanthurus* spp. were feeding on it. Since previous toxicological surveys and epidemiological reports had indicated an unusually large ciguatoxic fish population at Palmyra Island, it was decided to collect samples of some of the more dominant kinds of algae and to test them for toxicity. The samples consisted of *L. majuscula, Boodlea composita* (Harvey), *Turbinaria ornata* (Agardh), *Caulerpa serrulata* (Forskål), *Bryopsis pennata* var. *secunda* (Harvey), *Jania capillacea* Harvey, *Jania tenella* Kutzing, *Ceramium mansoni* Dawson, *Enteromorpha sp?, Homothamnion solutum* Bornet and Flahault, and *Centroceras clavulatum* (Agardh). The results of this study were reported by Habekost *et al.* (1955). Aqueous homogenized extracts were prepared and injected intraperitoneally into white mice. It was found that all the algal extracts were moderately toxic to mice, i.e., one or more of the mice injected died within a period of 36 hours, with the exceptions of *Enteromorpha* and *Hormothamnion* which were only weakly toxic. It is of significance that two other species of algae *Lithothrix aspergillus* Gray and *Corallina officinalis* var. *chilenus* (Harvey) collected from Corona Del Mar, Calif., were also tested in the same manner and found to be nontoxic to mice. This led the investigators to believe that the results in toxicity could hardly be explained by merely a toxic salt concentration in the algal extracts. It is of further significance that in a series of eight reef fishes taken in the immediate vicinity and at the same time where the Palmyra algae had been collected, all the fishes were toxic as follows: strongly toxic (one specimen, *Arothron hispidus*), moderately toxic (five specimens, *Acanthurus fuliginosus,*

Chaetodon auriga, Abudefduf septemfasciatus), and weakly toxic (two specimens, *Acanthurus fuliginosus, Abudefduf sordidus*). Moreover, specimens of *Lyngbya majuscula, B. pennata* var. *secunda*, and *Caulerpa serrulata* were found in the intestinal contents of these fishes.

Randall (1958) prepared a review of the problem of ciguatera with particular reference to a possible source of the poison. He discussed many of the earlier theories of the origin of ciguatoxin and presented what he has termed "a new hypothesis." His hypothesis, however, is not new, but is largely a modification and an assemblage of data which has been known to marine biotoxicologists for many years. Randall has provided, however, a valuable review which adds further support to some of the basic elements of the 18th century algal food chain theory. Randall cited certain observations, an abbreviated edition of which is as follows:

1. Ciguatoxic fishes are usually bottom-dwellers, but may be open-water forms. In any case they are shore fishes and are usually found at depths of less than 100 m.

2. Poisonous fishes are generally associated with coral reefs. Fishes that occur over sand, mud, or turtle grass bottoms are less apt to be toxic.

3. Most poisonous fishes attain moderate or large size.[13]

4. Fishes which cause ciguatera are carnivores or feeders on detritus, or benthic algae. None appear to be plankton-feeders as adults (with the exception of clupeoid fishes). *Selar crumenophthalmus* and *Decapturus pinnulatus* are among the most abundant food fishes in Tahiti and are said to be universally nonpoisonous.[14]

5. There appears to be a correlation between the amount of fish in the diet and the degree of toxicity in poisonous fishes that feed primarily on fishes and crustaceans.

6. The food habits of surgeonfishes and knowledge of their relative toxicity may shed light on the nature of the organism producing ciguatera toxin. *Ctenochaetus striatus* is an abundant reef fish, frequently strongly toxic, and is largely an herbivore. Similar examples are to be found in other species of surgeonfishes such as *Naso brevirostris* and *Acanthurus triostegus*.

7. Some gastropods and echinoids, thought to be mainly herbivores, are reported to cause a ciguateralike illness.[15]

8. The geographical distribution of ciguatoxic fishes is spotty. Fishes in one sector around an island may be safe to eat, whereas fishes from another location may be toxic.

---

[13] This is a generality that is only partially true. There is ample toxicological data to show that many ciguatoxic fishes are small species and are consequently less frequently eaten by man and therefore attract less attention as a public health menace.

[14] There are a significant number of plankton feeders that have been incriminated as ciguatoxic fishes, including *Selar crumenophthalmus* which Randall cites as "universally nonpoisonous" (*see* p. 87).

[15] Further substantiation of the possible occurrence of ciguatoxin in marine invertebrates is presented in the recent observations of McFarren *et al.* (1965), who have reported the occurrence of a ciguateralike poison in oysters, clams, and *Gymnodinium breve* cultures.

9. That poisonous fishes are found in one part of a reef and not another in the nearby vicinity suggests that reef fishes are generally nonmigratory.

10. A region that once supported a toxic fish population may eventually become a safe area, or the reverse may be true.

11. A poisonous fish population may suddenly develop after a severe storm, hurricane, earthquake, or manmade catastrophic event.

12. Areas of slight or intermittent freshwater drainage may support a poisonous fish population.

13. The majority of records of poisonous fishes are from island areas. The island environment appears to be more conducive to ciguatera because there is a greater percentage of reef areas.

On the basis of these observations, which have in most instances been well established, Randall presented the following conclusions:

1. Fishes become poisonous because of some factor in their environment.

2. The toxicity of fishes is associated with their food supply.

3. The basic poisonous organism is benthic.

4. Since obligate herbivorous and detritus-feeding fishes may be poisonous, the toxic organism would most likely be an alga, a fungus, a protozoan, or a bacterium. If an alga, it must be fine because ciguatoxic surgeonfishes such as *Acanthurus triostegus* cannot feed on coarse types.

5. Blue-green algae, members of the Cyanophyta, would be the most likely source for ciguatoxin since this same group is also involved in freshwater poisonings.

6. The most poisonous fishes are the large predacious species, especially those that feed on other fishes. Studies have shown that the toxin disappears slowly, if at all, in poisonous fishes. Therefore it appears that they would accumulate the toxin and the larger older fishes would be the most dangerous. When a predator eats an herbivore, detritus-feeding fish, invertebrate, or another carnivore that preys on such forms, he acquires in one short period the toxin accumulated by the animal over its lifetime.

7. The organisms producing ciguatoxin may be one of the first growing on new or denuded surfaces in tropical seas in normal ecological succession. The availability of new surfaces may be the result of storms, hurricanes, earthquakes, dumping of war matériel, garbage, dredging operations, blasting, or by the outflow of turbid fresh water which kills the normal marine flora, thus exposing surfaces for new algal growth after the fresh water is swept away. The reader should refer to the original article by Randall (1958) in order to obtain a complete appreciation of the algal food theory as the author has presented it. Certainly it is the most complete review of the subject of any that have appeared to date.

Further confirmation of Randall's suggestion regarding ecological succession and the resulting increase in growth of precursor ciguatoxic marine algae is to be found in the excellent report by Dawson (1959) on the "Changes in Palmyra Atoll and its Vegetation through the Activities of Man, 1913-1958."

Rayner (1972) has obtained impressive evidence for the existence of fat-soluble toxins in herbivorous fishes, which tends to further support the algal food chain theory.

Banner (1959) and Banner *et al.* (1960), on the other hand, have been unable to find any relationship between ciguatoxin as found in reef fishes and that found in *Lyngbya majuscula*. Banner based his conclusion on several factors:

1. There is a difference in solubilities between ciguatoxin and the poison found in *Lyngbya.*[16] The toxin from both *Lyngbya* and *Lutjanus bohar* can be extracted with concentrated ethanol. The algal toxin can be extracted with hot water or cold petroleum ether, but the fish toxin cannot.

2. Patch tests made from *Lyngbya* extracts applied to human skin produce erythema, but similarly prepared tests of *Lutjanus bohar* extracts do not.

3. The geographical incidence of toxic *Lyngbya* is not accompanied by a corresponding incidence of ciguatoxic fishes.

Starr *et al.* (1962) have found *Lyngbya* to be toxic to human cells in tissue cultures, and to possess antibacterial properties against *Pseudomonas fluorescens*, *Micrococus pyogenes* var. *aureus*, and *Mycobacterium smegmatis*, and antiviral activity against mouse meningopneumonitis virus, and pronounced antifungal activity against *Candida albicans*.

Halstead *et al.* (1963) prepared aqueous extracts of *Lyngbya majuscula* from Palmyra Island by low temperature precipitation from methanol, pretreated with silicic acid, and finally fractionated on a silicic acid column using ether and ether-methanol (gradient elution). The major amount of toxicity as shown by the frog nerve test (Hessel *et al.*, 1960) was found to be in the material removed from silicic acid by ether. The two fractions corresponded to the ciguatoxin fractions of the silicic acid column and were weakly toxic. No conclusions were reached concerning the nature of the algal poison until further chemical and pharmacological evaluations have been completed.

Helfrich and Banner (1963) demonstrated that it was possible to induce toxicity in nontoxic surgeonfish *Acanthurus xanthopterus* by feeding them on a diet of toxic *Lutjanus bohar*. They pointed out in their report that the crux of the food chain hypothesis rests on three assumptions: (1) that the toxin can be acquired in the diet of the fish, (2) that the toxin can be eaten without detrimental effect to the fish, and (3) that the ingested toxin can be stored within the body of the fish in a pharmacologically unchanged condition. Failure to substantiate these conditions would necessitate discarding the food chain theory. Four specimens of *A. xanthopterus* were used in the experiments. Two of them were fed toxic *L. bohar* flesh and the controls were fed nontoxic

---

[16] There is some indication that ciguatoxin may be a complex of poisons having both fat and water-soluble fractions. So the apparent differences in solubilities may be of little or no significance. For further discussion on the solubilities of ciguatoxin see Halstead and Ralls (1954), Ralls and Halstead (1955), Hashimoto (1956), Hessel (1961), Scheuer (1969), and Yasumoto and Scheuer (1969).

skipjack *Katsuwonus pelamis*. The fish were fed 10 g of flesh every other day for a maximum period of 18 months. No apparent harm was observed in the test fish. Toxicity tests were conducted by feeding samples of the flesh and viscera of the *A. xanthopterus* to mongooses. The results conclusively demonstrated that the surgeonfish did acquire toxicity.

Banner *et al.* (1963*a*) have shown that when ciguatoxic *Lutjanus bohar* are maintained on a nontoxic diet of commercial Hawaiian skipjack tuna and herring over a period of 18 months there is no marked loss in toxicity. They also reported a positive correlation between size and toxicity in *L. bohar*. They found that in a random sample of 437 fish from the Line Islands taken in 1959 and 1960, the toxicity in all degrees increases from about 25 percent in those below 1 kg in size to 100 percent in those over 5 kg. The smaller fish were found to be either nontoxic or weakly toxic, and the larger fish moderately or strongly toxic. The feeding activity of *L. bohar* was found to be irregular with little difference between daylight or night. The major food categories taken from 1,790 stomachs examined from Palmyra and Christmas Atolls, consisted of fish (predominantly acanthurids), crustaceans (mainly megalopa larvae of crabs), mollusks (mainly gastropods), and miscellaneous invertebrates (planktonic tunicate *Pyrosoma*). The study indicated that the acanthurids could be the chief dietary source of ciguatoxin, and that there was a positive correlation between the size of the predator and the size of the prey. It was suggested that only the larger snappers were able to eat the deep-bodied herbivores, which may partially account for the observation that only the larger fish are strongly toxic. They also did a preliminary study on the feeding habits of some of the acanthurids, particularly *Acanthurus triostegus* and *Ctenochaetus striatus*, but received no clues as to the source of the poison. The fish were found to eat almost all algae found on the reef, but they were usually in small quantities. The blue-green alga *Plectonema terebrans* (Borinet and Flahault) is suspected of being associated with fish poisoning in some parts of the Gilbert Islands and was also found to grow in abundance epiphytically on most of the algae eaten by the common acanthurids at Christmas Island (Cooper, 1964).

Cooper (1964) published a valuable field study on the ciguatera problem in the Gilbert Islands. In this report she discussed the evolution of toxicity in the Gilberts, the spotty distribution as found in the various islands, and native theories as to the origin of ciguatoxin. It is interesting to note that native fishermen repeatedly incriminated marine algae as the source of the poison. Natives also reported the dumping of war matériel, garbage, wrecks, bombings, and other catastrophic events as precipitating causes of fish poisoning.

In many atolls of French Polynesia, outbreaks of ciguatera fish poisoning have been reported within a few months or years after stresses had been observed on limited areas of coral reefs or lagoons. Bagnis (1972) gives some results of a study on the conditions in which ciguatoxications have appeared after human aggression on atoll benthos. In the light of this information Bagnis considers ciguatera fish poisoning as a reaction of coral reef life to different types of human interference.

Helfrich *et al.* (1968) and Helfrich and Banner (1968) completed an exhaustive study on the ecology of reef fishes in the Line Islands as it relates to the ciguatera problem. They analyzed the stomach contents of 2,276 specimens of *L. bohar* and found that fishes were present in the majority of cases, and that acanthurids were most dominant. Fishes were followed in frequency by crustaceans, mollusks, and miscellaneous invertebrates. Seasonal fluctuations in the relative volume of food organisms in the diet were noted, with fishes dominating in summer months (87 percent) and dropping to 30 percent in December, being replaced largely by crustaceans. A brief dietary survey of *Acanthurus triostegus, C. striatus*, and *C. cyanoguttatus* revealed that cyanophytes of the genera *Lyngbya* and *Plectonema* were commonly found in the stomachs along with a variety of other algae. The local distribution of toxic *Lutjanus bohar* around Palmyra and Christmas Islands did not show any clearly nontoxic areas in contrast to toxic areas as is observed in other archipelagoes.

The question has been raised regarding a possible relationship between ciguatoxications and radioactive substances in the ocean (Halstead, 1958). Studies conducted by Bartsch *et al.* (1959) and Helfrich (1960) failed to reveal any correlation between toxicity and radioactivity (gross beta and gamma activity) in the reef fishes of the Central Pacific. The study was based on a series of samples of muscle, liver, and gonads from ciguatoxic lutjanid fishes taken from the Marshall and Line Islands.

In summarizing the algal food chain theory, it can be stated that this theory is not of recent origin, but probably dates back to at least the 18th century. Most investigators are in agreement that the variable toxicity and spotty geographical distribution favor the idea that ciguatoxin arises somewhere in the environment of the fish. Since most ciguatoxic fishes are bottom-dwelling shore fishes (although not exclusively so), the bulk of their food is derived from a benthic organism of some type. Moreover, since many obligate herbivorous fishes are ciguatoxic, it appears that the poison originates with either a benthic marine alga or with a microorganism living in close association with it. The surgeonfishes, acanthurids, are typical of some of these obligate herbivorous reef fishes. Stomach content analyses of acanthurids have shown that some of the algal members of the Cyanophyta, such as *Lyngbya majuscula*, are present to a major extent in ciguatoxic speciments. The blue-green algae, both fresh and saltwater species, have been shown to be toxic. It is therefore believed that certain species of the Cyanophyta serve as at least one possible source of ciguatoxin. Probably other kinds of algae—including certain species of dinoflagellates—may on occasion also serve as a source of the poison (Banner, 1975; Scheuer, 1977). It has been shown that ciguatoxicity can be experimentally induced in nontoxic fish without apparent harm to the fish. There is no reason to believe that ciguatoxin can be transvectored from the alga via an herbivorous invertebrate (mollusk, echinoderm, etc.) to the fish. Zooxanthellae living as symbionts with a variety of other marine invertebrates, i.e., radiolarians, coelenterates, mollusks, bryozoans, worms, ascidians, etc., must be considered as potential sources of the poison. Carnivorous fishes become ciguatoxic by ingesting herbivorous reef fishes. It has not been determined whether ciguatoxin is a single poison or a complex of poisons, but it is probably a complex structure having several fractions. Larger predacious fishes are more likely to be toxic than smaller-sized fishes of the same species. The exact biogenesis of ciguatoxin is still unknown.

Jardin (1972) has suggested that ciguatoxic cycles in fish may be triggered by contamination by metallic or metalloid compounds, which appear in a highly toxic organic form.

MECHANISM OF INTOXICATION

Ciguatera poisoning results from the ingestion of any one of a large variety of shore or reef fishes which are usually subtropical or tropical in their distribution. Generally speaking, large-size predacious fishes are more likely to be ciguatoxic than smaller fishes. However, many species of small herbivorous fishes are also dangerous to eat. The most toxic part of the fish is usually the liver, followed by the intestines, then the testes, ovaries, and least of all the muscle. The biotoxin is believed to be derived from the food of the fish. Ciguatoxications are in no way associated with bacterial food poisoning. There is no evidence of a seasonal incidence in ciguatoxicity, but the spawning season in some of the large predacious fishes may be a more dangerous period than the other seasons of the year. There is no possible way to detect a ciguatoxic fish by its appearance, and the degree of freshness has no bearing on its toxicity.

MEDICAL ASPECTS

Clinical Characteristics

Ciguatera fish poisoning is often confused with other forms of gastrointestinal disturbances, particularly bacterial food poisonings and other types of fish biotoxications. Ciguatera fish poisoning results from the ingestion of certain kinds of warm water shore or reef fishes. The disease is caused by a specific poison, or complex of poisons, which affect primarily the gastrointestinal and nervous systems of man.

Ciguatera fish poisoning has been variously designated in medical reports as ciguatos, ciguabo, siguatera, Pacific type of ichthyosarcotoxism, and ichthyosarcotoxism. The proper clinical terms used to designate this disease are ciguatera fish poisoning, ciguatoxication, ciguatera type of ichthyosarcotoxism, or simply ciguatera.

Ciguatera fish poisoning in its simplest uncomplicated form develops within 3 to 5 hours after the fish is eaten. There is a sudden onset of abdominal pain followed by nausea, vomiting, and a watery diarrhea. The gastrointestinal symptoms will occur in about 40 to 75 percent of the cases. The victim feels weak, generally ill, and may experience muscle aches throughout the back and thighs in about 10 percent of the cases. Soon after, the victim complains of numbness and tingling in and about the mouth which then extends to the extremities (present in about 50 percent or more of the cases). Fever, headache, and rash are generally absent, and the patient has no desire for food. The acute symptoms usually subside in about 8 to 10 hours, and within 24 hours after onset most of the patient's symptoms will have completely subsided except for a feeling of weakness. However, the numbness and tingling may continue to a lesser extent for a period of 4 to 7 days (Bartsch *et al.*, 1959). The foregoing résumé is typical of the majority of uncomplicated ciguatoxications that are generally encountered by the practicing physician in an endemic ciguatoxic locality such as the Marshall Islands.

Ciguatera, like many other diseases, may vary greatly in its clinical manifestations depending upon the toxicity of the fish that is eaten, the individual's sensitivity to the poison, amount of fish ingested, and other factors. In a broader sense ciguatera fish poisoning may be characterized as follows: The onset of symptoms may vary from almost immediately to within a period of 30 hours after ingestion of the fish, but is usually within a period of 6 hours. The initial symptoms in some cases are gastrointestinal in nature, consisting of nausea, vomiting, watery diarrhea, metallic taste, abdominal cramps, and tenesmus, whereas in other patients the initial symptoms consist of tingling and numbness about the lips, tongue, and throat. This may be accompanied by a sensation of dryness of the mouth. The muscles of the mouth, cheeks, and jaws may become drawn and spastic with an accompanying sensation of numbness throughout. Generalized symptoms of headache, anxiety, malaise, prostration, dizziness, pallor, cyanosis, insomnia, chilly sensations, fever, profuse sweating, rapid weak pulse, weight loss, myalgia, and back and joint aches may be present in varying degrees, or one or more of these symptoms may be entirely absent. The victims usually complain of a feeling of profound exhaustion and weakness. The feeling of weakness may become progressively worse until the patient is unable to walk. Muscle pains are generally described as a dull, heavy ache, or cramping sensation, but on occasion may be sharp, shooting, and affect particularly the arms and legs. Victims complain of their teeth feeling loose and painful in their sockets. Visual disturbances consisting of blurring, temporary blindness, photophobia, and scotoma are common. Pupils are usually dilated and the reflexes diminished. Skin disorders are frequently reported that are usually initiated by an intense generalized pruritus, accompanied by erythema, and maculopapular eruptions, blisters, extensive areas of desquamation—particularly of the hands and feet—and occasionally ulceration. There may also be a loss of hair and nails.

In severe intoxications the neurotoxic components are especially pronounced. Paresthesias involve the extremities, and paradoxical sensory disturbances may be present in which the victim interprets cold as a "tingling, burning, dry-ice or electro-shock sensation," or hot objects may give a feeling of cold. In regard to the paradoxical sensory disturbance, Mann (1938) cited the classical case of a naval officer who was poisoned by an amberjack. Four weeks later he was observed subconsciously blowing on his ice cream, which was "burning his tongue," in order to cool it. Ataxia and generalized motor incoordination may become progressively worse. The reflexes may be diminished, muscular paralyses may develop, accompanied by clonic and tonic convulsions, muscular twitchings, tremors, dysphonia, dysphagia, coma, and death by respiratory paralysis. The limited morbidity statistics show a case fatality rate of about 12 percent. Death may occur within 10 minutes, but generally requires several days.

In severe ciguatoxications recovery is slow and may be very prolonged, with extreme weakness, sensory disturbances, and excessive weight loss being the last symptoms to disappear. Severe intoxications have occurred in which the patient survived, but complete recovery required a period of several years (Meyer-Ahrens, 1855; Steinbach, 1895; Gudger, 1930). Similar reports have been received from natives who have suffered from fish poisoning. Meyer-Ahrens (1855), Steinbach (1895), and Gudger (1930) stated that symptoms may be present for as long as 25 years. Individuals who have been severely intoxicated have reported that during periods of stress, fatigue, exposure or poor nutrition, there is a recurrence of the myalgia and joint aches similar to that suffered during the original acute period of the disease.

Some of the most violent ciguatoxications reported in medical literature are caused by the eating of moray eels. It was formerly believed that moray eel or gymnothorax poisoning was caused by a poison that differed from ciguatoxin (Halstead and Lively, 1954; Halstead, 1958, 1959, 1962; Helfrich, 1961). However, recent chemical studies indicate that moray eel poison and ciguatoxin are either closely related or identical in their chemical and pharmacological properties (Halstead *et al.*, 1963; Banner *et al.*, 1963*a*). The characteristics present in moray eel intoxications that appear to differ from the usual clinical picture of ciguatoxications are apparently due to quantitative differences in the poison rather than qualitative ones. Toxicological studies have shown that large tropical moray eels are often violently poisonous. Because of some of these clinical differences the following description of moray eel intoxications may be useful.

Symptoms of tingling and numbness about the lips, tongue, hands, and feet, with a feeling of heaviness in the legs, usually develop within 20 minutes to 3 hours after ingestion of the eel. These symptoms may be followed by nausea, vomiting, diarrhea, abdominal pain, malaise, metallic taste, sore throat, laryngeal paralysis, meningismus, laryngeal spasm, aphonia, excessive mucuous production, foaming at the mouth, intense perspiration, increased body temperature, crying out as if in pain, conjunctivitis, paralysis of the respiratory muscles, ataxia, general motor incoordination, trismus, violent clonic and tonic convulsions, abnormal deep and superficial reflexes, coma, and death. Sensory reactions to deep and superficial pain are usually normal. Characteristic signs of this form of intoxication appear to be the absence of thoracic respiration with pronounced abdominal breathing, profuse perspiration, violent clonic and tonic convulsions, purposeless movements, and the extended period of time in which areflexia is present. In some cases the reflexes may be absent for a period of 2 months or more. The mortality rate is considerably higher than in outbreaks usually produced by other types of ciguatoxic fishes. The acute symptoms generally subside within 10 days in the milder form. Khlentzos (1950) has reported sequelae consisting of dizziness, blurred vision, tremors of the hands, sensory changes in the legs, alopoecia, interosseous atrophy, muscular weakness, ulnar palsy, foot drop, radial weakness, deviation of the tongue, atrophy of the tongue, and aphonia. In most instances these sequelae gradually disappeared within a period of 2 months. In most cases reported by Khlentzos (1950), the laboratory results were relatively normal. The white blood cell count varied between 8,500 and 20,000. The polymorphonuclears varied between 48 and 92. The sedimentation rate was elevated, varying between 2 mm and 45 mm per hour: hematocrit varied between 36 and 52 percent; NPH from 15 to 72 mg per 100 ml initially and at no time during hospitalization did it much exceed 72 mg per 100 ml; blood sugar from 85 to 155 mg per 100 ml; carbon dioxide combining power from 39 to 51 volumes per 100 ml; plasma chlorides from 545 to 627 mg per 100 ml; urobilinogen on three patients was elevated for a few days about 10 days after hospitalization—values were 1:220, 1:160, and 1:160 and dropped to 1:20 in 2 to 4 days. The electrocardiograms were not significant except in three patients who had pulsus alternans, the tracings showing electrical dissociation. Clinical reports on moray eel biotoxications have been given by Khlentzos (1950) and Ralls and Halstead (1955). An attack of ciguatera does not impart an immunity to the poison, but may provide the victim with an increased sensitivity to the poison (Halstead, 1958; Randall, 1958; Helfrich, 1963).

Of the authors listed in this chapter, the following have discussed the clinical characteristics of ciguatera fish poisoning: Cloquet (1821), Chevallier and Duchesne (1851*a, b*), Meyer-Ahrens (1855), Chevallier (1856), Fonssagrives and de Méricourt (1861), Day (1871), Fonssagrives (1877), Savtschenko (1886), Blanchard (1890), Steinbach (1895), Coutière (1899), Pellegrin (1899), Taschenberg (1909), Mowbray (1916), Phisalix (1922), Hoffmann (1927, 1929*a, b*), Pawlowsky (1927), Gudger (1930), Matsuo (1934), Mann (1938), Gilman (1942), Hiyama (1943), Ross (1947), Arcisz (1951), Meyer (1953), Wheeler (1953), Fish and Cobb (1954), Halstead and Lively (1954), Dack (1956), Hashimoto (1956), Mills (1956), Halstead (1958, 1959, 1962), Randall (1958, 1961), Kudaka (1960), Helfrich (1960, 1961), Bartsch and McFarren (1962), Bouder *et al.* (1962), Robinson (1962), Banner *et al.* (1963*b*), De Sylva (1963), Prosvirov (1963), Cooper (1964), and McFarren *et al.* (1965).

Pathology

Studies on gross pathological changes caused by ciguatera intoxications have been of a superficial nature. There are no worthwhile data on micropathology. The few reports available show evidence of acute visceral congestion. The pathology of ciguatera has been discussed by Chisholm (1808), Chevallier and Duschesne (1851*a, b*), Fonssagrives and de Méricourt (1861), Savtschenko (1886), Blanchard (1890), Coutière (1899), Colby (1943), and Li (1970).

Treatment

The treatment of ciguatoxications is largely symptomatic. An attack of fish poisoning does not impart immunity, and there are no known specific antidotes. There is still no general agreement as to the exact chemical and pharmacological properties of ciguatoxin. There is an increasing amount of chemical evidence that the poison currently designated as ciguatoxin may be either a complex of poisons or several different poisons which may vary in composition according to the fish or the source of the poison. At the time of this writing, it is still a matter of conjecture, since laboratory studies have not progressed to the point of permitting any conclusions.

In general the treatment of ciguatera fish poisoning is directed toward eliminating the poison from the body, combating the physiological effects of the poison, and providing whatever supportive therapy is required. The stomach should be emptied by gastric lavage, emetics, or saline purges at the earliest sign of ciguatoxication. The hyperventilation and shock associated with the acute phases of poisoning in laboratory animals suggest a profound alteration of electrolyte balance and blood pH which would be amenable to calcium therapy. Some cases of severe ciguatoxication have responded well to intravenous 10 percent calcium gluconate, whereas other have not. It is recommended that calcium therapy be tried (Khlentzos, 1950; Ralls and Halstead, 1955; Banner *et al.* 1963*b*).

Preliminary pharmacological studies have shown that at least one of the components of ciguatoxin acts as an irreversible cholinesterase inhibitor (Li, 1965, 1970). The observed symptoms of both laboratory animals and humans correspond with known symptoms of anticholinesterase intoxication. Thus the action of ciguatoxin parallels that of some of the organophosphorus compounds present in insecticides. Banner (1965) and his associates have found the administration of the oxime protopam chloride an effective remedy in laboratory animals (Li, 1965). At the time of this writing, protopam chloride had not been reported in the treatment of human ciguatoxications.

Preliminary pharmacological studies have shown that the biotoxin of moray eels, groupers, snappers, and sharks display anticholinesterase activity (Li, 1965). However, the barracuda biotoxin obtained from the Line Islands did not reveal anticholinesterase activity (Banner, 1965), which is further evidence that ciguatoxin may vary in composition from one fish species to another. Neostigmine (prostigmine), a cholinergic drug having an anticholinesterase action, has been successfully used in the treatment of barracuda poisoning (Banner *et al.*, 1963*b*). Neostigmine corresponds in its action to parasympathetic stimulation and is a potent antagonist of atropine. The case report given by the aforementioned authors is of significance because if the barracuda ciguatoxin in this particular case had possessed anticholinesterase activity, then the treatment might have been fatal to the patient. Moreover, in those cases of ciguatoxication in which an anticholinesterase fraction of ciguatoxin is present the use of neostigmine *would be decidedly contraindicated* *and must not be employed.* It is recommended that extreme caution be used in employing any anticholinesterase drug in the routine therapy of ciguatera fish poisoning. Banner *et al.* (1965) have recommended the use of 2-PAM (2-pyridine aldoxime methochloride) for the treatment of ciguatoxications.

Supportive therapy consisting of intravenous glucose in normal saline and injections of vitamin B_6 have been used with varying degrees of success (Bouder *et al.*, 1962; Helfrich, 1963; Banner *et al.*, 1963*a*). Victims suffering from moray eel poisoning appear to be particularly susceptible to violent convulsions and may present difficult nursing problems. Rest, quiet, and sedation are essential because the convulsions may be precipitated by noise or commotion of almost any type. Paraldehyde and ether inhalations have been reported to be effective in controlling the convulsions. Nikethamide or one of the other respiratory stimulants are advisable in cases of respiratory depression. In patients where excessive production of mucus is a factor, aspiration and constant turning are essential. Atropine has been found to make the mucus more viscid and difficult to aspirate, and may be contraindicated. If laryngeal spasm is present, intubation and tracheotomy may be required. Oxygen inhalation may be necessary. If the pain is severe, opiates may be necessary. Morphine given in small divided doses has been recommended. During later stages of recovery cool showers may be useful in relieving the severe itching that is sometimes present. Fluids given to patients suffering from the paradoxical sensory disturbance (temperature sensation reversal) should be tepid rather than either hot or cold (Khlentzos, 1950; Halstead, 1958, 1959). Intravenous procaine hydrochloride has been used by Gross (1960) with reputed success in the treatment of a single case of barracuda poisoning. Megavitamin doses of vitamin C should be tried.

Natives living in ciguatoxic endemic Pacific islands recommend using alcoholic (ethyl alcohol) or aqueous extracts of the salpiglossid plant *Duboisia myoporoides* Brown (Pl. 47, fig. a). The plant is reputedly useful in relieving the symptoms of ciguatoxications in mild cases, but is reported to be contraindicated in severe cases (Banner, 1965). The juice from the leaves of the kamachiri or kinkimoku plant *Pithecellobium dulce* (Roxburgh) (Pl. 47, fig. b) is also used in the treatment of fish poisoning. A very popular treatment in many tropical Pacific islands is ingestion of the juice or roasted leaves and bark of the monpakoi, also called false tobacco, hamaroki, or meganeoki plant *Tournefortea (Messerschmidia) argentea* Linnaeus (Pl. 47, fig c) (Hiyama, 1943). Cooper (1964) states that the Gilbertese use te tarai *Euphorbia atoto* Forster and Forster (Pl. 47, fig. d). A few drops of the milky sap of the plant are squeezed into a small drinking coconut and the mixture drunk. One of the best known remedies is the fruit of te non, the Indian mulberry *Morinda citrifolia* Linnaeus (Pl. 47, fig. e). Three unripe fruits are crushed with three ripe fruits, crushed with coconut juice and the mixture drunk. The buds from the seedless breadfruit tree te bukiraro, *Artocarpus sp?* (Pl. 48, fig. a) are also used. Another remedy is made from the fruit of the unripe papaya

165
165
165
165
165
166

166 *Carica papaya* Linnaeus (Pl. 48, fig. b), boiled with rainwater and then taken. 166 The fruit of the saltbush te mao, *Scaevola taccada* (Gaertner) (Pl. 48, fig. c) is used by squeezing the ripe fruit into a drinking coconut. In the Ryukyus 166 the natives prepare a tea from the leaves of *Vitex trifolia* Linnaeus (Pl. 48, fig. d) (Kudaka, 1960). The efficacy of these and other native plants which are widely employed by the native island peoples is presently unknown, but is in the process of being evaluated at the University of Hawaii.

The following plants have been used in the West Indies in the treatment of fish poisoning (Dunlop, 1917): West Indian alder or buttonwood bush 167 *Conocarpus erectus* Linnaeus (Pl. 49, fig. a), rabbit weed or Lord Lavington 167 *Hyptis capitata* Jacquin (Pl. 49, fig. c), and running pop or love-in-a-mist 167 *Passiflora foetida* Linnaeus (Pl. 49, fig. d). Most of these plants are used in the preparation of teas which are taken internally.

Prevention

Despite numerous statements to the contrary by native fishermen, there is no reliable method of detecting a poisonous fish by its appearance. Natives have suggested that certain poisonous fishes can be detected by observing the color of their throat which supposedly is orange-colored streaked with thin brown lines when toxic. Others recommend opening the stomach of the fish and examining it for worms—if worms are present the fish is safe to eat. Pacific islanders will sometimes place bits of the flesh in the vicinity of ant hills and consider the fish toxic if the ants refuse to eat it. Still others place a portion of the viscera out in the sun to dry. If the flies refuse to settle on it, the fish is believed to be poisonous. Black-colored teeth and a bitter tasting liver are reputed to be evidences of toxicity in barracuda. If when cleaning a fish the tissue juices produce a burning or stinging sensation of the hands the fish is said to be toxic. Some say to cook the fish and then tear the flesh into bits. If small dark veins are present the fish is thought to be edible. Some of the Marshallese place live captured fish in brackish water pools filled with organic matter for several days in order to let the fish detoxify. Okinawans test for toxicity by tasting the slime from the fish's eye, believing that poisonous fishes will produce a tingling sensation. Red snapper whose pectoral fins extend beyond the snout are reputedly poisonous. All fishes from a specific location about an island are said to be toxic. The appearance of the gills is also thought to be an important characteristic in determining toxicity. It is believed by many in the West Indies and the Pacific islands that if a silver coin is cooked with the fish it will turn black if the fish is toxic. The Gilbertese believe that if grated coconut is cooked inside the fish it will turn green if the fish is toxic. Numerous other similar statements have appeared over the years among native groups, but they are based largely on superstition or tradition and are without scientific merit.

Several native methods, however, are used and are relatively reliable. Some natives will feed portions of the fish that is under evaluation to one of their pet dogs or cats. If the animal shows no sign of toxicity, then the fish is eaten by the family. Another method is for the person to eat a small piece of the fish and if after several hours no ill effects have been noted the entire fish is served to other members of the family. Many islanders are very careful about feeding fish that is likely to be toxic to their children. The fish is first eaten by older members of the family and if they continue after several hours, or preferably overnight, without ill effects, then the fish is served to the children (Cooper, 1964). These are test methods that one can rely upon with a reasonable degree of safety assuming that the fish that is to be eaten is done so with moderation. It must be kept in mind that toxicity is directly related to the amount of fish eaten and biotoxin consumed.

At the time of this writing no simple chemical field test method has been developed that can be used to detect a poisonous fish. However, Banner (1965) and his associates have been exploring several methods in this area that may ultimately prove to be of value. At the present time the most reliable method is to test the fish by feeding samples of the flesh to a test animal—preferably a kitten—which is highly sensitive to most fish ichthyosarcotoxins.

If a person is faced with a survival situation, there are several fundamental points that should be kept in mind. *Never eat the viscera*, i.e., liver, gonads, intestines, etc., of tropical marine fishes under any circumstances. The roe of most tropical fishes is potentially dangerous and should always be eliminated. *Unusually large predacious reef fishes* such as snapper, barracuda, grouper, and jack *should not be eaten*. They are most likely to be dangerous during the reproductive season as is evidenced by a distended belly filled with large ripe roe or testes. *Ordinary cooking procedures* such as baking, frying, or drying, *do not render a toxic fish safe to eat*. Boiling the fish and discarding the water several times may be helpful, but cannot be completely relied upon. Some poisons are water soluble and others are not. If it is impossible to boil the fish then cut it into small strips, let them soak in several changes of seawater for at least 30 minutes, and squeeze out as much of the juice as possible. This procedure does not guarantee safe eating, but you may have no other choice. Eat only small amounts of an unknown variety of any tropical fish. *Tropical moray eels should never be eaten.* Some species of tropical moray eels are violently toxic and may produce convulsions and quick death. If at all possible, try to capture open water fishes rather than those near a reef or near the entrance to a lagoon. If natives are available ask their advice about eating reef fishes. Certainly the native population is your best source of local advice if you are stranded on a remote tropical island.

PUBLIC HEALTH ASPECTS

Ciguatoxications are for the most part limited to subtropical and tropical insular areas. Since most outbreaks occur in scattered undeveloped insular areas having limited public health facilities, the problem has received only minimal attention by trained epidemiologists. There are no reliable public health records on the subject of ciguatera fish poisoning in any part of the world. The disease is not generally reported to such international groups as

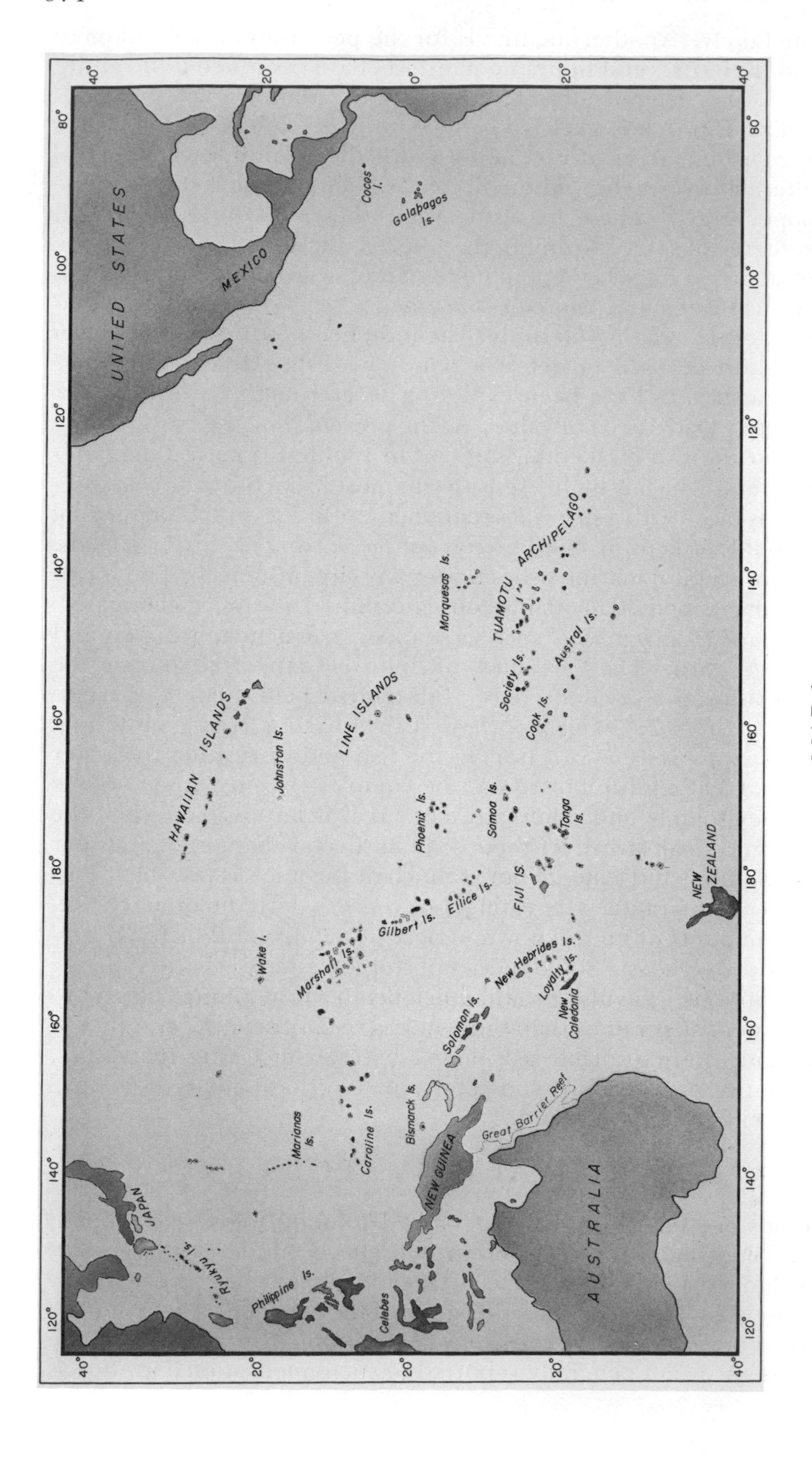

MAP 1
Principal island groups in the Pacific area where ciguatera is known to occur.
(R. kreuzinger)

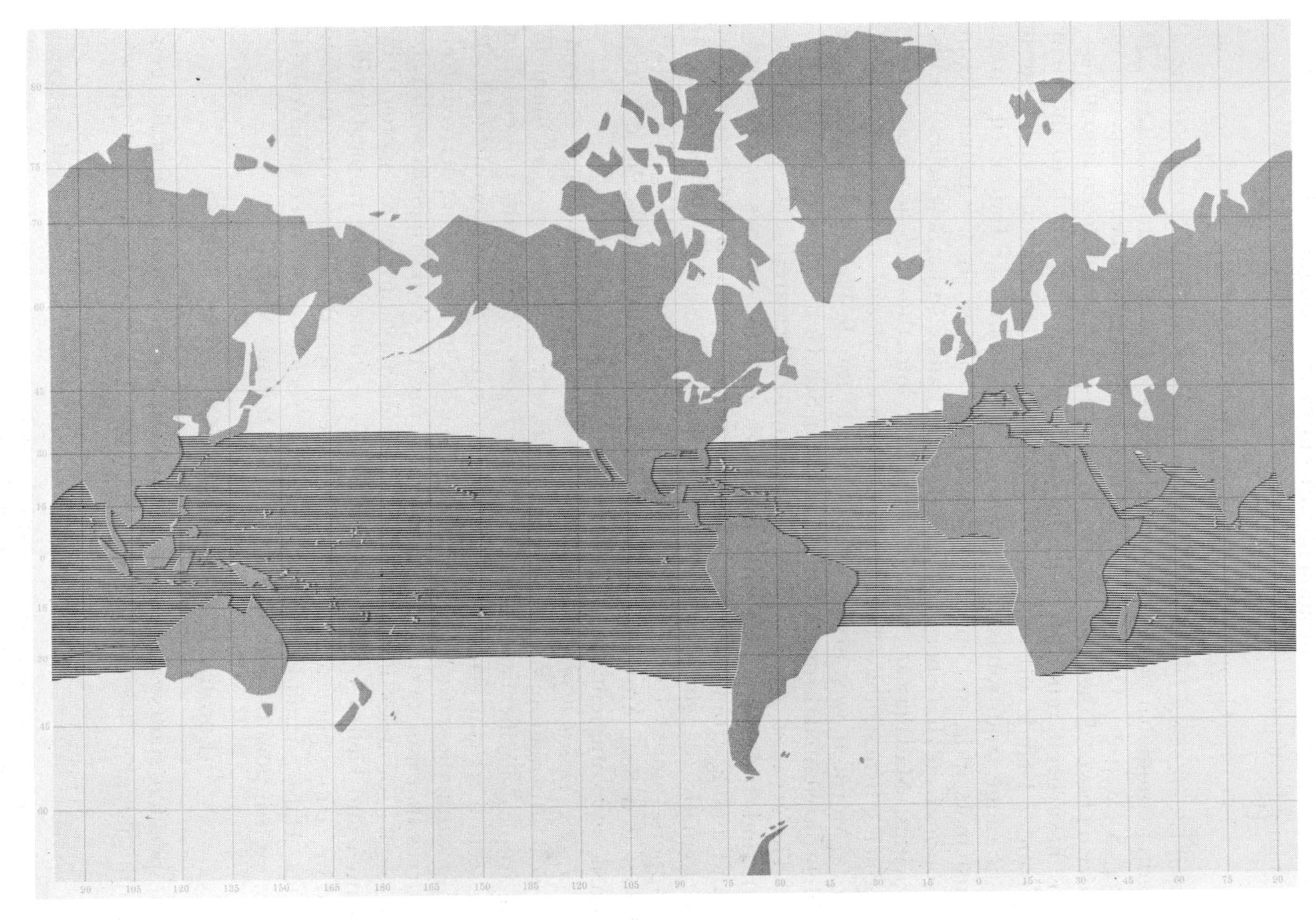

MAP 2

Map showing the approximate geographical distribution of ciguatoxic fishes. In general, ciguatoxic fishes are found in greatest abundance in subtropical-tropical insular areas of the West Indies, Pacific, and Indian Oceans. (L. Barlow)

the World Health Organization nor are outbreaks brought to the attention of very many local public health agencies.

Geography and Ecology of Ciguatera

General Geographical Distribution: Ciguatoxic fishes are largely restricted to a broad circumglobal belt which extends approximately from latitudes 35° N to 34° S (Map 2). Only in rare instances has ciguatera been suspected of occurring beyond these boundaries. In the eastern Pacific Ocean ciguatoxications have been reported from the Gulf of California south to the Juan Fernandez Islands, off the coast of Chile. Ciguatoxic fishes in the western Pacific range from southern Japan south throughout the Indo-Pacific area, Nanshan Islands, to Lord Howe Island and Taiwan. In the Indian Ocean poisonous fishes have been reported from Sri Lanka, India, Seychelles Islands, Mauritius, Madagascar, South Africa, and into the Red Sea. Ciguatera-producing fishes(?) have been reported from the Mediterranean area,[17] West Africa, Morocco, west to the Azores, south to Rio de Janeiro, Brazil, and are widely distributed throughout the Caribbean Sea, Gulf of Mexico, and Florida.

The greatest concentrations of ciguatoxic fishes are found in the Caribbean and Indo-Pacific regions. Although ciguatoxic fishes are primarily found among tropical islands, they may occur—particularly the ciguatoxic plectognaths (triggerfishes, filefishes, etc.), wrasses, moray eels, and lutjanid snappers—among tropical reefs along the continental areas, i.e., East Africa, India, Red Sea, Australia, Malaysia, Vietnam, United States, West Africa, Brazil, and continental Europe in the Mediterranean region.

Insular Distribution of Ciguatoxic fishes: According to all available epidemiological information, the greatest incidences of ciguatoxic fishes are found in the tropical Pacific Islands and the West Indies. The results presented in this section are concerned primarily with the insular distribution of ciguatoxic fishes within these two general localities. The data presented are derived from a survey that was conducted by Halstead during the years 1950 through 1957 among 64 countries of the world, supplemented by information gathered by Banner and Helfrich (1964) in their extensive studies in the tropical Pacific, and by various publications pertinent to the insular distribution of ciguatoxic fishes. Halstead and Cox (1973) have more recently conducted an investigation on fish poisoning in Mauritius, in the Indian Ocean.

Localized Distribution of Ciguatoxic Fishes

The peculiar spotty distribution of ciguatoxic fishes within a particular geographical region has been observed since the 17th century. Locke (1675) was

[17] There is considerable question as to whether or not ciguatera occurs in the Mediterranean Sea, but several piscine biotoxications have been reported which may possibly be ciguatera. The exact nature of these poisonings remains to be determined.

among the first to direct attention to the bizarre occurrence of these fishes, and stated:

> Those of the same species, size, shapes, color and taste, are one of them poyson, the other not in the least hurtful: And those that are, are so only to some of the company.

Since the time of Locke, numerous other authors have reported the scattered sporadic distribution of these fishes: Meyer-Ahrens (1855), Steinback (1895), Mowbray (1916), Phisalix (1922), Hoffmann (1929*a*), Matsuo (1934), O'Neill (1938), Gilman (1942), Hiyama (1943), Ross (1947), Arcisz (1950), Randall (1958), Cooper (1964), Helfrich *et al.* (1968), and Helfrich and Banner (1968).

Undoubtedly there are a variety of ecological factors that contribute to the spotty distribution of ciguatoxic fishes. Exactly what these factors are and how they interrelate no one knows, since very little research effort has been directed to the problem. There is also some evidence that these factors may vary somewhat with the geographical locality and the organisms involved.

Exact data on the zoogeography of poisonous fishes are almost totally lacking. Complete distributional data are not available for a single endemic island—to say nothing about extensive insular or continental areas. The geographical range of ciguatoxic fishes may occasionally extend beyond the previously mentioned latitudes. Ciguatoxic fishes do not appear to be restricted to coral-bearing regions as suggested by Hiyama (1943), Fish and Cobb (1954), and others. Previous studies serve to point up that because an area is presently free of ciguatoxic fishes is no assurance that this condition will continue to prevail. Areas that were once free of the problem have since become involved, and the reverse may occur. Some island areas appear to clear up after several years, whereas others seem to continue at about the same level of toxicity. Much additional field research is needed on the geographical and ecological aspects of the ciguatera problem, and the importance of these factors should not be minimized. Intelligent preventive measures and public health controls can be instituted only to the extent that the ecology of the disease is understood.

Incidence and Mortality Rate

Apparently there are no countries in which ciguatoxications are considered to be a reportable disease. Consequently statistics regarding outbreaks of this biotoxication are meager and grossly incomplete. In many areas where the incidence of ciguatera is highest the public health facilities for gathering the data are either minimal or nonexistent. In some places the natives are convinced that they have more effective home remedies than the sophisticated physician, and therefore refuse to seek medical aid. Hence numerous outbreaks, and particularly those of a mild character, pass by unnoticed by the local physician. The works of Bartsch *et al.* (1959) and Bartsch and McFarren (1962) are recommended to anyone interested in the epidemiological aspects of ciguatera fish poisoning.

Seasonal Variation in Toxicity: The spotty distribution and sudden occurrence of ciguatoxic fishes have suggested to some writers a seasonal fluctuation in toxicity. The fact that in some predacious species only the large mature specimens are toxic has also contributed to the idea that there is a definite relationship between gonadal development and toxicity. Arcisz (1950) has tabulated the occurrence of 26 outbreaks of fish poisoning caused by a variety of fishes in the West Indies and concluded that ciguatera is not seasonal. De Sylva (1963) came to similar conclusions in his studies on *Sphyraena barracuda*. Similar opinions have been expressed by Ross (1947), Belotte (1955), and Banner *et al.* (1963*a*). However, Bartsch *et al.* (1959), in their studies on the poisonous fishes of the Marshall Islands, noted a slight upswing in the incidence of intoxications during the spring months, but were unable to determine a distinct seasonal pattern. Although a seasonal variation has been established in most of the puffer fishes affected by gonadal activity, there is little evidence at present that a comparable phenomenon occurs in ciguatoxic fishes.

Ciguatoxicity Cycle in Fishes

Cooper (1964) has presented the idea that "there is a definite evolution of toxic conditions over the years" in some of the Pacific islands. Long-term fluctuations in toxicity have been recorded in the toxicity pattern of ciguatoxic reef fishes in the Line Islands, Marshall Islands, Johnston Island, Wake Island, Gilbert Islands, and the Hawaiian Islands. Cooper (1964) has suggested that this fluctuation in ciguatoxicity is a cyclic phenomenon which requires a period of not less than 8 years to conclude. During the initial phase only a few reef fishes are toxic, and during the maximum period almost all species are affected, but in the final phase only large eels, certain snappers, and groupers remain toxic.

Feeding experiments using 66 large toxic *Lutjanus bohar* which were maintained in tidal ponds of the Hawaii Marine Laboratory and fed nontoxic commercial Hawaiian skipjack and Puget Sound herring for a period of more than 30 months revealed no statistically significant decline in toxicity (Banner, 1965). Banner and his associates are thus of the opinion that the toxin is stored in the body of the fish and ciguatoxin is neither metabolized nor excreted for long periods of time. Whether there are environmental factors present in the normal reef biotope that might alter the detoxification period is not known. The time element may vary somewhat from one region to the next, but there is an increasing amount of field evidence which lends credence to Cooper's suggestion that the ciguatoxic condition may be a cyclic phenomenon—at least in some regions.

Relationship of Fish Size to Toxicity

Fonssagrives and de Méricourt (1861) and Rochas (1862) were among the first to direct attention to a possible correlation between the maturity of the fish (as indicated by its size) and its toxicity. They stated that *Lethrinus mambo* is edible until it attains a size of 70 to 80 cm, when it becomes violently poisonous. Savtschenko (1886) reported, "Many authors attribute the poisonous qualities of certain species to a particular age only; other fish of the same species, but younger, not being poisonous at all, and their meat safe for eating." Hiyama (1943) quoted the natives of Jaluit Island as saying that the larger specimens of barracuda, carangids, and lethrinids are most likely to be toxic. In one toxicological test, a 71 cm specimen of *Caranx melampygus* proved less toxic than a 90 cm one. Small specimens (30-40 cm) are widely sold for food at Saipan, but those of a meter or more in length are never eaten. Public health regulations instituted by the U.S. Navy regarding the eating of fishes by the military stationed at Saipan were largely based upon the size of certain species. It should be added that these beliefs regarding the toxicity of large-sized reef fishes, especially barracuda, carangids, and lutjanids, are widespread throughout the tropical Pacific. In his work on ciguatera in the Caribbean area, Arcisz (1950) stated: "Most of the outbreaks recorded in the literature have been due to large fish." There are exceptions, however, since Walker (1922) reported an outbreak caused by a small (1.1 kg) barracuda.

Although many opinions have been expressed on this subject, the only published laboratory studies of the relationship between the size of the fish and its toxicity are by Hessel *et al.* (1960) and Banner *et al.* (1963*a*). In both instances fish examined were *Lutjanus bohar* taken in the Line Islands. Hessel *et al.* found in a series of 190 specimens a low incidence of toxicity in fishes below 2.8 kg (only 18 percent toxic) but a sharp increase in toxicity in specimens weighing over that amount (69 percent toxic). Hessel's group used the frog nerve method of assay, and their studies indicated that there was not only a higher incidence of toxicity in the larger *L. bohar*, but also a greater concentration of poison within the fish. In a random sample of 437 *L. bohar* from the Line Islands taken in 1959 and 1960, Banner's group found that the toxicity in all degrees increased from about 25 percent in those fish weighing less than 1 kg to 100 percent toxicity in those fishes weighing over 5 kg. Banner's group assayed their fish by the standard mongoose feeding test.

Effects of Cooking

Although ciguatera intoxications occasionally result from eating raw fish, in most instances the fish have been cooked. Neither frying, baking, broiling, stewing, steaming, drying, salting, boiling, nor any other ordinary cooking procedure is known to significantly attenuate the toxin of ciguatera-producing fishes. Hiyama (1943) and Watanabe (1946) have noted that the eating of poisonous fishes which have been prepared in soup is a particularly common cause of the intoxication among certain native groups. Hiyama stated that there have been cases in which the members of the family that drank the soup, but did not eat the fish, were poisoned, and others that ate the fish, but did not drink the soup, were not intoxicated. Numerous other references might be cited in support of the observations of Hiyama and Watanabe, all illustrating the point that ordinary cooking procedures, salting, or drying cannot be relied upon to attenuate the poison.

Quarantine Regulations

Ancient mores regarding the eating of fishes are to be found among most ethnic groups. Taboos are particularly common among many of the island-dwelling peoples where fish poisoning has been a problem for centuries. Prohibitions regarding the eating of certain types of fishes appear in the Bible in Deuteronomy 14: 9-10. One of the earliest published accounts of a public health regulation governing the handling of ciguatoxic fishes is by Autenrieth (1833), who stated that the sale of fishes is prohibited in Haiti from May to October because of the frequency of intoxications during this period.

Public health regulations prohibiting the sale of poisonous fishes have been in existence in Cuba since 1842. Several versions and modifications of these Cuban public health regulations have appeared over the years, but one of the more comprehensive prohibitions is the one quoted by Hoffmann (1929*b*) which appeared as Presidential Decree No. 674 in the Sanitary Ordinance in Havana, Cuba, of July 6, 1914.

Coutière (1899) reported that the sale of the sardine *Harengula humeralis* was forbidden by the public health authorities in the Dominican Republic during the months of May through October.

In 1884, the fishermen of Mahebourg, Mauritius, petitioned the Governor to take action to forbid any selling of fishes outside the market walls, which had been permitted without the vendors obtaining a license to carry on such transactions (Jones, 1956). The fishermen presented a list of poisonous fishes which later constituted Ordinance No. 59 of 1884 and was approved by the Governor, continuing in Mauritius as of 1956 in the form of Government Notice No. 40 of 1935 (MP 1424/32).

During World War II poisonous fishes were frequently a problem to the military. Rear Admiral F. E. M. Whiting, USN, issued a directive, Order No. 14-45, listing the fishes poisonous at Saipan. Local military regulations required that all fishes that were to be used for food by American troops had to be inspected by the Fish Commissioner of the Military Government.

The most recent public health regulation regarding the sale and handling of poisonous fishes is that established by the Board of Health, Territory of Hawaii (Lee, 1954). An active public health surveillance program against ciguatoxic fishes is also operative in Japan.

A number of species of reef fishes suspected as toxic are prohibited from sale in the Suva, Fiji, fish markets (Banner and Helfrich, 1964).

The fact that public health organizations have attempted to regulate the handling and sale of ciguatoxic fishes is indicative of the need for protective regulations. However, without exception there are insufficient scientific facts regarding the epidemiology of the disease upon which to base these regulations. Laws should be established, but they must be backed by a fundamental knowledge of the biology of the disease, and the identity of the causative organisms, most of which are presently lacking.

TOXICOLOGY

Extraction Methods

Yasukawa (1935) was one of the first to attempt to extract ciguatoxin from the tissues of suspect reef fishes captured at Saipan. He prepared aqueous extracts from samples of the muscle, liver, gonads, and blood to which he added three parts of water and ground the mixture in a bowl. Since the initial collections were made in the field, the material was then preserved in chloroform and toluol, which were later removed at the laboratory, and the aqueous material tested by injecting it intraperitoneally into mice. Yasukawa failed to obtain any toxic effects, and none of the mice died. It now appears that the toxin was probably in the chloroform-toluol fraction which Yasukawa discarded and failed to test.

Macht (1936), Macht and Spencer (1941), and Macht, Brooks, and Spencer (1941) working with a collection of 65 species of North American fishes prepared tissue extracts by grinding up samples of fish muscle with an equal weight of sand, and to this mixture added physiological saline in the proportion of 2 ml to each gram of muscle. The homogenate was strained through muslin and the filtrate used for testing.

Sommer and Meyer (1937) in their studies on mussel and fish toxins used acid ethanol, neutral methanol, and methanol acidulated with hydrochloric acid for extractants.

Hiyama (1943) investigated several different methods of extracting fish poisons from reef fishes taken at Jaluit and Saipan islands. Ten g of raw fish muscle was left in 100 ml of absolute alcohol for 72 hours. The material was then filtered, and the alcohol boiled off. Distilled water was then added to make 100 ml and the extract injected subcutaneously into mice. A second method employed 200 ml of water to which was added 10 g of pulverized salted fish muscle and the mixture heated at 100° C for 2 hours. After filtering, it was concentrated to 15 ml and the extract tested. Another method involved 10 g of salted tissue placed in 200 ml of absolute alcohol and left at room temperature for 15 hours. Later it was filtered, 15 ml distilled water added, heated at 100° C until the alcohol was removed, concentrated to 2.7 ml, and tested. In his last method, Hiyama used 200 ml of water added to 10 g of salted tissue and left in a refrigerator at 10° C for 15 hours. After filtering, the filtrate was heated to 100° C for about 4 hours and concentrated to 13 ml and the extract was bioassayed. Hiyama believed that he obtained his best results using the absolute alcohol as the extractant. He concluded the poison was water soluble.

Kawakubo and Kikuchi (1942) prepared extracts from snappers (Lutjanidae) by grinding separate samples of raw and cooked muscle, viscera, and gonads with three parts of water, and then bioassayed the extracts.

Halstead and Bunker (1954*a*, *b*), Halstead and Ralls (1954), Ralls and Halstead (1955), Habekost *et al.* (1955), Goe and Halstead (1955), and Halstead and Schall (1955, 1956, 1958) prepared aqueous extracts, with various modifications, in most of their screening studies. Acidulated extracts were prepared by adding 2 ml of 0.1 *N* HCL for each gram in some of the tests, whereas in others acetic acid (pH 4.0) was used. In some extracts 2 ml of physiological saline were used for each gram of tissue, but later most of the testing was done by adding 2 ml of distilled water for each gram of tissue. In most instances a 7 g tissue sample was taken from each of the following: muscle, liver, gonads, and viscera. At first the material was boiled for 10 minutes and then centrifuged at 2,000 rpm for 25 minutes, but during subsequent experiments the heating and physiological and acidulated extracts were eliminated because they failed to show any significant advantages over the simpler distilled water technique. The extracts were then bioassayed.

Hashimoto (1956) prepared methanol extracts of a toxic barracuda (*Sphyraena picuda*) by using 5 g of minced tissue to which was added three parts of methanol acidulated with acetic acid to pH 3.8. The methanol was removed under vacuum and the residue washed three times with 5 ml portions of ether. The fat-free extract was then fed to cats, but without any toxic results. Later the fatty fraction, obtained from the ether-soluble fraction that had been dried, proved to be fatal to cats. He also prepared acetone extracts in the following manner: A methanol extract from 5 g of tissue was transferred into a separating funnel with a small portion of water and extracted four times with 15 ml portions of ether. After thorough washing with water, the ether fraction was filtered with a dried filter paper and dried by removing the ether. The oil substance obtained was then extracted three times with 3 ml portions of anhydrous acetone. The oily substance was fed to a cat which soon lapsed into a coma. Hashimoto next prepared extracts with the use of hot water. The acetone-soluble substance was then prepared as before, and extracted several times with hot water. The aliquots were then filtered into a clear solution with the use of barium carbonate. When dried the filtrate yielded a white amorphous mass mixed with a small amount of oil which proved to be toxic when fed to a cat. Hashimoto concluded that the barracuda poison was soluble in methanol, ether, acetone, and water, but insoluble in water acidulated with tartaric acid, and was heat stable.

Banner and Boroughs (1958) prepared muscle extracts from toxic *Lutjanus bohar* using various concentrations of ethanol. They concluded that ciguatoxin was soluble in 90 percent ethanol, and was detoxified in normal atmospheric oxygen, but retained its toxicity under nitrogen.

Banner *et al.* (1960) and Banner *et al.* (1961) later determined that the ethanol extraction method produced a substance that was toxic to test animals regardless of whether the fish was ciguatoxic or not. They found that ethanol extracts prepared from nonpoisonous Hawaiian tuna and lutjanids that were used as controls also gave toxic results. However, a more purified extract gave satisfactory differentiation between toxic fish and the control. They obtained 131 g of dried ground flesh from 6 highly toxic *Lutjanus bohar* which was extracted with 400 ml of 95 percent ethanol in a Soxhlet extractor for 24 hours and concentrated to 50 ml. The concentrate was then diluted with

150 ml of distilled water, yielding approximately a 25 percent alcoholic solution. The resulting solution was washed with three 100 ml portions of petroleum ether. It was found that the aqueous solution carried the toxin, and the petroleum ether extract was nontoxic. The aqueous alcoholic solution was then extracted with 300 ml of diethyl ether in three portions, which removed the bulk of the toxin. The ethereal extract was evaporated to dryness, yielding an average of 96.7 mg of residue containing the toxin. The residue was then placed in a peanut oil vehicle and injected. The petroleum ether extract and the aqueous alcoholic residue from both toxic and nontoxic fish would cause death in mice when injected at a level of 3.0 to 5.0 mg/g of mouse weight. However, the ethyl ether extract from toxic fish would produce kills at a level of 0.2 mg/g, or seven times the lethal level of the toxic fish. Their mouse intraperitoneal tests were supplemented by oral feeding tests using the mongoose.

Bartsch *et al.* (1959) and Bartsch and McFarren (1962) prepared tissue extracts from fishes taken in the Marshall Islands. They used 0.1 *N* HCL in the same manner that is used in testing for paralytic shellfish poison (Schantz *et al.*, 1958). They also prepared ether extracts by grinding 200 g of raw muscle and extracting three times with 100 ml portions of ether. The ether was decanted after each extraction, pooled, and evaporated at low heat. An oily liquid containing a small amount of suspended white material was obtained. The material was then bioassayed.

Hessel *et al.* (1960) and Hessel (1961, 1963) used acetone and ether soluble extracts of toxic *Lutjanus bohar* muscle that were emulsified into frog Ringer's solution by means of ultra sound, and tested the extract on the excised frog sciatic nerve bathed in the toxic emulsion and mounted on an electrode assembly to measure the action potential.

In their studies on the presence of a "ciguateralike poison" in oysters and clams, McFarren *et al.* (1965) have used ethyl ether extracts in obtaining suitable materials for testing.

Helfrich and Banner (1964) have experimented with the use of Tween 40[18] (polyoxyethylene sorbitan monopalmitate) and Tween 60 (polyoxyethylene sorbitan monostearate) as an emulsifier. The choice of these agents varies with the type of extract required because the two materials have slightly different solubility characteristics. They have recommended using either 80 g of Tween 40, or Tween 60 with 920 ml of normal saline solution. They then mix a variable amount of the ground tissue (1 to 80 mg) per 0.5 ml of the homogenizing Tween solution for testing purposes. This technique is still in the experimental stages of development but appears to offer considerable promise in simplifying screening procedures by the mouse intraperitoneal method.

Bioassay Methods

Numerous different kinds of test animals, and even plants, have been em-

[18] Tween is the proprietary name for the product produced by the Atlas Chemical Co., Wilmington, Del.

ployed in the bioassay of fish poisons. Probably some of the most ancient methods have been developed by native islanders who watched to see if flies, ants, or other insects avoided feeding on pieces of fish flesh that had been set out to dry. Other natives learned to test the edibility of a suspect fish by throwing a piece of the meat to a cat or dog and then observing them for a period of time to see whether the animal developed any symptoms. Another method is that reported by Cooper (1964) who noted that some of the Gilbertese adults would first partake of the fish to see if it was safe to eat. If no untoward effects resulted, they would then proceed to feed the fish to their children.

The earliest "scientific" attempts at developing bioassay methods for ichthyosarcotoxins were for poisonous puffers rather than ciguatoxic fishes. Fukushima (1716, cited by Suehiro, 1947) is said to have determined the toxicity of puffers by feeding them to prisoners. The Russian physician Savtschenko (1886) dissected puffers and fed them to dogs in order to determine the edibility of the fish. One of the most extensive bioassay experiments during the latter part of the 19th century was conducted by the French physician Rémy (1883) who dissected out various organs of Japanese puffers, and prepared simple aqueous extracts which were injected subcutaneously into dogs. A second series of tissue samples was fed to dogs, and the results of the two methods compared. In his work on Japanese poisonous puffers, Ishihara (1918) experimented with a variety of test animals including dragonflies, leaf chafers, loaches, carp, plaice, sculpin, newts, frogs, toads, snakes, turtles, sparrows, doves, chickens, rabbits, guinea pigs, mice, and rats. The crude toxins were administered by a variety of routes into the test animals, i.e., oral, intravenous, subcutaneous, intracutaneous, subdural, intracranial, intramedullary, anterior chamber of the eye, and conjunctiva. Numerous other workers have experimented with a variety of bioassay methods in testing for puffer poison and other marine biotoxins, but the preceding ones are illustrative of the development of some of these early assay methods that employed animals.

One of the most unique methods used in the evaluation of ichthyosarcotoxins is the one developed by Macht (1942) and Macht and Spencer (1941) who tested more than 65 different species of North American fishes for the presence of toxic substances. Aqueous extracts of fish muscle were tested by intraperitoneal injections into mice and on the growth of plants *Lupinus albus*. The increment or growth of the single straight root of the lupine seedling was measured after the plant had been grown for 24 hours in the dark at 15° C in the tissue extract to be tested. Growth of the root was then compared with controls which were grown in plant physiological or Shive's solution. The index of growth or phytotoxic index was expressed by the ratio of X/N or average growth in the unknown solution over the growth of the controls.

The bioassay method that has been used quite extensively in the testing of ciguatoxic fishes is a modification of the techniques employed by Sommer and Meyer (1937) in their studies on mussel and fish toxins. They used acid ethanol, neutral methanol, and methanol acidulated with hydrochloric acid for extractants. However, Hashimoto and Migita (1950) later recommended extraction with acetic acid and methanol because the hydrochloric acid destroyed certain poisons, e.g., puffer toxin. It was this same method that Hashimoto (1956) used in his studies on poisonous barracuda. Sommer and Meyer, and Hashimoto and Migita did all their testing by intraperitoneal injections on mice, whereas Hashimoto fed his barracuda extracts to cats.

Hiyama (1943) used mice, cats, and puppies for assaying his Marshall Island fishes. For his feeding experiments, weighed samples were taken of muscle, intestine, liver, gonads, and blood. Some of the samples were cooked, whereas others were fed raw. In some instances it was necessary to use salted and dried meat. His extractions were done with absolute alcohol, and the extracts injected subcutaneously into mice. Hiyama also fed ciguatoxic fishes to mice, but found the oral route to be unreliable. Many times a toxic fish when fed to a mouse would yield a negative result. Hiyama concluded that the extraction-injection method was the most reliable.

Kawakubo and Kikuchi (1942) tested the toxicity of lutjanid snappers on dogs, cats, mice, and frogs. The fish were fed raw, fried, or boiled, and aqueous extracts were prepared from the tissue and injected subcutaneously into dogs, cats, chickens, and the lymph sac of frogs. They concluded that man and dogs were more susceptible to ichthyosarcotoxins than either the cat or pig, and that the chicken reacted the least of all.

Halstead and Bunker (1954*a, b*), Halstead and Ralls (1954), Ralls and Halstead (1955), Habekost *et al.* (1955), Goe and Halstead (1955), Halstead and Schall (1955, 1956, 1958), and Halstead and his associates (several unpublished studies) assayed more than 25,000 fish tissue samples using a variety of test animals including crayfish, frogs, chickens, mice, hamsters, rats, cats, and dogs. Most of the tests employed the use of aqueous, homogenized, centrifuged extracts prepared from various fish tissues, i.e., musculature, liver, intestines, gonads, etc. The whole raw tissue and aqueous extracts were administered by oral, subcutaneous, intramuscular, intravenous, and intraperitoneal routes. As a result of these studies the authors concluded: (1) The most reliable method was to feed the test material to cats, but difficulties were encountered because ciguatoxin readily induces a vomiting reflex in cats. Moreover, cats are relatively expensive, at times difficult to obtain in adequate supply, and not suitable for mass screening of large numbers of reef fishes. (2) Mice were found to be the next most satisfactory test animal, and the intraperitoneal injection the most effective route of administration. Mice had the other advantages of being economical and readily available. Mouse feeding tests were found to be of little value because of the variation in physiological response of individual mice to a given test sample. Some mice would react to toxic fish samples, whereas others would not. (3) Rats and hamsters were found to be satisfactory test animals, but were more expensive than mice and did not appear to be any more sensitive to ciguatoxin. (4) Chickens and crayfish did not yield satisfactory results. Dogs were eliminated because of their excessive cost and legal problems associated with their procurement and maintenance. The studies conducted indicated a high correlation of results between mouse intraperitoneal injection and cat feedings, as well as published clinical records of human intoxications. The greatest difficulty encountered was the danger of false negatives and failure to critically observe the reactions of the mouse. The following criteria were established for recording the toxicological response in the mouse by the intraperitoneal injection method: *Negative*—when the mouse remains completely asymptomatic during the maximum test period of 36 hours. *Weakly Positive*—when the mouse shows definite symptoms such as lacrimation, diarrhea, ruffling of the hair, hypoactivity, inactivity, ataxia, etc., but the animal recovers. *Moderately Positive*—when the mouse develops ataxia, hypoactivity, paralysis, and dies within a period of 1 to 36 hours. *Strongly Positive*—when the mouse develops ataxia, followed by clonic or tonic convulsions of varying degrees, paradoxical respiration, respiratory paralysis, and death occurs within 1 hour.

Banner and Boroughs (1958), in their studies on ciguatoxic *Lutjanus bohar*, fed samples of the fish to cats, mongooses (*Herpestes mungo*), mice, and by intraperitoneal injection into mice. They found cats to be a satisfactory test animal, but not readily available. The mongoose compared quite favorably with the cat in its sensitivity to ciguatoxin. The reactions in the mongoose included failure to withdraw the foot when touched lightly, a progressive loss of motor coordination in first the back and then the forelegs, paralysis, Cheyne-Stokes breathing, coma, and death. The mongoose had the advantage of not vomiting the fish which was a constant problem in cats. In determining the lethal dose in the mongoose, no fish was found to be lethal under 5 percent of body weight, and no fish was used that did not cause death at 15 percent of body weight of the mongoose. Liquid extracts were fed by concentrating them and mixing them with dog food. Banner and Boroughs confirmed that the toxin was not destroyed by drying at 102-108° C for 24 hours, or by freezing for a period up to 6 months. They discarded the mouse feeding test because of insensitivity to the poison. They also discarded the mouse intraperitoneal tests using aqueous extracts because they felt that it was insensitive to small quantities of ciguatoxin. In a later report, Banner *et al.* (1960) classified the reactions of the mongoose in oral feeding tests on the following basis: 0, no reaction; 1, slight weakness and flexion of the forelimbs; 2, slight motor ataxia, more pronounced flexion of the forelimbs, and weakness of the hind limbs; 3, moderate motor ataxia with weakness and partial paralysis of limbs and body musculature; 4, acute motor ataxia and extreme weakness, capable of only limited movement, or coma; and 5, death. They found that reactions 3-5 were accompanied by excessive salivation and trembling just prior to death. They also test-fed a series of 37 organisms in an attempt to find a more suitable test animal, including protozoans, coelenterates, turbellarians, rotifers, crustaceans, insects, mollusks, echinoderms, fishes, amphibians, parakeets, pigeons, chicks, rabbits, guinea pigs, rats, and mice. Only crayfish, turtles, and mice reacted to ciguatoxin, but there were difficulties encountered in determining the amount of the sample ingested, or in evaluating the reactions they exhibited, and so they were discarded as a test animal. Banner *et al.* also pointed up that mongooses are not completely satisfactory test animals because they come from a wild population, are biologically variable, and consume too much fish to permit replicate testing necessary to establish an LD_{50}. They again reviewed the advisability of using intraperitoneal injections in mice and found that satisfactory results for a quantitative test could be obtained by using an alcoholic extract of fish flesh, subsequently purified by washing with petroleum ether and re-extracted with diethyl ether. Their mouse test was modified from the Halstead and Bunker technique in that they used 25 g of fish flesh in each case; cooled the container of the blender to assure temperatures below 40° C to prevent excessive precipitation of proteins; and the resulting slurry was centrifuged for 5 minutes, then decanted and recentrifuged at 2,000 rpm for 25 minutes. The diethyl ether extracts were prepared as previously outlined, and the extracts injected intraperitoneally into mice in a peanut oil vehicle. (*See also* Banner *et al.* (1961) for a discussion of a bioassay method for ciguatoxin.)

Hessel *et al.* (1960) and Hessel (1961, 1963) developed a bioassay method which they found to be more suitable for small-scale chemical fractionation techniques than either feeding or injection tests, and proved to be comparable in sensitivity to the cat feeding tests. It was observed that ciguatera toxin has an inhibitory effect on the action potential of excised frog sciatic nerve. An acetone and ether soluble extract of fish muscle was emulsified into frog Ringer's solution by means of ultra sound, and the excised nerve, mounted on an electrode assembly arranged for the detection of the action potential, was bathed in the emulsion for 55 minutes. At 55 minutes and every 5 minutes thereafter up to a total of 70 or 80 minutes, the action potential, generated under a stimulation voltage that is maximal for the untreated nerve, was measured by means of a low-level preamplifier and a cathode ray oscilloscope. A final potential appreciably below the standard for the nerve indicated a toxic emulsion. Hessel and associates also considered the percent retention of action potential in the frog nerve test to be a rough measure of the concentration of toxin in the muscle of the fish.

Granade, Cheng, and Doorenbos (1976) have developed a sensitive bioassay using larval brine shrimps *(Artemia salina)* which compares quite favorably with the more costly mongoose, cat, and mouse methods.

Critique of Bioassay Methods

The methods most commonly used in the bioassay of ciguatoxic fishes have been the cat, mongoose, and mouse feeding tests, mouse intraperitoneal injections, and the isolated frog sciatic nerve test. None of the present methods have proven to be completely ideal since they all have their limitations and undesirable characteristics.

In the past the extraction and bioassay methods have been based on the assumption by some workers that ciguatoxin was a water-soluble substance

(Yasukawa, 1935; Hiyama, 1943; Kawakubo and Kikuchi, 1942; Halstead and Bunker, 1954*a, b*; Halstead and Ralls, 1954; Ralls and Halstead, 1955; Habekost *et al.*, 1955; Goe and Halstead, 1955; Halstead and Schall, 1955, 1956, 1958), whereas others reported that the poison was a fat-soluble substance and could not be removed by a simple water extraction (Banner and Boroughs, 1958; Randall, 1958; Banner *et al.*, 1960; Hessel *et al.*, 1960; Bartsch and McFarren, 1962; McFarren *et al.*, 1965). Therefore most of the previous extraction and bioassay methods were directed to either a water-soluble or a fat-soluble substance, but not both. However, the dialysis studies of Halstead and Ralls (1954) on ciguatoxin from moray eel (*Gymnothorax sp?*),[19] red snapper (*Lutjanus vaigiensis*), and jack (*Caranx melampygus*) indicated that a water-soluble toxin, capable of passing through cellophane dialyzing tubing, appeared in the dialyzate and was toxic when injected intraperitoneally into mice. It was also noted that a significant portion of the poison remained in the residue in each instance. This was the first indication that possibly two or more fractions were present in the poison referred to as ciguatoxin. Further indications of the complex chemical nature of ciguatoxin were found in the solubility studies of Hashimoto (1956) on the poison of the barracuda *Sphyraena barracuda*. He found that the toxin had both fat-soluble and water-soluble properties, being soluble in methanol, ether, acetone, and water, but insoluble in acidulated water with tartaric acid. The silicic acid chromatographic studies of Hessel (1961, 1963) have shown that there are probably several toxic fractions that constitute ciguatoxin. Hessel's studies indicated that the most active fractions of ciguatoxin were probably fat-soluble substances, but the nature of some of the other fractions was not determined. Banner *et al.* (1963*a*) reported extracting a toxic substance from the surgeonfish *Ctenochaetus striatus* in aqueous alcohol which was not removed by solvent extraction with diethyl ether, whereas the poison from the red snapper (*L. bohar*) was a fat-soluble substance and readily soluble in diethyl ether (*see also* Banner *et al.*, 1961). In brief, the criticisms against the use of aqueous homogenized extracts because ciguatoxin is a fat-soluble poison are not fully justified. At present no one has been able to fully define the complex nature of ciguatoxin. There is an increasing amount of evidence which indicates that crude ciguatoxin contains both fat and water soluble fractions. There is also the possibility that the combination of these fractions may vary from one fish species, or even one specimen to the next. A satisfactory solution to the problem of preparing ciguatoxic extracts suitable for mouse intraperitoneal injections may be suggested by the method of Banner and Helfrich (1964) in using a suitable homogenizing agent such as Tween.

In summarizing the advantages of the various test animals there are certain factors which should be considered. Oral cat feeding tests, particularly in kittens, have proven to be a useful and sensitive bioassay for ciguatoxin. The disadvantages are the difficulty in procuring an adequate number of

[19] The species has since been determined as probably *Gymnothorax javanicus*.

cats, and that ciguatoxin elicits vomiting in cats to the extent that the desired toxic response cannot always be obtained. Dogs are useful, but prohibitive in cost for screening studies, and usually unobtainable in adequate numbers. The mongoose has been found to be a fairly satisfactory animal for oral feeding tests, but it is difficult to manage, biologically variable, usually unavailable, and prohibited by law in many regions. Feeding tests in rodents are generally unsatisfactory because of their low sensitivity to ciguatoxin by the oral route, and apparent variability in their reactions to the poison by this method. Mice are still the most satisfactory test animal for routine screening. They are economical, readily available, and easy to manage. The intraperitoneal method is recommended, but if simple aqueous homogenized extracts are used one cannot depend upon merely death times as an index of response (Goe and Halstead, 1955). The mouse must be carefully observed for such reactions as hypoactivity, hypersalivation, cyanosis, labored breathing, ataxia, and other evidence of neurotoxic effects. The greatest danger of the mouse intraperitoneal method when using aqueous extracts is that of false negatives. Negative reports are frequently the failure upon the part of the technician to critically observe the reaction of the mouse. Physiologically the most active part of crude ciguatoxin is believed to be in the fat-soluble fraction, which in simple aqueous homogenized extracts is present in only minimal quantities in the form of a milky suspension. The test is therefore less sensitive than some of the other more refined techniques that employ greater concentrations of the lipid fraction (Banner and Boroughs, 1958; Banner *et al.*, 1961; Banner *et al.*, 1961). However, some of these refined techniques also have their weakness in that they are testing merely the lipid fraction and nothing else. The question of false positives has been raised on several occasions, and some feel that the mouse will yield false positive reactions because of sensitivity to fish protein injected intraperitoneally as an aqueous homogenate. Tests conducted on more than 25,000 fish tissue samples representing hundreds of species of marine fishes have shown that a majority of these specimens and species *will not produce a toxic reaction* (Hessel *et al.*, 1960). Muscle samples taken from temperate zone fish species have consistently shown a negative response. These results tend to negate the criticism that the mouse will routinely yield false positives because of sensitivity to fish protein injected intraperitoneally as an aqueous homogenate. Exceptions may occur, but present evidence indicates that foreign protein reactions by the intraperitoneal route are a rarity. If the mouse intraperitoneal method is to be used in screening toxic fishes, a suitable emulsifier such as Tween (Banner and Helfrich, 1964) should be used so that both the water-soluble and fat-soluble fractions of crude ciguatoxin are being tested. The frog sciatic nerve test of Hessel *et al.* (1960) appears to be relatively sensitive but is time consuming and requires a skilled technician to operate it. It is therefore not suitable as a routine screening procedure. The bioassay technique of Granade, Cheng, and Doorenbos (1976) using larval brine shrimp *(Artemia salina)* offers considerable promise as both a sensitive and economical test for ciguatoxin.

PHARMACOLOGY

Despite the large amount of literature that has been written on ciguatera, only recently have reports been made on physiological and pharmacological properties of the poison. Banner *et al.* (1963*a*) and Banner *et al.* (1963*b*) tested semipurified extracts of ciguatoxin derived from barracuda *Sphyraena barracuda* and red snapper *Lutjanus bohar* on isolated toad sciatic nerve-muscle and guinea pig phrenic nerve-diaphragm preparations. They con-concluded that while the action potential of the nerve may be lost from long immersion in a solution of ciguatoxin the immediate effect was upon the nerve-muscle junction, and was similar in its effect to the end plate blocking action of *d*-tubocurarine, the depolarizing block induced by decamethonium, or by inhibiting the release of acetylcholine from nerve endings as caused by botulism toxin. The contractility of the muscle on direct stimulation was not impaired.

The most complete report on the pharmacology of ciguatoxin is by Li (1965). He used semipurified ciguatoxin extracted from a single batch of the musculature of *L. bohar*. The extract was diluted with 4.5 percent Tween 60 (polyoxyethylene sorbitan monostearate) in 0.9 percent saline solution as an emulsifier. The lethal dose (LD_{50}) by mouse intraperitoneal injections was approximately 12 mg/kg of body weight, or approximately an 0.7 percent solution of the poison. Intravenous injections of a nonlethal dose of ciguatoxin (0.5 μg/g) caused in an anesthetized rat a prompt fall in blood pressure and a slight increase in respiration. After a few minutes both returned to normal. No electrocardiographic changes were noted. Intravenous injection of a sublethal or lethal dose of ciguatoxin (2-3 μg/g) caused the blood pressure to increase slightly and remain below normal. Respiration was stimulated initially both in rate and depth, then decreased, and irregularities developed following a Cheyne-Stokes pattern. Electrocardiograms showed brachycardia and frequent ventricular blocks. Arrhythmic heartbeats with displacement of the QRS-T wave interval and omission of the T wave were noted. If a lethal dose was given, the respiration stopped after several hours, and the blood pressure remained only slightly lower than normal until death. Intravenous injection of a large lethal dose (4-6 μg/g) of ciguatoxin caused a sudden increase in blood pressure, the effects being similar to an injection of adrenaline. Brachycardia, ventricular block, and arrhythmic heartbeat always occurred. Prolonged ventricular blocks developed prior to cessation of respiration in a few rats. Changes in respiration associated with the fall in blood pressure usually led to complete respiratory arrest within a few minutes.

Intraperitoneal injections of sublethal or lethal doses of ciguatoxin on intact mice or rodents were similar in their effects to those caused by the administration of anticholinesterase. The symptoms consisted of initial restlessness and increasing abdominal distress, followed by muscular fasciculation, diarrhea, salivation, lacrimation, profuse sweating, and generalized weakness. In lethal doses there was respiratory distress, tremors, ataxia, dyspnea, and respiratory arrest. The heartbeat slowed, but continued after respiration had ceased.

Ciguatoxin applied to the eyes of rabbits resulted in miosis, but subsequent application of a 1 percent solution of atropine immediately counteracted the effect.

The anticholinesterase effect of ciguatoxin was demonstrated with the use of rabbit intestine suspended in Tyrode's solution to which was added acetylcholine chloride (0.1 μg/ml). The intestinal contractions were then observed. When the intestine was washed twice with Tyrode's solution, normal contractions resumed. If blood was first mixed with ciguatoxin and permitted to stand for 20 minutes before the acetylcholine chloride was added, contractions of greater amplitude occurred. The presence of ciguatoxin inhibited the hydrolytic effect of cholinesterase in blood upon acetylcholine. If the rabbit intestine was first treated with ciguatoxin for 2 minutes the contractions caused by the additional acetylcholine chloride were much greater than those elicited by acetylcholine chloride alone, but the ciguatoxin alone failed to produce any contractions. Human red blood cells treated with ciguatoxin from shark liver (*Carcharhinus menisorrah*) inhibited the cholinesterase enzyme activity as measured by the electrometric method of Michel.

The effects of drugs were also tested for their prophylactic and therapeutic activity. Atropine (up to 5 mg/kg) injected before or after an injection of ciguatoxin was effective in combating the muscarine effects of ciguatoxin. Atropine with magnesium sulphate (200 mg/kg) had the additional effect of eliminating the muscular paralysis and tremors due to the nicotinelike action of ciguatoxin. Atropine was able to prolong the life of the test animal, but failed to prevent death. Physostigmine injected with or without atropine had little protective action against ciguatoxin in mice and rats. Protopam chloride (2-formyl-1-methyl pyridinium chloride oxime), when administered with atopine, prevented death of both rats and mice injected with ciguatoxin. Protopam was found to be most effective when given in six successive doses (20 μg/g intravenously and 120 μg/g intramuscularly or 120 μg/g intraperitoneally for the initial dose, reduced by 50 percent in successive doses) at 45-minute intervals. The number of doses given varied according to the amount of ciguatoxin administered. The effects of ciguatoxin resembled those of the lipid-soluble organophosphorous compounds, but the antagonistic action of protopam was not as great as against diisopropyl fluorophosphate and paraxon. The mode of antagonistic action of protopam to ciguatoxin has not been determined. The cause of death from ciguatoxin is believed to be asphyxia.

Lethal doses of ciguatoxin (1/18 to 76.6 mg/kg) were administered to rats by Kosaki and Anderson (1968). Impaired respiration occurred first, followed by cessation; hypotension was observed, then bradycardia and arrhythmias. Partial blockage of nerve-muscle response was noted. In animals given non-lethal extracts, slightly irregular respiration and transient hypotension occurred. Similar pharmacological actions were observed by Ogura, Nara, and Yoshida (1968) in their comparative studies on ciguatoxin and tetrodotoxin.

In studies conducted by Ohshika (1971), ciguatoxin produced an initial negative and chronotropic effect on the isolated rat atria. The initial depression could be effectively blocked by atropine (1×10^{-7} g/ml), and partially blocked by hexamethonium or meicholinium-3 (2×10^{-5} g/ml). The positive inotropic and chronotropic action of ciguatoxin was reduced by pretreating the atria with MJ-1999 (1×10^{-6} g/ml) or guanethidine (1×10^{-6} g/ml). Abolition of the response to ciguatoxin could also be accomplished with a single dose of 2 mg/kg of reserpine given to the rat 24 hours before isolating the heart.

According to Setliff *et al.* (1971) and Rayner (1972), the pharmacological actions of ciguatoxin appear to be related to its direct effects on excitable membranes rather than to its previously reported anticholinesterase properties *in vitro* (Rayner, Kosaki, and Fellmeth, 1968). This toxin was found to increase the permeability of frog skin to Na^+ moving down a concentration gradient and to produce a tetrodotoxin-sensitive depolarization of frog muscle membranes. The maximal steady-state potential after adding ciguatoxin was about minus 50 mv/. This potential was the same as that produced by immersing the muscle in zero Ca^{++} Ringer (containing 1 mM EDTA) and in this medium, ciguatoxin had no further depolarizing effect. Kinetic studies indicated that ciguatoxin may be regarded as a competitive inhibitor of the membrane polarizing action of Ca^{++} ions $K_{Ca} = 3.3 \times 10^{-5}$ M; $K_{CT} = 2.9 \times 10^{-10}$M).

Rayner and Szekerczes (1973) found no inhibition of Na-K ATPase with the most highly purified extract of ciguatoxin at an estimated concentration as high as 10^{-4} M and with an erythrocyte membrane protein to toxin ratio of 0.66. They concluded that a lipid-soluble Na-K ATPase inhibitor may be present as a contaminant of partially purified extracts prepared by standard methods for ciguatoxin extraction. Ciguatoxin obtained from the flesh of the moray eel *(Gymnothorax javanicus)*, grouper *(Epinephelus fuscoguttatus)*, and the liver of the gray shark *(Carcharhinus menisorrah)* were similar in their effects to extracts from *L. bohar*.

CHEMISTRY

The term "ciguatoxin" has been applied to a poison which occurs in a wide variety of marine fishes. The broad phylogenetic distribution of the poison has contributed to the difficulty of developing suitable isolation procedures. The existence of the toxin has long been recognized; but investigations into the chemical and physical properties of the toxin are of recent origin.

Halstead and Ralls (1954) demonstrated that a considerable fraction of the toxin was present in the dialysate from the homogenized tissues of several species of ciguatoxic fishes. These included red snapper (*Lutjanus vaigiensis*) and an undetermined species of moray eel (*Gymnothorax*). It was noted that there was a difference in the percentage distribution between dialysate and nondialyzable fractions in the various species.

Scheuer *et al.* (1967) have proposed that the toxin responsible for ciguatera fish poisoning, isolated from the moray eel *Gymnothorax javanicus*, in pure form is a phospholipid, and has a minimum lethal dose of 0.26 g/g of mouse by intraperitoneal injection.

Procedures for the extraction and partial purification of ciguatoxin from *Lutjanus bohar* have been outlined by Hessel, Halstead, and Peckham (1960) and Hessel (1961). However, the most comprehensive studies as to the chemical nature of ciguatoxin have been published by Scheuer and his associates (1967, *see also* Mosher, 1966). After preparative thin-layer chromatography they obtained ciguatoxin as a transparent, light yellow viscous oil which could not be crystallized. The ciguatoxin obtained was relatively unstable and lost its toxicity upon contact with air, light, and chromatographic adsorbents (alumina, florisil, or silicic acid). Activity was lost to some extent even when the sample was stored in chloroform solution in the dark at -20° C. The fact that they were unable to obtain a stable crystalline derivative raised the question as to whether ciguatoxin is a single substance with a tendency to decompose or whether it consists of several closely related compounds. Combustion data obtained by ultramicro methods indicated an empirical formula for ciguatoxin of $C_{35}H_{65}NO_8$—if there is only one nitrogen atom. They concluded that ciguatoxin can be considered as a lipid containing a quaternary nitrogen atom, one or more hydroxyl groups, and a cyclopentanone moiety.

Yasumoto and Scheuer (1969) succeeded in isolating ciguatoxin from the liver of moray eels.

Kamiya and Hashimoto (1973) have isolated an emetic factor from the liver of the red snapper *Lutjanus bohar* from the Ryukyu Islands, which they termed *ciguaterin*. The substance was purified by column chromatography. The purified product was positive to ninhydrin, peptide, and Pauly reagents, but negative to Dragendorff and sugar reagents. It showed only a slight shoulder near 331 nm in the ultraviolet region. On acid hydrolysis, it afforded 14 amino acids, of which lysine, glutamic and aspartic acids were prominent. Total amino acid residues accounted for 40 percent of the solid. The results suggested that ciguaterin possesses a peptide moiety.

Chungue *et al.* (1977) isolated two toxins from a parrotfish *Scarus gibbus*, which appear to be confirmed by epidemiological and clinical observations. The fish taken from the Gambier Islands was extracted by the usual method adopted for ciguatoxin, and the crude toxin was fractionated by column chromatography using silicic acid and DEAE-cellulose. Two fat-soluble toxins were separated by DEAE-cellulose column chromatography. The polar one was similar to ciguatoxin and the other differed quite markedly. The latter toxin (designated SG-1) was further purified by repeated gel filtration on Sephadex LH-20 to give an oily substance with a lethality to mice of 0.03 mg/kg. There was a similarity in chemical and biological properties, but a dissimilarity in chromatographic behavior.

LITERATURE CITED

All Union Scientific Research Institute
1957 Fish Resources of the USSR. [In Russian] Maritime Fish Economy and Oceanography of USSR, Moscow.

Arćisz, W.
1950 Ciguatera: tropical fish poisoning. U.S. Fish Wildlife Serv., Spec. Sci. Rept., Fish. No. 27, 23 p.

Autenrieth, H. F.
1833 Ueber das gift der fische. C. F. Osiander, Tübingen. 287 p.

Bagnis, R. A.
1972 Ciguatera et intervention humaine sur les ecosystèmes coralliens en Polynésie Française, p. 597-600. *In* M. Ruivo [ed.], Marine pollution and sea life. Food and Agriculture Organization of the United Nations and Fishing News Ltd., London, England.

Bagnis, R. A., Berglund, F., Elias, P. S., van Esch, G. J., Halstead, B. W., and K. Kojima
1970 Problems of toxicants in marine food products, 1. Marine biotoxins. Bull. World Hlth. Org. 42: 69-88.

Banner, A. H.
1959 Fish poisoning reports wanted for University of Hawaii study. South Pacific Quart. Bull. 9(3): 31.
1965 Regarding the identity of *Lutjanus tuluttharmna*. (Personal communication, Mar. 16, 1965.)
1975 *In* O. A. Jones [eds.], Biology and geology of coral reefs. Academic Press, New York 3: 177-213.

Banner, A. H., and H. Boroughs
1958 Observations on toxins of poisonous fishes. Proc. Soc. Exp. Biol. Med. 98: 776-778.

Banner, A. H., and P. Helfrich
1964 The distribution of ciguatera in the tropical Pacific. Tech. Rept. No. 3. Final Rept. NIH Contr. SA-43-ph-3741. 48 p. (Hawaii Marine Lab., Univ. Hawaii.)

Banner, A. H.., Helfrich, P., Scheuer, P. J., and T. Yoshida
1963*a* Research on ciguatera in the tropical Pacific. Proc. Gulf Caribbean Fish. Inst. 16th Ann. Sess., p. 84-98.

Banner, A. H., Sasaki, S., Helfrich, P., Alender, C. B., and P. J. Scheuer
1961 Bioassay of ciguatera toxin. Nature 189(4760): 229-230.

Banner, A. H., Scheuer, P. J., Sasaki, S., Helfrich, P., and C. B. Alender
1960 Observations on ciguatera-type toxin in fish. Ann. N.Y. Acad. Sci. 90(3): 770-787.

Banner, A. H., Shaw, S. W., Alender, C. B., and P. Helfrich
1963*b* Fish intoxication. South Pacific Comm. Tech. Paper No. 141. 17 p.

Bartsch, A. F., Drachman, R. H., and E. F. McFarren
1959 Report of a survey of the fish poisoning problem in the Marshall Islands. Dept. Interior, Public Health Serv. 117 p., 17 figs.

Bartsch, A. F., and E. F. McFarren
1962 Fish poisoning: a problem in food intoxication. Pacific Sci. 16(1): 42-56.

Behre, A.
1949 Vergiftungen durch fischereierzeugnisse. Z. Lebensm. Untersuch. Forsch. 89(3): 299-312.

Belotte, J.
1955 Les poissons [empoisonnés] du lagon Tahitien. Méd. Trop. 15: 232-236.

Berg, L. S.
1947 Classification of fishes both recent and fossil. [In Russian & English] J. W. Edwards, Ann Arbor, Mich. 517 p.

Blanchard, R.
1890 Traité de zoologie médicale, vol. 2. J. B. Baillière et Fils, Paris, p. 638-695.

Bleeker, P.
1862-77 Atlas ichthyologique des Indes Orientales Néerlandaises, publié sous les auspices du gouvernement colonial néerlandais. 9 vols. Amsterdam. 416 pls.

Bouder, H., Cavallo, A., and M. J. Bouder
1962 Poissons vénéneux et ichtyosarcotoxisme. Bull. Inst. Océanogr. 59 (1240), 66 p., 2 figs.

Bréta, F.
1939 Contribution à l'étude des poissons vénéneux. Bull. Lab. Maritime Dinard 12(21): 5-52, 2 figs.

Bryant, E. O.
1912 The moon and poisonous fish. Nature 90(2246): 305.

BURROWS, G. M.
1815 An account of two cases of death from eating mussels; with some general observations on fish-poison. London Med. Reposit. 3(18): 445-476.

CATESBY, M.
1743 The natural history of Carolina, Florida and the Bahama Islands. 2 vols. London.

CAVALLO, A., and H. BOUDER
1961 Ichthyotoxisme, et problèmes économiques essais de prophylaxie. Arch. Inst. Pasteur Nouméa. 61 p.

CHEVALLIER, A.
1856 Sur des cas de l'empoisonnement par des poissons. J. Chim. Méd. Pharmacol. Toxicol., 4, 2: 85-86.

CHEVALLIER, A., and E. A. DUCHESNE
1851*a* Mémoire sur les empoisonnements par les huitres, le moules, les crabes, et par certains poissons de mer et de rivière. Ann. Hyg. Publ. (Paris) 45(11): 387-437.
1851*b* Mémoire sur les empoisonnements par les huitres, les moules, les crabes, et par certains poissons de mer et de rivière. Ann. Hyg. Publ. (Paris) 46(1): 108-147.

CHISHOLM, C.
1808 On the poison of fish. Edinburgh Med. Surg. J. 4(16): 393-422.
1821 Ueber einige gifte in den amerikanischen inseln und eben da wachsende gegenmittel aus dem pflanzenreiche. Isis con Oken, Jena., p. 533-536.

CHUNGUE, E., BAGNIS, R., FUSETANI, N., and Y. HASHIMOTO
1977 Isolation of two toxins from a parrotfish *Scarus gibbus*. Toxicon 15(1): 89-93.

CLARK, E.
1950 Fisherman beware! Fishing for poisonous plectognaths in the western Carolines. Res. Rev. NAVEXOS P-510, p. 1-6, 4 figs.

CLARK, E., and H. A. GOHAR
1953 The fishes of the Red Sea: order Plectognathi. Publ. Marine Biol. Sta. Al Ghardaqa (Red Sea) No. 8. 80 p., 315 pls., 22 figs.

CLELAND, J. B.
1942 Injuries and diseases in Australia attributable to animals. Med. J. Australia 2(14): 313-32.

CLOQUET, H.
1821 Ichthyque [poison], p. 550-554. *In* Dictionnaire des Sciences Naturelles. Vol. 22.

COLBY, M.
1943 Poisonous marine animals in the Gulf of Mexico. Trans. Texas Acad. Sci. 26: 62-69.

COOPER, M. J.
1964 Ciguatera and other marine poisoning in the Gilbert Islands. Pacific Sci. 18(4): 411-440, 11 figs.

CORRE, A.
1865 Note pour servir à l'histoire des poissons vénéneux. Arch. Méd. Nav. 3: 136-147.

COUTIÈRE, H. M.
1899 Poissons venimeux et poissons vénéneux. Thése Agrég. Carré et Naud, Paris. 221 p.

DACK, G. M.
1956 Food poisoning. Rev. ed. Univ. Chicago Press, Chicago. 251 p.

DAWSON, E. Y.
1959 Changes in Palmyra Atoll and its vegetation through the activities of man, 1913-1958. Pacific Nat. 1(2), 51 p., 23 figs.

DAWSON, E. Y., ALEEM, A. A., and B. W. HALSTEAD
1955 Marine algae from Palmyra Island with special reference to the feeding habits and toxicology of reef fishes. Allan Hancock Found. Publs., Occ. Pap. 17, 39 p., 13 figs.

DAY, F.
1865 The fishes of Malabar. Bernard Quaritch, London. 293 p., 20 pls.
1871 On fish as food, or the reputed origin of disease. Indian Med. Gaz. 6: 5-8, 26-29.
1878-88 The fishes of India. 2 vols. London. 816 p., 198 pls.
1880-84 The fishes of Great Britain and Ireland. 2 vols. Williams and Norgate, London. 179 pls.

DEMPSTER, G. O.
1949 Fish poisoning. Brit. Med. J. 1(4608): 775.

DE SYLVA, D.
1963 Systematics and life history of the great barracuda *Sphyraena barracuda* (Walbaum). Stud. Trop. Oceanogr. Miami, No. 1. 179 p., 36 figs.

DE SYLVA, D. P., and A. H. HINE
1972 Ciguatera—marine fish poisoning—a possible consequence of thermal pollution in tropical seas? *In* M. Ruivo [ed.], Marine pollution and sea life. Fishing News (Books) Ltd., Surrey. p. 594-597.

DUMÉRIL, A.
1866 Des poissons vénéneux. Ann. Soc. Linn. Dept. Maine-et-Loire 8: 1-17.

DUNLOP, W. R.
1917 Poisonous fishes in the West Indies. West Indies Bull. 16(2): 160-167.

FAUST, F. S.
1906 Die tierischen gifte. F. Vieweg und Sohn, Braunschweig. 248 p.
1924 Tierische gifte. Handbuch Exp. Pharm., Berlin. (NSA)
1927*a* Die tropischen intoxikationskrankheiten. II. Vergiftungen durch tierische gifte, p. 865-1004. *In* C. Mense [ed.], Handbuch der Tropenkrankheiten. 3d., ed., Vol. 2. J. A. Barth, Leipzig.
1927*b* Vergiftungen durch tierische gifte, p. 1745-1927. *In* G. von Bergmann and R. Staehelin [eds.], Handbuch der inneren medizin. Vol. 2. J. Springer, Berlin.

FISH, C. J., and M. C. COBB
1954 Noxious marine animals of the central and western Pacific Ocean. U.S. Fish Wildlife Serv., Res. Rept. No. 36, p. 14-23.

FONSSAGRIVES, J. B.
1877 Animaux toxicophores, p. 620-636; fig. 71. *In* J. B. Fonssagrives, Traité d'hygiene navale. J. B. Baillière et Fils, Paris.

FONSSAGRIVES, J. B., and L. DE MÉRICOURT
1861 Recherches sur les poissons toxicophores exotiques des pays chauds. Ann. Hyg. Publ. Méd. Légale, 2, 16(2): 326-359.

FORSTER, J. G.
1777 Voyage round the world, in His Britannic Majesty's sloop, Resolution, commanded by Capt. James Cook, during the years 1772, 3, 4, and 5. 2 vols. G. Robinson, London.

FORSTER, J. R.
1778 Observations made during a voyage round the world, on physical geography, natural history, and ethic philosophy. G. Robinson, London. 649 p.

FOSBERG, F. R., and M. H. SACHET
1953 Handbook for atoll research (second preliminary edition). Atoll Res. Bull. No. 17, Natl. Res. Council. 129 p.

FOWLER, H. W.
1904 A collection of fishes from Sumatra. J. Acad. Nat. Sci. Philadelphia 12: 497-560, 28 pls.
1928 The fishes of Oceania. Mem. Bernice P. Bishop Mus. Vol. 10. 466 p., 49 pls.
1931 Contributions to the biology of the Philippine Archipelago and adjacent regions. The fishes of the families Pseudochromiidae, Lobotidae, Pempheridae, Priacanthidae, Lutjanidae, Pomadasyidae, and Teraponidae, collected by the U.S. Bureau of Fisheries steamer *Albatross*, chiefly in Philippine seas and adjacent waters. U.S. Nat. Mus., Bull. 100, Vol. 11. 388 p., 28 figs.
1933 Contributions to the biology of the Philippine Archipelago and adjacent regions. The fishes of the families Banjosidae, Lethrinidae, Sparidae, Girellidae, Kyphosidae, Oplegnathidae, Gerridae, Mullidae, Emmelichthyidae, Sciaenidae, Sillaginidae, Arripidae, and Enoplosidae collected by the U.S. Bureau of Fisheries steamer *Albatross*, chiefly in Philippine seas and adjacent waters. U.S. Nat. Mus., Bull. 100, Vol. 12. 465 p., 32 figs.
1936 The marine fishes of west Africa. Bull. Am. Mus. Nat. Hist. Vol. 70, Part I & II. 1,493 p., 567 figs.
1938 The fishes of the George Vanderbilt South Pacific expedition, 1937. Monogr. No. 2. Acad. Nat. Sci. Philadelphia, Pa. 349 p., 12 pls.
1941 Contributions to the biology of the Philippine Archipelago and adjacent regions. The fishes of the groups Elasmobranchii, Holocephali, Isospondyli, and Ostarophysi obtained by the U.S. Bureau of Fisheries steamer *Albatross*, in 1907 to 1910, chiefly in the Philippine Islands and adjacent seas. U.S. Nat. Mus., Bull. 100, Vol. 13. 879 p., 30 figs.

(NSA)—Not seen by author.

Fowler, H. W.—Continued
1956 Fishes of the Red Sea and southern Arabia. Vol. I. Branchiostomida to Polynemida. Weizmann Science Press of Israel, Jerusalem. 240 p., 117 figs.
1959 Fishes of Fiji. Govt. of Fiji, Suva. 670 p., 243 figs.

Gilman, R. L.
1942 A review of fish poisoning in the Puerto Rico-Virgin Islands area. U.S. Nav. Med. Bull. 40(1): 19-27, 2 pls.

Goe, D. R., and B. W. Halstead
1955 A case of fish poisoning from *Caranx ignobilis* Forskål from Palmyra Island, with comments on the screening of toxic fishes. Copeia (3): 238-241.

Gomez, I. C.
1926 Algunos peces venenosos de la República Mexicana. Bol. Direc. Estud. Biol. 3(4): 66-68.

Gosline, W. A., and V. E. Brock
1960 Handbook of Hawaiian fishes. Univ. Hawaii Press, Honolulu. 372 p., 273 figs.

Granade, H. R., Cheng, P. C., and N. J. Doorenbos
1976 Ciguatera I. Brine shrimp *(Artemia salina* L.) larval assay for ciguatera toxins. J. Pharm. Sci. 65(9): 1414-1415.

Gross, H. F.
1960 By trial—by error. Barracuda poisoning. J. Florida Med. Assoc. 47: 172-173.

Gudger, E. W.
1930 Poisonous fishes and fish poisonings, with special reference to ciguatera in the West Indies. Am. J. Trop. Med. 10(1): 43-55.

Günther, A. C.
1859-70 Catalogue of the fishes in the British Museum. 8 vols. British Museum, London.

Habekost, R. C., Fraser, I. M., and B. W. Halstead
1955 Observations on toxic marine algae. J. Wash. Acad. Sci. 45(4): 101-103.

Halstead, B. W.
1951 Poisonous fish—a medical-military problem. Res. Rev. U.S. Nav. NAVEXOS P-510: 10-16, 8 figs.
1958 Poisonous fishes. Public Health Report 73(4): 302-312, 11 figs.
1959 Dangerous marine animals. Cornell Maritime Press, Cambridge, Md. 146 p., 86 figs., 1 pl.
1962 Biotoxications, allergies, and other disorders, p. 521-540. *In* G. Borgstrom [ed.], Fish as food. Vol. II. Nutrition, sanitation, and utilization. Academic Press, New York.

Halstead, B. W., and N. C. Bunker
1954*a* A survey of the poisonous fishes of the Phoenix Islands. Copeia (1): 1-11, 5 figs.
1954*b* A survey of the poisonous fishes of Johnston Island. Zoologica 39(2): 61-77. 1 fig.

Halstead, B. W., and K. W. Cox
1973 An investigation on fish poisoning in Mauritius. Proc. Roy. Soc. Arts Sci. Mauritius. 4(Pt. 2): 1-26, 2 pts.

Halstead, B. W., Hessel, D. W., and J. Suchy
1963 An investigation of ciguatera poison. Final Rept. June 1961-Sept. 1962. Army Chem. Corps Grant No. DA-CML-18-108-61-G-19fi 9 p. (World Life Res. Inst., Colton, Calif.)

Halstead, B. W., and W. M. Lively
1954 Poisonous fishes and ichthyosarcotoxism. Their relationship to the Armed Forces. U.S. Armed Forces Med. J. 5(2): 157-175, 9 figs.

Halstead, B. W., and R. J. Ralls
1954 Results of dialyzing some fish poisons. Science 119(3083): 160-161.

Halstead, B. W., and D. W. Schall
1955 A report on the poisonous fishes captured during the Woodrow G. Krieger expedition to the Galapagos Islands, p. 147-172, 1 pl. *In* Essays in the natural sciences in honor of Captain Allan Hancock. Univ. Southern Calif. Press, Los Angeles.
1956 A report on the poisonous fishes captured during the Woodrow G. Krieger expedition to Cocos Island. Pacific Sci. 10(1): 103-109, 1 fig.
1958 A report on the poisonous fishes of the Line Islands. Acta Trop. 15 (3): 193-233.

Harry, R. R.
1953 Ichthyological field data of Raroia Atoll, Tuamotu Archipelago. Atoll Res. Bull. No. 18, Natl. Res. Council, p. 47-169, 177.

HASHIMOTO, Y.
1956 A note on the poison of a barracuda, *Sphyraena picuda* Bloch & Schneider. Bull. Japan. Soc. Sci. Fish. 21(11): 1153-1157.

HASHIMOTO, Y., and M. MIGITA
1950 On the shellfish poisons. I. Inadequacy of acidulated alcohols with hydrochloric acid as solvent. Bull. Japan. Soc. Sci. Fish. 16(3): 77-85.

HASHIMOTO, Y., and T. NOGUCHI
1971 Occurrence of a tetrodotoxin-like substance in a goby *Gobius criniger*. Toxicon 9: 79-84, 1 fig.

HELFRICH, P.
1960 A study of the possible relationship between radioactivity and toxicity in fishes from the central Pacific. U.S. Atomic Energy Comm. TID-5748. 16 p. (Hawaii Marine Lab., Univ. Hawaii.)
1961 Fish poisoning in the tropical Pacific. Hawaii Marine Lab., Univ. Hawaii. 16 p., 7 figs.
1963 Fish poisoning in Hawaii. Hawaii Med. J. 22: 361-372.

HELFRICH, P., and A. H. BANNER
1963 Experimental induction of ciguatera toxicity in fish through diet. Nature 197(4871): 1025-1026.
1964 Regarding the use of Tween 40 and Tween 60. (Personal communication, May 1964.)
1968 Ciguatera fish poisoning, 2. General patterns of development in the Pacific. Occ. Pap. Bernice P. Bishop Mus. 23(14): 370-382.

HELFRICH, P., PIYAKARNCHANA, T., and P. S. MILES
1968 Ciguatera fish poisoning, 1. The ecology of ciguateric reef fishes in the Line Islands. Occ. Pap. Bernice P. Bishop Mus. 23(14): 305-369, 15 figs.

HERRE, A. W.
1924 Poisonous and worthless fishes. Philippine J. Sci. 25(4): 415-510.
1927 Fishery resources of the Philippines. Philippine Islands Bur. Sci. Pop. Bull. (3): 60-65.
1953 Checklist of Philippine fishes. U.S. Fish Wildlife Serv., Res. Rept. 20. 977 p.

HESSEL, D. W.
1961 Marine biotoxins. II. The extraction and partial purification of ciguatera toxin from *Lutjanus bohar* (Forskål). Toxicol. Appl. Pharmacol. 3(5): 574-583.
1963 Marine biotoxins. III. The extraction and partial purification of ciguatera toxin from *Lutjanus bohar* (Forskål); use of silicic acid chromatography, p. 203-209. *In* H. L. Keegan and W. V. Macfarlane [eds.], Venomous and poisonous animals and noxious plants of the Pacific region. Pergamon Press, New York.

HESSEL, D. W., HALSTEAD, B. W., and N. H. PECKHAM
1960 Marine biotoxins. I. Ciguatera poison: Some biolgoical and chemical aspects. Ann. N.Y. Acad. Sci. 91(3): 788-794, 4 figs.

HIATT, R. W., and D. W. STRASBURG
1960 Ecological relationships of the fish fauna on coral reefs of the Marshall Islands. Ecol. Monographs 30: 65-127.

HIYAMA, Y.
1943 Report of an investigation on poisonous fishes of the South Seas. [In Japanese] Nissan Fish. Exp. Sta. (Odawara, Japan). 137 p., 29 pls. 83 figs.

HOFFMANN, W. H.
1927 La ciguatera, die fischvergiftung von Cuba. Hamburg Univ., Abhandl. Gebiet Auslandsk. 26: 197-200.
1929*a* Los peces venenosos de Cuba y la ciguatera. Rev. Chilena Hist. Nat. 33: 28-30.
1929*b* La ciguatera, enfermedad producida por peces venenosos de Cuba. Invest. Progreso 3(11): 101-102.

HUTCHINS, D. E.
1912 The moon and poisonous fish. Nature 90: 382, 417.

ISHIHARA, F.
1918 Über die physiologischen wirkungen des fugutoxins. Mitt. Med. Fak. Univ. Tokyo 20: 375-426; Pls. 23-25.

JARDIN, C.
1972 Organo-minerals and ciguatera. FAO Nutr. Newslett. 11(3): 14-25.

Jones, J. D.
1956 Observations on fish poisoning in Mauritius. Proc. Roy. Soc. Arts Soc. Mauritius 1(4): 367-385.

Jordan, D. S.
1905 A guide to the study of fishes. 2 vols. Henry Holt & Co., New York. 427 figs.
1929 Poisonous fishes in Samoa. Am. Nat. 63(687): 382-384.

Jordan, D. S., and B. W. Evermann
1896-1900 The fishes of North and Middle America. 4 parts. U.S. Nat. Mus., Bull. 47. 392 pls.
1905 The aquatic resources of the Hawaiian Islands. U.S. Fish. Comm. Vol. 23. 574 p., 229 figs., 138 pls.
1908 American food and game fishes. Doubleday, Page & Co., New York. 572 p.

Jordan, D. S., Tanaka, S., and J. O. Snyder
1913 A catalogue of the fishes of Japan. J. Coll. Sci., Tokyo Univ. 33, 497 p., 396 figs.

Jouan, H.
1867 Notes sur quelques poissons nuisibles du Japon. Mém. Soc. Sci. Nat. Cherbourg 13: 142-144.

Joubin, L. M.
1929-38 Faune ichthyologique de l'Atlantique Nord. Andr. Fred Høst & Fils, Copenhague. Nos. 1-18.

Kamiya, H., and Y. Hashimoto
1973 Purification of ciguaterin from the liver of the red snapper *Lutjanus bohar*. Bull. Jap. Soc. Sci. Fish. 39(11): 1183-1187.

Kawakubo, Y., and K. Kikuchi
1942 Testing fish poisons on animals and report of a human case of fish poisoning in the South Seas. [In Japanese] Kaigun Igakukai Zasshi 31(8): 30-34, 3 pls.

Khlentzos, C. T.
1950 Seventeen cases of poisoning due to ingestion of an eel, *Gymnothorax flavimarginatus*. Am. J. Trop. Med. 30(5): 785-793.

Knox, G.
1888 Poisonous fishes and means for prevention of cases of poisoning by them. [In Russian] Voenno-Med. Zh. 161(3): 399-464, 9 figs.

Kosaki, T. I., and H. H. Anderson
1968 Marine toxins from the Pacific, 4. Pharmacology of ciguatoxin(s). Toxicon 6: 55-58, 1 fig.

Kudaka, K.
1960 A preliminary investigation of the poisonous fish of the Ryukyu Islands. [In Japanese] Ann. Rept. Ryukyu Fish Inst. 1960: 95-103.

Kumada, T.
1937 Marine fishes of the Pacific coast of Mexico. Nissan Fisheries Institute & Co., Odawara, Japan. 75 p., 102 pls.

Lacépède, B. G.
1799-1804 Histoire naturelle des poissons . . . dédiée au citoyen Lacépède. 14 vols. Paris.
1819 Histoire naturelle des poissons, par M. le Comte de Lacépède, suite et complément des oeuvres de Buffon. 5 vols. Paris.
1833-35 Naturgeschichte der Fische. 2d ed. 3 vols. Desmarest [ed.], Paris.

Lagler, K. F., Bardach, J. E., and R. R. Miller
1962 Ichthyology. Wiley and Sons, New York. 545 p.

Lagraulet, J., Bereziat, G., Cuzon, G., and J. Polonovski
1973 Étude comparative des acides gras extraits de poissons *Ctenochaetus striatus* provenant de zones toxicogènes et non toxicogènes du lagon de Tahiti. Bull. Soc. Path. Exot. 66(1): 235-239.

Larson, E., and L. Rothman
1967 Ciguatera poisoning by the horse-eye jack, *Caranx latus*, a carangid fish from the tropical Atlantic. Toxicon 5: 121-124.

Lee, R. K.
1954 Subsection C. Regulations pertaining to marine and freshwater life that is toxic or dangerous to man. Territory of Hawaii Public Health Regulations. Second Amendment to Chapter 4, Sect. 3.

Li, K. M.
1965 Ciguatera fish poison: a cholinesterase inhibitor. Science 147: 1580-1581.
1970 Ichthyosarcotoxins in fishes of the Pacific Ocean with special reference to the mechanism of action, p. 857-882. *In* B. W. Halstead, Poisonous and Venomous Marine Animals, vol. 3. U.S. Government Printing Office, Washington, D.C.

LIN, S. I., LIN, W. M., and H. C. YANG
1953 Poisonous fishes of Taiwan. China Fishery (10): 5-12, 20 figs.

LOCKE, J.
1675 An extract of a letter written to the publisher by Mr. J. L. about poisonous fish in one of the Bahama Islands. Phil. Trans. Roy. Soc. London 10: 312.

LONGLEY, W. H., and S. F. HILDEBRAND
1940 XIV. New genera and species of fishes from Tortugas, Florida. Carnegie Inst. Wash. Publ. 517, p. 223-285.
1941 Systematic catalogue of the fishes of Tortugas, Florida. Carnegie Inst. Wash. Publ. 535. 331 p., 34 pls.

MAAS, T. A.
1937 Gift-tiere. *In* W. Junk, [ed.], Tabulae biologicae. Vol. XIII. N. V. Van de Garde & Co., Drukkerij, Zaltbommel, Holland. 272 p.

McFARREN, E. F., TANABE, H., SILVA, F. J., CAMPBELL, J. E., and K. H. LEWIS
1965 The occurrence of ciguatera-like poison in oysters and clams. Toxicon 3(2): 111-123.

MACHT, D. I.
1936 The phytotoxic reactions of normal and pathological blood sera. Protoplasma 27: 1-8.
1942 An experimental appreciation of Leviticus XI. 9-12 and Deuteronomy XIV. 9-10. Hebrew Med. J. 2: 165-170.

MACHT, D. I., BROOKS, D. J., and E. C. SPENCER
1941 Physiological and toxicological effects of some fish muscle extracts. Am. J. Physiol. 133(2): 372.

MACHT, D. I., and E. C. SPENCER
1941 Physiological and toxicological effects of some fish muscle extracts. Proc. Soc. Exp. Biol. Med. 46(2): 228-233.

MANN, G.
1954 Vida de los peces en aguas Chilenas. Ministerio de Agricultura, Univ. Chile, Santiago. 342 p.

MANN, W. L.
1938 Fish poisoning in Culebra-Virgin Islands area. U.S. Nav. Med. Bull. 36(4): 631-634.

MATSUO, R.
1934 Study of the poisonous fishes at Jaluit Island. [In Japanese] Nanyo Gunto Chihobyo Chosa Ronbunshu 2: 309-326, 9 pls.

MEEK, S. E., and S. F. HILDEBRAND
1923 The marine fishes of Panama. Part I. Field Mus. Nat. Hist. Publ. 215 (Zoology) 15: 1-330, 24 pls.
1925 The marine fishes of Panama. Part II. Field Mus. Nat. Hist. Publ. 226 (Zoology) 15: 331-708; Pls. 25-71.
1928 The marine fishes of Panama. Part III. Field Mus. Nat. Hist. Publ. 249 (Zoology) 15: 709-1045; Pls. 72-102.

MEYER, K. F.
1953 Medical progress. Food poisoning. New Engl. J. Med. 249: 765-773, 804-812, 843-852.

MEYER-AHRENS, K. M.
1855 Von den giftigen fischen. Schweiz. Z. Med. Chir. Geburtsh. (3): 188-230; (4-5): 269-332.

MILLS, A. R.
1956 Poisonous fish in the South Pacific. J. Trop. Med. Hyg. 59: 99-103.

MOQUIN-TANDON, A.
1860 Eléments de zoologie médicale. Baillière et Fils, Paris.

MOSHER, H. S.
1966 Non-protein neurotoxins. Science 151 (3712): 860-861.

MOWBRAY, L. L.
1916 Fish poisoning (ichthyotoxismus). Bull. N.Y. Zool. Soc. 19(6): 1422-1423.

NIAUSSAT, P., GAK, J.-C., and J.-P. EHRHARDT
1969 Etat actuel de l'ichthyosarcotoxisme en Polynésie Française. Soc. Méd.-Chir. Hôp. 1969(6): 582-605.

NICHOLS, J. T., and P. BARTSCH
1945 Fishes and shells of the Pacific world. Macmillan Co., New York. 201 p., 83 figs., 9 pls.

NIELLY, M.
1881 Animaux et vegetaux nuisibles, p. 709-710. *In* M. Nielly, Elements de pathologie exotique. Delahaye and Lecrosnier, Paris.

NOGUCHI, T., KAO, H., and Y. HASHIMOTO
1971 Toxicity of the goby, *Gobius criniger*. Bull. Jap. Soc. Sci. Fish. 37(7): 642-647, 2 figs.

OGURA, Y., NARA, J., and T. YOSHIDA
1968 Comparative pharmacological actions of ciguatoxin and tetrodotoxin, a preliminary account. Toxicon 6: 131-140, 10 figs.

OHSHIKA, H.
1971 Marine toxins from the Pacific, 9. Some effects of ciguatoxin on isolated mammalian atria. Toxicon 9: 337-343, 1 fig.

O'NEILL, J. B.
1938 Food poisoning. U.S. Nav. Med. Bull. 36(4): 629-631.

ORFILA, M. P.
1817 A general system of toxicology. Carey & Son, Philadelphia, p. 380-384, 399.

PAWLOWSKY, E. N.
1927 Gifttiere und ihre giftigkeit. Gustav Fischer, Jena. 515 p., 170 figs.

PELLEGRIN, J.
1899 Les poissons vénéneux. M.D. Thèse 510. Fac. Méd. Paris. 113 p., 16 figs.

PHISALIX, M.
1922 Animaux venimeux et venins. 2 vols. Masson et Cie., Paris.

PROKHOROFF, P.
1884 On the poisonous character of certain lampreys. [In Russian] Vrach. St. Petersb. 5(4): 54-55.

PROSVIROV, E.
1963 Poisonous fishes. [In Russian] Kaliningrad Publ. Ofc., Moscow. 79 p.

RALLS, R. J., and B. W. HALSTEAD
1955 Moray eel poisoning and a preliminary report on the action of the toxin. Am. J. Trop. Med. Hyg. 4(1): 136-140.

RANDALL, J. E.
1958 A review of ciguatera, tropical fish poisoning, with a tentative explanation of its cause. Bull. Marine Sci. Gulf Caribbean 8(3): 236-267, 2 figs.
1961 Ciguatera: tropical fish poisoning Sea Frontiers 7(3): 130-139.
1964 A revision of the filefish genera *Amanses* and *Cantherhines*. Copeia (2): 331-361, 18 figs.

RANDALL, J. E., and V. E. BROCK
1960 Observations on the ecology of epinepheline and lutjanid fishes of the Society Islands, with emphasis on food habits. Trans. Am. Fish. Soc. 89(1): 9-16.

RAYNER, M. D.
1972 Mode of action of ciguatoxin. Fed. Proc. 31(3): 1139-1145, 7 figs.

RAYNER, M. D., KOSAKI, T. I., and E. L. FELLMETH
1968 Ciguatoxin: more than an anticholinesterase. Science 160(3823): 70-71, 1 fig.

RAYNER, M. D., and J. SZEKERCZES
1973 Ciguatoxin: effects on the sodium potassium activated adenosine triphosphatase of human erythrocyte ghosts. Toxicol. Appl. Pharmacol. 24: 489-496.

RÉMY, C.
1883 Sur les poissons toxiques du Japon. Compt. Rend. Soc. Biol., 7, 5: 3-28.

ROBINSON, M.
1962 Report of an outbreak of poisoning by wahoo *Acanthocybium solandri* at Puerto Vallarta, Mexico. (Personal communication, Jan. 11, 1962.)

ROCHAS, V. de
1862 La Nouvelle Calédonie et ses habitants. F. Sartorius, Paris, p. 64-65.

ROSS, S. G.
1947 Preliminary report on fish poisoning at Fanning Island (Central Pacific). Med. J. Australia 2: 617-621.

RUSSELL, F. E.
1952 Poisonous fishes. Engin. Sci. 15: 11-14, 2 figs.

SAVTSCHENKO, P.
1886 Atlas of the poisonous fishes. Description of the ravages produced by them in the human organism, and antidotes which may be employed. [In Russian and French] St. Petersburg. 53 p., 11 pls.

SCHANTZ, E. J., MCFARREN, E. F., SCHAFER, M. L., and K. H. LEWIS
1958 Purified shellfish poison for bioassay standardization. J. Assoc. Offic. Agric. Chemists 41(1): 160-168.

SCHEUER, P. J.
1969 The chemistry of some toxins isolated from marine organisms. Fort. Chem. Org. Natur. 27: 322-339.
1977 Marine toxins. Accts. Chem. Res. 10: 33-39.

SCHEUER, P. J. TAKAHASHI, W., TSUTSUMI, J., and T. YOSHIDA
1967 Ciguatoxin: isolation and chemical nature. Science 155(3767): 1267-1268.

SCHNACKENBECK, W.
1943 Fischgifte und fischvergiftungen. Muench. Med. Wochschr. 90(8): 149-151.

SCHULTZ, L. P.
1943 Fishes of the Phoenix and Samoan Islands collected in 1939 during the expedition of the U.S.S. *Bushnell*. U.S. Nat. Mus., Bull. 180, 316 p., 27 figs.

SCHULTZ, L. P.—Continued
1958 Review of the parrotfishes Family Scaridae. U.S. Nat. Mus. Bull. (214): 1-143, 31 figs., 27 pls.

SCHULTZ, L. P., CHAPMAN, W. M., LACHNER, E. A., and L. P. WOODS
1960 Fishes of the Marshall and Marianas Islands. Vol. 2. Families from Mulidae through Stromateidae. U.S. Nat. Mus., Bull. 202, 238 p.; pls. 75-123.

SCHULTZ, L. P., HERALD, E. S., LACHNER, E.A., WELANDER, A. D., and L. P. WOODS
1953 Fishes of the Marshall and Marianas Islands. Vol. I. Families from Asymmetrontidae through Siganidae. U.S. Nat. Mus., Bull. 202, 685 p., 74 pls., 89 figs.

SEALE, A.
1912*a* Some poisonous Philippine fishes. Philippine J. Sci., D, 7(4): 289-291 1 fig.
1912*b* Poisonous fishes of the Philippine Islands. Bull. Philippine Bur. Health (9): 3-9, 6 figs.

SETLIFF, J. A., RAYNER, M. D., and S. K. HONG
1971 Effect of ciguatoxin on sodium transport across the frog skin. Toxicol. Appl. Pharmacol. 18: 676-684.

SMITH, R. O.
1947 Survey of the fisheries of the former Japanese Mandated Islands. U.S. Fish Wildlife Serv., Fish Leaflet 273. 105 p., 50 figs.

SOMMER, H. W., and K. F. MEYER
1937 Paralytic shell-fish poisoning. Arch. Pathol. 24(5): 560-598, 2 figs.

STARR, T. J., DEIG, E. F., CHURCH, K. K., and M. B. ALLEN
1962 Antibacterial and antiviral activities of algal extracts studied by acridine orange staining. Texas Repts. Biol. Med. 20(2): 271-278, 6 figs.

STEINBACH, E.
1895 Bericht über die gesundheitsverhältnisse der eingeborenen der marshallinseln im jahre 1893/94 und bemerkung über fischgift. Mitt. Forschungslab. Gelehrt. Deutsch. Schutzgeb. 8(2): 157-171.

SUEHIRO, Y.
1947 Poison of globe-fish, p. 140-159. *In* Y. Suehiro, Practice of fish physiology. [In Japanese] Takeuchi Bookstore, Tokyo.

TANAKA, S.
1914 Some *Lutianid* fish supposed to be poisonous. Dobutsugaku Zasshi 26 (310): 412-413, 2 figs.

TASCHENBERG, E. O.
1909 Die giftigen tiere. Ferdinand Enke, Stuttgart. 325 p., 64 figs.

TITCOMB, M., and M. K. PUKUI
1952 Native use of fish in Hawaii. Suppl. J. Polynesian Soc., Mem. 29. 162 p., 120 figs.

VALENCIENNES, M. A.
1835-50 Ichthyologie des Iles Canaries, ou histoire naturelle des poissons. Paris. 109 p., 26 pls.

VAN ZANT, C. B.
1914 Fish and shellfish poisoning with illustrative cases. Colorado Med. 11: 453-459.

WALKER, F. D.
1922 Fish poisoning in the Virgin Islands. U.S. Nav. Med. Bull. 17(2): 193-202, 7 pls.

WATANABE, M.
1946 Case of poisoning by the reef fish *Lutjanus vaigiensis*. [In Japanese] Res. Inst. Nat. Res., Rept. No. 6. 9 p.

WEBER, M., and L. F. DE BEAUFORT
1911-61 The fishes of the Indo-Australian Archipelago. 11 vols. E. J. Brill, Leiden, Holland.

WHEELER, J. F.
1953 The problem of poisonous fishes. Mauritius-Seychelles Fish. Survey, Fish. Publ. 1(3): 44-48.

WHITLEY, G. P.
1943*a* Poisonous and harmful fishes. Council Sci. Ind. Res., Bull. No. 159. 28 p., 16 figs.; Pls. 1-3.
1943*b* The fishes of New Guinea. Australian Mus. Mag. 8(4): 141-144, 5 figs.
1954 Are hussars edible? Australian Mus. Mag. 11(6): 194-199, 5 figs.

YASUKAWA, T.
1935 Report of an investigation of poisonous fishes within the jurisdiction of the Saipan branch of the government-general. [In Japanese] Tokyo Imp. Univ. Contagious Disease Res. Inst. 27 p.

YASUMOTO, T., HASHIMOTO, Y., BAGNIS, R., RANDALL, J. E., and A. H. BANNER
1971 Toxicity of the surgeonfishes. Bull. Jap. Soc. Sci. Fish. 37(8): 724-734, 4 figs.

YASUMOTO, T., and P. J. SCHEUER
1969 Marine toxins from the Pacific, 8. Ciguatoxin from moray eel livers. Toxicon 7: 273-276.

Chapter XI—VERTEBRATES

Class Osteichthyes

Poisonous: CLUPEOTOXIC FISHES

Clupeotoxism is a form of ichthyosarcotoxism caused by fishes of the order Clupeiformes, which includes the families Clupeidae, the herrings; Engraulidae, the anchovies; Elopidae, the tarpons; Albulidae, the bonefishes; Pterothrissidae, the deepsea bonefishes; and Aleopcephalidae, the deepsea slickheads. However, the families most commonly incriminated in human clupcotoxications are members of the Clupeidae and Engraulidae. Our knowledge of the biotoxication is based entirely upon clinical reports scattered over a period of several centuries of time. Although clupeotoxism is not new to biotoxicologists, it has been only recently separated from ciguatera as a distinct clinical entity (Randall, 1958; Helfrich, 1961). At the moment this distinction is based largely upon the clinical symptomatology of the disease. Whether clupeotoxism is caused by a biotoxin chemically distinct from ciguatoxin or whether the clinical evidence represents a quantitative difference in the amount of the poison ingested has not been determined. The fact that most clupeiform fishes are largely plankton feeders and that most ciguatoxic fishes are mainly bottom feeders leads one to believe that there may be a basic difference in the biotoxin involved, but at the moment this is only an assumption. Clupeotoxism is a sporadic unpredictable public health problem of the tropical Atlantic Ocean, Caribbean Sea, and the tropical Pacific Ocean. To what extent this form of ichthyosarcotoxism occurs elsewhere is not known.

The following authors have published on the systematics and biology of clupeiform fishes: Günther (1859-70), Bleeker (1874), Day (1878-88), Jordan and Evermann (1896-1900, 1905), Jordan, Tanaka, and Snyder (1913), Schultz (1949), Herre (1953), Gosline and Brock (1960), Hildebrand (1963), and Marshall (1964).

It is noteworthy that to date no one has reported any experimental work on the toxicology, pharmacology, or chemistry of these violently poisonous fishes. Apparently, clupeiform fishes, which are generally valuable food fishes, become toxic only at sporadic intervals. The biotoxicity of these fishes appears to be a completely unpredictable phenomenon.

REPRESENTATIVE LIST OF FISHES REPORTED AS CLUPEOTOXIC[1]

Phylum CHORDATA

Class OSTEICHTHYES

Order CLUPEIFORMES (ISOSPONDYLI): Herrings, Anchovies, Etc.

Family CLUPEIDAE

132 *Clupanodon thrissa* (Linnaeus) (*see* Pl. 1, fig. b, Chapter X). Sprat, thread herring, tassart, shad (USA), dorokui (Japan).

DISTRIBUTION: Indo-Pacific, China, Japan, Korea.

SOURCES: Fonssagrives and de Méricourt (1861), Savtschenko (1886), Pawlowsky (1927), Hiyama (1943), Helfrich (1961).
OTHER NAMES: *Clupea thrissa, Meletta thrissa.*

132 *Harengula humeralis*[2] (Cuvier) (*see* Pl. 2, fig. a, Chapter X). Red-ear sardine, sardine, pilchard, sprat, whitebill, pincer (USA), sardina, sardina de ley (West Indies).

DISTRIBUTION: Florida, Bermuda, West Indies to Brazil.

SOURCES: Coutière (1899), Dunlop (1917), Bouder, Cavallo, and Bouder (1962).
OTHER NAME: *Clupea humeralis.*

132 *Harengula ovalis* (Bennett) (*see* Pl. 2, fig. b, Chapter X). Sardine (USA), tunsoy, sardina (Philippines).

DISTRIBUTION: Indo-Pacific, Red Sea.

SOURCES: Meyer-Ahrens (1855), Linstow (1894), Coutière (1899), Banner *et al.* (1963), Banner and Helfrich (1964).
OTHER NAMES: *Clupanodon venenosa, Clupea venenosa, Harengula punctata, Meletta venenosa.*

170 *Harengula zunasi* Bleeker (Pl. 1, fig. a). Sardine (USA), ooloo caoo, ulukau (Tonga).

---

[1] In many of the earlier clinical accounts of fish poisoning caused by clupeiform fishes, it is impossible to determine the nature of the biotoxication. Consequently, some of the clupeiform fishes have been listed in the ciguatera section as well as in this chapter.

[2] The term "yellow-billed sprat" is a confusing vernacular name both in the literature and in the field. The species that is sometimes referred to in the early literature as the "yellow-billed sprat" has now been determined as *Opisthonema oglinum*, which is best known as the Atlantic thread herring. However, the "yellow-billed sprat" currently referred to at St. Kitts and in some adjacent islands by local fishermen is actually the red-ear sardine (*Harengula humeralis*). Regardless of this confusion, there is documentation which indicates that both of these species have been involved in some violent intoxications and death in the West Indies.

Distribution: Indo-Pacific, China, Japan, Korea.

Sources: Bouder *et al.* (1962), Banner and Helfrich (1964).

Ilisha africana (Bloch) (Pl. 1, fig. b). Herring (USA), ilischa (USSR). *170*
Distribution: West Coast of Africa from Senegal to Gulf of Guinea.

Source: Prosvirov (1963).

Macrura ilisha (Buchanan-Hamilton) (*see* Pl. 3, fig. a, Chapter X). Hilsa, sablefish (USA), palasoh, pulla, oolum (India), nga-tha-louk (Burma), matabelo (Indonesia), zoboor, soboor, shad (Iran). *133*
Distribution: Persian Gulf, India, Sri Lanka, Burma, Vietnam.

Sources: Coutière (1899), Phisalix (1922), Hiyama (1943).
Other name: *Clupea ilisha.*

Nematolosa nasus (Bloch) (*see* Pl. 3, fig. b, Chapter X). Gizzard shad, hairback herring, thread-finned gizzard shad, long-finned gizzard shad (USA, Australia), suagan, kabasi (Philippines), noonah, muddu candai (India), pla kōk, pla kup (Thailand), zobi, goaff, pahlowa, gooag (Iran). *133*
Distribution: Indo-Pacific, India, Burma.

Sources: Autenrieth (1833), Knox (1888), Coutière (1899).
Other name: *Clupea nasus.*

Opisthonema oglinum (LeSueur) (*see* Pl. 4, fig. a, Chapter X). Atlantic thread herring, hairyback, bristle herring (USA), sargo, sargo de gato, sardinha large, tassart, sardinha sargo, caillen, yellow-billed sprat, machuelo (West Indies). *133*
Distribution: West Indies, occasionally to Cape Cod.

Sources: Fonssagrives and de Méricourt (1861), Savtschenko (1886), Phisalix (1922), Randall (1958).
Other names: *Clupea thrissa, Clupea thryssa, Megalops thrissa, Meletta theissa, Meletta thrissa.*

Sardinella fimbriata (Valenciennes) (Pl. 1, fig. c). Sardine (USA), lemuru (Indonesia), tamban (Philippines). *170*
Distribution: Indo-Pacific, India, China.

Sources: Savtschenko (1886), Phisalix (1922), Helfrich (1961).
Other names: *Clupea fimbriata, Clupea fimbricata, Spratella fimbriata.*

Sardinella perforata (Cantor) (*see* Pl. 4, fig. b, Chapter X). Sardine (USA), tembang, tamban (Malaysia), tamban (Philippines), lemuru (Indonesia). *133*
Distribution: Indo-Pacific, Arabia, Persian Gulf.

Sources: Coutière (1899), Phisalix (1922), Bouder *et al.* (1962).
Other names: *Clupea perforata, Harangula perferata, Clupeonia perforata.*

170 *Sardinella sindensis* (Day) (Pl. 1, fig. d). Sardine (USA), tamban, kasig (Philippines), pla lang keo (Thailand), charlay, mutthi, louar, lee-gur, poonduringa (India).

DISTRIBUTION: Indo-Pacific, India, Taiwan.

SOURCES: Savtschenko (1886), Phisalix (1922).
OTHER NAMES: *Clupea sindensis, Clupea venonosa* (see also *Harengula ovalis*).

Family ENGRAULIDAE

170 *Engraulis japonicus* Schlegel (Pl. 1, fig. e). Anchovy (USA), katakuchi-iwashi, seguro (Japan).

DISTRIBUTION: China, Japan, Korea, Formosa.

SOURCES: Vincent (1889), Coutière (1899), Phisalix (1922), Fish and Cobb (1954).

134 *Thrissina baelama* (Forskål) (*see* Pl. 5, Chapter X). Anchovy (USA), dumpilas, tigi (Philippines), trich (Indonesia).

DISTRIBUTION: Indo-Pacific, Red Sea, enters rivermouths.

SOURCES: Duméril (1866), Coutière (1899), Phisalix (1922).
OTHER NAMES: *Anchovia baelama, Engraulis baelama, E. bollama, Thrissocles baelama.*

BIOLOGY

Family CLUPEIDAE: This family is economically one of the most important fish groups in the world. They are used extensively for food by man, marine mammals, birds, turtles, and predatory fishes. They are cosmopolitan in their distribution, except for arctic and antarctic seas. Some species live in fresh water, whereas others enter rivers to spawn, but the majority are oceanic species—wanderers of the open sea. Herring are frequently found swimming in immense schools. Their vertical range is quite variable, from less than 20 fathoms to several hundred fathoms or more. The dentition of herring is small and feeble, and their food consists of planktonic organisms, algae, crustaceans, mollusks, fish eggs, small fishes, and a variety of other minute creatures.

Family ENGRAULIDAE: Anchovies are small clupeiform fishes present in temperate and warm seas but most abundant in tropical regions. Although generally found in the open sea, they enter bays and ascend rivers. They are characterized by their snout projecting beyond their mouth. The flesh of some species is of excellent flavor, and they are commercially canned in large numbers. Anchovies are valuable bait fishes and serve as important food fishes. They are mainly plankton feeders, their food consisting of crustaceans, foraminifera, and a variety of other minute organisms.

BIOGENESIS OF CLUPEOTOXIN

Clupeotoxin is believed to originate in the food web of the fish since all the fishes that have been incriminated in human biotoxications also have a reputation as being edible and in some instances are valuable food fishes. The incidence of toxic clupeiform fishes is completely sporadic and unpredictable. Nevertheless, they may become violently poisonous, approaching or possibly surpassing puffers in the rapidity with which they kill a person.

Many of the observations that have been reported regarding the sporadic and spotty distribution of ciguatoxic fishes apply equally as well for clupeotoxic fishes. However, several observations may be of significance relative to this particular group of fishes. According to D'Arras (1877), Vincent (1889), and others, the incidence of clupeotoxism is seasonal, occurring only during the warm summer months (July-September in northern latitudes)—at which time the fish move in close to shore and do their feeding. D'Arras (1877) quoted Father Montrouzier, one of the early naturalists who lived in New Caledonia, as saying that sardines became poisonous because they fed on a "green monad" (apparently a dinoflagellate) which discolored great areas of the ocean near Balade, New Caledonia, during certain seasons of the year. He further noted that toxic clupeiform fishes occurred only in the region about Balade during the time that the monads appeared, but they did not occur at such other places as Belep, about 45 miles to the north, where populations of green monads were absent. D'Arras further reported that the monads caused conjunctivitis, coryza, and erythema in humans coming in contact with them, which indicated that a toxic substance was present in these dinoflagellates.

Helfrich (1961) and Banner and Helfrich (1964) reported that one species of herring is reputedly toxic all the time in Fiji, but a second species is toxic only in October and November, and possibly through January, but mainly during the swarming of the palolo worm.

At present the origin of clupeotoxin has not been determined, but it is suspected that the poison probably differs in its origin from that of ciguatoxin since the fish transvectors are largely plankton feeders, whereas most ciguatoxic fishes are bottom feeders. The dinoflagellate observations of Father Montrouzier may be worthy of serious consideration in future investigations on the origin of clupeotoxin.

Randall (1958) has quoted De Sylva as suggesting that blooms of the planktonic blue-green alga *Skujaella* (*Trichodesmium* in older literature) may be the source of clupeotoxin.

MECHANISM OF INTOXICATION

Clupeotoxications result from eating clupeiform fishes, i.e., sardines, herring, tarpon, etc. Most poisonings have occurred in tropical island areas, and were caused by fishes that had been captured close to shore. The viscera are generally regarded as the most toxic part of the fish. According to some reports, tropical clupeiform fishes are most likely to be toxic during the warm summer months of the year. Some workers have associated toxicity in clupeiform fishes with the swarming of palolo worms or local dinoflagellate blooms. There is no possible way to detect a toxic clupeiform fish by its appearance, and the degree of freshness has no bearing on its toxicity.

MEDICAL ASPECTS

Clinical Characteristics

The symptoms and signs of clupeotoxism are distinct and usually violent. The first indication of a biotoxication is the sharp metallic taste which may be present immediately upon ingestion of the fish. This is soon followed by nausea, dryness of the mouth, vomiting, malaise, abdominal pain, and diarrhea. The gastrointestinal upset may be accompanied by a feeble pulse, tachycardia, chills, cold clammy skin, vertigo, a drop in blood pressure, cyanosis, and other evidences of a vascular collapse. Within a very short period of time, or concurrently, a variety of neurological disturbances rapidly ensue such as nervousness, dilated pupils, violent headaches, numbness, tingling, hypersalivation, muscular cramps, respiratory distress, progressive muscular paralysis, convulsions, coma, and death. Death may occur in less than 15 minutes. Ferguson (1823) claimed that the poison was so rapid in its action that natives have died while in the very act of eating the yellow-billed sprat *(Clupea thrissa)*[3]—part of the fish was still in the victim's mouth at the time of death. Pruritus and various types of skin eruptions including desquamation and ulceration have been reported in victims that have survived. There are no accurate statistics available regarding the mortality rate, but it is believed to be much higher than that of most other forms of ichthyosarcotoxism. The crude case fatality rate derived from scattered reports that have appeared in the literature show a rate of about 45 percent, but at least one clinician indicated that he did not know of any victim who had ingested a strongly toxic clupeiform fish and recovered (Banner and Helfrich, 1964).

The clinical characteristics of biotoxications caused by clupeiform fishes have been discussed by Ferguson (1823), Autenrieth (1833), Fonssagrives and de Méricourt (1861), Duméril (1866), D'Arras (1877), Linstow (1894), Coutière (1899), Phisalix (1922), Randall (1958), and Helfrich (1961).

Pathology

There is only a single autopsy report on a case of clupeotoxism. This report, cited by several authors (Fonssagrives and de Méricourt, 1861; D'Arras, 1877; Savtschenko, 1886; Blanchard, 1890; Linstow, 1894), is based on an examination of one of the victims in the outbreak on board the French warship the *Catinat.* Examination of the skin revealed cyanosis; the viscera were congested and gave off an "odor of gangrene." The lungs were congested. No other findings were reported.

Treatment

There is no information available regarding treatment. Treatment is symptomatic. See section on ciguatera, Chapter X, p. 370 on treatment.

---

[3] The yellow-billed sprat, usually termed *Clupea thrissa* in older literature relating to the West Indies, is believed to be *Opisthonema oglinum* (Randall, 1958; Hildebrand, 1963).

Prevention

Persons should avoid eating clupeiform fishes in inshore tropical island areas during the summer months unless there is positive local information available regarding their edibility. Poisonous clupeiform fishes cannot be detected by their appearance. Cooking procedures, salting, and drying will not prevent intoxication.

PUBLIC HEALTH ASPECTS

Geographical Distribution

General: Clupeotoxic fishes appear to be largely restricted to tropical insular areas, with the possible exception of the Mediterranean Sea (Autenrieth, 1833; Meyer-Ahrens, 1855; Coutière, 1899). Most of the reported outbreaks have occurred in the West Indies (Ferguson, 1823; Autenrieth, 1833; Fonssagrives and de Méricourt, 1861; D'Arras, 1877; Savtschenko, 1886; Marestang, 1888; Coutière, 1899), but are known to be present in the tropical Pacific Ocean (Fonssagrives and de Méricourt, 1861; Vincent, 1883; Savtschenko, 1886; Linstow, 1894; Coutière, 1899; Helfrich, 1961; Banner and Helfrich, 1964), and the Indian Ocean (Tennent, 1861; Banner and Helfrich, 1964).

Localized Distribution of Clupeotoxic Fishes: Clupeotoxic fishes resemble ciguatoxic fishes in their spotty distribution within a given locality. Clupeotoxic fishes differ from most ciguatoxic fishes in that they are frequently found in immense schools, whereas most ciguatoxic fishes travel singly or in relatively small groups. However, the toxicity of individual fishes varies greatly within a single school. One fish will be violently toxic, whereas another one swimming along side will be edible (D'Arras, 1877; Banner and Helfrich, 1964), There are insufficient data available to permit any conclusions concerning the nature or whereabouts of "toxic areas" for clupeotoxic fishes.

Incidence and Mortality Rate

Clupeotoxism does not appear to be as frequent in its occurrence as ciguatera. Since clupeotoxism is not a reportable disease, there are no reliable data available regarding either the incidence or mortality rate. Fatality rates are considered to be very high (Banner and Helfrich, 1964).

Seasonal Variation

Differing from the evidence about ciguatera, there is a fairly consistent opinion among most investigators that clupeotoxic fishes occur on a seasonal basis. In Sri Lanka, and possibly elsewhere in the Indian Ocean, clupeiform fishes have been reported toxic during the months of December and January. It is said to be May through October in Tahiti (Savtschenko, 1886); October through January in Fiji (Helfrich, 1961; Banner and Helfrich, 1964); and May through October in the West Indies (Coutière, 1899; Autenrieth, 1833). Reliable data on this subject are still lacking.

Relationship of Fish Size to Toxicity

The size of the fish does not appear to be a factor in clupeotoxism. However, gonadal development may affect toxicity in certain species of clupeoid fishes such as *Harengula humeralis* (Marestang, 1888) in the West Indies, and *Ilisha africana* in west Africa (Prosvirov, 1963).

Quarantine Regulations

The marketing of sardines has been prohibited in several localities where they were known to be toxic. According to Tennent (1861) certain species of sardines were not permitted to be sold during the toxic season in Ceylon (Sri Lanka). The Governor's order of February 1823 read thus:

> Whereas it appears by information conveyed to the Government that at three several periods at Trincomalie, death has been the consequence to several persons from eating the fish called Sardinia during the months of January and December, enacts that it shall not be lawful in that district to catch sardines during these months, under pain of fine and imprisonment.

It is believed that the law was directed to the sale of *Clupanodon thrissa*. Similar regulations have been enacted in Tahiti prohibiting the sale of the same species of fish during the months of May through October (Savtschenko, 1886). Public health authorities prohibited the sale of *Harengula humeralis* in Hispaniola during the months of May through October (Autenrieth, 1833; Coutière, 1899). Suspect species of all toxic fishes are not permitted to be sold in Suva, Fiji, where *Harengula ovalis* (sometimes referred to as *Clupea venenosa*) is a problem (Banner and Helfrich, 1964).

TOXICOLOGY

(Unknown)

PHARMACOLOGY

(Unknown)

CHEMISTRY

(Unknown)

LITERATURE CITED

Autenrieth, H. F.
1833 Ueber das gift der fische. C. F. Osiander, Tübingen. 287 p.

Banner, A. H., and P. Helfrich
1964 The distribution of ciguatera in the tropical Pacific. Tech. Rept. No. 3. Final Rept. NIH Contr. SA-43-ph-3741. 48 p. (Hawaii Marine Lab., Univ. Hawaii).

Banner, A. H., Helfrich, P., Scheuer, P. J., and T. Yoshida
1963 Research on ciguatera in the tropical Pacific. Proc. Gulf Caribbean Fish. Inst. 16th Sess. Contr. No. 207, p. 84-98.

Blanchard, R.
1890 Traité de zoologie médicale. Vol. 2. J. B. Baillière et Fils, Paris, p. 638-695.

Bleeker, P.
1874 Poissons de Madagascar et de l'Ile de la Réunion. E. J. Brill, Leiden. 104 p., 21 pls.

Bouder, H., Cavallo, A., and M. J. Bouder
1962 Poissons vénéneux et ictyosarcotoxisme. Bull. Inst. Océanogr. 59(1240), 66 p., 2 figs.

Coutière, H.
1899 Poissons venimeux et poissons vénéneux. Thèse Agrég. Carré et Naud, Paris. 221 p.

D'Arras, L.
1877 Essai sur les accidents causés par les poissons. Thèse No. 156, Fac. Méd., Paris. 69 p.

Day, F.
1878-88 The fishes of India. 2 vols. London. 816 p., 198 pls.

Duméril, A.
1866 Des poissons vénéneux. Ann. Soc. Linn. Dept. Maine-et-Loire 8: 1-17.

Dunlop, W. R.
1917 Poisonous fishes in the West Indies. West Indies Bull. 16(2): 160-167.

Ferguson, W.
1823 On the poisonous fishes of the Carribee Islands. Trans. Roy. Soc. Edinburgh 9(1): 65-79.

Fish, C. J., and M. C. Cobb
1954 Noxious marine animals of the central and western Pacific Ocean. U.S. Fish Wildlife Serv., Res. Rept. Rept. No. 36, p. 14-23.

Fonssagrives, J. B., and L. de Méricourt
1861 Recherches sur les poissons toxiphores exotiques des pays chauds. Ann. Hyg. Publ. Méd. legale, 2. 16(2): 326-359.

Gosline, W. A., and V. E. Brock
1960 Handbook of Hawaiian fishes. Univ. Hawaii Press, Honolulu. 372 p., 273 figs.

Günther, A.
1859-70 Catalogue of the fishes in the British Museum. 8 vols. London.

Helfrich, P.
1961 Fish poisoning in the tropical Pacific. Hawaii marine Lab., Univ. Hawaii. 16 p., 7 figs.

Herre, A. W.
1953 Check list of Philippine fishes. U.S. Fish Wildlife Serv., Res. Rept. 20. 977 p.

Hildebrand, S. F.
1963 *Family* Elopidae, p. 111-131, 3 figs.; *Family* Albulidae, p. 132-147, 3 figs.; *Family* Engraulidae, p. 152-249, 34 figs.; *Family* Clupeidae, p. 257-454, 55 figs. *In* H. B. Bigelow [ed.], Fishes of the western North Atlantic. Part 3. Soft-rayed bony fishes. Mem. Sears Found. Marine Res. No. 1. Yale Univ., New Haven.

Hiyama, Y.
1943 Report of an investigation on poisonous fishes of the South Seas. [In Japanese] Nissan Fish. Exp. Sta. (Odawara, Japan), 137 p., 29 pls., 83 figs.

Jordan, D. S., and B. W. Evermann
1896-1900 The fishes of North and Middle America. 4 parts. U.S. Nav. Mus. Bull. 47. 392 pls.
1905 The aquatic resources of the Hawaian Islands. Part I. The shore fishes. Bull. U.S. Fish Comm. Vol. 23. 574 p., 229 figs., 138 pls.

Jordan, D. S. Tanaka, S., and J. O. Snyder
1913 A catalogue of the fishes of Japan. J. Coll. Sci., Tokyo Univ. 33, 497 p., 396 figs.

KNOX, G.
1888 Poisonous fishes and means for prevention of cases of poisoning by them. [In Russian] Voenno-Med. Zh. 161(3): 399-464, 9 figs.

LINSTOW, O. V.
1894 Die giftthiere und ihre wirkung auf den Menschen. August Hirschwald, Berlin. 147 p., 54 figs.

MARESTANG, —.
1888 L'île de Saint Barthélemy. III. Histoire naturelle. Arch. Méd. Nav. 49: 161-190.

MARSHALL, T. C.
1964 Fishes of the Great Barrier Reef and coastal waters of Queensland. Angus and Robertson, Melbourne. 566 p., 72 pls.

MEYER-AHRENS, K. M.
1855 Von den giftigen fischen. Schweiz. Z. Med. Chir. Geburtsh. (3): 188-230; (4-5): 269-332.

PAWLOWSKY, K. M.
1927 Gifttiere und ihre giftigkeit. Gustav Fischer, Jena. 515 p., 171 figs.

PHISALIX, M.
1922 Animaux venimeux et venins. 2 vols. Masson et Cie., Paris.

PROSVIROV, E.
1963 Poisonous fishes. [In Russian] Kaliningrad Publ. Ofc., Moscow. 79 p.

RANDALL, J. E.
1958 A review of ciguatera, tropical fish poisoning, with a tentative explanation of its cause. Bull. Marine Sci. Gulf Caribbean 8(3): 236-267, 2 figs.

SAVTSCHENKO, P.
1886 Atlas of the poisonous fishes. Description of the ravages produced by them in the human organism, and antidotes which may be employed. [In Russian and French] St. Petersburg. 53 p., 10 pls.

SCHULTZ, L. P.
1949 A further contribution to the ichthyology of Venezuela. Proc. U.S. Nat. Mus. Vol. 99. 211 p., 20 figs., 3 pls.

TENNENT, J. E.
1861 Sketches of the natural history of Ceylon. Longman, Green, Longman and Roberts, London. 500 p.

VINCENT, L.
1883 Les poissons vénéneux du Japon. Arch. Méd. Nav., 39: 392-394.
1889 Contribution à la géographie médicale: le Japon. Arch Méd. Nav. 52: 107-122.

Chapter XII—VERTEBRATES

Class Osteichthyes

Poisonous: GEMPYLOTOXIC FISHES

The gempylids, escolars, or pelagic mackerels, are a small group of predacious oceanic fishes. They have a band-shaped body, large sharp teeth, and are distinguished from the true mackerels by the complete absence of a lateral keel or ridge on the caudal peduncle. Two dorsal fins are present, the first of which is spinous and longer than the second. Gempylids produce an oil which has a pronounced purgative effect.

The systematics of this group has been discussed by Jordan, Evermann, and Clark (1930), Tinker (1944), Breder (1948), Herre (1953), Okada (1955), Randall (1955), and Smith (1961).

REPRESENTATIVE LIST OF FISHES REPORTED AS GEMPYLOTOXIC

Phylum CHORDATA

Class OSTEICHTHYES

Order PERCIFORMES (PERCOMORPHI): Snake Mackerels

Family GEMPYLIDAE

Ruvettus pretiosus Cocco[1] (Pl. 1). Oilfish, castor oil fish, purgative fish (USA), te icka ne paka (Line Islands), walu (Hawaiian Islands), uravena, palu, vena, kuravena (Polynesia), ulago (Philippines), baramutsu (Japan), pesce ruvetto (Italy), ingwandarumi (Ryukyus), purgierfische (Germany), escolar (Spain, Portugal, West Indies). *172*

DISTRIBUTION: Tropical Atlantic and Indo-Pacific Oceans.

SOURCES: Waite (1897, 1899), Gudger (1925), Macht and Barba-Gose (1931*a*, *b*), Phillips and Brady (1953), Fish and Cobb (1954), Kudaka (1960), Cooper (1964).

BIOLOGY

Little seems to be known about the habits of most gempylids. Some species are said to attain large size, 75 kg or more, and a length of about 2 m. They

[1] Some taxonomists separate the Atlantic species from the Pacific one, calling the latter *Ruvettus tydemani* Weber.

are frequently found in schools at certain seasons, and appear to be primarily nocturnal in their feeding habits, since according to most reports they are captured only during dark, calm nights, usually in deep water, in some species up to 400 fathoms or more. They will bite on a variety of bait, but pompanos and flyingfishes are especially good.

MECHANISM OF INTOXICATION

Gempylid fish poisoning results from eating the flesh which contains an oil causing a purgative effect. The purgative oil is also present in the bones, and sucking on the rich oily bones may result in diarrhea.

MEDICAL ASPECTS

Clinical Characteristics

The diarrhea caused by ingestion of the oil contained in the flesh and bones of gempylid fishes develops rapidly, is pronounced, but generally without pain or cramping (Lowe, 1843-60; Krämer, 1901). No other untoward effects have been reported. Macht and Barba-Gose (1931*a*) found that emulsions injected intravenously are not toxic.

The outbreak reported by Cohen *et al.* (1946), and thought to have been caused by *Ruvettus pretiosus*, was a clear-cut outbreak of ciguatera poisoning and should not be confused with gempylid poisoning. Gempylotoxism has also been referred to as gempylid diarrhea (Halstead, 1958).

Pathology

None reported.

Treatment

Symptomatic, but generally not required.

Prevention

Gempylid poisoning is usually not a serious matter. Many native groups esteem these fishes despite their purgative effects. Persons should merely be aware of the laxative effects of the rather potent oil.

PUBLIC HEALTH ASPECTS

Gempylotoxism is not a public health problem.

TOXICOLOGY

According to Macht and Barba-Gose (1931*a, b*) the gempylid oil is not toxic in the usual sense of the term since it produces no intoxicating effects and is not irritating to the tissues. The oil does have a pronounced purgative effect which is discussed in the section on pharmacology.

PHARMACOLOGY

Macht and Barba-Gose (1931*a, b*) have published the only pharmacological evaluation of the oil of *Ruvettus pretiosus*. Administration of 10 to 15 ml of *Ruvettus* oil to a cat by gastric intubation resulted in an abundantly soft stool after 2 hours. The same effects were produced in dogs with 25 to 30 ml of the oil. The authors further devised a special method for quantitative comparative studies. They prepared a thick suspension of finely divided animal charcoal and introduced 0.1 ml of it by means of a gastric tube into rats that had been previously fed on a standard dry diet and allowed to fast a day before the experiment. The rats were then killed, usually after about 50 minutes, and the entire gastrointestinal tract excised. The distance traversed by the carbon suspension, observed through the wall, was measured and expressed as a percentage of the total length of the intestine. An equal quantity (0.1 ml) of the oil to be examined was given in the same manner together with the black emulsion. Comparative figures were then obtained as to the laxative effect and the normal controls.

Macht and Barba-Gose found that *Ruvettus* oil has a laxative effect equivalent to castor oil, but a different pharmacodynamic action. Purgation in castor oil is produced by its saponifiable fraction, ricinoleic acid, whereas it is the unsaponifiable fraction in *Ruvettus* oil. Isolated intestinal loop studies by Moreau's method revealed that *Ruvettus* oil is not irritating, causing only slight increase of intestinal secretion and mucus, and less congestion than castor oil. *Ruvettus* oil does not act as a purgative when given subcutaneously or intramuscularly. Intravenous injections of an emulsion produced only transient fall of blood pressure in dogs and cats and no effects on respiration.

CHEMISTRY

Chemical studies by Macht and Barba-Gose (1931*a, b*) revealed that the principal constituent of the saponifiable fraction of *Ruvettus* oil was oleic acid. The relatively low iodine numbers obtained from the various fractions studied indicated little unsaturation of an order higher than mono-ethylenic acid. The more important unsaponifiable portion of *Ruvettus* oil was fractionated, and its principal constituent was found to be the saturated compound cetyl alcohol. The most important unsaturated compound in the unsaponifiable portion of the oil proved to be oleyl alcohol. The comparative pharmacological results made with ethyl esters of oleic, hydroxyoleic acids, and with cetyl acetate and oleyl acetate show that the most active compound from the purgative standpoint proved to be cetyl acetate.

LITERATURE CITED

BREDER, C. M., JR.
1948 Field book of marine fishes of the Atlantic coast. Rev. ed. G. P. Putnam Sons, New York. 332 p., 16 pls.

COHEN, S. C., EMERT, J. T., and C. C. GOSS
1946 Poisoning by barracuda-like fish in the Marianas. U.S. Nav. Med. Bull. 46(2): 311-317, 2 figs.

COOPER, M. J.
1964 Ciguatera and other marine poisoning in the Gilbert Islands. Pacific Sci. 18(4): 441-440, 11 figs.

FISH, C. J., and M. C. COBB
1954 Noxious marine animals of the central and western Pacific Ocean. U.S. Fish Wildlife Serv., Res. Rept. No. 36, p. 14-23.

GUDGER, E. W.
1925 A new purgative, the oil of the "castor oil fish," *Ruvettus*. Boston Med. Surg. J. 192(3): 107-111, 1 fig.

HALSTEAD, B. W.
1958 Poisonous fishes. Public Health Repts. 73(4): 302-312, 11 figs.

HERRE, A. W.
1953 Check list of Philippine fishes. U.S. Fish Wildlife Serv., Res. Rept. 20. 977 p.

JORDAN, D. S., EVERMANN, B. W., and H. W. CLARK
1930 Check list of the fishes and fishlike vertebrates of North America and Middle America north of the northern boundary of Venezuela and Colombia. Rept. U.S. Comm. Fish. 1928. 670 p.

KRÄMER, A.
1901 Der purgierfisch der gilbertinseln. Globus 79(12): 181-183, 3 figs.

KUDAKA, K.
1960 A preliminary investigation of the poisonous fish of the Ryukyu Islands. [In Japanese] Ann. Rept. Ryukyu Fish Inst.

LOWE, R. T.
1843-60 A history of the fishes of Madeira. Bernard Quaritch, London. 196 p., 27 pls.

MACHT, D. I., and J. BARBA-GOSE
1931a Pharmacology of *Ruvettus pretiosus*, or "castor-oil fish." Proc. Soc. Exp. Biol. Med. 28(7): 772-774.
1931b Two new methods for pharmacological comparison of insoluble purgatives. J. Am. Pharm. Assoc. 20(6): 556-564, 3 figs.

OKADA, Y.
1955 Fishes of Japan. Maruzen Co., Tokyo. 434 p., 391 figs.

PHILLIPS, C., and W. H. BRADY
1953 Sea pests—poisonous or harmful sea life of Florida and the West Indies. Univ. Miami Press, Miami, Fla. 78 p. 38 figs., 7 pls.

RANDALL, J. E.
1955 Fishes of the Gilbert Islands. Atoll Res. Bull. No. 47. Pacific Sci. Bd., Natl. Res. Council. 243 p.

SMITH, J. L.
1961 The sea fishes of southern Africa. 4th ed. Central News Agency, South Africa. 580 p., 1,232 figs., 102 pls.

TINKER, S. W.
1944 Hawaiian fishes. Tongg Publ. Co., Honolulu. 404 p., 138 figs.

WAITE, E. R.
1897 The mammals, reptiles, and fishes of Funafuti. Mem. Australian Mus. 3: 199-201.
1899 The fishes of Funafuti (supplement). Mem. Australian Mus. 5(9): 540-544.

Chapter XIII—VERTEBRATES

Class Osteichthyes

Poisonous: SCOMBROTOXIC FISHES

Scombrotoxism, or scombroid poisoning, is caused largely by perciform fishes of the suborder Scombroidei, all of which are members of the single family Scombridae, the tunas and related species. One of the members of the order Beloniformes, family Scomberesocidae, the Japanese saury *Cololabis saira*, has also been incriminated in scombroid poisoning. Certain other dark-meated marine fishes are also suspected as potential causative agents of scombroid poisoning, but there are no specific data available as to their identity. Scombroid poisoning is generally caused by the improper preservation of scombroid fishes, which results in certain bacteria acting on histidine in the muscle of the fish, converting it to saurine. Ingestion of saurine by humans results in an allergiclike reaction known as scombrotoxism. This is the only form of ichthyosarcotoxism known wherein bacteria play an active role in toxin production within the body of the fish.

Tunas and related species are among man's most valuable fisheries resources, but under certain circumstances these fishes may produce serious untoward effects in humans. Fortunately in more civilized nations with modern refrigeration and canning methods, scombroids usually do not constitute a public health hazard.

The systematics of scombroids is a controversial subject about which most authors are not in complete agreement. The nomenclature adopted in this chapter is that provided in the latest taxonomic studies by Collette and Gibbs (1963*a, b*) which have been largely followed by the World Scientific Meeting on the Biology of Tunas and Related Species, Food and Agriculture Organization of the United Nations. The reader is referred to the following authors for further discussion on the systematics and biology of scombroid fishes: Kishinouye (1923), Joubin (1929-38), Hildebrand (1946), Beaufort and Chapman (1951), Rivas (1951), and Rosa (1963).

REPRESENTATIVE LIST OF FISHES REPORTED AS SCOMBROTOXIC

Phylum CHORDATA

Class OSTEICHTHYES

Order PERCIFORMES (PERCOMORPHI)

Suborder SCOMBROIDEI: Tunas and Related Species

Family SCOMBRIDAE

174 *Euthynnus pelamis* (Linnaeus) (Pl. 1, fig. a). Skipjack tuna (USA), barrilete (Latin America), katsuwo, mandara (Japan), tjakalang (Celebes), tulingan, puyan, gulisan (Philippines), bonita, watermelon (South Africa), tonno bonita (Italy), bonite (France), oceanic bonito (England), gaiado (Portugal), bonito (Spain).

DISTRIBUTION: Circumtropical in tropical and subtropical seas.

SOURCES: Henderson (1830), Fonssagrives (1877), Savtschenko (1886), Phisalix (1922), Halstead (1954), Helfrich (1961).
OTHER NAMES: *Katsuwonus pelamis, Katsuwonus vagans, Scomber pelamis, Scomber pelamys, Thunnus pelamys, Thynnus pelamys, Thynnus vagans.*

174 *Sarda sarda* (Bloch) (Pl. 1, fig. b). Atlantic bonito (USA), bonito, bonite (Canary Islands), sarda (Madeira), bonito (Spain), palamita (Italy), pelamid (England), palamide (Germany), pelamida, bonitou, bonite (France), katonkel, sarrao, bonito (South Africa), palamida (USSR), bonito (Latin America).

DISTRIBUTION: Tropical and temperate Atlantic Ocean, Mediterranean Sea.

SOURCES: Markov (1943), Behre (1949).
OTHER NAMES: *Pelamide sarda sarda, Pelamys sarda.*

175 *Scomber scombrus* Linnaeus (Pl. 2, fig. a). Atlantic mackerel (USA), common mackerel (England), maquereau commun (France), maccarello (Italy), gemeine makrele (Germany), sarda (Portugal), caballa (Spain), skumbria (USSR).

DISTRIBUTION: Tropical and subtropical Atlantic Ocean, Mediterranean Sea.

SOURCES: Niel (1815), Autenrieth (1833), Meyer-Ahrens (1855), Moquin-Tandon (1860), Fredericq (1924).
OTHER NAME: *Scomber scomber.*

175 *Scomberomorus regalis* (Bloch) (Pl. 2, fig. b). Cero (USA), sierra, pintado (West Indies).

DISTRIBUTION: East coast of America, Cape Cod to Brazil.

SOURCES: Autenrieth (1833), Gatewood (1909), Mann (1938), Gilman (1942).
OTHER NAMES: *Scomber regalis, Cybium acervum, Cybium acerrum, Cybium regale.*

176 *Thunnus alalunga* (Bonnaterre) (Pl. 3, fig. a). Albacore (USA), albacor (England), albacore (Australia), voador (Portugal), germon (France), albacora (Spain).

DISTRIBUTION: Circumtropical, Mediterranean Sea and Atlantic Ocean.

SOURCES: Autenrieth (1833), Meyer-Ahrens (1855), Knox (1888).
OTHER NAMES: *Scomber alalonga, Scomber alatunga.*

Thunnus thynnus (Linnaeus) (Pl. 3, fig. b). Bluefin tuna (USA), common tunny (England), thon commun (France), atun (Spain), gemeine thunfisch (Germany), rabilho (Portugal), tunny (South Africa), túnez (USSR). 176

Distribution: Cosmopolitan, mainly in subtropical and temperate seas.

Sources: Burrows (1815), Pappenheim (1858), Fonssagrives (1877), Phisalix (1922).

Other names: *Thynnus vulgares, Thynnus vulgaris, Thynnus thynnus, Scomber thynnus.*

BIOLOGY

Family SCOMBRIDAE: Scombroid fishes are characterized by their adaptation for swift locomotion, having a sharp profile anteriorly and a slender tail with a widely forked caudal. There is a series of detached finlets on the back behind the second dorsal fin and on the undersurface behind the anal fin, a feature which distinguishes them from most other fishes. Scombroids have two dorsal fins, of which the first is composed of spines, and the second of soft rays. The fins fit into grooves or depressions of the body, the bones of the head lie flat, and the gill covers fit tightly against the sides. The scales are usually small, thin, and metallic in appearance, offering a minimum of friction in the water. Scombroids are indeed the epitome of grace, form, and speed.

Tunas and related species are largely oceanic fishes, migrating great distances through the open seas. Mackerels are largely inhabitants of littoral waters. Scombroids are widely distributed throughout all temperate and tropical seas, and some are occasionally found in arctic and antarctic waters. They usually swim in large schools. Scombroids generally swim near the surface of the sea during and after the spawning season. During the warmer months they approach the shore but retire to deeper waters during the cold seasons. They follow the plankton, swimming in the deeper strata during the day, and rising to the surface at night. Some of the tunas may be found at depths of 200 m or more, whereas other scombroids seldom descend below 40 m.

Scombroids are predatory and are voracious feeders. Tunas are said to feed chiefly on plankton in the open sea, but they may also ingest a wide variety of moderate-sized fishes. Wahoo are particularly voracious and feed largely on such schooling species as sardines, anchovies, mackerels, etc. In general, scombroids feed on swimming species rather than on benthic forms. Scombroids generally spawn in the open sea during the warmer seasons.

BIOGENESIS OF SCOMBROTOXIN

A toxic substance has been known to occur under certain conditions in scombroid fishes since the latter part of the 18th century. For many years scombrotoxin was thought to be a metabolic product of toxigenic bacteria. With the discovery of large quantities of histamine in the musculature of spoiled scombroid fishes by Suzuki, Yoneyama, and Odake (1912), and the fact that scombroid poisoning clinically resembled a severe histamine reaction, the possibility of histamine as the etiological agent was seriously considered. Guggenheim (1914, cited by Kimata, 1961) reported that histamine was readily formed by the action of putrefactive bacteria acting on muscle histidine. It was later confirmed by Igarashi (1938) that the sharp peppery flavor of spoiled fish meat was due primarily to the presence of histamine. He also noticed that the pungency increased with the spoilage of the fish, and thus concluded that histamine production was the result of an autolytic breakdown of fish tissues. Pergola (1937) hypothesized that a special toxigenic bacterium produced the substance causing scombroid poisoning. Markov (1943) suggested that scombrotoxin was produced by putrefactive bacteria acting upon certain chemical precursors that were readily available in scombroid tissues, but he failed to substantiate his theory with laboratory evidence. Geiger, Courtney, and Schnakenberg (1944) investigated the matter further and provided bacteriological and chemical evidence that the histamine content was low in fresh scombroid fishes, but increased rapidly after death as a result of bacterial action. Histamine production was explained on the basis of decarboxylation of histidine due to bacteriological activity. However, they did not identify the bacteria responsible for this action. Autolytic studies were also conducted by Kimata and Kawai (1953*a, d, e*), and they reported that very little histamine was formed in aseptic scombroid muscle suspensions by the action of enzymes inherent in the muscle tissue. It was further observed that the production of histamine under aseptic conditions was influenced by environmental temperatures and pH. The optimum pH for histamine production at 35° C was between 3.5 and 4.5, and the maximum amount of histamine produced was 6.9 mg/100 g of meat. The optimum temperature for histamine production was between 40° and 45° C, but the maximum amount of histamine produced during autolysis under the best aseptic conditions did not exceed 10 to 15 mg/100 g of meat. It became obvious that these amounts were negligible compared to those produced by bacterial action. It was thus concluded that enzymes alone inherent in fish muscle tissue are not responsible for the histamine appearing during the spoilage of scombroid fishes. Moreover, in tissue autolysis of shark and white-meat fishes, production of histamine is either absent or very insignificant compared to what may occur in dark meat fishes under comparable conditions (Igarashi, 1939; Kimata, 1961).

With the increasing amount of evidence that bacteria had a direct bearing on histamine production, it was important to identify the strains of bacteria responsible for this activity, and to determine whether histamine was indeed the scombrotoxic agent. A large amount of research has been conducted on the subject, much of which goes beyond the scope of this present work. Consequently, only a cursory review of the subject that is directly pertinent to the problem of scombrotoxism will be presented.

Van Veen and Latuasan (1950) isolated from the skipjack tuna *Euthynnus pelamis* two unidentified strains of bacteria—some small Gram positive cocci and some small Gram positive rods—both of which grew anerobically and aerobically and were capable of producing histamine from the histidine of fish muscle. Kimata and Kawai (1953*b*) and Kimata and Tanaka (1954*a*) reported that the histamine productivity was caused by a single organism, and was not due to most other strains of putrefactive bacteria found in fishes. The isolated organism closely resembled *Proteus morgani*, but since some of its growth characteristics differed from the usual strains of *P. morgani* it was designated a new species, namely, *Achromobacter histamineum*. Studies have since shown that the bacterium is a strain of *P. morgani* which has the ability to decarboxylize histidine to histamine. However, other bacteria (*Salmonella, Shigella, Clostridium, Escherichia, Proteus*) have also been shown to be capable of producing histamine (Kimata, 1961).

The question of whether histamine is completely responsible for scombroid poisoning has been considered by several investigators. Feldberg and Schilf (1930), Aiiso (1954), Geiger (1955), Kimata (1961), Kawabata (1962), and others have pointed out that oral or enteral administration of large amounts of histamine are not toxic to most mammals, including man.

Geiger (1955) contended that previous experiments with histamine on mammals failed to support the belief that histamine was the cause of scombroid poisoning. He further stated:

> . . . instead they indicate that other toxins produced by certain toxin-forming microorganisms in the course of spoilage may be mainly responsible for toxic symptoms. The possibility that histamine may participate in cases of human fish poisoning cannot be entirely eliminated, because highly seasoned hot dishes prepared from spoiled fish, or simultaneous consumption of alcoholic beverages, may possibly facilitate the intestinal passage of histamine.

Kawabata, Ishizaka, and Miura (1955*c, d*) also believed that histamine alone did not account for the entire problem of scombroid poisoning. They believed that there was some other substance in addition to histamine which acted either independently or synergistically with histamine. In the course of their investigations they isolated a vagal stimulant from the Japanese saury *Cololabis saira* which was closely related to histamine in its chemical and pharmacological properties. This substance they termed "saurine." It is presently believed that scombrotoxin is a complex of substances containing histamine, saurine, and possibly other unidentified toxic products which are produced as a result of bacterial action on dark-meat fishes, mainly of the suborder Scombroidei. There are still questions relating to the scombrotoxic problem which continue to remain unanswered.

Some of the more pertinent references to the subject of fish spoilage as it relates to the histamine and scombrotoxic problem are as follows: Suzuki, Yoneyama, and Odake (1912), Koessler and Hanke (1919), Igarashi (1938, 1939, 1949), Legroux *et al.* (1946*a, b*), Amano and Bito (1951), Kimata and

Kawai (1953*a, b, c, d, e,* 1958, 1959), Kimata and Tanaka (1954*a, b*), Kimata, Kawai, and Tanaka (1954*a, b*), Miyaki and Hayashi (1954), Simidu and Hibiki (1954*a, b, c, d, e, f,* 1955*a, b, c, d*), Williams (1954, 1956, 1957, 1959), Kawabata *et al.* (1955*a, b, c, d, e*), Kawabata and Suzuki (1959*a, b*), Okazaki and Harada (1959), and Shifrine *et al.* (1959). Kimata (1961) has written an excellent comprehensive account of this subject.

MECHANISM OF INTOXICATION

Scombroid poisoning, scombrotoxism, is caused by the ingestion of spoiled scombroid fishes, tunas, and related species. Intoxication is actually caused by histamine, saurine, and possibly other toxic byproducts resulting from bacterial action on histidine, a normal muscle constituent of dark-meat fishes. The poisoning is not believed to be caused directly by bacterial toxins *per se*. Dark-meat fishes other than scombroid species may rarely be involved in this form of ichthyosarcotoxism. The development of the toxic products is due to inadequate preservation. The degree of freshness of the fish is an important consideration in scombroid poisoning.

MEDICAL ASPECTS

Clinical Characteristics

The symptoms of acute scombroid poisoning resemble those of histamine intoxication. The symptoms are characteristic and appear with almost monotonous consistency. Frequently, toxic scombroid flesh can be detected immediately upon tasting it. Victims state that it has a "sharp or peppery" taste. Symptoms, usually developing within a few minutes after ingestion of the toxin, consist of intense headache, dizziness, throbbing of the carotid and temporal vessels, epigastric pain, burning of the throat, cardiac palpitation, rapid and weak pulse, dryness of the mouth, thirst, inability to swallow, nausea, vomiting, diarrhea, and abdominal pain. Within a short time a generalized erythema and an urticarial eruption may develop covering the entire body, accompanied by a severe pruritus. The face of the individual becomes swollen and flushed, the eyes become injected, and coryza develops. In severe cases there may be bronchospasm, suffocation, and severe respiratory distress. Various other minor discomforts such as fever, chills, malaise, tremors, metallic taste, and cyanosis of the lips, gums, and tongue may occur. There is danger of shock, and deaths have been reported (Autenrieth, 1833; Stephenson, 1838). However, the acute symptoms are generally transient, lasting only 8 to 12 hours. Galiay (1845) reported an unusually severe case of 8 days' duration.

If a poorly preserved neurotoxic scombroid fish is eaten, there may be a combination of both ciguatera and scombroid types of symptoms. In rare instances the clinical picture may become even more complex, involving in addition to the two types of ichthyosarcotoxism one of the ordinary types of bacterial food poisoning.

Scombroid poisoning has also been termed "tjakalang poisoning" by Van Veen and Latuasan (1950) and Dack (1956). In Japan, scombroid poisoning has been referred to as "saurine poisoning" (Kimata, 1961; Kawabata, 1962).

The clinical characteristics of scombroid poisoning have been discussed by the following authors: Niel (1815), Henderson (1830), Autenrieth (1833), Stephenson (1838), Galiay (1845), Meyer-Ahrens (1855), Knox (1888), Phisalix (1922), Gilman (1942), Van Veen and Latuasan (1950), Halstead and Lively (1954), Halstead (1954, 1958, 1959, 1962, 1964), Kawabata *et al.* (1955*a*), Dack (1956), Sapin-Jaloustre (1957), Helfrich (1961), Bouder, Cavallo, and Bouder (1962), and Listick and Condit (1964).

Pathology

Not reported.

Treatment

The treatment of scombroid poisoning is directed toward combating the effects of scombrotoxin which is believed to consist mainly of histamine and the closely related substance saurine. The experimental studies of Legroux *et al.* (1946*a, b*) and Legroux, Bovet, and Levaditi (1947) have shown that experimental animals suffering from scombrotoxism quickly respond to antihistamine drugs.

In the event of poisoning, the stomach should be immediately evacuated. Epinephrine is recommended as a specific physiological antagonist to histamine (Goodman and Gilman, 1955). Epinephrine will prevent or counteract all symptoms if administered promptly. If absorption of scombrotoxin has occurred to any extent, antihistamine drugs should be used to control the subsequent reaction. Pieper (1958, cited by Bartsch, Drachman, and McFarren, 1959) found that patients suffering from scombroid poisoning recovered rapidly with epinephrine, cortisone, and intravenous benadryl. In the event of shock, intravenous fluids and plasma may be required. See also p. 370 on the treatment of ciguatera. The treatment of scombroid poisoning has been briefly discussed by Sapin-Jaloustre (1957), Halstead (1958, 1964), Helfrich (1961), and Bouder *et al.* (1962).

Prevention

In warmer climates scombroids should be promptly refrigerated or eaten soon after capture. It has been shown that the histamine content in some of the scombroid fishes increases from 0.09 mg/100 g of tissue to about 95 mg/100 g when kept at room temperature (20°-25° C) for about 10 hours (Geiger, 1944). Toxic scombroid flesh cannot always be detected by appearance. The toxin content may be very high with little or no evidence of putrefaction. Scombroid meat having a "sharp or peppery" taste should be discarded. Van Veen and Latuasan (1950) have noted that if the whole fish is available the gills should be examined. If there is the slightest evidence of putrefaction the fish should be discarded.

PUBLIC HEALTH ASPECTS

Scombrotoxism is a potential health hazard wherever scombroid and other dark-meat fishes are known to occur. However, with the advent of modern refrigeration and canning methods, the public health problem is greatly minimized in most areas. In communities where fresh fishes are an important part of the diet and adequate preservation methods are not always employed, scombrotoxism continues to be a public health problem. Numerous outbreaks of scombrotoxism are known to occur in many parts of the world where public health facilities are limited or nonexistent, but public health data on this subject are unavailable. The only published series of reported outbreaks of scombroid poisoning are by Kawabata *et al.* (1955*a*). These outbreaks occurred in Japan during the years 1951 through 1954. During this period there were at least 14 outbreaks involving 1,215 persons, but no deaths. In a second series of outbreaks of fish poisonings based on 25,000 outpatient and hospital patient unpublished records at Guam Memorial Hospital during the period of March 1, 1956, through February 28, 1958, there were 12 outbreaks of various types of fish poisonings involving 41 persons and 1 death (Pieper, 1958, cited by Bartsch *et al.*, 1959). However, one of the outbreaks caused by eating "mackerel" involved an estimated 15 to 20 persons who did not report into the hospital. The exact number of persons involved in scombroid poisoning could not be determined from the records, but it appears that scombroid poisoning is one of the common forms of fish poisoning in Guam. No deaths from scombroid poisoning were reported. A single outbreak of scombroid poisoning has been reported by Listick and Condit (1964) caused by eating a Spanish mackerel (species not designated) at Los Angeles, Calif. Seven persons were made ill, and all recovered.

TOXICOLOGY

Marcacci (1891) is said to have been the first to determine the toxicity of scombroid poison by testing tuna blood on laboratory animals. The blood was administered intravenously and intraperitoneally to dogs, producing a progressive paralysis. No convulsions were noted. He found his experimental results to be comparable to those of Mosso (1888) on eel serum. Temperatures of 60° C for 10 minutes or the addition of a few drops of HCl destroyed the toxin. Marcacci therefore believed that he was working with an "albuminoid" substance. His work was crude and too incomplete to draw any conclusion as to the relationship of this compound to the causative agent of scombroid poisoning. Kossel (1896) and Thompson (1900) obtained a toxic "protamine" from the milt of the mackerel *Scomber scombrus*.

Most of the serious toxicological investigations on scombroid fishes began with the work of Geiger *et al.* (1944). They prepared aqueous extracts from sardines, *Sardinops caerulea;* mackerel, *Pneumatophorus diego;* albacore,

Germo alalunga; tuna, *Thunnus thynnus;* and bonito, *Sarda chiliensis.*[1] The histamine content was determined by using isolated guinea pig intestine. They found that 0.1 ml of mackerel extract when added to 25 ml of a Ringer-Locke bath induced a strong contraction of the guinea pig intestine. Moreover, the contraction was comparable to that produced by 0.1 μg histamine in 25 ml solution. Ten ml of the fish extract were ashed and the residue dissolved in 10 ml of 5 percent aqueous HCl. After neutralization with Na_2CO_3 the solution proved to be nearly inactive, indicating that the contraction of the intestine was not due to cations (Mg or K) present in the extract. Atropine treatment of the intestine failed to inhibit the activity of the extract, which excluded the action of a parasympathomimetic substance. Concentrated extracts injected subcutaneously into guinea pigs resulted in dyspnea and death by asphyxia. The effects were characteristic of histamine toxicity. Injections of the concentrate intravenously into a rabbit in ether narcosis increased the blood pressure to an extent comparable to that produced by a histamine solution of similar activity. The concentrate injected intravenously into a cat in nembutal anesthesia had a strong hypotensive action characteristic of histamine. When the concentrated solution was rubbed on the surface of the skin it produced a strong localized urticarial reaction. The fish extract gave a very pronounced Pauly and Knoop reaction, which was stronger than could be accounted for by the biological activity of histamine alone. The authors suggested that perhaps other imidazole derivatives might be responsible for this difference. They further observed that the histamine content of different samples of the same species of fish varied greatly. This variation in histamine content was attributed to different degrees of autolysis.

Geiger (1944) later recommended using histamine detection as a useful method in determining the state of freshness of scombroid meat. This technique employed guinea pig ileum, suspended in Ringer-Locke solution, connected with a frontal writing lever of a kymograph. Contractions of the gut using a test extract were compared with a standard known histamine solution.

Legroux *et al.* (1946*a*) and Legroux *et al.* (1947) tested a sample of tuna flesh (species not stated) which had poisoned four out of five persons. The color of the flesh was pink, oily, firm, and without an odor of putrefaction. Bacteriological examination for human pathogens was negative. Aqueous extracts of the tuna meat injected intramuscularly into guinea pigs produced muscular tremors, convulsions, and death. Autopsies of the animals revealed congestion of the gastric and intestinal mucosa. Heating and filtration of the agent through a Chamberland filter did not detoxify the substance. Futher studies on guinea pigs revealed that if they were first administered

[1] The studies of Geiger *et al.* (1944) suggest that these species should be considered as potential scombrotoxic fishes.

antihistaminics, they were protected from the symptoms produced by subsequent injections of the toxic extract. Similar results were obtained using injections of a known histamine solution. Tests were also run using guinea pig ileum. Contractions comparable to those obtained from a known histamine solution were obtained from the tuna extract. The authors therefore concluded that histamine was present in the tuna flesh, and it was probably produced by bacteria as a result of decarboxylation of histidine.

Van Veen and Latuasan (1950) have pointed out that dogs are extremely sensitive to scombroid poison, and a sign of intoxication is the swollen appearance of the head. In their tests on the toxicity of scombroid poison they followed the technique of previous workers, injecting extracts intramuscularly into guinea pigs.

Aiiso (1954) demonstrated that 450-500 mg percent administered by mouth to human volunteers did not produce any symptoms. This was the equivalent amount of histamine found to be present in the dried seasoned saury *Cololabis saira* that caused the poisonings at Hiroshima, Japan, in 1953 (Kawabata *at al.* 1955*a*).

Geiger (1955) questioned that histamine is the primary toxic substance present in scombroid poisoning in view of the fact that histamine administered by either oral or enteral routes is nontoxic to most mammals, probably because it is detoxified during its passage through the intestinal wall. However, he further commented that if histamine is the primary factor in scombroid poisoning then either the detoxification mechanism was disturbed, or the conditions of the intestinal tract were such that histamine was absorbed at an increased speed so that detoxification could not keep up with the entry of histamine into the circulation. Geiger cited some earlier experiments on dogs in which it was found that after introduction of dilute chloroform, ethyl alcohol, or HCL into the gut, the absorption of intraduodenally administered histamine produced a blood pressure depression characteristic of histamine toxicity. He also found that histamine administered to guinea pigs by stomach tube did not kill the animals, but if the intestinal mucosa was damaged beforehand by administration of saponin for 2 consecutive days the same dose of histamine killed 6 of 10 animals within 60 to 100 minutes. Care had to be taken in administering the histamine because if any of it was spilled in the mouth of the guinea pig severe symptoms and even lethal shock followed. These results were consistent with the earlier findings of Feldberg and Schilf (1930) who found that histamine produced shock when it was absorbed from the oral mucosa, the tongue, or trachea, but did not affect the animal if the histamine was administered directly into an intact intestinal tract. In another group of experiments Geiger incubated fresh tuna meat after inoculation with an unidentified strain of marine bacteria for 56 hours at 37° C. The samples contained, at the end of the experiment, 190 to 210 mg of histamine per 100 g of fresh substance. This spoiled fish meat was fed to dogs, cats, rats, and mice. In addition, the guinea pigs received by gavage 10 ml of a concentrated aqueous extract containing 80 to 100 mg of histamine made from spoiled fish. "No symptoms of poisoning could be observed in these experiments, and all the animals survived." The author then concluded that there was a lack of supporting evidence that histamine is the primary cause of intoxication in scombroid poisoning, and suggested that other toxins produced by toxigenic microorganisms in the course of spoilage were responsible for the toxic symptoms, but finally admitted that histamine may be a factor possibly because of a temporary disturbance in the normal intestinal barrier.

Simidu and Hibiki (1955*d*) reported that poisoning by spoiled scombroid fishes usually occurs in humans when the histamine concentration reaches a level of about 100 mg/100 g of fish muscle.

Shifrine *et al.* (1959) found that if tuna muscle was first cooked and then spoiled for 3 to 4 days the material was toxic to chicks when fed in amounts of 30 percent of their diet, whereas spoiled uncooked muscle was innocuous. The toxicity was manifested by rapid weight loss and occasional deaths. Autoclaving did not alter the toxicity, but the toxic substance could be removed by water extraction following acetone extraction. The LD_{50} of histamine by intraperitoneal injection in mice is 13 g/kg (Stecher, 1960).

PHARMACOLOGY

The exact chemical nature of scombrotoxin is not known, but is believed to be a complex of several substances including histamine, saurine, and possibly other unidentified toxic chemical constituents.

The pharmacology of histamine has been studied extensively by numerous investigators since its discovery in 1907 by Windaus and Vogt. Since much of the research on this subject is beyond the scope of the present monograph, this discussion will be limited to a summary of the more pertinent actions of histamine and saurine. Histamine occurs in both plants and animals and is also found in formed blood elements (platelets and granulocytes), tissue mast cells, and elsewhere. Histamine is a normal constituent of many tissues, but irregular in its occurrence and varying in amounts present from one species and one tissue to the next. It is generally high in the lungs and skin but low in the muscles of most animals.

Histamine is a versatile and active compound affecting a variety of organs and tissues. However, its greatest effects are those manifested by the blood vessels, various smooth muscle structures, and glands. Histamine is the most potent capillary dilator known. Its action is directly upon the contractile mechanism which is independent of any innervation. The dilatation is accompanied by a marked increase in capillary permeability, which differs from most other vasodilators. Histamine will produce a seepage of fluid, plasma proteins, and even some red blood cells. Injection of histamine intradermally will result in a characteristic wheal indicative of the increased capillary permeability, but if it is injected intravenously all the capillaries of the body will be affected and there will be a measurable change in the plasma volume and hematocrit. There will also be a marked increase in the flow of lymph from the thoracic duct and other major lymphatic vessels.

The cutaneous reaction caused by the injection of histamine is characterized by a local reddening which appears promptly at the site of the injection, a wheal of localized edema which develops within 1 minute and obscures the original injection spot, and a scarlet halo surrounding the wheal which starts to develop in 30 seconds to a minute, attaining a diameter up to 5 cm during the next 5 minutes and then fading away in about 10 minutes. The condition is termed urticaria when crops of itching wheals occur surrounded by areas of erythema.

In most species histamine produces a vasoconstrictor effect when it is perfused through isolated arteries or through the vascular bed of isolated organs. One would expect that injection of histamine would result in a rise in blood pressure, but in man and most other animals there is a pronounced drop in blood pressure. This paradoxical effect is explained as being due to the ability of histamine to release the arterioles and capillaries from a normal state of tonic contraction which more than offsets the direct action upon the smooth muscle of the arterial vessel walls.

Histamine does not have a pronounced effect upon the heart except as caused by its vascular action. A compensatory tachycardia may be induced by the vasodilatation and fall of blood pressure and thereby produce an increased cardiac output and coronary blood flow.

Histamine contracts the smooth muscle of the uterus and gall bladder and stimulates secretion of many of the glands of external secretion. Probably the greatest effect is on the gastric glands.

Histamine is highly effective when administered parenterally, but ineffective by mouth. This is believed to be due to poor absorption, enzymatic destruction, and partly to conjugation to an inactive acetylated histamine. When injected, histamine disappears rapidly from the blood stream due to diffusion into the tissues, inactivation, conjugation, and excretion (Dragstedt, 1958).

The fact that histamine taken by mouth is generally nontoxic in humans suggested to Miyaki and Hayashi (1954) and Kawabata *et al.* (1955*a, b, e*) that scombrotoxism was caused by one or more other toxic substances working independently or in conjunction with histamine. Kawabata and his group isolated a second substance from toxic scombroid musculature which they believed contributed to scombrotoxications in humans. This substance was isolated from the spoiled flesh of the Japanese saury *Cololabis saira* and was termed saurine. Saurine was found to closely resemble acetylcholine and histamine in its pharmacological properties, but was chemically distinct from both of them. Previous Magnus tests had shown that isolated mouse and guinea pig intestine were sensitive to acetylcholine, but only guinea pig intestine was sensitive to histamine. The crude methanolic saurine extract contracted guinea pig intestine, but had no effect on mouse intestine, which eliminated acetylcholine as one of the active constituents. Further tests revealed that if crude saurine extracts were injected intracutaneously into rabbits they developed wheals that were quite comparable to those produced by histamine controls on the adjacent skin of the same rabbit. When the crude saurine extract was injected intravenously into guinea pigs, it produced a shock syndrome with symptoms of dyspnea, hiccup, convulsions, etc., comparable to those produced by histamine and acetylcholine. It was also found that antiallergic drugs such as adrenalin, ephedrine hydrochloride, and methyl ephedrine hydrochloride were effective in relaxing guinea pig intestine stimulated by saurine. As a result of these experiments the authors concluded that saurine closely resembled histamine in its pharmacological properties and that these allergiclike fish poisonings were attributable to the combined effects of histamine and saurine.

Kawabata *et al.* (1955*d*) later explored the suggestion of Miyaki and Hayashi (1954) that possibly other biogenic amines might enhance the effects of histamine in these allergiclike fish intoxications. Amines such as trimethylamine (TMA) or trimethylamine oxide (TMAO), phosphorylcholine, agmatine, etc., were suggested. However, Kawabata and his associates found that when TMA or TMAO were added to histamine or saurine there were no appreciable changes in the pharmacological effects as measured by the previously mentioned tests in guinea pigs, rabbits, and mice. They therefore concluded that these did not contribute significantly to this form of fish poisoning.

CHEMISTRY

Scombrotoxism is caused by an intoxicant that is not present to any toxicological extent in fresh dark-meat fishes but develops within a period of 3 to 4 days *post mortem.* Since the clinical and pharmacological effects of this biotoxin are almost identical to those produced by histamine intoxication most chemical research on this subject has been directed to the chemistry of histamine.

Histamine ($C_5H_9N_3$) is a base (β-iminazolylethylamine) and is obtained by the decarboxylation of free amino acid histidine. It is soluble in water, alcohol, hot chloroform, but only slightly soluble in ether. It acts as a bivalent radical in the formation of salts of which the phosphate and the hydrochloride are the most commonly employed. Solutions of the salts are stabilized to boiling when acidified, but not when alkalized. Histamine is adsorbed by charcoal, magnesium trisilicate, cotton acid succinate, and other agents. The fundamental action of histamine remains to be established (Dragstedt, 1958; Stecher, 1960).

The fact that oral histamine is generally nontoxic to humans, and the suggestion of Miyaki and Hayashi (1954) and others that allergiclike fish poisoning, scombrotoxism, might be due to the combined effects of histamine and some other biogenic amine, led Kawabata *et al.* (1955*a, b, c, d*) to investigate the spoilage problem to determine whether some other biotoxic agents were present. Their studies led them to conclude that scombrotoxism was caused by a combination of histamine and a second substance of unknown structure which they called saurine.

Kawabata *et al.* (1955*b*) isolated saurine from histamine present in spoiled Japanese saury *Cololabis saira* by one- and two-dimensional paper chromatography, which was confirmed by Magnus' method using guinea pig intestine. Saurine was differentiated from histamine by its R_f value of 0.1 with *n*-butanol-acetic acidwater (4:1:2) and by its reaction with diazo reagent. It is insoluble in ether, acetone, benzene or chloroform, and possibly ethanol. Saurine is easily extracted with 80 percent methanol at room temperature. It is stable to boiling for 1 hour in 60 percent methanol with 1 percent concentrated HCl. Its dialyzability, electrodialytic character, and precipitability with phosphotungstic acid suggest that saurine may be a basic substance with relatively low molecular weight.

It appears that the substance designated as scombrotoxin may be a combination of chemical constituents including histamine, saurine, and possibly other toxic products resulting from the *post-mortem* action of microorganisms on dark-meat fishes.

The chemical aspects of the scombrotoxic problem have also been discussed by Kimata and Kawai (1953*c*, 1958, 1959), Williams (1954, 1956, 1957, 1959), Kimata (1961), and Amano and Bito (1951).

LITERATURE CITED

AIISO, K.
1954 On the samma-sakuraboshi poisoning. [In Japanese] Shinryo Shitsu 6: 44. (NSA)

AMANO, K., and M. BITO
1951 Consequence of free amino acids generated from decomposing fish muscle. [In Japanese, English summary] Bull. Japan. Soc. Sci. Fish. 16(12): 10-16.

AUTENRIETH, H. F.
1833 Ueber das gift der fische. C. F. Osiander, Tübingen. 287 p.

BARTSCH, A. F., DRACHMAN, R. H., and E. F. MCFARREN
1959 Report of a survey of the fish poisoning problem in the Marshall Islands. Dept. Interior, Public Health Serv. 117 p., 17 figs.

BEAUFORT, L. F. DE, and W. M. CHAPMAN
1951 The fishes of the Indo-Australian archipelago. IX. Percomorphi (concluded), Blennoidea. E. J. Brill, Leiden. 484 p., 89 figs.

BEHRE, A.
1949 Vergiftungen durch fischereierzeugnisse. Z. Lebensm. Untersuch. Forsch. 89(3): 229-312.

BOUDER, H., CAVALLO, A., and M. J. BOUDER
1962 Poissons vénéneux et ichthyosarcotoxisms. Bull. Inst. Océanogr. 59 (1240), 66 p., 2 figs.

BURROWS, G. M.
1815 An account of two cases of death from eating mussels; with some general observations on fish-poison. London Med. Reposit. 3(18): 445-476.

COLLETTE, B. B., and R. H. GIBBS, JR.
1963*a* Preliminary field guide to the mackerel- and tuna-like fishes of the Indian Ocean (Scombridae). Smithsonian Institution, Washington, 48 p., 8 pls.
1963*b* A preliminary review of the fishes of the family Scombridae. FAO Fish. Rept. No. 6, 3: 977-978.

DACK, G. M.
1956 Food poisoning. Rev. ed. Univ. Chicago Press, Chicago. 251 p.

DRAGSTEDT, C. A.
1958 Histamine and antihistaminics, p. 619-625. *In* V. A. Drill [ed.], Pharmacology in medicine. 2d ed. McGraw-Hill Book Co., Inc., New York.

FELDBERG, W., and E. SCHILF
1930 Histamin: seine pharmakologie und bedeutung für die humoralphysiologie. Julius Springer, Berlin. 572 p., 86 figs.

FONSSAGRIVES, J. B.
1877 Animaux toxicophores, p. 621-636; Fig. 71. *In* J. B. Fonssagrives, Traité d'hygiene navale. J. B. Baillière et Fils, Paris.

FREDERICQ, L.
1924 Die sekretion von schutz- und nutzstoffen, p. 1-87. *In* H. Winterstein [ed.], Handbuch der vergleichenden physiologie. Vol. II. Gustav Fischer, Jena.

GALIAY, D. M.
1845 Empoisonnement accidental de plusieurs personnes, produit par un poisson de mer connu sous le nom de thon ou de *Scomber thynnus*. Bull. Gén. Thérap. 29: 204-211.

GATEWOOD, J. D.
1909 Naval hygiene. P. Blackiston's Son & Co., Philadelphia. 779 p.

GEIGER, E.
1944 Histamine content of unprocessed and canned fish. A tentative method of quantitative determination of spoilage. Food Res. 9(4): 293-297.
1955 Role of histamine in poisoning with spoiled fish. Science 121(3155): 865-866.

GEIGER, E., COURTNEY, G., and G. SCHNAKENBERG
1944 The content and formation of histamine in fish muscle. Arch. Biochem. 3(3): 311-319.

(NSA)—Not seen by author.

GILMAN, R. L.
1942 A review of fish poisoning in the Puerto Rico-Virgin Islands area. U.S. Nav. Med. Bull. 40(1): 19-27, 2 pls.

GOODMAN, L. S., and A. GILMAN
1955 The pharmacological basis of therapeutics. 2d. ed. MacMillan Co., New York. 1,831 p., 89 figs.

HALSTEAD, B. W.
1954 A note regarding the toxicity of the fishes of the skipjack family, Katsuwonidae. Calif. Fish Game 40(1): 61-63.
1958 Poisonous fishes. Public Health Rept. (U.S.) 73(4): 302-312, 11 figs.
1959 Dangerous marine animals. Cornell Maritime Press, Cambridge. 146 p. 88 figs.
1962 Biotoxications, allergies, and other disorders, p. 521-540. *In* G. Borgstrom [ed.], Fish as food. Vol. II. Nutrition, sanitation, and utilization. Academic Press, New York.
1964 Fish poisonings—their diagnosis, pharmacology, and treatment. Clin. Pharmacol. Therapy. 5(5): 615-627.

HALSTEAD, B. W., and W. M. LIVELY
1954 Poisonous fishes and ichthyosarcotoxism. Their relationship to the Armed Forces. U.S. Armed Forces Med. J. 5(2): 157-175, 9 figs.

HELFRICH, P.
1961 Fish poisoning in the tropical Pacific. Hawaii Marine Lab., Univ. Hawaii. 16 p., 7 figs.

HENDERSON, P. B.
1830 Case of poisoning from the bonito, *(Scomber pelamis)*. Edinburgh Med. J. 34(105): 317-318.

HILDEBRAND, S. F.
1946 A description catalog of the shore fishes of Peru. U.S. Nat. Mus. Bull. 189. 530 p., 95 figs.

IGARASHI, H.
1938 The pungent principles of fishes produced by decrease in freshness, Part I. J. Chem. Soc. Japan 59: 1258-1259.
1939 The pungent principles of fishes produced by decrease in freshness, Part II. Bull. Japan. Soc. Sci. Fish. 8: 158-160.
1949 On the ptomaine of fish. Pamphlet, Municipal Office, Hakodate City. 33 p.

JOUBIN, L. M.
1929-38 Faune ichthyologique de l'Atlantique nord. Nos. 1-18. Andr. Fred Høst & Fils, Copenhague.

KAWABATA, T.
1962 Problems involved in the research on fish and shellfish poisonings. Japan. J. Med. Sci. Biol. 15: 141-143.

KAWABATA, T., ISHIZAKA, K., and T. MIURA
1955*a* Studies on the food poisoning associated with putrefaction of marine products. I. Outbreaks of allergy-like food poisoning caused by "samma sakuraboshi" (dried seasoned saury) and canned seasoned mackerel. [In Japanese, English summary] Bull. Japan. Soc. Sci. Fish. 21(5): 335-340.
1955*b* Studies on the food poisoning associated with putrefaction of marine products. II. Causative toxic substance and some of its chemical properties. [In Japanese, English summary] Bull. Japan. Soc. Sci. Fish. 21(5): 341-348.
1955*c* Studies on the food poisoning associated with putrefaction of marine products. III. Physiological and pharmacological properties of newly isolated vagus-stimulant, named "saurine." [In Japanese, English summary] Bull. Japan. Soc. Sci. Fish. 21(5): 347-351.
1955*d* Studies on the food poisoning associated with putrefaction of marine products. IV. Epidemiology and the causative agents of the outbreaks of allergy-like food poisoning caused by cooked frigate-mackerel meat at Kawasaki and by "samma sakuraboshi" at Hamamatsu. [In Japanese, English summary] Bull. Japan. Soc. Sci. Fish. 21(11): 1167-1170.
1955*e* Studies on the food poisoning associated with putrefaction of marine products. V. Influence of trimethylamine and trimethylamine oxide added to histamine or "saurine" upon their own action tested on the contractility of intestine and uterus of guinea pigs. [In Japanese, English Summary] Bull. Japan. Soc. Sci. Fish. 21(11): 1172-1176. (NSA)

(NSA)—Not seen by author.

KAWABATA, T., and S. SUZUKI

1959*a* Studies on the food poisoning associated with putrefaction of marine products. VIII. Distribution of *l*-(—) histidine decarboxylase among *Proteus* organisms and the specificity of decarboxylating activity with washed cell suspension. [In Japanese, English summary] Bull. Japan. Soc. Sci. Fish. 25(6): 473-480.

1959*b* Studies on the food poisoning associated with putrefaction of marine products. IX. Factors affecting the formation of *l*(—)—histidine decarboxylase by *Proteus morganii*. Bull. Japan. Soc. Sci. Fish. 25(6): 481-487.

KIMATA, M.

1961 The histamine problem, p. 329-352. *In* G. Borgstrom [ed.], Fish as food. Vol. I. Production, biochemistry, and microbiology. Academic Press, New York.

KIMATA, M., and A. KAWAI

1953*a* The freshness of fish and the amount of histamine present in the meat. I. Mem. Res. Inst. Food Sci. Kyoto Univ. 6: 3-11.

1953*b* A new species of bacterium which produces large amounts of histamine on fish meats, found in spoiled fresh fish. Mem. Res. Inst. Food Sci. Kyoto Univ. 6: 1-2.

1953*c* The production of histamine by the action of bacteria causing the spoilage of fresh fish. I. Bull. Res. Inst. Fac. Sci. Kyoto Univ. 12: 29-33.

1953*d* Biochemistry of *Achromobacter histamineum*. I. Mem. Res. Food Sci. Kyoto Univ. 6: 3-11. (NSA)

1953*e* The freshness of fish and the amount of histamine present in the meat. II. Mem. Res. Inst. Food Sci. Kyoto Univ. 6: 12-22.

1958 Studies on the histamine formation of *Proteus morganii*. Mem. Coll. Agr. Kyoto Univ., Fish. Ser. Spec. issue, p. 92-99.

1959 Studies on the histamine formation of *Proteus morganii* (cont.). Mem. Res. Inst. Food Sci. Kyoto Univ. 18: 1-7.

KIMATA, M., KAWAI, A., and M. TANAKA

1954*a* The freshness of fish and the amount of histamine present in the meats. III. Mem. Res. Inst. Food Sci. Kyoto Univ. 7: 6-11.

1954*b* The freshness of fish and the amount of histamine present in the meats. IV. Mem. Res. Inst. Food Sci. Kyoto Univ. 8: 1-6.

KIMATA, M., and M. TANAKA

1954*a* On the bacteria causing spoilage of fresh fish, especially on their activity which can produce histamine. Mem. Res. Inst. Food Sci. Kyoto Univ. 7: 12-17.

1954*b* A study whether the bacteria having an activity which can produce a large amount of histamine, so-called histamine-former, are present or not, on the surface of fresh fish. Mem. Res. Inst. Food Sci. Kyoto Univ. 8: 7-16.

KISHINOUYE, K.

1923 Contributions to the comparative study of the so-called scombroid fishes. J. Coll. Agr., Imp. Univ. Tokyo 8(3), 473 p., 26 figs.

KNOX, G.

1888 Poisonous fishes and means for prevention of cases of poisoning by them. [In Russian] Voenno-Med. Zh. 161(3): 399-464, 9 figs.

KOESSLER, K. K., and M. T. HANKE

1919 Studies on proteinogenous amines. IV. The production of histamine from histidine by *Bacillus coli communis*. J. Biol. Chem. 39: 539-556. (NSA)

KOSSEL, H.

1896 Ueber die basischen stoffe des zellkerns. Hoppe-Seylers Z. Physiol. Chem. 22: 176-187.

LEGROUX, R., BOVET, D., and J. C. LEVADITI

1947 Présence d'histamine dans la chair d'un thon responsable d'une intoxication collective. Ann. Inst. Pasteur 73(1): 101-104.

LEGROUX, R., LEVADITI, J. C., BOUDIN, G., and D. BOVET

1946*a* Intoxications histaminiques collectives consécutives à l'ingestion de thon frais. Presse Méd. 29: 545.

1946*b* A propos des intoxications histaminiques collectives d'origine alimentaire. Presse Méd. 53: 743.

(NSA)—Not seen by author.

LISTICK, F. A., and P. K. CONDIT
1964 Scombroid poisoning—California. Morbidity and Mortality Weekly Rept., Communicable Disease Center 13(4): 30.

MANN, W. L.
1938 Fish poisoning in Culebra—Virgin Islands area. U.S. Nav. Med. Bull. 36(4): 631-634.

MARCACCI, A.
1891 Sur le pouvoir toxique du sang de thon. Arch. Ital. Biol. 16: 1.

MARKOV, S.
1943 Seefischvergiftung. Wien. Med. Wochschr. 93(26-27): 388.

MEYER-AHRENS, K. M.
1855 Von den giftigen fischen. Schweiz. Z. Med. Chir. Geburtsh. (3): 188-230; (4-5): 269-332.

MIYAKI, K., and M. HAYASHI
1954 Food poisoning caused by ordinary putrefaction. II. Detection of histamine and its synergistic factor in deteriorated dried mackerel pike. [In Japanese, English summary] J. Pharm. Soc. Japan 74: 1145-1148.

MOQUIN-TANDON, A.
1860 Eléments de zoologie médicale. Baillière et Fils, Paris.

MOSSO, A.
1888 Un venin dans le sang des murenides. Arch. ital. Biol. 10: 141-169.

NIEL, J. G.
1815 Observations sur les accidents qui proviennent de l'usage des poissons empoisonnés, etc. Ann. Clin. Rec. Period. (Montpellier) 36: 315-330.

OKAZAKI, H., and S. HARADA
1959 Studies on an outbreak of histamine poisoning caused by fresh fish in Kagawa prefecture. I. General view of the outbreak and histamine-producing ability of the isolated bacteria. [In Japanese, English summary] J. Japan. Vet. Med. Assoc. 12(6): 259-262.

PAPPENHEIM, L.
1858 Fleischnahrung, p. 592-595. *In* L. Pappenheim, Handbuch der Sanitäts-Polizei. August Hirschwald, Berlin.

PERGOLA, M.
1937 Bactéries toxigènes chez le poisson frais avec ou sans la présence de l'ichthyovenin. Boll. Sez. Ital. Soc. Intern. Microbiol. 9(5): 105-108. (NSA)

PHISALIX, M.
1922 Animaux venimeux et venins. 2 vols. Masson et Cie., Paris.

RIVAS, L. R.
1951 A preliminary review of the western north Atlantic fishes of the family Scombridae. Bull. Marine Sci. Gulf Caribbean 1(3): 209-230.

ROSA, H., JR., *ed.*
1963 Proceedings of the world scientific meeting on the biology of tunas and related species. 3 vols. FAO Fisheries Reports No. 6.

SAPIN-JALOUSTRE, H. and J.
1957 Une toxi-infection alimentaire peu connue: l'intoxication histaminique par le thon. Concours Méd. 79(22): 2705-2708.

SAVTSCHENKO, P.
1886 Atlas of the poisonous fishes. Description of the ravages produced by them in the human organism, and antidotes which may be employed. [In Russian and French] St. Petersburg. 53 p., 10 pls.

SHIFRINE, M., OUSTERHOUT, L. E., GRAU, C. R., and R. H. VAUGHN
1959 Toxicity to chicks of histamine formed during microbial spoilage of tuna. Appl. Microbiol. 7: 45-50.

SIMIDU, W., and S. HIBIKI
1954*a* Studies on putrefaction of aquatic products. XIII. Comparison on putrefaction of different kinds of fish (1). Bull. Japan. Soc. Sci. Fish. 20(4): 298-301.
1954*b* Studies on putrefaction of aquatic products. XIV. Comparison on putrefaction of different kinds of fish (2). [In Japanese, English summary] Bull. Japan. Soc. Sci. Fish. 20(4): 302-304. (NSA)
1954*c* Studies on putrefaction of aquatic products. XV. Comparison of putrefaction for round, fillet, minced, and denatured fishes. [In Japanese, English summary] Bull. Japan. Soc. Sci. Fish. 20(5): 388-391. (NSA)
1954*d* Studies on putrefaction of aquatic products. XVI. Consideration on difference in putrefaction for various kinds of fish (1). [In Japanese, English summary] Bull. Japan. Soc. Sci. Fish. 20(5): 392-395. (NSA)

(NSA)—Not seen by author.

SIMIDU, W., and S. HIBIKI—Continued

1954*e* Studies on putrefaction of aquatic products. XVII. Consideration of difference in putrefaction for various kinds of fish (2). [In Japanese, English summary] Bull. Japan. Soc. Sci. Fish 20(8): 717-719. (NSA)

1954*f* Studies on putrefaction of aquatic products. XII. On putrefaction of bloody muscle. Bull. Japan. Soc. Sci. Fish. 20(3): 216-218.

1955*a* Studies on putrefaction of aquatic products. XIX. Influence of certain substances upon histamine formation. [In Japanese, English summary] Bull. Japan. Soc. Sci. Fish. 20(9): 808-810. (NSA)

1955*b* Studies on the putrefaction of aquatic products. XX. Considerations of difference in putrefaction for various kinds of fish (3). Influence of autolysis on putrefaction. [In Japanese, English summary] Bull. Japan. Soc. Sci. Fish. 21(4): 267-279. (NSA)

1955*c* Studies on putrefaction of aquatic products. XXI. Considerations of difference in putrefaction for various kinds of fish (4). On difference of concentration of histidine between the interior and exterior cells. [In Japanese, English summary] Bull. Japan. Soc. Sci. Fish. 21(5): 357-360. (NSA)

1955*d* Studies on putrefaction of aquatic products. XXIII. On the critical concentration of poisoning for histamine. [In Japanese] Bull. Japan. Soc. Sci. Fish. 21(5): 365-367. (NSA)

STECHER, P. G.

1960 The Merck index of chemicals and drugs. An encyclopedia for chemists, pharmacists, physicians, and members of allied professions. 7th ed. Merck & Co., Inc., Rahway, N.J. 1,642 p.

STEPHENSON, J.

1838 Medical zoology and mineralogy. John Churchill, London. 350 p.

SUZUKI, U., YONEYAMA, C., and S. ODAKE

1912 Composition of "bonito" salt paste. [In Japanese] J. Coll. Agr. Tokyo Imp. Univ. 5: 33-41. (NSA)

THOMPSON, W. H.

1900 Die physiologische wirkung der protamine und ihrer spaltungsprodukte. Hoppe-Seylers Z. Physiol. Chem. 29: 1-19, 7 figs.

VAN VEEN, A. G., and H. E. LATUASAN

1950 Fish poison caused by histamine in Indonesia. Doc. Neerland. Indon. Morbis Trop. 2(1): 18-20.

WILLIAMS, D. W.

1954 Report on chemical indices of decomposition in fish (histamine). J. Assoc. Offic. Agr. Chemists 37(3): 567-572.

1956 Report on chemical indices of decomposition in fish (histamine). J. Assoc. Offic. Agr. chemists 39: 609-612.

1957 Report on decomposition in fish (histamine). J. Assoc. Offic. Agr. Chemists 40(2): 420-421.

1959 Report on chemical indices of decomposition in fish (histamine). J. Assoc. Offic. Agr. Chemists 42(2): 287-289.

(NSA)—Not seen by author.

Chapter XIV—VERTEBRATES

Class Osteichthyes

Poisonous: TETRODOTOXIC FISHES

Tetrodotoxications constitute one of the most violent forms of marine biotoxications. This type of fish poisoning is more commonly designated as tetraodon or puffer poisoning. The causative transvectors are members of the order Tetraodontiformes, formerly termed the Plectognathi, which includes the families Tetraodontidae, the puffers; Diodontidae, the porcupinefishes; Canthigasteridae, the sharp-nosed puffers; and Molidae, the molas or ocean sunfishes. The order also includes several other families such as the Triacanthodidae and Triacanthidae, the spikefishes; Balistidae, the triggerfishes; Monacanthidae and Aluteridae, the filefishes; Aracanidae and Ostraciontidae, the trunkfishes; and Triodontidae, the three-toothed puffers. Although most, if not all, of these families contain species which have been incriminated in human intoxications, the term tetrodotoxication (tetraodon or puffer poisoning) has been largely reserved for biotoxications caused by members of the families Tetraodontidae, Diodontidae, Canthigasteridae, and possibly the Molidae and Triodontidae.[1] The members of these families are believed to transvect tetrodotoxin. Some of the fishes of the other families within the order Tetraodontiformes may also transvect tetrodotoxin, but as yet there is insufficient laboratory or clinical evidence to warrant this conclusion. There is reason to believe that any of the tetraodontiform fishes may also transvect ciguatoxin, but this has yet to be established as fact. In the subtropical and tropical regions of the world, the tetrodotoxic fishes constitute a public health problem, but fortunately tetraodontiform fishes are readily distinguished by their peculiar morphology.

The phylogenetic relationships of tetraodontiform fishes are far from being satisfactorily understood despite the large amount of literature concerning the subject. Since there seems to be no general agreement regarding the correct taxonomic nomenclature of the group, we have for the purposes of this chapter adopted a modification of the terminology as proposed by Fraser-Brunner (1943, 1951), with the exception of the puffers of Japan and

---

[1] There has been some confusion regarding the proper spelling of "tetraodon" versus "tetrodon" poisoning. Both terms continue to be used. Since the proper spelling of the generic term for the fish is *Tetraodon*, then "tetraodon" poisoning is preferred. However, the designation of "tetraodontoxin" has never gained acceptance by chemists either in the United States or Japan (Halstead, 1953). Therefore, since "tetrodotoxin," which is a bastard term, continues to be widely used by scientific writers, the spelling "tetrodotoxication" is the most popular spelling for the clinical term.

adjacent regions in which we have largely followed Abe (1949, 1950, 1951, 1952, 1954). Breder and Clark (1947) have published a useful work on the phylogeny of the Plectognathi in which they discussed the more important publications dealing with the systematics and visceral anatomy of tetraodontoid fishes.

The following authors have published on the systematics and biology of tetraodontid fishes: Bleeker (1851, 1862-77), Day (1878-88), Jordan, Tanaka, and Snyder (1913), Meek and Hildebrand (1928), Fraser-Brunner (1943, 1951), Breder and Clark (1947), Abe (1949, 1950, 1951, 1952, 1954), Clark and Gohar (1953), Herre (1953), Tomiyama, Abe, and Tokioka (1958), Smith (1961), Smith and Smith (1963), and Marshall (1964).

Biotoxicologists owe a debt of gratitude to Japanese scientists for their careful investigations over a period of several centuries which have finally culminated in a more exacting knowledge of the biology, pharmacology, and chemistry of puffers and tetrodotoxin. The public health authorities in Japan are also to be congratulated for the thorough manner in which they have maintained their epidemiological records in marine biotoxications since the latter part of the 19th century. Certainly no other country can make similar claims toward maintaining such outstanding epidemiological performance. It is hoped that future investigations will be directed toward elucidating the biogenesis of tetrodotoxin. A major question that continues to confront biotoxicologists is why one puffer is toxic and the next specimen of the same species taken in another locality at the same season of the year is not.

REPRESENTATIVE LIST OF FISHES REPORTED AS TETRODOTOXIC

Phylum CHORDATA

Class OSTEICHTHYES

Order TETRAODONTIFORMES (PLECTOGNATHI): Puffers, Porcupinefishes, Molas, Etc.

Family CANTHIGASTERIDAE

179 *Canthigaster rivulatus* (Temminck and Schlegel) (Pl. 2, fig. a). Sharp-nosed puffer (USA), pu'u olai (Hawaii), luab (Marshall Islands), sue mimi (Samoa), boriring, butinga (Philippines), kitamukura, yokofugu, kinchakufugu, gobanfugu (Japan), kappa, plachee (India), zeehaan (Indonesia).

DISTRIBUTION: Indo-Pacific, Japan.

SOURCES: Schlegel (1850), Tahara (1894, 1896, 1897*a*), Taschenberg (1909), Itakura (1917), Yudkin (1944), Fish and Cobb (1954).

OTHER NAMES: *Tetrodon rivulatus, Tetrodon grammatocephalus, Spheroides grammatocephalus, Spheroides rivulatus.*

Family DIODONTIDAE

Chilomycterus affinis Günther (Pl. 1, fig. a). Porcupinefish, spiny puffer, balloonfish, tiger puffer, swelltoad, hairy boxfish, burrfish, globefish, cucumberfish, rabbitfish, spiny boxfish (USA), puerco espino, sorospin, erizo (Mexico), o'opu okala, oopuhue, torabuku (Hawaii), mojannur, japonke, mejangir (Marshall Islands), sou nichukew, soutu (Carolines), derudn (Palau), duto, boteting laut, botiting laot, buteteng laot (Philippines), ishigakifugu (Japan). *178*

Distribution: Southern coast of California, Galapagos Islands, westward to Hawaii and Japan.

Sources: Yamasaki (1914), Suehiro (1947), Halstead and Schall (1955).
Other name: *Chilomycterus californiensis.*

Chilomycterus antennatus (Cuvier) (Pl. 1, fig. b). Bridled burrfish (USA), sapo (Madeira), boiacu (Angola), spiny blaasop (South Africa). *178*

Distribution: Atlantic coast of tropical America, west and South Africa.

Source: Prosvirov (1963).

Chilomycterus atinga (Linnaeus) (Pl. 2, fig. b). Spotted burrfish (USA), atinga (West Indies). *179*

Distribution: Florida Keys, West Indies, Bermuda.

Sources: Coutière (1899), Maass (1937), Bréta (1939).
Other names: *Chilomycterus reticulatus, Diodon atinga, Diodon attinga.*

Chilomycterus orbicularis (Bloch) (Pl. 1, fig. c). Porcupinefish, spiny puffer, balloonfish, tiger puffer, swelltoad, hairy boxfish, burrfish, globefish, cucumberfish, rabbitfish, spiny boxfish (USA), puerco espino, sorospin, erizo (Mexico), o'opu okala, oopuhue, torabuku (Hawaii), mojannur, japonke, mejangir (Marshall Islands), sou nichukew, soutu (Carolines), derudn (Palau), duto, boteting laut, botiting laot, buteteng laot (Philippines), porcupinefish, spiny blaasop, penvisse (South Africa). *178*

Distribution: Indo-Pacific, westward to Cape of Good Hope.

Sources: Pellegrin (1899), Herre (1924), Maass (1937), Bréta (1939).

Chilomycterus tigrinus (Cuvier) (Pl. 3, fig. a). Burrfish (USA), katu peyttheya, mulla peytthai (Sri Lanka). *180*

Distribution: Indian Ocean.

Sources: Meyer-Ahrens (1855), Kobert (1902), Maass (1937).
Other name: *Diodon tigrinus.*

Diodon holacanthus Linnaeus (Pl. 4, fig. a). Balloonfish (USA), atinga (West Indies), puerco espino, sorospin, erizo (Mexico), baiacus de *181* espinhos (Brazil), o'opu okala, oopuhue, torabuku (Hawaii), mojannur, japonke, mejangir (Marshall Islands), sou nichukew, soutu (Carolines), derudn (Palau), duto, boteting laut, botiting laot, buteteng laot (Philippines), porcupinefish (Australia), katu peyttheya, mulla peytthai (Sri Lanka), harisembon (Japan), yu hu (China), porcupinefish, spiny blaasop, penvisse (South Africa), shokei'eiya, meshwaka (Egypt), pez erizo (Spain).

DISTRIBUTION: All warm seas.

SOURCES: Pawlowsky (1927), Maass (1937), Suehiro (1947), Fish and Cobb (1954).

OTHER NAMES: *Diodon holocanthus, Diodon bleekeri, Diodon maculatus, Diodon novemmaculatus, Paradiodon maculatus, Diodon spinosissimus.*

181 *Diodon hystrix* Linnaeus (Pl. 4, fig. b). Porcupinefish (USA), atinga (West Indies), puerco espino, sorospin, erizo (Mexico), baiacus de espinhos (Brazil), o'opu okala, oopuhue, torabuku (Hawaii), mojannur, japonke, mejangir (Marshall Islands), sou nichukew, soutu (Carolines), derudn (Palau), duto, boteting laut, botiting laot, buteteng laot (Philippines), porcupinefish (Australia), katu peyttheya, mullu peytthai (Sri Lanka), yu hu (China), porcupinefish, spiny blaasop, penvisse (South Africa), shokei'eiya, meshwaka (Egypt), pez erizo (Spain).

DISTRIBUTION: All tropical seas.

SOURCES: Coutière (1899), Herre (1924), Maass (1937), Bréta (1939), Fish and Cobb (1954), Halstead and Lively (1954), Helfrich (1961).

OTHER NAMES: *Diodon histrix, Tetrodon hystrix, Diodon punctatus.*

Family MOLIDAE

180 *Mola mola* (Linnaeus) (Pl. 3, fig. b). Ocean sunfish, headfish, king of the mackerels, round-tailed sunfish (USA), kahala, makua (Hawaii), sunfish (Australia), ukigi, manbo, kusabi-fugu (Japan), bahlul (Egypt), lune, poisson lune, lune de mer, mola, mole, motebut, muola (France), pez luna, muela de molina, muela, mula, mola, rodador, atlua, illargiarraia, bezedor (Spain), bezedor, lua, peixe lua, pendao, orelhao, roda, rodim rolim (Portugal), klumpfisch, mondfisch, meermond, schwimmender kopf (Germany).

DISTRIBUTION: Temperate and tropical seas.

SOURCES: Cloquet (1821), Coutière (1899), Herre (1924), Maass (1937), Prosvirov (1963).

OTHER NAMES: *Orthagoriscus mola, Orthagoriscus molla, Tetraodon mola, Tetrodon lunae.*

Family TETRAODONTIDAE

182 *Amblyrhynchotes honckeni* (Bloch) (Pl. 5, fig. a). Puffer, swelltoad, blower, toadfish, swellbelly, swellfish, swellingfish, blowfish, jugfish, rabbitfish

(USA), blazer, tinga tinga, ikan butak, ikan noginogi, ikan buntal, buntal pisang (Indonesia), ho t'um yu, kay-po-oy (China).

Distribution: Indonesia, China, South Africa.

Sources: Coutière (1899), Yamasaki (1914), Yudkin (1944), Smith (1950).
Other names: *Spheroides honckeni, Tetraodon honkenyi, Tetraodon houckenyi, Tetraodon houkenyi, Tetrodon honckeni, Tetrodon honckenyi, Tetrodon honkengi, Tetrodon houckenyi, Aptodactylus punctatus, Aplodactylus punctatus, Geneion maculatum, Gneion maculatum, Spheroides maculatum, Tetraodon maculatus, Tetrodon maculatum, Tetrodon sceleratus.*

Arothron aerostaticus (Jenyns) (Pl. 5, fig. b). Puffer, swelltoad, blower, toadfish, swellbelly, swellfish, swellingfish, blowfish, jugfish, rabbitfish (USA), toadfish, globefish, swellfish, blowfish, puffer (Australia), furube, fougno, fugu (Japan), oopuhue, maki maki, keke, akeke (Hawaii), luap, wat (Marshall Islands), te buni (Gilbert Islands), nipou, wata, nemata, sete, tiwadi (Carolines), botiti, botete, tikung, langiguihon, tinga tinga (Philippines), ikan buntal, buntal pisang (Malaysia), blazer, tinga tinga, ikan butak, ikan noginogi, ikan buntal, buntal pisang (Indonesia), ca noc vàng, ca noc, canoc hot mit (Vietnam), blaasop, tobies, aufblaser, tinga tinga (South Africa), dremma, deremah (Egypt). *182*

Distribution: Indo-Pacific, Australia, Japan, Red Sea, east Africa.

Sources: Yamasaki (1914), Clark and Gohar (1953).
Other names: *Tetraodon aerostaticus, Tetraodon aerostatus, Tetrodon aerostaticus.*

Arothron hispidus (Linnaeus) (Pl. 6, fig. a). Puffer, swelltoad, blower, toadfish, swellbelly, swellfish, swellingfish, blowfish, jugfish, rabbitfish (USA), oopuhue, maki maki, keke, akeke (Hawaii), luap, wat (Marshall Islands), te buni (Gilbert Islands), nipou, wata, nemata, sete, tiwadi (Carolines), botiti, botete, tikung langiguihon, tinga tinga (Philippines), ikan buntal, buntal pisang (Malaysia), blazer, tinga tinga, ikan butak, ikan noginogi, ikan buntal, buntal pisang (Indonesia), ca noc vàng, ca noc, canoc hot mit (Vietnam), blaasop, tobies, aufblaser, tinga tinga (South Africa), dremma, deremah (Egypt), yokoshimafugu (Japan). *183*

Distribution: Panama, Indo-Pacific, Japan, Australia, South Africa, Red Sea.

Sources: Coutière (1899), Phisalix (1922), Yudkin (1945), Dawson, Aleem, and Halstead (1955), Banner and Boroughs (1958), Helfrich (1961, 1963), Bartsch and McFarren (1962).
Other names: *Spheroides hispidus, Tetraodon hispidus, Tetrodon hispidus.*

Arothron meleagris (Lacépède) (Pl. 6, fig. b). Puffer, swelltoad, blower, toadfish, swellbelly, swellfish, swellingfish, blowfish, jugfish, rabbitfish (USA), oopuhue, maki maki, keke, akeke (Hawaii), luap, wat (Marshall *183*

Islands), te buni (Gilbert Islands), nipou, wata, nemata, sete, tiwadi (Carolines), botiti, botete, tikung, langiguihon, tinga tinga (Philippines), ikan buntal, buntal pisang (Malaysia), blazer, tinga tinga, ikan butak, ikan noginogi, ikan buntal, buntal pisang (Indonesia), ca noc vàng, ca noc, canoc hot mit (Vietnam), mizorefugu (Japan), tambor, botete (Mexico, Central America).

DISTRIBUTION: West coast of Central America and throughout the Indo-Pacific.

SOURCES: Yamasaki (1914), Hiyama (1943), Fish and Cobb (1954).
OTHER NAMES: *Tetraodon lacrymatus, Tetraodon meleagris, Tetrodon lacrymatus, Tetrodon meleagris.*

183 *Arothron nigropunctatus* (Bloch and Schneider) (Pl. 6, fig. c). Puffer, swelltoad, blower, toadfish, swellbelly, swellfish, swellingfish, blowfish, jugfish, rabbitfish (USA), oopuhue, maki maki, keke, akeke (Hawaii), luap, wat (Marshall Islands), te buni (Gilbert Islands), nipou, wata, nemata, sete, tiwadi (Carolines), botiti, botete, tikung, langiguihon, tinga tinga (Philippines), ikan buntal, buntal pisang (Malaysia), blazer, tinga tinga, ikan butak, ikan noginogi, ikan buntal, buntal pisang (Indonesia), ca noc vàng, ca noc, canoc hot mit (Vietnam), blaasop, tobies, aufblaser, tinga tinga (South Africa), kokutenfugu (Japan), toadfish, globefish, swellfish, blowfish, puffer (Australia), daghemeiah (Egypt).

DISTRIBUTION: Tropical Indo-Pacific, Japan, Australia, east coast of Africa, Red Sea.

SOURCES: Herre (1924), Hiyama (1943), Whitley (1953), Flaschentrager and Abdalla (1957), Halstead (1959).
OTHER NAMES: *Arothron diadematus, Ovoides nigropunctatus, Tetraodon nigropunctatus, Tetrodon nigropunctatus, Amblyrhynchotes diadematus.*

184 *Arothron reticularis* (Bloch and Schneider) (Pl. 7, fig. a). Puffer, swelltoad, blower, toadfish, swellbelly, swellfish, swellingfish, blowfish, jugfish, rabbitfish (USA), oopuhue, maki maki, keke, akeke (Hawaii), luap, wat (Marshall Islands), te buni (Gilbert Islands), nipou, wata, nemata, sete, tiwadi (Carolines), botiti, botete, tikung, langiguihon, tinga tinga (Philippines), ikan buntal, buntal pisang (Malaysia), blazer, tinga tinga, ikan butak, ikan noginogi, ikan buntal, buntal pisang (Indonesia), ca noc vàng, ca noc, canoc hot mit (Vietnam), blaasop, tobies, aufblaser, tinga tinga (South Africa), patoca, bengfula, kappa, kuddul mahcutchee, kadal-mahcutchee, plachee, pulli-plachee, nanju-pethai, nanju-meen, peyah, sootha, thonthe, khachaka, kend (India), pipa-machchi (Burma), toadfish, globefish, swellfish, blowfish, puffer (Australia).

DISTRIBUTION: Indo-Pacific, Australia, India.

SOURCES: Day (1878-88), Seale (1912), Herre (1924), Hiyama (1943).
OTHER NAMES: *Tetraodon reticularis, Tetrodon reticularis.*

Arothron setosus (Smith) (Pl. 5, fig. c). Puffer, swelltoad, blower, toadfish, swellbelly, swellfish, swellingfish, blowfish, jugfish, rabbitfish (USA), oopuhue, maki maki, keke, akeke (Hawaii), luap, wat (Marshall Islands), te buni (Gilbert Islands), nipou, wata, nemata, sete, tiwadi (Carolines), botiti, botete, tikung, langiguihon, tinga tinga (Philippines), ikan buntal, buntal pisang (Malaysia), blazer, tinga tinga, ikan butak, ikan noginogi, ikan buntal, buntal pisang (Indonesia), ca noc vàng, ca noc, canoc hot mit (Vietnam), tambor, botete (Mexico, Central America). *182*

Distribution: West coast of Mexico, Galapagos and Cocos Islands.

Sources: Halstead and Schall (1955, 1956).

Arothron stellatus (Bloch and Schneider) (Pl. 8). Puffer, swelltoad, blower, toadfish, swellbelly, swellfish, swellingfish, blowfish, jugfish, rabbitfish (USA), oopuhue, maki maki, keke, akeke (Hawaii), luap, wat (Marshall Islands), te buni (Gilbert Islands), nipou, wata, nemata, sete, tiwadi (Carolines), botiti, botete, tikung, langiguihon, tinga tinga (Philippines), ikan buntal, buntal pisang (Malaysia), blazer, tinga tinga, ikan butak, ikan noginogi, ikan buntal, buntal pisang (Indonesia), ca noc vàng, ca noc, canoc hot mit (Vietnam), shiroamifugu (Japan), blaasop, tobies, aufblaser, tinga tinga (South Africa), fahaka (Egypt). *185*

Distribution: Tropical Indo-Pacific, Japan, Australia, Red Sea, and east coast of Africa.

Sources: Day (1878-88), Buddle (1930), Bonde (1953).

Boesemanichthys firmamentum (Temminck and Schlegel) (Pl. 7, fig. b). Puffer, swelltoad, blower, toadfish, swellbelly, swellfish, swellingfish, blowfish, jugfish, rabbitfish (USA), toad, toadfish, globefish, swellfish, blowfish, puffer (Australia), hoshifugu (Japan). *184*

Distribution: Australia, New Zealand, Japan.

Sources: Rémy (1883), Coutière (1899), Phisalix (1922).
Other names: *Tetraodon firmamentum, Tetrodon firmamentum, Tetrodon firmamentus, Tetrodon gillbanksi.*

Chelonodon patoca (Hamilton-Buchanan) (Pl. 9, fig. a). Puffer, swelltoad, blower, toadfish, swellbelly, swellfish, swellingfish, blowfish, jugfish, rabbitfish (USA), oopuhue, maki maki, keke, akeke (Hawaii), luap, wat (Marshall Islands), te buni (Gilbert Islands), nipou, wata, nemata, sete, tiwadi (Carolines), botiti, botete, tikung, langiguihon, tinga tinga (Philippines), ikan buntal, buntal pisang (Malaysia), blazer, tinga tinga, ikan butak, ikan noginogi, ikan buntal, buntal pisang (Indonesia), ca noc vàng, ca noc, canoc hot mit (Vietnam), blaasop, tobies, aufblaser, tinga tinga (South Africa), pipa-machchi (Burma), patoca, bengfula, kappa, kuddul mahcutchee, kadal-mahcutchee, plachee, pulli-plachee, nanju-pethai, nanju-meen, peyah, sootha, thonthe, khachaka, kend (India). *186*

DISTRIBUTION: Indo-Pacific, Australia, China, India, southeast Africa.

SOURCES: Seale (1912), Moses (1922), Jones (1956).
OTHER NAMES: *Tetraodon patoca, Tetrodon patoca.*

186 *Ephippion guttifer* (Bennett) (Pl. 9, fig. b). Puffer (USA), odjonekpountane, kookoe, adanjame, awulenj, awule, ewude, okpo, eku (west Africa), tamboril (Spain).

DISTRIBUTION: West Africa, Canary Islands, Portugal, Spain, Mediterranean Sea.

SOURCES: Coutière (1899), Irvine (1947), Prosvirov (1963).
OTHER NAMES: *Gneion maculatum, Hemiconiatus gutifer, Tetrodon maculatum.*

186 *Fugu basilevskianus* (Basilewsky) (Pl. 9, fig. c). Puffer (USA), ho t'um yu, kay-po-oy (China), bok oh (Korea), koraifugu (Japan).

DISTRIBUTION: North China, northwest Korea.

SOURCES: Read (1939), Tani (1945), Suehiro (1947).
OTHER NAMES: *Sphaeroides basilevskianus, Sphaeroides basilewskianus, Spheroides basilewskianus, Sphoeroides basileurskianus, Sphoeroides basilewskianus.*

184 *Fugu chrysops* (Hilgendorf) (Pl. 7, fig. c). Puffer (USA), akemefugu (Japan).

DISTRIBUTION: Pacific coast of Japan.

SOURCES: Takahashi and Inoko (1892), Tanaka (1914), Tani (1945), Suehiro (1947), Fish and Cobb (1954).
OTHER NAMES: *Sphaeroides chrysops, Spheroides chrysops, Sphoeroides chrysops, Tetraodon chrysops, Tetrodon chrysops.*

187 *Fugu niphobles* (Jordan and Snyder) (Pl. 10, fig. a). Puffer (USA), kusafugu (Japan), ho t'um yu, kay-po-oy (China), botiti, botete, tikung, langiguihon, tinga tinga (Philippines).

DISTRIBUTION: Japan, China, Philippines.

SOURCES: Vincent (1889), Migita and Hashimoto (1951), Suyama and Uno (1957).
OTHER NAMES: *Sphaeroides niphobles, Spheroides niphobles, Sphoeroides niphobles, Tetrodon niphobles, Tetraodon nivaeus.*

188 *Fugu oblongus* (Bloch and Schneider) (Pl. 11, fig. a). Puffer, swelltoad, blower, toadfish, swellbelly, swellfish, swellingfish, blowfish, jugfish, rabbitfish (USA), oopuhue, maki maki, akeke (Hawaii), luap, wat (Marshall Islands), te buni (Gilbert Islands), nipou, wata, nemata, sete, tiwadi (Carolines), botiti, botete, tikung, langiguihon, tinga tinga (Philippines), ikan buntal, buntal pisang (Malaysia), blazer, tinga tinga, ikan butak, ikan noginogi, ikan buntal, buntal pisang (Indonesia), ca noc vàng, ca noc, canoc hot mit (Vietnam), blaasop, tobies, aufblaser, tinga tinga (South Africa), ho t'um yu, kay-po-oy (China), fugu (Japan), toadfish, globefish, swellfish, blowfish, puffer (Australia).

DISTRIBUTION: Indo-Pacific, southeast Africa, India, south China, Australia.

SOURCES: Day (1878-88), Pellegrin (1899), Yudkin (1944), Fish and Cobb (1954).
OTHER NAMES: *Tetraodon oblongus, Lagocephalus oblongus, Spheroides oblongus.*

Fugu ocellatus obscurus (Abe) (Pl. 11, fig. b). Puffer (USA), mefugu (Japan), ho t'um yu, kay-po-oy (China), bok oh (Korea), *188*

DISTRIBUTION: Rivers and seas of central and north China, west Korea; east China Sea and adjoining waters.

SOURCES: Autenrieth (1833), Coutière (1899), Suehiro (1947).
OTHER NAMES: *Diodon ocellatus, Spheroides ocellatus, Sphoeroides ocellatus, Tetraodon occelatus, Tetraodon ocellatus, Tetrodon ocellatus.*

Fugu ocellatus ocellatus (Linnaeus) (Pl. 11, fig. c). Puffer (USA), meganefugu (Japan), ho t'um yu, kap-po-oy (China), botiti, botete, tikung, langiguihon, tinga tinga (Philippines). *188*

DISTRIBUTION: China, Japan, Philippines.

SOURCES: Autenrieth (1833), Coutière (1899), Phisalix (1922), Tani (1945), Yudkin (1944), Suehiro (1947), Fish and Cobb (1954).
OTHER NAMES: *Diodon ocellatus, Spheroides ocellatus, Tetraodon occelatus, Tetraodon ocellatus, Tetrodon ocellatus.*

Fugu pardalis (Temminck and Schlegel) (Pl. 10, fig. b). Puffer (USA), higanfugu (Japan), ho t'um yu, kay-po-oy (China), bok oh (Korea). *187*

DISTRIBUTION: China, Japan.

SOURCES: Rémy (1883), Tanaka (1914), Suehiro (1947), Fish and Cobb (1954).
OTHER NAMES: *Higan fugu pardales, Sphaeroides pardalis, Spheroides pardalis, Sphoeroides pardalis, Tetraodon pardalis, Tetrodon pardalis, Tetrodon perdalis.*

Fugu poecilonotus (Temminck and Schlegel) (Pl. 11, fig. d). Puffer, swelltoad, blower, toadfish, swellbelly, swellfish, swellingfish, blowfish, jugfish, rabbitfish (USA), oopuhue, maki maki, keke, akeke (Hawaii), luap, wat (Marshall Islands), te buni (Gilbert Islands), nipou, wata, nemata, sete, tiwadi (Carolines), botiti, botete, tikung, langiguihon, tinga tinga (Philippines), ikan buntal, buntal pisang (Malaysia), blazer, tinga tinga, ikan butak, ikan noginogi, ikan buntal, buntal pisang (Indonesia), ca noc vàng, ca noc, canoc hot mit (Vietnam), blaasop, tobies, aufblaser, tinga tinga (South Africa), komonfugu (Japan), ho t'um yu, kay-po-oy (China), bok oh (Korea). *188*

DISTRIBUTION: Indo-Pacific, China, Korea, Japan.

SOURCES: Takahashi and Inoko (1892), Tani (1945), Suehiro (1947), Fish and Cobb (1954).
OTHER NAMES: *Sphaeroides alboplumbeus, Spheroides alboplumbeus, Sphoeroides alboplumbeus, Tetrodon alboplumbeus, Tetrodon poicilonotus, Tetrodon poecilonotus.*

189 *Fugu rubripes chinensis* (Abe) (Pl. 12, fig. a). Puffer (USA), fugu (Japan), ho t'um yu, kay-po-oy (China), bok oh (Korea).
DISTRIBUTION: East China Sea and adjoining waters.

SOURCES: Halstead and Bunker (1953), Halstead and Ralls (1954).
OTHER NAMES: *Sphoeroides rubripes chinensis, Tetrodon rubripes.*

189 *Fugu rubripes rubripes* (Temminck and Schlegel) (Pl. 12,. fig. b). Puffer (USA), torafugu (Japan), ho t'um yu, kay-po-oy (China), bok oh (Korea).
DISTRIBUTION: Japan, China, Korea.

SOURCES: Takahashi and Inoko (1892), Itakura (1917), Yudkin (1944), Suehiro (1947), Fish and Cobb (1954).
OTHER NAMES: *Sphaeroides rubripes, Spheroides rubripes, Sphoeroides rubripes, Tetrodon rubripes.*

189 *Fugu stictonotus* (Temminck and Schlegel) (Pl. 12, fig. c). Puffer (USA), gomafugu (Japan), ho t'um yu, kay-po-oy (China), bok oh (Korea).
DISTRIBUTION: Southern Korea, east China Sea and adjoining waters, Japan.

SOURCES: Rémy (1883), Yamasaki (1914), Tani (1945), Yudkin (1944), Suehiro (1947).
OTHER NAMES: *Sphaeroides stictonotus, Sphaeroides sticutonotus, Spheroides stochonatus, Spheroides stictonotus, Sphoeroides stictonotus, Tetrodon stictonotus, Tetrodon strictonotus.*

190 *Fugu vermicularis porphyreus* (Temminck and Schlegel) (Pl. 13, fig. a). Puffer (USA), mafugu (Japan), ho t'um yu, kay-po-oy (China), bok oh (Korea).
DISTRIBUTION: East China Sea and adjoining waters, Japan.

SOURCES: Rémy (1883), Itakura (1917), Tani (1945), Migita and Hashimoto (1951), Halstead and Bunker (1953).
OTHER NAMES: *Sphaeroides borealis, Sphaeroides porphyreus, Spheroides porphyreus, Sphoeroides porphyreus, Sphoeroides vermicularis porphyreus, Tetraodon porphyreus, Tetrodon porphyreus.*

190 *Fugu vermicularis radiatus* (Abe) (Pl. 13, fig. b). Puffer (USA), fugu (Japan), ho t'um yu, kay-po-oy (China), bok oh (Korea).
DISTRIBUTION: East China Sea and adjoining waters.

SOURCES: Hashimoto and Migita (1951), Halstead and Bunker (1953), Murtha, Stabile, and Wills (1958).

Other names: *Sphaeroides vermicularis radiatus, Spheroides radiatus, Sphoeroides vermicularis radiatus.*

Fugu vermicularis vermicularis (Temminck and Schlegel) (Pl. 13, fig. c). Puffer (USA), shosaifugu (Japan), ho t'um yu, kay-po-oy (China), bok oh (Korea). *190*

Distribution: East China Sea, Japan.

Sources: Rémy (1883), Kobert (1902), Maass (1937), Tani (1945), Halstead and Bunker (1953).

Other names: *Sphaeroides abotti, Spheroides vermicularis, Sphoeroides vermicularis, Sphoeroides vermicularis vermicularis, Tetraodon vermicularis, Tetraodon virmicularis, Tetrodon vermicularis.*

Fugu xanthopterus (Temminck and Schlegel) (Pl. 12, fig. d). Puffer (USA), shimafugu (Japan), ho t'um yu, kay-po-oy (China), bok oh (Korea). *189*

Distribution: China, Korea, southern Japan.

Sources: Rémy (1883), Coutière (1899), Read (1939), Tani (1945).

Other names: *Sphaeroides xanthopterus, Sphoeroides xanthopterus, Tetrodon xanthoperus, Tetrodon xanthopterus.*

Lagocephalus laevigatus inermis (Temminck and Schlegel) (Pl. 14, fig. a). Puffer (USA), toadfish, globefish, swellfish, blowfish, puffer (Australia), kanafugu (Japan), ho t'um yu, kay-po-oy (China), patoca, bengfula, kappa, kuddul mahcutchee, kadal-mahcutchee, plachee, pulli-plachee, nanju-pethai, nanju-meen, peyah, sootha, thonthe, khachaka, kend (India). *191*

Distribution: Australia, Japan, China Sea, Indian Ocean, east coast of Africa.

Sources: Coutière (1899), Tani (1945), Suehiro (1947), Migita and Hashimoto (1951).

Other names: *Lagocephalus laevigatus, Spheroides inermis, Tetraodon inermis, Tetrodon inermis, Tetrodon laevigatus.*

Lagocephalus laevigatus laevigatus (Linnaeus) (Pl. 14, fig. b). Smooth puffer (USA), baiacú, mamaiacú (Brazil), pompus, toade, tambor, tamboril (West Indies), kookoe, adanjme, awulenj, awule, ewude, okpo, eku, (West Africa). *191*

Distribution: Tropical Atlantic Ocean, Gulf of Mexico.

Sources: Pellegrin (1899), Bréta (1939), Yudkin (1944), Larson, Lalone, and Rivas (1960), Prosvirov (1963).

Other names: *Lagocephalus laevigatus, Lagocephalus levigatus, Lagocephalus loevigatus, Lagocephalus pachycephalus, Tetrodon laevigatus.*

Lagocephalus lagocephalus (Linnaeus) (Pl. 14, fig. c). Oceanic puffer (USA), baiacú, mamaiacú (Brazil), pompus, toade, tambor, tamboril (West Indies), kookoe, adanjme, awulenj, awule, ewude, okpo, eku (West Africa). *191*

DISTRIBUTION: Tropical Atlantic Ocean; Mediterranean Sea.

SOURCES: Phisalix (1922), Charnot (1945).
OTHER NAMES: *Lagocephalus lagocephalus, Spheroides lagocephalus, Tetrodon lagocephalus.*

192 *Lagocephalus lunaris* (Bloch and Schneider) (Pl. 15, fig. a). Puffer (USA), oopuhue, maki maki, keke, akeke (Hawaii), luap, wat (Marshall Islands), te buni (Gilbert Islands), nipou, wata, nemata, sete, tiwadi (Carolines), patoca, bengfula, kappa, kuddul mahcutchee, kadal-mahcutchee, plachee, pulli-plachee, nanju-pethai, nanju-meen, peyah, sootha, thonthe, khachaka, kend (India), qarrad (Red Sea), blaasop, tobies, aufblazer, tinga tinga (South Africa), toadfish, globefish, swellfish, blowfish, puffer (Australia), hơ t'um yu, kay-po-oy (China), sabafugu (Japan).

DISTRIBUTION: Indo-Pacific, India, Red Sea, south and east coast of Africa, Australia, China, Japan.

SOURCES: Rémy (1883), Yamasaki (1914), Migita and Hashimoto (1951), Whitley (1953), Kudaka (1960).
OTHER NAMES: *Pleuranacanthus lunaris, Sphaeroides lunaris, Spheroides lunaries, Spheroides lunaris, Tetrodon lunaris, Gastrophysus spadiceus, Sphaeroides spadiceus, Spheroides spadiceus, Tetrodon spadiceus, Lagocephalus spadiceus.*

192 *Lagocephalus oceanicus* Jordan and Evermann (Pl. 15, fig. b). This is believed by some workers to be a geographic variety of *Lagocephalus lagocephalus* and therefore might be termed *L. lagocephalus oceanicus* (Tyler, 1965). Puffer, swelltoad, blower, toadfish, swellbelly, swellfish, swellingfish, blowfish, jugfish, rabbitfish (USA), oopuhue, maki maki, keke, akeke (Hawaii), luap, wat (Marshall Islands), te buni (Gilbert Islands), nipou, wata, nemata, sete, tiwadi (Carolines), botiti, botete, tikung, langiguihon, tinga tinga (Philippines), ikan buntal, buntal pisang (Malaysia), blazer, tinga tinga, ikan butak, ikan noginogi, ikan buntal, buntal pisang (Indonesia), ca noc vàng, ca noc, canoc hot mit (Vietnam), blaasop, tobies, aufblaser, tinga tinga (South Africa), patoca, bengfula, kappa, kuddul mahcutchee, kadal-mahcutchee, plachee, pulli-plachee, nanju-pethai, nanju-meen, peyah, sootha, thonthe, khachaka, kend (India), ho t'um yu, kay-po-oy (China), kumasakafugu (Japan).

DISTRIBUTION: Tropical Pacific Ocean, Japan, South Africa.

SOURCES: Autenrieth (1833), Mazumdar (1915), Bonde (1953).
OTHER NAMES: *Lagocephalus lagocephalus, Lagocephalus lagocephalus oceanicus, Spheroides lagocephalus, Tetraodon lagocephalon, Tetrodon lagocephalus, Tetrodon pennanti.*

192 *Lagocephalus sceleratus* (Forster) (Pl. 15, fig. c). Puffer, swelltoad, blower, toadfish, swellbelly, swellfish, swellingfish, blowfish, jugfish, rabbitfish (USA), oopuhue, maki maki, keke, akeke (Hawaii), luap, wat (Marshall

Islands), te buni (Gilbert Islands), nipou, wata, nemata, sete, tiwadi (Carolines), botiti, botete, tikung, langiguihon, tinga tinga (Philippines), ikan buntal, buntal pisang (Malaysia), blazer, tinga tinga, ikan butak, ikan noginogi, ikan buntal, buntal pisang (Indonesia), ca noc vàng, ca noc, canoc hot mit (Vietnam), blaasop, tobies, aufblaser, tinga tinga (South Africa), patoca, bengfula, kappa, kuddul mahcutchee, kadal-machutchee, plachee, pulli-plachee, manju-pethai, nanju-meen, peyah, sootha, thonthe, khachaka, kend (India), senninfu (Japan).

Distribution: Indo-Pacific, southern Japan, Australia, east Africa.

Sources: Anderson (1776), Furtado (1903), Whitley (1953), Flaschentrager and Abdalla (1957), Kudaka (1960).
Other names: *Gastrophysus sceleratus, Lagocephalus scleratus, Pleuranacanthus sceleratus, Sphoeroides sceleratus, Spheroides sceleratus, Tetraodon argenteus, Tetrodon argentens, Tetrodon sceleratum, Tetrodon sceleratus, Tetrodon scelectarus, Tetrodon bicolor.*

Sphaeroides annulatus (Jenyns) (Pl. 16, fig. a). Puffer (USA), tambor, botete (Mexico, Central America). *193*

Distribution: California to Peru, Galapagos Islands.

Sources: Pellegrin (1899), Phisalix (1922), Gomez (1926), Goe and Halstead (1953), Lalone, DeVillez, and Larson (1963).
Other names: *Cheilichthys heraldi, Spheroides annulatus, Spheroides heraldi, Spheroides annulatus, Tetrodon geometricus, Tetrodon heraldi, Tetrodon herraldi.*

Sphaeroides maculatus (Bloch and Schneider) (Pl. 16, fig. b). Puffer (USA), pompus, toade, tambor, tamboril (West Indies). *193*

Distribution: Atlantic coast of United States to Guiana.

Sources: Taschenberg (1909), Yudkin (1944, 1945), Larson *et al.* (1960), Lalone *et al.* (1963).
Other names: *Spheroides maculatum, Spheroides maculatus, Sphoeroides maculatus, Tetrodon maculatus.*

Sphaeroides spengleri (Bloch) (Pl. 17, fig. a). Bandtail puffer (USA), pompus, toade, tambor, tamboril (West Indies), baiacú, mamaiacú (Brazil), tambor, botete (Mexico), kookoe, adanjme, awulenj, awule, ewude, okpo, eku (west Africa). *194*

Distribution: Tropical Atlantic Ocean, Gulf of Mexico, west coast of Africa, West Indies.

Sources: Coutière (1899), Gomez (1926), Mucciolo (1956), Prosvirov (1963).
Other names: *Leisomus marmoratus, Spheroides marmoratus, Spheroides spengleri, Sphoeroides marmoratus, Tetrodon marmoratus, Tetrodon spengler.*

Sphaeroides testudineus (Linnaeus) (Pl. 17, fig. b). Puffer (USA), pompus, toade, tambor, tamboril (West Indies), baiacú, mamaiacú (Brazil). *194*

DISTRIBUTION: Atlantic coast of United States, West Indies, Brazil.

SOURCES: Bréta (1939), Yudkin (1944), Larson *et al.* (1960).
OTHER NAMES: *Spheroides testudineus, Tetraodon bajacu, Tetraodon punctuatus, Tetrodon punctatus, Tetrodon testudinis, Tetrodon testudineus.*

194 *Tetraodon lineatus* Linnaeus (Pl. 17, fig. c). Puffer (USA), kookoe, adanjme, awulenj, awule, ewude, okpo, eku (west Africa), fahaka (Egypt).
DISTRIBUTION: Rivers of northern and western Africa.

SOURCES: Autenrieth (1833), Maass (1937), Yudkin (1944), Fish and Cobb (1954).
OTHER NAMES: *Tetrodon lineatus, Tetrodon physa, Tetrodon fahaka, Spheroides lineatus, Spheroides fahaka.*

194 *Torquigener hamiltoni* (Gray and Richardson) (Pl. 17, fig. d). Puffer, swelltoad, blower, toadfish, swellbelly, swellfish, swellingfish, blowfish, jugfish, rabbitfish (USA), oopuhue, maki maki, keke, akeke (Hawaii), luap, wat (Marshall Islands), botiti, botete, tikung, langiguihon, tinga tinga (Philippines), te buni (Gilbert Islands), nipou, wata, nemata, sete, tiwadi (Carolines), ikan buntal, buntal pisang (Malaysia), blazer, tinga tinga, ikan butak, ikan noginogi, ikan buntal, buntal pisang (Indonesia), ca noc vàng, ca noc, canoc hot mit (Vietnam), blaasop, tobies, aufblaser, tinga tinga (South Africa), toadfish, globefish, swellfish, blowfish, puffer (Australia).
DISTRIBUTION: Indo-Pacific, Australia.

SOURCES: Pellegrin (1899), Phisalix (1922), Whitley (1953), Fish and Cobb (1954).
OTHER NAMES: *Sphaeroides hamiltoni, Spheroides hamiltoni, Tetraodon hamiltoni, Tetrodon hamiltoni.*

BIOLOGY

Family CANTHIGASTERIDAE: The sharp-nosed puffers are small brilliantly colored marine fishes which abound in coral reefs and shoal areas. *Canthigaster amboinensis* is said to attain a length of 25 cm or more, but most of the species attain a maximum length of less than 13 cm. These attractive little fishes generally travel singly or in pairs, hovering almost motionless for a time and then suddenly darting in and out among the corals with a speed that is unusual for most puffers. When frightened or injured, canthigasterids have the typical pufferlike habit of inflating themselves with water and grating their teeth in a noisy manner.

Family DIODONTIDAE: The porcupinefishes are spiny marine fishes which are widely distributed in all warm seas. *Diodon hystrix* is said to attain a total length of 91 cm or more, but most diodontids range from 20 to 50 cm in length. Porcupinefishes inhabit coral reefs and shoal areas, generally traveling singly or in pairs. They also have the ability to inflate themselves until they are almost spherical in outline. During the inflation process the spines which are usually depressed are extended, giving the fish a formidable appearance. Porcupinefishes have been known to kill larger carnivorous fishes by inflating themselves and becoming stuck in the throat of the would-be captor. Diodontids are also capable of giving a severe bite. Darwin (1945) stated that diodons have been found alive in the stomachs of sharks, and in some instances have mortally wounded their captors by gnawing through the stomach walls and sides of the sharks and thereby escaping.

Family MOLIDAE: The Molidae, the largest of the tetraodontoid fishes, are members of the genus *Mola*. They attain a length of more than 3 m and a weight in excess of 3½ tons. However, *Ranzania*, another genus of this family, rarely exceeds a length of 61 cm. Sunfishes inhabit open seas and are generally seen singly or in pairs but seem to become gregarious at certain seasons when they band together in small schools which may consist of as many as a dozen individuals. The name "headfishes" is sometimes given to these grotesque creatures because they have the appearance of being composed of an enormous head to which small fins are attached. When swimming they are said to progress with a waving motion from side to side with the dorsal fin projecting from the water. The smaller specimens of *Mola mola* can move with surprising rapidity and at times have been observed jumping several feet clear of the surface of the water. During calm weather these fishes may be observed basking in the sun at the surface of the sea, lying more or less on their sides. Large basking specimens are generally slow moving, unable to move faster than a boat can row, and are easily overtaken and harpooned. They are said to emit a grunting or groaning noise when captured. Most of the basking specimens are believed to be sick and dying as a result of their heavy parasitic infestation. Young headfish are alert, active, and have occasionally been observed leaping out of water. Molas have been found to contain fishes in their stomachs which are known to live at depths of at least 100 fathoms; hence it is believed that at times molas may descend to considerable depths. According to Fraser-Brunner (1951), the inshore migrations of molas coincide with the invasions of medusae, salps, and ctenophores upon which they largely feed. However, specimens of molas taken inshore have usually been found to be feeding on littoral forms of crustacea, ophiuroids, mollusks, hydroids, ctenophores, corallines, and algae. *Mola mola* have been captured on hook and line with live anchovy bait in California waters. Relatively little is known about their breeding habits.

Family TETRAODONTIDAE: Some authors divide the puffer family Tetraodontidae into four separate families—Chonerhinidae, Colomesidae, Lagocephalidae, and Tetraodontidae. However, more recent studies suggest that these puffers should be combined into the single family Tetraodontidae. Tetraodontids inhabit a variety of ecosystems, marine, estuarine, and fresh water. Most lagocephalids range in size from about 25 to 50 cm, but *Lagocephalus lagocephalus* is said to attain a total length of 61 cm. Some of the other puffers attain even greater size. *Arothron stellatus* has been reported to reach a total length of 91 cm, but most tetraodontids range from about 20 to 40 cm.

Most puffers are considered to be shallow water fishes. However, it should be pointed out that some members inhabit relatively deep waters. *L. oceanicus* has been taken at 40 m or more, *L. lagocephalus inermis* at 90 m, *L. sceleratus* at 60 m, and *Sphaeroides oblongus* at 100 m. *Liosaccus cutaneus* is usually conceded to be a deep water species and *Boesmanichthys firmamentum* has been taken at a depth of 100 fathoms. Species of *Lagocephalus* have been taken hundreds of km from shore at depths of 4,000 fathoms, and are a common constituent in the stomachs of pelagic fishes such as tunas, wahoos, and other scombroids. Whether these deep water puffers are nontoxic or the poison is not transvectored by scombroids is not known. The Indo-Pacific tetraodontids living around coral reef areas tend to travel singly or in small groups. *Sphaeroides* species tend to be more gregarious and are sometimes observed in large groups, but apparently they do not school in the manner that some fishes do. Stomach analyses conducted in laboratories on *A. hispidus* indicate that this species is omnivorous in its eating habits, since fragments of corals, sponges, algae, mollusks, and fish are commonly found in their stomachs. At Tagus Cove, Isabela Island, during the Krieger Galapagos expedition (Halstead and Schall, 1955), it was observed that hundreds of *Sphaeroides annulatus* were attracted to the surface of the water with a night light. Specimens could then be readily captured by spear or dip net. When puffers are at rest they appear to hover almost motionless with only their pectorals fanning the water. The pectoral, dorsal, and anal fins are the chief locomotory organs, the tail being used principally as a rudder. Despite their reputation of being slow-moving fishes, puffers are capable of moving with surprising rapidity when they are frightened. They are likewise capable of inflicting severe bites. According to Gimlette (1923) and others, the human male genital organs are frequent targets of their attacks. They are quite vicious and readily snap at almost any bait offered them.

The inflating mechanism of puffers has been a subject of interest among anatomists for many years. The research that has been conducted on this interesting mechanism has been well presented by Breder and Clark (1947), to whom we are indebted for the following information. Inflation is employed only as a defense mechanism. If a puffer is frightened or annoyed it will gulp down its fluid medium, thus causing inflation. There is no evidence that puffers come to the surface in order to inflate themselves with air as is commonly believed. The inflating mechanism consists of the powerful muscles of the first branchiostegal ray, which depress a pad covering the ceratohyals, thus expanding the mouth cavity and drawing in water or air if the fish is out of water. The elevation of the ceratohyals forces the fluid into the saclike ventral diverticulum of the stomach from which it is partially separated by a sphincterlike ring. Fluid is retained in the diverticulum by a strong esophageal sphincter and the pylorus, and not by the flaplike breathing valves which are present in the mouth. The opercular valves prevent leakage during compression, but the sac can remain distended when the valves are held open or removed. The water or air in the sac is released by relaxation of the esophageal sphincter, permitting escape through the oral or opercular openings. Puffers make considerable noise during inflation by grinding their heavy jaw teeth together. Some species of puffers are covered with short prickly bristles which they appear to have the ability to withdraw or extend at will. Puffers have a distinctive offensive odor which is particularly noticeable when they are being dressed or dissected.

The genus *Colomesus* is comprised of a single species which inhabits the rivers of northern South America and the West Indies. Specimens have been taken from the mouth of the Amazon to the fringe of the Andes Mountains —a distance of more than 3,000 miles from the nearest salt water. The colomesids likewise have the ability to inflate themselves with water or air. Judging from the literature, relatively little is known regarding its habits.

The *Xenopterus* is a small genus of freshwater fishes which, with the possible exception of a single African species, is restricted to the rivers of southern Asia and Indonesia. These fishes are said to attain a total length of about 28 cm. According to Day (1878-88), xenopterids are very pugnacious and capable of inflicting serious wounds with their sharp beaklike jaws. Burmese natives claim that if a person should fall into the water where these fishes abound, they will attack in droves and kill the victim by their bites. Apparently little else has been reported on the habits of this interesting group.

BIOGENESIS OF TETRODOTOXIN

The origin of the poison in tetraodontiform fishes is still a question. Although the toxicity of tetraodontiform fishes appears to be more consistent than in most other types of poisonous fishes, it is nevertheless subject to considerable fluctuation. Toxic puffers are found in both salt and fresh water, in temperate and torrid latitudes. One puffer specimen of a given species in one locality may be poisonous whereas another specimen of the same species in another locality may be nontoxic. Moreover, the amount of poison varies somewhat from one specimen to the next of the same species in the same locality. Although it is recognized that the toxicity of puffers is influenced greatly by their reproductive cycle, the preceding observations have led some workers to believe that the food habits of the fish may also be a factor in the uptake or production of the poison. Therefore much of the material presented in the section on the "Biogenesis of Ciguatoxin," p. 354, may also be applicable to the study of the biogenesis of tetrodotoxin. There is also the possibility that puffers may serve as transvectors of ciguatoxin. Some of the previous reports regarding the origin of oral fish poisons that appear to be particularly pertinent to tetraodontiform fishes are therefore briefly mentioned in this section.

Forster (1778) attributed the toxicity of pargos and puffers to jellyfish (or plants) upon which he believed these fishes fed. Chisholm (1808) reviewed the popular theories of his period, in which fishermen suggested that puffers and other toxic fishes became poisonous from eating green sea moss (*Corallina opuntia*), medusae, holothurians, or from feeding on copper banks. Moreau de Jonnès (1819, 1821) was apparently the first to investigate these theories by conducting feeding experiments and by studying the habitat of poisonous fishes. His results showed that the cause was neither manchineel berries, small crustaceans, physalids, nor the proximity of copper banks. He concluded that the poison developed as a pathological state in certain fishes and represents a morbid alteration, which was not to be confused with putridity after death. In spite of these experiments, Günther (1880) stated quite boldly that

> the fishes, the flesh of which appears always to have poisonous properties, are *Clupea thrissa, Clupea venenosa*, and some species of *Scarus, Tetrodon*, and *Diodon* . . . . all or nearly all these fishes acquire their poisonous properties from their food which consists of medusae, corals, or decomposing substances.

Pellegrin (1899) pointed out that fishes may become poisonous from putrefaction, or that during spawning season "physiological alkaloids" are secreted in the viscera of tetraodontiform fishes.

Fonseca (1917) stated that the poisonous character of tetraodontids has been attributed to the ingestion of coelenterates, the presence of parasites, or an infectious disease that is sometimes prevalent in fishes. In ancient Egypt it was believed that the overflow of the Nile River influenced the toxicity of the puffer.

Pawlowsky (1927) wrote that "puffer poison is not a decomposition product, but a normal component of the living fish." Kariya (1914) is quoted as saying that "puffer poison is a hormone of the sex glands." In his discussion on puffers and other poisonous fishes, Larsen (1942) reported that Hawaiian puffers eat crabs and the shellfish *Rocellaria gigantea, Chlamys albolineatus*, and *Crepidula aculeata*. When *Rocellaria* is ingested by humans it is said to produce a choking sensation. In his review of Japanese research on poisonous puffers, Yudkin (1944) concluded that the poison was not a protamine, but probably an acyclic compound, the origin of which was entirely unknown. Numerous other workers have expressed their opinions regarding the origin of tetrodotoxin, but the aforementioned are representative of them.

It has been established, as discussed elsewhere in this monograph, that the reproductive cycle of puffers exerts a pronounced influence on the toxicity of puffers. They are generally most toxic immediately preceding and during the height of gonadal activity. The female puffer is considerably more toxic than the male. It has been found (Halstead, 1951) that two specimens of the same puffer species *Arothron meleagris* taken about the same time of the year, one from the Hawaiian Islands and the other from the Phoenix Islands, differed in toxicity. The Hawaiian specimen was innocuous, whereas the Phoenix Islands puffer was violently poisonous. Preliminary fish feeding experiments conducted by Halstead and his associates using California species of *Leptocottus armatus* and *Paralabrax clathratus* demonstrated that if they were fed poisonous puffer meat they became toxic within a period of 10 to 15 days with no apparent injury to the fish ingesting the material. Thus the poison can be transvectored from one fish to the next. The peculiar spotty distribution of poisonous puffers, i.e., a species being toxic in one area and not in another, leads one to believe that tetraodontiform fishes become poisonous as a result of their food habits. Like ciguatoxin the original substance may be a toxin or in the form of a nontoxic chemical constituent which is later metabolized into a toxic product. Dawson *et al.* (1955) found that the stomach contents of *A. hispidus* specimens examined contained sponge spicules, protozoan cysts, the algae *Caulerpa serrulata*, and a species of *Lyngbya*. In this same study of a random series of 107 samples of gut contents examined, it was found that more algae than any other type of food were present in poisonous fishes. "It was interesting to note that *Lyngbya* was detected in a large majority of the samples of poisonous fishes." The authors were unable to come to any definite conclusions because of the limited amount of material available for study. Habekost, Fraser, and Halstead (1955) likewise reported finding toxic substances in aqueous extracts obtained from a number of tropical and temperate marine algae. It is believed that toxic algae also may be a factor in toxin production in tetraodontiform fishes. However, a great deal of additional research is required on the food habits of puffers before the dietary role in the production of tetrodotoxin can be fully evaluated.

MECHANISM OF INTOXICATION

Puffer or tetraodon poisoning is caused by ingesting the flesh, viscera, or skin of toxic tetraodontiform fishes, i.e., sharp-nosed puffers, porcupinefishes, molas, and various other puffers. These fishes are most dangerous to eat immediately prior to and during the reproductive season of the fish since there is a distinct relationship between gonadal activity and toxicity. The skin, liver, ovaries, and intestines are the most toxic portions of the fish. The musculature of the fish is usually safer to eat than other parts of the fish, but at times even it may be toxic. The toxicity of the fish cannot be determined by its appearance or size since even small puffers may contain sufficient poison to be lethal.

MEDICAL ASPECTS

Clinical Characteristics

Despite the numerous reports in which diodontids and molids have been listed as poisonous, an exhaustive review of the literature on poisonous fishes fails to yield much data of a specific nature on the symptomatology caused by these two families of fishes. The following discussion is a résumé of the clinical characteristics of the cases of tetrodotoxications which have appeared in the literature to date. Careful analysis fails to reveal any significant symptomatic differences resulting from the ingestion of fishes of the families Canthigasteridae, Diodontidae, Molidae, and Tetraodontidae. It is therefore assumed that there is either a single poison, tetrodotoxin, or a group of closely related substances that are responsible for tetrodotoxications, and thus the subject will be dealt with as a single clinical entity.

Tetrodotoxication, or puffer poisoning, has been variously designated as ichthyotoxism, tetrodotoxism, ciguatera, fuguism, fuguismus, furube, tetrodon, tetraodon, botete, toado, balloonfish, globefish, blaasop, swellfish, fugu, fougou, ikan buntal, toby poisoning, and various other terms. Puffer poisoning is one of the primary types of ichthyosarcotoxism and is probably the most violent form known.

The onset and types of symptoms in puffer poisoning vary greatly, depending upon the person and the amount of poison ingested. However, symptoms of malaise, pallor, dizziness, paresthesias of the lips and tongue, and ataxia most frequently develop within 10 to 45 minutes after ingestion of the fish, but cases have been reported in which the symptoms did not develop for 3 hours or more. The paresthesias which the victim usually describes as a "tingling or prickling sensation" may subsequently involve the fingers and toes, then spread to other portions of the extremities, and gradually develop into severe numbness. In some cases the numbness may involve the entire body, in which instances the patients have stated that it felt as though their bodies were "floating." Hypersalivation, profuse sweating, extreme weakness, precordial pain, headache, subnormal temperatures, decreased blood pressure, and a rapid weak pulse usually appear early in the succession of symptoms. Gastrointestinal symptoms of nausea, vomiting, diarrhea, and epigastric pain are sometimes present early in the disease, whereas in other cases they are totally lacking. Contradictory statements appear in the literature relative to pupillary changes, but these differences can probably be resolved on the basis of the time at which the examination was made. Apparently the pupils are constricted during the initial stage and later become dilated. As the disease progresses the eyes become fixed and the pupillary and corneal reflexes are lost.

Shortly after the development of the paresthesias, respiratory symptoms become a prominent part of the clinical picture. Respiratory distress, increased rate of respiration, movements of the nostrils, and a diminution in depth of respiration are generally observed. Respiratory distress later becomes very pronounced, and the lips, extremities, and body become intensely cyanotic. Petechial hemorrhages involving extensive areas of the body, blistering, and subsequent desquamation have been reported (Fig. 1). Severe hematemesis has also been known to occur. Muscular twitching, tremor, and incoordination become progressively worse and finally terminate in an extensive muscular paralysis. The first areas to become paralyzed are usually the throat and larynx, resulting in aphonia, dysphagia, and later complete aphagia. The muscles of the extremities become paralyzed, and the patient is unable to move. As the end approaches, the eyes of the patient become fixed and glassy, and convulsions may occur. The victim may become comatose but in most instances retains consciousness, and the mental faculties remain acute until shortly before death.

Two interesting examples of the comatose condition in puffer poisoning have been cited by Akashi (1880). A gambler ostensibly died from eating fugu, and the body was placed in storage for officials to examine. About 7

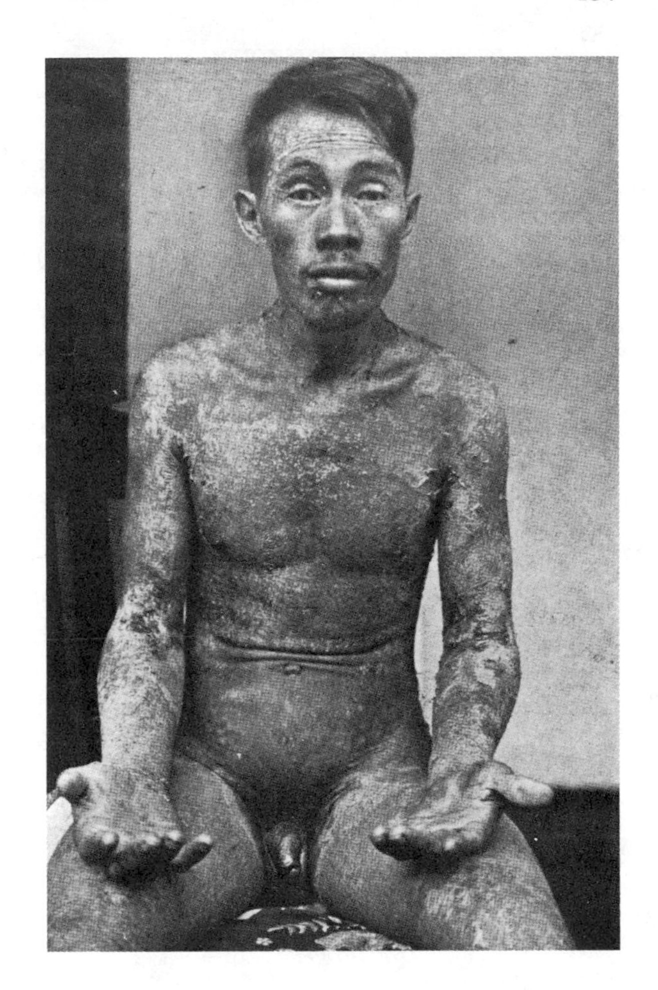

Figure 1.—Photograph of a 30-year-old Javanese victim of tetraodon poisoning. The causative fish was believed to have been *Arothron immaculatus*. This case is of particular interest because it is a good example of the intense desquamation that sometimes appears in puffer poisoning. The case developed as a severe dry redness of the skin, about 3 hours after digesting the fish, followed by swelling and infiltration. The penis and scrotum became very edematous. Shortly after, petechia developed involving the entire body, except at the elbows, knees, and tibial areas where there were extensive subcutaneous hemorrhages. About the third day large skin blisters developed, and about the ninth day the skin began to peel off. Various other systemic effects were present, but the patient fully recovered. (*See* Leber, 1927, for complete case history.)

days later the man became conscious and finally recovered. The victim claimed to have recalled the entire incident, and stated that he was afraid he would be buried alive. In the second case the victim was considered to be dead and was placed on a cart and shipped to a crematorium in a nearby town. The man recovered as he was being removed from the cart and walked away. This latter victim also claimed to have been aware of what was happening.

Death results from a progressive ascending paralysis involving the respiratory muscles. On the basis of the Japanese statistics (Kawabata, 1965), the mean case fatality rate is 59.4 percent. If death occurs, it generally takes place within the first 6 hours or within 24 hours at the latest. Burnett (1846) and Richardson (1893) reported cases in which death occurred within 17 minutes after ingestion of the fish. The prognosis is therefore said to be good if the patient survives for 24 hours. There are no data available regarding significant laboratory findings of tetraodon poisoning.

For the sake of convenience, Fukuda and Tani (1941) have divided the disease into degrees according to various stages of progression. These degrees are characterized as follows:

First Degree:	Oral paresthesias present, sometimes accompanied by gastrointestinal symptoms.
Second Degree:	Advanced paresthesias, motor paralysis of extremities, but reflexes still intact.
Third Degree:	Gross muscular incoordination, aphonia, dysphagia, respiratory distress, precordial pain, cyanosis, drop in blood pressure, but victim is conscious.
Fourth Degree:	Mental faculties become impaired, respiratory paralysis, extreme drop in blood pressure, heart continues to pulsate for a short period.

Some of the more complete case histories have been reported by Akashi (1880), Richardson (1893), Tahara (1896), Larsen (1925), Fukuda and Tani (1941), and Bonde (1948).

It has been said that the poison may be absorbed through the skin by simple contact with the fish. A classic example is that of Heckel, in New Caledonia, who spent 2 days dissecting a large puffer and suffered general malaise, intense cephalalgia, itching and rash on the face and hands, and rather serious gastrointestinal troubles (Boyé, 1911; Phisalix, 1922).

Pathology

In comparison with the large number of recorded deaths resulting from tetraodon poisoning, very little autopsy data are available. Since autopsy data on this subject are rarely given, the following cases are given in detail. Richardson (1861) was one of the first to report on the pathology of this disease. Two Dutch sailors died in 17 to 20 minutes as a result of eating the liver of a South African puffer. The autopsy findings, which were very brief, were as follows:

General: Appearance not unusual. Skin cyanotic in dependent parts, but not to a remarkable degree. Partly masticated food flowed from the mouth.
Thoracic and abdominal viscera: Stomach is externally negative; mucous membranes around cardiac orifice deep purple, highly congested, extending along the lesser curvature, gradually lessening as the lower end is approached. Heart contained a fibrinous clot in right ventricle, with small quantity of dark fluid blood; left ventricle moderately distended with dark fluid blood; muscular tissue appeared normal.
Musculature: Firm, florid, no abnormalities noted.

Bennett (1871) reported his findings on two Australian children, age 8 and 9 years, who died from eating puffer. Death times were not given. His findings were as follows:

General: Well nourished. Cadaverous lividity well marked on dependent parts of body.
Mouth: Pallid, tightly closed, lips livid externally, mucous membranes of tongue and mouth pallid.
Eyes: Pupils moderately dilated.
Lungs: Engorged with fluid and dark blood.

Heart: Somewhat contracted; contained a black clot and dark fluid blood.
Stomach: Veins of external surface dilated, engorged with black blood; mucous membranes at pylorus inflammed, rugae deep red, but intervening spaces paler.
Liver: Engorged with black blood.
Small Intestines: Mucosa congested, reddened patches with intervening areas of pallor.
Large Intestine: Not congested, fecal matter present.
Kidneys: Congested.
Brain: Cerebral veins engorged with dark blood.
Musculature: Unusually rigid.

The second body was similar to the first with the exception that the muscular rigidity was more intense and cyanosis of lips was more intense, contrasting with paleness of the mouth. The small intestine was more congested. The urinary bladder was empty.

Two deaths of Chinese, who died within 4 hours after eating puffer, were reported by Larsen (1925). His findings were as follows:

Lungs: Extreme pulmonary congestion with marked edema.
Heart: Normal in appearance, but no coagulation of blood within 12 hours after death.
Stomach: Full of food, walls definitely injected; no other signs of acute gastrointestinal irritation.
Liver: Light colored with scattered petechiae. Histological sections were similar to those of an ecclamptic liver.
Kidneys: Markedly injected. The appearance of the organs indicated a diffuse violent poison.

Duncan (1951) reported an Australian boy, age 11, who died within 2 hours from eating puffer. "Apart from some pulmonary edema and some congestion of the organs, no abnormality was detected. The routine histological sections did not assist in any way."

A few experimental studies have been conducted on the pathology of this disease in laboratory animals. Rémy (1883) found in his experiments with dogs that the salivary glands, pancreas, stomach, and small intestines were ecchymotic. The liver and kidneys were engorged with dark blood. No appreciable lesions were found in the nervous system. Shimazono (1914) published the most complete neuropathological report that has appeared to date. His autopsies were based on four rabbits which were injected subcutaneously once daily with aqueous solution of 0.0005 to 0.002 g of tetrodotoxin. Animals developed paralysis but recovered within a few days. Autopsies of rabbits showed:

Spinal Cord: Nissl's preparations of large ganglion cells of anterior horns, especially from the central group, were swollen, rounded, and Nissl's granules were undergoing dissolution. Similar changes in posterior horns and grey matter were more distinctive. Satellitosis was often pronounced, and accompanied by neurophagia. Larger glial cells of pia showed proliferation. The cell bodies stained deeply, especially nucleus, and had thick processes. Glial cells of medium size showed similar cytoplasmic swelling.

Degenerative changes included irregular nuclei, indistinct chromatin bodies, and pale cytoplasm that was occasionally granular, or showed several basophilic or fucinophilic bodies. Preamoeboid cells developed from medium-sized glial cells were also present. Proliferating glial cells were present. The proliferating glial cells and preamoeboid cells contained green bodies and black layers in the cytoplasm when stained with Marchi's method. Spinal cord changes of a rabbit surviving for about two months, with fixation one hour after death, showed glial proliferation and preamoeboid cells. The latter often contained Alzheimer's methyl blue granules, and only a few fucinophilic granules. Occasionally these cells showed marked degenerative changes with pronounced vacuolation. Eventually, nuclei showed degenerative changes, including basophilia with pyknosis.

Brain: Ganglionic cells showed milder changes than those of spinal cord. Distinctive changes were reported in the medulla and pons using Nissl's preparation of formalin fixed material. In the smaller ganglionic cells the nucleus was not separate from the cytoplasm, and vacuolation frequently occurred along the margin and central portion. Eventually the entire body became degenerated. Small perivascular hemorrhages were observed around the smaller vessels of the cerebrum and medulla. No cellular reaction was seen. In one instance, gitter cells were reported. Elzholz's bodies along with wallerian degeneration were found in the optic nerves. Glial cells revealed the same changes as in the spinal cord.

Peripheral Nerves: Major changes were not noted. Two cases showed nuclear degeneration of Schwann's cells with swollen, vacuolated, cytoplasm, containing Elzholz's bodies. In the region of the gasserian ganglion slight glial cell proliferation was noted.

Hori (1957) found that in acute tetrodotoxications in mice there was observed an intense congestion of the brain, lungs, liver, and kidney. Hemorrhaging and glomerular degeneration were prominent renal changes. In chronic administration of sublethal doses of tetrodotoxin in rabbits, rats, and mice, Hori (1958) reported finding the following:

Brain: Congestion of the cortices of the cerebrum and cerebellum, but no nerve cell changes.
Heart: Marked congestion, and atrophy of the heart muscle, particularly in mice.
Lungs: Marked congestion, and evidence of hemorrhage in rabbits.
Liver: Congestion, especially in the center of the lobes, and irregularity in size, pyknosis, and loss in nuclear detail. Staining of the sections was positive for peroxidase in rats.
Spleen: Strongly congested, with atrophy, necrosis, and thickening of the capsule.

In summary, the pathology of puffer poisoning in humans was characterized by pulmonary edema and generalized congestion of the viscera with localized changes in the gastric mucosa. No data were available regarding neuropathological changes in the human. Significant neuropathological alterations were seen in the brain and spinal cord of rabbits. These changes consisted of swollen cytons, dissolution of Nissl's granules, satellitosis, glial cell proliferation, and neurophagia. Other degenerative changes included pyknosis, granular cytoplasm, basophilic and fucinophilic bodies, vacuolation, and entire cell disintegration. Occasional small perivascular hemorrhages were reported. Usually there was no cellular reaction. The cerebral changes were generally milder than those in the spinal cord.

Treatment

The treatment of tetraodon poisoning is largely symptomatic. However, Golin and Larson (1969) found strychnine or pralidoxime combined with atropine to be an effective antidote on experimental animals. It is essential to remove the poison from the body as rapidly as possible. Ingestion of large quantitites of sodium bicarbonate solution has been recommended by Kimura (1927) as efficacious in attenuating the effects of the toxin, and as an emetic, although this same author found that intravenous injections of sodium bicarbonate were ineffective in rabbits and mice that had been intoxicated by fugu poison. Any of the routine emetics are helpful, but apomorphine is said to yield the best results (Fukuda and Tani, 1937*a*). Laxatives and enemas have also been found to be useful. As the disease progresses the physician must direct his main efforts toward combating respiratory distress and the accompanying drop in blood pressure. Iwakawa and Kimura (1922) and Kimura (1927) have reported success in reducing the untoward respiratory and vasomotor symptoms with the use of epinephrine and posterior pituitary extract. On the other hand, Imahasi (1928) found in his studies on mice that epinephrine, picrotoxin, physostigmine, and guanidine were not only useless but in some instances detrimental. Caffeine, hexeton, coramine, cardiazole, and glucose were found to be helpful occasionally. He found that he was able to prolong life and decrease the mortality rate in mice most effectively with the use of lobeline and sodium bicarbonate.

Yano (1938), without offering any experimental data, stated that his best results were obtained in frogs and rabbits by using coramine. Tyramine was said to be effective in combating the drop in blood pressure. Caffeine, lobeline, and sodium thiosulfate were recommended to be particularly effective when administered in conjunction with intravenous physiological saline.

Fukuda and Tani (1937*a*) have recommended intravenous lactate Ringer's solution. Artificial and faradic stimulation of the phrenic nerve have been suggested by Goertz (1875), Iwakawa and Kimura (1922), and Kaminishi (1942).

The experimental studies of Murtha *et al.* (1958) and Murtha (1960) on laboratory animals indicate that artificial respiration is the most important therapeutic measure. Although a large number of analeptics were tested singly and in combination with each other, only pentylenetrazol (Metrazol) appeared to be significantly effective. Although artificial respiration may prolong life for a brief period, there is no indication that it will prevent death in severe intoxications; nevertheless, it should be tried along with Metrazol.

Torda, Sinclair, and Ulyatt (1973) suggested the following regime for the treatment of puffer poisoning. Treatment should be aimed at 1) maintenance of airway and ventilation, 2) maintenance of adequate circulation and renal function, and 3) treatment of cardiac dysrhythmias.

On suspicion of puffer poisoning, vomiting should be induced, provided there is no hazard of aspiration of vomitus (in which case gastric lavage can be carried out after intubation with a cuffed endotracheal tube). Whether gastric lavage is indicated more than 3 hours after ingestion is uncertain. As gastric emptying may be slowed by tetrodotoxin, it is probably safer to do it than not. Where symptoms are confined to paraesthesia and weakness not affecting the muscles of respiration and deglutition, probably no effective treatment is possible. Light sedation may be desirable, and close observation must be maintained at least until symptoms begin to recede. In the presence of any difficulty in swallowing, oral intake must be suspended and an intravenous infusion set up to maintain hydration and prevent hypotension.

Any difficulty in dealing with saliva or respiratory secretions is an indication for endotracheal intubation, as is increasing dyspnea, a rising respiration rate, or progressive elevation of the arterial carbon dioxide tension. Torda and his associates suggest that ventilatory insufficiency is an indication for assisted or controlled ventilation rather than oxygen supplementation. The short natural course of puffer poisoning makes tracheotomy unnecessary. Their preference is for nasal intubation, with a cuffed polyvinyl chloride tube. As paralysis is likely to become complete, controlled ventilation is preferable to assisted, unless the ventilator used is able to "take over" automatically if the patient fails to trigger it. Adequate humidification of the inspired gases is necessary.

Fluid administration is regulated according to arterial blood pressure, central venous pressure, and urinary output. Since puffer poisoning results in vasodilation (as well as central cardiac depression in larger doses), it is rational to infuse a plasma expander until urine output exceeds 40 ml/hr. However, if the central venous pressure rises without restoration of urine output, inotropic agents such as isoprenaline or metaraminol are indicated. After urine output has been restored, 2 l of dextrose-saline will supply maintenance fluid and sodium for 24 hours. Potassium chloride 40 to 60 mEq per day should be added after it has been ascertained that the patient is not hyperkalaemic.

The electrocardiogram is monitored for bradycardia. The effect of isoprenaline on bradycardia or conduction disturbances appears to be uninvestigated. Complete atrioventricular dissociation may be an indication for insertion of a temporary pacemaker.

Prevention

From the viewpoint of preventing tetraodon poisoning, no one has ever improved on the ancient Mosaic sanitary laws: eliminate all scaleless fishes from the diet, and there is no opportunity to come in contact with tetraodontiform fishes. Although everyone recognizes that abstinence is the most effective solution, it does not seem to be acceptable in some regions. If a person is living in Japan and has a desire to eat fugu, he should purchase the fish from a first class authorized restaurant having a licensed puffer cook. It is important that the individual preparing the food have a thorough knowledge regarding the fish and its toxicity, and the relative toxicity of the various organs of the puffer. It is advisable to eliminate all the visceral organs and skin from the diet regardless of the species. Although the testes are usually nontoxic, it must be kept in mind that this organ is frequently confused with the violently toxic ovaries, particularly during the season of the year in which the reproductive organs are in their dormant state.

Cooking by frying, stewing, baking, boiling, etc., does not inactivate the toxin. Halstead and Bunker (1953) found that the toxin may still be present in lethal quantities even when passed through the commercial canning process. The toxin can be inactivated chemically by cooking the meat in a strong solution of sodium bicarbonate, ordinary baking soda, for a prolonged period of time. This latter technique also destroys the flavor of the fish and renders it useless for consumption.

Our knowledge regarding the toxicity of most puffer species, exclusive of Japanese forms, is meager. Many of the tropical tetraodontiform species have produced violent deaths at one time or another. Unless a species is definitely known to be nontoxic, which is questionable even under the best of circumstances, all tropical forms should be eliminated from the diet.

If it is a question of survival—either eat the fish or die—and the edibility of the fish is unknown, it is recommended that the fish be enviscerated promptly and only the musculature be used. Moreover, the meat should be cut or torn into small bits and soaked in water for a minimum of 3 or 4 hours. During this period the flesh should be kneaded while in the water and the water changed at frequent intervals. The toxin is water-soluble, and leaching will effectively remove it.

Immunity: Repeated injections or ingestion of puffer poison over prolonged periods of time will not produce an immunity. Fukuda (1937, 1951) stated that there are cases of persons who have habitually eaten puffer for a period of 31 years and then became intoxicated. There are other instances in which individuals have been poisoned several times and then finally died because of subsequent fugu poisoning.

Finally, it should be remembered that the toxicity of puffers varies according to species, geographical locality, season of the year, and the organ of the animal. The factors governing the toxicity of these animals are as yet unknown. There is no simple technique whereby the edibility of tetraodontiform fishes can be evaluated in the field. The consumption of these fishes almost always involves some risk.

PUBLIC HEALTH ASPECTS

Geographical Distribution

Poisonous tetraodontiform fishes are likely to be encountered wherever these fishes occur throughout their geographical range. The canthigasterids, or sharp nosed puffers, are largely restricted to the coral reefs and shoal areas

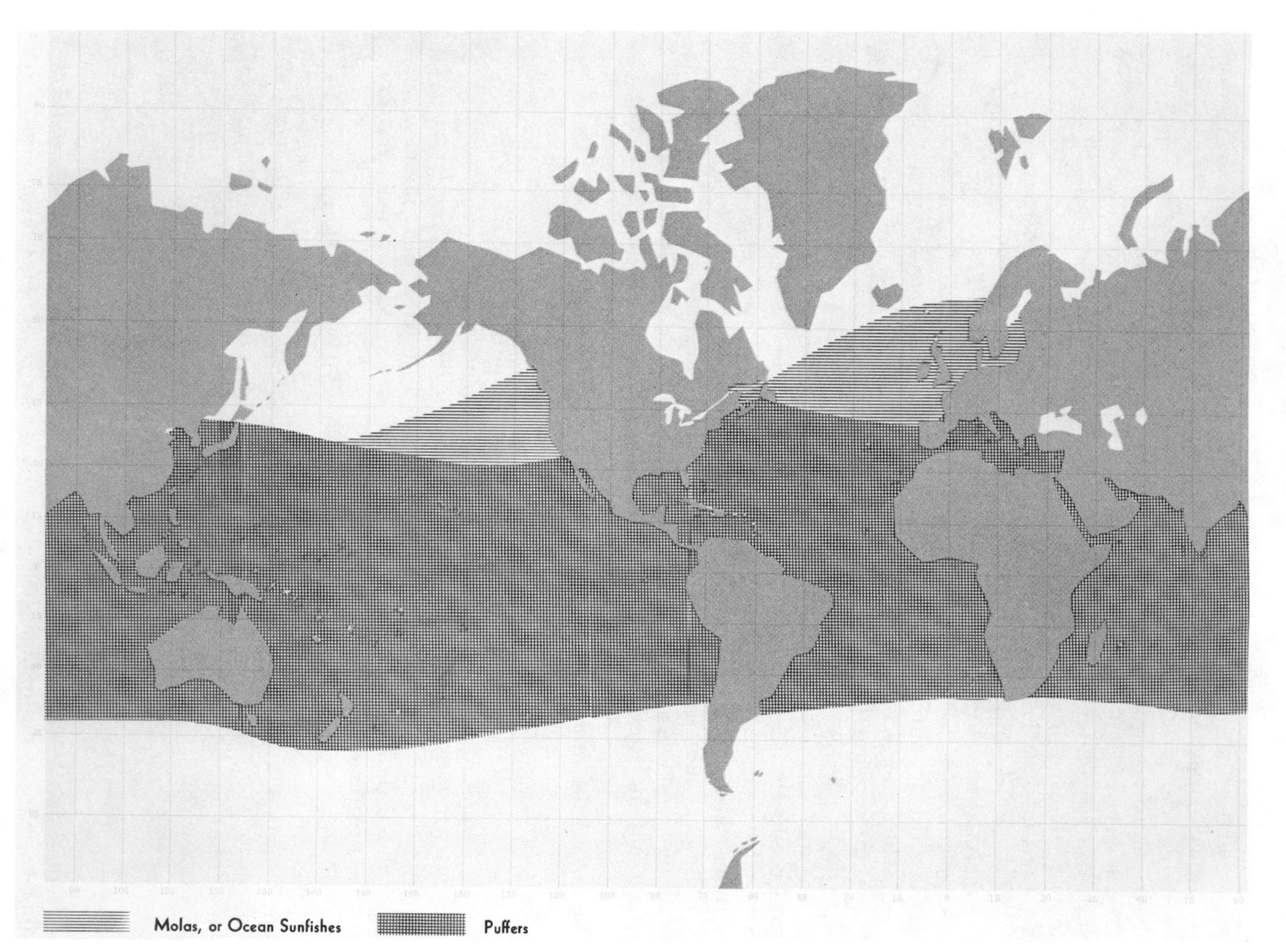

MAP 1

Map showing the world distribution of poisonous tetraodontiform fishes.

of tropical and semitropical seas. The diodontids, or porcupinefishes, are widely distributed in all warm seas. The tetraodontids have a somewhat greater distribution in that representatives may be found throughout a broad circumglobal belt which extends approximately from latitudes 47° N to 47° S. Molas are primarily inhabitants of the open seas and may range as far north as latitude 63° N (Map 1). Outbreaks of tetrodotoxications may be expected to occur wherever these fishes are to be found.

In Japan, where the most accurate epidemiological data are available, Ogura (1958) found that the areas having the greatest incidence of puffer poisoning were in northern Kyushu in the Fukuoka Prefecture and in the area surrounding the Inland Sea of Seto. Comparable information for other countries is not available.

Incidence and Mortality Rate

Global statistics regarding the incidence and mortality rate of tetrodotoxications are not available since this disease is not required to be reported in most countries. The public health aspects of this biotoxication are further complicated by the lack of adequate public health facilities in many of the localities where these fishes are found.

The most accurate statistics concerning the incidence of puffer poisoning in a specific locality are those from Japan. Fukuda and Tani (1937*a*) reported that fugu poisoning accounted for about 42 percent of the total food intoxications in Japan and was regarded as the most important single type of food poisoning in that country. The public health significance of fugu poisoning in Japan has also been emphasized by Ogonuki and Ide (1942) and other public health workers. A report on the incidence of fugu poisoning in Japan between the years 1910-49 by Fukuda (1951) is most illuminating. The highest number of deaths (470) occurred during 1947, which is almost three times as many deaths (175) as given for 1942, the prewar record year.

According to Fukuda the apparent increase in mortality statistics can be explained by the inauguration of a more efficient public health reporting system. Prior to World War II Japanese physicians were not required to specify puffer poisoning as the cause of death on the medical certificate; however, since World War II, Japanese public health laws have listed fugu poisoning as a reportable disease. A secondary reason cited for the increase in puffer poisoning during 1947 was the drastic postwar food shortages at that period when many people took chances which normally they would not have considered. The gradual decline in deaths from puffer poisoning during 1948-49 (386-246) is given as an indication of the validity of this conclusion. Fukuda (1951) further points out that in prewar times the relative percentage of puffer poisonings among other types of food poisonings was 44.6 percent, but after World War II its percentage declined to 21.6 percent. The relative decrease is explained by Fukuda on the basis that the total number of food poisonings reported of all types has been greatly increased because of recent public health laws requiring the reporting of all types of food intoxications.

Despite the relative decline of puffer poisoning among the various types of food intoxications, it continued to head the list as the number one cause of fatal food poisonings in Japan. Bacterial and mushroom intoxications are almost insignificant in comparison.

The statistics covering the period 1886-1963 were obtained from the Food and Sanitation Section, Bureau of Disease Prevention, Ministry of Public Health and Welfare of Japan by Kawabata (1965). They represent the most complete set of statistics available to us at this time. Excluding the 6 years 1943-48, for which the data are grossly incomplete, the case fatality rate was 59.4 percent.

A similar statistical study on outbreaks of puffer poison which is based on essentially the same statistics as those provided by the Ministry of Public Health and Welfare of Japan has been made by Ogura (1958).

Seasonal Incidence: The puffer season begins about October and continues through March, but fugu are of particularly fine flavor from November to February. The toxin content within the ovaries and liver also increases during this period, reaching its peak during the height of gonadal activity—the spawning season which takes place during May and June for most Japanese puffers. Fukuda and Tani (1937*c*) have shown that the greatest incidence of fugu intoxications also occurs between late fall and late spring.

Incidence of Intoxications According to Sex: Puffer poisoning is much more frequent in human males than in females. Fukuda and Tani (1937*c*) observed that during the years 1925-34 there was a total of 909 deaths from eating fugu. A total of 705 or 77.6 percent were men and 204 or 22.4 percent were women. Fukuda explained this condition on the basis of differences in eating habits of the two groups.

Causative Agents of Intoxications: In a survey of 136 cases of fugu poisoning by Japanese puffers, Fukuda (1951) found the species breakdown to be as follows:

Species	Cases
Fugu rubripes	41
Fugu vermicularis porphyreus	19
Fugu pardalis	13
Fugu vermicularis vermicularis	8
Fugu xanthopterus	4
Fugu stictonotus	2
Fugu poecilonotus	1
Unknown	48
Total	136

The principal factors involved are frequency with which a species is eaten and its toxicity.

Organs of the Fish Most Frequently Involved: Intoxication from eating puffer musculature in Japan is a rare occurrence but less common among cases of tetraodon poisoning elsewhere. In a series of 129 cases Fukuda (1951) found the incidence of poisoning according to the part of the fish eaten as follows:

Liver	64
Ovary	55
Skin	10
Total	129

Frequency of Poisoning According to Method of Cooking: Most cases of puffer poisoning in Japan occur as a result of eating cooked food usually in the form of soup or "chiri" (a type of puffer stew which contains meat, viscera, and skin). No cases are known to have occurred from eating "sashimi": sliced puffer meat.

Relationship of Fish Size to Toxicity: There is little evidence that fish size is of great significance in toxic tetraodontiform fishes once the fish has attained sexual maturity. A larger fish generally contains more of the poison because of its larger size, but on a per kilogram basis a large fish may not contain any more poison than a smaller one. However, there are insufficient data available at this time to permit any generalized conclusions concerning the relationship of size to toxicity in tetraodontiform fishes.

Quarantine Regulations

The biblical admonition prohibiting the eating of scaleless fishes, which would include all of the tetraodontiform fishes, has been previously discussed in the historical chapter. Public health regulations governing the sale of poisonous fishes, which included *Diodon attinga, D. hystrix, D. orbicularis, Tetrodon testudinis, T. laevigatus*, and *Paradiodon sp.*, have been in operation in Cuba since 1842 (Hoffmann, 1929*a, b*; Fonseca and Diaz, 1957). *D. hystrix* has been prohibited from being sold in Mauritius since 1884 (Jones, 1956). During World War II Rear Admiral F. E. M. Whiting, USN, prohibited the use of puffers and porcupines at Saipan. Public health regulations of the Board of Health, State of Hawaii, Section 2015, Revised Laws of Hawaii, 1945, prohibit the importation or sale of any poisonous fishes in Hawaii. However, for the most part there are few regulations governing the sale of tetraodontiform fishes in most countries other than in Japan.

The sale of puffers in Japan is governed largely by public health or police regulations within the individual prefectures rather than by the central government. Laws have existed since 1668 forbidding the sale of certain species (Rémy, 1883; Vincent, 1883). Fukuda (1937) has reviewed some of the laws which were in operation at that time. The Kanagawa, Osaka, Kyoto, Shiga, Ishikawa, and Kagawa Prefectures have prohibited by police regulation the sale, display, handling, or distribution of puffers. In Hokkaido, Aichi, Hiroshima, Tottori, and Ehime Prefectures, the sale or distribution of whole puffers is prohibited, but eviscerated fish may be freely marketed if there has been scientific confirmation that only the viscera of the species is toxic. In

Nagasaki Prefecture only fresh puffers are permitted to be sold. The Oita Prefecture prohibits the sale of puffer under the guise of any other type of fish. The consumer is required by law to be informed that he is purchasing fugu. In the Shizuoka Prefecture the governor issued a warning against eating globefish. Some of the prefectures have no regulations on this problem. In general, those prefectures which have enforced their regulations concerning the sale of puffers have effected a pronounced reduction in their mortality statistics.

The Tokyo Prefecture has instituted one of the better health programs regarding the sale of fugu in Japan. The law was termed the "Professional Globefish Treatment Regulation" and was instituted as Metropolitan Police Board Ordinance Nos. 13 and 15, dated June 1892. The law was poorly enforced until the program was reorganized, and two new laws were passed by the Tokyo Legislature as Regulation No. 43, "Regulations for handling fugu," dated April 5, 1949, and Regulation No. 95, "Detailed rules for executing the regulations for handling fugu."

Persons interested in becoming puffer cooks in the Tokyo Prefecture must pass the examination held by the prefectural governor, obtain his permission, and become licensed. As indicated in the preceding globefish regulations, there are specific requirements which the candidate must meet if he is to be certified. The Tokyo Metropolitan Office has published an official text entitled *Textbook for Fugu Cooks* (Fig. 2). This book contains all the necessary information one is required to know in order to pass the examination. In addition, instructions are available to cooks on the preparation and identification of puffers. The examination for fugu cooks in Tokyo is administered under exacting requirements and careful supervision. In addition to the written examination, a practical or oral examination is also given. The selling and handling of poisonous puffers is rigidly controlled by the Tokyo Metropolitan Office. All cooks and restaurants handling fugu must be licensed.

Economic Aspects

Tetraodontid fishes hold a unique position in commercial fisheries in that they are discarded as "trash species" by some people, whereas they are considered the most delectable of seafoods by others. In Japan the puffer commands a higher price than most other types of food fishes. Since the puffer fishery has been developed to a greater extent in Japan than any other country, the following discussion will be primarily concerned with the Japanese puffer industry.

Puffers are taken by Japanese fishermen largely as a byproduct of the croaker fisheries. There are no facilities that are devoted exclusively to the catching of puffers. Despite the relative importance of the industry in Japan, there are no figures available regarding the size of the annual catches. The largest number of puffers is captured by trawler in the east China Sea. However, these puffers are said to be inferior in quality. The best quality puffers are taken by hook and line in the vicinity of southwest Honshu and northern

Figure 2.—Cover of the Japanese *Textbook for Fugu Cooks*. This book is remarkable because it is the only known cookbook on the preparation of poisonous fishes for human consumption.
(Library, World Life Research Institute)

Table 1.—*Toxic Power of Globefish (by T. Fukuda and Others)*

Species	Vernacular Name	Ovary	Testes	Liver	Intestine	Skin	Meat	Blood
Spheroides alboplumbeus	KOMONFUGU	V	S	V	S	S	W	—
S. pardalis	HIGANFUGU	V	W	V	S	S	N	N
S. porphyreus	MAFUGU	V	N	V	S	S	N	—
S. vermicularis	SHOSAIFUGU	V	N	W	W	S	N	—
S. chrysops	AKAMEFUGU	S	N	S	W	S	N	N
S. niphobles	KUSAFUGU	S	N	S	W	S	N	—
S. rubripes	TORAFUGU	S	N	S	W	N	N	N
S. xanthropterus	SHIMAFUGU	W	N	S	W	N	N	—
S. stictonatus	GOMAFUGU	S	N	S	N	N	N	—
S. spadiceus	SABAFUGU	N	N	N	N	N	N	—

Legend: V: Virulent toxin
S: Strong toxin
W: Weak toxin
N: No toxin
—: No data available

Kyushu. Prior to 1953 large numbers of puffers were captured along the coasts of southern Korea, but the South Korean government has issued laws prohibiting Japanese fishermen to come within 16 miles of the Korean coast.

Upon capture the puffers are either iced or placed in saltwater tanks and shipped to the larger cities alive. Some of the more elite Tokyo restaurants import their puffers via air transportation from Shimonoseki. Practically all the better class restaurants deal directly with the fishermen, whereas others purchase their puffers through wholesale fish dealers at the Tokyo

Central Wholesale Fish Market. In the Tokyo area there are about 1,800 dealers who specialize in various seafood delicacies such as porgies, abalone, puffers, etc.

The fugu season in Japan begins about October and continues through March. Puffers are in their prime during the months of November through February, during which time they command their highest price. During 1977 puffers were selling in the wholesale fish markets of Tokyo on the average of 6,000 ¥ ($12.00) per pound. However, during the peak of the season "Tiger" puffer was selling for as much as 13,000 ¥ ($25.00) per pound. Beginning about the latter part of March, the fugu begins its reproductive period, at which time the gonads become greatly enlarged. The poison content is increased at this time, and the flesh is said to lose its delicate flavor. As the reproductive activities gradually subside the fine flavor returns.

There are four species of puffers which are most commonly eaten and command the highest prices in Japan. They are *Fugu rubripes, F. vermicularis vermicularis, F. vermicularis porphyreus*, and *F. pardalis*. All of these species are violently poisonous, but the toxicity varies with the individual specimen, sex, part of the body, and the season of the year. The fact that these fishes are poisonous is common knowledge among the Japanese. The eating of puffer in Japan might in some respects be considered a national habit, at least among the better class. Although fugu is said to resemble young chicken in taste, the flavor is not the sole attraction. It is said that imbibers seem to relish the physiological aftereffects which they receive from the attenuated poison present in the flesh. The reaction has been described as an exhilarating state consisting of sensations of warmth, flushing of the skin, mild paresthesias of the tongue and lips, and euphoria. Some claim that they never get this sensation, so there appears to be some variance in opinion regarding this matter.

Usually, only the musculature, fins, and testes of the puffer are eaten. The meat is generally served as "sashimi," sliced raw meat, which is relatively safe to eat. Fugu is also commonly eaten as "chiri," partially cooked fillets taken from a kettle containing skins, livers, intestines, or testes. This latter dish has been the cause of numerous fatal intoxications. However, an attempt is made to use only those species which are known to have nontoxic flesh and skin. The musculature and skin of *Fugu rubripes rubripes* are not known to be poisonous, but its visceral organs such as liver, ovaries, and gastrointestinal tract are highly toxic, whereas *F. vermicularis vermicularis, F. vermicularis porphyreus*, and *F. pardalis* are reported to have strongly toxic skin and sometimes weakly toxic musculature. Consequently, Fukuda (1951) has recommended that a person eat sparingly of the meat of these latter species. With the possible exception of *F. pardalis*, the testes of most Japanese puffers of commercial value are reputed to be safe for consumption. One of the major dangers of eating testes is the fact that they are all too frequently confused with ovaries.

Poisonous puffer ovaries can be eaten and are sold commercially in the

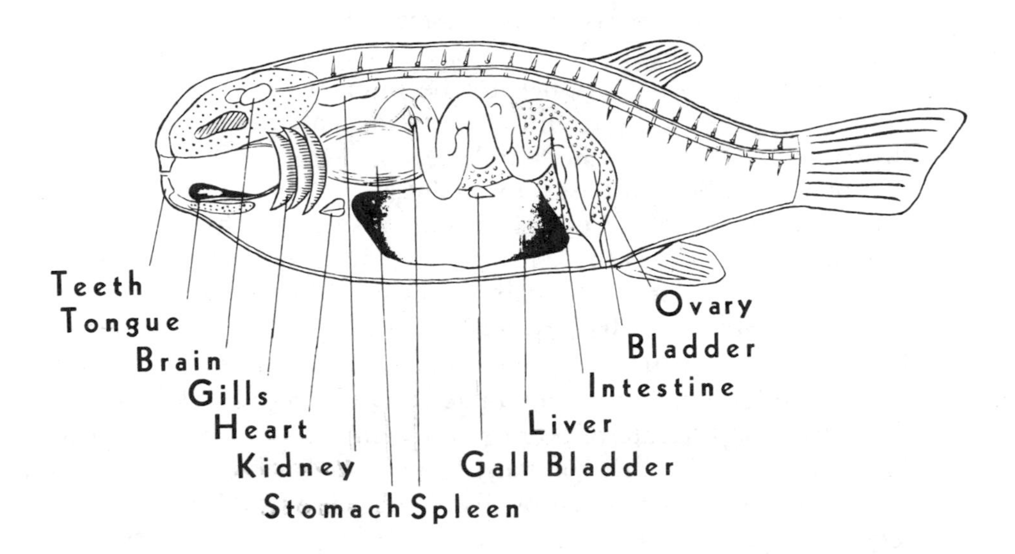

FIGURE 3.—Anatomy of the fugu (From Japanese *Textbook for Fugu Cooks*). The textbook provides this diagram showing the various organs of the fish. Cooks are required to learn the names and positions of these organs.

Ishikawa Prefecture, Japan, as pickled puffer roe. According to Migita and Hashimoto (1951) the poisonous roe of all species of puffer is generally used for this purpose. A variety of methods is used, but they basically follow the same general pattern. The roe is salted and then preserved in rice bran for a period of 3 or 4 years. The toxicity of the roe and rice bran gradually decreases during the pickling process. One study that was made revealed that about 40 percent of the toxicity disappeared within 7 months, and 70 percent is gone in 17 months. Because of the combined factors of reduction in toxicity and extreme saltiness, which prevents the individual from ingesting excessive amounts of the material, pickled roe is relatively safe to eat.

Fukuda (1951) states that skilled fugu cooks are able to prepare toxic puffer livers in such a manner as to render them fit for human consumption. The liver is boiled, mashed, and then boiled again in fresh water. This process is repeated several times until the cook believes all of the poison has been removed. Fukuda adds the disconcerting note that even the finest puffer cooks occasionally succumb to their own cooking. At best, people who indulge in this form of epicurean pastime must consider themselves as "living dangerously."

In their studies on puffer poisoning on the Asiatic continent, Fukuda and Tani (1942) found that the Koreans eat about the same species of puffer that are eaten in Japan. However, the Koreans prefer cooking their fish in bean paste soup with parsley. They believe that the parsley will neutralize the poison. The Chinese people prefer *Fugu ocellatus* to other species of puffers. The most desirable parts of the fish are the testes and liver which are broiled. Puffer meat is usually dried or partially dried and pickled in salt. The pickled puffers are then shipped to the interior of China where they are sold. There appears to be no information available concerning the incidence of puffer poisoning in China and Korea.

Puffer poison has been used as an analgesic and in the treatment of a number of disorders, e.g., arthritis, rheumatism, pruritus, impotence, enuresis, tetanus, pertussis, asthma, headaches, etc. Tetrodotoxin is said to be particularly efficacious as a muscular relaxant in disorders in which convulsions are dominant. The drug, prepared from the ovaries of Japanese puffers, has been manufactured by the Sankyo Company of Tokyo under the name of "tetrodotoxin." The injectable solution, packaged in 1 ml ampoules, sold in 1953 at the rate of $0.71 per 10 ampoules. The poison could also be obtained from this same company in powder form at a cost of $5.97 per gram. The solution is prepared by dissolving 1.6 g of the powder in distilled water containing 0.5 percent of carbolic acid as a preservative. This will make 1 liter of the solution. The lethal dosage of the powder is said to be 1 μg/g of mouse body weight.

According to Hess and Weinstock (1926), the liver of *Sphaeroides maculatus*, the Atlantic puffer, is unusually rich in vitamin D and is said to contain 5,700 international units per gram of liver, or to be 15 times as potent as cod-liver oil. However, puffers have not been used as a commercial source of vitamin D.

Sphaeroides maculatus was introduced to the fish markets along the Atlantic seaboard of the United States during World War II. The fish is said to be abundant and taken in large quantities in certain seasons by trap net and other fisheries (Yudkin, 1944). The fish has been eaten in various local communities for many years and is reputed to have exceptionally good flavor. The fish seems to be gaining in acceptance and appears on the market in limited quantities, both fresh and frozen. There is at least one report of a fatality from having eaten an unidentified species of puffer found along the Atlantic coast of the United States. A woman died March 23, 1963, at Homestead, Fla., from eating a puffer of unknown identification, possibly *S. maculatus*. Undoubtedly more deaths will occur in the United States as puffers grow in popularity and authorities fail to instruct people in the proper preparation of these fishes. Yudkin (1945) demonstrated that a cardioinhibitor was present in *S. maculatus*, and Larson *et al.* (1960) and Lalone *et al.* (1963) have published on the toxicity of several different species of puffers inhabiting the eastern seaboard of the United States. Public health authorities should be cognizant of the dangers of eating these fishes.

It is of interest to note that Reintjes and King (1953) found in a survey of 1,097 stomachs of yellowfin tuna *Neothunnus macropterus*, captured in the Central Pacific during 1950-51, that identified tetraodontids, diodontids, and molids as comprising 3.0 percent of the aggregate total volume of the stomach contents. Tetraodontoid fishes were present in 2.9 percent of the stomachs examined. Nakamura (1950) has reported similar findings. Why these poisons are not transvectored to scombroid fishes is not known.

Military Medicine Aspects

From the days of the ancient imperial Japanese army until modern times, puffers have been recognized as a health hazard to military personnel. Many a sailor or soldier in a foreign land has been tempted to eat an improperly prepared puffer in a moment of curiosity, jest, or emergency. The mortality statistics attest to the folly of such an act. According to reports received from reliable Japanese sources (Hiyama, 1952), more than 400 of their military persons succumbed to fish poisoning in Micronesia during World War II. There is reason to believe that a sizable number of these individuals had eaten improperly prepared tetraodontoid fishes since they are routinely used as food in Japan. It has been definitely established that a puffer species may be nontoxic in one area and violently poisonous in another. Unfamiliarity with the unusual characteristics of puffers can result in serious consequences on the part of the unwary gourmet. Numerous warnings have appeared in the military preventive medicine manuals and publications of various nations. The reader is referred to the following works for additional information along this line: Vincent (1889), Buddle (1930), Hiyama (1943), Whitley (1943), Halstead (1951), and Fish and Cobb (1954).

TOXICOLOGY

Rochas (1860, 1862) was among the first to employ experimental methods in studying the effects of puffer poison. He conducted several feeding experiments, demonstrating that the eggs of puffer could kill a cat within an hour. His tests indicated that the rest of the fish was relatively harmless. Rochas was also one of the first to perform autopsies on laboratory animals that had died of puffer poisoning (Rémy, 1883).

The experiments of Savtschenko (1882) showed that the poison was not confined to the eggs but could also be found in the liver and other viscera. He found that cooking did not destroy the toxin, which "affects exclusively the nervous system and with horrible swiftness."

Rémy's work (1883) was the most complete toxicological study attempted up to that time on puffer poisoning. Samples of various species of Japanese puffers were fed to cats and dogs, which were observed for a period of time for the development of symptoms. Rémy found five puffer species to be poisonous, *Tetrodon pardalis, T. rubripes, T. lineatus, T. vermicularis*, and *T. rivulatus*. From these studies he concluded that the poison in puffers lies exclusively in the gonads, and the toxicity of the fish is in proportion to its gonadal activity. Rémy also observed that the poison acted chiefly on the nervous centers, affecting the general and special "sensibilities." He considered death to be caused by cardiac paralysis and suffocation.

Miura and Takesaki (1887, 1889) attempted to determine the anatomical distribution of puffer poison. Alcoholic and aqueous extracts were prepared from the musculature, ovary, testes, liver, kidneys, heart, spleen, and skin of *T. rubripes*. The extracts were filtered, heated to dryness over a waterbath, and then dissolved in distilled water. The samples were tested by making subcutaneous injections into rabbits. These authors also concluded that only the ovary contained the poison.

One of the more interesting early toxicological investigations is that of Takahashi and Inoko (1890), in which they tested aqueous and alcoholic ovarian extracts of puffer on rabbits, frogs, pigeons, chickens, snakes, puffers, and various other species of marine and freshwater fishes. The authors prepared extracts from three species of puffers, *T. vermicularis, T. stictonotus*, and *T. pardalis*. They observed that injected puffer poison was innocuous to both toxic and nontoxic species of puffers but was usually toxic to most other fishes. In a later work (1892) the same authors conducted a series of experiments to show the localization of puffer poison. In this instance, Takahashi and Inoko injected aqueous extracts of the various puffer organs into frogs and rabbits. They found the frog to be an ideal test animal because of it sensitivity to the poison. Their study was important since it was the most complete one that had appeared up to that time. They demonstrated that the poison was present in organs other than the ovaries and confirmed the earlier observations of Rémy that the toxicity of the fish was influenced by gonadal activity. Takahashi (1897) later conducted a more exacting study in order to determine the physiological effects of puffer poison on various warmblooded animals, such as dogs, cats, rabbits, mice, doves, and chickens. However, this latter study was primarily concerned with the physiological effects on certain organ systems rather than with a toxicological screening technique.

Silvado (1911) attempted to evaluate the toxicity of three Brazilian puffers *T. marmoreus, T. punctatus*, and *T. testudineus*, by feeding samples of the liver, gonads, and musculature to chickens. As a result of these studies he concluded that all three species were toxic.

Fonseca (1917) prepared physiological saline extracts from the musculature, bile, liver, ovaries, testes, skin, and cutaneous mucus of the Brazilian puffers *Spheroides greeleyi* and *S. spengleri*. Tests were made by intravenous and intraperitoneal routes. The author came to the conclusion that there was much variability in the localization of the poison and questioned any relationship between the toxicity in the gonads and liver.

Ishihara (1918) used a broad spectrum of test animals in his pharmacological studies on puffer poison, viz., rabbits, guinea pigs, rats, mice, sparrows, doves, chickens, loaches, frogs, toads, eels, snakes, turtles, carp, plaice, sculpin, dragonflies, leaf chafers, newts, and puffers. The toxins were administered by various routes: oral, intravenous, subcutaneous, intracutaneous, intracardial, intramuscular, intraperitoneal, subdural, intracranial, intramedullary, anterior chamber of the eye, and conjunctiva. Most of his extracts were prepared with the use of physiological saline, but in some instances he used the powdered or commercial forms of the poison. He also concluded that tetraodon poison is innocuous when injected into other puffers.

Sommer and Meyer (1937) in their studies on mussel and fish toxins used acid ethanol, neutral methanol, and methanol acidulated with hydrochloric acid for extractants. Later Hashimoto and Migita (1950) demonstrated that methanol acidulated with hydrochloric acid "produces an artificial poison from some common components of shellfish." They also observed (1951), with puffer poison, that methanol acidulated with hydrochloric acid destroys tetrodotoxin. They recommended extraction with the use of methanol and acetic acid. This latter method is said to extract the toxin readily in large quantities and yields results similar to plain water extraction.

Macht and Spencer (1941) tested more than 65 different species of North American fishes for the presence of toxic substances. Tissue extracts were prepared by grinding up samples of fish muscle with an equal weight of sand, and to this mixture was added physiological saline in the proportion of 2 ml to each gram of muscle. The homogenate was strained through muslin, and the filtrate was then tested by injecting 1 to 3 ml intraperitoneally into white mice weighing from 20 to 25 g each. Macht also used a phytopharmacological technique for testing fish toxins, which employed the seedlings of *Lupinus albus*. The increment or growth of the single straight root of the seedling was measured after the plant had been grown for 24 hours in the dark at 15° C in the extract to be tested. Growth of the root was then compared with controls which were grown in plant-physiological (Shive) solution. The index of growth or phytotoxic index was expressed by the ratio of X/N or average growth in the unknown solution over the growth of the controls. They found that heating destroyed the toxicity of most fish muscle extracts with the exception of those from *Diodon hystrix* and *Sphaeroides maculatus*. These extracts could be heated to 60° C without markedly impairing their toxicity for plants.

The monumental toxicological works of Tani (1940, 1941, 1945) have probably made the greatest contribution to our understanding of the toxic nature of puffers. Tani attempted to correlate the anatomical distribution of puffer poison, size of fish, size of gonads and liver, and toxicity with the season of the year. The poison was extracted by boiling samples of the muscle, liver, ovary, testes, skin, and intestine in weak acetic acid for 30 minutes. Samples were taken from a wide variety of Japanese tetraodontoid fishes. The tissues were ground with a mortar and pestle. Water was then added to the homogenate and the mixture filtered by the addition of warm water for 30 minutes at 60° C. The residue was filtered again, washed with warm water, and then filtered several additional times. The filtrate was evaporated at a temperature which did not exceed 60° C. The residue was then dissolved in lukewarm water in sufficient quantity to equal the weight of the original tissue sample, so that 1 ml of the extract was equivalent to 1 g of the original tissue sample. The extracts were later injected into white laboratory mice weighing between 15 and 20 g.

Tani defined his mouse unit as the amount of toxin required per gram of mouse weight to kill a 15-20 g mouse. In other words, 1 ml of extract (representing 1 g of an organ) having a toxicity of 1,000 mouse units would be lethal to 1,000 g of mice or fifty 20-g mice. The minimum lethal dose for man, by Tani's standard, is about 200,000 units. Tani's mouse unit is somewhat cumbersome and difficult to use, but it is nevertheless one of the first attempts to evalute the toxicity of puffers on a quantitative basis.

The conclusions which Tani derived from these studies were as follows: In general, the ovary and liver were strongly toxic, intestines and skin were moderately toxic, and the musculature, blood, testes, and eyeballs were either weakly toxic or nontoxic. Puffers having black or blue colored integument had nontoxic skins. *Sphoeroides rubripes* and *S. porphyreus* had the greatest quantity of poison within their ovaries, whereas *S. porphyreus, S. pardalis*, and *S. rubripes* had the most toxic livers. In terms of the combined toxicity of ovary and liver, *S. porphyreus* and *S. rubripes* proved to be the most toxic species tested. *S. niphobles, S. alboplumbeus, S. pardalis*, and *S. vermicularis* were also strongly toxic species, but because of their small size the total amount of toxin present within the body of a single specimen was relatively small. *S. porphyreus* and *S. rubripes* were frequently less toxic on a per gram basis but in the aggregate were more toxic because of their large size. It is noteworthy that in those instances where large series of puffers were tested the toxicity varied considerably according to the specimen, species, season of the year, and locality. It was found that the toxicity of puffers generally began to increase about December and reached a peak about May or June, or just before the spawning season, and then rapidly subsided with the decline in gonadal activity. Tani demonstrated that postmortem changes in the body of the fish did not influence its toxicity. He also found that the parasitic copepods *Argulus* which are frequently present on the skin of puffers and generally considered by Japanese to be directly related to the toxicity of the fish, do not affect the toxicity and are nontoxic.

In his valuable work on the poisonous fishes of the South Seas, Hiyama (1943) used mice, cats, and puppies for his screening tests, which consisted of both feeding and extraction techniques. For his feeding experiments, weighed samples were taken of muscle, intestine, liver, gonads, and blood. Some of the samples were cooked while others were fed raw. In some instances it was necessary to use salted and dried meat. He tried a number of extraction techniques, but the one giving the most satisfactory results consisted of placing 10 g of raw flesh in 200 ml of absolute alcohol where it was left for 15 hours at room temperature (31°-32° C). After filtering, 15 ml of distilled water was added and the solution then heated for about 2 hours until it was concentrated to 2.7 ml. Hiyama found it difficult in cases of mild poisoning to detect the symptoms in mice, and also found a great deal of variation in their reactions to ingested poisonous flesh. He came to the conclusion that, since the results of the feeding techniques were dependent upon the amount of flesh ingested by the animals, the extraction injection method is preferable. Hiyama found the livers of *Tetraodon hispidus, T. meleagris*, and *T. nigropunctatus* to be toxic.

Kawakubo and Kikuchi (1942) found that man and dog were generally more susceptible to ichthyosarcotoxins than either the cat or pig, and the chicken reacted least of all.

Goe and Halstead (1953), Halstead and Bunker (1953, 1954*a*, *b*), and Dawson *et al.* (1955) have reported on methods used for the routine screening of fish poisons. Since most of these reports have been discussed in detail in Chapter X, "Ciguatoxic Fishes," only a few pertinent comments are needed. Goe and Halstead (1953) found the musculature and viscera of *Sphoeroides annulatus* from the Gulf of California to be toxic. In their toxicity studies on *S. rubripes chinensis*, *S. vermicularis porphyreus*, *S. vermicularis vermicularis*, and *S. vermicularis radiatus*, Halstead and Bunker (1953) found that the poison was usually not destroyed by the commercial canning process. This process included placing the open cans in a steam exhaust at 100° C for 10 minutes. The cans were then sealed and were placed in a steam retort under a pressure of 12.5 pounds per square inch at 116.6° C for 75 minutes. Aqueous extracts were prepared in the usual manner and injected intraperitoneally into laboratory mice.

Benson (1956) reported on the toxicity of *S. testudineus* that had been incriminated in human fatalities in southern Florida.

Flaschentrager and Abdalla (1957) investigated the toxicity of two Red Sea puffers *Amblyrhynchotes diadematus* and *Arothron hispidus*. Aqueous extracts were prepared from the tissues of the fishes and injected into the thoracic lymphsacs of frogs (*Bufo regularis*). The intestine, liver, muscle, gonads, and skin were found to be toxic.

Larson *et al.* (1960) prepared tissue extracts from the skin, muscle, liver, and gonads of *S. maculatus*, *S. testudineus*, *Lactophrys tricornis*, *L. trigonus*, *Lagocephalus laevigatus*, and *Chilomycterus schoepfi*. The species of *Lactophrys* were nontoxic, but the *Sphoeroides* and *Chilomycterus* were poisonous. All species were captured in Florida. Lalone *et al.* (1963) have presented a rather thorough analysis of the toxicity of *S. maculatus* in Florida.

Ogura (1963*a*, *b*) has developed a biological assay method for testing the crystalline tetrodotoxin by using a rat vagal nerve-stomach preparation. The quantity of tetrodotoxin present is estimated by determining its ability to inhibit the contraction of the stomach electrically stimulated through the vagus nerve. The preparation is placed in a 30 ml bath that is maintained at a temperature of 30° C and aerated with oxygen. The vagus nerve is stimulated for 3 minutes at 5 minute intervals. Each dose of tetrodotoxin to be evaluated is added to the bath 2 minutes prior to electrical stimulation. They found that when the results were plotted against the log dose a straight line was obtained between 1 and 10 mg/ml. Ogura found this method to be useful in making a biological estimation of minute quantities of tetrodotoxin.

Kao (1966) has provided a valuable review of the toxicology of tetrodotoxin.

Itokawa and Cooper (1969) found that tetrodotoxin (3×10^{-8} molar) promoted the release of thiamine from perfused rat and frog nerve preparations in a manner similar to other neuroactive drugs. The rats were injected with thiamine labeled with sulfur^{-35}; analysis of brain, spinal cord, and sciatic nerve homogenates revealed labeled thiamine in membrane, synaptosomes, and mitochondrial subfractions. However, on incubation of these fractions with tetrodotoxin, thiamine was released only from the membrane fragments.

In order to distinguish between solutions of tetrodotoxin and saxitoxin (*see* p. 70-71 on *Gonyaulax*), Waterfield and Evans (1972) suggested a simple method consisting of bio-assay of dilutions before and after heating to 100° C at pH 1.0 for 20-30 min. In strongly acid solutions, for example, tetrodotoxin rapidly lost toxicity at 100° C, while saxitoxin remained stable.

The acute cardiovascular, antiarrhythmic, and toxic effects of tetrodotoxin in unanesthetized dogs were studied by Duce, Feldman, and Smith (1972). Five unanesthetized dogs were given I.V. doses of 0.3, 0.6, 1.2, 2.4, and 4.8 μg/kg of tetrodotoxin infused over 5 minute periods at 65 minute intervals. In similar experiments 4 anesthetized dogs were given serial doses of saline. Emesis occurred in 3 or 4 dogs after each dose of tetrodotoxin, and the highest dose produced respiratory arrest and death in 4 of the 5 animals. The tetrodotoxin produced increases in heart rate and arterial pressure and decreases in rectal temperature. Central venous pressure and peripheral resistance were unaltered. Cardiac output, as determined by dye dilution, was increased after 0.6 μg/kg. In another series of experiments, 4 unanesthetized dogs with ventricular arrhythmias produced by two-stage coronary artery occlusion were given cumulative I.V. doses of 6-7 μg/kg of tetrodotoxin as repeated individual doses of 1-4 μg/kg infused over 5 minute periods at 65-minute intervals. Such treatment produced emesis in all 4 dogs, and respiratory arrest and death in 2 dogs. The incidence of abnormal ventricular beats was transiently reduced in 3 animals at the higher cumulative doses, but this effect appeared more related to simultaneously occurring increases in atrial rate than to suppression of ectopic impulse formation.

PHARMACOLOGY*

Action of Tetrodotoxin on the Nervous System

General: Death from puffer poisoning is nearly always preceded by deterioration of nervous activity in man and animals (Bennett, 1871; Fonseca, 1917; Itakura, 1917; Ishihara, 1918, 1924), and when death does not occur the first symptoms appearing are usually of nervous origin (Autenrieth, 1883; Bennett, 1871; Rémy, 1883; Itakura, 1917; Ishihara, 1918; Fukuda and Tani, 1937*a, b*). It is reported that muscular paralysis in humans may be so complete that a patient suffering from puffer poisoning may be buried alive (Akashi, 1880). This same author, as well as Leber (1927) and Fukuda and Tani (1937*a*), stated that even during this profound paralysis mental powers remained essentially normal. Should the poisoning end with death, the sensorium became cloudy only immediately preceding expiration. It is frequently noted that a degree of anesthesia accompanies paralysis (Autenrieth, 1833; Ishihara,

---

* The assistance of Dr. Don R. Goe in preparing this section is gratefully acknowledged. Dr. Goe investigated the pharmacological properties of tetrodotoxin as his Ph.D. thesis.

1918; Iwakawa and Kimura, 1922; Yano, 1937), although Fukuda (1951) stated that anesthesia became evident only at near lethal doses. The fact that animals in the lower phylogenetic series tolerate the toxin better than higher animals (Iwakawa and Kimura, 1922; Kawakubo and Kikuchi, 1942) suggested that the lethal attack of the toxin may be on the central nervous tissues, for these lower animals are less dependent for life on the physiological integrity of the central nervous system.

Ogura, Nara, and Yoshida (1968) compared the pharmacological actions of tetrodotoxin and ciguatoxin. They found them similar in many respects (*see* Pharmacology section in Chapter X, "Ciguatera Fishes") with the following exceptions: (a) in the case of tetrodotoxin, the oral toxicity is very low in comparison with the I.V. toxicity; (b) hypotension induced with tetrodotoxin is not antagonized with atropine; (c) pressor response to epinephrine is not suppressed with tetrodotoxin; (d) tetrodotoxin possesses ganglionic blocking action; (e) tetrodotoxin has no anticholinesterase activity; (f) tetrodotoxin has no influence on the acetylcholine-induced contracture in the isolated frog muscle; (g) tetrodotoxin is insoluble in organic solvents, except for acidic methanol.

Effects on the Central Nervous System: In harmony with the preceding observations, it has been repeatedly noted that one of the first effects of poisoning by fugu is the paralysis of central nervous tissues (Akashi, 1880; Takahashi and Inoko, 1890; Itakura, 1917; Ishihara, 1918, 1924; Iwakawa and Kimura, 1922; Fukuda and Tani, 1937*a, b*; Yano, 1937, 1938; Fukuda, 1951). Of these central nervous tissues the vital centers of the medulla have been given the most attention. Many investigators agree that one of the first bulb centers to be affected is the respiratory, and death is many times ascribed to the paralysis of this center (Takahashi and Inoko, 1890; Itakura, 1917; Ishihara, 1918; Iwakawa and Kimura, 1922; Yano, 1937; Fukuda, 1951; Murtha, 1960; Murtha *et al.*, 1958; Li, 1963). Many other studies have also demonstrated that there is a paralysis of the vasomotor center which is reflected in a progressive hypotension (Goertz, 1875; Takahashi and Inoko, 1890; Yamasaki, 1914; Ishihara, 1918; Iwakawa and Kimura, 1922; Yano, 1938; Nomiyama, 1942).

There are several reports that tetrodotoxin paralyzes the spinal cord (Takahashi and Inoko, 1890; Itakura, 1917; Ishihara, 1918, 1924; Iwakawa and Kimura, 1922). Whether this paralysis of the cord precedes, accompanies, or follows the paralysis of the vital medullary centers is not clearly shown. However, Iwakawa and Kimura (1922) stated that paralysis of the cord followed paralysis of higher centers. Both sensory and motor components of the cord were paralyzed (Ishihara, 1917, 1918; Itakura, 1917).

It has been determined that both longitudinal and transverse conduction of the cord are paralyzed (Takahashi and Inoko, 1890, 1892; Ishihara, 1918, 1924). Ishihara found that paralysis occurred in both areas almost simultaneously, but transverse conductive paralysis at times slightly preceded the other.

During the past decade several hundred articles have appeared on the use of tetrodotoxin as a neurophysiological probe, but since most of this literature goes beyond the scope of this present work no attempt has been made to review this data.

The reports of the effect of the toxin on other central nervous tissues are scant and inconclusive. Iwakawa and Kimura (1922) stated that it may have a narcotic effect on the brain. This they surmised from their studies concerning the activity of frogs following poisoning. Duncan (1951) has noted some sedative or narcotic effects on humans, and tetrodotoxin has been used as a sedative in the treatment of alkaloid addiction (Hsiang, 1939). The toxin seems then to exert some narcotic effects, but the data presented above are too inadequate to be conclusive. Itakura (1917) stated that, although tetrodotoxin may act as a local anesthetic and may reduce temperature, there is no evidence of a narcotic effect directly on the brain. Itakura arrived at these conclusions because he failed to note any change in peripheral sensory perceptions (by direct stimulation of an exposed nerve) with the injection of the toxin into the scapular region. A perceptible change in the respiratory patterns was used as an indication of sensory perception. The protocol did not state if the animals were anesthetized although they probably were not. It is of interest to note that, even under deep anesthesia, one can obtain a respiratory response to a painful stimulation; so this response appears to be reflex in nature, and it is therefore difficult to understand how this method could be used to determine the narcotic effects of the poison.

Vomiting is a sign that is consistently present in puffer poisoning. The cause of emesis in puffer poisoning was first studied from a pharmacological viewpoint by Nomiyama (1941, 1942). She concluded that the site of the emetic action was in the vomiting center of the medulla oblongata since morphine blocked emesis, but atropine did not. It was believed at that time that morphine directly antagonized the emetic drug action on the vomiting center. Hayama and Ogura (1958, 1963) investigated this earlier work by Nomiyama in view of the more modern concept that in drug induced vomiting the central emetic agent acts upon a specialized receptor site and not upon the vomiting center *per se.* Using crystalline tetrodotoxin on dogs, minimum effective dose of 0.7 g/kg by subcutaneous and intramuscular routes, or 0.3 g/kg intravenously, they found that if the chemoreceptor trigger zone was ablated the dogs failed to vomit. Chlorpromazine and hexamethonium failed to prevent the emetic response to tetrodotoxin whereas tetraethylammonium was effective. They believed that a nicotinic action in the vomiting mechanism of tetrodotoxin was suggested. Similar findings have been reported by Murtha *et al.* (1958), Murtha (1960), and Borison *et al.* (1963). The latter group of investigators have also shown that intracerebroventricular injections of tetrodotoxin, in doses from 1/250 to 1/500 of the intravenous requirement, caused volitional paralysis, hypothermia, and respiratory failure; but the pinna, corneal, and deep tendon reflexes remained active.

Effects on the Spinal Cord: The effects of the toxin as demonstrable in the activity of the cord itself have been discussed above. In this section will be considered only the effect of cord paralysis in the mediation of reflexes and sensory and motor conduction.

Subarachnoidal injections of tetrodotoxin into the cord of the rabbit results in complete flaccid paralysis of the posterior extremities, loss of the knee reflex, and incontinence of urine (Itakura, 1917). Concomitantly, there is anesthesia below the level of injection (Itakura, 1917). As the toxin spreads craniad, anesthesia follows; in the case of both sensory and motor nerves it can be demonstrated that it is the synapses within the cord rather than the nerves themselves that are paralyzed. Ishihara (1918) stated that both the posterior and anterior horns of the cord are paralyzed by the toxin. He came to these conclusions because he observed an interruption of the reflex arc. With no more detail than is given it is difficult to determine in which portion of the pathways the paralysis occurred. One cannot determine from Itakura's work (1917) if these portions of the cord were paralyzed by the toxin. Other investigators, however, as previously noted, indicate that there is a sensory and motor paralysis in the cord. It is not known whether the sensory or motor synapses are first paralyzed, nor whether this sensory and motor paralysis precedes or follows the cessation of cross transmission and longitudinal conduction in the cord.

Repeated subcutaneous applications of tetrodotoxin have been shown to cause definite microscopic changes in the cord (Shimazono, 1914). Whether or not these changes are correlated with altered physiological function is not stated.

Although the responses of animals treated with curare and those treated with tetrodotoxin (so that the cord is paralyzed) are similar, the mechanism action of the two has been shown by Ishihara (1918) to be quite different. Curare paralyzes the motor endplate only, whereas tetrodotoxin paralyzes the cord, the motor and the sensory nerves, and perhaps the motor endplate last of all (*see* section on "Effects on Motor Nerves," p. 482).

It is not clear as to whether the paralysis of the cord precedes or follows the paralysis of the peripheral nerve. Iwakawa and Kimura (1922) stated that there was convincing evidence which indicated that spinal paralysis occurred later than that of the motor nerves, and they believed that spinal paralysis may not be demonstrable with mild poisoning, whereas motor nerve paralysis was obvious. No experimental procedure or protocol is given, so it is difficult to evaluate these statements.

Effects on Peripheral Nerves

General: It has been reported that peripheral nerves can be anesthetized or paralyzed by tetrodotoxin as indicated in the discussion on respiration (Ito, 1889; Takahashi and Inoko, 1890, 1892; Itakura, 1917; Ishihara, 1918, 1924; Iwakawa and Kimura, 1922; Yano, 1937; Nomiyama, 1942; Fukuda, 1951; Matsumura and Yamamoto, 1954; Murtha *et al.*, 1958; Goe, 1958; Furukawa, Sasaoka, and Hosoya, 1959; Murtha, 1960; Dettbarn *et al.*, 1960; Cheymol *et al.*, 1961; Cheymol, Bourillet, and Ogura, 1962*a, b*; Narahashi *et al.*, 1964). Most authors agree that the first peripheral nerve to be paralyzed is the phrenic and that paralysis of the other nerves follows with a longer time or with a more concentrated dosage. It has been demonstrated, however, that both motor nerve and sensory nerve conduction is blocked by the toxin (Itakura, 1917; Ishihara, 1918; Yano, 1937; Fukuda, 1951; and others). Matsuo (1934) believed that sensory precedes motor interference. Ishihara (1918) has found that it takes longer to paralyze the nerves of coldblooded animals than it takes to paralyze the nerves of warmblooded animals. Ishihara immersed a nerve in a solution of the toxin and determined that only the immersed part was paralyzed. This is in agreement with the work of Yano (1937) and Suehiro (1948) who stated that the toxin does not travel along the nerves but is carried in the blood. Ishihara (1924), Matsumura and Yamamoto (1954), Kuriaki and Wada (1957), and Goe (1958) found that washing the nerve in saline or Ringer's solution restored the excitability after paralysis with the toxin. Itakura (1917) and Hashimoto (cited by Yano, 1937) noted slight reversible morphological changes in the axones of rabbits that had received prolonged (up to 14 days) injections of the toxin.

Effects on Motor Nerves: The early data concerning the curarelike action of tetrodotoxin on the neuromuscular junction are very confusing. Some investigators reported that the poison had an effect while others indicated that they were unable to demonstrate an effect. Ito (1889), Miura and Takesaki (1889), Takahashi and Inoko (1890), Itakura (1917), and Iwakawa and Kimura (1922) believed that the toxin had a curarelike effect on the motor endplates. Iwakawa and Kimura (1922) stated that heavy poisoning caused nearly simultaneous paralysis of the cord and of the motor nerve endings, while slight poisonings paralyzed the motor nerve endings, leaving the cord still reactive. However, they offered no experimental evidence to support their statements. Ishihara (1918) was unable to determine if the motor endplate was paralyzed or not, for he stated that the paralysis of the muscles occurred simultaneously with the paralysis of the peripheral motor nerves. The last statement concerning the time necessary for paralysis of the muscle and of the motor nerve is not in agreement with a statement earlier in his own work where he stated that the excitability of muscle outlasted the excitability of the nerve when both were immersed in a solution of the toxin. The statement also does not agree with the work of others. The report of Matsumura and Yamamoto (1954) has clarified at least a portion of the controversy. They found that tetrodotoxin in minimum doses (10^{-5}-3×10^{-5}) produced reversible blockage of neuromuscular transmission in the nerve-sartorius preparation of the toad owing to reduction of the muscle potential without significant changes in the endplate potential. This suggests that the toxin initially raises the threshold of the muscle fibers for propagated impulses. Increased doses of tetrodotoxin abolish generation of the endplate potential completely. Tetrodotoxin was also found to depress impulse propagation in the sciatic nerve fibers in a higher concentration than that at which it blocked synaptic transmission, initially affecting the potentials referable to the impulses propagated in A-fibers. This is in harmony with the earlier report of Itakura (1917) that direct application of the toxin to the nerve trunk of cold and warmblooded animals interrupts motor conduction. Similar findings have been reported by Goe (1958).

More recent pharmacological investigations on the effects of tetrodotoxin on the neuromuscular junction have provided still further insight into the problem. The findings are as follows: Furukawa *et al.* (1959) concluded that tetrodotoxin had a potent narcotic action on nerve and muscle without polarizing them but did not suppress the sensitivity of the endplate to acetylcholine even at much higher concentrations than is needed for the narcosis of nerve and muscle. In this regard, tetrodotoxin has been found to differ from most other anesthetics. They used a frog nerve sartorius preparation in their studies.

Fleisher, Killos, and Harrison (1960) reported that incubation of the isolated rat phrenic nerve-diaphragm in solutions containing as little as 0.06 μg of tetrodotoxin per milliliter produced neuromuscular block accompanied by marked inhibition of acetylcholine release. They concluded that the primary effect of tetrodotoxin on neuromuscular transmission was an inhibition of acetylcholine release. Similar findings have been reported by Cheymol *et al.* (1961) and Cheymol *et al.* (1962*a, b*), who further noted that in this respect tetrodotoxin is somewhat like cocaine.

On the other hand, Dettbarn *et al.* (1960) found that tetrodotoxin reversibly inhibited conduction in single nerve fiber preparation of frog at a concentration of 3×10^{-10} *M*. In the isolated electroplax of the electric eel *Electrophorus electricus*, higher concentrations of the poison blocked both transmission and conduction. They believed that tetrodotoxin was not a potent acetylcholinesterase inhibitor. They decided that the mechanism of the toxin in blocking transmission and conduction had not been established.

Narahashi *et al.* (1964) reported that tetrodotoxin, at very low concentration, blocked the action potential production through its selective inhibition of the sodium-carrying mechanism while keeping the potassium-carrying mechanism intact.

Kao (1966, 1972) has provided a valuable review of the similarities and contrasts between tetrodotoxin and saxitoxin in their significance in the study of the excitation phenomena of excitable cells. Saxitoxin and tetrodotoxin are unique in acting only on changes in sodium permeability normally associated with excitation.

Narahashi (1972) adds that tetrodotoxin is unique in its selective blocking action on the nerve membrane sodium conductance increase, the mechanism whereby an action potential is produced on stimulation. The effect is exerted at very low concentrations (10^{-7} to 10^{-8} *M*). Tetrodotoxin selectively blocks the ionic channels through which sodium ions normally flow during the rising phase of the action potential or during peak transient current under voltage-clamp conditions. Currents carried by ions other than sodium through this peak transient channel are also susceptible to tetrodotoxin. Tetrodotoxin has no rectifying action on the peak transient current. The effect of tetrodotoxin is exerted only from outside of the nerve membrane in cationic forms. The number of tetrodotoxin molecules bound to the nerve membrane is estimated to be 13 or less per square micron of the membrane. The resting sodium conductance of the nerve membrane is partially decreased by tetrodotoxin. Among several tetrodotoxin derivatives studied, deoxytetrodotoxin and tetrodonic acid are totally ineffective in blocking the action potential or peak transient current. Saxitoxin, which originally comes from the dinoflagellate, has essentially the same action as tetrodotoxin on the nerve membrane.

Effects on Sensory Nerves: It has been previously mentioned that sensory nerves are also paralyzed. The question arises, however, as to the resistance of sensory nerves to the toxin and whether they are paralyzed before or after the sensory receptors. Itakura (1917) determined that the toxin produced a greatly prolonged reaction time which is referable to anesthesia of the sensory nerves rather than to paralysis of the motor nerves. Ishihara (1918) demonstrated that with both man and animals, tetrodotoxin administration resulted in a dulling or total loss of sensitivity. Itakura (1917), Ishihara (1918), and Yano (1937) stated that the anesthetizing effect of the toxin is less than that of a dilute solution of cocaine, although the effect may be considerably longer lasting (*see also* Cheymol *et al.*, 1962*a, b*). Iwakawa and Kimura (1922), Fukuda and Tani (1937*a*), and Yano (1937) observed that the clinical use of the toxin frequently resulted in hypoesthesia of lips, tongue, and finger tips. He also reported that cats poisoned by tetrodotoxin exhibited hypoesthesia to pain in skin incisions. The clinical utility of tetrodotoxin is limited because its anesthetic properties are attained only with near lethal doses (Nomiyama, 1942; Fukuda, 1951). None of the above experiments was designed to indicate whether it was the sensory end organ or the sensory nerve which was affected. Also, there is no time correlation of the onset of this anesthesia with muscular paralysis. Iwakawa and Kimura (1922) demonstrated that if a frog's foot is immersed in a solution of the toxin, sensory perception is reduced. They believed this was due to paralysis of the end organ of sense, but from their experimental results it could just as well have been due to paralysis of the peripheral portion of the sensory nerve. Itakura (1917) reported anesthetic effects of the toxin in the spinal frog, on the cornea and nose of the rabbit, and with subcutaneous and intradermal injections in man.

Loewenstein, Terzuolo, and Washizu (1963) reported that in two receptors, the crustacean stretch receptor neuron and the pacinian corpuscle, tetrodotoxin was found to act selectively on the all-or-none component, blocking the action potential but leaving the generator potential unaffected.

Effects on Autonomic Nerves: It is generally conceded that fugu toxin paralyzes the autonomic nerves and that they are somewhat more resistant to the toxin than are the other peripheral nerves (Itakura, 1917; Ishihara, 1918, 1924). Ishihara stated, however, that large peripheral nerves may be more slowly paralyzed than the autonomics. This may be referable to the poor blood supply of these large nerves and therefore the relatively slow diffusion of the toxin into these nerves. It is also reported that the parasympathetics are paralyzed somewhat before the sympathetics (Ishihara, 1918, 1924; Kimura, 1927).

Itakura (1917) believed that tetrodotoxin has the ability to paralyze the autonomic ganglia, whereas Ishihara (1918, 1924) stated that it could not unless the poison was applied directly in large doses. Ishihara believed, however, that the toxin could paralyze the postganglionic fibers. Matsumura and Yamamoto (1954), using the isolated toad cervical ganglion, found that tetrodotoxin blocked ganglionic transmission, eliminating the postganglionic potentials and leaving the preganglionic potentials intact. Because adrenalin seemed to reverse the paralyzing effect on the sympathetics, Ishihara believed that the end apparatus of the adrenergic nerves was not paralyzed. Ishihara, by pilocarpine administration after paralysis of parasympathetic nerves, also claimed to have demonstrated that the end apparatus of these cholinergic nerves was not paralyzed by the toxin. Pilocarpine, however, does not act through the cholinesterase mechanisms but rather acts directly on the target organ. Ishihara also claimed that atropine paralyzed the terminal apparatus of these nerves, but here again, since atropine does not act through the cholinesterase mechanism, his conclusions are without experimental support.

Ogura and Hori (1961) administered a single subcutaneous or intraperitoneal injection of crystalline tetrodotoxin into pyloric ligated rats and found that this resulted in significant decreases in gastric secretory volume, free acid and total acidity, with increases in pH and gastric pepsin concentration. It was further observed that histamine alone in subcutaneous doses of 5 mg/100 g body weight caused no significant effect on gastric secretion. When tetrodotoxin was given prior to the administration of histamine, the usual tetrodotoxic effect on the secretion of gastric juice and acid were not seen. Physostigmine showed an effect similar to that of histamine. When adrenocorticotropic hormone was given in doses of 2 units per day intramuscularly for 5 days, the tetrodotoxic effect on gastric juice and acid remained unchanged, and pepsin concentration increased, maintaining the pepsinogen concentration in the gastric wall at the same level. Next, rats were adrenalectomized 4 days prior to the pyloric ligation, and tetrodotoxin administered. The rats were fed sodium chloride 10 g/liter and given an ordinary diet. The inhibitory effect of tetrodotoxin on gastric secretion observed in adrenal-intact rats was also observed in adrenalectomized rats. Insulin was shown to stimulate the secretion of gastric juice and acid, but this was prevented when tetrodotoxin was administered. From these results the authors concluded that the inhibitory effect of tetrodotoxin on the secretion of gastric juice might be caused by ganglionic blocking action since the poison has been shown to inhibit ganglionic function in animals.

Additional mention will be made of the autonomic nerves as they are involved in the response of other organs and tissues to tetrodotoxin.

Effects on the Neural Components of Respiration: Most investigators believe that the usual cause of death in tetrodotoxications is paralysis of the respiratory mechanism (Goertz, 1875; Miura and Takesaki, 1890; Takahashi and Inoko, 1890; Fonseca, 1917; Itakura, 1917; Ishihara, 1918, 1924; Iwakawa and Kimura, 1922; Fukuda and Tani, 1937*a*, *b*; Nomiyama, 1942; Fukuda, 1951; Goe, 1958; Murtha *et al.*, 1958; Murtha, 1960; Li, 1963; Borison *et al.*, 1963; and others). However, stimulation of respiration following administration of the toxin has been reported (Nomiyama, 1942). The fact that young animals withstand the toxin better than adults may be the same reason that young animals also withstand anoxia better than mature animals. There are, however, some differences in opinion as to just when paralysis of respiration occurs in the chain of events following puffer poisoning.

Several workers (Hayashi and Muto, 1901; Kaminishi, 1942; Fukuda,

1951) have indicated that paralysis of the peripheral nerves first appears. In the articles (Hayashi and Muto, 1901; Ishihara, 1918) reporting where these nerves were studied selectively, it was pointed out that the phrenic nerve failed first. Other investigators (Takahashi and Inoko, 1890, 1892; Ishihara, 1918; Yano, 1937; Nomiyama, 1942; Fukuda, 1951; Goe, 1958; Murtha *et al.,* 1958; Murtha, 1960; Borison *et al.,* 1963; Li, 1963) believed that it was a central paralysis which was ultimately responsible for death. Fukuda and Tani (1937*a*) stated that it was "narcosis of the respiratory muscles" that caused difficulty in breathing. Iwakawa and Kimura (1922) investigated this problem using a toxin dilution of moderate potency. They found, using the rabbit, that subcutaneous injection resulted in paralysis of the center while intravenous administration resulted in a depressed excitability of the phrenic nerve before paralysis of the center. They observed that after the center had become depressed the phrenic nerve many times regained its excitability (Takahashi, 1897, found the phrenic nerve excitable by electric stimulation after respiratory arrest). However, in using concentrated solutions with subcutaneous injections they found the nerve was the first to fail (results similar to those of Fukuda, 1951), and in using dilute doses with intravenous injection the bulbar center was the first to fail. It is also noted that if repeated dilute doses are given (Hayashi and Muto, 1901) the paralyses of the center and the nerves occur nearly simultaneously. Iwakawa and Kimura (1922) explained this simultaneous paralysis as a result of the slow absorption of the toxin. It should be noted that with very slow injection Ishihara (1918) obtained paralysis of nearly all nerves. However, the recent studies of Borison *et al.* (1963) and Li (1963) clearly indicate that the respiratory arrest is from central action of the toxin on the brain stem respiratory centers and is the primary cause of death.

Respiratory stasis may or may not be accompanied by convulsions (Takahashi and Inoko, 1890, 1892; Yamasaki, 1914; Iwakawa and Kimura, 1922; Yano, 1937). The presence or absence of convulsions depends on the animal species and the route of injection. The various factors involved in this situation have not been fully delineated.

When respiration fails, artificial respiration may prolong life (Itakura, 1917; Murtha *et al.,* 1958; Murtha, 1960) even when the blood pressure cannot be readily restored (Kaminishi, 1942). According to Aomura, Yen, and Oikawa (1930) artificial respiration cannot prevent death in heavy poisoning. Ishihara (1918) has demonstrated that artificial respiration can prevent death when an animal is injected with 1.8 times the lethal dose of the toxin. Murtha *et al.* (1958) found that cats receiving lethal doses of puffer poisoning usually recovered within 1.3 to 5.3 hours, if kept alive by continuous artificial respiration and that administration of Metrazol greatly shortened the period during which artificial respiration was necessary. It is probable that other lethal changes occur in addition to respiratory stasis and also contribute to the cause of death.

Ishihara (1918) determined that the response of the respiratory system to fugu toxin is very marked even at low dosages so that according to him it is the first measurable response to the toxin. Depending on a variety of factors (site of injection, dose, species) respiration may become slower and shallower (Yamasaki, 1914; Ishihara, 1918; Yano, 1937) or become increased in rate and amplitude (Nomiyama, 1942). That recovery is usually rapid following mild poisoning is evidence, as is discussed elsewhere, that the poison is seemingly detoxified by the body.

Effects on Vasomotor System: Several investigators (Goertz, 1875; Takahashi and Inoko, 1890, 1892; Hayashi and Muto, 1901; Ishihara, 1918; Iwakawa and Kimura, 1922; Fukuda and Tani, 1937*a*, *b*; Yano, 1938; Kaminishi, 1942; Nomiyama, 1942; Fukuda, 1951; Goe, 1958; Murtha *et al.*, 1958; Murtha, 1960; Tsukada, 1960; Li, 1963; and others) have reported that paralysis of the vasomotor system and concomitant hypotension accompanies or follows the paralysis of the respiratory center. Takahashi and Inoko (1890, 1892) believed that the effect was entirely on the vasomotor center. This is not in agreement with later investigations. However, Ishihara (1918, 1924) demonstrated that in mild poisoning only the medullary centers and spinal centers for vasomotor control are paralyzed and finally, with larger doses, the sympathetic ganglia are also paralyzed. Fukuda and Tani (1937*a*, *b*) stated that death from fugu toxin resulted from a paralysis of the blood vessels and a paralysis of the respiratory centers, though Ishihara (1918, 1942) claimed that the paralyzing effect was on the center or on the nerves and not on the smooth muscle of the vessels. Iwakawa and Kimura (1922), Yano (1937, 1938) and Nagayosi (1941, cited by Murtha, 1960), on the other hand, felt that the dilatation resulted from a paralysis of the blood vessel muscles as well as of the vasomotor centers. Nomiyama (1942) believed that there were two phases to the paralysis: first an anesthesia of the blood pressure and respiratory centers near the diencephalon resulting in an increase in both blood pressure and respiration, followed by an anesthesia of the medullary centers of blood pressure and respiration leading to a decrease in blood pressure and a depression of respiration. Ishihara (1918, 1924) determined, by the application of adrenalin, that the autonomic end apparatus was not paralyzed. Ishihara also demonstrated that the nerves to the vessels were paralyzed later, or by larger doses of the toxin, than the centers or ganglia. It is interesting to note that the ganglion, where acetylcholine is the conductor, is paralyzed while the end apparatus of the sympathetic nerves to the blood vessels of the mesentery, where sympathin is the chemical mediator, is not paralyzed. Li (1963) found that administration of a small sublethal dose of tetrodotoxin produced hypotension only, without respiratory or other apparent changes, thus indicating an independent action. Preliminary investigation showed that cardiac output as measured with a cardiometer was unaffected, which suggested that hypotension is attributable to direct action of puffer poison on peripheral blood vessels. Support of Li's hypothesis is found in the fact that further lowering of diastolic pressure occurred when the toxic fraction was given after injection of a ganglionic blocking agent, tetraethylammonium bromide. The administration of antihistamine diphenhydromine prior to injection of a sublethal dose of toxin prevented the fall in blood pressure, which indicated that the hypotension may be due to the histaminic action of tetrodotoxin. Li felt that the hypotension was unlikely to be due to autonomic effect because the pressure response to carotid occlusion or adrenalin was not affected by administration of the toxin, and atropine did not influence the hypotensive effect. It was further found that infusion of noradrenalin following injection of the toxin maintained the arterial blood pressure at normal levels for about 1 hour.

A variety of vasomotor responses to different concentrations of toxin have been noted. Kunisho (1934) found that tetrodotoxin in small doses acted initially like a stimulus to the vasodilatory center and later like a stimulus to the vasoconstrictor center. Larger doses (0.1 percent) caused a strong initial vasoconstriction. In dogs, Nomiyama (1942) observed that small doses of puffer toxin produced an initial rise in blood pressure, an increase in heart rate, and an increase in respiratory amplitude, followed by a drop in blood pressure and respiratory rate and further increase in respiratory amplitude. Higher concentrations produced an immediate lowering of blood pressure followed by a gradual increase in heart rate and respiratory amplitude. Ishihara (1918) using perfusion techniques noted no effect on the rate of flow of the perfusate with administration of the toxin, though he found that direct or intravenous application of a concentrated solution resulted in peripheral dilatation. Kobayashi (1944), however, found that while small doses produced no observable change in the size of certain cerebral vessels, medium to large doses produced a strong initial constriction followed immediately by a marked vasodilatation. The latter response he attributed to a direct vasodilatory action plus a passive reaction resulting from the overall effect on the peripheral blood pressure. Yano (1937) likewise noted vasodilatation resulting in an increased flow of perfusate in a rabbit's ear preparation after administration of tetrodotoxin. He found that adrenalin could reduce the magnitude and duration of the response to toxin, as could barium. The barium, however, was effective to a lesser degree. Because of the different sites of action (barium on smooth muscle, adrenalin at the nerve endings), Yano felt that the vasodilatation produced by tetrodotoxin must be the result of a paralysis of the nerve, followed by a paralysis of the smooth muscle of the blood vessels.

In a comparative study of the pharmacology of tetrodotoxin and saxitoxin, Kao (1972) determined that both toxins cause severe hypotension involving a direct relaxant effect of the toxins on the vascular muscle (at low doses) and also a blockade of vasomotor nerves (at high doses).

Effects on Cardiovascular System: The vasomotor component of this system has been previously discussed. Altered functions of the vasomotor system during puffer poisoning will be mentioned here only as necessary in explanations of other effects.

Most investigators agree that tetrodotoxications result in a drop in blood pressure (Takahashi and Inoko, 1890, 1892; Ishihara, 1918, 1924; Iwakawa and Kimura, 1922; Kimura, 1927; Yano, 1937, 1938; Kaminishi, 1942; Nomiyama, 1942; Fukuda, 1951; Li, 1963; etc.). Because this drop may be ameliorated by both adrenalin and barium (Yano, 1937), the hypotension may be referable to a paralysis of the vasomotor system and of the smooth muscles of the terminal arterial bed. Similar findings have been reported by Li (1963).

Several investigators (Ishihara, 1918, 1924; Kimura, 1927; Yano, 1937, 1938; Kaminishi, 1942; Yudkin, 1945) have demonstrated a cardioinhibitor in the fugu toxin, although Fukuda (1951) believed that "almost no harm" is done to the heart in fugu poisoning.

Since there appears to be little difference in the effect of the toxin on the hearts of coldblooded animals and on the hearts of warmblooded animals, in the following discussion the subject will be presented in general without reference to species differences. Yano (1938) stated that the toxin seemed to paralyze the cardiac muscle and the course of the impulse. Ishihara (1918, 1924) described the action of the toxin on the isolated heart of the toad. He found that the toxin stopped the heart beat, but since the heart beat returned to normal after washing with saline he claimed that the toxin had no permanent effect on the contractility of the heart muscle. By using a perfusion technique he concluded that the toxin had no effect on the activity of the sinus venosus. Using this same technique he reported an early auriculoventricular block followed later by sinoauricular block. In working with an *in vivo* heart, Ishihara frequently observed a transient tachycardia before the fibrillation, palpitation, and conduction block. This he assumed to be due to a specific stimulating action of the toxin. However, the tachycardia may have been reflexly induced as a result of the paralysis of the vasomotor centers and the rapidly dropping blood pressure. The fact that he occasionally observed tachycardia of the atria with ventricular stasis suggests that the auriculoventricular conduction system is more susceptible to the toxin than are the afferent and efferent nerves concerned with the carotid and aortic pressoreflexes. Nomiyama (1942) also noticed transient tachycardias in dogs as blood pressure began to fall. Yano (1938) denied that the suppression of the heart in amplitude and rhythm that led to systolic standstill was due secondarily to changes in blood pressure.

Kimura (1927) confirmed and extended Ishihara's observations. He distinguished three relatively specific effects of the toxin on the heart: (1) paralysis of the musculature, (2) paralysis of the impulse pathways, and (3) paralysis of the impulse center. He produced graphic records of complete ventricular diastolic stoppage with a relatively normal atrial rhythm. Kimura also determined that repetitive applications of the toxin produced less and less effect on the heart. From this he concluded that fugu toxin can act as a so-called "potential poison of Straub." Yudkin (1945) corroborated the data concerning ventricular diastolic stasis during atrial activity. Yudkin did not use the Japanese prepared tetrodotoxin but prepared his principles from fresh Atlantic northern puffer *Sphaeroides maculatus*. Yudkin felt that the activity of the toxin might lie in its anticholinesterase activity but he was unable to demonstrate such activity. (*See also* Tsukada, 1957, 1960; Murtha *et al.*, 1958; Murtha, 1960; for further comments regarding the cardioinhibitory effects of tetrodotoxin.)

Takahashi and Inoko (1890), Ishihara (1918, 1924), and Larson (1925) reported that tetrodotoxin markedly delayed the clotting of blood. Inaba (1935) found that coagulation was enhanced by small to medium doses and inhibited by large doses of toxin. These reactions could not be demonstrated *in vitro*. The observation that clotting is delayed by puffer poison may be related to the discovery by Heilbrunn *et al.* (1954) and Couillard (1953) that the ovaries of many animals (invertebrates and fishes, including puffers) contain heparin or heparinlike compounds. The identity of the toxin substances from puffer ovaries with these heparinlike compounds is not proven.

Bernstein (1969) investigated the effects of tetrodotoxin on various biologic parameters. His drug interaction studies demonstrated that tetrodotoxin can prevent or reverse ouabain-induced bradycardia, tachycardia, and arrhythmias. He also showed that reserpine can completely block the actions of tetrodotoxin, whereas propranolol and phenoxybenzamine tend to modify its various actions.

Action of Tetrodotoxin on Other Organs and Systems

Skeletal Muscle: It has been suggested by several investigators that tetrodotoxin may have a direct effect on skeletal muscle. Yamasaki and Kikkawa (1909), and Yamasaki (1914) reported that tetrodotoxin relieved the muscle spasm of tetanus, though this may have been more of an effect on the motor nerves than in the muscle itself. Takahashi (1897) found that tetrodotoxin-poisoned muscles retained reactivity to electric stimulation. Related to this is the finding of Yano (1937) that a poisoned muscle loses its excitability to direct stimuli less readily than to stimulation through its nerve. Itakura (1917) demonstrated a direct paralyzing effect of the toxin on skeletal muscle. A graphic record of the effect of the toxin resembles a typical fatigue curve but differs in being irreversible. Itakura further demonstrated that the effect is on the musculature itself and not in the motor endplate, as is that of curare. Iwakawa and Kimura (1922) reported that muscle excitability nearly always outlasts the excitability of the cord and of the motor nerves. Ishihara (1918, 1924) found similar relationships. Fukuda and Tani (1937*a*) remarked that the difficulties in breathing resulting from puffer poisoning are the result of the "narcosis" of the breathing muscles, but they offered no experimental evidence in explanation. More evidence of a direct effect comes from the work of Matsumura and Yamamoto (1954) who found that tetrodotoxin acts first and only on muscle fibers until the concentration reaches a certain level, when the motor endplate region becomes affected. The result, then, is first a raising of the threshold of the muscle fibers to a propagated impulse followed by a direct action on the endplate region. In addition to physiological effects, microscopic pathological changes in the form of slight degeneration of various striated fibers plus a hyperplasia of nuclei may be produced in poisoned muscle cells (Hasiguti, 1931). Murtha *et al.* (1958) and Murtha (1960) observed that the poison can affect the response of skeletal muscle to both indirect and direct excitation.

Smooth Muscle: Very little information is available concerning the direct action of puffer toxin on smooth muscle. Ishihara (1918) reported no paralyzing effect on the smooth muscle of an isolated intestinal strip, uterus, iris of the eye, or hair muscle. In 1924, Ishihara also reported no effect of the toxin on the stomach band on the toad. The remainder of this discussion will deal with the effects of the toxin on organs containing smooth muscle.

An increase in intestinal motility and vomiting are very commonly associated with puffer poisoning (Rémy, 1883; Takahashi and Inoko, 1890; Tahara, 1896; Takahashi, 1897; Ishihara, 1918, 1924; Kimura, 1927; Nomiyama, 1942). Takahashi and Inoko (1890) and Takahashi (1897) felt that the increased intestinal motility in puffer poisoned animals was referable to anoxia. Ishihara (1918, 1924), however, demonstrated that direct application of the toxin to a well oxygenated intestine caused motility. Kimura (1927) studied this problem in some detail. He noted that with mild poisoning there was an increase in tone and a decrease in movement of the intestine. With more heavy poisoning and a concomitant high grade of respiratory disturbance, there was an increase in movement as well as an increase in tone which Kimura felt was due to asphyxia. This is in disagreement with the work of Ishihara (1918, 1924) in which he attributed the increased motility directly to the toxin. Ogura (1957) believed the increased activity of the gut to be the result of paralysis of the sympathetics rather than stimulation of the parasympathetics or of altered activity of Auerbach's plexus.

With *in vitro* studies Ishihara (1918, 1924) demonstrated an increased tone and motility of the isolated gut of the rabbit. In investigating this same problem, Kimura (1927) determined that small doses resulted initially in a decreased tone and motility followed frequently by a rise of both, though seldom to normal levels. In heavy poisoning, however, the transitory flabbiness was followed by an increased tone and movement to supranormal levels. These were the usual results, and great variability was observed from one preparation to another. The increased activity was elicited by a concentration of the toxin so high that it probably could not explain the increased *in vivo* motility. Ishihara (1918) believed the flabbiness to be due to stimulation of the sympathetic nerves. Kimura (1927) used atropinized preparations and obtained no greater reduction of tone or movement of the gut, and also observed a loss of tone of the ileocecal sphincter. From these data he concluded that the relaxation of the intestine was caused by a vagal paralysis rather than by sympathetic stimulation as asserted by Ishihara. As to the nature of the increased tone and motility with heavy poisoning, Kimura agreed with Ishihara that the cause was probably sympathetic paralysis. Ogura (1957, 1958, 1959) reported that tetrodotoxin prepared from puffer ovarian extract exerted a parasympathomimetic effect on isolated guinea pig intestine, but crystalline tetrodotoxin did not show this property. Hamada (1957, 1960) found that crystalline tetrodotoxin inhibited the contraction of isolated guinea pig ileum produced by nicotine and serotonin, but failed to produce any inhibitory effect to contractions produced by acetylcholine and histamine. Tetrodotoxin inhibited the peristaltic action of isolated ileum, but did not alter the acetylcholine content in the stretched isolated intestinal wall muscle. The poison appeared to have a local anesthetic action on the intestine.

The effect of mild poisoning with puffer toxin on uterine smooth muscle is an increase in tone of the uterus *in situ* while the action on isolated tissue is extremely variable (Kimura, 1927). Ishihara (1918, 1924) reported that the toxin acts neither on the uterine nerves nor on the uterine muscle.

According to Ishihara (1918), urinary bladder smooth muscle is paralyzed after the bladder blood vessels are paralyzed. In view of the fact that these vessels are innervated by postganglionic sympathetic fibers, this is presumptive evidence that these nerves can be paralyzed by puffer poison.

Hamada (1960) reported that contraction of the rectus muscle of the frog was inhibited by cocaine or procaine but unaffected by tetrodotoxin. The relaxant effect of nicotine in the guinea pig's tracheal chain was abolished following pretreatment with hexamethonium, cocaine, and tetrodotoxin. The bronchoconstriction produced by physostigmine and nicotine in the guinea pig's tracheal chain was abolished after being treated with hexamethonium, cocaine, and tetrodotoxin. Puffer poison did not demonstrate atropinic, antihistaminic, or antiadrenaline action in the tracheal chain and did not affect the anaphylactic response of tracheal muscle by egg albumin.

Effect of Tetrodotoxin on the Eye: Considerable confusion exists as to the exact effects of tetrodotoxin on the eye. Rémy (1883), Miura and Takesaki (1889), Takahashi and Inoko (1890), and Ishihara (1918) reported that the pupil remained uniformly dilated (there was occasionally a slight transient narrowing) before respiratory arrest occurred. Iwakawa and Kimura (1922) stated that the pupils are almost always conspicuously constricted during this same period, and Yano (1937) found that tetrodotoxin applied to the cornea caused a slight pupillary constriction. Fukuda and Tani (1937*a*) observed that the toxin abolished or at least diminished the pupillary reflex.

Takahashi and Inoko (1890) reported that the pupillary dilatation which accompanied the onset of respiratory stasis is rapidly followed by constriction and later by a slow and somewhat incomplete dilation.

Iwakawa and Kimura (1922) claimed that direct application of the toxin to the eye, either *in vivo* or *in vitro*, produces no constriction of the pupil (*see* Yano, 1937). They concluded that the constriction of the pupil noted in fugu poisoning must be central in nature.

Iwakawa and Kimura (1922), however, demonstrated that direct application of adrenalin to the isolated eye caused pupillary dilatation when the eye was previously treated with tetrodotoxin. They also determined that topical application of adrenalin could prevent the pupillary constriction produced by the toxin. Atropine administration produced similar results to adrenalin administration. The mode of action of these drugs differed (atropine acts directly on the end organs, or target, while adrenalin acts through the terminal apparatus or mediator), but the effect in this condition would be expected to be essentially the same.

With topical application of the toxin, Ishihara (1918, 1924) studied the effect of the toxin on the sensory, oculomotor, and sympathetic nerves of the eye. He determined that the motor and sensory nerves were paralyzed nearly simultaneously and then later the sympathetic nerves were paralyzed. He stated that the oculomotor nerve was the first to regain activity and recovery and that the sympathetics to the eyes were paralyzed much later than other sympathetic nerves. He noted that the smooth muscle of the iris was not paralyzed by the toxin because it became active with the application of eserine. These results suggested that it was the nerve (pre- or post-ganglionic) or the ganglion that was paralyzed, because the application of eserine would be expected to reverse a paralysis in any of these locations. This is also presumptive evidence in favor of the hypothesis that there is no paralysis of the terminal conductive apparatus of the sympathetic nerves of the eye. Stimulation of the cervical sympathetics, however, failed to produce any response in pupil size. Experiments with nicotine indicated that the ganglion is paralyzed as well as the nerves, but it is probably paralyzed much more slowly than the nerve fibers. In his later work, Ishihara disagreed somewhat with his earlier statements in this regard. Topical application of the toxin to nerve trunks elsewhere in the body also paralyzed sympathetic fibers when they were present in the trunk. Without submitting supporting data, Ishihara stated that he believed the postganglionic fibers to be paralyzed before the pre-ganglionic fibers.

Additional experiments by Ishihara also indicated that the somatic fibers of the oculomotor nerves are paralyzed while the terminal conduction apparatus remains functional. This same relationship pertained to both parasympathetic nerves and nerve endings. Ishihara also reported that tetrodotoxin does not affect the intraocular pressure.

Takahashi (1897) thought that the effect of the toxin on the abolition of the corneal reflex was a result of the depression of respiration.

Iwakawa and Kimura (1922) reported eye movement and movement of the nictitating membrane due to the effects of tetrodotoxin, but these signs are difficult to interpret. Itakura (1917) studied the effect of the toxin on sensory functions of the eye. He injected and sprinkled the toxin into the eye and observed uniform paralysis of sensory function.

Fleisher *et al.* (1960) injected 50 μg of tetrodotoxin and saline into the anterior chamber of a rabbit's eye *in vivo* with a resulting prolonged mydriasis. They found that degeneration of past ganglionic adrenergic fibers to the eye prior to injection had no effect on the production of mydriasis with tetrodotoxin, which suggested no involvement of adrenergic pathways. Carbamylcholine (0.1 mg/kg) injected subcutaneously into a rabbit having mydriasis following intraocular injection of tetrodotoxin resulted in marked pupillary constriction, which indicated that tetrodotoxin has little or no effect on post-synaptic cholinergic receptors.

Effect on Adrenalin Discharge and Blood Sugar: Because adrenalin discharge is the most potent general physiological stimulus for the production of hyperglycemia, it is felt that the discharge of adrenalin and increased blood sugar levels might well represent evidence of the activation of a single mechanism.

Itakura (1917) observed that when tetrodotoxin was injected into rabbits the blood sugar content of the blood rose more or less without reaching a level sufficient to cause glycosuria. Tateyama (1942) also reported hyperglycemia following administration of fugu toxin. In addition, he noted a subsequent fall in blood sugar to hypoglycemic levels. On the other hand, without stating his experimental procedure, Kimura (1927) claimed that fugu toxin administration reduced the sugar content of the blood and noted no initial hyperglycemia. There are not sufficient data available to resolve this difference of opinion. It is interesting to note, however, that tetrodotoxin caused an increased secretion of adrenalin in the dog (Aomura *et al.*, 1930; Satake, 1954). This would seem to corroborate the work of Itakura (1917). On the other hand, Kimura (1927) claimed that the hypoglycemia is referable to a low adrenalin content of the adrenals. Kimura believed the hypoglycemic effect on rabbits was dependent on the nerves and was central in nature. Whether or not a direct effect exists is not known. It is possible that these variations in results may be due to species differences.

The value of administering adrenalin in order to reduce the toxic effects of puffer poison has been studied by several investigators (Ishihara, 1918; Iwakawa and Kimura, 1922; Kimura, 1927; Hashimoto and Migita, 1951). They reported that adrenalin administration weakened the effect of the toxin. By using intravenous injections of the toxin as well as subcutaneous injection, Kimura (1927) determined that the beneficial effects of adrenalin administration were probably due to its vasoconstrictor activity, because adrenalin administration had little effect on the course of the toxin in intravenous injections of the toxin. Since the toxin is rather rapidly detoxified in the body, any mechanism which delays the absorption of the toxin would weaken its effect.

Effects of Tetrodotoxin on Miscellaneous Organs and Tissues: Ishihara (1918) presented evidence that tetrodotoxin does not paralyze the nerve endings to the pilomotor muscle nor the muscle itself. He also found (1924) that the poison resulted in no increase in sweat secretion on the cat's paw, but rather a frequent decrease in sweat secretion which he attributed to paralysis of the medullary center and the cord. Because pilocarpine injection elicited sweat secretion, he reasoned that the nerve-end apparatus was not paralyzed. Kawakubo and Kikuchi (1942) reported no effect on salivation in cases of severe puffer poisoning, whereas Nomiyama (1942) found increased salivation as an early symptom of poisoning in the marmot.

Because tetrodotoxin produced hypoglycemia in some rabbits, which may be referable to carbohydrate storage, Kimura (1927) wondered if it might not also affect the CO_2 tension of the blood. It is not surprising that Kimura was unable to demonstrate such an effect since the two factors are unrelated.

It has been noted (Itakura, 1917) that tetrodotoxin ameliorates conditions of impotence; Itakura observed in rabbits erection of the penis which lasted for hours. The phenomenon also occurred in a spinal preparation. It was further demonstrated that such substances as yohimbine hydrochloride and magnesium sulfate when used to paralyze the lower part of the cord also caused erection. That this erection was "visceral" rather than "somatic" in nature is evident from the fact that the skeletal muscle was paralyzed and flaccid. The toxins, therefore, as Itakura reasoned, must work on certain structures in the lower section of the spinal cord related to autonomic or "visceral" erection. Itakura was confused in assuming the sympathetics to be primarily involved in erection. There is some question concerning explanation of the action of the toxin in causing erection because it is generally a paralyzing rather than a stimulating drug.

The physiological and pharmacological effects of puffer poison suggest that it might have anticholinesterase activity, but Yudkin (1945) who worked specifically on this matter was unable to demonstrate any anticholinesterase activity of the toxin with *in vitro* studies.

Itakura (1917) and Fukuda and Tani (1937*a*) have observed urinary incontinence resulting from tetrodotoxications. Inouye (1913) reported that large doses of tetrodotoxin caused a paralysis of micturition. He also noted that the urinary urge, produced in dogs by silver nitrate solutions injected into the bladder, appeared to cease after subcutaneous injection of one-half the lethal dose of tetrodotoxin. In the clinical use of the toxin, Inouye (1913) reported albuminuria, epithelial fragments, and frequent urination without urinary infection. Wada (cited by Yamasaki, 1914) found in the rabbit a slight atrophy of the kidney after repeated injections of the toxin. Itakura (1917) observed that tetrodotoxin in the rabbit caused albuminuria and cylinduria. He found no definite change in the kidney and stated that in human subjects with healthy kidneys he failed to note any renal changes as a result of therapeutic injection of the toxin.

Ogura (1959) reported that tetrodotoxin reduced the edema induced by 5-hydroxytryptamine, histamine, and albumin. He also concluded that tetrodotoxin did not directly affect the capillary permeability (Ogura and Nara, 1963).

Working with female rats, Ogura and Hori (1959) found that tetrodotoxin had no direct effect on the weight, suprarenals, uterus, thymus, and thyroid, but would increase the blood volume into these organs under certain circumstances. They also found that tetrodotoxin did not affect the basal metabolism of rats.

Ogura and Nakagima (1959) reported on disturbances of the righting reflex of cats due to the effects of tetrodotoxin.

Blankenship (1976) has provided an excellent review on the use of tetrodotoxin as a research tool dealing with the nature of the sodium channel in excitable membranes, the development of the quantal theory of synaptic transmission, and the evolution of excitatory-current channels in vertebrate muscle. Earlier reviews on the use of tetrodotoxin as a pharmacological tool have been prepared by Dettbarn (1971) and Gage (1971).

Fate of Tetrodotoxin in the Animal Body: According to Yano (1937) tetrodotoxin is not stored in the body (because accumulative effects are not noted) and is apparently readily metabolized by the body. Kimura (1927) determined that slow injections of the toxin increased the minimum lethal dose. This he attributed to detoxification in the body of the animal. The rate of this detoxification was such that a near lethal dose could be almost completely neutralized in an hour. Kimura next sought to determine if the poison was excreted, absorbed, or destroyed. By ligating the renal arteries he determined that excretion by the kidneys was not important. There was no difference in the minimum lethal dose for animals so treated compared with the control animals in which the renal arteries were not ligated.

Kimura also investigated the role of the liver in detoxification. Because the minimum lethal dose was not reduced by injecting the toxin into the portal circulation, he reasoned that the liver played no important role in destruction or fixation of the toxin. However, circulation time is so short in rabbits (the animals used in Kimura's studies) that little difference would be expected in the minimum lethal dose when it was injected by these two routes. Therefore, it is doubtful that Kimura's studies can be expected as ruling out the activity of the liver in detoxification of tetrodotoxin. In fact, Kabuki (1936, 1937) found that tetrodotoxin was more toxic when the liver was damaged and that it was less toxic when injected into the renal portal vein rather than into the marginal ear vein. Kabuki also found that perfusion of an isolated rabbit liver with tetrodotoxin resulted in the removal of the toxin from the solution. Kimura (1927) ruled out fixation by muscular tissue when he failed to reduce the minimum lethal dose by first passing the blood through the hind legs of rabbits. Naruke (1935), on the other hand, found that the toxicity of tetrodotoxin was suppressed in the rabbit when given with extracts of various tissues, viz., heart, brain, liver, and muscle.

With *in vitro* studies using blood, Kimura demonstrated that the tetrodotoxin was slowly inactivated. This represents, however, only a fraction of the ability of the animal's body to inactivate the toxin. Kariya (1914) assumed that blood alkalies destroyed the toxin, because he found that the toxin is destroyed by relatively strong alkali such as 0.6 percent sodium hydroxide. Although Kariya's data did not support his conclusion, it was demonstrated by Kimura that a bicarbonate solution of a titratable alkalinity similar to that of blood rather rapidly destroyed the activity of the toxin. Why the detoxifying activity of this bicarbonate solution was greater than that of the blood was not explained. Kimura also demonstrated that the active detoxifying principle in the blood was dialyzable and heat stable. It is now known that because of the blood buffer systems, titration is not a reliable method of determining blood alkalinity. It is not surprising that Kimura should have obtained a greater effect from the bicarbonate solution than from the blood because by his method of determining alkalinity, the bicarbonate solution was actually much more alkaline than the buffered blood.

CHEMISTRY

by Donovan A. Courville

The first recorded attempt to study the chemical and physical properties of the toxin of the puffer is that of Takahashi and Inoko (1889*a, b*). These workers noted some of the elementary properties, including solubilities, diffusability through a membrane, and failure to precipitate with the alkaloidal reagents. From the data obtained, it was concluded that the toxin was not an enzyme or an organic base.

The problem of isolation was taken up by Yashizawa (1894, cited by Suehiro, 1947) who isolated two crystalline toxic fractions. These were designated as "tetrodonin" and "fugamin."

More significant from this early period is the work of Tahara (1894, 1896, 1897*a, b,* 1910). Tahara also isolated two toxic fractions which he referred to as "tetrodonin" and "tetrodonic acid."[2] Later work indicated that tetrodonin was but an impurity containing small amounts of the true toxin and that the toxin itself had no true acid properties. Hence, these terms were abandoned in favor of the terms "tetrodotoxin" and "fugu toxin" which are more descriptive of the derivation of the toxin from the fish. These terms were in general usage until Yokoo (1950) isolated the toxin in relatively pure form. The term "spheroidine" was introduced to refer to this more potent product.

During the period roughly marked by the first half of the 20th century, other investigators have made contributions with reference to the chemical and physical properties of the toxin, and improvements in the methods of isolation were devised. Ishihara (1924) studied the properties of the toxin and concluded that the toxin molecule contained glucose as an ester. This conclusion was later challenged by Nagai and Ito (1939).[3] These workers concluded that all the nitrogen of the toxin was amino nitrogen; furthermore, approximately one-third of the nitrogen was also of amide character. Yokoo (1948*a, b, c,* 1950, 1952) described improved techniques of isolation including details of his isolation procedure for spheroidine. A series of papers by Tsuda and Kawamura (1950, 1951, 1952*a, b,* 1953), Tsuda and Shan-hai (1951), Tsuda and Umezawa (1951), and Tsuda *et al.* (1952) provide data which culminated in a tentative empirical formula for the toxin. In the presentation of his more recent work (Yokoo and Morosawa, 1955), Yokoo has shown an identity between his product (spheroidine) and that obtained by Tsuda. Of interest is the isolation by Kanayama (1943) of a toxic fraction from the puffer which contains phosphorus but no nitrogen.

Recent work on the chemistry of the toxin includes (1) attempts to isolate the toxic principle in a state of high purity and (2) attempts to determine the structural configuration. These aims have now been successfully attained through the researches of Tsuda, Goto, Woodward, and their coworkers.

---

[2] However, see Tsuda (1963) where this term is introduced to refer to a derivative of the toxin rather than to the toxin itself.

[3] More recent work has confirmed the position of Nagai and Ito.

Problem of Correlation of Data by Various Investigators

As one traces the early development of information relative to the chemistry of tetrodotoxin, it becomes apparent that certain discrepancies are to be found in the descriptions provided by various investigators. It is not always clear just what relationships exist between the various preparations obtained during this period of development. The fact that the same physiological and pharmacological properties seem to carry through the various preparations logically suggests that the variations are due largely to differences in the degree of purity. However, since different isolation methods have been employed, the possibility should not be overlooked that the preparations represented different hydrolysis products of an original more complex molecule which contained the active part of the structure. This problem was not completely clarified until the recent establishment of the structural relationship between the toxin and certain readily formed derivatives. The isolation by Kanayama (1943) of a toxic fraction containing phosphorus but no nitrogen suggested that the puffer may contain more than one toxic substance. To date this suspicion has not been confirmed.

Isolation Procedures

As is to be expected, the methods of isolation of the toxin of the puffer have undergone a process of evolution since the crude product was first obtained. These variations in isolation technique were introduced in an effort (a) to simplify the procedure, which at best is much involved, (b) to bypass certain steps which were suspected as highly destructive to the toxin itself, and (c) to eliminate the use of expensive processes. This review is limited to a brief description of the general procedures used by those investigators who employed their product for the study of the chemical or physical properties of the toxin. It is presumed that those workers who used their product for the study of pharmacological properties used either less pure products or used methods of isolation which did not vary appreciably from those given here. Because the procedures in many cases are highly involved and because the methods are now obsolete, no attempt is made here to describe the methods in sufficient detail to serve as isolation instructions.

The potency of various toxic fractions obtained in the processes of isolation is commonly expressed by the Japanese workers as a "toxicity of *x* gamma." The first expression adopted here should be interpreted to mean that *x* gamma of the product per gram of mouse weight are required to produce death in a 15- to 20-g mouse.

Fractionations by Takahashi and Yashizawa: Takahashi and Inoko (1889*a*, *b*, 1890, 1892) evidently secured their data on the toxin of the puffer using only extraction fractions and without the isolation of any product that might be referred to as even a crude toxin.

Yashizawa (1894, cited by Suehiro, 1947), following the leads provided by Takahashi, was able to isolate certain crude toxin fractions by dialysis through a cow bladder followed by removal of some of the impurities with lead acetate and alcohol which reagents did not precipitate the toxin. The essential features of the separation techniques were incorporated into those used by Tahara and hence no further detail is given here.

Tahara's Early Isolation Method (1897): Tahara's early method (1894, 1896, 1897*a, b*) involved, as the first step, the dialysis of a homogenate of fugu eggs using a cow bladder as membrane. The dialysate was concentrated by evaporation (presumably at ordinary pressure). Phosphoric acid and certain other impurities were removed by treatment with lead acetate, the lead being in turn removed by means of hydrogen sulfide. As the acidity increased during further concentrations, ammonium carbonate was added. When the product was reduced to a thick syrup, the toxin was precipitated in impure form upon treatment with alcohol. Addition of small amounts of water to the separated crystals resulted in the dissolution of the toxin, leaving behind certain water-insoluble materials which were removed by filtration. The small amount of chloride remaining was precipitated with silver hydroxide. On acidifying the filtrate, the toxin precipitated as the silver salt. The silver was removed with hydrogen sulfide. Evaporation of the filtrate from the silver sulfide yielded a light brown resinous material. This, on purification with charcoal and recrystallization, gave a crystalline material having toxic properties.

Tahara's Later Method (1910): Tahara later revised his method of isolation to eliminate the long process of dialysis and reconcentration and to reduce the amount of expensive reagents used. The ovaries used as the source of the toxin were washed to remove salt, homogenized with two volumes of water, heated for 3 hours on a steam bath, and the precipitated protein materials filtered off. The liquid was then reduced to about one-fourth volume by evaporation in weakly acid solution. Further protein which precipitated during the process was removed. Treatment with lead acetate served to remove phosphoric acid, chloride, and some other unidentified impurities. Addition of ammonia to the filtrate precipitated the toxin together with lead hydroxide and other extraneous materials. The isolated precipitate was washed with dilute ammonia, then with water, suspended in water and the lead removed by treatment with hydrogen sulfide. The filtrate was concentrated to a syrup under reduced pressure and the toxin precipitated by the addition of absolute alcohol. The separated toxin was dried over sulfuric acid in a desiccator. A still more toxic fraction was obtained by evaporation of the filtrate from the alcohol treatment to dryness *in vacuo* at low temperature.

These fractions were further purified by one or the other of two methods. The first method involved a reprecipitation of an aqueous solution of the toxin with alcohol and then ether. The second method was a repetition of the isolation procedure using lead acetate to precipitate impurities, followed by precipitation of the toxin with ammoniacal lead acetate and removal of the lead with hydrogen sulfide. Some samples were further treated with charcoal; others were not.

The toxicities of the fractions obtained were followed through the purification procedures. The most potent product required a dose of approximately 7 mg/kg of body weight to produce death in a period of 15 minutes. Since the toxicity values are based on death time and are not given in terms of minimum lethal dose, it is not possible to compare accurately the potencies of the various fractions. The minimum lethal dose for the rabbit was determined for only one fraction, and this was found to be between 3 and 4 gamma/g of body weight.

By treatment of the crude toxin solution with gold chloride, Tahara was able to remove an impurity having a very high nitrogen content. This base was named "tetronine." From the elementary analyses, an empirical formula $C_{11}H_{11}N_9O_2$ was hypothesized. The substance was not identified.

A further impurity separated out when a solution of the toxin was reduced to a small volume and permitted to stand for a period of time. The precipitate thus obtained had a sweet taste and the analytical values were in agreement with those of a simple sugar. This substance was believed to be a pentose, and Tahara gave it the name "tetrodopentose." This fraction was later identified by Tsuda and Kawamura (1950) as a mixture of mesoinositol and scillitol. These substances have elemental analyses identical to pentose.

The analytical data and properties of tetrodotoxin as given by Tahara are in terms of the product obtained after removal of tetronine and the mixture of mesoinositol and scillitol. Tahara's process of extracting tetrodotoxin was patented in the United States in 1913 (*see* Tahara, 1913).

Isolation by Ishihara: Ishihara (1924) evidently used the same method as described by Tahara (Ishihara, 1924; Nagai and Ito, 1939).

Isolation by Nagai and Ito (1939): These workers followed Tahara in using dialysis of an extract of the ovaries of fugu fish as the first step in the isolation. The dialysate was then concentrated to one-fifth volume under reduced pressure and the impurities were precipitated by successive treatment with mercuric sulfate, phosphotungstic acid, and silver nitrate. The excess of the heavy metals and phosphotungstic acid were removed with hydrogen sulfide and hot barium hydroxide respectively. The resulting solution was then evaporated under reduced pressure to give a slightly brown mass which was recognized as still containing some inorganic salt as impurity. The total nitrogen and amino nitrogen were followed during the process of purification and the ultimate purity of the product was assumed on the basis of approximate coincidence between total nitrogen and amino nitrogen.[4] The toxicity of the final product was given as 11.3 gamma.

Isolation by Hayashi and Miyagi (1946): Nagai (1954) refers to an isolation by Hayashi and Miyagi (1946) using aluminum oxide chromatography to yield a product with a minimum lethal dose of 1.0 gamma, but the details of the isolation were not available to him.

---

[4] Comparison of the toxicity values for the products with those later obtained on products of greater purity would seem to indicate that this assumption was not well founded.

Isolation Procedures Described by Yokoo: The isolation method for tetrodotoxin first reported by Yokoo (1948*a*) began with repeated extractions of fugu eggs with 0.5 percent formalin and concentration of the extracts under reduced pressure. The crude toxin was obtained from this concentrate in a manner similar to that described by Tahara (1910), using (1) lead acetate to remove impurities, (2) ammoniacal lead acetate to precipitate the toxin fraction, and (3) hydrogen sulfide to remove the lead. The filtrate was reduced to a small volume and the crude toxin precipitated with alcohol and ether.

The crude toxin thus obtained was carried through an involved process of purification in which phosphotungstic acid, phenylhydrazine, mercury picrate, and picrolonic acid were used as successive precipitating agents for impurities. The toxin was then precipitated with refianic acid and separated by filtration. The free toxin was released from the rufianate by treatment with wool and dilute sulfuric acid. The wool was removed by filtration and the sulfuric acid removed with lead acetate. The lead in turn was then removed with hydrogen sulfide. The filtrate from this process was reduced to a small volume *in vacuo* and the toxin fraction precipitated with methanol. The toxicity at this point in the isolation was 0.8 gamma.

Adsorption of the aqueous toxin on a chromatographic column of alumina, followed by elution with hot water, reduction of the volume *in vacuo*, and precipitation with methanol and ether gave a product with toxicity of 0.3 gamma. Repetition of the rufianic acid treatment and of the chromatographic separation eventually yielded a product with toxicity of 0.16 gamma.

In a subsequent paper (1948*a*), Yokoo described certain modifications and possible alternates in his technique. Techniques involving precipitations with ferric hydroxide, copper hydroxide, and silver hydroxide are described. The purity of the products obtained was not as good as by the previous method and evidently the procedures were abandoned.

In 1950, a further report appeared by Yokoo in which a simplified modification of his initial procedure is described. The procedure remained much the same through the step by which impurities were removed with picrolonic acid. Treatment with rufianic acid was eliminated. A chromatographic separation using a different preparation of alumina (AP8 by Japan Aluminum Co.) as adsorbing agent gave a toxic fraction with a toxicity of 0.09 gamma. Repeated recrystallizations from aqueous solution by methanol or methanol-ether mixtures resulted in a product with toxicity 0.01 gamma. This product was named spheroidine. Analysis: C, 40.40, 40.06; H, 5.54, 5.88; N, 11.12; and ash, 0.3 percent. The presence of sulfur in the ash was suspected. By the time Yokoo had made his 1952 report, he had secured larger samples of spheroidine and made certain corrections in the data. The toxicity was corrected to 0.013-0.014 gamma; the nitrogen content was corrected to 11.64 percent, and the suspicion that the ash contained sulfur was abandoned.

In a fourth report (1952), Yokoo described a still further simplified process of isolation. Chemical precipitations were omitted except for the initial precipitation with ammoniacal lead acetate. The chromatographic separation does not differ greatly from that previously described except that a different alumina of American production was used in the column. This product evidently proved more efficient than the Japanese varieties previously employed. This reference would seem to represent a description of the simplest isolation procedure of the toxin to date (Yokoo, 1952). This description given below starts with the product obtained by precipitation with lead acetate and removal of the lead with hydrogen sulfide, as described previously in this section.

Next a 5 × 35 cm. glass column is packed with 250 g. of alumina (Merck's alumina standardized for chromatographic adsorption according to Brockmann) and then the filling is washed by passing 250 cc. of water through. The solution of 20 g. of the above poison (M.L.D. 10 gamm/g.) in 400 cc. of water is passed through the previously washed column of alumina. The column is washed by passing 400 cc. of water through. The adsorbed poison is eluted with 3 liters of hot water. The eluted solution is acidified with acetic acid and then evaporated on the steam bath into a syrupy body. The residuary syrup is dissolved in 80 cc. of 80 vol. percent methanol, and filtered. The filtrate is again evaporated on the steam bath under reduced pressure into a syrupy body. To the syrup are added 2 cc. of water, and 20 cc. of methanol. To the resultant solution is added 80 cc. of ether. The precipitate of the poison is decanted and then dried over sulfuric acid in a vacuum desiccator. The yield is about 1,400 mg. The minimum lethal dose for the laboratory mouse is about 0.7 gamma/g. To the above yield of the dried poison is added 2 cc. of water and made alkaline with ammonia. The solution is left alone for an hour. Thus the poison is crystallized of itself. The white crystals of the poison are filtered with a suction and washed five times each with 1 cc. of water, two times each with 1 cc of methanol. The crystals are dissolved in 0.5 cc. of water with a small amount of acetic acid and again filtered and washed with 10 cc. of methanol. To the filtrate is added 30 cc of ether. The mixture is let stand for one day. Then it is again crystallized of itself. The white crystals of the poison are filtered with a suction and washed five times, each with 1 cc. of water, then two times each with 1 cc. of methanol. Next the crystals pass the same process once again to be recrystallized. The yield is about 16 mg.

Anal. Calcd, for $C_{12}H_{17}O_{10}N_3$; C, 39.65; H, 4.72; N, 11.57.

Found: C, 39.80, 39.49, 40.08; H, 5.23, 4.89, 5.48; N, 11.73, 12.07, 11.89.

Isolation by Tsuda and Kawamura (1950, 1952a, b, 1953): The crude toxin used by Tsuda and Kawamura (1950) was obtained from the Sankyo laboratory. The steps in its preparation by Sankyo laboratory are described as follows: The ovaries were extracted with hot water and the extract concentrated *in vacuo*. A precipitation with lead acetate in alkaline medium was carried out and the precipitate was redissolved in water, reprecipitated with 10 percent sulfuric acid, and centrifuged. The filtrate was again concentrated *in vacuo* and precipitated with lead acetate in methanolic ammonia and centrifuged. The filtrate was again concentrated *in vacuo* and precipitated with lead acetate in methanolic ammonia and centrifuged. The filtrate was again precipitated with alcoholic ammonia and the precipitate dissolved in water and precipitated with hydrogen sulfide. The precipitate of lead sulfide was filtered off and the filtrate evaporated to dryness *in vacuo* to give the crude poison.

Starting with this crude product, mesoinositol and scillitol were isolated from the portion insoluble in 75 percent alcohol (Tsuda and Kawamura, 1950). The filtrate, after removal of the mesoinositol and scillitol, was carried through a process of purification by means of precipitations with alcohol, lead acetate, and phosphotungstic acid. From the fractions referred to as D and F, which were least potent in terms of the toxin, glucose was isolated as the phenylosazone, the *p*-nitrophenylosazone, and the pentaacetate. Fraction E, which was most toxic, gave no osazone. This fraction was evidently regarded as still not representing the pure toxin.

Using one-dimensional paper chromatography (Tsuda and Kawamura, 1952*a*) with 90 percent phenol solution as the partitioning solvent, glycine, taurine, alanine, valine, leucine, and glucose were shown to be present as impurities in the crude toxin. The poisonous component had a very low R_f value, suggesting that a separation could be made by column chromatography. This procedure was attempted using an aqueous solution of the poison in a column composed of potato starch and celite. The column was developed by phenol-water until the solution reached the tip of the column. The column was then divided into 10 equal portions numbered 1 to 10 from top to bottom and each washed with ether and extracted with cold water. The extracts were evaporated *in vacuo*, the residues extracted with 70 percent ethanol without heat, and the alcoholic extracts evaporated to dryness *in vacuo*. Tests showed the toxin to be present only in parts 1 and 2. An unidentified reducing substance appeared in parts 8 and 10. The presence of glycine, taurine, alanine, valine, leucine, and phenylalanine was confirmed in the various parts. These same amino acids were found in the sulfuric acid hydrolysate of the ovaries of the globefish.

It was found that the amino acids and glucose could be separated from the toxin by means of adsorption chromatography using an active charcoal (Edokol). The toxin was eluted with 5 to 10 percent ethanol while the amino acids and glucose were eluted by water but not by the dilute alcohol.

The toxin thus obtained was further purified by development through a chromatopile composed of 500 filter papers (Tsuda and Kawamura, 1953). Papers 50 to 70 were allowed to adsorb aqueous toxin solution to the extent of 500 mg (potency 0.2 gamma) and the pile then developed with 90 percent phenol-water as the partitioning solvent. The papers at R_f 0.0 to 0.1 were removed, washed free of phenol with ether, and digested with cold water. This extract was evaporated under reduced pressure. The residue was taken up in dilute acetic acid and precipitated with alcohol to give, on recrystallization, crystals of the purified toxin with a potency of 0.01 to 0.12 gamma. Anal. Found: C, 41.88; H, 5.98; N, 11.81.

The toxin thus obtained was carried through a process of five recrystallizations to a constant toxicity of 0.010 to 0.009 gamma. The elemental analyses of the 5 products varied through the following ranges but not in order: C, 40.23-41.14; H, 5.25-5.88; N, 12.68-13.28.

Isolation by Shirota, Fujita, and Kawamura (1952): Starting with a sample of the toxin as produced by Tahara (minimum lethal dose 5 gamma) these workers attempted a further purification by chromatography using activated carbon and Hyflo Super Gel in the column. An aqueous solution of the toxin was poured into the column followed by distilled water. The refractive index and toxicity of the collected fractions were followed. It was noted that the toxin did not appear until the refractive index of the eluate approached that of distilled water. At this point a fraction was obtained containing the toxin. Concentration of this fraction yielded a product some ten times as potent by weight as the starting material (minimum lethal dose 0.43 gamma).

The purification by dialysis was studied quantitatively, using Tahara's toxin as starting material. By using either a cow bladder or cellophane, it was found that the MLD increased from 5 gamma to 4 gamma by the process.

Kanayama's Toxin and Its Isolation: Kanayama (1943) reported the isolation of a toxin from the puffer which contained phosphorus but no nitrogen. Other than the comment by Nagai (1954) that no other worker had observed phosphorus in puffer toxin, the literature is essentially silent on Kanayama's toxin. Many of the observed differences in properties of this toxin from the toxin commonly referred to as "tetrodotoxin" might be explained on the basis that again he was dealing with an impure sample of the same toxin. However, it is rather difficult to explain the absence of nitrogen on such a basis. It would seem that Kanayama isolated a toxic fraction from the puffer which is not the same as tetrodotoxin. If so, the presence of such a toxin has not been confirmed. The use of mercuric acetate instead of lead acetate would seem to be the essential point of difference in the isolation procedure.

A water extract of the puffer ovaries was reduced to about one-fifth volume under reduced pressure at 50° C. An equal volume of methyl alcohol was then added and the precipitate formed was filtered off. The volume of the filtrate was again reduced and phosphotungstic acid added. The precipitate formed was filtered off, and the excess phosphotungstic acid was removed by the addition of barium hydroxide. The volume was again reduced and further precipitation of impurities was accomplished by the use of mercuric acetate. After removal of the precipitate, the toxin was precipitated on addition of sodium carbonate and mercuric acetate. The addition of mercuric acetate was stopped when the precipitation of red-brown mercuric oxide began to form. The toxin was separated by centrifugation and washed with 50 percent methyl alcohol, dissolved in water, and the mercury removed with hydrogen sulfide. The mercury sulfide was filtered off and the precipitation with mercury acetate and sodium carbonate repeated with subsequent removal of the mercury.

The filtrate was evaporated to dryness *in vacuo* and the residue treated with methyl alcohol, leaving an insoluble residue containing taurine, and possibly other impurities. The filtrate contains the crude toxin. Further purification was accomplished by precipitation with pyridine and a final precipitation of the toxin with absolute alcohol.

Isolation Procedure by Nagai (1956): The earliest report on the isolation of the toxin using ion-exchange techniques is that by Nagai (1956). The toxin was extracted from the viscera with six volumes of hot water (60°-80°) for 5 hours. The residue showed but minor amounts of unextracted toxin following this procedure. A column of Amberlite IRC-50, converted to the sodium form, was used to absorb the toxin from which it was eluted with 0.5*N* HCl. The HCl was then removed by passage through a second column containing Amberlite IR-4B. The HCl-free solution was reduced in volume to a pulpy consistency and the toxin extracted several times with absolute alcohol. After evaporation of the alcohol, the crude toxin remained as a brown greasy mass which was taken up in water. This aqueous solution was then passed through a third column containing Amberlite IRC-50 in the hydrogen form. In this process, the remaining impurities are fixed to the resin while the liquid eluant contains the toxin. The eluate was reduced to a small volume and the toxin precipitated with 5 percent $HgCl_2$ solution. After refrigeration overnight, the Hg^{+2} ion was precipitated with H_2S and the Cl- ion was removed by passage through a column of Amberlite IR-4B. The toxin was then precipitated from the resulting solution as fine needle-shaped crystals with "pressure." Recrystallization was from hot 70 percent ethanol.

Elemental analyses of the products from different lots showed significant variations in the C, H, and N content even after several recrystallizations, suggesting that the toxin was not absolutely pure. However, the remaining impurities may have been largely or entirely in the nature of derivatives of the toxin, produced in the process of isolation. This possibility is supported by the report of Goto *et al.,* (1964*a*). Employing a method of isolation which, from the abstract of the report would seem to be based on the procedure of Nagai, the toxin obtained contained anhydroepitetrodotoxin as an impurity. Separation of the toxin from this derivative was accomplished by conversion to the picrate and reconstitution of the toxin by hydrolysis to the hydrobromide.

Isolation of Puffer Toxin in Quantities: With the recognition that the problem of the structure of puffer toxin would probably not be solved until a satisfactory method had been devised for obtaining the pure toxin in quantities, the development of such a process was undertaken by Cutter Laboratories (Berkeley, Calif.) under Army Chemical Corps contract No. CML-4564, Project No. 4-08-03-001 (1954). With the development of a process which seemed feasible for transfer from a laboratory scale to a production scale, a contract for quantity production of the toxin was granted to Loma Linda University. This work was under the direction of George J. Nelson.

The raw materials were obtained by arrangement with Tokiharu Abe of the University of Tokyo, Japan, who contracted in turn with certain commercial fisheries for the collection and storage of the liver and gonads of puffer. The raw materials were shipped under refrigeration at the direction of the chemical operations officer of the U.S. Army General Depot in Japan, and on arrival were kept in cold storage until used. The essential data relative to the isolation process are here briefly described.

The materials were worked up in batches of 100 pounds, the yield of toxin from each batch amounting to between 50 and 100 mg. The thawed material was ground in a power meat grinder and covered with 20 gallons of methanol. Concentrated sulfuric acid was added with stirring until the mixture was definitely acid to indicators. The mixture was allowed to settle for 2 or 3 days and the supernatant liquid drawn off. This was followed by a second similar extraction. The first extract was used as the starting point for the subsequent step while the second extract was used as the medium for the first extraction of the next batch.

Starting with 40 gallons of accumulated extract, an acetone precipitation of the toxin was carried out. Twenty-five gallons of acetone was introduced into a 35-gallon stainless steel tank equipped with a power stirrer and a bottom drain. One pound of Johns-Manville Cellite filter aid was mixed with the acetone and pumped through a 4-chamber filter press (15 by 15 inches) for a precoat. The system was arranged so that the solvent could be recirculated or discharged into a solvent storage tank as desired. Five gallons of the methanol extract was added to the acetone, a half-pound of Cellite added to the mixture, the pH adjusted to 6.5-7.0, and after 14 minutes the mixture was pumped through the filter. This was repeated batchwise until the 40 gallons of extract had been filtered free of the precipitate formed by the addition of the acetone.

The solid phase containing the toxin and the filter aid was removed from the filter frame and the residual acetone permitted to evaporate in the open air for 2 or 3 days. Extraction of the toxin from the filter aid was accomplished with demineralized water, or with eluate from the CS-101 ion-exchange columns used in a subsequent step and followed with demineralized water. The use of 8 to 10 gallons of extraction water removes 95 percent of the toxin. The pH of the extract was adjusted to 7.0 with saturated sodium carbonate solution, and the extract was stored for several days at 35° C. A gummy precipitate settled out which contained no toxin but interfered with the adsorption of the toxin on the ion-exchange column.

The supernatant from this precipitation was passed through three columns of CS-101 resin arranged in tandem. The columns were 3 inches in diameter and made of borosilicate glass. These were filled to a depth of 48 inches with the sodium form of the resin. The resin was washed by passing about 10 gallons of demineralized water through the columns at a rate of about 1½ liters per hour. Elution of the toxin from the resin was accomplished by means of 8 to 10 gallons of 1 *N* acetic acid solution. The toxin appeared when the pH of the effluent dropped below 7.4, at which point the eluant was collected in 3-liter fractions. The fractions were assayed for toxin content.

Fractions containing 150 or more mouse units per milliliter were combined for the subsequent steps while the less concentrated fractions were used as solvent for acetone precipitation as previously noted.

The eluate was reduced in volume at 40° C by means of a Rinco Rotating Vacuum Evaporator with suitable condensing apparatus kept at near minus 20° C. Evaporation was continued until there was no further evidence of deposition of volatile amines on the connecting tubes leading to the condensing flask. The unevaporated residue was dissolved in about 1 liter of water and the pH adjusted to 6.5.

The solution of the toxin was then pumped through two ion-exchange columns in series containing thoroughly washed XE-80 resin (16-50 mesh) in the sodium form. The columns were 2 inches in diameter and 60 inches high. Each column was separated into three sections by plastic discs provided with a groove for an 'O' ring. This 'O' ring held a nylon screen which served to prevent the resin particles from passing through the holes in the disc. In case the liquid front became irregular, this arrangement served to restore a front which was perpendicular to the axis of the column. Each segment of the column was filled to a depth of about 15 inches with the resin.

After the toxin solution had been put on the column, demineralized water was passed through until the effluent was clear. The toxin was then eluted with about 8 gallons of 0.08 *N* acetic acid. When the pH of the effluent dropped below 7.0, fractions were collected. Fractions containing over 200 mouse units per millimeter were combined and again concentrated to a thick syrup on the Rinco Rotary Evaporator. Fractions with lower concentration were combined with the next batch to be passed through the XE-89 system.

The residue from the vacuum evaporation was dissolved in a minimum of demineralized water. A pilot precipitation of the toxin was made on a 10-20 ml sample by careful adjustment of the pH to 8.2 with near saturated sodium carbonate. The toxin precipitated on overnight refrigeration at 35° C. This precipitate was used to seed the main portion similarly treated. If the toxin failed to precipitate after a few days, the whole batch was processed again through the XE-89 system. Separation of the crystallized toxin was by centrifugation. Final purification was by repeated recrystallization.

The resins used in the columns required regeneration after processing each batch. This was accomplished by a reverse flow through the columns, first with 5 percent sodium hydroxide solution followed by a solution consisting of 200 g of sodium chloride and 500 ml of 5 percent sodium hypochlorite dissolved in 5 gallons of water. Five gallons was required for regenerating the two XE-89 columns and 15 gallons for regenerating the three CS-101 columns. The columns were then washed with demineralized water to displace the regenerating solution, then with 5 percent sulfuric acid to convert the resin to the acid form. The acid was then displaced with demineralized water and the entire process repeated three times. Ordinary commercially available distilled water was found to be inadequate for use throughout the isolation procedure.

The toxin obtained by this process of isolation over a period of years (1957-64) was released to Dr. R. B. Woodward at Harvard University through the U.S. Army Chemical Research and Development Laboratories to be used in studies to determine the structure of the poison. The results of these studies are reported in a subsequent section of this chapter.

Properties of Tetrodotoxin (Spheroidine)

Physical Properties: Tahara (1910) described the purified toxin as a colorless, amorphous, hygroscopic powder, without odor or taste, dialyzable, very soluble in water, soluble with difficulty in ethanol, and insoluble in the other organic solvents.

Some of these properties had been observed by Takahashi and Inoko (1889*a*) and were recognized by Tahara at an earlier date (1894). From observations on purer samples of the toxin produced later, it would seem that some of these early conclusions were misconceptions due to the impurity of the products observed. More recent studies, for example, indicate that the toxin is very slightly soluble in water (Tsuda and Kawamura, 1952*b*) and has a degree of solubility in methyl alcohol (Yokoo, 1952). Later investigations also fail to indicate any hygroscopic property of the toxin as reported by the early workers.

Tahara (1897*a, b,* 1910) indicated that the toxin had no melting point and carbonized on heating. Yokoo (1950) and Tsuda and Kawamura (1952*b*) reported similar observations on the purified toxin. Ishihara (1924) on the other hand gave a melting point of 120° C for the toxin. Kanayama (1943) reported a melting point of 155° C for his isolated toxin, which is probably not identical with the tetrodotoxin of other workers.

Early reports described the toxin preparations as amorphous. Yokoo (1950) obtained the toxin in the form of crystal needles. Tsuda and Kawamura (1952*b*) used the terms "silky needles" and "prismatic crystals" in referring to the crystalline form.

Ishihara (1924) gave the specific rotation value of his product as $[\alpha]_D +17°$ to $+27°$. Nagai and Ito (1939) gave the specific rotation as $[\alpha]^{15}_D$ -22.85°. Tsuda and Kawamura (1953) gave it as $[\alpha]_D$-8.64°.

Nagai and Ito (1939) reported that adsorbing agents such as permutite, aluminas, kaolin, and fuller's earth "had no effect" on the toxin. Tahara (1910) reported that when he did not use charcoal in the process of purification, his product was more toxic than when it was used. Yokoo (1952) used alumina as the adsorbing agent in his chromatographic separation from which the toxin was eluted with hot water. Tsuda and Kawamura (1952*a*) used potato starch and celite as the adsorbing agent and noted also that the toxin could be adsorbed on activated charcoal (Edokol) and eluted with 5 to 10 percent ethanol.

Yokoo (1948*a*) reported that the toxin migrates toward the negative electrode during electrophoresis. Kanayama (1943) indicated that even at low pH conditions, the toxin migrated toward the positive electrode.

Precipitation Reactions: Tahara (1897*a*) reported that the toxin is not precipitated by gold chloride, platinum chloride, mercuric chloride, phosphotungstic acid, tannic acid, picric acid, Mayer's reagent, or iodine in potassium iodide. It is not precipitated by lead sugar [sic], lead acetate in acid solution, or any of the alkaloid or albumin precipitating agents, but is precipitated by ammoniacal lead acetate (Tahara, 1910). Nagai and Ito (1939) observed that the toxin could be purified of certain impurities by precipitation with mercuric sulfate, phosphotungstic acid, or silver nitrate without simultaneous precipitation of the toxin. Yokoo (1948*a*) reported that the toxin could not be precipitated with phosphotungstic acid, mercurous picrate, picrolonic acid, flavianic acid, Reinecke's salt, laudaniric acid, phenylhydrazine, nitranilic acid, or quinine but that it could be precipitated with rufianic acid, copper hydroxide, ferric hydroxide, and silver hydroxide.

Stability of Tetrodotoxin: The relative stability of the toxin to cold mineral acids and its instability to warm mineral acids and to alkalis was recognized by Tahara (1910). It was his conclusion that boiling with dilute acetic acid had little or no destructive effect and that boiling with glacial acetic acid produced partial destruction only after long periods. He likewise was of the opinion that boiling in water for short periods did not destroy the toxin.

Ishihara (1924) reported the toxin stable in boiling water for three hours. A loss of one-half of the activity was noted at 4 hours and loss of 80 percent of the activity at 6 hours. It was further noted that the toxin was stable to cold concentrated mineral acids but that boiling 1 hour with one-tenth volume of concentrated hydrochloric acid was sufficient to give complete destruction. Shirota *et al.*, (1952) reported that a loss in toxicity was experienced when the toxin was boiled for 30 minutes in 10 percent acetic acid and that if heated in water for 30 minutes at 100° C, 90° C, and 70° C, the toxicity was reduced by 20 percent, 10 percent, and 5 percent respectively.

Tahara (1910) cautioned that in the precipitation of the toxin with ammoniacal lead acetate, the concentration of free ammonia should not exceed 0.34 percent because of the sensitivity of the toxin to alkali. The basis for taking this concentration as the maximum tolerable was not given.

Ishihara (1924) observed no change in activity on storage at -15° C for 12 hours or from exposure of the toxin to sunlight for 20 days. It was also found stable to the action of trypsin, ptyalin, emulsion, invertin, and bile but totally destroyed by 0.2 percent pepsin solution in 0.5 percent hydrochloric acid in 4 hours. The toxicity was lost on treatment with 10 percent calcium chloride solution of 10 percent iodine trichloride solution. Salts in general had no effect.

Yokoo (1950) found that the toxicity was destroyed by the action of 2 percent sodium hydroxide treatment in the cold for 90 minutes, but that part of the toxicity was restored by retreatment with acid. It was on this basis that Yokoo concluded that the molecule contained a lactone ring which was necessary for the toxicity of the molecule and that the loss of activity was due to the opening of this ring by alkali and the restoration of the toxicity in part was due to a reversibility of this ring rupture. This concept of a lactone ring was later abandoned (Yokoo and Morosawa, 1955) in the light of a more detailed study of this reversible loss of toxicity. It was found that if a sample of the toxin with toxicity 0.01 gamma were treated with dilute acetic acid, the potency dropped rapidly to a toxicity value of 0.017-0.021 gamma and then decreased gradually to still lower toxicity. If the solution was made alkaline at the point of the rapid drop, all toxicity was immediately lost. If this was then reacidified, the potency again returned to the value of 0.017 but did not return to the original toxicity value.

Studies Prior to 1956 on the Constitution of the Tetrodotoxin Molecule

Studies on Reactive Groups in the Tetrodotoxin Molecule: Many qualitative tests have been carried out in attempts to determine the chemical nature of tetrodotoxin. In some cases the same test has been conducted by more than a single worker with conflicting results. This is perhaps to be expected since, until the isolation of spheroidine by Yokoo (1950), the products used contained variant impurities depending on the method of isolation and purification. Where conflicting results are reported, it would seem that the more recent results of Yokoo and Tsuda would be more valid.

The reported results of qualitative tests made on the toxin are given in Table 2.

Tetrodotoxin forms crystalline salts with hydrochloric and tartaric acids, both being hygroscopic (Tsuda, Kawamura, and Hayatsu, 1960). The salts are soluble in water, the free base precipitating on neutralization. The free base can also be obtained by dissolving in an acetic acid-ethanol mixture and precipitating by the addition of ether. According to Tsuda's report, the toxin does not form a hydrate. It would seem clear from Woodward's report (Woodward, 1964) that the toxin is a hydrate. One mole of nitrogen is

TABLE 2.—*Qualitative Tests used on Tetrodotoxin*

TEST	GROUPS TESTED	INVESTIGATORS	DATE	RESULTS
Fehling's test	Reducing groups	Tahara	1910	+
		Shirota *et al.*	1952	Weak
		Tsuda & Kawamura	1953	—
Silver mirror test	Reducing groups	Tahara	1910	Strong
		Ishihara	1924	+
		Yokoo	1948*b*	+
		Tsuda & Kawamura	1953	Weak
Potassium ferricyanide test	Reducing groups	Tahara	1910	—
Trommer test	Glucose	Ishihara	1924	+
Barfoed test	Reducing carbohydrates	Ishihara	1924	+

TABLE 2.—*Qualitative Tests used on Tetrodotoxin*—Continued

TEST	GROUPS TESTED	INVESTIGATORS	DATE	RESULTS
Molisch test	Carbohydrates	Ishihara	1924	Weak before but clear after hydrolysis
		Tsuda & Kawamura	1953	+
Osazone test	Reducing carbohydrates	Ishihara	1924	— before but + after hydrolysis
Dinitrophenylhydrazine test	Reducing carbohydrates	Yokoo	1948*b*	—
Orcinol test	Pentose	Ishihara	1924	+
Mucic acid test	Galactose	Ishihara	1924	—
Naphthoresorcinol test	Glucuronic acid	Ishihara	1924	—
Not stated	Mosite	Ishihara	1924	—
Not stated	Formose	Ishihara	1924	—
Lauth's test	Aromatic rings	Nagai & Ito	1939	—
Ninhydrin test	Amino acids	Nagai & Ito	1939	+
		Tsuda & Kawamura	1953	—
		Shirota *et al.*	1952	Weak
Nitrous acid test	Primary amines	Nagai & Ito	1939	All N is amino 30 percent also amide
		Shirota *et al.*	1952	No change in toxicity with HNO_2 treatment
Biuret test	Proteins	Nagai & Ito	1939	—
		Tsuda & Kawamura	1953	—
		Tahara	1910	—
		Ishihara	1924	—
Millon's test	Tyrosin	Nagai & Ito	1939	—
		Ishihara	1924	—
Murexide test	Purines	Nagai & Ito	1939	—
		Ishihara	1924	—
Sakaguchi's test	Guanidine group	Nagai & Ito	1939	—
		Tsuda & Kawamura	1953	—
Diazo reaction	Aromatic amines	Ishihara	1924	—
Hypobromite test	Primary amines	Nagai & Ito	1939	—
Tarugi Lenci reaction	Amines, amides	Tsuda & Kawamura	1953	Weak
Pauly reaction	Pyrimidines and imidazols	Tsuda & Kawamura	1953	—
Ehrlich reaction	Indoles	Tsuda & Kawamura	1953	—
Xanthoproteic acid test	Tyrosin	Tsuda & Kawamura	1953	—

TABLE 2.—*Qualitative Tests used on Tetrodotoxin*—Continued

TEST	GROUPS TESTED	INVESTIGATORS	DATE	RESULTS
Sulfur test	Sulfur	Ishihara	1924	+
		Nagai	1954	+ (from impurity)
		Yokoo	1948*b*	?
		Yokoo	1950	—
Bromine water	Double bonds	Nagai & Ito	1939	—
Acetic anhydride	Alcohol and amine groups	Nagai & Ito	1939	—
		Yokoo	1948*b*	+
Benzoyl chloride	Alcohol groups	Nagai & Ito	1939	—
Iversen test	Phosphorus	Tsuda	1953	—
Beilstein test	Halogens	Tsuda	1953	—
Ferric chloride	Phenols, enols, oximes, etc.	Nagai & Ito	1939	—
Permanganate	Reducing groups	Nagai & Ito	1939	+

released on treatment with sodium hypobromite but not with sodium nitrite and acetic acid. Treatment with periodic acid for 1 hour at 25° C resulted in an uptake of 2 moles; after 48 hours, 3.2 moles. These results were taken to prove the presence of an α-glycol group in the toxin molecule.

Efforts prior to 1956 to Establish the Empirical Formula of the Toxin: Tahara (1910) suggested the empirical formula $C_{16}H_{31}NO_{16}$ based on the analysis of his purest product. Evidently no molecular or equivalent determinations were made.

Yokoo (1948*a*) arrived at the formula $C_4H_7O_3N$. This formula was fortified with a molecular weight determination by the cryoscopic method to give a value of 116. Analysis of derivatives seemed to support the formula also. However, later work (1950) would seem to show this early data to be in error, unless Yokoo was dealing with some smaller fragment of the toxin which still retained its toxicity. More probable is the assumption that the molecular weight determination was in error and that the correct formula approached $C_4H_7O_3N_3$. Based on the analysis of spheroidine, Yokoo (1952) suggested an empirical formula of $C_{12}H_{17}O_{10}N_3$. The molecular weight by the cryoscopic method was found to be 335. At that time, Yokoo believed the toxin to contain a lactone group and an amine group. Titration of what was considered to be a lactone group gave equivalent weight values of 364 and 350. Titration of the latter group gave equivalent weight values of 365 and 355. As previously noted, Yokoo later abandoned the concept of a lactone group being present and regarded this titration as representing a change from the hydrochloride of the amine to the free amine (Yokoo and Morosawa, 1955). This concept was also shown later to be in error (Woodward, 1964).

Tsuda and Kawamura (1953) made equivalent weight determinations on a series of preparations representing the products of consecutive recrystallizations. The toxicity and elemental analysis were followed through the series of recrystallizations. The final product of crystallization was assumed to be pure toxin on the basis of constancy of toxicity. Tsuda did not suggest an empirical formula based on the results of his analysis. It is apparent that his data are not in satisfactory agreement with the formula suggested by Yokoo($C_{12}H_{17}O_{10}N_3$)(*see* Table 3), but are in good agreement with values calculated for $C_{11}H_{19}O_8N_3$. It is possible that this hesitancy may have risen from the fact that essentially all the molecular and equivalent weight determinations have resulted in values more in harmony with the formula suggested by Yokoo. The molecular weight for $C_{11}H_{19}O_8N_3$ is only 321.

From their study on the reversible loss of toxicity in alkali, Yokoo and and Morosawa (1955) conceived the idea that there were two forms of the toxin with different toxicity values and that the change from one form to the other was reversibly brought about by alkali and acid. This concept was con-

TABLE 3.—*Analytical Data of Tsuda and Yokoo for Tetrodotoxin (Spheroidine)*

Author	Description of sample	Percent			Molecular and equivalent weights	Toxicity in gamma/gm mouse
		C	H	N		
Tsuda & Kawamura, 1952*b*	Product prior to final series of crystallizations	41.88	5.98	11.81	—	0.01–0.012
Tsuda & Kawamura, 1953	Product after recrystallization to constant toxicity	40.23	5.41	12.85	363	0.009–0.01
		41.14	5.36	12.65	352	
		40.92	5.88	13.00	355	
Yokoo, 1950	First isolation of spheroidine	40.40	5.54	[a] 11.12	—	[b] 0.01–0.011
		40.06	5.88	—	—	
Yokoo, 1952	Product described in 1952	39.80	5.23	11.73	350–364E.W	—
		39.49	4.89	12.07	335M.W	—
		40.08	5.48	11.89	—	—
Yokoo & Morosawa, 1955	Form I	([c])	([c])	([c])	([c])	0.01
Yokoo & Morosawa, 1955	Form II	40.90	5.84	13.04	—	[d] 0.02
	Calculated for $C_{12}H_{17}O_{10}N_3$	39.65	4.72	11.57	363	—
	Calculated for $C_{11}H_{19}O_8N_3$	41.1	5.92	13.08	321	—

[a] This value was refined to 11.64 in 1952.
[b] This value was refined to 0.013–0.014 in 1952.
[c] No data given but presumed to be that of the 1950 product.
[d] When the method of Tsuda and Kawamura was employed in the assay, this value was 0.01 and identical to that for Tsuda's purified product.

firmed by the actual isolation of two forms of the toxin. If the toxin were crystallized from a small amount of dilute acetic acid by the additions of methanol or ethanol and much ether, the toxin crystallized with a toxicity of 0.02 gamma. This form is referred to as "II." If II were dissolved in a small amount of dilute acetic acid and crystallized by the addition of ethanol only and kept in the refrigerator, form "I" was obtained. Crystallizations at slightly elevated temperatures gave a product with toxicity values between those of forms I and II.

The absorption spectra and the paper chromatograms for the two forms were found to be identical. The analytical data for the two forms, however, differed. Data obtained from form II approximated closely those obtained by Tsuda and Kawamura on their purified product (*see* Table 3). Form II and the purified toxin of Tsuda and Kawamura also gave identical toxicity values when both were assayed by the method of the latter workers (0.01 gamma). Yokoo and Morosawa, however, obtained a toxicity value for form II of 0.02 gamma. The discrepancy between the values in the two cases suggests that care must be used in attempting to correlate the products of various investigators in terms of the toxicity values since evidently small differences in procedure may give somewhat different results.

For convenience of comparison, the analytical data and toxicity values of the toxin as prepared by Tsuda and Kawamura, and Yokoo are brought together in tabular form in Table 3.

Studies prior to 1956 on the Hydrolysis Products of Tetrodotoxin: Tahara was the first investigator to report results from the study of hydrolysis products of tetrodotoxin (1910). Since it is now apparent that his product was far from pure, it is not possible to determine whether the hydrolytic products isolated were from tetrodotoxin or from some impurity in his product. A brief outline of his results is presented with no attempt at evaluation.

The toxin was hydrolyzed by boiling with 20 parts of 10 percent hydrochloric acid for 3 hours. The products of hydrolysis were evaporated to near dryness in order to remove the hydrochloric acid. The residue was taken up in water and filtered through animal charcoal. Precipitation with gold chloride gave two gold salts. The first was brick red and the other yellow, melting at 181° d and 184° d respectively. Only the latter was subjected to analysis and this was found to contain 44.65 percent gold. After removal of the gold with hydrogen sulfide, Tahara obtained a colorless crystalline substance, easily soluble in water but only slightly soluble in alcohol. It liberated ammonia on heating with potassium hydroxide. The hydrochloride of the base did not melt even at 200° C. It formed a yellow crystalline picrate on treatment with picric acid. The picrate had no melting point. The filtrate from the gold chloride precipitation was evaporated to dryness, the residue taken up in a very small amount of water, and on addition of alcohol, a white powdery precipitate was obtained which was very hygroscopic. This product had acid properties, decomposing barium carbonate readily with liberation of carbon dioxide. It was precipitable with silver nitrate but not with mercurous chloride. It formed a gelatinous precipitate with copper acetate which became granular when warmed. This copper salt was readily soluble in dilute acetic acid but not in water. Analysis gave the following data: C, 19.85; H, 2.78; Cu, 38.22; N, 0.0. The products described were not identified.

Tahara (1910) attempted an alkaline hydrolysis but was unable to obtain any recognizable products from the experiment.

Ishihara (1924) reported the presence of glucose among the hydrolytic products of the toxin. The probability that this came from the hydrolysis of an impurity has been noted previously.

Hayashi and Miyagi (1946, cited by Nagai, 1954) reported that the hydrolysate of the toxin showed a positive reaction for glucosamine and gave a positive Sakaguchi reaction. Nagai (1954) considered the results to be due to impurities in the toxin sample.

Studies prior to 1956 on the Chemical Structure of the Toxin: Tahara (1910) concluded from his study that tetrodotoxin was not an alkaloid or a protein. The first conclusion was evidently based on the failure of alkaloid precipitants to precipitate the toxin; the second on the low nitrogen content. More recent determinations of the nitrogen content show that the values obtained by Tahara were too low. However, the later data on nitrogen content still show less nitrogen than is found in proteins.

Ishihara (1924) was convinced that the toxin contained glucose which was liberated on hydrolysis and that since the toxin was relatively stable to cold acid, the glucose must be linked as an ester rather than as glucoside. The presence of glucose as part of the toxin molecule was challenged by Nagai and Ito (1939) on the basis of ready removal of carbohydrate without involving any hydrolytic procedures. It is of course possible to explain the results obtained by Ishihara on the basis that his glucose was obtained from the hydrolysis of an impurity in his toxin preparation rather than from the toxin. This point has been eventually clarified in favor of the contention of Nagai and Ito. With the establishment of the structural formula of the toxin, it has become clear that it is quite out of the question that glucose could have been formed by hydrolysis of the toxin.

Tsuda and Kawamura (1953) regarded the toxin molecule as an acid amide of a polyhydroxyacyl compound. These investigators were not able to get a positive Fehling's test on their highly purified product but still obtained a strong positive Molisch test. Whether this resulted from carbohydrate impurity in what they regarded as a highly purified product, or whether the toxin molecule is capable of degradation to a furfural derivative is not apparent.

Nagai and Ito (1939) concluded that all of the nitrogen of the toxin was amino nitrogen but that approximately one-third was also amide nitrogen. Shirota *et al.* (1952), however, reported that no change in toxicity occurred when the toxin was treated with nitrous acid.

Tsuda and Kawamura (1953) interpreted the result from spectrophotometric studies as indicating the presence of OH (or NH) and $CONH_2$ (or CN) groups in the molecule. Two double-bonded linkages are indicated which were assumed to be the double bonds of the carbonyl groups of the amides. Absence of absorption in the region 5.7-5.9μ was taken as evidence that the molecule does not contain any ester, lactone, lactam, ketone, or aldehyde type carbonyl. Yokoo obtained infrared spectra from both forms I and II of the toxin which were quite identical to that obtained by Tsuda. The paper chromatographs of the two products were also identical.

Yokoo and Morosawa (1955) suspected a possible relation of the toxin to the pyrrolidones. The compound 3-oxy-2-pyrrolidone was synthesized and the toxicity studied. It was found to be of an order entirely different from that of tetrodotoxin (MLD, 20 mg). On the basis of similarity of pharmacological action of the toxin to muscarine, a number of amino aldehydes related to muscarine were also synthesized and the toxicity values examined. They were all found to be very weak toxins in comparison to tetrodotoxin.

Oxidation with periodic acid for 1 hour at 25° C resulted in an uptake of 2 moles; after 48 hours, 3.2 moles. The presence of an α-glycol group was deduced from these results (Tsuda *et al.*, 1960).

Biochemistry of Tetrodotoxin: While tetrodotoxin is among the most toxic substances known among the nonproteins, little is known relative to the mechanism of its action. A few observations have been made relative to the biochemistry of the toxin, some of which may be involved in the mechanism of its action.

Kuriaki and Nagano (1957) noted that tissue respiration and dehydrogenase activity in the brain of the cock and mouse were inhibited by low concentrations but stimulated by higher concentrations of the toxin. Of the enzymes studied, choline acetylase was the most sensitive to the poison. The action resembles that of botulinus toxin which has been assumed to attack the end of the cholinergic nerve fibres, reducing the production of acetylcholine. The concentrations of the toxin used were regarded as readily attainable in cases of human poisoning. Higher concentrations were found to depress cytochrome oxidase and acetylcholine esterase, but no effect was noted on dehydrogenases or on the yellow enzyme.

The effects of the toxin on glucose uptake and glycogenesis were studied by Kuriaki and Wada (1959). Glucose uptake was determined by the difference in glucose concentration of the medium before and after incubation at 37° C with a specified weight of diaphragm for 30 minutes. Determinations in the presence of and in the absence of tetrodotoxin indicated that the glycogen production was doubled in the presence of the poison at a concentration of 1×10^{-4}. At concentrations of toxin which were ineffective in increasing the production of glycogen over the control, the addition of insulin resulted in a significant increase of glycogen formation over that produced by the action of insulin alone. Epinephrine depresses glycogenesis at a concentration of 1×10^{-4}. Epinephrine at this concentration in the presence of insulin also depresses glycogenesis. Insulin acts synergistically with tetrodotoxin and norepinephrine but antagonistically with epinephrine.

The effect of crystalline tetrodotoxin on various enzyme systems in animal tissues has been studied by Nakazawa and Ogura (1961) and Ogura and Fujimoto (1963). It has been previously demonstrated that the oxygen uptake (using a glucose substrate) of brain cortex slices increased remarkably with the addition of 0.02 to 0.1 *M* KCl. This phenomenon has been termed the "potassium effect," and is considered to be related to the control mechanism on carbohydrate metabolism and the permeability of membranes in brain cortex cells. They found that tetrodotoxin showed no measurable influence on the potassium effect in guinea pig brain cortex. Moreover, no effects were noted in pyruvate oxidation using homogenates of liver and brain when tetrodotoxin was used at concentrations of 10^{-5} g/ml. Tests were also conducted in an effort to determine the action of tetrodotoxin on glutamate oxidation, on cytochrome C oxidase activity, on adenosine triphosphatase activity in homogenates and mitochondria of liver and brain, and on carbonic anhydrase of erythrocytes at concentrations of 10^{-4} g/ml. The results of these studies indicated that tetrodotoxin at concentrations of less than 10^{-4} g/ml showed little or no effects on the respiratory enzyme system and related enzymes *in vitro*. The MLD for mice was 1.2×10^{-8} g per gram of body weight.

The influence of crystalline tetrodotoxin on the movement of aldolase from excited rat diaphragm has been investigated by Ogura and Fujimoto (1963). It has been found that aldolase is released from rat diaphragm in a normal potassium medium at 37° C. This aldolase efflux is increased by high K concentration in the bathing solution which leads to a marked reduction in membrane polarization. It has been suggested that the magnitude of aldolase efflux can be taken as an index of membrane permeability. They found that tetrodotoxin failed to affect the aldolase efflux from guinea pig muscle and suggested that the toxin might have more effect on sodium influx across muscle membrane than the potassium efflux (*see also* Ogura, 1963*b*).

The studies of Narahashi *et al.* (1960) suggested that tetrodotoxin blocks conduction of nerve and muscle through its rather selective inhibition of the sodium-carrying mechanism. In order to verify this hypothesis Narahashi *et al.* (1964) observed sodium and potassium currents in the lobster giant axons treated with tetrodotoxin by means of the sucrose-gap voltage clamp technique. They found that tetrodotoxin at concentrations of 1×10^{-7} to 5×10^{-9} g/ml blocked the action potential but had no effect on the resting potential. Partial or complete recovery might have occurred on washing with normal sodium. The increase in sodium conductance normally occurring upon depolarization was effectively suppressed when the action potential was blocked after tetrodotoxication, while the delayed increase in potassium conductance underwent no change. They concluded that tetrodotoxin, at very low concentration, blocks the action potential production through its selective inhibition of the sodium-carrying mechanism while keeping the potassium-carrying mechanisms intact.

Kao (1964) believes the available evidence indicates that the toxin exerts its effects by way of a potent axonal blocking action. This concept was based on observations (Kao and Fuhrman, 1963) to the effect that the toxin "can abolish propagated action potentials in desheathed frog sciatic nerve in a concentration as low as 0.003 μM (about 1 μg/liter) and in this concentration range produce profound physiological derangement in the whole animal."

Recent Studies on the Structure of Tetrodotoxin

With the availability of methods for obtaining the pure toxin in larger quantities in the United States, the problems of establishing the formula was undertaken by R. B. Woodward at Harvard University under contract with the U.S. Army Chemical Research and Development Laboratories. Efforts in this same direction were being carried out simultaneously by Tsuda and coworkers at the University of Tokyo and by Goto and coworkers at Nagoya University in Japan. Concomitantly a study was in progress at Stanford University under the direction of H. S. Mosher and F. A. Fuhrman on a topic which was not even suspected to be related to the studies of Tsuda and Woodward. These studies involved the toxin present in the California newt *Taricha torosa* (*see* Figs. 22 and 23). It was only after this study had been in progress for some time that the suspicion grew to a conviction that the toxin in *T. torosa* was identical or nearly identical to tetrodotoxin. The eventual interlocking of the results of these four projects conducted in areas so far apart, and in the latter case on such a widely different phylogenetic group of animals, makes it desirable to treat this material under headings distinct from the information at hand up through 1955. The reports of the four groups of workers will be reviewed separately.

Studies on Tetrodotoxin by Tsuda and Coworkers: As a result of attempts to acetylate tetrodotoxin, the isolation of two acetates was reported. These were believed to be formed, not from the toxin itself, but from degradation products that were formed in the process of acetylation (Tsuda *et al.*, 1958, 1960). One was a diacetate, difficult to crystallize in the free state, but which formed a picrate (mp 199°-201°) and a hydrochloride salt (mp 219°-221°). The results of elemental analysis of the latter suggested the formula $C_{15}H_{20}O_9N_2 \cdot HCl$; $[\alpha]_D + 16.7 (c = 1.7, EtOH)$; U.V $^{EtOH}_{max}$ 235-238 mμ. The other was a monoacetate, (mp 150°-151° C). Results from elemental analysis were in satisfactory agreement with the formula $C_9H_{13}O_4N$; MW 205. Both acetates were regarded as having been formed by loss of carbon and nitrogen atoms.

Reduction of the diacetate with sodium borohydride gave acetaldehyde and a secondary amine, C_5H_9ON (Tsuda *et al.*, 1960). This amine formed a dibenzoate (mp 146°-148°) and a N, O-ditosylate (mp 132°-134°). Elemental analysis of the latter favored the formula $C_{19}H_{21}O_5NS_2$. Hydrogenation of the ditosylate yielded a dihydride derivative (mp 122°); $[\alpha]_D$-48 (c=0.9, $CHCl_3$). Reduction with lithium aluminum hydride resulted in substitution of the O-tosyl group with hydrogen and the formation of a simple secondary amine $C_9H_{11}N$. This amine was identified as dl-β-methylpyrrolidine by comparison of its N-tosylate (mp 72°-73°) with the tosylate of the synthetic compound. The melting points were the same and no change in melting point was observed when the two were mixed. It was concluded that β-methylpyrrolidine forms an integral part of the skeleton of the molecule of tetrodotoxin.

A series of degradation products of tetrodotoxin resulting from a 10-hour heating with potassium hydroxide in 60 percent ethanol has been described by Kawamura (1960). From the acidified dark red solution (pH4-5) thus obtained, a substance was extracted with butanol that on elemental analysis gave an empirical formula $C_9H_{12}O_2N_3Cl$. This product constituted about 30 percent of the yield and melted at 280° C. This base is hereafter referred to as the C-9 base (*see* reports by Goto *et al.*, 1962*a*; Kishi *et al.*, 1964, for further reference to this C-9 base). Oxalic acid was also isolated to the extent of about 20 percent of the mixture. The hydrochloride salt of the C-9 base was purified by sublimation. This salt was readily convertible back to the free base on treatment with potassium hydroxide. A dark green coloration resulted from dissolving the C-9 base in excess sodium hydroxide and addition of ferric chloride.

Two absorption maximums were observed at 235 and 262 mμ in the ultraviolet spectrum of an ethanol solution of the C-9 base (Kawamura, 1960). The maximum at 262 mμ was displaced to 279 mμ in 10 percent alcoholic sodium hydroxide, indicating that an enol-OH group was present in the structure. The infrared spectrum revealed bands for OH and NH besides characteristic bands at 1658, 1637, and 650-720 cm^{-1}, typical of aromatic amines, and at 1580-1588, typical of the C=C group. Comparison of the infrared spectrum with that for 2-amino-pyrimidine suggested a close relationship to this structure. One mole of hydrogen was absorbed on hydrogenation of the base using platinum in ethanol as catalyst. The dihydro-derivative thus formed melted at 275°-278° C. The ultraviolet and infrared spectra of the dihydro-derivative showed maximal characteristics of an aromatic ring, indicating that the reduction had occurred at the point of the enol-OH structure and not in the ring system. The dihydro-base was acetylated to give a triacetyl-derivative (mp 197°-199°) that was saponifiable in alkali to restore the original base. The infrared spectrum of the acetylated derivative revealed bands suggestive of the presence of ester and amide groups. These results were interpreted to indicate that the C-9 base had, in addition to the NH_2 and enol-OH groups, another OH group not linked directly to the aromatic ring.

To obtain further information relative to the structure of the aromatic ring (Kawamura, 1960), the C-9 base was oxidized in alkaline potassium permanganate at 60°-70° C. A product of oxidation was isolated which melted at 327° C with decomposition. Elemental analysis gave a formula $C_6H_5O_4N_3$. This product formed a dimethyl ester $C_8H_9O_4N_3$ (mp 185°-186°). Pyrolysis of the oxidation product resulted in the release of carbon dioxide with concomitant formation of a substance $C_4H_5N_3$ (mp 123°-124° C). The oxidation product must then have been a dicarboxylic acid derivative of 2-amino-pyrimidine since the infrared spectrum of the dimethyl ester revealed two ester-carbonyl bonds. The 4,5-dicarboxylic acid derivative of 2-amino-pyrimidine was synthesized and shown by ultraviolet and infrared spectra and melting points to be identical to that of the oxidation product of the C-9 base. The incomplete formula for the base was hypothesized to be that shown in Figure 4.

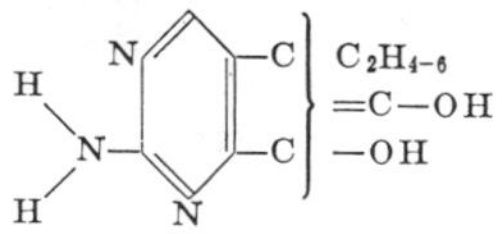

FIGURE 4.—Incomplete formula hypothesized for the C-9 base.

On the basis of a reexamination of the various analytical experiments, the empirical formula for the C-9 base was refined to $C_9H_9O_2N_3$ in a subsequent report (Tsuda *et al.*, 1962*a*, *b*). In this same report, the more exact structure for the base was established by the following steps. The methyl ether of the C-9 base was chlorinated with sulfuryl chloride to replace any ring hydroxyl with Cl. The Cl was then reduced to H with palladium and H_2 to yield a product $C_{10}H_{11}ON_3$. This was shown to be identical to the synthesized 2-amino-8-methoxy-6-methylquinazoline. The C-9 base is thus 2-amino-6-hydroxymethyl-8-hydroxyquinazoline (Fig. 5). The synthesis of the methyl ether of the base is described in a separate paper (Tsuda *et al.*, 1962*a*, *b*).

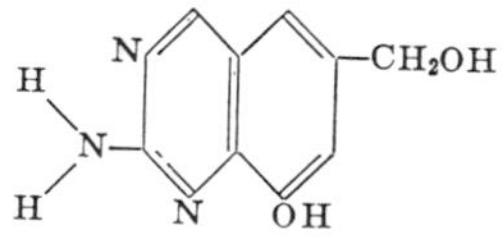

FIGURE 5.—The established structure of the C-9 base.

Still further confirmation of the structure of the C-9 base resulted from reduction with hydrochloric acid and red phosphorus (Tsuda *et al.*, 1962*a*, *b*). A crystalline substance was obtained which proved to be the hydriodic acid salt of a base that could not be obtained in pure form, either as the salt or the free base. Treatment of the impure product with potassium ferricyanide and subsequent purification of the product by chromatography on aluminum oxide yielded a yellow crystalline product (mp 232°-234° C) with formula $C_9H_9N_3$. The ultraviolet spectrum showed maximum absorption bands very similar to those observed from 2-amino-quinazoline. Since it could be expected that the hydriodic acid-red phosphorus reduction would eliminate the oxygen atoms and since the previous investigations had led to a hydroxymethyl derivative of quinazoline, it was presumed that the base obtained by the reduction was 2-amino-6-methylquinazoline. This derivative was synthesized and found to be identical in properties to the reduction product from the toxin. Tsuda (1963) reported the confirmation of the structure of the C-9 base by Goto and coworkers (Goto, Kishi, and Hirata, 1962*a*, *b*) based on resonance spectra of protons of the base as compared with those of related materials.

The find was also reported (Tsuda, 1963) that tetrodotoxin readily loses a molecule of water under mild conditions to yield a substance ($C_{11}H_{19}O_9N_3$) with amino acid properties. This derivative was called tetrodonic acid ($[\alpha]_D + 10.1$, c=1.09, 2 percent HCl). Tetrodonic acid forms a hydrohalogen acid salt with HBr ($C_{11}H_{17}O_8N_3 \cdot HBr$) with molecular weight and elemental analysis agreeing with this formula. This salt was usable for X-ray diffraction analysis studies to determine the structural framework of the toxin molecule.

The space constants for the orthorhombic modification of the tetrodonic acid crystal were found to be: a=20.231 A; b=10.590 A; c=6.813 A; density= 1.821 g/cm³. From the constants it was calculated that the elementary space cell contained four molecules. From the cell volume, the density and the number of molecules per cell, the molecular weight could be calculated as 400.2, confirming the empirical formula of the hydrobromide salt of tetrodonic acid as $C_{11}H_{17}O_8N_3 \cdot HBr$. The systematic gaps observed indicated a space grouping of the type $P2_12_12_1$.

On the basis of the Weissenberg moving film photographs and the chemical and physical data relative to the groups present in the molecule, it was concluded that the spatial structure of the tetrodonic acid molecule was as shown in Figure 6.

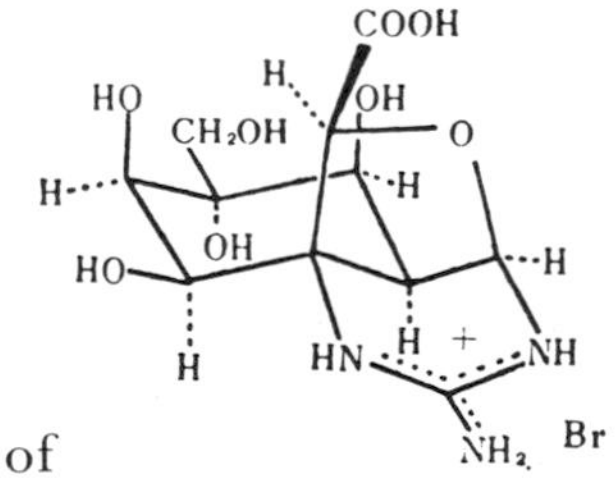

FIGURE 6.—Spatial structure of the hydrobromide salt of tetrodonic acid (Tsuda, 1963).

Information relative to the position of the bromide ion was obtained by 3-dimensional Patterson syntheses (regular and modified types). Studies on numerous 3-dimensional Fourier syntheses, with special attention to the location of the bromide ion, led to a model of the molecular structure. In Figure 7 is shown the calculated projection of the 3-dimensional electron density along the c-axis of the crystal.

On the basis of the infrared spectrum and the pK_a (8.3), it was concluded that tetrodotoxin itself contains a lactam group (Tsuda, 1963). There is no absorption band for an aldehyde group visible. The hydroxymethyl side-chain on the cyclohexane ring was confirmed by the proton resonance spectrum and by the results of periodic acid oxidation. Structures I, II, or III of

Figure 8 were regarded as possible for the tetrodotoxin molecule. The possibility was entertained that the molecule exists as a dimer with a C_{22} formula.

Evidently Tsuda was not prepared to indicate the nature of the change involved in the formation of tetrodonic acid by loss of water. However, an amino-derivative of the anhydro-form was prepared by uniting two molecules of the toxin through the N-atom of ammonia. He hypothesized that the toxin molecule was of similar configuration through an O-atom (Tsuda, 1963).

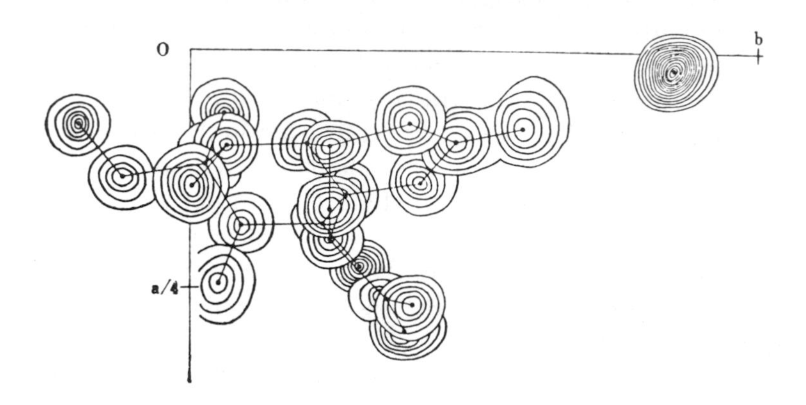

FIGURE 7.—Projection of 3-dimensional Fourier synthesis along the c axis (Tsuda, 1963).

Subsequently, the toxin was purified by conversion to the diacetylanhydrotoxin and reconversion to the toxin by hydrolysis in 5 percent hydrochloric acid (Tsuda *et al.*, 1964*a*, *b*). On the basis of elemental analysis of the product thus purified, the previously suggested empirical formula $C_{12}H_{19}O_9N_3$ was revised to $(C_{11}H_{17}O_8N_3)_n$ where *n* might be 1 or 2. Except for the problem of the value of *n*, Tsuda was now in agreement with Goto and with Woodward; however, in the latter's opinion, there existed one-half mole of water of crystallization. In this report, Tsuda provided the results of attempts to determine whether or not there occurred any inversion of H atoms at C-9 or at $C\text{-}4_a$ in the interconversion from one derivative to another. The various derivatives were treated with deuterium oxide and the magnetic resonance spectra were examined before and after conversion. It was Tsuda's conclusion that inversion does occur at C-9 in the enolization when tetrodotoxin is converted to tetrodonic acid but that the configuration at $C\text{-}4_a$ remains unchanged through all interconversion. (For numbering of the C atoms, *see* the structure of Fig. 9.)

Recognizing that the carboxyl group of the tetrodonic acid did not occur in the toxin itself, Tsuda concluded that the toxin structure must involve either a lactam, a lactone, or an ortho-acid linkage. Of these three, the only one which appeared to agree with the now known behavior of the toxin was the latter. The structure as given in Figure 9 was regarded as the correct one for the toxin.

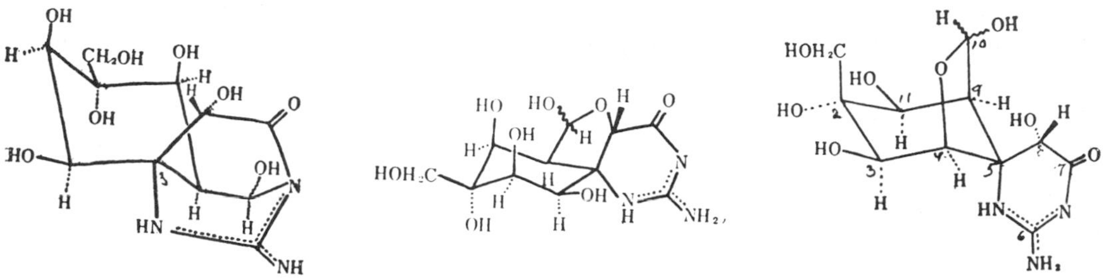

I II III

FIGURE 8.—Structure regarded as possible for the tetrodotoxin molecule (Tsuda *et al.*, 1963).

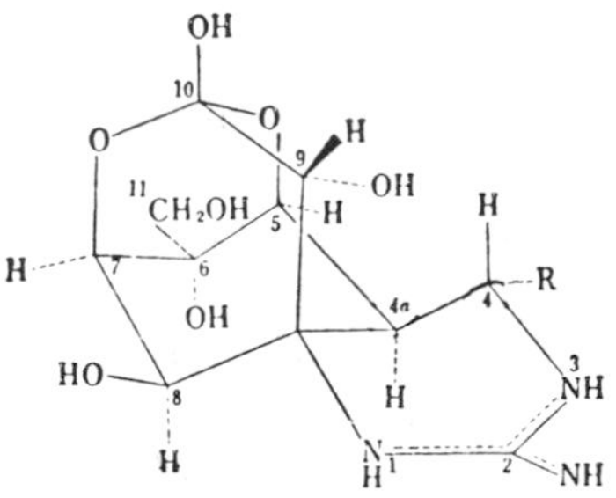

I : R=OH VI : R=NH$_2$
VII : R=OCH$_3$ VIII : R=H

FIGURE 9.—The formula of tetrodotoxin as concluded by Tsuda *et al.* (1964*b*).

Studies on Tetrodotoxin by Goto and Coworkers: Following the report of Kawamura (1960) in which a partial structure of the C-9 base had been suggested, Goto and coworkers (Goto *et al.*, 1962*a*, *b*; Kishi *et al.*, 1964) undertook the problem of determining a more complete formula for this derivative of the toxin. The empirical formula suggested by Kawamura ($C_9H_{11}N_3O_2$) was altered to $C_9H_9N_3O_2$ based on analytical data from the hydrochloride salt. Evidence was presented for identifying the base as 2-amino-4-methyl-8-hydroquinazoline.

Hydrolysis of the toxin in concentrated sulfuric acid at room temperature over a period of 3 days gave a C-8 base (Goto *et al.*, 1962*b*; Kishi *et al.*, 1964). The hydrochloride of this base on treatment with acetic anhydride formed a diacetate. Elemental analysis of the diacetate provided data agreeing with the formula [$C_8H_5ON_3(COCH_3)_2$]. Identification of the C-8 base as 2-amino-4-methyl-6-hydroxyquinazoline was possible by comparison of properties with the synthetic product.

On the basis of elemental analyses and molecular weight determinations of the toxin and the above derivatives, the empirical formula for the toxin was altered from that suggested by Tsuda ($C_{12}H_{19}O_9N_3$) to $C_{11}H_{17}O_8N_3$ (Goto *et al.*, 1963*a*). Certain properties of the toxin were reported. It has no melting point but decomposes at 220° C. It shows no absorption in the ultraviolet range. Guanidine is liberated by permanganate oxidation. It is a monoacidic base with pK_a 8.3 attributed to the guanidine group. Two moles of periodate are absorbed in 0.1*N* sulfuric acid at 5° C with liberation of formaldehyde. Hydrolysis by heating in water gave an acid ($C_{11}H_{19}O_9N_3$) in 45 percent yield. This acid is referred to as tetrodoic acid but is presumed to be the same as the tetrodonic acid of Tsuda. The acid gave a negative Sakaguchi reaction, indicating that it is not a mono-substituted one. The spectrum of the acid, as the hydrochloric salt, in deuterium oxide solution indicated the presence of eight protons which could not be exchanged by deuterium. Treatment of the acid with 1 mole of periodic acid liberated formaldehyde, just as with the toxin itself. This was accomplished without destruction of the acid carbonyl group but there was a coincident introduction of an α-ketol group. Absorption of a second mole of periodic acid altered the structure still further with a presumed introduction of a hemiacetal group. The acid, like the toxin, could be converted to the C-9 base by heating with aqueous barium hydroxide. On the basis of the observed properties, the structures of the acid derivatives and of the toxin were hypothesized. The structure of the toxin was at this time regarded as having one or the other of the formulas I or II in Figure 10 (Goto *et al.*, 1963*a*). The derivatives of the toxin on which this study was based are further amplified by Kishi *et al.* (1964).

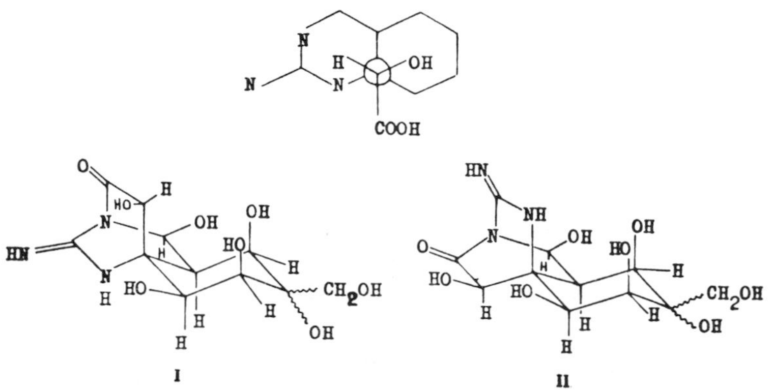

N N N H OH COOH

O HO H N OH OH HN HO H N H HO CH2OH H H H OH

I

HN NH N OH OH O HO H HO H HO HO CH2OH H H H OH H

II

FIGURE 10

Decision in favor of I (Goto *et al.*, 1963*b*) was made on the basis of properties of the anhydrotetrodoic acid prepared by the action of 5 percent barium hydroxide solution at room temperature on the toxin in an atmosphere of nitrogen. This anhydro-derivative absorbed 1 mole of aqueous bromine (BrOH). Absorption spectra showed the presence of a γ-lactone which could be opened by neutralization to pH 7.5. X-ray analysis showed the position of the bromine atom at C-4 in the structure shown below. The OH group must then have been added at the other end of the double bond (C-4) with the coincident formation of lactone with the OH of C-6. The reactions involved were hypothesized as shown in structures III and IV of Figure 11.

Since the toxin has no acid function and shows no absorption in the ultraviolet, neither the carboxyl group nor the double bond of the anhydroacid

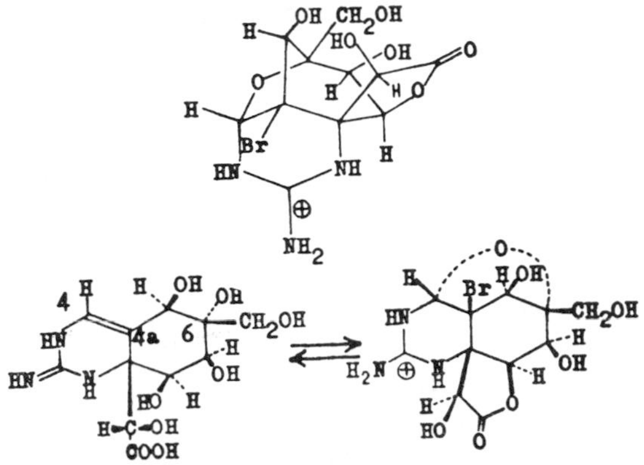

III IV

FIGURE 11

are present in the toxin molecule. On the basis of spectral data, it was hypothesized that only four structures for the toxin were possible. These are shown as V to VIII in Figure 12.

Of the four structures, VIII was ruled out on the basis that no absorption band for lactone was evident between 1,700 and 1,800 cm^{-1} in the infrared

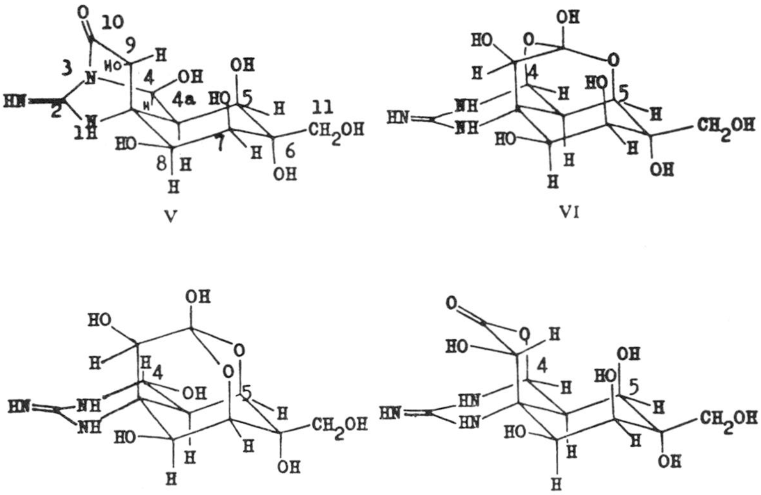

V VI

VII VIII

FIGURE 12

spectrum. Numbers VI and VII were ruled out since the orthoester formula does not account for the low pK_a 8.3. Hence the decision was in favor of VI.

In the meantime, Tsuda had reported further data which called for a reconsideration of formula VI as an orthoester (Tsuda *et al.*, 1964*a*). In this report, the possibility was entertained that the toxin existed as a dimer and that the configuration at C-9 of the bromolactone and tetrodoic acid was an epimeric configuration from that previously proposed. In the reconsideration of the earlier conclusions (Goto *et al.*, 1964*a, b, c, d, e, f*), the inversion at C-9 was believed to have occurred either in the formation of the anhydrotetrodoic acid or in the hydrolysis of the toxin to the acid. Proof that the inversion occurred during the hydrolysis of the toxin to the acid was obtained by comparison of the nuclearmagnetic resonance spectra of the anhydroacid and that of the additional product resulting from treatment with aqueous bromine, carried out first in water and then in deuterium oxide.

Evidence from the titration curve of the toxin, purified through the picrate, eliminated any possibility that the toxin existed as a dimer (Goto *et al.*, 1964*a*). The same was found to hold true for the amino derivative (Goto *et al.*, 1964*b*). If the toxin were a dibasic acid or a diacidic base, it could be presumed that the pK values would differ significantly. No such difference was observed. The monomeric form of the toxin was also confirmed by Woodward and Gougoutas (1964) by means of crystalographic analysis of the toxin.

Evidence suggesting an orthoester structure of the toxin was obtained by treatment with 1 mole of periodic acid in 0.1 *N* sulfuric acid at 0° (Goto *et al.*, 1963*a*). During this treatment, formaldehyde was liberated and a carbonyl group appeared. A reversible titration of this group was possible, suggesting the presence of a lactone. After titration with sodium hydroxide, a carboxyl absorption band at 1,600 cm^{-1} was present but no lactone band between 1,700 and 1,800 cm^{-1}. Back titration showed the group to have a pK_a 3.35 corresponding to a free carboxyl group. Assuming the carboxyl to be the same as that in tetrodoic acid, this carboxyl could form a lactone with hydroxyl groups in gamma and delta positions relative to the carboxyl group. Since a gamma lactone was evident in the bromoderivative (Tomie *et al.*, 1963), it could be presumed that a similar situation exists in the toxin also. The appearance of a weak absorption band at 1,747-1,750 cm^{-1} by the salts of the toxin (sulfate, hydrobromide, and picrate) was regarded as explainable in terms of tautomerism between an orthoester and lactone forms. This tautomerism is represented by formulas IX and X as shown in Figure 13.

Since tetrodotoxin and its crystalline hydrobromide show no carbonyl absorption, the suggestion is that the toxin exists completely in the form of the orthester.

A further derivative containing the orthoester group as acetate ester was produced by the acetylation of tetrodotoxin with acetic anhydride and *p*-toluene sulfonic acid to yield tetraacetylanhydroepitetrodotoxin *p*-toluene sulfonate (Goto *et al.*, 1964*b*). This tetraacetate could be converted to a penta-

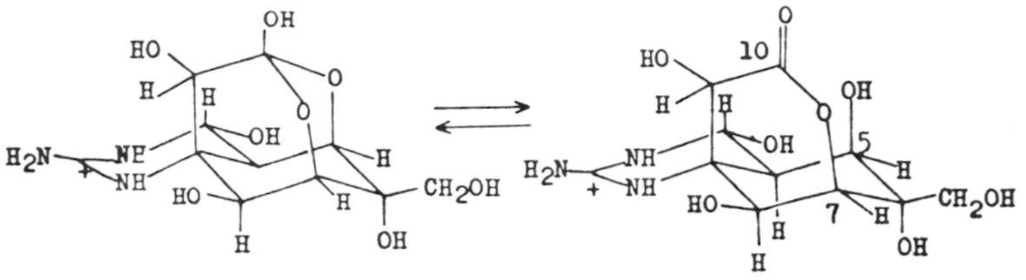

IX X

FIGURE 13

acetate on treatment with acetic anhydride and pyridine. Hydrolysis of the pentaacetate with aqueous ammonia gave a diacetyl derivative. If tetrodotoxin was acetylated with acetic anhydride and pyridine, small amounts of a hexaacetate were obtained. The formulas deduced for the tetra-, penta-, and hexaacetate were as shown in structures XI, XII, and XIII respectively in Figure 14.

In the course of a study on the IR-spectra of a number of derivatives of tetrodotoxin and other model compounds containing the guanidinium group, the interesting observation was made (Takahashi, Goto, and Hirata, 1964*a, b*) that while all of these showed two strong bands in the region 1,700-1,500 cm^{-1}, the wave length differences for the two bands differed in such a manner that the compounds could be divided into two groups. In one group, the difference was of the order of 60 cm^{-1} or less, while in the other group the differences were of the order of 80 cm^{-1} or more. In the first group, the alpha substituent to the guanidinium is hydroxyl or amino or bromo. In the second, the alpha substituent is a strongly electron-attracting group such as a carbonyl double bond or a sterically-strained system such as a five-membered ether ring.

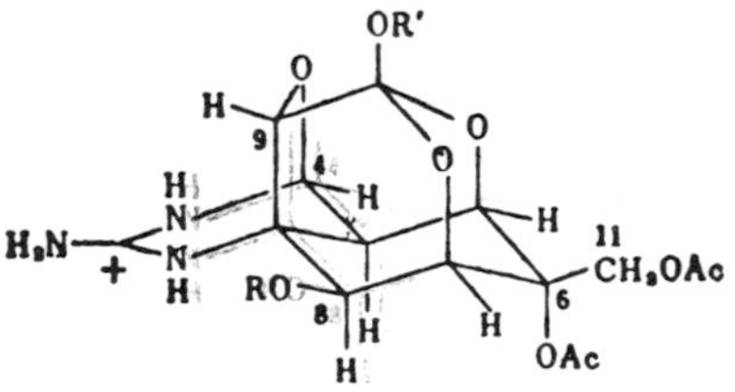

XI 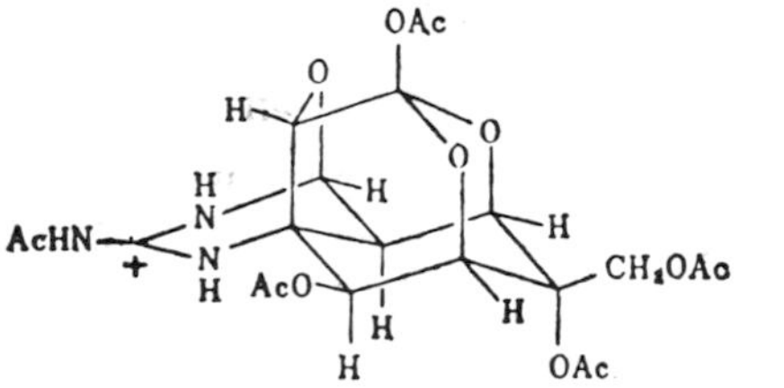 XII

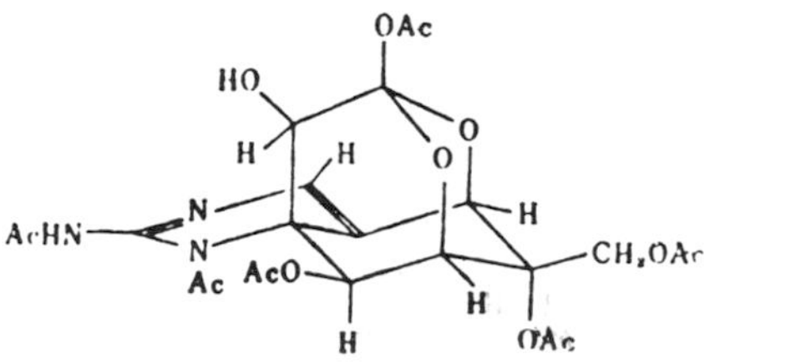

XIII

FIGURE 14

Studies on Tetrodotoxin by Woodward and Coworkers: Utilizing the isolated toxin provided by Nelson at Loma Linda University, a series of studies was conducted by Woodward and associates at Harvard University during the period 1960-64. An examination of the work that had been done up to this time led Woodward to conclude that the failure to establish the empirical formula with certainty was due largely to the tenacity with which the toxin retained solvents, particularly water. This phenomenon was probably due to the large number of OH and NH groups present. Previous attempts had only defined the probable limits of the formula as lying in the area $C_{10\text{-}12}H_{15\text{-}19}O_{8\text{-}10}N_3$.

While the near equivalency of the number of C atoms to the sum of oxygen and nitrogen atoms permitted the conclusion that the structure was a most unusual one, the possibility that one or more bonds were satisfied between N and O seemed negligible. This conclusion logically follows from the observation that guanidine had been observed among the products of vigorous oxidation. It seemed highly probable that the guanidine unit was present in intact form in the toxin molecule.

In a previous connection, it was pointed out that Tsuda had isolated a decomposition product in the form of a base with empirical formula $C_9H_9O_2N_3$ which has been referred to as the C-9 base. This base formed an acetyl derivative $C_{15}H_{15}O_5N_3$. An examination of the infrared and nuclear magnetic resonance spectra led to hypothesizing a series of possible structures for this C-9 base as derivatives of quinazoline. All of these proposed structures could be eliminated on the basis of known properties except two. Decision between these two was possible through the formation and isolation of a crystalline copper chelate derivative. The structural formula of the C-9 base and its triacetate was concluded to be that shown in Figure 15. This was the same conclusion reached by Tsuda and coworkers independently.

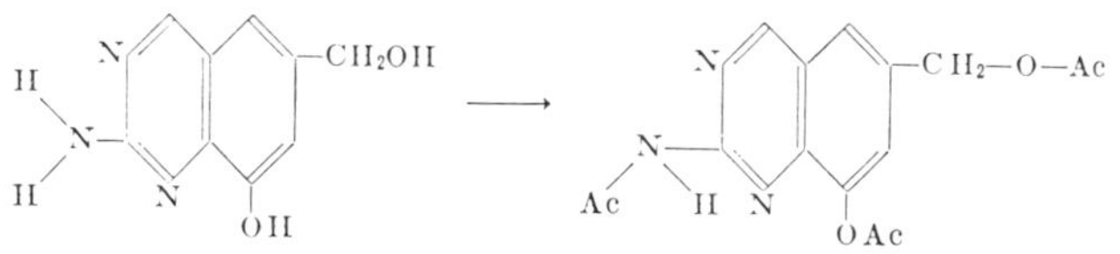

FIGURE 15.—Hypothesized formulas for the C-9 base and its triacetate as deduced by Tsuda and Woodward.

This coincidence permitted the conclusion with certainty that the toxin was degraded by alkali to a quinazoline ring system which contained the guanidine group. Tsuda, at this point, regarded as relatively certain that the original toxin molecule also contained this quinazoline ring system. Woodward, on the other hand, regarded the evidence as incomplete. The possibility remained that the quinazoline derivative was only an artifact produced from an open chain precursor in the process of degradation. It was recognized that the probability was large that the quinazoline ring system *was* present in the toxin molecule since several workers had obtained other quinazoline derivatives by different methods of degradation, including one from acid treatment rather than by alkali.

Because of unwillingness of the investigators to accept this conclusion short of unequivocal proof, attempts were made to complete the evidence. Since quinazoline contains eight C atoms (which number is a major fraction of the C atoms of the toxin), this probability would appear confirmed if it could be shown that the toxin contained a carbocyclic ring as well as the ring containing the guanidine group. The evidence would be even more convincing if the empirical formula could be established with certainty.

Results obtained from examination of the degradation products of the toxin to the C-9 base in the absence of protons proved inconclusive. The establishment of the empirical formula as $C_{11}H_{17}O_8N_3 \pm yH_2O$ was accomplished by mass spectrometric examination of the mixture of products produced by prolonged acetylation of tetrodotoxin under mild conditions. From the crude peracetylated toxin, Inayama isolated a heptaacetylanhydrotetrodotoxin and an octaacetylanhydrotetrodotoxin, the emipirical formulas being established beyond question by analysis and mass spectroscopy. The data obtained and the interpretations resulting from these data are provided in Table 4 as taken from Woodward's report.

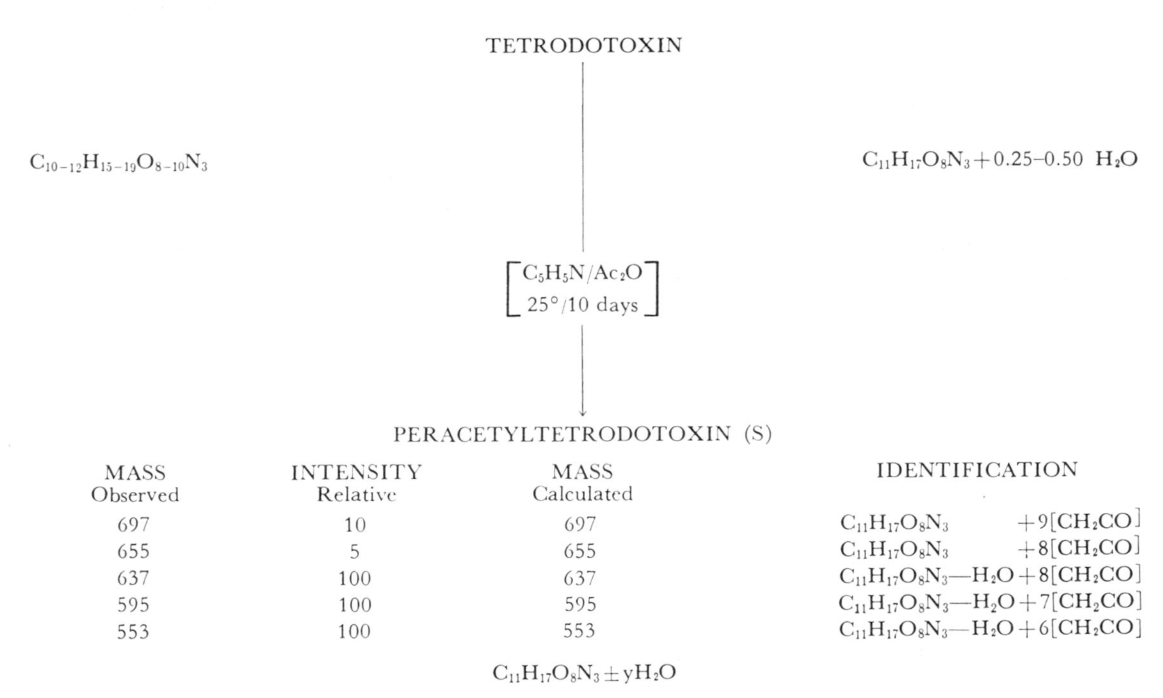

MASS Observed	INTENSITY Relative	MASS Calculated	IDENTIFICATION
697	10	697	$C_{11}H_{17}O_8N_3 + 9[CH_2CO]$
655	5	655	$C_{11}H_{17}O_8N_3 + 8[CH_2CO]$
637	100	637	$C_{11}H_{17}O_8N_3 - H_2O + 8[CH_2CO]$
595	100	595	$C_{11}H_{17}O_8N_3 - H_2O + 7[CH_2CO]$
553	100	553	$C_{11}H_{17}O_8N_3 - H_2O + 6[CH_2CO]$

$C_{11}H_{17}O_8N_3 \pm yH_2O$

TABLE 4.—Data and interpretations from mass spectrometric examination of products produced by prolonged acetylation of tetrodotoxin.

Examination of these data indicate that they are interpretable in terms of a heptaacetyl and an octaacetyl derivative of a molecule having the empirical formula $C_{11}H_{17}O_8N_3 \pm yH_2O$. Magnetic resonance spectra provided further confirmation of this formula as well as information relative to the number and positions of the C—H bonds. This evidence pointed to the presence of eight C—H bonds in the heptaacetate, only one of which could be present as CH_2, while the octaacetate contained only seven. The eighth acetyl group could be considered formed from the heptaacetate by the difference in partial configurations shown in Figure 16 where n is greater than, or equal to, O.

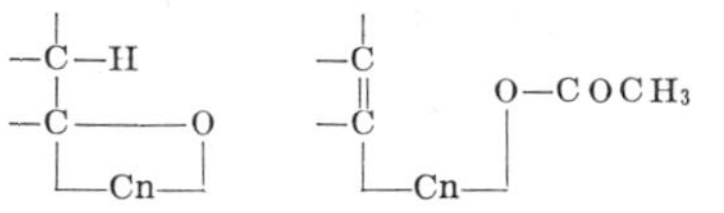

FIGURE 16.—Showing the probable shift in configuration in the formation of the octa-acetate from the heptaacetate.

However, since the two structures called for eight and seven C—H groups respectively in the total molecule (not including those from the acetate groups), it was not possible to deduce with certainty the number of C—H bonds in the toxin molecule itself. This point was of considerable importance in establishing the presence of a carbocyclic ring, which in turn was critical to establishing whether the toxin molecule contained the quinazoline ring system.

This problem was solved by the preparation by Woodward and Gougoutas (1964) of a crystalline derivative of the toxin by the action of hydrogen chloride in the presence of acetone and methanol. The composition of this derivative as O-methyl-O′, O″-isopropylidinetetrodotoxin hydrochloride monohydrate was established beyond question to be that shown in Figure 17. This was accomplished by measurement of its density and the determination of its unit cell dimensions through X-ray crystallographic methods.

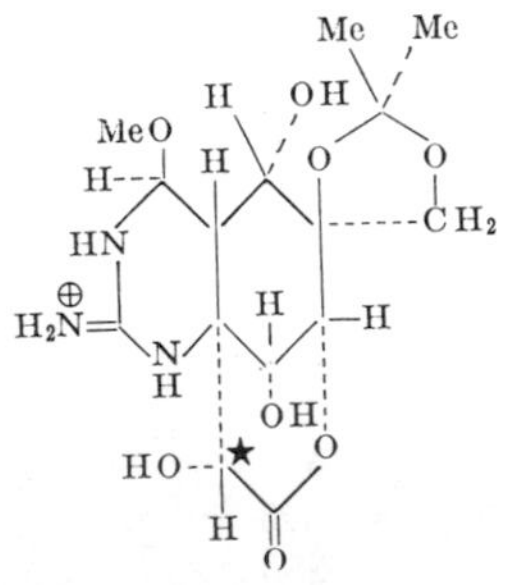

FIGURE 17.—The structure of the O-methyl-O′, O″-isopropylidinetetrodotoxin hydrochloride monohydrate derivative of the toxin.

This determination also provided further confirmation of the empirical formula $C_{11}H_{17}O_8N_3 \pm yH_2O$ for the toxin. The derivative showed the same nuclear magnetic resonance spectrum as did the toxin, indicating that no major change had occurred in the structure of the toxin in the formation of the derivative. It could now be stated with certainty that the toxin molecule contained eight C—H bonds, two of which are adjacent and attached respectively "to carbon bound only to other carbon atoms and to carbon bearing two electronegative atoms." From these data, it was possible to conclude beyond question that the toxin molecule contains a carbocyclic ring and hence that the quinazoline ring system must be present in the toxin molecule. Thus the inference of Tsuda was found to be correct, though based on inadequate evidence to constitute final proof.

With the establishment of the fact that the toxin molecule must contain the quinazoline nucleus, it was now possible to begin the process of developing a formula which would account for the information available relative to the structures of the toxin and its known derivatives. Among the derivatives to be given major attention were the various quinazolines obtained by various workers by acid and alkali degradation of the toxin. Of these, two were of special significance. These are shown in Figure 18.

Figure 18.—The structure of two quinazolines obtained by degradation of tetrodotoxin in base and acid respectively.

An examination of these formulas indicates that C-6 probably contains both an OH and a CH_2OH group and that C-8 also contains an OH group. The presence of CH_2OH on C-6 was provided confirmation by the observation that formaldehyde was formed as one of the degradation products from periodate oxidation. The nuclear magnetic resonance studies had indicated that only one HC—(CC) grouping is present and that it is adjacent to a (HC—ON or O) group. Thus the valences of all the atoms are satisfied except for carbons labelled 8 and a in Figure 18. The principal group still unaccounted for is a glyoxylic acid residue which must then have been derived from a precursor group on C-8 or C-a. The decision in favor of C-a was based on the formulas of the quinazolines shown in Figure 18. The degradation in basic solution could be feasibly accounted for in terms of structures with the glyoxylic acid residue in either portion C-8 or C-a. In acid solution it could occur only if the glyoxylic acid residue was at C-a. The mechanism by which this could occur is given in Figure 19 as taken from Woodward's report.

Since it was now certain that the toxin molecule did not contain a free carboxyl group and that the glyoxylic acid residue must exist in some combined form, the simplest solution lay in the assumption that the carboxyl group is combined with one of the numerous OH groups to form a lactone. This assumption was in agreement with the infrared spectra of the toxin hydrochloride and its O-methyl-O′, O″-isopropylidine derivative which had revealed

DEGRADATION BY ACID

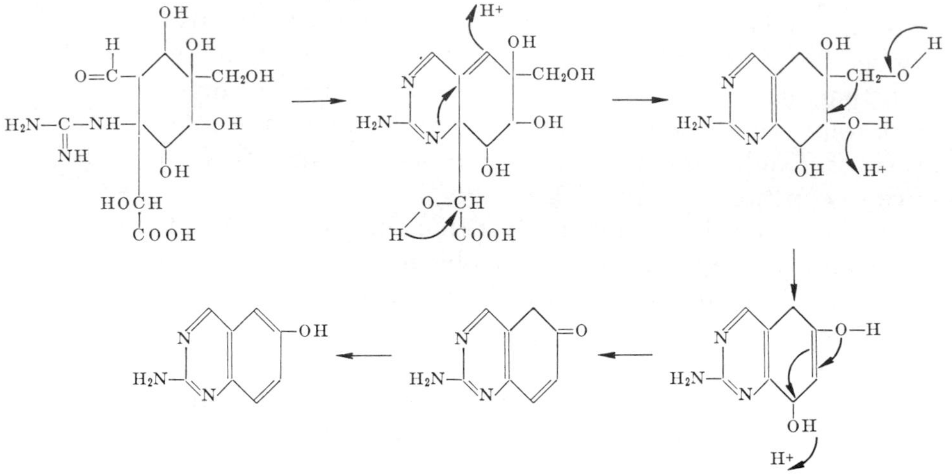

FIGURE 19.—Showing the intermediate steps by which the quinazoline degradation product from the toxin in acid solution could be formed.

bands at 5.71μ that do not change their positions on deuteration. Evidence from X-ray crystallographic analysis of the O-methyl-O′, O″-isopropylidine derivative provided the basis for concluding that the lactone linkage was with the OH of C-7. The structure of the hydrochloride salt is clearly that shown in Figure 20.

It might be presumed that the toxin is the deprotonated species of the hydrochloride at the guanidine nitrogen. Two facts stand against this structure for the toxin. Tetrodotoxin certainly does not contain a lactone group as indicated by the absence of bands characteristic of such a group. Further, the bands associated with the guanidinium system do not change on conversion to the free base. It must be concluded that the deprotonation must occur at some other site. The increase of pK_a observed in solvents of low dielectric constant indicates that the deprotonation occurs rather from an hydroxyl group. But no hydroxyl group of the molecule represented in Figure 20 could be expected to dissociate with a pK_a of 8.5.

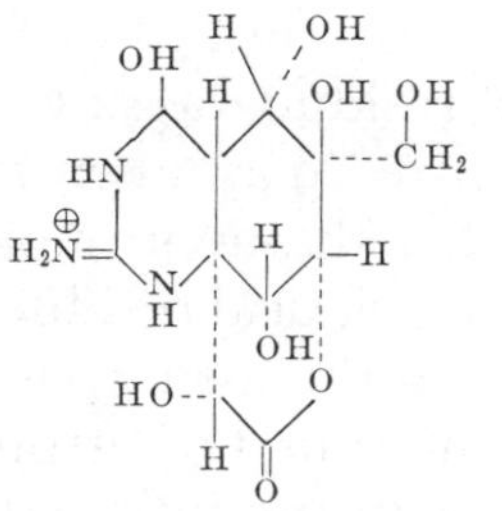

FIGURE 20.—Structure of tetrodotoxin hydrochloride.

A solution to the dilemma appears when it is recognized that the free base is a zwitterion, and that one of the hydroxyl groups has been added to the lactone grouping to form a hemilactal system. A three-dimensional representation of the toxin hydrochloride (Fig. 21) indicates that the OH on C-5 is properly located for formation of such a hemilactal.

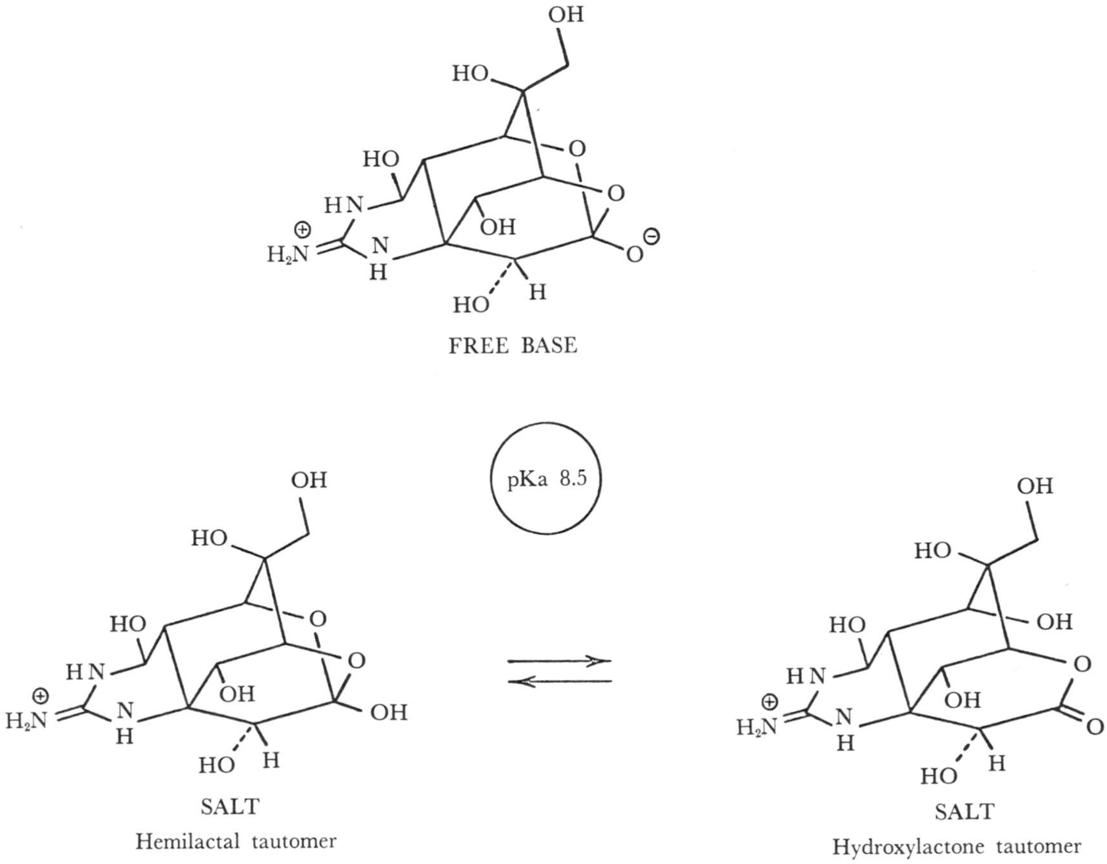

FREE BASE

pKa 8.5

SALT
Hemilactal tautomer

SALT
Hydroxylactone tautomer

FIGURE 21.—Showing the three-dimensional formula of tetrodotoxin as a zwitterion and its conversion to the salt in two tautomeric forms in equilibrium.

A number of puzzling relationships between the structures of the toxin and its derivatives now become apparent. Woodward (1964) states:[5]

The rigidity of the entire skeleton of the open chain tautomer is undoubtedly a factor which favors the establishment of equilibrium with the hemilactal form (Fig. 21) which, it is interesting to note, incorporates the feature of a dioxaadamantine nucleus. It is also worthy of note at this time that the lactone band in the infrared spectrum of (noncrystalline) tetrodotoxin hydrochloride is relatively weak, and the implication is

[5] The figure numbers as given in this quotation from Woodward (1964) have been changed to correspond with the figures as given in this book.

clear that in this material both open-chain and hemilactal ring chain tautomers co-exist. By contrast, the crystalline O-methyl-O′, O″-isopropylidinetetrodotoxin hydrochloride exists entirely in the lactone form (Fig. 21). However, this same substance is equilibrated to a greater or lesser degree with the isomeric hemilactal form, depending upon the environment. In deuterium oxide solution the carbonyl band in the infrared spectrum of this hydrochloride has all but vanished, and it may be estimated roughly that 90 percent of the molecules are in the hemilactal form in that medium.

If the already rigid system of the tetrodotoxin molecule is further modified by introduction of yet another element of constraint in the form of a new ring, the equilibrium just discussed between hydroxylactone and hemilactal forms is strongly displaced in favor of the hemilactal tautomer. These are the circumstances which obtain in the so-called *anhydro* series. When tetrodotoxin was treated with hydrogen chloride in acetone for a short time, a beautifully crystalline O, O′-isopropylidinetetrodotoxin hydrochloride is produced. This substance was easily formulated (Fig. 17), since it was clear that the carbinolamine hydroxyl group should share the ready capacity for ether formation which is characteristic of its class, and since the only hydroxyl group with which ether formation is possible is that at C. 11. Infrared spectra of the new hydrochloride, either as crystalline solid or in a variety of solvents, display no lactone band whatsoever, and it is clear that the derivative exists in the hemilactal form under all conditions. It is of much interest that heptaacetylanhydrotetrodotoxin is also a member of this *anhydro* series . . . . Its full structure is entirely in accord with its detailed nuclear magnetic resonance spectrum. In particular, it may be noted that with the change in configuration at C. 4, the carbon-hydrogen bond at that position makes a dihedral angle with the adjacent bridge-head carbon-hydrogen bond of approximately 90°, and that consequently, the coupling constant is very nearly 0. Further, double resonance experiments show the expected coupling between the resonances associated with the hydrogen atoms at C. 7 and C. 10. A final point of interest is the observation of the expected 1,3 coupling between the equatorial hydrogens at C. 5 and C. 7 with the absence of coupling between the axial hydrogens at C. 8 and C. 10. It may also be noted that the complete analysis of the nuclear magnetic resonance spectrum of this acetyl derivative permitted the complete derivation of the tetrodotoxin structure (Fig. 20), except for the stereochemistry at C. 6, without recourse to the independent evidence from our X-ray crystallographic study of O-methyl-O′, O″-isopropylidene-tetrodotoxin hydrochloride. Finally, we may note briefly that the full structure of octaacetylanhydrotetrodotoxin, which may be deduced readily from the relationship already adumbrated, is also in full accord with the detailed analysis of its nuclear magnetic resonance spectrum.

Of more than usual interest is the unique nature of the hemilactal function, not having been observed before in the structure of any organic molecule. If groups which are normally noninteracting are oppositely positioned in a rigid structure, the possibility is not excluded that these groups may enter into combinations which are not observed in simpler systems.

Tetrodotoxin and Tarichatoxin in the California Newt: The discovery of a toxin in the California newt, believed to be identical to tetrodotoxin in a widely different species (the Japanese puffer), stands as one of the most remarkable biotoxicological discoveries of recent times. The story has its beginning in the early 1930s when V. C. Twitty transferred from Yale University to Stanford University where he had been conducting research on the eastern salamander *Ambystoma punctatum*. In searching for a suitable substitute, he was referred to the California newt *Taricha torosa* (Figs. 22, 23). In the course of his studies, Twitty noted that when the eye vesicles of *T. torosa* were transplanted into an embryo of *A. tigrinum*, the host became paralyzed (Twitty and Elliott, 1934). From this and other observations, Twitty concluded that the embryos of *Taricha* contained a toxic substance. Efforts to isolate or identify the toxin were unsuccessful (Twitty and Johnson, 1934; Twitty, 1937).

The problem was taken up again by Brown and Mosher (1963) and Mosher *et al.*, (1964). Isolation of the toxin was eventually attained, the last step in the purification being a precipitation by the addition of ether to a solution of the impure toxin in 5 percent ethanolic acetic acid. The microcrystals obtained had a toxicity of c. 7,000 mouse units per milligram, thus placing the toxin in the category of the more potent fish toxins. A more satisfactory method of final purification was later developed by passing carbon dioxide through an aqueous suspension of the toxin until the toxin dissolved. The insoluble materials were then removable by centrifugation. The toxin solution, saturated with carbon dioxide was then let stand in a closed container over 0.01 percent solution of ammonium hydroxide. Microcrystals of the toxin formed on the surface of the solution. Mosher's group termed the poison "tarichatoxin."

During the course of the studies, the suspicion grew that the toxin from *T. torosa* was the same as that from the Japanese puffer, though such an equation seemed most unlikely in view of the wide biological differences between the two sources. The identity was believed to be established by showing coincidence of numerous properties. Among these properties common to tarichatoxin and tetrodotoxin are the following: (1) a toxicity of the same high order; (2) similar LD_{50} values in mice; (3) the toxicity from parenteral administration is some 40 to 50 times that from oral administration; (4) decomposition on heating without any true melting point; (5) solubility only in acid reagents; (6) coincidence of properties of the heptaacetates in terms of melting points, mass spectra, infrared spectra, magnetic-resonance spectra, and optical rotation; (7) coincidence of properties of the diacetates; (8) similar chromatographic behavior by thin-layer chromatography in four different solvent systems; (9) failure of administration of tetrodotoxin to *Taricha* to produce any toxic symptoms; and (10) the same empirical formula as deduced from the nuclear-magnetic-resonance spectra (*see also* Fuhrman *et al.*, 1963; Kao and Fuhrman, 1963; Buchwald *et al.*, 1964).

In 1969 Fuhrman *et al.* reported that they found tetrodotoxin from the eggs of puffer and newts to be an aminoperhydroquinazoline compound ($C_{11}H_{17}N_3O_8$) of proven structure. It was extremely toxic ($LD_{50}=10^{\circ}$ g/kg) and blocked conduction in nerves by selectively preventing the increase in conductance to sodium ions that normally accompanies excitation.

More recently, Kim, Brown, Mosher, and Fuhrman (1975) have reported the occurrence of tetrodotoxin in the skin of *Atelopus* frogs of Costa Rica.

The discovery that such widely different species are capable of developing the same toxin is most remarkable and poses some very interesting questions regarding the biogenesis of the poison. The story of this discovery was reviewed in the July, 1964, issue of *The Sciences* of the New York Academy of Sciences.

FIGURE 22.—California newt, *Taricha torosa* (Rathke), male (right, with rough skin) and female (left, with smooth skin). Average adult length about 15 cm. Egg cluster in foreground. Size of cluster about 2 cm in diameter. Tarichatoxin, which is believed to be identical to tetrodotoxin, is found most concentrated in the eggs or developing embryos of the newt. (Courtesy H. S. Mosher)

FIGURE 23.—Egg cluster of *Taricha torosa* attached to a twig in a stream. Diameter of individual clusters about 2 cm. Embryos appear as dark dots.

(Courtesy H. S. Mosher)

LITERATURE CITED

ABE, T.

1949 Taxonomic studies on the puffers (Tetraodontidae, Teleostei) from Japan and adjacent regions. V. Synopsis of the puffers from Japan and adjacent regions. Bull. Biogeogr. Soc. Japan 14(13): 89-140.

1950 Taxonomic studies on the puffers (Tetraodontidae, Teleostei) from Japan and adjacent regions. VI. Variation of pectoral fin. Japan. J. Ichthyol. 1(3): 198-206.

1951 Taxonomic studies on the puffers (Tetraodontidae, Teleostei) from Japan and adjacent regions. VI. Variation of pectoral fin, pt. 2. Japan. J. Ichthyol. 1(4): 272-283.

1952 Taxonomic studies on the puffers (Tetraodontidae, Teleostei) from Japan and adjacent regions. VII. Concluding remarks, with the introduction of two new genera, *Fugu* and *Boesemanichthys*. Japan. J. Ichthyol. 2(1): 35-44; (2): 93-97; (3): 117-127.

1954 Taxonomic studies on the puffers from Japan and adjacent regions. Corrigenda and addenda. Japan. J. Ichthyol. 3(3-5): 121-128, 1 fig., 1 pl.

AKASHI, T.

1880 Experiences with fugu poisoning. [In Japanese] Iji Shimbum (27): 19-23.

ANDERSON, W.

1776 An account of some poisonous fish in the South Seas. Phil. Trans. Roy. Soc. London 66: 544-552.

AOMURA, T., YEN, T. J., and K. OIKAWA

1930 Einfluss des tetrodotoxins auf die epinephrinabgabe beim hunde. Tohoku J. Exp. Med. 15(1-2): 36-49.

AUTENRIETH, H. F.

1833 Ueber das gift der fische. C. F. Osiander, Tübingen. 287 p.

BANNER, A. H., and H. BOROUGHS

1958 Observations on toxins of poisonous fishes. Proc. Soc. Exp. Biol. Med. 98: 776-778.

BARTSCH, A. F., and E. F. MCFARREN

1962 Fish poisoning: a problem in food intoxication. Pacific Sci.16(1): 42-56.

BENNETT, G.

1871 On the "toad fish" (*Tetraodon hamiltoni*) of New South Wales. N.S. Wales Med. Gaz. 1: 176-181, 1 pl.

BENSON, J.

1956 Tetraodon (blowfish) poisoning. A report of two fatalities. J. Forensic Sci. 1(4): 119-125.

BERNSTEIN, M. E.

1969 Pharmacologic effects of tetrodotoxin: cardiovascular and antiarrhythmic activities. Toxicon 7: 287-302, 10 figs.

BLAKENSHIP, J. E.

1976 Tetrodotoxin: from poison to powerful tool. Persp. Biol. Med. 19: 509-526.

BLEEKER, P.

1851 Over drie nieuwe soorten van tetraodon van den Indischen Archipel. Nat. Tijdschr. Neder-Indie 1: 96-971

1862-77 Atlas ichthyologique des Indes Orientales Néerlandaises, publié sous les auspices du gouvernement colonial néerlandais. 9 vols. Amsterdam.

BONDE, C. VON

1948 Mussel and fish poisoning. I. Mussel poisoning. S. African Med. J. 22: 760-761.

1953 The toxicity of the blaasop or toby. S. African Med. J. 27(33): 692-694.

BORISON, H. L., MCCARTHY, L. E., CLARK, W. G., and N. RADHAKRISHNAN

1963 Vomiting, hypothermia, and respiratory paralysis due to tetrodotoxin (puffer poison) in the cat. Appl. Pharmacol. 5(3): 350-357.

BOYÉ, L.

1911 Intoxications et empoisonnements, p. 387. *In* C. Grall and A. Clarac [eds.], Traité de pathologique, exotique, clinique, et thérapeutique. Paris.

BREDER, C. M., JR., and E. CLARK
1947 A contribution to the visceral anatomy, development, and relationships of the Plectognathi. Bull. Am. Mus. Nat. Hist. 88(5): 287-320, 8 figs.; Pls. 11-14.

BRÉTA, F.
1939 Contribution à l'étude des poissons vénéneux. Bull. Lab. Maritime Dinard 12(21): 5-52, 2 figs.

BROWN, M. S., and H. S. MOSHER
1963 Tarichatoxin: isolation and purification. Science 140(3564): 295-296.

BUCHWALD, H. D., DURHAM, L., FISCHER, H. G., HARADA, R., MOSHER, H. S., KAO, C. Y., and F. A. FUHRMAN
1964 Identity of tarichatoxin and tetrodotoxin. Science 143(3605): 474-475.

BUDDLE, R.
1930 Some common poisonous fishes found in Singapore waters. J. Roy. Nav. Med. Serv. 16(2): 102-111, 8 figs.

BURNETT, W.
1846 On the effects produced by poisonous fish on the human frame. Proc. Roy. Soc. London 5: 609.

CHARNOT, A.
1945 La toxicologie au Maroc. Mém. Soc. Sci. Nat. Maroc. 1945(47): 86-97.

CHEYMOL, J., BOURILLET, F., and Y. OGURA
1962*a* Influence de la tétrodotoxine cristallisée et de la cocaine sur la libération d'acétylcholine au niveau des terminaisons nerveuses motrices. Méd. Exp. 6: 79-87.
1962*b* Action de quelques paralysants neuromusculaires sur la libération de l'acétylcholine au niveau des terminaisons nerveuses motrices. Arch. Intern. Pharmacodyn. 139(1-2): 187-197, 2 figs.

CHEYMOL, J., KOBAYASHI, T., BOURILLET, F. and L. TÉTREAULT
1961 Sur l'action paralysante neuromusculaire de la tétrodotoxine. Arch. Intern. Pharmacodyn. 134(1-2): 28-53. 13 figs.

CHISHOLM, C.
1808 On the poison of fish. Edinburgh Med. Surg. J. 4(16): 393-422.

CLARK, E.
1950*a* Reef fish studies in the South Pacific. Final Field Rept. Pac. Sci. Bd., Natl. Res. Council. 24 p.
1950*b* Fisherman beware! Fishing for poisonous plectognaths in the western Carolines. Res. Rev. NAVEXOS P—510, p. 1-6.

CLARK, E., and H. A. GOHAR
1953 The fishes of the Red Sea: order Plectognathi. Publ. Marine Biol. Sta. Al Ghardaqa (Red Sea) No. 8, 80 p., 5 pls., 22 figs.

CLOQUET, H.
1821 Ichthyque [poison], p. 550-554. *In* Dictionnaire des sciences naturelles. Vol. 22.

COUILLARD, P.
1953 An antimitotic substance in the ovary of the common puffer, *Sphaeroides maculatus*. Biol. Bull. 105(2): 372.

COUTIÈRE, H.
1899 Poissons venimeux et poissons vénéneux. Thèse Agrég. Carré et Naud, Paris. 221 p.

DARWIN, C. R.
1945 The voyage of the Beagle. Reprinted E. P. Dutton & Co., Inc., New York. 496 p.

DAWSON, E. Y., ALEEM, A. A., and B. W. HALSTEAD
1955 Marine algae from Palmyra Island with special reference to the feeding habits and toxicology of reef fishes. Allan Hancock Found. Publs., Occ. Pap. 17, 39 p., 13 figs.

DAY, F.
1878-88 The fishes of India. 2 vols. London, 816 p., 198 pls.

DETTBARN, W. D.
1971 Mechanism of action of tetrodotoxin (TTX) and saxitoxin (STX). *In* L. L. Simpson [ed.] Neuropoisons, their pathophysiological actions. Plenum Press, New York. p. 169-186.

DETTBARN, W. D., HIGMAN, H., ROSENBERG, P., and D. NACHMANSOHN
1960 Rapid and reversible block of electrical activity by powerful marine biotoxins. Science 132(3422): 300-301.

DUCE, B. R., FELDMAN, H. S., and E. R. SMITH
1972 Acute cardiovascular, antirrhythmic and toxic effects of tetrodotoxin (TTX) in unanesthetized dogs. Toxicol. Appl. Pharmacol. 23: 701-712.

DUNCAN, C.
1951 A case of toadfish poisoning. Med. J. Australia 2(20): 673-675.

FISH, C. J., and M. C. COBB
1954 Noxious marine animals of the central and western Pacific Ocean. U.S. Fish Wildlife Serv., Res. Rept. No. 36, p. 14-23.

FLASCHENTRAGER, B., and M. M. ABDALLA
1957 Some toxic fishes of the Red Sea. Alexandria Med. J. 3(2): 177-188.

FLEISHER, J. H., KILLOS, P. J., and C. S. HARRISON
1960 Effects of puffer poison on neuromuscular transmission. Federal Proc. 19(1): 264.

FONSECA, O. O. DA
1917 Sobre os peixes venenosos. Brazil Med. 31: 90-91, 97-99.

FONSECA, R. C., and A. A. DIAZ
1957 Recordativo sobre la ciguatera. Mar y Pesca, p. 32-35.

FORSTER, J. R.
1778 Observations made during a voyage round the world, on physical geography, natural history, and ethic philosophy. G. Robinson, London. 649 p.

FRASER-BRUNNER, A.
1943 Notes on the plectognath fishes. VIII. The classification of the suborder *Tetraodontoidea*, with a synopsis of the genera. Ann. Mag. Nat. Hist., 11, 10: 1-18.
1951 The ocean sunfishes (family Molidae). Bull. Brit. Mus. (Nat. Hist.) 1(6): 87-121, 18 figs.

FUHRMAN, F. A., FUHRMAN, G. J., DULL, D. L. and H. S. MOSHER
1969 Toxins from eggs of fishes and amphibia. Agric. Food Chem. 17(3): 417-424, 6 figs.

FUHRMAN, F. A., FUHRMAN, G. J., and J. S. ROSEEN
1970 Toxic effects produced by extracts of eggs of the cabezon *Scorpaenichthys marmoratus*. Toxicon 8: 55-61, 2 figs.

FUHRMAN, F. A., KAO, C. Y., MOSHER, H. S., and M. S. BROWN
1963 Tarichatoxin—a potent neurotoxin from the California newt. Proc. West. Pharmacol. Soc. 6: 31-32, 1 fig.

FUKUDA, T.
1937 Puffer poison and the method of prevention. [In Japanese] Nippon Iji Shimpo 762: 1417-1421.
1951 Violent increase of cases of puffer poisoning. [In Japanese] Clinics and Studies 29(2).

FUKUDA, T., and I. TANI
1937*a* Records of puffer poisonings. Rept. 1. [In Japanese] Kyushu Univ. Med. News 11(1): 7-13.
1937*b* Records of puffer poisonings. Rept. 2. [In Japanese] Iji Eisei 7(26): 905-907.
1937*c* Statistische beobachtung über die fuguvergiftung. Japan. J. Med. Sci., IV, 10: 48-50.
1941 Records of puffer poisonings. Rept. 3. [In Japanese] Nippon Igaku oyobi Kenko Hoken (3258): 7-13.
1942 The puffer of the continent. [In Japanese] Nippon Igaku oyobi Kenko Hoken (3308): 13-17.

FURTADO, J. A.
1903 Pesquizas ichthyologicas na Bahia do Rio de Janeiro. These, Fac. Med. Rio de Janeiro, p. 131-142.

FURUKAWA, T., SASAOKA, T., and Y. HOSOYA
1959 Effects of tetrodotoxin on the neuromuscular junction. Japan. J. Physiol. 9(2): 143-152.

GAGE, P. W.
1971 Tetrodotoxin and saxitoxin as pharmacological tools. *In* L. L. Simpson [ed.], Neuropoisons, their pathophysiological actions. Plenum Press, New York, p. 187-212.

GIMLETTE, J. D.
1923 Malay poisons and charm cures. 2d ed. J. & A. Churchill, London. 260 p.

GOE, D. R.
1958 Some neurophysiological effects of an extract from the gulf puffer, *Sphoeroides annulatus* Jenyns. Doctoral Dissertation. Univ. Southern Calif. 86 p.

GOE, D. R., and B. W. HALSTEAD
1953 A preliminary report of the toxicity of the gulf puffer, *Sphoeroides annulatus*. Calif. Fish Game 39(2): 229-232, 1 fig.

GOERTZ, A.
1875 Ueber in Japan vorkommende fisch- und lack-vergiftungen. Mitt. Deut. Ges. Natur. Volk. Ostasiens (Yokohama) 8: 23-26.

GOLIN, S., and E. LARSON
1969 An antidotal study on the skin extract of the puffer fish *Spheroides maculatus*. Toxicon 7: 49-53.

GOMEZ, I. C.
1962 Algunos peces venenosos de la República Mexicana. Bol. Direc. Estud. Biol. 3(4): 66-68.

GOTO, T., KISHI, Y., and Y. HIRATA
1962*a* Structure of the C_9-base, an alkaline degradation product of tetrodotoxin. Bull. Chem. Soc. Japan 35(6): 1045-1046.
1962*b* Structure of the C_8-base, an acid degradation product of tetrodotoxin. Bull. Chem. Soc. Japan 35(7): 1244-1245.

GOTO, T., KISHI, Y., TAKAHASHI, S., and Y. HIRATA
1963*a* The structure of tetrodotoxin. Tetrahedron Letters (30): 2105-2113.
1963*b* The structure and stereochemistry of tetrodotoxin. Tetrahedron Letters (30): 2115-2118.
1964*a* The structures of tetrodotoxin and anhydroepitetrodotoxin. [In Japanese, English summary] J. Chem. Soc. Japan 85: 661-666.
1964*b* Acetylation of tetrodotoxin. [In Japanese, English summary] J. Chem. Soc. Japan 85: 667-671.
1964*c* Further studies on the structure of tetrodotoxin. Tetrahedron Letters (14): 779-786.
1964*d* Tetrodotoxin, III. A revised molecular formula for tetrodotoxin. Bull. Chem. Soc. Japan 37(2): 283-284.
1964*e* Aminodesoxytetrodotoxin. Tetrahedron Letters (27): 1831-1834.
1964*f* Extraction and purification of tetrodotoxin. [In Japanese, English summary] J. Chem. Soc. Japan 85: 508-511.

GÜNTHER, A.
1880 An introduction to the study of fishes. Adam & Charles Black, Edinburgh. 720 p.

HABEKOST, R. C., FRASER, I. M., and B. W. HALSTEAD
1955 Observations on toxic marine algae, J. Wash. Acad. Sci 45(4): 101-103.

HALSTEAD, B. W.
1951 Poisonous fish—a medical military problem. Res. Rev. U.S. Nav. NAVEXOS P—510: 10-16, 8 figs.
1953 Some general considerations of the problem of poisonous fishes and ichthysarcotoxism. Copeia (1): 31-33.
1959 Dangerous marine animals. Cornell Maritime Press, Cambridge. 146 p. 88 figs.

HALSTEAD, B. W., and N. C. BUNKER
1953 The effect of the commercial canning process upon puffer poison. Calif. Fish Game 39(2): 219-228, 1 fig.
1954*a* A survey of the poisonous fishes of the Phoenix Islands. Copeia (1): 1-11, 5 figs.
1954*b* A survey of the poisonous fishes of Johnston Island. Zoologica 39(2): 61-77, 1 fig.

HALSTEAD, B. W., and W. M. LIVELY
1954 Poisonous fishes and ichthyosarcotoxism. Their relationship to the Armed Forces. U.S. Armed Forces Med. J. 5(2): 157-175, 9 figs.

HALSTEAD, B. W., and R. J. RALLS
1954 Results of dialyzing some fish poisons. Science 119(3083): 160-161.

HALSTEAD, B. W., and D. S. SCHALL
1955 A report on the poisonous fishes captured during the Woodrow G. Krieger expedition to the Galapagos Islands, p. 147-172, 1 pl. *In* Essays in the natural sciences in honor of Captain Allan Hancock. Univ. Southern Calif. Press, Los Angeles.
1956 A report on the poisonous fishes captured during the Woodrow G. Krieger expedition to Cocos Island. Pacific Sci. 10(1): 103-109, 1 fig.

HAMADA, J.
1957 Etude pharmacologique de la tétrodotoxine cristallisée sur la fibre musculaire lisse. Première partie —L'action sur l'intestin isolé de certains animaux. [In Japanese] Ann. Rept. Inst. Food Microbiol. 10: 73-77.
1960 The effect of tetrodotoxin crystals on the smooth muscle. II. The effect of tetrodotoxin and ganglionic active drugs in the isolated intestine and bronchial muscle. [In Japanese, English summary] J. Chiba Med. Soc. 36(4): 1358-1368.

HASHIMOTO, Y., and M. MIGITA
1950 On the shellfish poisons. I. Inadequacy of acidulated alcohols with hydrochloric acid as solvent. Bull. Japan. Soc. Sci. Fish. 16(3): 77-85.
1951 On a method of quantitative analysis for fugu (puffer) toxin. [In Japanese] Bull. Japan Soc. Sci. Fish. 16(8): 341-346, 1 fig.

Hasiguti, M.
1931 On the changes in the striated muscles caused by globe-fish toxin. [In Japanese] Trans. Japan. Pathol. Soc. 21: 678-679.

Hayama, T., and Y. Ogura
1958 Emetic action of tetrodotoxin (preliminary note). Ann. Rept. Inst. Food Microbiol. 11: 77-78.
1963 Site of emetic action of tetrodotoxin in dog. J. Pharmacol. Exp. Therap. 139(1): 94-96.

Hayashi, H., and K. Muto
1901 Ueber athemversuche mit einigen giften. Arch. Exp. Pathol. Pharmakol. 47: 209-230.

Heilbrunn, L. V., Chaet, A. B., Dunn, A., and W. L. Wilson
1954 Antimitotic substances from ovaries. Biol. Bull. 106(2): 158-168, 1 fig.

Helfrich, P.
1961 Fish poisoning in the tropical Pacific. Hawaii Marine Lab., Univ. Hawaii. 16 p., 7 figs.
1963 Fish poisoning in Hawaii. Hawaii Med. J. 22: 361-372.

Herre, A. W.
1924 Poisonous and worthless fishes. Philippine J. Sci. 25(4): 415-510, 2 pls.
1953 Checklist of Philippine fishes. U.S. Fish Wildlife Serv., Res. Rept. 20. 977 p.

Hess, A. F., and M. Weinstock
1926 Puffer fish oil; a very potent antirachitic; its elaboration by fish deprived of sunlight. Proc. Soc. Exp. Biol. Med. 23: 407-408.

Hiyama, Y.
1943 Report of an investigation on poisonous fishes of the South Seas. [In Japanese] Nissan Fish. Exp. Sta. (Odawara, Japan). 137 p., 29 pls., 83 figs.
1952 Regarding tetrodotoxic fish. (Personal communication, 1952).

Hoffmann, W. H.
1929*a* Los peces venenosos de Cuba y la ciguatera. Rev. Chilena Hist. Nat. 33: 28-30.
1929*b* La ciguatera, enfermedad producida por peces venenosos de Cuba. Invest. Progreso 3(11): 101-102.

Hori, H.
1957 Etude histo-patholigique de la tétrodotoxine cristallisée. I. L'observation de la toxicose aigüe chez la souris. [In Japanese, English summary] Ann. Rept. Food Microbiol. 10: 71-72.
1958 Studies on histopathologic effects of crystalline tetrodotoxin. II. Observations in rabbits, rats and mice chronically treated with tetrodotoxin. [In Japanese, English summary] Ann. Rept. Inst. Food Microbiol. 11: 71-76.

Hsiang, N. S.
1939 A new method of treatment of alkaloid addicts by tetrodotoxin. [In Japanese] Manṣhu Igaku Zasshi 30; 639-647.

Imahasi, T.
1928 Über den einfluss einiger pharmaka auf die tödliche vergiftung durch tetrodotoxin. Okayama Igakkai Zasshi 40(12): 2452-2464.

Inaba, M.
1935 Effects of tetrodotoxin on blood coagulation. [In Japanese] Okayama Igakkai Zasshi 47: 3348-3361.

Inouye, S.
1913 Influence of tetrodotoxin on micturition and especially on enuresis nocturna. [In Japanese] Japan. Z. Dermatol. Urol. 13: 79-86.

Irvine, F. R.
1947 The fishes and fisheries of the Gold Coast. Crown Agents for the Colonies, London. 352 p., 217 figs.

Ishihara, F.
1918 Über die physiologischen wirkungen des fugutoxins. Mitt. Med. Fak. Univ. Tokyo 20: 375-426; Pls. 23-25.
1924 Studien über das fugotoxin. Arch. Exp. Pathol. Pharmakol. 103: 209-218.

Itakura, T.
1917 Zur kenntnis der pharmakologischen wirkung des tetrodotoxins. Mitt. Med. Fak. Kaiserl., Univ. Tokyo 17: 455-538, 11 figs.

Ito, J.
1889 A review of articles on fugutoxin. [In Japanese] Z. Tokyo Med. Ges. 3: 533-534.

Itokawa, Y., and J. R. Cooper
1969 Thiamine release from nerve membranes by tetrodotoxin. Science 166: 759-761, 2 figs.

IWAKAWA, K., and S. KIMURA
1922 Experimentelle untersuchungen über die wirkung des tetrodontoxins ("fugu-gift"). Arch. Exp. Pathol. Pharmakol. 93(4-6): 305-331.

JONES, S.
1956 Some deaths due to fish poisoning (ichthyosarcotoxism) in India. Indian J. Med. Res. 44(2): 353-360, 2 figs.

JORDAN, D. S., TANAKA, S., and J. O. SNYDER
1913 A catalogue of the fishes of Japan. J. Coll. Sci., Tokyo Univ. 33, 497 p., 396 figs.

KABUKI, E.
1936 Über die wirkung des tetrodotoxins auf blutdruck und atmung bei leberschädung. Japan. J. Med. Sci. (IV, Pharm.) 8: 109.
1937 Über die tetrodotoxindurchströmung der leber. [In Japanese] Chiba Igakukaai Zasshi 15: 2538-2543.

KAMINISHI, J.
1942 Experimental study of the principles of emergency treatment against puffer poison. [In Japanese] Nippon Igaku oyobi Kenko Hoko (3296): 13-15.

KANAYAMA, S.
1943 Purification and several chemical characteristics of puffer poison. [In Japanese] Fukuoka Acta Med. 36(4): 395-401, 1 fig.

KAO, C. Y.
1964 Tetrodotoxin: mechanism of action. Science 144: 319.
1966 Tetrodotoxin, saxitoxin, and their significance in the study of excitation phenomena. Pharm. Rev. 18(2): 997-1049.
1972 Pharmacology of tetrodotoxin and saxitoxin. Fed. Proc. 31(3): 1117-1123, 3 figs.

KAO, C. Y., and F. A. FUHRMAN
1963 Pharmacological studies on tarichatoxin, a potent neurotoxin. J. Pharmacol. Exp. Therap. 140: 31-40.

KARIYA, S.
1914 Experimentelle untersuchung über das tetrodongift. Mitt. Med. Ges. Tokyo 28(5): 1.

KAWABATA, T.
1965 Regarding outbreaks of puffer poisoning in Japan during the years 1952-63. (Personal communication, May 1, 1965.)

KAWAKUBO, Y., and K. KIKUCHI
1942 Testing fish poisons on animals and report of a human case of fish poisoning in the South Seas. [In Japanese] Kaigun Igakukai Zasshi 31(8): 30-34, 3 pls.

KAWAMURA, M.
1960 Untersuchungen der eierstock-extrakte von kugelfischen. XII. Über tetrodotoxin. Chem. Pharm. Bull. 8(3): 262-265.

KIM, Y. H., BROWN, G. B., MOSHER, H. S., and FUHRMAN
1975 Tetrodotoxin: Occurrence in atelopid frogs of Costa Rica. Science 189: 151-152.

KIMURA, S.
1927 Zur kenntnis der wirkung des tetrodongiftes. Tohoku J. Exp. Med. 9: 41-65, 7 figs.

KISHI, Y., TAGUCHI, H., GOTO, T., and Y. HIRATA
1964 Structure of C_9-base, C_8-base, and oxy-C_8-base, alkaline and acid degradation products of tetrodotoxin and its derivatives. [In Japanese, English summary] J. Chem. Soc. Japan 85: 564-572.

KOBAYASHI, T.
1944 Über den einfluss des fugugiftes auf die hirngefässe. Japan J. Med. Sci. (IV, Pharm.) 16(2): 68-69.

KOBERT, R.
1902 Ueber giftfische und fischgifte. Med. Woche (19): 199-201; (20): 209-212; (21): 221-225.

KUDAKA, K.
1960 A preliminary investigation of the poisonous fish of the Ryuku Islands. [In Japanese] Ann. Rept. Ryuku Fish Inst. p. 95-103.

KUNISHO, K.
1934 The reaction of curare, tetrodotoxin, strychnine, veratrin, and sodium nitrite on the peripheral blood vessels. [In Japanese] Okayama Igakkai Zasshi 47: 531-546.

KURIAKI, K., and H. NAGANO
1957 Susceptibility of certain enzymes of the central nervous system to tetrodotoxin. Brit. J. Pharmacol. 12(4): 393-396.

KURIAKI, K., and I. WADA
1957 Effect of tetrodotoxin on the mammalian neuro-muscular system. Japan. J. Pharmacol. 7(1): 35-37.
1959 Effect of tetrodotoxin, epinephrine and nor-epinephrine on glucose uptake of the rat diaphragm. Japan. J. Pharmacol. 8(2): 170-172.

LALONE, R. C., DEVILLEZ, E. D., and E. LARSON
1963 An assay of the toxicity of the Atlantic puffer fish, *Spheroides maculatus*. Toxicon 1: 159-164.

LARSEN, N. P.
1925 Fish poisoning. Queen's Hospital Bull. 2(1): 1-3, 1 fig.
1942 Tetrodon poisoning in Hawaii. Proc. 6th Pacific Sci. Congr. 5: 417-421.

LARSON, E., LALONE, R. C., and L. R. RIVAS
1960 Comparative toxicity of the Atlantic puffer fishes of the genera *Spheroides, Lactophrys, Lagocephelus* and *Chilomycterus*. Federation Proc. 19(1): 388.

LEBER, A.
1927 Über tetrodonvergiftung. Arb. Trop. Grenzgebiete 26: 641-643, 2 pls.

LI, K. M.
1963 Action of puffer fish poison. Nature 200(4908): 791.

LOEWENSTEIN, W. R., TERZUOLO, C. A., and Y. WASHIZU
1963 Separation of transducer and impulse-generating processes in sensory receptors. Science 142(3596): 1180-1181.

MAASS, T. A.
1937 Gift-tiere. *In* W. Junk [ed.], Tabulae biologicae. Vol. XIII. N. V. Van de Garde & Co's Drukkerij, Zaltbommel, Holland. 272 p.

MACHT, D. I., and E. C. SPENCER
1941 Physiological and toxicological effects of some fish muscle extracts. Proc. Soc. Exp. Biol. Med. 46(2): 228-233.

MARSHALL, T. C.
1964 Fishes of the Great Barrier Reef and coastal waters of Queensland. Angus and Robertson, Melbourne. 566 p., 72 pls.

MATSUMURA, M., and S. YAMAMOTO
1954 The effect of tetrodotoxin on the neuro-muscular junction and peripheral nerve of the toad. Japan. J. Pharmacol. 4(1): 62-68, 7 figs.

MATSUO, R.
1934 Study of the poisonous fishes at Jaluit Island. [In Japanese] Nanyo Gunto Chihobyo Chosa Ronbunsha 2: 309-326, 9 pls.

MAZUMDAR, S. K.
1915 A case of fish poisoning. Indian Med. Gaz. 50: 218-219, 1 fig.

MEEK, S. E., and S. F. HILDEBRAND
1928 The marine fishes of Panama. Part III. Field Mus. Nat. Hist. Publ. 249 (Zoology) 15: 709-1045; Pls. 72-102.

MEYER-AHRENS, K. M.
1855 Von den giftigen fischen. Schweiz. Z. Med. Chir. Geburtsh. (3): 188-230; (4-5); 269-332.

MIGITA, M., and Y. HASHIMOTO
1951 On the puffer roe pickled in salt and rice-bran. [In Japanese] Nippon Sui San Gakkatshi 16(8): 335-340.

MIURA, M., and H. TAKESAKI
1887 Investigation on the localization of puffer poison. [In Japanese] Tokyo Med. Assoc. Mag. 3(8): 451-456; (9): 514-518.
1889 Die ermittelung des tetradon-giftes und seine experimentelle untersuchung. [In Japanese] Z. Tokyo Med. Ges. 3(10): 514-518.
1890 Zur localisation des tetrodon-giftes. Arch. Pathol. Anat. Physiol. 122: 92-99.

MOREAU DE JONNÈS, A.
1819 Recherches sur les poissons toxicofères des Indes occidentales. Bull. Sci. Soc. Philomath. Paris, 3,6: 136.
1821 Sur les poissons toxicophores des Indes occidentales. Nouveau J. Méd. Chir., Pharm. 11: 356-389.

MOSES, S. T.
1922 A statistical account of the fish-supply of Madras. Madras Fish. Bull. 15: 161-161.

MOSHER, H. S., FUHRMAN, F. A., BUCHWALD, H. D., and H. G. FISCHER
1964 Tarichatoxin—tetrodotoxin: a potent neurotoxin. Science 144(3622): 1100-1110.

Mucciolo, P.
1956 Peixes venenosos e alimentação. Rev. Brasil. Med. 13(6): 486-487.

Murtha, E. F.
1960 Pharmacological study of poisons from shellfish and puffer fish. Ann. N.Y. Acad. Sci. 90(3): 820-836, 9 figs.

Murtha, E. F., Stabile, D. E., and J. H. Wills
1958 Some pharmacological effects of puffer poison. J. Pharmacol. Exp. Therap. 122(2): 247-254.

Nagai, J.
1954 Chemistry of fugu (Tetraodontidae) —poison and biochemistry of its intoxication. [In Japanese] Fukuoka Acta Med. 45: 1-12.
1956 Isolierung des kugelfischgiftes mit ionenaustauscher. Hoppe-Seylers Z. Physiol. 306: 104-106.

Nagai, J., and T. Ito
1939 On the chemical study of fugu (*Spheroides*) poison. J. Biochem. (Tokyo) 30(2): 235-238.

Nakamura, H.
1950 The food habits of yellowfin tuna, *Neothunnus macropterus* (Schlegel), from the Celebes Sea. U.S. Fish Wildlife Serv., Spec. Sci. Rept., Fish. 23. 8 p.

Nakazawa, T., and Y. Ogura
1961 Effect of crystalline tetrodotoxin on various enzymes and enzyme systems in animal tissues. Ann. Rept. Food Microbiol. 13: 59-61.

Narahashi, T.
1972 Mechanism of action of tetrodotoxin and saxitoxin on excitable membranes. Fed. Proc. 31(3): 1124-1132, 13 figs.

Narahashi, T., Deguchi, T., Urakawa, N., and Y. Ohkubo
1960 Stabilization and rectification of muscle fiber membrane by tetrodotoxin. Am. J. Physiol. 198: 934-938.

Narahashi, T., Moore, J. W., and W. R. Scott
1964 Tetrodotoxin blockage of sodium conductance increase in the lobster giant axons. J. Gen. Physiol. 47(5): 965-974.

Naruke, T.
1935 The study of anti-tetrodotoxin. [In Japanese] Folia Pharm. Japon. 18: 117-118.

Nomiyama, S.
1941 Analgesic and emetic action from the application of puffer poison. [In Japanese] J. Japan. Pharm. 31: 148-150.
1942 The pharmacological study of puffer poison. [In Japanese] Nippon Yakubutsugaku Zasshi 35(4): 458-496.

Ogonuki, H., and M. Ide
1942 Outbreaks of food poisoning in Japan. Proc. Pacific Sci. Congr. Pacific Sci. Assoc., 6th, Berkeley 1939, 5: 423-427.

Ogura, Y.
1957 Sur l'existence de l'acétylcholine dans les extraits fluides crues d'ovaire du tétrodon (fugu). [In Japanese, French summary] Ann. Rept. Inst. Food Microbiol. 10: 84-86.
1958 Some recent advances in the knowledge of tetrodotoxin. [In Japanese] Kiso Igakus, Seitai No Kagaku 9(4): 281-287.
1959 The effect of tetrodotoxin on oedema formation in the hind-paw of rat induced by 5-hydroxy-tryptamine, histamine and albumin (egg white). [In Japanese, English summary] Ann. Rept. Food Microbiol. 12: 105-109.
1963*a* The biological estimation of crystalline tetrodotoxin. III. On the isolated stomach vagal nerve preparation of rat. [In Japanese, English summary] Ann. Rept. Inst. Food Microbiol. 15-93-96.
1963*b* Analyse du phénomène de potentialisation de la contraction maximale provoquée par de faibles doses de tétrodotoxine. [In Japanese, French summary] Ann. Rept. Inst. Food Microbiol. 15: 97-100.

Ogura, Y., and K. Fujimoto
1963 Influence of crystalline tetrodotoxin on the movement of aldolase from excited rat diaphragm. [In Japanese, English summary] Ann. Rept. Inst. Food Microbiol. 16: 57-60.

Ogura, Y., and H. Hori
1959 L'influence de la tétrodotoxine cristallisée sur le poids de l'organo et la courbe pondérale. [In Japanese, French summary] Ann. Rept. Inst. Food Microbiol. 12: 100-104.
1961 Effect of crystalline tetrodotoxin on gastric secretory activity. [In Japanese] Folia Pharmacol. Japon. 57: 274-279.

Ogura, Y., and S. Nakagima
1959 L'action de la tétrodotoxine cristallisée sur les réflexes de redressement. [In Japanese, French summary] Ann. Rept. Inst. Food Microbiol. 12: 115-118.

Ogura, Y., and J. Nara
1963 Influence of crystalline tetrodotoxin on quantitative measurement of Evans blue space in the tissue of a rat. [In Japanese, English summary] Ann. Rept. Inst. Food Microbiol. 16: 61-63.

Ogura, Y., Nara, J., and T. Yoshida
1968 Comparative pharmacological actions of ciguatoxin and tetrodotoxin, a preliminary account. Toxicon 6: 131-140, 10 figs.

Pawlowsky, E. N.
1927 Gifttiere und ihre giftigkeit. Gustav Fischer, Jena. 515 p., 170 figs.

Pellegrin, J.
1899 Les poissons vénéneux. M.D. Thèse 510. Fac. Méd. Paris. 113 p., 16 figs.

Phisalix, M.
1922 Animaux venimeux et venins. 2 vols. Masson et Cie., Paris.

Prosvirov, E.
1963 Poisonous fishes. [In Russian] Kaliningrad Publ. Ofc., Moscow. 79 p.

Read, B. E.
1939 Chinese materia medica. Fish Drugs. Peiping Nat. Hist. Bull. 136 p., 190 figs.

Reintjes, J. W., and J. E. King
1953 Food of the yellowfin tuna in the central Pacific. U.S. Fish Wildlife Serv., Fish. Bull. 81, p. 98.

Rémy, C.
1883 Sur les poissons toxiques du Japon. Compt. Rend. Soc. Biol., 7, 5: 3-28.

Richardson, B. W.
1893 Fish-poisoning and the disease "siguatera." Asclepiad, London 10: 38-42.

Richardson, J.
1861 On the poisonous effect of a small portion of the liver of a *Diodon* inhabiting the seas of southern Africa. J. Linn. Soc. London 5: 213-216.

Rochas, V. de
1860 Essai sur la topographie hygiénique et médicale de la Nouvelle-Calédonie. M.D. Thèse 250. Fac. Méd. Paris. 35 p.
1862 La Nouvelle Calédonie et ses habitants. F. Sartorius, Paris. p. 64-65.

Satake, Y.
1954 Secretion of adrenaline and symphathins. Tohoku J. Exp. Med. 60: 121.

Savtschenko, P. N.
1882 A case of poisoning by fish. [In Russian] Medits. Pribav. Morsk. Sborniku, St. Petersburg (9): 55-61.

Schlegel, H.
1850 Pisces, p. 323. *In* P. F. Siebold, Fauna Japonica, sive descriptio animalium quae in itinere per Japoniam suscepto annis 1823-1830 collegit, notis, observationibus et adumbrationibus illustravit P. F. de Siebold. Conjunctis studiis C. J. Temminck, et H. Schlegel pro vertebratis atque W. de Haan pro invertebratis elaborata. 6 vols. in 4. Lugduni Batavorum, Leyden.

Seale, A.
1912 Poisonous fishes of the Philippine Islands. Bull. Philippine Bur. Health (9): 3-9, 6 figs.

Shimazono, J.
1914 Verhalten der nervensubstanz bei verschiedenen vergiftungen. Arch. Psychiat., Berlin 53(3): 1065-1068; Pls. 33-42.

Shirota, N., Fujita, K., and M. Kawamura
1952 Studies on the globefish poison. [In Japanese] Ann. Rept. Takamine Lab. 4: 45-49.

Silvado, J.
1911 Peixes nocivos da Bahia do Rio de Janeiro. Imprensa Nacional, Rio de Janeiro. 24 p., 20 pls.

Smith, J. L.
1950 The sea fishes of southern Africa. Central News Agency, Cape Town, S. Africa. 550 p., 103 pls.
1961 The sea fishes of southern Africa. 4th ed. Central News Agency, South Africa. 580 p., 1,232 figs., 102 pls.

Smith, J. L., and M. M. Smith
1963 The fishes of Seychelles. Rhodes Univ., Grahamstown. 215 p., 98 pls.

Sommer, W. W., and K. F. Meyer
1937 Paralytic shell-fish poisoning. Arch. Pathol. 24(5): 560-598, 2 figs.

Suehiro, Y.
1947 Poison of globe-fish, p. 140-159. *In* Y. Suehiro, Practice of fish physiology. [In Japanese] Takeuchi Bookstore, Tokyo.

Suyama, M., and Y. Uno
1957 Puffer toxin during the embryonic development of puffer, *Fugu (Fugu) niphobles* (J. et S.). Bull. Japan Soc. Sci. Fish. 23(7-8): 438-441, 1 pl.

TAHARA, Y.
1894 Report of discovery of puffer toxin. [In Japanese] Chugai Iji Shimpo (344): 4-9.
1896 Ueber die giftigen bestandtheile des tetrodon. Congr. Internat. Hyg. Demog., C. R. 1894 (Budapest) 8(4): 198-207.
1897*a* Discovery of puffer poison. [In Japanese] Dobutsugaka Zasshi 6(69): 268-275.
1897*b* Report on puffer poison. [In Japanese] Yakugaku Zasshi (328): 587-625.
1910 Über das tetrodongift. Biochem. Z. 30: 255-275.
1913 Tetrodotoxin and process of extracting the same. U.S. Patent No. 1,058,643. Patented Apr. 8, 1913. Commissioner of Patents, Washington, D.C. 2 p.

TAKAHASHI, D.
1897 The puffer poison. [In Japanese] Dobutsugaku Zasshi 5(56): 227-230; (57): 260-273; (60): 363-372.

TAKAHASHI, D., GOTO, T., and Y. HIRATA
1964*a* Structures of anhydrotetrodoic acid and bromoanhydrotetrodoic lactone. [In Japanese, English summay] J. Chem. Soc. Japan 85: 579-586.
1964*b* IR-spectra of guanidinium group in tetrodotoxin derivatives. [In Japanese, English summary] J. Chem. Soc. Japan 85: 586-589.

TAKAHASHI, D., and Y. INOKO
1889*a* Untersuchungen über das fugugift. Centralbl. Med. Wiss. 27(29): 529-530.
1889*b* Chemische untersuchungen über das fugugift. Centralbl. Med. Wiss. 27(49): 881-882.
1890 Experimentelle untersuchungen über das fugugift. Arch. Exp. Pathol. Pharmakol. 26: 401-418, 455-457.
1892 Localization of poison in the body of tetrodon. Sei-i-kwai Med. J. 11(5): 46-50; (6): 81-82.

TANAKA, S.
1914 The poisonous fish. [In Japanese] Dobutsugaku Zasshi 26(313): 516-517.

TANI, I.
1940 Seasonal changes and individual differences of puffer poison. [In Japanese] Nippon Yakubutsugaku Zasshi 29(1-2): 1-3.
1941 Poison of *Sphaeroides ocellatus* (Osbeck) and *Sphaeroides chrysops* (Hilgendorf). [In Japanese] Nippon Yakubutsugaku Zasshi 31(2): 1-2.
1945 Toxicological studies on Japanese puffers. [In Japanese] Teikoku Tosho Kabushiki Kaisha 2(3), 103 p., 22 figs.

TASCHENBERG, E. O.
1909 Die giftigen tiere. Ferdinand Enke, Stuttgart. 325 p., 64 figs.

TATEYAMA, T.
1942 Biochemical study of the puffer poison. [In Japanese] Fukuoka Acta Med. 35(9): 92-104.

TOMIE, T., FURUSAKI, A., KASAMI, K., YASUOKA N., MIYAKE, K., HAISA, M., and I. NITTA
1963 The crystal and molecular structure of bromoanhydrotetrodoic lactone hydrobromide. A derivative of tetrodotoxin. Tetrahedron Letters (30): 2101-2104.

TOMIYAMA, I., ABE, T., and T. TOKIOKA
1958 Encyclopaedia zoologica illustrated in colours. Vol. II. [In Japanese] Hokuryn-kan Publishing Co., Ltd., Tokyo. 392 p., 133 figs.

TORDA, T. A., SINCLAIR, E., and D. B. ULYATT
1973 Puffer fish (tetrodotoxin) poisoning: clinical record and suggested management. Med. J. Austr. 1: 599-602.

TSUDA, K.
1963 Die konstitution und konfiguration der tetrodonsäure. Chem. Pharm. Bull. (Japan) 11(11): 1473-1475.

TSUDA, K., HAYATSU, R., UMEZAWA, B., and T. NAKAMURA
1952 The constituents of the ovaries of globe-fish. V. Preparation of a hydrocarbon, $C_{54}H_{82}$, from cholesteryl tosylate, and the identification of the hydrocarbon, $C_{54}H_{82}$, obtained from various sources. J. Pharm. Soc. Japan 72(2): 182-186, 1 fig.

TSUDA, K., IKUMA, S., KAWAMURA, M., TACHIKAWA, R., BABA, Y., and T. MIYADERA
1962*a* Über die struktur der C_9-base, die

TSUDA, K., IKUMA, S., *et al.*—Continued
sich durch behandlung mit alkalilauge aus tetrodotoxin gewinnen lässt. Chem. Pharm. Bull. (Japan) 10(2): 247-249.
1962*a* Über tetrodotoxin. IV. Mitteilung. Die struktur der C_9-base, die sich durch einwirkung der alkalilauge aus tetrodotoxin gewinnen lässt. Chem. Pharm. Bull. (Japan) 10(9): 856-865.

TSUDA, K., IKUMA, S., KAWAMURA, M., TACHIKAWA, R., and T. MIYADERA
1962*a* Über tetrodotoxin. V. Mitteilung. Synthese des C_9-Base-methyläthers. Chem. Pharm. Bull. (Japan) 10(9): 865-867.
1962*a* Über tetrodotoxin. VI. Mitteilung. 2-Amino-6methylchinazolin als abbau produkt von tetrodotoxin. Chem. Pharm. Bull. (Japan) 10(9): 860-870.

TSUDA, K., and M. KAWAMURA
1950 The constituents of the ovaries of globefish. I. The isolation of mesoinositol and scillitol from the ovaries. J. Pharm. Soc. Japan 70(7-8): 432-435, 2 figs.
1951 The constituents of the ovaries of globefish. IV. Reducing sugar isolated during purification of the poisonous component. J. Pharm. Soc. Japan 7(4): 282-284.
1952*a* The constituents of the ovaries of globefish. VI. Purification of globefish poison by chromatography. J. Pharm. Soc. Japan 72(2): 187-190, 3 figs.
1952*b* The constituents of the ovaries of globefish. VII. Purification of tetrodotoxin by chromatography. J. Pharm. Soc. Japan 72(6): 771-773, 2 figs.
1953 The constituents of the ovaries of globefish. VIII. Studies on tetrodotoxin. Pharm. Bull. 1(2): 112-113, 1 fig.

TSUDA, K., KAWAMURA, M., and R. HAYATSU
1958 On the constituents of tetrodotoxin, Chem. Pharm. Bull. 6(2): 225-226.
1960 Untersuchungen der eierstockextrakte von kugelfischen. XI. Über tetrodotoxin. Chem. Pharm. Bull. (Japan) 8(3): 257-261.

TSUDA, K., and H. SHAN-HAI
1951 The constituents of the ovaries of globefish. III. Fatty acids that compose the fatty oil. J. Pharm. Soc. Japan 71(4): 279-282.

TSUDA, K., TACHIKAWA, R., SAKAI, K., AMAKASU, O., KAWAMURA, M., and S. IKUMA
1963 Die konstitution and konfiguration der tetrodonsäure. Chem. Pharm. Bull. 11(11): 1473-1475.

TSUDA, K., TACHIKAWA, R., SAKAI, K., TAMMURA, C., AMAKASU, O., KAWAMURA, M., and S. IKUMA
1964*a* On the structure of tetrodotoxin. Chem. Pharm. Bull. (Japan) 12(5): 642-645.
1964*b* Über die konstitution und konfiguration des anhydrotetrodotoxins. Chem. Pharm. Bull. (Japan) 12(5): 632-642.

TSUDA, K., and B. UMEZAWA
1951 The constituents of the ovaries of globefish. II. Steroids that constitute the fatty oil. J. Pharm. Soc. Japan 71(4): 273-278, 3 figs.

TSUKADA, O.
1957 Sur le mécanisme de la bradycardie provoquée par la tétrodotoxine cristallisée. [In Japanese, French summary] Ann. Rept. Inst. Food Microbiol. 10: 78-83.
1960 Effect of tetrodotoxin crystals on the circulatory reflex and experimental arythmia. [In Japanese, English summary] J. Chiba Med. Soc. 36(4): 1369-1379.

TWITTY, V. C.
1937 Experiments on the phenomenon of paralysis produced by a toxin occurring in *Triturus* embryos. J. Exper. Zool. 76(1): 67-104, 8 figs.

TWITTY, V. C., and H. A. ELLIOTT
1934 The relative growth of the amphibian eye, studied by means of transplantation. J. Exper. Zool. 68(2): 247-291, 22 figs.

TWITTY, V. C., and H. H. JOHNSON
1934 Motor inhibition in *Amblystomia* produced by *Triturus* transplants. Science 80(2064): 78-79.

TYLER, J. C.
1965 Comments regarding pufferfish. (Personal communication, Dec. 20, 1965.)

VINCENT, L.
1883 Les poissons vénéneux du Japon. Arch. Méd. Nav., Paris 39: 392-394.
1889 Contribution à la géographie médicale: le Japon. Arch. Méd. Nav., Paris 52: 107-122.

WATERFIELD, C. J., and M. H. EVANS
1972 A method for distinguishing tetrodotoxin from saxitoxin, by comparing their relative stabilities when heated in acid solutions. Experientia 28: 670-671.

WHITLEY, G. P.
1943 Poisonous and harmful fishes. Council Sci. Ind. Res., Bull. No. 159. 28 p., 16 figs.; Pls. 1-3.
1953 Toadfish poisoning. Australian Mus. Mag. 11(2): 60-65, 6 figs.

WOODWARD, R. B.
1964 The structure of tetrodotoxin. Pure Appl. Chem. 9(1): 49-74.

WOODWARD, R. B., and J. Z. GOUGOUTAS
1964 The structure of tetrodotoxin. (Personal communication, 1964.)

YAMASAKI, S.
1914 Hepatotoxin. [In Japanese] Osaka Med. Assoc. Mag. 13(4): 443-463.

YAMASAKI, S., and T. KIKKAWA
1909 The application of tetrodotoxin for leprosy and other diseases. [In Japanese] Osaka Med. Assoc. Mag. (10): 1140-1141.

YANO, I.
1937 The pharmacological study of tetrodotoxin. [In Japanese] Fukuoka Med. Coll. J. 30(9): 1169-1784, 9 figs.
1938 An experimental study on the globe fish (furu). Japan. Soc. Int. Med., 33d Ann. Mtg. (5): 99-101.

YOKOO, A.
1948*a* Chemical studies on tetrodotoxin. Rept. I. [In Japanese] Rept. Inst. Physics Chem. 24(3): 136-139.
1948*b* Chemical studies on tetrodotoxin. Rept. 2. [In Japanese] Hiroshima Igaku 1(2): 52-53.
1948*c* Studies on a toxin of the globe-fish. [In Japanese, English summary] Bull. Tokyo Inst. Technol. 13(8): 8-12.
1950 Chemical studies on tetrodotoxin. Rept. III. Isolation of spheroidine. [In Japanese] J. Chem. Soc. Japan 71(11): 591-592.
1952 Studies on toxin of a globe-fish. IV. Proc. Japan Acad. 28(4): 200-202.

YOKOO, A., and S. MOROSAWA
1955 Studies on the toxin of a globe-fish. Rept. V. Comparison with tetrodotoxin. J. Pharmacol. Soc. Japan 75(2): 235-236, 3 figs.

YUDKIN, W. H.
1944 Tetrodon poisoning. Bull. Bingham Oceanogr. Coll. 9(1): 1-18, 1 fig.
1945 The occurrence of a cardio-inhibitor in the ovaries of the puffer, *Spheroides maculatus*. J. Cell. Comp. Physiol. 25(2): 85-95, 3 figs.

Chapter XV—VERTEBRATES

Class Osteichthyes

Poisonous: ICHTHYOOTOXIC FISHES

Ichthyootoxism is one of the lesser known forms of fish poisoning. Ichthyootoxic fishes constitute a group of organisms that produce a poison which is generally restricted to the gonads of the fish. There is a definite relationship between gonadal activity and toxin production. The musculature and other parts of the fish are usually edible. Fishes in this group are mainly freshwater, but others are anadromous, brackish water, or marine species. A cursory review of some of the freshwater forms has been included in this section in order to complete the overall picture of ichthyootoxications. Many of the fishes involved in ichthyootoxism are phylogenetically unrelated.

The biology and systematics of most of the fishes incriminated as ichthyootoxic have been discussed by Günther (1859-70), Day (1878-88), Garman (1895), Jordan and Evermann (1896-1900), Berg (1947, 1962-65), Palombi and Santarelli (1961), and Svetovidov (1963). The nomenclature as proposed by Berg has largely been followed in dealing with this group of fishes.

The meager amount of information on the occurrence of ichthyootoxins suggests that the biotoxin is probably widespread in numerous phylogenetically unrelated fresh and saltwater fish species. The problem offers a wealth of virgin subject matter that is in need of further study. Future investigations may turn up some very interesting problems relative to the biogenesis of the poison.

REPRESENTATIVE LIST OF FISHES REPORTED AS ICHTHYOOTOXIC[1]

Phylum CHORDATA

Class OSTEICHTHYES

Order ACIPENSERIFORMES: Sturgeons, Etc.

Family ACIPENSERIDAE

Acipenser güldenstädti Brandt (Pl. 1, fig. a). Russian sturgeon (USA, England), russkii osetr (USSR). *196*

DISTRIBUTION: Persian and Siberian Rivers, Caspian Sea, Danube.

SOURCES: Danilewsky (1885), Arustamoff (1891, 1898), Lisunoff (1892).

---

[1] Some of the accounts of poisonings produced by the fishes listed in this section are too vague to classify with any degree of finality, but from the evidence presented it is assumed that they are of the ichthyootoxic variety.

196 *Acipenser sturio* Linnaeus (Pl. 1, fig. b). Sturgeon (USA, England), esturgeon (France), stor (Germany), storione (Italy), storge (Scandinavia).

DISTRIBUTION: Both coasts of the Atlantic, Mediterranean Sea, rivers of Europe and Russia.

SOURCES: Sengbusch (1844), Koch (1857), Schreiber (1884), Konstansoff (1904, 1906), Konstansoff and Manoiloff (1914), Phisalix (1922), Fredericq (1924), Maass (1937).

196 *Huso huso* (Linnaeus) (Pl. 2). Sturgeon (USA, England), beluga (USSR), wiz, wyz (Poland), morun (Rumania), hausen (Germany), kyrpy (Kazan).

DISTRIBUTION: Black Sea, Sea of Azov, Caspian Sea, Mediterranean Sea, and rivers that drain into these seas.

SOURCES: Sengbusch (1844), Koch (1857), Schreiber (1884), Arustamoff (1891, 1898), Lisunoff (1892), Konstansoff (1904, 1906), Konstansoff and Manoiloff (1914), Phisalix (1922).
OTHER NAME: *Acipenser huso.*

Order LEPISOSTEIFORMES (GINGLYMODI): Gars

Family LEPISOSTEIDAE

197 *Lepisosteus tristoechus* (Bloch and Schneider) (Pl. 3). Alligator gar (USA), manjuarí (Cuba).

DISTRIBUTION: Rivers of Cuba and bays and coastal waters of the Gulf of Mexico.

SOURCES: Coutière (1899), Colby (1943).
OTHER NAMES: *Lepidosteus tristoechus, Lepisosteus tristoechus* (some authors consider *L. tristoechus* to be a synonym of *L. spatula*).

Order CLUPEIFORMES (ISOSPONDYLI): Clupeiform Fishes

Suborder SALMONOIDEI: Salmons, Trouts, Whitefishes, Etc.

Family SALMONIDAE

197 *Salmo salar* Linnaeus (Pl. 4, fig. a). Atlantic salmon (USA).

DISTRIBUTION: North Atlantic, ascending all suitable rivers in northern Europe and regions north of Cape Cod.

SOURCES: Autenrieth (1833), Chevallier and Duchesne (1851*a, b*), Knox (1888).

197 *Stenodus leucichthys* (Güldenstädt) (Pl. 4, fig. b). Inconnu, whitefish (USA, England), belorybitsa (USSR).

DISTRIBUTION: Volga River, rivers of Siberia from Arctic Ocean.

SOURCES: Arustamoff (1891, 1898), Lisunoff (1892).

Suborder ESOCOIDEI (HAPLOMI): Pikes

Family ESOCIDAE

Esox lucius Linnaeus (Pl. 5). Northern pike (USA), hecht (Germany), brochet (France), luccio (Italy), gadda (Sweden), shchuka (USSR). *197*

DISTRIBUTION: Europe, northern Asia, and North America.

SOURCES: Autenrieth (1833), Chevallier and Duchesne (1851*a, b*), Linstow (1894), Kobert (1902), Faust (1906, 1924, 1927*a, b*), Taschenberg (1909), Pawlowsky (1927), Romano (1940).

Order CYPRINIFORMES (OSTARIOPHYSI): Minnows, Carps, Catfishes, Etc.

Suborder CYPRINOIDEI (EVENTOGNATHI): Minnows, Carps

Family CYPRINIDAE

Abramis brama (Linnaeus) (Pl. 6, fig. a). Bream (USA, England), breme, brax (France), brachsen, bley (Germany), leshch (USSR). *198*

DISTRIBUTION: Europe above the Pyrenees and Alps to the Volga River and Caspian Sea.

SOURCES: Autenrieth (1833), Chevallier and Duchesne (1851*a, b*), Coutière (1899), Taschenberg (1909), Phisalix (1922), Pawlowsky (1927).
OTHER NAMES: *Cyprinus brama, Cyprinus farenus.*

Barbus barbus (Linnaeus) (Pl. 6, fig. b). Barbel (USA, England), brzana (Poland), mrena (Bulgaria), barbe (Germany), usach, marena (USSR). *198*

DISTRIBUTION: Northern and central Europe.

SOURCES: Schlegel (1801), Chevallier and Duchesne (1851*a, b*), Knox (1888), Coutière (1899), Kobert (1902), Phisalix (1922), Giunio (1948).
OTHER NAMES: *Barbus barba, Barbus fluviatilis, Cyprinus barbus.*

Cyprinus carpio Linnaeus (Pl. 6, fig. c). Carp (USA, England), carpe (France), karpfen (Germany), karp, sazan (USSR). *198*

DISTRIBUTION: Europe, Russia.

SOURCES: Autenrieth (1833), Chevallier and Duchesne (1851*a, b*), Phisalix (1922).

Diptychus dybowski Kessler (Pl. 7, fig. a). Naked osman (USA, England), golyi osman, kokcha, alabugá (USSR). *199*

DISTRIBUTION: Central Asia.

SOURCE: Knox (1888).
OTHER NAME: *Schizothorax przewalski.*

199 *Schizothorax argentatus* Kessler (Pl. 7, fig. b). Balkhash marinka, snow trout, fine scaled carp (USA, England), balkhashskaya marinka (USSR), asla, dinnawah, adoee, loh-one (India).
DISTRIBUTION: Central Asia.

SOURCES: Knox (1888), Albahary (1912).
OTHER NAME: *Schizothorax argenteus.*

199 *Schizothorax intermedius* McClelland (Pl. 7, fig. c). Common marinka, snow trout, fine scaled carp (USA, England), marinka obyknovennaya (USSR), asla, dinnawah, adoee, loh-one (India).
DISTRIBUTION: Central Asia.

SOURCES: Knox (1888), Hiyama (1943).

199 *Tinca tinca* (Linnaeus) (Pl. 7, fig. d). Tench (USA, England), tanche (France), schleie (Germany), tenca (Spain), lin (USSR).
DISTRIBUTION: Europe.

SOURCES: Autenrieth (1833), Faust (1906, 1924, 1927*a, b*), Phisalix (1922), Pawlowsky (1927), Giunio (1948).
OTHER NAMES: *Cyprinus tinca, Cyprinus tinco, Tanche vulgaire, Tinca vulgaris.*

Suborder SILUROIDEI (NEMATOGNATHI): Catfishes

Family AGENEIOSIDAE

200 *Ageneiosus armatus* Lacépède (Pl. 8 fig. a). Sheatfish, horned wel (USA).
DISTRIBUTION: Surinam.

SOURCES: Kolb (1826), Autenrieth (1833), Knox (1888), Coutière (1899), Pawlowsky (1927).
OTHER NAME: *Silurus militaris.*

Family ARIIDAE

200 *Bagre marinus* (Mitchill) (Pl. 8, fig. b). Sea catfish (USA), bagre marina (Latin America).
DISTRIBUTION: Seas of tropical America.

SOURCES: Autenrieth (1833), Knox (1888), Coutière (1899), Pawlowsky (1927).
OTHER NAMES: *Silurus bagre, Silurus bagrus.*

Family ICTALURIDAE

200 *Ictalurus catus* (Linnaeus) (Pl. 8, fig. c). White catfish (USA).
DISTRIBUTION: Delaware River to Texas.

SOURCES: Macht and Spencer (1941), Macht (1942), Taft (1945).
OTHER NAME: *Ameirus catus.*

Pseudobagrus aurantiacus (Temminck and Schlegel) (Pl. 8, fig. d). Catfish (USA), gibachi, gingyo, gigyo, gigyu (Japan). *200*

DISTRIBUTION: Japan.

SOURCES: Coutière (1899), Phisalix (1922), Pawlowsky (1927), Hiyama (1943).
OTHER NAME: *Bagrus aurantiacus.*

Family SILURIDAE

Parasilurus asotus (Linnaeus) (Pl. 9, fig. a). Mudfish, catfish (USA), namazu (Japan), amurskissom (USSR). *201*

DISTRIBUTION: Island waters of Japan, Korea, Manchuria, and China.

SOURCES: Knox (1888), Coutière (1899), Phisălix (1922), Pawlowsky (1927).
OTHER NAMES: *Silurus asotus, Silurus japonicus.*

Silurus glanis Linnaeus (Pl. 9, fig. b). Sheatfish (USA, England), wels (Germany), som (USSR). *201*

DISTRIBUTION: Europe.

SOURCES: Autenrieth (1833), Knox (1888).

Order GADIFORMES (ANACANTHINI): Codfishes and Hakes

Family GADIDAE

Lota lota (Linnaeus) (Pl. 10). Burbot, eelpout (USA, England), lotte (France), quappe, aalraupe, treische (Germany), nalim (USSR), mietus (Poland). *201*

DISTRIBUTION: Fresh waters of northern and central Europe.

SOURCES: Brandt and Ratzeburg (1829-33), Chevallier and Duchesne (1851*a, b*), Coutière (1899), Pawlowsky (1927), Giunio (1948).
OTHER NAMES: *Gadus lota, Lota vulgaris, Lotta vulgaris.*

Order CYPRINODONTIFORMES (MICROCYPRINI, CYPRINODONTES): Killifishes, Topminnows, Etc.

Family CYPRINODONTIDAE

Aphanius calaritanus (Cuvier and Valenciennes) (Pl. 11, fig. a). Killifish, dogpike (USA), solinarka, nono (Yugoslavia), hundshecht (Germany). *202*

DISTRIBUTION: Southern Europe and north Africa. (Pawlowsky lists from Sumatra but this appears to be in error.)

SOURCES: Coutière (1899), Pawlowsky (1927), Giunio (1948).
OTHER NAMES: *Cyprinodon calaritanus, Cyprinodon calarinatus, Lebias calaritana.*

Fundulus diaphanus (LeSueur) (Pl. 11, fig. b). Banded killifish (USA), freshwater mummichog (Canada). *202*

DISTRIBUTION: Streams and lakes, Maritime Provinces, Canada, south to North Carolina.

SOURCE: White, Medcof, and Day (1965).

Order PERCIFORMES (PERCOMORPHI): Perchlike Fishes, Etc.

Suborder COTTOIDEI (CATAPHRACTI; SCLEROPAREI; LORICATI): Sculpins, Etc.

Family COTTIDAE

202 *Scorpaenichthys marmoratus* (Ayres) (Pl. 12). Cabezon (USA).

DISTRIBUTION: Pacific coast of North America.

SOURCES: Walford (1931), Schultz (1938), Hubbs and Wick (1951), Roedel (1953), Halstead (1974).

Suborder BLENNIODEI (JUGULARES, in part): Blennies, Pricklefish, Etc.

Family STICHAEIDAE

202 *Stichaeus grigorjewi* Herzenstein (Pl. 13). Japanese prickleback, blenny (USA), nagazuka (Japan).

DISTRIBUTION: Tohoku, Hokkaido, Japan; Korea.

SOURCES: Sakai *et al.* (1962), Asano and Itoh (1962), Hatano *et al.* (1964).
OTHER NAME: *Dinogunellus grigorjewi.*

BIOLOGY

Order ACIPENSERIFORMES

Family ACIPENSERIDAE: Sturgeons are a group of freshwater or anadromous fishes having elongate fusiform bodies covered with five rows of bony scutes. The head is covered with bony shields, and the snout is elongate, conical, or spatulate. The jaws are toothless. They are found in Europe, northern Asia, and North America. There are 4 genera and about 25 species. Sturgeons interbreed easily. Anadromous sturgeons have seasonal forms. The spring form ascends the rivers in spring, and spawns in spring or summer of the same year. The winter form ascends the rivers usually in autumn and spawns in the following year. They are clumsy, sluggish bottom feeders. Sturgeons feed on a variety of foods including fishes, mollusks, crustaceans, worms, etc. They may attain large size, up to about 30 m, with a weight of 1300 kg or more. Their flesh is coarse and beefy, but they are nevertheless of great economic importance in some regions. The fecundity rate of sturgeons is exceedingly high. A single fish may produce more than 7.5 million eggs in a single season. Spawning occurs usually in the central part of the river in a rocky area. The roe of sturgeons is used commercially in the preparation of caviar. Sturgeons are said to be long lived, attaining an age of more than 100 years.

Order LEPISOSTEIFORMES (GINGLYMODI)

Family LEPISOSTEIDAE: The gars are characterized by having an elongate body and a long snout covered with ganoid scales. The dorsal and anal fins are located in the posterior portion of the body near the caudal fin. They are found in North and Central America and on the island of Cuba. Gars are largely freshwater fishes, inhabiting streams, lakes, and bayous but may occur in brackish water. They are sluggish swimmers and voracious in their eating habits, feeding on crayfish and small fishes, and quite destructive to other fishes in general. The flesh of gars is usually considered to be of poor quality. Most species move whenever possible to coastal areas to spawn. Gars may attain large size, 6 m or more in length. The eggs of two species *Lepisosteus spatula* and *L. tristoechus* have been reported to be toxic to humans. It is suspected that the eggs of other species may also be poisonous to eat.

Order CLUPEIFORMES (ISOSPONDYLI)

Suborder SALMONOIDEI

Family SALMONIDAE: The salmons and trouts are freshwater and anadromous fishes of the Northern Hemisphere. They are a group of outstanding fishes because of their beauty, activity, gaminess, and quality of food. There are less than 100 species. They are characterized by an elongate, or moderately elongate, body covered with cycloid scales, by a naked head, and by the presence of an adipose fin. The salmonids are a commercially important group. They are active and voracious feeders, living largely on smaller fishes and a variety of invertebrates. The roe of some European salmonids has been reported as toxic to humans.

Suborder ESOCOIDEI (HAPLOMI)

Family ESOCIDAE: The pikes are a small family having one genus and about a half dozen species. They are limited to the fresh waters of North America, Asia, and Europe. The pikes have a long slender body and a large mouth armed with strong sharp teeth. All the pikes are voracious carnivores and consume large numbers of fishes. Some species attain large size, 1.5 m, and a weight of 35 kg of more. Pikes are good eating and generally of considerable economic value. On occasion, the roe of some species has been found to be toxic to humans during the reproductive season.

Order CYPRINIFORMES (OSTARIOPHYSI)

Suborder CYPRINOIDEI (EVENTOGNATHI)

Family CYPRINIDAE: This is the largest of all fish families, containing more than 2,000 species. Minnows are found scattered throughout the north temperate zone except the Arctic Circle and extend only to a limited extent south of the Tropic of Cancer in the Western Hemisphere. They are particularly abundant in the rivers of southern Asia and tropical Africa. The family is characterized by highly specialized pharyngeal teeth. With few exceptions, cyprinids are small and relatively feeble fishes. They constitute most of the food of the predatory river fishes. The diet of cyprinids varies greatly according to the species; some are strictly herbivorous, whereas others are carnivorous. Most cyprinids are freshwater fish; only a few species are able to tolerate any degree of salinity. Some species are found in cold waters of melting snows (*Schizothorax*, etc.), whereas others are found in stagnant ponds. Many of the cyprinids found in the clear swift streams of eastern Europe and central Asia contain toxic roe during the reproductive season.

Suborder SILUROIDEI (NEMATOGNATHI)

Family AGENEIOSIDAE: This is a small family of South American catfishes which inhabits the streams and rivers of the Amazon Basin northward to Surinam. Little is known regarding their habits. The flesh of *Ageneiosus armatus* is said to be generally of fine flavor, but during the mating season the fish is reputedly poisonous to eat, apparently because of its toxic roe.

Family ARIIDAE: Most marine catfishes are members of the family Ariidae. It is a large family with numerous species. Ariid catfishes inhabit tropical and subtropical regions throughout the world. They are sleek silvery fishes covered with smooth skin, but the head is protected by a coat of mail pierced by a central fontanelle. They have four to six barbels around the mouth. The dorsal and pectoral spines of some species are very sharp and can inflict a painful venomous wound. They are more active than many of their freshwater counterparts and spend little time resting on the bottom but may frequently be seen in schooling formation. Ariid catfishes are generally tough having coarse flesh but are used as foodfishes in some regions. Sea catfishes are sometimes seen swarming in shallow water, in sandy bays, or in the vicinity of rivermouths. The roe of some species is reported to be toxic.

Family ICTALURIDAE: The North American catfishes of this family are typical catfishes having four pectoral fins possessing strong spines. They range from Canada to Guatemala. All are freshwater fishes. Among the smaller ictalurids are the madtoms which possess venomous dorsal and pectoral spines. They are able to inflict very painful wounds. Voracious bottom feeders, they will eat almost anything they can swallow. Some of the ictalurids are among the most esteemed food fishes of the catfish clan. Ictalurids may attain a weight of 75 kg or more.

Family SILURIDAE: The members of this catfish family are characterized by a naked body, a long anal fin, and the absence of an adipose fin. They are largely freshwater fishes widespread throughout Europe and Asia. Some species of this family are sometimes found in brackish water, particularly in the southern seas of the USSR. They feed mainly on other fishes. *Siluris glanis* is one of the largest river fishes in Europe, attaining a weight of 200 kg of more. Some of the silurids are commercially valuable in certain parts of Europe. The roe of some species is reported poisonous to eat.

Order GADIFORMES (ANACANTHINI)

Family GADIDAE: The codfishes are among the most valuable of all food fishes. They are found in all seas of the Northern Hemisphere, and a few species penetrate into the Southern Hemisphere. One species is a freshwater fish. Codfishes among other things are characterized by one, two, or three dorsal fins without any spines. An unpaired barbel is usually present under the chin. The members of this family are largely deepwater fishes but are also found close in to shore. They tend to be voracious eaters and will feed on almost anything that is available. They are prolific breeders; a single fish may produce more than 9 million eggs in a single season. They are similar to salmon in that, unlike most fishes, they spawn in cooling water. The liver produces an oil of considerable economic value. Some species may attain a weight of about 75 kg. The freshwater *Lota* of Europe and Asia is reported to have toxic roe.

Order CYPRINODONTIFORMES (MICROCYPRINI, CYPRINODONTES)

Family CYPRINODONTIDAE: The killifishes are a family of small fishes, most of which are freshwater fish, and range from New England south to Argentina and to Asia and Africa. Among other things they are characterized by a small protractile mouth. There are about 200 species. The largest species rarely reaches a length of 30 cm. Most of them are found in swamps, ditches, reservoirs, lakes or small ponds. Some species are herbivorous, but many of them are insectivorous. Because of their insectivorous habits, some species are used quite extensively in mosquito abatement. *Fundulus* inhabits both fresh water and salt water. The marine species are usually found in shallow bays or protected areas close to shore. Cyprinodonts are generally too small to be considered of much food value.

Order PERCIFORMES (PERCOMORPHI)

Suborder COTTOIDEI (CATAPHRACTI, SCLEROPAREI, LORICATI)

Family COTTIDAE: The sculpins constitute a large fish family of the northern seas. They differ from the Scorpaenidae, from which they may have been derived, because of their greater number of vertebrae, relatively feeble spinous dorsal and ventral fins, and their scales. A few species possess scales, but most of them are covered with a naked skin, small spines, or platelets. Cottids are largely marine bottom fishes inhabiting shallow waters along coastal areas, but a few are found in deep water, and several species live in fresh water. They feed on a variety of invertebrates and small fishes. None of them is considered to be of much economic value. The roe of at least one species *Scorpaenichthys marmoratus* has been found to be toxic to humans.

Suborder BLENNIODEI (JUGULARES, in part)

Family STICHAEIDAE: The pricklebacks are a group of blennylike fishes having moderately elongated bodies covered with overlapping scales. The rays of the dorsal fin are entirely spiny in most species. There are more than 50 species in the family, and all of them live in the Northern Hemisphere in cold marine waters. Most of them are shallow water inhabitants, but a few species are found in deep water. Pricklebacks feed on small fishes and various invertebrates. The family is of little or no economic importance in most regions, but *Stichaeus grigorjewi* is used in the manufacture of boiled fish paste (Kamaboko) and seasoned crushed meat (Soboro) in northern Japan. The fish is generally eviscerated prior to use, but due to periodic labor shortages the viscera are not always removed. If the viscera are not removed during the reproductive season of the fish, human intoxications sometimes occur from eating the roe.

BIOGENESIS OF ICHTHYOOTOXIN

There is no information available concerning the origin of the poison in the roe of ichthyootoxic fishes. The occurrence and concentration of the poison in the roe of the fish appear to be directly related to the reproductive cycle of the fish, but whether other factors such as the food of the fish are also involved in the production of the biotoxin is not known.

MECHANISM OF INTOXICATION

Ichthyootoxism results from ingestion of the roe of various salt and freshwater fishes of Europe, Asia, and North America, and to a lesser extent in the torrid zone. The roe of many freshwater and estuarine fishes of eastern Europe and Asia are dangerous to eat during the reproductive season of the year (usually March through June). The most dangerous types of ichthyootoxic fishes are members of the genera *Barbus, Schizothorax, Tinca* (Europe and Asia), and *Stichaeus* (Japan). These fishes have caused innumerable intoxications in Europe and Asia. Although cooking is said to destroy most ichthyootoxins, it cannot be relied upon as a completely safe procedure since the poison in some fishes appears to be resistant to heat. The musculature and other parts of ichthyootoxic fishes are generally safe to eat during the reproductive season.

MEDICAL ASPECTS

Clinical Characteristics

Signs and symptoms develop soon after ingestion of the roe and consist of abdominal pain, nausea, vomiting, diarrhea, dizziness, headache, fever, bitter taste, dryness of the mouth, intense thirst, sensation of constriction of the chest, cold sweats, rapid irregular pulse, low blood pressure, cyanosis, pupillary dilatation, syncope, chills, dysphagia, and tinnitus. In severe cases there may be muscular cramps, paralysis, convulsions, coma, and death. *Barbus* roe usually does not cause death, but fatalities have resulted from eating *Schizothorax* roe (Knox, 1888). Deaths have also occurred from eating the roe of the Japanese prickleback *Stichaeus* (Asano and Itoh, 1962). The victims generally recover within 3 to 5 days if they are given reasonable supportive therapy, but it may take longer. Clinical accounts of ichthyoo-toxism have been reported by Autenrieth (1833), Chevallier and Duchesne (1851*a, b*), Coutière (1899), McCrudden (1921), Phisalix (1922), Pawlowsky (1927), Hubbs and Wick (1951), Sakai *et al.* (1962), Asano and Itoh (1962), etc.

Pathology

(Unknown)

Treatment

The stomach of the patient should be promptly evacuated. There is no known antidote. The remainder of the treatment is symptomatic. Respiratory and cardiac stimulants may be required. Ichthyootoxins of the type included in this section are believed to be chemically distinct from either ciguatoxin or tetrodotoxin. However, the therapy recommended in these other types of ichthyosarcotoxism may prove helpful and should be considered. There are no reliable therapeutic data available.

Prevention

Avoid eating the roe of any fish during the reproductive season of the year. This advice is particularly pertinent to the freshwater and brackishwater fishes of Europe and Asia. The eating of fish eggs of a little-known variety is not without danger of intoxication. Cooking cannot be relied upon to prevent poisoning. Some ichthyootoxins appear to be destroyed by heat, but others do not. Mice and cats are usually sensitive to ichthyootoxins and can be used as test animals by feeding them a generous sample of the roe before human consumption.

PUBLIC HEALTH ASPECTS

The problem of ichthyootoxic fishes is a public health problem in some parts of Europe and Asia. However, since the biotoxication is not reportable in most regions, little is known regarding the incidence of outbreaks or the over-all mortality rate.

TOXICOLOGY

Knox (1888) appears to have been one of the very few to evaluate the toxicity of freshwater fish roe. He fed mice *Schizothorax argentatus* roe which had been preserved in alcohol for 6 months. The eggs, which were apparently washed in fresh water before the experiment, killed mice within 30 hours after in-gestion.

McCrudden (1921) prepared aqueous extracts from fresh pike and barbel roe and injected them intravenously into rabbits. Both fish extracts proved to be toxic, but the pike extracts were the more poisonous. He also fed fresh roe to dogs and cats but produced only mild evidence of intoxication. Un-fortunately most of his fresh roe was obtained during the nontoxic period in the fall and winter months, so it is quite likely that his oral tests might have shown more dramatic results if the roe had been taken during the peak of the reproductive season. Intravenous injections of the aqueous extracts of pike roe in rabbits caused respiratory distress, convulsions, and death in a period of about 1 hour. No significant differences were observed in the results obtained from desiccated roe with that obtained from the fresh roe. After using various routes of administration, i.e., subcutaneous, oral, intraperitoneal, McCrudden (1921) concluded that he obtained his best results using the intravenous route. He was unable to detect any qualitative differences in the toxic reaction obtained between pike and barbel roe. He found that boiling destroyed the toxicity of the poison.

Hubbs and Wick (1951) tested the roe of a large *Scorpaenichthys marmoratus* captured near La Jolla, Calif., by feeding an aqueous homogenate orally to 12 rats and 2 guinea pigs. Four rats and one guinea pig died, and all the animals exhibited diarrhea and nasal discharge. The intensity of the intoxication appeared to diminish as the dosage was decreased. There was no evidence that the flesh of this fish was toxic.

Asano and Itoh (1962) tested the roe of *Stichaeus (Dinogunellus) grigorjewi* by feeding dried roe to chickens. One portion of the roe was air dried at room temperature, and a second portion was autoclaved at 120° C for 30 min. The two chickens that were fed the air dried material died, but those that were fed the autoclaved roe survived. Saline extracts were prepared from another batch of roe, and similar extracts were prepared from some autoclaved roe. The extracts were injected intraperitoneally into mice. The mice receiving the autoclaved material survived, whereas the mice receiving the untreated extracts died. In general, the usual extraction methods employed for testing paralytic shellfish poison gave inconsistent results. When bioassayed, some of the mice survived, whereas others did not. Another series of toxicity tests was conducted using filtrates of *Stichaeus* roe that had been deproteinized by various methods. The roe poison appeared to be precipitated with other nontoxic proteins. There was no evidence of toxicity when these deproteinized filtrates were injected intraperitoneally into mice. It was found that the roe extracts had antigenic properties. Injections of formalin and phenol toxoids of the ichthyootoxin could prolong the survival time of rats. However, antigen production was found to be incomplete. There was no evidence of toxicity in the egg capsule. The poison was not dialyzable. Separation of the albumin and globulin fractions indicated that the poison was located in the globulin fraction. Autopsies of intoxicated rats showed the occurrence of fatty degeneration of the liver. The mechanism by which this degeneration was produced was not known.

Hatano *et al.* (1964) prepared acetone extracts from the roe of *S. grigorjewi* and injected them intraperitoneally into mice. This resulted in ruffling of the hair, severe paralysis, and death within 24 hours.

Fuhrman *et al.* (1969) found that the eggs of the marbled sculpin, or cabezon (*Scorpaenichthys marmoratus*), contain a toxin that inhibits growth of cells in tissue culture and produces necrosis in the liver and spleen. The toxin appeared to be a protein or was attached to protein in the eggs. Intraperitoneal injection of sterile extracts of cabezon toxin resulted in increased numbers of total white cells (Fuhrman, Fuhrman, and Roseen, 1970). At the same time the number of circulating mononucleated cells fell markedly. The authors concluded that the effects of cabezon toxin on the numbers and types of white blood cells in peripheral blood appeared to reflect changes produced by the toxin on the liver and spleen.

PHARMACOLOGY

There is very little information regarding the pharmacological properties of ichthyootoxins. McCrudden (1921) suggested that the vomiting might be caused by the local irritating effects of the poison on the gastric mucosa. He believed that the primary effects of the poison were on the central nervous system, basing his conclusion on the observation that in rabbits there was a rapidly progressing paralysis which affected primarily the respiratory center. Sensory paralysis appeared prior to motor paralysis, and death occurred through respiratory paralysis.

CHEMISTRY

by Donovan A. Courville

While the presence of a toxic substance in the roe of certain fishes has long been recognized, the chemical nature of this toxin has remained obscure. Studies were initiated by McCrudden (1921) using the roe of pike and barbel. It was found that the toxic substance in the roe of the two species was apparently the same, except perhaps in the degree of toxicity. A number of properties of the toxin were observed. Among these were solubility in water, instability to heating, precipitation with lead acetate, colloidal iron or saturated ammonium sulfate, and its high molecular weight as indicated by its failure to dialyze through a semipermeable membrane. These properties suggested that the toxic substance was a protein of the albumin class. During all treatments, the toxin remained bound to the protein.

However, since it was known that vitellin of bird's eggs has toxic properties and belongs to the globulin fraction, the possibility was entertained that the toxin belonged to a similar fraction of fish roe to which the term "ichthulin" had been applied (McCrudden, 1921). To check this possibility, the albumin and globulin fractions of fish roe were separated and tested separately for toxic properties. The toxic properties were found to be lodged in the albumin fraction while the globulin fraction contained only traces of the toxic material. The improbability that the globulin fraction contained the toxin was confirmed by the observation that a solution of the toxin in dilute salt solution failed to reprecipitate on removal of the salt by dialysis. This behavior did not coincide with that to be expected of a globulin.

Studies on the chemical nature of the toxic material of fish roe were revived by Asano and Itoh (1962) using the roe of the prickleback *Stichaeus (Dinogunellus) grigorjewi* as source material. As a result of these investigations, the conclusion of McCrudden that the toxin belonged to the class of toxalbumins was disallowed, and evidence was presented to indicate that the protein was a globulin rather than an albumin. Since the toxin "has a similar nature to ichthulin," the term dinogunellin was introduced to refer to the protein complex.

Important to the clarification of the chemical nature of the toxin was the observation (Asano and Itoh, 1962) that the toxicity was associated with a lipid moiety in conjugation with protein and that the toxicity was retained by the lipid fraction on hydrolysis. Direct extraction of the homogenized roe with various fat solvents, with subsequent evaporation of the solvent, uniformly revealed that the lipid residue produced the toxic effects on injection into animals. This observation would seem to be at variance with that of McCrudden who reported that the toxin remained with the protein throughout all fractionation operations.

The observation that the toxin is extractable with ethanol with considerable loss of toxicity, and the inconsistency of results with variations in the extraction procedures, suggested that the toxicity is not present as a lipoprotein as found in fish roe but is rather an artifact produced in the process of isolation, possibly by a process of dehydration, possibly by hydrolysis of the lipoprotein. However, the antigenic properties of the toxin seemed to demand recognition that at least part of the toxin was of the nature of a lipid-protein complex.

In an attempt to determine the nature of the protein to which the lipid is conjugated, it was noted that the toxin is soluble in dilute sodium chloride solutions and is precipitated on removal of the salt by dialysis. This led to the conclusion that the protein is a globulin, a conclusion which was the reverse of that by McCrudden who found that the toxin did not reprecipitate on removal of the salt from a saline solution of the toxin. Recognizing the discrepancy from McCrudden's observations, Asano and Itoh (1962) repeated the experiment using the same method of fractionation reported by McCrudden. The results were the same. It is thus difficult to determine from the reports whether the discrepancy is to be explained by the difference in source materials or whether it is due to experimental error. The lipoprotein fraction of roe was isolated by Chargaff's method in the belief that the discrepancy might be the result of conjugation of the protein to a nonprotein moiety of lipid nature such as is true of vitellin in birds' eggs. This fraction proved to contain the toxin, thus confirming the nature of the toxin as a lipoprotein. Preliminary analysis of the lipoprotein showed that it contained about 80 percent protein and 20 percent lipid. The more exact identity of both parts of the toxin molecule remained obscure. On the basis of the work by Kaneda *et al.* (1955) confirming the toxicity of autoxidation products of unsaturated lipids in marine animal oils, Asano and Itoh (1962) assumed that the inherent poisonous principle of the roe is a lipoprotein "dinogunellin" and that the toxicity of the lipid is probably due to secondary changes, possibly the formation of unsaturated fatty acid peroxide or hydrolysis of the lipoprotein, the toxicity remaining in both the hydrolyzed and combined states of the lipid. It was also the observation of Asano and Itoh that the toxin was not as easily destroyed by heat as thought by McCrudden, 30 minutes heating at 120° C in an autoclave being necessary for complete destruction.

A lipid-soluble toxin has also been observed in the liver of certain fish (Sakai *et al.*, 1962). This toxin, while of a lipid nature, seems certainly to be distinct from that found in the roe of *Stichaeus (Dinogunellus) grigorjewi*.

Further analysis of the lipid moiety of the toxin in the roe of *S. grigorjewi* (Hatano *et al.*, 1964) revealed the presence of glycerol, choline, phosphorous, fatty acids, and amino acids, suggesting relation to lecithins in further conjugation with amino acids. The infrared spectrum (KBr) showed characteristic absorption at 1730, 1675, 1645, and 1570 cm^{-1} and rich discrete absorption in the 900 cm^{-1} to 1500 cm^{-1} region.

LITERATURE CITED

ALBAHARY, J. M.
1912 Sur quelque poissons toxiques à laitances vénéneuses. Bull. Suisse Pêche Piscicult. 13: 194-196.

ARUSTAMOFF, M.
1891 Ueber die natur des fischgiftes. Zentr. Bakteriol. Parasitenk., Abt. I, 10(4): 113-119.
1898 On the nature of fish poison. Vrach. (36): 1055-1057, 10 figs.

ASANO, M., and M. ITOH
1962 Toxicity of a lipoprotein and lipids from the roe of a blenny *Dinogunellus grigorjewi* Herzenstein. Tohoku J. Agr. Res. 13(2): 151-167.

AUTENRIETH, H. S.
1833 Ueber das gift der fische. C. F. Osiander, Tübingen. 287 p.

BERG, L. S.
1947 Classification of fishes both recent and fossil. [In Russian & English] J. W. Edwards, Ann Arbor, Mich.
1962-65 Freshwater fishes of the U.S.S.R. and adjacent countries. 3 vols. 4th ed. [Engl. transl.] IPST Cat. Nos. 471-473. Israel Prog. Sci. Transl., Jersualem.

BRANDT, J. F., and J. T. RATZEBURG
1829-33 Pisces. Vol. 2. Medizinische Zoologie. Berlin. 55 p.

CHEVALLIER, A., and E. A. DUCHESNE
1851*a* Mémoire sur les empoisonnements par les huitres, les moules, les crabes, et par certains poissons de mer et de rivière. Ann. Hyg. Publ. (Paris) 45(11): 387-437.
1851*b* Mémoire sur les empoisonnements par les huitres, les moules, les crabes, et par certains poissons de mer et de rivière. Ann. Hyg. Publ. (Paris) 46(1): 108-147.

COLBY, M.
1943 Poisonous marine animals in the Gulf of Mexico. Trans. Texas Acad. Sci. 26: 62-69.

COUTIÈRE, H.
1899 Poissons venimeux et poissons vénéneux. Thèse Agrég. Carré et Naud, Paris. 221 p.

(NSA)—Not seen by author.

DANILEWSKY, K.
1885 The case of poisoning by fish poison. [In Russian] Vrach. (50): 854-855.

DAY, F.
1878-88 The fishes of India. 2 vols. London. 816 p., 198 pls.

FAUST, E. S.
1906 Die tierischen gifte. F. Vieweg und Sohn, Braunschweig. 248 p.
1924 Tierische gifte. Handbuch Exp. Pharm., Berlin. (NSA)
1927*a* Die tropischen intoxikationskrankheiten. II. Vergiftungen durch tierische gifte, p. 865-1004. *In* C. Mense [ed.], Handbuch der tropenkrankheiten. 3rd ed., Vol. 2. J. A. Barth, Leipzig.
1927*b* Vergiftungen durch tierische gifte, p. 1745-1927. *In* G. von Bergmann and R. Staehelin [ed.], Handbuch der inneren medizin. Vol. 2. J. Springer, Berlin.

FREDERICQ, L.
1924 Die sekretion von schutz- und nutzstoffen, p. 1-87. *In* H. Winterstein [ed.], Handbuch der vergleichenden physiologie. Vol. II. Gustav Fischer, Jena.

FUHRMAN, F. A., FUHRMAN, G. J., DULL, D. L., and H. S. MOSHER
1969 Toxins from eggs of fishes and amphibia. Agric. Food Chem. 17(3): 417-424, 6 figs.

FUHRMAN, F. A., FUHRMAN, G. J., and J. S. ROSEEN
1970 Toxic effects produced by extracts of eggs of the cabezon *Scorpaenichthys marmoratus*. Toxicon 8: 55-61.

GARMAN, S.
1895 The cyprinodonts. Mem. Mus. Comp. Zool. Harvard Coll. 19(1), 179 p., 12 pls.

GIUNIO, P.
1948 Otrovne ribe. Higijena (Belgrade) 1(7-12): 282-318, 20 figs.

GÜNTHER, A.
1859-70 Catalogue of the fishes in the British Museum. 8 vols. London.
1880 An introduction to the study of fishes. Adam & Charles Black, Edinburgh. 720 p.

Halstead, B. W.
1974 Marine biotoxicology, p. 212-239. *In* F. Coulston and F. Korte [eds.], Environmental quality and safety, vol. 3. Academic Press, Inc., New York.

Hatano, M., Zama, K., Takama, K., Sakai, M., and H. Igarashi
1964 Toxic substance of the roe of northern blenny. I. On the extraction of toxic substance and its chemical properties. [In Japanese, English summary] Bull. Fac. Fish., Hokkaido Univ. 15(2): 138-146.

Hiyama, Y.
1943 Report of an investigation on poisonous fishes of the South Seas. [In Japanese] Nissan Fish. Exp. Sta. (Odawara, Japan). 137 p., 29 pls., 83 figs.

Hubbs, C. L., and A. N. Wick
1951 Toxicity of the roe of the cabezon, *Scorpaenichthys marmoratus*. Calif. Fish Game 37(2): 195-196.

Jordan, D. S., and B. W. Evermann
1896-1900 The fishes of North and Middle America. 4 parts. U.S. Nat. Mus., Bull. 47. 392 pls.

Kaneda, T., Ishii, S., Sakai, H., and K. Arai
1955 Bull. Tokai Regional Fish. Research Lab. No. 12 (Contr. A, No. 20). (NSA)

Knox, G.
1888 Poisonous fishes and means for prevention of cases of poisoning by them. [In Russian] Voenno-Med. Zh. 161(3): 399-464, 9 figs.

Kobert, R.
1902 Ueber giftfische und fischgifte. Med. Woche (19): 199-201; (20): 209-212; (21): 221-225.

Koch, T.
1857 Ueber das fischgift. Bericht an das medicinal-department des ministeriums des innern. Med. Zeitung Russlands 14(45): 337-341.

Kolb, J. N.
1826 Bromatologie oder uebersicht der bekanntesten Nahrungsmittel der bewohner der verschiedenen welttheile, naturhistorisch und mit hinweisung auf ihren diatatischen und pharmakody-namischen werth. Part I. Gelehrten Buchhandlung, Hadamar, p. 236-317.

Konstansoff, S. W.
1904 On the nature of fish poison. [In Russian] Arkh. Biol. Nauk. 10(5): 474-507.
1906 Ueber das wesen des fischgiftes. Zentr. Bakteriol. Parasitenk., Abt. I. 38: 542-557.

Konstansoff, S. W., and E. O. Manoiloff
1914 Ueber die einwirkung des verdauungsferments auf das sogenannte fischgift. Wien. Klin. Wochschr. 27(25): 883-886.

Linstow, O. V.
1894 Die giftthiere und ihre wirkung auf den menschen. August Hirschwald, Berlin. 147 p., 54 figs.

Lisunoff, S. A.
1892 On the poisoning caused by salted red fish (sturgeon). [In Russian] Russ. Med. (23-24): 358-360; (25-26): 378-380.

Maass, T. A.
1937 Gift-tiere. *In* W. Junk [ed.], Tabulae biologicae. Vol. XIII. N. V. Van de Garde & Co's Drukkerij, Zaltbommel, Holland. 272 p.

McCrudden, F. H.
1921 Pharmakologische und chemische studien über barben- und hechtrogen. Arch. Exp. Pathol. Pharmakol. 91: 46-80.

Macht, D. I.
1942 An experimental appreciation of Leviticus XI: 92-12 and Deuteronomy XIV: 9-10. Hebrew Med. J. 2: 165-170.

Macht, D. I., and E. C. Spencer
1941 Physiological and toxicological effects of some fish muscle extracts. Proc. Soc. Exp. Biol. Med. 46(2): 228-233.

Palombi, A., and M. Santarelli
1961 Gli animali commestibili dei mari d'Italia. 2d ed. Ulrico Hoepli, Milano, Italy. 435 p., 3 pls., 280 figs.

Pawlowsky, E. N.
1927 Gifttiere und ihre giftigkeit. Gustav Fischer, Jena. 515 p., 170 figs.

Phisalix, M.
1922 Animaux venimeux et venins. 2 vols. Masson et Cie., Paris.

Roedel, P. M.
1953 Common ocean fishes of the California coast. Calif. Dept. Fish Game, Fish Bull. 91. 184 p., 175 figs.

(NSA)—Not seen by author.

ROMANO, S.
1940 Animali velenosi della fauna Italiana. Natura 31(4): 137-167, 7 figs.

SAKAI, M., SHINANO, H., KIMURA, T., EME, Y., SAKA, M., and I. HAYASHI
1962 Poisoning produced by the ovaries of *Stichaeus grigorjewi* Herzenstein. [In Japanese] Shokuhin Eisei Kenkyu 12(7): 53-68.

SCHLEGEL, J. H.
1801 Barbeneier. Neue materialien für die staatsarzneiwissenschaften. Jena, Pt. 2. p. 150-153.

SCHREIBER, J.
1884 Ueber fischvergiftung. Berlin. Klin. Wochschr. (11-12): 161-163, 183-185.

SCHULTZ, L. P.
1938 Treasures of the Pacific. Nat. Geogr. Mag. 74(4): 466, 485.

SENGBUSCH, E.
1844 Über das fischgift, mit besonderer berücksichtigung der in Russland vorgekommenen vergiftungen durch gesalzene fische. Med. Zeitung Russlands (St. Petersburg) (48-52): 377-414.

SVETOVIDOV, A. N.
1963 Fauna of U.S.S.R. Clupeidae. Vol. II. No. I. [Engl. transl.] IPST Cat. No. 126. Israel Prog. Sci. Transl., Jerusalem. 428 p., 53 pls.

TAFT, C. H.
1945 Poisonous marine animals. Texas Rept. Biol. Med. 3(3): 339-352.

TASCHENBERG, E. O.
1909 Die giftigen tiere. Ferdinand Enke, Stuttgart. 325 p., 64 figs.

WALFORD, L. A.
1931 Handbook of common commercial and game fishes of California. Calif. Dept. Fish Game, Fish Bull. 37. 120 p.

WHITE, H. C., MEDCOF, J. C., and L. R. DAY
1965 Are killifish poisonous? J. Fish. Res. Board Can. 22(2): 635-638.

Class Osteichthyes

Poisonous: ICHTHYOHEMOTOXIC FISHES

Although the subject of ichthyohemotoxins, or fish serum poisons, is actually concerned with a variety of phylogenetically unrelated fishes, most of the research has been concerned with toxins derived from three types of eels—morays, congers, and anguillids—which are members of the single order Anguilliformes (Apodes). The members of this group are characterized by an eellike body, abdominal pelvic fins (when present), and an air bladder connected with the intestine by a duct. There are no spines in the fins. The scales, if present, are cycloid. The vertebrae are numerous. Gill openings are narrow or slitlike. The dorsal and anal fins are very long, and usually confluent posteriorly (Berg, 1947).

In reviewing the research on ichthyohemotoxins it is noteworthy that most reports deal with the immunological, toxicological, and chemical aspects, and that the usual array of clinical reports are absent.

The biology and systematics of this group have been discussed by Günther (1880), Jordan and Evermann (1896-1900), Joubin (1929-38), Jordan, Evermann, and Clark (1930), Fowler (1936), Ginsberg (1951), Okada (1955), Luther and Fiedler (1961), Palombi and Santarelli (1961), Berg (1962-65), and Riedl (1963).

General reviews on eel serum poison have been written by Blanchard (1890), Mitchell (1906), Taschenberg (1909), Scott (1921), Phisalix (1922, 1931), Pawlowsky (1927), Anon. (1928, 1947), Guillaume (1931), Engelsen (1922), Moru (1934), Gessner (1938), Brocq-Rousseu and Fabre (1942), Giunio (1948), Ralls and Halstead (1955), and Halstead (1964).

LIST OF FISHES REPORTED AS ICHTHYOHEMOTOXIC

Phylum CHORDATA

Class OSTEICHTHYES

Order ANGUILLIFORMES (APODES): EELS

Family ANGUILLIDAE

Anguilla anguilla (Linnaeus) (Pl. 1). Common European eel (USA), gemeine aal (Germany), anguille (France), enguia (Portugal), anguilla (Spain), ugor (USSR). *204*

Distribution: Europe, fresh and salt water.

Sources: Springfeld (1889), Pellegrin (1899), Phisalix (1922), Giunio (1948).
Other names: *Anguilla fluviatilis, Anguilla vulgaris.*

204 *Anguilla japonica* Temminck and Schlegel (Pl. 2, fig. a). Japanese eel (USA), unagi (Japan), nuchaunagi (Okinawa).
DISTRIBUTION: Pacific coast and southward from middle of Japan, southwestern Korea and China, fresh and salt water.

SOURCE: Inoko (1892).

204 *Anguilla rostrata* (LeSueur) (Pl. 2, fig. b). American eel (USA), anguilla (Latin America).
DISTRIBUTION: Atlantic coast of the United States, from Maine to Mexico.

SOURCES: Macht and Spencer (1941), Macht (1942), Taft (1945).
OTHER NAMES: *Anguilla chrisypa, Anguilla chryspa.*

Family CONGRIDAE

205 *Ariosoma balearica* (De la Roche) (Pl. 3, fig. a). Spanish conger eel (USA), madre del sofio, congre, congre de sacre, congre dols (Spain).
DISTRIBUTION: Tropical Atlantic and Mediterranean.

SOURCE: Bellecci and Polara (1907).
OTHER NAME: *Congromuraena balearica.*

205 *Conger conger* (Linnaeus) (Pl. 3, fig. b). Conger eel (USA, England), le congre, le congre commun (France), grongo vulgare (Italy), congro (Portugal), der meeraal (Germany), congrio safio, congrio (Spain), gôbo paguête, dyeye (west Africa).
DISTRIBUTION: Atlantic Ocean, Mediterranean.

SOURCES: Mosso (1888), Pellegrin (1899), Bellecci and Polara (1907), Philsalix (1922), Giunio (1948).
OTHER NAME: *Conger vulgaris.*

Family MURAENIDAE

205 *Gymnothorax moringa* (Cuvier) (Pl. 3, fig. c). Spotted moray (USA).
DISTRIBUTION: Atlantic Ocean.

SOURCE: Kobert (1902).
OTHER NAME: *Muraena moringa.*

135, 258 *Muraena helena* (Linnaeus) (Pl. 7, fig. b, Chapter X). Moray eel (USA), moray, murry (England), murena, morena (Italy), morena (Spain) muräne (Germany).
DISTRIBUTION: Eastern Atlantic Ocean and Mediterranean Sea.

SOURCES: Blanchard (1890), Bellecci and Polara (1907), Coutière (1907), Kopaczewski (1917*a, b, c, d, e, f, g, h,* 1918), Camus and Gley (1919), Brocq-Rousseu and Fabre (1942), Giunio (1948), Halstead (1974).
OTHER NAMES: *Gymnothorax helena, Muraena helina, Muraena halena.*

Family OPHICHTHYIDAE

Echelus myrus (Linnaeus) (Pl. 4, fig. a). Worm eel (USA, England), grongo muro, miro (Italy), congrio pintado, congre pintat (Spain), le myre commun (France). *206*

DISTRIBUTION: Mediterranean Sea, Gulf of Gascogne.

SOURCES: Mosso (1888), Bellecci and Polara (1907), Giunio (1948).
OTHER NAMES: *Conger myrus, Myrus vulgaris.*

Ophichthus rufus (Rafinesque) (Pl. 4, fig. b). Spanish snake eel (USA, England), le serpent de mer, l'ophisure serpent (France), ophisurus serpente (Italy), culebrita roja marina (Spain). *206*

DISTRIBUTION: Mediterranean Sea.

SOURCE: Bellecci and Polara (1907).
OTHER NAMES: *Ophichthys hispanus, Ophisurus hispanus.*

Oxystomus serpens (Linnaeus) (Pl. 4, fig. c). Spanish snake eel (USA, England), le serpent de mer, l'ophisure serpent (France), ophisurus serpente (Italy). *206*

DISTRIBUTION: Eastern Atlantic Ocean and Mediterranean Sea.

SOURCES: Bellecci and Polara (1907), Giunio (1948).
OTHER NAMES: *Ophichthys serpens, Ophisurus serpens, Muraena serpens.*

BIOLOGY

Family ANGUILLIDAE: Freshwater eels are widely distributed throughout the world, with the exception of the Arctic, Antarctic, west coast of Africa, Pacific coast of America, and Atlantic coast of South America, from which areas they are absent. Anguillids are primarily freshwater and brackishwater fishes, but migrate to the depths of the sea for the purpose of breeding. When the common freshwater eel of Europe attains a length of about 30 cm, it changes from its usually yellow color to a silvery appearance. "Silver eels" make their way down to the mouths of rivers. Migration from estuaries and into the open sea takes place during late summer and autumn. Then starts the extended journey in the deep waters of the central Atlantic, to an area known as the Sargasso Sea. Reproductive activities take place within this area. The young leaf-shaped eels, which are termed leptocephalids, slowly make their way to Europe. It is estimated that this migration takes about 3 years (Russell and Yonge, 1944). During this period the leptocephalids gradually change into the true eellike appearance. Upon reaching the shores of Europe they ascend the rivers as typical little eels, known as "elvers." They remain in fresh water for a period of 5 to 20 years and finally return to the sea to spawn and die. The American eel has a similar life history, but the leptocephalids are believed to require only about a year for their migration. Eels are reputed to be among the more voracious of carnivorous fishes. Their bill of fare consists of a large variety of bony fishes, shrimp, and crayfish. They are powerful, rapid swimmers and somewhat nocturnal in their habits. During the day they may hide in crevices, under rocks, under logs, or burrow in the mud. Freshwater eels are generally considered to be excellent food fishes. The females, larger than the males, may attain a length of about 1 m.

Family CONGRIDAE: Conger eels are similar in appearance to anguillids but are without scales, have a somewhat different mouth, and have the dorsal fin beginning nearer the head. They are widely distributed in marine, brackish or fresh water, in warm and temperate zones. The principal species, *Conger conger,* with which this chapter is concerned, is largely marine. Congers prefer deep water where the bottom is rocky or sandy areas surrounded by rocks, although its bathymetric range extends from tidal zones to 50 fathoms or more. They may prefer lobster pots to secure their contents and prey upon flatfishes, pilchard, hake, herring, crabs, and a large variety of other types of marine animals. Congers may attain a length of more than 2 m and a weight of more than 40 kg.

Family MURAENIDAE: See Chapter X, p. 349.

Family OPHICHTHYIDAE: Snake eels are found in temperate and tropical seas, inhabiting shallow water, coral reefs, and sandy shores. A few species are deepwater forms, whereas others extend into estuaries, rivers, and streams. Snake eels have the ability to burrow rapidly into the sand with their sharp-pointed tails. Their food consists of small fishes and crustaceans. Although their teeth are relatively small, they can inflict a painful bite.

MECHANISM OF INTOXICATION

Ichthyohemotoxins are largely parenteral poisons, although there are a few instances on record where individuals have become intoxicated due to ingestion of massive quantities of the toxin. Most of the species listed in this section are considered to be good food fishes. However, ingestion of fresh serum from ichthyohemotoxic eels may result in severe, if not fatal, consequences. Under normal circumstances the fishes listed in this group should not be considered as a public health hazard.

MEDICAL ASPECTS

Clinical Aspects

Very little is known regarding the symptomatology of ichthyohemotoxism in humans. Our information is based on the observations of such early authors as Kobert (1902), Mosso (1889), Springfeld (1889), Steindorff (1914), and Giunio (1948).

Fish serum intoxications may be of two types: (1) Systemic, a form that results from drinking fresh, uncooked, fish blood. The symptoms consist of diarrhea, bloody stools, nausea, vomiting, frothing at the mouth, skin eruptions, cyanosis, apathy, irregular pulse, weakness, paresthesias, paralysis, respiratory distress, and possibly death. (2) Topical, a form that produces severe inflammatory response when raw eel serum is placed in the eye or on the tongue. Steindorff (1914) has termed the condition "conjuctivitis ichthyotoxica." Oral symptoms consist of burning, redness of the mucosa, and hypersalivation. Ocular symptoms are severe burning and redness of the conjunctivae which develop within 5 to 20 minutes after contamination. Lacrimation and swelling of the eyelids are usually present. A sensation as though a foreign body was present in the eye may persist for several days. Repeated inoculations of eel serum in the eye gradually result in an immunity with progressing decrease in the severity of symptoms with each subsequent inoculation.

Pathology

(Unknown)

Treatment

Symptomatic. There is no specific antidote. Repeated subcutaneous injections of eel serum will produce an immunity in laboratory animals.

Prevention

Care should be taken in the handling of fresh eel serum. Raw eel serum or blood should never be ingested. Cooking is said to destroy the toxic properties of the serum.

PUBLIC HEALTH ASPECTS

Ichthyohemotoxism generally is not considered to be a public health problem.

TOXICOLOGY

Mosso (1888) initiated research on the toxicology of ichthyohemotoxins. He had noted earlier in his work that if a few drops of eel blood, *Anguilla anguilla, Conger conger, Echelus myrus*, or *Muraena helena*, were placed on the tongue there was at first a salt taste, then burning, hypersalivation, and a feeling of tightening of the throat. It was also observed that boiling destroyed the toxicity. Blood was collected by cutting the tail off a fresh fish which had been hung up to drain. The serum was obtained from the ice-cooled coagulated blood. Two volumes of 0.75 percent NaCl were mixed with each volume of whole blood, and this was allowed to stand for a period of time. The remaining corpuscles were removed by centrifugation and the clear serum utilized. Most of his experiments consisted of administering parenteral injections of the diluted serum into dogs, rabbits, guinea pigs, pigeons, and frogs.

Autopsies were conducted on most of the animals killed, but little was found other than visceral congestion and failure of the blood to clot. Mosso termed fish serum poison "ichthyotoxicum."

Mosso further determined that heating of *Anguilla* and *Muraena* sera to 100° C resulted in loss of its bitter, burning taste and its toxicity. Drying did not destroy its toxicity. Biliary salts were not present in the serum. Parenteral administration of fish serum was toxic, but it was harmless when ingested. He thought the toxin to be an albuminoid substance.

Experiments similar to those of Mosso were conducted by Springfeld (1889) with the serum of *A. anguilla* on rabbits. He reported that ichthyohemotoxism in rabbits occurred in two stages, first excitation and then relaxation. In both instances there was pronounced stimulation of the respiratory center. His results were essentially the same as those of Mosso, with the exception that Springfeld did not find his rabbits to be completely anesthetized or paralyzed.

Marcacci (1891) experimented with the serum of tuna. Serum was injected intraperitoneally and intravenously into dogs. He noted that the symptoms consisted of progressive paralysis without convulsions, and finally death. Marcacci found that heating the serum to 60° C for 10 minutes destroyed its toxic properties and that the addition of a few drops of HCl had the same effect. He also believed that he was dealing with an albuminoid substance which was similar in action to that studied by Mosso in eels.

In his work with the serum of *Petromyzon marinus*, Cavazzani (1892) observed that its toxic properties were similar to those possessed by the sera of *Muraena* and *Conger*. Cavazzani tested the serum of *Petromyzon* by injecting it subcutaneously and intraperitoneally into frogs, rabbits, and dogs. Frogs developed hypoactivity, decreased sensitivity, weakness, exaggerated reflexes, paralysis, and death. Rabbits showed increased respiratory and cardiac rates, muscular paralysis, and weakness. Comparable symptoms were present in dogs. Cavazzani concluded that the ichthyohemotoxin of *Petromyzon* was closely allied to that of *Muraena*.

Inoko (1892) studied the toxicity of the Japanese eel *A. japonica*. Subcutaneous and intraperitoneal injections of the serum into dogs and rabbits produced respiratory distress, drop in blood pressure, hemolysis, failure of blood to clot, muscular paralysis, and death. Heating the serum to 100° C or the addition of anhydrous alcohol destroyed the toxicity. Inoko believed the toxic substance to be a toxalbumen. The serum of *Congromuraena anago*, a Japanese conger, was also tested but found to be nontoxic.

Maglieri (1897) worked on the minimal lethal dose of *Anguilla* serum for rabbits. Serum diluted with 0.75 percent NaCl in the ratio of 1:3 was administered intravenously. Tests on a series of 17 rabbits showed an MLD of 0.020 to 0.025 ml/kg of the undiluted serum. Subcutaneous injections in a series of 10 rabbits resulted in an MLD of 0.40 to 0.45 ml of undiluted serum per kg. Intraperitoneal tests on a series of eight rabbits showed an MLD of 0.20 to 0.25 ml/kg.

Wehrmann (1897) determined that the toxicity of eel serum varied according to the season of the year and locality from which the fish were obtained. In a series of tests he found that the lethal dosage varied from 0.1 to 0.2 ml of the serum by intraperitoneal injections on guinea pigs. Wehrmann recommended the intraperitoneal route since subcutaneous injections resulted in a local inflammatory reaction and failed to give a consistent lethal dosage.

Pettit (1898) studied the pathological changes produced by eel serum on the kidneys of rabbits. Pathological findings consisted of degenerative changes in the tubular epithelium of the kidney. However, his findings were inconclusive since such changes are commonly observed in surgically removed and rapidly-fixed kidney specimens.

Scofone and Buffa (1900) tested *Anguilla* serum by intramuscular injections on the goldfish *Carassius auratus.* Dosages of serum sufficient to kill a 27-pound dog failed to produce any symptoms in a 50-g fish.

Gley (1904) tested the blood of *Torpedo marmorata* and found it to be toxic when administered intravenously to dogs, rabbits, guinea pigs, and frogs. Symptoms consisted of excitation, diarrhea, fall in blood pressure, muscular paralysis, and death due to respiratory paralysis. The serum also produced hemolysis and a slowing of the clotting mechanism. Heating of the serum to 57° C for 15 minutes resulted in a loss of toxicity.

Steindorff (1914) conducted extensive experiments on the toxic effects of *Anguilla* serum on the eyes of humans and laboratory animals—monkeys, goats, cats, rabbits, dogs, doves, chickens, cockatoos, and frogs. Inoculations of the eye with eel serum generally resulted in an inflammatory reaction, as evidenced by redness of the conjunctiva, lacrimation, chemosis, swelling of the eyelids, cloudiness of the cornea, and hyperemia. Histological sections of the skin surface adjacent to the conjunctiva of an eye which had been inoculated with eel serum were examined for pathological changes. It was found that the connective tissue layer was widely separated from the epithelial layer by hemorrhage and edema. The tissues showed evidence of an inflammatory reaction with infiltrations of lymphocytes and polymorphonuclear leucocytes. Intravenous injections of eel serum into rabbits produced miosis within 10 to 15 minutes, and the vessels of the iris became dilated. After 7 to 8 hours the pupils returned to their normal size. Repeated injections of eel serum in a rabbit gradually produced an immunity with a progressive decrease in the severity of the ocular symptoms with each subsequent injection. Irradiation of eel serum for 15 minutes with a mercury ultraviolet light was found to destroy the miotic properties of eel serum but did not alter its immunological properties. Intravenous injections of eel serum in a cat failed to produce any pupillary changes.

In his testing of *Muraena* serum on various laboratory animals, Kopaczewski (1917*a*, 1918) calculated the per kg lethal dose of undiluted serum for guinea pigs at 0.2 ml, 0.15 ml for the rabbit, and 0.1 ml for the dog. Symptoms generally consisted of weakness, collapse, respiratory distress, and convulsions. He also determined (1917*d*, *e*) that its toxicity could be retained for a period of 30 days or more if it was placed in sealed ampules and stored in the dark. Sunlight was found to destroy its toxicity. Freezing appeared to have no deleterious effect on toxicity, but heating to 60°-65° C destroyed it.

Moray serum could be dried without any appreciable loss in toxicity. Kopaczewski (1917*g*) also demonstrated that ultraviolet rays with wave length above 300 mu do not reduce the toxic properties in 4 hours and 30 minutes. However, ultraviolet rays with wave lengths of 224-300 μ were found to destroy its toxic characteristics within 90 minutes.

Eel serum lyses *Staphylococcus aureus* after prolonged contact (Kopaczewski, 1917*b*, 1918) and inactivates rabies virus *in vitro* (Phisalix, 1926).

Buglia (1919*a*, *b*, *c*) investigated the toxicity of the serum of anguillid eels during their leptocephalous stage. Aqueous extracts of the bodies of leptocephalids were found to have a hemolytic action on defibrinated beef blood, an action analogous to that of adult eel blood. This analogy was further confirmed by the fact that the factors which modify the action of leptocephalid extracts also modify adult eel serum. Extracts from the skin or mucous secretions from leptocephalid eels also have hemolytic properties. Buglia further reported finding leptocephalid extracts and adult eel serum as having an accelerating action on the coagulation *in vitro* of dog blood.

Summarizations of the toxicological investigations of previous workers have been published by Phisalix (1922), Pawlowsky (1927), Maass (1937), Brocq-Rousseu and Fabre (1942), and Ralls and Halstead (1955).

PHARMACOLOGY

Some of the more pronounced toxic actions of eel serum are those affecting the nervous system (Mosso, 1888; Inoko, 1892; Camus and Gley, 1898*b*, *d*; Gley, 1907; Kopaczewski, 1918; Mitomo, 1927). However, the manifestation of nervous system involvement was found to vary with the dosage (Camus and Gley, 1898*a*; Kopaczewski, 1918; Mitomo, 1927). Relatively strong doses resulted immediately in convulsions and respiratory arrest, whereas weaker doses produced various paralytic symptoms which sometimes terminated in death.

When paralysis occurred the primary involvement appeared to be central (Mitomo, 1927), for even after complete paralysis, the motor nerves and the skeletal muscles had essentially normal excitability. Later there ensued a cord paralysis which abolished all lower reflex activity. The neurological effects noted appeared to be referable first to a direct action of the toxin on central nervous system tissues and secondly to a failure of circulation which was concomitant with a respiratory stasis.

In the terminal stages of the paralytic form, the skeletal muscles sometimes underwent fibrillar twitching (Camus and Gley, 1898*d*; Mitomo, 1927) which was believed to be the result of direct stimulation of the motor nerve endings (Mitomo, 1927), an effect which could be blocked by curare. Fibrillation could also be initiated by sensory stimulation (Camus and Gley, 1898*b*). In relatively mild poisoning, the hind quarters were affected to a much greater extent than the forelimbs (Mosso, 1888; Camus and Gley, 1898*b*). In some cases there was a spastic extensor paralysis (Mosso, 1888).

Autonomic effects included pupillary constriction (transient dilatation in dogs), increase in intestinal motility, urination, salivation, and lacrimation (Mosso, 1888; Camus and Gley, 1898*d*).

Death from poisoning by eel serum was almost always the result of respiratory depression. It is generally agreed that the depression was central and not peripheral (Mosso, 1888; Inoko, 1892; Camus and Gley, 1898*d*). Mosso (1888) reported that respiration may end in inspiratory stasis, suggesting that there had been a specific poisoning of the expiratory center. Except with maximal lethal doses, the respiratory nerves and muscles were still excitable at death (Mosso, 1888). Cavazzani (1892) reported profound respiratory arrhythmias resulting from eel serum poisoning. With moderately lethal doses, artificial respiration was found to be effective (Mosso, 1888).

Eel serum toxin was also said to have a depressing effect on circulation (Mosso, 1888; Inoko, 1892; Kopaczewski, 1918; Mitomo, 1927), mainly through its action on the medullary centers (Mosso, 1888; Inoko, 1892). There was some indication that, at least in massive doses, the heart was also affected directly (Mitomo, 1927). In general, the most important cardiovascular effect was the lowering of blood pressure (Mosso, 1888; Inoko, 1892; Mitomo, 1927). According to Mitomo (1927), small doses of eel poison sometimes produced a transient increase in blood pressure as a result of the direct stimulation of the heart and thus an increase in amplitude and a decrease in rate of contraction. Later, with larger initial doses, there was a fall in blood pressure which reflected a shortening of diastole in particular and an increase in heart rate. According to Bardier (1898), intravenous injection of eel serum into the rabbit caused initially a marked bradycardia followed by a complete but transient arrhythmia. Recovery of the normal rhythm was said to be accomplished relatively soon after the period of arrhythmia. Counteracting the effects on the heart were those on the blood vessels, where small doses caused a slight dilatation and larger doses vasoconstriction. Jaques (1955) found that small doses of eel serum produced vasoconstriction of long duration in the isolated perfused rabbit ear.

Another interesting vascular effect was the inhibition or retardation of blood coagulation (Mosso, 1888; Inoko, 1892) which was said to be the result of the production of a "thrombase" by the liver of the poisoned animal (Delezenne, 1897; Coutière, 1899). Inhibition was so complete that the blood in an animal that had been dead from 1 to 2 days would still be unclotted or incompletely clotted. Retraction did not occur in either case. Hemolysis sometimes occurred on addition of eel serum to the blood of dogs and rabbits in a manner similar to the hemolysis resulting from treatment of blood with urea (Inoko, 1892).

Smooth muscle, in general, was stimulated by eel serum, and poisoning in the intact individual usually resulted in an increase in intestinal motility, accompanied by urination and pupillary constriction (Mosso, 1888; Gley, 1907; Mitomo, 1927; Jaques, 1955). According to Jaques (1955), a potent, nondialyzable substance from eel serum produced slow contractions in isolated perfused guinea pig ileum. The contractions persisted after treatment with atropine and neoantergan in doses sufficient to suppress strong acetylcholine or histamine contractions, and they remained unaffected after tryptamine desensitization.

When introduced into the gut, eel serum produced no toxic manifestations. In studying detoxification by acid and alkali, Buglia and Barbieri (1922) found that the amount of acidity or alkalinity required to destroy or attenuate the toxic action of an extract of whole eel larvae was of the same order of magnitude as the acidity and alkalinity present in the mammalian stomach and intestine respectively.

Apparent contradictions in the literature regarding the effects of eel serum poison are believed due to the differences in dosages that were administered.

Immunological Aspects

Native Immunity: Animals vary a great deal in their response to eel serum. Unlike the reactions in other susceptible animals, the erythrocytes of marmots (Camus and Gley, 1905*a, b*) and cats (Camus and Gley, 1911) were not readily hemolyzed *in vivo* or *in vitro* with the toxin. Erythrocytes of newborn rabbits were also very resistant to hemolysis (Camus and Gley, 1899*a, b*). Pigeons were also included in this group since their erythrocytes seemed to possess greater relative resistance than did those in the intact animal (Tchistovitch, 1899). Although the general toxicity of eel serum for the frog (*Rana temporaria* and *R. esculenta*), toad (*Bufo vulgaris*), tortoise (*Testudo groeca*), and bat (*Vespertilio murinus*) is not known with certainty, all of these animals do possess resistant erythrocytes (Mosso, 1888; Camus and Gley, 1899*a, b*; Tchistovitch, 1899; Mitomo, 1927).

Natural immunity was reported to be associated with resistance of the cells to injury and not to the presence of natural serum antitoxins (Camus and Gley, 1898*c*, 1899*a, b*). Normal canine serum, nevertheless, apparently had a mild antitoxic effect *in vitro* (Héricourt and Richet, 1897*b*). Eel serum was also shown to contain antihuman blood group O agglutinins (Grubb, 1949, 1950) and has been produced commercially for clinical laboratory and investigational use.

Acquired Tolerance: Early investigators claimed that susceptible animals receiving heated eel serum (Phisalix, 1896; Wehrmann, 1897; Camus and Gley, 1898*d*), water-diluted eel serum (Wehrmann, 1897), alcohol precipitate of eel serum redissolved in water (Phisalix, 1896), heated hedgehog serum (Camus and Gley, 1898*b, d*), and peptone (Clerc and Loeper, 1902*a, b*) were refractory within several hours and remained so for about 3 days to otherwise lethal doses of fresh eel serum. The tolerance for the toxin was admittedly very light and evanescent. Similar claims have not been advanced by later investigators. The phenomenon is reminiscent of "tachyphylaxis" described by Cesa-Bianchi (1911) and Champy and Gley (1911).

Acquired Cellular Immunity: Washed erythrocytes and isolated viable intestine and heart from susceptible animals subjected to prolonged immunization with small increasing doses of native eel serum were more resistant to the toxin than the same tissues from nonimmunized animals (Camus and Gley, 1898*c, d,* 1899*b*; Tchistovitch, 1899; Mitomo, 1927). This phenomenon apparently represented an acquired cellular resistance to the toxin. Whether it actually involved residual free antibodies, sessile antibodies, structural alterations in the cell membrane, a selection of resistant cells, or some other mechanism was not known. The appearance of cytologic resistance was not strictly correlated with the concentration and activity of humoral antibodies (Camus and Gley, 1899*b*; Tchistovitch, 1899; Mitomo, 1927).

Acquired Humoral Immunity: The following animals were readily immunized with native eel serum:

Rabbits . . . . .(Maglieri, 1897; Wehrmann, 1897; Camus and Gley, 1898*b*; Tchistovitch, 1899; Kopaczewski, 1917*f,* 1918).
Dogs . . . . (Héricourt and Richet, 1897*a, b*; Tchistovitch, 1899; Mitomo, 1927).
Goats . . . . . (Tchistovitch, 1899).

Rabbits were also readily immunized with nontoxic heated serum (58° C for 15 minutes) (Camus and Gley, 1898*c, d*) or with formalin treated serum (Puccinelli, 1931; Stefanopoulo, 1931). Fowl and pigeons could only be weakly immunized (Tchistovitch, 1899). Guinea pigs were immunized with extreme difficulty (Phisalix, 1897; Tchistovitch, 1899; Kopaczewski, 1917*h,* 1918).

Successfully immunized animals were protected against 10, 15, 20 and even 60 lethal doses of the native eel serum (Maglieri, 1897; Camus and Gley, 1898*b, d*; Tchistovitch, 1899; Kopaczewski, 1917*f, h,* 1918; Mitomo, 1927; Puccinelli, 1931; Stefanopoulo, 1931) except when the animal was injected under the occipitoatloidian membrane (Gley, 1907).

Passive immunity could be conferred upon susceptible animals through the transfer of immune serum (Héricourt and Richet, 1897*a, b*; Maglieri, 1897; Wehrmann, 1897; Kopaczewski, 1917*h,* 1918). Antibodies passed through the placenta of immunized pregnant female rabbits and thus conferred passive protection upon their offspring (Camus and Gley, 1899*a, b*).

A specific antitoxin was demonstrated (1) by injection of a toxin-immune serum mixture into animals (Wehrmann, 1897; Héricourt and Richet, 1897*b*; Camus and Gley, 1898*d*; Tchistovitch, 1899; Doerr and Raubitschek, 1908; Kopaczewski, 1917*h,* 1918; Stefanopoulo, 1931); (2) by the preservation of susceptible erythrocytes exposed to the neutralized toxin *in vitro* (Camus and Gley, 1898*a, c, d*; Tchistovitch, 1899;) or *in vivo* (Camus and Gley, 1898*d*); and (3) under suitable conditions by flocculation of the complex in the test tube using the serum of immunized rabbits, guinea pigs, dogs, and goats (Tchistovitch, 1899; Puccinelli, 1931—*see* Kopaczewski, 1918 to the contrary). Antitoxin has also been demonstrated by lack of action of the toxin-immune serum mixture on isolated, surviving rabbit intestine (Mitomo, 1927).

Cross-protection and neutralization tests have shown that the toxins of congers, eels, and morays are closely related (Camus and Gley, 1910, 1919). More distant antigenic relationships have been found among the sera of the eel and moray (Kopaczewski, 1917*f*, *h*, 1918), mixed snake venom (*Crotalus*, asp, *Bothrops*) (Calmette, 1895, cited by Wehrmann, 1897), and the serum of the asp (Wehrmann, 1897). Finally, diptheria antiserum in large quantities has been shown to neutralize the ichthyohemotoxin *in vitro* and *in vivo* (Wehrmann, 1897).

The administration of antivenomous serum (*Crotalus*, cobra, and *Bothrops* venoms) to eels reduced the toxicity of their blood (Wehrmann, 1897).

Hypersensitive Reactions: Guinea pigs and rabbits became highly sensitized with small doses of native eel serum (Kopaczewski, 1918), heated serum (Doerr and Raubitschek, 1908), acid inactivated and neutralized eel serum (Doerr and Raubitschek, 1908), or formalin treated serum (Puccinelli, 1931). Shocking doses of the toxoid or native toxin induced a fatal anaphylaxis when administered 3 weeks after the initial dose.

Conclusions: It is clear that the sera of eels and the toxins which they contain are capable of eliciting a number of immunological phenomena. Other than the natural resistance of the tissues of some animals, it has been observed that both susceptible and resistant animals respond to the antigenic stimulus which the ichthyohemotoxin provides with the formation of specific antibodies. Under appropriate conditions the antitoxin neutralized the toxin and apparently flocculated it *in vitro*. Serological studies suggested a close antigenic relation among the toxins of the various eels and the venoms of a number of animals including the moray, *Crotalus*, *Bothrops*, and asp. There was evidence that susceptible animals may have acquired a rapid tolerance to eel serum through prior administration of certain biological agents. The development of cellular immunity as well as hypersensitivity to the toxin has also been reported following repeated exposure of experimental animals to the serum.

CHEMISTRY

by DONOVAN A. COURVILLE

The name ichthyotoxin was applied by A. Mosso (1888) to the toxic principle in the serum of the anguilla, moray, and conger eels. It has been assumed from the time of the earliest studies that the toxins from the various types of eels are identical, such assumption evidently being based on the similarity of observed symptoms. The principal question occupying the attention of early researchers on this subject has been whether the various symptoms observed are all due to a single substance or to a plurality of substances in eel serum.

The question seems to have been raised first by Camus and Gley (1905*a*, *b*, 1912), who pointed out evidence for and against the concept of plurality of physiologically active substances. In 1917, the question was revived by

Kopaczewski (1917*d*) who observed that the hemolytic properties of moray serum were destroyed on heating for 15 minutes at 75° C., while the toxic properties persisted after such treatment though at a much decreased potency. From these observations, it was concluded that the hemolytic properties were due to a substance different from that which produced the toxic symptoms. Lumière (1929) interpreted the report of Camus and Gley as stating that the symptoms were produced by the same substance. This conclusion was challenged on the basis of inconclusive evidence. Camus and Gley (1929) replied to the criticism of Lumière by pointing out that they had made no conclusions in view of the apparent inconsistency of the evidence. In the meantime, Buglia (1921) presented evidence indicating that the hemolytic properties of eel serum resulted from the presence of salts of higher fatty acids and that the toxic properties certainly could not be related exclusively to this group of substances. Further evidence pointing to different substances as responsible for the hemolytic and toxic properties was reported by Kopaczewski (1918) who found that if an alcoholic extract of the serum was evaporated to dryness under vacuum and the residue redissolved in water, the resulting solution had no toxic properties, but the hemolytic properties still persisted.

A. Mosso (1888) reported the toxin to be insoluble in hot alcohol but readily soluble in water and in the serum which contains it. Phisalix (1922) observed that the toxin remained in solution after treatment with an equal volume of absolute alcohol. A precipitate formed on such treatment, and the hemolytic principle seemed to be distributed between the precipitate and the filtrate. U. Mosso (1889) observed that the toxin was not precipitated by long passage of carbon dioxide into the serum nor by magnesium or ammonium sulfates. This same worker found the toxin to be nondialyzable.

The stability of the toxin, or perhaps more correctly its instability, has been the subject of numerous studies. A. Mosso (1888) noted that the toxicity and the characteristic burning taste of the serum disappeared simultaneously on heating to 100° C, and the toxin was thus regarded as responsible for this peculiar taste. U. Mosso extended this relationship by noting that any treatment which caused a loss of toxicity also resulted in loss of the characteristic taste. The toxin was found to be destroyed on heating to 70° C, at which temperature the albumin fraction of the serum coagulates (Mosso, 1889). Camus and Gley (1898*b, d*) observed considerable loss of toxicity on heating to 58° C for 15 minutes. Similar observations are recorded by Kopaczewski (1918) and Phisalix (1896).

Buglia (1919*c*) later challenged the heat labile nature of the toxin as previously reported. He found that if the coagulum formed by heating the serum were triturated with sand or subjected to a limited tryptic digestion, the toxic properties reappeared in solution. It was concluded that the toxin is combined with albumin in the serum because the serum coagulated on heating to suggest the erroneous conclusion that the heat has destroyed the toxin. The apparent validity of the conclusion was confirmed by the same worker with the observation that while the toxin was not dialyzable before such treatment it became dialyzable afterward. A similar concept of conjugation of the toxin with an albumin had been previously entertained by U. Mosso (1889), though he was evidently not clear on the nature of the change brought about by heat.

As suggested by A. Mosso (1888) and confirmed by Camus and Gley (1898*d*) and by Kopaczewski (1917*d*, 1918), the toxicity is maintained essentially without loss through the process of rapid drying under vacuum and resolution. Kopaczewski further noted that considerable loss of toxicity occurred on storage of the serum in sealed capsules over a period of 30 days. Complete destruction of the toxin was reported by U. Mosso (1889) on treatment of 0.1 ml of serum with 0.0024 ml of hydrochloric acid (D. 1.19), larger amounts of weaker acids being necessary to produce the same result. Addition of an equivalent amount of alkali to the acid-treated serum caused restoration of the toxicity. On the other hand the corresponding destruction produced by alkali could not be reversed by neutralization. Salts had no effect even in saturated solution. The results of Buglia and Barbieri (1922) on the effects of acid on the toxin are discussed in another connection.

Maglieri (1897) found that the toxin was destroyed on 24-hour storage with a trace of tricresol.

Kopaczewski (1917*d*, 1918) destroyed the toxin by exposure to the action of sunlight for 48 hours. Ultraviolet light from a Zeiss ultraviolet lamp (wave length 300-400 μ) yielded no observable destruction, while exposure to ultraviolet light from a Cooper-Hewitt quartz lamp (8,000 candle power, wave length 224-300 μ) resulted in partial destruction. Exposure to X-rays had no effect.

The ichthyohemotoxin does not survive the action of certain of the digestive enzymes. A. Mosso, in his earliest study (1888), noted that the toxin was destroyed by passage through the digestive tract and hence was harmless when ingested. The toxicity was apparent only when injected directly into the blood or into the small intestine. U. Mosso (1889) observed destruction of the toxin by gastric juice but found that the acid of the digestive fluid was sufficient to account for the results. Wehrmann (1897) found that the bile was also capable of destroying the toxin. Buglia (1919*c*) was able to liberate the toxin from its conjugated protein by the limited action of trypsin; but this action was always accompanied by heavy loss of toxicity, suggesting an action of the enzyme on the toxin as well as on the conjugated protein.

Buglia and Barbieri (1922), believing the previously reported data to be inconclusive as to whether enzymes or acids caused the destruction by gastric juice, studied this point in detail. Samples of serum were treated with various amounts of acid, left standing for one-half hour, and neutralized. The toxicity was then compared with the toxicity of the original serum. The data are somewhat difficult to interpret. It was found that as the amount of acid was increased up to a certain point, less and less toxin survived the treatment. As the amount of acid increased beyond this point, the toxic properties began to increase and continued until a toxicity was reached approaching that of the original serum. These workers concluded that the observed destruction by gastric juice was due to the acid rather than to enzymes.

Bénech (1899) studied the moray to determine whether other portions of the eel than the serum might also contain the toxin. Extracts of tissues from which all the blood had been washed were still found toxic to the rabbit on injection. If the extracting liquid used was equal to the volume of the tissue extracted, death resulted in a few days on injection of 14 ml. Again it is tacitly assumed, presumably on the basis of similarity of toxic symptoms, that the toxin in the flesh of the eel is the same as that in the serum. However, Bénech found the toxin from the tissues to be readily precipitated with ammonium sulfate somewhat short of saturation. This property is different from that reported by U. Mosso (1889) for the toxin of eel serum.

If the precipitate formed by treating the aqueous extract of the tissues with ammonium sulfate was redissolved and dialyzed free of ammonium sulfate, the nondialyzable fraction had the following properties as further reported by Bénech: A precipitate readily formed on treatment with saturated ammonium sulfate, less efficiently with sodium chloride, and not at all by magnesium sulfate. Acids tended to produce a slight turbidity which dissolved in excess of the reagent. Bases had no action. Alcohol formed an abundant coagulum. The solution gives a positive biuret reaction in the cold and a positive Millon reaction. Precipitates formed on treatment of the solution with Tanret's, Esbach's, and Mehu's reagents and with tannin and platinum chloride. When the extract was evaporated to dryness *in vacuo*, it left a white powder of disagreeable taste and bitter aftertaste. Analysis of this powder gave C, 50.31, 50.04; H, 7.95, 7.81; N, 15.60, 16.61.

No information is yet available as to the nature of the eel toxin. Several theories have been proposed as to the nature of the protein to which it is evidently bound. A. Mosso (1888) regarded it as an albuminoid. U. Mosso (1889) considered it an albumin. He observed that the protein had no enzymic activity. Based on the appearance of the toxin in the fraction containing the "serines," he concluded the toxin was a "serine."

Several observations have been made on the composition of eel serum as distinguished from mammalian sera (Drilhon, 1953, 1954, 1955). These observations may or may not have a significant relation to the problem of the nature of the toxin molecule. Drilhon observed that the sera of the conger and moray eels (as with many other fishes) contain large amounts of lipids with a larger fraction of the protein belonging to the μg-globulin type. Fontaine and Callamand (1949) noted that the serum of eels also contains an unusually active lipolytic enzyme and a high antigonadotropic activity toward mammals.

Methods have been devised for purifying the toxin from anguilla, conger, and moray eels based on procedures for fractionation of proteins (Ghiretti and Rocca, 1963). The method giving the best results used various concentrations of ammonium sulfate in the initial precipitation step. The fraction which precipitated with ammonium sulfate between 25 and 35 percent saturation contained all the toxin but only 4 percent of the protein. Redissolving the precipitate thus obtained in water, dialyzing against 0.15 *M* NaCl, adjustment of the pH to 5.5, and dilution to the starting serum volume gave a product with toxicity equal to that of the original serum. A further precipitation with 25 percent ethyl alcohol gave an additional fourfold purification which contained nearly all of the toxin. However, of the many effects shown by whole serum, all had now disappeared except that on the nervous system and eventual death. The suggestion is the same as that revealed by earlier work, i.e., that the eel serum contains more than a single toxic material.

LITERATURE CITED

ANONYMOUS

1928 Poisonous fish. Fish Gaz. (London) 96(2654): 164.

1947 Eel's blood is poisonous. Ciba Symp. 9(8): 759.

BARDIER, E.

1898 Action cardiaque du sérum d'anguille. Compt. Rend. Soc. Biol., 10, 5(8): 548.

BELLECCI, A., and G. POLARA

1907 Sulla tossicità del siero di sangue di alcune specie di Murenoidi. Arch. Farm. Sper. 6: 598-622.

BÉNECH, E.

1899 Toxalbumine retirée de la chair d'anguille de rivière. Compt. Rend. Acad. Sci. 128(13): 833-836.

BERG, L. S.

1947 Classification of fishes both recent and fossil. [In Russian & English] J. W. Edwards, Ann Arbor, Mich.

1962-65 Freshwater fishes of the U.S.S.R. and adjacent countries. 3 vols. 4th ed. [Engl. transl.] IPST Cat. Nos. 471-473, Israel Prog. Sci. Transl., Jerusalem.

BLANCHARD, R.

1890 Traité de zoologie medicale. Vol. 2. J. B. Baillière et Fils, Paris. p. 638-695.

BROCQ-ROUSSEU, D., and R., FABRE

1942 Les toxalbumines. Hermann & Cie., Paris. 72 p.

BUGLIA, G.

1919*a* Sur l'action toxique exercée sur le sang par les extraits aqueux du corps des jeunes anguilles encore transparentes (cieche). Arch. Ital. Biol. 69: 119-133, 3 figs.

1919*b* Sur la toxicité des extraits aqueux du corps des jeunes anguilles encore transparentes (cieche). Arch. Ital. Biol. 69: 185-202.

1919*c* Ricerche sulla natura del veleno dell'anguilla. I. L'ittiotossico è termostabile. Atti Accad. Naz. Lincei, Rend., Classe Sci. Fis., Mat., Nat. 28: 54-58.

1921 Ricerche sulla natura del veleno dell'anguilla. VII. Della sostanza che emolizza il sangue. Atti Soc. Toscana Sci. Nat. Pisa, Proc. Verbali Mem, 34: 87-105.

BUGLIA, G., and G. BARBIERI

1922 Perchè il veleno dell'anguilla introdotto per via gastrica non è tossico. Arch. Sci. Biol. (Naples) 3: 26-38.

CAMUS, L., and E. GLEY

1898*a* De la toxicité du sérum d'anguille pour des animaux d'espèce différente (lapin, cobaye, hérisson). Compt. Rend. Soc. Biol., 10, 50: 129-130.

1898*b* De l'action destructive d'un sérum sanguin sur les globules rouges d'une autre espèce animale. Immunisation contre cette action. Compte Rend. Acad. Sci. 126: 428-431.

1898*c* Sur le mécanisme de l'immunisation contre l'action globulicide de sérum d'anguille. Compte. Rend. Acad. Sci. 127(6): 330-332.

1898*d* Recherches sur l'action physiolgique du sérum d'anguille. Contribution à l'étude de l'immunité naturelle et acquise. Arch. Intern. Pharmacodyn. 5: 247-305.

1899*a* Expériences concernant l'état refractaire au sérum d'anguille. Immunité cytologique. Compt. Rend. Acad. Sci. 129: 231-233.

1899*b* Nouvelles recherches sur l'immunité contre le sérum d'anguille. Contribution à l'étude de l'immunité naturelle. Ann. Inst. Pasteur 13: 779-787.

1905*a* Action hématolytique et toxicité générale du sérum d'anguille pour la marmotte. Compte. Rend. Acad. Sci. 140(26): 1717-1718.

1905*b* Comparison entre l'action hématolytique et la toxicité du sérum d'anguille chez la marmotte (*Arctomys marmota*). Arch. Exp. Pharmacol. Therap. 15: 159-169.

1910 Recherches sur l'immunisation contre les sérums toxiques. J. Physiol. Pathol. Gén. 12: 781-795.

1911 De l'action du sérum d'anguille sur le chat. Compt. Rend. Soc. Biol. 71: 158-159.

1912 Sur le mécanisme de l'action hémolytique du sérum d'anguille. Compt. Rend. Acad. Sci. 154: 1630-1633.

CAMUS, L., and E. GLEY—Continued

1919 Immunisation croisée. Action réciproque du sérum d'anguille ou du sérum de murène sur des animaux immunisés contre l'une ou l'autre de ces ichthyotoxines. Compt. Rend. Soc. Biol. 82: 1240-1241.

1929 La relation entre la toxicité du sérum d'anguille et son action globulicide. Compt. Rend. Soc. Biol. 101: 866-867.

CAVAZZANI, E.

1892 L'ittiotossico nel *Petromyzon marinus*. Giorn. Accad. Med. Torino 40(3): 872-876.

CESA-BIANCHI, D.

1911 Sull'azione reciproca degli estratti dei diversi organi, Pathologica 3: 344-350.

CHAMPY, C., and E. GLEY

1911 La tachyphylaxie croisée. Compt. Rend. Soc. Biol. 71: 430-432.

CLERC, A., and M. LOEPER

1902*a* Influence des injections intraveineuses de peptone sur l'intoxication par le sérum d'anguille. Compt. Rend. Soc. Biol. 54: 1061-1862.

1902*b* Formule hemoleucocytaire de l'intoxication par le sérum d'anguille. Compt. Rend. Soc. Biol. 54: 1062-1064.

COUTIÈRE, H.

1899 Poissons venimeux et poissons vénéneux. Thèse Agrég. Carré et Naud, Paris. 221 p.

1907 Sur le prétendu appareil venimeux de la Murène hélène. Bull. Soc. Philomath. Paris 9(9): 229-234, 3 figs.

DELEZENNE, C.

1897 De l'action du sérum d'anguille sur la coagulation du sang. Compt. Rend. Soc. Biol. 4: 42-43.

DOERR, R., and H. RAUBITSCHEK

1908 Toxin und anaphylaktisierende substanz des aalserums. Berlin. Klin. Wochschr. 45(33): 1525-1528.

DRILHON, A.

1953 Etude de quelques diagrammes électrophorétiques de plasma de poisson. Compt. Rend. Acad. Sci. 237: 1779-1781.

1954 Etude des lipoprotéides sériques chez quelques poissons au moyen de l'électrophorèse sur papier. Compt. Rend. Acad. Sci. 238: 940-942.

1955 Euglobines chez les poissons. Compt. Rend. Soc. Biol. 149: 2124-2126.

ENGELSEN, H.

1922 Om giftfisk og giftige fisk. Nord. Hyg. Tidskrift 3: 316-325.

FONTAINE, M., and O. CALLAMAND

1949 La lipase sérique chez un cyclostome (*Petromyzone marinus* L.) et divers poissons téléostéens. Bull. Inst. Oceanogr. Monaco N. 943.

FOWLER, H. W.

1936 The marine fishes of west Africa. Bull. Am. Mus. Nat. Hist. Vol. 70. Part. I & II. 1,493 p., 567 figs.

GESSNER, O.

1938 Tierische gifte, p. 61-66, 81-83. *In* Huebner and J. Schuller [eds.], Handbuch der experimentellen pharmakologie. Vol. 6. Julius Springer, Berlin.

GHIRETTI, F., and E. ROCCA

1963 Some experiments on ichthyotoxin, p. 211-216. *In* H. L. Keegan and W. V. MacFarlane [eds.], Venomous and poisonous animals and noxious plants of the Pacific region. Pergamon Press, New York.

GINSBERG, I.

1951 The eels of the northern Gulf Coast of the United States and some related species. Texas J. Sci. 3(3): 431-485, 16 figs.

GIUNIO, P.

1948 Otrovne ribe. Higijena (Belgrade) 1(7-12): 282-318, 20 figs.

GLEY, E.

1904 Recherches sur le sang des sélaciens. Action toxique du sérum de torpillo (*Torpedo marmorata*). Compt. Rend. Acad. Sci. 138: 1547-1549.

1907 De l'action des ichtyotoxines sur le système nerveux des animaux immunisés contre ces substances. Compt. Rend. Acad. Sci. 145(24): 1210-1212.

GRUBB, R.

1949 Some aspects of the complexity of the human ABO blood groups. Acta Pathol. Microbiol. Scand. 84: 1-72.

1950 Quelques aspects de la complexité des groupes ABO. Rev. Hematol. 5: 268-275.

GUILLAUME, A.

1931 Les poissons venimeux. Rev. Sci. 69: 428-434, 16 figs.

GÜNTHER, A. C.
1880 An introduction to the study of fishes. Adam & Charles Black, Edinburgh. 720 p.

HALSTEAD, B. W.
1964 Fish poisonings—their diagnosis, pharmacology, and treatment. Clin. Pharmacol. Therap. 5(5): 615-627.
1974 Marine biotoxicology, p. 212-239. *In* F. Coulston and F. Korte [eds.], Environmental quality and safety, vol. 3. Academic Press, New York.

HÉRICOURT, J., and C. RICHET
1897*a* Action locale du sérum d'anguille. Sérothérapie contre les effets toxique du sérum d'anguille. Compt. Rend. Soc. Biol. 49: 74.
1897*b* Sérothérapie in vitro dans l'intoxication par le sang d'anguille. Compt. Rend. Soc. Biol. 49: 367-369.

INOKO, D.
1892 Poison existing in the blood of *Anguilla japonica* Sieb. s. *A. bostoniensis* Les. [In Japanese] Tokyo Iji Shinshi Wochschr. (754): 1-2.

JAQUES, R.
1955 A substance from eel serum producing slow contractions. Nature (London) 175(4448): 212.

JORDAN, D. S., and B. W. EVERMANN
1896-1900 The fishes of North and Middle America. 4 parts, U.S. Nat. Mus,, Bull. 47. 392 pls.

JORDAN, D. S., EVERMANN, B. W., and H. W. CLARK
1930 Check list of the fishes and fishlike vertebrates of North and Middle America north of the northern boundary of Venezuela and Colombia. Rept. U.S. Comm. Fish. 1928. 670 p.

JOUBIN, L. M.
1929-38 Faune ichthyologique de l'Atlantique nord. Nos. 1-18. Andr. Fred Høst & Fils, Copenhague.

KOBERT, R.
1902 Ueber giftfische und fischgifte. Med. Woche (19): 199-201; (20): 209-212; (21): 221-225.

KOPACZEWSKI, W.
1917*a* Recherches sur le sérum de la murène (*Muraena helena* L.). I. La toxicité du sérum de murène. Compt. Rend. Acad. Sci. 164: 963-974.
1917*b* Recherches sur le sérum de la murène (*Muraena helena* L.). II. L'action physiologique du sérum. Compt. Rend. Acad. Sci. 165: 37-39.
1917*c* Sur le venin de la murène (*Muraena helena* L.). Compt. Rend. Acad. Sci. 165: 513-515.
1917*d* Recherches sur le sérum de la muréne (*Muraena helena* L.). III. La toxicité et les propriétés physiques du sérum. Compt. Rend. Acad. Sci. 165: 600-602
1917*e* Recherches sur le sérum de la murène *(Muraena helena)*. IV. L'équilibre moléculaire et la toxicité du sérum. Compt. Rend. Acad. Sci. 165: 725-727.
1917*f* Recherches sur le sérum de la murène (*Muraena helena* L.). V. Sur le mécanisme de la toxicité du sérum de murène. Compt. Rend. Acad. Sci. 165: 803-806.
1917*g* Influence des radiations lumineuses sur la toxicité du sérum de la murène Compt. Rend. Soc. Biol. 80: 884-885,
1917*h* Essais d'immunisation contre la toxicité du sérum de la murène. Compt. Rend. Soc. Biol. 80: 886-888.
1918 Recherches sur le sérum de la murène (*Muraena helena*). Ann. Inst. Pasteur 32(12): 584-612.

LUMIÈRE, A.
1929 La toxicité des sérums d'anguille, de murène et de certains autres poissons est-elle due à leur pouvoir globulicide? Compt. Rend. Soc. Biol. 100(14): 1209-1210.

LUTHER, W., and K. FIEDLER
1961 Die unterwasserfauna der mittelmeerküsten. Paul Parey, Hamburg. 253 p., 46 pls.

MAASS, T. A.
1937 Gift-tiere. *In* W. Junk [ed.], Tabulae biologicae. Vol. XIII. N. V. Van de Garde & Co's Drukkerij, Zaltbommel, Holland. 272 p.

MACHT, D. I.
1942 An experimental appreciation of Leviticus XI: 9-12 and Deuteronomy XIV: 9-10. Hebrew Med. J. 2: 165-170.

MACHT, D. I., and E. C. SPENCER
1941 Physiological and toxicological effects of some fish muscle extracts. Proc. Soc. Exp. Biol. Med. 46(2): 228-233.

MAGLIERI, C.
1897 Sull'azione tossica, immunissante e battericida del siero de sangue di anguilla. Ann. Igiene 7: 191-214.

MARCACCI, A.
1891 Sur le pouvoir toxique du sang de thon. Arch. Ital. Biol. 16: i.

MITCHELL, C. A.
1906 The toxine of eel's blood. Knowl. Sci. News 29: 459.
MITOMO, Y.
1927 Studien über das aalserum. I. Mitteilung. Die pharmakologische wirkung des aalserums. Tohoku J. Exp. Med. 8: 284-326.
MORU, J.
1934 Contribution à l'étude de la toxicité des animaux marins. Thèse, Fac. Méd., Paris. 47 p.
MOSSO, A.
1888 Un venin dans le sang des murènides. Arch. Ital. Biol. 10: 141-169.
MOSSO, U.
1889 Recherches sur la nature de venin qui se trouve dans le sang de l'anguille. Arch. Ital. Biol. 12: 229-236,
OKADA, Y.
1955 Fishes of Japan. Maruzen Co., Tokyo. 434 p., 391 figs.
PALOMBI, A., and M. SANTARELLI
1961 Gli animali commestibili dei mari d'Italia. 2d ed. Ulrico Hoepli, Milano, Italy. 435 p., 3 pls., 280 figs.
PAWLOWSKY, E. N.
1927 Gifttiere und ihre giftigkeit. Gustav Fischer, Jena. 515 p., 170 figs.
PELLEGRIN, J.
1899 Les poissons vénéneux. M. D. Thèse 510. Fac. Méd. Paris. 113 p., 16 figs.
PETTIT, A.
1898 Altérations rénales consécutives à l'injection de sérum d'anguille. Compt. Rend. Soc. Biol. 50: 320-322.
PHISALIX, C.
1896 Propriétés immunisantes du sérum d'anguille contre le venin de vipère. Compt. Rend. Soc. Biol. 3: 1128-1130.
1897 Venins et animaux venimeux dans la série animale. Rev. Sci. (Sér. 4) 8: 97-104, 195-201, 329-335.
PHISALIX, M.
1922 Animaux venimeux et venins. 2 vols. Masson et Cie., Paris.
1926 Immunité naturelle de l'anguille vis-à-vis du virus rabique et action rabicide de son sérum. Bull. Mus. Hist. Nat., Paris. p. 89-91.
1931 Le venin de quelques marins. Notes sta. océanogr. salammbo, Tunis (22): 3-15.
PUCCINELLI, E.
1931 Sulla proprietà antigeni del siero di anguilla formolato. Pathologica 23: 531-535.
RALLS, R. J., and B. W. HALSTEAD
1955 Moray eel poisoning and a preliminary report on the action of the toxin. Am. J. Trop. Med.Hyg. 4(1): 136-140.
RIEDL, R.
1963 Fauna und flora der adria. Ein systematischer meeresführer für biologen und naturfreunde. Paul Parey, Hamburg. 640 p., 17 figs., 122 pls.
RUSSELL, F. S., and C. M. YONGE
1944 The seas. Our knowledge of life in the sea and how it is gained. 2d ed. Frederick Warne & Co. Ltd., London. 379 p., 65 figs., 127 pls.
SCOFONE, L., and E. BUFFA
1900 Action du sérum de quelques animaux sur les poissons. Recherches expérimentales. Arch. Ital. Biol. 33: 367-372.
SCOTT, H.H.
1921 Vegetal and fish poisoning in the tropics, p. 790-798, 5 figs., 2 pls. *In* W. Byam and R. G. Archibald [eds.], The practice of medicine in the tropics. Vol. I. Henry Frowde and Hodder & Stoughton, London.
SPRINGFELD, A.
1889 Über die giftige wirkung des blutserums des gemeinen fluss-aales, *Anguilla vulgaris*. Thesis, Greiswald. 35 p.
STEFANOPOULO, G. J,
1931 Sur les propriétés antigènes des ichthyotoxines formolées. Compt. Rend. Sco. Biol. 106 (11): 917-920.
STEINDORFF, E.
1914 Experimentelle untersuchungen über die wirkung des aalserums auf das menschliche und tierische auge. Graefes Arch. Ophthalmol. 88: 158-183.
TAFT, C. H.
1945 Poisonous marine animals. Texas Rept. Biol. Med. 3(3): 339-352.
TASCHENBERG, E. O.
1909 Die giftigen tiere. Ferdinand Enke, Stuttgart. 325 p., 64 figs.
TCHISTOVITCH, T.
1899 Etudes sur l'immunisation contre le sérum d'anguilles. Ann. Inst. Pasteur 13(5): 406-425.
WEHRMANN, C.
1897 Recherches sur les proprietés toxiques et antitoxiques du sang et de la bile des anguilles et des vipères. Ann. Inst. Pasteur 11: 810-828.

Class Osteichthyes

Poisonous: MISCELLANEOUS ORAL FISH BIOTOXICATIONS

Ichthyoallyeinotoxic Fishes

Ichthyoallyeinotoxism, or hallucinogenic fish poisoning, is caused by ingesting certain types of reef fishes known to occur in the tropical Pacific and Indian Oceans. The biotoxication may result from eating either the head or flesh of the fish.

Since most ichthyoallyeinotoxic fishes have been incriminated as transvectors of ciguatoxin, the reader is referred to the section on ciguatoxic fishes (p. 325) for information regarding their taxonomy.

LIST OF FISHES REPORTED AS ICHTHYOALLYEINOTOXIC

Phylum CHORDATA

Class OSTEICHTHYES

Order PERCIFORMES (PERCOMORPHI): Perchlike Fishes

Family ACANTHURIDAE

Acanthurus triostegus sandvicensis Streets (Pl. 1, fig. a). Convict surgeonfish, tang (USA), manini (Hawaii). *208*

DISTRIBUTION: Hawaii.

SOURCES: Helfrich and Banner (1960), Helfrich (1961, 1963), Bouder, Cavallo, and Bouder (1962).

Family KYPHOSIDAE

Kyphosus cinerascens (Forskål) (Pl. 1, fig. b). Sea chub (USA), nenue, manaloa (Hawaii), bluefish, rudderfish, chub (South Africa). *208*

DISTRIBUTION: Indo-Pacific.

SOURCES: Helfrich and Banner (1960), Helfrich (1961, 1963), Bouder *et al.* (1962).

Kyphosus vaigiensis (Quoy and Gaimard) (Pl. 1, fig. c). Sea chub (USA), brass bream (South Africa). *208*

DISTRIBUTION: Indo-Pacific.

SOURCE: Helfrich and Banner (1960).

Family MUGILIDAE

148 *Mugil cephalus* Linnaeus (Pl. 27, fig. b, Chapter X). Common mullet (USA), ama (Hawaii), haarder, mullet, flathead mullet (South Africa).

DISTRIBUTION: Cosmopolitan.

SOURCES: Helfrich and Banner (1960), Helfrich (1961, 1963), Bouder *et al.*. (1962).

209 *Neomyxus chaptalli* (Eydoux and Souleyet) (Pl. 2). Mullet (USA).

DISTRIBUTION: Indo-Pacific.

SOURCES: Helfrich and Banner (1960), Helfrich (1961, 1963), Bouder *et al.* (1962).

Family MULLIDAE

148 *Mulloidichthys samoensis* (Günther) (Pl. 28, fig. b, Chapter X). Surmullet, goatfish (USA), weke, weke'a'a, weke-ula (Hawaii), baybao, tubac, tuyo (Philippines), jome (Marshall Islands), tebaweina (Gilbert Islands), afolu i'a sina (Samoa), gold-striped goatfish (Australia).

DISTRIBUTION: Indo-Pacific.

Sources: Helfrich (1961, 1963), Bouder *et al.* (1962).

149 *Upeneus arge* Jordan and Evermann (Pl. 29, fig. b, Chapter X). Surmullet, goatfish (USA), weke, weke pueo (Hawaii), baybao, tubac, tuyo (Philippines), jome (Marshall Islands), tebaweina (Gilbert Islands), afolu i'a sina (Samoa), gold-striped goatfish (Australia).

DISTRIBUTION: Indo-Pacific.

SOURCES: Jordan, Evermann, and Tanaka (1927), Titcomb and Pukui (1952), Randall (1958), Helfrich (1961, 1963), Bouder *et al.* (1962).

Family POMACENTRIDAE

149 *Abudefduf septemfasciatus* (Cuvier) (Pl. 30, Chapter X). Sergeant major, damselfish (USA), maomao (Hawaii), ulavapua, alala saga, mutu (Samoa), bakej (Marshall Islands), tebukibuki (Gilbert Islands), palata (Philippines), damselfish (Australia).

DISTRIBUTION: Indo-Pacific, Australia, east Africa, China.

SOURCE: Cooper (1964).

Family SERRANIDAE

210 *Epinephelus corallicola* (Cuvier and Valenciennes)(Pl. 3). Grouper (USA), gatala (Samoa), rero (Tahiti), baraka, kugtung (Philippines), coral rockcod (Australia), vieille (Seychelles).

DISTRIBUTION: Tropical Indo-Pacific.

SOURCE: Cooper (1964).

Family SIGANIDAE

Siganus argenteus (Quoy and Gaimard)(Pl. 39, fig. d, Chapter X). Rabbitfish (USA), baliwis, palit (Philippines), cordonnier (Mauritius). *158*

DISTRIBUTION: Indo-Pacific.

SOURCE: Halstead and Cox (1973).
OTHER NAME: *Teuthis rostrata.*

Siganus corallinus (Valenciennes) (Pl. 5, fig. a, Chapter XXIII). Rabbitfish (USA), baliwis, palit (Philippines), cordonnier (Mauritius). *262*

DISTRIBUTION: Indo-Pacific.

SOURCE: Halstead and Cox (1973).
OTHER NAME: *Teuthis corallina.*

Siganus oramin (Schneider) (Pl. 36, fig. b, Chapter X). Rabbitfish (USA), ellok, mole (Marshall Islands), lopauulu (Samoa), baliwis, palit (Philippines), gold-lined spinefoot (Australia), rabbitfish, slimy, spiny (South Africa). *155*

DISTRIBUTION: Indo-Pacific, east Africa, Saudi Arabia.

SOURCE: Wheeler (1953).
OTHER NAME: *Teuthis oxamin.*

Siganus rivulatus (Forskål). Rabbitfish (USA), baliwis, palit (Philippines), cordonnier (Mauritius).

DISTRIBUTION: Indo-Pacific.

SOURCE: Halstead and Cox (1973).
OTHER NAME: *Teuthis rivulata.*

BIOLOGY

Family ACANTHURIDAE: See p. 350 for a discussion on the biology of this family.

Family KYPHOSIDAE: The rudderfishes are characterized by the absence of molars, the front of the jaws being occupied by incisors, which are often serrated, loosely attached, and movable. There are numerous species, and they are largely inhabitants of warmer seas, although a few species are found in temperate waters. Rudderfish are usually found in shallow water around rocks, reefs, or tidepool areas. They tend to be herbivorous in their eating habits.

Family MUGILIDAE: See p. 351 for a discussion on the biology of this family.

Family MULLIDAE: See p. 352.

Family POMACENTRIDAE: See p. 352.

Family SERRANIDAE: See p. 352.

Family SIGANIDAE: See p. 353.

MECHANISM OF INTOXICATION

Ichthyoallyeinotoxism may result from eating the flesh or the head of any of the incriminated reef fishes. The poison is reputedly concentrated in the head of the fish which is believed to be the most dangerous part of the fish to eat. The biotoxication is quite sporadic and completely unpredictable in its occurrence. The poison is not destroyed by any ordinary cooking procedure.

MEDICAL ASPECTS

Clinical Characteristics

The poison affects primarily the central nervous system. The symptoms may develop within minutes to 2 hours after ingestion, persisting up to 24 hours or more. Symptoms consist of dizziness, loss of equilibrium, lack of motor coordination, hallucinations, and mental depression. A common complaint of the victim is that "someone is sitting on my chest," or there is a sensation of a tight constriction around the chest. The conviction that they are going to die or other frightening nightmares are a characteristic part of the clinical picture. Other complaints consist of itching, burning of the throat, muscular weakness, and rare abdominal distress. No fatalities have been reported, and in comparison with other forms of ichthyosarcotoxism, hallucinogenic fish poisoning is relatively mild.

Pathology

(Unknown)

Treatment

The stomach should be immediately evacuated. The remainder of the treatment is symptomatic. There are no known specific antidotes.

Prevention

Caution should be exercised in eating those species of reef fishes that have been incriminated as causative agents of ichthyoallyeinotoxism. When possible, local natives should be consulted before eating these fishes in tropical areas. It is advisable not to eat the head of a tropical reef fish. An hallucinogenic fish cannot be detected by its appearance.

PUBLIC HEALTH ASPECTS

Hallucinogenic fishes may be a public health problem in some tropical areas but there is no information available on this subject.

TOXICOLOGY

The only recorded attempt to test for ichthyoallyeinotoxin is by Helfrich and Banner (1960). Specimens of *Mugil cephalus, Mulloidichthys samoensis*, and *Upeneus arge* captured in Hawaii were fed to humans, cats, and mongooses but without any evidence of ill effects.

PHARMACOLOGY

(Unknown)

CHEMISTRY

(Unknown)

Ichthyohepatotoxic Fishes

The livers of certain species of fishes, generally considered safe to eat, are sometimes found to be toxic to humans. Most of the outbreaks of ichthyohepatotoxism have been reported from Japan. Very little is known regarding the incidence of the biotoxication, the species of fishes involved, or the precise nature of the poison. However, hypervitaminosis A is believed to be an important factor in the biotoxication.

The biology and systematics of the Japanese fishes incriminated as ichthyohepatotoxic have been discussed by Jordan, Tanaka, and Snyder (1913), Okada (1955), and Tomiyama, Abe, and Tokioka (1958).

LIST OF FISHES REPORTED AS ICHTHYOHEPATOTOXIC

Phylum CHORDATA

Class OSTEICHTHYES

Order PERCIFORMES (PERCOMORPHI): Perchlike Fishes

Family SCOMBRIDAE

Scomberomorus niphonius (Cuvier and Valenciennes) (Pl. 4, fig a). Japanese mackerel (USA), sawara (Japan). *210*

Distribution: Japan, Korea, northern China.

Source: Mizuta *et al.* (1957).

Family SERRANIDAE

Stereolepis ischinagi (Hilgendorf) (Pl. 4, fig. b). Seabass (USA), ishinagi, medai (Japan), karasu (Korea). *210*

Distribution: Japan, Korea.

Source: Abe and Kinumai (1957).

Family SPARIDAE

Petrus rupestris (Valenciennes) (Pl. 4, fig. c). Porgy (USA), steenbras (South Africa). *210*

Distribution: South Africa.

Source: Smith (1961).

Family TRICHODONTIDAE

210 *Arctoscopus japonicus* (Steindachner) (Pl. 4, fig. d). Japanese sandfish (USA), hatahata, kaminariuwo (Japan).

DISTRIBUTION: Japan, Korea, east coast of USSR, Alaska.

SOURCE: Mizuta *et al.* (1957).

BIOLOGY

Family SCOMBRIDAE: The sawara *Scomberomorus niphonius* is generally distributed from southern Hokkaido to Honshu, Shikoku, Kyushu, Korea, and northern China. It lives in water of 10° C to 20° C. This species has a wide vertical migratory range, living near the surface in the summer but moving down to deeper water in the winter. The spawning season is during April and May. For a general discussion on the biology of Scombridae, see p. 352.

Family SERRANIDAE: The ischinagi *Stereolepis ischinagi* inhabits the rocky bottoms at a depth of 500 m or more. The species is common in the waters of the middle and northern part of Honshu and is particularly abundant in Hokkaido. The fish come close to shore for spawning during May and June. For a general discussion on the biology of the Serranidae, see p. 352.

Family SPARIDAE: *Petrus rupestris* is found among reefs or rocks in deep water and occasionally enters deep estuaries. It is an excellent game fish and is capable of inflicting serious wounds with its powerful jaws and teeth. Large specimens may attack people in the water. It is an aggressive carnivore. This fish attains a length of about 1.7 m and a weight of more than 70 kg. For a general discussion on the biology of the Sparidae, see p. 353.

Family TRICHODONTIDAE: The Japanese sandfish *Arctoscopus japonicus* ranges along the two coasts of northern Honshu northward to Kamchatka and Alaska. They are particularly common in the Akita and Yamagata Prefectures. They are usually found on muddy or sandy bottoms at a depth of about 150 m, but by the end of November they begin to migrate to vegetated shore areas of about 1 m deep for the purpose of spawning, and after spawning they return to deeper water. They feed mainly on copepods. The sandfish is an important food fish in the Akita Prefecture.

MECHANISM OF INTOXICATION

Intoxication is caused by eating the liver of the fish. It is not known if fish liver poisoning is a seasonal phenomena. All of the reported Japanese outbreaks occurred during October and November. It is believed that the livers of the incriminated Japanese fishes are sometimes safe to eat, but there are no reliable data regarding this aspect of the problem. Ordinary cooking procedures apparently do not destroy the poison.

MEDICAL ASPECTS

Clinical Characteristics

Symptoms appear within 30 minutes to 12 hours, usually appearing in about 1 hour and attaining a maximum intensity within 7 hours after ingestion of the fish liver. The initial symptoms consist of nausea, vomiting, fever, and headache. The headache may be very severe and is said to be intensely aggravated by the slightest movement of the body, head, or eyes. A mild diarrhea may be present, but abdominal pain is generally absent. The face of the victim usually becomes flushed and edematous. A macular rash having large patchy erythematous areas develops shortly. Within 3 to 6 days desquamation appears. Large areas of skin may peel off around the nose, mouth, head, neck, and upper extremities and gradually extend over the entire body. Epilation may result. Desquamation may continue for about 30 days. Vesicular formation of the oral mucosa and bleeding from the lips have been reported. Orbital pain, joint aches, and palpitation with a rapid pulse rate may be present. Victims have complained of a slippery sensation of the tip of the tongue. Most of the more acute symptoms disappear in about 3 to 4 days. Residual symptoms consist of chapping of the lips, stomatitis, and a mild hepatic dysfunction. No other sequelae have been reported. Recovery is usually uneventful. No fatalities have been reported. The liver may be enlarged but no evidence of jaundice has been noted. Evidence of liver dysfunction may be present.

A summary of clinical findings have been written by Abe and Kinumai (1957) and Mizuta *et al.* (1957).

There is some clinical evidence that suggests excessive vitamin A intake may be a factor in ichthyohepatotoxism.

Pathology

(Unknown)

Treatment

Symptomatic.

Prevention

Care should be exercised in eating the liver of any fish. In general, the liver is one of the most dangerous parts of a fish to eat. If a fish is poisonous, a greater concentration of the poison is likely to be found in the liver than almost any other part of the fish. Cooking does not destroy the poison. Most outbreaks of ichthyohepatotoxism have resulted from eating fish livers that had been sautéed, or as part of a soup. A toxic liver cannot be determined by its appearance. It is recommended that fish livers be eliminated from the diet unless there is reliable information which indicates that they are safe to eat.

PUBLIC HEALTH ASPECTS

Insufficient data available.

TOXICOLOGY

(Unknown)

PHARMACOLOGY

(Unknown)

CHEMISTRY

There is insufficient data available at the present time to permit any firm conclusions regarding the chemical nature of ichthyohepatotoxins. Abe and Kinumai (1957) believe that the poisoning is due to ingestion of excessive amounts of vitamin A. However, Mizuta *et al.* (1957) believe the biotoxication is the result of a histaminelike substance. There is no experimental evidence to back either suggestion.

Ichthyosarcotoxism of Unknown Etiology

Periodically, reports have appeared in scientific literature of outbreaks of ichthyosarcotoxism of unknown etiology in man and animals. Very little is known regarding the clinical characteristics of these outbreaks, and the poisons have not been studied. *Myoxocephalus (Cottus) scorpius* Linnaeus (Pl. 5, fig. a), which inhabits the North Atlantic Ocean, has been reported by Bøje (1939) to have caused intoxication in dogs feeding on this fish. This same author has reported similar intoxications from eating the Greenland halibut *Reinhardtius hippoglossoides* (Walbaum) (Pl. 5, fig. b), which is found in the Arctic Atlantic Ocean. Various other temperate zone fishes have been incriminated in the Mediterranean Sea, Europe, and North America, but most of the offending species have been listed with a query in Chapter X on ciguatoxic fishes.

211

211

Minamata Disease

The term "Minamata disease" is derived from a series of outbreaks of neurological disorders resulting from the ingestion of shellfish and fish taken in Minamata Bay, Kyushu, Japan. The disease was first reported in 1953 and was caused by the ingestion of marine organisms; such organisms served as transvectors of methylmercury derived from industrial wastes that were being dumped in Minamata Bay. A second outbreak occurred during 1964-65 at the mouth of the Agano River, Niigata, Japan, involving reportedly 120 persons, of which 6 died. The actual number of persons in Japan involved with Minamata disease is presently not known, since many of these are asymptomatic, but there have been estimates as high as 100,000 persons.

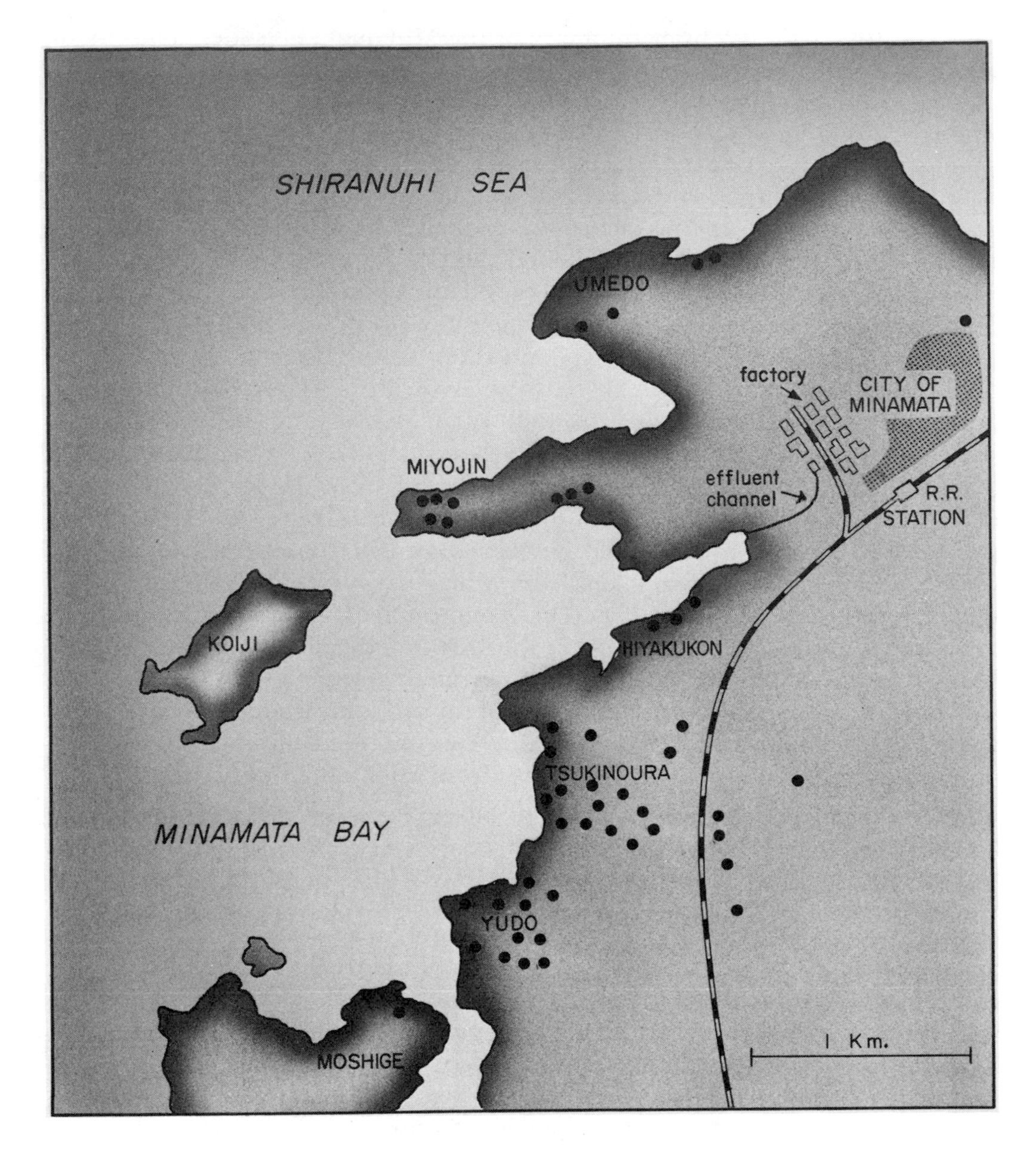

Figure 1.—Map showing the distribution of the first 52 cases of Minamata disease (1953-56). During 1957-60 there were an additional 31 cases. Some of the later outbreaks occurred in villages north and south of the areas shown on this map. The original 52 victims were studied quite intensively, and by 1960 it was found that only 3 were able to return to work within 6 months; 9 were "improved"; 23 were listed as stationary and disabled; 17 died, giving a 33 percent case fatality.

(R. Kreuzinger, after Kurland, Faro, and Siedler)

The following is a brief summary of the Minamata episode. The city of Minamata is located on the southwest coast of Kyushu, a few miles inland from the Shiranuhi Sea (Fig. 1). It has a population of about 40,000 people, most of whom are dependent upon local industries for their employment. A large chemical and fertilizer plant which had been producing various chemical products in the region for several decades began the production of vinyl chloride in 1949. The effluent from this factory was chanelled into the Shiranuhi Sea until 1950, when a new channel was opened directly into nearby Minamata Bay. The bay is small and is not generally used for commercial fishing, but was utilized by many of the local families as a source of seafood. With the discharge of the factory effluent into the bay and the resulting toxicity of the marine organisms, a series of outbreaks ensued involving a reported total of 121 persons during the years 1953 through 1970, of which 46 died. In each instance the victims had eaten crabs, shellfish, or fish taken from Minamata Bay that had been contaminated by the toxic effluent from the factory. Furthermore, it was observed that cats and fish-eating birds were also affected. It was found that the disease could be reproduced in laboratory animals that were fed on fish or shellfish taken from Minamata Bay. The causative agent was believed to be an organic mercury compound. There was a correlation between the quantity of mercury in the fish tested and the amount of fish required to produce the disorder in experimental animals. Unusually large amounts of mercury were detected in the mud from Minamata Bay (up to 2010 ppm). Shellfish collected from the bay contained significant quantities of mercury (more than 100 ppm). Water from Minamata Bay was abnormally high in mercury (1.6 to 3.6 ppb). The organs of humans and cats dying spontaneously of Minamata disease and of cats fed shellfish experimentally showed at autopsy elevated values in mercury content. Field investigations revealed that the mercury probably came from the effluent of the local chemical factory since large quantities of mercuric chloride were utilized as a catalytic agent in the production of vinyl chloride. Since the mercury compound in the seafood could not be extracted with organic acid or basic solvents, it was thought that a simple alkyl mercury compound may have combined to form a protein complex which was released during the process of digestion. It was later suggested that the toxic agent was an organic mercurial in combination with vinyl having the possible formula of $(CH_2\text{-}CH)_2 \cdot 3HgCl_2$.

It is now known that anaerobic bacteria *(Methanobacterium omelianskii)* present in detritus and sediments on the bottom of the ocean, bays, lakes, or rivers, are able to convert any form of mercury into methyl- and dimethylmercury. The aquatic food chain is the primary mechanism by which mercury is concentrated. At each trophic level less mercury is excreted than ingested. Consequently there is proportionately more mercury in algae than in the surrounding water, more still in fish that feed on algae, and so on. Regardless of the mercurial pollutant involved at the beginning of the conversion process, methylmercury $(CH_3\text{-}Hg^{+})$ is the principal toxic form found in shellfish and fish (Wallace *et al.*, 1971). Normal background levels of mercury from supposedly uncontaminated sources have been recorded as follows: Air—0.003 to 0.009 ppb; soil—0.01 to 15.0 ppm; sea water—0.03 to 2.0 ppb; human tissues (i.e., liver, kidney, etc.)—0.023 to 2.75 ppm; whole blood—0 to 0.05 ppm; hair—0.2 to 6.0 ppm; marine plants—0.023 to 0.037 ppm; marine fish—usually below 0.10 to 0.15 ppm (Wallace *et al.*, 1971).

Minamata disease is a severe illness characterized by widespread involvement of the central nervous system. There is cellular degeneration, particularly of the granular layer of the cerebellum, and lesser involvement of the basal ganglia, hypothalmus, mid-brain, and cerebral cortex. The onset of the disease may be either acute or subacute, beginning with a progressive numbness of the distal parts of the extremities, or the lips and tongue. This is followed by an ataxic gait, loss of motor coordination of the hands, dysarthria, dysphagia, deafness, and blurring of vision, associated with constriction of the visual fields. Voluntary movements were limited in most victims examined, but muscle atrophy was rarely apparent. Spasticity and rigidity were often present (Fig. 2). Muscle stretch reflexes were usually preserved or became hyperactive, and extensor plantar responses were occasionally elicited during later stages. Hypoesthesia was sometimes present.

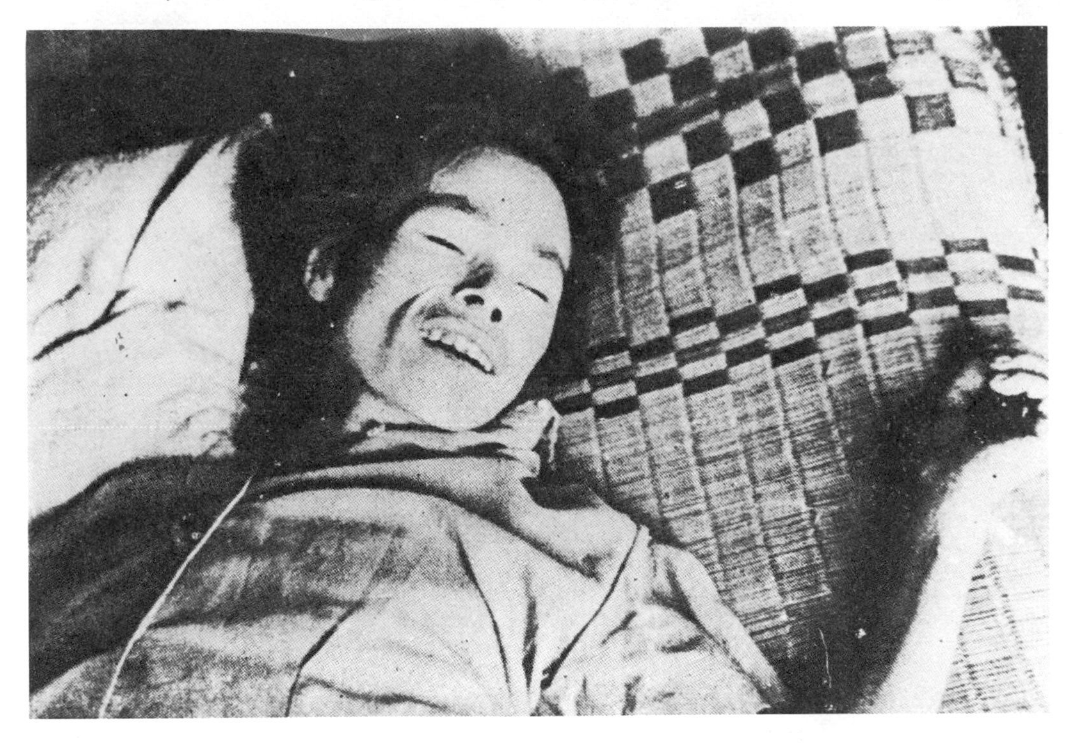

Figure 2.— Victim of Minamata disease showing the "childish" facial appearance which is one of the more obvious clinical signs of this intoxication.

(From Katsuki *et al.*, 1958)

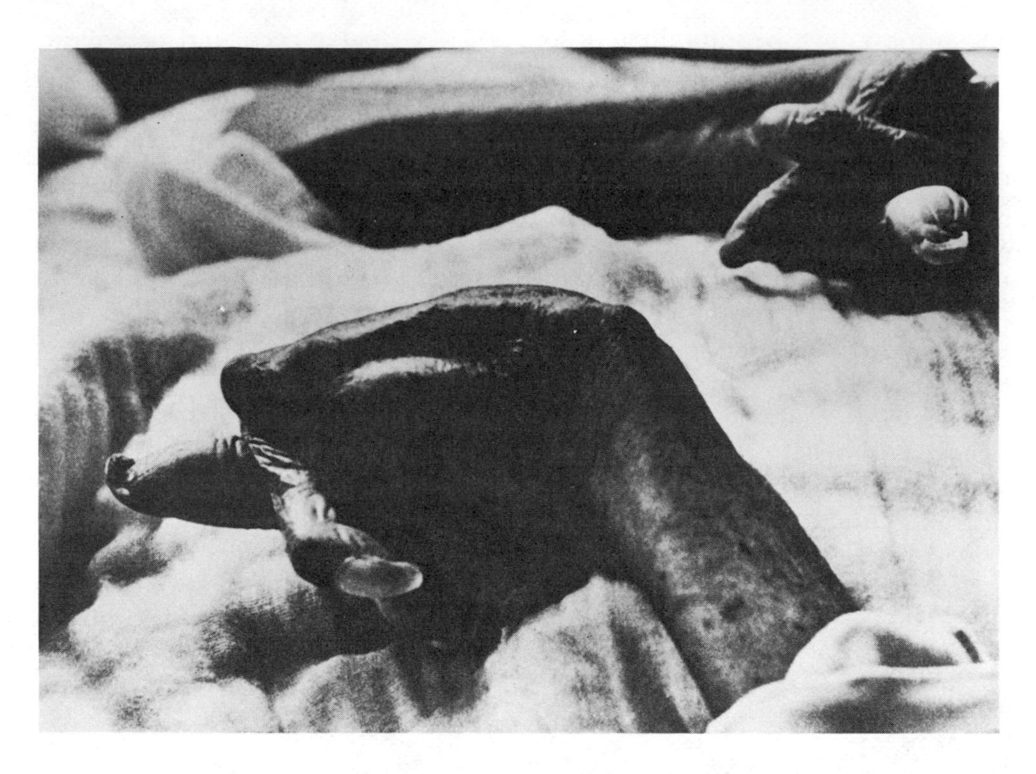

FIGURE 3.—Hands of a Minamata victim. The muscles of the upper and lower extremities become atrophied and finally fixed to such an extent that they become useless appendages. (Photo by W. Eugene Smith, courtesy of H. Inoue)

Insomnia, agitation, hypomania, and loss of emotional control were frequently observed. Stupor and coma were present in some patients. Generalized convulsions occurred in only a few cases. In severe cases generalized muscle wasting developed. Most of the victims had abnormal involuntary movements, including choreoathetosis, myoclonus, coarse resting tremors, and action tremors. Intellectual impairment was sometimes noted. Death was usually due to intercurrent infection, aspiration pneumonia, or inanition. The duration of the disease in fatal cases ranged from 26 days to 4 years.

The treatment of methylmercury poisoning is essentially symptomatic. BAL (British anti-lewisite, dimercaprol), the classical antidote for other forms of mercury poisoning, has only limited usefulness in Minamata disease, and may even aggravate the induced nervous system symptoms. Recent clinical studies have suggested using EDTA and N-acetyl-DL-penicillamine as useful ligands in chelating out methylmercury (Hojo, Sugiura, and Tarraka, 1975; Aaseth, 1975, 1976; Aaseth, Wannag, and Norseth, 1976).

One of the more serious aspects of Minamata disease are the congenital effects. Infants born to mothers exposed to methylmercury develop mental retardation and cerebral palsy with convulsions. These effects have been found in infants born of asymptomatic mothers. Thus, alkylmercurials are able to pass both the blood, brain, and placental barriers. Fetal intoxication is possible for as long as 3-4 years (Wallace *et al.*, 1971; Berglund *et al.*, 1971).

In brief, the alkylmercurials are especially hazardous because of their extreme toxicity, their ability to penetrate epithelial barriers, and their predilection for the central nervous system. Methylmercury, unlike inorganic mercurial compounds, is difficult to detect and a long-lasting poison. Methylmercury is bound by hemoglobin in red blood cells and circulates in this form for months, being excreted at the rate of about 1 percent per day, mostly in the feces. The biological half-life of methylmercury in man is 70-74 days and 200 days in fish. It passes readily into the central nervous system where it selectively and irreversibly damages the cells of the granular layer in the cerebellum, basal ganglia, cerebral cortex, hypothalmus, and the calcarine cortex. Methylmercury quickly passes the placental barrier and accumulates in the fetal brain and blood, building up to 30 percent higher erythrocyte levels than in the mother. It is especially injurious to the central nervous system of infants and children. The symptoms of methylmercury poisoning appear several weeks or months, and possibly years, after ingestion or absorption of a toxic dose (Kurland, Faro, and Siedler, 1960; Eyl, 1971).

The mercury content in the organs of some of the first Minamata victims that died was as high as 71 ppm in their livers, 144 ppm in their kidneys, 21 ppm in their brains, and 70 ppm in their hair (Kurland, Faro, and Siedler, 1960). Blood levels for mercury content, unfortunately, were not determined in the victims during the initial Minamata epidemic, but blood studies were conducted in the Niigata outbreaks. It was estimated that the methylmercury levels in the fish and shellfish eaten by the victims was from 10 to 20 ppm. An important element in the clinical evaluation of mercury poisoning is the determination of the level of methylmercury in the body in relationship to the clinical effects produced. Studies thus far have indicated that the levels in the blood and in the hair may be the best indices available in determining the extent of the intoxication and central nervous system involvement. Japanese studies indicated that toxic symptoms were present when blood levels reached 0.5 to 1 ppm of Hg in whole blood. In the Niigata cases the lowest blood level at which symptoms occurred was of the order of 0.2 ppm. Poisoning was generally evident when mercury levels in the hair were about 200 ppm, although in some cases the levels may have been considerably lower (Berglund *et al.*, 1971; HEW, 1971).

In 1959 legal action was taken against the Shin Nihon Chisso Company, whose chemical factory in Minamata was allegedly responsible for the mercury outbreaks. The victims' relatives were reported to have received compensation at the rate of $800 for each death, and for those permanently crippled, $280 for each adult per year, and $80 for each child per year (Ui, 1972*b*). According to recent reports (1973) the battle for adequate compensation for the Minamata victims still continues.

The toxic effects of mercury have been under consideration by the Joint Food and Agriculture Organization of the United Nations/World Health

Organization Codex Alimentarious Commission in order to establish an *acceptable daily intake* (ADI) for the international health community. After reviewing an enormous amount of data on the toxicity of mercury and other heavy metals the Joint FAO/WHO Expert Committee on Food Additives (FAO/WHO, 1972; cited by Lu, 1973, 1974) concluded that it was inappropriate for them to establish ADI's for heavy metals such as mercury, lead, and cadmium. The reader is referred to the original report for the reasons for their conclusions. However, the Committee did recommend the establishment of a *provisional tolerable weekly intake* of 0.3 mg of total mercury per person, of which no more than 0.2 mg should be present as methylmercury, CH_3Hg^+ (expressed as mercury). These amounts are equivalent to 0.005 mg and 0.0033 mg respectively, per kg of body weight.

The recommended limits of fisheries products contaminated with methylmercury for human consumption in the United States, Canada, Australia, and a number of other countries is 0.5 ppm. Sweden and Japan have established their limits as 1.0 ppm, a figure which is considered to be too high by many authorities.

The Minamata outbreaks should be of great concern to agencies dealing with the relationship of water pollution, marine organisms, and the resulting public health implications. Greater attention should be focused on those environmental factors influencing the ecology of marine life which may have been responsible for specific conditions conducive to the development of a highly specific intoxicating agent. Moreover, it should be noted that contaminants were picked up by marine organisms from the water and benthos, transvectored without apparent harm to themselves, but were lethal to the humans who ate them.

For further information on Minamata disease the reader is referred to the writings of the following authors: Kitamura *et al.* (1957*a, b*), McAlpine and Araki (1958), Katsuki *et al.,* (1958), Anon. (1959), Takeuchi *et al.* (1959, 1962), Kurland, Faro, and Siedler (1960), Kiyoura (1960), Tokuomi *et al.* (1961*a, b*), Takeuchi (1961), Hirakawa, Inoue, and Uchida (1961), Uchida, Hirakawa, and Inoue (1961*a, b*), Irukayama *et al.* (1962), Kawabata (1962), and Halstead (1970).

A number of excellent technical reviews on mercury pollution in the environment have recently been published, viz., Miller and Berg (1969), Larsson (1970), Nelson *et al.* (1970), Pecora (1970), Goldwater (1971), HEW (1971), Montague and Montague (1971), Wallace *et al.* (1971), D'Itri (1972), Lambou (1972), and Ruivo (1972). Particular attention is called to the "red book" on *Minamata Disease* by the Study Group on Minamata Disease, Kumamota University, by Kutsuna *et al.,* (1968). This is by far the most comprehensive analysis of the Minamata outbreak. Another outstanding report on the mercury problem is by the Swedish Expert Group entitled *Methylmercury in Fish*, by Berglund *et al.* (1971). The tragedy of the Minamata victims has been graphically depicted by Kuwabara (1970) in his excellent 222-page pictorial work entitled *Minamata Disease*, published in Japanese with an English summary. Attention is also directed to the untiring efforts of Jun Ui (1972*a, b*) of Tokyo University and his associates of the Jishu-Citizen's Movement in attempting to curb the inroads of pollution in Japan. Excellent and comprehensive bibliographies on mercury contamination have been prepared by Dow Chemical Company, 3 volumes (1970), and by Jardin (1972) of the Food and Agriculture Organization. A superb pictorial publication dealing with the devastating results of this intoxication, entitled *Minamata*, has been published by Smith and Smith (1975). Anyone concerned with the social implications of this disease should not fail to see this moving pictorial account.

LITERATURE CITED

AASETH, J.

1975 The effect of N-acetyl-homocysteine and its thiolactone on the distribution and excretion of mercury in methyl mercuric chloride injected mice. Acta Pharmacol. Toxicol. 36: 193-202.

1976 Mobilization of methylmercury *in vivo* and *in vitro* using N-acetyl-DL-penicillamine and other complexing agents. Acta Pharmacol. Toxicol. 39: 289-301.

AASETH, J., WANNAG, A., and T. NORSETH

1976 The effect of N-acetylated DL-penicillamine and DL-homocysteine thiolactone on the mercury distribution in adult rats, rat foetuses and maccoca monkeys after exposure to methyl mercuric chloride. Acta Pharmacol. Toxicol. 39: 302-311.

ABE, H., and M. KINUMAI

1957 Food poisoning caused by eating the livers of *Stereolepis ichinagi.* [In Japanese] Nichi Ju Kaichi 10: 125-127.

ANONYMOUS

1959 A view on the theory of organic mercury as the causal substance of "Minamata disease." (Unpublished data.)

BERGLUND, F., *et al.*

1971 Methyl mercury in fish: a toxicologic-epidemiologic evaluation of risks. Report from an expert group. Nat. Inst. Publ. Health, Stockholm, Sweden. 364 p.

BØJE, O.

1939 Toxin in the flesh of the Greenland shark. Meddel. Grønland 125(5): 1-16.

BOUDER, H., CAVALLO, A., and M. J. BOUDER

1962 Poissons vénéneux et ichtysarcotoxisme. Bull. Inst. Océanogr. 59 (1240), 66 p., 2 figs.

COOPER, M. J.

1964 Ciguatera and other marine poisoning in the Gilbert Islands. Pacific Sci. 18(4): 411-440, 11 figs.

D'ITRI, F. M.

1972 The environmental mercury problem. CRC Press, Cleveland, Ohio. 136 p.

DOW CHEMICAL COMPANY

1970 Environmental aspects of mercury usage. 3 vols. Dow Chemical Co., Midland, Michigan.

EYL, T. B.

1971 Methyl mercury poisoning in fish and human beings. Clin. Toxicol. 4(2): 291-296.

FRIBERG, L. T.

1972 Mercury in the environment. CRC Press, Cleveland, Ohio. 300 p.

GOLDWATER, L. J.

1971 Mercury in the environment. Sci. Am. 224(5): 15-21, 8 figs.

HALSTEAD, B. W.

1970 Toxicity of marine organisms caused by pollutants. FAO Technical Conference on marine pollution and its effects on living resources and fishing. 9-18 December 1970. FIR: MP/70/R-6 (WM/A8087): 1-21. Also in Ruivo (1972).

HALSTEAD, B. W., and K. W. COX

1973 An investigation on fish poisoning in Mauritius. Proc. Roy. Soc. Arts Sci. Mauritius 4(2): 1-26, 2 figs.

HARTUNG, R., and B. D. DINMAN

1974 Environmental mercury contamination. Ann Arbor Science Publ., Ann Arbor, 345 p.

HELFRICH, P.

1961 Fish poisoning in the tropical Pacific. Hawaii Marine Lab., Univ. Hawaii. 16 p., 7 figs.

1963 Fish poisoning in Hawaii. Hawaii Med. J. 22: 361-372.

HELFRICH, P., and A. H. BANNER

1960 Hallucinatory mullet poisoning. J. Trop. Med. Hyg. 1960: 1-4.

H.E.W.

1971 A report on mercury hazards in the environment. Mercurial Pesticide Registration Review Panel, U.S. Govt. Dept. Health, Educ. & Welfare, Wash., D.C. 85 p.

HIRAKAWA, K., INOUE, T., and M. UCHIDA

1961 The non-volatile mercury compound in the toxic shellfish *Hormomya mutabilis*. Kumamoto Med. J. 14(4): 192.

HOJO, S., SUGIURA, Y., and H. TARRAKA

1975 Specific ability of sulfur-ligands to remove mercury-203-labeled organomercury from hemoglobin in comparison with nitrogen-ligands. Fac. Pharm. Sci. (Kyoto) 24(10): 684-688.

IRUKAYAMA, K., KONDO, T., FUJIKI, M., and F. KAI
1962 Report II. Comparison of the mercury compound in the shellfish from Minamata Bay with various mercury compounds experimentally accumulated in the control shellfish. [In Japanese, English summary] Japan. J. Hyg. 16: 467-475.

JARDIN, C.
1972 Mercury in the environment and its related toxicological aspects—a selected bibliography. FAO, Rome (Microfiche), No. 16221.

JONES, H. R.
1971 Mercury pollution control. Noyes Data Corp., Park Ridge, N. J. 249 p. (Pollution Control Review No. 1)

JORDAN, D. S. EVERMANN, B. W., and S. TANAKA
1927 Notes on new or rare fishes from Hawaii. Proc. Calif. Acad. Sci., 4, 16(20): 674.

JORDAN, D. S., TANAKA, S., and J. O. SNYDER
1913 A catalogue of the fishes of Japan. J. Coll. Sci., Tokyo Univ. 33. 497 p., 396 figs.

KATSUKI, S., FURUKAWA, K., ISHIDA, S., and N. TANAKA
1958 A study on metabolism in the so-called Minamata disease with special reference to metal metabolism. Kyushu J. Med. Sci. 9: 102: 109, 1 fig.

KAWABATA, T.
1962 Fish-borne food poisoning in Japan, p. 467-479. Vol. II. Nutrition, sanitation, and utilization. Academic Press, New York.

KITAMURA, S.,, MIYATA, C., DATE, S., MISUMI, H., MINAMOTO, H., NOGUCHI, Y., TOMITA, M., UEDA, K., KOJIMA, T., KURIMOTO, S., and R. NAKAGAWA
1957*a* Epidemiological investigation of the unknown central nervous disorder in the Minamata district. [In Japanese] Kuamoto Med. J. 31: 1-61.

KITAMURA, S., MIYATA, C., DATE, S., MISUMI, H., HONDA, S., TOMITA, M., UEDA, K., KOJIMA, T., and K. FUKAGAWA
1957*b* Epidemiological investigation of the unknown central nervous disorder in the Minamata district. [In Japanese] Kuamoto Med. J. 31: 238-341.

KIYOURA, R.
1960 Study on the highly toxic substances extracted from the fishes in the Minamata Bay. (Unpublished data.)

KURLAND, L., FARO, S., and H. SIEDLER
1960 Minamata disease. World Neurol. 1(5): 370-395, 9 figs.

KUTSUNA, M., *et al.*
1968 Minamata disease. Study group of Minamata disease. Kumamoto Univ., Kumamoto. 338 p., illus.

KUWABARA, S.
1970 Minamata disease 1960-1970. Tokyo, Japan. 222 p.

LAMBOU, V. W.
1972 Report on the problem of mercury emissions into the environment of the United States. U.S. Environmental Protection Agency, Wash., D.C. (Unpublished.)

LARSSON, J. E.
1970 Environmental mercury research in Sweden. Swedish Environmental Protection Board, Stockholm. 71 p., 3 figs.

LU, F. C.
1973 Toxicological evaluation of food additives and pesticide residues: The role of WHO, in conjunction with FAO. WHO Chron. 27(2): 43-48.
1974 Mercury as a food contaminant. WHO Chron. 28(1): 8-11.

MCALPINE, D., and S. ARAKI
1958 Minamata disease. An unusual neurological disorder caused by contaminated fish. Lancet 19 8: 629-631.

MILLER, M. W., and G. G. BERG, *eds.*
1969 Chemical fallout; current research on persistent pesticides. C. C. Thomas, Publisher, Springfield, Illinois. 531 p.

MIZUTA, M., ITA, T., MURAKAMI, T., and M. MIZOBE
1957 Mass poisoning from the liver of sawara and iwashikujira. [In Japanese] Nihon Iji Shimpo (1710): 27-34.

MONTAGUE, K., and P. MONTAGUE
1971 Mercury. Sierra Club, San Francisco, Calif. 158 p.

NELSON, N.
1970 Hazards of mercury. Study Group on mercury hazards, U.S. Govt. Dept. Health, Educ. & Welfare, Wash., D.C., 97 p.

OKADA, Y.
1955 Fishes of Japan. Maruzen Co., Tokyo. 434 p., 391 figs.

PECORA, W. T., *director*
1970 Mercury in the environment. U.S. Govt. Printing Off., Wash., D.C. 67 p. (U.S. Geological Survey Professional Paper 713.)

RANDALL, J. E.
1958 A review of ciguatera, tropical fish poisoning, with a tentative explanation of its cause. Bull. Marine Sci. Gulf Caribbean 8(3): 236-267, 2 figs.

RUIVO, M., *ed.*
1972 Marine pollution and sea life. Food and Agriculture Organization of the United Nations and Fishing News (Books) Ltd., London, England. 627 p.

SMITH J. L.
1961 The sea fishes of southern Africa. 4th ed. Central News Agency, South Africa. 580 p., 1,232 figs., 102 pls.

SMITH, W. E., and A. M. SMITH
1975 Minamata. Holt, Rinehart and Winston, New York. 192 p., illus.

TAKEUCHI, T.
1961 A pathological study of Minamata disease in Japan. Symp. Geogr. Neurol., VII Intern. Congr. Neurol. Rome. 24 p.

TAKEUCHI, T., KAMBARA, T., MORIKAWA, N., MATSUMOTO, H., SHIRAISHI, Y., and H. ITO
1959 Pathological observations of the Minamata disease. Acta Pathol. Jap. 9: 769-783. (NSA)

TAKEUCHI, T., MORIKAWA, N., MATSUMOTO, H., and Y. SHIRAISHI
1962 A pathological study of Minamata disease in Japan. Acta Neuropathol. 2: 40-57, 17 figs.

(NSA)—Not seen by author.

TITCOMB, N., and M. K. PUKUI
1952 Native use of fish in Hawaii. Suppl. J. Polynesian Soc., Mem. 29. 162 p., 120 figs.

TOKUOMI, H., OKAJIMA, T., KANAI, J., TSUNODA, M., ICHIYASU, Y., MISUMI, H., SHIMOMURA, K., and M. TAKABA
1961*a* Minamata disease. World Neurol. 2(6): 536-545, 6 figs.
1961*b* Minamata disease—an unusual neurological disorder occurring in Minamata, Japan. Kumamoto Med. J. 14(2): 47-64, 15 figs.

TOMIYAMA, I., ABE, T., and T. TOKIOKA
1958 Encyclopaedia zoologica, illustrated in colours. Vol. II. [In Japanese] Hokuryn-kan Publ. Co., Ltd., Tokyo. 392 p., 133 figs.

UCHIDA, M., HIRAKAWA, K., and T. INOUE
1961*a* Biochemical studies on Minamata disease. III. Relationships between the causal agent of the disease and the mercury compound in the shellfish with reference to their chemical behaviors. Kumamoto Med. J. 14(4): 171-179, 2 figs.
1961*b* Biochemical studies on Minamata disease. IV. Isolation and chemical identification of the mercury compound in the toxic shellfish with special reference to the causal agent of the disease. Kumamoto Med. J. 14(4): 181-187, 2 figs.

UI, J., *ed.*
1972*a* Polluted Japan. Tishu-Koza, Tokyo, Japan. 78 p., illus.

UI, J.
1972*b* The singularities of Japanese pollution. Jap. Quart. 19(3): (11p.)

WALLACE, R. A., FULKERSON, W., SHULTS W. D., and W. S. LYON
1971 Mercury in the environment: the human element. Oak Ridge National Laboratory, Oak Ridge, Tennessee, 61 p.

WHEELER, J. F.
1953 The problem of poisonous fishes. Mauritius-Seychelles Fish. Survey, Fish Publ. 1(3): 44-48.

Chapter XVIII—VERTEBRATES

Class Chondrichthyes

Venomous: SHARKS, RAYS, CHIMAERAS

The class Chondrichthyes comprises the sharks, rays, skates, chimaeras, and their relatives. The members of this group are characterized by their highly developed jaws and fins, and by a general body structure developed along the familiar piscine plan. The skeleton is cartilaginous. Fertilization in most species is of the internal type.

The class CHONDRICHTHYES, according to Lagler, Bardach, and Miller (1962), is divided into the following major groups:

Subclass ELASMOBRANCHII: Sharks, skates, and rays. This large subclass is further divided into the orders Squaliformes (or Pleurotremata) which includes the various families of sharks; and the Rajiformes (Hypotremata or Batoidei) which includes the sawfishes, skates, rays, mantas, and their relatives.

Subclass HOLOCEPHALI: Chimaeras. This subclass contains the order Chimaeriformes and a single family.

Chondrichthyes includes both poisonous and venomous fish species, but only the ones that possess a venom apparatus will be discussed in this chapter.

Venomous Sharks

Sharks are a group of fishes characterized by a cartilaginous skeleton, primitive ribs, and a complete braincase or chondrocranium. The gill openings are elongate and lateral or partially so; the gill pouches are supported by cartilaginous arches. Median and paired fins are present and supported by cartilaginous rods at their bases from which radiate unsegmented horny rays. The paired fins are attached to simple cartilaginous girdles. Portions of the pelvic fins in the males are modified to form a copulatory intromittent organ. The nostrils are paired and ventral in location. The mouth is ventral and equipped with well-developed jaws, the upper part of which is not fused with the cranium.

Venomous sharks are limited to those species possessing dorsal fin spines, namely, certain members of the families Heterodontidae, Squalidae, and Dalatiidae. Within these families there are at least 11 genera known to possess dorsal spines. However, only the two species *Heterodontus francisci* and *Squalus acanthias* are known to have venomous spines. The remaining species, according to the literature, have not been studied by venomologists. Some of the dalatiid sharks have only a single rudimentary fin spine, or the spines are entirely absent. Whether or not the dorsal fin spine of dalatiid sharks is an actual venom apparatus is not known.

The biology and systematics of the sharks listed in this section have been discussed by Günther (1859-70), Jordan and Evermann (1896-1900), Garman (1913), Jordan, Tanaka, and Snyder (1913), Joubin (1929-38), Bigelow and Schroeder (1948), Gilbert (1963), Soljan (1963).

REPRESENTATIVE LIST OF SHARKS REPORTED AS VENOMOUS

Phylum CHORDATA

Class CHONDRICHTHYES

Order SQUALIFORMES: Sharks

Family HETERODONTIDAE

214 *Heterodontus francisci* (Girard) (Pl. 1, fig. a). Hornshark (USA).

Distribution: California coast from Point Conception to Lower California and into the Gulf of California.

Sources: Evans (1943), Halstead (1956, 1959), Halstead and Mitchell (1963).
Other name: *Gyropleurodus francisci.*

Family SQUALIDAE

214 *Squalus acanthias* Linnaeus (Pl. 1, fig. b). Spiny dogfish (USA, England), spinarolo (Italy), koliuciaia akula (USSR), dornhund (Germany), aguillat (France), galhudo (Portugal), ferrón (Spain), aburazuno (Japan), galludo, melga (Latin America).

Distribution: Atlantic and Pacific Oceans.

Sources: Coutière (1899), Evans (1920, 1923, 1943), Phisalix (1922), Gudger (1943), Halstead (1956, 1959), Halstead and Mitchell (1963).
Other names: *Acanthias americanus, Acanthias vulgaris, Spinax acanthias, Squalus suckleyi.*

BIOLOGY

Family HETERODONTIDAE: Heterodontids are Pacific sharks, ranging from north temperate to south temperate latitudes. *Heterodontus* is a common inshore species, particularly in areas where kelp is found. Horned sharks are thought to migrate from shallow to deeper waters at certain seasons of the year. They may be found at depths of 85 fathoms or more. Their food consists of mollusks, crabs, and other hardshelled invertebrates. They are oviparous, laying a horny spiral-shaped egg case.

Family SQUALIDAE: Squalids are widely distributed throughout subarctic, temperate, tropical, and subantarctic seas.

Most dogfish are somewhat sluggish in their movements and erratic in their migrations, traveling singly or in schools. Their bathymetric range extends from the surface to depths of 100 fathoms or more. They are not pelagic, preferring relatively shallow protected bays. The migration of squalids seems to be governed by thermal changes, preferring water temperatures from 7° to 15° C. Squalids are viviparous, giving birth to their young from late summer through the winter in some regions but earlier in others. Dogfish are voracious and include a variety of fishes in their diet: capelin, herring, menhaden, mackerel, hake, cod, haddock. They also feed on coelenterates, mollusks, crustaceans, and worms. Squalids have been used considerably as fertilizer and a source of vitamins A and D. Dogfish are economically important because of the damage they do to fishing gear.

MORPHOLOGY OF THE VENOM APPARATUS

The following material is taken from an unpublished thesis prepared by Don R. Goe (1950) in partial fulfillment for the degree of Master of Arts, Department of Biology, Walla Walla College. The histological research was done at the School of Tropical and Preventive Medicine, Loma Linda, Calif., under the direction of Bruce W. Halstead.

Materials and Methods: The specimen material described in this section was obtained from the Pacific dogfish *Squalus acanthias* from Ship Harbor, Fidalgo Island, Skagit County, Wash., during June through August 1949. The sharks were caught on set-lines baited with fish in water 2 to 6 fathoms deep over a sandy bottom sparsely covered with eel grass. The stings were immediately severed from the captured fish and preserved in 5 percent formalin. Decalcification of the stings was accomplished by the phloroglucin-nitric acid method according to Carleton (1926). The process was carried out under a vacuum of 500-600 mm of mercury, thereby reducing the necessary time from several days to less than an hour. Following decalcification, the stings were washed in running water for 12 hours, then put through an embedding process with low-viscosity nitrocellulose according to the technique of Koneff and Lyons (1937). The nitrocellulose blocks were hardened in successive changes of chloroform and kept in 80 percent ethyl alcohol. Sections were cut 15μ thick with a sliding microtome and stained with triosin and Harris' hematoxylin. Every fifth section was stained and mounted on a glass slide with permount.

Gross Anatomy

Squalus acanthias has a sharp strong spine immediately anterior to each dorsal fin (Figs. 1a, b; 2 a, b). The length of the exposed portion of the anterior spine is approximately one-third the height of the fin. The length of the exposed portion of the posterior spine is approximately one-half to two-thirds the height of the fin. The anterior spine is only slightly curved anteroposteriorly, while the posterior spine is more curved and in lateral view somewhat sabre-

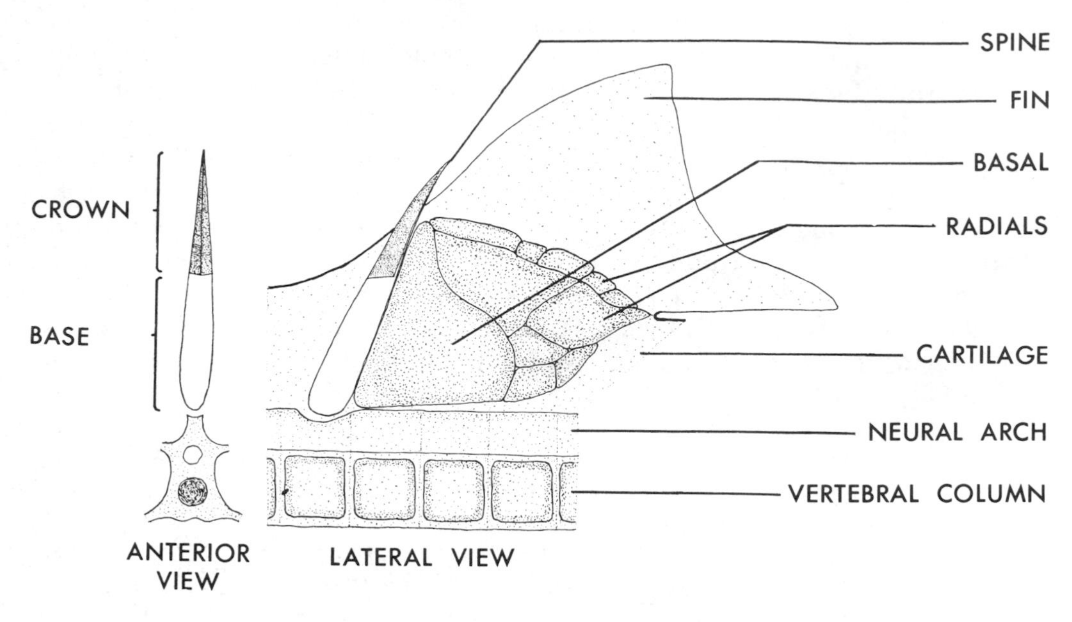

Figure 1a.—First dorsal fin and spine of *Squalus acanthias*. (Courtesy D. Goe)

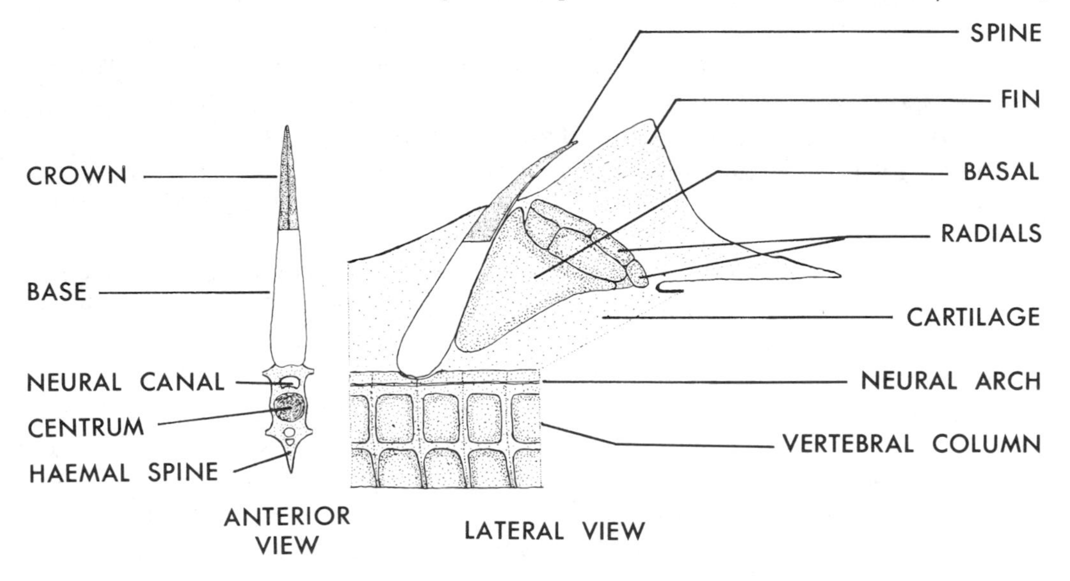

Figure 1b.—Second dorsal fin and spine of *Squalus acanthias*. (Courtesy D. Goe)

shaped. The spine itself is roughly triangular in cross section with the apex directed anteriorly. The two sides are slightly convex; the longitudinally grooved posterior aspect of the spine forms the base of the triangle.

The spine is grooved only in its exposed portion, the groove becoming shallower toward the tip. When the fresh spine is examined macroscopically, the shallow groove is seen to be occupied by a glistening white substance which extends for a variable distance toward the tip. The tissue of this grooved region contains the venom gland and is described in detail in the following section.

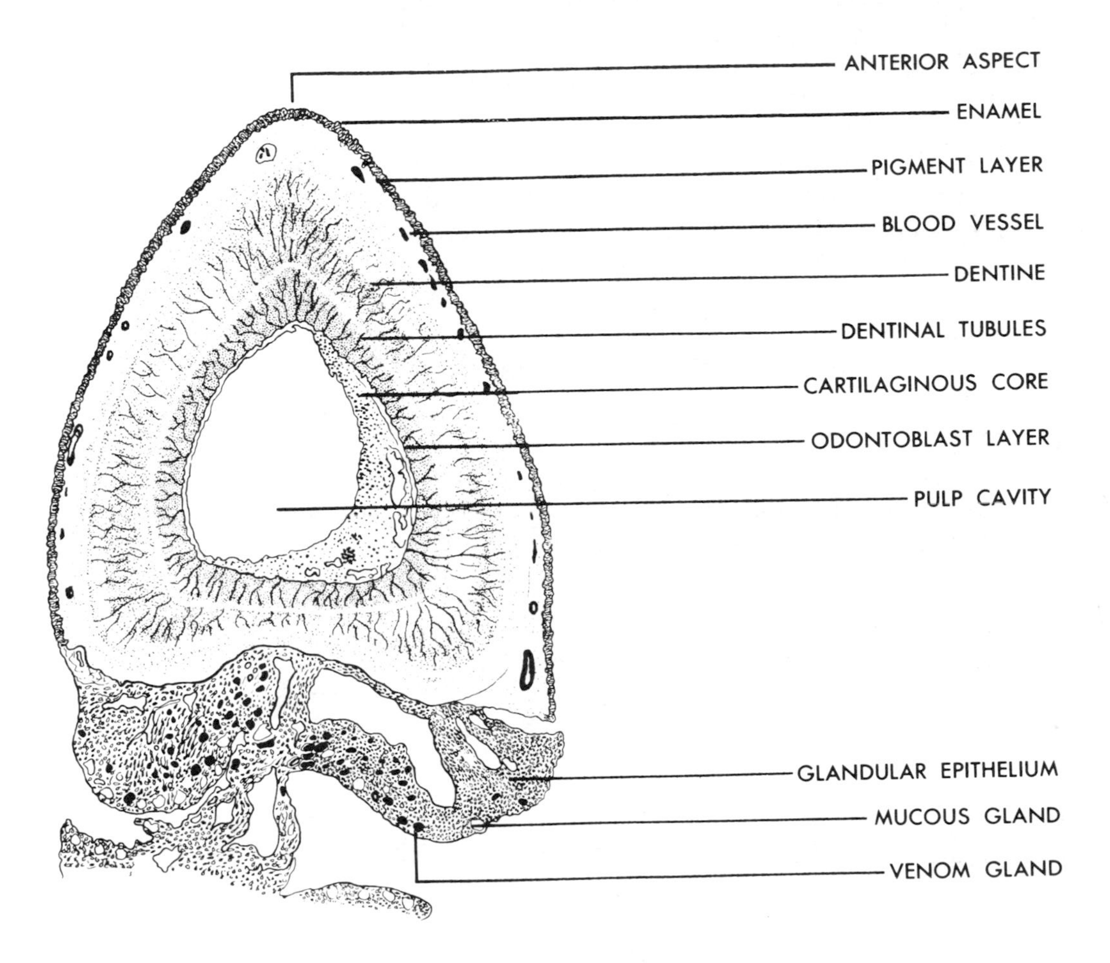

FIGURE 2a.—Drawing of cross section of the dorsal sting of *Squalus acanthias* at the level of the middle third. (Courtesy D. Goe)

The spine is buried in the integument for more than half its length. The embedded portion is designated as the root or base and the remainder the crown or spine proper. The crown is hollow, the hollow space being occupied by a cartilaginous ray which forms the basal portion of the spine and articulates with the vertebral column.

The walls of the spine are made of dentine which in the crown consists of a double layer. The surface of the exposed portion of the spine, except for the posterior grooved aspect, is covered with a layer of enamel. A more or less compact layer of pigment separates the enamel from the dentine (Daniel, 1934).

For approximately two-thirds of its total length, the spine is adjoined posteriorly to the basal cartilage. The articulating surface on the basal portion of the spine rests in a depression in the neurapophysis of the vertebra. A close-fitting capsule of fibrous connective tissue binds the articulating cartilage to the vertebral column and to the supporting basal cartilage immediately posterior to the spine.

Microscopic Anatomy

The spine in cross section in its basal third is pointed at the anterior margin and rounded at the posterior margin. The anterior half is covered with a thin layer of collagenous connective tissue. Under the collagen fibers is a narrow layer of dentine bordered internally by another band of collagenous connective tissue. This layer of connective tissue surrounds a central core of relatively undifferentiated hyaline cartilage. Around the periphery of this central core may be seen a narrow chain of islets of dentine. Scattered about this narrow band of dentine are numerous odontoblasts.

At the level of the middle third, the spine in cross section is trigonal, the anterior aspect forming the apex (Figs. 2a; 3; 4). Small rounded projections of the wall occur at either posterolateral angle of the spine. When observed under low magnification, the walls of the spine are seen to be composed externally of a layer of enamel roughly one-tenth the width of the wall. This enamel covers only the anterior and lateral portions of the distal half of

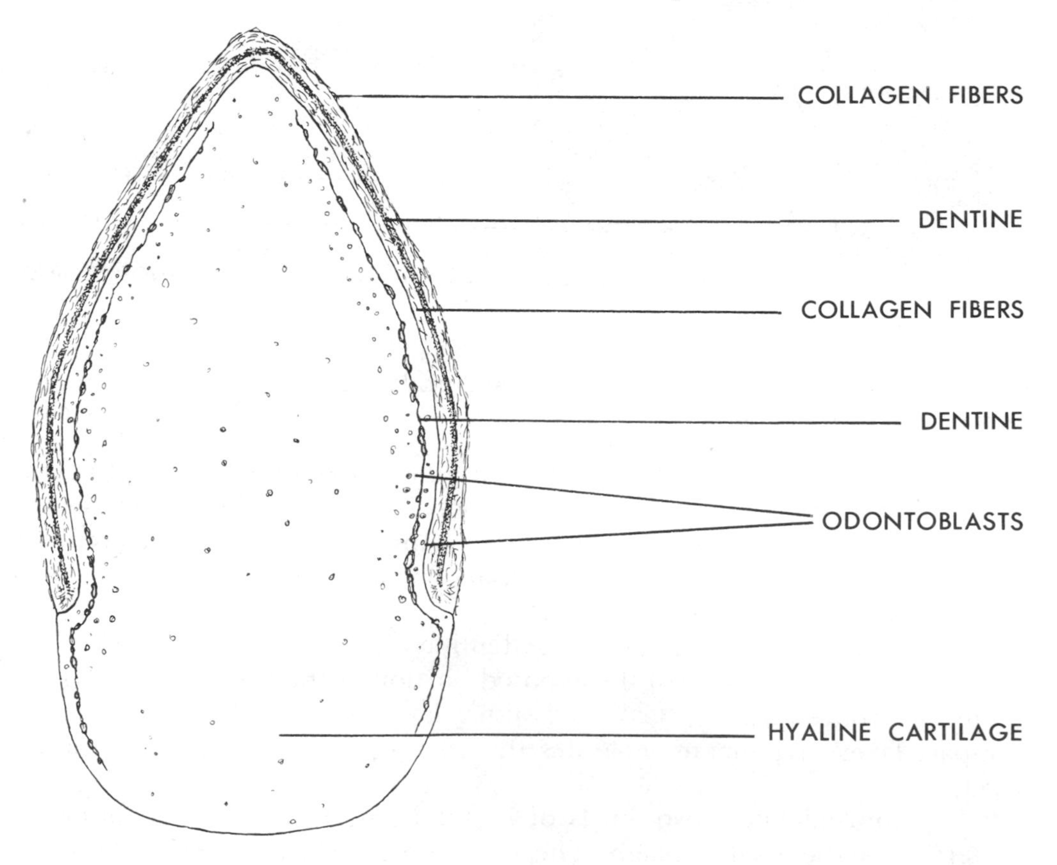

FIGURE 2b.— Drawing of cross section of the dorsal sting of *Squalus acanthias* at the level of the basal third. (Courtesy D. Goe)

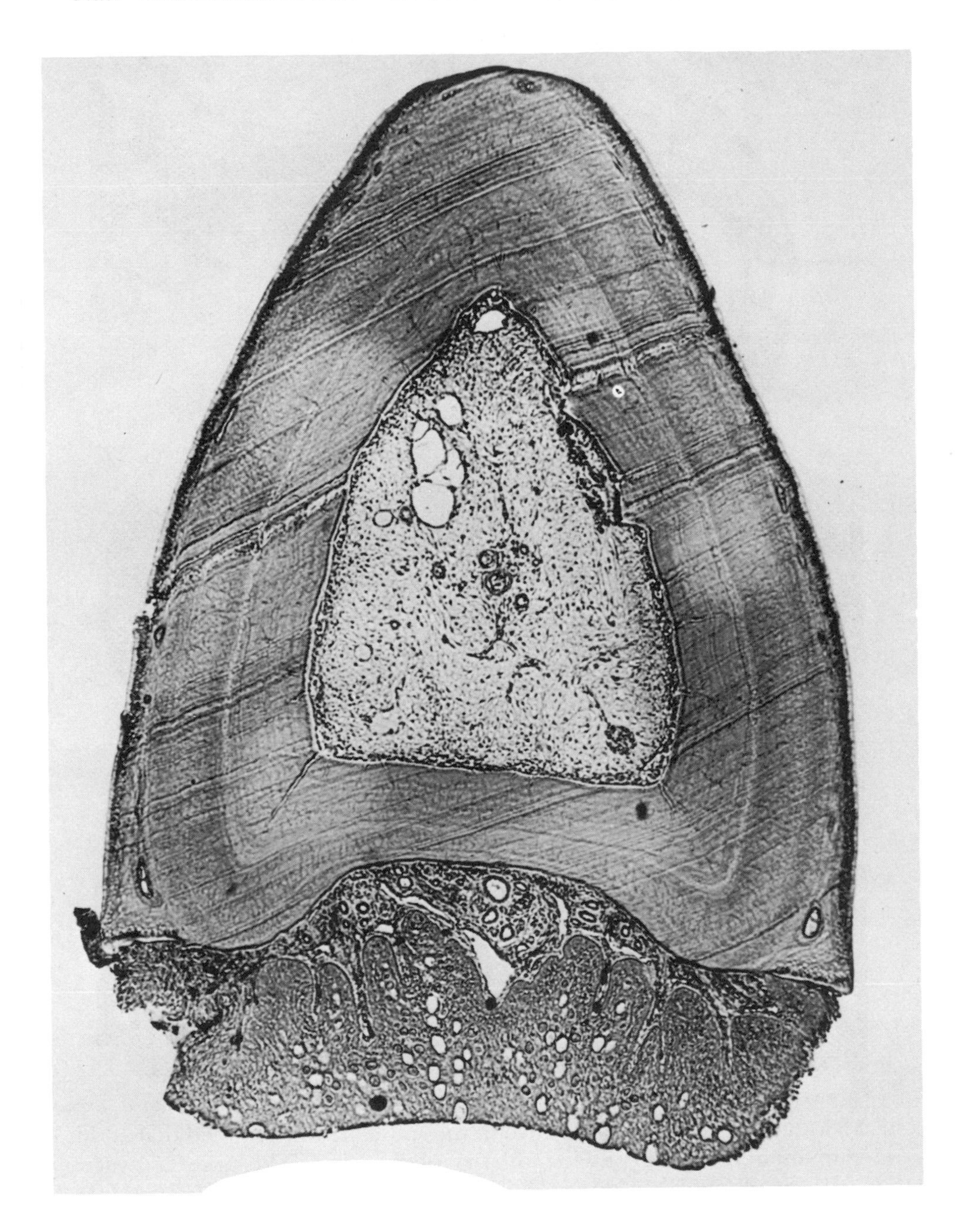

FIGURE 3.—Photomicrograph of cross section of the dorsal sting of *Squalus acanthias* taken at the level of the distal third of the spine. It will be noted that at this level the integumentary sheath of the spine is not in contact with the integument of the fin. The venom-producing cells are the large dark vacuolated ones. × 40. (Courtesy D. Goe)

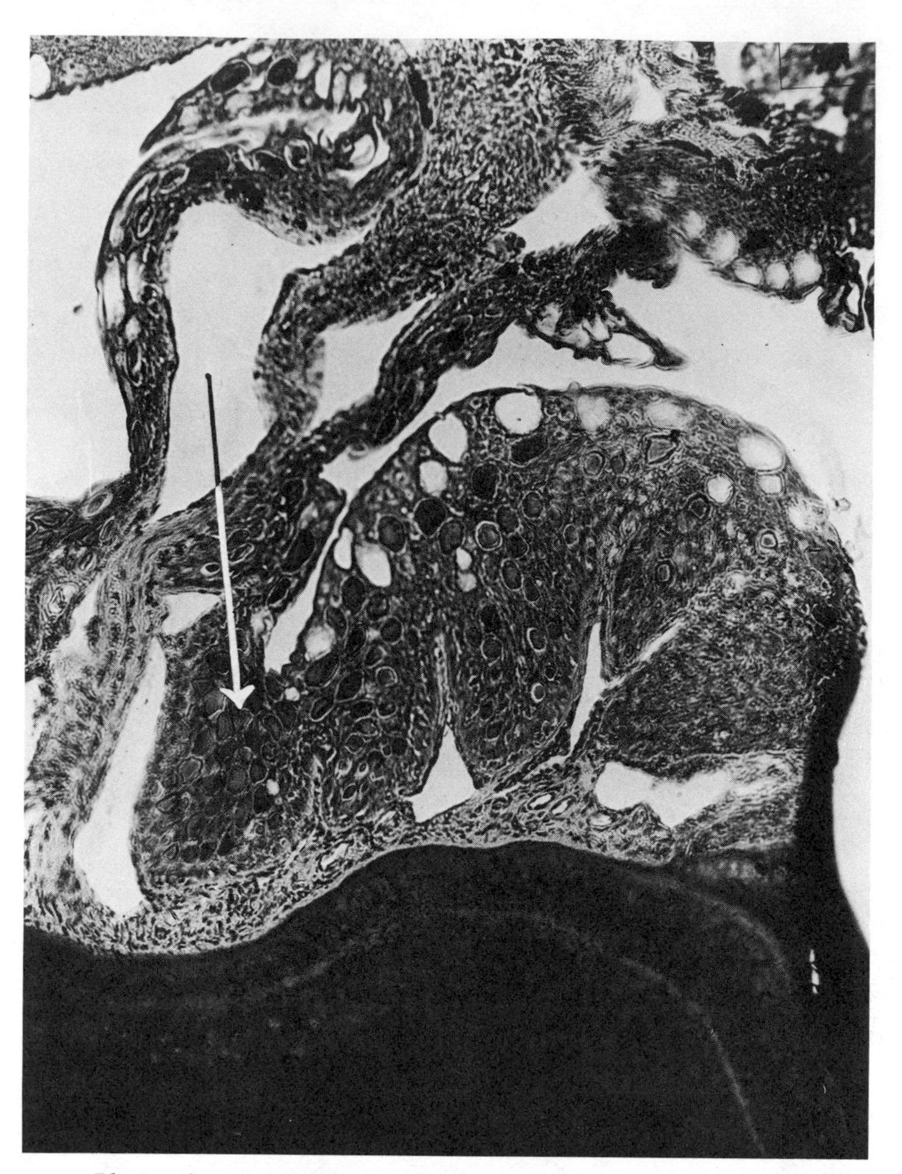

FIGURE 4.—Photomicrograph of cross section of the dorsal sting of *Squalus acanthias* taken at a lower level of the middle third of the spine. It will be noted that at this level the integument of the spine connects with that of the spine. The large dark vacuolated venom cells are abundant in this section (see arrow). × 140. (Courtesy D. Goe)

the spine, no enamel being found on the posterior aspect or on the basal half. A narrow compact layer of pigment cells lies between the enamel and the dentine. Along the inner margin of this outer layer of dentine are scattered blood vessels. The dentine appears to have been laid down in concentric layers of varying thickness. Dentine constitutes the bulk of the spine. Canaliculi or dentinal tubules are found in the innermost layer of dentine. These tubules begin on the inner surface of the dentine next to the cartilaginous core, extend radially to the limit of the dentine, and become attenuate as they pass through the dentine. They give off minute side branches which anastomose with similar branches from other tubules. The cartilaginous core at this level does not occupy all the space in the hollow spine. Around the wall of this empty space, the pulp cavity, is an odontoblast layer. The portion of the core which persists is composed of older hyaline cartilage than that found in the basal third. Between the posterior aspect of the spine and the basal cartilage of the fin skeleton is a collagenous connective tissue binding the spine to the cartilage in the region of the middle third.

The distal third of the spine in cross section is an elongate trigone. At either posterolateral angle are small rounded thickenings or projections similar to those seen in the middle third. Externally the anterolateral walls of the spine are covered with a thin layer of enamel which is absent on the posterior wall. Immediately beneath the enamel and separating it from the dentine layer is the pigment layer mentioned in the description of the middle third. Cross sections of vascular channels occur in this outer layer of dentine. Here also is noted the concentric layering of the dentine. The cartilaginous core is absent from this region. Only small strands of connective tissue can be seen around the periphery of the hollow space. The depression on the posterior aspect of the spine is occupied by typical dermis of principally collagenous fibers. Cross sections of vascular channels lined with endothelium occur in the collagenous matrix.

The distribution of the integumentary sheath of *Squalus acanthias* differs quite radically from that of stingrays. Stingray stings show a relatively even distribution of the integument around the spine, whereas in *Squalus* the sheath is bunched up within the interdentate depression and present in only a thin layer or completely absent on the anterolateral aspects of the spine. The integument is comprised of two layers, an outer epidermis and an inner dermis. The inner dermis consists of areolar connective tissue and vascular channels. The epidermis consists of modified stratified squamous epithelium with both polygonal and spindle-shaped cells arranged with a smooth surface and variable-sized papillary down-growths. The basal cells are more nearly columnar in type, and the more superficial cells are spindle-shaped and arranged with the long axis parallel to the basement membrane. Scattered throughout the superficial portion of the epithelium are large vacuoles which measure up to 200μ or more in diameter. Two types of glandular cells are found within the epithelium. As one approaches the superficial portion of the epithelium there are polygonal-shaped, clear, finely granular cells having slightly pycnotic nuclei. These cells appear to be of a mucin type. There are also oval-shaped cells having a homogenous brown or green-staining material with accumulations of finely granular material. These latter cells are believed to be the venom-producing cells. Venom production is apparently by a holocrine type of secretion.

MECHANISM OF INTOXICATION

Wounds are inflicted by the two dorsal stings which are located adjacent to the anterior margins of each of the two dorsal fins. The spines are most frequently encountered by humans in attempting to remove the fish from a hook and line or from a net in which the fish has become entangled. Skin divers occasionally incur severe wounds because of carelessness in handling spiny dogfish that have been speared.

MEDICAL ASPECTS

Clinical Characteristics

The symptoms consist of immediate, intense, stabbing pain which may continue for a period of hours. The pain may be accompanied and followed by a generalized erythema and severe swelling of the affected part. Tenderness of the affected part may continue for several days. According to Coutière (1899), dogfish stings may be fatal. The only clinical reports are by Evans (1920, 1923, 1943).

Pathology

(Unknown)

Treatment

Wounds produced by spined sharks are usually of the puncture wound variety. Since shark spines do not have an enveloping integumentary sheath and the bulk of the glandular tissue is located near the base of the spine, it would be a rare instance for the glandular tissue to become embedded in the wound of the victim. In most instances effects resulting directly from the action of the venom are of minor concern. Nevertheless, it is advisable to irrigate the wound with saltwater and do whatever debridement may be necessary if the tissues have been lacerated. If there is little or no bleeding, then moderate bleeding should be encouraged. The pain is usually mild in comparison with most stingray stings, but opiates may be needed. The extremity should be submerged in hot water for a period of 30 minutes or more at as high a temperature as the victim can tolerate without doing further injury. The addition of sodium chloride or magnesium sulfate to the water is optional. Suturing may be required. Antitetanus agents should be administered. Secondary infections from shark spines may sometimes occur, and antibiotic therapy may be needed. Elevation of the injured limb is recommended.

Prevention

Care should be taken in handling spiny dogfish because they have the ability to give a sudden lunge with their body which can easily drive the dorsal spines into the legs or hands of the unwary fisherman.

PUBLIC HEALTH ASPECTS

Shark stings are not a significant public health problem.

TOXICOLOGY

There are no specific laboratory data available, but the venom is believed to be relatively mild in its toxicity.

PHARMACOLOGY

(Unknown)

CHEMISTRY

(Unknown)

Venomous Stingrays

The order Rajiformes, which includes the skates and rays, consists of a group of fishes having the gills confined to the ventral surface of their entire body. The body proper is elongate, depressed, and almost elliptical in cross section. The pectoral fins have wide bases which are greatly extended and attached along the side of the body from the tip of the snout to the anterior margin of the pelvic fins. The anal fin is absent.

Rays may be both poisonous to eat and venomous, but only the venomous members of the Rajiformes will be discussed in this section. The suborder Myliobatoidea (Bigelow and Schroeder, 1953) includes the ray families Dasyatidae (stingrays or whiprays), Potamotrygonidae (river rays), Gymnuridae (butterfly rays), Urolophidae (round stingrays), Myliobatidae (eagle rays or bat rays), Rhinopteridae (cow-nosed rays), and Mobulidae (devil rays or mantas).

Stingrays probably constitute the most important single group of venomous fishes. The reports of stingray attacks in literature and the testimonials of victims attest to the frequency in which man departs from the stingray habitat "sadder but wiser" for his experience.

The systematics of rays pertinent to this chapter have been discussed by Günther (1859-70), Jordan and Evermann (1896-1900), Garman (1913), Fowler (1928, 1936, 1941), Bigelow and Schroeder (1953), Marshall (1964), and others.

REPRESENTATIVE LIST OF STINGRAYS REPORTED AS VENOMOUS

Phylum CHORDATA

Class CHONDRICHTHYES

Order RAJIFORMES: Rays

Family DASYATIDAE

215 *Dasyatis akajei* (Müller and Henle) (Pl. 2, fig. a). Stingray (USA), akaei (Japan), shao yang yü, hai yao yü (China), hai yoe uh (Korea).
DISTRIBUTION: Japan, China seas, Korea.

SOURCES: Fish and Cobb (1954), Okada (1955).
OTHER NAMES: *Trigon acajei, Trigon acajei, Trigon pastinaca.*

215 *Dasyatis brucco* (Bonaparte) (Pl. 2, fig. b). Stingray (USA), trigone, brucco (Italy), brucko (Germany), pastinague, brune (France).
DISTRIBUTION: Mediterranean Sea.

SOURCES: Porta (1905), Romano (1940), Maretić (1968).
OTHER NAME: *Trygon brucco.*

215 *Dasyatis dipterurus* (Jordan and Gilbert) (Pl. 2, fig. c). Diamond stingray (USA).
DISTRIBUTION: San Diego Bay southward.

SOURCES: Halstead and Modglin (1950), Russell (1953*a*), Halstead (1956, 1959).
OTHER NAMES: *Amphotistius dipterurus, Dasyatis dipterura.*

216 *Dasyatis longus* (Garman) (Pl. 3, fig. a). Stingray (USA).
DISTRIBUTION: Gulf of California to Panama, Galapagos Islands.

SOURCE: Russell *et al.* (1958*a*).

216 *Dasyatis pastinaca* (Linnaeus) (Pl. 3, fig. b). Stingray (USA), stingray, fire flare (England), gemeine stechrochen (Germany), pastenague commune (France), uge, ugo (Portugal), pastinaga (Spain), pastinaca (Italy).
DISTRIBUTION: Northeastern Atlantic Ocean, Mediterranean Sea, Indian Ocean.

SOURCES: Coutière (1899), Porta (1905), Evans (1916, 1921, 1923, 1943), Phisalix (1922), Romano (1940), Gudger (1943), Schnackenbeck (1943), Fleury (1950), Halstead and Modglin (1950), Halstead and Bunker (1953), Ocampo, Halstead, and Modglin (1953), Fish and Cobb (1954), Halstead (1956, 1959), Maretić (1968).
OTHER NAMES: *Dasyatis pastinacus, Pastinaca marina, Raia pastinaca, Trygon pastinaca, Trygon pastinacus, Trygon vulgaris.*

Dasyatis sayi (LeSueur) (Pl. 4, fig. a). Bluntnose stingray (USA), raya (Latin America). *129, 217*

DISTRIBUTION: Western Atlantic from New Jersey to southern Brazil.

SOURCES: Gudger (1914), Phillips and Brady (1953).
OTHER NAME: *Raia pastinaca.*

Dasyatis sephen (Forskål) (Pl. 4, fig. b). Stingray (USA), pagi, pantikan (Philippines), ikan pari (Malaysia), buki (Marshall Islands), fei (Caroline Islands), tenkee (India). *217*

DISTRIBUTION: Indo-Pacific, Australia, Indian Ocean, Red Sea.

SOURCES: Tweedie (1941), Fish and Cobb (1954).
OTHER NAME: *Trygon sephen.*

Dasyatis violacea (Bonaparte) (Pl. 5). Violet or pelagic stingray (USA), ferracia violacea (Italy), violette stechrochen (Germany), pastinague violette (France), pez prelado (Spain). *218*

DISTRIBUTION: Mediterranean Sea.

SOURCES: Coutière (1899), Porta (1905), Evans (1916), Phisalix (1922), Romano (1940), Gudger (1943), Halstead and Modglin (1950), Halstead and Bunker (1953).
OTHER NAMES: *Trigon violaceus, Trygon violacea.*

Taeniura lymma (Forskål) (Pl. 6, fig. b). Stingray (USA), pagi, pantikan (Philippines), ikan pari (Malaysia), buki (Marshall Islands), fei (Caroline Islands), tenkee (India). *219*

DISTRIBUTION: Indo-Pacific, Australia, Indian Ocean, Red Sea.

SOURCES: Smith (1950), Fish and Cobb (1954).
OTHER NAME: *Trygon ornatus.*

Urogymnus africanus (Schneider) (Pl. 6, fig. a). Stingray (USA), thorny ray (Australia), pagi, pantikan (Philippines), ikan pari (Malaysia), buki (Marshall Islands), fei (Caroline Islands), tenkee (India). *219*

DISTRIBUTION: Indo-Pacific, Australia, Indian Ocean, Red Sea.

SOURCE: Gimlette (1923).
OTHER NAME: *Urogymnus asperimus.*

Family GYMNURIDAE

Gymnura marmorata (Cooper) (Pl. 7, fig. a). California butterfly ray (USA), raya (Mexico). *220*

DISTRIBUTION: Point Conception, Calif., south at least to Mazatlan, Mexico.

SOURCES: Halstead and Modglin (1950), Campbell (1951), Russell (1953*a*), Halstead (1956, 1959).
OTHER NAME: *Pteroplatea marmorata.*

Family MYLIOBATIDAE

129, 220 *Aetobatus narinari* (Euphrasen) (Pl. 7, fig. b). Spotted eagle ray (USA), hihimanu, ihimanu (Hawaii), abuki (Marshall Islands), pagi manok, paol, banogan, taligmanok (Philippines), pari lang (Malayasia), eeltenkee, currooway tiriki (India), bonnet skate, spotted ray, keppierog, pylstert (South Africa), impogo, kombedo (west Africa).

DISTRIBUTION: Tropical and warm-temperate belts of the Atlantic, Red Sea, Indo-Pacific.

SOURCES: Coutière (1899), Gudger (1914, 1943), Phisalix (1922), Gimlette (1923), Tweedie (1941), Evans (1943), Halstead and Bunker (1953), Ocampo *et al.* (1953), Phillips and Brady (1953), Fish and Cobb (1954), Halstead (1956, 1959).

OTHER NAMES: *Aetobatis narinari, Stoasodon narinari.*

221 *Myliobatis aquila* (Linnaeus) (Pl. 8, fig. a). Eagle ray (USA), alderrochen (Germany), colombo (Italy), mourine (France), ratão (Portugal), cucho (Spain).

DISTRIBUTION: Eastern Atlantic, Mediterranean Sea.

SOURCES: Coutière (1899), Porta (1905), Phisalix (1922), Fleury (1950), Maretić (1968).

OTHER NAMES: *Aetobatus aquila, Myliobates aquila, Raja aquila.*

221 *Myliobatis californicus* Gill (Pl. 8, fig. b). California bat stingray (USA), raya (Mexico).

DISTRIBUTION: Oregon to Magdalena Bay, Lower California.

SOURCES: Halstead and Modglin (1950), Halstead and Bunker (1953), Russell (1953*a*), Halstead (1956, 1959).

OTHER NAMES: *Aetobatus californicus, Holorhinus californicus.*

222 *Pteromylaeus punctatus* Macleay and Macleay (Pl. 9). Eagle ray (USA).

DISTRIBUTION: Admiralty Islands.

SOURCE: Fish and Cobb (1954).

OTHER NAME: *Myliobatus punctatus.*

Family POTAMOTRYGONIDAE[1]

222 *Potamotrygon hystrix* (Müller and Henle) (Pl. 10, fig. a). Freshwater stingray (USA), raya de rio, raya fluvial (Latin America), liba spari (Surinam).

DISTRIBUTION: Freshwater rivers of Argentina, Uruguay, Paraguay, Brazil, and the Guianas.

SOURCES: Coutière (1899), Phisalix (1922), Gudger (1943), Castex and Loza (1964).

OTHER NAME: *Trygon hystrix.*

---

[1] The family Potamotrygonidae is comprised of freshwater stingrays found in the streams and rivers of South America. Some of these species are encountered in coastal brackish waters in the vicinity of river

Potamotrygon magdalenae (Duméril). Freshwater stingray (USA), raya de rio, raya fluvial (Latin America), liba spari (Surinam).
DISTRIBUTION: Magdalena River, Colombia.

SOURCES: Coutière (1899), Phisalix (1922), Castex and Loza (1964), Castex (1967).
OTHER NAME: *Taeniura magdalenae*.

Potamotrygon motoro (Müller and Henle) (Pl. 10, fig. b). Freshwater stingray (USA), raya de rio, raya fluvial (Latin America), liba spari (Surinam). *222*
DISTRIBUTION: Freshwater rivers of Argentina, Paraguay, Uruguay, and Brazil.

SOURCES: Vellard (1931, 1932), Halstead and Bunker (1953), Holloway, Bunker, and Halstead (1953), Castex (1967).
OTHER NAMES: *Paratrygon motoro, Potamotrygon garrapa, Taeniura dumerili, Taeniura mülleri, Trigon garappa.*

Family RHINOPTERIDAE

Rhinoptera bonasus (Mitchill) (Pl. 11). Cownose ray (USA), raya (Latin America). *223*
DISTRIBUTION: Coastal western Atlantic from southern New England to Brazil.

SOURCE: Phillips and Brady (1953).
OTHER NAME: *Rhinoptera quadriloba.*

Rhinoptera marginata (Geoffroy Saint-Hilaire). Cownose ray (USA).
DISTRIBUTION: Mediterranean Sea.

SOURCES: Romano (1940), Maretić (1968).

Family UROLOPHIDAE

Urolophus halleri Cooper (Pl. 12, fig. a). Round stingray (USA). *223*
DISTRIBUTION: Point Conception, Calif., to Panama Bay.

SOURCES: Halstead and Modglin (1950), Campbell (1951), Halstead and Bunker (1953), Russell (1953*a, b*), Russell and van Harreveld (1954).
OTHER NAME: *Urobatis halleri.*

---

mouths, particularly *Potamotrygon motoro* and *P. brumi.* However, since these are primarily freshwater fishes, only a brief résumé of the group is included in this monograph. There are 16 species presently recognized (according to Castex, 1964) which are members of the genus *Potamotrygon*, namely, *P. motoro* (Müller and Henle), *P. hystrix* (Müller and Henle), *P. strongylopterus* (Schomburgk), *P. magdalenae* (Dumèril), *P. brachyurus* (Günther), *P. reticulatus* (Günther), *P. signatus* Garman, *P. humerosus* Garman, *P. scobina* Garman, *P. circularis* Garman, *P. laticeps* Garman, *P. brumi* Devicenzi, *P. pauckei* Castex, *P. labradori* Castex, Maciel, and Achenbach, *P. falkneri* Castex and Maciel, and *P. schumacheri* Castex. In addition, there is a species of another genus, *Disceus thayeri* Garman (about which little appears to be known) and two other species of questionable validity, *Elipesurus spinicauda* Schomburgk and *Paratrygon aiereba* (Walbaum). All of the potamotrygonids are venomous and potentially dangerous to man. Countries having one or more of the freshwater stingray species are Argentina, Paraguay, Brazil, Colombia, Guianas, probably Peru, and several other northern South American countries. For a comprehensive discussion of this subject, see Castex (1963) and Castex and Loza (1964).

223 *Urolophus jamaicensis* (Cuvier) (Pl. 12, fig. b). Yellow stingray (USA), raya (Latin America).

DISTRIBUTION: Western tropical Atlantic from southern Caribbean to Florida.

SOURCES: Coutière (1899), Phisalix (1922), Phillips and Brady (1953).
OTHER NAMES: *Urobatis sloani, Urolophus torpedinus.*

BIOLOGY

Family DASYATIDAE: The dasyatids are a group of stingrays characterized by the absence of a caudal fin; the outer anterior margins of pectorals continuous along sides of the head, without separate cephalic fins or rostral lobes; eyes and spiracles on top of head; disc not more than 1.3 times as broad as long; length of tail from center of cloaca to tip longer than breadth of disc; no distinct dorsal fin; and the pelvis not bearing a slender median process directed forward. Dasyatids are most commonly observed lying on the bottom of the flats of bays, shoal lagoons, and river mouths, or on patches of sand between coral areas. At times they are active swimmers and are capable of progressing rapidly by undulating motions of the margins of their pectorals. At times they may be seen swimming at the surface. A habit that makes them particularly dangerous to man is the frequency in which they are found completely or partially buried in the sand or mud with only their tails, eyes, and spiracles exposed. They feed on worms, mollusks, or crustaceans which they excavate from the bottom with the use of their pectoral fins. They also feed on smaller fishes. Stingrays often fall prey to sharks, and it is not unusual to find a shark with a stingray sting embedded in its mouth (Gudger, 1946). Water is taken in through the spiracles and expelled through the gill openings. Dasyatids occur in greatest abundance along shallow coastal tropical waters, but two species have been recorded at a depth of 60 fathoms or more. They attain a size of more than 2 m across the disc. They are ovoviviparous.

Family GYMNURIDAE: Gymnurids differ from dasyatids in that the disc is more than 1.5 times as broad as long, length of tail from center of cloaca to tip is considerably shorter than breadth of disc, and the tail is with or without a small dorsal fin near its midlength. The butterfly rays are closely related to the dasyatids, and some authors include them within the Dasyatidae. However, they differ so widely in the shortness of the tail and shape of the disc in addition to several other characteristics that most authors prefer to consider them as a separate family. They attain a size of about 2 m across the disc. Their habits are similar to those of the dasyatids. They are ovoviviparous.

Family MYLIOBATIDAE: The eagle rays are very similar in appearance to the devil rays (Mobulidae), but they differ in that the anterior subdivisions of the pectorals form either one soft fleshy lobe which extends forward below the front of the head or two such lobes joined together basally. They have a single subrostral lobe or fin. Eagle rays spend much of their time cruising over the bottom or near the surface of the water with a flying-like motion. They feed largely on crustacea and mollusks which they excavate from the sand or mud with the use of their large pectoral fins. They prefer warm temperate or tropical waters. They are most often found in shallow water, but they have been taken at depths of 60 fathoms. Large specimens attain a width of 1.5 m or more. They are ovoviviparous.

Family POTAMOTRYGONIDAE: The freshwater stingrays differ from the dasyatids in that the pelvis bears a long slender median process which is directed forward, and they are largely confined to fresh or brackish water near river mouths. They are similar to dasyatids in their habits, spending most of their time completely or partially buried in the mud. Since they are generally found in turbid water, they are difficult to detect and constitute a serious hazard to anyone wading in streams inhabited by them. They feed largely on small invertebrates and fishes. Potamotrygonids are usually small in size, less than 1 m in length. They are ovoviviparous.

Family RHINOPTERIDAE: The cownose rays differ from the eagle rays by possessing a pair of subrostral lobes or fins. The morphological differences between the Rhinopteridae and Myliobatidae are so slight that some authors include them in a single family. They are similar in their habits to the myliobatids, feeding largely on bivalve mollusks and crustaceans. They attain a size of about 2 m or more in width. They are ovoviviparous.

Family UROLOPHIDAE: The round stingrays can be distinguished from other rays by the presence of a well-developed caudal fin with cartilaginous radial supports. They are most commonly found buried in the mud like their relatives the dasyatids. They feed on worms and crustaceans which they dislodge with their pectoral fins. Urolophids are for the most part inhabitants of shallow waters but have been taken at depths of 40 fathoms or more. Round stingrays are small in size, usually less than 1 m in length. They are a real hazard to fishermen and swimmers because of their very powerful muscular armed tails. They are ovoviviparous.

MORPHOLOGY OF THE VENOM APPARATUS

The venom apparatus of a stingray (Fig. 5) consists of the caudal appendage, the retroserrate dentinal spine with its enveloping integumentary sheath and associated venom glands, and the cuneiform area of the integument with which the sting is in contact when at rest. The term "sting" refers to the vasodentinal spine together with its enveloping integumentary sheath and associated venom glands. The term "spine" refers only to the vasodentinal portion of the sting.

Types of Stingray Caudal Appendages

The ability of a stingray to sting varies according to the structure of its caudal appendage and the placement of the spine on the caudal appendage. The various types of caudal appendages in relationship to the stinging ability of the ray have been classified into four general categories by Halstead and

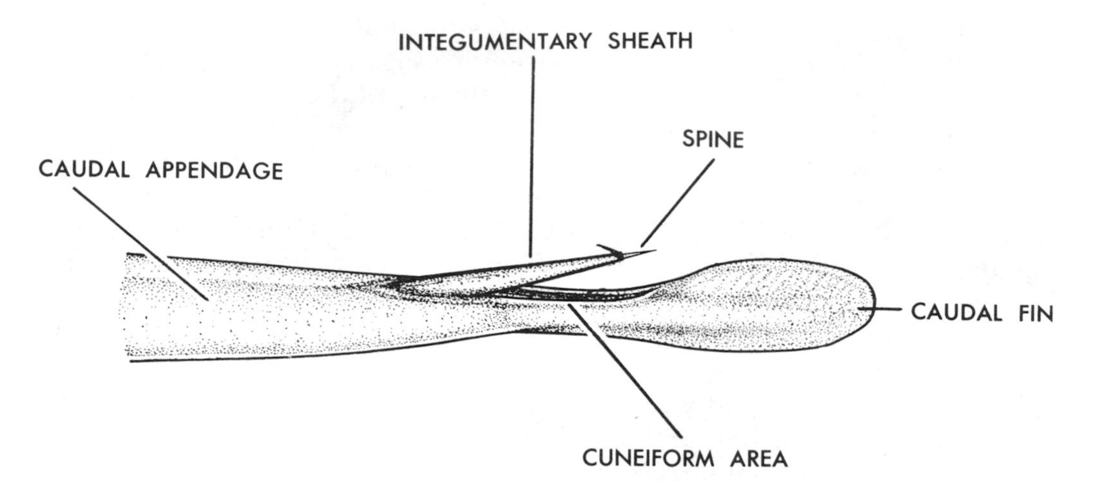

FIGURE 5.—Caudal appendage of a stingray showing the gross morphology of the venom apparatus. (R. Kreuzinger)

Bunker (1953). One or more of these types may be found within a single family of stingrays. This classification is not intended to be of phylogenetic significance but is more directly concerned with the adaptability of the caudal appendage as a striking organ in the various stingray species. This classification is as follows:

1. Gymnurid type (Fig. 6): Caudal appendage cylindrical and tapering in cross section, greatly reduced in size. Spine is small, may be feeble, about 2.5 cm or less in length, usually placed in the middle or proximal third of tail, with the exception of some of the potamotrygonids, whose spines may be placed further out on tail. Dorsal and caudal fins are variable, present as cutaneous folds or indistinct. The cuneiform area is usually moderately or poorly developed. Stingrays having the gymnurid type of caudal appendage are not well adapted for stinging and seldom inflict casualties in man. Examples: *Gymnura*, the butterfly rays, and *Potamotrygon brumi*, a freshwater stingray.

2. Myliobatid type (Fig. 6): Caudal appendage cylindrical and tapering in cross section but terminating in a long whiplike tail. Vertical caudal fin fold is absent or indistinct. A single dorsal fin is present immediately proximal to the origin of the sting. The spine is generally located on the proximal portion of the basal third of the tail. The spine is moderate to large in size, usually about 5 to 12 cm in length. The cuneiform area is moderately developed. Stingrays having this type of caudal appendage are capable of inflicting serious stings depending upon how the attack is incurred. Nevertheless, the striking ability of the ray is comparatively limited. Examples: *Myliobatis* and *Aetobatus*, the eagle rays; *Rhinoptera*, the cownose rays; *Mobula*, the devil rays.

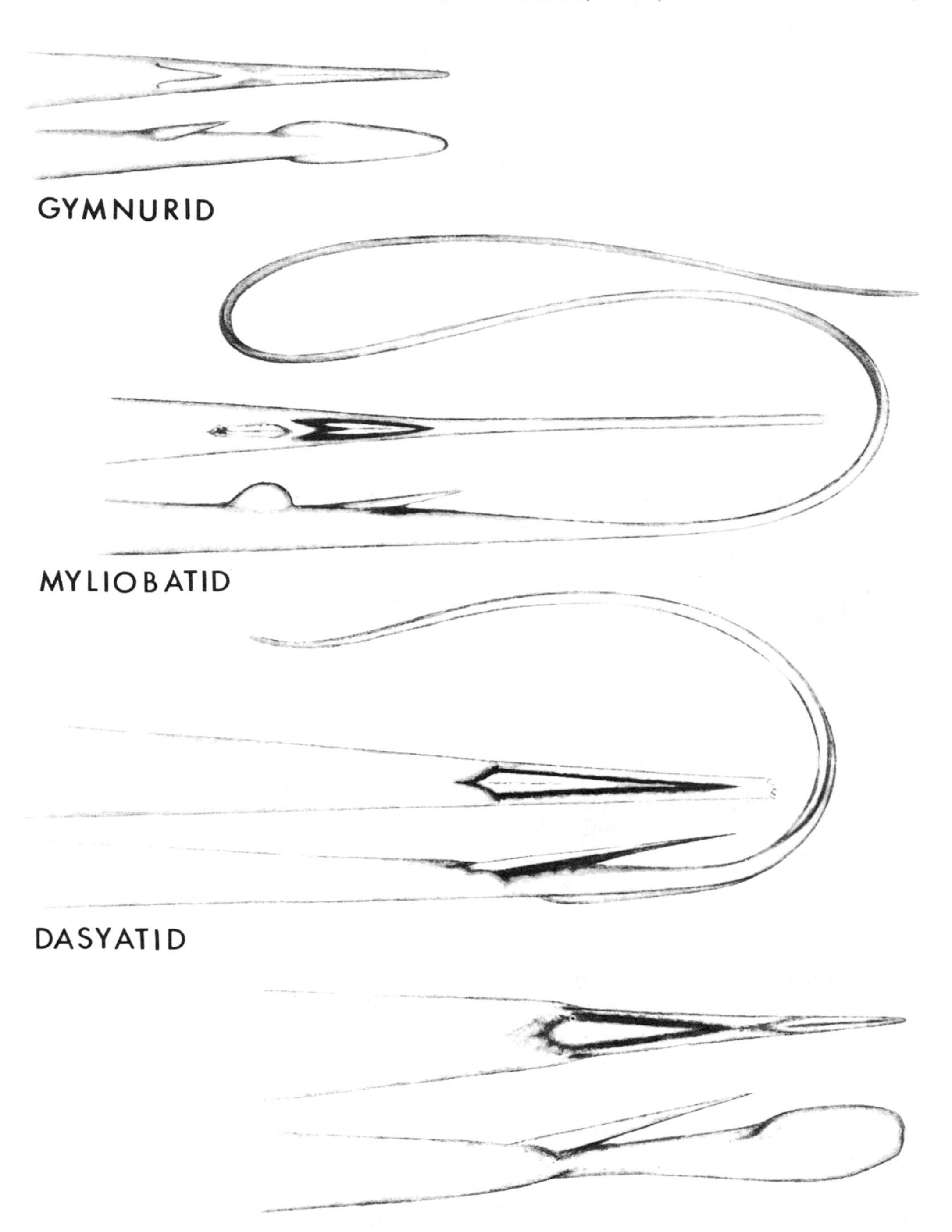

FIGURE 6.—Drawing showing the various types of stingray caudal appendages. This classification is not intended to be of phylogenetic significance but is concerned with the adaptability of the caudal appendage as a striking organ in the various stingray species. (Mrs. R. Kreuzinger)

3. Dasyatid type (Fig. 6): Caudal appendage depressed proximal to sting, becoming cylindrical and tapering in cross section distally and finally terminating in a long whiplike tail. Dorsal fin is absent. Sting is moderate to large, may attain a length of 37 cm or more in some species, and is usually located in the distal portion of the basal or middle third of the tail. Vertical caudal fin fold is indistinct in some species but pronounced in others. Cuneiform area is usually poorly developed. Stingrays having this type of caudal appendage generally have a slightly greater striking advantage than that of the myliobatid rays. Example: *Dasyatis*, the true stingray.

4. Urolophid type (Fig. 6): Caudal appendage relatively short, muscular, depressed proximal to sting, not produced as a whiplike structure, becoming compressed distal to sting and usually forming a more or less distinct caudal fin. Dorsal fin may be present or absent. The bodies of the lateral caudal muscle fasciculi diminish abruptly in size as they approach the origin of the sting and continue distally as tendinous attachments. The sting is usually located in middle or distal third of tail, moderate in size, usually about 4 cm or more in length; cuneiform area is moderately to well developed, forming a distinct depression in some species. This type of caudal appendage is a powerful muscular organ having a well-developed cuneiform area and a distally placed sting, all of which provide the ray with a highly developed striking organ. Examples: *Urolophus*, the round stingray and *Potamotrygon*, most species of freshwater stingrays.

Gross Anatomy

The venom apparatus of stingrays consists of a bilaterally retroserrate spine and its integumentary sheath (Fig. 7). The spine is an elongate tapering structure that ends in an acute-sagittate tip. The spine is composed of an inner core of vasodentine covered by a thin outer layer of enamel. It is firmly anchored in a dense collagenous network of the dermis on the dorsum of the caudal appendage. The dorsal surface of the spine is marked by a number of shallow longitudinal furrows. These furrows are usually more pronounced on the basal portion of the spine and disappear distally. The serrate edges of the spine are termed the dentate margins. Medial to each dentate margin is a longitudinal groove, the ventrolateral-glandular groove. The grooves are separated from each other by the median ventral ridge of the spine. Contained within the grooves of an "unsheathed" or traumatized sting is a strip of gray tissue. This tissue consists of the portion of the dermis in the integumentary sheath where the blood vessels are concentrated and the basal cells of the epidermis are located. The ventrolateral-glandular grooves protect the vascular structures essential to the maintenance and regeneration of the epidermis.

The retrorse teeth of the spine are hidden in an intact sting. They become visible only after the sting has been traumatized. Pressure exerted on the epidermis of the sting against the teeth as they enter the flesh of the victim concurrently results in: (1) damage to the stingray epidermis, (2) further

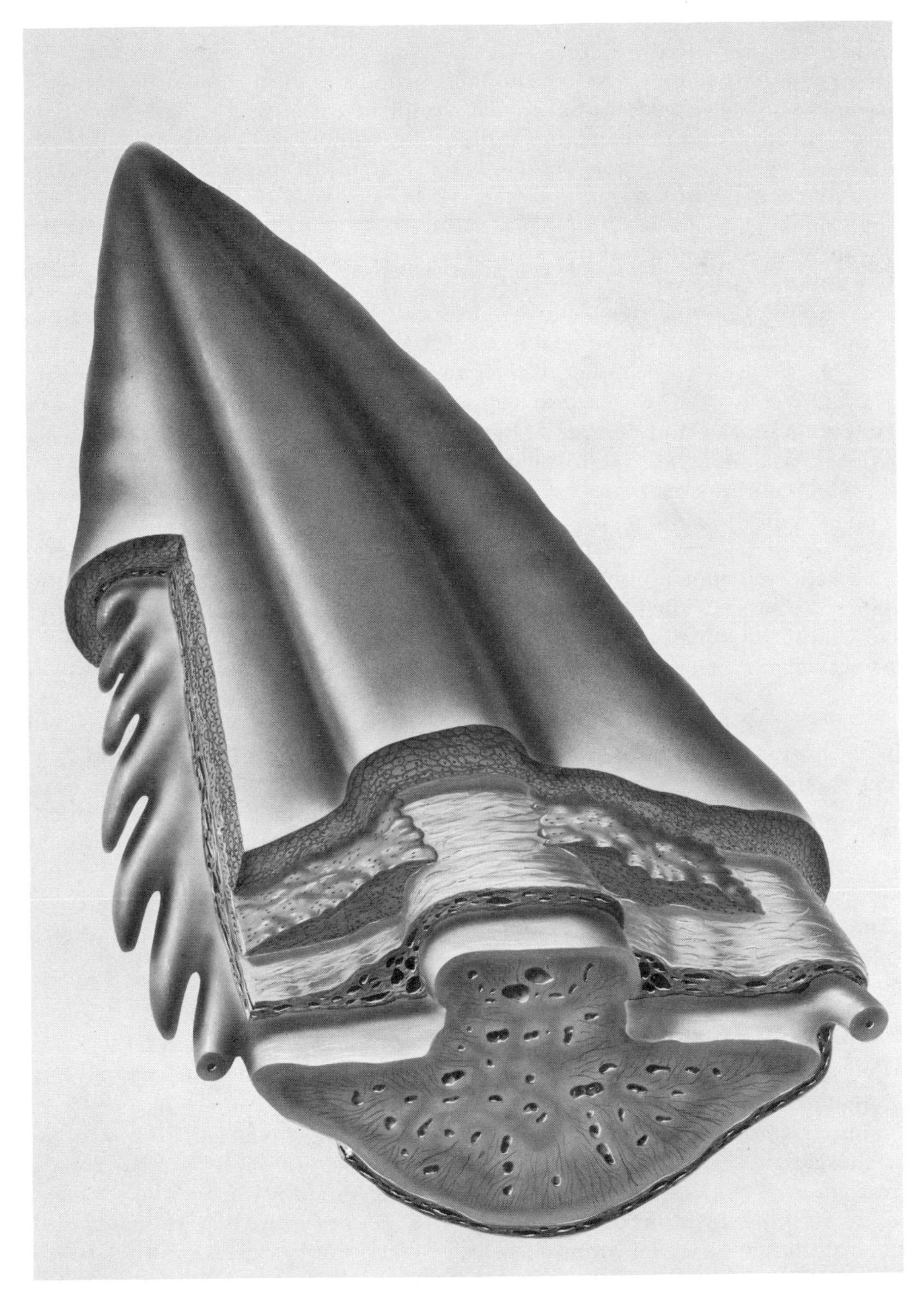

FIGURE 7.—Drawing showing the gross anatomy of a typical stingray sting. (L. Barlow)

traumatization of the recipient's wound. The two processes thus produce an ideal setting for the release and absorption of the venom. A damaged sting frequently has not lost its venomous properties even though the integumentary sheath appears to have been completely removed and the spine is bare. In an examination of more than 180 traumatized stingray stings of *Myliobatis californicus*, it was observed that a strip of epidermis usually continued to remain within the protection of the ventrolateral-glandular grooves (Fig. 8a, b). Moreover, histological sections seem to indicate that there is greater venom production per unit mass of epidermis in a traumatized sting than in an intact one.

The thickened wedge-shaped portion of the integument on the dorsum of the caudal appendage ventral to the sting is known as the cuneiform area (Fig. 5). Toxicological studies of the cuneiform integument indicate that the glandular cells of this area also secrete venom. Thus, the sting would be bathed in mucus and venom originating from the ventrolateral-glandular groove tissue and the cuneiform area.

Some authors have suggested that the spine is replaced yearly, but there is no evidence to support this idea. It is more likely that replacement occurs when needed. Gudger (1943), Bigelow and Schroeder (1953), and others have reported that multiple spines may be present (Fig. 9). When several spines are present the oldest is usually the most posterior one in position on the tail, but this may vary from one specimen to the next.

Comparative Gross Anatomy of Stingray Spines: There is very little information available as to how the morphology of the spines of one stingray species compares with that of another. The following data which have been gleaned from several sources provide some insight as to the amount of variation that may occur among stingray species. It should be kept in mind, however, that there is some variation in the size and shape of the spine from one specimen to the next. Unfortunately our data are too meager at this time to draw any conclusions as to what might be considered a standard description of a spine for a given stingray species. There is some evidence that there does exist a morphological pattern that is species specific. It will require much additional research before these patterns can be fully evaluated. The spine of *M. californicus* was arbitrarily used as a point of reference.

Family DASYATIDAE

Dasyatis akajei (Müller and Henle) (Fig. 10a, b). Mature spine slender and elongate. Spines from five specimens were examined, ranging in size from 65 mm by 5 mm to 88 mm by 5 mm. The spines were similar in size and shape to those of *M. californicus*. Dorsum of spine is finely to moderately furrowed and generally smooth in the distal third. The median ventral ridge of the spine is quite pronounced, resulting in a deeper ventrolateral-glandular groove than is observed in most spines of comparable size from other stingray species. Sizes of the rays are unknown, but all were reported to be adult specimens. Collected by T. Abe in Tokyo, Japan.

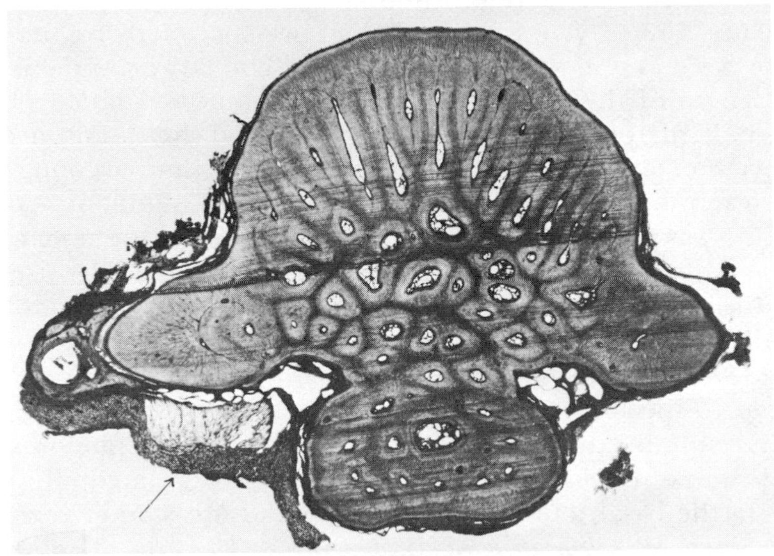

FIGURE 8a.— Photomicrograph of traumatized sting of *Myliobatis californicus*. Note that the integumentary sheath has been completely removed except for the portion lying within the protection of the lower left ventrolateral-glandular groove. The arrow points to a large amount of glandular tissue which continues to remain active within this groove. A sting of this type is still capable of inflicting a serious envenomation. × 20.

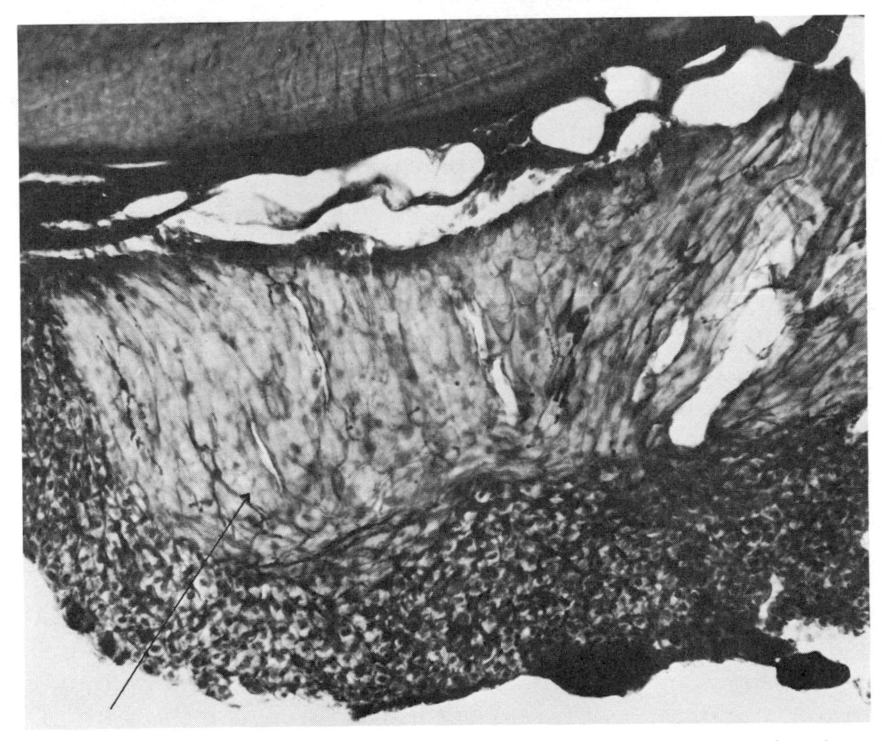

FIGURE 8b.—Enlargement of the glandular triangle of the above sting showing venom-producing cells (see arrow). × 90. (From Halstead, Ocampo, and Modglin)

Dasyatis brevicaudata (Hutton) (Fig. 11a, b). Mature spine slender and very elongate. Spines were examined from two specimens having lengths of 240 mm by 12 mm and 370 mm by 14 mm. The largest spine is about 5.2 times the length of the longest *M. californicus* spine examined. Dorsum of spine is deeply furrowed, becoming smooth toward the distal third of spine. Retrorse teeth extend from basal sixth to sagittal tip and are long and stout. The ventrolateral-glandular grooves were deep and highly developed. The shorter spine was taken from a ray having a disc width of 1.2 m, captured at Heron Island, Great Barrier Reef, Australia, by J. Booth. The size and exact locality of the other spine was unknown, but the ray was believed to have been captured near Sydney, Australia. Gift of G. Whitley.

Dasyatis brevis (Garman) (Fig. 12a, b). Mature spine slender and elongate. Spines from 10 specimens were examined, varying in length from 60 mm by 4 mm to 125 mm by 10 mm in the longest. The largest spine was about 1.7 times the length of the largest *M. californicus* spine examined. Retrorse teeth extend from the basal fourth to sagittate tip and are stouter than those observed in either *M. californicus* or *D. dipterurus*. Dorsum of spine is usually deeply furrowed. Furrows vary in number and pattern, smooth out and gradually disappear in the distal third of spine. Ventrolateral-glandular grooves are deep and well developed. Exact size of rays was unknown, but reported to be in excess of 1 m in length. Captured in the Gulf of California near Guaymas, Baja California, by J. Fitch.

Dasyatis dipterurus Jordan and Gilbert) (Fig. 13a, b). Mature spine slender, elongate, generally longer than *M. californicus* spines. Numerous spines were examined. Typical example measured 85 mm by 4 mm. Retrorse teeth, which extend from basal third to sagittate tip are similar to *M. californicus* but slightly more robust. Dorsum of spine is variably marked, sometimes with a single median furrow which gradually disappears in the terminal third. Ventrolateral-glandular grooves are deep and well developed. Size of rays was unknown but were taken from adult specimens. Captured near Mission Bay, San Diego, Calif.

Dasyatis pastinaca (Linnaeus) (Fig. 14a, b). Mature spine slender, elongate, similar in appearance to those of *M. aquila*. In one series the spines varied in length from 50 mm to 120 mm. The width was not given. Retrorse teeth extending from the basal third to the sagittate tip are apparently similar to those of *D. dipterurus*. Ventrolateral-glandular grooves are well developed. Description is based on data published by Fleury (1950). The specimen illustrated was collected at Venice, Italy. It measured 56 mm by 4 mm. Courtesy of Smithsonian Institution.

Taeniura lymma (Forskål) (Fig. 15a, b). Mature spines elongate and slender. Spines from four specimens were examined. Each tail had two spines on it. Spines ranged in size from 50 mm by 3 mm to 71 mm by 5 mm. Similar in length to those of *M. californicus* but usually greater in breadth. Dorsum of spines is rather deeply furrowed, but smooth in the distal third. Retrorse teeth extend from middle third to sagittate tip and are slightly more robust

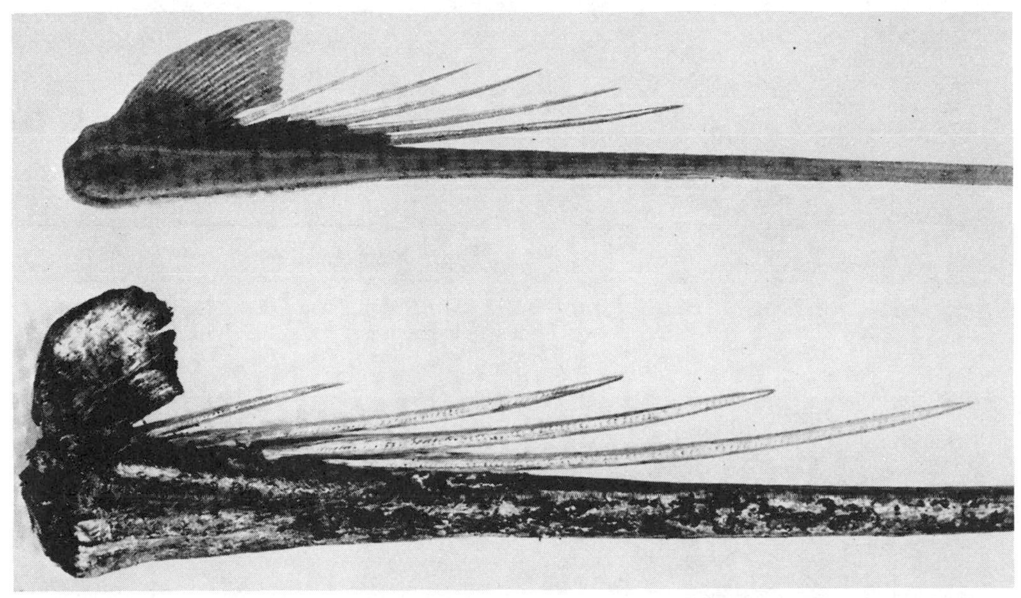

FIGURE 9.—Photographs of the caudal appendage of *Aetobatus narinari* showing examples of multi-spined specimens. It is not unusual to find more than one spine on the tail of many species of stingrays. There is no evidence that these spines are replaced annually as some authors have suggested. Multiple spines are also frequently found on some of the freshwater potamotrygonid stingrays. (From Gudger)

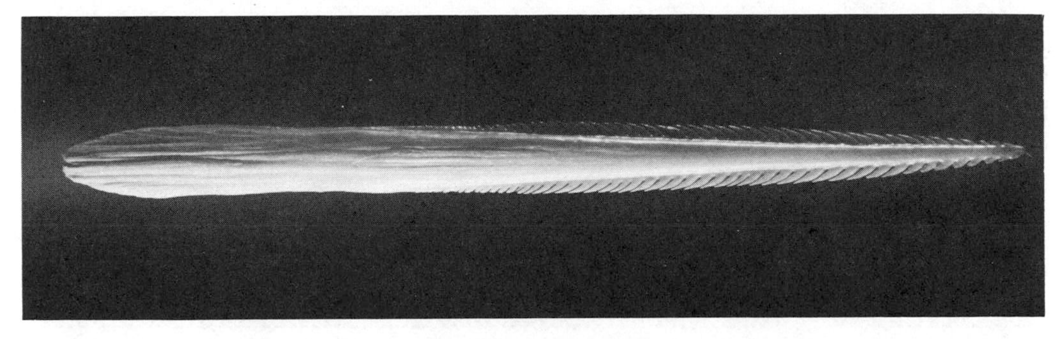

FIGURE 10a.—Dorsal view of the spine of *Dasyatis akajei*. Length 8.5 cm.

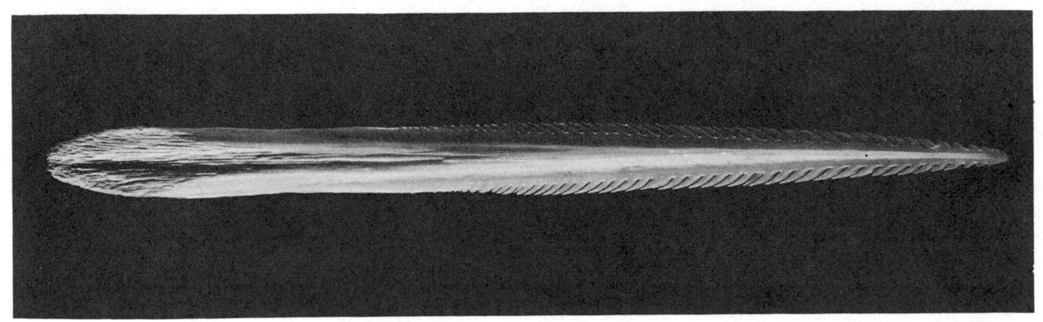

FIGURE 10b.—Ventral view of the spine of *D. akajei*. (R. Kreuzinger)

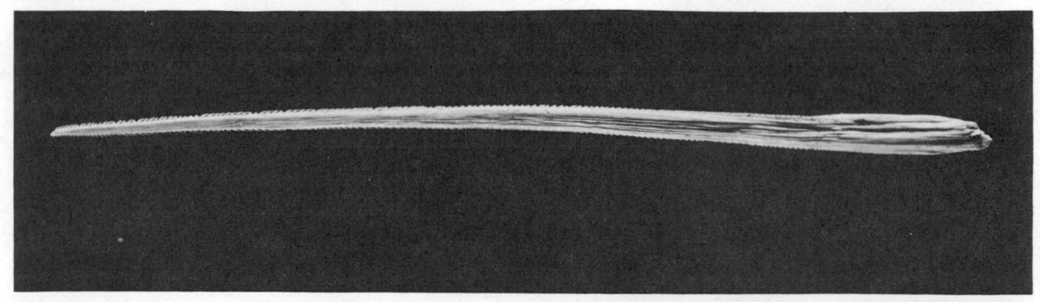

FIGURE 11a.—Dorsal view of the spine of *Dasyatis brevicaudata*. Tip is damaged. Length 37 cm.

FIGURE 11b.—Ventral view of the spine of *D. brevicaudata*. (R. Kreuzinger)

FIGURE 12a.—Dorsal view of the spine of *Dasyatis brevis*. Length 12 cm.

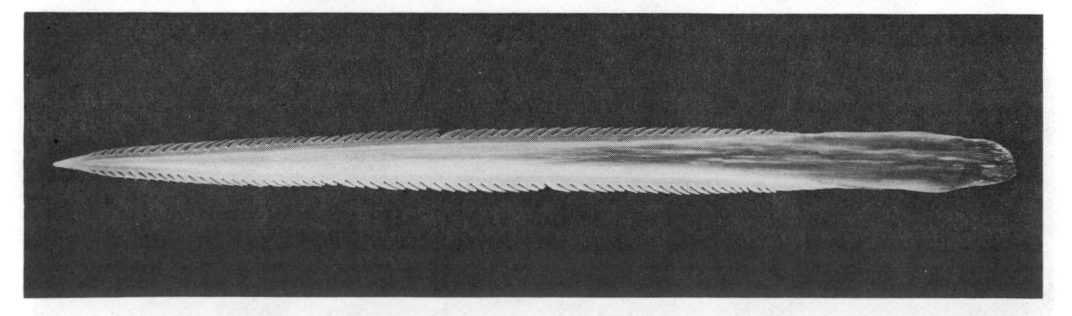

FIGURE 12b.—Ventral view of the spine of *D. brevis*. (R. Kreuzinger)

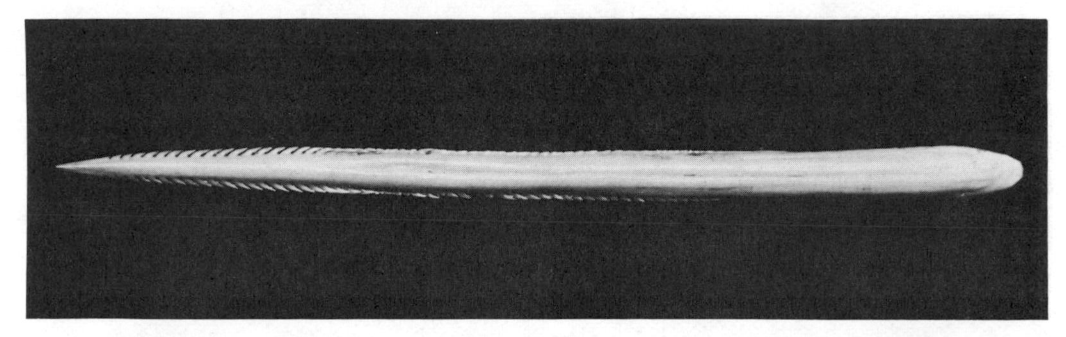

FIGURE 13a.—Dorsal view of the spine of *Dasyatis dipterurus*. Length 8.5 cm.

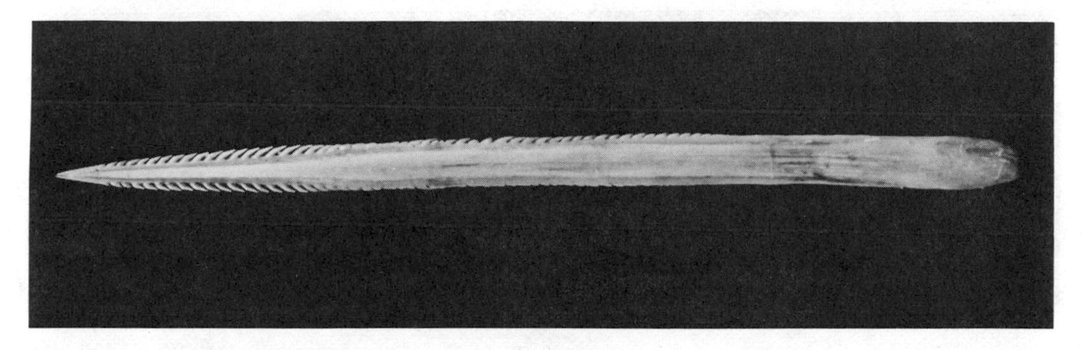

FIGURE 13b.—Ventral view of the spine of *D. dipterurus*.

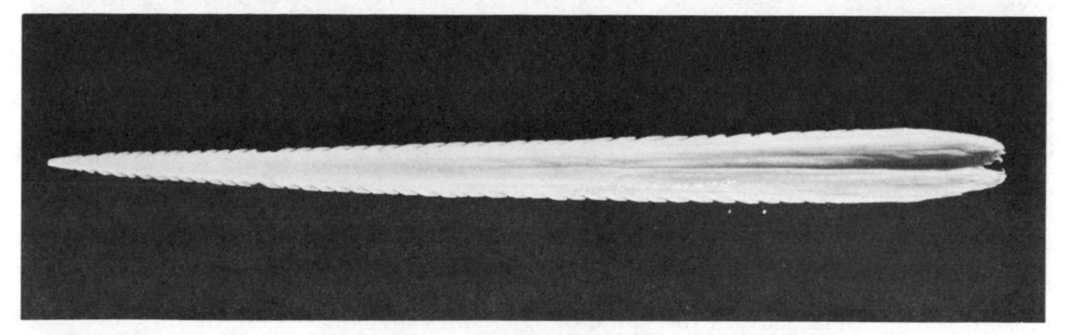

FIGURE 14a.—Dorsal view of the spine of *Dasyatis pastinaca*. Length 10 cm.

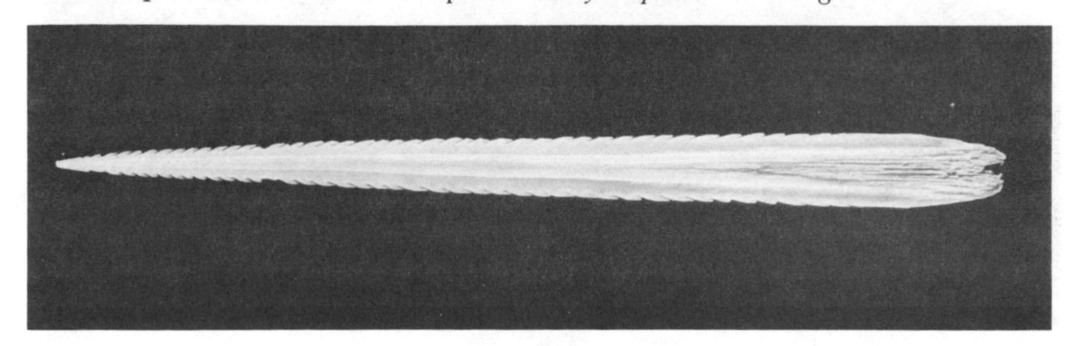

FIGURE 14b.—Ventral view of the spine of *D. pastinaca*. (R. Kreuzinger)

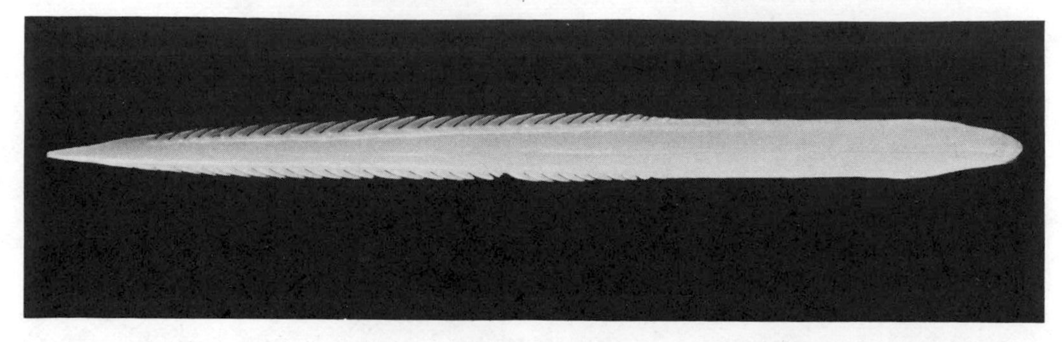

FIGURE 15a.—Dorsal view of the spine of *Taeniura lymma*. Length 7 cm.

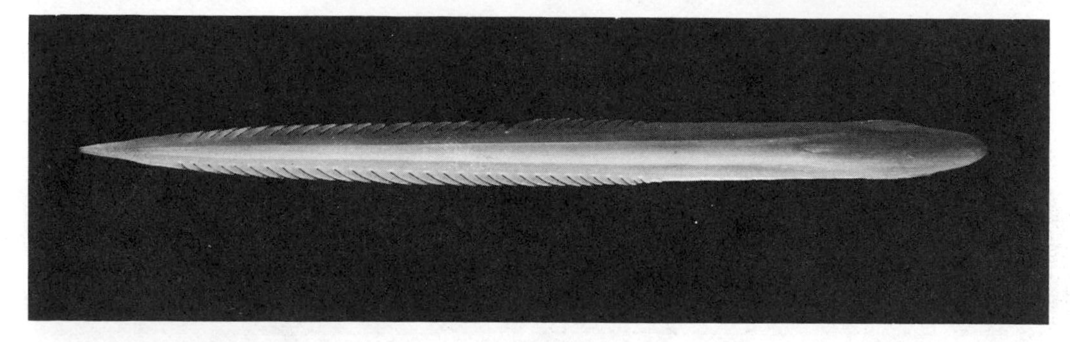

FIGURE 15b.—Ventral view of the spine of *T. lymma*. (R. Kreuzinger)

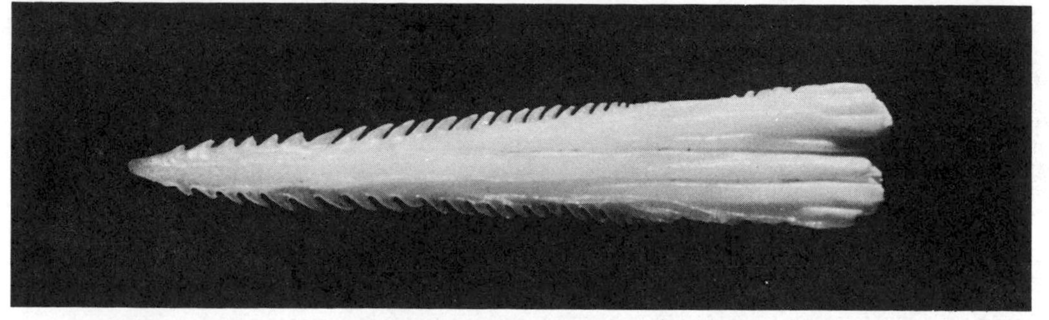

FIGURE 16a.—Dorsal view of the spine of *Gymnura marmorata*. Length 1.7 cm.

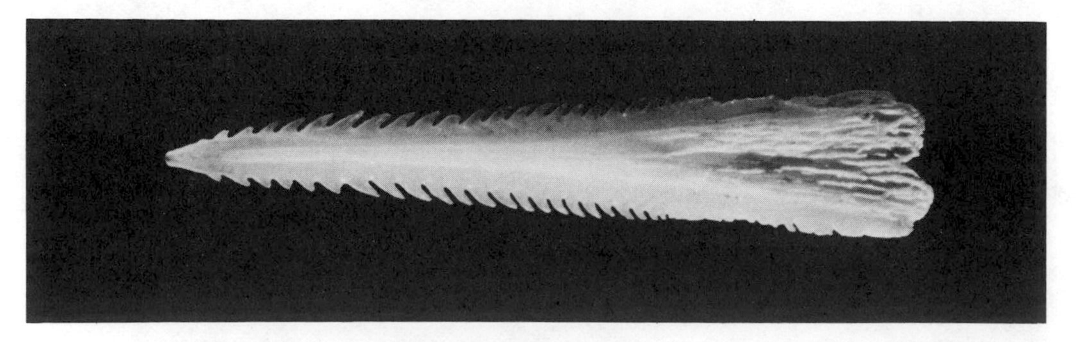

FIGURE 16b.—Ventral view of the spine of *G. marmorata*. (R. Kreuzinger)

than those of *M. californicus*. Ventrolateral-glandular grooves are deep and well developed. Size of rays was unknown but reported to be adult specimens. Collected at Townsville, Queensland, Australia, by G. Coates.

Family GYMNURIDAE

Gymnura marmorata (Cooper) (Fig. 16a, b). Mature spine short, usually about one-third the length of *M. californicus* spines. Two young rays measuring 26 cm and 34 cm in width had spines that measured 8 mm by 2 mm and 17 mm by 4 mm respectively. Mature spines of several large rays were examined, but the measurements were lost. Retrorse teeth, which extend from middle third of shaft to sagittate tip are well developed, relatively stout, and rather blunt in mature spines. Dorsum of spine is irregularly marked with furrows which terminate distally at the level of middle third of spine. Ventrolateral-glandular grooves are moderately developed. Specimens taken at Scripps Pier, La Jolla, Calif., by C. Limbaugh.

Family MYLIOBATIDAE

Aetobatus narinari (Euphrasen) (Fig. 17a, b). Mature spine elongate and slender, except in the basal fourth which is expanded giving the base of the spine a wedge-shaped appearance. Two spines were examined, one from an adult ray (exact size unknown) and the other from a young ray. The largest spine measured 120 mm by 6 mm along the shaft, and 12 mm in width at the base. The smaller spine measured 60 mm by 5 mm, and 5 mm. Retrorse teeth extend from the basal sixth to the sagittate tip. Teeth beginning about the end of the basal third are short and stout, whereas the teeth near the base are less robust and shorter. Dorsum of spine is rugose throughout its length. Ventrolateral-glandular grooves are only moderately developed. Locality from which ray was obtained is unknown.

Myliobatis aquila (Linnaeus) (Fig. 18a, b). Mature spine closely resembles that of *M californicus* but tends to be slightly broader at the base. Spines from five specimens were examined, ranging in size from 50 mm by 5 mm to 80 mm by 6 mm. Fleury (1950) reported a spine 220 mm from a ray having a span of 1.05 m. Retrorse teeth extend from basal fourth to the sagittate tip and are generally irregular in their placement, comparatively elongate and slender. Dorsum of spine is irregularly marked with shallow furrows, becoming smooth in the distal third. Ventrolateral-glandular grooves are deep and well developed. Size of rays is unknown but reported to be adult. Collected from Paso de la Patria, Corrientes, Argentina, by M. N. Castex.

Myliobatis californicus Gill (Fig. 19a, b). Mature spine elongate, varying in length from 40 mm by 3 mm to 77 mm by 5 mm. (Spines were removed from fish measuring from 49.6 cm to 122.0 cm in total length, based on a series of 180 spines.) Retrorse teeth are elongate, slender, and acute, extending from proximal third to sagittate tip. Dorsum of spine is marked with a variable number of irregular furrows which seldom extend beyond the junction of the middle and distal third of spine. A single deep median furrow is frequently

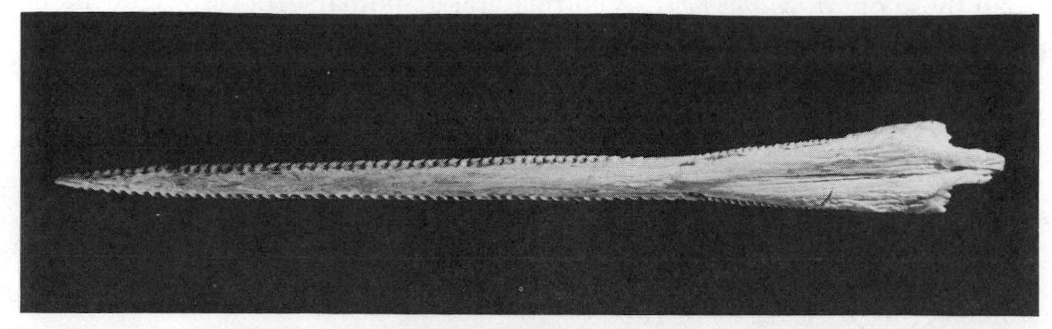

FIGURE 17a.—Dorsal view of the spine of *Aetobatus narinari*. Length 12.5 cm.

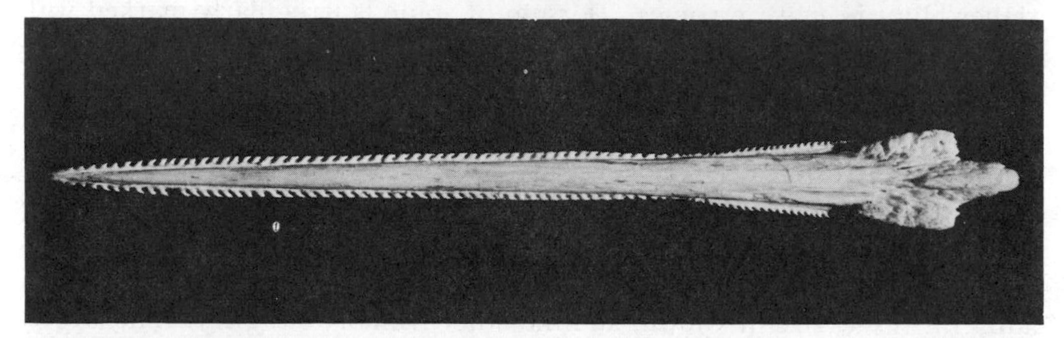

FIGURE 17b.—Ventral view of spine of *A. narinari*.

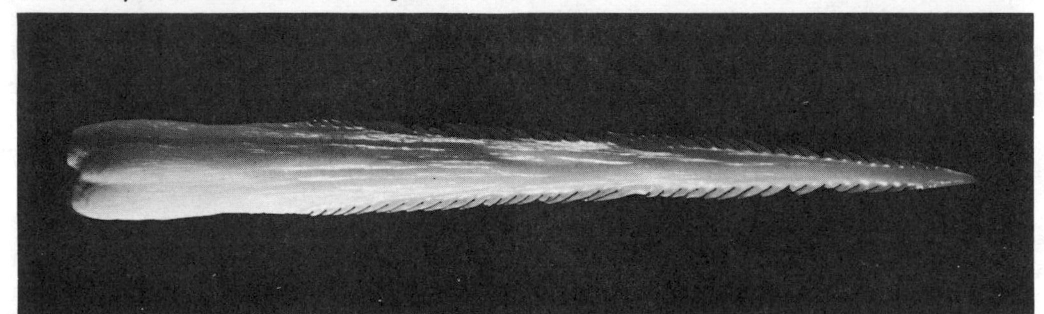

FIGURE 18a.—Dorsal view of the spine of *Myliobatis aquila*. Length 7.5 cm.

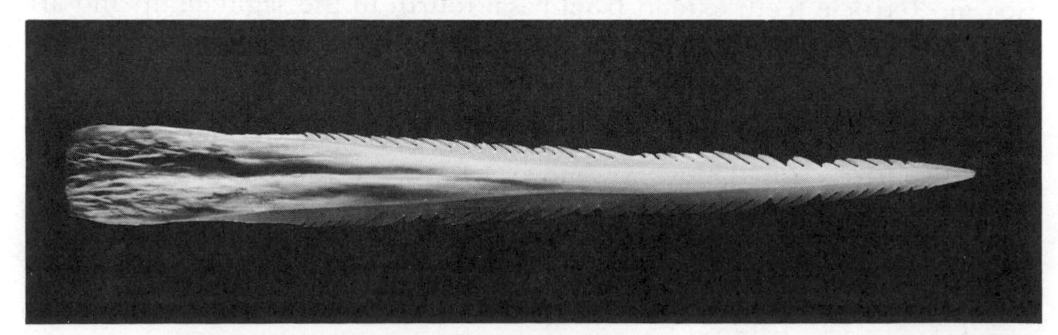

FIGURE 18b.—Ventral view of spine of *M. Aquila*. (R. Kreuzinger)

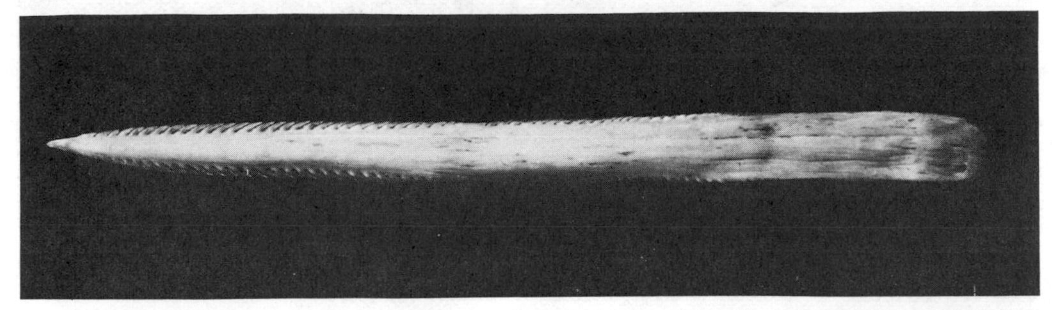

FIGURE 19a.—Dorsal view of the spine of *Myliobatis californicus*. Length 7 cm.

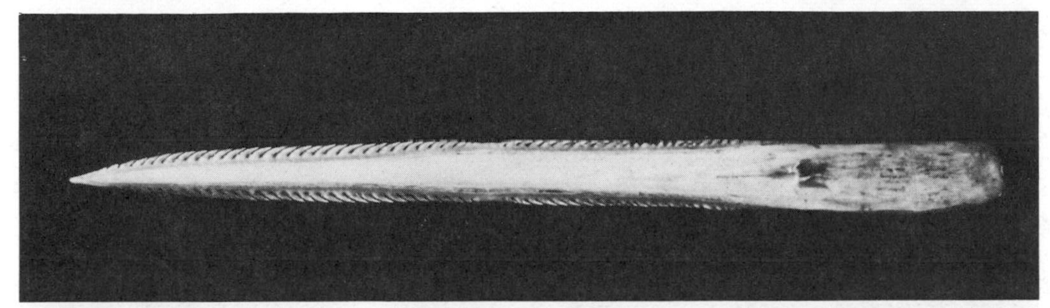

FIGURE 19b.—Ventral view of same spine.

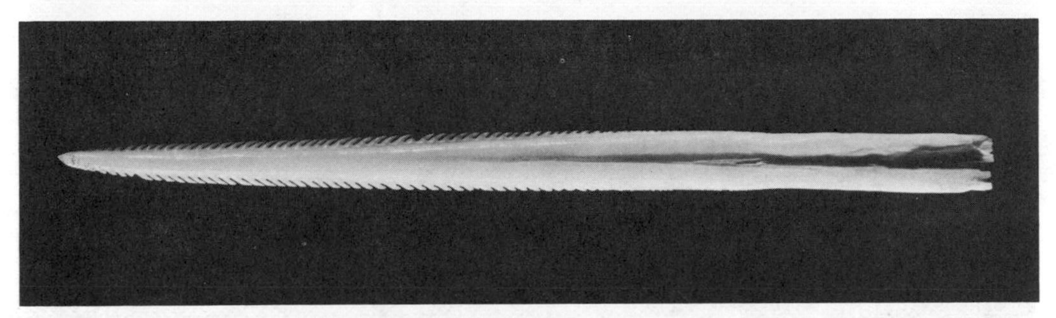

FIGURE 20a.—Dorsal view of the spine of *Potamotrygon motoro*. Length 7.5 cm.

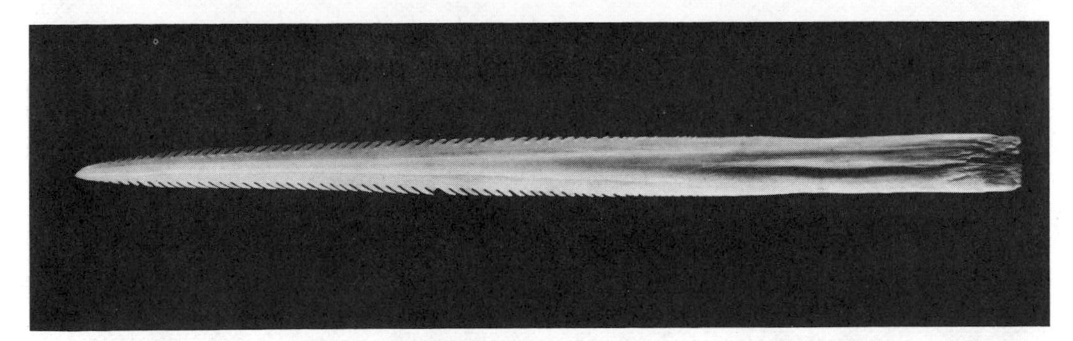

FIGURE 20b.—Ventral view of same spine. (R. Kreuzinger)

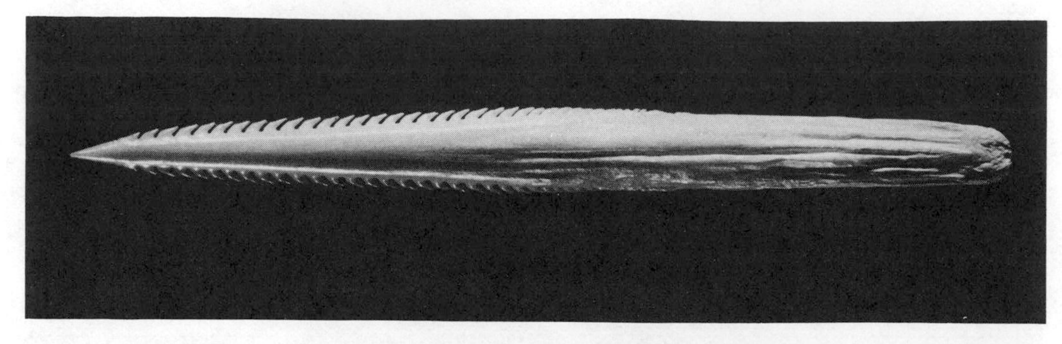

FIGURE 21a.—Dorsal view of the spine of *Rhinoptera steindachneri*. Length 7.2 cm.

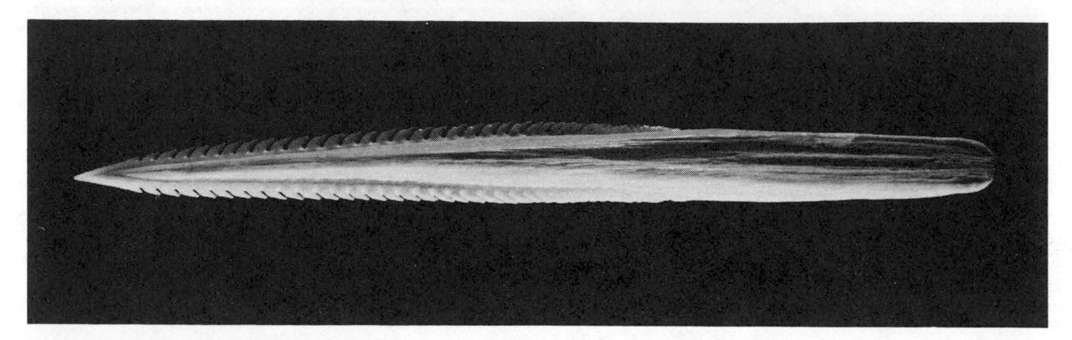

FIGURE 21b.—Ventral view of same spine.

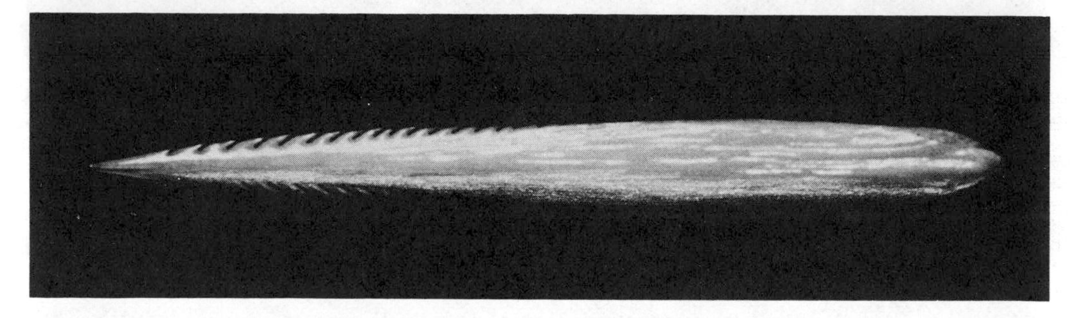

FIGURE 22a.—Dorsal view of the spine of *Urolophus halleri*. Length 4 cm.

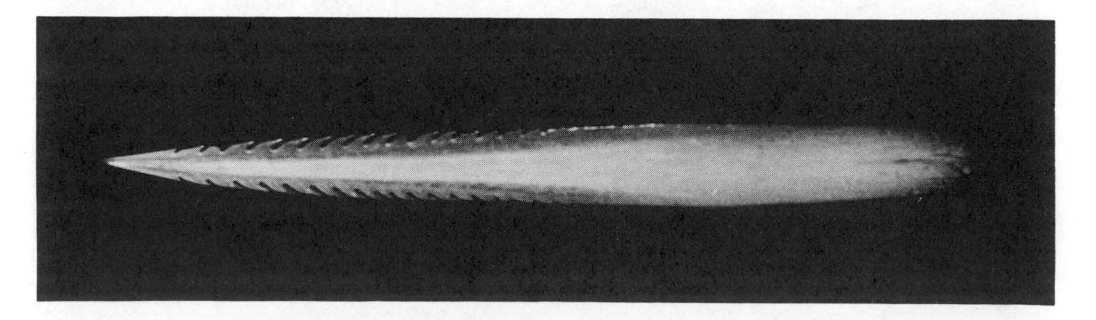

FIGURE 22b.—Ventral view of same spine. (R. Kreuzinger)

present on the dorsum of the spine. Ventrolateral-glandular grooves are moderately developed. Collected from San Francisco Bay, Burlingame, Calif., by E. S. Herald.

Family POTAMOTRYGONIDAE

Potamotrygon motoro (Müller and Henle) (Fig. 20a, b). Mature spine slender and elongate. Spines from two specimens were examined. One ray had two spines present. Spines varied from 55 mm by 5 mm to 80 mm by 5 mm. Spines closely resemble those of a dasyatid ray. Retrorse teeth extend from the middle third to the sagittate tip and are slender and acute. Dorsum of spine is marked with a single deep furrow which extends more than half way up the shaft of the spine. Ventrolateral-glandular grooves are deep and well developed. Size of rays is unknown, but reported to be adults. Specimens collected from Paso de la Patria, Corrientes, Argentina, by M. N. Castex.

Family RHINOPTERIDAE

Rhinoptera steindachneri Evermann and Jenkins (Fig. 21a, b). Mature spines elongate, but comparatively broad and stout. Spines from five specimens were examined ranging in size from 50 mm by 4 mm to 68 mm by 6 mm. Dorsum of spine has a single median furrow which extends only about half way up the shaft of the spine; remainder of dorsum is smooth. Retrorse teeth extend from basal or middle third to sagittate tip; teeth are stouter than in most of the myliobatid or dasyatid rays. Ventral median ridge is broad and relatively low, resulting in a comparatively shallow ventrolateral-glandular groove. Size of rays is unknown but said to have been adult. Specimens collected near Guaymas, Gulf of California, Mexico.

Family UROLOPHIDAE

Urolophus halleri Cooper (Fig. 22a, b). Mature spine short, usually about one-half to three-fifths the length of a *M. californicus* spine. Several hundred spines were examined, most of them measuring between 28 mm by 3 mm to 41 mm by 4 mm. All spines were taken from adult fish. Retrorse teeth similar to those of *D. dipterurus* extend from middle third to tip of spine. Dorsum of spine is marked with numerous fine irregular longitudinal depressions which gradually disappear as they approach the terminal third of spine. Deep ventrolateral-glandular grooves appear to be most highly developed in this species. Specimens collected from Mission Bay, San Diego, Calif., by G. Kuhn.

Microscopic Anatomy

The following discussion of the spine and dermis is based largely on the sting of *Myliobatis californicus*. In the stingrays studied thus far it appears there is little variation from one species to the next in the histological structure of the spine and dermis, but there are very significant species differences in the epidermis. For the purposes of this presentation, *M. californicus* has been arbitrarily selected as a point of reference.

Spine: The spine is a vasodentinal structure perforated by numerous dentinal canals. The dentine canals vary considerably in size, are filled with a finely fibrillar, moderately cellular connective tissue, and contain one to four small blood vessels. Each canal has an associated elaborate system of fine branching lines suggesting a canalicular system. The dorsal and ventrolateral aspects of the spine are completely covered by a thin layer of enamel or vitrodentine, which is absent on the median aspect of the ventrolateral-glandular grooves and the median ventral ridge. The retrorsed marginal teeth are attached or immediately adjacent to the horizontal limb of the spine depending on the cut of the section. Specific differences in the shape of the spine are mainly due to variations in length and width of the median ventral ridge, the presence or absence of a dorsal groove, and variations in the length and width of the dentate margins. The shape of the spine varies somewhat from one species to the next, but the basic histological structure is quite similar.

Dermis: The dermis of all the species examined is a moderately dense fibrous connective tissue containing varying numbers of chromatophores filled with brown granular pigment. The thickest and most vascular portion of the dermis is located within the "glandular triangle." The dermis is thick in *Aetobatus narinari, M. californicus, Dasyatis dipterurus*; moderately thick in *Gymnura marmorata*; but greatly reduced in *Urolophus halleri*. Scattered throughout the dermis but most heavily concentrated in the glandular triangle are variable-sized cellular clumps having a definite acinar arrangement circumscribed by bands of collagenous connective tissue (Fig. 23). The cells within these clumps are polygonal with colorless or pink cytoplasm and have centrally-placed, dark blue, spherical nuclei. These cells originally were thought to be glandular in nature and associated with a duct system of some type. However, more recent studies indicate that the dermis contains only nerves and vascular structures. Histochemical studies are needed which will reveal the nature and function of these acini. These polygonal cells were similarly abundant in all of the species but *G. marmorata* and *U. halleri* in which they were greatly reduced in number. The chromatophores were fewer and diffusely scattered in the dermis of *D. dipterurus* and *M. californicus*, more numerous and slightly concentrated in the superficial dermis of *U. halleri*, and most numerous and densely concentrated immediately beneath the basement membrane in *G. marmorata* and *A. narinari*. This peculiar distribution of chromatophores in the latter two species resulted in a continuous black or brown line completely encircling the spine.

Comparative Microscopic Anatomy of the Epidermis: The greatest species variation in the histology of stingray stings is found in the epidermal layer of the integumentary sheath. There appears to be specific histological patterns which may ultimately prove useful in studying the phylogeny of these rays. However, our histological data are still too meager to permit the development of many valid conclusions regarding the morphological pattern of a given species at this time.

Family DASYATIDAE

Dasyatis akajei: No histological data available.

Dasyatis brevicaudata: No histological data available.

Dasyatis brevis: No histological data available.

Dasyatis dipterurus (Fig. 24). The epidermis is stratified and rests on a distinct basement membrane. The basement membrane in the region of the median ventral ridge is thrown into deep folds between which are thin cores of dermis. The epidermis is thickest in the dorsal region and thinnest in the ventral. There is a single basal layer of basophilic columnar cells and an intermediate layer of slightly acidophilic polygonal cells with centrally placed spherical nuclei. Within this zone are irregular yellow-brown areas in which cellular outlines and nuclei are observed. These areas apparently indicate

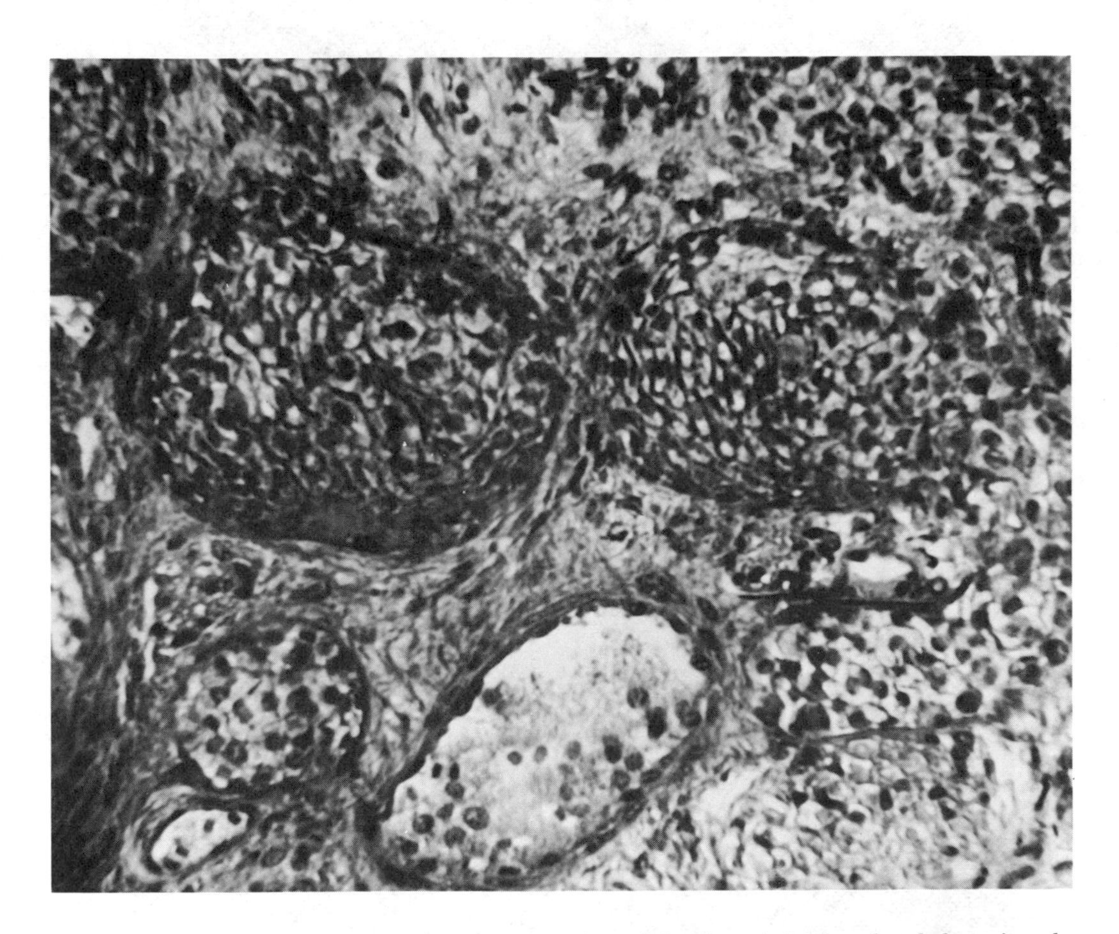

FIGURE 23.—Photomicrograph of a cross section of the dermis in the glandular triangle of *Myliobatis californicus* showing the "acini" surrounded by bands of collagenous connective tissue. It was originally thought that these acini contained glandular structures. However, recent studies indicate that the dermis contains only nerves and vascular structures. Hematoxylin and triosin stain. × 380.

(From Halstead, Ocampo, and Modglin)

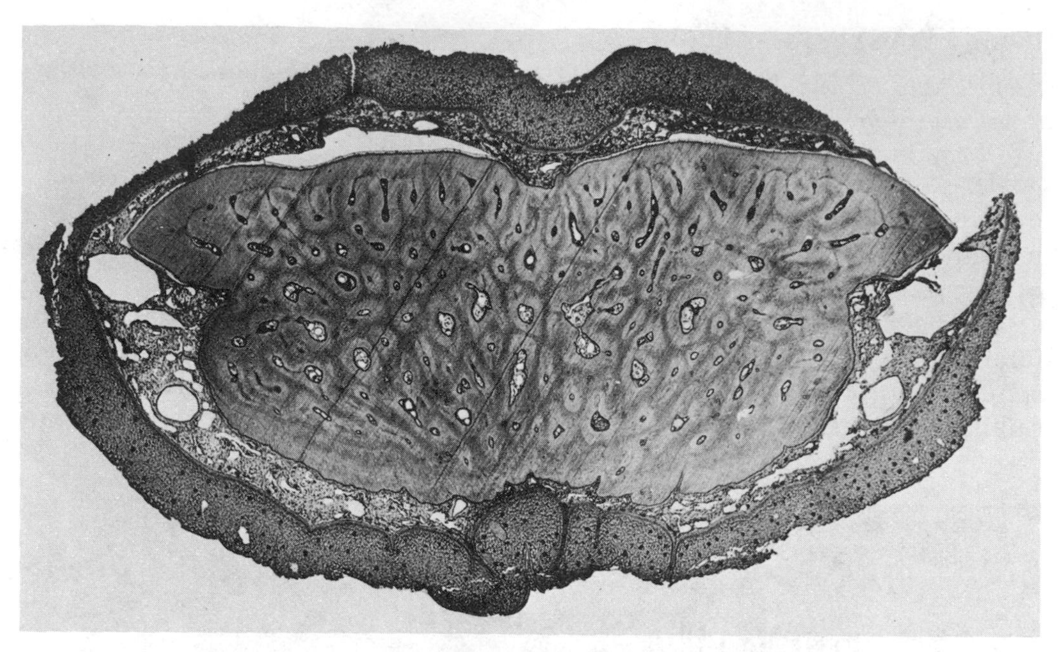

FIGURE 24a.—Photomicrograph of a cross section of a mature intact sting of *Dasyatis dipterurus*. Hematoxylin and triosin stain. × 60.

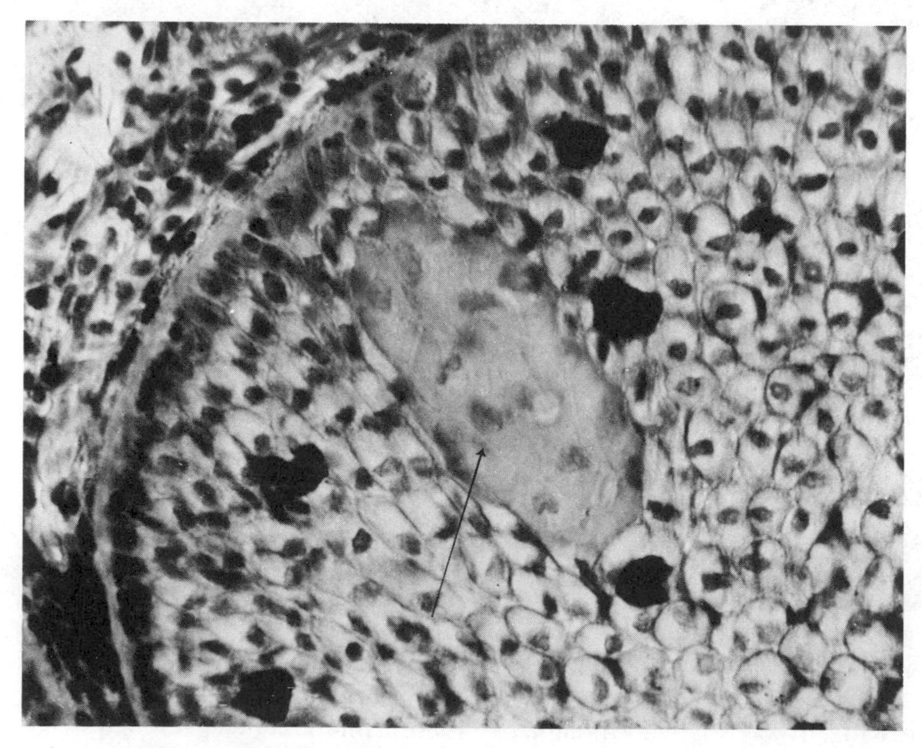

FIGURE 24b.—Enlargement of glandular epithelium of *Dasyatis dipterurus*. Arrow points to area of secretory activity. Hematoxylin and triosin stain. × 900.
(From Halstead, Ocampo, and Modglin)

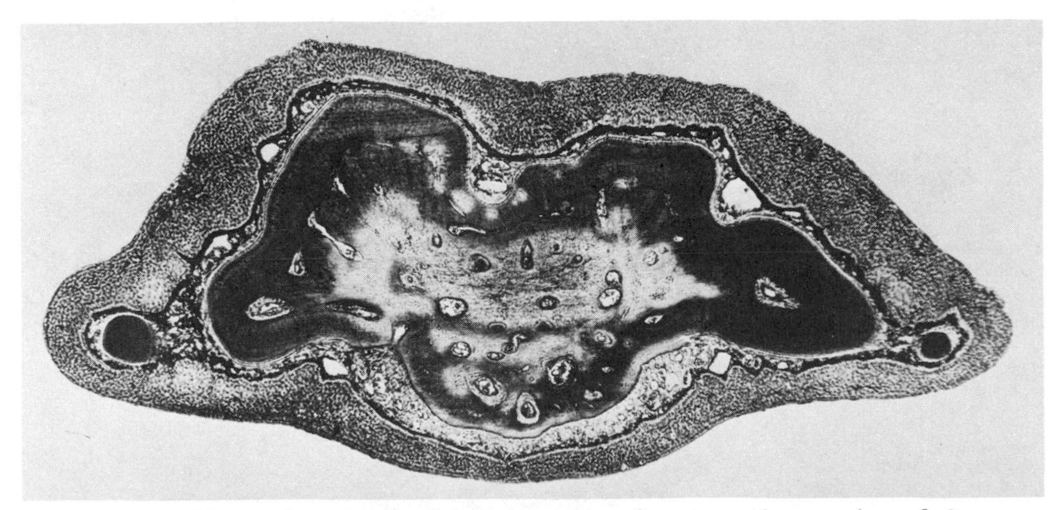

FIGURE 25a.—Photomicrograph of a cross section of a mature intact sting of *Gymnura marmorata*. Hematoxylin and triosin stain. × 75.

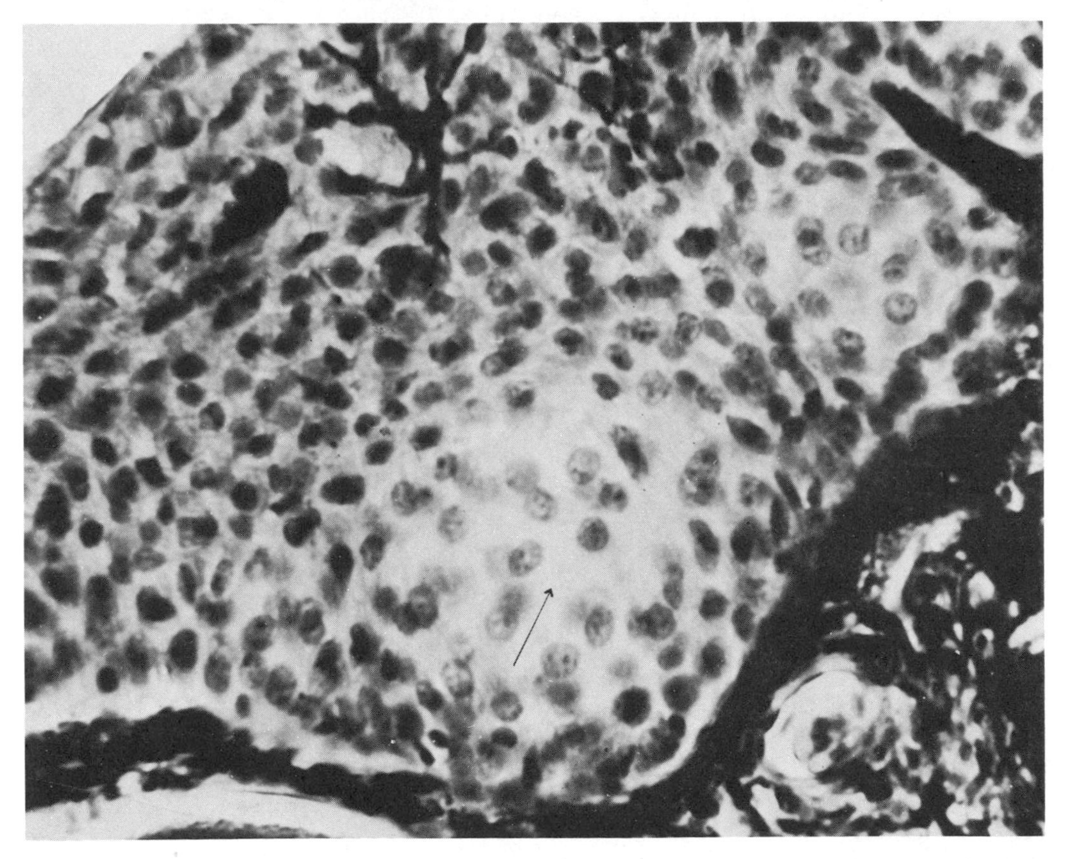

FIGURE 25b.—Enlargement of glandular epithelium of *Gymnura marmorata*. Arrow points to area of secretory activity. Hematoxylin and triosin stain. × 750.
(From Halstead, Ocampo, and Modglin)

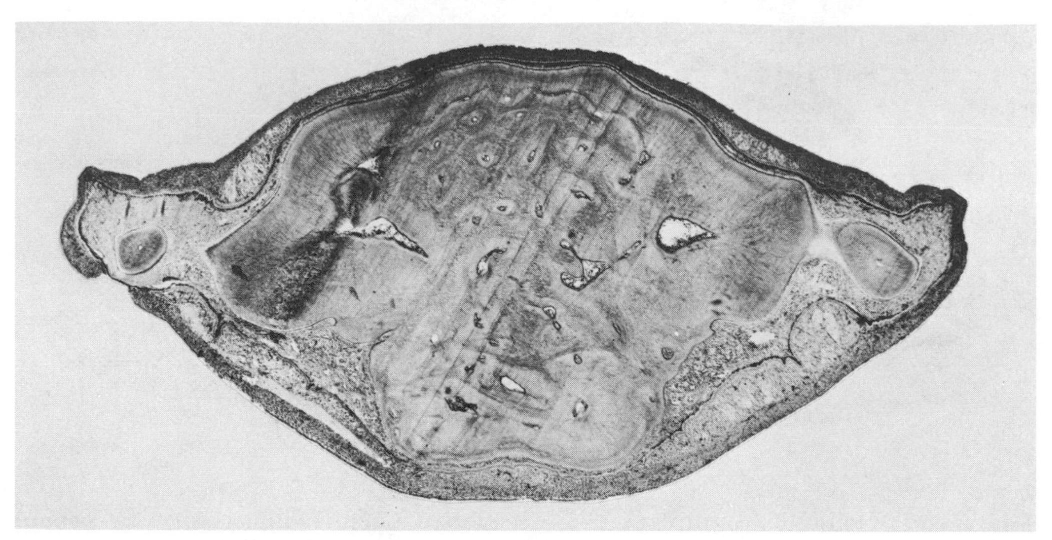

FIGURE 26a.—Photomicrograph of a cross section of a mature sting of *Aetobatus narinari*. Hematoxylin and triosin stain. × 90.

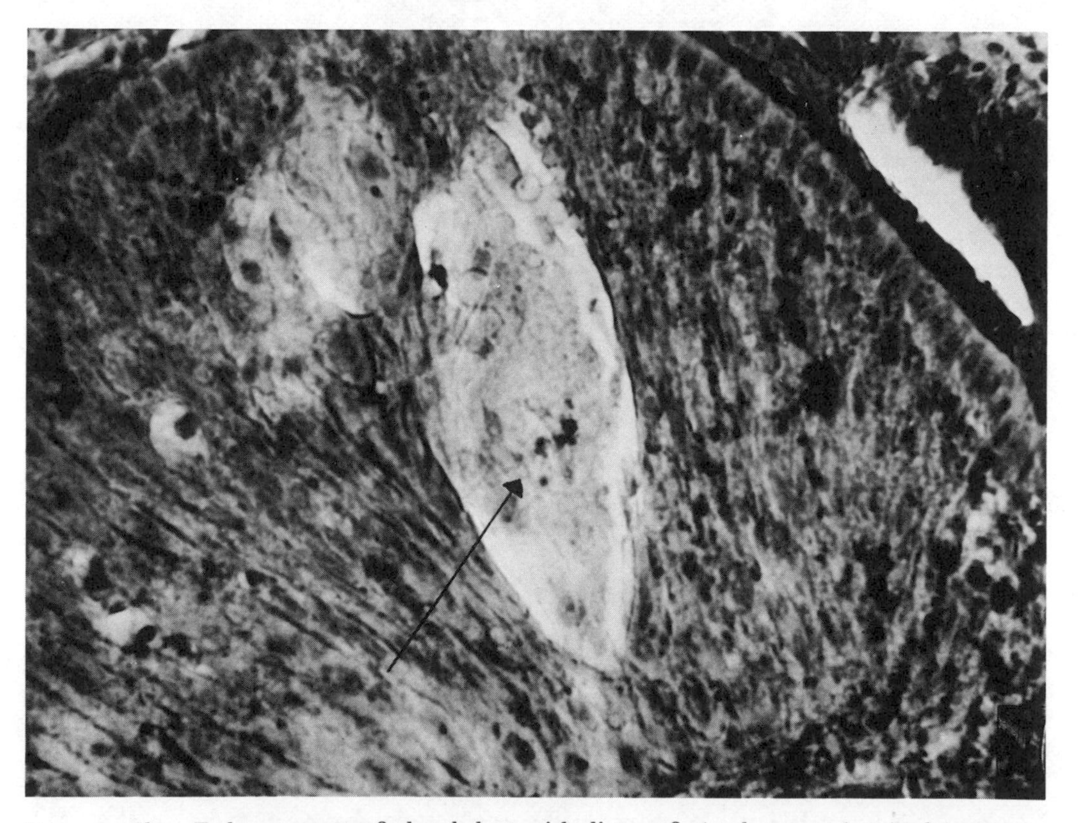

FIGURE 26b.—Enlargement of glandular epithelium of *Aetobatus narinari* taken in the area of the glandular triangle. Arrow points to area where cells are undergoing a holocrine type of lysis. Hematoxylin and triosin stain. × 550.

(From Halstead, Ocampo, and Modglin)

some type of secretory activity (Fig. 24b). The peripheral layer is comprised of basophilic columnar cells with basally placed, irregular, somewhat pyknotic nuclei and clear cytoplasm. The cells of this region are rounded and do not have the appearance of the stratified squamous epithelial cells as observed in *A. narinari* and *M. californicus*. Chromatophores, moderate in number, are diffusely scattered and show little branching. A definite cuticle is present.

Taeniura lymma: No histological data available.

Family GYMNURIDAE

Gymnura marmorata (Fig. 25a, b). The epidermis of this species is similar to that of *M. californicus*. However, the outlines of the epidermal cells are less distinct, and the areas of glandular activity appear to be small and largely restricted to the glandular triangle. The circular clear areas with basophilic amorphous material and chromatophores are less numerous than in *M. californicus*.

Family MYLIOBATIDAE

Aetobatus narinari (Fig. 26a). The epidermis is stratified and rests upon a distinct basement membrane. Projecting into the epidermis for a short distance are small unbranching dermal papillae. The dorsolateral and ventral portions of the epidermis are divided into three distinct zones of about equal width. There is a basal zone, an intermediate secretory zone, and an outer peripheral zone. The basal zone is comprised of the usual layer of cuboidal or columnar cells four to six cells thick, having pink granular cytoplasm with blue-staining nuclei that are ovoid to elliptical in shape. The cells of the intermediate zone are definitely lysed, indicative of a holocrine type of secretion (Fig. 26b). The secretion within this zone is pink, granular, and amorphous. Scattered throughout the secretion are nuclei which tend to be pyknotic or to have undergone karyolysis and karyorrhexis. The peripheral zone is a layer of stratified squamous epithelial cells four to ten cells thick. Many of these cells have an almost water-clear cytoplasm. Chromatophores are few, show little tendency to branching, and are diffusely scattered. The remainder of the epidermis is similar in morphology and thickness to that of *M. californicus*, with the single exception that the circular clear areas with basophilic amorphous material were not observed in the sections studied.

Myliobatis aquila. Judging from Fleury's (1950) description, the histology of the epidermis of this species is similar to that of *M. californicus*.

Myliobatis californicus (Figs. 27b, 28b). The epidermis is stratified and rests upon a distinct basement membrane. There is a progressive and uniform decrease in thickness from about 40 cells in the glandular triangle to about 20 cells on the dorsal surface. Many stellate and twig-shaped chromatophores filled with brown granular pigment are uniformly scattered throughout the epidermis. The basal layer is a single sheet of columnar cells with dark blue oval nuclei. The peripheral cells become progressively more ovoid in shape. The cells of the basal one-fourth to one-third are slightly basophilic, whereas the remaining cells are acidophilic with spherical blue nuclei. Frequently there are observed circumscribed patches in the basal half of the epidermis which give evidence of secretory activity (Fig. 27b). In some areas these secretory patches may extend from the basal layer to the periphery of the epidermis, whereas in other cases the epidermis is divided into three distinct zones similar to that observed in *A. narinari*. It is believed that venom production occurs in these secretory areas. The cells within these patches are distended and elongate, with clear spaces around the nuclei having pink or yellow cytoplasm concentrated toward the pole. The clear areas within the cells are traversed by fine pink strands suggestive of coagulated protein. Some of the cells at the outer margin of the patches have undergone lysis resulting in large areas of pink, homogeneous, amorphous secretion. In other secretions (Fig. 28b), the epidermis of the glandular triangle shows an antemortem autolysis of the intermediate and basal cells resulting in large areas of homogeneous, light pink material containing the nuclei of the disrupted cells. Intercellular bridges are present in the outer one-half of the epidermis. Numerous large, clear, circular areas partially filled with slightly basophilic amorphous material are scattered throughout the epidermis but are most concentrated in the glandular triangle. The outer layer of cells is a peculiar stratified squamous type, and their peripheral cytoplasm forms a distinct pink line suggestive of either a striated border or cuticle.

Family POTAMOTRYGONIDAE

Potamotrygon motoro (Fig. 29a, b). The histology of the epidermis of this species closely resembles that of *M. californicus*.

Family RHINOPTERIDAE

Rhinoptera steindachneri. No histological data available.

Family UROLOPHIDAE

Urolophus halleri (Fig. 30a, b, c). The epidermis is stratified and rests upon a distinct basement membrane. The distribution and cellular structure of the epidermis differs widely from any of the aforementioned species. The epidermis in the dorsal region of the sting at the thickest point is about 20 cells in depth and in most places considerably less, whereas the ventral epidermis is more than 45 cells in depth at the thickest point. Numerous simple and branching dermal processes project deeply into the thicker areas of the epidermis. The dorsolateral and ventral regions of the epidermis of this species are likewise divided into three distinct zones. There are a thin basal zone of columnar cells, a remarkably broad intermediate zone of large oval cells having peripherally-placed oval vesicular nuclei and colorless cytoplasm (Fig. 30b, c), and a thin peripheral zone of modified, stratified, squamous epithelium. Some of the cells in the outer margin of the patches have undergone lysis resulting in large areas of pink, granular, amorphous secretion. These cells are believed to contain the venom. A definite cuticle is

FIGURE 27a.—Photomicrograph of a cross section of a mature intact sting of *Myliobatis californicus*. Taken at the level of the middle third of sting. At this level the osseous spine is flattened and the glandular triangle broadened. Hematoxylin and triosin stain. × 90.

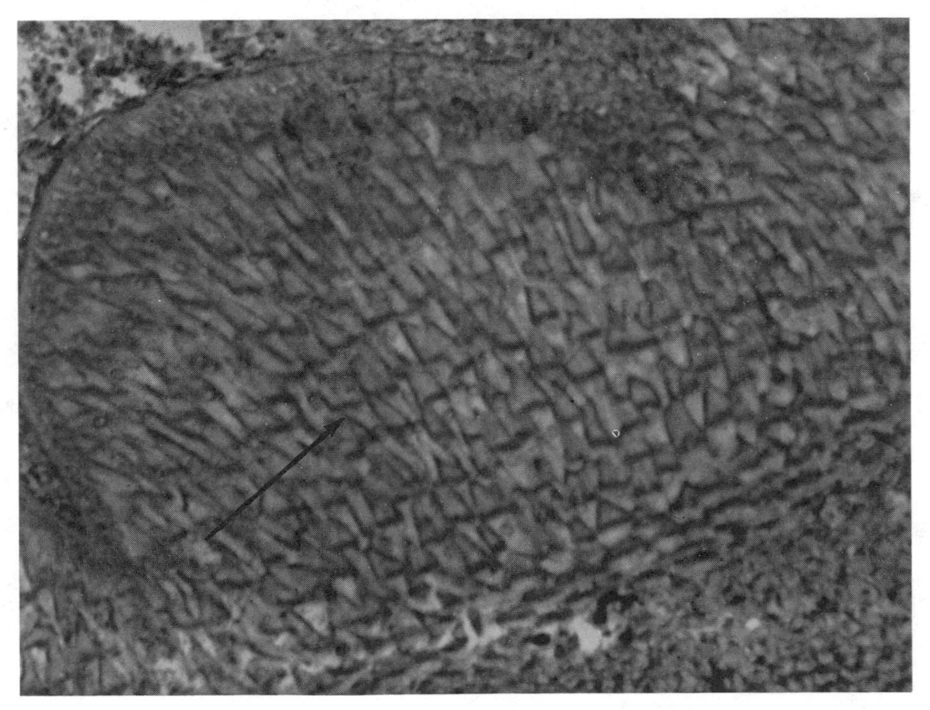

FIGURE 27b.—Enlargement of glandular triangle of *M. californicus*. Arrow points to area of greatest secretory activity. × 1000. (From Halstead, Ocampo, and Modglin)

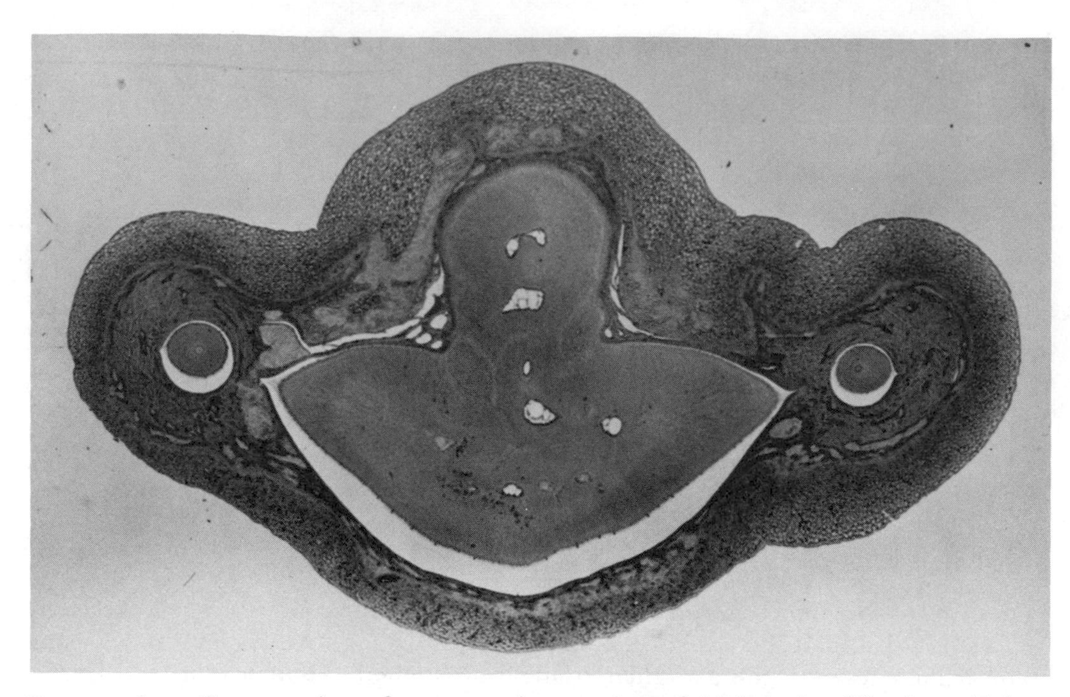

FIGURE 28a.—Cross section of a mature intact sting of *Myliobatis californicus*. Taken at the level of the distal third of sting. At this level the osseous spine and the glandular triangle are more trigonal in shape. Hematoxylin and triosin stain. × 90.

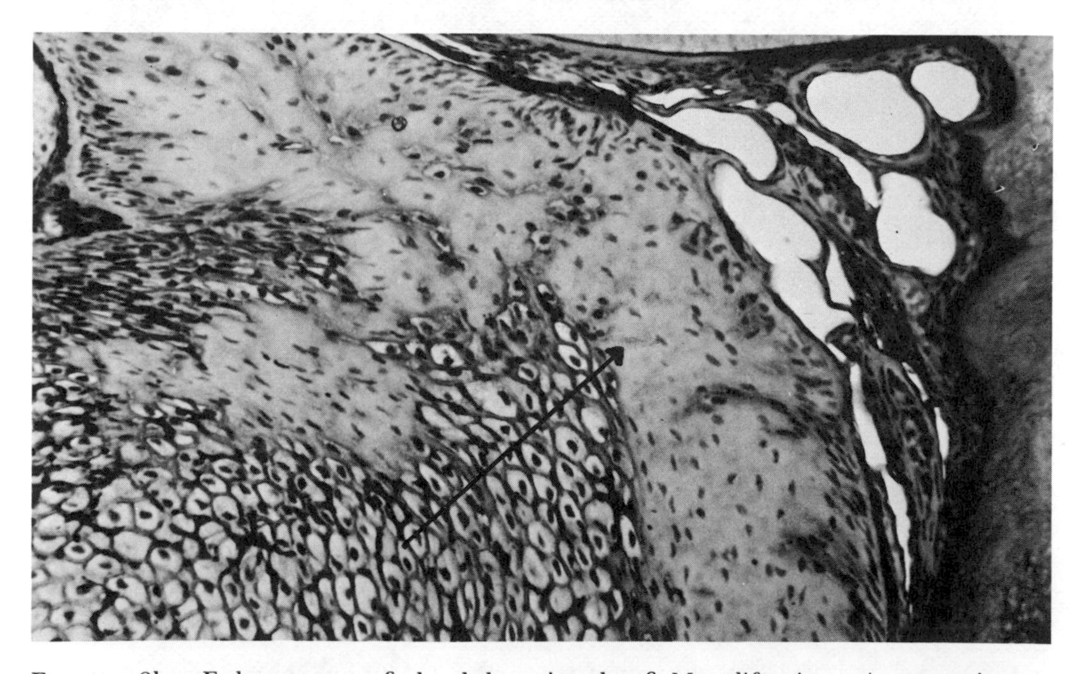

FIGURE 28b.—Enlargement of glandular triangle of *M. californicus*. Arrow points to area of lysis. Hematoxylin and triosin stain. × 220.

(From Halstead, Ocampo, and Modglin)

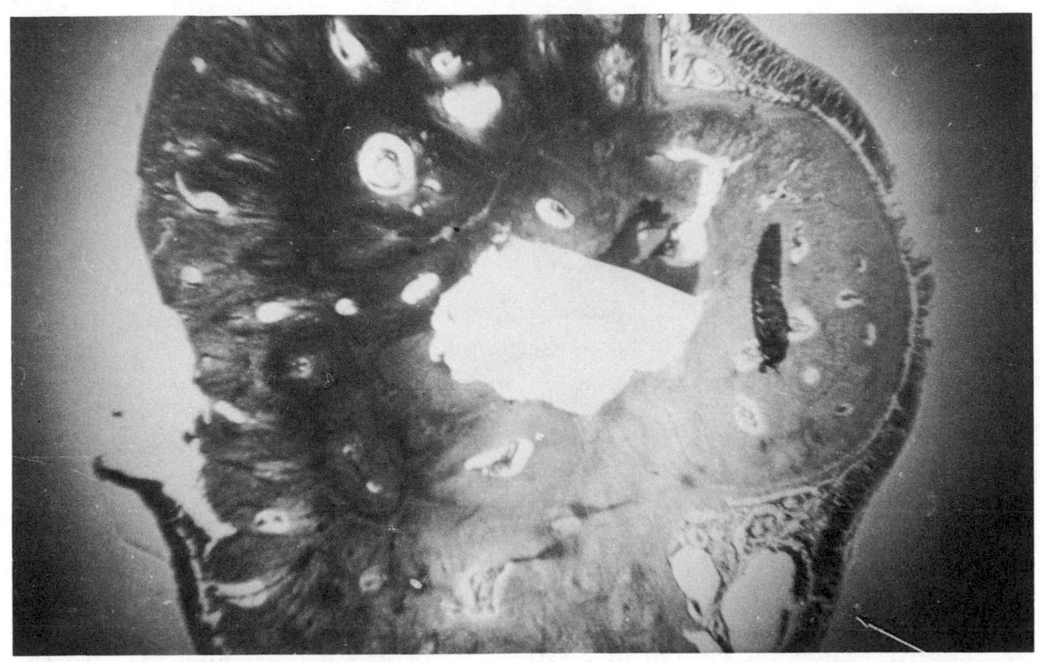

FIGURE 29a.—Photomicrograph of a cross section of a mature sting of *Potamotrygon motoro*. The large cavity in the center of the spine is an artifact due to a tear in the section. The integumentary sheath in this particular specimen is relatively thin. Arrow points to glandular triangle. Approx. × 30.

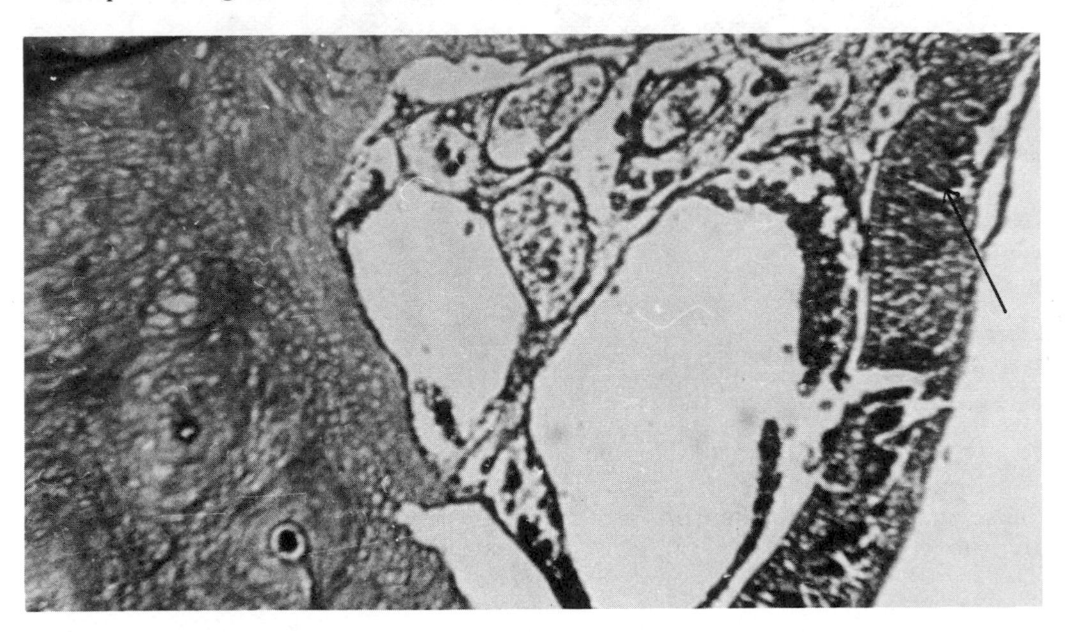

FIGURE 29b.—Enlargement of the glandular triangle of *P. motoro*. The histological details in this section are indistinct. Arrow points to glandular area. Approx. × 90. (Courtesy M. N. Castex)

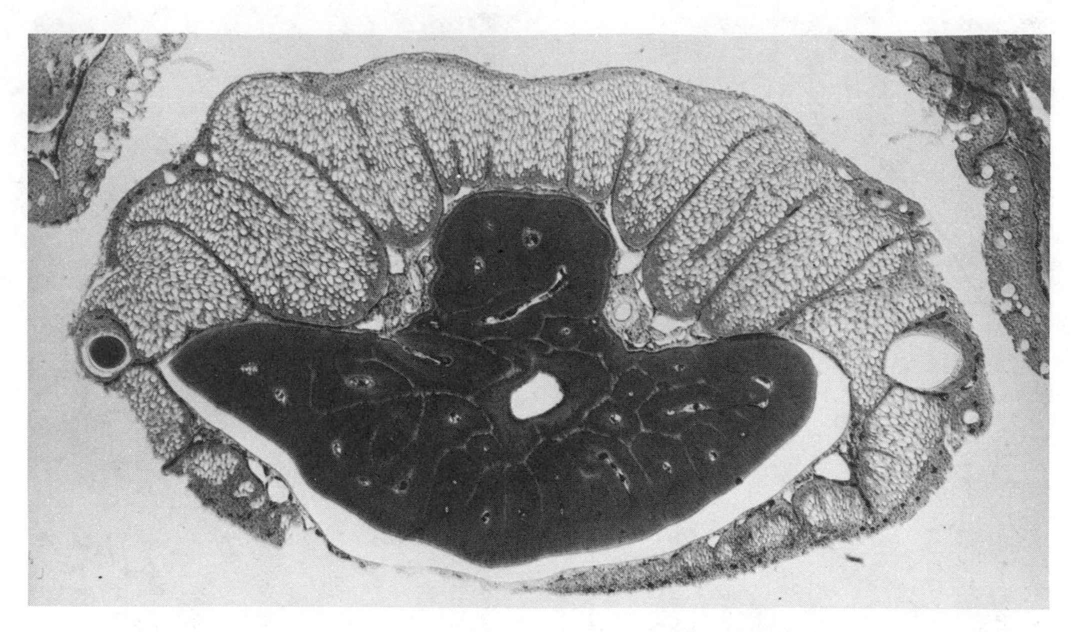

Figure 30a.—Photomicrograph of a cross section of a mature sting of *Urolophus halleri*. Note the thick dense layer of glandular epithelium which almost surrounds the entire spine. Hematoxylin and triosin. × 75.

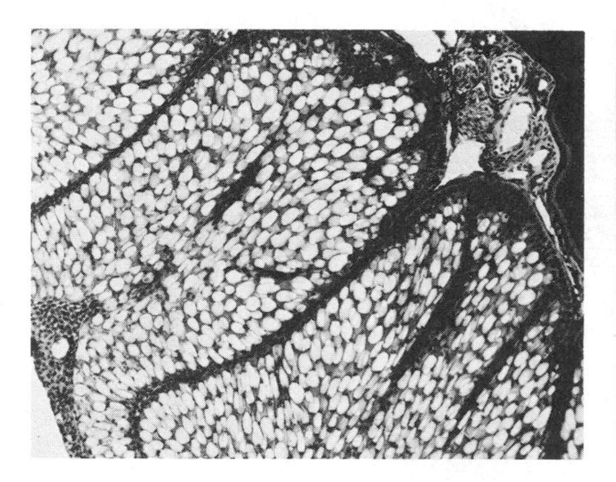

Figure 30b.—Enlargement of the glandular triangle of *Urolophus halleri*. Note the dense mass of large glandular cells. Hematoxylin and triosin. × 125.

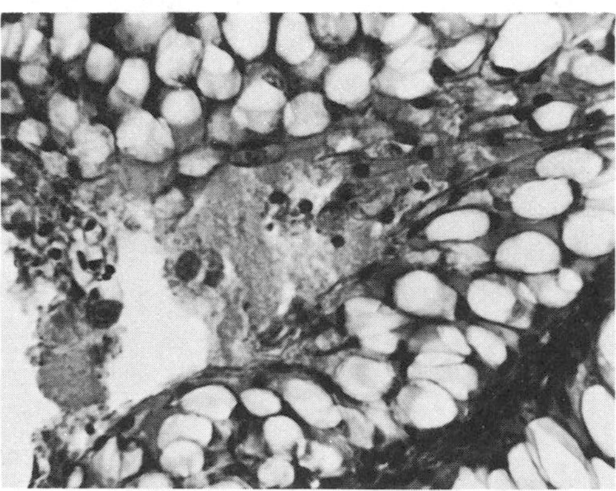

Figure 30c.—Enlargement of glandular epithelium of *Urolophus halleri* showing evidence of secretory activity. Hematoxylin and triosin stain. × 450.
(From Halstead,, Ocampo, and Modglin)

present. Chromatophores are few, most of them unbranching and diffusely scattered. Circular clear areas containing amorphous material are present but few in number.

Discussion on the Structure of the Venom Apparatus: Evans' early works (1916, 1921) on the stingray venom apparatus erroneously depicted a scheme of canals, follicles, nipples, and secretory filaments in the ventrolateral-glandular grooves of the sting of *D. pastinaca* (Fig. 31). However, Evans (1923) later rectified these errors with more accurate figures, photomicrographs, and this statement:

> The micro-photographs show clearly the points which I have endeavored to describe, and clearly do not give any support to the observations of Porta. At the same time they effectively nullify the views I held in 1916 as to the significance of the canals, nipple-shaped projections and filamentous processes.

It is clear from a study of Evans' photomicrographs that he sectioned traumatized stings, hence the epithelium in the glandular triangles which he has illustrated is fragmentary. Furthermore, the "lateral flaps" which he describes are apparently nothing more than badly torn areolar connective tissue. No "alveolar"[2] connective tissue is apparent in the glandular triangle of any of the species which we have examined thus far. The "alveoli" which Evans mentions are probably vascular structures.

More recently Fleury (1950) and Castex and Loza (1964) reiterated Porta's (1905) suggestion that the venom is produced by glandular acini located in the dermis and that the venom was discharged from these acini through a network of ducts. As Russell (1965) has already noted, these "ducts" described by Fleury closely resemble vascular channels. However, in a subsequent communication from Castex (1964), he indicated that he has discarded his previous views on this matter and now believes that venom production occurs in the epidermal layer. Castex further concurs with Evans (1923), Halstead *et al.* (1955), and Russell (1965) in that the "venom ducts" are vascular channels.

It is believed that the venom is produced largely by the glandular epithelial cells which are concentrated in the ventrolateral-glandular grooves and the ventral portion of the integumentary sheath. Histological evidence indicates that venom production in the species studied, at least in the epidermis, is the result of a holocrine type of lysis. To what extent, if any, the acini within the dermis of the glandular triangles contribute to venom production is not known. Earlier toxicological studies by Holloway *et al.* (1953) on *U. halleri* also indicate that the ventrolateral-glandular groove of the sting is the primary site of venom production.

MECHANISM OF INTOXICATION

Venom is secreted and introduced into the body of the victim by the sting or venom apparatus. Campbell (1951) has published the only critical analysis

[2] The term "alveolar" as published in an earlier report on *M. californicus* by Halstead and Modglin (1950) should be corrected to read "areolar connective tissue."

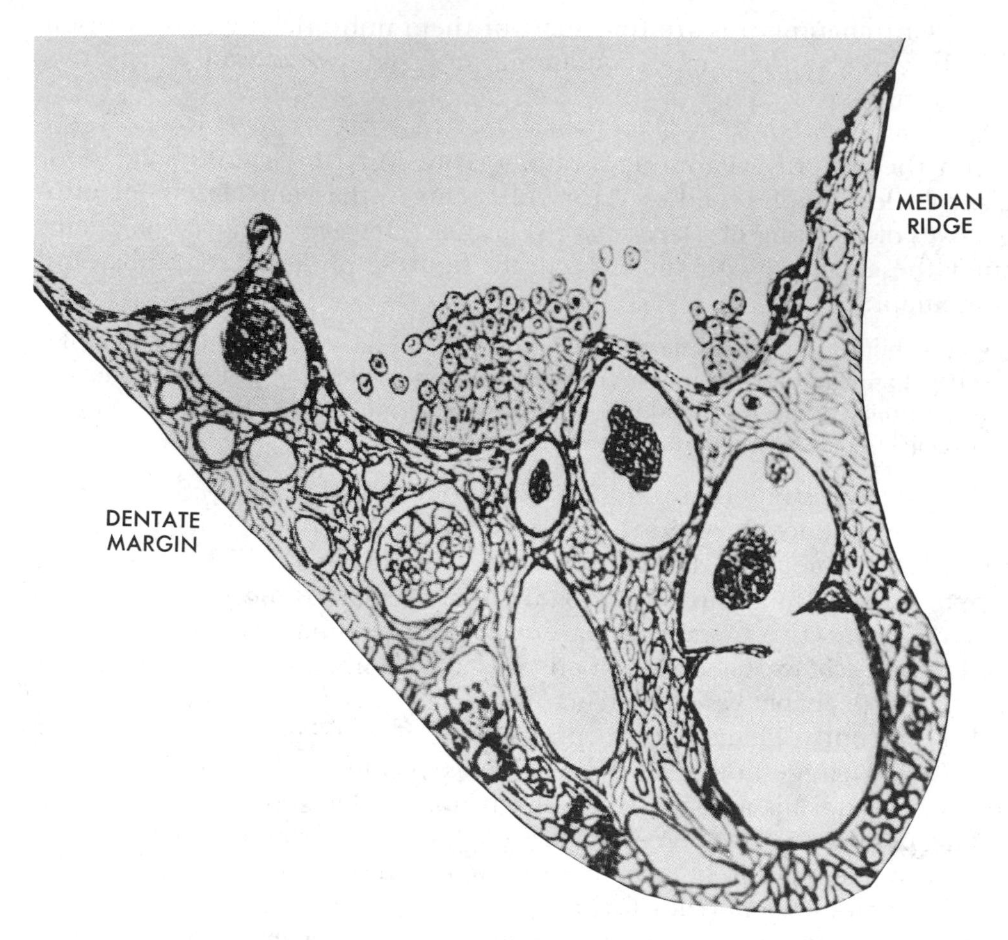

FIGURE 31.—Drawing of a longitudinal section of the sting of *Dasyatis pastinaca*. (From Evans)

of the stinging action of stingrays that has been found to date. "Though this fish [*Urolophus*] *halleri* is greatly feared by bathers, in the aquarium it proved very reluctant to sting." He found that in order to evoke the stinging response, it was necessary to pin the fish to the bottom of the tank by the sudden application of a large rubber suction disc on the end of a stick (Fig. 32). This technique produced repeated thrusts with the sting. In rays in which the spinal cord had been severed, the stinging reaction followed stimulation of the caudal one-third of the pectoral fins and the entire surface of the pelvic fins. When the trunk was stimulated there was no response. His experiments revealed that the stinging was always directed to the side by a bending of the tail toward the evoking stimulus. The tip of the tail took no active part in the stinging action. Halstead and Bunker (1953) reported observing the stinging action of *U. halleri* during the capture of a large number of specimens.

The stimulus consisted of poking the back of the fish with a blunt stick. It was noted that for a quick stinging response the stimulus had to be applied to the posterior third of the ray's body. The stinging reaction time rapidly decreased as the application of stimuli progressed in the anterior direction. The stinging response in the series of rays tested was found to be consistent but varied with the type of stimulus used. When a stick was applied to the posterior third of the back, the ray immediately arched the tail through a vertical plane over its back and struck with precision at the probe. There were no horizontal movements or general lashing of the caudal appendage but a very precise thrust. When a more diffuse stimulus was applied to the posterior third of the ray's body, both vertical and horizontal movements were observed. At a later date a second series of urobatids were captured, and it was noted that both the vertical and the horizontal type of thrust as observed by Campbell were used.

The sharp, arrowlike tip and backward-pointing teeth of the stingray spine make an effective weapon. The spine is so constructed that penetration of the victim's flesh is accomplished with ease, but the barb is difficult to remove and lacerates the tissues as it is withdrawn. The sheath of the sting is easily damaged. When the sting penetrates the flesh, the sheath is torn, the dentate margins of the sting are exposed, and venom is released. Laceration of the tissues aids in the absorption and distribution of the toxin, thus producing a violent tissue reaction. The stinging action of rays has also been discussed by Russell (1953*b*) and Russell and Lewis (1956).

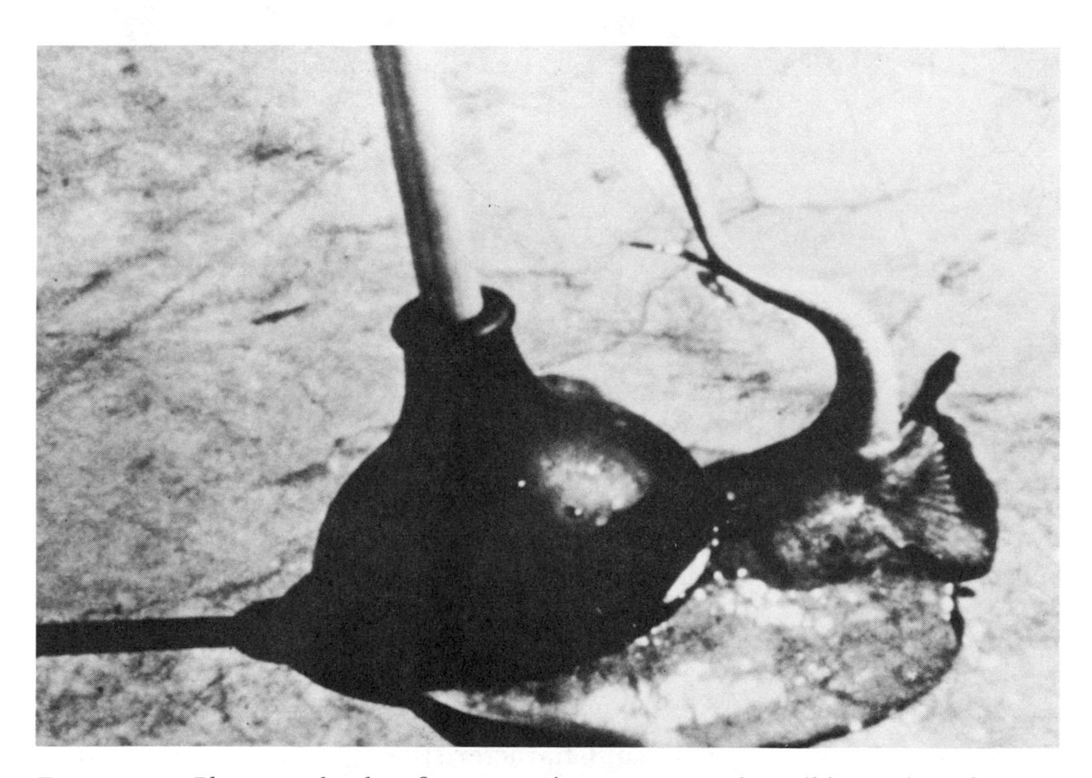

FIGURE 32.—Photograph taken from a movie sequence on the striking action of *Urolophus halleri*. The striking position of the caudal appendage can be clearly seen as the rubber plunger presses against the back of the ray. (Courtesy P. Saunders)

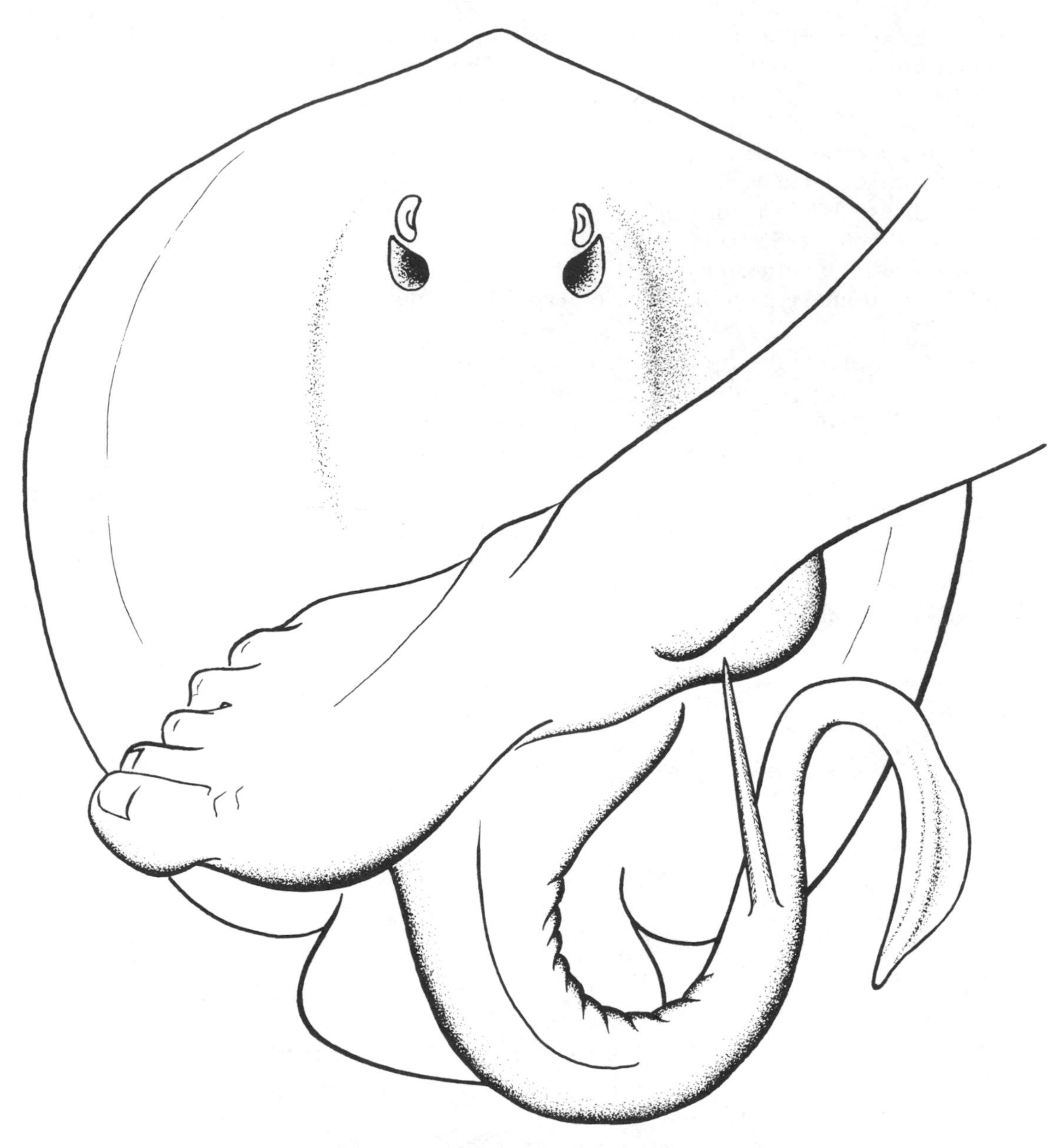

FIGURE 33.—Drawing showing the method by which a stingray usually inflicts its sting. (R. Kreuzinger)

MEDICAL ASPECTS

Clinical Characteristics

Pain is the predominant symptom and usually develops immediately or within a period of 10 minutes following the attack. Russell (1953*b*, 1959, 1961*a, b,* 1965), Russell *et al.* (1958*b*), Russell and Lewis (1956), who have made an extensive study of the clinical characteristics of stingray attacks, have found that the onset of pain usually embraces an area of approximately 10 cm in diameter about the wound. The pain generally becomes more severe during the first 30 minutes and may radiate to include the entire extremity. The pain usually attains its maximum intensity in less than 90 minutes and although gradually diminishing may continue for 6 to 48 hours. The pain is variously described as sharp, shooting, spasmodic, or throbbing in character. Potamotrygonids or freshwater stingrays are reputed to cause extremely painful wounds (Schultz, 1944; Castex *et al.*, 1963; Castex and Loza, 1964). More generalized symptoms of fall in blood pressure, vomiting, diarrhea, sweating, arrythmia, muscular paralysis, and death have been reported. Russell (1953*b*, 1965) stated that he has never observed either a spastic or flaccid paralysis in a stingray victim and suggests that the "spasms" may be due to "contractures initiated as flexion reflexes stimulated by the intense pain." This is further substantiated by the fact that medications directed at alleviating the pain also relieved the muscular spasm. The nausea, faintness, vertigo, and bradycardia so often experienced by the victim within the first few minutes after the accident are believed to be attributable to the primary shock caused by the extreme pain. Russell believes that the fall in systemic arterial pressure is due to peripheral vasodilatation, which occurs when small amounts of stingray toxin are experimentally administered intravenously, and that mechanisms related to this fall may accentuate or even precipitate a transient cerebral anoxis and shock. Deaths resulting from stingray wounds have been reported by Mordecai (1860), Crevaux and Le Janne (1882), Gudger (1943), Wright-Smith (1945), Russell (1953*a*, *b*, 1959), and Russell *et al.*, 1958*a* (Figs. 34; 35).

FIGURE 34.—Drawing illustrating probable contact between 11-year-old boy and *Dasyatis americana* at Point Bolivar, Texas. Victim died 2½ days later. He was stung July 4, 1954. (Courtesy F. Russell)

FIGURE 35.—Photograph of *Dasyatis longus* which killed a 12-year-old boy near San Felipe, Baja California. Victim was stung October 12, 1955. (Courtesy F. Russell)

Stingray wounds are either of the laceration or puncture type (Fig. 36). Penetration of the skin and underlying structure is usually accomplished without serious damage to the surrounding tissues. However, withdrawal of the sting may result in extensive tissue damage due to the retrorse serrations of the spine. Russell cited one case in which a sting having a width of 5 mm produced a wound 3.5 cm in length. Cleland (1942) tells of a fisherman stung by a giant 3 m Australian stingray, which drove its sting through the lower third of the man's leg between the tibia and fibula leaving a wound 17.7 cm on the lateral aspect of the leg and 10 cm long on the medial side. The lacerated edges of a stingray wound bleed freely but not abnormally so. Swelling in the vicinity of the wound is a constant finding. Crevaux and Le Janne (1882) reported a case in which the edema extended from the right heel to the epigastrium. The edema may persist despite elevation of the extremity which suggests pathological changes in the locally affected tissues. Russell believes that much of the localized edema can be attributed to lymphatic obstruction, precipitated by inflammation, and to damage to the

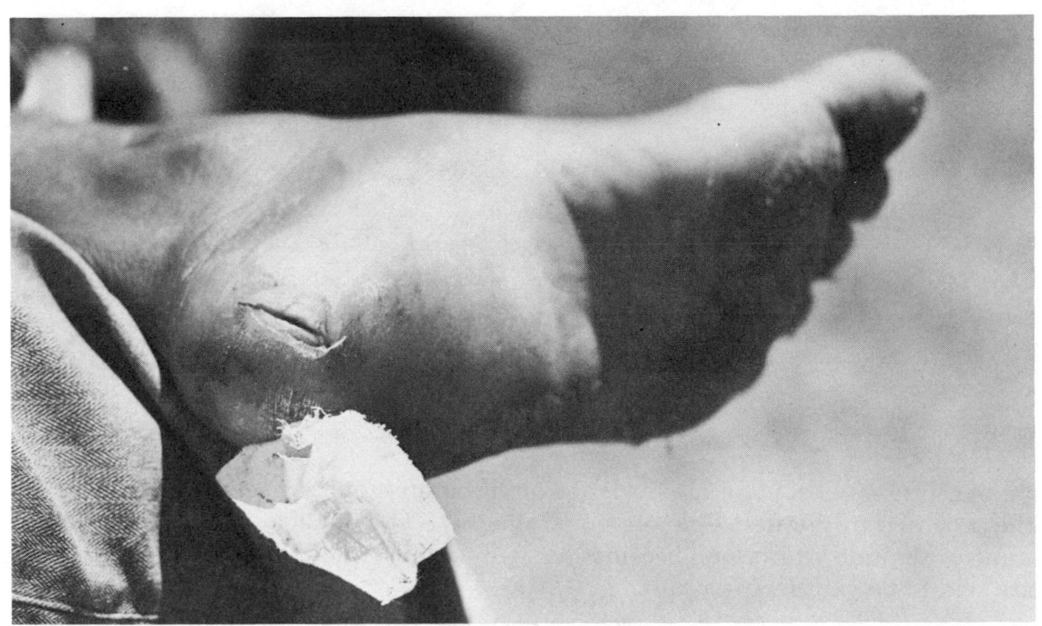

FIGURE 36.—Stingray wound inflicted by *Aetobatus narinari*. Vera Cruz, Mexico. (From Halstead and Bunker)

lymphatics and supporting structures. These changes are said to be due to the direct action of the venom. The area about the wound at first has an ashy appearance, later becomes cyanotic and then reddened. Tissue necrosis in the vicinity of the wound frequently occurs, suggesting proteolytic properties of the venom. Although stingray injuries occur most frequently about the ankle joint and foot as a result of stepping on the ray, Ronka and Roe (1945), Wright-Smith (1945), and Russell *et al.* (1958*a*) have reported instances in which the wounds occurred in the chest about the heart (Fig. 34), and Cadzow (1960) cited a puncture wound of the liver by a stingray spine.

Castex *et al.* (1963) and Castex and Loza (1964) have prepared an excellent review of the clinical aspects of attacks from freshwater stingrays of South America. They have termed the clinical entity of freshwater stingray attacks the "paratrygonic syndrome," in which there are only minimal systemic symptoms, but the lesions frequently progress to a state of chronic ulceration and gangrene (Fig. 37a, b).

Pathology

No specific pathological studies have been reported on marine stingray attacks.

Treatment

The following recommendations are based on the clinical investigations of 1,725 cases of stingray attacks over a period of 15 years by Russell (1953*b*, 1965) and his associates, who have had more first-hand experience in this field than any other group.

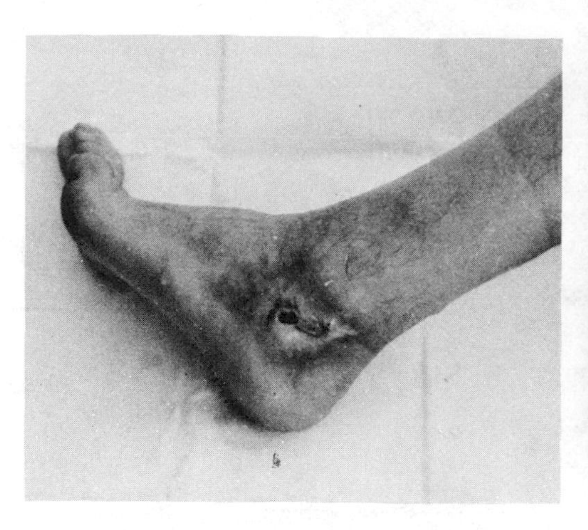

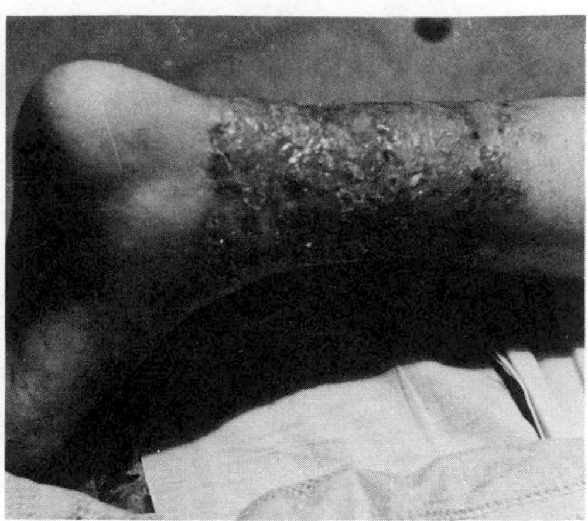

FIGURE 37a.—(Left) A typical early lesion of the paratrygonic syndrome resulting from the sting of a freshwater potamotrygonid stingray. The initial lesion is a laceration or puncture wound which soon becomes necrotic and ulcerated. Potamotrygonid stings are said to be extremely painful.

FIGURE 37b.—(Right) If improperly treated the chronic skin ulcer in the paratrygonic syndrome gradually becomes gangrenous, and the leg may have to be amputated.
(Courtesy M. M. Castex)

Efforts of treatment should be promptly and vigorously instituted. The treatment is directed toward alleviating pain, combating the effects of the venom, and preventing secondary infection. Successful results are largely dependent upon the rapidity with which treatment is instituted. The victim should immediately irrigate the wound with the cold saltwater at hand. This procedure facilitates removal of the venom, and the cold water tends to act as a vasoconstrictor thus reducing the amount of absorption of the poison while serving as a mild anesthetic agent. A tourniquet may be applied immediately above the stab site but must be released every few minutes in order to preserve circulation. The wound should be explored carefully for evidence of pieces of the sting's integumentary sheath. All pieces of integumentary sheath must be completely removed, or envenomation will continue and the results of the treatment will be greatly impaired. As soon as the wound has been thoroughly cleansed the injured member should be soaked in hot water. The water should be maintained at as high a temperature as the patient can tolerate without producing further injury to the tissues. Soaking should continue for a period of 30 to 90 minutes. The addition of magnesium sulfate to the water is sometimes desirable because of its mild anesthetic properties. The addition of other anesthetic and antiseptic agents is optional. Following the soaking procedure the wound should be debrided, cleansed, and closed with dermal sutures. The use of antitetanus agents is recommended. Antibiotic agents may be required. The use of intramuscular or intravenous demerol has been found effective in controlling the pain. The primary shock

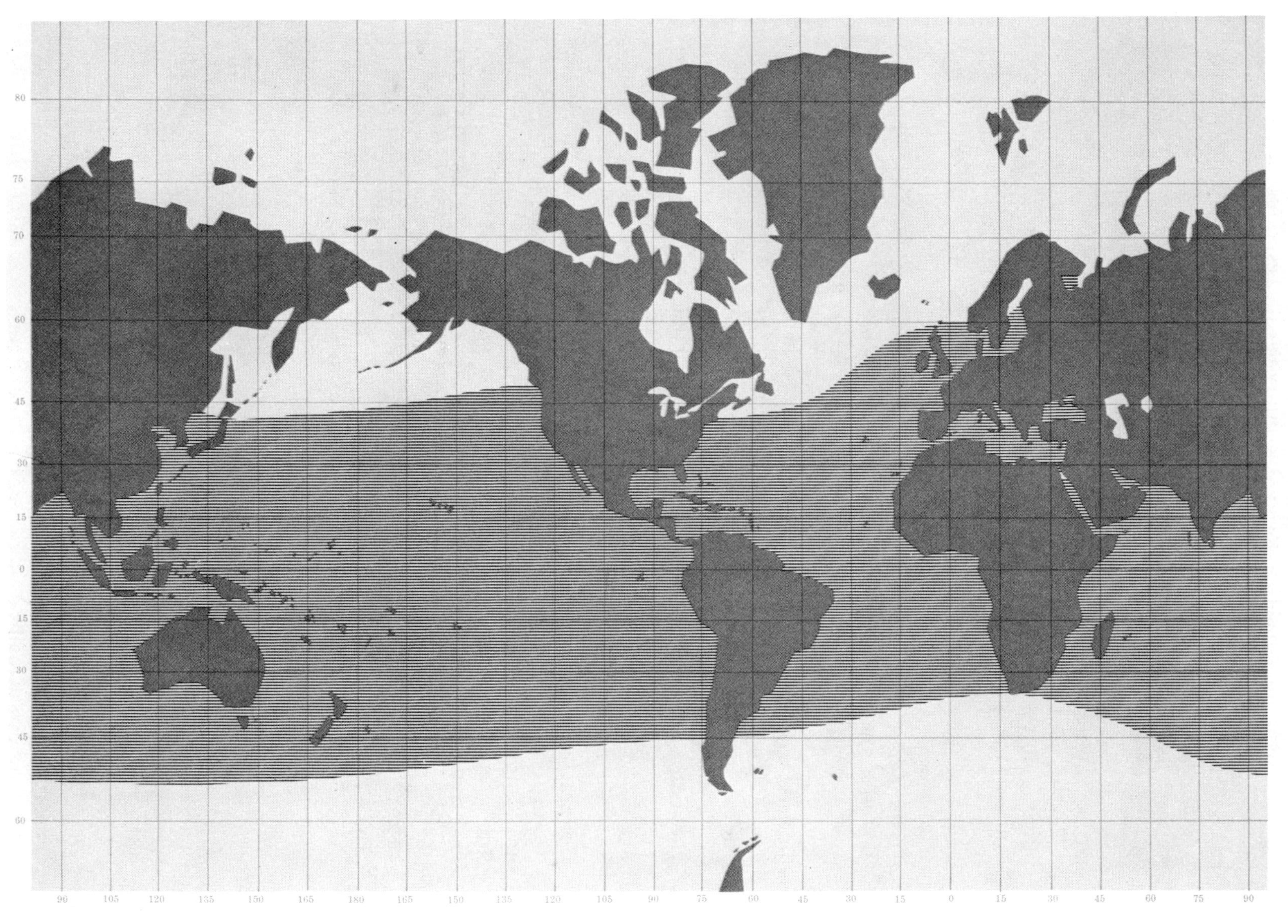

MAP 1

Map showing the approximate geographical distribution of stingrays. It will be noted that stingrays are primarily inhabitants of the warmer latitudes. (L. Barlow)

so often seen immediately following the injury usually responds satisfactorily to routine supportive measures. However, the secondary shock resulting directly from the action of the venom on the cardiovascular system may require immediate and vigorous therapy. Treatment should be directed toward maintaining cardiovascular tone and preventing of any further complicating factors. Elevation of the injured member is advisable.

The use of potassium permanganate, ammonia, and cryotherapy (Mullins, Wilson, and Best, 1957) is not only useless but may even have adverse effects. They are not recommended for the treatment of stingray stings. For further reading on the treatment of stingray attacks see Evans (1943), Halstead and Bunker (1953), Russell and Lewis (1956), Halstead (1959), Russell (1959), and Halstead and Mitchell (1963).

Prevention

It should be kept in mind that stingrays commonly lie almost completely buried in the upper layer of a sandy or muddy bottom. Stingrays are therefore a hazard to anyone wading in water inhabited by them. As Schultz (1944) and others have pointed out, the chief danger is in stepping on one that is buried. This danger can be largely eliminated by shuffling one's feet along the bottom. Usually the body of the ray is pinned down by the weight of the victim, thereby permitting the beast to make a successful strike. Pushing one's feet along the bottom eliminates this danger and at the same time routs the stingray from its lair. It is also recommended that a stick be used to probe along the bottom in order to rid the area of lurking rays.

PUBLIC HEALTH ASPECTS

Stingrays are tropical and subtropical fishes, only a few species range into warm temperate latitudes. Stingrays constitute a moderate public health problem in coastal areas, particularly during the warmer months of the year when beach swimming and wading are popular pastimes (Map 1). There are no accurate statistics as to the worldwide incidence of stingray attacks since it is not a reportable item. Russell (1959, 1961*a*) has estimated that there are approximately 750 stingray attacks each year in North America. Over a period of 15 years, Russell (1965) collected data on 1,725 cases of stingray injury, most of them having occurred in the United States.

The bathymetric distribution of stingrays along beach areas is noteworthy. In a series of 1,000 specimens of *Urolophus halleri* that were captured along Southern California beaches and then released by Russell (1953*a*, 1955, 1965) and his workers between 1951 to 1958, it was found that only 14 rays were retaken at a distance greater than 25 km from their original point of capture. There were 807 of the rays recaptured within 5 km of the point at which they were tagged. None of the rays was found over 15 fathoms deep or more than 3 km from shore.

TOXICOLOGY

Vellard (1931, 1932) was the first to demonstrate by toxicological research that stingrays actually produce a venom, although the earlier histological studies of Porta (1905), Pawlowsky (1907), and Evans (1916, 1921, 1923) had provided ample morphological evidence that their stings were venom-producing organs. Vellard worked with the South American freshwater stringray *Potamotrygon motoro*. He obtained crude extracts by scraping the ventrolateral-glandular grooves of the sting with a scalpel and triturating the material in both distilled water and physiological saline solution. The solution gave an acid reaction with litmus and was highly toxic when injected into animals. Tests were made by injecting crude extracts and by stabbing the animals with intact stings. The venom was tested on dogs, rabbits, mice, pigeons, reptiles, and batrachians. Dogs usually survived the crude extract injections but frequently died when wounded with the intact sting. Intravenous and intramedullary routes of administration were found to be the most efficacious. Vellard found that the venom from one-fourth of a sting killed a 30 g mouse within 10 to 20 minutes. Boiling, strong acids, and alcohol were said to destroy the action of the venom. Solutions of the venom stored at 5° C were reported to lose their potency within a few days.

Fleury (1950) conducted a series of toxicological experiments on guinea pigs in which he stuck them with spines from living specimens of *Myliobatis aquila* and *Dasyatis pastinaca*. The live rays were placed in aquaria, and the guinea pigs were strapped to boards. The sting of the ray was then grasped with a pair of forceps, and the guinea pigs were stabbed in the thigh with the stings. Several different tests were made using stings that were intact and stings in which the integumentary sheath had been completely removed. Fleury also injected a guinea pig that had been sensitized to stingray venom by previous injection. In order to determine the site of venom production he prepared extracts from the "glandular region" of the integumentary sheath and injected the crude venom by needle. The guinea pigs injected with the intact sting showed evidence of pain, developed dyspnea, rapid shallow breathing, paralysis, agitation, and clonic convulsions, but gradually recovered after several hours. The guinea pigs that had been stabbed with the sting in which the integumentary sheath had been completely removed showed evidence of pain but did not exhibit any of the other pathological effects. The guinea pig that had been previously sensitized developed only rapid breathing when it was injected. The guinea pigs that received injections of the crude glandular venom extract showed evidence of intense pain and all the pathological signs exhibited by the animals injected by the intact stings. Fleury was unable to detect any significant differences in the results obtained from the stings of *M. aquila* and those obtained from *D. pastinaca* and thus concluded that the venoms are similar in their toxicological properties. He felt that this experimental evidence substantiated his conclusion that a venom is produced in the glandular region of the integumentary sheath of the sting.

Working with a series of 125 specimens of *Urolophus halleri* taken from Seal Beach, Calif. and another series of 12 specimens from Mission Bay, Calif., Holloway *et al.* (1953) demonstrated that this species also produces a venom. Material for extracts was obtained by scraping the integumentary tissue from the ventrolateral-glandular grooves of the spine, cuneiform area, caudal integument from other than the cuneiform area, and slime associated with the sting. Distilled water extracts were prepared and injected into white mice. Toxic symptoms appeared to consist initially of markedly increased abdominal breathing and motor hyperactivity which at first seemed an exaggeration of purposeful movements but later reflected an apparent inhibition of self-control. Motor ataxia followed, and the mouse sometimes exhibited running motions for a few moments while lying incapacitated on its side. Dyspnea became marked, and paradoxical breathing with ultimately complete paralysis of the intercostal muscles was often seen. The tail and ears exhibited varying degrees of cyanosis. Violent convulsions soon began, and within seconds a marked clonus of the extremities was observed. Reflexes appeared to be hyperactive. The agonal period was approximately 10 seconds. Complete respiratory arrest occurred just before all motions ceased. Inactivation of the toxin, probably by autolysis, was inhibited but not prevented by temperatures as low as -7° C and was complete in about 72 hours. The toxic principle was found to be concentrated in the epithelium lining the ventrolateral-glandular grooves of the sting. Integument taken from the caudal appendage was nontoxic. Only one sample from the cuneiform area proved to be toxic and was regarded as of questionable significance. The authors concluded that the toxic substance acted as a convulsant capable of causing death in mice by respiratory arrest.

The toxicology of stingray venom has been most extensively investigated by Russell and his group, who have done most of their research on the California round stingray *U. halleri*. Some of the most important papers on this subject are: Russell and van Harreveld (1954), Russell *et al.* (1957), Russell *et al.* (1958*a*), Russell *et al.* (1958*b*), and Russell and Long (1960). A complete listing of their articles relating to toxicology appears in Russell (1965). Their findings can be summarized as follows: The crude venom loses its toxicity within 4 to 18 hours by standing at room temperature. It is more stable at lower temperatures or in 20 to 40 percent glycerol. Most of the toxicity is lost in lyophilization. The crude venom has been tested on frogs, mice, rats, cats, and monkeys. The details as to the pharmacological properties of stingray venom are discussed in the following section. The LD_{99} for stingray venom has been calculated as 28 mg dried crude venom per kg mice (Russell *et al.*, 1958*b*). Studies have also been conducted on the venoms of *D. dipterurus* and *M. californicus*, but they appear to be toxicologically similar to the venom of *U. halleri* (Russell *et al.*, 1958*b*). A small amount of crude venom was obtained from *Gymnura marmorata*, but the amount was too meager and the action of the venom too weak to permit analysis.

Castex *et al.* (1963) have conducted a large series of experiments on the toxicology of South American freshwater stingrays. He has also studied experimentally the pathological effects of the lesions produced in laboratory animals. Since the bulk of this work is concerned with freshwater stingrays, it is beyond the scope of this present volume. Readers having further interest in this subject should read Castex's original report.

PHARMACOLOGY

The most comprehensive studies on the pharmacological properties of stingray venom have been conducted by Russell and his associates working primarily with the venom of *Urolophus halleri*.

Cardiovascular system. Stingray venom has a deleterious effect on the vertebrate cardiovascular apparatus. The action on the blood vessels appears to be diphasic. Low concentrations of the venom give rise to simple peripheral vasodilatation or vasoconstriction. With massive doses the venom causes vasoconstriction without a preliminary period of dilatation. The most obvious effect, and perhaps the more important one, is that of vasoconstriction. This effect has been observed in all the blood vessels examined. Some of the most serious effects were those directly upon the heart. The most consistent change seen in the electrocardiographic pattern of cats that were injected with small amounts of the venom was bradycardia with an increase in the PR interval giving a first, second, or third degree atrioventricular block. The second degree block was usually followed by sinus arrest. Reversal of the small dose effect occurred within 30 seconds following the end of the injection. When cats were given larger amounts of the venom they showed almost immediate ST, T wave change indicative of ischemia in addition to the PR interval change and, in some animals, true muscle injury. High concentrations of the venom caused marked vasoconstriction of the large arteries and veins as well as the arterioles. The direct effects on the heart muscle are quite drastic. The venom produces changes in the heart rate and amplitude of systole and may cause complete, irreversible, cardiac standstill. It appears that stingray venom affects the normal pacemaker. The new rhythm evoked following cardiac standstill is frequently irregular and is believed to be elaborated outside the sino-atrial node (Russell and van Harreveld, 1954; Russell *et al.*, 1957).

Respiratory system. Stingray venom depresses respiration. Although part of the respiratory depression is secondary to the cardiovascular changes, the venom may have a direct effect on the respiratory centers of the medulla.

Miscellaneous effects. Stingray venom produces many changes in the behavior of animals. Some of these changes can be attributed to the direct effects of the venom on the central nervous system. In mammals the venom occasionally produces convulsive seizures, but the mechanism of these seizures is not apparent. They may be due in part to cardiovascular failure. Seizure patterns were not reported in electroencephalograms from anesthetized animals (Russell *et al.*, 1958*a*). The venom does not seem to have a deleterious effect on neuromuscular transmission (Russell and Long, 1960; Russell and Bohr, 1962). When the venom is injected into the lateral ventricles of mammals, it produces transient apathy, astasia, and licking motions (Russell and Bohr, 1962). Mice injected with lethal doses of venom developed hyperkinesis, prostration, marked dyspnea, blanching of the ears and retina, and exophthalmos. These signs were followed by complete atonia, gasping respiratory movements, coma, and death. A similar syndrome was observed in cats and monkeys including ataxia, dilated pupils, increased salivation, micturation, defecation, marked atonia, cyanosis, and hypoactive or absent deep and superficial reflexes. One monkey exhibited a tonic-clonic generalized motor seizure accompanied by hypersalivation, twitching of the head, and marked dilatation of the pupils (Russell *et al.*, 1958*a*; Russell, 1965).

CHEMISTRY

There is very little information available regarding the chemistry of stingray venoms. The most specific data are those provided by Russell *et al.* (1958*a, b*) and Russell (1965). The freshly prepared water extract of crude venom prepared from *Urolophus halleri* is described as clear, colorless, or faintly gray in color. The pH was 6.76. The crude extract loses its toxicity within 4 to 18 hours upon standing at room temperature but is more stable at lower temperatures or in 20 to 40 percent glycerol. The venom will not tolerate lyophilization. The total protein content was found to be approximately 30 percent, total nitrogen 3 percent, and total carbohydrate 3 percent. Ten amino acids have been found to be present. With the use of disc electrophoresis they have identified 15 fractions in extracts from the venomous integumentary sheath of *U. halleri*. Extracts prepared from sponges that had been stabbed with fresh stings were found to contain 10 fractions. Further studies on these extracts using gel filtration (Sephadex G 100 and G 200) suggested that the toxic protein, or proteins, may have a molecular weight in excess of 100,000. The fraction of the toxin having the greatest lethality was found to have two or three distinct bands when subjected to disc electrophoresis. Crude venom extracts have been shown to contain serotonin, 5-nucleotidase, and phosphodiesterase. Protease and phospholipase activity were absent.

Venomous Chimaeras

The subclass Holocephali includes the chimaeras, elephantfishes, or ratfishes, as they are variously termed. They are a group of cartilaginous fishes having a single external opening on either side covered by an opercular fold with cartilaginous supports leading to a common branchial chamber into which the true gill clefts open. Externally, chimaeras are more or less compressed laterally, tapering posteriorly to a slender tail. The snout is rounded or conical, extended as a long pointed beak, or bearing a curious hoe-shaped proboscis. There are two dorsal fins. The first dorsal is triangular, usually higher than the second, and edged anteriorly by a strong-pointed bony spine which serves as a venom apparatus.

Our knowledge regarding venomous chimaeras is also scanty and is limited to two species, *Chimaera monstrosa* of the North Atlantic Ocean and *Hydrolagus colliei* of the Pacific Ocean. Evans (1923, 1943) has made brief reference to the dorsal spine of *H. affinis* but did not describe the venom apparatus if such is present.

Some of the more important works on the systematics and biology of chimaeras are by Jordan and Evermann (1896-1900), Joubin (1929-38), Whitley (1940), Fowler (1941), Bigelow and Schroeder (1953), and Marshall (1964).

LIST OF CHIMAERAS REPORTED AS VENOMOUS

Phylum CHORDATA

Class CHONDRICHTHYES

Subclass HOLOCEPHALI

Order CHIMAERIFORMES (CHIMAERAE): Chimaeras

Family CHIMAERIDAE

Chimaera monstrosa Linnaeus (Pl. 13, fig. a). Chimaera (USA, England), chimera (Italy), seeroche (Germany), chimère (France), peixe rato (Portugal), quimera (Spain), khimery (USSR). *224*

DISTRIBUTION: North Atlantic from Norway and Iceland to Cuba, the Azores, Morocco, and the Mediterranean Sea; also South Africa.

SOURCES: Evans (1923, 1943), Halstead and Bunker (1952), Phillips and Brady (1953), Halstead (1956, 1959), Castex (1967).

Hydrolagus affinis (Capello) (Pl. 13, fig. b). Chimaera (USA, England), chimera (Italy), seeroche (Germany), chimère (France), peixe rato (Portugal), quimera (Spain), khimery (USSR). *224*

DISTRIBUTION: Atlantic Ocean from Cape Cod to Portugal.

SOURCES: Evans (1923, 1943), Halstead and Bunker (1952), Russell (1965).
OTHER NAME: *Chimaera affinis*.

Hydrolagus colliei (Lay and Bennett) (Pl. 13, fig. c). Ratfish (USA). *224*

DISTRIBUTION: Pacific coast of North America.

SOURCES: Halstead and Bunker (1952), Phillips and Brady (1953), Halstead and Mitchell (1963), Russell (1965).

BIOLOGY

Family CHIMAERIDAE: Chimaeras have a wide geographical range, extending from north temperate to south temperate waters. However, the specific range of the individual members of this order indicates a definite preference for cooler temperatures. The depth range extends from the surface down to 1,400 fathoms or more. The vertical distribution of *Hydrolagus colliei* differs from most other chimaeras. Goode and Bean (1895) reported that the Pacific ratfish may be seen swimming at the surface of the water in southeastern Alaska and British Columbia, but it generally inhabits deeper water southward. The usual depth range for this species is given as 5 to 60 fathoms, while most other chimaeras inhabit depths of 100 to 1,900 fathoms or more. Chimaeras are weak swimmers, propelling themselves largely by undulations of the posterior part of their body, second dorsal fin, and caudal. The large pectorals serve both as propelling and balancing organs. Chimaeras are somewhat nocturnal in their habits. They are not vigorous, struggle little, and die soon after being removed from the water. They have well-developed dental plates and can inflict a nasty wound. Dean (1906) has expressed the opinion that fishermen fear the jaws of the ratfish more than they do the formidable dorsal spine. He mentioned one case in which a ratfish bit a piece of flesh from the hand of a fisherman. Chimaeras tend to be omnivorous in their eating habits, the bulk of their diet consisting of small invertebrates, mollusks, a variety of crustaceans, and small fishes. They are said to bite on almost any bait and are readily taken by hook and line. Fertilization is internal and is effected by the pelvic claspers. All chimaeras are oviparous, laying their eggs in large horny capsules. Chimaeras attain a length of about 2 m.

MORPHOLOGY OF THE VENOM APPARATUS

The following morphological description is based largely on the findings of Halstead and Bunker (1952) on *Hydrolagus colliei* which were taken in 50 fathoms off Roberts Point, Blaine, Wash.

Gross Anatomy

The venom apparatus of *Hydrolagus colliei* consists of the dorsal spine, the glandular epithelium of the spine and connecting membrane, and the enveloping integumentary sheath.

The mature spine is elongate, tapers to an acute point, and is composed of a cartilaginous core covered by a sheath of vasodentine. This structure is generally considered to be a modified denticle (Goodrich, 1909). A lateral view of the spine reveals that it curves gracefully backward with the bend becoming slightly more marked in the terminal one-fifth (Fig. 38). The anterior margin of the spine is keeled and entire, while the two posterolateral margins are serrate with the teeth pointing in a downward direction. The anterior keel of the spine is translucent and colorless; the remainder is dark brown except for light areas where the pigment layer has been worn away. The anterolateral surface of the spine is marked by a number of fine ridges diminishing in size as they approach the posterior margins.

In cross section (Fig. 39) the spine is roughly trigonal in outline with a pronounced anterior keel and a shallow posterior depression, termed the

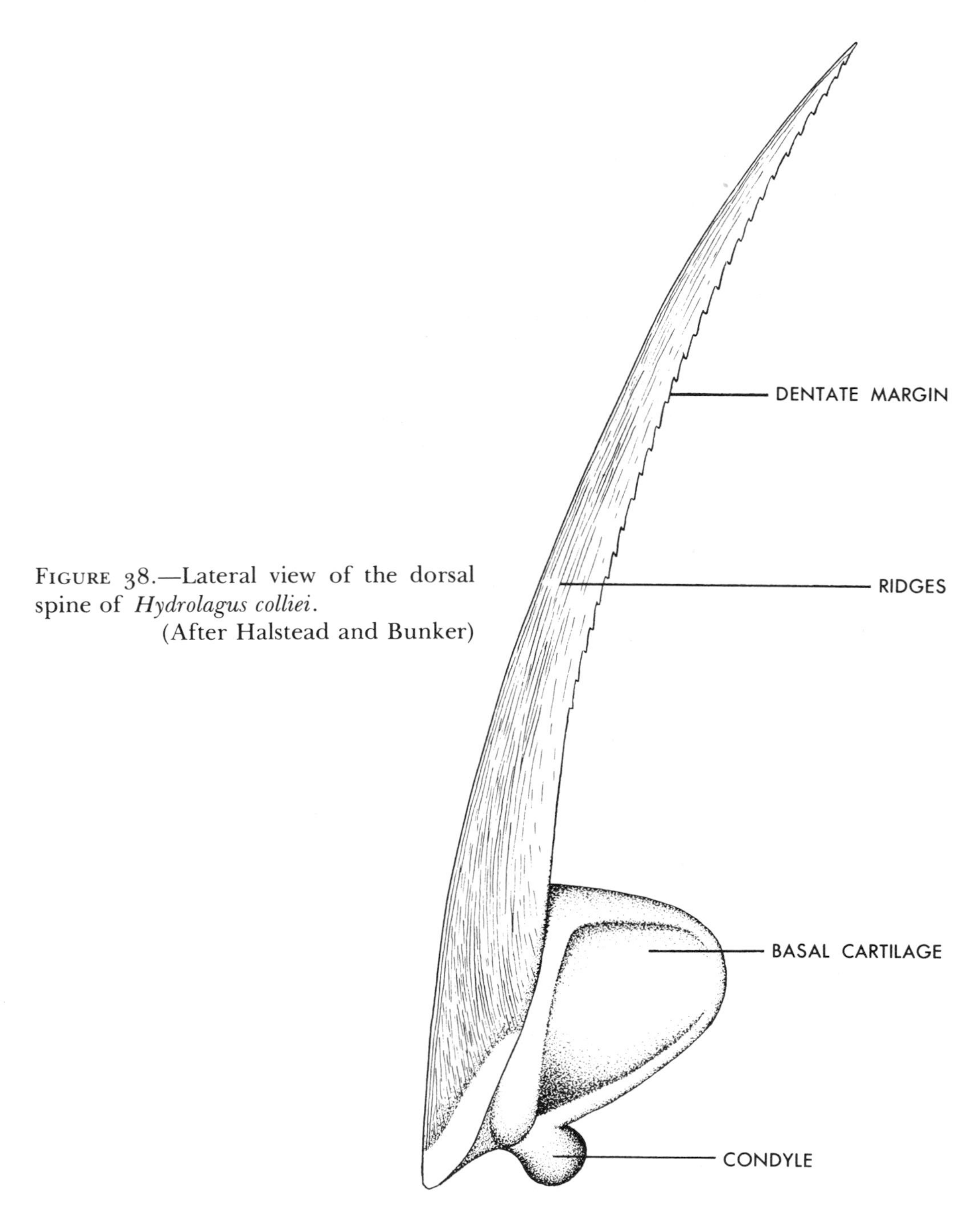

Figure 38.—Lateral view of the dorsal spine of *Hydrolagus colliei*.
(After Halstead and Bunker)

interdentate depression. Lying within the depression is a strip of soft, grayish, sparsely pigmented tissue. Running in a vertical direction along the midline of this tissue is a dark line marking the junction where the spine was formerly connected with the dorsal fin. In most of our specimens the dorsal spine is connected with the anterior margin of the dorsal fin only at the

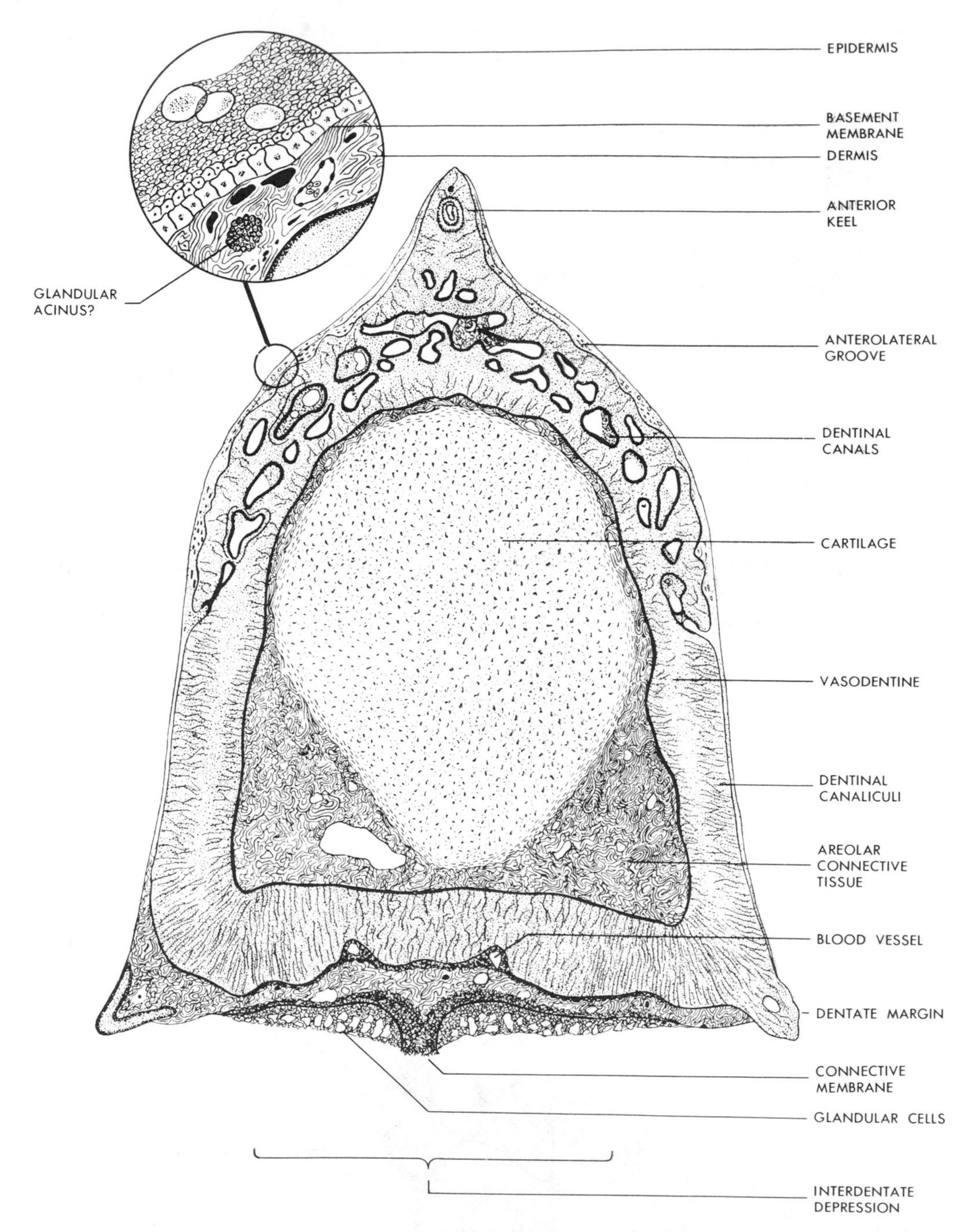

FIGURE 39.—Cross section of the dorsal sting of *Hydrolagus colliei*.
(After Halstead and Bunker)

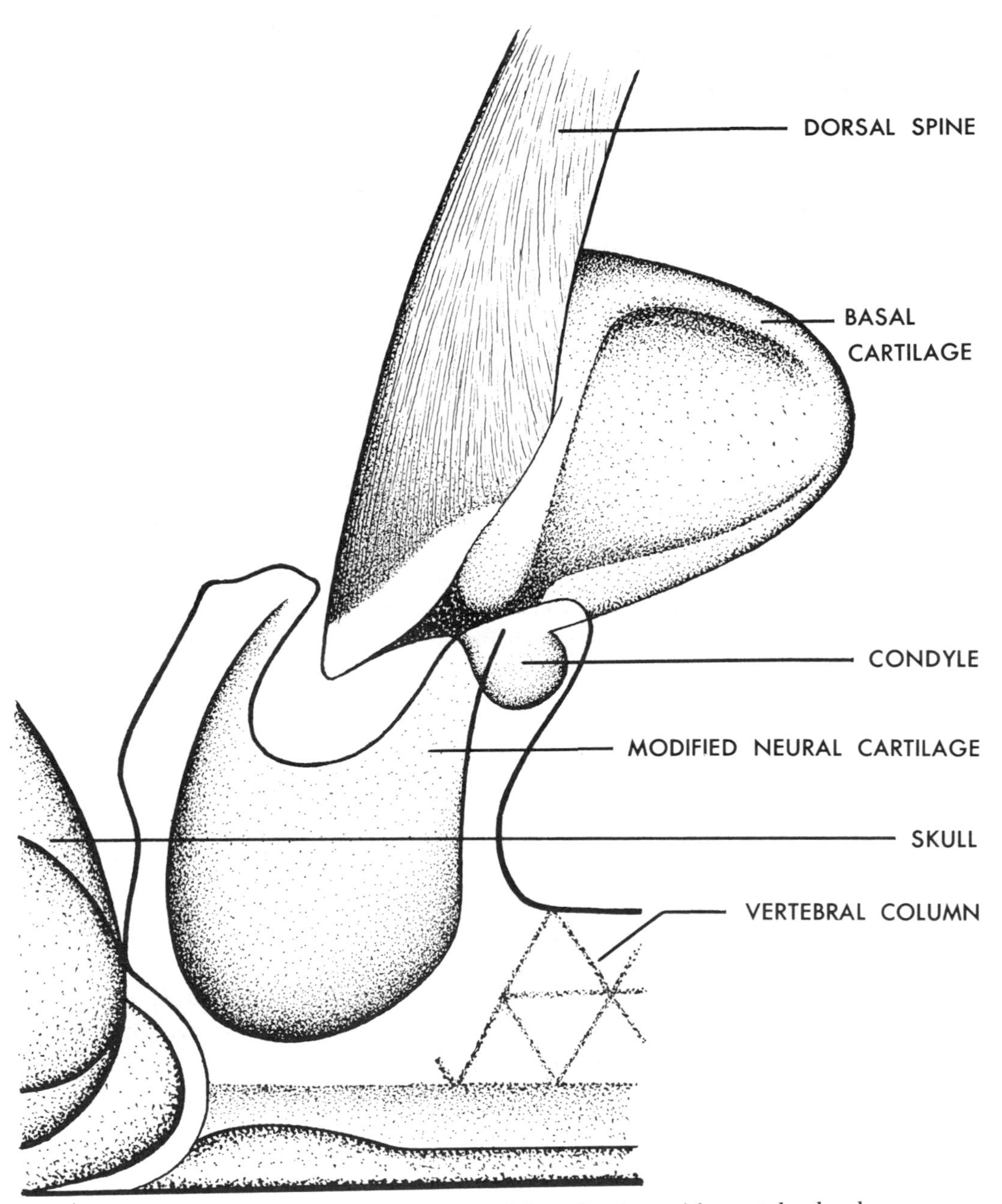

FIGURE 40.—Lateral view of articulation of dorsal spine with vertebral column. (After Halstead and Bunker)

proximal one-half of the spine, the terminal portion having been torn away. The remaining connecting membrane between the dorsal spine and fin is thickened and glistening white in appearance. Careful examination of this area fails to reveal the deep grooves which Evans (1923) observed in *Chimaera monstrosa*. However, when the dorsal fin is depressed and the spine is resting against the fin, the connecting membrane forms a fold which covers most of the dentate margin of the proximal one-half of the spine. Judging from the anatomical evidence available, it appears that this membrane is an integral part of the venom apparatus of the dorsal spine. Apparently one of the functions of this membrane is to bathe the dentate margins of the spine with secretions.

The base of the dorsal spine is situated directly above the origin of the pectoral fin. If the flesh at the base of the spine is dissected away, it will be seen that the spine is comprised of two portions—a hard dentinal part extending almost the entire length of the spine and a short basal portion composed of cartilage. Further observation reveals that the vasodentine portion is only an outer sheath enveloping an inner core of cartilage. This core is an elongate, superior extension of the cartilaginous base of the spine which extends up into the sheath for approximately six-sevenths of its total length. These anatomical relationships can best be demonstrated by drying the spine. In the dried state, the dentinal sheath can be lifted off the rest of the spine, thus exposing the core.

It will now be observed that the cartilaginous portion of the spine is comprised of three parts: an elongated tapering extension or inner core, the articulating base, and a posterior wedge-shaped projection. Since the superior extension was described in the previous paragraph, it will need no further discussion. Continuing to view the spine from the side, one will note that the articulating basal portion is further subdivided by a transverse notch into an anterior V-shaped process and a posterior condyle (Figs. 38, 40). Immediately above the condyle and arising from the posterior aspect of the shaft of the core is a greatly expanded, compressed, wedge-shaped piece of cartilage—the basal supporting cartilage for the dorsal fin. The basal cartilage serves as an attachment for the insertion of the depressor muscles of the spine and a portion of the abductor muscles for the dorsal fin rays (Fig. 41).

The dorsal spine is supported by a highly modified, cartilaginous, neural spine of the anterior portion of the spinal column. The neural spine is greatly elevated and strongly compressed medially to form a moderately expanded anterior process and a broadly expanded, winglike, posterior process. These two processes are separated above by a deep notch. The posterior side of the posterior neural process is depressed so as to form a fossa of moderate depth. The fossa is subdivided into two lateral depressions by a low, vertical median ridge. The depressor muscles of the dorsal spine, one on each side, have their origin in these lateral depressions. Situated on the dorsum of the posterior neural process is the fossa in which the condyle of the dorsal spine articulates.

There are six muscles, bilaterally distributed, which control the movement of the dorsal fin and spine (Fig. 41). These muscles consist of a pair of erectors, a pair of depressors, and a pair of abductors. The erector muscle has its origin in the depression between the anterior and posterior neural processes. This wedge-shaped muscle passes upward and backward over the anterior and dorsal surface of the posterior neural process and inserts obliquely along the basal one-fifth of the dorsal spine and aids the abductor

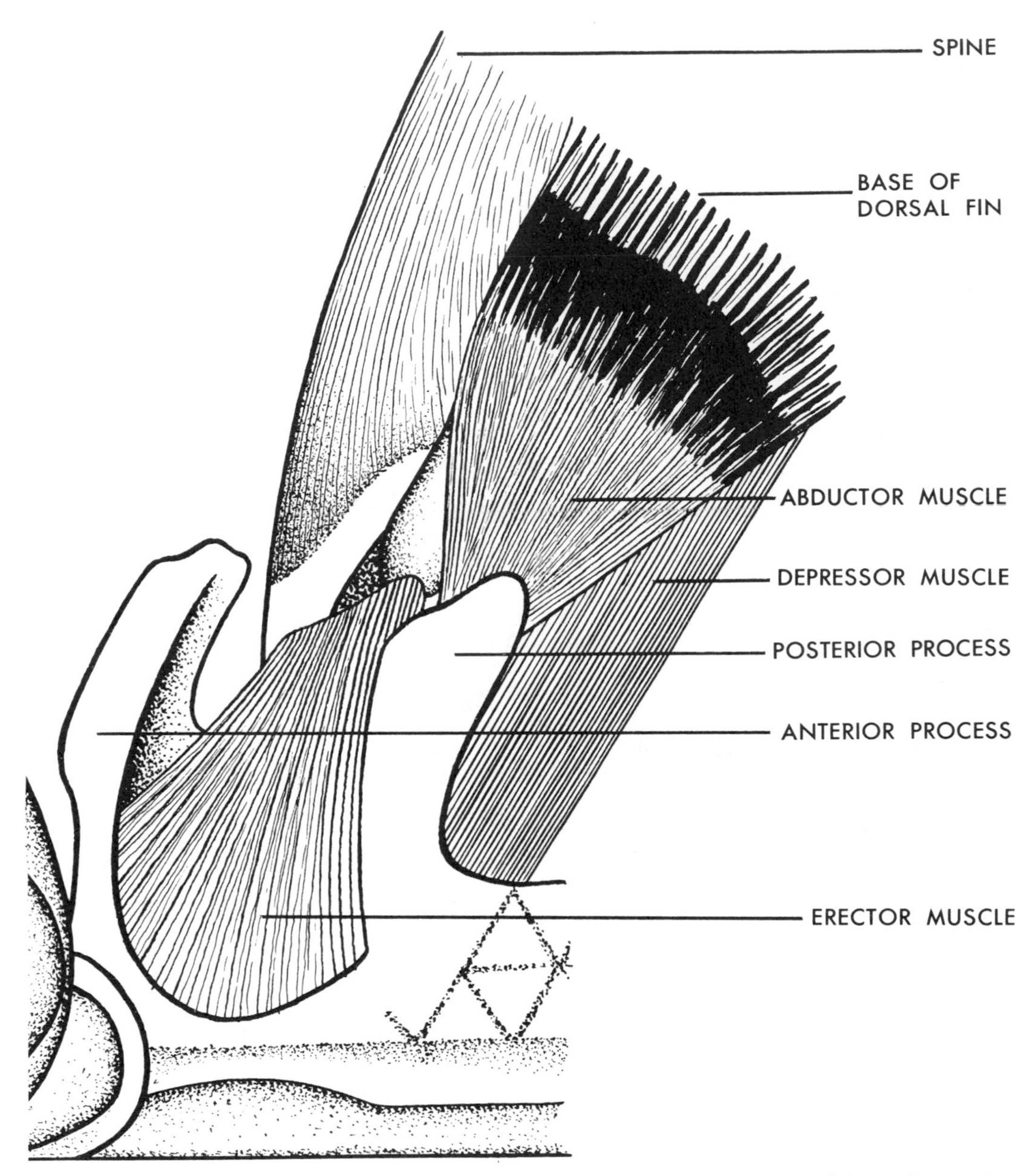

FIGURE 41.—Lateral view of base of dorsal spine showing origins and insertions of muscles. (After Halstead and Bunker)

muscle in spreading the dorsal fin rays. The depressor muscle originates in the posterolateral depression of the posterior neural process and inserts on the inferoposterior margin of the basal cartilage of the dorsal spine. Its function is to depress this spine. The abductor originates on the dorsum of the posterior neural process just lateral to the articular fossa and from the side of the basal cartilage. The muscle extends upward and, fanning out in a

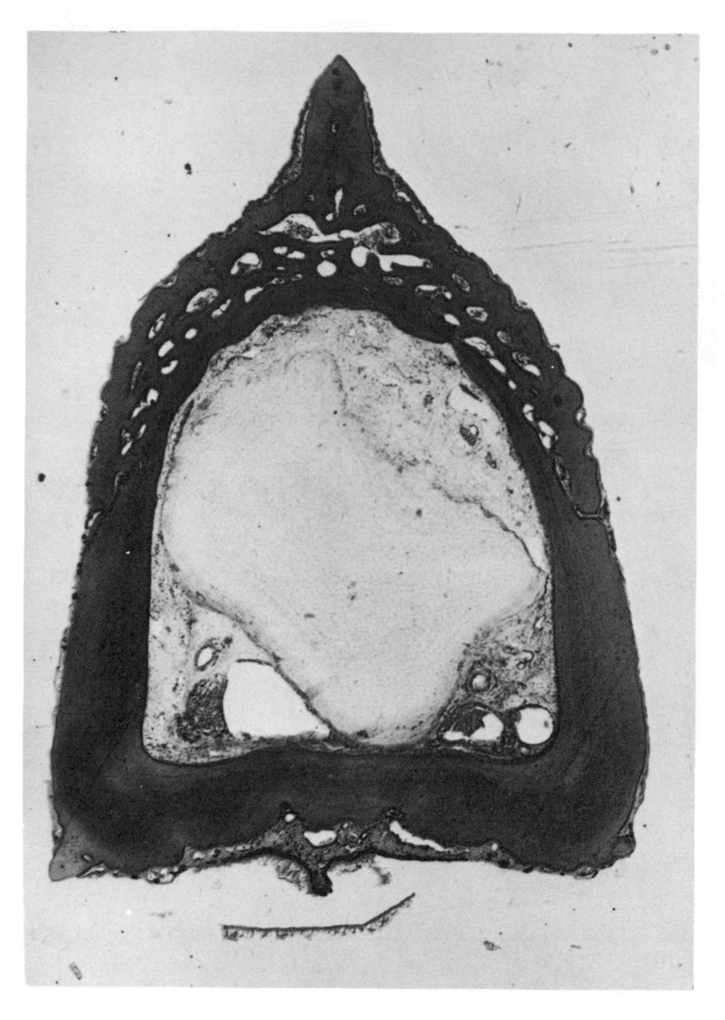

FIGURE 42.—Photomicrograph of a cross section of a mature dorsal sting of *Hydrolagus colliei*. Most dorsal stings have been traumatized, and it is extremely difficult to find one that is intact. It will be observed that most of the integumentary sheath of this sting has been worn away but portions of it remain in the anterolateral grooves and in the interdentate depression. Hematoxylin and triosin stain. × 30. (E. Rich)

posterior direction, inserts to the base of each of the dorsal fin rays by a number of small slips. The function of this muscle is to spread or abduct the dorsal fin rays. In the use of the term abduction, the dorsal spine is used as the axis or point of reference. When in the depressed position, the dorsal fin rests in a longitudinal groove along the back. When referring to the depression, the term "dorsal fin groove" will be used.

Microscopic Anatomy

The anterior angle of the spine is more acute than the two posterior angles and is heavily keeled (Figs. 39, 42). The anterolateral surfaces are convex and marked by a number of grooves. The posterior side is slightly indented to form the broad, shallow, interdentate depression. Cross sections of the marginal serrations can be seen at the posterolateral angles. The serrated

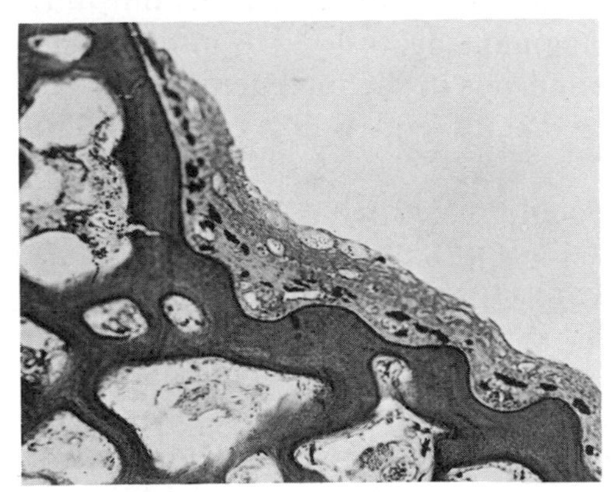

FIGURE 43a.—Photomicrograph showing glandular epithelium in the region of the anterolateral grooves of the dorsal sting of *Hydrolagus colliei*. × 210.

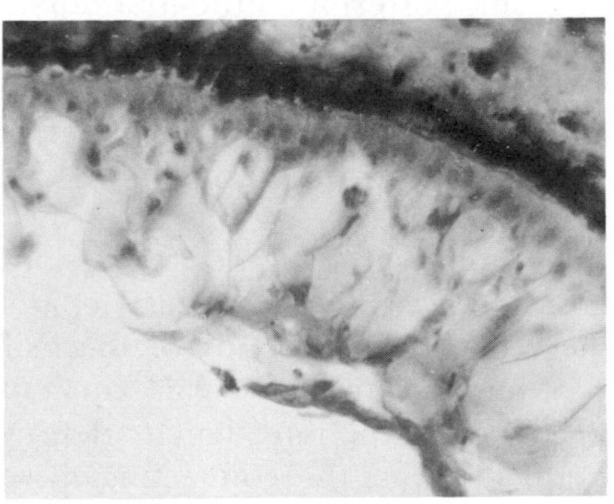

FIGURE 43b.—Glandular epithelium in the region of the anterolateral grooves of the dorsal sting of *H. colliei*. × 600.

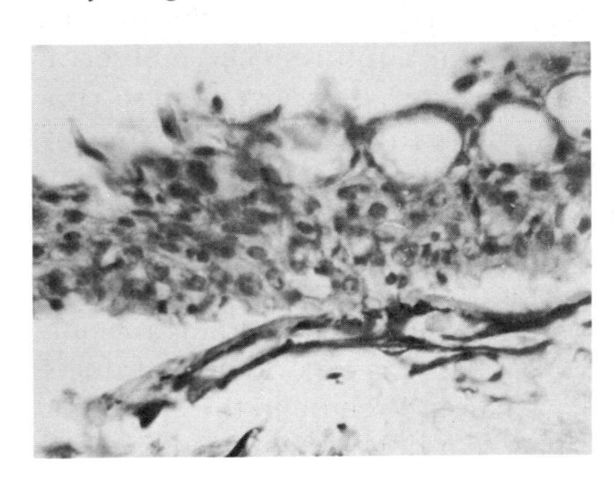

FIGURE 43c.—Glandular epithelium in the region of the dorsal fin groove of *H. colliei*. Hematoxylin and triosin stain. × 600. (From Halstead and Bunker)

portion of the spine is designated as the "dentate margin." This general configuration will be observed throughout most of the sections taken from the distal three-fourths of the spine. As one approaches the basal one-fourth, the anterior angle becomes obtuse and the posterior side of the spine becomes convex. At a lower level, the basal supporting cartilage comes into view.

The spine is covered externally by a thin layer of integument, consisting of two layers—an outer epidermis and an inner dermis. The epidermis, comprised of squamous epithelium, is avascular, and rests on an acellular basement membrane. Underlying the epidermis is the vascularized layer of fibrous connective tissue, the dermis. In the uninjured spine, this integument completely envelops the spine except along the middle of the interdentate depression where the dorsal fin joins the spine. In sections where the spine is still connected with the soft dorsal, it will be observed that the integument is continuous with that of the dorsal fin. The epithelium is most concentrated in the anterolateral grooves just posterior to the anterior keel of the spine (Fig. 43a, b, c) and in the region of the interdentate depression. Posterior to the first dorsal ray, the epithelial layer thins out until it is only one or two cells in thickness. Pigment granules are interspersed throughout the epithelium but are most numerous in the interdentate depression. Mucous cells may also be seen scattered at infrequent intervals throughout this same layer.

In examining sections from the spines of young fish 250 mm in length or less, one receives the impression that the epithelium in these two concentrated areas is the ordinary epidermal variety such as may be found elsewhere covering the body of the fish. However, examination of the spines of older specimens reveals concentrations of fully developed glandular epithelium in the interdentate depression, the epidermal layer of the anterolateral grooves, the connecting membrane between the spine and the first dorsal ray, and in the epithelial lining of the groove on the back which receives the dorsal fin when in the depressed position. Furthermore, it will be noted that the cells of this glandular epithelium are in the process of secretory activity as evidenced by vacuolation, increased cytoplasmic area, and phantom nuclei. Some of the more superficial cells in the region of the interdentate depression appear to be undergoing dissolution or rupture. These observations strongly suggest a holocrine type of secretion; moreover, these concentrated glandular patches are not seen elsewhere in the epidermis covering the dorsal fin or in the integument of the back.

By carefully examining the dermis located in the anterolateral grooves and the interdentate depression, one will see an occasional clump of what appears to be glandular tissue. It was formerly believed that these "acini" were glandular tissue. However, recent studies indicate that they are only nerves and vascular structures in the dermis. There is no evidence of ducts. The venom is probably secreted by the glandular epithelium in the epidermal portion of the integument of the spine and the adjoining connecting membrane of the dorsal fin.

Goodrich (1909) stated that the spine does not contain an enamel layer. Likewise, Halstead and Bunker (1952) have been unable to demonstrate that such a layer exists. There is however a light blue line that is occasionally observed around the periphery of the vasodentine. If this line is compared with the enamel line of the dorsal spine of *Squalus acanthus*, it will be seen that there is considerable difference both in structure and width.

In cross sections taken at the level of the middle third of the spine, it will be observed that the vasodentine layer forms a broad wall which completely encircles the inner core of cartilage. Subsequent sections taken from lower levels reveal that this dentinal wall thins out posteriorly until it gradually disappears. The anterior half of this wall, including the keel, contains many dentinal canals. These canals are occupied mostly by areolar connective tissue and are heavily vascularized with small blood vessels which in some cases are filled with nucleated erythrocytes. Some of these dentinal canals open into the small grooves on the anterolateral portion of the spine. The inner edge of the dentinal wall of young spines shows a light-staining area which is noncalcified vasodentine.

The odontoblast layer adjoins the inner margin of the dentinal wall and consists of a row of darkly-stained cells with large nuclei, one or two cells in thickness. The odontoblasts produce the vasodentine. When the process of vasodentine formation has been completed, the odontoblasts cease to function and apparently disappear, because in older spines this layer is not detected. The remainder of the tissue between the odontoblast layer and the cartilaginous core is comprised of loose areolar connective tissue interspersed with small blood vessels containing nucleated blood cells. The core itself is composed of embryonic hyalline cartilage. This cartilage continues about six-sevenths of the distance up the length of the spine. In the terminal portion the cavity of the spine is filled with loose areolar connective tissue.

MECHANISM OF INTOXICATION

Envenomations from chimaeras are the result of wounds inflicted by their single, sharp, dorsal spines.

MEDICAL ASPECTS

Clinical Characteristics

Wounds inflicted by chimaeras are usually of the puncture wound variety. There is very little information available as to the clinical characteristics. Evans (1943) has reported that the wounds can be painful and serious. Chimaera wounds are generally less severe than those produced by stingrays.

Pathology

(Unknown)

Treatment

Chimaera wounds should be treated in a manner similar to those produced by stingrays (*see* p. 655). However, in the case of chimaera wounds there is less opportunity for the venomous glandular epithelium to become embedded in the wound of the victim because of the anatomical arrangement of the venom apparatus.

Prevention

One must exercise caution in handling live chimaeras. They can deliver a nasty bite, and the single dorsal spine is extremely sharp.

PUBLIC HEALTH ASPECTS

Chimaera envenomations are not a public health problem.

TOXICOLOGY

Our knowledge of the toxicology of chimaera venom is based on the single report by Halstead and Bunker (1952) on *Hydrolagus colliei*.

An intradermal scratch test performed with the use of the dorsal spine of *H. colliei* on the volar surface of the left forearm of the author resulted in a zone of ischemia having raised borders surrounded by a 2 cm zone of erythema, which appeared in 3 to 4 minutes. In about 5 minutes there was a mild dull ache in the area around the scratch which lasted for about 10 minutes. The eruption disappeared in about 30 minutes. A similar test was performed on the same arm using the dorsal spine of *Squalus acanthias*, but no reaction was produced. It is believed that the reason for the difference is that the glandular epithelium in *S. acanthias* is situated further down the shaft of the spine, whereas in *H. colliei* the glandular epithelium is located nearer the tip. In other words, to achieve comparable results one would have to insert the *S. acanthias* spine much deeper into the skin.

Aqueous extracts were prepared from the glandular epithelium of the interdentate depression of the dorsal spine of *H. colliei*. A control test was made by preparing similar extracts from the skin taken from the back of the same fish. The extracts were injected intraperitoneally into laboratory mice. The extracts from the glandular epithelium of the dorsal spine proved to be toxic, resulting in immediate agitation, dragging of the hind limbs, jerking motions, inactivity, and death of the mice on the third day. The controls remained asymptomatic.

PHARMACOLOGY

(Unknown)

CHEMISTRY

(Unknown)

LITERATURE CITED

AUTENRIETH, H. F.
1833 Ueber das gift der fische. C. F. Osiander, Tübingen. 287 p.

BIGELOW, H. B., and W. C. SCHROEDER
1948 Fishes of the western North Atlantic. Vol. I, pt. I. Mem. Sears Found. Marine Res. 576 p., 106 figs.
1953 Fishes of the western North Atlantic. Vol. I, pt. II. Mem. Sears Found. Marine Res. 588 p., 127 figs.

BOTTARD, A.
1889 Les poissons venimeux. Octave Doin, Paris. 198 p.

CADZOW, W. H.
1960 Puncture wound of the liver by stingray spines. Med. J. Australia 47(1): 936-937.

CAMPBELL, B.
1951 The locomotor behavior of spinal elasmobranchs with an analysis of stinging in *Urobatis*. Copeia (4): 277-284, 3 figs.

CARLETON, H. M.
1926 Histological technique. Oxford Univ. Press, London. 398 p.

CASTEX, M. N.
1963 La raya fluvial. Notas histórico-geográficas. Librería y Editorial Castellri. Sante Fe, Argentina. 120 p., 26 figs.
1964 Stingray venom apparatus. (Personal communication, November, 1964.)
1967 Freshwater venomous rays, p. 167-176. *In* F. E. Russell and P. R. Saunders [eds.], Animal toxins. Pergamon Press, New York.

CASTEX, M. N., and F. LOZA
1964 Etiológia de la enfermedad paratrygónica: estudio anatómico, histológico y funcional del aparato agresor de la raya fluvial americana (gen. *Potamotrygon)*. Rev. Assoc. Med. Arg. 78(6): 314-324.

CASTEX, M. N., MACIEL, I., PEDACE, E. A., and F. LOZA
1963 La enfermedad paratryconica Librería y Editorial Castellri, Sante Fe, Argentina. 184 p., 64 figs.

CLELAND, J. B.
1942 Injuries and diseases in Australia attributable to animals. Med. J. Australia 2(4): 313-320.

COUTIÈRE, H.
1899 Poissons venimeux et poissons vénéneux. Thèse Agrég. Carré et Naud, Paris. 221 p.

CREVAUX, J., and E. LE JANNE
1882 Recit du troisième voyage dans l'Amérique équatoriale. Arch. Méd. Nav. 37: 50-53.

DANIEL, J. F.
1934 The elasmobranch fishes. University of California Press, Berkeley. 332 p.

DEAN, B.
1906 Chimaeroid fishes and their development. Carnegie Inst. Wash. Publ. 32. 194 p., 11 pls., 144 figs.

EVANS, H. M.
1916 The poison-organ of the sting-ray *(Trygon pastinaca)*. Proc. Zool. Soc. London 29(1): 431-440, 7 figs.
1920 The poison of the spiny dog-fish. Brit. Med. J. 1(3087): 287-288, 5 figs.
1921 The poison organs and venoms of venomous fish. Brit. Med. J. 2(3174): 690-692, 2 figs.
1923 The defensive spines of fishes, living and fossil, and the glandular structure in connection therewith, with observations on the nature of fish venoms. Phil. Trans. Roy. Soc. London, B, 212: 1-33, 3 pls., 14 figs.
1943 Sting-fish and seafarer. Faber and Faber, Ltd., London. 180 p., 31 figs.

FISH, C. J., and M. C. COBB
1954 Noxious marine animals of the central and western Pacific Ocean. U. S. Fish Wildlife Serv., Res. Rept. No. 36, p. 14-23.

FLEURY, R.
1950 L'appareil venimeux des sélaciens trygoniformes (anatomie—histologie—physiologie). Mém. Soc. Zool. France (1): 1-38.

FOWLER, H. W.
1928 The fishes of Oceania. Mem. Bernice P. Bishop Mus. Vol. 10. 466 p., 49 pls.
1936 The marine fishes of west Africa. Bull. Am. Mus. Nat. Hist. Vol. 70, part I and II. 1,493 p., 567 figs.
1941 Contributions to the biology of the Philippine Archipelago and adjacent regions. The fishes of the groups Elasmobranchii, Holocephali, Isospondyli, and Ostarophysi obtained by the U.S. Bureau of Fisheries steamer "Albatross," in 1907 to 1910, chiefly in the Philippine Islands and adjacent seas. U.S. Nat. Mus., Bull. 100, Vol. 13. 879 p., 30 figs.

FREDERICQ, L.
1924 Die sekretion von schutz- und nutzstoffen, p. 1-87. *In* H. Winterstein [ed.], Handbuch der vergleichenden physiologie. Vol. II. G. Fischer, Jena.

GARMAN, S.
1913 The plagiostomia (sharks, skates, and rays). Mem. Mus. Comp. Zool. Harvard College. Vol. 36 514 p., 75 pls.

GILBERT, P. W., *ed.*
1963 Sharks and survival. D. C. Heath and Co., Boston. 578 p.

GIMLETTE, J. D.
1923 Malay poisons and charm cures. 2d ed. J. and A. Churchill, London. 260 p.

GOE, D. R.
1950 The gross and microscopic anatomy of the dorsal fin spine of the Pacific dogfish *Squalus suckleyi* (Girard) 1854. M.A. thesis, Walla Walla College, Wash. 18 p., 10 figs.

GOODE, G. B., and T. H. BEAN
1895 Oceanic ichthyology. U. S. Nat. Mus., Spec. Bull. 553 p., 415 figs.

GOODRICH, F. S.
1909 Vertebrata craniata. 518 p., 515 figs. *In* R. Lankester [ed.], A treatise on zoology. Part IX. A. and C. Black, London.

GUDGER, E. W.
1914 History of the spotted eagle ray, *Aetobatus narinari*, with a study of its external structures. Carnegie Inst. Wash., Papers Tortugas Lab. 6: 241-323, 10 pls., 19 figs.
1943 Is the sting ray's sting poisonous? A historical résumé showing the development of our knowledge that it is poisonous. Bull. Hist. Med. 14: 467-504, 12 figs.
1946 Does the sting ray strike and poison fishes? Sci. Monthly 63: 110-116, 4 figs.

GÜNTHER, A.
1859-70 Catalogue of the fishes in the British Museum. 8 vols. British Museum, London.

HALSTEAD, B. W.
1956 Animal phyla known to contain poisonous marine animals, p. 9-27. *In* E. E. Buckley and N. Porges [eds.], Venoms. Am. Assoc. Adv. Sci., Washington, D.C.
1958 Poisonous fishes. Public Health Rept. (U.S.) 73(4): 302-312, 11 figs.
1959 Dangerous marine animals. Cornell Maritime Press, Cambridge. 146 p., 88 figs.

HALSTEAD, B. W., and N. C. BUNKER
1952 The venom apparatus of the ratfish, *Hydrolagus colliei*. Copeia (3): 128-138, 1 pl., 4 figs.
1953 Stingray attacks and their treatment. Am. J. Trop. Med. Hyg. 2(1): 115-128, 12 figs,

HALSTEAD, B. W., and L. R. MITCHELL
1963 A review of the venomous fishes of the Pacific area, p. 173-202, 16 figs. *In* H. L. Keegan and W. V. MacFarlane [eds.], Venomous and poisonous animals and noxious plants of the Pacific region. Pergamon Press, New York.

HALSTEAD, B. W., and F. R. MODGLIN
1950 A preliminary report on the venom apparatus of the bat-ray, *Holorhinus californicus*. Copeia (3): 165-175, 6 figs.

HALSTEAD, B. W., OCAMPO, R. R., and F. R. MODGLIN
1955 A study on the comparative anatomy of the venom apparatus of certain North American stingrays. J. Morphol. 97(1): 1-21, 4 pls.

HOLLOWAY, J. E., BUNKER, N. C., and B. W. HALSTEAD
1953 The venom of *Urobatis halleri* (Cooper), the round stingray. Calif. Fish Game 39(1): 77-82, 1 fig.

JORDAN, D. S., and B. W. EVERMANN
1896-1900 The fishes of North and Middle America. 4 parts. U.S. Nat. Mus., Bull. 47. 392 pls.

JORDAN, D. S., TANAKA, S., and J. O. SNYDER
1913 A catalogue of the fishes of Japan. J. Coll. Sci., Tokyo Univ. 33. 497 p., 396 figs.

JOUBIN, L. M.
1929-38 Faune ichthyologique de l'Atlantique nord. Nos. 1-18 A. F. Høst et Fils, Copenhagen.

KONEFF, A. A., and W. R. LYONS
1937 Rapid embedding with hot low-viscosity nitrocellulose. Stain Technol. 12(2): 57-59.

LAGLER, K. F., BARDACH, J. E., and R. R. MILLER
1962 Ichthyology. John Wiley and Sons, New York. 545 p., 14 figs.

Maretić, Z.
1968 Venomous animals and their toxins. [In Croatian] Pro Medico 3: 93-114.

Marshall, T. C.
1964 Fishes of the Great Barrier Reef and coastal waters of Queensland. Angus and Robertson, Melbourne. 566 p., 72 pls.

Mordecai, E. R.
1860 Wound inflicted by a stingray. New Orleans Med. News Hosp. Gaz. 7: 679-680.

Mullins, J. F., Wilson, C. J., and W. C. Best
1957 Cryotherapy in the treatment of stingray wounds. Southern Med. J. 50(4): 533-535.

Ocampo, R. R., Halstead, B. W., and F. R. Modglin
1953 The microscopic anatomy of the caudal appendage of the spotted eagleray, *Aetobatus narinari* (Euphrasèn), with special reference to the venom apparatus. Anat. Rec. 115(1): 87-99, 3 pls.

Okada, Y.
1955 Fishes of Japan. Maruzen Co., Tokyo. 434 p., 391 figs.

Pawlowsky, E. N.
1907 Contribution to the structure of the epidermis and its glands in venomous fish. [In Russian] Trav. Soc. Imp. Nat., St. Petersburg 38(1): 265-282, 1 pl., 1 fig.

Phillips, C., and W. H. Brady
1953 Sea Pests—Poisonous or harmful sea life of Florida and the West Indies. Univ. Miami Press, Miami, Fla. 78 p., 38 figs., 7 pls.

Phisalix, M.
1922 Animaux venimeux et venins. 2 vols. Masson et Cie., Paris.
1931 Le venin de quelques poissons marins. Notes Sta. Océanogr. Salammbo, Tunis (22): 3-151.

Porta, A.
1905 Ricerche anatomiche sull'apparecchio velenifero di alcuni pesci. Anat. Anz. 26: 232-247, 2 pls., 5 figs.

Romano, S.
1940 Animali velenosi della fauna Italiana. Natura 31(4): 137-167, 7 figs.

(NSA)—Not seen by author.

Ronka, E. K., and W. F. Roe
1945 Cardiac wound caused by the spine of the stingray (suborder—Masticura). Mil. Surg. 97: 135-137, 2 figs.

Russell, F. E.
1953*a* The stingray. Eng. Sci. 17(3): 15-18, 6 figs.
1953*b* Stingray injuries: a review and discussion of their treatment. Am. J. Med. Sci. 226: 611-622, 3 figs.
1955 Multiple caudal spines in the round stingray, *Urobatis halleri*. Calif. Fish Game 41: 213-217. (NSA)
1959 Stingray injuries. Public Health Repts. 74(10): 855-859.
1961*a* Injuries by venomous animals in the United States. J. Am. Med. Assoc. 177(13): 903-907.
1961*b* Stechrochenverletzungen und ihre ursachen. Arch. Fischereiwiss. 13(1/2): 73-79, 5 figs.
1965 Marine toxins and venomous and poisonous marine animals, p. 255-384, 20 figs. *In* F. S. Russell [ed.], Advances in marine biology. Vol. 3. Academic Press, New York.
1967 Comparative pharmacology of some animal toxins. Fed. Proc. 26(4): 1206-1224.

Russell, F. E., Barritt, W. C., and M. D. Fairchild
1957 Electrocardiographic patterns evoked by venom of the stingray, Proc. Soc. Exp. Biol. Med. 96: 634-635.

Russell, F. E., Fairchild, M. D., and J. Michaelson
1958*b* Some properties of the venom of the stingray. Med. Arts Sci. 12(2): 78-86, 3 figs.

Russell, F. E., and V. C. Bohr
1962 Intraventricular injection of venom. Toxicol. Appl. Pharmacol. 4: 165-173. (NSA)

Russell, F. E. and A. van Harreveld
1954 Cardiovascular effects of the venom of the round stingray, *Urobatis halleri*. Arch. Internat. Physiol. 62(3): 322-333, 6 figs.

Russell, F. E., and R. D. Lewis
1956 Evaluation of the current status of therapy for stingray injuries, p. 43-53, 3 figs. *In* E. E. Buckley and N. Porges [eds.], Venoms. Am. Assoc. Adv. Sci., Washington, D.C.

RUSSELL, F. E., and T. E. LONG
1960 Effects of venoms on neuromuscular transmission, p. 101-116, 3 figs. *In* H. R. Viets [ed.], Myasthenia gravis. Charles C. Thomas, Springfield, Ill.

RUSSELL, F. E, PANOS, T. C., KANG, L. W., WARNER, W. M., and T. C. COLKET
1958*a* Studies on the mechanism of death from stingray venom. A report of two fatal cases. Am. J. Med. Sci. 235: 566-584.

SCHNACKENBECK, W.
1943 Fischgifte und fischvergiftungen. Münch. Med. Wochschr. 90(8): 149-151.

SCHULTZ, L. P,
1944 The stingarees, much feared demons of the seas, U.S. Nav. Med. Bull. 42(3): 750-754, 3 figs.

SMITH, J. L.
1950 The sea fishes of southern Africa. Central News Agency, Cape Town, S. Africa. 550 p., 103 pls.

SOLJAN, T.
1963 Fauna and flora of the Adriatic. Vol. I. Fishes of the Adriatic. [Eng. Transl.] Nolit Publ. Hse., Belgrade, Yugoslavia. 428 p.

TWEEDIE, M. W.
1941 Poisonous animals of Malaya. Malaya Publishing House, Singapore. 90 p., 29 figs.

VELLARD, J.
1931 Venin des raies *(Taeniura)* du Rio Araguaya (Brésil). Compt. Rend. Acad. Sci. 192: 1279-1281.
1932 Poissons venimeux du Rio Araguaya. Mem. Soc. Zool. France 29: 513-539.

WHITLEY, G. P.
1940 The fishes of Australia. Part I. The sharks, rays, devil-fish, and other primitive fishes of Australia and New Zealand. Royal Zoological Society of New South Wales, Sydney. 280 p., 303 figs.

WRIGHT-SMITH, R. J.
1945 Case of fatal stabbing by stingray. Med. J. Australia 2: 466-467, 2 figs.

Chapter XIX—VERTEBRATES

Class Osteichthyes

Venomous: CATFISHES

The suborder Siluroidei (Nematognathi) includes a group of fishes having a wide variety of sizes and shapes. Their body shape may vary from short to greatly elongate and even eellike. The head is extremely variable, sometimes very large, wide or depressed, again very small. The mouth is not protractile, but the lips are sometimes greatly developed, usually with long barbels, generally with at least one pair from rudimentary maxillaries, often one or more pairs about the chin, and sometimes one from each pair of nostrils. A unique characteristic of this group is the Weberian apparatus consisting of the fusion of the four anterior vertebrae in association with a chain of small bones which link the airbladder with the perilymph-filled spaces surrounding the inner ear. The skin of these fishes is thick and slimy or has bony plates. No true scales are ever present. About 1,000 species are included within this group; most are found in the freshwater streams of the tropics. A few species are marine.

Since most venomous catfishes are freshwater fishes and therefore beyond the scope of this present work, only brief reference will be made to them. However, the freshwater species that have been incriminated in human envenomations will be listed and certain aspects of their venomology discussed.

The systematics of the venomous catfishes listed below have been discussed by Hamilton-Buchanan (1822), Day (1878-88), Eigenmann and Eigenmann (1890), Eigenmann (1912), Jordan, Tanaka, and Snyder (1913), Weber and de Beaufort (1913), Hubbs and Raney (1944), Schultz (1944), Gosline (1945), Hubbs and Lagler (1947), Tomiyama, Abe, and Tokioka (1958), and Marshall (1964).

REPRESENTATIVE LIST OF CATFISHES REPORTED AS VENOMOUS

Phylum CHORDATA

Class OSTEICHTHYES

Order CYPRINIFORMES (OSTARIOPHYSI)

Suborder SILUROIDEI (NEMATOGNATHI): Catfishes

Family ARIIDAE

226 *Arius heudeloti* Cuvier and Valenciennes (Pl. 1). Sea catfish (USA), dakak, konko, kong, poissons chat (west Africa).
DISTRIBUTION: West Africa.

SOURCES: Citterio (1926), Halstead, Kuninobu, and Hebard (1953).

200 *Bagre marinus* (Mitchill) (Pl. 8, Chapter XV). Sea catfish (USA), bagre marino (Latin America).
DISTRIBUTION: East coast of America from Cape Cod to Brazil.

SOURCES: Colby (1943), Halstead (1959).
OTHER NAMES: *Aelrulichthys marinus, Bagre marina.*

226 *Galeichthys feliceps* (Valenciennes) (Pl. 2, fig. a). Sea catfish (USA), dakak, konko, kong, poissons chat (west Africa).
DISTRIBUTION: Azores, west Africa, south Africa.

SOURCE: Smith (1961).
OTHER NAME: *Tachysurus feliceps.*

226 *Galeichthys felis* (Linnaeus) (Pl. 2, fig. b). Sea catfish (USA), bagre marino (Latin America).
DISTRIBUTION: Cape Cod to Gulf of Mexico.

SOURCES: Colby (1943), Halstead *et al.* (1953).
OTHER NAME: *Hexanematichthys felis.*

227 *Genidens genidens* (Cuvier and Valenciennes) (Pl. 3, fig. a). Catfish (USA), mandi (Brazil).
DISTRIBUTION: Brazil.

SOURCE: Vellard (1932).
OTHER NAME: *Genidens granulosus.*

227 *Netuma barbus* (Lacépède) (Pl. 3, fig. b). Sea catfish (USA), bagre marino (Latin America), machoiran (French Guiana).
DISTRIBUTION: East coast of South America from Argentina to the Guianas.

SOURCES: Phisalix (1922), Maass (1937).
OTHER NAME: *Bagrus barbatus.*

227 *Osteogeniosus militaris* (Linnaeus) (Pl. 4, fig. a). Catfish (USA), pla hua on (Thailand), poné kelí tí (India).
DISTRIBUTION: Java, Borneo, Sumatra, India.

SOURCES: Phisalix (1922), Maass (1937), Fish and Cobb (1954).
OTHER NAMES: *Arius militaris, Siluris militaris.*

Selenaspis herzbergi (Bloch) (Pl. 4, fig. b). Catfish (USA), bagre (Latin America). *227*

Distribution: British Guiana to Venezuela.

Sources: Phisalix (1922), Maass (1937).
Other name: *Arius herzbergi*.

Family BAGRIDAE

Liobagrus reini Hilgendorf (Pl. 5, fig. a). Catfish (USA), akaza, hinamazu (Japan). *228*

Distribution: Freshwater streams of Japan.

Source: Tange (1955*b*).

Pseudobagrus aurantiacus (Temminck and Schlegel) (Pl. 5, fig. b). Bagrid catfish (USA), gibachi, gingyo, gigyo, gigyu (Japan). *228*

Distribution: Japan.

Sources: Tange (1953, 1955*c*).

Family CLARIIDAE

Clarias batrachus (Linnaeus) (Pl. 5, fig. c). Labyrinthic catfish (USA), ikan keli (Malaysia), pla duk dam (Thailand). *228*

Distribution: Indo-Pacific, India.

Sources: Taschenberg (1909), Evans (1943), Fernando (1965).
Other name: *Clarias magus*.

Family DORADIDAE

Centrochir crocodili (Humboldt) (Pl. 6, fig. a). Doradid armored catfish (USA), bagre (Latin America). *229*

Distribution: Magdalene Basin, Colombia.

Sources: Phisalix (1922), Maass (1937).
Other name: *Doras crocodili*.

Pterodoras granulosus (Valenciennes) (Pl. 6, fig. b). Doradid armored catfish (USA), bagre (Latin America). *229*

Distribution: Amazon and Paraguay Rivers.

Sources: Phisalix (1922), Maass (1937), Halstead *et al.* (1953).
Other name: *Doras maculatus*.

Family HETEROPNEUSTIDAE[1]

Heteropneustes fossilis (Bloch) (Pl. 6, fig. c). Catfish (USA), pla cheet (Thailand), chelu meenu (India). *229*

Distribution: Freshwater streams of the Indo-Pacific, India.

---

[1] According to some authors this group is placed under the Saccobranchidae.

SOURCES: Bhimachar (1944), Halstead (1959), Fernando and Fernando (1960).
OTHER NAME: *Saccobranchus fossilis.*

Family ICTALURIDAE

230 *Noturus flavus* Rafinesque (Pl. 7, fig. a). Stonecat (USA).
DISTRIBUTION: Great Lakes region, westward to Montana and Wyoming and south to Texas.

SOURCES: Reed (1907), Phisalix (1922).

230 *Noturus furiosus* Jordan and Meek (Pl. 7, fig. b). Carolina madtom (USA).
DISTRIBUTION: Eastern North Carolina, Neuse, Tar, and Little Rivers.

SOURCES: Phisalix (1922), Evans (1943).
OTHER NAME: *Schilbeodes furiosus.*

231 *Noturus mollis* (Mitchill) (Pl. 8, fig. a). Tadpole madtom (USA).
DISTRIBUTION: Eastern and central North America river drainages.

SOURCES: Reed (1907), Phisalix (1922).
OTHER NAME: *Schilbeodes mollis.*

231 *Noturus nocturnus* Jordan and Gilbert (Pl. 8, fig. b). Freckled madtom (USA).
DISTRIBUTION: Central North American drainages.

SOURCES: Reed (1907), Maass (1937).
OTHER NAME: *Schilbeodes nocturnus.*

Family PIMELODIDAE

231 *Pimelodus clarias* (Bloch) (Pl. 8, fig. c). Pimelodid catfish (USA), bagre (Latin America).
DISTRIBUTION: Rio da Prata to Panama.

SOURCES: Phisalix (1922), Maass (1937).
OTHER NAMES: *Pimelodus maculatus, Silurus clarias.*

Family PLOTOSIDAE

232 *Cnidoglanis megastoma* (Richardson) (Pl. 9, fig. a). Plotosid catfish (USA), estuary catfish (Australia).
DISTRIBUTION: Australia.

SOURCES: Cleland (1912), Whitley (1943).
OTHER NAME: *Plotosus megastomus.*

232 *Paraplotosus albilabris* (Valenciennes) (Pl. 9 fig. b). Plotosid catfish (USA), patuna, sumbilang (Philippines), white-lipped catfish (Australia), ikan sembilang (Malaysia).

Distribution: Indo-Pacific.

Sources: Evans (1943), Le Mare (1952).

Plotosus canius (Hamilton-Buchanan) (Pl. 9, fig. c). Plotosid catfish (USA), irung-kelletee (India). *232*

Distribution: Indian Ocean, brackish waters and rivers.

Sources: Pawlowsky (1913, 1914), Phisalix (1922), Buddle (1930), Maass (1937), Le Mare (1952).
Other names: *Plotosus limbatus, Plotosus unicolor.*

Plotosus lineatus (Thunberg) (Pl. 9 fig. d). Oriental catfish (USA), striped catfish (Australia), ikan sembilang (Malaysia), pla duk tale (Thailand), patuna, sumbilang (Philippines), gonzui, gigi (Japan), barbel-eel, barber, nkunga (South Africa). *232*

Distribution: Indo-Pacific (marine).

Sources: Pawlowsky (1913, 1914), Phisalix (1922), Le Gac (1936), Evans (1943), Bhimachar (1944), Le Mare (1952), Halstead and Smith (1954), Marshall (1964).
Other names: *Plotosus anguillaris, Plotosus arab, Plotosus castaneus, Plotosus ikapor.*

Tandanus bostocki Whitley (Pl. 10, fig. a). Plotosid catfish (USA), catfish (Australia). *233*

Distribution: West Australia, New South Wales, north Australia, south Australia, Victoria, Tasmania.

Source: Cleland (1912).
Other names: *Cnidoglanis bostocki, Plotosus bostocki.*

Family SILURIDAE

Chrysichthys cranchi (Leach) (Pl. 10, fig. b). Eurasian catfish (USA), kokuni, manoro (west Africa). *233*

Distribution: Rivers of west Africa.

Source: Taschenberg (1909).
Other name: *Bagrus nigritus.*

Parasilurus asotus (Linnaeus) (Pl. 10, fig. c). Mudfish (USA), namazu (Japan), ijou (China). *233*

Distribution: Japan, Korea, Manchuria, and China.

Sources: Maass (1937), Fish and Cobb (1954).
Other name: *Silurus japonicus.*

Synodontis batensoda Rüppell (Pl. 10, fig. d). Upside-down fish (USA, England). *233*

Distribution: Nile River, Egypt.

Source: Evans (1943).

BIOLOGY

Family ARIIDAE: The ariid catfishes are a large group of subtropical and tropical marine fishes which are worldwide in distribution. They are active fishes and, unlike their freshwater counterparts, are constantly on the move, frequently in large schools. They resemble most other catfishes in appearance but differ from other species in that the anterior and posterior nostrils are close together, the latter covered by a valve. Sea catfish have an interesting habit in which the male catfish incubates their eggs by placing them in his mouth. The male places up to 50 eggs or more in his mouth for a period of 2 months. After the eggs are hatched out the young fish remain in the mouth for an additional period of 2 weeks.

Family BAGRIDAE: The bagrid catfishes are distributed throughout certain parts of Africa, Asia Minor, southern and eastern Asia, and Japan. They resemble the South American pimelodids from which they differ anatomically. Bagrids are largely freshwater inhabitants. Some species grow to large size. They are carnivorous in their eating habits and voracious feeders.

Family CLARIIDAE: Clarid catfishes are distributed throughout the fresh waters of Africa, Madagascar, southern and eastern Asia, including the Malay Archipelago, and the Philippines. The clarids differ from other catfishes by having an auxiliary breathing apparatus situated in a pocket that extends back and upward from the gill cavity. The treelike breathing organs are attached to the second and fourth gill arches and permit clarids to live out of water much longer than most other catfishes. Most clarids lack an adipose fin and a dorsal spine. The pectoral spines are the only fin spines present. The anal fin is usually very long.

Family DORADIDAE: The doradid catfishes are limited to the fresh waters of South America. The skull is generally broad, strongly ossified, and produced posteriorly as a flat bony plate extending as far back as the base of the dorsal fin. The body is thick-set and sometimes tadpole-shaped. The sides of the body may be covered with strong, spiny plates or may be naked. The dorsal and pectoral spines are well developed and capable of inflicting painful wounds (Fig. 1).

Family HETEROPNEUSTIDAE: This family is sometimes termed the Saccobranchidae. Heteropneustid catfishes are characterized by a long air sac extending back from the branchial cavity. Members of this family are limited in their distribution to the fresh waters of Asia. *Heteropneustes fossilis* is reported to inflict stings which may result in death.

Family ICTALURIDAE: Ictalurids are typical freshwater North American catfishes ranging from Canada to Guatemala. Some of the best known venomous species in this family are the madtoms *(Schilbeodes)*, which are capable of inflicting painful wounds with their pectoral and dorsal stings.

Family PIMELODIDAE: The pimelodids range from Mexico southward throughout most of South America. This is the largest single family of South American catfishes. Typically, they have long barbels, an adipose fin, and spines in front of the dorsal and pectoral fins.

Family PLOTOSIDAE: The plotosids are a group of elongate, torpedo-shaped catfishes, whose skin is naked. The plotosids are largely marine fishes, although a few species are freshwater. They are found along the coasts of east Africa, south and east Asia, Japan, Philippines, Malay Archipelago, and northern Australia. *Plotosus lineatus* is one of the most dangerous venomous fishes known. Envenomations from this fish may be fatal.

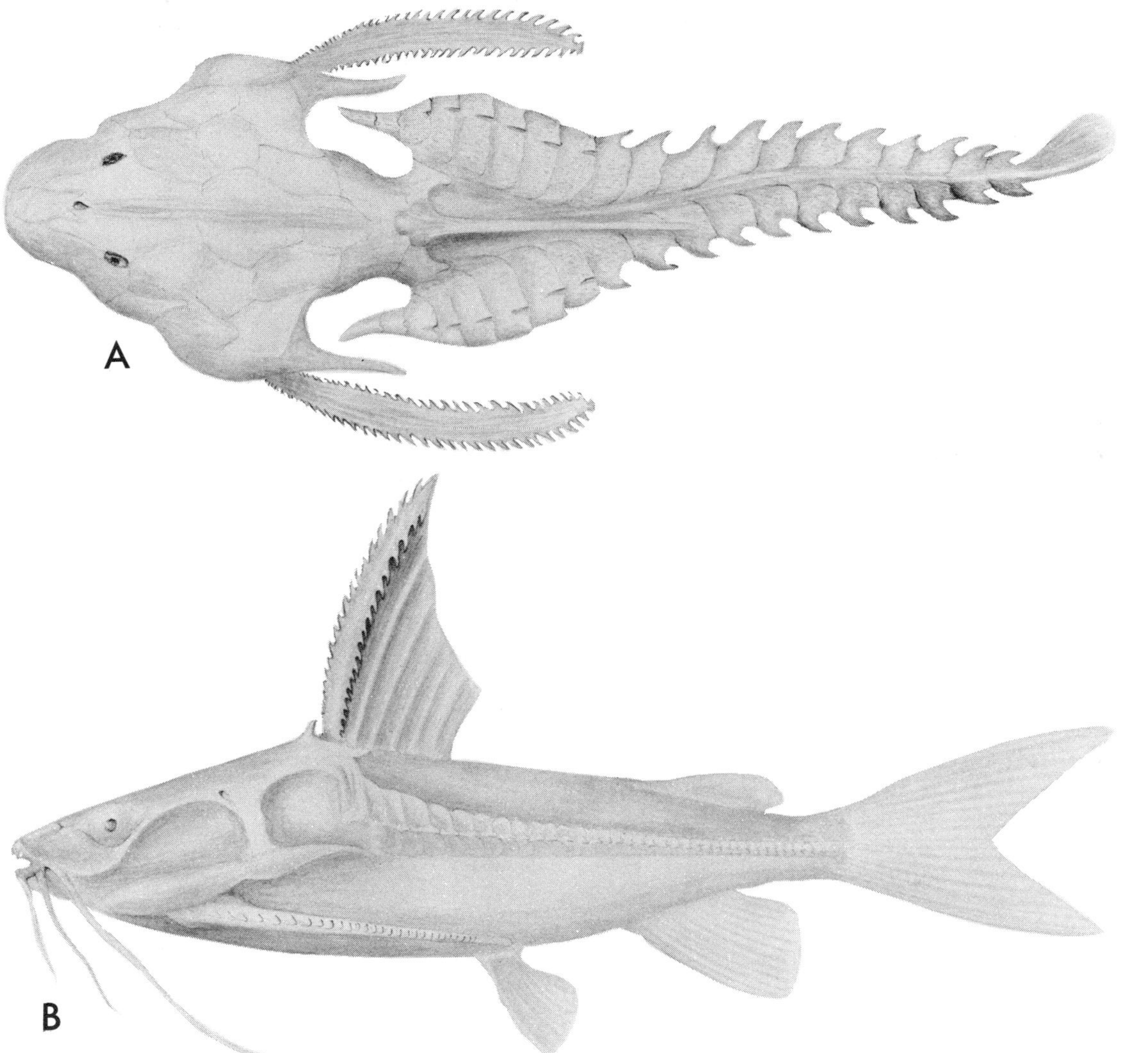

FIGURE 1.—Doradid catfishes are noted for their heavy armor and highly developed dorsal and pectoral spines. a. Dorsal view of *Platydoras costatus* (Linnaeus) showing heavy armored skin consisting of spiny scutes or plates, and the heavily armed pectoral spines. b. Side view of *Centrochir crocodili* (Humboldt) showing dorsal spine. Note the large retrorse teeth along the margins of the dorsal spine. The spines depicted in these drawings are representative of this group of catfishes. In some species of doradid catfishes the osseous spines are said to be accompanied by venom glands, and together they represent a formidable traumagenic device capable of inflicting painful wounds.

(K. Fogassy, after Eigenmann)

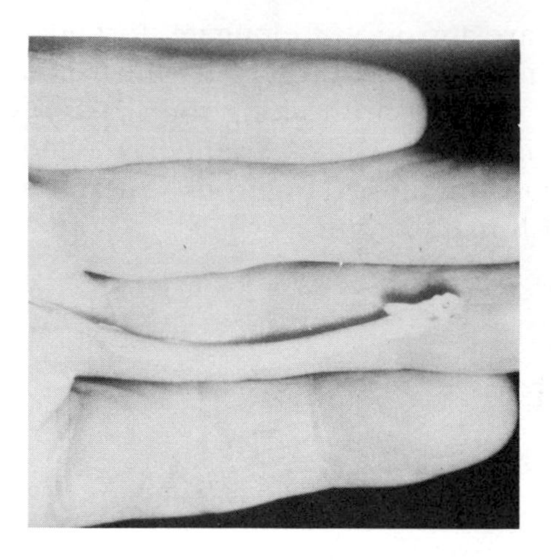

FIGURE 2.—This small Amazonian catfish which is commonly called the candirú *(Urinophilus sp?)* has the remarkable ability to penetrate the human urethra. Once the candirú has entered the urethra it is almost impossible to remove without surgical intervention.
(J. Tashijan)

Family SILURIDAE: Silurids are Old World catfishes having an entirely naked skin; they resemble clarids by the dorsal spine being either greatly reduced or entirely absent, but differ by the dorsal fin being very short or entirely absent. The adipose fin is also absent. The anal fin is very elongate. They are freshwater inhabitants, found in Europe, Africa, and parts of Asia. *Silurus glanis* is the largest known catfish, attaining a length of about 4 m and a weight of more than 275 kg.

Catfishes vary greatly in their adaptations to a large variety of ecological conditions. In addition to the usual array of forms common to North America, there is a Syrian catfish *Euglyptosternum* having a plaited structure on the thorax permitting the fish to maintain its position in rapid currents. The South American astroblepids are equipped with flattened suctorial lips which permit them to scale the vertical sides of cataracts. One of the few known vertebrate "parasites" is the pygidid catfish *Urinophilus* (Fig. 2) which is known to penetrate the urethral orifice of mammals, including man (Gudger, 1929). *Malapterus*, an Egyptian catfish, is provided with electrical organs capable of giving a severe jolt. Some siluroids live in the clear cold waters of mountain streams, whereas others are found in stagnant ponds, marshes, and lakes, or may burrow in the mud. *Plotosus lineatus*, a marine species, is usually found lurking by schools among tall seaweed. Most catfishes are carnivorous bottom feeders ingesting almost anything they can swallow. They are generally regarded as food fishes, but marine species have tough and flavorless flesh. They are oviparous in their development.

MORPHOLOGY OF THE VENOM APPARATUS

The venom organs of only two species of marine catfishes have been studied to any extent, viz., *Galeichthys felis* by Halstead *et al.* (1953) and *Plotosus lineatus* by Pawlowsky (1914) and Bhimachar (1944). The venom organs of several species of freshwater catfishes have been studied in detail, but since they are beyond the scope of the present work only brief reference will be made to them.

Gross Anatomy

GALEICHTHYS FELIS: The venom apparatus of *Galeichthys felis* includes the dorsal and pectoral stings and the axillary venom glands. The dorsal and pectoral spines are comprised of modified or coalesced soft rays which have become ossified and are, as in most catfishes, so constructed that they can be locked in the extended position at the will of the fish. Thus, *G. felis* is equipped with a formidable and efficient defensive weapon.

Dorsal spine (described in the upright position, Fig. 3): The mature dorsal spine is a stoutly-elongate, compressed, tapered, slightly arched, osseous structure bearing a series of retrorse dentations along the anterior and posterior surfaces and having an acute sagittate tip. Examination of an intact sting will reveal that the osseous spine is completely enveloped by a thin layer of sparsely pigmented skin, the integumentary sheath, which is continuous with that of the soft-rayed portion of the fin. There is no external evidence of a venom gland, nor does the integumentary sheath have the swollen appearance such as is observed in the stings of *Noturus mollis*. The dorsal spine terminates distally in a soft tip, the so-called spurious ray, which has been described by Reed (1924*a*), viz.,

> Growth of the spine results from the gradual coossification of the anterior branches of the spurious ray with the tip of the spine in the form of caps, thus adding to its length and producing successively new points . . . Due to the constant process and method of adding new parts to the end of the spine, it is always soft, so that in mechanical contact with solid objects its newer parts are mutilated.

The anterior denticulate margin of the dorsal spine gives it a somewhat keeled appearance. There is no evidence of a groove along the anterior margin of the spine. The retrorse, anterior dentations are most pronounced in the upper third of the spine and gradually alter in form until they appear as a series of distinct notches on the basal third of the spine. The anterolateral surfaces of the dorsal spine are marked by numerous, short, irregular, shallow, anastomosing, longitudinal furrows. Located along the posterior margin of the spine is a shallow groove which gradually deepens and broadens as it approaches its point of origin near the base of the spine. Lying within this groove are the posterior dentations which gradually increase in size at about the level of the junctions of the basal and middle thirds of the spine and then abruptly terminate near the inferior margin of the distal opening of the central canal. The posterior surface of the sagittate tip, superior to the first posterior dentation, is flattened and set at an angle and contains the distal opening of the central canal of the spine. From a posterior view, the spine gradually broadens until the basal third is approached and then abruptly expands into the triangular-shaped basal articulation. At the upper apex of the triangular base is located the proximal opening of the central canal. By inserting a fine wire through the proximal opening of the central canal, one will observe that the canal extends throughout the entire length of the spine,

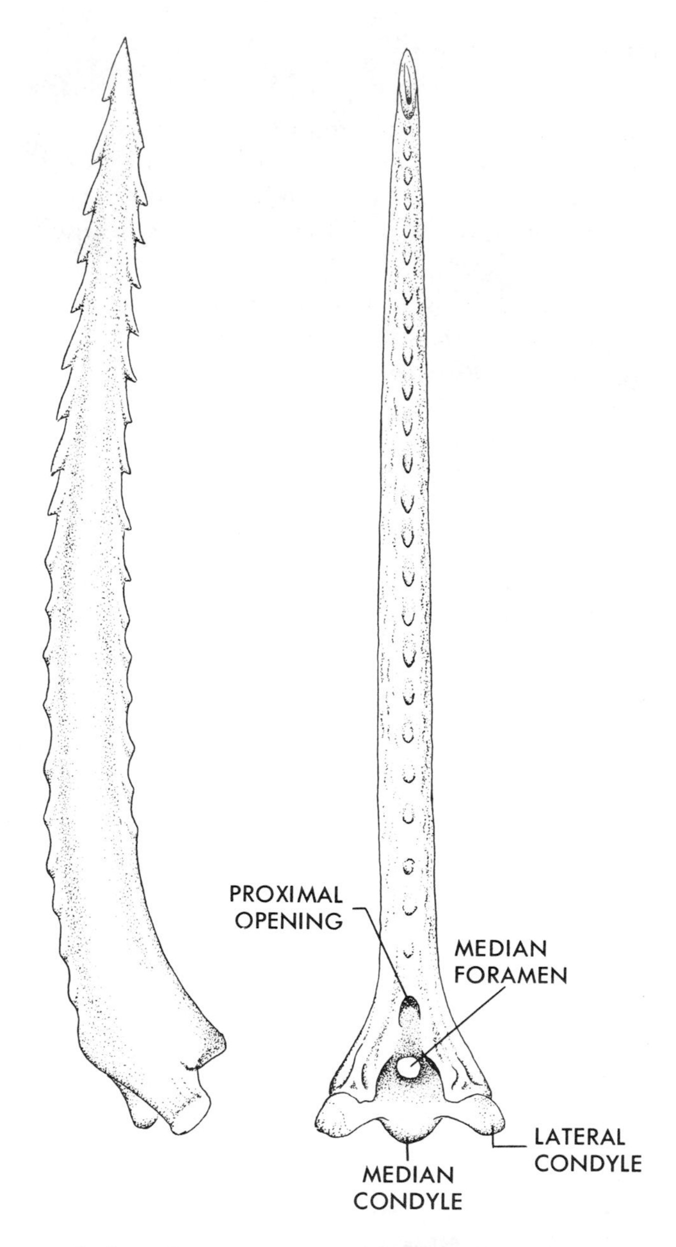

FIGURE 3.—Dorsal spine of *Galeichthys felis*. (Left) Lateral view of dorsal spines showing the retrorse dentations along the anterior (left side) and posterior margins of the spine. (Right) Posterior view showing the proximal opening and median foramen at the base of spine. (R. Kreuzinger, after Halstead, Kuninobu, and Hebard)

and the wire can be removed from the distal opening. Immediately beneath the proximal opening of the central canal is a median foramen which receives the ringlike articulation of the second interneural spine. The articulating portion of the base of the dorsal spine is comprised of three condyles, a large, median, and small lateral condyle on either side.

The locking device of the dorsal spine (Fig. 4) is a complicated mechanism consisting of the highly modified first and second interneurals and their associated dorsal spines. The upper portion of these interneurals is greatly expanded and coosified so as to form a stirrup-shaped shield, the so-called buckler or dorsal plate. However, the lower portion of the interneurals are separate and similar to those of other teleosts. From a ventral view it will be observed that the central aperture of the buckler is divided by the shafts of the interneurals into two small anterior and two large posterior foramina. Only the large posterior foramina can be seen from a dorsal view since the anterior foramina are hidden from above by the head of the buckler. Situated between the two posterior foramina is a large, wedge-shaped condyle. The anterior and major portion of this condyle belongs to the first interneural spine, but the posterior component is part of the second interneural. Inserted between the head of the buckler and the condyle is a sagittiform bone, the modified first dorsal spine. The anterior foramina convey the tendons of the erector muscles of the first dorsal spine, whereas the posterior foramina receive the latero-

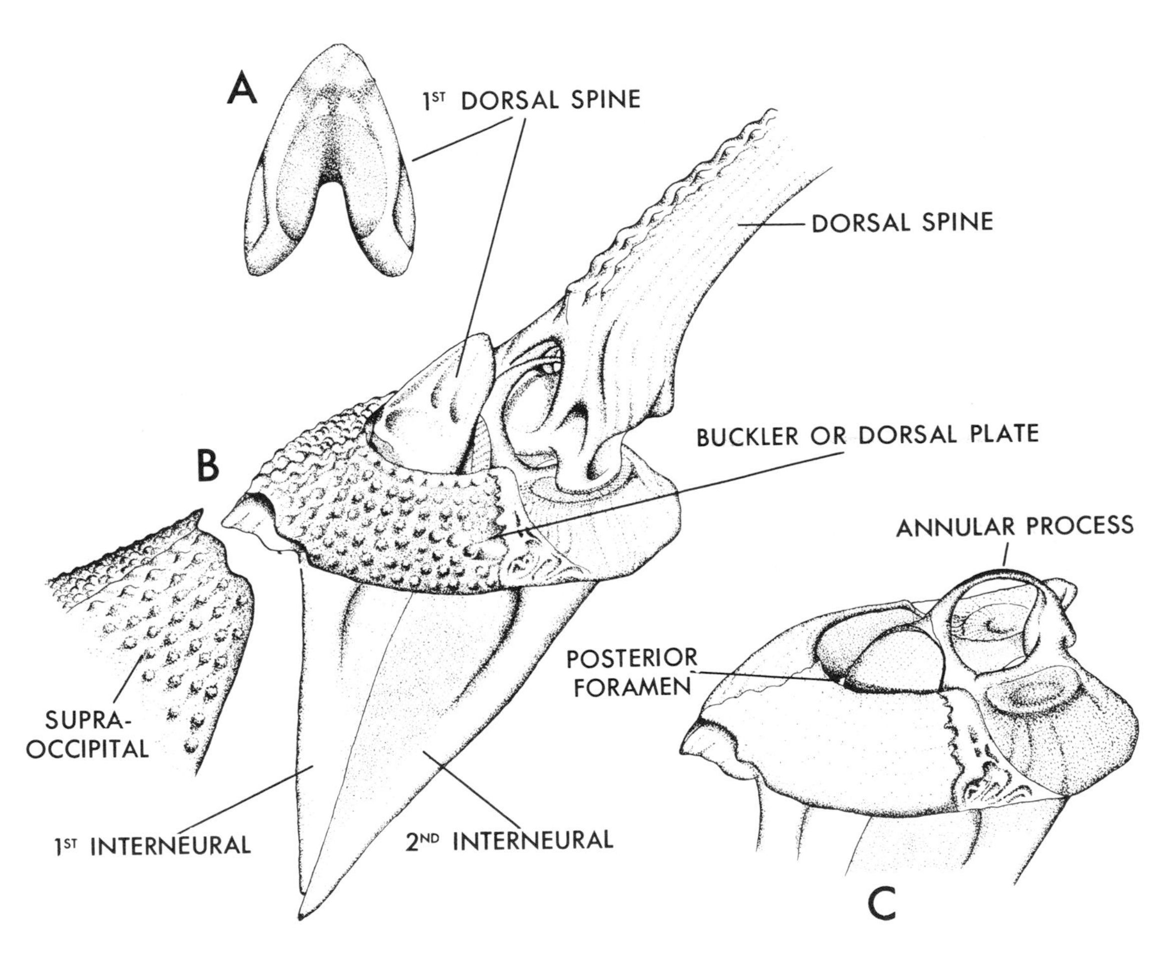

Figure 4.—Locking device of the dorsal spine of *Galeichthys felis*. a. Modified first dorsal spine, posterior aspect. b. Anterolateral view of the locking device of dorsal spine showing its relationship to the supraoccipital of the skull. c. Modified head of the first and second interneurals with the dorsal spine removed.

(R. Kreuzinger, after Halstead, Kuninobu, and Hebard)

inferior tips of the first dorsal spine and convey the tendons of the erector muscles of the second dorsal spine. The erector muscles of the first dorsal spine have their insertion along the inferior margin of the first dorsal spine and pull in a downward and forward direction. The erector muscles of the second dorsal spine insert about the level of the median foramen on the antero-inferior surface of the base of the second dorsal spine and likewise pull in a downward and forward direction. The depressor muscles of the second dorsal spine insert on the spurlike tuberosities on the posterior surface of the base of the spine. The first dorsal spine is apparently without a depressor muscle of its own. The first and second dorsal spines are connected to each other by ligamentous attachments so that the movements of one affect the other. Depression of the dorsal spines is probably further aided by the depressor muscles of the soft rays since the spines are attached to the rays by a connecting membrane.

The locking process is accomplished by bringing the roughened posterior surface of the sagittiform first dorsal spine in contact with the roughened anterior surface of the cuneiform condyle of the first interneural. Thus, contraction of the erector muscles in a downward and forward direction results in both erection and locking of the spine.

The head of the second interneural supports three articulations, two laterally placed broad concavities and a median articular ring. This median articulation consists of a deep semilunar notch which is closed by a curved rod-like extension of bone termed the annular process, giving the joint a ring-like appearance. This particular joint is referred to as the annular articulation of the second interneural spine. The annular process is inserted through the anteroposterior foramen at the base of the dorsal spine. Thus, the dorsal spine is interlocked with the head of the second interneural like a link in a chain. The median condyle of the base of the dorsal spine articulates within this ring in an arc of about 90°. The lateral concavities articulate with the lateral condyles of the dorsal spine.

Pectoral Spine (described in the horizontal position, Fig. 5): The shaft of the pectoral spine is similar to the dorsal spine in its general morphology, the possession of a spurious ray, and an integumentary sheath. There is also present a central canal with a distal and proximal opening. However, the basal third of the pectoral spine differs considerably from that of the dorsal spine. There are three articulating condyles present on the base of the pectoral spine. The superior condyle is greatly enlarged, kidney-shaped, and set at an angle, giving the base of the spine a somewhat twisted appearance. The median condyle is relatively small and lies along the same plane as the superior condyle. The superior and inferior condyles are separated posteriorly by a broad V-shaped groove and anteriorly by the median condyle and a small semilunar notch. Immediately posterior to the median condyle and cutting across the base of the superior condyle at right angles is a large semilunar notch which receives the scapular condyle of the pectoral girdle.

As previously mentioned by Starks (1930), "the scapular, coracoid, and the

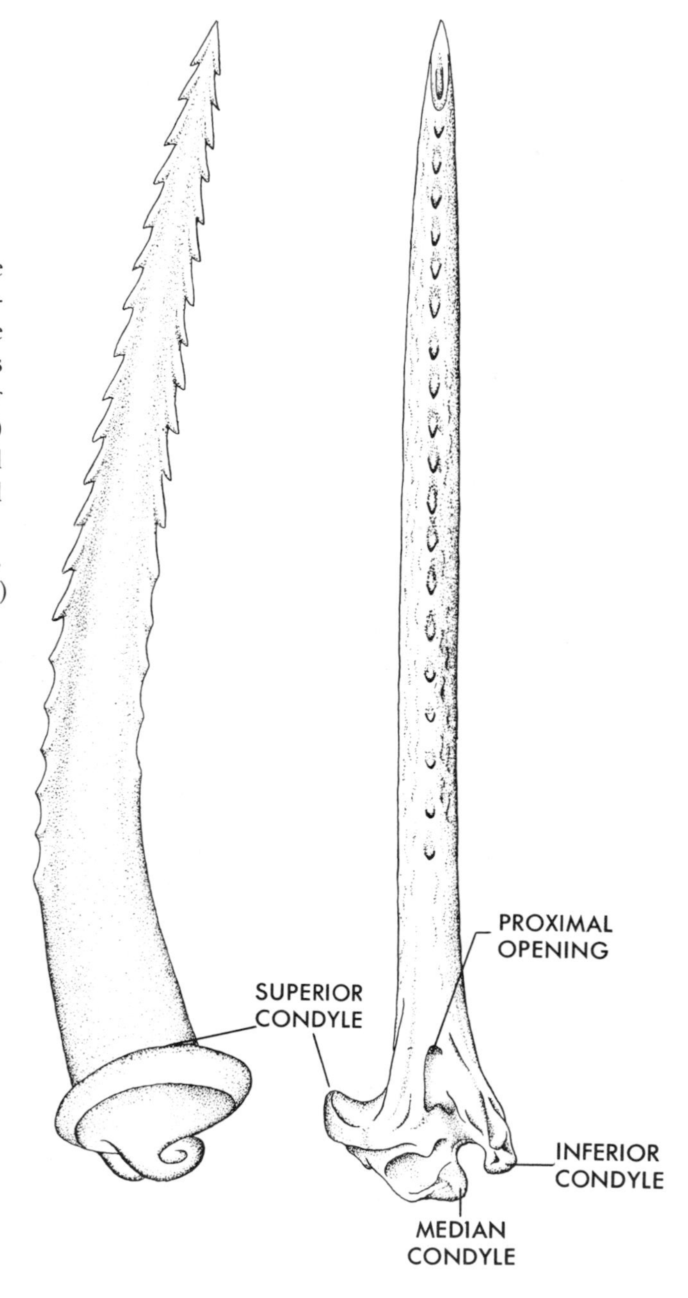

Figure 5.—Right pectoral spine of *Galeichthys felis*. (Left) Superior view of right pectoral spine showing the retrorse dentations along the anterior and posterior margins of the spine. (Right) Posterior view of right pectoral spine showing the proximal opening at the base.
(R. Kreuzinger, after Halstead, Kuninobu, and Hebard)

mesocoracoid are ankylosed into a single complex bone. The coracoid is closely attached to the cleithrum, leaving no interosseous space between." The base of the pectoral spine is received by a gnarled opening in the cleithrum and articulates with a large condyle in the scapular region (Fig. 6). Just posterior to the scapular condyle is a small pedicle of bone with which the first actinost articulates. On the underside of the arched portion of the cleithrum which passes over the pectoral joint is a deep groove in which the superior

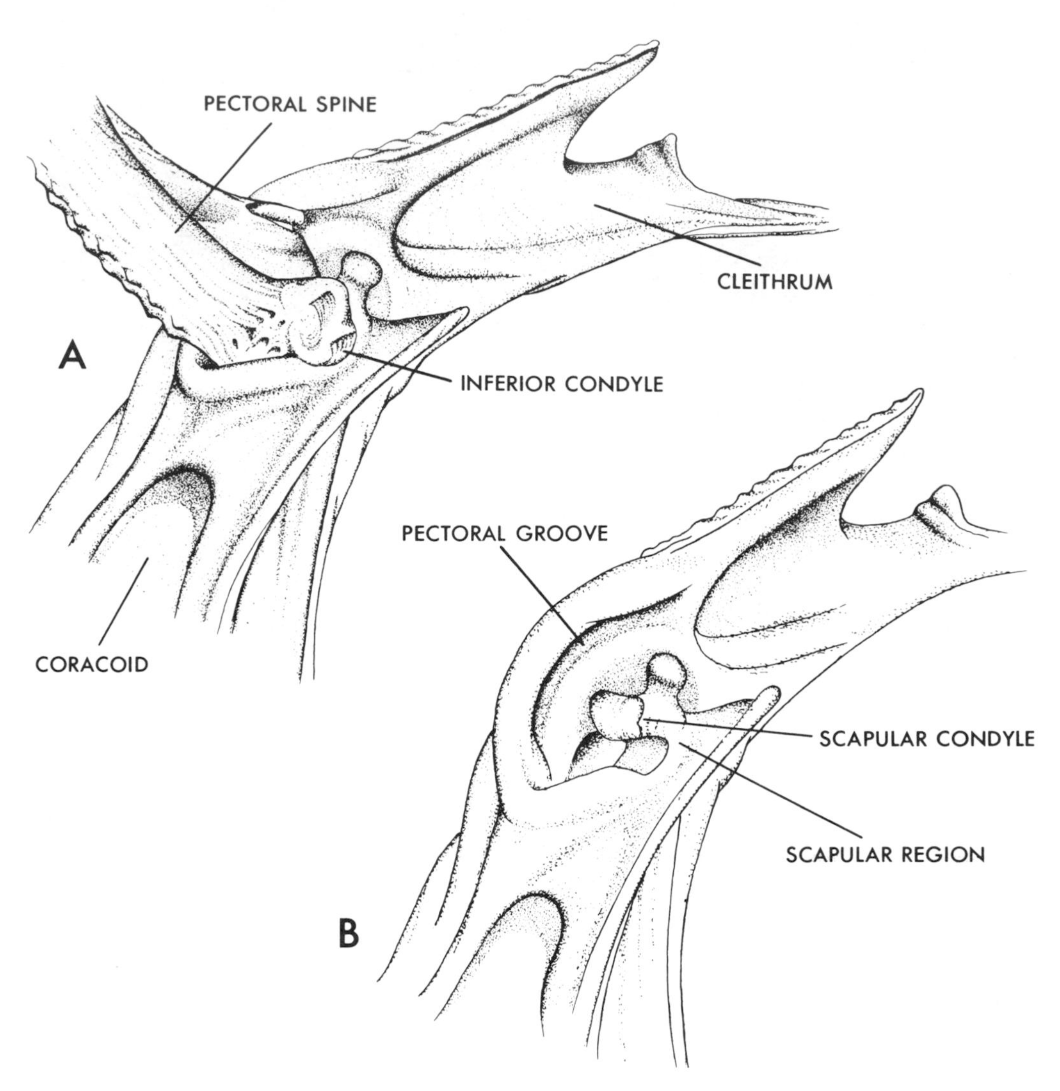

FIGURE 6.—Locking device of the left pectoral spine of *Galeichthys felis*. a. Pectoral spine in articulation with the cleithrum. b. Pectoral spine removed, showing the pectoral groove of the cleithrum.

(R. Kreuzinger, after Halstead, Kuninobu, and Hebard)

condyle of the pectoral glides back and forth. The surfaces of both the condyle and the groove are roughened, and the two structures are closely fitted; hence all that is needed to lock the spine is a slight change in horizontal pitch.

The muscular system of the pectoral spine is most remarkable and complex in its arrangement. The movements of the pectoral spine are controlled by four muscles. There is a large abductor muscle which originates on the inner aspect of the coracoid, enters a foramen of the cleithrum, then passes inferior to the tendon of the extensor spine, and inserts on the superior-medial aspect of the superior condyle of the pectoral spine. The large abductor muscle originates on the outer aspect of the coracoid, passes inferior to the belly of

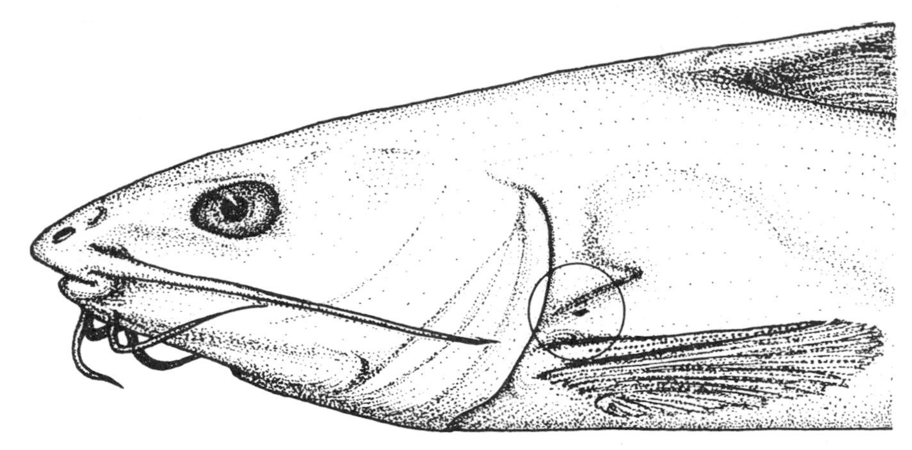

FIGURE 7.—Lateral view of head and shoulder region of *Galeichthys felis*. Circle is drawn around the posthumeral region to designate the small axillary pore which opens to the axillary gland. (From Halstead, Kuninobu, and Hebard)

the abductor muscles of the pectoral rays, and inserts on the posteromedial aspect of the base of the superior condyle of the pectoral spine. The extensor[2] and flexor muscles are relatively small and may be easily overlooked. The extensor muscle is located within the body of the cleithrum and can be observed only by making a large bone flap in the region of the arch over the pectoral joint. The extensor muscle crosses over the tendon of the abductor muscle and inserts just above the abductor tendon on the superior-medial aspect of the superior condyle of the pectoral spine. Adjacent to the anterior margin of the cleithrum inferior to the pectoral joint is the origin of the small flexor muscle. This muscle inserts on the inferior condyle of the pectoral fin. The functions of these muscles are indicated by their respective names. However, the principal movement of the pectoral spine takes place in the horizontal plane and is controlled by the abductor and adductor muscles. Movements of flexion and extension are very limited and appear to be concerned primarily with the locking action.

Axillary Gland: The axillary pore, outlet of the axillary gland, is located just below the vertical center of the posthumeral process of the cleithrum (Fig. 7). In *G. felis*, the pore is small, having an ovoid or slitlike opening which can be contained about three times in the diameter of the pupil of the eye. There is no external evidence of the axillary gland.

If the axillary gland is approached via the visceral cavity, it will be observed that the long axis lies in a vertical plane at right angles to the posthumeral process. The axillary gland is triangular in shape and situated so that the posthumeral process crosses over the upper half of the gland. In a fish with a standard length of 75 mm, the axillary gland measures about 2.0 mm by

[2] The terms extension and flexion are used in reference to the central axis of the body. The movements of the pectoral spine in the dorsal or upward direction would be extension whereas flexion would be in the ventral direction.

2.8 mm. The gland is bounded laterally by the posthumeral process and skin, medially by the peritoneum and the visceral cavity, postero-inferiorly by the abdominal muscles, and anteriorly by the adductor muscles of the pectoral rays. The gland is weakly attached to the adjacent structures by strands of loose areolar connective tissue.

PLOTOSUS LINEATUS: The following morphological description of the venom apparatus of *P. lineatus* is based on the writings of Pawlowsky (1914), Bhimachar (1944), and Tange (1955*a*). Tange has published the most complete description thus far. The venom apparatus of *P. lineatus* is comprised of the dorsal and pectoral stings and the axillary gland.

Dorsal Spine (described in the upright position, Fig. 8): The mature dorsal spine is a stout, moderately elongate, compressed, tapered, almost straight, osseous structure bearing a series of retrorse dentations along the anterior and posterior surfaces and ending in an acute sagittate tip. The serrations are more prominent on the anterior side of the spine extending from the basal third of the spine to the sagittate tip but are reduced in number and size along the posterior margin.

Examination of the intact sting will reveal that the osseous spine is completely enveloped by a thin layer of skin, the integumentary sheath, which is continuous with that of the soft-rayed portion of the fin. There is no external evidence of a venom gland.

The dorsal spine terminates distally in a soft tip, the so-called spurious ray, which has been previously described for *G. felis* and other catfishes. Evidence of the growth process of the dorsal spine by the continual addition of caps of spurious rays is seen in the mature portion of the spine by a series of shallow oblique grooves or markings which have been described in detail by Tange (1955*a*).

Judging from Tange's (1955*a*) report, the base of the spine is similar to that of *G. felis*. The shaft of the spine, which is called the central canal, is hollow and is similar to that of *G. felis*. There is no evidence of any connection between the central canal and the venom glands located along the outer shaft of the dorsal spine. The spine in cross section is spindle-shaped.

The venom glands lie beneath the integumentary sheath and appear as two glandular plate-like masses, one on either side of the shaft of the spine. The glands extend almost the entire length of the spine, from the tip to the proximal base, and cover most of the width of the spine.

Pectoral Spine: There are very little descriptive data available regarding the pectoral spine of *Plotosus lineatus*, but apparently the shaft of the pectoral spine is similar to the dorsal spine in its general morphology and in the possession of a spurious ray and integumentary sheath. There is also present a central canal with a distal and proximal opening. The venom glands are grossly similar in morphology to those of the dorsal sting. In one specimen (total length of fish 243 mm) described by Tange (1955*a*), the pectoral sting glands measured 12 mm long, about 1 mm wide, and 0.15 mm average in thickness. However, the venom glands were thickest in the distal third of the spine.

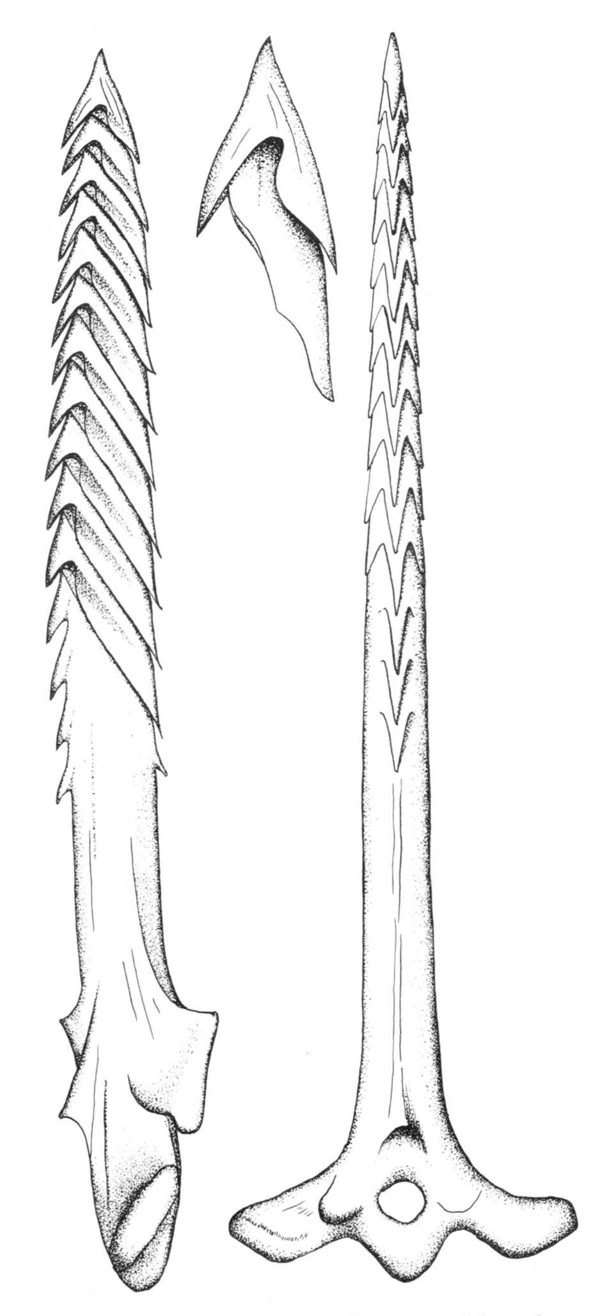

Figure 8.—Dorsal spine of *Plotosus lineatus.* (Left) Lateral view of dorsal spine showing the retrorse dentations along the anterior (left side) and posterior margins of the spine. The soft tip or so-called spurious ray has been separated from the main shaft of the spine to show its structure. It will be noted that the serrations of the spine are formed by the accumulated additions of these spurious rays. (Right) Posterior view of the dorsal spine showing the proximal opening and the median foramen at the base of the spine. (R. Kreuzinger)

The base of the spine has not been described but apparently resembles that of *G. felis.*

Axillary Gland: Tange (1955*a*) has described the axillary gland of *P. lineatus* as being somewhat chestnut-shaped and situated near the base of the pectoral spine. The dimensions of the gland in one fish (total length of fish 243 mm) were 2.5 mm by 1.8 mm. The axillary gland is surrounded on the outside with a thick capsule of connective tissue. There is an excretory canal which opens on the dorsal aspect of the pectoral spine base as a small pore. Venom is believed to be supplied to the outside of the pectoral stings through this axillary pore. Thus the pectoral stings are equipped with their own fin glands and are further supplied with venom from the axillary glands.

Microscopic Anatomy

GALEICHTHYES FELIS: Dorsal sting: Since the microscopic anatomy of the dorsal sting does not appear to differ significantly from that of the pectoral sting, only the latter will be described.

Pectoral Sting: Examination of cross or oblique sections taken from the middle third of the pectoral sting (Fig. 9a-c) will reveal that the sting may be divided into three distinct zones: a peripheral integumentary sheath, an intermediate osseous portion, and a central canal. The structure is ovoidal in outline, possessing anterior and posterior denticles which give it a bi-keeled appearance. The integumentary sheath is continuous with the integument covering the soft rays and the body of the fish. The sheath is comprised of a relatively thick outer layer of epidermis and a thin layer of dermis which are separated from each other by a basement membrane and a distinct pigment line. With the exception of areas where glandular epithelium is concentrated, stratified squamous epithelium, about five to six cells in thickness, is the principal component of the epidermal layer. Scattered throughout the stratified squamous epithelium are large mucous cells (termed clavate cells by Reed, 1907, and others), which are identical in morphology with those found elsewhere in the integument. The mucous cells are generally ovoid to round in outline, measure about 24μ to 52μ in their greatest diameter, and are binucleate. The nuclei are basophilic in their staining reaction, and the cytoplasm is clear and vacuolated. The glandular cells, which comprise the venom gland, are best observed in tissue sections taken from the middle and distal thirds and are most concentrated at the anterolateral and posterolateral margins of the sting where they are sometimes clumped two or three cells deep. At the level of the middle third of the sting, the venom gland appears as a cellular sheet wedged between the pigment layer and the stratified squamous epithelium. Although the mucous cells appear to merge with the glandular cells, it will be observed that the mucous cells are generally more spherical, are less likely to be collected in sheets or clumps, possess a very light-staining or clear cytoplasm, and are more polygonal in outline. Mucous cells stain a pale green with Masson's trichrome, while glandular cells stain red. Both mucous and glandular cells are binucleate. The glandular cells range in size from about 12μ to 28μ, and many of them have a ragged or greatly vacuolated appearance, apparently having discharged their cytoplasmic secretions so that only the nuclei and a few vacuoles remain. The dermis, lying immediately below the epidermis and in direct contact with the epidermal basement membrane, consists of a thin layer of dense areolar connective tissue interspersed with minute vascular channels which completely invests the osseous portion of the sting.

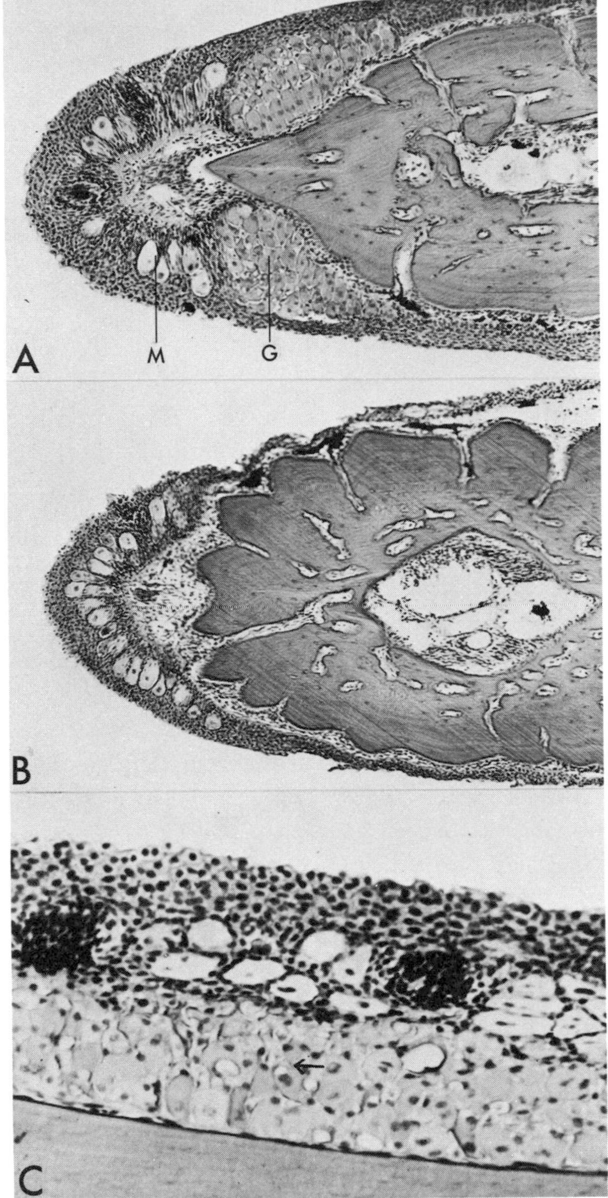

Figure 9.—Photomicrographs of the pectoral sting of *Galeichthys felis*. a. Oblique section of the pectoral sting taken at the level of the middle third. M: mucous cells; G: glandular cells (venom gland). The glandular cells range from 12 to 28μ in diameter. b. Cross section of the pectoral sting taken at the level of the basal third. It will be observed that the venom gland has disappeared and only the mucous cells remain. c. Median longitudinal section of the pectoral sting at the level of the basal third. The arrow is located in the dense glandular area (venom gland). Clear mucous cells are sparsely scattered in the stratified squamous epithelium above. (From Halstead, Kuninobu, and Hebard)

The osseous portion of the sting, the pectoral spine, is more than twice the thickness of the integumentary sheath. Scattered throughout the spine are a variable number of small, irregular, branching, osseous canals occupied by loose areolar connective tissue and small blood vessels. Some of the canals lie in a transverse plane, extending from the dermis of the integumentary sheath to the central canal. Interspersed throughout the osseous substance are small lacunae which contain osteocytes.

The central canal of the pectoral spine is lined with loose areolar connective tissue and a variable number of small nerve trunks. Scattered throughout the mid-portion of the canal are blood vessels and areolar connective tissue. Hence, one would conclude that glandular activity is limited to the integumentary sheath and the primary function of the central canal is to convey vascular and nervous structures rather than venom as one might be tempted to rationalize.

As the basal articulation of the pectoral sting is approached, it will be observed that the glandular tissue gradually diminishes in size and finally disappears (Fig. 9b). Sections taken near the base show an integument which is similar to that found elsewhere on the body, viz., containing scattered mucous cells but no glandular cells. The osseous portion of the sting becomes broader and the canals more numerous. The only significant change noted in the central canal is the progressive enlargement and decrease in numbers of the nerve trunks. Cross sections taken from the distal end of a traumatized sting generally show very little evidence of stratified squamous epithelium or mucous cells. Glandular cells which are heaped up in relatively large ovoid masses on either side of the spine are the chief cellular component of the epidermal layer of this region. However, intact specimens are similar in their morphology to sections taken from the level of the middle third. Although reduced in size, the dermis is still evident. The pectoral spine becomes triangular in outline, the central canal gradually decreases in size, and the osseous canals are fewer in number as the distal tip of the spine is approached. In cross sections taken through the distal opening of the central canal, the canal appears as a deep depression opening on the posterior aspect of the spine. At the extreme tip the central canal disappears completely, and only one or two small osseous canals remain.

Median longitudinal sections of the pectoral spine reveal that the integumentary sheath extends throughout the length of the spine. The line of demarcation between the mucous cells and the venom gland is very distinct in sections taken at this plane (Figs. 9c; 10a). The glandular cells are congregated along the anterior margin of the spine in the form of a cellular sheet with the greatest concentration of the cells near the tip and in the shallow depressions between denticles. As one proceeds toward the base, the gland gradually thins out and finally disappears as the basal third is approached. The venom gland along the posterior margin of the spine is relatively thin and irregular in its distribution. Tissue sections cut at the median longitudinal plane also clearly reveal the central canal extended throughout the length of the spine. The remaining structures do not appreciably differ in their morphology from the description previously given of the cross sections of the sting.

There is no evidence of any glandular ducts. In general, the basal glandular cells of the epidermis are columnar in their arrangement, having two ovoid to spherical basophilic nuclei and an eosinophilic cytoplasm. Continuing outward the cells appear to become enlarged and more spherical and subsequently undergo a holocrine type of lysis. Thus the venom-producing

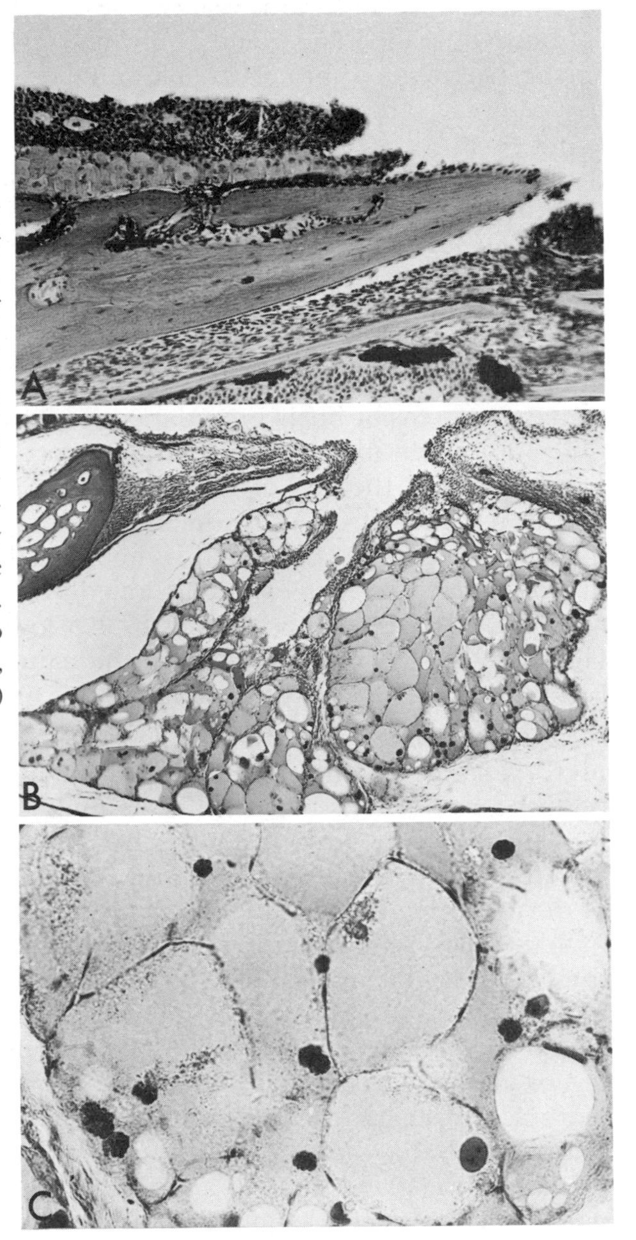

Figure 10.—Photomicrographs of dorsal sting and axillary gland of *Galeichthys felis*. a. Median longitudinal section through the top of dorsal sting. It will be observed that the mucous and glandular cells are similar in their appearance and distribution to those observed in the pectoral sting in Figure 9c. b. Cross section through the axillary pore and duct of the axillary gland. c. Enlarged view of the secretory cells of the axillary gland. These cells vary in size from 45 to 210μ in diameter. (From Halstead, Kuninobu, and Hebard)

cells are destroyed at the time the secretion is released. Since the venom glands extend out to the distal extremity of the sting, insertion of the tip into the flesh of the victim would also damage cells and thereby release venom into the wound. The amount of venom received by the person would depend to a large extent upon the depth to which the sting penetrated.

Axillary Gland (Fig. 10b, c): Examination of longitudinal tissue sections of the axillary gland will reveal that it is enclosed within a capsule of fibrous connective tissue. The gland is divided into three or four lobes which are further subdivided into a variable number of lobules by fibrous connective tissue septa. The lobules are composed of large secretory cells which are ovoid to polygonal in outline and measure 45μ to 210μ in their greatest diameter. The secretory cells are further characterized by having two large, granular, basophilic nuclei and a greatly distended, finely granular, eosinophilic, vacuolated cytoplasm. Each of the secretory cells is surrounded by a layer of fibrous connective tissue one cell in thickness. There is no evidence of a basal undifferentiated cell layer such as is generally observed in a holocrine type of gland. In some of the sections there are observed interlobar and interlobular ducts lined by a modified type of squamous epithelium and containing areas of finely granular, amorphous secretion. The axillary gland is a compound tubular gland having a merocrine type of secretion.

Because of the lack of fresh specimens, Halstead *et al.* (1953) were unable to test the secretions of the axillary gland of this species. Since the histological structure of this gland is similar to that which Reed (1907) observed in *Schilbeodes* and *Noturus*, it is assumed that the secretions are toxic. Similar conclusions have been arrived at by Pawlowsky (1914) and Tange (1955*a*). However, one cannot be certain of the nature of the secretions until toxicological tests have been conducted.

PLOTOSUS LINEATUS: Dorsal and Pectoral Stings: The dorsal (Fig. 11) and pectoral stings resemble each other in their microscopic anatomy. Examination of cross sections taken from the middle third of a fin sting reveals that it is divided into three distinct zones: a peripheral integumentary sheath, an intermediate osseous portion, and a central canal. The cellular composition of the integumentary sheath, which contains the venom glands, is comparable to that of *G. felis*. However, it appears to vary markedly in the distribution of the glandular elements. The venom glands of *P. lineatus* are much larger than in *G. felis* (Fig. 9) and surround most of the osseous portion of the spine except in the anterior and posterior portions where the glandular cells thin out or abruptly terminate just short of the anterior and posterior denticular margins of the spine. The glandular cells are described as large cylindrical or ovoid cells that measure about 50μ to 60μ by 25μ to 35μ. In the peripheral portion of the gland the cells are polygonal in shape. The cells are arranged perpendicular to the longitudinal axis of the spine, appearing as a cellular sheet several cells in depth. The cylindrical cells are in close proximity to the osseous spine, whereas the ovoid and polygonal cells are more superficial in their distribution. The glandular cells are acidophilic in their staining and have a vacuolated cytoplasm but are less vacuolated than the glandular cells of most other venomous fish species. The largest vacuole seen consumed about one-third the cell volume. The cytoplasm of the intact cells is finely granulated. The nuclei of the glandular cells are poor in chromatin, rounded, eccentric in position, and quite small in relation to the size of the cell, measuring 6μ to 10μ. There are two to three nuclei for each cell. The nucleoli

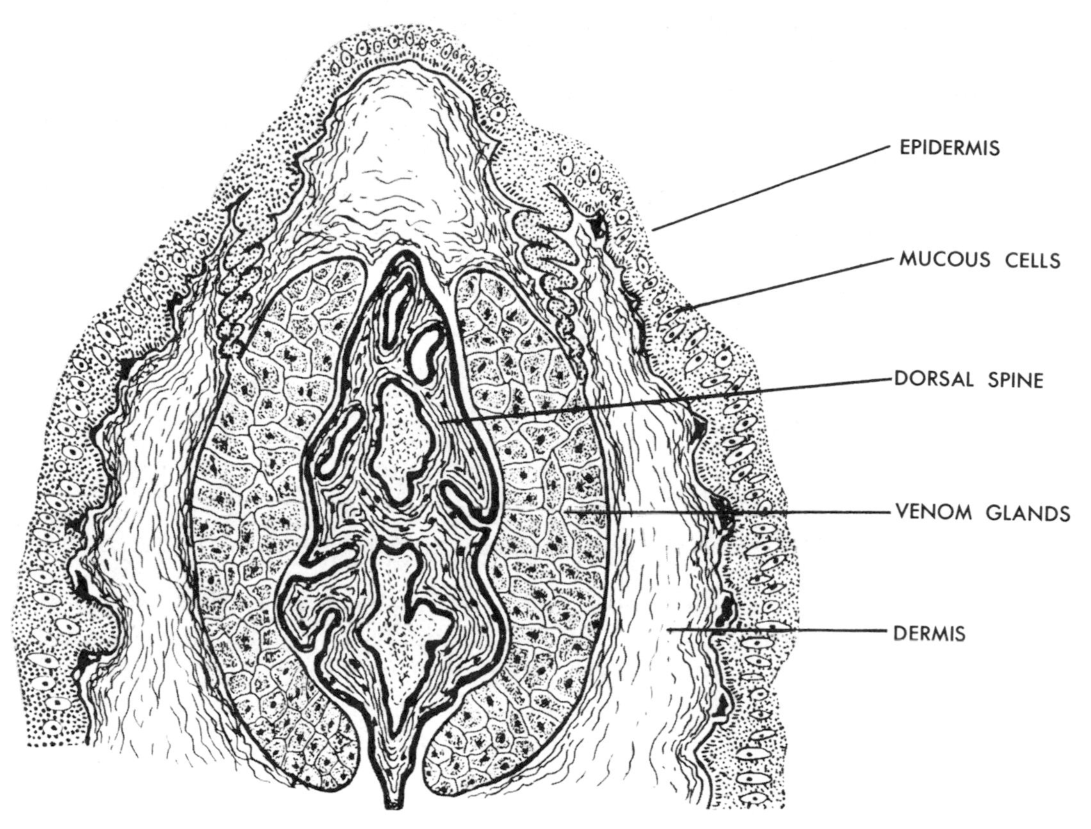

Figure 11.—Drawing of a cross section of the dorsal sting of *Plotosus lineatus*. Note the large plate-like arrangement of the venom glands which cover most of the surface of the dorsal spine. × 125 (approx.) (R. Kreuzinger, after Pawlowsky)

are clearly defined. In between the glandular cells are flattened supporting cells which have a size of 6μ to 10μ by 2μ to 3μ. Their nuclei are much richer in chromatin than those of the glandular cells. There is no evidence of excretory ducts.

Scattered throughout the integument, exclusive of the principal venom gland, there is an abundance of dilated cells about 50μ to 60μ by 30μ to 40μ in size. The cells are for the most part rounded, oval, or almost cylindrical in shape. Their cytoplasm is acidophilic and resembles the cells of the venom glands. The nuclei are rounded or ovoid, but richer in chromatin than those of the venom gland cells. The nuclei measure 7μ to 8μ by 6μ to 7μ. Animal experiments with this portion of the integumentary sheath seem to indicate that this portion of the integument may also produce a toxic secretion.

The rest of the integument resembles that described for *G. felis*.

Axillary Gland: The axillary gland is surrounded on the outside by a thick capsule of connective tissue which penetrates into the gland and divides the glandular parenchyma into several sections or pockets. The axillary gland is comprised of very large glandular cells which measure 70μ to 130μ by 60μ

to 80μ. They are generally round or ovoid in shape. The cytoplasm stains intensely acidophilic and resembles that found in the cells of the venom glands of the fin stings. The cytoplasm of these cells are also vacuolated and in some instances the vacuoles may consume two-thirds of the cell volume. The nuclei of these cells are round or ovoid and quite small in comparison to the size of the cell, measuring about 12.5μ by 16μ. They usually appear to be rich in chromatin. The nucleoli are clearly defined. Supporting cells are situated between the glandular cells. These cells are spindle-shaped, flattened, and have an average dimension of about 5μ to 10μ by 6μ. Their cytoplasm does not have a homogenous appearance but is finely granular and basophilic in its staining. The nuclei are not rich in chromatin.

In the center of each glandular pocket there appears to be a definite lumen which apparently leads to the excretory duct. The main excretory duct is quite short, its walls composed of integument, and empties on the dorsal aspect of the pectoral fin base as a small pore.

Venom Apparatus of Freshwater Catfishes

Since the morphology of the venom organs of freshwater catfishes is beyond the scope of the present work, the reader's attention is directed to the following publications for further information on this subject: Reed (1907, 1924*a, b*), Pawlowsky (1913), Phisalix (1922), Bhimachar (1944), and Tange (1955*b, c*). Histological sections of the stings of several representative species *(Ictalurus melas, I. nebulosus, I. punctatus, Noturus insignis, N. miurus, N. mollis)* of freshwater catfishes are illustrated (Figs. 12-18).

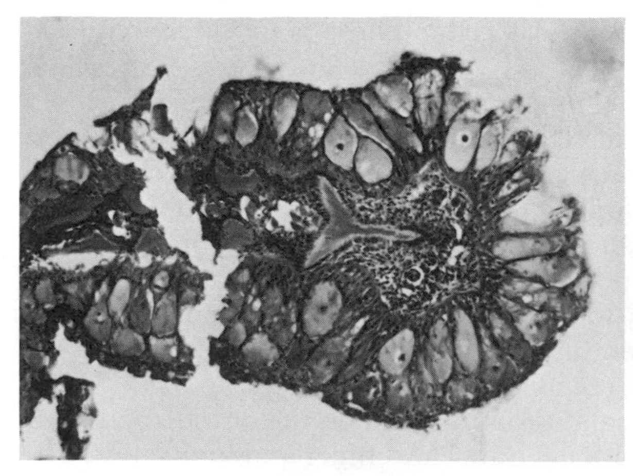

FIGURE 12.—Photomicrograph of the pectoral sting of *Ictalurus melas* (Rafinesque). Cross section taken at the level of the middle third of the spine. Note that most of the outer layer of stratified squamous epithelium which contains the mucous cells is absent in this section. The large cells seen in this photomicrograph are the glandular cells which comprise the venom gland. Hematoxylin and triosin stains. ×175. Sections prepared by M. Chitwood and B. W. Halstead. (E. Rich)

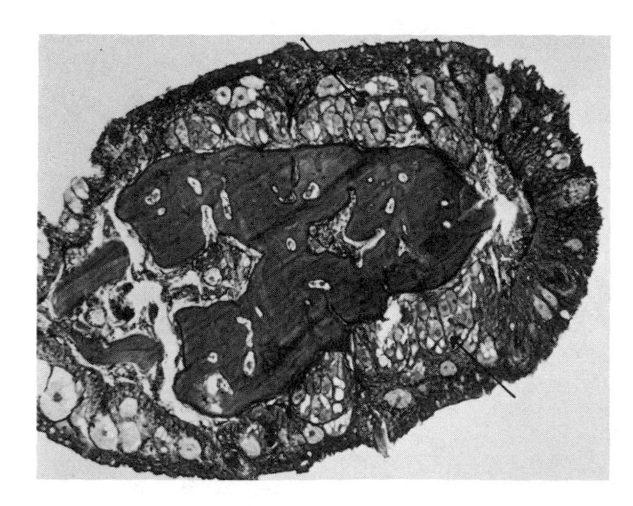

Figure 13.—Photomicrograph of the pectoral sting of *Ictalurus nebulosus* (LeSueur). Cross section is taken at the level of the middle third of the spine. Arrows point to clusters of glandular cells which comprise the venom gland. Note that the venom gland encompasses most of the surface of the trigonal-shaped center mass of pectoral spine. Mucous cells are situated in the outer layer of stratified squamous epithelium. Hematoxylin and triosin stains. × 95. Sections prepared by M. Chitwood and B. W. Halstead. (E. Rich)

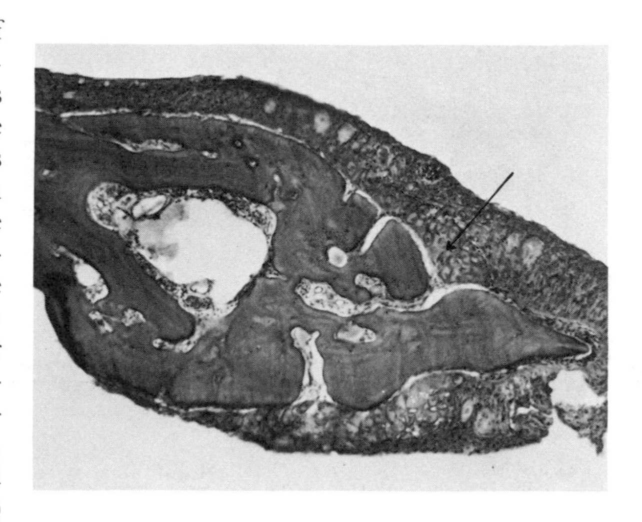

Figure 14.—Photomicrograph of the pectoral sting of *Ictalurus punctatus* (Rafinesque). Cross section is taken at the level of the middle third of the spine. Arrow points to a cluster of glandular cells which comprise the venom gland. Note that the venom gland in this particular section does not appear to be as extensive as found in *I. nebulosus*. The larger cells in the outer layer of stratified squamous epithelium are mucous cells. Hematoxylin and triosin stains. × 140. Sections prepared by M. Chitwood and B. W. Halstead. (E. Rich)

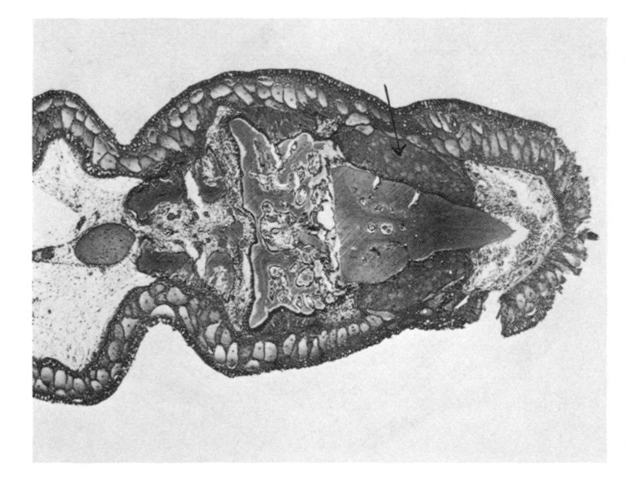

Figure 15.—Photomicrograph of the pectoral sting of *Noturus insignis* (Richardson). Cross section is taken at the level of the middle third of the spine. Arrow points to clearly defined mass of glandular cells which comprise the venom gland. The larger cells in the outer layer of stratified squamous epithelium are mucous cells. Hematoxylin and triosin stains. × 55. Sections prepared by M. Chitwood and B. W. Halstead. (E. Rich)

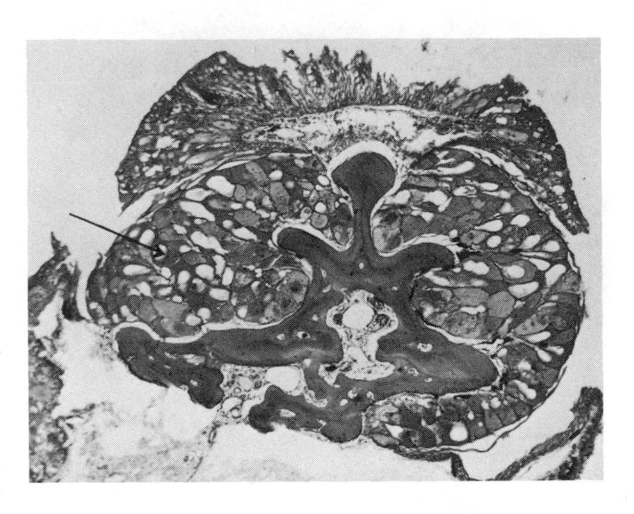

FIGURE 16.—Photomicrograph of the dorsal sting of *Noturus miurus* Jordan. Cross section is taken at the level of the middle third of the spine. Note that part of the superficial layer of stratified squamous epithelium (top) has been traumatized. Arrow points to one of the large masses of glandular cells which comprise the venom glands on either side of the dorsal spine. Epithelium along the sides of the sting is largely absent. Hematoxylin and triosin stains. × 63. Sections prepared by M. Chitwood and B. W. Halstead. (E. Rich)

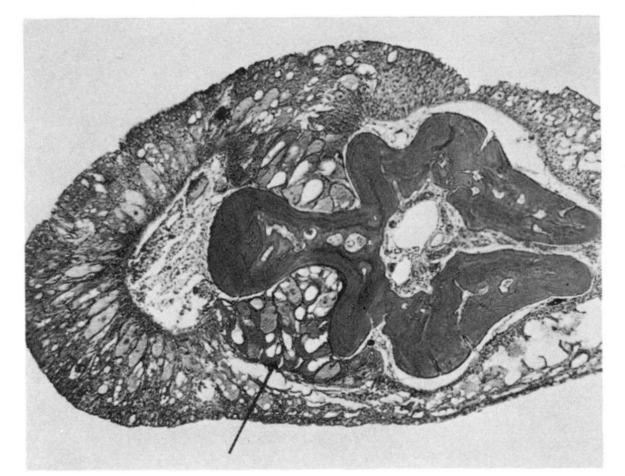

FIGURE 17.—Photomicrograph of the pectoral sting of *Noturus miurus* (Jordan). Cross section is taken at the level of the middle third of the spine. Note that the venom glands are limited in this section to either side of the anterior aspects of the pectoral spine (see arrow). The epithelial layer is well-developed and the mucous glands are clearly evident. Hematoxylin and triosin stains. × 68. Sections prepared by M. Chitwood and B. W. Halstead. (E. Rich)

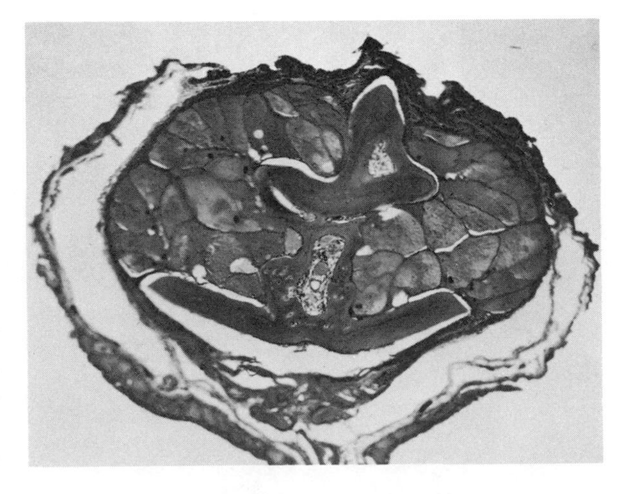

FIGURE 18.—Photomicrograph of the pectoral sting of *Noturus mollis* (Mitchill). Cross section is taken at the level of the middle third of the spine. The large masses of glandular cells on either side of the spine comprise the venom glands. The outer layer of stratified squamous epithelium is almost completely absent in this section. Hematoxylin and triosin. × 63. Sections prepared by M. Chitwood and B. W. Halstead. (E. Rich)

MECHANISM OF INTOXICATION

Envenomations from catfishes are generally incurred as a result of handling the fish—such as removing the fish from the net or taking a hook out of its mouth. Some species of catfishes may display aggressive behavior (Fig. 19). The fin spines of catfishes are particularly dangerous because when the fish becomes agitated or is handled, the dorsal and pectoral spines are firmly locked into a rigid extended position. In most of the venomous species the

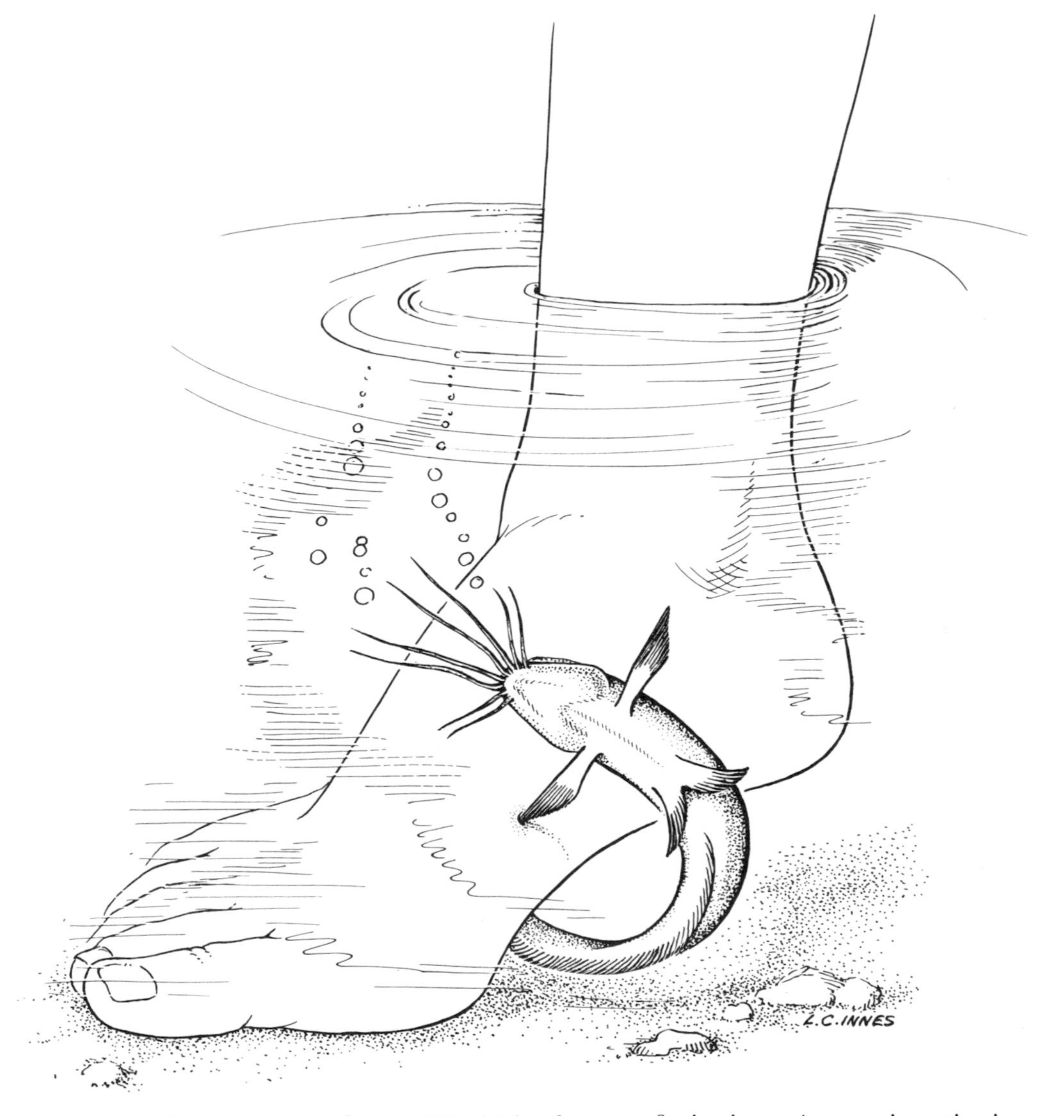

FIGURE 19.—*Heteropneustes fossilis* (Bloch) in the act of stinging. Aggressive stinging behavior such as this is unusual among catfishes since most species tend to be passive stingers. However, *H. fossilis* may demonstrate aggressive behavior when threatened. (L. C. Innes, after Fernando)

spines are enveloped in an integumentary sheath containing the venom glands. The fin spines are very sharp and will readily penetrate through the skin of the victim. The spines of some species are also equipped with a series of sharp retrorse teeth capable of producing a severe laceration of the victim's flesh, thus facilitating absorption of the venom and subsequent secondary infection. During the process of inflicting the wound, the delicate integumentary sheath of the sting is damaged, the venom glands are exposed, and the toxic secretions of the venom glands thus gain entrance into the traumatized flesh of the victim.

MEDICAL ASPECTS

Clinical Characteristics

The severity of the symptoms incident to an attack varies with the species of catfish and the amount of venom received. The pain is generally described as an instantaneous stinging, throbbing, or scalding sensation which may be localized or radiate up the affected limb. In wounds produced by species of *Schilbeodes, Noturus*, and *Arius*, the pain usually subsides within 30 minutes or less. Some of the tropical species of *Plotosus, Selenaspis, Sciades, Pimelodus, Pterodoras, Netuma, Osteogeniosus*, etc., are capable of producing violent pain which may last for 48 hours or more. The area about the wound becomes ischemic immediately after being stung. The pallor about the wound is soon followed by a cyanotic appearance and then by redness and swelling. In extreme cases there may be a massive edema involving the entire limb, accompanied by lymphadenopathy, numbness, and gangrene of the area about the wound. Primary shock may be present and manifested by such symptoms as faintness, weakness, nausea, loss of consciousness, cold clammy skin, rapid weak pulse, low blood pressure, and respiratory distress. Improperly treated cases frequently result in secondary bacterial infections of the wound. According to Phisalix (1922) and Bhimachar (1944), tetanus may be an additional complication which should be considered. Wounds produced by some of the less venomous catfishes, e.g., *Schilbeodes* and *Noturus*, are generally mild in nature and cause no further discomfort after the pain subsides. However, the more dangerous species such as *Plotosus* will frequently produce wounds which may take weeks to heal. In Herre's case (1949), he was unable to use the thumb, which had been stung by a *Plotosus*, for five and a half months. Deaths have resulted from the stings of *Plotosus lineatus, Pterodoras granulosus*, and *Pimelodus clarias*.

The symptoms resulting from wounds by *G. felis* apparently are similar to those produced by *Noturus* and *Schilbeodes*. In August 1949, while collecting fishes at Boca del Rio, Vera Cruz, Mexico, the author was jabbed in the right thumb by the dorsal sting of a small specimen measuring less than 150 mm in total length. On another occasion a similar wound was received from the pectoral sting of this same species. In each instance there was an immediate, intense stinging and throbbing sensation which continued for about 20 minutes and then gradually subsided. There was an ischemic zone around the wound and beyond that a zone of erythema. No other symptoms were noted and no treatment taken. According to the native fishermen at Boca del Rio, the larger specimens of *G. felis* can produce very severe lacerations and extremely painful wounds. No deaths were reported.

The painful nature of wounds resulting from catfish stings has been discussed by numerous authors, for example: Phisalix (1922), Pawlowsky (1927), Buddle (1930), Le Gac (1936), Bhimachar (1944), Herre (1949), and Halstead (1959).

Pathology

(Unknown)

Treatment

The treatment of catfish envenomations is similar to that of stingray stings. See p. 655 for details.

Prevention

Most catfish stings can be prevented by avoiding contact with the dorsal and pectoral stings of the fish. When handling a catfish, one should grasp the animal firmly with one hand immediately behind the dorsal and pectoral fins. If the specimen is too small to handle in this manner, it is better to grasp the fish with a pair of pliers or a large pair of forceps. If the fish is *Plotosus lineatus*, a person would show even greater judgment by leaving it completely alone.

PUBLIC HEALTH ASPECTS

Catfish envenomations are not reported to be a public health problem.

TOXICOLOGY

Data regarding the toxic effects of catfish venom upon laboratory animals appear to be the work of two persons, Toyoshima (1918) and Bhimachar (1944).

Toyoshima (1918) obtained the venom by macerating dorsal and pectoral spines removed from *Plotosus lineatus*, a marine catfish taken at Okinoshima, Kochi Prefecture, Japan. Physiological saline and glycerine were found to yield the best results. His extracts were crude, using as a standard lethal dose 0.1 ml, the amount of extract sufficient to kill a 10 to 12 g white rat within a period of 3 hours by subcutaneous injection. Glycerine gave a yield about four times as potent as physiological saline extracts. His test animals consisted of beavers, dogs, rabbits, white rats, sparrows, frogs, snakes, newts, carp, loaches, chickens, and pigeons.

Intravenous injections resulted in the immediate development of symptoms consisting of muscular spasm, respiratory distress, and death. The same amount of extract injected intraperitoneally produced the same symptoms within a period of 10 to 20 minutes. It was noted that in rabbits there was also mydriasis and exophthalmos. Most animals also showed a decrease in body temperature of 1 to 2 degrees.

Using the "standard poisonous liquor," Toyoshima attempted to determine the minimal lethal dosage for various types of laboratory animals. The following specimens were used: 25 white rats, 14 beavers, 10 rabbits, 9 black rats, 1 pigeon, 6 sparrows, 11 chickens, 5 frogs, 5 snakes, 11 newts, 5 carp, and 19 loaches. The minimal lethal dosages for these animals, the amount which killed the animals within a period of 5 days, is given in Table 1.

Studies of the effect of temperature upon the action of the toxin revealed that the amount of toxin at 37° C sufficient to kill a newt was inactive when cooled to temperatures of 7 to 8° C.

Toxic symptoms were produced when the toxin was administered by subcutaneous, intravenous, intramuscular, intraperitoneal, intestinal, vaginal, intraspinal, and epineurial routes but were inactive when taken orally. Subcutaneous injection of the toxin produced an area of cyanosis surrounded by a zone of erythema, which lasted for about 2 days and then developed into dry gangrene. Rabbits which had died from poisoning showed a generalized visceral congestion. Toyoshima concluded that the primary site of action of the toxin was in the brain.

Plototoxin was found to have both neurotoxic and hemotoxic properties. Utilizing the blood corpuscles of various animals, Toyoshima observed that the toxin produced hemolysis in humans, rabbits, carp, dogs, beavers, rats, eels, chickens, pigs, sparrows, cats, and tortoises but had no effect upon bovine, toad, sheep or pigeon red blood cells. The toxin produced hemolysis in nonresistant species without the aid of lecithin, but the addition of lecithin failed to produce hemolysis in resistant species. Tests conducted on the red blood cells of dogs and rabbits at temperatures of 0° to 1° C did not produce hemolysis, but at 37° C, hemolysis occurred. Pigeons and sheep appear to be more resistant to the toxic effects of plototoxin than most other animals that were tested, requiring large amounts of the toxin to produce death.

Toyoshima (1918) also studied the effects of various physical agents upon plototoxin. Heating the toxin at 100° C for 2 hours or at 145° C for 30 seconds destroyed its hemolytic properties. No destruction was noted when the toxin was kept at -10° C to -15° C for 5 days. Toxin kept at -191° C in liquid air overnight failed to show any alteration in its toxic properties. However, the death time appeared to be slightly lengthened. Exposure of the poison to sunlight for a period of 30 days showed no destruction, but 100 days' exposure reduced the toxicity to one-half. The greatest destruction of hemolytic properties was produced by red light rays, but the greatest damage to its neurotoxic properties was obtained from the violet light rays. Exposure of the toxin to ultraviolet light for 40 minutes reduced its neurotoxic activity by about one-half, but there was no alteration in its hemolytic characteristics even when exposed for 60 minutes. The potency of the toxin was greatly reduced when exposed

TABLE 1.—*MLD's for animals injected with plototoxin. (From Toyoshima)*

Animal	Route of Administration and Amount (ml)		
	Subcutaneous	Intravenous	Intramuscular
Rat (white)	2. 5	—	—
Beaver	3. 5	0. 12	—
Rabbit	4. 0	0. 3	—
Sparrow	—	—	4. 3
Rat (black)	—	0. 28	—
Chicken	—	—	?
Pigeon	—	—	?
Carp	—	—	1. 1
Frog	7. 0	—	—
Loach	—	—	9. 9
Snake	10. 5	—	—
Newt	46. 2	—	—

to X-rays for 40 to 60 minutes and almost entirely destroyed by 80 minutes of exposure. The neurotoxic properties were also reduced by one-half when the toxin was placed in an electric shaker and vibrated at 190 times per minute, but the hemolytic activity was only slightly altered. Placing the toxin in a U-tube and charging with 5 milliamperes of electricity for 3 hours resulted in only a slight decrease in its neurotoxicity and no effect upon hemolysis.

The effects of various acids were also tried, viz., hydrochloric, acetic, and lactic acids. Sodium hydroxide, ammonia, calcium chloride, iodine trichloride, Lugol's solution, hydrogen peroxide, hydrogen sulfide, oxygen, hydrogen, carbonic acid, gastric juice, pancreatin, saliva, urea, lecithin, alcohol, neurin, chloroform, adrenalin, and animal sera were also tested in order to determine their effects upon plototoxin. Hydrochloric was found to be the only acid that was destructive to the poison. The toxin was resistant to ammonia but slightly weakened when exposed to 1/10*N* sodium hydroxide. Calcium chloride and iodide trichloride were ineffective, but if Lugol's solution was mixed with equal parts of the toxin, the poison was destroyed. Hydrogen peroxide was also found to be very destructive. Although hydrogen sulfide destroyed the hemolytic properties, oxygen, hydrogen, and carbonic acid gases were all ineffective. Pancreatin and gastric juices completely destroyed plototoxin over a period of several hours. Saliva had no deleterious effects. Urea appeared to decrease its neurotoxicity but increase its hemolytic properties. Addition of alcohol resulted in complete destruction of the poison. Lecithin, neurin, and adrenalin were ineffective. Chloroform weakened both fractions of the toxin. The sera of humans, sheep, horses, cattle, and chickens failed to show evidence of any natural-occurring antibodies for plototoxin.

With the use of homogenized rabbit brain tissue, saline, and plototoxin incubated at 37° C for 3 hours, Toyoshima was able to absorb the neurotoxic fraction or have it combine with the nerve cells, leaving a clear supernatant liquid which he believed to be an almost hemolytic fraction. As a result of this experiment he finally termed the combined toxin "plototoxin," the neurotoxic fraction "plotospasmin," and the hemolytic substance "plotolysin."

Various dialysis experiments were also conducted using as a filter membrane sheep-skin paper, carp air-bladder, collodium, and the urinary bladder of cattle. Three ml of the standard toxin was placed in small bags made from each of the aforementioned substances and placed in water for a period of time. Comparison was later made with the liquids inside and outside of the bags by condensing the liquid outside of the bag to 3 ml or the same amount as inside of the bag. It was found that the neurotoxic fraction passed readily through the carp air-bladder and sheep-skin paper but little or none at all through the collodium and urinary bladder. However, the hemolytic fraction failed to pass through the membranes or passed through in only small amounts. The toxin also failed to pass through a Chamberland filter.

Injection of sublethal doses of plototoxin in rabbits produced a leukopenia within 15 minutes, but the leukocyte count was restored to normal within a period of 4 hours. A leukocidin test proved to be negative. Neither plotospasmin nor plotolysin inhibited the activity of immune hemolytic and agglutinating antibodies *in vitro* or *in vivo*. Crude plototoxin reduced but did not completely inactivate serum complement. Crude plototoxin did not suppress the growth nor kill common pathogenic microorganisms suspended in peptone water such as *Salmonella typhosa, Escherichia coli*, and *Micrococcus pyogenes* var. *albus*.

Normal serum antitoxins were not detected in any of the animal sera studied including those from the pigeon and *Plotosus*. The sera from patients with tuberculosis, cancer, leprosy, and syphilis did not affect the hemolytic power of plotolysin. Both plotospasmin and plotolysin were found to be antigenic and gave rise on repeated injection to antitoxins in low concentration and of moderate effectiveness. Plotospasmin apparently elicited a better antibody response than plotolysin. Beavers sensitized with crude plototoxin responded to shock injections with mild anaphylactic symptoms. The response may be related either to the toxins or to extraneous substances. Toyoshima's claim that the antitoxins were found in the serum albumin fraction should be reexamined in the light of present knowledge that antibodies are confined to globulins. Plotospasmin treated with heat failed to combine with its specific antibody. Plotolysin treated either with heat or hydrogen sulfide also failed to combine with its specific antibody. Both plotospasmin and plotolysin, however, were unaffected in this respect by exposure to X-rays or hydrogen peroxide. Both plotospasmin and plotolysis lost their antigenicity when treated with heat but only partially lost it when treated with hydrogen peroxide. Erythrocytes from susceptible animals immunized against plotolysin retained their susceptibility to lysis. The hemolysins of eel serum, hemolytic micrococci, and plototoxin were found to be antigenically unrelated. The rate at which antitoxin neutralized plotospasmin was observed to be directly proportional to the antibody concentration. When it did occur, neutralization required at least 2 hours' contact. Antiplotolysin was believed to compete with susceptible erythrocytes for the plotolysin and may effectively have neutralized the lytic activity when added simultaneously or even after the plotolysin had acted upon the erythrocytes for as long as 30 minutes. More antiplotolysin was required to neutralize the lysin as the time of exposure to erythrocytes was increased. The antiserum to plototoxin was also found to be capable of inducing mild anaphylactic symptoms in animals. The resistance of antiplotolysin to heat was much stronger than that of antiplotospasmin. Neither was found to lose its activity through dessication. Toyoshima demonstrated that injections of immune serum would alleviate the toxic symptoms in laboratory animals or prolong their death time.

Bhimachar (1944) reported finding both neurotoxic and hemolytic properties in the venom from *Heteropneustes fossilis*, a freshwater Indian catfish. Subcutaneous injections of glycerinated *Heteropneustes* venom in frogs resulted in hypoactivity, tonic convulsions, collapse, and death within 15 to 20 minutes after injection. Washed bovine red blood cells inoculated with *Heteropneustes* toxin showed marked hemolysis.

PHARMACOLOGY

(Unknown)

CHEMISTRY

(Unknown)

LITERATURE CITED

BHIMACHAR, B. W.
1944 Poison glands in the pectoral spines of two catfishes—*Heteropneustes fossilis* (Bloch) and *Plotosus arab* (Forsk), with remarks on the nature of their venom. Proc. Indian Acad. Sci., B, 19: 65-70, 3 figs.

BUDDLE, R.
1930 Some common poisonous fishes found in Singapore waters. J. Roy. Nav. Med. Serv. 16(2): 102-111, 8 figs.

CITTERIO, V.
1926 L'apparato vulnerante di *Arius heudeloti* (C. e V.). Atti Soc. Ital. Sci. Nat. 17: 24-27.

CLELAND, J. B.
1912 Injuries and diseases of man in Australia attributable to animals (except insects). Australasian Med. Gaz. 32(11): 269-274; (12): 297-299.

COLBY, M.
1943 Poisonous marine animals in the Gulf of Mexico. Trans. Texas Acad. Sci. 26: 62-69.

DAY, F.
1878-88 The fishes of India. 2 vols., illus. London.

EIGENMANN, C. H.
1912 The freshwater fishes of British Guiana, including a study of the ecological grouping of species and the relation of the fauna of the plateau to that of the lowlands. Mem. Carnegie Mus. Vol. V. 578 p., 103 pls.

EIGENMANN, C. H., and R. S. EIGENMANN
1890 A revision of the South American Nematognathi or cat-fishes. California Academy of Sciences, San Francisco. 508 p.

EVANS, H. M.
1943 Sting-fish and seafarer. Faber and Faber, Ltd., London. 180 p., 31 figs.

FERNANDO, C. H.
1965 A preliminary study of the defensive spines of some Malayan freshwater fishes. Bull. Fish. Res. Stn., Ceylon 17(2): 169-176.

FERNANDO, C. H., and A. FERNANDO
1960 The defensive spines of the freshwater fishes of Ceylon. Ceylon J. Sci. (Biol. Sci.) 3: 133-141.

FISH, C. J., and M. C. COBB
1954 Noxious marine animals of the central and western Pacific Ocean. U.S. Fish Wildlife Serv., Res. Rept. No. 36, p. 14-23.

GOSLINE, W. A.
1945 Catálogo dos nematognatos de agua-doce da America do sul e central. Bol. Mus. Nac. (Rio de Janeiro), New Ser. No. 33. 138 p.

GUDGER, E. W.
1929 Nicholas Pike and his unpublished paintings of the fishes of Mauritius, western Indian Ocean, with an index to the fishes. Bull. Am. Mus. Nat. Hist. 58(9): 489-530.

HALSTEAD, B. W.
1959 Dangerous marine animals. Cornell Maritime Press, Cambridge. 146 p., 88 figs.

HALSTEAD, B. W., KUNINOBU, L. S., and H. G. HEBARD
1953 Catfish stings and the venom apparatus of the Mexican catfish, *Galeichthys felis* (Linnaeus). Trans. Am. Microscop. Soc. 72(4): 297-314, 6 figs.

HALSTEAD, B. W., and R. L. SMITH
1954 Presence of an axillary venom gland in the oriental catfish *Plotosus lineatus*. Copeia (2): 153-154.

HAMILTON-BUCHANAN, F.
1822 An account of the fishes found in the river Ganges and its branches. 2 vols. A. Constable and Co., Edinburgh. 405 p., 59 pls.

HERRE, A. W.
1949 A case of poisoning by a stinging catfish in the Philippines. Copeia (3): 222.

HUBBS, C., and K. F. LAGLER
1947 Fishes of the Great Lakes region. Bull. Cranbrook Inst. Sci. 26: 1-186, 276 figs.

HUBBS, C. L., and E. C. RANEY
1944 Systematic notes on North American siluroid fishes of the genus *Schilbeodes*. Occ. Pap. Mus. Zool., Univ. Michigan (487): 1-26, 1 pl.

JORDAN, D. S., TANAKA, S., and J. O. SNYDER
1913 A catalogue of the fishes of Japan. J. Coll. Sci., Tokyo Univ. 33. 497 p., 396 figs.

LE GAC, P.
1936 Accidents consécutifs à la piqûre d'un poisson venimeux, le *Plotosus lineatus*. Bull. Soc. Madagascar 29: 925-927.

LE MARE, D. W.
1952 Poisonous Malayan fish. Med. J. Malaya 7(1): 1-8, 1 pl.

MAASS, T. A.
1937 Gift-tiere. *In* W. Junk [ed.], Tabulae biologicae. Vol. XIII: N. V. Van de Garde & Co's Drukkerij, Zaltbommel, Holland. 272 p.

MARSHALL, T. C.
1964 Fishes of the Great Barrier Reef and coastal waters of Queensland. Angus and Robertson, Melbourne. 566 p., 72 pls.

PAWLOWSKY, E. N.
1913 Sur la structure des glandes à venin de certains poissons et en particulier de celles de *Plotosus*. Compt. Rend. Soc. Biol. 74(18): 1033-1036, 1 fig.
1914 Über den bau der giftdrüsen bei *Plotosus* und anderen fischen. Zool. Jahrb. 38: 427-442, 3 pls., 4 figs.
1927 Gifttiere und ihre giftigkeit. Gustav Fischer, Jena. 515 p., 170 figs.

PHISALIX, M.
1922 Animaux venimeux et venins. 2 vols. Masson et Cie., Paris.

REED, H. D.
1907 The poison glands of *Noturus* and *Schilbeodes*. Am. Naturalist 41: 553-566, 5 figs.
1924*a* The morphology and growth of the spines of siluroid fishes. J. Morphol. 38(3): 431-451, 14 figs.
1924*b* The morphology of the dermal glands in Nematognathus fishes. Z. Morphol. Anthropol. 24: 227-264, 32 figs.

SCHULTZ, L. P.
1944 The stingarees, much feared demons of the seas. U.S. Nav. Med. Bull. 42(3): 750-754, 3 figs.

SMITH, J. L.
1961 The sea fishes of southern Africa. 4th ed. Central News Agency, South Africa. 580 p., 102 pls., 1,232 figs.

STARKS, E. C.
1930 The primary shoulder girdle of the bony fishes. Stanford Univ. Publ., Biol. Ser. 6(2): 1-92, 38 figs.

TANGE, Y.
1953 Beitrag zur kenntnis der morphologie des giftapparates bei den japanischen fischen nebst bemerkungen über dessen giftigkeit. I. Über das vorkommen des giftapparates bei den japanischen knochenfischen. Yokohama Med. Bull. 4(2): 120-128, 2 figs.
1955*a* Beitrag zur kenntnis der morphologie des giftapparates bei den japanischen fischen. XV. Über den giftapparat bei *Plotosus anguillaris* (Lacépède). Yokohama Med. Bull 6(6): 424-437, 7 figs.
1955*b* Beitrag zur kenntnis der morphologie des giftapparates bei den japanischen fischen. XIII. Über den giftapparat bei *Liobagrus reinii* (Hilgendorf). Yokohama Med. Bull. 6(4): 255-265, 5 figs.
1955*c* Beitrag zur kenntnis der morphologie des giftapparates bei den japanischen fischen. XIV. Über den giftapparat bei *Pseudobagrus aurantiacus* (Temminck et Schlegel). Yokohama Med. Bull. 6(5): 345-354, 6 figs.

TASCHENBERG, E. O.
1909 Die giftigen tiere. F. Enke, Stuttgart. 325 p., 64 figs.

TOMIYAMA, I., ABE, T., and T. TOKIOKA
1958 Encyclopedia zoologica illustrated in colours. Vol. II. [In Japanese] Hokuryn-kan Publ. Co., Ltd., Tokyo. 392 p., 133 figs.

TOYOSHIMA, T.
1918 Serological study of toxin of the fish *Plotosus anguillaris* Lacépède. [In Japanese] J. Japan Protoz. Soc. 6(1-5): 45-270.

VELLARD, J.
1932 Poissons venimeux du Rio Araguaya. Mém. Soc. Zool. France 29: 513-539.

WEBER, M., and L. F. DE BEAUFORT
1913 The fishes of the Indo-Australian Archipelago. II. Malacopterygii. Myctophoidea, Ostariophysi: I Siluroidea. E. J. Brill, Leiden. 404 p., 151 figs.

WHITLEY, G. P.
1943 The fishes of New Guinea. Australian Mus. Mag. 8(4): 141-144, 5 figs.

Chapter XX—VERTEBRATES

Class Osteichthyes

Venomous: WEEVERFISHES

The family Trachinidae, weevers, are a group of small marine fishes confined to the eastern Atlantic and Mediterranean coasts. One species is found along the coast of Chile, but according to Delfin (1901), Gill (1907), and Berg (1947), there is considerable question regarding the validity of this Chilean species. Weevers are generally not considered to be of commercial importance, although they are deemed a great delicacy in France. According to Irvine (1947), *Trachinus radiatus* of the Gold Coast is edible, but local fishermen state that it is not. Weeverfishes are among the most dangerous of the venomous fishes, and their stings are extremely painful and may be fatal.

LIST OF WEEVERFISHES REPORTED AS VENOMOUS

Phylum CHORDATA

Class OSTEICHTHYES

Order PERCIFORMES (PERCOMORPHI)

Suborder PERCOIDEI: Perchlike fishes

Family TRACHINIDAE

Trachinus araneus Cuvier (Pl. 1, fig. a). Araneus weeverfish (USA), dragonfish, weever (England), aragno (Italy), mitteländische petermännchen (Germany), aragna (France), aranhuco (Portugal), arana (Spain). *236*

DISTRIBUTION: Portugal, Mediterranean Sea, southward along the Atlantic coast of North Africa.

SOURCES: Ulmer (1865), Gressin (1884), Bottard (1889*a, b*), Coutière (1899), Briot (1903), Porta (1905), Pawlowsky (1911, 1927), Phisalix (1922), Maretić (1962, 1966), Halstead and Carscallen (1964).

OTHER NAMES: *Trachinus araenea, Trachinus araneus, Trachinus aranius.*

236 *Trachinus draco* Linnaeus (Pl. 1, fig. b). Greater weeverfish (USA), greater weeverfish, sea-cat, stringbull, dragonfish (England), dragone (Italy), gemeine queise (Germany), aragna (France), aranha grande (Portugal), arana (Spain).

DISTRIBUTION: Norway, British Isles, southward to Mediterranean Sea, coasts of North Africa, Black Sea.

SOURCES: Byerley (1849), Gressin (1884), Bottard (1889*a, b*), Pohl (1893), Coutière (1899), Phisalix (1899), Briot (1902, 1903, 1904), Porta (1905), Evans (1906, 1907*a, b*, 1916, 1923), Pawlowsky (1906, 1907, 1911, 1929), Phisalix (1922), Kazda (1931), De Marco (1936, 1937, 1938), Russell and Emery (1960), Maretić (1962, 1966), Skeie (1962*a, b, c, d, e*), Legeleux (1965).
OTHER NAMES: *Draco marinus, Dragono marinus sive viva, Trachinus major.*

236 *Trachinus radiatus* Cuvier (Pl. 1, fig. c). Weeverfish (USA, England), vive à tête rayonnée (France), aranha (Portugal), arana (Spain), trachino raggiato (Italy), straklen-petermännchen (Germany).

DISTRIBUTION: Mediterranean Sea, southward along the west coast of Africa.

SOURCES: Gressin (1884), Bottard (1889*a, b*), Pohl (1893), Coutière (1899), Porta (1905), Pawlowsky (1911, 1927), Phisalix (1922), Russell and Emery (1960), Maretić (1962, 1966), Halstead and Carscallen (1964).

236 *Trachinus vipera* Cuvier (Pl. 1, fig. d). Lesser weeverfish (USA), adder pike, black fin, little weever, lesser weever (England), trachino vipera (Italy), viperqueise (Germany), petite vive (France).

DISTRIBUTION: North Sea, southward along the coasts of Europe, Mediterranean Sea.

SOURCES: Byerley (1849), Gressin (1884), Bottard (1885, 1889*a, b*), Coutière (1899), Phisalix (1899), Briot (1903), Porta (1905), Pawlowsky (1911, 1927, 1929), Phisalix (1922), Russell and Emery (1960), Maretić (1962), Halstead and Carscallen (1964), Legeleux (1965).

BIOLOGY

Family TRACHINIDAE: Weevers are small marine fishes, the largest of which, *Trachinus draco*, is not known to exceed a total length of 46 cm. Trachinids are primarily inhabitants of flat, sandy, or muddy bays. *T. draco* is said to be more common in deeper waters, whereas the lesser weever, *T. vipera*, is found in shallower areas and has a more southern distribution. During the spring and early summer weevers migrate into shallow areas to spawn. According to Day (1880-84), weevers are generally abundant in areas inhabited by shrimp. The food of weevers consists essentially of squid, crustaceans, annelids, and small fishes. Weevers will bury themselves in the soft sand or mud with only their heads partially exposed, darting out with great rapidity at their prey. Despite their sedentary habits, weevers can move with swiftness and are said to be able to strike an object with their opercular stings with unerring accuracy. According to Byerley (1849), weevers are a terror to fishermen working in shallow, sandy areas in which these fish are concealed. When trachinids are at rest, their dorsal fin is depressed, but when provoked their fin is instantly erected and their opercular expanded. According to Bottard (1889*a, b*) and Gressin (1884), the opercular spine may be abducted to an angle of 35° to 40° with the longitudinal axis of the body. The slightest touch on the body of the fish is said to cause them to strike with severity and accuracy.

Weevers can remain alive for an extended period of time after being removed from the water. Spawning occurs during the later spring or early summer. The food of weevers consists of invertebrates and small fishes. Weevers are fished commercially in some parts of Europe. The catch in Denmark alone amounted to 1,044 tons in 1957 (Skeie, 1962*b*). It is claimed that the venomous nature of the fish has greatly hampered the development of the industry. The flesh of the weever is white, firm, and of excellent flavor.

MORPHOLOGY OF THE VENOM APPARATUS

The bulk of the morphological studies on weeverfishes have been conducted on the venom organs of *Trachinus draco* and to a lesser extent on *T. vipera*. The anatomy of the venom organs of weeverfishes has been described by Allman (1840), Byerley (1849), Gressin (1884), Parker (1888), Bottard (1889*a, b*), Porta (1905), Pawlowsky (1906, 1907), Evans (1907*a, b*, 1923), Halstead and Modglin (1958), Russell and Emery (1960), and Skeie (1962*b*).

Gross Anatomy

The venom apparatus of *Trachinus draco* consists of five dorsal spines, two opercular spines located one on either side of the head, and their enveloping integumentary sheaths (Fig. 1). Since the spines are covered by a thin-walled integumentary sheath, all of the spines are clearly evident even in nontraumatized specimens.

Opercular Spines (Figs. 2-3): This particular description is based on the right opercular spine in the anatomical position. Removal of the thin integumentary sheath reveals a broad, compressed, "dagger-like" opercular spine which terminates in an acute tip. Attached to the superior and inferior margins of the spine, one on either margin, are two flattened, pyriform, glistening white, soft, spongy masses—the venom glands. The base of each of these masses is partially protected by being situated within two conic cavities, one on either side, which are formed by the junction of the operculum and the base of the spine. The masses taper rapidly as the tip of the spine is approached. Microscopic sections of these masses indicate that they are comprised of glandular tissue. The integumentary sheath surrounding the opercular spine is divided in such a way as to form a separate cul-de-sac

FIGURE 1.–Head of the weeverfish *Trachinus draco* showing the venomous dorsal and opercular stings. (See arrows). It will be observed that the dorsal stings have been extended in the position that is customarily used for striking a victim. (K. Fogassy)

for each of the glandular masses. Further dissection fails to reveal any gross evidence of excretory ducts or special muscles associated with the venom glands.

The operculum of *T. draco* is a triangular-shaped, shieldlike bone with the apex extending downward (Figs. 2-3). Extending horizontally and backward from the upper anterior portion of the operculum is the stout opercular spine. The basal one-third of the spine is attached to the remainder of the operculum, whereas the distal two-thirds is free and lies superficial to the supero-posterior portion of the operculum. In the larger specimen the free portion of the spine measured 1 mm in length. The superolateral and inferolateral margins of the free portion of the spine are deeply grooved for the reception of the venom glands. Aside from the grooves, the remainder of the spine is solid. The remainder of the operculum is unarmed.

Movements of the opercular spine are controlled primarily by three muscles. Abduction of the opercle is by contraction of the large "cheek" or opercular muscle and of a smaller muscle which originates on the distal tip of the pterotic bone and inserts along the upper margin of the process of the operculum superior to the base of the opercular spine. Adduction is by contraction of a muscle which appears to originate on the inferior aspect of the pterotic bone and inserts on the inner aspect at the base of the opercular spine.

Dorsal Spines (Figs. 4-5): This description is based on extended dorsal spines in their anatomical position. The six or seven dorsal spines are enclosed within thin-walled integumentary sheaths from which protrude their needle-sharp tips. Each of the spines are connected by a thin interspinous membrane. Removal of the integumentary sheaths reveals a strip of slender, elongate, fusiform, whitish, spongy tissue, lying within the distal portion of each of the anterolateral-glandular grooves of the spines. The tissue within these grooves is the glandular epithelium or venom-producing portion of the sting. The dorsal stings of *Trachinus* are grossly similar to those of *Scorpaena*. Table 1 shows a comparison between the length of each spine and the length of their associated strip of glandular tissue. The spines were measured from their articulating base to distal tip. The glandular tissue was measured within the groove from the proximal origin to the distal extremity. There is no evidence

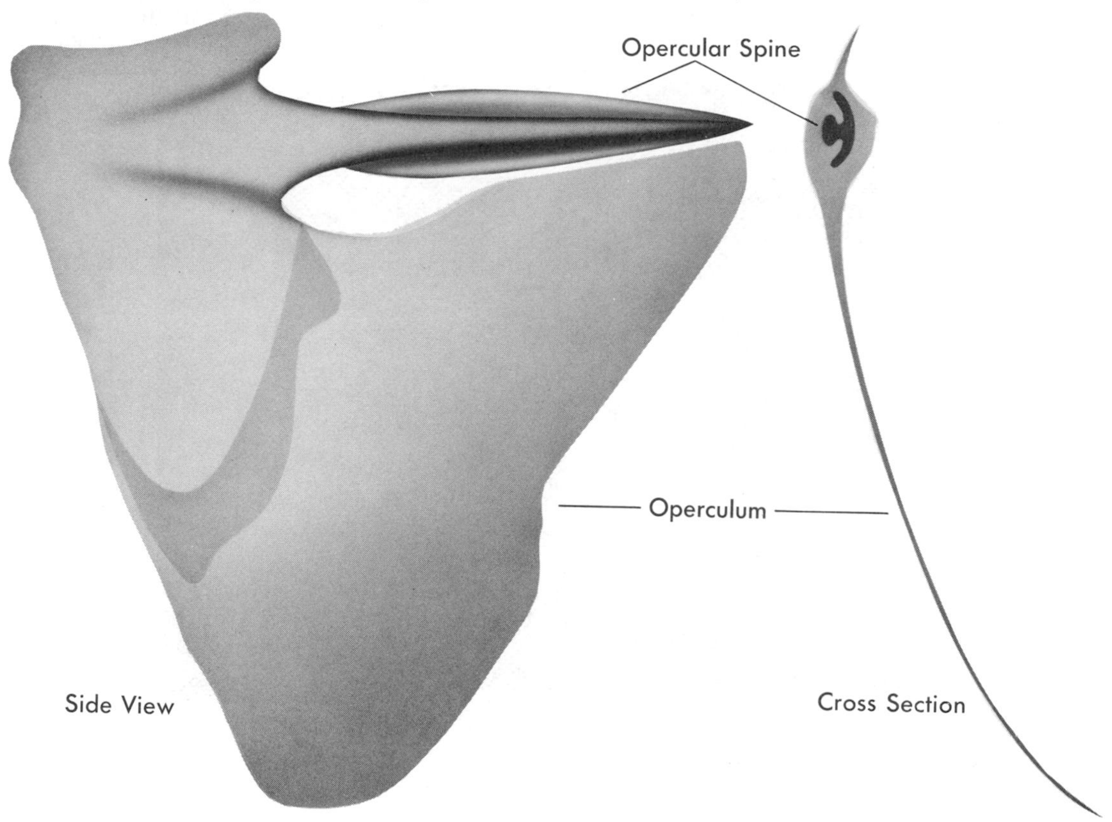

FIGURE 2.—Left bony operculum of the weeverfish *Trachinus*. Note the deeply-grooved daggerlike spine on the upper portion of the operculum. (R. Kreuzinger)

of excretory ducts. The glands extend to the distal tip of the spine except in those instances in which the spine has been traumatized. These particular measurements were taken from a fish having a standard length of 210 mm. The first three dorsal stings are the most highly developed as venom organs. The fourth, fifth, sixth, and seventh (when present) are progressively less developed and more slender. Glandular grooves and venom glands are usually absent or greatly reduced in the last dorsal spine.

The second dorsal spine may be considered as representative of the others. The spine is relatively stout, solid, and straight and terminates in an acute tip. In cross section, the spine is trigonal in outline throughout most of its length. A pronounced anteromedian ridge is present. The anterolateral gooves, one on either side, extend throughout the length of the spine. The grooves are broader and deeper in the distal third of the spine.

An anterior view of the spine shows that the base is broadened and comprised of two lateral condyles, one on either side of the median foramen through which the ringlike process of the supporting interneural spine passes. It is noted that the anteromedian ridge arises from a point immediately over the median foramen in dorsal spine I, arises from the right lateral condyle in dorsal spine II, but arises from the left lateral condyle in dorsal spine III. This alternating pattern is followed throughout the first dorsal fin.

FIGURE 3.—Lateral view of the left opercular sting of *Trachinus draco*. The integumentary sheath has been removed to show the underlying opercular spine and venom glands which are seen as whitened triangular masses lying above and below the basal portion of the shaft of the opercular spine. The length of the opercular spine is 25 mm.
(R. Kreuzinger and B. W. Halstead)

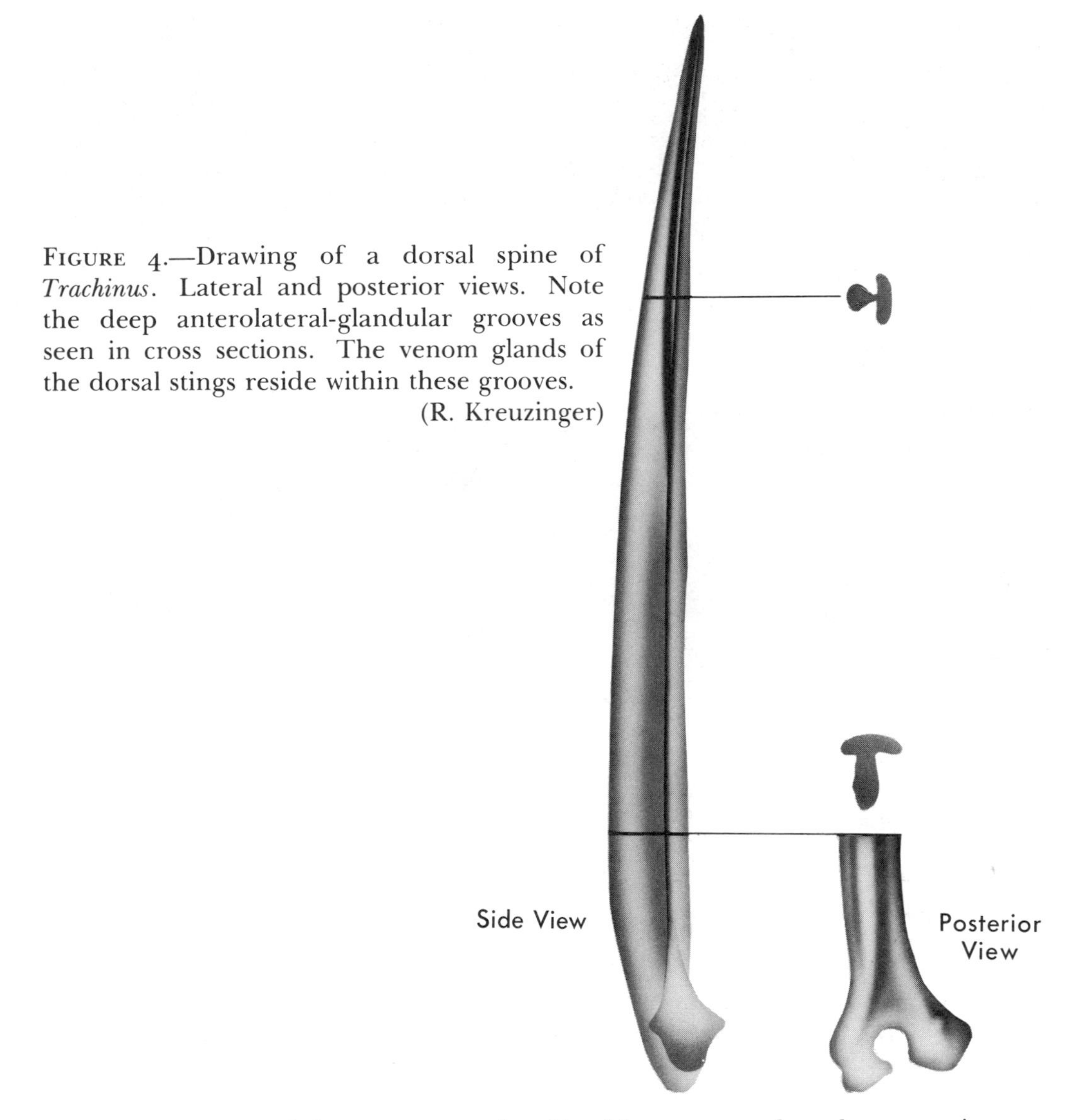

FIGURE 4.—Drawing of a dorsal spine of *Trachinus*. Lateral and posterior views. Note the deep anterolateral-glandular grooves as seen in cross sections. The venom glands of the dorsal stings reside within these grooves.
(R. Kreuzinger)

The anterior condylar processes in *Trachinus* are reduced, appearing as mere tuberosities.

A posterior view of the spine reveals that the shaft is rounded near the base, becoming flattened as the tip is approached. Extending backward from each of the lateral condyles are the posterior condylar processes.

Erection of the dorsal spines is accomplished by contraction of a pair of bilaterally distributed muscles which originate from the anterior aspect of each of the supporting interneural spines. Tendinous attachments also arise from the adjacent neural spines. The erector muscles ascend vertically, pass anterior to the head of the interneural spine, and insert on the anterior condylar tuberosity. The dorsal fin is depressed by a pair of bilaterally distributed muscles originating from the posterior aspect of each of the sup-

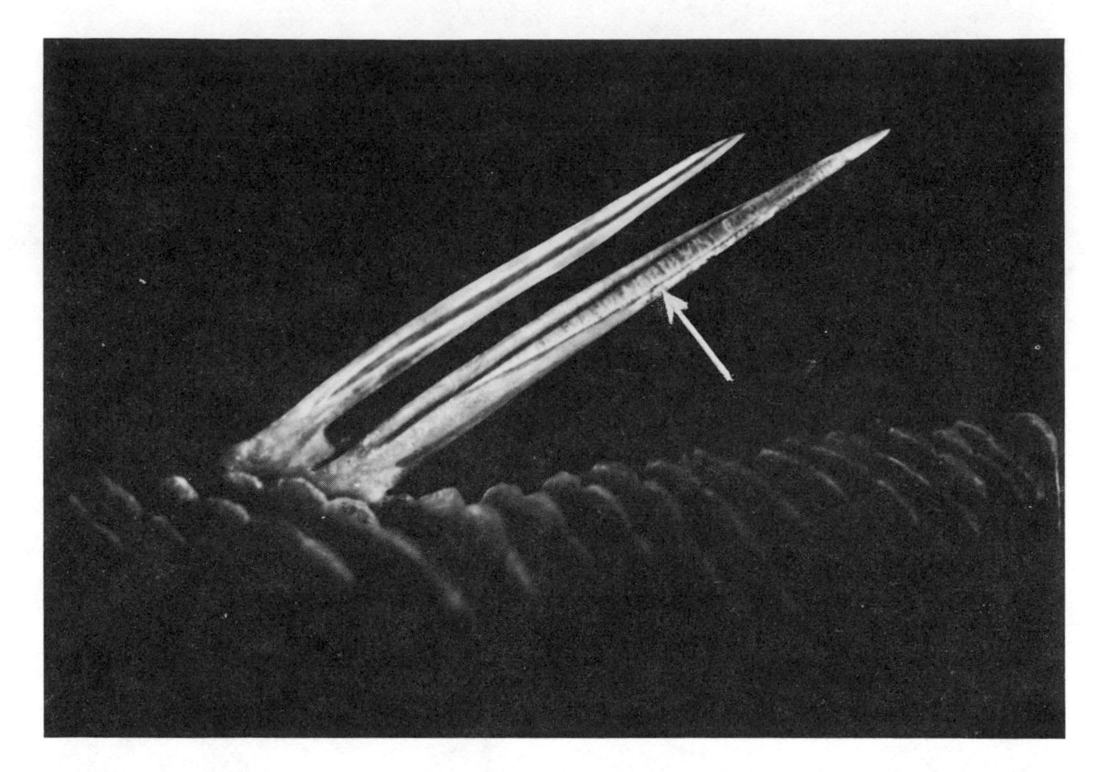

FIGURE 5.—Lateral view of dorsal spines of *Trachinus draco*. Only the first two dorsal spines are seen in this photograph. The integumentary sheaths have been removed from both spines. Note the fusiform venom gland lying within the anterolateral-glandular groove of the second spine (see arrow). The venom gland in the first spine has been completely removed and only the groove is seen. The length of the second dorsal spine is 20 mm. (R. Kreuzinger and B. Halstead)

porting interneural spines. Tendinous attachments also arise from the adjacent neural spines. The depressor muscles ascend vertically, pass posterior to the head of the interneurals, and insert on the posterior condylar processes.

Microscopic Anatomy

Opercular Stings (Fig. 6a, b): The operculum of *Trachinus draco* and *T. vipera* in cross section appears as an elongate, crescent-shaped structure. The section in the vicinity of the opercular spine may be divided into three principal zones: a peripheral layer of integument, a dentinal portion, and a glandular zone. The integumentary sheath appears in this section as two distinct but continuous layers between which are sandwiched the glandular and dentinal structures. The integument is comprised of a comparatively thin peripheral layer of epidermis and a thick inner layer of dermis. The epidermis on the outer side of the operculum is comprised of a layer of stratified squamous epithelium five or six cells in thickness, whereas the layer of epidermis serving as the inner lining of the operculum is comprised of a thin layer of stratified squamous epithelium about two cells in thickness. In either case the epithelial

TABLE 1.—*Comparison Between Length of Spines and Length of Associated Glandular Tissue in* Trachinus draco

Dorsal Spine	1	2	3	4	5	6	7
Spine (mm)	12	18	18	14	8	6	4
Gl. Tissue (mm)	5	8	9	6	3	2	?

cells rest upon a distinct basement membrane which separates them from the underlying dermis. Scattered throughout the epidermis are occasional large secretory or mucous cells having a clear cytoplasmic border. The dermis is comprised of a thick layer of dense, fibrous, connective tissue. Interspersed throughout the dermis are vascular structures containing nucleated red blood cells and occasional clumps of pigment cells. Situated in about the center of the crescent-shaped operculum and surrounded by dermis is the opercular spine. The spine is separated from the dermis by a broad space. Frequently there is observed within this space patches of pink, amorphous, finely granular secretion. The lining of the surrounding dermis

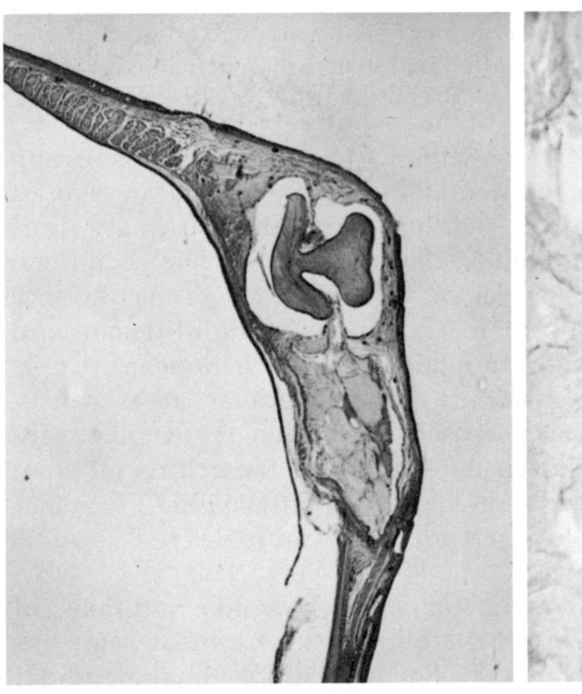

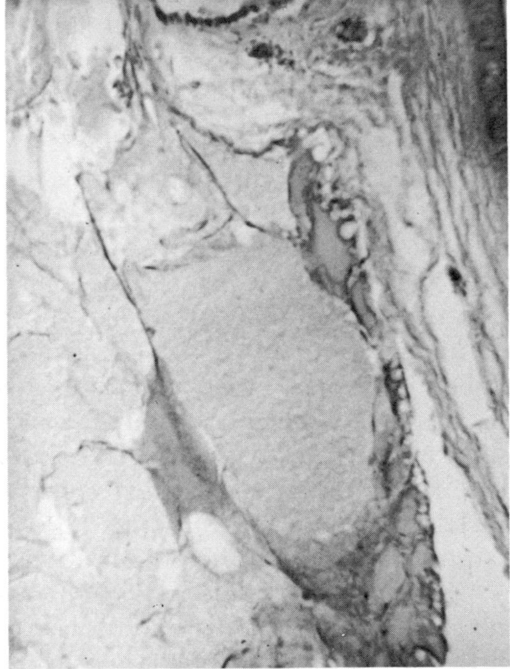

FIGURE 6a.—(Left) Photomicrograph of a cross section of the opercular sting of *Trachinus vipera*. The "T"-shaped dentinal spine is clearly seen in the upper portion of the operculum, and below is seen a large venom gland. × 15. b. (Right) Enlargement of the inferior opercular venom gland of *T. vipera* shown at left. Hematoxylin and triosin stains. Sections prepared by B. Halstead and M. Chitwood. × 190.

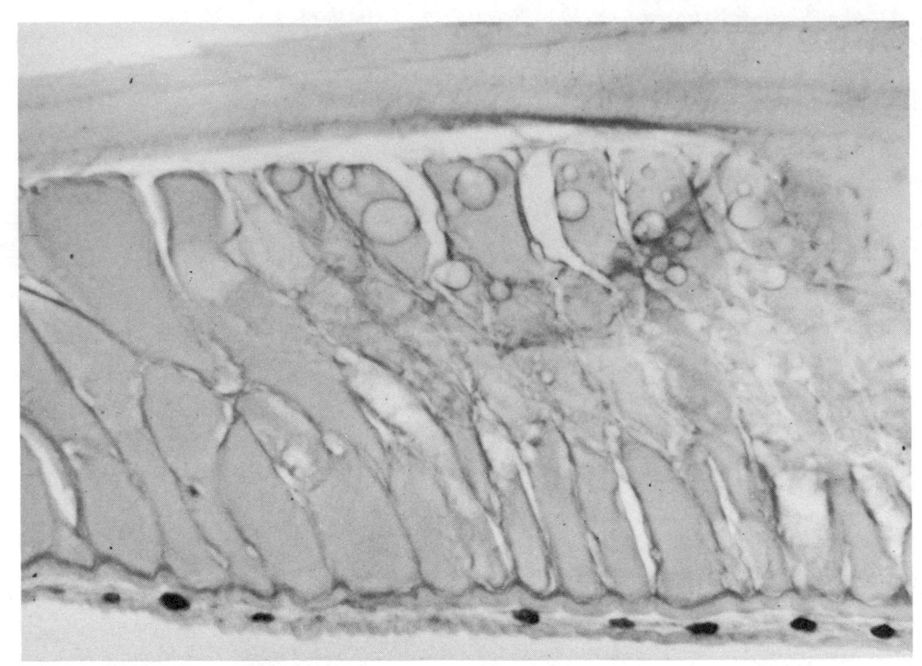

FIGURE 7.—Longitudinal section of the inferior opercular venom gland of *Trachinus vipera*. Only a small segment of the adjacent dentinal structure is seen above. Hematoxylin and triosin stains. Sections prepared by B. Halstead and M. Chitwood. × 210.

consists of a strip of simple cuboidal epithelium. The spine in cross section appears as a solid inverted "T"-shaped dentinal structure marked by a series of fine concentric rings. Diffusely scattered throughout the matrix are short tubular segments suggestive of a fine canalicular mechanism. The peripheral layer of the spine consists of a thin layer of vitreodentine. Projecting into each of the glandular grooves of the spine is a thin extension of dermis and glandular tissue from the adjacent glandular areas which appears deeply embedded within the surrounding dermis. The glandular areas consist of masses of large, polygonal, variable-sized, distended, light pink cells filled with finely granular secretion. In most instances the cell membranes have ruptured, releasing their contents. One or two dark-staining vesicular nuclei are frequently present. The largest cell observed measured 280μ by 140μ in its greatest dimensions.

Longitudinal sections (Fig. 7) reveal that the dentine-like substance of the spine is solid, and there is observed the minute short tubular segments diffusely scattered throughout the matrix. The largest polygonal glandular cells in one longitudinal section measured 172μ by 40μ, but they varied considerably in their size and morphology. In general, these cells appear as large columnar secretory cells arranged perpendicular to the axis of the spine upon which they rest. Pigment cells are particularly prominent in longitudinal sections. Otherwise the histological details are similar to those previously described.

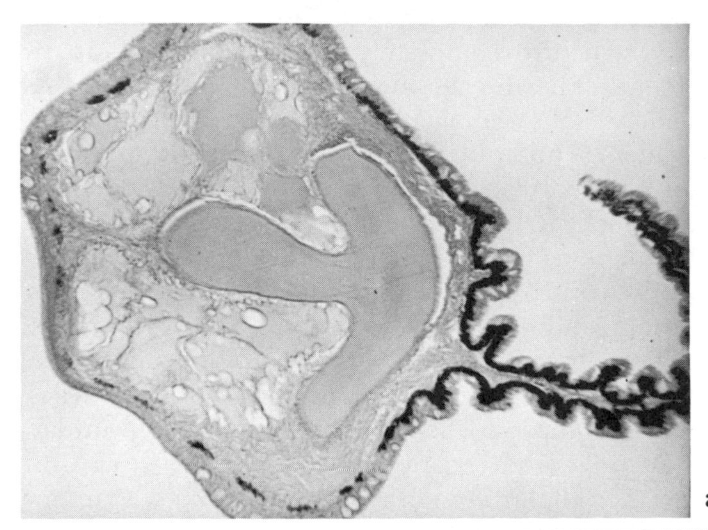

a

b

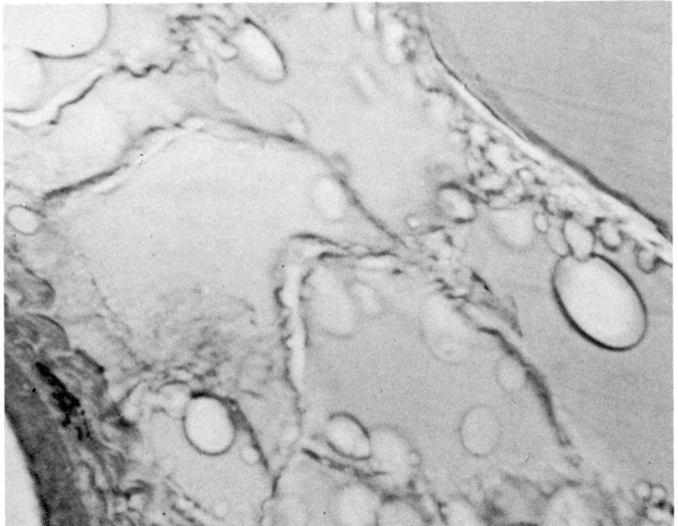

FIGURE 8a.—(Top) Photomicrograph of a cross section of a dorsal sting of *Trachinus vipera*. Note the relatively massive venom glands situated on either side of the anteromedian ridge of the spine. × 110. b. (Center) Enlargement of one of the venom glands seen in the photomicrograph above. × 350. c. (Bottom) Longitudinal section of a dorsal sting of *T. vipera*. × 250. Hematoxylin and triosin stains. Sections prepared by B. Halstead and M. Chitwood.

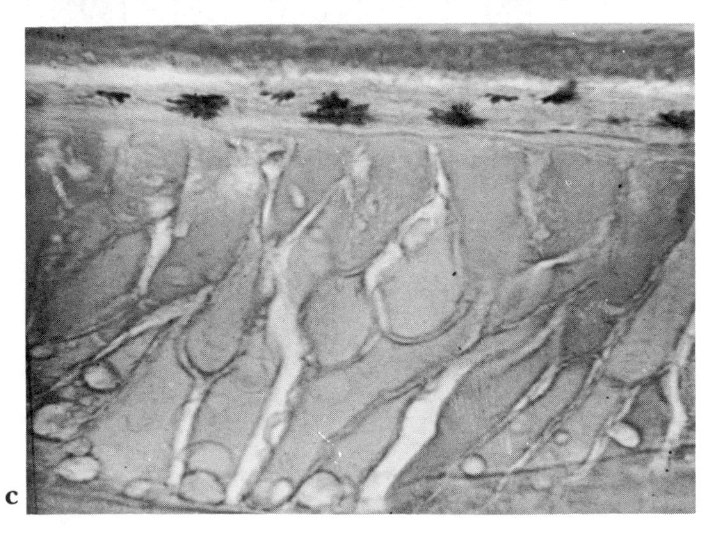

c

Dorsal Stings (Fig. 8a, b): A dorsal sting in cross section is essentially an inverted "T"-shaped dentinal structure completely invested by a comparatively thin layer of integument. The dorsal sting is similar to that of the operculum. There is no evidence of a canalicular mechanism such as was observed in the opercular spine. The epidermis is comprised of stratified squamous epithelium about five or six cells in thickness. Immediately beneath the basement membrane upon which the epithelial cells rest is an almost continuous thick layer of pigment. The dermis, as in the opercular sting, is comprised of dense, fibrous, connective tissue. The glandular tissue is situated within the anterolateral-glandular grooves in such a way as to form a glandular triangle similar to that observed in other fishes but does not have the pinnate arrangement as in *Scorpaena*. The glandular cells are large, polygonal, variable-sized, pink-staining, vacuolated, having a finely granular cytoplasm. One of the largest cells observed measured 220μ by 84μ in its greatest dimensions. Some of the cellular membranes appear to have undergone dissolution thereby releasing their contents.

In longitudinal sections (Fig. 8c) it is observed that the large polygonal glandular cells are arranged perpendicular to the axis of the spine upon which they rest. The largest glandular cells observed measured 200μ by 36μ. Again, there is no evidence of the minute tubular segments that were observed in the opercular spine. The remaining histological details are similar to those previously described for the opercular spine.

Pawlowsky (1906) described strands of connective tissue interlacing the glandular cells which they believe serve as "supporting" structures. According to Pawlowsky, these "supporting" cells are more evident in *Scorpaena* than in *Trachinus*, but Halstead and Modglin (1958) were unable to demonstrate them in either genera. Although Bottard (1889*a, b*) made references to excretory ducts and "tubular glands" in *Trachinus*, all other authors deny their existence. It is the consensus of opinion that venom production is the result of a holocrine type of secretion.

MECHANISM OF INTOXICATION

Envenomations are produced by the opercular and dorsal stings of weeverfishes. Stings from *Trachinus draco* are usually incurred by fishermen attempting to remove the fish from their nets since *T. draco* are generally encountered in deeper waters. However, some of the other species of weevers are found in shallow bays buried just beneath the sand with their head and dorsal and opercular stings protruding. They may dart out rapidly and strike an object with unerring accuracy with their opercular stings. When a weever is provoked, the dorsal fin is instantly erected and the gill covers expanded. The weever is thus prepared for immediate action and is capable of inflicting a serious sting. Swimming close to a sandy bottom inhabited by weeverfishes may be an invitation to a painful experience. One must proceed with considerable caution when wading or swimming in shallow sandy bays in which weevers are likely to occur.

MEDICAL ASPECTS

Clinical Characteristics

Weever wounds generally produce instantaneous pain described as a burning, stabbing, or "crushing" sensation, which initially is confined to the area about the wound and then gradually radiates throughout the affected limb. The pain becomes progressively more severe until it attains an excruciating peak, usually within a period of 30 minutes or less. The severity of the pain is such that the victim frequently screams in agony, thrashes about wildly, and may lose consciousness. Bourienne (1782) reported a case in which the pain was so violent that the individual finally amputated his finger and thereby obtained prompt relief. Evans (1943) mentioned fishermen hammering their fingers with a thole pin, and others wrapping the wound with vinegar-soaked paper and lighting it in a desperate hope of deadening pain. In most reported instances, morphine fails to give relief. If untreated, the pain usually subsides within a period of 2 to 24 hours. There may develop a sensation of urgency to urinate shortly after receiving the sting. Tingling followed by numbness subsequently develops in the area about the wound. Injection of the venom produces at first an area of ischemia about the wound which is soon followed by redness, heat, and swelling. The edema is progressive, and within a half hour or more the entire limb is involved. Movement of the affected part is greatly restricted. The swelling may continue for a period of 10 days or longer. These initial symptoms of pain and inflammation may be associated with or followed by headache, fever, chills, delirium, nausea, vomiting, syncope, sweating, cyanosis, joint aches, ankylosis, aphonia, torpidity, cardiac palpitation, bradycardia, pronounced psychic depression, respiratory distress, and clonic or tonic convulsions. Deaths from weever stings have been reported by Ulmer (1865), Evans (1943), Halstead (1957), Halstead and Modglin (1958), Russell and Emery (1960), Skeie (1962*b, e*), and others. Secondary infections commonly occur in improperly treated cases. Lymphangitis, lymphadenitis, pyogenesis, necrosis, and sloughing of the tissues about the wound are not infrequent. Gangrene may develop, necessitating amputation. Primary shock is a usual complication which must be guarded against. The recovery period is extremely variable depending upon the individual, amount of venom injected, species of weever, season of the year, and possibly the locality, but the period usually varies from a few days to several months. Bottard (1885) reported a case which was complicated by a severe secondary infection which resulted in atrophy of the index finger, peripheral neuritis, and ankylosis. More than 3 years later the patient was still unable to touch an object with the affected finger without severe pain. De Marco (1936) presented an interesting case of a 15-year-old boy in which a weever sting on one of his fingers appeared to have stimulated a generalized epileptiform attack. He concluded that the venom acted as a cortical sensory-motor stimulant. See Halstead (1957) for a clinical account of an envenomation by the weeverfish *Trachinus vipera.*

Pathology

(Unknown)

Treatment

The treatment of weever envenomations is similar to that used for stingray stings (see p. 655 for complete details). In addition, it should be noted that Maretić (1957) has found intravenous calcium gluconate to be effective in alleviating the pain associated with weever stings. Local injections of procaine may be helpful in less severe envenomations. Intravenous meperidine is recommended for the relief of severe pain which continues beyond the first hour following the injury. At present there is no specific antivenin available although this aspect of treatment has been investigated by Skeie (1962*a*).

Prevention

Weever stings are generally contracted either from handling the fish or by accidentally stepping on the fish while wading in waters inhabited by them. One must use extreme caution in handling a weever in order to avoid coming in contact with either the dorsal or opercular stings. Weeverfish tenaciously cling to life even after being removed from water over a period of time. Thus the careless handling of a "dead fish" may result in a severe sting. When wading in bays inhabited by weevers one must be especially careful not to step on the dorsal stings of a partially buried fish. Diving or swimming close to a sandy bottom containing weevers may be a very dangerous procedure and should be avoided.

PUBLIC HEALTH ASPECTS

Weeverfish envenomations are not generally considered to be a public health problem although they are of concern to fishermen, peddlers, and the consumer because of their venomous spines. Weeverfishes are limited in their distribution to the North Sea, the eastern Atlantic from the British Isles to tropical west Africa, Mediterranean, and the Black Seas (see Map). The overall incidence of weever stings in Europe and Africa is not known, but probably compares quite favorably to the incidence of scorpionfish stings for those areas. Skeie (1962*b*, *e*) has estimated that in Denmark there are probably about 10 stings each year, but overall incidence for Europe and Africa probably involves several hundred persons each year. Fortunately many of the stings are relatively minor in nature and seldom come to the attention of the physician. Skeie (1962*b*) has suggested that the danger of envenomations from the stings of this fish has probably hampered the development of commercial fisheries for weevers. Some cities in France have passed police regulations to the effect that it is illegal to sell weevers on the commercial market without first removing the opercular and dorsal stings.

TOXICOLOGY

Winkler (1868, cited by Coutière, 1899) was one of the first to attempt to

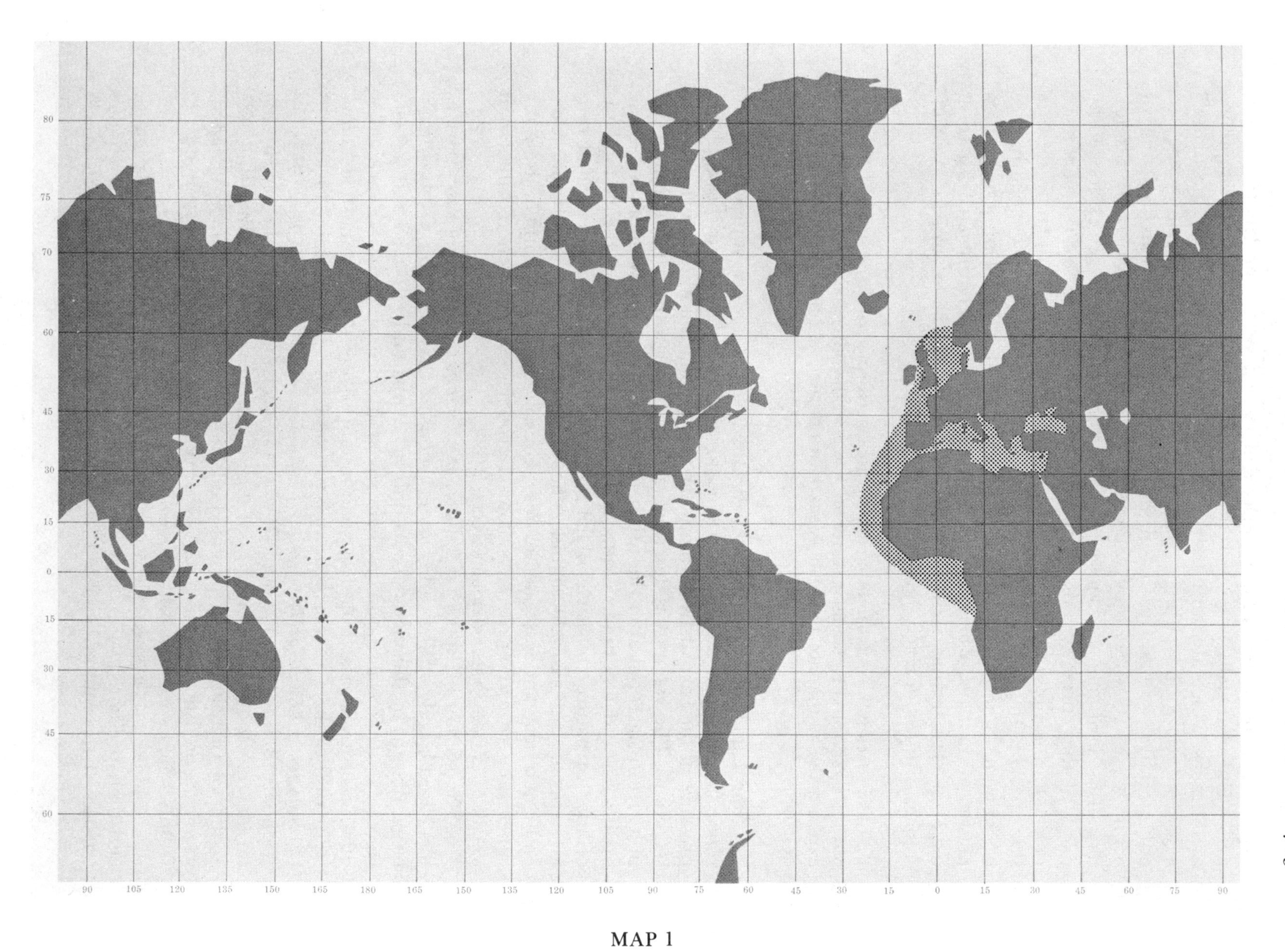

MAP 1

Map showing the geographical distribution of weeverfishes. *Trachinus draco* also occurs in the Black Sea. (L. Barlow)

experimentally demonstrate the effects of weever venom on laboratory animals. Unfortunately, he prepared his extracts from the dorsal interspinous membranes, which are not a part of the venom apparatus, and "inoculated" a dog and two young starlings. As might be expected, the experiment yielded only negative results. The first productive experimental research was that of Schmidt (1874), whose observations were published in an obscure Scandinavian medical journal. Consequently, his observations were overlooked by many of the workers of his time. Schmidt attempted to test the venom of *Trachinus draco* by introducing small bits of opercular glandular tissue in a skin incision. In some instances the opercular sting itself was used to introduce the venom. Some of the venom tissue had been preserved in brandy for a period of 14 days, whereas other experiments were conducted with fresh material. His techniques were crude, the results variable, and in some instances of questionable value. The author stated:

> Results prove on the one hand that inoculations performed on frogs exerted "venomous" effects from the glandular secretion, while on the other hand, judging from the experiments on the rabbits, it can by no means be assumed that the effects are very destructive.

The experimental techniques of Gressin (1884) were somewhat more refined than those of Schmidt. Gressin obtained the venom by applying pressure at the base of fresh opercular venom glands and then collected the secretion. The venom was described as being clear with a slightly bluish tint in its fresh state, but after the fish had been dead for a few hours it became opalescent and less fluid. Under low power magnification the venom was cellular in nature consisting of large, refractile, lymphlike, nucleated cells interspersed in a plasmalike liquid. The venom was coagulated by strong acids, bases, and heat, and was water soluble. Gressin made no attempt to draw any conclusions regarding the physiological properties of the venom other than to call attention to its noxious effects.

The section of Bottard's *Les Poissons Venimeux* concerning weeverfishes was based largely on the earlier observations of Gressin (1884) from which a large amount of the material was quoted. Apparently Bottard and Gressin were friendly collaborators in their research since they quoted each other quite extensively and in at least one instance from the unpublished manuscript of the other. In this regard it is noted that Gressin dedicated his famous *Contribution à l'Etude de l'Appareil à Venin chez les Poissons du Genre "Vive"* to Alphonse Bottard (among others), who was interning at the time at l'Hôpital du Havre. Bottard's description of the appearance of the venom is the same as that of Gressin, but he provided additional experimental data regarding its effects on fish and guinea pigs. The venom was apparently obtained from freshly killed specimens of *T. draco*.

Pohl (1893) reviewed the work of Gressin (1884) at considerable length and then presented original experimental data. In addition to the characteristics of the venom described by Gressin, Pohl found that the poison was destroyed by treatment with ethyl alcohol, ether, or chloroform. Stings that had been dried usually lost their toxic properties within a period of 24 hours. Pohl stated that he tested more than 50 frogs and never observed a convulsion nor an increase in reflex excitability. Moreover, the excitability of peripheral nerves and muscles was believed to be unaltered. The principal symptoms which he observed consisted of paralysis, decreased cutaneous sensitivity, hypnosis, and circulatory disturbances. He considered that the toxin acted primarily on the heart and was most rapidly effective when applied directly to the organ. A small amount of glandular tissue applied in the vicinity of the conus arteriosus of an exposed frog heart resulted in a decrease in cardiac rate from 44 to 9 beats per minute within a period of 17 minutes. The decrease in heart rate was associated with diastolic stasis and a loss in the excitability of the heart muscle. The progression of events was: retardation of the beat with strong diastole and systole, then because of defective filling of the ventricle, the diastolic tension was decreased and the energy of systolic contraction lagged considerably. A period of 2 to 3 hours followed in which there was a slow uniform cardiac rate, gradually becoming less intense while the rhythm was maintained and sinking gradually to only a very superficial wave which could hardly be considered a systolic contraction. Finally the heart stopped in diastole. The heart muscle was then only locally excitable. Pohl administered venom with atropine, camphor, caffeine, helleboreine, and hydrastinine. Atropine had no noticeable effect in the course of intoxication. Caffeine, helleboreine, and hydrastinine did not appear to alter the effects of the venom on the heart. Pohl decided that the frog was not a suitable laboratory animal for the evaluation of the action of the venom on man. The author suggested that possibly the disparity between his experimental results and those of Gressin was on the basis of the two venoms that were used. Gressin captured his fish at Le Havre in May, whereas Pohl obtained his at Trieste in September. It was suggested that differences in locality and time of year may have had a bearing on the toxicity of the venoms.

Phisalix (1899) prepared glycerinated extracts from the opercular and dorsal stings of *T. vipera* and *T. draco* and injected them subcutaneously into the thighs of guinea pigs. The injections produced evidence of immediate pain, paralysis, localized swelling, and death. Because of the accompanying inflammatory reaction, Phisalix was of the opinion that the manifestations were due to both the venom and a secondary bacterial infection.

The extensive work of Coutière (1899) does not contain any original experimental data on the venom of *Trachinus*. It is, however, an excellent review of the early literature on this subject.

Research on the venom of *Trachinus* reached its acme during the first three decades after the turn of the century. Most outstanding among the researchers of this period were Briot, Evans, Pawlowsky. Briot (1902) was the first to attempt to evaluate the immunological properties of *Trachinus* venom. He found that if one drop of washed red blood cells (type not stated) suspended in 2 cc of normal saline was added to 0.2 cc of animal serum (type not stated) heated for 1 hour at 60° C with 10 to 20 drops of weever venom, hemolysis was completed in all tubes to which venom had been added within a period of an hour and a half. He further observed that if the serum was not heated, the venom failed to produce hemolysis. From these experiments he concluded that there was an "anti-hemolysin" present in normal unheated serum which was capable of protecting the red blood cells against the hemolytic action of the venom. Experiments also demonstrated that the hemolytic activity of weever venom was intact even after it had been heated for 1 hour at 75° C. If the venom was heated for 20 minutes at 100° C, the hemolytic properties of the venom were attenuated but nevertheless present. Calmette's snake anti-venin failed to counteract the action of weever venom. Weever venom did not seem to alter the coagulation time of blood. In another series of experiments, Briot (1902) attempted to show the effects of *T. draco* venom when injected intravenously into various laboratory animals. The venom was obtained from the opercular and dorsal stings of 420 weevers taken during the months of May, June, and September. The stings were ground in a mortar, extracted with glycerine, and then filtered. It was found that if 0.5 cc were injected into the vein the rabbit died before the injection had been completed. Venom in 0.2 cc or 0.1 cc amounts produced death within several minutes. Briot concluded that death was due to a respiratory paralysis. He was not certain as to whether the poison acted upon the respiratory center or the muscles but assumed that it was the latter. If the venom was heated to 100° C for 30 minutes, filtered, and then administered, it was observed that frogs would tolerate four or five times the unheated lethal dose of venom. A rabbit given 5 cc intravenously of the attenuated venom demonstrated only a passing prostration, but the following day the entire right side of the head and right shoulder were edematous. The eye became gangrenous. The rabbit died on the 25th day. Thus it was found that heating of the venom at 100° C for 30 minutes was not sufficient to completely destroy it. If the venom was heated for 1 hour it was completely inactivated. The venom was also destroyed by calcium hypochlorite and gold chloride in dilutions of 1 to 1,000. In a third series of experiments, Briot (1902) demonstrated the antigenic properties of *Trachinus* venom in rabbits. With these studies Briot found that it was possible to actively immunize a rabbit with two subcutaneous sublethal doses of weever venom. Animals which had been immunized were able to tolerate doses of *Trachinus* venom three or four times greater than the lethal dosage used in the controls. It was also observed that serum taken from immunized animals could be used to passively immunize other rabbits. Briot's 1903 paper is a compilation of his earlier work together with an elaboration on certain details not previously mentioned. Venom heated at 100° C for 30 minutes was toxic but attenuated in its activity. If heated at 100° C for 1 hour, the venom was completely destroyed. Venom was said to lose its antigenic properties if heated. Briot also pointed out that passive immunity usually lasted for a period of three and a half months, but the period varied in individual animals. In still another report, Briot (1903) conducted a comparative study on the relative venom content in opercular and dorsal stings. The author ground up the opercular and dorsal stings from 150 weevers, prepared glycerinated extracts of the venom, and tested them on fish, frogs, and rabbits. Briot concluded that the dorsal spines contained considerably less venom and were less dangerous than opercular stings. However, the studies were not conducted on a quantitative basis, so no conclusions were drawn concerning the relative virulence on a weight-for-weight basis. Briot (1904) later attempted to show the presence of a kinase in weever venom and the ability of the enzyme to activate an "inactive pancreatic sugar." Six test tubes were taken labeled A through F. One cc of "inactive pancreatic sugar" and a small cube of coagulated egg white were placed in each of the tubes. In addition, 0.1 cc of venom was placed in tube B, 0.3 cc of venom in tube C, and 0.5 cc of venom in tube D. To tube E was added 0.3 cc of glycerine, and tube F received 0.5 cc of glycerinated weever venom which had been heated to 100° C for a half hour. Digestion of the egg white occurred only in tubes B, C, and D, being progressively more rapid in the tubes having the higher concentrations. The remainder of the tubes were negative. From these experiments the author concluded that a kinase was present in weever venom. Briot failed to complete the experiment by showing the effects of the venom on egg white without the presence of the "inactive pancreatic sugar."

Evans (1907*a*, *b*) was the next to investigate the toxicological properties of weever venom. The venom was collected by extracting it with a syringe directly from the opercular spines of freshly caught *T. draco* specimens. Evans likewise made a series of toxicity tests on goldfish, frogs, mice, and guinea pigs. In his experiments on rabbits and cats he found that variable quantities of venom injected intravenously usually resulted in an immediate marked fall in blood pressure, abnormally strong contractions of the heart persisting for about 10 seconds, an additional slight fall of 5 seconds duration, a marked rise in blood pressure momentarily to a normal level, and finally a deep drop which initially is rather rapid and then gradually declines until death occurs. In his hemolytic experiments, Evans used defibrinated blood of pigeon, guinea pig, sheep, ox, dog, horse, and man. The reader is referred to the original report for the complete details of the experiments. However, Evans found that weever venom produced hemolysis of the red blood cells from all of the animals tested without serum being present. The addition of the serum of the same animal did not inhibit but rather increased the activity of the venom. In most cases heating the serum for an hour to 62° C diminished its activating power. Minimal doses of the venom could be activated by the sera of other animals in the case of man, or the corpuscles were rendered more susceptible to the poison. Evans believed that the poison was an amboceptor which united with the "endocomplement" of the blood cells. The results obtained by Evans were in contradiction with those of Briot (1904). Evans (1923, 1943) later suggested that the discrepancy was probably on the basis of the variation in methods used in preparing the venom and in the freshness of the poison. Therefore, Evans repeated his experiments using the techniques as given by Briot. The results were similar to those of the French worker with the exception that Evans observed a slight hemolysis in the tube containing only glycerinated venom and red blood cells. The normal serum acted as an "antihaemolysin," and the heated serum acted as a "complement." However, with the unfiltered venom there was complete and rapid hemolysis. Further study led Evans to the conclusion that the addition of glycerine and subsequent filtration altered both the virulence and character of the venom. If heated serum was added to filtered glycerinated venom, the amount of hemolysis was increased and virulence was decreased. Unheated serum added to glycerinated venom gave variable results. In some instances hemolysis was increased, remained the same, or in still others was decreased. Heated serum added to glycerinated venom passed through a Berkfeld filter markedly increased hemolysis. Unheated serum added to glycerinated venom passed through a Berkfeld filter resulted in a total absence of hemolysis. Evans attempted to explain these results by assuming the presence of a complement-like body in addition to an amboceptor in fresh venom. The complement seemed to pass through the filter less readily than the amboceptor, and its passage apparently could be increased by adding salt solution. He also observed that lecithin was able to complement the venom amboceptor in the cases in which heated serum increased the hemolysis. Finally, if venom extracted direct by syringe from the venom gland was dried *in vacuo*, and then after several days was dissolved in normal saline and filtered, it was possible to obtain a solution capable of hemolyzing washed red blood cells without the addition of either heated or unheated serum.

Phisalix (1922) wrote a lengthy but incomplete review of the early research on the physiological effects of weever venom. No original experimental data were presented. The most complete review of the early literature on weever venom is that by Pawlowsky (1927).

De Marco (1936, 1937) also published on the toxicological effects of weever venom. His interest in the problem originated from a case of "reflex epilepsy" which was precipitated accidentally by "stimulating the skin in an attempt to remove a weever sting" from a patient. De Marco believed that the weever venom caused an increased excitation of the cortical motor-sensory centers of the brain, thereby resulting in an epileptiform attack. In his experimental work he prepared glycerine extracts and venom direct from the glands of *T. draco*. Various dosages of extracts were then injected into the dorsal lymph sacs of frogs. There was produced an immediate increase of reflex excitability, followed by tonic and clonic convulsions, paralysis, loss of position and spinal reflexes, and finally death. He observed an increase in central nervous system permeability to potassium following injection of the venom into frogs. De Marco (1938) demonstrated a more rapid exhaustion of the frog's gastrocnemius in the presence of the venom but no deleterious effects that might be attributed to the direct action of the venom on the muscle. Maretić (1957) found that guinea pigs stabbed with the stings of weevers became restless and noisy and later developed a paresis of the hind legs. One animal developed tachycardia, but aside from this he did not observe any significant systemic effects as a result of the poisoning.

Russell and Emery (1960) analyzed venom extracts prepared from *T. draco* and *T. vipera* and found that the lethal fraction of the toxin was nondialyzable. Details of their findings of some of the pharmacological properties of weever venom will be discussed in greater detail in the following section. Carlisle (1962) reported that the dialyzable fraction of weever venom produced the pain characteristic of the whole venom. The nondialyzable fraction failed to provoke any pain but did produce a rise in pulse rate and respiratory distress. His pain experiments were conducted by injecting the extracts into himself.

Skeie (1962*d*) extracted weever venom using a micropipette to which was attached a vacuum of 1/5 atmosphere of pressure. The pipette was inserted into the venom grooves of the stings of the fishes, and the venom was sucked into a test tube reservoir. With the use of this technique Skeie extracted the venom from 600 weevers into 40 ml of physiological saline, making a total quantity of 60 ml of venom solution. He found the MLD_{100}(LD_{100}) in two mice to be 0.0004 ml. On the basis of this determination he calculated that each weever contained sufficient venom to kill 250 mice weighing 17 g each. In a series of 19 batches of weever venom that were tested upon mice, he found the average venom content to vary from 6-1,066 LD_{100} per weever, or from 640-2,500 LD_{100} per ml of venom. Since the opercular glands of the weever are considerably larger than the dorsal fin glands, they contained the greatest quantity of toxin. Injections of weever venom into the skin of guinea pigs produced a localized reaction which remained for about 48 hours. Weever venom was also tested on monolayer cultures using chick fibroblasts, human embryonic kidney cells, *Rhesus* monkey kidney cells, and human amnion cells. It was observed that even very small concentrations of toxin produced growth and cellular changes consisting of vacuolation and profuse granulation. The use of chick fibroblasts provided a sensitive method by which to measure the toxicity of weever venom. Various methods of storing weever venom were investigated, and it was found that the venom rapidly loses its toxicity over a relatively short period of time. Temperature was observed to be an important factor. Storage in glycerin, albumin, and various enzyme inhibitors was generally ineffective. The results from freeze drying were unsatisfactory. After trying various methods of stabilizing the toxin, he revealed that the best results were obtained by extracting and centrifuging (at low temperatures) the venom after capture of the fish. The toxin was then filtered under sterile conditions through Seitz EKS pads under pressure, 15 percent glycerin added, and the toxin frozen quickly in a mixture of alcohol and carbon dioxide ice and stored at -60° C. Skeie found that the toxin would remain unchanged under these conditions for a period of more than 2 years.

Skeie's (1962*a*) investigations on the chemical nature of *T. draco* venom revealed that only part of the protein in the crude venom possessed toxic activity and that it was possible to eliminate some of the nontoxic proteins. However, he did not purify the remaining toxic fraction. He confirmed the earlier work of Briot (1902) and others, that weever venom possesses antigenic properties which could be enhanced with the addition of aluminum hydroxide.

In his particular series of experiments he was able to detoxify weever venom with the use of formalin, but the antigenic properties were considerably decreased. Skeie suggested that under certain experimental conditions it may be possible to detoxify the venom and still retain its antigenic activity, which would be a significant point in the development of a useful vaccine.

Skeie (1962*c*) injected *T. draco* venom intravenously, subcutaneously, and intraperitoneally into white mice. After injection of 1 LD intravenously into a mouse, it ran in circles, developed opisthotonus, tonic and clonic spasms of the extremities, respiratory distress, frothy discharge from the nose and mouth, and died within seconds to minutes. Injections of sublethal doses of weever toxin resulted in an ischemic necrosis of the ears and tip of the tail, and in about 4 to 5 days they fell off. In addition, large patches of hair fell out of the back and sides of the mouse. Tolerance to weever venom was found to vary from one animal species to the next and according to the route of administration. Four times more toxin was needed to kill mice by intraperitoneal than by intravenous injection, and 16 times more by subcutaneous injection. The symptoms also varied somewhat with the route of administration. Large areas of necrosis developed at the site of injection by subcutaneous and intramuscular administration of the venom. However, these necrotic areas did not occur after repeated administration of the venom. These results were interpreted as evidence of an increasing immunity to the poison.

PHARMACOLOGY

Most of the earlier researches on the action of weever venom were concerned with the toxicological effects. Probably Russell and Emery (1960) were the first to conduct what might be considered a pharmacological analysis of the poison. Assays indicated that the lethal portion of the venom was in the nondialyzable fraction. They found that extracts of the venom of *Trachinus draco* had no deleterious effects on mammalian neuromuscular transmission. There was some shortening of the muscle and a very gradual depression of both the directly and indirectly elicited contractions when large amounts of the extract were added to the nerve-muscle bath, but there was no evidence of a differential that could be attributed to changes at the neuromuscular junction. The venom produced in cats a precipitous fall in systemic arterial pressure concomitant with changes in the pulse, pulmonary arterial, venous, and cisternal pressures, cardiac rate, respiration, and the electrocardiogram and electroencephalogram. The findings were similar to those produced by stingray venom. The electrocardiograms indicated that the venom could produce both changes in rhythm and injury to the heart muscle. The fall in pulmonary arterial pressure suggested either a failure of the heart to maintain an effective stroke volume or a decrease in pulmonary resistance. Studies of pulmonary artery blood flow using a gated sine-wave electromagnetic flowmeter indicated that the blood flow in this vessel was reduced during the period of decreased pressure. The findings of lowered pulmonary arterial pressure and flow, a decrease in heart rate with various degrees of auriculoventricular block, and evidence of heart muscle injury were interpreted as an indication that some degree of cardiac failure was probably responsible for the fall in systemic arterial pressure and the rise in venous and cisternal pressures. Smaller amounts of weever venom produced transient vasoconstriction or vasodilation, depending upon the quantity injected. The authors concluded that there was little or no deleterious effect upon the heart, although there were changes in cardiac and respiratory rates. The depression in central nervous system activity observed following the intravenous injection of lethal doses of the venom was attributed to ischemic anemia produced by the lowered systemic arterial pressure. The wave pattern was typical of that produced during cardiovascular failure from any of a number of causes. It was suggested that the venom might have a direct effect on central nervous system activity, but these effects were not defined. The rate and depth of respiration following injection of weever venom were subject to considerable variation. Small doses of the venom produced a slight increase in rate. Slightly larger doses of venom resulted in an increase or decrease in rate, and lethal doses usually caused a decrease in rate leading to complete cessation of respiration. The evidence suggested that a part of the respiratory crisis could be attributed to the changes in the cardiovascular system. It was found that neither continuous artificial respiration nor direct stimulation of the diaphragm significantly altered the cardiovascular crisis, nor did these measures save the life of the animal.

Skeie (1962*c*) concluded from his physiological studies on weever venom that heart failure was the primary cause of death. Electrocardiograms taken during the various experiments on cats showed pronounced ischemic changes in the myocardium which resulted in a terminal ventricular flutter. Autopsies of the animals revealed a severe stasis in the lungs and liver together with pulmonary edema. He believed that the fall in blood pressure after injection of the venom was probably due to cardiac failure and the effect of hyaluronidase with a reduced circulating blood volume. Injections of weever venom markedly affected the depth and frequency of respiration. Large doses of venom produced a transient apnea which was believed to be caused by CO_2 accumulations due to tissue ischemia and the hemolysis with loss of O_2 transport. Whether there was a direct neural effect on the respiratory organs was not determined.

CHEMISTRY

Fresh weever venom is clear, gray in color, and has a fishy taste and ammoniacal odor. The lethal portion of the venom is reported to be in the nondialyzable fraction. Russell and Emery (1960) reported an elemental and proximate chemical analysis as given in Table 2.

Carlisle (1962) investigated the chemistry of the venom of *Trachinus vipera* and found that it contained a dialyzable and a nondialyzable fraction. Analysis of the dialyzable fraction by ion-exchange column and paper chromatography revealed two spots, one of which was identified as 5-hydroxy-

TABLE 2.—*Analysis of venom extracts from the opercular and dorsal stings of* Trachinus *(From Russell and Emery, 1960)*

Elemental analysis	Per cent by weight	Proximate analysis	Per cent
Carbon	22. 8	Moisture	41. 5
Hydrogen	3. 5	Protein	21. 7
Nitrogen	6. 9	Lipids (ether soluble)	3. 5
Phosphorus	2. 2	Inorganic matter	12. 9
Sulfur	0. 0	Carbohydrate (calculated)	20. 4

tryptamine. The other substance did not appear to be an indole derivative but gave a violet color which faded to brown with ninhydrin. When these materials were eluted and injected under the skin of a person, the 5-hydroxytryptamine produced the characteristic pain, whereas the other substance caused no pain but the wheal and flare of a histamine releaser. The two substances together produced the same effect when injected as the unfractionated dialyzable material. He concluded that the chief pain-producing substance of the weeverfish venom was 5-hydroxytryptamine, and that it was associated with a substance of low molecular weight which acted as a histamine releaser. The nondialyzable fraction was subjected to paper electrophoresis in a hanging strip apparatus and consisted of two albumins and a mucopolysaccharide. The albumins were eluted as one extract and the polysaccharide as another and injected into goby fish *Gobius ruthensparri* for bioassay. Both fractions were lethal in doses corresponding to the lethal dose of the unfractionated venom. The whole venom was assayed for anticoagulation properties against a standard preparation of heparin in which 0.0077 mg represented 1 I.U., and it was found that 156 μg of venom had less anticoagulant activity than 1.9 μg of heparin. Carlisle concluded that the mucopolysaccharide of the venom is not a heparinlike material—a conclusion consonant with the lack of sulphur in the sample of venom analyzed by Russell and Emery (1960). Thus, the polysaccharide in weever venom does not appear to serve an anticoagulant function such as is sometimes observed in the mucopolysaccharides in other venoms.

Haavaldsen and Fonnum (1963) isolated three protein fractions by electrophoresis from the venom of *T. draco*. They obtained two spots on a paper chromatogram which were identified as histamine and a catecholamine. Photofluorimetric studies demonstrated the presence of adrenaline and noradrenaline in large concentrations. The venom showed considerable cholinesterase activity but did not reveal the presence of 5-hydroxytryptamine, lecinthinase, or phosphodiesterase. It was concluded that *T. draco* does not contain 5-hydroxytryptamine in large quantities such as that found in the toxin of *T. vipera.*

LITERATURE CITED

Allman, G. J.
1840 On the stinging property of the lesser weever-fish *(Trachinus vipera)*. Ann. Mag. Nat. Hist. 6: 161-165, 1 fig.

Berg, L. S.
1947 Classification of fishes both recent and fossil. [In Russian & English] J. W. Edwards, Ann Arbor, Mich.

Bottard, A.
1885 Note sur la piqûre de la vive. Compt. Rend. Soc. Biol. 2: 23-26.
1889*a* L'appareil à venin des poissons. Compt. Rend. Acad. Sci. 108: 534-537.
1889*b* Les poissons venimeux. Octave Doin, Paris. 198 p.

Bourienne
1782 Mémoire sur les effets de la piqûre des arêtes de la vive. J. Med. Mil. Paris 1: 371-382.

Briot, A.
1902 Sur l'action du venin de la vive *(Trachinus draco)*. Compt. Rend. Soc. Biol. 54: 1169-1171.
1903 Etudes sur le venin de la vive *(Trachinus draco)*. J. Physiol. Pathol. Gen. Paris 6: 271-282.
1904 Sur l'existence d'une kinase dans le venin de la vive *(Trachinus draco)*. Compt. Rend. Soc. Biol. 56: 1113-1114.

Byerley, I.
1849 On the *Trachinus draco*, or otter-pike, sting fish, or weever. Proc. Lit. Phil. Soc. Liverpool (5): 156-168.

Carlisle, D. B.
1962 On the venom of the lesser weever-fish, *Trachinus vipera*. J. Marine Biol. Assoc. U.K. 42: 155-162.

Coutière, H.
1899 Poissons venimeux et poissons vénéneux. Thèse Agrég. Carré et Naud, Paris.

Day, F.
1880-84 The fishes of Great Britain and Ireland. 2 vols., illus. Williams and Norgate, London.

Delfin, F. T.
1901 Catálogo de los peces de Chile. Rev. Chilena Hist. Nat. 5: 82.

De Marco, R.
1936 Sull'azione neurotossica del veleno di tracine e sull'epilessia umana riflessa. Riv. Patol. Nerv. Ment. 47(1): 204-208.
1937 Effetti del veleno di *"Trachinus"* e di quello di *"Scorpaena"* sulla attività neuromuscolare della rana. Arch. Fisiol. 37(3): 398-404.
1938 Effetti del veleno di *"Trachinus"* sulla capacita di lavoro del gastrocnemio di rana. Riv. Biol. 25: 225-234, 4 figs.

Evans, H. M.
1906 Note on the treatment of weever sting. Brit. Med. J. 2: 23.
1907*a* Observations on the poisoned spines of the weever fish *(Trachinus draco)*. Brit. Med. J. 1: 73-76, 3 figs.
1907*b* Observations on the poisoned spines of the weever fish, *Trachinus draco*. Trans. Norfolk Norwich Nat. Soc. 8: 355-368.
1916 The poison-organ of the sting-ray *(Trygon pastinaca)*. Proc. Zool. Soc. London 29(I): 431-440, 7 figs.
1923 The defensive spines of fishes, living and fossil, and the glandular structure in connection therewith, with observations on the nature of fish venoms. Phil. Trans. Roy. Soc. London, B, 212: 1-33, 3 pls., 14 figs.
1943 Sting-fish and seafarer. Faber and Faber, Ltd., London. 180 p., 31 figs.

Gill, T.
1907 Life histories of toadfishes (Batrachoidida), compared with those of weevers *(Trachinus)* and stargazers (Uranoscopids). Smithsonian Misc. Coll. 48(1697): 388-427, 21 figs.

Gressin, L.
1884 Contributions à l'étude de l'appareil à vénin chez les poissons du genre "vive" *(Trachinus)*. Thèse No. 289, Fac. Méd. Paris. 50 p.

Haavaldsen, R., and F. Fonnum
1963 Weever venom. Nature 199(4890): 288-287.

HALSTEAD, B. W.
1957 Weever stings and their medical managements. U.S. Armed Forces Med. J. 8(10): 1441-1451, 8 figs.
HALSTEAD, B. W., and L. J. CARSCALLEN
1964 Marine organisms, p. 336-343. *In* P. L. Altman and D. S. Dittmer [eds.], Biology data book. Fed. Am. Soc. Exp. Biol., Washington, D.C.
HALSTEAD, B. W., and R. F. MODGLIN
1958 Weeverfish stings and the venom apparatus of weevers. Z. Tropenmed. Parasitol. 9(2): 129-146, 12 figs.
IRVINE, F. R.
1947 The fishes and fisheries of the Gold Coast. Crown Agents for the Colonies, London. 352 p., 217 figs.
JOUBIN, L. M.
1929-38 Faune ichthyologique de l'Atlantique nord. Nos. 1-18. A. F. Høst et Fils, Copenhagen.
KAZDA, F.
1931 Finergangran nach stich durch *Trachinus draco* (petermännchen). Arch. Klin. Chir. 166: 546-548, 1 fig.
LEGELEUX, G.
1965 Contribution à l'étude des poissons toxicophores. Imprimerie Douladoure, Toulouse, France. 124 p.
MARETIĆ, Z.
1957 Erfahrungen mit stichen von giftfischen. Acta Tropica 14(2): 157-161, 2 figs.
1962 Venomous animals. Liječnički Vjesnik 84(12): 1233-1252, 8 figs.
1966 On venomous animals of the Adriatic Sea [In Croatian] Morsko Ribarst. 18: 168-173, 7 figs.
PARKER, W. N.
1888 On the poison-organs of *Trachinus*. Proc. Zool. Soc. London (3): 359-367, 1 pl., 2 figs.
PAWLOWSKY, E. N.
1906 Microscopic structure of the poison glands of *Scorpaena porcus* and *Trachinus draco*. [In Russian] Trav. Soc. Imp. Nat., St. Petersburg 37(1): 316-337, 1 pl.
1907 Contribution to the structure of the epidermis and its glands in venomous fish. [In Russian] Trav. Soc. Imp. Nat., St. Petersburg 38(1): 265-282, 1 pl., 1 fig.
1911 Contribution to the question of the structure of the poison glands of certain fish of the Scorpaenidae family. [In Russian] Trav. Soc. Imp. Nat., St. Petersburg 41(1): 317-328.
1927 Gifttiere und ihre giftigkeit. G. Fischer, Jena. 515 p., 170 figs.
1929 Poisons and poison-producing organs in the animal kingdom. [In Russian] Russ. J. Trop. Med. 7: 4-11.
PHISALIX, C.
1899 Expériences sur le venin des vives (*Trachinus vipera* et *Tr. draco*). Bull. Mus. Hist. Nat. 5(5): 256-258.
PHISALIX, M.
1922 Animaux venimeux et venins. 2 vols. Masson et Cie., Paris.
POHL, J.
1893 Beitrag zur lehre von den fischgiften. Prag. Med. Wochschr. 18(4): 31-33.
POLL, M.
1947 Poissons marins. Faune de Belgique. Mus. Roy. Hist. Nat. Belgique, Brussels. 452 p., 267 figs.
1951 Expédition océanographique Belge dans les eaux cotières Africaines de l'Atlantique sud (1948-49). Résultats scientifiques. Vol. IV, No. 1. Poissons. I—Généralités. Institut Royal des Sciences Naturelles de Belgique, Brussels. 154 p., 13 pls.
1954 Expédition océanographique Belge dans les eaux cotières Africaines de l'Atlantique sud (1948-49). Résultats scientifiques. Vol. IV, No. 3a. Poissons. IV—Téléostéens acanthoptérygiens, pt. 1. Institut Royal des Sciences Naturelles de Belgique, Brussels. 390 p., 9 pls.
1959 Expédition océanographique Belge dans les eaux cotières Africaines de l'Atlantique sud (1948-1949). Résultats scientifiques. Vol. IV, No. 3b. Poissons. V—Téléostéens acanthoptérygiens, pt. 2. Institut Royal des Sciences Naturelles de Belgique, Brussels. 417 p., 7 pls.
PORTA, A.
1905 Ricerche anatomiche sull'apparecchio velenifero di alcuni pesci. Anat. Anz. 26: 232-247, 5 figs., 2 pls.

RUSSELL, F. E., and J. A. EMERY
1960 Venom of the weevers *Trachinus draco* and *Trachinus vipera*. Ann. N.Y. Acad. Sci. 90(3): 805-819, 4 figs.

SCHMIDT, F. T.
1874 Om Fjärsingens stik og giftredskaber. Nord. Med. Arkiv. 6(2): 1-20.

SKEIE, E.
1962*a* Weeverfish toxin. Acta Pathol. Microbiol. Scand. 56: 229-238.
1962*b* The venom organs of the weeverfish *(Trachinus draco)*. Medd. Dan. Fisk. Havund. 3(10): 327-338, 13 figs.
1962*c* Toxin of the weeverfish *(Trachinus draco)*. Experimental studies on animals. Acta Pharmacol. Toxical 19: 107-120, 8 figs.
1962*d* Weeverfish toxin. Extraction methods, toxicity determinations and stability examinations. Acta Pathol. Microbiol. Scand. 55: 167-174.
1962*e* Problemer vedrørende den giftige fisk fjaesing i Danmark. Nord. Med. 67(14): 429-460.

SMITT, F. A.
1892-95 Skandinaviens fiskar. 3 vols. P. A. Norstedt & Söners, Stockholm. 1,239 p., 53 pls., 380 figs.

SOLJAN, T.
1963 Fauna and flora of the Adriatic. Vol. I. Fishes of the Adriatic [Engl. Transl.] Nolit Publ. Hse., Belgrade, Yugoslavia. 428 p.

ULMER,—
1865 Tod durch den stich eines *Trachinus draco*. Alg. Mil. Arztl. Ztg., Vienna 6(42): 329-332.

Chapter XXI—VERTEBRATES

Class Osteichthyes

Venomous: SCORPIONFISHES

Members of the family Scorpaenidae, the scorpionfishes, are widely distributed throughout all tropical and temperate seas. A few species are also found in arctic waters. Many scorpaenids attain large size and are valuable food fishes, whereas others are relatively small and of no commercial value. Some species are extremely venomous.

Considerable confusion exists in the literature concerning the taxonomic status of scorpaenids, and the situation is unlikely to improve until a complete world revision of this important group of fishes is accomplished. According to McCulloch (1929-30), Whitley (1932), and Smith (1950), the group should be divided into two distinct families: the Scorpaenidae—scorpionfishes proper, and Synancejidae—the stonefishes. However, Fowler (1928) places them in a single family, Scorpaenidae, which is the classification adopted in this present work. For additional information concerning the systematics and identifying characteristics of the Scorpaenidae, the reader is referred to the works of Fowler (1928), Joubin (1929-38), Matsubara (1943), Lozano y Rey (1947, 1952), Poll (1947, 1951, 1954, 1959), Smith (1950, 1957, 1958, 1961), Herre (1952, 1953), Ginsburg (1953), Phillips (1957), Marshall (1964), Eschmeyer (1965, 1969), Schultz, Woods, and Lachner (1966), Eschmeyer and Rao (1973), and Eschmeyer and Randall (1975).

REPRESENTATIVE LIST OF SCORPIONFISHES REPORTED AS VENOMOUS

Phylum CHORDATA

Class OSTEICHTHYES

Order PERCIFORMES (PERCOMORPHI)

Suborder COTTOIDEI (CATAPHRACTI, SCLEROPAREI, LORICATI): Scorpionfishes, Sculpins, Etc.

Family SCORPAENIDAE

Apistus carinatus (Bloch and Schneider) (Pl. 1). Scorpionfish (USA), bull-rout, sulky, waspfish, cobbler, rockcod (Australia), hirekasago, hachi (Japan), ikan-semaram (Malaysia). *238*

DISTRIBUTION: Indo-Pacific, Australia, Japan, China, India.

SOURCES: Pawlowsky (1913, 1914, 1927), Tweedie (1941), Tange (1953*a, c, d*), Halstead and Mitchell (1963), Russell (1965).

OTHER NAMES: *Apistus alatus, A. cottoides, A. evolans, A. venenans*. (Herre, 1953, believes *A. alatus* Cuvier and Valenciennes to be a valid species.)

238 *Brachirus brachypterus* (Cuvier) (Pl. 2, fig. a). Lionfish, turkeyfish, featherfish, short-finned scorpionfish (USA), laong, lalung, laiuñg, laoung (Philippines), butterflyfish (South Africa), zebrafire fish, butterfly cod (Australia).

DISTRIBUTION: Indo-Pacific, Australia, Indian Ocean.

SOURCES: Heere (1927), Halstead and Mitchell (1963), Russell (1965).
OTHER NAME: *Dendrochirus brachypterus.*

240 *Brachirus zebra* (Quoy and Gaimard) (Pl. 4, fig. a). Lionfish, turkeyfish, featherfish, short-finned scorpionfish (USA), laong, lalung, laiuñg, laoung (Philippines), butterflyfish (South Africa), zebrafire fish, butterfly cod (Australia).

DISTRIBUTION: Indo-Pacific, Polynesia to east Africa.

SOURCES: Blanchard (1890), Coutière (1899), Pawlowsky (1911), Tange (1953*b*), Halstead and Mitchell (1963), Marshall (1964).
OTHER NAMES: *Dendrochirus zebra, Pterois zebra.*

239 *Centropogon australis* (White) (Pl. 3, fig. a). Scorpionfish (USA), scorpionfish, waspfish, fortesque (Australia).

DISTRIBUTION: New South Wales and Queensland, Australia.

SOURCES: Cleland (1912), Scott (1921), Whitley (1943, 1963), Halstead and Mitchell (1963), Marshall (1964), Russell (1965).

239 *Erosa erosa* (Langsdorf) (Pl. 3, fig. b). Scorpionfish (USA), daruma-okoze (Japan).

DISTRIBUTION: Japan.

SOURCES: Tange (1954*a*), Halstead and Mitchell (1963).
OTHER NAME: *Synanceia erosa.*

238 *Gymnapistes marmoratus* (Cuvier and Valenciennes) (Pl. 2, fig. b). Scorpionfish (USA), southern Australian cobbler (Australia).

DISTRIBUTION: Northern territory and South Australia.

SOURCES: Whitley (1943), Halstead and Mitchell (1963).

240 *Inimicus filamentosus* (Cuvier and Valenciennes) (PL. 4, fig. b). Devil scorpionfish (USA), lupo, lopo, gatas-gatasan (Philippines), lepou Malaysia), padang, priaman (Sumatra).

DISTRIBUTION: Indo-Pacific.

SOURCES: Coutière (1899), Pawlowsky (1909, 1911), Faust (1924), Halstead and Mitchell (1963).
OTHER NAME: *Pelor filamentosus.*

241 *Inimicus japonicus* (Cuvier and Valenciennes) (Pl. 5). Devil scorpionfish (USA), oniokoze (Japan).

DISTRIBUTION: Japan.

Sources: Coutière (1899), Taschenberg (1909), Pawlowsky (1909, 1911, 1927), Halstead and Mitchell (1963), Russell (1965).
Other name: *Pelor japonicum.*

Minous adamsi Richardson (Pl. 6, fig. a). Scorpionfish (USA), himeokoze (Japan). *242*
Distribution: Japan, China.

Sources: Tange (1953*a, c, d*), Halstead and Mitchell (1963).

Notesthes robusta (Günther) (Pl. 7, fig. a). Scorpionfish (USA), bullrout, scorpionfish (Australia). *243*
Distribution: New South Wales and Queensland, Australia.

Sources: Cleland (1912), Kesteven (1914), Scott (1921), Duhig (1929), Whitley (1943, 1963), Cilento (1944), Halstead and Mitchell (1963), Russell (1965), Cameron and Endean (1966).

Pterois antennata (Bloch) (Pl. 8). Zebrafish, lionfish, tigerfish, scorpionfish, turkeyfish, featherfish, firefish (USA), sausaulele (Samoa), e si ĕd, smú ĕ, o ses, lóĕ, kó kĕl tá ŏt (Palau), lepu pangantien, ikan, sowanggi, ikan bambou (Malaysia), lalung, lalong, laiuñg, lalouñg, lallong, lolong (Philippines), purrooah, cheeb-ta-ta-dah, kurrum, flying dragon, toombi (India), fini maja, rathu gini maha, saval min, striped butterflyfish, gon maha (Sri Lanka), navire (Réunion), devilfish, butterflyfish, fireworksfish, sting-fish (South Africa), mino-kasago, yamanokami (Japan). *244*
Distribution: Indo-Pacific, Indian Ocean, China.

Sources: Coutière (1899), Pawlowsky (1911), Tange (1953*b*), Halstead and Mitchell (1963), Russell (1965).
Other names: *Scorpaena antennata, Pterois artemata.*

Pterois russelli Bennett (Pl. 7, fig. d). Zebrafish, lionfish, tigerfish, scorpionfish, turkeyfish, featherfish, firefish (USA), sausaulele (Samoa), e si ĕd, smú ĕ, o ses, lóĕ, kó kĕl tá ŏt (Palau), lepu pangantien, ikan, sowanggi, ikan bambou (Malaysia), lalung, lalong, laiuñg, lalouñg, lallong, lolong (Philippines), purrooah, cheeb-ta-ta-dah, kurrum, flying dragon, toombi (India), fini maja, rathu gini maha, saval min, striped butterflyfish, gon maha (Sri Lanka), navire (Réunion), devilfish, butterflyfish, fireworksfish, stingfish (South Africa), mino-kasago, yamanokami (Japan). *243*
Distribution: Indo-Pacific, Australia, Indian Ocean.

Sources: Pawlowsky (1911, 1914, 1927), Tweedie (1941), Tange (1953*a, b, c, d*), Halstead and Mitchell (1963), Russell (1965).
Other name: *Pterois lunulata* (*See* Smith, 1957).

Pterois volitans (Linnaeus) (Pl. 7, fig. b, c). Zebrafish, lionfish, tigerfish, scorpionfish, turkeyfish, featherfish, firefish (USA), sausaulele (Samoa), e si ĕd, smú ĕ, o ses, lóĕ, kó kĕl tá ŏt (Palau), lepu pangantien, ikan, *243* sowanggi, ikan bambou (Malaysia), lalung, lalong, laiuñg, lalouñg, lallong, lolong (Philippines), purrooah, cheeb-ta-ta-dah, kurrum, flying dragon toombi (India), fini maja, rathu gini maha, saval min, striped butterflyfish, gon maha (Sri Lanka), navire(Réunion), devilfish, butterflyfish, fireworks-fish, stingfish (South Africa), mino-kasago, yamanokami (Japan).

DISTRIBUTION: Indo-Pacific, Australia, Japan, China, Indian Ocean, Red Sea.

SOURCES: Schnee (1908), Herre (1927), Whitley (1949), Tange (1953*b*), Halstead, Chitwood, and Modglin (1955*b*), Saunders (1959*a, b, c,* 1960*a*), Atz (1962*b*), Halstead and Mitchell (1963), Russell (1965).

OTHER NAMES: *Pterois miles, P. muricata. Pterois volitans* is subject to considerable growth changes involving the length of pectorals, free portions of pectoral rays, head spination, and the length of the ocular tentacle. Smith (1957) believes that *P. miles* is merely the adult form of *P. volitans.* Klausewitz (1957) has retained *P. miles* as a valid species. In general, there is still much confusion regarding the systematics and diagnoses of the various members of the subfamily Pteroinae. It appears that all of the members of this subfamily are equipped with well-developed venom organs in their dorsal and anal spines.

242 *Scorpaena grandicornis* Cuvier and Valenciennes (Pl. 6, fig. b). Lionfish, long-horned scorpionfish (USA), rascacio (West Indies), mangangá (Brazil).

DISTRIBUTION: Florida Keys, West Indies, Panama, Brazil.

SOURCES: Blanchard (1890), Coutière (1899), Scott (1921), Pawlowsky (1927), Bayley (1940).

244 *Scorpaena guttata* Girard (Pl. 9, fig. a). California scorpionfish, sculpin (USA).

DISTRIBUTION: Central California south into the Gulf of California.

SOURCES: Halstead, *et al.* (1955*a*), Saunders (1959*a*), Halstead and Mitchell (1963), Taylor (1963), Legeleux (1965), Russell (1965).

245 *Scorpaena plumieri* Bloch (Pl. 10, fig. a). Spotted scorpionfish, sculpin (USA), rascassio, rascacio (West Indies), escorpião (Brazil).

DISTRIBUTION: Atlantic coast from Massachusetts to the West Indies and Brazil.

SOURCES: Coutière (1899), Scott (1921), Bayley (1940), Colby (1943), Phillips and Brady (1953), Halstead and Carscallen (1964).

OTHER NAMES: *Scorpaena bufo, S. ginsburgi.*

244 *Scorpaena porcus* Linnaeus (Pl. 9, fig. b). Scorpionfish (USA), scorpena nera (Italy), hogfish, sea scorpion (England), braune drachenkopf (Germany), rascasse (France), escorpena (Spain), rascasso (Portugal), ciernomorskii morskoi ersch (USSR).

DISTRIBUTION: Atlantic coast of Europe from the English Channel to the Canary Islands, French Morocco, Mediterranean Sea, Black Sea.

Sources: Pohl (1893), Sacchi (1895*a, b*), Coutière (1899), Cecca (1902), Scott (1921), Tuma (1927), De Marco (1937), Lumière and Meyer (1938), Romano (1940), Halstead *et al.* (1955*b*), Maretić (1962, 1966, 1968), Russell (1965).

Scorpaena scrofa Linnaeus (Pl. 10, fig. b). Scorpionfish (USA), scorpena nera (Italy), grossen drachenkopf (Germany), rascasse rouge (France), rascasso (Portugal), rascacio (Spain). *245*

Distribution: West coast of France south to Cabo Blanco, Northwest Africa and Mediterranean Sea.

Sources: Bottard (1889*a, b*), Blanchard (1890), Dunbar-Brunton (1896), Coutière (1899), Cecca (1902), Pawlowsky (1906, 1907, 1911, 1927), Taschenberg (1909), Engelsen (1922), Faust (1924), Maretić (1962, 1968).

Scorpaenopsis gibbosa (Bloch and Schneider) (Pl. 11). Scorpionfish (USA), oni-kasago, rokubu (Japan), nohu, omaka (Hawaii), scorpionfish, stingfish (South Africa), ikan satan, ikan-sowangi, bezard (Malaysia), scorpionfish, waspfish (Australia). *246*

Distribution: Indo-Pacific, Indian Ocean.

Sources: Halstead and Mitchell (1963), Russell (1965).
Other names: *Scorpaena nesogallica, S. mesogallica.*

Sebastes auriculatus Girard (Pl. 12, fig. d). Brown rockfish (USA). *247*

Distribution: Pacific coast of America, from Cape Mendocino to Cedros Island.

Sources: Roche and Halstead (1972), Roche (1973).
Other name: *Auctospina auriculata.*

Sebasticus marmoratus (Cuvier and Valenciennes) (Pl. 12, fig. a). Scorpionfish (USA), kasago, hachi-kasago (Japan). *247*

Distribution: Indo-Pacific, China Sea, Japan.

Sources: Pawlowsky (1911, 1927), Tange (1953*a*), Halstead and Mitchell (1963), Russell (1965).
Other name: *Sebastes marmoratus.*

Sebastodes inermis (Cuvier) (Pl. 12, fig. b). Japanese stingfish (USA), mebaru (Japan). *247*

Distribution: Japan.

Sources: Pawlowsky (1927), Tange (1953*a, c*), Halstead and Mitchell (1963), Legeleux (1965).
Other name: *Sebastes inermis.*

Sebastodes joyneri (Günther) (Pl. 12, fig. c). Joyner stingfish (USA), togottomebaru (Japan). *247*

Distribution: Japan.

Sources: Pawlowsky (1911, 1927), Halstead and Mitchell (1963).
Other name: *Sebastes joyneri.*

248, 249 *Synanceja horrida* (Linnaeus)[1] (Pl. 13, Pl. 14). Stonefish (USA), stonefish, star-gazer, warty-ghoul, dornorn (Australia), ikan-satan, devilfish (Java), ikan-lepu, goblinfish, lepou, ikan-swanggi (Malaysia), crapaud de mer (Martinique), laffe (Mauritius), lumpfish, sumalapao, lupu, gatasan, laoung (Philippines), no'u, nufu, nuhu, joju, gofu, toadfish (Polynesia), sherova (Portuguese East Africa), rapau de mer (Réunion), stonefish, devilfish, stingfish (South Africa).

DISTRIBUTION: Indo-Pacific, Australia, China, India.

SOURCES: Schnee (1911), Duhig (1929), Whitley and Boardman (1929), Whitley (1943), Cilento (1944), Saunders (1958, 1959*a, b,* 1960*a*), Wiener (1959*a, b*), Saunders and Tökés (1961), Atz (1962*a*), Whitley (1963), Legeleux (1965), Deakins and Saunders (1967), Russell (1967).

250 *Synanceja verrucosa* Bloch and Schneider (Pl. 15, figs. a, b). Stonefish (USA), stonefish, stargazer, warty-ghould, dornorn (Australia), ikan-satan, devilfish´(Java), ikan-lepu, goblinfish, lepou, ikan-swanggi (Malaysia), crapaud de mer (Martinique), laffe (Mauritius), lumpfish, sumalapao, lupu, gatasan, laoung (Philippines), no'u, nufu, nuhu, joju, gofu, toadfish (Polynesia), synancée (New Caledonia), sherova (Mozambique), rapau de mer (Réunion), stonefish, devilfish, stingfish (South Africa).

DISTRIBUTION: Indo-Pacific, Australia, Indian Ocean, Red Sea.

SOURCES: Le Juge (1871), Sacchi (1895*a*), Schnee (1911), Cleland (1912), Faust (1924), Whitley (1943, 1963), Gail and Rageau (1956), Saunders (1959*a, c*), Atz (1962*a*), Halstead and Mitchell (1963), Russell (1965), Bagnis (1968).

OTHER NAMES: *Scorpaena brachion, Synanceia brachio, Synanceichthys verrucosus, S. emmydrichthys, S. vulcans.*

BIOLOGY

Family SCORPAENIDAE: The family of scorpionfishes includes several hundred species, most of which are found in temperate marine waters, although the bulk of the venomous members are tropical inhabitants. Because of their complex anatomical features many of the species are extremely difficult to identify. However, all of the members of this family have at least one anatomical characteristic in common and that is the presence of a bony plate, or stay, which extends across the cheek from the eye to the gill cover. Consequently, scorpionfishes are sometimes referred to as the "mailcheeked fishes." Although varying somewhat in their habitats, scorpionfishes are for the most part bottom fishes, preferring rocky coastal areas—hence the term rockfishes. Many of the venomous species are found in or around coral reefs or kelp beds with which they are able to blend in quite favorably because of their protective coloration. Some species bury themselves in the sand, and most of the dangerous ones have the habit of lying motionless for long periods

[1] The spelling of "*Synanceja*" is said to be a misprint for "*Synanceia,*" which is therefore said to be the correct spelling (Eschmeyer and Rao, 1973).

of time. A few species are deep-water inhabitants and may be taken at depths of several hundred fathoms or more. Some of the temperate zone scorpaenids are considered to be of considerable commercial value. They are carnivorous fishes, feeding on invertebrates and other fishes. Most scorpionfishes bear living young.

The maroon-colored zebrafish, *Pterois*, is among the most beautiful and ornate of coral reef fishes. It is generally found in shallow water, hovering about in a crevice or at times swimming unconcernedly in the open. *Pterois* have been called turkeyfish because of their interesting habit of slowly swimming about, spreading their fanlike pectorals and lacy dorsal fins, like a turkey gobbler displaying its plumes. Zebrafish are frequently observed swimming in pairs and are apparently fearless in their movements. Acceptance of the invitation to reach out and grab one of these majestic fish results in an extremely painful experience, because hidden beneath the "lace" are the needlesharp fin stings of this fish. *Pterois volitans* is known to display a defensive behavior (Fig. 1) which may result in an extremely painful experience for the uninitiated diver who fails to recognize its tactics. This interesting behavior has been well described by Steinitz (1959) and Atz (1962*a, b*). *Pterois* reacts quickly to an object moving closely toward it from the side by confronting the object with its dorsal stings. This is accomplished by rotating its body about 90° from the usual upright position or by rotating 180° starting from the lateral-side-up position. With a rapid darting motion

Figure 1.—Under the proper circumstances the zebrafish *Pterois* reacts quickly to an object that is moving toward it. The dorsal spines are fully erected, and the fish rotates its body in order to confront the object with its dorsal spines. The fish inflicts its sting with a rapid darting motion as shown in the accompanying photograph. (From J. W. Atz, courtesy New York Zoological Society)

the fish thereupon inflicts its sting. The immediate stinging process is inflicted with little or no pain, but this painless injection soon blossoms into extreme torture (Atz, 1962*b*).

Members of the genus *Scorpaena* are for the most part shallow-water bottom dwellers, found in bays, along sandy beaches, rocky coastlines, or coral reefs. Their habit of concealing themselves in crevices, among debris, under rocks, in seaweed, together with their protective coloration which blends them almost perfectly into their surrounding environment, makes them difficult to see. Scorpaenids are generally captured by hook and line, and in many regions they are a popular and important food fish. When they are removed from the water they have the defensive habit of erecting their spinous dorsal fin and flaring out their armed gill covers, pectoral, pelvic, and anal fins. The pectoral fins, although dangerous in appearance, are unarmed. Hinton (1962) and Breder (1963) have reported observations on the defensive behavior of *Scorpaena guttata* and *S. plumieri* respectively. Whenever an object came in close proximity of *S. guttata* the dorsal spines were immediately erected, and the fish moved swiftly toward the object so as to deliver a sharp blow to the object with the side of its head or with its dorsal sting. In the case of *S. plumieri* it was observed that the fish was generally quiescent, but when touched with a stick, it would settle down tightly on the sand and sometimes arch its back. The head of the fish was directed slightly downward and the opercula expanded. If the intrusion continued the fish would suddenly change its stance and expose large yellow patches on the pectoral fins (Figs. 2, 3). Prior to this, all of the exposed surface of the fish was somewhat drab, but with the change in stance, the pectorals would be suddenly flipped over, displaying the brightly colored undersurface. The pale dots near the base of the pectorals became a bright iridescent blue, and the interradial light-colored patches along the margin to mid-part of the fins were bright yellow. The dark area around the blue spots became an intense black and the fin margin gray. If the provocation continued, the fish would repeatedly ram or butt the intruding object.

Stonefishes, members of the genus *Synanceja*, are shallow-water dwellers, commonly found in tidepools and shoal reef areas. *Synanceja* has the habit of lying motionless in coral crevices, under rocks, in holes, or buried in sand and mud. They appear to be fearless and completely disinterested in the careless intruder, swimming sluggishly about at infrequent intervals. Whitley and Boardman (1929) reported seeing mollusks and prawns crawling over the head and across the mouth of a stonefish with no observable movement or apparent interest on the part of the fish.

Because of a thick coating of slime, irregular wartlike texture of the skin, and the habit of burying itself in the sand, stonefishes frequently become coated with bits of coral debris, mud, and algae. Coutière (1899) reported that specimens have been observed with algae growing on their skin. The phenomenal ability of the stonefish to camouflage itself makes it a real menace to bathers and those wading about in areas inhabited by this fish. Interesting and valuable accounts of its habits have been written by Saville-Kent (1900),

Figure 2.—When at rest scorpaenids are commonly observed concealing themselves in crevices, among debris, under rocks or in seaweed along sandy beaches, rocky coastlines, or coral reefs. The *Scorpaena porcus* shown in this drawing is at rest with the vertical fins in the retracted position. (From *Fish Resources of the USSR*)

Figure 3.—When scorpionfishes become excited and develop aggressive behavior, their vertical and pectoral fins are fully extended in the manner shown in this illustration. When an object moves toward the fish, it moves quickly toward the invader, striking it with either its head or dorsal stings.

(From *Fish Resources of the USSR*)

Whitley (1932), and Roughley (1947). Endean (1961) has observed that stonefish will not move even when approached and that the dorsal stings are not erected until the fish is actually touched. If the fish is picked up and returned to the water it will usually not attempt to swim off but will settle on the substratum immediately below the point where it is dropped. The stonefish makes a shallow depression by scooping up sand or mud with its pectoral fins until it is piled up around the sides of the body. The algae which frequently cover the skin of the fish are the same color and the same types as those found in the immediate surroundings. The algae further contribute to camouflaging the fish. Apparently the sticky milky fluid produced by the wartlike protuberances of the skin forms a film on the skin of the fish to which the algae adhere. If the fish is approached while swimming, it will point its dorsal stings toward the intruder. Stonefishes can survive long periods, 24 hours or more, out of water provided their surrounding are kept moist. Stonefishes may attain a length of 38 cm and a weight of 1.5 kg or more. This is undoubtedly one of the most dangerous venomous fishes known.

MORPHOLOGY OF THE VENOM APPARATUS

Morphologically the venom organs of scorpionfishes appear to fall into three distinct general types as exemplified by the genera (1) *Pterois*, (2) *Scorpaena*, (3) *Synanceja*. Although this classification is an arbitrary one and has recognized limitations, it has proven to be useful in dealing with such a morphologically diversified group of fishes. The bulk of the detailed anatomical discussions that follow in these pages will deal with a single species that has been selected to represent each group, viz., *Pterois volitans* (Linnaeus), *Scorpaena guttata* Girard, and *Synanceja horrida* (Linnaeus). A commentary regarding other related species has been added whenever possible. The venom organs of scorpionfishes are among the most highly developed in any of the fishes. A morphological comparison of these basic types appears in Table 1 and Fig. 4. Since the venom organs of most genera and species of scorpionfishes have never been adequately studied, the present classification can only be considered as tentative.

Gross Anatomy

PTEROIS TYPE (Fig. 4):

The venom apparatus of *Pterois volitans* includes 13 dorsal spines, three anal spines, two pelvic spines, their associated venom glands, and their enveloping integumentary sheaths. The sheaths are gaily covered, having bands of white alternating with dark red to black. The light-colored areas are relatively transparent in preserved specimens, and through them can be observed on either side of the spine a slender, elongate, fusiform strand of gray or pinkish tissue lying within each of the glandular grooves. This tissue is the glandular epithelium or venom-producing portion of the sting. Table 2 shows a comparison between the length of each spine and that of its associated strip of glandular tissue. The spines were measured from their articulating base to distal tip. After the glandular tissue was exposed by removing the outer layer of the integumentary sheath, it was measured within the groove from the proximal origin to the distal extremity. Measurements were taken from a fish having a standard length of 120 mm.

The following descriptions are of the spines after removal of the integumentary sheaths, thereby revealing the dentinal structure of the sting.

Dorsal Spines (Fig. 5): This description is based on dorsal spine IV which is representative of the others. The spine is described in the upright position. A lateral view of the fourth dorsal spine of a mature fish will reveal it to be elongate, slender, and almost straight for its entire length except near the proximal and distal portions which are slightly curved craniad. The acute distal tip of the spine is trigonal in outline. The anterolateral-glandular grooves originate superolaterally to the lateral condyles at the base of the dorsal spine. The anterolateral-glandular grooves appear as deep channels extending the entire length of the shaft.

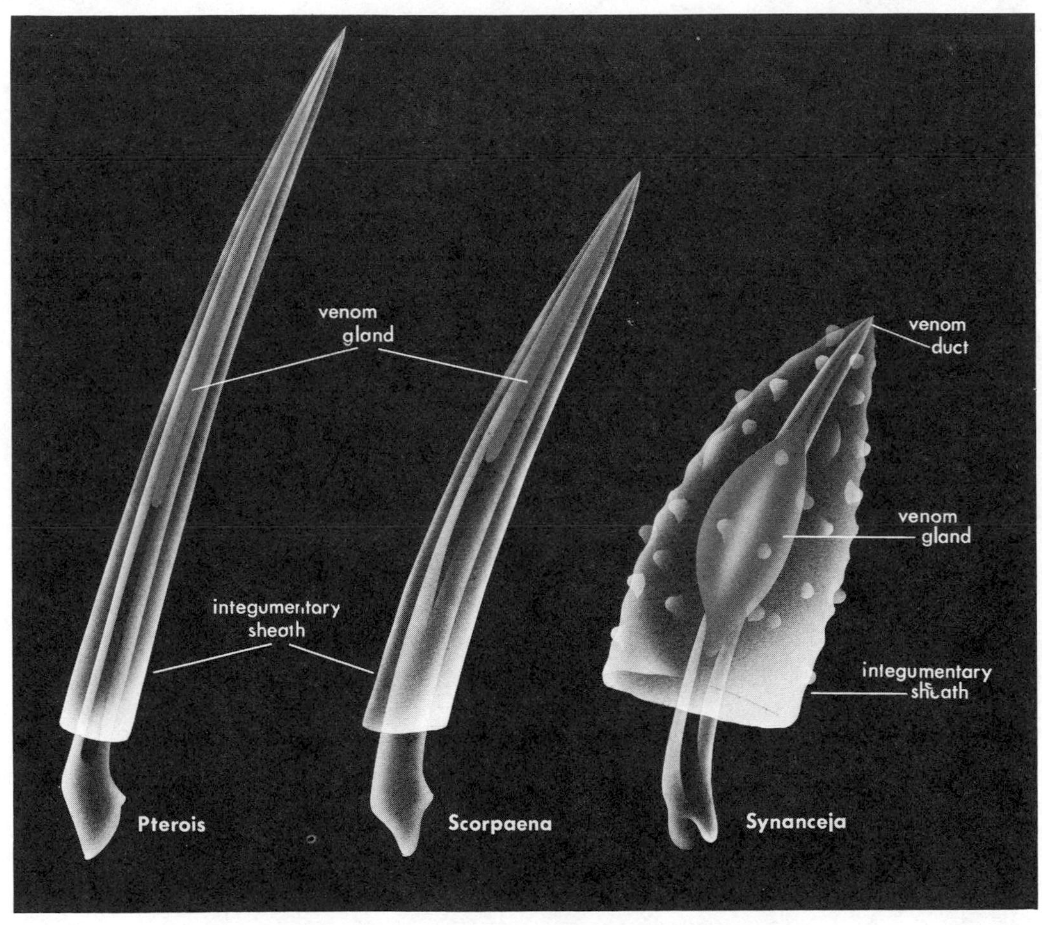

FIGURE 4.—This drawing graphically illustrates some of the more obvious morphological differences in the stings of *Pterois, Scorpaena*, and *Synanceja* which are considered to be typical examples of the three basic types of scorpionfish venom organs. (R. Kreuzinger)

TABLE 1.—*A Comparison of the Venom Organs of* Pterois, Scorpaena, *and* Synanceja *(Halstead and Mitchell, 1963*; see also *Fig. 4)*

STRUCTURE	*PTEROIS*	*SCORPAENA*	*SYNANCEJA*
Fin spines	elongate, slender	moderately long, heavy	short, stout
Integumentary sheath	thin	moderately thick	very thick
Venom glands	small-sized, well-developed	moderate-sized, well-developed	very large, highly developed
Venom duct	not evident or poorly developed	not evident or poorly developed	well developed

An anterior view of the spine reveals that approximately the basal one-sixth of the anteromedian ridge is furrowed by a shallow anteromedian groove which originates at the median foramen at the base of the spine. The anteromedian ridge extends throughout the remainder of the length of the shaft and finally terminates in a sharp point at the distal tip.

A posterior view of the spine reveals that the posterolateral ridges are separated by a groove, the posteromedian groove, of variable depth throughout the entire length of the spine.

Anal Spines: This description is based on the third anal spine which is morphologically similar to the others. The spine is described in the extended anatomical position. A lateral view of the anal spine shows it as comparatively short, stout, and slightly curved craniad, terminating in an acute trigonal tip. The anterolateral-glandular grooves originate below the lateral condyles of the spine and are narrow and shallow at the point of origin, but broaden and deepen as they approach the distal tip of the spine.

An anterior view reveals that the anteromedian groove is absent in the third anal spine; instead, the anteromedian ridge originates from the base of the left lateral condyle, thus producing an elongate depression on the right side inferior to the base of the right lateral condyle. The second anal spine differs slightly in that the anteromedian ridge originates largely from the base

TABLE 2.—*A Comparison Between the Length of the Spine of* Pterois volitans *and the Length of its Glandular Tissue (Halstead* et al., *1955b)*

Dorsal spine	1	2	3	4	5	6	7	8	9	10	11	12	13
Spine (mm)	23	29	39	42	45	49	46	48	45	40	12	11	15
Glandular tissue (mm)	16	18	26	31	30	30	30	31	30	29	5	5	5
Anal spine	1	2	3										
Spine (mm)	8	15	20										
Glandular tissue (mm)	4	8	8										
Ventral spine	Rt.	Lt.											
Spine (mm)	21	21											
Glandular tissue (mm)	11	11											

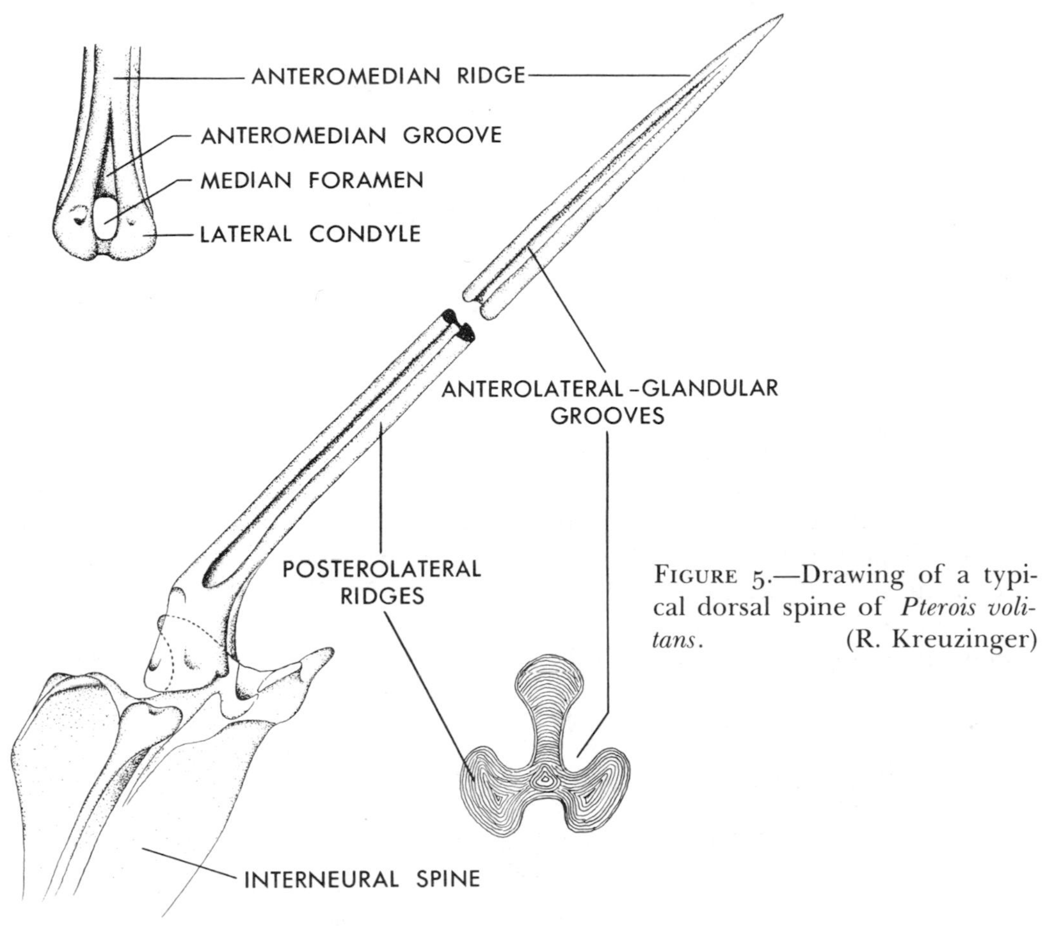

FIGURE 5.—Drawing of a typical dorsal spine of *Pterois volitans*. (R. Kreuzinger)

of the right lateral condyle. The origin of the anteromedian ridge of the first anal spine is equally distributed between the bases of the two lateral condyles, and the anteromedian groove is greatly reduced.

The posterior aspect of the spine is marked by a broad, shallow groove which extends from the median foramen to the distal tip.

Ventral Spines: This description is based on the right ventral spine described in the extended anatomical position. A superior view of the spine reveals that it is short, stout, and slightly curved craniad, terminating in an acute trigonal tip. The superior and inferior glandular grooves, which are the equivalent of the anterolateral-glandular grooves of the dorsal and anal spines, originate at the base of their respective condyles, and extend as deep grooves for the entire length of the spine. The anteromedian ridge extends from the median notch at the base of the spine to the distal tip. The posterior aspect of the spine is slightly ridged.

SCORPAENA TYPE (Fig. 4):

The venom apparatus of *Scorpaena guttata* includes 12 dorsal spines, three anal spines, two pelvic spines, their associated venom glands, and their

TABLE 3.—*A comparison Between the Length of the Spine of* Scorpaena guttata *and the Length of its Glandular Tissue (Halstead* et al.,*1955a)*

Dorsal spine	1	2	3	4	5	6	7	8	9	10	11	12
Spine (mm)	25	35	46	48	47	45	41	39	34	29	23	31
Glandular tissue (mm)	15	20	26	29	28	21	18	18	15	13	9	12
Anal spine	1	2	3									
Spine (mm)	24	41	35									
Glandular tissue (mm)	17	23	17									
Ventral spine	Rt.	Lt.										
Spine (mm)	34	34										
Glandular tissue (mm)	15	20										

enveloping integumentary sheaths. The dorsal spines, like the anal spines, are connected by deeply incised interspinous membranes. If the integumentary sheath is removed, a slender, elongate, fusiform strand of gray or pinkish tissue can be observed lying within the glandular grooves on either side of the spine. The tissue within these grooves is the glandular epithelium or venom-producing portion of the sting. Table 3 shows a comparison between the length of each spine and the length of its associated strip of glan-

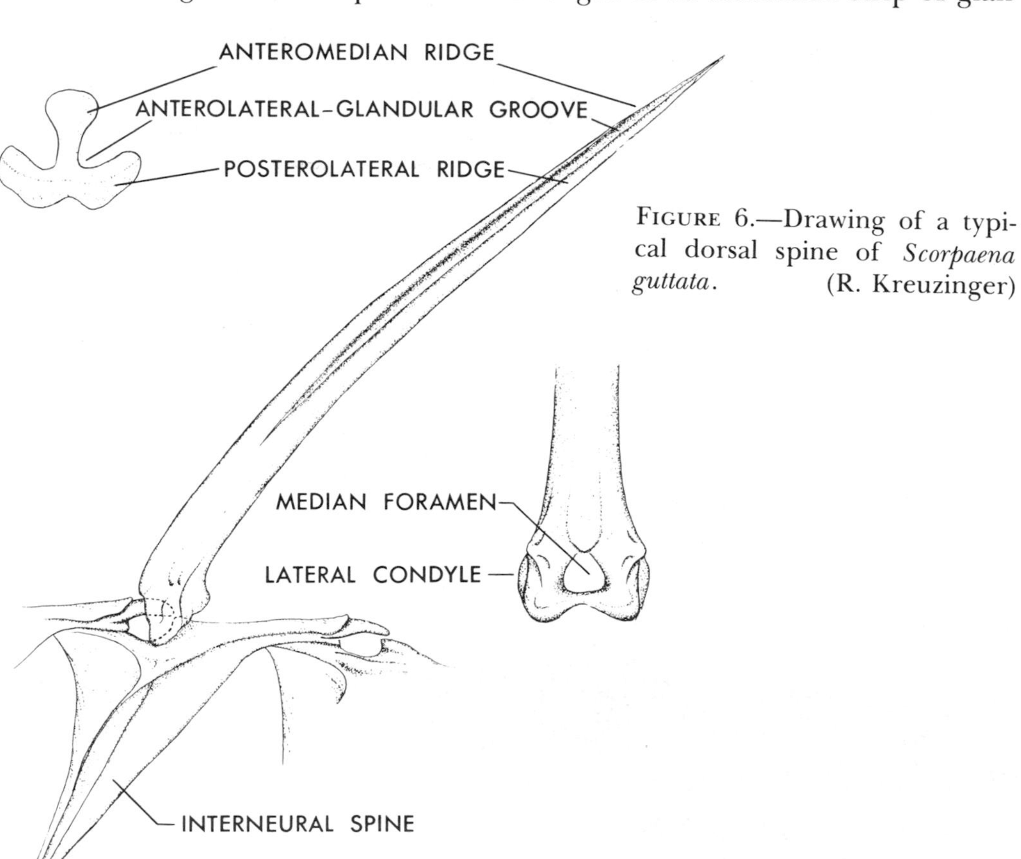

FIGURE 6.—Drawing of a typical dorsal spine of *Scorpaena guttata*. (R. Kreuzinger)

dular tissue. The spines were measured from their articulating base to the distal tip. The glandular tissue was measured within the groove from the proximal origin to the distal tip. Measurements were taken from a single fish having a standard length of 255 mm.

The following descriptions of the spines were made after removing the integumentary sheath, thereby revealing the dentinal structure of the stings.

Dorsal Spines (Fig. 6): This description is based on dorsal spine IV which is representative of the others. The spine is described in the upright position. A lateral view of the fourth dorsal spine of a mature fish shows it as elongate, slender, and slightly arched craniad. The acute tip of the spine is trigonal in outline. The anterolateral-glandular grooves originate in the middle third of the spine as shallow grooves which become progressively deeper and broader as they approach the distal end of the spine. An anteromedian ridge extends throughout the length of the spine. At the base of the spine between the lateral condyles is a median foramen through which the ringlike process of the interneural spine passes.

A posterior view of the spine shows that the posterolateral ridges are separated by a groove, the posteromedian groove, which is fairly pronounced near the base, becoming progressively more shallow and finally disappearing as the distal third of the spine is approached.

Anal Spines: This description is based on anal spine II which is morphologically similar to the others. The spine is described in the extended anatomical position. A lateral view of the second anal spine reveals that it is relatively long, stout, and slightly curved craniad, terminating in an acute trigonal tip. The anterolateral-glandular grooves originate just distal to the lateral condyles of the spine. The grooves are narrow and relatively shallow in the basal one-half of the spine, but broaden and deepen as they approach the distal tip. As in the dorsal spine, a small median foramen is present through which the process of the interhaemal spine passes. The posteromedian groove is most pronounced in the proximal half of the spine, becoming less distinct as the distal tip is approached.

Pelvic Spines: This description is based on the left pelvic spine described in the extended anatomical position. A superior view of the spine reveals that it is of moderate length, stout, and slightly curved craniad in the distal third. The superior and inferior glandular grooves, which are the equivalent of the anterolateral-glandular grooves of the dorsal and anal spines, originate at the base of their respective condyles. The grooves are relatively shallow in the proximal half of the spine but become deeper and broader distally. The anteromedian ridge, which is rather acute, extends for the entire length of the spine. A definite posteromedian groove is present. The spine is otherwise similar to those previously described for this species.

Opercular Spines: Examination of the opercular spines fails to reveal any evidence of glandular tissue. It will be observed under microscopic examination, that the spines are rounded and without glandular grooves (Fig. 18).

SYNANCEJA TYPE (Fig. 4):

The venom apparatus of *Synanceja horrida* includes 13 dorsal spines, three anal spines, two pelvic spines, their associated venom glands, and their enveloping integumentary sheaths. The integumentary sheaths are dusky or mottled with deep brown, and are fleshy and tuberculate. The distal acute tips of the dorsal spines can be observed protruding slightly through their sheaths. However, the anal and pelvic spines are completely embedded within their thick sheaths and require dissection before they can be observed.

If the dorsal fin is erected, the first three dorsal spines are in an almost vertical position, but the remainder of the dorsal spines appear to be restricted in their movement by the interspinous membranes and lie at an inclined plane. Removal of the integumentary sheaths from the dorsal spines reveals a pair of remarkably large, grayish-brown, fusiform glands, one on either side of each spine. The glands of the first three dorsal spines are located about the middle third of the shaft of the spine; however, the glands on the remainder of the dorsal spines are more distally situated.

TABLE 4.—*A comparison Between the Length of the Spine of* Synanceja horrida *and the Length of Its Glandular Tissue (Halstead* et al., *1956)*

Dorsal spine	1	2	3	4	5	6	7	8	9	10	11	12	13
Spine (mm)	36.4	46.4	44.1	39.2	37	39.8	39.8	40.8	40.7	40.9	40.2	39.5	42.3
Glandular tissue (mm)	12.8	15.5	13.7	10.0	9.0	11.7	10.5	11.8	11.4	9.5	10.5	10.5	10.7

Anal spine	1	2	3
Spine (mm)	12.5	18.3	21.5
Glandular tissue (mm)	3.8	3.4	3.1

Pelvic spine	Rt.	Lt.
Spine (mm)	32.0	29.0
Glandular tissue (mm)	3.4	4.0

The glands of the anal spines are similar in their morphology but smaller in size. The gland of the first anal spine is situated at about the middle third of the spine, but the other two glands are more distally located.

On the superior and inferior aspect of each pelvic spine is located a minute gland, similar in form, but very much smaller than the other fin glands.

All the glands appear to be attached to the spines by relatively tough strands of connective tissue. The proximal ends of the glands terminate as a thin, glistening band of tough, fibrous, connective tissue. The distal ends of the glands terminate in ductlike structures lying within the glandular grooves. These ducts extend to the distal tips of the spines. When pressure was placed on the glands it was observed that fluid could be made to pass through the ducts to the distal tips of the spines.

The relative size of the dorsal, anal, and pelvic glands, and their relationship to the length of their corresponding spines are given in Table 4. Each

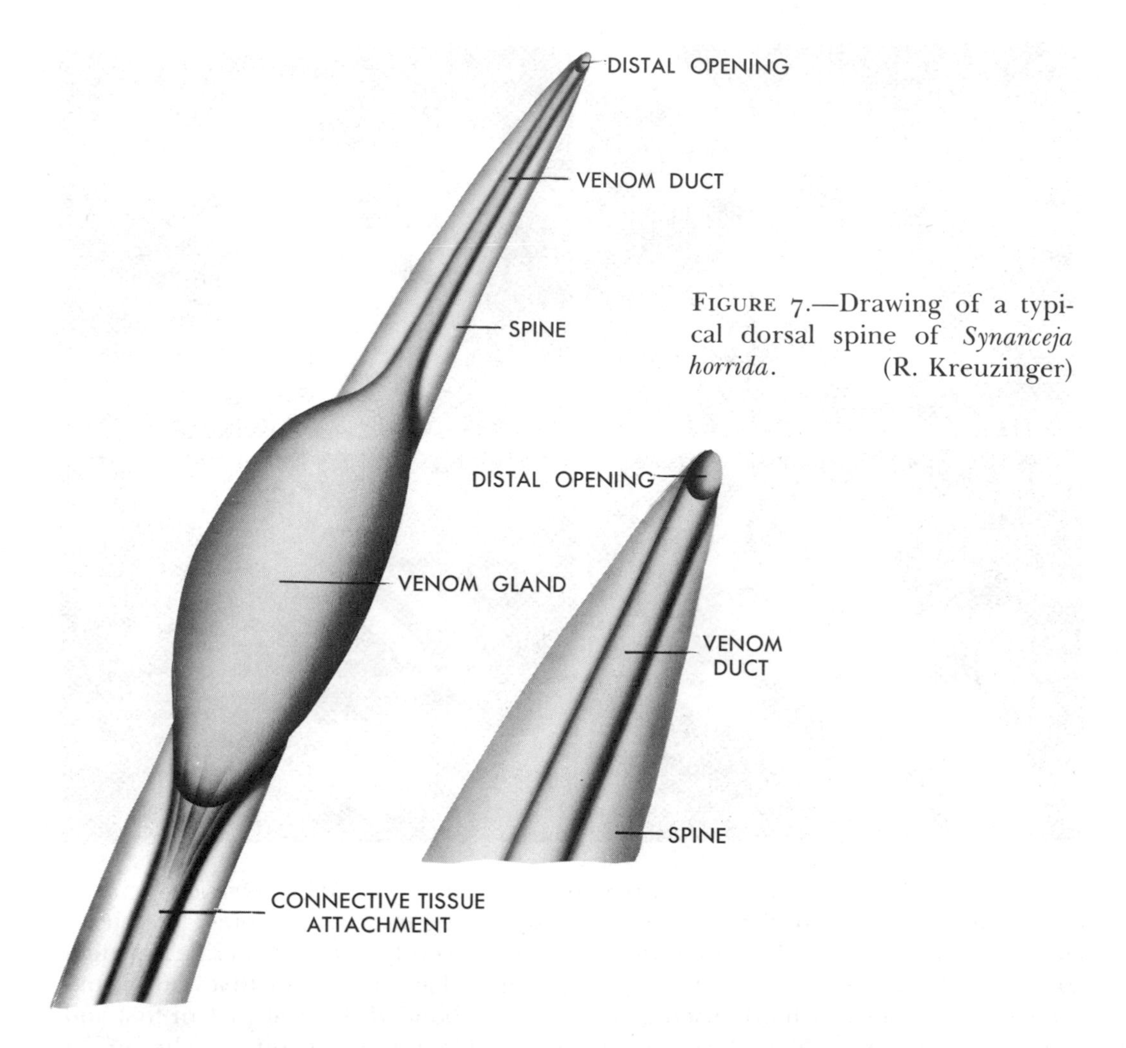

FIGURE 7.—Drawing of a typical dorsal spine of *Synanceja horrida*. (R. Kreuzinger)

spine was measured from its articulating base to its distal tip. In order to measure the venom glands it was necessary to remove their integumentary sheaths. The glands were measured while attached to the glandular grooves of the spine. The width of the spine was measured at the level of the midpoint of the attached venom gland. The width of the venom gland was taken at its widest point. Measurements were taken from a specimen of *S. horrida* having a standard length of 265 mm.

The following descriptions of the spines were made by removing the integumentary sheaths, thereby revealing the dentinal structure of the stings.

Dorsal Spines (Figs. 7; 8; 9): This description is based on dorsal spine II which is representative of the others. The spine is described in the upright anatomical position. A lateral view of the second dorsal spine of a mature fish will reveal it to be elongate, moderately stout, and almost straight for its entire length except near the proximal third, at which point it curves craniad. The acute distal tip of the spine is trigonal in outline. From an anterior view

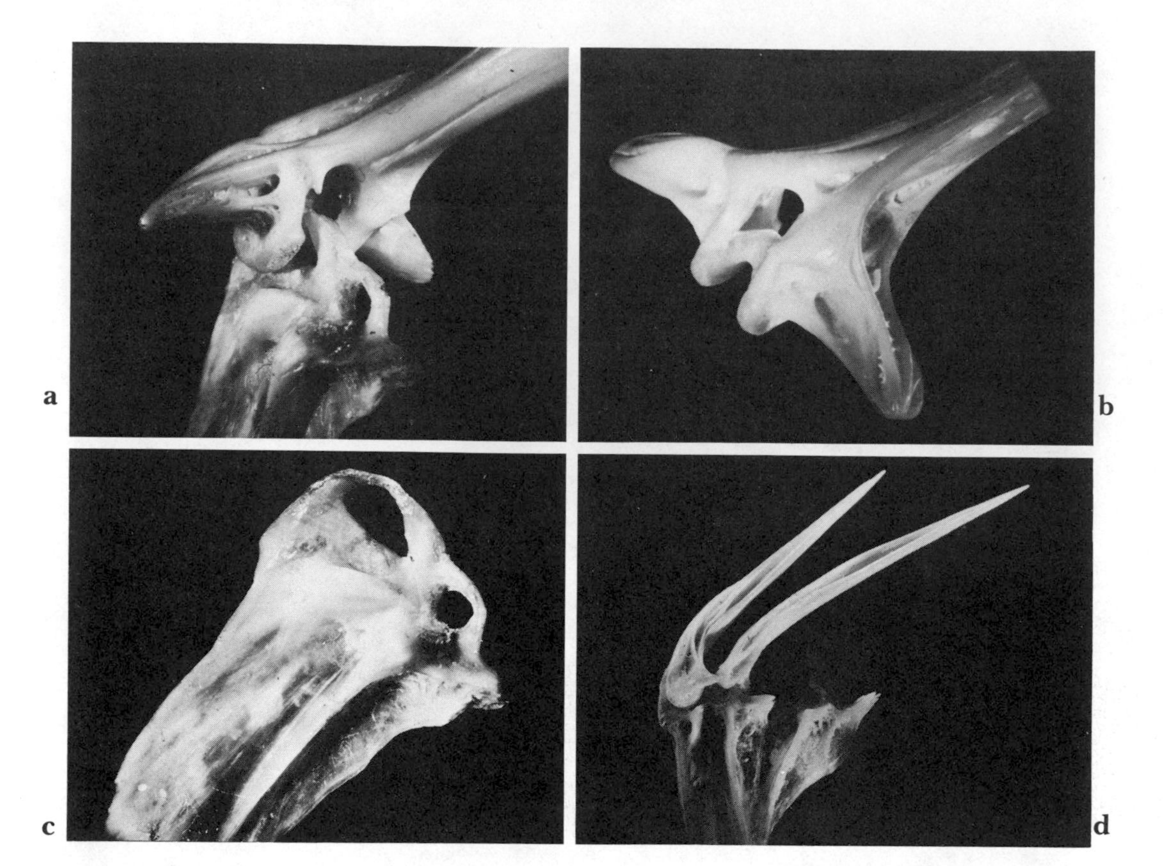

FIGURE 8.—a. Posterolateral view of base of first dorsal spine of *Synanceja horrida* in articulation with the annular process of the supporting interneural spine. b. Anterolateral view of base of first dorsal spine. Note lateral condyles inferolateral to median foramen through which the annular ring passes. c. Lateral view of first and second interneural spines which are fused to form a single bone. d. Lateral view of first and second dorsal spines. Note deep grooves on sides along shaft of spines. The venom glands normally lie along these anterolateral-glandular grooves. Length of first spine 36.4 mm. (From Halstead, Chitwood, and Modglin)

it will be seen that the base of the spine is comprised of two articular lateral condyles, one on either side, which are separated by a median foramen. Passing through the median foramen, in an anterior-posterior direction, is the annular or ringlike process of the interneural spine with which the dorsal spine articulates. Originating distal to the lateral condyles of the dorsal spine and extending throughout its entire length, one on either side, are the two anterolateral-glandular grooves. The basal third of the glandular grooves is relatively shallow and narrow, but the middle third deepens and broadens abruptly and then gradually narrows in the distal third. The main body of the venom glands is attached to the middle third of the dorsal spine.

The proximal portion of the venom gland lies within and is attached to the basal third of the anterolateral-glandular grooves by strands of connective

FIGURE 9.—Photograph of anterior view of first dorsal sting of *Synanceja horrida*. The integumentary sheath has been peeled back to show the large venom glands situated on either side of spine shaft. Length of spine 36.4. (R. Kreuzinger)

tissue. The duct of the gland lies within the protection of the distal third of the groove. Separating the two anterolateral-glandular grooves is a heavy, flat-to-rounded ridge, the anteromedian ridge, which arises superior to the median foramen, extends throughout the entire length of the spine, and terminates as a sharp point in the distal tip. The posterior aspect of the spine is rounded.

Anal Spines (Fig. 10a): These spines are described in the extended anatomical position. A lateral view of the anal spines reveals that they are relatively short and stout. The first spine is straight, but the second and third curve slightly backward near the distal tip. The first spine is anteriorly and posteriorly flattened, whereas the second and third spines tend to be rounded. The distal tips of the anal spines are acute but not as sharp as those of the dorsal spines. The lateral condyles, median foramen, and anteromedian ridge are similar in their morphology to that described for the dorsal spines. The anterolateral-glandular grooves originate distal to the lateral condyles of the spine. The grooves are narrower and shallower than those observed in the dorsal spine. With the few foregoing exceptions, the anal spines are similar to those of the dorsal fin.

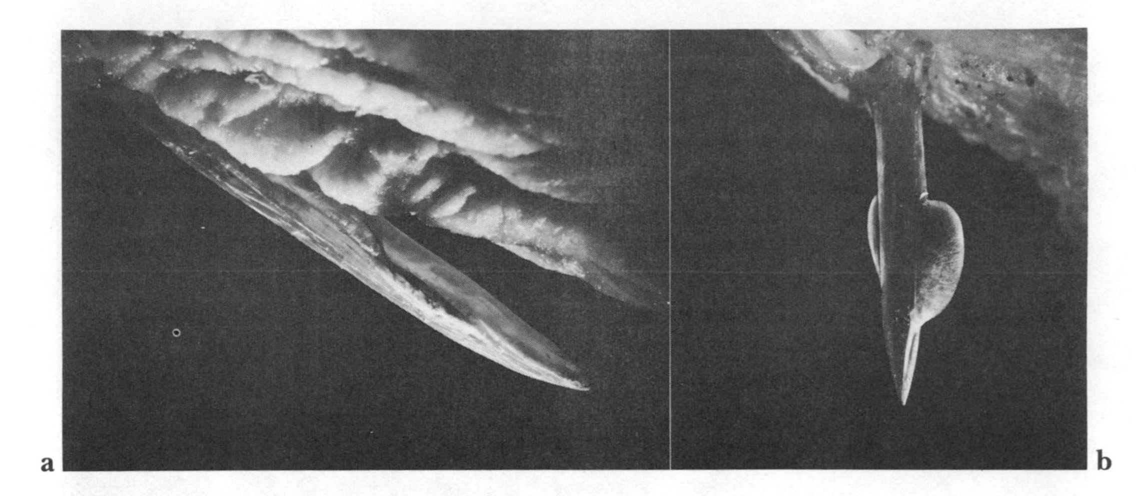

FIGURE 10.—a. Anterior view of first anal sting of *Synanceja horrida*. The integumentary sheath has been removed to show venom glands situated on either side of spine shaft. Length of spine 12.5 mm. b. Anterior view of right pelvic sting. The integumentary sheath has been removed to show venom glands situated on either side of spine shaft. Note that the glands are considerably smaller than those found in any of the other stings. Length of spine 29 mm.

(From Halstead, Chitwood, and Modglin)

Pelvic Spines (Fig. 10b): This description is based on the right pelvic spine described in the extended anatomical position. A superior view of the spine reveals that it is moderate in length, stout, and slightly curved craniad, terminating in an acute trigonal tip. The superior and inferior glandular grooves, which are the equivalent of the anterolateral-glandular grooves of the dorsal and anal spines, originate at about the middle of the length of the spine and continue as narrow, shallow grooves to the acute distal tip. The articular base of the spine is similar to that previously described for the dorsal and anal spines, with the exception that the superior condyle terminates posteriorly in a hooklike process. The anal spines are similar to those of the dorsal and anal fins.

Microscopic Anatomy

PTEROIS TYPE:

Since the microscopic anatomy is basically the same, this description will deal largely with the dorsal stings of *Pterois volitans* with brief reference to certain exceptions which are found in the anal and ventral ones.

Dorsal Stings: A dorsal sting (Figs. 11a, b; 12) in cross section is essentially an inverted T-shaped dentinal structure completely invested by a layer of integument. The spine is composed of a solid dentinelike substance marked by a series of numerous, concentric, fine rings. The peripheral layer of the spine consists of a very thin layer of vitreodentine. Sections taken from the basal and middle thirds of the spine reveal that posteriorly the posterolateral

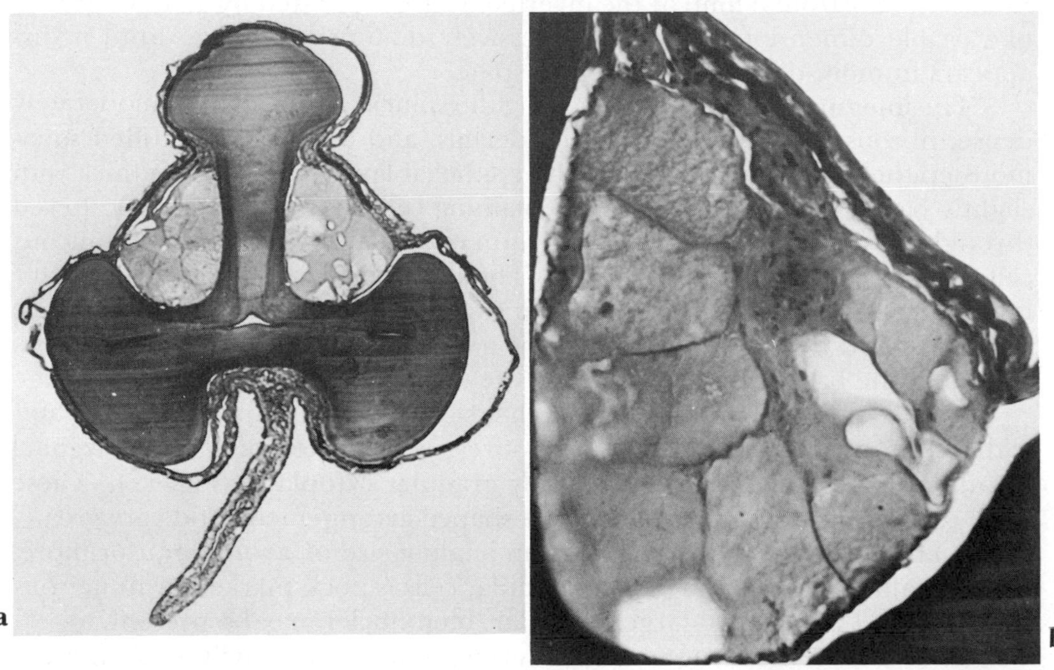

FIGURE 11.—a. (Left) Photomicrograph of a cross section of the middle third of the 7th dorsal sting of *Pterois volitans*. Note that the integumentary sheath completely surrounds the sting. The venom glands appear as heart-shaped clusters of large polygonal glandular cells on either side of the anteromedian ridge. b.(Right) Enlargement of the glandular triangle of section shown in Fig. 11a. × 300. Hematoxylin and eosin stains. × 100. Preparations by B. W. Halstead and M. J. Chitwood. (E. Rich)

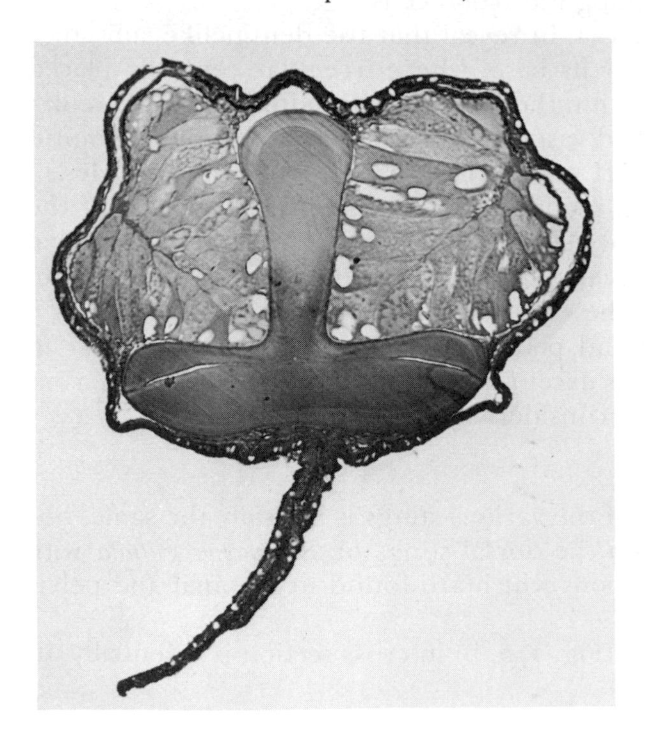

FIGURE 12.—Photomicrograph of a cross section of the distal third of the 9th dorsal sting of *Pterois volitans*. Note the increase in the glandular tissue at this level as compared with that seen in Fig. 11a. The tail-like posterior extension of the integumentary sheath is the interspinous membrane. × 80. Hematoxylin and eosin stains. Preparations by B. W. Halstead and M. J. Chitwood. (E. Rich)

ridges, the horizontal limb of the inverted T, are separated by a broad groove of variable dimensions, which progressively diminishes in size until it disappears in more distal sections of the spine.

The integumentary sheath investing the spine is a thin layer of moderately dense, fibrous, connective tissue, the dermis, and a layer of stratified squamous epithelium, the epidermis. The epithelial layer is 4 to 6 cells thick with slightly basophilic basal cells. The remaining cells are acidophilic. Scattered throughout this epithelial layer are numerous large secretory or mucous cells having a clear cytoplasmic area. The peripheral cytoplasm of the outer layer of cells forms a distinct pink line suggestive of a striated border. The epidermis and dermis are separated by a continuous thick layer of pigment adjoining the basement membrane of the epidermis.

Located in the dermal layer, on either side of the anteromedian ridge and within the anterolateral-glandular grooves, is a cluster of large polygonal glandular cells with pinkish-gray, finely granular cytoplasm (Fig. 12). These large cells tend to have a pinnate, heart-shaped arrangement, and vary greatly in size and morphology. Some of them attain a size of $270\mu \times 75\mu$ or more. Occasionally the cytoplasm of the glandular cells is pock-marked by numerous vacuoles, and one or two large, vesicular, blue nuclei may be present.

Also scattered throughout the dermis, but most heavily concentrated in the glandular triangles and the posteromedian groove, are cellular clumps of variable size having acinarlike arrangements and circumscribed by bands of collagenous connective tissue. The cells within these clumps are polygonal with colorless or pink cytoplasm, and have centrally placed, dark blue, spherical nuclei. Interspersed throughout the dermis are numerous vascular channels of variable size containing red blood cells.

Longitudinal sections (Fig. 13a, b) reveal that the dentinelike substance of the spine is solid except near its base, where irregular centrally placed marrow cavities are present. Within these cavities are adipose tissue, areolar connective tissue, and blood-filled vascular channels. In a complete longitudinal section of dorsal spine VIII, measuring 56 mm from the condyles to the distal tip, it was found that glandular tissue extended throughout the distal 35 mm of the anterolateral-glandular grooves. The largest polygonal glandular cells in one longitudinal section measured 120μ x 75μ. The remaining histological details are the same as previously described.

Cross sections of the anal and pelvic stings (Fig. 14a, b) are similar to those obtained from dorsal stings except that both the dermis and epidermis of the integumentary sheath are considerably thicker.

SCORPAENA TYPE:

Since the microscopic anatomy of the various stings is basically the same, this description will deal largely with the dorsal stings of *Scorpaena guttata* with brief reference to certain exceptions which are found in the anal and pelvic stings.

Dorsal Stings: A dorsal sting (Fig. 15a, b) in cross section is essentially an

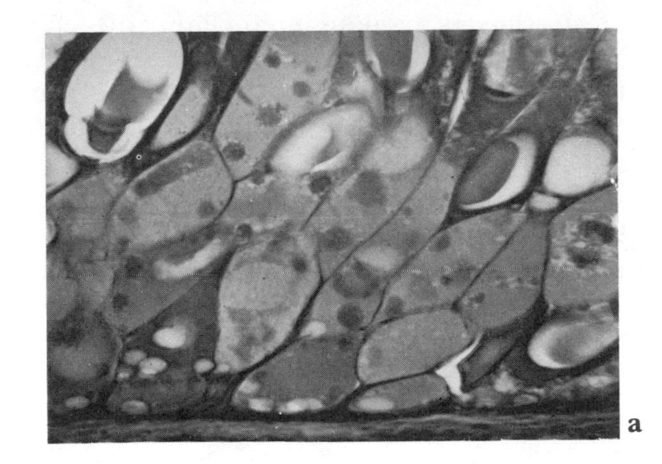

Figure 13.—a. (Top) Longitudinal section of the glandular tissue in the anterolateral-glandular groove at the middle third of the 8th dorsal sting of *Pterois volitans.* × 175. b. (Bottom) Longitudinal section of the glandular tissue of the same sting as above taken at the level of the distal tip. Note the segments of the integumentary sheath along the upper margin of the sting. × 60. Hematoxylin and eosin stains. Preparations by B. W. Halstead and M. J. Chitwood. (E. Rich)

inverted T-shaped dentinal structure completely invested by a layer of integument. The spine is composed of a solid dentinelike substance marked by a series of numerous, concentric, fine rings. The peripheral layer of the spine consists of a thin layer of vitreodentine. Sections taken from the basal and middle thirds of the spine reveal that posteriorly the posterolateral ridges, the horizontal limbs of the inverted T, are separated by a distinct posteromedian groove, which progressively diminishes in size until it completely disappears in more distal sections of the spine.

The integumentary sheath investing the spine is a comparatively thick layer of moderately dense fibrous connective tissue, the dermis, and a layer of stratified squamous epithelium, the epidermis. The epithelial layer is 8 to 10 cells thick with slighly basophilic basal cells. The remaining cells tend to be acidophilic. Scattered throughout this epithelial layer are numerous large secretory or mucous cells having a clear cytoplasmic area. The peripheral cytoplasm of the outer layer of epithelium forms a distinct striated border which in some areas appears remarkably thick. The epidermis and dermis are separated by a continuous, relatively thick layer of pigment cells adjoining the basement membrane of the epidermis.

Located in the dermal layer, on either side of the anteromedian ridge and within the anterolateral-glandular grooves, is a cardiform cluster of large polygonal glandular cells with pink, finely granular cytoplasm (Fig. 15a).

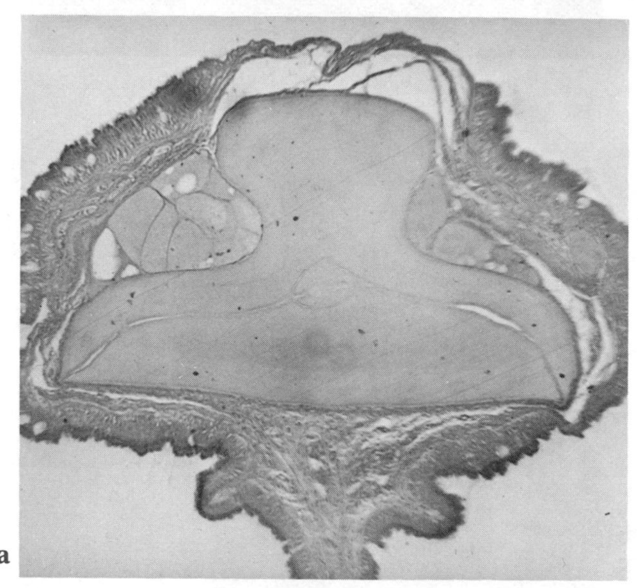

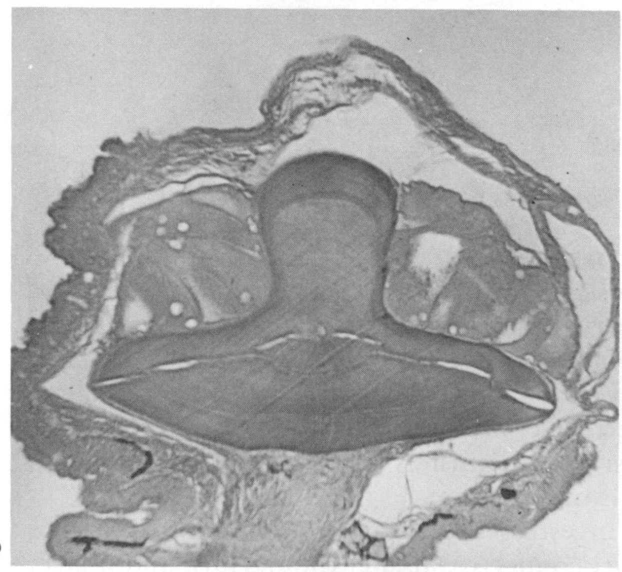

FIGURE 14.—a. (Left) Cross section of the middle third of the 3rd anal sting of *Pterois volitans*. × 100. b. (Bottom) Cross section of the middle third of left pelvic sting of *Pterois volitans*. × 100. Hematoxylin and eosin stains. Preparations by B. W. Halstead and M. J. Chitwood. (E. Rich)

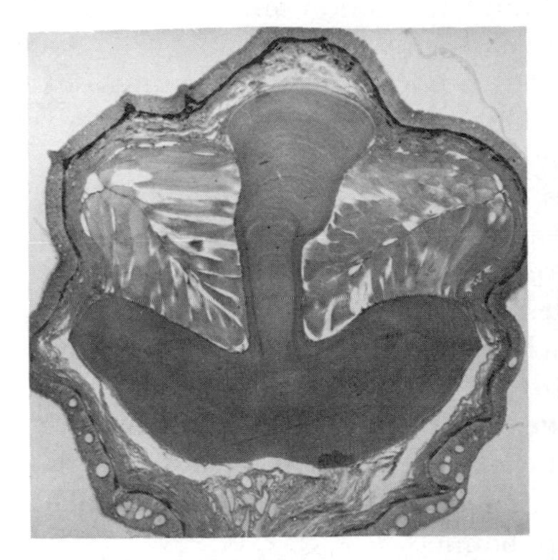

a

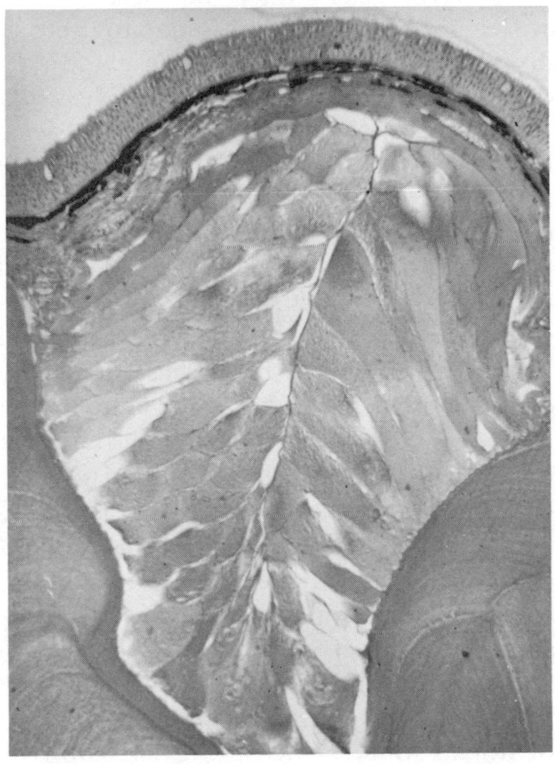

b

FIGURE 15.—Photomicrographs of cross sections of the dorsal stings of *Scorpaena guttata*. a. (Above, left) Cross section of the middle third of 3rd dorsal sting. Note the pinnate heart-shaped arrangement of the large glandular cells. × 30. b. (Above, right) Enlargement of the glandular triangle of section shown in preceding figure. × 100. c. (Below, left) Cross section of the middle third of 8th dorsal sting. × 30. d. (Below, right) Cross section of the distal third of 10th dorsal sting. × 30. Hematoxylin and eosin stains. Preparations by B. W. Halstead and M. J. Chitwood. (E. Rich)

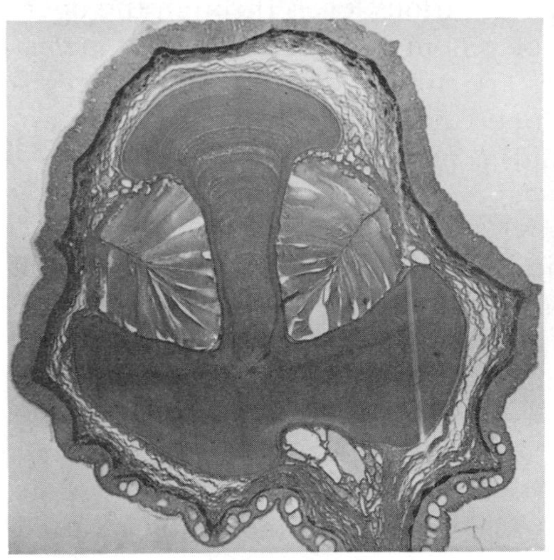

c

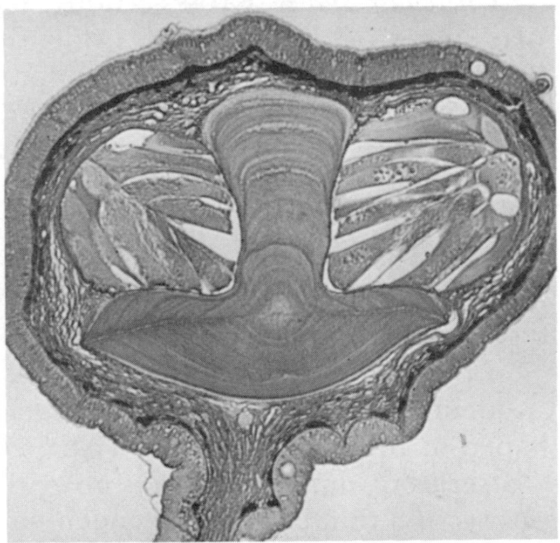

d

These large cells tend to be arranged in a pinnate fashion and vary greatly in size and morphology, some attaining a size of $275\mu \times 105\mu$ or more. Occasionally the cytoplasm of the glandular cells is pock-marked by numerous vacuoles, and one or more large, vesicular, blue nuclei may be present.

Also scattered throughout the dermis, but most heavily concentrated in the vicinity of the glandular triangles and the posteromedian groove, are cellular clumps of variable size having acinarlike arrangements and circumscribed by bands of collagenous connective tissue. The cells within these clumps are polygonal with colorless or pink cytoplasm and have centrally placed, dark blue, spherical nuclei. Throughout the dermis are numerous vascular channels of variable size containing red blood cells.

Longitudinal sections (Fig. 16a, b) reveal that the dentinelike substance of the spine is solid except near the base of the spine where irregular, central marrow cavities are present. Within these cavities are adipose tissue, areolar connective tissue, and blood-filled vascular channels.

It will be observed that the glandular cells are divided into two principal zones. The cells within the inner zone appear as large columnar secretory cells resting upon the spine and are arranged perpendicularly to the axis of the spine upon which they rest. The outer zone is a series of layers of similar columnar cells cut in cross section and apparently arranged at right angles to the inner zone. The largest polygonal glandular cells in one longitudinal section measured 285μ x 105μ. In a complete longitudinal section of dorsal spine I, measuring 26.6 mm from the condyles to the distal tip, it was found that glandular tissue extends throughout the distal 12 mm of the anterolateral-glandular grooves. The remaining histological details are similar to those described above for the dorsal sting.

Anal and Pelvic Stings: Cross sections of the anal and pelvic stings (Fig. 17a, b) are similar to those obtained from dorsal stings with the exception that both the dermis and epidermis of the integumentary sheath are considerably thicker.

Opercular Spines: Cross sections taken at various levels through the opercular spines fail to reveal any evidence of venom glands in *Scorpaena guttata* (Fig. 18). However, its opercular spines are surrounded by a thick integumentary sheath. Whether or not the opercular spines of this species are representative of all species of scorpaenids is not known.

SYNANCEJA TYPE:

Since the microscopic anatomy is basically the same, this description will deal largely with the dorsal stings of *Synanceja horrida* and with brief reference to certain exceptions found in the anal and ventral stings.

Dorsal Stings (Figs. 19a, b, c, d; 20a, b): Cross sections of the sting show a vasodentinal spine covered by an integument. The spine is somewhat goblet-shaped at the base but more T-shaped distally. The solid spine is marked by a series of numerous, concentric, fine rings. The periphery of the spine consists of a thin layer of vitreodentine. The horizontal limb of the inverted T-shaped spine is slightly rounded and without a groove.

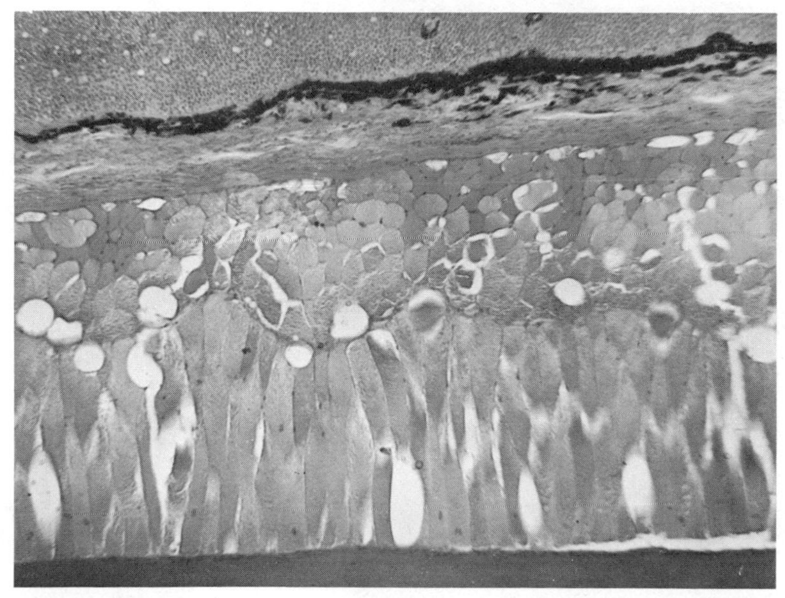

Figure 16a.—Photomicrographs of longitudinal sections of the first dorsal sting of *Scorpaena guttata*. Longitudinal section through the middle third of first dorsal sting. × 100. (E. Rich)

Figure 16b.—Longitudinal section through the distal tip of the first dorsal sting. × 30. Hematoxylin and eosin stains. Preparations by B. W. Halstead and M. J. Chitwood. (E. Rich)

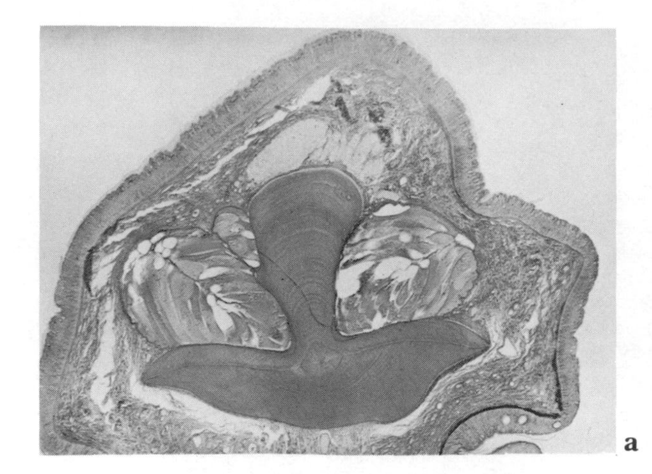

FIGURE 17a.—Cross section through the distal third of the 3rd anal sting of *Scorpaena guttata*. × 60.

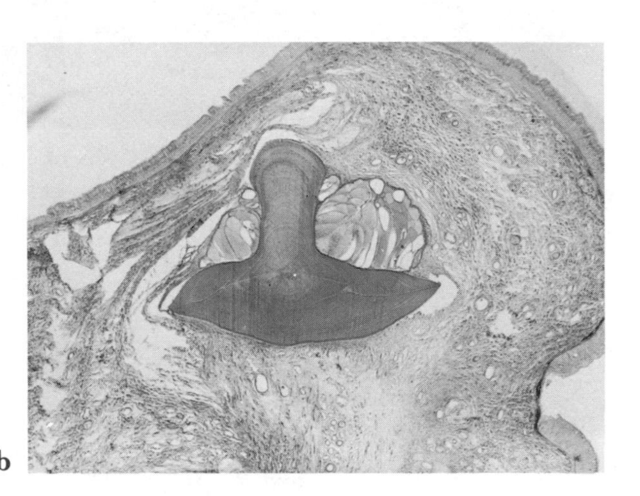

FIGURE 17b.—Cross section through the distal third of the left pelvic sting of *Scorpaena guttata* × 60. Hematoxylin and eosin stains. Preparations by B. W. Halstead and M. J. Chitwood. (E. Rich)

Packed between the vertical and horizontal limbs of the T, in distal sections of the spine, is a circular mass of round and oval-shaped, closely packed, variable-sized, slightly granular, tan bodies. They have definite outlines suggesting cell membranes. Although nuclei are not present, there are "ghostlike" membranes suggesting nuclear membranes. These bodies are enclosed within this area by a thin sheet of dense fibrous connective tissue which extends between the distal tips of the vertical and horizontal limbs of the T-shaped dentinal structure, thus forming a closed duct. Similar areas at more proximal levels of the sting have expanded in all directions, resulting in large areas of these bodies. These clusters of bodies are the areas of venom production. Many of the bodies are filled with brown granules of about 15μ in their greatest dimension. Scattered throughout the bodies are irregular patches of yellowish, homogeneous, amorphous secretion. Most of the bodies at lower levels range in color from red to tan with hematoxylin and triosin stains. Peripherally, these bodies are less regular and frequently occur in clumps that are traversed by many fine-cell, membrane-like lines that give the clump a mosaic appearance.

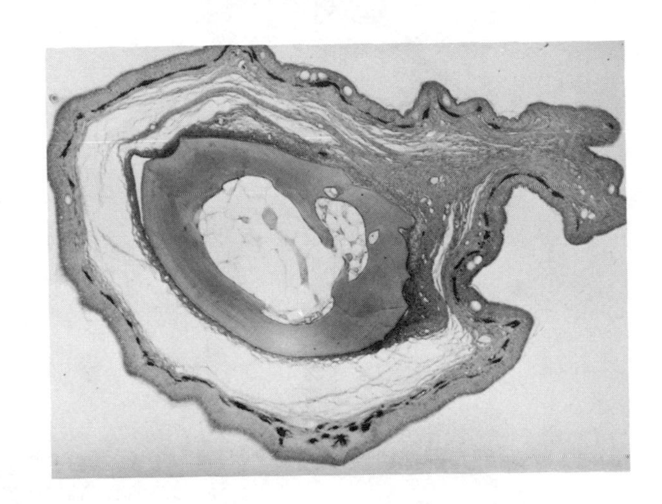

FIGURE 18.—Cross section through the distal third of an opercular spine of *Scorpaena guttata*. There is no evidence of glandular tissue, but the spine is surrounded by a thick integumentary sheath. None of the opercular spines examined in this species have shown any evidence of venom glands. Whether or not the opercular spines of this species are representative of all species of scorpaenids is not known. Hematoxylin and eosin stains. Preparations by B. W. Halstead and M. J. Chitwood. (E. Rich)

The preceding histological description is based on the work of Halstead, Chitwood, and Modglin (1956) who were limited in their studies to formalin-preserved specimens. More recently Endean (1961) has described the microscopic anatomy of *Synanceja* working with fresh material that had been preserved in Bouin's fluid, Zenker's fluid, and 5 percent formalin in sea-water. Endean used hematoxylin and eosin and Mallory's dye mixture. He described the venom gland as follows:

A layer of densely fibrous connective tissue, 12-15μ in thickness, formed the boundary of each gland. The fibrillar material was collagenous and it enclosed a mosaic of gland cells and granule-filled compartments. Often, the cells were subcircular in outline and such cells had an average diameter of 36μ with a range from 28-52μ. Frequently, however, when the cells were grouped in clusters, they each had a pentagonal or hexagonal outline, owing to adjacent cells abutting on one another.

Eosin dyed the cytoplasm of the cells pink to red. Mallory's dye mixture dyed them purplish to reddish. Presumably, the behaviour of the cells to these dyes is influenced by the stage each had reached in its secretory cycle. A positive PAS reaction was given by the cytoplasm and it exhibited β-metachromasy with toluidine blue. Much of this metachromasy was due to RNA. With Sudan black B the cells gave a faint reaction for lipid. When spherical, the cell nuclei averaged 6μ in diameter. However, some nuclei were ellipsoidal and in some of the smaller cells nuclei undergoing division were observed. Nuclei were dyed faintly by haematoxylin and scarcely at all by Mallory's dye mixture. They were dyed by orcein and they were clearly revealed by the Feulgen technique.

The granule-filled compartments (Fig. 20a, b) were roughly circular in outline and they had an average diameter of 65μ. The walls of the compartments appeared to possess very fine branching fibres and minute pores but the lodging of large numbers of granules against these walls made observations of their structure difficult. Frequently, the walls of adjacent compartments had fused, possibly owing to the action of the fixatives used, and it was thought, initially, that the granules were enclosed in a

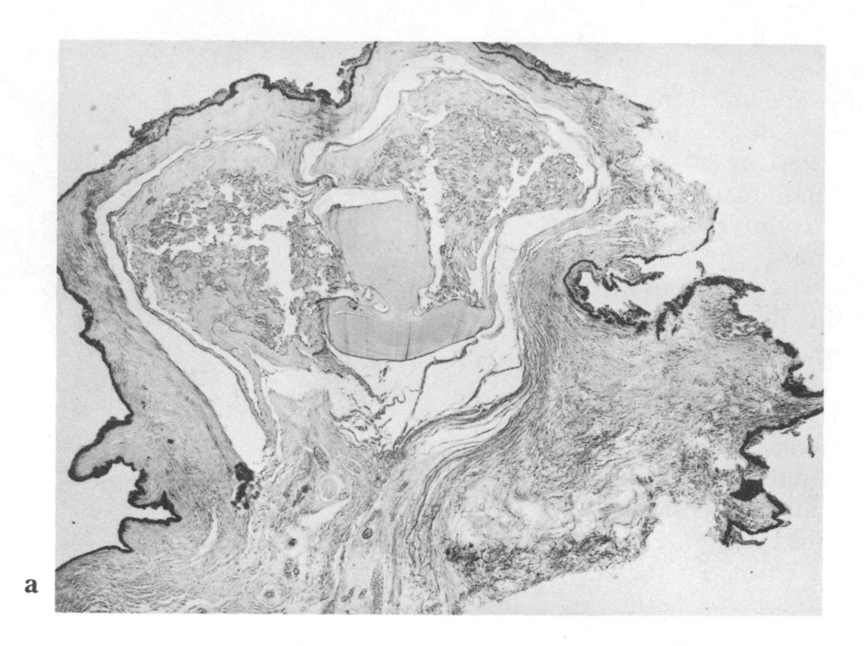

FIGURE 19.—Photomicrographs of the dorsal sting of *Synanceja horrida*. a. (Top) Cross section through the middle third of the 2nd dorsal sting. Note the thick integumentary sheath surrounding the spine, and the extensive glandular area containing the large clusters of venom granules in front of and on either side of the dorsal spine. × 20. b. (Right) Enlargement of the glandular area showing venom granules. × 50.

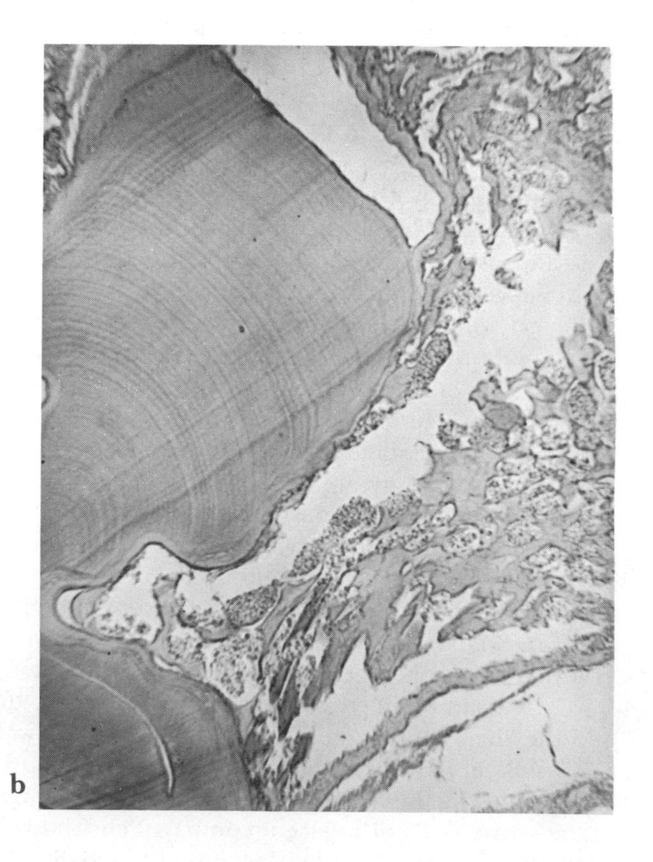

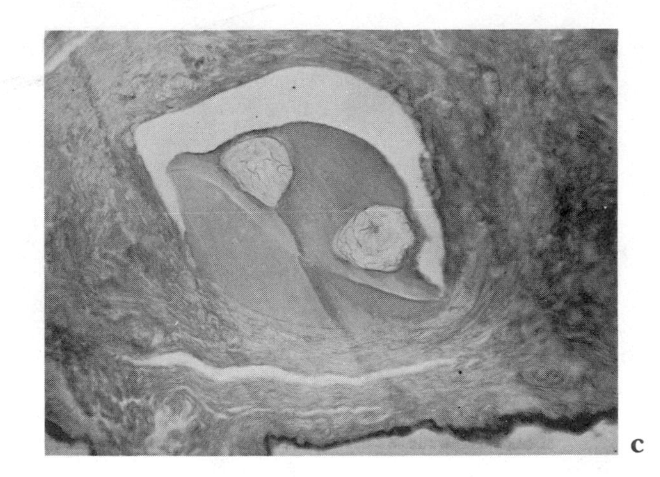

Figure 19c.—Cross section of the distal tip of the 5th dorsal sting. Situated within the excretory ducts, on either side of the anteromedian ridge, are collections of venom granules. × 22.

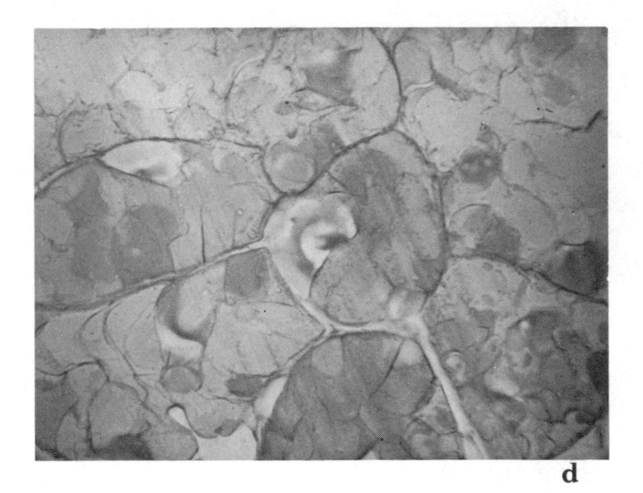

Figure 19d.—Longitudinal section through the venom gland of the 6th dorsal sting. Endean(1961) has suggested that the patches of yellowish, homogeneous, amorphous secretion seen in this section are artefacts due to inadequate fixation. The pink-stained areas are believed to be degenerating gland cells and the amorphous secretion probably represents disrupted venom granules. × 75. Hematoxylin and eosin stains. Preparations by B. W. Halstead and M. J. Chitwood.

(E. Rich)

meshwork which ramified throughout the gland. However, close study of both longitudinal and transverse sections revealed that the granules were enclosed in spherical compartments which abutted on gland cells, other compartments, and, when at the periphery of the gland, on the collagenous walls of the gland.

Collagenous material was not present in the walls of the compartments as these were not dyed by the aniline blue component of Mallory's dye mixture. Although they were stained heavily with Verhoeff's and Gomori's elastic tissue stains, variable results were obtained when the Taenzer-Unna orcein method for elastic tissue was employed. Bouin-fixed material gave a positive result. Zenker-fixed material gave

FIGURE 20a.—Photomicrograph of secretion of venom gland of *Synanceja horrida (S. trachynis)* showing gland cells and granule-containing compartments as described by Endean (1961). The gland was fixed in Zenker's fluid and the section was dyed with Mallory's dye mixture. Appx. × 80.

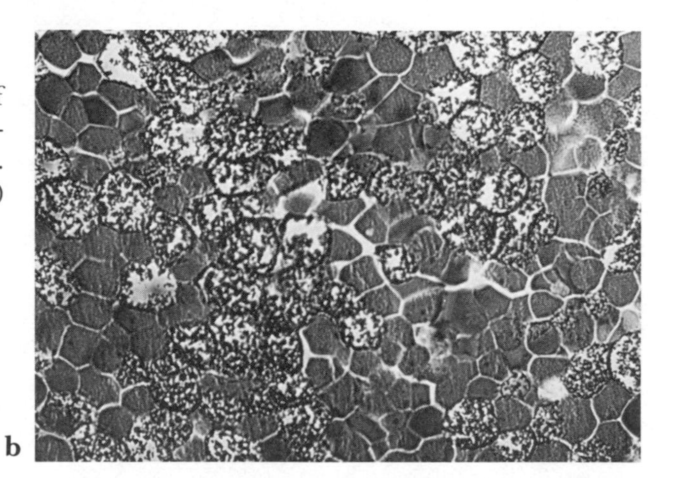

FIGURE 20b.—Enlargement of gland cells and granule-containing compartments. Appx. × 80. (Courtesy R. Endean)

a negative result, the dye being removed from the tissue by 70% alcohol. Negative results were obtained when Weigert's elastic tissue stain was employed.

The granules present in the meshwork were often irregular in shape, but when sub-spherical they had diameters ranging from 1.5 to 3.5μ. They were strongly eosinophilic, were dyed orange by Mallory's dye mixture, and were dyed black by Verhoeff's elastic tissue stain. They gave a positive reaction only for protein with the histochemical reagents used. Counts of granules in each of ten randomly selected compartments apparent in a 10-μ section gave figures which ranged from 72 to 264.

It was established that the granules were produced by the gland cells. Many of the larger (greater than 35μ in diameter) gland cells contained vacuoles with diameters ranging from 3 to 6μ. In these vacuoles, granules often occurred in small clusters. The granules were dyed orange by Mallory's dye mixture whilst the rest of the cell was dyed reddish to purplish. Also, the vacuolar contents were dyed strongly by Verhoeff's elastic tissue stain. Although the nucleus of each cell was coloured by this stain, the rest of the cell was not coloured. Some cells were packed with vacuoles and it was difficult to distinguish these cells in sections from compartments in which the granules were tightly packed. It was evident that compartments containing granules are the end products of the development of gland cells. The cytoplasm of the vacuole-packed cell disintegrates and the vacuoles break down to liberate their contained granules although the liberation of the granules from the vacuoles was not detected in the sections examined. Possibly the fixatives used disrupted the vacuoles of the gland cells which had reached a late stage in their development. The gland cell walls form the boundaries of the compartments. Judging by the reactions of the boundaries of the compartments to stains, particularly elastic tissue stains, the chemical constitution of the gland cell walls must be modified during their later states of development.

Some glands appeared to contain more granule-filled compartments than other glands but otherwise no marked differences were observed in the structure of the glands examined. Moreover, the structure was essentially the same from distal to proximal ends of the same gland.

A thin layer of dense fibrous connective tissue is in contact with most of the spine's surface. It spans the areas of ovoid bodies and only partially covers the parts of the spine that form the pockets containing these bodies.

Venom production is the result of a holocrine type of secretion and the secretory products are in the form of the previously described venom granules.

The dermis of the integumentary sheath is a dense, collagenous, connective tissue which contains varying numbers of small blood vessels, nerves, and melanophores. The junction between dermis and epidermis is marked by a continuous line of black granular pigment in sections through the base, but at the tip this line almost encircles the sting. The epidermis is a stratified squamous epithelium of varying thickness. It appears to be of two equally thick layers. There is an inner layer of epithelium such as is normally observed in fishes and an outer layer with ghostlike cellular characteristics suggesting a transition between parakeratotic and hyperparakeratotic change, probably representing a protective adaptation to the fish's peculiar environment. Scattered somewhat irregularly throughout this layer are large secretory or mucous cells having a clear cytoplasmic area. In some areas it is seen that the peripheral cytoplasm of the outer layer of epithelium forms a distinct striated border.

Anal Stings (Fig. 21a): The microscopic anatomy of the anal stings is similar to that of the dorsal stings, but there are a few noteworthy exceptions. Distal cross sections show that the anal spines are modified for the purpose of conveying venom, but they tend to be more goblet- than T-shaped in outline. Also, the grooves are narrow and greatly reduced in area in comparison with those of the dorsal stings. The areas of venom production are likewise reduced, and the integumentary sheaths are relatively thicker than those of the dorsal stings.

Pelvic Stings (Fig. 21b): The microscopic anatomy of the pelvic stings is similar to that of the anal stings.

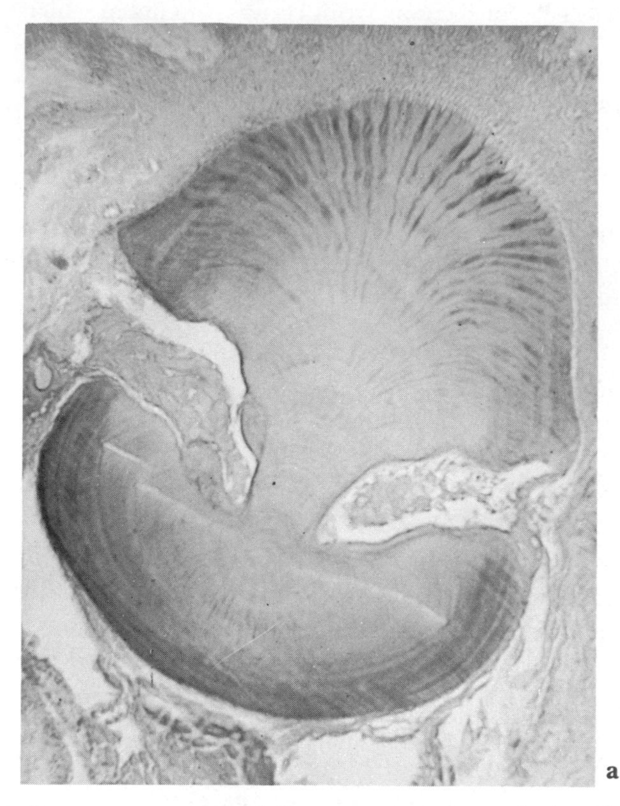

FIGURE 21—a. (Top) Cross section through the distal third of the 3rd anal sting. Note coagulated venom in anterolateral-glandular grooves of the spine. × 100. b. (Below) Cross section through the distal third of left pelvic sting. Note venom granules in anterolateral-glandular grooves of the spine. Hematoxylin and eosin stains. Preparations by B. W. Halstead and M. J. Chitwood. (E. Rich)

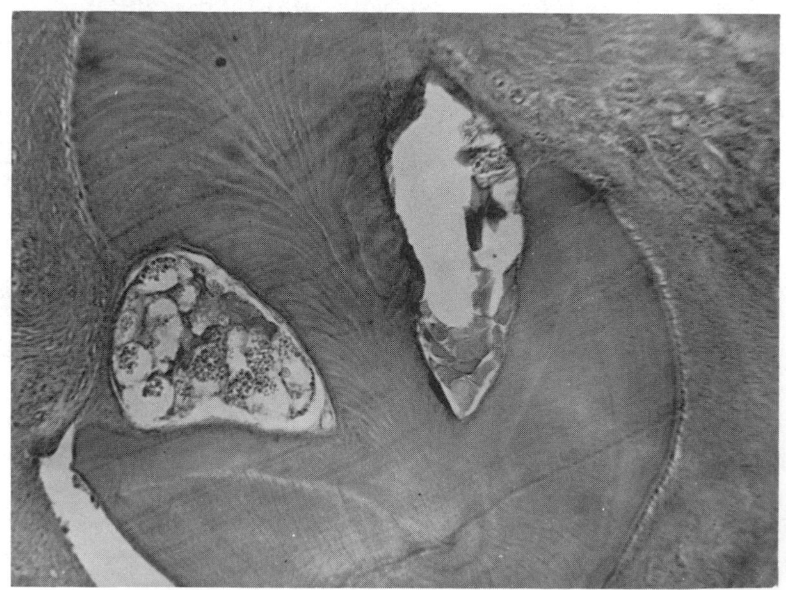

Comments on the Venom Apparatus of Other Scorpaenid Species

Pterois Type: Apparently the only other scorpaenid species having venom organs similar to *Pterois volitans* that has been studied in detail is *P. lunulata*, which has been described by Pawlowsky (1914, 1927) and Tange (1953*b*). Pawlowsky (1914) found that a dorsal sting of *P. lunulata* measuring 44 mm in length contained a venom gland 33 mm in length, or about three-fourths the total length of the sting. But according to Tange (1953*a, b, c, d*), the venom glands of the sting correspond to about one-fourth the length of the sting. However, the findings of Halstead *et al.* (1955*b*) on *P. volitans* agreed with those of Pawlowsky. It was observed that the venom glands of *Pterois* were fusiform in shape and that the attenuated proximal ends of the glands in the longer dorsal stings extended to within one-half to one-fourth of the proximal end of the stings. The complete extension of the gland is usually not seen without the aid of a dissecting microscope. Particular care must be exercised in removing the integumentary sheath or portions of the gland are removed and the gland appears shorter than it actually is. It is believed that these may be some of the reasons why Tange (1953*b*) came to the conclusion that the venom glands were only one-fourth the length of the spines. According to Tange the glandular cells of *P. lunulata* measure $50 \times 20\mu$, whereas the glandular cells of *P. volitans* are considerably larger. Pawlowsky (1906, 1907, 1911, 1927) described large, flattened elements which he termed supporting cells. However, neither Tange (1953*a, b, c, d*) nor Halstead *et al.* (1953*b*) was able to corroborate Pawlowsky's findings. There is no evidence of a lumen or glandular duct through which the venom is secreted in the *Pterois* type. Venom production appears to be the result of a holocrine type of secretion.

Scorpaena Type: Scorpionfishes believed to have a venom apparatus similar to *Scorpaena guttata* include the following: *Apistus evolans* (Pawlowsky, 1914; Tange, 1953*c*), *Erosa erosa* (Tange, 1954*a*), *Inimicus didactylus* (Pawlowsky, 1909), *I. filamentosus* (Pawlowsky, 1909, 1911), *I. japonicus* (Fig. 22) (Pawlowsky, 1909, 1911, 1927; Tange, 1954*b*), *Minous adamsi*[2] (Tange, 1953*a, c, d*), *M. inermis* (Tange, 1953*d*), *S. scrofa* (Bottard, 1889*a, b*; Sacchi, 1895*a, b*), *S. porcus* (Bottard, 1889*a, b*; Sacchi, 1895*a, b*; Pawlowsky, 1906), *S. ustulata* (Sacchi, 1895*a, b*), and *Sebastes marinus* (Pawlowsky, 1909). The following species have also been examined and are said to have a venom apparatus similar to the preceding species, although complete anatomical descriptions are lacking: *Hypodytes rubripinnis* (Pawlowsky, 1914; Tange, 1953*c*), *Scorpaena mauritiana* (Bottard, 1889*a, b*), *S. mesogallica* (Bottard, 1889*a, b*), *S. neglecta* (Pawlowsky, 1911, 1927), *Scorpaenopsis gibbosa* (Bottard, 1889*a, b*), *Sebasticus marmoratus* (Pawlowsky, 1911, 1927), *Sebastodes inermis* (Pawlowsky, 1927; Tange, 1953*a, c*), and *S. joyneri* (Pawlowsky, 1911). Moreover, there is some

---

[2] Halstead and Mitchell (1963) had suggested that *Erosa*, *Minous*, and *Inimicus* had venom organs of the *Synanceja* type. However, further study indicates that they more closely resemble those of *Scorpaena* because of the absence of a well-developed excretory duct.

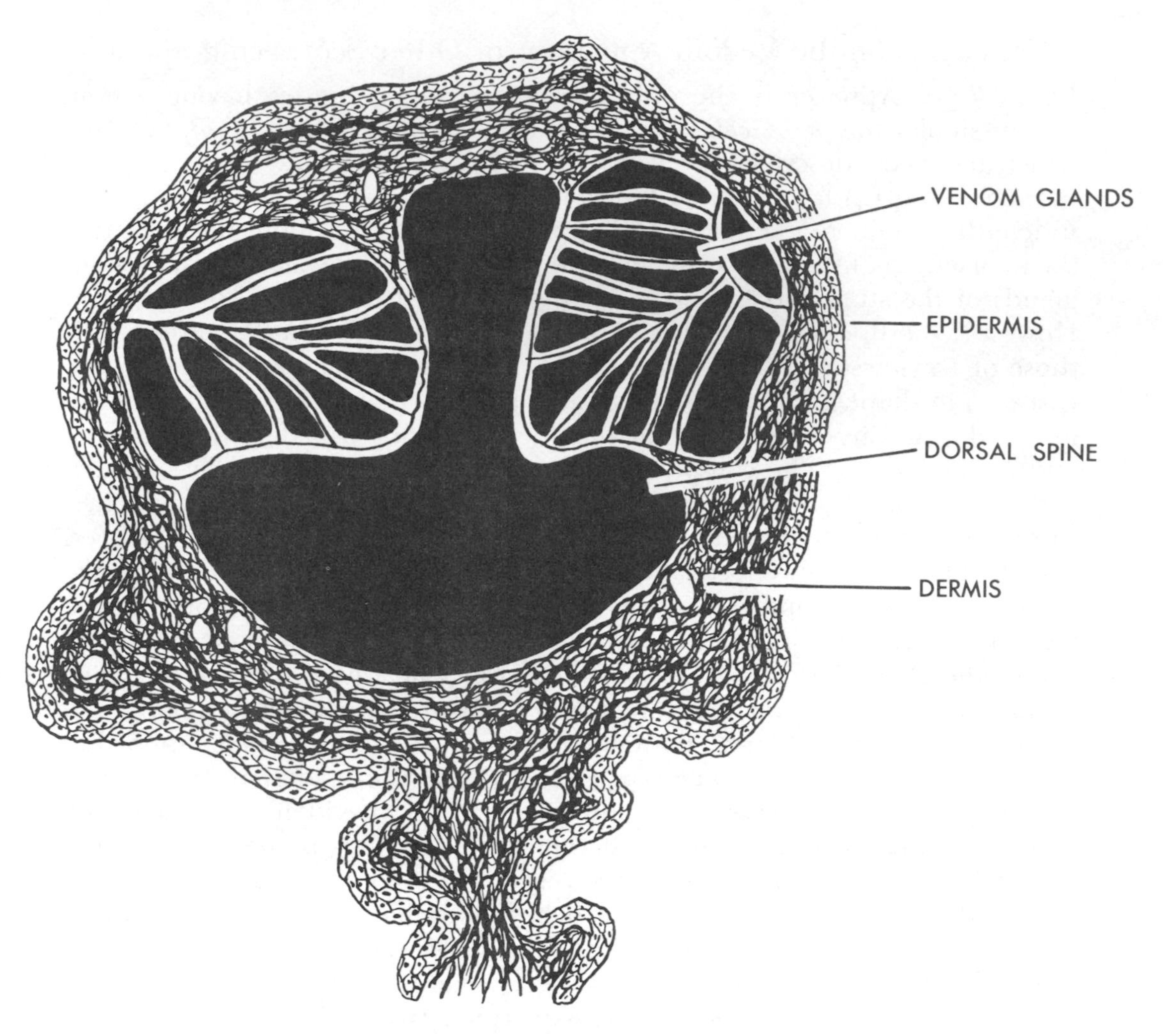

FIGURE 22.—Drawing of a cross section through a dorsal sting of *Inimicus (Pelor) japonicus*. The close resemblance to the morphological arrangement of the venom glands of *Scorpaena guttata* is clearly seen in this drawing. (R. Kreuzinger, after Pawlowsky)

controversy as to whether or not venom organs are actually present in *Sebasticus marmoratus* (Pawlowsky, 1909, 1911, 1927; Tange, 1953*a*) and *Sebastes marinus* (Bottard, 1889*a, b*; Schnackenbeck, 1943).

Cameron and Endean (1966) have described the venom apparatus of *Notesthes robusta* which is said to consist of 15 dorsal spines, three anal spines, the single paired pelvic spines, together with their enveloping integumentary sheaths and their underlying venom glands. Each venom gland is composed of a double row of large gland cells derived from a central core of epidermal cells in the interior of the gland, and each gland cell is surrounded and supported by epidermal cells. Proteinaceous venom granules are elaborated in cytoplasmic vacuoles in the gland cells. The cytoplasm subsequently disintegrates and the granules remain in fluid-filled sacs which are pushed to the periphery of the gland. Partial regeneration of venom glands occurs within 6 days. A bridge of epidermal cells connecting a regenerating venom gland with the epidermis of its surrounding integumentary sheath was demonstrated.

Roche and Halstead (1972) have described the venom organs of the following species of California rockfishes: *Sebastes auriculatus, S. constellatus, S. dalli, S. eos, S. goodei, S. helvomaculatus, S. levis, S. maliger, S. miniatus, S. nebulosus, S. paucispinus, S. semicinctus*, and *S. serriceps*. The venom apparatus of the *Sebastes* examined consisted of 13 dorsal spines, three anal spines, two pelvic spines, their associated musculature, venom glands, and integumentary sheaths. The gross anatomical features closely resembled that of *Scorpaena* species. However, the regular pinnate, or two-layered arrangement, and the cardiform cross-sectional outline described for *Scorpaena* and *Pterois* were not found in any species of *Sebastes* that were examined histologically. The irregular arrangement of venom gland cells in the California *Sebastes* were similar to that described for the European *S. marinus* and the Japanese *Sebastodes joyneri* (Pawlowsky, 1909, 1911) and *S. inermis* (Tange, 1954*c*). The reader is referred to the original article by Roche and Halstead (1972) for more detailed descriptions and illustrations of the *Sebastes* venom organs.

Venom organs are said to be absent in *Sebastodes taczanowski* (Pawlowsky, 1909) and *Sebastichthys pachycephalus* (Tange, 1953*a*). Further anatomical research will have to be done to decide whether or not venom organs are present in other scorpionfishes of the above genera.

There is no evidence of excretory ducts associated with the venom glands in *Scorpaena guttata* nor does the venom appear to flow freely between the layer of glandular cells and the sheath such as was described by Bottard (1889*b*) for *S. mauritiana* and *S. mesogallica*. Pawlowsky (1906, 1907, 1911, 1927) described large flattened elements in the scorpionfishes which he examined, and Tange (1953*c*) reported similar findings for *Apistus evolans*, but Halstead *et al.* (1955*a*) failed to find such structures in *S. guttata*. Pawlowsky (1906) also described the "cuticula of the epidermis folding in the groove at the tip of the distal tip of the spine, thus forming an excretory duct" in the scorpaenids which he examined, but most other workers have failed to corroborate this. However, Sacchi (1895*a, b*) and Tuma (1927) indicated that the venom glands of some scorpaenids which they studied contained a central excretory duct. No duct has been described for *S. guttata*. Venom production appears to be the result of a holocrine type of secretion.

Synanceja Type: It was previously believed that there were several other genera of scorpionfishes that possessed the *Synanceja* type of venom apparatus (Halstead and Mitchell, 1963). However, subsequent study indicates that the highly developed type of venom apparatus found in *S. horrida, S. verrucosa*, and possibly one or two other species characterized by the presence of an excretory duct is restricted to the members of the genus *Synanceja*. Unfortunately our knowledge of the morphology of the venom organs of some of the other closely related species of scorpionfishes is much too meager at present to permit a final decision on this matter.

Endean's (1961) findings on the histology of the venom organs of *S. trachynis* (generally accepted as a synonym of *S. horrida*) are similar to those of Halstead *et al.* (1956). However, he has noted that formalin fixation may cause shrinkage and vacuolization of glandular cells, disruption of the walls of compartments, and result in the granules being transformed into an amorphous mass (Fig. 20a, b). Venom production in *Synanceja* appears to be of the holocrine type.

MECHANISM OF INTOXICATION

Envenomations by scorpionfishes are most frequently produced by their dorsal stings and less commonly by their pelvic or anal stings. Most scorpionfish stings are due to careless handling of the fish in removing them from a net or hook, and jabbings which occur in the process. Contacts with *Pterois* are sometimes made by uninitiated persons who are attracted to them because of their beautiful, soft, lacelike appearance, but soon find to their chagrin and painful sorrow that the fish is quite deceptive in its appearance. Stonefish *(Synanceja)* wounds generally result from an unwary victim stepping on the dorsal stings of a fish that is partially buried in the sand or by reaching in a crevice and thereby being stuck by the stings of a well-camouflaged specimen. It should be kept in mind that some species of scorpionfishes are quite aggressive in their attitude and will strike out with their dorsal fins if one reaches out a hand within striking distance of their spines (Fig. 1). Serious wounds have also resulted from the careless handling of a dead specimen.

MEDICAL ASPECTS

Clinical Characteristics

The introduction of *Pterois* venom into a wound produces almost immediately an intense, sharp, shooting, or throbbing pain which generally radiates from the affected part. The pain may continue for several hours or longer, depending upon the amount of poison received. The general area about the wound may become reddened, swollen, and hot, and the specific area in the immediate vicinity of the wound become cyanotic. Subsequently, there may occur sloughing of the tissues about the wound and gangrene. Secondary bacterial infections of the wound as a result of inadequate medical care are not infrequent. Cardiac failure, delirium, convulsions, and nervous disturbances have been reported. Primary shock may be present, manifested by such symptoms as faintness, weakness, nausea, loss of consciousness, cold clammy skin, rapid weak pulse, low blood pressure, and respiratory distress. Deaths resulting from zebrafish stings have been reported by Schnee (1908), Faust (1924), Herre (1952), Atz (1962*a, b*), and Whitley (1963). Saunders (1960*b*) published one of the few clinical reports on a sting by *Pterois* in which there were serious local and systemic effects.

Stings from other species of scorpionfishes, i.e., *Scorpaena, Sebastes,*

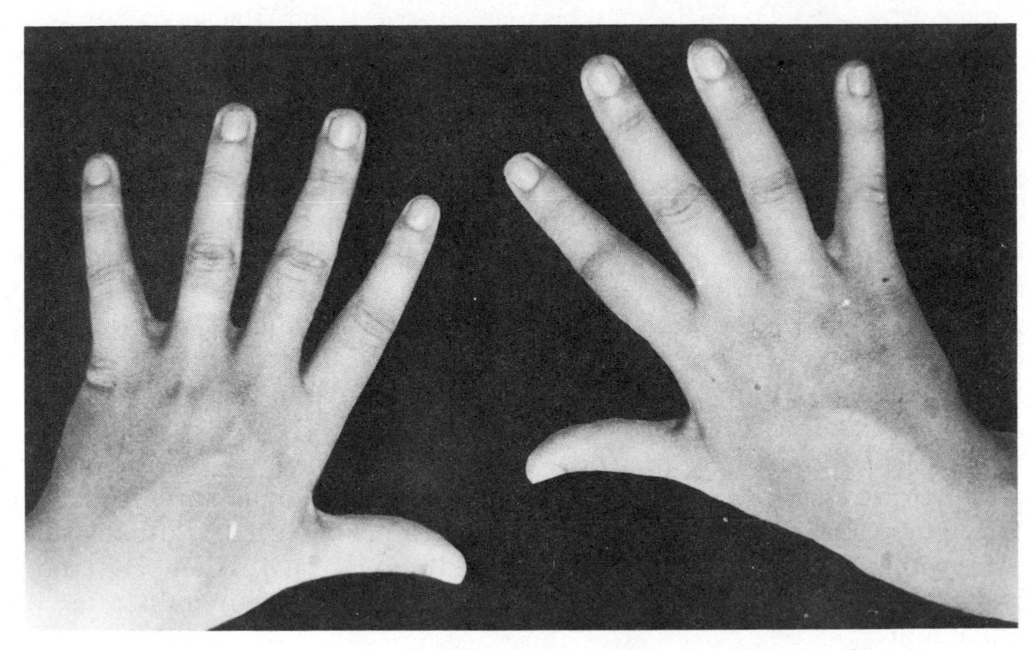

Figure 23.—Results of a sting inflicted by *Scorpaena porcus* one hour after envenomation. Note swelling of right index finger. Victim was a 19-year-old girl. Taken near Pula, Yugoslavia. (Courtesy of Z. Maretić)

Apistus, Minous, Erosa, Hypodytes, etc., vary somewhat from one species to the next, but generally the introduction of scorpaenid venom immediately produces an intense throbbing pain. Within a few minutes the area about the wound becomes ischemic and then cyanotic. The pain becomes progressively more severe and may radiate to the groin or axilla. The intensity of the pain may be comparable to that produced by renal colic and may continue for several hours. Within a short period of time the affected part becomes swollen, erythematous, and indurated (Fig. 23). Profuse perspiration, pallor, dyspnea, restlessness, nausea, vomiting, diarrhea, loss of consciousness, and extreme tachycardia are commonly present. Abscesses, necrosis, and sloughing of the tissues about the wound have been reported. Bayley (1940) and Colby (1943) state that a maculopapular or scarlatiniform rash over the body may occur. Cecca (1902) cited a case which resulted in peripheral neuritis, paralysis, and muscular atrophy due to a sting by *Scorpaena nera*. Secondary bacterial infections, tetanus, and primary shock are frequent complications which must be considered. According to Blanchard (1890), Coutière (1899), Scott (1921), and Colby (1943), scorpionfish stings may result in death.

The puncture wounds from stonefish *(Synanceja)* are disproportionately small in comparison with the violent pain that they produce. The pain, which is generally described as instantaneous, intense, sharp, or burning, radiates within a few minutes from the wound site, involving the entire leg, groin, and abdomen, or if in the upper extremities, the axilla, shoulder, neck, and head. The pain may become so severe that the victim thrashes about, rolling on the ground and screaming in agony, and may lose consciousness. The area about the wound becomes numb, and the skin, even at some distance from the site of the injury, becomes painful to touch—a condition which may continue for many days. In some instances, complete paralysis of the limb may ensue.

Localized symptoms consist of ischemia within the immediate vicinity of the wound, bordered by an area of redness. Swelling develops rapidly and is extensive, involving the entire appendage, usually to such an extent that movement of the affected part is impaired. Nausea, vomiting, and diarrhea are frequently present, together with a feeling of generalized weakness, profuse sweating, delirium, headache, irregular pulse, and decreased body temperature.

In severe cases there may be a feeling of constriction of the chest, respiratory distress, convulsions, and even death. According to surveys made in Portuguese East Africa by Smith (1951), death, if it occurs, usually takes place within the first 6 hours. However, Le Juge (1871), Taschenberg (1909), and others report that death may occur days and even months after the initial injury.

In less serious cases the ischemic area about the wound gradually becomes cyanotic, abscessed, ulcerated, and within a few days, gangrenous, if proper treatment has not been instituted. Large blisters containing a clear exudate may develop over the affected appendage. Lymphangitis, lymphadenitis, joint aches, fever, and secondary bacterial infections are common. The edema and pain subside slowly and in some instances may continue for two or more months. The return of sensation to the affected part is also gradual, and sometimes it never completely returns. The venom is said to have a pronounced adverse effect on general health and vitality.

Bottard (1889*a, b*), Schnee (1911), Gallagher (1928), Bonne and Neuhaus (1936), Smith (1951) have published valuable clinical case histories on stonefish stings. The general medical aspects have been discussed by Blanchard (1890), Coutière (1899), Cleland (1912, 1932), Scott (1921), Engelsen (1922), Faust (1924), Whitley and Boardman (1929), Tweedie (1941), Ralph (1943), Whitley (1943, 1963), Cilento (1944), Halstead *et al.* (1956), Wiener (1958), Saunders (1960*a*), Jüptner (1960), Halstead and Mitchell (1963), Russell (1965, 1971), and Roche (1973).

Pathology

(Unknown)

Treatment

The treatment of scorpionfish stings is directed toward alleviating the pain, combating the effects of the venom, and preventing secondary infection. Most scorpionfish stings are of the puncture-wound variety, usually small in diameter, and difficult to irrigate, but whenever possible they should be

a

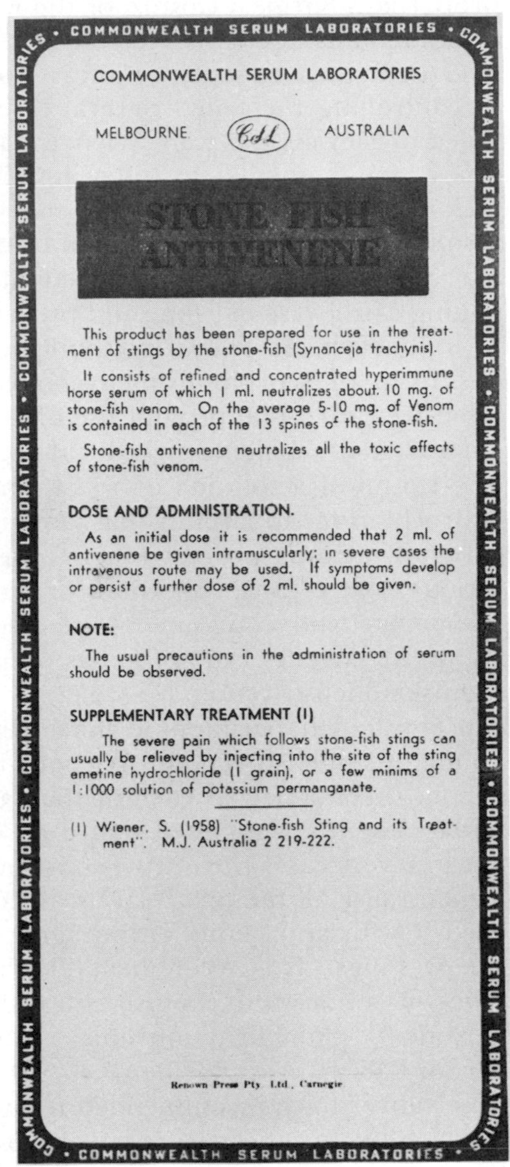

• COMMONWEALTH SERUM LABORATORIES •

COMMONWEALTH SERUM LABORATORIES

MELBOURNE (CSL) AUSTRALIA

STONE FISH ANTIVENENE

This product has been prepared for use in the treatment of stings by the stone-fish (Synanceja trachynis).

It consists of refined and concentrated hyperimmune horse serum of which 1 ml. neutralizes about 10 mg. of stone-fish venom. On the average 5-10 mg. of Venom is contained in each of the 13 spines of the stone-fish.

Stone-fish antivenene neutralizes all the toxic effects of stone-fish venom.

DOSE AND ADMINISTRATION.

As an initial dose it is recommended that 2 ml. of antivenene be given intramuscularly; in severe cases the intravenous route may be used. If symptoms develop or persist a further dose of 2 ml. should be given.

NOTE:

The usual precautions in the administration of serum should be observed.

SUPPLEMENTARY TREATMENT (1)

The severe pain which follows stone-fish stings can usually be relieved by injecting into the site of the sting emetine hydrochloride (1 grain), or a few minims of a 1:1000 solution of potassium permanganate.

(1) Wiener, S. (1958) "Stone-fish Sting and its Treatment". M.J. Australia 2 219-222.

Renown Press Pty. Ltd., Carnegie

b

FIGURE 24.—a. (Above) Stonefish antivenin produced by the Commonwealth Serum Laboratories, Melbourne, Australia. b. (Right) Literature describing the treatment of stonefish stings and the use of stonefish antivenin. (Courtesy S. Wiener)

thoroughly cleansed in order to remove as much of the venom as possible. Bleeding should be encouraged. A tourniquet may be applied immediately above the stab site but must be released every few minutes in order to preserve circulation. The injured member should be soaked in hot water as soon as possible. The water should be maintained at as high a temperature as the patient can tolerate without producing further injury to the tissue. Soaking should continue for a period of 30 to 90 minutes. The addition of magnesium sulfate to the water is sometimes desirable because of its mild anesthetic properties. Surgical closure of the wound is usually not required. The use of antitetanus agents is recommended. Antibiotic agents may be required. The use of intramuscular or intravenous demerol may or may not be effective in controlling the pain. Several cases have been cited in which morphine failed to alleviate the pain (Wiener, 1958). The primary shock which is sometimes seen immediately following the injury will probably respond satisfactorily to routine supportive measures. However, the secondary shock resulting directly from the action of the venom on the cardiovascular system may require vigorous therapy. Treatment should be directed toward maintaining cardiovascular tone and preventing any further complicating factors.

Stonefish *(Synanceja)* poisoning is a particularly severe and very dangerous form of scorpionfish envenomation and usually requires immediate intensive care. The prompt use of hot water is recommended. Injection of emetine hydrochloride directly into the wound has been found to be of value (0.5-1.0 ml of a solution of 1g of emetine hydrochloride per ml). Emetine hydrochloride apparently has an antagonistic action against the venom, generally affords prompt relief from the pain, and is said to prevent more serious toxic effects. Injections of 0.1 to 0.5 ml of 5 percent potassium permanganate and congo red have also been recommended (Wiener, 1958, 1959*a, b*). Wiener (1959*a, b*) has prepared a high potency antivenin for use against stonefish venom (Fig. 24a, b). The antivenin consists of refined and concentrated hyperimmune horse serum of which 1 ml neutralizes about 10 mg of stonefish venom. Wiener found that on the average 5 to 10 mg of venom is contained in each of the 13 dorsal spines of the stonefish. As an initial dose it is suggested that 2 ml of antivenin be given intramuscularly, but in severe cases the intravenous route may be used. This antivenin is said to neutralize all the toxic effects of stonefish venom. Russell (1965) has also used it with apparently encouraging results in the treatment of *Scorpaena guttata* stings. It is recommended that stonefish antivenin be tried in other types of serious scorpionfish stings that do not respond to other forms of treatment. Stonefish antivenin is available through the Commonwealth Serum Laboratories, Melbourne, Australia.

Natives have recommended using *Abrus precatorius* and the sap of mangrove trees for the treatment of stonefish stings (Gail and Rageau, 1956; Wiener, 1958).

A scorpionfish antivenin has also been developed by Dr. Z. Maretić, Medical Center, Pulu, Yugoslavia.

Prevention

One must use extreme care in handling scorpionfishes, particularly when removing them from a hook or net. They may suddenly flare their gill covers and erect their dorsal spines which can easily penetrate a heavy glove. Placing one's hands in crevices or holes inhabited by these fishes must be done with caution because if the hand is brought close to them, some scorpionfishes will erect their dorsal stings and fearlessly strike the approaching object, thereby producing envenomation. When wading in shallow coral areas inhabited by stonefishes, one should be aware of the fact that these fishes will not dart from their hiding place in the sand but rather will erect their dorsal stings and wait for the unwary victim to step on them. The dorsal stings are very stout and sharp and quite capable of penetrating a thin rubber-soled sneaker or fin. Whenever possible the individual should drag his feet and be careful where he steps. Don't succumb to the temptation of trying to grab a beautiful, lacy, slow-swimming zebrafish. There is venom in those lovely fins that will long be remembered if one succeeds in catching the prey.

PUBLIC HEALTH ASPECTS

Scorpionfish envenomations are generally not considered to be a public health problem. However, in certain localities such as the New Hebrides and other regions where stonefish and some of the more venomous species of scorpionfishes commonly occur they may constitute a public menace (Wiener, 1958). Scorpionfishes are among the most widely distributed venomous fishes known to biotoxicologists (Map 1). Russell (1961, 1965) estimates that there are about 300 scorpionfish envenomations in the United States each year. There are no records available for other areas.

One of the most notable scorpionfish envenomations occurred when one of the aquanauts, Commander Scott Carpenter, was stung by *Scorpaena guttata* when emerging from the hatchway of the U.S. Navy's Sealab II at a depth of 68 m, off La Jolla, Calif., in October, 1965 (Carpenter, 1965). Groping in the dark with his hands, he was stung by the dorsal stings of a scorpionfish on the right index finger. Scorpionfish seemed to be attracted to lighted areas around Sealab II and later became so abundant as to constitute a hazard to the divers. Several other divers were also stung before the project was completed. Fortunately, the Sealab II divers were stung by *S. guttata* rather than some of the more dangerous tropical species, or the results might have been disastrous (Clarke, Flechsig, and Grigg, 1967).

TOXICOLOGY

Research on the physiological effects of the venom of *Scorpaena* appears to have started with Pohl (1893) who worked on the venom of *Trachinus draco* and *Scorpaena porcus*. He came to the conclusion that *Trachinus* and *Scorpaena* venom have a similar cardiac inhibitory action on frogs, but *Scorpaena* venom is less potent and more variable in its effects. Dunbar-Brunton (1896) also worked with the venom derived from these same two species of fishes. Subcutaneous injections of the venom into guinea pigs produced evidence of pain, incontinence, motor paralysis of the limbs, clonic and tonic convulsions, respiratory distress, and death. Briot (1904, 1905) prepared venom extracts from one of the European scorpaenids by macerating the dorsal and opercular stings in glycerin and physiological saline. (He gave only the common name of hogfish, which generally refers to *S. porcus*.) The extracts were tested by subcutaneous injections into frogs, rabbits, and

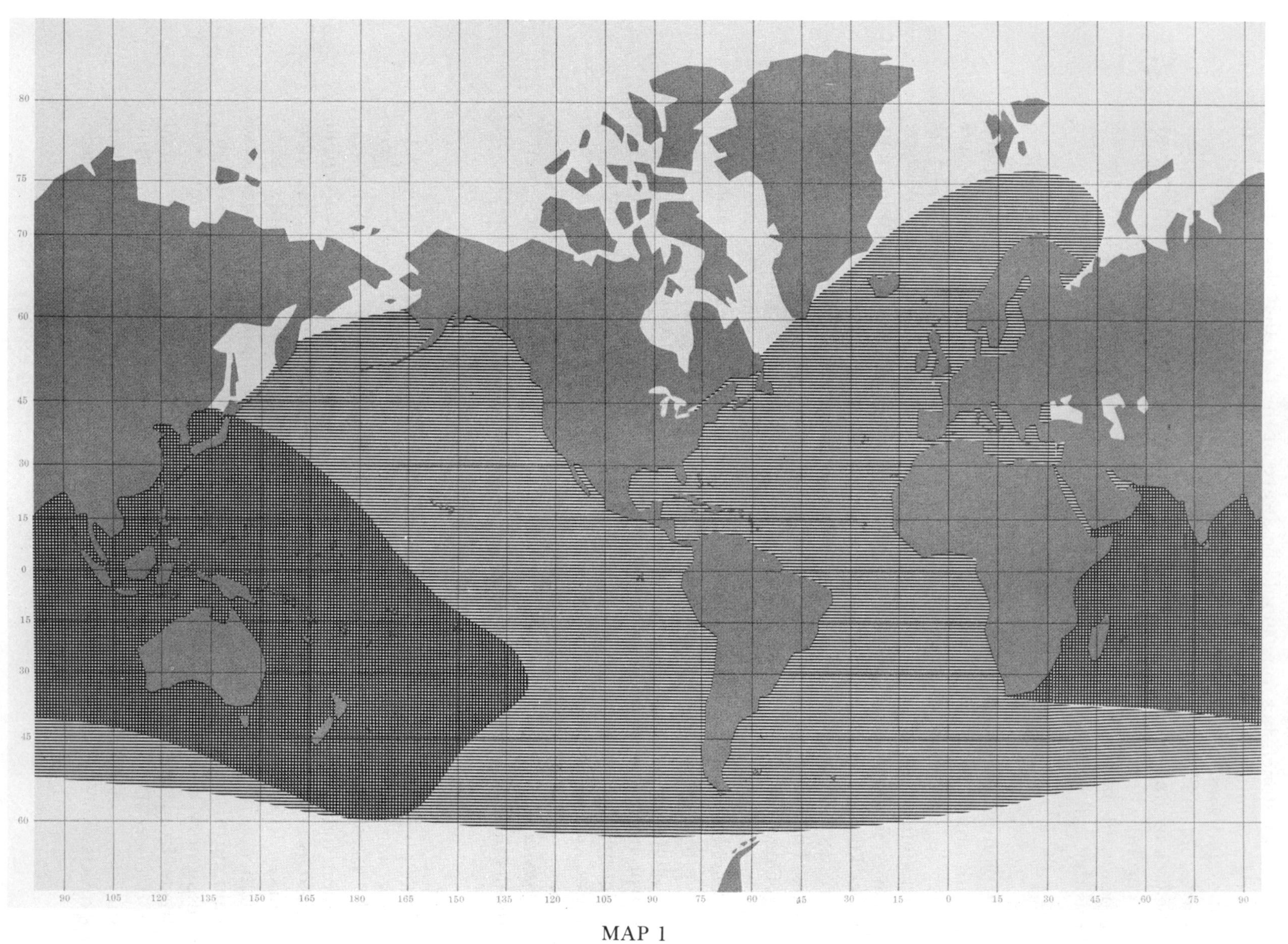

MAP 1

Map showing the geographical distribution of venomous scorpionfishes. The darker area indicates the distribution of the stonefish *Synanceja*. (L. Barlow)

white rats. Since the injections produced only a mild transitory paralysis in the frogs, Briot concluded that the European hogfish did not produce a venom and that injuries resulting from contact with their spines were purely traumatic in nature. Phisalix (1931) conducted similar experiments using venom from the European hogfish *S. porcus* on sparrows, pigeons, and guinea pigs. The effects on the pigeons and sparrows were mild and transitory, but larger doses in guinea pigs produced stupor, trembling, decreased body temperature, paralysis of the hind legs, convulsions, respiratory paralysis, and finally death. Similar results have been obtained by De Marco (1937) who studied the effects of the venom of *S. porcus* on frogs. Lumière and Meyer (1938) also repeated Briot's experiments, using the marine fishes *Gobius lota* and *Crenilabrus operculatus* and guinea pigs as test animals. Their findings agreed with those of Phisalix and De Marco, and they also came to the definite conclusion that *S. porcus* does produce a venom.

The first scorpaenid venom that was studied to any great extent was that of *Synanceja*, the stonefish. According to Bottard (1889*a, b*) and Duhig and Jones (1928*a, b*), the venom obtained from living specimens was a clear, bluish, watery liquid, which became opalescent and cloudy soon after the fish died. Microscopically, it had the appearance of an albuminous liquid in which were contained large, highly refractile, rounded cells having single, small, centrally placed nuclei. When tested by litmus paper the venom gave a neutral to weakly acid reaction. The venom was said to be readily coagulated by nitric acid, alcohol, ammonia, water, and heat. Dried venom did not dissolve in either concentrated alcohol or normal saline.

Bottard (1889*a, b*) was the first to attempt to evaluate the toxicological effects produced by stonefish venom. The venom was aspirated from the dorsal fin glands and then tested by injecting it subcutaneously into dogs, frogs, and himself. Injections into dogs produced signs of pain, incontinence, tremors, ruffling of the hair, and evidence of anorexia and thirst. Frogs developed a decreased response to stimuli and motor paralysis of the hind limbs and died within a period of 3 hours. Injections of the venom into human skin resulted in a painful tingling sensation, necrosis, and gangrene, but the wound eventually healed.

Duhig and Jones (1928*a, b*) and Duhig (1929) studied *Synanceja horrida* venom and its effects on guinea pigs. They concluded that tissue extracts derived from the dorsal fin glands and injected subcutaneously into guinea pigs exerted a toxic action on the voluntary and involuntary muscles. Injection of the venom produced, within a few minutes, marked irritability, evidence of pain, anorexia, tremors, clonic and tonic convulsions, motor paralysis of the limbs, loss of corneal reflexes, and respiratory distress. They were unable to determine for certain if the venom was a neurotoxin or a myotoxin but believed it to be the former. *Synanceja* venom also had a hemolytic action which was capable of producing lysis of guinea pig, sheep, and human red blood cells without the assistance of lecithin or complement. Fresh unheated serum, when added to cells of the same species, inhibited hemolysis of guinea pig, sheep, and human red blood cells by this venom. In the case of guinea pig serum, a very high dilution was sufficient to protect against a minimum hemolytic dose. Human serum, besides protecting against lysis of cells of its own species, also protected sheep cells. Guinea pig serum heated to 55° C for 20 minutes inhibited hemolysis, but the same serum heated to 60° C for half an hour did not completely inhibit hemolysis. Lecithin was found to reactivate venom, the cytolytic power of which had been inhibited by fresh serum.

Gail and Rageau (1956) injected the glandular contents of a single dorsal sting from *S. verrucosa* into a frog which died in extensor paralysis in about 4 hours. Injections of the venom into one rat resulted in defecation, respiratory distress, hypothermia, spasmatic contractions of the abdominal muscles, marked weakness in the hind limbs, and death in about 7 hours. An autopsy revealed blood in the thorax. A second rat was injected, and it developed similar signs and died in about 16 hours. A dog was injected with the venom from the glandular contents of five dorsal stings, and it developed agonal signs within 45 seconds, i.e., vocalization, trismus, convulsions, relaxation of the sphincters, respiratory distress, and loss of consciousness. Death occurred within 1 minute of the injection and was attributed to cardiac collapse and respiratory paralysis.

Wiener (1958) found that if emetine hydrochloride was mixed with the venom of *S. horrida (trachynis)* before being injected intravenously into mice the emetine protected the mice against a lethal dose of the venom. Sublethal doses of stonefish venom also protected the mice against lethal doses of emetine, thus indicating an antagonistic action between these two substances. Wiener further observed that stonefish venom was heat labile. It was coagulated and destroyed by boiling and lost its toxicity in less than 30 minutes at a temperature of 52° C. The effect of heat suggested that stonefish venom was an albumin. This conclusion was further supported by his findings that the venom was a good antigen and readily stimulated the production of neutralizing antibodies in rabbits.

Wiener (1959*a*) reported that fresh stonefish venom was opalescent and had a pH of about 6.0. Lyophilized venom stored in a desiccator over silica gel for 3 months did not show any appreciable loss of toxicity. The venom gave all the reactions of protein. It was precipitated by mineral acids, alcohol, and picric acid. No reducing sugars were present upon hydrolysis, and a copious precipitate was produced by boiling. The toxicity of the venom was destroyed almost instantly at a pH below 4. About 50 percent of the toxicity was destroyed at pH 8.6 in about 3 hours and was almost completely destroyed after 24 hours. The venom was stable between pH 7.0 and 7.6 and could be stored at 4° C for 24 hours with only slight reduction in toxicity. However, when the venom was stored for 48 hours there was a reduction in toxicity by 50 percent, and there was a complete loss in toxicity after 4 or 5 days. Freezing and thawing increased the rate of detoxification, and the adverse effect of pH changes on stability were enhanced when the temperature was increased. The average yield of venom from a functional dorsal sting of *S. horrida* was 0.03 ml or 5.1 to 9.8 mg of the dried venom. The LD_{50} in mice by venous injection was 0.005 to 0.01 mg/15 g of mouse weight; the subcutaneous LD_{50} was 0.04 to 0.06 mg; and the intraperitoneal LD_{50} was 0.02 to 0.03 mg. Lethal injections of the venom produced muscular incoordination, paralysis of the hind limbs, irregular respiration, coma, and convulsions. These signs were rapidly followed by prostration, cessation of respiratory movements, and cardiac arrest. Autopsy revealed the lungs to be hemorrhagic and filled with a gelatinous frothy fluid. In guinea pigs the venom caused necrosis at the injection site, muscular weakness, respiratory depression, coma, and death. Autopsies of the guinea pigs revealed that the lungs were emphysematous, and the fluid was bloodstained. Similar results were attained in rabbits. Intravenous injection of 3.5 mg of venom into an 8-kg dog produced cardiac and respiratory arrest in less than 2 minutes. Smaller amounts of the venom caused a rise in blood pressure followed by a fall, accompanied by a tripled pulse pressure and respiratory distress. Intravenous injection of 0.5 mg of venom into a chicken resulted in a precipitous drop in blood pressure from about 200 mm of mercury to almost 0. The blood pressure returned to normal during the next minute but then fell again to 80 mm of mercury where it remained for about 10 minutes. A second injection of 1 mg of the venom killed the bird in 4 minutes. Wiener further observed that the toxicity of stonefish venom could be destroyed by the addition of emetine hydrochloride and a variety of buffered dyes and other substances (pH 7.4), such as methyl violet, phloxine, congo red, pontamine sky blue, hydrogen peroxide iodine, potassium permanganate, and sodium oleate. However, studies on mice revealed that the protective action of both potassium permanganate and congo red was the result of their direct action on the venom since once the dyes entered the circulation of the animal they failed to demonstrate any protective effect. Wiener also found that his venom solution was only weakly hemolytic and concluded that since there was an absence of hemolysis in the blood of animals that had died from the effects of stonefish venom, it was unlikely that hemolysis played a significant role in stonefish envenomations.

Wiener (1958, 1959*a, b*) also investigated the antigenic properties of stonefish venom. He found that the venom was antigenic in the mouse, rabbit, and horse. Both precipitating and neutralizing antibodies were produced in the rabbit and horse. The antivenin produced in the horse was refined and concentrated so that 1 ml neutralized 10 mg of venom. Antigen-antibody reactions in agar diffusions produced two well-defined lines, a fact which suggested that the stonefish venom contained at least two antigenic components.

Saunders (1958, 1959*a, b, c,* 1960*a*) and Saunders and Tökés (1961) have also made significant contributions to our knowledge of the toxicology of stonefish venoms *(S. horrida* and *S. verrucosa)*. They found the fresh venom of *S. horrida* to be a clear, colorless fluid with a pH of 6.8, a nitrogen content of

2 percent, and a protein content (biuret determination) of about 13 percent. The intravenous LD_{50} in mice was approximately 200μg of protein per kg of body weight. They therefore estimated that the extracts of the stings from one fish would contain 10,000 to 25,000 LD_{50} for mice. The lethal fraction was nondialyzable. Lyophilized or glycerol-treated extracts were tested and found to retain 50 to 100 percent of their toxicity when stored for 1 year at -20° C. Extracts prepared from the dorsal stings of a single fish contained about 3,000 LD_{50} for mice. The mean intravenous lethal dose/kg in rabbits was approximately one-tenth the LD_{50} for mice. This amount produced ataxia, paralysis of the limbs and neck, convulsions, respiratory arrest, and death in rabbits. Although respiratory arrest was a terminal event, artificial respiration failed to prolong the life of the animal. Starch gel electrophoresis of the venom revealed that it contained 7 to 10 bands. However, the lethal fraction could be recovered from only one of these bands. The lethal fraction had an approximate mouse LD_{50} of 15 μg of nitrogen/kg of body weight. Re-electrophoresis of this lethal fraction gave only a single band. The venoms of *S. horrida* and *S. verrucosa* were believed to be physiologically identical. Death followed intravenous injections into mice in about 30 seconds to 30 minutes, depending upon the dose. Animals that lived beyond the 30-minute period usually survived indefinitely. The signs included ataxia, circling movements, and partial or complete paralysis of the limbs. The animal then appeared to recover briefly, but this was soon followed by rolling or rapid running movements, convulsion, respiratory arrest, and death. The auricles and usually the ventricles were still beating when the chest was opened within 1 minute after respiration had ceased. Some of the other pharmacological properties of *Synanceja* venom described by Saunders and his associates are discussed in the following section.

Venom extracts were prepared from the stings of *Pterois volitans* (Saunders, 1959*a*; Saunders and Taylor, 1959). Bioassay of the extracts was found to produce symptoms that were similar to those caused by stonefish venom. The only apparent difference was a more pronounced skeletal muscular weakness, i.e., the animal's head was often seen to be resting on the floor of the cage prior to death. The mean LD_{50} of freshly prepared extracts for mice by intravenous injection was about 1 mg of protein/kg. Storage of lyophilized or glycerol-treated extracts was similar to that for stonefish venom. The active fraction of the venom was nondialyzable.

Venom extracts were also prepared from the stings of *S. guttata* and tested on mice (Saunders, 1959*a*). These extracts were found to produce toxic effects similar to those produced by *Synanceja* and *Pterois* venoms, and storage durability was similar.

Russell (1971) has published a valuable review on the toxicology of scorpionfish venoms.

PHARMACOLOGY

The pharmacological properties of scorpionfish venom have been studied by

Wiener (1958, 1959*a*), Saunders (1958, 1959*a, b, c,* 1960*a*), Saunders and Taylor (1959), Austin, Cairncross, and McCallum (1961), and Saunders *et al.* (1962). Most of these investigations have been concerned with the venoms of *Synanceja horrida, S. verrucosa*, and to a lesser extent *Pterois volitans* (Saunders, 1959*a*; Saunders and Taylor, 1959). The primary effect on the cardiovascular system was the production of a marked hypotension accompanied by an increase in respiratory rate. The hypotension was believed to be due primarily to a peripheral vasodilatation, although a decrease in force of cardiac contractions was observed soon after the blood pressure began to fall. The location of the vascular bed in which vasodilatation occurred was not determined. These effects were transient in sublethal doses. However, larger doses caused more marked changes in arterial pressure, respiratory rate, and alterations in the electrocardiogram. Lethal doses caused a marked accentuation of these changes, including first degree atrioventricular block, ventricular fibrillation, and respiratory arrest. Death resulted in some of the experimental rabbits in 90 seconds. According to Saunders *et al.* (1962), adrenergic blockade was excluded as a mechanism of action of the venom, and evidence of a lack of effect on autonomic ganglionic transmission was obtained. The pressor responses to carotid artery occlusion, anoxia, and the depressor responses to central stimulation of the severed right vagus and depressor nerves were not affected by hypotensive doses of stonefish venom.

The findings of Austin *et al.* (1961) were for the most part consistent with those of Wiener and Saunders. However, they have made several additional observations worthy of note. They observed that the intravenous administration of stonefish venom to rabbits resulted in hypotension, respiratory arrest, and muscular paralysis. Their experimental evidence indicated that the systemic effects were due to the potent myotoxic properties of the venom which resulted in paralysis of cardiac, involuntary, and skeletal muscle. Paralysis developed with the appearance of a conduction block in the muscle. They suggested that the conduction block was due to a slow depolarization of the muscle and that the cause of death was due to paralysis of the diaphragm. Russell (1965) commented on the findings of Austin *et al.* (1961) thus:

> Their most important finding was that records of the action potential from the phrenic nerve indicated that following injection of the venom, respiratory movements were affected before efferent respiratory center activity, and also, before conduction in the phrenic nerve was impaired. This would indicate that stonefish venom either blocked neuromuscular conduction across the phrenic nerve-diaphragm junction or paralyzed the diaphragmatic musculature.

The pharmacological activity of *P. volitans* and *Scorpaena guttata* is similar to that of stonefish venom (Saunders, 1959*a, b, c*; Saunders and Taylor, 1959).

Russell (1965) has noted the remarkable similarity in the pharmacological properties of the venoms of *Urolophus, Trachinus, Pterois, Scorpaena*, and *Synanceja*. He believes that the venoms are probably closely related chemically.

Russell (1971) has also published a useful review on the pharmacology of scorpionfish venom.

CHEMISTRY

Very little is known about the chemistry of scorpionfish venoms. Chemical research has been directed primarily to the venom of a single species, namely, *Synanceja horrida*. Studies thus far indicate that the lethal component or components of these venoms are nondialyzable (Saunders, 1959*a, b*), possess antigenic properties (Wiener, 1958, 1959*a, b*), and are believed to be protein in nature. Further studies by Saunders and Tökés (1961) have shown that the fraction lethal to mice was insoluble in deionized water but was increasingly soluble with increasing ionic strength. The lethal fraction was most stable around neutrality, and activity was lost at pH 5.6 or less and at pH 10.6. The lethal fraction was found to be heat labile but could be stored for long periods in the lyophilized or glycerol-treated state at low temperatures (-20° C). Starch-gel electrophoresis in phosphate buffer at pH 8 and ionic strength of 0.076-0.1 generally revealed the presence of 7 or 8 components or even 10 components in some samples. Material lethal to mice could be recovered from only one of these bands, and the active fraction eluted from the gel was about twice as potent as the original venom. Re-electrophoresis of this active fraction gave only a single band.

No structural studies have been reported on the chemistry of scorpionfish venoms.

LITERATURE CITED

Atz, J.
1962a Stonefishes—the world's most venomous fishes. Aquarist & Pondkeeper 27(8): 146-148, 5 figs.
1962b The flamboyant zebrafish. Animal kingdom 55(2): 34-40.

Austin, L., Cairncross, K. D., and I. A. McCallum
1961 Some pharmacological actions of the venom of the stonefish *"Synanceja horrida."* Arch. Intern. Pharmacodyn. 131(3-4): 339-347, 7 figs.

Bagnis, R.
1968 A propos de 51 cas de piqûres venimeuses par la "rascasse" tropicale *Synanceja verrucosa* dans les Iles de la Société et des Tuamotus. Méd. Trop. 28(5): 612-620.

Bayley, H. H.
1940 Injuries caused by scorpion fish. Trans. Roy. Soc. Trop. Med. Hyg. 34(2): 227-230, 1 pl.

Blanchard, R.
1890 Traité de zoologie médicale. Vol. 32. J. B. Baillière et Fils, Paris. p. 638-695.

Bonne, W. M., and K. Neuhaus
1936 Vergiftiging door de steek van een visch. Geneesk. Tijdschr. Nederlandsch-Indie 76: 2404-2405, 1 fig.

Bottard, A.
1889a L'appareil à venin des poissons. Compt. Rend. Acad. Sci. 108: 534-537.
1889b Les poissons venimeux. Octave Doin, Paris. 198 p.

Breder, C. M., Jr.
1963 Defensive behavior and venom in *Scorpaena* and *Dactylopterus*. Copeia (4): 698-700, 2 figs.

Briot, A.
1904 La rascasse a-t-elle un venin? Compt. Rend. Soc. Biol. 57: 666-667.
1905 Sur l'action soi-distant venimeuse de la rascasse. Compt. Rend. Assoc. Franc. Adv. Sci. 33: 904.

Cameron, A. M., and R. Endean
1966 The venom apparatus of the scorpion fish *Notesthes robusta*. Toxicon 4: 111-121, 8 figs.

Carpenter, S.
1965 A venomous lash from a scorpionfish. 200 feet down, the next U.S. frontier. Life Mag. 59(16): 100-106, figs.

Cecca, R.
1902 Sugli effetti tossici delle punture di alcuni pesci. Clin. Med. Ital. 41: 82-87.

Cilento, R.
1944 Some poisonous plants, sea and land animals of Australia and New Guinea (notes for medical, nursing and ambulance staffs). W. R. Smith and Paterson Pty. Ltd., Brisbane, Australia. 37 p.

Clarke, T. A., Flechsig, A. O., and R. W. Grigg
1967 Ecological studies during project Sealab II. Sci. 157 (3795): 1381-1389, 7 figs.

Cleland, J. B.
1912 Injuries and diseases of man in Australia attributable to animals (except insects). Australasian Med. Gaz. 32(11): 269-274; (12): 297-299.
1932 Injuries and diseases in Australia attributable to animals (other than insects). Med. J. Australia, 4, 1932: 1257-166.

Colby, M.
1943 Poisonous marine animals in the Gulf of Mexico. Trans. Texas Acad. Sci. 26: 62-69.

Coutière, H.
1899 Poissons venimeux et poissons vénéneux. Thèse Agrég. Carré et Naud, Paris. 221 p.

Deakins, D. E., and P. R. Saunders
1967 Purification of the lethal fraction of the venom of the stonefish *Synanceja horrida* (Linnaeus). Toxicon 4: 257-262, 1 fig.

De Marco, R.
1937 Effetti del veleno di *"Trachinus"* e di quello di *"Scorpaena"* sulla attività neuromuscolare della rana. Arch. Fisiol. 37(3): 398-404.

DUHIG, J. V.
1929 The nature of the venom of "*Synanceja horrida*" (the stonefish). Z. Immunitätsforsch. 62(3-4): 185-189, 2 figs.

DUHIG, J. V., and G. JONES
1928*a* The venom apparatus of the stone fish (*Synanceja horrida*). Mem. Queensland Mus. 9(2): 136-150, 8 figs.
1928*b* Haemotoxin of the venom of *Synanceja horrida*. Australian J. Exp. Biol. 5(2): 173-179.

DUNBAR-BRUNTON, J.
1896 The poison-bearing fishes, *Trachinus draco* and *Scorpaena scrofa*: the effects of the poison on man and animals and its nature. Lancet 2(3809): 600-602.

ENDEAN, R.
1961 A study of the distribution, habitat, behaviour, venom apparatus, and venom of the stone-fish. Australian J. Marine Freshwater Res. 12(2): 177-190, 3 pls., 3 figs.

ENGELSEN, H.
1922 Om giftfisk og giftige fisk. Nord. Hyg. Tidskrift 3: 316-325.

ESCHMEYER, W. M.
1965 Three new scorpionfishes of the genera *Pontinus, Phenacoscorpius*, and *Idiastion* from the Western Atlantic Ocean. Bull. Mar. Sci. 15(3): 521-534, 5 figs.
1969 A systematic review of the scorpionfishes of the Atlantic Ocean (Pisces: Scorpaenidae). Calif. Acad. Sci. Occ. Pap. No. 79. 130 p., 13 figs.

ESCHMEYER, W. N., and J. E. RANDALL
1975 The scorpaenid fishes of the Hawaiian Islands, including new species and new records (Pisces: Scorpaenidae). Proc. Calif. Acad. Sci. 40(11): 265-334, 25 figs.

ESCHMEYER, W. N., and K. V. R. RAO
1973 Two new stonefishes (Pisces: Scorpaenidae) from the Indo-West Pacific, with a synopsis of the subfamily Synanceiinae. Proc. Calif. Acad. Sci. 39(18): 337-382, 13 figs.

FAUST, E. S.
1924 Tierische gifte. Handbuch Exp. Pharm., Berlin.

FOWLER, H. W.
1928 The fishes of Oceania. Mem. Bernice P. Bishop Mus. 10, 466 p., 49 pls.

GAIL, R., and J. RAGEAU
1956 Premières sur un poisson marin venimeux de la Nouvelle-Calédonie: la synancée (*Synanceia verrucosa* Bloch). Bull. Soc. Pathol. Exotique 49(5): 846-854.

GALLAGHER, M. J.
1928 Clinical notes on a case of poisoning by stone-fish venom. Mem. Queensland Mus. 9(2): 148-150.

GINSBURG, I.
1953 Western Atlantic scorpionfishes. Smithsonian Inst. Misc. Collections 121(8), 103 p., 6 figs.

HALSTEAD, B. W., and L. J. CARSCALLEN
1964 Marine organisms, p. 336-343. *In* P. L. Altman and D. S. Dittmer [eds.], Biology data book. Fed. Am. Soc. Exp. Biol., Washington, D.C.

HALSTEAD, B. W., CHITWOOD, M. J., and F. R. MODGLIN
1955*a* The venom apparatus of the California scorpionfish, *Scorpaena guttata* Girard. Trans. Am. Microscop. Soc. 74(2): 145-158, 15 figs.
1955*b* The anatomy of the venom apparatus of the zebrafish, *Pterois volitans* (Linnaeus). Anat. Rec. 122(3): 317-333, 16 figs.
1956 Stonefish stings, and the venom apparatus of *Synanceja horrida* (Linnaeus). Trans. Am. Microscop. Soc. 75(4): 381-397, 12 figs.

HALSTEAD, B. W., and L. R. MITCHELL
1963 A review of the venomous fishes of the Pacific area, p. 173-202, 16 figs. *In* H. L. Keegan and W. V. MacFarlane [eds.], Venomous and poisonous animals and noxious plants of the Pacific region. Pergamon Press, New York.

HERRE, A. W.
1927 Fishery resources of the Philippines. Philippine Islands Bur. Sci. Pop. Bull. (3): 60-65.
1952 A review of the scorpaenoid fishes of the Philippines and adjacent seas. Philippine J. Sci. 80(4): 381-482.
1953 Check list of Philippine fishes. U.S. Fish Wildlife Serv., Res. Rept. 20. 977 p.

HINTON, S.
1962 Unusual defense movements in *Scorpaena plumieri mystes*. Copeia (4): 842.

Joubin, L. M.
1929-38 Faune ichthyologique de l'Atlantique nord. Nos. 1-18. A. F. Høst et Fils, Copenhague.

Jüptner, H.
1960 Verletzungen durch steinfische *(Synanceja trachynis)*. Z. Tropenmed. Parasitol. 11: 475-477.

Kesteven, L.
1914 The venom of the fish *Notesthes robusta*. Proc. Linnean Soc. NSW 39(1): 91-92.

Klausewitz, W.
1957 Feuerfische der Gattungen *Dendrochirus* und *Pterois*. Aquar.-Terrar.-Z. (12): 319-323, 5 figs.

Legeleux, G.
1965 Contribution à l'étude des poissons toxicophores. Imprimerie Douladoure, Toulouse, France. 124 p.

Le Juge, E.
1871 Note sur le *Synanceia brachio* (Cuvier). Soc. Roy. Arts Sci. Maurice 5: 19-24.

Lozano y Rey, L.
1947 Peces ganoideos y fisóstomos. Mem. Real Acad. Cienc., Madrid. Vol. 11, 839 p., 20 pls., 189 figs.
1952 Los peces fluviales de España. Ministerio de Agricultura, Madrid. 251 p., 31 figs.

Lumière, A., and P. Meyer
1938 La rascasse a-t-elle un venin? Compt. Rend. Soc. Biol. 127(4): 328-330.

McCulloch, A. R.
1929-30 A check-list of the fishes recorded from Australia. Mem. 5. Australian Mus., Sydney. 534 p.

Maretić, Z.
1962 Venomous animals. Lijčnički Vjesnik 84(12): 1233-1252.
1966 On venomous animals of the Adriatic Sea. [In Croatian] Morsko Ribarst. 18: 168-173, 7 figs.
1968 Venomous animals and their toxins. [In Croatian] Pro Medico 3: 93-114.

Marshall, T. C.
1964 Fishes of the Great Barrier Reef and coastal waters of Queensland. Angus and Robertson, Melbourne. 566 p., 72 pls.

Matsubara, K.
1943 Studies on the scorpaenoid fishes of Japan. Trans. Sigenkagaku Kenkyusyo (1-2), 486 p., 4 pls.

Pawlowsky, E. N.
1906 Microscopic structure of the poison glands of *Scorpaena porcus* and *Trachinus draco*. [In Russian] Trav. Soc. Imp. Nat., St. Petersburg 37(1): 316-337, 1 pl.
1907 Contribution to the structure of the epidermis and its glands in venomous fish. [In Russian] Trav. Soc. Imp. Nat., St. Petersburg 38(1): 265-282, 1 pl., 1 fig.
1909 Contribution to the question of poisonous skin glands of certain fishes. [In Russian] Trav. Soc. Imp. Nat., St. Petersburg 40(1): 109-126, 5 figs.
1911 Contribution to the question of the structure of the poison glands of certain fish of the Scorpaenidae family. [In Russian] Trav. Soc. Imp. Nat., St. Petersburg 41(1): 317-328.
1913 Sur la structure des glandes à venin de certains poissons et en particulier de celles de *Plotosus*. Compt. Rend. Soc. Biol. 74(18): 1033-1036, 1 fig.
1914 Über den bau der giftdrüsen bei *Plotosus* und anderen fischen. Zool. Jahrb. 38: 427-442, 3 pls., 4 figs.
1927 Gifttiere und ihre giftigkeit. G. Fischer, Jena. 515 p., 170 figs.

Phillips, J. B.
1957 A review of the rockfishes of California (family Scorpaenidae). Calif. Dept. Fish and Game, Fish Bulletin No. 104. 158 p., 66 figs.

Phillips, C., and W. H. Brady
1953 Sea pests—poisonous or harmful sea life of Florida and the West Indies. Univ. Miami Press, Miami. 78 p., 7 pls., 38 figs.

Phisalix, M.
1931 Le venin de quelques poissons marins. Notes Sta. Océanogr. Salammbo, Tunis (22): 3-15.

Pohl, J.
1893 Beitrag zur lehre von den fischgiften. Prag. Med. Wochschr. 18(4): 31-33.

Poll, M.
1947 Poissons marins. Faune de Belgique. Mus. Roy. Hist. Nat. Belgique, Brussels. 452 p., 267 figs.
1951 Expédition océanographique Belge dans les eaux côtières africaines de l'Atlantique sud (1948-49). Résultats

POLL, M.—Continued
scientifiques. Vol. IV, No. 1. Poissons. I—Généralités. Institut Royal des Sciences Naturelles de Belgique, Brussels. 154 p., 13 pls.

1954 Expédition océanographique Belge dans les eaux côtières africaines de l'Atlantique sud (1948-49). Résultats scientifiques. Vol. IV, No. 3a. Poissons. IV—Téléostéens acanthoptérygiens (1st pt.). Institut Royal des Sciences Naturelles de Belgique, Brussels. 390 p., 9 pls.

1959 Expédition océanographique Belge dans les eaux côtières africaines de l'Atlantique sud (1948-49). Résultats scientifiques. Vol. IV, No. 3b. Poissons. V—Tèlèostèens acanthoptérygiens (2d pt.). Institut Royal des Sciences Naturelles de Belgique, Brussels. 417 p., 7 pls.

RALPH, C. C.
1943 Poison of the stonefish. Victorian Naturalist 60: 77.

ROCHE, E. T.
1973 Venomous marine fishes of California. Calif. Dept. Fish and Game, Marine Resources Leaflet No. 4. 11 p., 9 figs.

ROCHE, E. T., and B. W. HALSTEAD
1972 The venom apparatus of California rockfishes (family Scorpaenidae). Calif. Dept. Fish and Game, Fish Bulletin 156. 49 p., 45 figs.

ROMANO, S.
1940 Animali velenosi della fauna Italiana. Natura 31(4): 137-167, 7 figs.

ROUGHLEY, T. C.
1947 Wonders of the Great Barrier Reef. Charles Scribner's Sons, New York. 282 p.

RUSSELL, F. E.
1961 Injuries by venomous animals in the United States. J. Am. Med. Assoc. 177(13): 903-907.

1965 Marine toxins and venomous and poisonous marine animals. Adv. Marine Biol. 3: 255-385.

1967 Comparative pharmacology of some animal toxins. Fed. proc. 26(4): 1206-1224.

1971 Pharmacology of toxins of marine organisms. *In* H. Rašková [ed.], Internat. Encycl. Pharm. Therap., Sect. 71, Pharmacology and toxicology of naturally occurring toxins. Vol. 2 (Part III): 3-114.

SACCHI, M.
1895*a* Sulla struttura degli organi del veleno della Scorpena. I. Spine delle pinne impari. Boll. Mus. Zool. Genova (30): 1-10.

1895*b* Sulla struttura degli organi del veleno della Scorpena. II. Spine delle pinne pari. Boll. Mus. Zool. Genova (36): 1-4, 1 pl.

SAUNDERS, P. R.
1958 Venom of the stonefish *(Synanceja horrida)*. Federation Proc. 17(1): 1612.

1959*a* Venoms of scorpionfishes. Proc. Western Pharmacol. Soc. 2: 47-54.

1959*b* Venom of the stonefish *Synanceja horrida* (Linnaeus). Arch. Intern. Pharmacodyn. 123: 195-205.

1959*c* Venom of the stonefish *Synanceja verrucosa*. Science 129(3344): 272-274, 1 fig.

1960*a* Pharmacological and chemical studies of the venom of the stonefish (Genus *Synanceja)* and other scorpionfishes. Ann. N.Y. Acad. Sci. 90(3): 784-804, 2 figs.

1960*b* Sting by a venomous lionfish. U.S. Armed Forces Med. J. 11(2): 224-227.

SAUNDERS, P. R., ROTHMAN, S., MEDRANO, V. A., and H. P. CHIN
1962 Cardiovascular actions of venom of the stonefish *Synanceja horrida*. Am. J. Physiol. 203(3): 429-432, 4 figs.

SAUNDERS, P. R., and P. B. TAYLOR
1959 Venom of the lionfish *Pterois volitans*. Am. J. Physiol. 197(2): 437-440, 3 figs.

SAUNDERS, P. R., and L. TÖKÉS
1961 Purification and properties of the lethal fraction of the venom of the stonefish *Synanceja horrida* (Linnaeus). Biochim. Biophys. Acta 52: 527-532.

SAVILLE-KENT, W.
1900 The Great Barrier Reef of Australia. Its products and potentialities. 2d ed. W. H. Allen and Co., Ltd., London. 387 p., 61 pls.

SCHNACKENBECK, W.
1943 Fischgifte und fischvergiftungen. Münch. Med. Wochschr. 90(8): 149-151.

Schnee, S.
1908 Vorläufige mitteilungen über eine beobachtete vergiftung durch den feuerfisch *(Pterois)*. Arch. Schiffs. Tropenhyg. 12: 166-167.
1911 Drei fälle von verletzung durch den giftigen fisch Synanceia (Nufu). Arch. Schiffs Tropenhyg. 15: 312-316.

Schultz, L. P., Woods, L. P., and E. A. Lachner
1966 Fishes of the Marshall and Marianas Islands, vol. 3. U.S. Nat. Mus., Bull. 202. 176 p., 24 pls., 156 figs.

Scott, H. H.
1921 Vegetal and fish poisoning in the tropics, p. 790-798, 5 figs., 2 pls. *In* W. Byam and R. G. Archibald [eds.], The practice of medicine in the tropics. Vol. I. H. Frowde and Hodder and Stoughton, London.

Smith, J. L.
1950 The sea fishes of southern Africa. Central News Agency, Cape Town, S. Africa. 550 p., 103 pls.
1951 A case of poisoning by the stonefish, *Synanceja verrucosa*. Copeia (3): 207-210.
1957 The fishes of the family Scorpaenidae in the Western Indian Ocean. Part II. The subfamilies Pteroinae, Apistinae, Setarchinae and Sebastinae. Rhodes Univ., Dept. Ichthyology, Ichthyological Bull. (5): 75-87, 2 pls., 4 figs.
1958 Fishes of the families Tetrarogidae, Caracanthidae and Synanciidae, from the Western Indian Ocean with further notes on Scorpaenid fishes. Rhodes Univ., Dept. Ichthyology, Ichthyological Bull. (12): 167-181, 2 pls.
1961 The sea fishes of southern Africa. 4th ed. Central News Agency, South Africa. 580 p., 102 pls., 1,232 figs.

Steinitz, H.
1959 Observations on *Pterois volitans* (L.) and its venom. Copeia (2): 158-160.

Tange, Y.
1953*a* Beitrag zur kenntnis der morphologie des giftapparates bei den japanischen fischen, nebst bemerkungen über dessen giftigkeit. I. Über das vorkommen des giftapparates bei den japanischen knochenfischen. Yokohama Med. Bull. 4(2): 120-128, 2 figs.
1953*b* Beitrag zur kenntnis der morphologie des giftapparates bei den japanischen fischen, nebst bemerkungen über dessen giftigkeit. II. Über den giftapparat bei *Pterois lunulata* Temminck et Schlegel. Yokohama Med. Bull. 4(3): 178-184, 3 figs.
1953*c* Beitrag zur kenntnis der morphologie des giftapparates bei den japanischen fischen, nebst bemerkungen über dessen giftigkeit. III. Über den giftapparat bei *Apistus evolans* Jordan et Starks. Yokohama Med. Bull. 4(5): 318-324, 3 figs.
1953*d* Beitrag zur kenntnis der morphologie des giftapparates bei den japanischen fischen, nebst bemerkungen über dessen giftigkeit. IV. Über den giftapparat bei *Minous adamsii* Richardson. Yokohama Med. Bull. 4(6): 374-380, 4 figs.
1954*a* Beitrag zur kenntnis der morphologie des giftapparates bei den japanischen fischen, nebst bemerkungen über dessen giftigkeit. VI. Über den giftapparat bei *Erosa erosa* (Langsdorf). Yokohama Med. Bull. 5(2): 118-124, 3 figs.
1954*b* Beitrag zur kenntnis der morphologie des giftapparates bei den japanischen fischen, nebst bemerkungen über dessen giftigkeit. VII. Über den giftapparat bei *Inimicus japonicus* (Cuvier et Valenciennes). Yokohama Med. Bull. 5(3): 234-242, 5 figs.
1954*c* Beitrag zur kenntnis der morphologie des giftapparates bei den japanischen fischen, nebst bemerkungen über dessen giftigkeit. IX. Über den giftapparat bei *Sebastodes inermis* (Cuvier et Valenciennes). Yokohama Med. Bull. 5(6): 429-434, 3 figs.

Taschenberg, E. O.
1909 Die giftigen tiere. F. Enke, Stuttgart. 325 p., 64 figs.

Taylor, P. B.
1963 The venom and ecology of the California *Scorpaena guttata* Girard. Ph.D. dissertation, Univ. Calif., San Diego. 138 p.

TUMA, V.

1927 Les glandes vénéneuses chez la rascasse *(Scorpaena porcus* L.). Trav. Inst. Histol. Embryol. Fac. Med. Univ. Charles, Prague, p. 153-154, 3 figs.

TWEEDIE, M. W.

1941 Poisonous animals of Malaya. Malaya Publishing House, Singapore. 90 p., 29 figs.

WHITLEY, G. P.

1932 The chinaman fish. Australian Mus. Mag. 4(11): 394-396, 2 figs.

1935 Some fishes of the Sydney district. Australian Mus. Mag. 5(9): 291-304, 35 figs.

1943 Poisonous and harmful fishes. Council Sci. Ind. Res., Bull. No. 159. 28 p., 16 figs.; Pls. 1-3.

1949 Solvol fish book. Boylan and Co., Sydney. 20 p., 3 pls.

1963 Dangerous Australian fishes, p. 41-64, 2 pls. *In* J. W. Evans [chmn], Proc. first international convention on life saving techniques. Part III. Scientific Section. Suppl. Bull. Post Grad. Comm. Med., Univ. Sydney.

WHITLEY, G. P., and W. BOARDMAN

1929 Quaint creatures of a coral isle. Australian Mus. Mag. 3(11): 366-374, 13 figs.

WIENER, S.

1958 Stonefish sting and its treatment. Med. J. Australia 45(7): 218-222, 2 figs.

1959*a* Observations on the venom of the stone fish *(Synanceja trachynis)*. Med. J. Australia 1(19): 620-627.

1959*b* The production and assay of stonefish antivenene. Med. J. Australia 2: 715-719, 11 figs.

Chapter XXII—VERTEBRATES

Class Osteichthyes

Venomous: TOADFISHES

The Batrachoididae, or toadfishes, are a group of small bottom fishes which inhabit the warmer waters of the coasts of America, Europe, Africa, and India. Toadfishes are of little commercial value and are not generally considered as food fishes, although they are eaten in some countries. The flesh is said to be fine flavored, but the fishes are small and bony. According to Taschenberg (1909), the liver of some of the batrachoids is poisonous to eat. However, the greatest interest of these fishes to biotoxicologists is their unique and highly developed venom organs.

The systematics and identifying characteristics of batrachoid fishes have been discussed by Day (1865, 1878-88), Jordan and Evermann (1896-1900), Gilbert and Starks (1904), Bean and Weed (1910), Meek and Hildebrand (1923), Jordan, Evermann, and Clark (1930), Hubbs and Schultz (1939), Fowler (1936, 1941), Smith (1952, 1961), Mendis (1954), Poll (1959), Marshall (1964), Cervigon (1966), and Collette (1966).[1]

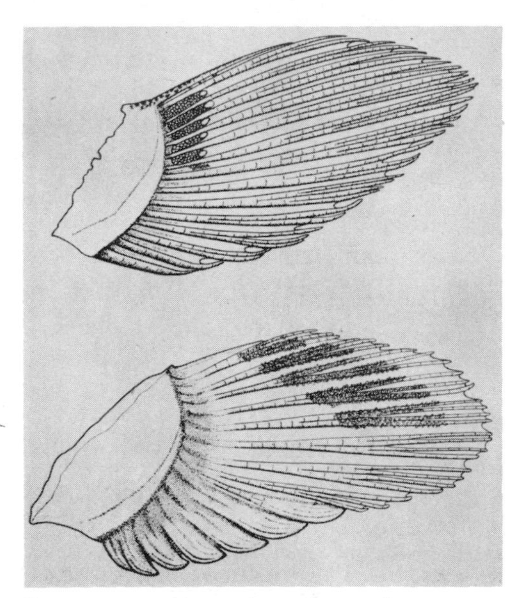

FIGURE 1.—Posterior view of pectoral fins of *Thalassophryne* showing the distribution of the pectoral fin glands. Upper is *T. dowi*; lower, *T. maculosa*. (From Collette)

[1] Collete (1966) has resurrected the genus *Daector* for the eastern Pacific species of venomous toadfishes: *reticulata* Günther, *gerringi* Rendahl, and *dowi* Jordan and Gilbert. He states, "The most significant character that separates the the three eastern Pacific species *(Daector)* of Thalassophryninae from the six western Atlantic *(Thalassophryne)* is the presence of discrete glands between bases of the uppermost pectoral fin rays." In actuality it is not a matter of the presence or absence of these glands, but rather it concerns their anatomical position on the pectoral fins (*see* Fig. 1). In the eastern Pacific species the glands are situated between the bases of the upper pectoral fin rays, whereas in the western Atlantic species the same glands are situated more distally on the upper pectoral fin rays. Merely the differences in position of these glands on the pectoral fins, in the number of precaudal and caudal vertebrae, and in the number of second dorsal and anal fin rays appear to be insufficient to warrant generic status for *Daector*. Therefore Bean and Weed (1910) have been followed in this monograph for the nomenclature of the Thalassophryninae. Some toadfishes are apparently equipped with both a venom apparatus and ichthyocrinotoxic organs (axillary and pectoral glands). However, the ichthyocrinotoxic nature of the batrachoid fishes has not been fully determined.

REPRESENTATIVE LIST OF BATRACHOID FISHES REPORTED AS VENOMOUS

Phylum CHORDATA

Class OSTEICHTHYES

Order BATRACHOIDIFORMES (HAPLODOCI): Toadfishes

Family BATRACHOIDIDAE

252 *Barchatus cirrhosus* (Klunzinger) (Pl. 1, fig. a). Toadfish (USA).
DISTRIBUTION: Red Sea.

SOURCES: Demoreau (1908), Phisalix (1922), Smith (1952), Halstead (1959), Halstead and Carscallen (1964).
OTHER NAME: *Batrachus cirrhosus.*

252 *Batrachoides didactylus* (Bloch) (Pl. 1, fig. b). Toadfish (USA), sapo (Spain).
DISTRIBUTION: Mediterranean Sea and nearby Atlantic coasts.

SOURCES: Gill (1907), Halstead (1959), Halstead and Carscallen (1964).
OTHER NAMES: *Batrachus didactylus, B. liberiensis.*

252 *Batrachoides grunniens* (Bloch) (Pl. 1, fig. c). Toadfish (USA), munda (Sri Lanka).
DISTRIBUTION: Coasts of Sri Lanka, India, Burma, Malaysia.

SOURCES: Cantor (1839), Calmette (1908), Taschenberg (1909), Dunlop (1917), Castellani and Chalmers (1919), Scott (1921), Halstead (1956, 1959), Halstead and Carscallen (1964).
OTHER NAME: *Batrachus grunniens.*

256 *Halophryne diemensis* (LeSueur) (Pl. 5, fig. a). Toadfish (USA), banded frogfish (Australia).
DISTRIBUTION: Australia, New Guinea.

SOURCE: Marshall (1964).

252 *Marcgravichthys cryptocentrus* (Valenciennes) (Pl. 1, fig. d). Toadfish (USA), niguim, niguin, sapo bocon (Brazil).
DISTRIBUTION: Panama to Brazil.

SOURCE: Demoreau (1908).
OTHER NAMES: *Amphichthys cryptocentrus, Batrachus cryptocentrus.*

253 *Opsanus tau* (Linnaeus) (Pl. 2, fig. a). Oyster toadfish (USA), sapo (West Indies).
DISTRIBUTION: Atlantic coast from Massachusetts to West Indies.

SOURCES: Wallace (1893), Gill (1907), Reed (1907), Calmette (1908), Taschen-

berg (1909), Gudger (1910), Castellani and Chalmers (1919), Phisalix (1922), Taft (1945), Phillips and Brady (1953), Halstead (1959), Halstead and Carscallen (1964).
OTHER NAME: *Batrachus tau.*

Thalassophryne amazonica Steindachner (Pl. 2, fig. b). Brazilian toadfish (USA), pocomon, niquim, niquin, sapo (Brazil). *253*
DISTRIBUTION: Amazon river and its tributaries, Brazil. Primarily a freshwater species, but may be taken in brackish water at the river mouth.

SOURCES: Pawlowsky (1927), Froes (1933*a, b*).

Thalassophryne dowi Jordan and Gilbert (Pl. 2, fig. c). Dow's toadfish (USA), sapo (Latin America). *253*
DISTRIBUTION: Pacific coast of Central America from Punta Arenas, Costa Rica, to Panama.

SOURCES: Gill (1907), Bean and Weed (1910), Halstead (1956), Halstead and Mitchell (1963).

Thalassophryne maculosa Günther (Pl. 3, fig. a, b). Toadfish (USA), sapo (West Indies). *254*
DISTRIBUTION: Caribbean Sea.

SOURCES: Günther (1864), Fayrer (1878), Linstow (1894), Coutière (1899), Kobert (1902), Gill (1907), Gonsalves (1907), Calmette (1908), Demoreau (1908), Taschenberg (1909), Silvado (1911), Pryor (1918), Castellani and Chalmers (1919), Engelsen (1922), Phisalix (1922), Faust (1924), Pawlowsky (1927), Froes (1932, 1933*a, b*).

Thalassophryne montevidensis Berg (Pl. 3, fig. c). Toadfish (USA), sapo (Buenos Aires). *254*
DISTRIBUTION: Buenos Aires.

SOURCES: Bean and Weed (1910), Pawlowsky (1927).
OTHER NAME: *Thalassothia montevidensis.*

Thalassophryne nattereri Steindachner (Pl. 4, fig. a). Toadfish (USA), pocomon, niquim, niquin, sapo (Brazil). *255*
DISTRIBUTION: Baía, Brazil.

SOURCE: Froes (1933*a, b*).
OTHER NAMES: *Thalassophryne branneri, T. maculosa* (Pellegrin, *non* Günther).

Thalassophryne punctata Steindachner (Pl. 4, fig. b). Toadfish (USA), niquim, niquin, sapo (Brazil). *255*
DISTRIBUTION: Baía, Brazil.

SOURCES: Pawlowsky (1927), Froes (1933*a, b*).

255 *Thalassophryne reticulata* Günther (Pl. 4, fig. c). Toadfish (USA), sapo (Latin America).

DISTRIBUTION: Pacific coast of Central America.

SOURCES: Günther (1864), Woodward (1864), Day (1871, 1878-88), Fayrer (1878), Savtschenko (1886), Bottard (1889), Linstow (1894), Kobert (1894, 1902, 1905), Coutière (1899), Gill (1907), Reed (1907), Calmette (1908), Demoreau (1908), Taschenberg (1909), Bean and Weed (1910), Pawlowsky (1911, 1927), Silvado (1911), Hilzheimer and Haempel (1913), Pryor (1918), Castellani and Chalmers (1919), Scott (1921), Engelsen (1922), Phisalix (1922), Faust (1924), Froes (1932, 1933*a*, *b*), Norman (1937), Giunio (1948), Halstead (1959), Halstead and Mitchell (1963), Halstead and Carscallen (1964).

BIOLOGY

Family BATRACHOIDIDAE: Batrachoid fishes, with their broad depressed heads and large mouths, are somewhat repulsive in appearance. Most toadfishes are marine shore forms, but some are estuarine or entirely fluviatile, ascending rivers for great distances. The geographical distribution of toadfishes is shown on Map 1. They appear to enjoy turbid water. Regardless of the type of water in which they are found, batrachoids are primarily bottom fishes. They hide in crevices and burrows, under rocks and debris, or among seaweed or lie almost completely buried under a few centimeters of sand or mud. Froes (1932, 1933*a*) stated that the Brazilian species of *Thalassophryne* has the habit of covering itself with a thin layer of sand or mud, but with careful observation one can usually detect the outline and protruding eyes of the fish as one wades along in the clear shallow water of sandy beaches. Toadfishes are quite hardy and are able to live for several hours after being removed from the water. According to Goode (1884), the bottom temperature of the water frequented by these fishes would appear to range from 10° C to 32° C. During the winter months toadfishes tend to migrate to deeper water where they remain in a somewhat torpid condition. They also are experts at camouflage. Their ability to change color to lighter or darker shades at will and their mottled pattern make these fishes difficult to see.

Most toadfishes tend to be somewhat sluggish in their movements, but when going after food they can dart out with surprising rapidity. They are somewhat omnivorous in habit but seem to prefer, among other things, crabs, mollusks, worms, and small fishes. Toadfishes are said to be quite vicious and will snap at almost anything upon the slightest provocation. Although they are not capable of producing a severe wound, they can inflict a severe bite. When they are disturbed or their dorsum is touched, they immediately erect their dorsal spines and flare out their opercular spines in defiance. Toadfishes do not school, but they are gregarious and tend to congregate together. For additional information regarding the habits of these interesting fishes the reader is referred to the works of Gill (1907) and Gudger (1910).

MORPHOLOGY OF THE VENOM APPARATUS

Gross Anatomy

The venom apparatus of *Thalassophryne dowi* consists of two dorsal spines, two opercular spines which are located one on either side of the head (Fig. 2), their associated venom glands, and their enveloping integumentary sheaths. In non-traumatized specimens the spines are almost completely hidden by their sheaths, which can be easily retracted to expose their acute tips. The following gross anatomical descriptions are based on three specimens having a standard length of 123 to 230 mm. The fish were taken at Punta de Juan, Panama, July 2, 1953, during the Woodrow G. Krieger Panama Expedition of the School of Tropical and Preventive Medicine, Loma Linda, Calif.

Opercular Stings (Figs. 3-4): This particular description is based on the right opercular spine in the anatomical position. Removal of the integument covering the operculum reveals that the opercular spine, with the exception of the extreme distal tip, is encased within a glistening, whitish, pyriform mass. The broad rounded portion of this mass is situated at the base of the spine and tapers rapidly as the tip of the spine is approached. Attached to this mass

FIGURE 2.—The head of the toadfish *Thalassophryne dowi* showing the location of the opercular and dorsal stings (see arrows). The venom organs of the toadfishes are among the most highly developed of any group of fishes.
(R. Kreuzinger)

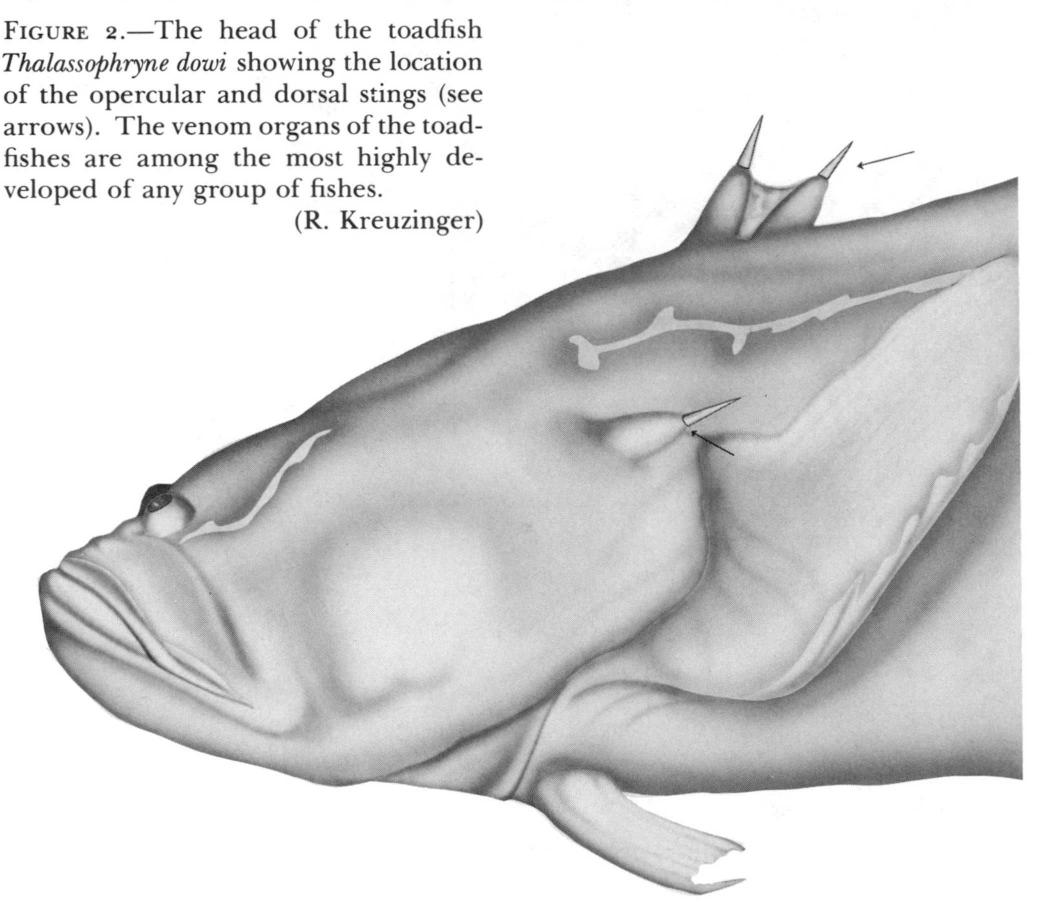

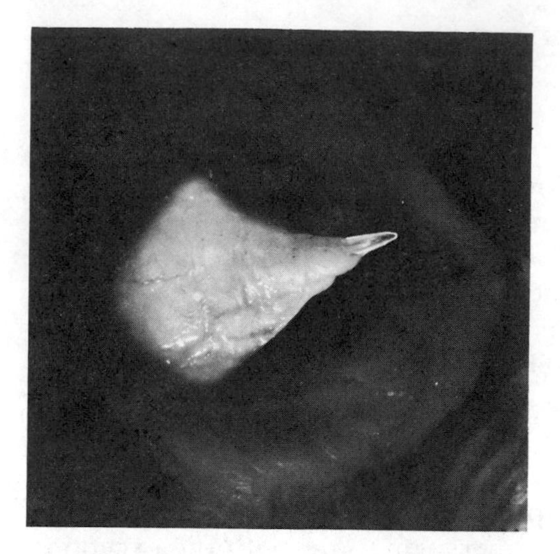

FIGURE 3.—Photograph of the left opercular sting of *Thalassophryne dowi*. The integumentary sheath has been removed to expose the venom gland. The distal opening of the venom canal is barely visible. Length of spine from the base of the gland to the tip of spine is 10 mm.
(R. Kreuzinger)

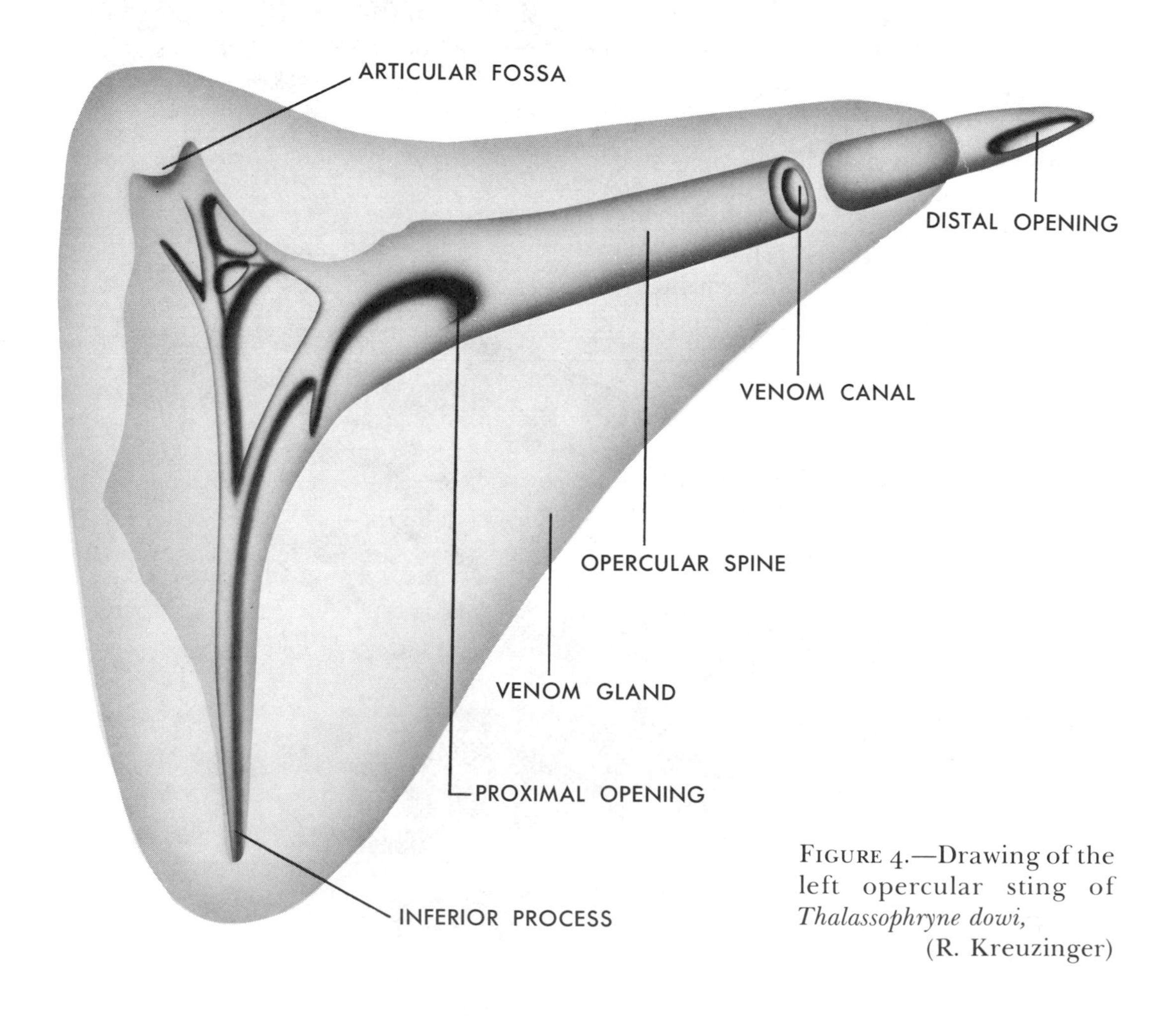

FIGURE 4.—Drawing of the left opercular sting of *Thalassophryne dowi*,
(R. Kreuzinger)

FIGURE 5.—Photograph of the dorsal stings of *Thalassophryne dowi*. The integumentary sheath has been removed to expose the large venom glands which are seen at the base of the two dorsal spines. Length of first dorsal spine from the base of the venom gland to the tip of the spine is 11 mm. (R. Kreuzinger)

are loose connective and adipose tissues. By careful dissection it is observed that the mass is situated within a special compartment that is separated from the pre-opercular muscle by a thin sheet of connective tissue attached to the posterior margin of the muscle. The pyriform mass consists of a tough saclike outer covering of connective tissue containing a soft, granular, gelatinouslike substance having the appearance of fine tapioca. Microscopic sections of the mass indicate that it is the opercular venom gland. Dissection of the gland fails to reveal any gross evidence of ducts. Apparently the hollow spine serves as the glandular duct.

The operculum of *T. dowi* is highly specialized as a defensive organ for the introduction of venom. The bone is roughly ┌-shaped. The horizontal limb of the operculum is a slender hollow bone which curves slightly lateral and terminates in an acute tip. Located on the lateral aspect of the tip of the spine is the ovoid distal opening of the venom canal, whereas the proximal opening is located on the inferior aspect at the base of the spine. The inferior limb or process of the opercle is ridged on the lateral aspect and flattened on the medial side but is not developed as a spine. The articular fossa is located superior to the junction of the horizontal and inferior processes of the opercle.

Movements of the opercular spine are controlled by two muscles. Abduction of the opercle is by contraction of a large fan-shaped muscle which originates on the superior aspect of the frontal and supra-occipital bones and inserts on the roughened superior aspect of the articular fossa of the opercle. Adduction of the opercle is by contraction of a muscle which originates on the posterolateral projection of the supra-occipital and inserts on the superior aspect of the base of the opercular spine.

Dorsal Stings (Figs. 5-6): This description is based on extended dorsal spines in their anatomical position. The two dorsal spines are enclosed within a single integumentary sheath but separated from each other by a thin sheet of connective tissue. Removal of the sheath reveals that each of the dorsal spines is further enveloped by loose connective tissue and has an individual venom gland similar in shape and structure to those previously described for the opercular stings.

The dorsal spines are likewise highly specialized as defensive organs for the introduction of venom. The first dorsal spine is about the same length as that of the opercle, but the second dorsal spine is slightly longer. The dorsal spines are slender and hollow, curve craniad slightly, and terminate in acute tips. Located on the anterior aspect of the tip of the spine is the ovoid distal opening. The proximal opening is situated on the anterior aspect of the base. From above downward, the spines gradually broaden and then in the proximal third rather abruptly expand into the triangular-shaped basal articulations. Immediately beneath the proximal opening of each spine is the median notch which articulates with the interneural spine. On either side of the median notch are the lateral condyles which likewise articulate with the head of the supporting interneural spine. Projecting from the anterior aspect of each of the lateral condyles are two spinelike processes termed the anterior condylar processes. Superior and lateral to the lateral condyles are two smaller processes, one on either side. These lateral processes tend to flare out in a posterolateral direction and are termed the posterior condylar processes. The posterior condylar processes are less developed than the anterior condylar processes. Moreover, the anterior condylar processes are well developed in the first dorsal spine and reduced in the second dorsal spine.

The two dorsal spines are erected and depressed as a single unit since they are surrounded by a single sheath and joined together by connective tissue. Erection of the spines is accomplished by contraction of two pairs of bilaterally distributed muscles. The first muscle pair originates from the parapophyses of the first vertebra, ascends vertically past the body of the vertebra, and inserts on the distal tip of the anterior condylar processes of the first dorsal spine. The second muscle pair originates from the parapophyses of the third vertebra and inserts on the anterior condylar processes of the second dorsal spine. Depression of the first dorsal spines appears to be controlled by three sets of muscles. The first muscle pair originates from the parapophyses of the first vertebra, ascends vertically past the body of the vertebra, and inserts on the posterior condylar processes. The second muscle pair originates from the sides of the third vertebra and inserts on the posterior condylar processes of the second dorsal spine. The paired dorsal division of the epaxial muscles also appears to aid in the depression of the dorsal spines to which they are joined by strong connective tissue attachments.

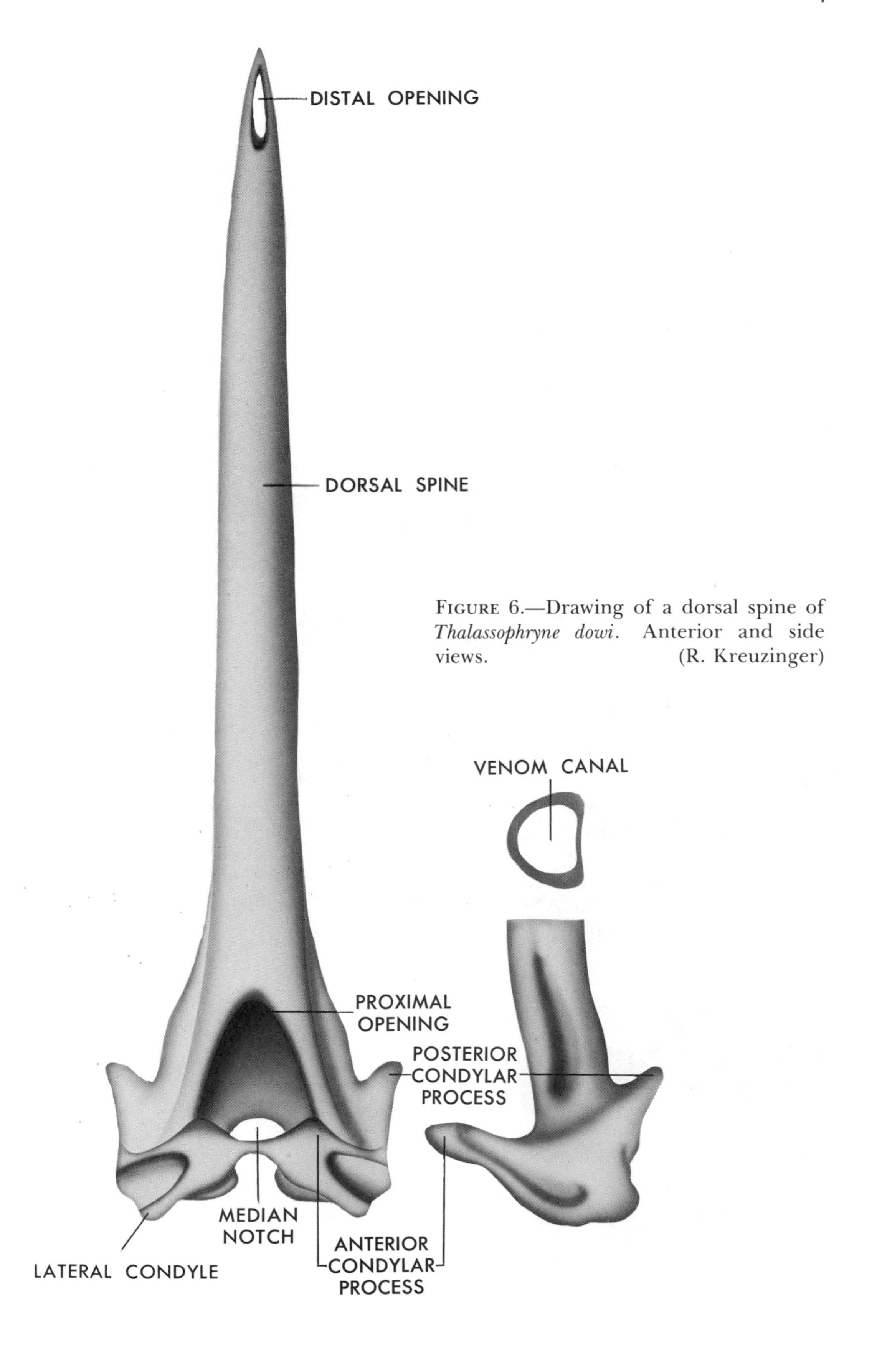

Figure 6.—Drawing of a dorsal spine of *Thalassophryne dowi*. Anterior and side views. (R. Kreuzinger)

Microscopic Anatomy[2]

Opercular Stings (Fig. 7): The opercular sting of *Thalassophryne dowi* in cross section may be divided into four principal zones: a peripheral integumentary sheath, a glandular zone, a dentinal portion, and a central canal. The integumentary sheath is comprised of a thin outer layer of epidermis and a thick inner layer of dermis. These two layers are separated from each other by a distinct basement membrane and an interrupted pigment layer. The epidermis is comprised of stratified squamous epithelium three to four cells in thickness. Scattered throughout the epidermis are occasional large secretory or mucous cells having a clear cytoplasmic area. The peripheral cytoplasm of the outer layer of cells forms a distinct pink line suggestive of a striated border. The dermis is comprised of a thick layer of dense, fibrous, connective tissue. Scattered throughout the dermis are vascular structures containing nucleated red blood cells. The glandular zone is crescent-shaped and contains strands of areolar connective tissue, large, distended, light pink-stained, polygonal cells filled with finely granular secretion and vascular channels. In some instances the polygonal cells appear to have undergone complete lysis and there remains only areas of amorphous secretion. Some of the larger areas measure 300μ or more in diameter. Situated at one side within the crescent of the glandular area is the opercular spine. The spine in cross section appears as an ovoid continuous ring of a dentinelike substance marked by a series of fine concentric rings. The periphery of the spine consists of a fine layer of vitreodentine. The central canal of the spine is lined with a thin layer of areolar connective tissue and minute vascular channels filled with red blood cells. The remainder of the central canal is filled with large, polygonal-shaped, glandular cells and concentrations of light pink-stained, finely granular, amorphous secretion.

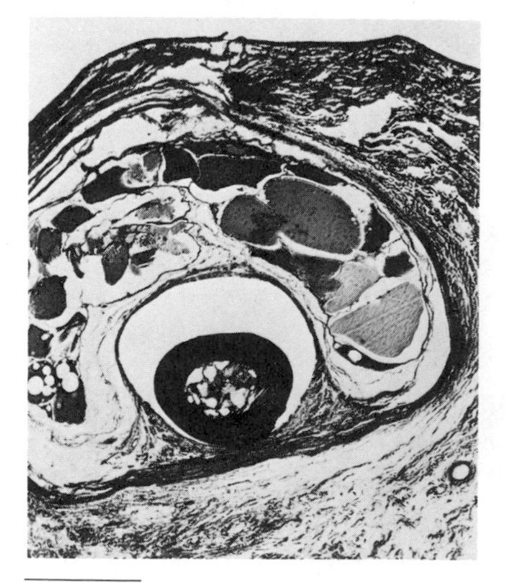

FIGURE 7.—Cross section of the opercular sting of *Thalassophryne dowi*. Note the horseshoe-shaped venom gland surrounding the hollow opercular spine. Coagulated venom is present in the venom canal. × 15. Hematoxylin and triosin stains. Preparations by B. W. Halstead and M. J. Chitwood. (E. Rich)

[2] The histological preparations were stained with hematoxylin and triosin as previously described by Halstead, Ocampo, and Modglin (1955).

Dorsal Stings (Fig. 8a-d): The dorsal stings are essentially the same in their morphology as those of the opercular. However, two minor differences are noted, viz., the dentinal portions of the stings are centrally located within each of the two glandular zones, and the glandular areas appear to be slightly larger. Basal sections reveal that the two glandular areas are separated by a fibrous tissue system derived from the dermis. Because of the similarity of the dorsal stings to those of the opercula, no further description of these stings is necessary.

Longitudinal sections (Fig. 8d) through the dorsal stings show that the glandular cells are elongate, polygonal, and arranged in interrupted layers through which are scattered areas of amorphous secretion. The glandular cells measure 375μ by 60μ or more in size. These sections fail to reveal sufficient additional anatomical detail to warrant further description at this time.

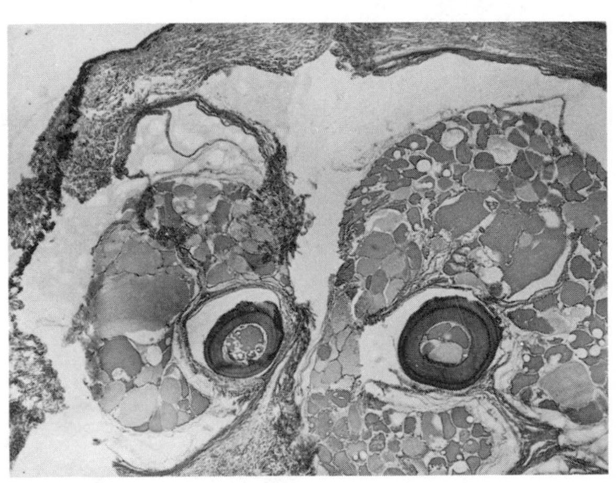

FIGURE 8a.—Cross section through the middle third of the dorsal stings of *Thalassophryne dowi.* Note that each of the spines is almost completely surrounded by large venom glands. × 20.

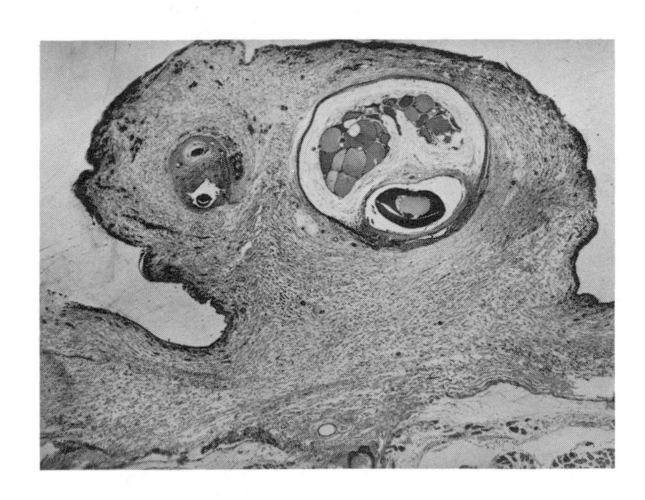

FIGURE 8b.—Cross section through the distal third of the dorsal stings of *Thalassophryne dowi.* Note the decrease in the amount of glandular area and the increase in the thickness of the integumentary sheath surrounding the stings. × 20.

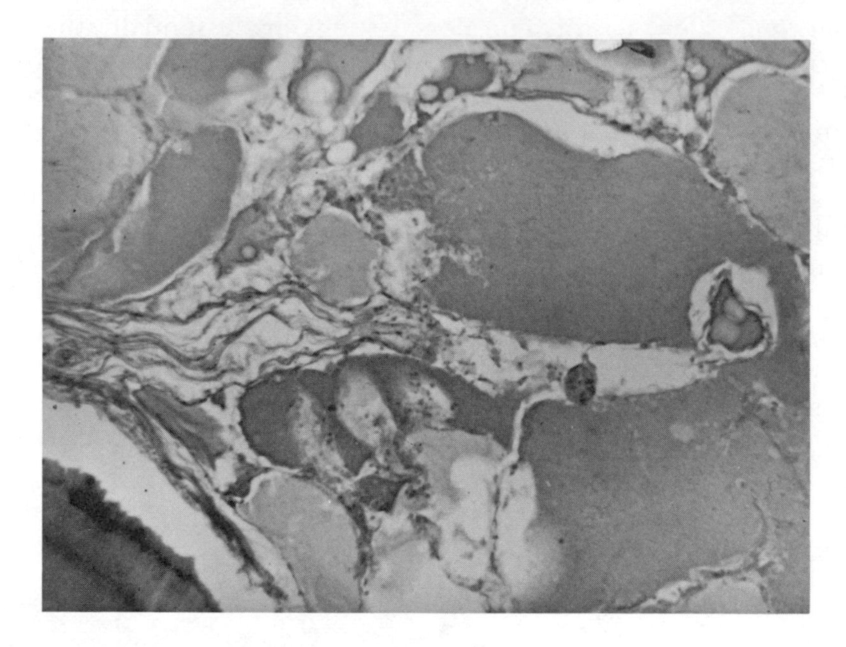

FIGURE 8c.—Enlargement of a cross section of one of the venom glands shown in Fig. 8a. × 115. (E. Rich)

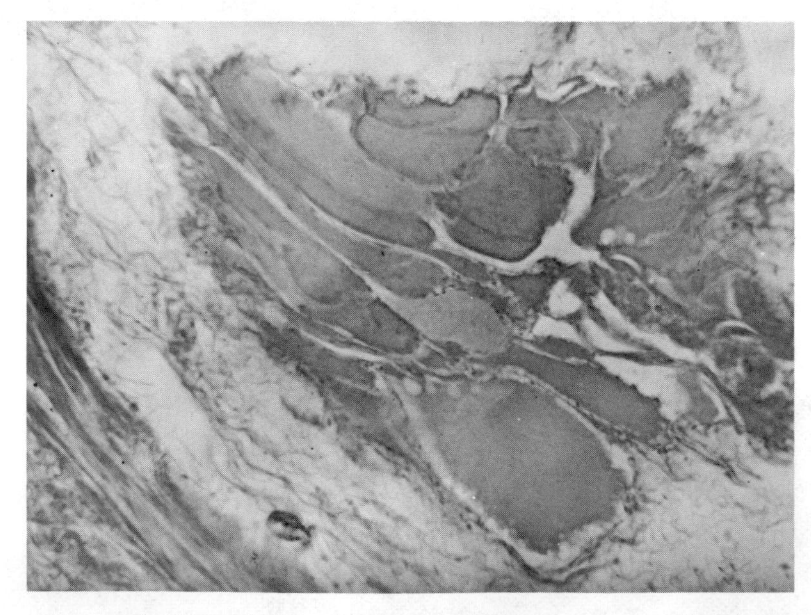

FIGURE 8d.—Longitudinal section through the venom gland of a dorsal sting. × 115. Hematoxylin and triosin stains. Preparations by B. W. Halstead and M. J. Chitwood. (E. Rich)

Comments on the Venom Apparatus of Other Batrachoid Species

The opercular and dorsal spines of *Porichthys notatus* Girard, which ranges from Panama to Puget Sound, are solid and do not display the remarkable adaptive structures for the conveyance of venom as does *Thalassophryne*. Moreover, dissection of the integumentary sheaths of *P. notatus* fails to reveal any gross evidence of venom glands in either the opercular or dorsal spines. Microscopic sections of the opercular and dorsal spines also fail to demonstrate the presence of venom glands. One may therefore conclude that despite certain superficial resemblances to *Thalassophryne* there are no venom organs present in the spines of *P. notatus*.

Opsanus tau has been incriminated as a venomous species by numerous authors, but according to Nigrelli (1944) and Taft (1945), this fish does not possess venomous spines. The latter author states that he has been stuck numerous times by *O. tau* without toxic effects.

No anatomical information is available concerning the presence of venom organs in *Barchatus, Batrachoides*, and *Marcgravichthys*.

The nature of the axillary and pectoral glands in batrachoid fishes as described by Wallace (1893) and Collette (1966) are presently unknown, but these glands are believed to have an ichthyocrinotoxic function. (See Chapter XXIV for further information on ichthyocrinotoxic fishes.)

MECHANISM OF INTOXICATION

Wounds are generally contracted from stepping on toadfishes which are partially buried in sand or mud. In some instances wounds have been received from the opercular spines as a result of careless handling.

MEDICAL ASPECTS

Clinical Characteristics

Little is known about the clinical characteristics of toadfish stings. According to Froes (1932, 1933*a, b*), who is one of the few available sources of clinical data on this problem, wounds produced by these fishes are extremely painful. The pain develops rapidly, is radiating and intense. Some have described the pain as being similar to that resulting from a scorpion sting. The pain is soon followed by swelling, redness, and heat. Secondary bacterial contamination may result if the wound is not properly treated. Froes (1932) cited a case in which the patient remained ill as a direct effect of the sting for a period of about five months and a mild ankylosis of the affected toe continued to persist. According to Dow (1865), batrachoid wounds are not generally serious. No fatalities have been recorded in the literature.

Pathology

(Unknown)

Treatment

The treatment of toadfish stings is similar to that used for scorpionfish stings. There is no antivenin available for toadfish envenomations. See p. 782 for information on the treatment of scorpionfish stings.

Prevention

One should shuffle one's feet when wading in shallow, muddy, sandy bays or other coastal areas inhabited by toadfishes. They remain buried in the mud and when disturbed will quickly erect their sharp dorsal stings. These stings can readily penetrate a thin-soled flipper but might encounter some difficulty in penetrating very deeply through a sneaker because of the short length of the spines. When handling toadfishes, one must be careful to avoid contact with the pungent opercular and dorsal spines.

PUBLIC HEALTH ASPECTS

Toadfishes generally are not considered to be of great public health significance.

TOXICOLOGY

Froes (1933*a*, *b*) appears to be the only investigator who has given even casual study to the venom of *Thalassophryne*. Froes stated that the venom can easily be ejected from the apertures of the spines by applying pressure at their base. The venom is described as a slightly acid, clear, albuminous fluid containing cellular constituents. Guinea pigs injected subcutaneously with the dorsal spines of a Brazilian *Thalassophryne* died about 12 hours later with mydriasis, ascites, paralysis of the hind limbs, and local necrosis about the injection site. Subcutaneous injections of dorsal spine venom into a chick resulted in clonic and tonic convulsions and death within 1 minute. No hemolysis or other alterations of the blood were noted. The authors concluded that the venom of *Thalassophryne* had both local proteolytic and general neurotoxic properties. Froes (1933*b*) stated that a more comprehensive report on the physiological properties of *Thalassophryne* venom was to appear in a separate report, but careful search fails to locate any additional information concerning the nature of this venom by Froes or others.

PHARMACOLOGY

(Unknown)

CHEMISTRY

(Unknown)

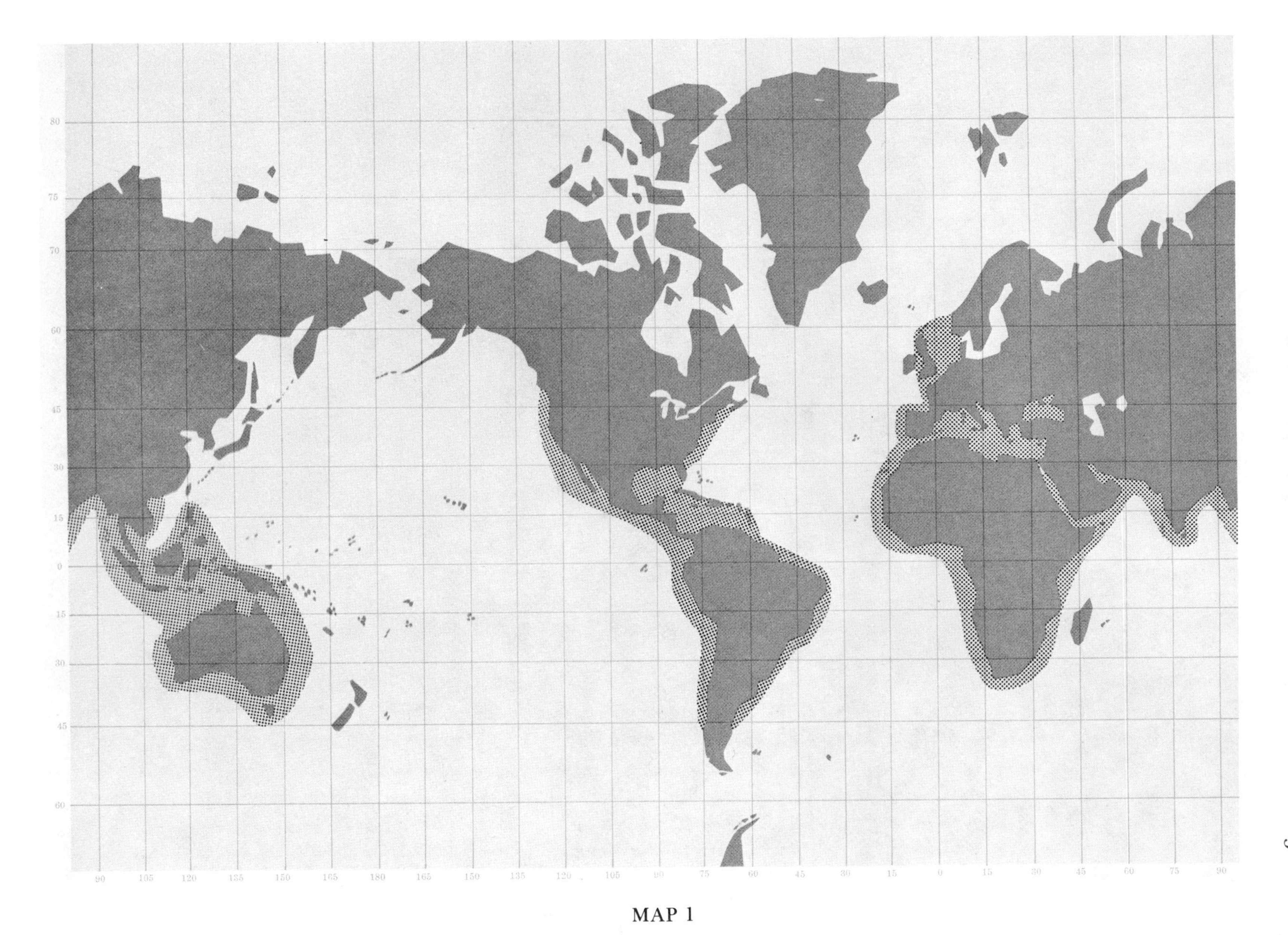

MAP 1

Map showing the geographical distribution of toadfishes. (L. Barlow)

LITERATURE CITED

BEAN, B. A., and A. C. WEED
1910 A review of the venomous toadfishes. Proc. U.S. Nat. Mus. 38(1764): 511-526, 4 pls., 8 figs.

BOTTARD, A.
1889 Les poissons venimeux. Octave Doin, Paris. 198 p.

CALMETTE, A.
1908 Venoms. Venomous animals and antivenomous serum-therapeutics. [Engl. Transl.] J. Bale, Sons and Danielsson, Ltd., London. 403 p., 123 figs.

CAMERON, A. M.
1969 The venom apparatus and the epidermal glands of some Queensland teleosts. Ph.D. thesis, Univ. Queensland. 270 p., 81 figs.

CANTOR, T. E.
1839 Notes respecting some Indian fishes, collected, figured, and described, etc. J. Roy. Asiatic Soc. Bengal 5: 165-172.
1849 Catalogue of Malayan fishes. J. Roy. Asiatic Soc. Bengal 18: 983-1042.

CASTELLANI, A., and A. J. CHALMERS
1919 Venomous animals: Protozoa to Arthropoda, p. 203-241; Figs. 17-26. *In* A. Castellani and A. J. Chalmers, Manual of tropical Medicine. 3d ed. Wm. Wood Co., New York.

CERVIGON, M. F.
1966 Los peces marinos de Venezuela, vol. 2, p. 856-865, 3 figs. Estación de Investigaciones Marinas de Margarita, Fundación La Salle de Ciencias Naturales, Caracas.

COLLETTE, B. B.
1966 A review of the venomous toadfishes, subfamily Thalassophryninae. Copeia (4): 846-864, 12 figs.

COUTIÈRE, H.
1899 Poissons venimeux et poison vénéneux. Thèse Agrég. Carré et Naud, Paris. 221 p.

DAY, F.
1865 The fishes of Malabar. Bernard Quaritch, London. 293 p., 20 pls.
1871 On fish as food, or the reputed origin of disease. Indian Med. Gaz. 6: 5-8, 26-29.
1878-88 The fishes of India. 2 vols. F. Day, London.

DEMOREAU, G.
1908 Contribution à l'étude des piqûres de poissons au cours des accidents du travail, p. 6-10. Thesis, Paris.

DOW, J. M.
1865 (Letter to Dr. Günther regarding *Thalassophryne* and its poisonous power.) Proc. Zool. Soc. London. 1865: 677.

DUNLOP, W. R.
1917 Poisonous fishes in the West Indies. West Indies Bull. 16(2): 160-167.

ENGELSEN, H.
1922 Om giftfisk og giftige fisk. Nord. Hyg. Tidskrift 3: 316-325.

FAUST, E. S.
1924 Tierische gifte: fische, Pisces, p. 1841-1854. *In* Handbuch der Experimentellen Pharmakologie, Berlin. Vol. 2.

FAYRER, J.
1878 Venomous animals. Edinburgh Med. J. 23: 105-109.

FOWLER, H. W.
1936 The marine fishes of West Africa. Bull. Am. Mus. Nat. Hist. 70(11): 1075-1077, 1 fig.
1941 Contributions to the biology of the Philippine Archipelago and adjacent regions. The fishes of the groups Elasmobranchii, Holocephali, Isospondyli, and Ostarophysi obtained by the U.S. Bureau of Fisheries steamer "Albatross," in 1907 to 1910, chiefly in the Philippine Islands and adjacent seas. U.S. Nat. Mus., Bull. 100. Vol. 13. 879 p., 30 figs.

FROES, H. P.
1932 Sur un poisson toxiphore brésilien: le "niquim" *Thalassophryna maculosa*. Rev. Sud-Am. Med. Chir. 3(11): 871-878, 3 figs.
1933*a* Peixes toxiforos do Brasil. Bahia Med. (4): 69-75.
1933*b* Studies on venomous fishes of tropical countries. J. Trop. Med. Hyg. 36: 134-135, 2 figs.

GILBERT, C. H., and E. C. STARKS
1904 The fishes of Panama Bay. Mem., Calif. Acad. Sci., 4,304 p., 33 pls.

Gill, T.
1907 Life histories of toadfishes (Batrachoidida), compared with those of weevers *(Trachinus)* and stargazers (Uranoscopids). Smithsonian Misc. Coll. 48(1697): 388-427, 21 figs.

Giunio, P.
1948 Otrovne ribe. Higijena (Belgrade) 1(7-12): 282-318, 20 figs.

Gonsalves, A. D.
1907 Peixes venenosos da Bahia. Gaz. Med. Bahia 38(10): 441-452.

Goode, G. B.
1884 The fisheries and fishery industries of the United States, Sect. I, p. 249. U.S. Government Printing Office, Washington, D.C.

Gudger, E. W.
1910 Habits and life history of the toadfish *(Opsanus tau)*. Bull. U.S. Bur. Fish. 1908, (28): 1071-1109; Pls. 107-113.

Gunther, A.
1864 On a poison organ in a genus of batrachoid fishes. Proc. Zool. Soc. London (1): 155-158, 2 figs.
1869 An account of the fishes of Central America, based on collections made by Capt. J. M. Dow, F. Godman, Esq., and O. Salvin, Esq. Trans. Zool. Soc., London 6: 377-494, 24 pls.
1880 An introduction to the study of fishes. A. and C. Black, Edinburgh. 720 p., 321 figs.

Halstead, B. W.
1956 Animal phyla known to contain poisonous marine animals, p. 9-27. *In* E. E. Buckley and N. Porges [eds.], Venoms. Am. Assoc. Adv. Sci., Washington, D.C.
1959 Dangerous marine animals. Cornell Maritime Press, Cambridge. 146 p., 88 figs.

Halstead, B. W., and L. J. Carscallen
1964 Marine organisms, p. 336-343. *In* P. L. Altman and D. S. Dittmer [eds.], Biology data book. Fed. Am. Soc. Exp. Biol., Washington, D.C.

Halstead, B. W., and L. R. Mitchell
1963 A review of the venomous fishes of the Pacific area, p. 173-202, 16 figs. *In* H. L. Keegan and W. V. MacFarlane [eds.], Venomous and poisonous animals and noxious plants of the Pacific region. Pergamon Press, New York.

Halstead, B. W., Ocampo, P. R., and F. R. Modglin
1955 A study on the comparative anatomy of the venom apparatus of certain North American stingrays. J. Morphol. 97(1): 1-21, 4 pls.

Hilzheimer, M., and O. Haempel
1913 Biologie der wirbeltiere. F. Enke, Stuttgart. 756 p.

Hubbs, C. L., and L. P. Schultz
1939 A revision of the toadfishes referred to *Porichthys* and related genera. Proc. U.S. Nat. Mus. 86(3060): 473-496.

Jordan, D. S., and B. W. Evermann
1896-1900 The fishes of North and Middle America. 4 parts. U.S. Nat. Mus., Bull. 47. 392 pls.

Jordan, D. S., Evermann, B. W., and H. W. Clark
1930 Check list of the fishes and fishlike vertebrates of North and Middle America north of the northern boundary of Venezuela and Colombia. Rept. U.S. Comm. Fish. 1928. 670 p.

Kobert, R.
1894 Compendium der praktischen toxikologie, p. 90-93. F. Enke, Stuttgart. (NSA)
1902 Ueber giftfische und fischgifte. Med. Woche (19): 199-201; (20): 209-212; (21): 221-225.
1905 Ueber giftfische und fischgifte. F. Enke, Stuttgart. 36 p., 11 figs.

Linstow, O. V.
1894 Die giftthiere und ihre wirkung auf den menschen. A. Hirschwald, Berlin. 147 p., 54 figs.

Marshall, T. C.
1964 Fishes of the Great Barrier Reef and coastal waters of Queensland. Angus and Robertson, Melbourne. 566 p., 72 pls.

Meek, S. E., and S. F. Hildebrand
1923 The marine fishes of Panama. Part I. Field Mus. Nat. Hist. Publ. 215 (Zoology), 15: 1-330, 24 pls.

Mendis, A. S.
1954 Fishes of Ceylon. Fish. Res. Sta., Dept. Fisheries, Ceylon. 222 p., 8 figs.

(NSA) Not seen by author.

NIGRELLI, R. F.
1944 Fish may be poisonous, too. Animal Kingdom 47(5): 122-124,. 1 fig.
NORMAN, J. R.
1937 Illustrated guide to the fish gallery. Brit. Mus. Nat. Hist., London. 175 p.
PAWLOWSKY, E. N.
1911 Contribution to the question of the structure of the poison glands of certain fish of the Scorpaenidae family. [In Russian] Trav. Soc. Imp. Nat., St. Petersburg 41(1): 317-328.
1927 Gifttiere und ihre giftigkeit. G. Fischer, Jena. 515 p., 170 figs.
PELLEGRIN, J.
1899 Les poissons vénéneux. M.D. Thèse 510, Fac. Méd., Paris.
1908 Les poissons d'eau douce de la Guyane Française. Rev. Colon. (67): 557-591.
PHILLIPS, C., and W. H. BRADY
1953 Sea pests—poisonous or harmful sea life of Florida and the West Indies. Univ. Miami Press, Miami, Fla. 78 p., 38 figs., 7 pls.
PHISALIX, M.
1922 Animaux venimeux et venins. 2 vols. Masson et Cie., Paris.
POLL, M.
1959 Order Batrachoidiformes, p. 331-335, 2 figs. *In* M. Poll, Expédition océanographique Belge dans les eaux côtières africaines de l'Atlantique sud (1948-1949). Vol. IV, part 3B. Poissons. V. Téléostéens acanthoptérygiens. 2d part. Inst. Roy. Sci. Nat. Belg., Brussels.
PRYOR, J. C.
1918 Food, p. 153-155; Marine animal life dangerous to man, p. 309-318. *In* J. C. Pryor, Naval hygiene. P. Blakiston's Son and Co., Philadelphia.
REED, H. D.
1907 The poison glands of *Noturus* and *Schilbeodes*. Am. Naturalist 41: 553-566, 5 figs.
SAVTSCHENKO, P.
1886 Atlas of the poisonous fishes. Description of the ravages produced by them in the human organism, and antidotes which may be employed. [In Russian and French] St. Petersburg. 53 p., 10 pls.
SCOTT, H. H.
1921 Vegetal and fish poisoning in the tropics, p. 790-798, 5 figs., 2 pls. *In* W. Byam and R. G. Archibald [eds.], The practice of medicine in the tropics. Vol. I. H. Frowde and Hodder and Stoughton, London.
SILVADO, J.
1911 Peixes nocivos da Bahia do Rio de Janeiro. Imprensa Nacional, Rio de Janeiro. 24 p., 20 pls.
SMITH, J. L.
1952 The fishes of the family Batrachoididae from South and east Africa. Ann. Mag. Nat. Hist. 5(12): 313, 3 figs.
1961 The sea fishes of southern Africa. 4th ed. Central News Agency, South Africa. 580 p., 102 pls., 1,232 figs.
TAFT, C. H.
1945 Poisonous marine animals. Texas Rept. Biol. Med. 3(3): 339-352.
TASCHENBERG, E. O.
1909 Die giftigen tiere. F. Enke, Stuttgart. 325 p., 64 figs.
WALLACE, L. B.
1893 The structure and development of the axillary gland of *Batrachus*. J. Morphol. 8: 563-568, 1 pl.
WOODWARD, H.
1864 Discovery of poison organs in fishes. Intellect. Observ. 5: 253-257.

Chapter XXIII—VERTEBRATES

Class Osteichthyes

Venomous: MISCELLANEOUS FISHES

The families presented in this chapter comprise a variety of fishes which are reported to have a venom apparatus. For the most part, little is known about either the anatomical structure of their venom organs or the nature of their venom. The families are arranged phylogenetically by order. Some fishes whose venomous nature were not sufficiently documented for the purposes of this publication have been omitted. The reader is referred to the more exhaustive descriptions by Halstead (1970) with the hope that further investigation in this greatly neglected aspect of piscine venomology will be encouraged.

Order ANGUILIFORMES (APODES)

Suborder ANGUILLOIDEI

Family MURAENIDAE: Moray Eels

The meager amount of research that has been conducted on so-called "venomous eels" is concerned with the European moray eel *Muraena helena* (Linnaeus) (Pl. 1), which is found in the eastern Atlantic Ocean and Mediterranean Sea. Other species have been listed as venomous by modern authors, but their work is largely presumptive. They apparently have assumed that since *M. helena* is reputed to have a venom apparatus, all morays have one. Although there are no published data to support these conclusions, recent studies indicate that some morays may secrete a poison from their palatine mucosa.

258

The reputation of the moray eel as a venomous fish seems to have originated during the second century B.C. with the writings of Nicander (cited by Coutière, 1899), who stated that fear of this fish caused fishermen who were unlucky enough to capture a moray to throw themselves into the sea "because its bite poisons with a venom like that of the viper." Deipnosophistes of Athens (ca. A.D. 400) is also said to have discussed the venomous prowess of the moray (Fig. 1).

Du Tertre (1667) reported that the muraenid eels and congers of the Antilles inflicted a dangerous bite. Corre (1865) stated that people of the French West Indies feared the bite of *Gymnothorax moringa*. These early

FIGURE 1.—The mouth of a moray eel with its powerful jaws and teeth is a fearsome sight to behold. Early writers believed that the moray eel possessed hollow fangs similar to those of a rattlesnake, but modern anatomical research has shown that this is not true. The larger canine teeth are solid and quite capable of inflicting an extremely painful bite. (M. Shirao)

reports served as a stimulus to Bottard (1889) to determine if *M. helena* actually possessed a venom apparatus. Bottard concluded from his anatomical research that a true venom apparatus was present, consisting of the palatine teeth and a secretory gland supposedly situated between the palatine bone and the mucosa. Coutière (1899) quoted with obvious reserve the opinion of Bottard, but in a later note (1907) stated that there was no anatomical evidence of a venom gland in relation to the palatine teeth. The problem was further complicated by Porta (1905) who disagreed with both Bottard and Coutière. Porta said that Galasso later became convinced, after studying Porta's histological preparations, that a venom apparatus was present in *M. helena*. Porta further stated that the venom glands of *M. helena* were homologous to the intermaxillary glands of amphibians and were not the same as those described by Bottard. With this conflicting state of affairs, Pawlowsky (1909) attempted to settle the problem by working with fresh material. He contended that Porta's description was too schematic and that his illustrations failed to support his conclusions. Pawlowsky prepared serial sections of the palate of *M. helena*. Study of these sections failed to reveal any glands of the types described by either Bottard or Porta. Pawlowsky, however, suggested that perhaps a venom was secreted by the palatine mucosa. Pawlowsky's suggestion appears to have been substantiated by the work of Kopaczewski (1917) who demonstrated that saline extracts prepared from the palatine mucosa were toxic to guinea pigs.

There was apparently no further research until the unpublished work in 1956 of Halstead and Teel, who were able to obtain fresh specimens of

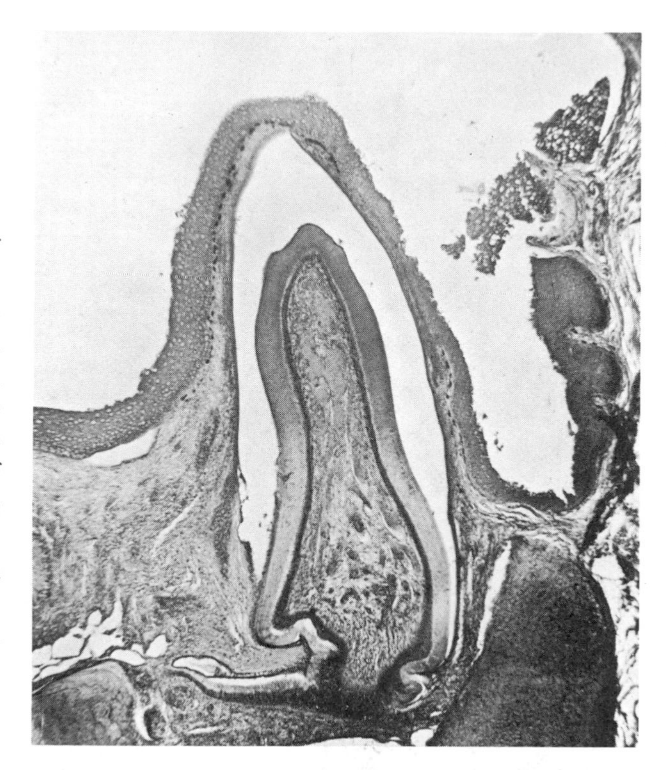

Figure 2.—Sagittal section through one of the vomerine teeth of *Muraena helena*. This tooth is covered by a layer of palatine mucosa. Note that the tooth is not hollow in the sense that it contains a venom canal which opens to the surface such as has been described by some of the early writers. Thus there is no evidence that the tooth of the moray eel is similar to the fangs of a crotalid rattlesnake. × 36. Hematoxylin and triosin stains.

Preparation by R. Ocampo.

Muraena helena from Madeira, Portugal. Careful gross examination of the premaxillary, maxillary, inner maxillary, vomer, and dentary teeth failed to reveal any so-called hollow teeth (Fig. 2). A few of the teeth did appear to have a mucosal sheath or evidence of the remnants of a sheath around the base of the teeth. There was no evidence of a venom duct such as one might expect to find in venomous snakes. Routine histological sections were prepared using hematoxylin and triosin stains. These sections consistently showed a thick layer of oral mucosa comprised of typical stratified squamous epithelium in which were included numerous mucous cells (Fig. 3). Underlying the epithelium was a layer of supporting connective tissue. There was no evidence of any specialized glandular structures which might be interpreted as venom glands. However, this does not preclude the possibility that the palatine mucosa may produce toxic secretions.

The toxicological experiments of Kopaczewski (1917) and Phisalix (1931) are inconclusive, and neither prove nor disprove the presence of a venom apparatus, since either an ichthyosarcotoxin or an ichthyohemotoxin could have produced the same results. It is quite likely that their toxicological results were due to a serum toxin rather than to an ichthyoacanthotoxin (Ghiretti and Rocca, 1963). See also the sections on ciguatoxic eels, Chap-

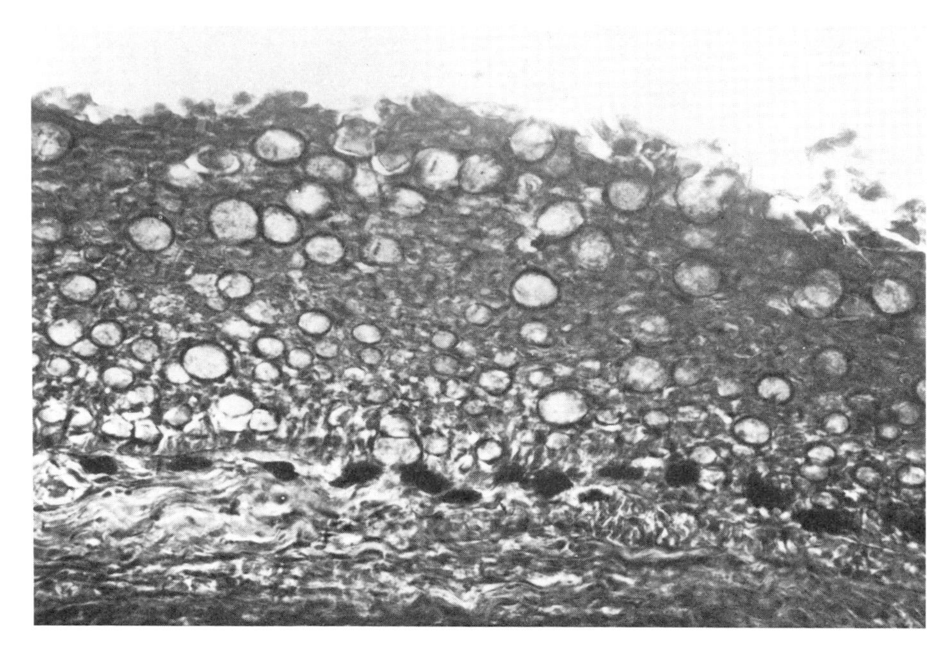

FIGURE 3.—Photomicrograph of the oral mucosa of upper jaw of *Muraena helena*. Numerous mucous cells are seen scattered throughout the layer of stratified squamous epithelium. There is no morphological evidence of the type of venom apparatus ascribed to it by Bottard (1889) and others. However, this does not preclude the possibility that the palatine mucosa does not secrete a toxic substance. If a toxin is present in the mucosal secretions, then a crude mechanism of envenomation takes place by means of the trauma inflicted by the teeth. × 290. Hematoxylin and triosin stains. Preparation by R. Ocampo.

ter X, and ichthyohemotoxic eels, Chapter XVI. Moray eels are also reported to have toxic skin secretions (*see* Chapter XXIV, p. 885) on ichthyocrinotoxic fishes.

Order PERCIFORMES (PERCOMORPHI)

Suborder PERCOIDEI

Family CARANGIDAE: Jacks, Scads, Pompanos

Carangids are a group of medium- or large-sized, metallic-, silvery-, or golden-colored fishes which are usually adapted for rapid swimming. The group, containing both oceanic and coastal members, is widely distributed in all warm and some temperate seas. A particular characteristic of this group is the presence of two separate spines in front of the anal fins. There are no published data regarding the venom apparatus of carangids. Some articles merely list various species of carangids as possessing venomous spines with no supportive morphological data.

Scomberoides sanctipetri became of venomological interest because of reports from fishermen that they sometimes received extremely painful stings after having been jabbed by the anal spines of the fish. *Scomberoides sanctipetri* has also been reported to use its anal spines to sting other fishes which they capture for food. The venom organs of carangids is an area worthy of further study.

The systematics of this group have been studied by Fowler (1928, 1936), Joubin (1929-38), Jordan *et al.* (1930), Weber and Beaufort (1931), Smith (1950), Schultz *et al.* (1953), and Herre (1953).

LIST OF CARANGIDS REPORTED AS VENOMOUS

Family Carangidae

Caranx hippos (Linnaeus) (Pl. 13, fig. a, Chapter X). Crevalle, jack (USA), crevalle (West Indies). *139*

Distribution: Tropical Atlantic.

Sources: Kobert (1902), Taschenberg (1909), Engelsen (1922).

Oligoplites saurus (Pl. 2, fig. a). Leatherjacket, runner, zapatero, quiebra, sauteur (USA, West Indies). *259*

Distribution: Both coasts of tropical America.

Source: Phillips and Brady (1953).

Scomberoides sanctipetri (Cuvier) (Pl. 2, fig. b). Leatherback, lae, runner (USA). *259*

Distribution: Indo-Pacific.

Source: Halstead *et al.* (1972).

Selar crumenophthalmus (Bloch) (Pl. 2, fig. c). Big-eyed scad (USA), chicharro (West Indies), atule (Samoa), matang baka, bunutan, ataloy (Philippines), mijinga, korkofanha (West Africa). *259*

Distribution: Circumtropical.

Sources: Kobert (1902), Taschenberg (1909), Engelsen (1922).
Other names: *Caranx crumenopthalmus, Trachurops crumenophthalmus, Trachurus crumenophthalma.*

Trachurus trachurus (Linnaeus) (Pl. 2, fig. d). Saurel, horse mackerel (USA, England), saurel (France), suro (Italy), stöcker (Germany), jurel (Spain). *259*

Distribution: Mediterranean Sea and north and tropical Atlantic Ocean.

Sources: Taschenberg (1909), Hilzheimer and Haempel (1913), Engelsen (1922).
Other name: *Caranx trachurus.*

BIOLOGY

Family CARANGIDAE: This family includes the pompano, jack, crevally, and ulua, which are a large group of swift-swimming oceanic fishes. They are particularly common in the vicinity of coral reefs and islands. Some of the Pacific species are especially noted for their long migrations. Carangids are carnivorous in their eating habits. Many are desirable game fishes, and most of them are valued as food. The only species of carangid in which the venom organs have been studied in detail is the leatherback *Scomberoides sanctipetri* (Halstead *et al.*, 1972). Leatherbacks are considered good eating by some persons but are of minor commercial importance as a table fish. Occasionally the Hawaiian tuna fishermen use its tough, silvery skin for making feathered lures for hooking tuna.

In Hawaii, where the habits of *S. sanctipetri* have been studied to some extent, it has been found that this species predominantly inhabits inshore areas and is most frequently encountered in sheltered bays and harbors having mud and sand bottoms. They are often captured in brackish water near the mouths of streams and have been observed upstream for a considerable distance. Adult lae rarely venture beyond the 25-fathom curve or more than several miles offshore. Local fishermen catch them in relatively large numbers on baited hooks and artificial lures.

Little information is available concerning the life history of *S. sanctipetri*. Spawning occurs from early June to late November or early December. The adults attain a size of about 63 cm in total length. They are a vigorous, fast swimming predator feeding predominantly on smaller schooling fishes such as *Stolephorus purpureus, Pranesus insularium*, and *Kuhlia sandwicensis*.

Juvenile lae up to 50 mm in standard length are unusual in that they selectively feed upon nehu scales, *S. purpureus*. Juvenile lae are extremely aggressive toward nehu and frequently contribute extensively to the bait-fish mortalities in the bait wells aboard ship. When several nehu are introduced into an aquarium containing one or more lae, the attacks begin immediately and continue until the nehu are dead and lying motionless on the bottom, usually within 30 minutes to an hour. On close examination of the nehu after removal from the aquarium it is seen that the scales and much of the flesh had been removed from the back and sides of the head and the anterior back. Occasionally, the flesh of the back had been removed to the extent that many of the small bony structures were exposed.

MORPHOLOGY OF THE VENOM APPARATUS

Gross Anatomy

The following description of the venom apparatus of *Scomberoides sanctipetri* is based on data from Halstead *et al.* (1972).

The venom apparatus of *S. sanctipetri* consists of seven dorsal spines and two anal spines, their associated musculature, venom glands, and en-

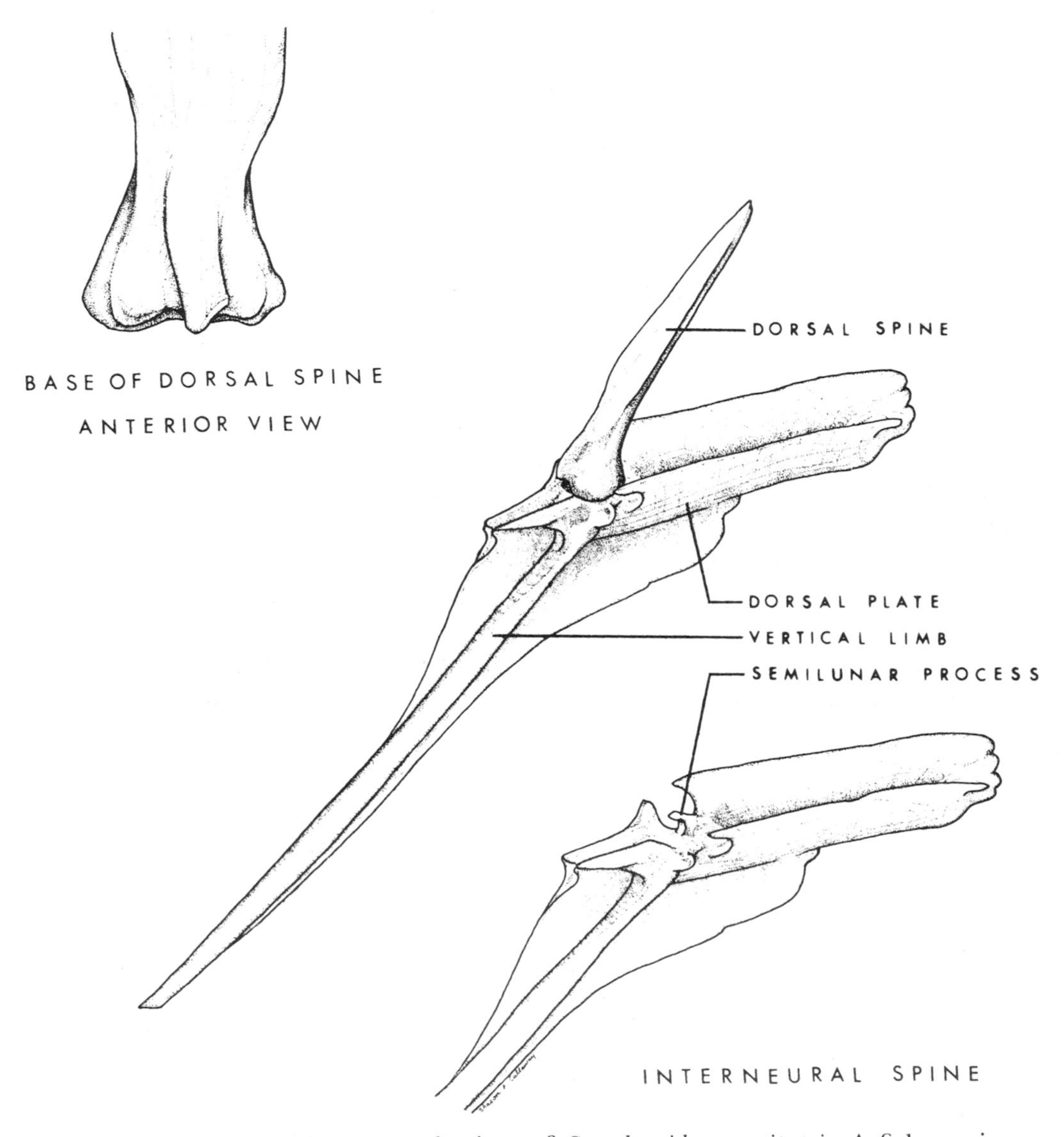

FIGURE 4.—Dorsal and interneural spines of *Scomberoides sanctipetri*. A fish specimen having a standard length of 28 cm was found to have a 4th dorsal spine 8 mm in length. The 4th is the longest of the dorsal spines. (S. K. Calloway)

veloping integumentary sheaths. There is a single dorsal spine separate from the other spines and longer and more slender than the others. This spine is adherent to the first dorsal ray and could not effectively serve as a venom organ. A similar situation exists in the anal spines. The third anal spine is similar in size to the first detached anal spines but is adherent to the first ray and not capable of serving as a venom organ.

Dorsal Spine (Fig. 4): This description is based on dorsal spine IV which is representative of the others. The spine is described in the upright or extended position. A lateral view of the mature dorsal spine reveals that it is relatively short and anteroposteriorly flattened and does not have the usual pronounced trigonal shape that is found in the venomous stings of most other fishes and the anal spines of this species. The flattened shape of the dorsal spine is due to the absence of the customary anteromedian ridge in this species. The anterior surface of the dorsal spine is smooth, having no grooves or ridges. The posterior side of the spine contains a shallow median groove, the posteromedian groove. On either side of the groove is a low-lying ridge. The posterolateral ridges form the median borders of the shallow posterolateral grooves which extend throughout the length on either side of the shaft. Thus, the posterior surface of the shaft of the spine is marked by a single shallow posteromedian groove and two very shallow posterolateral grooves. As will be shown later, the venom glands lie within these posterolateral grooves. These grooves apparently play a minor role in assisting in the introduction of venom during the envenomation process. The base of the dorsal spine is slightly enlarged and expanded into two asymmetrical lateral articular condyles. Between the condyles is a small foramen which receives the semilunar process of the head of the associated interneural spine.

The dorsal spine rests on and articulates with an interneural spine. The following is a description of the interneural spine of dorsal spine IV,

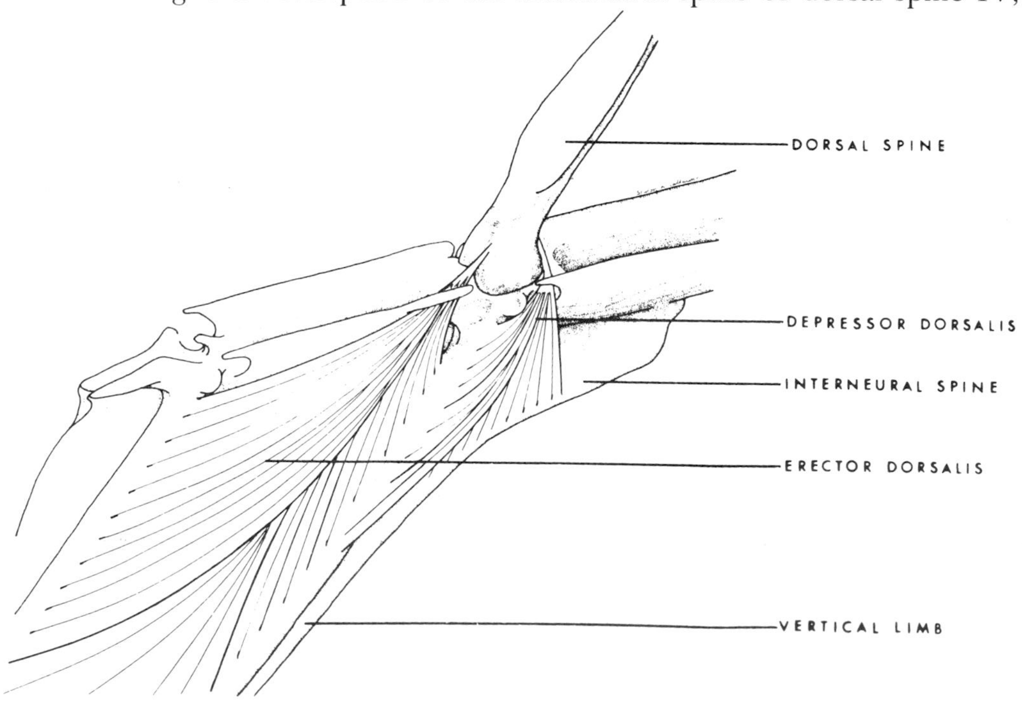

FIGURE 5.—Erector and depressor muscles of the dorsal spines of *Scomberoides sancti-petri*. (S. K. Calloway)

which is representative of the others. The interneural spine is roughly ┌-shaped. However, the dorsal or horizontal limb of the interneural forms an angle of about 125° with that of the descending or vertical limb. The dorsal or horizontal limb is referred to as the dorsal plate of the interneural. The dorsal plate is expanded into an elongate trough-like bone which receives the dorsal spine when it is fully depressed. The width of the dorsal plate is about one-quarter of its length. The head of the vertical limb of the interneural spine is only slightly expanded and modified into two articular areas, one on either side, which receive the lateral condyles of the dorsal spine. Arising from the floor of the trough immediately posterior to the articular area is a small semilunar bone or process which arches upward and forward and passes through the foramen situated between the lateral condyles of the dorsal spine. The sides of the trough-like dorsal plate are notched in the vicinity of the semilunar process. These notched areas provide the passageways for the depressor dorsalis muscles of the dorsal spine (Fig. 5). The vertical limb of the interneural terminates in a sharp point. Projecting forward from the anterior aspect along the median line of the vertical limb is a thin flat sheet of membranous bone which gradually thins out into a cartilagenous sheet and connects with the preceding interneural spine. Extending beneath the dorsal plate of the interneural spine is a relatively large triangular-shaped sheet of membranous bone which also thins out into a cartilagenous sheet and connects with the interneural spine behind it. The posterior extremity of the trough-like dorsal plate of the interneural spine narrows and then joins the anterior aspect of the head of the adjacent interneural spine. This narrow area provides the passageway for the erector dorsalis muscle.

Erection and depression of the individual dorsal spines appear to be controlled largely by two sets of muscles (Fig. 5). Removal of the epaxial muscle mass and the thin layer of fascia covering the interneural spines permits examination of the erector and depressor dorsalis muscles. A set of these muscles is situated on either side of each dorsal spine and the associated interneural spine. An erector dorsalis muscle of each dorsal spine originates by fascial attachments from each side of the membranous bone which projects from the vertical limb of the interneural spine. From each side of the interneural spine the erector dorsal muscle passes up between the passageway which was formed by the posterior tip of the dorsal plate of the preceding interneural and the head of the connecting vertical limb and inserts on the anterior aspect of each of the lateral condyles of the dorsal spine. A depressor dorsalis muscle originates by fascial attachments from each side of the membranous bone which extends from the posterior aspect of the vertical limb of the interneural spine. From each side of the interneural spine a depressor dorsalis muscle passes upward through the passageway provided by each of the notched areas of the trough-like dorsal plate of the interneural and inserts by tendinous attachments to the posterior aspect of each of the lateral condyles of the dorsal spine. The protractor, retractor, and superficial inclinator dorsalis muscles appear to be undeveloped in this fish species.

The venom glands are not grossly visible within the intact dorsal sting of this species.

Anal Spines (Fig. 6): This description is based on anal spines I and II. The spines are described in the anatomical extended position. A lateral view of a mature anal spine I reveals that it is an elongate, stout, almost straight shaft which terminates in a sharp tip. The length of anal I is about one-third longer than the length of most of the dorsal spines and slightly shorter than anal spine II. The base of anal spine I has a swollen or bulbous appearance posteriorly. The base is expanded into two symmetrical articular condyles separated by a narrow deep slot which can be seen when viewed in the vertical plane. The general shape of the shaft in cross section is roundish and smooth,

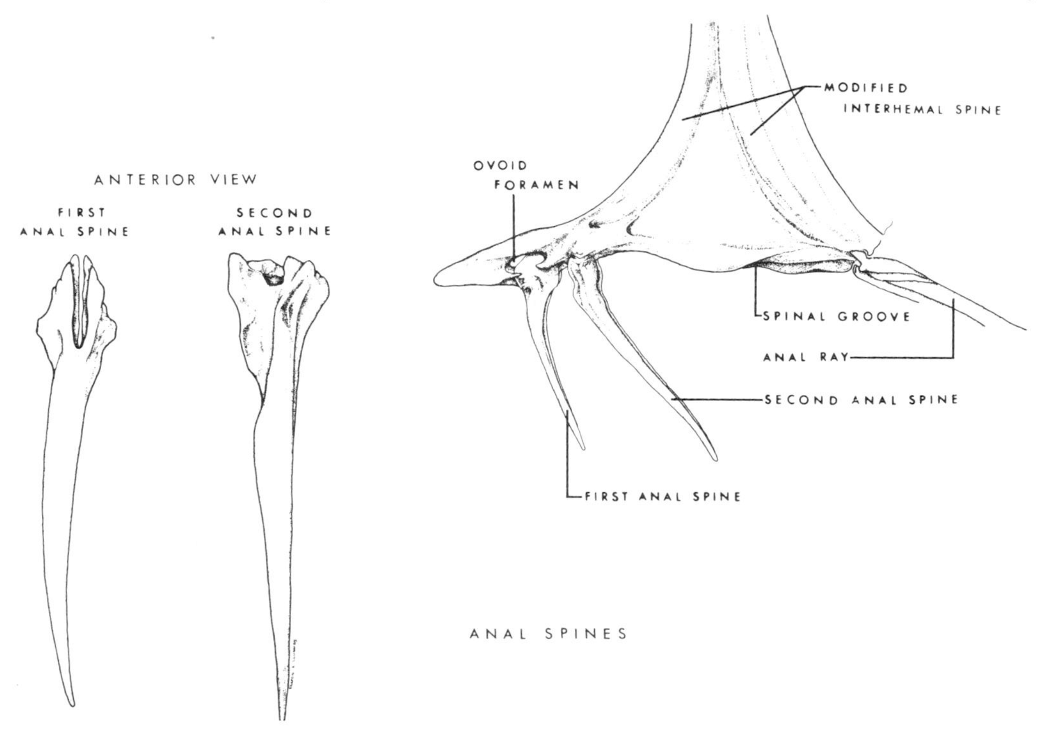

FIGURE 6.—Anal spines of *Scomberoides sanctipetri*. A fish specimen having a standard length of 28 cm had a 2nd anal spine 12 mm in length. (S. K. Calloway)

without ridges or grooves. The back side of the spine contains a shallow posteromedian groove which receives the interspinous membrane. An anterior view of the spine reveals that the shaft of the spine is bent slightly to the right and the base is bulbous, but less pronounced than when observed laterally. When viewed from the back or anteriorly, it is seen that the lateral articular condyles are separated by a narrow deep median slot which extends down into the base for a distance of about one-fifth the length of the spine. This slot receives the median articular crest of the highly modified ventral plate of the associated first interhemal spine.

Anal spine II is slightly longer and heavier in construction than anal spine I. From a lateral view the base of anal spine II is only slightly expanded. The anterior side of the anal spine is almost straight. The posterior side of the spine is also straight except near the base where it bulges slightly. When viewed anteriorly it will be observed that the base of the spine is divided into two lateral articular condyles which are separated by a median foramen. This foramen receives the annular ring of the ventral plate of the supporting interhemal spine. The anterior surface of the left condyle and lower portion of the shaft of the anal spine is grooved to receive the shaft of anal spine I when the two spines are depressed. When anal spines I and II are fully extended, they form a 45° angle with each other.

The first of three interhemal spines of the anal fin (Fig. 6) of this fish are heavy bodied and modified as support structures for the remarkably well-developed anal spines I and II. From a lateral viewpoint the first interhemal spine has a plow-share appearance with the anterior tip or head of the spine forming the share. The side of the anterior tip is penetrated by a large ovoid foramen and joins with the vertical limb in making a sweeping arc which ascends steeply upward and backward to connect with the hemal spine of the associated vertebrae. The vertical limbs of the first and second interhemal spines fuse to form a strong support structure for the first two anal spines. A broad triangular-shaped membranous bone extends backward from the fused vertical limb to attach to the vertical limb of the modified third interhemal spine. Thus the vertical limbs of the first two interhemal spines and the vertical limb of the third interhemal spine form the ascending borders of the triangular membranous bone. Removal of the anal spines reveals that the ventral heads of the interhemal spines are highly modified in order to provide the necessary articulations and a reception area for the anal spines when they are depressed. From a ventral view it will be seen that the anterior tip or head of the first interhemal spine is sagittaform. Descending between the lateral barbs of the arrow and forming the lower border of the ovate foramen is a compressed crest-like bone which articulates within the deep median slot in the base of the first anal spine. The posterior margin of the crest passes backward and upward and finally merges with the vertical plate of the second interhemal spine. Descending from the ventral plate of the second interhemal spine is the annular ring which passes through the medial foramen of the second anal spine. The annular process in this case forms a complete ring. On either side of the ring the vertical plate provides a small articular depression which receives the lateral condyles of the second anal spine. On the basis of the bony structure of the spine it it evident that the locking mechanism of the first and second anal spines operates on a frictional basis employing the use of muscles and tendons rather than bony structures. The posterior portion of the ventral head of the second interhemal spine and the ventral plate of the third interhemal spine are grooved or canaliculate and thus provide a deep protective area for the anal spine when they are depressed.

The erector and depressor muscles of anal spines I and II are situated

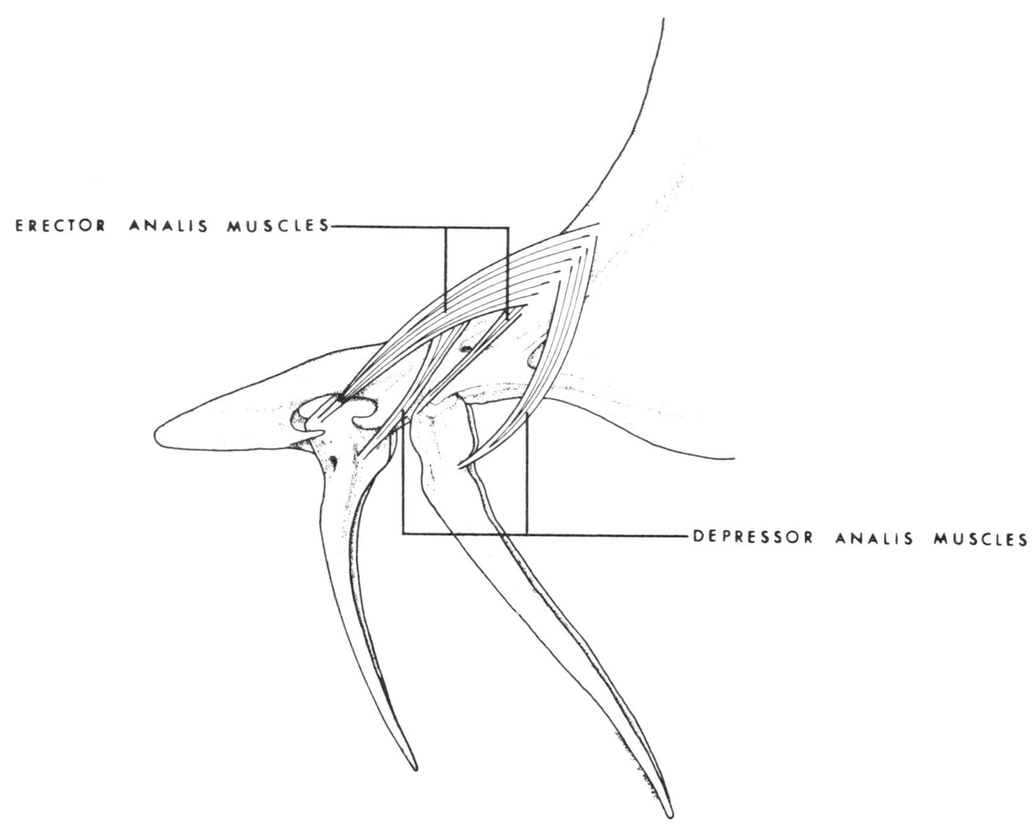

FIGURE 7.—Erector and depressor muscles of the anal spines of *Scomberoides sanctipetri*. Left side. (S. K. Calloway)

deep to the great lateral hypaxial muscles (Fig. 7). The erector and depressor muscles of the anal spines are covered by a sheath of fascia which must be removed in order to expose the muscles beneath. The erector analis muscles of the first anal spine are relatively large muscles and have their origin, one on either side, along the sides of the sweeping arc of the first interhemal spine. The tendons of these first erector muscles pass downward and posterior to the barb-like projections of the sagittaform ventral head of the first interhemal bone and insert on the anterior aspect of each of the lateral condyles at the base of anal spine I. The depressor analis muscles of the first anal spine are much shorter and smaller-bodied muscles than the erectors. The depressors, situated one on either side, have their origin along the lower sides of the vertical limb of the first interhemal spine. They pass almost vertically downward. Their tendons pass through a shallow notch in the sides of the ventral head of the first interhemal and insert on the posterior aspect of the lateral condyles. The erector analis muscle of the second anal spine is situated just posterior to the depressor analis of anal spine I. The two muscles lie immediately adjacent to each other. The erector analis muscles of the second anal spine originate, one on either side from the vertical limb of the second interhemal spine. The muscles pass downward and slightly forward,

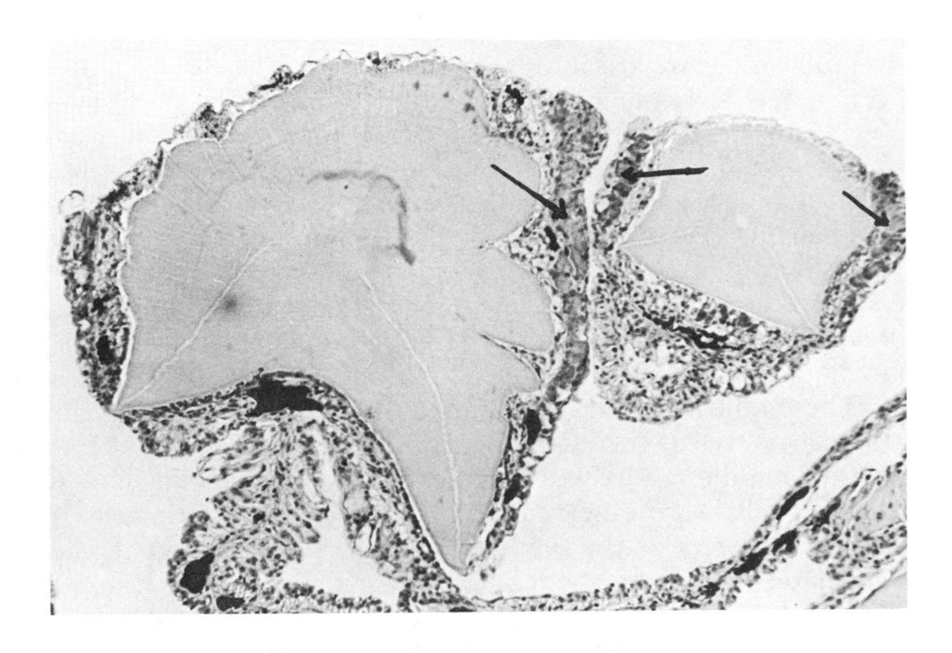

Figure 8.—Photomicrograph of cross sections of the 4th and 5th dorsal stings of *Scomberoides sanctipetri*. Note the large venom-producing cells along the sides of the spines (see arrows). The 4th dorsal sting (the smaller section) is at the level of the distal third, and the 5th dorsal sting at the level of the middle third. × 100. Hematoxylin and triosin stains. (P. C. Engen)

and the tendons pass through the previously mentioned shallow notch in the sides of the ventral head of the first interhemal and insert on the anterior aspect of each of the lateral condyles of the second anal spine. The depressor analis muscles of the second anal spine are almost as large as the first erector analis muscles. These depressor muscles originate on the lower sides of the triangular membranous bone of the second interhemal spine and pass downward and slightly forward. The tendons insert on the posterior aspect of each of the lateral condyles.

Microscopic Anatomy

Dorsal Sting (Fig. 8): A dorsal sting in cross section appears at the level of the distal or middle third as a somewhat rectangular bony structure which is completely invested by a layer of integument. The spine is more rectangular out toward the tip, becoming more irregular in configuration as one progresses toward the base. The spine is composed of acellular bone since no enclosed osteocytes are present in the bony matrix. The marrow spaces in the middle and distal thirds are small and few in number, but vascular channels are fairly numerous. A prominent layer of osteoblasts is present adjacent to the bone. There is no evidence of well-developed anterolateral glandular grooves such as are found in the scorpionfishes.

The integumentary sheath investing the spine consists of a dermis of moderately dense fibrous connective tissue and an epidermis of stratified squamous epithelium. Immediately subjacent to the epithelium is a very dense layer of connective tissue. From this layer dense collagenous fiber bundles can be seen extending inward for varying distances. In some areas, particularly where the dermis is thin, these fibers can be observed extending through the entire thickness of the connective tissue and into the bone as Sharpey's fibers. These fiber bundles are prominent in non-glandular regions and usually very inconspicuous in the areas underlying the glandular cells.

The surface epithelium of the integumentary sheath is quite thin, consisting of about four to five layers of stratified squamous epithelial cells. Scattered among the superficial epithelial cells are numerous mucous secretory cells. These mucous cells are prominent in the epithelium covering the body of the fish as well as the integumentary sheath of the sting. However, on the lateral, middle, and distal surfaces of the dorsal and anal stings, large glandular cells were observed which were confined primarily to the basal and middle layers of the epithelium. These cells are generally rectangular in shape, measuring about 7-10μ by 15-20μ. The nuclei are large, vesicular, and lie flattened against the internal surface of the cell membrane. It is these cells that are believed to be the venom-producing structures and they appear to be completely filled with a secretion leaving only a thin and ill-defined layer of cytoplasm at the periphery. This secretory material stained in a polychromatophilic manner, the color varying from a pink to a bluish-purple. It was difficult to determine the distribution of the cells on a particular sting because the stings were apparently traumatized during the capture of the fish. As shown in the photomicrographs the loss of epithelium was especially great on the anterior surface of the spine.

Anal Stings (Figs. 9, 10): The distal third of the anal sting is more or less rectangular in shape in cross section. The anterior edge of the anal spine is rounded, whereas the posterior margin is depressed to form a shallow trough to provide an attachment for the interspinous membrane. The remainder of the spine is similar to that described for the dorsal spine. There are no glandular grooves. The anal spine is completely invested by an integumentary sheath. The dermis and epithelium of the integumentary sheath are similar to that described for the dorsal stings. The venom cells are well developed in the anal sting and similar in size and appearance to those described for the dorsal sting.

Comments

The presence of venom organs in carangid fishes is of particular biotoxicological interest because most venomous fishes are sedentary in their habits. However, the carangid *Scomberoides sanctipetri* is not a sedentary fish, but is a fast-swimming, mid-water and surface, shore species. The venom apparatus of *Scomberoides* differs quite markedly from all other venomous perciform fishes. Whether these anatomical differences are found in other species of carangids which are suspected as being venomous is not known. In

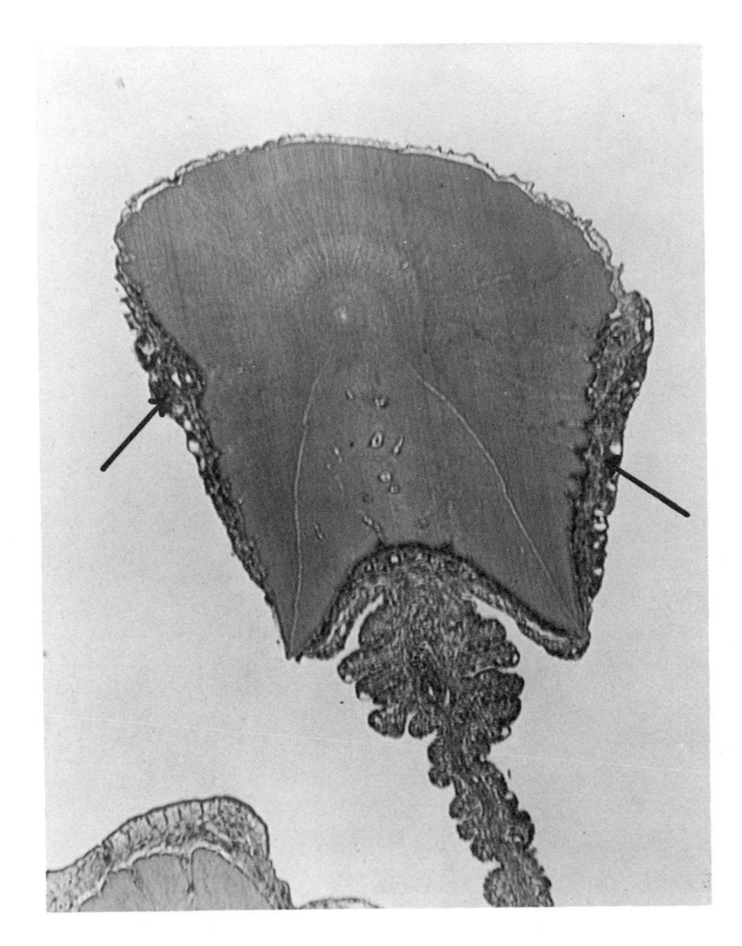

FIGURE 9.—Cross section of the distal third of the first anal sting of *Scomberoides sancti-petri*. Arrows point to the glandular areas. The integumentary sheath along the anterior margin of the spine has been severely traumatized. × 16. Hematoxylin and triosin stains. (P. C. Engen)

most perciform fishes the dorsal stings are the most highly developed as far as the venom apparatus is concerned, but such is not the case in *Scomberoides*. The dorsal stings of *Scomberoides* are relatively short and do not appear to be as well supplied with venom glands as the anal stings. The anal stings are highly developed, stouter, and about one-third longer than the longest dorsal sting. The musculature is correspondingly better developed. In most other venomous perciform fishes the anal stings are either poorly developed, or the venom glands are completely absent. *Scomberoides* also differs from other perciformes in that the anal spines are equipped with a frictional locking device which resembles that found in the fin spines of catfishes and balistid triggerfishes. Histologically the venom glands of *Scomberoides* more closely resemble those of catfishes than any of the other perciform fishes. In the case of most catfish the venom gland appears as a

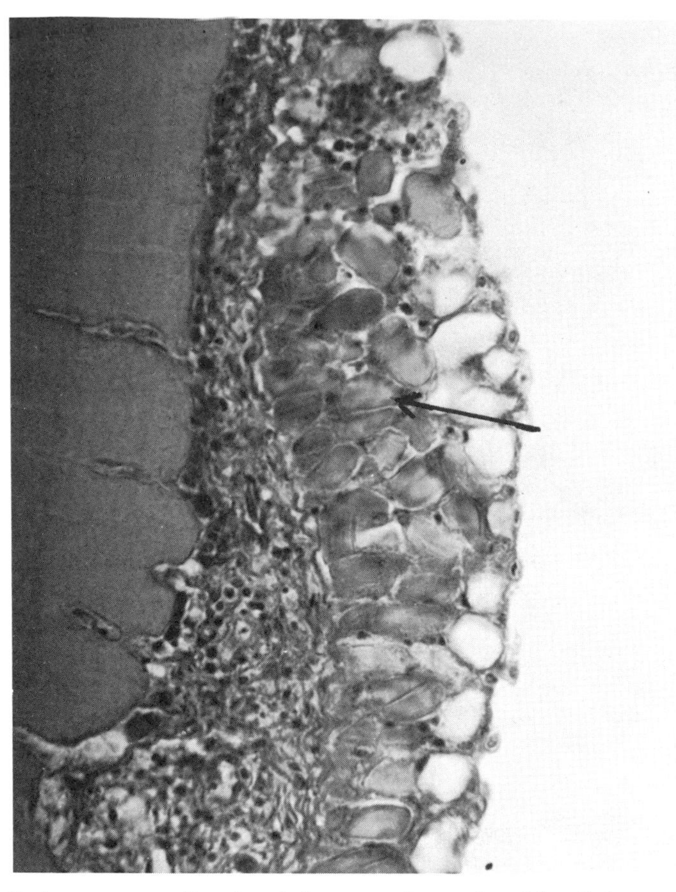

FIGURE 10.—Enlargement of a glandular area from the side of the second anal spine at the level of the distal third. Arrow points to venom-producing glandular cells. × 160. Hematoxylin and triosin stains. (P. C. Engen)

cellular sheet wedged between the pigment layer and the stratified epithelium along the side of the fin spine (Halstead, 1970), such as is found in *Scomberoides.*

Venom production is believed to be the result of a holocrine type of secretion.

MECHANISM OF ENVENOMATION

The use which *Scomberoides sanctipetri* makes of its venomous dorsal and anal stings has not been fully determined. Many Hawaiian tuna fishermen believe that the lae purposely stab the nehu with their venomous anal spines. In making contact with the nehu, the lae makes short rapid attacks with its dorsal and anal spines fully erect. However, other observations have shown that the lae makes contact by means of its mouth on the head and upper body of the nehu. These attacks are so rapid and of such short duration that accurate observations are difficult without the aid of high-speed photography. Since adult lae retain the venomous quality of their dorsal and anal spines, it is possible that these spines are employed both offensively and defensively, but no further information seems to be available to substantiate this. Because of the large size and locking mechanism of the anal stings it is apparent that they are the most dangerous ones. When lae are removed from the water, both the dorsal and anal stings are erected, but only the anal stings are locked into place.

MEDICAL ASPECTS

Clinical Characteristics

Hawaiian fishermen that have been stung by either dorsal or anal stings of an adult lae report that the resulting pain is quite intense and may last up to four hours. Accidental injury from one of the venomous stings of a juvenile specimen feels much like a mild bee sting and may last for 20 to 30 minutes. No delayed reactions have been reported. Morphological evidence indicates that the anal stings probably inflict the most serious wounds.

Treatment

The treatment of *Scomberoides* stings is similar to that used for scorpionfish stings. (*See* p. 782.)

Prevention

Care should be taken in handling leatherbacks to avoid contact with the dorsal and anal stings.

TOXICOLOGY

(Unknown)

PHARMACOLOGY

(Unknown)

CHEMISTRY

(Unknown)

Order PERCIFORMES (PERCOMORPHI)

Suborder PERCOIDEI

Family SCATOPHAGIDAE: Scats

The scats or spadefishes are a small family of perchlike Indo-Pacific fishes, of about three species, all less than 35 cm in length. They are primarily marine, but commonly invade brackish and freshwater streams.

Richardson (1844-48) was one of the first to observe that scats were able to inflict painful wounds, and Maass (1937) reported scats as venomous.

Whitley (1943) commented that the aborigines of the Northern Territory, Australia, refer to these fish as "poison bone" and published an account of an envenomation by *Selenotoca multifasciata*. Marshall (1964) reported that scats are able to inflict more painful wounds than do other allied species. Scats have also been listed as venomous by Halstead and Mitchell (1963) and Halstead (1970). However, Cameron (1969) and Cameron and Endean (1970) were the first to describe the venom apparatus of the scatophagids.

The systematics of this family has been studied by Fowler (1928), Weber and Beaufort (1936), Heere (1953), Marshall (1964), and Munro (1967).

LIST OF SCATS REPORTED AS VENOMOUS

Family SCATOPHAGIDAE

260 *Scatophagus argus* (Linnaeus) (Pl. 3, fig a). Spotted scat, spadefish (USA), spotted butterfish (Australia), kipar (Indonesia).

DISTRIBUTION: Indo-Pacific.

SOURCES: Maass (1937), Marshall (1964), Cameron (1969), Cameron and Endean (1970), Halstead (1970).

260 *Selenotoca multifasciata* (Richardson) (Pl. 3, fig. b). Many-banded scat, spadefish (USA), striped butterfish (Australia).

DISTRIBUTION: East, north and western coasts of Australia, New Guinea.

SOURCES: Richardson (1844-48), Whitley (1943), Marshall (1964), Cameron (1969), Cameron and Endean (1970).

BIOLOGY

Little appears to be known of the habits of scats in nature. They inhabit inshore coastal areas, particularly around sewer outlets of coastal cities, and have become a favorite aquarium fish because of the ability of the young to adapt readily to freshwater. The young feed upon algae. Scats are frequently observed feeding around sewer outlets, apparently ingesting excrement and refuse—hence the name of the genus *Scatophagus* (eaters of dung). Scats tend to be soft-bodied, but are nevertheless valued as food by some people. They are difficult to breed in captivity and are said to be oviparous.

MORPHOLOGY OF THE VENOM APPARATUS

Gross Anatomy

Our knowledge of the venom organs of scatophagid fishes is based largely on the studies of Cameron (1969) and Cameron and Endean (1970) in which they studied *Scatophagus argus* and *Selenotoca multifasciata*. The gross anatomical descriptions are by B. W. Halstead.

The venom apparatus of scatophagids is comprised of 11 or 12 dorsal spines, two anal spines, a pair of ventral spines, their associated musculature, venom glands, and enveloping integumentary sheaths. All of the spines are equipped with venom glands. However, the venom glands appear to be largely confined to juveniles and gradually disappear as the fish matures. The dorsal spines, like the anal spines, are connected by a moderately incised interspinous membrane. The integumentary sheaths of scatophagids are scaled and rugose on the proximal portion of the sting. The venom glands appear as a slender, elongate, fusiform strand of gray or pinkish tissue, lying within the anterolateral glandular grooves on either side of the spine. The venom glands are difficult to detect in preserved specimens. The length of the glands vary from spine to spine and with each fish depending upon its maturity. In the specimens studied the venom glands usually do not extend to the distal tip of the spine. The ratio of venom gland length to spine length in a series of 69 glands from six specimens of *S. argus* was found to have a mean value of 0.20 with a range of 0.01 to 0.45. The longest gland measured was associated with the fifth dorsal spine and was 10.36 mm in length, taken from a fish specimen having a standard length of 125 mm. There is no evidence of an excretory duct. The distal terminations of venom glands varied in their distance from the distal ends of the spines bearing them. This distance ranged from 10 to 3230μ.

Dorsal Spines (Fig. 11): This description is based on dorsal spine IV which is representative of the others. The spine is described in the upright anatomical position. From the lateral aspect the spine is elongate but somewhat stoutish. The proximal fourth is slightly curved with a craniad convexity and tapered distally to a very sharp point. The acute tip of the spine is trigonal in shape. Two distinct anterolateral-glandular grooves are present, one on either side, which originate near the base of the spine. The grooves are best developed in the vicinity of the middle third of the spine and gradually disappear in the distal fourth, but do not extend to the tip of the spine. The glandular grooves are separated by an anteromedial ridge. The base of the spine is expanded into two somewhat asymmetrical lateral condyles. No median foramen is present. A posterior view of the spine shows that the posterolateral ridges are separated by a distinct posteromedian groove which extends throughout the length of the spine.

Each dorsal spine rests on and articulates with an interneural spine which is generally ┌ -shaped. The dorsal or horizontal limb of the interneural is expanded into a flattened, platelike structure called the dorsal plate. The upper surface of the dorsal plate is modified to receive the lateral condyles of the articulating dorsal spine. The annular process, which is highly developed in some of the scorpionfishes, is absent or reduced in scatophagids. A central ridge is present on the median plate. For the most part the interneurals of scatophagids are not atypical of those found in most other fishes.

The dorsal spine musculature has not been described.

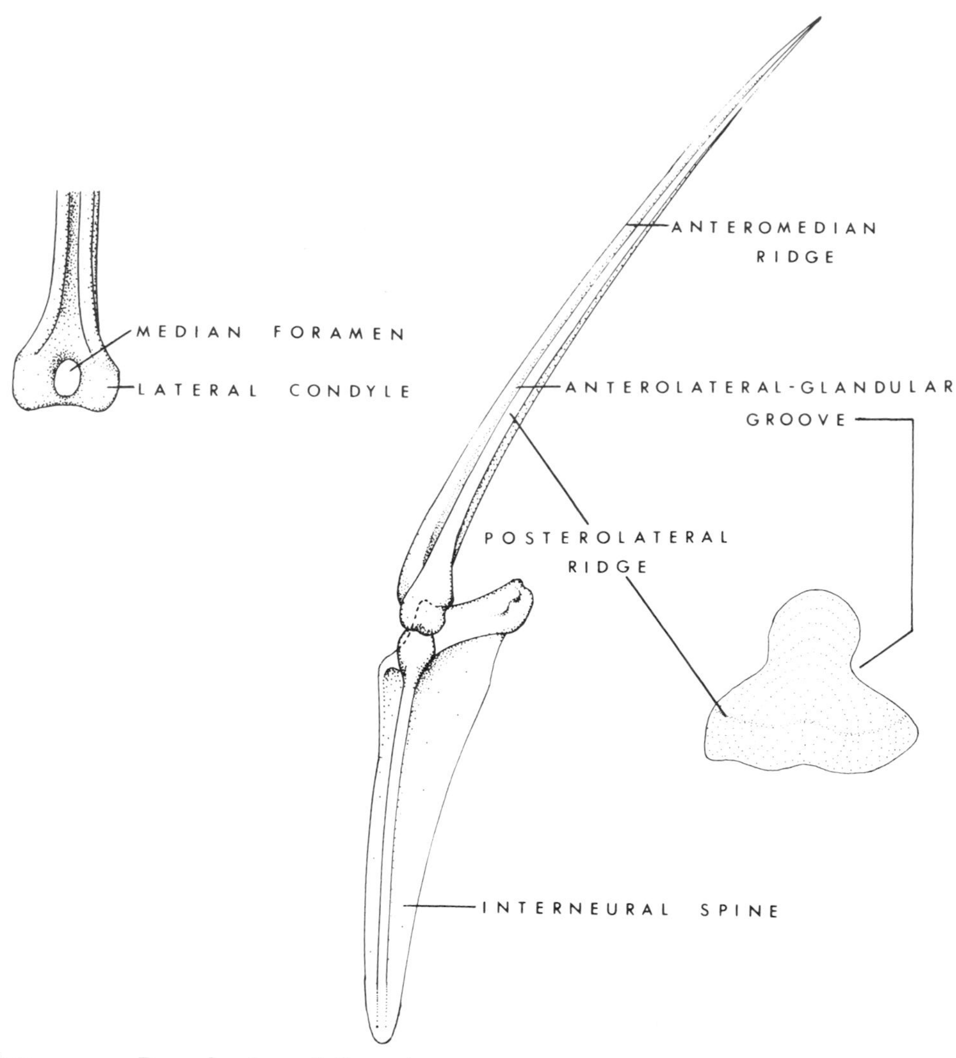

FIGURE 11.—Dorsal spine of *Scatophagus argus*. The dorsal spines of *Selenotoca multifasciata* are similar in morphology to those of *S. argus*. × 6. (P. Mote)

Anal spines: This description is based on anal spine II which is described in the dependent anatomical position. The basic configuration of the anal spines is similar to that of the dorsal spines. The anal spines, however, are stouter. The anterolateral-glandular grooves are distinct, but as in the case of the dorsal spines do not extend to the distal tip. The posteromedian groove extends to the distal tip of the spine. The base of the spine is expanded into two almost symmetrical lateral condyles which are separated by a well-developed median foramen.

The four anal spines are supported by and articulate with corresponding interhemal spines. The first two interhemal spines are heavy bodied and fused together to form a plow-shaped structure. Removal of anal II spine reveals that the ventral head of the associated interhemal spine is equipped with a well-developed annular ring which passes through the median foramen of the anal spine with which it articulates. The musculature of the anal spines has not been described.

Pelvic Spines: The pelvic girdle of scatophagids are thoracic in position and are equipped with a small scaly axillary process. This description is based on the left pelvic spine in the extended anatomical position.

The shaft of the spine is elongate, moderately stout, trigonal in shape, and terminates in an acute tip. The superior and inferior glandular grooves (the equivalent of the anterolateral-glandular grooves of the dorsal and anal spines) are well-developed for most of the length of the spine, but do not extend to the tip. The lateral condyles are almost symmetrical and are separated by a median foramen. The posteromedian groove is well-developed and extends throughout the length of the spine.

The pelvic girdle of scatophagids is comprised of two bones, the basipterygii, which fuse together along the midline and provide support for the two pelvic spines and the pelvic rays. The pelvic girdle is in the shape of an isosceles triangle with the apex directed anteriorly. The length of the pelvic girdle is only slightly longer than the pelvic spines. The greatest width of the pelvic girdle is about half the length of the pelvic spines. The lateral sides of the pelvic girdle are stiffened by the longitudinal suprapelvic and subpelvic keels. On the posterior end of the pelvic girdle along the medial line, a small spine projects from each basipterygium which fuse together to form the notched postpelvic process. A subpelvic process is also present. The lateroposterior angle of each basipterygium is equipped with a specialized condylar area having an annular ring which passes through and articulates with the median notch of the pelvic spine. The remainder of the posterior surface of the pelvic girdle provides a condylar surface which articulates with the pelvic rays. The musculature of the pelvic girdle has not been described.

Microscopic Anatomy

The microscopic anatomy of the dorsal, anal, and pelvic stings of *Scatophagus argus* and *Selenotoca multifasciata* are quite similar, so the following description will deal primarily with the anatomical features of the dorsal sting of *S. argus*.

Dorsal Stings (Figs. 12a, b): A dorsal sting in cross section is essentially an inverted T-shaped acellular bony structure invested by a layer of integument. Cross sections reveal indentations representing the two anterolateral-glandular grooves and the posteromedian groove. The integumentary sheath investing the spine consists of a thin layer of dermal collagenous connective tissue, a basement membrane and an epidermis of stratified epithelium. Dermal scales are present in the proximal parts of the integumentary sheath but do not extend to the distal tip of the spine. Aggregations of gland cells are present in the epidermis filling the anterolateral-glandular grooves.

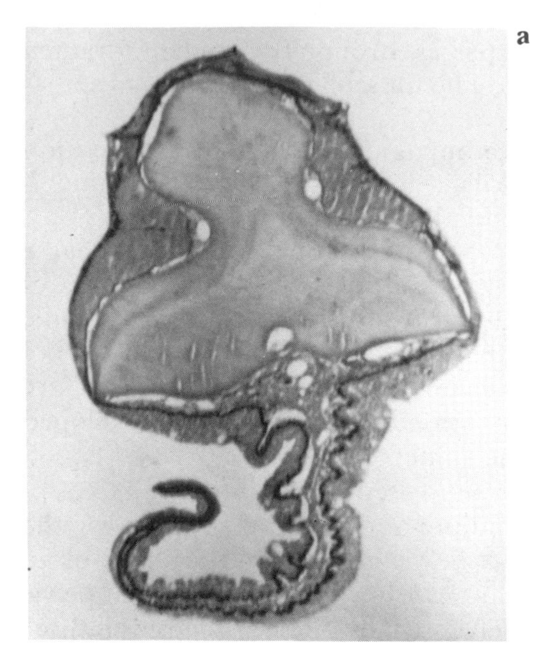

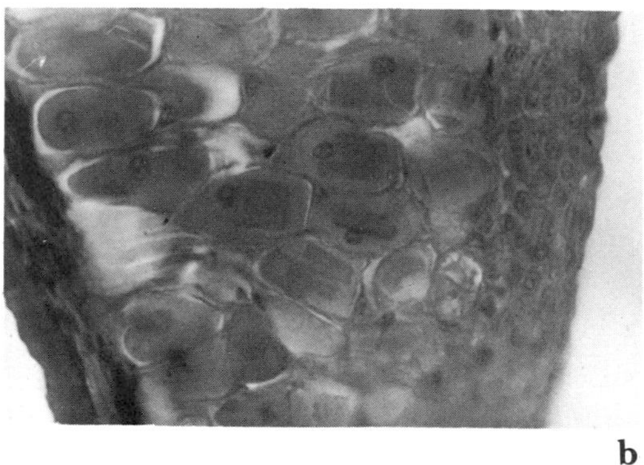

FIGURE 12a.—Cross section through the middle third of the 5th dorsal spine of *Scatophagus argus*. × 17.5. Hematoxylin and eosin.
FIGURE 12b.—Cross section through the glandular triangle at the level of the middle third of the 9th dorsal spine of *S. argus*. × 168. Hematoxylin and eosin.
(Courtesy A. Cameron)

These are considered to be the venom glands. These aggregations of venom gland cells are not encapsulated by dermal connective tissue capsules as are the venom glands of scorpionfishes, but their location within the glandular grooves is similar to that of scorpaenids. The venom glands of scatophagids are merely thickenings of the epidermal portion of the integumentary sheath with concentrations of clavate (venom) cells in the region of the glandular grooves. This is similar to the glandular arrangement found in the catfish *Galeichthys felis*. Measurements made of gland width and depth within the glandular grooves of the spine indicated that the venom glands are wedge-shaped in cross section and possess tapered ends, the greatest cross-sectional area being about midway in the length of the gland. The extent of the venom glands near the base of the spine is limited by the encroachment of scales of the integumentary sheath. In a series of six specimens of *S. argus* ranging from 62 to 296 mm in standard length, the shortest gland length for a total of 69 dorsal spine glands measured 340μ in length from a fish specimen of 125 mm, and the longest gland measured 10,360μ from a fish 296 mm in standard length. The venom gland cells are irregularly shaped but tend to be columnar in the deepest portion of the gland, abutting on the glandular groove of the spine. There is no basal layer of germinal cells separating the venom-producing cells from the basement membrane of the epidermis in the deeper portion of the glandular groove but a basal layer is present elsewhere. Each venom cell possesses a spherical nucleus which is centrally situated. The cytoplasm appears homogeneous and no granular secretory products were observed. However, angular rod-like bodies occurred in some of the venom cells. They colored with dyes as did the venom cell cytoplasm but were more intense. They yielded a strongly positive reaction for protein with mercuric bromphenol blue reagent but were PAS negative. The mature venom cell cytoplasm colored red with eosin, yellow-green with Lillie's allochrome stain, orange to red with Mallory's triple stain, deep blue with mercuric bromphenol blue reagent and yielded a weakly positive PAS reaction. Stages in the development of venom cells from small, undifferentiated epidermal cells possessing basophilic cytoplasm are apparent in some regions of the glands, particularly at their proximal and distal ends. As the epidermal cells increase in size they lose their affinity for hematoxylin and progressively acquire an affinity for eosin. Irregularly shaped gland cells with a granule-packed cytoplasm are distributed sparsely in the epidermis peripheral to the venom glands of some spines. The cytoplasm of these cells respond to dyes similar to the venom cells but are less intense in coloration. In some sections spherical to pear-shaped goblet cells are present in the epidermis.

The arrangement of the venom glands of *S. multifasciata* is essentially the same as that of *S. argus*.

MECHANISM OF INTOXICATION

Envenomations by scatophagids are most frequently produced by their dorsal stings and occasionally by their pelvic and anal stings. Most scatophagid stings are due to careless handling of the fish. Wounds may also result from the careless handling of a dead specimen.

MEDICAL ASPECTS

Clinical Characteristics

Scatophagid stings cause painful wounds. The pain has been described as an intense, shooting or throbbing sensation, "electric shock-like." The pain may radiate up the arm and down the side of the body. Smaller specimens are said to inflict more painful wounds than larger ones. Little else seems to be known concerning the clinical characteristics of scatophagid wounds.

Pathology

(Unknown)

Treatment

The treatment of scatophagid wound envenomations is similar to that of scorpionfish stings. (*See* p. 782.)

TOXICOLOGY

(Unknown)

PHARMACOLOGY

(Unknown)

CHEMISTRY

(Unknown)

Order PERCIFORMES (PERCOMORPHI)

Suborder PERCOIDEI

Family URANOSCOPIDAE: Stargazers

The stargazers are a group of small, carnivorous, bottom-dwelling, marine fishes. They are characterized by having a large cuboid head, an almost vertical mouth with fringed lips, and an elongate, conic, subcompressed body. Their eyes are small and anteriorly situated on the flat upper surface of the head. The family contains less than a dozen species, most of which are about 40 cm or less in length. Despite the small number of species, representatives of the family Uranoscopidae are distributed throughout the Mediterranean Sea and the warmer parts of the Pacific, Atlantic, and Indian Oceans. Stargazers are of medical importance by virtue of their venomous cleithral or shoulder spines. Since uranoscopids spend a considerable portion of their time buried in the sand or mud with only their eyes and a portion of their mouth protruding, they present a menace to the person entering their environment.

The systematics of uranoscopids have been discussed by Bottard (1889), Herre (1953), Smith (1961), Marshall (1964), and others.

LIST OF STARGAZERS REPORTED AS VENOMOUS

Family URANOSCOPIDAE

261 *Uranoscopus japonicus* Houttuyn (Pl. 4, fig. a). Stargazer (USA), mishima pufferfish, mishimaokoze, osen, ushisakanbo, ginu (Japan).

DISTRIBUTION: Southern Japan, southern Korea, China, Philippines, Singapore.

SOURCE: Maass (1937).

261 *Uranoscopus scaber* Linnaeus (Pl. 4, fig. b). Stargazer (USA), mishima pufferfish, mishimaokoze, osen, ushisakanbo, ginu (Japan), gemeine sternseher (Germany), uranoscopo, scabro, lucerna (Italy), uranoscope, rat, miou (France), rata (Spain), besmek (Yugoslavia), bufo, aranhuco (Portugal).

DISTRIBUTION: Eastern Atlantic and Mediterranean.

SOURCES: Bottard (1889), Coutière (1899), Porta (1905), Taschenberg (1909), Hilzheimer and Haempel (1913), Engelsen (1922), Phisalix (1922), Maass (1937), Giunio (1948), Maretić (1968).

MORPHOLOGY OF THE VENOM APPARATUS

Gross Anatomy

The venom apparatus of *Uranoscopus scaber* is comprised of the two cleithral or shoulder spines (situated one on either side), their associated venom glands, and their enveloping integumentary sheaths (Fig. 13a, b). There is no evidence of any venomous fin spines.

The cleithral sting appears as a sharp conical spine which generally protrudes to a greater or lesser extent through the enveloping integumentary sheath. The cleithral spine protrudes conspicuously from along the super-posterior margin of the operculum at a point which is superior to the upper margin of the pectoral fin. The distance between the base of the spine and

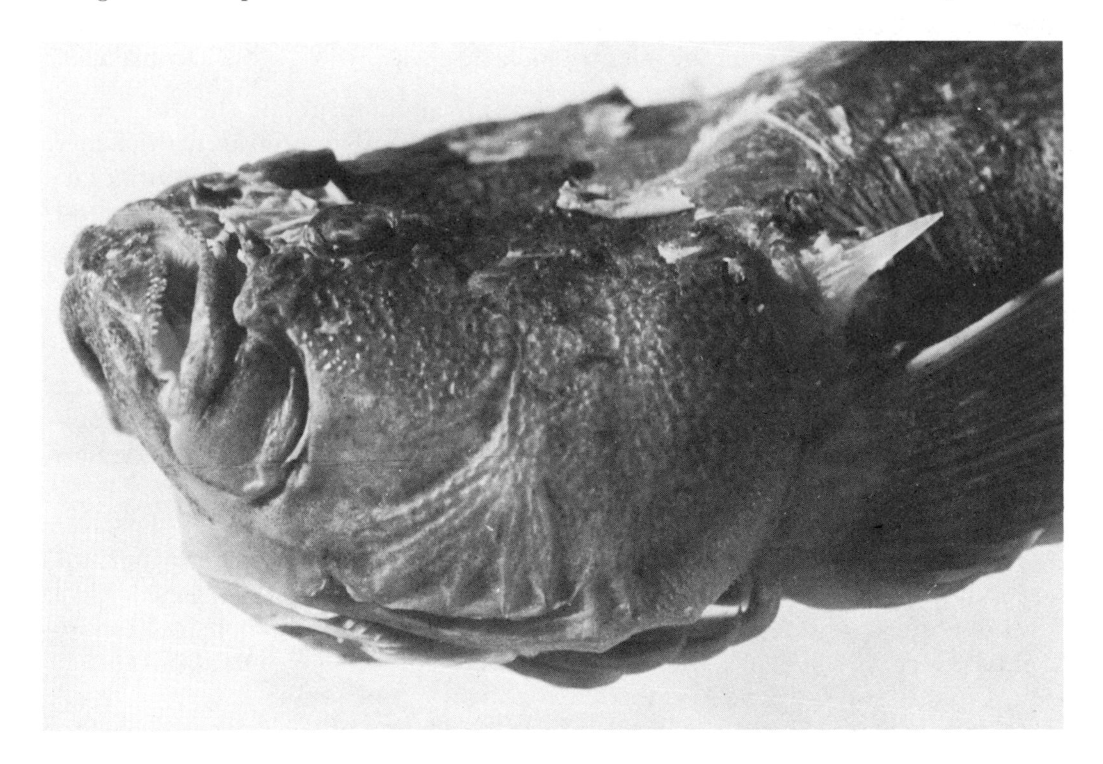

FIGURE 13a.—The head of *Uranoscopus scaber* showing the left cleithral spine. The integumentary sheath has been removed to expose the spine. Length of spine 13 mm.

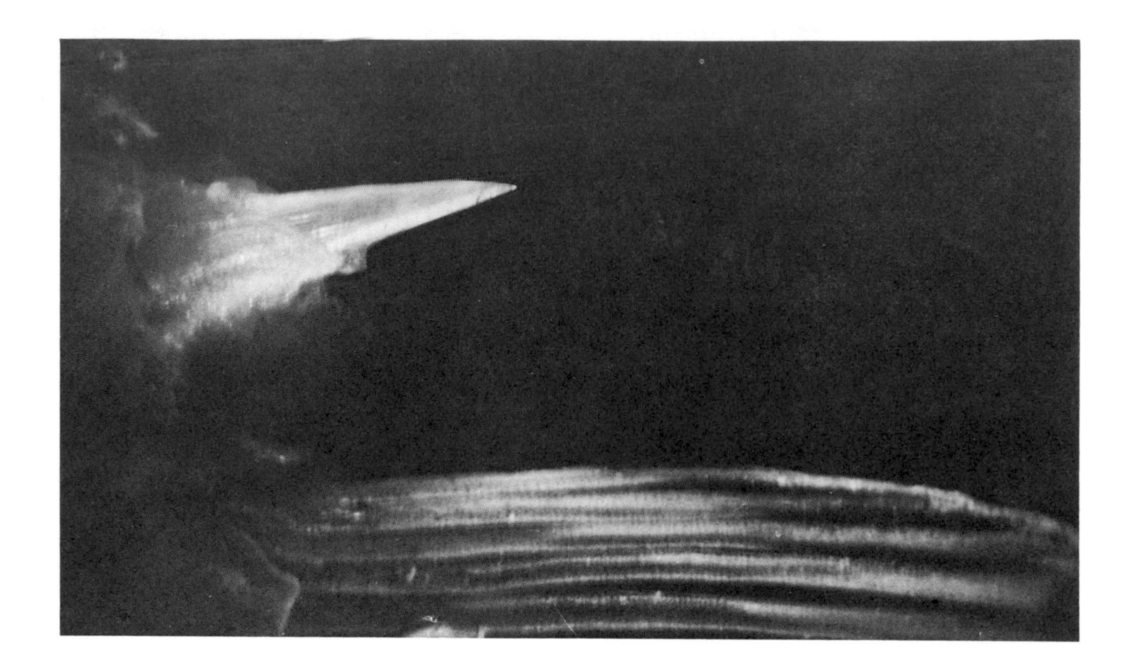

FIGURE 13b.—Close-up of cleithral spine showing the venom gland which surrounds the base of the spine. Length of spine 13 mm. (R. Kreuzinger)

the upper margin of the pectoral fin is equivalent to about the transverse diameter of the orbit. The integumentary sheath is only moderately thickened and mottled gray or brown in appearance. Removal of the integumentary sheath by careful dissection (Fig. 13a, b) reveals a thin covering of loose fascia and a whitish granular spongy tissue which is very thinly spread over the outer aspect of the cleithral spine, becoming thickened along the upper and lower margins of the spine, gradually spreading into a thickened mass which lies between the spine and the body of the fish. This spongy mass is most concentrated beneath the base of the cleithral spine and gradually fans out for a short distance over the upper pectoral area of the body. Subsequent histological studies reveal that this spongy mass constitutes the venom gland of the cleithral sting.

Without the opportunity of working with living specimens of *U. scaber*, it is difficult to ascertain the mechanism of action of the venom apparatus, but judging from the anatomical relationships it appears that the cleithral sting is almost rigid and relatively limited in its use as an offensive striking organ. Certainly it lacks the maneuverability of the venom apparatus as found in stingrays, weevers, catfishes, toadfishes, and others.

Cleithral Spine (Figs. 14; 15a, b): The bony portion of the venom apparatus of *Uranoscopus* consists of the superoposterior prolongation of the cleithrum. The shaft of the cleithrum is a long, slender, almost straight bone, which in the anatomical position extends forward and below where it attaches to the

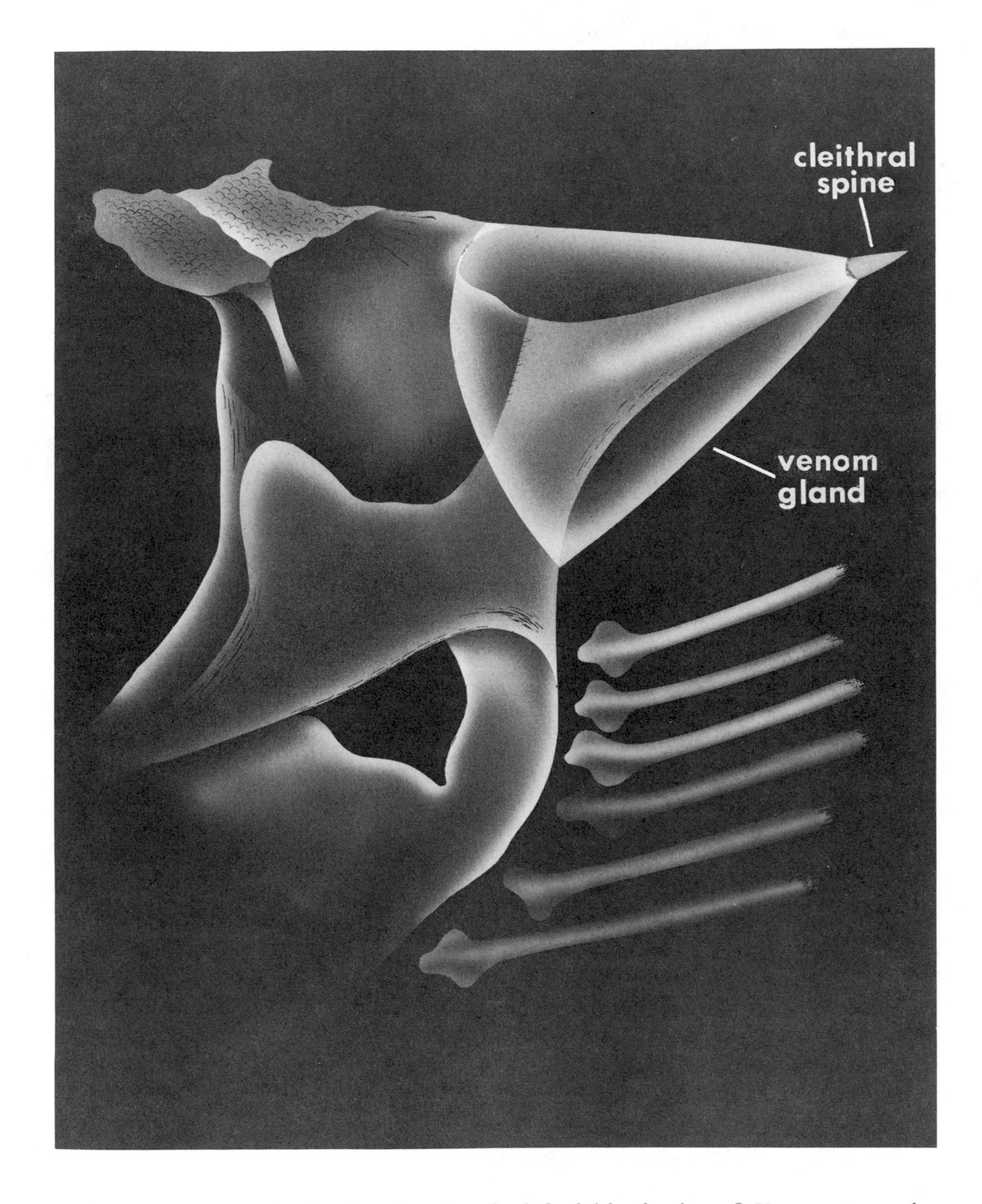

Figure 14.—Phantom drawing showing the left cleithral spine of *Uranoscopus scaber* and its anatomical relationship to the venom gland. (R. Kreuzinger)

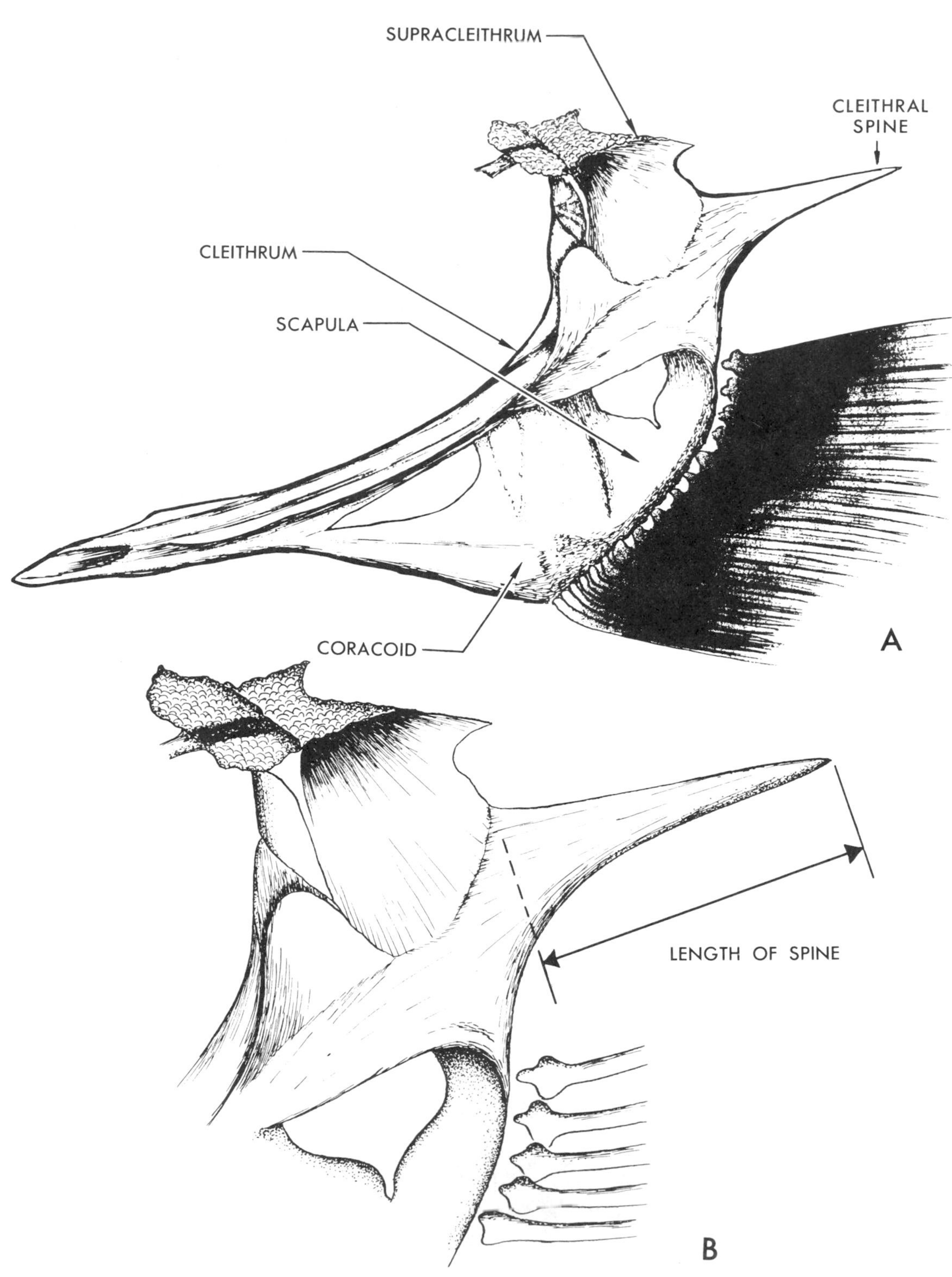

FIGURE 15.—a. Drawing showing the left pectoral girdle of *Uranoscopus scaber* and the cleithral spine. b. Drawing showing the portion of the cleithral spine which was measured in determining its length in these particular studies. (R. Kreuzinger)

adjacent cleithrum of the opposite side at a point along the midventral line immediately anterior to the pelvic girdle. The upper end of the cleithrum is expanded along the vertical plane and terminates as a sharp, slender, elongate, conical spine. The base of the spine is somewhat compressed. The inferior margin of the spine contains a pronounced but shallow groove which rapidly disappears near the middle third of the spine. There is only a slight suggestion of a tiny groove at the base along the superior margin of the spine. The surface of the outer aspect of the base of the spine is somewhat irregularly grooved and rugose in texture. The surface of the inner aspect of the base of the spine is relatively smooth.

Microscopic Anatomy

The cleithral spine consists of a free distal portion and a proximal portion covered by skin which surrounds the spine as a sheath. The material composing the spine is an acellular cementumlike material which shows concentric growth laminae. The distal half tends to be smooth surfaced and nearly round. The proximal part becomes flattened and grooved lateromedially. The grooves deepen and become more irregular near the juncture of the spine and the cleithrum. Some of the grooves eventually penetrate the interior of the spine to form canals. The canals enlarge within and are continuous with cancellous spaces in the cleithrum.

The skin which ensheaths the spine and extends from it is composed of stratified cuboidal or polyhedral cells, interspersed with unicellular mucous glands (Figs. 16-18). The basal germinal cells are columnar. The external surface cells are slightly compressed polyhedral cells. The free surface of the external cells may be concave or convex. This free surface is crossed by parallel ridges which upon superficial examination resemble microvilli. By focusing up and down, one can see that they are ridges which sometimes anastomose. The ridges may run in different directions on different cells, and on any one cell the ridges may be straight or appear to be arranged like lines on a contour map or a fingerprint.

The connective tissue of the dermis is relatively thick and is separated from the epidermis by a prominent basement membrane. Within large spaces or pockets of the dermis is a gelatinous material containing what appears to be scattered fibrocytes. The dermis is somewhat loosely attached to the spine. At the surface of attachment there is a palisade of columnar cells which produces the hard tissue of the spine where development is in progress.

Deep to the dermis, the connective tissue becomes a periosteum of the bone. The bone is largely, but not entirely, acellular. The large spaces within the bone previously mentioned also contain a gelatinous material as well as delicate-walled blood vessels. The canals do not contain ducts for the passage of a specific secretion. Macrophages are abundant in many of the cancellous spaces. The gelatinous material appears to be a breakdown product of cells by holocrine secretion or a denaturation of the connective tissue fibers themselves. This gelatinous material is believed to be the venom.

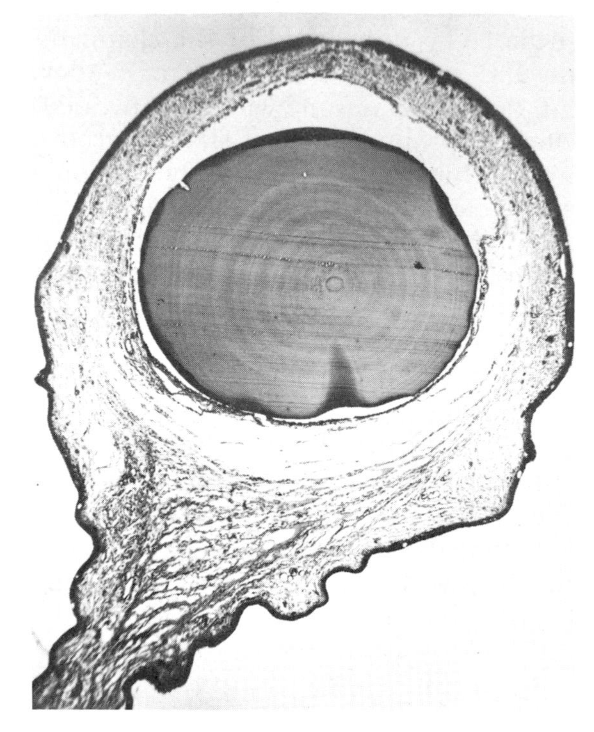

FIGURE 16.—(Left) Cross section through the distal third of the cleithral sting of *Uranoscopus scaber*. The cleithral spine is shown surrounded by the integumentary sheath. × 20. Hematoxylin and triosin stains.

(From Halstead and Dalgleish)

FIGURE 17.—(Below) Section of the integumentary sheath of the cleithral spine showing the large mucous cells of the epidermis. Arrow points to a pocket on the dermis containing a gelatinous material and scattered fibrocytes which are not clearly visible in this photomicrograph. The venom is believed to be contained in the gelatinous material. × 100. Hematoxylin and triosin.

(From Halstead and Dalgleish)

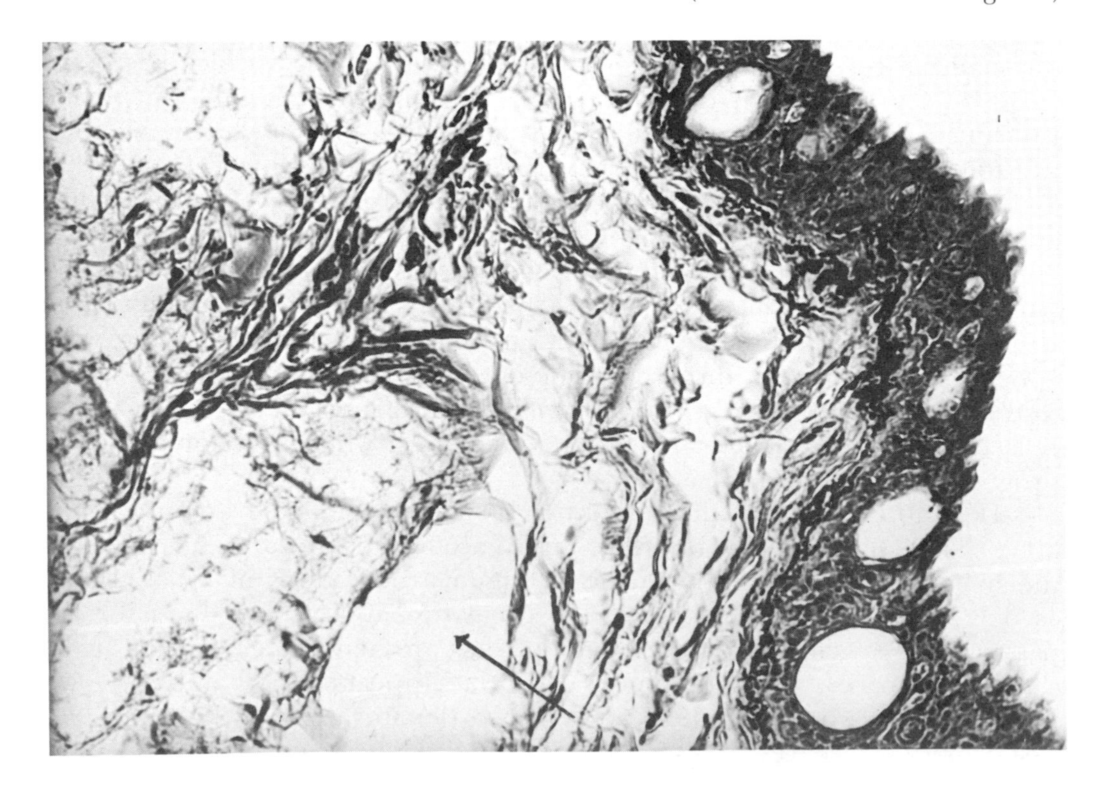

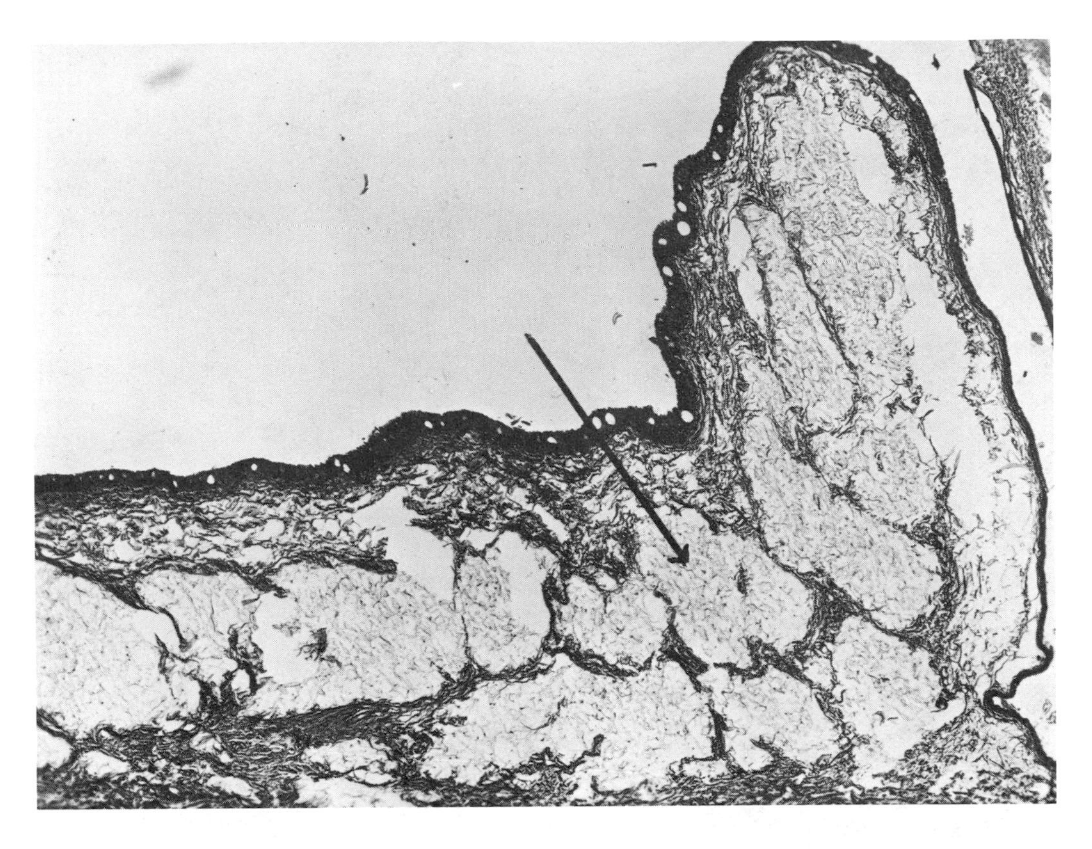

Figure 18.—Cross section of the integumentary sheath of the cleithral spine taken near the base of the spine. Arrow points to one of the dermal pockets which contains the gelatinous material (venom). × 30. Hematoxylin and triosin stains.
(From Halstead and Dalgleish)

Between bones, there is often dense, ligamentous, connective tissue as well as tendons of muscle insertions. A short distance from the spine, a gland-like arrangement of hypertrophied cells occurs within the framework of some of the loose connective tissue bundles (Fig. 19a, b). The cells, centrally, are epitheloid in appearance but are reduced in size and roundness of shape as they approach the peripheral areas of aggregation. They appear to be derived from fibrocytes which have enlarged among the fibers of collagen. No blood vessels can be noted within the cluster. It is postulated that modified fibrocytes either produce the venom continuously, or hypertrophied fibrocytes produce a holocrine secretion which is collected within dermal pockets or cancellous spaces of the bone near the base of the spine. An enzyme may also be responsible for what appears to be a dissolving of fibers of the connective tissue to make room for the pockets of gelatinous material. There are seen what appear to be developmental stages in this breakdown. Near the base of the spine and running between muscles are rather large vascular channels with very thin walls which can very easily be ruptured when the sheath is stripped from the spine as it enters the flesh of the victim.

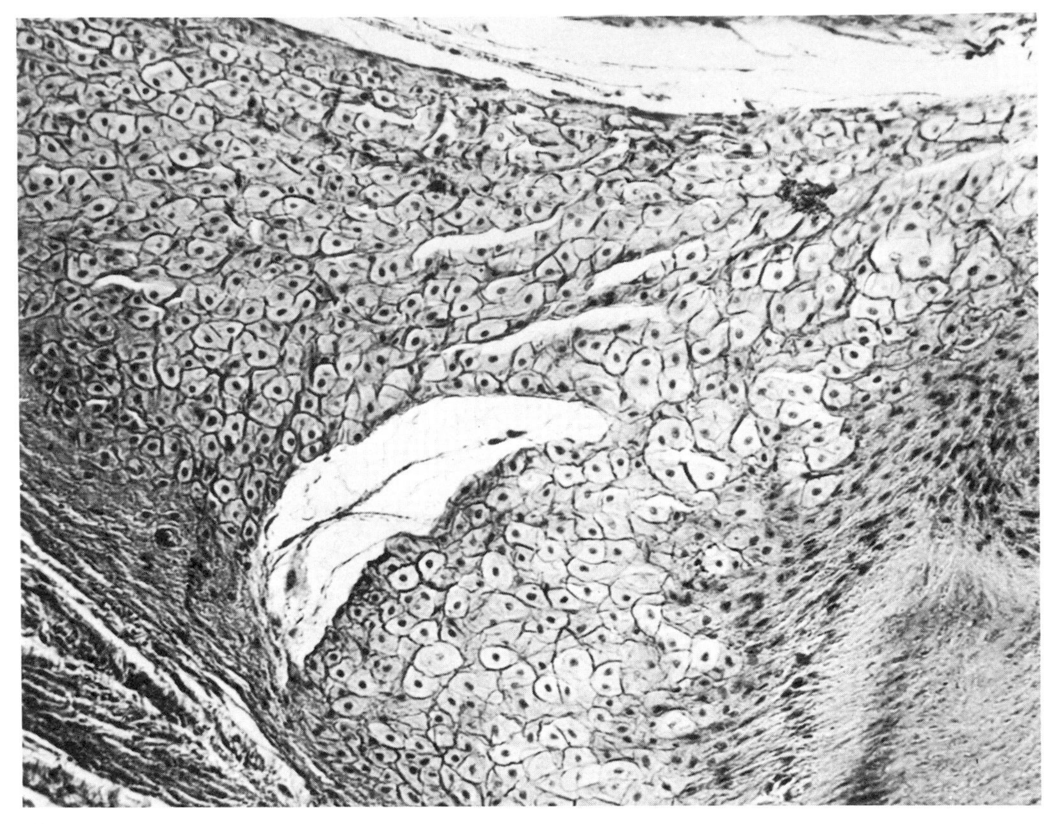

a

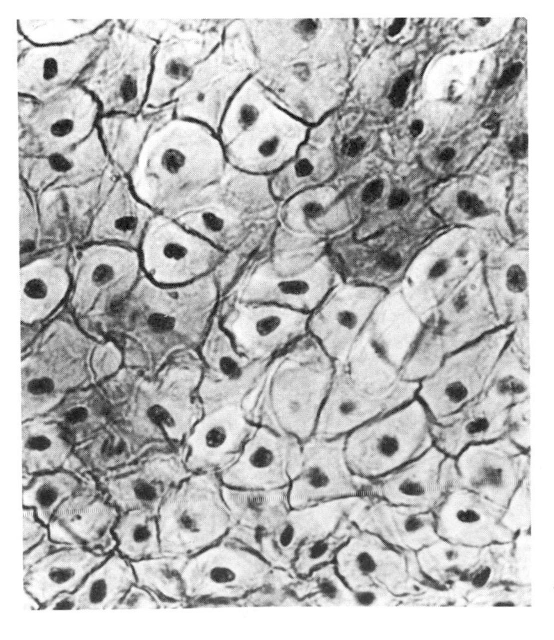

b

FIGURE 19.—a. (Above) A section through a large dermal pocket containing hypertrophied cells. The section was taken near the base of the cleithral spine. The hypertrophied cells are believed to be the producers of the venom or gelatinous material. × 110. b. (Left) Enlargement of the hypertrophied cells of the venom gland. × 500. Hematoxylin and triosin stains.

(From Halstead and Dalgleish)

Discussion

The venom apparatus of *Uranoscopus scaber* is grossly in agreement with the brief description given by Bottard (1889). The cleithral venom apparatus of *U. scaber* superficially more closely resembles the opercular venom apparatus of *Thalassophryne* than any other group of fishes studied to date. However, *U. scaber* differs from *Thalassophryne* structurally. One significant difference is that the cleithral spine of *U. scaber* is not hollow as is the opercular spine of *Thalassophryne*. Venom conduction in *U. scaber* is accomplished by means of a very shallow groove which extends along the inferior margin of the cleithral spine. Conduction of the venom to the tip of the cleithral spine is aided by the integumentary sheath which surrounds the venom gland and the spine. Thus the structural arrangement suggests that when the spine penetrates the flesh, the sheath of skin surrounding the spine is stripped away and the gelatinous glandular material with blood and other tissue fluids are forced into the wound through the grooves of the spine which provide a passageway between the skin of the victim and the surface of the spine. It appears that the venom is located in the gelatinous material of the dermal pockets and in cancellous spaces of the bone continuous with canals and grooves of the cleithral spine.

MECHANISM OF INTOXICATION

Stings from stargazers are most likely to occur as a result of careless handling of the fish when removing them from nets or hooks. Stargazers inflict their stings with the use of their cleithral or shoulder spines.

MEDICAL ASPECTS

Nothing is known about the clinical characteristics of stargazer envenomations. It is assumed that they are similar to the stings of *Scorpaena* or *Thalassophryne*. According to Halstead and Dalgleish (1967), fatalities have been reported in the Mediterranean area, apparently due to *Uranoscopus scaber*, but there are no details concerning the circumstances. References in the literature to the venomous properties of stargazers are scanty and largely of a general or nonspecific type. The following authors make mention of the venomous properties of this group: Bottard (1889), Coutière (1899), Porta (1905), Taschenberg (1909), Hilzheimer and Haempel (1913), Engelsen (1922), Phisalix (1922), Nobre (1928), Maass (1937), Giunio (1948), and Halstead (1959). Treatment is similar to that used for scorpionfishes (*see* p. 782).

There is no information available on the pharmacology or chemistry of uranoscopid venoms.

Order PERCIFORMES (PERCOMORPHI)

Suborder SIGANOIDEI (AMPHACANTHINI)

Family SIGANIDAE: Rabbitfishes

Siganids are a group of spiny-rayed fishes which closely resemble the acanthurids. They differ from all other fishes in that the first and last rays of the ventral fins are spinous. Rabbitfish are of moderate size, usually valued as food, and abound about rocks and reefs from Polynesia to the Red Sea.

The systematics of siganids has been discussed by Fowler (1928), Schultz *et al.* (1953), Herre (1953), Smith (1961), and Marshall (1964).

REPRESENTATIVE LIST OF SIGANIDS REPORTED AS VENOMOUS

Family SIGANIDAE[1]

262 *Siganus corallinus* (Valenciennes) (Pl. 5, fig. a). Rabbitfish (USA), batawayi, dangie, layap (Philippines), cordonnier, rabbitfish (Mauritius, Seychelles).

DISTRIBUTION: Indo-Pacific, Philippines to the Seychelles.

SOURCE: Halstead *et al.* (1971).
OTHER NAME: *Teuthis corallina.*

262 *Siganus fuscescens* (Houttuyn) (Pl. 5, fig. b). Rabbitfish (USA), nuga (Fiji), mole, mere (Marshall Islands), batawayi, dangie, layap (Philippines).

DISTRIBUTION: Indo-Pacific.

SOURCES: Amemiya (1921), Maass (1937), Fish and Cobb (1954).
OTHER NAME: *Teuthis fuscescens.*

262 *Siganus lineatus* (Valenciennes) (Pl. 5, fig. c). Rabbitfish (USA), golden-lined spinefoot (Australia), nuga (Fiji), mole, mermer (Marshall Islands), batawayi, dangie, layap (Philippines).

DISTRIBUTION: Indo-Pacific, Australia, Japan.

SOURCES: Coutière (1899), Taschenberg (1909), Engelsen (1922), Halstead *et al.* (1971).

263 *Siganus puellus* (Schlegel) (Pl. 6, fig. a). Rabbitfish (USA), nuga (Fiji), mole, mermer (Marshall Islands), batawayi, dangie, layap (Philippines).

DISTRIBUTION: Indo-Pacific.

SOURCES: Fish and Cobb (1954), Halstead *et al.* (1971).

---

[1] According to some authors, this family should be called the Teuthididae. However, because of the continual nomenclature confusion regarding *Teuthis* versus *Siganus* and the general usage and acceptance of the latter, Woodland (1972) has proposed that *Teuthis* be suppressed and that *Siganus* and *Siganidae* be retained. Woodland's recommendations are herewith followed.

Siganus punctatus (Schneider) (Pl. 6, fig. b). Rabbitfish (USA), nuga (Fiji), mole, mermer (Marshall Islands), batawayi, dangie, layap (Philippines). *263*

DISTRIBUTION: Indo-Pacific.

SOURCE: Halstead *et al.* (1971).
OTHER NAME: *Teuthis punctata.*

Siganus rostratus (Valenciennes) (Pl. 7, fig. a). Rabbitfish (USA), nuga (Fiji), mole, mermer (Marshall Islands), batawayi, dangie, layap (Philippines). *264*

DISTRIBUTION: Indo-Pacific.

SOURCE: Halstead *et al.* (1971).

Siganus vulpinus (Schlegel and Müller) (Pl. 7, fig. b). Rabbitfish (USA), nuga (Fiji), mole, mermer (Marshall Islands), batawayi, dangie, layap (Philippines). *264*

DISTRIBUTION: Indo-Pacific.

SOURCE: Halstead *et al.* (1971).
OTHER NAMES: *Teuthis vulpina, Lo vulpinus.*

MORPHOLOGY OF THE VENOM APPARATUS

Gross Anatomy

The venom apparatus of *Siganus* consists of 13 dorsal spines, 7 anal spines, 4 pelvic spines, their associated musculature, venom glands, and enveloping integumentary sheaths. The dorsal spines and the anal spines are connected with each other by deeply incised interspinous membranes. The pelvic fins differ from other spiny-rayed fishes in that each pelvic fin is equipped with an inner and outer well-developed pelvic spine. The following gross anatomical description is based on the studies of Amemiya (1921) and Tange (1955*a*) on *Siganus fuscescens* and by Halstead *et al.* (1971) on *S. vulpinus*. Although most of the details are based on *S. vulpinus*, the spines of most of the siganids closely resemble each other.

Dorsal Spines (Fig. 20, 21, 22): This description is based on dorsal spines I, VIII, and XIII of *Siganus vulpinus* which are representative of the others. The spine is described in the upright or extended position. A lateral view of the mature dorsal spine reveals that it is an elongate, rather stout, almost straight shaft which terminates in an acute tip. The base of the dorsal spine is enlarged and expanded into two asymmetrical condyles. Between the condyles is a small foramen which increases in size from the first to the 13th dorsal spine. The general shape of the shaft of the spine in cross section is trigonal, but varies greatly in morphology and diameter depending upon the particular dorsal spine and the level of the spine that is examined. There is a single anteromedian ridge at the level of the distal third, but in prog-

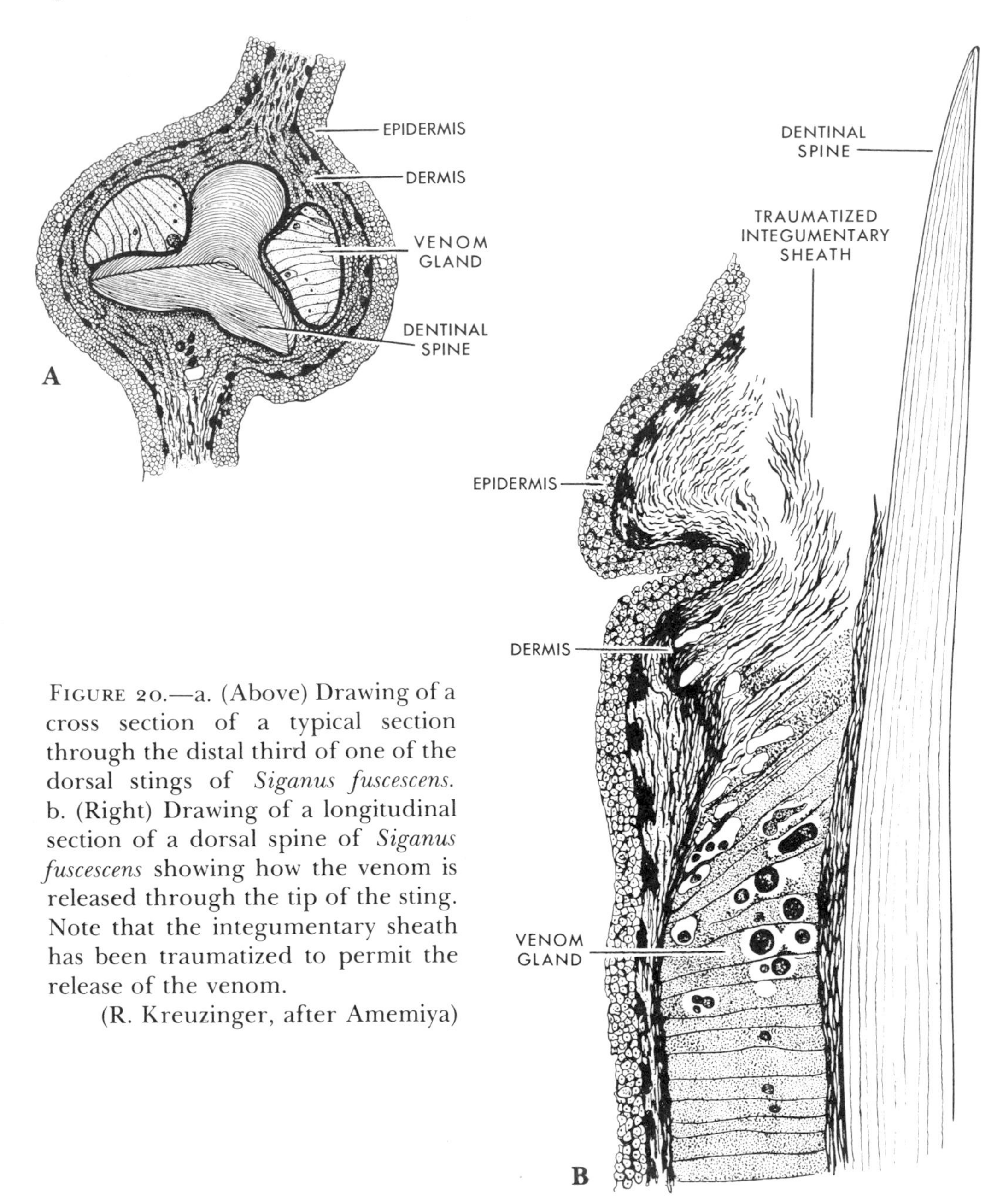

FIGURE 20.—a. (Above) Drawing of a cross section of a typical section through the distal third of one of the dorsal stings of *Siganus fuscescens*. b. (Right) Drawing of a longitudinal section of a dorsal spine of *Siganus fuscescens* showing how the venom is released through the tip of the sting. Note that the integumentary sheath has been traumatized to permit the release of the venom.
(R. Kreuzinger, after Amemiya)

ressing toward the proximal portion of the shaft the anteromedian ridge veers to the right or left and terminates in the corresponding lateral condyle at the base of the spine. Beginning with the first dorsal spine and proceeding to the last dorsal spine, the positioning of the anteromedian ridge on the dorsal spines follows an alternating pattern. This is a characteristic feature of the arrangement of the dorsal spines of all siganids. The anterolateral-glandular grooves in which the venom glands are situated are extremely variable in teuthids. In most venomous fishes there are two distinct anterolateral-glandular grooves—one on either side of the anteromedian ridge which may or may not extend throughout the entire length of the spine. In the siganids the anterolateral-glandular grooves extend throughout the length of the spines, but vary in their depth and width. There is also a variable number of secondary grooves, most of which originate in the proximal third of the spine and extend up along the middle third of the shaft for a variable distance. Despite the small size of the secondary grooves, some of these grooves contain fully developed venom glands. At the base of the spine the two articular lateral condyles are separated by a Λ-shaped articular depression which receives the articular base of the annular process of the head of the associated interneural spine. Situated immediately above the apex of the Λ-shaped articular depression is a small foramen which receives the spurlike annular process.

The dorsal spine rests on and articulates with an interneural spine. The following is a description of the interneural of the dorsal spine VIII which is more or less representative of the others. The interneural spine is ┌-shaped. The dorsal or horizontal limb of the interneural is expanded into a flattened platelike structure which will be referred to as the dorsal plate of the inter-

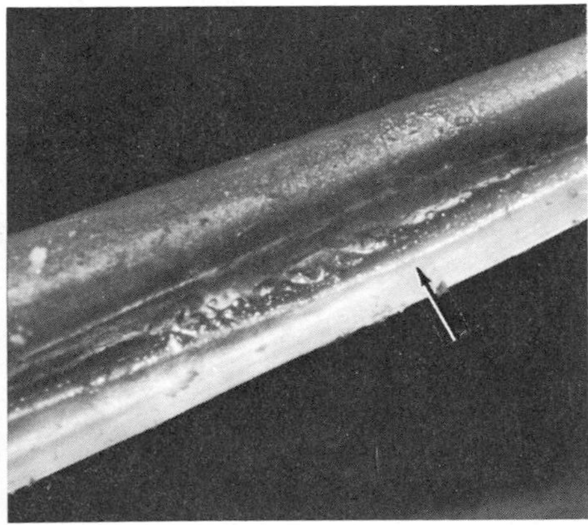

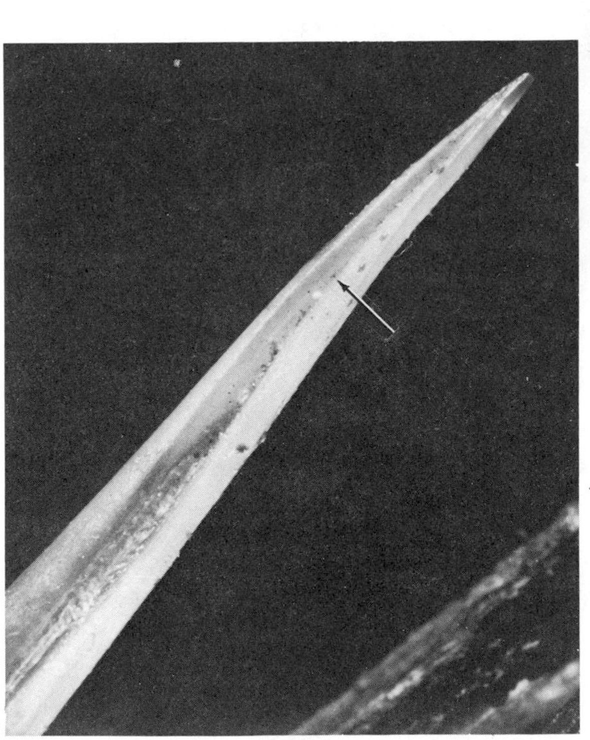

Figure 21.—a. (Left) Photograph showing a close-up of the distal third of the fifth dorsal sting of *Siganus fuscescens*. The integumentary sheath has been removed to expose the venom gland which appears as a strip of tissue lying within the anterolateral-glandular groove. Length of spine appearing in photo is 15 mm. b. (Above) Close-up of venom gland lying within the anterolateral-glandular groove. (R. Kreuzinger)

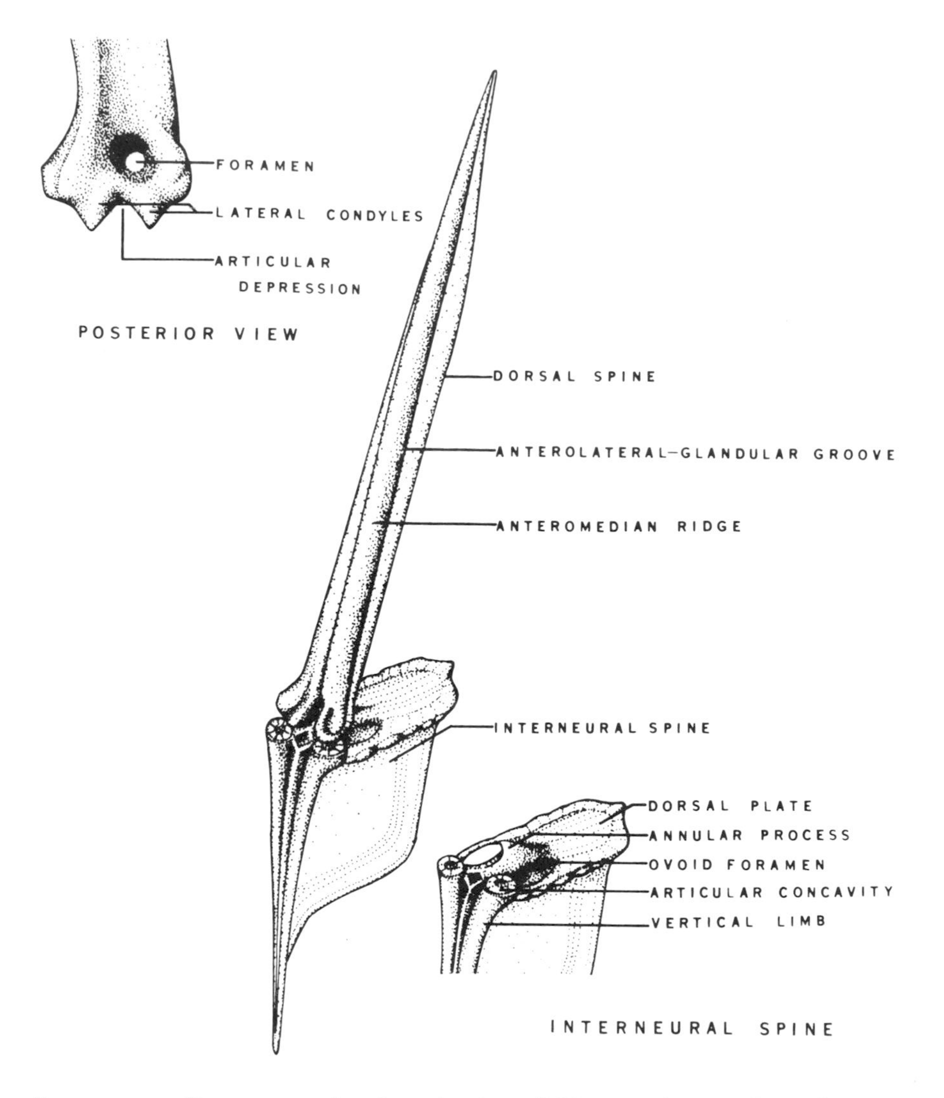

Figure 22.—Representative dorsal spine of *Siganus*. Approximately × 5. (P. Mote)

neural. The width of the dorsal plate is about 1.5 of its length. The upper surface of the dorsal plate is modified to receive the lateral condyles of the adjoining dorsal spine and provides passageways for the depressor dorsalis muscles. The posterior half of the upper surface of the dorsal plate is flattened. The anterior half of the upper surface of the dorsal plate supports three articular surfaces. There is a median articular process which appears as a spurlike bone that extends into the foramen at the base of the dorsal spine. With the exception of the first two dorsal spines, the annular process does not form a complete ring such as is found in the catfish *Galeichthys* (Halstead, Kuninobu, and Hebard, 1953). The semilunar-shaped basal portion of the annular process provides a smooth articular surface which is received by and articulates with the Λ -shaped articular notch at the base of the dorsal spine. Situated on either side of the annular process is a large ovoid foramen. The foramen in its greatest diameter is about one-third the length of the dorsal plate. These foramina provide passageways for the depressor dorsalis muscles of the dorsal spine. Immediately anterior to each foramen is a small articular concavity. These concavities receive the

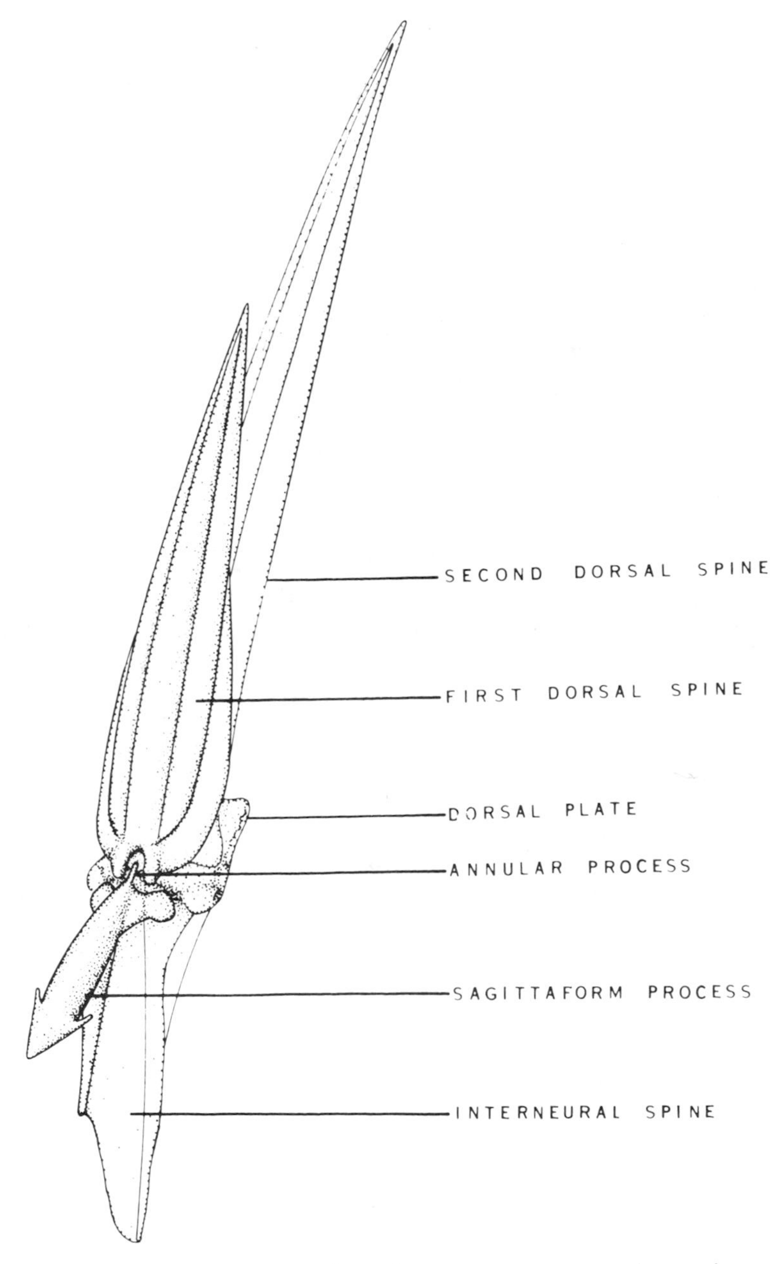

Figure 23.—First dorsal spine of *Siganus* showing the sagittaform process of the dorsal plate of the first interneural spine. × 6. (P. Mote)

lateral condyles of the dorsal spine. The vertical limb of the interneural terminates ventrally in a rather sharp point. From an anterior view it will be seen that the vertical limb is trigonal in shape. The anterior surface of this vertical limb is provided with a shallow median depression which extends for most of the length of the limb. This depression is so arranged as to receive the posterior margin of the adjoining interneural median plate. Extending beneath the dorsal limb of the interneural and posterior to the vertical limb is a flat sheet of rhomboid-shaped bone, the median plate of the interneural. The posterior aspect of the vertical limb and the anterolateral aspect of the median plate provide surfaces for the origin of the depressor dorsalis muscles of the dorsal spine.

Most of the interneural spines follow a similar morphological pattern with the exception of the first interneural spine. The dorsal plate of the first interneural spine has a sharp strongly-developed arrow-shaped horizontal prolongation which is herewith termed the sagittaform process (Fig. 23). This process appears to be unique with the family Siganidae. Although the sagittaform process is enveloped by a thick integumentary covering, there is no microscopic evidence of glandular cells. The exact function of this process is unknown, but it nevertheless presents a formidable appearance. It is noteworthy that the annular processes of the first two interneural spines are highly developed and form complete annular rings such as is found in *Galeichthys*, the catfish.

Erection and depression of the individual dorsal spine are controlled largely by three sets of muscles (Fig. 24).[2] Removal of the epaxial muscle mass and the thin layer of fascia covering the interneural spines permits examination of the erector and depressor dorsalis muscles. A set of these muscles are situated on either side of each dorsal spine. The origin and distribution of these muscles become apparent when two adjacent dorsal spines and their associated interneurals are examined, e.g., dorsal VII and VIII. The erector dorsalis muscle of dorsal VIII originates from the latero-posterior aspect of the median plate of the preceding interneural and the anterior aspect of the vertical limb of the VIII interneural spine. The erector dorsalis of dorsal VIII then passes through an opening formed by the posterior margin of the interneural dorsal plate of VII and anterior margin of the interneural dorsal plate of VIII. The erector dorsalis muscle then inserts on the anterior aspect of the base of the dorsal VIII spine. The depressor dorsalis originates from the lateroanterior aspect of the median plate and the posterior aspect of the vertical limb of the interneural. The depressor dorsalis passes through the foramen of the dorsal plate of the interneural and inserts on the posterior aspect of the base of the dorsal spine. A similar set of muscles is found on both sides of each dorsal spine. A third set of muscles which control the erection and depression of the dorsal fin consists

---

[2] No attempt has been made to describe the superficial inclinator dorsalis muscles which bend or incline the fin spines to the side.

of the protractor dorsalis and the retractor dorsalis muscles. The protractor dorsalis originates on the dorsoposterior aspect of the skull where it is divided into two muscle bundles by the supraoccipital bone, passes backward and upward, and inserts by a series of tendinous attachments to the anterior aspect of the base of the first dorsal spine. Contact with the other dorsal spines is accomplished by an interspinous ligament which is attached to the lateral aspect of the base of each of the dorsal spines and rays and finally terminates in the retractor dorsalis. The retractor dorsalis originates in the hypural bone of the caudal and inserts on the posterior aspect of the base of the last dorsal ray and the same interspinous ligament previously discussed. The protractor dorsalis appears to be a major force in the erection of the dorsal fin and is much larger in size than the retractor dorsalis.

The venom glands of the dorsal spines appear as a slender, elongate, fusiform strip of pinkish tissue lying within the distal third of the antero-lateral-glandular groove. These glands can be seen through the translucent integumentary sheath which surrounds the spine, but become more readily apparent when the integumentary sheath is removed.

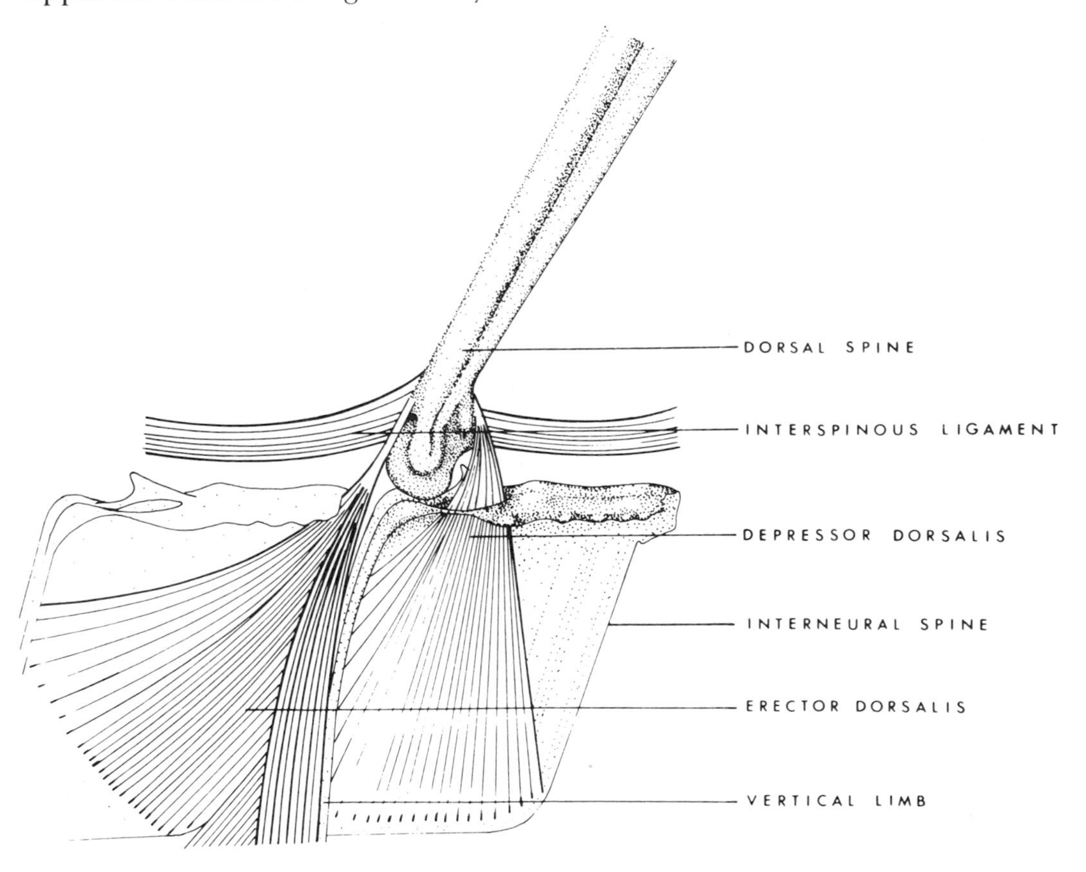

Figure 24.—Drawing showing the muscles controlling the vertical movements of the dorsal spine of *Siganus*. Approx. × 8. (P. Mote)

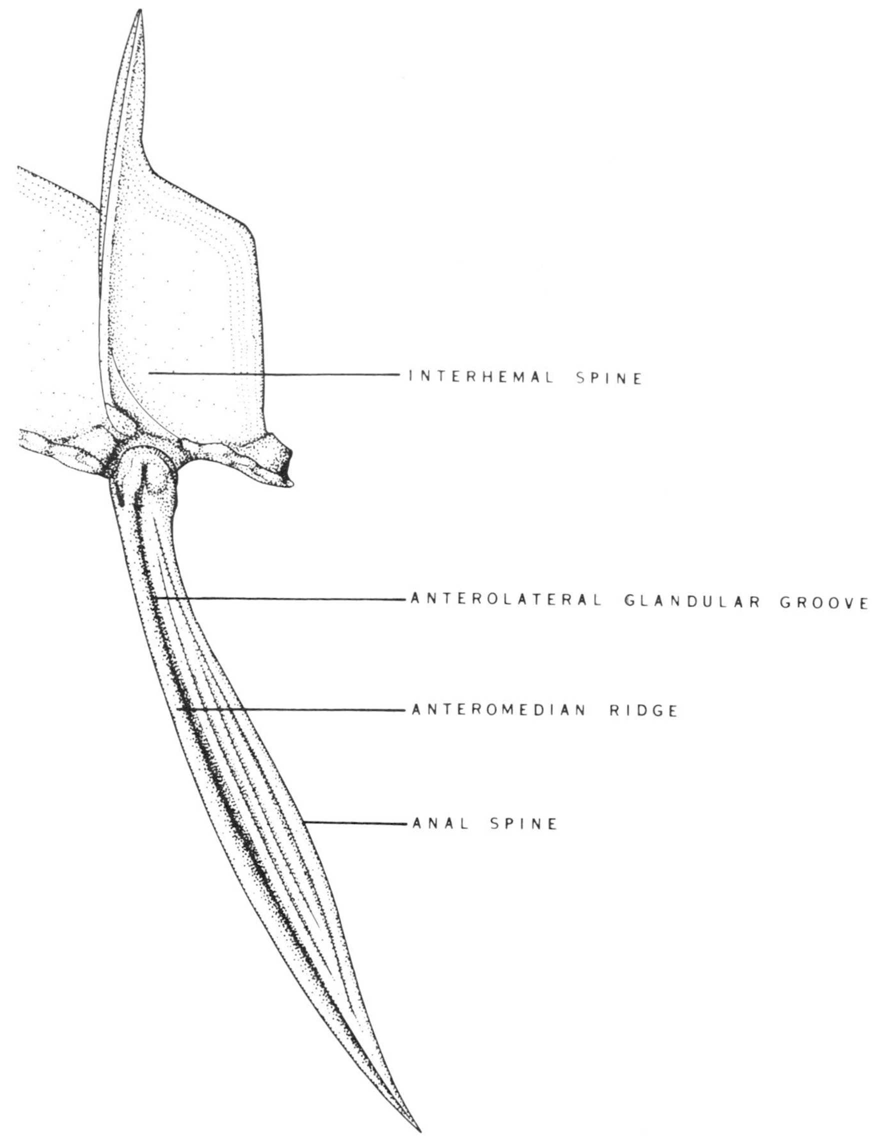

FIGURE 25.—Representative anal spine of *Siganus*. × 5. (P. Mote)

Anal Spines (Fig. 25): This description is based on anal spines III, IV, and V. The spine is described in the extended position. A lateral view of a mature anal spine reveals that it is an elongate, stout, slightly arched shaft which terminates in a very sharp tip. The base of the anal spine is enlarged and expanded into two asymmetrical condyles. Between the condyles is a minute foramen. The general shape of the shaft in cross section is trigonal, but because of the acute angle of the leading edge of the spine it has an almost sabre-like appearance from a side view. The anteromedian ridge of the anal spines is more pronounced than any found in the dorsal spines. The anteromedian ridge of the anal spine, as in the dorsal spine, veers off to the left or right and terminates in the corresponding lateral condyle at the base of the spine. The same alternating pattern of the positioning of the anteromedian ridge as seen in the dorsal spines is also present in the anal spines. The anterolateral-glandular grooves of the anal spines are well developed and somewhat less irregular than those of the dorsal spines. Secondary grooves, varying in number, are most pronounced along the middle third of the shaft of the spine. At the base of the spine the two articular lateral condyles are separated by a Λ -shaped articular depression which receives the articulating annular process of the head of the associated interhemal spine. The annular process and the articular arrangement of the anal spine closely resemble that of the dorsal spines.

Pelvic Spine (Fig. 26): The lateral or outer pelvic spine is an elongate relatively slender shaft, terminating in an acute tip. Although the spine is somewhat trigonal in cross section, it is dorso-ventrally flattened and has a broad base. Situated on the inferior aspect of the shaft are two grooves,

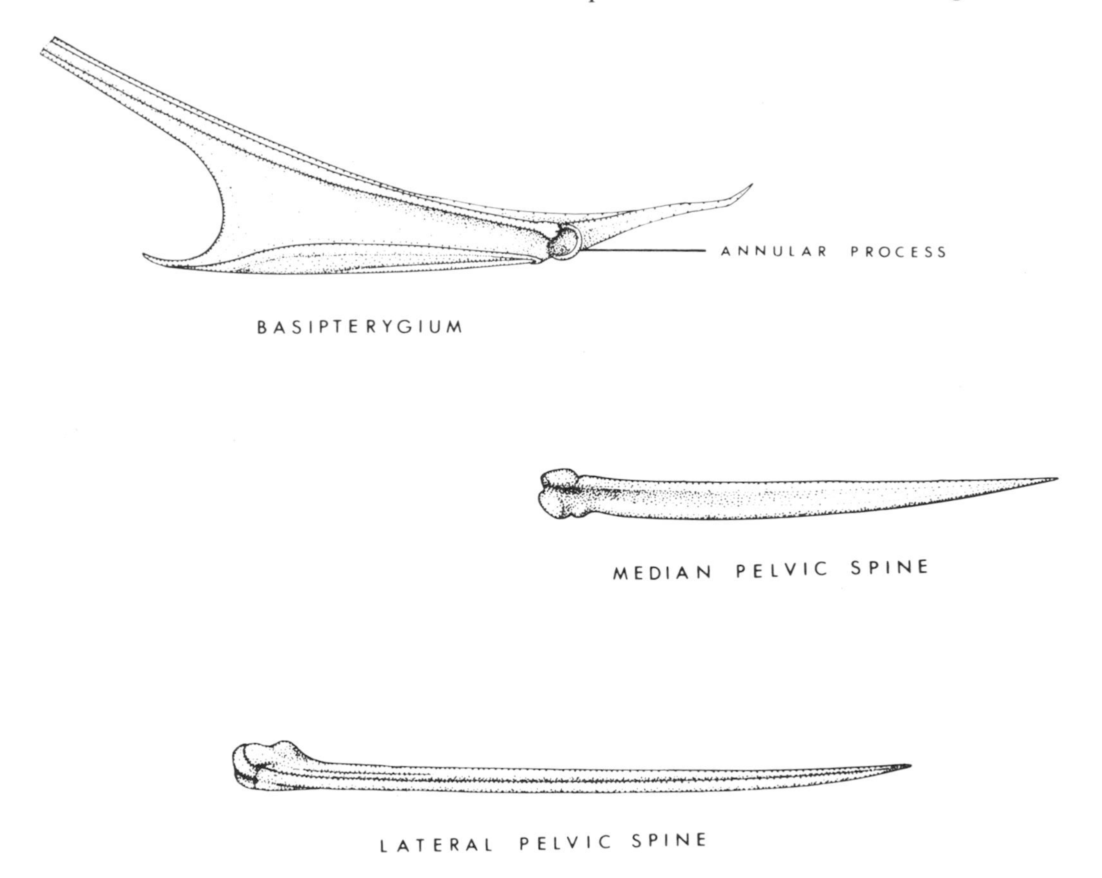

Figure 26.—Pelvic spines and basipterygium of the pelvic girdle of *Siganus*. × 4. (P. Mote)

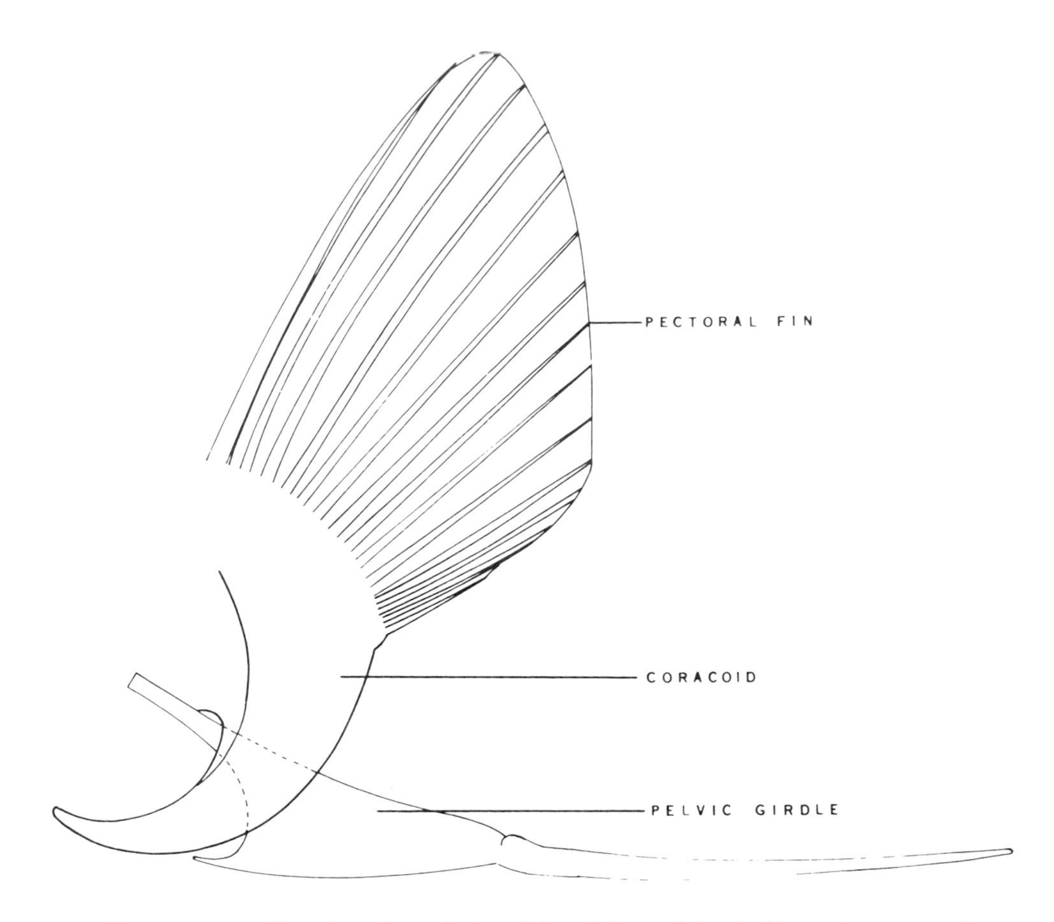

FIGURE 27.—Showing the relationship of the pelvic girdle to the pectoral girdle of *Siganus*. × 3. (P. Mote)

one on either side of the spine. These grooves are the equivalent of the anterolateral glandular grooves of the dorsal and anal spines. The grooves are separated by a pronounced ridge, similar to that of the anteromedian ridge of the dorsal and anal spines. Removal of the integumentary sheath reveals a slender, elongate fusiform strand of venom glands. The base of the lateral pelvic spine is broad and depressed. At the base of the spine the two articular condyles are separated by a notch which receives and articulates with the annular process at the basipterygium. The median or inner pelvic spine is similar to the outer pelvic spine but is more slender, and its base possesses a simple articular surface. The pelvic fin is attached to the basipterygium which makes up the pelvic girdle. From a side view the basipterygium is shaped somewhat like the prow of a boat. The basipterygium is comprised of two bones which form either half of the pelvic girdle. The ventral base of the basipterygium is horizontally expanded to form a flat keel-like projection. Extending forward and upward from the base is the vertical element of the basipterygium. The upper margin of the vertical element gradually flares outward proceeding from the anterior tip along the median line, then downward and backward, and finally fuses with the posterior margin of the horizontal keel. The two halves of the vertical element of the pelvic girdle fuse along the median plane. However, because of the flaring arrangement of the posterior portion of the horizontal base of the basipterigium, only the anterior half is fused along the median plane. The posterior portion of the base provides a hollow receptacle which houses the origin of one of the abductor muscles of the pelvic fin (Figs. 27; 28a, b). The abductor muscles housed within the basipterygial receptacle insert on the ventral aspect of the base of the median condyle of the lateral or outer pelvic spine and on the ventral aspect of the base of the median or inner pelvic spine. Another bundle of abductor muscles originates within the V-shaped depression that is formed by the horizontal keel and the upper margin of the vertical element of the basipterygium. This latter bundle of the abductor muscle inserts on the ventral aspect of the lateral condyle of the pelvic spine. The function of abduction appears to be handled primarily through the outer pelvic spine. The adductor muscles of the pelvic fin originate on the

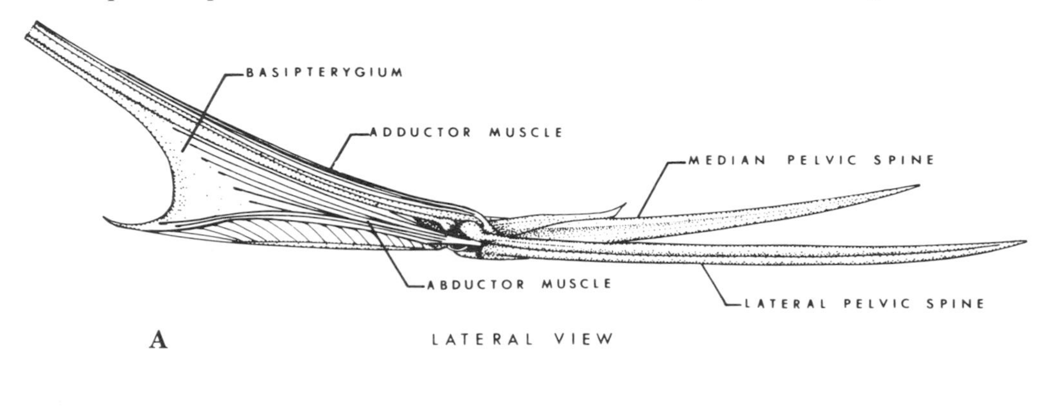

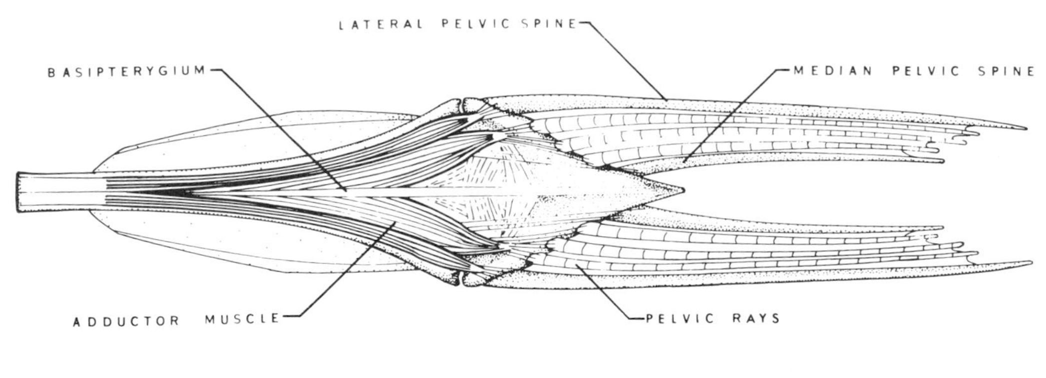

FIGURE 28.—Pelvic fin muscles of *Siganus*. × 4. (P. Mote)

a. Lateral view.

b. Dorsal view.

dorsal surface of the vertical element of the basipterygium. The adductor muscle inserts on the dorsal aspect of the base of the pelvic spines and rays.

Spines of Miscellaneous Siganids: It was found that the gross morphology of the dorsal, anal, and pelvic spines of *Siganus argenteus, S. corallinus, S. doliatus, S. fuscescens, S. lineatus, S. puellus*, and *S. punctatus* is similar to that found in *S. vulpinus*.

Microscopic Anatomy

Dorsal Sting: This histological description is based primarily on the spines of *Siganus vulpinus* and *S. punctatus* (Fig. 29a, b, c). The histological structure of each of the dorsal spines is basically the same, making a single description adequate. The only differences are in the gross shape of the spines, and this does not alter the basic histology.

Grooves indenting the bone are quite prominent and serve as channels through which blood vessels and nerves reach the distal portion of the spine and house the venom glands themselves. The bony spine itself consists of dense acellular bone with centrally placed marrow spaces except in the apical region where the bone is solid. Small blood vessels are prominent in at least

a

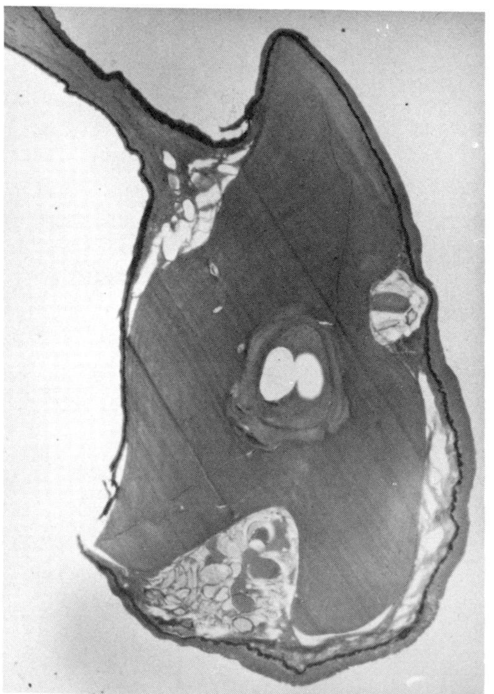

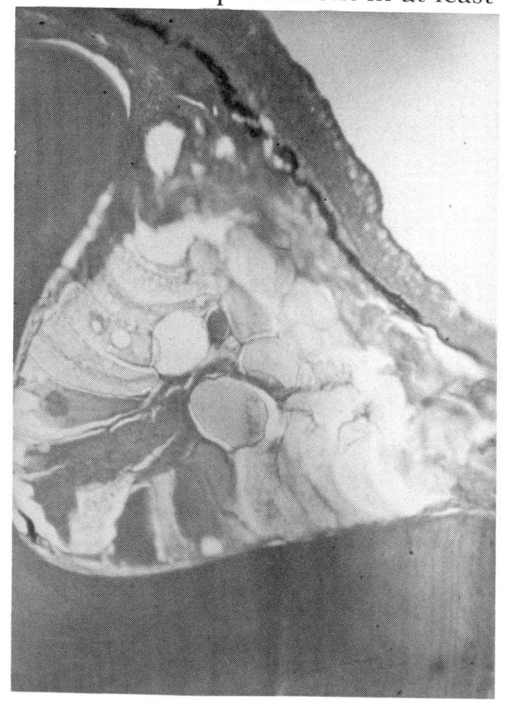

b

FIGURE 29.—a. (Left) Photomicrograph of the cross section of the distal third of the 4th dorsal sting of *Siganus punctatus*. Note that the spine is canaliculated, but there is no evidence of venom within any of these canals. Venom glands are situated within the anterolateral-glandular grooves. × 45. b. (Right) Enlargement of glandular triangle of a cross section of the distal third of the 4th dorsal sting. Yellowish deposits of venom are clearly evident in this section. × 175. (E. Rich)

FIGURE 29c.—Longitudinal section of the 5th dorsal sting of *Siganus punctatus*. Note the similarity of the morphology of the glandular cells to those found in *Pterois volitans*. × 250. Hematoxylin and triosin stains. Preparations by B. W. Halstead and M. Chitwood. (E. Rich)

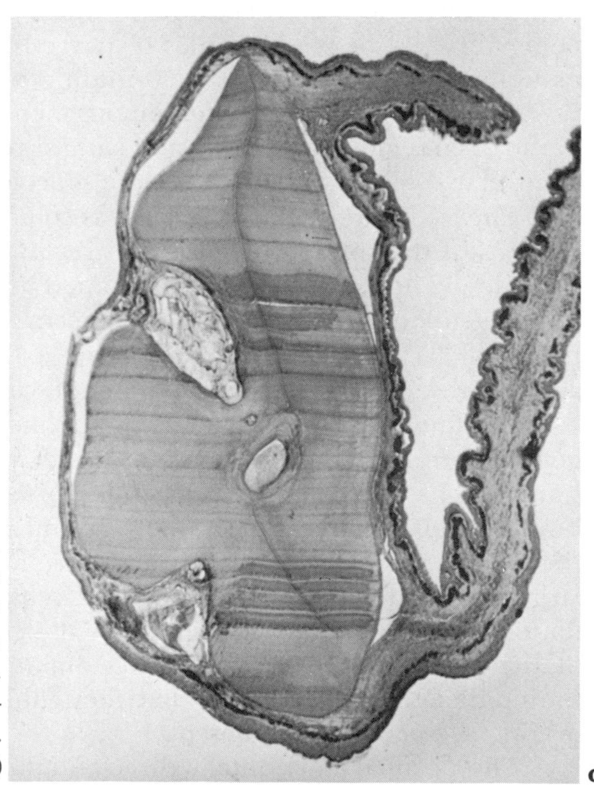

FIGURE 29d.—Cross section taken through the distal third of the pelvic sting of *Siganus punctatus*. Note the presence of small venom glands in anterolateral-glandular grooves. ×60. Hematoxylin and triosin stains. Preparations by B. W. Halstead and M. Chitwood. (E. Rich)

one of the marrow spaces. Branches from these vessels reach the surface through canals which open into the deep portion of the bony grooves as well as the base of the interspinous membrane. The bone surrounding the marrow spaces has undergone considerable resorption followed by redeposition. Sharpey's fibers are conspicuous in the regions of membrane attachment and grooves. Osteoblasts appear as very flattened cells on the bony surfaces.

The integumentary sheath investing the spine is composed of an outer layer of stratified squamous epithelium and a rather dense underlying connective tissue dermis. The thickness of the epithelium varies depending on the amount of exposure it receives. Since the spines are arranged alternately in two rows, some surfaces are more protected than others. The medial surface tends to be shielded by the spines of the other row; the cells here are about three to four cells deep. The epithelium of the unprotected surfaces is about twice as thick with six to eight layers of cells. The epithelium of the interspinous membranes is uniformly thin on both sides consisting of two to three layers of cells. Mucous secreting cells are rather scarce. The surface appears finely sculptured.

Supporting the epithelium is a very dense layer consisting of two components. The basement membrane forms the outer layer. The inner layer is a relatively thick collagenous compact stratum containing scattered nuclei.

The remainder of the supporting connective tissue is less dense. The pigment layer is located near the periphery of this region, the melanophores being very distinctive. Bundles of nerve fibers are quite prominent. Elastic tissue appears to be absent surrounding the spine proper. Orcein staining reveals elastic fibers in the interspinous membrane only. Especially prominent in the basal region of the sting and lying superficial to even the more proximal glandular cells are islands of a loose mucoid type of connective tissue.

The toxin-producing cells exist as columns in the anterolateral-glandular grooves of the spine, which in cross section are referred to as the glandular triangles. These cell columns are limited to the distal half of the spine. They may be subdivided by bands of connective tissue, but a single cell group is the usual arrangement. These cells are extremely large polygonal structures and represent bound accumulation of cell manufacture. Apparently the cells become greatly enlarged as the toxin is produced; this material undergoes maturation and/or degenerative changes as evidenced by changes in staining characteristics and granule formation. Cells are noted that are homogenous in appearance, finely granular, and coarsely granular. Finally, there is what appears to be a coalescence of these coarse granules into a few large irregular globules. These large cells are surrounded by a capsule of flattened cells which are thought to be the germinal cells. The ultimate origin of these cells was not determined. Staining characteristics were similar to the epithelium, and it is felt that they must be of epithelial or neural crest origin. A duct system is absent.

The peculiar horizontally directed anterior spine is covered throughout its length by body tissue except for the apical tip. Marrow spaces appear to be limited to the proximal third of the spine. Well-defined grooves are absent, and no venom glands are associated with this structure. The spine is surrounded by dense connective tissue. The surface epithelium and superficial dermis are the same as that covering the spines. The same type of loose mucoid tissue noted in the spines is present dorsally and dorsolaterally. Scales are absent dorsally to the spinc, but begin just laterally to it beneath the pigment layer. Deep to the scales and lateral to the spine are upward of two dozen very dense connective tissue layers forming a practically solid mat. The fiber bundles forming these layers are arranged at right angles to each other. One layer has its fibers oriented along the long axis of the body. The fibers of the adjacent layers run in a dorsoventral direction. A third fiber grouping, limited largely to the outer half, passes inward at intervals at right angles to these two layers. These fiber bundles may pass through a number of layers or be limited to connecting two adjacent layers. Underlying this dense and well-organized structure is striated muscle.

Anal Stings: The histological features of the anal stings are almost identical to the dorsal spines.

Pelvic Stings (Fig. 29d): Except for gross differences in the shape of these spines as noted above, the pelvic stings are similar to the other spines, and the histology remains the same. The grooves equivalent to the anterolateral grooves of the dorsal and anal spines contain the venom glands.

Stings of Miscellaneous Siganids: The microscopic anatomy of the dorsal, anal, and pelvic stings of *Siganus argenteus, S. doliatus, S. fuscescens, S. lineatus, S. puellus*, and *S. punctatus* is similar to that of *S. vulpinus*.

MEDICAL ASPECTS

There is very little specific information available regarding the clinical aspects of rabbitfish stings. Judging by the few reports that have been obtained, rabbitfish stings are comparable to those produced by *Scorpaena*. They are said to be quite painful. The treatment is similar to that of scorpionfish stings (*see* p. 782).

Order PERCIFORMES (PERCOMORPHI)

Suborder ACANTHUROIDEI (TEUTHIDOIDEA)

Family ACANTHURIDAE: Surgeonfishes

Surgeonfishes, or tangs, are a group of shore fishes inhabiting warm seas. The principal genus *Acanthurus* is characterized by the presence of a sharp, lancelike, movable spine which is located on each side of the caudal peduncle. Surgeonfishes are particularly common in surge channels and shoal areas in the vicinity of coral patch reefs. Most surgeonfishes are small to moderate in size. Surgeonfishes have been repeatedly incriminated as venomous fishes since the time of Corre (1865). However, very little is known regarding the nature of their venom apparatus. Recent studies indicate that a definite but feeble venom apparatus is present in some species of acanthurids.

The biology and systematics of this group have been studied by Jordan *et al.* (1930), Fowler (1936), Smith (1950), Schultz *et al.* (1953), Herre (1953), and Marshall (1964).

REPRESENTATIVE LIST OF SURGEONFISHES REPORTED AS VENOMOUS

Family ACANTHURIDAE

136 *Acanthurus chirurgus* (Bloch) (Pl. 10, fig. a, Chapter X). Doctorfish, surgeonfish (USA).

DISTRIBUTION: Tropical Atlantic Ocean.

SOURCES: Corre (1865) Maass (1937), Phillips and Brady (1953).
OTHER NAMES: *Acanthurus hepatus, Acanthurus phlebotomus.*

137 *Acanthurus triostegus* (Linnaeus) (Pl. 11, fig. c, Chapter X). Convict, surgeonfish, tang (USA), manini (Hawaii), kuban (Marshall Islands), koinawa (Gilbert Islands), shimahagi (Japan), five-banded surgeonfish (Australia),

DISTRIBUTION: Indo-Pacific.

SOURCES: Pawlowsky (1919), Maass (1937), Evans (1943).
OTHER NAMES: *Acanthurus triostegus, Teuthis triostegus.*

138 *Ctenochaetus striatus* (Quoy and Gaimard) (Pl. 12, fig. a, Chapter X). Surgeonfish (USA), kala (Hawaii), maito (Tahiti), te ribabui (Gilbert Islands), sazanamihagi (Japan), labahita, indangan, mangadlit, yaput (Philippines), vanaki, pidja, abila (Indonesia), imim, ael, bir, diepro, tiebro (Marshall Islands), pone, palagi, nanife, ili ilia (Samoa), balagi (Fiji).

DISTRIBUTION: Indo-Pacific, to the Red Sea, excluding Hawaii.

SOURCES: Harry (1953), Randall (1955).
OTHER NAME: *Ctenochaetus strigosus.*

136 *Ctenochaetus strigosus* (Bennett) (Pl. 10, fig. c, Chapter X). Surgeonfish, tang (USA), kala (Hawaii).

DISTRIBUTION: Hawaiian Islands, Johnston Island.

SOURCES: Harry (1953), Randall (1955).

265 *Prionurus microlepidotus* Lacépède (Pl. 8). Surgeonfish (USA).

DISTRIBUTION: Japan.

SOURCE: Tange (1955*b*).
OTHER NAME: *Xesurus scalprum.*

MORPHOLOGY OF THE VENOM APPARATUS

Gross Anatomy

According to Tange (1955*b*), the venom apparatus of the Japanese surgeonfish *Prionurus microlepidotus* (formerly *Xesurus scalprum*) consists of nine dorsal spines, three anal spines, two pelvic spines, their associated venom glands, and enveloping integumentary sheaths. No reference is made to the caudal peduncular spines.

The dorsal spines are of moderate length, slightly arched craniad, thickened proximally, but taper off rapidly to an acute distal tip. The spine in cross section is trigonal or T-shaped in outline. The anterolateral-glandular grooves, separated by an anteromedian ridge, of *Prionurus* are less developed than in some of the scorpaenids. The grooves more closely resemble those of *Siganus* but are more symmetrical in their development. The anterolateral-glandular grooves extend for almost the entire length of the spine, being deepest on the proximal end but becoming progressively more shallow as they approach the distal tip. The length of the fin spine varies somewhat with each fin and according to its position in the fin.

The venom glands lie within the bony protection of the anterolateral-glandular grooves. The glands are elongate, fusiform, and extend from approximately the junction of the middle and distal third to almost the acute tip of the spine. The glands are completely enveloped by the integumentary sheath of the interspinous membranes.

Microscopic Anatomy

The dorsal sting of *Prionurus* in cross section is essentially an inverted T-shaped dentinal structure completely invested by a layer of integument (Fig. 30a, b). The spine is composed of a solid, dentinelike substance which closely resembles that described for *Pterois* and some of the other scorpaenids. The integumentary sheath investing the spine is a thin layer of moderately dense, fibrous, connective tissue, the dermis, and a layer of stratified squamous epithelium, the epidermis. The epithelium resembles that described for *Pterois*. In most fish species studied to date the venom glands are completely surrounded by a connective tissue capsule, but according to Tange (1955*b*) the venom glands in younger specimens may be only partially surrounded by connective tissue. In younger specimens the connective tissue may be partially or almost completely absent along the outer or anterolateral margin of the glandular triangle but generally present along the medial aspect. The glandular cells situated in the dermal layer vary in size, arrangement, and appearance of their cytoplasm. There are usually observed about 10 to 40 large glandular cells variously clumped together. These cells may be round, ovoid, or polygonal in shape and range in size from 23 to 33μ by 12 to

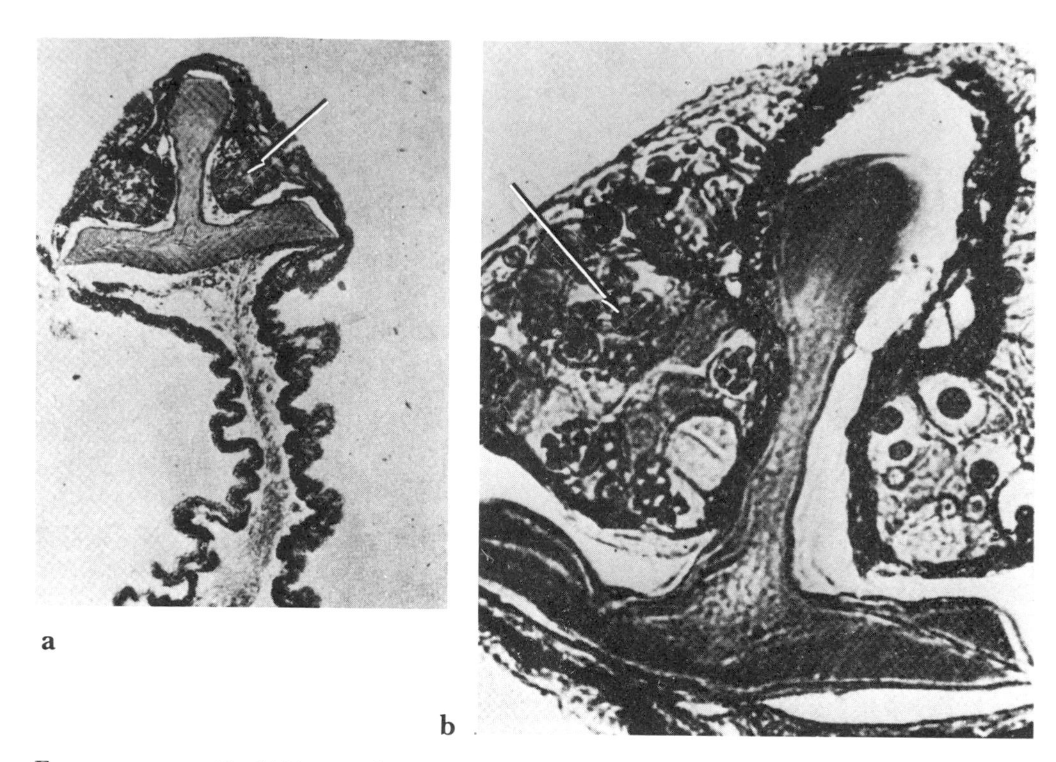

FIGURE 30.—a. (Left)Photomicrograph of a cross section through the 3rd dorsal sting of *Prionurus microlepidotus* (formerly *Xesurus scalprum*). The venom glands are seen as two dark masses, one on either side of the anteromedian ridge of the spine. The dentinal spine is the T-shaped structure. × 100. b. (Right) Enlargement of the above cross section showing greater detail of the venom glands. × 400. Hematoxylin and eosin stains. (From Tange)

18μ. The cytoplasm of the glandular cells stain acidophilic with hematoxylin and eosin. The cells are vacuolated, and the vacuoles are filled with round colloidal bodies of various sizes and shapes. The nuclei of the glandular cells are eccentrically situated and poor in chromatin. Situated between the glandular cells are occasionally seen flat or ovoid supporting cells. No glandular ducts were observed. Venom production appears to be by a holocrine type of secretion. In general, the venom glands of *Prionurus microlepidotus* are much smaller than those observed in other kinds of venomous fishes.

Comments on the Caudal Spine Venom Organs of Surgeonfishes

There are no anatomical data available on the caudal spine venom organs of acanthurids. Clinical reports by Harry (1953) and Randall (1959) strongly suggest that the caudal spines of at least some species of acanthurids constitute a venom apparatus. Superficial examination of the caudal spines of *Acanthurus achilles, A. aliala, A. lineatus, A. mata, A. nigricans, A. olivaceus,* and *A. xanthopterus* reveals that they are well developed and are enveloped in an integumentary sheath (Figs. 31; 32). In some species of acanthurids the

FIGURE 31.—Cross section through one of the caudal stings of *Acanthurus xanthopterus*. Note that the spine is completely surrounded by a very extensive integumentary sheath, but there is very little evidence of secretory activity in this particular section. The sheath is lined by a thin layer of stratified squamous epithelium similar to that found elsewhere on the body of the fish. Clinical evidence strongly suggests that some species of surgeonfishes have venomous caudal spines, but present morphological data are too inadequate to permit any conclusions at this time. × 30. Hematoxylin and triosin stains. Preparations by B. Halstead and M. Chitwood. (R. Kreuzinger)

integumentary sheath is very thin and poorly developed. However, in *A. mata* and in several of the other larger species of surgeonfishes the integumentary sheaths are relatively thick and are probably capable of secretory activity. Sections prepared from the caudal spines of *Acanthurus xanthopterus* and *A. triostegus* fail to reveal much evidence of secretory activity in the integumentary lining of the caudal spine (Fig. 31). A much larger series of specimens would have to be examined before any conclusions are made. The caudal spines of the members of the genera *Acanthurus, Ctenochaetus, Paracanthurus,* and *Zebrasoma* are movable and can be extended, or abducted at right angles to the body or adducted at the will of the fish. When adducted the caudal spine resides in a deep, elongate, fusiform depression which contains the bulk of the spine. The caudal spine depression is lined with epithelium which is believed to secrete both mucus and venom. Thus the caudal spine would be constantly bathed in venom when at rest.

When the caudal spine is viewed in its normal anatomical position on the side of the caudal peduncle of the fish, it appears as an acutely fusiform

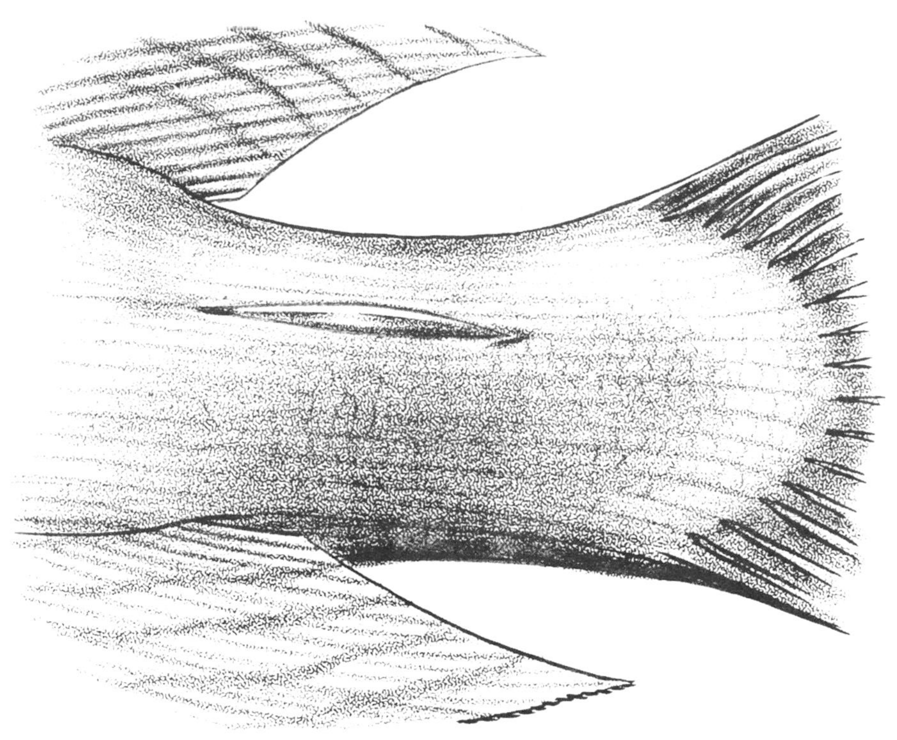

Figure 32.—a. (Above) Tail of a surgeonfish *Acanthurus* showing the caudal spine in the contracted position. b. (Below) Tail of *Acanthurus* showing the caudal spine in the extended position. Surgeonfishes can inflict painful wounds with their caudal spines. (From Halstead)

FIGURE 33.—a. (Above) Tail of *Acanthurus glaucopareius* showing the caudal spine in a partially extended position. b. (Below) Tail of an *Acanthurus* showing the caudal spine in an extended position. (Courtesy D. Ollis)

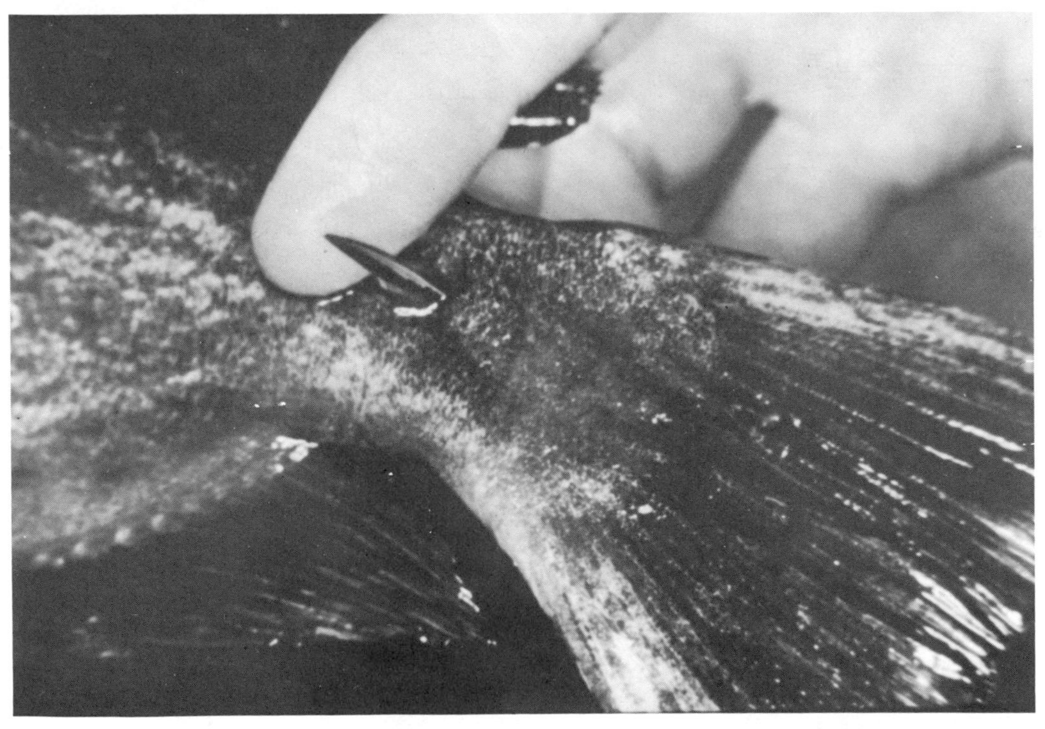

dentinal structure. The anterior and posterior tips of the spine each terminate in a sharp point. If the spine is cut in cross section at the level of the middle third, it is found to be compressed, having knifelike margins. When the spine is in the extended position, the anterior or leading edge provides a sharp cutting device, and the posterior margin is almost as sharp. If the spine is dissected out from the body of the fish, it will be observed that the inner cutting edge of the spine flares out at the level of the middle third of the spine and continues to almost the posterior tip (Fig. 34a, b). This flaring forms the base of the spine to which the muscular attachments that regulate the movement of the spine are attached. There are two groups of muscles which govern the movement of the caudal spine: an anterior and a posterior pair (Fig. 34c). These muscles have been described by Souché (1935) for

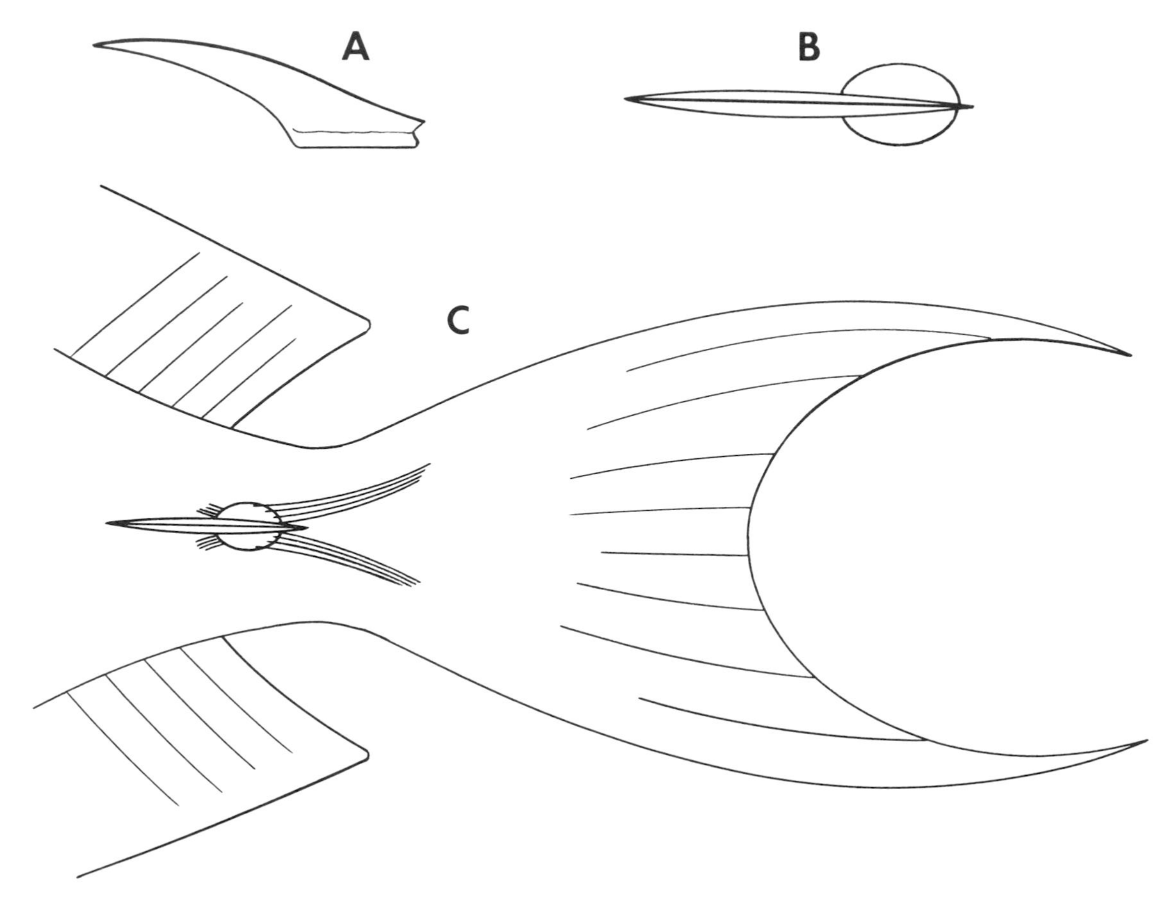

FIGURE 34.—a. Side view of *Acanthurus* caudal spine which has been removed from the fish. Anterior tip of spine is pointing to the reader's left. b. Caudal spine of *Acanthurus* showing the expanded base of spine at the posterior third of the spine, viewed from above. c. Caudal spine of *Acanthurus chirurgus* showing muscular attachments which govern the abduction and adduction of the caudal spine. Note attachment of the two pairs of muscles to the base of the spine. The short anterior pair of muscles act as adductors and the long posterior pair serve as abductors. (R. Kreuzinger, after Souché)

Acanthurus chirurgus (see also Monod, 1959). The two anterior muscles have their origin on the underside of the anterior margin of the base of the caudal spine and insert on the lateral aspect of the last caudal vertebra. This anterior pair of muscles acts as the adductor muscles for the caudal spine and are said to be the weaker and the shorter of the two muscle pairs. The posterior pair of muscles originate on the dorsal aspect along the posterior margin of the base of the caudal spine and insert by means of two long thin tendinous attachments which insert on the base of the upper and lower fin rays. The posterior muscles act as the abductor muscles of the caudal spine. Apparently these muscles are quite evident in *A. chirurgus*, but they are much less evident in some of the other species of acanthurids. This is a subject which is in need of further study since the exact mechanism of action of these caudal-spine muscles is not fully understood.

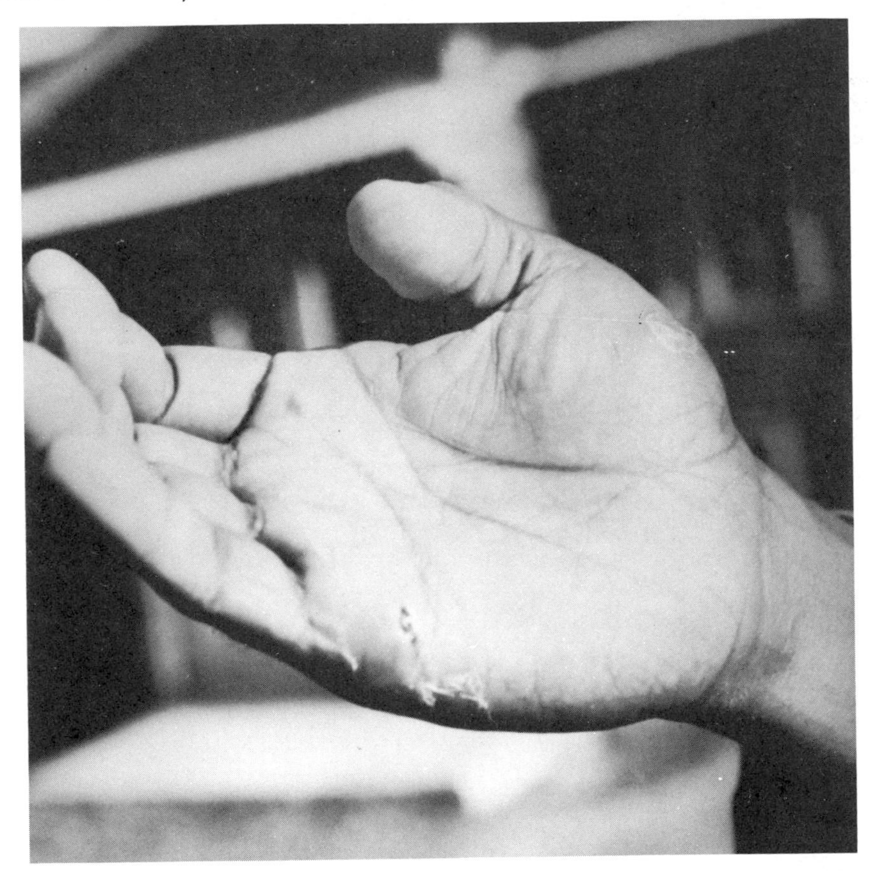

FIGURE 35.—Surgeonfish wound inflicted by *Acanthurus olivaceus*. The victim was attempting to remove the surgeonfish from the shaft of a spear when he was struck by one of the caudal spines. Bleeding was immediate and profuse. Severe swelling began about 1 hour later and involved most of his hand. Drainage of the wound and swelling continued for about 10 days. The wound healed in about 3 weeks. Treatment consisted of soaking the wound daily in warm water containing epsom salts and applying a bacteriacidal cream to the wound. (Courtesy G. Mote)

MECHANISM OF INTOXICATION

Envenomations from surgeonfishes may be produced by either their fin spines or by the spines on the caudal peduncle. Wounds are most frequently encountered when one attempts to remove the fish from nets or hooks. Attacks have been known to occur to a person when wading in a confined area containing a large number of surgeonfishes. Upon becoming excited surgeonfishes extend their caudal spines and slash the legs of the victim with their tails, producing deep and painful wounds.

MEDICAL ASPECTS

Clinical Characteristics

Surgeonfishes produce puncture wounds or cuts (Fig. 35). The associated pain is described as an immediate, intense, stinging or throbbing sensation which soon radiates out to involve the entire appendage. This is followed shortly by severe swelling. The acute pain generally subsides after about 12 hours, but the pain may continue to some extent for at least a week. A feeling of nausea has been reported (Harry, 1953; Randall, 1959).

Pathology

(Unknown)

Treatment

The treatment of surgeonfish stings is similar to that of scorpionfish envenomations. (*See* p. 782).

PUBLIC HEALTH ASPECTS

Surgeonfishes are not a significant public health problem.

PHARMACOLOGY

(Unknown)

CHEMISTRY

(Unknown)

LITERATURE CITED

AMEMIYA, I.
1921 On the structure of the poison sting of *Aigo (Siganus fuscescens)*. [In Japanese] Suisan Gakkai Ho 3(3): 196-204, 1 pl.

BOTTARD, A.
1889 Les poissons venimeux. Octave Doin, Paris. 198 p.

CAMERON, A. M.
1969 The venom apparatus and the epidermal glands of some Queensland teleosts. Ph.D. Thesis, University of Queensland, St. Lucia, Queensland, Australia.

CAMERON, A. M., and R. ENDEAN
1970 Venom glands in scatophagid fish. Toxicon 8: 171-178, 8 figs.

CASTELLANI, A., and A. J. CHALMERS
1919 Venomous animals: Protozoa to Arthropoda, p. 203-241; Figs. 17-26. *In* A. Castellani and A. J. Chalmers, Manual of tropical medicine. 3d ed. W. Wood Co., New York.

CORRE, A.
1865 Note pour servir à l'histoire des poissons vénéneux. Arch. Méd. Nav. 3: 136-147.

COUTIÈRE, H.
1899 Poissons venimeux et poissons vénéneux. Thèse Agrég. Carré et Naud, Paris. 221 p.
1907 Sur le prétendu appareil venimeux de la Murène Hélène. Bull. Soc. Philomath. Paris 9(9): 229-234, 3 figs.

DU TERTRE, R. P.
1667 Histoire naturelle des Antilles habitées par les Français. 4 vols. Paris.

ENGELSEN, H.
1922 Om giftfisk og giftige fisk. Nord. Hyg. Tidskrift 3: 316-325.

EVANS, H. M.
1943 Sting-fish and seafarer. Faber and Faber, Ltd., London. 180 p., 31 figs.

FISH, C. J., and M. C. COBB
1954 Noxious marine animals of the central and western Pacific Ocean. U.S. Fish Wildlife Serv., Res. Rept. No. 36, p. 14-23.

FOWLER, H. W.
1928 The fishes of Oceania. Mem. Bernice P. Bishop Mus. Vol. 10. 466 p., 49 pls.
1936 The marine fishes of West Africa. Bull. Am. Mus. Nat. Hist. Vol. 70, Parts I and II. 1493 p., 567 figs.

FREDERICQ, L.
1924 Die sekretion von schutz- und nutzstoffen, p. 1-87. *In* H. Winterstein [ed.], Handbuch der vergleichenden physiologie. Vol. II. G. Fischer, Jena.

GHIRETTI, F., and E. ROCCA
1963 Some experiments on ichthyotoxin, p. 211-216. *In* H. L. Keegan and W. V. MacFarlane [eds.], Venomous and poisonous animals and noxious plants of the Pacific region. Pergamon Press, New York.

GIMLETTE, J. D.
1923 Malay poisons and charm cures. 2d ed. J. and A. Churchill, London. 260 p.

GIUNIO, P.
1948 Otrovne ribe. Higijena (Belgrade) 1(7-12): 282-318, 20 figs.

HALSTEAD, B. W.
1959 Dangerous marine animals. Cornell Maritime Press, Cambridge. 146 p., 88 figs.
1970 Poisonous and venomous marine animals of the world. Vol. 3. U.S. Government Printing Office, Washington, D.C.

HALSTEAD, B. W., and A. E. DALGLEISH
1967 The venom apparatus of the European stargazer, *Uranoscopus scaber* Linnaeus, p. 177-186, 11 figs. *In* F. E. Russell and P. R. Saunders [eds.], Animal toxins. Pergamon Press, New York.

HALSTEAD, B. W., DANIELSON, D. D., BALDWIN, W. J., and P. C. ENGEN
1972 Morphology of the venom apparatus of the leatherback fish *Scomberoides sanctipetri* (Cuvier). Toxicon 10: 249-258, 5 figs.

HALSTEAD, B. W., and L. R. MITCHELL
1963 A review of the venomous fishes of the Pacific area, p. 173-202, 16 figs. *In* H. L. Keegan and W. V. MacFarlane [eds.], Venomous and poisonous animals and noxious plants of the Pacific region. Pergamon Press, New York.

HALSTEAD, B. W., KUNINOBU, L. S., and H. G. HEBARD
1953 Catfish stings and the venom apparatus of the Mexican catfish, *Galeichthys felis* (Linnaeus). Trans. Am. Microscop. Soc. 72(4): 297-314, 6 figs.

HALSTEAD, B. W., ENGEN, P. C., and D. D. DANIELSON
1971 Morphology of the venom organs of the rabbitfishes (family Teuthidae), p. 121-157, 29 figs. *In* A. de Vries and E. Kochva [eds.], Toxins of animals and plant origin, vol. 1. Gordon and Breach Science Publishers, New York.

HARRY, R. R.
1953 Ichthyological field data of Raroia Atoll, Tuamotu Archipelago. Atoll Res. Bull, Natl. Res. Council (18): 47-169, 177.

HERRE, A. W.
1953 Check list of Philippine fishes. U.S. Fish Wildlife Serv., Res. Rept. 20. 977 p.

HILZHEIMER, M., and O. HAEMPEL
1913 Biologie der wirbeltiere. F. Enke, Stuttgart. 756 p.

JORDAN, D. S., EVERMANN, B. W., and H. W. CLARK
1930 Check list of the fishes and fishlike vertebrates of North and Middle America north of the northern boundary of Venezuela and Colombia. Rept. U.S. Comm. Fish. 1928. 670 p.

JOUBIN, L. M.
1929-38 Faune ichthyologique de l'Atlantique nord. Nos. 1-18. A. F. Høst et Fils, Copenhague.

KOBERT, R.
1894 Compendium der praktischen toxikologie, p. 90-93. F. Enke, Stuttgart. (NSA)
1902 Ueber giftfische und fischgifte. Med. Woch (19): 199-201; (20): 209-212; (21): 221-225.

KOPACZEWSKI, W.
1917 Sur le venin de la murène (*Muraena helena* L.) Compt. Rend. Acad. Sci. 165: 513-515.

MAASS, T. A.
1937 Gift-tiere. *In* W. Junk [ed.], Tabulae biologicae. Vol. XIII. N. V. Van de Garde & Co's Drukkerij, Zaltbommel, Holland. 272 p.

MARETIĆ, Z.
1968 Venomous animals and their venoms. [In Croatian] Pro Medico 3: 93-114.

MARSHALL, T. C.
1964 Fishes of the Great Barrier Reef and Coastal waters of Queensland. Angus and Robertson, Melbourne. 566 p., 72 pls.

MONOD, T.
1959 Notes sur l'épine latéro-caudale et la queue de l'*Acanthurus monroviae*. Bull. Inst. Française Afrique Noire 21, Ser. A, (2): 710-734, 32 figs.

MUNRO, I. S. R.
1967 The fishes of New Guinea. Department of Agriculture, Stock, and Fisheries, Port Moresby, New Guinea. 651 p., 78 pls.

NOBRE, A. F.
1928 Animals venenosos de Portugal. Porto 1: 1-10, 39-55.

PAWLOWSKY, E. N.
1909 Contribution to the question of poisonous skin glands of certain fishes. [In Russian] Trav. Soc. Imp. Nat., St. Petersburg. 40(1): 109-126, 5 figs.

PHILLIPS, C., and W. H. BRADY
1953 Sea pests—poisonous or harmful sea life of Florida and the West Indies. Univ. Miami Press, Miami, Fla. 78 p., 38 figs., 7 pls.

PHISALIX, M.
1922 Animaux venimeux et venins. 2 vols. Masson et Cie., Paris.
1931 Le venin de quelques poissons marins. Notes Sta. Océanogr. Salammbo, Tunis (22): 3-15.

PORTA, A.
1905 Ricerche anatomiche sull'apparecchio velenifero di alcuni pesci. Anat. Anz. 26: 232-247, 5 figs., 2 pls.

(NSA) Not seen by author.

Randall, J. E.
1955 A revision of the surgeon fish genus *Ctenochaetus*, family Acanthuridae, with descriptions of five new species. Zoologica 40(4): 149-166, 2 figs.
1959 Report of a caudal-spine wound from the surgeonfish *Acanthurus lineatus* in the Society Islands. Wasmann J. Biol. 17(2): 245-248.

Richardson, J.
1844-48 Ichthyology of the voyage of H.M.S. *Erebus* and *Terror*. London. 139 p., 60 pls.

Schultz, L. P., Herald, E. S., Lachner, E. A., Welander, A. D., and L. P. Woods
1953 Fishes of the Marshall and Marianas Islands. Vol. I. Families from Asymmetrontidae through Siganidae. U.S. Nat. Mus., Bull. 202. 685 p., 74 pls., 89 figs.

Smith, J. L.
1950 The sea fishes of southern Africa. Central News Agency, Cape Town, S. Africa. 550 p., 103 pls.
1961 The sea fishes of southern Africa. 4th ed. Central News Agency, South Africa. 580 p., 102 pls., 1,232 figs.

Souché, G.
1935 Contribution à l'étude des épines de l'*Acanthurus chirurgus* Bl. Bull. Sta. Biol. d'Arcachon, Univ. Bordeaux 32: 31-38, 7 figs.

Tange, Y.
1955*a* Beitrag zur kenntnis der morphologie des giftapparates bei den japanischen fischen, nebst bemerkungen über dessen giftigkeit. XI. Über den giftapparat bei *Siganus fuscescens* (Houttuyn). Yokohama Med. Bull. 6(2): 115-122, 4 figs.
1955*b* Beitrag zur kenntnis der morphologie des giftapparates bei den japanischen fischen. XII. Über den giftapparat bei *Xesurus scalprum* (Cuvier et Valenciennes). Yokohama Med. Bull. 6(3): 171-178, 4 figs.

Taschenberg, E. O.
1909 Die giftigen tiere. F. Enke, Stuttgart. 325 p., 64 figs.

Weber, M., and L. F. de Beaufort
1931 The fishes of the Indo-Australian Archipelago. VI. Perciformes (continued). Families: Serranidae, Theraponidae, Sillaginidae, Emmelichthyidae, Bathyclupeidae, Coryphaenidae, Carangidae, Rachycentridae, Pomatomidae, Lactariidae, Menidae, Leiognathidae, Mullidae. E. J. Brill, Leiden. 448 p., 81 figs.
1936 The fishes of the Indo-Australian Archipelago. VII. Perciformes (continued). Families: Chaetodontidae, Toxotidae, Monodactylidae, Pempheridae, Kyphosidae, Lutjanidae, Lobotidae, Sparidae, Nandidae, Sciaenidae, Malacanthidae, Cepolidae. E. J. Brill, Leiden. 607 p., 106 figs.

Whitley, G. P.
1943 Poisonous and harmful fishes. Council Sci. Ind. Res., Bull. No. 159, 28 p., 16 figs.; Pls. 1-3.

Woodland, D. J.
1972 Proposal that the genus name *Teuthis* Linnaeus (Pisces) be suppressed. Bull. Zool. Nomencl. 29(4): 190-193.

Chapter XXIV—VERTEBRATES

Class Osteichthyes

ICHTHYOCRINOTOXIC FISHES

The fishes included in this chapter comprise a recently established major category of ichthyotoxic fishes. Ichthyocrinotoxic fishes produce their poisons by glandular secretion but lack a venom apparatus, i.e., the glandular structures are not associated with spines, teeth, or some other mechanical traumagenic device. The toxic glandular contents are merely secreted into the water. There is no mechanism whereby the poison can be injected into a victim. Our present knowledge of this group of fishes is meager, but the role of these fishes in the marine ecosystem may ultimately prove to be of great significance. With the exception of the few preliminary studies reported in this chapter, ichthyocrinotoxic fishes have been largely overlooked by marine biotoxicologists. The number and phylogenetic distribution of ichthyocrinotoxic fishes is unknown, but they are probably widely distributed among piscine organisms.

The poison glands of ichthyocrinotoxic fishes undoubtedly assist in the defensive mechanism of the fish as warning or repellent substances. However, these glands may exert some of their most profound effects in the marine ecosystem by secreting toxic metabolites into the ocean environment.

The reasons for the succession of planktonic organisms in a biocenose, having apparently stabilized nutritional and biochemical conditions, have long been a subject of interest and speculation. Although there is probably a complex of factors, it is believed that the changes may be due, at least in part, to growth-inhibiting or growth-promoting biochemical substances produced by one of more members of the community. The succession of organisms in a marine community has been explained on the basis of antibiosis[1] (Hardy, 1956; Nigrelli, 1958, 1962).

Succession phenomena in marine biocenoses have been studied intensively by Hardy (1956) who proposed the theory of animal exclusion. The theory states that the distributional relationship between marine plants and animals is due to a modification in the vertical migrational behavior of animals in relationship to concentrations of plants. Metabolic products of the organ-

[1] The term "antibiosis" as used here includes any biochemical interaction that results in the reduction of a population of any susceptible plant or animal species either by a killing action or by interfering with metabolic processes essential for growth or reproduction (Nigrelli, 1962).

isms involved are suspected of being responsible for the exclusion effect. Marine biotoxicological investigations have shown that toxic metabolites are produced by a great variety of marine organisms, i.e., algae, invertebrates, fishes, etc. Thus the toxic products of one group of organisms probably affect the number and distribution of associated plants and animals and thereby determine the character of the biocenose. It is in this area of animal exclusion that ichthyocrinotoxic fishes may possibly play their most significant role in the ocean environment. Thus crinotoxic processes in marine organisms may serve as regulatory mechanisms in population dynamics. The toxic glandular secretions of some ichthyocrinotoxic fishes are known to be lethal to other adult fishes and are probably highly toxic to larval forms. The extent to which ichthyocrinotoxic fishes control the development of other fishes has not been determined. The relationship of skin toxins, or ichthyocrinotoxins, i.e., poisons produced by glandular elements in the skin of fishes, to the various types of ichthyosarcotoxins (particularly cyclostome poison and tetrodotoxin) is unknown. At present this is an area of biochemical oceanography about which we know little and which is in need of exploration.

The biology and systematics of the ichthyocrinotoxic fishes mentioned in this chapter have been discussed by Jordan, Evermann, and Clark (1930), Schultz *et al.,* (1953), Gosline and Brock (1960), and others.

REPRESENTATIVE LIST OF FISHES REPORTED AS ICHTHYOCRINOTOXIC

Phylum CHORDATA

Class AGNATHA

Order MYXINIFORMES (MARSIPOBRANCHII): Hagfishes, Lampreys[2]

Family MYXINIDAE

122 *Myxine glutinosa* Linnaeus (Pl. 1, fig. a, Chapter VIII). Atlantic hagfish (USA), borer, hagfish, poison ramper (England), piral eller pilal (Sweden).

DISTRIBUTION: North Atlantic.

SOURCE: Engelsen (1922).

Family PETROMYZONTIDAE

123 *Lampetra fluviatilis* (Linnaeus) (Pl. 2, fig. b, Chapter VIII). River lamprey (USA), neva lamprey, rechnaya minoga, nevskaya minoga (USSR).

DISTRIBUTION: Baltic and North Sea basins, westward to Ireland and France; enters the rivers from the sea.

SOURCES: Engelsen (1922), Pawlowsky (1927), Maass (1937).

---

[2] The lampreys and hagfishes are reputed to contain toxic substances in their blood, skin, and possibly their flesh. The chemical nature of these poisons is largely unknown. Consequently, the relationships of these several poisons to each other are presently not understood. Only those references pertaining to the toxic skin secretions of cyclostomes have been included in this chapter. (*See also* Chapter VIII for additional information on toxic cyclostomes.)

Petromyzon marinus Linnaeus (Pl. 1, fig. b, Chapter VIII). Sea lamprey, large nine-eyes (USA, England), meerneunauge, seelamprete (Germany), lamproie (France), pricke (Austria), lampreda (Italy), hafs-nejnoga (Sweden), morskaya minoga (USSR). *122*

DISTRIBUTION: Coasts and rivers of both sides of the Atlantic, rivers of the Mediterranean.

SOURCES: Engelsen (1922), Fredericq (1924), Pawlowsky (1927).

Class OSTEICHTHYES

Order CYPRINIFORMES (OSTARIOPHYSI): Minnows, Catfishes, etc.[3]

Order ANGUILLIFORMES (APODES): Eels

Family MURAENIDAE

Muraena helena (Linnaeus) (Pl. 1, Chapter XXIII). Moray eel (USA), moray, murry (England), murène (France), murena, morena (Italy), mourena (Spain), muräne (Germany). *135, 258*

DISTRIBUTION: Eastern Atlantic Ocean and Mediterranean Sea.

SOURCE: Maass (1937).

Order PERCIFORMES (PERCOMORPHI): Perchlike Fishes

Family SERRANIDAE[4]

Grammistes sexlineatus (Thunberg) (Pl. 1, fig. a). Golden striped bass (USA). *268*

DISTRIBUTION: Indo-Pacific.

SOURCES: Liguori *et al.* (1963), Randall *et al.* (1971).[5]

Pogonoperca punctata Cuvier and Valenciennes.

DISTRIBUTION: Indo-Pacific.

SOURCES: Hashimoto and Kamiya (1969), Randall *et al.* (1971).

---

[3] Many species of catfishes possess an axillary gland located near the base of the pectoral spine. The gland secretes a toxic substance. It is believed that the axillary gland is part of the pectoral venom apparatus. However, the axillary secretion is released into the surrounding environment, and, although it may contribute to an envenomation inflicted by the pectoral sting, it is quite possible that the axillary gland may also have a crinotoxic function. Unfortunately, this is a problem which has never received any attention. See Chapter XIX on venomous catfishes for further information on the axillary glands of catfishes. Since there is no definite information available on this subject, no particular species of catfishes are listed in this section.

[4] According to some authors the grammistids have now been placed in a separate family, the Grammistidae. (*See* Randall *et al.*, 1971).

[5] Randall *et al.* (1971) have incriminated an additional group of ichthyocrinotoxic fishes, which include the following grammistid species: *Aulacoephalus temminicki* Bleeker, *Diploprion bifasciatum* Cuvier and Valenciennes, *Grammistops ocellatus* Schultz, *Rypticus bicolor* (Valenciennes), *R. subbifrenatus* Gill, and *R. randalli* Courtenay.

268 *Rypticus saponaceus* (Bloch and Schneider) (Pl. 1, fig. b). Soapfish (USA), jaboncillo (West Indies).

DISTRIBUTION: Tropical and subtropical Atlantic.

SOURCE: Maretzki and Castillo (1967), Randall *et al.* (1971).

Order TETRAODONTIFORMES (PLECTOGNATHI): Triggerfishes, Puffers, Trunkfishes, Etc.

Family DIODONTIDAE

181 *Diodon hystrix* Linnaeus (Pl. 4, fig. b, Chapter XIV). Porcupinefish (USA), atinga (West Indies), puerco espino, sorospin, erizo (Mexico), baiacus de espinhos (Brazil), o'opu okala, oopuhue, torabuku (Hawaii), mojannur, japonke, mejangir (Marshall Islands), sou nichukew, soutu (Carolines), derudn (Palau), duto boteting laut, botiting laot, buteteng laot (Philippines), porcupinefish (Australia), katu peyttheya, mullu peytthai (Sri Lanka), yu hu (China), porcupinefish, spiny blaasop, penvisse (South Africa), shokei'eiya, meshwaka (Egypt), pez erizo (Spain).

DISTRIBUTION: All tropical seas.

SOURCE: Eger (1963).

Family OSTRACIONTIDAE

163, 269 *Ostracion meleagris* Shaw (Pl. 4). Trunkfish (USA), pahu (Hawaii), bill (Marshall Islands), obuluk, tabayong (Philippines), umisuzume, suzume-fugu (Japan), cowfish, boxfish (Australia), cowfish, trunkfish, boxfish, oskop, seekoei, seevarkie (South Africa).

DISTRIBUTION: Indo-Pacific.

SOURCES: Brock (1956), Thomson (1963), Eger (1963).
OTHER NAME: *Ostracion lentiginosus.*

268 *Rhinesomus bicaudalis* (Linnaeus) (Pl. 3). Spotted trunkfish (USA), chapin (West Indies).

DISTRIBUTION: West Indies to Florida.

SOURCES: Brown (1945).
OTHER NAME: *Lactophrys bicaudalis.*

Family TETRAODONTIDAE

183 *Arothron hispidus* (Linnaeus) (Pl. 6, fig. a, Chapter XIV). Puffer, swelltoad, blower, toadfish, swellbelly, swellfish, swellingfish, blowfish, jugfish, rabbitfish (USA), oopuhue, maki maki, keke, akeke (Hawaii), luap, wat (Marshall Islands), te buni (Gilbert Islands), nipou, wata, nemata, sete, tiwadi (Carolines), botiti, botete, tikung, langiguihon, tinga tinga (Philippines), ikan buntal, buntal pisang (Malaysia), blazer, tinga tinga, ikan butak, ikan nogi-nogi, ikan buntal, buntal pisang (Indonesia), ca noc vàng, ca noc, canoc hot mit (Vietnam), blaasop, tobies, aufblaser, tinga tinga (South Africa), dremma, deremah (Egypt), yokoshimafugu (Japan).

DISTRIBUTION: Panama, Indo-Pacific, Japan, Australia, South Africa, Red Sea.

SOURCES: Flaschentrager and Abdallah (1957), Eger (1963), Thomson (1963).

Arothron meleagris (Lacépède) (Pl. 6, fig. b, Chapter XIV). Puffer, swelltoad, blower, toadfish, swellbelly, swellfish, swellingfish, blowfish, jugfish, rabbitfish, (USA), oopuhue, maki maki, keke akeke (Hawaii), luap, wat (Marshall Islands), te buni (Gilbert Islands), nipou, wata, nemata, sete, tiwadi (Carolines), botiti, botete, tikung, langiguihon, tinga tinga, ikan butak, ikan noginogi, ikan buntal, buntal pisang (Indonesia), ca noc vàng, ca noc, canoc hot mit (Vietnam), mizorefugu (Japan), tambor, botete (Mexico, Central America). *183*

Distribution: West coast of Central America and throughout the Indo-Pacific.

Source: Eger (1963).

Arothron nigropunctatus (Bloch and Schneider) (Pl. 6, fig. c, Chapter XIV). Puffer, swelltoad, blower, toadfish, swellbelly, swellfish, swellingfish, blowfish, jugfish, rabbitfish (USA), oopuhue, maki maki, keke, akeke (Hawaii), luap, wat (Marshall Islands), te buni (Gilbert Islands), nipou, wata, nemata, sete, tiwadi (Carolines), botiti, botete, tikung, langiguihon, tinga tinga (Philippines), ikan buntal, buntal pisang (Malaysia), blazer, tinga tinga, ikan butak, ikan noginogi, ikan buntal, buntal pisang (Indonesia), ca noc vàng, ca noc, canoc hot mit (Vietnam), blaasop, topies, aufblaser, tinga tinga (South Africa), kokutenfugu (Japan), toadfish, globefish, swellfish, blowfish, puffer (Australia), daghemeiah (Egypt). *183*

Distribution: Indo-Pacific, Japan, Australia, east Africa, Red Sea.

Source: Flaschentrager and Abdallah (1957).
Other names: *Arothron diadematus, Amblyrhynchotes diadematus.*

Fugu pardalis (Temminck and Schlegel) (Pl. 10, fig. b, Chapter XIV). Puffer (USA), higanfugu (Japan), ho t'um yu, kay-po-oy (China), bok oh (Korea). *187*

Distribution: China, Japan.

Sources: Tani (1945), Macomber (1956).
Other name: *Sphoeroides pardalis.*

Fugu poecilonotus (Temminck and Schlegel) (Pl. 11, fig. d, Chapter XIV). Puffer, swelltoad, blower, toadfish, swellbelly, swellfish, swellingfish, blowfish, jugfish, rabbitfish (USA), oopuhue, maki maki, keke, akeke (Hawaii), luap, wat (Marshall Islands), te buni (Gilbert Islands), nipou, wata, nemata, sete, tiwadi (Carolines), botiti, botete, tikung, langiguihon, tinga tinga (Philippines), ikan buntal, buntal pisang (Malaysia), blazer, tinga tinga, ikan butak, ikan noginogi, ikan buntal, buntal pisang (Indonesia), ca noc vàng, ca noc, canoc hot mit (Vietnam), blaasop, tobies, aufblaser, tinga tinga (South Africa), komonfugu (Japan), ho t'um yu, kay-po-oy (China), bok oh (Korea). *188*

Distribution: Indo-Pacific, China, Korea, Japan.

Source: Tani (1945).
Other name: *Sphoeroides alboplumbeus.*

189 *Fugu stictonotus* (Temminck and Schlegel) (Pl. 12, fig. c, Chapter XIV). Puffer (USA), gomafugu (Japan), ho t'um yu, kay-po-oy (China), bok oh (Korea).
DISTRIBUTION: Southern Korea, east China Sea and Japan.

SOURCE: Tani (1945).
OTHER NAME: *Sphoeroides stictonotus*.

193 *Sphaeroides maculatus* (Bloch and Schneider) (Pl. 16, fig. b, Chapter XIV). Puffer (USA), pompus, toade, tamboril (West Indies).
DISTRIBUTION: Atlantic coast of United States to Guiana.

SOURCES: Larson, Lalone, and Rivas (1961), Lalone, DeVillez, and Larson (1963).

Order BATRACHOIDIFORMES (HAPLODOCI): Toadfishes, etc.

Family BATRACHOIDIDAE

253 *Opsanus tau* (Linnaeus) (Pl. 2, fig. a, Chapter XXII). Oyster toadfish (USA), sapo (West Indies).
DISTRIBUTION: Atlantic coast of United States, Massachusetts to West Indies.

SOURCES: Wallace (1893), Collette (1966).
OTHER NAME: *Batrachus tau*.

253 *Thalassophryne dowi* Jordan and Gilbert (Pl. 2, fig. c, Chapter XXII). Dow's toadfish (USA), sapo (West Indies, Latin America).
DISTRIBUTION: Pacific coast of Central America, Costa Rica to Panama.

SOURCE: Collette (1966).
OTHER NAME: *Daector dowi*.

254 *Thalassophryne maculosa* Günther (Pl. 3, fig. a, b, Chapter XXII). Toadfish (USA), sapo (West Indies).
DISTRIBUTION: Caribbean Sea.

SOURCE; Collette (1966).

BIOLOGY

Order MYXINIFORMES

Family MYXINIDAE: The hagfishes are exclusively marine fishes and are only occasionally encountered in brackish waters. They are very sensitive to salinity, generally requiring an optimum range of 3.2 to 3.4 percent and water temperatures of less than 13° C. These tolerances tend to confine hagfishes to depths of 15 to 20 fathoms except during the colder seasons of the year. Hags prefer a clay, mud, or sandy bottom, where they seem to spend most of their time. Representatives of this group inhabit temperate and subtropical waters of the Atlantic and Pacific Oceans. Since the hag is somewhat of a scavenger in its habits, its food consists of dead, disabled, or netted fishes. Polychaete worms form the main item of diet in some hagfishes. Hags are sometimes hauled on board ship still clinging to the side of a fish which they had just victimized. Their habit of attacking netted fishes has given them a bad reputation among haddock fishermen. Hagfishes are blind and locate most of their food by scent. They spawn throughout most the year. The eggs are less than 30 in number, and generally they are deposited attached to a fixed object on the bottom of the sea. Hags do not pass through a larval stage. The skin of the hagfish is richly supplied with large mucous cells. A large hagfish is said to be capable of filling a 2-gallon bucket with slime.

Family PETROMYZONTIDAE: The members of the family Petromyzontidae are inhabitants of marine and fresh waters of the Northern Hemisphere. Some species are marine and anadromous, whereas others are confined to fresh waters. The habits of *Petromyzon marinus* are representative of the marine and anadromous species. Since lampreys are seldom seen in the open sea, little is known about this phase of their life. When encountered in the sea, they are usually close to the land or in estuaries or other comparatively shallow-water areas. Lampreys have been taken at 547 fathoms off Nantucket, Mass. Life history studies indicate that lampreys tolerate a wide range of temperatures and salinities.

Adult lampreys are parasitic, feeding on the blood of other fishes. They attach themselves to the body of a host fish by sucking with the oral disk; by means of their horny teeth, they rasp through the skin and scales and then suck the blood until the host is dry. Gage and Gage-Day (1927) have shown that the buccal secretion of lampreys contains an anticoagulant which facilitates the flow of blood. Mackerel, shad, cod, pollack, salmon, basking sharks, herring, swordfish, hake, sturgeon, and eels are favorite prey of lampreys. Larval lampreys feed largely on microscopic organisms.

P. marinus, an anadromous species, seeks a river having a gravelly bottom with rapid running water for spawning beds and muddy or soft sandy bottom in quiet water for the larvae. In the New England states, migration up the rivers usually takes place during the spring and early summer months. The parents die soon after spawning. The larval lampreys remain in the parent stream, living in mud burrows, for a period of 3 to 5 years. When they reach a length of 10 to 13 cm they migrate to the sea, usually during the autumn.

Order ANGUILLIFORMES (APODES)

Family MURAENIDAE: Moray eels are a group of savage, moderate-sized marine fishes. Their bathymetric range extends from intertidal reef flats to depths of several hundred meters. They are found throughout temperate and tropical seas. Moray eels inhabit a large variety of biotopes, in surge channels, coralline ridges, interislet channels, reef flats, and lagoon patch reefs. They are nocturnal in their habits, hiding in crevices or holes and under rocks or coral during the day and coming out at night. With the aid of a light one can observe morays wriggling over a reef flat at night in large numbers. They may often be seen thrusting their heads out of coral holes, their mouths slowly opening and closing, waiting for some unwary victim to pass by their lair. They are able to strike with great rapidity and ferocity. Their long, fang-like, depressible teeth are exceedingly sharp and can inflict serious lacerations. Their powerful muscular development, tough leathery skin, and dangerous jaws make them formidable animals. Some of the larger morays attain a length of 3 m or more and may constitute a real hazard to divers, particularly when poking one's hands into holes and crevices. Morays are carnivorous and predacious. They can readily be lured out by placing dead fish in front of their lair. Natives use spears, hooks, lines, snares, and traps in capturing them. The flesh of some species is used by certain peoples as food and is said to be agreeable, but oily and not readily digestible. Some species are violently poisonous to eat.

Order PERCIFORMES (PERCOMORPHI)

Family SERRANIDAE: Sea bass or groupers are robust, carnivorous, predacious shore fishes of tropical and temperate waters. A variety of biotopes are inhabited by this large family: coral reefs, rocks, sandy areas, and kelp. A few are found in fairly deep water, but most live in shallow water. Some attain a weight of 300 kg or more. They are usually considered good food fishes although some seabass are poisonous to eat.

Order TETRAODONTIFORMES (PLECTOGNATHI)

Family DIODONTIDAE: The porcupinefishes are spiny marine fishes widely distributed in all warm seas. *Diodon hystrix* attains a total length of 91 cm or more, but most diodontids range from 20 to 50 cm in length. Porcupinefishes inhabit coral reefs and shoal areas, generally traveling singly or in pairs. They also have the ability to inflate themselves until they are almost spherical in outline. During the inflation process the spines, which are usually depressed, are extended, giving the fish a formidable appearance. Porcupinefishes have been known to kill larger carnivorous fishes by inflating themselves and becoming stuck in the throat of the would-be captor. Diodontids are also capable of giving a severe bite. Darwin (1945) states that diodontids have been found alive in the stomachs of sharks; in some instances they have mortally wounded their captors by gnawing through the stomach walls and sides of the sharks and thus escaped.

Family OSTRACIONTIDAE: Trunkfishes are especially peculiar in construction. The body is enveloped within a bony box, comprised of six-sided scutes, leaving openings only for the jaws, fins, and tail. They live in tropical seas and are frequently seen swimming slowly among the corals in shallow water. Many of them are brilliantly colored. The "exoskeleton," which serves them well for protection, is used by primitive peoples as a container in which the trunkfish is cooked over a fire. Little is known of the habits of trunkfishes.

Family TETRAODONTIDAE: Some authors divide the puffer family Tetraodontidae into four separate families—Chonerhinidae, Colomesidae, Lagocephalidae, and Tetraodontidae. However, more recent studies (Tyler, 1965) suggest that these puffers should be combined into the single family Tetraodontidae. Tetraodontids inhabit a variety of ecosystems, i.e., marine, estuarine, and freshwater. Most lagocephalids range in size from about 25 to 50 cm, but *Lagocephalus lagocephalus* may attain a total length of 61 cm. Some of the other puffers attain even greater size. *Arothron stellatus* has been known to reach a total length of 91 cm, but most tetraodontids range from about 20 to 40 cm. Most puffers are considered shallow-water fishes. However, some members do inhabit relatively deep waters. *L. oceanicus* has been taken at 40 m or more, *L. lagocephalus inermis* at 90 m, *L. sceleratus* at 60 m, and *Sphaeroides oblongus* at 100 m. *Liosaccus cutaneus* is usually conceded to be a deepwater species, and *Boesmanichthys firmamentum* has been taken at a depth of 100 fathoms. Species of *Lagocephalus* have been taken hundreds of km from shore at depths of 4,000 fathoms and are a common constituent in the stomachs of pelagic fishes such as tunas, wahoos, and other scombroids. Whether these deepwater puffers are nontoxic or the poison is not transvectored by scombroids is not known. The Indo-Pacific tetraodontids living around coral reef areas tend to travel singly or in small groups. *Sphaeroides* species tend to be more gregarious and are sometimes observed in large groups, but apparently they do not school in the manner that some fishes do. Stomach analyses on *A. hispidus* indicate that this species is omnivorous in its eating habits, since fragments of corals, sponges, algae, mollusks, and fish are commonly found in their stomachs. At Tagus Cove, Isabela Island, during the Krieger Galapagos expedition (Halstead and Schall, 1955), it was observed that hundreds of *Sphaeroides annulatus* were attracted to the surface of the water with a night light. Specimens could then be readily captured by spear or dip net. When puffers are at rest they appear to hover almost motionless with only their pectorals beating the water. The pectoral, dorsal, and anal fins are the chief locomotory organs, the tail being used principally as a rudder. Despite their reputation of being slow-moving fishes, puffers are capable of moving with surprising rapidity when they are frightened. They are likewise capable of inflicting severe bites. According to Gimlette (1923) and others, the human genital organs are frequent targets of their attacks. Puffers are quite vicious and readily snap at almost any bait offered them.

The inflating mechanism of puffers has been a subject of interest among anatomists for many years. The research conducted on this interesting mechanism has been well presented by Breder and Clark (1947), to whom we are indebted for the following information. Inflation is employed only as a defense mechanism. If a puffer is frightened or annoyed, it will gulp down its fluid medium, thus causing inflation. There is no evidence that puffers come to the surface in order to inflate themselves with air as is commonly believed. The inflating mechanism consists of the powerful muscles of the first branchiostegal ray which depress a pad covering the ceratohyals, thus expanding the mouth cavity and drawing in water or air if the fish is out of water. The elevation of the ceratohyals forces the fluid into the saclike ventral diverticulum of the stomach, from which it is partially separated by a sphincterlike ring. Fluid is retained in the diverticulum by a strong esophageal sphincter and by the pylorus, and not by the flaplike breathing valves which are present in the mouth. The opercular valves prevent leakage during compression, but the sac can remain distended when the valves are held open or removed. The water or air in the sac is released by relaxation of the esophageal sphincter, permitting escape through the oral or opercular openings. Puffers make considerable noise during inflation by grinding their heavy jaw teeth together. Some species of puffers are covered with short prickly bristles which they withdraw or extend, apparently at will.

Puffers have a distinctive offensive odor which is particularly noticeable when they are being dressed or dissected.

The genus *Colomesus* is comprised of a single species which inhabits the rivers of northern South America and the West Indies. Specimens have been taken from the mouth of the Amazon to the fringe of the Andes Mountains —a distance of more than 5,000 km from the nearest salt water. The colomesids likewise have the ability to inflate themselves with water or air. Judging from the literature, relatively little is known regarding the habits of this genus.

Xenopterus is a small genus of freshwater fishes which, with the possible exception of a single African species, is restricted to the rivers of southern Asia and Indonesia. These fishes attain a total length of about 28 cm. According to Day (1878-88), xenopterids are very pugnacious and capable of inflicting serious wounds with their sharp beaklike jaws. Burmese natives claim that if a person should fall into the water where these fishes abound, they will attack in droves and kill the victim by their bites. Apparently little else has been reported on the habits of this interesting group.

Order BATRACHOIDIFORMES (HAPLODOCI)

Family BATRACHOIDIDAE: Batrachoid fishes, with their broad, depressed heads and large mouths, are somewhat repulsive in appearance. Most toadfishes are marine shore forms, but some are estuarine or entirely fluviatile, ascending rivers for great distances. The geographical distribution of toadfishes is shown on the map (Chapter XXII, p. 813). They appear to enjoy turbid water. Regardless of the type of water in which they are found, batrachoids are primarily bottom fishes. They hide in crevices and burrows, under rocks and debris, or among seaweed or lie almost completely buried under a few centimeters of sand or mud. The Brazilian species of *Thalassophryne* has the habit of covering itself with a thin layer of sand or mud, but with careful observation one can usually detect the outline and protruding eyes of the fish as one wades along in the clear shallow water of sandy beaches. Toadfishes are quite hardy and are able to live for several hours after being removed from the water. The bottom temperature of the water frequented by these fishes would appear to range from 10° C to 32° C. During the winter months toadfishes tend to migrate to deeper water where they remain in a somewhat torpid condition. They also are experts at camouflage. Their ability to change color to lighter or darker shades at will and their mottled pattern make these fishes difficult to see.

Most toadfishes tend to be somewhat sluggish in their movements, but after having food they can dart out with surprising rapidity. They are somewhat omnivorous in habit but seem to prefer, among other things, crabs, mollusks, worms, and small fishes. Toadfishes are said to be quite vicious and will snap at almost anything upon the slightest provocation. Although they are not capable of producing a severe wound, they can inflict a bite that is not readily forgotten. When they are disturbed or their dorsum touched, they immediately erect their dorsal spines and flare out their opercular spines in defiance. Toadfishes do not school, but they are gregarious.

MORPHOLOGY OF THE POISON ORGANS

The poison glands of ichthyocrinotoxic fishes have been studied only to a limited extent. Our knowledge of these glandular structures was formerly restricted to three species: *Myxine glutinosa, Ostracion meleagris*, and *Arothron hispidus*, but more recently Randall *et al.* (1971) and Aida *et al.* (1973) have described the crinotoxic glands of some of the grammistid soapfish, and Hashimoto, Shiomi, and Aida (1974) have reported the occurrence of a skin toxin in gobies (*Gobiodon spp.*).

Myxine glutinosa: The following résumé on the histology of the skin and slime glands of cyclostomes is based on the excellent histological descriptions of the skin of *M. glutinosa* by Schreiner (1916, 1918) and Blackstad (1963). Since their descriptions go far beyond the scope of the material pertinent to this chapter, the reader is referred to the original articles for further details on this subject.

The poison glands of cyclostomes seem to be confined to the skin. The skin of cyclostomes is comprised of three main layers: the epidermis (90μ), the dermis (90μ) with collagen bundles, and a hypodermis (240μ) rich in fat cells. The epidermis consists of five different types of cells: undifferentiated basal cells, large mucous cells, small mucous cells, thread cells, and sensory cells (Figs. 1-3).

The undifferentiated cells and small mucous cells are present in greatest abundance, and thread cells are more numerous than large mucous cells. The specialized cells are derived from the undifferentiated cells, consequently transitional cell forms are also present. The undifferentiated cells are found in greatest abundance in the basal half of the epidermis, most of them in contact with the basement membrane. The undifferentiated cells are elongate, with the outer ends pointed and the basal ends flattened against the basement membrane. The cytoplasm of these cells is distinctly divided into an endo- and ectoplasm. The endoplasm, located mainly in the infra- and supranuclear regions, contains granules like mitochondria and centrioles.

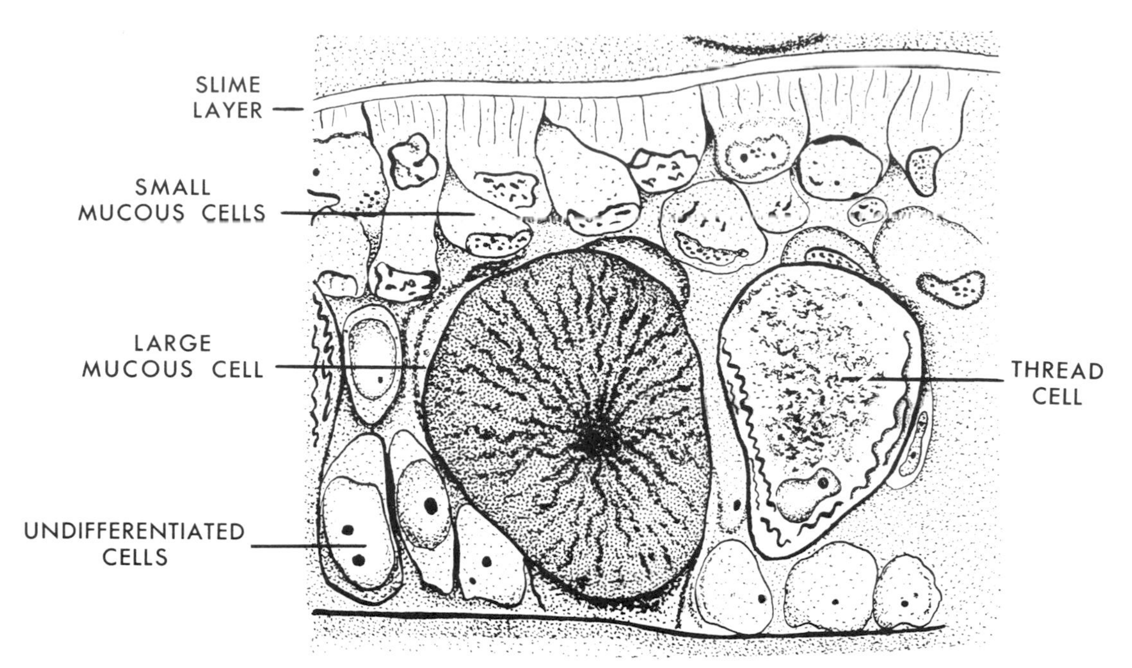

FIGURE 1.—Section of the epidermis of *Myxine glutinosa* showing four of the most common cell types. It is not certain as to which of these cells produce toxic secretions, but it is suspected that the mucous and thread cells are involved. Approx. × 580. (R. Kreuzinger, modified from Schreiner)

The small mucous cells are elongate and occupy most of the outer half of the epidermis. These cells have a cup-shaped nucleus, and many of the cells reach the surface of the epithelium. The loose endoplasm above the nucleus is distended with secretion. A thin rim of ectoplasm forms the lateral and basal boundaries of the cell. The apical ectoplasm is pierced by numerous fine canaliculi which lead from the endoplasm to the surface and allow the secretion to flow out of the cell. These cells are about 9μ in diameter. Granules of varying size and stainability are present in the endoplasm.

The large mucous cells are about 45μ in diameter, five to six times the diameter of the undifferentiated cells and the small mucous cells, and are generally found in the basal half of the epidermis. Mucous secretion occupies the major part of each cell, but in the center there is a nucleus closely connected with radiant strands in the adjacent area while finer strands pervade peripherally.

The thread cells are one of the most conspicuous cell types in the epidermis (Fig. 2). They are about the same size as the large mucous cells but are elongated and mainly situated in the middle of the epidermis. The upper end of the cell approaches the epidermal surface where their contents are emptied. These cells have a large central mass of dense granules, a slightly flattened nucleus, and peripherally spiral threads lying in a thin aqueous matrix.

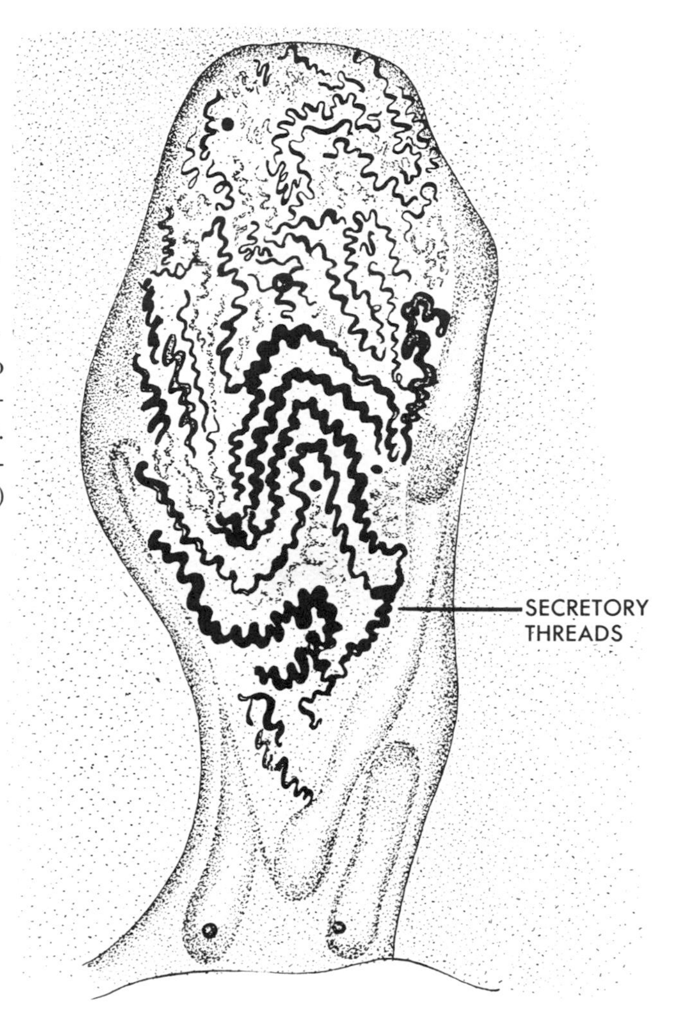

FIGURE 2.—Drawing of an enlarged view of a thread cell from *Myxine glutinosa* showing the dominant secretory threads of the cell. This cell is one of the most conspicuous cells of the epidermis and is believed to be one of the cells that produce toxic secretions. Approx. × 1270. (R. kreuzinger, modified from Schreiner)

The sensory cells are found scattered over the entire body of the fish but are most concentrated on the tentacles. Sensory cells are found ramifying among epidermal cells.

The dermis consists mainly of bundles of collagen fibers which cross each other but stay parallel to the basement membrane of the epidermis.

The subcutaneous tissue consists of a few layers of fat cells, 60 to 100μ in diameter, and strands of connective tissue, nerves, and blood vessels.

There appears to be no specific information available concerning the exact source of the toxic skin secretions of cyclostomes. The thread cells of hagfishes (Van Oosten, 1957) are modified mucous cells. They secrete a spirally coiled mucous thread which can be shot out and unwound for a considerable distance. It is thought that the thread cells and possibly some of the mucous cells produce the toxins present in the slime of these fishes. However, this is conjecture since little is known regarding the exact chemical nature of their secretions. It is not known whether the slime glands are involved in the

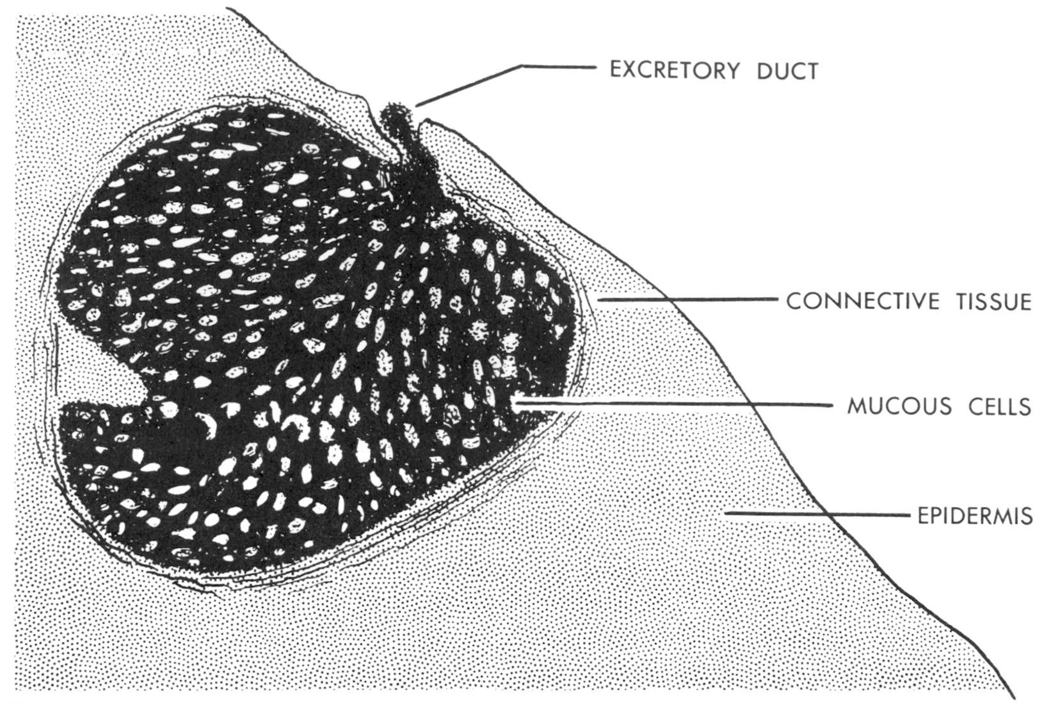

FIGURE 3.—Drawing of a slime gland taken from the caudal region of *Myxine glutinosa*. This drawing is based on a photomicrograph of the surface of a gelatin block stained with thionin. The mucous cells are stained dark and the thread cells appear as unstained oval structures. Slime is seen within the excretory duct. The slime glands are suspected as secreting toxic substances. Approx. × 20.

(R. Kreuzinger, modified from Blackstad)

production of toxic secretions. Slime glands (Fig. 3) are situated bilaterally in the epidermis and are segmentally arranged from the head to the tail of the fish.

Ostracion meleagris: The following résumé is based largely on the histological investigations of Thomson (1963).[6]

The skin of *Ostracion*, the boxfish or trunkfish, contains both hard and soft parts. The hard bony dermal carapace extends throughout the body trunk and the head. The soft skin covers only the fleshy lips, the base of the fins, and the caudal peduncle. There is an infolding of the skin at the base of the fins and at the junction of the trunk carapace with the caudal peduncle. Observations of the living fish revealed that when the animal is disturbed, copious amounts of foamy mucous secretions are released in these areas of infolding at the base of the fins and caudal peduncle. The epidermis about the lips, parts of the carapace, and caudal peduncle are partitioned by dermal septa into shallow pockets having a reticulated appearance externally. The epidermis around the fin bases and parts of the carapace is quite irregular,

[6] Thomson's (1963) work was based on *Ostracion lentiginosus* which is now considered by some authors to be a synonym of *O. meleagris*.

owing to the infoldings, ridges, and protuberances which thus effectively increase the secretory surface area. There was no macroscopic evidence of axillary glands or pockets of "jelly" such as have been mentioned by Brown (1945) for *Rhinesomus bicaudalis.*

Skin

The skin (Fig. 4) of the boxfish is comprised of three main layers: the outer epidermis, the dermis, and subcutaneous tissue. Since skin secretions are limited to the epidermal layer, the other layers will not be described. The epidermis consists of stratified squamous epithelium and is sharply delineated from the dermis by a layer of melanophores. There are at least three distinct cell types in the epidermal layer: undifferentiated basal cells, mucous cells, and club cells. Undoubtedly sensory cells and possibly other types are present, but no attempt has been made to describe them.

The undifferentiated basal cells have been described as forming the "matrix" of the epidermis. The specialized cells are derived from these undifferentiated cells. They are distinguished by their large, usually spherical, nuclei which are about 4μ in diameter. The polyhedral cell outlines are irregular and the cytoplasm is small compared to the large nucleus. In the basal layer of the epidermis the cells tend to be columnar, and the nuclei are fusiform. Elsewhere the undifferentiated cells are more polyhedral or polygonal in outline, and the nuclei are spherical. On the surface of the epidermis there is a single layer of squamous cells with serrated edges which forms the external boundary of the stratified epithelium.

The mucous cells are about 12 to 16μ in diameter and are present in greatest abundance at or near the surface of the epithelium. Their nuclei are spindle-shaped, peripheral, and the intact cells are full of mucus. In some sections mucous cells can be seen discharging their contents (Fig. 5).

The club cells are large (25 to 50μ), variable in shape, and the most conspicuous in the epidermis (Fig. 6). Club cells characteristically discharge their secretion while in the middle layers of the epidermis. The resulting cell is then pushed to the surface of the epidermis and discarded (Rabl, 1931; Andrew, 1959). Club cells are a common constituent of teleost epidermis. Their secretion seems to be the merocrine type. The toxic secretions are probably produced by the club cells. They appear to be a modified mucous cell but differ from the other mucous cells in their staining affinities.

Thomson found that the mucous cells stained positively with Mallory's triple (blue), periodic acid leucofuchsin Shiff's method (PAS) (red violet), and Gomori's aldehyde fuchsin (purple), which indicated their mucoid nature (Figs. 6-7). However, the secretions in the club cells stained negatively with Mallory's triple (yellow-orange to red) and PAS (pale pink). On the other hand, the club cell secretions were stained by toluidine blue and mucicarmine which are also mucous stains. Thomson concluded that both the mucous cells and the club cells contain mucins, but they are of distinctly different types.

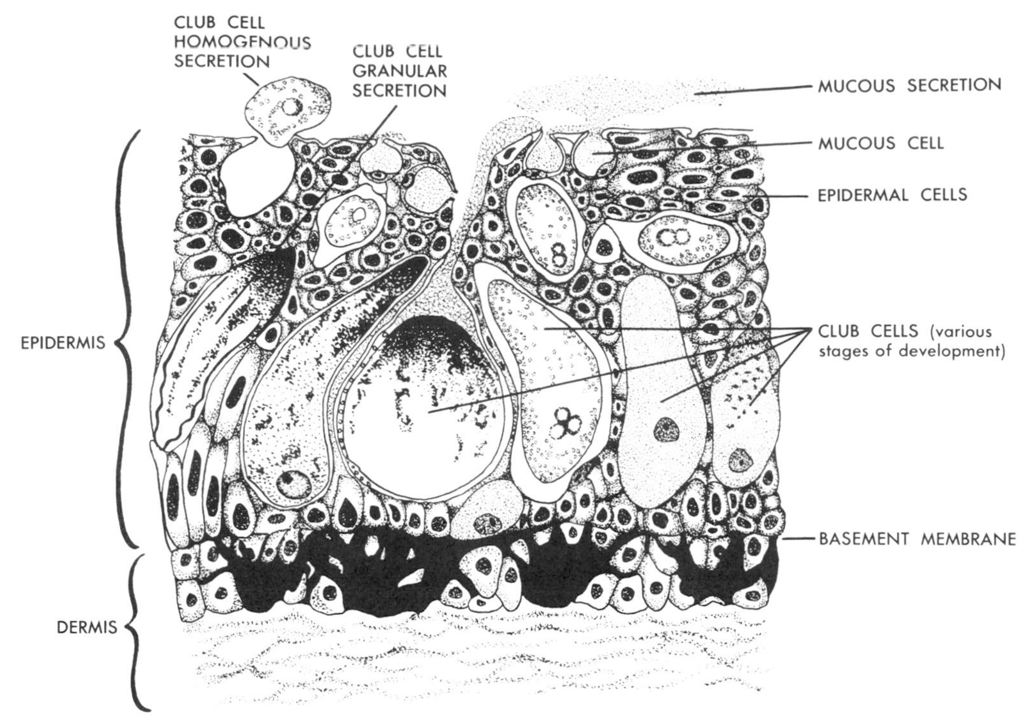

FIGURE 4.—(Above) Drawing of a skin section of *Ostracion meleagris* showing some of the more common cell types. The toxic secretions are probably produced by the club cells. (R. Kreuzinger, modified from Thomson)

FIGURE 5.—(Below) Photomicrograph of the caudal skin of *Ostracion meleagris* showing club cell glands. Note the dermal septum separating the glands and the dark secretory product being extruded by the gland on the right. Heidenhain's hematoxylin stain. × 440. (Courtesy D. A. Thomson)

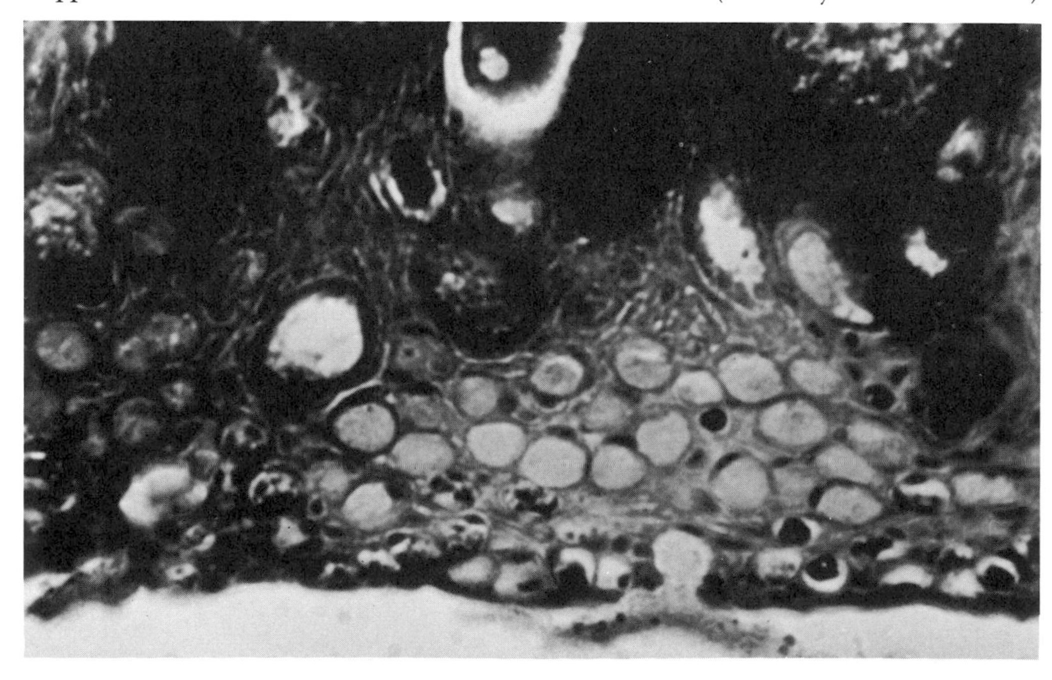

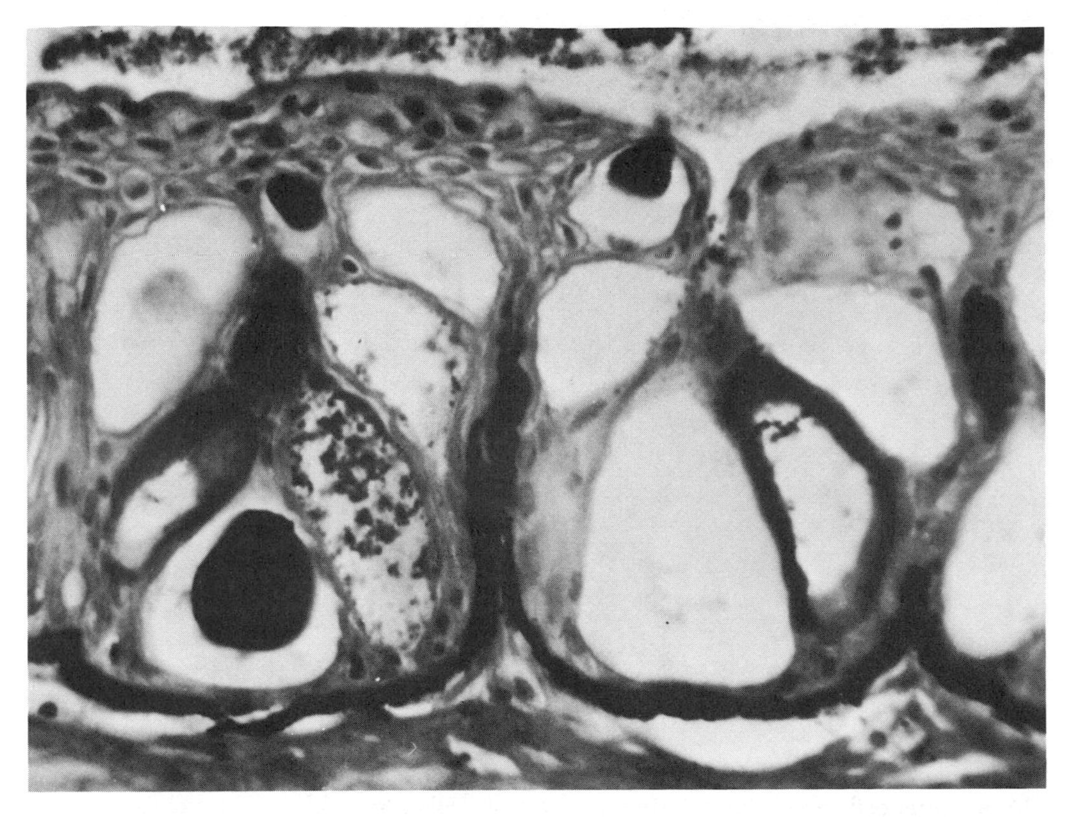

Figure 6.—Section of the skin of *Ostracion meleagris* showing one of the mucous cells extruding its contents. × 360. Heidenhain's hematoxylin stain.
(Courtesy D. A. Thomson)

Randall *et al.* (1971) and Aida *et al.* (1973) have studied the histology of the skin of a number of different Pacific grammistid fishes, particularly *Grammistes sexlineatus, Rypticus bicolor, Diploprin bifasciatus*, and *Aulacocephalus temmincki*. They found that there were two types of mucous cells in the epidermis and well developed mucous glands in the dermis under the scales in the skin of both *G. sexlineatus* and *R. bicolor*. They designated the mucous cells as Type I and II. Type I is the usual type found in the epidermis of most fishes. The mucous cells were relatively small and arranged along the upper layer of the epidermis. Moreover, Type I was positive to PAS reaction and negative to Sudan Black B staining. On the other hand, Type II were large cells, more deeply situated, and were negative to the PAS reaction and positive to Sudan Black B staining. The authors believed that the Type II mucous cells are found only in grammistid fishes. The dermal mucous glands were found to have a duct leading to the surface of the skin, and the glands were composed of small compartments. The material in these glands was PAS negative and is strongly stained with Sudan Black B. The skin toxin of the grammistids studied was also found to stain black with Sudan Black B.

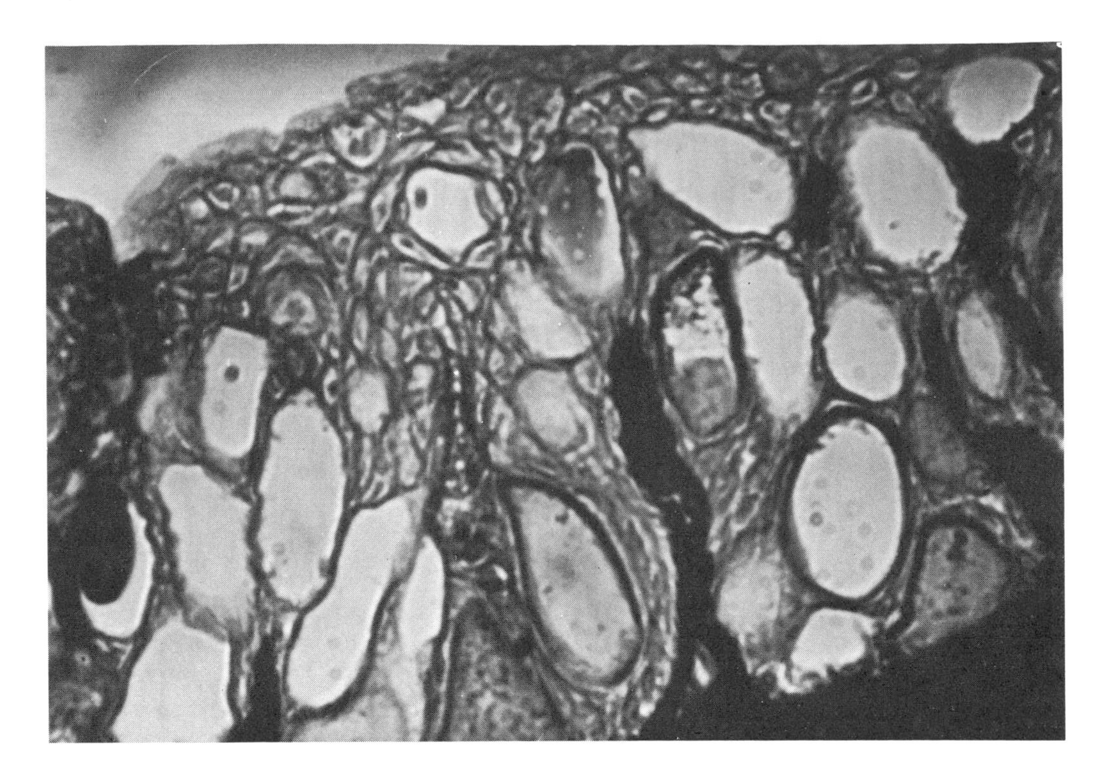

FIGURE 7.—Section of the skin of *Ostracion meleagris* stained with Mallory's triple stain. The mucous cells are stained blue and the club cells are reddish. The club cells are believed to produce toxic secretions. × 360. (Courtesy D. A. Thomson)

Labial Glands

The labial glands of *Ostracion* were also studied and histological sections prepared of the entire lips, including the jaws and epithelial lining of the buccal cavity and esophagus. The most conspicuous structures in the anterior portion of the buccal cavity of *Ostracion* are the labial villi attached to the ventral floor of the vestibular region between the external lips and the teeth (Figs. 8-9). The labial villi are part of an extensive elaboration of the mucous membrane of the oral cavity and have been termed "labial glands" by Thomson (1963). The labial glands have been classified as the compound acinous type.

The labial glands of *Ostracion* consist not only of dorsal and ventral villi, but of glandular pockets which extend deep into the dermis of the lips through an infolding of the lip epidermis, thereby effectively forming a duct that empties outside of the buccal cavity. Glandular pockets are found in both dorsal and ventral portions of the lip and extend in an anterior to posterior direction with the duct opening at the surface of the lip epidermis. Other pockets are found deep in the epithelial lining of the buccal cavity with ducts emptying into the oral cavity behind the teeth (Fig. 10a, b).

The labial villi are contained in large pockets formed by connective tissue

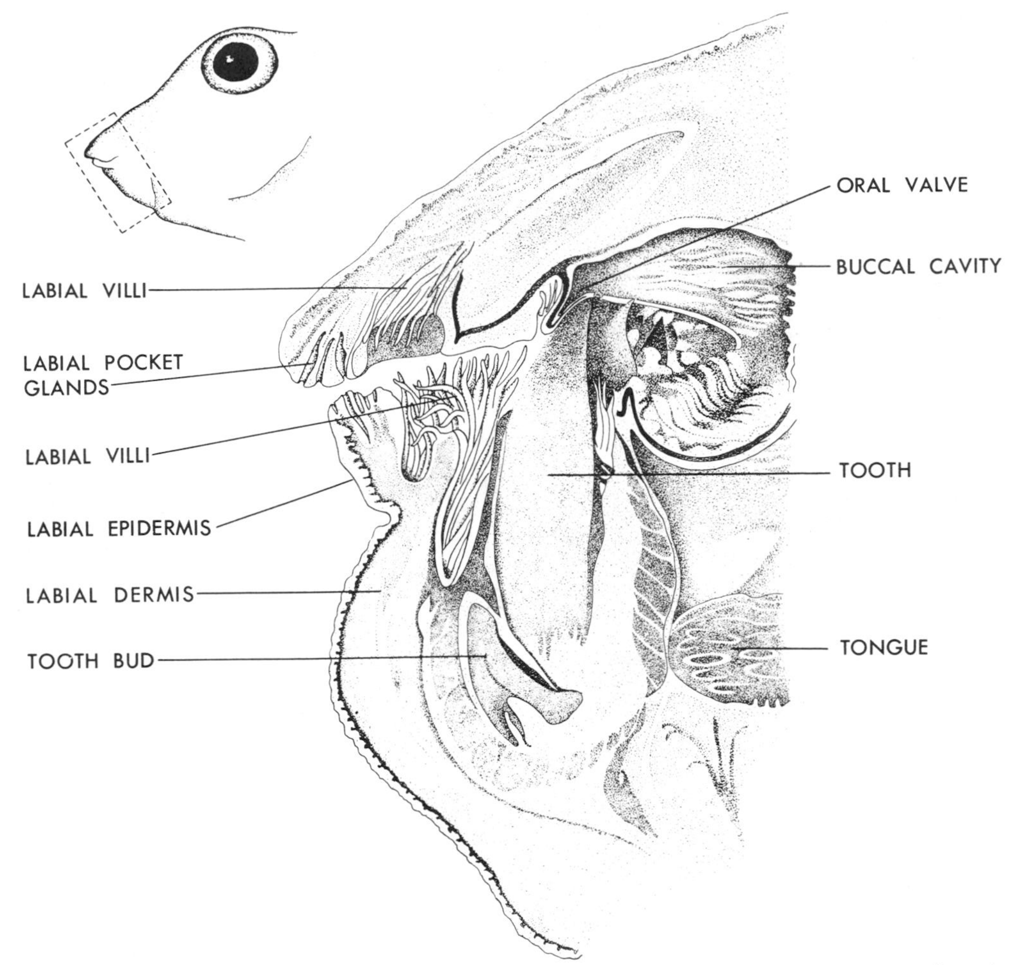

Figure 8.—(Above) Drawing of a sagittal section of the lips of *Ostracion meleagris* showing the location of the labial villi and glands. (R. Kreuzinger, after Thomson)

Figure 9.—(Below) Drawing of a cross section of the mouth of *Ostracion meleagris* through the dorsal teeth showing the arrangement of the labial villi around the teeth. (R. Kreuzinger, after Thomson)

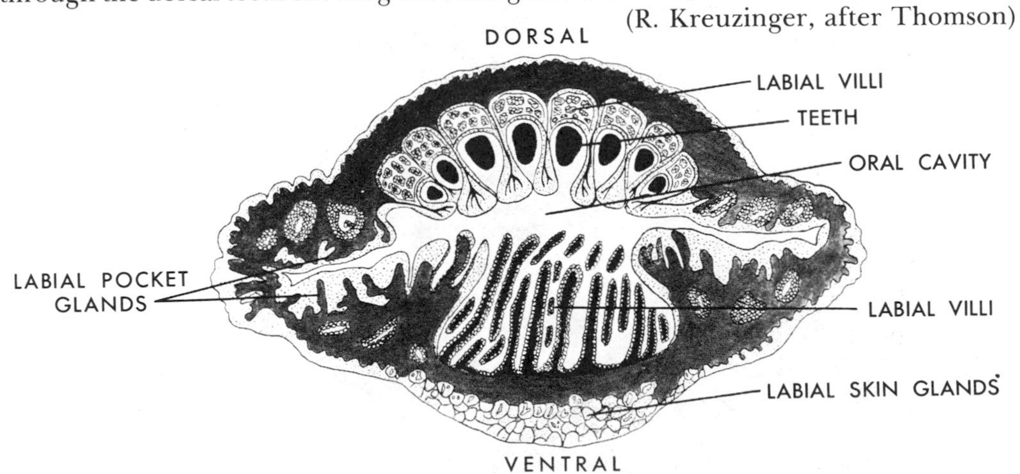

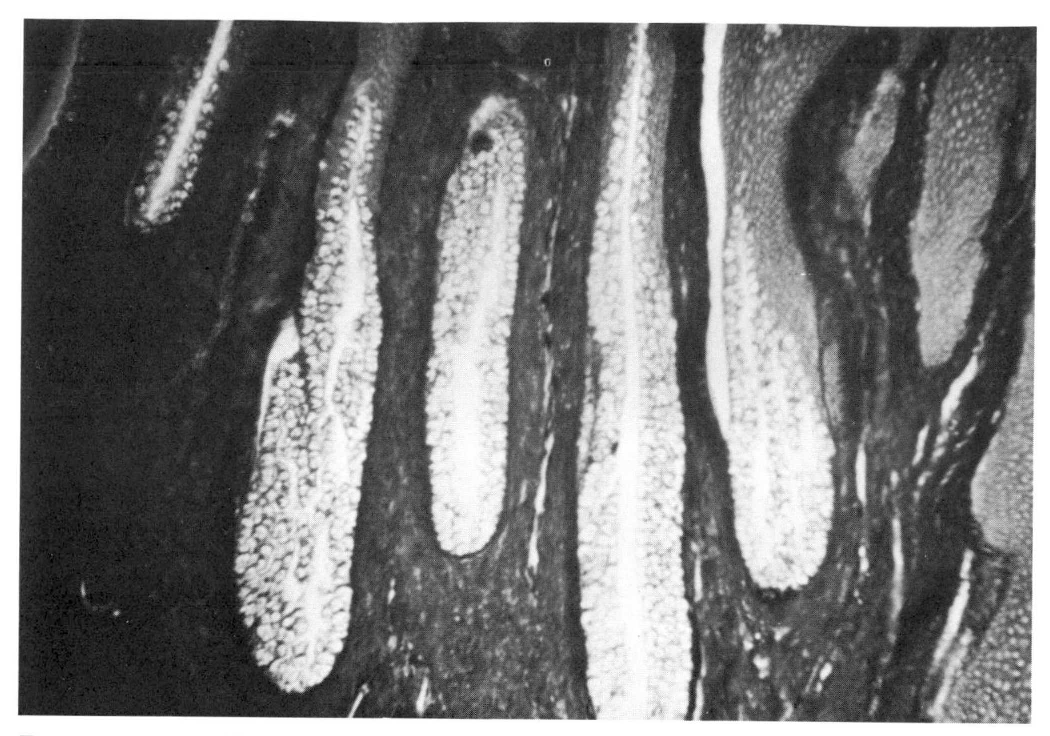

FIGURE 10.—a. (Above) Photomicrograph of a sagittal section of the ventral lips of *Ostracion meleagris* showing labial pocket glands and their pseudoducts. × 90. Mallory's triple stain. b. (Below) A cross section of the ventral lips of *Ostracion meleagris* showing the labial pocket glands. × 400. Gomori's fuchsin-trichrome stain.

(Courtesy D. A. Thomson)

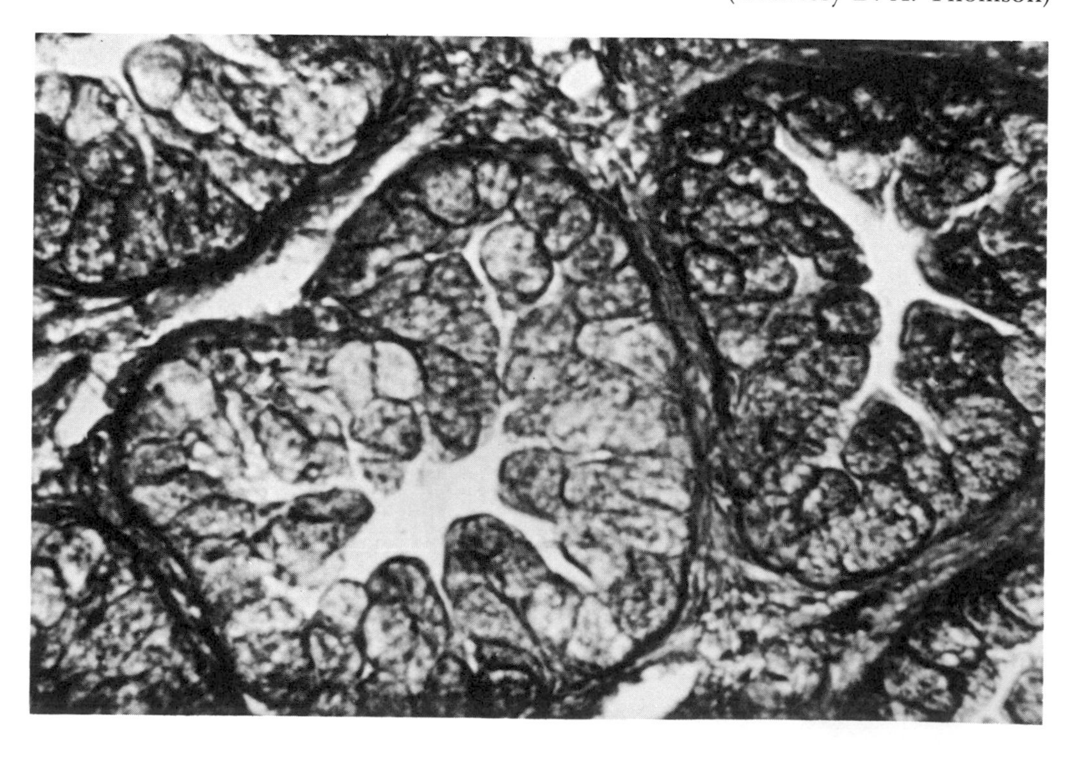

Table 1.—*Staining Reactions of Secretory Products in Boxfish Epidermal Gland Cells After Bouin's Fixation (Thomson, 1963)*

Stains	COLOR OF SECRETORY PRODUCTS			
	Mucous Cells	CLUB CELLS		Labial Cells
		Homogenous Secretion	Coarse Granules	
Harris hematoxylin triosin	light red	red	pale pink	clear
Heindenhain's iron hematoxylin	clear	black	black	clear
Mallory's triple stain	blue	orange, red, or blue	dark red to violet	clear
Gomori's aldehyde fuchsin trichrome	deep purple	orange, red, or green	deep purple	purple granules
PAS	deep red-violet	pale pink	pale pink	red-violet granules
Toluidine blue	clear	deep aqua	blue to light aqua	clear
Mucicarmine	clear	pink to red	pink	....
Sudan black B	clear	gray	very light gray	....

septa both above the dorsal row of teeth and below the ventral row. There is a pocket opposite each tooth, and these have been referred to as "dental pockets" (Fig. 9).

The labial gland cells extend throughout the mucous membrane of the buccal cavity and are found lining the folds and papillae of the palate but are not present in the esophagus.

The epidermis of the lips is thicker and more compact than that of the rest of the skin. The epidermal cells of the lips consist of the undifferentiated basal cells, mucous cells, club cells, and "labial cells."[7] They are similar in appearance to those previously described for other skin areas but are nevertheless distinctive. The undifferentiated basal cells predominate, although mucous and club cells are found in the upper layers and in the glandular pockets. The staining affinities of the epidermal cells of the lips differ somewhat from those previously described for the other skin cells. These differences in staining reactions appear in Table 1.

[7] "Labial cells" are modified mucous cells which differ somewhat in their staining affinities from other mucous cells and are most prominent in the labial glands.

Experimental studies by Thomson (1963, 1964) suggested that the labial glands of *Ostracion* produce a toxic secretion and therefore should be considered as poison glands. Since the labial glands are not associated with a venom apparatus, they can be classified as ichthyocrinotoxic structures. It has also been suggested that perhaps two or more different toxins are being secreted by these various cell types since their staining affinities are not the same. However, the differences (if any) have not been substantiated.

Arothron hispidus: The following résumé of the histology of the skin of *Arothron hispidus* is based largely on the reports of Rosen (1913*a, b*) and Eger (1963).

The skin of *A. hispidus* is comprised of three main layers: the epidermis, the dermis, and the subcutaneous tissue. A prominent feature of the integument is the presence of numerous cartilaginous spines or prickles which are distributed over the body but are most concentrated in the ventral region of the fish. Since the dermis and subcutaneous tissues are not concerned with poison production, no attempt will be made to describe them.

The prickles of puffers are dermal in origin. The base of the prickles are deeply imbedded in the dermis, and the shaft of the prickles passes through a canal of loose connective tissue to the surface of the skin. All of the prickles appeared to have a covering of epidermis, and in no case was there evidence that the prickle had pierced through the epidermis. The prickle, with its associated epidermis, usually does not project beyond the margin of the adjacent skin surface but lies in pockets formed by invaginations of the epidermis and dermis (Fig. 11a). The distal third of the prickle is enveloped by a thick, whitish mass of glandular tissue which is generally seen

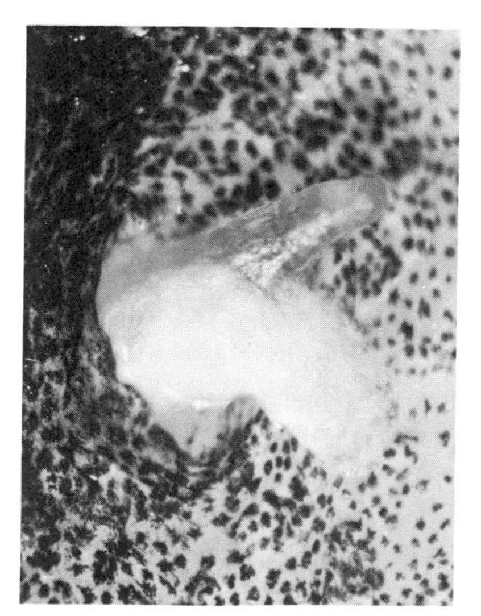

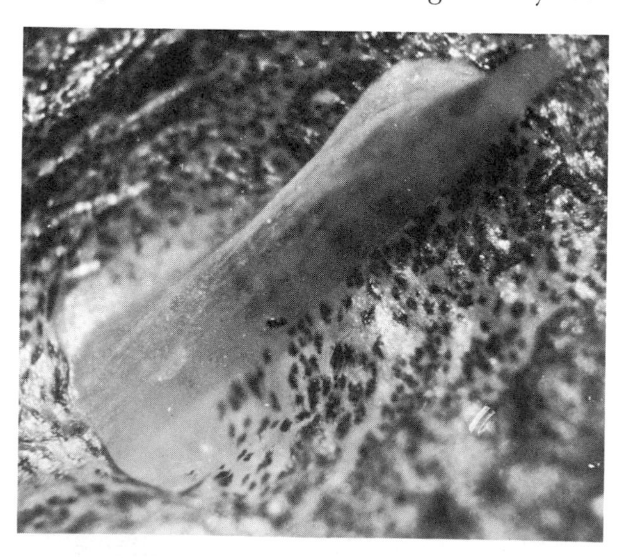

FIGURE 11.—a. (Left) Photograph of a close-up of the skin of *Arothron hispidus* showing a partially extended spine with the associated glandular mass of tissue and integumentary invagination. × 50. b. (Right) A fully extended spine with the associated glandular mass of tissue compressed. × 50. (Courtesy W. H. Eger)

resting in the pocket. When the puffer is disturbed, air or water is taken into the expansible stomach, pushing against the ventral body wall and inflating the fish. As a result of this action the prickles are extended. The glandular tissue mass enveloping the spine becomes stretched and compressed along the sides of the spine (Figs. 11b; 12a, b).

The epidermis of *Arothron* is comprised of stratified squamous epithelium. There are at least three distinct cell types in the epidermis, viz., undifferentiated basal cells, mucous cells, and club cells.[8] Probably sensory cells and possibly other types are present, but no attempt has been made to describe them. Mucous cells are numerous and prominent throughout the epidermis of *Arothron*.

There are noticeable differences in the epidermis in the immediate vicinity of the spines in comparison with the epidermis elsewhere. The epidermis near the base of the spine becomes thinner and invaginates around the spine to form a depression or pocket. The epidermis continues around the distal tip of the spine where it gradually increases in thickness. There are few undifferentiated basal cells in the epidermis in the vicinity of the spines, and mature mucous cells can be seen several layers deep in some sections. In general, the mucous cells around the spines are larger than those in the nonspinous epidermal areas. Club cells are commonly found in the thin epidermal depressions around the spines and in the epidermis enveloping the shaft of the spine. Club cells are infrequently found in the nonspinous epidermis. The club cells may be distinguished from mucous glands by their homogeneous cell contents as contrasted to the more granular, vacuolated contents of the mucous glands, by the more rounded nucleus as contrasted to the flattened nucleus of the mucous cells, and by their characteristic staining affinities. Eger's (1963) results were similar to those reported by Thomson (1963) for the mucous and club cells in *Ostracion* (see Table 1).

Eger (1963) concluded that toxin is produced primarily by the club cells of the epidermis. Because of the association of the glandular epithelium with the spines or prickles of puffers, one might be tempted to conclude that these structures constitute a true venom apparatus. However, in comparing the cartilagenous prickles of a puffer with the highly developed dentinal venom apparatus of elasmobranchs, catfishes, scorpionfishes, weevers, or any other known venomous fish, the marked differences between these structures become very apparent. The prickle apparatus of puffers should not be considered as a true venom organ, but rather as a specialized ichthyocrinotoxic structure.

Batrachoid fishes: Certain members of the toadfish family Batrachoididae contain pectoral fin and axillary glands. These glands are thought to serve as crinotoxic structures since they are not directly associated with a traumagenic device, but appear to have some of the general morphological characteristics

---

[8] Eger (1963) refers to the club cells as "serous glands," but later (1966) stated that these "serous glands" are probably analogous in function to the "club cells" of Rabl (1931) and Thomson (1963).

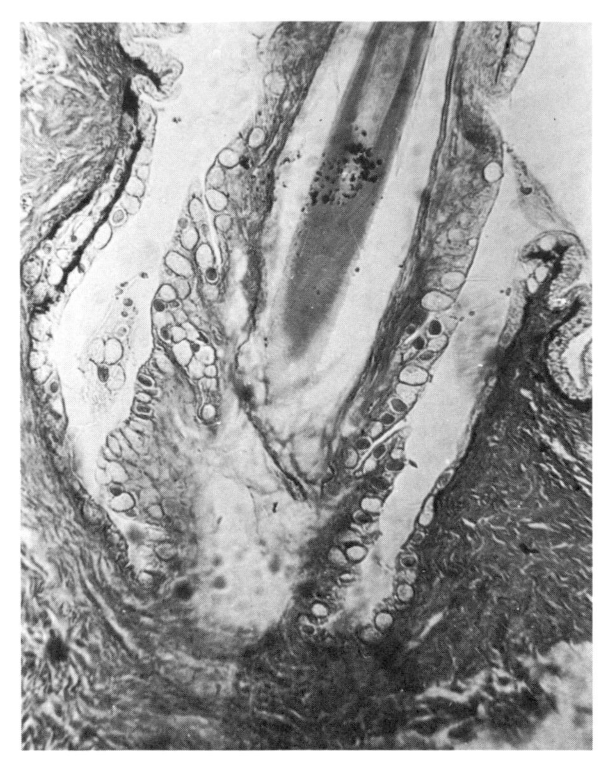

FIGURE 12.—a. (Left) A sagittal section through a skin prickle of *Arothron hispidus* showing the orientation of integumentary invaginations and tissue layers. × 100. Harris' hematoxylin. b. (Opposite page) A cross section through a skin prickle of *Arothron hispidus* showing the associated serous glands (dark stained) and the larger mucous glands. The prickle appears to have been cut at a slight angle. × 100. Gomori's aldehyde fuchsin and trichrome stains.
(Courtesy W. H. Eger)

of crinotoxic organs. Wallace (1893) has given a brief morphological description of the axillary glands of *Opsanus tau*, and a further discussion of the pectoral fin and axillary glands of batrachoid fishes in general appears in Collette's (1966) review of this fish family.

The pectoral fin glands of batrachoid fishes are described as discrete glands on the upper pectoral fin rays (Fig. 13). Unfortunately there are no histological descriptions of these glands available.

The axillary glands of *Opsanus tau* consist of a pouchlike sac situated in the pectoral axilla (Fig. 14a). When the toadfish becomes excited, the pectoral fins are thrown outward and forward until the large foramen is opened to its fullest extent. Whether the axillary glands are especially active at this time has not been determined. If the axillary glands are dissected out from the underlying loose connective tissue, it is found to be a pouch-shaped sac having a tough fibrous wall lined by a layer of epidermal tissue. This tissue is thrown into minute folds having a wrinkled appearance. The cavity is divided into two to five chambers caused by separate invaginations of the skin. Wallace (1893) shows an enlarged view of one of the folds which demonstrates the six different kinds of cells found in the epidermal lining (Fig. 14b, c). His description of these cells is as follows:

(a) The superficial layer is characterized by small, deeply-staining nuclei closely crowded together and is extremely delicate, often torn off in sections, leaving (b), the polygonal cells on the surface. These are much larger than the preceding, the nuclei

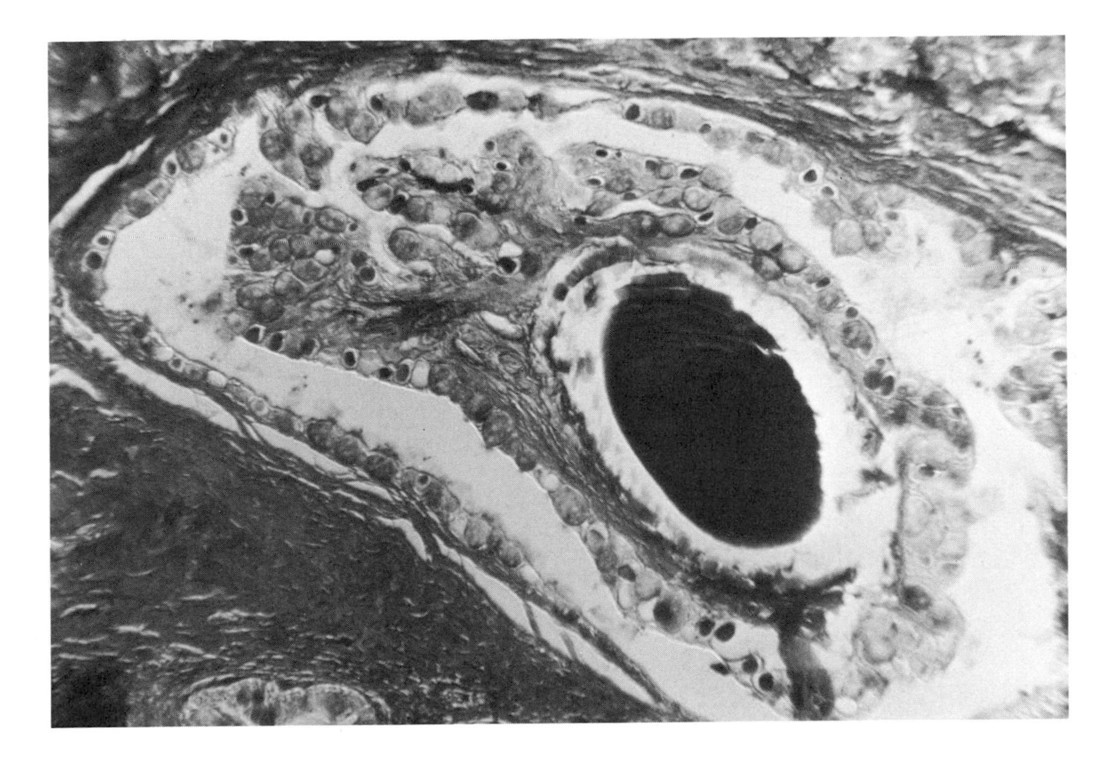

measuring 5μ in diameter and the entire cell having an average diameter of 12μ. The nucleus takes a faint stain and has a distinct but not conspicuous nucleolus. (c) Beneath these, and differing from them only in form, are the elongated cells which are sometimes ovate, sometimes spindle-shaped, and as the cell increases in length the nucleus is seen nearer the upper end rather than at the center. The cells are 35μ in length and the round nucleus has a diameter of 7μ. (d) Resting on the basement-membrane and arranged with comparative regularity, are the columnar cells, 20μ in length and containing a nucleus 7μ in diameter. Neither these nor the elongated cells have a distinct nucleolus. The outline of the cell-wall is sometimes rounded at the top, but more often sends a prolongation into the interstices of the elongated cells lying above. (e) At regular intervals in the polygonal layer of cells and projecting through the superficial layer are the mucous cells, usually so full of well-stained mucin that little idea can be had of their general character beyond their shape and striated contents. If, however, they are carefully taken from living tissue and freshly stained with methyl green, the structure becomes clear. Globate or oval in form, they attain a length of 30μ and have a large, bright nucleus at the base 10μ in diameter. The protoplasm is vacuolated and sends protoplasmic strands through an aperture at the top, the vacuoles, no doubt being filled with transparent mucin. In dead cells found in the secretion of the gland, the exact form of the aperture can be distinctly made out and is usually oval with a perfectly smooth edge. (f) The huge clavate cells vary somewhat in shape according to their position, the majority being rounded at the top with a tapering base. In sections the contents of the cell seem to be an evenly stained, homogeneous substance while in specimens macerated for twenty-four hours in nitric acid or Haller's Fluid distinct spherules are abundant, which are probably dissolved or broken up by long treatment.

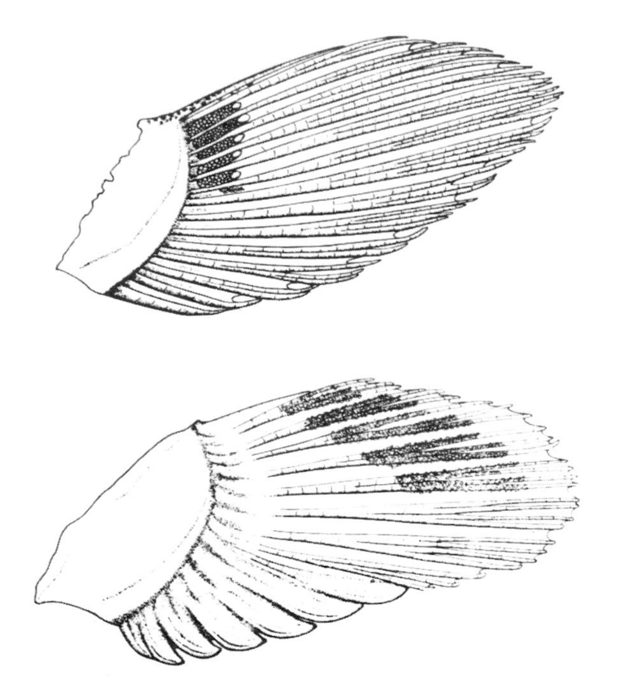

FIGURE 13.—Drawing of the pectoral fins of *Thalassophryne* showing the position of the pectoral fin glands on two different species of toadfishes. Upper: *Thalassophryne dowi* showing discrete glands between the bases of the upper pectoral fin rays. Lower: *T. maculosa* showing the glandular tissue scattered medially on the upper pectoral fin rays. The toxicity of these glandular structures has not been determined, but it is believed that they may serve as crinotoxic structures. (From Collette)

Ichthyocrinotoxic Organs of Other Fishes: Axillary glands are also found in some catfishes, but they appear to be a part of the pectoral venom apparatus since they are directly associated with the traumagenic pectoral spine (*see* Chapter XIX, venomous catfishes, p. 679). Whether the catfish axillary gland also has a crinotoxic function is not known. There is no information available regarding the ichthyocrinotoxic structures of other fishes.

MECHANISM OF INTOXICATION

The means by which persons become intoxicated by ichthyocrinotoxic fishes apparently varies from one species to the next. Very little has been reported regarding the actual mechanism by which these poisons are encountered by man. The slime of some species of cyclostomes is toxic to ingest and may produce an inflammatory reaction if brought in contact with the mucous membranes of humans. The skin of certain species of moray eels and puffers is poisonous to eat. Dermal contact with the slime of *Rypticus saponaceus*, and probably other species of ichthyocrinotoxic fishes, may produce a dermatitis. Clark and Gohar (1953), Brock (1956), Thomson (1964), and others have observed that when trunkfishes *(Ostracion)* are alarmed or under stress they secrete a foamy toxic substance from their mouths and at the base of their fins (Plate 3). This substance is toxic to fish and mice (via intraperitoneal injection) and is said to be toxic when ingested by humans (Brown, 1945).

269

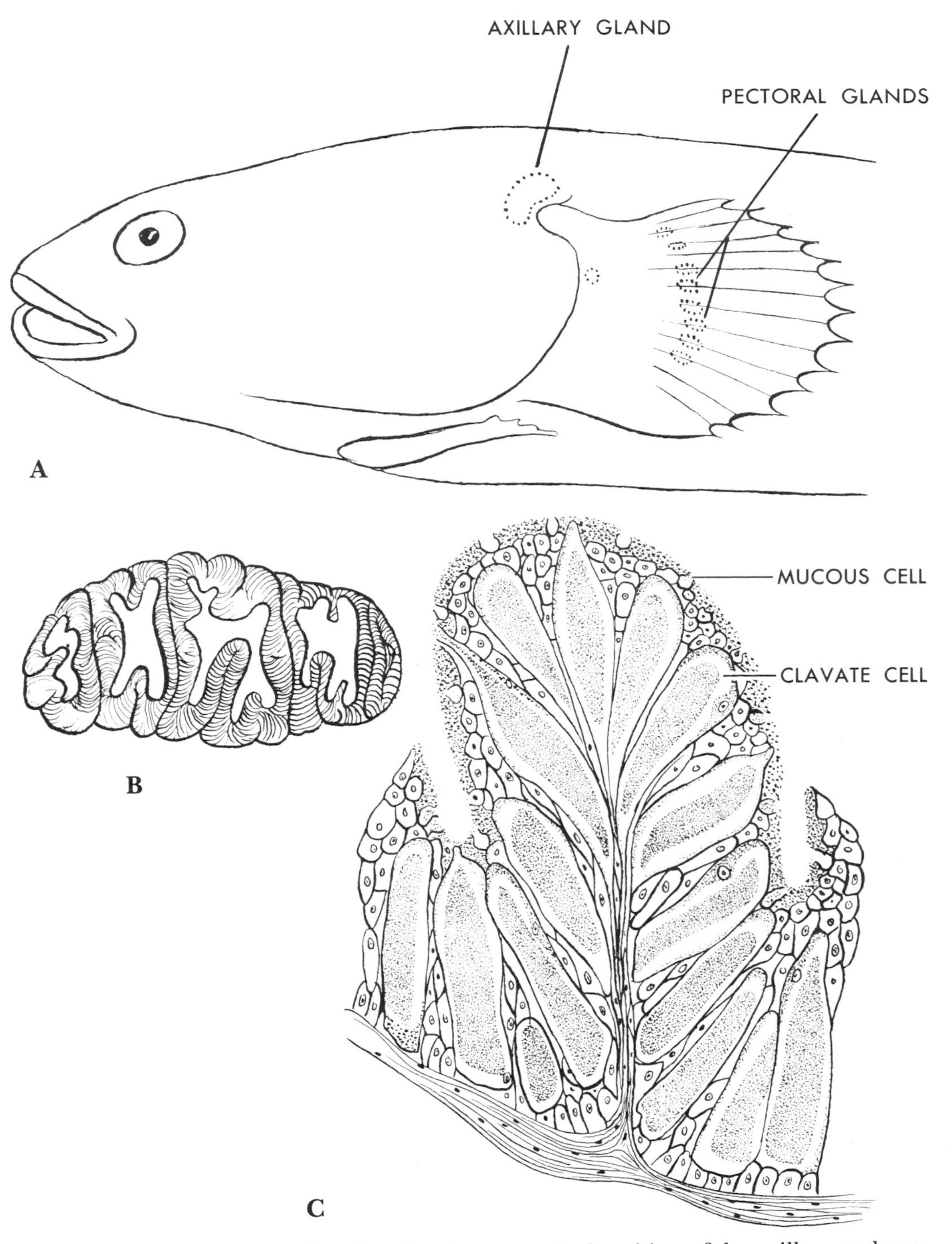

FIGURE 14.—a. (Top) Drawing showing the anatomical position of the axillary and pectoral fin glands of the toadfish *Opsanus tau*. Arrows point to the respective glandular areas. b. (Left) Cross section of the axillary gland of *O. tau* showing the chambers produced by the epidermal folds. c. (Right) Enlargement of one of the epidermal folds showing the various types of cellular structures. (R. Kreuzinger, after Wallace)

MEDICAL ASPECTS

Clinical Characteristics

In the case of cyclostome poisoning it is not certain whether intoxication is due to poisons that are present primarily in the slime, flesh, or both. The symptoms consist of nausea, vomiting, dysenteric diarrhea, tenesmus, abdominal pain, and weakness. Symptoms usually begin to appear within a few hours after ingestion of the fish. Recovery generally takes place within a period of several days. See also Chapter VIII on cyclostome poisoning.

Randall (1963) reported an encounter with the slime of *Rypticus saponaceus*. After spearing the fish he placed it in the front part of his swimming trunks. Within a few minutes he noticed a burning sensation of the urethra and promptly removed the fish. The burning pain was comparable to urethral contact with a strong soap. The irritation disappeared shortly thereafter. There is no information available concerning the toxic effects of other ichthyocrinotoxic fishes other than puffers.

Brown (1945) claims that ingestion of trunkfishes (Ostraciontidae) may result in "unsteadiness of gait similar to drunkenness and the effects may be serious." It is not known whether Brown (1945) was referring to the effects of ostracitoxin or ciguatoxin.

The toxicity of the skin of puffers is well known to public health officials familiar with tetrodotoxications in Japan. Fukuda (1951) found that in a series of 129 cases of puffer poisoning, 10 were due to eating the skin of the fish. The symptoms of puffer poisoning have been discussed at length on p. 461 in Chapter XIV (tetrodotoxic fishes).

Randall *et al.* (1971) have reported that, when the mucus of grammistid fishes comes in contact with the tongue, it has an immediate unpleasant bitter taste and produces a slight stinging sensation. Moreover, when the fish are confined to a small volume of water and agitated, the unpleasant taste may be detected in the water.

Pathology

(Unknown)

Treatment

Symptomatic. See also p. 461, Chapter XIV on the treatment of puffer poisoning.

Prevention

Cyclostomes should be deslimed before eating. Hagfishes and lampreys should be covered with salt and left in a concentrated brine solution for several hours prior to cooking. Care should be taken in handling ichthyocrinotoxic fishes. Fishes having toxic slime may produce skin irritations. The skin of puffers should be removed prior to eating since the skin is one of the most toxic parts of the fish.

PUBLIC HEALTH ASPECTS

The toxicity of the skin of puffers is well known to public health officials familiar with tetrodotoxications. The public health aspects of puffer poisoning have been discussed at length in the section on tetrodotoxic fishes (Chapter XIV). There is no information available regarding the public health aspects of other forms of ichthyocrinotoxications.

TOXICOLOGY

For information on the toxicology of cyclostome poisons, see Chapter VIII.

Liguori *et al.* (1963) have shown that the slime of the golden striped bass *Grammistes sexlineatus* has both antibiotic and toxic properties. Tactile stimulation causes the fish to secrete copious amounts of a white viscid mucus through its skin. Tests were conducted by collecting the mucus, suspending it in distilled water, and lyophilizing it. It was found that both aqueous and 95 percent ethanol extracts of the lyophilized material inhibited the growth of *Escherichia coli* when assayed by the paper disk method on plate cultures at 37° C. The zone of inhibition increased with concentration. The ethanol extractable material was found to be lethal to the topminnow *Fundulus heteroclitus* when immersed in a solution of 70 or more parts per million, killing the fish in 1 hour or less. The symptoms were said to be primarily neurological but were not described. Tests conducted on sea urchin eggs and sperm were quite striking. Various concentrations were capable of immobilizing sperm motility and caused severe cytolysis of unfertilized eggs. Animalization was produced after 5 minutes and 1 hour exposure of the 2-cell stage in solutions of the toxic mucus of 100 parts and 50 parts per million. Six ninhydrin areas were present on paper chromatograms of the ethanol extract, but the eluates of these failed to show either the antibiotic or the toxic activity, possibly because of the low concentrations.

Maretzki and Castillo (1967) have investigated the toxic properties of the slime of the soapfish, *Rypticus saponaceus*. They found that if a single specimen of soapfish was placed in a tank or live well with other fishes, these others all died, apparently due to the toxic slime of the soapfish. They collected the slime by placing the soapfish in a small volume of sea water and then agitating the container. Under these conditions the fish became very slimy to the touch, and the water turned foamy and opalescent. The soapfish itself survived the treatment if it was washed with water before returning it to fresh sea water. Intraperitoneal injection of the crude toxic solution in mice resulted in motor unrest, convulsions, and death. However, they found that the most convenient routine assay procedure was with the use of the guppy, *Lebistes reticulatus*. If guppies were placed in fresh sea water, they survived without any evidence of ill effects for at least 24 hours but were extremely sensitive to relatively small concentrations of sea water containing soapfish toxin. Immediately after the addition of soapfish toxin the guppies underwent a period of motor excitation or hyperactivity. Gradually the guppies became quiet and rested at the bottom of the tank. If the concentration of the toxin was low, the guppies slowly regained their strength and within 1 or 2 hours returned to normal activity. With higher concentrations the guppies lost their sense of balance, swam to the surface occasionally, and tumbled back; death occurred within 20 to 30 minutes. With even more concentrated solutions guppies lost their sense of balance almost immediately, and death followed in 3 to 5 minutes. Small reef fishes exposed to the toxin exhibited identical signs. Solutions of the toxins at concentrations which were initially lethal to a group of guppies had a much less pronounced effect on a subsequent lot of fish placed in the same tank. It appeared that the toxin was ingested, absorbed, or somehow inactivated by the fish. But even if no fish was placed in the tank, the toxin gradually lost its activity.

Clark and Gohar (1953), Brock (1956), and Thomson (1964) reported that boxfish of the genus *Ostracion* produced a toxic mucous secretion of the skin when placed under stress. The toxic properties of the slime of the boxfish, *Ostracion meleagris* have been investigated by Thomson (1964). He found that placing a freshly captured and highly excited boxfish in an aquarium with other fishes often resulted in the death of all the other fish inhabitants within a few minutes. It was also observed that boxfish out of water secreted a copious watery mucus which foamed profusely on agitation. The toxic nature of this mucus was demonstrated by rinsing the skin secretions of the boxfish into aquaria containing other reef fishes. The signs exhibited by fishes so exposed were initial irritability and gasping, followed by quiescence, decreased rate of opercular movements, loss of equilibrium and locomotion, sporadic convulsions, and death. No recovery occurred once the initial symptoms appeared. A standard bioassay was developed by adding a portion of the total rinse volume to 100 ml of sea water containing four to six newborn sailfin mollies, *Mollienesia latipinna*, 10 to 12 mm long. The mean survival time was used as an index of toxicity.

The boxfish toxin was extracted by pooling the filtered aqueous rinses of six to eight boxfishes, centrifuging at 16,000 rpm, and dialyzing against tapwater. The clear supernatant fluid retained its original toxicity, but the precipitate and dialyzate were nontoxic. Heating the toxic supernatant fluid produced further precipitation but no loss in toxicity. Moreover, boiling to dryness caused no appreciable loss in toxicity of the residue. The crude toxin residue was soluble in water, methanol, ethanol, acetone, and chloroform, but insoluble in diethyl ether and benzene. The toxin was stable in acid (pH 2.0) and mildly basic solutions (pH 11.0) but was rapidly detoxified by strong bases (KOH). There was some evidence of bacterial detoxification since the toxicity of fresh secretions gradually decreased at refrigeration temperatures, whereas boiled solutions could be maintained for months at the same temperatures with little loss in toxicity. Repeated extraction of the dried residue with acetone or chloroform and diethyl ether resulted in a particular substance capable of producing stable foams in aqueous solutions. This material was toxic to fish at concentrations of 1:1 million. Approximately

50 to 100 mg of this crude toxin (ostracitoxin) could be obtained at one time from the skin secretions of a single adult boxfish. Ostracitoxin could be detected only in the epidermal mucous secretions of living, distressed boxfish. Aqueous and ethanolic extracts of the skin, viscera, and muscles of freshly killed boxfish were surprisingly nontoxic. It was believed that the poison was "activated" during the process of secretion. It was further noted that boxfish differ from puffers in that boxfish are sensitive to their own poison. If fresh boxfish slime was injected intramuscularly into a boxfish, there was an immediate loss of balance, and death occurred within a few minutes. On the other hand, puffers are immune to tetrodotoxin, even though other fishes are sensitive to the poison (Larson *et al.,* 1960; Eger, 1963).

Ostracitoxin caused desensitization of sea anemone and hydroid tentacles, inhibition of cleavage of sea urchin blastomeres and was generally toxic to invertebrates. Fishes were found to be highly susceptible to ostracitoxin and were rapidly killed when immersed in sea water containing this poison. Intraperitoneal injections of ostracitoxin into mice caused ataxia, labored breathing, coma, and death. The MLD in mice was 0.2 mg/g of mouse. Sublethal injections of the poison into mice caused signs of poisoning, but recovery was complete, which was unlike its irreversible action on fishes.

Ostracitoxin caused hemolysis of vertebrate red blood cells *in vitro* at concentrations as low as 1.0 part per million.

The toxicology of puffer poison appears in Chapter XIV, on p. 473.

Hashimoto and Kamiya (1969) investigated the toxin found in the skin of the Indo-Pacific soapfish, *Pogonoperca punctata.* Encapsulated skin extracts that were force-fed to cats or injected subcutaneously induced ciguatera-like effects. The cats developed vomiting, diarrhea, hypersalivation, loss of motor activity, coma, and paralysis within a few days. Extracts of the muscle tissue of the fish were nontoxic on the basis of 10 percent of body weight. Intraperitoneal injections in mice resulted in paralysis of the hind limbs, respiratory distress, and death. The MLD was 320 μg/gm, which was equivalent to 70 mg of the raw skin. Death time was 24 to 48 hours. When the Japanese ricefish, or medaka *(Oryzias latipes),* was placed in a solution of the grammistid toxin, the fish showed motor excitation, loss of equilibrium, and death. The MLD to the fish was 0.78 mg per 50 ml (or 15.6 ppm). Hemolytic activity was demonstrated using a 2 percent rabbit blood cell suspension. The hemolytic activity was detected from extracts made from the liver. Extracts from the flesh and other viscera did not display such activity. The toxin was only partially extractable with water, physiological saline, 10 percent acetic acid, or hot ethanol, but effectively with hot 70 percent ethanol. It was soluble in *n*-butanol and *n*-amyl alcohol but insoluble in diethyl ether, petroleum ether, or chloroform. In an alkaline medium (pH 12) the toxin was heat labile. About 80 percent of the hemolytic activity was lost on heating in a bath of boiling water for 15 minutes. No marked loss of activity was observed when heated on the acidic side (pH2).

Randall *et al.* (1971) developed a taste test in order to perform a quick screening method for grammistid toxin. They found that placing a small amount of the mucus of soapfishes on one's tongue resulted in an immediate unpleasant bitter taste and a slight stinging sensation. They also found that, when the fish was confined to a small volume of water and agitated, the taste could be detected in the water. This taste test provided a rapid method of initial screening for grammistid toxin. The water in which the fish is confined also forms a long-lasting foam like a soap solution when shaken vigorously.

Randall and his associates also tested the ichthyotoxic activity of grammistid toxin derived from *Grammistes sexlineatus, Aulacocephalus temmincki*, and *Diploprion bifasciatum* on nongrammistid fishes in a manner similar to that carried out by Hashimoto and Kamiya (1969) on *Pogonoperca punctata*.

When a live *Grammistes sexlineatus* 93 mm in standard length (SL) was kept in 150 ml of distilled water for 1.5 minutes, the water became turbid and foamy. Five individuals of *Oryzias latipes* placed in this water died in 11 minutes on the average.

One *Diploprion bifasciatum* 158 mm in standard length was agitated in 800 ml of seawater for 5 minutes. Five adults of *O. latipes* were then placed in this water. They showed immediate motor excitation and died within 15 minutes. Although a freshwater species, *O. latipes* can ordinarily survive in seawater for at least a day.

An individual of *Aulacocephalus temmincki* about 200 mm in standard length was kept under agitation in 400 ml of seawater for 3 minutes. One *O. latipes* was killed in 81 minutes when placed in this water, and a girellid fish *(Girella punctata)* succumbed in 63 minutes in the same water.

Tests were also made of the effect of partially purified toxin from the mucus and skin of the following soapfishes on live *O. latipes: P. punctata, G. sexlineatus, D. bifasciatum, A. temmincki, G. ocellatus*, and *R. saponaceus*. The tests were run with the same *n*-butanolic extracts as used for the investigation of hemolytic activity (see below). One saponin unit of toxin (determined from the hemolytic activity equivalent to 1 mg of standard saponin) from each of the six species was dissolved in separate containers containing 50 ml of distilled water at 20° C. Five live medaka were placed in each container and the death time of the fish recorded. The results are shown in Table 2. They indicate that the relationship between hemolytic and ichthyotoxic activity is similar regardless of the species of soapfish.

TABLE 2.—*The Death Time of Medaka* (Oryzias latipes) *in Solution of Skin Toxin from Grammistid Fishes.*

Species	*Pogonoperca punctata*	*Grammistes sexlineatus*	*Diploprion bifasciatum*	*Aulacocephalus temmincki*	*Grammistops ocellatus*	*Rypticus saponaceus*
Death time (min.)	68, 80, 84 84, 100	61, 64, 68 78, 84	84, 88, 90 91, 95	68, 71, 71 72, 111	47, 95, 102 116, 120	72, 95, 103 125, 130
Average	83	69	89	79	96	105

Effect of a Soapfish on a Predator: On December 4, 1968, it was noted that a *Grammistes sexlineatus* 21 mm in standard length was eaten during the night by another fish of the same species 38 mm in standard length in an aquarium at the Eniwetok Marine Biological Laboratory, Eniwetok Atoll, Marshall Islands. Since grammistids do not die in water in which they have been provoked into liberating skin toxin which kills other fishes in the same container, it is presumed that they are tolerant of their own toxin. They would not, therefore, be expected to be deterred by the toxin in feeding. As mentioned previously, it has been hypothesized that the skin toxin of grammistids is repelling to predators. To test this, the 38-mm *G. sexlineatus* was placed in a large aquarium at Eniwetok which contained a lionfish *(Pterois volitans)* about 150 mm in standard length which was routinely fed small live fishes as food. Before the little grammistid had reached the bottom of the aquarium, the lionfish was already in slow pursuit. It soon seized the soapfish in its jaws, but immediately expelled it. The lionfish made two expelling movements of its mouth after the soapfish was released, suggesting that a lingering bad taste was being experienced. No further attempt was made by the *Pterois* to eat the *Grammistes.*

Hemolytic Activity: The skin of grammistids causes hemolysis of rabbit blood cells. Hemolytic activity was tested for six species of these fishes. For *Pogonoperca punctata, Grammistes sexlineatus*, and *Grammistops ocellatus* the test solutions were prepared by first grinding up the skin without removing the mucus and washing with distilled water. The residue was then extracted three times with hot 70 percent ethanol for 30 minutes. The three extracts were combined and concentrated *in vacuo*, acidified with acetic acid, and defatted with diethyl ether. The residue was then dissolved in distilled water. Some of this solution was used for testing, and some was extracted three times with *n*-butanol; the three *n*-butanol layers were combined, the residue dissolved in distilled water, and this also used for testing.

The test solutions for *Diploprion bifasciatum, Aulacocephalus temmincki*, and *Rypticus saponaceus* were made from the mucus and then the skin with the external mucus removed. The mucus was carefully scraped off the skin, suspended in 50 ml of distilled water and shaken three times with an equal volume of *n*-butanol. The *n*-butanol layers were combined and evaporated to dryness under reduced pressure. The residue was dissolved in distilled water and acidified with acetic acid. The skin freed from slime was treated in the same manner as the first three species mentioned.

Hemolytic activity was determined with a 2 percent rabbit blood cell suspension. The results (Table 3) are expressed in terms of saponin units (S.U.). One saponin unit is the hemolytic activity equivalent of 1 mg of standard saponin.

For the three species in which a separate test was made on the mucus, it is evident that much more toxin is found in the mucus than the skin alone. Considerable individual variation is evident within a species, undoubtedly reflecting in part the treatment of the individual fishes prior to their death.

TABLE 3.—*The Hemolytic Activity of Skin and Mucous Toxins of Grammistid Fishes*

Species	No.	Skin			Mucus	
		Hot 70% EtOH Ext.		n-BuOH Ext.	n-BuOH Ext.	
		S.U./g skin	S.U./mg	S.U./mg	S.U./g mucus	S.U./mg
Pogonoperca punctata	3	17.4	0.75	0.78		
	4	51.8	0.99	1.70		
	5	3.5	0.10	0.21		
	6	8.0	0.26	0.53		
	7	19.2	0.41	0.58		
	8	15.6	0.27	0.50		
	9	19.5	0.22	0.91		
	10	20.5	0.23	0.94		
Grammistes sexlineatus	6	31	0.36	2.3		
	7	25	0.31	0.86		
Diploprion bifasciatum	1	0.22	0.004		1.2	0.15
	2	0.18	0.003	0.02	0.8	0.05
	3	0	0.003		1.8	0.09
Aulacocephalus temmincki	1	0.14	0.002		1.0	0.03
	2	0.30	0.004		1.9	0.08
	3	0.21	0.003	0.16	3.2	0.08
Grammistops ocellatus	1	1.67	0.15	0.60	—	—
Rypticus saponaceus	1	0.21	0.04	0.15	1.6	0.04

Nevertheless, it seems evident that more toxin may be found per unit of skin in *Grammistes sexlineatus* and *Pogonoperca punctata* than in the other species.

The same procedure for testing hemolytic activity of *G. sexlineatus* and *P. punctata* was carried out for a specimen of the pseudogrammid fish *Suttonia lineata* from Hawaii 75 mm in standard length. No hemolysis of the rabbit blood cells was detectable for this species.

Hashimoto, Shiomi, and Aida (1974) isolated a toxin from the epidermis of gobies *Gobiodon spp.* which produced a stinging and bitter taste, toxic to a fish, and had hemolytic activity.

PHARMACOLOGY

No pharmacological studies have been reported dealing specifically with ichthyocrinotoxins, with the single exception of tetrodotoxin. See Chapter XIV on the pharmacology of puffer poison.

CHEMISTRY

The chemistry of the toxin from the slime of the soapfish, *Rypticus saponaceus*, has been studied by Maretzki and Castillo (1967). Soapfish toxin is non-dialyzable and insoluble in solutions of low ionic strength. Crude toxin solutions at neutral pH undergo rapid inactivation but are stable at pH 3-4. The toxic principle cannot be extracted with ethyl ether, chloroform, or methyl acetate, but is soluble in water saturated *n*-butanol. Crude toxin solutions subjected to 65° C for 2 hours retained a high toxicity level. These studies suggested that the toxin is, or is associated with, a protein or a polypeptide.

Randall *et al.* (1971) utilized thin-layer chromatography preparations obtained by *n*-butanol extraction for determination of hemolytic activity. The preparations resulted in several components, each of which was positive to Dragendorff reagent. The toxic components were located on the cellulose plate by spraying with 2 percent rabbit blood cell suspension. The spots having hemolytic activity were almost the same as those positive to Dragendorff reagent. They found that the patterns of the chromatograms for the different species were similar to each other.

Preparations of skin toxin of three grammistid fishes, *Grammistes sexlineatus, Pogonoperca punctata*, and *Rypticus saponaceus* were further purified by silicic acid column chromatography by Randall and his associates. They also attempted to purify the skin toxin from *Diploprion bifasciatum*, but they were able to obtain only a preparation of low purity. Although there were variations in the relative amounts of the various amino acids, the four dominant acids, namely, leucine, isoleucine, phenylolanine, and glycine, were the same for all four species of fish. There were differences in the composition of the amino acids, ultraviolet and infrared spectra, but they were unable to determine to what extent the differences may have been due to variations in purity rather than the chemical structure of the toxin. Hashimoto and Oshima (1972) fractionated grammistins A, B, and C from *Pogonoperca punctata* by means of countercurrent distribution and Sephadex LH-20 column chromatography. They were different from each other in both hemolytic activity and amino acid composition. These components are peculiar polypeptides having an unknown moiety positive to Dragendorff reagent.

Hashimoto, Shiomi, and Aida (1974) extracted the toxin of *Gobiodon quinquestrigatus* with ethanol and purified by precipitation at 25 percent saturation of NaCl, countercurrent distribution and Sephadex LH-20 column chromatography. Thin layer chromatography preparations still gave three spots, each positive to Dragendorff reagent, ninhydrin, and each having hemolytic activity. The toxin was easily dialyzable through a cellophane membrane and showed an absorption band at around 260 nm. The amino acid composition was similar to that of the 1-butanolic extract from the mucus of the soapfish *Pogonoperca punctata* and the toxin was assumed like grammistin to be a mixture of peptides containing a tertiary or quaternary amine moiety. When the whole fish or epidermis was stored without preheating, the toxin disappeared rapidly, probably due to enzymatic degradation.

LITERATURE CITED

Aida, K., Hibiya, T., Mitsuura, N., Kamiya, H., and Y. Hashimoto
1973 Structure of the skin of the soapfish *Pogonoperca punctata*. Bull. Japanese Soc. Sci. Fish. 39(12): 1351.

Andrew, W.
1959 Textbook of comparative histology. Oxford Univ. Press, New York, 652 p.

Blackstad, T. W.
1963 The skin and the slime glands, p. 195-230, 17 figs. *In* A. Brodal and R. Fänge [eds.], The biology of myxine. Universitetsforlaget, Oslo.

Breder, C. M., Jr., and E. Clark
1947 A contribution to the visceral anatomy, development, and relationships of the Plectognathi. Bull. Am. Mus. Nat. Hist. 88(5): 287-320, 8 figs.; Pls. 11-14.

Brock, V. E.
1956 Possible production of substances poisonous to fishes by the boxfish, *Ostracion lentiginosus* Schneider. Copeia (3): 195-196.

Brown, H. H.
1945 Fish poisoning, p. 34-37. *In* The fisheries of the Windward & Leeward Islands. Develop. Welfare West Indies, Bull. No. 20.

Clark, E., and H. A. Gohar
1953 The fishes of the Red Sea: order Plectognathi. Publ. Marine Biol. Sta. Al Ghardaqa (Red Sea) No. 8, 80 p., 5 pls., 22 figs.

Collette, B. B.
1966 A review of the venomous toadfishes, subfamily Thalassophryninae. Copeia (4): 846-864, 12 figs.

Darwin, C. R.
1945 The voyage of the *Beagle*. Rep. ed. E. P. Dutton & Co., New York. 496 p.

Day, F.
1878-88 The fishes of India. 2 vols., illus. London.

Eger, W. H.
1963 An exotoxin produced by the puffer, *Arothron hispidus*, with notes on the toxicity of other plectognath fishes. M.S. Thesis. Univ. Hawaii. 88 p.

Engelsen, H.
1922 Om giftfisk og giftige fisk. Nord. Hyg. Tidskrift 3: 316-325.

Flaschentrager, B., and M. M. Abdallah
1957 Some toxic fishes of the Red Sea. Alexandria Med. J. 3(2): 177-188.

Fredericq, L.
1924 Die sekretion von schutz- und nutzstoffen, p. 1-87. *In* H. Winterstein [ed.], Handbuch der vergleichenden physiologie. Vol. II. G. Fischer, Jena.

Fukuda, T.
1951 Violent increase of cases of puffer poisoning. [In Japanese] Clinics and Studies 29(2).

Gage, S. H., and M. Gage-Day
1927 The anti-coagulating action of the secretion of the buccal glands of the lampreys *(Petromyzon, Lampetra,* and *Entosphenus)*. Science 66(1708): 282-284, 2 figs.

Gimlette, J. D.
1923 Malay poisons and charm cures. 2d ed. J. and A. Churchill, London. 260 p.

Gosline, W. A., and V. E. Brock
1960 Handbook of Hawaiian fishes. Univ. Hawaii Press, Honolulu. 372 p., 273 figs.

Halstead, B. W., and D. W. Schall
1955 A report on the poisonous fishes captured during the Woodrow G. Kreiger expedition to the Galapagos Islands, p. 147-172, 1 pl. *In* Essays in the natural sciences in honor of Captain Allan Hancock. Univ. Southern Calif. Press, Los Angeles.

Hardy, A. C.
1956 The new naturalist. The open sea, its natural history. The world of plankton. Collins, London. 335 p.

Hashimoto, Y., and H. Kamiya
1969 Occurrence of a toxic substance in the skin of a sea bass *Pogonoperca punctata*. Toxicon 7(1): 65-70, 2 figs.

Hashimoto, K., and Y. Oshima
1972 Separation of grammistins A, B, and C from a soapfish *Pogonoperca punctata*. Toxicon 10(3): 279-284, 4 figs.

Hashimoto, Y., Shiomi, K., and K. Aida
1974 Occurrence of a skin toxin in coral-gobies *Gobiodon spp.* Toxicon 12: 523-528.

Jordan, D. S., Evermann, B. W., and H. W. Clark
1930 Check list of the fishes and fishlike vertebrates of North and Middle America north of the northern boundary of Venezuela and Colombia. Rept. U.S. Comm. Fish. 1928. 670 p.

Lalone, R. C., DeVillez, E. D., and E. Larson
1963 An assay of the toxicity of the Atlantic puffer fish, *Spheroides maculatus*. Toxicon 1: 159-164.

Larson, E., Lalone, R. C., and L. R. Rivas
1960 Comparative toxicity of the Atlantic puffer fishes of the genera *Spheroides, Lactophrys, Lagocephalus* and *Chilomycterus*. Federation Proc. 19(1): 388.

Liguori, V. R., Ruggieri, G. D., Baslow, M. H., Stempien, M. F., Jr., and R. F. Nigrelli
1963 Antibiotic and toxic activity of the mucus of the Pacific golden striped bass *Grammistes sexlineatus*. Am. Zool. 3(4): 546.

Maass, T. A.
1937 Gift-tiere. *In* W. Junk [ed.], Tabulae biologicae. Vol. XIII. N. V. Van de Garde and Co's Drukkerij, Zaltbommel, Holland. 272 p.

Macomber, R. D.
1956 An observation on pufferfish toxin. J. Wash. Acad. Sci. 46(3): 85.

Maretzki, A., and J. del Castillo
1967 A toxin secreted by the soapfish *Rypticus saponaceus*. Toxicon 4: 245-250, 1 fig.

Nigrelli, R. F.
1958 Dutchman's baccy juice or growth-promoting and growth-inhibiting substances of marine origin. Trans. N. Y. Acad. Sci., II, 20(3): 248-262.
1962 Antimicrobial substances from marine organisms. Introduction: The role of antibiosis in the sea. Trans. N. Y. Acad. Sci., II, 24(5): 496-497.

Pawlowsky, E. N.
1927 Gifttiere und ihre giftigkeit. G. Fischer, Jena. 515 p., 170 figs.

Rabl, H.
1931 I. Integument der anamnier, p. 271-345. *In* L. Bolk, E. Goppert, E. Kallium, and W. Lubosch [eds.], Handbuch der vergleichenden anatomie der wirbeltiere. Urban and Schwarzenberg, Berlin.

Randall, J. E.
1963 Regarding the toxic effects of the slime of the soapfish *Rypticus saponaceus*. (Personal communication, Sept. 19, 1963.)

Randall, J. E., Aida, K., Hibiya, T., Mitsuura, R., Kamiya, H., and Y. Hashimoto
1971 Grammistin, the skin toxin of soapfishes, and its significance in the classification of the Grammistidae. Publ. Seto mar. Biol. Lab. 19(2/3): 157-190.

Rosen, N.
1913*a* Studies on the plectognaths. The integument. Arkiv Zool. Stockholm 8(10): 1-29.
1913*b* Studies on the plectognaths. The body muscles. Arkiv Zool. Stockholm 8(18): 1-14.

Schreiner, K. E.
1916 Zur kenntnis der zellgranula. Untersuchungen über den feineren bau der haut von *Myxine glutinosa*. I. 1st part. Arch. Mikr. Anat. (Abt. I) 89: 79-188.
1918 Zur kenntnis der zellgranula. Untersuchungen über den feineren bau der haut von *Myxine glutinosa*. I. 2d part. Arch. Mikr. Anat. (Abt. I) 92: 1-63.

Schultz, L. P., Herald, E. S., Lachner, E. A., Welander, A. D., and L. P. Woods
1953 Fishes of the Marshall and Marianas Islands. Vol. I. Families from Asymmetrontidae through Siganidae. U.S. Nat. Mus. Bull. 202. 685 p., 74 pls., 89 figs.

Tani, I.
1945 Toxicological studies on Japanese puffers. [In Japanese] Teikoku Tosho Kabushiki Kaisha 2(3), 103 p., 22 figs.

Thomson, D. A.
1963 A histological study and bioassay of the toxic stress secretion of the boxfish, *Ostracion lentiginosus*. Ph.D. Thesis. Univ. Hawaii. 194 p.
1964 Ostracitoxin: an ichthyotoxic stress secretion of the boxfish, *Ostracion lentiginosus*. Science 146(3641): 244-245.

Tyler, J. C.
1965 Regarding the nomenclature of tetraodontid fishes. (Personal communication, December, 1965.)

VAN OOSTEN, J.
1957 The skin and scales, p. 207-244, 10 figs. *In* M. E. Brown [ed.], The physiology of fishes, vol. I. Metabolism. Academic Press Inc., New York.

WALLACE, L. B.
1893 The structure and development of the axillary gland of Batrachus. J. Morphol. 8: 563-568, 1 pl.

WOODLAND, D. J.
1973 Information regarding the correct nomenclature of rabbitfishes. (Personal communication.)

Chapter XXV—VERTEBRATES

Class Reptilia

SEA TURTLES, SEA SNAKES

The class Reptilia is comprised of a group of poikilothermal vertebrates characterized by dry cornified skin with scales or scutes. Typically, there are four limbs, each with five clawed toes which are reduced or absent in some. The skeleton is bony, having one occipital condyle. Reptiles inhabit terrestrial, freshwater, or marine environments in warm, temperate, and tropical regions. There are about 6,000 species.

The living members of the class Reptilia are divided into four orders:

CHELONIA (TESTUDINATA): Turtles, tortoises, and terrapins.

RHYNCHOCEPHALIA: Beaked headed reptiles. Only one species remains, *Sphenodon punctatum*. The other members are known only from fossil remains.

SQUAMATA: Lizards and snakes.

CROCODILIA: Alligators, caimans, gavials, and crocodiles.

Only two of these orders are of interest to marine biotoxicologists: Chelonia, which includes marine turtles that are sometimes deadly poisonous to eat; and Squamata, which includes the venomous sea snakes.

Poisonous Sea Turtles

Reptiles of the order Chelonia (Testudinata) are characterized by a broad body incased in a bony shell comprised of a rounded dorsal carapace and a flat ventral plastron, joined at the sides and covered by polygonal laminae (scutes, scales) or leathery skin. The jaws are edentulous and equipped with horny sheaths. The quadrate bone is united to the skull. The ribs are fused to the shell, and the sternum is absent. All turtles (tortoises, terrapins) are oviparous in their reproduction. Although there are about 265 species in this order, only five marine species of turtles have been reported as poisonous to man.

The biology and systematics of turtles have been discussed by Stejneger (1907), Rooij (1915), Stejneger and Barbour (1933), Carr (1952), Caldwell (1960), Wermuth and Mertens (1961), Nikol'skii (1963), and others.

REPRESENTATIVE LIST OF MARINE TURTLES REPORTED AS POISONOUS

Phylum CHORDATA

Class REPTILIA

Order CHELONIA: Turtles

Family CHELONIIDAE

272 *Chelonia mydas* (Linnaeus) (Pl. 1, fig. a). Green turtle, rock turtle, meat turtle, sand turtle, right turtle, milk turtle (USA), tortuga (Latin America), gal kasbava, mas kasbava, vali kasvava, perr amai, pal amai (Sri Lanka).

DISTRIBUTION: All tropical and subtropical oceans.

SOURCES: Taylor (1921), Loveridge (1945), Halstead (1959).
OTHER NAMES: *Chelonia japonica, Chelonia virgata, Testudo mydas.*

272 *Eretmochelys imbricata* (Linnaeus) (Pl. 1, fig. b). Hawksbill turtle, scute turtle, shell turtle, comb turtle, spectacled turtle, fowl turtle (USA), tortuga (Latin America), pothu kasbava, lelli kasvava, pana kasvava, kanadi kasbava, kukulu kasbava, alunk amai (Sri Lanka), caret (France).

DISTRIBUTION: All tropical and subtropical oceans.

SOURCES: Bierdrager (1936), Deraniyagala (1939), Loveridge (1945), Carr (1952), Halstead (1959), Cooper (1964), Hashimoto, Konosu, and Yasumoto (1967).

Family DERMOCHELIDAE

272 *Dermochelys coriacea* (Linnaeus) (Pl. 1, fig. c). Leathery turtle, leatherback turtle, trunk turtle, harp turtle, luth turtle, ridge turtle, three-ridged turtle, bat turtle, oil turtle, ship turtle, boat turtle, seven-banded turtle (USA, Australia), tortuga (Latin America), dhara kasbava, thun dhara kasbava, vavul kasbava, thel kasbava, navu kasbava, mavalla, dhom amai, yelu vari, amai (Sri Lanka).

DISTRIBUTION: Largely circumtropical but occasionally taken in temperate seas off the coasts of North and South America, Mediterranean area, British Isles, and Japan.

SOURCES: Deraniyagala (1939), Halstead (1959).

BIOLOGY

Family CHELONIIDAE: The green sea turtle *Chelonia* usually inhabits water less than 25 m in depth and prefers areas sheltered by reefs where it feeds on algae. It is also common in bays and lagoons. Occasionally *Chelonia* will make its way into freshwater lakes. Green turtles are sometimes seen basking on reefs and beaches of islands uninhabited by man. They are omnivorous but primarily vegetarian, feeding upon *Cymodocea, Thalassia, Zostera, Halophila,* and other algae. When kept in captivity they seem to show a preference for a diet of meat and fish. Green turtles nest between the latitudes 30° north and 30° south of the equator. They will migrate considerable distances, leaving their usual haunts to get to their breeding grounds. The nest site is usually selected on a beach having loose sand within reach of the waves. When the exact spot is selected, the loose sand is brushed away with the front flippers, but the actual digging is done with the hind ones. About 60 to 190 eggs may be laid at a time. Upon completion of laying, the turtle covers her nest completely with sand. She obliterates her tracks by throwing sand over her back with the front flippers as she moves away. The entire nesting process requires about 2 hours. The breeding season seems to be from July to November in Sri Lanka, but October to mid-February in Australia. This species is considered one of the more valuable turtles for use as food.

The hawksbill turtle *Eretmochelys* is generally found close to land in tropical and subtropical oceans. Seldom does it enter lagoons. Although usually considered carnivorous, it is omnivorous and at times may subsist entirely upon algae. The breeding range is between 25° north and 25° south of the equator. Eggs are laid on sandy beaches in a manner similar to that used by *Chelonia*. As many as 115 eggs or more are laid at a time. The egg-laying season extends from November to February in some areas, but seems to take place during April to June in others. *Eretmochelys* is of commercial importance because of its overlapping scutes which are utilized in the manufacturing of jewelry, etc.

Family DERMOCHELIDAE: The leather turtle *Dermochelys*, said to be the swiftest and the largest of living chelonians, usually inhabits relatively deep water near the edge of the Continental Shelf. Newly hatched leatherbacks head directly for the open ocean and do not return to shallow water until they are ready for egg laying. An adult may attain a weight of more than 780 kg. Their food consists of algae, crustaceans, and fishes. *Dermochelys* is believed to lay eggs three or four times a year, which in Sri Lanka takes place during May to June. The eggs are laid on sandy beaches at night. Often several females will deposit their eggs in close proximity to each other.

For additional information regarding the biology of sea turtles, see Deraniyagala (1939), Carr (1952), and Caldwell (1960).

BIOGENESIS OF CHELONITOXIN

The origin of turtle poison (chelonitoxin) is unknown, but most investigators who have studied the problem appear to be rather consistent in their opinion that the toxin is derived from poisonous marine algae eaten by turtles (Deraniyagala, 1939; Loveridge, 1945; Romeyn and Haneveld, 1956; Pillai *et al.*, 1962). Turtle poisoning has a remarkable resemblance to ciguatera in its sporadicity and spotty geographical distribution; a species of turtle may be safe to eat in one locality but deadly in another. These observations lend support to the idea that turtle poison is derived from the food of the animal.

MECHANISM OF INTOXICATION

Chelonitoxications result from the ingestion of the flesh, fat, viscera, or blood of various species of tropical sea turtle. Toxicity in turtles is sporadic and may occur at any time of the year. The degree of freshness of the turtle meat has no bearing on the toxicity of the organism.

MEDICAL ASPECTS

Clinical Characteristics

The symptoms of chelonitoxication vary with the amount of flesh ingested and the person. Symptoms generally develop within a few hours to several days after eating the turtle. In one large outbreak involving 100 persons, most of the victims developed symptoms about 12 hours after eating the turtle. The initial signs and symptoms usually consist of nausea, vomiting, diarrhea, facial tachycardia, pallor, severe epigastric pain, sweating, coldness of the extremities, and vertigo. There is frequently reported an acute stomatitis consisting of a dry, burning sensation of the lips, tongue, lining of the mouth, and throat. Some victims complained of a sensation of tightness of the chest. The victim frequently becomes lethargic and unresponsive. Swallowing become very difficult, and hypersalivation is pronounced. The oral symptoms may be slow to develop but become increasingly severe after several days. The tongue develops a white coating, the breath becomes foul, and later the tongue may become covered with multiple pinhead-size, reddened pustular papules. The pustules may persist for several months, whereas in some instances they break down into ulcers. Desquamation of the skin over most of the body has been reported (Cooper, 1964). Some victims develop a severe hepatomegaly with right upper quadrant tenderness. The conjunctivae become icteric. Headaches and a feeling of "heaviness of the head" are frequently reported. Deep reflexes may be diminished. Somnolence is one of the more pronounced symptoms present in severe intoxications and is usually indicative of an unfavorable prognosis. At first the victim is difficult to awaken and then gradually becomes comatose, which is followed rapidly by death. The symptoms presented are typical of a hepatorenal death. The overall case fatality rate on reported outbreaks is about 28 percent (Halstead, 1970).

The clinical characteristics of marine turtle poisoning have been discussed by Bierdrager (1936), Deraniyagala (1939), Kinugasa and Suzuki (1940), Romeyn and Haneveld (1956), Pillai *et al.* (1962), and others.

Pathology

The few autopsy reports that are available cite pathological findings consistent with an alimentary toxicosis. Aside from the oral findings described in the

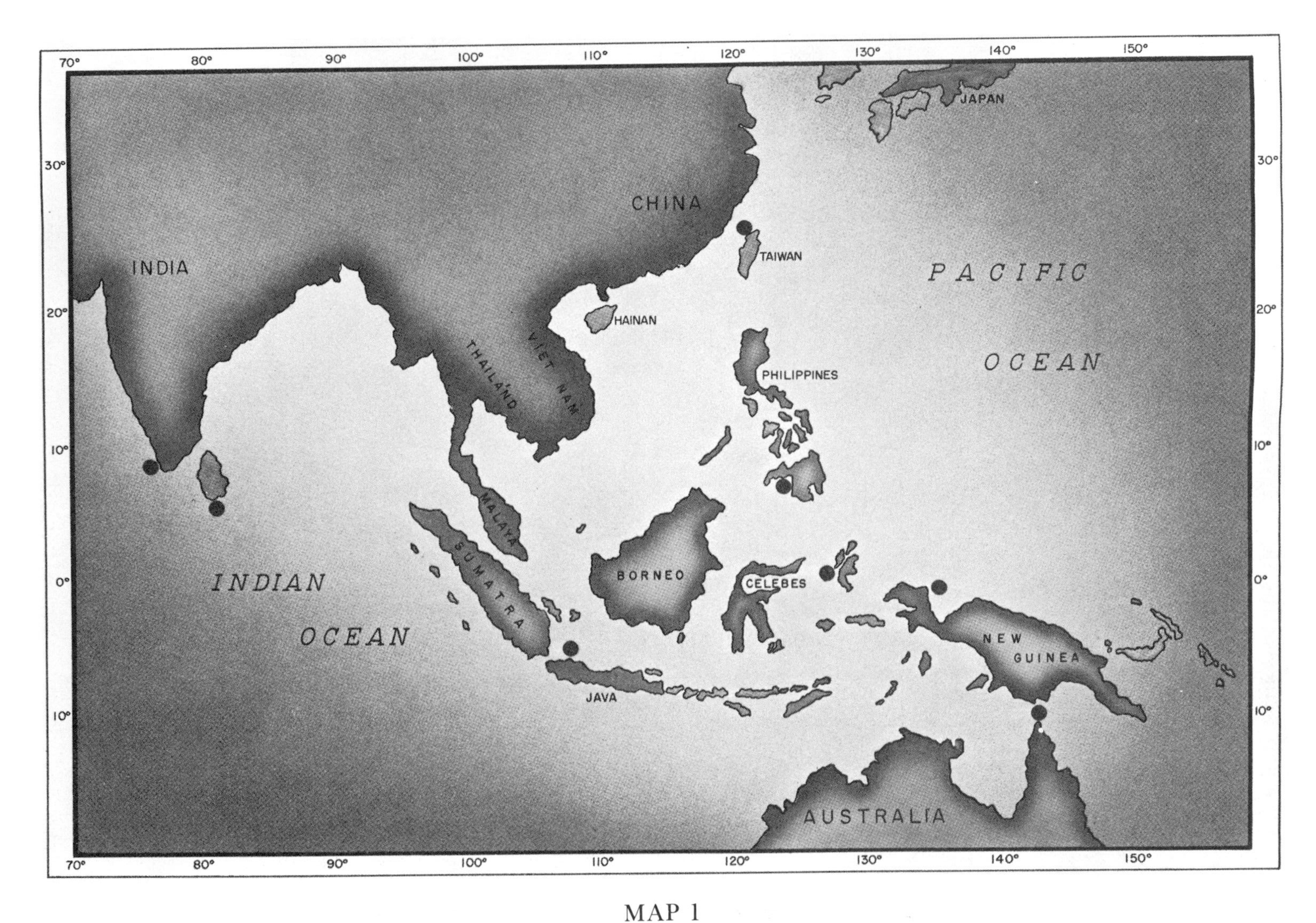

MAP 1
Map showing areas where major outbreaks of sea turtle biotoxications have occurred.
(R. Kreuzinger, from various sources)

preceding clinical section, the esophagus may show ulceration of the mucosa. The serosal surface of the stomach in some cases has revealed dilatation of the blood vessels along the lesser curvature, with scattered ecchymoses along the vessels. The mucosa was generally rough and eroded, but without marked ulcer formation. Pathological changes are quite marked in the intestines. The small intestine showed dilated blood vessels, scattered regions of congestion, and irregular scattered areas of ecchymoses. Most of the changes appear in the proximal portion of the small intestine, but some are observed in the ileocecal region. Aside from dilatation of the coronary arteries, there are no remarkable changes reported for the heart. Liver pathology varies somewhat, but generally the liver is enlarged and the capsule irregular and tense, pale or yellow in color, or else soft and friable. The liver most commonly shows evidence of a massive fatty degeneration, centrilobular congestion, and necrosis of the adjacent liver cells. In some instances the fatty changes are so great as to give the appearance of solid fatty tissue. The kidneys may also vary in appearance from cloudy swelling, fatty degeneration, or necrosis of the tubules.

Autopsy reports on marine turtle intoxications have been published by Bierdrager (1936), Kinugasa and Suzuki (1940), Romeyn and Haneveld (1956), and Pillai *et al.* (1962).

Treatment

There are no known antidotes for chelonitoxin. The treatment is symptomatic.

Prevention

There are no reliable external characteristics which differentiate a poisonous sea turtle from a nonpoisonous one. According to Bierdrager (1936), the natives on the northern coast of west New Guinea claim that toxic turtles have a long neck, black tongue, and black under the chin. However, apparently these natives do not have much confidence in their ability to recognize the edible turtles from the poisonous ones because they test the turtle by feeding parts of the flesh to cats and dogs before they partake of it. If one is going to eat sea turtles in a region where they are reputed to be toxic, the safest procedure is first to feed a sample of the flesh to a dog or cat and then wait at least 24 hours before eating the turtle. Natives frequently succumb to turtle poisoning; nevertheless, one should always check with the local inhabitants concerning the safety of eating turtle meat. If there is the slightest suspicion regarding the edibility of a marine turtle, the turtle should be discarded.

PUBLIC HEALTH ASPECTS

Geographical Distribution

Aside from the somewhat vague report of two cases of marine turtle poisoning in the Windward Islands, West Indies, most of the outbreaks have taken place in the Indo-Pacific and particularly in the Malay Archipelago, southern

Sri Lanka, and southern India (*see* Map 1). It is noteworthy that the distribution of poisonous turtles does not coincide with the general zoogeographical distribution of sea turtles, but the toxic individuals appear to be confined to rather specific localities in the Indo-Pacific region. Areas that are considered to be especially dangerous are the Moluccas Islands; Japen Island, west New Guinea; Quilon, India; and Pantura, Sri Lanka. Numerous outbreaks of turtle poisoning have occurred over the years along the coasts of Mindanao, Philippine Islands, but there are no specific locality data available.

Incidence and Mortality Rate

The overall incidence of chelonitoxications is unknown, but it is quite apparent that the incidence is much higher than the few reports cited indicate. Like ciguatera, most outbreaks of turtle poisoning occur in areas where public health facilities are either minimal or nonexistent. In regions where poisonous sea turtles are endemic they do constitute a definite public health problem to the local populace and must be reckoned with by the transient visitor. The reported outbreaks show a total of 365 persons poisoned, of which 103 persons died, giving a case fatality of about 28 percent.

Seasonal Incidence

It is believed that sea turtle poisoning may occur during any season of the year, but the most dangerous months, apparently, are April through August.

Ecology

Most investigators who have studied the problem of sea turtle poisoning to any extent are convinced that the poison is derived from toxic marine algae. However, ecological and food web studies have not been conducted on this subject (*see* section on Biogenesis of Chelonitoxin, p. 919).

TOXICOLOGY

There appear to be only three toxicological reports in which attempts were made to test the toxic properties of turtle flesh, viscera, or blood experimentally on laboratory animals. Marcacci (1891) injected blood from a large sea turtle (species not given) intravenously and intraperitoneally into dogs. The injections resulted in a progressive paralysis unaccompanied by convulsions. Marcacci's report is too brief to be of much value.

Loisel (1904) prepared alcoholic extracts from the eggs of the European freshwater turtle *Testudo pusilla*. Intravenous injections of the extracts into the ear veins of rabbits resulted in tonic convulsions and death. Loisel did not investigate the toxicity of sea turtles.

Kinugasa and Suzuki (1940) prepared alcoholic extracts from the flesh of a large sea turtle of an unidentified species which had poisoned 57 persons at Koryu, Taiwan. Seven of the victims died. The extracts were injected subcutaneously into two guinea pigs, one mouse, and six frogs. Both the guinea pigs vomited several times about 15 minutes after injection of 20 ml subcutaneously, but none of them died. The mouse was given 2 ml subcutaneously but showed no reaction. The following day the same mouse was given 10 ml of the extract. About 30 minutes after the injection the mouse became restless, rubbed its face with its forepaws, and then vomited a small amount of mucus and undigested food. The mouse showed no further signs but died 30 hours after the injection. Injections of the extracts in the frogs showed stimulation of the vagus nerve and slowing of heart action with small doses (0.5 ml), paralysis of the vagus nerve and increased heart rate with larger doses (1.0), and cardiac arrest with larger injections (10-15 ml).

PHARMACOLOGY

(Unknown)

CHEMISTRY

(Unknown)

Venomous Sea Snakes

Snakes are characterized by a body covering of horny epidermal scales or shields. The quadrate bone is movable with the skull. Vertebrae are usually of the procoelous type. Snakes may be either oviparous or ovoviviparous. They are further modified by the absence of limbs, feet, ear openings, sternum, or urinary bladder. The mandibles are joined anteriorly by ligaments. Their eyes are immobile, covered by transparent scales, and without lids. The tongue is slender, bifid, and protrusible.

The only serpents pertinent to this monograph are the sea snakes of the family Hydrophiidae. Sea snakes have a body more or less compressed posteriorly and a strongly compressed, paddle-shaped tail. Their nostrils, with the exception of *Laticauda*, are situated on the upper surface of the snout and have watertight, valvelike closures. The tongue in sea snakes is short, and only the cleft portion is protrusible. The loreal shield is usually absent. Venom fangs are present, and the maxillary teeth are grooved.

The biology and systematics of sea snakes have been discussed by such authors as Wall (1921), Smith (1926, 1935, 1943), Ditmars (1931), Volsøe (1939), Loveridge (1945), Kinghorn (1961), Taylor (1965), and Dunson (1975).

REPRESENTATIVE LIST OF SEA SNAKES REPORTED TO CAUSE HUMAN ENVENOMATIONS

Phylum CHORDATA

Class REPTILIA

Order SQUAMATA: Lizards and Snakes

Family HYDROPHIIDAE

Enhydrina schistosa (Daudin) (Pl. 2, fig. a). Sea snake (USA, Australia), hogly pattee, valakadyen (India), oelar lempeh (Java), ular laut (Malaysia). *273*

Distribution: North coast of Australia, Vietnam, to the Persian Gulf.

Sources: Rogers (1903), Bokma (1942), Barme (1958), Barme and Detrait (1959), Carey and Wright (1960*a, b*), Reid, Thean, and Martin (1963), and Reid (1975).

Hydrophis cyanocinctus Daudin (Pl. 2, fig. b). Sea snake (USA). *273*

Distribution: Indonesia, Japan, to the Persian Gulf.

Sources: Swaroop and Grab (1954), Reid *et al.* (1963), Werler and Keegan (1963), Homma *et al.* (1964).

Hydrophis obscuris Daudin (Pl. 3, fig. a). Sea snake (USA), chittul, banded sea snake (England). *274*

Distribution: Burma, to the east coast of India.

Sources: Fayrer (1872), Kermorgant (1902), Klemmer (1963).
Other name: *Hydrophis chloris.*

Hydrophis ornatus (Gray) (Pl. 3, fig. b). Spotted sea snake (USA, Australia). *274*

Distribution: Northern Australia.

Sources: Smith (1926), Klemmer (1963).
Other name: *Hydrophis ocellatus.*

Hydrophis spiralis (Shaw) (Pl. 4, fig. a). Banded sea snake (USA), chittul (India). *275*

Distribution: Malay Archipelago, to the Persian Gulf.

Sources: Reid and Lim (1957), Klemmer (1963), Reid *et al.* (1963).

Kerilia jerdoni Gray (Pl. 4, fig. b). Sea snake (USA). *275*

Distribution: Strait of Malacca, to the east coast of the Indian Peninsula, Sri Lanka, Mergui Archipelago.

Sources: Reid and Lim (1957), Klemmer (1963), Reid *et al.* (1963).

Lapemis hardwicki Gray (Pl. 4, fig. c). Sea snake (USA, Australia), oelar lempeh (Indonesia). *275*

Distribution: Southern Japan, to the Mergui Archipelago, coast of northern Australia.

Sources: Reid and Lim (1957), Barme (1958), Barme and Detrait (1959), Klemmer (1963), Halstead *et al.* (1977).

276 *Microcephalophis gracilis* (Shaw) (Pl. 5). Graceful sea snake (USA, Australia).

DISTRIBUTION: Southern China, northern Australia, to the Persian Gulf.

SOURCES: Reid (1956*c*), Reid and Lim (1957), Klemmer (1963).
OTHER NAME: *Hydrophis gracilis.*

277 *Pelamis platurus* (Linnaeus) (Pl. 6). Yellow-bellied sea snake (USA, Australia), kullunder (India).

DISTRIBUTION: Indo-Pacific area, extending from southern Siberia to Tasmania; from the west coast of Central America westward to the east coast of Africa. This species is the most widely distributed of any of the sea snakes.

SOURCES: Fayrer (1872), Wall (1921), Swaroop and Grab (1954), Klemmer (1963).
OTHER NAMES: *Hydrophis bicolor, Hydrus platurus, Pelamis bicolor, Pelamydrus platurus.*

278 *Thalassophina viperina* (Schmidt) (Pl. 7, fig. a). Sea snake (USA).

DISTRIBUTION: Southern China, Malay Archipelago, to the Persian Gulf.

SOURCES: Kermorgant (1902), Klemmer (1963).
OTHER NAME: *Hydrophis nigra.*

278 *Thalassophis anomalus* (Schmidt) (Pl. 7, fig. b). Sea snake (USA).

DISTRIBUTION: Moluccas, Borneo, Java, east Sumatra, Gulf of Thailand.

SOURCES: Fossen (1940), Klemmer (1963).

BIOLOGY

Sea snakes are aquatic inhabitants of the tropical Pacific and Indian Oceans, ranging for the most part from the Samoan Islands westward to the East Coast of Africa, Japan to the Persian Gulf, along the coasts of Asia, through Indo-Australian seas to Australia. One species, *Pelamis platurus*, has an enormous geographical range extending from the west coast of Latin America across the Pacific and Indian Oceans to the East Coast of Africa, and from Southern Siberia to Tasmania (Map 2). With the exception of a single freshwater species, *Hydrophis semperi*, which lives in freshwater Lake Bombon (also called Lake Taal), Luzon, Philippine Islands, all are marine. *Laticauda crockeri* is limited to the brackish water of Lake Tungano, Rennel Island, Solomon Islands.

Sea snakes generally prefer sheltered coastal waters and are particularly fond of river mouths. Around shore, sea snakes may inhabit rock crevices,

Figure 1.—Diver pursuing a large sea snake, probably *Aipysurus laevis* at Heron Island, Great Barrier Reef, Queensland, Australia. Coral reefs are a favorite habitat of sea snakes. This sea snake is venomous, but there are no human fatalities recorded. This is one of the largest of the sea snakes. (R. Taylor, courtesy *Mondo Sommerso*)

tree roots, coral boulders, or pilings (Fig. 1). In regions where sea snakes are plentiful, more than 100 snakes may be taken by fishermen in a single net haul. A factor governing the distribution of most sea snakes seems to be the depth of water in which they feed. The depth must be shallow enough for them to go to the bottom to feed and to rise to the surface for air. However, Herre (1942) claimed that he observed "various species of sea snakes 100 to 150 miles from land." *P. platurus* is a notable example of a sea snake that may at times be found several hundred km out in the open sea. Some of the members of the Laticaudinae are said to be semiterrestrial and never venture far from the tidal zone to which they return to feed (Fish and Cobb, 1954). Sea snakes may sometimes appear swimming together in large groups. Lowe (1932, cited by Werler and Keegan, 1963) encountered the largest concentration of sea snakes ever reported while en route between the Malay Peninsula and Sumatra. A mass of Stoke's sea snakes *(Astrotia stokesi)* was about "10 feet wide and fully 60 miles long." The group was estimated to consist of millions of individuals and was thought to be either migrating or breeding.

With their compressed, oarlike tails, sea snakes are well adapted for

FIGURE 2.—Diver attempting to capture a sea snake *Aipysurus laevis* using a simple noose on the end of a spear. The snake measured 1.5 m. Photo taken at Heron Island, Great Barrier Reef. (B. Cropp, courtesy *Mondo Sommerso*)

locomotion in their marine environment. Instead of the imbricate position of the scales, such as is found in terrestrial snakes, most of the scales of hydrophiids are juxtaposed and hexagonal in position, which further demonstrates adaptation for an aquatic existence. Swimming is accomplished by lateral undulatory movements of the body. Volsøe (1939) pointed out that sea snakes have the remarkable ability to move backward or forward in the water with equal rapidity, but when placed on land, they are awkward and move about with difficulty. Sea snakes are able to float, lying motionless for long periods of time. Volsøe believed that this is accomplished with the use of a large fat body which covers most of the visceral organs, thereby reducing the specific gravity of the snake.

A second organ contributing to their floating ability is an extensive lung which apparently retains a portion of the air at all times. Only the right lung is fully developed, whereas the left lung is vestigial or absent. However, the right lung is very long, extending in some species to the vent. The lung, particularly the posterior portion, is comprised of a thick wall of smooth muscle having a shiny appearance. Volsøe suggested that the posterior portion seemed to have little or no respiratory function and probably served as a hydrostatic organ similar to a fish's bladder. However, there is a difference in that no gas-secreting or gas-absorbing organs have been found in sea snakes, so the hydrostatic function is a rather simple process of inhaling and exhaling air, thereby regulating the buoyancy of the snake. There is no evidence that sea snakes are able to utilize oxygen from the water such as is done by fishes. In addition to the true lung, sea snakes have a well-developed tracheal lung extending from behind the head and reaching to the true lung. This arrangement permits a more complete utilization of the inspired air. The air is taken in through the small nostrils situated upon the dorsum of the upper jaw. Closure of the nostrils is effected by membranous valves in association with special muscles. Moreover, the usual notch in the rostral shield, which permits passage of the protrusible tongue and is always open in land snakes, is provided with a special mechanism for closing. The means by which the snake is able to open its mouth for catching and swallowing a fish underwater is not known. Sea snakes are reported to be able to remain submerged for hours.

Sea snakes capture their food underwater, usually swallowing the fishes head first. A considerable portion of time is spent feeding on or near the bottom, around rocks, or in crevices where they capture eels and other small fishes which are promptly killed with a vigorous bite of their venomous jaws (Klawe, 1964).

The breeding season varies with the locality. Breeding areas are generally located in sheltered, rocky, or cavernous limestone locations. In most regions breeding probably takes place during March through May. The Hydrophiidae are said to be ovoviviparous, and the Laticaudinae are oviparous. The gestation period is about 8 months (Bergman, 1949).

Sea snakes, according to Wall (1921), do not hiss but make a gurgling sound. Günther (1864) claimed that some of the larger species may attain a maximum length of 3 m, but the average is about 1.2 m (*see* Stone and Wall, 1913; Smith, 1926; Werler and Keegan, 1963). Sea snakes are delicate creatures and do not tolerate captivity for very long periods of time. Dunson (1975) has published the most comprehensive report on the biology of sea snakes.

Behavior of Sea Snakes: The disposition of sea snakes is a subject of controversy and one of great practical import to those coming in contact with them. Cantor (1841) quoted Schlegel (1837) as saying that sea snakes are more docile than most venomous terrestrial reptiles. Cantor disagreed with this view, because in his experiences with sea snakes in the Bay of Bengal and the Gangetic estuary, he found them to be "ferocious." The statement is sometimes made in the literature that "sea snakes are not known to attack bathers" (Wall, 1921; Smith, 1943). However, Fayrer (1872) cited the case of a ship's captain who was bathing by his ship at Moulmein, Burma, and was fatally bitten by a sea snake. Rogers (1903) believed that deaths from sea snake bites in Madras were most common among oyster fishermen who were working in shallow water and accidentally stepped on them. Becke (1909) also reported a case in which a native diving for clam shells in Torres Straits was suddenly attacked by a small *Pelamis platurus* which coiled itself around

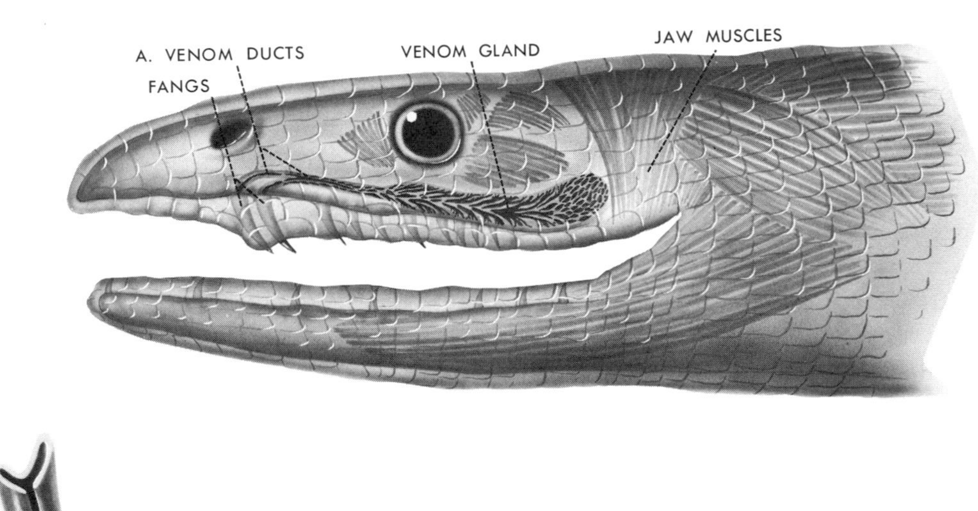

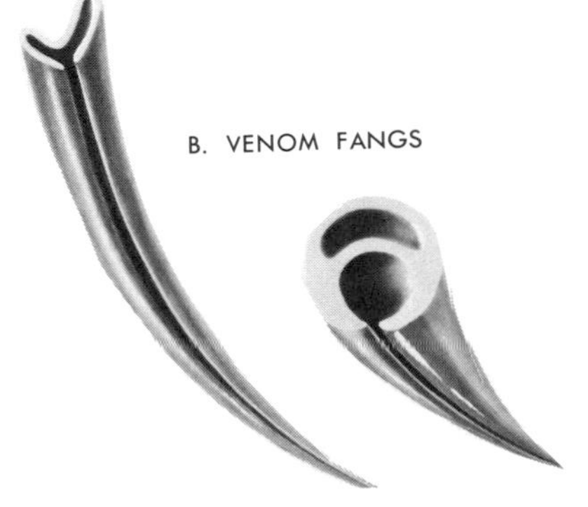

FIGURE 3.—Semidiagrammatic drawing of the venom apparatus of sea snakes.
(L. Barlow, after West)

his forearm and bit him on his forefinger. When provoked, sea snakes will bite viciously (Fitzsimons, 1919; Ditmars, 1931). Wall (1921) claimed that he was unable to get *Enhydrina schistosa* to bite and found it to be of a very gentle nature. On the other hand, Bokma (1942) reported a fatality resulting from the bite of *E. schistosa*, and Tweedie (1941), Le Mare (1952), and others are of the opinion that this species is probably the cause of most sea snake bites and fatalities. Smith and Hindle (1931) stated that *Laticauda colubrina* is exceedingly gentle and can be readily handled without fear. *Hydrophis elegans, H. klossi, H. melanosoma, H. ocellatus, H. spiralis, Kerilia jerdoni*, and *Lapemis hardwicki* are also reputed to be harmless because of their weak striking movement or because of their docility (Gray, 1930). The difference in the gapes of some sea-snake species is also thought to be a striking factor. The gape of *K. jerdoni* is only about 45°, whereas in *E. schistosa* and *H. cyanocinctus* it is said to be equal to that of a cobra (Fayrer, 1872; Fairley, 1934). Herre (1942), who had extensive field experience in the Philippine Islands, believes that most sea snakes tend to be gentle and slow to bite. He found *L. semifasciata* particularly docile, but *H. cyanocinctus* quite vicious. Sea snakes are most likely to be vicious and aggressive during their breeding season. Heatwole (1975) has also reviewed the subject of sea-snake attacks on divers. One can only conclude from these varied opinions that the docility of a sea snake varies with the species, the season of the year, and the manner in which it is approached. One should not become overconfident and take advantage of their so-called gentle nature which is accompanied by a set of fangs and an extremely virulent venom.

MORPHOLOGY OF THE VENOM APPARATUS

The venom apparatus of sea snakes is comprised of two or more functional fangs, their venom glands, and associated musculature (Figs. 3; 4a, b; 5a, b; 6; 7a, b; 8; 9; 10). In some species there are usually only two functional fangs at a time, one on either maxillary bone, but in other species there may be two or three functional fangs on either side. In addition, there may be three or four replacement fangs clustered together on each maxillary bone. The functional fangs are solidly anchored to the maxillary bone, whereas the reserve fangs have a softened immature base and are easily dislodged during dissection. The replacement fangs may be found in various stages of development. The most mature replacement fang is usually situated in the anterior position and medial to the functional fang. The replacement fangs remain in a depressed position until becoming functional. Gradually the oldest replacement fang moves to the anterior position and becomes erect; the base then gradually ankyloses, permitting the fang to be fully operational. During this replacement period, two or more functional fangs may be present for a short period of time, but eventually only one functional fang is present. Thus the transition is made in such a manner that a fully functional fang is present at all times. The other replacement fangs are clustered together immediately

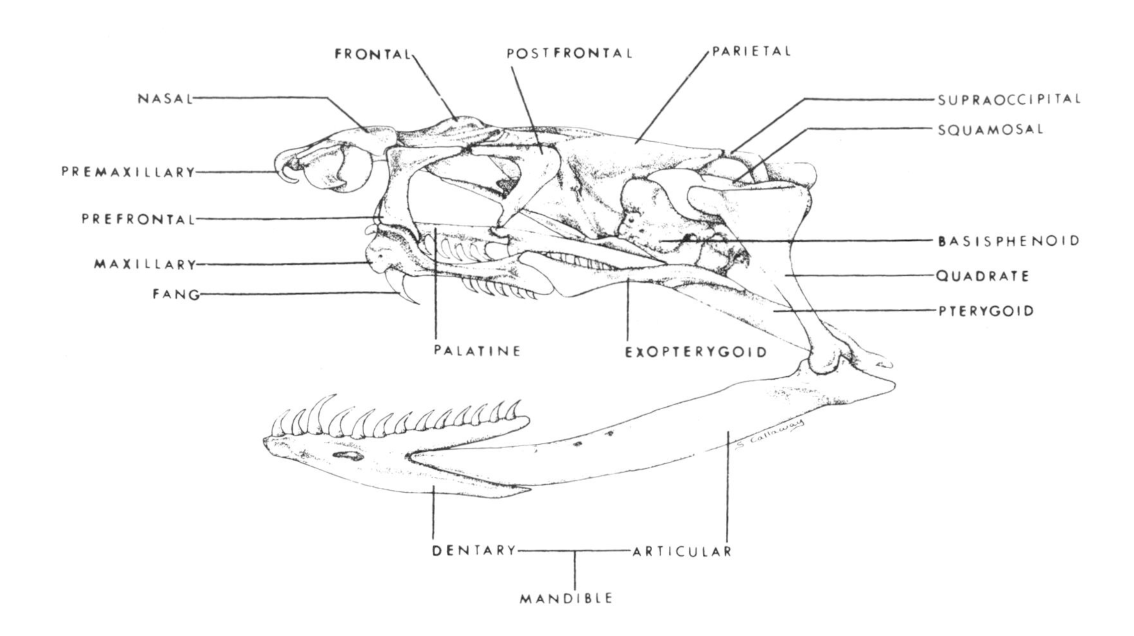

FIGURE 4a.—Lateral view of skull of *Lapemis hardwicki*, showing bony relationships of the fangs. × 2.7. (S. K. Callaway)

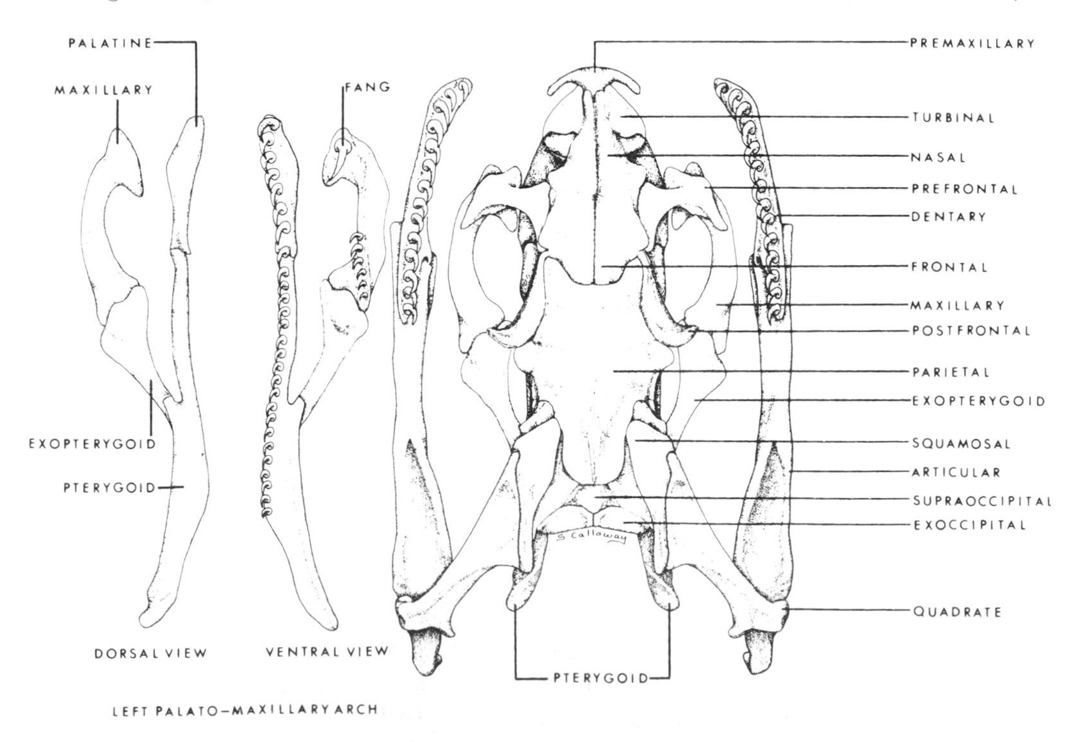

FIGURE 4b.—Dorsal view of skull of *Lapemis hardwicki*. × 2.7. (S. K. Callaway)

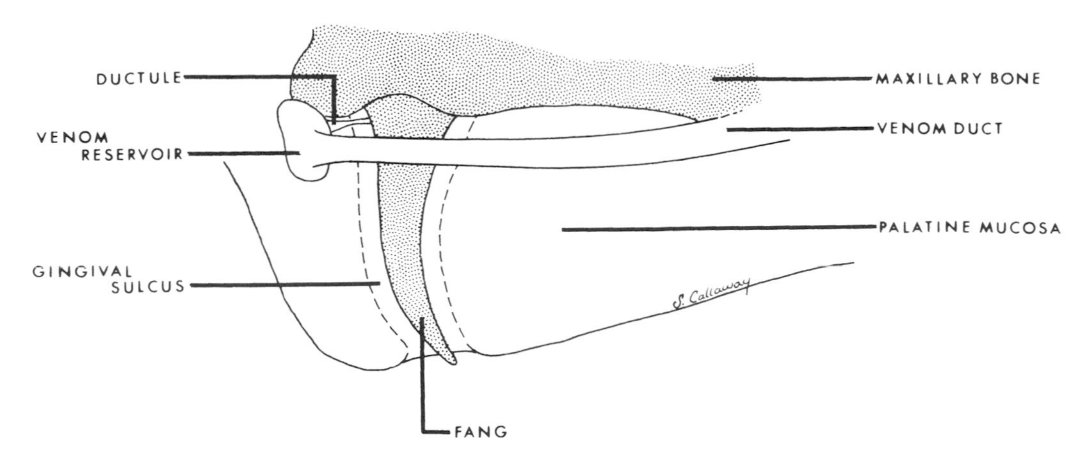

Figure 5a.—Showing the relationship of fang to the venom duct and palatine mucosa. Lateral view. × 17.5. (S. K. Callaway)

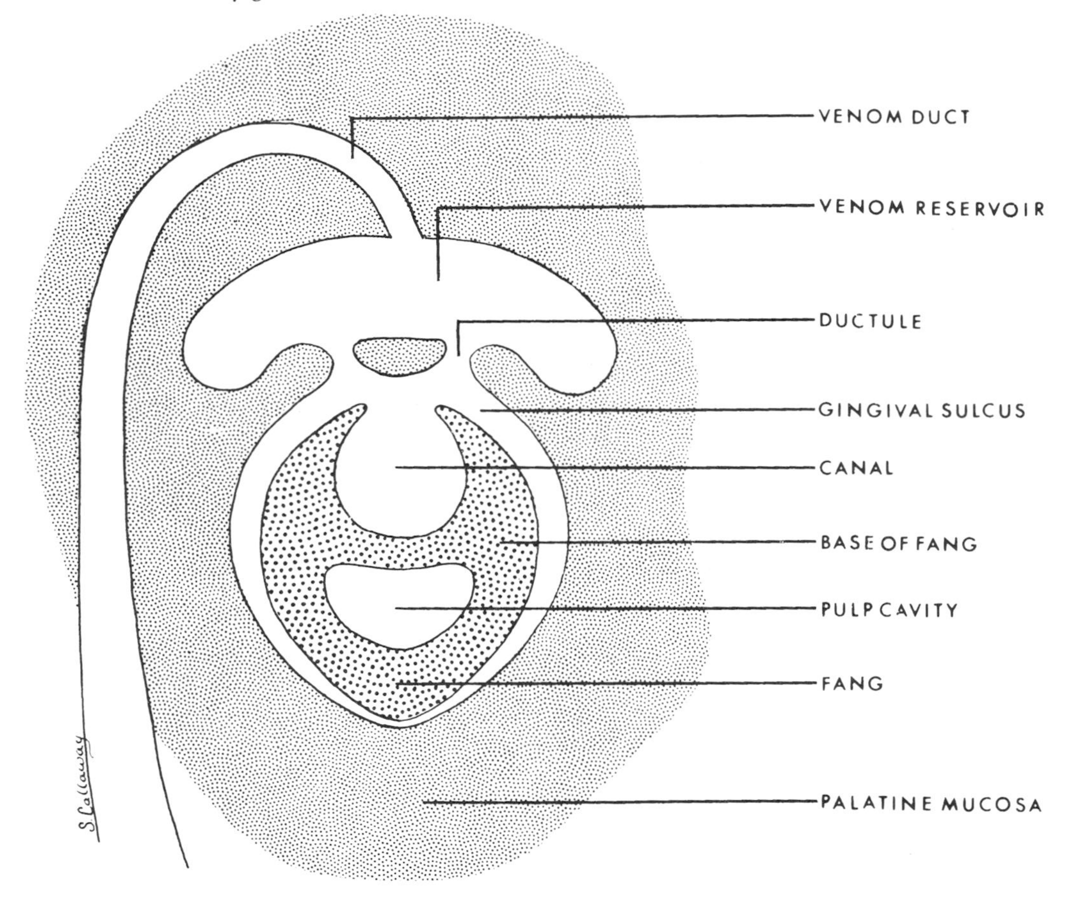

Figure 5b.—Cross section through the fang showing the relationship of the fang to venom duct and palatine mucosa. × 43. (S. K. Callaway)

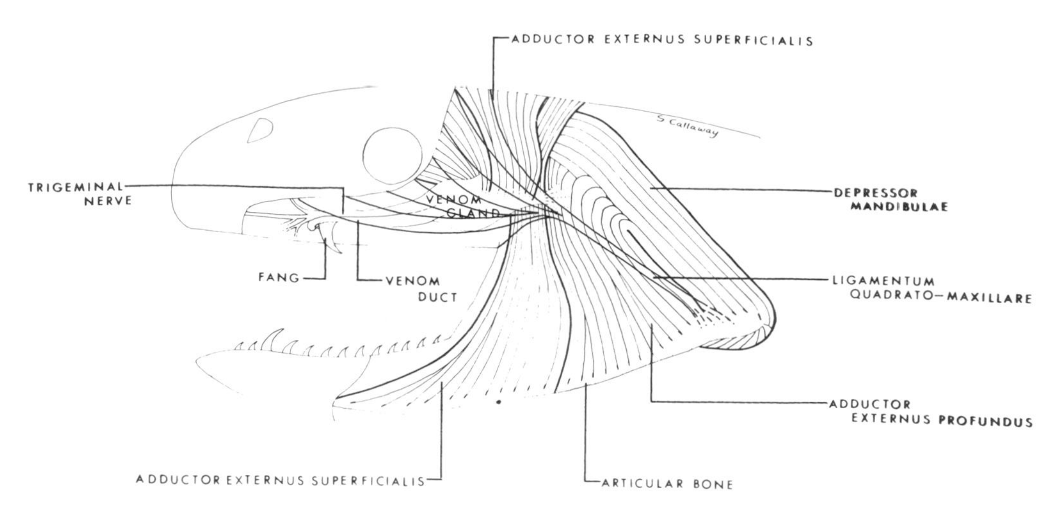

FIGURE 6.—Lateral view of the head of *Lapemis hardwicki*, showing cranial musculature associated with the venom apparatus. × 3. (S. K. Callaway)

posterior to the first replacement fang and the functional fang. The degree of hardness of the base of each of these replacement fangs varies with their degree of maturity, the younger the fang the softer the base. The frequency with which fangs are replaced in sea snakes is not known. The functional fang is partially hidden, and the replacement fangs are completely covered by fleshy lobes of palatine mucosa. Posterior to the functional fang(s) is an interspace about equal to that of the diameter of the eye; then begins a series of smaller maxillary teeth.

The following description of the venom apparatus is based on the studies of Halstead, Engen, and Tu (1976) on *Lapemis hardwicki*, which appears to be the only sea snake whose venom apparatus has been studied in detail.

Fang Structure (Figs. 4a, b; 5a, b; 7c): A functional fang of *L. hardwicki* is a moderately stout, conic, gracefully arched dentinal structure which terminates in an almost needle-sharp tip. The base, or pedestal, of the fang is only slightly expanded where it engages the socket of the maxillary bone. From an anterior view it will be observed that the fang is laterally compressed, and the fang is grooved throughout most if its length. The upper end of the groove at the base of the fang has a V-shaped opening. The groove is tightly closed throughout most of its length so as to form a canal except where it reaches the tip and expands slightly into a somewhat elliptical opening, the discharge orifice. The length of the mature fang in *L. hardwicki* seldom exceeds 4 mm. Further details as to the structure of the fang are dealt with in the section on microscopic anatomy.

Venom Gland (Figs. 5a, b; 6): The main body of the venom gland lies superior to the maxillary bone and extends between the lower posterior margin of the eye and then passes posteriorly beneath the adductor externus superficialis muscle. The lower portion of the adductor externus superficialis

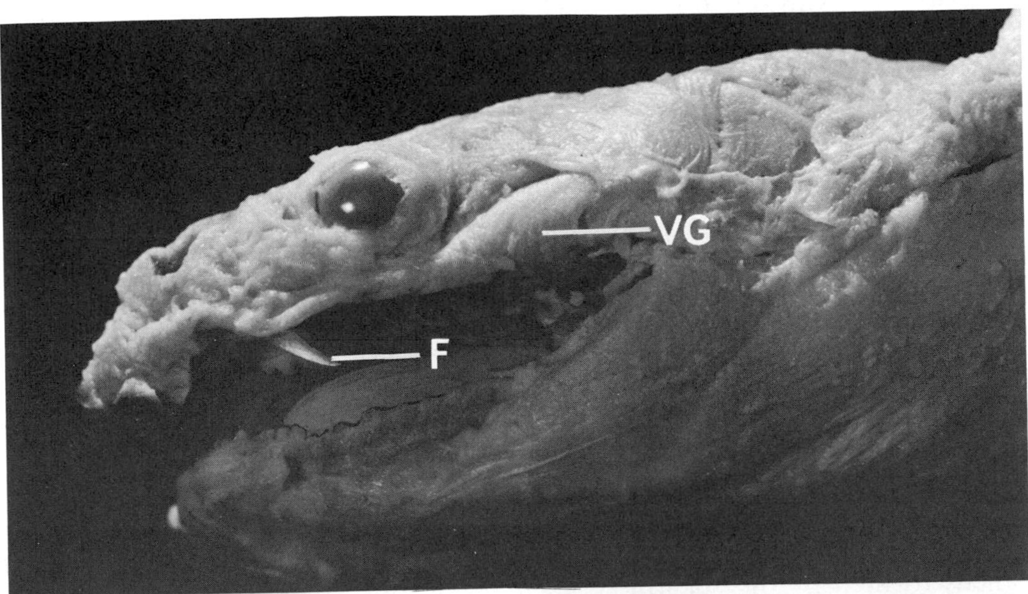

Figure 7a.—Head of *Lapemis hardwicki*. × 2. (E. Roche)

Figure 7b.—Skull of *Lapemis hardwicki*. (E. Roche)

Figure 7c.—Close-up showing fang of *Lapemis hardwicki*. × 5. (E. Roche)

muscle divides into two lobes in such a way as to border the upper margin of the body of the venom gland, whereas the other lobe covers the posterior portion of the venom gland. The posterior portion of the venom gland is attached to the adductor externus superficialis muscle by strands of muscle and connective tissue. The width of the venom gland is about equal to the diameter of the eye. The venom gland appears as a distinct ovate structure encapsulated by a strong sheath of dense connective tissue. The venom gland begins to taper at about the level of the postorbial margin of the eye, from which point it continues anteriorly as the venom duct. It will be shown histologically that as the venom gland tapers along the supralabial ridge and merges into the venom duct in the suborbital region it incorporates the so-called accessory venom gland.

Venom Duct (Figs. 5a, b; 6): The venom duct extends anteriorly from the body of the venom gland beginning at about the level of the posterior margin of the eye, passing between the maxillary bone and the orbit to the fangs. The venom duct empties into an expanded venom reservoir from which extends one or more small ductules which finally empty into the gingival sulcus formed by the sheathlike palatine mucosa which surrounds the base or pedestal of the fang, the entrance to the venom canal of the fang, and usually most of the fang. The venom finally passes through the canal and makes its exit through the discharge orifice of the fang.

The trigeminal nerve enters the capsule of the venom gland from the medial aspect at about the level of the postorbital margin of the eye, then passes anteriorly and emerges from the capsule in the vicinity of the cervix of the gland, passing adjacent and superior to the venom gland. The gross morphology of the trigeminal nerve and the venom duct emerging from the capsule of the venom gland give the appearance of two venom ducts. However, more critical examination of these structures reveals that the inferior

FIGURE 8.—*Aipysurus laevis*. Heron Island, Great Barrier Reef. (Julie Booth)

venom duct terminates in the sheath at the base of the fang, whereas the superior trigeminal nerve continues anteriorly giving off several branches in the nasal region.

Ligaments and Musculature (Fig. 6): There are two primary ligaments associated with the capsule of the venom gland. The largest of these is the ligamentum quadrato-maxillare. This ligament extends from the upper part of the maxilla, passes backward beneath the orbital region adjacent to the venom duct and attaches to the capsule of the venom gland at the cervix. The ligament emerges beneath the outer fleshy portion of the adductor externus superficialis muscle and then passes backward and slightly downward to connect with the aponeurosis covering the quadrato-mandibular articulation. The second ligament, known as the ligamentum internum quadrato-glandulare is not as well developed as the ligamentum quadrato-maxillare. This second ligament extends from the upper and inner aspect of the venom gland upward and backward and fuses with an aponeurosis covering the quadrato-squamosal articulation. Both of these ligaments are comparatively weak and are not as well developed as those found in other members of the Proteroglypha such as the Viperidae. There does not appear to be any evidence of the ligamentum quadrato-maxillare anterior. If it is present, it is greatly reduced in size.

The adductor externus superficialis muscle originates from the posterior edge of the postfrontal bone and the adjacent portion of the parietal bone. As the muscle extends downward and backward toward the venom gland, it divides into two parts. There is an outer fleshy part which overlies the posterior portion of the venom gland and attaches to it by strands of connective tissue. The deeper portion of the muscle passes medial and posterior to the venom gland and then passes downward, helping to form the angle of the mouth, at which point the muscle fans out and inserts along the ventral surface of the articular.

The adductor externus profundus muscle originates in the area of the quadrato-squamosal articulation and along the face of the quadrate bone, passes downward, backward, and deep to the adductor externus superficialis, and inserts on the ventral surface of the articular adjacent and posterior to the adductor externus superficialis.

The depressor mandibulae muscle originates on the posterior face of the quadrate bone and inserts on the retroarticular process.

The other head muscles have not been described because they do not appear to serve any significant function in the venom apparatus of *L. hardwicki.*

The prime function of the adductor muscles, as their name implies, is to draw the lower jaw upward and close the mouth. However, it appears that one of the important special functions of the adductor externus superficialis is in controlling the ejection of the venom by exerting pressure on both sides of the gland. It is doubtful that any of the other adductor or cranial muscles play a very significant role in venom ejection.

West (1895; *see also* Halstead, 1970) described "two large cushions of

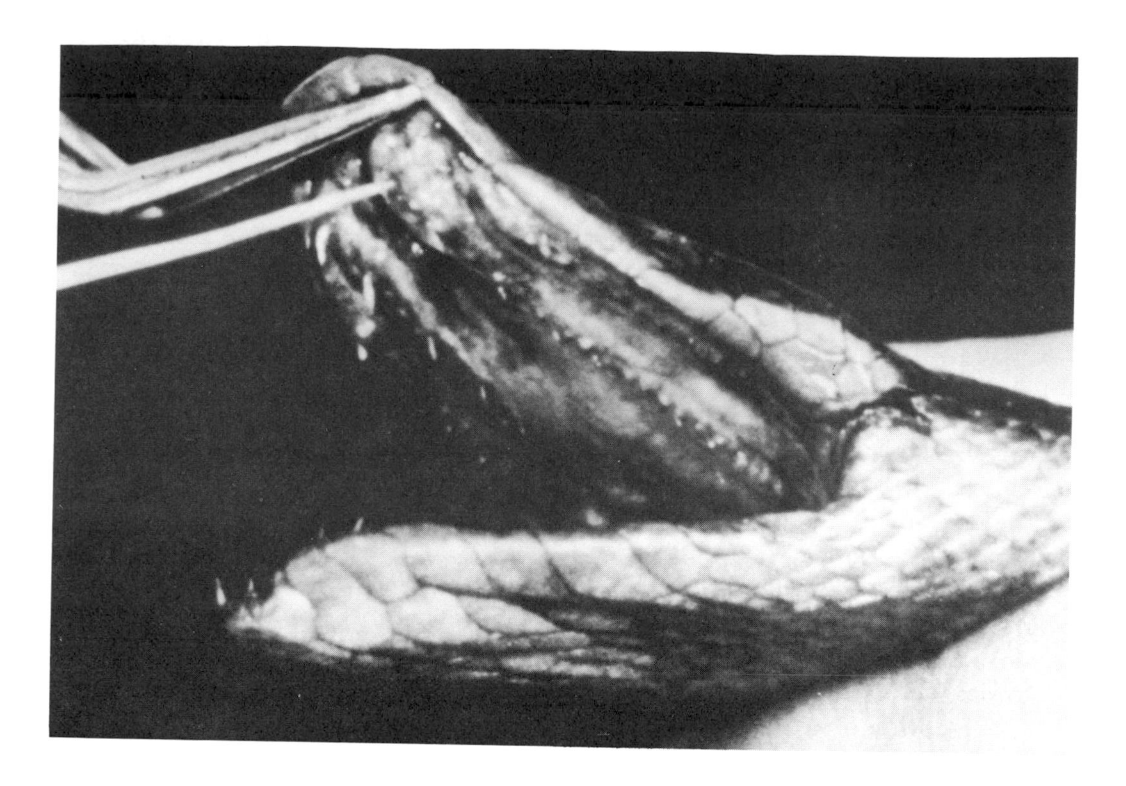

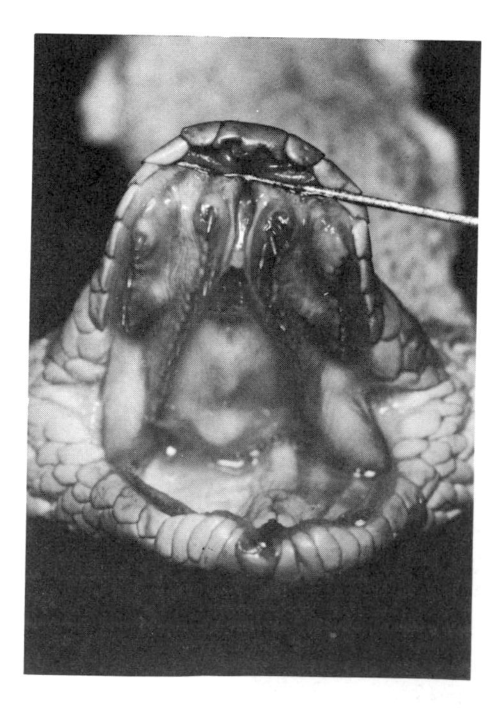

FIGURE 9 (Above).—Open mouth of *Pelamis platurus* showing teeth and fangs. The needle points to one of the fangs. Despite the small size of the fangs of this snake, they are capable of inflicting a fatal envenomation in humans. (Courtesy P. Saunders)

FIGURE 10 (Left).—Open mouth of the sea snake *Astrotia stokesi* showing teeth and fangs. This snake is venomous, but has not been incriminated in human envenomations possibly because of the small size of the fangs. Nevertheless, this species should be regarded as a dangerous reptile. (K. Gillett)

muscular tissue," one in front of each fang. According to West the ejection of venom into the fangs is controlled by these muscles. There is reportedly a vertical slit between these muscles through which the venom duct passes just as it empties into the mucosal folds surrounding the base of the fangs. West provides a rather elaborate description of this muscular venom-control mechanism. However, examination of the cranial muscles of *L. hardwicki* fails to reveal any gross or histological evidence of these muscles.

Background articles that were found to be helpful in the study of the venom apparatus of *L. hardwicki* include the studies of the cranial muscles of various species of snakes by Haas (1952), Kochva (1958), Boltt and Ewer (1964), Bellairs (1969), and on cephalic glands by Rosenberg (1967), Kochva and Gans (1970), Kochva and Wollberg (1970), and Burns and Pickwell (1972).

Microscopic Anatomy

The venom gland of *Lapemis hardwicki* is situated immediately posterior to the orbit just beneath the skin and tapers along the supralabial ridge. The venom gland is surrounded by dermal connective tissue. Inferiorly, the venom gland is separated from the oral mucosa by numerous small clusters of mucous glands which empty individually into the oral cavity by a small duct (Fig. 11).

The venom gland and its duct system can be divided into four distinct components: 1) the venom gland proper, 2) the elongate duct system, 3) a venom reservoir, and 4) the ductules connecting the venom reservoir with the gingival sulcus at the base of the fang.

The venom gland is somewhat pyriform in shape. It is completely surrounded by a thick capsule of dense connective tissue. Muscle fibers are attached to the dorsal surface of the capsule. Trabeculae from the capsule can be seen penetrating the gland. The connective tissue stroma tends to be quite loose, irregular, and abundant. The connective tissue is increased in amount in the central area of the gland which contains the main body of the tubules forming the venom gland (Fig. 12). The parenchyma appears to be organized as a compound tubular gland with the terminal ends becoming small ducts which join others to form finally the single main venom duct. The tubules in the central portion of the gland are much enlarged, apparently dilated by the accumulated secretion.

The venom gland cells vary in shape from cuboidal to columnar. This variation in shape is apparently due to the degree of distention of the tubules. The nuclei are large and rounded to ovoid in shape. The chromatin material appears as small granules lying against the nuclear membrane as well as being dispersed in the nuclear sap. Nucleoli were not evident. The cytoplasm appears quite uniform and somewhat granular.

The lumen of the tubules contains two types of material. The nature of this material was better demonstrated in frozen sections than those that were double embedded. The material appeared to be in the form of a finely granular precipitate. In addition, degenerating cells were observed. These

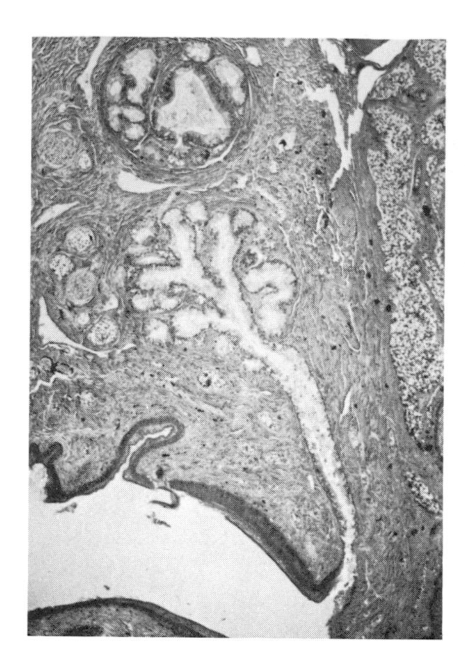

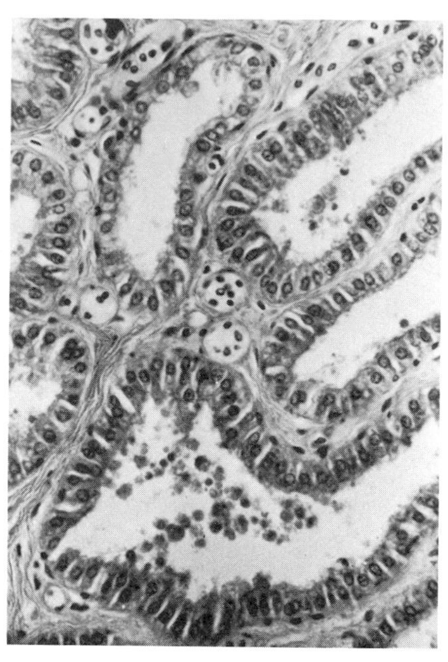

Figure 11 (Left).—Cross section through one of the acini of the supralabial glands emptying into the oral cavity. This gland is located lateral to the maxillary bone just beneath the labial scales and extends posteriorly to about the mid-section of the venom gland. The cells are of a mucous-secreting type. × 45. Hematoxylin and triosin. (P. Engen)

Figure 12(Right).—Cross section of venom gland showing some of the compound tubular glands having cuboidal to columnar cells. Note degenerating cells in the lumen of the tubules. × 285. Hematoxylin and triosin. (P. Engen)

were usually in the form of clumps and loose aggregations presumably shed from the lining surface epithelium. These bodies tended to be round, variable in size, and rather dense. Nuclear material was often present which demonstrated both pyknosis and karyorrhexis. Vacuoles were also frequently present.

As the tubules approach the central region of the gland, they form small ducts. This transition can be readily recognized when stained with the PAS technique because the apical ends of the lining cells of the small ducts become quite strongly PAS positive. This reaction is absent in the tubular cells in the central region of the gland.

The main venom duct is accompanied by blood vessels and a branch of the trigeminal nerve, all of these structures being enclosed in a definite connective tissue sheath (Figs. 13, 14). The duct and its associated accessory venom glands are further isolated by a separate connective tissue layer of their own. The accessory venom glands appear as numerous small glands which empty their secretions into the lumen of the venom ducts (Fig. 15).

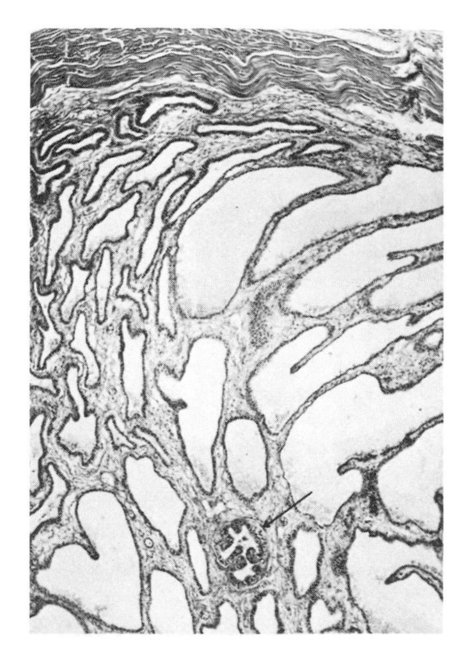

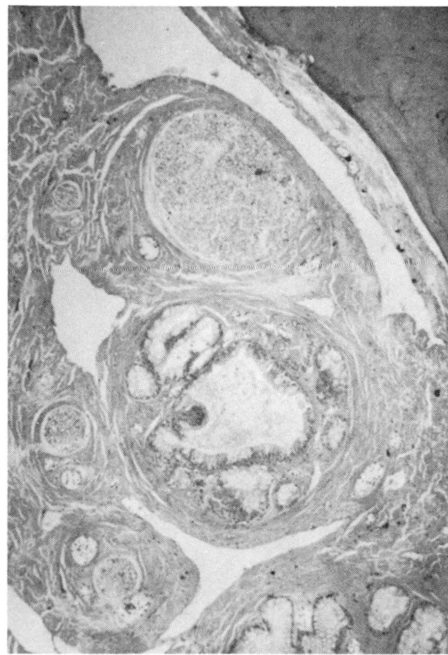

FIGURE 13 (Left).—Cross section of venom gland showing the beginning of venom duct (*see* arrow). × 45. Hematoxylin and triosin. (P. Engen)

FIGURE 14 (Right).—Cross section through the posterior third of venom duct showing the relationship to the trigeminal nerve (above) to the venom duct (below). Note secretions present in the lumen of duct. × 170. Hematoxylin and triosin. (P. Engen)

These glands appear to be of a mucous nature. Two cell types are present in the accessory venom gland. One is a rather typical columnar cell. The nuclei of these cells tend to be round or ovoid, but their contours are often irregular. PAS positive granules are present in the apical half of the cytoplasm. The staining intensity and number of these granules vary, indicating perhaps the activity of these cells at the time of the staining. The second type of cell observed was a mucous secreting cell type. These mucous cells are quite numerous and are the major cell type. The cells lining the main venom duct are of a stratified or pseudostratified columnar type. The nuclei are ovoid and the apical cytoplasm of the cells adjacent to the lumen contain PAS positive granules. Interspersed among these cells, singly, but usually in groups, are the mucous secreting type of cells. These mucous cells are absent at each end of the duct, but appear to be the dominant cell type in the mid-portion of the venom duct.

Near the termination of the venom duct, the accessory glands disappear and the lumen narrows. The distal end of the venom duct finally terminates by expanding into a sac-like venom reservoir. The venom reservoir is a flattened semilunar structure lying anterior and inferior to the base of the fang (Figs. 16-18). It appears to be a single sac, which could be easily com-

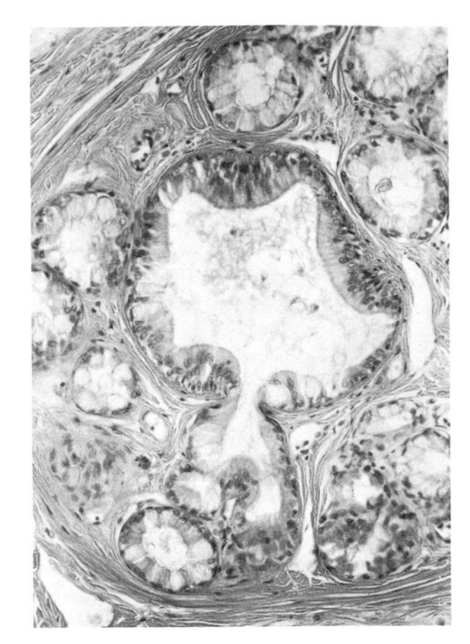

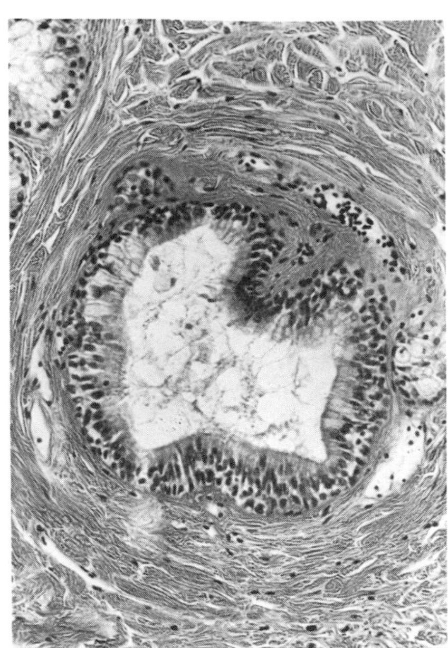

FIGURE 15(Left).—Cross section through the middle third of venom duct surrounded by accessory venom glands which are comprised of mucous cells. Note one of these glands is shown emptying into the duct. × 182. Hematoxylin and triosin. (P. Engen)

FIGURE 16 (Right).—Cross section of distal third of venom gland approaching the termination of the duct. Note the absence of accessory venom glands. × 182. Hematoxylin and triosin. (P. Engen)

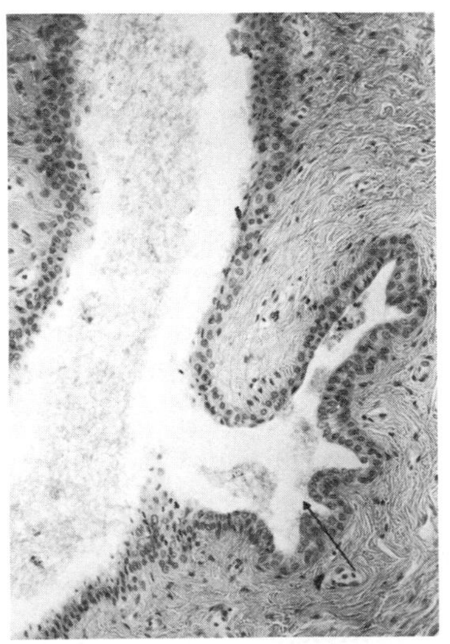

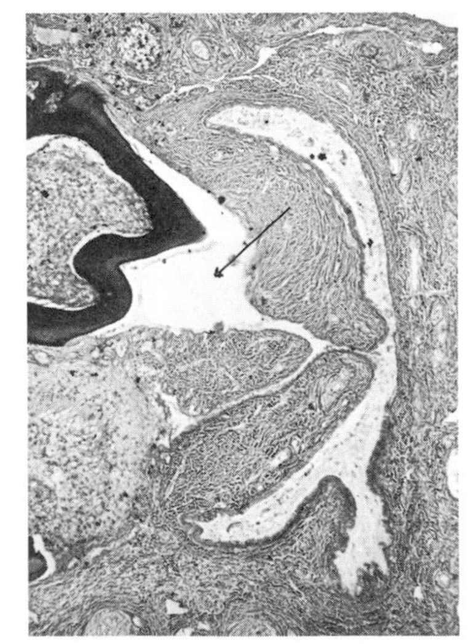

FIGURE 17(Left).—Cross section showing venom duct emptying into venom reservoir. Arrow points to venom duct. × 170. Hematoxylin and triosin. (P. Engen)

FIGURE 18 (Right).—Section showing venom duct emptying into venom reservoir near the base of the fang. Note connecting ductule between venom reservoir and gingival sulcus. Arrow points to sulcus. × 55. Hematoxylin and triosin. (P. Engen)

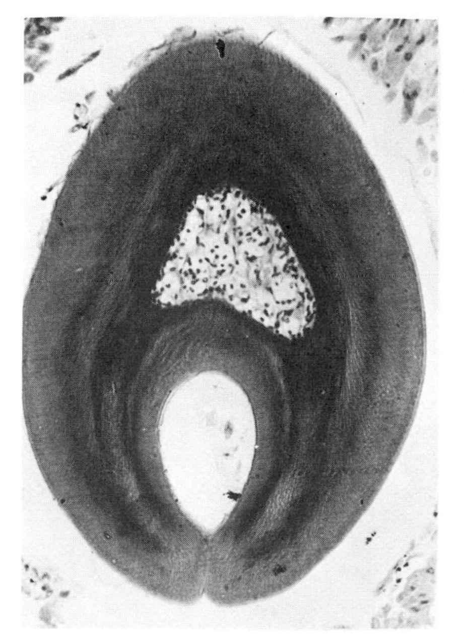

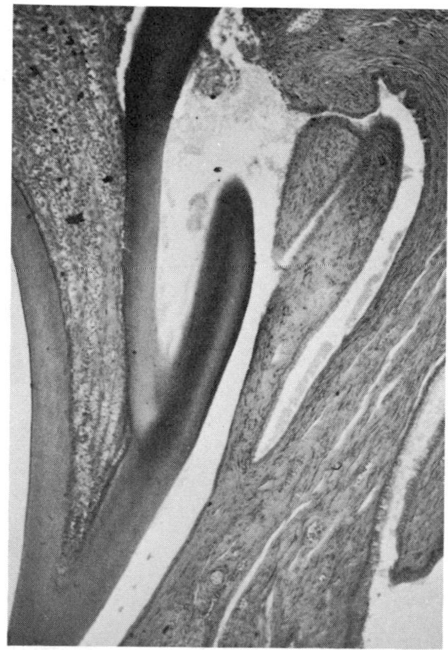

FIGURE 19 (Left).—Cross section through upper part of fang showing the pulp cavity (upper) and venom canal (lower orifice). Note pulp cavity is filled with odontoblasts. × 182. Hematoxylin and triosin. (P. Engen)

FIGURE 20 (Right).—Sagittal section of fang showing venom reservoir with connecting ductules emptying into gingival sulcus. × 45. Hematoxylin and triosin. (P. Engen)

pressed by any object pressed against the base of the fang. The venom reservoir is lined by a stratified cuboidal epithelium 2-3 cells in thickness and surrounded by dense connective tissue. No muscle elements were found to be associated with this reservoir. Moreover, except for the muscle fibers attached to the dorsal surface of the capsule of the venom gland proper the entire duct structure is devoid of muscular connections.

Leading from the venom reservoir to the basal region of the fang which is surrounded by the gingival sulcus is one or more short ductules. The epithelium lining the sulcus is similar to that lining the reservoir, namely cuboidal and stratified squamous cell types. A cross section of the upper part of the fang reveals the pulp cavity filled with odontoblasts and an ovate incompletely closed venom canal (Fig. 19). A sagittal section of the upper part of the fang reveals the venom reservoir with its connecting ductules emptying into the gingival sulcus (Fig. 20).

MECHANISM OF INTOXICATION

Sea-snake envenomations are usually encountered by fishermen handling their nets, sorting fish, stepping accidentally on them, wading, swimming, collecting shells, or accidentally bumping into them in some other manner (Fig. 21). About two-thirds of the sea-snake bites occur on the lower limbs,

FIGURE 21.—Sea snakes are accidentally taken in fishing nets and, in the process of sorting the fish and removing the reptiles from the nets, the fishermen are frequently bitten. Sea snakes which are normally docile become quite irritable because of the rough handling which they receive in the netting process. (Courtesy H. A. Reid)

i.e., foot, leg, toe, or thigh. The remainder of the bites are on the finger, hand, or forearm. The species most frequently causing human envenomations are *Enhydrina schistosa, Hydrophis spiralis, H. cyanocinctus, Kerilia jerdoni,* and *Pelamis platurus*, but other species have been incriminated. Sea snakes inflict their wounds with the use of fangs which are reduced in size but are of the elapine or cobra type. In comparison with other venomous snakes, their dentition is relatively feeble but nevertheless fully developed for venom conduction. Although some species are docile, they are all venomous and potentially dangerous.

MEDICAL ASPECTS

Clinical Characteristics

The latent period from the bite to the onset of symptoms usually varies from 5 minutes to 8 hours. It is generally believed that the length of the latent period varies according to the amount of venom injected. The larger the dose of venom is, the shorter the latent period and the period in which death may occur. Reid and Lim (1957), however, have found that this correlation may not always hold true in sea-snake bites. Moreover, no correlation was found between the latent period and the recovery time in either trivial or severe cases.

Symptoms usually begin mildly and slowly become progressively worse. Localized symptoms at the site of the bite are minor. After the initial prick there is no pain or reaction at the site of the bite (Fig. 22). The initial generalized symptoms in some victims consist of a mild euphoria, whereas in others there is malaise and anxiety. A sensation of a thickening of the tongue and a generalized feeling of aching or stiffness of the muscles may gradually develop. Speaking and swallowing become increasingly difficult. A general motor weakness may ensue which progresses into a distinct muscular paralysis beginning with the legs and ascending rapidly, within an hour or two, involving the trunk, arm, and neck muscles. The paralysis is usually flaccid with decreased or absent tendon reflexes, but may be spastic with an initial hyperreflexia. A bilateral painless swelling of the parotid glands may develop. Nausea, vomiting, and nasal regurgitation may be present. The pulse be-

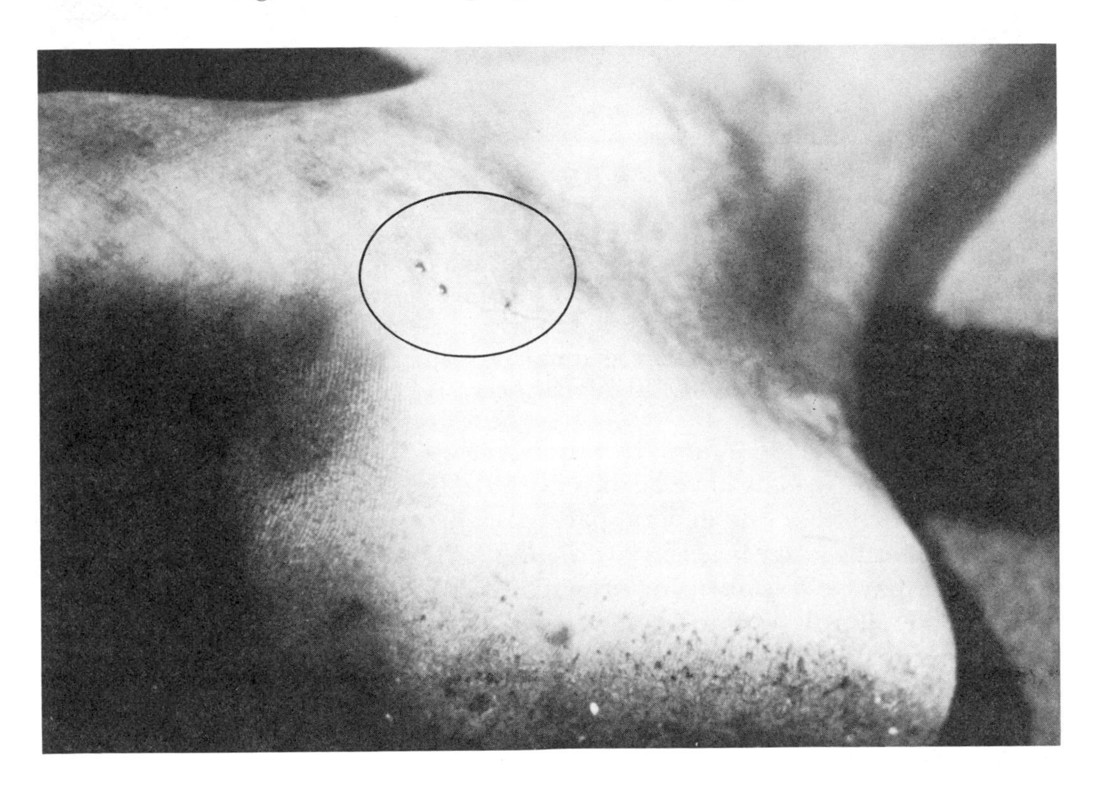

FIGURE 22.—Sea-snake bite on the instep of the foot. The fang marks appear in this case as three dots (circle). Sea-snake bites are usually inconspicuous, painless, and without swelling. The identity of the sea snake was not known, but was suspected as *Enhydrina schistosa*. Bite occurred near Penang, Malaya. (Courtesy H. Reid)

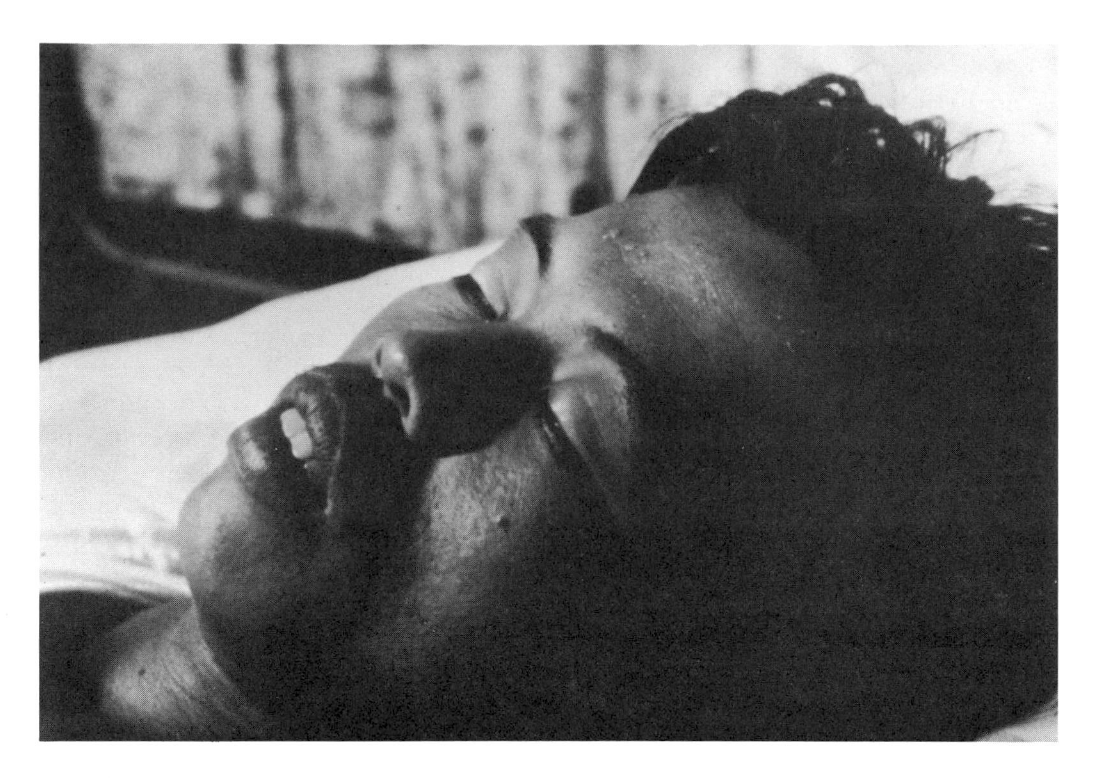

FIGURE 23.—Close-up of sea-snake victim. Sweating, especially on the forehead, is evident in severe cases, particularly when respiratory failure is imminent. Death occurs in about 10 percent of the untreated cases and is most commonly due to respiratory failure from direct muscle damage to respiratory muscles, or inhalation of vomitus or of secretions. Death may also be due to hyperkalemic cardiac arrest, the excess potassium being released from damaged muscle cells. After several days death may also occur from renal failure due to tubular necrosis. (Courtesy H. A. Reid)

comes weak and irregular, and the pupils dilated. Muscle twitchings, twisting movements, and spasms have been reported. Bulbar paralysis may precede, accompany, or follow the general paralysis. Jaw stiffness due to trismus, an important feature, may be easily overlooked. Ptosis is an early and important sign which may be mistaken for drowsiness. However, the patient is usually mentally alert until respiratory failure is advanced. Dryness of the throat, burning sensation, thirst, and a general feeling of coldness and increased sweating are common early symptoms. Dysarthria and dysphagia, usually due to trismus, are frequently present. Ocular and facial paralysis may occur. There is no clinical evidence of cerebral, cerebellar, or extrapyramidal involvement, and sphincter disturbance is unusual. Secondary shock is frequently present. As the paralysis becomes more severe, even passive movements of the appendages may be painful. Deep reflexes are diminished.

Albuminuria and hemoglobinuria are common. Myoglobinuria becomes evident about 3 to 6 hours after the bite. The urine becomes dusky yellow

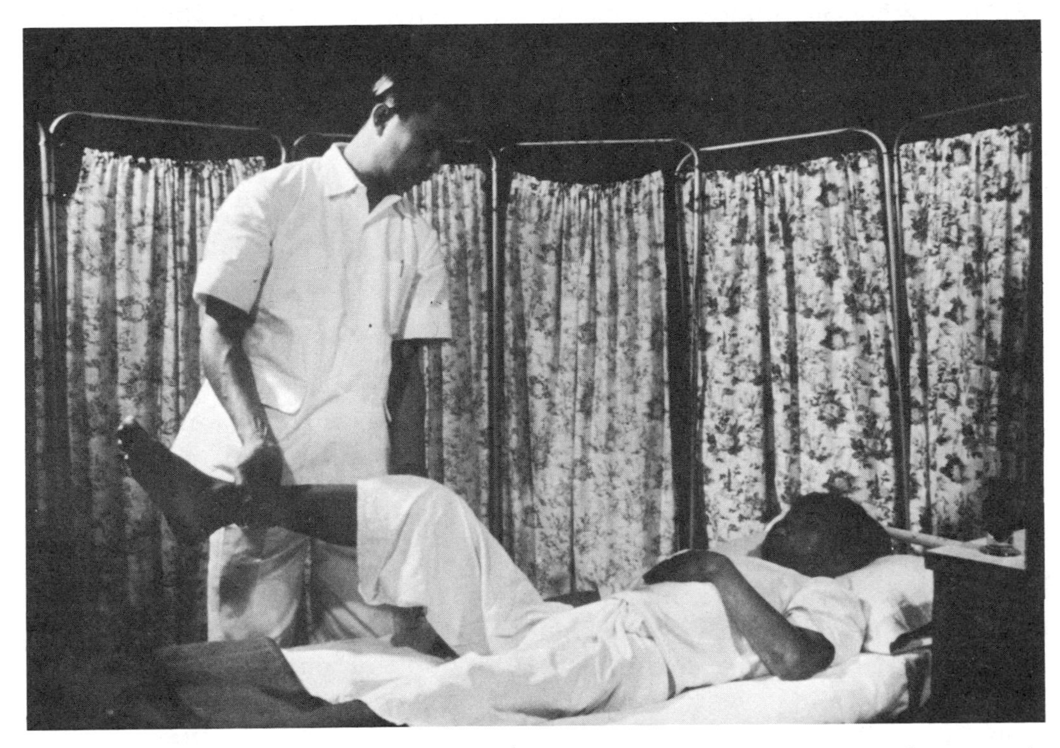

FIGURE 24 (above).—Flexing the limbs is extremely painful because sea-snake venom is myotoxic and directly attacks skeletal muscle. Flexing the limbs is a useful clinical diagnostic procedure in determining sea-snake envenomations.

FIGURE 25 (below).—The face of this victim show ptosis and a pseudotrismus. Difficulty in opening the mouth is due to a spasm of the undamaged jaw muscles which protect the damaged muscles against being stretched. A beaker containing the red-brown myoglobinuric urine of the patient is seen on the bedside table.

(Courtesy H. Reid)

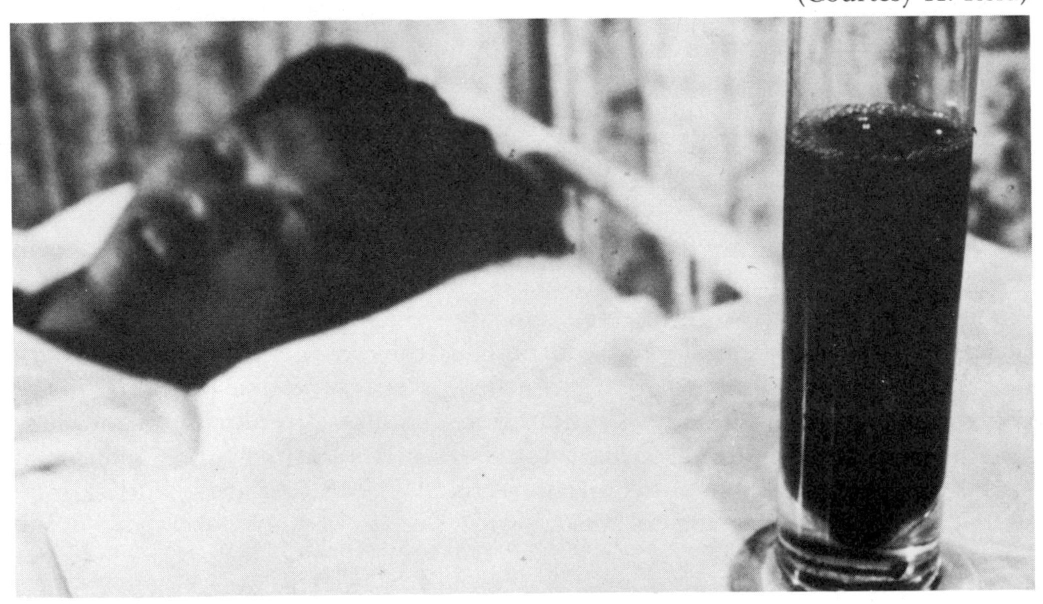

with positive protein and occult blood tests. This precedes by about 1 hour the red-brown color of myoglobinuria (Fig. 25). Myoglobinuria can be distinguished from hemoglobinuria by a hand spectroscope using fluorescent lighting or by electrophoresis (Reid, 1959, 1963*a*). The cerebrospinal fluid is normal. The blood picture is normal apart from a leucocytosis and from hemoconcentration.

There is no clinical evidence of direct cardiovascular involvement. Failing vision is considered to be a terminal sign. The generalized pains resulting from muscle movements and the myoglobinuria are said to be the outstanding clinical signs of sea-snake envenomation. In fatal cases, respiratory paralysis with terminal hypertension and cyanosis usher in death. In some cases little or no peripheral paralysis may be evident, and the victim dies from bulbar paralysis. Death may take place within several hours or several days after the bite. There are no accurate statistics on the mortality rate of sea-snake poisoning, but the case fatality rate is estimated to be about 3 percent (Reid, 1956*a*, 1957).

The prognosis of a case of sea-snake envenomation is based on an evaluation of the following signs: multiple toothmarks suggest a high venom dose, but single marks do not necessarily indicate a small dose. Vomiting, ptosis, weakness of the external eye muscles, dilatation of pupils with a sluggish light reaction, and a leucocytosis exceeding 20,000 per mm^3 are serious signs. If the victim survives the first week, recovery may be predicted, although it may require weeks or months for complete recovery. The total number of days myoglobin is visible in the urine indicates the approximate number of weeks of expected illness (Reid, 1959).

The clinical characteristics of sea-snake bites have been discussed by Rogers (1903), Becke (1909), Fitzsimmons (1919), Kopstein (1930), Bokma (1942), Reid (1956*a, b, c*, 1957, 1959, 1963*a*), Reid and Lim (1957), Keegan (1958), and Reid *et al.* (1963).

Pathology

Data on the pathological effects of sea-snake venom are meager, and the reports somewhat conflicting. Rogers (1903) was one of the first to investigate the subject, and he found little noteworthy after he had autopsied various laboratory animals that had been injected with sea-snake venom.

Lamb and Hunter (1907) investigated the histological changes in the nervous system resulting from injections of venom from *Enhydrina valakadien* (synonym of *E. schistosa*) into monkeys. Five monkeys were used in this particular series and were injected subcutaneously with 0.05 to 0.12 mg/kg of venom. Death time varied according to the dosage from 1.75 to 6.75 hours. The monkeys were then autopsied. Representative portions of the central and peripheral nervous system were fixed in perchloride of mercury and stained with thionin according to Nissl's method and with hematoxylin and eosin, whereas other portions were fixed in Müller's fluid and stained with Congo red according to Busch's and Donaggio's method. The nerve endings were fixed and stained in osmic acid. The investigators found that the venom had a widespread action on the tissues of the entire nervous system, including ganglion cells and nerve fibers. Widespread degenerative changes were found in the ganglion cells throughout the nervous system. These changes were particularly evident in sections of the anterior horns of the cord. Degenerative changes were marked in the cervical, dorsal, and lumbosacral regions of the cord.

The cellular degenerative changes were described as follows: The cells stained deeply, showing a central nucleus and nucleolus, but the cytoplasm was devoid of granules (ghost cells) and presented an ill-defined reticulum which stained less intensely than the Nissl granules. Fragmented granules were still present in some cells, but the granules were generally at the periphery. Vacuolation was seen only in a small proportion of the cells. Degenerative changes were less marked in the motor nuclei of the pons and medulla but were nevertheless present. Changes were also observed in sections from the 3rd, 5th, 6th, 7th, and 12th nuclei. The large motor cells of the motor cortex showed changes similar to those of the anterior horn cells. The smaller cells of the cortex seemed even more degenerated. Donaggio's stain revealed definite degenerative changes in both central and peripheral nerve fibers. Axis cylinders were seen to be wavy in outline in some areas but swollen in others so as to form spindlelike thickenings. There was also observed patchy staining of the myelin. In some areas the myelin and axis cylinder were stained with the same degree of intensity so that the two could not be distinguished from each other. Similar degenerative changes were observed in the myelin in the nerve terminals. The authors concluded that the degenerative changes produced by sea-snake venom closely resembled those produced by cobra and krait venom.

Nauck (1929) killed two guinea pigs with the venom of *Pelamis platurus* and failed to find either gross or microscopic changes at the site of the bite, but the capillaries of the brain, lungs, liver, and kidneys were congested. Bokma (1942) reported the autopsy findings of a Javanese fisherman who had died 7½ hours after being bitten by *Enhydrina schistosa*. His gross findings were that of acute visceral congestion together with a microscopic hyperemia of the viscera, portal infiltration of leucocytes in the liver, and necrosis and infiltration around the fang tracks. The nervous system was not examined. The blood showed no hemolysis or bilirubin despite hemoglobinuria in life. Reid (1956*b, c*) reported on the autopsy findings of several cases resulting from sea-snake bites. His findings were essentially similar to those of Bokma (1942). Examination of the nervous system failed to reveal any significant pathology.

Marsden and Reid (1961) and Reid (1963*b*) found that the principal pathological lesion in sea-snake poisoning in man is necrosis of skeletal muscle. The toxin appears to select individual muscle fibers, leaving a healthy fiber next to a necrosed one and affects only one or more segments of varying lengths in a fiber, usually with an abrupt transition to normal

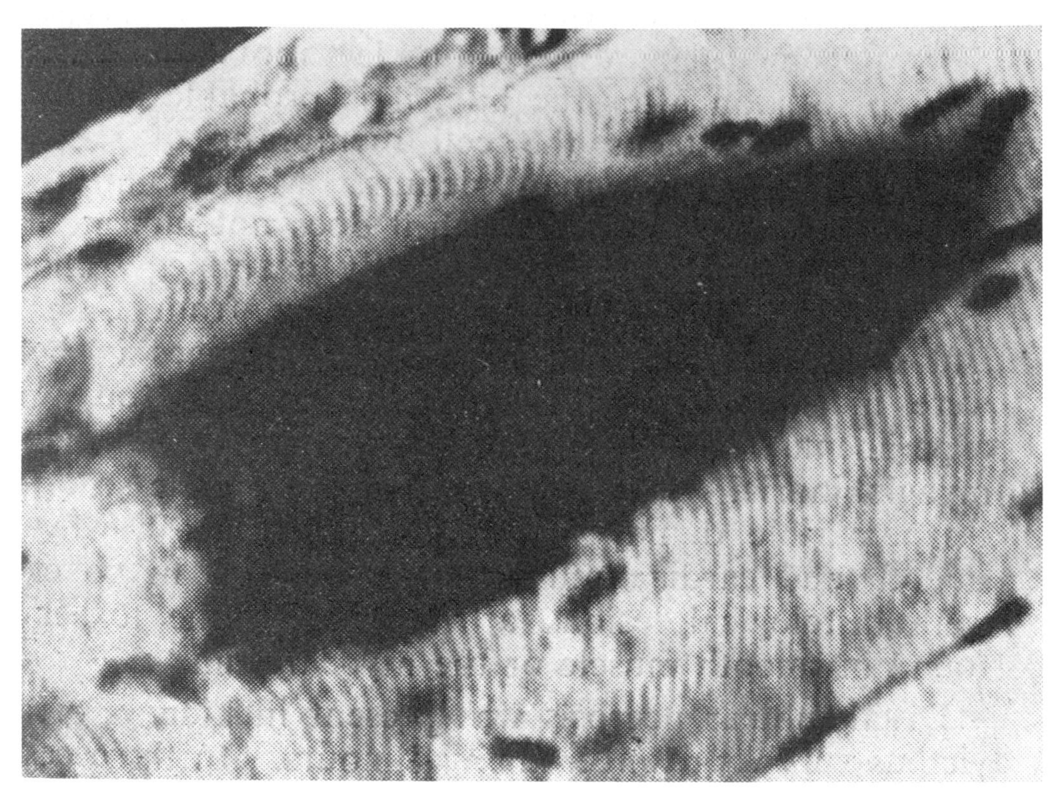

FIGURE 26.—Early necrosis of a muscle fiber caused by a sea-snake envenomation. × 400. (From Marsden and Reid)

muscle (Fig. 26). The number of fibers with focal necroses varies from muscle to muscle in the same patient. Apparently, the affected segment of a muscle fiber first becomes swollen and the sarcoplasm undergoes a coagulation necrosis. Initially the nuclei and swollen myofibrils can still be seen, but the disks of the myofibrils soon lose their differential staining and both nuclei and myofibrils disappear as they become fused into an amorphous hyaline mass. Gradually the necrotic material contracts, leaving an empty space beneath the sarcolemma which remains intact but slightly collapsed. Soluble products, i.e., myoglobin, enzymes, and electrolytes, are rapidly absorbed into the bloodstream. The insoluble products are phagocytosed by histiocytes and macrophages. Regeneration and repair is surprisingly rapid, beginning 1 or 2 weeks after the bite with multiplication of surviving muscle nuclei. Thus the muscle fiber is regenerated within its original sarcolemmal sheath and repair is remarkably complete since very few muscle fibers are entirely necrosed.

Renal damage often results from sea-snake poisoning as with other types of myonecrosis. The whole kidney is engorged, but the boundary zone is intensely congested and distinct. The major damage is in the tubules. There is extensive necrosis of the epithelium of the loop of Henle, the second convoluted tubule, and the collecting tubules. The lumen, filled with desquamated cells of granular and amorphous debris, forms casts in the distal and collecting tubules.

Marsden and Reid concluded that the myonecrosis developed rapidly and was clinically apparent within 1 hour of the bite. These lesions explain one of the chief symptoms of sea-snake envenomations, namely, muscle movement pains, stiffness, and the resulting paralysis. Myoglobinuria is evidence of the myonecrosis. Distal tubular necrosis was present in the kidneys of all victims surviving longer than 48 hours. Acute renal failure appeared to be the primary cause of death in these late cases. The authors felt that the myoglobin was the main cause of the renal damage, but raised the question as to whether sea-snake venom was nephrotoxic. This conclusion was questioned since no renal damage was observed in patients who had died within 24 hours of the bite. The authors noted some pathological differences between sea-snake poisoning in man and that in animals. Myonecrosis and renal damage has not been previously reported in goats and rabbits experimentally injected with sea-snake venom (Carey and Wright, 1960*a, b*). However, Marsden and Reid injected *E. schistosa* venom into dogs and obtained widespread myonecrosis. Histological changes in the nervous system and smooth and cardiac muscle were notably absent.

Treatment

The following recommendations are based on the extensive clinical experiences and researches of Reid (1956*a, b*, 1957, 1959, 1961, 1963*a, b*), Reid and Lim (1957), Russell (1962), and Russell *et al.* (1966) on snake-bite poisoning in general and sea-snake envenomations in particular. The work by Boys and Smith (1959) will also be found useful.

Sea-snake poisoning is a medical emergency requiring immediate attention and the exercise of considerable judgment. Tragic consequences may result from delayed or inadequate treatment. Before any first aid or therapeutic measures are instituted, it is always important to determine if envenomation has occurred. Needless treatment can cause discomfort and deleterious results. A sea snake may bite without injecting venom. The fangs and teeth of a sea snake are generally small and may not have penetrated the skin sufficiently to have resulted in envenomation. M'Kenzie (1820), Day (1869), and Reid (1956*a, b, c*, 1957, 1959) have observed that the majority of sea-snake bites do not result in envenomation. It should be kept in mind that some persons become emotionally disturbed even when bitten by nonvenomous snakes. These emotions may give rise to hysteria, disorientation, withdrawal, faintness, dizziness, rapid respiration, hyperventilation, rapid pulse, and even primary shock.

Diagnosis of Sea Snake Bite: The diagnosis of sea-snake bite is based on the following criteria:

1. A history of the victim being in the ocean, along the seashore, river mouth, sorting nets, wading, swimming, etc. The opportunity for contact with a sea snake must be clearly established.

2. Absence of pain. The victim usually does not complain of pain after the initial prick or bite.

3. Presence of definite fangmarks at the site of the bite. The patient is sometimes unaware of the bite until the fangmarks have been seen. Sea-snake bites are characterized by multiple pinhead-sized puncture wounds (Fig. 22). Fangmarks are most frequently seen around the ankle, foot, hand, leg, or arm. Sea-snake fangmarks are small, rarely exceeding 4 mm in length, and usually appear as circular dots as though made by a hypodermic needle. The fangmarks (the anterior two or four teeth of the upper jaw of the snake) and the posterior maxillary teeth vary in number according to the species of snake and the extent to which the teeth have made contact with the skin of the victim. The anterior fangmarks vary from 1 to 4 dots, and the posterior maxillary marks may vary from 3 to 20 dots. Reid (1956*a*) most frequently encountered four fangmarks. The fangmarks are usually larger than those of the maxillary teeth. Fangs and teeth may break off and remain embedded in the skin. Fishermen will sometimes draw a hair over the site to extract the teeth, thereby confirming that a bite has been made.

4. Positive identification of the snake. It is important to determine if the snake has a flat, paddlelike tail typical of a sea snake. Whenever possible, the offending snake should be killed by a sharp blow on the head and retained for positive identification by a knowledgeable person. Sea-snake bites are sometimes mistaken for sea-urchin stings, fish stings, etc., which can readily be distinguished by the associated pain at the site of the wound. Sea-snake bites are usually painless.

5. Development of the symptoms of snakebite envenomation. Important diagnostic symptoms are painful muscle movements, paralysis of the legs, trismus, and ptosis. Usually symptoms will develop within 1 hour after the bite. If there is no evidence of poisoning within 6 hours after the bite then one may conclude that either there has been no envenomation or it was not a sea-snake bite.

General: If treatment is to be effective it must be instituted immediately following the bite. First aid and treatment should generally be directed toward 1) removing as much venom as possible from the wound, 2) retarding absorption of the venom, 3) neutralizing the venom, 4) mitigating the effects produced by the venom, and 5) preventing complications, including secondary infections.

First Aid: Absorption of sea-snake venom is rapid. In most instances the venom is absorbed before first aid can be administered. Suction is of value only if it can be applied within the first few minutes following the bite. Incision and suction are said to be of little value in sea-snake envenomations. It is generally advisable to leave the bite alone. The affected limb should be immediately immobilized, and *all exertion must be avoided.* The patient should lie down and keep the immobilized part below the level of the heart. A tourniquet should be applied tight enough to occlude the superficial venous and lymphatic return. Apply the tourniquet to the thigh in leg bites or above the elbow in upper-limb bites. It should be released for 90 seconds every 10 minutes. A tourniquet is of little value if applied later than 30 minutes following the bite, and it should not be used for more than 4 hours. The tourniquet should be removed as soon as antivenin therapy has been started. Some workers believe that the tourniquet is of little or no value in sea-snake bites. If sea-snake antivenin, or a polyvalent antivenin containing a krait (elapidae) fraction is available, it should be administered intramuscularly either in the buttocks or at some other site distant from the bite. The antivenin should be given only after the appropriate skin or conjunctival test has been made (*see* p. 955 for details on how the test is made). There appears to be an increasing number of physicians who are recommending injecting the antivenin in adequate quantities intravenously without concern for side reactions. They believe that any hypersensitivity can be controlled with the use of adrenalin. Usually one unit (vial or ampule) is sufficient until the patient can be transported to a physician. Keep the patient warm. He should not be given alcoholic beverages but may be given water, coffee, or tea. Keep calm and reassure the patient with encouraging words and actions. Transport the patient immediately to the nearest doctor or hospital, but do not make him walk or exert himself in any way. If at all possible place the offending dead snake in a container and give it to the physician for identification.

Therapy: Sea-snake envenomation is a medical emergency requiring immediate attention. A delay in instituting proper medical treatment can lead to consequences far more tragic than might be incurred in an ordinary traumatic injury. The first step is to determine if envenomation has actually occurred. In most instances, by the time the patient reaches the physician, the 1-hour test period since the bite was made will have been exceeded so there should be some clinical evidence of poisoning if any is to develop. If the patient has been carefully examined and there is no evidence of intoxication, the patient needs only reassurance and observation for a period of an hour or two. If after a period of observation there is no evidence of toxemia, the patient can be released.

If a period of more than 1 hour has elapsed and there is definite evidence of intoxication, antivenin therapy should be instituted immediately upon arrival at the hospital or physician's office. At this time, suction and incision are useless. There is now available a sea-snake antivenin, which is prepared from the venom of *Enhydrina schistosa* and can be obtained from the Commonwealth Serum Laboratories, Melbourne, Australia, and the Snake and Venom Research Laboratories, Penang, Malaysia. This antivenin is concentrated and purified so as to minimize hypersensitivity reactions. If sea-snake antivenin is not available, use a polyvalent antiserum containing a krait (elapidae) fraction. A skin or conjunctival test should be done prior to injection of the antivenin. In persons with a history of extensive allergies, the antivenin should be injected with caution even in the presence of a negative skin test.

Follow the instructions given in the brochure that accompanies the antivenin, or if instructions are not available, the procedure given below: The antivenin can be given either intramuscularly or intravenously. The intramuscular route is said to be safer, but less effective. Again, physicians are increasingly recommending injecting the antivenin intravenously without concern for hypersensitivity. Russell *et al.* (1966) recommends that a portion of the first ampule should be injected subcutaneously proximal to the bite, surrounding the wound, or in advance of the swelling. Antivenin should never be injected into a finger or toe. Avoid injecting large amounts of the antivenin into the injured part. A second portion of the antivenin should be injected intramuscularly into a large muscle mass at some distance from the bite. The last portion should be given intravenously. Intravenous antivenin is indicated with patients in shock. When given intravenously the antivenin should be given by the drip method over a period of 1 hour. The antivenin can be added to the physiological saline solution and given in a continuous drip. Subsequent doses can then be added to the saline solution. The incidence of sensitivity reactions is said to be lessened when the antivenin is combined with hyaluronidase (10 ml of antivenin with 1,000 units of hyaluronidase) and given intramuscularly. Reid (1956*a, b, c,* 1957) recommends an injection of 250 mg of cortisone intramuscularly prior to administering the antivenin but not at the same time that the antivenin is given. If hypersensitivity reactions occur, they can usually be controlled with adrenalin subcutaneously with or without intravenous antihistaminic drugs. If the patient shows a positive reaction to the initial sensitivity test, the antivenin will have to be administered in graded doses in accordance with standard medical procedures for desensitization. One ampule of sea-snake antivenin neutralizes 10 mg of *Enhydrina schistosa* venom and the venom of other common sea-snake species. The minimum effective dosage is 1 ampule. In severe poisoning as evidenced by ptosis, weakness of external eye muscles, dilatation of pupils with sluggish light reaction, and leucocytosis exceeding 20,000, 3 to 4 or more ampules are required. Children generally respond satisfactorily to smaller doses of antivenin than that required by adults, which is contrary to previous textbook statements. If envenomation is present, antivenin therapy should be given as soon as possible, but it may be successful even 8 or more hours after the bite has occurred. Parenteral fluids should always be given following a severe envenomation. Cryotherapy is not recommended.

Supportive measures such as blood transfusions, plasma, vasopressor drugs, antibiotics, antitetanus agents, oxygen, etc., may be required. Corticosteroids are the drugs of choice in combating delayed allergic reactions provoked by the venom or the horse serum. Neurological disorders such as cerebral, meningeal, myelitic, radicular, and neuritic disorders (particularly of the fifth and sixth cervical roots) have been reported (Reid, 1957). Delayed reactions usually do not appear until 3 to 7 days after administration of the antivenin.

Sensitivity Tests:[1] A sensitivity test must be carried out on all victims of sea-snake poisoning before horse serum antivenin is administered. Directions for these tests will usually be found in the package containing the antivenin. In the absence of specific instructions, follow these steps:

A. Inject 0.10 ml of a 1:10 dilution of the horse serum or antivenin intracutaneously on the inner surface of the forearm. Use the specific hypodermic needle provided for the test. If one is not provided, use a short, 27-gauge needle. If the test is done correctly, a wheal will be raised at the site of the injection. The wheal is white at first but if the test is positive the area about the point of injection will become red within 10 to 15 minutes. If any local or systemic allergic manifestations develop within 20 minutes of the test, *do not* give antivenin. Leave this decision to the physician.

If the patient develops a severe reaction to the test (restlessness, flushing, sneezing, urticaria, swelling of the eyelids and lips, respiratory distress, or cyanosis), inject 0.3 to 0.5 ml of 1:1,000 adrenalin subcutaneously and observe the patient closely. Be prepared to administer artificial respiration. A cardiac stimulant may also be needed if shock develops.

B. An alternative to the skin test is the eye test. One or two drops of a 1:10 solution of the horse serum or antivenin are placed on the conjunctiva of one eye. If the test is positive, redness of the conjunctiva will develop within a few minutes. If the reaction is very severe, it should be controlled by depositing a drop or two of 1:1,000 adrenalin directly on the conjunctiva.

C. If a serum sensitivity test is positive, desensitization should be carried out before administering antivenin.

Prevention

Although sea snakes are generally considered to be docile and at times reluctant to bite, some species are aggressive and have no biting inhibitions. It should be kept in mind that even the "harmless" snakes are fully equipped with a venom apparatus and a potentially lethal poison. Sea snakes thus merit respect and thoughtful consideration. Sea snakes may occasionally bite a bather. Reid (1959) has statistically estimated that one sea-snake bite occurs per 270,000 man-bathing-hours in an endemic area such as Penang, Malaysia. The most dangerous areas in which to swim are river mouths. It is probable that accidents are more likely to occur in a river mouth where the sea snakes are more numerous and the water more turbid. The turbidity of the water and resulting poor visibility for the sea snake may contribute to their becoming fearless and less discriminating in their biting habits. When possible, avoid swimming in a river mouth. When wading in an area

---

[1] The author is indebted to F. E. Russell for the use of this material.

inhabited by sea snakes, shuffle your feet. This will prevent stepping on sea snakes. The majority of sea-snake bites have taken place among native fishermen removing fish from their nets. A person should be extremely careful in handling a net haul containing sea snakes. It is advisable to first remove the snakes with the use of a hooked stick or a wire before attempting to handle the fish. Sea snakes are occasionally captured while fishing with a hook and line. Do not attempt to remove the hook from the mouth of the snake. Cut the line, and let the snake drop into the water. Do not handle the snake.

PUBLIC HEALTH ASPECTS

Sea snakes are among the most abundant and venomous of the world's reptiles (Map 2). Poisonings may occur wherever sea snakes are found, but the incidence is highest along the coasts of southeast Asia, Persian Gulf, and the Malay Archipelago. The total number of snakebite deaths[2] in the world (excluding China, USSR, and central Europe) is estimated to be between 30,000 to 40,000 annually, and of this number about 30,000 occur in Asia alone (Swaroop and Grab, 1954; Russell, 1963). Accurate statistics regarding the overall incidence of sea-snake bites are not available since registration of data of this type is almost nonexistent throughout many of the localities in which sea snakes are endemic. Local superstitions and customs in some of these regions further prevent any attempt to obtain even reasonably accurate figures.

The only report on the incidence of sea-snake bites is by Reid and Lim (1957), who conducted a survey among the fishing villages in northwest Malaysia. They found that in 17 villages having a total population of about 40,100 persons, there had been 144 sea-snake bites during the period 1944 (or before) through 1955. An analysis of these reported cases showed that 55 were trivial, 48 severe, and 41 fatal.

The results of the survey were obtained by questioning native fishermen and showing examples of various types of sea snakes. The investigators were fully aware of the weaknesses and difficulties entailed in a study of this type. Many of the natives were reluctant to discuss the subject of sea-snake bites on a personal basis because they believed that talk about sea-snake bites offended the snakes and would change their timidity to aggression. They also believed that if one talks about his bite he will be bitten again, and if a pregnant woman hears of a sea-snake bite she will die. Obviously superstitions of this type are not helpful in obtaining a complete record. Because of the reluctance of the fishermen to disclose bites, the number forgotten, and the relatively small proportions of persons interviewed in the larger villages, Reid and Lim concluded that their statistics were a gross underestimate of the actual number. They believed that the total incidence of sea-snake bites

[2] This includes all species of venomous snakes, both marine and terrestrial.

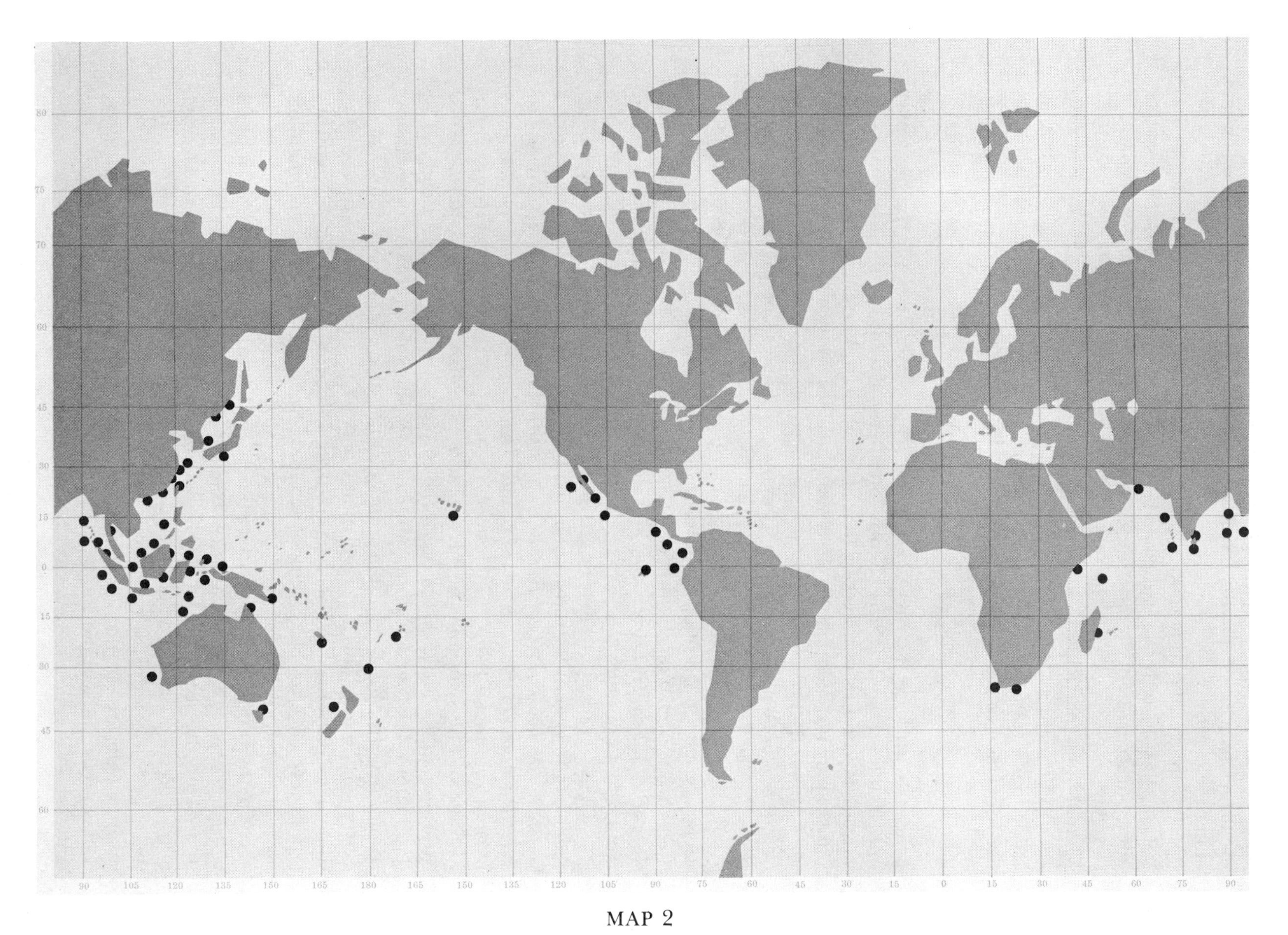

MAP 2

Map showing the geographical distribution of the sea snake *Pelamis platurus*. This is the most widely distributed of the sea snakes and has been incriminated in fatal human envenomations. (L. Barlow)

in Malaysian fishing villages is about 150 persons each year. The incidence in villages near river mouths is believed to be about twice that of other areas along the coast.

It is interesting to note Reid's (unpublished) observation that during the period 1955-59, only two bathers in Penang Island waters died from sea-snake bites, but 21 bathers on the same beaches died from accidental drowning.

Barme (1963) stated that sea snakes are abundant along the coasts of Vietnam and particularly in South Vietnam. Sea snake bites are common among Vietnamese fishermen, and there are "numerous" fatalities each year, but reliable statistics are unknown because of superstitions which forbid the fishermen to discuss their snake bites with anyone.

TOXICOLOGY

The early work on the toxicology of sea-snake venom was for the most part quite crude. Russell (1801, cited by Stephenson, 1838) tested the toxicity of various species of sea snakes by permitting them to bite fowl and in some instances pigeons. The bite from *Hydrophis cyanocinctus* caused death in a fowl in 8 minutes, and *H. nigrocinctus* produced death in a fowl in 7 minutes. *Enhydrina schistosa* killed a fowl in 5 minutes, and in tests on pigeons the venom proved to be 10 times as potent as that from a cobra.

Cantor (1841) conducted a series of experiments similar to those of Russell, using in addition to fowl a tortoise *(Trionyx gangeticus)*, a snake *(Coluber catenularis)*, and a puffer *(Tetraodon patoca)*. The hydrophiids tested were *E. schistosa, H. nicrocinctus,* and *H. cyanocinctus*. The death time varied from 7 to 30 minutes.

The largest reported series in which sea snakes were tested by having them bite fowl and dogs was published by Fayrer (1872). The snakes tested were *E. schistosa, H. cyanocinctus, H. obscurus, H. lapemoides, H. stricticollis, Pelamis platurus, Lapemis hardwicki,* and *Thalassophina viperina*. Death times in the fowl ranged from 1½ minutes to 7 hours and in the dogs from 1 hour to 8 days. The symptoms in the fowl usually consisted of muscular spasms, drooping appearance, dilatation of the pupils, convulsions, and death. The dogs showed hypersalivation, muscular spasms, involuntary evacuations, respiratory distress, convulsions, and death. The venom was found to be extremely potent, but the small fangs and weak musculature of the jaws sometimes prevented the sea snakes from inflicting a fatal bite. Noguchi's (1909) reference to the toxicity of sea-snake venom is largely a commentary upon the work of Fayrer.

The first of a series of more critical toxicological analyses on hydrophiid venom began with the research of Rogers (1903). The venom was collected by making the sea-snake bite on a watchglass covered with a thin layer of gutta percha tissue stretched tightly across it. The venom was ejected onto the glass as clear drops free from saliva. The material was then dried over calcium chloride or strong sulfuric acid and stored until needed in dry, well-corked vials without loss of potency. *E. schistosa* was found to yield the largest amount of venom of the species tested. In some of the smaller species Rogers was unable to extract any venom. In the dried form, the venom appeared as white, shining scales, freely soluble in water or physiological saline, dissolving as a colorless liquid without the yellow tinge that was found in some of the other snake venoms. Thirteen specimens of *E. schistosa* were found to yield an average of 9.4 mg of venom (dried weight) per bite, which was about one twenty-fifth as much venom ejected by a cobra bite. The largest amount of venom obtained from a single bite was 13.0 mg. Upon boiling, the venom became slightly opalescent. The toxicity of the venom was largely destroyed after boiling for 1 minute. Varying amounts of the venom were injected subcutaneously into rabbits, rats, birds, and fishes. Symptoms generally consisted of signs of drowsiness, muscular weakness, ataxia, respiratory distress, paralysis, convulsions, and death. See Table 1 for the minimal lethal dosages obtained.

Table 1. —*MLD's of sea snake venoms for various animals (From Rogers, 1903)*

ANIMAL	*E. schistosa*	*T. viperina*	*H. cyanocinctus*	*M. cantoris*	*P. platurus*	COBRA*
White rat	0.07 mg/kg	....	....	....	....	0.5 mg/kg
Rabbit	0.04 mg/kg	....	....	....	....	0.7 mg/kg
Pigeon	0.05 mg/kg	0.5 mg/kg	0.5 mg/kg	0.5 mg/kg	0.075 mg/kg	0.5 mg/kg "birds"
Fowl	0.04 mg/kg	....	....	....	....	
Mudfish (*Saccobranchus fossilis*)	0.5 mg/kg	0.75 mg/kg	1.0 mg/kg	0.75 mg/kg	0.25 mg/kg	25.0 mg/kg

* Note the potency of sea-snake venom in comparison with that of cobra venom. Rogers (1903) stated that in some specimens of *E. schistosa* the venom was found to be from 10 to 20 times as potent as cobra venom.

Taking the minimal lethal dose of *Enhydrina* venom as 0.05 mg/kg for warm-blooded animals, Rogers estimated that the minimal lethal dosage for a 70-kg man would be 3.5 mg or about one-third of the average amount of venom ejected by a fresh adult sea snake. Sea-snake venom was also tested on frogs and terrestrial non-venomous snakes, but the experiments were too incomplete to be conclusive.

Another series of experiments on the minimum lethal dosages of *E. schistosa* and *L. curtus* venom was conducted by Fraser and Elliott (1905). The venom collected from six specimens of *E. schistosa*, averaging about 76 cm in length, amounted to about 30 to 40 mg (due to an accident the exact amount was unknown). Eight specimens of *L. curtus* (size of snakes not given) yielded 22.0 mg of venom or 2.75 mg of venom per snake. The venom was collected by removing the glands from the snake. The venom was expressed from the larger glands onto a watchglass and dried in a dessicator over sulfuric acid. The smaller glands were dried whole in a similar manner. The material was then stored in sealed vials. Later, the small glands were opened, macerated in successive small quantities of thymol water, filtered, and evaporated to dryness over sulfuric acid under vacuum. The material obtained was a yellow, brittle, scaly product, soluble in water, having much the same appearance as the venom expressed from the fresh glands. The venom was dissolved in Ringer's solution for injection purposes. By this process, 60 dried glands of *E. schistosa* weighing 1.09 g yielded 280 mg of crude venom which was used in determining the following MLD's:

ANIMALS	*E. schistosa*	*L. curtus*	COBRA
White rats	0.09 mg/kg	0.6 mg/kg	0.5 mg/kg
Rabbits	0.06 mg/kg	...	0.6 mg/kg
Cats	0.2 mg/kg	...	1.0 mg/kg

Fraser and Elliot found the potency of the dried venom to be a little less than half that of fresh venom. Animals given Calmette's cobra antivenin were found to have only a "feeble" amount of protection from the sea-snake venom. These findings agree with Rogers (1903), who stated that Calmette's serum is "of no use against the poison of the Hydrophiidae."

Smith and Hindle (1931) worked with the venom of *Laticauda colubrina*. The venom was "milked" from three snakes. The crystals of the dried venom were pale yellow in color and weighed 10.2 mg. The yield of a single adult snake was estimated to be 5.1 mg. The MLD for mice was found to be 0.113 mg/kg body weight, whereas cobra venom had an MLD of 0.25 mg/kg. An intraperitoneal injection of 0.51 mg/kg in an eel caused death in 100 minutes. A second eel was injected with 0.1275 mg/kg and died during the night. A black wrasse injected intraperitoneally with 0.51 mg/kg died in 51 minutes. The authors concluded that the bite of *L. colubrina* is not likely to be fatal to a healthy adult human unless the yield of venom is greater than the comparatively small quantity which they obtained.

One of the most extensive reports on the toxicology of sea-snake venoms is by Reid (1956*c*). Reid collected the venom from a large series of sea snakes comprising ten different species. The venom was collected by making the snake bite on a plastic spoon covered with a plastic sheet tautly drawn over and tied below with a string. The spoons were then placed in a silica gel dessicator in the dark for 2 or more days until the venom was dried. The dried venom was then removed from the spoon, placed in a vial, and weighed. The maximum dried-venom yield per snake was as follows: *E. schistosa*, 54.4 mg; *H. cyanocinctus*, 21.3 mg; *H. spiralis*, 10.8; *K. jerdoni*, 5.61 mg; and *L. hardwicki*, 5.4 mg. It was found that a spontaneous bite gave a greater yield than by applying pressure over the venom gland. The specimens of sea snakes were assumed to be adult specimens and varied in length from 40 to 100 cm or more.

The largest venom yield was usually taken on the first milking, but at times this would vary. After the maximum yield had been obtained, subsequent yields were substantially smaller. The yield from *H. klossi*, which has a very small head, was repeatedly small and believed to be of negligible danger to man.

Reid compiled a table based on the findings of various investigators as to

Table 2.—*MLD (mg/kg) of sea-snake venom for various animals (modified from Reid, 1956c)*

SEA SNAKE SPECIES	MOUSE	RAT	GUINEA PIG	RABBIT	PIGEON	CHICKEN	CAT	DOG	MUDFISH
Enhydrina schistosa	0.1–0.15 s.c. 0.13 i.v.	0.07 s.c. 0.09 s.c.	0.09 i.v.	0.02 s.c. 0.06 s.c. 0.05 s.c. 0.08 i.v.	0.05 s.c.	0.04 s.c.	0.25 s.c.	0.035 s.c.	0.5 s.c.
Hydrophis cyanocinctus					0.5 s.c.				1.0 s.c.
Pelamis platurus					0.075 s.c.				0.25 s.c.
Praescutata viperina					0.5 s.c.				0.75 s.c.
Microcephalophis cantoris					0.5 s.c.				0.75 s.c.
Lapemis curtus		0.6 s.c.							
Laticauda colubrina	0.25 s.c.								

the minimum lethal dosage of different species of sea snakes (Table 2). He suggested that since sea-snake venom is largely neurotoxic without "thrombase," it is unlikely that the route of administration would significantly affect the results. The toxicity of *E. schistosa* venom compared favorably with that of the three most poisonous land snakes (*see* Table 3). It is interesting to note that *E. schistosa* is about twice as toxic as the tiger snake *Notechis scutatus*, 4 times as toxic as the death adder *Acanthophis antarcticus*, and 10 times as toxic as the common cobra *Naja naja*.

Reid also investigated the rate of absorption of *E. schistosa* venom. Several dogs were placed under ethyl chloride anesthesia, and the tails were bitten by different specimens of *E. schistosa*. He observed that in each case, after one to four strong jaw contractions of the snake, fangmarks could be seen on the dorsum of the dog's tail. In six of the dogs, the tail was amputated 8 cm above the bite by a single hatchet blow, at intervals timed by a stopwatch. Bleeding was immediately arrested with the use of a tourniquet. The four dogs whose tails were amputated in 1 to 2 minutes showed no ill effects, but the dog which had the tail amputated in 3 minutes died in 27 minutes, and the last dog whose tail was amputated in 4 minutes died in 30 minutes. In the four control cases in which there were no amputations, one dog died in 34 minutes and the others showed no ill effects. The study showed that a lethal venom dose can be absorbed within 3 minutes or less. One noteworthy

Table 3.—*MLD in mg dried venom per kg weight of various animals (subcutaneous injection) (modified from Reid, 1956c)*

SPECIES	MOUSE	RAT	RABBIT	PIGEON	CAT	DOG
Enhydrina schistosa	0.13	0.07	0.02	0.05	0.25	0.035
Naja naja		0.8	0.25	0.5	10.0	1.0
Notechis scutatus	0.25	0.4	0.045		0.1	
Acanthophis antarcticus	0.7	0.4	0.15		0.5	

finding was that three of the four controls did not develop any symptoms despite well-marked jaw contractions and fangmarks. These experiments showed that sea snakes may bite without injecting any significant amounts of venom.

Tu (1959) investigated the toxicity of the venom of *L. semifasciata* of Taiwan on rabbits, mice, guinea pigs, and frogs. The results were as follows: the LD_{50} by subcutaneous injection for mice was 0.338 (± 0.013) mg/kg; for guinea pigs 0.0897 (±0.0028) mg/kg; and for rabbits 0.0495 (±0.0027). The LD_{50} by intravenous injection for mice was 0.211 (±0.012) mg/kg; for guinea pigs 0.0631 (±0.0023); and for rabbits 0.0486 (±0.0025) mg/kg. Thus the frogs were found to be the most tolerant to this venom and the rabbits the most susceptible. Tu *et al.* (1962) studied the toxicity of the venom of *L. colubrina* of Taiwan on rats, rabbits, and frogs. They found the LD_{50} for rats varied from 0.378 to 0.45 mg/kg. The MLD for rabbits was 0.11 mg to 0.13 mg/kg. Details of their study were not given.

Carey and Wright (1960*a*) reported that the LD_{50} of *E. schistosa* venom for a variety of laboratory animals was in the region of 50 to 100 μg/kg or about four times the toxicity of cobra venom. The yield per bite for captive snakes was up to 55 mg of dried venom. It was estimated that one bite may contain up to 15 LD_{50} for humans. The lethal dose in man is believed to be about 3.5 mg (Carey and Wright, 1960*b*). Sea-snake venom resembles cobra venom in some of its properties but differs in others. The toxic fraction of *E. schistosa* venom was electropositive at pH 9.8, whereas cobra venom was electronegative at that pH. Cobra and *E. schistosa* venoms gave characteristic but different immuno-electrophoretic patterns with their specific antisera at pH 6.8, and there was only a slight cross-precipitation between these venoms and antisera. Cobra antivenom gave only a low degree of neutralization of *E. schistosa* venom by the *in vitro* method of testing. Cobra venom had a high resistance to heat, but *E. schistosa* venom fell off very rapidly with heating. After 2 minutes of heating at 100° C, only 14 percent of the toxicity remained, and no toxicity remained after 30 minutes at that temperature. The researchers obtained low titers in potency tests on their sera and believed that it was because the toxic fraction of the venom is a poor antigen which may be attributed to its low molecular weight and possibly to its very strong electropositive character. *In vitro* testing revealed that 1 ml of their most potent antiserum would neutralize the toxicity of 0.2 mg of venom. Therefore, on that basis, 75 ml would be required to neutralize the toxicity of the venom of one average bite. The authors believed that specific *E. schistosa* antivenom would not equally protect against *H. cyanocinctus* and *H. spiralis* venoms, but there was sufficient cross-neutralization to indicate that a polyvalent antivenom against the three common dangerous sea snakes would be a useful lifesaving therapeutic agent.

Barme (1963) gave the MLD (dried venom) by the intravenous route on 20 g mice for various species of sea snakes as follows:

Lapemis hardwicki	4.0 μg
Enhydrina schistosa	2.5 μg
Hydrophis cyanocinctus	7.0 μg
Kolpophis annandalei	11.0 μg
Microcephalophis gracilis	2.5 μg
Hydrophis fasciatus	3.5 μg
Thalassophina viperina	7.0 μg
Pelamis platurus	10.0 μg

These levels are considerably higher than those of venoms from land snakes evaluated under the same conditions. The venom of the water viper *Ancistrodon piscivorius* had an MLD of 150 μg, and the venoms of the Egyptian cobra *Naja naja* and the bamboo snake *Trimeresurus grammineus* had an MLD of 20 μg each. Barme found that there was an antigenic similarity existing between the venoms of different species of the Hydrophiidae. He prepared a specific antivenin in a horse with the venom of *Lapemis*. At the end of 6 months the horse furnished a serum in which 1 ml neutralized 400 μg of homologous venom (LD_{100} in mice). The same serum that neutralized 400 μg of the venom of *Lapemis* neutralized 150 μg (60 LD_{100}) of the venom of *Enhydrina* and 250 μg (35 LD_{100}) of the venom of *H. cyanocinctus*. The antivenin was tested on a human case with good results.

Homma *et al.* (1964, 1965) have recently shown that the venoms of *H. cyanocinctus, L. semifasciata*, and *L. laticaudata* captured near Amami Oshima, Ryukyu Islands, are fatal when injected intramuscularly into mice. rats, and guinea pigs. They found that the survival in these animals could be prolonged by neutralizing the venom *in vitro* with various agents such as edathamil calcium disodium, tetracycline, and sodium citrate. Okonogi, Hattori, and Igarashi (1967) have experimentally immunized rabbits against the venom of *L. semifasciata* using formalized venom as the immunogen.

Tamiya and Arai (1966) isolated two toxic fractions from the sea snake *Laticauda semifasciata* which they designated as erabutoxin *a* and *b*. Both of the toxins had LD_{50} values of 0.15 μg/g body weight for mice and 0.07 μg/g for rats.

Cheymol *et al.* (1967) studied the venom taken from sea snakes *Enhydrina schistosa, Hydrophis cyanocinctus* and *Lapemis hardwicki* collected along the coast of Vietnam. Toxicity tests were conducted using lots of 10 Swiss mice weighing 18-22 g each. The venom was injected intravenously. They calculated the LD_{50} for each of the snake species as follows: *E. schistosa*, 0.35 mg/kg; *H. cyanocinctus*, 0.66 mg/kg; *L. hardwicki*, 0.44 mg/kg.

Toxicity studies on the venom of the sea snake *Laticauda laticaudata* and *L. semifasciata* have been conducted by Vick (1968; *see* Halstead, 1970) in the mouse, dog, monkey, and chimpanzee. A total of 400 mice, 100 dogs, 20 monkeys, and 2 chimpanzees were used. The intravenous LD_{100} dose was first established in the mouse. This data provided a basis from which to operate and was used as a guide in the dog studies. Dogs were anesthetized with sodium pentobarbital, 30 mg/kg, and recorded for both physiological and pharmacological changes following intravenous challenge with venom. The parameters monitored in the dog were cortical electrical activity (EEG), respiration, heart rate, arterial and venous blood pressure, total blood histamine, and serum catecholamines. All animals were followed from time of injection until death. Chart 1 shows a typical response of the dog to 0.05 mg/kg of *L. laticaudata* venom. Chart 2 shows a typical response of the dog to 0.05 mg/kg of *L. semifasciata* venom. An interesting component of the response of the dog to the sea-snake venom was the relative lack of physiological changes during the first 5-10 minutes after challenge. Of the 38 animal venoms studied to date these two sea-snake venoms were the only venoms which did not produce a precipitous fall in blood pressure and a marked decrease in heart rate immediately after injection. From this data it appears that the venom of the sea snake has a unique and singular effect—that of producing a block of the nerve impulse transmission at the myoneural junction of the diaphragm. Death seems to quickly follow this interruption in respiratory function. Stimulation of the phrenic nerve at time of death does not produce a perceivable response in the diaphragmatic musculature. Direct stimulation of the diaphragm does, however, produce a contracture. The exact mechanism of the block is as yet unknown but does appear to be "curare-like."

Studies in the monkey and chimpanzees closely paralleled those in the dog with the important exception that all venoms were given intramuscularly to unanesthetized preparations. This was to mimic more closely those changes in vital functions that might be expected following an actual envenomation. Parameters were monitored using indwelling catheters and telemetry transmitter devices. The changes observed in the primates were identical to those observed in the dog. As might be expected the onset of action was somewhat slower because of the intramuscular route of administration.

No significant change in blood histamine or serum epinephrine, norepinephrine levels were noted with either venom.

The actual lethal doses for each venom are as follows:

	L. laticaudata	*L. semifasciata*
I.V. Mouse	0.06 mg/kg	0.08 mg/kg
I.V. Dog	0.05 mg/kg	0.01 mg/kg
I.V. Monkey	0.08 mg/kg	0.10 mg/kg
I.V. Chimpanzee	0.10 mg/kg	0.10 mg/kg

Levey (1969) extracted venom from the sea snake *Laticauda colubrina* taken at Pulau Sudon, near Singapore. The LD_{50} was determined at 0.45 mg/kg using the method of Litchfield and Wilcoxon.

Tu and Ganthavorn (1969) conducted toxicity tests on the following sea snakes taken from various localities: *Enhydrina schistosa* and *Hydrophis cyanocinctus* from the Strait of Malacca, *Lapemis hardwicki* from the Gulf of Thailand, S. Thailand, and *Pelamis platurus* from N. Formosa. Toxicity tests were made

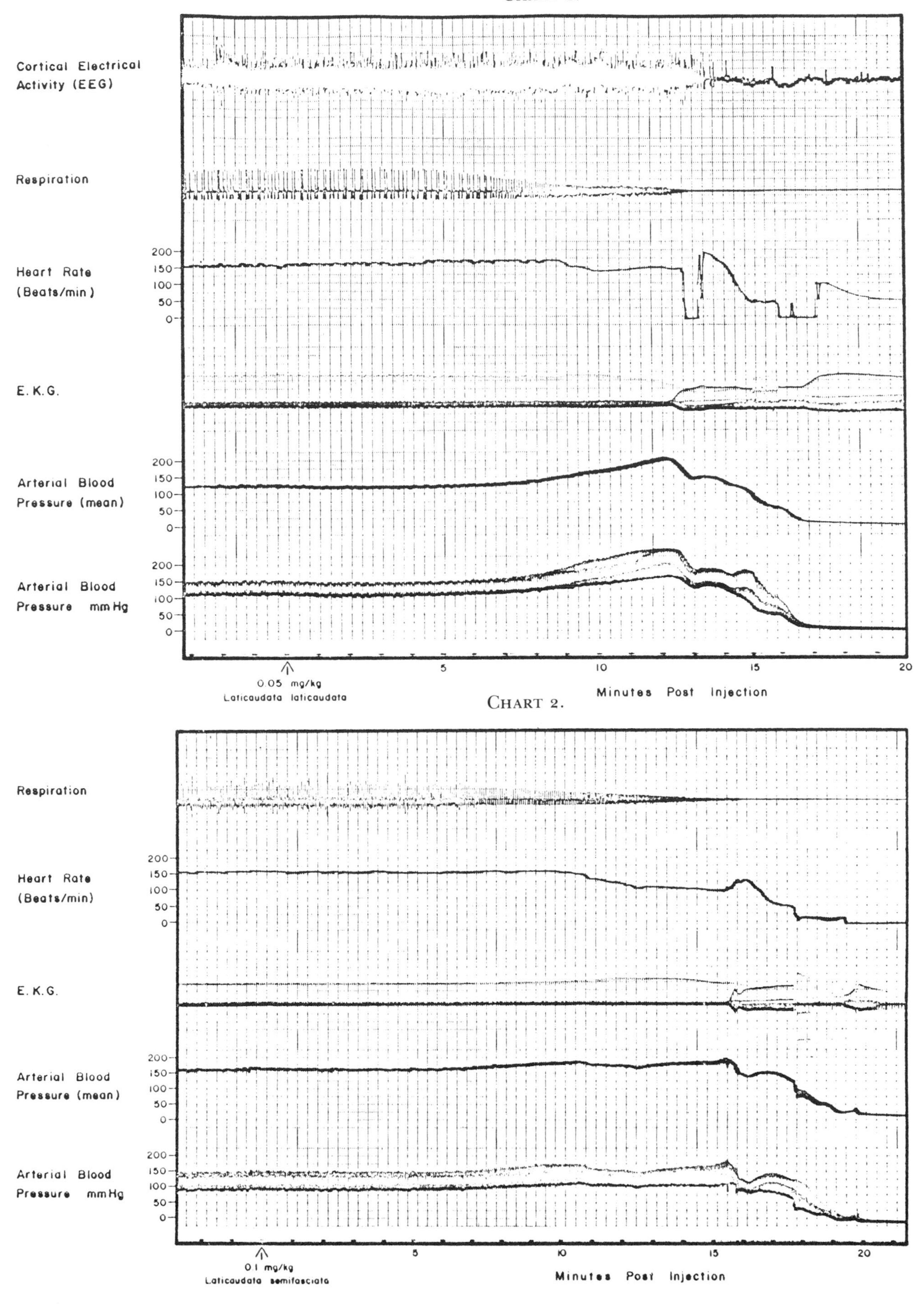
CHART 1.
Cortical Electrical Activity (EEG)
Respiration
Heart Rate (Beats/min)
200
150
100
50
0
E.K.G.
Arterial Blood Pressure (mean)
200
150
100
50
0
Arterial Blood Pressure mmHg
200
150
100
50
0
5
10
15
20
0.05 mg/kg
Laticaudata laticaudata
Minutes Post Injection
CHART 2.
Respiration
Heart Rate (Beats/min)
200
150
100
50
0
E.K.G.
Arterial Blood Pressure (mean)
200
150
100
50
0
Arterial Blood Pressure mmHg
200
150
100
50
0
5
10
15
20
0.1 mg/kg
Laticaudata semifasciata
Minutes Post Injection

TABLE 4.—*Yield and Toxicity of* Lapemis hardwicki.

Year Collected	Number of Snakes	Venom Yield (mg/snake)	Toxicity in Mice LD_{50} (mg/kg—IV)
1967	68	5.2	0.71
1969	4305	2.4	1.4

by injecting 0.25 ml of venom of varying concentrations intravenously into Swiss white mice weighing 20 g each. The number of mice that died in 24 hours was observed. Six dosage levels of five mice each were used and the LD_{50} value was calculated by the method of Litchfield and Wilcoxon. The LD_{50}'s determined were as follows: *E. schistosa*—0.09 mg/kg; *H. cyanocinctus*—0.35 mg/kg; *P. platurus*—0.18 mg/kg; *L. hardwicki*—0.71 mg/kg. It will be observed that the venom of *E. schistosa* was the most toxic and that of *L. hardwicki* the least toxic.

Tu and Hong (1971) collected specimens of *Lapemis hardwicki* which were captured in the Gulf of Thailand and off the east coast of Kra Isthmus, Thailand, during July, 1967, and July-August, 1969. The toxicity tests were done by injecting 0.2 ml of venom of varying concentrations intravenously into Swiss white mice weighing 20 g each. Five mice were used at each of the six dosage levels. The number of mice which died was observed after 24 hours. The toxicity was determined statistically with the method of Litchfield and Wilcoxon and expressed as the lethal dosage 50 percent value. The yield and toxicity resulting from the collections of *Lapemis hardwicki* taken during 1967 and 1969 are summarized in Table 4.

In 1969, a large number of *L. hardwicki* were collected, including many small-sized snakes, which probably accounts for the smaller yield of venom in the 1969 group in comparison with the 1967 group.

Tu and Toom (1971) isolated the toxic component from the venom of *Enhydrina schistosa* captured in the Strait of Malacca during the summer of 1967. The toxin was isolated by passage of the venom through a carboxylmethyl cellulose column. The preparation was shown to be homogeneous by zonal electrophoresis, ultracentrifugation, isoelectric focusing, and gel filtration. Toxicity was determined by injecting 0.25 ml of toxin dissolved in 0.85 percent NaCl solution into the tail vein of 20 g Swiss mice. Ten to fifteen mice were injected at each of 15 dosage levels for each toxicity determined. The median LD_{50} was determined by the method of Litchfield and Wilcoxon and was found to be 0.44 μg/g of body weight.

PHARMACOLOGY

Reports on studies that deal specifically with the pharmacological effects of sea-snake venom on various organ systems are the product of only a few authors. Some of the first reports on the pharmacology of sea-snake venom are by Rogers (1903, 1904*a, b, c,* 1905). He worked with the venom of *Enhydrina bengalensis* (a synonym of *E. schistosa*) from specimens taken near

Puri, India. The venom was collected by having the snakes bite through a thin layer of gutta percha tissue stretched tightly over a watchglass. The venom was then dried over calcium chloride or concentrated sulfuric acid and finally transferred to a tightly stoppered vial for use in testing.

Varying dilutions of *Enhydrina* venom dissolved in physiological saline were used to test its effect on pigeon and human red blood cells. It was found that cobra venom produced a comparatively rapid hemolysis of pigeon cells (7 minutes), but the sea-snake venom produced a very gradual hemolysis over a prolonged period of time (24 hours or more). Human red blood cells were slightly more resistant and required a longer period of time to hemolyze. Similar results were obtained from the venom of the sea snakes *Disteira cyanocincta* (a synonym of *Hydrophis cyanocinctus*) and *Hydrophis cantoris* (a synonym of *Microcephalophis cantoris*). Rogers concluded cobra venom was 1,000 times more potent in its hemolytic properties than *Enhydrina* venom.

It was found that cobra venom had the ability to destroy the clotting mechanism of blood. However, Rogers found that, when tested under the same conditions, the *Enhydrina* venom failed to inhibit the clotting mechanism. From his studies on the blood, he concluded that sea snakes produce their lethal effects by acting directly on the nervous system.

Initially, Rogers (1903) lacked adequate physiological recording equipment for his experiments. Nevertheless, he attempted to determine the primary site of the neurotoxic effects by injecting rabbits intravenously with *Enhydrina* venom in doses of 1 mg/kg of body weight or about 20 times the MLD. He then carefully observed the results. He concluded that the primary effect of the venom was a paralyzing action on the respiratory center and that cardiac failure was secondary to that of the respiratory paralysis. In his later studies, Rogers (1904*a, b, c,* 1905) was able to confirm his earlier observations. He decided that the primary action of the poison was directly on the respiratory center, quickly followed by paralysis of the end plates of the phrenic nerves. He felt that *Enhydrina* venom resembled cobra venom and curare in that it had a marked action in paralyzing the motor end plates but did not perceptibly alter nerve conduction. He further decided that the action on the reflex functions of the spinal cord were slight and secondary to the effects on respiration.

Rogers also tested the affinity of *Enhydrina* venom for nerve tissue by emulsifying fresh pigeon-brain tissue (cerebrum and cerebellum) with minimum lethal doses of the venom. The emulsion was then permitted to stand for varying periods of time. He used as a control similar solutions of venom to which no brain tissue had been added. He found that the emulsified samples were much less toxic than the controls, even though they contained equal amounts of the venom. He believed that the grey matter was effective in fixing the poison.

Fraser and Elliot (1904, 1905) studied the pharmacology of the venom of *E. valakadien* (a synonym of *E. schistosa*) and *Enhydrus curtus* (a synonym of *Lapemis curtus*). The venom was obtained by removing the venom glands from the snake, expressing the venom in a watchglass, and then placing the material in a desiccator to dry over sulfuric acid. The tests were conducted by dissolving the dried venom in Ringer's solution in various proportions.

The venom was tested on the red blood cells of rats to determine if there was any evidence of hemolysis, but there was none. Perfusion experiments were conducted using frog heart to determine if sea-snake venom had a direct effect on the heart. The results were negative except for very strong solutions (1:5,000) which produced tonic contractions of the frog ventricle. It had been previously found that cobra venom produced constriction of the blood vessels even in dilute quantities (1:10,000,000). Similar tests with *Enhydrina* venom were negative. Experiments with the mammalian heart exposed *in situ* using anesthetized cats and rabbits failed to reveal any evidence of direct action on the vagal cardioinhibitory center. This was found to be in striking contrast to cobra venom which acts as a powerful stimulant on the cardioinhibitory center and heart muscle. There was also no evidence of direct action on the vasomotor center. Fraser and Elliot concluded, as did Rogers (1904*a*), that the primary action of the venom was directly on the respiratory center of the brain. They also found that there was paralysis of the motor end plates.

The pharmacological properties of sea-snake venom have more recently been studied by Tu (1959), using the venom of *Laticauda semifasciata*. He found that in rabbits the venom produced a pronounced respiratory inhibition and a gradual rise in blood pressure with small doses, a transient fall in blood pressure in moderate doses, and a rapid rise of blood pressure in large doses. There were no remarkable cardiac changes revealed by *in situ* experiments. The venom induced a stimulating action on the isolated frog heart but much less stimulation in rabbit heart. The coronary outflow of rabbit heart was slightly increased. The venom produced a vasodilatation of the hind-leg vessel of frogs in small doses but vasoconstriction after dilatation in larger doses. The venom constricted the ear vessels of rabbits after a transient dilatation. The venom stimulated contraction of rabbit and frog smooth muscles of the intestinal tract and uterus. It was found to have both a curarelike action and a direct paralytic action on isolated frog gastrocnemius. There was no paralytic action on either the sensory or the motor nerve fibers of the sciatic nerve of the frog. A slight analgesic effect in mice was observed.

Carey and Wright (1961) concluded from their studies on the venom of *Enhydrina schistosa* that the paralyzing effect of the venom on isolated rat phrenic nerve diaphragm preparation indisputably demonstrated a peripheral action. The initial effect of the venom was at the neuromuscular junction, but with larger doses there was permanent muscle damage. When the whole venom was injected directly into the medulla of rabbits, death resulted immediately even with small doses. However, if equivalent doses of a fraction containing lecithinase activity was used, similar effects on the medulla were observed, although the fraction was nontoxic when injected intraperitoneally. On the other hand, they found that the neurotoxic fraction, which acted similarly to the whole venom when injected intraperitoneally, had no effect on the medulla.

The pharmacological effects of *E. schistosa* venom on nerve and muscle have been reviewed by Meldrum (1965).

Tamiya and Arai (1966) demonstrated with frog-muscle preparations that the toxic fractions erabutoxin a and b from *Laticauda semifasciata* acted on the postsynaptic membrane to block neuromuscular transmission.

Cheymol *et al.* (1967) studied the effects of the venom of *Lapemis hardwicki* on intact rats and cats and on isolated nerve-muscle preparations. They found that the toxin produced a neuromuscular blocking action which resulted in paralysis of the peripheral nerves without affecting the muscular fibers or the nerve itself. They believed that the site of action was at the specific receptors of the postsynaptic membrane which was blocked irreversibly by the toxin.

Intravenous administration of 0.40 mg/kg of *Laticauda colubrina* venom to an anesthetized dog produced respiratory arrest, cardiovascular collapse, and death (Levey, 1969). The symptoms of envenomation were suggestive of medullary paralysis, but were nevertheless inconclusive. Aqueous extracts of the plant *Clinacanthus nutans* which is used in southeast Asia as a sea-snake antidote were found to be ineffective in prolonging the survival time of mice injected by lethal doses of the venom.

Sato and Tamiya (1968) determined that when erabutoxin b from *Laticauda semifasciata* was iodinated at pH 9.6, the histidine residue at position 26 was preferentially modified to form the di-ido derivative without loss of toxicity. Their findings provided histochemical evidence which supported the view that ^{131}I-labeled erabutoxin b binds to the end plates of the mouse diaphragm. Further studies were conducted by Sato, Abe, and Tamiya (1970) using ^{131}I-labeled erabutoxin b that had been incubated with isolated mouse diaphragms, which resulted in radioautograms showing that the radio-isotope was concentrated at the end plates. The results of these studies suggested that erabutoxin b binds specifically to the end plates *in vitro* as well as *in vivo* and that the bound toxin is not removed by washing.

Tu and Ganthavorn (1969) studied the immunologic properties of the sea snakes *Enhydrina schistosa, Hydrophis cyanocinctus, Lapemis hardwicki*, and *Pelamis platurus*. Antivenin of *E. schistosa* was found to neutralize effectively the venoms of each of the four species.

Tu, Passey, and Toom (1970) isolated phospholipase A from the venom of the sea snake *Laticauda semifasciata*. The enzyme isolated exhibited hemolytic activity which was greatly intensified by the addition of lecithin. The purified phospholipase A was nontoxic, nonhemorrhagic, and exhibited only slight myolitic activity.

Cheymol *et al.* (1967) studied the paralyzing action of the venom of *Laticauda semifasciata* and of its neurotoxins (erabutoxins a and b) *in situ* in cat and rat neuromuscular preparations and on isolated preparations. The crude venom and the toxins caused a slowly developing paralysis which was spontaneously reversible. The erabutoxins had a similar but more potent paralyzing activity than the venom. The paralysis was peripheral and blocked the end-plate receptors without affecting the muscle fibers or the acetylcholine output at nerve endings. They concluded that the venom and the toxins do not act by simple competition with acetylcholine as does d-tubocurarine.

Sato, Ogahara, and Tamiya (1972) obtained antisera by injecting the formaldehyde-treated erabutoxin a or b in rabbits. Each 0.1 ml of the best preparations of the anti-erabutoxin a- and b-sera neutralized the toxicity of 25 μg of both erabutoxins. On double diffusion in agarose gel, the antisera formed a single continuous precipitation line with erabutoxins a and b and diiodoerabutoxin b. The antisera did not precipitate, however, with laticotoxin a of *Laticauda laticaudata* or with cobrotoxin of *Naja naja atra*. The radioimmunoassay experiments showed that the affinity of the antibodies to laticotoxin a or cobrotoxin was less than one-hundredth of that of erabutoxins.

CHEMISTRY

by DONOVAN A. COURVILLE

There are very little published data available regarding the chemistry of sea-snake venom. Carey and Wright (1960*a, b*) have shown that the venom of *Enhydrina schistosa* contains at least three antigenic fractions. The neurotoxic component was the most electropositive (pH 6.8 in agar gel). It would pass through a cellulose but not nylon dialysis sac. They found that after repeating the cellulose sac dialysis against large volumes of distilled water, about 1 percent of the initial toxicity remained inside the sac. It was suggested that this may be due to a different toxic component or to adsorption of traces of dialyzable component onto nondialyzable material. Immunoelectrophoresis of the cellulose sac dialysate using antiserum to the whole venom revealed only one main precipitation line. However, further studies showed, by varying the relative concentrations of venom and antiserum in the immunoelectrophoresis, that it consisted of at least two antigenic components. Placing the dialysate on a carboxymethyl-cellulose column and eluting it with pH 6.6 buffers of graded molarities from 0.06 to 0.08 *M* resulted in further purification of the toxic component. They recovered 80 to 100 percent of the toxicity of the initial sample in the eluant in one sharply-defined zone. Immunoelectrophoresis of the sample using antiserum to the whole venom gave a single precipitation line only over a range of relative concentrations of venom and antiserum. When whole venom was mixed with excess antiserum prepared against the toxic fraction only, the supernatant was nontoxic but still contained antigenic material as demonstrated by immunoelectrophoresis.

Carey and Wright (1962) isolated the toxic factor of the venom of *E. schistosa* by dialysis and column chromatography. They found that the leci-

FIGURE 27.—Crystals of laticaudatoxin from the venom of the sea snake *Laticauda semifasciata*. This particular fraction has been termed "erabutoxin *a*" by Tamiya and Arai (1966). × 150.

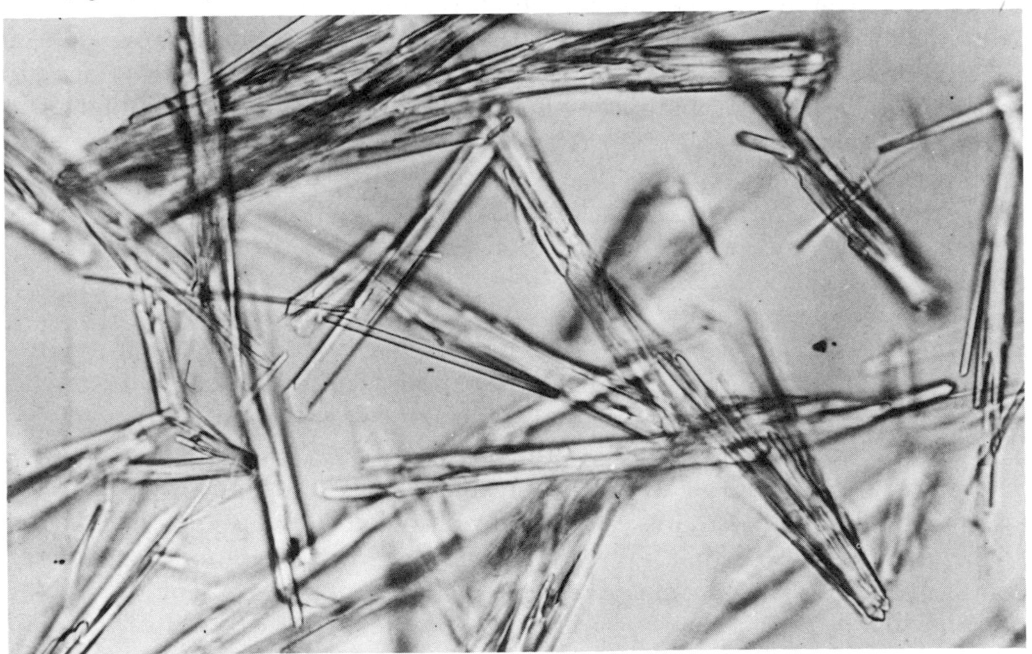

FIGURE 28.—Crystals of laticaudatoxin from the venom of the sea snake *Laticauda semifasciata*. This particular fraction has been termed "erabutoxin *b*" by Tamiya and Arai (1966). The term "erabutoxin" was derived from the Japanese vernacular name of the sea snake "erabu-umihebi." × 200. (Courtesy N. Tamiya)

thinase activity of the whole venom was not associated with the toxic fraction. Antiserum to the toxic factor was able to neutralize the toxicity of large doses of the whole venom.

Proceeding from earlier evidence that the venom of sea snakes contains toxic components of a protein nature, Tamiya and Arai (1966) undertook the problem of isolating and further characterizing the venom from specimens of *L. semifasciata*. Two toxic proteins were isolated, to which were given the names "erabutoxin *a*" (Fig. 27) and "erabutoxin *b*" (Fig. 28) after the vernacular Japanese name of the snake "erabu-umihebi."

The secretion from the parotid glands of the snakes was dried over sodium hydroxide pellets at -20° C, suspended in 0.01 M phosphate buffer at pH 6.4, and centrifuged at 4000 rpm. This process was repeated on the resuspended residue, the combined supernatants were dialyzed against the same buffer, the centrifugation was repeated, and the centrifugate was chromatographed on a CM-cellulose column equilibrated to the same buffer. The concentration of the sodium chloride in the eluant was increased by increments, and the fractions were assayed for protein content. A small-scale experiment showed that the major fraction of the toxicity (95 percent) was nearly equally distributed in the second and third of three peaks. The toxin in the second peak was assigned the name erabutoxin *a*; that in the third peak erabutoxin *b*.

The combined eluants of erabutoxin *a* from a series of such separations were concentrated at freezing temperature, redialyzed, and concentrated to a small volume. Addition of saturated ammonium sulphate to yield a 30 percent saturation resulted in the formation of crystaline needles (Fig. 27) on standing at 4° C. Erabutoxin *b* (Fig. 28) was crystallized by a similar procedure. Solutions of each toxin revealed that no other protein components were present. Electrophoresis of a solution of the crude toxin revealed two components which at pH 4.3 moved toward the cathode. These two components corresponded to erabutoxins *a* and *b*. Other protein components moved toward the anode.

Both toxins were found to be unstable to heat, with losses of 70 percent of the toxicity in 5 and 10 minutes boiling respectively. The molecular weights were determined by ultracentrifugation methods and found to be the same at 7430, a figure in satisfactory agreement with results from amino acid analysis.

The amino acid composition of the two toxins was determined by analysis of the acid hydrolysate in an automatic analyser. The tryptophane content, lost by the acid hydrolysis, was estimated by ultraviolet absorption. The results of the analysis showed that both toxins contain 61 or 62 amino acids, the remaining question being related to the number of tryptophane molecules present (one or two). Alanine and methionine were found absent. The only difference detectable from the analyses was that one aspartic acid of erabutoxin *a* is replaced by histidine in erabutoxin *b*.

Tu, Passey, and Toom (1970) isolated phospholipase A from the venom of the sea snake *Laticauda semifasciata* on CM-cellulose followed by chromatography on DEAE-cellulose. The preparation, which was shown to be homogeneous by zonal electrophoresis, ultracentrifugation, isoelectric focusing, and gel filtration, had a specific activity seven times that of crude venom. By using ovolecithin of known composition in the 1 and 2 positions, the enzyme was shown to be specific for the 2 position, liberating mainly unsaturated fatty acids. The molecular weight of the enzyme, determined by both Sephadex gel filtration and by a combination of sedimentation and diffusion coefficients from ultracentrifugation was 11,000.

Ibrahim (1970) studied the action of phospholipase A in heated sea-snake *(Enhydrina schistosa)* venom on purified phospholipids, and the phospholipids of plasma and intact muscle cells. The enzyme required activation by deoxycholate in order to achieve maximal hydrolysis of purified lecithin, phosphatidylserine, and the phospholipids of human plasma. The hydrolysis of purified phosphatidylethanolamine is inhibited by deoxycholate. The choline, serine, and ethanolamine phosphatides of the intact muscle cell were also hydrolyzed. The author suggested that perhaps the myotoxic action of sea-snake venom may reside in the ability of its phospholipase A fraction to directly attack the three-di-acyl phospholipid species of the intact muscle cell membrane. He further suggested that blood (erythrocytes and plasma) is a poor substrate for the phospholipase A of sea-snake venom and no hemolysis is produced. Because of the lack of specificity of the sea-snake enzyme for blood phospholipids, the competition between blood and muscle substrate may be less with sea-snake venom than with cobra venom. Eventually the circulating enzyme of sea-snake venom may reach the muscle cells and attack membrane phospholipids producing myolysis and myoglobinuria. This may account for the latent period between sea-snake bites and the development of symptoms, since myoglobinuria occurs several hours after the bite.

Tu, Hong, and Solie (1971) found that the venom of *Laticauda semifasciata* contained two toxins. These toxins were isolated and purified by means of Sephadex gel filtration and CM-cellulose column chromatography. The purity of the toxins was established using isoelectric focusing, electrophoresis, sedimentation velocity, and sedimentation equilibrium. Both toxins were isolated in crystalline form. The lethality of the purified toxins increased from five to sixfold when compared to the original venom. The molecular weight as determined by amino acid composition, gel filtration, and by sedimentation equilibrium was approximately 6800 for both toxins. The amino acid composition of the two toxins was quite similar. Toxin *a* contained 62 amino acid residues and toxin *b* contained 61 amino acid residues. No free sulfhydryl groups were detectable. End-group analysis showed both toxins to be a single polypeptide chain with arginine at the N terminal and aspartic acid (asparagine)) at the C terminal. The isoelectric points were determined by isoelectric focusing to be 9.15 for toxin *a* and 9.34 for toxin *b*. The toxicity of the toxins was unaltered by heating to 100° C for 30 minutes or by exposure to pH extremes from 1 to 11. The toxicity was completely lost when the tryptophan residue in the purified toxins was modified with N-bromosuccinimide. However, no change in antigenicity was observed for the N-bromosuccinimide-modified toxins. No significant change in the toxicity was observed when the lysine and arginine residues of the purified toxins were modified.

Tu and Hong (1971) isolated the toxic fraction from the venom of *Lapemis hardwicki* by means of a four-step purification by column chromatography. The toxin was proven homogenous by zonal electrophoresis, analytical ultracentrifugation, and isoelectric focusing. The toxin was also isolated in crystal form. The lethality of the purified toxin increased 23-fold when compared to the original venom. The molecular weight as determined by amino acid composition and by sedimentation equilibrium was aproximately 6800. The toxin consisted of 61 amino acid residues and did not contain a free sulfhydryl group. End group analysis indicated that there was a histidine at the NH_2-terminal and an aspartic acid or asparagine at the COOH-terminal. The isoelectric point as determined by isoelectric focusing was 9.85.

The toxic component from the venom of the sea snake *Enhydrina schistosa* was isolated by Tu and Toom (1971) by passage of the venom through a carboxymethyl cellulose column. The preparation was shown to be homogeneous by zonal electrophoresis, ultracentrifugation, isoelectric focusing, and gel filtration. The molecular weight of the toxin as determined by sedimentation velocity and diffusion experiments was found to be 7200. The minimum molecular weight calculated from amino acid analysis was 7000. The isoelectric point, determined by isoelectric focusing was 9.2. Although hyaluronidase, alkaline phosphatase, phosphodiesterase, phospholipase A, acetylcholinesterase, DNase, leucine amino peptidase, and clotting activity could all be detected in unfractionated venom, no enzymatic activity could be found in the purified toxin. The compound N-bromosuccinimide completely destroyed the toxicity of the preparation which, under the conditions employed, strongly suggested that tryptophan is necessary for toxicity.

Tu and Passey (1972) isolated phospholipase A from venom of *Laticauda semifasciata* by column chromatography using a combination of DEAE- and CM-cellulose. Homogeneity was established by ultracentrifugation, isoelectric focusing, electrophoresis on cellulose acetate, and gel filtration. Molecular weight was determined by the following two independent methods: a combination of s(1.93×10^{-13} sec) and D($1.41 \times 10^{-6}cm^2/sec$) and Sephadex gel chromatography. The molecular weight was 11,000 by combination of s and D and 10,700 from the gel filtration method. Among the several phospholipids tested, only L-α-lecithin was hydrolyzed, indicating that the enzyme was very specific. The enzyme exhibited hemolytic activity which was greatly intensified by the addition of lecithin. The isolated phospholipase A had relatively little activity on mouse embryo cells and very weak myolytic activity.

Raymond and Tu (1972) reported that the single tyrosine residue contained in the neurotoxin from the venom of *Lapemis hardwicki* was found to be essential for the toxic action through iodination and nitration. It was suggested that tyrosine seems to be "masked" or "buried" in the protein structure. Their work indicated that there were certain chemical similarities between cobra and sea-snake venoms.

Sato, Ogahara, and Tamiya (1972) treated erabutoxins a and b obtained from *Laticauda semifasciata* with 2 percent (w/v) formaldehyde. It was found that erabutoxin b lost toxicity, changed in its ORD and CD properties and was modified on the amino, guanidyl, and hydroxyphenyl groups of the amino residues. The modified erabutoxin b was monodispersed on ultracentrifuge and had a molecular weight of 68,000—about ten times that of the the original toxin.

Liu *et al.* (1973) separated three toxic fractions from the crude venom of *Hydrophis cyanocinctus* by column chromatography. The most abundant fraction which they termed "hydrophitoxin a" was studied for amino acid composition and was found to contain 19 of the common amino acids (including asparagine and glutamine). Phenylalanine was absent. The total number of amino acid residues per molecule was estimated to be 59. Methionine was found to occupy the N-terminal position and asparagine was at the C-terminal end. They concluded that from these preliminary results "hydrophotoxin a" appeared to be similar in molecular size and structure to several other sea-snake venoms that had been previously reported by other workers.

Tu (1974) lists the species of sea snakes from which the venoms have been purified; these include *Lapemis hardwicki, Enhydrina schistosa, Laticauda laticaudata, L. colubrina*, and *L. semifasciata*. The terms "*laticotoxin*" and "*erabutoxin*" have been applied to some of these toxins which appear to be chemically identical (*see also* Uwatoko, *et al.*, 1966*a;* Seto *et al.*, 1970).

Tamiya (1975) published a detailed chemical analysis on the amino acid sequence of sea-snake neurotoxins. This data should be examined by anyone interested in this topic.

LITERATURE CITED

BARME, M.

1958 Contribution à l'étude des serpents marins venimeux *Hydrophiidae* du Viet-Nam. Bull. Soc. Pathol. Exot. 51: 258-265.

1963 Venomous sea snakes of Viet Nam and their venoms, p. 373-378. *In* H. L. Keegan and W. V. MacFarlane [eds.], Venomous and poisonous animals and noxious plants of the Pacific region. Macmillan Co., New York.

BARME, M., and J. DETRAIT

1959 Etude de la composition des venins des Hydrophiides. Compt. Rend. Séan. Acad. Sci. 248: 312-315.

BECKE, L.

1909 Venomous sea-snakes, p. 305-310. *In* L. Becke, 'Neath Austral skies. J. Milne, London.

BELLAIRS, A.

1969 The life of reptiles, 2 vols. Weidenfeld and Nicolson, London.

BERGMAN, R. A.

1949 The anatomy of *Lapemis hardwicki* Gray. Proc. Konink. Ned. Akad. Wetenschap. 52(8): 882-898.

BIERDRAGER, J.

1936 Een geval van massale schildpadvergiftiging in Nw. Guinee. Ned. Tijdschr. Geneesk. 76(18): 1933-1944, 1 fig.

BOKMA, H.

1942 Casuïstische mededeeling. Nog eens een beet van een zeeslang. Ned. Tijdschr. Geneesk. 82: 87.

BOLTT, R. E., and R. F. EWER

1964 The functional anatomy of the head of the puff adder, *Bitis arietans* (Merr.). J. Morphol. 114: 83-106, 12 figs.

BOYS, F., and H. M. SMITH

1959 Poisonous amphibians and reptiles—recognitions, and bite treatment. C. C. Thomas, Springfield, Ill. 149 p., 29 figs.

BURNS, B., and B. V. PICKWELL

1972 Cephalic glands in sea snakes *(Pelamis, Hydrophis*, and *Laticauda)*. Copeia 1972(3): 547-559, 9 figs.

CALDWELL, D. K., *coord.*

1960 Sea turtles of the United States. U.S. Fish Wildlife Serv., Fish Leaflet 492. 20 p., 28 figs.

CANTOR, T.

1841 Observations upon pelagic serpents. Trans. Zool. Soc. London 2: 303-313; Pls. 56-57.

CAREY, J. E., and E. A. WRIGHT

1960*a* Isolation of the neurotoxic component of the venom of the sea snake, *Enhydrina schistosa*. Nature 185: 103-104.

1960*b* The toxicity and immunological properties of some sea-snake venoms with particular reference to that of *Enhydrina schistosa*. Trans. Roy. Soc., Trop. Med. Hyg. 54(1): 50-67, 7 figs.

1961 The site of action of the venom of the sea snake *Enhydrina schistosa*. Trans. Roy. Soc. Trop. Med. Hyg. 55(1): 153-160, 2 figs.

1962 Studies on the fractions of the venom of the sea snake *Enhydrina schistosa*. Aust. J. Exp. Biol. 40: 427-435.

CARR, A.

1952 Handbook of turtles. The turtles of the United States, Canada, and Baja California. Cornell Univ. Press, Ithaca, N.Y. 542 p., 82 pls., 37 figs.

CHEYMOL, J., BARME, M., BOURILLET, F., and M. ROCH-ARVEILLER

1967 Action neuromusculaire de trois venins d'hydrophiidés. Toxicon 5: 111-119, 8 figs.

COOPER, M. J.

1964 Ciguatera and other marine poisoning in the Gilbert Islands. Pacific Sci. 18(4): 411-440, 11 figs.

DAY, F.

1869 On the bite of the sea-snake. Indian Med. Gaz. 4: 92.

DERANIYAGALA, P. E.

1939 The tetrapod reptiles of Ceylon. Vol.. I. Testudinates and crocodilians. Ceylon J. Sci., Colombo Mus. Nat. Hist. Ser. 412 p., 24 pls., 137 figs.

Ditmars, R. L.
1931 Snakes of the world. Macmillan Co., New York. 207 p.
Dunson, W. A.
1975 The biology of sea snakes. University Park Press, Baltimore, 530 p.
Fairley, N. H.
1934 Snake bite: its mechanism and modern treatment. Proc. Roy. Soc. Med. 27: 1083-1094.
Fayrer, J.
1872 The thanatophidia of India, being a description of the venomous snakes of the Indian Peninsula, with an account of the influence of their poison on life; and a series of experiments. J. and A. Churchill, London. 156 p., 31 pls.
Fish, C. J., and M. C. Cobb
1954 Noxious marine animals of the central and western Pacific Ocean. U.S. Fish Wildlife Serv., Res. Rept. No. 36, p. 14-23.
Fitzsimons, F. W.
1919 The snakes of South Africa. Their venom and the treatment of snake bite. 3d ed. T. Maskew Miller, Cape Town. 547 p., 26 pls., 152 figs.
Fossen, A.
1940 Vergiftiging door den beet van zeeslangen. Ned. Tijdschr. Geneesk. 80: 1164-1166.
Fraser, T. R., and R. F. Elliot
1904 Contributions to the study of the action of sea-snake venoms—Part I. Proc. Roy. Soc. (London) 74(498): 104-109.
1905 Contributions to the study of the action of sea-snake venoms—Part I. Venoms of *Enhydrina valakadien* and *Enhydris curtus.* Phil. Trans. Roy. Soc. London, Ser. B, 197: 249-279.
Gray, M. E.
1930 Note on sea snakes. Australian Nat. 8: 88.
Günther, A.
1864 On a poison organ in a genus of batrachoid fishes. Proc. Zool. Soc. London (1): 155-158, 2 figs.
Haas, G.
1952 The head muscles of the genus *Causus* (Ophidia, Solenoglypha) and some remarks on the origin of the Solenoglypha. Proc. Zool. Soc. London, 122: 573-592, 13 figs.
Halstead, B. W.
1959 Dangerous marine animals. Cornell Maritime Press, Cambridge. 146 p., 88 figs.
1970 Poisonous and venomous marine animals of the world, vol. 3. Vertebrates. U.S. Govt. Printing Office, Washington, D.C.
Halstead, B. W., Engen, P. C., and A. T. Tu
1977 The venom and venom apparatus of the sea snake *Lapemis hardwicki* Gray. Linnean Society of London (Zoology). [In press]
Hashimoto, Y., Konosu, S., and T. Yasumoto
1967 Investigation on toxic marine animals in the Ryukyu and Amami Islands, 4. A survey on turtle poisoning. [In Japanese] Tech. Rept. Lab. Mar. Biochem. Fac. Agric., Univ. Tokyo.
Heatwole, H.
1975 Attacks by sea snakes on divers. *In* W. A. Dunson [ed.], The biology of sea snakes. University Park Press, Baltimore, p. 503-516.
Herre, A. W.
1942 Notes on Philippine sea-snakes. Copeia (1): 7-9.
Homma, M., Abe, R., Okonogi, T., Kosuga, T., and S. Mishima
1965 Studies on Habu snake and Erabu sea-snake venoms. Outlines of biological toxicities of the two snake venoms, and inhibitory actions of tannic acid on them. [In Japanese, English summary] Nihon Saikingaku-zasshi 20(6): 281-289, 4 figs.
Homma, M., Okonogi, T., and S. Mishima
1964 Studies on sea snake venom. 1) Biological toxicities of venoms possessed by three species of sea snakes captured in coastal waters of Amami Oshima. Gunma J. Med. Sci. 13(4): 283-296, 6 figs.
Ibrahim, S. A.
1970 A study on sea-snake venom phospholipase A. Toxicon 8: 221-224 2 figs.
Keegan, H. L.
1958 Some venomous animals of the Far East. 406th Medical General Laboratory, Camp Zama, Japan. 33 p., 60 figs.

KERMORGANT, A.
1902 Les serpents de mer et leur venin. Ann. Hyg. Med. Colon. 5: 431-435, 6 figs.

KINGHORN, J. R.
1961 The snakes of Australia. Rep. ed. Angus and Robertson, London. 197 p., 137 figs.

KINUGASA, M., and W. SUZUKI
1940 Über untersuchungen der ursache der massenhaften vergiftung nach dem genuss von fleisch einer an der küste von koryo in der präfektus sintiku gefangenen seeschildkröte. [In Japanese, German summary] Taiwan Igohkai Zassi 39(74): 66-74.

KLAWE, W. L.
1964 Food of the black-and-yellow sea snake, *Pelamis platurus*, from Ecuadorian coastal waters. Copeia (4): 712-713.

KLEMMER, K.
1963 Liste der rezenten giftschlangen: Elapidae, Hydropheidae, Viperidae und Crotalidae, p. 255-464, 37 pls. *In* Behringwerke-Mitteilungen [ed.], Die giftschlangen der erde. N. G. Elwert Universitäts- und Verlags-Buchhandlung, Marburg.

KOCHVA, E. T.
1958 The head muscles of *Vipera palaestinae* and their relation to the venom gland. J. Morphol. 102: 23-53, 13 figs.

KOCHVA, E., and C. GANS
1970 Salivary glands of snakes. Clin. Toxicol. 3(3): 363-387, 37 figs.

KOCHVA, E., and M. WOLLBERG
1970 The salivary glands of Aparallactinae (Colubridae) and the venom glands of *Elaps* (Elapidae) in relation to the taxonomic status of this genus. Zool. J. Linnean Soc. 49(3): 217-224, 4 pls.

KOPSTEIN, F.
1930 Die giftschlangen Javas und ihre bedeutung für den menschen. Z. Morphol. Oekol. Tiere 19: 339, 361-363.

LAMB, G., and W. K. HUNTER
1907 On the action of venoms of different species of poisonous snakes on the nervous system. VI. Venom of *Enhydrina valakadien*. Lancet 2: 1017-1019.

LE MARE, D. W.
1952 Poisonous Malayan fish. Med. J. Malaya 7(1): 1-8, 1 pl.

LEVEY, H. A.
1969 Toxicity of the venom of the sea-snake, *Laticauda colubrina*, with observations on a Malay "Folk Cure." Toxicon 6: 269-276.

LIU, C.-S., HUBER, G. S., LIN, C.-S., and R. Q. BLACKWELL
1973 Fractionation of toxins from *Hydrophis cyanocinctus* venom and determination of amino acid composition and end groups of hydrophitoxin *a*. Toxicon 11: 73-79, 4 figs.

LOISEL, G.
1904 Substances toxiques extraites des oeufs de tortue et de poule. Compte. Rend. Soc. Biol. 57: 133-135.

LOVERIDGE, A.
1945 Reptiles of the Pacific world. Macmillan Co., New York. 259 p., 7 pls.

MARCACCI, A.
1891 Sur le pouvoir toxique du sang de thon. Arch. Ital. Biol. 16: i.

MARSDEN, A. T., and H. A. REID
1961 Pathology of sea-snake poisoning. Brit. Med. J. 1: 1290-1293, 5 figs.

MELDRUM, B.S.
1965 The actions of snake venoms on nerve and muscle. The pharmacology of phospholipase A and of polypeptide toxins. Pharm. Rev. 17(4): 393-445.

M'KENZIE,—
1820 An account of venomous sea snakes, on the coast of Madras. Asiatick Res. 13: 329-336.

NAUCK, E. G.
1929 Untersuchungen über das gift einer seeschlange *(Hydrus platurus)* des Pazifischen Ozeans. Arch. Schiffs. Tropenhyg. 33: 167-170.

NIKOL'SKII, A. M.
1963 Fauna of Russia and adjacent countries. Reptiles. Vol. I. Chelonia and Sauria. [Engl. transl.] IPST Cat. No. 838. Israel Prog. Sci. Transl., Jerusalem. 352 p., 9 pls., 69 figs.

NOGUCHI, H.
1909 Snake venoms. An investigation of venomous snakes with special reference to the phenomena of their venoms. Carnegie Inst. Wash. Publ. No. 111. 315 p., 33 pls.

Okonogi, T., Hattori, Z., and I. Igarashi
1967 Experimental studies on immunization against sea-snake venom. [In Japanese, English summary] Nihon Saikingaku-zasshi 22(3): 173-177, 2 figs.

Pillai, V. K., Nair, M. B., Ravindranathan, K., and C. S. Pitchumoni
1962 Food poisoning due to turtle flesh (a study of 130 cases). J. Assoc. Phys. India 10(4): 181-187.

Raymond, M. L., and A. T. Tu
1972 Role of tyrosine in sea-snake neurotoxin. Biochim. Biophys. Acta 285: 498-502, 2 figs.

Reid, H. A.
1956*a* Sea-snake bites. Brit. Med. J. 2: 73-85, 4 figs.
1956*b* Three fatal cases of sea-snake bite, p. 367-371. *In* E. E. Buckley and N. Porges [eds.], Venoms. Am. Assoc. Adv. Sci., Washington, D.C.
1956*c* Sea-snake bite research. Trans. Roy. Soc. Trop. Med. Hyg. 50(6): 517-542.
1957 Antivenene reaction following accidental sea-snake bite. Brit. Med. J. 2: 26-29.
1959 Sea-snake bite and poisoning. Practitioner 183: 530-534.
1961 Diagnosis, prognosis, and treatment of sea-snake bite. Lancet 2: 399-402.
1963*a* Treatment of snake-bite poisoning. Brit. Med. J. 1: 1675.
1963*b* Snakebite in Malaya, p. 355-362. *In* H. L. Keegan and W. V. Macfarlane [eds.], Venomous and poisonous animals and noxious plants of the Pacific region. Macmillan Co., New York.
1975 Epidemiology and clinical aspects of sea snake bites. *In* W. A. Dunson [ed.], The biology of sea snakes. University Park Press, Baltimore, p. 418-462.

Reid, H. A., and K. J. Lim
1957 Sea-snake bite. A survey of fishing villages in northwest Malaya. Brit. Med. J. 1957: 1266-1272.

Reid, H. A., Thean, P. C., and W. J. Martin
1963 Specific antivenene and prednisone in viper-bite poisoning: controlled trial. Brit. Med. J. 2: 1378-1380.

Rogers, L.
1903 On the physiological action of the poison of the Hydrophidae. Proc. Roy. Soc. London 71: 481-496.
1904*a* On the physiological action of the poison of the Hydrophidae. Part II —Action on the circulatory, respiratory, and nervous systems. Proc. Roy. Soc. London 72: 305-319.
1904*b* The physiological action and antidotes of snake venoms with a practical method of treatment of snake bites. Lancet 1: 349-355.
1904*c* On the physiological action and antidotes of colubrine and viperine snake venoms. Proc. Roy. Soc. London 72(485): 419-423.
1905 The physiological action and antidotes of colubrine and viperine snake venoms. Phil. Trans. Roy. Soc. London, Ser. B, 197: 123-191.

Romeyn, T., and G. T. Haneveld
1956 Vergiftigingen door het eten van schildpadvless *(Eretmochelys imbricata)* op Nederlands Nieuw-Guinea. Ned. Tijdschr. Geneesk. 100: 1156-1159.

Rooij, N. de
1915 The reptiles of the Indo-Australian Archipelago. I. Lacertilia, Chelonia, Emydosauria. E. J. Brill, Leiden. 384 p., 132 figs.

Rosenberg, H. I.
1967 Histology, histochemistry, and emptying mechanism of the venom glands of some elapid snakes. J. Morphol. 123: 13: 156, 26 figs.

Russell, F. E.
1962 Snake venom poisoning, p. 197-210B. *In* G. M. Piersol [ed.], Cyclopedia of medicine, surgery and specialities. Vol. 2. F. A. Davis Co., Philadelphia.
1963 Research review: snake venom and antivenins. Med. News 1963: 4 p.

Russell, F. E., Quilligan, J. J., Jr., Rao, S. J., and F. A. Shannon
1966 Snakebite. J. Am. Med. Assoc. 195(7): 596-597.

Sato, S., Abe, T., and N. Tamiya
1970 Binding of iodinated erabutoxin *b*, a sea snake toxin, to the endplates of the mouse diaphragm. Toxicon 8: 313-314, 1 fig.

Sato, S., and N. Tamiya
1968 The primary structure of erabutoxins. [In Japanese] Abstr. Symp. Protein Structures 19: 13.

SATO, S., OGAHARA, H., and N. TAMIYA
1972 Immunochemistry of erabutoxins. Toxicon 10: 239-243, 4 figs.

SETO, A., SATO, S., and N. TAMIYA
1970 The properties and modification of tryptophan in sea snake toxin, erabutoxin *a*. Biochim. Biophys. Acta 214: 483-489.

SMITH, M. A.
1926 Monograph of the sea-snakes (Hydrophiidae). 1964 Rep. ed. British Museum, London, 130 p., 2 pls., 35 figs.
1935 The sea snakes (Hydrophiidae). DANA Rept. No. 8, Copenhagen. 6 p.
1943 Serpentes, p. 379-441. The fauna of British India, Ceylon, and Burma. Reptilia and Amphibia, vol. 3. Taylor and Francis, London.

SMITH, M., and E. HINDLE
1931 Experiments with the venom of *Laticauda, Pseudechis* and *Trimeresurus* species. Trans. Roy. Soc. Trop. Med. Hyg. 25(2): 115-120.

STEJNEGER, L.
1907 Herpetology of Japan and adjacent territory. Bull. U.S. Nat. Mus. 58: 400-442.

STEJNEGER, L., and T. BARBOUR
1933 A check list of North American amphibians and reptiles. 3d ed. Harvard Univ. Press, Cambridge. 185 p.

STEPHENSON, J.
1838 Medical zoology and mineralogy. J. Churchill, London. 350 p.

STONE, F. S., and F. WALL
1913 An unusually large sea-snake *(Distira brugmansi)*. J. Bombay Nat. Hist. Soc. 22: 403-404.

SWAROOP, S., and B. GRAB
1954 Snakebite mortality in the world. Bull. World Health Organ. 10: 35-76.

TAMIYA, N.
1975 Sea-snake venoms and toxins. *In* W. A. Dunson [ed.]. The biology of sea snakes. University Park Press, Baltimore, p. 385-415.

TAMIYA, N., and H. ARAI
1966 Studies on sea-snake venoms. Biochem. J. 99: 624-630, 5 figs.

TAYLOR, E. H.
1921 Amphibians and turtles of the Philippine Islands. Pacific Islands Bur. Sci. Publ. No. 15. Bureau of Printing, Manila. 193 p.
1965 The serpents of Thailand and adjacent waters. Univ. Kansas Sci. Bull. 45(9): 609-1096.

TU, A. T.
1974 Distribution of sea snakes in southeast Asia and the Far East and chemistry of venoms of three species. *In* H. T. Hummand and E. Lane [eds.], Bioactive compounds from the sea. Vol. 1. Marine Science. Marcel Dekker, New York. p. 207-230.

TU, A. T., and S. GANTHAVORN
1969 Immunological properties and neutralization of sea-snake venoms from Southeast Asia. Am. J. Trop. Med. Hyg. 18(1): 151-154, 4 figs.

TU, A. T., and B.-S. HONG
1971 Isolation and characterization of the toxic component of *Enhydrina schistosa* (common sea snake) venom. J. Biol. Chem. 246(4): 1012-1016, 4 figs.

TU, A. T., HONG, B.-S., and T. N. SOLIE
1971 Characterization and chemical modifications of toxins isolated from the venoms of the sea snake, *Laticauda semifasciata*, from Philippines. Biochemistry 10(8): 1295-1304, 9 figs.

TU, T., LIN, M. J., YANG, H. M., LIN, H. J., and C. N. CHEN
1962 Toxicological studies on the venom of a sea snake, *Laticauda colubrina* (Schneider). [In Japanese] J. Formosan Med. Assoc. 61(12): 122.

TU, A. T., and R. B. PASSEY
1972 Phospholipase A from sea snake venom and its biological properties, p. 419-436, 8 figs. *In* A. de Vries and E. Kochva [eds.], Toxins of animal and plant origin. Gordon and Breach Science Publ., London.

TU, A. T., PASSEY, R. B., and P. M. TOOM
1970 Isolation and characterization of phospholipase A from sea snake, *Laticauda semifasciata* venom. Arch. Biochem. Biophys. 140: 96-106.

TU, T.
1959 Toxicological studies on the venom of a sea snake, *Laticauda semifasciata* (Reinwardt) in Formosan waters. J. Formosan Med. Assoc. 58(4): 182-203, 10 figs.

TU, A. T., and P. M. TOOM
1971 Isolation and characterization of the toxic component of *Enhydrina schistosa* (common sea snake) venom. J. Biol. Chem. 246(4): 1012-1016, 4 figs.

TWEEDIE, M. W.
1941 Poisonous animals of Malaya. Malaya Publishing House, Singapore. 90 p., 29 figs.

UWATOKO, Y., NOMURA, Y., KOJIMA, K., and F. OBO
1966*a* (Title not available). Acta Med. Univ. Kagoshima 8: 141. (NSA)
1966*b* Ibid. 8: 151.

VOLSØE, H.
1939 The sea snakes of the Iranian Gulf and the Gulf of Oman, with a summary of the biology of the sea snakes. E. Munksgaard, Copenhagen. Danish Sci. Investig. Iran (Pt. I). 45 p.

WALL, F.
1921 *Ophidia taprobanica;* or the snakes of Ceylon. H. R. Cottle, Govt. Printer, Colombo, Ceylon. 581 p.

WERLER, J. E., and H. L. KEEGAN
1963 Venomous snakes of the Pacific area, p. 219-326. *In* H. L. Keegan and W. V. MacFarlane [eds.], Venomous and poisonous animals and noxious plants of the Pacific region. Macmillan and Co., New York.

WERMUTH, H., and R. MERTENS
1961 Schildkröten-krokodile brückenechsen. G. Fischer, Jena. 442 p., 270 figs.

WEST, G. S.
1895 On the buccal glands and teeth of certain poisonous snakes. Proc. Zool. Soc. London 52: 812-826, 3 pls.

(NSA)—Not seen by author.

Chapter XXVI—VERTEBRATES

Class Mammalia

Poisonous: WHALES, POLAR BEARS, WALRUSES, SEALS

The species presented in this chapter comprise a small diverse group of poisonous marine mammals. Most of the species involved inhabit cold temperate or arctic waters.

Mammals are characterized by a body that is usually covered with hair and a skin containing various types of glands. The skull possesses two occipital condyles. The jaws usually have differentiated teeth which are contained in sockets. The limbs are variously adapted for walking, climbing, burrowing, swimming, or flying. The feet have claws, nails, or hoofs. The heart is four-chambered, with only a left aortic arch. The lungs are large and elastic. There is a diaphragm between the thoracic and abdominal cavities. The male has a penis and fertilization is internal. The eggs are small or minute and usually retained in a uterus for development. The females have mammary glands that secrete milk to nourish the young. Body temperature is regulated. It is estimated that there are more than 4,000 species of living mammals.

Although Walker *et al.* (1964) list 19 orders of mammals for the class Mammalia in their monumental work *Mammals of the World*, there are only three orders of concern to the marine zootoxicologists. These three orders are:

CETACEA: Whales, dolphins, and porpoises.
CARNIVORA: Carnivores, particularly the polar bear.
PINNIPEDIA: Seals, sea lions, and walruses.

Poisonous Whales, Dolphins, and Porpoises

Mammals of the order Cetacea are characterized by their spindle-shaped body form. The head is long, often pointed, and joined directly to the body. Some species have a fleshy dorsal fin. The flippers or fore limbs are broad and paddle-like; the digits are embedded and have no claws. There are no hind limbs. The tail is long and ends in two broad, transverse, fleshy flukes notched in the midline. The teeth, when present, are all alike and lack enamel. The nostrils are on top of the head. The ear openings are minute. The body surface is smooth, without hairs except for a few on the muzzle. There are no skin glands except mammary and conjunctival glands. A thick layer of fat (blubber) under the skin affords insulation.

The living cetaceans comprise 38 genera and about 90 species, distributed among eight families. They occur in all seas of the world and in certain rivers and lakes. There are three marine species which are reportedly poisonous to eat.

The biology and systematics of the cetaceans mentioned in this chapter have been discussed by Tate (1947), Troughton (1947), Norman and Fraser (1948), Bee and Hall (1956), Tomilin (1957), Hall and Kelson (1959), Walker *et al.* (1964), Daugherty (1965), and Harrison and King (1965). Hershkovitz' (1966) *Catalog of Living Whales* was found to be especially useful.

REPRESENTATIVE LIST OF CETACEANS REPORTED AS POISONOUS

Phylum CHORDATA

Class MAMMALIA

Order CETACEA: Whales, Dolphins, Porpoises

Family BALAENOPTERIDAE

280 *Balaenoptera borealis* Lesson (Pl. 1, fig. a). Sei whale (USA), agalagitakg (Aleutians), iwashikujira, kaguo-kuzira (Japan), seiwal (Germany), rorqual de Rudolf (France), seihval (Norway), saidianoi kit (USSR), sandereydur (Iceland).

DISTRIBUTION: Atlantic Ocean, coast of Labrador southward to Campeche; Pacific Ocean, Bering Sea southward to Baja California.

SOURCE: Mizuta *et al.* (1957).

Family MONODONTIDAE

280 *Delphinapterus leucas* (Pallas) (Pl. 1, fig. b). White whale, beluga (USA), sisuak (Arctic Alaska), baleine blanche (France), huitingar (Iceland), hvitfisk (Sweden, Norway), belukha (USSR).

DISTRIBUTION: Arctic and Subarctic seas.

SOURCE: Stefansson (1924, 1944).

Family PHYSETERIDAE

280 *Physeter catodon* Linnaeus. (Pl. 1, fig. c). Sperm whale (USA), kegutilik (Greenland), pottwal (Germany), cachalot (France), makko kum kujira (Japan), kashalot (USSR).

DISTRIBUTION: Polar, temperate, and tropical seas.

SOURCE: Sahashi (1933).

BIOLOGY

Family BALAENOPTERIDAE: The fin back whales are the largest of living animals. The largest member of the group, the blue or sulphur-bottom whale,

Sibbaldus musculus (Linnaeus), attains a length of 30 m and a weight of 112,500 kg. This family is composed of three genera and six species and occurs in all oceans. This is one of two families of baleen whales in which the embryonic teeth are replaced by baleen plates in the adult animals. Fin back whales are frequently called "rorquals," which refers to "a whale having folds or pleats." The rorquals are equipped with longitudinal furrows, usually 10 to 100 in number and 2.5 to 5 cm deep, which are present on the throat and chest. These furrows increase the capacity of the mouth when opened. The members of this family are the fastest swimmers of the baleen whales, some of them attaining speeds up to 48 km/hr. They usually travel singly or in pairs, but several hundred individuals may congregate when food is abundant. Their food consists largely of euphausiid shrimp, copepods, amphipods, and other zooplankton. Some species even include fishes and penguins in their diet. The zooplankton are captured by gulping and swallowing or skimming. When skimming, the whale swims through the zooplankton with its mouth open and its head above the surface of the water. When a mouthful of organisms has been filtered from the water by the baleen plates, the whale dives, closes its mouth, and swallows the plankton. Rorquals breed and give birth in the warmer waters within their range. The larger species give birth to a single calf every other year, but the smaller forms breed more frequently. Several of the members of this family are hunted commercially for their oil and meat. Whales are considered to be among the most healthy of all living mammals since evidence of pathology is seldom observed (Slijper, 1962).

Family MONODONTIDAE: The family of white whales consists of only two genera, each having a single species, *Delphinapterus leucas*, the white whale, and *Monodon monoceros*, the narwhal. Both species are found in arctic seas but ascend rivers. Ingestion of the flesh of the white whale has caused fatalities (Stefansson, 1924, 1944). There is no information available concerning the edibility of the narwhal. The white whale attains a length of more than 4 m and a weight of about 900 kg. The body shape of the white whale resembles that of the members of the Delphinidae. The snout is blunt and there is no beak. There are no external grooves on the throat. White whales usually live in schools, sometimes consisting of more than 100 individuals. They migrate in response to the shifting pack ice and rigorous winters. The white whale can swim for hours at a speed of 9 km per hour and can remain underwater for periods of 15 minutes. They emit various sounds which are probably produced by the emission of a stream of bubbles rather than by the voicebox. They feed mainly on benthic organisms, cephalopods, crustaceans, and fishes. The gestation period is about 14 months, and the calf is about 1.5 m long at birth. White whales are of economic importance and are hunted mainly for their skins which are sometimes sold as "porpoise leather" (Walker *et al.*, 1964).

Family PHYSETERIDAE: The family of sperm whales consists of two genera and two species which inhabit all oceans. The sperm whale *Physeter catodon* attains a large size up to 20 m and more than 55 metric tons. The other member of the family, the pygmy sperm whale *Kocia breviceps*, is small and attains a length of about 4 m and a weight of about 320 kg. The characteristic features of *Physeter* are its tremendous barrel-shaped head and the underslung jaw. The sperm whale is said to be the only cetacean with a gullet large enough to swallow a man. It sometimes lifts its head out of the water to look and listen. When necessary, it can swim up to 12 knots. They usually travel in groups up to 20 individuals, but large schools may number in the hundreds. The gestation period in *Physeter* is about 16 months. They feed on squid, cuttlefish, fishes, and elasmobranchs. *Physeter* is hunted primarily for the oil and spermaceti, used for making candles and ointments. Ambergris is a substance that is unique to the sperm whale and is believed to be formed from solid wastes coalescing around a matrix of indigestible matter. The meat of the sperm whale is usually discarded by pelagic whalers, but some of the Pacific coast stations freeze the meat as food for fur-bearing animals or treat it to yield oil and meat meal.

For information regarding the biology of cetaceans, the excellent monograph by Slijper (1962) entitled *Whales* is recommended.

MECHANISM OF INTOXICATION

Sei whale poisoning is caused by eating the liver of *Balaenoptera borealis*. Cooking or degree of freshness of the liver does not affect the course of the biotoxication. Asiatic porpoise poisoning is caused by eating the liver, eyes, blood, viscera, and flesh of *Neophocaena phocaenoides*. Most of the biotoxications have been caused by eating porpoises taken from the Yangtze River in the vicinity of Yangchow (Kiangtu), China. Porpoises taken near the mouth of the river are thought to be safe to eat. The spring of the year is the most dangerous period. Ingestion of the flesh of *Delphinapterus leucas* may cause human fatalities. There are no specific clinical data available concerning human biotoxications from the sperm whale *Physeter catodon* other than vague reports that the oil and possibly the flesh are toxic to eat.

MEDICAL ASPECTS

Clinical Characteristics

Sei Whale Poisoning: Intoxications have been reported from eating the liver of the Sei whale *Balaenoptera borealis*. The signs and symptoms of Sei whale liver poisoning develop within 24 hours after ingestion and consist of severe occipital headaches, neck pain, flushing and swelling of the face, nausea, vomiting, abdominal pain, diarrhea, fever, chills, photophobia, epiphora, and an erratic blood pressure. After a day or two the patient's lips become dry, and desquamation develops around the mouth, gradually spreading to the cheeks, forehead, and neck. The desquamation usually does not involve the entire body.

Laboratory findings vary somewhat from one case to the next, but the following findings have been reported. There is usually an increase in urinary urobilinogen, negative diazo reaction, and negative urinary protein tests. Blood tests show normal values for the sedimentation rate. The protein concentration is normal. The albumin-globulin ratio may be decreased, and the albumin concentration is decreased. An increase in globulin concentration may occur. The cephalin-cholesterol flocculation test is usually positive to a medium degree, and the serum cholinesterase activity is decreased. The serum alkalin-phosphatase and phenol turbidity tests may be positive. A slight increase in nonprotein nitrogen concentration has been reported, although jaundice has not been observed. In general, there is definite evidence of liver impairment, but it is usually mild. Acute symptoms almost always subside within a period of 2 days, but the desquamation may continue for a longer period of time.

The chemical and pharmacological properties of Sei whale poison are unknown, but it is believed to be a histaminelike substance since methanol extracts of the poison gave a marked histaminelike reaction on isolated rabbit intestine. Japanese physicians have termed this biotoxication as "iwashi-kujira liver poisoning." The only clinical report on Sei whale liver poisoning is by Mizuta *et al.* (1957).

Asiatic Porpoise Poisoning: This biotoxication is caused by eating the viscera or flesh of the Asiatic porpoise *Neophocaena phocaenoides*. The signs and symptoms consist of abdominal pain, abdominal distention, swelling and numbness of the tongue, loss of vision, cyanosis, a sensation of numbness of various areas of the skin, hypersalivation with the saliva having a greenish tinge, and finally muscular paralysis. Death may be rapid; the case fatality rate is said to be very high. The only clinical reports available are by Macgowan (1884, 1886, 1887). Most of these biotoxications occur during the spring of the year.

White Whale Poisoning: Ingestion of the flesh may cause human fatalities (Stefansson, 1924, 1944). Clinical characteristics are unknown.

Sperm Whale Poisoning: There is no information available concerning the clinical characteristics of sperm whale biotoxications.

Pathology

(Unknown)

Treatment

Sei Whale Poisoning: If the poison is a histaminelike substance, the use of epinephrine should be considered since it is a specific physiological antagonist to histamine. Antihistamines should prove of value. The remainder of the treatment is symptomatic.

Asiatic Porpoise Poisoning: There is no specific antidote. The treatment is symptomatic. The natives claim that aqueous extracts of the plants *Mirabilis jalapa* Linnaeus, *Mimosa corniculata* Loureiro, Chinese olive *(Canarium sp.)*, and camphor are effective antidotes.

White Whale Poisoning: Treatment is symptomatic.
Sperm Whale Poisoning: Treatment is symptomatic.

Prevention

The liver of the Sei whale *Balaenoptera borealis* should never be eaten. The viscera and flesh of the Asiatic porpoise *Neophocaena phocaenoides* should not be eaten, particularly when taken in the spring of the year from the Yangtze River in the vicinity of Yangchow (Kiangtu), China. The liver, blood, eyes, and flesh of the upper back are especially dangerous to eat. If the porpoise must be eaten, it is advisable to boil the meat and discard the broth. Natives prepare the porpoise by cutting out the dangerous parts of the animal, washing it thoroughly with fresh water, and then boiling for 8 hours with either Chinese olives or sugar cane. If the flesh is boiled, the water should always by discarded. One should always check first with the local inhabitants to determine the edibility of the Asiatic porpoise. Natives believe that the farther up the river the porpoise is captured the more dangerous it is to eat. The flesh of the white whale *Delphinapterus leucas* should never be eaten.

TOXICOLOGY

There are no toxicological data available on Sei whale, white whale, or Asiatic porpoise poisons.

Sahashi (1933) found that when sperm whale oil is fed to rats with their normal diet for a period of 3 weeks, they develop seborrhea and their growth is retarded. He also observed that if 2 ml of oleic alcohol or its acetate (oleic alcohol is one of the principal constituents of sperm whale oil) were injected intraperitoneally into rats, they died. The oleic alcohol killed the rats in 10 to 40 hours, and the oleic acetate killed them within 6 days. Whether or not oleic acid is the principal toxic substance in sperm whale oil and meat is not known. The entire problem needs further study.

PHARMACOLOGY

(Unknown)

CHEMISTRY

(Unknown)

Poisonous Polar Bears

The carnivores are a group of small to large mammals, having four or five toes, claws, mobile limbs, complete and separate radius and ulna, tibia and fibula, small incisors, and canines as slender fangs. The only marine carnivore toxic to man is the polar bear *Thalarctos maritimus* (Phipps). Most mammalogists are of the opinion that there is but a single species.

The systematics and biology of the polar bear are discussed by Bee and Hall (1956), Hall and Kelson (1959), Ognev (1962), and Walker *et al.* (1964). An excellent bibliography on polar bears has been published by the Arctic Institute of North America (1966).

LIST OF POLAR BEARS REPORTED AS POISONOUS

Phylum CHORDATA

Class MAMMALIA

Order CARNIVORA: Carnivores

Family URSIDAE

Thalarctos maritimus (Phipps)[1] (Pl. 2). Polar bear, water bear, white bear, ice bear (USA, England), eisbär (Germany), isbjørnen (Sweden), ours polaire, ours blanc (France), belyi medved' (USSR), nanuk (Arctic Alaska). *281*

DISTRIBUTION: Arctic, circumpolar.

SOURCES: Kane (1856), Jackson (1899), Krogh and Krogh (1913), Stefansson (1921, 1944), Köhl (1929), Doutt (1940), Anon. (1942), Sutton (1942), Rodahl and Moore (1943), Rodahl (1949*a*, *b*), Russell (1967).
OTHER NAME: *Ursus maritimus*.

BIOLOGY

The polar bear is an inhabitant of the ice pack of the Arctic Ocean, where icebergs and broken pan ice are interspersed with stretches of open water. With the ever-shifting margin of the polar cap and the open ocean, the polar bear migrates according to the season of the year. Seldom does it venture any great distance inland from the sea. The world population of polar bears is estimated to be more than 10,000 (Harington, 1964). It is a remarkable and powerful swimmer. Peary is said to have seen the tracks of a polar bear along the course of a lead covered with young ice, "about 200 miles from land." Individual bears have actually been observed swimming in the open sea "more than 40 miles from shore" (Anthony, 1917). Either the forepaws alone or all four feet may be used in swimming. The soles of the feet are covered with fur overshoes which lend protection and provide sure footing on the slippery ice. The polar bear enjoys a wide variety of foods—whale meat, birds, fish, tundra vegetation, or almost anything else that is available—but it is particularly fond of seals and young walruses. Its sight is probably better than that of most bears, and its sense of smell is extraordinary. Mating occurs in June or early July, and the cubs are born about January. Polar bears attain large size, the males up to a length of 3.3 m and a weight of more than 450 kg. They have occasionally been known to stalk and kill humans during the

[1] Miller and Kellogg (1955) list three different species of polar bears, but Hall and Kelson (1959) recognize a single species and five subspecies. Walker *et al.* (1964) recognize only a single species.

winter when food was scarce but are generally considered to be less offensive during the summer months. Man is the primary predator of the polar bear and hunts them largely for their skins. There is growing concern regarding the depletion of the polar bear population and the threat of extinction. International legislation governing their harvesting is presently under study by the various countries involved, and rigid regulations concerning the hunting of these magnificent animals are already in force in most arctic areas. Intensive studies on the biology of polar bears have also begun (Mikhailov, 1961; Harington, 1964; and Norderhaug, 1965).

BIOGENESIS OF POLAR BEAR POISON

The biogenesis of the poison present in polar bear liver has not been completely established, but it is believed that the toxic properties of the liver and kidneys are due primarily to the presence of excessive quantities of vitamin A. The vitamin A apparently is accumulated through the diet of the bear—seals, walruses, fishes, etc., whose bodies contain small quantities of vitamin A.

MECHANISM OF INTOXICATION

Polar bear poisoning is caused by eating moderate to large quantities of the liver or kidneys. The intoxication is believed to be due to ingestion of excessive quantities of vitamin A which is present in polar bear liver and kidneys.

MEDICAL ASPECTS

Clinical Characteristics

The symptoms usually begin about 2 to 5 hours after ingestion of the material. The predominant symptoms are intense throbbing or dull frontal headaches, nausea, vomiting, diarrhea, abdominal pain, dizziness, drowsiness, irritability, weakness, muscle cramps, visual disturbances, and collapse. The headaches may become intensified during the first 8 hours and cause insomnia since they are aggravated by lying down. Gradually the headache lessens in severity and may disappear by the following day. Numerous cases have been cited in which desquamation occurred in various parts of the body, particularly the face, arms, legs, and feet. Tonic and clonic convulsions may be present. Sutton (1942) reported a case in which photosensitization apparently was a prominent symptom. If fatalities do occur from these intoxications, they are very rare. The amount of liver ingested appears to have a direct bearing upon the severity of the symptoms produced. Eskimos believe that ingestion of polar bear liver may result in depigmentation of the skin. However, investigation of these cases fails to substantiate this belief (Stefansson, 1921; Halstead, 1966).

The clinical characteristics of polar bear poisoning have been discussed by Krogh and Krogh (1913), Köhl (1929), Bøje (1939), Doutt (1940), Anon. (1942), Sutton (1942), Rodahl (1949*a*), Becker and Klotzsche (1955), Jeghers and Marraro (1958), and others.

Jeghers and Marraro (1958) have reviewed the clinical characteristics of acute hypervitaminosis A. The disorder consists primarily of central nervous system manifestations due to an abrupt and marked elevation of spinal fluid pressure. Symptoms develop when a single large dose of vitamin A is taken of 200,000 International Units (I.U.) for infants or one to several million units for adults. The symptoms develop within hours after ingestion and consist of violent headaches, nausea, vomiting, drowsiness, polyarthralgia, exophthalmos (rarely), etc., all of which are supposedly related to a marked increase in spinal fluid pressure. In addition, generalized or localized desquamation of the skin about the lips or exposed surfaces are usually evident by the second day. In acute intoxications, manifestations are said to result from direct toxic effects on specific tissues and are not dependent on supersaturation of liver stores.

Pathology

(Unknown)

Treatment

Symptomatic. Emetics and laxatives promptly administered are sometimes useful in relieving the severity of the symptoms. The clinical manifestations gradually disappear after ingestion of the toxic material has discontinued. The reduction of cerebro-spinal fluid pressure by lumbar puncture has been suggested (Jeghers and Marraro, 1958).

Prevention

There is no reliable method of detecting toxic polar bear liver or kidneys by merely visual examination. The age of the bear seems to have no bearing on the edibility of the meat. The liver from cub bears has been known to cause intoxications (Kane, 1856). In general, polar bear liver and kidneys should be eaten only in small quantities (less than 230 g, or about a half pound), and the meat should be thoroughly cooked. Cooking is said to attenuate the toxin.

PUBLIC HEALTH ASPECTS

Polar bear poisoning is of minor public health importance since most Eskimos usually discard polar bear liver. Intoxication is most likely to occur among uninformed travelers in the arctic.

TOXICOLOGY

Kane (1856) was one of the first to investigate the toxicity of polar bear liver. He fed bear liver to his men and ate it himself on several occasions. He found that the toxicity varied from one bear to the next. He too fed the liver to dogs but did not observe any ill effects.

The first attempt to study the toxicity of polar bear liver in the laboratory was by Vaughan Harley (cited by Jackson, 1899), who prepared aqueous, ethereal, and alcoholic extracts of liver tissue. The extracts were injected intraperitoneally and subcutaneously into dogs, guinea pigs, and mice. Subcutaneous injections of ether extracts resulted in death of the mice within 3 days. All the other tests were negative.

Stefansson (1921) conducted a series of human experiments in which he and his men experimentally ate the liver of polar bears. He too found that the toxicity of the liver varied with the individual bear.

Rodahl and Moore (1943) studied the toxicity and biochemical constituents obtained from polar bear and bearded seal *(Phoca barbata)* livers taken from northeast Greenland. The livers had been preserved in brine. Chemical tests revealed that the liver samples were very high in vitamin A. The first specimen was from a 2-year-old female bear and contained about 18,000 I.U. of vitamin A per gram of wet liver. A second specimen was taken from a 4-year-old male and also contained 18,000 I.U./g. However, the specimen from a third bear had only 13,000 I.U./g. A single specimen of bearded seal liver contained 13,000 I.U./g. The high vitamin A content suggested that the toxicity might be due to hypervitaminosis A. In making their toxicity test, they used the liver from the 2-year-old female bear, which at the time of testing had dropped to 10,000 I.U. of vitamin A per gram. The rats were reluctant to eat the liver, but the experimenters managed to feed one rat 33.1 g of bear liver over a period of 22 days, which amounted to an average of 15,000 I.U. of vitamin A per day. The rat became anemic, and the hind legs were paralyzed. When the rat became moribund, it was killed. Autopsy revealed profuse internal hemorrhages, particularly under the skin, which are typical of hypervitaminosis A. Hemorrhages were also found in the pericardium. One rat died accidentally, and the three other rats were unaffected. The authors concluded that the toxicity in polar bear liver poisoning was due to hypervitaminosis A. They estimated that it would require about 7,500,000 I.U. of vitamin A to cause acute hypervitaminosis A in man. This amount would be present in 375 g of bear liver containing 20,000 I.U. of vitamin A per gram, which is not an excessive portion to be eaten by a person at a single meal.

The most comprehensive study on the toxicity of polar bear liver that has appeared to date is by Rodahl (1949*a, b*). He obtained two wet tissue specimens of bear liver from northeast Greenland, containing 26,700 and 21,900 I.U. per gram of vitamin A respectively. Feeding tests were conducted with rats. Ingestion of 0.5 to 0.6 g of polar bear liver proved to be lethal. He further found that polar bear liver from which the vitamin A had been removed was nontoxic. "Bear liver, bear liver oil containing all its vitamin A, and purified vitamin A concentrates, had identical effects when given in equivalent amounts with regard to the vitamin A content." Equivalent amounts of bear liver oil in which the vitamin A had been removed had no deleterious effects. Moreover, the symptoms increased with increasing amounts of vitamin A. The author therefore concluded that the toxicity of polar bear liver was due to vitamin A. It is interesting to note that Rodahl also found in this same study that the livers of the snow hare and walrus contained only small quantities of vitamin A, but the liver of the Greenland fox, which is known to be poisonous, contained 12,000 I.U. of vitamin A per gram of liver. Rodahl's (1949*a*) report is of great value to anyone interested in this subject because he provides very detailed studies of the pathological effects of polar bear liver poisoning in rats. Russell (1967) determined the vitamin A content in the livers of two polar bears. One was an older bear which had caused "illness in 3 dogs which ate of it" and was found to contain 10,400 μg/g (34,600 I.U./g), and the other sample was taken from a younger bear and had 9,215 μg/g (30,700 I.U./g).

PHARMACOLOGY

The following summary of the pharmacology of vitamin A is based on a review by Severinghaus (1965). The functions of vitamin A concern the combination of retinine with proteins to form the light-sensitive pigments in the retina. Vitamin A also stabilizes the ratio between the glycoproteins or mucoproteins and the keratoproteins in epithelial structures throughout the body. Some investigators have suggested that the very labile sulfhydryl groups in proteins are stabilized by combination with vitamin A, which may be its general mechanism of action. The absorption of vitamin A follows the same general pathways for lipids, which require the availability of fats and bile salts. Transportation within the body is by free alcohol and esters with fatty acids, which are intimately associated with proteins. Absorption from the digestive tract is dependent upon adequate protein content. The metabolic end products of vitamin A have not been identified. Little is known about excretion, but the urinary losses are negligible. Vitamin A is stored in the liver, kidneys, and fat depots of the body. Circulating vitamin A may be greatly reduced during various disease processes, with apparent sequestration in tissues. The action may be followed by release of the vitamin into the bloodstream during convalescence, even without further intake. The minimum daily requirement has been given as 1,500 to 4,000 I.U.

CHEMISTRY

by Donovan A. Courville

Although the question of the toxicity of polar bear liver has not been completely resolved to everyone's satisfaction, the preponderance of evidence thus far is that an intake of excessive doses of vitamin A is a major factor (Rodahl and Moore, 1943; Rodahl, 1949*a, b*).

Vitamin A has as its precursors three orange plant pigments known as α-, β, and γ-carotenes. The structures of α-carotene and vitamin A are shown below. Only in α-carotene are the two halves of the molecule alike. Since the vitamin activity is lodged in a structure represented by half of α-carotene, only this form is capable of yielding two molecules of the vitamin on oxidative rupture of the central double bond. This oxidative rupture occurs only in animals. The principal site is in the intestine though it occurs also in the liver. An aldehyde probably forms as the first step in this rupture. If so, it is quickly reduced to the alcohol form as shown below. This is esterified with a long-chain fatty acid in the liver where it is stored. Certain species of fishes and marine mammals store large amounts of the vitamin in the liver, making the liver oils good sources of the vitamin. Since animals, including humans, must depend heavily on the carotene precursors in plants for their supply of vitamin A, more than the theoretical amount of carotene must be ingested to secure the calculated amount needed to meet daily nutritional requirements.

CH_3 CH_3 — CH_3 — CH_3 — CH_3 — CH_3 — CH_3 CH_3

$-CH{=}CH{-}C{=}CH{-}CH{=}CH{-}C{=}CH{-}CH{=}CH{-}CH{=}C{-}CH{=}CH{-}CH{=}C{-}CH{=}CH-$

$-CH_3$ — CH_3-

α-carotene

↓

CH_3 CH_3 — CH_3 — CH_3

2 $-CH{=}CH{-}C{=}CH{-}CH{=}CH{-}C{=}CH{-}CH_2OH$

$-CH_3$

vitamin A

In both vitamin A and its carotene precursors, there is a theoretical possibility of a cis- or trans-arrangement at each of the double bonds in the side chain. The major proportion of naturally-occurring carotenes are of the trans-configuration. One or more of the spatial arrangements must be isomerized to the cis-form in the animal body. Some marine crustacea contain the vitamin with a cis-arrangement at one of the double bonds of the chain. A similar form is a short-lived intermediate in the changes involved in the phenomenon of sight in humans and many other animals (West and Todd, 1961).

Poisonous Walruses and Seals

The pinnipeds are a group of small to large marine mammals having a spindle-shaped body and limbs modified into flippers for aquatic locomotion. The toes are included in webs, and the tail is very short. The males are usually larger than the females. The livers of walruses and certain species of seals may at times be poisonous to eat.

The systematics and biology of walruses and seals have been discussed by Ognev (1962), Troughton (1947), Bee and Hall (1956), Fay (1957), Scheffer (1958), Hall and Kelson (1959), Walker *et al.* (1964), Daugherty (1965), Harrison and King (1965).

REPRESENTATIVE LIST OF WALRUSES AND SEALS REPORTED AS POISONOUS

Phylum CHORDATA

Class MAMMALIA

Order PINNIPEDIA

Family ODOBENIDAE

Odobenus rosmarus (Linnaeus).[2] (Pl. 3, fig. a). Walrus (USA, England), avik (Arctic Alaska), kelyuch (Arctic USSR). *282*

DISTRIBUTION: Arctic Ocean—northeastern coast of Siberia, the northwestern coast of Alaska, and north to northwest Greenland and Ellesmere Island.

SOURCE: Fay (1960).

Family PHOCIDAE

Erignathus barbatus (Erxleben).[3] (Pl. 3, fig. b). Bearded seal (USA), oogruk, ugruk (Arctic Alaska), lachtak (Arctic Eurasia). *282*

DISTRIBUTION: Inhabits the edge of the ice along the coasts and islands of North America and northern Eurasia.

SOURCES: Krogh and Krogh (1913), Rodahl and Moore (1943), Rodahl (1949*a*).
OTHER NAME: *Phoca barbata.*

Pusa hispida (Schreber) (Pl. 3, fig. c). Ringed seal, Caspian seal (England, Canada, USA), baikal (USSR), seehund (Germany). *282*

DISTRIBUTION: Circumboreal near the edge of the ice, to the North Pole.

SOURCE: Rodahl (1949*a*).
OTHER NAME: *Phoca hispida*.

BIOLOGY

Family ODOBENIDAE: Walruses inhabit the open waters of the Arctic Ocean near the edge of the polar ice. They migrate south in the winter with the advance of the ice and move north in the spring as the ice retreats. Walruses are frequently observed riding the ice floes during their migration. Walruses have a thick swollen body, rounded head and muzzle, a short neck, and a tough wrinkled skin. There is no tail. They have a conspicuous mus-

[2] There are two subspecies recognized: *Odobenus rosmarus divergens* (Illiger), the Pacific form, and *O. rosmarus rosmarus* (Linnaeus), the Atlantic form.

[3] There are two subspecies that are recognized by some authors: *Erignathus barbatus barbatus* (Erxleben), the Eurasian form, and *E. barbatus nauticus* (Pallas), the Pacific form.

tache consisting of large bristles that are richly supplied with blood vessels and nerves. They are especially characterized by their large protruding ivory tusks. The bulls attain a length of about 3.7 m and a weight of more than a metric ton. The cows are about one-third the size of the bulls. They are generally found in mixed herds consisting of bulls, cows, and calves and numbering 100 or more individuals. Walruses can swim about 24 km per hour. They may be observed using their tusks to haul themselves out of the water. Walruses make a bellowing noise and at times may sound like a herd of trumpeting elephants. They seem to have poor senses of smell and hearing but fairly good eyesight. Walruses feed on the bottom and are thought to use their bristles and possibly their tusks to forage for food, which consists mainly of mollusks and other marine life.

Rogue walruses, individuals which remain separate from the rest of the herd, may feed on seals, narwhal, white whales, or other dead cetaceans. It is the rogue walruses that are most likely to have toxic livers (Fay, 1960). The rogue is a solitary bull and feeds almost exclusively on vertebrates. The rogue has a characteristic appearance. It is relatively lean and slender with shoulders and forelimbs that appear unusually large and powerfully developed. The chin, neck, and breast are impregnated with oil from frequent contact with seal blubber; the oxidized oil imparts an amber color to these parts. The tusks are unusually long, slender, and sharp-pointed; their surfaces are covered with scratches. Rogues are said to constitute less than 0.1 percent of the walrus population. Eskimos claim that rogues develop from calves who are separated from their mothers in the first year or two of life and are not sufficiently familiar with bottom feeding techniques to sustain themselves. Consequently only a few of the stronger bulls are able to survive by scavenging and preying on whatever vertebrates they are able to capture. Actually no one knows exactly how rogues originate in the normal herd.

Walruses are polygamous. The gestation period is about 12 months, and most of the births take place from April to June. Eskimos use every part of the walrus, either for food, shelter, or boats, and the tusks are used extensively in ivory carvings.

Family PHOCIDAE: The family of true seals consists of 13 genera and 18 species which are widely distributed throughout the coastal and oceanic waters of polar, temperate, and tropical regions of the world.

Bearded seals *(Erignathus)* are solitary creatures, living alone on ice floes not far from land, except during the mating season. They derive their name from the beardlike tuft or stout white bristles growing down on each side of their muzzles. Their food consists largely of crustaceans and mollusks. Fishes are sometimes eaten but are apparently less desirable. Bearded seals seek their food on the bottom of the sea and may dive to great depths. The single pup is born about late March. The adult male attains a length of about 3 m and a weight exceeding 360 kg. The flesh is said to be tough, coarse, and most appreciated by the Eskimos when it is decomposed and frozen. The liver is toxic to eat, apparently because of the excessive vitamin A content (Rodahl and Moore, 1943).

The Australian sea lion *(Neophoca)* is the largest of its kind, inhabiting rocky coastal areas along the southern shore of Australia. For the most part, these seals are nonmigratory, usually remaining in the immediate environs of their birthplace throughout their life. They are generally docile, except during the mating season when they become somewhat ill-tempered. Their food consists mainly of penguins, which are available in abundance, and fishes. They also have the interesting habit of ingesting stones, apparently used as an aid in digesting their food. The breeding season is from October to early December, during which time the community spends most of its time inshore. The harems are relatively small, consisting of one to four females to each male. The females give birth to a single pup. The male attains a maximum length of about 3.7 m. The flesh of some specimens has been reported to be highly toxic (Leigh, 1839).

MECHANISM OF INTOXICATION

Walrus and seal poisoning is usually the result of ingesting the livers of the rogue bull walrus *(Odobenus)* or the bearded seal *(Erignathus)* and the flesh of the Australian sea lion *(Neophoca)*. Cooking may partially attenuate the poison but apparently does not completely destroy it. Intoxication is believed to be due to the intake of excessive quantities of vitamin A present in the liver. The nature of the poison involved in the flesh of the Australian sea lion is unknown.

MEDICAL ASPECTS

Clinical Characteristics

The symptoms of walrus and seal liver poisoning are believed to be similar to those of polar bear liver poisoning. See p. 990.

Pathology

(Unknown)

Treatment

See polar bear liver poisoning, p. 991.

Prevention

Avoid eating the livers of rogue bull walruses and bearded seals and the flesh of the Australian sea lions. See also polar bear liver poisoning, p. 991.

PUBLIC HEALTH ASPECTS

Walrus and seal poisoning is usually not a public health problem since natives are generally acquainted with the toxicity of these organisms and avoid eating them. These intoxications are most likely to occur among uninformed arctic travelers.

TOXICOLOGY

Rodahl and Moore (1943) tested the vitamin A content and toxicity of the liver of the bearded seal *(Erignathus)* and found one sample to contain 13,000 I.U. of vitamin A per gram of liver tissue. They believed that seal liver intoxication was due to hypervitaminosis A just as in polar bear liver poisoning. See p. 991.

There is no information available concerning the nature of the poison reported to be present in the flesh of the Australian sea lion *Neophoca*.

PHARMACOLOGY

Vitamin A is believed to be the causative agent of the intoxication. See p. 993 on the pharmacology of vitamin A in polar bear liver poisoning.

CHEMISTRY

See p. 993 on the chemistry of vitamin A in polar bear liver poisoning.

LITERATURE CITED

ANONYMOUS
1942 Is polar bear liver poisonous? J. Am. Med. Assoc. 118(4): 337.

ANTHONY, H. E., *ed.*
1917 Animals of America. Mammals of America. Garden City Publ. Co., Inc., Garden City, N.Y. 335 p.

BECKER, W. and C. KLOTZSCHE
1955 Die hypervitaminose A. Ärtztl. Woch. 24: 545-550.

BEE, J. W. and E. R. HALL
1956 Mammals of northern Alaska on the arctic slope. Mus. Nat. Hist., Univ. Kansas, Lawrence, Kansas. 309 p., 127 figs.

BØJE, O.
1939 Toxin in the flesh of the Greenland shark. Medd. Grønland 125(5): 1-16.

DAUGHERTY, A. E.
1965 Marine mammals of California. Calif. Dept. Fish Game. Sacramento, Calif. 87 p.

DOUTT, J. L.
1940 Toxicity of polar bear liver. J. Mammal. 21: 356-357.

FAY, F. H.
1957 History and present status of the Pacific walrus population. Trans. 22d North Am. Wildlife Conf., Wildlife Manag. Inst.: 431-445.
1960 Carnivorous walrus, some arctic, zoonoses. Arctic 13(2): 111-1221.

HALL, E. R., and K. R. KELSON
1959 The mammals of North America. 2 vols. Ronald Press Co., New York.

HALSTEAD, B.W.
1966 Field report on the epidemiology of biotoxications from eating the livers of polar bears and other arctic mammals. (Unpublished data.)

HARINGTON, C. R.
1964 Polar bears and their present status. Can. Audubon Mag. 26(1): 3-10.

HARRISON, R. J., and J. E. KING
1965 Marine mammals. Hutchinson and Co., Ltd., London. 192 p.

HERSHKOVITZ, P.
1966 Catalog of living whales. U.S. Nat. Mus. Bull. No. 246. 259 p.

JACKSON, F. G.
1899 A thousand days in the Arctic. Harper and Brothers, London. 940 p.

JEGHERS, H., and H. MARRARO
1958 Hypervitaminosis A: its broadening spectrum. Am. J. Clin. Nutr. 6(4): 335-339.

KANE, E. K.
1856 Arctic explorations: the second Grinnell expedition in search of Sir John Franklin, 1853, '54, '55. 2 vols. Childs and Peterson, Philadelphia.

KÖHL, H.
1929 Kann die leber der fleischfresser giftig sein? Z. Fleisch-Milchhyg. 40(3): 45-48.

KROGH, A., and M. KROGH
1913 A study of the diet and metabolism of Eskimos undertaken in 1908 on an expedition to Greenland. Medd. Grønland 51(1): 11.

LEIGH, W. H.
1839 Reconnoitering voyages and travels, with adventures in the new colonies of South Australia, p. 164. Smith, Elder and Co., London.

MACGOWAN, D. J.
1884 Porpoise poison. China Insp. Customs Med. Rept. 27: 12.
1886 Poisonous fish and fish-poisoning in China. Chinese Rec. Mission J. 17(2): 45-49; (4): 139-140.
1887 Poisonous fish in China. Bull. U.S. Fish. Comm. 6: 130-131.

MIKHAILOV, S. V.
1961 The polar bear. [In Russian] Okhota Okhot. Khoz. 7(7): 24.

MILLER, G. S., JR., and R. KELLOGG
1955 List of North American recent mammals. U.S. Mus. Bull. No. 205. 954 p.

MIZUTA, M., ITO, T., MURAKAMI, T., and M. MIZOBE
1957 Mass poisoning from the liver of sawara and iwashikujira. [In Japanese] Jap. Med. J. (1710): 27-34.

NORDERHAUG, M.
1965 The polar bear, conservation problem of the Arctic Ocean. [In Swedish] Sver. Natur. Årsbok: 100-106.

NORMAN, J. R., and F. C. FRASER
1948 Giant fishes, whales, and dolphins. New ed. Putnam, London. 376 p., 8 pls.

OGNEV, S. I.
1962 Mammals of USSR and adjacent countries. Vol. III. Carnivora (Fissipedia and Pinnipedia). [Engl. transl.] IPST Cat. No. 231. Israel Prog. Sci. Transl., Jerusalem. 641 p., 299 figs.

RODAHL, K.
1949*a* The toxic effect of polar bear liver. Norsk Polarinstitutt, Skrifter No. 92. J. Dybwad, Oslo. 90 p., 18 pls.
1949*b* Toxicity of polar bear liver. Nature 164(4169): 530-531.

RODAHL, K., and T. MOORE
1943 The vitamin A content and toxicity of bear and seal liver. Biochem. J. 37: 166-168.

RUSSELL, F. E.
1967 Vitamin A content of polar bear liver. Toxicon 5: 61-62.

SAHASHI, Y.
1933 Nutritive value of sperm whale oil and finback whale oil. Sci. Papers, Inst. Phys. Chem. Res. (Tokyo) 20(416): 245-253, 17 figs.

SCHEFFER, V. B.
1958 Seals, sea lions, and walruses; a review of the Pinnipedia. Stanford Univ. Press, Stanford, Calif. 179 p., 15 figs., 32 pls.

SEVERINGHAUS, E. L.
1965 The vitamins, p. 1005-1023. *In* J. R. DiPalma [ed.], Drill's pharmacology in medicine. McGraw-Hill Book Co., Inc., New York.

SLIJPER, E. J.
1962 Whales. Basic Books, Inc., New York. 475 p., 229 figs.

STEFANSSON, V.
1921 The friendly Arctic, the story of five years in polar regions. Macmillan Co., New York. 784 p.
1924 My life with the Eskimo. Macmillan Co., New York. 538 p.
1944 Arctic manual. Macmillan Co., New York. 556 p.

SUTTON, R. J., JR.
1942 Is polar bear liver poisonous? J. Am. Med. Assoc. 118: 1026.

TATE, G. H.
1947 Mammals of eastern Asia. Macmillan Co., New York. 366 p., 79 figs.

TOMILIN, A. G.
1957 Fauna of the USSR and adjacent lands. Vol. 9. Cetacea. [In Russian] Academy of Sciences of the USSR, Moscow. 756 p., 12 pls., 146 figs.

TROUGHTON, E.
1947 Furred animals of Australia. C. Scribner's Sons, New York. 374 p., 25 pls.

WALKER, E. P., WARNICK, F., LANGE, K. I., UIBLE, H. E., HAMLET, S. E., DAVIS, M. A., and P. F. WRIGHT
1964 Mammals of the world. 3 vols. Johns Hopkins Press, Baltimore.

WEST, E. S., and W. R. TODD
1961 Textbook of biochemistry. Macmillan Co., New York. 1423 p.

Glossary

The following is a list of technical terms most commonly encountered in marine biotoxicology. Definitions of many of these terms are not generally available in current dictionaries. Whenever possible, the literature citation where the term was first proposed or discussed is included.

ABALONE VISCERA POISON—A form of intoxication resulting from the ingestion of the viscera of certain Japanese abalone of the mollusk family Haliotidae.

ACIPENSERIN—A toxic substance obtained from the gonads of sturgeon, *Acipenser* (Kurajeff, 1901).

ACTINOCONGESTIN—A synonym of congestin, a poison obtained from the stinging tentacles of sea anemones (Richet, 1908). *See also Congestin.*

ACTINOTOXIN—The term applied by Richet, Perret, and Portier (1902) to the crude poison obtained from alcoholic extracts of the tentacles of sea anemones.

APPROXIMATE LETHAL DOSE (ALD)—The approximate lethal dose required to kill a test animal. This term is generally unacceptable.

ALEXIPHARMAC—An ancient and largely obsolete term used to designate an antidote used against poisons. The term was originally derived from Nicander's poem *Alexipharmaca* which discussed poisons and their treatment (Nicander, 275-135 B.C).

AMPHIPORINE—The toxic substance obtained from tissue extracts prepared from the nemertean worm *Amphiporus lactifloreus* and other species (Bacq, 1937). Amphiporine is said to resemble acetylcholine and nicotine in its pharmacological properties. Its chemical nature is unknown.

APLYSIN—The neurotoxin obtained from the digestive gland of the sea hare *Aplysia* (Winkler and Tilton, 1962).

ASARI POISON—A synonym for venerupin shellfish poison (Akiba, 1943). *See also Venerupin.*

ASTEROTOXIN—The poison found in starfishes, members of the echinoderm class Asteroidea (Halstead, 1965).

BIOGENESIS—In natural products and biotoxicology, the manner in which chemical compounds are synthesized by living organisms—also termed biosynthesis.

BIOSYNTHESIS—The manner in which chemical compounds are synthesized by living organisms—also termed biogenesis.

BIOTOXICOLOGY—The science of poisons produced by living things, their cause, effects, nature, detection, and treatment of intoxications produced by them.

BIOTOXINS—Poisons derived from either plants or animals.

BIOTOXICATION—Intoxication resulting from plant or animal poisons.

BONELLININ—A toxic aqueous substance obtained from the proboscis worm *Bonellia*. Originally termed "bonellin," but because of a conflict in nomenclature with "bonelline," a green pigment, has been changed to "bonellinin" (Nigrelli, Stempien, Ruggieri, Liguori, and Cecil, 1967).

BRANDAROPURPURINE—A form of purpurine, now termed murexine, obtained from the median zone of the hypobranchial gland of *Murex brandaris* (Erspamer, 1946, 1947). *See also Murexine.*

CALLISTIN SHELLFISH POISONING—A form of shellfish poisoning caused by eating *Callista brevisiphonata* obtained in the vicinity of Mori, Hokkaido, Japan. Originally described by Asano, Takayangi, and Furuhara (1950), the intoxication is believed to be due to a choline which is present in large quantities in the ovaries of the shellfish.

CEPHALOTOXIN—The poison found in the salivary glands of the cephalopods, believed to be an active protein (Ghiretti, 1959, 1960). *See also Eledoisin.*

CHELONITOXIN—The poison present in the flesh, fat, blood, and viscera of certain species of marine turtles. The poison is believed to be derived from marine algae that are eaten by turtles. The term comes from *Chelonia*, the order of turtles. The chemical and pharmacological properties of the poison are unknown (Halstead, 1965).

CIGUABO—A corruption of the term ciguatera.

CIGUATERA—One of the forms of ichthyosarcotoxism caused by eating the flesh or viscera of various species of subtropical and tropical marine shore fishes. The intoxication is characterized by a gastrointestinal upset, paresthesias, muscular weakness, paralysis, and various other neurological disturbances. The exact chemical nature of ciguatoxin has not been fully elucidated, although Scheuer (*see* Mosher, 1966) has proposed the empirical formula of $C_{28}H_{52}NO_5Cl$. Whether this particular substance accounts for the complete clinical syndrome of ciguatera poisoning remains to be seen. The clinical syndrome was first described by Parra (1787).

CIGUATERATOXIN—The term used by Scheuer to designate the poison found in ciguatoxic fishes (Mosher, 1966). Ciguateratoxin is a synonym of ciguatoxin.

CIGUATERIN—The term proposed by Hashimoto (1967) for a water-soluble toxic fraction found in ciguatoxic fishes.

CIGUATOS—A corruption of the term ciguatera.

CIGUATOXIN—The term proposed by Halstead (1964) to designate the poison present in ciguatoxic fishes. *See also Ciguatera.*

CIGUATOXICATION—A clinical designation for ciguatera poisoning (Halstead, 1965).

COMPLETE LETHAL DOSE (CLD)—Complete lethal dose required to kill a test animal. This term is generally unacceptable.

CLUPEOID FISH POISONING—One of the principal types of ichthyosarcotoxism which is usually caused by a clupeoid fish (generally one of the tropical herrings) (Desportes, 1770; Randall, 1958; Helfrich, 1961). *See also Clupeotoxism.*

CLUPEOTOXISM—The clinical designation for clupeoid poisoning. The first reference to this form of fish poisoning was by Desportes (1770) in the West Indies, but the earliest case history is that by Chisholm (1808). *See also Clupeoid Fish Poisoning.*

CNIDAE—A term generally used interchangeably with nematocysts.

CNIDOCIL—The triggerlike hair situated on the outer surface of the cnidoblast which when stimulated discharges the contents of the nematocysts.

CNIDOBLAST—The interstitial cell from which the nematocyst develops.

CNIDOCYST—A synonym of nematocyst.

CNIDOM—The nematocyst pattern or spectrum of a particular coelenterate. The types of nematocysts present in a particular coelenterate are believed to be sufficiently specific and consistent to be of taxonomic significance (Cutress, 1955).

CONGESTIN—A term proposed by Richet (1903*a*, *b*) to designate a toxic substance derived from the tentacles of sea anemones. When injected into dogs, congestin causes an intense congestion of the splanchnic vessels.

CUNEIFORM AREA — The thickened wedge-shaped area on the dorsum of the caudal appendage ventral to the proximal portion of the batoidean sting (Halstead and Modglin, 1950). The glandular epithelium of the cuneiform area is believed to produce venom.

CYPRININ—The toxic substance obtained from the milt of the carp *Cyprinus carpio* (Miescher, 1874).

DINOGUNELLIN—The toxic lipoprotein found in the roe of a Japanese blenny *Stichaeus (Dinogunellus) grigorjewi* (Asano and Itoh, 1962). *See also Lipostichaerin.*

ECTYONIN—The specific term proposed by Nigrelli, Jakowska, and Calventi (1959) for an antimicrobial substance obtained from the sponge *Microciona prolifera*. Preliminary tests indicated that ectyonin was nontoxic for mice and fish. The term derived from the name of the sponge subfamily Ectyoninae.

ELEDOISIN—An endecapeptide isolated by Anastasi and Erspamer (1962) from the posterior salivary glands of the octopus *Eledone*. Eledoisin is said to have powerful vasodilator and hypotensive effects on most animals. It was originally termed moschatin, but the name was changed due to the confusion resulting from the plant alkaloid moschatin from *Achillea moschata*. *See also Cephalotoxin.*

ENVENOMATION—To sting or impregnate an organism with a toxin by means of a venom apparatus. *See also Stinging.*

EPTATRETIN—A term proposed by Jensen (1963) to designate a potent cardiostimulant obtained from the branchial heart of the Pacific hagfish *Eptatretus stouti.* The substance is reported to be a highly unstable aromatic amine. Its chemical structure has not been fully defined, but it is not a catecholamine or some other commonly occurring biochemical.

ERABUTOXIN—The active toxic principle obtained from the venom of the sea snake *Laticauda semifasciata* (Tamiya, 1966). The term is derived from the Japanese vernacular name of the sea snake, "Erabu-umihebi."

ERINACEOPURPURINE—A term used by Erspamer (1946, 1947) to designate the form of purpurine, now called murexine, obtained from the median zone of the hypobranchial gland of *Murex erinaceus. See also Murexine.*

EZOWASURE-GAI POISONING—The Japanese term for callistin shellfish poisoning caused by gastropods of the genus *Callista* (Kawabata, 1962). *See also Callistin Shellfish Poisoning.*

FUGAMIN—A term proposed by Yashizawa (1894; cited by Suehiro, 1947) to designate one of two crystalline substances which he isolated from puffer poison. The chemical nature of the substance is unknown. The term is now obsolete.

FUGU TOXIN—The Japanese term for puffer poison.

GLANDULAR TRIANGLE—A term proposed by Porta (1905) to designate a cross section of the ventrolateral-glandular grooves of the stingray's sting, but which was later expanded to include cross sections of the glandular grooves of other fishes regardless of the anatomical position of the groove.

GONYAULAX POISON—The toxic principle produced by certain members of the dinoflagellate genus *Gonyaulax*, sometimes termed paralytic shellfish poison, mussel poison, clam poison, mytilotoxin, or saxitoxin.

HALITOXIN—The term applied to the crude toxic aqueous extract obtained from *Haliclona viridis* (Baslow and Turlapaty, 1969).

HALLUCINOGENIC FISH POISONING—A form of ichthyosarcotoxism caused by eating the head or flesh of certain species of tropical marine shore fishes. Surgeonfish, chub, mullet, damsel, grouper, and goatfish have been incriminated. These fish may produce hallucinations or frightening dreams when eaten. The term "hallucinatory fishes" has been used by some authors, but "hallucinogenic" is preferred (Helfrich and Banner, 1960). *See also Ichthyoallyeinotoxin.*

HOLOTHURIGENIN—A term proposed by Matsuno and Yamanouchi (1961) to designate the triterpenoid sapogenin isolated from the sea cucumber *Holothuria vagabunda.* Holothurigenin is thought to be a trihydroxylactonediene belonging to the triterpenoid series. It has been found to have the empirical formula $C_{30}H_{44}O_5$.

HOLOTHURIN—A term used to designate the toxic mixture of steroid glycosides obtained from holothurians or sea cucumbers. In some species, holothurin is found to be most concentrated in the Organs of Cuvier, but may be found in other parts of the organism as well. The term has been independently proposed by Yamanouchi (1942, 1943*a, b,* 1955) and Nigrelli (1952).

HOLOTOXIN—A steroidal glycoside obtained from the sea cucumber *Stichopus japonicus* which exhibits antifungal activity against *Trichophyton, Candida,* etc. (Shimada, 1969).

HYPNOTOXIN—A term proposed by Portier and Richet (1902) to designate a toxic substance obtained from the tentacles of the hydrozoan *Physalia.* The drug characteristically produces a central nervous system depression affecting both motor and sensory elements.

ICHTHULIN—A term applied to the globulin portion of fish eggs. It is analogous to the vitellin of bird eggs. The origin of the term is unknown. The ichthulin of some fish species is toxic to warm-blooded animals (McCrudden, 1921; Asano and Itoh, 1962).

ICHTHYISMUS—An obsolete clinical term formerly used to designate fish poisoning. *See Ichthyotoxism.*

ICHTHYOACANTHOTOXIN — The poison secreted by the venom apparatus of fishes. Synonymous with the term "fish venom."

ICHTHYOACANTHOTOXISM — The clinical term proposed by Halstead (1953) to designate an intoxication resulting from injuries produced by the stings, spines, or "teeth" of venomous fishes.

ICHTHYOALLYEINOTOXIN—The generic term proposed by Halstead in this monograph to designate the poison produced by hallucinogenic fishes. The poison when ingested by humans generally produces frightening dreams. *See also Hallucinogenic Fish Poisoning.*

ICHTHYOCRINOTOXIN—The term first proposed by Halstead in this monograph to designate the poison produced by a glandular secretion of fishes which is not associated with a venom apparatus. Ichthyocrinotoxic glands are generally located in the skin of the fish, and the glandular contents are released directly into the water. The secretions from these glands are toxic and usually lethal to other fishes.

ICHTHYOHEMOTOXIN—The toxic substance present in the blood of certain species of marine fishes (Halstead, 1964).

ICHTHYOHEMOTOXISM—The clinical term used to designate fish blood poisoning (Halstead, 1964).

ICHTHYOOTOXIN—A toxic substance derived from the roe of fishes. This particular term is reserved for those fish poisons which appear to be limited in their anatomical distribution to the roe (Halstead, 1964). This term should not be confused with ichthyosarcotoxin. Ichthyosarcotoxin may also be found in the roe of fishes, but it is found in other parts of the fish as well, whereas ichthyootoxin is limited to the roe.

ICHTHYOOTOXISM—The clinical term used to designate intoxications resulting from the ingestion of the toxic roe of fishes (Halstead, 1964).

ICHTHYOSARCEPHIALTILEPSIS — The clinical term proposed by Helfrich and Banner (1961) for hallucinogenic (hallucinatory) fish poisoning. The term for obvious reasons has not been generally adopted.

ICHTHYOSARCOTOXIN — The poison found in the flesh of poisonous fishes. This term does not include those poisons produced by bacterial contamination.

ICHTHYOSARCOTOXISM — The clinical term proposed by Halstead (1953) to designate an intoxication resulting from the ingestion of the flesh, i.e., in the broadest sense, musculature, viscera, or gonads of poisonous fishes.

ICHTHYOSISMUS—An obsolete clinical term formerly used to designate fish poisoning.

ICHTHYOTOXICOLOGY — The science which treats poisons derived from fishes, their effects, nature, antidotes, and recognition.

ICHTHYOTOXICUM—A term proposed by Mosso (1889) to designate the toxic substance found in eel serum. The term is obsolete.

ICHTHYOTOXIN—A general term used to designate any type of poison derived from fishes.

ICHTHYOTOXISM—A general term used to designate any form of intoxication produced by a fish.

ICHTHYOVENIN—A toxic substance reputedly produced by a special toxigenic bacteria acting on the flesh of inadequately preserved scombroid fishes. This term was introduced by Pergola (1937). Ichthyovenin is now believed to be histamine in combination with saurine. The term ichthyovenin is now obsolete.

INTEGUMENTARY SHEATH—The integument which envelops the spine or bony portion of the venomous sting of a fish.

IRUKANDJI STINGS—A type of jellyfish sting originally described by Flecker (1945, 1952). It is a well-defined clinical syndrome caused by the carybdeid jellyfish *Carukia barnesi.* The term "Irukandji" refers to the name of the local aboriginal tribe whose area was roughly the same as that in which the stingings have taken place, namely, the vicinity of Cairns, Queensland, Australia (Barnes, 1964).

IWASHIKUJIRA LIVER POISONING—The Japanese clinical term for sei whale liver poisoning which is caused by eating the liver of *Balaenoptera borealis.* The nature of the poison is unknown, but is believed to be a histamine-like substance.

LATICOTOXIN—The toxic principle obtained from the venom of the sea snake *Laticauda laticaudata* (Tamiya, 1966).

LETHAL DOSE—This term is seldom used in modern toxicological literature. It is largely replaced by the more acceptable term of Minimum Lethal Dose (MLD).

LETHAL DOSE$_{50}$ (LD$_{50}$)—The amount of poison required to kill 50 percent of the animals tested.

LIPOSTICHAERIN—The term proposed by Asano and Itoh (1966) to replace their former term "Dinogunellin" because of the revision of the generic name of the fish from *Dinogunellus* to *Stichaeus*. Lipostichaerin is said to be analogous to "lipovitellin" in hen egg yolk. Lipostichaerin has been fractionated into α-, β-, and γ-lipostichaerins. α-, β-, and γ-stichaerins represent the protein moieties of the corresponding lipostichaerins. *See also Dinogunellin.*

MACULOTOXIN—A potent neurotoxin found in the posterior salivary glands of the blue-ringed octopus *Octopus maculosus* (Freeman and Turner, 1971), now considered to be identical with tetrodotoxin. *See Tetrodotoxin.*

MARINBUFAGIN—A cardiotonic steroid found in the poison from the parotid gland of the toad *Bufo marinus* (Pataki and Meyer, 1955; Siperstein, Murray, and Titus, 1957). This toad is not included in this work since it is generally considered to be a freshwater species.

MARINOBUFOTOXIN—A toxic substance obtained from the parotid gland of the toad *Bufo marinus* (Pataki and Meyer, 1955; Siperstein *et al.*, 1957). This toad is not included in this work since it is generally considered to be a freshwater species.

MEDUSO-CONGESTIN—A term proposed by Dujarric de la Rivière (1915) to designate a toxic substance derived from the tentacles of the jellyfish *Rhizostoma cuvieri.* Injection of the substance into laboratory animals results in intense visceral congestion. Meduso-congestin is believed to be identical with the substance which Richet (1903*a, b*) termed congestin. *See also Congestin.*

MIMI POISONING—A form of intoxication caused by the ingestion of Asiatic horseshoe crabs and their eggs.

MINIMAL LETHAL DOSE (MLD)—The smallest amount of poison required to kill a test animal.

MITHRIDATE—A term used to designate the polyvalent antidotes for poisons used during the ancient, medieval, and early modern periods. The term is derived from Mithridates, King of Pontus (120-63 B.C.), who experimented with poisons and originated various antidotes for poisons.

MOSCHATIN—*See Eledoisin.*

MUREX POISON—The toxic principle found in the hypobranchial gland of the mollusk *Murex*. A synonym of murexine. *See also Murexine.*

MUREXINE—A term proposed by Erspamer and Dordoni (1947) to designate the neurotoxic principle derived from the median zone of the hypobranchial gland of *Murex* and other prosobranch mollusks. According to Erspamer and Benati (1953), the substance has been defined chemically as β-imidazolyl-(4)-acrylcholine. The substance is believed to be the same as Dubois' purpurine. *See also Purpurine.*

MYTILOCONGESTIN—A term proposed by Richet (1905) to designate a toxic substance derived from *Mytilus edulis*. Upon injection, the substance produces congestion and hemorrhage as does congestin. The relationship of this substance to Brieger's mytilotoxin is not exactly known. The chemical nature of the poison has not been defined.

MYTILOTOXIN—A term proposed by Brieger (1885, 1888, 1889) to designate the neurotoxin derived from toxic mussels of the genus *Mytilus*. It is sometimes used interchangeably with paralytic shellfish poison, mussel poison, clam poison, or saxitoxin.

NEMATOCYST—One of the minute stinging capsules of coelenterates.

NEMATOCYTE—A synonym for cnidoblast.

NEMERTINE—A nerve stimulant extracted from the tissues of nemertean worms *Lineus lacteus* and *L. longissimus* (Bacq, 1937). The chemical and pharmacological properties of nemertine are unknown.

NEREISTOXIN—A neurotoxic substance first isolated from the Japanese polychaete annelid *Lumbriconereis heteropoda* (Nitta, 1934). The empirical formula of $C_5H_9NS_2$ has been proposed for nereistoxin.

OSTRACIN—The term originally proposed by Thomson (1963) to designate the ichthyocrinotoxin produced by fishes of the genus *Ostracion*, but was later amended by the author to ostracitoxin, which is the preferred designation. *See also Ostracitoxin.*

OSTRACITOXIN—The term proposed by Thomson (1964) to designate the ichthyocrinotoxin produced by fishes of the genus *Ostracion*. The poison is heat-stable, nondialyzable, hemolytic, and nonprotein in nature. The chemical structure of the poison is being studied by Scheuer and his group at the University of Hawaii.

PAHUTOXIN—A poison isolated from the mucous secretions of fishes of the family Ostraciontidae, the boxfishes. *See also Ostracitoxin* (Thomson, 1964; Boylan and Scheuer, 1967).

PALYTOXIN—A poison isolated from the sea anemone *Palythoa vestitus* by Paul Scheuer at the University of Hawaii. The chemical nature of the poison had not been reported at the time of this writing.

PARALYTIC SHELLFISH POISON—The toxic principle found in poisonous mussels, clams, and other shellfish originally derived from toxic dinoflagellates. A term that is sometimes used interchangeably with mussel poison, clam poison, dinoflagellate poison, or saxitoxin. It is believed to have a molecular formula of $C_{10}H_{17}N_7O_4 \cdot 2HCl$. *See also* other terms listed above.

PARATRYGONICA—A term applied by Castex *et al.* (1963) to the clinical entity of envenomations from freshwater stingrays. It is more generally referred to as the *paratrygonic syndrome.*

PELOMETOXIN—The name applied by Markoff (1939) to the nonprotein, heat-stable, toxic compound found in scombroid fishes; also termed scombrotoxin or saurine—a histamine-like substance. Pelometoxin is seldom used.

PHYSALAEMIN—A polypeptide obtained from the South American amphibian *Physalaemus fuscumaculatus* having pharmacological properties similar to octopodan eledoisin (Erspamer, Bertaccini, and Cei, 1962).

PLOTOLYSIN—The hemotoxic fraction of the catfish poison plototoxin. *See also Plototoxin.*

PLOTOSPASMIN—The neurotoxic fraction of the catfish poison plototoxin. *See also Plototoxin.*

PLOTOTOXIN—A term proposed by Toyoshima (1918) to designate the poison derived from the catfish *Plotosus lineatus*. The toxin is said to be comprised of a hemotoxic fraction, plotolysin, and a neurotoxic fraction, plotospasmin.

POISON—Any substance which when ingested, injected, absorbed, or applied to the body in relatively small quantities, by its chemical action may cause damage to structure or disturbance of function.

PURPURINE—A term proposed by Dubois (1903) to designate the neurotoxic principle derived from the median zone of the hypobranchial or purple gland of the gastropods *Murex* and *Purpura*. According to later authors, the substance is believed to be an ester or a mixture of esters of choline, which has been termed as "murexine" by Erspamer and Dordoni (1947). *See also Murexine.*

SALMIN—A term proposed by Miescher (1874) to designate a toxic substance obtained from the milt of salmon.

SAURINE POISONING—*See Scombroid Poisoning.*

SAXITOXIN—A term used by Schuett and Rapoport (1962) to designate the toxin obtained from poisonous mussels *(Mytilus)*, clams *(Saxidomus)*, and plankton *(Gonyaulax)*. Saxitoxin has been used interchangeably with paralytic shellfish poison or clam poison or gonyaulax poison. Saxitoxin is said to have a molecular formula of $C_{10}H_{17}N_7O_4 \cdot 2HCl$.

SCOMBROID POISONING—A form of ichthyosarcotoxism caused by the improper preservation of scombroid and other dark-meated marine fishes, which results in certain bacteria acting on histidine in the muscle of the fish converting it to histamine, saurine, and possibly other toxic substances. Ingestion of this toxic material results in an allergiclike reaction. *See also Scombrotoxin.*

SCOMBROTOXIN—The term used by Halstead (1964) to designate the toxic substance found in improperly-preserved scombroid and other dark-meated marine fishes. The exact chemical nature of scombrotoxin is not known, but it is believed to be a complex of several substances, including histamine, saurine, and possibly other toxic chemical constituents.

SIGUATERA—A misspelling of the term ciguatera.

SPHEROIDINE—A term proposed by Yokoo (1950) to designate a toxic fraction from puffer poison. Spheroidine was believed to have the empirical formula $C_{12}H_{17}O_{13}N_3$. *See also Tetrodotoxin.*

SPINE—The bony portion of the sting exclusive of the integumentary sheath, venom gland, and other soft tissues.

STING—The osseous spine, integumentary sheath, and accompanying venom glands. The integumentary sheath generally includes the venom glands which are associated with it. This particular definition is in reference to the stings of fishes only. Generally speaking, the term "sting" refers to the complete venom apparatus of the organism.

STINGING—The act of introducing venom into the flesh of a victim by means of a venom apparatus. *See also Envenomation.*

SUBERITIN—A term proposed by Richet (1906) to designate a toxic substance derived from the marine sponge *Suberites domunculus* which when injected into dogs produces intestinal hemorrhages and respiratory distress. The chemical nature of the poison has not been defined.

SYGUATERA—One of several spellings of the term ciguatera (Savtschenko, 1886).

TARICHATOXIN—The potent neurotoxin which occurs in the California newt *Taricha torosa* (Rathke). The presence of a paralyzing toxic substance in *T. torosa* was first described by Twitty and Elliott (1934), Twitty and Johnson (1934), and Twitty (1937). However, the term "tarichatoxin" was first applied to this newt poison by Brown and Mosher (1963). The poison is now believed to be identical to tetrodotoxin (Mosher *et al.*, 1964; Woodward, 1964). *See also Tetrodotoxin.*

TETRAMINE—A term proposed by Ackermann, Holtz, and Reinwein (1923) for $N(CH_3)_4OH$, tetramethylammonium hydroxide, which they isolated as a toxic fraction from the sea anemone *Actinia equina.*

TETRAODONTOXIN—This term was proposed by Halstead (1953) as a substitute for "tetrodotoxin" since the name of the toxin is derived from the generic name of the fish which is *Tetraodon*. However, tetraodontoxin has never found acceptance, and since tetrodotoxin is so firmly entrenched in the literature, the term tetraodontoxin has been discarded.

TETRODOIC ACID—A term proposed by Kishi, Goto, and Hirata (1964) for a product resulting from tetrodotoxin. *See also Tetrodotoxin.*

TETRODONIC ACID—A term proposed by Tahara (1894) for what he believed to be the acid fraction of puffer poison. It was later learned that the toxin itself did not have acid properties and the term was abandoned. Tsuda *et al.* (1963) resurrected the term to designate a derivative of tetrodotoxin having the empirical formula $C_{11}H_{19}O_9N_3$ and possessing properties of an amino acid.

TETRODONIN—The term proposed by Yashizawa (1894; cited by Suehiro, 1947), to designate one of two toxic crystalline substances which he isolated from puffer poison. The chemical nature of the substance is unknown. This term was subsequently adopted by Tahara (1894) for one of two fractions which he isolated from puffer poison, but it was later shown to be an impurity which contained only a small amount of the pure poison. Tahara (1910) hypothesized the empirical formula of tetrodonin as $C_{11}H_{11}N_9O_2$.

TETRODOPENTOSE—A term proposed by Tahara (1897) to designate an impurity separated out from puffer poison which he believed to be pentose. This substance was later identified by Tsuda and Kawamura (1950) as a mixture of mesoinositol and scillitol.

TETRODOTOXIN—The term proposed by Tahara (1897) for the pure crystalline puffer poison. It is also called fugu toxin. Although improperly spelled, the term "tetrodotoxin" has become widely adopted and is now deeply entrenched in biotoxicological literature. It is the pure crystalline puffer poison derived originally from fish of the genus *Tetraodon*. The empirical formula is believed to be $C_{11}H_{17}O_8N_3 \pm YH_2O$ (Woodward, 1964). *See also Tetraodontoxin.*

TETRONIN—The term proposed by Tahara (1910) to designate a nitrogenous base obtained from crude puffer poison. The base was thought to have the empirical formula of $C_{11}H_{11}N_9O_2$. The substance was never clinically identified.

THALASSIN—A term proposed by Richet (1902) to designate a toxic compound derived from the tentacles of *Anemonia sulcata*. The substance when injected into dogs was found to produce allergic symptoms.

THERIAC—An ancient antidote for a poison. Originally derived from Mithridate, it was later modified by Andromachus the Elder (ca. 60 B.C.) and comprised of 70 pulverized drugs mixed with honey. Such a mixture was given by Andromachus to Emperor Nero. The drug gained great popularity during ancient times and was used in the treatment of many different types of poisonings.

TJAKALANG POISONING—A synonym of scombroid poisoning. *See Scombroid Poisoning.*

TOMEA—The Tahitian term for scombroid poisoning.

TOXICOLOGY—The science which treats of poisons (regardless of their origins), their effects, their detection, and treatment of the conditions which they produce.

TOXIGENESIS—The process of poison production.

TOXIN—Any poisonous substances of microbic, mineral, vegetable, or animal origin. There is some disagreement in the use of this term since, in the older and strictest sense, toxins are more or less unstable, do not cause symptoms of intoxications until after a period of incubation, and are antigenic. However, in modern usage the terms "toxin," "toxicology," and "toxic" have been widely adopted to designate inorganic and organic poisonous substances having a vast array of chemical and biological properties.

TOXINOLOGY—A term proposed by the International Society of Toxinology to designate biological poisons or biotoxins. Because of the confusion that exists between the terms "toxicology" and "toxinology," the term "biotoxicology" is recommended since it clearly designates the biological origin of the poison, whereas "toxinology" does not.

TRANSVECTOR—An organism which serves as a purveyor or transmitter of a poison which is not generated within its body but is obtained from another source. A typical example would be the mussel *Mytilus* which serves as a transvector of paralytic shellfish poison derived from the dinoflagellate *Gonyaulax* (Halstead, 1965).

TRITURUS EMBRYONIC TOXIN—*See Tarichatoxin.*

TRUNCULOPURPURINE—A term used by Erspamer (1946, 1947) to designate the form of purpurine, now called murexine, obtained from the median zone of the hypobranchial gland of *Murex trunculus.*

VENERUPIN—The toxic principle believed to be an amine found in certain Japanese pelecypods which were formerly placed under the genus *Venerupis* (Akiba, 1943; Togashi, 1943). This toxin is entirely distinct from paralytic shellfish poison found in other types of bivalves. The exact chemical nature of the toxin is unknown. The poison is now believed to originate with a dinoflagellate *(Prorocentrum sp?)*, which is ingested by the shellfish in a manner similar to paralytic shellfish poisoning (Nakazima, 1963).

VENERUPIN SHELLFISH POISONING—An organotropic form of shellfish poisoning resulting from the ingestion of oysters *(Crassostrea gigas)* and/or asari *(Tapes semidecussata)* from certain restricted areas of Japan. It was first described by Akiba (1943) and Togashi (1943). The exact chemical nature of the poison is unknown but is believed to be an amine. *See also Venerupin.*

VENOM—The poison secreted by a venom apparatus of an animal. Venoms are usually a large molecular protein or are in association with a protein which may serve as a carrier, but it is becoming increasingly obvious that there may be many exceptions to this generalization.

VENOM APPARATUS—The traumagenic device, venom gland, and accessory organs directly concerned with the production and transmission of a venom.

VENOMOUS ANIMAL—An animal that is equipped with a traumagenic device, i.e., a spine, tooth, nematocyst, etc., and a poison or venom gland, and associated accessory organs capable of introducing the venom into the flesh of the victim and thereby producing an envenomation.

WHELK POISONING—A form of intoxication resulting from the ingestion of whelks, members of the mollusk family Buccinidae. The poison is localized in the salivary gland of the mollusk, and the principal ingredient is believed to be tetramine.

LITERATURE CITED

Ackermann, D., Holtz, F., and H. Reinwein
1923 Reindarstellung und konstitutionsermittelung des tetramins, eines giftes aus *Aktinia equina*. Z. Biol. 79: 113-120.

Akiba, T.
1943 Study of poisons of *Venerupis semidecussata* and *Ostrea gigas*. [In Japanese] Japan Iji Shimpo 1078: 1077-1082, 1 fig.

Anastasi, A., and V. Erspamer
1962 Occurrence and some properties of eledoisin in extracts of posterior salivary glands of *Eledone*. Brit. J. Pharmacol. 19(2): 326-336, 4 figs.

Asano, M., and M. Itoh
1962 Toxicity of a lipoprotein and lipids from the roe of a blenny *Dinogunellus grigorjewi* Herzenstein. Tohoku J. Agr. Res. 13(2): 151-167.
1966 Lipoproteins (Lipostichaerins) in the roe of blenny *Strichaeus grigorjewi* Herzenstein. Tohoku J. Agr. Res. 16(4): 299-316.

Asano, M., Takayangi, F., and Y. Furuhara
1950 Studies on the toxic substances in marine animals. II. Shellfish-poisoning from *Callista brevisiphonata* Carpenter occurred in the vicinity of Mori, Kayabe county. [In Japanese, English summary] Bull. Fac. Fish., Hokkaido Univ. (7): 26-36, 3 figs.

Bacq, Z. M.
1937 L' "amphiporine" et la "némertine" poisons des vers némertiens. Arch. Intern. Physiol. (Liége) 44(2): 190-204.

Barnes, J. H.
1964 Cause and effects of irukandji stingings. Med. J. Australia 1: 897-904, 3 figs.

Baslow, M. H., and P. Turlapaty
1969 *In vivo* antitumor activity and other pharmacological properties of halitoxin obtained from the sponge *Haliclona viridis*. Proc. West. Pharmacol. Soc.

Boylan, D. B., and P. J. Scheuer
1967 Pahutoxin; a fish poison. Science 155: 52-56.

Brieger, L.
1885 Ueber basische produkte in der miesmuschel. Deut. Med. Wochschr. 11(53): 907-908.
1888 Zur kenntniss des tetanin und des mytilotoxin. Arch. Pathol. Anat. Physiol. 112: 549-551.
1889 Beitrag zur kenntniss der zusamensetzung des mytilotoxins nebst einer uebersicht der bisher in ihren haupteigenschafter bekannten ptomaine und toxine. Arch. Pathol. Anat. Physiol., 9, 115(5): 483-492.

Brown, M. S., and H. S. Mosher
1963 Tarichatoxin: isolation and purification. Science 140(3564): 295-296.

Castex, N. M., Maciel, I., Pedace, E. A., and F. Loza
1963 La enfermedad paratrygónica. Inst. Biol. Col. Immaculada Concepcion, Santa Fe, Argentina.

Chisholm, C.
1808 On the poison of fish. Edinburgh Med. Surg. J. 4(16): 393-422.

Cutress, C. E.
1955 An interpretation of the structure and distribution of Cnidae in Anthozoa. Systematic Zool. 4(3): 120-137, 10 figs.

Desportes, J. B.
1770 Histoire des maladies de S. Domingue, Vol. I, p. 108-110. Lejay, Paris.

Dubois, R.
1903 Sur le venin de la glande à pourpre des murex. Compt. Rend. Soc. Biol. 55: 81.

Dujarric de la Rivière, R.
1915 Sur l'existence d'une médusocongestine. Compt. Rend. Soc. Biol. 78: 596-600.

Erspamer, V.
1946 Sulle reazioni colorate degli estratti de ghiandola ipobranchiale di *Murex trunculus, Murex brandaris* e *Murex erinaceus*. Ric. Sci. 16(7): 938 940.
1947 Ricerche chimiche e farmacologiche sugli estratti di ghiandola ipobranchiale di *Murex trunculus, Murex brandaris* e *Tritonalia erinacea*. II.

ERSPAMER, V.—Continued
Reasioni chimiche colorate degli estratti. Arch. Intern. Pharmacodyn. 74(2): 113-150, 2 figs.

ERSPAMER, V. and O. BENATI
1953 Identification of murexine as β [imidazolyl-(4)]-acryl-choline. Science 117(3033): 161-162.

ERSPAMER, V., BERTACCINI, G., and J. M. CEI
1962 Occurrence of an eledoisin-like polypeptide (physalaemin) in skin extracts of *Physalaemus fuscumaculatus.* Experientia 18: 562-563.

ERSPAMER, V., and F. DORDONI
1947 Ricerche chimiche e farmacologiche sugli estratti di ghiandola ipobranchiale di *Murex trunculus, Murex brandaris* e *Tritonalia erinacea.* III. Presenze negli estratti di un nuovo derivato della colina o di una colina omologa: la murexina. Arch. Intern. Pharmacodyn. 74(3-4): 263-285, 9 figs.

FLECKER, H.
1945 Injuries by unknown agents to bathers in North Queensland. Med. J. Australia 1: 417.
1952 Irukandji sting to North Queensland bathers without production of weals but with severe general symptoms. Med. J. Australia 2: 89-91.

FREEMAN, S. E., and R. J. TURNER
1970 Maculotoxin, a potent toxin secreted by *Octopus maculosus.* Hoyle Toxicol. Appl. Pharmacol. 16(3): 681-690, 4 figs.

GHIRETTI, F.
1959 Cephalotoxin: the crab-paralysing agent of the posterior salivary glands of cephalopods. Nature 183: 1192-1193.
1960 Toxicity of octopus saliva against crustacea. Ann. N.Y. Acad. Sci. 90(3): 726-741, 11 figs.

HALSTEAD, B. W.
1953 Some general considerations of the problem of poisonous fishes and ichthyosarcotoxism. Copeia (1): 31-33.
1964 Fish poisonings—their diagnosis, pharmacology, and treatment. Clin. Pharmacol. Therap. 5(5): 615-627.
1965 Poisonous and venomous marine animals of the world. Vol. I. Invertebrates. U.S. Government Printing Office, Washington, D.C.

HALSTEAD, B. W., and F. R. MODGLIN
1950 A preliminary report on the venom apparatus of the bat-ray, *Holorhinus californicus.* Copeia (3): 165-175, 6 figs.

HASHIMOTO, Y.
1967 Report on the second conference, Japan-U.S. cooperative studies on ciguatera in the tropical and subtropical Pacific. (Unpublished data.)

HELFRICH, P.
1961 Fish poisoning in the tropical Pacific. Hawaii Marine Lab., Univ. Hawaii. 16 p., 7 figs.

HELFRICH, P., and A. H. BANNER
1960 Hallucinatory mullet poisoning. J. Trop. Med. Hyg. 1960: 1-4.

JENSEN, D.
1963 Eptatretin: a potent cardioactive agent from the branchial heart of the Pacific hagfish, *Eptatretus stoutii.* Comp. Biochem. Physiol. 10: 129-151.

KAWABATA, T.
1962 Fish-borne food poisoning in Japan, p. 467-479. *In* G. Borgstrom [ed.], Fish as food. Vol. II. Nutrition, sanitation, and utilization. Academic Press, New York.

KISHI, Y., GOTO, T., and Y. HIRATA
1964 Structure of tetrodoic acid, a hydrolysis product of tetrodotoxin. [In Japanese, English summary] Nippon Kagaku Zasshi 85: 572-579.

KURAJEFF, D.
1901 Ueber das protamin aus den spermatozoen des *Accipenser stellatus.* Hoppe-Seylers Z. Physiol. Chem. 32: 197-200.

MCCRUDDEN, F. H.
1921 Pharmakologische und chemische studien über barben- und hechtrogen. Arch. Exp. Pathol. Pharmakol. 91: 46-80.

MARKOFF, W. N.
1939 Zum problem der seefisch-fäulnis. Zentr. Bakteriol. Parisitenk., Abt. II, 101: 157-171.

MATSUNO, T., and T. YAMANOUCHI
1961 A new triterpenoid sapogenin of animal origin (sea cucumber). Nature 191(4783): 75-76.

(NSA) Not seen by author.

MIESCHER, F.
1874 Das protamin, eine neue organische base aus den samenfäden des rheinlachses. Ber. Deut. Chem. Ges. (Berlin) 7: 376-379. (NSA)

MOSHER, H.S.
1966 Chemistry (C). Non-protein neurotoxins. Science 51(3712): 860-861.

MOSHER, H. S., FUHRMAN, F. A., BUCHWALD, H. D., and H. G. FISCHER
1964 Tarichatoxin-tetrodotoxin: a potent neurotoxin. Science 144(3622): 1100-1110.

MOSSO, U.
1889 Recherches sur la nature du venin qui se trouve dans le sang de l'anguille. Arch. Ital. Biol. 12: 229-236.

NAKAZIMA, M.
1963 Summary on the cause of shellfish poisoning in Hamana Lake. (Personal communication, May 23, 1963.)

NIGRELLI, R. F.
1952 The effects of holothurin on fish, and mice with sarcoma 180. Zoologica 37: 89-90.

NIGRELLI, R. F., JAKOWSKA, S., and I. CALVENTI
1959 Ectyonin, an antimicrobial agent from the sponge, *Microciona prolifera* Verrill. Zoologica 44(4): 173-176.

NIGRELLI, R. F., STEMPIEN, M. F., JR., RUGGIERI, G. D., LIGUORI, V. R., and J. T. CECIL
1967 Substances of potential biomedical importance from marine organisms. Fed. proc. 26: 1197-1205.

NITTA, S.
1934 Ueber nereistoxin, einen giftigen bestandteil von *Lumbriconereis heteropoda* Marenz (Eunicidae). [In Japanese] J. Pharmacol. Soc. Japan 54: 648-652.

PARRA, A.
1787 Descripción de diferentes piezas de historia natural, las más del ramo marítimo, p. 1-135, 13 pls. Havana, Cuba.

PATAKI, S., and K. MEYER
1955 Helv. Chim. Acta 38: 1631. (NSA)

PERGOLA, M.
1937 Matiéres toxigénes chez le poisson frais avec ou sans la présence de l'ichthyovenin. Boll. Sez. Ital. Soc. Intern. Microbiol. 9(5): 105-108. (NSA)

PORTA, A.
1905 Ricerche anatomiche sull'apparecchio velenifero di alcuni pesci. Anat. Anz. 26: 232-247, 2 pls., 5 figs.

PORTIER, P., and C. RICHET
1902 Sur les effets physiologiques du poison des filaments pecheurs et des tentacules des coelentérés (Hypnotoxine). Compt. Rend. Acad. Sci. 134-247-248.

RANDALL, J. E.
1958 A review of ciguatera, tropical fish poisoning, with a tentative explanation of its cause. Bull. Marine Sci. Gulf Caribbean 8(3): 236-267, 2 figs.

RICHET, C.
1902 Du poison pruritogéne et urticant contenu dans les tentacules des actinies. Compt. Rend. Soc. Biol. 54: 1438-1440.
1903*a* Des poisons contenus dans les tentacules des actinies (congestine et thalassine). Compt. Rend. Soc. Biol. 55: 246-248.
1903*b* De la thalassine, considerée comme antitoxine cristallisée. Compt. Rend. Soc. Biol. 55: 1071-1073.
1905 De l'action de la congestine (virus des actinies) sur les lapins et de ses effets anaphylactiques. Compt. Rend. Soc. Biol. 58: 109-112.
1906 De l'action toxique de la subéritine (extrait aqueux de *Suberities domuncula*). Compt. Rend. Soc. Biol. 61: 598-600.
1908 De la substance anaphylactisante ou toxogenine. Compt. Rend. Soc. Biol. 64: 846-848.

RICHET, C., PERRET, A., and P. PORTIER
1902 Des propriétés chimiques et physiologiques de poison des actinies (actinotoxine). Compt. Rend. Soc. Biol. 54: 788-790.

SAVTSCHENKO, P.
1886 Atlas of the poisonous fishes. Description of the ravages produced by them in the human organism, and antidotes which may be employed. [In Russian] St. Petersburg. 53 p., 10 pls.

(NSA) Not seen by author.

SCHUETT, W., and H. RAPOPORT
1962 Saxitoxin, the paralytic shellfish poison. Degradation to a pyrrolopyrimidine. J. Am. Chem. Soc. 84: 2266.

SHIMADA, S.
1969 Antifungal steroid glycoside from sea cucumber. Science 163: 1462.

SIPERSTEIN, M. D., MURRAY, A. W., and E. TITUS
1957 Biosynthesis of cardiotonic sterols from cholesterol in the toad, *Bufo marinus*. Arch. Biochem. Biophys. 67(1): 154-160.

SUEHIRO, Y.
1947 Poison of globe-fish, p. 140-159. *In* Y. Suehiro, Practice of fish physiology. [In Japanese] Takeuchi Bookstore, Tokyo.

TAHARA, Y.
1894 Report of discovery of puffer toxin. [In Japanese] Chugai Iji Shimpo (344): 4-9.
1897 Report on puffer poison. [In Japanese] Yakugaku Zasshi (328): 587-625.
1910 Über das tetrodongift. Biochem. Z. 30: 255-275.

TAMIYA, N.
1966 (Personal communication, April, 1966.)

THOMSON, D. A.
1963 A histological study and bioassay of the toxic stress secretion of the boxfish *Ostracion lentiginosus*. Ph.D. Thesis. Univ. Hawaii. 194 p.
1964 Ostracitoxin: an ichthyotoxic stress secretion of the boxfish, *Ostracion lentiginosus*. Science 146(3641): 244-245.

TOGASHI, M.
1943 Clinical study of the poisoning by *Venerupis semidecussata*. [In Japanese] Japan Iji Shimpo (May). (NSA)

TOYOSHIMA, T.
1918 Serological study of toxin of the fish *Plotosus anguillaris* Lacépède. [In Japanese] J. Japan Protoz. Soc. 6(1-5): 45-270.

TSUDA, K., TACHIKAWA, R., SAKAI, K., AMAKASU, O., KAWAMURA, M., and S. IKUMA
1963 Die konstitution und konfiguration der tetrodonsäure. Chem. Pharm. Bull. 11(11): 1473-1475.

TSUDA, K., and M. KAWAMURA
1950 The constituents of the ovaries of globefish. I. The isolation of meso-inositol and scillitol from the ovaries. J. Pharm. Soc. Japan 70(7-8): 432-435, 2 figs.

TWITTY, V. C.
1937 Experiments on the phenomenon of paralysis produced by a toxin occurring in *Triturus* embryos. J. Exp. Zool. 76(1): 67-104, 8 figs.

TWITTY, V. C., and H. A. ELLIOTT
1934 The relative growth of the amphibian eye, studied by means of transplantation. J. Exp. Zool. 68(2): 247-291, 22 figs.

TWITTY, V. C., and H. H. JOHNSON
1934 Motor induction in *Amblystoma* produced by *Triturus* transplants. Science 80(2064): 78-79.

WINKLER, L. R., and B. E. TILTON
1962 Predation on the California sea hare *Aplysis californica* Cooper, by the solitary great green sea anemone, *Anthopleura xanthogrammica* (Brandt), and the effect of sea hare toxin and acetylcholine on anemone muscle. Pacific Sci. 16(3): 286-290, 2 figs.

WOODWARD, R. B.
1964 The structure of tetrodotoxin. Pure Appl. Chem. 9(1): 49-74.

YAMANOUCHI, T.
1942 Study of poisons contained in holothurians. [In Japanese] Teikoku Gakushiin Hokoku (17): 73.
1943*a* On the poison contained in *Holothuria vagabunda*. [In Japanese] Folia Pharmacol. Japan. 38(2): 115.
1943*b* Distribution of poison in the body of *Holothuria vagabunda*. [In Japanese] Zool. Mag. (Tokyo) 55(12): 87-88.
1955 On the poisonous substance concontained in holothurians. Publ. Seto Marine Biol.. Lab. 4(2-3): 183-203, 2 figs.

YOKOO, A.
1950 Chemical studies on tetrodotoxin. Rept. III. Isolation of spheroidine. [In Japanese] Nippon Kagaku Zasshi 71(11): 590-592.

(NSA) Not seen by author.

Personal Name Index

Only the most important sources, current researchers, and historical names are included. The reader is referred to the "Literature Cited" at the end of each chapter for a full listing of names occurring in the text.

General Index

Page references to species occurring in the plates are listed in boldface. Popular names of species are listed under that species name, e.g., "textile (cone)" appears under the heading "Cone shells."

Phylum Protozoa

PLATES

CHAPTER II

II: PLATE 1

FIGURE a.—*Gymnodinium breve* Davis. × 1,350.

FIGURE b.—*Gonyaulax catenella* Whedon and Kofoid. × 1,350.

FIGURE c.—*Gonyaulax tamarensis* Lebour. × 1,125.

FIGURE d.—*Pyrodinium phoneus* Woloszynska and Conrad. × 1,650.

(R. Kreuzinger, after various sources)

FIGURE e.—*Prymnesium parvum* Carter. × 5,000. (H. Vester, after Reich)

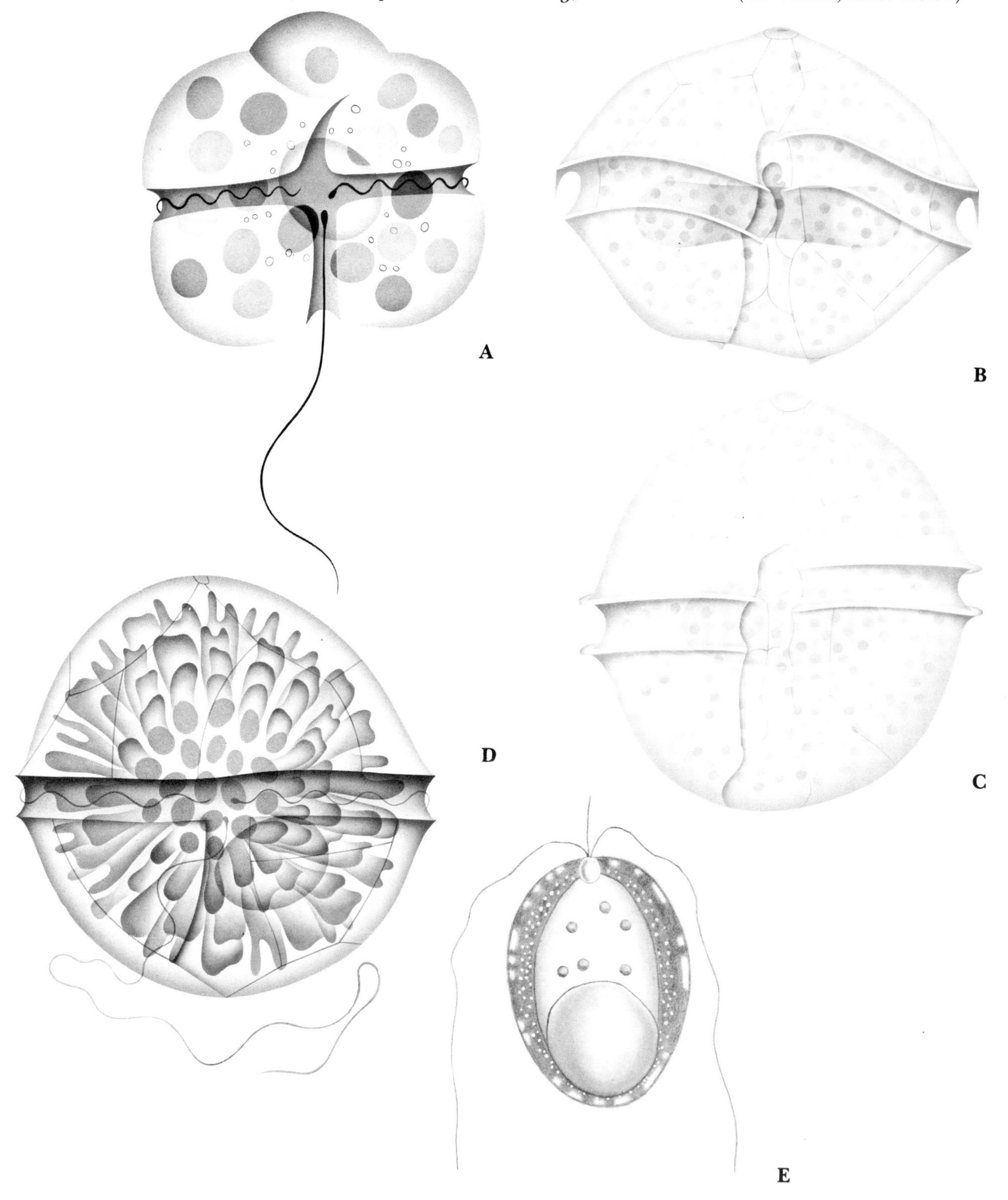

II: PLATE 2

FIGURE a.—*Gonyaulax tamarensis* in a typical chain formation. Taken from culture. × 440.
(Photo by H. Sommer, courtesy of K. F. Meyer, George Williams Hooper Foundation)

FIGURE b.—Photomicrograph showing predation of *Gonyaulax tamarensis* by *Favella* sp. *G. tamarensis* can be seen in a semi-digested state within the body of the predator. × 350. (Courtesy of A. Prakash, Fisheries Research Board of Canada)

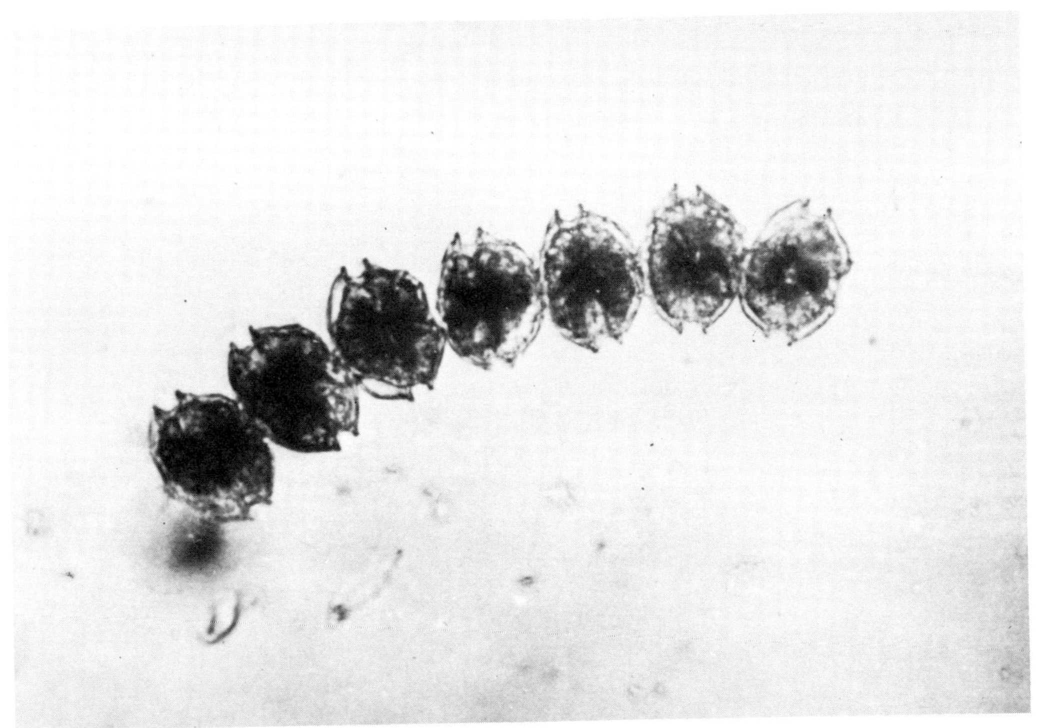

A

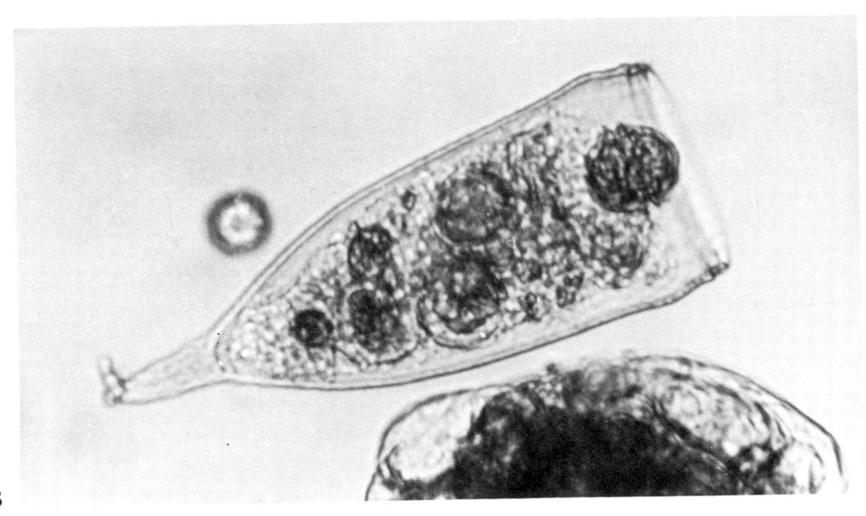

B

II: PLATE 3

FIGURE a.—*Spisula solidissima* (Dillwyn). × 0.4.
FIGURE b.—*Mya arenaria* Linnaeus. × 0.2.
FIGURE c.—*Mytilus californianus* Conrad. × 0.3.
FIGURE d.—*Mytilus edulis* Linnaeus. × 0.9.
FIGURE e.—*Modiolus modiolus* (Linnaeus). × 0.4.
FIGURE f.—*Protothaca staminea* (Conrad). × 0.8.
FIGURE g.—*Saxidomus giganteus* (Deshayes). × 0.7.
FIGURE h.—*Saxidomus nuttalli* Conrad. × 0.3. (R. Kreuzinger, courtesy of S. S. Berry)

II: PLATE 4

FIGURE a.—Internal anatomy of the mussel *Mytilus* showing parts of the body of the shellfish likely to contain paralytic shellfish poison.

FIGURE b.—Internal anatomy of the clam *Saxidomus* showing parts of the body of the shellfish likely to contain paralytic shellfish poison. (R. Kreuzinger)

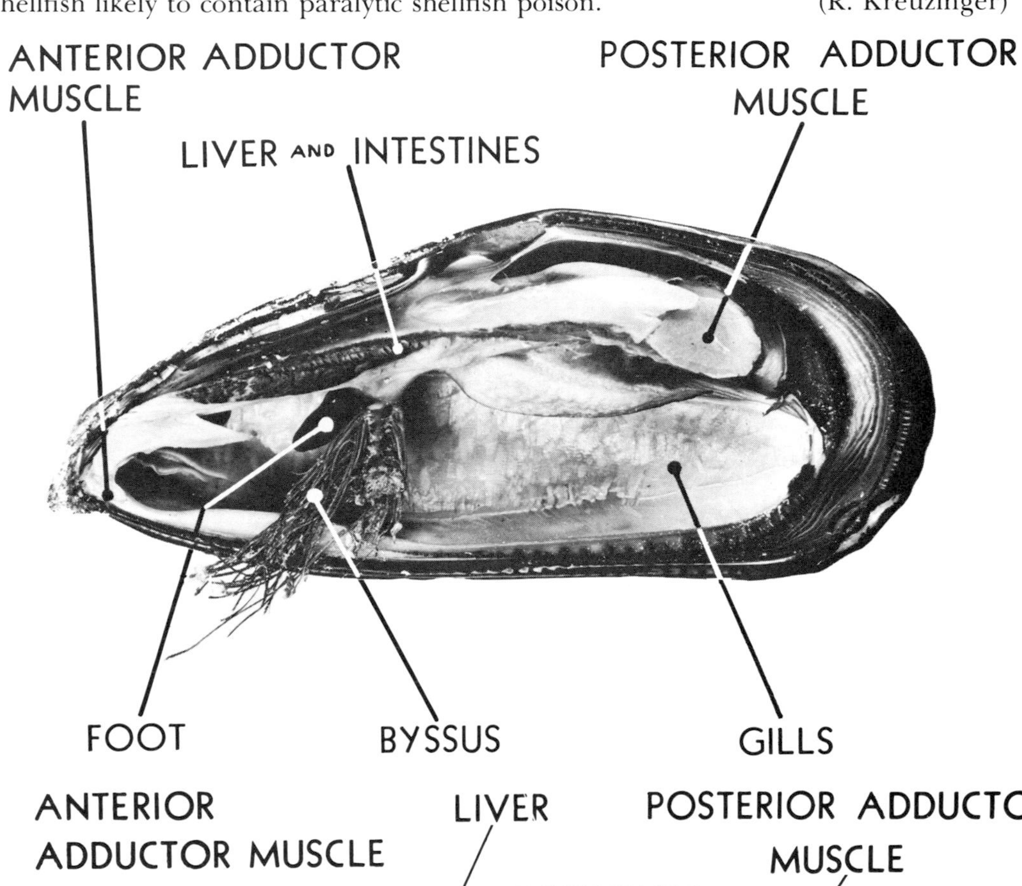

A

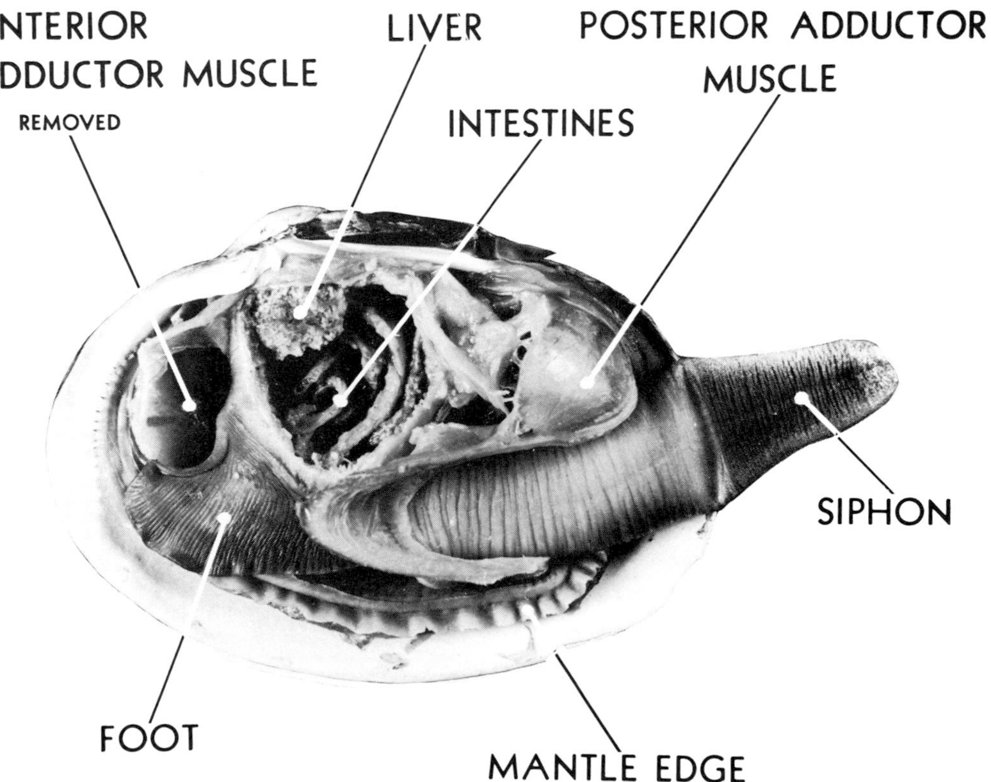

B

Phylum Porifera

PLATES

CHAPTER III

III: PLATE 1

FIGURE a.—*Fibulia nolitangere* (Duchassaing and Michelotti). Produces a stinging sensation and swelling of the hands when touched. Specimen from Maiden Cay, Jamaica, collection of Peabody Museum of Natural History, Yale University. × 0.7.

(Courtesy of W. D. Hartman)

FIGURE b.—*Haliclona viridis* (Duchassaing and Michelotti), living specimen. Taken at Harrington Sound, Bermuda. (G. Lower)

FIGURE c.—*Tedania ignis* (Duchassaing and Michelotti) in its natural habitat, off the coast of Florida, showing various growth formations. Contact with this species is said to produce an almost instantaneous stinging sensation which may result in a serious dermatitis. (R. Straughan)

A

B

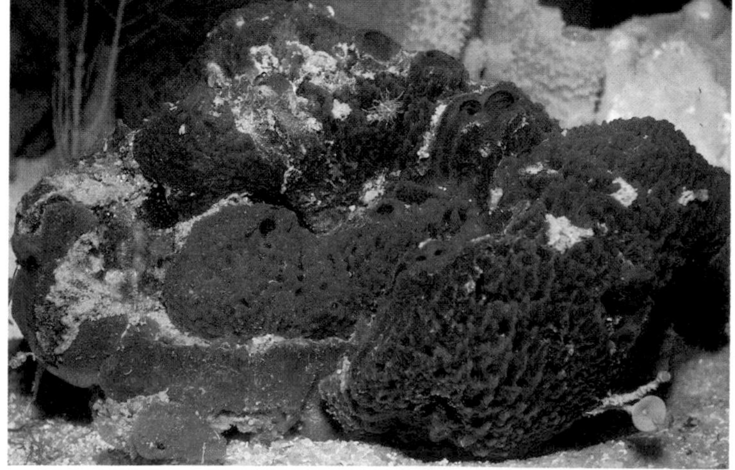

C

III: PLATE 2

Microciona prolifera (Ellis and Solander). This sponge is the cause of a contact dermatitis known as red moss or sponge poisoning which occurs among the fishermen in Northeastern United States. Symptoms consist of redness, stiffness of the finger joints, and swelling of the hands. If not treated properly the dermatitis may spread and can become serious. Forms a cluster having a height of about 150 mm. Taken at Woods Hole, Mass. (G. Lower)

Phylum Coelenterata

PLATES

CHAPTER IV

IV: PLATE 1

Millepora alcicornis Linnaeus. This particular example has fused branches and demonstrates a platelike growth. Taken from the West Indies. × 0.1. (From Boschma)

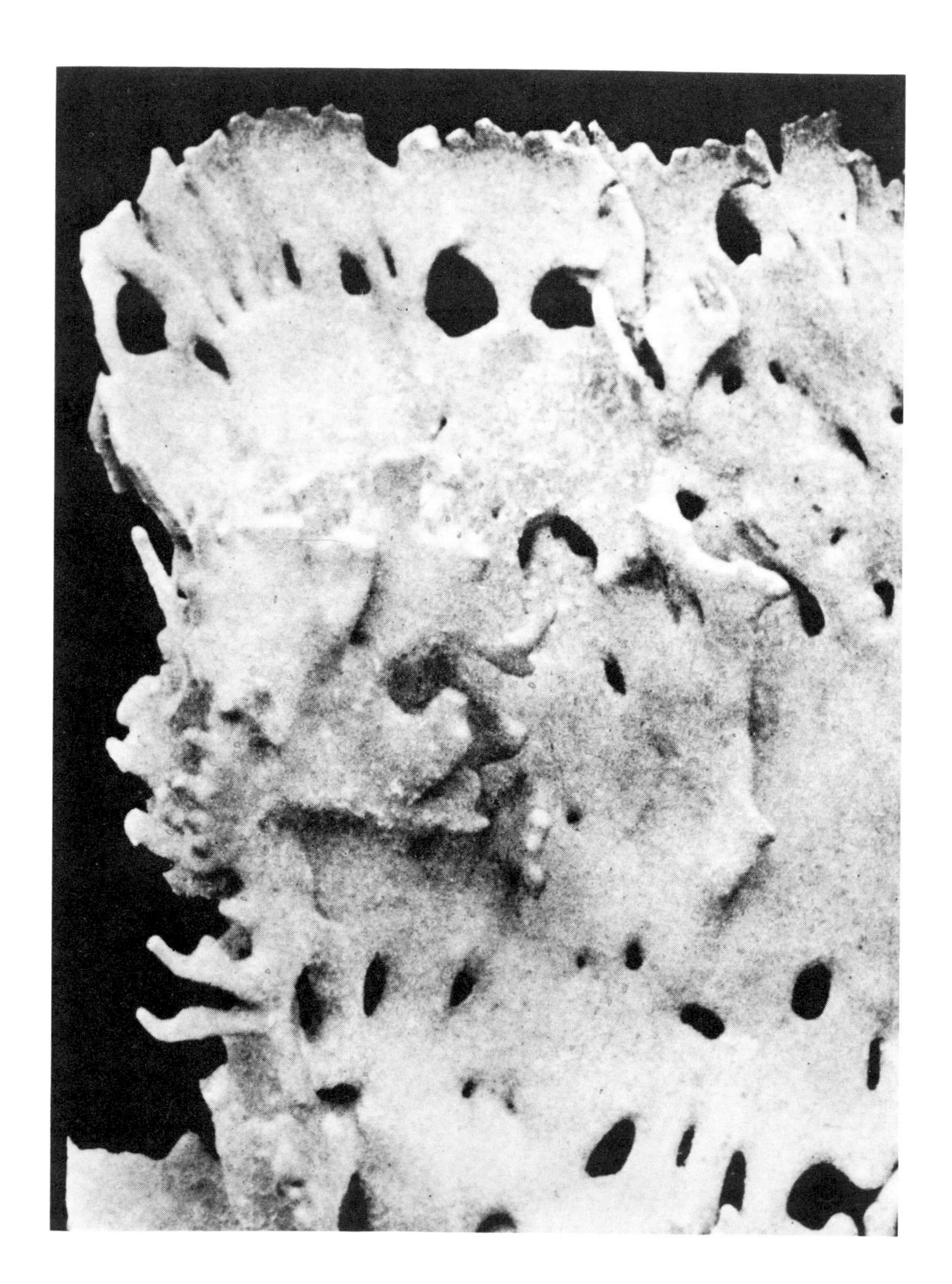

IV: PLATE 2

Millepora dichotoma Forskål. Taken on the School of Tropical and Preventive Medicine Red Sea Expedition at Al Ghardaqa, Egypt.

(D. Ollis, courtesy of Marine Biological Station, University of Cairo)

IV: PLATE 3

Gonionemus vertens (Agassiz). Diameter of bell up to 15 mm. Living specimen taken in an aquarium at Friday Harbor, Wash. This species is generally considered to be innocuous, but Aznaurian (1964) reported 37 persons stung while swimming near Vladivostok, USSR. The initial sting was like that of a nettle, but six persons subsequently died. The causative agent was said to be *Gonionemus*, but the species was believed to be *G. vertens*. (Photo by G. E. MacGinitie, courtesy of P. Saunders)

IV: PLATE 4

Olindioides formosa Goto. × 0.6. (R. Kreuzinger, after Goto)

IV: PLATE 5

FIGURE a.—*Physalia utriculus* (La Martinière). Attains a length of 13 cm. Tentacles may hang down a distance of 12 m. Drawn from specimen which stung the author at Saipan, Mariana Islands. (M. Shirao)

FIGURE b.—*Cyanea capillata* (Linnaeus). Diameter of bell 20 cm. Drawn from specimen collected in northern California. (M. Shirao)

FIGURE c.—*Chrysaora quinquecirrha* (Desor). Diameter of bell 25 cm. (M. Shirao, after Mayer)

IV: PLATE 6

FIGURE a.—*Physalia physalis* (Linnaeus). Taken in an aquarium at Miami, Fla. Closeup showing the float and upper portion of tentacles.

FIGURE b.—Closeup of same specimen showing upper portion of tentacles.

FIGURE c.—Closeup of same specimen showing lower portion of tentacles.

(R. Straughan)

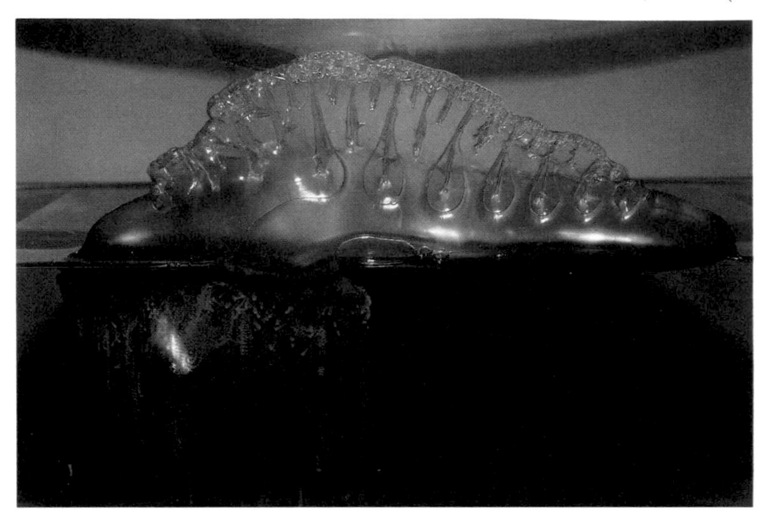

A

B

C

IV: PLATE 7

Physalia utriculus (La Martinière). Taken in an aquarium at Sydney, Australia. Attains a length of 13 cm. Tentacles may hang down a distance of 12 m. Its sting is reputedly less intense than that of *P. physalis*. (K. Gillett)

IV: PLATE 8

Closeup of fishing tentacles and smaller polyps of *Physalia utriculus* (La Martinière). Taken at Sydney, New South Wales, Australia. (Courtesy of Justice F. Myers)

IV: PLATE 9

Sting caused by the Portuguese man-o-war *Physalia physalis* (Linnaeus). On January 24, 1966, the victim was diving off Isla de Providencia, about 455 km north of Panama, and surfaced directly under a large *Physalia*. The creature was about 25 cm in length and had long tentacles which were estimated to be hanging down in the water for a depth of more than 6 m. He was stung on the upper chest and shoulder. The pain was immediate and of extreme intensity. Fortunately he surfaced near his dugout canoe and his partner assisted him in removing the *Physalia* and helped him into the boat. His partner was also severely stung while removing the animal from his shoulder. The pain reached its greatest intensity in the boat on the way back to town during the first hour after contact. Witnesses reported that the victim was carried screaming through town to one of the local physicians. The doctor, who had never encountered a case of this type before, gave him an injection of a sedative and applied a white salve of an unknown type. The victim was delirious with pain most of the night. This photograph was taken the following day at which time the victim appeared to be rational, but had developed the large skin blisters shown in the picture. After a day or two he became seriously ill and was reported to have developed "lockjaw." Since there was no tetanus vaccine available on the island, he was transported to the neighboring island of San Andres, about 77 km to the south. On February 19, he was still reported to be in very serious condition, but on February 25 he was reported to have been taken off the critical list. Unfortunately there is no further information concerning the outcome of this case, but it is believed that his subsequent recovery was uneventful.

(Photograph and legend material courtesy S. Anderson)

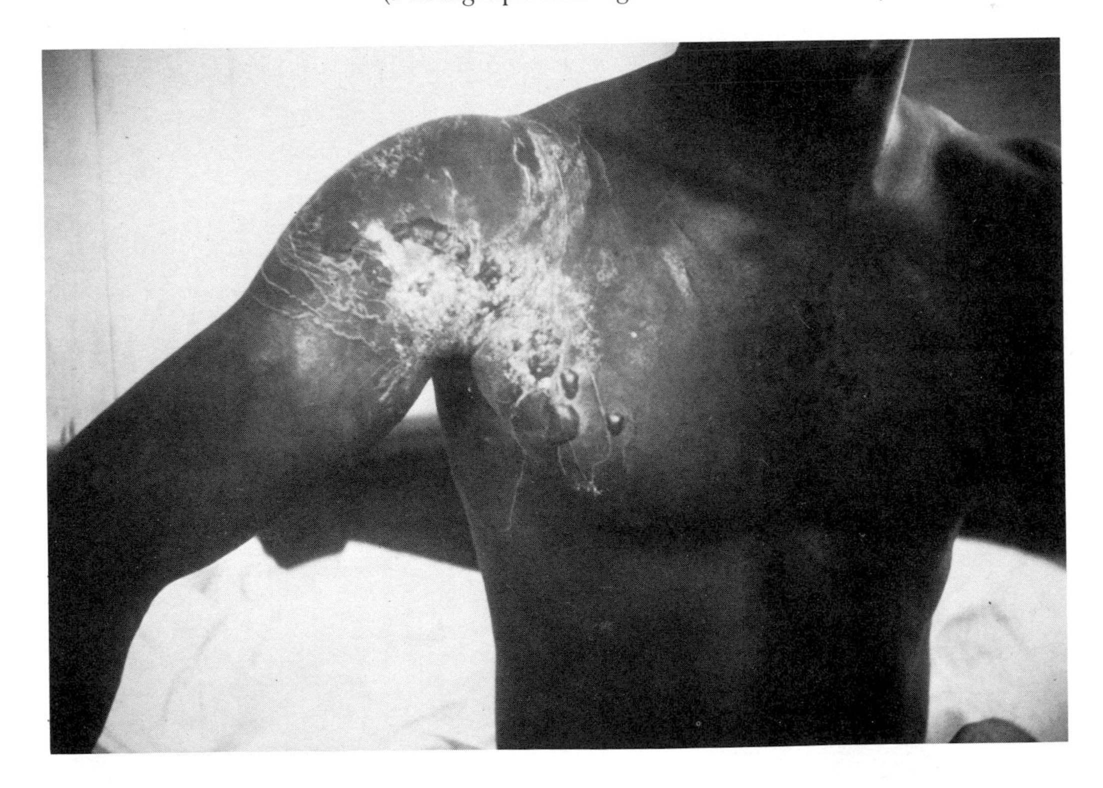

IV: PLATE 10

Single branch of living specimen of *Aglaophenia cupressina* Lamouroux. (K. Gillett)

IV: PLATE 11

FIGURE a.—*Lytocarpus philippinus* (Kirchenpauer). Stings from this hydroid are similar to *Pennaria* and *Aglaophenia*, but generally more painful and may be serious. Note that this large colony is growing from a single main stalk. Height about 200 mm. Taken at Heron Island, Great Barrier Reef, Australia. (A. Power)

FIGURE b.—Photomicrograph of nematocysts of *Lytocarpus philippinus*. The dactylozooid (large ovoid structure, upper right) contains a number of elongate fingerlike undischarged nematocysts. Extending from the lower end of the dactylozooid are a number of discharged nematocysts. The thread tubes are somewhat indistinct since part of them run off the edge of the picture (upper left). Five loose undischarged elongate nematocysts are seen scattered about the lower portion of the picture. ×400. (K. Gillett)

A

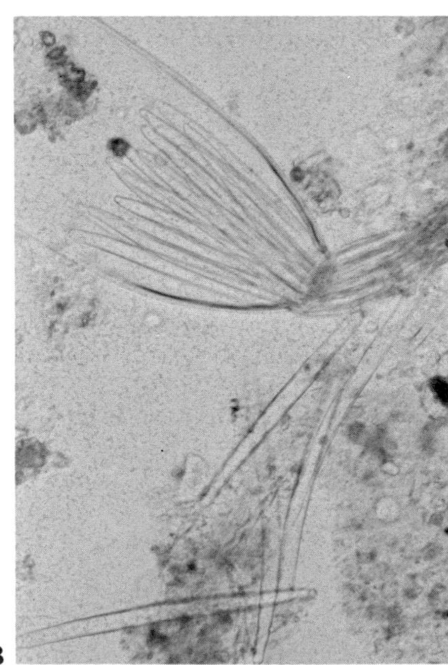

B

IV: PLATE 12

Rhisophysa eysenhardti Gegenbaur. × 1. (R. Kreuzinger, after Lens and Riemsdijk)

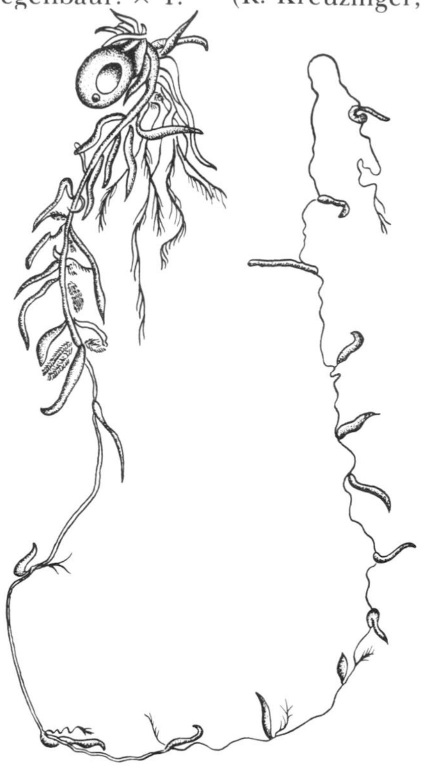

IV: PLATE 13

Carybdea rastoni Haacke. Taken in an aquarium. × 1. (Courtesy of R. V. Southcott)

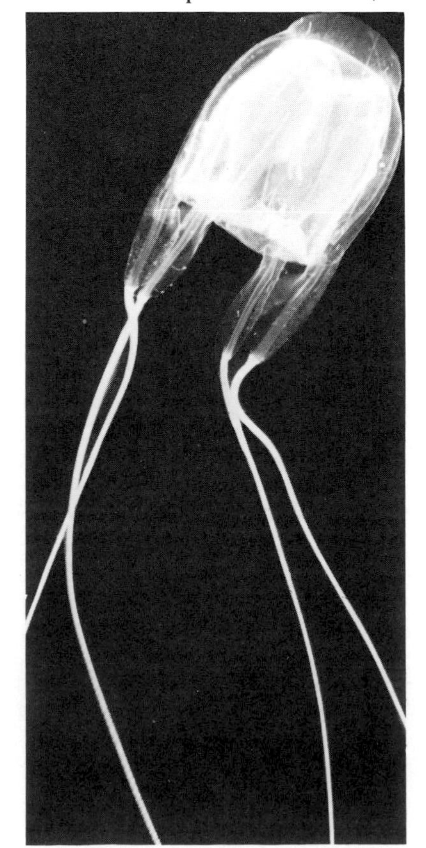

IV: PLATE 14

FIGURE a.—*Carukia barnesi* Southcott, carybdeid jellyfish which causes Irukandji stings along the coast of North Queensland, Australia. This jellyfish causes a violent self-limiting, nonfatal sting. The causative agent of these Irukandji jellyfish stings has only recently been identified and was named in honor of Dr. Jack Barnes, the renowned Australian physician who has contributed so much to our knowledge of coelenterate stings. (Photo courtesy J. H. Barnes)

FIGURE b.—Photomicrographs of nematocysts of *Carukia barnesi*.

1. Discharged tentacular nematocyst.

2-3. Undischarged tentacular nematocysts. (From R. V. Southcott)

FIGURE c.—*Carukia barnesi* (on left) compared with two immature specimens of *Carybdea rastoni* Haacke (on right).

FIGURE d.—Drawing of *Carukia barnesi* Southcott. Side view of entire adult. Only one tentacle is shown full length. (From R. V. Southcott)

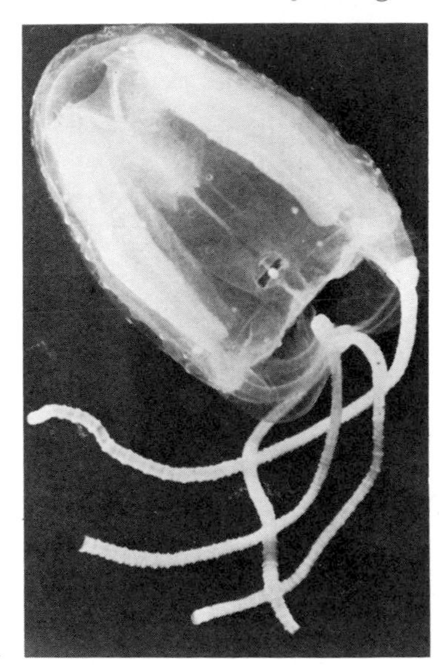

A

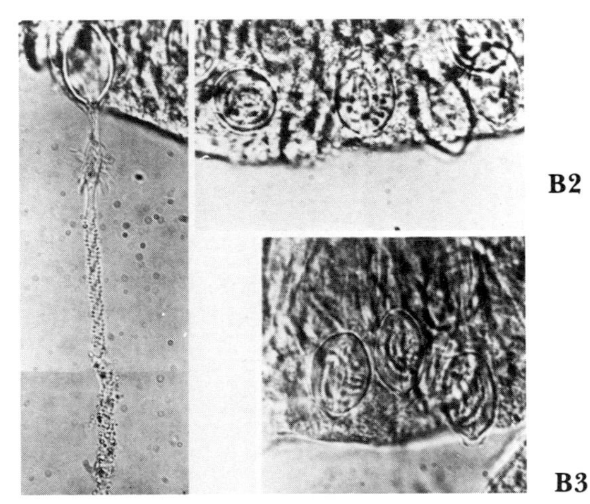

B1 B2 B3

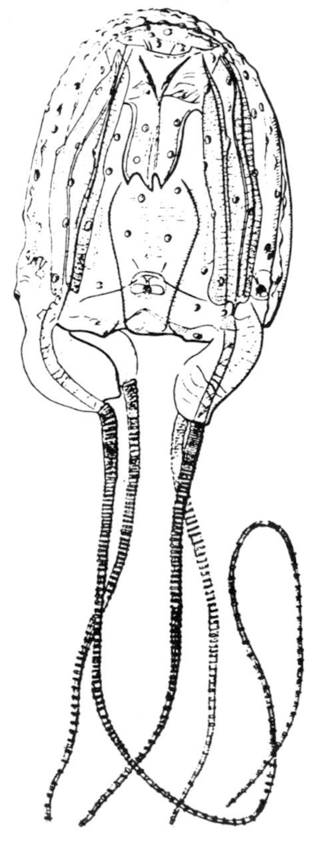

D

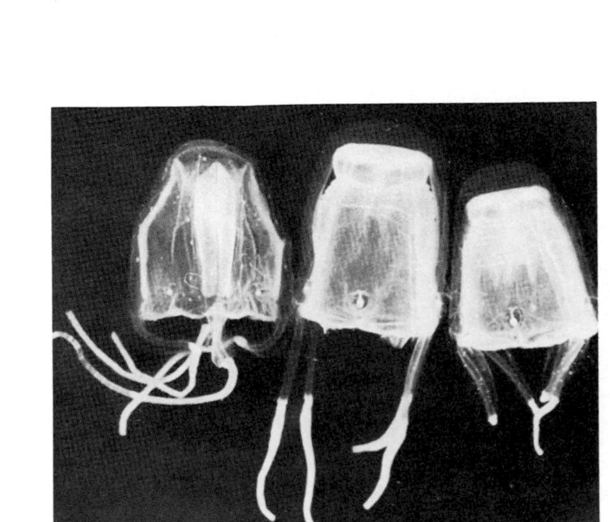

C

IV: PLATE 15

FIGURE a.—*Catostylus mosaicus* (Quoy and Gaimard). This is the commonest large jellyfish found along the eastern Australian coastline. It is largely an estuarine form. Sydney, New South Wales, Australia. (Courtesy of I. Bennett)

FIGURES b-c.—Close up of *Catostylus mosaicus*. Note the typical cross mark on the bell. The stinging ability of this jellyfish seems to vary with the season of the year. At times it is handled with impunity, whereas during its breeding season it may cause severe stings, intense pain, cyanosis, and difficulty in respiration. (Courtesy of E. Pope)

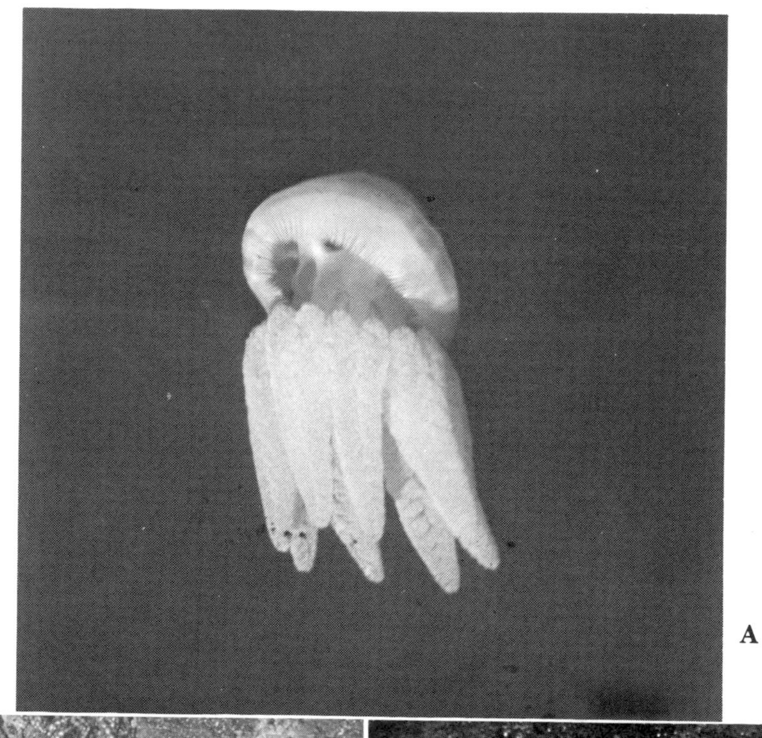

A

B C

IV: PLATE 16

Chironex fleckeri Southcott. This is probably the most dangerous venomous marine animal known. It may cause death within a period of seconds to several minutes. Stringent protective measures in the form of adequate diving clothing, neoprene-coated nylon, must be worn to avoid contact with the tentacles. All exposed portions of the body must be covered. The extreme rapidity of action of the poison frequently precludes opportunity for treatment of the victim. Living specimen taken near Cairns, Queensland, Australia. (Courtesy of K. Gillett)

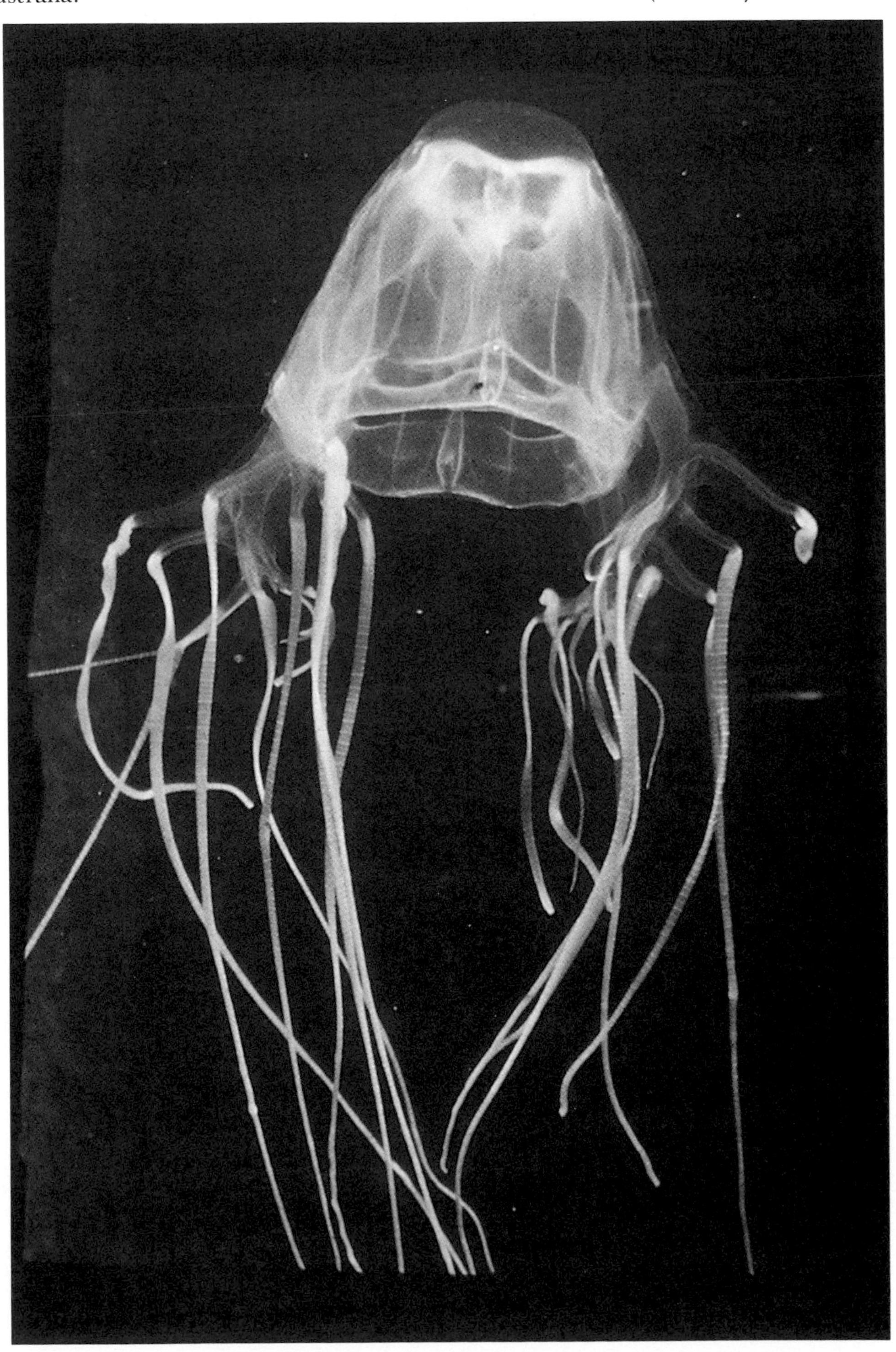

IV: PLATE 17

Morphology of the deadly sea wasp, *Chironex fleckeri* Southcott:

FIGURE a.—Side view of *Chironex*. × 0.5 (approx.).

FIGURE b.—View of *Chironex* from below. Tentacles have been removed. × 0.5 (approx.). (From R. V. Southcott)

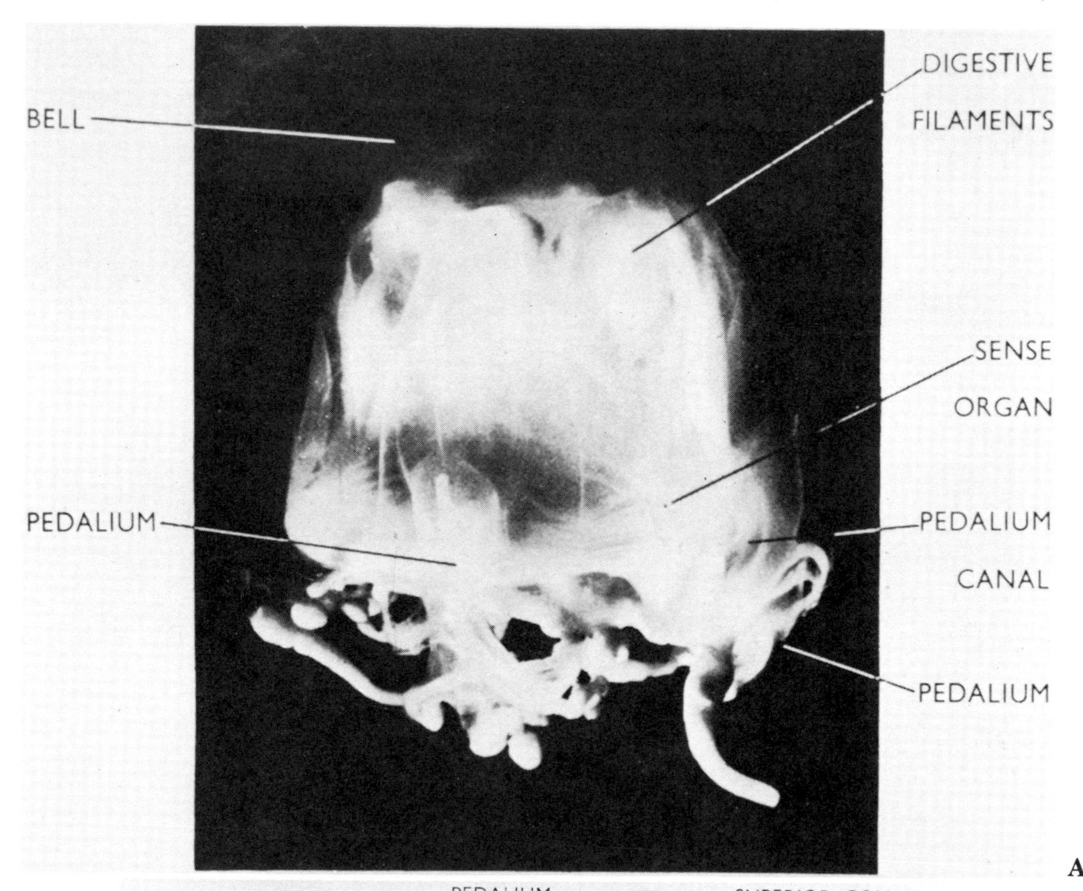

A

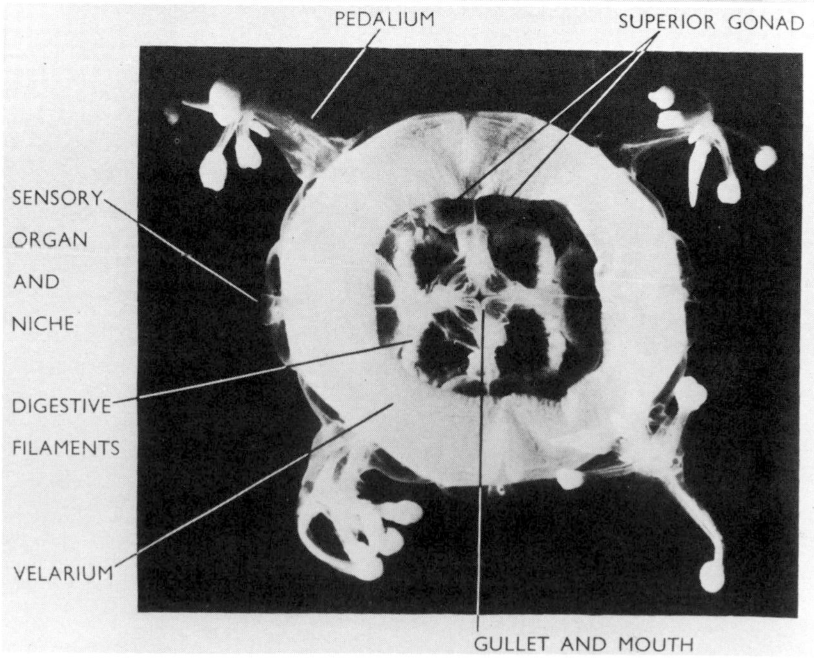

B

IV: PLATE 18

FIGURE a.—Photograph of arm of a 4-year-old male fatally stung by *Chironex fleckeri* at Halfmoon Bay, about 16 km north of Cairns, Queensland, Australia, on January 31, 1966. The child was being supported in approximately one meter of water by his father, who was teaching him how to swim. The child screamed that he had been stung and was immediately carried to the beach where he was vigorously rubbed with wet sand. When severe pain persisted, the child was taken by car to the Cairns Base Hospital where he collapsed while waiting in the casualty department. Mouth-to-mouth resuscitation and external cardiac massage were applied, followed by tracheal intubation, oxygen under intermittent positive pressure, intracardiac epinephrine, and cardiac massage through the open chest. The response was only transient. The time between the injury and collapse was about 35 minutes. Methyl alcohol was not applied until after arrival at the hospital. Nematocysts recovered from the skin lesions were identical with those of *C. fleckeri*. (Courtesy of J. H. Barnes)

FIGURE b.—Photograph of a 10-year-old boy fatally stung by *Chironex fleckeri* at Bucasia Beach, about 20 km north of Mackay, Queensland, Australia, on February 12, 1963. The boy collapsed on the beach immediately after being assisted from the water. Mouth-to-mouth resuscitation and external cardiac massage were promptly applied but without any apparent benefit. The victim was severely stung on the arms, legs, and trunk. The child died less than 10 minutes after being stung. Nematocysts recovered from the skin lesions were identical with those of *C. fleckeri*.
(Courtesy of J. H. Barnes)

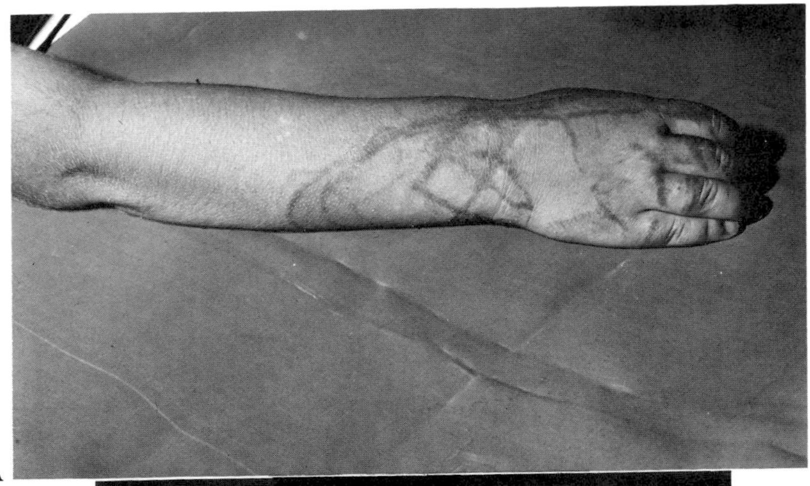

A

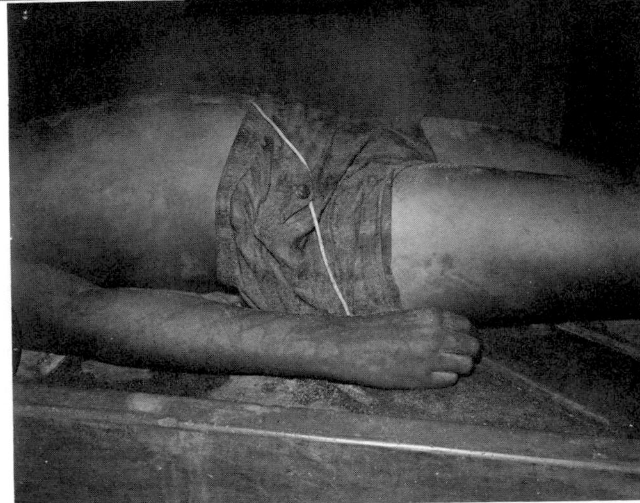

B

IV: PLATE 19

Chiropsalmus quadrigatus Haeckel. Living specimen swimming in aquarium at Bells Beach near Cairns, Queensland, Australia. Specimen collected by G. Rowell. The unpaired tentacle on each pedalium is usually tinged with lavender, and the first tentacles may be orange or yellow; the remainder are whitish. Under natural conditions, the body of the jellyfish is almost invisible under water. Height of body about 45 mm. There are five tentacles to each pedalium. This cubomedusae is sometimes confused with *Chironex fleckeri* which it closely resembles. Stings from this species may be very severe, but are generally less dangerous than *C. fleckeri*.

(Photo and legend courtesy of J. H. Barnes)

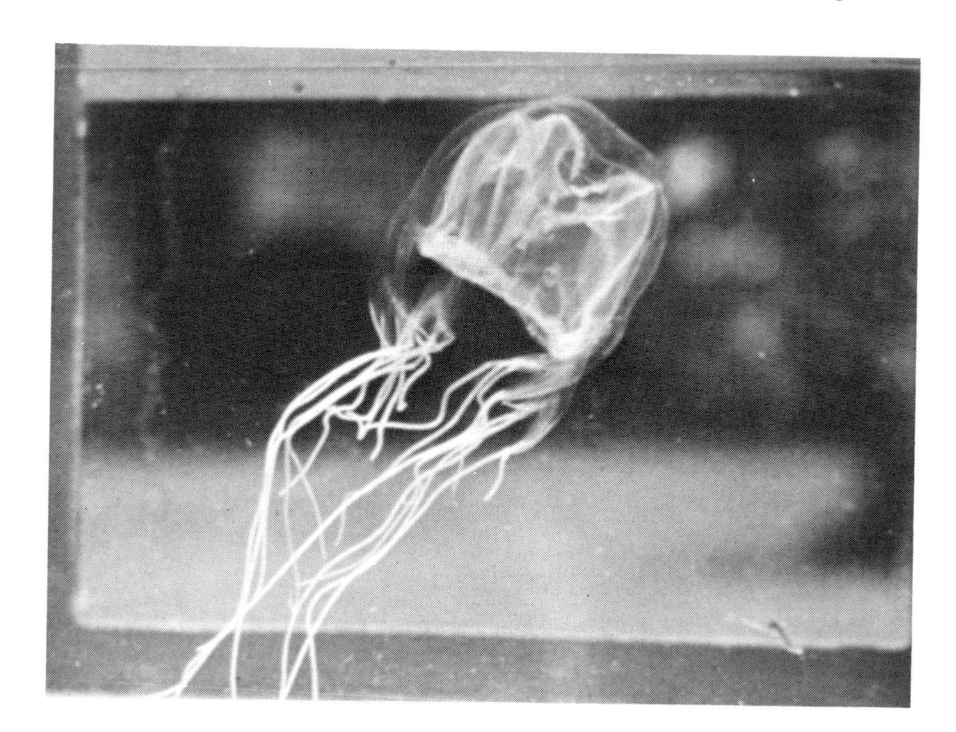

IV: PLATE 20

Chiropsalmus quadrumanus (Müller). × 0.3. Stings are believed to be similar to those produced by *C. quadrigatus*. (R. Kreuzinger, after Mayer)

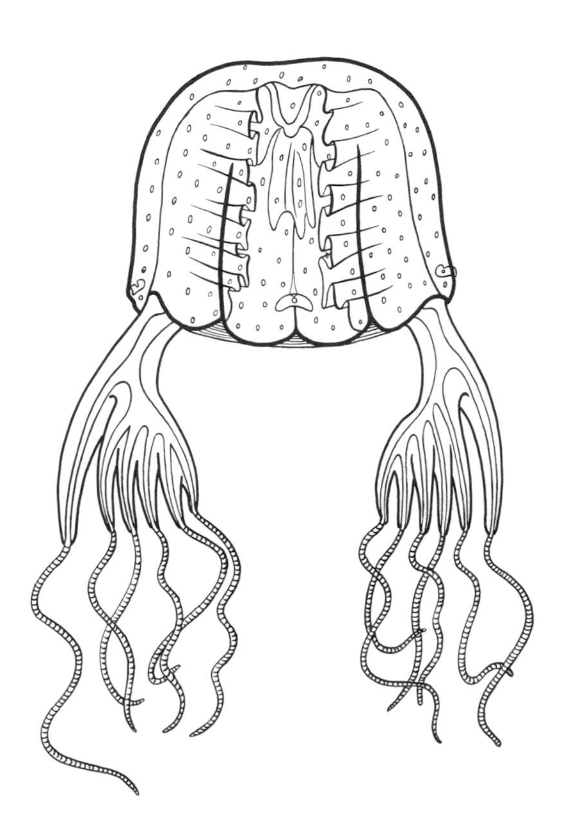

IV: PLATE 21

Cyanea capillata (Linnaeus) taken at the Woods Hole Marine Biological Laboratory, Massachusetts. Diameter of bell 22 cm. Photograph of living specimen. (G. Lower)

IV: PLATE 22

FIGURE a.—*Lychnorhiza lucerna* Haeckel. × 0.5. This jellyfish produces a urticarial lesion that is accompanied by an intense burning sensation, dizziness, vomiting, dyspnea, and agitation. (R. Kreuzinger, after Mayer)

FIGURE b.—*Nausithoë punctata* Kölliker, juvenile, or Scyphostoma larval form. × 0.6. The larval form which was formerly and erroneously described as a separate species, *Stephanoscyphus* and *S. corniformis*, has been termed "stinging alga" because of its algallike appearance during this stage of its growth when it is attached to the ocean floor. It can produce intense pain and itching. *(See also* Fig. c for adult form.)

(T. Komai)

FIGURE c.—*Nausithoë punctata* Kölliker. Diameter 15 mm. (R. Kreuzinger, after Mayer)

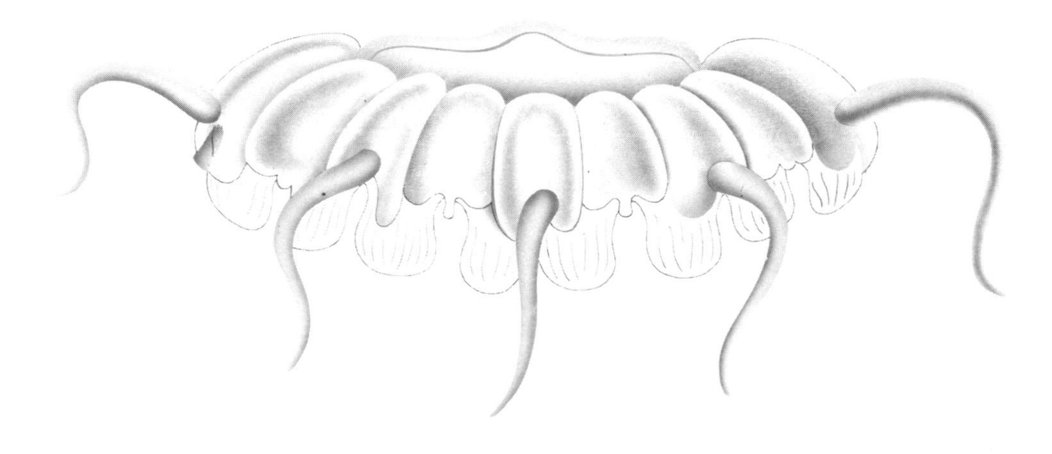

IV: PLATE 23

FIGURE a.—Sting inflicted by *"Stephanoscyphus racemosus,"* which is the larval stage of *Nausithoë punctata* Kölliker. The victim accidentally brushed his hand against the coelenterate and received an immediate and intense stinging sensation. Within a few minutes his hand became swollen and remained so for a period of several days. The skin in the area of the sting developed a maculopapular reaction and within a few days the skin became ulcerated. Several different lotions of an unknown variety were applied but the lesions appeared to be resistant to treatment. The swelling continued for more than 2 weeks, and the ulceration gradually healed after about 3 weeks. During most of this period, the victim complained of a painful pruritus. (D. Ludwig)

FIGURE b.—Cluster of *"Stephanoscyphus racemosus"* which stung the Palauan native in Fig. a. The cluster measured about 15 cm in diameter. (G. Mote)

FIGURE c.—Close-up of above. Taken at Koror, Palau, Western Caroline Islands. (G. Mote)

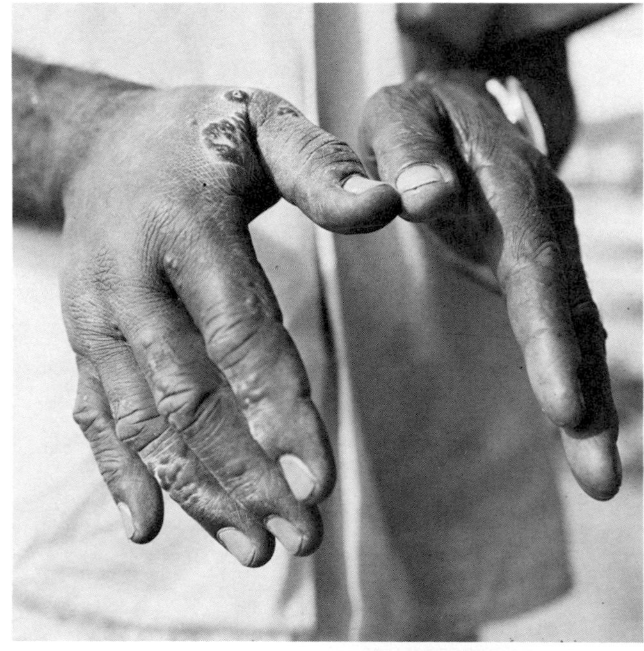

IV: PLATE 24

Pelagia noctiluca (Forskål). ×1.5. A wide range in color variation is quite typical of this species. (R. Kreuzinger, after Mayer)

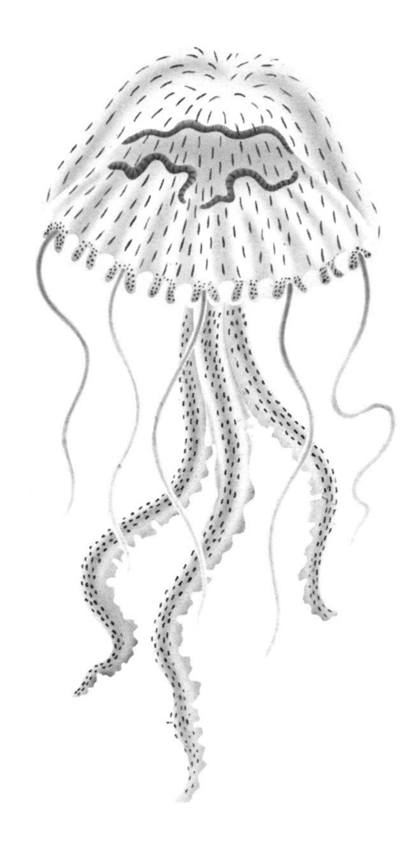

IV: PLATE 25

Rhizostoma pulmo (Macri), living specimen. Taken in an aquarium at the Zoological Station, Naples, Italy. Diameter of bell up to 600 mm. Stings are generally mild, consisting of local effects, urticaria, pain, and itching. Severe general symptoms have been reported on rare occasions. Excessive mucous production is common.
(Courtesy P. Giacomelli)

IV: PLATE 26

FIGURES A-C.—*Acropora palmata* (Lamarck). Large branch measuring about 45 cm in height, and close-up showing skeletal details. (Courtesy of Smithsonian Institution)

IV: PLATE 27

FIGURE a.—*Actinia equina* Linnaeus. Diameter up to 7 cm.

FIGURE b.—*Anemonia sulcata* (Pennant). Diameter up to 20 cm.

FIGURE c.—*Adamsia palliata* (Bohadsch). Diameter up to 4 cm.

FIGURE d.—*Sagartia elegans* (Dalyell). Diameter up to 8 cm.

(H. Baerg, after Stephenson)

IV: PLATE 28

FIGURE a.—*Actinia equina* Linnaeus. Photograph of living specimen taken in an aquarium at the Zoological Station, Naples, Italy. Diameter up to 7 cm. Stings may result in swelling, erythema, itching, sloughing of the tissue, and ulceration of the affected area. The severity of the sting appears to vary with the person.
(Courtesy of P. Giacomelli)

FIGURE b.—*Actinodendron plumosum* Haddon. Photograph of living specimen taken in Aquarium de Noumea, New Caledonia. Diameter up to 30 cm. Stings are similar to that produced by a nettle, but may be even more severe with the effects lasting more than a week. (Courtesy of J. Catalla)

FIGURE c.—*Anemonia sulcata* (Pennant), living specimen taken in an aquarium at the Zoological Station, Naples, Italy. Diameter up to 20 cm. Stings are similar to *A. equina*.
(Courtesy of P. Giacomelli)

FIGURE d.—*Anemonia sulcata*, taken in a tidal pool at low tide, at Roscoff, Britanny, France. (Courtesy of R. Buchsbaum)

A

B

C

D

IV: PLATE 29

FIGURES a.-b.—Photograph of a 14-year-old boy who was stung on the upper lip by *Anemonia sulcata.*

FIGURES c.-d.—Skin lesion within a few hours after envenomation. The lesion was reddened, painful, swollen, and pruritic.

FIGURES e.-f.—Skin lesion several days later. Most of the initial pain had disappeared, but the lesion was still hypersensitive and pruritic. Taken near Pula, Yugoslavia.

(Courtesy Z. Maretić)

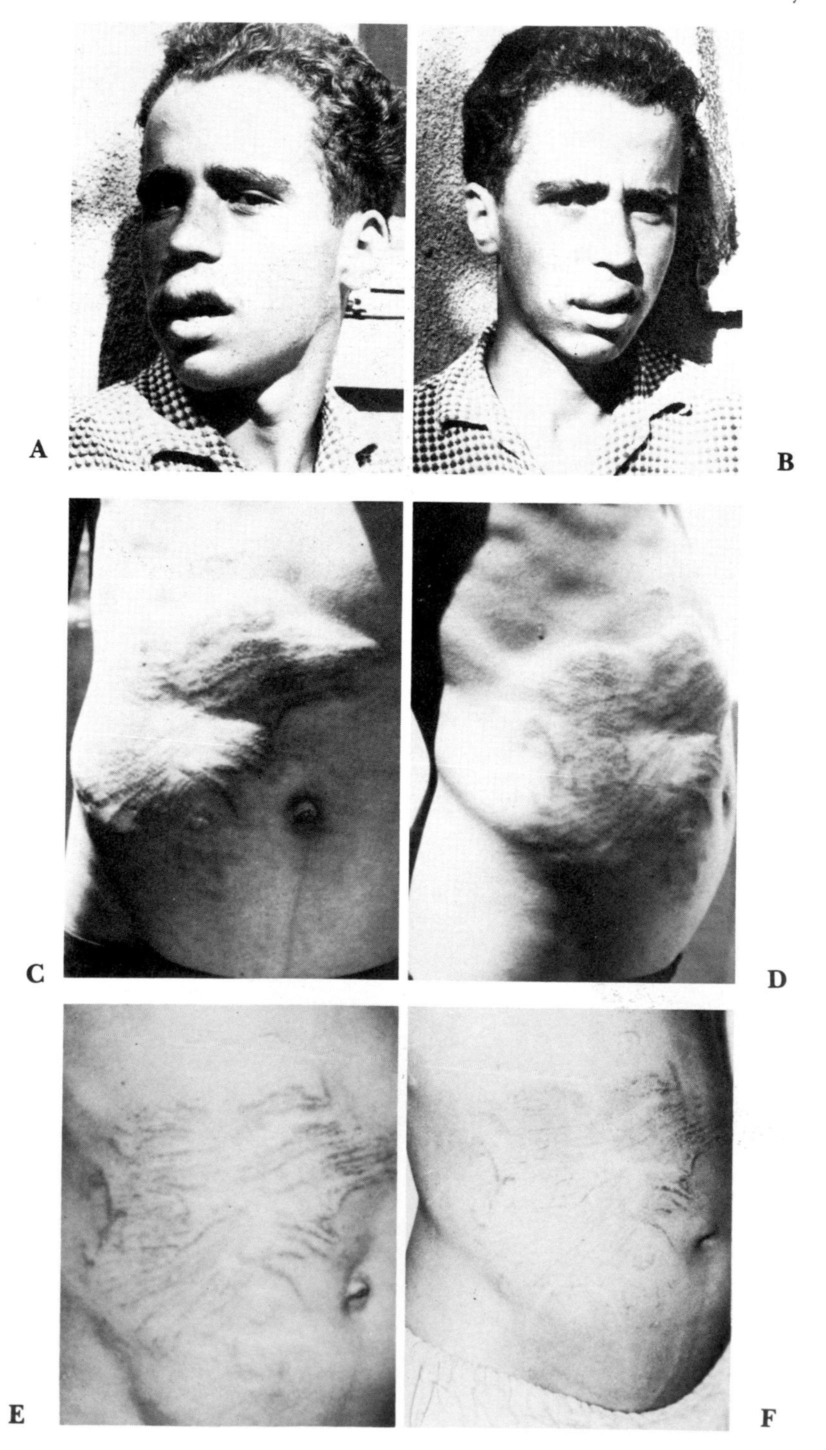

IV: PLATE 30

FIGURE a.—*Actinodendron plumosum* Haddon. × 0.5. (From Saville-Kent)

FIGURE b.—Sting produced by a "hell's fire sea anemone." The exact identification of this sea anemone is unknown, but it is believed to have been *Actinodendron plumosum* Haddon. Taken at Koror, Palau, Western Caroline Islands. (D. Ludwig)

FIGURE c.—*Renilla muelleri* Kölliker. Rachis about 5 cm in width. (R. Knabenbauer)

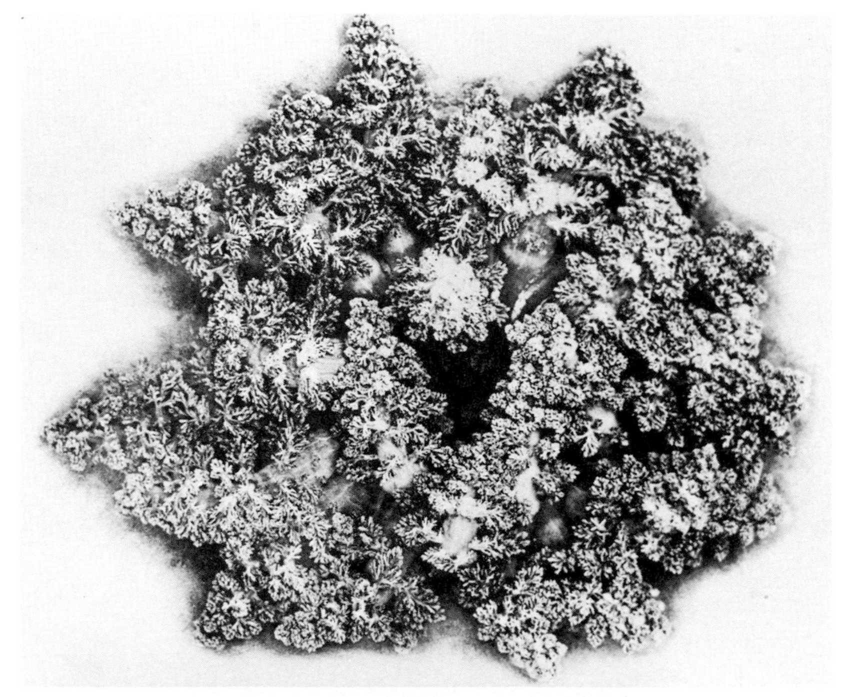

A

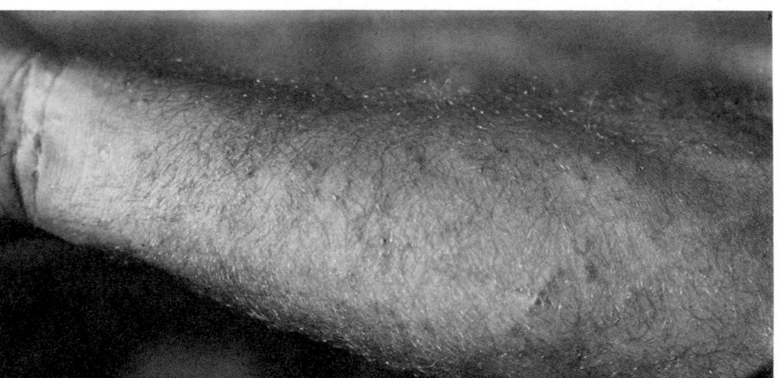

B

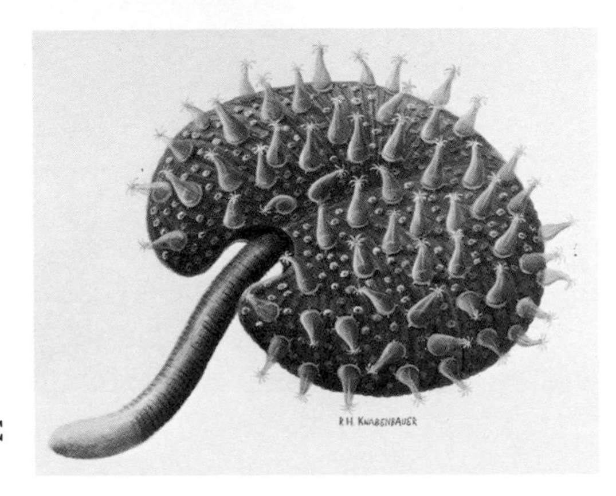

C

IV: PLATE 31

Closeup showing beds of *Rhodactis howesi* Kent. The size of the anemone can be determined by comparing it with the man's hand. (Courtesy of E. J. Martin)

A

B

IV: PLATE 32

FIGURE a.—*Alicia costae* (Panceri). Tentacles are extended. Living specimen taken at the Zoological Station, Naples, Italy. Diameter about 7 cm. (Courtesy of P. Giacomelli)

FIGURE b.—*Palythoa tuberculosa* Esper. ×0.5 approx. (S. A. Reed, courtesy of P. Scheuer)

IV: PLATE 33

FIGURE a.—*Lebrunia danae* (Duchassaing and Michelotti). May produce a mild urticarial rash. Diameter up to 5 cm. Taken near La Paguera, Puerto Rico.

(Courtesy of C. Cutress)

FIGURE b.—*Triactis producta* Klunzinger.

1. Taken in the vicinity of the Marine Biological Laboratory, Eilat, Israel. Daytime form. Note tentacles are contracted. Diameter of base about 4 cm.
2. Nighttime form. Note tentacles are in extended position.
3. Envenomation inflicted by *T. producta* involving volar surface of left wrist. The sting of *Triactis* produces an initial stinging or burning sensation, followed by redness, edema, and pain which increases in intensity. There is usually no ulceration.

(D. Masry)

A

B1

B2

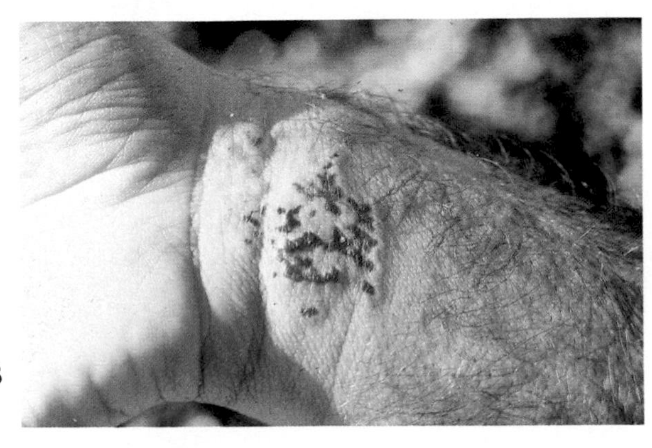

B3

IV: PLATE 34

FIGURE a.—*Adamsia palliata* (Bohadsch). Photograph of living specimen taken in an aquarium at the Zoological Station, Naples, Italy. Stings similar to those produced by *Actinia equina*.

FIGURE b.—*Calliactis parasitica* (Couch). Photograph of living specimen taken in an aquarium at the Zoological Station, Naples, Italy. Diameter up to 10 cm. Stings similar to those of *Actinia equina*. (Courtesy of P. Giacomelli)

FIGURE c.—*Corynactis australis* Haddon and Duerden. Photograph of living specimen taken in natural habitat near Sydney, New South Wales, Australia.

FIGURE d.—*Corynactis australis*, close-up of preceding specimen. Diameter up to 2 cm. Note the white-tipped tentacles which appear as swollen knobs. Stings from this anemone are painful and cause itching. (Courtesy of Justice F. Myers)

A

B

C

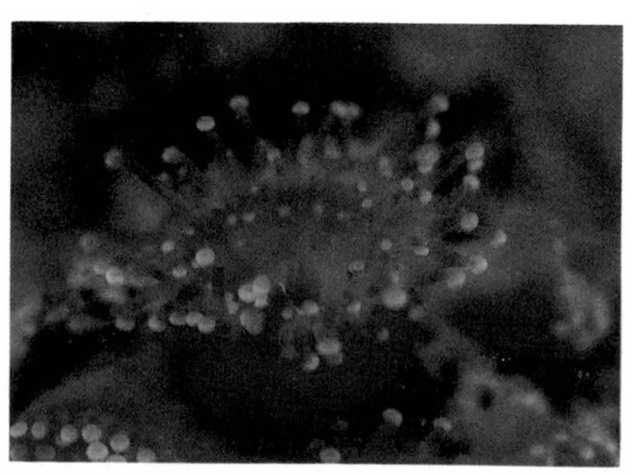

D

IV: PLATE 35

FIGURE a.—*Radianthus paumotensis* (Dana). Diameter up to 8 cm or more. Not dangerous to handle, but is reputed to be poisonous to eat when raw.

FIGURE b.—*Alicia costae* (Panceri). Diameter about 7 cm. May inflict a mild urticarial rash.

FIGURE c.—*Physobrachia douglasi* Kent. Diameter about 5 cm. May inflict a mild urticarial rash. (H. Baerg)

IV: PLATE 36

Palythoa australiensis Carlgren. Diameter of colony about 6 cm. This member of the Zoanthidae is a coral without a skeleton. This particular specimen was taken on the reef at Heron Island, Great Barrier Reef, Australia. (B. W. Halstead)

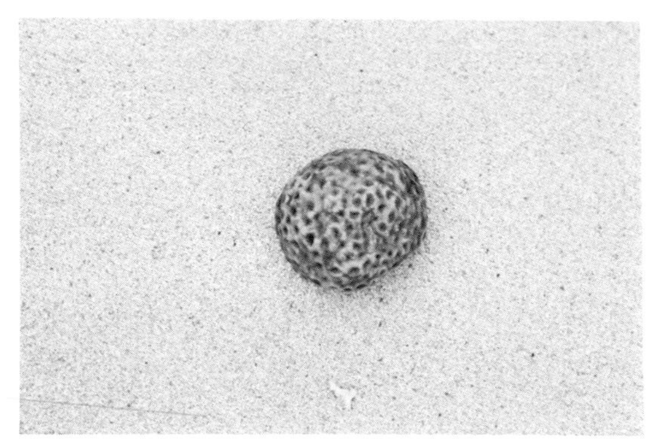

IV: PLATE 37 (opposite page)

Morphological types of nematocysts (according to Weill, 1934).

FIGURE a.—*Desmonemes*, tube threadlike, closed, forming a corkscrewlike coil. Also called *volvents*.

FIGURE b.—*Anacrophoric rhopalonemes*, tube club-shaped, without an apical appendix.

FIGURE c.—*Acrophoric rhopalonemes*, tube club-shaped, with an apical appendix.

FIGURE d.—*Holotrichous isorhizas*, tube spiny throughout.

FIGURE e.—*Atrichous isorhizas*, tube devoid of spines.

FIGURE f.—*Basitrichous isorhizas*, tube spiny at base only.

FIGURE g.—*Homotrichous anisorhizas*, spiny throughout, spines equal.

FIGURE h.—*Heterotrichous anisorhizas*, spiny throughout, spines larger at base.

FIGURE i.—*Microbasic mastigophores*, tube continued beyond the butt; butt not more than three times the capsule length. A stinging type of major medical importance to man. Found in *Chironex, Carybdea, Chiropsalmus, Chirodropus*, and others.

FIGURE j.—*Macrobasic mastigophores*, tube continued beyond the butt; butt four or more times the capsule length.

FIGURE k.—*Microbasic amastigophores*, no tube beyond the butt; butt not more than three times the capsule length.

FIGURE l.—*Macrobasic amastigophores*, no tube beyond the butt; butt four or more times the capsule length.

FIGURE m.—*Homotrichous microbasic euryteles*, butt short, spines on butt of equal size.

FIGURE n.—*Heterotrichous microbasic euryteles*, butt short, spines on butt of unequal size.

FIGURE o.—*Telotrichous macrobasic euryteles*, butt long, with distal spines only.

FIGURE p.—*Merotrichous macrobasic euryteles*, butt long, spines elsewhere than to the ends.

FIGURE q.—*Stenoteles*, or *penetrants*, undischarged and discharged. A stinging type of major medical importance to man. Found in *Physalia* and others.

(R. Kreuzinger, after Mayer, and Hyman)

FIGURE r.—Discharged microbasic mastigophores of *Carybdea rastoni*.

FIGURE s.—Discharged microbasic mastigophores of *Chiropsalmus quadrigatus*.

FIGURE t.—Discharged microbasic mastigophores of *Chironex fleckeri*.

FIGURE u.—Undischarged microbasic mastigophores of *Chirodropus sp.*

(H. Vester, after Southcott)

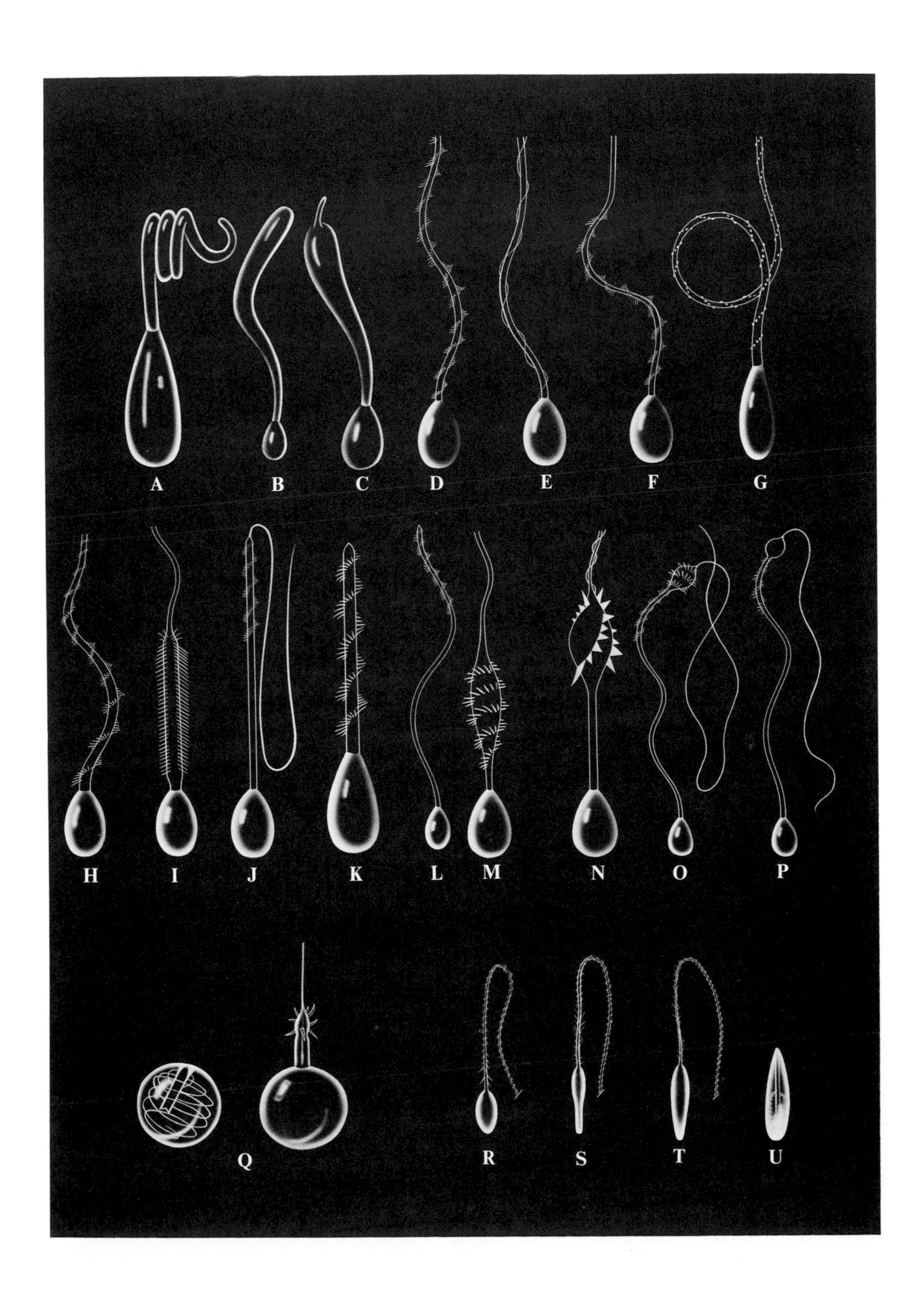
A
B
C
D
E
F
G
H
I
J
K
L
M
N
O
P
Q
R
S
T
U

IV: PLATE 38

Morphology of a typical coelenterate nematocyst.

FIGURE a.—Semidiagrammatic drawing of an undischarged coelenterate nematocyst, greatly enlarged. Note that in the undischarged nematocyst the coiled thread tube which conveys the venom is in the inverted position.

FIGURE b.—The same nematocyst after discharge. The operculum is open and the thread tube has undergone eversion, or an unfolding of itself. The venom, which originates in the capsule, is conveyed to the victim through the thread tube.

(R. Kreuzinger, from Halstead, courtesy of Cornell Maritime Press)

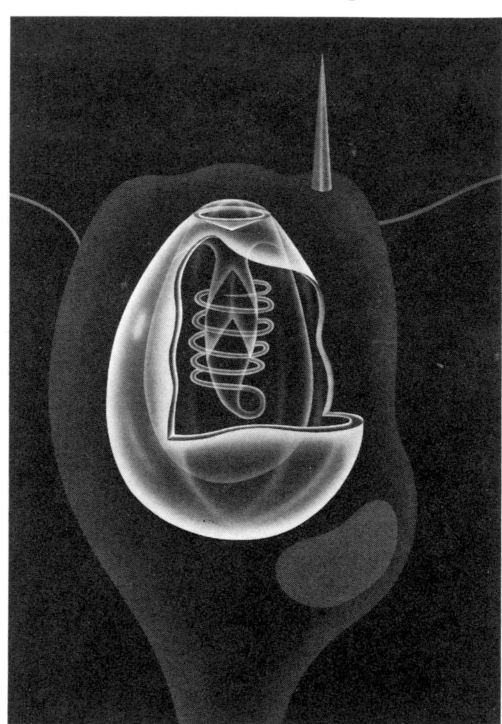

A

B

IV: PLATE 39

Photomicrograph of the nematocysts of *Physalia physalis*.

FIGURE a.—Isolated nematocysts of *Physalia*. Although separated from the tentacles of *Physalia*, they are still reactive and capable of producing a painful sting. Note how the inner thread tube is tightly coiled inside the capsule. When the nematocyst reacts normally, the thread tube is extruded with sufficient force to penetrate a surgical glove. *Physalia* has two general sizes of nematocysts; the smaller is about 9μ in diameter, and the larger is about 27μ in diameter. Both sizes of nematocysts are seen in this photomicrograph. Shadow of micrometer appears on the right side. × 640.

(Courtesy of C. E. Lane)

FIGURE b.—Discharged nematocysts of *Physalia*. Note discharged thread tubes extending from the nematocyst capsules. × 320. (Courtesy of J. C. Fardon)

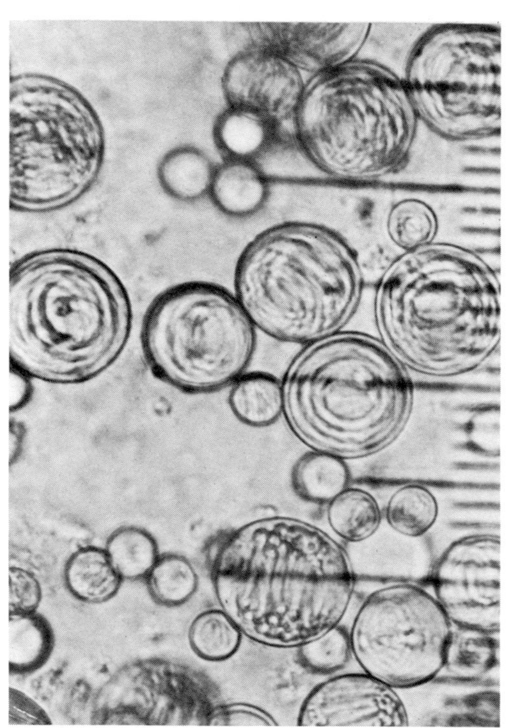

A

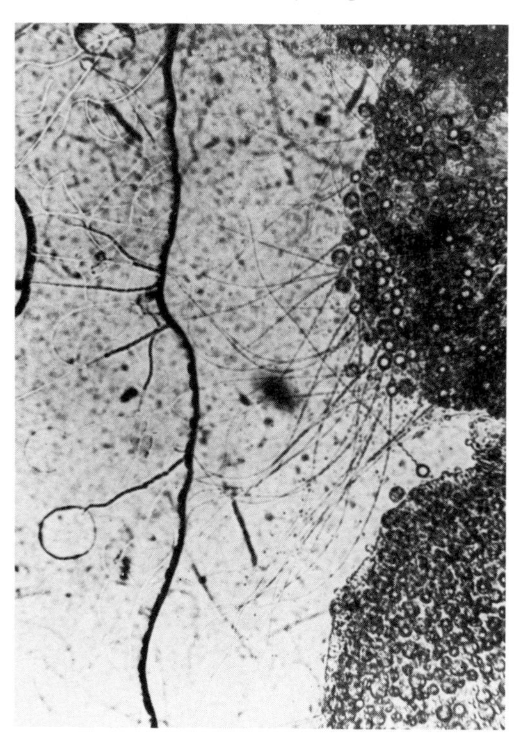

B

IV: PLATE 40

Photomicrographs of the nematocysts of *Physalia physalis*.

FIGURE a.—Phase-contrast photomicrograph of discharging nematocysts. Note opercular area of large nematocyst about to discharge (lower left). × 800.

FIGURE b.—Later stage of discharging nematocyst. Note the thread tube length is already many times the diameter of the parent capsule, but is far from completely extended. Again note the two size groups of nematocysts. × 500.

FIGURE c.—Photomicrograph showing portions of two intertwined thread tubes; one from a large capsule and the more tortuous from a small capsule. × 1,280.

(Courtesy of C. E. Lane)

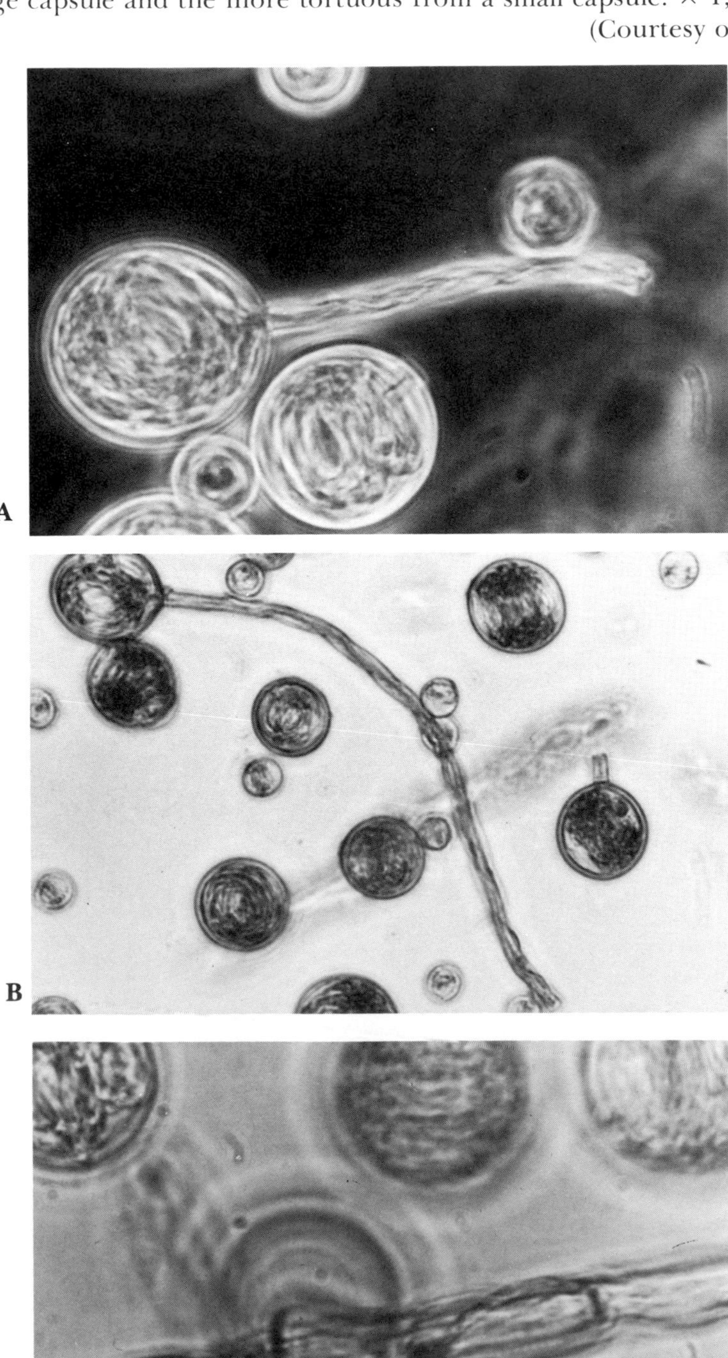

IV: PLATE 41

FIGURE a.—Photomicrograph of nematocysts of *Chironex fleckeri*, unfixed and unstained, undischarged, from living tentacle. The elongate nematocyst is a microbasic mastigophoric type, size 70 to 85μ in length. This is the principal stinging type and can be readily recovered from the skin of victims. The large oval capsule is less plentiful and functions mainly as a glutinant. The small circular capsules are believed to be immature forms.

FIGURE b.—Photomicrograph of a large microbasic mastigophore of *Chironex fleckeri*. Note the heavily barbed thread tube which is longitudinally arranged in three loops. The "corkscrew" near the larger pole is the proximal portion of the butt of the thread tube. The two small nematocysts in this picture have discharged their thread tubes which usually form a twisted skein within the capsule.

FIGURE c.—Discharged microbasic mastigophores of *Chironex fleckeri*. Note the long fine spines on the butt, and the three spirals of short stout barbs on the shaft of the thread tube.

FIGURE d.—A remarkable photomicrograph of microbasic mastigophores of *Chironex fleckeri* taken immediately after discharge. Note the droplets of venom deposited near the openings of the thread tubes of two of the nematocysts, and the apparent increase in volume of venom over that of the nematocyst capsule.

(Photomicrographs and legend courtesy of J. H. Barnes)

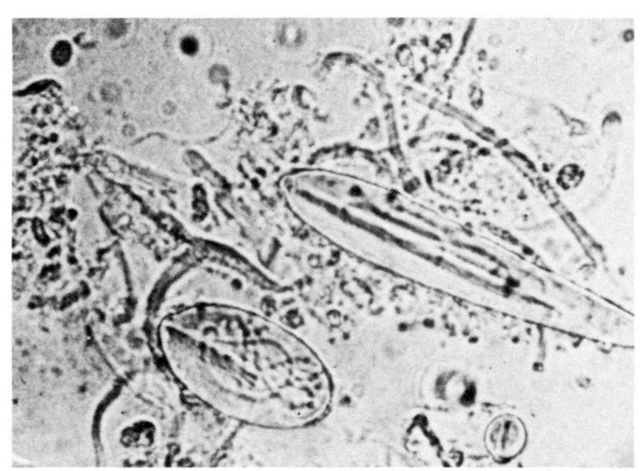
A

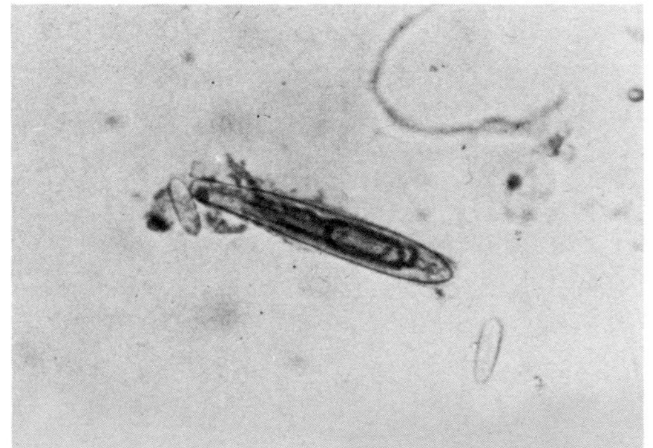
B

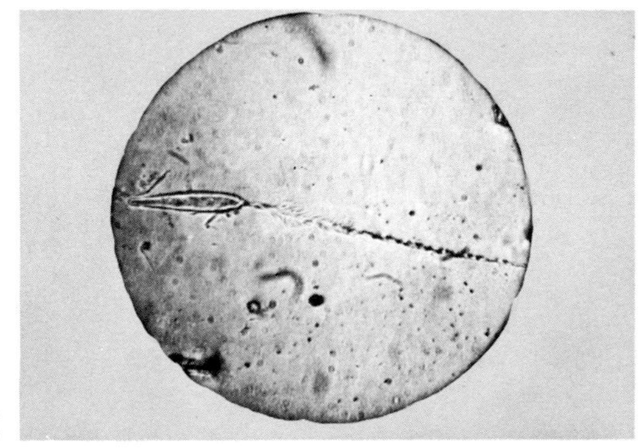
C

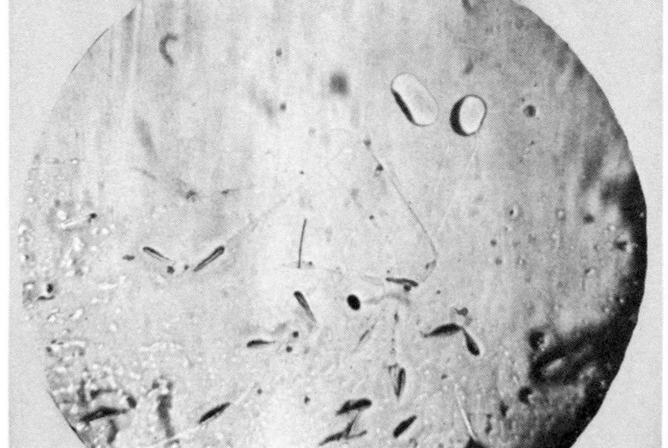
D

IV: PLATE 42

Photomicrograph of nematocysts, unfixed and unstained, taken from a tentacle of a living specimen of *Chironex fleckeri*. The large, elongate, undischarged microbasic mastigophores are particularly evident in this view and are believed to be the ones most dangerous to man. Size 70 to 85μ in length. The ovoid nematocysts are believed to serve as glutinants. (K. Gillett)

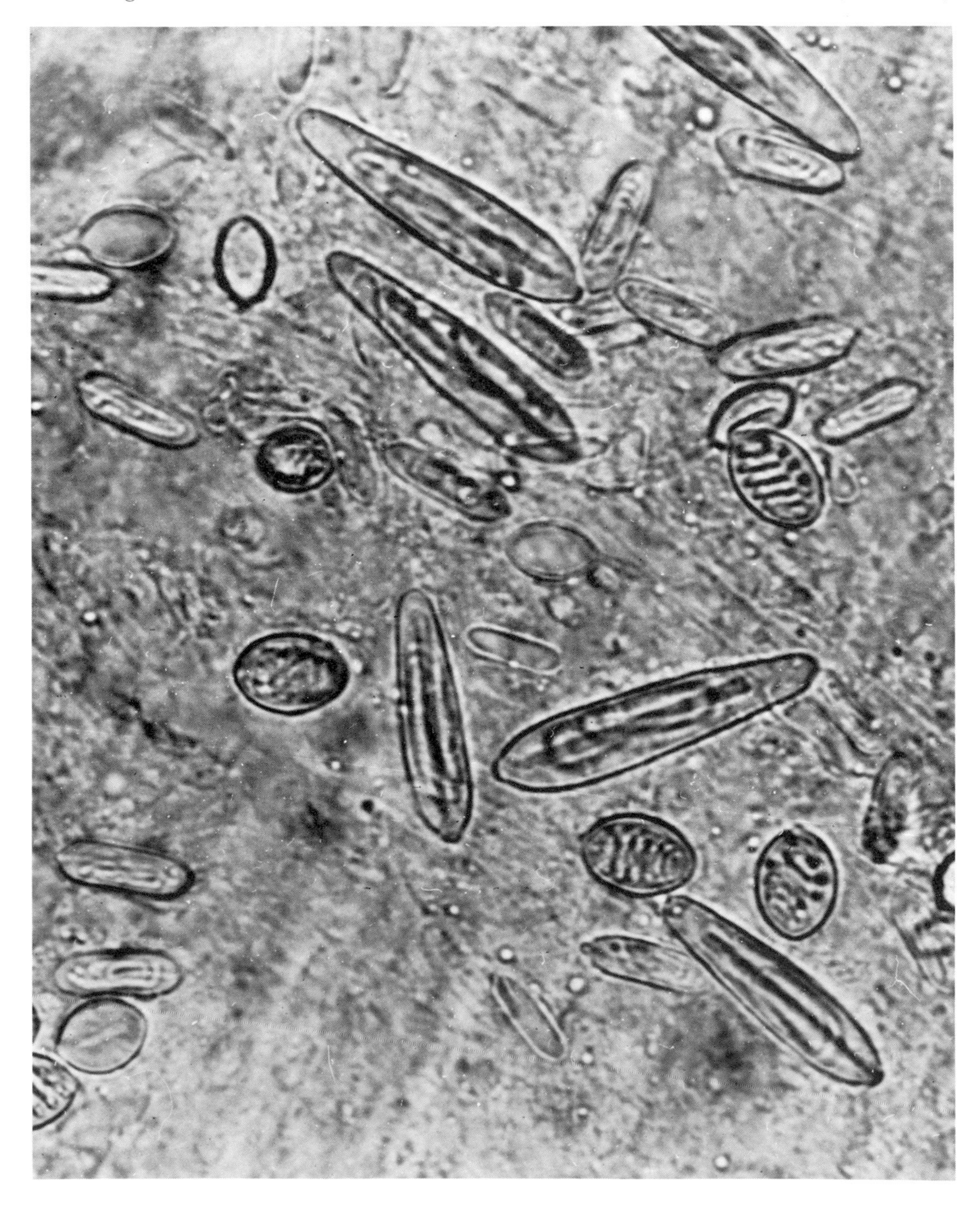

IV: PLATE 43

FIGURE a.—Stings produced by the hydroid *"Halecium beani."* Described as "erythematopapular lesions with faint morbilliform eruption." (From De Oreo)

FIGURE b.—Sting produced by a small specimen of the jellyfish *Pelagia noctiluca*. (Photo by W. Stephenson, courtesy of D. Squires)

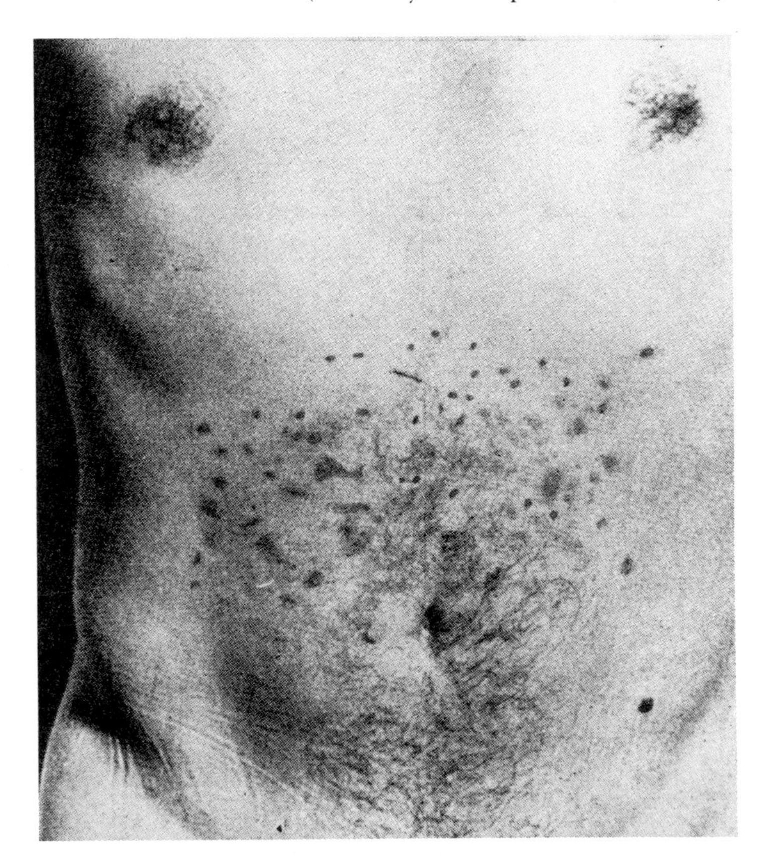

A

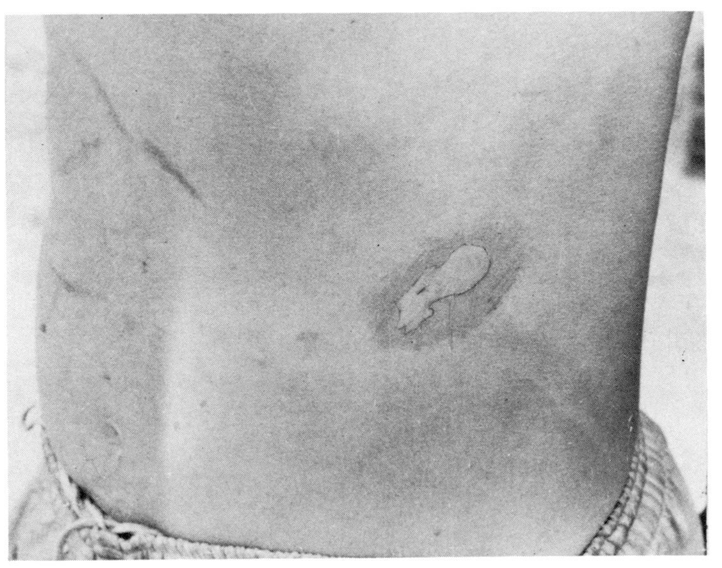

B

IV: PLATE 44

FIGURE a.—Skin lesions produced by the Portuguese man-o-war *Physalia physalis*. The victim, Robert P. L. Straughan, a professional diver and underwater photographer, collided with a large specimen of *Physalia* while coming up from a dive off the coast of southern Florida. The tentacles of the jellyfish made full contact with Straughan's back. The almost instantaneous pain was described as intense beyond description. The envenomation caused extreme difficulty in respiration, rupturing of blood vessels on his back, trismus, and a more or less general muscular paralysis to the extent that he almost drowned. The intense pain continued for about 5 hours, and then gradually subsided. The scars lasted for about a year, and his back continued to be sensitive to sunburn for an even longer period of time. (Courtesy of R. Straughan)

FIGURE b.—Sting caused by *Physalia utriculus*. The lesions produced by the Pacific form of *Physalia*, as seen in this photograph, differ markedly from that generally caused by the Atlantic species *P. physalis*, as seen in Figure a. *P. utriculus* generally has a single tentacle which makes a linear lesion. The "baring effect," or interrupted markings of the linear lesion, are due to the contractions of the tentacle. The lateral extensions of the lesions as seen in the upper part are believed to be due to rubbing or scratching by the victim. The Pacific species of *Physalia* may occur with two or three long tentacles which would increase the number of linear lesions accordingly, but in general the markings as seen in this photograph are diagnostic for *P. utriculus*. Stingings occurred near Cairns, Australia. (Courtesy of J. H. Barnes and D. J. Lee)

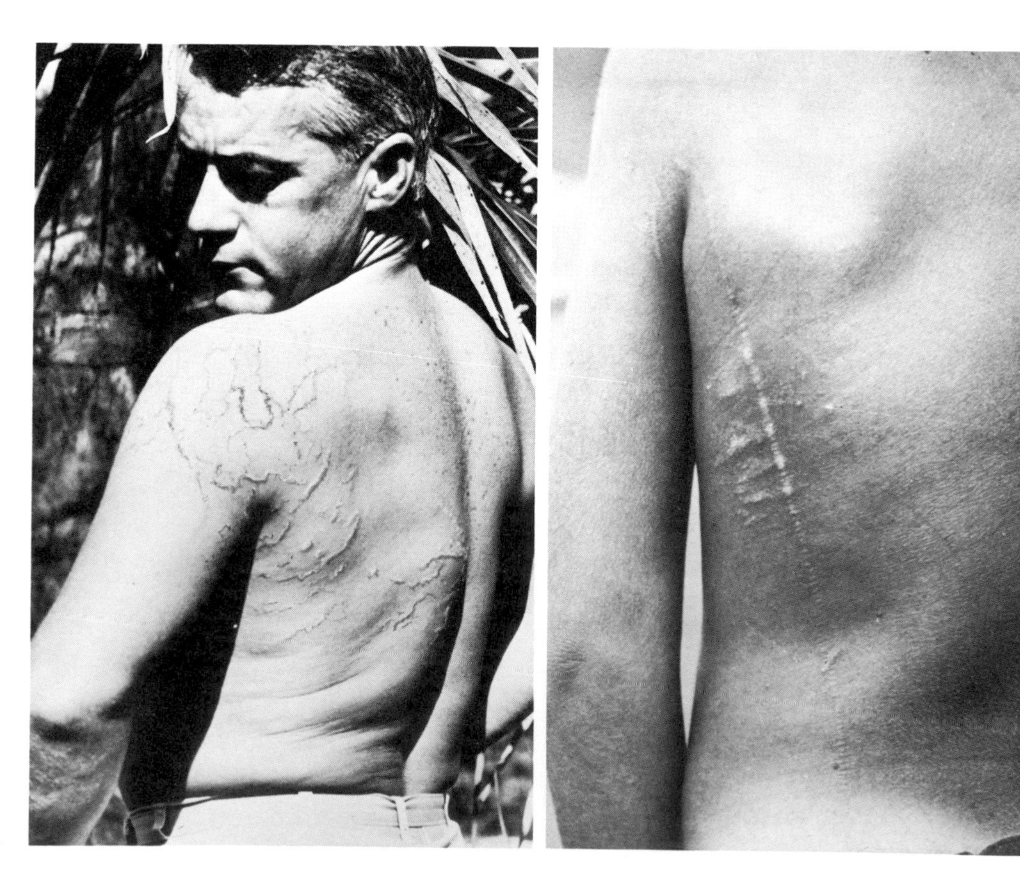

A B

IV: PLATE 45

FIGURE a.—Sting inflicted by *Millepora dichotoma*. Taken at the Marine Biological Station, University of Cairo, at Al Ghardaqa, Red Sea. Lesion consists of a mildly painful reddened urticarialike rash which gradually subsided after about 6 hours. (D. Ollis)

FIGURE b.—Experimental sting produced by *Carybdea rastoni* on forearm of human volunteer. The central dull wheal is 24 hours old. It is in three parts corresponding to the contacting tentacles of the cubomedusa. The middle of these three dull wheals is from two tentacles. On either side of the old wheal are fresh wheals 13 minutes old. The one on the left is from the combined effects of four tentacles. The long wheal to the immediate right of the dull wheal is from the effect of two tentacles, and the interrupted wheal on the extreme right is from a single tentacle. (Copyright R. V. Southcott)

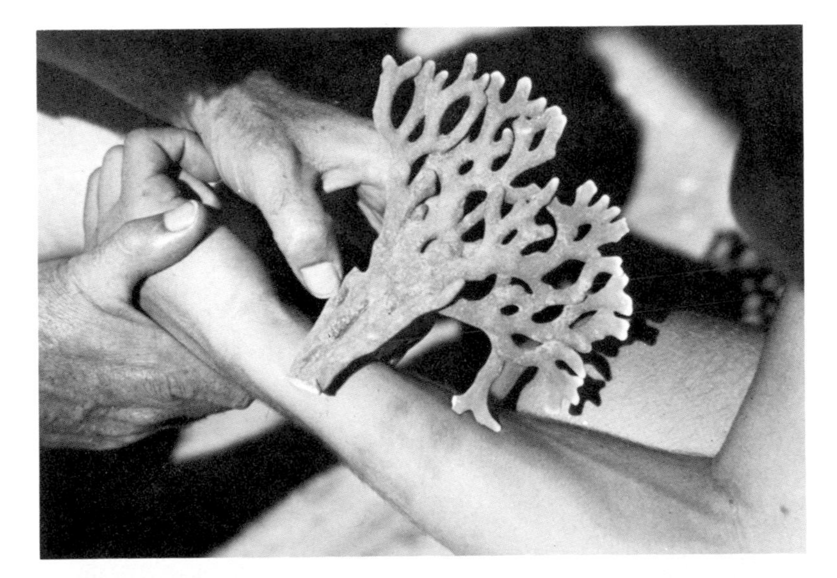

A

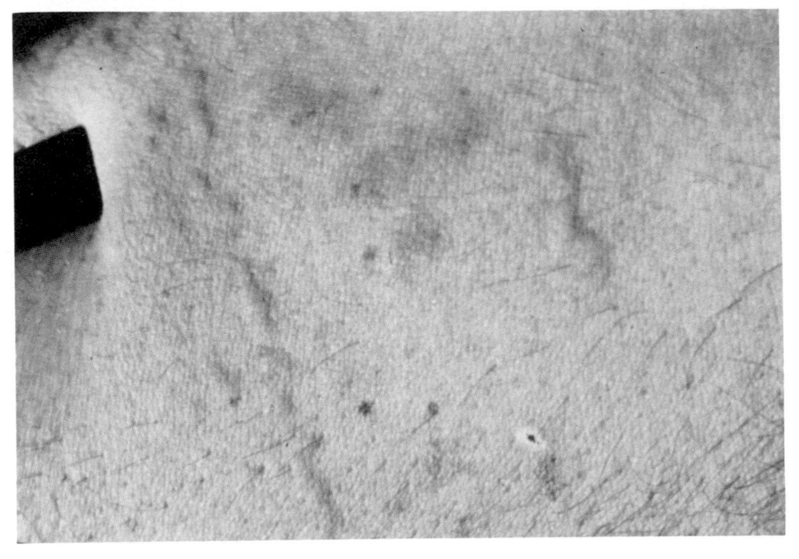

B

IV: PLATE 46

Sting inflicted by *Cyanea capillata* on small boy. Photographed 1 hour after stinging. The sharp angulations and undulations characteristic of *Cyanea* stings are well demonstrated in this case. This bizarre effect is caused by the filamentous tentacles which tend to attach initially at widely separated points. Subsequent movement of victim and water determine the final pattern of attachment. Nematocysts on *Cyanea* tentacles are grouped in minute clumps closely set in a regular mosaic; thus stings appear fairly uniform in intensity. A zig-zag pattern of tiny papules may be visible within a few minutes after stinging, but are soon lost in the erythematous flare that follows. Pain is comparatively mild, and often described as burning rather than stinging. Symptoms usually subside within 20 minutes. The persistence of vivid reddish streaks for several days after symptoms have disappeared is of diagnostic significance. Transient vesication and mild pigmentation have been observed on tender skins. Permanent scarring is generally minimal or absent unless secondary infections have been caused by scratching or inadequate therapy. The clinical diagnosis was supported by examination of the nematocysts from the tentacles. Cairns, Queensland, Australia.

(Photo and legend courtesy of J. H. Barnes)

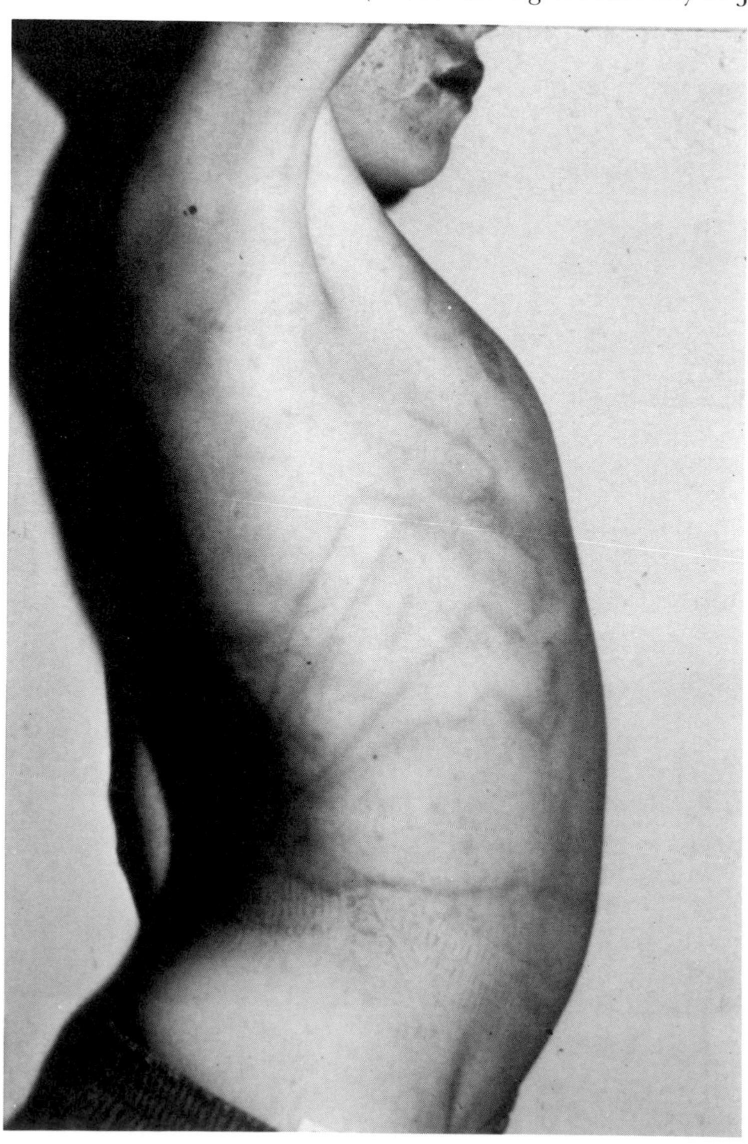

IV: PLATE 47

Near-fatal sting inflicted by *Chironex fleckeri* on 13-year-old white male. Despite early removal of tentacles, the skin injury was severe and painful, necessitating repeated injections of pethidine during the first 12 hours. The large indurated wheals, evident soon after stinging, had subsided when this picture was taken about 12 hours later, and the areas of tentacle contact appear as seared lines with an erythematous halo. Vesication soon commenced on the left hip and forearm. A few minutes after having been stung, the victim showed evidence of circulatory failure with pronounced cyanosis of the skin and lips. He felt faint and dizzy, but noticed no difficulty in breathing. All the sting areas were flooded with methyl alcohol (rubbing alcohol), whereupon the tentacles "shrivelled and fell away like magic." The boy's condition then improved rapidly, with pain the only remaining symptom. The survival of this boy was attributed to the immediate application of methyl alcohol. The medusa was seen but not captured. However, the description given was that of a large specimen of *Chironex fleckeri*, and this was further supported by an examination of tentacle fragments which were consistent with such an identification. Cairns, Queensland, Australia

(Photo and legend courtesy of J. H. Barnes)

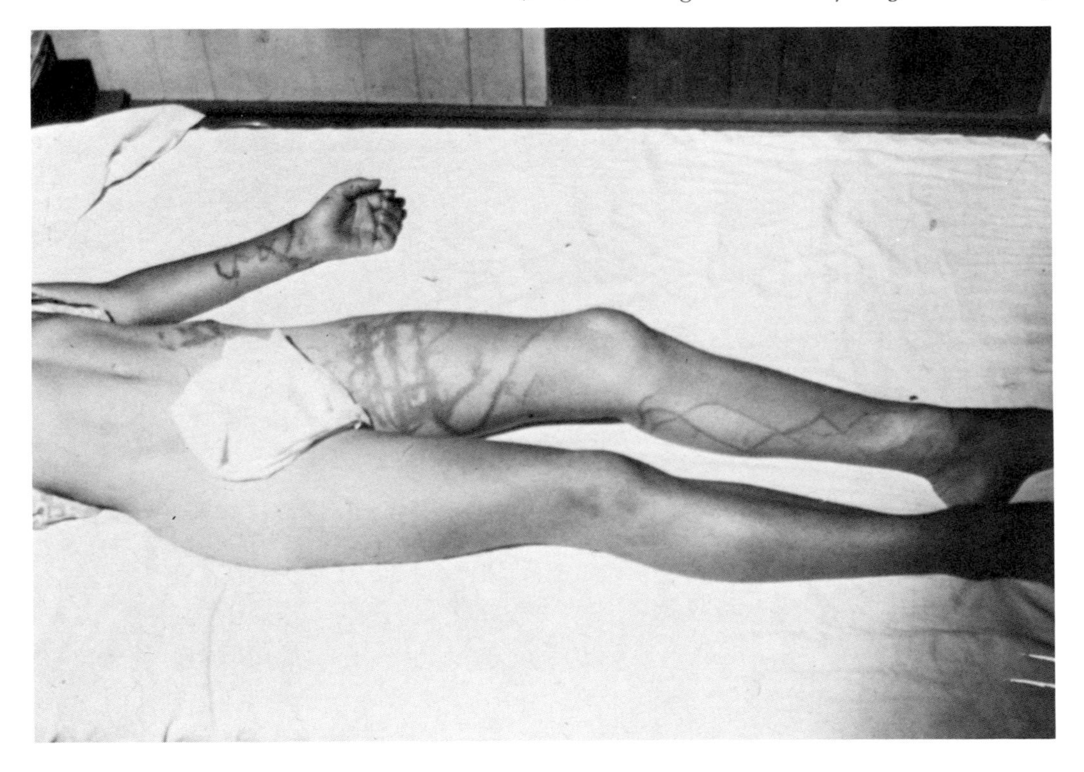

IV: PLATE 48

Sting believed inflicted by *Chironex fleckeri* on a 24-year-old Chinese male. Photographed 17 days after contact. Victim was standing in clear water when he felt a pain like "hot needles" on his chest and both arms, near water level. He looked around, but was unable to locate the medusa. On reaching the beach about 2 minutes later, he rubbed all painful areas with dry sand, and removed a length of tentacle from his left foot—this being the only tentacle still adhering. Discomfort reached a maximum soon after stinging, easing 25 minutes later without further treatment. Wheals reached .6 cm in height on the left arm, and a long vesicle developed at the site of the ankle injury. No systematic effects were observed. Aside from itching for the first few days, recovery was uneventful.

Measurements along skin markings indicate that about 5 meters of tentacles contacted the victim, but with the exception of small areas on the arms and left ankle such contact was transient and sustained contact did not occur. The diffuse markings on the chest and abdomen resulted from tentacles rolling across the skin, distributing a low concentration of nematocysts over a wide area. Tentacles attached themselves near the left elbow and wrist, but were removed before exerting their full effects. A characteristic "ladder pattern" is visible on healed lesions in these areas. Cairns, Queensland, Australia. (Photo and legend courtesy of J. H. Barnes)

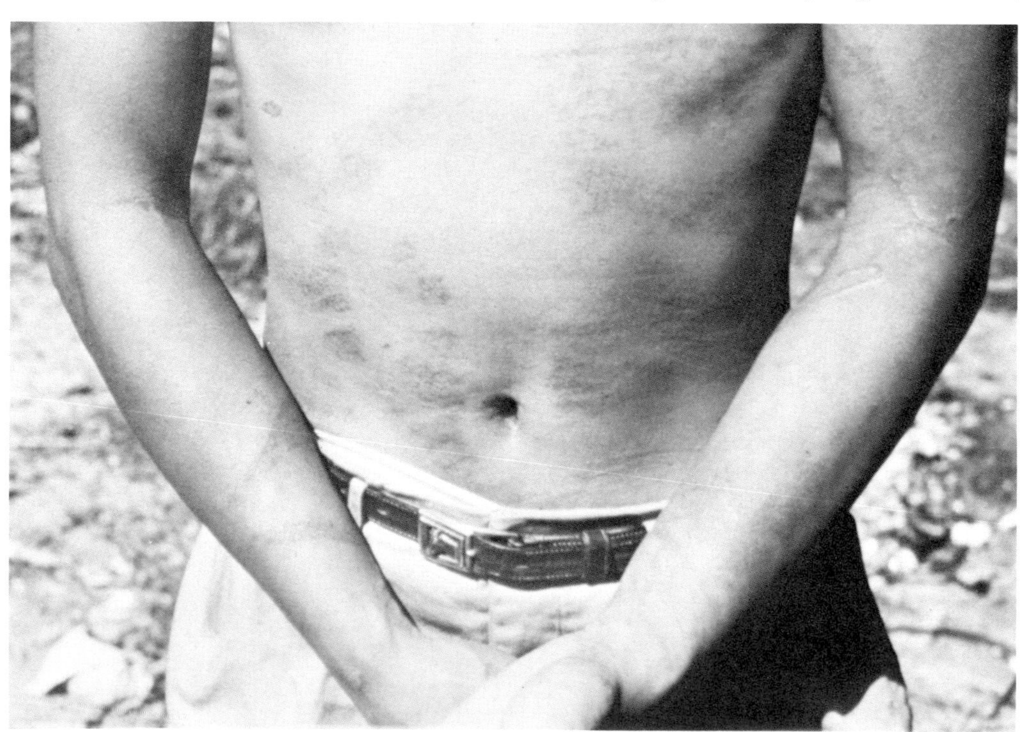

IV: PLATE 49

FIGURE a.—Sting inflicted by *Chiropsalmus quadrigatus* on leg of 8-year-old boy while swimming at Machants Beach, Cairns, Queensland, Australia. Initial pain was severe, but gradually subsided after about 40 minutes. Two hours later the stings were visible as slightly raised lines of erythema which developed a few small watery vesicles during the next 3 days. The photograph was taken 11 days after the stinging at which time healing was almost complete under the dry scabs, and the only complaint was periodic itching. The injury is typical of stings produced by medium-sized *C. quadrigatus*. Examination of tentacle fragments recovered from the skin confirmed the clinical diagnosis.

FIGURE b.—Sting on lower leg and ankle probably caused by *Chiropsalmus quadrigatus*. Photograph taken 7 days after injury shows the irregular pattern of multiple tentacles, necrosis, and desiccation of superficial skin marking the lines of contact. Each line is made up of a regular series of transverse bars, corresponding to the raised rings of nematocyst tentacles of *Chironex fleckeri* and *Chiropsalmus quadrigatus*. Since both species were present on the day of this stinging, either might have been responsible for the injury. Nematocyst studies were inconclusive. Clinically the effects were typical of *C. quadrigatus*, and lacked the intensity of *Chironex* stings of similar magnitude. The victim experienced immediate and severe pain, which moderated after the application of dilute formalin solution. Slight pain persisted for 24 hours, being replaced by an itch which was still intermittently troublesome 8 days later. Whealing, erythema, edema, and minimal vesication were evident for 2 days. After desquamation of superficial epithelium some brownish pigmentation remained, gradually disappearing over several months. No permanent scars remained. Cairns, Queensland, Australia.

(Photos and legends courtesy of J. H. Barnes)

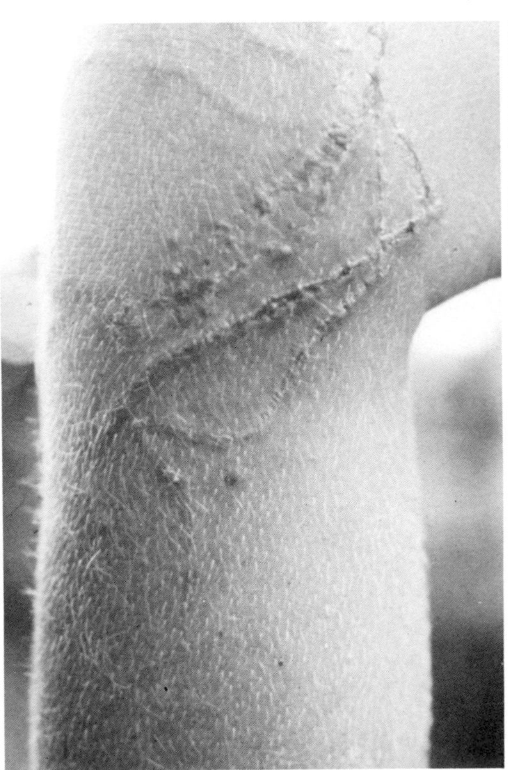

A

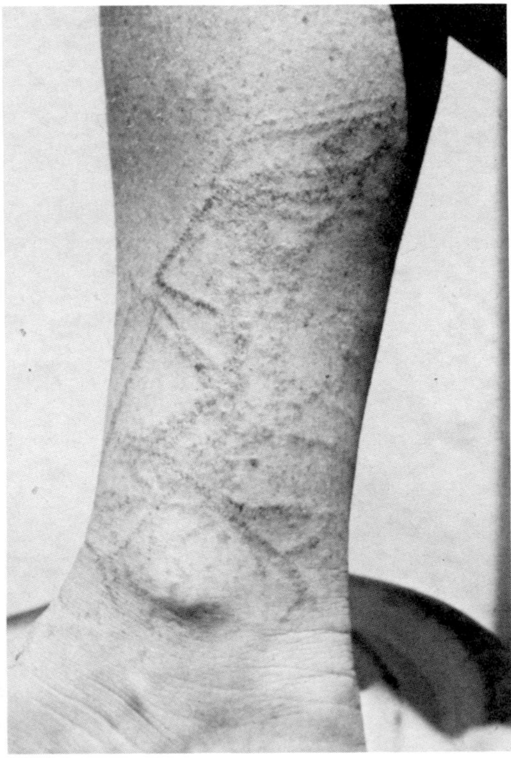

B

IV: PLATE 50

Human skin section from a fatal case of stinging by *Chironex fleckeri*. On the surface of the human skin (upper left center) can be seen the collapsed capsules of the nematocyst. Immediately beneath the capsules can be seen the trajectory where the nematocyst threads have penetrated through the horny layer of the skin down to the malpighian layer. Arrows point to paths of penetration by nematocysts. Masson's Stain. Paraffin. × 1,600. (Courtesy of D. J. Lee)

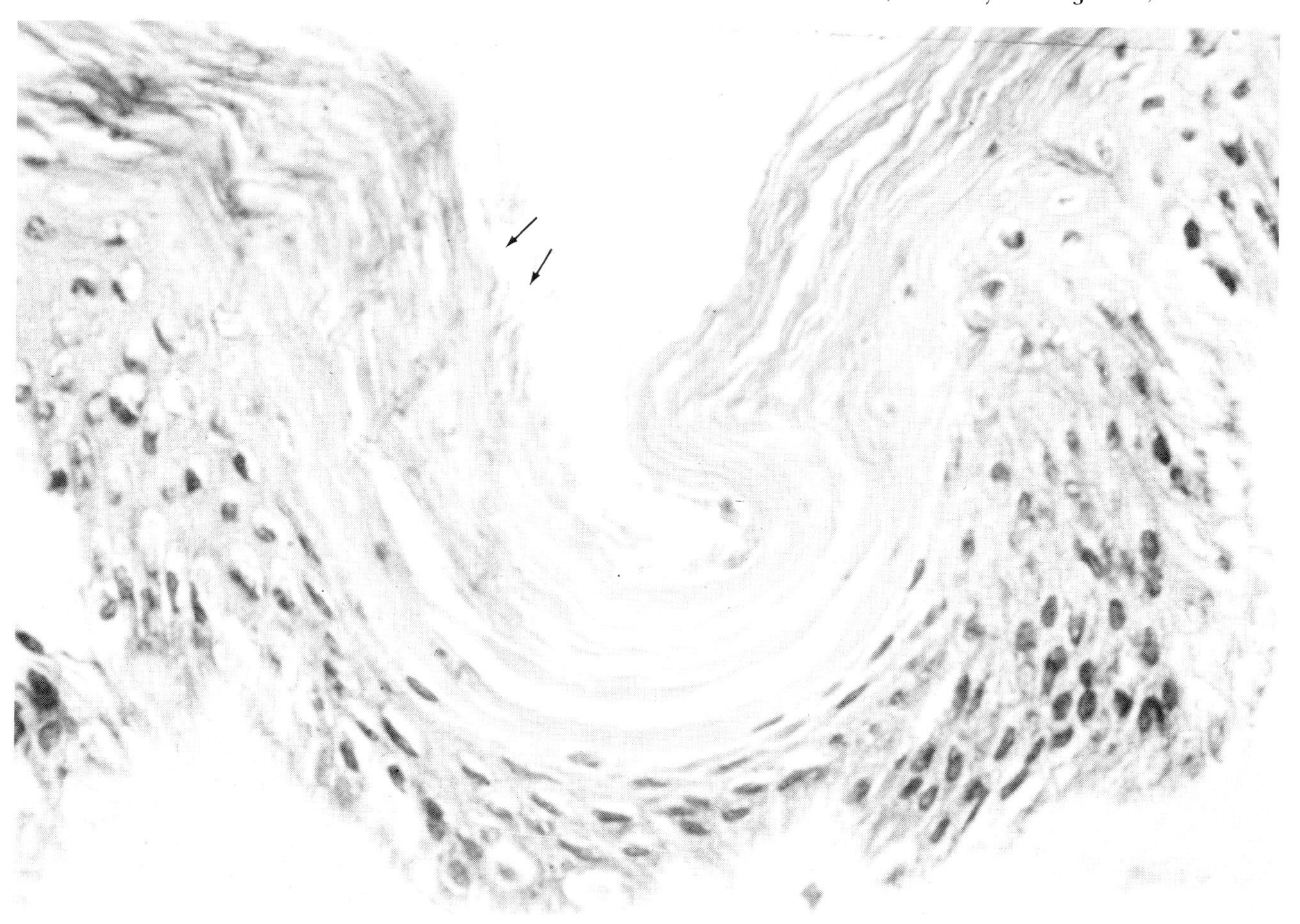

Phylum Echinodermata

PLATES

CHAPTER V

V: PLATE 1

FIGURE a.—*Acanthaster planci* is commonly found among corals which the starfish voraciously feeds upon. The rays, or arms, vary from 13 to 16 in number. The color of this starfish varies from gray, or brown, to a deep red, generally blending with its surroundings. Note the manner in which this venomous starfish has denuded the adjacent coral. Taken at Heron Island, Great Barrier Reef, Australia.

FIGURE b.—Closeup of spines of *Acanthaster planci*. Aboral surface. Spines may attain a length of 6 cm or more. The venom is produced by large numbers of acidophilic glandular cells which are distributed throughout the epidermis of the integumentary covering of the spine. (K. Gillett)

A

B

V: PLATE 2

FIGURE a.— Photomicrograph of a longitudinal section of the tip of an aboral spine of *Acanthaster planci* (Linnaeus). The epidermis, dermis, and underlying endoskeletal supportive structure of the spine are clearly delineated. × 82.5.

FIGURE b.—Cross section of the same spine. Arrow is pointing to indentation of the epidermal layer at which point is located an intraepithelial venom gland. × 82.5.

FIGURE c.—Enlargement showing, at tip of arrow, a typical intraepithelial gland comprised of acidophilic glandular cells which are believed to secrete the venom.

FIGURE d.—Enlargement showing at tip of arrow basophilic mucous glands. To the left is a darkened cellular mass, an intraepithelial venom gland. × 390.

Spines obtained from specimen taken at Heron Island, Great Barrier Reef, Australia, by K. Gillett. (E. Rich)

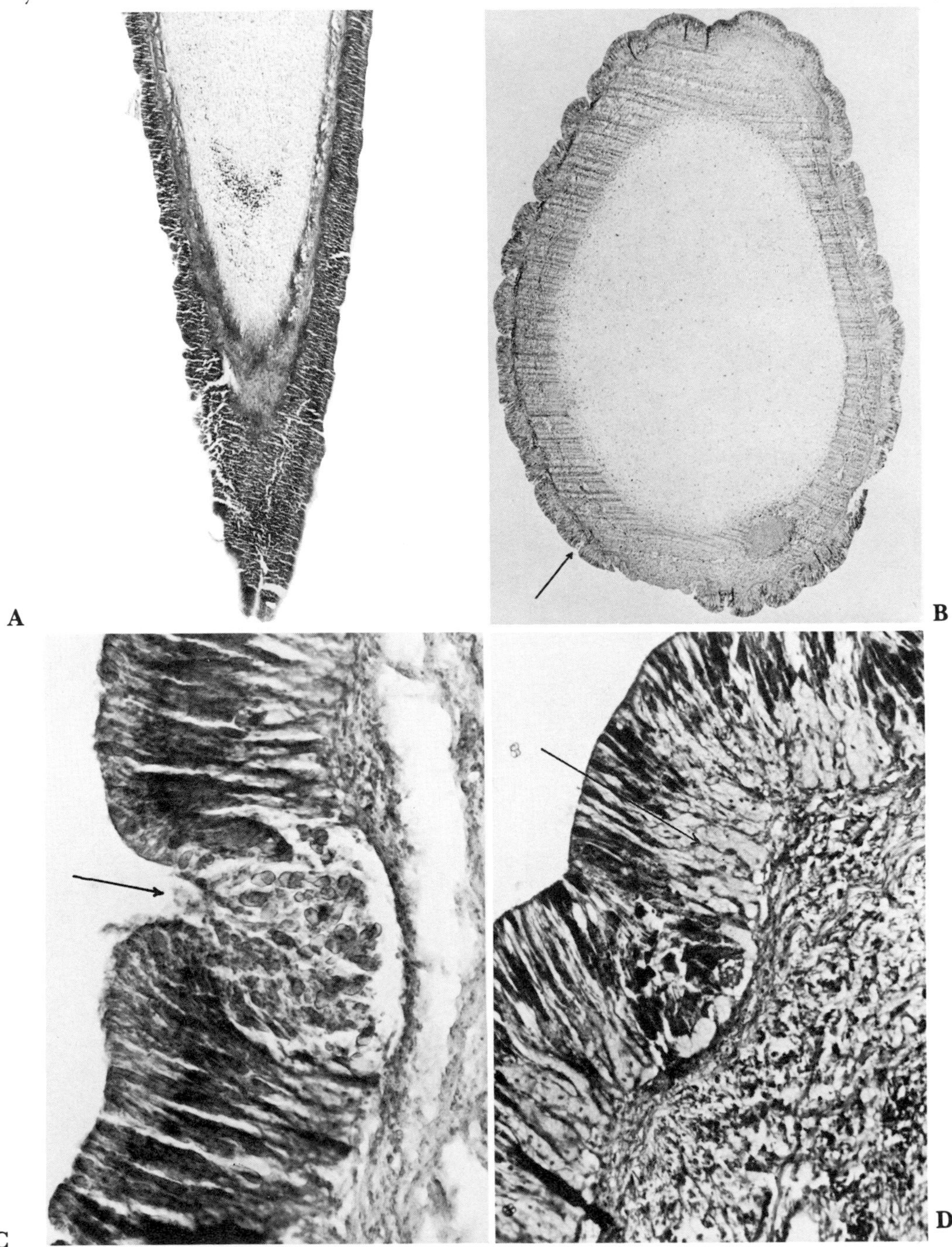

V: PLATE 3

FIGURE a.—*Aphelasteria japonica* (Bell). Diameter 8 cm.

FIGURE b.—*Asterias amurensis* Lütken. Diameter 30 cm.

FIGURE c.—*Asterina pectinifera* Müller and Troschel. Diameter 6 cm.

FIGURE d.—*Astropecten scoparius* Valenciennes. Diameter 8.5 cm. A toxic substance, asterotoxin, has been obtained from the alimentary tissues of the Japanese starfish species *A. japonica, A. amurensis,* and *A. scoparius*. Asterotoxin has been shown to have a cardioinhibitory effect on oyster heart. A toxic saponin has been isolated from *A. pectinifera* that is lethal to fish. The effects of these poisons on humans are unknown. Photographs of living specimens taken from Tokyo Bay, Japan. (Y. Hiyama)

V: PLATE 4

Solaster papposus (Linnaeus). This starfish produces a poison that causes convulsions, paralysis, and death in cats and other laboratory animals. The effects of the toxin on humans are unknown. Diameter 36 cm. Photographed at Plymouth, England.
(Courtesy of D. P. Wilson)

V: PLATE 5

Figure a.—*Diadema antillarum* Philippi. This West Indian species resembles very closely its Pacific counterparts. The needlelike spines of *Diadema* are hollow and said to contain a poison. Injuries from these spines are extremely painful, and because of the friable nature of the spines, they are difficult to remove. Diameter of test up to 11 cm. Living specimen taken at the Miami Seaquarium, Miami, Fla.
(Courtesy A. Mueller, Miami Seaquarium)

Figure b.—*Diadema setosum* (Leske). A characteristic feature of this Pacific form is the isolated blue spots on the apical plates of the test. These spots are visible to a greater or lesser extent in all of the specimens shown in this series. A dangerous species to handle. Diameter of test up to 10 cm. Specimen taken on the reef at Heron Island, Great Barrier Reef, Australia. (A. Powers)

Figure c.—Juvenile form of *Diadema setosum*. Note the banded appearance of the spines which is characteristic of the young of this species. Banding is sometimes observed in adults, but spines are generally black, reddish, or greenish. Taken in aquarium at Sydney, New South Wales, Australia. (A. Healy)

A

B

C

V: PLATE 6

FIGURE a.—*Echinothrix calamaris* (Pallas). The spines of the genus *Echinothrix* are hollow, reputed to contain a poison, and as dangerous as those of *Diadema*. Spines are generally shorter than in *Diadema*. Stings from the members of this genus are extremely painful. The spines vary in color from the banded pattern that is seen in the juvenile forms to white, green, purple, red, or black in the adult. Living specimen taken in Tokyo Bay, Japan. Diameter of test up to 13 cm or more.

(Courtesy of Y. Hiyama)

FIGURE b.—*Paracentrotus lividus* Lamarck. The globiferous pedicellariae of this European sea urchin are venomous, and the gonads contain a poison which produces paralyses, convulsions, and death in experimental studies on rabbits. The effects of this poison on man are unknown. Diameter of test 7 cm. The color in this species varies from a deep violet to green. Living specimen taken in the aquarium at the Zoological Station, Naples, Italy. (Courtesy of P. Giacomelli)

A

B

V: PLATE 7

FIGURE a.—*Araesoma thetidis* (Clark). This Australian sea urchin possesses sharp secondary aboral spines which are encased in a venom sac and are capable of inflicting painful stings. They are generally captured in deep water trawls. When first captured, they may appear melon-shaped and bristling with spines, but soon deflate themselves by losing a large amount of water at which time they assume a flattened appearance such as is generally seen in preserved specimens. The spines are very fragile and break off readily in the wound. This specimen was taken in 70 fathoms of water off Coogee, New South Wales, Australia. Diameter of test up to 20 cm.
(Photograph by Justice F. Myers, courtesy of E. Pope)

FIGURE b.—*Araeosoma thetidis*, preserved specimen. Taken off the coast of New Zealand.
(Courtesy of H. B. Fell and M. D. King)

A

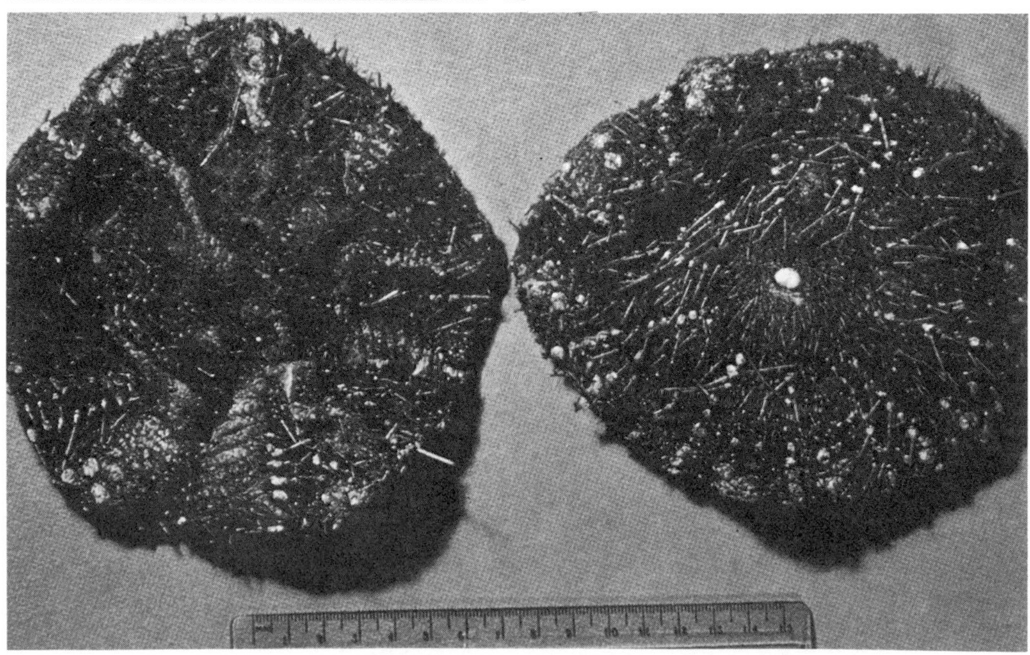

B

V: PLATE 8

FIGURE a.—Tip of a secondary aboral spine of *Araeosoma thetidis* (Clark) ensheathed in venom sac. When the spine penetrates the flesh, the venom sac is ruptured and the venom released in the wound. Spines may attain a length of 35 mm, but are generally shorter.

FIGURE b.—Tip of a primary oral spine of *Phormosoma bursarium* Agassiz ensheathed in venom sac. (R. Kreuzinger, after Mortensen)

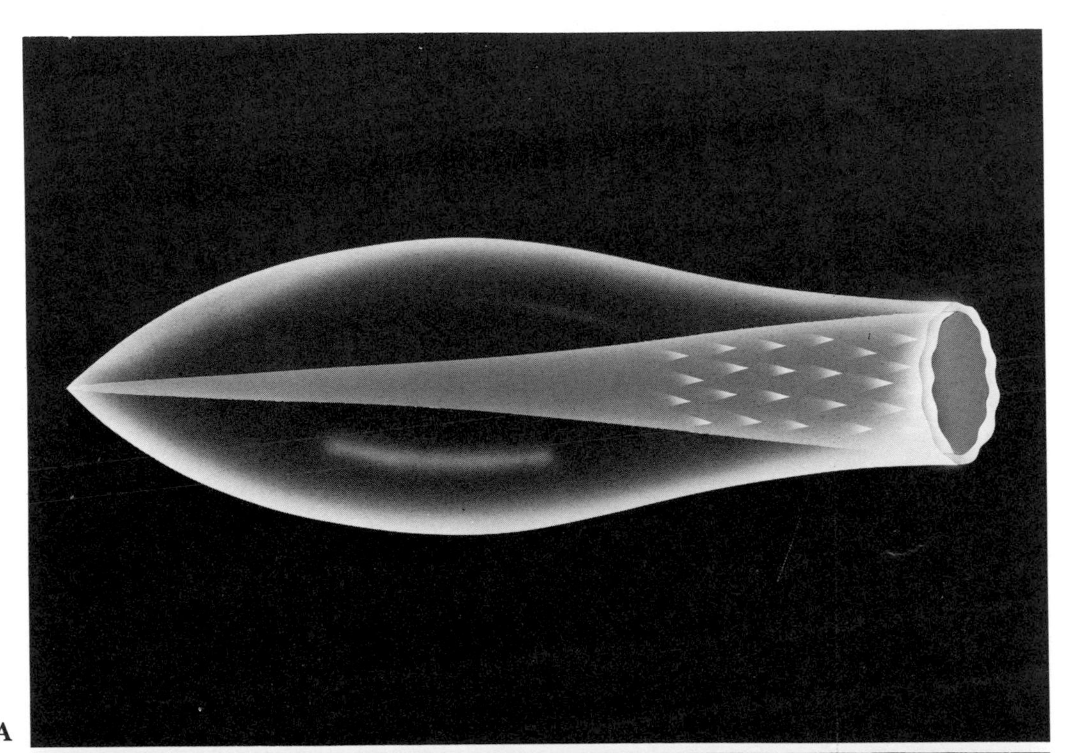

A

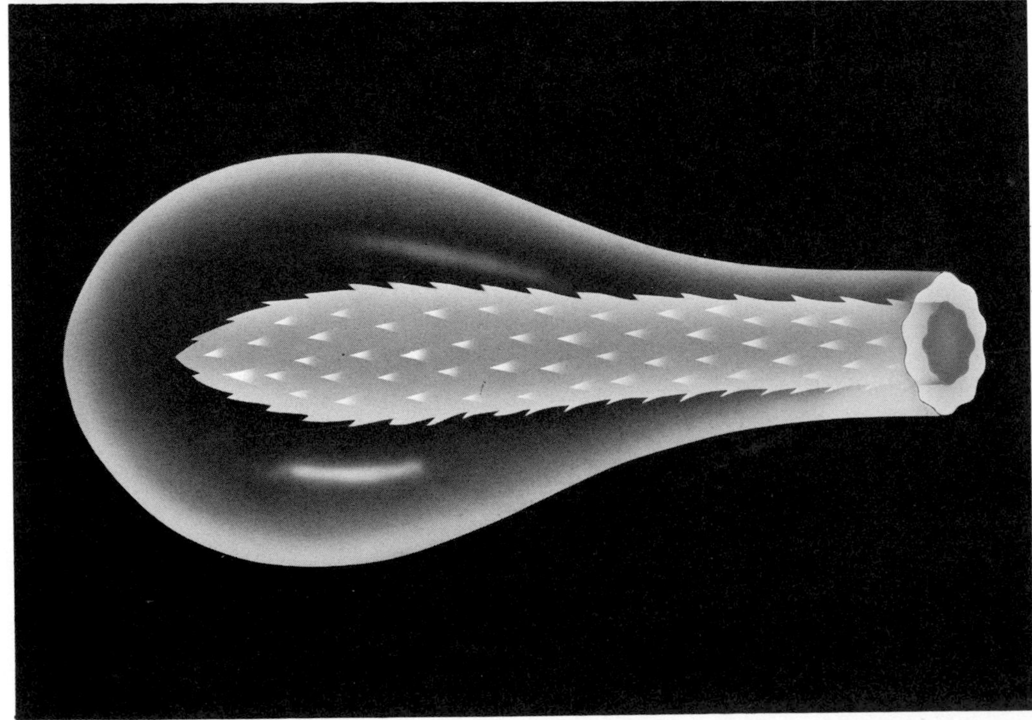

B

V: PLATE 9

Phormosoma bursarium Agassiz. The primary oral spines of this tropical Pacific sea urchin are ensheathed in venom sacs. This is a dangerous species to handle. Diameter of test 16 cm. Oral view. (From Mortensen)

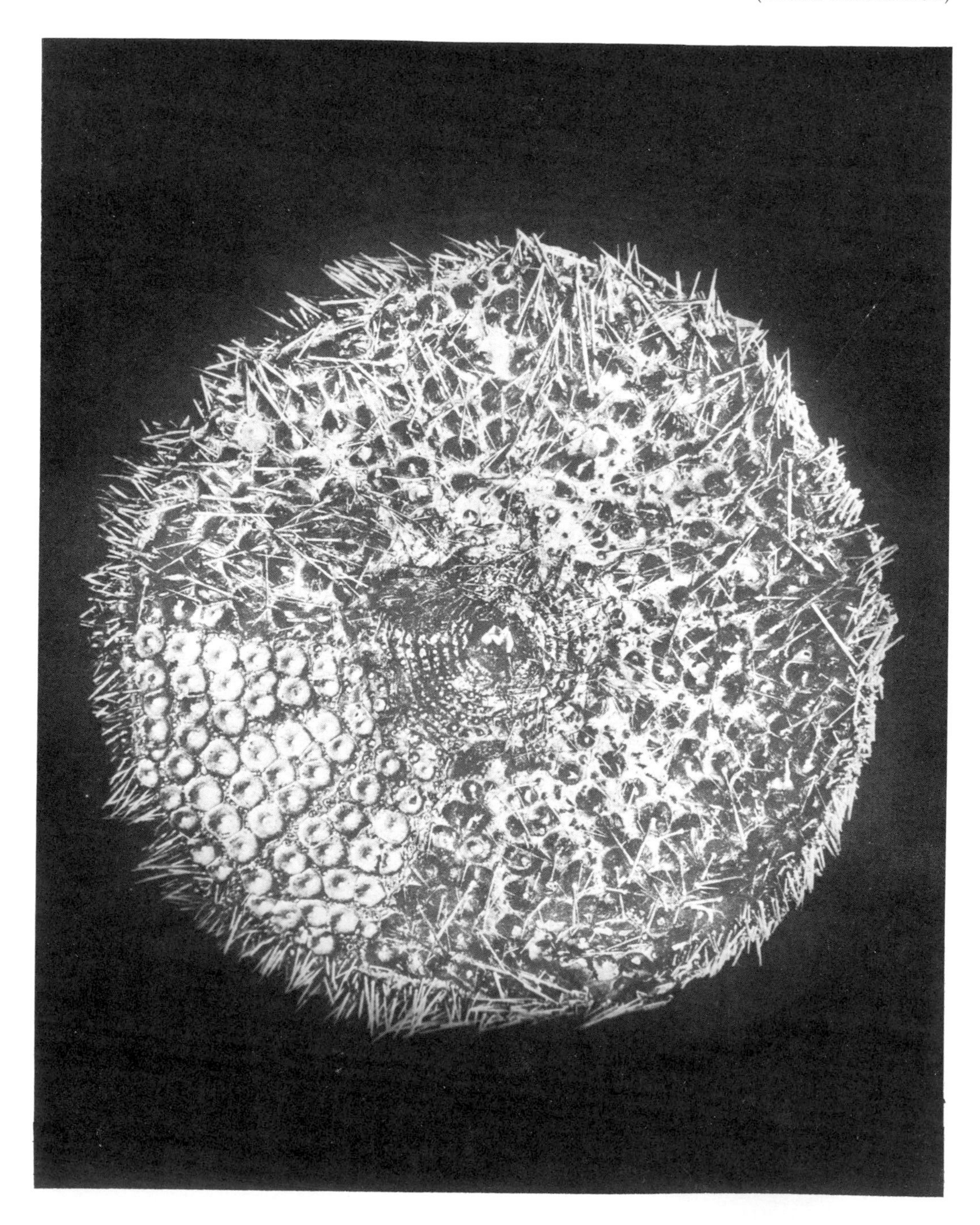

V: PLATE 10

FIGURES a and d.—*Toxopneustes pileolus* (Lamarck). The short spines, in this and other species of this family, frequently appear in life as covered by a brilliantly colored dense bed of small flowers—the venomous pedicellariae. The globiferous pedicellariae are highly developed in this group and vary in color—purple, yellow, red, green, or white. Stings from this Indo-Pacific sea urchin are painful and can be fatal. The poison has a direct action on the nervous system. Preserved specimens. (a) Oral side, from Misaki, Japan; (d) Aboral side, from Mauritius. Diameter of test 13 cm.

FIGURES b and c.—*Toxopneustes elegans* Döderlein. The venomous globiferous pedicellariae are highly developed in all the members of this genus. A dangerous species to handle. Preserved specimens from Kagoshima, Japan. Diameter of test 7.5 cm. (b) Aboral side; (c) Oral side. (From Mortensen)

B

A

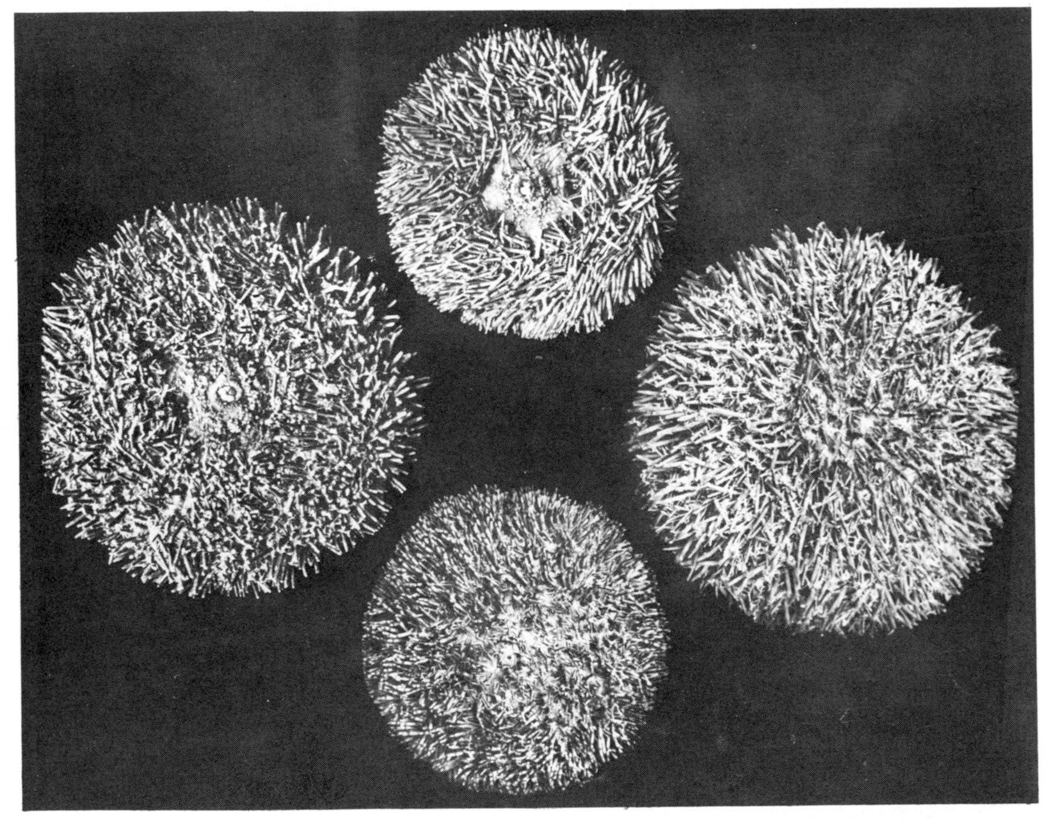

D

C

V: PLATE 11

FIGURE a.—*Toxopneustes pileolus* (Lamarck). Diameter of test about 13 cm. Pedicellariae extended.

FIGURE b.—Closeup showing some of the spines and pedicellariae. Note the triangular-shaped venomous globiferous pedicellariae with the jaws wide open. Taken in Sydney Harbor, New South Wales, Australia.

(Photos by C. V. Turner, courtesy of E. Pope, Australian Museum)

V: PLATE 12

FIGURE a.—*Toxopneustes elegans* Döderlein. The venomous globiferous pedicellariae appear as small flowerlike structures extending slightly beyond the spines. The tube feet appear as elongate, slender, translucent structures extending beyond the pedicellariae. The pedicellariae of this species can inflict a painful sting which may be lethal. Diameter of test 7.5 cm. Living specimen taken from Tokyo Bay, Japan. (T. Iwago)

FIGURE b.—*Tripneustes gratilla* (Linnaeus). This species is believed by some to be identical with *T. ventricosus* in which case the gonads are suspect as being poisonous to eat during certain periods of the year. Living specimen taken near Sydney, New South Wales, Australia. Diameter of test 10 cm. (Courtesy of I. Bennett)

FIGURES c-d.—*Tripneustes ventricosus* (Lamarck). Ingestion of the ova of this West Indian sea urchin may cause severe gastrointestinal disturbances, pain, nausea, vomiting, and diarrhea in some persons. However, the ova of this species are commonly eaten in the West Indies, where it is called the "white sea egg" or "edible sea egg." The ova are eaten during the period from September through April. The nature of the poison—if such is present—is not known, and it has been suggested that these intoxications are on an allergic basis. Living specimen taken at the Bermuda Biological Station, side and dorsal views. (G. Lower)

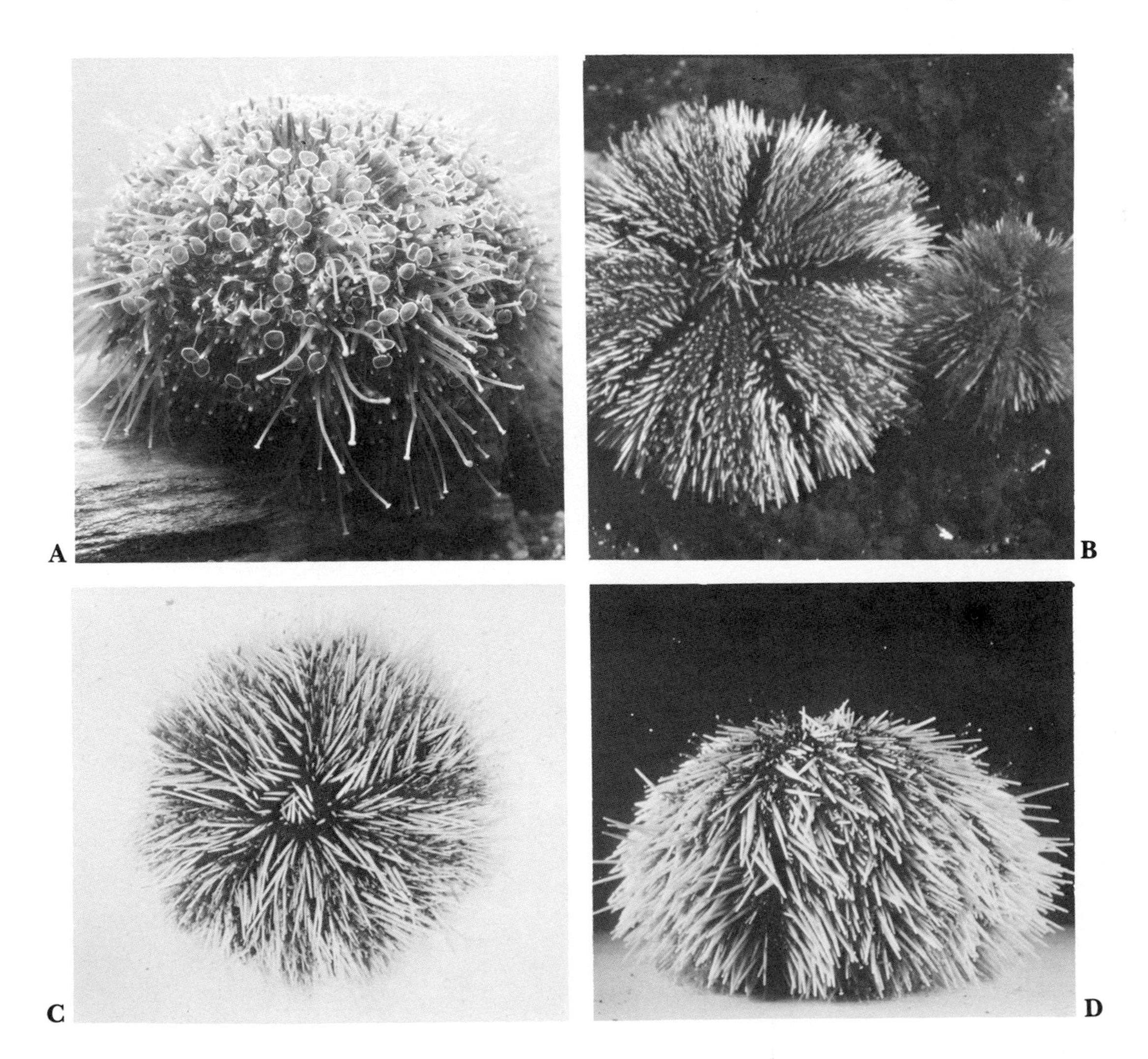

V: PLATE 13

FIGURE a.—(Top) Sea urchin bite inflicted by *Toxopneustes roseus* (Agassiz). This sea urchin was handled at some length by one person without ill effects, but when handed over to a second person, he was promptly bitten. Note welts near base of little finger. The initial bites consisted of three white, painful marks that spread into a single welt about 1 cm in diameter within minutes. The marks had diminished by the following morning, but three red puncture marks surrounded by an area of tenderness remained. The sea urchin was captured on Cruise 16 of the R/V Anton Bruun approximately 100 m west of El Viejo Island, about 5 km offshore from Valdivia, Bahia de Santa Elena, Ecuador. Collected by Sylvia A. Earle and John Hall.

(Photo and legend courtesy S. Earle, Cape Haze Marine Laboratory)

FIGURE a.—(Bottom) Close-up photograph showing the venomous pedicellariae of the sea urchin *Toxopneustes roseus* (Agassiz). This sea urchin is a member of one of the more dangerous genera. The photos in Figures b and c reveal that there is an apparent difference between daytime and nighttime activity.

FIGURE b.—This photograph was taken in the daytime and reveals that the pedicellariae were open and active, whereas in FIGURE c, the pedicellariae at nighttime were closed and inactive. This sea urchin can inflict a painful sting with these pedicellariae. Taken in the Gulf of California by the California Academy of Sciences Sea of Cortez Expedition. (Courtesy D. Wobber)

V: PLATE 14

Photograph of a 6-year-old girl who was stung by the sea urchin *Arbacia lixula* (Linnaeus). The child developed chills and fever. The wound area became edematous and reddened. There were no spines imbedded in the skin. It was believed that some of the lesions were inflicted by the venomous pedicellariae of the urchin. Taken near Pula, Yugoslavia. (Courtesy Z. Maretić)

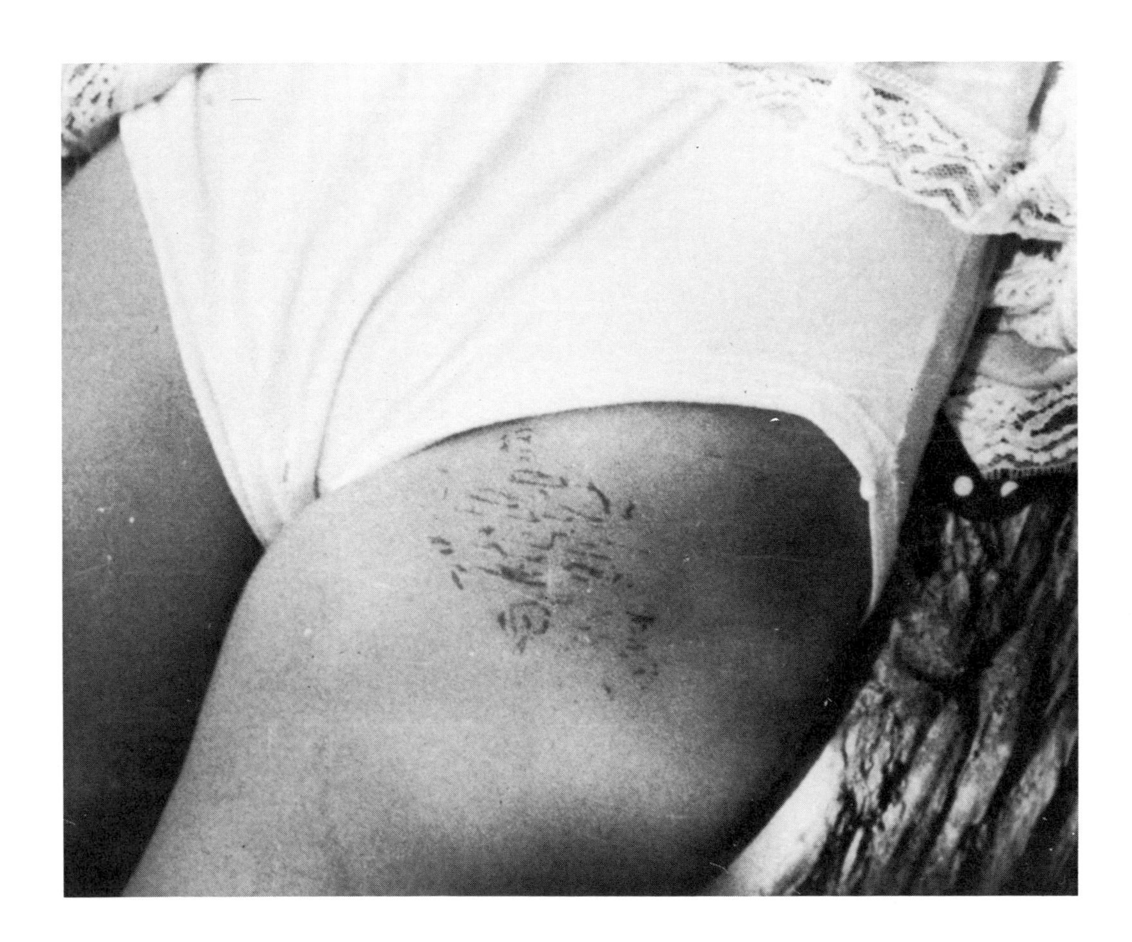

V: PLATE 15 (following page)

Examples of basic types of sea urchin pedicellariae.

FIGURE a.—Globiferous pedicellaria, venomous type, from *Salmacis bicolor* Agassiz. × 87.

FIGURE b.—Tridentate pedicellaria from *Echinus multidentatus* Clark. Left, head, × 25; right, valve, × 45.

FIGURE c.—Triphyllous pedicellaria from *Salenia unicolor* Mortensen. Left, valve, × 340; right, head, × 110.

FIGURE d.—Ophiocephalous pedicellaria from *Hygrosoma petersi* (Agassiz). Left, valve, × 55; right, head, × 32.

FIGURE e.—Dactylous pedicellaria from *Areosoma paucispinum* Clark. Left, head, × 20; right, valve, × 22. (R. Kreuzinger, after Mortensen)

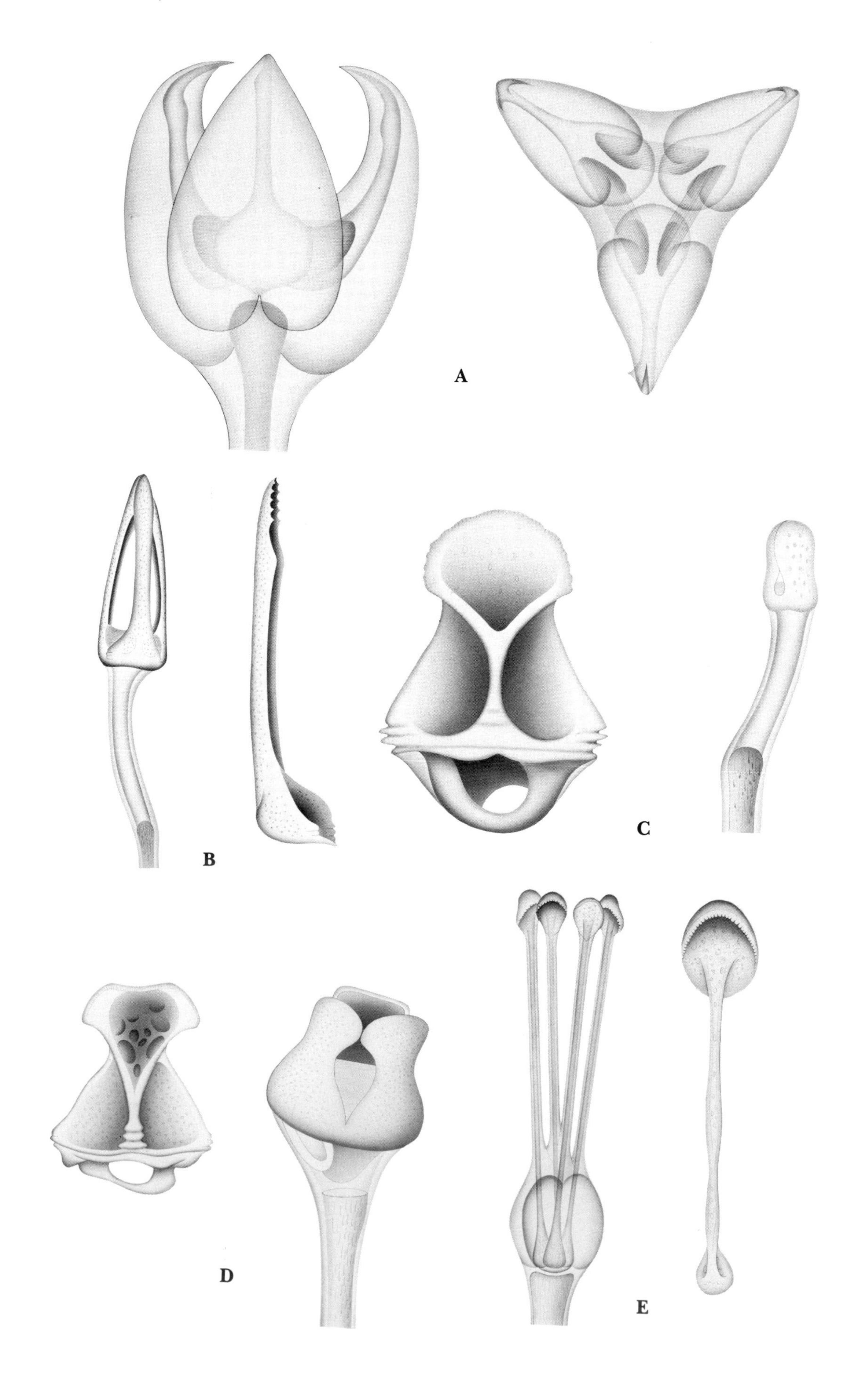
A
B
C
D
E

V: PLATE 16

Examples of venomous globiferous pedicellariae of sea urchins.

FIGURE a.—*Echinus esculentus* Linnaeus. Left, head, × 15; right, valve, × 11.

FIGURE b.—*Heterocentrotus mammillatus*(Linnaeus). Left, head, small form, × 33; right, valve, large form, × 14.

FIGURE c.—*Lytechinus anamesus* Clark. Left, head, turned down so as to look at it directly from above. Note stalk glands on the supporting stalk of the pedicellaria, × 21; right, valve, × 24.

FIGURE d.—*Paracentrotus lividus* (Lamarck). Left, head, × 31; right, valve, front and side view, × 36.

FIGURE e.—*Psammechinus microtuberculatus* (Blainville). Valve, × 42.

FIGURE f.—*Sphaerechinus granularis* (Lamarck). Head, × 21. Note stalk glands.

FIGURE g.—*Strongylocentrotus sachalinicus* Döderlein. Left, head, × 14; right, valve, × 25.

FIGURE h.—*Toxopneustes pileolus*(Lamarck). Left, head, looking down at it directly from above, × 4; right, valve, × 8.

FIGURE i.—*Toxopneustes roseus* (Agassiz). Head of large globiferous pedicellaria, × 6. Note stalk glands. (R. Kreuzinger, after Mortensen)

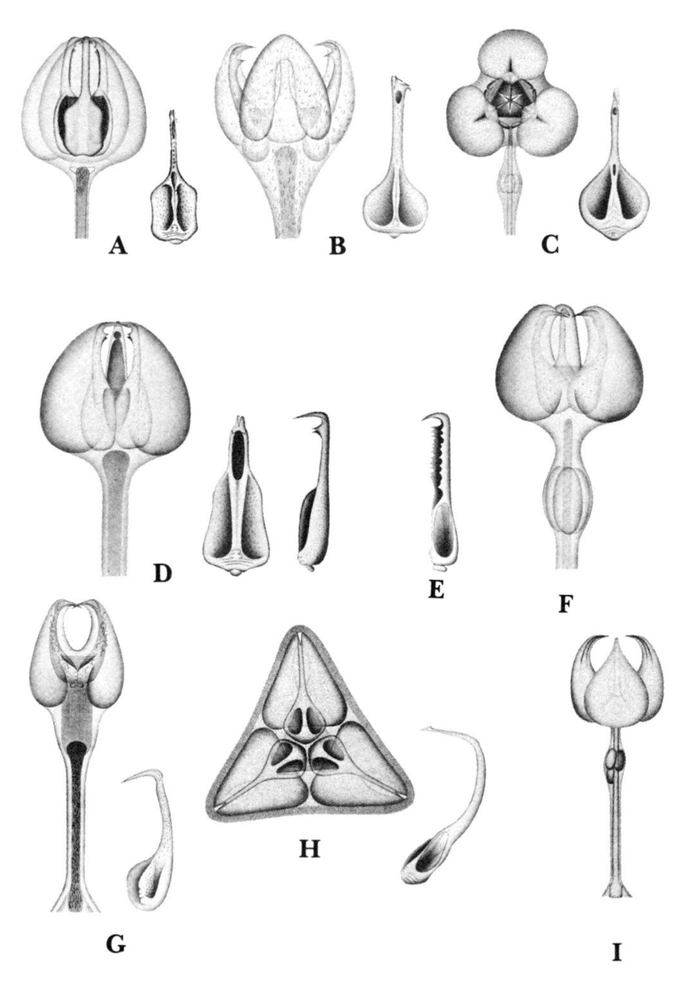

V: PLATE 17

FIGURE a.—Living globiferous pedicellariae on the test of *Sphaerechinus granularis*. Spines have been removed to show the pedicellariae. The head, stalk, and stalk glands are clearly visible. × 7.

FIGURE b.—Living globiferous pedicellaria removed from the test of *Sphaerechinus granularis*. × 12.

FIGURE c.—The same pedicellaria with the jaws open. Note the fang where the venom is released appearing at the extreme tip of each valve or jaw. × 12.

FIGURE d.—Enlargement of the previous pedicellaria with the jaws open. Specimens taken at the aquarium of the Zoological Station at Naples, Italy. × 24.

(Courtesy of P. Giacomelli)

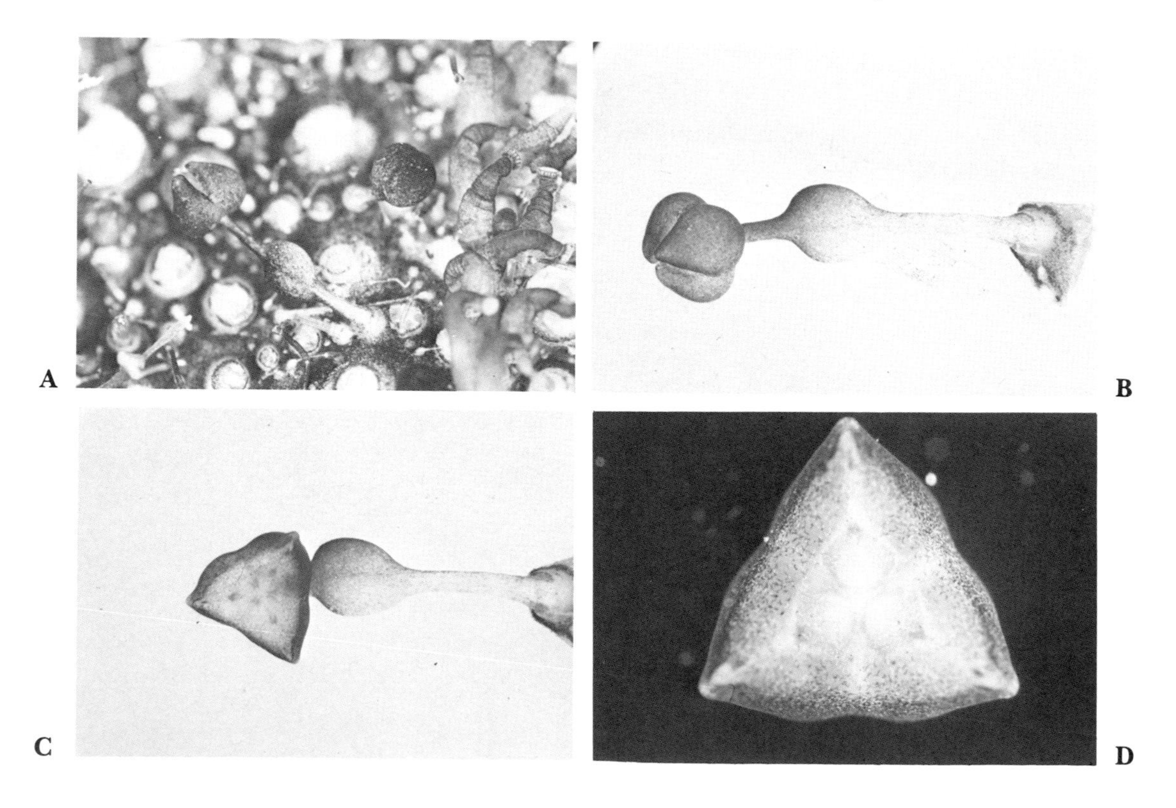

V: PLATE 18

FIGURE a.—*Actinopyga agassizi* (Selenka). Living specimen taken on the reef at Heron Island, Great Barrier Reef, Australia.

FIGURE b.—*Stichopus chloronotus* Brandt. Living specimen taken on the reef at Heron Island, Great Barrier Reef, Australia. Length 30 cm. (K. Gillett)

FIGURE c.—*Holothuria tubulosa* Gmelin. Living specimen taken at the aquarium of the Zoological Station at Naples, Italy. Length 25 cm. (Courtesy of P. Giacomelli)

A

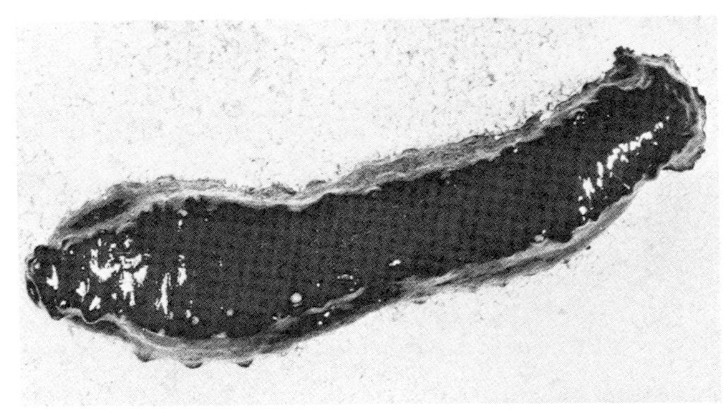

B

C

V: PLATE 19

Cucumaria echinata Von Marenzeller. Japan. Length 8 cm.
(H. Baerg, after Kuroda and Uchida)

V: PLATE 20 (opposite page)

FIGURE a.—*Afrocucumis africana* (Semper). Length 8 cm.
(After Kuroda and Uchida; Utinomi)

FIGURE b.—*Stichopus japonicus* Selenka. Length 40 cm. Japan. (After Utinomi)

FIGURE c.—*Holothuria vagabunda* Selenka. Length 20 cm. (After Kuroda and Uchida)

FIGURE d.—*Paracaudina chilensis* var. (Von Marenzeller). Length 20 cm. Japan.

FIGURE e.—*Thelenota ananas* (Jaeger). Length 75 cm. Japan. (After Utinomi)

FIGURE f.—*Holothuria monocaria* Lesson. Length 40 cm. Japan.

FIGURE g.—*Stichopus japonicus* Selenka. Length 30 cm. Japan.
(After Fisheries Society of Japan)

FIGURE h.—*Pentacta australis* (Ludwig). Length 8 cm. Japan. (After Utinomi)

FIGURE i.—*Holothuria impatiens* (Forskål). Length 40 cm. Australia. (After Clark)

FIGURE j.—*Cucumaria japonica* Semper. Length 20 cm. Japan.
(After Fisheries Society of Japan)

All the sea cucumbers appearing on this plate have been found to contain holothurin, a toxic saponin. Visceral liquid ejected by some sea cucumbers may produce a skin rash or blindness if brought in contact with the eyes. Although certain of these cucumber species are commonly eaten in some areas, the ingestion of them should be considered with caution. (Drawing by H. Baerg, after the above sources)

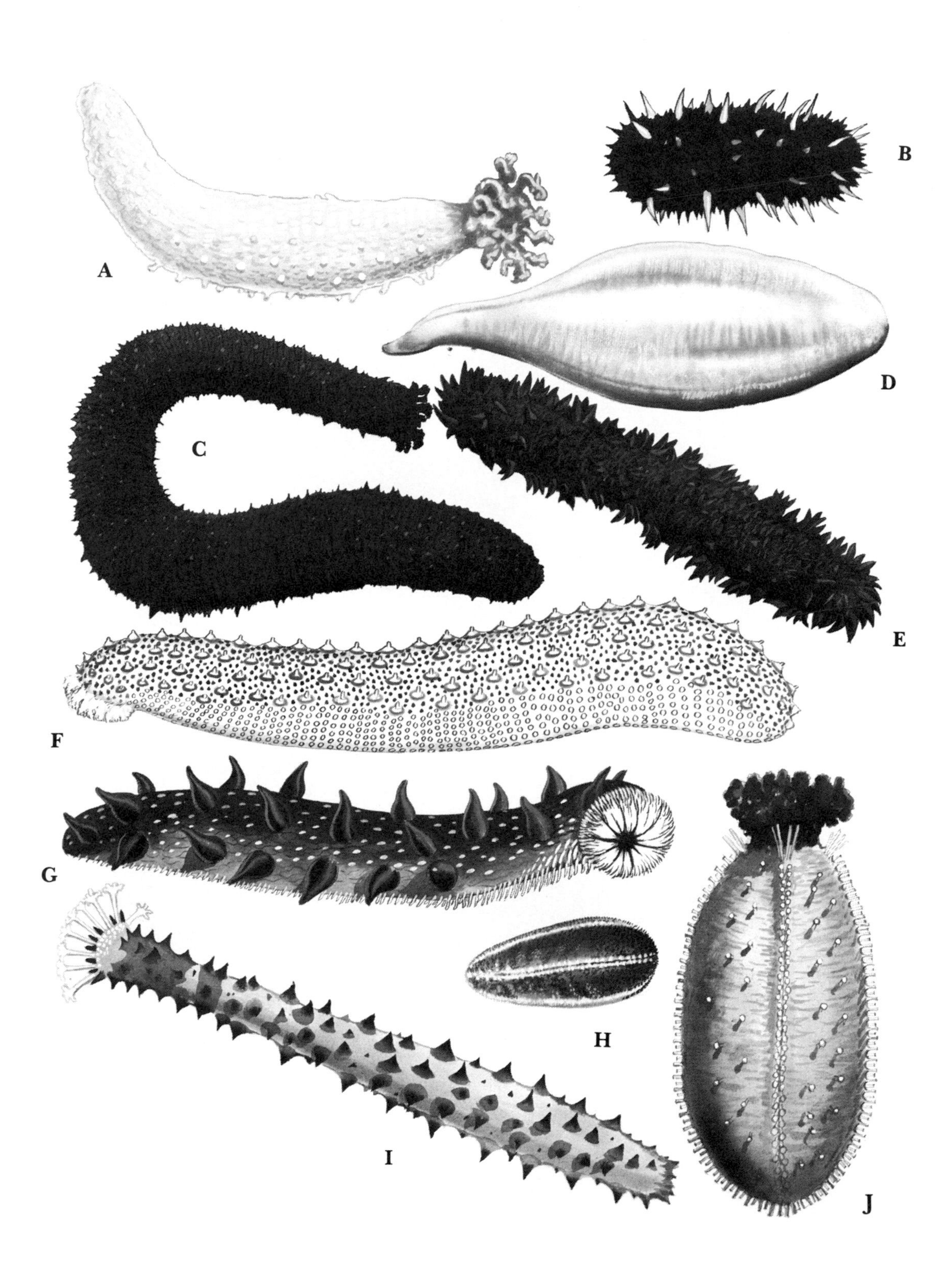
A
B
C
D
E
F
G
H
I
J

V: PLATE 21

FIGURE a.—*Holothuria argus* (Jaeger). Living specimen taken at Heron Island, Great Barrier Reef, Australia. Length 30 cm or more. (K. Gillett)

FIGURE b.—*Euapta lappa* (Müller). Living specimen taken near Miami, Fla. (R. Straughan)

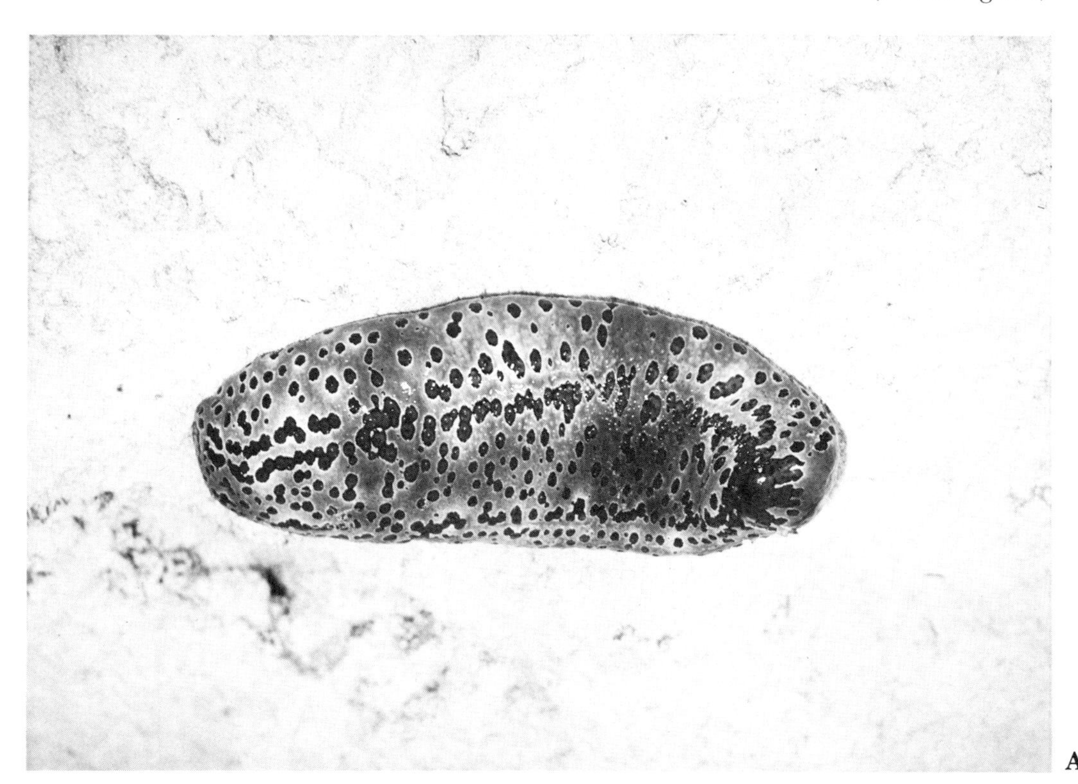

A

B

Phylum Mollusca

PLATES

CHAPTER VI

VI: PLATE 1

FIGURE a.—*Neptunea antiqua* (Linnaeus). Whelk poison is concentrated in the salivary glands. Although this particular species is commonly eaten in Europe and has not been reported in human intoxications, it is known to contain the poison and is viewed with suspicion. Intoxication is by ingestion. Attains a length of about 10 cm.

(From Forbes and Hanley)

FIGURE b.—*Neptunea arthritica* (Bernardi). This and the related Japanese whelk, *N. intersculpta*, have been involved in large numbers of human oral intoxications in Hokkaido. Symptoms consist largely of intense headaches, dizziness, weakness, gastrointestinal upset, and visual disturbance. Length about 10 cm.

(From Kuroda and Uchida)

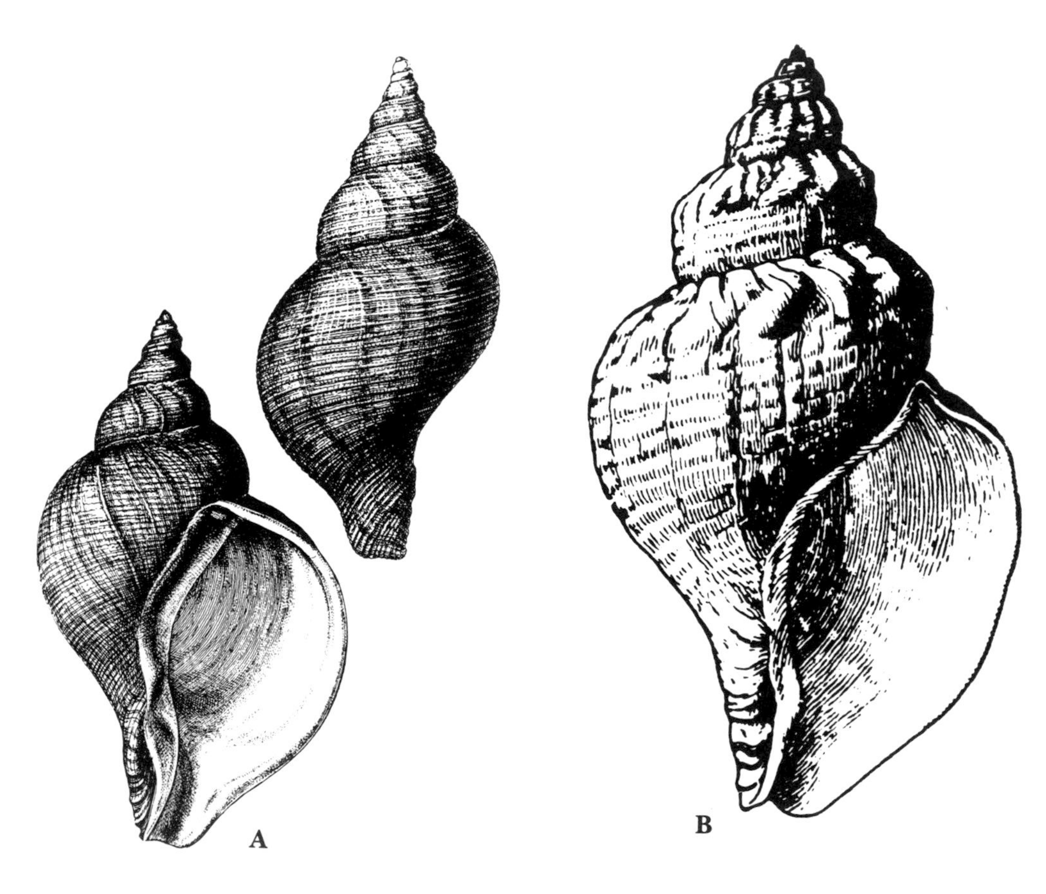

VI: PLATE 2

FIGURES a-b.—*Conus aulicus* Linnaeus. One of the more dangerous species of cone shells. This species is believed capable of inflicting a fatal sting. Most cone shell stings are similar in their clinical characteristics. See text for details. Attains a length of 15 cm. Drawings of specimen collected in the Marshall Islands.

FIGURES c-d.—*Conus geographus* Linnaeus. This is probably the most dangerous species of cone shell and has been involved in more fatal and serious human stings than any other member of this group. The venom apparatus is well developed and capable of delivering a relatively large quantity of venom. Attains a length of 13 cm or more. Drawing of specimen collected in the Marshall Islands. (S. Arita)

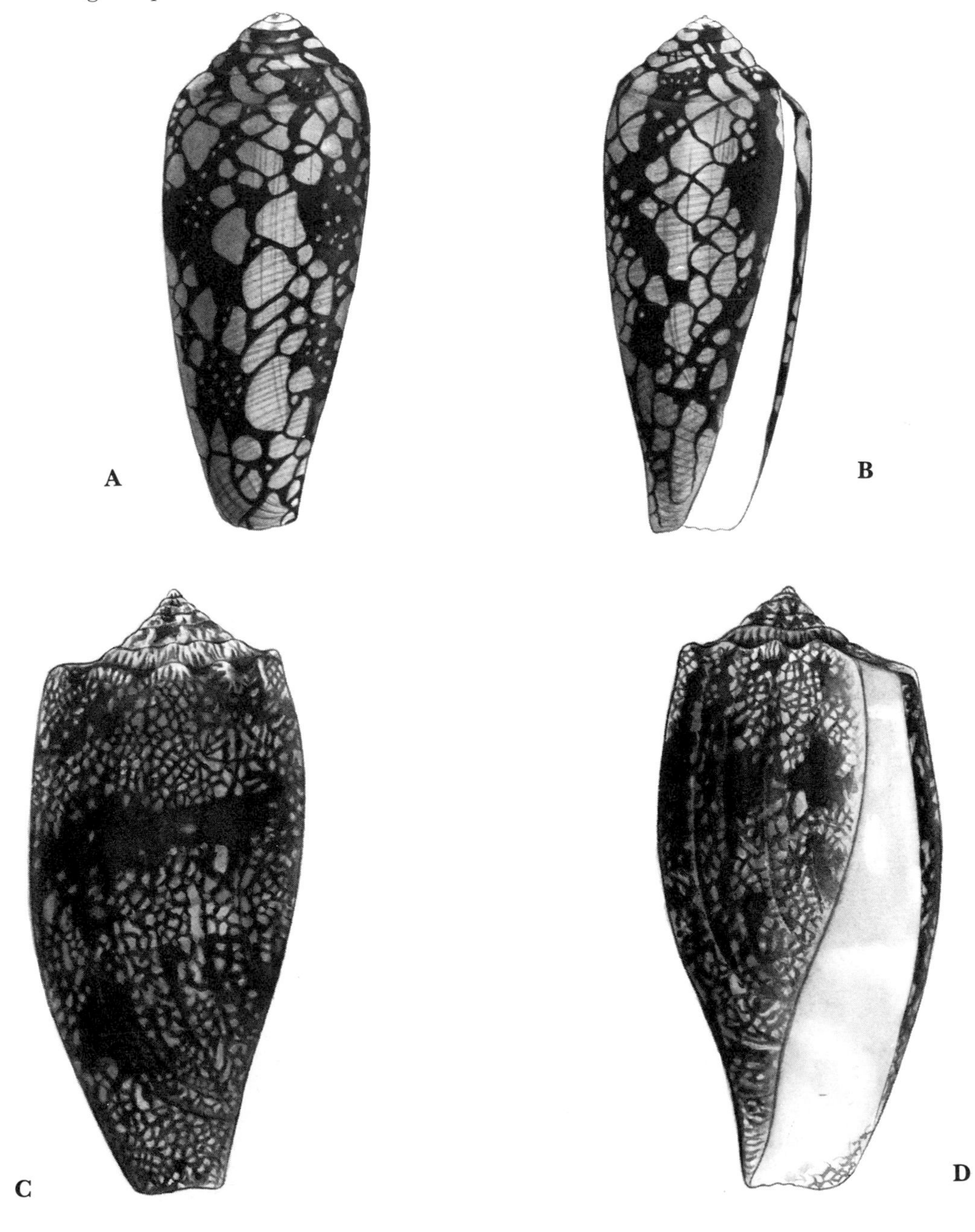

VI: PLATE 3

FIGURES a-b.—*Conus marmoreus* Linnaeus. This species has stung humans and is a dangerous species, although no human fatalities have been reported. The venom apparatus is large and well developed. Attains a length of 10 cm. Drawing of specimen collected in the Marshall Islands. (S. Arita)

FIGURE c.—*Conus striatus* Linnaeus. This is an aggressive and dangerous species, and has caused human fatalities. It has a large and well-developed venom apparatus. Attains a length of 10 cm. Drawing of specimen collected in the Marshall Islands. (K. Tomita)

FIGURE d.—*Conus gloria-maris* Hwass. This is a very rare and highly prized species. Fewer than 100 specimens are said to have ever been captured. Although its venom apparatus has not been studied and no human intoxications have been reported, it is a potentially dangerous species. Attains a length of 12.5 cm. Drawing of specimen collected in the Moluccas. (H. Baerg)

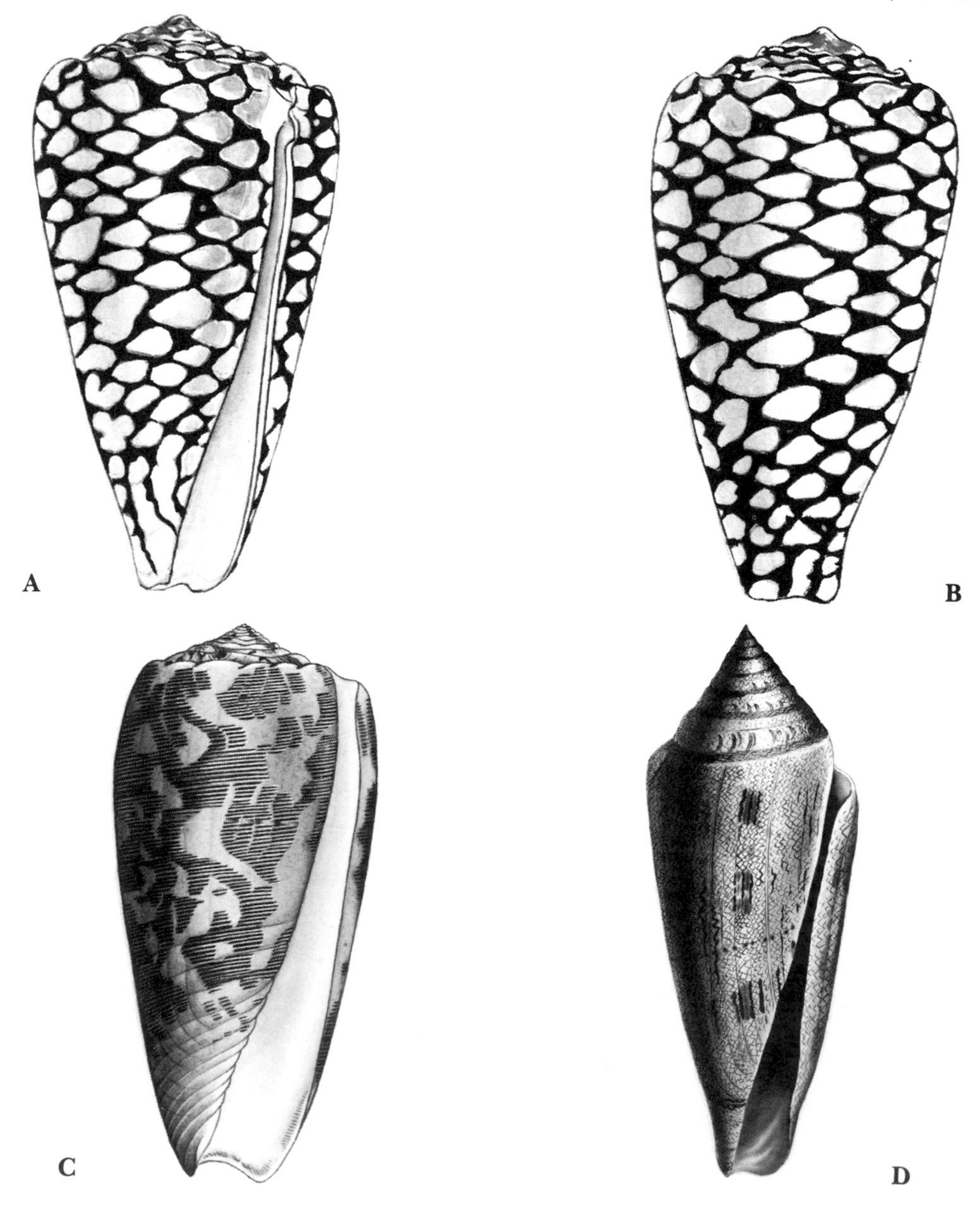

VI: PLATE 4

Conus magus Linnaeus. Length 6 cm. The specimen shown was collected at Jaluit, Marshall Islands, by J. Fitch. This cone shell has a well-developed venom apparatus and is similar in structure to that of other members of this family. The morphology of the venom apparatus of *C. magus* has been described in detail by Endean and Duchemin (1967). Endean and Izatt (1965) have shown that the venom of *C. magus* is more toxic to mice than any other cone shell venoms that have been studied to date. Amounts as low as 0.2 mg (wet wt.) of *C. magus* venom per kg are lethal to mice. Venom extracted from posterior regions of the venom duct of *C. magus* was lethal to mice and produced a contracture of isolated skeletal and smooth musculature of the rat in the absence of electrical stimulation. The effect was sustained and the musculature became paralyzed in the contracted state. Posterior duct venom increased the strength of contraction of the musculature of the ventricle of the toad but decreased its rate of contraction. Cardiac musculature was not paralyzed by the venom at concentrations tested. The effects produced by the venom in skeletal, smooth, or cardiac musculature could be reversed by washing out the venom. The MLD or LD_{50} were not determined. There is no information available concerning the chemical characteristics of the venom.

(Photo by R. Kreuzinger, courtesy S. S. Berry)

VI: PLATE 5

FIGURES a-b.—*Conus textile* Linnaeus. One of the more dangerous cone shells. Human fatalities have been caused by this species. Attains a length of 10 cm.

FIGURES c-d.—*Conus tulipa* Linnaeus. This is believed by some authorities to be the most venomous of the cone shells. Although no human fatalities have been reported, it is an aggressive stinger. Attains a length of about 7.5 cm. (K. Tomita)

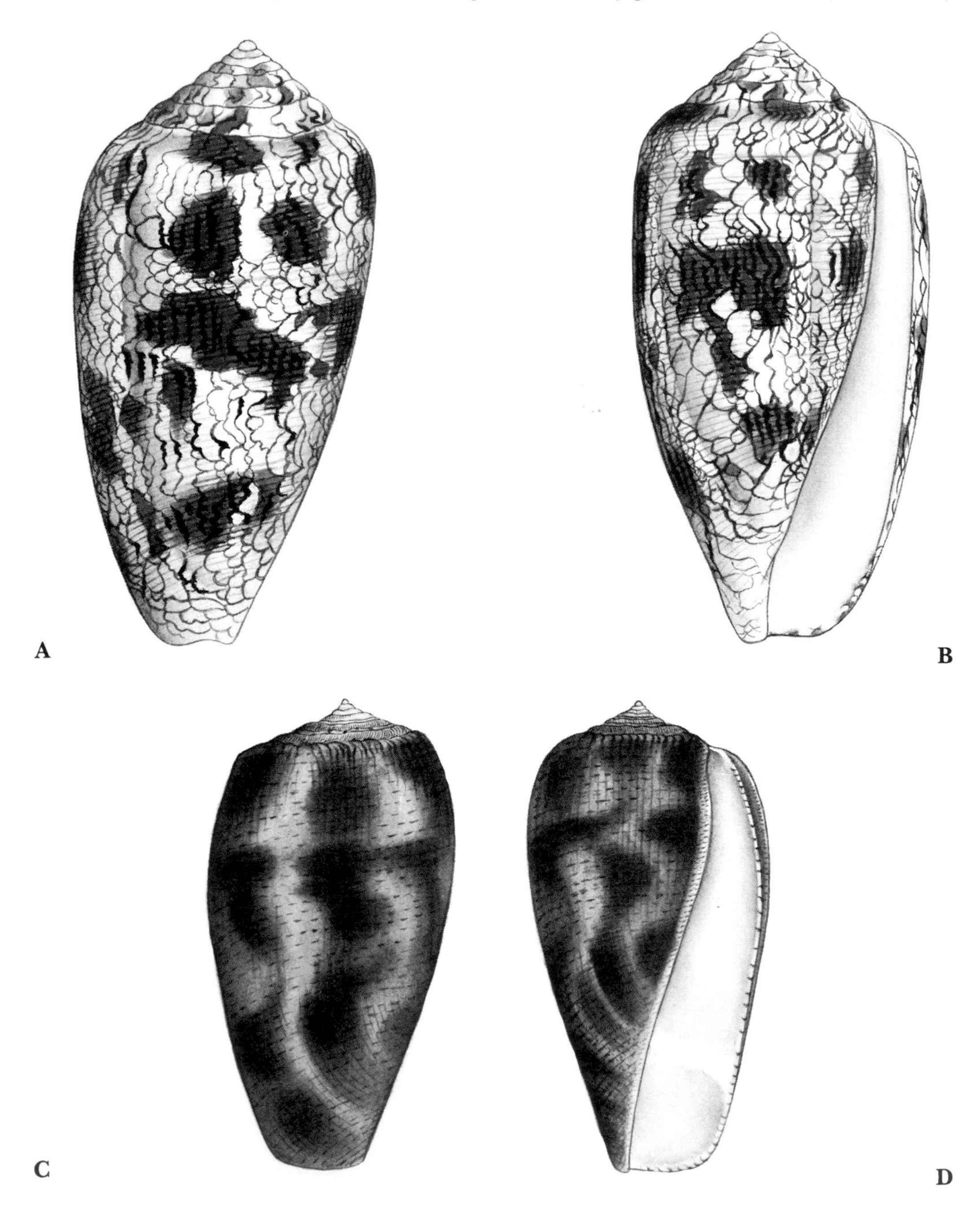

VI: PLATE 6

FIGURE a.—*Conus marmoreus*, living specimen taken at New Caledonia. (E. Pope)

FIGURE b.—*Conus tulipa* Linnaeus. One of the most dangerous species of the cone shells. Living specimen taken at Heron Island, Great Barrier Reef, Australia. Length 7·5 cm. (K. Gillett)

A

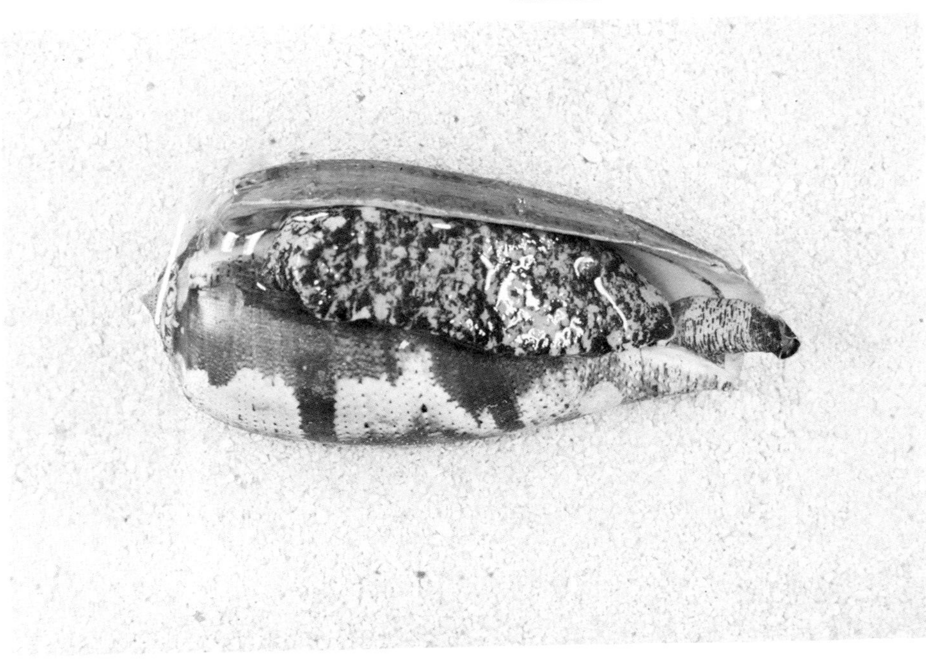

B

VI: PLATE 7

FIGURE a.—*Conus textile*, closeup, showing extended siphon and tubular proboscis in the foreground. Extending from either side at the base of the proboscis can be seen the tentacles. The process of stinging is accomplished with the use of the muscular proboscis.

FIGURE b.—*Conus textile*, closeup, showing the cone shell in the act of striking a top shell *Tegula*. The radular teeth originate in the radular sheath, are passed through the pharynx (a single tooth at a time), and then into the proboscis. A radular tooth is used as a harpoonlike device which is grasped by the proboscis until the victim has been impaled. In this shot, which was taken from an actual movie sequence of the stinging process, one can see the radular tooth extending from the extreme tip of the tubular proboscis and penetrating into the body of the *Tegula*. Note the siphon to the right and below the extended proboscis. (Courtesy of P. Saunders)

FIGURE c.—*Conus textile*, same specimen as above in process of stinging a top shell *Tegula*. The proboscis is also used to grasp the victim. Taken at Guam, Mariana Islands. (Photo by E. Pankow, courtesy of P. Saunders)

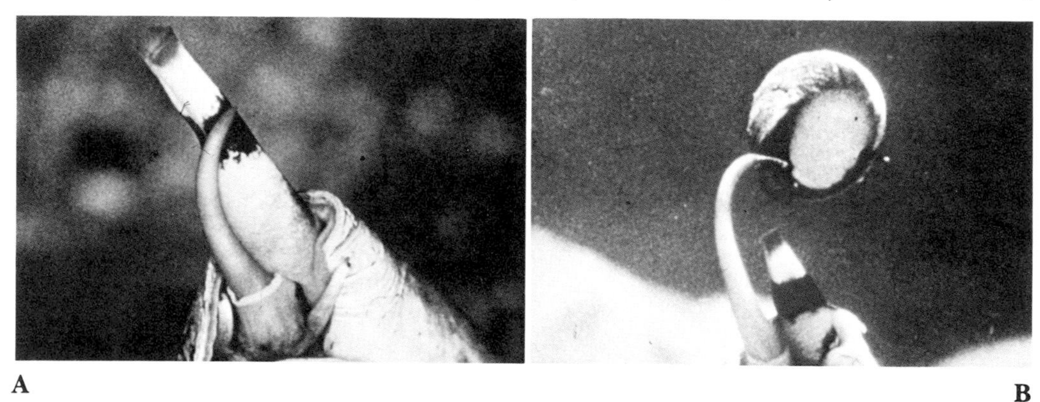

A B

C

VI: PLATE 8

Figure a.—Needle pointing to radular tooth of *Conus striatus* projecting from the opercular area of cottoid fish *Clinocottus analis*. Experimental sting made in an aquarium.

Figure b.—Base of *Conus geographus* radular tooth showing the ligamentous attachment. Note opening at the base through which the venom enters the hollow tooth. Length of shaft about 8 mm, diameter 0.2 mm.

Figure c.—Anterior tip of radular teeth of *Conus striatus*. This particular species has one of the most striking and elaborately designed harpoonlike teeth of any of the cones. Length of shaft of mature tooth about 7 mm, diameter 0.4 mm.

(Courtesy of P. Saunders)

Figure d.—Anterior tip of radular tooth of *Conus textile*. Taken from fresh living specimen. Note that there appears to be a thin membrane surrounding the tooth. Length of shaft of mature tooth about 10 mm, diameter 0.12 mm. Specimen taken at Heron Island, Great Barrier Reef, Australia. (K. Gillett)

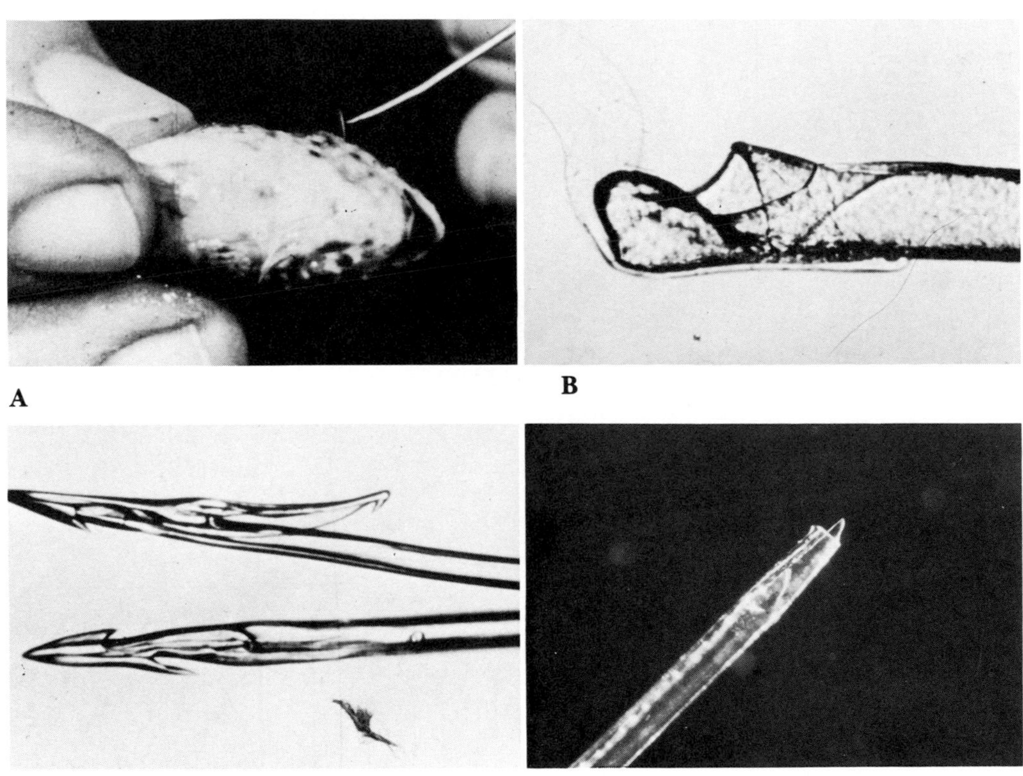

VI: PLATE 9

FIGURE a.—Dissection of *Conus californicus* showing general anatomical relationships of the venom apparatus. Shell has been removed and the mollusk is essentially intact. × 2.5.

FIGURE b.—Dissection of *Conus textile* showing general anatomical relationships of the venom apparatus. Shell has been removed and the mollusk is essentially intact. × 2. (R. Kreuzinger, dissection by B. W. Halstead)

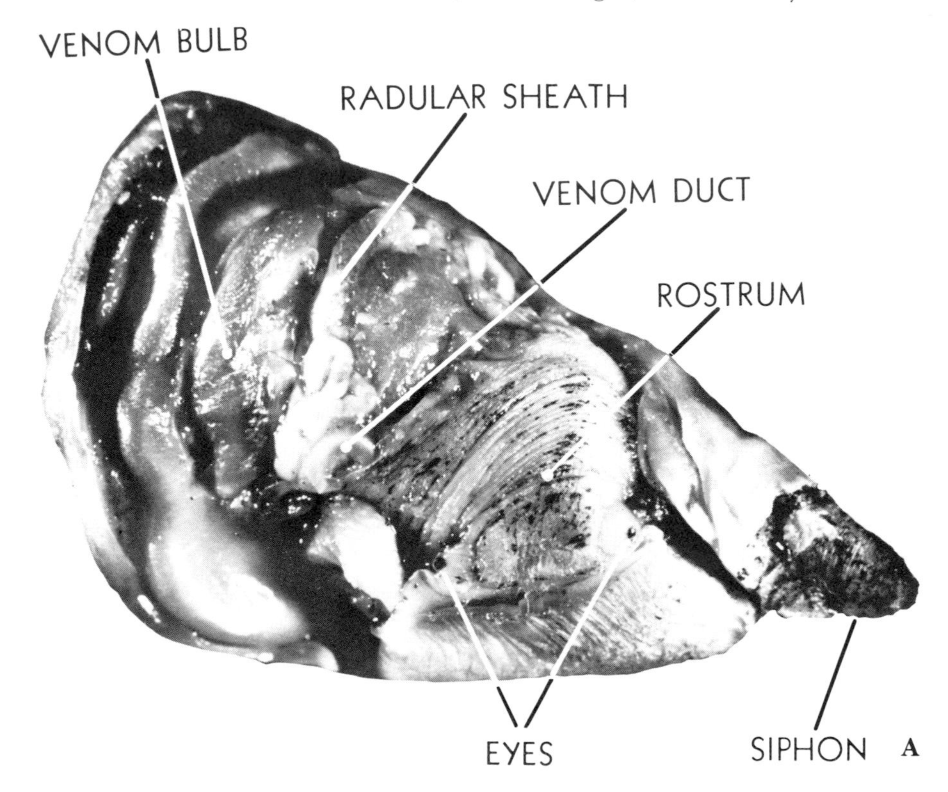

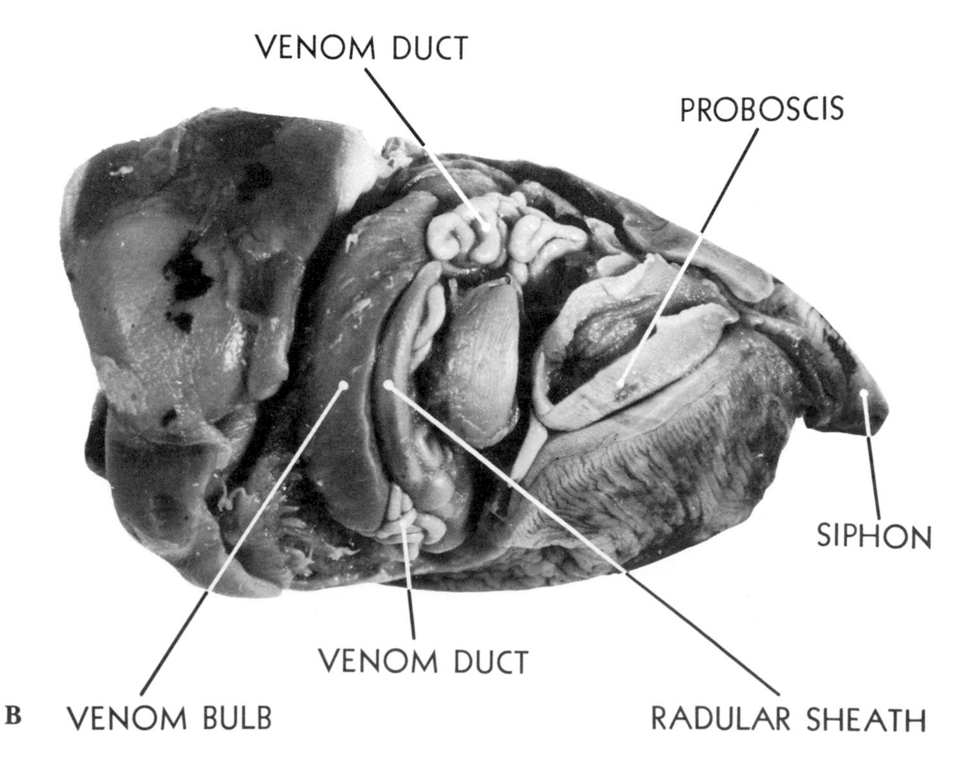

VI: PLATE 10

FIGURE a.—Dissection of *Conus geographus* to expose the relative anatomical position of the venom apparatus in the body of the mollusk. Note that a radular tooth has been released and is no longer held by the proboscis, but lies in the buccal cavity. × 1.2.

FIGURE b.—Venom apparatus of *Conus textile* dissected out from body of the mollusk. × 1.6. (R. Kreuzinger, dissection by B. W. Halstead)

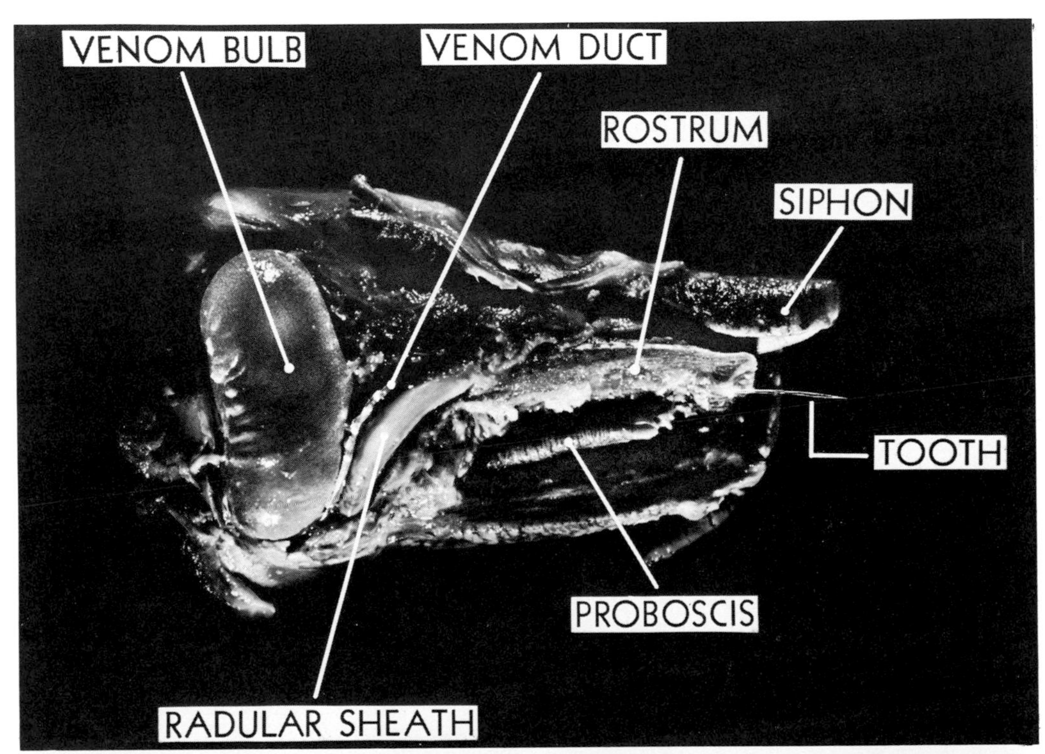

A

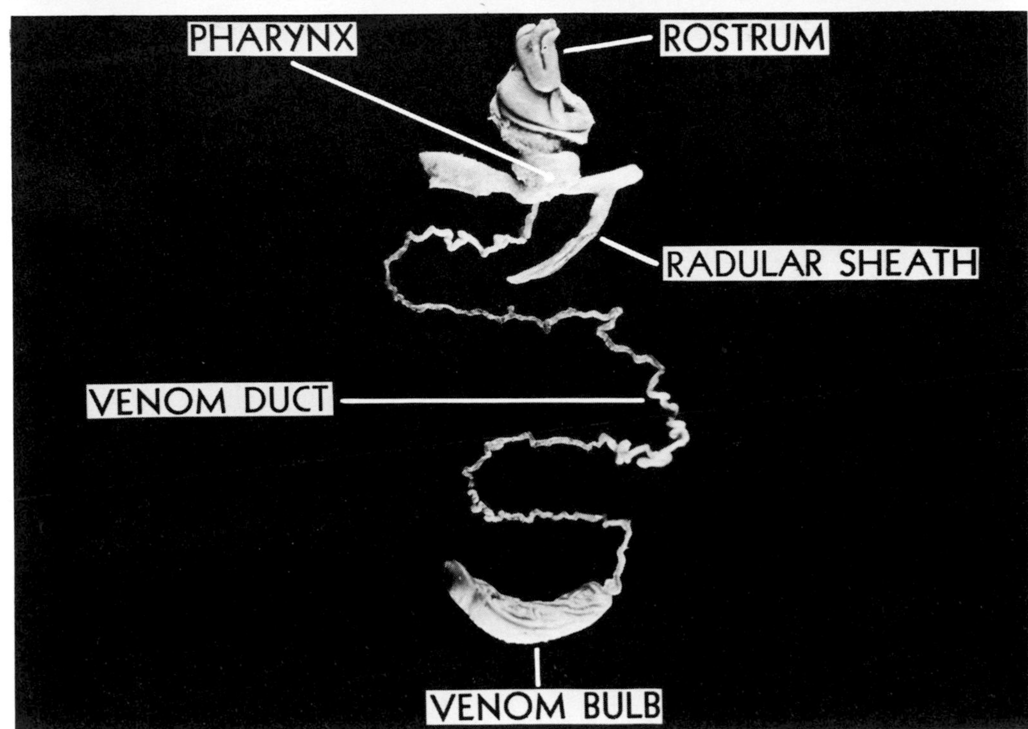

B

VI: PLATE 11

FIGURE a.—Radular tooth of *Conus striatus*, anterior tip, showing elaborate structural design. × 48.

FIGURE b.—Radular tooth of *Conus striatus*. Entire tooth with tooth ligament attached. × 8.

FIGURE c.—Radular tooth of *Conus textile*. Tooth ligament is not shown. × 15.

(R. Kreuzinger, preparations by B. W. Halstead)

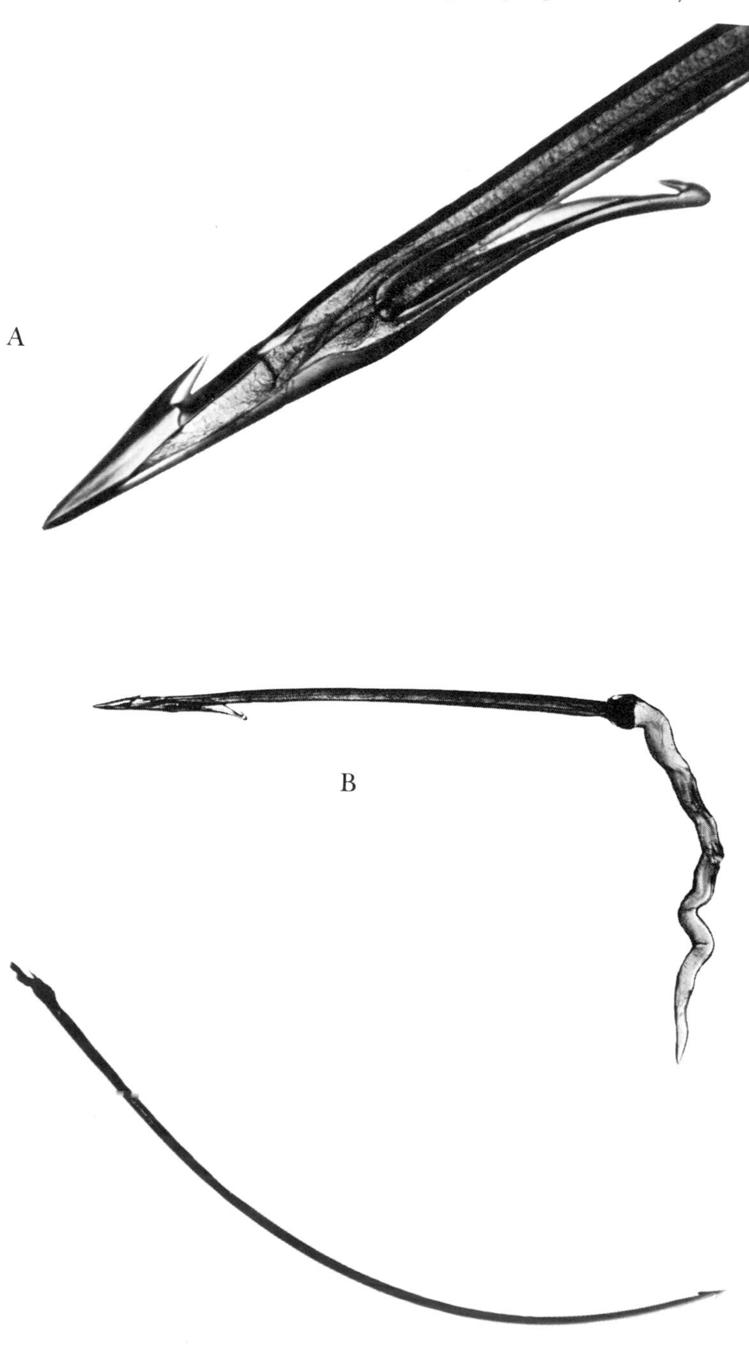

VI: PLATE 12

FIGURE a.—Photomicrograph of cross section of venom bulb of *Conus californicus*. Note the two thick layers of smooth muscle separated by connective tissue septum and very thin epithelial lining. There is no evidence of venom production in this organ. Consequently the venom bulb is believed to serve largely as a pressure regulating organ aiding in charging the hollow radular teeth with venom and facilitating propulsion of the teeth into the pharynx and proboscis. × 30.

FIGURE b.—Enlargement of previous section of venom bulb of *C. californicus* showing the epithelial lining and adjacent smooth muscle. × 220.

FIGURE c.—Photomicrograph of cross section of venom bulb of *Conus textile*. Note that it follows the same morphological pattern as in *Conus californicus*. This pattern has been found to be quite consistent in all the species of *Conus* examined thus far. × 21.

FIGURE d.—Enlargement of previous section of venom bulb of *Conus textile*. × 80.

(R. Kreuzinger, preparations by B. W. Halstead)

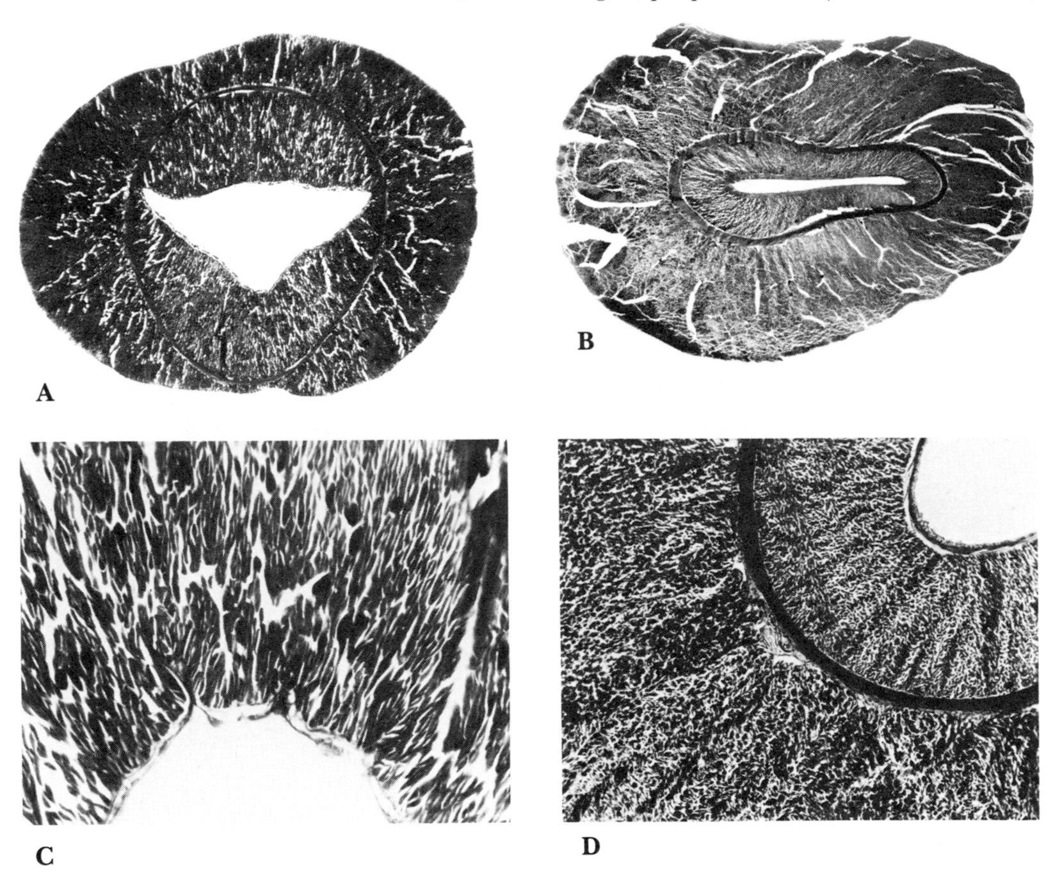

VI: PLATE 13

FIGURE a.—Photomicrograph of cross section of venom duct of *Conus californicus*. Note glandular epithelium with columns of spherical bodies, and amorphous secretion in the lumen. This morphological pattern has been found to be quite consistent in all the species of *Conus* examined thus far. × 105.

FIGURE b.—Enlargement of previous section of venom duct of *Conus californicus*. × 330. (R. Kreuzinger., preparations by B. W. Halstead)

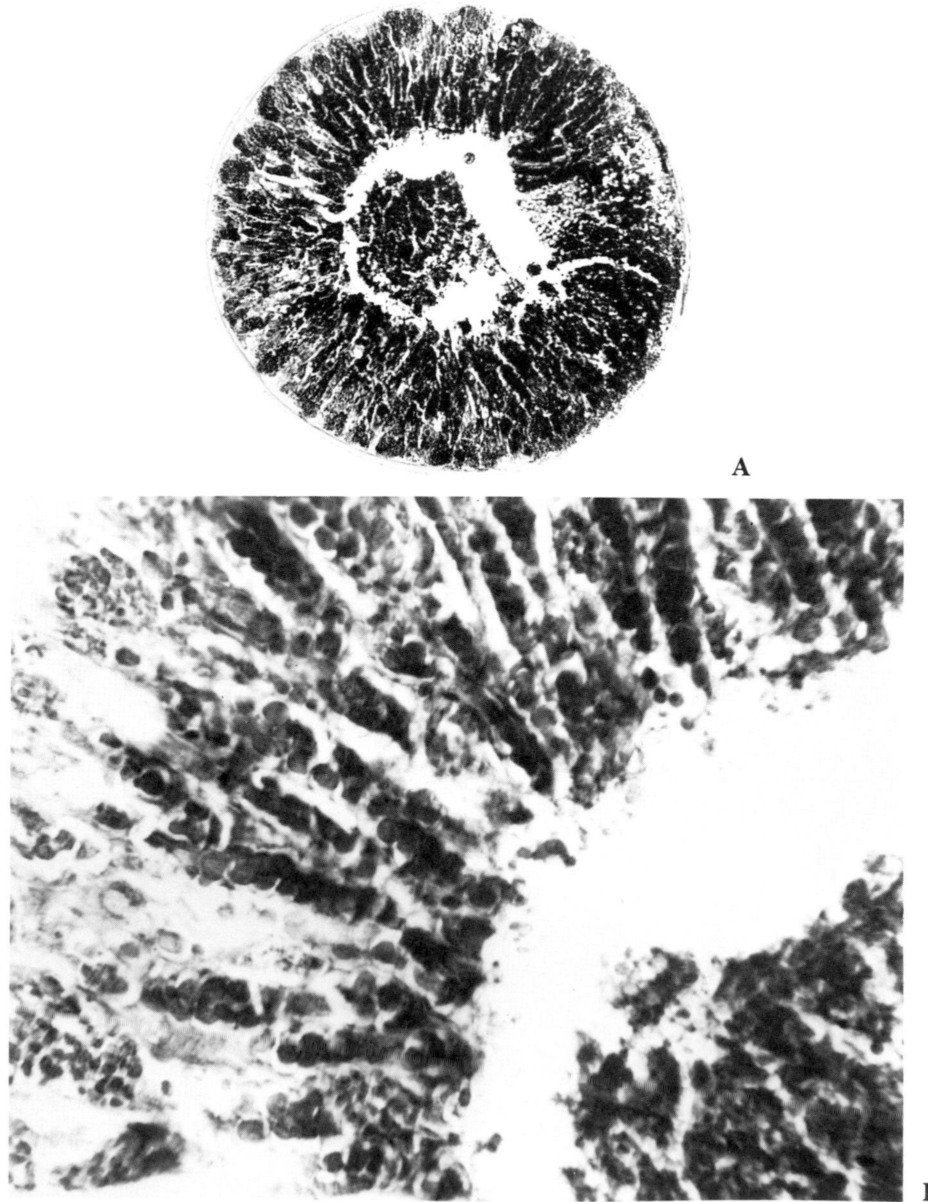

VI: PLATE 14

FIGURE a.—Photomicrograph of cross section through proximal third of long arm of radular sheath of *Conus californicus*. Spherical bodies are cross sections of radular teeth. × 82.5.

FIGURE b.—Photomicrograph of cross section through middle third of long arm of radular sheath of *Conus californicus*. The convoluted structure of the teeth is quite evident in this section. × 82.5.

FIGURE c.—Cross section through distal third of long arm of radular sheath of *Conus californicus*. Note longitudinal fold and cross sections of the radular teeth. × 73.

FIGURE d.—Cross section through middle third of long arm of radular sheath of *Conus textile*. × 33. (R. Kreuzinger, preparations by B. W. Halstead)

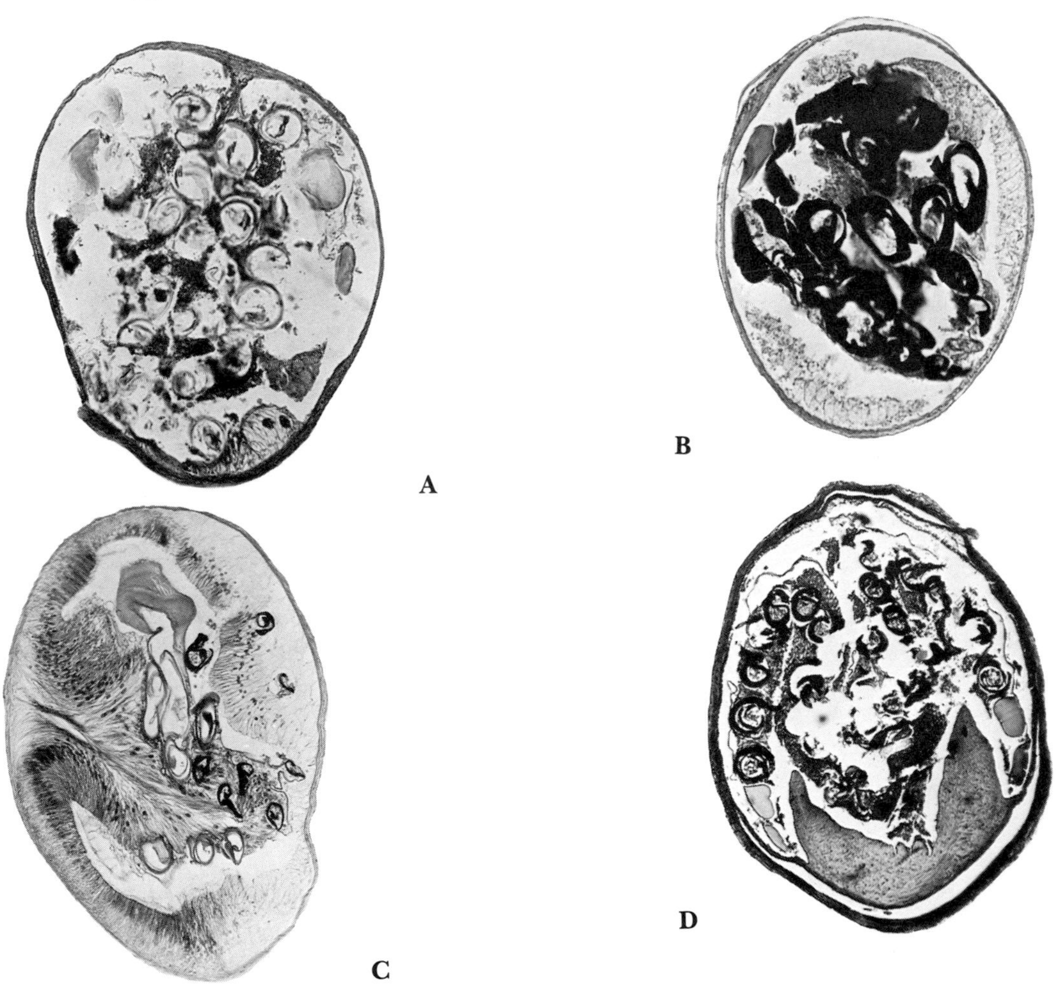

(Below) Radular sheath showing various levels from which cross sections in Figures a-d were taken.

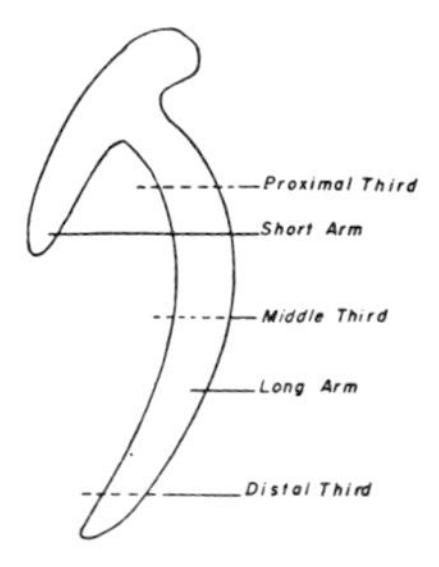

VI: PLATE 15

FIGURE a.—Photomicrograph of longitudinal section of the pharynx of *Conus californicus*. Note arrow pointing to entrance of venom duct posterior to pharyngeal muscle. Base of proboscis is seen to the right of pharyngeal muscle bundle. × 22.

FIGURE b.—Longitudinal section of pharynx of *Conus californicus* with arrow pointing to entrance of radular sheath anterior to pharyngeal muscle through which the radular teeth are passed. Opening of venom duct is not shown on this section. Base of proboscis appears to the right of the pharyngeal muscle mass. × 28.

FIGURE c.—Ciliated pseudostratified epithelium of pharynx of *Conus californicus*. The cilia serve to propel the radular teeth through the lumen of the pharynx and into the proboscis. × 300.

FIGURE d.—Longitudinal section of proboscis of *Conus californicus*. The radular tooth is finally released to the tip of the proboscis where it is firmly grasped ready to plunge into the victim. Pharyngeal muscles do not appear in this section. × 15.

(R. Kreuzinger, preparations by B. W. Halstead)

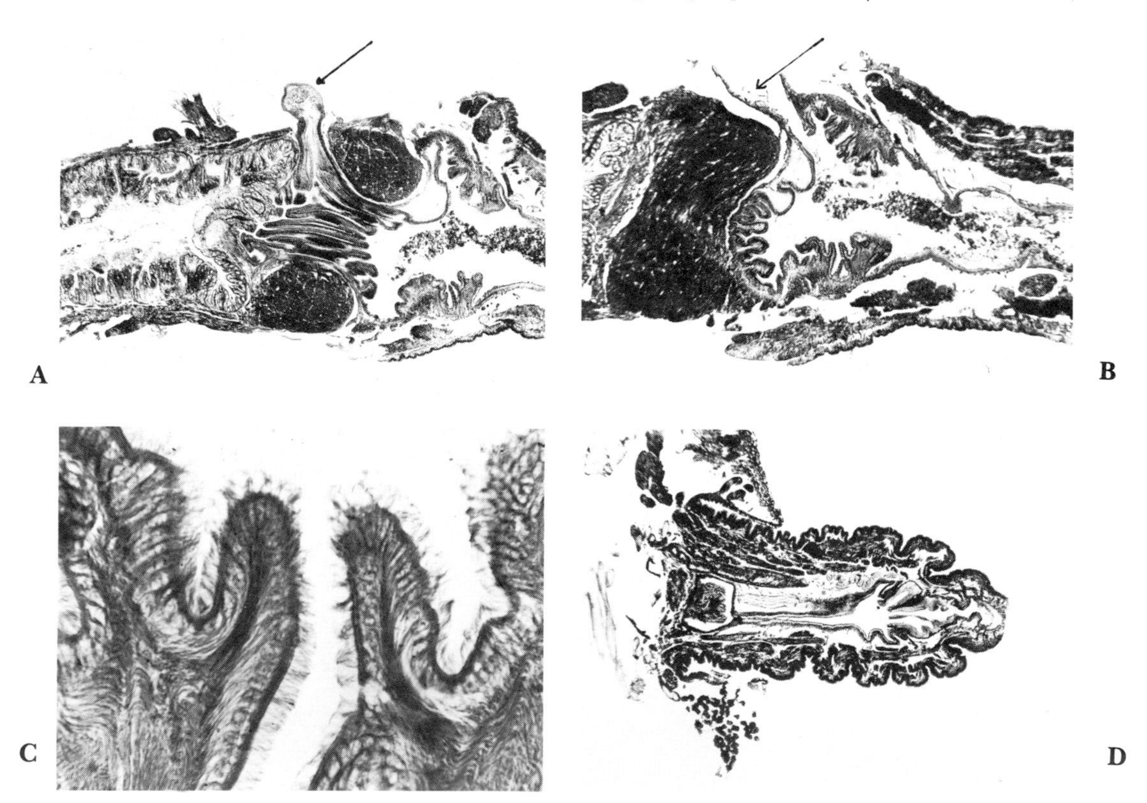

VI: PLATE 16

Semidiagrammatic drawing of typical tooth of *Conus*. The radular teeth of cone shells are comprised of a flat sheet of chitin rolled into a hollow tubelike structure. The venom is apparently injected into the base and along the open shaft of the tooth under pressure from the venom bulb and duct. Thus when the radular tooth is released it is fully charged with venom. Left: tip of tooth. Right: base of tooth with tooth ligament attached. Greatly enlarged. (R. Kreuzinger)

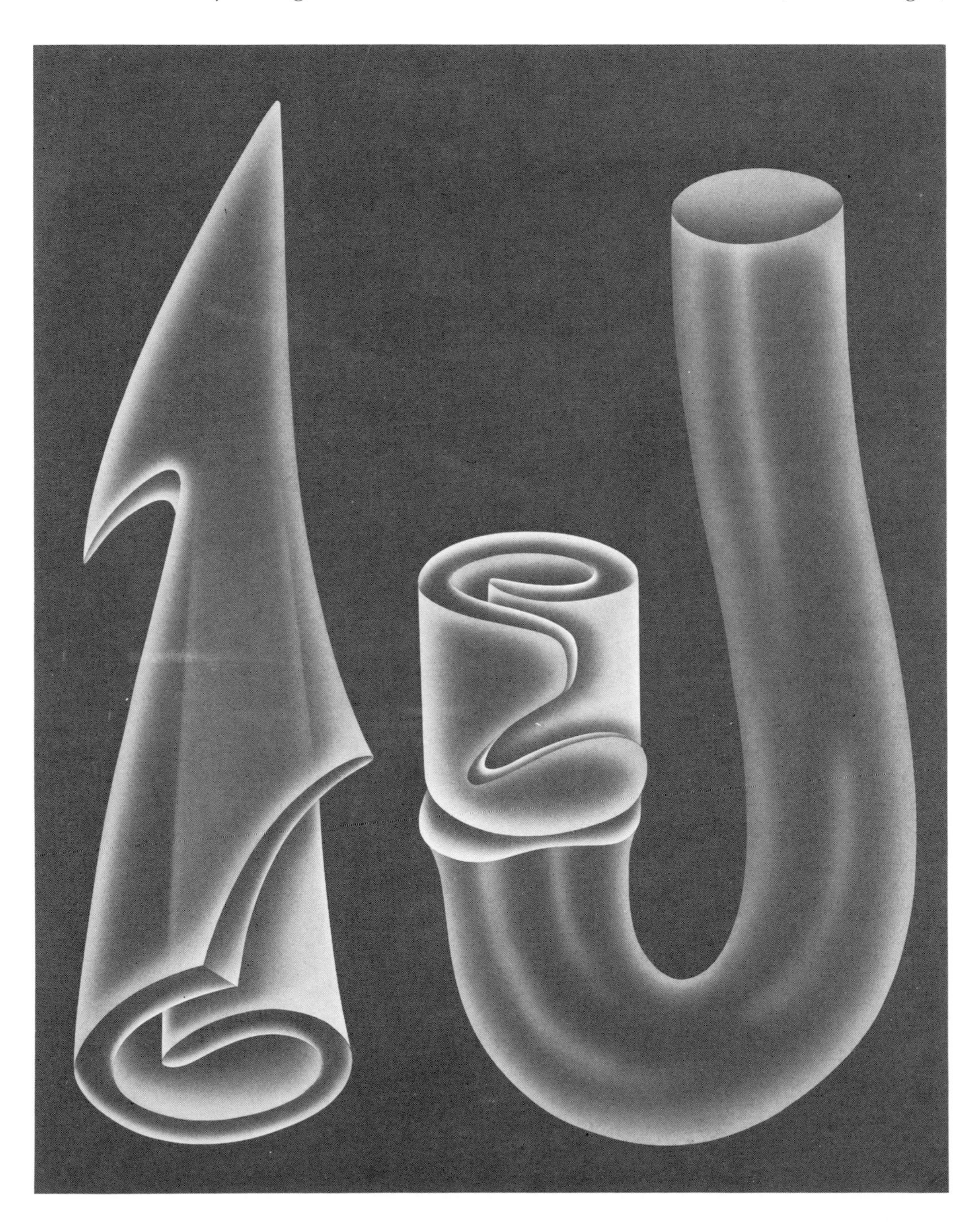

VI: PLATE 17

Haliotis discus Reeve. This is one of two species of Japanese abalones which is known to cause abalone viscera poisoning in humans. The disease is caused by ingesting the viscera of the abalone. Symptoms consist of burning and stinging sensations accompanied by a severe urticariallike reaction, erythema, itching, edema, and ulceration of the skin. Skin lesions are limited to those portions of the body exposed to sunlight. A second Japanese abalone *H. sieboldi* Reeve, which is not illustrated, has also caused this form of intoxication. Length 15 cm.

(N. Hastings, courtesy of S. S. Berry)

VI: PLATE 18

FIGURE a.—*Murex brandaris* Linnaeus. Length 7.5 cm.

FIGURE b.—*Murex trunculus* Linnaeus. The two European species of *Murex* illustrated on this plate produce a toxic substance in their hypobranchial gland known as murexine. The poison is believed to affect humans only if administered parenterally. Both *M. brandaris* and *M. trunculus* are eaten in certain parts of Europe. However, *M. brandaris* is said to be poisonous to eat at certain times, and has reputedly caused death. Specimens photographed at the Zoological Station at Naples, Italy. Length 5 cm.

A

B

VI: PLATE 19

FIGURE a.—*Thais lapillus* Linnaeus. A toxic substance similar to, or identical with, murexine has been extracted from the purple gland of this mollusk. The poison has been termed "purpurine." Its effects on humans are unknown. Length 5 cm.

FIGURE b.—*Livona pica* Linnaeus. This mollusk has a long history of being toxic to eat at certain unpredictable times. No laboratory studies have been reported on its toxicity, and it is a popular seafood in the West Indies. This shell is an algal feeder and it is possible that under some circumstances it may pick up toxic substances in its diet. Length 10 cm.

FIGURE c.—*Dosinia japonica* (Reeve). This is one of three species of Japanese bivalves known to contain venerupin poison. This particular species has not been reported in human intoxications, but should be considered a potentially dangerous species to eat if taken in the Schizuoka or Kanagawa Prefectures of Japan during a period when venerupin poisoning is known to occur. Length 7 cm.

FIGURE d.—*Tapes semidecussata* Reeve. One of the causative agents of venerupin poisoning in Japan. Symptoms consist of severe gastrointestinal upset, anemia, halitosis, jaundice, and other evidences of liver damage. The intoxication may be fatal. Length 5 cm. (R. Kreuzinger, courtesy of S. S. Berry)

VI: PLATE 20

FIGURE a.—*Aplysia californica* Cooper. A toxic substance termed aplysin has been isolated from the digestive gland of this sea hare (nudibranch). The poison acts as a cholinergic agent affecting blood pressure, respiration, and the sympathetic nervous system of injected laboratory animals. Effects on humans are unknown. Length 38 cm. Living specimen photographed in an aquarium at Oakland, Calif. (R. Ames)

FIGURE b.—*Aplysia depilans* Linnaeus, living specimen. Photographed at the Zoological Station at Naples, Italy. Length 25 cm.

FIGURE c.—*Aplysia punctata* Cuvier. Living specimen taken at the Zoological Station at Naples. This species is said to secrete an irritant capable of producing urticaria, inflammation, and alopecia. Attains a length of 20 cm. (Courtesy of P. Giacomelli)

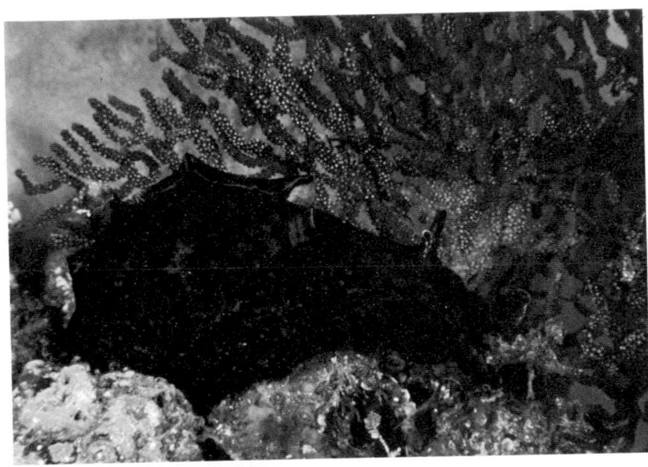

A

B

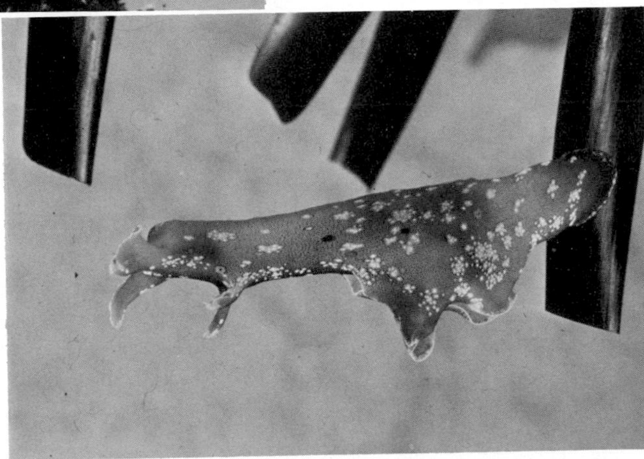

C

VI: PLATE 21

Creseis acicula Rang. Little is known regarding the toxicity of pteropods. Certain arctic mammals reputedly derive their source of poison from the ingestion of toxic pteropods, but this has not been demonstrated for a certainty. This particular species has caused "stingings" in Florida, resulting in a maculopapular skin rash. The mechanism by which the rash is produced is not known. Length 33 mm.

(H. Baerg, after Rang)

VI: PLATE 22

Callista brevisiphonata Carpenter. The causative agent of callistin shellfish poisoning in Japan. Ingestion of these shellfish produce a violent urticarial reaction, accompanied by a severe gastrointestinal upset, paralysis and numbness of the throat, mouth, and tongue. No deaths have been reported. The poison is said to be concentrated in the ovaries of the shellfish. Length 8 cm. (Courtesy of Smithsonian Institution)

VI: PLATE 23

Crassostrea gigas (Thunberg). This is the only oyster that has been incriminated in venerupin poisoning in Japan. Length to 25 cm or more.
(R. Kreuzinger, courtesy of S. S. Berry)

VI: PLATE 24

Tridacna gigas (Linnaeus). Length of valves up to 137 cm. (D. Ludwig)

VI: PLATE 25

Sepia officinalis Linnaeus. (From Jatta)

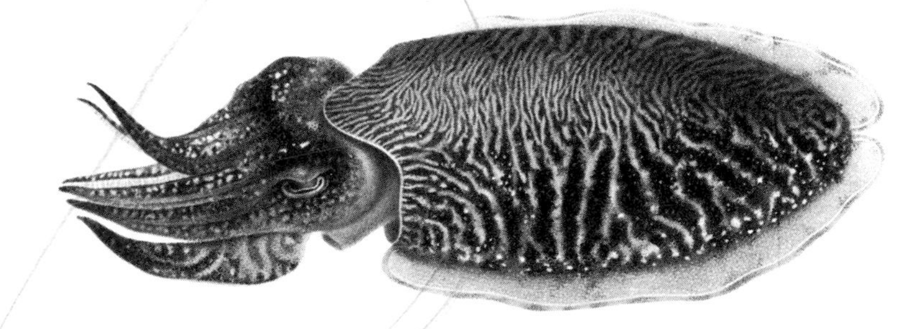

VI: PLATE 26

Ommastrephes sloani pacificus Steenstrup. This common Japanese squid has been incriminated in hundreds of oral intoxications involving several thousands of persons in the Niigata, Yamagata, and Nagano Prefectures of Japan. The symptoms were similar to those caused by *Octopus dofleini*. Bacteriological tests were negative for all the ordinary types of food pathogens. The nature of the poison is unknown. Length about 30 cm. Drawing made from actual specimen in Japan. (S. Arita)

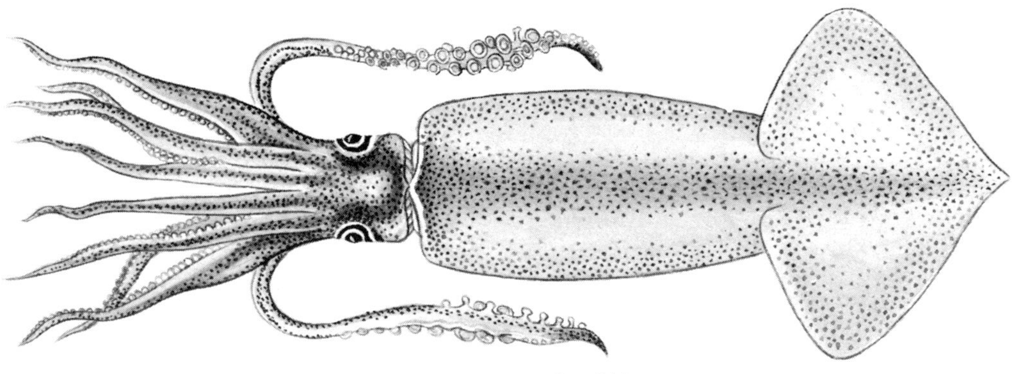

VI: PLATE 27

Eledone moschata (Lamarck). An active endecapeptide fraction of cephalotoxin, known as eledoisin, has been isolated from the salivary secretions of this octopus. Eledoisin is said to have powerful vasodilator and hypotensive effects. The effects of the poison on humans are unknown. Length 40 cm. Taken in an aquarium at the Zoological Station at Naples, Italy. (Courtesy of P. Giacomelli)

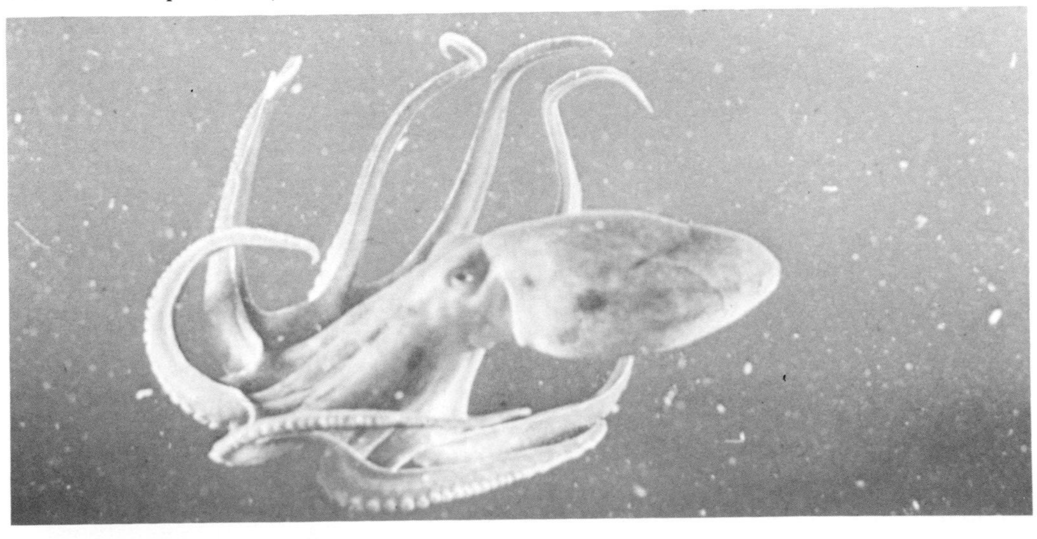

VI: PLATE 28

FIGURE a.—*Octopus* cf. *apollyon* Berry taken in the Vancouver Public Aquarium. The color and pattern are subject to considerable fluctuation in life, depending to a large extent upon the background in which the octopus is found. May attain a span of more than 9 m. (D. Hanna, courtesy of Vancouver Public Aquarium)

FIGURE b.—Octopus bites believed to be inflicted by *O. apollyon*. The bites were encountered while removing the octopus from an aquarium at the Marine Science Center, Oregon State University, Newport Oregon. (Courtesy of Ken Hilderbrand, Oregon State University)

A

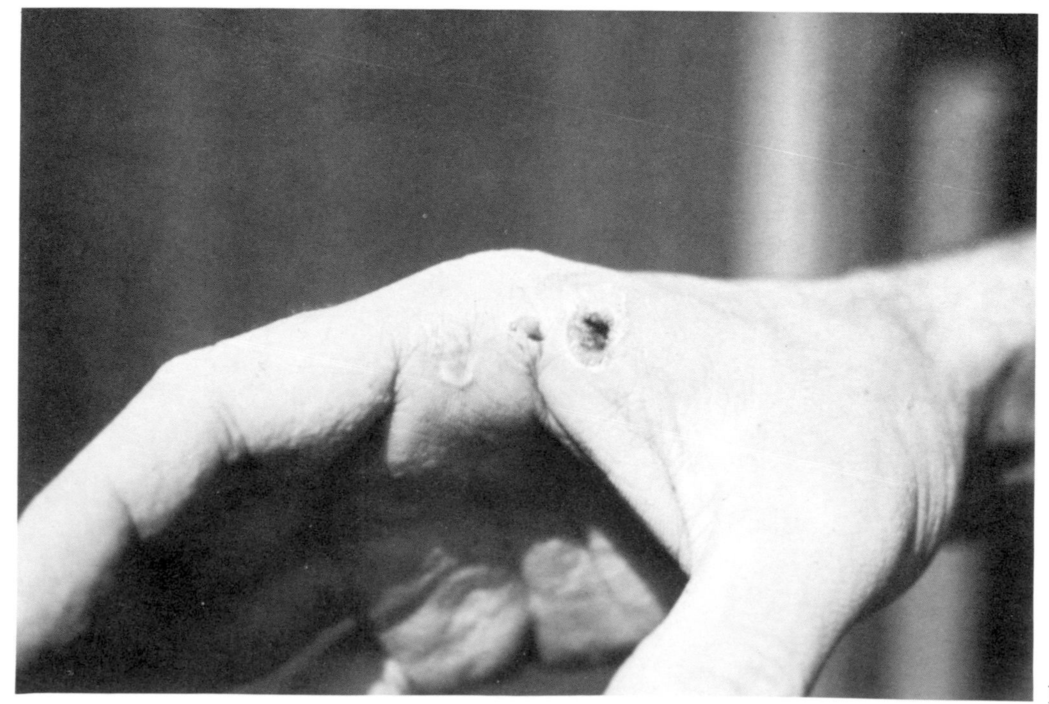

B

VI: PLATE 29

Octopus maculosus Hoyle, living specimen, taken at Sydney, Australia. This is the only species of octopus which has been incriminated in a human fatality due to a bite. The specimen which caused the death of an Australian skindiver was small, about 20 cm in total spread. The victim had placed the octopus on his shoulders and was bitten on the upper back near the spinal column. Symptoms consisted of dryness of the mouth, difficulty in swallowing, vomiting, ataxia, bleeding from the wound, weakness, respiratory distress, loss of consciousness, and death within a period of 2 hours after having been given artificial respiration and treated in an iron lung. The episode occurred at East Point, about 5 km from Darwin, Northern Territory, Australia.
(K. Gillett)

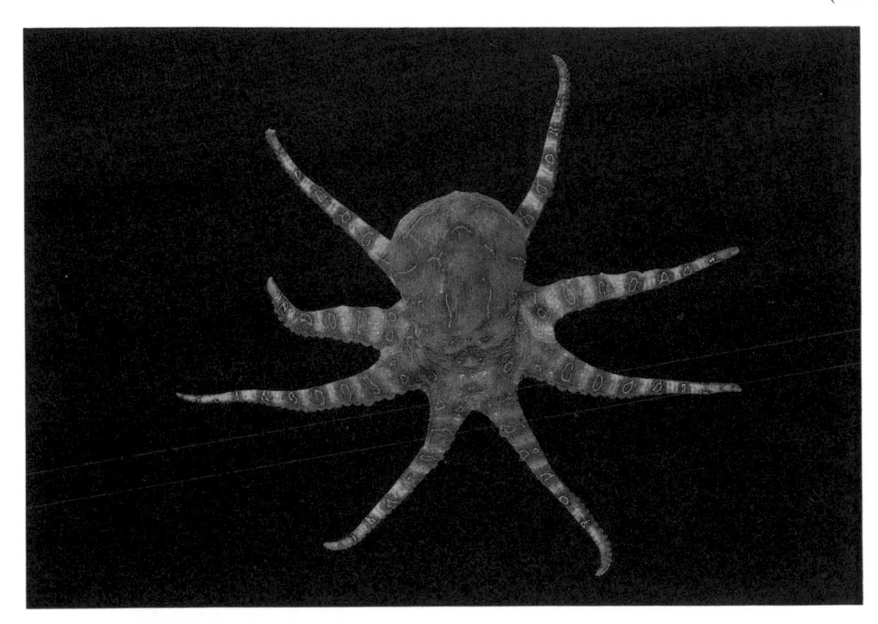

Closeup of a tentacle.

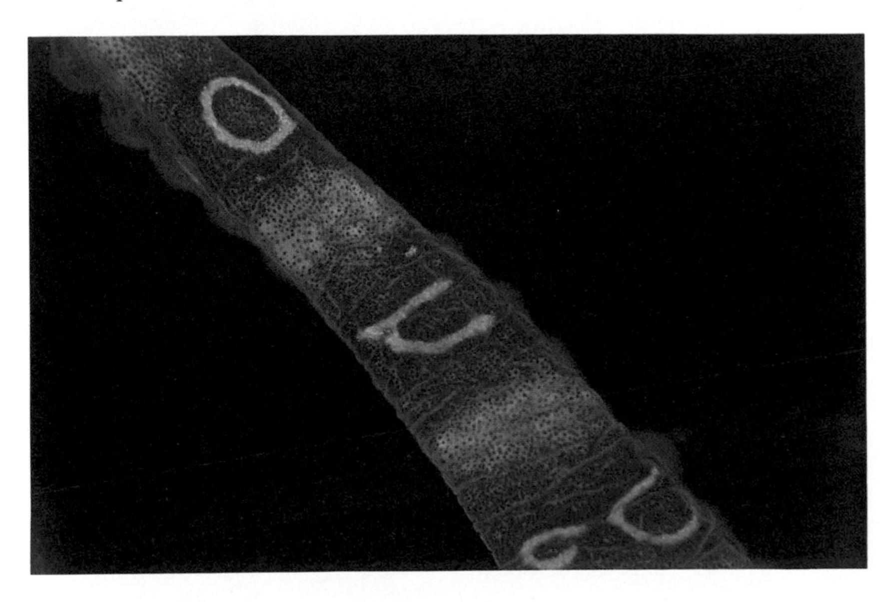

VI: PLATE 30

FIGURE a.—*Octopus macropus* Risso, living specimen, taken in an aquarium at the Zoological Station at Naples, Italy. (Courtesy of P. Giacomelli)

FIGURE b.—*Octopus vulgaris* Lamarck. This species has been used extensively in Europe in research on cephalotoxin. It has not been reported to have bitten humans. This same species also occurs in Japan where it has produced oral intoxications similar to those caused by *O. dofleini*. Photograph of living specimen taken at the aquarium of the Zoological Station, Naples, Italy. Length about 80 cm.

FIGURE c.—Same specimen as above showing the undersurface of the web, with buccal mass and jaws at center. (Courtesy of P. Giacomelli)

A

B

C

Phyla Platyhelminthes, etc.

PLATES

CHAPTER VII

VII: PLATE 1

FIGURE a.—*Leptoplana tremellaris* Oersted. Length 22 mm.

FIGURE b.—*Stylochus neapolitanus* (Delle Chiaje). Length 25 mm.

FIGURE c.—*Thysanozoon brocchi* Grube. Length 60 mm.

The European turbellarians appearing on this plate secrete a toxic substance believed to be produced by their rhabdoids and certain poison glands. Aside from the general toxic effects of the poison on laboratory animals, little is known about the nature of the poison or its effects on humans. (H. Baerg, after Lang)

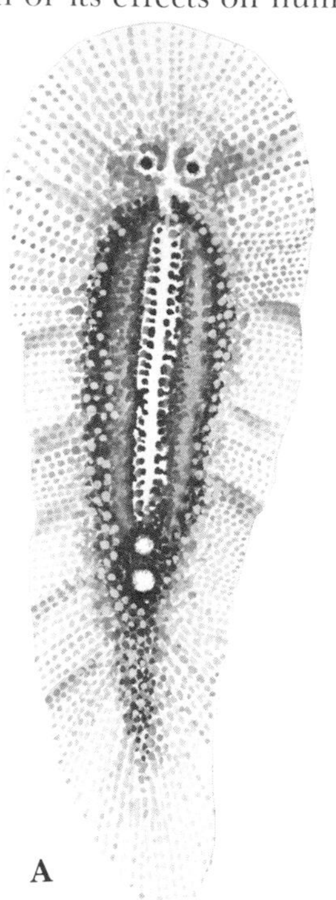

A

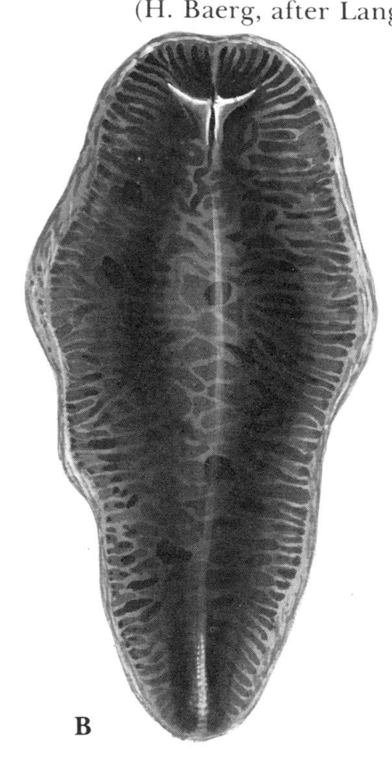

B

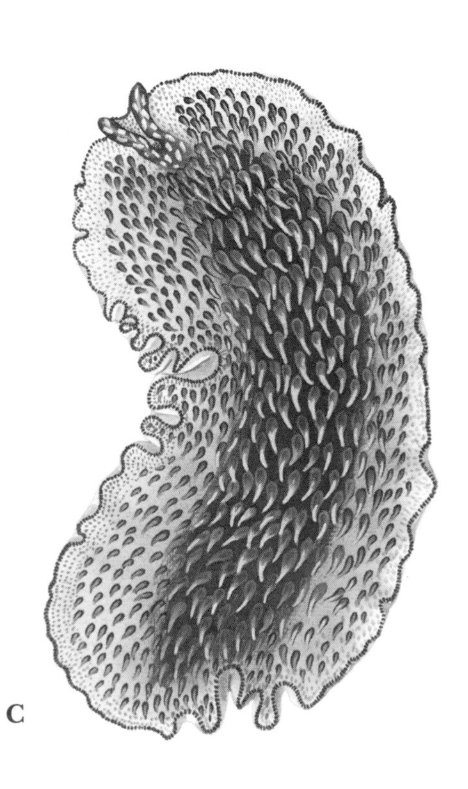

C

VII: PLATE 2

FIGURE a.—*Bdelloura candida* (Girard). Length 15 mm.

FIGURE b.—*Procerodes lobata* (Schmidt). The appearance of this triclad worm is variable as shown in the drawing. Length 7 mm. Extracts prepared from the tissues of these European triclad worms are toxic to laboratory animals, but the effects of the poison on humans are not known. (From Wilhelmi)

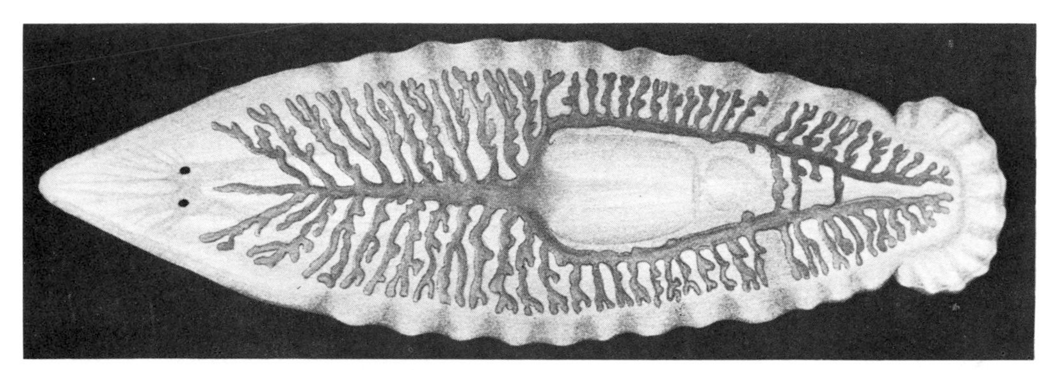

A

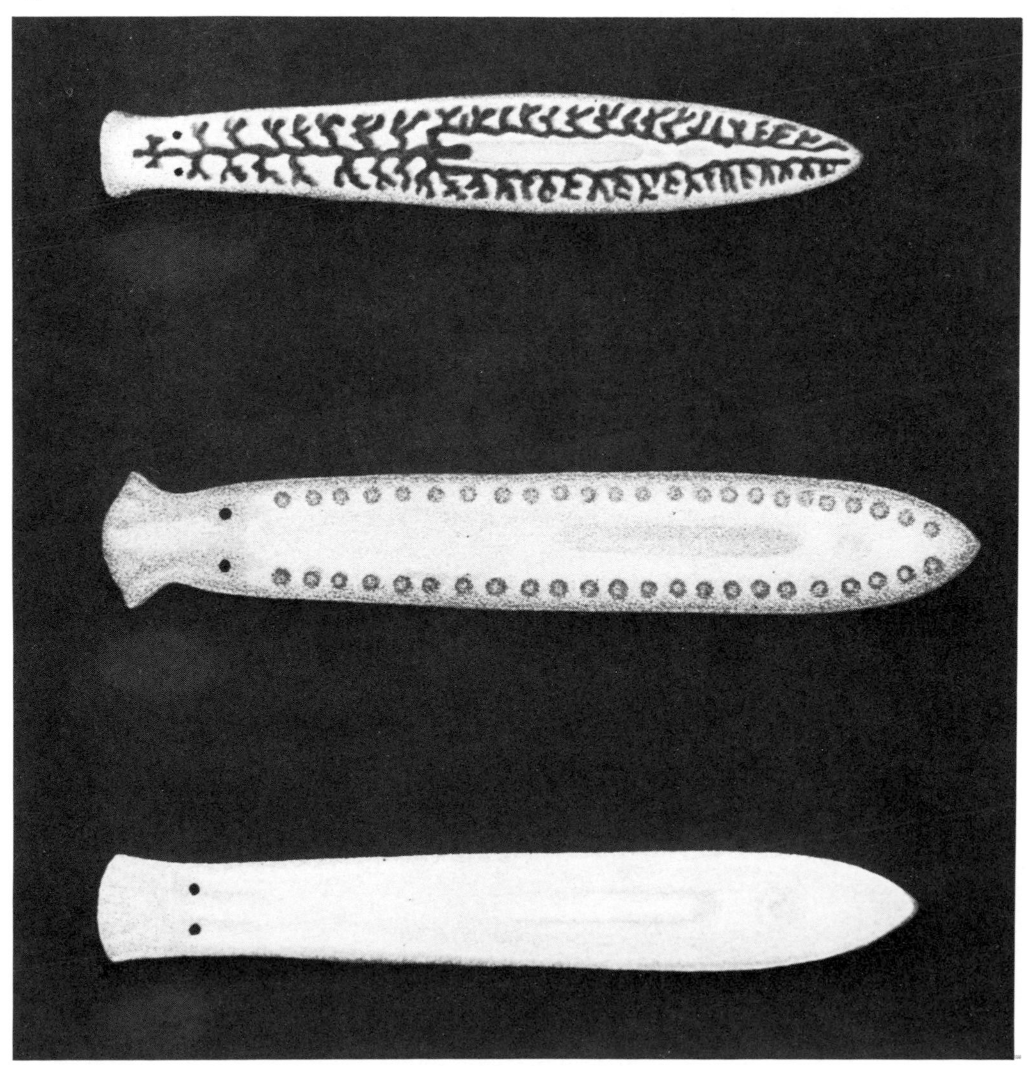

B

VII: PLATE 3

FIGURE a.—*Amphiporus lactifloreus* (Johnston). Length 10 cm. (After McIntosh)

FIGURE b.—*Drepanophorus crassus* (Quatrefages). Length 14 cm. Two toxic substances have been extracted from the tissues of these nemertean worms, "amphiporine," and a less toxic material, "nemertine." These poisons are known only by their effects on laboratory animals. (H. Baerg, after Lang)

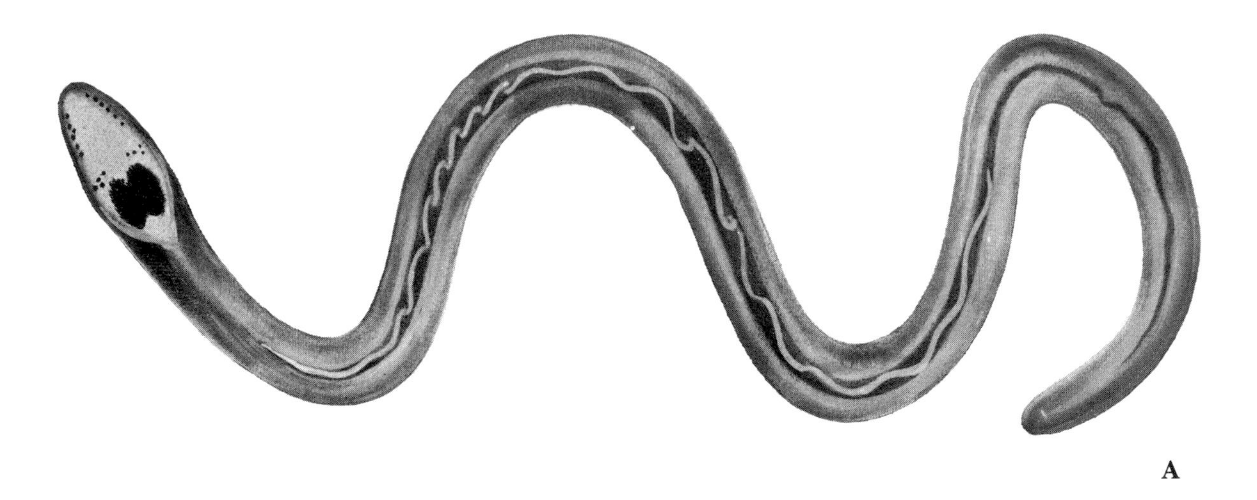

A

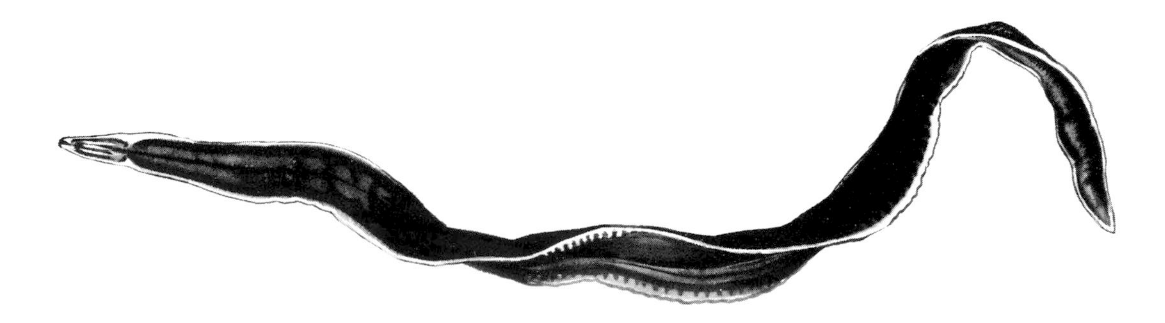

B

VII: PLATE 4

Structure of a nemertean worm in relationship to its venom apparatus.

FIGURE a.—Phantom drawing showing the stylet or venom apparatus and internal anatomy of a hoplonemertine worm. Note worm with proboscis and stylet apparatus in the extended position (upper right). (R. Kreuzinger, modified from Barnes)

FIGURE b.—Stereogram showing the general anatomy of a typical hoplonemertine worm with particular reference to the relationship of the various body cavities and the proboscis to which is attached the stylet apparatus. (R. Kreuzinger, modified from Hyman)

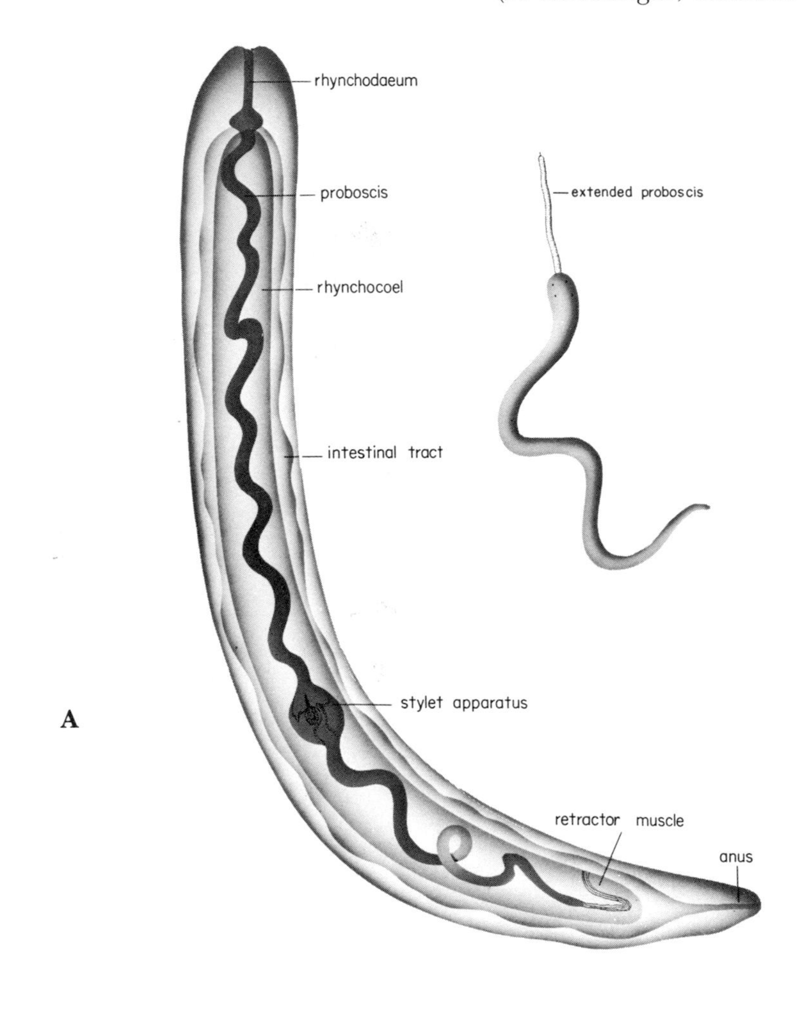

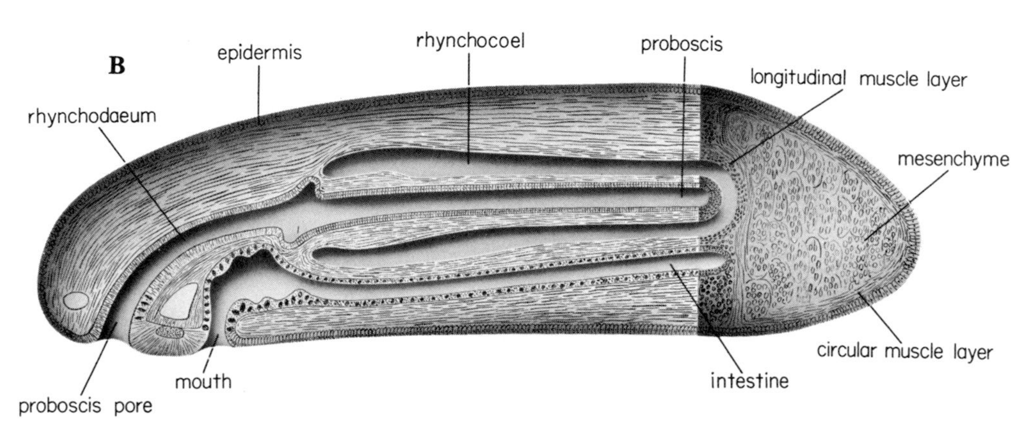

VII: PLATE 5

FIGURE a.—*Chloeia flava* (Pallas). Living specimen taken near Sydney, New South Wales, Australia. Length 80 mm. (K. Gillett)

FIGURE b.—*Chloeia viridis* Schmarda. Living specimen taken at the Miami Seaquarium. Length 70 mm. (Courtesy of A. C. Mueller, Miami Seaquarium)

These worms possess elongate pungent setae which project from the parapodia. The setae can readily penetrate gloves and may produce a severe stinging sensation. These worms should be handled with care.

A

B

VII: PLATE 6

Eurythoë complanata (Pallas). This West Indian annelid possesses hollow pungent parapoidal setae which are reported to be venomous. Stings produced by the setae of this worm may result in an intense inflammatory reaction of the skin, consisting of redness, swelling, burning sensation, numbness, and itching. The setae can pass readily through gloves so these worms should be handled with great care. Length 100 mm or more.

FIGURE a.—Preserved specimen of *Eurythoë complanata*. West Indies. Note that the setae are in the retracted position during which time it may be confused with any other innocuous polychaete worm.

FIGURE b.—Preserved specimen of *Eurythoë complanata* taken from same locality. Closeup showing the parapoidal setae in the extended position. Bristle worms are difficult to handle when the setae have been extended.

(Courtesy of Smithsonian Institution)

FIGURE c.—Living specimen of *Eurythoë complanata* taken near Sydney, New South Wales, Australia. Setae are retracted. During life, these worms may display a magnificent irridescence. (Photo courtesy of Justice F. Myers)

FIGURE d.—Living specimen of *Eurythoë complanata* taken near Sydney. Setae are extended.

FIGURE e.—Closeup showing the extended setae. (E. Pope)

VII: PLATE 7

FIGURE a.—*Hermodice carunculata* (Pallas). This West Indian bristle worm is one of the most dangerous of the group. It is said to be capable of inflicting a "paralyzing effect" with its hollow, venom-filled, pungent setae. Length 30 cm or more.

(From Mullin)

FIGURE b.—*Hermodice carunculata* living specimen taken on the reef. Locality unknown.

(World Life Research Institute)

A

B

VII: PLATE 8

Glycera dibranchiata Ehlers. The jaws are at the anterior end of the proboscis, which can be extended or retracted with explosive rapidity. Blood worms live in burrows in mud in the intertidal zone and below low-water areas. They are used extensively in the United States and Canada as bait worms. They can inflict a painful bite with their venomous jaws. Living specimen from Maine. Length up to 37 cm.

(Courtesy P. Saunders)

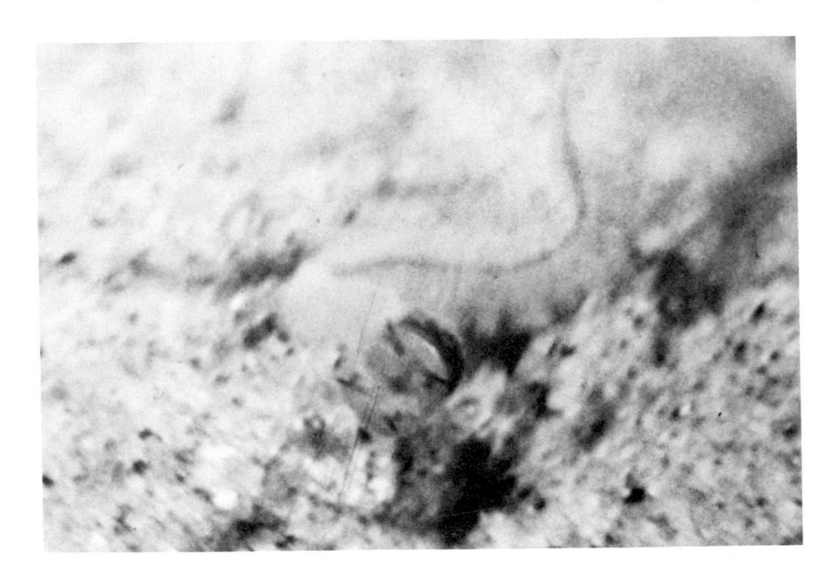

VII: PLATE 9

Lumbriconereis heteropoda Marenzeller. This worm is commonly used for fish bait in Japan. Nereistoxin, a neurotoxic substance, has been isolated from the tissues of this worm. The poison is toxic to warm-blooded animals, but its effects on man are not known. Note the flies which have died as a result of feeding on the dead bodies of these worms. Length 50 cm or more. (From Okaichi and Hashimoto)

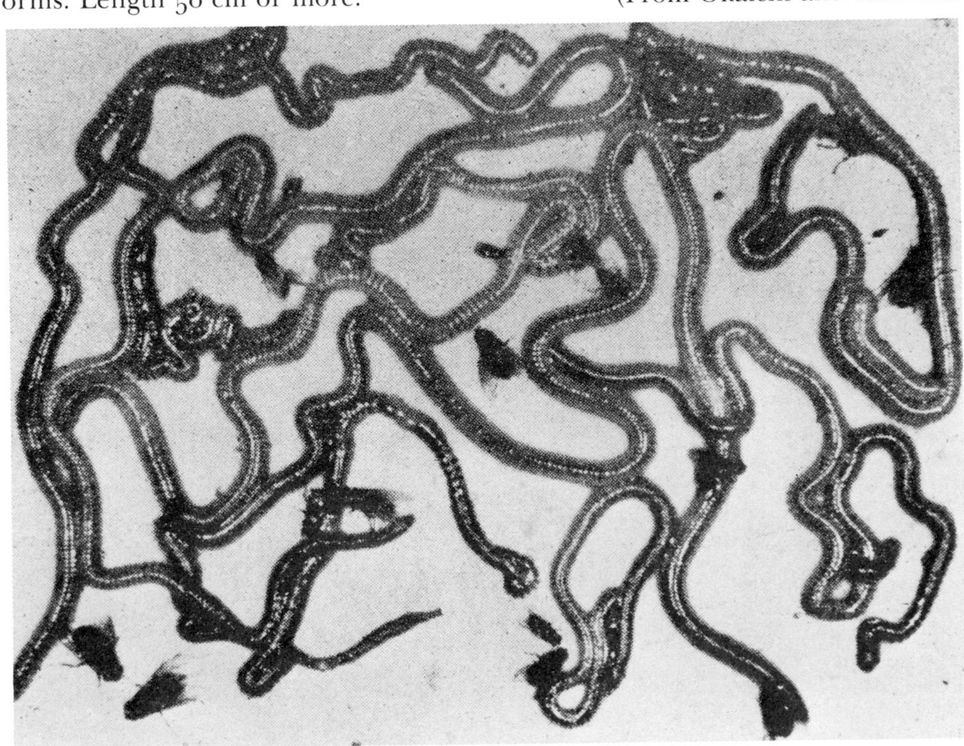

VII: PLATE 10

FIGURE a.—*Carcinoscorpius rotundicauda* (Latreille). The flesh, viscera, and eggs of this Asiatic horseshoe crab may be extremely toxic during the reproductive season of the year. Ingestion may result in dizziness, headache, gastrointestinal upset, weakness, paresthesias, paralysis, loss of consciousness, and death. There is no known antidote. Total length of adult 33 cm. 1A.—Adult male, lateral view, preserved specimen from Singapore. From the collection of the Osborn Zoological Laboratory, Yale University. 2A.—Dorsal view of above specimen; 3A.—Ventral view of above specimen.

FIGURE b.—*Tachypleus gigas* (Müller). The flesh, viscera, and eggs of this Asiatic horseshoe crab may be toxic during the reproductive season of the year. Ingestion can cause death. Total length of adult about 50 cm. 1B.—Adult female, lateral view, preserved specimen from Singapore. From the collection of the Osborn Zoological Laboratory, Yale University. 2B.—Dorsal view of above specimen. 3B.—Ventral view of above specimen. (Courtesy of T. H. Waterman)

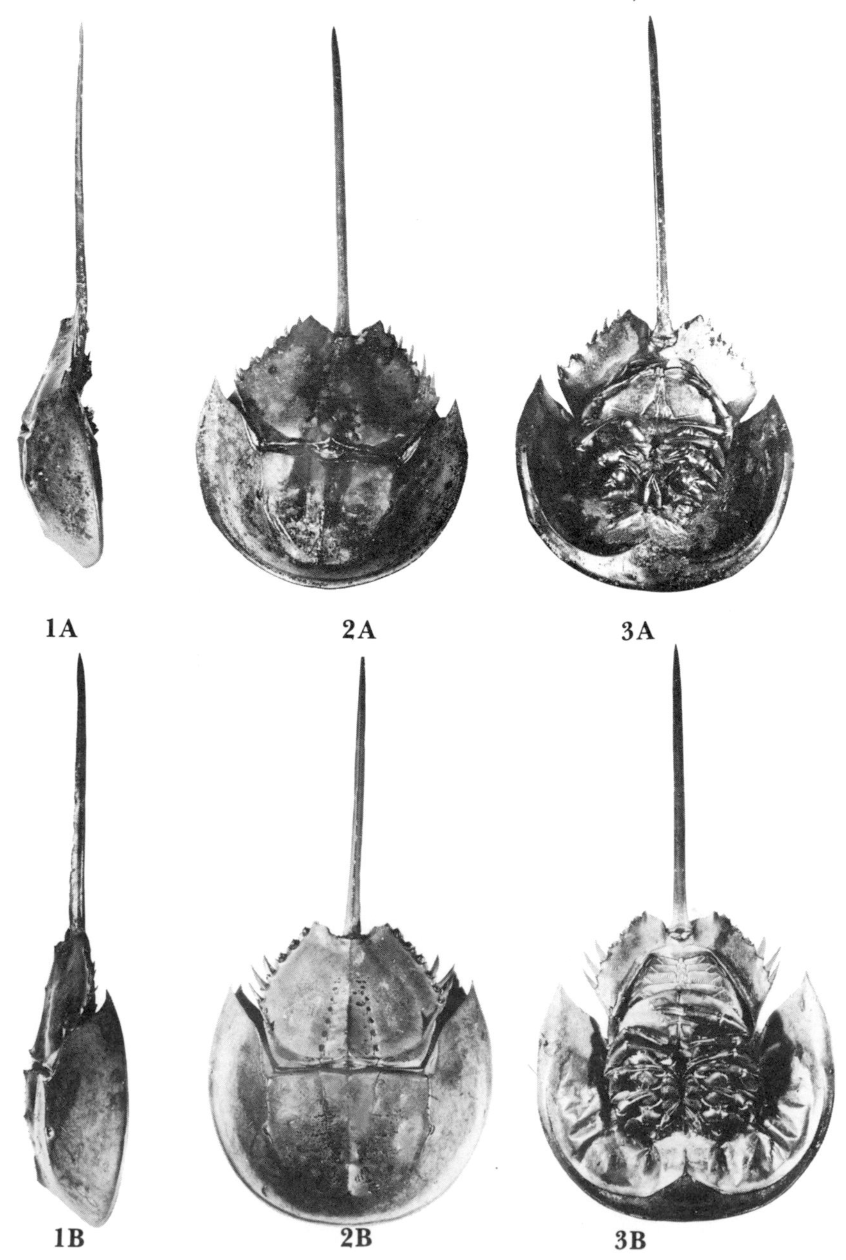

VII: PLATE 11

FIGURE a.—*Carpilius convexus* (Forskål). Width of carapace 60 mm.

FIGURE b.—*Carpilius maculatus* (Linnaeus). Width of carapace 140 mm. (From Guinot)

FIGURE c.—*Demania alcalai* Garth. Length of carapace 58.5 mm.

FIGURE d.—*Demania toxica* Garth. Length of carapace 43.1 mm. (From Garth)

FIGURE e.—*Lophozozymus pictor* (Fabricius). Width of carapace 90 mm. (From Teh and Gardiner)

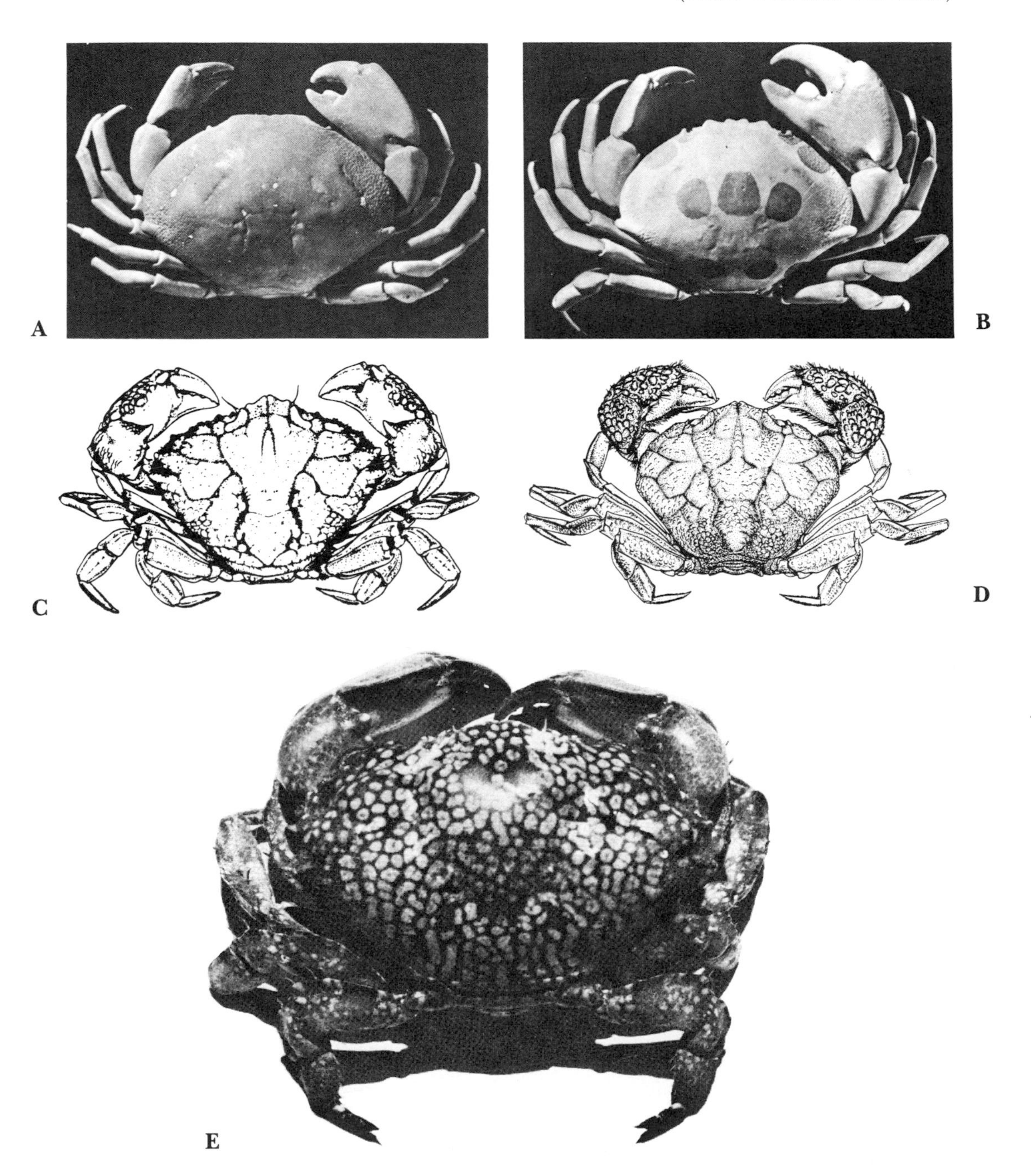

VII: PLATE 12

Crab parasitized by the barnacle, *Sacculina carcini* Thompson. The barnacle appears as a tumor mass located on the underside of the crab at the junction between the thorax and the abdomen. Rootlike appendages extend from the tumor mass throughout the body and appendages of the crab. A toxic substance is produced by the barnacle which inhibits the growth of the crab. (R. Kreuzinger, after Parker and Haswell)

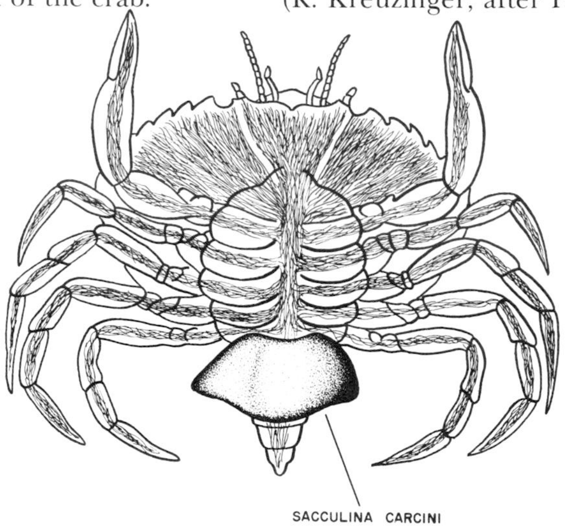

VII: PLATE 13

Birgus latro (Linnaeus). Length of carapace 11 cm. (M. Shirao)

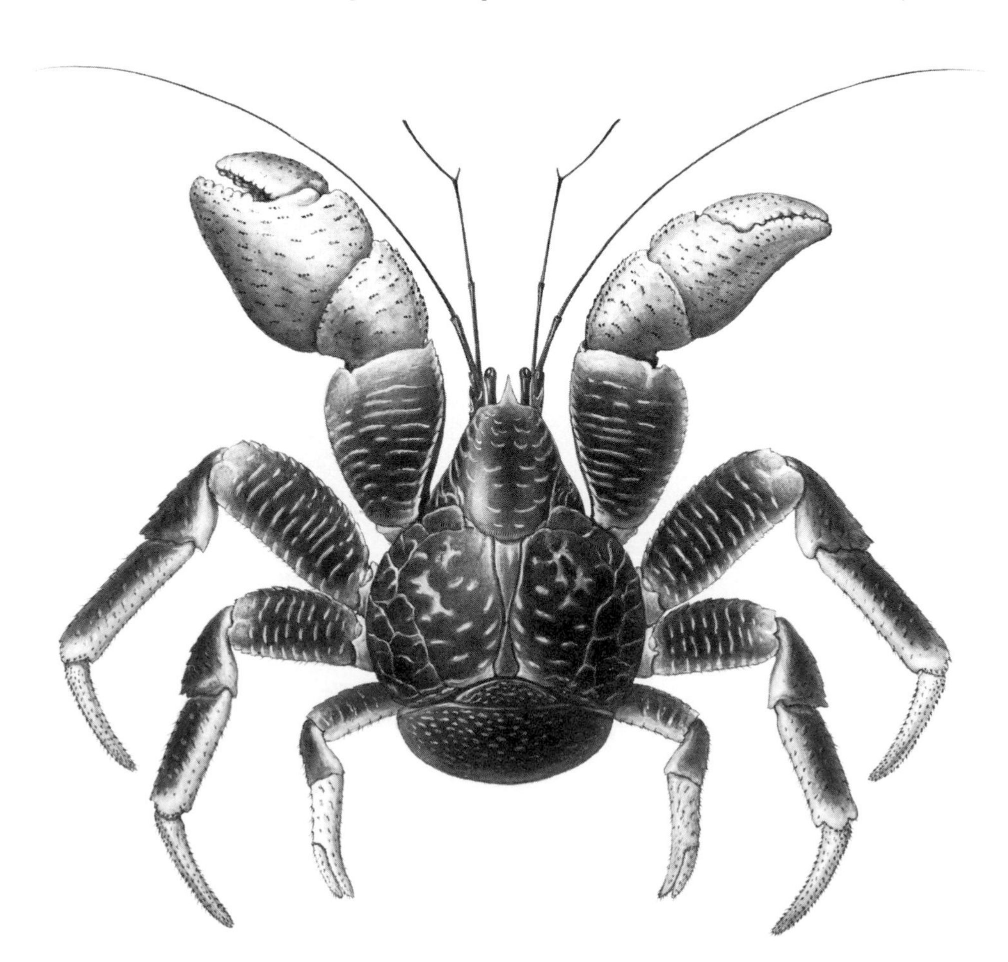

VII: PLATE 14

FIGURE a.—*Zozymus aeneus* (Linnaeus). Width of carapace 6 cm.

FIGURE b.—*Platypodia granulosa* (Rüppell). Width of carapace 3.7 cm.

FIGURE c.—*Atergatis floridus* (Linnaeus). Width of carapace 4.5 cm.

FIGURE d.—*Eriphia sebana* (Shaw and Nodder). Width of carapace 3.5 cm.

(M. Shirao)

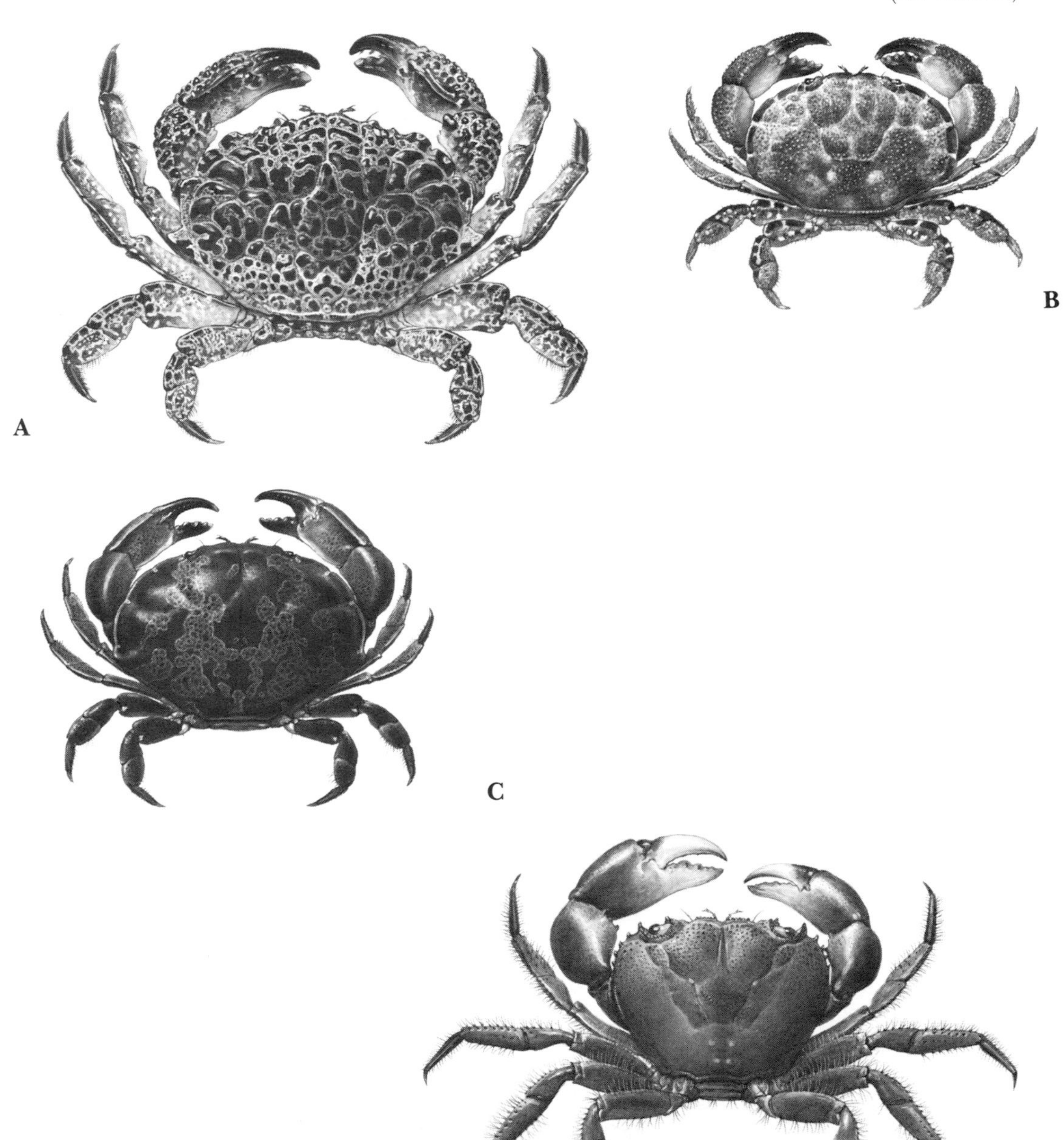

VII: PLATE 15

Bonellia viridis Rolando. Tissue extracts contain a substance that is toxic to a variety of laboratory animals. The effects of the poison on man are unknown. Length 10 cm or more. Photograph is of living specimen taken at the aquarium of the Zoological Station at Naples, Italy. (Courtesy of P. Giacomelli)

VII: PLATE 16

Alcyonidium gelatinosum (Linnaeus). (After Osburn)

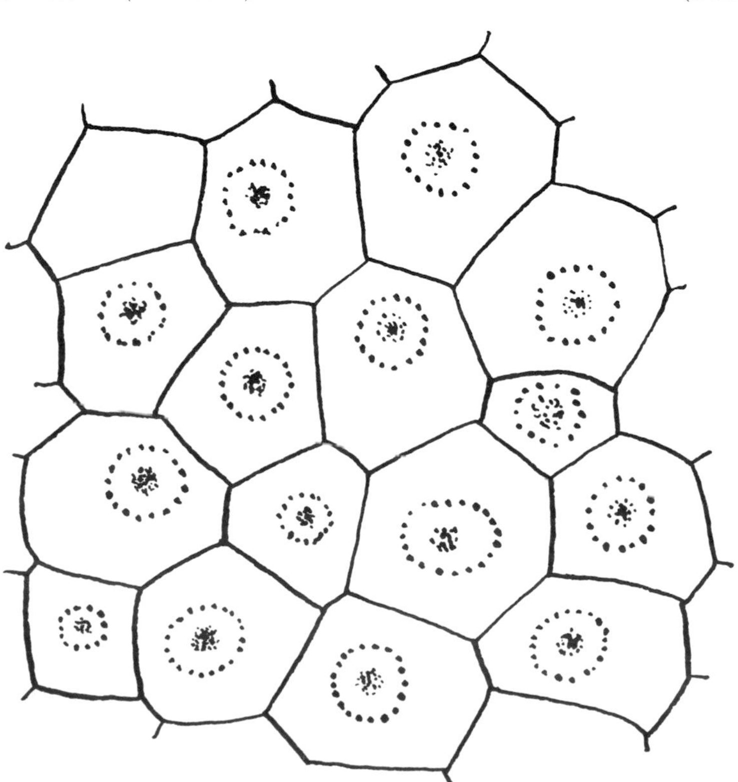

Class Agnatha

PLATES

CHAPTER VIII

VIII: PLATE 1

FIGURE a.—*Myxine glutinosa* Linnaeus. Length up to 79 cm. Side view and buccal cavity. (R. Kreuzinger after Bigelow and Schroeder)

FIGURE b.—*Petromyzon marinus* Linnaeus. Oral disc (lower center). Length up to about 84 cm. (R. Kreuzinger after Bigelow and Schroeder)

The species shown in this plate have been incriminated in cyclostome poisoning in humans.

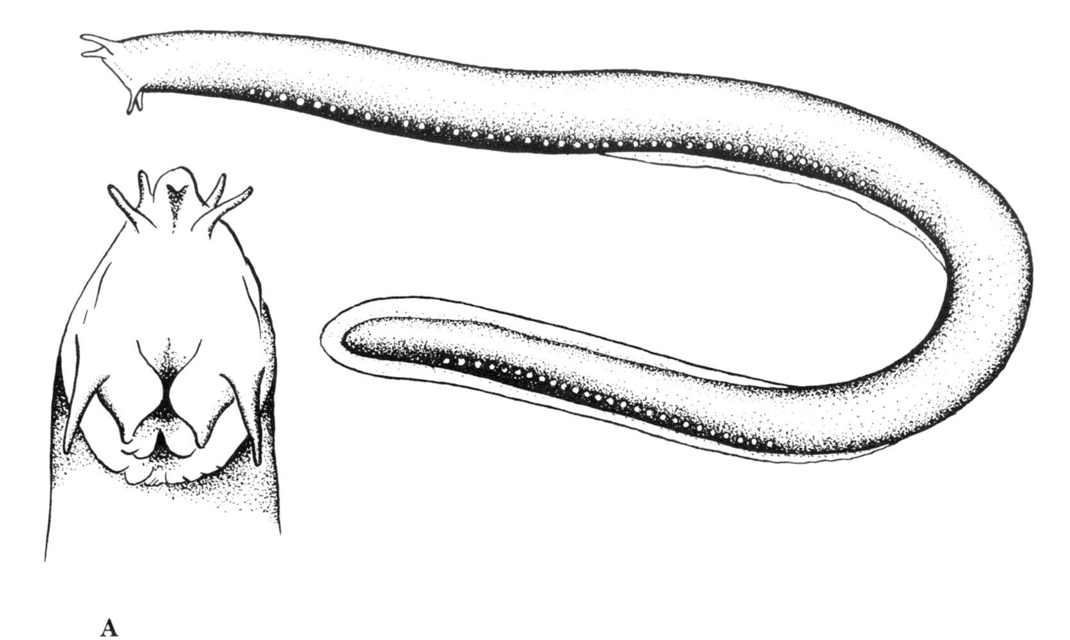

A

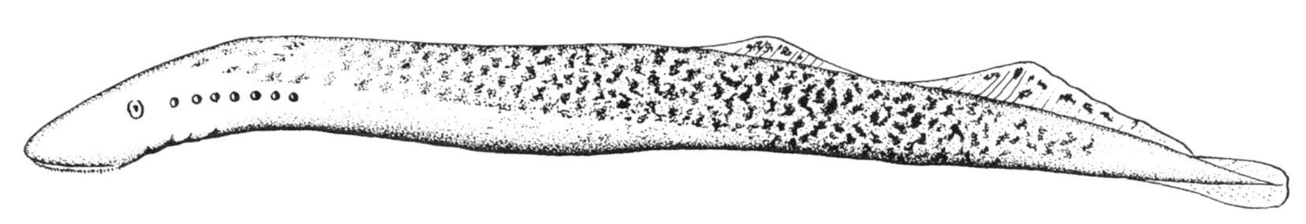

B

VIII: PLATE 2

FIGURE a.—*Caspiomyzon wagneri* (Kessler). Length up to 55 cm or more.

FIGURE b.—*Lampetra fluviatilis* (Linnaeus). Length up to 36 cm or more.
(From Berg)

The species shown in this plate have been incriminated in cyclostome poisoning in humans.

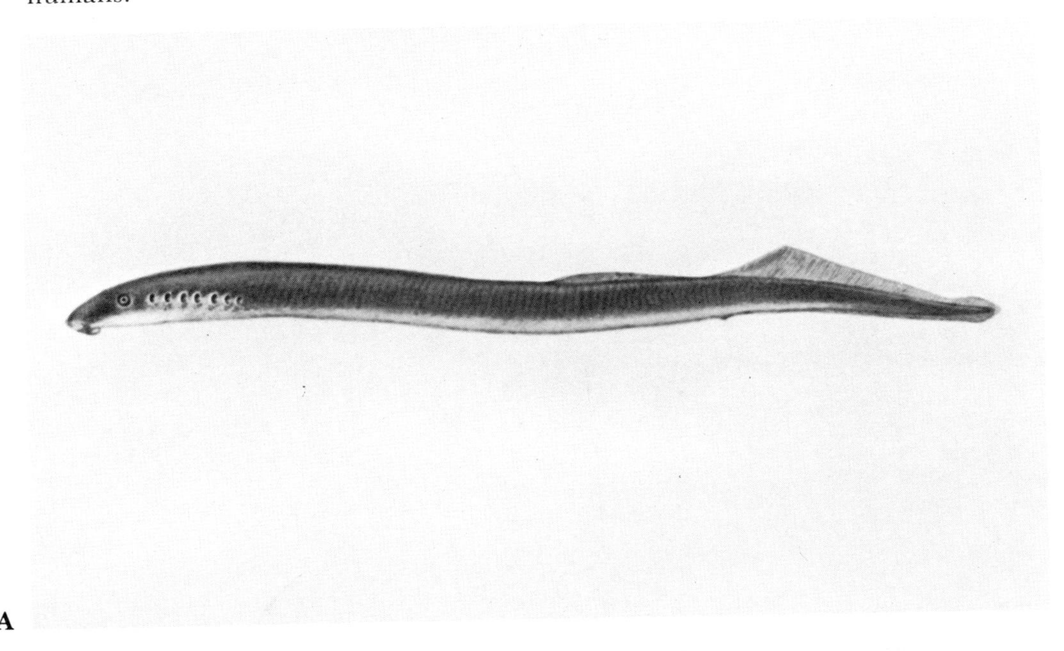

A

B

Class Chondrichthyes–Poisonous Sharks, etc.

PLATES

CHAPTER IX

IX: PLATE 1

FIGURE a.—*Carcharhinus melanopterus* (Quoy and Gaimard). Length 2 m.
FIGURE b.—*Carcharhinus amblyrhynchos* (Bleeker). Length 1.3 m.
FIGURE c.—*Galeorhinus galeus* (Linnaeus). Length 2 m.
FIGURE d.—*Prionace glauca* (Linnaeus). Length 4 m or more. (M. Shirao)
The sharks appearing on this plate have been incriminated in elasmobranch poisoning.

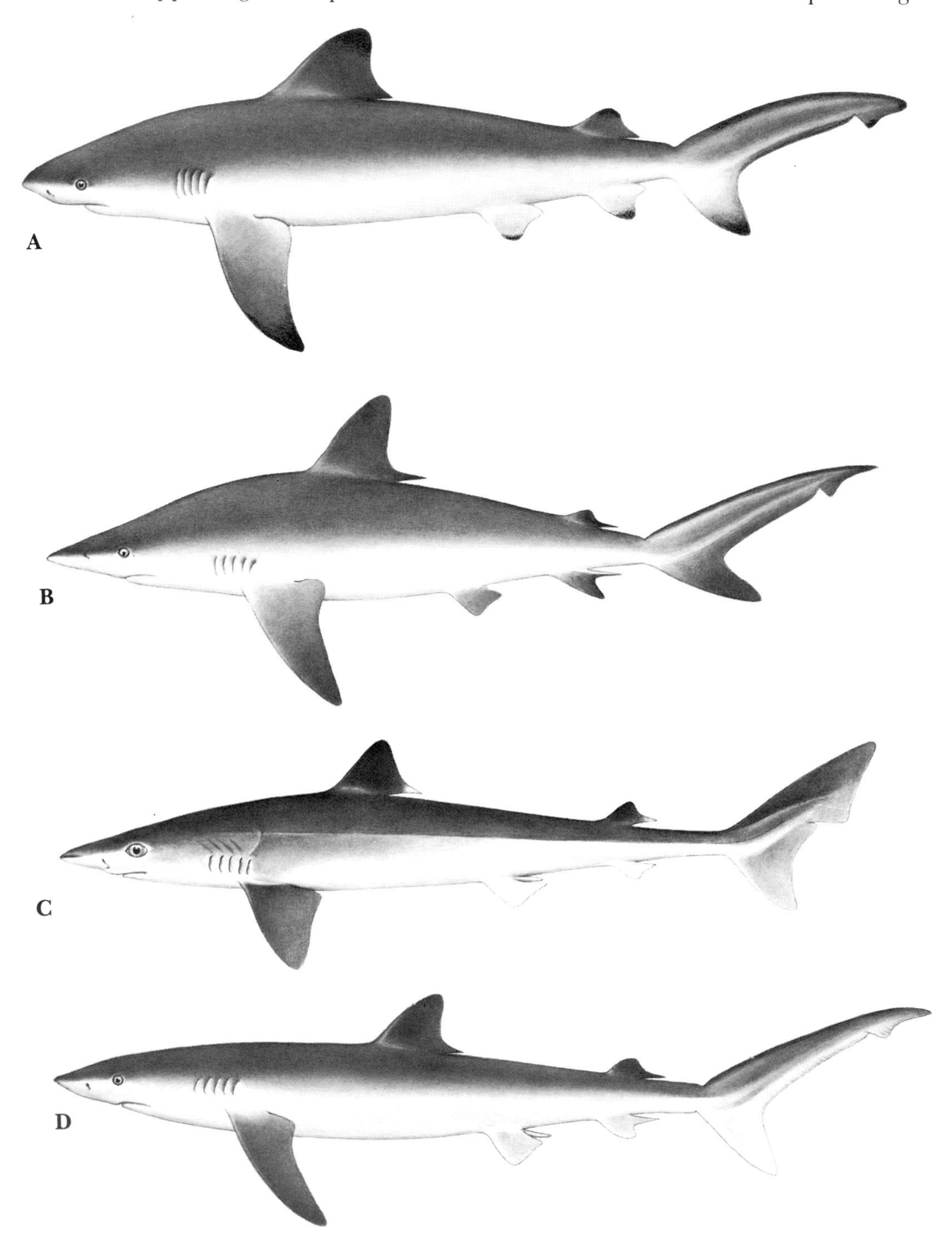

IX: PLATE 2

FIGURE a.—*Somniosus microcephalus* (Bloch and Schneider).

FIGURE b.—*Hexanchus griseus* (Bonnaterre). Length 5 m.

FIGURE c.—*Carcharodon carcharias* (Linnaeus). Length up to 11 m or more. (M. Shirao)

The sharks appearing on this plate have been incriminated in elasmobranch poisoning.

IX: PLATE 3

FIGURE a.—*Scyliorhinus caniculus* Linnaeus. Length 80 cm.

FIGURE b.—*Sphyrna zygaena* (Linnaeus). Length 4 m. (T. Kumada)

IX: PLATE 4

FIGURE a.—*Dasyatis sayi* (LeSueur). Width 1 m.

FIGURE b.—*Aetobatus narinari* (Euphrasen). Width 2 m or more. (M. Shirao)

FIGURE c.—*Raja batis* Linnaeus. Length 2.5 m, width 1.5 m.
(M. Shirao after Andriyashev)

FIGURE d.—*Torpedo marmorata* Risso. Length 70 cm.
(M. Shirao after Luther and Fiedler)

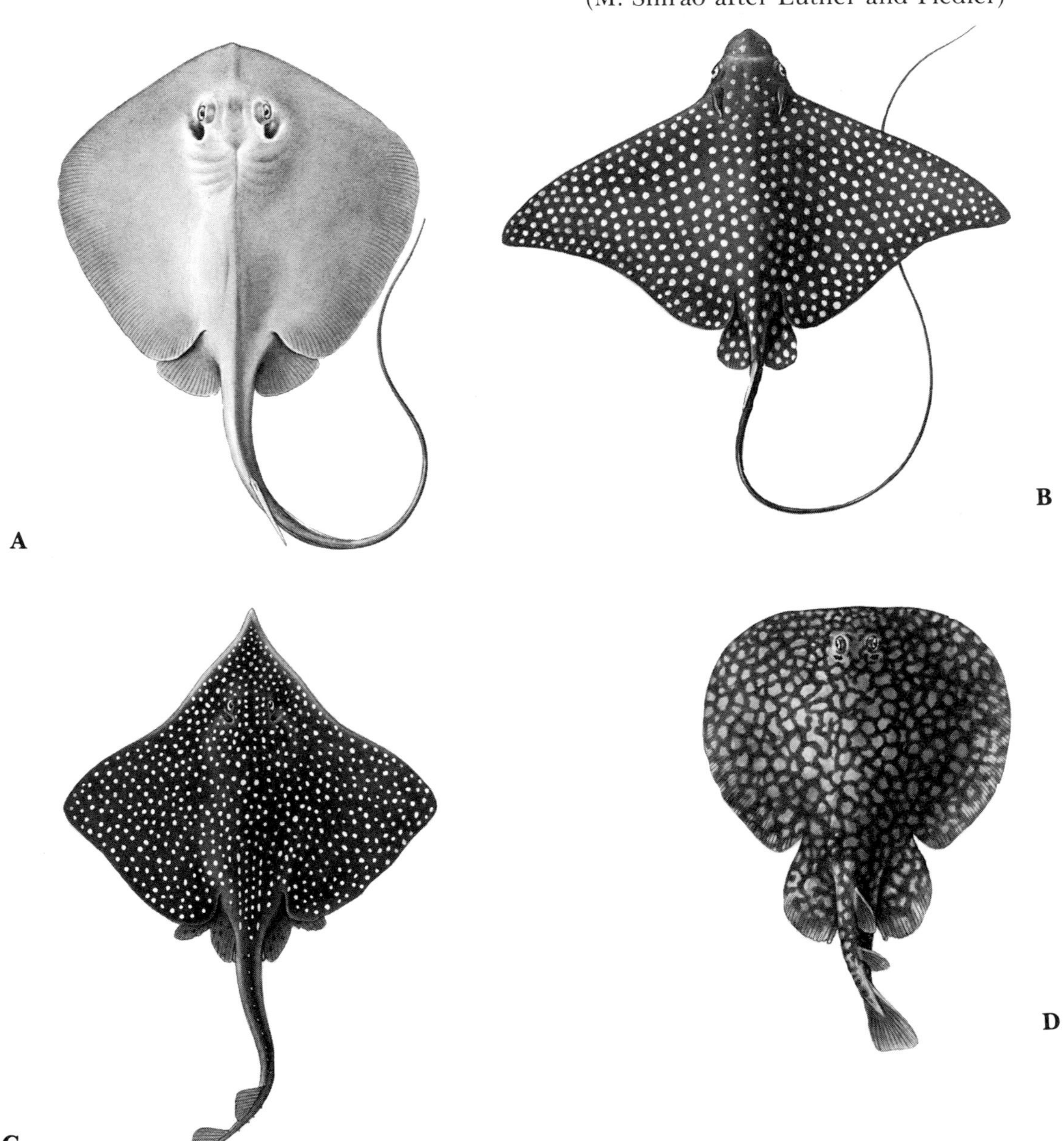

IX: PLATE 5

FIGURE a.—*Chimaera monstrosa* Linnaeus. Female. Length up to 1 m.

FIGURE b.—*Hydrolagus colliei* (Lay and Bennett). Female. Length 1 m. (M. Shirao)

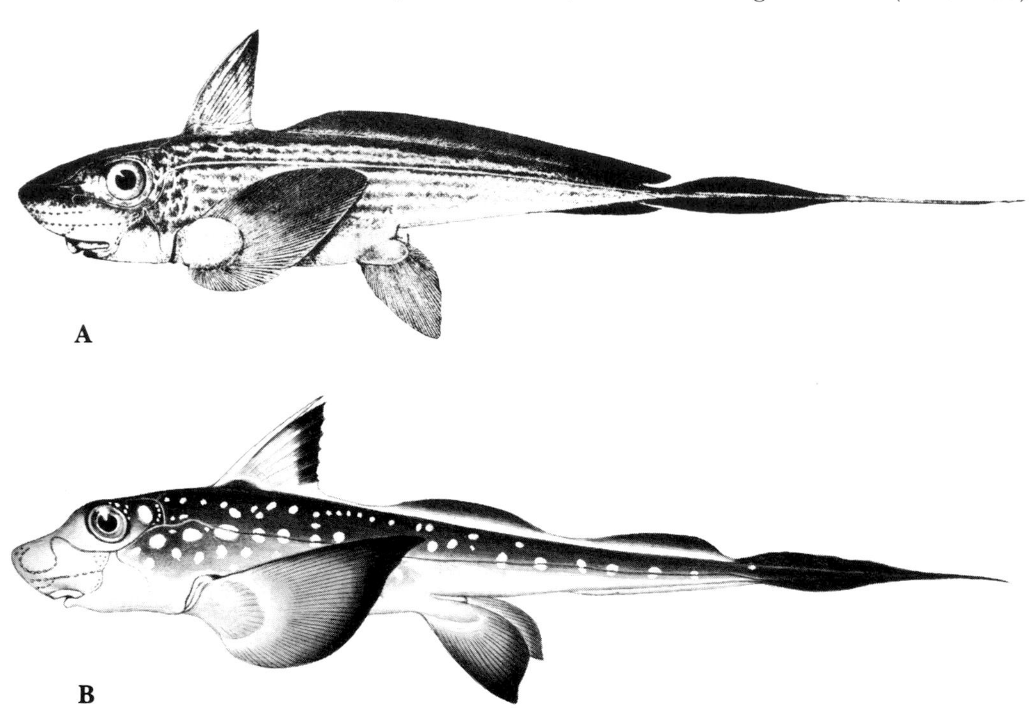

Class Osteichthyes–Ciguatoxic Fishes

PLATES

CHAPTER X

X: PLATE 1

FIGURE a.—*Albula vulpes* (Linnaeus). Length 1 m.

FIGURE b.—*Clupanodon thrissa* (Linnaeus). Length 25 cm. (S. Arita)

X: PLATE 2

FIGURE a.—*Harengula humeralis* (Cuvier). Length 41 cm. (From Rivas)

FIGURE b.—*Harengula ovalis* (Bennett). Length 15 cm. (From Day)

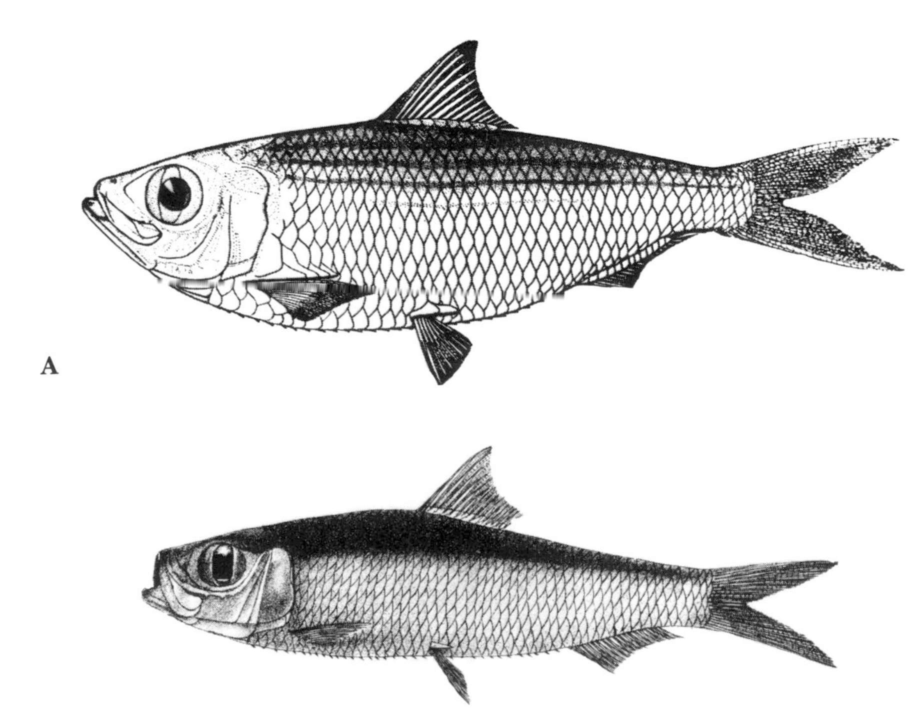

X: PLATE 3

FIGURE a.—*Macrura ilisha* (Buchanan-Hamilton). Length 25 cm.

FIGURE b.—*Nematolosa nasus* (Bloch). Length 20 cm. (From Day)

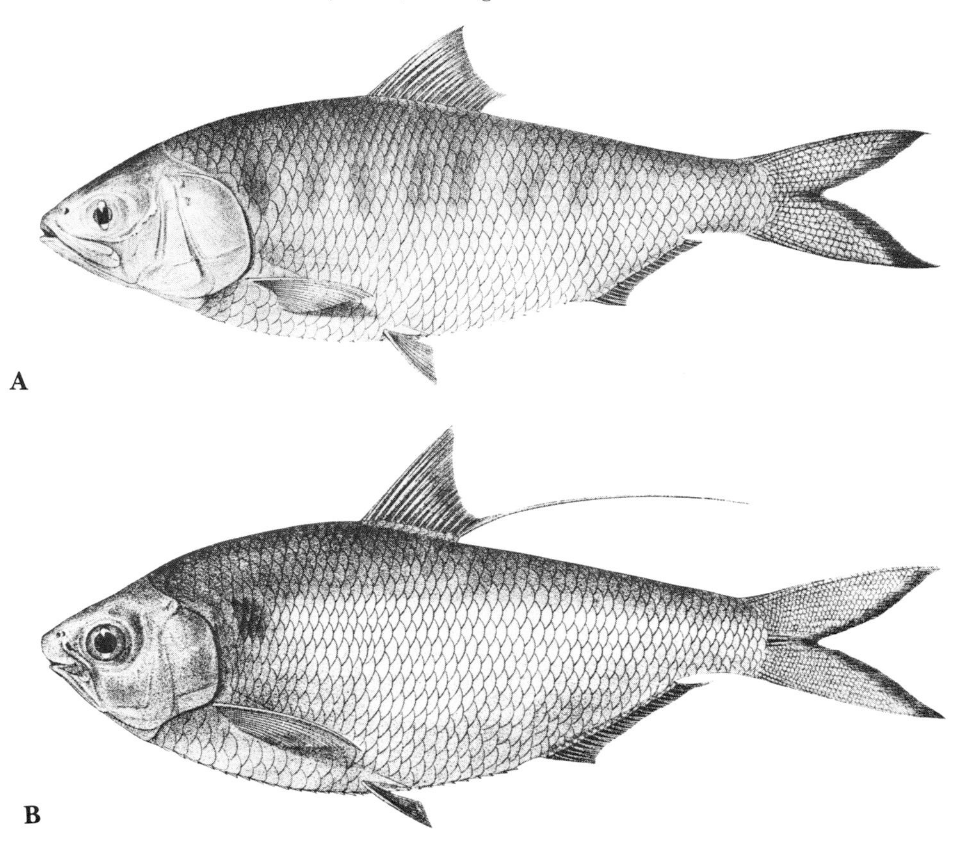

X: PLATE 4

FIGURE a.—*Opisthonema oglinum* (LeSueur). Length 25 cm. (From Hildebrand)

FIGURE b.—*Sardinella perforata* (Cantor). Length 18 cm. (From Bleeker)

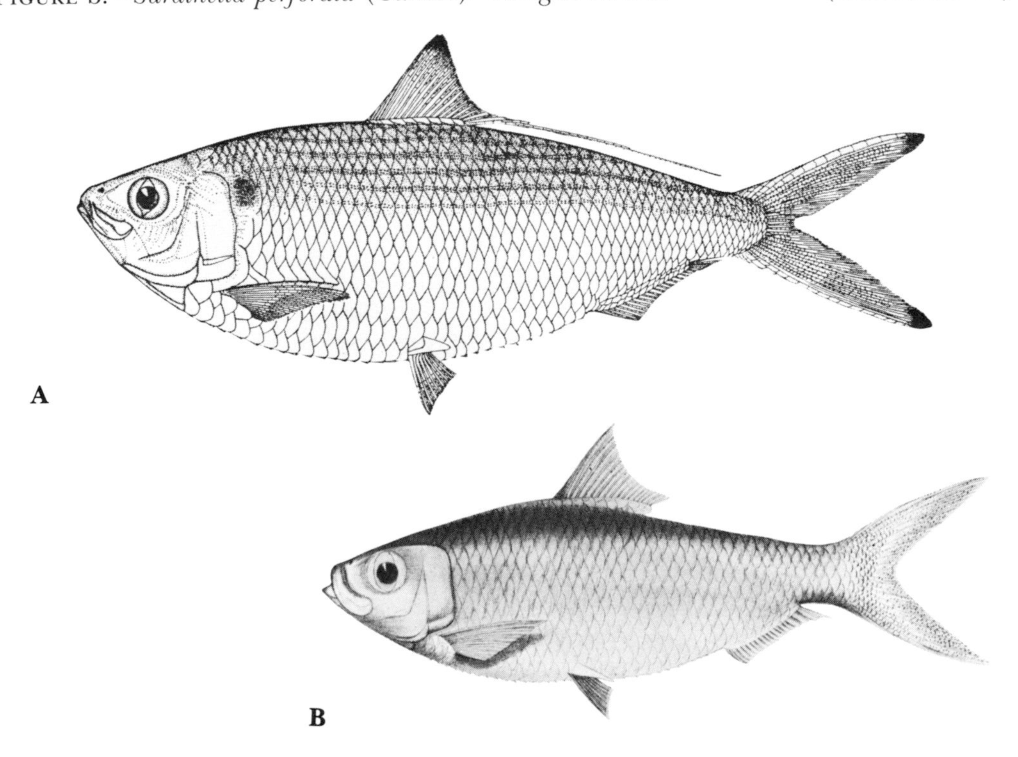

X: PLATE 5

Thrissina baelama (Forskål). Length 12 cm. (From Day)

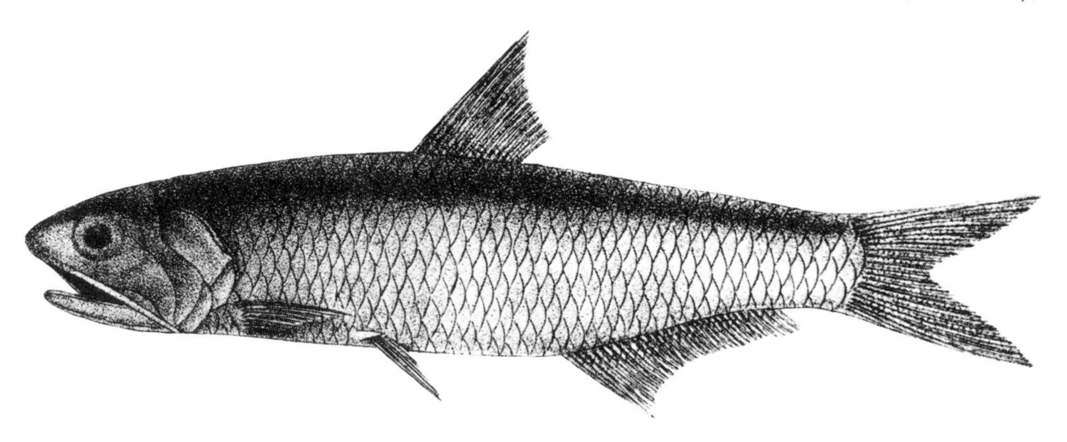

X: PLATE 6

FIGURE a.—*Gymnothorax flavimarginatus* (Rüppell). Length 1.5 m. (M. Shirao)
FIGURE b.—*Gymnothorax javanicus* (Bleeker). Length 1.5 m.
FIGURE c.—*Gymnothorax meleagris* (Shaw and Nodder). Length 1 m. (T. Kumada)

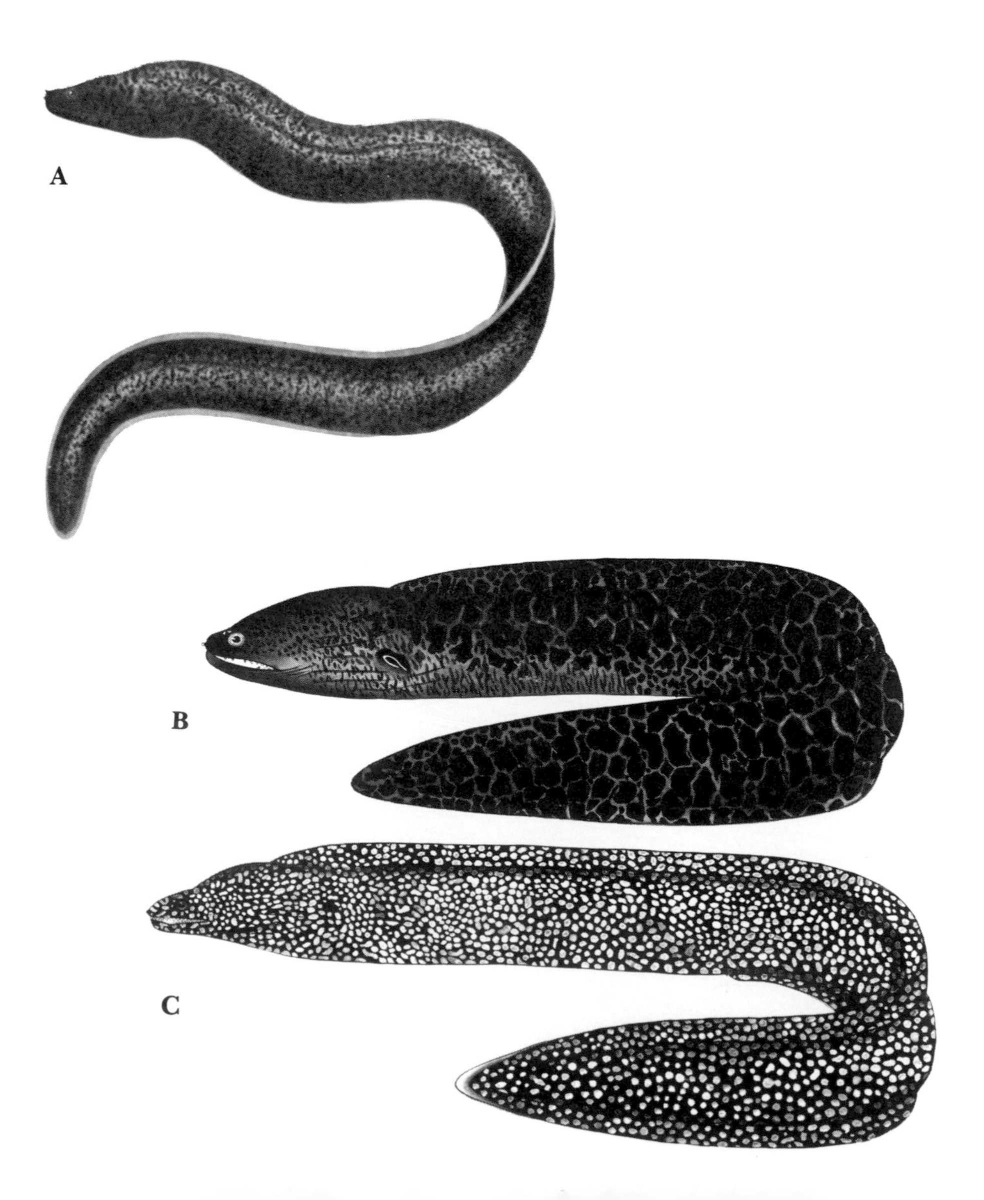

X: PLATE 7

FIGURE a.—*Gymnothorax undulatus* (Lacépède). Length 1.5m. (T. Kumada)

FIGURE b.—*Muraena helena* Linnaeus. Living specimen taken in an aquarium, Zoological Station, Naples, Italy. Length 1.5 m. (P. Giacomelli)

A

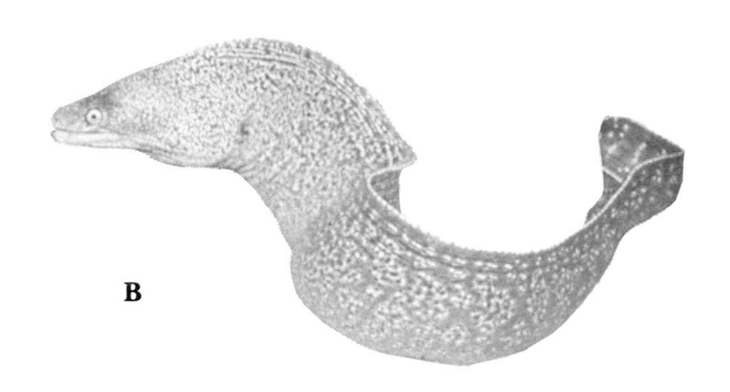

B

X: PLATE 8

FIGURE a.—*Strongylura caribbaea* (LeSueur). Length 38 cm. (Courtesy J. Randall)

FIGURE b.—*Hemiramphus brasiliensis* (Linnaeus). Length 38 cm. (From Jordan and Evermann)

A

B

X: PLATE 9

Myripristis murdjan (Forskål). Length 30 cm. (T. Kumada)

X: PLATE 10

FIGURE a.—*Acanthurus chirurgus* (Bloch). Length 22 cm. Pale color phase. Florida.

FIGURE b.—*Acanthurus lineatus* (Linnaeus). Length 18 cm. Onotoa, Gilbert Islands.

FIGURE c.—*Ctenochaetus strigosus* (Bennett). Length 13 cm. Hawaiian Islands.

(Courtesy J. Randall)

A

B

C

X: PLATE 11

FIGURE a.—*Acanthurus nigrofuscus* (Forskål). Length 15 cm. (S. Arita)

FIGURE b.—*Acanthurus olivaceus* Bloch and Schneider. Length 16 cm. (K. Tomita)

FIGURE c.—*Acanthurus triostegus* (Linnaeus). Length 26 cm. (S. Arita)

FIGURE d.—*Acanthurus xanthopterus* Cuvier and Valenciennes. Length 35 cm. (G. Coates)

A

B

C

D

X: PLATE 12

FIGURE a.—*Ctenochaetus striatus* (Quoy and Gaimard). Length 19 cm. (K. Tomita)

FIGURE b.—*Naso lituratus* Bloch and Schneider. Length 38 cm. (S. Arita)

FIGURE c.—*Zebrasoma veliferum* (Bloch). Length 30 cm. (T. Kumada)

A

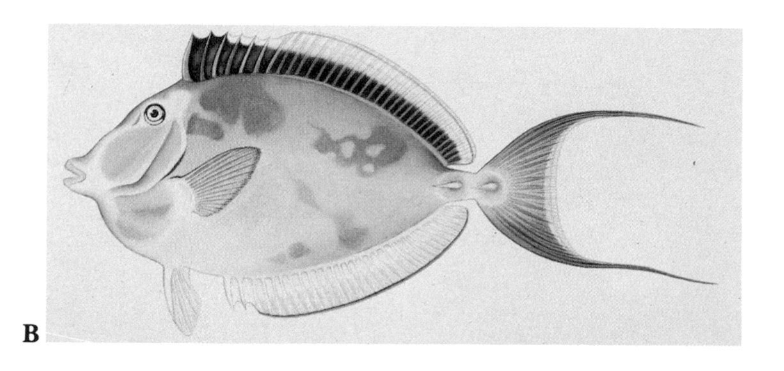

B

C

X: PLATE 13

FIGURE a.—*Caranx crysos* (Mitchill). Length 44 cm. (From Evermann and Marsh)

FIGURE b.—*Caranx hippos* (Linnaeus). Length 75 cm. (T. Kumada)

FIGURE c.—*Caranx ignobilis* (Forskål). Length 55 cm. (From Jordan and Evermann)

FIGURE d.—*Caranx latus* Agassiz. Length 90 cm. (Courtesy J. Randall)

FIGURE e.—*Caranx lugubris* Poey. Length 60 cm. (S. Arita)

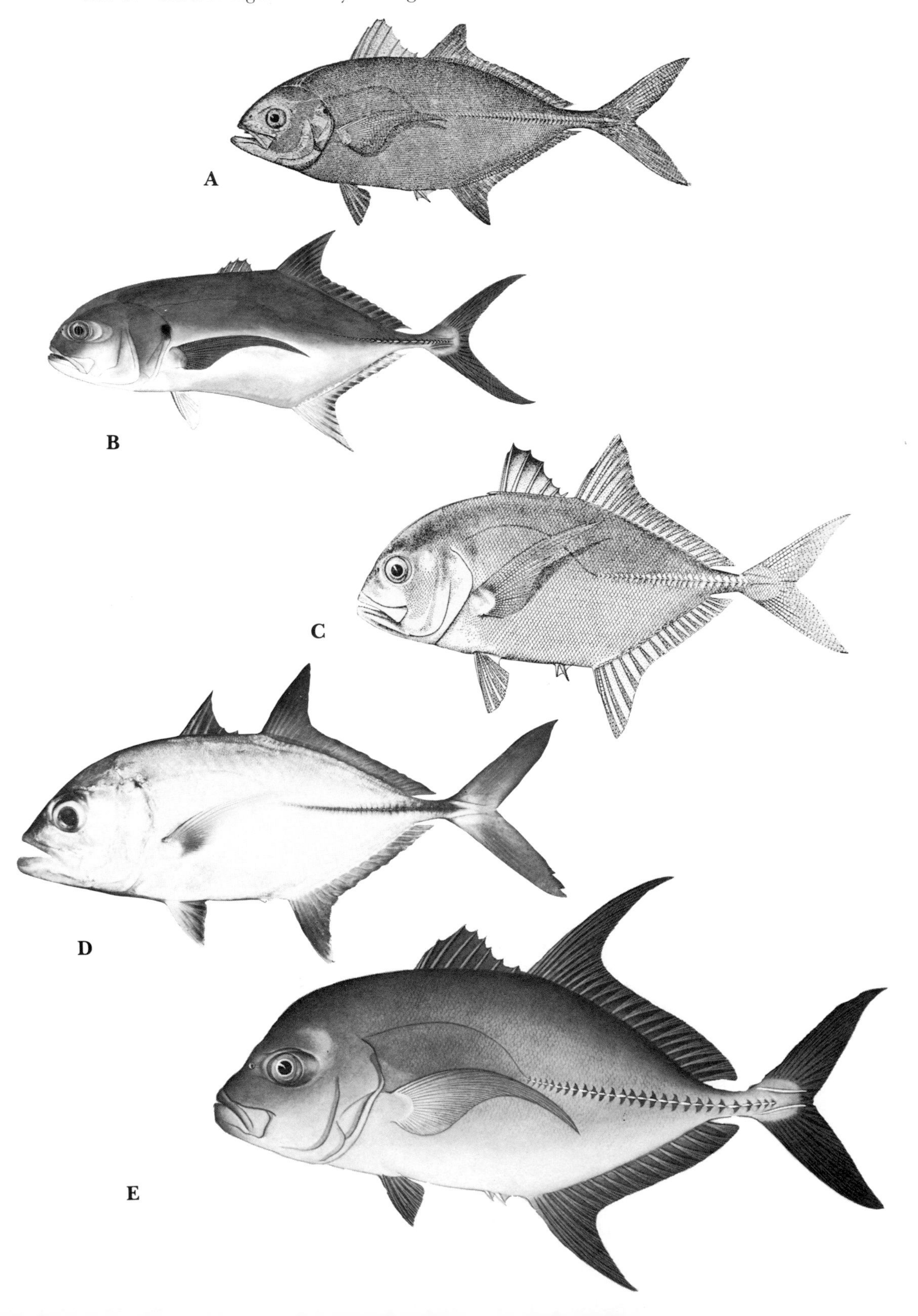

X: PLATE 14

FIGURE a.—*Caranx melampygus* Cuvier. Length 65 cm. (K. Tomita)

FIGURE b.—*Caranx sexfasciatus* Quoy and Gaimard. Length 50 cm.

FIGURE c.—*Elagatis bipinnulatus* (Quoy and Gaimard). Length 65 cm. (T. Kumada)

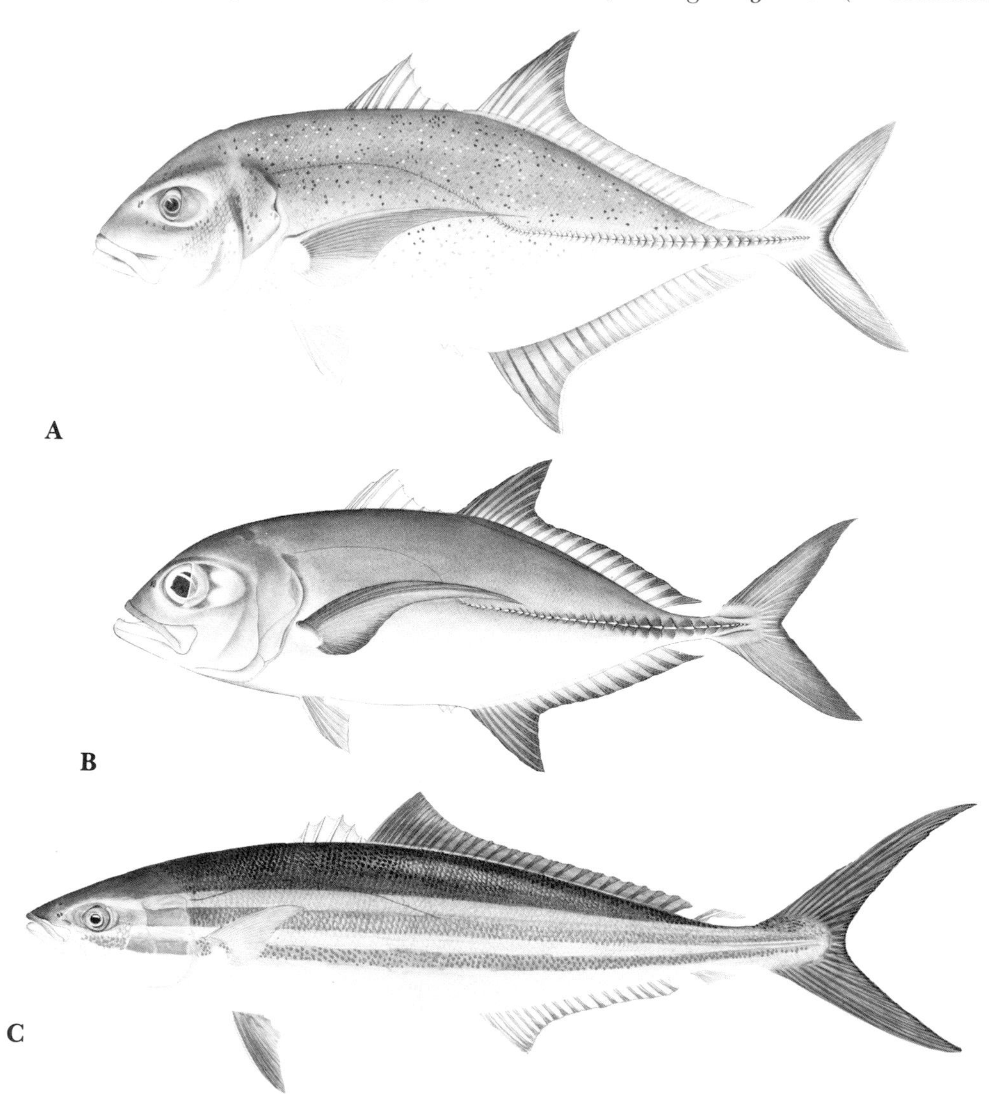

X: PLATE 15

Selar crumenophthalmus (Bloch). Length 60 cm. (T. Kumada)

X: PLATE 16

FIGURE a.—*Seriola dumerili* (Risso). Length 90 cm. (From Palombi and Santarelli)

FIGURE b.—*Seriola falcata* Valenciennes. Length 30 cm. (Courtesy Smithsonian Institution)

X: PLATE 17

FIGURE a-b.—*Coryphaena hippurus* Linnaeus. Length 1.2m.
a. Adult male. (From Walford)
b. Adult female. (From Hiyama)

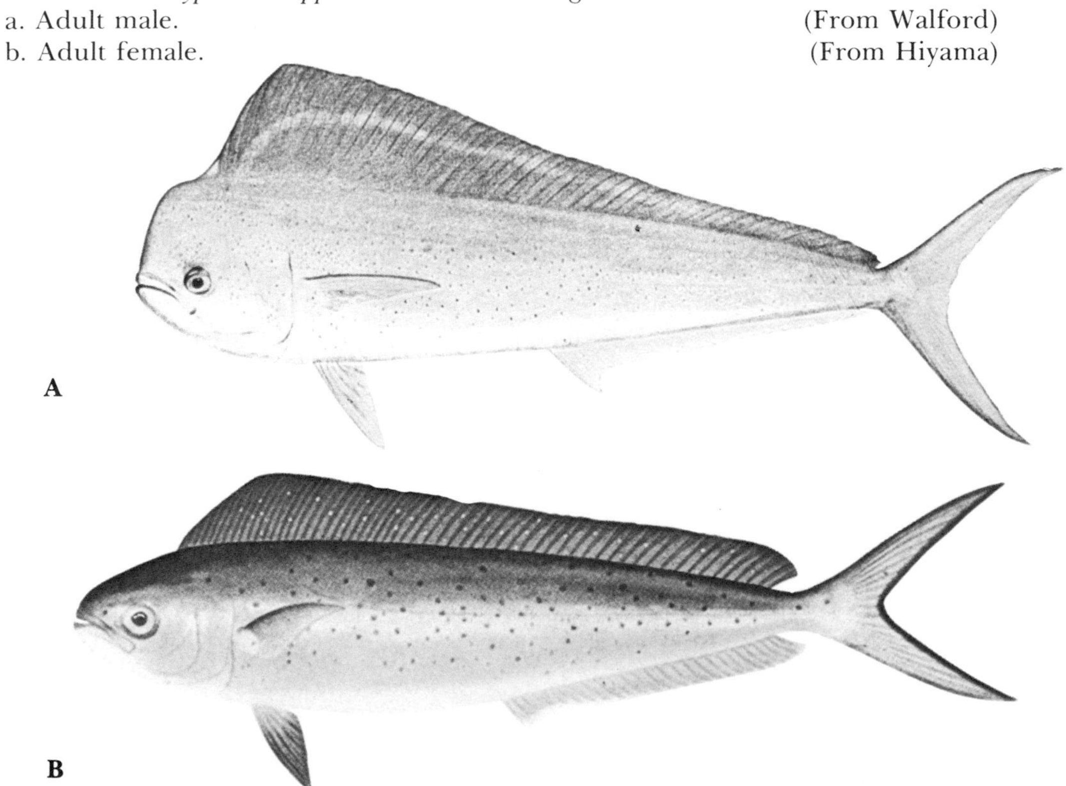

X: PLATE 18

Ctenogobius criniger (Cuvier and Valenciennes). Length 10 cm.
(From Jordan and Seale)

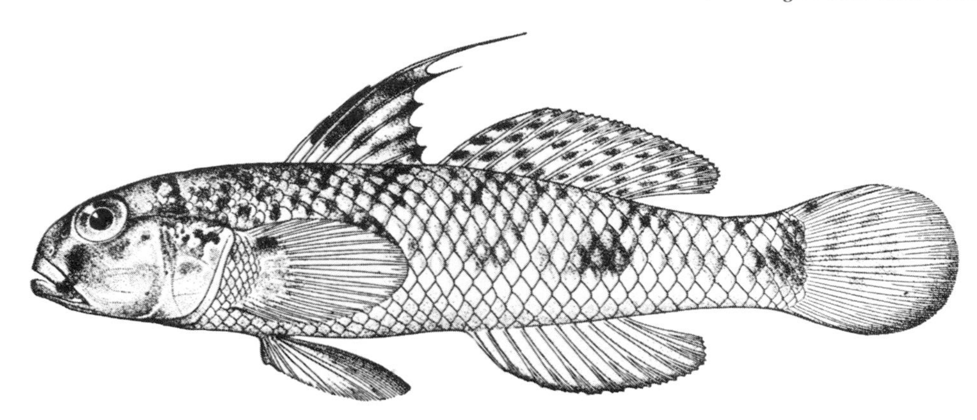

X: PLATE 19

Coris gaimardi (Quoy and Gaimard). Length 30 cm. (T. Kumada)

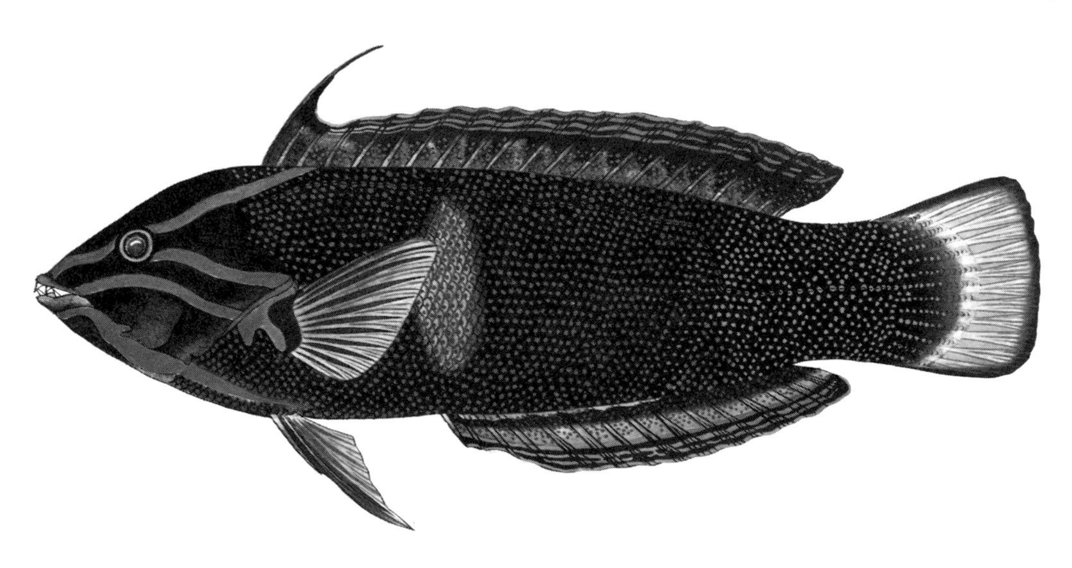

X: PLATE 20

Epibulus insidiator (Pallas). Length 30 cm. (T. Kumada)

X: PLATE 21

FIGURE a.—*Aprion virescens* Cuvier and Valenciennes. Length 72 cm. (T. Kumada)

FIGURE b.—*Gnathodentex aureolineatus* (Lacépède). Length 25 cm. (K. Tomita)

FIGURE c.—*Gymnocranius griseus* (Temminck and Schlegel). Length 35 cm. (From Hiyama)

A

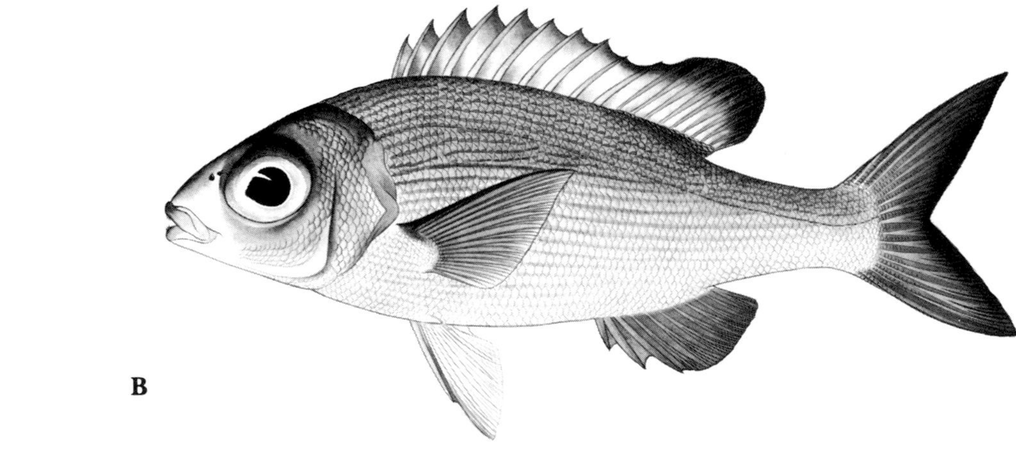

B

C

X: PLATE 22

FIGURE a.—*Lethrinus harak* (Forskål). Length 30 cm. (From Hiyama)

FIGURE b.—*Lethrinus miniatus* (Forster). Length 45 cm.

FIGURE c.—*Lethrinus variegatus* Cuvier and Valenciennes. Length 35 cm. (T. Kumada)

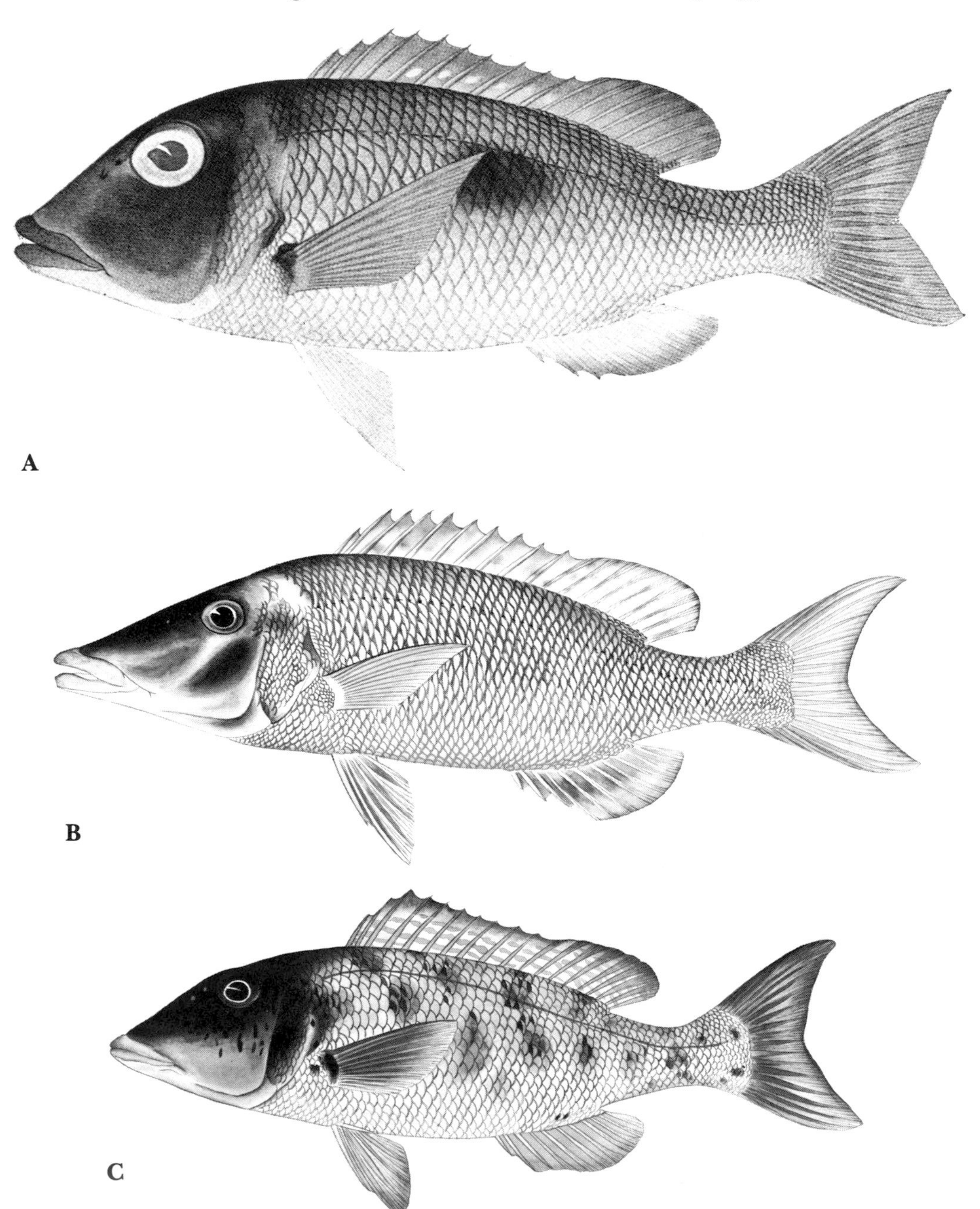

X: PLATE 23

FIGURE a.—*Lutjanus argentimaculatus* (Forskål). Length 50 cm. (T. Kumada)

FIGURE b.—*Lutjanus aya* (Bloch). Length 76 cm. (S. Arita, after Evermann and Marsh)

FIGURE c.—*Lutjanus bohar* (Forskål). Length 90 cm. (M. Shirao)

FIGURE d.—*Lutjanus gibbus* (Forskål). Length 40 cm. (K. Tomita)

FIGURE e.—*Lutjanus janthinuropterus* (Bleeker). Length 60 cm. (T. Kumada)

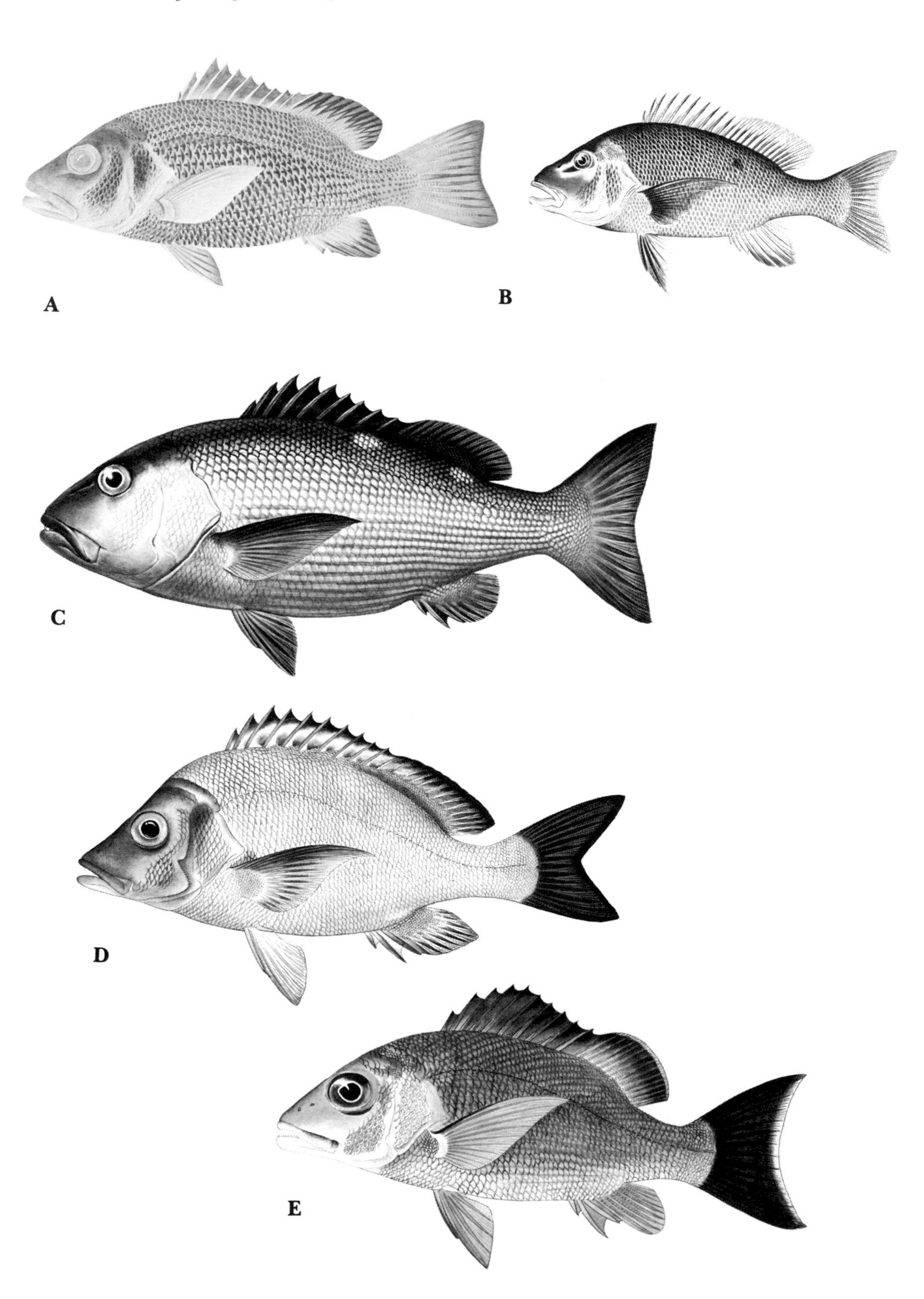

X: PLATE 24

Figure a.—*Lutjanus jocu* (Bloch and Schneider). Length 75 cm.
(S. Arita, after Evermann and Marsh)

Figure b.—*Lutjanus kasmira* (Forskål). Length 25 cm. (From Hiyama)

Figure c.—*Lutjanus monostigmus* (Cuvier and Valenciennes). Length 30 cm.
(K. Tomita)

X: PLATE 25

FIGURE a.—*Lutjanus nematophorus* Bleeker. Length 30 cm. (G. Coates)

FIGURE b.—*Lutjanus semicinctus* Quoy and Gaimard. Length 30 cm. (T. Kumada)

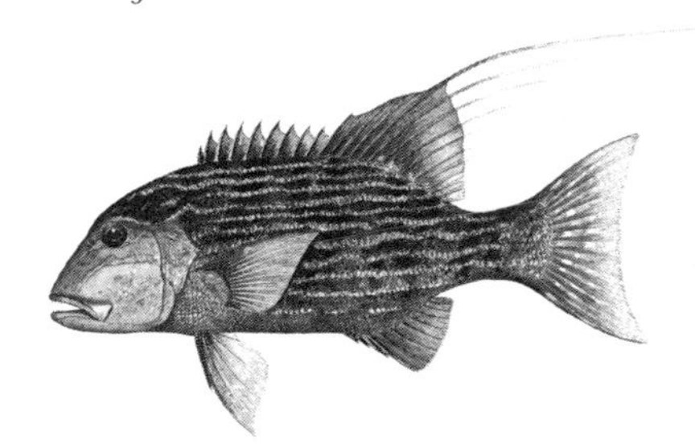

A

B

X: PLATE 26

FIGURE a.—*Lutjanus vaigiensis* (Quoy and Gaimard). Length 50 cm. (K. Tomita)

FIGURE b.—*Monotaxis grandoculis* (Forskål). Length 32 cm. (S. Arita)

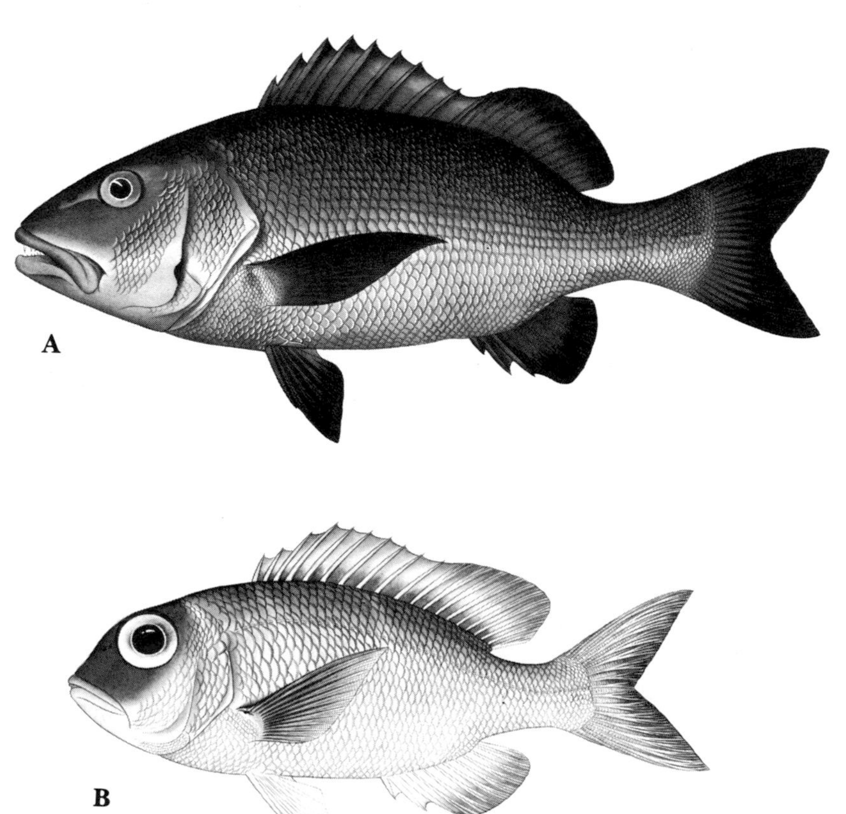

A

B

X: PLATE 27

FIGURE a.—*Chelon vaigiensis* (Quoy and Gaimard). Length 32 cm. (From Marshall)

FIGURE b.—*Mugil cephalus* Linnaeus. Length 30 cm. (From Hiyama)

X: PLATE 28

FIGURE a.—*Mulloidichthys auriflamma* (Forskål). Length 35 cm.

FIGURE b.—*Mulloidichthys samoensis* (Günther). Length 30 cm.

(From Jordan and Evermann)

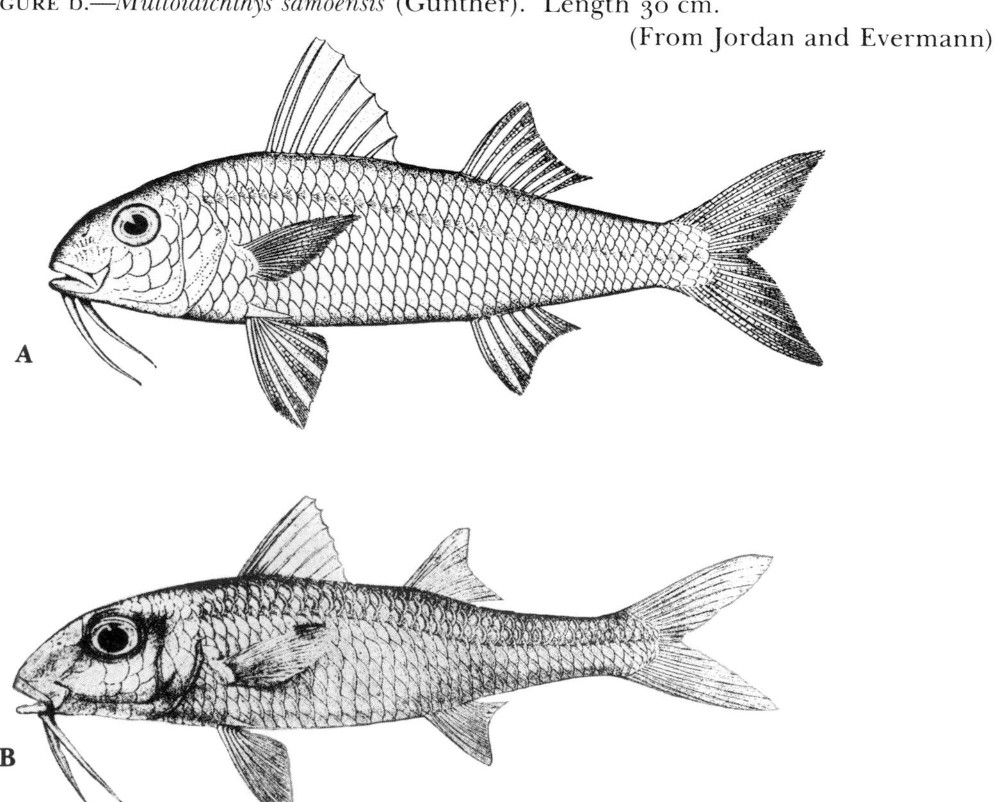

X: PLATE 29

Figure a.—*Parupeneus chryserydros* (Lacépède). Length 30 cm.
(From Jordan and Evermann)

Figure b.—*Upeneus arge* Jordan and Evermann. Length 32 cm. Barbels are not shown.
(From Jordan and Evermann)

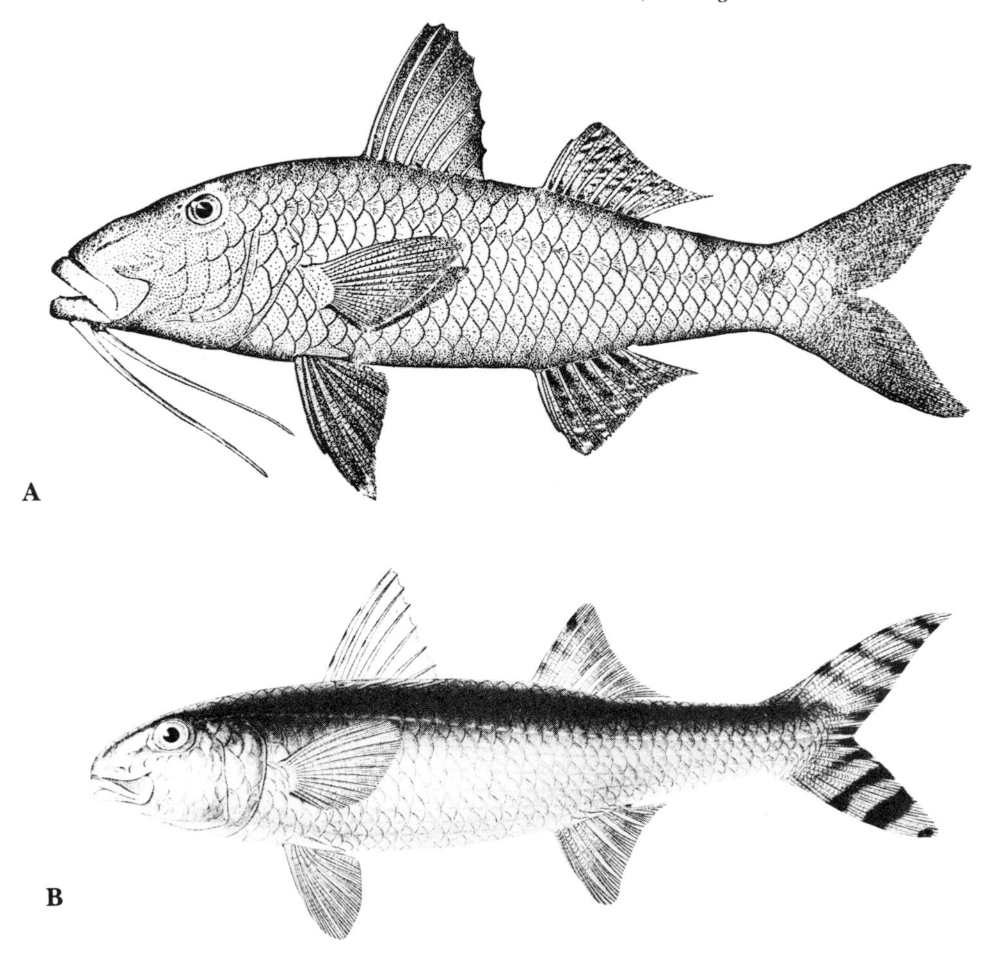

X: PLATE 30

Abudefduf septemfasciatus (Cuvier). Length 18 cm. (From Day)

X: PLATE 31

Figure a.—*Chlorurus pulchellus* (Rüppell). Length 90 cm. (From Smith)

Figure b.—*Scarus coeruleus* (Bloch). Length 90 cm. (S. Arita, after Evermann and Marsh)

Figure c.—*Scarus ghobban* Forskål. Length 1.1 m. (From Smith)

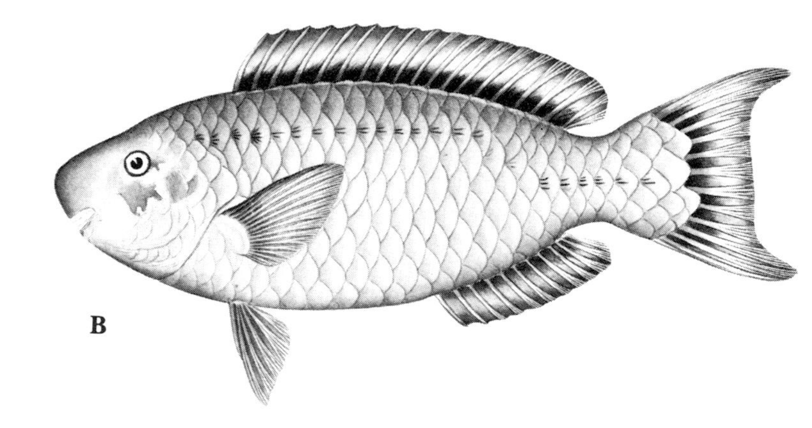

X: PLATE 32

Figure a.—*Scarus microrhinos* Bleeker. Length 33 cm. (T. Kumada)

Figure b.—*Scarus vetula* Bloch and Schneider. Length 30 cm. (S. Arita, after Evermann and Marsh)

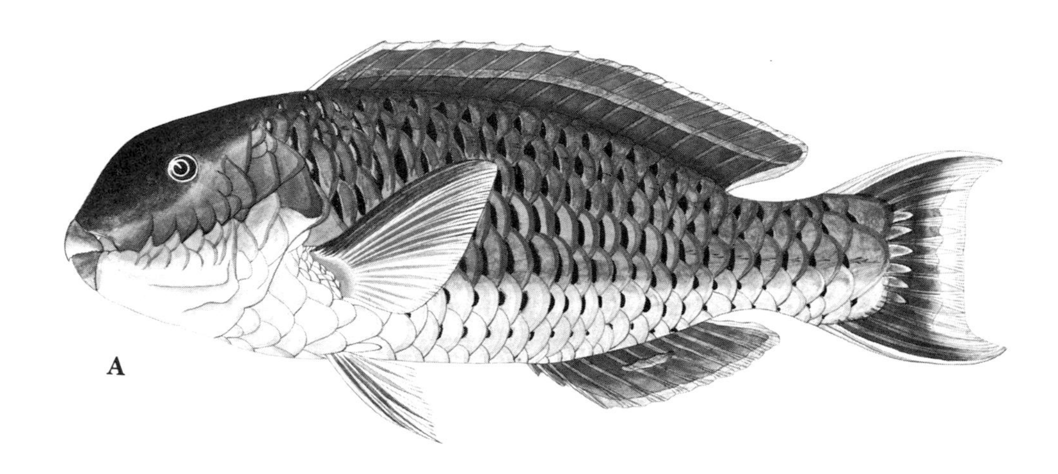

A

B

X: PLATE 33

FIGURE a.—*Acanthocybium solandri* (Cuvier). Length 2 m. (From Walford)

FIGURE b.—*Scomberomorus cavalla* (Cuvier). Length 1.5 m. (From Goode)

FIGURE c.—*Cephalopholis fulvus* (Linnaeus). Length 25 cm. (From Jordan and Evermann)

A

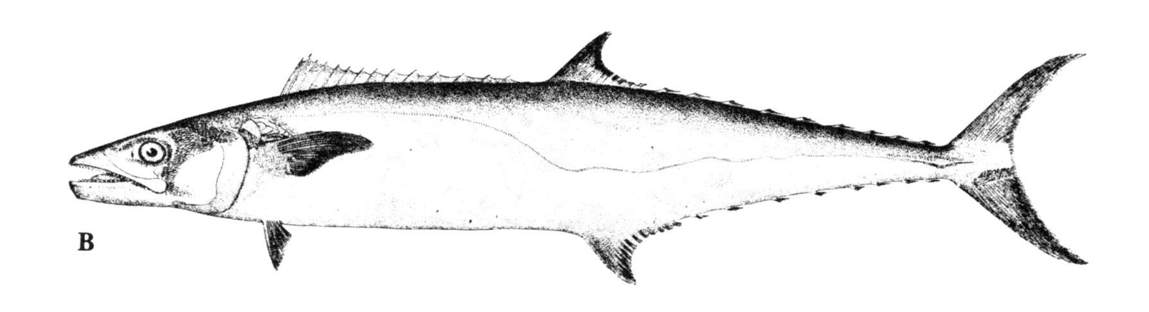

B

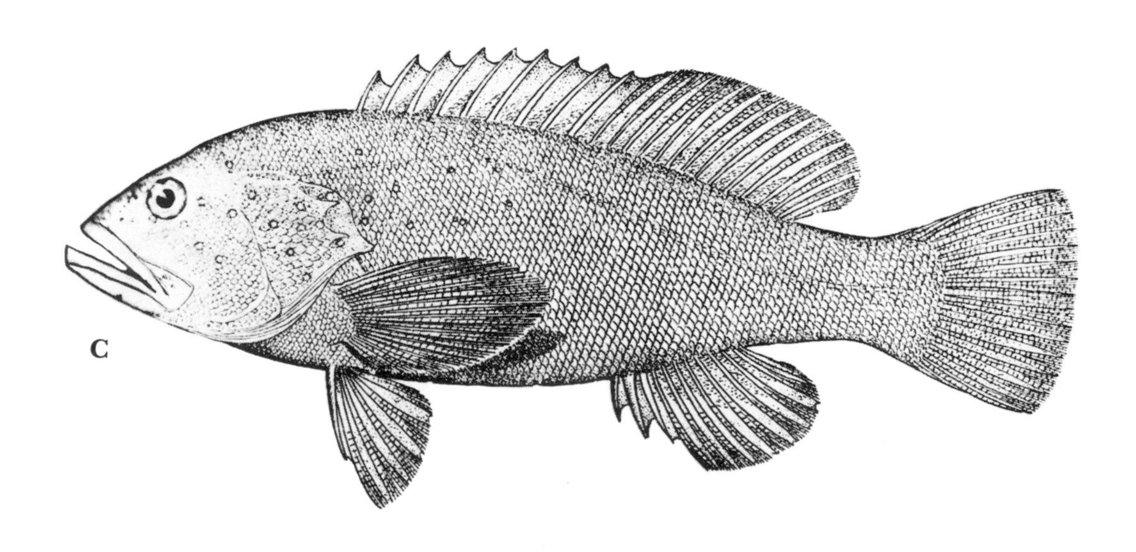

C

X: PLATE 34

FIGURE a.—*Anyperodon leucogrammicus* (Cuvier and Valenciennes). Length 40 cm. (From Marshall)

FIGURE b.—*Cephalopholis argus* Bloch and Schneider. Length 50 cm. (K. Tomita)

FIGURE c.—*Cephalopholis leopardus* (Lacépède). Length 50 cm. (T. Kumada)

FIGURE d.—*Cephalopholis miniatus* (Forskål). Length 25 cm. (From Marshall)

A

B

C

D

X: PLATE 35

FIGURE a.—*Epinephelus adscensionis* (Osbeck). Length 38 cm.
(S. Arita, after Evermann and Marsh)

FIGURE b.—*Epinephelus fuscoguttatus* (Forskål). Length 60 cm. (From Day)

FIGURE c.—*Epinephelus guttatus* (Linnaeus). Length 45 cm.
(S. Arita, after Evermann and Marsh)

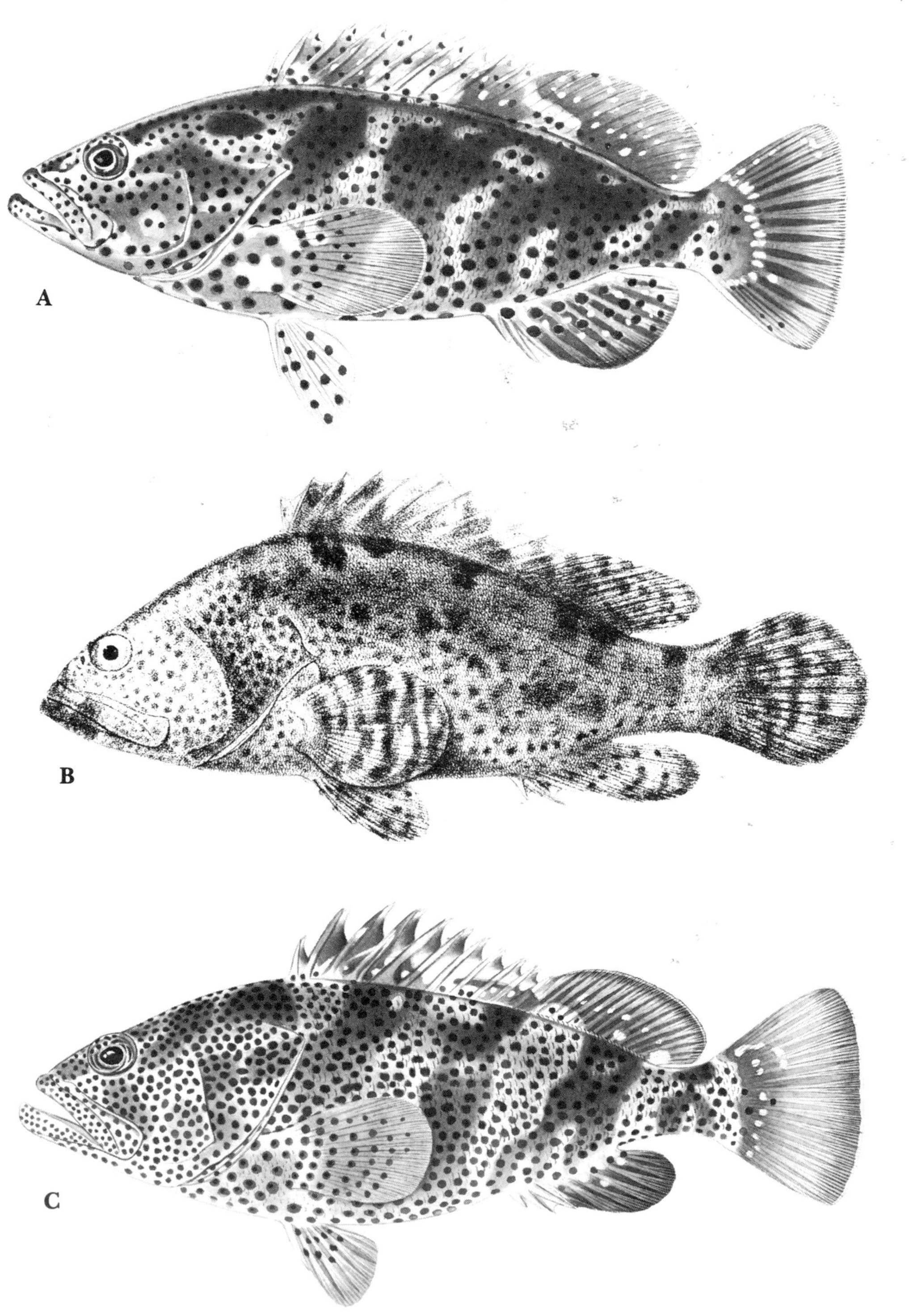

X: PLATE 36

FIGURE a.—*Mycteroperca venenosa* (Linnaeus). Length 90 cm. (From Jordan and Evermann)

FIGURE b.—*Siganus oramin* (Schneider). Length 30 cm. (From Day)

A

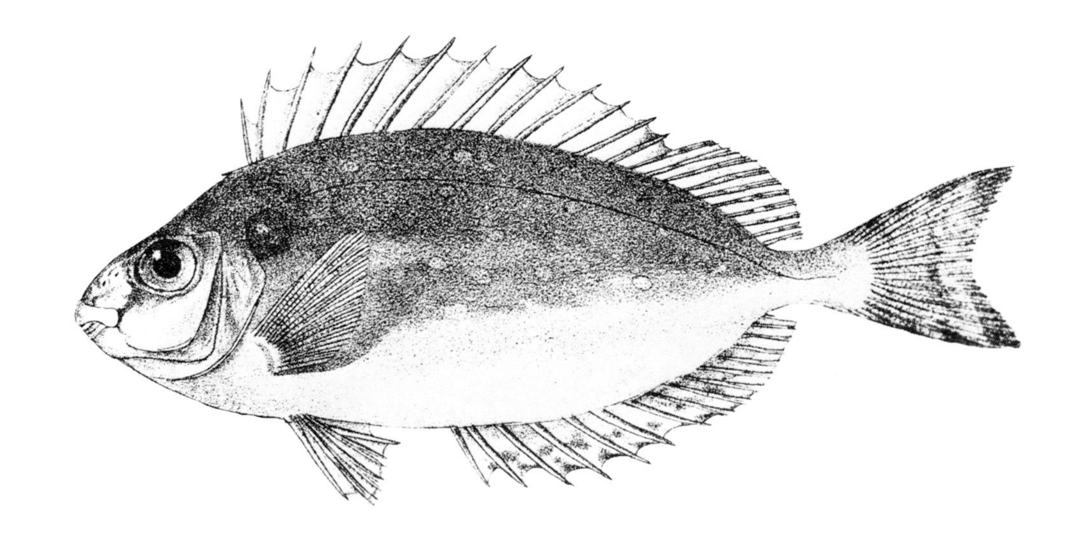

B

X: PLATE 37

FIGURE a.—*Epinephelus tauvina* (Forskål). Length 2.1 m. (From Marshall)
FIGURE b.—*Plectropomus leopardus* (Lacépède). Length 35 cm. (From T. Kumada)
FIGURE c.—*Plectropomus maculatus* (Bloch). Length 55 cm. (From Marshall)

X: PLATE 38

FIGURE a.—*Plectropomus oligacanthus* Bleeker. Length 55 cm.

FIGURE b.—*Plectropomus truncatus* Fowler and Bean. length 52 cm. (T. Kumada)

FIGURE c.—*Variola louti* (Forskål). Length 60 cm. (S. Arita)

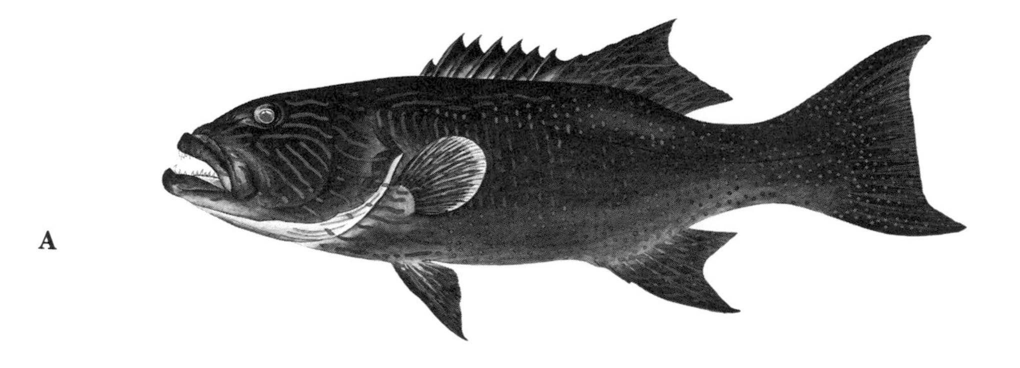

A

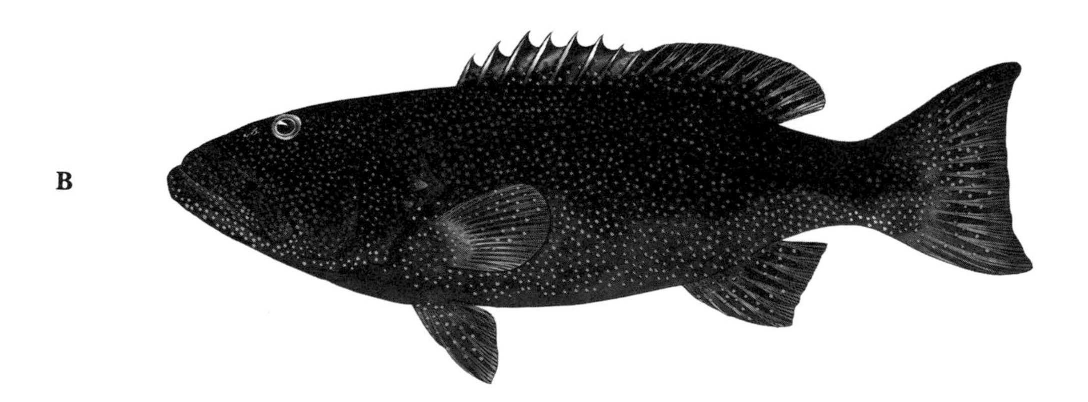

B

C

X: PLATE 39

FIGURE a.—*Siganus fuscescens* (Houttuyn). Length 25 cm. (S. Arita)

FIGURE b.—*Siganus lineatus* (Valenciennes). Length 30 cm. (G. Coates)

FIGURE c.—*Siganus puellus* (Schlegel). Length 27 cm. (T. Kumada)

FIGURE d.—*Siganus argenteus* (Quoy and Gaimard). Length 35 cm. (From Günther)

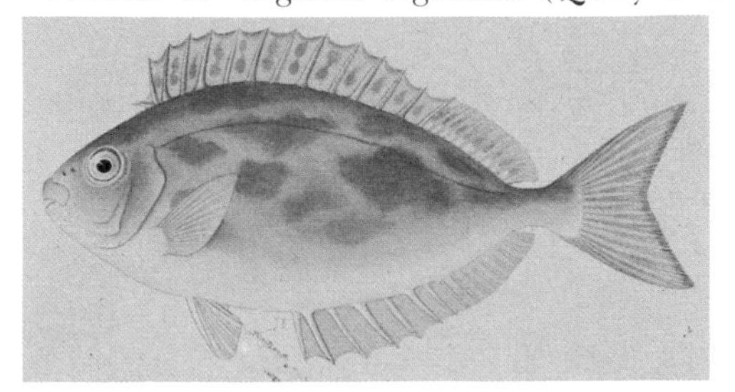

A

B

C

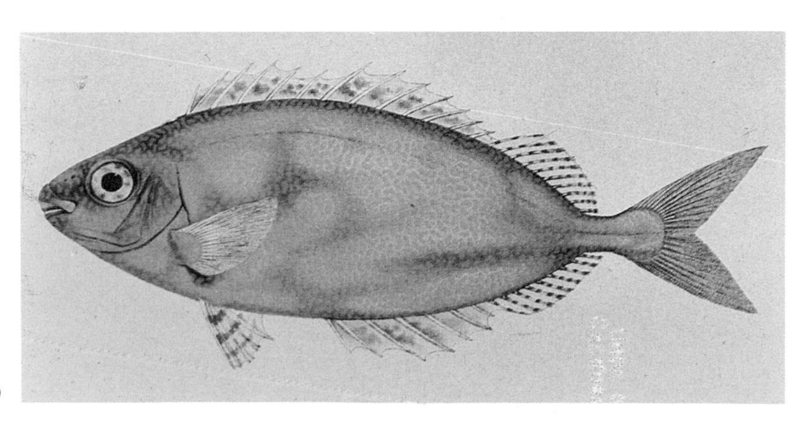

D

X: PLATE 40

FIGURE a.—*Sparus sarba* Forskål. Length 35 cm. (From Hiyama)

FIGURE b.—*Pagellus erythrinus* (Linnaeus). Length 40 cm. (From Joubin)

FIGURE c.—*Pagrus pagrus* (Linnaeus). Length 38 cm. (From Day)

X: PLATE 41

Sphyraena barracuda (Walbaum). Length 1.6 m. (T. Kumada)

X: PLATE 42

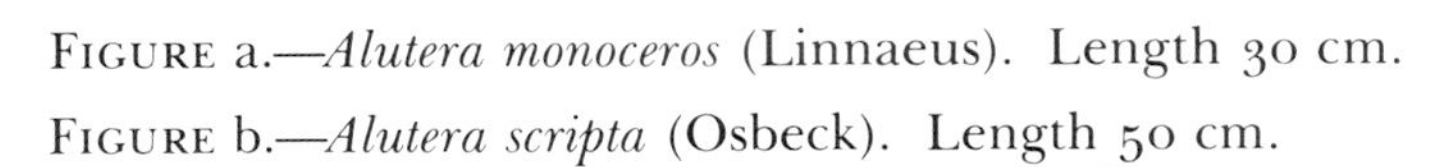

FIGURE a.—*Alutera monoceros* (Linnaeus). Length 30 cm. (From Hiyama)

FIGURE b.—*Alutera scripta* (Osbeck). Length 50 cm. (T. Kumada)

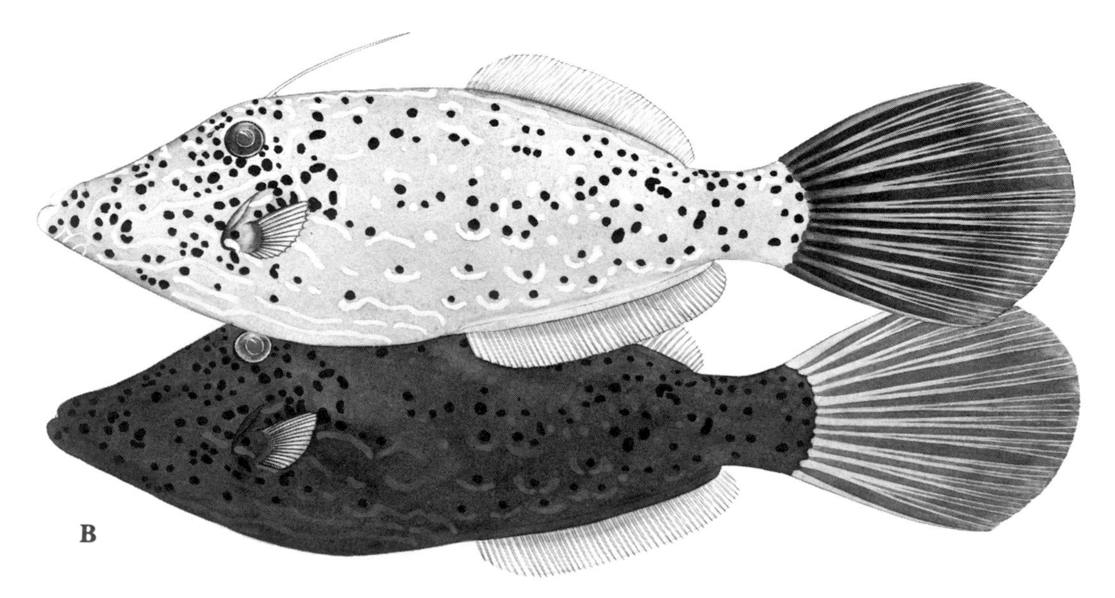

X: PLATE 43

FIGURE a.—*Balistapus undulatus* (Mungo Park). Length 35 cm. (From Marshall)

FIGURE b.—*Balistes vetula* Linnaeus. Length 38 cm. (From Evermann and Marsh)

FIGURE c.—*Balistoides conspicillum* Bloch and Schneider. Length 35 cm. (T. Kumada)

X: PLATE 44

FIGURE a.—*Odonus niger* (Rüppell). Length 30 cm. (K. Tomita)

FIGURE b.—*Pseudobalistes flavimarginatus* (Rüppell). Length 30 cm. (S. Arita)

FIGURE c.—*Amanses sandwichiensis* (Quoy and Gaimard). Length 25 cm. (From Herre)

X: PLATE 45

FIGURE a.—*Lactoria cornuta* (Linnaeus). Length 40 cm. (From Tomiyama and Abe)

FIGURE b.—*Stephanolepis setifer* (Bennett). Length 11 cm. (From Fraser-Brunner)

FIGURE c.—*Acanthostracion quadricornis* (Linnaeus). Length 25 cm. (From Evermann and Marsh)

FIGURE d.—*Ostracion meleagris* Shaw. Length 20 cm. (From Jordan and Evermann)

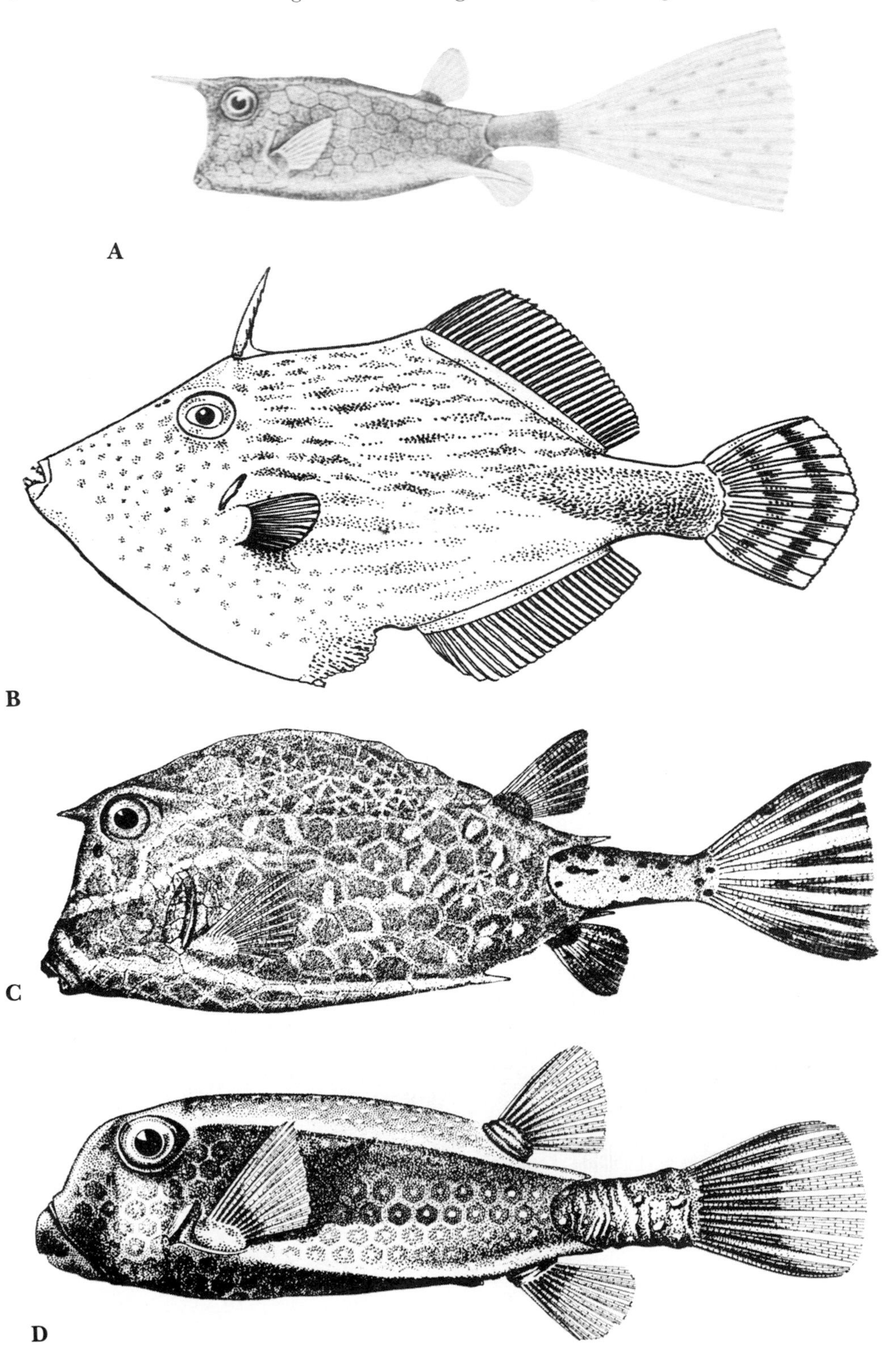

X: PLATE 46

FIGURE a.—*Histrio histrio* (Linnaeus). Length 20 cm. (S. Arita)

FIGURE b.—*Lophiomus setigerus* (Vahl). Length 60 cm. (T. Kumada)

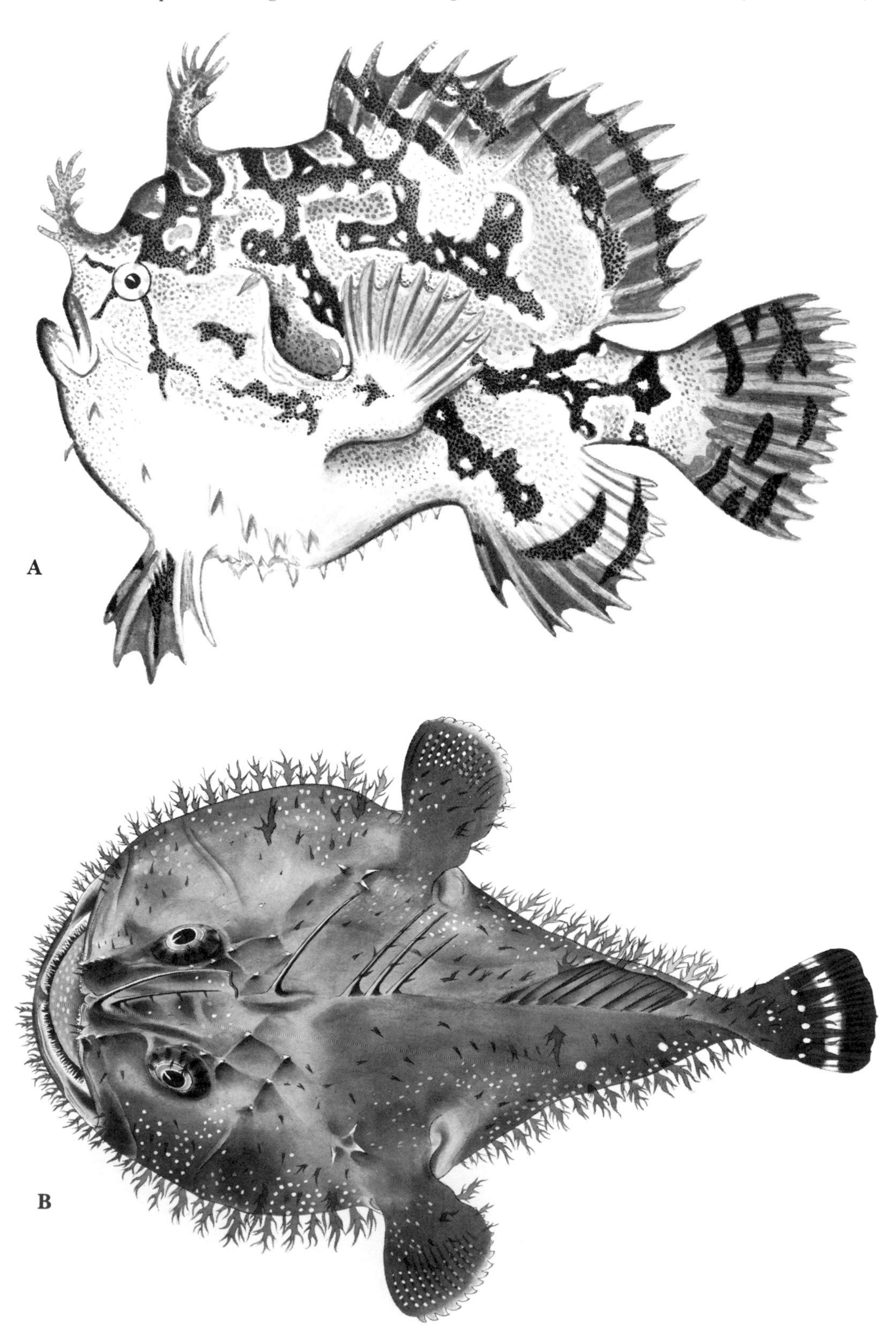

X: PLATE 47

Plants used by Pacific Islanders for the treatment of fish poisoning.

FIGURE a.—*Duboisia myoporoides* Brown.

FIGURE b.—*Pithecellobium dulce* (Roxburgh).

FIGURE c.—*Tournefortea (Messerschmidia) argentea* Linnaeus. Taken at Wake Island.

FIGURE d.—*Euphorbia atoto* Forster and Forster.

FIGURE e.—*Morinda citrifolia* Linnaeus. Taken at Pago Point, Guam, Mariana Islands.
(Courtesy R. Morgan, San Diego Natural History Museum)

X: PLATE 48

Plants used by Pacific Islanders for the treatment of fish poisoning.

FIGURE a.—*Artocarpus* sp? Guam, Mariana Islands.

FIGURE b.—*Carica papaya* Linnaeus. Guam, Mariana Islands. Tree showing cluster of fruit.

FIGURE c.—*Scaevola taccada* (Gaertner). Guam, Mariana Islands. Tree.
(Courtesy R. Morgan, San Diego Natural History Museum)

FIGURE d.—*Vitex trifolia* Linnaeus. Guam, Mariana Islands.
(Courtesy Smithsonian Institution)

X: PLATE 49

Plants used in the West Indies for the treatment of fish poisoning.

Figure a.—*Conocarpus erectus* Linnaeus. Dade County, Florida.

Figure b.—Herbarium sheet of *C. erectus*. Specimen collected from Lee County, Fla., University of Miami Herbarium.

Figure c.—*Hypis capitata* Jacquin. Specimen collected at Fort Myers, Fla. University of Miami Herbarium.

Figure d.—*Passiflora foetida* Linnaeus. (Courtesy T. R. Alexander)

A B C

D

Class Osteichthyes–Clupeotoxic Fishes

PLATE

CHAPTER XI

XI: PLATE 1

FIGURE a.—*Harengula zunasi* Bleeker. Length 15 cm. (From Okada)

FIGURE b.—*Ilisha africana* (Bloch). Length 12 cm. (Mrs. R. Kreuzinger, after Prosvirov)

FIGURE c.—*Sardinella fimbriata* (Valenciennes). Length 15 cm. (From Day)

FIGURE d.—*Sardinella sindensis* (Day). Length 20 cm. (From Day)

FIGURE e.—*Engraulis japonicus* Schlegel. Length 13 cm. (S. Arita)

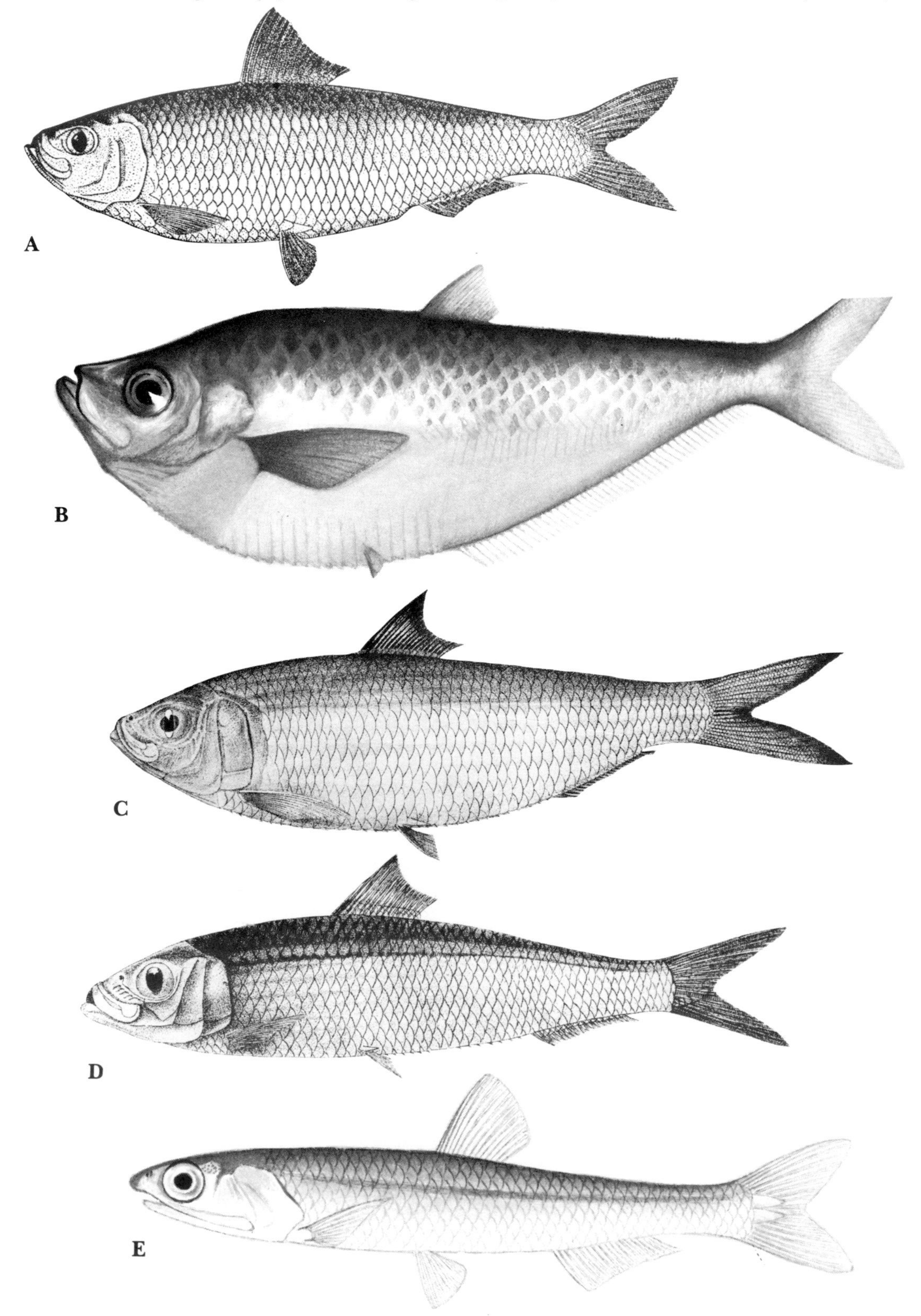

Class Osteichthyes–Gempylotoxic Fishes

PLATE

CHAPTER XII

XII: PLATE 1

Ruvettus pretiosus Cocco. Length 1.3 m. (From Hiyama)

Class Osteichthyes–Scombrotoxic Fishes

PLATES

CHAPTER XIII

XIII: PLATE 1

FIGURE a.—*Euthynnus pelamis* (Linnaeus). Length 1 m. (T. Kumada)

FIGURE b.—*Sarda sarda* (Bloch). length 91 cm.

(Courtesy National Geographic Society)

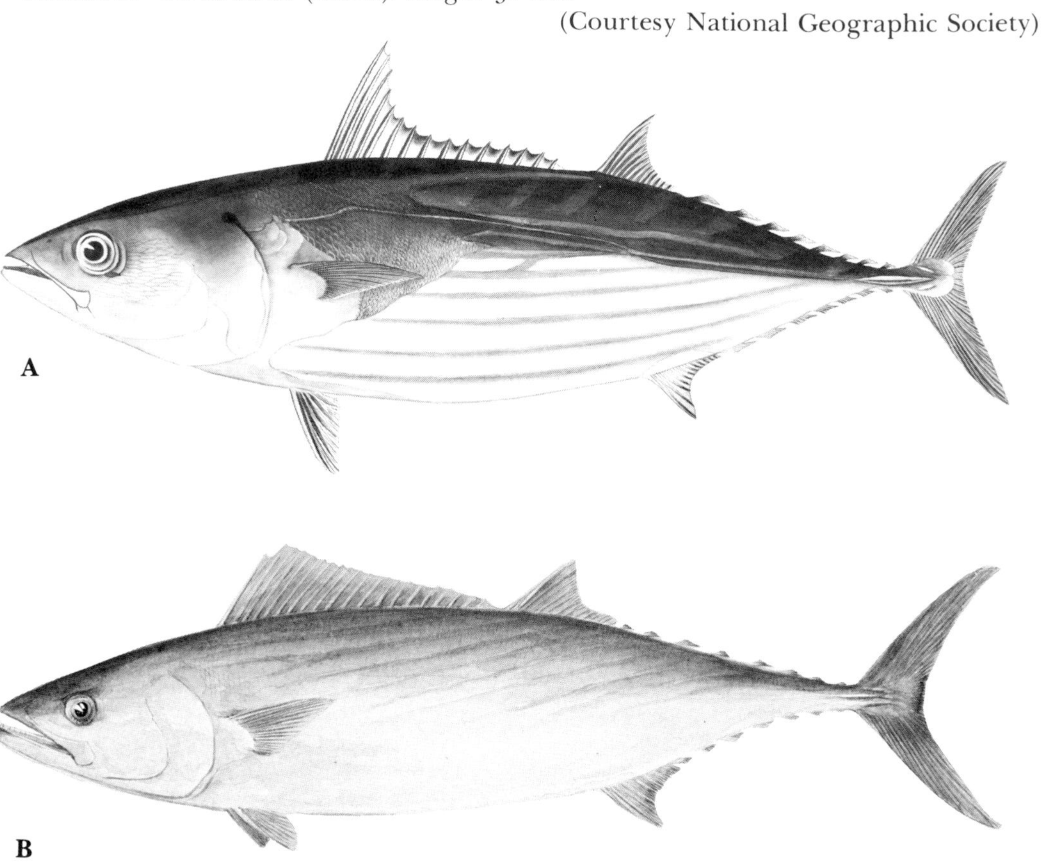

XIII: PLATE 2

FIGURE a.—*Scomber scombrus* Linnaeus. Length 61 cm.

FIGURE b.—*Scomberomorus regalis* (Bloch). Length 1.8 m.

(Courtesy National Geographic Society)

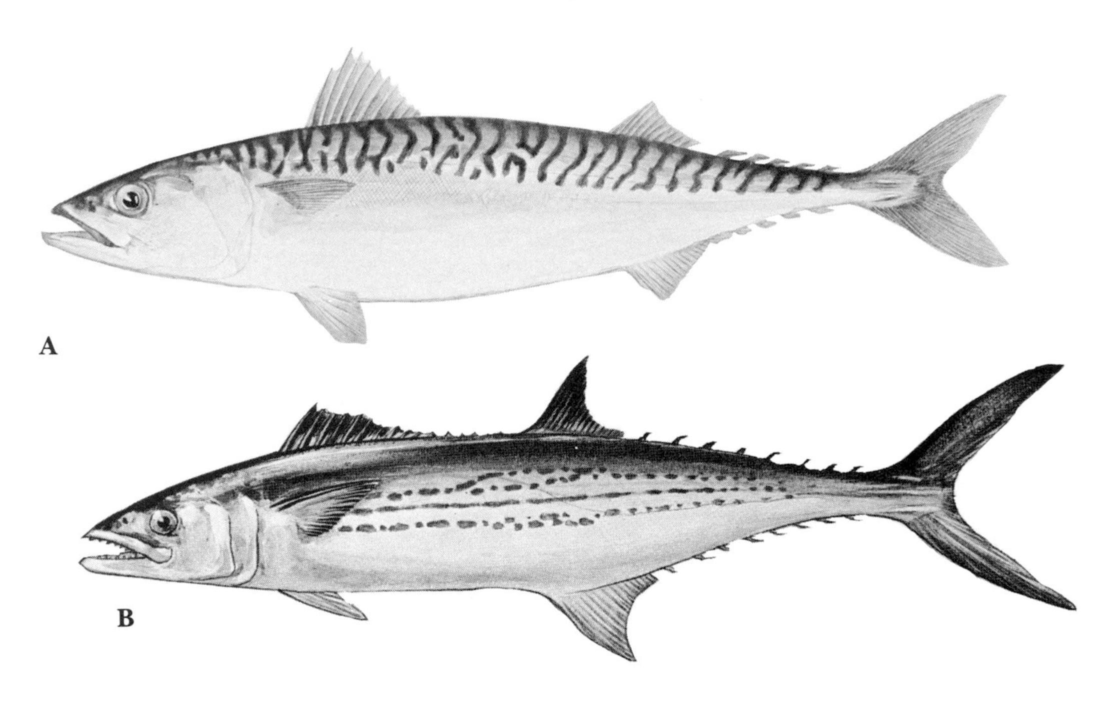

XIII: PLATE 3

FIGURE a.—*Thunnus alalunga* (Bonnaterre). Length 1.3 m.

FIGURE b.—*Thunnus thynnus* (Linnaeus). Length 3 m. (From Hiyama)

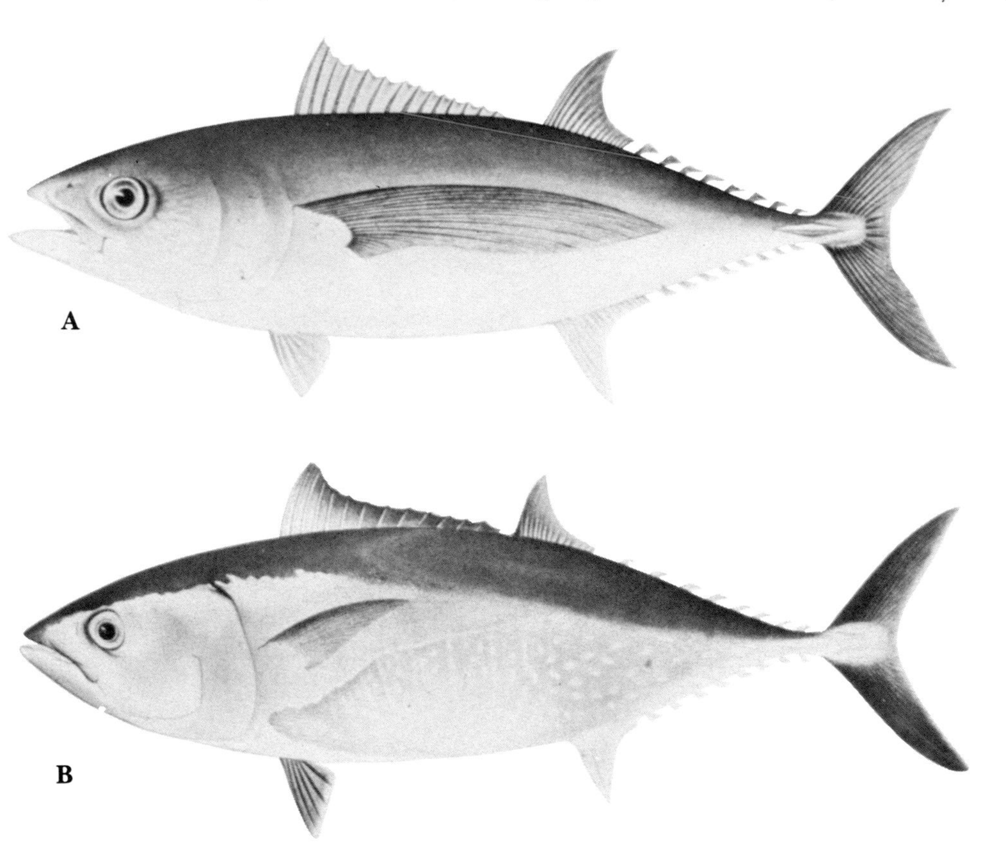

Class Osteichthyes–Tetrodotoxic Fishes

PLATES

CHAPTER XIV

XIV: PLATE 1

FIGURE a.—*Chilomycterus affinis* Günther. Length 17 cm. (M. Shirao)

FIGURE b.—*Chilomycterus antennatus* (Cuvier). Length 50 cm. (From Prosvirov)

FIGURE c.—*Chilomycterus orbicularis* (Bloch). Length 26 cm. (S. Arita)

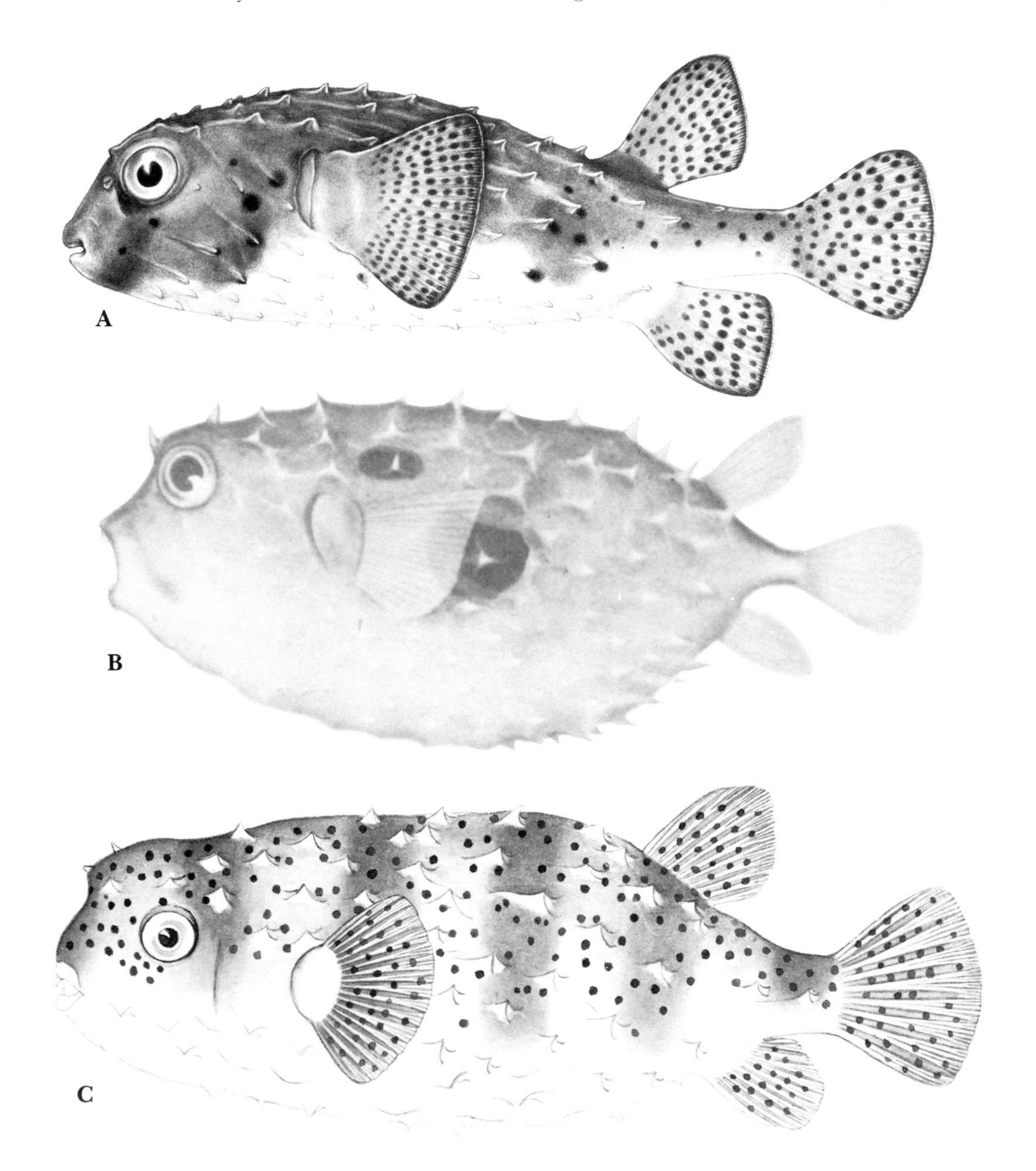

XIV: PLATE 2

FIGURE a.—*Canthigaster rivulatus* (Temminck and Schlegel). Length 10 cm. (M. Shirao)

FIGURE b.—*Chilomycterus atinga* (Linnaeus). Length 20 cm. (M. Shirao)

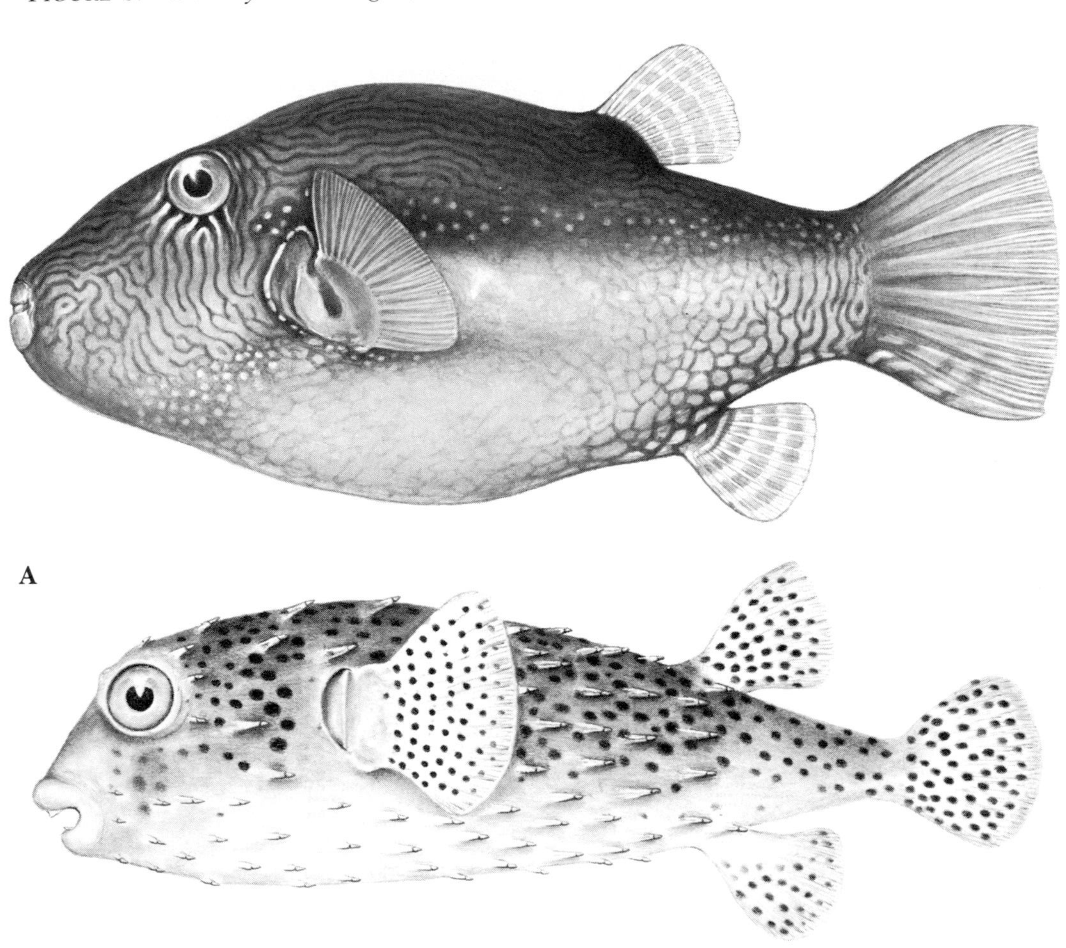

XIV: PLATE 3

FIGURE a.—*Chilomycterus tigrinus* (Cuvier). Length 20 cm.

FIGURE b.—*Mola mola* (Linnaeus). Length over 3 m. (M. Shirao)

XIV: PLATE 4

FIGURE a.—*Diodon holacanthus* Linnaeus. Length 50 cm.

FIGURE b.—*Diodon hystrix* Linnaeus. Length 90 cm. (M. Shirao)

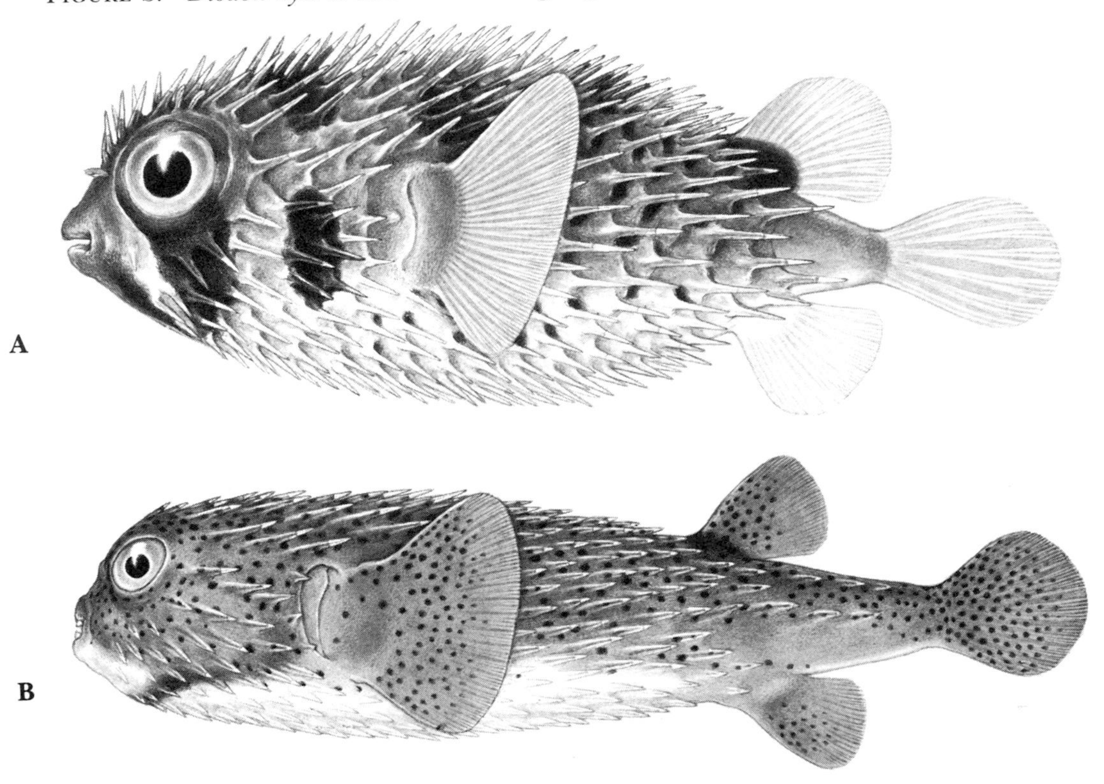

XIV: PLATE 5

FIGURE a.—*Amblyrhynchotes honckeni* (Bloch). Length 30 cm.

FIGURE b.—*Arothron aerostaticus* (Jenyns). Length 54 cm. (M. Shirao)

FIGURE c.—*Arothron setosus* (Smith). Length 30 cm. (From Jordan and Evermann)

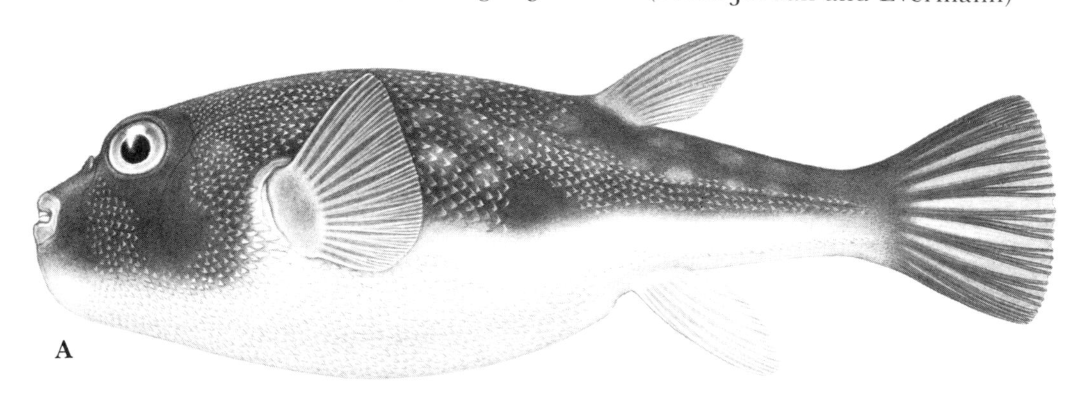

A

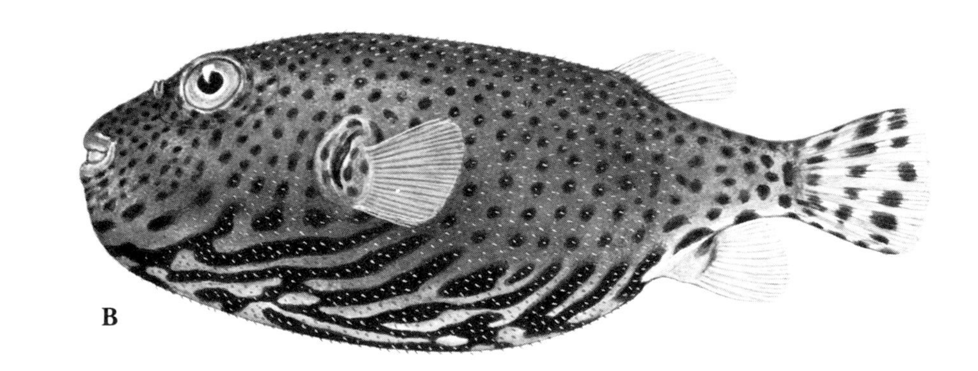

B

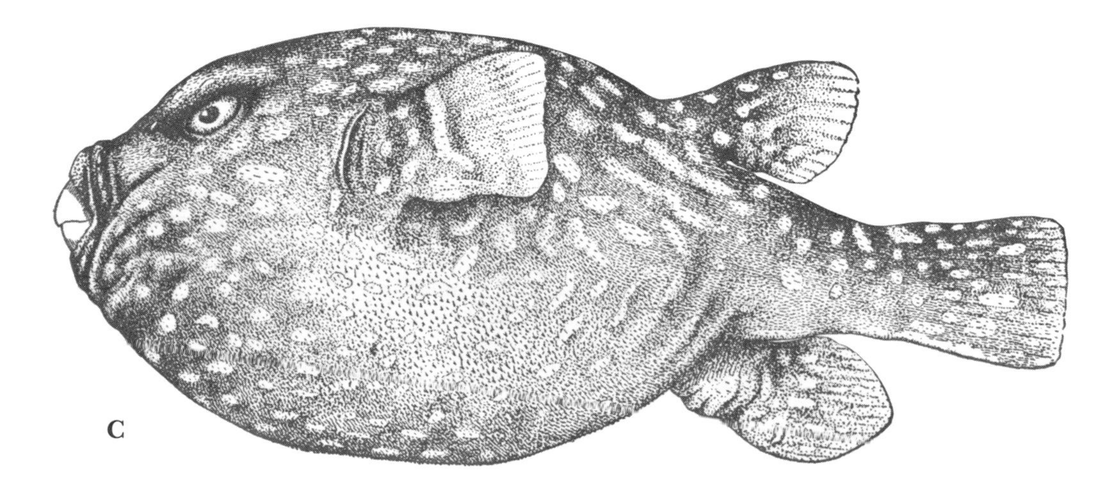

C

XIV: PLATE 6

FIGURE a.—*Arothron hispidus* (Linnaeus). Length 53 cm.

FIGURE b.—*Arothron meleagris* (Lacépède). Length 32 cm.

FIGURE c.—*Arothron nigropunctatus* (Bloch and Schneider). Length 25 cm. (S. Arita)

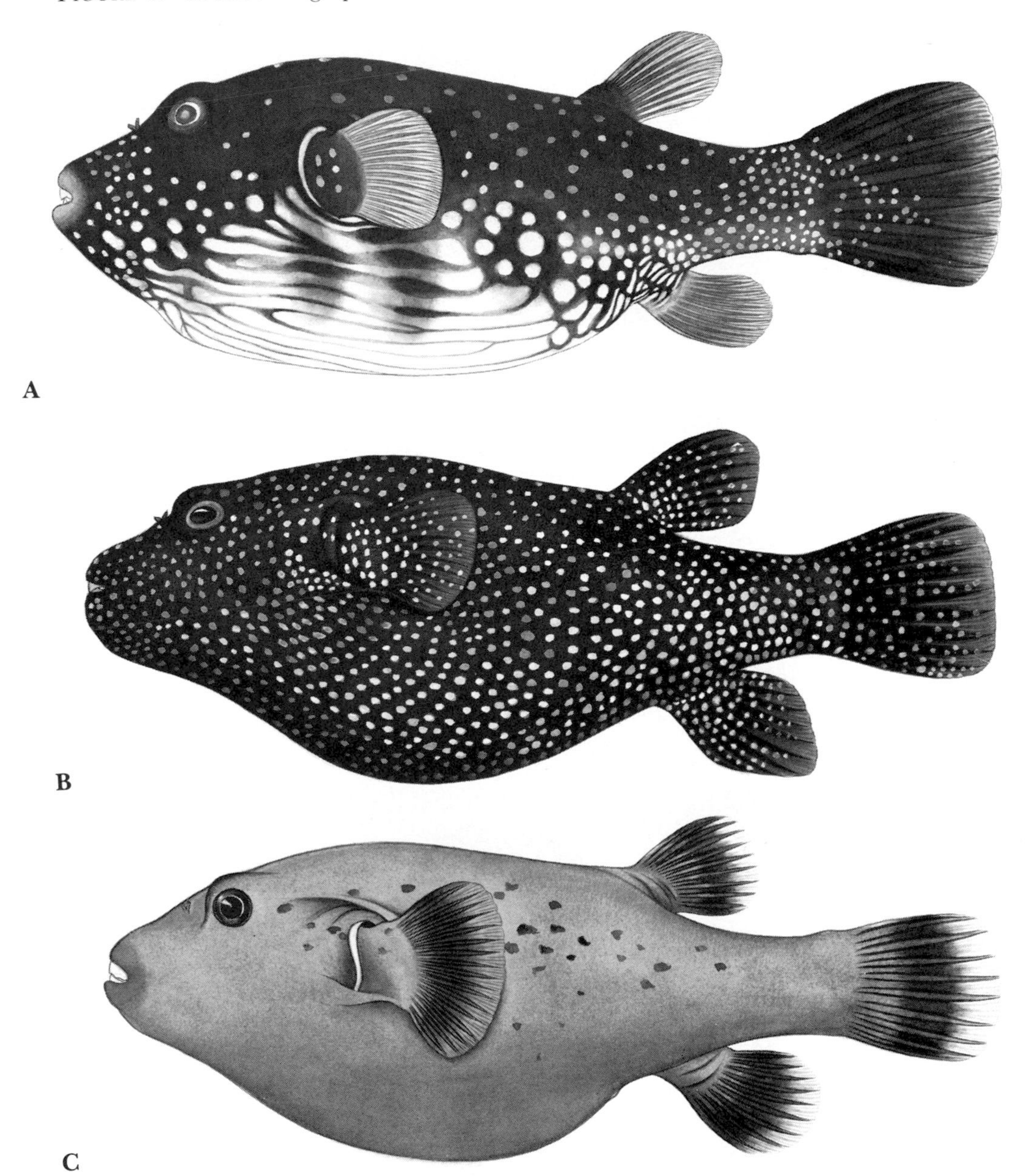

XIV: PLATE 7

FIGURE a.—*Arothron reticularis* (Bloch and Schneider). Length 42 cm.

FIGURE b.—*Boesemanichthys firmamentum* (Temminck and Schlegel). Length 35 cm. 1. Adult; 2. Juvenile (6 cm).

FIGURE c.—*Fugu chrysops* (Hilgendorf). Length 20 cm. (M. Shirao)

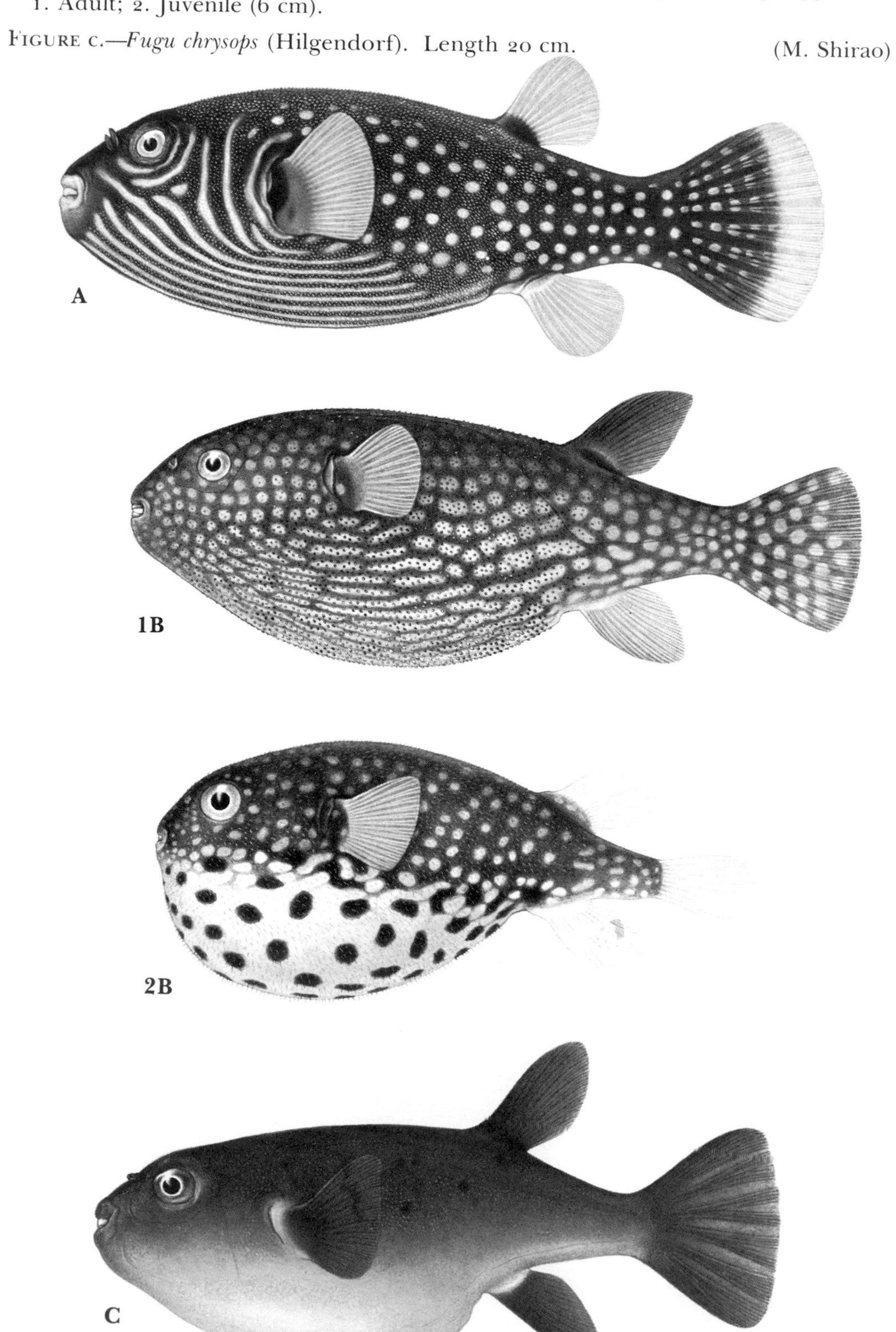

XIV: PLATE 8

Arothron stellatus (Bloch and Schneider). Length 90 cm.
a. Adult; b. Juvenile (approx. 10 cm). (M. Shirao)

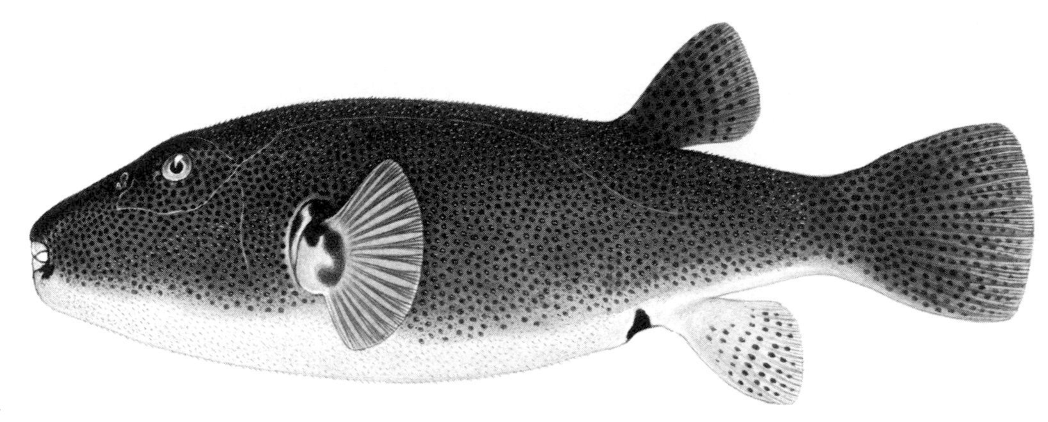

A

B

XIV: PLATE 9

FIGURE a.—*Chelonodon patoca* (Hamilton-Buchanan). Length 33 cm.

FIGURE b.—*Ephippion guttifer* (Bennett). Length 70 cm.

FIGURE c.—*Fugu basilevskianus* (Basilewsky). Length 35 cm. (M. Shirao)

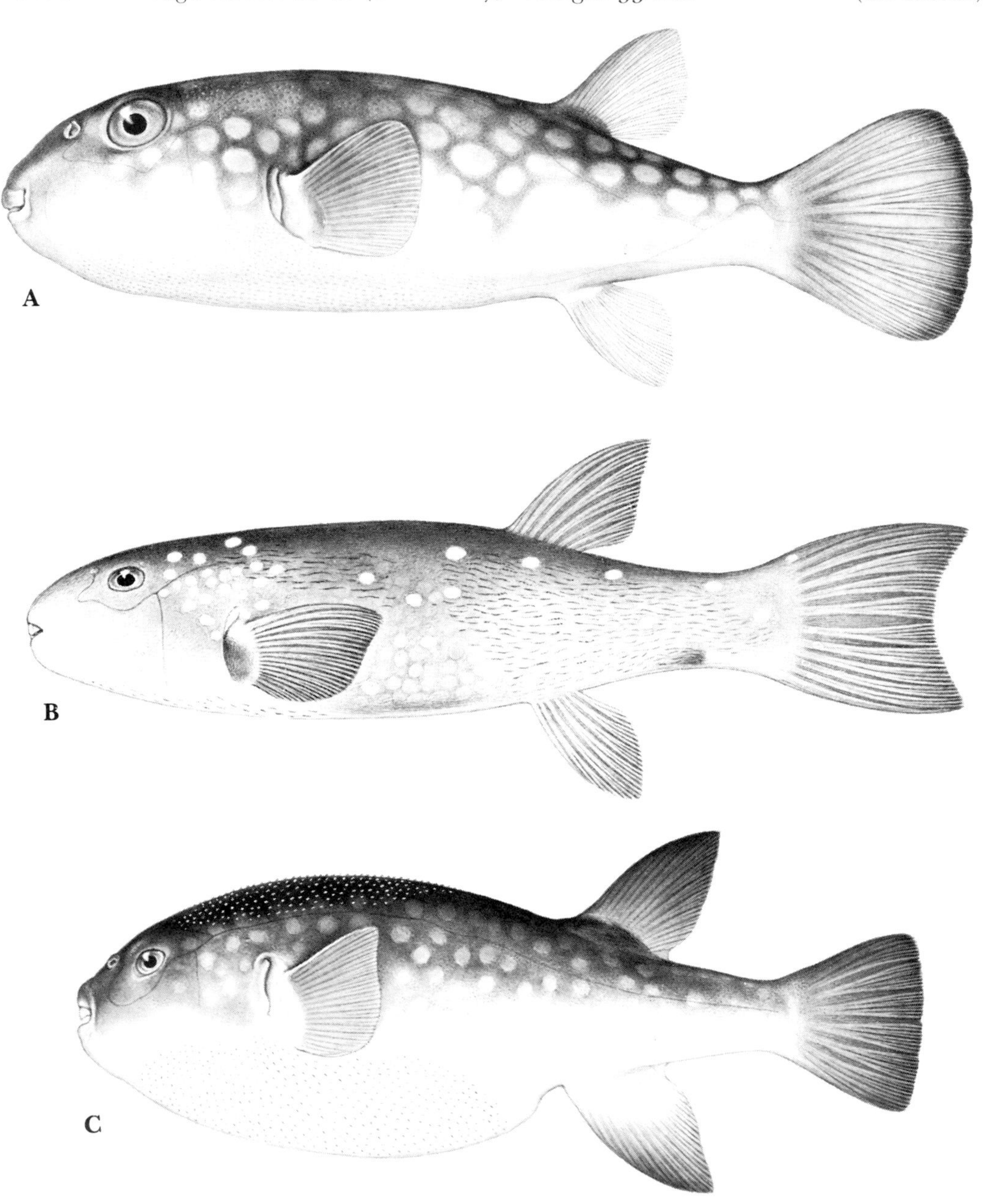

XIV: PLATE 10

FIGURE a.—*Fugu niphobles* (Jordan and Snyder). Length 16 cm.

FIGURE b.—*Fugu pardalis* (Temminck and Schlegel). Length 36 cm. (M. Shirao)

XIV: PLATE 11

FIGURE a.—*Fugu oblongus* (Bloch and Schneider). Length 26 cm.

FIGURE b.—*Fugu ocellatus obscurus* (Abe). Length 12 cm.

FIGURE c.—*Fugu ocellatus ocellatus* (Linnaeus). Length 28 cm.

FIGURE d.—*Fugu poecilonotus* (Temminck and Schlegel). Length 25 cm. (M. Shirao)

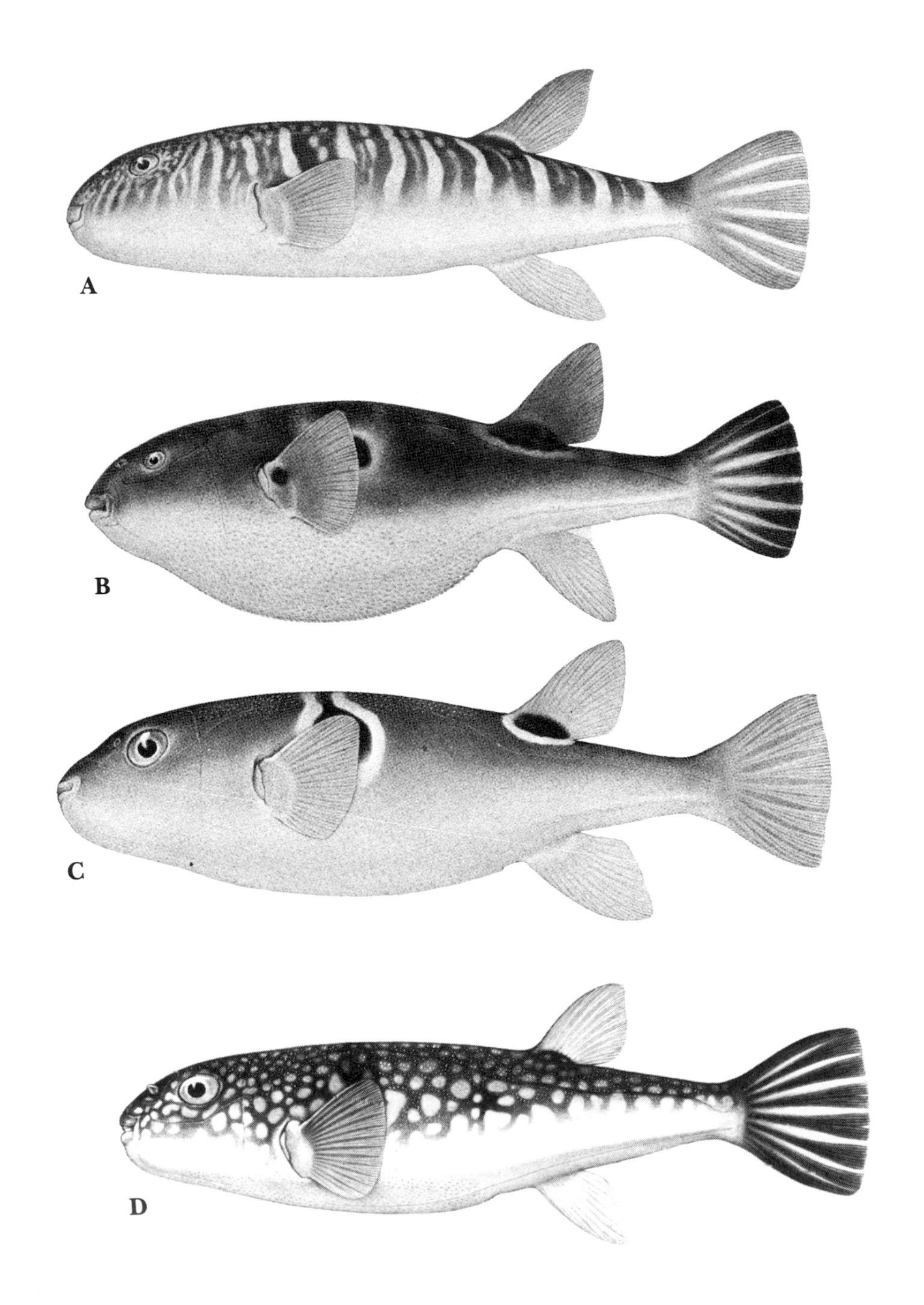

XIV: PLATE 12

FIGURE a.—*Fugu rubripes chinensis* (Abe). Length 55 cm. This species differs from *F. rubripes rubripes* in having a black anal fin, and the irregular black spots behind the ocellated humeral blotch are not as conspicuous. Otherwise the color is similar to *F. rubripes rubripes*.

FIGURE b.—*Fugu rubripes rubripes* (Temminck and Schlegel). Length 45 cm.

FIGURE c.—*Fugu stictonotus* (Temminck and Schlegel). Length 40 cm.

FIGURE d.—*Fugu xanthopterus* (Temminck and Schlegel). Length 60 cm. (M. Shirao)

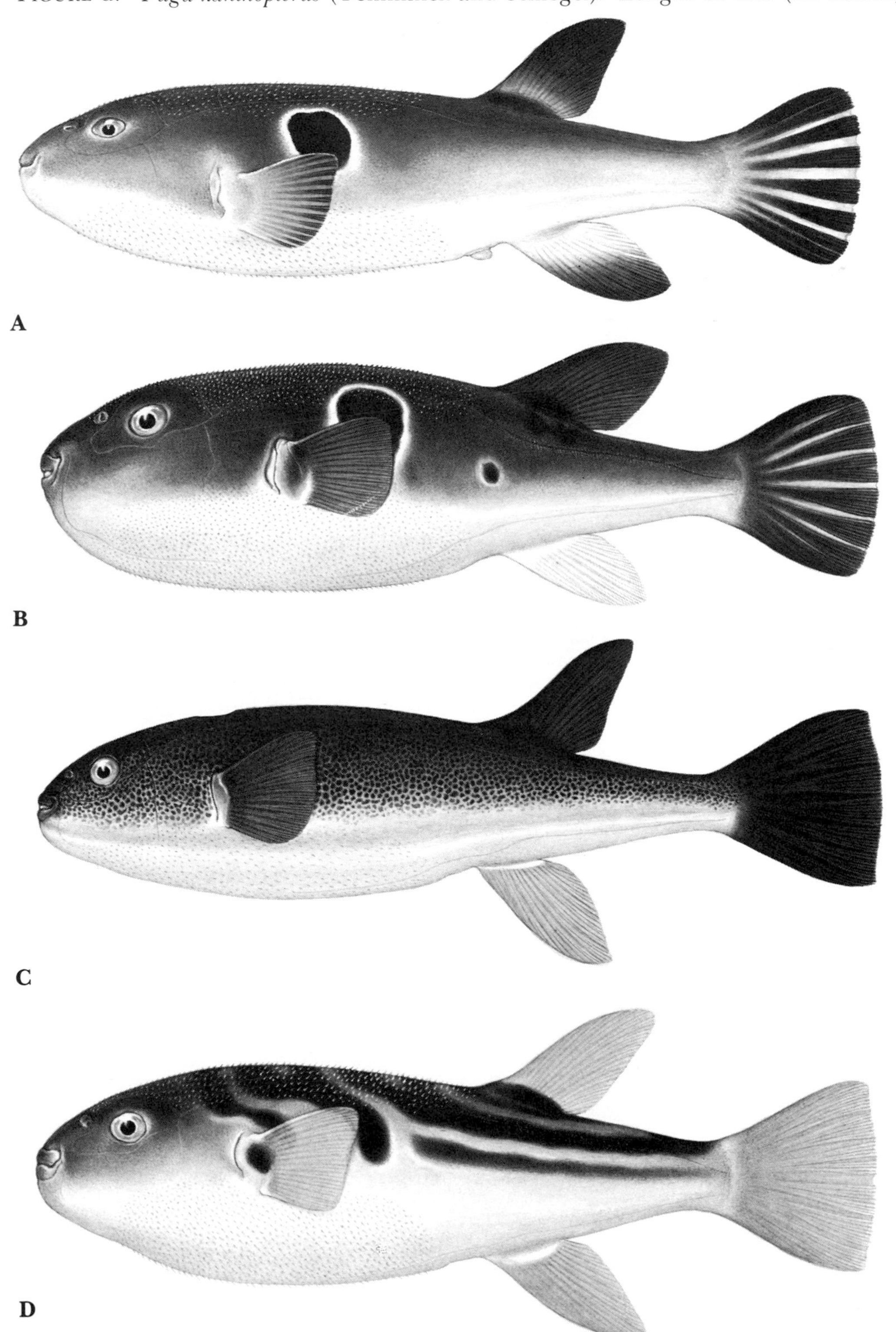

XIV: PLATE 13

FIGURE a.—*Fugu vermicularis porphyreus* (Temminck and Schlegel). Length 47 cm.
1. Adult; 2. Juvenile (approx. 5 cm).

FIGURE b.—*Fugu vermicularis radiatus* (Abe). Length 25 cm.

FIGURE c.—*Fugu vermicularis vermicularis* (Temminck and Schlegel). Length 33 cm.
(M. Shirao)

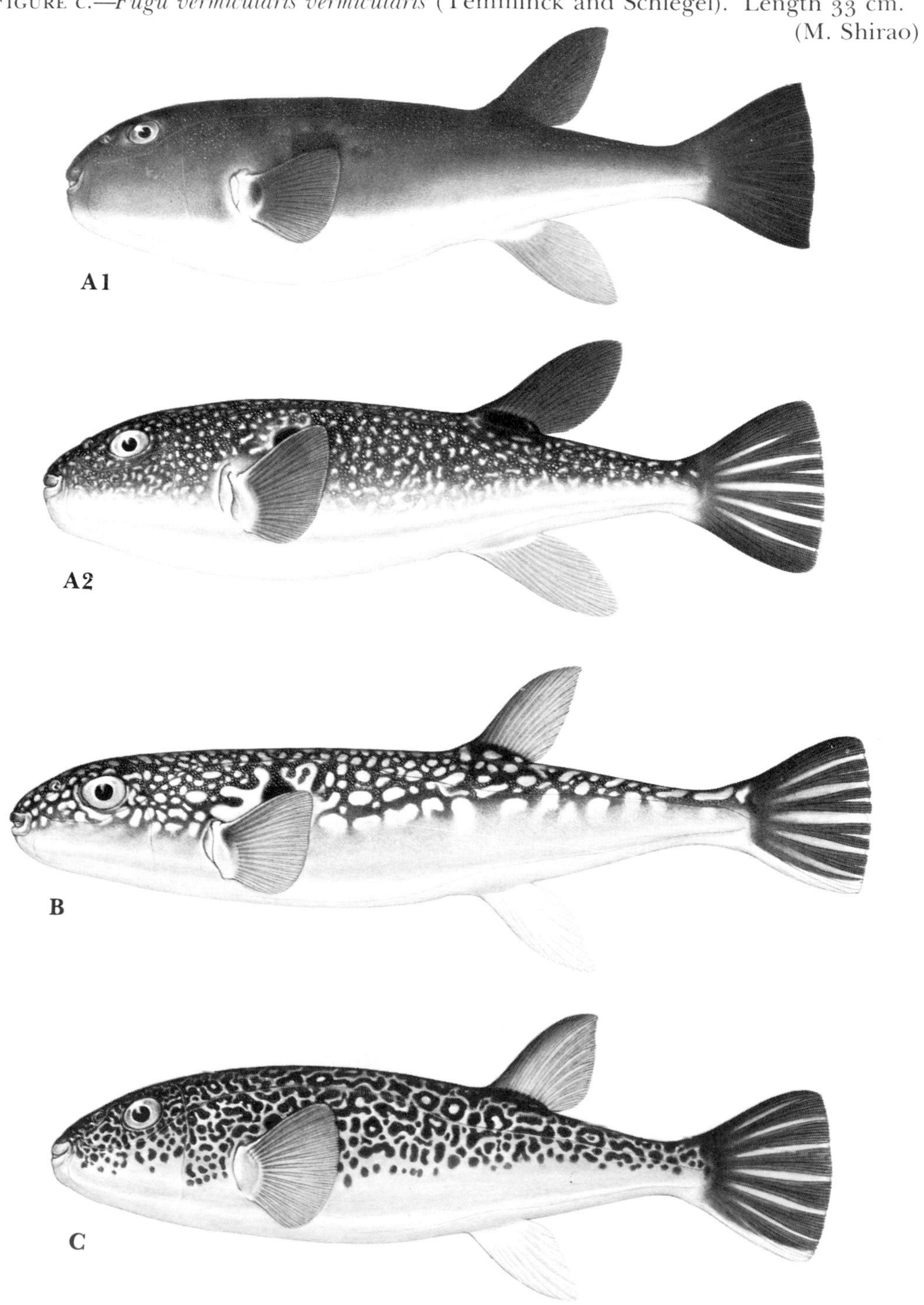

XIV: PLATE 14

FIGURE a.—*Lagocephalus laevigatus inermis* (Temminck and Schlegel). Length 60 cm. (M. Shirao)

FIGURE b.—*Lagocephalus laevigatus laevigatus* (Linnaeus). Length 60 cm. (From Prosvirov)

FIGURE c.—*Lagocephalus lagocephalus* (Linnaeus). Length 40 cm. (From Joubin)

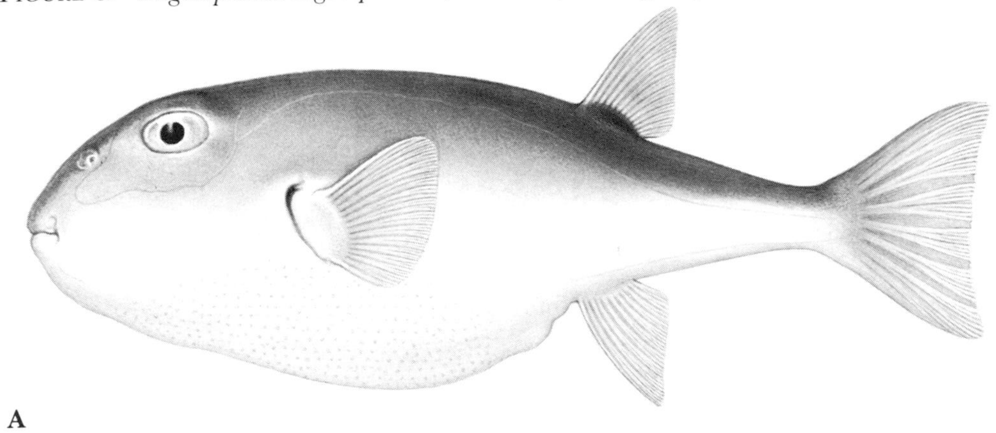

A

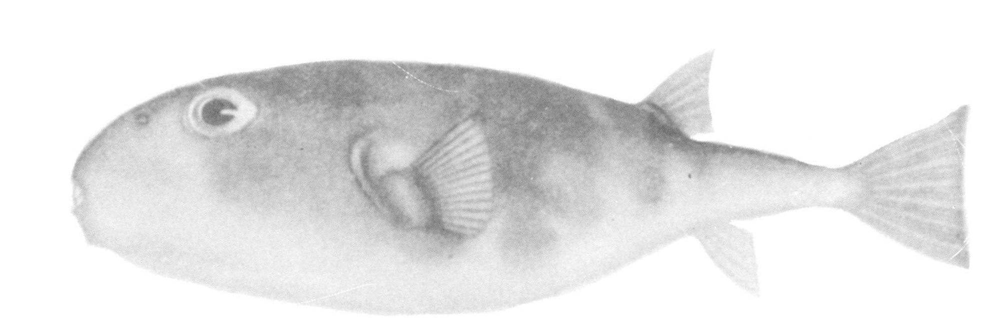

B

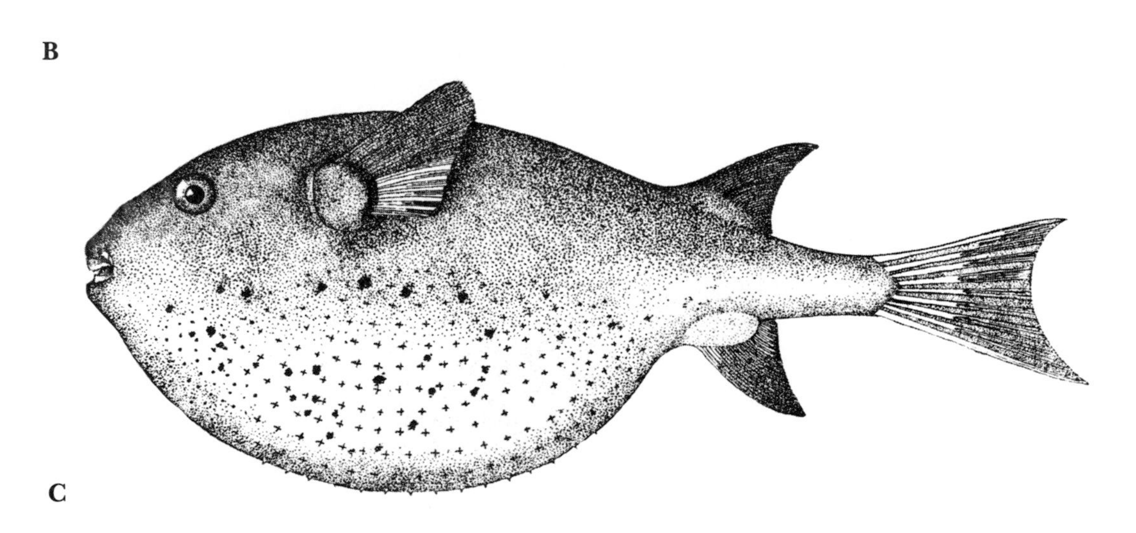

C

XIV: PLATE 15

FIGURE a.—*Lagocephalus lunaris* (Bloch and Schneider). Length 30 cm.

FIGURE b.—*Lagocephalus oceanicus* Jordan and Evermann. Length 40 cm.

FIGURE c.—*Lagocephalus sceleratus* (Forster). Length 75 cm. This species is famous in the annals of marine biotoxicology because it is the fish that almost terminated Captain Cook's second world voyage at New Caledonia on September 7, 1774. Ingestion of this fish made Cook and several members of his crew violently ill. (M. Shirao)

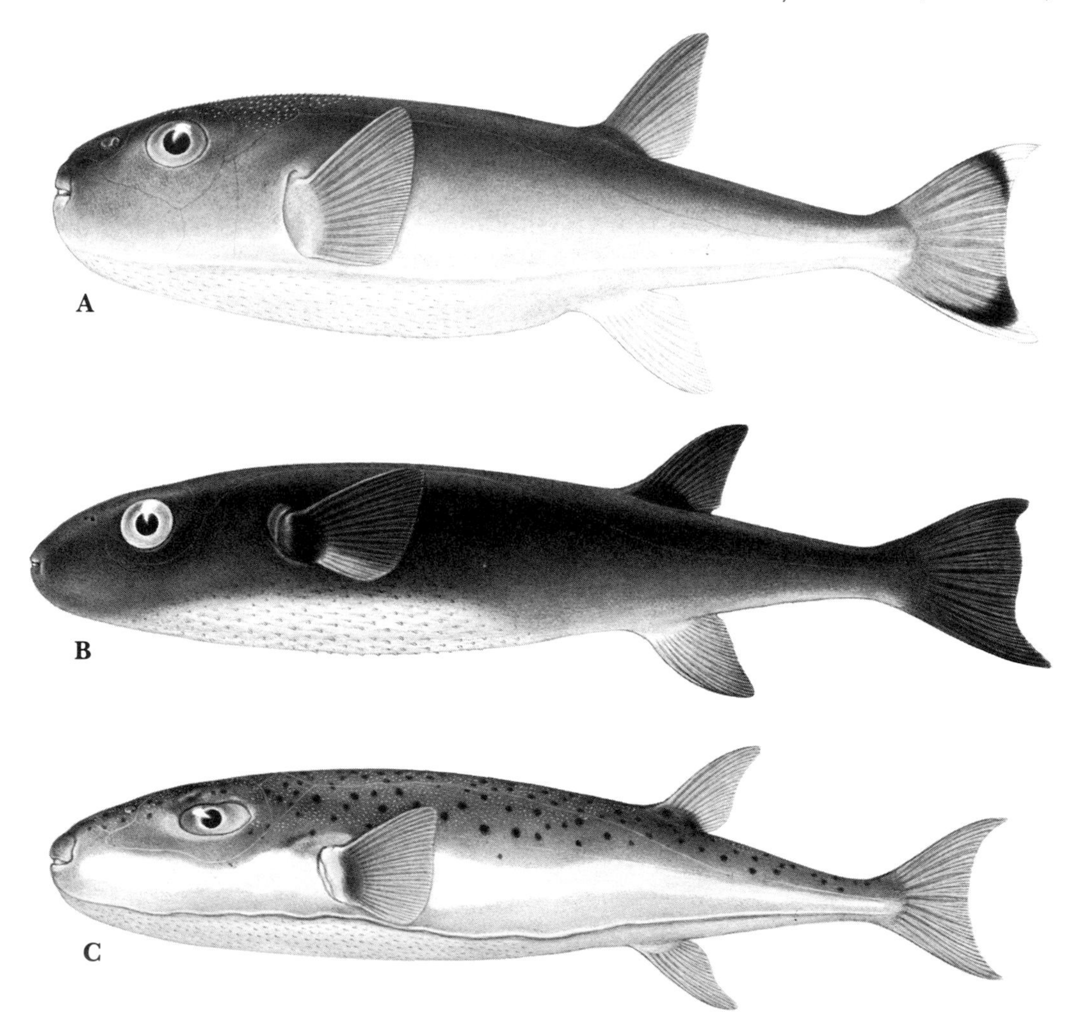

XIV: PLATE 16

FIGURE a.—*Sphaeroides annulatus* (Jenyns). Length 28 cm. (T. Kumada)

FIGURE b.—*Sphaeroides maculatus* (Bloch and Schneider). Length cm. (M. Shirao)

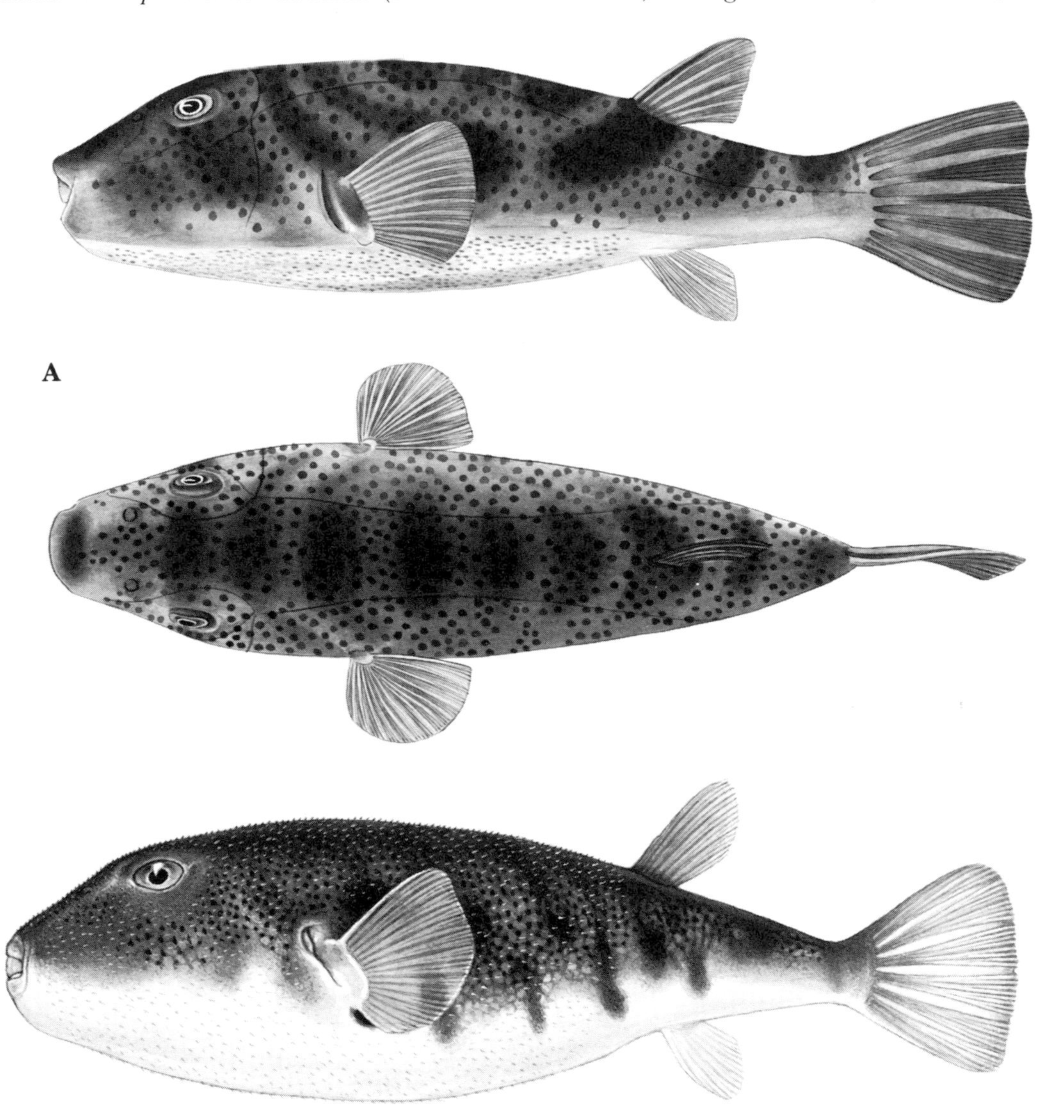

B

XIV: PLATE 17

FIGURE a.—*Sphaeroides spengleri* (Bloch). Length 20 cm.

FIGURE b.—*Sphaeroides testudineus* (Linnaeus). Length 21 cm.

FIGURE c.—*Tetraodon lineatus* Linnaeus. Length 45 cm.

FIGURE d.—*Torquigener hamiltoni* (Gray and Richardson). Length 9 cm. (M. Shirao)

Class Osteichthyes–Ichthyootoxic Fishes

PLATES

CHAPTER XV

XV: PLATE 1

FIGURE a.—*Acipenser güldenstädti* Brandt. Length 1.4 m.

FIGURE b.—*Acipenser sturio* Linnaeus. Length 3.0 m.

(From *Fish Resources of the USSR*)

XIV: PLATE 2

Huso huso (Linnaeus). Length 1.8 m. This fish attains enormous size. One specimen captured near the estuary of the Ural measured over 8 m and weighed more than 1,000 kg. (From *Fish Resources of the USSR)*

XV: PLATE 3

Lepisosteus tristroechus (Bloch and Schneider). Length 3 m.
(From Jordan and Evermann)

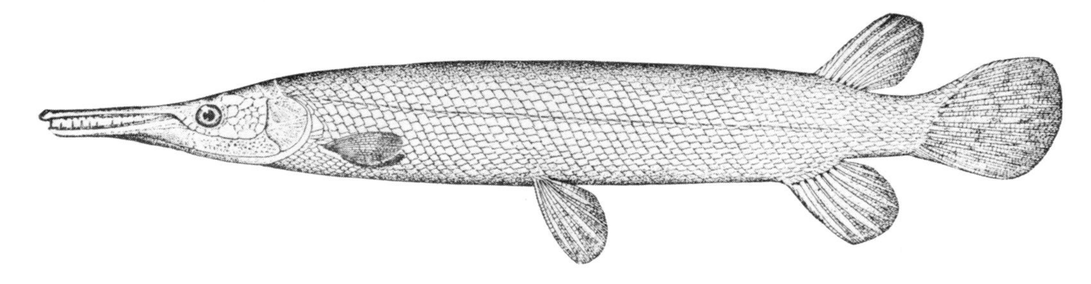

XV: PLATE 4

FIGURE a.—*Salmo salar* Linnaeus. Length 1.4 m.

FIGURE b.—*Stenodus leucichthys* (Güldenstädt). Length 71 cm.
(From *Fish Resources of the USSR*)

XV: PLATE 5

Esox lucius Linnaeus. Length 34 cm. (From *Fish Resources of the USSR*)

XV: PLATE 6

FIGURE a.—*Abramis brama* (Linnaeus). Length 35 cm.

FIGURE b.—*Barbus barbus* (Linnaeus). Length 85 cm. (M. Shirao)

FIGURE c.—*Cyprinus carpio* Linnaeus. Length 1 m. (From *Fish Resources of the USSR*)

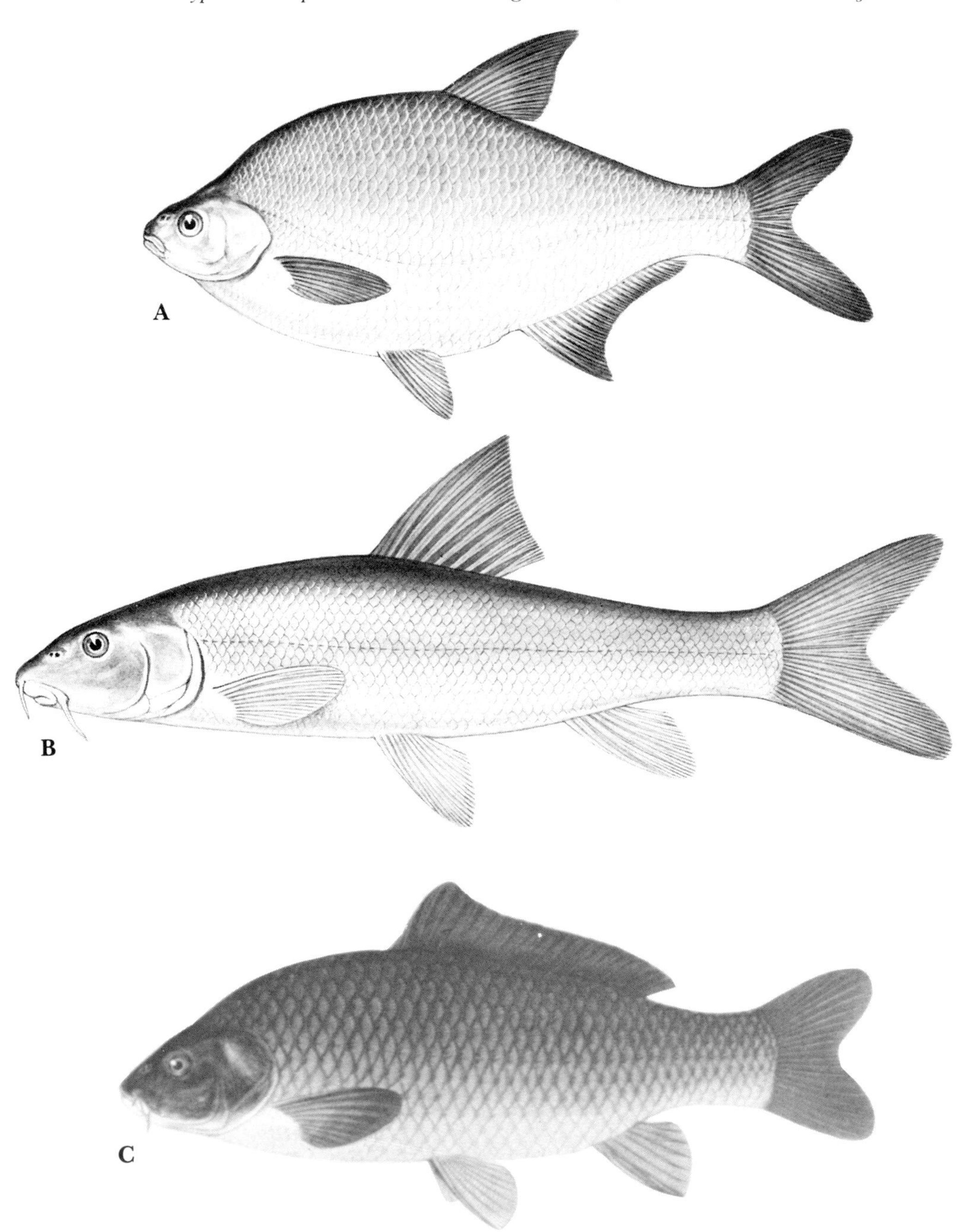

XV: PLATE 7

FIGURE a.—*Diptychus dybowski* Kessler. Length 42 cm.

FIGURE b.—*Schizothorax argentatus* Kessler. Length 80 cm. (From *Fish Resources of the USSR*)

FIGURE c.—*Schizothorax intermedius* McClelland. Length 47 cm. (From Day)

FIGURE d.—*Tinca tinca* (Linnaeus). Length 60 cm. (M. Shirao)

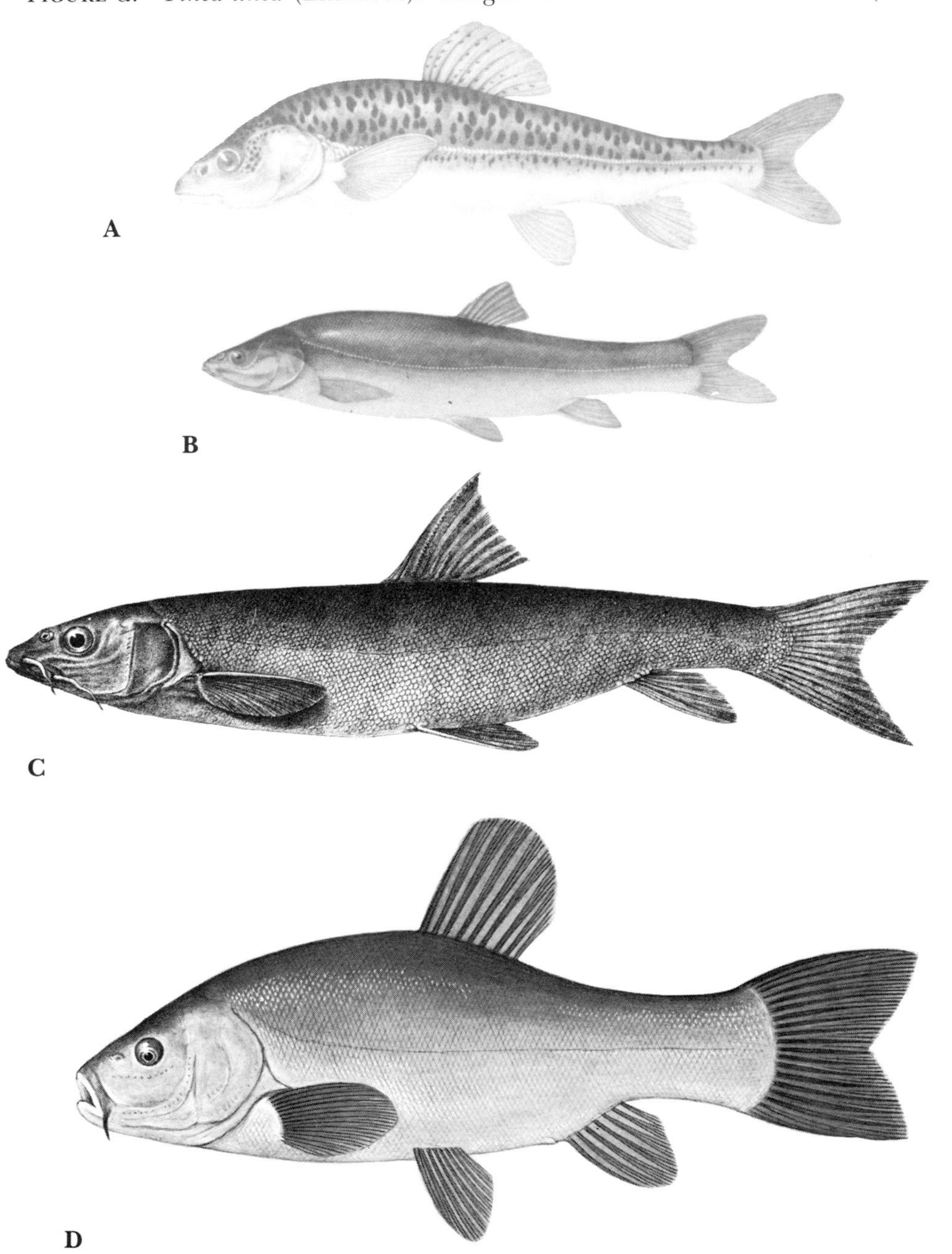

XV: PLATE 8

FIGURE a.—*Ageneiosus armatus* Lacépède. Length 30 cm. (From Bloch)

FIGURE b.—*Bagre marinus* (Mitchill). Length 35 cm. (From Goode)

FIGURE c.—*Ictalurus catus* (Linnaeus). Length 60 cm. (From Fowler)

FIGURE d.—*Pseudobagrus aurantiacus* (Temminck and Schlegel). Length 21 cm. (S. Arita)

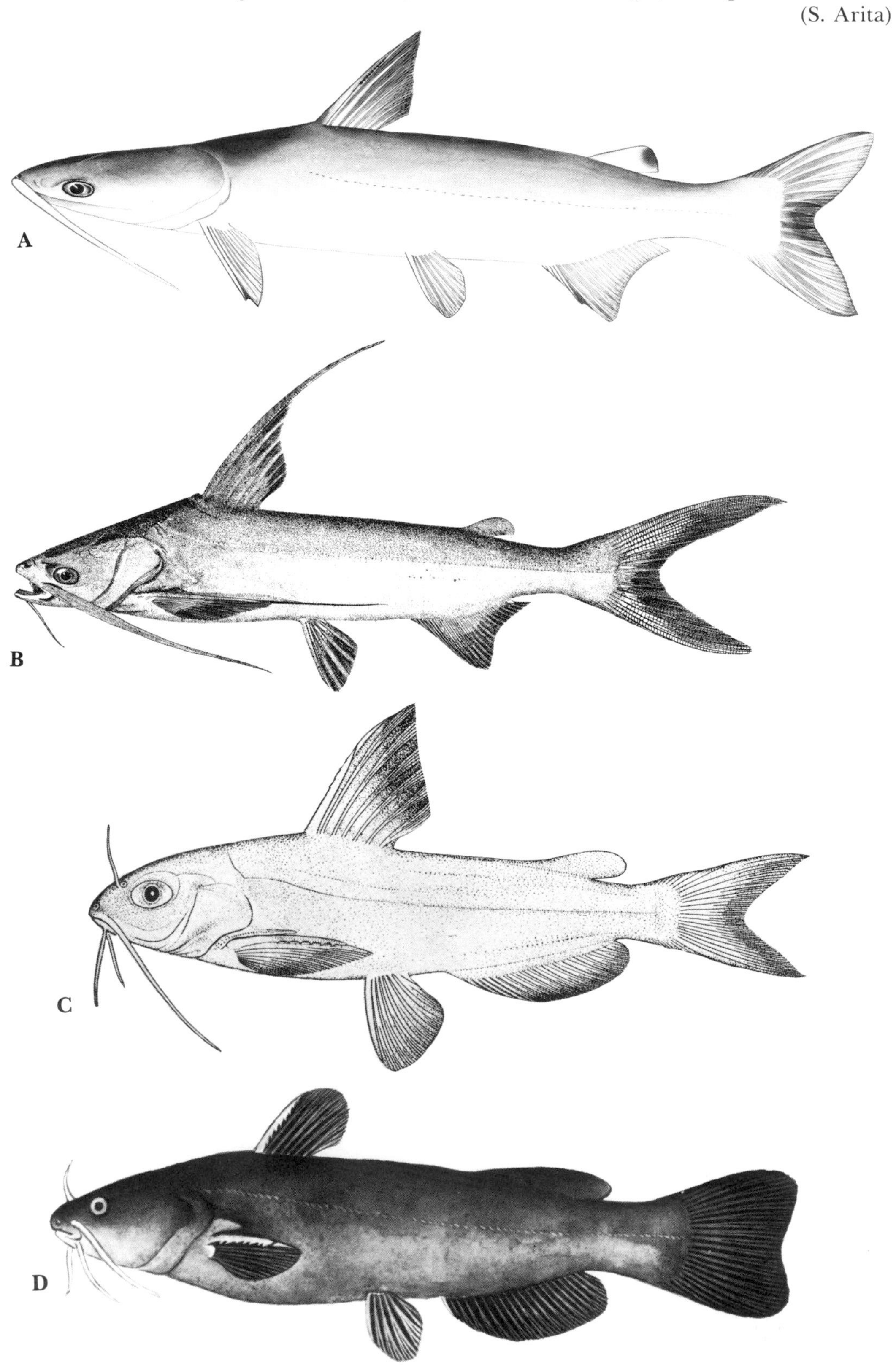

XV: PLATE 9

FIGURE a.—*Parasilurus asotus* (Linnaeus). Length 40 cm. (S. Arita)

FIGURE b.—*Silurus glanis* Linnaeus. Length 3.5 m. (From *Fish Resources of the USSR*)

XV: PLATE 10

Lota lota (Linnaeus). Length 1 m. (From *Fish Resources of the USSR*)

XV: PLATE 11

FIGURE a.—*Aphanius calaritanus* (Cuvier and Valenciennes). Length 5 cm. Female (Left); Male (Right) (From Boulenger)

FIGURE b.—*Fundulus diaphanus* (LeSueur). Length 5 cm. (From Jordan and Evermann)

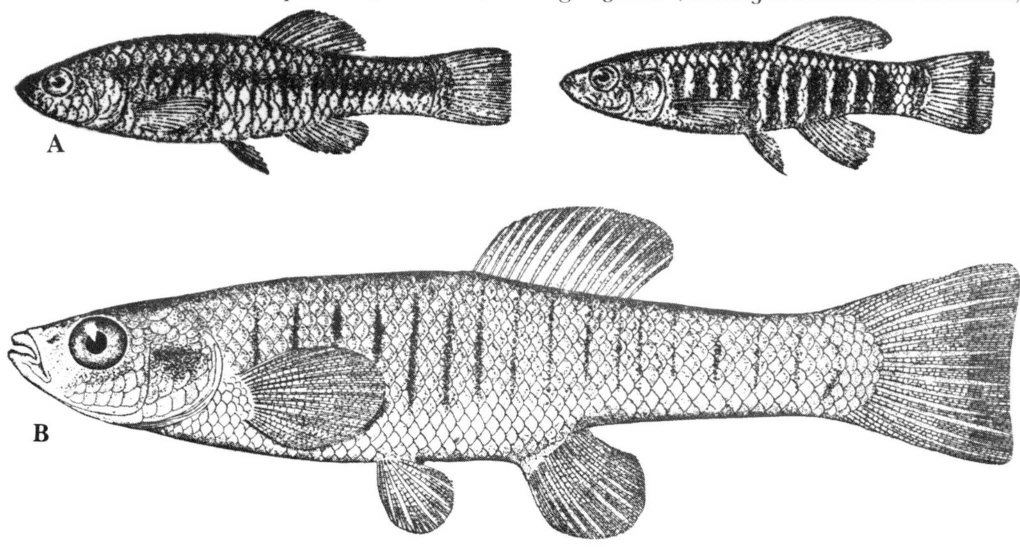

XV: PLATE 12

Scorpaenichthys marmoratus (Ayres). Length 75 cm. (M. Shirao)

XV: PLATE 13

Stichaeus grigorjewi Herzenstein. Length 50 cm. (From Tomiyama, Abe, and Tokioka)

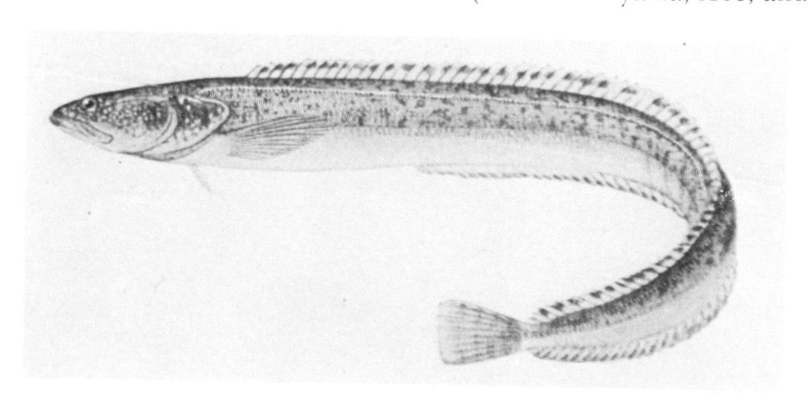

Class Osteichthyes–Ichthyohemotoxic Fishes

PLATES

CHAPTER XVI

XVI: PLATE 1

Anguilla anguilla (Linnaeus). Length 1 m. (From *Fish Resources of the USSR*)

XVI: PLATE 2

FIGURE a.—*Anguilla japonica* Temminck and Schlegel. Length 1 m. (S. Arita)

FIGURE b.—*Anguilla rostrata* (LeSueur). Length 1 m. (From Jordan and Evermann)

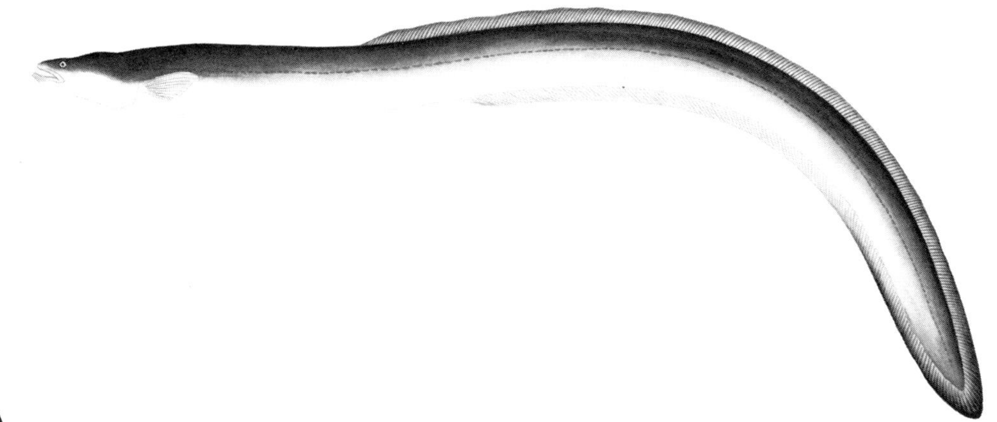

A

B

XVI: PLATE 3

FIGURE a.—*Ariosoma balearica* (De la Roche). Length 40 cm. (From Lozano y Rey)

FIGURE b.—*Conger conger* (Linnaeus). Length 2 m. (From Goode)

FIGURE c.—*Gymnothorax moringa* (Cuvier). Length 1 m. (From Evermann and Marsh)

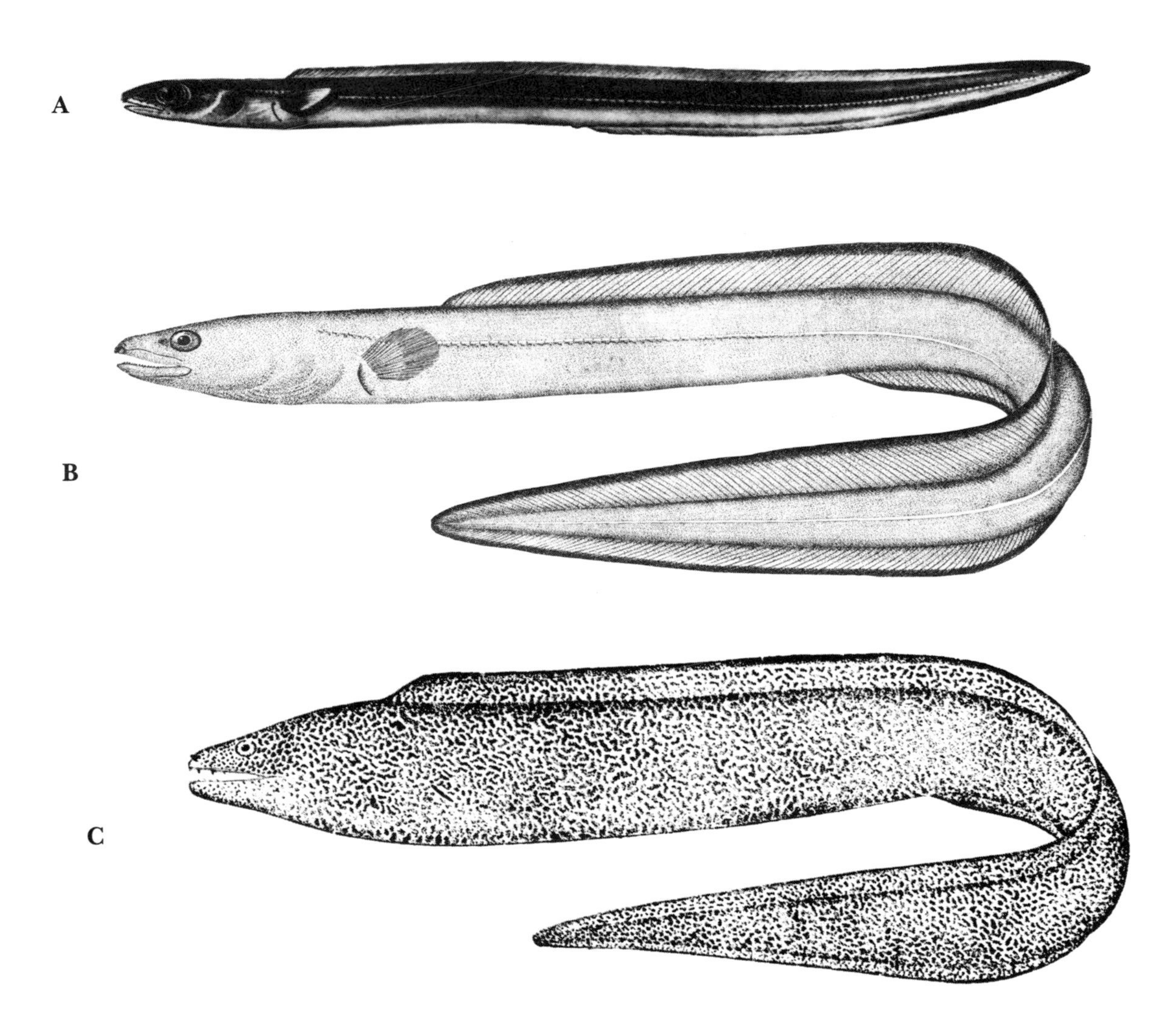

XVI: PLATE 4

Figure a.—*Echelus myrus* (Linnaeus). Length 75 cm. (From Lozano y Rey)

Figure b.—*Ophichthus rufus* (Rafinesque). Length 60 cm. (K. Fogassy)

Figure c.—*Oxystomus serpens* (Linnaeus). Length 2 m. (From *Mondo Sommerso*)

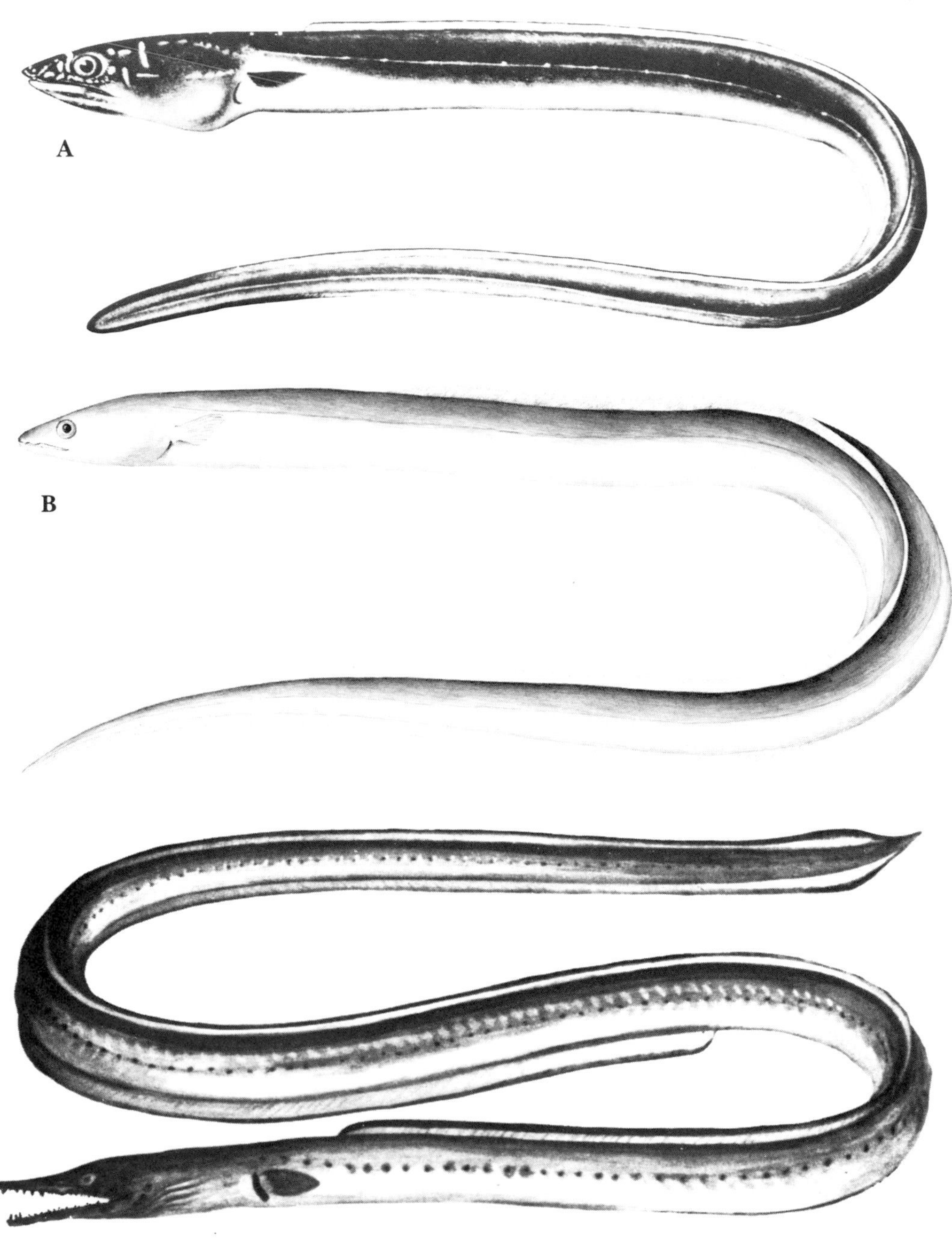

Class Osteichthyes–Poisonous Miscellaneous Fishes

PLATES

CHAPTER XVII

XVII: PLATE 1

FIGURE a.—*Acanthurus triostegus sandvicensis* Streets. Length 20 cm. (Courtesy J. Randall)

FIGURE b.—*Kyphosus cinerascens* (Forskål). Length 50 cm. (From Bleeker)

FIGURE c.—*Kyphosus vaigiensis* (Quoy and Gaimard). Length 40 cm. (From Grant)

XVII: PLATE 2

Neomyxus chaptalli (Eydoux and Souleyet). Length 12 cm. (From Jordan and Evermann)

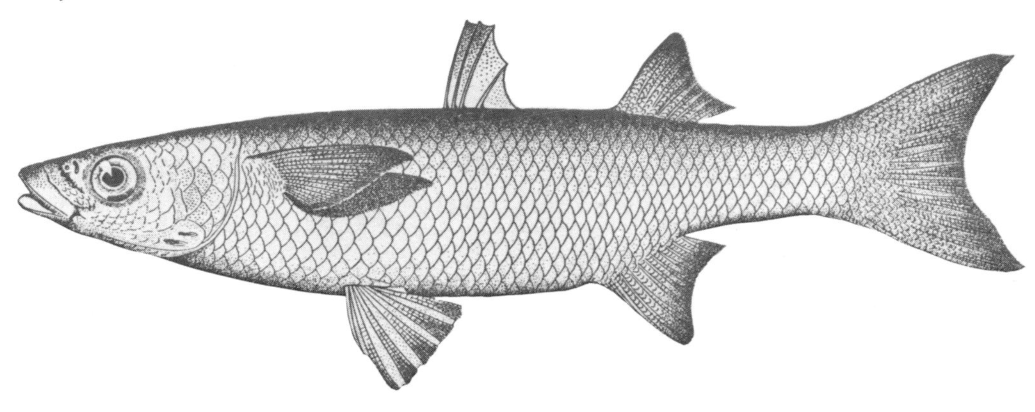

XVII: PLATE 3

Epiniphelus corallicola (Cuvier and Valenciennes). Length 40 cm. (From Günther)

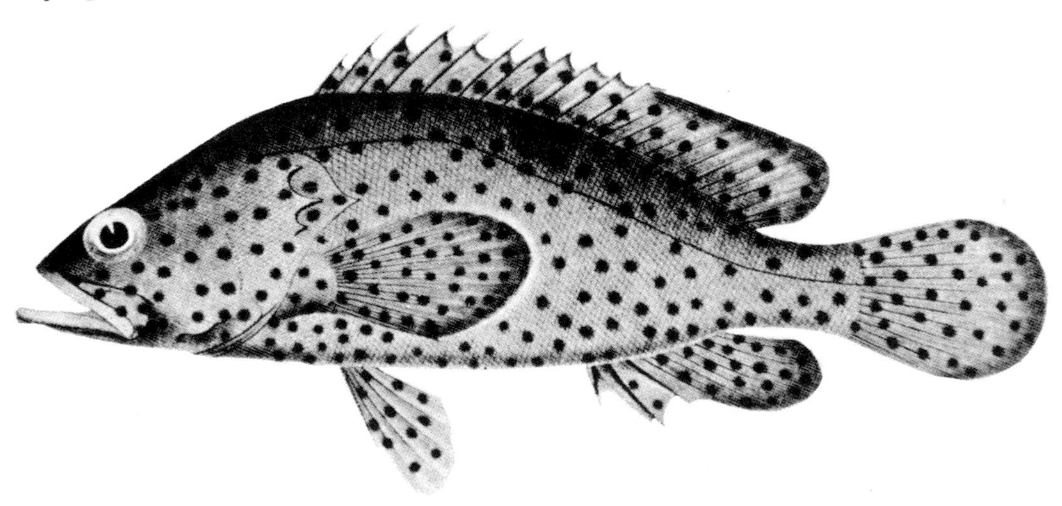

XVII: PLATE 4

FIGURE a.—*Scomberomorus niphonius* (Cuvier and Valenciennes). Length 1 m.

FIGURE b.—*Stereolepis ischinagi* (Hilgendorf). Length 2 m. (From Hiyama)

FIGURE c.—*Petrus rupestris* (Valenciennes). Length 1.7 m. This species is said to be very dangerous since it attains a large size, 70 kg or more, and is equipped with large powerful jaws and teeth. It is reputed to be an aggressive carnivore and capable of inflicting a serious wound. The liver is toxic. (K. Fogassy, after Smith)

FIGURE d.—*Arctoscopus japonicus* (Steindachner). Length 26 cm. (From Hiyama)

XVII: PLATE 5

FIGURE a.—*Myoxocephalus scorpius* Linnaeus. Length 21 cm.

FIGURE b.—*Reinhardtius hippoglossoides* (Walbaum). Length 38 cm.

(From *Fish Resources of the USSR*)

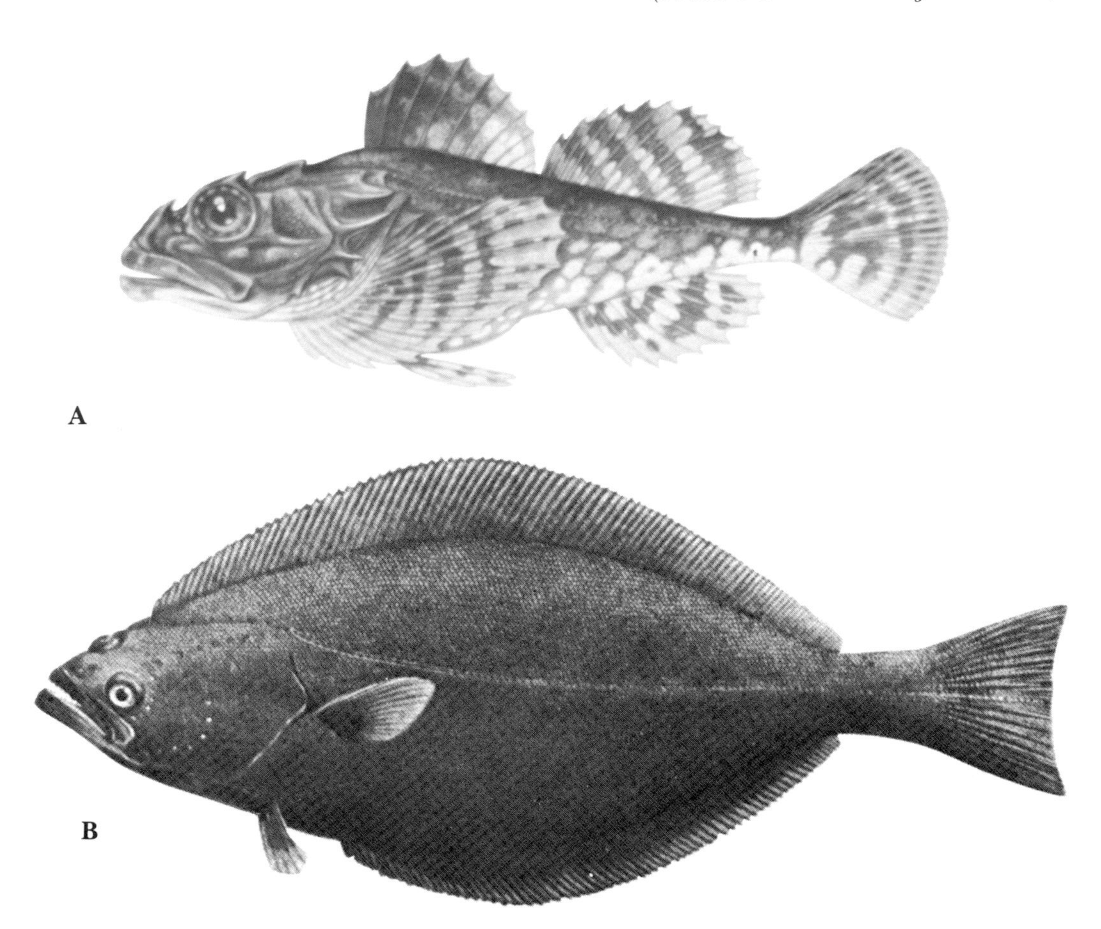

Class Chondrichthyes–Venomous Sharks, etc.

PLATES

CHAPTER XVIII

XVIII: PLATE 1

FIGURE a.—*Heterodontus francisci* (Girard). Length 1.2.

FIGURE b.—*Squalus acanthias* Linnaeus. Length 1 m. (M. Shirao)

XVIII: PLATE 2

FIGURE a.—*Dasyatis akajei* (Müller and Henle). Length 1 m. (M. Shirao)

FIGURE b.—*Dasyatis brucco* (Bonaparte). Length 1.4 m. (R. Kreuzinger)

FIGURE c.—*Dasyatis dipterurus* (Jordan and Gilbert). Length 2 m. (M. Shirao)

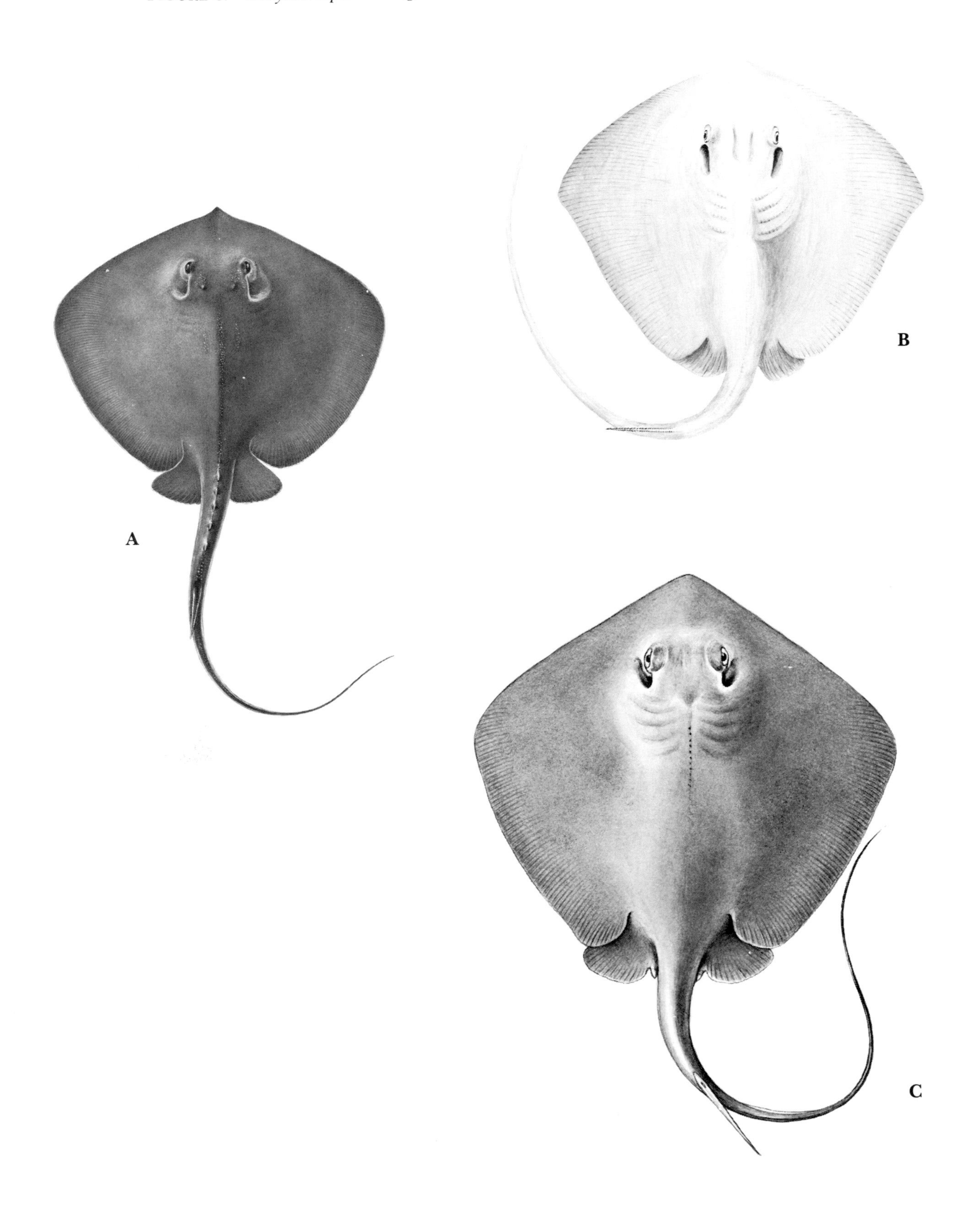

XVIII: PLATE 3

FIGURE a.—*Dasyatis longus* (Garman). Length 1 m. (From Garman)

FIGURE b.—*Dasyatis pastinaca* (Linnaeus). Length 2.5 m. (From *Fish Resources of the USSR*)

XVIII: PLATE 4

FIGURE a.—*Dasyatis sayi* (LeSueur). Width of disc 1 m. (R. Kreuzinger)

FIGURE b.—*Dasyatis sephen* (Forskål). Width of disc 50 cm. (G. Coates)

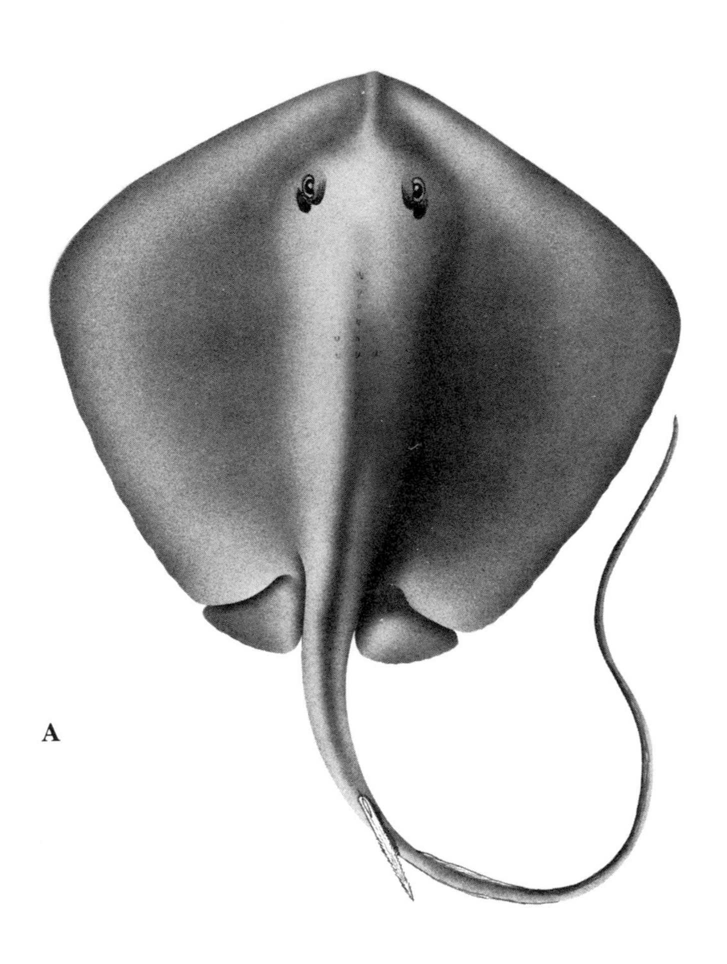

A

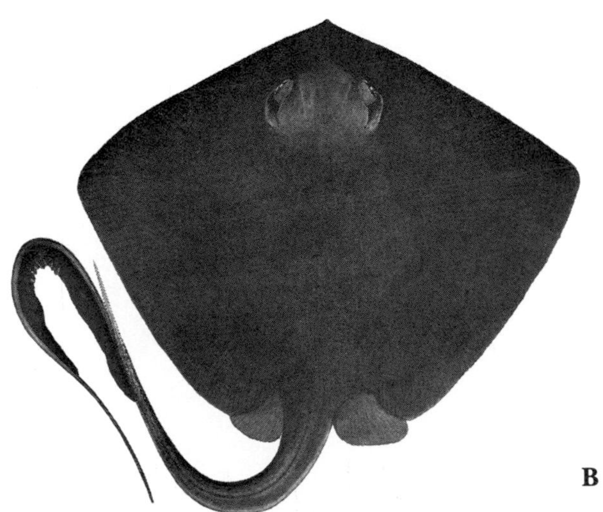

B

XVIII: PLATE 5

Specimens of pelagic stingrays which were taken aboard the *Alcyon* of the Caribbean Fisheries Development Project, FAO, in the western Atlantic, east of the Lesser Antilles. A fatality was attributed (1967) to a pelagic ray which was taken aboard a long-line tuna boat operating from St. Maarten. A crewman who was stationed at the rail when the fish was taken aboard was struck in the cardiac region and he died within 5 minutes. The pelagic ray was believed to have been *Dasyatis violacea* (Bonaparte). This is believed to be the only species of its genus that is found near the surface of the water in either the Atlantic or Pacific Oceans.

(Legend and photo courtesy of W. F. Rathjen)

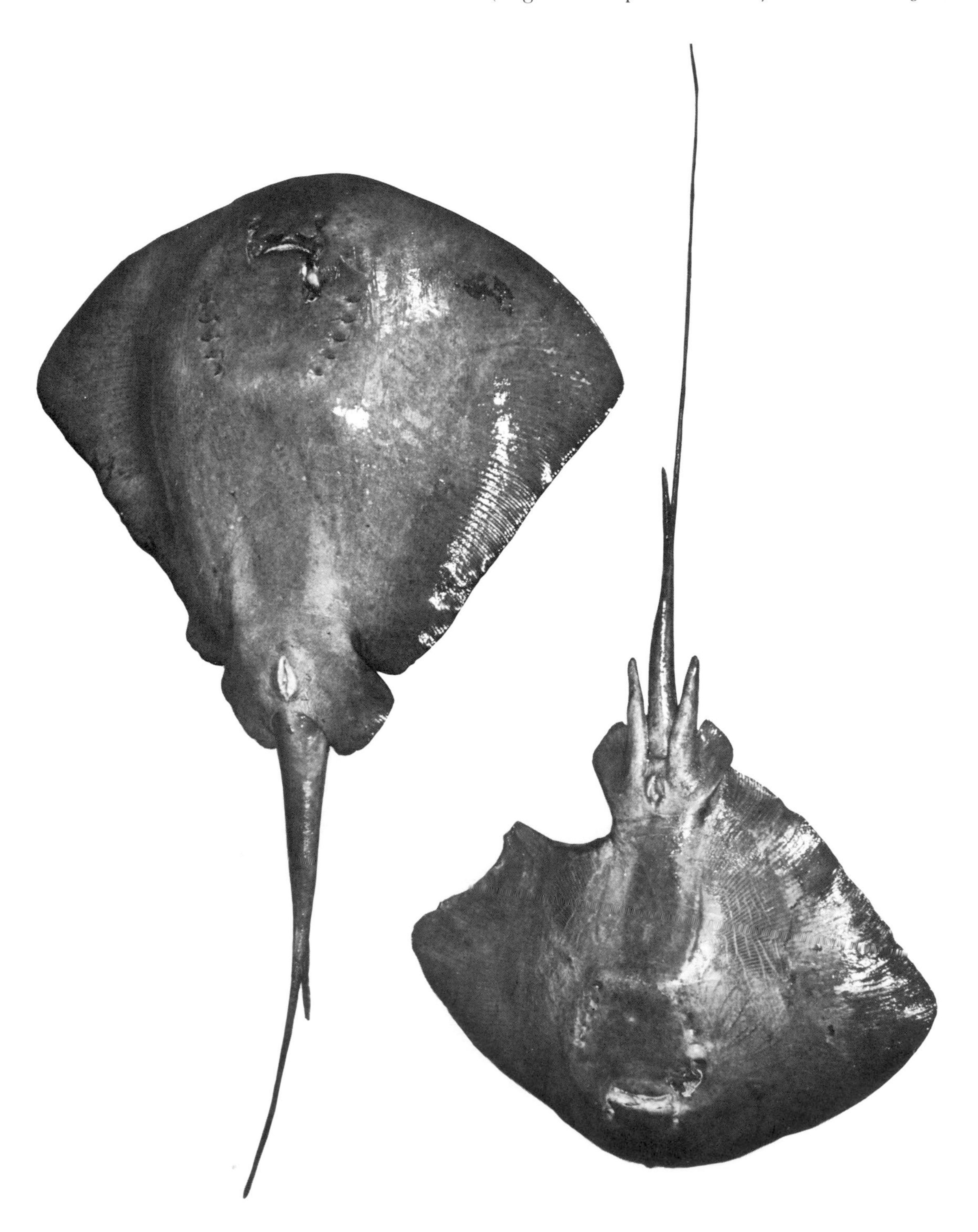

XVIII: PLATE 6

FIGURE a.—*Urogymnus africanus* (Schneider). Width of disc 60 cm. This species has been listed as venomous, but it does not possess a dorsal sting so its ability to produce envenomation is unlikely unless the osseous tubercules which cover the body are bathed by toxic skin secretions. (From Day)

FIGURE b.—*Taeniura lymma* (Forskål). Width of disc 18 cm. (G. Coates)

XVIII: PLATE 7

FIGURE a.—*Gymnura marmorata* (Cooper). Width of disc 220 cm.

FIGURE b.—*Aetobatus narinari* (Euphrasen). Width 2 m. (S. Arita)

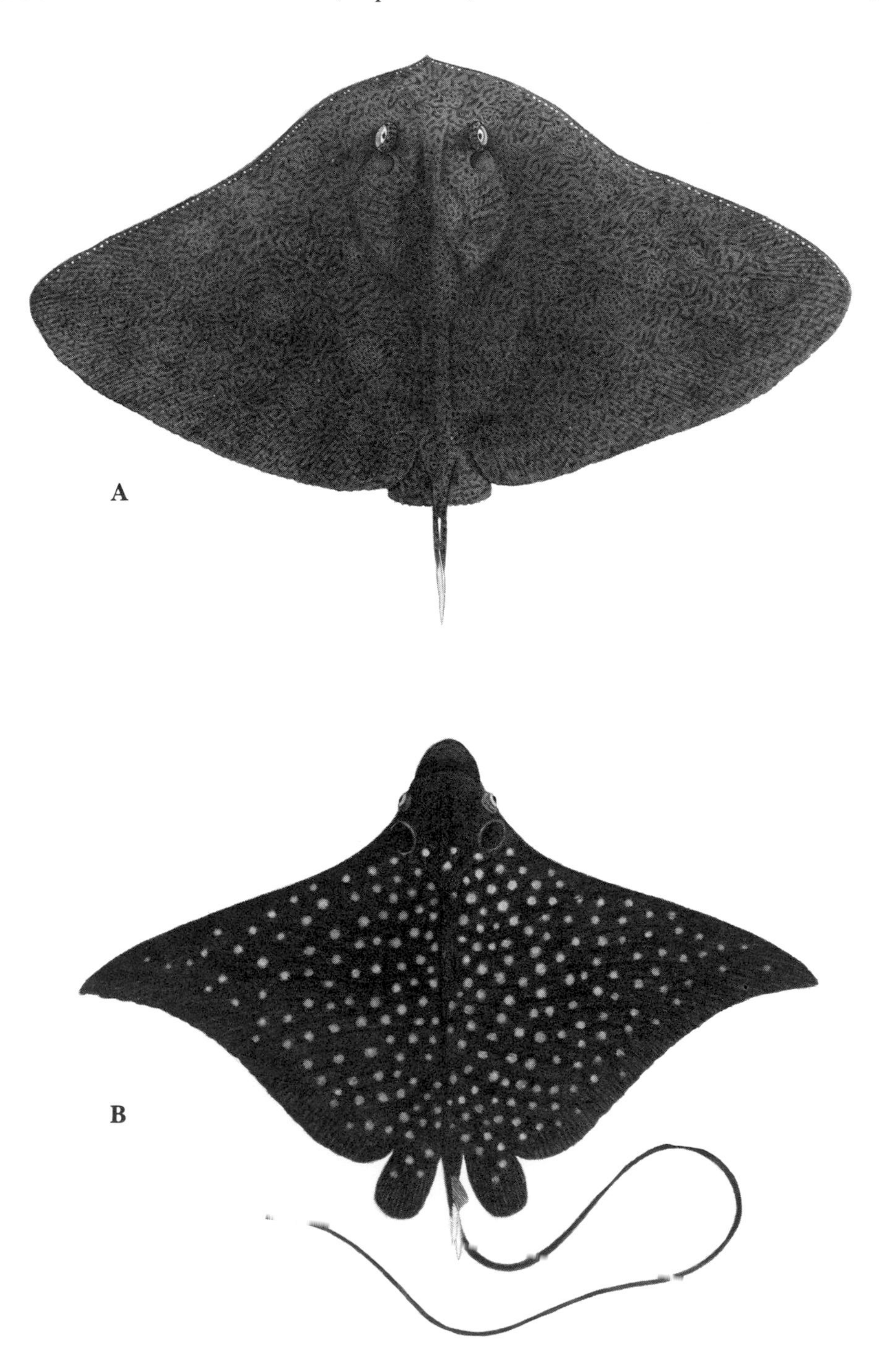

XVIII: PLATE 8

FIGURE a.—*Myliobatis aquila* (Linnaeus). Width 1 m. (R. Kreuzinger)

FIGURE b.—*Myliobatis californicus* Gill. Width 1.2 m. (M. Shirao)

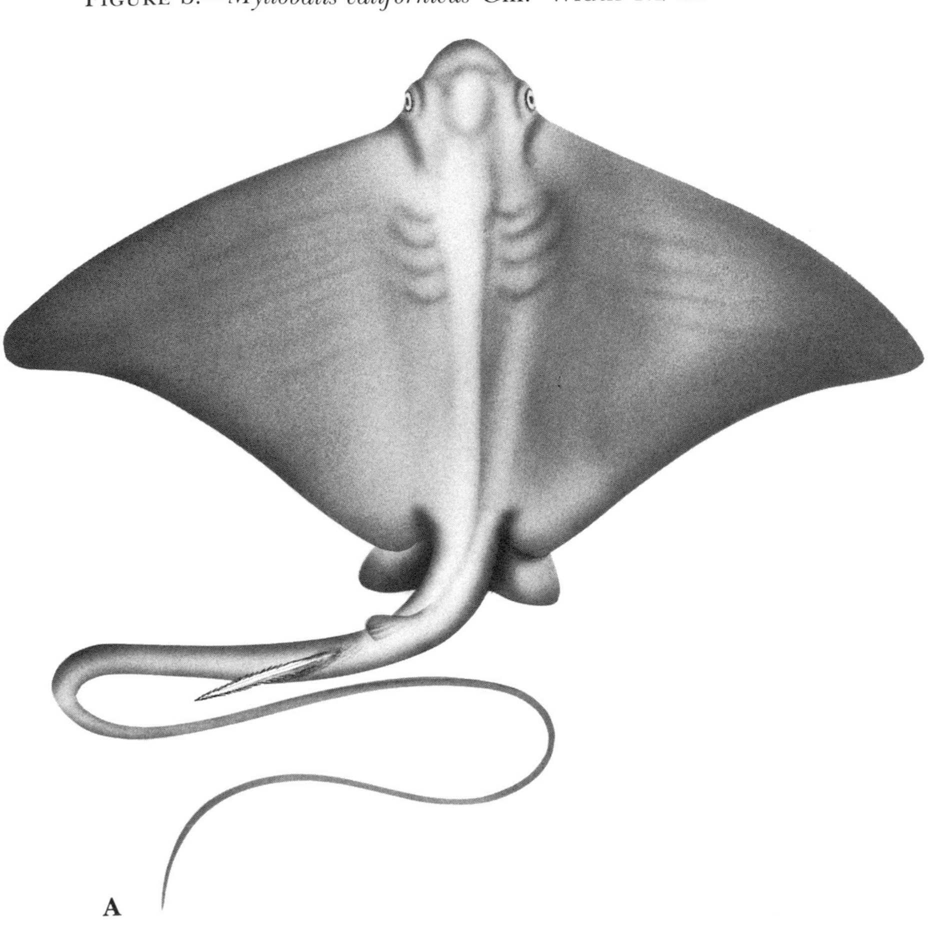

A

B

XVIII: PLATE 9

Pteromylaeus punctatus Macleay and Macleay. Width 1 m.

(K. Fogassy, after Whitley)

XVIII: PLATE 10

FIGURE a.—*Potamotrygon hystrix* (Müller and Henle). Width of disc 50 cm.

FIGURE b.—*Potamotrygon motoro* (Müller and Henle). Width of disc 54 cm.

(M. Shirao, courtesy of M. Castex)

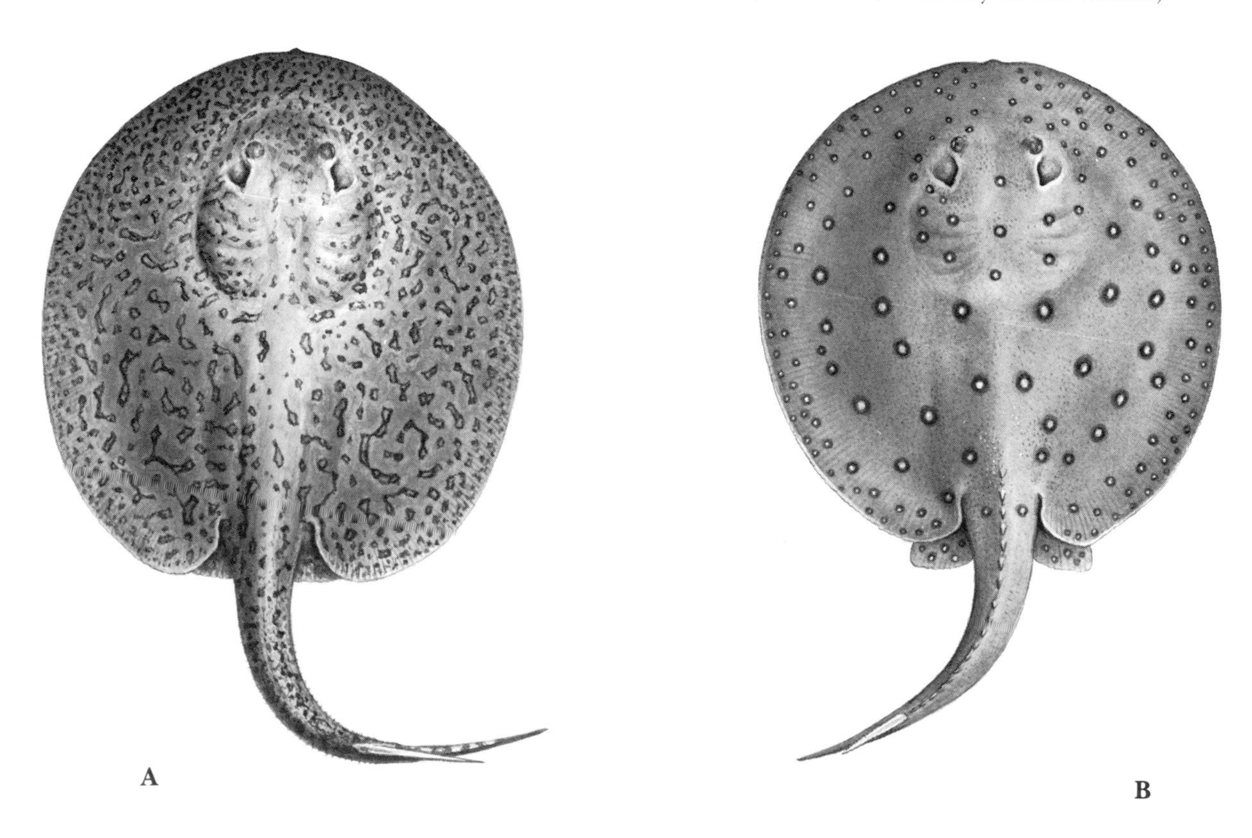

XVIII: PLATE 11

Rhinoptera bonasus (Mitchill). Width 2 m. (K. Fogassy)

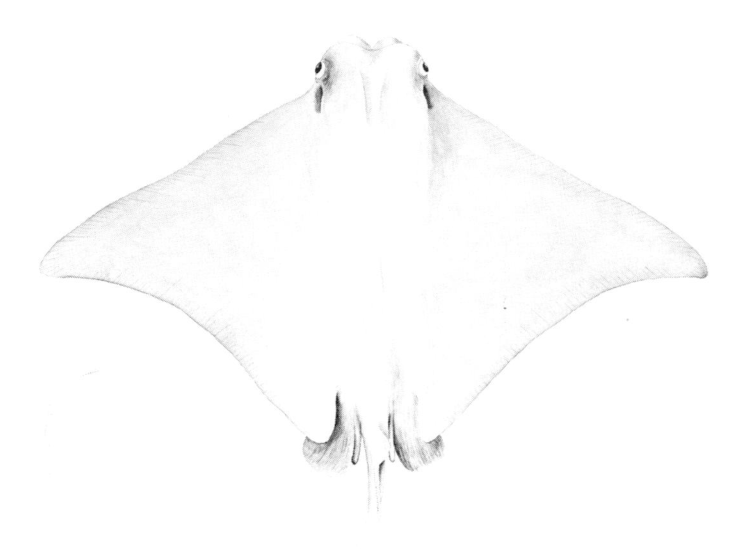

XVIII: PLATE 12

FIGURE a.—*Urolophus halleri* Cooper. Length 50 cm. (M. Shirao)

FIGURE b.—*Urolophus jamaicensis* (Cuvier). Length 67 cm. (K. Fogassy)

XVIII: PLATE 13

FIGURE a.—*Chimaera monstrosa* Linnaeus. Female. Length up to 1 m. (M. Shirao)

FIGURE b.—*Hydrolagus affinis* (Capello). Female. Length 1.2 m. (From Good and Bean)

FIGURE c.—*Hydrolagus colliei* (Lay and Bennett). Length up to 1 m. 1. Male; 2 Female. (M. Shirao)

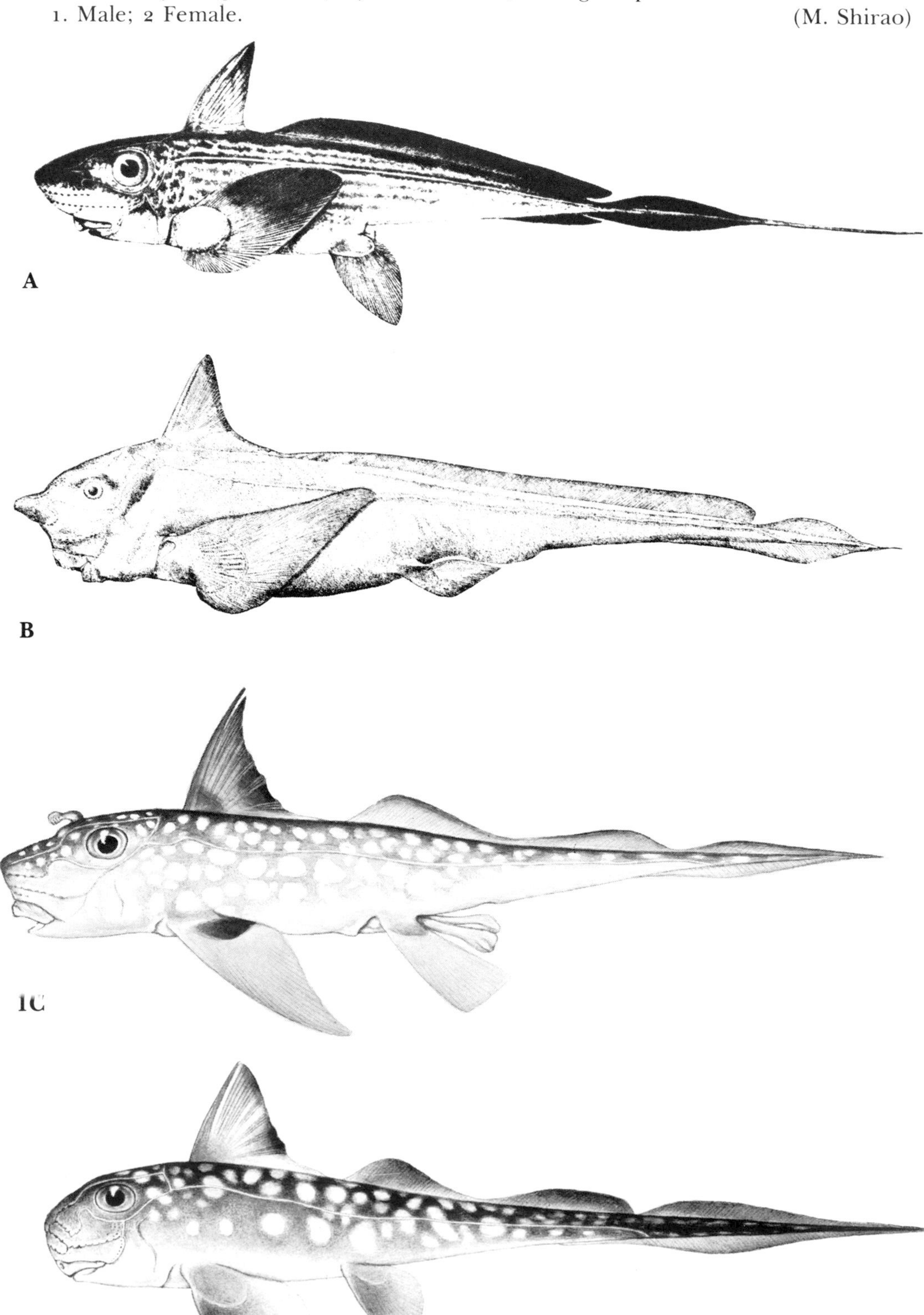

Class Osteichthyes–Catfishes

PLATES
CHAPTER XIX

XIX: PLATE 1

Arius heudeloti Cuvier and Valenciennes. Length 54 cm. (From Prosvirov)

XIX: PLATE 2

FIGURE a.—*Galeichthys feliceps* (Valenciennes). Length 40 cm. (From Poll)

FIGURE b.—*Galeichthys felis* (Linnaeus). Length 40 cm. (From Goode)

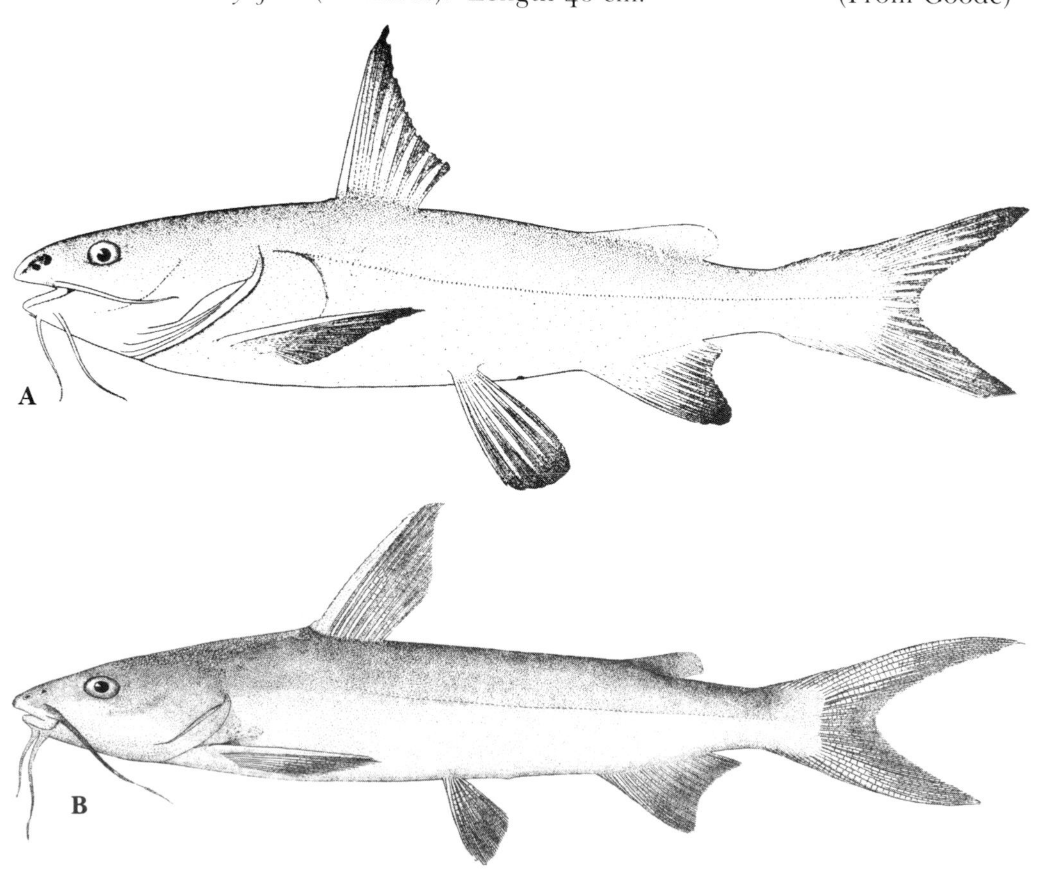

XIX: PLATE 3

FIGURE a.—*Genidens genidens* (Cuvier and Valenciennes). Length 40 cm. (R. Kreuzinger)

FIGURE b.—*Netuma barbus* (Lacépède). Length 40 cm. (From Castelnau)

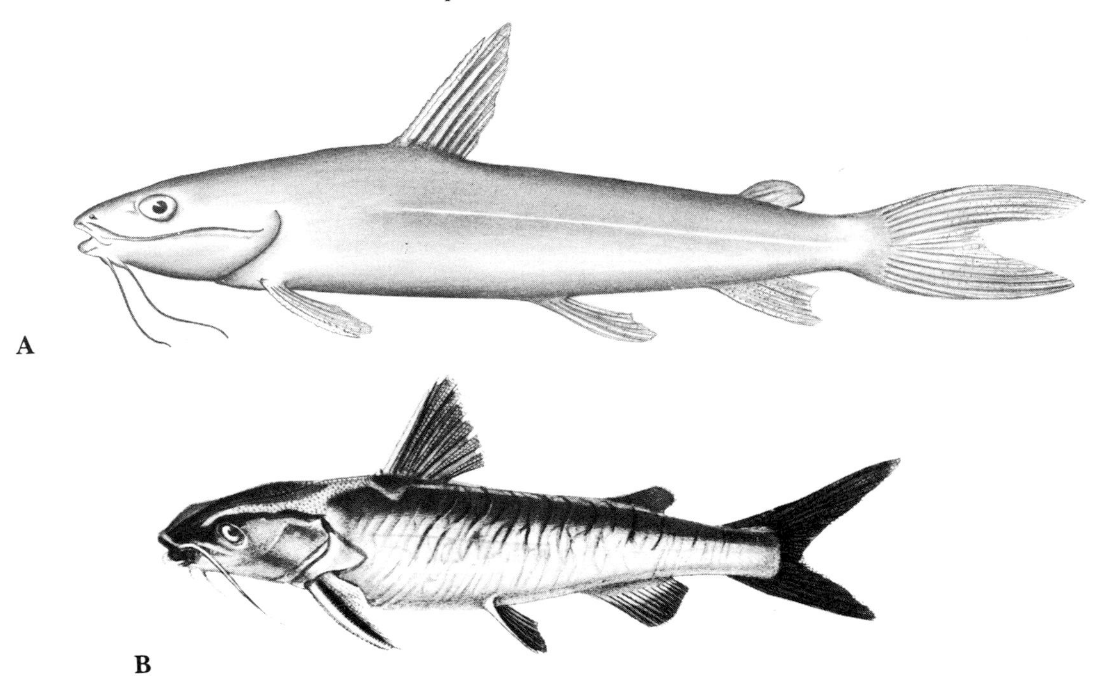

XIX: PLATE 4

FIGURE a.—*Osteogeniosus militaris* (Linnaeus). Length 30 cm. (From Day)

FIGURE b.—*Selenaspis herzbergi* (Bloch). Length 35 cm. (From Eigenmann)

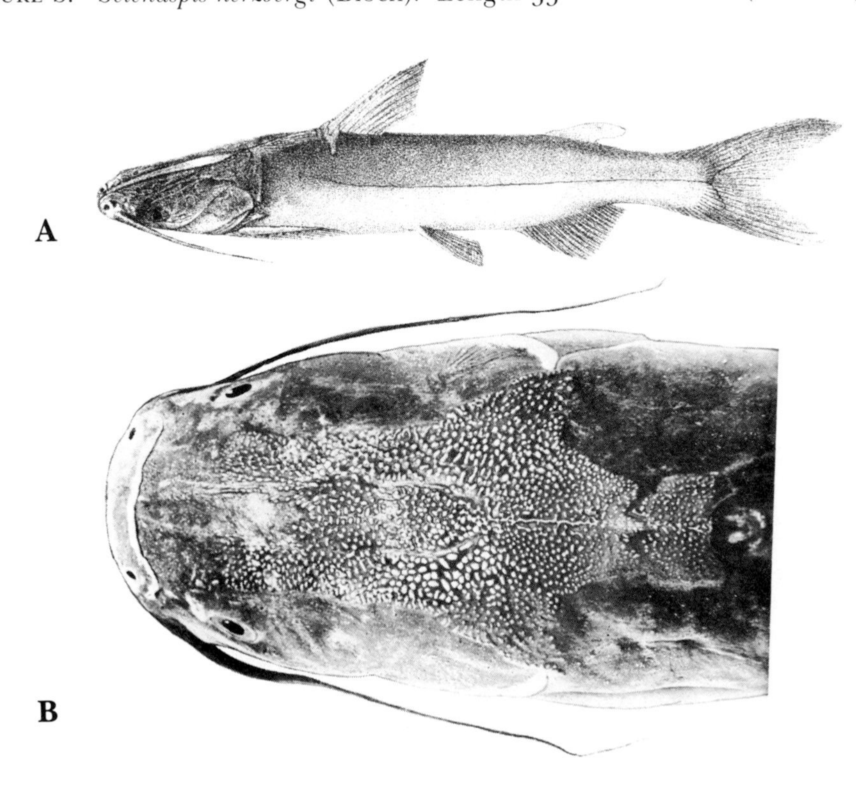

XIX: PLATE 5

FIGURE a.—*Liobagrus reini* Hilgendorf. Length 10 cm.
(From Tomiyama, Abe, and Tokioka)

FIGURE b.—*Pseudobagrus aurantiacus* (Temminck and Schlegel). Length 21 cm.
(S. Arita)

FIGURE c.—*Clarias batrachus* (Linnaeus). Length 30 cm. (From Day)

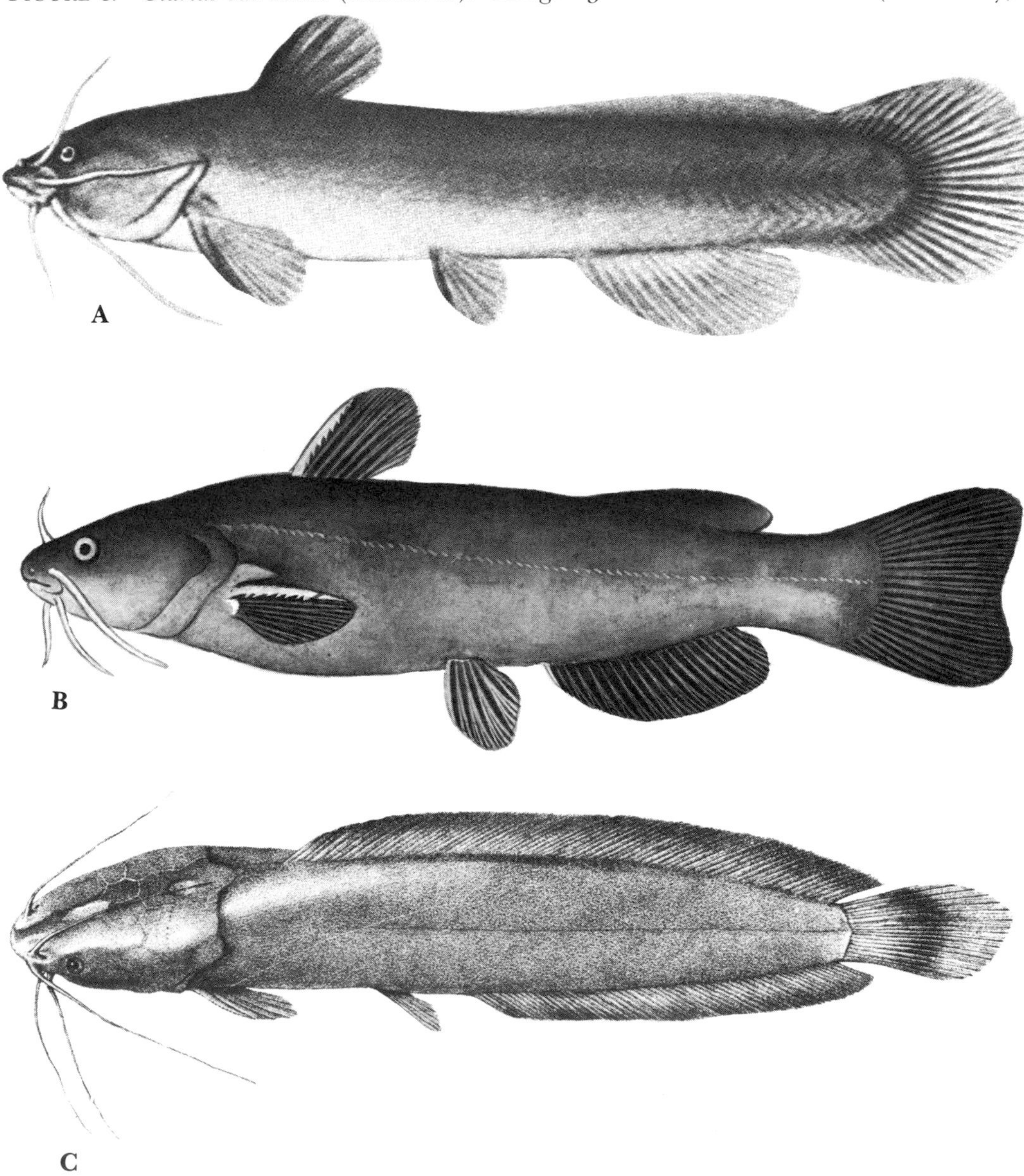

XIX: PLATE 6

FIGURE a.—*Centrochir crocodili* (Humboldt). Length 15 cm.

FIGURE b.—*Pterodoras granulosus* (Valenciennes). Length 39 cm.
1. Sideview. 2. Dorsal view of head. (From Eigenmann)

FIGURE c.—*Heteropneustes fossilis* (Bloch). Length 25 cm. (From Day)

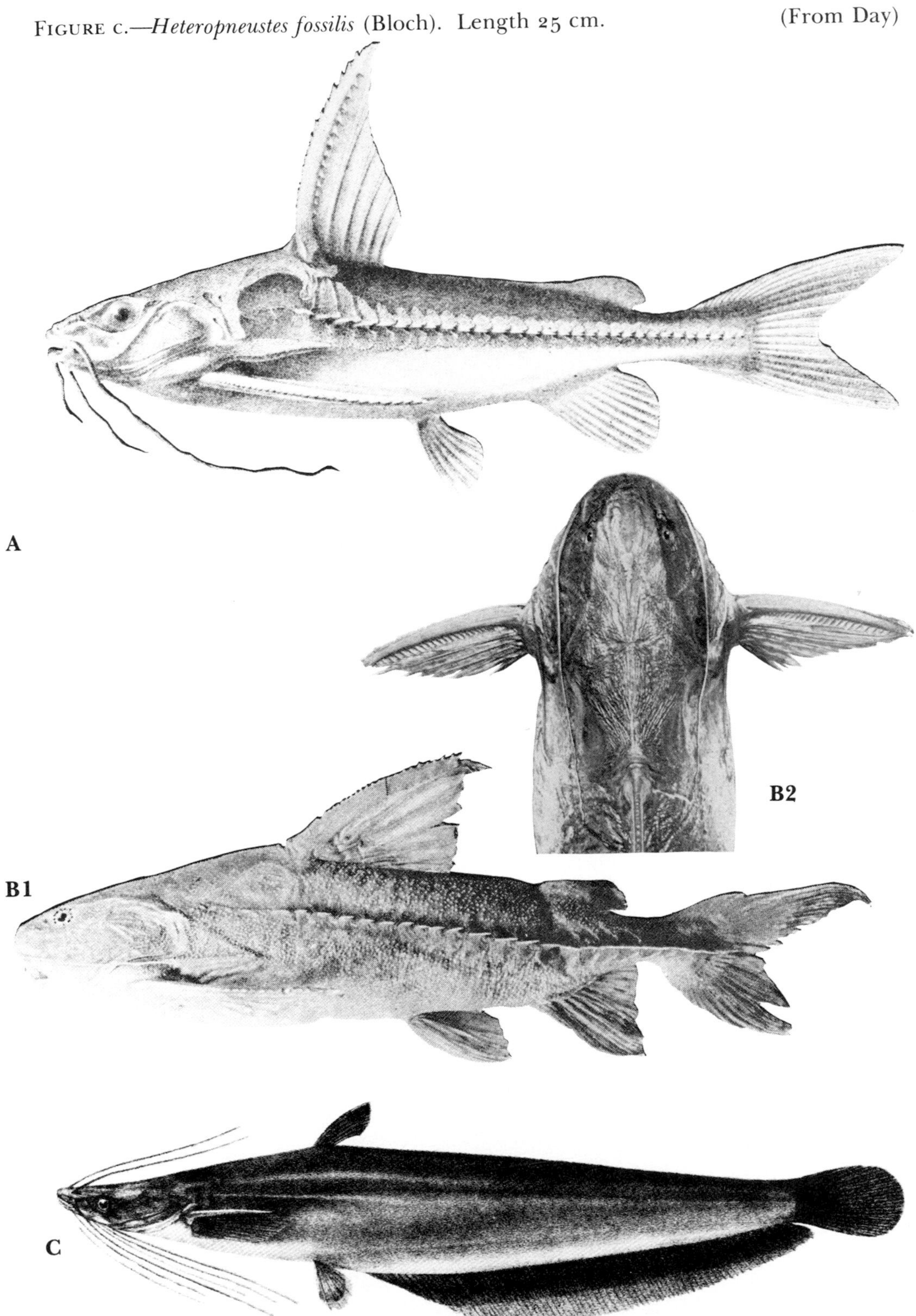

XIX: PLATE 7

FIGURE a.—*Noturus flavus* Rafinesque. Length 30 cm. (M. Shirao)

FIGURE b.—*Noturus furiosus* Jordan and Meek. Length 12 cm. (From Jordan)

A

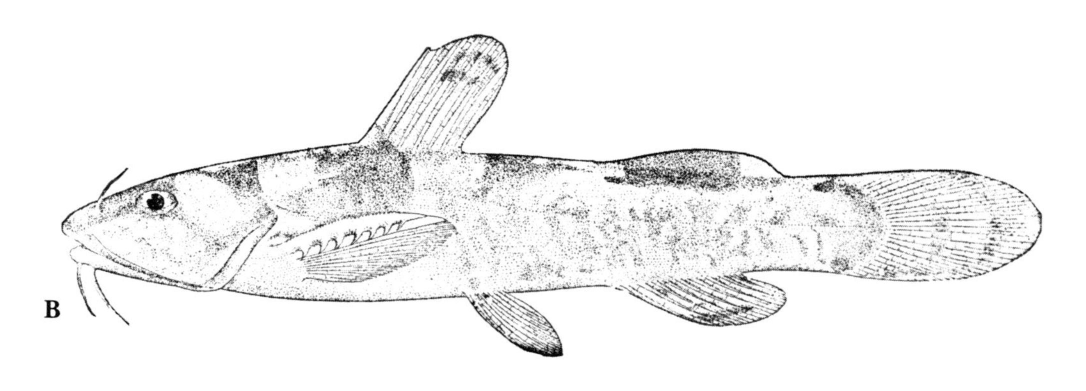

B

XIX: PLATE 8

FIGURE a.—*Noturus mollis* (Mitchill). Length 10 cm. (M. Shirao)

FIGURE b.—*Noturus nocturnus* Jordan and Gilbert. Length 8 cm. (From Evermann and Kendall)

FIGURE c.—*Pimelodus clarias* (Bloch). Length 28 cm. This species is sometimes found with large spots on its head and body. (From Schultz)

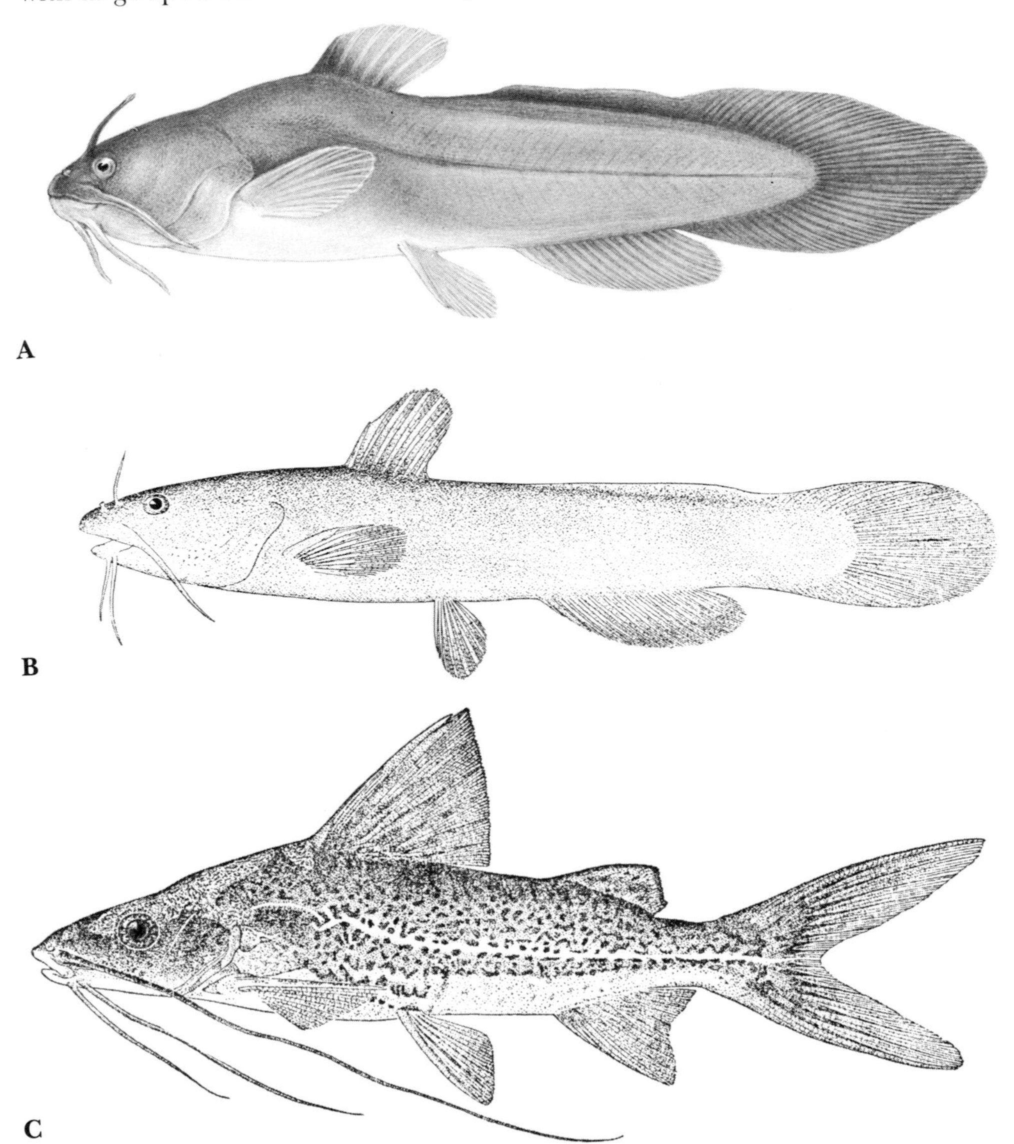

XIX: PLATE 9

FIGURE a.—*Cnidoglanis megastoma* (Richardson). Length 90 cm.

FIGURE b.—*Paraplotosus albilabris* (Valenciennes). Length 35 cm. (From Richardson)

FIGURE c.—*Plotosus canius* (Hamilton-Buchanan). Length 90 cm. (From Day)

FIGURE d.—*Plotosus lineatus* (Thunberg). Length 30 cm. (M. Shirao)

XIX: PLATE 10

FIGURE a.—*Tandanus bostocki* Whitley. Length 38 cm.

FIGURE b.—*Chrysichthys cranchi* (Leach). Length 1 m. (From Boulenger)

FIGURE c.—*Parasilurus asotus* (Linnaeus). Length 40 cm. (S. Arita)

FIGURE d.—*Synodontis batensoda* Rüppell. Length 24 cm. (From Boulenger)

A

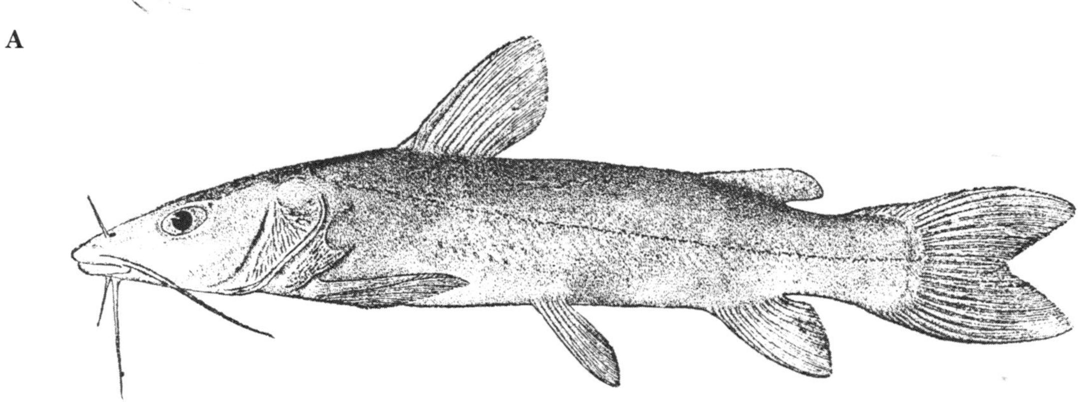

B

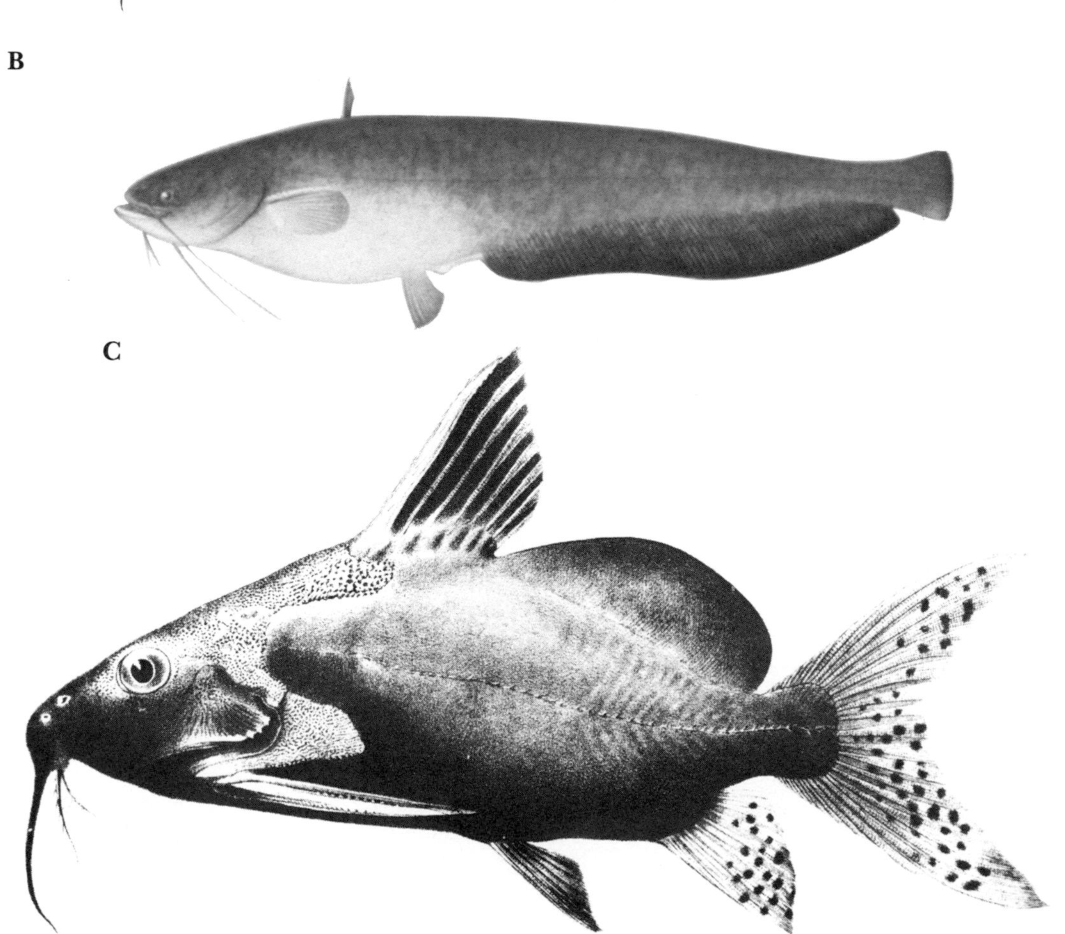

Class Osteichthyes–Weeverfishes

PLATE
CHAPTER XX

XX: PLATE 1

FIGURE a.—*Trachinus araneus* Cuvier. Length 32 cm. (From Joubin)
FIGURE b.—*Trachinus draco* Linnaeus. Length 45. (M. Shirao)
FIGURE c.—*Trachinus radiatus* Cuvier. Length 26 cm. (From Joubin)
FIGURE d.—*Trachinus vipera* Cuvier. Length 15 cm. (M. Shirao)

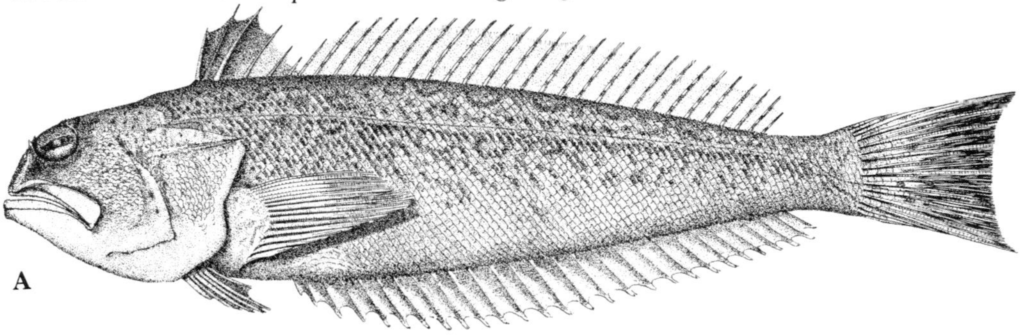

A

B

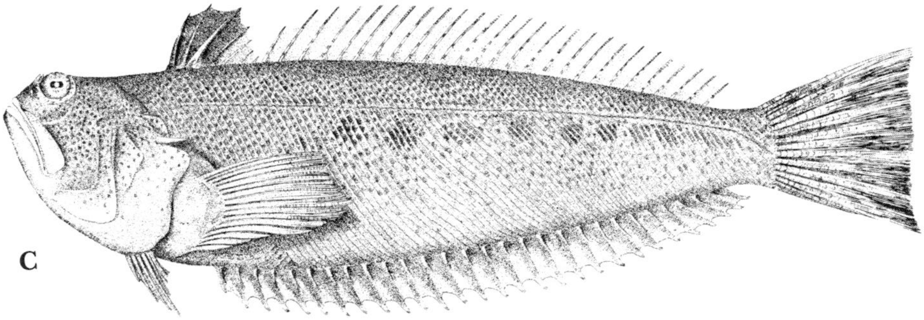

C

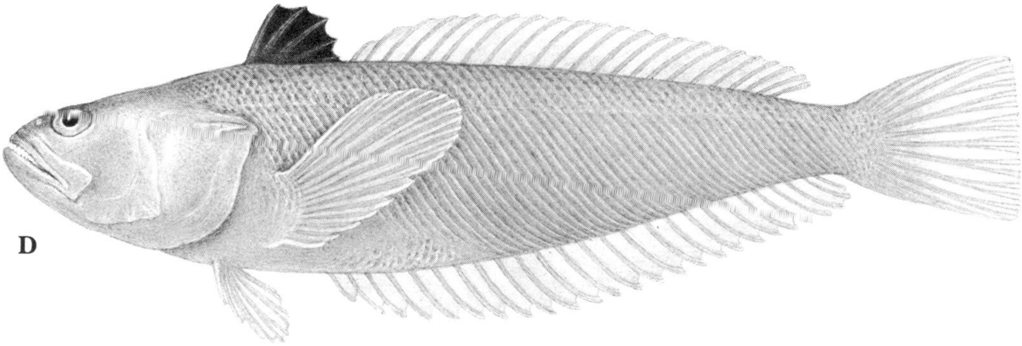

D

Class Osteichthyes–Scorpionfishes

PLATES

CHAPTER XXI

XXI: PLATE 1

Apistus carinatus (Bloch and Schneider). Length 16 cm. (S. Arita)

XXI: PLATE 2

Figure a.—*Brachirus brachypterus* (Cuvier). Length 15 cm. (From Shirao)

Figure b.—*Gymnapistes marmoratus* (Cuvier and Valenciennes). Length 20 cm. The spines of the head and dorsal vertical fins of *G. marmoratus* are said to be venomous and can cause excruciating pain. (From Waite)

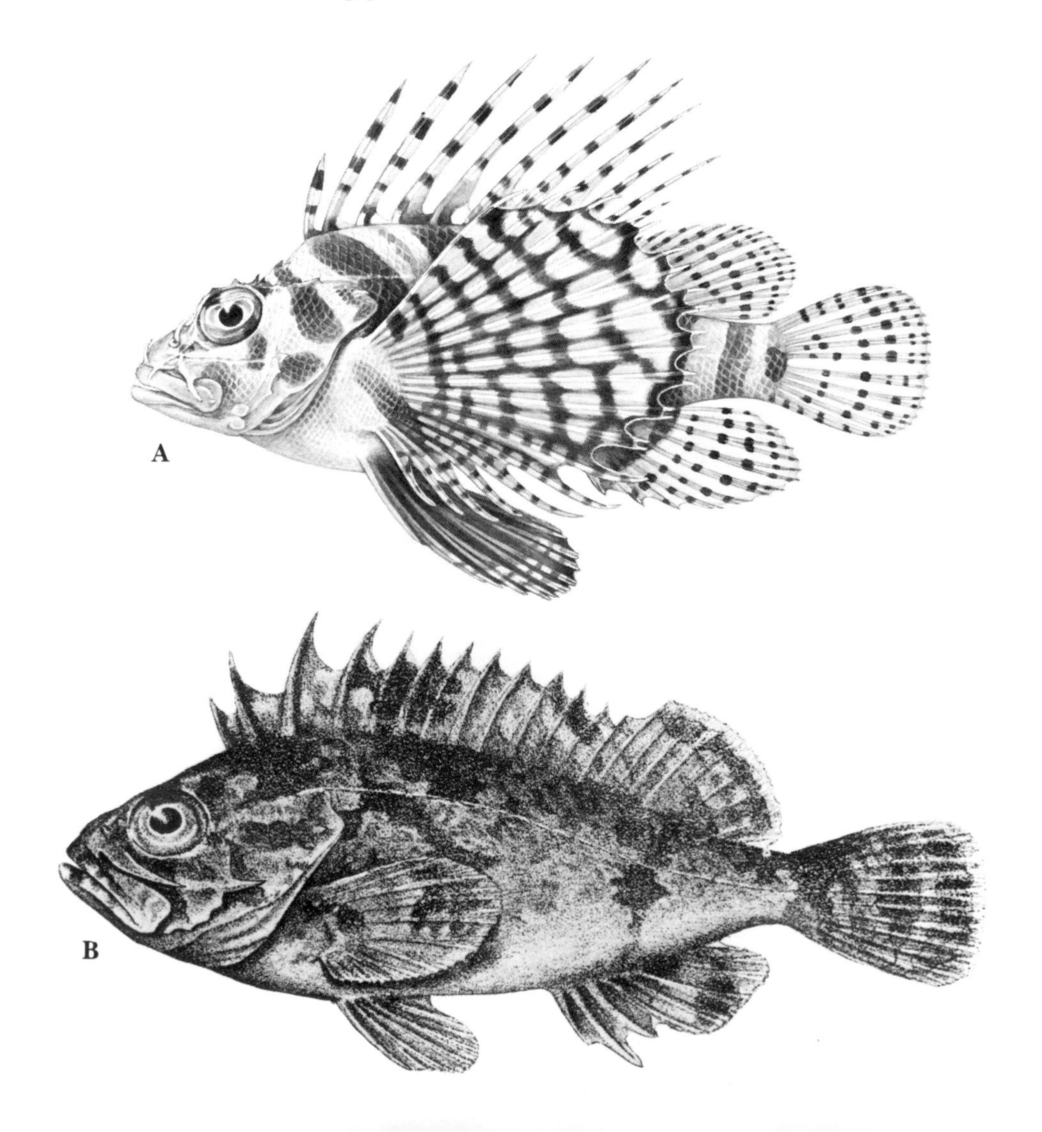

XXI: PLATE 3

FIGURE a.—*Centropogon australis* (White). Length 15 cm. Living specimen taken near Sydney, Queensland, Australia.

FIGURE b.—*Erosa erosa* (Langsdorf). Length 15 cm. (From Tomiyama and Abe)

A

B

XXI: PLATE 4

FIGURE a.—*Brachirus zebra* (Quoy and Gaimard). Length 12 cm. Taken at the Waikiki Aquarium, University of Hawaii, Honolulu, Hawaii. (Courtesy R. Ames)

FIGURE b.—*Inimicus filamentosus* (Cuvier and Valenciennes). Length 16 cm. (From Smith)

A

B

XXI: PLATE 5

Inimicus japonicus (Cuvier and Valenciennes). Length 20 cm. This illustration depicts some of the many color variations of this venomous fish. (M. Shirao)

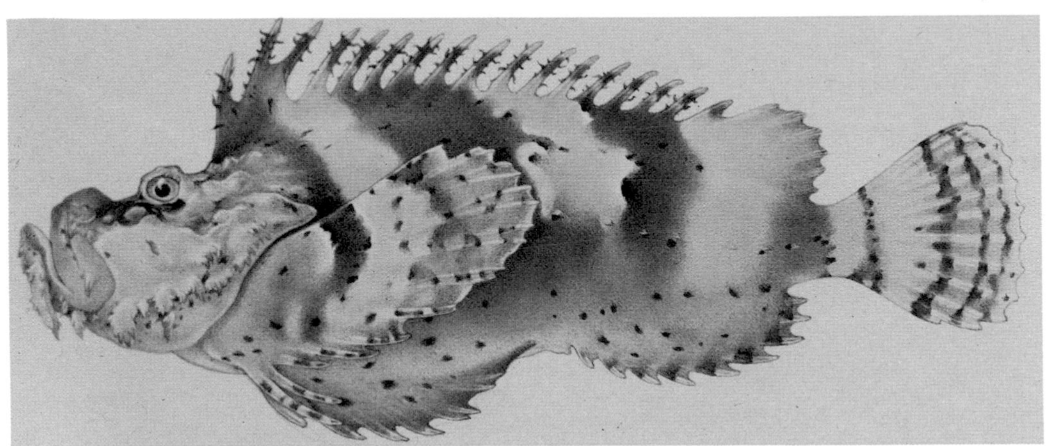

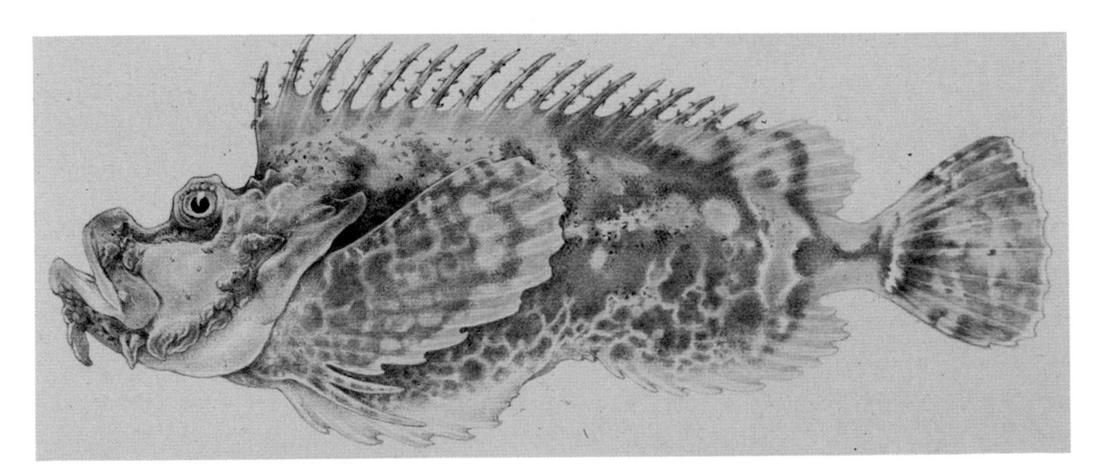

XXI: PLATE 6

FIGURE a.—*Minous adamsi* Richardson. Length 11 cm. (From Jordan and Starks)

FIGURE b.—*Scorpaena grandicornis* Cuvier and Valenciennes. Length 30 cm. (From Evermann and Marsh)

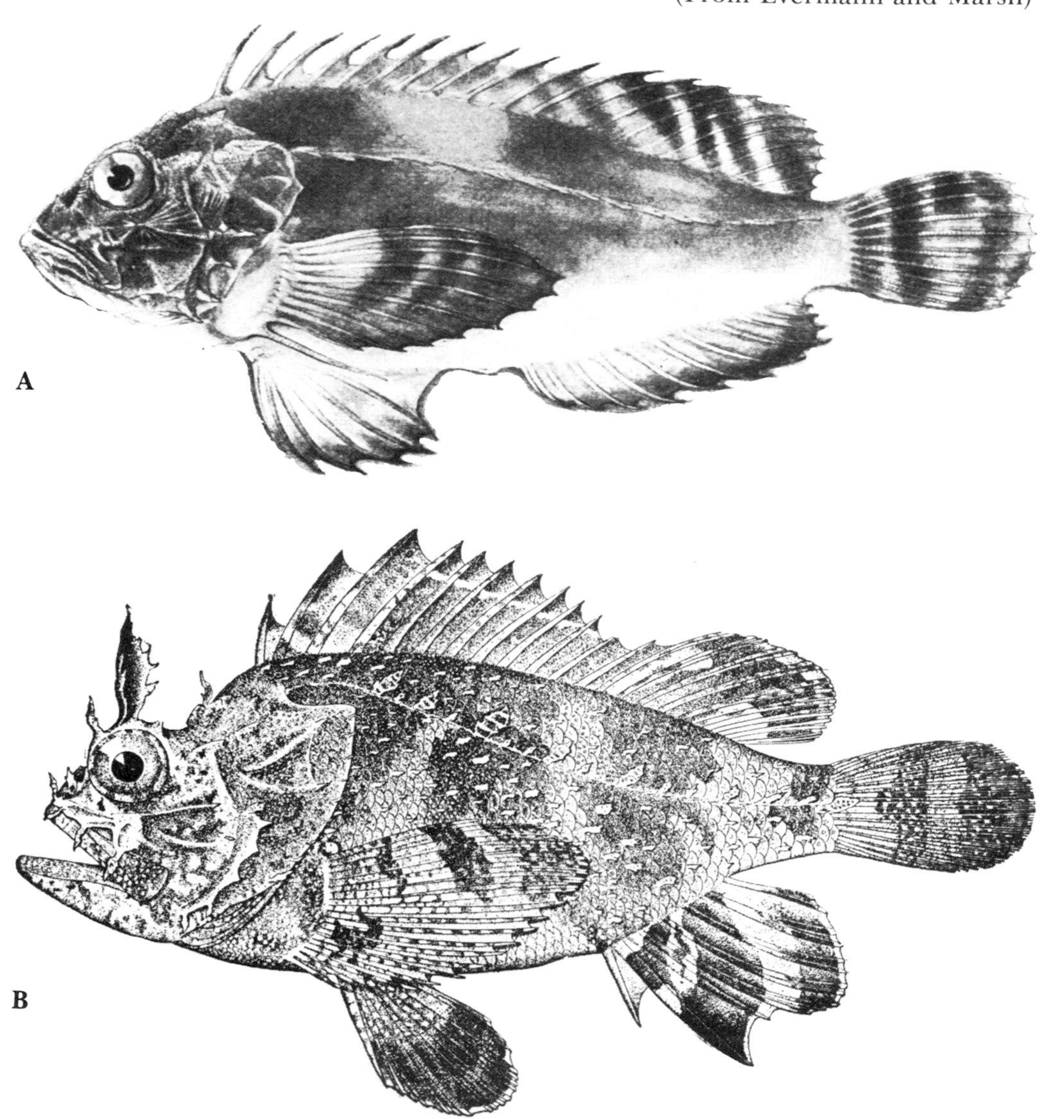

XXI: PLATE 7

FIGURE a.—*Notesthes robusta* (Günther). Length 25 cm. (S. Arita, after Bleeker)

FIGURE b.—*Pterois volitans* (Linnaeus). Length 28 cm. (M. Shirao)

FIGURE c.—*Pterois volitans*. This is the *P. miles* form which is now believed to be the adult form of *P. volitans*. (G. Coates)

FIGURE d.—*Pterois russelli* Bennett. Length 35 cm. (S. Arita)

FIGURE e.—*Pterois lunulata* Schlegel. Length 35 cm. This is the original illustration of this species which is now believed to be a synonym of *P. russelli*.
(From Temminck and Schlegel)

A

B

C

D

E

XXI: PLATE 8

Pterois antennata (Bloch). Length 15 cm. This is one of the most spectacular of the reef fishes, but hidden within these lacy fins are venomous spines. Taken near Heron Island, Great Barrier Reef, Queensland, Australia. (A. Powers)

XXI: PLATE 9

FIGURE a.—*Scorpaena guttata* Girard. Length 43 cm. (M. Shirao)

FIGURE b.—*Scorpaena porcus* Linnaeus. Length 30 cm. (From *Fish Resources of the USSR*)

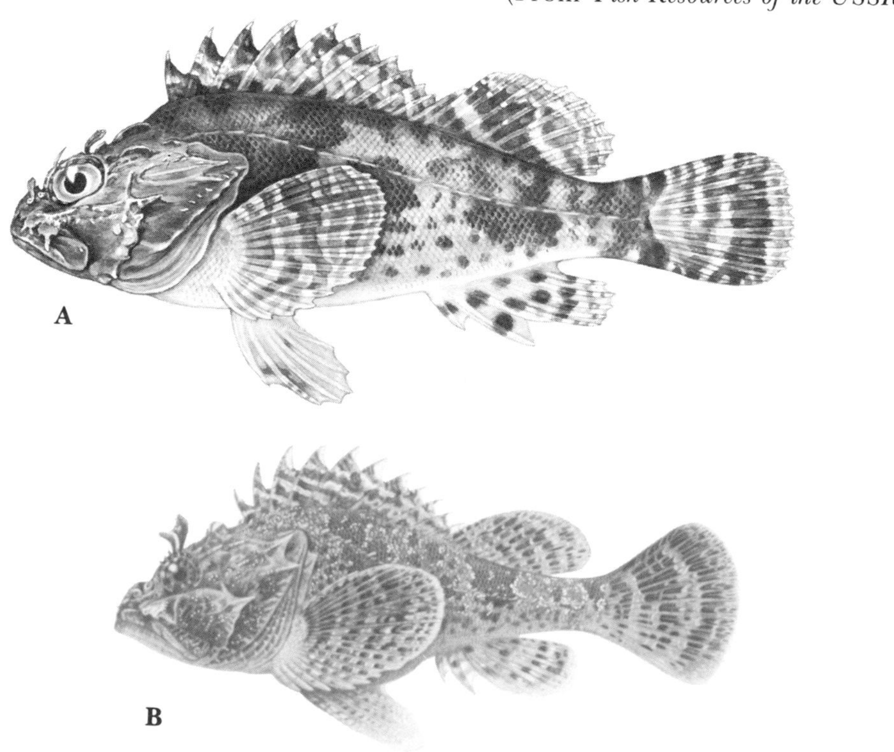

XXI: PLATE 10

Figure a.—*Scorpaena plumieri* Bloch. Length 30 cm. Living specimen taken in an aquarium. (Courtesy Miami Seaquarium)

FIGURE b.—*Scorpaena scrofa* Linnaeus. Length 50 cm. This is one of the species of venomous fishes best known to European writers of antiquity. Taken at the aquarium, Zoological Station, Naples, Italy. (P. Giacomelli)

A

B

XXI: PLATE 11

Photograph showing a representative habitat in which *Scorpaenopsis gibbosa* (Bloch and Schneider) is encountered. Scorpaenids are difficult to detect because of their habit of camouflaging themselves and concealment in crevices. They can inflict painful wounds with their dorsal spines. This species attains a length of more than 20 cm. Taken near Heron Island, Great Barrier Reef, Queensland, Australia.

(A. Powers)

XXI: PLATE 12

Figure a.—*Sebasticus marmoratus* (Cuvier and Valenciennes). Length 15 cm. (S. Arita)

Figure b.—*Sebastodes inermis* (Cuvier and Valenciennes). Length 20 cm. (From Hiyama)

Figure c.—*Sebastodes joyneri* (Günther). Length 20 cm. (S. Arita)

Figure d.—*Sebastodes auriculatus* Girard. Length 43 cm.
(Courtesy of J. B. Phillips, State of California Department of Fish and Game)

XXI: PLATE 13

Synanceja horrida (Linnaeus). Although toadfishes *Thalassophryne* probably have the most highly developed venom apparatus of any of the fishes, the venom glands of *Synanceja horrida* are the largest found in fishes. The stonefishes *Synanceja* are some of the most dangerous venomous fishes known. Their stings are extremely painful, and they have reportedly caused death. Length 61 cm. (S. Arita)

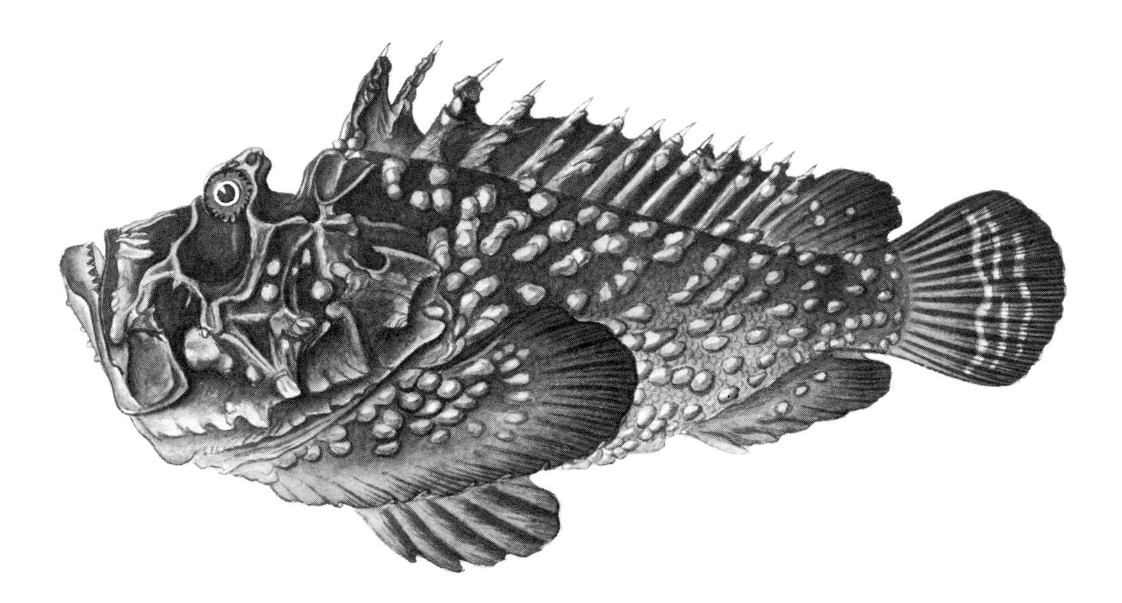

XXI: PLATE 14

FIGURE a.—Side view of *Synanceja horrida* (Linnaeus) showing a specimen in which two of the dorsal spines have had their integumentary sheaths removed in order to reveal the spines and the attached venom glands. Note their very large size in comparision with the shaft of the spine. The fish is about 30 cm in length.

FIGURE b.—Head of *Synanceja horrida* showing the ensheathed venom glands, one on either side of the shaft of the dorsal spine. During the act of stinging, the spines penetrate the flesh of the victim and in so doing the integumentary sheaths are compressed toward the base of each spine. Pressure is exerted on the integumentary sheaths and thus on the venom glands which are attached to the sides of the spines. The venom glands are covered with a dense, collagenous, connective tissue; consequently pressure exerted on the venom gland forces the venom to be expelled through the venom ducts which open at the distal tip of the spine. It is necessary for a spine to penetrate the flesh to a depth of 0.6 cm to 1.0 cm before sufficient pressure is exerted on the venom glands to expel the venom. The dorsal spines of *Synanceja* are sharp, stout, and can readily pass through a rubber sneaker or sandal. Stings from these fish may be extremely painful, and fatalities have been reported. Taken at Heron Island, Great Barrier Reef, Queensland, Australia. (K. Gillett)

A

B

XXI: PLATE 15

FIGURE a.—Side view of *Synanceja verrucosa* Bloch and Schneider. Length 30 cm. (K. Gillett)

FIGURE b.—*Synanceja verrucosa* Bloch and Schneider. Length about 30 cm. This specimen is shown buried in the sand. When covered with sand, these fish can easily be stepped on. Taken at Palau, Western Caroline Islands. (D. Ludwig, World Life Research Institute)

A

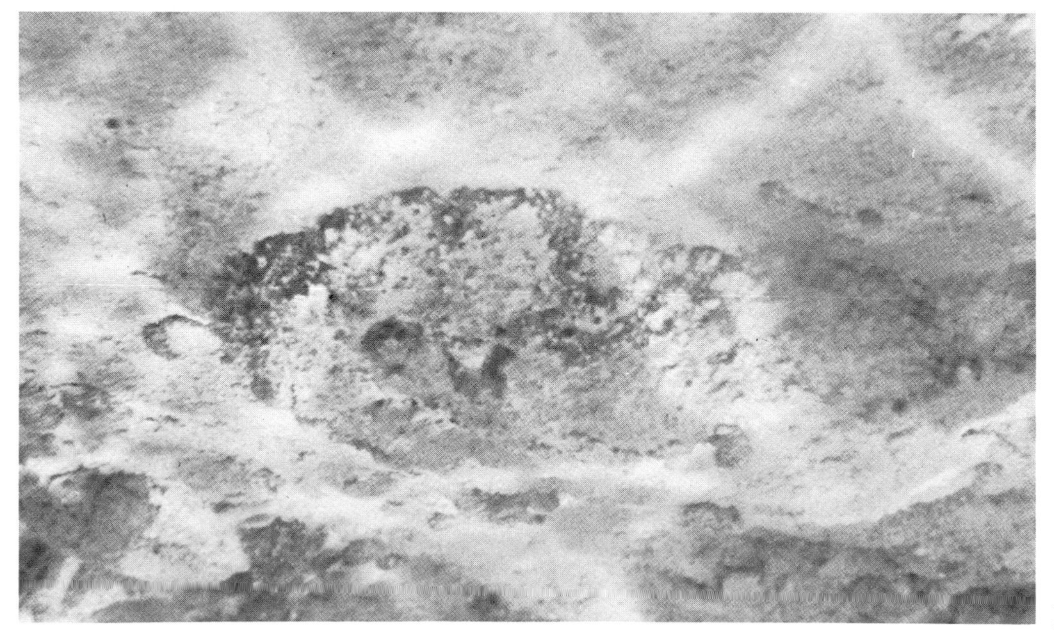

B

Class Osteichthyes–Toadfishes

PLATES

CHAPTER XXII

XXII: PLATE 1

FIGURE a.—*Barchatus cirrhosus* (Klunzinger). Length 25 cm. (From Klunzinger)

FIGURE b.—*Batrachoides didactylus* (Bloch). Length 19 cm.

1. Side view. (From Smitt)
2. Dorsal view. (From Pol)

FIGURE c.—*Batrachoides grunniens* (Bloch). Length 20 cm. (From Blegvad)

FIGURE d.—*Marcgravichthys cryptocentrus* (Valenciennes). Length 28 cm. (From Cervigon)

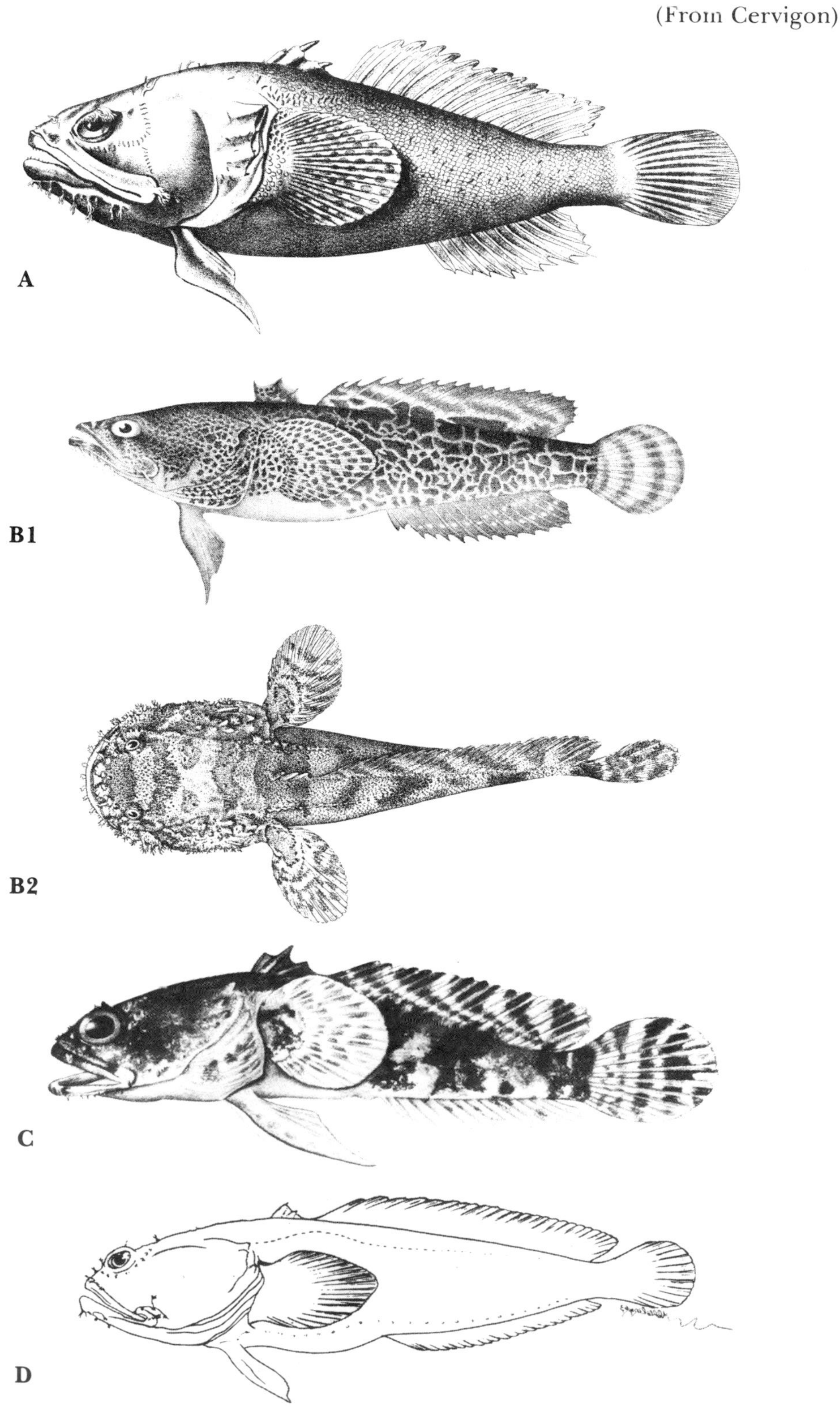

XXII: PLATE 2

FIGURE a.—*Opsanus tau* (Linnaeus). Length 30 cm. (From Storer)

FIGURE b.—*Thalassophryne amazonica* Steindachner. Length 9 cm. (From Collette)

FIGURE c.—*Thalassophryne dowi* Jordan and Gilbert. Length 13 cm. (M. Shirao)

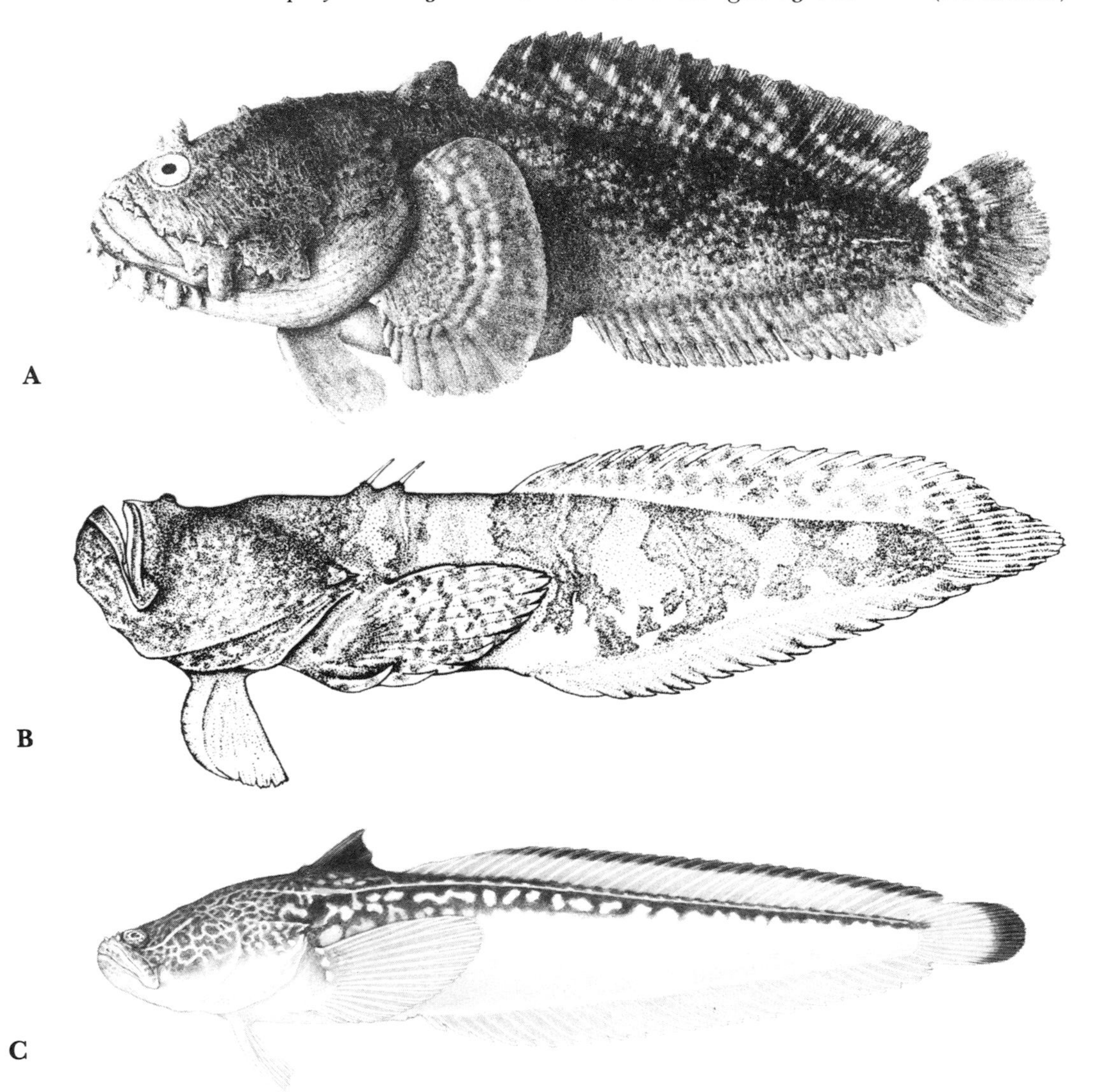

XXII: PLATE 3

FIGURE a, b.—*Thalassophryne maculosa* Günther. Length 14 cm. Two specimens are illustrated showing color pattern variations.

FIGURE c.—*Thalassophryne montevidensis* Berg. Length 15 cm. (From Collette)

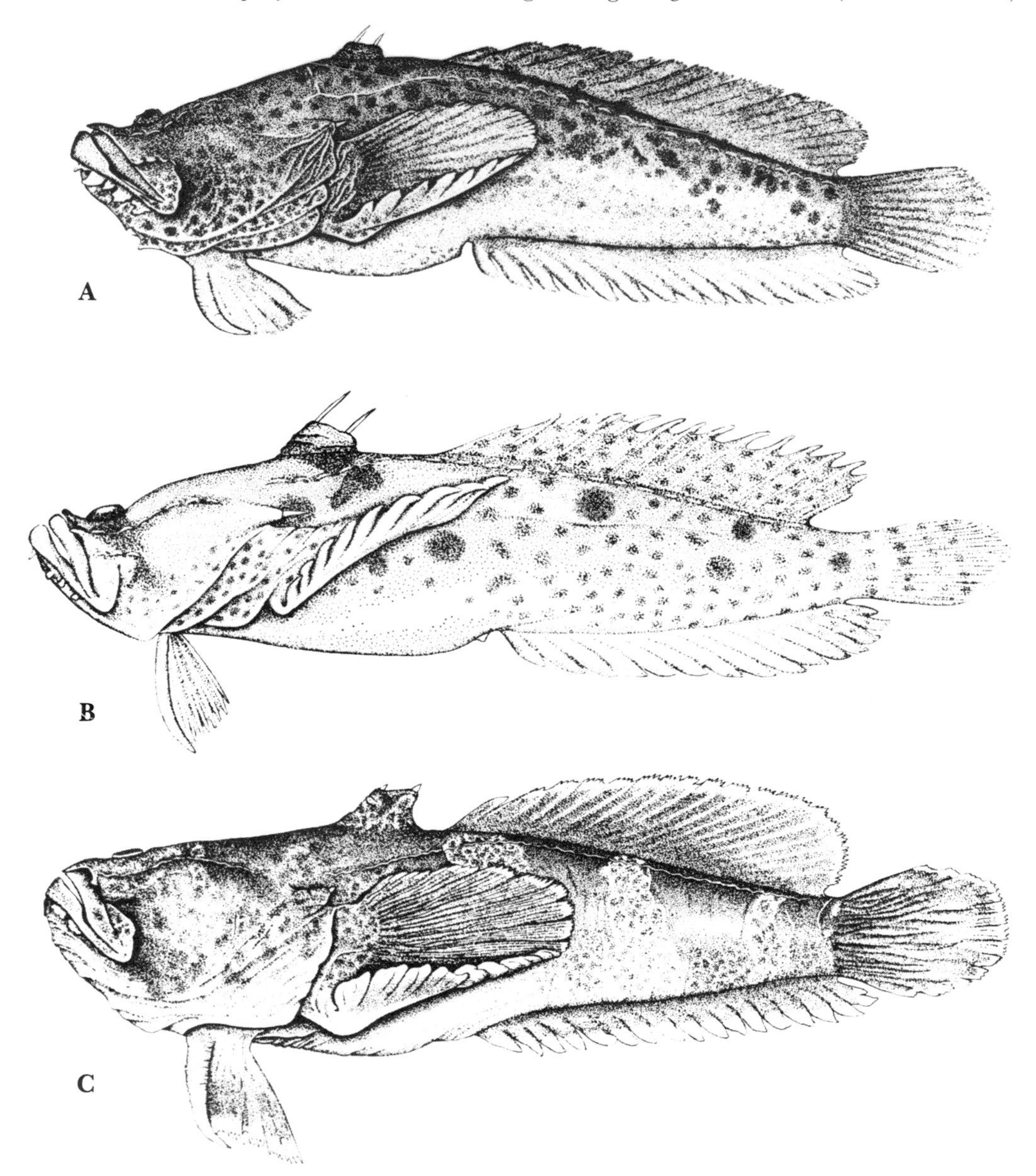

XXII: PLATE 4

FIGURE a.—*Thalassophryne nattereri* Steindachner. Length 12 cm. (From Collette)

FIGURE b.—*Thalassophryne punctata* Steindachner. Length 15 cm. (Courtesy Harvard Museum of Comparative Zoology)

FIGURE c.—*Thalassophryne reticulata* Günther. Length 25 cm. (M. Shirao)

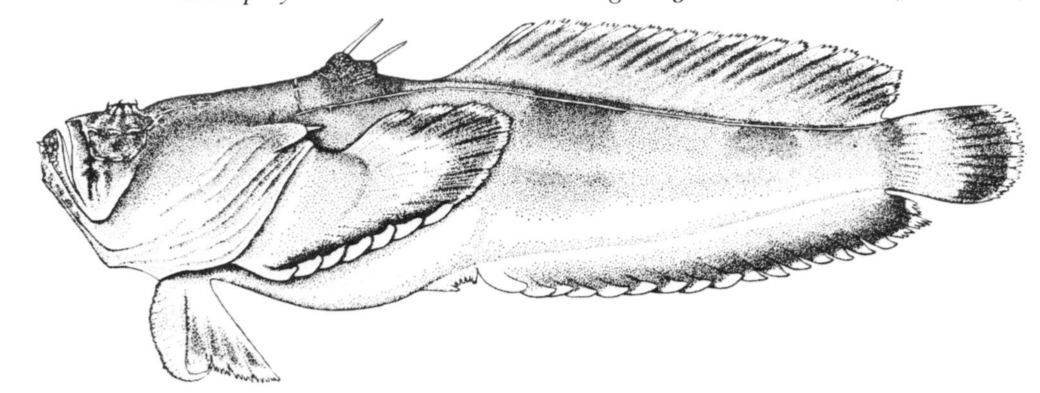

A

B

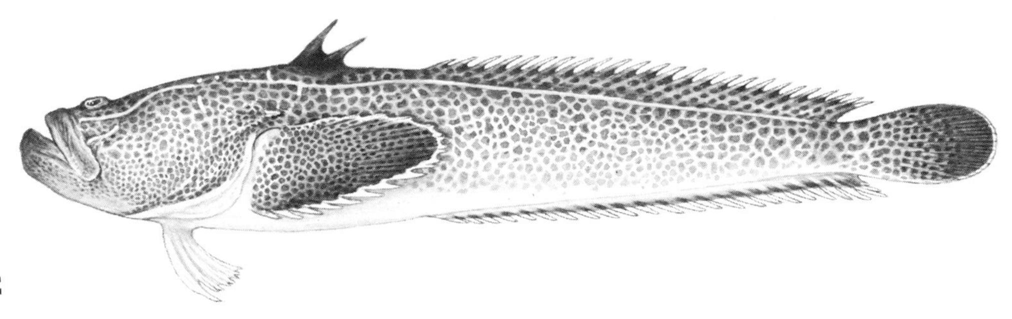

C

XXII: PLATE 5

FIGURE a.—*Halophryne diemensis* (LeSueur). Length 25 cm. This fish is greatly feared by the natives of the Palm Islands, North Queensland, Australia, where it is known locally as a "stonefish." *Halophryne* is not related to the true stonefish, which is a scorpaenid and a member of the genus *Synanceja*. However, Cameron (1969) examined the dorsal fin spines of *H. diemensis* and found no histological evidence of venom glands. (J. Booth)

FIGURE b.—*Opsanus beta* (Goode and Bean). Length 30 cm. This is a close relative of *O. tau* which is reputed to have venomous dorsal spines. It is not known whether this particular species has venomous spines or not. (Courtesy Miami Seaquarium)

A

B

Class Osteichthyes–Venomous Miscellaneous Fishes

PLATES

CHAPTER XXIII

XXIII: PLATE 1

Muraena helena Linnaeus. Length 1.5 m. This European moray eel has long been ascribed as possessing hollow fangs and fully developed venom glands. However, modern anatomical research fails to corroborate these suppositions. The canine teeth are solid, and there is no evidence of venom glands of the type that is found in venomous snakes and other fishes. According to some workers the palatine mucosa does secrete a toxic substance. *See* text, p. 819, fig. 2. (World Life Research Institute)

XXIII: PLATE 2

FIGURE a.—*Oligoplites saurus* (Bloch). Length 26 cm.

FIGURE b.—*Scomberoides sanctipetri* (Cuvier). Length 65 cm. (C. T. Conley)

FIGURE c.—*Selar crumenophthalmus* (Bloch). Length 60 cm. (From Kumada)

FIGURE d.—*Trachurus trachurus* (Linnaeus). Length 30 cm. (From Jordan and Evermann)

The anal spines of these carangids are reputed to be venomous, but there are no supporting morphological data.

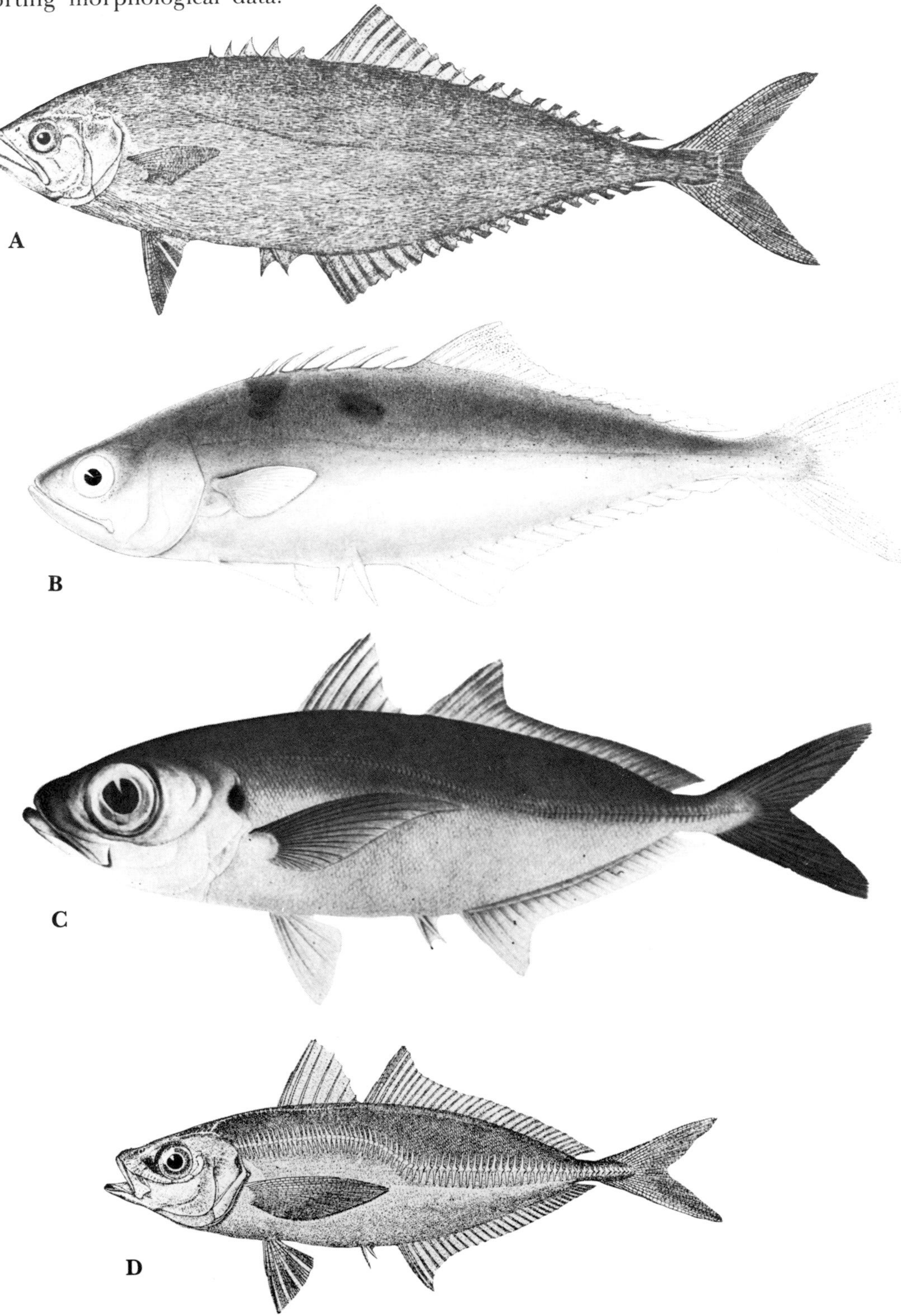

XXIII: PLATE 3

FIGURE a.—*Scatophagus argus* (Linnaeus). Length 15 cm. (T. Kumada)

FIGURE b.—*Selenotoca multifasciata* (Richardson). Length 40 cm. (From Richardson)

The dorsal and anal spines are capable of inflicting painful wounds.

A

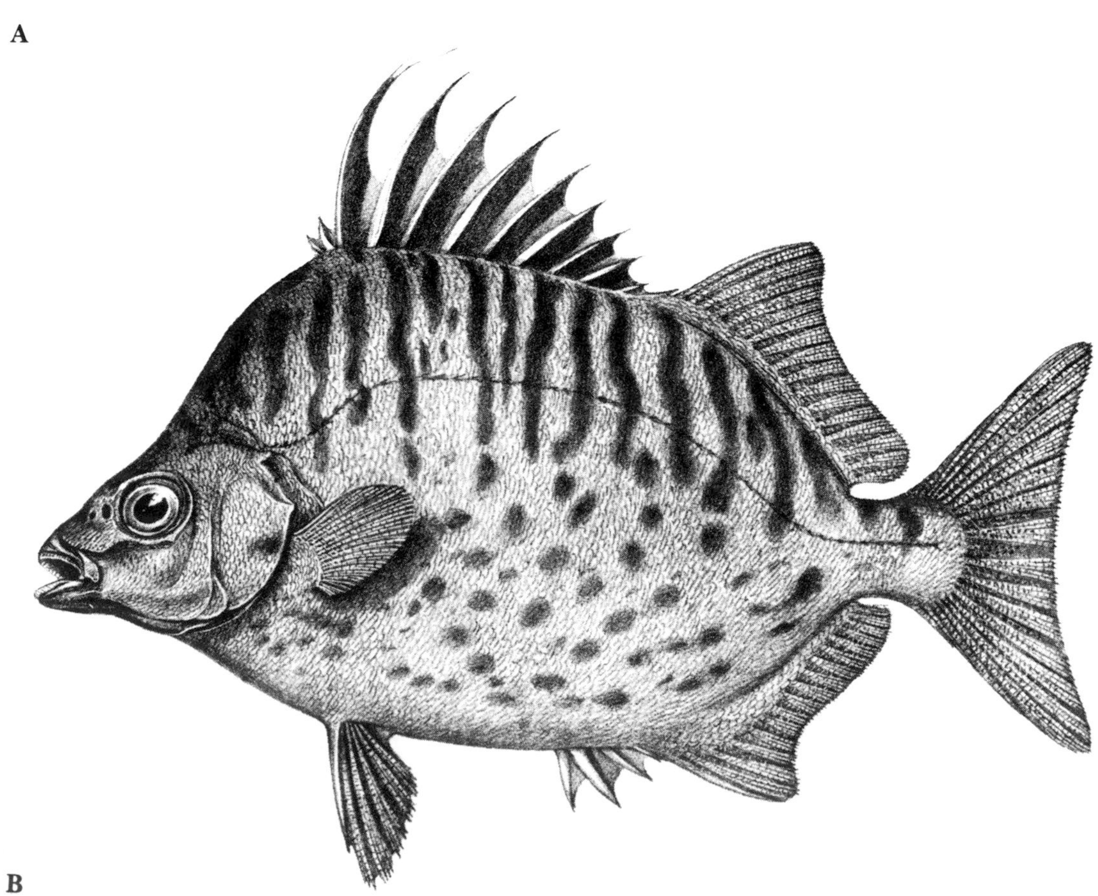

B

XXIII: PLATE 4

FIGURE a.—*Uranoscopus japonicus* Houttuyn. Length 25 cm.
(From Temminck and Schlegel)

FIGURE b.—*Uranoscopus scaber* Linnaeus. Length 15 cm.
1. Dorsal view; 2. Side view. (M. Shirao)

The cleithral or shoulder spines are venomous. Wounds from these fishes can be very painful, and unsubstantiated fatalities have been reported.

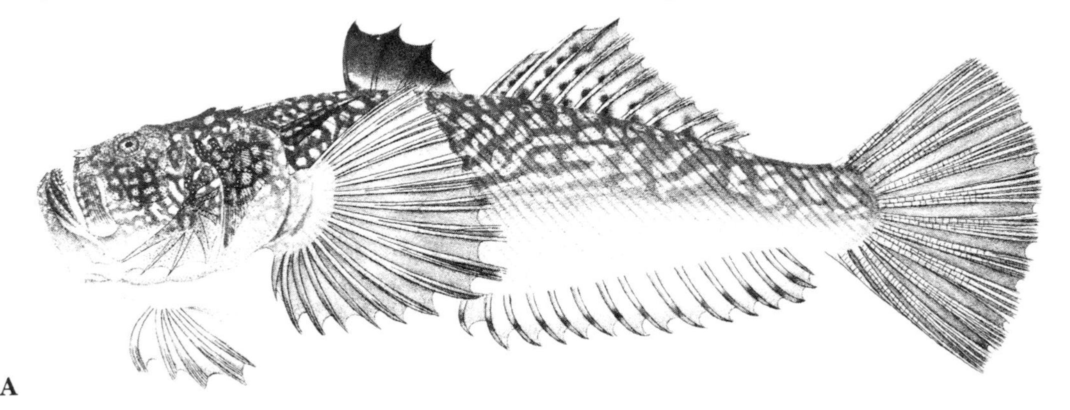

A

1B

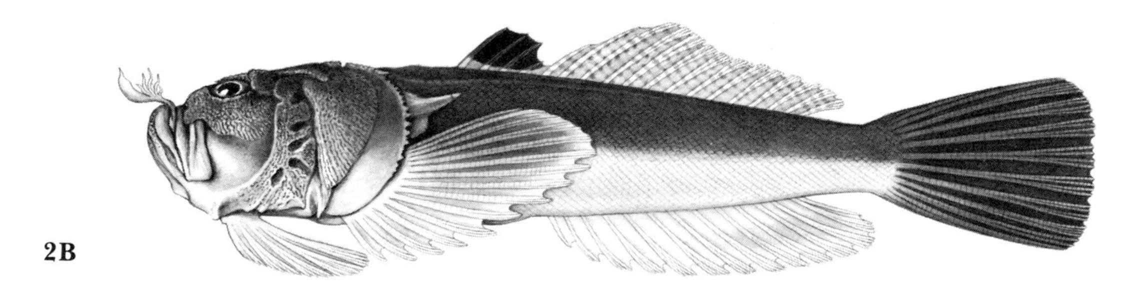

2B

XXIII: PLATE 5

FIGURE a.—*Siganus corallinus* (Valenciennes). Length 25 cm. (C. T. Conley)
FIGURE b.—*Siganus fuscescens* (Houttuyn). Length 25 cm. (S. Arita)
FIGURE c.—*Siganus lineatus* (Valenciennes). Length 30 cm. (G. Coates)

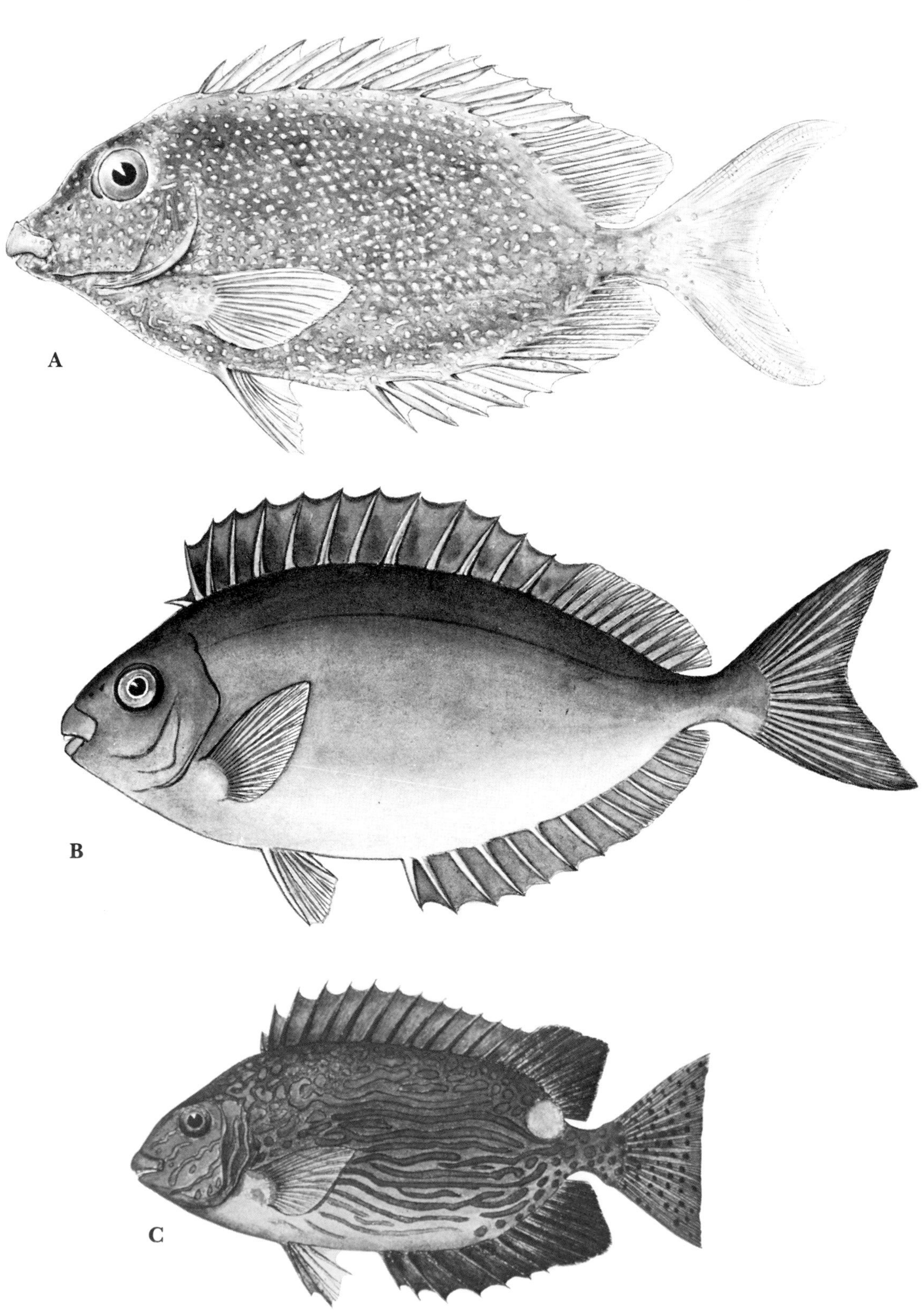

XXIII: PLATE 6

FIGURE a.—*Siganus puellus* (Schlegel). Length 27 cm. (T. Kumada)

FIGURE b.—*Siganus punctatus* (Schneider). Length 32 cm. (Courtesy Smithsonian Institution)

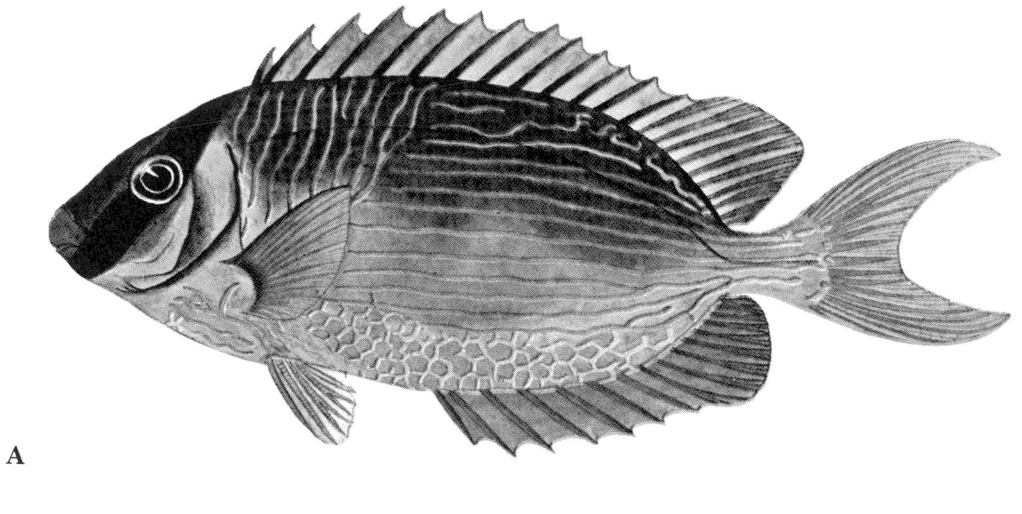

A

B

XXIII: PLATE 7

FIGURE a.—*Siganus rostratus* (Valenciennes). Length 30 cm.

FIGURE b.—*Siganus vulpinus* (Schlegel and Müller). Length 20 cm. (P. Mote)

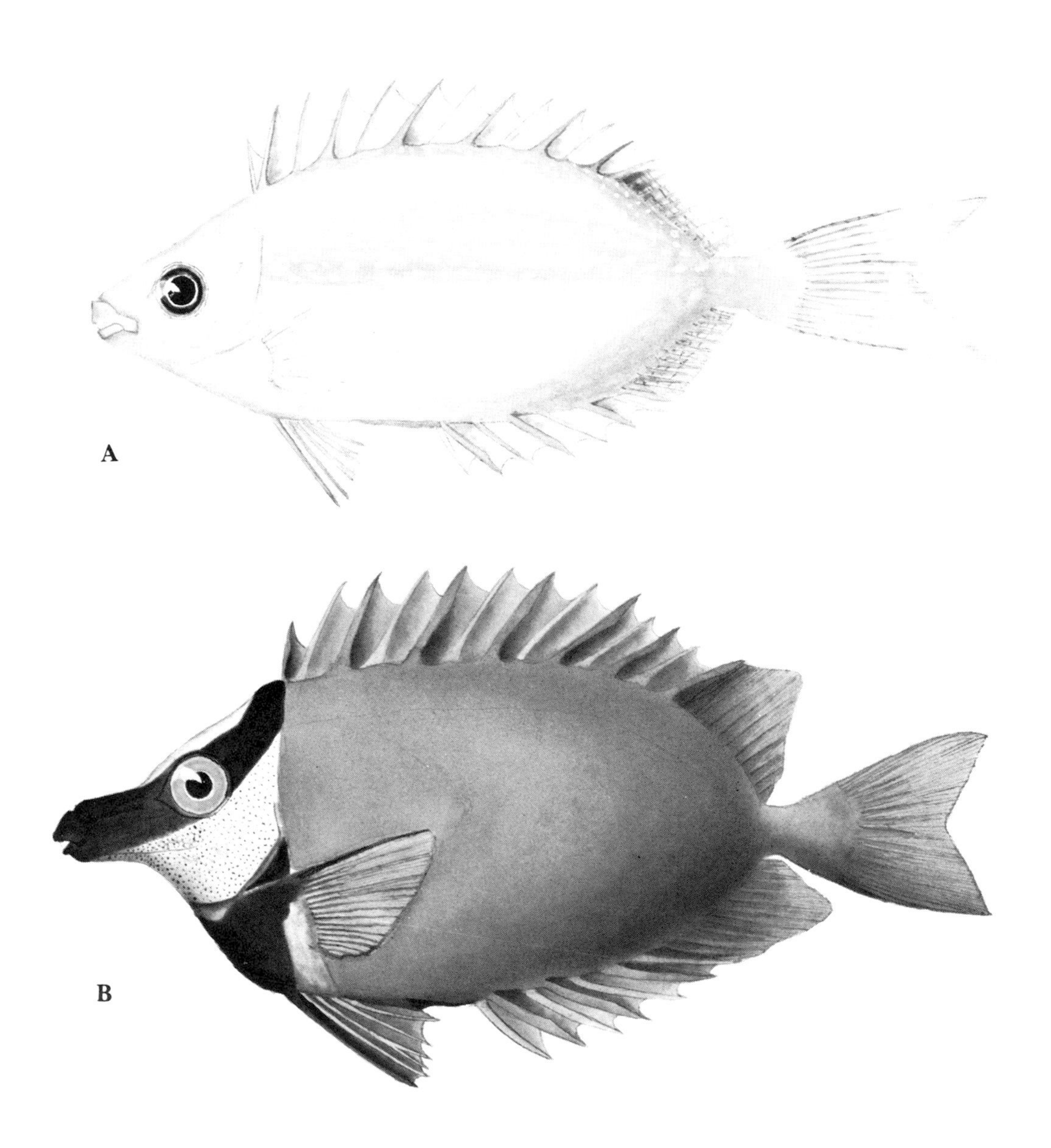

XXIII: PLATE 8

Prionurus microlepidotus Lacépède. Length 30 cm. (From Okada)

Class Osteichthyes–Ichthyocrinotoxic Fishes

PLATES

CHAPTER XXIV

XXIV: PLATE 1

Figure a.—*Grammistes sexlineatus* Thunberg). During stress, this fish can secrete copious amounts of a white viscid mucus through its skin. This secretion has been shown to possess both antibiotic and toxic properties. The secretion is toxic to fishes, but it is not known whether it is poisonous to man. Length 15 cm.

(From Jordan and Seale)

Figure b.—*Rypticus saponaceus* (Bloch and Schneider). Length 45 cm. (From Bean)

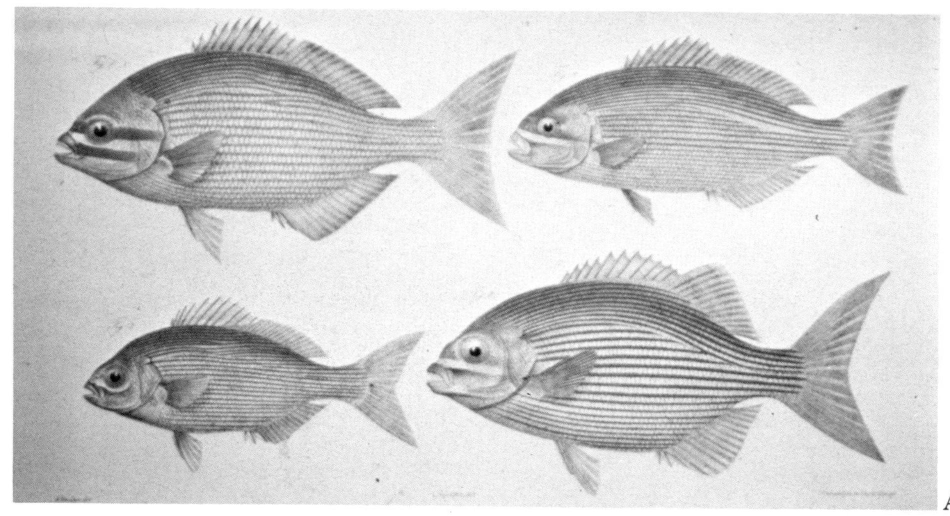

A

B

XXIV: PLATE 2

Rhinesomus bicaudalis (Linnaeus). Length 30 cm. (From Evermann and Marsh)

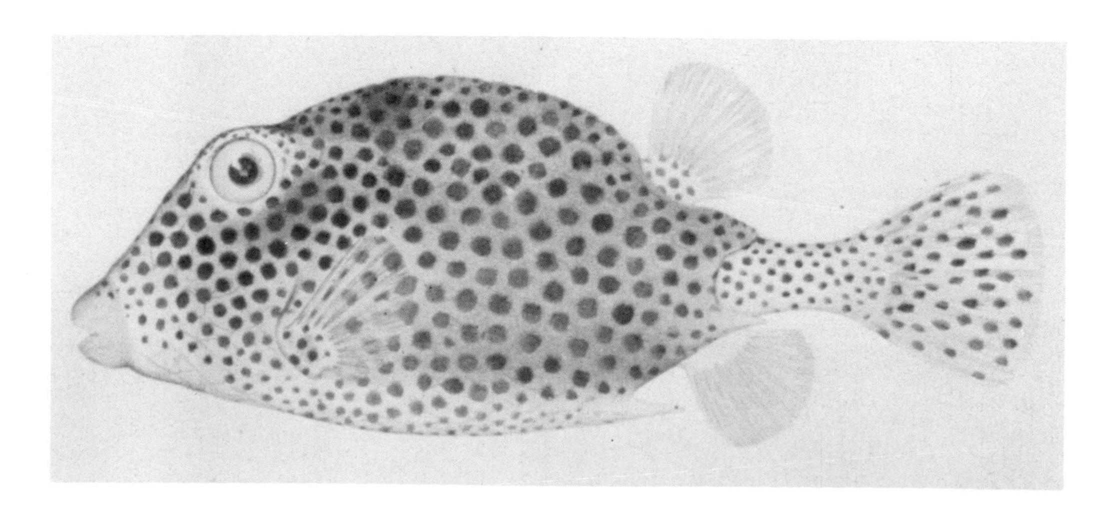

XXIV: PLATE 3

FIGURE a.—*Ostracion meleagris* Shaw. Length 20 cm. (From Jordan and Evermann)

FIGURE b.—*O. meleagris* during the stress reaction in which the fish secretes a foamy toxic substance (ostracitoxin) from the mouth and at the base of its fins. This substance has been found to be toxic to other aquarium fishes and mice. Ostracitoxin is said to be toxic to man when ingested. (Courtesy of D. A. Thomson)

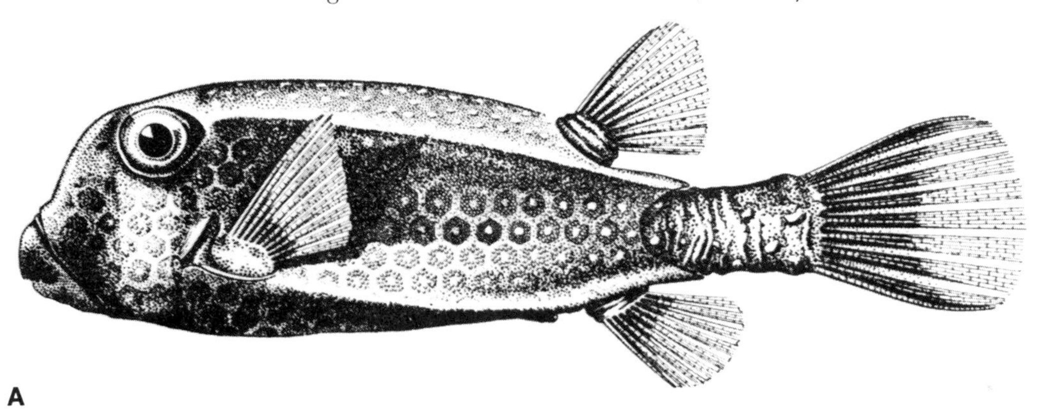

A

B

Class Reptilia

PLATES

CHAPTER XXV

XXV: PLATE 1

FIGURE a.—*Chelonia mydas* (Linnaeus). May attain a weight of over 250 kg, and a carapace length of about 120 cm. The flesh of this turtle may be poisonous to eat in some localities. (H. Baerg)

FIGURE b.—*Eretmochelys imbricata* (Linnaeus). May attain a weight of about 125 kg, and a carapace length of about 85 cm. The flesh of this turtle may be poisonous to eat in some localities. (Mrs. R. Kreuzinger)

FIGURE c.—*Dermochelys coriacea* (Linnaeus). May attain a weight of over 250 kg and a carapace length of about 120 cm. The flesh of this turtle may be poisonous to eat in some localities. However, in most regions this turtle is highly esteemed for its fine flavor. The toxicity of marine turtles is believed to be due to their feeding on toxic plants. Heron Island, Queensland, Australia. (H. Baerg)

XXV: PLATE 2

FIGURE a.—*Enhydrina schistosa* (Daudin). Length 1.2 m. This species is said to be one of the more common species of sea snakes and is the most frequent cause of human fatalities from sea-snake bites. There are varying opinions regarding its aggressiveness, but it is certain that when provoked *E. schistosa* can inflict a fatal envenomation. This species must be treated with respect.

(R. H. Knabenbauer)

FIGURE b.—*Hydrophis cyanocinctus* Daudin. Length 1.8 m. This species has caused human fatalities. (R. H. Knabenbauer)

A

B

XXV: PLATE 3

FIGURE a.—*Hydrophis obscuris* Daudin. Length 1.2 m. This species is reported to have caused human fatalities.

(F. Shultz, after Fayrer, courtesy U.S. Armed Forces Institute of Pathology)

FIGURE b.—*Hydrophis ornatus* (Gray). Length 95 cm. This species has caused human fatalities.

(M. Shirao)

A

B

XXV: PLATE 4

FIGURE a.—*Hydrophis spiralis* (Shaw). Length 1.8. This species has caused human fatalities.

FIGURE b.—*Kerilia jerdoni* Gray. Length 90 cm. This species has been incriminated in human envenomations and may cause fatalities.

(F. Shultz, after Fayrer, courtesy U.S. Armed Forces Institute of Pathology)

FIGURE c.—*Lapemis hardwicki* Gray. Length 85 cm. This species has caused human fatalities. (M. Shirao)

A

B

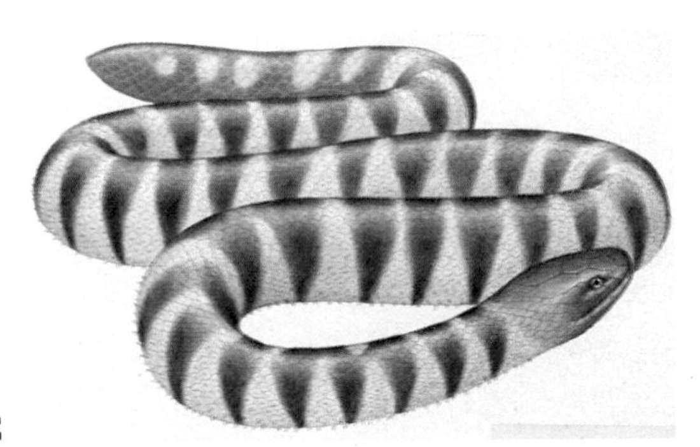

C

XXV: PLATE 5

Microcephalophis gracilis (Shaw). Length 1 m. This species has caused human fatalities. (M. Shirao)

XXV: PLATE 6

Pelamis platurus (Linnaeus). Length about 85 cm. Living specimen taken near Keelung, Taiwan. This species has caused human fatalities. (M. Shirao)

XXV: PLATE 7

FIGURE a.—*Thalassophina viperina* (Schmidt). Length 92 cm. This species may cause human fatalities.

(F. Shultz, after Fayrer, courtesy U.S. Armed Forces Institute of Pathology)

FIGURE b.—*Thalassophis anomalus* Schmidt. Length 80 cm. This species has caused human fatalities.

(M. Shirao)

A

B

Class Mammalia

PLATES

CHAPTER XXVI

XXVI: PLATE 1

FIGURE a.—*Balaenoptera borealis* Lesson. Length 18 m. The liver of this whale is reported to be poisonous on occasion, causing severe headaches, gastrointestinal upset, flushing of the face, facial edema, desquamation of the skin, and liver impairment. The nature of the poison is unknown. (M. Shirao)

FIGURE b.—*Delphinapterus leucas* (Pallas). Length 5.5 m. Ingestion of the flesh of this whale has been reported to cause human fatalities.

FIGURE c.—*Physeter catodon* Linnaeus. Length 18 m. The flesh and oil of this whale are reported to be toxic in some areas. (Mrs. R. Kreuzinger)

XXVI: PLATE 2

Thalarctos maritimus (Phipps). Length 2.5 m. Polar bears may attain a weight of 720 kg. Although the polar bear is one of the most widely known of the arctic mammals, there is surprisingly little information available concerning the biology of this magnificent mammal. Man is its greatest enemy and unless conservation laws are rigidly enforced, this animal will ultimately face extinction. The liver and kidneys are toxic to eat. Ingestion of these parts may cause gastrointestinal upset, drowsiness, dizziness, weakness, visual disturbances, desquamation, and convulsions. Fatalities are thought to be rare. The primary cause of the intoxication is believed to be due largely to hypervitaminosis A. (H. Baerg)

XXVI: PLATE 3

FIGURE a.—*Odobenus rosmarus* (Linnaeus). Length 3.5 m. Attains a weight of about 1,300 kg. A large bull walrus taken at Walrus Island, Bristol Bay, Alaska.

FIGURE b.—Herd of walruses at Walrus Island, Bristol Bay, Alaska.

(Courtesy of U.S. Fish and Wildlife Service)

A

B

XXVI: PLATE 4

FIGURE a.—*Erignathus barbatus* (Erxleben). Length 2.7 m. The liver of the bearded seal may be poisonous to eat. The toxicity is believed to be due to hypervitaminosis A. (R. Wallen, courtesy Alaska Department of Fish and Game)

FIGURE b.—*Pusa hispida* (Schreber). Length 1.4 m. This seal is probably the most common and important pinniped in the economy of the arctic natives. However, the liver of some of the older specimens may at times be toxic to eat. (Courtesy V. Scheffer)

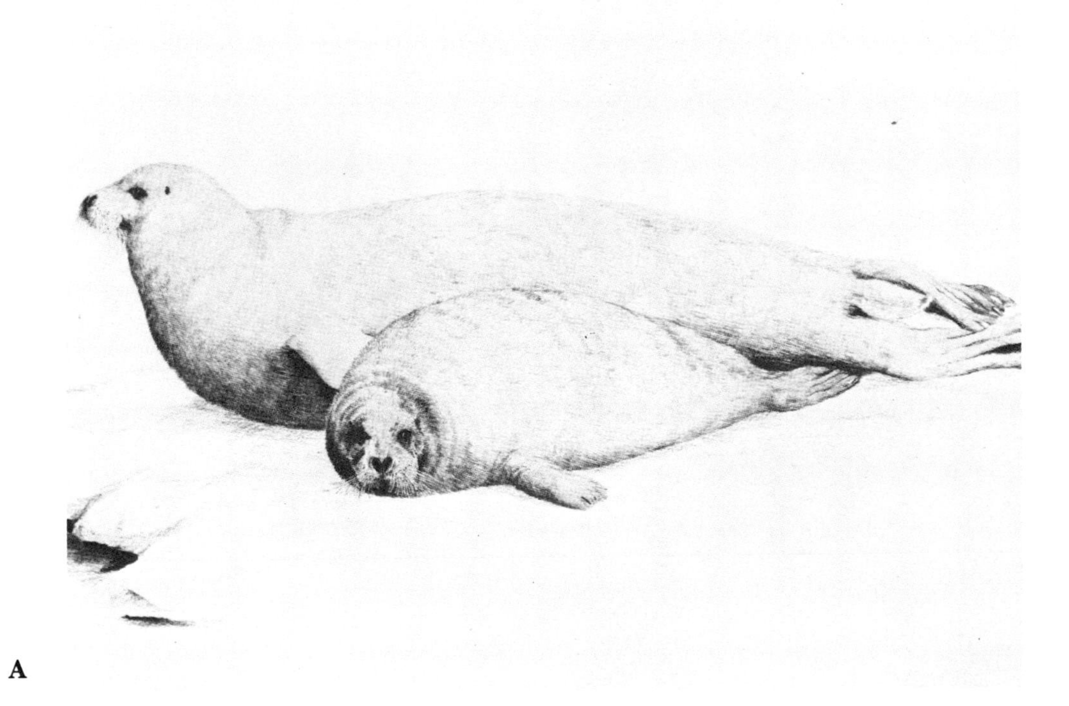

A

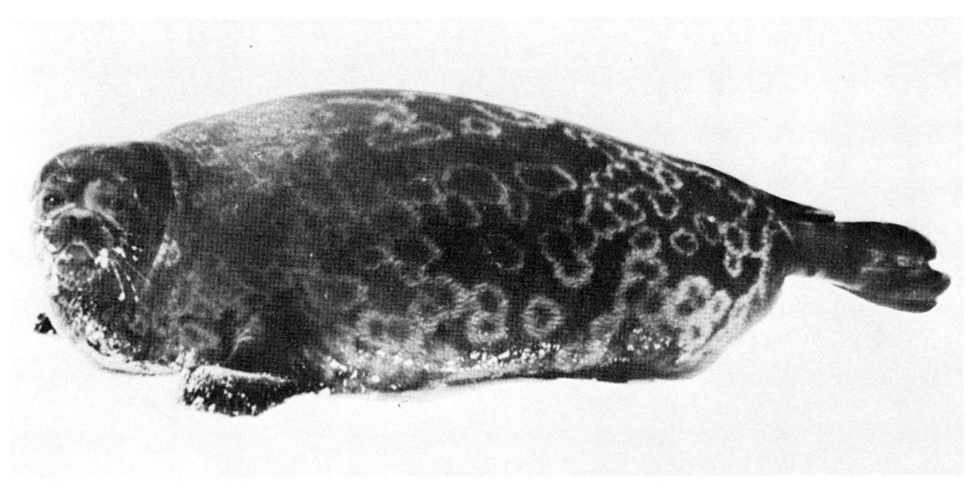

B